Recommended Dietary Allowances (RDA) and Adequate Intakes (AI) for Vitamins

Age (yr)	Thiamin RDA (mg/day)	Riboflavin RDA (mg/day)	Niacin RDA (mg/day)[a]	Biotin AI (µg/day)	Pantothenic acid AI (mg/day)	Vitamin B_6 RDA (mg/day)	Folate RDA (µg/day)[b]	Vitamin B_{12} RDA (µg/day)	Choline AI (mg/day)	Vitamin C RDA (mg/day)	Vitamin A [illegible]/day)[c]	[illegible])[d]	[illegible])[e]	[illegible]	
Infants															
0–0.5	0.2	0.3	2	5	1.7	0.1	65	0.4	125	40	400	[illegible]	[illegible]	2.0	
0.5–1	0.3	0.4	4	6	1.8	0.3	80	0.5	150	50	500	5	5	2.5	
Children															
1–3	0.5	0.5	6	8	2	0.5	150	0.9	200	15	300	5	6	30	
4–8	0.6	0.6	8	12	3	0.6	200	1.2	250	25	400	5	7	55	
Males															
9–13	0.9	0.9	12	20	4	1.0	300	1.8	375	45	600	5	11	60	
14–18	1.2	1.3	16	25	5	1.3	400	2.4	550	75	900	5	15	75	
19–30	1.2	1.3	16	30	5	1.3	400	2.4	550	90	900	5	15	120	
31–50	1.2	1.3	16	30	5	1.3	400	2.4	550	90	900	5	15	120	
51–70	1.2	1.3	16	30	5	1.7	400	2.4	550	90	900	10	15	120	
>70	1.2	1.3	16	30	5	1.7	400	2.4	550	90	900	15	15	120	
Females															
9–13	0.9	0.9	12	20	4	1.0	300	1.8	375	45	600	5	11	60	
14–18	1.0	1.0	14	25	5	1.2	400	2.4	400	65	700	5	15	75	
19–30	1.1	1.1	14	30	5	1.3	400	2.4	425	75	700	5	15	90	
31–50	1.1	1.1	14	30	5	1.3	400	2.4	425	75	700	5	15	90	
51–70	1.1	1.1	14	30	5	1.5	400	2.4	425	75	700	10	15	90	
>70	1.1	1.1	14	30	5	1.5	400	2.4	425	75	700	15	15	90	
Pregnancy															
≤18	1.4	1.4	18	30	6	1.9	600	2.6	450	80	750	5	15	75	
19–30	1.4	1.4	18	30	6	1.9	600	2.6	450	85	770	5	15	90	
31–50	1.4	1.4	18	30	6	1.9	600	2.6	450	85	770	5	15	90	
Lactation															
≤18	1.4	1.6	17	35	7	2.0	500	2.8	550	115	1200	5	19	75	
19–30	1.4	1.6	17	35	7	2.0	500	2.8	550	120	1300	5	19	90	
31–50	1.4	1.6	17	35	7	2.0	500	2.8	550	120	1300	5	19	90	

NOTE: For all nutrients, values for infants are AI. The glossary on the inside back cover defines units of nutrient measure.

[a]Niacin recommendations are expressed as niacin equivalents (NE), except for recommendations for infants younger than 6 months, which are expressed as preformed niacin.

[b]Folate recommendations are expressed as dietary folate equivalents (DFE).

[c]Vitamin A recommendations are expressed as retinol activity equivalents (RAE).

[d]Vitamin D recommendations are expressed as cholecalciferol and assume an absence of adequate exposure to sunlight.

[e]Vitamin E recommendations are expressed as α-tocopherol.

Recommended Dietary Allowances (RDA) and Adequate Intakes (AI) for Minerals

Age (yr)	Sodium AI (mg/day)	Chloride AI (mg/day)	Potassium AI (mg/day)	Calcium AI (mg/day)	Phosphorus RDA (mg/day)	Magnesium RDA (mg/day)	Iron RDA (mg/day)	Zinc RDA (mg/day)	Iodine RDA (µg/day)	Selenium RDA (µg/day)	Copper RDA (µg/day)	Manganese AI (mg/day)	Fluoride AI (mg/day)	Chromium AI (µg/day)	Molybdenum RDA (µg/day)
Infants															
0–0.5	120	180	400	210	100	30	0.27	2	110	15	200	0.003	0.01	0.2	2
0.5–1	370	570	700	270	275	75	11	3	130	20	220	0.6	0.5	5.5	3
Children															
1–3	1000	1500	3000	500	460	80	7	3	90	20	340	1.2	0.7	11	17
4–8	1200	1900	3800	800	500	130	10	5	90	30	440	1.5	1.0	15	22
Males															
9–13	1500	2300	4500	1300	1250	240	8	8	120	40	700	1.9	2	25	34
14–18	1500	2300	4700	1300	1250	410	11	11	150	55	890	2.2	3	35	43
19–30	1500	2300	4700	1000	700	400	8	11	150	55	900	2.3	4	35	45
31–50	1500	2300	4700	1000	700	420	8	11	150	55	900	2.3	4	35	45
51–70	1300	2000	4700	1200	700	420	8	11	150	55	900	2.3	4	30	45
>70	1200	1800	4700	1200	700	420	8	11	150	55	900	2.3	4	30	45
Females															
9–13	1500	2300	4500	1300	1250	240	8	8	120	40	700	1.6	2	21	34
14–18	1500	2300	4700	1300	1250	360	15	9	150	55	890	1.6	3	24	43
19–30	1500	2300	4700	1000	700	310	18	8	150	55	900	1.8	3	25	45
31–50	1500	2300	4700	1000	700	320	18	8	150	55	900	1.8	3	25	45
51–70	1300	2000	4700	1200	700	320	8	8	150	55	900	1.8	3	20	45
>70	1200	1800	4700	1200	700	320	8	8	150	55	900	1.8	3	20	45
Pregnancy															
≤18	1500	2300	4700	1300	1250	400	27	12	220	60	1000	2.0	3	29	50
19–30	1500	2300	4700	1000	700	350	27	11	220	60	1000	2.0	3	30	50
31–50	1500	2300	4700	1000	700	360	27	11	220	60	1000	2.0	3	30	50
Lactation															
≤18	1500	2300	5100	1300	1250	360	10	14	290	70	1300	2.6	3	44	50
19–30	1500	2300	5100	1000	700	310	9	12	290	70	1300	2.6	3	45	50
31–50	1500	2300	5100	1000	700	320	9	12	290	70	1300	2.6	3	45	50

Tolerable Upper Intake Levels (UL) for Vitamins

Age (yr)	Niacin (mg/day)[a]	Vitamin B_6 (mg/day)	Folate (μg/day)[a]	Choline (mg/day)	Vitamin C (mg/day)	Vitamin A (μg/day)[b]	Vitamin D (μg/day)	Vitamin E (mg/day)[c]
Infants								
0–0.5	—	—	—	—	—	600	25	—
0.5–1	—	—	—	—	—	600	25	—
Children								
1–3	10	30	300	1000	400	600	50	200
4–8	15	40	400	1000	650	900	50	300
9–13	20	60	600	2000	1200	1700	50	600
Adolescents								
14–18	30	80	800	3000	1800	2800	50	800
Adults								
19–70	35	100	1000	3500	2000	3000	50	1000
>70	35	100	1000	3500	2000	3000	50	1000
Pregnancy								
≤18	30	80	800	3000	1800	2800	50	800
19–50	35	100	1000	3500	2000	3000	50	1000
Lactation								
≤18	30	80	800	3000	1800	2800	50	800
19–50	35	100	1000	3500	2000	3000	50	1000

[a]The UL for niacin and folate apply to synthetic forms obtained from supplements, fortified foods, or a combination of the two.

[b]The UL for vitamin A applies to the preformed vitamin only.

[c]The UL for vitamin E applies to any form of supplemental α-tocopherol, fortified foods, or a combination of the two.

Tolerable Upper Intake Levels (UL) for Minerals

Age (yr)	Sodium (mg/day)	Chloride (mg/day)	Calcium (mg/day)	Phosphorus (mg/day)	Magnesium (mg/day)[d]	Iron (mg/day)[b]	Zinc (mg/day)	Iodine (μg/day)	Selenium (μg/day)	Copper (μg/day)	Manganese (mg/day)	Fluoride (mg/day)	Molybdenum (μg/day)	Boron (mg/day)	Nickel (mg/day)	Vanadium (mg/d
Infants																
0–0.5	—[e]	—[e]	—	—	—	40	4	—	45	—	—	0.7	—	—	—	—
0.5–1	—[e]	—[e]	—	—	—	40	5	—	60	—	—	0.9	—	—	—	—
Children																
1–3	1500	2300	2500	3000	65	40	7	200	90	1000	2	1.3	300	3	0.2	—
4–8	1900	2900	2500	3000	110	40	12	300	150	3000	3	2.2	600	6	0.3	—
9–13	2200	3400	2500	4000	350	40	23	600	280	5000	6	10	1100	11	0.6	—
Adolescents																
14–18	2300	3600	2500	4000	350	45	34	900	400	8000	9	10	1700	17	1.0	—
Adults																
19–70	2300	3600	2500	4000	350	45	40	1100	400	10,000	11	10	2000	20	1.0	1.8
>70	2300	3600	2500	3000	350	45	40	1100	400	10,000	11	10	2000	20	1.0	1.8
Pregnancy																
≤18	2300	3600	2500	3500	350	45	34	900	400	8000	9	10	1700	17	1.0	—
19–50	2300	3600	2500	3500	350	45	40	1100	400	10,000	11	10	2000	20	1.0	—
Lactation																
≤18	2300	3600	2500	4000	350	45	34	900	400	8000	9	10	1700	17	1.0	—
19–50	2300	3600	2500	4000	350	45	40	1100	400	10,000	11	10	2000	20	1.0	—

[d]The UL for magnesium applies to synthetic forms obtained from supplements or drugs only.

[e]Source of intake should be from human milk (or formula) and food only.

NOTE: An Upper Limit was not established for vitamins and minerals not listed and for those age groups listed with a dash (—) because of a lack of data, not because these nutrients are safe to consume at any level of intake. All nutrients can have adverse effects when intakes are excessive.

SOURCE: Adapted with permission from the *Dietary Reference Intakes* series, National Academy Press. Courtesy of the National Academy Press, Washington, D.C.

NUTRITION THERAPY
AND
PATHOPHYSIOLOGY

NUTRITION THERAPY AND PATHOPHYSIOLOGY

Marcia Nahikian Nelms,
Southeast Missouri State University

Kathy Sucher
San Jose State University

Karen Lacey
University of Wisconsin, Green Bay

Pamela Goyan Kittler
Food, Culture and Nutrition Consultant, Sunnyvale, CA

R. Gerald Nelms
Southern Illinois University

Annalynn Skipper
Author and Consultant

Melissa Hansen-Petrik
The University of Tennessee-Knoxville

Christina Lee Frazier
Southeast Missouri State University

Robert D. Lee
Central Michigan University

Thomas J. Pujol
Southeast Missouri State University

Mildred Mattfeldt-Beman
Saint Louis University

Sara Long
Southern Illinois University

Maria Karalis
Abbott Renal Care Abbott Park, IL

Jessie M. Pavlinac
Oregon Health & Science University

Jordi Goldstein-Fuchs
University of California, San Francisco,

Roschelle A. Heuberger
Central Michigan University

Ethan A. Bergman
Central Washington University

Nancy S. Buergel
Central Washington University

Deborah A. Cohen
Southeast Missouri State University

Cade Fields-Gardner
The Cutting Edge

Elaina Jurecki
Kaiser Permanente Medical Center, Northern California,

Joyce Wong
Kaiser Permanente Medical Center, Northern California

Australia • Brazil • Canada • Mexico • Singapore • Spain
United Kingdom • United States

Australia • Brazil • Canada • Mexico • Singapore • Spain
United Kingdom • United States

Nutrition Therapy and Pathophysiology

Marcia Nelms, Kathryn Sucher, Sara Long

Executive Editor: Peter Marshall
Development Editor: Elizabeth Howe
Assistant Editor: Elesha Feldman
Editorial Assistant: Lauren Vogelbaum
Technology Project Manager: Donna Kelley
Marketing Manager: Jennifer Somerville
Marketing Communications Manager: Jessica Perry
Project Manager, Editorial Production: Cheryll Linthicum
Creative Director: Rob Hugel
Art Director: Lee Friedman
Project Coordination, Copyediting, and Electronic Composition:
Pre-Press Company, Inc.
Print Buyer: Rebecca Cross
Permissions Editor: Roberta Broyer
Text Designer: Lisa Devenish
Illustrations and Photo Research: Pre-Press Company, Inc.
Cover Designer: Ross Carron
Cover Image: Photodisc/Getty Images
Cover Printer: C&C Offset Printing Co., Ltd.
Compositor: Pre-Press Company, Inc.
Printer: C&C Offset Printing Co., Ltd.

Printed in China

3 4 5 6 7 10 09 08 07

For more information about our products, contact us at:
Thomson Learning Academic Resource Center
1-800-423-0563
For permission to use material from this text or product, submit a request online at.
Any additional questions about permissions can be submitted by e-mail to thomsonrights@thomson.com.

Thomson Higher Education
10 Davis Drive
Belmont, CA 94002-3098
USA

Library of Congress Control Number: 2006923284

ISBN-13: 978-0-534-62154-4
ISBN-10: 0-534-62154-6

DEDICATION

For our colleagues in
nutrition and dietetics

For our students: past,
present and future.

For Jerry, Taylor, and Emory
Marcia Nahikian-Nelms

For my supportive
and loving husband,
Peter and my son, Alexander
Kathryn Sucher

JKR . . . your love
and friendship make life
a joyous experience
Sara Long

BRIEF TABLE OF CONTENTS

PART I

THE ROLE OF NUTRITION THERAPY IN HEALTH CARE

1 Health Care Systems and Reimbursement 1

2 The Role of the Dietitian in the Healthcare System 29

PART II

THE NUTRITION CARE PROCESS

3 The Nutrition Care Process 39

4 Complementary And Alternative Medicine 65

5 Assessment of Nutrition Status and Risk 101

6 Documentation of Nutrition Care 137

7 Methods of Nutrition Support 149

PART III

INTRODUCTION TO PATHOPHYSIOLOGY

8 Fluid and Electrolyte Balance 181

9 Acid-Base Balance 203

10 Cellular & Physiological Response to Injury 219

11 Genomics 237

12 Immunology 259

13 Pharmacology 297

PART IV

NUTRITION THERAPY

14 Energy Balance and Body Weight 323

15 Diseases of the Cardiovascular System 371

16 Diseases of the Upper Gastrointestinal Tract 421

17 Diseases of the Lower Gastrointestinal Tract 457

18 Diseases of the Hepatobiliary System: Liver, Gallbladder, Exocrine Pancreas 509

19 Diseases of the Endocrine System 549

20 Diseases of the Renal System 609

21 Diseases of Hematological System 651

22 Diseases of the Neurological System 687

23 Diseases of the Respiratory System 715

24 Neoplastic Disease 751

25 Metabolic Stress 785

26 HIV and AIDS 805

27 Diseases of the Musculoskeletal System 843

28 Metabolic Disorders 881

REFERENCES R-1

APPENDICES A-1

GLOSSARY GL-1

INDEX I-1

TABLE OF CONTENTS

PART 1

1

Health Care Systems and Reimbursement 1

Introduction 1

Health Care Facilities in the United States 4

Preventive and Primary Health Care Services 4

Secondary and Tertiary Care 5

Restorative Care 5

Long-Term Care 5

Financing the Health Care Industry 5

Private Insurance 6

Traditional Fee-for-Service Plans 6

Group Contract Insurance 6

Public Insurance 7

The Medicare Program 9

The Medicaid Program 12

The State Children's Health Insurance Program 12

The Uninsured 13

Demographic Trends and Health Care 14

The Need for Health Care Reform 15

The High Cost of Health Care 15

Efforts at Cost Containment 15

Equity and Access as Issues in Health Care 18

Racial and Ethnic Disparities in Health 18

Health Care Reform in the United States 24

Nutrition as a Component of Health Care Reform 24

The Cost-Effectiveness of Nutrition Services 26

Conclusion 26

2

The Role of the Dietitian in the Health Care System 29

Introduction 29

The Registered Dietitian in Clinical Practice 30

The Role of the Clinical Dietitian 30

The Clinical Nutrition Team 30

Other Health Professionals—Interdisciplinary Teams 30

Members of the Health Care Team 31

Developing Critical Thinking Skills and Professional Competencies 33

Definition of Critical Thinking 33

Components of Critical Thinking 34

Specific Knowledge Base 34

Experience 34

Competencies 34

Attitudes 36

Standards 36

Levels of Clinical Reasoning 37

Conclusion 37

PART 2

3

The Nutrition Care Process 39

Introduction 39

Framework for Nutrition Care: Evaluation of Nutritional Status 39

Key Concepts: Nutritional Status 41

Purpose of Providing Nutrition Care 41

Key concepts: Nutrition Care 41

ADA's Standardized Nutrition Care Process (NCP) and Model Promotes High-Quality, Individualized Nutrition Care 42

Brief History of ADA's NCP 42

Standardized Nutrition Diagnostic Language 42

Improved Quality of Care 44

Critical Thinking 45

Key Concepts: ADA's Standardized Nutrition Care Process 45

Big Picture of Nutrition Care: The Model 46

Central Core 46

Two Outer Rings 47

Supportive Systems: Screening and Referral System and Outcomes Management System 47
Key Concepts: Nutrition Care Process and Model 48

Steps of the NCP 48
Step 1 Nutrition Assessment 48
Obtain and Verify Appropriate Data 48
Cluster and Organize Assessment Data 48
Evaluate the Data Using Reliable Standards 49
Key Concepts: NCP Step 1 Nutrition Assessment 49
Step 2 Nutrition Diagnosis 49
Example of Nutrition Diagnoses 55
Criteria for Evaluating PES Statements 56
Relationship of Nutrition Diagnosis to the Other Steps of the NCP 56
Key Concepts: NCP Step 2 Nutrition Diagnosis 58
Step 3 Nutrition Intervention 58
Sub-Step 3a: Plan 58
Sub-Step 3b: Implement the Nutrition Intervention 59
Key Concepts: NCP Step 3 Nutrition Intervention 59
Step 4 Nutrition Monitoring and Evaluation 59
Sub-Step 4a: Monitor Progress 59
Sub-Step 4b: Measure Outcomes 60
Sub-Step 4c: Evaluate Outcomes 61
Key Concepts: NCP Step 4 Nutrition Monitoring and Evaluation 61

Documentation 62

Conclusion 63

4

Complementary and Alternative Medicine 65

Introduction 65
Who Chooses CAM? 66
CAM Rationale 67
Biomedical Response 71

Alternative Medical Systems 73
Ayurvedic Medicine 73
Chiropractic Medicine 74
Homeopathic Medicine 74
Naturopathic Medicine 77
Osteopathic Medicine 77
Traditional Chinese Medicine/Acupuncture 78

Complementary Therapies 79
Chelation Therapy 79
Folk Healing 79
Natural Products 84
Dietary Therapies 91
Vitamin/Mineral Supplements and Megavitamin Therapy 92
Mind-Body Therapies 92

Medical Pluralism in Practice 95

Conclusion 95

5

Assessment of Nutrition Status and Risk: The First Step in the Nutrition Care Process 101

Introduction 101
Nutritional Status and Nutritional Risk 102

An Overview: Nutrition Assessment and Screening 102
Subjective and Objective Data Collection 103
Dietary Information 103
Psychosocial Information 104
Information Regarding Education, Learning and Motivation 105
Tools for Data Collection 105

Dietary Assessment Methods 108
Twenty-Four-Hour Recall 108
Food Record/Food Diary 108
Food Frequency 109
Observation of Food Intake/ "Calorie Count" 112

Analysis of Food Intake 112
Analysis Based on USDA's MyPyramid 112
Analysis Based on Exchanges/Carbohydrate Counting 112
Specific Nutrient Analysis 112
Computerized Dietary Analysis 112

Evaluation and Interpretation of Dietary Analysis Information 112
USDA's MyPyramid 112
U.S. Dietary Guidelines for Americans 113
Daily Values/Dietary Reference Intakes 113

Nutritional Physical Examination 114

Anthropometric/Body Composition Assessment 114
Anthropometrics 114
Height/Stature 114
Weight 115
Interpretation of Height and Weight: Infants and Children 116
Growth Charts 116
Body Mass Index 116
Interpretation of Height and Weight: Adults 116
Usual Body Weight 116
Percent Usual Body Weight and Percent Weight Change 116
Desirable or "Ideal" Body Weight 116
Healthy Body Weight 117
Body Mass Index (BMI) 117
Frame Size 117
Waist-Hip Ratio and Waist Circumference 117

Body Composition 117
- *Skinfold Measurements 118*
- *Bioelectrical Impedance Analysis (BIA) 119*
- *Hydrodensitometry – Underwater Weighing 120*
- *Near Infrared Interactance (NIR) 120*
- *Dual Energy X-Ray Absorptiometry (DEXA) 120*

Biochemical Assessment 120

Protein Assessment 121
- *Somatic Protein Assessment 121*
- *Visceral Protein Assessment 122*

Immunocompetence 125
- *Delayed Cutaneous Hypersensitivity (DCH) 125*
- *Total Lymphocyte Count (TLC) 125*

Hematological Assessment 126
- *Hemoglobin (Hgb) 126*
- *Hematocrit (Hct) 126*
- *Mean Corpuscular Volume (MCV) 126*
- *Mean Corpuscular Hemoglobin (MCH) 126*
- *Mean Corpuscular Hemoglobin Concentration (MCHC) 126*
- *Ferritin 126*
- *Transferrin Saturation 126*
- *Protoporphyrin 126*
- *Serum Folate 126*
- *Serum B12 126*

Vitamin and Mineral Assessment 127

Other Labs for with Clinical Significance 127

Functional Assessment 127

Energy and Protein Requirements 128

Measurement of Energy Requirements 130

Measurement of Protein Requirements 131

Estimation of Energy Requirements 131
- *Energy Requirements Based on DRI 131*
- *Activity Factor 132*
- *Stress Factors 132*

Estimation of Protein Requirements 133
- *RDA for Protein 133*
- *Protein Requirements in Metabolic Stress, Trauma, and Disease 133*
- *Protein-Kilocalorie Ratio 133*

Interpretation of Assessment Data 133

Conclusion 133

Documentation 137

An Overview: Writing in the Profession 137

The Functions, Context, Parts, and Processes of Writing 137
- *Rhetorical Norms 138*
- *Levels of Discourse 138*
- *Steps in the Writing Process 138*

Charting: Documentation of the Nutrition Care Process 139

Standardized Language and Medical Abbreviations 140

Medical Records 140

Organization of Nutrition Documentation 140
- *SOAP 140*
- *IER Notes 142*
- *Focus Notes 143*
- *PIE Notes 143*
- *PES, or Problem, Etiology, Signs/Symptoms Statements 143*
- *Assessment, Diagnosis, Intervention, Monitoring/Evaluation (ADIM) 144*
- *Charting by Exception 145*

Keeping a Personal Medical Notebook 145

Guidelines for All Charting 145
- *Confidentiality 146*

Writing for Non-Medical Audiences 147

Instructional Materials for Patients, their Families, and the Public 147
- *Tips for Writing Instructional Materials 147*

Reporting Your Own Research 147

Conclusions: Your Ethos—Establishing Expertise 147

Methods of Nutrition Support 149

Introduction 149

Oral Diets 149

Texture Modifications 150

Oral Supplements 151

Appetite Stimulants 153

Specialized Nutrition Support (SNS) 153

Enteral Nutrition 154
- *Indications 154*
- *Gastrointestinal Access 154*
- *Formulas 157*
- *Feeding Techniques 161*
- *Equipment 161*
- *Putting It All Together: Determination of the Nutrition Prescription 161*
- *Complications 164*
- *Monitoring for Complications 165*

Parenteral Nutrition 167
- *Indications 167*
- *Venous Access 167*
- *Solutions 169*
- *Putting It All Together: Determination of the Nutrition Prescription 172*
- *Administration Techniques 172*

Patient Monitoring 172
Complications 172

Conclusion 176

PART 3

8

Fluid and Electrolyte Balance 181

Introduction 181

Normal Anatomy and Physiology of Fluids and Electrolytes 181

Total Body Water 181
Movement of Fluid between Blood and Interstitial Spaces 182
Movement between Extracellular Fluid and Intracellular Fluid 182
Total Body Water Balance 183
Fluid Intake 183
Fluid Output 184
Fluid Requirements 185

Body Solutes 185

Types of Solutes 185
Distribution of Electrolytes 186
Movement of Solutes 186
Electrolyte Requirements 186

Physiological Regulation of Fluid and Electrolytes 187

Thirst Mechanism 187
Renal Function 187
Hormonal Influence: Renin-Angiotension-Aldosterone System (RAAS) 187
Electrolyte Regulation 187
Sodium 187
Potassium 187
Calcium and Phosphorus 187

Disorders of Fluid Balance 188

Alterations in Volume 189
Hypovolemia 189
Hypervolemia 191
Alterations in Osmolality 192
Sodium Imbalances 192
Hyponatremia 192
Hypernatremia 194
Potassium Imbalances 194
Hypokalemia 194
Hyperkalemia 195
Calcium Imbalance 195
Hypocalcemia 196
Hypercalcemia 196
Phosphorus Imbalance 196
Hypophosphatemia 196
Hyperphosphatemia 197
Magnesium Imbalances 197
Hypomagnesemia 197
Hypermagnesemia 197

Conclusion 198

9

Acid-Base Balance 203

Introduction 203

Acids 203
Bases 204
Buffers 204
pH 204
Terms Describing pH 205

Regulation of Acid-Base Balance 205

Chemical Buffering 205
Other Chemical Buffer Systems 206
Respiratory Regulatory Control 206
Renal Regulatory Control 207
Other Renal Regulatory Controls 207
Effect of Acid and Base on Electrolyte Balance 208

Assessment of Acid-Base Balance 208

Acid-Base Disorders 208

Respiratory Acidosis 209
Etiology 210
Pathophysiology 210
Clinical Manifestations 210
Treatment 210
Respiratory Alkalosis 211
Etiology 211
Pathophysiology 211
Clinical Manifestations 211
Treatment 211
Metabolic Acidosis 211
Etiology 212
Pathophysiology 212
Clinical Manifestations 213
Treatment 213
Metabolic Alkalosis 213
Etiology 213
Pathophysiology 214
Clinical Manifestations 214
Treatment 214
Mixed Acid-Base Disorders 214
Assessment of Acid-Base Disorders 214

Conclusion 215

10

Cellular and Physiological Response to Injury 219

Introduction 219
Defining Disease and Pathophysiology 219
Disease Process: Epidemiology, Etiology, Pathogenesis, Clinical Manifestations, Outcome 219
Cellular Injury 220
Mechanisms of Cellular Injury 221
Cellular Response to Injury 221
Cellular Alterations in Size 222
Cellular Injury from Infection 223
Inflammation 226
Cellular Death 230
Conclusion 231

11

Nutrigenomics 237

Introduction 237
An Overview of the Structure and Function of Genetic Material 241
Deoxyribonucleic Acid (DNA) and Genome Structure 241
Translating the Message from DNA to Protein 242
The Genetic Code 242
Intervening Sequences 243
Transcription and Translation 243
Genetic Variation 245
Inheritance 245
Single Nucleotide Polymorphisms 247
Other Polymorphisms 248
Epigenetic Regulation 248
DNA Methylation 248
Histone Modification 249
The Epigenotype 249
Dietary Regulation and Measurement of Gene Expression 250
Dietary Components Influence Gene Expression 250
Measuring Gene Expression 251
Nutrigenomics in Disease 251
Cancer 251
From Single Gene Inherited Cancers to Gene-Nutrient Interactions 251
Variations in Xenobiotic Metabolism Influence Risk 251
MTHFR and ADH Polymorphisms Interact with Dietary Folate and Alcohol 252
Fruits and Vegetables 252
Obesity and Diabetes 253
Obesity 253
Developmental Origins of Adult Disease 254
Diabetes 254
Cardiovascular Disease 255
Individual Variation in Response to Environmental Influences 255
Dietary Modification is Effective in Monogenic Disease 256
Dietary Fats Interact with Various Genotypes to Influence Outcomes 256
Nutrigenomics and the Practice of Dietetics 256
Individual Testing in the Marketplace 257
Evolving Knowledge and Practice Requirements for Dietitians 257
Conclusion 257

12

Immunology 259

Introduction 259
Natural Resistance 260
Antigens: The Key to Recognizing Pathogens and Altered Cells 261
Antigens and Immunogens 261
Characteristics of an Antigen 262
Immune System Overview 262
Functions and Requirements of the Immune System 262
Divisions of the Immune System 262
Cells of the Immune System 263
Origin of Cells of the Immune System 263
Cells Derived from the Myeloid Stem Cell 264
Cells Derived from the Lymphoid Stem Cell 268
Organs of the Immune System 269
Central Lymphoid Organs 269
Peripheral (Secondary) Lymphoid Organs 269
Soluble Mediators 270
Complement 271
Cytokines 271
Antigen Recognition Molecules 272
Major Histocompatibility Complex 272
Antibody 273
B Cell Receptor 274
T Cell Receptor 274
Interaction between the T Cell Receptor and the MHC Antigens 274
Immune Response: Attacking Pathogens 274
Modes of Attack 275

Progression of the Immune Response: Putting the Parts Together 276
Immunity to Infection: Attacking Specific Types of Pathogens 278

Attacking Altered and Foreign Cells: Tumors and Transplants 278
Tumor Immunology 278
Transplantation Immunology 279

Immunization 281
Passive Immunization 281
Active Immunization 281
Types of Vaccines 282

Immunodeficiency 282
Malnutrition and Immunodeficiency 283
Inherited Immunodeficiencies 283
Acquired Immunodeficiencies 284

Tolerance 284
Auto Tolerance 284
Induced Tolerance 284
Central Tolerance 284
Peripheral Tolerance 285

Attack on Harmless Antigens: When the Immunological System Causes Harm 285
Hypersensitivity 285
Autoimmunity 290

Conclusion 293

13

Pharmacology 297

Introduction 297
Role of Nutrition Therapy in Pharmacotherapy 301

Drug Mechanisms 301

Administration of Drugs 302
Pharmacokinetics: Absorption of Drugs 303
Pharmacokinetics: Distribution of Drugs 303
Pharmacokinetics: Metabolism of Drugs 303
Pharmacokinetics: Excretion of Drugs 304
Alterations in Drug Pharmacokinetics 304
Altered GI Absorption 304
Altered Distribution 305
Altered Metabolism 305
Altered Urinary Excretion 305

How Do Food and Drugs Interact? 305
Effect of Nutrition on Drug Action 306
Effect of Nutrition on Drug Dissolution 306
Effect of Nutrition on Drug Absorption 306
Effect of Nutrition on Drug Metabolism 306
Effect of Nutrition on Drug Excretion 307
Nutritional Complications Secondary to Pharmacotherapy 307
Drug Consequence: Effect on Nutrient Ingestion 307
Drug Consequence: Effect on Nutrient Absorption 307
Drug Consequence: Effect on Nutrient Metabolism 309
Drug Consequence: Effect on Nutrient Excretion 309
At-Risk Populations 309
Drug-Nutrient Interactions in the Elderly 309
Drug-Nutrient Interactions in HIV and AIDS 309
Drug-Nutrient Interactions in Nutrition Support 310

Nutrition Therapy 311
Nutrition Implications 311
Nutrition Interventions 311

Conclusion 312

PART 4

14

Energy Balance and Body Weight 323

Introduction 323

Energy Balance 324
Energy Intake 324
Energy Expenditure 324
Resting Energy Expenditure 325
Thermic Effect of Food 325
Physical Activity Energy Expenditure 326
Estimating Energy Requirements 326
Equations 326
Indirect Calorimetry 329
Doubly Labeled Water 330
Direct Calorimetry 330

Regulation of Energy Balance 330
The Adipocyte and Adipose Tissue 331

Body Composition, Obesity, and Overweight 332
Body Fat Distribution 332

Epidemiology of Overweightand Obesity 335
Overweight and Obesity in the United States 335
Overweight and Obesity in Canada 337
Overweight and Obesityin Europe 338
Effects of Race, Ethnicity, Socioeconomic Status, and Age 339

Adverse Health Consequences of Overweight and Obesity 341

Etiology of Obesity 342
Medical Disorders and Medical Treatments 342
Genetics and Body Weight 343
Obesigenic Environment 345
Energy Expenditure 347

Treatment of Overweight and Obesity 348
Assessment 348
Management 349
Nutrition Therapy 350
Physical Activity 352
Behavior Therapy 352
Pharmacologic Treatment 352
Surgery 354
Eating Disorders 355
Etiology of Eating Disorders 357
Anorexia Nervosa 358
Health Complications of Anorexia Nervosa 358
Bulimia Nervosa 360
Health Complications of Bulimia Nervosa 361
Eating Disorders Not Otherwise Specified 361
Nutrition Therapy for Eating Disorders 362
Nutrition Therapy for Anorexia Nervosa 362
Nutrition Therapy for Bulimia Nervosa 363
Nutrition Therapy for Eating Disorders Not Otherwise Specified 364
Conclusion 364

15

Diseases of the Cardiovascular System 371

Introduction 371
Anatomy and Physiology of the Cardiovascular System 371
The Heart 371
Electrical Activity of the Heart 372
Cardiac Cycle 373
Cardiac Function 373
Regulation of Blood Pressure 374
Hypertension 376
Epidemiology 376
Etiology 376
Pathophysiology 377
Treatment 378
Nutrition Therapy 378
Nutrition Implications 378
Nutrition Interventions 378
Weight Loss 378
Sodium 380
Developing the Nutrition Therapy Prescription 383
Atherosclerosis 383
Definition 383
Epidemiology 385
Etiology 386
Family History 386
Age and Sex 386
Obesity 387
Dyslipidemia 387
Hypertension 388
Physical Inactivity 389
Atherogenic Diet 389
Diabetes Mellitus 389
Impaired Fasting Glucose and Metabolic Syndrome 390
Cigarette Smoke 391
Pathophysiology 391
Clinical Manifestations 392
Treatment 392
Nutrition Therapy 392
Nutrition Implications 392
Nutrition Interventions 394
Ischemic Heart Disease 401
Definition 401
Epidemiology 402
Etiology 403
Pathophysiology 403
Clinical Manifestations 406
Diagnosis 406
Treatment 406
Nutrition Therapy 407
Nutrition Implications 407
Nutrition Interventions 407
Peripheral Arterial Disease 408
Definition 408
Epidemiology 408
Pathophysiology 408
Clinical Manifestations and Diagnosis 409
Heart Failure 410
Definition 410
Epidemiology 410
Etiology 410
Pathophysiology 410
Clinical Manifestations 413
Treatment 414
Nutrition Therapy 414
Nutrition Implications 414
Nutrition Interventions 414
Conclusion 415

16

Diseases of the Upper Gastrointestinal Tract 421

Introduction 421
Normal Anatomy and Physiology of the Upper Gastrointestinal Tract 421
Motility, Secretion, Digestion and Absorption 422

Anatomy and Physiology of the Oral Cavity 423
Oral Cavity Motility 423
Oral Cavity Secretions 423
Normal Anatomy and Physiology of the Esophagus 423
Normal Anatomy and Physiology of the Stomach 424
Gastric Motility 424
Gastric Secretions 424
Gastric Digestion and Absorption 427
Pathophysiology of the Oral Cavity 428
Oral Disease 428
Dental Caries 428
Inflammatory Conditions of the Oral Cavity 429
Conditions Resulting in Altered Salivary Gland Function 429
Surgical Procedures for the Oral Cavity 430
Impaired Taste: Dysgeusia/Ageusia 430
Nutrition Therapy for Pathophysiology of the Oral Cavity 430
Nutritional Implications 430
Nutrition Intervention 431
Pathophysiology of the Esophagus 433
Gastroesophageal Reflux Disease (GERD) 433
Barrett's Esophagus – A Complication of GERD 435
Nutrition Therapy - Gastroesophageal Reflux Disease 436
Nutrition Implications 436
Nutrition Assessment 436
Nutrition Interventions 436
Dysphagia 436
Nutrition Therapy 437
Achalasia 441
Nutrition Therapy 442
Hiatal Hernia 442
Pathophysiology of the Stomach 442
Indigestion 442
Nausea and Vomiting 442
Nutrition Therapy 444
Gastritis 445
Peptic Ulcer Disease 445
Nutrition Therapy 447
Gastric Surgery 448
Vagotomy 448
Gastroduodenostomy (Billroth I); Gastrojejunostomy (Billroth II); Roux en Y Procedure 448
Nutrition Therapy for Gastric Surgery 448
Other Conditions of Gastric Pathophysiology 450
Stress Ulcers 450
Zollinger-Ellison Syndrome 451
Conclusion 451

17

Diseases of the Lower Gastrointestinal Tract 457

Introduction 457
Normal Anatomy and Physiology of the Lower Gastrointestinal Tract 457
The Small Intestine 457
Anatomy 457
Motility 458
Secretions 459
Digestion 460
Absorption 460
The Large Intestine 460
Anatomy 460
Motility 463
Secretions 463
Digestion and Absorption 463
Pathophysiology of the Lower Gastrointestinal Tract 466
Diarrhea 467
Definition 467
Epidemiology 467
Etiology 467
Clinical Manifestations 472
Diagnosis 472
Treatment 472
Nutrition Therapy 472
Constipation 474
Definition 474
Epidemiology 475
Etiology 475
Diagnosis 475
Treatment 475
Nutrition Therapy 476
Malabsorption 477
Definition 477
Etiology 477
Pathophysiology 477
Treatment 480
Nutrition Therapy 480
Celiac Disease 481
Definition 481
Epidemiology 481
Etiology 482
Pathophysiology 482
Clinical Manifestations 482
Diagnosis 482
Prognosis and Treatment 482
Nutrition Therapy 483

Irritable Bowel Syndrome 484
Definition 484
Epidemiology 485
Etiology 485
Pathophysiology 485
Clinical Manifestations 485
Treatment 485
Nutrition Therapy 486
Inflammatory Bowel Disease 488
Definition 488
Epidemiology 488
Etiology 489
Pathophysiology 490
Clinical Manifestations 490
Treatment 492
Nutrition Therapy 494
Diverticulosis/Diverticulitis 496
Definition 496
Epidemiology 496
Etiology 496
Pathophysiology 496
Clinical Manifestations 497
Treatment 497
Nutrition Therapy 497
Common Surgical Interventions for the Lower GI Tract 498
Ileostomy and Colostomy 498
Short Bowel Syndrome 499
Definition 499
Epidemiology 499
Etiology 499
Pathophysiology 499
Treatment 500
Nutrition Therapy 500
Bacterial Overgrowth 503
Definition 503
Pathophysiology 503
Clinical Manifestations 503
Treatment 503
Nutrition Therapy 503
Conclusion 503

18

Diseases of the Hepatobiliary: Liver, Gallbladder, Exocrine Pancreas 509

Introduction 509
Normal Anatomy and Physiology of the Liver 509
Anatomy 509
Functions of the Liver 512
Bile 512
Enterohepatic Circulation 512
Jaundice 513
Pertinent Laboratory Values and Procedures 513
Use of Tests 514
Pathophysiology of the Liver 514
Alcoholism and Malnutrition 514
Diagnosis and Epidemiology of Chronic Alcoholism 514
Alcohol Withdrawal Syndrome 516
Metabolism of Alcohol 516
Fatty Liver 517
Mechanisms of Malabsorption in the Alcoholic 518
Nutrition Implications of Alcoholism 519
Hepatitis 525
Definition and Epidemiology 525
Clinical Manifestations 525
Nutrition Therapy 525
Alcoholic Hepatitis 526
Treatment and Nutrition Therapy 526
Cirrhosis 526
Definition 526
Etiology 526
Clinical Manifestations 526
Complications 527
Liver Transplant 533
Nutrition Therapy 533
Cystic Fibrosis-Associated Liver Disease (CFALD) 533
Epidemiology 533
Etiology 533
Clinical Manifestations and Treatment 533
Nutrition Therapy 534
The Gallbladder 535
Normal Function of the Gallbladder 535
Cholelithiasis (Gallstones) 535
Epidemiology 535
Etiology 536
Complications 536
Roentgenography of the Biliary System 536
Treatment 537
Nutrition Therapy 537
The Pancreas 539
Normal Anatomy and Physiology of the Pancreas 539
Pancreatitis 539
Definition and Clinical Manifestations 539
Etiology and Pathogenesis 539
Nutrition Therapy 540
Conclusion 543

19

Diseases of the Endocrine System 549

Introduction 549

Normal Anatomy and Physiology of the Endocrine System 550

Classification of Hormones 550

Endocrine Function 553

Pituitary Gland 553

Thyroid Gland 553

Adrenal Glands 553

Endocrine Pancreas 554

Endocrine Control of Energy Metabolism 559

Endocrine Disorders 559

Thyroid Disorders 564

Hypothyroidism 564

Hyperthyroidism 566

Diabetes Mellitus 570

Definition 570

Management 570

Type 1 Diabetes Mellitus (T1DM) 570

Epidemiology 572

Etiology 572

Pathophysiology and Clinical Manifestations 575

Diagnosis 577

Laboratory Measurements 577

Treatment 580

Nutrition Therapy 587

Type 2 Diabetes Mellitus (T2DM) 591

Epidemiology 591

Etiology 591

Pathophysiology 591

Clinical Manifestations 593

Treatment 593

Nutrition Therapy 596

Gestational Diabetes Mellitus (GDM) 597

Definition 597

Epidemiology 597

Etiology 597

Pathophysiology 597

Clinical Manifestations 597

Diagnosis 598

Treatment 598

Nutrition Therapy 601

Hypoglycemia 601

Definition 601

Etiology 601

Pathophysiology 602

Clinical Manifestations 603

Treatment 603

Nutrition Therapy 603

Conclusion 604

20

Diseases of the Renal System 609

Introduction 609

The Kidneys 609

Normal Anatomy 609

Normal Physiology 611

Diagnostic Procedures 612

Nephrotic Syndrome 613

Definition 613

Epidemiology 614

Etiology 614

Clinical Manifestations 614

Treatment 615

Nutrition Therapy 615

Chronic Kidney Disease 615

Definition 616

Epidemiology 617

Etiology 617

Pathophysiology 617

Treatment 619

Nutrition Therapy 623

CKD Stages 1 and 2 623

CKD Stages 3 and 4 624

Evaluation/Outcome Measurement 625

CKD Stage 5 625

Nutrition Therapy for Comorbid Conditions and Complications 638

Cardiovascular Disease 638

Secondary Hyperparathyroidism (SHPT) 638

Anemia 637

Medicare Coverage for Medical Nutrition Therapy 639

Nutritional Requirements of the Post Transplant Patient 640

Acute Renal Failure 643

Definition 643

Epidemiology and Etiology 643

Pathophysiology 643

Clinical Manifestations 643

Treatment 644

Nutrition Therapy 644

Nephrolithiasis 645

Definition 645

Epidemiology 645

Treatment 645
Nutrition Therapy 646
Conclusion 646

21

Diseases of the Hematological System 651

Introduction 651
Blood Composition 652
Anatomy and Physiology of the Hematological Systems 652
The Cells of the Hematological Systems 652
The Development of the Hematological Cells 653
Hemoglobin 655
Homeostatic Control of the Hematological Systems 657
Blood Clotting 657
Factors Affecting Hemostasis 658
Summary 658
Nutritional Anemias 658
Microcytic Anemias: Iron Deficiency and Functional Anemia 659
Definition 659
Epidemiology 661
Etiology 661
Pathophysiology 662
Clinical Manifestations 666
Treatment 667
Nutrition Therapy 668
Megaloblastic Anemias 670
Definition 670
Epidemiology 670
Etiology 670
Pathophysiology 672
Clinical Manifestations 672
Treatment 673
Nutrition Therapy 673
Hemochromatosis 674
Definition 674
Epidemiology 674
Etiology 674
Pathophysiology 675
Clinical Manifestations 675
Treatment 675
Nutrition Therapy 675
Nutrition Implications 675
Nutrition Interventions 675
Hemoglobinopathies: Non-Nutritional Anemias 676
Sickle Cell Anemia 676
Nutrition Therapy 676
Thalassemia 676
Nutrition Therapy 676
Polycythemia 676
Nutrition Therapy 676
Hemolytic Anemia 678
Nutrition Therapy 678
Anemia of Prematurity 678
Nutrition Therapy 678
Aplastic Anemia 678
Nutrition Therapy 678
Other Rare Anemias 678
Nutritional Implications of Non-Nutritional Anemias 678
Nutritional Interventions for Non-Nutritional Anemias 678
Clotting and Bleeding Disorders 679
Hemophilia 679
Nutrition Therapy 679
Hemorrhagic Disease of the Newborn 679
Nutrition Therapy 679
Thrombosis 679
Nutrition Therapy 679
Diseases Involving WBC Types and Bone Marrow Failure Requiring Bone Marrow Transplant (BMT) 679
Definition 679
Epidemiology 681
Etiology 681
Pathophysiology 681
Clinical Manifestations 681
Treatment 681
Nutrition Therapy 682
Nutrition Implications 682
Nutrition Interventions 682
Conclusion 683

22

Diseases of the Neurological System 687

Introduction 687
Normal Anatomy and Physiology of the Nervous System 687
Central Nervous System (CNS) 689
Neurological Disorders 690
Epilepsy and Seizure Disorders 691

Stroke and Aneurysm 694

Progressive Neurological Disorders 697

Parkinson's Disease 698

Definition 698

Epidemiology 698

Etiology 699

Pathophysiology 699

Clinical Manifestations 699

Diagnosis 699

Treatment 699

Nutrition Therapy 700

Amyotrophic Lateral Sclerosis 701

Definition 701

Epidemiology 701

Etiology 701

Pathophysiology 701

Clinical Manifestations 702

Treatment 702

Nutrition Therapy 702

Guillain-Barré 702

Definition 702

Epidemiology 702

Etiology 702

Pathophysiology 702

Clinical Manifestations 702

Treatment 702

Nutrition Therapy 702

Myasthenia Gravis 703

Definition 703

Epidemiology 703

Etiology 703

Pathophysiology 703

Clinical Manifestations 703

Treatment 703

Nutrition Therapy 703

Multiple Sclerosis 703

Definition 703

Epidemiology 703

Etiology 703

Pathophysiology 704

Clinical Manifestations 704

Treatment 704

Nutrition Therapy 704

Alzheimer's Disease and Other Forms of Dementia 705

Definition 705

Epidemiology 705

Etiology 705

Pathophysiology 706

Clinical Manifestations 707

Treatment 707

Nutrition Therapy 707

Neurotrauma and Spinal Cord Injury 707

Traumatic Brain Injury 707

Definition 709

Epidemiology 709

Etiology 709

Pathophysiology 709

Clinical Manifestations 709

Treatment 709

Nutrition Therapy 709

Spinal Cord Injury 710

Definition 710

Epidemiology 710

Etiology 710

Pathophysiology 710

Treatment 710

Nutrition Therapy 711

Conclusion 711

23

Diseases of the Respiratory System 715

Introduction 715

Normal Anatomy and Physiology of the Respiratory System 715

Measures of Pulmonary Function 717

Nutrition and Pulmonary Health 719

Asthma 720

Definition 720

Epidemiology 720

Etiology 720

Pathophysiology 720

Clinical Manifestations 721

Treatment 721

Nutrition Therapy 721

Nutrition Implications 721

Nutrition Interventions 722

Brochopulmonary Dysplasia 722

Definition 722

Etiology 722

Treatment 723

Nutrition Therapy 723

Nutrition Implications 723

Nutrition Interventions 723

Chronic Obstructive Pulmonary Disease 725

Definition 725

Epidemiology 726

Etiology 726

Pathophysiology: Chronic Bronchitis 726

Clinical Manifestations: Chronic Bronchitis 726
Pathophysiology: Emphysema 727
Clinical Manifestations: Emphysema 727
Treatment 727
Nutrition Therapy 727
Energy and Macronutrient Needs 729
Vitamins and Minerals 729
Feeding Strategies 730

Cystic Fibrosis 731
Definition 731
Epidemiology 731
Etiology 732
Pathophysiology 732
Nutrition Therapy 732
Nutrition Implications 732
Nutrition Interventions 733

Pneumonia 737
Definition 737
Epidemiology 737
Etiology 737
Nutrition Implications 737
Aspiration Pneumonia 737
Patients with Tracheostomies 739

Respiratory Failure 740
Definition 740
Nutrition Therapy 741
Nutrition Implications 741
Nutrition Interventions 741

Transplantation 743
Definition and Epidemiology 743
Pathophysiology 743
Nutrition Therapy 744

Upper Respiratory Infection 744
Definition and Epidemiology 744
Pathophysiology 744
Clinical Manifestations 745
Nutrition Implications 745

Conclusion 745

24

Neoplastic Disease 751

Introduction 751

Definition 751

Epidemiology 752

Etiology 752
Cancer Screening and Prevention 754

Pathophysiology 754
Diagnosis 756

Clinical Manifestations 758

Treatment 758
Surgery 759
Cancer Diagnoses Requiring Surgery for Treatment 759
Chemotherapy 762
Radiation 763
Other Therapies 765

Nutrition Therapy 766
Nutrition Implications 766
Cachexia 766
Abnormalities in Carbohydrate, Protein and Lipid Metabolism 767
Nutritional Implications of Cancer Treatment 767
Nutrition Interventions 768
Nutrition Assessment 768
Determining Nutrient Requirements 770
Nausea and Vomiting 771
Early Satiety 773
Mucositis 774
Diarrhea 775
Dysguesia 776
Xerostomia 776
Anorexia 777
Nutrition Support 778
Home Nutrition Support 779

Conclusion 779

25

Metabolic Stress 785

Introduction 785

Physiological Response to Starvation 785

Physiological Response to Stress 786
Definition 786
Epidemiology 787
Etiology 787
Clinical Manifestations 787
Pathophysiology 788
Nutrition Therapy 789
Nutrition Assessment 789
Nutrition Interventions 790

Burns 794
Definition 794
Epidemiology 794
Etiology 794
Clinical Manifestations 794

Pathophysiology 795
Treatment 795
Nutrition Therapy 795
Nutrition Implications 795
Nutrition Assessment 796
Nutrition Interventions 796

Surgery 797
Definition 797
Epidemiology 797
Etiology 797
Clinical Manifestations 797
Nutrition Therapy 799
Nutrition Implications 799
Nutrition Interventions 799

Sepsis, Systemic Inflammatory Response Syndrome (SIRS), and Multi-organ Distress Syndrome (MODS) 799
Definition 799
Epidemiology 799
Etiology 799
Clinical Manifestations 801
Pathophysiology 801
Treatment 801
Nutrition Therapy 801
Nutrition Implications 801

Conclusion 801

26

HIV and AIDS 805

Introduction and Epidemiology 805

Normal Anatomy and Physiology of the Immune System 806

Etiology 807

Pathophysiology 807

Diagnosis 809

Clinical Manifestations 812

Treatment 814
Anti-HIV Therapies 814
Prevention and Treatment of Opportunistic Disease 816

Nutrition Therapy 816
Nutrition Implications 817
Nutrition Assessment 818
Physical Assessment 818
Biochemical Assessment 822
Medical History Assessment 828
Dietary Evaluation 828
Nutrition Interventions 828
Macronutrient Therapy 830
Micronutrient Therapy 831
Non-Nutrient Therapy for Nutritional Status Maintenance 834
Nutrition Care Plan 836
Implementation 837
Evaluation/Outcome Measurement 837

Conclusion 838

27

Diseases of the Musculoskeletal System 843

Introduction 843

Normal Anatomy and Physiology of the Skeletal System 843
Cartilage 844
Bone 844
The Cells of Osseous Tissue 845
Skeletal Growth and Development 845
Cortical and Trabecular Bone 846
Hormonal Control of Bone Metabolism 848

Osteoporosis 849
Diagnosis 851
Epidemiology 852
Health and Economic Impact of Fractures 853
Etiology 854
Prevention 855
Calcium 855
Vitamin D 858
Physical Activity 859
Cigarette Smoking 859
Alcohol 860
Other Nutrients and Food Components 860
Medical Management 861
Pharmacologic Prevention and Treatment 861

Paget Disease 862

Rickets and Osteomalacia 863
Rickets 863
Epidemiology, Etiology, and Clinical Manifestations 863
Prevention 863
Treatment 864
Osteomalacia 864
Etiology and Clinical Manifestations 864
Treatment 864

Arthritic Conditions 865
Definition and Epidemiology 865
Osteoarthritis 865
Epidemiology, Etiology, and Clinical Manifestations 865
Treatment 866

Rheumatoid Arthritis 867
Epidemiology, Etiology, and Clinical Manifestations 867
Treatment 868
Diet and Rheumatoid Arthritis 868
Gout 870
Epidemiology and Etiology 870
Clinical Manifestations 870
Treatment 870
Fibromyalgia 870
Definition and Epidemiology 870
Etiology 871
Diagnosis and Clinical Manifestations 871
Treatment 871
Diet and Fibromyalgia 873
Conclusion 873

28

Metabolic Disorders 881

Introduction and Definition 881
History 881
Epidemiology and Inheritance 882
Pathophysiology of Impaired Metabolism 882
Diagnosis / Newborn Screening 883
Clinical Manifestations 883
Treatment 885
Acute Therapy 885
Chronic Therapy 885
Restriction of Precursors 885
Replacement of the End Products 885
Providing Alternate Substrates for Metabolism 885
Use of Scavenger Drugs to Remove Toxic By-Products 885
Supplementation of Vitamins or Other Cofactors 885
Amino Acid Disorders 886
Epidemiology, Etiology and Clinical Manifestations 886
Phenylketonuria 886
Nutrition Interventions 887
Nutritional Concerns 890
Adjunct Therapies 892
Urea Cycle Disorders 892
Epidemiology, Etiology and Clinical Manifestations 892
Acute Treatment 894
Nutrition Interventions 894
Nutritional Concerns 895
Adjunct Therapies 895
Mitochondrial Disorders 896
Etiology and Clinical Manifestations 896
Nutrition Interventions 897
Adjunct Therapies 897
Disorders Related to Vitamin Metabolism 897
Etiology and Clinical Manifestations 897
Nutrition Interventions 897
Nutritional Concerns 898
Disorders of Carbohydrate Metabolism 898
Galactosemia 899
Epidemiology, Etiology and Clinical Manifestations 899
Nutrition Interventions 900
Nutritional Concerns 900
Adjunct Therapies 901
Hereditary Fructose Intolerance 901
Etiology and Clinical Manifestations 901
Nutrition Interventions 902
Nutritional Concerns 902
Adjunct Therapies 902
Glycogen Storage Diseases 902
Epidemiology, Etiology and Clinical Manifestations 902
Nutrition Interventions 904
Nutritional Concerns 905
Adjunct Therapies 905
Disorders of Fat Metabolism 905
Etiology and Clinical Manifestations 905
Nutrition Interventions 907
Nutritional Concerns 908
Adjunct Therapies 908
Conclusion 909

References R1
Appendix A—General Information A-1
Appendix B—Nutrition Assessment B-1
Appendix C—Parental and Enteral Formulas C-1
Appendix D—Nutrient Tables D-1
Appendix E—Menu Planning E--1
Appendix F—Contemporary and Alternative Medicine Tables F-1
Appendix G—Answers to Case Study Questions G-1
Appendix H—Answers to End-of-Chapter Questions H-1
Glossary GL-1
Index I-1

PREFACE

Students often tell us that it is during their nutrition therapy course that much of their coursework—sometimes previously disjointed—comes together for them. That is in part due to the nature of nutrition science. The science of nutrition incorporates principles of biological, chemical, psychological, and social sciences and it may be difficult for students to grasp how these diverse fields connect to become the integrated pieces of the clinical puzzle. The American Dietetic Association defines medical nutrition therapy as: "an essential component of comprehensive health care services. Individuals with a variety of conditions and illnesses can improve their health and quality of life by receiving medical nutrition therapy. MNT can improve consumers' health and well-being, and increase productivity and satisfaction levels through decreased doctor visits, hospitalizations and reduced prescription drug use" (ADA 2006). Nutrition therapy forces the student and practitioner to rely on his or her academic preparation and continuing education in order to face the 21st century challenges of nutrition and medical care.

The authors of this text are educators, clinicians, and researchers. Our purpose in creating this text is to assist students to navigate the realm of clinical nutrition. Most of us look to primary reference texts as the cornerstone of our practice. Many names come to mind—Zeman's *Clinical Nutrition*; Harrison's *Book of Internal Medicine*; or *The Merck Manual of Diagnosis and Therapy*. The primary goal of this text is to not only provide the reference material needed to understand clinical nutrition practice, but provide it in such a way that the learning environment will support the student's development of critical thinking, clinical reasoning, and decision making skills.

What makes this text different from other clinical nutrition texts? The clinical environment evolves as a result of the impacting forces of research, health care funding, evidence-based nutrition practice, and development of the nutrition care process, standardized language, and standardized nutrition diagnoses. To meet the demands of these evolving forces, this text includes an overview of health care systems, the role of the registered dietitian as a member of the health care team, nutrition diagnoses, guidelines for documentation and other professional writings, complementary and alternative medicine, and nutrigenomics. Incorporation of the newest framework for nutrition therapy will provide students and practitioners with a background in the most current research on the integration of evidence-based practice within the context of the nutrition care process.

Second, the text provides the foundation chapters for nutrition therapy practice: a comprehensive review of the physiology required to integrate nutrition therapy as a component of medical care. Foundation chapters cover physiological response to injury, fluid and electrolyte balance, pharmacology, genetics, and the immune system. These chapters focus specifically on the application of each of these topics to clinical nutrition practice.

The text is organized using a systems approach consistent with other medical texts. Each nutrition therapy chapter discusses the normal structure and function of a body system, explains how the disease process interrupts normal functioning, and then describes appropriate medical and nutrition interventions. In addition, each nutrition therapy chapter includes a case study, an overview of the nutrition care process for one of the featured disorders, a table describing drug-nutrient interactions specific to each disorder, and an interview with a current clinical practitioner. This approach allows any health care professional to benefit from this text.

Though every effort has been made to address the most recent research and the most common clinical and medical practices, this text has the same limitation that any medical textbook will have: new diagnoses, new drugs, new treatments, and a new understanding of the relationship between nutrition and disease which will inevitably continue to be cultivated after its publication. Thus, this book strives to educate students about not only the facts and theories that comprise current medical knowledge, but also the process of skill development that empowers students to grow in expertise within their field. As practitioners of the future utilize the nutrition care process, it will be refined even as their knowledge of disease and its treatment evolves.

As clinical practitioners and current dietetic educators, we have experienced the need not only for a different approach to the clinical nutrition text, but also a reference for clinical practitioners. We believe that this text fills both voids.

ACKNOWLEDGMENTS

We have had significant assistance in the development and writing of this text. We would like to first thank Sandra Witte, PhD, RD and Peter Marshall for their insight and wisdom in the early stages of this book.

We are grateful for our contributing authors for their expertise, persistence and dedication in the development for each of their chapters:

Ethan Bergman, PhD, RD
Nancy Buergel, PhD, RD
Deborah Cohen, MMSC, RD
Cade Fields-Gardner, MS, RD
Christina Frazier, PhD
Jordi Goldstein-Fuchs DSc, RD
Roschelle Heuberger, PhD, RD
Robert Lee,PhD, RD
Paula Hansen, PhD, RD
Elaina Jurecki, MS, RD
Maria Karalis, MBA, RD
Pamela Kittler, MS
Karen Lacey, MS, RD
Millie Matfeldt-Beman, PhD, RD
Jessie M. Pavlinac, MS, RD, CSR, LD
Thomas J. Pujol, EdD, FACSM
Annalynn Skipper, PhD, RD, FADA
Joshua Tucker, MS
Joyce Wong, MS, RD

We believe the practitioner interviews will assist students in understanding the role of the clinical dietitian within the health care practice and serve as significant role models. We would like to thank the following individuals for their gracious consent for interviews within this text:

Mary Ellen Beindorff, RD, LD
Shelly Case, BSc, RD
Jordan Davidson, RD, LD, CNSD
Margaret Davis, MBA, RD
Nancy Duhaime, MS, RD, LD
Kathleen Huntington, MS, RD, LD
Marianne Hutton, RD, CDE
Kelly Leonard, MS, RD
Sandra Luthringer, RD, LDN
Eileen MacKusick, MS, RD
La Paula Sakai, MS, RD, CNSD
Kathryn Sikorski, RD, CDE
Valerie Simler, MS, RD, CDE
Linda White, RD, CDE
Jill Weisenberger, MS, RD, CDE

We would like to thank the many reviewers whose insights and suggestions proved invaluable during the writing process:

Judith Ashley, PhD, RSPH, RD
University of Nevada Reno

William J. Banz, PhD, RD
Southern Illinois University

Charlotte Baumgart PhD, RD, CDN
D'Youville College

Mallory Boylan, PhD
Texas Tech University

Lauren Bronich-Hall, MS, RD, LDN
Towson University

Jennifer L. Bueche, PhD, RD, CDN
SUNY College at Oneonta

Jerrilynn D. Burrowes, PhD, RD
C.W. Post Campus of Long Island University

Nancy H. Burzminski, EdD, RD, LD
Kent State University

Jayne L. Byrne, MS, RD, LD
The College of St. Benedict/St. John's University

Christina Campbell, PhD, RD, LN
Montana State University

Dina Chapman, MS, RD, LD, CDE
Barnes-Jewish College of Nursing and Allied Health

Cathy Cunningham, PhD, RD
Tennessee Technological University

Wendy Cunningham PhD, RD, ETT
California State University Sacramento

Julie Davis, MS, RD
Benedictine University

Brenda M. Davy, PhD, RD
Virginia Tech University

Mary L. Dundas, PhD, RD
Idaho State University

Miriam Edlefsen, PhD, RD
Washington State University

Kelly Eiden, MS, RD, LD
Barnes-Jewish College of Nursing and Allied Health

Stephanie L. England, MS, RD, LDN
University of Tennessee at Chattanooga

Jamie Erskine, PhD, RD
University of Northern Colorado

Erin E. Francfort, MHE, RD, LD
Idaho State University

Susan Fredstrom, PhD, RD, CNSD
Minnesota State University, Mankato

Teresa Fung, ScD, RD
Simmons College

Mary Kathryn Gould MS, RD, LD
Marshall University

Virginia B. Gray, PhD, RD
Mississippi State University

Linda D. Griffith, PhD, RD, CNSD
Huntsville, AL (unaffiliated)

Janet K. Grommet, PhD, RD
Brooklyn College, City University of New York

Elizabeth J. Guthrie, MS, RD, LD
The Ohio State University

Theresa L. Han-Markey, MS, RD
University of Michigan

Susan Edgar Helm, PhD, RD
Pepperdine University

Gina Jarman Hill, PhD, RD
Texas Christian University

Tawni Holmes PhD, RD, LD
University of Central Oklahoma

Debra Geary Hook, MPH, RD
California State University, San Bernardino

Norman Hord, PhD, MPH, RD
Michigan State University

Paula Inserra, PhD, RD
Virginia State University

Kendra K. Kattelmann, PhD, RD, LN
South Dakota State University

Jay Keller, MS, RD
Idaho State University

Danita S. Kelley, PhD, RD
Western Kentucky University

Anne Kendall, PhD, RD
University of Florida

Karla Kennedy-Hagan, PhD, RD, LDN
Eastern Illinois University

Jacqueline D. Lee, PhD, RD
California State University, Long Beach

Anne B. Marietta, PhD, RD, LD
Southeast Missouri State University

Allison Marshall, MS, RD, CDN
Hunter College, City University of New York

Sharon L. McWhinney, PhD, RD, LD
Prairie View A & M University

Mark S. Meskin, PhD
California State Polytechnic University, Pomona

Donna H. Mueller, PhD, RD, FADA.
Drexel University

Sharon M. Nickols-Richardson, PhD, RD
Virginia Polytechnic Institute and State University

Carol E. O'Neil, PhD, RD
Louisiana State University

Susan Polasek, MA, RD, LD
University of Texas at Austin

Tonia Reinhard, MS, RD
Wayne State University

Tania Rivera, MS, RD, LD/N
Florida International University

Nina L. Roofe, MS, RD, LD, CLC
University of Central Arkansas

Andrew Rorschach, PhD, RD
University of Houston

Mary Sand, MS, RD
Iowa State University

Kelly A. Tappenden, PhD, RD
University of Illinois at Urbana-Champaign

Martha L. Taylor, PhD, RD
University of North Carolina Greensboro

Julie Poh Thurlow, DrPH, RD
University of Wisconsin, Madison

Dr. Wilfred H. Turnbull
Life University (Marietta, GA)

Jane B. Uzcategui, MS, RD, CNSD
California State University Los Angeles

Mardell A. Wilson, EdD, RD, LD
Illinois State University

Joy Winzerling, PhD, RD
University of Arizona

Gloria Young, EdD, RD
Virginia State University

Linda O. Young, MS, RD, LMNT
University of Nebraska Lincoln

Finally, we acknowledge and appreciate the significant editorial assistance throughout the development and writing of this book. This book could not have been accomplished without Elesha Feldman and Elizabeth Howe. Thank you.

NUTRITION THERAPY

AND

PATHOPHYSIOLOGY

1

Health Care Systems and Reimbursement

Marcia Nelms, Ph.D., R.D.

Southeast Missouri State University

CHAPTER OUTLINE

Introduction

Health Care Facilities in the United States

Preventive and Primary Health Care Services • Secondary and Tertiary Care • Restorative Care • Long-Term Care

Financing the Health Care Industry

Private Insurance • Public Insurance • The Uninsured • Demographic Trends and Health Care • The Need for Health Care Reform • Health Care Reform in the United States

Introduction

Every day, people throughout the world rely on medical services for both routine and emergency health care. The accessibility, availability, and quality of those services vary widely. For example, comparing the experiences of patients admitted for a simple nonemergency surgical procedure in different areas of the world demonstrates these significant differences. Three patients—Mr. B, Mrs. C, and Mr. A—were diagnosed with **cholecystitis** and advised by their physicians to undergo an elective **cholecystectomy.**

Mr. B is a 55-year-old man living in St. Louis, Missouri. Most of the cost of his surgery will be covered by private insurance. This private insurance is purchased through his retirement plan and is managed through a preferred provider organization (PPO). Eighty percent of the cost of this surgery will be covered by his insurance as long as he uses a surgeon and hospital that are members of his PPO. Mr. B can utilize any hospital and surgeon he chooses, but if those providers are outside the PPO network he will have to pay significantly more for the surgery.

Mrs. C is a 60-year-old female who lives in Toronto, Canada. Her primary physician has referred her to the National Health Service to schedule her surgery. Mrs. C will have no expense for this surgery because her health care is provided through Canada's national health plan. Canada's

cholecystitis—an inflammation of the gallbladder, usually due to a gallstone

cholecystectomy—the procedure for removing the gallbladder surgically

Parts of this chapter are based on *Community Nutrition in Action: An Entrepreneurial Approach,* 4E, by Marie A. Boyle and David H. Holben.

BOX 1.1 HISTORY OF HEALTH CARE IN THE UNITED STATES

1900s

- Surgery is now common, especially for removing tumors, infected tonsils, and appendectomies.
- Doctors are not expected to provide free services to hospital patients.
- Railroads are leading industry to develop employee health programs.

1910s

- American hospitals are now modern scientific institutions, using antiseptics and methods to relieve pain.
- Reformers argue for health insurance but opposition comes from physicians.
- America enters World War I.

1920s

- Growing cultural influence of the medical profession—physicians' incomes are higher and prestige is established.
- General Motors signs a contract with Metropolitan Life to insure 180,000 workers.
- Penicillin is discovered, but 20 years will pass before it is used to combat infection and disease.

1930s

- The Depression changes priorities, with greater emphasis on unemployment insurance and "old age" benefits.
- The Social Security Act is passed, omitting health insurance.
- Against the advice of insurance professionals, Blue Cross begins offering private coverage for hospital care in dozens of states.

1940s

- Penicillin comes into use.
- President Truman offers a national health program plan, proposing a single system that would include all of American society. Truman's plan is denounced by the American Medical Association (AMA) and is called a Communist plot by a House subcommittee.

1950s

- At the start of the decade, national health care expenditures are 4.5% of the Gross National Product.
- Many legislative proposals are made for different approaches to hospital insurance, but none succeed.
- Many more medications are available now to treat a range of diseases, including infections, glaucoma, and arthritis, and new vaccines become available that prevent dreaded childhood diseases, including polio. The first successful organ transplant is performed.

1960s

- Over 700 insurance companies selling health insurance.
- In the 1950s, the price of hospital care doubled. Now, in the early 1960s, those outside the workplace, especially the elderly, have difficulty affording insurance.
- President Lyndon Johnson signs Medicare and Medicaid into law.

health care system is financed through public funds and is delivered through private hospitals and physicians. Their services are reimbursed through the national health care program (Anonymous August 7, 2002a., August 7, 2002b).

Finally, consider Mr. A, a 71-year-old retired businessman living in Tokyo, Japan. His recommended surgery will be provided through the Japanese health care system. All citizens of Japan must join a health care system either through their employer or through the national program. Individuals pay approximately 20% of their health care costs, while all other expenses are paid through the national insurance program (Anonymous August 7, 2003c).

Even this brief comparison reveals that the United States health care system is unique. Box 1.1 describes the history of health care in the United States (U.S.). The first factor that differentiates the U.S. system from that of other countries is reliance on the private sector not only to provide health care but also to primarily fund health care services (Angell 1999). In the U.S., the majority of health care services are purchased through private employers. Employers pay more than 80% of health insurance premiums (Iglehart 1999). Of course, employees ultimately pay the cost through a reduction in salary and their own monetary contributions (insurance premiums). The other providers of health care coverage are government-based programs, such as Medicare and Medicaid, which are only available to certain portions of the population. Participation in these government-based programs is dependent upon age and/or income.

While the U.S. has been extremely successful in designing and developing one of the most technologically advanced health care systems in the world, much less effort has been focused on assuring consistent access to care for all individuals. Access and cost of health care in the U.S. continue to be important factors that differentiate the U.S. from other countries. The cost of health care in the U.S. was estimated to be $1.7 trillion in 2004 (Lundberg 2005), one-seventh of the total U.S. economy and larger than the gross national products of most countries of the world (Lundberg 2005). This cost exceeds that of any other country when measured either as a percentage of the gross domestic product (GDP)

1970s

- President Richard Nixon renames prepaid group health care plans as health maintenance organizations (HMOs), with legislation that provides federal endorsement, certification, and assistance.
- The number of women entering the medical profession rises dramatically. In 1970, 9% of medical students are women; by the end of the decade, the proportion exceeds 25%.
- Healthcare costs are escalating rapidly, partially due to unexpectedly high Medicare expenditures, rapid inflation in the economy, expansion of hospital expenses and profits, and changes in medical care, including greater use of technology, medications, and conservative approaches to treatment. American medicine is now seen as in crisis.

1980s

- Corporations begin to integrate hospital systems.
- Under President Reagan, Medicare shifts to payment by diagnosis (DRG) instead of by treatment. Private plans quickly follow this model.
- "Capitation" payments to doctors become more common.

1990s

- Health care costs rise at double the rate of inflation.
- Expansion of managed care helps to moderate increases in health care costs.
- Federal health care reform legislation fails again to pass in the U.S. Congress.
- By the end of the decade there are 44 million Americans, 16% of the nation, with no health insurance at all.
- By June 1990, 139,765 people in the U.S. have HIV/AIDS, with a 60% mortality rate.
- Health care costs are on the rise again.
- Medicare is viewed by some as unsustainable under the present structure and must be "rescued." Changing demographics of the workplace lead many to believe the employer-based system of insurance can't last.

2000+

- The Family and Medical Leave Act allows workers to take up to 12 weeks of unpaid leave to care for seriously ill family members, newborn or adoptive children, or their own serious health problems without fear of losing their jobs.
- The five-year, $24 billion State Children's Health Insurance Program (SCHIP) provides health care coverage for up to 5 million children.
- Kennedy-Kassebaum Health Insurance Portability and Accountability Act helps individuals keep health insurance when they change jobs, guarantees renewability of coverage, and ensures access to health insurance for small businesses.
- In 2003, the Human Genome Project identified 20,000–25,000 genes in human DNA and determined the sequences of 3 billion chemical base pairs. More chromosomes continue to be identified.

Reprinted from: Public Broadcasting Service [dated 2000; cited 2005 October 25]. Available from: www.pbs.org/healthcarecrisis/history.htm.

or per capita. Despite these expenditures, over 45 million Americans are without health insurance. This means approximately 16% of the U.S. population has no health insurance and many more have very limited coverage (Rhoades, Brown, and Vistnes 1998).

The US Census 2004 recently assessed health systems throughout the world using the following as benchmarks: provision of good health measured in life expectancy, responsiveness to expectations of the population, and fairness of individuals' financial contribution toward their health care. This report indicated Japan, Australia, France, Sweden, and Spain had the longest life expectancy. U.S. life expectancy at 70 years was ranked as twenty-fourth throughout the world. Next, WHO judged responsiveness in health care by a nation's respect for the dignity of individuals, the confidentiality of health records, prompt attention in emergencies, and choice of provider. The U.S., Switzerland, Luxembourg, Denmark, and Germany were ranked highest for these qualities. Financial fairness was measured by equal distribution of health cost faced by each household. Top-ranked countries providing financial fairness were Colombia, Luxembourg, Belgium, Djibouti, and Denmark; the U.S. ranked fifty-fourth (Anonymous 2005).

Attempting to reduce costs and to improve the quality of care is a daunting challenge for any health care program. It is the staggering cost and fragmented access to medical care that challenge the future of health care in the U.S., where certain trends have emerged as a result of these issues. Fewer patients are hospitalized now than 20 years ago, but when they are hospitalized, these patients require a much higher level of care. Attempting to decrease the length of hospital stays has increased the need for more home care and skilled nursing facilities. As the U.S. population continues to age, there will be an increased need for long-term care: approximately 35 million people over the age of 65 were counted in the year 2000. Finally, the U.S. health care system has historically been an acute care system—a traditional model that primarily addresses current health problems rather than focusing on care that could prevent health problems. It is essential to address preventive care in order to begin to

TABLE 1.1

Levels of Health Care

Level of Health Care	Focus	Example(s)
Preventive Health Care	To prevent illness and chronic disease	Public health programs providing health education to reduce risk of transmission of HIV
Primary Health Care	Treatment of acute illness and general health maintenance provided by a physician or other health care team member (nurse, physician's assistant, or nurse practitioner)	Physician's office visit for treatment of ear infection, strep throat, or for yearly physical examination
Secondary Health Care	Medical care with a specialist upon referral from the patient's primary health care provider	Appointment with endocrinologist for diagnosis and treatment of diabetes mellitus; admission to hospital for surgical procedure
Tertiary Health Care	Medical care from specialists who have access to highly technical services	Treatment within a bone marrow transplant unit
Restorative Care	Rehabilitation from acute and chronic illness	Transfer to rehabilitation facility for comprehensive care after a cerebrovascular accident (stroke)
Long-Term Care	Provision of ongoing care for those individuals who are dependent for all levels of health care	Care for patient with Alzheimer's disease

decrease the costs of the long-term care associated with chronic illnesses (Hampl, Anderson, and Mullis 2002).

Where do nutrition services fit within our current health care picture? First, nutrition is one of the cornerstones of preventive health care. The focus of the U.S. health care system has begun to shift toward preventive health care, and as this trend continues it will be crucial that nutrition remain at the forefront of this effort. Secondly, nutrition therapy remains an essential component of medical treatments and research indicates its importance will continue to be recognized. The provision of nutrition therapy is affected by health care financing. The registered dietitian (RD) is the credentialed professional who provides nutrition therapy. As providers of nutrition therapy, dietitians must understand how health care is organized and financed in order to effectively participate in the U.S. health care system.

Health Care Facilities in the United States

The provision of medical care, including nutrition therapy, can occur in many different environments. The current health care system in the U.S. provides six levels of health care (see Table 1.1). Levels of care describe the scope of services and settings where health care is offered to clients as well as the type of provider for those services. Each level of care creates different requirements and opportunities for the dietitian.

Preventive and Primary Health Care Services

Preventive and primary health care focus on prevention of acute or chronic illness and general health maintenance, respectively. Examples of preventive services include immunizations, screening tests such as mammograms and **colonoscopy,** and even annual physical examinations that assist in early identification of disease. Preventive health care can also include education about the ways lifestyle choices affect health and influence the development or course of disease (Stone et al. 2002). Nutrition is a crucial element of preventive health care because it is a significant component of disease prevention.

Preventive services can be offered at several types of facilities. These include city, county, and state health departments and clinics; schools; physician's offices; occupational health services; professional organizations such as The American Dietetic Association (ADA); nonprofit or volunteer agencies such as the American Heart Association or the American Cancer Society; and many other community-based organizations. Dietitians working in preventive health provide nutrition expertise to program design, policy development, and evaluation for all levels of health education and promotion.

colonoscopy—a procedure for evaluating the lining of the colon using a long, flexible tubular video probe that is inserted into the rectum

Secondary and Tertiary Care

Secondary and tertiary levels of health care involve the diagnosis and treatment of illness and encompass all areas of **acute care**. Traditionally, secondary care and tertiary care are provided in hospitals. But these services can include treatment provided in physicians' offices, treatment in outpatient facilities such as same-day surgical units, urgent care, and emergency room care. Tertiary care typically has been designated as medical care requiring expert, technical intervention that is provided in highly specialized facilities such as intensive care units, burn units, and bone marrow transplant facilities.

There are several different classifications of hospitals. These include (see Box 1.2):

- Public nonprofit
- Private nonprofit
- Private profit
- Veterans's and military

Nutrition therapy is a crucial component of secondary and tertiary health care. Nutrition therapy is a fundamental element of treatment for any number of diseases. In fact, for some conditions such as **celiac disease**, nutrition therapy is the only treatment. The most common cardiac diagnoses—hypertension, atherosclerosis, myocardial infarction, and congestive heart failure—all require nutrition therapy as a significant part of treatment. Furthermore, nutrition therapy is an important tool in recovery and maintenance of nutritional status when disease or injury places an individual at nutritional risk that could impair recovery.

BOX 1.2 CLASSIFICATIONS OF HOSPITALS

Health Care Systems	Definition
Public Not-for-Profit	The focus is on the community, not on a group of shareholders looking to make a profit on the services. At not-for-profit, community-based hospitals, profitable services indirectly fund unprofitable—yet necessary—services like trauma and women's health.
Private Not-for-Profit	They are owned and managed by communities, religious institutions, district health councils, or their hospital boards.
Private Profit	For-profit hospitals are motivated to offer services that typically generate surpluses or high profits, like cardiac care, surgery and orthopedics. Investor-owned (for-profit) health care organizations also have a big financial incentive not only to avoid caring for uninsured and underinsured patients, but also to avoid locating their facilities in poorer geographic areas.
Veteran's and Military	This system provides health care to veterans of U.S. military service and operates 158 hospitals, 840 ambulatory care and community-based outpatient clinics, 133 nursing homes, 206 community-based outpatient psychiatric clinics, and 57 regional benefits offices.

Source: www.ohiohealth.com/aboutus/whoweare/notforprofit/notforprofit.htm; www.web.net/~ocsco/private_for_profit_sub.shtml; www.whitehouse.gov/omb/budget/fy2005/va.html

Restorative Care

The goal of restorative care is to assist individuals recovering from an acute or chronic illness to return to and maintain their optimal level of functioning. Restorative care occurs in rehabilitation facilities and long-term care facilities, and through outpatient clinics and home health services. As in other levels of health care, nutrition services are an important facet of care. Recovery from disease and injury require comprehensive nutrition services. Utilizing the special expertise of the RD for nutrition assessment, nutrition diagnosis, intervention, monitoring, and evaluation for the unique requirements of each individual will increase the likelihood of positive outcomes for health care.

Long-Term Care

Long-term care health care provides maintenance, custodial, and health services for individuals who are chronically ill or disabled. Diagnoses include musculoskeletal diseases such as osteoporosis, neurological diseases such as Parkinson's disease, and Alzheimer's and dementia disorders, as well as circulatory and respiratory diseases. Patients who require long-term care are at potential nutritional risk due to increased nutritional requirements and decreased nutritional intake. Furthermore, nutrition assessment, evaluation, and monitoring are required components for accreditation of long-term care facilities through the Centers for Medicare and Medicaid Services, formerly the Health Care Financing Administration.

Financing the Health Care Industry

Understanding an individual's access to health care requires an awareness of the current financial organization of health care. If nutrition services are to be an integral component of all levels of health care, providers need to comprehend how

acute care—medical treatment rendered to people whose illnesses or medical problems are short term or don't require long-term continuing care; acute care facilities are hospitals that mainly treat people with short-term health problems

celiac disease—a disease caused by both genetic and autoimmune factors; patients' exposure to gluten results in damage to the intestinal mucosa

FIGURE 1.1 Categories of Health Insurance

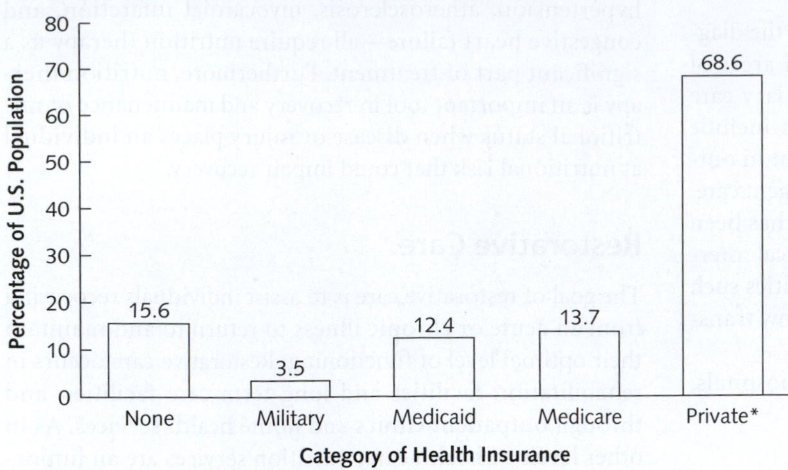

Categories of Health Insurance and Percentage of U.S. Population Enrolled

Categories of health insurance and percentage of U.S. population that has each type. Of 288 million people in the United States, 243 million had health insurance in 2003.

Percentages do not add to 100 because a person can be covered by more than one type of health insurance during the year.

*About 88 percent of people covered by private insurance are covered through an employer.

Source: U.S. Census Bureau.

M. Boyle and D. Holben, *Community Nutrition in Action*, 4e, copyright © 2006, p. 278

these services fit into the overall structure of health care financing in the U.S. The pluralistic system of health care in the U.S. includes many components: private insurance, group insurance, Medicare, Medicaid, workers' compensation, the Veterans Health Administration medical care system, Department of Defense hospitals and clinics, the Public Health Service's Indian Health Service, state and local public health programs, and the Department of Justice's Federal Bureau of Prisons. Currently, the system is structured around the provision of **health insurance**. In 2003, 84.4% of the U.S. population was insured, and 15.6% was not (US Census 2004). Some people choose not to have health insurance because they can pay for their health care; however, many Americans are forced to live without health insurance because they cannot afford it. The uninsured will be discussed later in this chapter. In the U.S., there are two general categories of health insurance: private and public. Approximately 68.6% of the U.S. population has private insurance, and 29.6% is covered by public health insurance provided by the government (see Figure 1.1).

Private Insurance

More Americans carry private insurance than are covered under governmental health programs. The following sections discuss a variety of plans within this privatized system. Private insurance can be in the form of traditional fee-for-service insurance or **group contract insurance**.

Traditional Fee-for-Service Plans Traditional fee-for-service plans include a billing system in which the provider of care charges a fee for each service rendered. This type of insurance is provided by commercial insurance companies such as Blue Cross/Blue Shield, not-for-profit organizations, and independent employee health plans. Traditional fee-for-service plans account for only about 10% of insurance coverage today. Critics of fee-for-service plans claim they encourage physicians to provide more services than are necessary (Stern 1989). Proponents of fee-for-service systems prefer the greater flexibility and unrestricted access to physicians, tests, hospitals, and treatments.

Group Contract Insurance In the latter part of the twentieth century, the nation's private health care system went through a major transition from the traditional unmanaged fee-for-service system to a predominantly **managed-care system**, represented by managed care organizations: **health maintenance organizations** (HMOs) and **preferred provider organizations** (PPOs). Both are prepaid group practice plans that offer health care services through groups of medical practitioners. The presumed goal of managed care is an

health insurance—financial protection against health care costs associated with treatment of disease or accidental injury

group contract insurance—health insurance offered through businesses, union trusts, or other groups and associations

managed-care system—a health care approach in which insurers try to limit the use of health services, reduce costs, or both; these health plans are subject to utilization review (UR), which aims to prevent unnecessary treatment by requiring enrollees to obtain approval for nonemergency hospital care, denying payment for wasteful treatment, and monitoring severely ill patients to ensure that they get cost-effective care

health maintenance organization (HMO)—a health plan that provides comprehensive medical services to its members for a fixed, prepaid premium; members must use participating providers

preferred provider organization (PPO)—a type of insurance in which the managed care company pays a higher percentage of the costs when a preferred (in-plan) provider is used; the participating providers have agreed to provide their services at negotiated discount fees

FIGURE 1.2 HMO Enrollments

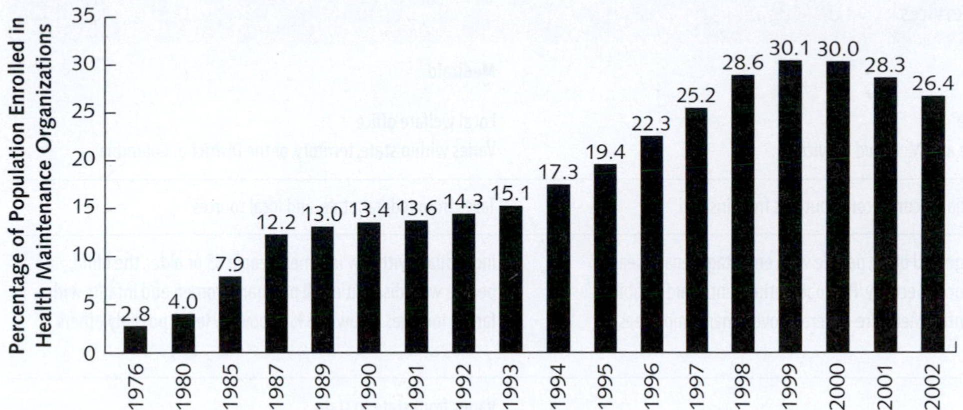

Source: M. Boyle and D. Holben, *Community Nutrition in Action,* 4e, copyright © 2006, p. 280

improved quality of care with decreased costs. In 2002, almost 76.1 million Americans were enrolled in managed-care plans. This large group of individuals also includes people with public insurance in a managed-care plan (12.8 million and 5.4 million Medicaid and Medicare beneficiaries, respectively) (Health, United States, National Center for Health Statistics 2003). Figure 1.2 shows HMO enrollments since the mid-seventies. The number of HMOs peaked in 1999 and is now declining. Considering job-based coverage, in 1988, only 27% of employees were enrolled in a managed-care plan, and this figure increased to 54%, 73%, and 86% in 1993, 1996, and 1998, respectively. In 2004, 95% of employees were enrolled, with the majority of workers with job-based coverage belonging to a PPO plan (55%), followed by HMOs (25%), and point of service (POS) plans (15%).

By law, employers with 25 or more employees must offer their employees HMO membership as an alternative to traditional fee-for-service health insurance plans. In HMOs, physicians practice as a group, sharing facilities and medical records. Physicians may either be salaried or provide contractual services. Reimbursement for physician services within the HMO model is often based on **capitation**. Capitation is a payment system that is based on a fixed amount per enrollee. No additional reimbursement is provided if the enrollee's care is more costly than the fixed amount. There are four general models of HMOs:

- *Staff model*: The HMO owns and operates its own facility, which is equipped for laboratory, pharmacy, and X-ray services; it hires its own physicians and other health care providers including RDs. Kaiser Permanente is an example of a staff HMO model.
- *Group model*: The HMO contracts with one or more multispecialty group practices that provide health care services exclusively to its members.
- *Network model*: Much like the group model, the HMO contracts with multiple group practices, hospitals, and other providers to provide services to its members, but in a nonexclusive arrangement.
- *Independent practice association (IPA)*: A decentralized model—or HMO without walls—in which the HMO contracts with individual physicians to care for plan members in their own private offices for a discounted fee. Physicians are free to contract with more than one plan and may provide care on a fee-for-service basis as well.

The HMO idea—a fixed cost to the consumer, with health care insurer and health care provider roles combined (Stern et al. 1989)—is viewed as a more cost-effective way of practicing medicine than the traditional fee-for-service systems. HMOs should have a greater stake in your wellness than most fee-for-service models because profit margins are higher if you stay healthy (Wolfe 1985). Prepaid group health plans emphasize health promotion because they provide health care services at a preset cost. By keeping people healthy, HMOs theoretically avoid the need for lengthy hospitalizations and costly services. Research does seem to indicate enrollees of HMOs are hospitalized less frequently than patients of fee-for-service physicians (Stern et al. 1989). Some managed care organizations are forming ancillary health care providers, which may include networks for RDs. RDs can obtain provider numbers from these managed care organizations, which enhances coverage for nutrition services, builds referrals, and promotes recognition of the value of nutrition therapy.

Public Insurance

The federal Centers for Medicare and Medicaid Services (CMS) of the Department of Health and Human Services

capitation—a payment or fee of a fixed amount per person

TABLE 1.2

Comparison of Medicare and Medicaid Services

	Medicare	Medicaid
Administration	Social Security Office Centers for Medicare and Medicaid Services	Local welfare office Varies within state, territory, or the District of Columbia
Financing	Trust funds from Social Security, contributions from insured	Taxes from federal, state, and local sources
Eligibility	People 65 years of age and older, people with end-stage renal disease, people eligible for Social Security/Railroad Retirement Board disability programs for 24 months, Medicare-covered government employees, possibly others	Individuals with low incomes, people 65 or older, the blind, people with disabilities, all pregnant women and infants with family incomes below 133% of poverty level, possibly others
Benefits	Same in all states **Hospital insurance (Part A)** *helps* pay for inpatient hospital care, skilled nursing facility care, home health care, hospice care. **Medical insurance (Part B)** *helps* pay for physicians' services, outpatient hospital services, home health visits, diagnostic X-ray, laboratory, and other tests; necessary ambulance services, other medical services and supplies, outpatient physical or occupational therapy and speech pathology; partial coverage of mental health treatment, kidney dialysis, medical nutrition therapy services for people with diabetes or kidney disease, and certain preventive services.*	Varies from state to state **Hospital services:** inpatient and outpatient hospital services; other laboratory and X-ray services; physician services; screening, diagnosis; and treatment of children; home health care services **Medical services:** many states pay for dental care; health clinic services; eye care and glasses; prescribed medications; and other diagnostic; rehabilitative; and preventive services; including nutrition services.
Typical Exclusions	Regular dental care and dentures, routine physical exams and related tests, eyeglasses, hearing aids and examinations to prescribe and fit them, most prescription drugs, nursing home care (except skilled nursing care), custodial care, immunizations (except for pneumonia, influenza, and hepatitis B), cosmetic surgery	Varies from state to state
Premium Costs (2005)	Part A: none if eligible, or $206–$375/month Part B: $78.20/month	None (federal government contributes 50% to 80% to states to cover eligible persons)

*Medicare beneficiaries who have both Part A and Part B can choose to get their benefits through a variety of risk-based plans (e.g., HMOs, PPOs), known as the Medicare Advantage Plan, which may expand coverage. An additional premium may apply. Certain recipients are covered for bone mass measurements, colorectal cancer screening, diabetes self-management training and supplies, glaucoma screening, mammogram screening, Pap test and pelvic examination, prostate cancer screening, and certain vaccinations. The Medicare Modernization Act of 2003 expanded coverage. For more information, visit www.medicare.gov.

Source: Adapted from U.S. Department of Health and Human Services, 2002 *Guide to Health Insurance for People with Medicare* (Washington, D.C.: U.S. Department of Health and Human Services, 2002); and Centers for Medicare and Medicaid Services, *Your Medicare Benefits* (Baltimore, MD: U.S. Department of Health and Human Services, 2004).

Source: This table was taken from Boyle/Holben's *Community Nutrition in Action* 4e, table 9.1, page 282

medicare—federal health insurance program, administered by the Centers for Medicare and Medicaid Services, for individuals over the age of 65, persons with disabilities, and those persons with end-stage renal disease

medicaid—entitlement program that pays for medical assistance for certain individuals and families with low incomes and resources

are responsible for administering **Medicare, Medicaid, State Children's Health Insurance (SCHIP),** and several other health-related programs, including the **Health Insurance Portability and Accountability Act** (**HIPAA**) of 1996) and Clinical Laboratory and Improvement Amendments (CLIA). The two major public health insurance plans in the United States are Medicare and Medicaid. A comparison of their features is provided in Table 1.2. **Workers' compensation**, which pays benefits to workers who have been injured on the job, is another public-sector health benefit program. SCHIP provides health coverage to uninsured children whose families earn too much money to qualify for Medicaid but too little to afford private coverage (HHS 2002). Health care services are also provided by the Department of Veterans Affairs (VA), the Public Health Service (including the **Indian Health Service**), the Department of Defense (including the **Civilian Health and Medical Program of the Uniformed Services, (CHAMPUS)**, public hospitals, community health centers, and state and local public health programs (Shi and Singh 2001a).

The Medicare Program In 2003, over 39 million individuals were enrolled in Medicare. This program was established in 1965 by Title XVIII of the Social Security Act and is administered by the CMS. The Social Security Administration provides information about program eligibility and handles enrollment (CMS 2005). Medicare is designed to assist the following:

- People 65 years of age or older;
- People of any age with end-stage renal disease;
- People eligible for Social Security or Railroad Retirement Board disability benefits up to 24 months;
- Individuals receiving or eligible to receive retirement benefits from Social Security or Railroad Retirement Boards; and
- People who had Medicare-covered government employment prior to retirement.

Recipients of Medicare benefits are offered the Original Medicare Plan or a Medicare Advantage Plan (which provides additional benefits). Basically, Medicare consists of two separate parts: hospital insurance (Part A) and medical insurance (Part B). No monthly premium is required for Medicare Part A if a person or his or her spouse is entitled to benefits under either Social Security or the Railroad Retirement System, or has worked a sufficient period of time in federal, state, or local government to be insured, because premiums would have been paid through payroll taxes while the individual or spouse was working (CMS 2005). Those who do not meet these qualifications (those who have fewer than 40 quarters of Medicare-covered employment) may purchase Part A coverage if they are at least age 65 and meet certain requirements (CMS 2005).

Medicare Part A. Medicare Part A provides hospital insurance benefits that include inpatient hospital care, care at a skilled nursing facility, and some home health care. Deductible and **coinsurance** fees apply. Hospital inpatient charges are reimbursed according to a **prospective payment system (PPS)** known as **diagnosis-related groups (DRGs)** (discussed in detail later in this chapter).

Since 1983, the government has shifted a larger portion of health care costs to Medicare beneficiaries through larger deductibles, greater use of services with coinsurance, and use of services not covered by Medicare.

Medicare Part B. Medicare Part B is an optional medical insurance program financed through premiums paid by enrollees and contributions from federal funds. Part B provides supplementary medical insurance benefits for eligible physician services, outpatient hospital services, certain home health services, and durable medical equipment.

As of 2002, Medicare pays RDs who enroll in the Medicare program as providers, regardless of whether they provide **medical nutrition therapy (MNT)** services in an independent practice setting, a hospital outpatient department, or any other setting. However, Medicare does not pay RDs for services provided to patients in an inpatient stay in a hospital or in a skilled nursing facility (CMS, 2002; Michael 2001; Ochs 2002). Enrolled Medicare MNT

state children's health insurance program (SCHIP)—federal children's health insurance initiative that allows each state to offer health insurance for children up to age 19 who are not already insured

health insurance portability and accountability act (HIPAA)—legislation that guarantees that people who lose their group health insurance will have access to individual insurance, regardless of preexisting medical problems; the law also allows employees to secure health insurance from their new employer when they switch jobs even if they have a preexisting medical condition

workers' compensation—insurance coverage that compensates employees for work-related injuries or disabilities, which employers are required to provide by state law

indian Health Service—an agency within the Department of Health and Human Services that operates a comprehensive health service delivery system for American Indians and Alaska Natives

civilian health and medical program of the uniformed services (CHAMPUS)—the health plan that serves the dependents of active-duty military personnel and retired military personnel and their dependents

coinsurance—a cost-sharing requirement under some health insurance policies in which the insured person pays some of the costs of covered services

prospective payment system (PPS)—system that pays hospitals a fixed sum per case according to a schedule of diagnosis-related groups

diagnosis-related groups (DRGs)—groups developed by Medicare that classify a patient's illness(es) according to principal diagnosis and treatment requirements for the purpose of establishing payment rates

medical nutrition therapy (MNT)—nutritional diagnostic, therapy, and counseling services for the purpose of disease management that are furnished by a registered dietitian or nutrition professional pursuant to a referral by a physician

TABLE 1.3

State Licensure or Certification for Registered Dietitians and Nutritionists

Licensing *— statutes include an explicitly defined scope of practice, and performance of the profession is illegal without first obtaining a license from the state.*
Statutory certification *— limits use of particular titles to persons meeting predetermined requirements, while persons not certified can still practice the occupation or profession.*
Registration *— the least restrictive form of state regulation. As with certification, unregistered persons are permitted to practice the profession. Typically, exams are not given and enforcement of the registration requirement is minimal*

State	Type of Licensure, certification or registration	State	Type of Licensure, certification or registration
Alabama	Licensing of dietitian/nutritionist	Montana	Licensing of nutritionist/dietitian title protection
Alaska	Licensing of dietitian/nutritionist	Nebraska	Licensing of medical nutrition therapist
Arkansas	Licensing of dietitian	Nevada	Certification of dietitian
California	Registration of dietitian	New Hampshire	Licensing of dietitian
Connecticut	Certification of dietitian	New Mexico	Licensing of dietitian/nutritionist
Delaware	Certification of dietitian/nutritionist	New York	Certification of dietitian/nutritionist
District of Columbia	Licensing of dietitian/nutritionist	North Carolina	Licensing of dietitian/nutritionist
Florida	Licensing of dietitian/nutritionist/nutrition counselors	North Dakota	Licensing of dietitian/nutritionist
Georgia	Licensing of dietitian	Ohio	Licensing of dietitian
Hawaii	Certification of dietitian	Oklahoma	Licensing of dietitian
Idaho	Licensing of dietitian	Pennsylvania	Licensing of dietitian/nutritionist
Illinois	Licensing of dietitian/nutrition counselors	Puerto Rico	Licensing of dietitian/nutritionist
Iowa	Licensing of dietitian	Rhode Island	Licensing of dietitian/nutritionist
Kansas	Licensing of dietitian	South Dakota	Licensing of dietitian/nutritionist
Kentucky	Licensing of dietitian, certification of nutritionist	Tennessee	Licensing of dietitian/nutritionist
Louisiana	Licensing of dietitian/nutritionist	Texas	Certification of dietitian
Maine	Licensing of dietitian and dietetic technician	Utah	Certification of dietitian
Maryland	Licensing of dietitian/nutritionist	Vermont	Certification of dietitian
Massachusetts	Licensing of dietitian/nutritionist	Virginia	Certification of dietitian/nutritionist
Minnesota	Licensing of dietitian/nutritionist	Washington	Certification of dietitian
Mississippi	Licensing of dietitian/nutritionist title protection	West Virginia	Licensing of dietitian
Missouri	Certification of dietitian	Wisconsin	Certification of dietitian

Source: Commission on Dietetic Registration. Chicago (IL): The American Dietetic Association. 2005 – [cited 2005 October 25]. Available from: www.cdrnet.org/certifications/licensure/index.htm.

providers are able to bill Medicare for MNT services provided to Medicare beneficiaries with type 1 diabetes, type 2 diabetes, gestational diabetes, nondialysis kidney disease, and post-kidney-transplant status using specified MNT **current procedural terminology (CMT) codes.**

A physician's referral for MNT is required. This has proven to be an obstacle to reimbursement for nutrition services. Data appear to indicate that physicians may not be aware of this third-party reimbursement for MNT benefits, or that they are simply not referring patients for these specific nutrition therapies (Smith 2005). Recognized providers of MNT include RDs or nutrition professionals who meet the following qualifications: BS degree or higher from a program in nutrition and dietetics, at least 900 hours of practice experience, and licensed or certified. State licensure is required in 46 states (see Table 1.3). Licensure or certification of dietetics practitioners assists the consumer in obtaining nutritional care only from qualified professionals. The use of the title *dietitian* or *nutritionist* is specific to only those appropriately credentialed professionals.

Currently, there are approximately 7,000 RDs who have enrolled as providers of MNT for third-party reimbursement (ADA 2005). Actual use of current Medicare benefits, however, is much lower than what was predicted when these benefits were initially proposed. Statistics indicate that only 211,000 individuals received MNT, whereas over 8.6 million may be eligible for it (Smith et al. 2005).

current procedural terminology (CMT) codes—coding system established by the Centers for Medicare and Medicaid Services for identifying medical care interventions

FIGURE 1.3 Medicare Savings after Three Years of Reimbursement for Nutrition Services

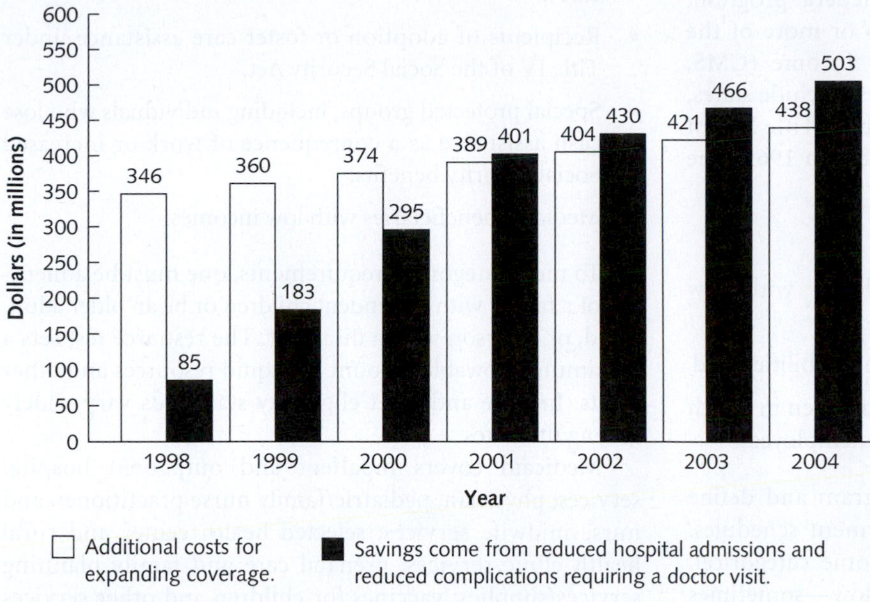

Source: M. Boyle and D. Holben, *Community Nutrition in Action*, 4e, copyright © 2006, p. 300

Pursuing third-party reimbursement for nutrition services remains a major objective of The American Dietetic Association (ADA). An ADA-financed independent study projected the cost of extending coverage of medical nutrition therapy to all Medicare beneficiaries under Medicare Part B to be less than $370 million over 7 years, when savings are considered. Savings would be greater than costs after the third year of enactment (see Figure 1.3) (Community Nutrition Institute 1997). For example, if coverage had begun in 1998, in 2001 an additional cost to Medicare Part B of $389 million would have been offset by a reduction in cost to Part A of $401 million, resulting in a net savings of $11 million. Savings to the Medicare program come from fewer hospital admissions and fewer complications requiring a physician's visit. Data used in the study were particularly significant for persons with diabetes and cardiovascular disease. Spending for diabetes and cardiovascular disease accounts for about 60% of annual Medicare spending (Monsen 1997). In the long run, the program would save more in medical expenses than it costs to operate.

The most recent changes in Medicare include an initial preventive physical examination, which may identify needs for MNT as diagnoses are made. Another program, **Voluntary Chronic Care Improvement Programs (CCIP)**, authorized and currently being tested, will provide another avenue for providing medical nutrition therapy that will be covered by Medicare reimbursement. Integration of coverage within Medicare modernization will be crucial to the future of the dietetics profession (Smith et al. 2005).

Coverage Gaps. Other types of health care, in addition to certain nutrition therapies, lack coverage by Medicare. The two most notable gaps in Medicare coverage have been prescription drug coverage and skilled nursing/long-term institutional care. Traditionally, most prescription drugs are not covered at all under the Medicare program. Only 100 days of skilled nursing/long-term care are covered annually by Medicare Part A. Thereafter, patients or their families must either pay costs themselves or "spend down" their assets in order to reduce their net worth and be eligible for Medicaid coverage of long-term care. However, in December 2003, President George W. Bush signed into law the Medicare Prescription Drug, Improvement, and Modernization Act of 2003 (Medicare Modernization Act) (CMS, *Medicare: A Brief Summary*, 2005). The Medicare Modernization Act provides optional coverage to Medicare recipients, including drug discount cards/prescription drug plans and other preventive benefits (wellness physical exam, cardiovascular disease blood screening, and diabetes screening for those at risk) in addition to those preventive benefits already covered (cancer screening, bone mass measurements, and vaccinations) (HHS 2004). For those in the Original Medicare Plan, a Medigap policy may be purchased if the individual participates in both Medicare Part A and Part B. A Medigap policy is a supplemental insurance policy sold by private insurance companies to help pay the deductible, coinsurance fees, prescription drug costs, and certain services not covered by Medicare. Alternatively, participants may choose the Medicare Advantage Plan, which generally provides greater benefits than the Original Medicare Plan (CMS, *Medicare and You* 2005, 2004).

For additional benefits, Medicare recipients often explore other options. They may continue insurance coverage through a current or former employer. Individuals may also choose to purchase nursing home or long-term care policies, which pay cash amounts for each day of covered nursing home or at-home care. Finally, individuals may qualify for full Medicaid (see the next section) benefits or at least to receive some state assistance in paying Medicare costs.

voluntary chronic care improvement programs (CCIP)—programs designed to improve the quality of care and life for people living with chronic illnesses, development and testing of which were authorized by the Medicare Modernization Act of 2003 (MMA); chronic illnesses account for a significant share of Medicare expenditures

The Medicaid Program Medicaid, an entitlement program insuring almost 36 million individuals (US Census 2004), was established as a joint state and federal program, with the federal government paying 50% or more of the costs depending on a state's per capita income (CMS, *Medicaid: A Brief Summary*, 2005). ("States" include states, U.S. territories, and the District of Columbia.) Title XIX of the Social Security Act established Medicaid in 1965. The program helps to finance medical care for:

- Eligible persons with low incomes;
- Certain pregnant women and children with low incomes;
- Older adults, the blind, and people with disabilities; and
- Members of families with dependent children in which one parent is absent, incapacitated, or unemployed.

Individual states administer the program and define eligibility, benefits and services, and payment schedules. Typically, one must meet three criteria: income, categorical, and resource. Income must often be below—sometimes significantly below—**federal poverty guidelines**. Because states administer the program, an individual may qualify for Medicaid in one state but not in another. Generally, those eligible for Medicaid include the following:

- Those eligible for **Temporary Assistance for Needy Families (TANF)** or **Supplemental Security Income (SSI)**.
- Children under 6 years old living in a household at or below 133% of the poverty guidelines and all children born after September 30, 1983, who are under age 19 and living in a household with a total income at or below the poverty guidelines.
- Pregnant women (eligible only for services related to pregnancy/complications, delivery, and postpartum care).
- Recipients of adoption or foster care assistance under Title IV of the Social Security Act.
- Special protected groups, including individuals who lose cash assistance as a consequence of work or increased Social Security benefits.
- Medicare beneficiaries with low incomes.

To meet categorical requirements, one must be a member of a family with dependent children or be an older adult, blind, or a person with a disability. The resource test sets a maximum allowable amount for liquid resources and other assets. Income and asset eligibility standards vary widely among the states.

Medicaid covers inpatient and outpatient hospital services; physician, pediatric/family nurse practitioner, and nurse-midwife services; selected health center and rural health clinic services; prenatal care and family planning services/supplies; vaccines for children and other services for those under 21 years; laboratory and X-ray services; and skilled nursing home and home health services, among others. Some states include other benefits, such as prescription drug coverage, dental services, and nutrition services, but there is significant variability among states (CMS 2005; Stollman 1995).

Medicaid covers less than half of those below the poverty line (Institute of Medicine 2000). The American Medical Association has recommended that Medicaid be expanded to provide acute-care coverage for all persons below the poverty line (Ahluwalia 1990). Although this would increase cost of services provided, it would improve access to health services and potentially decrease health care costs in the long run.

The State Children's Health Insurance Program President William J. Clinton signed into law the Balanced Budget Act of 1997, which included Title XXI, the State Children's Health Insurance Program (SCHIP). SCHIP was the largest single expansion of health insurance coverage for children in more than 30 years (CMS 2005). At the time, nearly 11 million American children—1 in 7—were uninsured. In fact, from 1988 to 1998 the proportion of children insured through Medicaid increased from 15.6% to 19.8%, while the percentage of children without health insurance increased from 13.1% to 15.4%. This increase was attributed to fewer children being covered by employer-sponsored health insurance (CMS 2005). The SCHIP initiative was designed to reach these children, many of whom were part of working families with incomes too high to qualify for Medicaid but too low to afford private health insurance. For example, in 2004, in most states, uninsured children under the age of 19 whose families earned up to $36,200 per year (family of

federal poverty guidelines—guidelines published annually by the Department of Health and Human Services to define "poverty" for legislative purposes; updates to the poverty guidelines (poverty line) can be found at http://aspe.hhs.gov/poverty/index.shtml

temporary assistance for needy families (TANF)—program that provides assistance and work opportunities to needy families by granting states the federal funds and wide flexibility to develop and implement their own welfare programs. It was formerly known as the welfare programs Aid to Families with Dependent Children (AFDC) and the Job Opportunities and Basic Skills Training (JOBS) programs

supplemental security income (SSI)—a federal income supplement program designed to help aged, blind, and disabled people who have little or no income that provides cash to meet basic needs for food, clothing, and shelter

four) were eligible (CMS 2005). States are able to use part of their federal funds to expand outreach and ensure all children eligible for Medicaid and the new SCHIP program are enrolled. The initiative is a partnership between the federal and state governments that helps provide children with health coverage. Because Medicaid allows states flexibility in determining eligibility, states currently cover children whose family incomes range from below the poverty level (as defined by government poverty guidelines) to as high as 300% of the poverty level income. Funds for the program became available to states in 1997. States receive federal matching funds only for actual expenditures to insure children. In 2003, almost 6 million children were covered by SCHIP (CMS 2005). Figure 1.4 summarizes the enrollment in SCHIP since 1999.

Under the program, states have flexibility in targeting eligible uninsured children. States may choose to expand their Medicaid programs, design new child health insurance programs, or create a combination of both types of programs. States choosing a new children's health insurance program may offer one of the following benchmark plans: the standard Blue Cross/Blue Shield Preferred Provider Option offered by the Federal Employees Health Benefit Program, a health benefit plan offered by the state to its employees, or the HMO benefit plan with the largest commercial enrollment in the state. A state may also choose to offer the equivalent of one of the benchmark plans. To qualify as "equivalent," a state's plan's value must be at least equal to the benchmark plan's, and it must include inpatient and outpatient hospital services, physicians' surgical and medical services, laboratory and X-ray services, and well baby/child care services, including immunizations. In addition, a benchmark-equivalent plan must include benefits similar to the benchmark plan coverage of prescription drugs, mental health services, vision care, and hearing-related care. States choosing the Medicaid option must offer the full benefit package offered to Medicaid recipients.

FIGURE 1.4 SCHIP Enrollment, 1999–2003

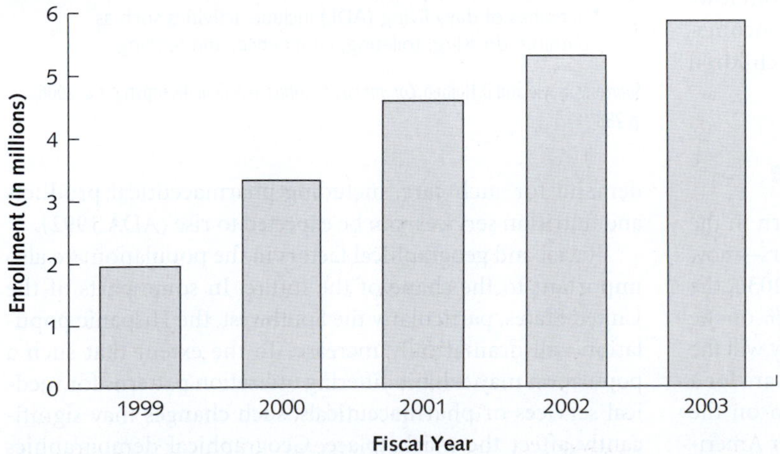

Source: M. Boyle and D. Holben, *Community Nutrition in Action,* 4e, copyright © 2006, p. 286

The Uninsured

In theory, health care coverage is available to virtually all U.S. citizens through one of four routes: Medicare for the elderly and people with disabilities; Medicaid for low-income women and children, some low-income men, and people with certain disabilities; employer-subsidized coverage at the workplace; or self-purchased coverage for those ineligible for the previous three options (Consumers Union 1998). Yet an estimated 45 million people (15.6% of the population) live in the U.S. with no insurance coverage at all, and perhaps an even larger number of people have coverage that is inadequate for covering the costs associated with a major illness (Shi 2001; US Census 2004).

In 1987, 12.9% of the population was uninsured, and the proportion of uninsured continued to increase until it peaked in 1998 at 16.3%. The current rate of lack of insurance represents an increase from a rate that fell to 14.2% in 2000. Although the 2003 data represent a percentage increase over 2002 of those without coverage, the number of people insured actually increased.

Who, then, are the uninsured? Statistics show they are not the elderly, who have Medicare, or the very poor, who have Medicaid. Instead, those who lack coverage are primarily people in the middle (for example, the working poor and those who work for small businesses). More than half are in families with incomes less than 200% of the poverty level (Schroeder 2001). They also include the self-employed, those who work part-time, seasonal workers, the unemployed, full-time workers whose employers offer unaffordable insurance or none at all, and early retirees—aged 55 through 64—who retired from companies that either offered no health insurance or discontinued it after these individuals retired (US Census 2001). These persons are classified further as the employed uninsured and nonworking uninsured. In 2001, included among the uninsured were 9.2 million children (US Census 2003; Strunk and Cunningham 2004).

Strunk and Cunningham (2004) report that one out of every seven Americans had difficulty accessing the medical care that they needed during 2001. A closer look reveals that the number of people covered by employment-based health insurance also decreased during that year (Strunk and Cunningham 2004). Health insurance premiums increased 59% from 2001 to 2004, with employee contributions increasing from 49% to 57% over that same period (Gabel et al. 2004).

In 2001, 9.2 million children under 19 years of age (12.1%) were uninsured (US Census 2001). The proportion of children without insurance did not change (11.4% of all children) from 2002 to 2003; however, impoverished children continued to be the children most likely not to be insured (19.2%). Statistics concerning

FIGURE 1.5 Percentage of Persons in the United States without Health Care Coverage, by Age and Race or Ethnic Origin, 2003.

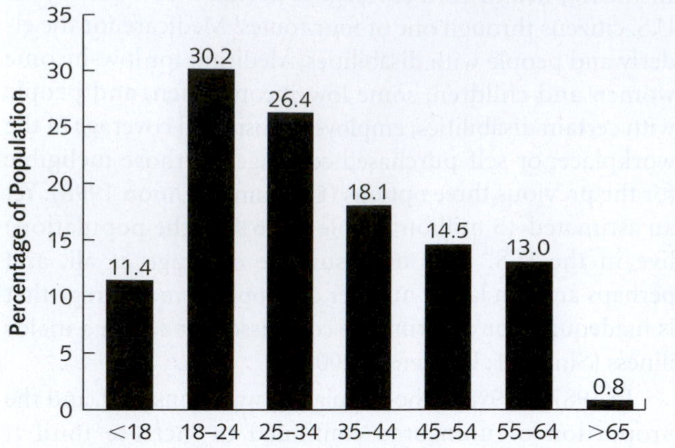

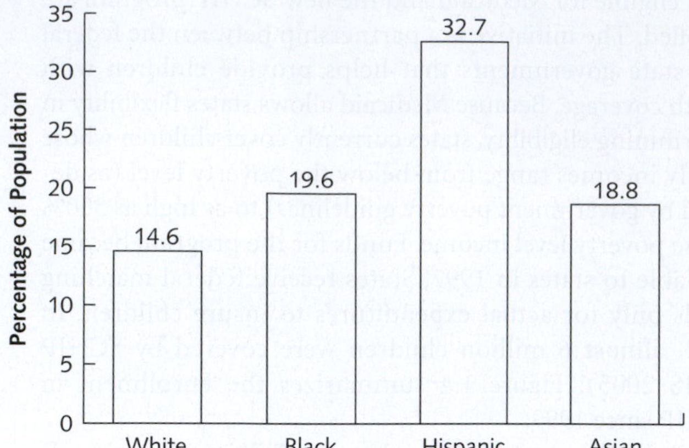

Source: M. Boyle and D. Holben, *Community Nutrition in Action,* 4e, copyright © 2006, p. 287

race but disregarding age reveal that 19.6% of blacks, 18.8% of Asians, and 32.7% of Hispanics lacked insurance in 2003; these figures were unchanged from 2002 (see Figure 1.5). For 2001–2003, 27.5% of American Indian and Alaska natives did not have health insurance; this percentage was also unchanged (US Census 2005).

When those without health insurance do get sick, they often wind up using the most expensive treatment available—hospital emergency room care—or they delay getting treatment and later require more expensive and prolonged medical services. These costs are eventually shifted to the people who are insured by the increase in overall medical costs. All community members, including the employed and nonworking uninsured, the homeless, and others, should be able to obtain medical care when it is needed. However, cost was a barrier for health care access for one out of every seven individuals living in the U.S. in 2003. On the other hand, between 2001 and 2003, access to needed medical care improved. During this time, the percentage of those who had no insurance and a low income fell from 16.4% to 13.2%. In fact, unmet medical needs of children from low-income households decreased to such a degree that income-related differences in access to health care for children disappeared (Strunk and Cunningham 2004).

FIGURE 1.6 Number of Elderly Needing Long-Term Care, 1990 and 2030.

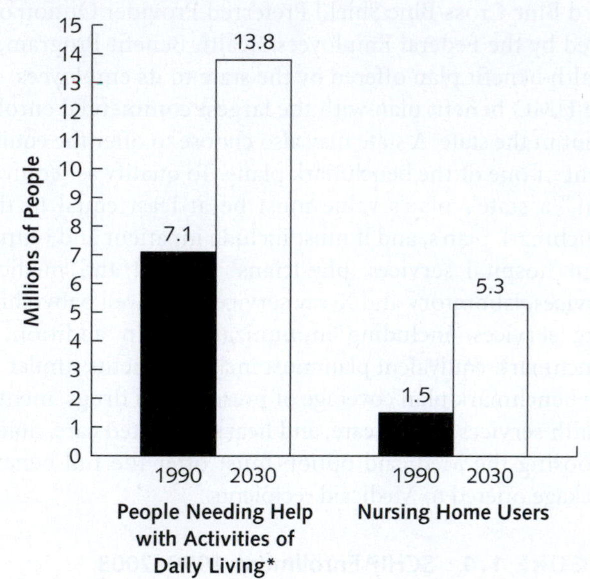

*Activities of daily living (ADL) include activities such as bathing, dressing, toileting, continence, and feeding.

Source: M. Boyle and D. Holben, *Community Nutrition in Action,* 4e, copyright © 2006, p. 288

Demographic Trends and Health Care

Between 1946 and 1964, 78 million babies were born in the United States; these individuals—the baby boomers—now make up one-third of the population. By the year 2030, the baby boom will become a senior boom, with 21% of the population over 65 years of age (ADA 1992). Not only will the elderly be greater in number, but they may require care for a greater number of years, placing a heavier burden on the long-term care system (see Figure 1.6). Because older Americans consume a disproportionate amount of medical care, the demand for such care, including pharmaceutical products and nutrition services, can be expected to rise (ADA 1992).

Racial and geographical factors in the population are also important to the shape of the future. In some parts of the United States, particularly the Southwest, the Hispanic population will dramatically increase. To the extent that such a population may exhibit differing utilization patterns for medical services or pharmaceuticals, such changes may significantly affect the marketplace. Geographical demographics will also be important, especially if the population drift from the Northeast to the Southwest and the Sun Belt continues (Institute of Medicine 2001).

The Need for Health Care Reform

To rate the success of a particular health care system, three crucial variables should be considered: cost, quality, and access (Wolfe 1985). At the bottom end of the rating scale is no health care system. As we have seen, millions of Americans cannot afford to buy into or gain meaningful, ongoing access to any health care at all. At the other end of the scale is high-quality, reasonably priced, accessible health care. On such a rating scale, how does the U.S. health care system fare?

Before you respond, imagine how you would react in the following situation (Fox 2000): You are the decision maker in a large corporation, and I approach you and try to sell you a product. I say I want to sell you a key piece of equipment that meets the following specifications:

- It will cost you $3,200 per employee per year.
- It will consume up to half of each profit dollar and will rise in price by 15% to 30% annually.
- There is a tremendous unexplained variation in the characteristics of this product depending on who uses it.
- There is no way to measure its quality in terms of appropriateness, reliability, or outcome.
- And you'll just have to take my word for it when I tell you that we adhere to the highest professional standards.

Would you buy this product? Many believe the current U.S. health care system fits this description. Not only is it expensive, but we don't necessarily know what we are paying for or whether what we are paying for is worth its price (Fox 2000).

The term *health care reform* refers to current efforts undertaken to ensure that everyone in the U.S. has access to affordable, quality health care. Among the challenges for health care reform are how to make health care accessible to everyone, how to contain costs, how to provide nursing home care to those who need it, and how to ensure that Medicare and Medicaid can serve all who are eligible.

Cost, access, and quality are interrelated; manipulating one has an astounding impact on the others. Consider the three candidates for a cholecystectomy from the beginning of this chapter and their varying health care options. Some people argue that we should abandon free enterprise and turn the system over to the government, as has been done in other countries, including Canada. Critics of government-run health care systems say such systems appear promising at first but soon bog down in bureaucracy, unable to keep pace with advances in medical technology. Some critics point to the Canadians who travel to the United States to purchase treatment out of their own pockets rather than wait in line for Canadian health care (Brodsky 1992). Is there a way to reduce health care costs and increase access to the system without sacrificing quality?

Health care policy makers are studying alternative models of delivery and financing, in hopes that approaches that have been successful in other nations might be applied to the U.S. system (Neel 1992). Per capita health spending in the United States exceeds that of other industrialized countries by huge margins (Reinhardt, Hussey, and Anderson 2004). As pointed out earlier, the U.S. health care system appears both to have higher costs (see Figure 1.7) and to offer less access than the systems of other industrialized nations. The following sections consider health care costs, equity, and access for different population groups.

The High Cost of Health Care In 2002 Americans spent almost $1.6 trillion for health care (CMS 2004). Figure 1.8 tracks the rise in U.S. health care costs since 1960. In fact, since that year, health care expenditures have increased over 800%. These statistics are indicative of health care inflation, an increase in the volume and costs of care in the U.S. over time. This growth, which is expected to continue, is a result of various factors, including an aging population, increased demand (fostered in part by more consumer awareness of health issues), and continuing advances in medicine that make it possible to offer more treatment options than ever before (Shi and Singh 2001a).

A major contributor to health care expenditures in the U.S. is the administrative cost of the insurance process itself. Yet another factor contributing to the cost of U.S. health care is the phenomenon of ever-rising professional liability costs. Some people say the U.S. has become a litigious society. For example, an obstetrician-gynecologist reported that in 2002 his professional liability insurance premium was $23,000. It then increased to $47,000 in 2003 and finally to $84,000 in 2004. Patient safety and litigation remain at the forefront of the medical malpractice crisis, and in order to help curtail the cost of liability insurance, reforms are necessary (Sage 2004).

Efforts at Cost Containment Efforts to control soaring health care costs cover a broad spectrum: slowing hospital construction, modifying hospital and physician reimbursement mechanisms, reducing the length of hospital stays, increasing copayments and deductibles for insured employees and Medicare recipients, changing eligibility requirements for Medicaid, reducing unnecessary surgery by requiring patients to obtain second opinions, restricting access to new technology, encouraging alternative delivery systems, and emphasizing prevention (ADA 1991). Generic drugs have also been utilized to help contain the costs of health care (Abramson et al. 2004).

The recent cost containment effort in the U.S. is actually a fierce competition among third-party payers (government, insurance companies, and employers) to control their own costs. This effort has been characterized by three trends (Fox 1991):

1. There has been a movement away from traditional fee-for-service health care to newer models of managed care, evident in the enrollments in HMOs and PPOs.

FIGURE 1.7 Total Health Expenditures as a Percentage of Gross Domestic Product (GDP), 1970–2001

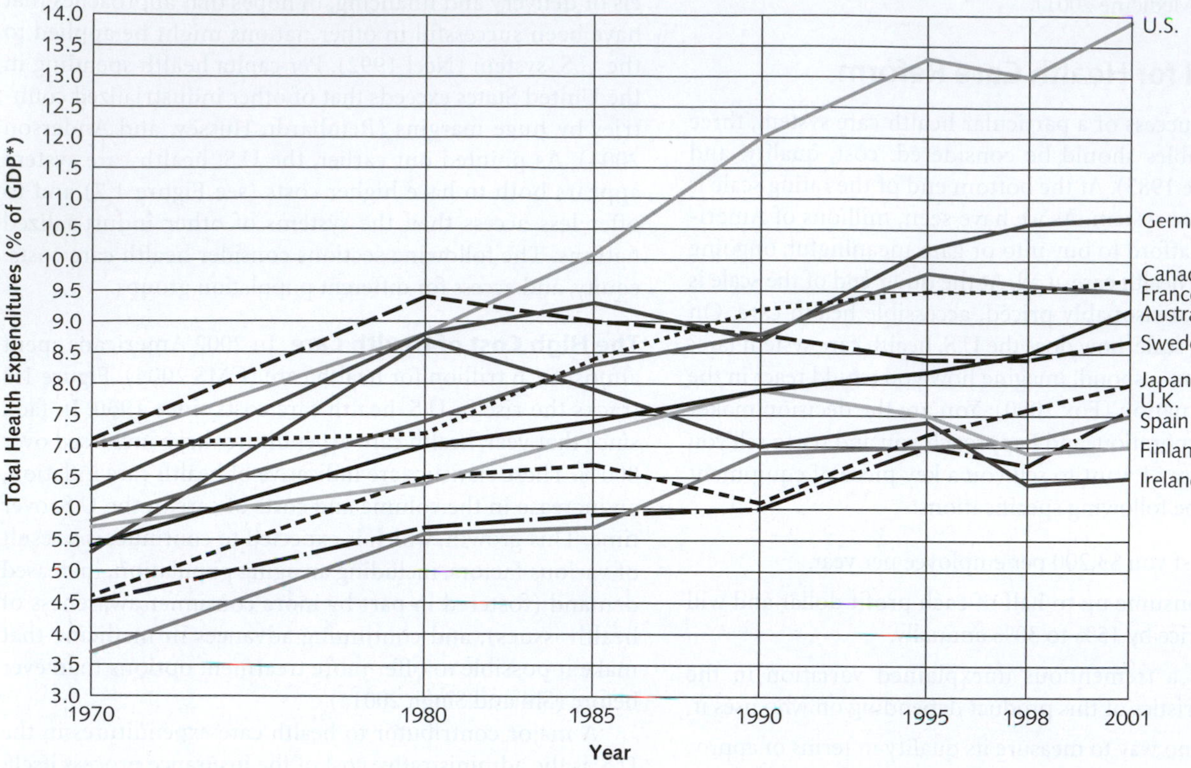

*Gross domestic product (GDP) represents the total value of a nation's output, income, or expenditures produced within its borders. GDP is more specific than gross national product (GNP), the total retail market value of all goods and services.

Source: National Center for Health Statistics, Health, United States, 2001 (Hyattsville, MD: National Center for Health Statistics, 2001). M. Boyle and D. Holben, *Community Nutrition in Action*, 4e, copyright © 2006, p. 290

2. As a growing portion of their profits are siphoned off into health care coverage, companies are increasingly attempting to manage the health care of their employees themselves in order to reduce expenditures. In an effort to avoid **cost shifting**, many businesses are moving to self-insured health plans. Such plans allow companies to determine the covered benefits and to assume the risks involved themselves (Shi and Singh 2001b).
3. The payers (government, insurance companies, and employers) are actively setting reimbursement restrictions and limitations.

The largest components of U.S. health care expenditures are hospital care (31%) and physician and clinical services (25%), as shown in Figure 1.9. Therefore, efforts to contain costs have largely been aimed at providers of those services.

cost shifting—a much-criticized aspect of the existing health care system in which hospitals and other providers bill indemnity (fee-for-service) insurers at higher rates to recover the costs of charity care and to make up for discounts given to HMOs, PPOs, Medicare, and Medicaid

One example of cost containment is the prospective payment system (PPS) that the federal government implemented as a result of the 1983 Social Security Act Amendments. The purpose of the PPS was to change the behavior of health care providers by changing the incentives under which care is provided and reimbursed. Prospective payment means knowing the amount of payment in advance. The PPS uses diagnosis-related groups (DRGs) as a basis for reimbursement. Patients are classified according to their principal diagnosis, secondary diagnosis, sex, age, and surgical procedures for which they are admitted to the hospital.

The DRG approach is based on a system of classifying hospital admissions. The system begins with the ninth edition of International Classification of Diseases: Clinical Modifications (ICD-9-CM), which contains approximately 10,000 possible reasons (organized into 23 major categories) for a hospital admission. The 23 categories are subdivided into 490 DRGs. The average cost per discharge is determined by state, region (rural or urban), number of hospital beds per facility, and other factors (Shi and Singh 2001c). All DRGs have been assigned a relative weight that reflects the cost of caring for a patient in the particular category.

FIGURE 1.8 National Health Expenditures (Billions of Dollars), 1960–2002

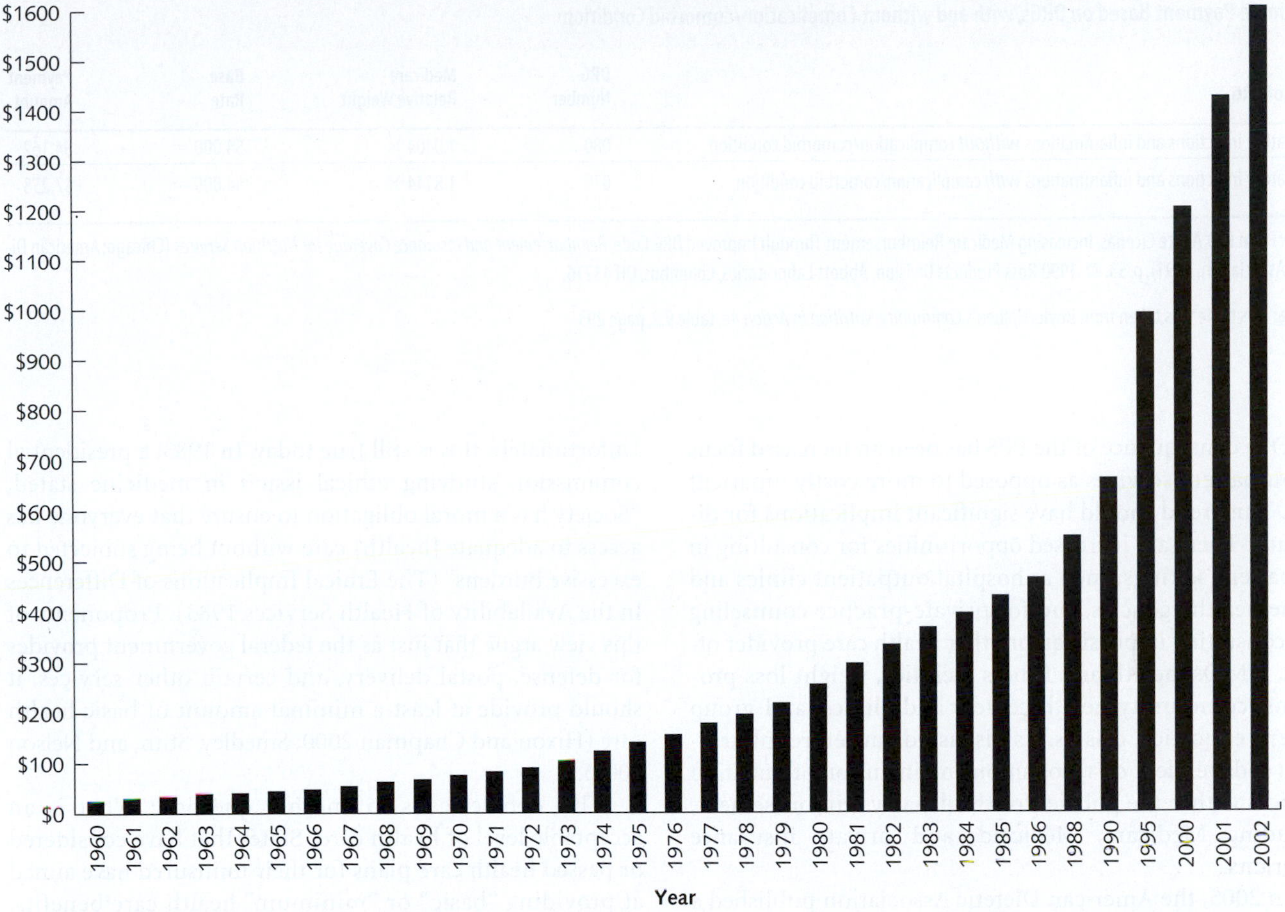

$1600
$1500
$1400
$1300
$1200
$1100
$1000
$900
$800
$700
$600
$500
$400
$300
$200
$100
$0
1960
1961
1962
1963
1964
1965
1966
1967
1968
1969
1970
1971
1972
1973
1974
1975
1976
1977
1978
1979
1980
1981
1982
1983
1984
1985
1986
1988
1990
1995
2000
2001
2002
Year

Source: Adapted from *Source Book of Health Insurance Data* (Washington, D.C.: Health Insurance Association of America, 2003). M. Boyle and D. Holben, *Community Nutrition in Action*, 4e, copyright © 2006, p. 291

FIGURE 1.9 National Health Care Costs of Hospital Care, Physician, and Clinical Services

Program Administration and Net Cost 7%
Other Spending 20%
Prescription Drugs 10%
Nursing Home Care 7%
Hospital Care 31%
Physician and Clinical Services 25%

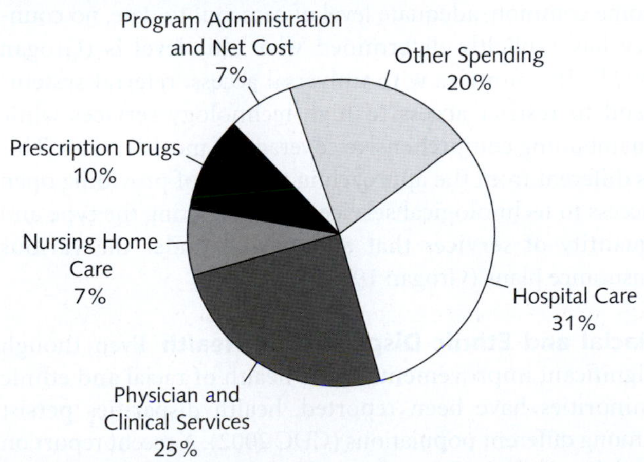

Source: M. Boyle and D. Holben, *Community Nutrition in Action*, 4e, copyright © 2006, p. 291

Table 1.4 shows a sample payment based on DRGs. Note that a patient with a complication or comorbidity (e.g., with malnutrition) is assigned a higher relative weight, reflecting the need for more intensive services. Assuring that the comorbidity condition is coded increases the hospital payment.

As mentioned earlier, beginning in 2002, RDs can enroll as providers for medical nutrition therapy with Medicare. The provision of medical nutrition therapy is coded using the specific procedural terminology codes for medical nutrition therapy. These codes can be used to document and report care that may be reimbursable through private insurance companies, depending on each company's specific coverage allowances. Billing for MNT is not available for inpatient (hospitalized) nutrition care, because nutrition services are part of routine hospital care and are not billed separately. Table 1.5 lists DRGs used to identify nutrition-related primary diagnoses, and Table 1.6 describes the MNT CPT codes.

TABLE 1.4

Sample Payment Based on DRGs, with and without Complication/Comorbid Condition

Name of DRG	DRG Number	Medicare Relative Weight	Base Rate	Payment Amount
Respiratory infections and inflammations *without* complication/comorbid condition	080	1.0404 ×	$4,000 =	$4,162
Respiratory infections and inflammations *with* complication/comorbid condition	079	1.8144 ×	$4,000 =	$7,258

Source: From D. D'Abate Cicenas, Increasing Medicare Reimbursement Through Improved DRG Code, *Reimbursement and Insurance Coverage for Nutrition Services* (Chicago: American Dietetic Association, 1991), p. 53. © 1990 Ross Products Division, Abbott Laboratories, Columbus, OH 43216.

Source: This table was taken from Boyle/Holben's *Community Nutrition in Action* 4e, table 9.2, page 293

One consequence of the PPS has been an increased focus on outpatient services as opposed to more costly inpatient care. This trend should have significant implications for dietitians. There are increased opportunities for consulting in outpatient settings, such as hospital outpatient clinics and home health agencies, and for private-practice counseling and consulting in physician or other health care provider offices, HMOs, health and fitness facilities, weight loss programs, community health centers and clinics, and group patient education classes. As discussed earlier, reimbursement is dependent on recognition and inclusion of nutrition services within the policies for third-party reimbursement, including Medicare, Medicaid, and private insurance programs.

In 2005, the American Dietetic Association published a list of official diagnostic terms for nutrition-related problems (see Chapter 3). This nutrition diagnostic terminology will assist in measuring outcomes and will improve the consistency and quality of care. Furthermore, they will assist dietitians to be more competitive and provide data for research (ADA 2006). (See Chapter 3 for more information on nutrition diagnostic terms.)

Equity and Access as Issues in Health Care Is health care a basic right? The majority of U.S. citizens polled (54%) say providing health insurance to the uninsured should be a top legislative priority (Gallop 2005; CNN 2005; *USA Today* 2005). In reality, as deVise wrote over 30 years ago, health care may be more of a privilege than a right:

> If you are either very poor, blind, disabled, over 65, male, female, white, or live in a middle- or upper-class neighborhood in a large urban center, you belong to a privileged class of health care recipients, and your chances of survival are good. . . . But, if you are none of these, if you are only average poor, under 65, female, black, or live in a low-income urban neighborhood, small town, or rural area, you are a disenfranchised citizen as far as health care rights go, and your chances of survival are not good. (deVise 1973)

Unfortunately, this is still true today. In 1983, a presidential commission studying ethical issues in medicine stated, "Society has a moral obligation to ensure that everyone has access to adequate [health] care without being subjected to excessive burdens" (The Ethical Implications of Differences in the Availability of Health Services 1983). Proponents of this view argue that just as the federal government provides for defense, postal delivery, and certain other services, it should provide at least a minimal amount of basic health care (Hixon and Chapman 2000; Smedley, Stith, and Nelson 2002).

This debate leads to another question: What is an acceptable level of health care? States that have considered or passed health care plans for their uninsured have aimed at providing "basic" or "minimum" health care benefits, unlike the "comprehensive benefits" offered through the national health plans of other industrialized countries.

Providing comprehensive benefits, of course, does not necessarily mean providing unlimited care. The right to health care in Britain, Germany, and Canada does not mean the right to all treatments. Although most services provided in these countries are covered, the extent to which services are offered varies substantially across countries. In reality, equity in health care means a commitment to providing some common, adequate level of care, but to date, no country has explicitly determined what this level is (Grogan 1992). In countries with universal access, referral systems tend to restrict access to high-technology services while maintaining comprehensive coverage of most services. This is different from the approach in the U.S. of providing open access to technological services but restricting the type and quantity of services that are covered under the various insurance plans (Grogan 1992).

Racial and Ethnic Disparities in Health Even though significant improvements in the health of racial and ethnic minorities have been reported, health disparities persist among different populations (CDC 2002). A recent report on racial and ethnic disparities presents national trends in race- and ethnicity-specific rates for 17 health status indicators

TABLE 1.5

DRGs that Diagnose Nutrition-Related Primary Diagnoses

Condition	ICD-9 International Code	DRG Code Example	A Sample of Nutrition Diagnosis Code that may present in these conditions
Achalasia	530.3	#183	NI-1.4 Inadequate Energy Intake
Acquired immune deficiency syndrome		#423	NI-1.1 Hypermetabolism
Acrodermatitis enteropathica	686.8	#278	
Addison's disease	255.4	#301	
Alcohol dependence	305	#436	
Alcoholic liver disease	571.3	#202	NC-2.1 Changes in ability to absorb or metabolize nutrients
Altered calcium metabolism	275.4	#297	
Alzheimer's disease	331.0	#429	
Amputation	84.17	#114	
Amyotrophic lateral sclerosis	335.20	#12	
Anal procedures (e.g. hemorrhoidectomy)	49.46	#157	
Anorexia nervosa	307.1	#428	NB-2.2 excessive exercise
Appendectomy/complications	47	#164	
Ascites	789.5	#464	
Atherosclerosis, ischemic heart disease	414	#132	
Bacterial endocarditis	421	#126	
Biliary atresia	751.61	#208	
Biliary cirrhosis	571.6	#207	
Bowel surgery		#148	
Brain tumor		#10	
Bronchial asthma		#96	
Burns	949	#457	NI-5.1 Increased nutrient needs
Cancer	e.g., leukemia 209.9 or lymphoma 202.8	#403	
Cerebral palsy	343.9	#35	
Cerebrovascular disease	436	#131	NI-51.1 excessive fat intake
Chemotherapy		#410	
Cholecystectomy	574.0	#195	
Cholecystitis	574.0	#208	
Chronic obstructive pulmonary disease	496	#88	
Cleft palate	749	#52 surgical	
Colostomy	46.10	#149	
Coma	780.0	#27	
Congenital heart disease		#137	
Congestive failure or shock	428	#127	
Constipation	564	#183	
Cor pulmonale	416.9	#78	
Coronary bypass/cardiac catheterization	36.10	#106	
Craniotomy		#1	
Cushing's syndrome	255.0	#301	
Cystic fibrosis	277	#298	
Cystinosis	270	#299	

(continued on the following page)

TABLE 1.5 *(continued)*

DRGs that Diagnose Nutrition-Related Primary Diagnoses

Condition	ICD-9 International Code	DRG Code Example	A Sample of Nutrition Diagnosis Code that may present in these conditions
Dental difficulties/oral disorders:		#185	NC-1.2 Impaired ability to manipulate or masticate food for swallowing
Broken or wired jaw	802.2		
Edentulism	525.1		
Mouth ulcers	528.9		
Tongue disorders	529.9		
Depression	311	#426	
Dermatitis herpetiformis	692.9	#284	
Diabetes insipidus	253.5	#301	
Diarrhea	558.9	#183	
Diverticular disease	562.11	#183	NI-53.4 Inadequate fiber intake
Down's syndrome	758	#429	
Dyspepsia	536.8	#183	
Electrolyte imbalances	K+, Ca++, Mg++, Na+ 275.6	#297	NI-55.1 Inadequate mineral intake NI-55.2 Excessive mineral intake
Enteritis or Crohn's disease	555.9	#179	
Epilepsy	345.9	#25	NC-2.3 Food medication interaction
Failure to thrive	783.4	#298	NI-1.4 Inadequate energy intake
Fever, unknown origin	780.6	#419	
Food allergy/allergic reactions	493	#447	
Asthma		#97	
Food poisoning	005.9	#183	NB-3.1 Intake of food and/or fluids intentionally or unintentionally
Fracture, hip	821	#210 surgical #74 medical #236 medical	
Gastrectomy or vagotomy	43.89	#155	
Gastric bypass or stapling	44.31	#288	NI-54.1 Inadequate vitamin intake
Gastric retention	536.8	#183	
Gastritis/gastroenteritis	535.5	#183	
Gestational diabetes	648.8	#372	
Glomerulonephritis, acute	580.9	#332	
Glomerulonephritis, chronic	582.9	#332	
Gluten-induced enteropathy	479.9	#183	
Goiter	240.9	#301	
Gout	274.9	#245	
Hartnup disease	270	#299	
Heart valve disorder		#136	
Heartburn/esophagitis	553.3	#183	
Hepatic cirrhosis	571.5	#202	
Hepatic encephalopathy &/or coma	572.2	#206	
Hepatitis	573.3	#206	
Hirschsprung's disease	751.3	#190	

(continued on the following page)

TABLE 1.5 *(continued)*

Homocystinuria	270.4	#299	
Huntington's chorea	333.4	#12	
Hyperinsulinism	251.1	#301	
Hyperlipidproteinemias	272.4	#299	
Hyperthyroidism	242.9	#301	
Hypoglycemia	251.2	#297	NI-53.4 Inconsistent carbohydrate intake
Hypothyroidism	244.9	#301	
Hysterectomy, abdominal	68.	#353	
Ileitis	558.9	#183	
Ileostomy, permanent	46.23	#149	
Inborn errors of CHO metabolism		#299	
Infectious mononucleosis	075	#421	
Infertility	628.9	#369	
Insulin-dependent diabetes mellitus	250.01	#295	
Intestinal fistula	569.81	#189	
Intestinal lipodystrophy (Whipple's disease)		#188	
Iron-deficiency anemia	280.9	#395	NI-55.1 Inadequate mineral intake
Irritable colon	564.1	#183	
Jaundice	782.4	#464	
Jejunoileal bypass	45.91	#288	
Ketoacidosis &/or coma	250.11	#29	
Kwashiorkor, severe PCM	260	#297	
Lactose malabsorption	371.3	#183	
Liver transplant		#191	
Low-birth-weight infant	765.1	#385	
Malabsorption syndrome	579.9	#183	
Malnutrition, mild degree	263.1	#297	
Malnutrition, moderate degree	263	#297	NI-5.2 Evident protein–energy malnutrition
Malnutrition, other PCM	263.8	#297	
Maple syrup urine disease	270.3	#299	
Meniere's syndrome		#73	
Multiple sclerosis	340	#13	
Muscular dystrophy	359.1	#256	
Myasthenia gravis	358	#12	
Myocardial infarction	410.9	#121	
Necrotizing enterocolitis		#184	
Nephritis	583.9	#332	
Nephrosclerosis	403.9	#332	
Nephrotic syndrome	581.9	#326	
Neurological trauma/spinal cord injury	952.9	#9	
Noninsulin-dependent diabetes mellitus	250.00	#294	
Nutritional anemias	281	#395	
Nutritional marasmus	261	#297	
Obesity	278.0	#297	
Osteoarthritis	715.9	#245	

(continued on the following page)

TABLE 1.5 *(continued)*

DRGs that Diagnose Nutrition-Related Primary Diagnoses

Condition	ICD-9 International Code	DRG Code Example	A Sample of Nutrition Diagnosis Code that may present in these conditions
Osteomalacia	268.2	#245	
Osteoprorosis	733	#245	
Other severe PCM/nutritional edema	262	#297	
Pancreatic insufficiency	577.8	#204	
Pancreatitis	577.0	#204	
Parathyroid disorders	252	#301	
Parkinson's disease	332	#12	NB-2.6 Self-feeding difficulty
Peptic ulcer/complications	533.9	#176	
Periodontal disease	523.9	#185	
Peripheral vascular disease	443.9	#130	
Pernicious vomiting	536.2	#183	
Phenylketonuria	270.1	#299	
Pneumonia bronchitis		#90	
Poliomyelitis	045.9	#20	
Postrenal transplantation	55.69	#302	
Prader-Willie syndrome	759.8	#385	
Pregnancy	650	#373	
Pregnancy-induced hypertention	642.4	#372	
Pressure ulcer	707.0	#271	NI-52.1 Inadequate protein intake
Psychosis		#430	
Pulmonary embolus	415.1	#78	
Pyelonephritis.chronic	590.0	#320	
Radiotherapy		#409	NC-1.1 Swallowing difficulty
Renal dialysis: hemodialysis	39.95	#317	
Renal dialysis: peritoneal	54.98	#317	
Renal failure, acute	584.9	#316	
Renal failure, chronic	585	#316	
Respiratory failure, acute	799.1	#87	
Retinal cataract surgery		#36	
Rheumatiac fever	390	#241	
Rheumatoid arthritis	714	#242	
Scleroderma		#241	
Sickle-cell anemia	282.6	#395	
Skin disorders		#284	
Acne	706.1	#284	
Chronic urticaria	708.9	#284	
Infantile eczema	692.9	#284	
Psoriasis	696.1	#283	
Systemic lupus erythematosus	710	#241	
Tonsillectomy/adenoidectomy	28.3	#57-58	
Total hip arthroplasty	81.59	#209	
Tropical sprue	579.1	#183	
Tuberculosis	011.9	#80	

(continued on the following page)

TABLE 1.5 *(continued)*

Typoid fever	002	#423	
Tyrosinemia	270.4	#299	
Ulcerative colitis/ileostomy	556	#149	
Underweight or general debility	269.9	#297	
Urolithiasis	592.9	#323	
Vitamin deficiencies		#297	
Cheilosis	266.0	#297	NI-54.1 Inadequate vitamin intake
Scurvy	267	#297	
Xerophthalmia	264.7	#47	
Vitamin D-resistant rickets	275.3	#299	
Wilson's disease	275.1	#299	
Zollinger-Ellison syndrome	251.5	#176	

Source: Escott-Stump, Sylvia. *Nutrition & Diagnosis-Related Care,* 5th ed. Lippincott Williams & Wilkins. Table B-11, page 774. American Dietetic Association. Nutrition Diagnosis: A Critical Step in the Nutrition Care Process. Chicago IL: ADA, 2006, pp. 23–31.

TABLE 1.6

Medicare Medical Nutrition Therapy (MNT) Current Procedural Terminology (CPT) Codes

CPT Code	Descriptions	Fee Schedule (CMS Reimbursement 80%; Client Is Responsible for 20%)
97802	MNT – Initial individual face-to-face patient assessment and intervention – 15 minutes	$17.92 – 15 minutes
97803	Reassessment; individual face-to face patient – 15 minutes	$17.92 – 15 minutes
97804	Group (2 or more individuals), 30 minutes	$7.09 – each group participant – 30 minutes
G0270	MNT; reassessment following initial assessment for changes in diagnosis, condition or treatment in the same calendar year as the initial assessment; individual face-to-face patient – 15 minutes	$17.92 – 15 minutes
G0271	MNT; reassessment following initial assessment for changes in diagnosis, condition or treatment in the same calendar year as the initial assessment; group – 30 minutes	$7.09 – each group participant – 30 minutes

Source: Centers for Medicare and Medicaid Services. 2002. Department of Health and Human Services Program Memorandum, Transmittal: A-020115.

during the 1990s. All racial and ethnic groups experienced improvements in rates for 10 of the 17 indicators. At the same time, the report shows that despite these overall improvements, in some areas the disparities for ethnic and racial minorities remained the same or even increased (Staveteig and Wigton 2002).

The report is part of the U.S. Department of Health and Human Services's Healthy People initiative—an effort to set health goals for each decade and then measure the progress made toward achieving them (HHS 2000). The indicators reflect various aspects of health and include infant mortality, teen births, prenatal care, and low birthweight. Also included are indicators for death rates for all causes and for heart disease, stroke, lung and breast cancer, suicide, homicide, motor vehicle crashes, and work-related injuries.

Infectious diseases such as tuberculosis and syphilis are additionally included. The percentage of children in poverty and the percentage of the population living in communities with poor air quality round out the set of measures developed to allow comparisons among national, state, and local areas on a broad set of health indicators.

All racial and ethnic groups experienced improvement in rates for 10 of the indicators: prenatal care; infant mortality; teen births; death rates for heart disease, homicide, motor vehicle crashes, and work-related injuries; tuberculosis case rate; syphilis case rate; and poor air quality (Institute of Medicine 2002; Monheit and Vistnes 2000). For five more indicators—total death rate and death rates for stroke, lung cancer, breast cancer, and suicide—there was improvement in rates for all groups except American Indians and Alaska natives. The percentage of children under 18 years old living in poverty improved for all groups except Asian or Pacific Islanders, while the percentage of low-birthweight infants improved only for black non-Hispanics.

One of the goals of the Healthy People initiative is to reduce disparities in health. However, for about half of the indicators, disparities improved only slightly, and disparities actually widened substantially for deaths due to work-related injuries, to motor vehicle crashes, and to suicide. "In many ways, Americans of all ages and in every racial and ethnic group have better health today," former Surgeon General David Satcher said. "But our work isn't done until all infants have the same chance to thrive, all mothers have equal access to prenatal care, and all Americans are equally protected from cancer, heart disease, and stroke "(Satcher 2002). Whereas the goals of Healthy People 2000 aimed at reducing disparities, the Healthy People 2010 plan aims at elimination of disparities in health among all population groups, with special emphasis on six areas: infant mortality, child and adult immunizations, human immunodeficiency virus/acquired immunodeficiency syndrome, cardiovascular disease, breast and cervical cancer screening and management, and diabetes complications (HHS 1998).

Health Care Reform in the United States

Practically all industrialized countries except the U.S. have national health care programs (American College of Physicians 1990). Coverage is generally universal (everyone is eligible regardless of health status) and uniform (everyone is entitled to the same benefits). Costs are paid entirely from tax revenues or by some combination of individual and employer premiums and government subsidization.

The concept of government-sponsored comprehensive health care is not new to the U.S. (Ahluwalia 1990). In 1934, President Franklin D. Roosevelt strongly supported national health insurance and considered including it with old age and unemployment insurance in the Social Security Act of 1935. Fearing national health insurance might jeopardize passage of the Social Security Act, however, he decided to drop the proposal. Years later, through the efforts of Presidents John F. Kennedy and Lyndon B. Johnson, Congress enacted the Social Security Amendments of 1965, which created Medicare (Title XVIII) and Medicaid (Title XIX). This legislation provided a form of public insurance for individuals without private insurance, but still made no provision for national health insurance.

Now, four decades later, increased health care costs and decreased patient satisfaction with the health care available in the U.S. have prompted consideration of a new approach to health care. Rather than proposing comprehensive reform of health care, many commentators suggest incremental reforms that address broad issues, such as health insurance, physician malpractice insurance/litigation, and incentives to induce businesses to include health promotion initiatives in their insurance plans (Johnson and Coulston 1995). Changes in medical education are also being discussed. Because **allopathic** (conventional) medicine has its roots in treatment of acute disease, medical education emphasizes training physicians in acute care and the treatment of chronic disease. Chronic diseases are the most prevalent problem in health care today and require a coordinated management team, including RDs, to address the complex issues involved (Holman 2004). Again, reform undoubtedly needs to include an increased focus on prevention of chronic disease, and nutrition therapy is a vital component of preventive care. Consumers are most concerned about prevention of disease and treating disease with both conventional and complementary and alternative methods. (Chapter 2 will focus on complementary and alternative remedies.)

allopathic—referring to modern or conventional medicine

Health care reform for the U.S. raises a formidable list of issues, including overall cost containment, universal access, emphasis on prevention, and reduction in administrative superstructure and costs (Omenn 1993). These issues require difficult decisions addressing several key questions (Division of Government Affairs 1991):

- Who should be covered?
- How can coverage be increased to reach all people?
- What services should be included in basic health care packages?
- Should health care cover both acute problems and prevention?
- Who should decide what constitutes preventive services?
- Who will pay for this coverage—consumers, employers, government?
- Where will government get the money to pay for it?
- How can health care costs be reduced or contained?
- What are the advantages and disadvantages of managed competition versus single-payer systems?

While the government remains undecided on what kind of health care system is needed and on how to pay for it, health care reform is evolving at an accelerating rate without legislation. The health care industry's determined efforts to curb growth of costs while increasing access to services has transformed the traditional approach to health care in the U.S. into one emphasizing a managed-care approach (Chima and Pollack 2002).

Nutrition as a Component of Health Care Reform

Community health care systems must include provision of nutrition services in order to preserve health and prevent disease. Realizing the importance of nutrition in overall health, even in the face of a nontraditional health care approach such as managed care, the ADA believes nutrition therapy is an essential component of disease management and health care, and that qualified RDs must provide it (Chima and Pollack 2002). Gro Harlem Brundtland, MD, MPH, director-general of the WHO, has said, "Nutrition is a cornerstone that affects and defines the health of all people, rich and poor. It paves the way for us to grow, develop, work, play, resist infection and aspire to realization of our fullest potential as individuals and societies. Putting first things first, we must realize that resources allocated to preventing and eliminating disease will be effective only if the underlying causes of malnutrition—and their consequences—are successfully addressed. This is the 'gold standard': health and human rights. It makes for both good science and good sense, economically and ethically" (Brundtland 2003).

One cannot have good health without proper nutrition. Conversely, poor nutrition contributes substantially to infant mortality; retarded growth and development of children; premature death, illness, and disability in adults; and frailty in the elderly, causing unnecessary pain and suffering, reduced productivity in the workplace, and increased health care costs.

Many believe nutrition services are the foundation of cost-effective prevention and are essential to halting the spiraling cost of health care. The ADA has advocated the inclusion of provision for nutrition services in any health care reform legislation (ADA 1995). In addition, health care reform legislation should recognize the RD as the nutrition expert of the health care team, whose scope of practice includes the following (ADA 1996):

- Nutrition assessment, for the purpose of determining individual and community needs and making appropriate nutrient intake recommendations for maintenance, recovery, or improvement of health.
- Nutrition counseling and education of individuals, families, community groups, and health professionals.
- Research and development of appropriate nutrition practice guidelines.
- Administration through management of time, finances, personnel, protocols, and programs.
- Consultation with patients, clients, and other health professionals.
- Evaluation of the effectiveness of nutrition counseling/education and community nutrition programs.

The Cost-Effectiveness of Nutrition Services The cost-effectiveness of nutrition services has been well documented (ADA 1995, 1996; Pastors 2003). Registered dietitians need to compete successfully for a fair share of the health care dollar. To do so, they must document the demand for and effectiveness of nutrition services so that they can market those services to health care officials, providers, payers, and the public.

Obviously, no payer in the health care system wants additional costs. For a new technology or service, including nutrition services, to be a reimbursable benefit, it must prove its cost-effectiveness. Only services that have a proven impact on quality of patient care will be funded. As Simko and Conklin have stated, no expenditure of resources is justified for a service that fails to achieve its intended outcome (Simko and Conklin 1989). Cost-effectiveness studies compare the costs of providing health care against a desirable change in patient health outcomes (for example, a reduction in serum cholesterol in a patient with hypercholesterolemia) (Simko and Conklin 1989).

In an effort to enhance the quality, efficiency, and effectiveness of the health care system, policy makers are urging physicians and other health professionals to develop practice guidelines or protocols that clearly specify appropriate care and acceptable limits of care for each disease state or condition. Care delivered according to a protocol has been linked with positive outcomes for the patient or client (American College of Physicians 1990). Examples of outcomes include measures of control (serum lipid profiles, glycolated hemoglobin), quality of life, dietary intake, and patient satisfaction. The ADA has developed a variety of client protocols that define the minimum number of office visits and activities required for successful nutrition intervention and the outcomes that can be expected from the dietetics professional implementing the protocol and, more recently, a series of evidence-based practice guidelines (Inman-Felton, Smith, and Johnson 1997). Nutrition protocols serve as frameworks to help practitioners in the assessment, development, and evaluation of nutrition interventions. Developing standardized protocols of care (practice guidelines) for nutrition intervention is considered a prerequisite for achieving payment for nutrition services and expanding current levels of third-party reimbursement (Gould 1991).

Documentation of specific outcomes of nutrition intervention—clinical data, laboratory measures, anthropometric measures, and dietary intake data—is also necessary. Figure 1.10 shows examples of outcome measures of nutrition intervention in burn injury, prenatal care, diabetes, and obesity (Splett 1991).

When one is developing protocols for a clinical practice setting, using an evidence-based approach will undoubtedly yield the best data with which to answer practice questions related to the protocol. The contribution of nutrition to

FIGURE 1.10 Measurable Outcomes of Nutrition Intervention

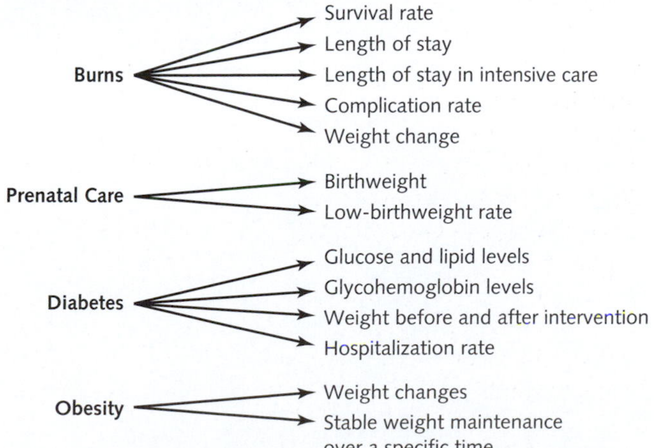

Source: From R. Gould, The next rung on the ladder: achieving and expanding reimbursement for nutrition services. Copyright © 1991 The American Dietetic Association. Reprinted by permission from *Journal of the American Dietetic Association* 91 (1991): 1383.

preventing disease, prolonging life, and promoting health is well recognized. Accumulated evidence shows that when nutrition services are integrated into health care, diet and nutritional status change. Positive outcomes of this integration include (ADA 1995; Gray and Gray 2002):

- Birthweight of infants born to high-risk mothers improves.
- Prevalence of iron-deficiency anemia is reduced.
- Serum cholesterol and risk of heart attacks are reduced.
- Glucose tolerance in persons with diabetes improves.
- Blood pressure in hypertensive patients is lowered.

Benefits of providing nutrition services far outweigh costs of providing those services. Research demonstrates the cost-effectiveness of nutrition therapy. For example:

- Oxford Health Plan operated a pilot nutrition screening program with the Medicare population in New York between 1991 and 1993. The program saved $10 for every $1 spent on nutrition counseling for these at-risk elderly patients. Monthly costs for Medicare claims alone tumbled from $66,000 before the nutrition program to $45,000 afterwards. As a result, the health plan continued its use of nutrition screenings.
- The Lewin Group documented an 8.6% reduction in hospital utilization and a 16.9% reduction in physician visits associated with nutrition therapy for patients with cardiovascular disease (Johnson 1999).
- The Lewin Group additionally documented a 9.5% reduction in hospital utilization and a 23.5% reduction in physician visits when nutrition therapy was provided to persons with diabetes mellitus (Johnson 1999).
- The University of California at Irvine demonstrated that lipid drug eligibility was obviated in 34 of 67 subjects as a result of nutrition intervention. The estimated annual cost savings from the avoidance of lipid medication was $60,652 (Sikland 1998).
- Pfizer Corporation projected $728,772 in annual savings from reduced cardiac claims of their employees from an on-site nutrition/exercise intervention program (Pfizer Corporation).
- The U.S. Department of Defense saved $3.1 million in the first year of a nutrition therapy program utilizing RDs who counseled 636,222 patients with cardiovascular disease, diabetes, and renal disease (Lewin Group Inc 1998).

The American Dietetic Association has taken significant steps to define nutrition services within our health care systems. The goal of this nutrition care process is to more "accurately describe the spectrum of nutrition care that can be provided by dietetics professionals" (Lacey 2003, p. 1063). By establishing standardized language, diagnoses and interventions built on the foundation of evidence-based research, nutrition therapy will gain increased recognition within our health care system.

Conclusion

In summary, nutrition therapy is an integral component of cost-effective medical treatment. It can reduce health costs by improving patient outcomes and reducing recovery time. Coverage of appropriate nutrition therapy, when medically necessary, should be included in any basic health care benefit package.

WEB LINKS

Agency for Health Care Research and Quality: An arm of the U.S. Department of Health and Human Services.

http://www.ahcpr.gov

American Dietetic Association: Updates on medical nutrition therapy information and resources.

http://www.eatright.org/Member/Login.cfm?TargetPage=/Member/83_12954.cfm&CFID=9131387&CFTOKEN=12357899

America's Health Insurance Plans: A trade group for HMOs and PPOs.

http://www.ahip.org

CDC's Office of Minority Health: Minority health resources and training from the Centers for Disease Control and Prevention.

http://www.cdc.gov/omh/default.htm

Centers for Medicare and Medicaid Services: The federal agency that administers Medicare, Medicaid, SCHIP, HIPAA, and CLIA.

http://www.cms.hhs.gov

Department of Health and Human Services: The principal agency for protecting the health of all Americans. Provides essential human services, especially for those who are least able to help themselves.

http://www.hhs.gov

Health Pages: Issues report cards on major managed-care plans.

http://www.thehealthpages.com

Indian Health Service: A U.S. federal agency.

http://www.ihs.gov

Joint Commission on Accreditation of Healthcare Organizations (JCAHO): The primary accreditation organization that evaluates hospitals and outpatient clinics.

http://www.jcaho.org

Medicare: Official U.S. government site for people with Medicare.

http://www.medicare.gov

National Committee for Quality Assurance (NCQA): Evaluates managed-care plans in terms of patient records, complaints, equipment, and personnel.

http://www.ncqa.org

National Institutes of Health: Comprises 27 institutes and centers. Mission is to uncover new knowledge that will lead to better health for everyone.

http://www.nih.gov

Office of Minority Health: Useful publications—such as "Closing the Gap"—and links to related sites.

http://www.omhrc.gov

Social Security Administration: The best site for information on the Medicare and Medicaid programs.

http://www.ssa.gov

END-OF-CHAPTER QUESTIONS

1. What are the main characteristics of the United State's health care system?
2. Explain the difference between the following private health insurance coverage: fee for service, health maintenance organization (HMO), and preferred provider organization (PPO).
3. Which governmental agency oversees administering Medicare/Medicaid services?
4. List four differences between Medicare and Medicaid eligibility and services.
5. What is the history of diagnostic-related groups (DRGs), and how do you think it has impacted health care coverage in the U.S.?
6. How does "capitation" affect the amount a health care provider/agency is paid for client care?
7. How does Medicare reimburse for inpatient hospital charges? Can Medicare reimburse inpatient nutrition support?
8. Can registered dietitians become Medicare program providers? Which nutrition therapies does Medicare currently cover?
9. How might the development of nutrition diagnoses and standardized language support reimbursement for nutrition services?
10. List five reasons why health care in the United States needs to be reformed.

2 Role of the Dietitian in the Health Care System

Kathryn Sucher, Sc.D., R.D.

San Jose State University

Sandra Witte, Ph.D., R.D.

California State University, Fresno

CHAPTER OUTLINE

The Registered Dietitian in Clinical Practice

The Role of the Clinical Dietitian • The Clinical Nutrition Team

Other Health Professionals—Interdisciplinary Teams

Members of the Health Care Team

Developing Critical Thinking Skills and Professional Competencies

Definition of Critical Thinking • Components of Critical Thinking • Levels of Clinical Reasoning

Introduction

The connection between diet and health has long been recognized—for example, the relationship between specific foods and the development of scurvy was discovered in the mid-1700s (Beeuwkes 1948). However, the profession of dietetics was first defined in 1899 by the American Home Economics Association as "individuals with knowledge of food who provide diet therapy for the medical profession." After 1917, dietitians were affiliated with the American Dietetic Association (ADA) (Cassell 1990). At that time, dietitians worked primarily in hospitals or in programs providing food assistance. Over time, the clinical dietitian's role in the hospital became the provision of specialized care and modification of diets to treat various medical conditions.

In the early 1970s, after high levels of malnutrition in hospitalized patients were reported (Butterworth 1974) and new and improved procedures for delivering enteral and parenteral nutrition were developed, clinical dietitians began to take a more involved role in screening patients and monitoring their needs for adequate nutrition support. In addition, as research pointed to the role of diet in the development of chronic disease, clinical dietitians began to provide primary and secondary disease prevention for such diseases as atherosclerosis, cancer, and type 2 diabetes mellitus (Winterfeldt, Bogle, and Ebro 2005). The information provided in this chapter is meant to help you understand your contribution to the care of a patient as part of the heath care team and how to apply the critical thinking skills that are necessary for the nutrition care process.

TABLE 2.1

Responsibilities and Tasks of Clinical Nutrition Team Members

Clinical Nutrition Team Member	Responsibilities	Major Tasks
Clinical Nutrition Manager	Directing the activities of clinical dietitians, dietetic technicians, and dietetic assistants	Hiring, evaluating, and training employees; reviewing productivity reports, writing job descriptions, scheduling employees, developing policies and procedures, designing performance standards, and developing and implementing goals and objectives of the department (Digh and Dowdy 1994)
Clinical Dietitian (RD)	Providing nutritional care for patients	Nutritional screening/assessment of patients to determine the presence of or risks of developing malnutrition, development of nutrition care plans, and delivery of counseling and education
Dietetic Technician (DTR)	Assisting the clinical dietitian	Gathering data for nutritional screening; assigning a level of risk for malnutrition according to predetermined criteria; administering nourishment and dietary supplements for patients and monitoring tolerance; and providing information to help patients select menus and giving simple diet instructions
Dietetic Assistant/Diet Clerk	Assisting the clinical dietitian and/or dietetic technician in some routine aspects of nutritional care	Processing diet orders, checking menus against standards, setting up standard nourishment, tallying special food requests; distributing and collecting patient menus; and distributing and collecting trays; may be involved in evaluating patient food satisfaction and helping to gather food records used to evaluate nutrient records

The Registered Dietitian in Clinical Practice

The Role of the Clinical Dietitian

Today, the practice of clinical nutrition is called *nutrition therapy* (Commission on Accreditation for Dietetic Education 2002). Clinical dietitians are the educated and trained professionals who can best deliver nutrition therapy, which includes nutritional assessment and care. Nutrition therapy is usually referred to as the nutrition care process. The nutrition care process, as explained in Chapter 3, consists of four steps: (1) nutrition assessment, (2) nutrition diagnosis, (3) nutrition intervention, and (4) nutrition monitoring and evaluation (Lacey and Pritchett 2003).

The Clinical Nutrition Team

Depending on the health care facility, nutrition therapy services may be organized along different lines. The manager of the services may have the title of chief clinical manager or clinical nutrition manager. They often report to the director of nutrition service, who commonly supervises the clinical nutrition manager and food service manager/directors. In turn, inpatient and outpatient clinical dietitians usually report to the clinical manager. Other important personnel in nutrition therapy services are registered dietetic technicians (DTR), who assist the dietitians in the nutritional screening of patients, in addition to other duties, and dietary assistants/diet clerks who are often responsible for processing diet orders, checking menus, and so forth. Table 2.1 provides detailed job specifications for clinical nutrition team members.

Clinical dietitians' services may be provided to general patient care units, such as those on a general medical or surgical floor, or may be based on a medical specialization, such as treatment of patients in intensive care units (e.g., burn/trauma unit or pediatric/neonatal intensive care units). In addition, clinical dietitians may be certified in a medical specialty and become diabetes educators, lactation consultants, or nutrition support specialists. Advanced nutrition therapy practice certifications and their requirements are listed in Table 2.2.

Other Health Professionals—Interdisciplinary Teams

In the health care setting, individuals from different disciplines communicate with each other regularly in order to best care for their patients (Wagner 2000). Dietitians are integral members of the patient's health care team and collaborate with physicians, pharmacists, nurses, speech pathologists, occupational therapists, social workers, and many others when providing nutritional treatment. Dietitians must know the

TABLE 2.2

Advanced Dietetic Practice Certifications Requirements

Specialty	Certifying Organization	Requirements
Pediatric Specialist (CSP)	American Dietetic Association/ Commission on Dietetic Registration (www.eatright.org)	Current RD, and Three years minimum length of RD status, and 4,000 hours of pediatric practice within the last five years, and Successful completion of the Board Certification as a Specialist in Dietetics examination.
Renal Specialist (CSR)	American Dietetic Association/ Commission on Dietetic Registration (www.eatright.org)	Current RD, and Three years minimum length of RD status, and 4,000 hours of renal practice within the last five years, and Successful completion of the Board Certification as a Specialist in Dietetics examination.
Diabetes Educator (CDE)	National Certification Board for Diabetes Education (www.ncbde.org)	A *minimum* of two years (to the day) of professional practice experience in diabetes self-management training, and A *minimum* of 1,000 hours of diabetes self-management training experience, and Current employment in a defined role as a diabetes educator with a minimum of four hours per week, or its equivalent, at the time of application, and Successful completion of the Certified Diabetes Educator Examination.
Nutrition Support (CNSD)	National Board of Nutrition Support Certification (www.ptcny.com/clients/NBNSC)	It is recommended that candidates have at least two years of experience in specialized nutrition support (parenteral and enteral nutrition). Successful completion of the Certification Examination for Nutrition Support Dietitians.
Lactation Consultant (IBCLC)	The International Board of Lactation Consultant Examiners (www.iblce.org/)	Completed comprehensive continuing education in lactation, and Have had extensive practical experience providing breastfeeding counseling, and Passed a certification examination.

roles of the other team members in order to be effective. Table 2.3 covers the education and training requirements for health professionals with whom a dietetic student should be familiar when first starting to practice dietetics.

Members of the Health Care Team

The practice of medicine by **medical doctors** includes the diagnosis, treatment, correction, advisement, or prescription for any human disease, ailment, injury, infirmity, deformity, pain or other condition, physical or mental, real or imaginary (American Medical Association 2006). All physicians in the United States (U.S.) have advanced training and certification in a specialized area of medicine or surgery. Table 2.4 lists the recognized board specialties and subspecialties. Nutritionally, doctors are responsible for ordering nutrition support and writing diet orders. Dietitians must consult with physicians in order to start or modify a patient's nutritional order.

The largest group of health care workers in the United States is **nurses**. They assist individuals in the performance of activities contributing to health or recovery from injury or illness, are responsible for assisting patients in carrying out therapeutic plans initiated by physicians and other health professionals, and assist other members of the

medical doctor—a health professional who has earned a post-bachelor degree of doctor of medicine (MD) or doctor of osteopathy (DO) and who has passed a licensing examination

nurse—a health care worker who has earned at least an associate's degree in nursing, has been licensed by the state, and assists patients in activities related to maintaining or recovering health

TABLE 2.3

Education and Certification Requirements of Selected Members of the Health Care Team

Health Profession	Education	Degree Initials	Credentialing	Web Resources
Medical Doctor	Four-year post-bachelor degree plus residency	MD	State licensure exam	American Medical Association (www.ama-assn.org)
Osteopathic Doctor	Four-year post-bachelor degree plus residency	DO	State licensure exam	American Osteopathic Association (www.DO-Online.org)
Nurse	Two- or four-year degree	AA (2-year) BSN (4-year)	State licensure exam (RN)	National League for Nursing Accrediting Commission (NLNAC) (www.nlnac.org)
Pharmacist	Six-year post-secondary education	PharmD	State licensure exam	American Pharmacists Association (www.aphanet.org)
Occupational Therapist	Master's degree required starting in 2007	MOT, MS, or MA	National exam for registration (OTR)	American Occupational Therapist Assn. (www.aota.org)
Speech Language Pathologist	Master's degree plus a clinical fellowship	MS or MA	National exam for Certificate of Clinical Competence (CCC)	American Speech-Language-Hearing Association (ASHA) (www.asha.org/default.htm)
Social Worker	Bachelor's degree or master's degree	BSW or MSW	State licensing, certification, or registration	The National Association of Social Workers (www.naswdc.org)
Health Educator	Bachelor's degree	BS or BA	Voluntary credentialing	American Association for Health Education (www.aahperd.org)

medical team (Potter and Perry 1997). Since they provide care for 24 hours a day, 7 days a week, nurses are commonly responsible for documenting a patient's food intake as well as notifying the dietitian if a patient has inadequate intake.

licensed pharmacist—a licensed health professional with a doctorate of pharmacy (PharmD) who compounds and dispenses medications, checks laboratory results for therapeutic drug levels, and reviews risk for drug interactions

occupational therapist—a health professional who has obtained a bachelor's degree and passed a national registration exam, who helps individuals with mentally, physically, developmentally, or emotionally disabling conditions improve their ability to perform tasks in their daily living and working environments

speech-language pathologist—a health professional who has earned a master's degree and passed a national examination, who assesses, diagnoses, treats, and helps to prevent speech, language, cognitive, communication, voice, swallowing, fluency, and other related disorders

A **licensed pharmacist** compounds and dispenses medications following prescriptions issued by physicians, dentists, or other authorized medical practitioners. In addition, they monitor laboratory results for therapeutic drug levels as well as electrolyte levels for patients receiving parenteral nutrition, and review risks for drug-drug and drug-nutrient interactions (American Association of Colleges of Pharmacy 2006). Pharmacists are commonly responsible for compounding parenteral nutrition support solutions.

Occupational therapists (OTs) work with individuals who have conditions that are mentally, physically, developmentally, or emotionally disabling to help improve their ability to perform tasks in their daily living and working environments. They assist clients in performing activities of all types, ranging from using a computer to caring for daily needs such as dressing, cooking, and eating (U.S. Department of Labor, Bureau of Labor Statistics 2005). Occupational therapists often work with patients with swallowing disorders and clients with physical disabilities to provide special instructions on eating and using adaptive feeding devices.

Speech-language pathologists, sometimes called speech therapists, assess, diagnose, treat, and help to prevent speech,

TABLE 2.4

American Board of Medical Specialties

Specialty Board	
Allergy & Immunology	Orthopedic Surgery
Anesthesiology	Otolaryngology
Colon & Rectal Surgery	Pathology
Dermatology	Pediatrics
Emergency Medicine	Physical Medicine & Rehabilitation
Family Practice	Plastic Surgery
Internal Medicine [1]	Preventive Medicine
Medical Genetics	Psychiatry & Neurology
Neurological Surgery	Radiology
Nuclear Medicine	Surgery
Obstetrics & Gynecology	Thoracic Surgery
Ophthalmology	Urology

*Subspecialties of Internal Medicine include: Cardiovascular Disease, Clinical & Laboratory Immunology, Critical Care Medicine, Endocrinology, Diabetes & Metabolism, Gastroenterology, Geriatric Medicine, Hematology, Infectious Disease, Medical Oncology, Nephrology, Pulmonary Disease, Rheumatology, and Sports Medicine.

(This list was effective March 2002. American Board of Medical Specialties® www.abms.org/)

[1] The subspecialties are only noted for Internal Medicine.

language, cognitive, communication, voice, swallowing, fluency, and other related disorders. Speech-language pathologists working in a health center provide clinical services to individuals with swallowing disorders, and they work closely with physicians, nurses, and dietitians to help assess the need for and to provide nutrition support (U.S. Department of Labor, Bureau of Labor Statistics 2005).

Medical and public health **social workers** provide persons, families, or vulnerable populations with the psychosocial support needed to cope with chronic, acute, or terminal illnesses. They also advise family caregivers, counsel patients, and help plan for patients' needs after discharge by arranging community and financial resources to cover medical needs and costs (U.S. Department of Labor, Bureau of Labor Statistics 2005). Social workers help patients obtain food-related services and access to food.

In health care settings, **health educators** teach patients about medical procedures, operations, services, and therapeutic regimens; promote self-care; and instruct individuals about how to protect, promote, or maintain their health and reduce risky behaviors, such as smoking cigarettes. They address topics such as disease prevention/health promotion, exercise, nutrition, pregnancy, stress, substance abuse, and violence (Coalitions of National Health Education Organizations 2005).

Developing Critical Thinking Skills and Professional Competencies

The 2001 Medicare benefit legislation defined medical nutrition therapy as "nutritional diagnostic, therapy, and counseling services for the purpose of disease management, which are furnished by a registered dietitian or nutrition professional" (Department of Health and Human Services 2001). As mentioned previously, clinical dietitians are educated and trained professionals, and are considered to be the members of the health care team best able to deliver accurate and appropriate clinical judgments in order to provide appropriate nutritional care. Completing the didactic program in dietetics is your first step to becoming a registered dietitian (RD). The next step, supervised practice, allows you to apply your education in the clinical setting.

As a dietetics student, you will acquire a great deal of knowledge during your didactic education. However, students usually do not have the opportunity to apply their knowledge other than through hypothetical disease case assignments. When a student enters the clinical environment, usually as a dietetic intern, he or she quickly finds that providing nutrition care requires more than mastery of a textbook. The textbook gives you information about the nutrition care process, or a medical condition, its diagnosis, and dietary treatment, but it does not integrate the diagnosis or treatment with the patient's own experiences, symptoms, behaviors, values, social perspectives, and other medical problems.

To provide optimal nutritional care, all of the aspects of a patient's life must be considered. To do this, the practitioner must be able to think critically in order to solve problems and develop a path that leads to the best solution for a client's needs. Dietetic educators know that the dietitians' problem-solving skills, along with their critical thinking skills, evolve with experience and practice. Thus, the path to becoming an RD requires both education and practice.

Definition of Critical Thinking

The act of thinking involves using the mind to organize and integrate information, identify relationships, make inferences, form conclusions, and make decisions. When thinking

social worker—a professional with at least a bachelor's degree in social work who provides persons, families, or vulnerable populations with psychosocial support, advises family caregivers, counsels patients, and helps plan for patients' needs after discharge

health educator—an individual with a bachelor's degree who educates patients about medical practices, self-care, and health promotion/disease prevention

critically, one also challenges assumptions, creates alternatives, and makes informed decisions. In 1990, a group of 46 expert critical thinkers convened the Delphi Consensus Group and defined critical thinking as:

> . . . purposeful, self-regulatory judgment which results in interpretation, analysis, evaluation, and inference as well as explanation of the evidential, conceptual, methodological, criteriological or contextual considerations upon which that judgment is based. (*The Delphi Report* 1990)

Critical thinking skills are very important, but few students or practitioners understand their application in the nutrition care process. The following section will outline the components of critical thinking and their applications as well as the broader implications of critical thinking for the American Dietetic Association's Standards of Professional Practice for Dietetics (1998).

Components of Critical Thinking

The dietitian who effectively uses critical thinking skills will make clinical judgments that result in effective nutritional care. Five components have been identified as essential in critical thinking: specific knowledge base, experience, competence, attitudes, and standards (Kataoka-Yahiro and Saylor 1994).

Specific Knowledge Base The first component of critical thinking is the dietitian's knowledge about nutrition and its role in health and disease. RDs will all have a minimum level of knowledge based on the Standards of Education of the American Dietetic Association set by the Commission on Accreditation for Dietetic Education (2002). Most dietitians exceed the minimum standards, depending on the programs from which they have graduated, the continuing education choices they make, and the advanced degrees they pursue. The dietitian's knowledge base includes information and theories related to communications, physical and biological sciences, social sciences, research, food, nutrition, management, and health care systems.

The dietitian's knowledge base is also continually changing and expanding. Learning is a lifelong process, and dietitians engage in continuing education throughout their careers. Nutrition is a developing science, and as new information becomes available, dietitians must apply the new developments to practice. The Standards of Professional Practice of the ADA require dietetic professionals to reflect on their practice in order to anticipate and react to change and remain effective practitioners (American Dietetic Association 1998).

Experience The second component of critical thinking is experience in dietetics practice. Dietetics students in nutrition therapy courses often feel overwhelmed by all the information they are expected to know. Though they can effectively apply the information to "mock patients" in contrived case studies, students do not think themselves capable of applying what they have learned to patients in the "real" clinical setting. This occurs in part because dietitians do not learn from textbooks alone; they also learn by observing, listening to patients, interacting with other health care professionals, and reflecting on the situations that arise.

Dietitians are required to complete a supervised practice experience (dietetic internship or coordinated program) before they can be eligible to write the exam for registration. This period of supervised practice allows the dietetic intern to gain experience in the clinical environment without the risk of causing serious harm to patients. Real patients with real problems provide the most effective learning experiences by stimulating the dietetic intern's intellectual curiosity and promoting retention of the information.

As an illustration, consider a skill that you developed in the past and now may take for granted, the skill of driving a car. Sometime around the age of 15 or 16, you had the opportunity to drive a car after years of observing someone else drive a car. You most likely had to complete a driver education course that included learning about all the legal aspects of driving. Then you probably got behind the wheel with the driver training instructor or one of your parents. No doubt that first ride was a little frightening, and you may have made some decisions that could be improved upon. Now compare your performance that first time behind the wheel with the way you drive today. The difference is that now you have had experience driving and making related decisions.

Competencies The third component of critical thinking involves the cognitive processes that a dietitian goes through to make clinical judgments. These processes are essentially the same as those used by physicians and other health care professionals and are referred to as medical problem solving (Elstein, Shulman, and Sprafka 1978).

In addition to having knowledge and skills related to nutritional care, you must also have the ability to identify problems and make decisions regarding the most appropriate solutions. These competencies or abilities include the scientific method, problem solving, decision making, and diagnostic reasoning.

Scientific Method The basic steps in the scientific method are:

- Identify the phenomenon.
- Collect data about the phenomenon.
- Formulate a hypothesis to explain the phenomenon.
- Test the hypothesis through experimentation.
- Evaluate the hypothesis.

For an example of the application of the scientific method to clinical practice, see Box 2.1.

BOX 2.1 CLINICAL APPLICATIONS—THE SCIENTIFIC METHOD IN PRACTICE

The dietitian is alerted that an elderly patient is not eating most of the food on his trays at mealtime (identification of phenomenon). The dietitian checks the medical record and sees that the patient lost his wife six months ago and has no family to visit him. The dietitian visits the patient, and she or he learns that the patient wears dentures. By observing and asking questions, the dietitian determines that the patient is experiencing a great deal of discomfort with them (collection of data about the phenomenon). The dietitian suspects that the cause of the patient's poor intake might be that he cannot chew the foods due to pain (formulation of a hypothesis to explain the phenomenon); hence, the dietitian requests that soft, easily chewed foods be served to the patient (test of the hypothesis through experimentation). The next day, the nurse reports that the patient's intake has been 100% for the last three meals (evaluation of the hypothesis).

Evidence-Based Dietetics Practice As defined by the American Dietetic Association, "**evidence-based dietetics practice** is the incorporation of systematically reviewed scientific evidence into food and nutrition practice decisions. It integrates professional expertise and judgment with client, customer and community values and evaluates outcomes" (American Dietetic Association Evidence Analysis Library 2006).

Changes in nutrition therapy recommendations are inevitable because of new developments in science and medicine, including ongoing research in nutrition therapy. Some of what you learn in school today will be outdated by the time you finish your internship and become a registered dietitian. As a dietitian, you must be able to critically review research findings by utilizing the research methodology skills you learned during your dietetic education. Dietetics practice should not be based on tradition but on evidenced-based research (see Box 2.2 for an example). This process is not unique to dietetics, and it is rapidly becoming the standard for all health care professions.

The ADA has been instrumental in the development of the Evidence Analysis Library and ADA Evidence Based Guidelines which are posted online (www.adaevidencelibrary.com) for its members. As defined by ADA, a *guideline* is a "systematically developed statement based on scientific evidence to assist practitioner and patient decisions about appropriate health care for specific clinical circumstances" (American Dietetic Association Evidence Analysis Library 2006). Guidelines are tied to the evidence-based library, which is updated as new research is published.

Problem Solving The process of problem solving involves obtaining information about the problem and then using the information to effectively solve the problem. For instance, suppose you walk into a room and flip the light switch, but the light does not go on. To determine the source of the problem, you would probably check the light bulb first. If the light bulb is not burned out, then other possible sources of the problem need to be checked. A similar process is used to determine a patient's nutritional problem. The practitioner can assume that there is a problem when the patient's nutritional status is not optimal; the patient's nutrition-related information is then collected in order to find "clues" that point to the solution (see Box 2.3 for an example).

BOX 2.2 CLINICAL APPLICATIONS—EVOLVING STANDARDS OF PRACTICE

Imagine you are an RD working in a facility where the standard of practice is that the initial diet order for all postoperative patients is clear liquids. However, you have just attended the local dietetic district meeting, and the speaker mentioned that most abdominal postoperative patients can tolerate a regular house diet and do not necessarily need to be on clear liquids. What do you do? A search of the medical literature reveals two research articles that support the use of regular diets postoperatively (Jeffery et al. 1996; Pearl et al. 2002). You should critically analyze each article and summarize the evidence. If the evidence supports use of regular diets postoperatively, this information should be presented to the appropriate staff members at your facility for discussion.

BOX 2.3 CLINICAL APPLICATIONS—PROBLEM SOLVING

Consider the client who presents with abdominal cramps, diarrhea, and flatulence throughout the day. Assessment information includes: 50-year-old Asian-American female, postmenopausal; family medical history includes a 75-year-old mother who developed osteoporosis resulting in a broken hip; and her 24-hour diet recall indicates she recently started drinking an 8-oz. glass of 1% fat milk at every meal and sometimes before going to bed, because her doctor had told her to consume more calcium in her diet. Although additional information remains to be collected to rule out other possible problems, one plausible explanation for the client's abdominal symptoms could be excessive intake of the milk sugar lactose.

evidence-based dietetics practice—dietetics practice in which systematically reviewed scientific evidence is used to make food and nutrition practice decisions

Decision Making Making a decision involves making a choice. The activities involved in decision making include:

- Identify and define a problem or situation.
- Assess all options for solving the problem.
- Weigh each option against a set of criteria.
- Test possible options.
- Consider the consequences of the decision (examine the positive and negative aspects of each option).
- Make a final decision.

The activities do not necessarily take place in a particular sequence. The clinician is usually moving back and forth and considering things simultaneously. The outcome of this process is a decision that is informed and supported by evidence and reasoning. Continuing the previous example from Box 2.3, if it is determined that a patient has lactose intolerance, options for solving this problem could include: discontinue the use of milk but take a calcium supplement; drink smaller quantities of milk more frequently; or continue to consume milk with every meal but use products containing lactase.

Diagnostic Reasoning Diagnostic reasoning is defined as a series of clinical judgments that result in an informal judgment or a formal diagnosis (Carnevali and Thomas 1993). Physicians are responsible for making a patient's medical diagnosis. However, dietitians continually use the process of diagnostic reasoning to make judgments about a patient's progress and/or response to nutrition therapy.

For example, a patient with protein-energy malnutrition (PEM) will manifest specific signs and symptoms associated with this condition. Once nutrition intervention has begun, the dietitian continues to observe anthropometric and laboratory values and compare them with those common to PEM. This diagnostic reasoning process allows the dietitian to make clinical inferences about the patient's progress.

Attitudes The fourth component of critical thinking is related to attitudes. Attitudes reflect the dietitian's values and should ensure that clinical judgment is made fairly and responsibly. Table 2.5 summarizes the attitudes necessary for effective critical thinking. In Box 2.2, concerning standards for postoperative feeding, the dietitian's attitude includes integrity, thinking independently, and risk taking. His or her questioning of a standard not based on the latest scientific evidence (integrity) may change a traditional practice of postoperative feeding (thinking independently); it takes courage to take action even when change is based on solid research that improves client care (risk taking).

Standards The final component of critical thinking is standards, both intellectual and professional (see Table 2.6). Application of intellectual standards involves a rigorous approach to critical thinking and ensures that clinical decisions are sound.

TABLE 2.5

Critical Thinking Attitudes

Attitude	Application
Confidence	When you are confident, the patient is more trusting of your competence.
Thinking independently	Consider all viewpoints, and base your decision on your own conclusions about the issue.
Fairness	Listen to both sides of a discussion; weigh all facts.
Responsibility and authority	Ask for help when you need it. Follow established Standards of Practice.
Risk taking	If you have reason to question the judgment of others, do so.
Discipline	Be thorough at all times. Follow established procedures.
Perseverance	Be determined to find the most effective solution. Don't settle for quick solutions.
Creativity	Look for different options when outcomes are not as expected.
Curiosity	Always ask, "*Why?*" Find out as much as you can before making a judgment.
Integrity	Question and test your personal knowledge and beliefs. Be willing to admit inconsistencies in your beliefs.
Humility	Admit your limitations. Be willing to rethink a situation and seek additional knowledge.

Source: R. Paul (1993). The art of redesigning instruction. In J. Willsen & A. Blinker (Eds.), *Critical Thinking: how to prepare students for a rapidly changing world.* Santa Rosa, CA: Foundation for Critical Thinking.

In the client with diarrhea presented earlier (Box 2.3) these standards should be used so a nutritional diagnosis can be determined and a treatment plan developed. The RD should seek to ensure that the dietary information obtained is adequate, that any confusing statements made by the client are clarified, and that the nutritional diagnosis is plausible and consistent with the assessment data collected.

Professional standards for critical thinking include ethical standards, criteria-based evaluation of outcomes, and standards for professional responsibility. The ADA has established both a Code of Ethics and Standards of Professional Practice, and both include the need for measurable, evidence-based evaluation of outcomes. For example, one

TABLE 2.6

Intellectual Standards that Universally Apply to Critical Thinking	
• Accurate	• Fair
• Adequate	• Logical
• Broad	• Plausible
• Clear	• Precise
• Complete	• Relevant
• Consistent	• Significant
• Deep	• Specific

Source: R. Paul (1993). The art of redesigning instruction. In J. Willsen & A. Blinker (Eds.), *Critical Thinking: how to prepare students for a rapidly changing world.* Santa Rosa, CA: Foundation for Critical Thinking.

principle in the Code of Ethics is, "The dietetic practitioner practices dietetics based on scientific principles and current information" (American Dietetic Association 1999). While the Standards of Professional Practice are broader statements to help guide the practice of dietetics, they do require that dietitians continuously improve their knowledge and skills, and that they regularly evaluate the quality of their practice, revising it if necessary (American Dietetic Association 1998). An important link between the Standards of Professional Practice and the Code of Ethics is that **outcomes research** should be a consequence of regular evaluation of practice quality. If abdominal postoperative diets are changed from clear liquid diets to regular house diets, data should be collected on patient outcomes. Did this change improve, worsen, or have no effect on surgical outcome? Published research on the outcomes would be added to the evidenced-based research library, possibly resulting in the release of updated clinical guidelines for postoperative feeding. In the classroom, critical thinking typically is used for exams and assignments, but for the practitioner, critical thinking leads to high levels of clinical reasoning that could influence the practice of dietetics.

Levels of Clinical Reasoning

When a dietetic student completes his or her internship and then passes the Dietetic Registration Examination, he or she is considered to have entry-level competence. As the new dietitian develops professionally and moves beyond entry level, he or she becomes more proficient and develops expertise in his or her area of practice.

For an entry-level dietitian, critical thinking may be at a basic level. The dietitian has only limited experience and relies on the facts and sets of rules or principles to make decisions. These facts and principles are perceived as absolutely correct because they are established by the authorities. There may be little or no adaptation of the principles to the patient's own unique needs. As the dietitian becomes more experienced, she or he begins to examine alternatives independently and systematically, disconnecting from the authorities. The dietitian is better prepared to anticipate possible outcomes and identify a broader range of options. It is evident that there is more than one solution to a problem and that the patient's own unique needs will influence which solutions are viable.

The highest level of critical thinking involves analysis of the entire situation: the person, the illness, the meaning of the illness to that person, the person's lifestyle, the family's needs, the social influences, and the physical environment in which the person lives (Kataoka-Yahiro and Saylor 1994). At this level, the dietitian is acting in support of the patient, the principles of nutrition therapy, and the professional standards that underlie the discipline of dietetics. A specific characteristic of a dietitian at the highest level of critical thinking is accountability for decisions and continuous quality of care assessment.

Conclusion

During your nutrition therapy course, you will be required to complete assignments that will help you apply much of the information presented in this text. You may find some assignments so overwhelming that you do not know where to start. Do not despair; after you have completed the first such assignment, the subsequent ones will become easier, and you will gain confidence just as the inexperienced driver does with daily practice driving a car. As with driving, after you start your supervised practice, your fears will decrease and your competence will grow with each new patient.

outcomes research—evaluation of care that focuses on the status of participants after receiving care

WEB LINKS

The American Dietetic Association (ADA): Information and resources for professionals, dietetic students, and the public. Members can access the *Journal of the American Dietetic Association*, Evidence Analysis Library, Nutrition Diagnoses Resources and the *Nutrition Care Manual.*

http://www.eatright.org

National Certification Board for Diabetes Education (NCBDE): Information on eligibility requirements and the written examination for diabetes educators.

http://www.ncbde.org

National Board of Nutrition Support Certification (NBNSC): Information on eligibility requirements and the written examination for nutrition support dietitians.

http://www.ptcny.com/clients/NBNSC

The International Board of Lactation Consultant Examiners (IBLCE): Information on eligibility requirements and the written examination for lactation consultants.

http://www.iblce.org

Occupational Outlook Handbook: Handbook on career information published by the U.S. Department of Labor, Bureau of Labor Statistics.

http://www.bls.gov/oco/home.htm

The Critical Thinking Community: The Foundation and Center for Critical Thinking provides resources on critical thinking for instruction in primary and secondary schools, colleges, and universities.

http://www.criticalthinking.org

END-OF-CHAPTER QUESTIONS

1. Identify members of the clinical nutrition care team. What are the major tasks performed by the clinical dietitian and the chief clinical manager?
2. What are the five components needed for critical thinking skills? Why is supervised practice a requirement for becoming a registered dietitian?
3. Why is continuing education necessary for the practice of dietetics?
4. What are the components of medical problem solving? How does evidence-based dietetics practice contribute to critical thinking skills?
5. A new friend finds out you are nutrition major and asks your advice about overeating late in the day. She tells you that she has no time to eat lunch and wants to save money, but then she eats too much when she gets home. Suggest three possible solutions. What are the possible consequences (both positive and negative) of each solution? How could your attitude affect each solution?
6. Why is outcomes research necessary for the advancement of dietetics practice?

3

The Nutrition Care Process

Karen Lacey, M.S., R.D., C.D.

University of Wisconsin–Green Bay

CHAPTER OUTLINE

Framework for Nutrition Care: Evaluation of Nutritional Status

Purpose of Providing Nutrition Care

ADA's Standardized Nutrition Care Process (NCP)

Steps of the NCP

Step 1: Nutrition Assessment • Step 2: Nutrition Diagnosis • Step 3: Nutrition Intervention • Step 4: Nutrition Monitoring and Evaluation

Introduction

A person's state of health is a continuum that can span from (1) being totally healthy and resistant to disease, to (2) having an acute illness, to (3) living with a chronic disease or condition that significantly alters one's capacity to function well, and finally to (4) having a terminal illness. Regardless of the state of health, adequate nutrition is important; poor nutrition may lead to a variety of health problems and may even make them worse. Table 3.1 illustrates how the focus of providing nutrition care is different for various states of health. An example of a primary prevention strategy may be promoting appropriate caloric balance and physical activity to prevent undesirable weight gain, and thus maintain health, whereas **nutrition intervention** in patients with chronic diseases is intended to help reduce complications and restore nutritional balance. It is important to determine both a person's health status and nutritional status, because these guide the type of nutrition intervention provided. The purpose of this chapter is to describe how dietetics professionals can provide quality nutrition care to individuals for the purpose of improving both nutrition and health status.

Framework for Nutrition Care: Evaluation of Nutritional Status

The adequacy of a person's nutrient intake is determined by (1) evaluating the amounts and types of nutrients that a person consumes, and then (2) comparing those findings to nutrient requirements needed at various stages throughout the continuum of growth, health, and illness. If one consumes adequate amounts and types of nutrients to support and optimize a given health state, the balance between nutrient intake and nutrient requirements is considered to

nutrition intervention—a specific set of activities and associated materials used to address a (nutrition-related) problem (Lacey and Pritchett 2003)

TABLE 3.1

Health State and Focus of Nutrition Interventions

Health State	Focus of Nutrition Intervention
Resistant and Resilient	Primary prevention strategies to maintain health
Stage of Susceptibility (at risk)	Primary prevention strategies to promote health and reduce risk
Presymptomatic Disease (subclinical)	Early identification and intervention to prevent or delay progression
Clinical Condition (physical or pathological change)	Diagnosis and treatment to reduce severity and duration, and restore or improve health
Chronic Condition, Disease or Disability (permanently diminished or altered capacity)	Disease management to reduce complications, accommodate limitations, and enable optimal functioning
Terminal Illness	Palliative care/comfort care to relieve discomfort and maintain dignity

Source: Adapted from Conceptual Framework for a Standardized Nutrition Language by P Splett and members of ADA's Standardized Language Task Force 2004.

be "good." However, if there is an inadequate or excessive intake of nutrients, or the form of nutrients is not well utilized by the body, a nutrient "imbalance" is present. Nutrient imbalance can result in significant health consequences. An excess of kilocalories (kcal) and undesirable eating patterns are associated with the progression of a number of chronic diseases such as obesity, diabetes mellitus, coronary artery disease, and hypertension. Inadequate intake of kcal and certain nutrients such as protein, on the other hand, can contribute to a compromised immune system and poor wound healing.

Nonetheless, evaluating nutrient intake alone does not describe the broader picture of nutritional status. Even though nutrient balance implies that one is consuming all of the necessary nutrients in their appropriate amounts, assessing nutritional status is not merely a simple equation of intake compared to needs. A person's nutritional status implies that a number of internal and external factors are also present that support optimum nutritional health. A wide variety of factors influence a person's ability to maintain optimum nutritional status (see Table 3.2) **Human biological factors** such as age, sex, physiological stages, illness, and physical and functional abilities determine nutrient requirements. For example, a mother who is breastfeeding needs to consume more kcal and protein compared to a nonbreastfeeding mother. Kcal and protein needs are also increased following major surgery. Furthermore, the form of nutrient may need to be altered depending on the degree of organ function. A person who has had a large portion of the small intestine removed may not be able to digest large molecular nutrients such as triglycerides and would benefit from specialized nutrient forms such as short- or medium-chain triglycerides. **Lifestyle factors** including attitudes, knowledge, and behaviors influence the type of choices that one makes about food and physical activity. For instance, understanding which foods contain saturated fats and cholesterol can influence what type and amount of meats and spreads one consumes. **Food and nutrient factors** describe the nutrients that are available for use by the body. Obtaining accurate information about a person's dietary intake is essential to evaluating nutritional status. **Environmental factors** such as social and

human biological factors—conditions that determine a person's nutrient requirements; one's age, gender, and stage of growth and development are used to estimate kcal and nutrient needs; illnesses that alter organ function or metabolism influence not only the amount of nutrients required but also the form of nutrients that the body needs and can tolerate

lifestyle factors—a person's knowledge, attitudes/beliefs, and behavior patterns directly impact the choices that are made regarding food and physical activity; assessment of these factors provides information about a person's ability and/or readiness to make behavior changes

food and nutrient factors—the amount and type of foods and nutrients that are consumed and therefore made available to the body

environmental factors—social and economic factors (wages, transportation, etc.) that impact both lifestyle choices and the consumption of food and nutrients; other external factors such as food safety and sanitation determine the quality of food that is consumed, and food availability/access contributes to the amount and type of food consumed

TABLE 3.2

Factors Affecting Nutritional Status

1. Human Biology Factors (determine nutrient requirements—normal, increased, decreased, change in form, etc.)
   a. Biological factors (age, sex, genetics)
   b. Physiological phases (growth, pregnancy, lactation, aging)
   c. Pathological factors (disease, trauma, altered organ function or metabolism)
2. Lifestyle Factors (determine food, physical activity, and related choices)
   a. Attitudes/beliefs
   b. Knowledge
   c. Behaviors
3. Food and Nutrient Factors (determine the type and amount of nutrients available for use by the body)
   a. Intake/composition
   b. Quantity
   c. Quality
   d. Feeding route and form
4. Environmental Factors (external influences that impact consumption and lifestyle)
   a. Social (cultural food practices and beliefs, parenting, peer influences)
   b. Economic (household finances, economy of the community/country)
   c. Food safety and sanitation (contaminated or unwholesome food, unsafe food handling)
   d. Food availability/access
5. System Factors (external influences that impact on delivery and services)
   a. Health care system
   b. Educational system
   c. Food supply system (industry, agriculture, institutions)

Source: Adapted from "Conceptual Framework for a Standardized Nutrition Language by P Splett and members of ADA's Standardized Language Task Force. 2004.

cultural food preferences and practices are external influences that impact both food consumption and lifestyle choices. For example, people frequently consume more food than usual at a social event where food is served. It is also common that adults prefer the types of foods that were typically consumed in the household while growing up as a child. Finally, **system factors** such as the health system, educational system, and food supply system impact the delivery of food, nutrition, and health services. A family whose income is near or at the poverty level and that has limited access to health care will likely purchase fewer fresh foods and may use the services of urgent care more frequently.

Key Concepts: Nutritional Status

- Adequacy of nutrient intake is important but does not completely describe nutritional status.
- Determination of a person's nutritional status is dependant on a wide variety of factors (biological, pathological, behavioral, cognitive, environmental, and systems).

Purpose of Providing Nutrition Care

The purpose of providing nutrition care is to restore a state of nutritional balance by influencing whatever factors are contributing to the imbalance or altered state of nutritional status. Because of the wide variety of and interaction among the many variables noted above, identifying the underlying causes of a nutritional status imbalance can be a complex process. If a person's caloric intake is less than desired, it is important to determine which, if any, of the following are contributing to the cause of this problem: a disease condition that is increasing the nutrient needs, a lack of knowledge as to how many kcal are in certain foods, a lack of resources (money, food preparation skills, transportation), or a cultural belief about limiting the intake of certain foods. The type of intervention and/or education that will be provided depends significantly on the underlying cause of this problem.

Registered dietitians (RDs) are highly trained health professionals who are best prepared to provide nutrition care. The American Dietetic Association's Commission on Accreditation of Dietetic Education (CADE) has identified eight different subject areas as part of the foundation knowledge and skills required of dietetics professionals. These include: (1) communication, (2) physiological and biological sciences, (3) social sciences, (4) research, (5) food, (6) nutrition, (7) management, and (8) health care systems. Academic preparation in all of these areas assists in the ability to understand and evaluate the degree to which any of the factors in Table 3.2 influence the nutritional balance equation.

Key Concepts: Nutrition Care

- Providing nutrition care can influence and change the factors that contribute to an imbalance in nutritional status and thus restore an improved state of nutritional health.
- Registered dietitians are highly qualified to provide nutrition care.

system factors—external factors (health care, education, and food supply systems) that influence the type of services that are available to individuals and how these services are delivered

ADA's Standardized Nutrition Care Process (NCP) and Model Promotes High-Quality, Individualized Nutrition Care

A Brief History of ADA's NCP

Dietetics professionals have always conducted **nutrition assessments** and provided some type of nutrition care. In the past, this was commonly described as a process of assessment, planning, implementation, and evaluation. However, a variety of methods were used to assess a patient's nutritional status and provide nutrition care. A nutrition assessment typically consisted of obtaining extensive data from the anthropometric, biochemical, clinical, and dietary intake (ABCD) model. The dietetics professional reviewed and recorded data that described all of the conditions that could be associated with nutrition. A nutrition assessment frequently resulted in the dietetics professional describing nutritional imbalance as mild, moderate, or severe. Much data was collected, but it was not necessarily linked directly to specific nutrition problems. Nutrition care, in the form of education or provision of modified diets or nutrition support, was frequently implemented to restore a state of nutritional balance. Similar types of nutrition care were given to clients with similar health needs. Patients with chronic renal failure generally required modification of protein, fluid, and electrolytes such as sodium, potassium, and phosphorus. Patients who desired weight loss were advised to decrease caloric intake and increase physical activity. However, the processes that were used among professionals were not necessarily similar. The language used to describe the care was not consistent, and nutrition problems were not always named. Specific signs and symptoms that resulted from a nutrition problem were not always clearly identified. Documentation did not necessarily link the assessment data to a problem, the etiology of a problem was commonly assumed but not always articulated, and specific goals and outcomes may or may not have been established. Hence, a great deal of variation in practice occurred.

In early 2002, dietetics professionals identified the need to create a more standardized method of providing nutrition care in order to improve both the quality of care and the likelihood of producing positive outcomes. A member task force was appointed by ADA's House of Delegates to address this important professional issue, and in March 2003, the House of Delegates adopted the Nutrition Care Process and Model for "implementation and dissemination to the dietetic profession and the Association for the enhancement of the practice of dietetics" (Lacey and Pritchett 2003). ADA's **Nutrition Care Process** is defined as "a systematic problem solving method that dietetics professionals use to critically think and make decisions to address nutrition related problems and provide safe, effective, high quality nutrition care" (Lacey and Pritchett 2003). This NCP consists of four distinct but interrelated and connected steps: (1) nutrition assessment, (2) **nutrition diagnosis**, (3) nutrition intervention, and (4) **nutrition monitoring and evaluation** (Lacey and Pritchett 2003). The second step, nutrition diagnosis, is the newest addition to the nutrition care process.

Standardized Nutrition Diagnostic Language Standardized language refers to a uniform terminology that is used to describe practice. The lack of a standardized nutrition language and common terminology has made it very difficult for dietetics professionals to communicate with each other and other health professionals. Many other health professionals, including physicians, nurses, and physical therapists, had developed standardized terminology; however, none existed for nutrition care. Because of this lack of agreement for nutrition language, there was no easy way to classify, measure, and report on the outcomes of nutrition interventions in various patient populations. The lack of specific uniform terminology used in dietetics practice made it impossible to gather data needed for research, education, and reimbursement justification via outcomes analysis. Most notably missing was language that described the specific nutrition problems. Therefore, a Standardized Language Task Force was formed in May of 2003, immediately following the adoption of the NCP, to develop standardized language for nutrition.

Sixty-two terms, now known as the Nutrition Diagnostic Terminology, were approved by the ADA's board of directors and House of Delegates in May 2005 (see Table 3.3). Dietetics professionals can now use these terms to clearly describe specific types of nutrition problems that contribute to a person's nutritional imbalance. Nutrition diagnoses give

nutrition assessment—a systematic process of obtaining, verifying, and interpreting data in order to make decisions about the nature and cause of nutrition-related problems (Lacey and Pritchett 2003)

nutrition care process (NCP)—a systematic problem solving method developed by the ADA that dietetic professionals use to think critically, make decisions addressing nutrition-related problems, and provide safe, effective, high-quality nutrition care (Lacey and Pritchett 2003)

nutrition diagnosis—the identification and descriptive labeling of an actual occurrence of a nutrition problem that dietetics professionals are responsible for treating independently (Lacey and Pritchett 2003)

nutrition monitoring and evaluation—an active commitment to measuring and recording the appropriate outcome indicators relevant to a nutrition diagnosis in order to determine the degree to which progress is being made and whether or not the client's goals are being met (Lacey and Pritchett 2003)

standardized language—a uniform terminology that is used to describe practice

TABLE 3.3

Nutrition Diagnostic Terminology

INTAKE	NI
Defined as "actual problems related to intake of energy, nutrients, fluids, bioactive substances through oral diet or nutrition support"	
Caloric Energy Balance (1)	
Defined as "actual or estimated changes in energy (kcal)"	
❑ Hypermetabolism (Increased energy needs)	NI-1.1
❑ Increased energy expenditure	NI-1.2
❑ Hypometabolism (Decreased energy needs)	NI-1.3
❑ Inadequate energy intake	NI-1.4
❑ Excessive energy intake	NI-1.5
Oral or Nutrition Support Intake (2)	
Defined as "actual or estimated food and beverage intake from oral diet or nutrition support compared with patient goal"	
❑ Inadequate oral food/ beverage intake	NI-2.1
❑ Excessive oral food/ beverage intake	NI-2.2
❑ Inadequate intake from enteral/parenteral nutrition infusion	NI-2.3
❑ Excessive intake from enteral/parenteral nutrition	NI-2.4
❑ Inappropriate infusion of enteral/parenteral nutrition (use with caution)	NI-2.5
Fluid Intake (3)	
Defined as "actual or estimated fluid intake compared against patient goal"	
❑ Inadequate fluid intake	NI-3.1
❑ Excessive fluid intake	NI-3.2
Bioactive Substances (4)	
Defined as "actual or observed intake of bioactive substances, including single or multiple functional food components, ingredients, dietary supplements, alcohol"	
❑ Inadequate bioactive substance intake	NI-4.1
❑ Excessive bioactive substance intake	NI-4.2
❑ Excessive alcohol intake	NI-4.3
Nutrient (5)	
Defined as "actual or estimated intake of specific nutrient groups or single nutrients as compared with desired levels"	
❑ Increased nutrient needs (specify)	NI-5.1
❑ Evident protein- energy malnutrition	NI-5.2
❑ Inadequate protein- energy intake	NI-5.3
❑ Decreased nutrient needs (specify)	NI-5.4
❑ Imbalance of nutrients	NI-5.5
Fat and Cholesterol (51)	
❑ Inadequate fat intake	NI-51.1
❑ Excessive fat intake	NI-51.2
❑ Inappropriate intake of food fats (specify)	NI-51.3
Protein (52)	
❑ Inadequate protein intake	NI-52.1
❑ Excessive protein intake	NI-52.2
❑ Inappropriate intake of amino acids (specify)	NI-52.3
Carbohydrate and Fiber Intake (53)	
❑ Inadequate carbohydrate intake	NI-53.1
❑ Excessive carbohydrate intake	NI-53.2
❑ Inappropriate intake of types of carbohydrate (specify)	NI-53.3
❑ Inconsistent carbohydrate intake	NI-53.4
❑ Inadequate fiber intake	NI-53.5
❑ Excessive fiber intake	NI-53.6
Vitamin Intake (54)	
❑ Inadequate vitamin intake (specify)	NI-54.1
❑ Excessive vitamin intake (specify)	NI-54.2
❑ A	❑ C
❑ Thiamin	❑ D
❑ Riboflavin	❑ E
❑ Niacin	❑ K
❑ Folate	❑ Other ________
Mineral Intake (55)	
❑ Inadequate mineral intake (specify)	NI-55.1
❑ Calcium	❑ Iron
❑ Potassium	❑ Zinc
❑ Other ____________	
❑ Excessive mineral intake (specify)	NI-55.2
❑ Calcium	❑ Iron
❑ Potassium	❑ Zinc
Other ____________	

(continued on the following page)

TABLE 3.3 *(continued)*

Nutrition Diagnostic Terminology

CLINICAL	NC	BEHAVIORAL-ENVIRONMENTAL	NB
Defined as "nutritional findings/problems identified as related to medical or physical conditions"		Defined as "nutritional findings/problems identified as related to knowledge, attitudes/beliefs, physical environment, or food supply and safety"	
Functional (1)		Knowledge and Beliefs (1)	
Defined as "change in physical or mechanical functioning that interferes with or prevents desired nutritional consequences"		Defined as "actual knowledge and beliefs as observed or documented"	
❑ Swallowing difficulty	NC-1.1	❑ Food and nutrition-related knowledge deficit	NB-1.1
❑ Chewing (masticatory) difficulty	NC-1.2	❑ Harmful beliefs/attitudes about food or nutrition-related topics (use with caution)	NB-1.2
❑ Breastfeeding difficulty	NC-1.3	❑ Not ready for diet/lifestyle change	NB-1.3
❑ Altered GI function	NC-1.4	❑ Self-monitoring deficit	NB-1.4
Biochemical (2)		❑ Disordered eating pattern	NB-1.5
Defined as "change in capacity to metabolize nutrients as a result of medications, surgery, or as indicated by altered lab values"		❑ Limited adherence to nutrition-related recommendations	NB-1.6
❑ Impaired nutrient utilization	NC-2.1	❑ Undesirable food choices	NB-1.7
❑ Altered nutrition-related laboratory values	NC-2.2	Physical Activity and Function (2)	
❑ Food-medication interaction	NC-2.3	Defined as "actual physical activity, self-care, and quality of life problems as reported, observed, or documented"	
Weight (3)		❑ Physical inactivity	NB-2.1
Defined as "chronic weight or changed weight status when compared with usual or desired body weight"		❑ Excessive exercise	NB-2.2
❑ Underweight	NC-3.1	❑ Inability or lack of desire to manage self-care	NB-2.3
❑ Involuntary weight loss	NC-3.2	❑ Impaired ability to prepare foods/meals	NB-2.4
❑ Overweight/obesity	NC-3.3	❑ Poor nutrition quality of life	NB-2.5
❑ Involuntary weight gain	NC-3.4	❑ Self-feeding difficulty	NB-2.6
		Food Safety and Access (3)	
		Defined as "actual problems with food access or food safety"	
		❑ Intake of unsafe food	NB-3.1
		❑ Limited access to food	NB-3.2

Source: From Nutrition Diagnosis: A Critical Step in the Nutrition Care Process © 2006 American Dietetic Association. Reproduced with permission.

purpose and focus to the assessment step. They are the missing link in nutrition care and a critical step in the nutrition care process. The standardized nutrition diagnostic terminology now allows dietetics professionals to make explicit that which has been implicit in the past.

Improved Quality of Care

The NCP is a standardized process—not standardized care. A standardized process refers to a consistent structure and framework used to provide nutrition care, whereas standardized care implies that all clients receive the same care. When professionals use a systematic process with standardized language, there is less variation of practice and a higher degree of predictability in terms of outcomes. The Institute of Medicine defines quality as "the degree to which health services for individuals and populations increase the likelihood of desired health outcomes and are consistent with current professional knowledge" (Institute of Medicine 2001). Dietetics professionals' quality performance can be assessed by measuring clients' outcomes (end results of intervention and treatment) *or* the degree to which providers adhere to an accepted care process. Clients and patients want service that results in positive outcomes. Health care administrators, payers, and the government require cost-effective, high-quality service based on current evidence-based practice. Use of ADA's NCP is the means by which dietetics professionals greatly increase their potential to provide high-quality nutrition care to individuals and groups. It combines both the *process of care* (the steps of the nutrition care process in a systematic and consistent manner) and the *content of care* (incorporation of evidence-based practice guides) to produce improved quality of care and improved nutritional status. Figure 3.1 illustrates how combining the process of care with the content of care improves quality.

Critical Thinking

The NCP enhances the quality of care provided by dietetics professionals and challenges the dietetics professional to use good critical thinking skills. Critical thinking integrates facts, informed opinions, active listening, and observations; it is creative and rational, and it requires the ability to conceptualize. Each step of the NCP identifies unique and specific types of critical thinking that, when applied, improve the likelihood that the process is being implemented in an effective manner. These specific critical thinking skills are described in Table 3.4.

Key Concepts: ADA's Standardized Nutrition Care Process

The four steps of the Nutrition Care Process are:

- Nutrition Assessment
- Nutrition Diagnosis
- Nutrition Intervention
- Nutrition Monitoring and Evaluation

By using the Nutrition Care Process, dietetics professionals can demonstrate that nutrition care improves outcomes because it:

- Is a systematic method used to make decisions to provide safe and effective care
- Provides a common language for documenting and communicating the impact of nutrition care

FIGURE 3.1 Demonstrating Quality

Source: Adapted from Slide #8 of ADA's Nutrition Care Process and Model: Providing Quality Nutrition Care in a Variety of Settings Power Point Prepared by ADA's Nutrition care Process Task Force, 2004

TABLE 3.4

Critical Thinking Used in the Nutrition Care Process

Nutrition Assessment	Nutrition Diagnosis	Nutrition Intervention	Nutrition Monitoring and Evaluation
Observe for nonverbal and verbal cues to prompt effective interviewing methods. Determine appropriate data to collect. Select assessment tools and procedures. Apply assessment tools in valid and reliable ways. Distinguish relevant from irrelevant data. Distinguish important from unimportant data. Validate the data. Organize and categorize the data in a meaningful framework that relates to nutrition problems. Determine when a problem requires consultation with or referral to another provider.	Find patterns and relationships among the data and possible causes. Make inferences ("If this continues to occur, then this is likely to happen"). State the problem clearly and singularly. Suspend judgment (be objective and factual). Make interdisciplinary connections. Rule in /rule out specific diagnoses. Prioritize the relative importance of problems.	Set and prioritize goals. Define the nutrition prescription or basic plan. Make interdisciplinary connections. Initiate behavioral and other interventions. Match intervention strategies with client needs, diagnoses, and values. Choose from among alternatives to determine a course of action. Specify the time and frequency of care.	Select appropriate indicators/measures. Use appropriate reference standards for comparison. Define where patient/client is now in terms of expected outcomes. Explain variance from expected outcomes. Determine factors that help or hinder progress. Decide between discharge or continuation of nutrition care.

Source: Adapted from Lacey K & Pritchett E. Nutrition Care Process and Model: ADA adopts road map to quality care and outcomes management. *J Amer Diet Assoc.* 2003; 103:1061–1072.

- Relies on an evidence-based approach.
- Uses specific critical thinking skills for each step.

Big Picture of Nutrition Care: The Model

The provision of nutrition care does not occur in a vacuum. The Nutrition Care Model in Figure 3.2 is a visual representation that reflects key concepts of each step of the NCP and illustrates the greater context within which nutrition care is provided. The model also identifies other systems that influence and impact the quality of care. It depicts the overlapping relationships of these components and how they interact to result in the best quality nutrition care possible.

Central Core

Central to providing nutrition care is the relationship between the client and the dietetics professional or team of dietetics professionals. The client's previous experiences and readiness for change as well as the ability of the dietetics professional to establish trust, demonstrate empathy, and communicate effectively with the client, influence this relationship. If a person believes that changing the intake of saturated fat will decrease his or her risk of cardiovascular disease and has had previous nutrition counseling that was helpful, that person is more likely to want to meet again with

FIGURE 3.2 Nutrition Care Process and Model

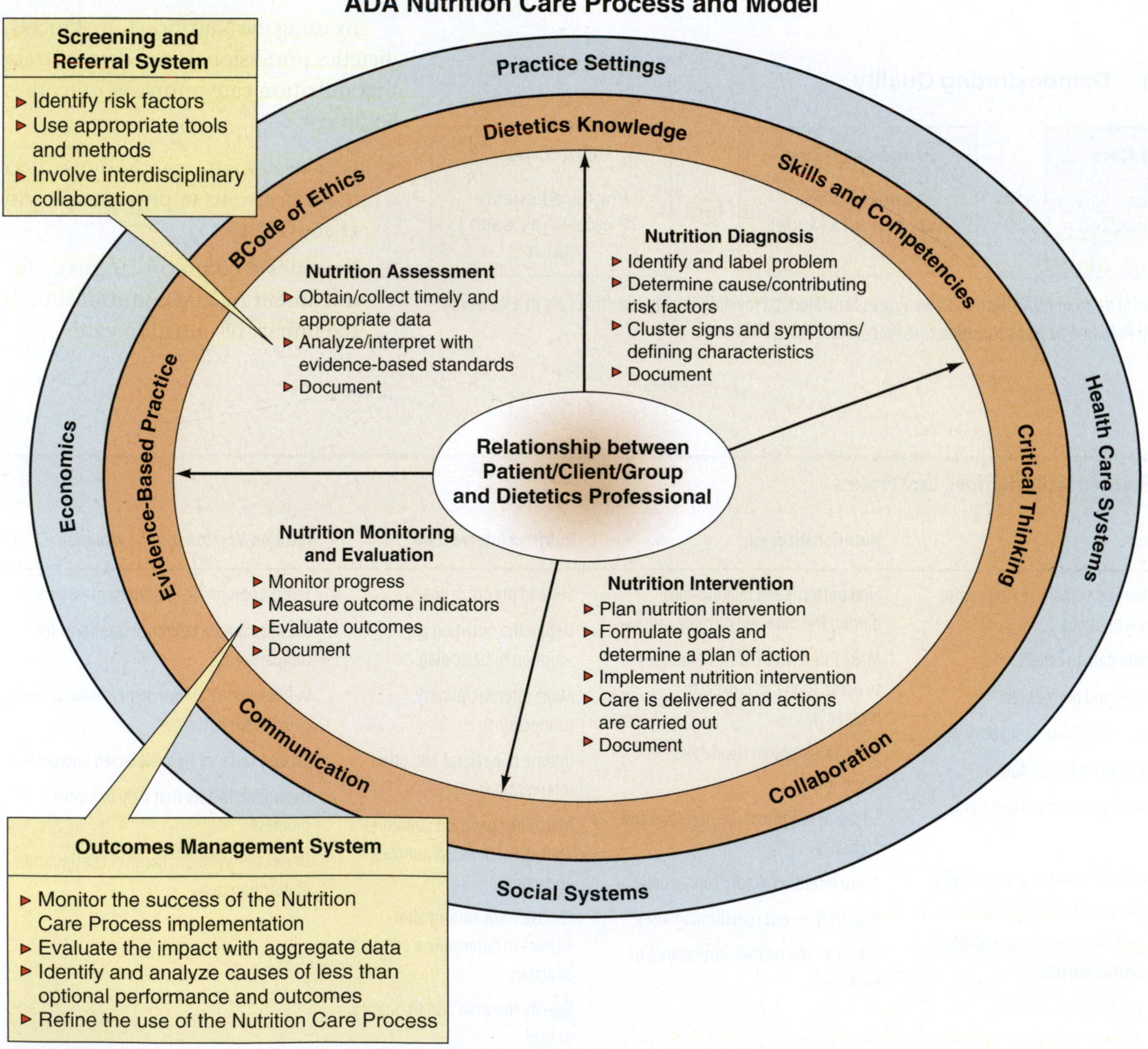

Source: Reprinted from Journal of the American Dietetic Association, Vol. 103, Lacey K and Pritchett E, *Nutrition Care Process and Model: ADA adopts road map to quality care and outcomes management,* pages 1061–1072, Copyright 2003, with permission from American Dietetic Association.

a dietitian; in contrast, an individual who believes that lifestyle behavior changes will have little to no impact on the risk of disease will probably be less receptive. It is important for the dietetic professional to establish trust and be able to communicate effectively with others, and it is essential that the client be actively involved in the care whenever possible and if culturally acceptable. This means that the client is aware of the purpose of care and participates in the decision making process of goal setting and intervention selection. This central core reinforces the importance of providing care that is individualized and client-focused.

Two Outer Rings

The *outermost ring* of the model identifies environmental factors, including practice settings, health care systems, social systems, and economics, that have an impact on the ability of the client to receive and benefit from the interventions of nutrition care. Dietetics professionals need to assess these factors and be able to evaluate the degree to which they may either be positive or negative influences on the outcomes of care. A health care plan that allows for up to three outpatient nutrition counseling sessions per year, where the client pays only a small portion of the cost of the nutrition counseling, is a much more positive external influence than a health care plan in which the client has to pay for the entire visit. The *inner adjoining ring* recognizes the strengths that dietetics professionals bring to the nutrition care process. These include professional knowledge/skills and competencies, code of ethics, evidence-based practice, and skills of critical thinking, collaboration, and communication. The dietetics professional must develop appropriate entry-level knowledge and skills through didactic course work and supervised practice prior to passing the National Registration Examination. In addition, he or she must obtain continuing education to meet individual learning needs in order to continue to practice as a credentialed professional. Providing nutrition care that is based on sound scientific evidence increases the likelihood that there will be a positive outcome for the client. Nutrition care also requires a great deal of collaboration with other health care professionals and community services.

Supportive Systems: Screening and Referral System and Outcomes Management System

The two supportive systems, the screening and referral system and the outcomes management system, although essential to providing effective and efficient nutrition care, are not considered steps of the nutrition care process itself primarily because they may not be accomplished solely by dietetics professionals. A **screening and referral system** identifies those individuals or groups who would benefit from nutrition care provided by dietetics professionals. Screening may be completed by nurses, by clients themselves, or through physician referral. Regardless of whether dietetics professionals are actively involved in conducting the screening process, they are still accountable for providing input into the development of appropriate screening parameters to ensure that the screening process asks the right questions. They should also evaluate how effective the screening process is in terms of correctly identifying the clients who require nutrition care. Screening parameters need to be tailored to the population and to the nutrition care services provided. A referral process may also ensure that the client has an identifiable method of being linked to dietetics professionals who will ultimately provide the nutrition care or medical nutrition therapy that is necessary. For example, using the DETERMINE check list (see Chapter 5) at elderly congregate meal sites can identify those clients who are at risk and could then be seen by the dietitian employed at a senior center.

The other system supporting the NCP is the **outcomes management system**. An outcomes management system is used to evaluate the effectiveness and efficiency of the entire NCP process (assessment, diagnosis, interventions, outcomes, costs, and other factors) when nutrition care is provided to a number of patients. Outcomes management is different from the fourth step, nutrition monitoring and evaluation, which refers to the evaluation of a single patient's/client's progress in achieving goals and desired outcomes. Outcomes management links care processes and resource utilization with outcomes. Health care organizations use complex information management systems to manage resources and track performance. Selected information documented throughout the nutrition care process is entered into these central information management systems and structured databases. Relevant aggregate data, data from a number of individual sources that have been summed together to create a larger whole, are collected and analyzed in a timely manner. Performance can be adjusted based on this analysis in order to improve outcomes. For example, data collected over time might reveal that less than 50% of clients seen in an outpatient setting received follow-up appointments, and that of those 50%, less than half met desired outcome goals. These data would then be used to more closely

screening and referral system—a supportive system within the Nutrition Care Process and Model that helps identify those persons who would benefit from nutrition care (Lacey and Pritchett 2003)

outcomes management system—a system that evaluates the effectiveness and efficiency of the entire NCP: assessment, diagnosis, implementation, cost, and other factors; it links care processes and resource utilization with outcomes (Lacey and Pritchett 2003)

examine the system of access and record keeping as well as the type of interventions used to provide care. Such an analysis can assist in the creation of policies for increasing the number of patients who receive follow-up care and in better achieving expected outcomes. When nutrition services have systems in place that can measure and evaluate data from many clients (aggregate data), these data can then be combined with data from other nutrition care providers and be part of evidence-based research that demonstrates the benefit and effectiveness of nutrition care.

Key Concepts: Nutrition Care Process and Model

- Nutrition Care is provided within the context of a larger model that includes a central core focused on individual care and positive relationships.
- Both external (environmental) and internal (resources of dietetics professionals) factors influence the type of nutrition care provided.
- The steps of the Nutrition Care Process are supported by two other systems, the screening and referral system and the outcomes management system. Dietetics professionals participate in both of these systems, but may not have sole responsibility for accomplishing the tasks they perform.

Steps of the NCP

Step 1: Nutrition Assessment

The first step of the NCP provides important information that helps determine a person's nutritional status. It is initiated by the referral and/or screening of individuals or groups for nutritional risk factors. A nutrition assessment is a very systematic process of obtaining, verifying, and interpreting data in order to make decisions about the nature and cause of nutrition-related problems (see Chapter 5). It is an ongoing, dynamic process that involves not only initial data collection, but also continual reassessment and analysis of a client's needs and condition. A nutrition assessment is not in and of itself a measure of a dietetics professional's level of productivity. A nutrition assessment provides data to accurately describe nutrition problems and facilitate the formulation of a nutrition diagnosis at the next step of the NCP. Assessment data also provides a means to reevaluate the nutrition problem as part of nutrition monitoring and evaluation, the fourth step in the NCP.

Nutrition assessment focuses on understanding the wide variety of factors (human biological, lifestyle, food and nutrient, environment, and system) that influence a person's nutritional status, as noted previously in Table 3.2. These data provide information about the types of nutrition problems that exist as well as their likely causes. Data gathered during the assessment step are also used to describe the severity of the problem. For example, if the nutrition problem is NB 1.7, "undesirable food choices," further clarification of the specific type of food that is undesirable, such as *32 ounces of high-fructose fruit drinks a day*, would be used to describe and quantify that problem. These data will then determine what types of outcomes are desired. In this case, the amount and type of beverages consumed would be tracked over time. Once a problem has been accurately defined and quantified, client goals can be established. If a client is consuming too much high-fructose fruit drink, the goal might be to consume 4 ounces of 100% fruit juice in place of the fruit drinks and substitute water for the remainder of the fluid needs in the day.

Obtain and Verify Appropriate Data The specific type of data gathered in the assessment can vary depending on a number of factors such as practice settings or the individual's/group's present health status. Dietitians who serve clients at a Women, Infants, and Children (WIC) clinic will obtain anthropometric data on head circumference and height and weight plotted on growth charts in order to assess the development of children. Dietitians in outpatient clinics will obtain height and weight measurements for adults and may also gather information about body fatness using skinfold thickness or bioelectrical impedance analysis. Recommended practices, as indicated in ADA's Evidence-Based Guides for Practice, may also influence the type of data collected in a nutrition assessment. Lipid profiles for patients with type 2 diabetes and cardiovascular diseases would be appropriate, whereas BUN, creatinine, and serum phosphorus should be evaluated when providing nutrition care to patients with renal disease.

Whether an initial assessment or a reassessment is being conducted influences the type of data collected. A thorough, detailed diet history is valuable during an initial assessment, but a brief investigation of a specific type of nutrient such as fiber might be more valuable in follow-up visits, especially if inadequate intake of fiber was one of the nutrition problems identified during the initial appointment. The dietitian needs to know what type of data is most appropriate to collect and to be able to determine whether those data are valid and accurate. For example, a stated weight may or may not represent the current weight of a client. Accurate and valid diet history information is dependant on the ability of the dietitian to establish a trusting and nonthreatening relationship with the client. In all cases, the data that are reviewed should be related to the types of nutrition problems likely to be encountered.

Cluster and Organize Assessment Data Nutrition assessment data that includes anthropometric, biochemical, clinical, and dietary intake data should be clustered and organized in a meaningful way that relates to specific types of nutrition problems. Clustering and organizing the data can reveal possible nutrition problem domains and/or classifications from which a specific nutrition diagnostic statement can then be formulated. Table 3.5 uses the factors affecting nutritional status (as illustrated in Table 3.2) as a means of organizing and clustering the assessment data according to possible types of nutrition diagnosis. The dietetics professional will examine anthropometric data with the intended purpose of ruling in or ruling out the possibility of weight classification problems: underweight, involuntary weight loss or gain, or overweight/obesity. Specific data from a dietary intake assessment reveal important information about the extent of possible intake nutrition diagnoses found in the Intake domain. Information gathered in an interview that reveals a person's knowledge and beliefs about health and nutrition allows the dietetics professional to rule in or rule out possible problems in the Knowledge and Behavior classification of the Behavioral-Environmental domain. This structure, when used to guide nutrition assessment, provides focus and purpose to the gathering of data. Each piece of assessment data helps answer a question regarding the possible impact it may have on the presence, severity, and cause of a specific nutrition problem. Using an organized structure that focuses on nutrition problem areas assists dietetics professionals to think critically about the meaning of the data and logically move into the next steps of the NCP.

Evaluate the Data Using Reliable Standards It is not only important that data be linked to specific types of problems; it is equally important that the information obtained in a nutrition assessment be compared to reliable standards or ideal goals. It is essential to use current and scientifically valid standards in the determination of whether a nutrition problem actually exists, and if so, to what degree. A nutrition diagnosis is made only after the data gathered are evaluated and compared to an established reference standard or recommendation (taking into account individual physiological needs). For example, calculating accurate estimated energy needs requires the use of appropriate physical activity factors. A reliable standard to use when evaluating the intake of kcal from fat, carbohydrates, and protein would be the Institute of Medicine's 2002 Acceptable Macronutrient Distribution Ranges for healthful diets (AMDR), whereas the ideal goals established as part of Adult Treatment Panel III (ATP III) would be used to evaluate and compare intake of saturated fats and monounsaturated fats of clients with cardiovascular risk or disease (NCEP 2002).

Key Concepts: NCP Step 1, Nutrition Assessment

- Nutrition assessment should ensure that appropriate and reliable data are collected for use in determining the existence of specific nutrition problems.
- Organizing and categorizing data according to possible nutrition diagnoses improves the efficiency and effectiveness of nutrition assessment.

Step 2: Nutrition Diagnosis

Nutrition diagnosis, the second step of the NCP, consists of the identification and descriptive labeling of an actual occurrence of a nutrition problem that dietetics professionals are responsible for treating independently. Nutrition diagnosis is the heretofore missing link between nutrition assessment and nutrition intervention. A nutrition diagnosis should not be confused with a medical diagnosis; the distinction between the two is extremely important. A medical diagnosis is the art of distinguishing one disease from another and describes the nature of that disease. A disease is further defined as any deviation from or interruption of the normal structure or function of a part, organ, or body system. Treatment of a disease involves the management and care of a patient for the purpose of combating the disorder (Dorland 2003). Many diseases have profound effects on a person's nutritional balance. The alteration of normal structure and function of organs can result in changes in nutrient intake, losses, requirements, and/or utilization. In some cases, nutrition therapy may be one of the most important ways of treating and managing the disease.

A nutrition diagnosis, in contrast to a medical diagnosis, is written in terms of a client problem for which nutrition-related activities provide the primary intervention. The goal of nutrition care is to improve the nutritional status of a client/patient by impacting the underlying cause of the nutritional problem. Nutrition diagnoses and care focus on nutrition issues that may be consequences of or contribute to diseases. Nutrition diagnoses also address behaviors that impact food choices.

Nutrition diagnostic terms are grouped into three domains: Intake, Clinical, and Behavioral-Environmental. The **Intake domain** contains nutrition problems that are related to the intake of energy, nutrients, fluids, and bioactive substances through oral diet or nutrition support. Labels such as inadequate, excessive, or inappropriate are used to describe the specific nutrient or substance that is altered. The **Clinical domain** contains nutrition problems that are related to medical or physical conditions. These include problems in swallowing, chewing, digestion, absorption, and maintaining

intake domain—domain that contains standardized nutrition diagnostic terms that describe actual problems related to the intake of energy, nutrients fluids, and bioactive substances through oral diet or nutrition support (enteral or parenteral nutrition) (Nutrition Diagnosis, ADA 2006)

clinical domain—Domain that contains standardized nutrition diagnostic terms that describe nutritional problems that relate to medical or physical conditions (Nutrition Diagnosis, ADA 2006)

TABLE 3.5

Organized Nutrition Assessment

Each piece of nutrition assessment data is collected for a specific purpose. It helps answer the following types of questions:

1. What is this data telling us about a person's nutritional status?
2. What possible nutrition diagnosis/es might this data provide evidence for?
3. What additional data— necessary to validate the presence of the suspected nutrition diagnoses?

These tables illustrate how different types of assessment data are associated with the various factors affecting nutritional status and the exploration of specific nutrition problems. As dietetics professionals collect data, they should simultaneously be thinking about the "why" (factors of nutritional status) and the "what" (possible nutrition diagnoses).

Factors Affecting Nutritional Status (Why):
Human Biology—determines nutrient requirements (normal, increased, decreased); amounts and types

Biology

- Age
- Sex
- Genetic endowment
- Weight

Physiological phases

- Growth
- Pregnancy
- Lactation
- Aging

Pathological

- Metabolism
- Trauma
- Neoplasm
- Infection
- Chronic disease

Physical/Functional

- Physical activity
- Physical and mental impairments affecting adl

Types of Nutrition Assessment Data

Anthropometric Data

- Height
- Weight (current, ideal, usual)
- BMI
- Waist-to-hip ratio
- Head circumference
- Skin-fold thickness
- Other girth measurements
- Growth
- Weight change

Biochemical /Laboratory Data

- Albumin, prealbumin
- Hemoglobin/hematocrit
- Electrolytes
- Glucose
- Lipid panel
- Micronutrient level indicators
- Other disease or body system function indicators (e.g. creatinine, BUN, CRP)
- Genetic (phenotype)

Medical/Health Hhistory

- Chief complaint/current concern
- Present/past illness
- Current health
- Surgeries, injuries
- Family history of disease

Nutrition-Focused Physical Examination

- Overall musculature
- Hair, skin, nails
- Eyes
- Oral (tongue, gums, lips, mucous membranes)
- Chewing/swallowing abilities
- Physical limitations and adaptations
- Bowel habits

Physical Activity and Exercise

- Activity patterns; intensity, frequency and duration

Possible Nutrition Diagnoses (What)

Weight (3)

Defined as "chronic weight or changed weight status when compared with usual or desired body weight"

❑ Underweight	NC-3.1
❑ Involuntary weight loss	NC-3.2
❑ Overweight/obesity	NC-3.3
❑ Involuntary weight gain	NC-3.4

Nutrient (5)

Defined as "actual or estimated intake of specific nutrient groups or single nutrients as compared with desired levels"

❑ Increased nutrient needs *(specify)* ________	NI-5.1
❑ Evident protein- energy malnutrition	NI-5.2
❑ Decreased nutrient needs *(specify)* ________	NI-5.4

(continued on the following page)

TABLE 3.5 *(continued)*

Biochemical (2)

Defined as "change in capacity to metabolize nutrients as a result of medications, surgery, or as indicated by altered lab values"

- ❑ Impaired nutrient utilization NC-2.1
- ❑ Altered nutrition-related laboratory values NC-2.2
- ❑ Food-medication interaction NC-2.3

Caloric Energy Balance (1)

Defined as "actual or estimated changes in energy (kcal)"

- ❑ Hypermetabolism *(Increased energy needs)* NI-1.1
- ❑ Increased energy expenditure NI-1.2
- ❑ Hypometabolism *(Decreased energy needs)* NI-1.3

Functional (1)

Defined as "change in physical or mechanical functioning that interferes with or prevents desired nutritional consequences"

- ❑ Swallowing difficulty NC-1.1
- ❑ Chewing (masticatory) difficulty NC-1.2
- ❑ Breastfeeding difficulty NC-1.3
- ❑ Altered GI function NC-1.4

Physical Activity and Function (2)

Defined as "actual physical activity, self-care, and quality of life problems as reported, observed, or documented"

- ❑ Physical inactivity NB-2.1
- ❑ Excessive exercise NB-2.2
- ❑ Inability or lack of desire to manage self-care NB-2.3
- ❑ Impaired ability to prepare foods/meals NB-2.4
- ❑ Poor nutrition quality of life NB-2.5
- ❑ Self-feeding difficulty NB-2.6

Factors Affecting Nutritional Status (Why):
Lifestyle— determines food, physical activity and related choices

Behavior (actions)
Knowledge
Attitudes/beliefs

Types of Nutrition Assessment Data

Personal History

Age
Gender
Occupation
Role in family
Education level

Nutrition and Health Awareness and Management

Knowledge of nutrition and dietary recommendations
Attitude toward food and eating
Self-management
Concerns, goals, priorities, motivation
Past MNT, education

Medical/Health History

Mental/emotional health
Cognition, perception

(continued on the following page)

TABLE 3.5 *(continued)*

Organized Nutrition Assessment

Possible Nutrition Diagnoses (What)

Knowledge and Beliefs (1)

Defined as "actual knowledge and beliefs as observed or documented"

❑ Food and nutrition-related knowledge deficit	NB-1.1
❑ Harmful beliefs/attitudes about food or nutrition-related topics	NB-1.2
❑ Not ready for diet/lifestyle change	NB-1.3
❑ Self-monitoring deficit	NB-1.4
❑ Disordered eating pattern	NB-1.5
❑ Limited adherence to nutrition-related recommendations	NB-1.6
❑ Undesirable food choices	NB-1.7

Physical Activity and Function (2)

Defined as "actual physical activity, self-care, and quality of life problems as reported, observed, or documented"

❑ Inability or lack of desire to manage self-care	NB-2.3
❑ Impaired ability to prepare foods/meals	NB-2.4
❑ Poor nutrition quality of life	NB-2.5

Factors of Nutritional Status (Why):
Food and Nutrient Factors—nutrients available for use by the body

Intake/Consumption

Quantity
Quality

Feeding route and form

Types of Nutrition Assessment Data

Dietary/Nutrition History

Food consumption

Food intake (usual, 24-hour, food frequency, food diary)
Meal and snack patterns
Appetite
Food allergies
Food preferences
Food-related cultural/religious practices

Medication/supplement history

Medication usage
Prescriptions
OTC
Herbal supplements
Dietary supplements
Illegal drugs

Possible Nutrition Diagnoses (What)

Caloric Energy Balance (1)

Defined as "actual or estimated changes in energy (kcal)"

❑ Inadequate energy intake	NI-1.4
❑ Excessive energy intake	NI-1.5

Oral or Nutrition Support Intake (2)

Defined as "actual or estimated food and beverage intake from oral diet or nutrition support compared with patient goal"

❑ Inadequate oral food/beverage intake	NI-2.1
❑ Excessive oral food/beverage intake	NI-2.2
❑ Inadequate intake from enteral/parenteral nutrition infusion	NI-2.3
❑ Excessive intake from enteral/parenteral nutrition	NI-2.4
❑ Inappropriate infusion of enteral/parenteral nutrition	NI-2.5

Biochemical (2)

Defined as "change in capacity to metabolize nutrients as a result of medications, surgery, or as indicated by altered lab values"

❑ Food-medication interaction	NC-2.3

Fluid Intake (3)

Defined as "actual or estimated fluid intake compared against patient goal"

❑ Inadequate fluid intake	NI-3.1
❑ Excessive fluid intake	NI-3.2

(continued on the following page)

TABLE 3.5 *(continued)*

Bioactive Substances (4)

Defined as "actual or observed intake of bioactive substances, including single or multiple functional food components, ingredients, dietary supplements, alcohol"

- ❑ Inadequate bioactive substance intake NI-4.1
- ❑ Excessive bioactive substance intake NI-4.2
- ❑ Excessive alcohol intake NI-4.3

Nutrient (5)

Defined as "actual or estimated intake of specific nutrient groups or single nutrients as compared with desired levels"

- ❑ Inadequate protein-energy intake NI-5.3
- ❑ Imbalance of nutrients NI-5.5

Fat and Cholesterol (51)

- ❑ Inadequate fat intake NI-51.1
- ❑ Excessive fat intake NI-51.2
- ❑ Inappropriate intake of food fats *(specify)* ________ NI-51.3

Protein (52)

- ❑ Inadequate protein intake NI-52.1
- ❑ Excessive protein intake NI-52.2
- ❑ Inappropriate intake of amino acids *(specify)* NI-52.3

Carbohydrate and Fiber Intake (53)

- ❑ Inadequate carbohydrate intake NI-53.1
- ❑ Excessive carbohydrate intake NI-53.2
- ❑ Inappropriate intake of types of carbohydrate *(specify)* ________ NI-53.3
- ❑ Inconsistent carbohydrate intake NI-53.4
- ❑ Inadequate fiber intake NI-53.5
- ❑ Excessive fiber intake NI-53.6

Vitamin Intake (54)

- ❑ Inadequate vitamin intake *(specify)* ________ NI-54.1
- ❑ Excessive vitamin intake *(specify)* ________ NI-54.2
  - ❑ A ❑ C
  - ❑ Thiamin ❑ D
  - ❑ Riboflavin ❑ E
  - ❑ Niacin ❑ K
  - ❑ Folate ❑ Other ________

Mineral Intake (55)

- ❑ Inadequate mineral intake *(specify)* NI-55.1
  - ❑ Calcium ❑ Iron
  - ❑ Potassium ❑ Zinc
  - ❑ Other ________
- ❑ Excessive mineral intake *(specify)* NI-55.2
  - ❑ Calcium ❑ Iron
  - ❑ Potassium ❑ Zinc
  - ❑ Other ________

Knowledge and Beliefs (1)

Defined as "actual knowledge and beliefs as observed or documented"

- ❑ Disordered eating pattern NB-1.5
- ❑ Undesirable food choices NB-1.7

Factors of Nutritional Status (Why):
Environmental Factors—external influences that impact consumption and lifestyle

Social
- Cultural food practices and beliefs
- Parenting,
- Peer influences

Economic
- Household finances
- Economy of community/country

Food safety and sanitation
- Contaminated or unwholesome food
- Unsafe food handling

Food availability/access

Types of Nutrition Assessment Data

Social History
- Economic status
- Housing situation
- Social and medical support
- History of recent crisis/stress
- Social isolation/connection
- Lifestyle
- Cultural/ethnic identity

Food availability
- Food planning, purchasing and preparation abilities
- Food safety practices
- Food/nutrition program utilization
- Food insecurity

(continued on the following page)

TABLE 3.5 *(continued)*

Organized Nutrition Assessment

Possible Nutrition Diagnoses (What)

Knowledge and Beliefs (1)

Defined as "actual knowledge and beliefs as observed or documented"

❑ Harmful beliefs/attitudes about food or nutrition related topics (use with caution)	NB-1.2
❑ Not ready for diet/lifestyle change	NB-1.3
❑ Self-monitoring deficit	NB-1.4
❑ Limited adherence to nutrition-related recommendations	NB-1.6
❑ Undesirable food choices	NB-1.7

Physical Activity and Function (2)

Defined as "actual physical activity, self-care, and quality of life problems as reported, observed, or documented"

❑ Inability or lack of desire to manage self-care	NB-2.3
❑ Impaired ability to prepare foods/meals	NB-2.4
❑ Poor nutrition quality of life	NB-2.5

Food Safety and Access (3)

Defined as "actual problems with food access or food safety"

❑ Intake of unsafe food	NB-3.1
❑ Limited access to food	NB-3.2

Source: Adapted from "Determinants of Nutritional Status" by P Splett and members of ADA's Standardized Language Task Force. 2004 and Nutrition Diagnosis: A Critical Step in the Nutrition Care Process. American Dietetic Association. 2005

appropriate weight. The **Behavioral-Environmental domain** includes problems that are related to knowledge, attitudes/beliefs, physical environment or access to food, and food safety. Within each of these domains, nutrition problems are further grouped according to classifications and subclassifications (refer to Table 3.5).

Each nutrition diagnostic term has a term number and a standard definition. For example, "inadequate protein intake" is defined as "lower intake of protein-containing foods or substances compared to established reference standards or recommendations based upon physiological needs" (ADA, Nutrition Diagnoses 2006). The use of standard definitions helps dietetics professionals use the language consistently within the profession. In addition to the numerical coding and standard definition, the American Dietetic Association has published a reference sheet for each diagnostic term that also identifies possible etiologies and signs and symptoms commonly associated with that nutrition problem. These reference sheets provide tools that the practitioner may use to examine the appropriate data and ask key questions when determining whether a nutrition diagnosis is present or not (see Box 3.1). As the dietetics professional gathers nutrition assessment data in order to determine whether or not a patient actually has a nutrition diagnosis of "inadequate protein intake," he or she should obtain information from the diet and client history that will provide evidence of the problem. In this case, data describing the amount of protein that is consumed would be essential for determining how far below the recommendation the patient's protein intake actually is. Other data from the assessment might provide clues about the cause of the problem, such as physiological reasons for increased need, lack of access to food, knowledge deficit, or psychological causes. By using these reference sheets, dietetics professionals can be assured that the diagnostic terminology is used consistently and accurately (ADA, Nutrition Diagnosis 2006).

behavioral-environmental domain—domain that contains standardized nutrition diagnostic terms that describe nutrition problems related to knowledge, attitudes/beliefs, physical environment, access to food, and food safety (Nutrition Diagnosis, ADA, 2006)

PES—problem, etiology, and signs and symptoms; the format used in the NCP to write a nutrition diagnosis; it clarifies a specific nutrition problem and logically links the nutrition diagnosis to nutrition intervention and to monitoring and evaluation

Most nutrition diagnoses are written in a **PES** (problem, etiology, signs/symptoms) format that lists the problem, its cause, and appropriate defining characteristics. The problem (P) is also referred to as the diagnostic label. It describes in a general way an alteration in the client's nutritional status. Words like excessive, inadequate, and inappropriate are frequently found in these labels. The related factors or etiology (E) are those factors that contribute to the cause or existence of a particular problem. Finally, the signs and symptoms (S) are the defining characteristics obtained from the subjective and objective nutrition assessment data. These data provide evidence that a problem exists and describe the severity of the problem. When these three parts are used to form the

BOX 3.1 SAMPLE NUTRITION DIAGNOSTIC TERM REFERENCE SHEET

INTAKE DOMAIN: Protein

INADEQUATE PROTEIN INTAKE (NI-52.1)

Definition

Lower intake of protein containing foods or substances compared to established reference standards or recommendations based upon physiological needs

Etiology (Cause/Contributing Risk Factors)

Factors gathered during the nutrition assessment process that contribute to the existence of or the maintenance of pathophysiological, psychosocial, situational, developmental, cultural, and/or environmental problems.

- Physiologic causes, e.g., increased nutrient needs due to prolonged catabolic illness, malabsorption, age, or condition
- Lack of access to food, e.g., economic constraints, cultural or religious practices, restricting food given to elderly and/or children
- Food- and nutrition-related knowledge deficit
- Psychological causes, e.g., depression or disordered eating

Signs/Symptoms (Defining Characteristics)

A typical cluster of subjective and objective signs and symptoms gathered during the nutrition assessment process that provide evidence that a problem exists; quantify the problem and describe its severity.

Nutrition Assessment Category	Potential Indicators of This Nutrition Diagnosis (one or more must be present)
Biochemical Data	
Anthropometric Measurements	
Physical Examination Findings	
Diet History	Report or observation of Insufficient intake of protein to meet requirements Cultural or religious practices that limit protein intake Economic constraints that limit food availability Prolonged adherence to a very low-protein weight loss diet
Client History	Conditions associated with a diagnosis or treatment of, e.g., severe protein malabsorptionsuch as bowel resection

National Academy of Sciences, Institute of Medicine. Dietary Reference Intakes for Energy, Carbohydrate, Fiber, Fat, Fatty Acids, Cholesterol, Protein, and Amino Acids. Washington DC: National Academy Press; 2002.

Edition: 2006

nutrition diagnostic statement, it is generally stated in the following way: the problem (P) *related to* the etiology (E) *as evidenced by* the signs and symptoms (S).

Example of Nutrition Diagnoses

- "Inadequate energy intake (P) *related to* changes in taste and appetite (E) *as evidenced by* average daily kcal intake 50% less than estimated recommendations (S)"
- "Involuntary weight loss (P) *related to* inadequate energy intake (E) *as evidenced by* eight pounds weight loss within four weeks (S)"

Let's examine how these diagnoses were made. A comprehensive nutrition assessment reveals the following data:

- Client is undergoing chemotherapy for cancer treatment (client's medical history).
- Complaints of meats tasting bitter and most beverages too sweet (food/nutrient intake history).
- Client states, "I have no appetite and no desire to eat" (food/nutrient intake history).
- Three-day food records reveal average kcal approximately 50% of estimated needs (dietary intake data compared to estimated needs).
- Weight loss of eight pounds since last outpatient visit one month ago (anthropometric data).

In order to evaluate the above nutrition assessment data, the dietetics professional applies a number of the critical thinking skills listed in Table 3.4. These include finding patterns and relationships between the data and possible causes, stating each problem clearly and singularly, ruling in/ruling out specific diagnoses, and prioritizing the importance of the diagnoses.

From the relationships that exist among the assessment data just noted, "inadequate energy intake" and "involuntary weight loss" are selected as relevant nutrition problems. It is essential to focus on problems for which nutrition interventions will be the primary treatment. Once the appropriate problems have been selected, the next step is to describe accurately the signs and symptoms. The signs and symptoms are used to validate and confirm the existence of problems. They also indicate the severity of the problems, answering the question, "How much?" or "How do I know?"

Finally, after validating the problem by identifying the appropriate signs and symptoms, the etiology or cause of the problem is explored. To determine the etiology, related factors and additional data from the assessment are reviewed. It is important to seek the answer to the question, "Why does this problem exist?" and explore all possibilities. It may even be necessary to frequently ask the question "Why?" to uncover the underlying root cause of the nutrition problem. To summarize:

- The problem is the "What?"
- The etiology is the "Why?"
- The signs/symptoms are the "How do I know?" or "How severe is the problem?"

In the present example, two important points about etiology are illustrated. First, even though medical diagnoses (cancer) and/or medical treatment (chemotherapy) contribute to nutrition problems, they should not be used as the primary etiology. Instead, it is best that a nutrition-related cause be part of the etiology. This is consistent with the guiding principle that distinguishes a nutrition diagnosis from other diagnoses. First, *a nutrition diagnosis is written in terms of a client problem for which nutrition-related activities provide the primary intervention.* Second, nutrition diagnostic terminology is always used to identify the nutrition problem (P). This language can also be used as etiology language. Behaviors and patterns of food and nutrient intakes that are undesirable (problems in and of themselves) can produce other problems such as changes in anthropometric, biochemical, or clinical findings. In the present example, inadequate caloric intake is the primary cause of unintentional weight loss.

Traditionally, nutrition care has been driven by diet orders associated with certain disease conditions, such as diet orders for a "renal diet," a "diabetic diet," or a "weight-loss diet." With the advent of the standardized nutrition language and nutrition diagnoses, nutrition care can and should be driven by the extent of a nutrition problem rather than solely by a diet order or medical condition. Medical conditions affect a person's ability to consume, digest, metabolize, and utilize nutrients. They also affect nutrient needs and requirements. However, the specific type of nutrition intervention can now be determined by the nutrition diagnosis. For example, instead of providing nutrition care/education as a result of a diet order for a diabetic diet or renal diet, the dietitian will now carefully assess the nutritional status of each patient to specifically identify what, if any, nutrition problems (diagnoses) exist. For example, a patient with type 2 diabetes could conceivably have inappropriate carbohydrate intake, undesirable food choices, or a self-monitoring deficit. A complete assessment may reveal, however, the absence of any nutritional problems at all. In the case of a patient with chronic renal disease, there could be problems such as excessive potassium intake or excessive fluid intake. Again, a complete assessment may show there are no problems. Another scenario might be two patients who present with a similar nutrition diagnosis, such as involuntary weight loss. Using the nutrition diagnoses to clarify and identify specific nutrition problems may reveal no nutrition problems at the present time, different nutrition problems for patients with similar medical diagnoses, or similar nutrition problems for patients with different medical conditions.

Criteria for Evaluating PES Statements Since the intent of nutrition diagnoses is to describe those problems for which nutrition intervention is the primary treatment, it is important to develop PES statements that accurately reflect that intent. The following questions were developed by ADA's Standardized Language Task Force to ensure that nutrition diagnoses are well written and accurately represent the nutrition problems:

- Can the dietetics professional impact, improve, or resolve the nutrition problem?
- Can an intervention reduce the significance of the signs and symptoms?
- Is the etiology truly the root cause?
- Is there an intervention that will address the root cause, thus increasing the likelihood that a positive change will result?
- Does the assessment data used to identify the nutrition diagnosis support and link to the diagnostic statement, etiology, and signs and symptoms?
- Are the signs and symptoms that are used to describe the problem specific enough to be measured?
- Are the problems clearly and singularly stated?

Box 3.2 demonstrates how these criteria are used to evaluate and refine PES statements for a client with diabetes.

Relationship of Nutrition Diagnosis to the Other Steps of the NCP Figure 3.3 illustrates the relationship of the nutrition diagnosis to the other steps of the NCP. As stated previously, a nutrition diagnosis is the missing link between nutrition assessment and nutrition intervention. An accurate nutrition diagnosis is generated from a focused nutrition assessment and sets the stage for the next two steps of the NCP: Step 3, nutrition intervention, and Step 4, nutrition monitoring and evaluation.

The signs and symptoms or defining characteristics represent data obtained from the nutrition assessment in Step 1. These data appropriately describe the particular problem by quantifying and qualifying how that specific problem is present at that point in time. If the problem is an energy intake imbalance (either NI 1.4 or NI 1.5), then a measurement of kcal (% of estimated caloric needs, average intake over time, an amount less than or more than desired, etc.) best describes the energy problem; if the problem is one of weight, then an appropriate anthropometric measurement

BOX 3.2 EVALUATING A NUTRITION DIAGNOSIS

When data is obtained from a nutrition assessment, there will be a number of findings that can provide clues to the presence of a particular nutrition diagnosis. The dietetics professional needs to distinguish among (1) data that is associated with a nutrition problem and/or may be a consequence of that problem, (2) data that will be used to document the specific signs and symptoms that describe and quantify that problem, and (3) data that will provide insight into the root cause of the problem.

Which of the following nutrition diagnoses is preferred and why?

A. Inconsistent carbohydrate intake related to not following a diabetic diet as evidenced by elevated A1c level of 10.5

B. Inconsistent carbohydrate intake related to inability to read labels correctly for carbohydrate content and lack of knowledge about amount of grams/carbohydrate units as evidenced by carbohydrate units in three meals of 1, 6, and 3

Can the dietetics professional impact, improve, or resolve the nutrition problem? In both examples, the nutrition problem "inconsistent carbohydrate intake" is one that can be improved or resolved.

Can an intervention reduce the significance of the signs and symptoms? In the case of example A, it is not clear that a change in carbohydrate intake alone will improve the A1c. There could be other factors that are impacting on the A1c such as medication, illness, and so on, whereas the signs and symptoms in B are more descriptive of the problem itself.

Is the etiology truly the *root* cause? Even though not following a diet plan is likely contributing to the inconsistent carbohydrate intake, there needs to be further understanding as to the reasons why a meal plan is not being followed. In other words, asking "why" uncovers the real reason for not following the plan and is more clearly stated in example B.

Is there an intervention that will address the root cause, thus increasing the likelihood that a positive change will result? Developing an intervention using example A might lead prematurely to a more traditional diet education of a diabetic diet, whereas addressing the real reason for not being able to follow a meal plan gives both the provider and the client a more realistic and specific plan for education. Focusing on the two topics in B should increase the likelihood that a positive change will occur compared to an education plan that is more general.

Does the assessment data used to identify the nutrition diagnosis support and link to the diagnostic statement, etiology, and signs and symptoms? Even though an elevated A1c provides a clue that there may be a problem with carbohydrate intake, it does not specifically describe the nutrition problem itself. This is an example of data that may be associated with or a consequence of a nutrition problem but that does not describe and quantify the specific problem.

Are the signs and symptoms used to describe the problem specific enough to be measured? In both cases, the signs and symptoms can be measured; however, improvement in carbohydrate units can be determined within a shorter time frame than can the A1c. Furthermore, changes in meal patterns can be expected in direct response to the intervention, whereas changes in A1c are influenced by more variables and may not occur directly in response to the nutrition education provided.

Are the problems clearly and singularly stated? In the case of example A, there are really two different types of problems embedded in this diagnosis: inconsistent carbohydrate intake and altered nutrition-related laboratory values. Example B describes a single problem in a straightforward manner, allowing the dietitian to deal with one problem at a time.

Therefore, after applying the criteria to each of the examples noted above, example B is the preferred nutrition diagnosis. It states the problem clearly and singularly, and it provides a quantifiable description of the signs and symptoms from which specific goals can be established and measured. It also provides clear direction for a specific intervention targeted at the root cause of the problem.

(BMI, relative weight, weight change over time, etc.) should be used to describe the weight problem. Data from the assessment also provide information used to determine the etiology.

These signs and symptoms then become the basis for setting ideal and measurable goals as part of Step 3, nutrition intervention. These signs and symptoms are also the **outcome measures** that are used to monitor and evaluate progress toward goals as part of Step 4, nutrition monitoring and evaluation. If kcal are inadequate by 50%, as in the previous example of a nutrition diagnosis, a desired goal might be to meet 75% of estimated caloric needs within 2 days. A kcal count or food record could be used to track and evaluate that outcome.

Finally, nutrition interventions as part of Step 3 should be logically linked to the cause of problems. If the root cause of inadequate intake is taste alteration and decreased appetite, interventions need to be linked to ways to enhance appetite and improve the taste of foods before a change in caloric intake will occur.

outcome measures—data used to evaluate the success of interventions; includes direct nutrition outcomes, clinical and health status outcomes, patient/client-centered outcomes, and health care utilization and cost outcomes

FIGURE 3.3 Relationship of the Nutrition Diagnosis to the Other Steps of the NCP

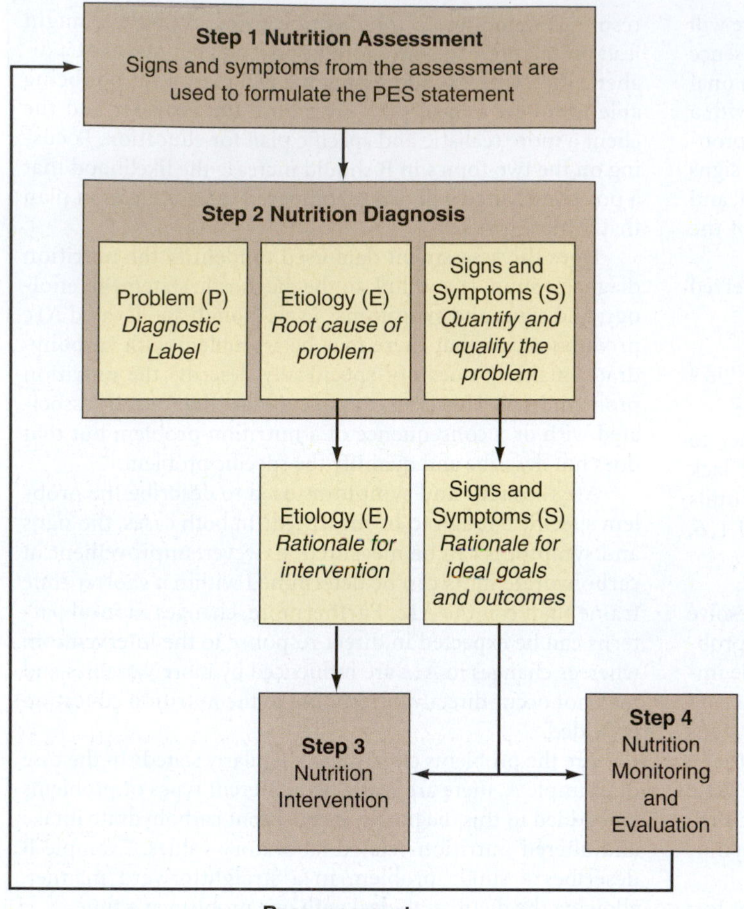

Source: Lacey K and Pritchett E, Nutrition Care Process and Model: ADA adopts road map to quality care and outcomes management. J Amer Diet Assoc. 2003;103:1061-1072.

When nutrition diagnoses are written as separate and distinct problems, even though one problem may actually cause another, the dietetics professional is able to prioritize which problem should be addressed first as part of Step 3, nutrition intervention. For example, when there is both an intake problem and a weight problem, caloric intake needs to change before a change in weight can be expected.

Key Concepts: NCP Step 2, Nutrition Diagnosis

- Nutrition diagnosis is not the same as medical diagnosis. It describes a problem for which nutrition-related activities provide the primary intervention.
- The desired format for writing a nutrition diagnosis is PES (problem, etiology, and signs and symptoms).
- Critical thinking skills such as finding patterns and relationships, stating problems clearly and singularly, and ruling in/ruling out certain diagnoses are essential to making accurate nutrition diagnoses.
- Accurate nutrition diagnoses set the stage for quality nutrition intervention (Step 3 of NCP) and nutrition monitoring and evaluation (Step 4 of NCP).

Step 3: Nutrition Intervention

The third step of the NCP, nutrition intervention, involves both planning and implementing. It is a specific set of activities and associated materials used to address the problem. Nutrition interventions are purposefully planned actions designed with the intent of changing a nutrition-related behavior, risk factor, environmental condition, or aspect of health status for an individual, target group, or the community at large (Lacey and Pritchett 2003). Dietetics professionals work collaboratively not only with other health care professionals, but more importantly with the client, family, or caregiver to create a realistic plan that has a good probability of positively influencing the diagnosed problem. This client-driven process is key to successful nutrition intervention, distinguishing it from previous planning steps that may or may not have involved the client to this degree.

Sub-Step 3a: Plan

Prioritize the Nutrition Diagnoses A nutrition assessment will likely result in the identification and labeling of multiple nutrition diagnoses; therefore, before any action can be taken, it is essential to *prioritize the diagnoses* that are identified in Step 2, nutrition diagnosis. The ranking of nutrition diagnoses permits dietetics professionals to arrange the problems in the order of their importance and urgency for the client. Once they have been sorted for safety, then another prioritization can be done based on such things as anticipated early response to an intervention, client preference of a behavior change, or the impact of one problem on another. Using the earlier example, it makes sense to first address the primary problem of energy intake before one expects to intervene on the secondary problem of weight:

- "Inadequate energy intake (P) *related to* changes in taste and appetite (E) *as evidenced by* average daily kcal intake 50% less than estimated recommendations (S)"
- "Involuntary weight loss (P) *related to* inadequate energy intake (E) *as evidenced by* eight pounds weight loss within four weeks (S)"

Identify Goals After having prioritized the diagnoses, it is necessary to *identify ideal goals and patient-focused* *expected outcomes.* **Ideal goals** are science-based values intended to control or improve specific health conditions. ADA's Evidence-Based Guides for Practice and other practice guides provide resources to assist dietetics professionals in selecting the appropriate goals for patients (NGC 2006). Consuming less than 7% of kcal from saturated fat is an example of an evidence-based ideal goal for the nutrition treatment of hyperlipidemia (NCEP 2002). This is the desired level of saturated fat that is associated with the least amount of cardiac risk. **Expected outcomes** are the desired change(s) to be achieved over time as a result of nutrition intervention. (Refer also to the section on nutrition monitoring and evaluation, sub-step 4b.) Expected outcomes are based on the nutrition diagnosis; for example, decreasing the intake of saturated fat by a specific amount or percentage of kcal is an expected outcome. Expected outcomes can also be defined in terms of a specific behavior that will result in the change in amount of saturated fat consumed; for example, the patient will substitute olive oil for solid margarine as the preferred spread on most breads. Expected outcomes should be written in observable and measurable terms that are clear and concise. They should be client-centered and realistically tailored to the client's circumstances and expectations for treatment. Interventions are then planned that will help a patient to meet these goals and outcomes.

Plan the Nutrition Intervention Finally, as part of the planning sub-step of Step 3, appropriate interventions need to be selected. All interventions must be based on scientific principles and rationales and must be grounded in quality research and evidence-based interventions when available. Once again, ADA's Evidence-Based Guides for Practice and other practice guides are invaluable resources for both identifying science-based ideal goals and selecting appropriate interventions. These guides link external scientific evidence regarding nutrition care to a specific health problem, thus giving dietetic professionals the confidence that they are making the best decisions when providing nutrition care. The use of evidence-based guides does not replace the expertise and judgment of dietetics professionals, but these tools enhance the value of dietetics professionals and increase the likelihood that a desired outcome will occur.

Sub-Step 3b: Implement the Nutrition Intervention Implementation is the action phase of the nutrition care process. It is during this phase that dietetics professionals communicate the plan of action to the client and other professionals. Dietetics professionals may directly carry out the intervention or may delegate or coordinate care provided by others. Once again, the central core of the Nutrition Care Model (relationship between patient/client/group and dietetics professional) recognizes that the client needs to be involved in this decision-making and action step of nutrition care.

Key Concepts: NCP Step 3, Nutrition Intervention

- First and foremost is the need to prioritize the nutrition diagnoses.
- Ideal goals and expected outcomes need to be identified prior to implementing nutrition interventions.
- Interventions are derived from accurate diagnoses and largely driven by client involvement.
- ADA's Evidence-Based Guides for Practice provide dietetics professionals with tools that promote quality service and demonstrate effectiveness of care.

Step 4: Nutrition Monitoring and Evaluation

The purpose of monitoring and evaluation is to determine the degree to which progress is being made and whether or not the client's goals or desired outcomes of nutrition care are being met. It is much more than merely "watching" what is happening. It requires an active commitment to measuring and recording the appropriate outcome indicators relevant to the nutrition diagnosis' signs and symptoms. Progress should be (1) monitored, (2) measured, and (3) evaluated on a planned schedule. Systematic use of each of these components provides consistency in practice, adds value, and demonstrates effectiveness of care.

Sub-Step 4a: Monitor Progress Monitoring refers specifically to determining that the goals and outcomes that are anticipated by the client and the dietetics professional are indeed occurring. Specific activities that are associated with this level of monitoring include:

- Determining whether the intervention is being implemented as planned
- Checking the client's understanding and attainment of goals
- Determining if changes in the client's condition are occurring (Lacey and Pritchett 2003)

Monitoring in this way may require gathering additional information about possible reasons for any lack of progress. Revision of a nutrition diagnosis and/or a change in plan may occur as a result of obtaining additional information. Using the NCP, therefore, may involve performing its steps more than once during the course of nutrition treatment.

ideal goals—science-based values intended to control or improve specific health conditions

expected outcomes—the desired change(s) to be achieved over time as a result of nutrition intervention

BOX 3.3 CLINICAL APPLICATIONS—NUTRITION THERAPY FOR CARDIOVASCULAR DISEASE

AJ is a 55-year-old male who works in sales. He has just returned from an annual health physical with his primary care physician. Following the usual exam, the physician initiated a referral to the health care system's outpatient dietitian. Below is pertinent assessment data that the dietitian obtained from both the patient chart and interview.

Step 1: Nutrition Assessment

Past medical and family history: Positive family history for premature heart disease; recent BP reading of 140/ 80

Laboratory data: LDL 130 mg/dL, TG 200 mg/dL, TC 200 mg/dL

Anthropometric data: 5'8", 185#, BMI 28.3

Physical activity: little to no exercise, states he is too busy at work, travels weekly by car and occasionally on plane for business, always takes the elevator to his fourth floor office and parks in the closest parking lot

Diet history summary (typical intake and usual amounts of key nutrients):

Average daily kcal = 3200 kcal (estimated needs based on adjusted body weight of 161# and Mifflin-St. Jeor formula and physical activity factor of 1.4 = 2155 kcal)

Saturated fat = 10% of total kcal (approximately 36 g)

- 7–8 oz portions of beef or chicken = 13 g SFA
- Chocolate cake/frosting and ice cream = 8 g SFA
- 3 oz bologna sandwich on white bread or fast-food cheeseburger = 6–8 g SFA
- 3 T butter = 15 g SFA
- 1 c whole milk = 4 g SFA

Sodium intake = 3500–4500 mg

- Cheeseburger = 600 mg
- 3 oz bologna = 226 mg
- Large french fries = 300 mg
- 2 c canned soup = 1600 mg
- 1–2 tsp salt added to foods = 1200–2400 mg

Key Concepts:

1. Assessment data is clustered and organized according to possible nutrition diagnoses.
2. Wherever possible, amounts of key nutrients are estimated.
3. Appropriate standards are applied to evaluate the data.

Step 2: Nutrition Diagnosis

P: Excessive energy intake
E: Daily intake of whole-fat dairy products, desserts, and large meat portions
S: Average caloric intake @ 150% in excess of estimated needs
P: Excessive saturated fat intake
E: Daily intake of whole-fat dairy products and large meat portions
S: Saturated fat intake @ 10% of total kcal
P: Excessive sodium intake
E: Daily consumption of convenience and fast foods
S: Typical daily sodium intake of 3500–4500 mg
P: Physical inactivity
E: Client perception of being too busy and frequent business travel
S: Little to no regular exercise and very sedentary lifestyle
P: Overweight
E: Excessive caloric intake and physical inactivity
S: BMI 28.3

Key Concepts:

1. Two different problems have similar causes; therefore, intervention can address more than one problem simultaneously.
2. One or two problems can actually be the cause of another problem. Therefore, it is necessary to address both the intake problem and physical inactivity before a change in weight will occur.

Sub-Step 4b: Measure Outcomes Measuring outcomes means that data are collected over time. This sub-step is critical for the dietetics professional to incorporate into practice. Interventions have too often been planned and acted upon with little regard for what has really happened as a result of the action taken.

The key to measuring outcomes is knowing what needs to be measured. The NCP provides clear examples of the types of outcomes to be measured (Lacey and Pritchett 2003). These include (among others):

- Direct nutrition outcomes such as knowledge gained, behavior changes, food or nutrient intake changes, and improved nutritional status
- Clinical and health status outcomes such as laboratory values, anthropometry and body composition, blood pressure, and risk factor profile
- Patient/client-centered outcomes such as quality of life, satisfaction, self-efficacy, and self-management
- Health care utilization and cost outcomes such as medication changes, special procedures, and planned/unplanned health care visits

The NCP now provides a clear and direct way to collect outcomes data regardless of the practice setting. Establishing a nutrition diagnosis that clearly describes the signs and symptoms establishes the type of outcome to be measured

Step 3: Nutrition Intervention

Establish Goals:

These are determined by consulting evidence-based practice guides and discussing expectations with the client. They are measurable and realistic and establish the type of outcome to be tracked over time.

1. Average daily caloric intake will be no more than 110% of estimated needs (approximately 2200 kcal; reduction of 1000 kcal/day).
2. Saturated fat intake will be 7% or less of total kcal.
3. Average daily sodium intake will be at or under 2400 mg.
4. Daily physical activity will increase by 2000 steps weekly to goal of 10,000/day.
5. Weight loss over time will average 1–2 pounds per week.

Implementation:

1. Assist client in making alternative food choices to lower total fat and saturated fat intake:
   - Smaller and leaner meat portions (chicken and fish).
   - Lower-fat dairy products (low-fat ice cream, 1% milk, etc.).
   - Plant stanols and/or olive oil in place of butter.
2. Assist client in making alternative food choices to lower daily intake of sodium:
   - Use a variety of seasonings at table in place of salt; provide with examples of recipes.
   - Order fast food hamburger and fresh vegetable or fruit salad in place of cheeseburger and french fries.
   - Decrease frequency of use of canned soups.
3. Assist client in increasing physical activity daily.
   - Arrange for client to obtain a step meter and instruct on its use and record keeping.

Key Concepts:

1. Client is actively involved in establishing realistic behavioral goals
2. Ideal goals are derived from the specific data obtained from the nutrition assessment; for example, current SFA intake is 10% of kcal; the ideal goal based on ATPIII is 7%.
3. Interactions are directly related to the etiology from the PES statements.

Step 4: Nutrition Monitoring and Evaluation

Monitor progress:

Dietetics professional may contact the client to provide support and clarify any questions regarding the plan. This is done in order to determine if the plan is being implemented and whether or not the client fully understands the information provided.

Measure Outcomes:

Direct nutrition outcomes:

- Behavior changes related to decreasing portion sizes, use of low-fat dairy foods and plant stanols, use of alternative seasonings in place of salt, food choices when eating at fast food establishments, and physical activity
- Intake changes of total kcal, total fat, SFA and sodium

Clinical and health status outcomes:

- Blood pressure, LDL, and BMI

Patient/client-centered outcomes:

- Satisfaction
- Self-management (food records, physical activity records)

Evaluate outcomes:

Baseline data will be compared to changes in the above outcome data that is tracked over time. Progress will be discussed with the client and any problems or barriers that are identified will be used to revise PES statements, modify interventions, and/or establish new goals.

and thus provides baseline data from which to begin measuring. A variety of documents and tools can be used to measure and track data, including electronic charting, coding systems, and spreadsheets.

Sub-Step 4c: Evaluate Outcomes It is not enough to just measure outcomes, however. Outcomes need to also be evaluated to determine what, if any, changes have occurred as a result of the nutrition intervention. Such an evaluation requires comparing the current findings with the previous signs and symptoms. Outcome evaluation may also involve additional data collection in order to explore why a change has *not* occurred as expected. Additional or revised nutrition diagnoses may also be needed. New goals and new interventions thus may further modify the nutrition care process.

Key Concepts: NCP Step 4, Nutrition Monitoring and Evaluation

- This step requires an active commitment to measuring and recording changes in the client's condition as they relate to the nutrition diagnosis' signs and symptoms.
- Progress should be monitored, measured, and evaluated on a planned schedule.

BOX 3.4 CLINICAL APPLICATIONS—ENTERAL NUTRITION FOLLOWING MOTOR VEHICLE ACCIDENT (MVA)

AT is a 25-year-old female in previously good health who was involved in a MVA. She was admitted to the hospital for extensive oral surgery resulting in a wired jaw. Despite numerous attempts to consume oral supplements, she was only able to meet 20% of her estimated kcal and protein needs. The physician writes the following order to begin tube feeding:

Use of hospital's high-protein, high-kcal tube feeding at goal of 80 mL/hr.

This formula provides 0.0616 g protein/mL, 1.5 kcal/mL and 75.8% free water. The standard hospital protocol for water flushes of tube feeding is 55 mL water q 8 hours.

According to the hospital's nutrition care protocol, the dietitian was contacted to complete a nutrition assessment.

Step 1: Nutrition Assessment

Anthropometric data: 5′5″, 125# (56.8 kg) BMI 20.8
Laboratory data: all WNL
Estimated kcal and protein needs: Based on actual body weight and Mifflin-St. Jeor formula for REE, physical activity factor of 1.3 and injury factor of 1.1. Protein needs based on 1.0–1.2 g/kg. Estimated needs = 1900 kcal and 57–68 grams protein

Estimated fluid needs based on 1 mL/kcal = 1900 mL

24 hours of high-protein, high-kcal tube feeding at goal rate of 80 mL/hr will provide: 2880 kcal, 118 g protein, and 1455 mL free water. Total TF water flushes provide an additional 150 mL water.

Step 2: Nutrition Diagnoses

P: Inadequate protein-energy intake
E: Inability to take nutrition orally secondary to oral surgery and wired jaw
S: Client only meeting 20% of estimated needs
P: Excessive intake from enteral/parenteral nutrition
E: Use of high-protein/high-kcal tube feeding at rate of 80 mL/hr
S: TF exceeding estimated needs of kcal and protein by 50% (2880 kcal and 118 g protein per order compared to estimated needs of 1900 kcal and 57–68 g protein)
P: Inadequate fluid intake
E: Use of concentrated tube feeding
S: 24-hour fluid intake 84% of estimated needs (1600 mL compared to estimated needs of 1900 mL)

Step 3: Nutrition Intervention

Goals:

1. Increase fluid to 1900 mL within 24 hours.
2. Decrease kcal and protein to 95–105% of estimated needs within 24 hours.

Intervention:

Recommend change tube feeding to following:

- Standard tube feeding providing 1 kcal/mL, 0.0366 g protein/mL, 83.3% free water.
- Begin tube feeding at 25 mL/hr for 8 hours, then increase to 50 mL next 8 hours, and if tolerated, increase to goal of 80 mL within 24 hours.
- Increase water flushes to 100 cc q 8 hours.
- 24 hours of this tube feeding will provide: 1920 kcal, 70 g protein, and 1900 cc free water.

Step 4: Nutrition Monitoring and Evaluation

- Verify that the tube feeding is being provided at desired rate.
- Record nutrients that are being provided.
- Compare nutrition provided to estimated needs and make changes and recommendation as appropriate.

- Direct nutrition outcomes, clinical and health status outcomes, patient/client-centered outcomes, and health-care utilization outcomes are the types of outcomes to be measured.
- Data from this step can be used to create an outcomes management system and can contribute to the body of evidenced-based research.

Documentation

Documentation (see Chapter 6) is an ongoing process that supports all of the steps of the NCP. The standardized language that is now part of the NCP improves both the written and oral communication among members of the health care team as well as communication with the patient. It allows dietetics professionals to more clearly name and document clinical judgments concerning nutrition problems (Hakel-Smith and Lewis 2004). Documentation should be relevant, accurate, and timely. A variety of charting formats have been used by dietetics professionals to communicate nutrition care. These formats include Subjective, Objective, Assessment, and Plan (SOAP); focus notes; and Problem, Intervention, and Evaluation (PIE). More recently, a newer form of charting based on the steps of the NCP has been introduced. This new form is the ADIM format (Assessment, Diagnosis, Intervention, and Monitoring). Electronic medical records are also becoming more widely used.

As dietetics professionals implement ADA's NCP and standardized language, more efficient and effective methods of documentation will surely evolve. Regardless of the specific format used, "when the systematic steps of the nutrition care process or the nutrition practitioner's clinical judgments are consistently defined and documented with standardized terms, this information can be collected, compared, and aggregated and therefore used to identify the most effective treatment" (Hakel-Smith, Lewis, and Eskridge 2005).

Conclusion

In Step 1, nutrition assessment, adequate and appropriate information is obtained in order to identify the nutrition problem and formulate a complete nutrition diagnosis in Step 2, nutrition diagnosis. A complete nutrition diagnosis includes both an etiology and accurate signs and symptoms. Step 3, nutrition intervention, establishes ideal goals and desired outcomes for which interventions likely to provide positive results are planned. In Step 4, nutrition monitoring and evaluation, appropriate indicators are measured over time to track progress toward desired goals. This completes the cycle of the NCP.

As dietetics practitioners incorporate the NCP into their practices, they recognize that the use of the standardized process along with the standardized nutrition diagnostic terminology changes both the way they think and the way they chart. Early adaptors remark, "You're changing the way you're thinking, you're changing the way you're charting—it's a huge change. There are no shortcuts you can take. . . . It's certainly worthwhile" (Mathieu, Foust, and Ouellette 2005).

WEB LINKS

American Dietetic Association—Quality Management:
To find the latest information on the nutrition care process published by the ADA, click on the "Practice" link; on the page that appears, click on "Quality Management." (member only access).

http://www.eatright.org

American Dietetic Association Evidence Analysis Library:
This website, available to ADA members only, summarizes the latest research and evidence-based guidelines related to nutrition and dietetics practice.

http://www.adaevidencelibrary.com

END-OF-CHAPTER QUESTIONS

1. List the internal and external factors that influence a person's ability to maintain optimal nutritional health.
2. What is the purpose of providing nutrition care?
3. List the four steps of ADA's Standardized Nutritional Care Process (NCP). Briefly describe each step.
4. Why is it important to have standardized nutrition diagnostic terminology in the practice of dietetics?
5. List and briefly describe the three domains of nutrition diagnostic terms.
6. Explain what P, E, and S are in the nutrition diagnosis.
7. Write an example of a PES nutrition diagnosis.
8. How does the nutrition diagnosis relate to the other steps in the Nutrition Care Process?
9. How does a nutrition diagnosis differ from a medical diagnosis?
10. Describe the criteria used to evaluate the quality of PES statements.
11. List and briefly describe the four types of outcome measures that can be monitored and evaluated in the NCP.
12. What is meant by "outcomes management system"? Why is it important in dietetic practice?

4

Complementary and Alternative Medicine

Pamela Goyan Kittler, M.S.

Food, Culture, and Nutrition Consultant, Sunnyvale, California

CHAPTER OUTLINE

Alternative Medical Systems

Ayurvedic Medicine • Chiropractic Medicine • Homeopathic Medicine • Naturopathic Medicine • Osteopathic Medicine • Traditional Chinese Medicine/Acupuncture

Complementary Therapies

Chelation Therapy • Folk Healing • Natural Products • Dietary Therapies • Vitamin/Mineral Supplements and Megavitamin Therapy • Mind-Body Therapies

Medical Pluralism in Practice

Introduction

Medical pluralism describes the consecutive or concurrent use of multiple health care systems and therapies by clients (Clark 1983). Historically, medical pluralism was the norm. In the nineteenth-century United States, medical practitioners of all types were unregulated and most learned their profession as apprentices, including "regular" or allopathic physicians.[1] Other doctors of homeopathy, osteopathy, naturopathy, and chiropractic, as well as religious and folk healers, were considered equally skilled (Baer 2001).[2] It wasn't until the acceptance of germ theory at the turn of the twentieth century and the subsequent transformation of **allopathic medicine** into biomedicine (see Box 4.1) that a single, conventional medical system became dominant, relegating other practices to the fringes of health care acceptability. According to biomedical standards, unconventional healing practices and products were judged ineffective, unscientific, and even dangerous.

Yet medical pluralism never completely disappeared. In recent decades, **complementary medicine** and **alternative medicine** have become increasingly popular. Complementary and alternative medicines are usually grouped together under the acronym *CAM*. It is estimated that in 2002, 36% of the U.S. population, nearly 103 million adults, used CAM (excluding prayer) practices or products (see

medical pluralism—the consecutive or concurrent use of multiple health care systems and therapies by clients

allopathic medicine—modern or conventional biomedicine

complementary medicine—unconventional modalities used by clients in addition to conventional biomedicine; may involve practitioner, but often self-prescribed

alternative medicine—unconventional therapeutic systems used by clients in place of or parallel to conventional biomedicine; typically administered by trained practitioner

BOX 4.1 CLINICAL APPLICATIONS—BIOMEDICINE

Biomedicine is the dominant, conventional medical system used in most western nations. It is based on the principles of the natural sciences, including anatomy, physiology, and biochemistry. Controlled experimentation and reproducible results are fundamental to biomedical theory. Mastery over nature is essential to biomedical practice. The practitioner fights infection, conquers disease, and kills pain with knowledge, skills, and technology. In the biomedical model, the body is analogous to a machine. When problems occur in the "equipment," diagnosis is achieved through numerical assessment of the "broken" parts. Data that fall within a standard are considered healthy; data outside the established parameters indicate illness. Symptoms that cannot be confirmed numerically are typically discounted as psychosomatic in origin. A duality between body and mind is assumed, and spirituality is completely beyond the scope of biomedicine.

Figure 4.1) (Barnes, Powell-Griner, McFann, and Nathin 2004). Consumer spending on CAM more than tripled during the last decade, rising from $11 billion annually to approximately $40 billion (Eisenberg, Kessler, Foster, Norlock, Calkins, and Delbanco 1993; Medstat PULSE Survey 2002). The market for dietary supplements alone now exceeds $17 billion each year (United States Food and Drug Administration 2002). Continued growth of CAM is predicted. By 2010, it is expected that the number of alternative medical practitioners will increase—per capita—by 88% compared to a 16% increase in biomedical physicians (Cooper and Stoflet 1996).

This chapter discusses which clients choose CAM and why, and provides an introduction to popular CAM modalities and theories. The practices and products included are presented as alternative systems and complementary therapies, though there is little consensus among researchers on what constitutes CAM, or how it should be categorized. While some CAM systems and therapies do not include a specific dietary component, they are included because there is significant crossover among CAM practitioners, and nutritional therapy (especially dietary supplements) may be offered as an added service. The emphasis, however, is on those aspects of CAM practices and products that may affect conventional medical nutrition therapy.

FIGURE 4.1 CAM Use by U.S. Adults—2002

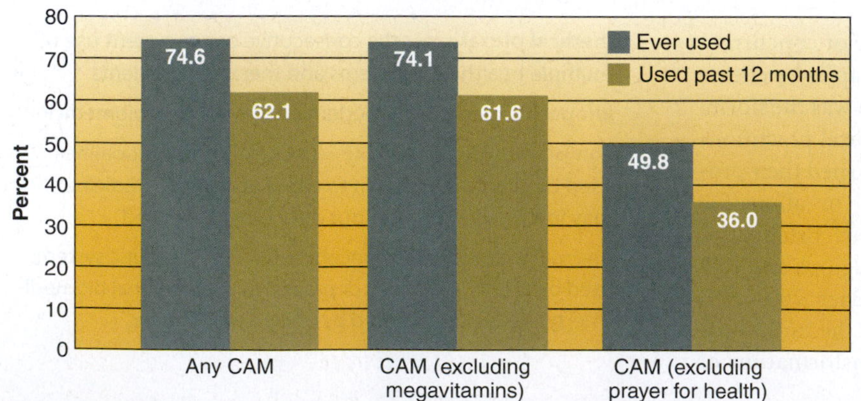

Source: National Center for Complementary and Alternative Medicine, NIH, DDHHS.

Who Chooses CAM?

An analysis of data collected in the 2002 National Health Interview Survey (NHIS) (Barnes et al. 2004) presents the most comprehensive profile of the typical CAM client to date: She is between the ages of 30 and 50, is well educated (bachelor's, master's, PhD, or professional degree), has a family income exceeding $75,000, and lives in an urban area in one of the Pacific states (see Table 4.1). She is most likely Asian or white if CAM is defined as all modalities excluding prayer. The study confirmed earlier data that CAM (excluding prayer) is used most by upper middle class women who are well-informed health care consumers (Eisenberg, Davis, Ettner, Appel, Wilkey, Van Rompay, and Kessler 1998; Murray and Robel 1992; Bausell, Lee, and Berman 2001).[3]

Each CAM client, however, varies from the collective profile, and data from the NHIS study identify trends in CAM preferences among subgroups (see Fig 4.2). When prayer specifically for health was included in the definition, the number of adults using CAM jumped dramatically to 62%. Prayer was the most commonly used CAM by African Americans and Latinos;[4] it was also a favorite of the poor, elders, persons living in the South, and men. Asians, who were least likely to use prayer, were most likely to employ CAM therapies such as meditation, megavitamins, and natural products. They were also most likely to use alternative medical systems such as homeopathy and naturopathy. These therapies and systems were reported as common in Pacific states, possibly due in part to high population concentrations of Asians in that region. Whites were most likely to use manipulative and body-based therapies (such as chiropractic and massage), which were also shown to be popular in the West and Midwest. Energy healing therapies (see the Mind-Body Therapies section) were not widely used by respondents, but were most commonly listed by persons living in the West and Northeast (Barnes et al. 2002).[5]

Figures on Native Americans were not included in the NHIS study, and data are limited. Native Americans were more

FIGURE 4.2 CAM Use by Race/Ethnicity

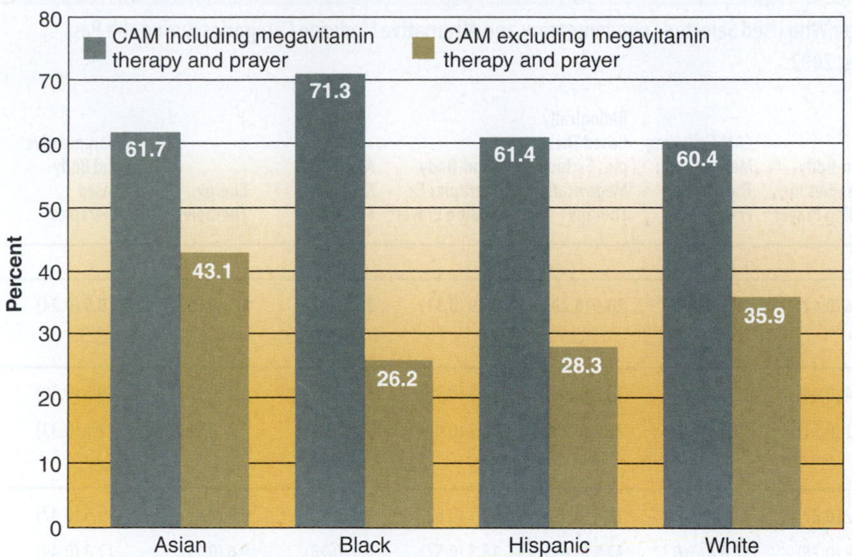

Source: National Center for Complementary and Alternative Medicine, NIH, DDHHS.

likely than any other demographic group to use CAM, according to a survey of 100,000 households conducted in 2001 (Medstat PULSE Survey 2002). Another study found that traditional folk medicine was popular with many urban Indians, although it is also widely believed that it is more commonly practiced in rural areas (Fuchs and Bashshur 1975). Use of CAM by more traditional, less acculturated U.S. minorities, rural populations, or recent immigrant arrivals should not be assumed, however. Studies report that the use of folk healers in some ethnic groups increases with education and income level and that acculturation is not associated with greater use of biomedicine (Marks, Solis, Richardson, Collins, Birba, and Hisserich 1987; Sawyers and Eaton 1992; Solis, Marks, Garcia, and Shelton 1990).

Health conditions most commonly treated with CAM are also detailed in the NHIS report (see Figure 4.3) (Barnes et al. 2004). Neck or back pain and head or chest colds were cited most often. Other ailments, such as headache, depression, stomach problems, high cholesterol, hypertension, and menopause, were reported by less than 5% of CAM clients. Yet because these data are calculated for the total of all CAM users in the survey, they underrepresent the percentage in each category. For example, studies show 54% of headache and neck pain sufferers and 46% of symptomatic menopausal women seek help from CAM (Wolsko, Eisenberg, Davis, Kessler, and Phillips 2003; Keenan, Mark, Fugh-Berman, Browne, Kaezmarczyk, and Hunter 2003). Further, some life-threatening illnesses were not considered in the NHIS findings. CAM care among cancer patients is estimated to range between 31% and 69%, increasing to as much as 80% if spiritual and psychotherapeutic practices are included (Richardson, Sanders, Palmer, Greisinger, and Singletary 2000; Kao and Devine 2000; Nagel, Hoyer, and Katenkamp 2004; Lee, Lin, Wrensch, Adler, and Eisenberg 2000). Recent data on HIV/AIDS patients also found frequent use of CAM strategies, ranging from 36% to over 67% (Fote-Ardah 2003; Gore-Felton, Vosvick, Power, Koopman, Ashton, Bachmann, Israelski, and Spiegel 2003; Kirksey, Goodroad, Kemppainen, Holzemer, Bunch, Corless, Eller, Nicholas, Nokes, and Bain 2002). The fact that many people who try CAM suffer from intractable pain or debilitating conditions correlates with data showing CAM users also make more frequent visits to conventional physicians and are more likely to have been hospitalized within the past year than are clients who do not use CAM (Barnes et al. 2004; Astin 1998; Druss and Rosenbeck 1999).

CAM Rationale

Research consistently demonstrates that adults choose CAM primarily because they believe it helps improve their health when combined with biomedicine (Barnes et al. 2004; Medstat PULSE Survey 2002). According to clients and practitioners, CAM offers a **holistic** approach (see Box 4.2) and is considerate of the body, mind, and spirit. Providers focus on optimal health and seek to enhance natural healing processes when illness occurs. They are especially attentive to client needs, let clients tell their stories, and use "touch and love" to empower self-healing. Responsiveness and caring typify the practitioner-client exchange; time is provided for full discussion of symptoms and concerns. Diagnosis is highly individualized for each client and often includes numerous or complex therapeutic prescriptions (Eisenberg 2002; Moura, Warber, and James 2002; Barrett et al. 2004; Burg 1996).

Some clients choose CAM out of curiosity; others turn to CAM if biomedical care is inconvenient or inaccessible (Murray and Robel, 1992). Many clients choose CAM when they feel that biomedical treatment has failed, especially in the management of chronic pain or the treatment of a terminal illness. CAM addresses quality of life issues and provides users with a sense of control over their situations (Barnes et al. 2004;

holistic medicine—a health care approach that considers the physical, environmental, mental, emotional, social, and spiritual aspects of human experience and aims to achieve functioning, balance, and well-being (American Holistic Medical Association 2004)

TABLE 4.1

Age-Adjusted Percentages of Adults 18 Years and Over Who Used Selected Complementary and Alternative Medicine Categories during the Past 12 Months, by Selected Characteristics: United States, 2002

Selected Characteristic	*Any use of:* CAM Including Megavitamin Therapy and Prayer[1]	Biologically Based Therapies Including Megavitamin Therapy[2]	Mind-Body Therapies Including Prayer[3]	CAM Excluding Megavitamin Therapy and Prayer[4]	Biologically Based Therapies Excluding Megavitamin Therapy[5]	Mind-Body Therapies Excluding Prayer[6]	Alternative Medical Systems[7]	Energy Therapies	Manipulative and Body-Based Therapies[8]
Percents (standard error)									
Total[9,10]	62.1 (0.40)	21.9 (0.30)	52.6 (0.42)	35.1 (0.38)	20.6 (0.29)	16.9 (0.31)	2.7 (0.12)	0.5 (0.05)	10.9 (0.24)
Sex[10]									
Male	54.1 (0.54)	19.6 (0.41)	43.4 (0.54)	30.2 (0.49)	18.2 (0.40)	12.5 (0.36)	2.2 (0.15)	0.3 (0.06)	9.5 (0.30)
Female	69.3 (0.49)	24.1 (0.40)	61.1 (0.51)	39.7 (0.50)	22.9 (0.39)	21.1 (0.42)	3.2 (0.17)	0.7 (0.08)	12.2 (0.33)
Age									
18–29 years	53.5 (0.84)	19.6 (0.63)	44.2 (0.87)	32.9 (0.80)	18.8 (0.62)	17.7 (0.62)	2.3 (0.25)	0.4 (0.09)	9.5 (0.47)
30–39 years	60.7 (0.75)	23.2 (0.64)	49.8 (0.75)	37.8 (0.76)	22.1 (0.63)	18.3 (0.57)	3.3 (0.28)	0.6 (0.11)	12.8 (0.49)
40–49 years	64.1 (0.68)	24.7 (0.64)	53.3 (0.75)	39.4 (0.73)	23.3 (0.62)	18.9 (0.59)	3.2 (0.25)	0.7 (0.12)	13.0 (0.51)
50–59 years	66.1 (0.85)	26.2 (0.72)	56.1 (0.90)	39.6 (0.82)	24.7 (0.71)	19.6 (0.67)	3.3 (0.29)	0.8 (0.16)	11.3 (0.52)
60–69 years	64.8 (0.97)	21.3 (0.81)	56.3 (1.04)	32.6 (0.93)	19.6 (0.79)	14.4 (0.70)	2.1 (0.29)	*0.4 (0.13)	9.8 (0.62)
70–84 years	68.6 (0.94)	15.3 (0.68)	63.3 (1.00)	25.1 (0.85)	13.3 (0.63)	9.4 (0.58)	1.4 (0.22)	*0.1 (0.06)	7.7 (0.52)
85 years and over	70.3 (2.05)	9.1 (1.35)	66.0 (2.16)	14.9 (1.58)	8.4 (1.32)	6.4 (1.14)	*0.9 (0.33)	*0.3 (0.18)	2.1 (0.52)
Race[10]									
White, single race	60.4 (0.44)	22.3 (0.33)	50.1 (0.46)	35.9 (0.42)	20.9 (0.32)	17.0 (0.35)	2.8 (0.13)	0.5 (0.06)	12.0 (0.28)
Black or African-American, single race	71.3 (0.98)	16.5 (0.71)	68.3 (0.98)	26.2 (0.85)	15.2 (0.68)	14.7 (0.69)	1.4 (0.22)	*0.3 (0.11)	4.4 (0.37)
Asian, single race	61.7 (1.94)	29.5 (1.87)	48.1 (1.99)	43.1 (2.03)	28.9 (1.83)	20.9 (1.67)	4.5 (0.74)	*0.6 (0.27)	7.2 (0.90)
Hispanic or Latino origin[10,11]									
Hispanic or Latino	61.4 (0.94)	20.6 (0.74)	55.1 (0.98)	28.3 (0.86)	19.8 (0.73)	10.9 (0.57)	2.4 (0.28)	*0.4 (0.14)	5.8 (0.43)
Not Hispanic or Latino	62.3 (0.43)	22.3 (0.32)	52.4 (0.45)	36.1 (0.40)	20.9 (0.31)	17.7 (0.33)	2.8 (0.12)	0.6 (0.05)	11.6 (0.26)
Education[10]									
Less than high school	57.4 (0.88)	12.5 (0.57)	52.0 (0.89)	20.8 (0.72)	11.7 (0.55)	8.0 (0.46)	1.3 (0.19)	*0.2 (0.06)	5.1 (0.40)
High school graduate/ GED[12] recipient	58.3 (0.68)	17.8 (0.47)	49.6 (0.70)	29.5 (0.61)	16.8 (0.46)	12.4 (0.46)	1.6 (0.16)	0.3 (0.08)	9.4 (0.39)
Some college— no degree	64.7 (0.76)	24.1 (0.64)	54.8 (0.81)	38.8 (0.77)	22.6 (0.63)	19.1 (0.60)	2.7 (0.23)	0.7 (0.12)	12.5 (0.54)
Associate of arts degree	64.1 (1.18)	24.6 (1.01)	53.8 (1.24)	39.8 (1.14)	23.1 (0.99)	20.2 (0.92)	3.0 (0.37)	*0.5 (0.17)	12.6 (0.79)
Bachelor of arts or science degree	66.7 (0.82)	29.8 (0.80)	54.9 (0.89)	45.9 (0.89)	27.7 (0.78)	25.0 (0.79)	4.6 (0.37)	0.9 (0.17)	15.3 (0.65)
Masters, doctorate, professional degree	65.5 (1.92)	31.5 (1.45)	52.7 (1.81)	48.8 (1.87)	29.8 (1.44)	26.5 (1.55)	5.2 (0.79)	*1.6 (0.67)	12.8 (0.78)
Family income[10,13]									
Less than $20,000	64.9 (0.84)	18.9 (0.65)	58.8 (0.84)	29.6 (0.78)	18.0 (0.64)	14.8 (0.58)	2.4 (0.23)	0.4 (0.12)	6.7 (0.38)
$20,000 or more	61.6 (0.44)	23.1 (0.34)	51.2 (0.46)	37.0 (0.43)	21.6 (0.34)	17.9 (0.35)	2.9 (0.14)	0.6 (0.06)	12.1 (0.28)
$20,000–$34,999	63.5 (0.80)	21.1 (0.70)	55.3 (0.82)	34.1 (0.83)	19.9 (0.67)	16.9 (0.66)	2.0 (0.25)	0.5 (0.15)	10.0 (0.53)
$35,000–$54,999	62.8 (0.83)	22.6 (0.72)	52.8 (0.86)	36.6 (0.84)	21.2 (0.68)	17.9 (0.64)	2.9 (0.28)	0.6 (0.11)	11.8 (0.55)

(continued on the following page)

TABLE 4.1 *(continued)*

$55,000–$74,999	60.9 (1.09)	22.7 (0.84)	50.1 (1.12)	37.4 (1.04)	21.2 (0.81)	18.2 (0.84)	2.4 (0.26)	0.4 (0.13)	11.0 (0.65)
$75,000 or more	61.9 (0.94)	27.1 (0.85)	48.7 (0.97)	43.3 (0.94)	25.6 (0.84)	20.7 (0.74)	4.0 (0.33)	0.7 (0.12)	15.2 (0.66)
Poverty status[10,14]									
Poor	65.5 (1.10)	17.9 (0.81)	60.8 (1.13)	28.2 (1.02)	17.0 (0.81)	14.1 (0.79)	2.0 (0.29)	*0.3 (0.13)	5.9 (0.52)
Near poor	64.3 (0.91)	19.1 (0.68)	57.1 (0.98)	30.4 (0.83)	18.3 (0.67)	14.7 (0.63)	1.9 (0.25)	*0.4 (0.13)	7.7 (0.52)
Not poor	62.6 (0.49)	24.7 (0.41)	51.2 (0.52)	39.8 (0.49)	23.2 (0.40)	19.5 (0.42)	3.2 (0.17)	0.6 (0.07)	13.1 (0.33)
Health insurance[15]									
Under 65 years:									
Private	61.4 (0.47)	24.6 (0.40)	50.0 (0.49)	39.4 (0.48)	23.2 (0.39)	19.3 (0.38)	3.0 (0.15)	0.6 (0.07)	13.1 (0.33)
Public	65.1 (1.21)	17.9 (0.88)	59.8 (1.22)	31.1 (1.10)	16.5 (0.85)	18.0 (0.92)	2.3 (0.36)	*0.4 (0.20)	7.3 (0.64)
Uninsured	57.7 (1.00)	21.1 (0.74)	49.5 (1.01)	31.2 (0.89)	20.4 (0.74)	14.7 (0.69)	3.1 (0.34)	0.7 (0.15)	8.0 (0.49)
65 years and over:									
Private	68.2 (0.96)	16.0 (0.72)	61.9 (1.04)	27.2 (0.86)	14.0 (0.67)	10.6 (0.59)	1.4 (0.23)	*0.2 (0.09)	9.4 (0.54)
Public	65.9 (1.18)	14.6 (0.83)	61.1 (1.26)	21.3 (1.00)	13.4 (0.81)	8.4 (0.70)	1.3 (0.26)	*0.1 (0.07)	4.5 (0.55)
Uninsured	74.4 (8.33)	18.2 (4.64)	73.2 (8.31)	19.7 (4.73)	18.2 (4.64)	*3.0 (1.52)	*0.7 (0.74)	*—	*0.7 (0.74)
Marital status[10]									
Never Married	60.2 (1.01)	21.0 (0.76)	52.0 (1.04)	33.0 (0.90)	19.7 (0.73)	18.0 (0.74)	2.6 (0.28)	0.7 (0.16)	9.4 (0.53)
Married	62.4 (0.55)	21.8 (0.43)	52.7 (0.57)	35.0 (0.51)	20.5 (0.42)	15.6 (0.39)	2.7 (0.17)	0.4 (0.08)	11.1 (0.32)
Cohabiting	59.4 (1.86)	25.9 (1.47)	47.7 (1.91)	37.9 (1.87)	24.6 (1.46)	20.4 (1.50)	2.9 (0.46)	*1.3 (0.44)	11.1 (1.15)
Divorced or Separated	65.4 (1.20)	23.5 (0.94)	57.5 (1.21)	38.8 (1.15)	22.2 (0.93)	22.1 (1.00)	2.6 (0.22)	0.6 (0.11)	11.1 (0.70)
Widowed	72.8 (2.39)	22.6 (3.90)	65.5 (2.52)	33.9 (4.05)	21.0 (3.87)	18.5 (3.68)	*2.0 (0.86)	*0.1 (0.07)	8.4 (1.86)
Urban/rural[10]									
Urban	62.6 (0.43)	22.9 (0.34)	53.2 (0.44)	36.0 (0.41)	21.5 (0.33)	18.0 (0.33)	2.9 (0.14)	0.6 (0.06)	10.8 (0.27)
Rural	60.4 (0.80)	19.3 (0.55)	50.9 (0.86)	32.6 (0.76)	18.3 (0.54)	13.9 (0.60)	2.1 (0.21)	0.4 (0.09)	11.1 (0.48)
Place of residence[10]									
MSA,[16] central city	63.5 (0.66)	22.5 (0.55)	55.3 (0.68)	34.9 (0.67)	21.1 (0.54)	18.3 (0.55)	3.1 (0.23)	0.6 (0.09)	9.9 (0.41)
MSA,[16] not central city	61.2 (0.52)	23.2 (0.42)	50.9 (0.55)	36.5 (0.49)	21.8 (0.41)	17.4 (0.40)	2.7 (0.15)	0.6 (0.07)	11.1 (0.32)
Not MSA[16]	62.1 (1.09)	18.2 (0.66)	53.1 (1.17)	31.9 (0.97)	17.2 (0.63)	13.9 (0.76)	2.1 (0.24)	0.3 (0.07)	11.6 (0.63)
Region[10]									
Northeast	57.9 (0.91)	22.6 (0.70)	46.9 (0.91)	35.7 (0.84)	21.1 (0.69)	16.9 (0.69)	3.1 (0.27)	0.7 (0.12)	10.9 (0.53)
Midwest	61.4 (0.80)	20.9 (0.60)	52.0 (0.82)	37.0 (0.77)	19.7 (0.57)	18.2 (0.59)	2.2 (0.20)	0.5 (0.10)	13.2 (0.57)
South	64.6 (0.65)	19.3 (0.45)	57.2 (0.66)	29.9 (0.61)	18.0 (0.44)	14.0 (0.45)	1.9 (0.15)	0.3 (0.07)	7.9 (0.33)
West	62.1 (0.91)	27.7 (0.70)	50.3 (1.08)	42.2 (0.82)	26.4 (0.69)	21.1 (0.82)	4.6 (0.36)	0.8 (0.13)	13.8 (0.55)
Pacific States[17]	64.0 (1.08)	27.7 (0.86)	52.4 (1.22)	43.0 (1.00)	26.4 (0.86)	22.4 (0.98)	4.8 (0.47)	0.8 (0.16)	13.3 (0.65)
Body weight status[10,18]									
Underweight	62.0 (2.55)	18.4 (2.00)	55.1 (2.57)	33.6 (2.38)	17.6 (1.96)	20.4 (2.08)	3.0 (0.74)	*0.5 (0.25)	8.9 (1.38)
Healthy weight	62.7 (0.60)	23.3 (0.49)	53.2 (0.61)	37.2 (0.57)	21.9 (0.47)	19.5 (0.49)	3.4 (0.21)	0.7 (0.10)	11.6 (0.39)
Overweight	60.1 (0.64)	21.9 (0.50)	49.6 (0.66)	34.8 (0.58)	20.6 (0.50)	15.8 (0.44)	2.6 (0.18)	0.5 (0.09)	11.2 (0.38)
Obese	64.6 (0.73)	21.1 (0.56)	56.3 (0.75)	33.4 (0.71)	19.8 (0.55)	15.3 (0.54)	1.9 (0.17)	0.4 (0.09)	10.3 (0.46)
Lifetime cigarette smoking status[10,19]									
Current smoker	57.2 (0.81)	19.7 (0.56)	47.6 (0.78)	32.9 (0.70)	18.7 (0.55)	16.8 (0.55)	2.0 (0.17)	0.5 (0.10)	9.2 (0.42)
Former smoker	66.6 (0.81)	27.0 (0.78)	55.6 (0.87)	41.9 (0.88)	25.3 (0.76)	21.1 (0.77)	4.0 (0.32)	0.8 (0.14)	13.6 (0.60)
Never smoker	62.8 (0.50)	21.2 (0.38)	54.3 (0.53)	34.1 (0.46)	20.0 (0.37)	16.1 (0.37)	2.6 (0.15)	0.5 (0.06)	10.7 (0.30)

(continued on the following page)

TABLE 4.1 *(continued)*

Age-Adjusted Percentages of Adults 18 Years and Over Who Used Selected Complementary and Alternative Medicine Categories during the Past 12 Months, by Selected Characteristics: United States, 2002

Selected Characteristic	*Any use of:* CAM Including Megavitamin Therapy and Prayer[1]	Biologically Based Therapies Including Megavitamin Therapy[2]	Mind-Body Therapies Including Prayer[3]	CAM Excluding Megavitamin Therapy and Prayer[4]	Biologically Based Therapies Excluding Megavitamin Therapy[5]	Mind-Body Therapies Excluding Prayer[6]	Alternative Medical Systems[7]	Energy Therapies	Manipulative and Body-Based Therapies[8]
Lifetime alcohol drinking status[10,20]									
Lifetime abstainer	61.6 (0.79)	14.9 (0.54)	56.9 (0.82)	24.3 (0.66)	14.0 (0.52)	10.8 (0.47)	1.5 (0.18)	*0.2 (0.06)	6.1 (0.33)
Former drinker	69.2 (0.96)	20.5 (0.82)	62.3 (0.99)	33.4 (0.97)	19.0 (0.79)	16.6 (0.74)	2.3 (0.27)	0.5 (0.13)	9.4 (0.57)
Current infrequent/light drinker	62.2 (0.56)	24.3 (0.45)	51.6 (0.58)	39.7 (0.55)	23.0 (0.46)	19.6 (0.45)	3.1 (0.18)	0.7 (0.08)	13.3 (0.37)
Current moderate/heavier drinker	57.0 (0.83)	25.5 (0.65)	43.5 (0.84)	38.5 (0.76)	24.0 (0.64)	18.4 (0.64)	3.4 (0.28)	0.6 (0.11)	12.1 (0.51)
Hospitalized in the last year[10]									
Yes	75.9 (0.97)	22.1 (0.91)	70.4 (1.04)	37.4 (1.14)	20.5 (0.89)	19.5 (0.91)	3.1 (0.40)	*0.5 (0.16)	11.2 (0.71)
No	60.6 (0.42)	22.0 (0.31)	50.8 (0.44)	34.9 (0.39)	20.7 (0.30)	16.7 (0.32)	2.7 (0.12)	0.5 (0.05)	10.9 (0.25)

* Estimates preceded by an asterisk have a relative standard error of greater than 30% and should be used with caution as they do not meet the standard of reliability or precision.

1 CAM including megavitamins and prayer includes acupuncture; ayurveda; homeopathic treatment; naturopathy; chelation therapy; folk medicine; nonvitamin, nonmineral, natural products; diet-based therapies; megavitamin therapy; chiropractic care; massage; biofeedback; meditation; guided imagery; progressive relaxation; deep breathing exercises; hypnosis; yoga; tai chi; qi gong; prayer for health reasons; and energy healing therapy/Reiki.

2 Biologically based therapies including megavitamin therapy includes chelation therapy; folk medicine; nonvitamin, nonmineral, natural products; diet-based therapies; and megavitamin therapy.

3 Mind body therapies including prayer includes biofeedback; meditation; guided imagery; progressive relaxation; deep breathing exercises; hypnosis; yoga; tai chi; qi gong; and prayer for health reasons.

4 CAM excluding megavitamins and prayer includes acupuncture; ayurveda; homeopathic treatment; naturopathy; chelation therapy; folk medicine; nonvitamin, nonmineral, natural products; diet-based therapies; chiropractic care; massage; biofeedback; meditation; guided imagery; progressive relaxation; deep breathing exercises; hypnosis; yoga; tai chi; qi gong; and energy healing therapy/Reiki.

5 Biologically based therapies excluding megavitamin therapy includes chelation therapy; folk medicine; nonvitamin, nonmineral, natural products; and diet-based therapies.

6 Mind-body therapies excluding prayer includes biofeedback; meditation; guided imagery; progressive relaxation; deep breathing exercises; hypnosis; yoga; tai chi; and qi gong.

7 Alternative medical systems includes acupuncture; ayurveda; homeopathic treatment; and naturopathy.

8 Manipulative and body-based therapies includes chiropractic care and massage.

9 Total includes other races not shown separately and persons with unknown education, family income, poverty status, health insurance status, marital status, body weight status, lifetime smoking status, alcohol consumption status, and hospitalization status.

10 Estimates were age adjusted to the year 2000 U.S. standard population using four age groups: 18–24 years, 25–44 years, 45–64 years, and 65 years and over.

11 Person of Hispanic or Latino origin may be of any race or combination of races. Similarly, the category "Not Hispanic or Latino" refers to all persons who are not of Hispanic or Latino origin, regardless of race.

12 GED is General Education Development high school equivalency diploma.

13 The categories "Less than $20,000" and "$20,000 or more" include both persons reporting dollar amounts and persons reporting only that their incomes were within one of these two categories. The indented categories include only those persons who reported dollar amounts.

14 Poverty status is based on family income and family size using the Census Bureau's poverty thresholds for 2001. "Poor" persons are defined as below the poverty threshold. "Near poor" persons have incomes of 100% to less than 200% of the poverty threshold. "Not poor" persons have incomes that are 200% of the poverty threshold or greater.

15 Classification of health insurance coverage is based on a hierarchy of mutually exclusive categories. Persons with more than one type of health insurance were assigned to the first appropriate category in the hierarchy. Persons under age 65 years and those age 65 years and over were classified separately due to the prominence of Medicare coverage in the older population. The category "Uninsured" includes persons who had no coverage as well as those who had only Indian Health Service coverage or had only a private plan that paid for one type of service such as accidents or dental care. Estimates are age-adjusted to the 2000 U.S. standard population using three age groups: 18–24 years, 25–44 years, and 45–64 years for persons under age 65, and two age groups: 65–74 years and 75 years and over for persons aged 65 years and over.

16 MSA is metropolitan statistical area.

17 Pacific states includes California, Oregon, Washington, Alaska, and Hawaii.

18 Body weight status was based on Body Mass Index (BMI) using self-reported height and weight. The formula for BMI is kilograms/meters^2. Underweight is defined as a BMI of less than 18.5; healthy weight is defined as a BMI of at least 18.5 and less than 25; overweight, but not obese, is defined as a BMI of at least 25 and less than 30; and obese is defined as a BMI of 30 or more.

TABLE 4.1 *(continued)*

19 Lifetime cigarette smoking status: Current smoker: smoked at least 100 cigarettes in lifetime and currently smoked cigarettes every day or some days; Former smoker: smoked at least 100 cigarettes in lifetime but did not currently smoke; Never smoker: never smoked at all or smoked less than 100 cigarettes in lifetime.

20 Lifetime alcohol drinking status: Lifetime abstainer is less than 12 drinks in lifetime; former drinker is 12 or more drinks in lifetime, but no drinks in past year; current infrequent/light drinker is defined as at least 12 drinks in lifetime and 1–11 drinks in past year (infrequent) or 3 drinks or fewer per week, on average (light); current moderate/heavier is defined as at least 12 drinks in lifetime and more than 3 drinks per week up to 14 drinks per week, on average for men and more than 3 drinks per week up to 7 drinks per week on average for women (moderate) or more than 14 drinks per week on average for men and more than 7 drinks per week on average for women (heavier).

NOTES: CAM is complementary and alternative medicine. The denominators for statistics shown exclude persons with unknown CAM information.

Source: Barnes P, Powell-Griner E, McFann K, Nahin R. *CDC Advance Data Report #343*. Complementary and Alternative Medicine Use Among Adults: United States, 2004. May 27, 2004

DATA SOURCE: National Health Interview Survey, 2002.

FIGURE 4.3 Disease/Condition for Which CAM Is Most Frequently Used*

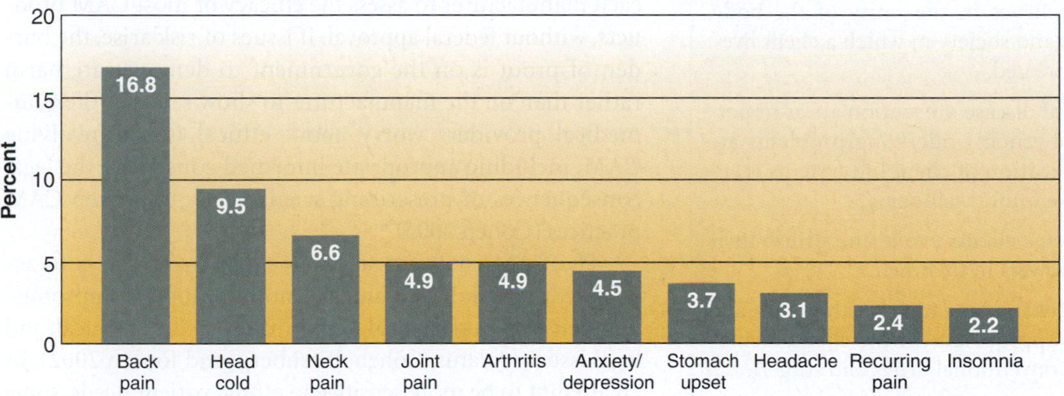

*These figures exclude the use of megavitamin therapy and prayer.

Source: National Center for Complementary and Alternative Medicine, NIH, DDHHS.

Gore-Felton et al. 2003; Li, Verhoef, Best, Otley and Hilsden, 2005). Congruence with personal beliefs and values, maintenance of ethnic identity, and culturally appropriate care are also cited as reasons some people select CAM (Murray and Robel 1992; Astin 1998; Kim and Chan 2004). Conversely, it has been suggested that women, minorities, and members of lower socioeconomic classes may use CAM to challenge the authority of the biomedical system (Baer 2001). Of least concern to consumers appears to be cost. Few choose CAM because it is less expensive, and in some situations, CAM practices and products can be more expensive than biomedical approaches, especially when insurance coverage of CAM is limited.

A significant finding in the NHIS report regarding CAM use was that only 12% of clients employed practitioner-based CAM. This supports earlier suggestions that a large majority of CAM consumers self-prescribe (Barnes et al. 2004; Eisenberg et al. 1998). Further, it has been reported that CAM clients frequently do not mention their use of CAM to their biomedical physicians (Eisenberg et al. 1998; Kao and Devine 2000; Lee et al. 2000).[6] The possibility of inappropriate CAM use that is unsupervised by any health care provider increases the potential for serious negative consequences due to erroneous application or adverse therapeutic interactions.

Biomedical Response

The response of the biomedical community to growing public use of CAM encompasses numerous, disparate positions. On the one hand, many conventional allied health professionals, especially nurses, have embraced CAM (Baer 2001), and there is evidence some physicians are becoming more accepting of CAM therapies.[7] Referrals to CAM practitioners have grown (Barnes et al. 2004; Medstat PULSE Survey 2002), and there is increased research activity on the efficacy of CAM (Barnes and Abbott, 1999). The U.S. government has demonstrated strong interest in CAM, particularly regarding the potential economic savings CAM modalities may offer in treating chronic conditions (Baer 2001). In 1992, the Office of Alternative Medicine (OAM) was established by the National Institutes of Health (NIH) to investigate and evaluate "unconventional" medical practices. In 1998, the National Center for Complementary and Alternative Medicine (NCCAM) was founded as an independent component of NIH, "dedicated to exploring complementary and alternative healing practices in the context of rigorous science; training complementary and alternative medicine (CAM) researchers; and disseminating authoritative information to the public and professionals"

BOX 4.2 CLINICAL APPLICATIONS—HOLISTIC MEDICINE

Holistic healing, as defined by the American Holistic Medical Association, considers the physical, environmental, mental, emotional, social, and spiritual aspects of human experience in order to attain the highest level of functioning, balance, and well-being (American Holistic Medical Association 2004). Holistic healing follows 10 principles:

1. Optimal health, the goal of holistic medicine, is not the absence of illness, but the highest state of being a person can achieve, regardless of health status.
2. Unconditional love is the most powerful medicine, especially when providers meet clients with grace, kindness, and acceptance.
3. Holistic medicine believes in the unity of the body, mind, spirit, culture, and society in which a client lives. The whole person is treated.
4. Health promotion and disease prevention are as important as symptomatic relief. Underlying problems are identified and modification of client life systems is encouraged to maximize future well-being.
5. Holistic medicine helps clients evoke and utilize their own innate healing powers in treatment.
6. Holistic medicine seeks to integrate all therapies and systems that may be appropriate for the client, including other CAM and conventional drugs and surgeries.
7. Holistic medicine is relationship-centered. The desires, opinions, and goals of both the patient and the provider are included in the ideal healing partnership.
8. Who the client is as a person is as important to treatment as what illness is presented.
9. Providers should lead clients through example.
10. The joy and pain of life, including birth, illness, and death, are learning experiences for both client and provider.

(National Center for Complementary and Alternative Medicine 2002). On the other hand, many physicians are uncertain about the competence of CAM providers and worry about offering unrealistic hope of a cure through CAM practices (Konefal 2002). Some believe CAM is quackery, with "implausible, dishonest, expensive, and sometimes dangerous claims" (Atwood 2003) by providers who prey on a naïve public. There have even been calls for defunding the NCCAM.

integrative medicine—the combination of conventional biomedicine with CAM modalities that have been scientifically proven to be safe and effective

A majority of biomedical providers likely fall somewhere between ardent support and outright rejection of CAM. Many question how CAM fits into an evidenced-based practice, arguing that studies on CAM approaches lack rigor and produce a double standard of validation for conventional biomedicine and CAM. Treatment effectiveness studies are limited in assessing the individualized therapies, use of multiple CAM products, variable practitioner techniques, and hard-to-measure outcomes typical of a holistic approach (Institute of Medicine of the National Academies 2005). Further, minimal government oversight of CAM training standards, practice guidelines, and products is also of concern to many biomedical providers. CAM practitioners are largely self-regulated, which results in variable education requirements and skill levels. And it is the responsibility of each manufacturer to assess the efficacy of most CAM products, without federal approval. If issues of risk arise, the burden of proof is on the government to demonstrate harm rather than on the manufacturer to show safety. Other biomedical providers worry about ethical issues involving CAM, including appropriate informed consent or the legal consequences of prescribing scientifically unproven CAM practices (Cohen 2005).[8]

The CAM challenge for biomedical providers is to acknowledge client beliefs and autonomy without compromising their own personal values regarding medical, ethical, and legal issues (Adams, Cohen, Eisenberg, and Jonsen 2002). In an attempt to be more sensitive to ethnic patient needs, some health centers have formed therapeutic alliances with folk healers; a few medical schools have entered into similar strategic partnerships with CAM practitioners as part of their training programs. This biomedical approach to medical pluralism has the potential to encourage "cooperation, research and open communication and respect between practitioners despite the possible existence of honest disagreement, and preserves the integrity of each of the treatment systems involved" (Kaptchuk and Millar 2005). A more hands-on method is **integrative medicine**, which blends conventional biomedicine with CAM modalities that have been scientifically proven to be safe and effective. Examples include physicians and nurses who provide CAM in private practice or hospital settings and independent integrative medical centers. Many health maintenance organizations and insurance companies now cover certain CAM practices and products. However, many CAM practitioners find that the attitudes and beliefs of biomedical health care providers are the largest impediment to successful collaboration (Barrett et al. 2004; Burg 1996; Curlin, Roach, Gorawara-Bhat, Lantos, and Chin 2005). Other CAM proponents caution that though well-intended, attempts at partnering and integration are really cases of the dominant biomedical system co-opting CAM for its own purposes, maintaining the conventional hierarchy in which the physician is superior to all other providers, and vetting CAM according to biomedical standards (Baer 2001). Whether the biomedical community ultimately accepts or spurns CAM,

BOX 4.3 NEW RESEARCH: CAM RESEARCH AND REGULATION

A report by the Institute of Medicine of the National Academies makes numerous recommendations regarding the research, regulation, and integration of CAM in the U.S. The report acknowledges that CAM can be difficult to validate using traditional scientific models, but states as its core message:

> That the same principles and standards of evidence of treatment effectiveness apply to all treatments, whether currently labeled as conventional medicine or CAM. Implementing this recommendation requires that investigators use and develop as necessary common methods, measures, and standards for the generation and interpretation of evidence necessary for making decisions about the use of CAM and conventional therapies. (Institute of Medicine of the National Academies 2005)

Other recommendations in the report include stronger regulation of supplements, from "seed to shelf," more comparative research between different CAM practices and products as well as between CAM and biomedical approaches, better education about CAM therapies and techniques for biomedical providers, greater encouragement of CAM practitioners to create training standards and practice guidelines for their professions, and development of more effective models of integrated care.

and vice versa, is irrelevant to many health care clients, who view medical pluralism as a matter of consumer choice and a right to self-help.

Alternative Medical Systems

Alternative medical systems operate parallel to the conventional biomedical system. They are typically holistic, practitioner-based therapies that are client-centered. Practitioners usually obtain academic training in accordance with established professional standards at U.S. or foreign schools, and many, such as doctors of chiropractic, naturopathy, osteopathy, homeopathy, and "oriental" medicine, are licensed in some states.

Ayurvedic Medicine

Ayurvedic is an ancient Asian Indian medical system created to promote a long and active life. Ayurvedic doctors, called *vaidyas*, traditionally complete a five-year course of study at government-sponsored schools in India. Graduate degrees (MS and PhD) in Ayurvedic are also offered at a few U.S. colleges of alternative medicine. In Ayurvedic, health is defined as balance of body, mind, and soul with respect to the natural, social, and spiritual worlds. The practitioner is more interested in who a client is than in what symptoms are presented. Personal tastes, work habits, temperament, family situation, and lifestyle are considered. Examination of the face (particularly the eyes) and the pulse provide further information. Character and constitution are considered predetermined at birth. Harmony between the client and the universe is achieved through diet, **botanical remedies,** exercise, and meditation.

Throughout life, the body experiences universal tendencies (also called laws of nature): creation (*sattwa*), maintenance (*rajus*), and dissolution (*tamas*) (Tirtha 1998). It also incorporates the natural elements of air/wind (*vata*), fire (*pitta*), and water and earth (*kapha*). These *doshas*, as they are known, constitute each body. Vata is the energy and activity of a body and mind, corresponding to the nerve force, circulatory system, and respiration. Pitta is metabolism, the intake of food and transformation into body substances (including ovum and sperm), corresponding to the digestive and endocrine processes. Kapha is the connective and protective structures, corresponding to bones, muscle, tendons, mucus, and synovial fluid. Strength, endurance, and creativity come when vata is balanced; when pitta is in balance, digestion is comfortable and there is a feeling of contentment. Physical and emotional stability comes from kapha balance.

Each body is dominated by one of the three doshas, presenting three specific body types (Goldber 2002). The vata type is found in persons who are slender, with prominent features or protruding veins, and a warm, vivacious, sometimes volatile personality. They are often energetic with erratic eating and sleeping patterns. They can be anxious and tend to have problems with insomnia, constipation, and menstrual cramping. The pitta type is seen in persons of medium build and fair complexion, with a quick intelligence, passionate temperament, and regular eating and sleeping habits. They are prone to heavy perspiration, indigestion, ulcers, hemorrhoids, and acne. The kapha type is observed in persons who are heavyset and strong, with thick hair and cool, oily skin. They eat slowly and sleep soundly. The kapha type is relaxed, tolerant, and affectionate. They may be obstinate, or procrastinate, and are subject to obesity, high cholesterol levels, and allergy or sinus problems.

ayurvedic medicine—an Asian Indian medical system based on ancient Sanskrit texts; the name comes from "ayur," meaning "longevity," and "vedic," meaning "knowledge or science"

botanical remedies—a comprehensive term covering all therapeutic parts of plants, including roots, bark, stems, gums, sap, leaves, and flowers; sometimes the whole plant is used because it is thought that the components work synergistically to produce a more effective cure

Illness is caused by aggravation of the doshas, and improper diet is a primary cause of this stress. Food is digested by *agnis* (fires) that produce juices and wastes. Too much or too little waste is a symptom of disease. Indigestible foods are especially harmful because they remain in the body, decompose, and produce toxins that require cleansing through purging, enemas, nasal douching, or bloodletting. Foods are classified according to which dosha they enhance or inhibit. In addition, foods are categorized according to their universal tendencies (sattwic, rajasic, or tamasic) as well as whether they are considered heat producing or cold producing, which is dependent not on temperature, but on how they affect the body. This classification system varies regionally in India; for example, lentils and peas are hot in western India, but cold in northern India (Achaya, 1994). Further, the hot-cold quality of a food can be altered by preparation. A cold food can be heated by the addition of chile peppers, for instance, and a hot food can be cooled by soaking it in cold water or adding yogurt (Ramakrishna and Weiss 1992). Thus, mangoes increase kapha, decrease vata and pitta, and are sattwic and hot. Chicken increases pitta and kapha, decreases vata, and is tamasic and cold. Dietary prescriptions account for imbalances leading to illness, as well as age of the client, gender, and seasonal variations. While sweet, sour, salty, and roasted dishes are preferred when digestion is strongest during summer and the monsoons, these foods are avoided during winter. Numerous botanical products are used in addition to diet to relieve symptoms and restore balance during illness (see Table 4.2). Research has proved the efficacy of many traditional botanical preparations (Pari and Saravanan 2004; Jagtap, Shirke and Phadke 2004; Virdi et al. 2003); however, a few may be toxic and several cases of lead poisoning from Ayurvedic remedies have been reported (Baliga et al. 2004; Centers for Disease Control and Prevention 2004).

chiropractic medicine—a medical system focusing on nonsurgical, drug-free care through manipulation of the spine and optimization of nerve function

subluxation—in chiropractic, a misalignment of the vertebrae

hydrotherapy—the use of immersion in water, steam baths, and saunas, colonic irrigation, and hot or cold compresses to treat health conditions; it can include therapeutic additives such as botanicals, minerals, or essential oils

homeopathic medicine—a medical system based on the idea that like cures like, in which minimal (usually diluted) doses of substances known to cause certain symptoms are used to treat those symptoms

Chiropractic Medicine

Chiropractic theory emerged at the turn of the twentieth century. Chiropractors are the third largest group of health care providers in the United States (U.S.), following physicians and dentists. They study for three to four years at a chiropractic college (admission requirements are similar to traditional medical school) and receive a Doctor of Chiropactic (DC) degree upon completion of the curriculum. Chiropractors are licensed in all 50 states and the District of Columbia (Cooper and Stoflet 1996).

The focus of chiropractic is on nonsurgical, drug-free care through manipulation of the spine and optimization of nerve function. **Subluxation,** or misalignment, of the vertebrae causes tension and impediment of nerve function, resulting in numerous health problems. The "innate intelligence" (Tirtha 1998) that regulates the vital processes of the body can be restored when subluxations are removed and the brain can communicate with the rest of the body without interference. The majority of clients consult chiropractors for musculoskeletal pain (Hawk, Long, and Boulanger 2001), though a small number with other conditions such as ulcers, hypertension, asthma, and addiction also use chiropractic care. Many chiropractors use only spinal manipulation to treat clients. Studies have found, however, that 60 to 80% of practitioners also provide diet and exercise counseling (Newman, Downes, Tseng, McProud, and Newman 1989; Hawk, Long, Perillo, and Boulanger 2004). Further, many chiropractors use other therapies, including acupuncture (see the Traditional Chinese Medicine section), **hydrotherapy,** vitamin/mineral supplements, botanical products, fasting, and colonic irrigation.

Homeopathic Medicine

Doctors of Homeopathy (DHt) complete a two- to three-year course of study at a school specializing in **homeopathy** or other alternative medical practice, such as naturopathy. Certification by one of several homeopathic organizations is available after completion of a recognized program. There are no nationally accepted education and practice standards, however, and homeopathy is mostly offered by doctors of chiropractics, osteopathy, naturopathy and a few biomedical specialists. Three states require homeopathic licensing for biomedical physicians: Arizona, Connecticut, and Nevada. Three other states, California, Minnesota, and Rhode Island, permit homeopaths unlicensed in other professions to practice. In the remaining states, homeopaths are subject to laws regarding who may and may not practice medicine.

The meaning of the word, *homeopathy*, explains its principle: "similar to disease," or "similar to suffering" (Carlston 2003).[9] It advocates remedies to stimulate the body's own healing powers according to four tenets. The first is "like cures like," meaning that substances that cause certain symptoms in a person are used to treat those symptoms. Symptoms are not considered negative, nor are they the disease.

TABLE 4.2

Selected Ayurvedic Remedies (*see also* Traditional Chinese Medicine Remedies; Folk Remedies, & Natural Products)

Remedy	DOSHA	Common CAM Use	Cautions*
Adhosa, vasaka (Malabar nut) *Adhatoda vasika*	Pitta, kappa (decrease) Vaya (increase)	Kappa disorders; expectorant, diuretic; treat diabetes (esp. thirst, wasting); treat nausea, vomiting, diarrhea, dysentery; treat urinary tract infections; treat asthma, cough, bronchitis, tuberculosis; treat rheumatism, neuralgia; treat epilepsy, hysteria	Should be used with caution by pregnant/lactating women
Akarakara (pellitory) *Anacyclus pyrethrum*	Vata, kappa (decrease) Pitta (increase)	Stimulant; nerve tonic; treat epilepsy, paralysis; treat diabetes; treat rheumatism; treat bowel conditions	Should not be used by pregnant/lactating women or by persons with gastrointestinal disorders (e.g., gastrointestinal reflux disorder, gastric ulcers, colitis, diverticulitis); may cause nausea, vomiting, diarrhea
Amalaki, amla (Indian gooseberry) *Emblica officinalis*	Vata, pitta (decrease) Kappa (increase)	Pitta diseases; strengthen blood, treat anemia, hemorrhaging; heart tonic; treat diabetes; prevent/treat cancer; treat urinary tract infections, prostate problems; cleanse intestines, colon; treat bowel disorders (esp. constipation, colitis, hemorrhoids); treat liver "weakness"; treat osteoporosis; treat insomnia, irritability; treat balding, premature graying of hair	May cause diarrhea
Arjun (arjuna) *Terminalia arjuna*	Vata, pitta, kappa (balanced)	Prevent/treat heart disease (e.g., endocarditis, mitral regurgitation, pericarditis, angina); treat hypertension; treat liver ailments (esp. cirrhosis); treat digestive disorders, diarrhea; promote urine; treat bone fractures; aphrodisiac	No adverse affects have been reported; other *teminalia* species may be hepatotoxic, nephrotoxic
Ashwagandha (winter cherry) *Withania somnifera*	Vata, kappa (decrease) Pitta (increase when used in excess)	General tonic; reduce stress; slow aging (esp. debilities—fatigue, sexual dysfunction, weak eyes, insomnia, memory loss, Alzheimer's disease); improve immune system; treat HIV/AIDS; prevent/treat cancer; treat multiple sclerosis; treat infertility, menstrual problems, restore hormonal balance in women; treat anemia; treat cough, difficulty breathing; treat arthritis, rheumatism, paralysis; treat alcoholism	No adverse affects have been reported
Brahmi (gotu kola) *Centella asiatica*	Vata, pitta, kappa (balanced)	Promote longevity, improve immune system; reduce stress; improve memory, intelligence; purify blood, adrenal; treat hypertension; treat HIV/AIDS; treat bowel disorders; treat rheumatism; rejuvenate brain cells, nerves; treat convulsions, epilepsy, tetanus; treat nervous disorders, dementia	Should not be used by pregnant/lactating women or women trying to conceive; should not be used by persons with epilepsy; may potentiate sedatives, alcohol; may interfere with cholesterol-lowering medications and hypoglycemic drugs; excessive doses may cause nausea, dizziness, photosensitivity
Eranda, vatari (castor oil plant) *Ricinus communis*	Vata (decrease) Pitta, kappa (increase)	Purgative; promote absorption of nutrients; treat colic, dyspepsia, belching, vomiting, dysentery, infantile diarrhea, constipation, hemorrhoids; treat urinary tract infections; treat kidney stones; treat enlarged spleen, liver, jaundice; treat headache, backache, sciatica; treat arthritis, rheumatism; promote menstruation	Should not be used by pregnant/lactating women or persons with kidney disease, bladder or intestinal infections; bile duct problems, or jaundice; prolonged use can promote hyperaldosteronism or inhibit intestinal motility; may potentiate corticosteroids; beans highly toxic
Gokshura (puncture vine) *Tribulis terrestris*	Vata, pitta, kappa (balanced)	Treat diabetes; treat urinary tract infections; promote kidney health, treat kidney problems (esp. stones, glomerulonephritis); treat back pain, neuropathy; treat rheumatism, sciatica; treat cough, difficulty breathing; treat impotence, boost testosterone levels (improve stamina, increase lean muscle)	Should not be used by pregnant/lactating women or men with enlarged prostate, prostate cancer, or testicular cancer

(continued on the following page)

TABLE 4.2 *(continued)*

Selected Ayurvedic Remedies (*see also* Traditional Chinese Medicine Remedies; Folk Remedies, & Natural Products)

Remedy	DOSHA	Common CAM Use	Cautions*
Gudmar, sarpadarushtrika *Gymnema sylvere*	Pitta, kappa (decrease) Vaya (increase)	Stimulate circulation; liver tonic; treat diabetes (esp. improve pancreatic function); promote urination; treat cough, fever	May potentiate hypoglycemic therapies (e.g., insulin, acarbose, metforman)
Guggula, guggul, (bedellium) *Commiphora mukal*	Kappa, vata (decrease) Pitta (increase)	Reduce serum cholesterol levels; treat diabetes; treat gout; treat arthritis, rheumatism; treat whooping cough, bronchitis; weight loss; regulate menstruation, treat endometriosis; treat dyspepsia, hemorrhoids; treat abscesses, edema	Should not be used by pregnant/lactating women or persons with liver disease, inflammatory bowel disease, or diarrhea
Haridra, gauri (turmeric) *Curcuma longa*	Kappa (decrease) Vaya, pitta (increase)	Regulate metabolism, improve protein digestion; improve intestinal flora; treat anorexia, dyspepsia, flatulence, inflammatory bowel syndrome (IBS), Crohn's disease, hemorrhoids; purify blood toxins; treat diabetes; treat cervical, colorectal, lung, prostate cancers; treat hepatitis, jaundice; treat urinary tract infections; treat arthritis; treat asthma, cough, loosen mucus; treat amenorrhea; treat chronic pain	Should not be used by pregnant/lactating women or persons with liver disorders; may potentiate anti-coagulant drugs; may cause nausea
Kumari (aloe vera) *Aloe* spp.	Juice, small doses: Vata, pitta, kappa (balanced) Powder, large doses: Pitta, vata (decrease) Kappa (increase)	Stomach tonic; purgative; promote fat/sugar digestion; treat colic, gastrointestinal reflux disorder, peptic ulcers, diverticulitis, diarrhea, dysentery, hemorrhoids; treat jaundice, hepatitis; treat breast, cervical, lung cancers; treat amenorrhea, alleviate symptoms of menopause	Should not be used by pregnant/lactating women or persons with inflammatory bowel syndrome, ulcerative colitis, Crohn's disease, appendicitis, or stomach pain; may cause serious potassium depletion if taken with diuretics or corticosteroids; may cause cramping, diarrhea
Nimb (neem) *Azadirachta india*	Pitta, kappa (decrease) Vaya (increase)	Purify blood toxins; cleanse liver; reduce blood glucose levels; weight loss; prevent/treat cancer; treat prostate problems; treat arthritis, rheumatism, muscle pain; treat nausea, vomiting; peptic ulcers; treat parasites; treat cough, bronchitis; contraception	Should not be used by pregnant/lactating women or by women trying to conceive
Sarpa-gandha (snakeroot) *Rauwolfia serpentina*	Pitta, kappa (decrease) Vata (increase)	Treat hypertension; treat bowel disorders, dysentery; treat central nervous system problems; treat hypochondria, violent mental disorders	Should not be used by pregnant/lactating women, persons taking MAO inhibitors, or persons with allergies, asthma, gallstones, ulcers, ulcerative colitis, heart disease, pheochromocytoma, kidney disease; Parkinsonism; epilepsy; or depression; may potentiate antihypertensive therapies; may cause drowsiness, dizziness; nausea, vomiting, diarrhea; arrhythmia; loss of sexual interest
Shilajit (mineral pitch, fulvic acid, "sweat of the rock")	Vata, pitta, kappa (balanced); large dose: Pitta (increase)	Treat HIV/AIDS; treat diabetes; rejuvenate kidneys, treat jaundice, kidney stones; treat gall stones; treat peptic ulcers; treat urinary tract infections; treat hemorrhoids, parasites; weight loss; treat menstrual disorders, sexual dysfunction; treat asthma; treat epilepsy, mental disorders	Should not be taken by persons with high uric acid levels or with fever

*Adverse side effects and/or interactions may occur even if not indicated.

Instead, symptoms indicate how the body is trying to restore itself to health. Treatment is for the fundamental, underlying imbalances. Second is "provings," the experimental analysis of substances in healthy persons to determine effects. Third, traditional homeopathic medicine promotes the use of single therapeutic substances because the effects of interactions could be harmful. However, modern homeopathic medicine often features a mixture of multiple substances. Finally, a minimal dose achieved through a series of dilutions is prescribed.

Homeopathic medicine is considered beneficial for headaches, respiratory problems, diseases of the digestive system, ankle sprains, and postoperative care (Kleignen, Knipschield, and ter Riet 1991). It is also used for management of colds and flu, diabetes, earaches, premenstrual problems, and other conditions. Substances used in homeopathy, such as mercury and belladonna, can be toxic in large quantities, but are not usually problematic in homeopathic doses. However, over-the-counter sales of homeopathic remedies present opportunities for misuse, since many consumers believe that if a little is good, more is better—the antithesis of homeopathic theory (Hawk et al. 2004).

Naturopathic Medicine

Naturopathic medicine was formulated in the nineteenth century, but draws on many ancient healing systems. Naturopathic doctors (ND's) attend four-year colleges that include many biomedical disciplines as well as training in other CAM modalities such as botanical medicine, homeopathy, hydrotherapy, and acupuncture. They are licensed in 13 states. Naturopathic medicine concentrates on primary care, and thus provides advice on keeping healthy as well as the diagnosis and treatment of both acute and chronic conditions.

The six principles of naturopathic medicine address the physical, mental, social, and spiritual well-being of the client: (1) utilize the natural healing powers of the body; (2) treat the cause, not the symptoms; (3) do no harm by using natural therapies; (4) treat the whole person; (5) educate and empower clients to adopt healthy lifestyles; (6) prevention is the best treatment. Illness is considered due to environmental toxins, common food allergies, inadequate vitamin and mineral intake, too much sugar, fat, and gluten in the diet, candidiasis (yeast infection), and dysbiosis (harmful biotic imbalance in the gut), in addition to spinal misalignment and energy problems. Further, emotional, social, and spiritual disharmony can also cause symptoms. Nutritional therapy, based on whole foods and dietary supplements, is the foundation of naturopathic health maintenance and healing. Most naturopaths use diet and lifestyle changes as the primary means of treatment, and offer additional, specialized services in one or more other CAM modalities (Tirtha 1998).

Detoxification through fasting, colonic irrigation, or a controlled diet is often the first recommendation in naturopathic medicine. Analysis of the underlying problems then leads to an individualized nutritional therapy, usually emphasizing a reduction in sugar, refined carbohydrates, sodium, protein intake from red meats, and alcohol. An increase in chicken, fish, legumes, whole grains, vegetables, and fruits is advocated. Naturopathic doctors may recommend macrobiotic diets, elimination diets to determine food allergies, and probiotics (see the Natural Products section) for some clients. Nutritional supplements, including vitamins, minerals, amino acids, and enzymes, are commonly recommended. One study found that naturopathic doctors were more likely than biomedical family physicians to prescribe medication, typically botanical remedies (Boon, Stewart, Kennard, and Guimond, 2003), such as Ayurvedic and traditional Chinese patent medicines, as well as specifically naturopathic formulations.

Osteopathic Medicine

Osteopathic medicine is the oldest alternative medical system originating in the U.S., dating to the early nineteenth century. Practitioners, who are doctors of osteopathy (DO's), complete a four-year training program and then several years of post-graduate study in their area of specialization. There is significant overlap in osteopathic and biomedical education, and osteopathic physicians are licensed to prescribe medications and perform surgery in all 50 states. Unlike biomedical doctors, osteopaths focus on musculoskeletal tension and restriction as the underlying cause of many health conditions. Osteopathic manipulative treatment (OMT) is the distinguishing practice.

Although disorders common to the musculoskeletal system, such as sports injuries, neck and back pain, arthritis, and headache, are frequently treated by doctors of osteopathic medicine, their focus is on primary care, considering the condition of the whole client instead of a single ailment. Analysis of posture, gait, and range of motion, physical symmetry, and inspection of soft tissue for hardening, tenderness, reflex activity, and fluid retention are osteopathic diagnostic tools (Tirtha 1998). For example, muscle stiffness and restriction in the upper body can indicate heart disease or gastrointestinal problems, including liver, pancreatic, and bowel dysfunction. Disorders can be cured through restoration of mobility and suppleness through OMT, including postural correction, diaphragmatic breathing, muscle relaxation techniques, and cranial manipulation. In addition, osteopathic physicians are trained in many CAM techniques, including diet therapy, botanical medicine, spirituality, acupuncture, and homeopathy (Saxon, Tunnicliff, Browkaw, and Raess 2004).

naturopathic medicine—a medical system based on the concept of vitalism, which defines life as an autonomous force that cannot be explained by physical or chemical processes; primary treatments are detoxification and nutritional therapy

osteopathic medicine—a medical system similar to biomedicine but distinguished by a focus on musculoskeletal tension and restriction as causes for conditions; uses osteopathic manipulative treatment

Traditional Chinese Medicine/Acupuncture

Traditional Chinese Medicine (TCM) is an ancient holistic medical system. It seeks to balance the vital forces of the body in order to maintain health.[10] Botanical medicine, nutritional therapy, acupuncture, massage, and exercise are used to restore harmony if illness occurs. In China, a physician is trained for five years and specializes either in botanical medicine or acupuncture, because the curriculum in each is vast and difficult to master. In the U.S., practitioners are likely to combine both in a three- to four-year program (Flaws and Sionneau 2001). Graduates receive a master's of science degree (MS), a master's of Traditional Chinese Medicine (MSTCM), master's of Acupuncture (MAc), or master's of acupuncture and oriental medicine (MAcOM). Over 50 U.S. colleges offer masters programs approved by the national Accreditation Commission for Acupuncture and Oriental Medicine. Some schools offer nonstandardized - doctoral programs (DOM or OMD), and others provide entry-level credentialing programs approved by the National Certification Commission for Acupuncture and Oriental Medicine, including a diplomat of acupuncture (DipAc). Credentials are sometimes obtained by other alternative health care providers to expand their services. Many (though not all) states require licensing for any practitioner of acupuncture or TCM.

In TCM, the vital forces of the body reflect the natural elements of the universe: fire, earth, metal, water, and wood. Each is equated with various body organs and with a secretion. There are also associations with the emotions, seasons, tastes, colors, directions, times of day, and other phenomena (Sheikh and Sheikh 1989; Ots 1990; Anderson 1987). These ancient classifications are further refined through the application of **yin/yang** principles. Yin represents all that is static, feminine, cold, dark, wet, soft, and mysterious in life, while yang is all that is active, masculine, warm, bright, dry, hard, and steadfast. Yin does not exist without yang, and vice versa. Further, health is influenced by life energies, such as **qi**, which travels along invisible channels known as *meridians* found along the surface of the body. Qi is associated with blood (sometimes called the physical manifestation of qi), and often moves with circulation. When qi is stagnant, deficient, or excessive, illness can occur. Other energies that must be balanced include *jing* (sexual energy or primordial essence) and *shen* (spiritual energy or the essence of higher consciousness) (Flaws and Sionneau 2001).

The heart, spleen, kidneys, lungs, and liver are yin organs, while the small intestines, stomach, large intestines, bladder, and gallbladder are yang organs. Each yin and yang organ corresponds to an element (Tirtha 1998). The function of each organ is interrelated to all others. Just as fire can consume wood, and water can extinguish fire, for example, the heart controls the liver, and kidneys direct the heart. A deficiency or excess in one, usually due to a deficiency or excess of yin or yang, can cause a domino effect of symptoms throughout the body. Diagnosis is highly individualized, made through close examination of the client, with special attention to palpitation of pulses, evaluation of the tongue, and an exhaustive history. Through this process a medical pattern is detected, in contrast to determining a specific disease or condition based on symptoms or laboratory testing. It is the medical pattern that determines the appropriate intervention, not the illness. Thus, it is said in Chinese medicine: "*Yi bing tong zhi, tong bing yi zhi* (Different diseases, same treatment; same diseases, different treatments)" (Saxon et al. 2004).

Diet is a primary therapy in TCM. A proper balance of yin and yang foods is considered essential to physical, emotional, and spiritual well-being. Classification of which foods are yin (or cold), and which are yang (or hot), varies regionally and may change with acculturation (Koo 1984; Kittler and Sucher 2004). In general, foods that are low in kilocalories, raw, boiled, or steamed, soothing, and green or white in color are yin; those that are high in kilocalories, cooked in oil, irritating to the mouth, and red, orange, or yellow in color are considered yang. Typical yin foods are most vegetables, fruits, and legumes. Chicken, duck, and honey are sometimes considered yin. Yang items are usually red meats, alcohol, seasonings such as chile peppers, onions, garlic, and ginger, and some produce, including tomatoes, eggplant, and persimmons. A yin food can be made yang through the addition of heat (cooking in oil or spicing), and a yang food can be cooled (through boiling, for example). Rice, noodles, and other Chinese staples are usually placed in a neutral category (Ots 1990; Ludman and Newman 1984). A healthy person may eat additional yang foods to balance the cold of winter, or yin foods to achieve harmony in summer. As people age, the body cools, and more yang foods can be helpful. Other conditions that are caused by too much yin, and thus respond to eating more yang foods, include pregnancy and childbirth, colds, flu, nausea, anemia, frequent urination, shortness of breath, weakness, and unexplained weight loss. Conditions due to excessive yang, which improve with an increase in yin food intake, include constipation, diarrhea, hemorrhoids, coughing, sore throat, fever, skin problems, conjunctivitis, earaches, and hypertension.

traditional chinese medicine—an ancient holistic medical system based on the concept that health is maintained by keeping the body's vital forces in balance

yin/yang—a philosophy with roots in Taoism, the way of nature; yin and yang are the fundamental duality of the universe, opposite and interacting principles of dark (yin) and light (yang).

qi—in Traditional Chinese Medicine, the fundamental essence or life force.

Treatment of a medical pattern determined by a TCM practitioner frequently requires multiple medications composed of natural products, including plants, animals, and minerals (see Table 4.3). For instance, ginseng may be used to fortify qi, and antelope horn can help cool too much yang in the liver (Molony 1998). Formulary mixtures of 5 to 10 substances are common. Most TCM remedies are prepared as decoctions. The client owns the prescription and can reuse it when symptoms occur or share it with family and friends. Studies have confirmed the efficacy of some Chinese remedies in treatment of cancer, hepatitis, and diabetes, among other conditions (Yin, Zhou, Jie, Xing, and Zhang 2004; Zhang et al. 2004; Lo, Tu, Liu, and Lin 2004). Other research has warned that traditional cures can be toxic (Chan et al. 1994; Pak, Esrason, and Wu 2004). The preparation of Chinese botanical medicines and formularies is unregulated, and studies have found imported patent medicines are often adulterated with undeclared pharmaceuticals (such as ephedrine and methyltestosterone) or heavy metals (including arsenic, mercury, and lead). Formularies with mislabeled or unlisted substances have led to hepatitis, renal failure, and death (Ko 1998; Vanherweghem 1994; Shad, Chinn, and Brann 1999).

Practitioners of TCM may also use therapies other than diet and Chinese botanical remedies. **Acupuncture** may be one of these. Qi travels along 12 meridians associated with specific organs (see Figure 4.4 and Box 4.4). When qi is restricted or out of balance, the flow can be enhanced through the stimulation of acupoints with hair-thin needles inserted just under the skin along the meridians. Unlike biomedicine, acupuncture does not provide "instant results" nor "discrete effects on a single symptom, organ or system" (Rothfeld and Levert 2002). Instead, the goal is to reestablish equilibrium without any side effects. Massage and exercise therapy, such as *qigong* (a practice combining movement and meditation—*tai chi* is its best known form), are also used to stimulate qi. Biomedical studies suggest that acupuncture and other qi therapies are effective in some conditions but that the placebo effect may play a role (Berman et al. 2004; Melchart et al. 2005).

Complementary Therapies

Although alternative medical systems are sometimes used in combination with biomedicine, complementary therapies are much more likely to be added. Some therapies are recommended by professional or lay experts; others are recommended by friends or family. Many are self-prescribed, often after an Internet search for information on specific conditions or symptoms (Walji, Sagaram, Meric-Bernstam, Johnson, and Bernstam 2004). Most complementary therapies are available without practitioner oversight, which increases their convenience and decreases their cost.

Chelation Therapy

Chelation therapy is used in conventional medicine to treat lead and other heavy metal poisoning. It is increasingly popular as a complementary practice used for the treatment of cardiovascular disease or cancer, and for antiaging purposes. It comes in two forms: intravenous chelation and oral chelation. Both feature mixtures of EDTA (ethylene-diamine-tetra-acetic acid) and various vitamins and minerals (especially antioxidants such as vitamin C and selenium). Other substances, such as the anticoagulant heparin (in intravenous solutions) and garlic or gingko biloba (in oral preparations) are sometimes included. In intraveneous treatments, provided by medical doctors (MDs), the EDTA is infused slowly over a period of three or four hours, with a total of 20 or more sessions required. Oral chelation is not considered as effective, but is often recommended as a follow-up to intravenous treatment and is sometimes self-administered. Nutritional counseling emphasizing whole, low-fat foods, reductions in caffeine and alcohol intake, smoking cessation, stress reduction, and exercise completes the therapeutic program.

Theoretically, chelation removes calcium build-up (in plaques) from arteries, which reduces atherosclerosis and restores circulation. Practitioners consider it a viable alternative to angioplasty and bypass surgery that is available for a tenth of the cost. Though most research suggests that chelation therapy is not usually harmful and may be beneficial in some conditions other than heavy metal toxicity (Buss, Torti, and Torti 2003), its efficacy in heart disease is unconfirmed (Knudtson et al. 2002). Some practitioners offer an accelerated intravenous treatment, infusing the EDTA solution in just 60 to 90 minutes. Side effects from an incorrect dose or increased rate of administration have been reported, including kidney damage and chelation of needed minerals (Neri, Sabah, and Samra 1993).

Folk Healing

Folk medical systems, including traditional healing practices and home remedies, are among the most commonly used complementary therapies. They are often the initial treatment for nonacute conditions and may be continued even if biomedical care is sought. In the U.S. common forms of folk healing are practiced by some African-Americans, Asians,

acupuncture—a traditional chinese medicine treatment in which hair-thin needles are inserted just under the skin at certain points on the body, in an effort to restore qi equilibrium

chelation therapy—the introduction of EDTA (ethylene-diamine-tetra-acetic acid) into the body to bind with and remove metal ions

TABLE 4.3

Selected Traditional Chinese Medicine Remedies: Often Used in Patent Formularies (*see also* Ayurvedic Remedies; Folk Remedies, & Natural Products)

Remedy	Property	Common CAM Use	Cautions*
Bai zhu, cang zhu *Atractylodes* spp.	Yang (warm)	Promote digestion; treat anorexia; treat diabetes; treat addiction to sweet, fatty foods; weight loss; treat goiter; treat hypertension; stimulate immune system in HIV/AIDS; treat arthritis; treat night blindness	No adverse effects reported
Ban xia (pinella) *Pinella ternatae*	Yang (warm)	Treat nausea, vomiting, diarrhea, abdominal bloating; treat esophageal cancer; treat stroke, dizziness, headache	Dietary supplements containing ephedrine were banned by the FDA in 2003, but the ban was overturned by a federal judge in 2005: interpretation and application of the ruling is pending/Ban did NOT apply to traditional Chinese herbal remedies; should not be used by pregnant/lactating women; toxic when consumed raw
Chai hu (hare's ear) *Bupleurum chinense*	Yin (cool)	Improve immune system; sedative; reduce serum cholesterol and triglyceride levels; treat colds, fever; treat asthma, bronchitis; treat anorexia, dyspepsia, diarrhea, constipation, colitis; hemorrhoids; weight loss; treat hypertension; treat bone cancer; liver tonic, treat liver ailments (e.g., hepatitis, cirrhosis); treat gallstones, inflammation of the gallblader; prevent kidney problems; treat premenstrual syndrome	Should be used with caution by pregnant/lactating women; excessive doses may cause dizziness, vomiting, diarrhea
Chen pi (mandarin, tangerine) *Citrus reticulata*	Yang (warm)	Regulate qi; improve immune system; treat HIV/AIDS; treat anorexia, dyspepsia, nausea, vomiting, peptic ulcer, flatulence, diarrhea; promote urination; prevent/treat congestion, allergies, asthma; treat stress, insomnia	Should be used with caution by pregnant/lactating women and by women with menstrual problems
Dan shen (salvia) *Salvia miltiorrhiza*	Yin (cold)	Heart, blood tonic; treat angina, arrhythmias, atherosclerosis; treat strokes; treat amenorrhea, endometriosis, fibrocystic disease; treat abdominal masses; treat chronic fatigue syndrome, chronic pain, insomnia	Should not be taken by pregnant/lactating women; should not be taken by women with breast cancer; may potentiate anticoagulants (e.g., aspirin) and nonsteroidal anti-inflammatory drugs; prolonged use may be harmful
Dong Quai (angelica) *Angelica sinensis*	Yang (warm)	Heart tonic; purify blood toxins; treat anemia; manage hypertension; treat esophageal, liver cancers; treat peptic ulcers, colitis; treat menstrual problems; treat arthritis; stimulate mucus-clearing cough, treat respiratory infections; treat constipation; treat headache	Should not be used by pregnant/lactating women; should not be used by women with heavy menstrual flow; may potentiate anticoagulants (e.g., aspirin, warfarin); may cause diarrhea, rash; prolonged or excessive doses may cause photosensitivity and changes in blood pressure, respiration
Fu zi (aconite, monkshood) *Aconitum carmichaeli*	Yang (hot)	Diuretic; improve kidney, spleen function; treat metabolic problems; heart tonic; treat pain	Should not be used by pregnant/lactating women; highly toxic when improperly prepared; may cause mouth tingling, tongue numbness, nausea, vomiting, stomach pain, respiratory distress, arrythmia, death
Gao teng (cat's claw, gambir) *Uncaria rhynchophylla*	Yin (cool)	Sedative; treat hypertension; treat liver ailments; treat tremors, seizures, convulsions; treat HIV/AIDS; treat fungal infections, herpes; purify blood toxins in pregnancy, birth; treat arthritis, chronic pain	Should be used with caution by pregnant/lactating women; may potentiate sedatives, anesthesia; excessive doses or prolonged use may cause nausea, diarrhea, swollen feet; stomach pain, kidney damage
Huang qi (milk vetch) *Astragalus membranaceus*	Yang (warm)	Spleen, blood, qi tonic; improve immune system; treat HIV/AIDS; treat heart disease; treat diabetes; weight loss; treat hyperthyroidism; adjunct to chemotherapy, treat melanoma, bladder, bone, colorectal, endometrial, kidney, liver, lung, ovarian cancers; treat edema; promote urination; treat diarrhea; treat colds; increase stamina, reduce fatigue	Should be used with caution by pregnant/lactating women; may interfere with immunosuppressive therapies; may cause bloating, flatulence

(continued on the following page)

TABLE 4.3 *(continued)*

Huang qin (Chinese, Baikal skullcap) *Scutellaria baicalensis*	Yin (cold)	Treat hypertension; treat cough, respiratory infections, asthma, allergies; treat gallstones; treat jaundice; treat irritability, fever, thirst; treat circulatory problems associated with diabetes; treat bone, liver cancers; treat threatened miscarriage; treat prostate problems	Should be used with caution by pregnant/lactating women; should not be used by persons with stomach or spleen disorders; may potentiate anticoagulant drugs; may interfere with carbohydrate metabolism in persons with diabetes
Ling zhi (reishi) *Ganoderma lucidum*	Yang (hot)	Regulate qi; improve immune system; treat HIV/AIDS; reduce serum cholesterol and triglyceride levels; treat hypertension; treat hepatitis; prevent/treat cancer, esp. cervical, colorectal, kidney, liver cancers; treat fibrocystic disease; treat neuralgia; increase stamina, reduce fatigue, stress	Should be used with caution by pregnant/lactating women; may potentiate anticoagulant and antihypotensive therapies; excessive doses or prolonged use may cause dry mouth, dizziness, nose bleeds, nausea, vomiting, itching
Long dan cao (gentiana, bitter root) *Gentiana scabra*	Yin (cold)	Treat hypertension; regulate sugar metabolism, treat hypoglycemia; treat anorexia, flatulence, nausea, vomiting, diarrhea; treat urinary tract infections; treat jaundice; treat genital pain, stimulate menstruation	No adverse effects reported
Ma huang (ephedra) *Ephedra sinica*	Yang (warm)	Analgesic; weight-loss; treat fatigue; improve physical performance; treat asthma, bronchitis; treat multiple sclerosis; treat night sweats	Dietary supplements containing ephedrine were banned by the FDA in 2003, but the ban was overturned by a federal judge in 2005: interpretation and application of the ruling is pending/Ban did NOT apply to traditional Chinese herbal remedies. Should not be used by pregnant/lactating women; should not be used by persons with heart disease, hypertension, diabetes, thyroid problems, glaucoma or enlarged prostate; may potentiate MAO inhibitors and antidepressants; prolonged use or excessive doeses may cause nausea, irritability, insomnia, tremors, seizures, respiratory distress, arrhythmia, cardiac arrest, stroke, death
Sheng di huang (raw)/shu di huang (steamed); rhemannia; Chinese foxglove *Rehmannia glutinosa*	Yin (cold when raw); yang (warm when steamed)	Diuretic; improve immune system; heart, kidney, liver tonic; treat arrhythmia; treat hypertension; treat diabetes, regulate sugar metabolism, treat hypoglycemia; treat goiter, hyperthyroidism; nourish liver, treat hepatitis; clear blocked bile; treat frequent urination; treat hemorrhage, menstrual problems, nosebleed; treat irritability, dizziness, insomnia	Should not be used by pregnant/lactating women; may cause diarrhea
Wu wei zi *Schizandra chinensis*	Yang (warm)	General tonic; improve immune system; treat arrhythmia; treat diabetes (esp. wasting, thirst); treat asthma, coughs; treat insomnia, fatigue; treat liver ailments; treat night sweats; treat premature ejaculation	No adverse effects reported
Xi ku cao (selfheal, all heal) *Prunella vulgaris*	Yin (cold)	Improve the immune system; treat HIV/AIDS; treat hypertension; treat cancer; treat goiter; treat headache, dizziness	No adverse effects reported
Ze xie (water plantain) *Alisma plantago-aquaticae*	Yin (cold)	Diuretic; treat kidney problems (esp. stones); promote urination; treat diarrhea, dysentery; treat diabetes; treat abdominal bloating; treat pelvic infections, herpes, vaginal discharge	Should be used with caution by pregnant/lactating women; may cause bloating, flatulence; mild itching, sneezing or severe hives, respiratory distress may occur in sensitive individuals
Zhi mu *Anemarrhena asphodeloidis*	Yin (cold)	Clear heat; treat diabetes, treat hypertension; treat HIV/AIDS; treat urinary tract infection; treat anorexia; treat cough, infection, inflammation	Should not be taken by persons with diarrhea

*Adverse side effects and/or interactions may occur even if not indicated.

BOX 4.4

CLINICAL APPLICATIONS: CUPPING, COINING, AND MOXIBUSTION

Other therapies that are designed to improve the flow of qi along the meridians or to extract toxins are cupping, coining, and moxibustion. In *cupping,* a heated cup or a cup with a wad of burning paper in it is placed upside down on the skin, creating a light suction. It leaves small round marks. In *coining,* coins or spoons dipped in tiger balm or other ointment are rubbed across the skin, resulting in streaky red marks. In *moxibustion,* a small bundle of herbs, or the tips of lit cigarettes, are burned on the skin. These practices are used mostly by Southeast Asians, but moxibustion is sometimes used in Traditional Chinese Medicine for yin conditions.

FIGURE 4.4 Qi Travels Along 12 Acupuncture Meridians Associated with Specific Organs

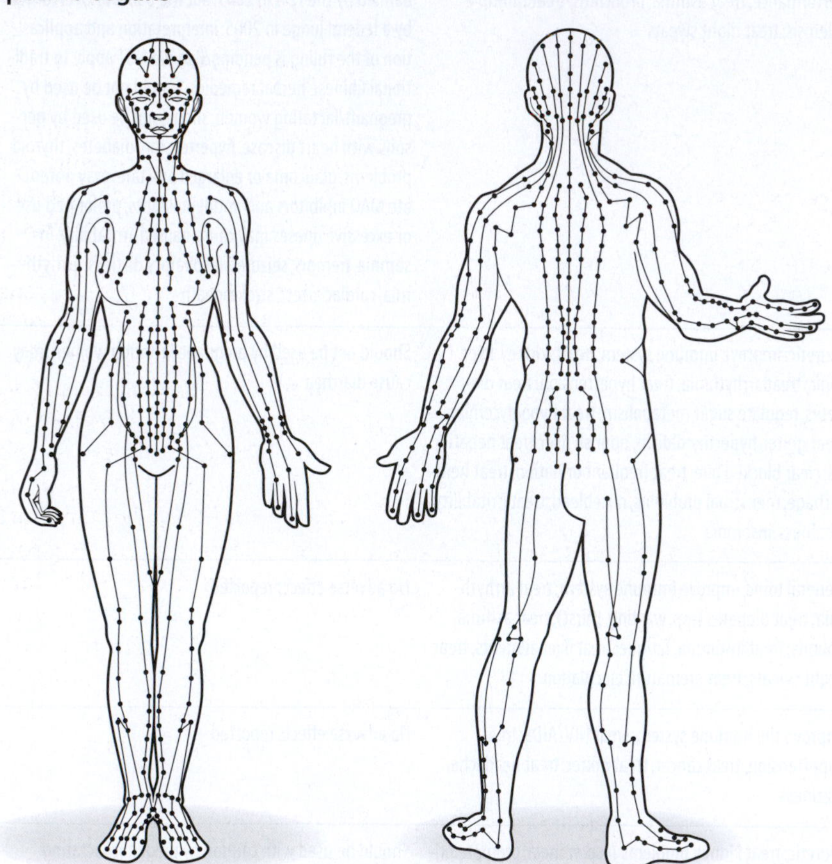

Source: Reprinted with permission from *Alternative Medicine: The Definitive Guide* by Burton Goldburg. Copyright © 2002 by AlternativeMedicine.com, Celestial Arts, Berkeley, CA. www.tenspeed.com

Latinos, Middle Easterners, Native Americans, and residents of certain rural communities. Only the briefest summary on these folk therapies can be included here; information on specific practices is available in other sources (Koo 1984; Spector 2003; Purnell and Paulanka 2002).

Though diverse, folk medical systems share certain common concepts. First and foremost is the idea that illness may be due to a variety of causes (Helman 1990). In biomedical care, a common attitude is that the patient is responsible for his or her health status. The smoker develops emphysema, the alcoholic suffers from cirrhosis of the liver, and the driver is injured when he or she fails to use a seat belt. Blaming the patient is uncommon in folk healing, however. The patient is rarely accountable for sickness. Instead, illness is attributed to outside forces: natural causes, social causes, and supernatural causes.

Natural causes can include weather, smoke, toxins, or pollution. Some Arabs, Chinese, Italians, Filipinos, and Mexicans believe that illness is due to "wind" or "bad air" that enters through body orifices, pores, or especially wounds. In general, a person who is out of balance with his or her environment suffers physical symptoms, a natural cause of illness found in humoral systems, such as Ayurvedic and Traditional Chinese Medicine (see Box 4.5 and the Alternative Medical Systems section). This harmony with nature can also be seen in the **hot-cold** systems found in parts of Latin America, the Middle East, and the Philippines. The definition of foods as being hot or cold varies, but generally depends on their characteristics, such as taste, color, how they are prepared, or their proximity to the sun during growth. Some Middle Easterners believe that consumption of hot or cold foods can cause the body to shift from hot to cold or vice versa. Illness can be due to eating incompatible hot-cold foods together or to overconsumption of a food in one category (Batmanglij 2000; Lipson and Meleis 1983). Some Mexicans and Filipinos classify an illness as either hot or cold, to be treated by a diet rich in the foods of the opposite category (Maduro 1983; Orque 1983). A person's astrology, which predetermines his or her health status at birth, is considered to be another natural cause of sickness.

Social causes of sickness occur through interpersonal conflict within a community. Enemies are blamed for ensuing symptoms. Throughout much of Africa, Asia, Europe, and the Middle East, an envious person may harm another with the "evil eye" (staring with malevolent intent). Conjury may be employed by a person to inflict illness or injury on someone they dislike. Practitioners who use magical charms, substances, chants, and curses are

BOX 4.5 HISTORICAL DEVELOPMENTS: HUMORAL MEDICINE

Humoral medicine is the basis of the ancient Greek healing system. It identified four characteristics of the natural world: air-cold, earth-dry, fire-hot, and water-moist. These were associated with four bodily humors and organs: blood and heart (hot and moist), phlegm and brain (cold and moist), yellow/green bile and liver (hot and dry), and black bile and spleen (cold and dry). The humors were affected by diet, lifestyle, and climate; illness resulted from an imbalance in the humors. For example, an early winter could increase black bile, or *melaina-chole* in Greek (root of the English "melancholy"), causing moodiness. Remedies included foods or hygienic practices of the opposite category, or a change of location. However, sometimes physical interventions, such as bloodletting and emetics, were required to restore humor harmony. Over the centuries, Greek humoral medicine spread to the Middle East and southern Europe. From there it spread throughout Latin America and the Philippines, where hot-cold principles of health and diet are still used today. Examples of older systems that incorporate humoral theories are Ayurvedic medicine and Traditional Chinese Medicine.

numerous, including "herb doctors," "rootworkers," *voodoo* (or *hoodoo*) doctors, *brujos* and *brujas* (Spanish for "witch"), sorcerers, "underworld men," *goofuhdus* men, and "conjure men."[11] Native American conjury often features natural phenomena, such as a snakebite or lightning, to strike a victim. A Latino brujo may cause sickness through contagious magic, such as placing a spell using a bit of hair or fingernail clippings from the targeted person. In turn, magic is needed to cure a person who has been "witched," "hexed," "mojoed," or "rooted." Botanical treatments are common, as are charms and incantations.

Supernatural causes of illness are many, primarily due to the intervention of gods, spirits, or the ghosts of ancestors (there is sometimes an overlap between illnesses due to supernatural causes and those due to social causes). The will of God, for example, is a factor for many Jews, Christians, and Muslims, with illness considered punishment for religious transgressions or part of the unknowable plan for humanity. Prayer is the most common form of CAM practiced in the U.S. (see Figure 4.5). Some Amish and Mennonite Pennsylvania Dutch also turn to *powwowing* (unrelated to Native American healing) or *Brauche*, which includes the laying on of hands, charms, blessings, spells, botanical remedies, and special teas to ameliorate symptoms (Hostetler 1976). In some Christian congregations, trained spiritualists use prayer to channel the healing powers of God.[12] For some people of African, Asian, Latino, Middle Eastern, Native American, or Pacific Islander heritage, it is malevolent spirits that cause illness. Spirit possession takes place when an evil spirit lodges within a person and causes irrational behavior. Sickness due to *soul loss* from spirit possession or extreme stress is common, and results in malaise, depression, weight loss, and sometimes death. Even the ghost of a Southeast Asian ancestor, who normally provides protection to the living, can turn on a person and make him or her ill if the victim has ignored or insulted the spirit. In situations involving supernatural causes, spiritual specialists are needed to cure the client. Native American *shamens*, Hmong *neng*, Mexican *curenderos*, Caribbean *espiritos* or *santeros*, and voodoo priests use ceremonial invocations to communicate with the spirits, saints, or gods to promote healing. Charms, spells, and botanical remedies are also common.

Folk medical systems also share the concept of culturally defined ailments. These folk illnesses are thought to be unique to each cultural group and are recognized by whatever symptoms, complaints, and disorders that group sanctions as sickness. For instance, a Pennsylvania Dutch infant with *livergrown* is irritable and colicky. Some African Americans may develop "high blood" not associated with hypertension, which occurs when too much blood migrates to a certain part of the body due to excessive consumption of rich foods or red-colored foods (such as beets, carrots, grape juice, and red meat—particularly pork).[13] "Low blood" can be caused by eating too many astringent and acidic foods, such as pickles and vinegar, and not enough meat (Jackson 1981). *Empacho* occurs when a wad of food gets stuck in the stomach of a Mexican American, and *pasmo*, a type of paralysis, happens when an imbalance of hot and cold causes digestive problems in Puerto Ricans (Lipson and Meleis 1983; Freidenberg, Mulvihill, and Caraballo 1993). A Korean American with indigestion, poor appetite or weight gain, stomach or chest pain, and hypertension suffers from *hwabyung* (Pang 1994). Typically, clients feel that a culturally defined illness is best cured by a folk healer.

Finally, home remedies are also used regularly in most folk medical systems. Some are recommended by traditional healers or by community experts, especially elder women skilled in health care. Others are considered conventional wisdom that is passed down in families from generation to generation. These home remedies include special foods to improve vitality: pork liver soup in China, thick eggnogs in Puerto Rico, chicken soup in Eastern Europe, and

hot-cold—a classification system that evolved from humoral medicine; unlike yin/yang, it is applied principally to diet, and sometimes to illness, but not to all of nature; to maintain health, hot foods must be balanced with cold foods, and hot or cold illnesses are treated with ample foods of the opposite category; it is sometimes combined with other classifications, such as "cool," "heavy or light," or "acidic or nonacidic"

FIGURE 4.5 10 Most Common CAM Therapies—2002

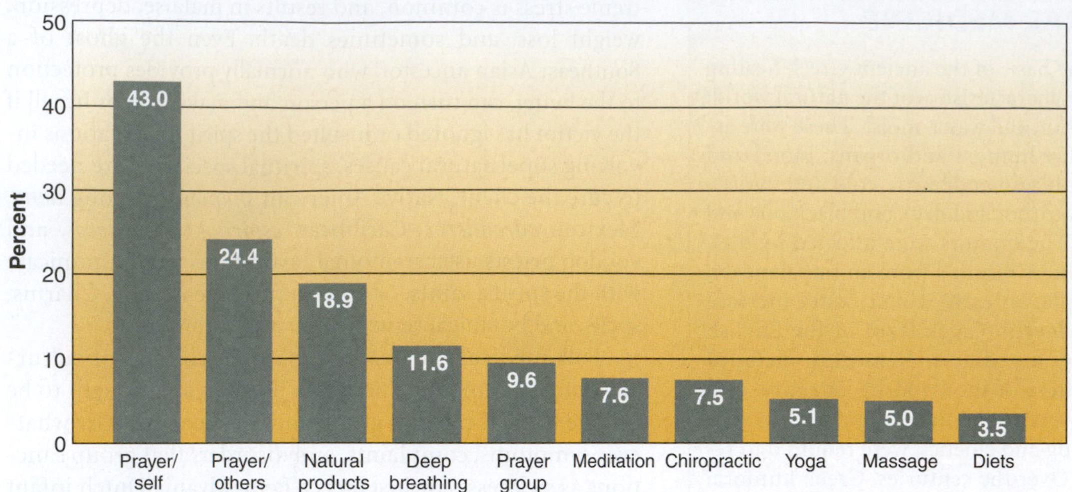

Source: National Center for Complementary and Alternative Medicine, NIH, DDHHS

for the Navajo, blue cornmeal. Other foods are consumed or avoided in the belief that like cures like or like causes like. For instance, some Italians drink red wine to boost their blood, and some American women eat gelatin (traditionally made from cow hooves) to improve their fingernails. Asian Indians may eat walnuts to boost brainpower, and Vietnamese may eat gelatinized tiger bones for overall strength. Some women in the U.S. refuse to eat strawberries during pregnancy for fear they will cause birthmarks on their babies. Botanicals of many types are frequently self-prescribed in folk healing (see Table 4.4). Research has documented effectiveness of these traditional natural products: Native American black cohosh, (Fugate and Church 2004), African American yellowroot (Okunade, Hufford, Richardson, Peterson, and Clark 1994), Brazilian guava leaf (Qian and Nihorimbere 2004), Mexican cactus pear (Wolfram, Kritz, Efthimiou, Stomatopoulos and Sinzinger 2002), Filipino licorice root (Dhingra, Parle and Kulkarni 2004), and Polynesian *noni* (Hornick, Myers, Adowska-Krowicka, Anthony, and Woltering 2003).[14] However, the potential adverse effects of inappropriate dosage and drug interactions of some home remedies are serious (Pierson 2003; Williamson 2001; Bielory 2004).[15]

Natural Products

Natural products are second only to prayer in CAM use. Included in the category are all nonvitamin/nonmineral supplements, particularly herbs and other botanicals. Other types of supplements include animal-based products such as glucosamine, enzymes, hormones, proteins/amino acids, and inorganic substances such as colloidal silver. Functional foods, probiotics, and prebiotics are also classified as natural products. As in other CAM modalities, there is no accepted definition of the category, and there is some overlap with folk healing. Data on usage varies, depending in part on what products are included. The 2002 NHIS CAM study reported nearly 19% of the general population had used natural products during the past year (Barnes et al. 2004). Figures from a survey of American households in 2000 found slightly higher numbers (Medstat PULSE Survey 2000, 2003). A member survey by a large health maintenance organization determined that nearly a third of adult clients used nonvitamin/nonmineral supplements (Schaffer, Gordon, Jensen, and Avins 2003). Most data suggest that, as with other CAM practices, women are more likely than men to use natural products (Gunther, Patterson, Kristal, Stratton, and White 2004; Millen, Dodd, and Subar 2004). However, use of performance-enhancing natural products is thought to be more prevalent in adolescent boys and young men (Perkin, Wilson, Schuster, Rodriguez, and Allen-Chabot 2002; Bell, Dorsch, McCreary, and Hovey 2004). Some research reports non-Hispanic whites as the most likely consumers, while other data found use highest among Native Americans (see Figure 4.6) (Bielory 2004; Medstat PULSE Survey 2000, 2003; Schaffer et al. 2003). Although natural products have been associated with better educated and wealthier clients, one study of low-income, rural patients found that 56% used **herbal remedies** (Gunther et al. 2004; Planta, Gunderson & Petitt 2000). Use is most prevalent in the South and West (Gunther et al. 2004).

Natural products are used to maintain health, prevent disease, treat pain, lose weight, reduce stress, induce sleep, and improve strength, stamina, speed, and mental acuity. They are taken as whole foods, teas, cold beverages and beverage

herbal remedies—technically, preparations made from leafy plants without woody stems; "herbal" is frequently used interchangeably with "botanical"

TABLE 4.4

Selected Folk Remedies (*see also* Ayurvedic Remedies; Traditional Chinese Medicine Remedies, & Natural Products)

Remedy	Preparation	Common CAM Use	Cautions*
Bearberry (Manzanita) *Arctostaphylos uva-ursi*	Leaf, stem tea, infusion or extract	Diuretic; treat diabetes; treat urinary tract infections, kidney problems (esp. stones); treat bronchitis	Should not be used by pregnant/lactating women; prolonged use or excessive doses toxic
Bitter root (Dogbane) *Apocynum* spp.	Root, fruit decoction or extract	Purgative; contraceptive/abortive; treat cardiovascular problems; treat kidney problems; treat liver ailments, gallstones; treat gout; treat edema; headache	May cause nausea; may increase heart rate and arterial blood pressure; excessive doses toxic
Black Cohosh *Cimicifuga racemosa*	Root decoction or extract	Treat menstrual problems; ameliorate menopause symptoms; ease labor; treat hypertension; treat kidney problems; treat arthritis; treat diarrhea; treat cough	Should not be used by pregnant/lactating women; should be used with caution by children, adolescents, and women with history of breast cancer or undergoing chemotherapy for breast cancer; prolonged use or excessive doses may cause mild nausea, vomiting, headaches, hypotension; dizziness; mastalgia, weight gain
Black Nightshade (Zhoa ia) *Solanum nigrum*	Leaf juice	Promote sleep; treat pain; ameliorate menopause symptoms; treat toothache, sore throats; treat colds, cough	Excessive doses may cause nausea, vomiting, disorientation; cardiac arrhythmia; respiratory depression; death
Bloodroot, Red Root, Red Puccoon *Sanguinaria canadensis*	Root juice	Emetic; stomach "cleansing"; treat dyspepsia, peptic ulcers; treat "weak" blood; treat kidney problems; stimulate mucus-clearing cough, treat asthma, croup, whooping cough, tuberculosis; treat liver ailments; treat arthritis, rheumatism; treat skin cancers	Topical applications can be caustic; excess doses may cause nausea, vomiting, dizziness, tremors, hypotension, shock, coma, death
Burdock *Arctium lappa*	Root juice, tea or extract	Prevent/treat cancer; improve immune system in HIV/AIDS: treat diabetes; reduce serum cholesterol levels; cleanse blood toxins; treat prostate cancer; treat kidney problems (esp. stones); treat hemorrhoids; treat back pain; treat gout; treat venereal diseases; treat asthma; treat acne	May interfere with absorption of some medications; may reduce need for insulin in type I diabetes
Guava Leaf *Psidium guajava*	Leaf tea or decoctions	General antioxidant; reduce serum cholesterol levels; reduce blood glucose levels; treat digestive tract disorders, diarrhea, dysentery	May be contraindicated for persons with heart conditions
Hawthorn *Crategus oxyacantha*	Root decoction; berry tea or extract	Diuretic; prevent/treat cardiovascular conditions (e.g., angina, arrhythmia, congestive heart failure); hypertension; reduce serum cholesterol levels; treat blood disorders, insomnia, sore throat	Should be used with caution by persons using beta-blockers; may potentiate digitalis; may cause nausea, headache
Kava Kava *Piper methysticum*	Root, rhizome tea, decoction, extract, powder additive for beverages	Sedative; euphoric; reduce stress, anxiety; treat urinary tract infections; treat bronchitis, asthma; treat venereal diseases; treat headache, backache; treat obsessive compulsive disorder	Should not be used by pregnant/lactating women or persons being treated for depression, hypertension or Parkinsonism; may be hepatotoxic (esp. when consumed with alcohol, Echinacea, or aspirin); may potentiate antiepileptic drugs; may cause central nervous system depression when taken with valerian or chamomile; may cause rash, drowsiness
Licorice Root *Glycyrrhiza glabra*	Root juice, tea, extract	Purgative; treat dyspepsia, gastric ulcers; stimulate endocrine system (esp. in HIV/AIDS); treat sore throat; treat tuberculosis, cough; liquefy mucus in cystic fibrosis; treat liver ailments; treat kidney tumors; treat arthritis, rheumatism; treat lupus erythematosis; menstrual problems	Should be used with caution by persons with heart or renal disease; may increase sensitivity to digitalis; may interfere with hypertension drugs; may potentiate insulin, corticosteroids, MAO inhibitors; prolonged use or excessive doses may cause hypokalemia, hypertension, headache, dizziness, edema

(continued on the following page)

TABLE 4.4 *(continued)*

Selected Folk Remedies (*see also* Ayurvedic Remedies; Traditional Chinese Medicine Remedies, & Natural Products)

Remedy	Preparation	Common CAM Use	Cautions*
Mandrake (Mayapple) *Podophyllum peltatum*	Root decoction, extract, resin	General tonic; purgative; emetic; treat treat anorexia; stomach problems, dyspepsia, constipation; treat urinary tract infections, incontinence; treat liver ailments (esp. hepatitis); lung conditions; treat rheumatism, arthritis	Should not be used by pregnant/lactating women (even topically—may cause fetal abnormalities or miscarriage); should not be used by children; excessive doses may cause rash, irritation, nausea, vomiting, renal failure, hepatotoxicity, cerebrotoxicity
Mango *Mangifera indica*	Leaf tea, decoction; fresh fruit; bark extract (Vimang)	Antioxidant; improve immune system, treat flu; treat diabetes; treat hypertension; treat liver ailments	Persons allergic to poison ivy or oak may also be allergic to mango sap
Mistletoe *Phoradendron leucarpum*	Leaf tea or extract	Treat cancer (esp. breast, ovarian, prostate); treat hypertension; treat cardiovascular disease; treat blood conditions, hemorrhaging, stomach disorders, diarrhea; ease anxiety, panic disorder	Should not be used by pregnant/lactating women or by children; may suppress immune system; may potentiate hypertension drugs and sedatives; berries highly toxic
Morning Glory *Ipomoea* spp.	Root tea or decoction	Purgative; treat diabetes; treat diarrhea; treat kidney problems; treat urinary tract infections; treat menstrual cramps; treat epilepsy, hysteria	Should not be used by pregnant/lactating women; may cause nausea, diarrhea
Noni (Indian Mulberry) *Morinda citirolia*	Fruit juice; leaf tea; bark extract	Improve immune system; prevent/treat cancer; treat diabetes; treat hypertension; treat anorexia, stomach problems, parasites; treat liver ailments; treat urinary tract infections; treat tuberculosis; treat edema; treat sore throat; treat eye problems	Should be used with caution by persons limiting potassium intake; may cause constipation; may turn urine pink
Prickly Pear Cactus *Opuntia* spp.	Fruit juice; pad extract; root decoction	Diuretic; treat diabetes; treat hypertension; reduce serum cholesterol levels; treat kidney problems (esp. stones); treat urinary tract infections; treat hangovers	Should be used with caution by pregnant/lactating women; some persons may experience allergic reactions, e.g., rash, hives, shortness of breath, and chest pain
Poke (Fitolaca) *Phytolacca americana*	Root, shoots decoction	Purgative; improve immune system; treat cancer; treat inflammation, fungal infections (esp. in HIV/AIDS); treat stomach problems; treat liver ailments; treat kidney problems; improve "weak" blood; treat prostate cancer; treat lung conditions; croup; treat arthritis; bursitis; rheumatism; treat toothache	Should not be used by pregnant/lactating women or by children or by persons using antidepressants or oral contraceptives; may be hepatotoxic; improper preparation or excessive doses may cause nausea, vomiting; berries toxic
Raspberry *Rubus* spp.	Leaf tea	Treat anorexia, diarrhea, stomach problems; treat hemorrhaging, anemia; menstrual problems; induce vomiting; induce labor	Should not be used by pregnant women during the first trimester
Willow *Salix* spp.	Bark decoction or extract	Treat inflammation, fever; treat diarrhea; treat osteoarthritis, rheumatism; aphrodisiac; treat premature ejaculation; treat headache, chronic pain; treat bedwetting	Should not be used by pregnant/lactating women or persons with sensitivity to salicylates; should not be given to children under age 16 with flu-like symptoms (to prevent Reyes syndrome); should be used with caution by persons being treated for diabetes, hemophilia, asthma, peptic ulcers, gout; may potentiate anticoagulant and antiplatelet therapies; excessive doses may cause rash, nausea, vomiting, kidney inflammation, tinnitus
Yellowroot *Xanthorhiza simplicissma*	Root tea	Improve the immune system; treat diabetes; treat hypertension; treat liver ailments (e.g., jaundice); treat stomach problems, dysentery	Should not be used by pregnant/lactating women or by infants; may interfere with B-vitamin metabolism

*Adverse side effects and/or interactions may occur even if not indicated.

FIGURE 4.6 Percentage of U.S. Households Using Herbal Supplements—2000

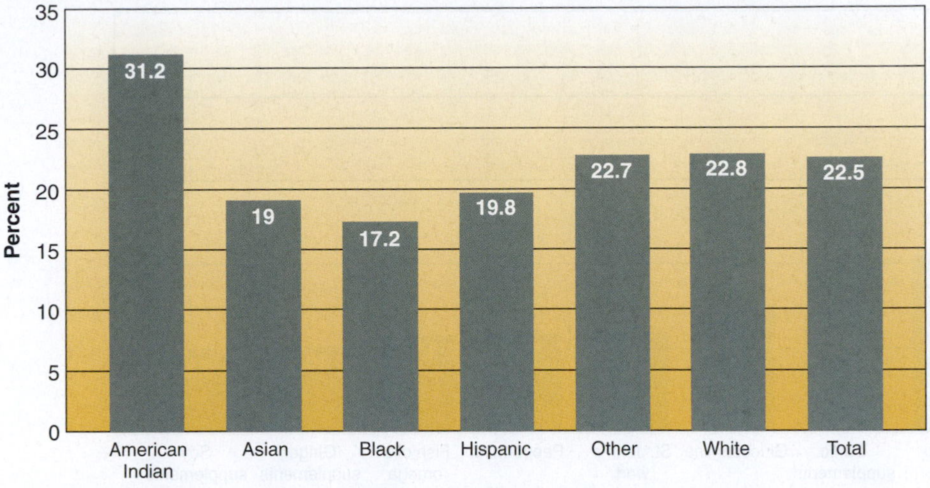

Source: http://www.medstat.com/healthcare/alternative5.asp. Medstat PULSE Survey, 2001.

> **BOX 4.6 CLINICAL APPLICATIONS: PREPARATION METHODS FOR BOTANICAL REMEDIES**
>
> Botanical remedies are often prepared in liquid form as teas, infusions, and decoctions (Debusk 2001). A *tea* is made by steeping fresh or dried botanicals briefly in hot water, usually for only a few minutes, before straining. An *infusion* is prepared by steeping the fresh or dried botanicals for up to 15 minutes in order to extract more of the active ingredients. A *decoction* is even more concentrated; the fresh or dried botanical is boiled in water for up to an hour. Teas and infusions are most often used for soft plant parts, such as leaves and flowers, while decoctions are used most often for hard plant parts, such as roots, stems, bark, and berries. Botanical powders and *extracts* (alcohol or glycerol is used as the solvent) are also sold as beverage additives. Some home remedies call for preparing botanicals in wine or whiskey.

supplements, nutritional bars, injections, tablets, capsules, powders, and suppositories (see Box 4.6). According to data from the NHIS report, echinacea was consumed most often, followed by ginseng, gingko biloba, and garlic pills (see Figure 4.7). Other studies have also found these products popular (Barnes et al. 2004; Medstat PULSE Survey, 2000, 2003; Gunther et al. 2004. Bilberries, evening primrose oil, goldenseal, grape seed extract, peppermint, saw palmetto, St. John's wort, and valerian are other examples of commonly used botanicals (see Table 4.5). In addition to single remedies, botanical compounds, featuring a blend of herbs (and sometimes other substances) are promoted for a variety of ailments.

Glucosamine is the most popular animal-based product; chondroitin, fish oil, shark cartilage, bee pollen, and deer antler velvet are other favorites.[16] Glandulars, which are organ extracts (e.g., heart, liver, mammary, pituitary, thyroid), are also animal products thought to support corresponding human organs. Hormonal supplements include melatonin for sleep problems, anxiety, depressed immune function, and other conditions; L-carnitine (derived from lysine) for serum cholesterol and triglyceride reduction; and dehydroepiandrosterone (DHEA) used to improve mood and memory, boost immunity, and slow aging. [17] Lactase (to mitigate symptoms of lactose intolerance), galactosidase (to reduce flatulence from vegetable and legume intake), pancreatic enzymes (to improve digestion), renal enzymes (to lower blood pressure), bromelain and papain (from pineapple and papaya, respectively, for digestion) and coenzyme Q_{10} (to prevent or cure hypertension, cardiovascular disease, diabetes, cancer, and other conditions) are popular enzyme and coenzyme supplements. Individual amino acids, amino acid mixtures, and protein powders are used primarily as performance enhancers, although they are also promoted to prevent heart disease, build immunity, treat cancer, improve thyroid function, and alleviate insomnia and depression.

Functional foods, probiotics, and prebiotics constitute a new and growing category of food-based natural products. Functional foods are those with ingredients that promote health or mitigate disease apart from the effect of established nutrient function. A 1999 survey found that 95% of respondents believe "certain foods have benefits that go beyond basic nutrition" (Perkin et al. 2002). The most popular are green tea and berries (sources of antioxidant phenols), broccoli (with isothiocyanates, glucosinolates, and other anticarcinogens), tomatoes (a source of the antioxidant lycopene), and soybeans (with isoflavone and other phytoestrogens). Most are used primarily for general health promotion and disease prevention, though soybean consumption is sometimes recommended to ameliorate

FIGURE 4.7 Top 10 Natural Products Used by Adult CAM Users—2002

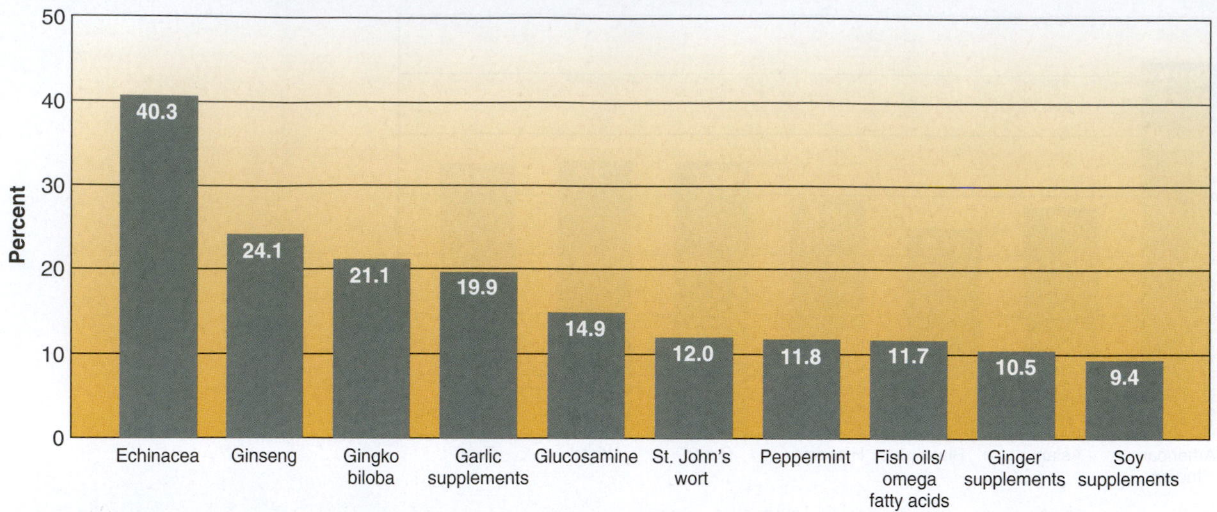

Source: National Center for Complementary and Alternative Medicine, NIH, DDHHS.

menopausal symptoms. **Probiotics**[18] are foods or products that include enough live bacteria (such as lactobacilli and bifidobacteria) to beneficially alter the microflora of the gut. This improves gut function, may reduce diarrhea and constipation, improve nutrient absorption, manage ulcerative colitis, and possibly lower cholesterol, enhance immunity, and prevent cancer. Yogurt is best known of the probiotic foods, and is sometimes inoculated with extra cultures to increase the type and number of microorganisms. Acidophilus milk, kefir, some fermented vegetables (such as cabbage) and cereals, and even salami, can also be cultured as probiotics. Studies suggest that the bacteria need not even be live to exert healthful effects (Schrezenmeir & de Vrese 2001). **Prebiotics** are the nondigestible oligosaccharides found in foods such as bananas, onions, leeks, Jerusalem artichokes, chicory, and honey. Human breast milk also contains prebiotics. These carbohydrates are thought to stimulate the growth and activity of beneficial bacteria in the gut. The therapeutic substances identified in functional foods, the bacteria in probiotics, and the oligosaccharides in prebiotics are also available in nonfood forms, such as capsules, powders, and suppositories.

Many natural products have proved effective for certain conditions, and food-based natural products are particularly promising (Covington 2004; Isolauri 2001; Saavedra, Abi-Hanna, Moore, and Yolken 2004; Tesch 2003). However, research has questioned or refuted some claims (Sleivert et al. 2003; Turner, Bauer, Woelkart, Hulsey, and Gangemi 2005; Van Hasselt, Gashe, and Ahmad 2004).[19] Beyond questions of efficacy, there are several potential problems with natural product consumption. Quality and content can vary from the label information, for example (Garrard, Harms, Eberly, and Matiak 2003). Mild toxicity to some substances has been reported by regional poison centers (Robinson, Griffith, Nahata, Mahan, and Casavant 2004; Yang, Dennehy, and Tsourounis 2003). More serious reactions can occur from interactions with prescription drugs or with other natural products. Therapeutic doses of garlic may act synergistically with fish oil to inhibit blood platelet aggregation, and there can be an additive effect between green tea and aspirin or other anticoagulants. Valerian may enhance depressants, such as barbiturates. St. John's wort can inhibit metabolism of numerous drugs, including birth control products, cyclosporin, warfarin, and digoxin (McKenna, Jones, and Hughes 2002). Gingko biloba reduces the effectiveness of some drugs, such as certain antacids and antianxiety medications, while potentiating others, including anticoagulants, antidepressants, and antipsychotics (Bressler 2005). Black cohosh may alter the response of cells to chemotherapy in patients under treatment for breast cancer (Rockwell, Liu, and Higgins 2005). Consumers may also experience allergic reactions to natural products (Bielory 2004) or poisoning due to interactions between two or more products (Bromley, Hughes, Leong, and Buckley 2005), and some researchers suggest autoimmunity diseases may be triggered by use of immune-boosting herbs in persons predisposed to such disorders (Lee and Werth 2004). Further, natural products can be adulterated with pesticides, heavy metals (such as mercury), or prescription drugs (such as warfarin or alprazolam) (Colson and De Broe 2005).

probiotics—foods or products containing live bacteria in quantities known to beneficially alter the microflora of the gut

prebiotics—foods or products containing nondigestible oligosaccharides and inulin, which are thought to stimulate the growth and activity of beneficial bacteria in the gut

TABLE 4.5

Selected Natural Products (*see also* Ayurvedic Remedies; Traditional Chinese Medicine Remedies, & Folk Remedies)

Product	CAM Dose	Common CAM Use	Cautions*
Bilberry *Vaccinium myrtillus*	80–160 mg	Prevent diabetic retinopathy; treat cataracts, macular degeneration; improve night vision; treat infections and inflammation; treat diarrhea, dyspepsia; treat mouth, throat problems; treat varicose veins	May cause allergic reaction or diarrhea; may interfere with iron absorption; prolonged use or excessive doses may be toxic
L-Carnitine	1000–6000 mg	Improve immune system; facilitate metabolism, protect heart in HIV/AIDS; prevent cardiovascular disease; treat angina; treat arrythmia; treat congestive heart failure; improve physical endurance; improve athletic performance; treat chronic fatigue syndrome; reduce memory loss; treat sports injuries; increase fat metabolism, weight loss	High doses may cause nausea, vomiting and diarrhea.
Chondroitin	400–600 mg	Treat osteoarthritis	May potentiate anticoagulant drugs (e.g., aspirin); excessive doses may cause nausea, diarrhea
Coenzyme Q_{10}	100–400 mg	Treat angina; treat arrhythmia; treat congestive heart failure; prevent heart disease; reduce hypertension; reduce serum cholesterol levels; reduce cancer risk; alleviate fibromylagia symptoms; increase energy levels, improve stamina in HIV/AIDS; treat chronic fatigue syndrome; treat fibrocystic disease; treat Parkinsonism; weight loss	Should be avoided by pregnant/lactating women; excessive doses may cause anorexia, diarrhea, fatigue, twitching
DHEA (dehydroepiandrosterone)	5–200 mg	Improve immune systerm; slow aging; treat chronic fatigue syndrome: alleviate fibromylagia symptoms; prevent muscle wasting in HIV/AIDS; treat lupus erythematosis; promote weight loss	Should not be taken by pregnant/lactating women or persons with ovarian, adrenal or thyroid tumors; may increase risk of breast, ovarian, liver cancer; may decrease serum HDL cholesterol levels; may precipitate mania in mood disorders; large doses may cause acne, facial hair on women, lowering of the voice-should be taken under supervision of health provider
Echinacea *Echinacea purpurea, E. angustifolia*	200–600 mg	Prevent/treat colds, flu; improve immune system; treat HIV/AIDS; treat chronic respiratory infections; treat urinary tract infections; maintain prostate health; treat chronic fatigue syndrome; treat fungal infections; prevent cancer	Should not be used by pregnant/lactating women, persons allergic to sunflower-family plants, or persons with certain systemic diseases, such as AIDS, tuberculosis, diabetes, multiple sclerosis, leukemia, and lupus erythematosis; may cause rash; may be hepatotoxic or nephrotoxic when combined with kava, salicylate (in herbs or aspirin), or hepatotoxic drugs such as anabolic steroids; prolonged use may be toxic
Evening Primrose Oil *Oenothera biennis*	2500–3000 mg	Treat diabetic neuropathy; treat impotency and female infertility; alleviate premenstrual syndrome and menopause symptoms; treat osteoarthritis; treat attention deficit hyperactivity disorder; treat memory loss; improve athletic performance; treat symptoms of alcohol withdrawal; treat skin disorders	Should not be used by persons being treated for epilepsy or schizophrenia (may cause seizures); may cause headaches, nausea; prolonged use may suppress immune system
Garlic *Allium sativum*	1200–2000 mg or (4 g fresh)	Prevent/treat colds, flu, sore throat; reduce hypertension; lower serum cholesterol levels; inhibit platelet aggregation; reduce risk of colon, esophageal, lung, stomach cancers; stimulate immune system in HIV/AIDS; stimulate mucus-clearing cough, treat coughs, bronchitis treat fungal infections; improve nails	May potentiate antihypertensive, hypoglycemic and anticoagulant drugs (e.g., aspirin); may interfere with certain protease inhibiters; large doses may cause nausea, heartburn, flatulence, diarrhea, body odor

(continued on the following page)

TABLE 4.5 *(continued)*

Selected Natural Products (*see also* Ayurvedic Remedies; Traditional Chinese Medicine Remedies, & Folk Remedies)			
Product	**CAM Dose**	**Common CAM Use**	**Cautions***
Gingko Biloba *Gingko biloba*	120–240 mg (extract)	Treat diabetic neuropathy; inhibit platelet aggregation; improve circulation; reduce macular degeneration, cataracts; improve hearing loss; treat involuntary ejaculation, impotence; treat depression, anxiety; treat headache, dizziness; treat Alzheimer's disease, reduce memory loss; optimize brain function	Should not be used by persons being treated for epilepsy or schizophrenia (may cause seizures); may increase blood pressure when taken with certain diuretics; may potentiate certain antidepressant, antipsychotic, and anticoagulant therapies, (e.g., aspirin, warfarin); may interfere with hypoglycemic drugs, antianxiety drugs and some antacids; excessive doses may cause nausea, headache or rash
Ginseng *Panax ginseng,* *P. quinquefolius*	300–2000 mg	Improve immune system; reduce stress and fatigue; improve physical, mental performance; reduce memory loss; regulate sugar metabolism, reduce blood glucose levels in diabetes, treat hypoglycemia; treat impotence and male infertility	Should not be used by persons being treated for acute infections or heart arrhythmia; should be used with caution by pregnant/lactating women and persons receiving immunosuppressive therapies; may potentiate anticoagulant, corticosteroid, hypoglycemic and estrogen drugs, also MAO inhibitors, NSAIDS, and stimulants, such as caffeine and Ritalin; may interfere with calcium channel blockers, opiates, and antipsychotic drugs; excessive doses may cause nausea, headache, insomnia, rash, breast tenderness
Glucosamine	900 mg/100 lb. of body weight	Treat osteoarthritis; alleviate back and joint pain	Should not be taken by persons allergic to shellfish; may increase insulin resistance; may interact with diuretics; excessive doses may cause nausea, diarrhea
Goldenseal *Hydrastis canadensis*	125–650 mg (extract)	Improve immune system; prevent/treat colds, flu; treat HIV/AIDS; inhibit lung cancer growth; treat urinary tract infections; treat chronic fatigue syndrome; treat fungal infections; treat diarrhea	Should not be used by pregnant/lactating women; should not be used by persons with high blood pressure, heart disease, or glaucoma; may potentiate other natural products; prolonged use or excessive doses may cause mouth irritation, nausea, vomiting, diarrhea, nosebleed, and lethargy
Grape Seed Extract *Vitis vinifera, V. coignetiae*	50–300 mg	General antioxidant; prevent/treat cancer; treat HIV/AIDS; lower serum cholesterol levels; reduce LDL oxidation; ameliorate fibromylagia symptoms; treat allergies; treat eczema, psoriasis; treat macular degeneration, cataracts, other vision problems	May interfere with hypocholesterolemic drugs; may potentiate anticoagulant drugs; may act synergistically with vitamin C
Melatonin	1–3 mg	Improve immune system, slow aging; treat cancer, adjunct to chemotherapy, radiation; treat seasonal affective disorder; treat insomnia	May cause excessive drowsiness when combined with sedatives, antihistamines, narcotic pain relievers; may interfere with corticosteroid drugs; may stimulate autoimmunity conditions
Peppermint *Mentha piperita*	450–2500 mg 3-4 cups (tea)	Relieve symptoms of irritable bowl syndrome, diverticulits, morning sickness; improve digestion; treat nausea, vomiting, diarrhea; dissolve gallstones; treat allergies, asthma; treat headache, chronic pain	Should be used with caution by pregnant women in the last trimester and persons with hiatal hernia; large doses may cause heartburn, muscle tremor or rash
St. John's Wort *Hypericum perforatum*	900–4000 mg	Treat depression, anxiety, stress; prevent/treat infections, inhibit growth of HIV/AIDS; reduce serum cholesterol levels; treat breast cancer (prevent infiltration of chest wall); treat premenstrual syndrome; treat chronic fatigue syndrome; alleviate fibromylagia symptoms; reduce memory loss; weight loss	Should be used with caution by pregnant/lactating women; adverse interactions with numerous over-the-counter and prescription drugs, e.g., cold and flu medications, antibiotics, MAO inhibiters, oral contraceptives and protease inhibitors; may interfere with chemotherapy; may precipitate mania in mood disorders; may increase sunburn damage

(continued on the following page)

TABLE 4.5 *(continued)*

Saw Palmetto *Serenoa repens,* *Sabal serrulata*	160–320 mg	Improve immune system; treat prostate problems, e.g., cancer; treat impotence; slow aging	Should not be used by pregnant/lactating women
Valerian *Valeriana* spp.	400–3000 mg	Treat flu; treat stress; treat insomnia; ease anxiety, panic disorder, obsessive compulsive disorder; treat headache; treat alcoholism	Should not be used by pregnant/lactating women; may potentiate other sedatives, e.g., alcohol, barbiturates; may impair driving and operation of machinery; prolonged use may cause headaches, irritability, insomnia, arrythmia

*Adverse side effects and/or interactions may occur even if not indicated.

Natural products are regulated by the 1994 Dietary Supplement and Education Act (DSHEA). The Act defines dietary supplements as neither foods nor drugs but in a separate category, and thus not subject to federal monitoring by the Food and Drug Administration (FDA). Safety evaluation, efficacy testing, and quality control is left up to manufacturers. Many in the industry have adopted uniform manufacturing standards; however, variation in potency is still common. The American Herbal Products Association has developed a numerical rating system for botanical safety: (1) safe when consumed appropriately, (2) restricted for certain uses, (3) use only under supervision of an expert qualified in the appropriate use of this product, and (4) insufficient data to make a safety claim. The FDA has the authority to protect the public from harmful natural products, but the government has the burden of proving a product is unsafe. Manufacturers may make statements regarding the structure and function of a product, but no claims regarding its use to prevent or cure specific illnesses and conditions can be stated. Ultimately, it is up to the consumer to make informed choices regarding natural product selection and use.[20]

Dietary Therapies

Nearly all people believe that a good diet is important in maintaining health and that a poor diet can contribute to disease. Yet diet quality has many definitions. Traditionally, Americans eat three "square" meals, with plentiful protein, starch, and a side of vegetable. Some Italians believe foods are heavy or light, wet or dry, and acid or nonacid. A wet meal, often with soup, is needed once a week to cleanse out the body system. Puerto Ricans, who classify foods as hot, cool, or cold, and heavy or light, may balance hot and cold at each meal, but eat heavy foods (such as starches) during the day and light foods (such as soups) in the evening. Filipinos may try to balance hot and cold ingredients in each dish. Many Middle Easterners believe an ample diet of fresh foods (canned and frozen items are avoided) is needed to maintain health.

Dietary regimes that include or exclude certain categories of foods are appealing to people who are trying to achieve specific health goals. Weight-loss diets are particularly popular: it is estimated that at any given time, 25 to 33% of Americans are eating to lose or control weight (Calorie Control Council National Consumer Survey 2004; The NPD Foodworld /NPD Group 2004). Some of the common approaches include the extremes of eating grapefruit or cabbage soup at every meal, low-carbohydrate diets (including Atkins, South Beach, and Zone) and low-fat diets (such as Weight Watchers and Jenny Craig). Many dieters tailor fad diets to meet their own tastes, picking and choosing food products marketed for current diet trends. Women are more likely than men to diet, and dieters have a higher household income than nondieters (Yin 2001).

Advocates of some diets claim disease prevention. Macrobiotics (based on a Japanese diet of whole grains, miso soup, and vegetables) is promoted as a way to avoid cancer and other illnesses. The Ornish and Pritikin diets (very low-fat, high-complex carbohydrate) were developed to reverse cardiovascular disease. Other specific dietary programs emphasize general health promotion. Organic foods, vegetarian diets (excludes red meat, may include some animal products, such as chicken or fish, eggs and dairy products), vegan diets (exclude all animal products), fruitarian diets (fruit and sometimes grains, seeds, nuts), and raw foodism (uncooked, unheated, unprocessed, organic vegan items) are just some examples.[21] It is also notable that many individuals customize their diets to account for personal food sensitivities or allergies (Tirtha 1998).

Many diets are followed for only short periods, but people who choose their diet based on strongly held convictions may resist modifications contrary to their food beliefs. Health benefits may be just one of many reasons a person adopts a specific diet; other factors can include ethnicity, religion, social or political concerns, and ethical considerations (Koo 1984). Studies on the advantages and disadvantages of dietary therapies are frequently inconclusive. For example, conclusions are conflicted regarding vegetarian diets and longevity, though it has been shown that vegetarianism may reduce the risk of cardiovascular disease and certain cancers (Singh, Sabate, and Fraser 2003; Willet 2003). There is also general agreement that the potential for nutritional deficiencies in vegetarian diets exists (i.e., B_{12}, iron, calcium, and omega-3 fatty acids), especially in children, adolescents, and individuals under conditions of high metabolic demand (Waldmann, Koschizke, Leitzmann, and Hahn 2003; Davis and Kris-Etherton 2003). More extreme regimes can present additional threats of nutritional insufficiency and adverse health consequences.

Vitamin/Mineral Supplements and Megavitamin Therapy

A comparison of data from the 1987, 1992, and 2000 National Health Interview Surveys found that the daily intake of multivitamin/mineral supplements has increased dramatically over the period, from 17 to 27% of the total population (Gunther et al. 2004). Single vitamin and mineral supplementation, such as vitamins A, C, and E, and calcium, also went up. Demographic information showed that women were slightly more likely than men to use vitamin/mineral supplements, and that as age and income increased, so did use. Whites took vitamins/minerals more often than blacks and Hispanics (other ethnic groups were not listed). Supplementation was most common in the West, followed by the Northeast, and was least common in the Midwest and South.

The NHIS reports did not investigate vitamin/mineral dosage among users. However, the 2002 NHIS CAM report found that approximately 3% of respondents used megavitamin therapy, taking supplements in excess of the Recommended Dietary Allowance (RDA) (Barnes et al. 2004). Though there is no consistent definition, the term *megavitamin therapy* usually encompasses megamineral intake as well, often in megadoses of over 10 times the RDA. It is sometimes also called *orthomolecular medicine*, a system that uses vitamin, mineral, and enzyme supplements to address the individual biochemical differences and needs of each client. Most practitioners of orthomolecular medicine are medical doctors (MDs), who typically prescribe a regimen of injections followed by tablets taken several times daily. However, many megavitamin consumers self-diagnose or rely on the advice of supplement salespeople in health food and other stores.

Over-the-counter megavitamin therapy is used frequently for minor complaints. For instance, vitamin C or zinc is taken for colds and chromium picolinate for carbohydrate cravings or to increase metabolism. Combinations of multiple vitamins and minerals are suggested for more serious conditions. Megadoses of vitamins A and C, copper, selenium, and zinc are suggested for osteoarthritis, for instance. Diabetes is sometimes self-treated with high amounts of vitamins B_{12}, C, and E, biotin, chromium picolinate, and zinc. Patent mixtures of vitamins and minerals are marketed for specific health problems; for example, thyroid stimulating compounds are marketed for hypothyroidism. Megavitamin therapy is especially associated with psychiatric conditions, including B_6, magnesium and zinc for autism, and B_3, B_6, B_{12}, C, folic acid, chromium, selenium, and zinc for depression. Megavitamin therapy is also used for children with behavioral disorders or developmental delays.

Proponents argue that megavitamins can remedy nutrient deficiencies that damage DNA, improve enzyme to coenzyme binding in numerous genetic disorders and in certain diseases, and reduce oxidant leakage from decaying mitochondria, thus slowing aging (Ames 2003). Advocates believe that megavitamin therapy is relatively inexpensive and safe: deaths from supplement overdose are rare. While the toxicity of vitamins C and E, chromium (trivalent forms), and beta-carotene is low in most persons, adverse effects from very high doses of vitamins A and D, niacin, pyridoxine, and selenium have been reported (Hathcock 1997). Efficacy is often unproven, and little research on long-term use of megavitamin therapy has been reported. Nutritional imbalances are possible, as reported for excess intake of zinc and deficiencies of copper (Igic, Lee, Harper, and Foach 2002). Hepatotoxicity and carcinogenicity from beta-carotene intake in smokers and drinkers suggest that even supplements presumed safe in the majority population may be dangerous for some clients (Leo and Lieber 1999). Further, megadose side effects are not uncommon. Headaches, insomnia, nausea, constipation or diarrhea, anorexia, mood changes, kidney stones, and allergic reactions (from dermatitis to anaphylactic shock) are possible (see Table 4.6).

Mind-Body Therapies

In the biomedical model, the physical reality of the body is separate from the psychological realm of the mind. Specialists treat one or the other, and psychosomatic illness is disparaged as "all in a client's head."[22] In many CAM systems, the body and mind are unified. The fundamental idea is that psychological status is manifested as physical symptoms, and therefore the diseases of the body cannot be cured without addressing underlying emotional or spiritual needs. Mind-body therapies try to maximize the innate healing power of psychophysiological connections through strengthening the conscious life force. These practices are typically combined with other CAM strategies, particularly nutritional therapies and dietary supplements (Tirtha 1998).

Numerous modalities are considered mind-body therapies, and research suggests some may be effective in certain conditions but ineffective in others (Astin, Shapiro, Eisenberg, and Forys 2003; Krucoff et al. 2005). Some modify how the brain operates in order to change physiological responses and actions. The best known is *biofeedback*, where a client can learn to consciously control bodily functions that are normally unconscious, such as blood pressure and heart rate, in order to optimize well-being. Biofeedback is also used for insomnia, headaches, asthma, and gastrointestinal disorders, including dyspepsia, irritable bowel syndrome, constipation, colitis, and eating disorders. *Hypnotherapy* addresses the unconscious through hypnotic suggestion to treat overeating, smoking, substance abuse,

TABLE 4.6

Selected Individual Vitamin and Mineral Supplements

Supplement	CAM Dose	Common CAM Use	Cautions*
Fat-soluble vitamins			
Vitamin A	2,000–25,000 IU	Improve immune system; prevent infection; prevent/treat cancer, cardiovascular disease; treat osteoarthritis; treat colds, flu	Toxic at high doses: total vitamin A intake may exceed 100,000 IU when combined with other sources of A (food and supplement) for acute infections; high intake may be associated with osteoporosis; may interfere with anticoagulants and anticonvulsants
Beta-carotene	15–100 mg	General antioxidant; reduce oxidation of LDL cholesterol; reduce cardiovascular disease risk; reduce cancer risk (esp. cervical cancer); protect lungs in cystic fibrosis	Hepatotoxic when combined with alcohol intake; may promote pulmonary cancer when used by smokers who drink; may interfere with prescription drugs
Vitamin E	800–1200 IU	General antioxidant; improve immune system; prevent infection; improve glucose tolerance; reduce oxidation of LDL and increase HDL cholesterol; improve circulation; reduce colon cancer risk; protect lungs in cystic fibrosis; reduce pain in osteoarthritis; treat depression, Alzheimer's, memory loss	May interfere with anticoagulants; high doses may suppress immune system
Water-soluble vitamins			
Ascorbic acid/Vitamin C	1000–20,000 mg	Improve immune system, prevent infection; shorten duration of colds, flu; prevent/treat cancer; improve iron absorption; reduce hypertension; reduce serum cholesterol levels and increase glutathione levels; reduce oxidation of LDL cholesterol; reduce pain of angina; improve glucose tolerance, reduce diabetic vascular damage; treat fungal infections, treat HIV/AIDS; protect lungs in cystic fibrosis; treat osteoarthritis; treat depression	May enhance iron absorption (increasing oxidative cellular stress) and decrease copper absorption; may increase risk of kidney stones; may increase in vitro conversion of amygdalin (natural laetrile) to cyanide; high doses can cause diarrhea
Biotin	300–16,000 mcg	Improve glucose metabolism, reduce blood glucose; prevent cracking, peeling nails	No adverse effects reported for oral intake
Cobalamin/Vitamin B_{12}	100–2000 mcg	Reduce plasma homocysteine levels and treat cardiovascular disease; treat anemia; treat diabetic neuropathy; improve brain function in HIV/AIDS, treat psychosis, depression, Alzheimer's, memory loss	No adverse effects for oral intake: B-vitamins are interdependent, excess of one may cause deficiency of others
Folic Acid	800–50,000 mcg	Reduce plasma homocysteine levels; reduce risk of cervical, colon cancer; treat anemia; treat depression, insomnia, irritability, dementia	Megadoses may inhibit cobalamine absorption; should not be used by persons with epilepsy
Niacin/Vitamin B_3	25–1000 mg	Reduce serum cholesterol, LDL, triglyceride levels, increase HDL levels; treat depression, mania, anxiety, dementia, memory loss	Megadoses may impair glucose tolerance; may increase plasma homocysteine; can induce hyperuricemia; should not be used by persons taking high dose aspirin or uricosuric drugs, or those with liver dysfunction, diabetes, or who abuse alcohol; can cause transient flushing, cramps, nausea, diarrhea
Pantothenic acid	50–1000 mg	Prevent infections; treat hypertension; treat depression, irritability; increase longevity	May reduce thiamin absorption and produce deficiency symptoms; can cause diarrhea
Pyridoxine/Vitamin B_6	25–1800 mg	Improve immune system; improve glucose tolerance; reduce homocysteine levels; treat artherosclerosis; reduce cervical cancer risk; alleviate premenstrual syndrome; treat morning sickness; treat depression, autism	High doses may reduce folate levels; can interfere with medications for Parkinsonism; may cause neuropathy; may cause rash
Riboflavin/Vitamin B_2	2–400 mg	Reduce frequency, severity of migraines; treat depression; improve immune system	No adverse effects for oral intake: B-vitamins are interdependent, excess of one may cause deficiency of others
Thiamin/Vitamin B_1	9–100 mg	Treat psychosis, anxiety, depression, irritability, Alzheimer's disease, memory loss	No adverse effects for oral intake: B-vitamins are interdependent, excess of one may cause deficiency of others

(continued on the following page)

TABLE 4.6 *(continued)*

Selected Individual Vitamin and Mineral Supplements

Supplement	CAM Dose	Common CAM Use	Cautions*
Minerals			
Boron	1–9 mg	Treat arthritis; improve bone density	Total intake may include additional amounts found in other supplements or foods; toxic in large doses: diarrhea, vomiting, death
Chromium	300–1000 mcg	Lower blood glucose levels; improve glucose tolerance; reduce LDL and increase HDL cholesterol levels; increase metabolism/weight loss; increase lean muscle, maintain muscle mass in HIV/AIDS	May cause rash; may cause renal or liver damage in large doses; may accumulate in body tissues causing oxidative damage; may be mutagenic; alters serotonin, dopamine, and norepinephrine metabolism in brain, can contribute to mood changes; may potentiate antidepressants
Copper	2–6 mg	Improve immune system; prevent infections; reduce risk of cardiovascular disease; prevent/treat cancer; treat osteoarthritis; treat anemia	Large doses may impair memory, cause depression, insomnia, depress immune system, cause oxidative tissue damage
Magnesium	400 mg	Improve pancreatic function, glucose tolerance; reduce birth defects, spontaneous abortion in pregnant women with diabetes; reduce diabetic retinopathy; relieve angina; treat hypertension; reduce cancer risk; treat autism, attention deficit hyperactivity disorder	Excessive doses may cause diarrhea; extremely large doses (usually due to antacid or Epsom salt abuse) may cause shock, coma, or cardiopulmonary arrest; should not be used by persons with poor kidney function
Potassium	200–500 mg	Improve glucose tolerance; prevent/treat hypertension; prevent muscle cramps	May interact with certain prescription and over-the-counter drugs (e.g., nonsteroidal anti-inflammatory drugs—NSAIDS); can cause hyperkalemia resulting in heart arrythmias, death
Selenium	50–400 mcg	Improve immune systerm; prevent infection; reduce risk of cardiovascular, cerebrovascular disease; reduce oxidation of LDL cholesterol; prevent/treat cancer (esp. prostate cancer); treat HIV/AIDS; treat osteoarthritis; protect lungs in cystic fibrosis; detoxify body of heavy metals; improve hair, nails, skin	Large doses may cause rash, irritability, gastrointestinal problems, impair immune system; may be hepatotoxic; may interact with lipid-lowering drugs
Zinc	10–75 mg	Improve immune system; prevent/treat colds, flu; prevent diabetes, reduce blood glucose levels; prevent/treat cancer; treat HIV/AIDS; treat osteoarthritis; treat prostate problems; treat autism; treat fungal infections; treat acne	May cause stomach upset; large doses inhibit copper and iron absorption; may suppress immunity; may affect absorption levels of prescription drugs

*Adverse side effects and/or interactions may occur even if not indicated.

insomnia, and other conditions. Some mind-body modalities concentrate on the emotional state of a person to enhance the immune response, such as *guided imagery* (promotion of positive thoughts about healing expressed through vision, hearing, smell, taste, and tactile sensation) and *neuro-linguistic programming* (NLP)—changing patterns of verbal and nonverbal negative expressions about healing to positive patterns. Many mind-body practices, including meditation and *breathwork* (diaphragmatic breathing to release negative emotions and tension and promote complete relaxation), promote relaxation for general health. Practices that improve the mind-body energy fields are commonly referred to as *energy therapies*. An example is *reiki*, which in Japanese means "free passage of universal life force." Reiki is used by 200,000 practitioners worldwide to manipulate the energy field around a client. When the practitioner places his or her hands above the client, or gently touches energy centers and pathways on the body, the client draws energy as needed to revitalize and heal (Tirtha 1998). *Reflexology*, another energy therapy is based on the concept that all parts of the body, including organs, are reflected in the hands and feet. Diagnosis is made through examination of the hand and foot reflex zones, and precise pressure is exerted to increase energy flow and promote healing. The Chinese practices of acupuncture, qigong, and tai chi are other examples of energy therapies (see the Traditional Chinese Medicine section).

Yoga, an ancient therapy with roots in Ayurvedic medicine (see the Alternative Medical Systems section), has achieved popularity beyond its Asian Indian origins. Yoga promotes the integration of physical, mental, and spiritual energies through exercise, detoxification, and purification.

The physical postures, known as *asana*, can be meditative or therapeutic. Breath control is used to increase energy flow throughout the body. Fasting, enemas, nasal cleansing, and eye cleaning are part of the complete practice. Yoga can reduce stress, increase strength and flexibility, ameliorate muscle pain, reduce blood pressure and pulse rate, and possibly improve glycemic control and nerve function in people with diabetes (Bharshankar, Bharshankar, Deshpande, Kaore, and Gosavi 2003; Malhotra, Singh, Tandon, Madhu, Prasad, and Sharma 2002; Williams et al. 2005). Other conditions that may benefit from yoga include headache, chronic back pain, arthritis, insomnia, addictions, and cancer.

TABLE 4.7

Factors in Risk-Benefit Analysis of Complementary and Alternative Medical Versus Conventional Biomedical Treatment
Severity and acuteness of illness
Curability with conventional treatment
Degree of invasiveness, associated toxicities, and side effects of conventional treatment
Quality of evidence of safety and efficacy of the desired CAM treatment
Degree of understanding of the risks and benefits of CAM treatment
Knowledge and voluntary acceptance of those risks by the patient
Persistence of the patient's intention to use CAM treatment

Source: Adams KE, Cohen MH, Eisenberg D & Jonsen AR. 2002. Ethical considerations of complementary and alternative medical therapies in conventional medical settings. *Annals of Internal Medicine,* 137: 660-664. P. 661

Medical Pluralism in Practice

The popularity and variety of CAM practices and products underscores the reality of medical pluralism. Acknowledged or not, many clients are combining conventional care with other treatments. Effective biomedical therapy may depend on working with a client to develop CAM-inclusive approaches.

It is beneficial for biomedical providers to keep an open mind about CAM, recognizing that CAM care may be effective for a client or may be addressing emotional, social, or spiritual needs that are unmet in most conventional clinical settings. An attitude of acceptance can encourage clients to share their CAM use with clinicians. Assessment should include not only what therapies are used but whether an alternative practitioner is employed or if the client has self-prescribed. The whole spectrum of CAM should be considered, and it is important not to overlook unrelated conditions. A woman with diabetes may be using CAM for depression, an older man with cardiovascular disease may be using it to help with his enlarged prostate, and the parent of an obese child may be using it to deal with behavioral problems. Intake information for all dietary regimes, natural products, over-the-counter products, patent medicines and vitamin/mineral supplements should be obtained.

Putting CAM use into clinical perspective can help the practitioner determine if it is medically, ethically, and legally responsible to endorse. A risk-benefit analysis of the CAM employed by a client (see Table 4.7) can clarify whether CAM is appropriate to use within the context of an individual treatment plan. Further, each CAM therapy and product can be classified as "(1) the medical evidence supports both safety and efficacy: recommend; (2) the medical evidence supports safety, but evidence regarding efficacy is inconclusive: accept but monitor; (3) the medical evidence supports efficacy, but evidence regarding safety is inconclusive: accept but monitor; and (4) the medical evidence indicates either serious risk or inefficacy: avoid and discourage" (Cohen 2005). Sometimes there will be limited scientific evidence regarding safety or efficacy of a CAM modality. Explaining that a practice or product is unproven as yet, and describing how CAM is largely unregulated, can help clients to make informed choices.

Biomedical providers have the opportunity to provide valuable oversight of client care, coordinating various practices and products in equal partnership with alternative practitioners or the client. Learning why a client uses CAM and what she or he hopes to achieve with the treatment assists providers to identify and discuss all therapeutic alternatives that are appropriate to the client's goals. Recommendations of suitable CAM modalities and guidance in selecting practitioners and products (see Table 4.8) are other useful strategies. Education about how some CAM therapies may potentiate or interfere with prescription drugs can increase intervention success and prevent serious adverse interactions. Acknowledgement of CAM benefits and rejection of only those practices dangerous to clients can help clinicians in building an effective therapeutic relationship with the client. A balance of biomedical and CAM approaches can fulfill the potential of medical pluralism to meet all client care needs.

Conclusion

Developing a plan for health care is a daunting task for even the savviest consumer. With the increased prevalence of the use of complementary and alternative medicine, patients need biomedical practitioners who are familiar with CAM and have developed the skills necessary to assist them as they make important decisions about their medical care. Through careful study of the information provided in this chapter—an overview of why clients choose CAM and an introduction to popular CAM modalities and theories, especially those aspects that may affect conventional nutrition therapy—the future practitioner can become better prepared to assist these clients.

TABLE 4.8

Guidelines for Client Selection of Complementary and Alternative Medicine

How to Choose Appropriate CAM Practices and Practitioners

Develop a list of personal goals for care, including physical, emotional, social, or spiritual needs.

Read books or articles online about various CAM approaches—learn as much as possible about the general principals of complementary and alternative medicine.

Find specific CAM practices that meet personal health care goals.

Look for negative reports on selected CAM practices: consider all information thoughtfully (CAM is often unregulated by the federal government and it is the consumer who must determine which CAM is effective and safe for personal use).

Choose a practitioner who will work in partnership with conventional health care providers.

Check that practitioners are trained in the therapies they provide: confirm credentials and licenses appropriate for each practice.

Select a practitioner who is easy to talk to and listens well. A good relationship is essential to healing. Discuss personal health care goals.

Discuss all biomedical care with the CAM practitioner, including use of prescription drugs and over-the-counter medications. Treatment for one condition can affect treatment of another problem.

How to Choose Appropriate CAM Products

Read books or online references regarding product claims. Look for evidence beyond advertising and personal stories about effectiveness. Look for clinical trials with a large number of participants (hundreds or thousands). Examine negative information about the product or therapy reported by some researchers, practitioners, or users, and weigh the evidence.

Become familiar with safe concentrations and dosage recommendations for product ingredients, including toxicity reports and upper safety levels for vitamins and minerals.

Avoid products with exaggerated claims. If it treats a multitude of complaints or promises a miracle cure, the product should be suspect; if it sounds too good to be true, it probably is.

Natural isn't always safe. If a product is effective against a symptom or disease, it is a drug and may be harmful if misused.

Ask for advice from trusted CAM practitioners or conventional health care providers who do not profit from sales of the product when in doubt.

How to Purchase Appropriate CAM Products: Check the Label

Check the active ingredients list: verify vitamins or minerals needed are included, and look for scientific name of botanicals desired (different plants may have same common name)—also confirm parts used to prepare product are those known for their therapeutic value.

Check concentration and daily dosage: Is the concentration of a botanical appropriate to achieve its benefits? Is the total daily dosage appropriate for expected results? Too little (insufficient concentration or dose) can reduce the effect, and too much (too concentrated or recommended dose too high, such as some megadoses) may result in serious side effects, including toxicity, even with natural products.

Check expiration date (loss of potency may occur after the date).

Find the lot number, as well as the name, address, phone number or website of manufacturer in case problems develop.

Look for indications of quality: check for USP insignia guaranteeing vitamin or mineral potency and purity as tested by the U.S. Pharmacopeia (USP) and/or the USP-NF insignia stating that herbal ingredients in the product meet the standards for safety and quality set forth in the USP National Formulary (NF) and/or the NNFA GMP seal certifying that a supplement is voluntarily produced with good manufacturing practices (GMP) according to the National Nutritional Foods Association (NNFA).

Store according to label instructions to maintain potency.

How to Coordinate CAM and Conventional Health Care

Select a biomedical health care provider who will work in partnership with CAM practitioners.

Discuss personal health care goals.

Share all CAM used with conventional health care provider. Products or practices used for one condition can affect care for another problem.

Discuss how CAM therapies can be used along with conventional care to meet goals.

Determine if CAM products have potentially harmful interactions with prescription drugs or over-the-counter medications.

PRACTITIONER INTERVIEW

Gretchen K. Vannice, MS, RD *Research Coordinator, Nordic Naturals, Inc (Watsonville, California)*

Mentoring Chair, ADA Practice Group—Nutrition in Complementary Care

It is very important that dietetic students and RDs know about CAM because nutrition therapy is one of its modalities and, if we are to be food and nutrition experts, we have to be able to accurately and effectively answer questions about the responsible use of nutrient/food supplements. The supplement industry has grown tremendously in recent years, mostly because of marketing, not science. Some supplements have efficacy, some don't; and through our dietetic education and training we have to know the difference. We need the skills to be able to tell our clients what is hype and what isn't, plus possible consequences of supplement use.

Many RDs and clients take the attitude: don't ask—don't tell. So it's important to keep an open mind when assessing a client and to ask them if they are using dietary supplements or herbs on a regular basis. For any supplement, you need to assess the dosage along with any contraindications with current therapies, the cost, and the evidence of efficacy—is the supplement appropriate for the condition and does it have a proven safety profile? I have found that some clients are emotionally attached to a supplement, and if it is not harmful, you probably should not recommend stopping its use. Prioritize the supplements to them: are they harmful, are they useful, or are they neither useful nor harmful? We cannot be experts on all the supplements out there, so don't try to pretend. If you don't know, do some research and get back to the client later.

Additional resources on CAM are available through the ADA practice group—Nutrition in Complementary Care (http://www.complementarynutrition.org), ADA's fastest growing practice group. My two other favorite resources are the National Center for Complementary and Alternative Medicine at http://nccam.nih.gov and Natural Medicines Comprehensive Database at http://www.naturaldatabase.com, a site run by pharmacists. The second one has a fee to use, but the information is objective, and I find it very useful.

The public attitude on CAM has shifted tremendously. CAM is now considered mainstream and not "hippy dippy." RDs have a huge opportunity to establish themselves as the "go to" experts between the supplement industry and the consumer. We have the education to provide expertise based on research. As a student, hone your skills to critically interpret research published in reputable journals so that the public will look to us for advice on nutrition and herbal supplements.

CASE STUDY 2

Introduction

Paula Thompson is a 55-year-old female who presents without having a menstrual cycle for five months. Her chief complaints include extreme fatigue, poor sleep, hot flashes and night sweats, periods of depression and anxiousness, and aching joints and muscles. Her gynecologist has diagnosed her with probable onset of menopause. Paula feels very strongly that she doesn't want to begin taking a prescription medication even though she is struggling to deal with her multiple symptoms. She has actively pursued alternative and complementary treatments to assist in the reduction of symptoms.

Nutrition Assessment

Ht. 5'3" Wt. 185# UBW 165# All labs normal.

Interview with the client indicates the following treatments:

- For aching joints and muscles: Peruvian bark, glucosamine sulfate (600 mg t.i.d.), flaxseed and evening primrose oil (1 capsule each evening);
- For her periods of depression: St. John's wort 4 g/day;
- For hot flashes and night sweats: dong quai (2 g of root t.i.d.) and black cohosh (40 mg daily).

Additionally, she is receiving acupuncture to assist with her muscle and joint pain. Ms. Thompson is also on Lipitor (80 mg/day) and Lisinopril 10 mg/day.

Nutrition history indicates that she is additionally attempting to reduce sugar intake and "nightshade" foods to assist in control of her menopausal symptoms.

Questions

1. Describe each of the medications and supplements that Ms. Thompson is currently taking.
2. What is the current evidence supporting the efficacy of each of these medications/supplements in the treatment of menopausal symptoms?
3. What is acupuncture? What is the rationale for its use in the treatment of aching joints and muscles?
4. As a registered dietitian, what advice and recommendations can you give Ms. Thompson to assist her in her decision making with regard to her use of complementary and alternative medicine?

WEB LINKS

ABC Homeopathy: Explanations of homeopathic theories, their use at home, short list of homeopaths, and online remedy finder. Can search by condition or remedy.

http://www.abchomeopathy.com

Acupuncture.com: Gateway to Chinese Medicine, Health, and Wellness. Extensive information on acupuncture and other Traditional Chinese Medicine (TCM) techniques. Includes online newsletter, "Ask the Doctor" page, referrals to practitioners, and a section for TCM students. Also contains consumer and TCM-professional information on the causes and treatment of different conditions/syndromes. Very user friendly.

http://www.acupuncture.com

American Holistic Medicine Association: Site dedicated to allopathic physicians interested in broadening their approach to healing. Explanation of holistic principles and advice to clients on finding a holistic healer.

http://www.holisticmedicine.org

Ayurvedic.com: Of special interest to health care professionals are the online resources for food and nutrition, including extensive food guidelines according to body type. Hosted by the Ayurvedic Institute.

http://www.ayurveda.com

Chiroweb.com: An online chiropractic community with information for practitioners, students, and consumers. Includes health articles, the ChiroFind feature, and "Ask a Chiropractic Doctor" section.

http://www.chiroweb.com

ConsumerLab.com: Independent testing facility posts results on vitamin, mineral, botanical, and other supplement evaluations. Includes Natural Products Encyclopedia. Subscription only: one-year, two-year, three-year, or 30-day access to single review.

http://www.consumerlab.com/index.asp

Dictionary of Chinese Herbs: Listing of TCM botanicals, with English, pinyin, Latin, and Chinese names. Also shows how the product is pronounced in Cantonese, Japanese, and Korean. Brief data on properties, meridians entered, action, indications, medical use, cautions, and dosage. Also includes information on acupuncture, acupressure, and qigong.

http://alternativehealting.org/chinese_herbs_dictionary.htm

Functional Foods for Health: Run by the University of Illinois, this site includes research and regulatory updates on functional foods. Online newletter is of particular interest. Consumer section includes downloadable handouts. Natural Products Alert database (NAPRALERT) of world literature describing ethnomedical and traditional uses, chemistry, and pharmacology of botanical, animal, and microbial extracts is available for purchase.

http://www.ag.uiuc.edu/~ffh

HerbMed: Interactive site providing information on 75 herbs without cost; data on all other botanicals require a subscription. Extensively referenced with links to each citation. Includes evidence for efficacy, evidence for activity, safety data, formulas and blends, traditional and folk use, and other information (pictures, cultivation, links). Daily subscriptions available.

http://www.herbmed.org/index.asp

Himalaya Herbal Healthcare: Includes Herb Finder feature for traditional Ayurvedic botanicals. Lists English, Sanskrit, Hindi, and Latin names as well as a brief overview of history, habitat, clinical constituents, pharmacology, toxicology, indications, and references (primarily from Indian scientific journals).

http://www.himalayahealthcare.com/index.htm

Memorial Sloan-Kettering Cancer Center: About Herbs, Botanicals and Other Products: Professional information on herbs, botanicals, and other products with research citations is available for the health care provider. Includes brief clinical summary, mechanisms of action, purported uses (not limited to cancer treatment), adverse reactions, and a literature review. Consumer information also available.

http://www.mskcc.org/mskcc/html/11570.cfm

National Center for Complementary and Alternative Medicine (NCCAM): Federal clearinghouse for information on CAM. Includes recent research, advisories and alerts, training opportunities, clinical trials, and resources for professionals and consumers (some in Spanish).

http://nccam.nih.gov

Natural Medicines Comprehensive Database: Online, evidence-based monographs on botanicals prepared by pharmacists for use by health care professionals. Safety, efficacy, and mechanics of action explained. Of special note are the list of alternative names for the product, interactions with other herbs and supplements, and interactions with food. Recommendations on approaching clients with data are offered. Extensive citations. Printable patient handouts also available. Subscription only: monthly, annual, two-year, three-year.

http://www.naturaldatabase.com

Naturopathy: Site run by the American Naturopathic Medical Association. Online newsletter includes numerous articles on the naturopathic approach to nutrition and diet.

http://www.anma.com/mon64.html

The Osteopathic Home Page: Information about the practice of osteopathy. Numerous articles on specific treatment topics of interest.

http://www.osteohome.com/MainPages/research.html

Quackwatch: Nonprofit site dedicated to combating health-related frauds, myths, fads, and fallacies from the allopathic perspective. Large section on alternative and complementary methods.

http://www.quackwatch.org

END-OF-CHAPTER QUESTIONS

1. What practices are included in complementary and alternative medicine (CAM), and what reasons have been reported to explain why individuals choose to use CAM?
2. What is "integrative medicine"?
3. List the alternative medical systems described in this chapter. Briefly describe one system that you have heard of and two that you have not. Pick one of the three and describe dietary recommendations included in the system.
4. List the alternative therapies described in this chapter. Briefly describe one therapy that you have heard of and two that you have not. Pick one of the three and describe dietary recommendations included in the therapy.
5. What is meant by hot-cold foods and yin/yang foods? If you had a client that told you they couldn't eat any yin foods because of their health condition, what would this mean?
6. Go to a grocery, pharmacy, or health food store and describe two nutritional therapies that you could have purchased. Pick one; has any sound nutritional/medical research been published that demonstrated its effectiveness? Describe how you conducted the literature search. If you found a published article, include the reference and the findings of the study.
7. Have you or someone you know ever used CAM? Why?
8. Do you think it is important for practitioners to know about CAM? Why or why not?

NOTES

[1] The term was used in the nineteenth century to differentiate "regular" doctors from those using homeopathy, which uses remedies that produce the same effect as the disease symptoms.

[2] The earliest advocates for the importance of diet in health in the United States were religious medical practitioners, including Sylvester Graham (1795–1851), a Presbyterian minister who emphasized good hygiene, rest, temperance, and a vegetarian diet high in unrefined starches to prevent illness, and Seventh-Day Adventists Ellen G. White (1827–1915) and John Harvey Kellogg (1852–1945), who opened health sanitariums throughout the nation promoting moderation, lacto-ovo vegetarianism, and avoidance of strong spices, caffeine, alcohol, and tobacco.

[3] One study questions the findings that indicate CAM use is highest among well-to-do consumers, suggesting that lower socioeconomic groups are undersampled in most surveys. Their data suggest that, similar to the general population, 85% of poor, urban patients use CAM diet, exercise, and prayer practices; more expensive therapies are used less often (Rhee, Garg, and Hershey 2004).

[4] Although nearly two-thirds of Hispanic respondents in a survey on alternative medicine employed CAM practices such as prayer and dietary supplements, most had more confidence in biomedical providers and drugs than in CAM practitioners and products (Mikhail, Wali, and Ziment 2004).

[5] A study of pediatric CAM use found that only 2% of parents reported consulting alternative providers (most often chiropractors and clergy or religious practitioners) for their children and adolescents. Herbal remedies and spiritual healing were the most common therapies (Yussman, Ryan, Auinger, and Weitzman 2004). Other studies suggest the numbers of children using CAM are much higher, from 12 to 33% (Guenther, Mendoza, Crouch, Moyer-Mileur, and Junkins 2005; Lin, Bioteau, Ferrari, and Berde 2004; Pitetti, Singh, Hornyak, Garcia, and Herr 2001).

[6] A study of prostate cancer patients found that treating MDs believed only 4% of clients used CAM, whereas 37% of clients reported using therapies such as natural remedies, megavitamins, chiropractic, massage, relaxation, and special diets (Nagel, Hoyer, and Katenkamp 2004).

[7] Biomedical CAM supporters, such as Andrew Weil, MD and Deepak Chopra, MD, have popularized modalities such as dietary supplements, functional foods, relaxation techniques, meditation, and yoga.

[8] A small survey of CAM organizations found that only 57% had an informed consent policy, and only 16% required their practitioners to obtain informed consent (Caspi and Holexa, 2005).

[9] Homeopathy is well respected in Europe. Homeopathic hospitals and clinics are part of the national health care system in Great Britain.

[10] Traditionally, a Chinese physician was paid when his client was healthy; if the client became ill, payment stopped.

[11] Practitioners of conjury are thought to get their powers from evil spirits or the devil, although some receive a calling from God or train under an expert. They are often distinguished from normal humans by several traits, such as dwarfism, the ability to see ghosts and spirits, being born on Christmas, being a seventh son, or having certain birthmarks.

[12] The International Order of St. Luke the Physician, Christian Healing Ministries, and the School of Pastoral Care are examples of interdenominational organizations that train clergy and laypersons in spiritual healing practices.

[13] Hypertension can be confused with high blood, and occasionally an African-American client may self-prescribe pickles and other salty items to counteract it. Pregnancy is considered a high blood period by a few black women; hence, they may avoid red meats.

[14] Recent research found that an extract from the fruit of the prickly pear cactus (*Opuntia ficus indica*) is a very effective hangover remedy (Morgan, Kori, and Thomas 2002).

[15] Folk remedies are sometimes exploited by commercial manufacturers without proof of efficacy or safety. The African cactus *Hoodia gordonii* is an example. Consumed by the Kalahari Desert tribes to stave off hunger and thirst, it is aggressively promoted (in pill and extract form) as an appetite suppressant and weight loss aid based on traditional use and a single study in rats (MacLean and Luo 2004).

[16] Although plant sources are available, most glucosamine is derived from crab shells, and chondroitin sulfate is made from cow, pig, or chicken cartilage.

[17] DHEA, a weak androgen hormone made from wild yams, enjoys a special exemption from the Anabolic Steroid Control Act of 2004, passed to address abuse of the muscle-building substances. Inclusion of DHEA was advocated by health officials and sports organizations, such as Major League Baseball, but lobbyists for the supplement industry fought successfully against it. Sales of DHEA were estimated to be $47 million in 2003 (Kornblut and Wilson 2005).

[18] Yogurt is sometimes used as a vaginal douche in the theory that probiotic bacteria will suppress the harmful organisms that cause yeast infections and urinary tract infections. Preliminary evidence suggests it may be effective (Wiese, McPherson, Odden, and Shlipak 2004).

[19] Despite its popularity as a cold remedy, recent research suggests *Echinacea angustifolia* does not prevent rhinovirus infections nor ease symptoms (Turner et al. 2005). Further studies on other echinacea species and on *E. angustifolia* in different dosages may be more conclusive on its efficacy (Debusk 2001).

[20] In much of Europe, botanicals are considered drugs, prescribed by medical doctors and purchased at pharmacies. The German Commission E Monographs (Blumenthal and Klein 1998) describe pharmacologic properties, indications, contraindications, dosages, side effects and toxicity for hundreds of botanicals.

[21] Raw foodists believe that heating destroys natural enzymes necessary for maximum nutrient utilization and allows dangerous levels of toxins to accumulate.

[22] Although some conventional health care providers deny any psychophysiological connections, others recognize the role of chronic stress in the development of conditions such as hypertension and heart disease and prescribe mind-body therapies, including exercise and relaxation techniques.

5

Assessment of Nutrition Status and Risk

Marcia Nelms, Ph.D., R.D.

Southeast Missouri State University

CHAPTER OUTLINE

Nutrition Assessment and Screening

Dietary Assessment Methods

Twenty-Four-Hour Recall • Food Frequency • Observation of Food Intake/ "Calorie Count"

Analysis of Food Intake

Analysis Based on USDA's MyPyramid • Analysis Based on Exchanges/Carbohydrate Counting • Specific Nutrient Analysis • Computerized Dietary Analysis

Evaluation and Interpretation of Dietary Analysis Information

USDA's MyPyramid • U.S. Dietary Guidelines for Americans • Daily Values/Dietary Reference Intakes

Nutritional Physical Examination

Anthropometric/Body Composition Assessment

Anthropometrics • Interpretation of Height and Weight • Body Composition

Biochemical Assessment

Protein Assessment • Immunocompetence • Hematological Assessment • Vitamin and Mineral Assessment • Other Labs with Clinical Significance

Functional Assessment

Energy and Protein Requirements

Measurement of Energy Requirements • Measurement of Protein Requirements • Estimation of Energy Requirements • Estimation of Protein Requirements

Interpretation of Assessment Data

Introduction

Just as the physical examination is the cornerstone of medical assessment, **nutrition assessment** provides the foundation for the nutrition care process. It is from information gathered in the nutrition assessment that a nutrition diagnosis can be determined. After interventions have been put into place, nutrition assessment data serve as benchmarks with which to measure the effectiveness of treatment.

nutrition assessment—analysis of an individual's nutrition status incorporating both subjective and objective data, including information on diet, psychosocial parameters, education, and motivation

Methods for nutrition assessment will change with the population, the nutrition diagnosis, and the desired outcomes for the nutrition therapy. The type of assessment required for the healthy individual will correlate with goals for a healthy population. For example, assessment used in screening a healthy population for disease risk will focus on those factors necessary for prevention of disease. In contrast, if assessment is planned for an at-risk population, data gathered may focus on those factors that confirm a diagnosis of malnutrition.

There is no one test that measures nutritional status. That is why assessment draws from many indices to provide a complete picture of nutritional health. It is through experience that a clinician can weigh results of multiple measures to critically evaluate the nutritional status of an individual or a population.

Nutrition assessment is defined as "an evaluation of the nutritional status of individuals or populations through measurements of food and nutrient intake and evaluation of nutrition-related health indicators" (Lee and Nieman 2003, p. 3). This chapter provides essential information the clinician will need in order to identify and use the appropriate nutrition assessment techniques and interpret the results for evaluation of nutritional status.

Nutritional Status and Nutritional Risk

Determination of *nutritional status* involves evaluating indices that reflect the body's nutrient stores. Nutritional status is altered when stores of energy, protein, water, vitamins, or minerals fluctuate as a result of either increased need, increased utilization, or altered intake. Historically, vitamin, mineral, and other nutrient deficiencies had the largest impact on nutritional status. Today, in most developed countries, concern for nutritional status focuses on the effect of excessive intake of nutrients and energy.

Development of nutritional deficiencies is a progressive process. The body works hard to maintain homeostasis, but at some point inadequacy or excess of a particular nutrient interrupts homeostasis and results in a physiological change. This process is exemplified by examining the development of iron deficiency. When dietary iron is inadequate, the body will initially release iron, primarily from ferritin, from stores to maintain adequate hemoglobin levels. Transferrin levels will also increase to ensure adequate transport. It is only when iron stores are low that there is inadequate iron available to maintain hemoglobin, hematocrit, and mean corpuscular value. At this point, physiological changes may be noted. These may include shortness of breath, paleness of the skin, increased heart rate, and fatigue. Changes in iron status are assessed and confirmed with numerous measures of dietary intake, biochemical levels, and clinical signs.

Determination of *nutritional risk* involves an attempt to project potential nutritional problems based on the client's current health status. Certain factors increase or decrease a client's nutritional risk; for example, a diagnosis of pancreatic cancer places an individual at a higher nutritional risk than admission for cholestectomy. It is understood that because of such a diagnosis and the likely treatment, nutritional changes and/or problems are probable. Significantly, most nutrition problems seen in the hospitalized population are the result of disease or its treatment. The patient will most likely have an increased requirement for certain nutrients and/or an inability to consume enough nutrients or metabolize the ones that can be digested and absorbed. Therefore, knowing the pathophysiology, treatment, and clinical course of a disease or diagnosis allows one to assess the nutritional risk of an individual and ultimately determine the nutritional diagnosis.

anthropometry—the study of the measurement of size and shape of the human body and its constituents (fat, lean tissue, and bone)

nutrition screening—the process of gathering data known to correlate with nutritional risk in order to identify individuals who are at risk

An Overview: Nutrition Assessment and Screening

As a component of the nutritional care process, the nutrition assessment consists of gathering data in the following areas: medical and social history; dietary history; subjective global assessment and the nutrition-focused physical examination; **anthropometrics** and body composition; biochemical data; potential drug-nutrient interactions; clinical symptoms; and estimation of energy, protein, and fluid requirements (American Dietetic Association 2000; Standing Committee on the Scientific Evaluation of Dietary Reference Intakes 2002, 2004). Assessment data from these areas may be both subjective and objective in nature. The assessment process then moves toward analysis of these data so that a summary of current and potential nutritional problems can be determined.

While it is not possible, nor even necessary, to complete a full nutritional assessment of every patient admitted to a clinic or hospital, it is essential to have a system in place that can quickly identify those patients who are at risk for nutritional problems. **Nutrition screening** is the process of gathering key pieces of information that have been correlated to nutrition risk. A recent study outlined easily available information from the medical record that was strongly correlated with nutrition risk (Brugler et al. 2005). Nutrition screening can be performed by dietetic technicians or other trained personnel, which allows for a more efficient and cost-effective collection and identification of at-risk patients. A dietitian can then perform a full nutrition assessment. The Joint Commission for Accreditation of Hospital Organizations requires that all patients receive nutrition screening within 48 hours of their admission to a hospital (JCAHO 2004).

Subjective and Objective Data Collection

Types of data include both subjective and objective information. Subjective data include information, usually obtained during interviews, coming directly from the patient, family members, or significant others. Thus, subjective data would include the client's perception of his or her medical condition, dietary intake, lifestyle conditions, current medications or supplement intake, and family medical history. Subjective data also include the interviewer's observations.

Objective data include information obtained from a verifiable source such as the current medical record and previous medical histories. These data could include anthropometric measurements, calculations such as estimation of energy and protein requirements, biochemical data, and the results of any medical intervention. The organization and content of the medical record will vary from institution to institution. See Tables 5.1 and 5.2 for examples of subjective and objective data, respectively.

Dietary Information A crucial skill necessary for conducting a nutrition assessment is the development and use of appropriate interviewing skills. The environment where the interview occurs, the rapport between interviewer and client, and the types of questions and the manner in which

TABLE 5.1

Subjective Nutrition Assessment Information

Subjective Information	
Diet-related eating habits and feeding abilities	Food allergies/aversion
Use and fit of dentures	Vitamin, mineral and nutrient supplement intake
Appetite and digestion problems	Complementary/alternative nutritional therapy
Nausea, vomiting, constipation, diarrhea, heartburn, etc.	Nutritional history and family nutritional history
Recent weight change	Presence of hunger
Any previous nutrition or dietary treatment that the patient describes	Method of obtaining foods/nutrient (e.g., Meals on Wheels)
Usual pattern of food intake	Ethnicity and religion (degree of observance)
Lifestyle/Psychosocial/Emotional	
Economic situation/income	Interaction with/between other family members or caretakers
Food insecurity: inability to purchase/prepare/store appropriate and acceptable food	Support systems
Living or eating alone	Coping mechanisms
Health promotion and exercise practices	Occupation
Smoking	
Medically Related	
Personal and family medical history (especially for diseases with nutritional implications, e.g., type 2 diabetes)	Medications, previous to admission (include prescription medications, over-the-counter [OTC] medications [antacids, laxatives, etc.], and any CAM medications)
Other physical problems	
Learning and Motivation Related	
Ability to communicate in English (speaking, comprehending, reading, and writing)	Educational level
Learning style/problem-solving abilities	Communication patterns
Patient's comments about previous prescribed diets/medical treatment and compliance issues	Attention span
Perception of health status/reasons for seeking health care	Long-term and recent memory
Desire to improve health or be involved in their own treatment or treatment decisions	Readiness to learn

TABLE 5.2

Objective Nutrition Assessment Information

Anthropometric data	
Age	Gender
Height	Weight
BMI	Weight change in 1 month, 6 months
% Usual Body Weight	
Biochemical lab results that are of nutritional relevance	
Visceral protein assessment	
Albumin	Transferrin
Prealbumin	Total protein
Immunocompetence	
Total lymphocyte count	
Hematological assessment	***Lipid assessment***
Hemoglobin	Total cholesterol
Hematocrit	HDL-C
Mean corpuscular volume	LDL-C
Mean corpuscular hemoglobin Concentration	Triglycerides
Mean corpuscular hemoglobin	
Total iron binding capacity	
Electrolytes	***Other***
Sodium	Blood glucose
Potassium	Glycated hemoglobin
Chloride	Blood urea nitrogen
Calcium	Serum creatinine
Phosphorus	Urinary protein
Clinical Findings	
Diagnosis	Medications ordered or received in the facility
Physical assessment	
Treatment orders (including diet orders)	Procedures to be administered (e.g., surgery)
Dietary Information	
Current intake that has been observed (not subjective)	Analysis of diet adequacy

they are posed directly affect the quality and accuracy of the information that is obtained. Table 5.3 lists some basic suggestions for conducting an effective interview.

During the patient interview, information regarding appetite and GI function should be obtained. This includes evaluation of the ability to chew, use and fit of dentures, swallowing ability, nausea, vomiting, constipation, diarrhea, and heartburn, or any other symptoms that might interfere with ability to maintain adequate nutritional intake.

TABLE 5.3

A Successful Patient Interview

Maintains an environment that is private and assures confidentiality.
Establishes good patient rapport.
Respects religious, cultural, and familial values and needs.
Provides for attentive listening skills.
Structures questions that are both open and neutral.
Avoids closed and leading questions.

A thorough diet history can identify the patient's usual pattern of intake, food preferences including ethnic, cultural and religious influences, and the patient's use of alcohol, complementary and alternative medicine, and vitamin, mineral, herbal or other types of supplements (see Chapter 2). Any previous nutrition education or nutrition therapy should be evaluated. Questions should also address food allergies or other food intolerances.

During this interview process, it is important to determine the availability of resources to both purchase and prepare food. This could include identification of kitchen facilities, food preparation skills, access to grocery stores, and any financial or social assistance (such as Meals on Wheels) that the client utilizes.

Psychosocial Information Many social factors—including socioeconomic status, social support systems, interactions with other people, and lifestyle—impact nutritional status. It is crucial to identify these factors during the interview, because they will impact planning and execution of nutrition education and intervention.

Economic situations directly impact nutritional status. Obviously, nutrition education has little meaning for clients who do not have access to adequate food for themselves or their families. (Box 5.1 explores assessment of food insecurity.) Food insecurity is defined as not having access to adequate food to support an active, healthy life at all times (Nord, Andrews, and Carlson 2003). It is estimated that 11% of households in the United States were food insecure at least some of the time during the year 2001, meaning they did not always have access to enough food for active, healthy lives for all household members because they lacked sufficient money or other resources for food. "The prevalence of food insecurity rose from 10.7% in 2001 to 11.1% in 2002, and the prevalence of food insecurity with hunger rose from 3.3% to 3.5%" (Nord, Andrews, and Carlson 2003).

Support systems and interaction with family members or caretakers are crucial factors when designing nutrition education and interventions. For example, if the client eats alone most of the time, appetite may be adversely affected. Lifestyle habits such as smoking, exercise, occupation (if still

BOX 5.1 CLINICAL APPLICATIONS

Assessing Food Access of Clients and Patients

According to the American Dietetic Association, "negative nutritional and non-nutritional outcomes have been associated with food insecurity in adults, adolescents, and children, including poor dietary intake and nutritional status, poor health, increased risk for the development of chronic diseases, poor psychological and cognitive functioning, and substandard academic achievement" (Holben 2006). Therefore, as part of the Nutrition Care Process (American Dietetic Association 2006), a comprehensive nutrition assessment includes obtaining adequate information to identify nutrition-related problems. Three domains have been utilized to cluster nutrition diagnoses and problems—intake, clinical, and behavioral-environmental (American Dietetic Association 2006). The behavioral-environmental cluster includes "actual problems with food access," diagnosis NB-3.2, "Limited access to food" (American Dietetic Association 2006).

Limited access to food can be evidenced by a variety of signs and symptoms. Regardless, it may hinder purchasing of food and prevent compliance to a prescribed diet (Holben 2006). As part of a food and nutrition history, food availability–related information should be obtained, including factors such as food planning, purchasing, preparation abilities and limitations, food safety practices, food/nutrition program utilization, and food insecurity (American Dietetic Association 2006).

Knowing and understanding the culture of your community will assist you to ask appropriate questions related to food access. The interview may include questions related to the following (Boeing and Holben 2003; Holben 2006; Holben and Myles 2004):

- Money for dietary prescriptions and/or medications
- Availability of a refrigerator/freezer, utilities, and transportation
- Participation in food assistance programs
- Gardening practices
- Other means of acquiring foods (hunting/fishing)
- Unintentional weight loss
- Quality of the diet
- Nutrition education need regarding meal planning and purchasing, label reading, and food safety

It has also been suggested that dietitians screen clients for lack of access to food due to resource constraints by using a single-item food sufficiency question: "Which of the following statements best describes the food eaten in your household?: (1) Enough of the kinds of food we want to eat, (2) Enough but not always the kinds of food we want to eat, (3) Sometimes not enough to eat; or 4) Often not enough to eat" (Kaiser 2005).

Source: David H. Holben, PhD, RD, LD; Associate Professor, Food, Nutrition, and Hospitality; Director, Didactic Program in Dietetics; Ohio University, Athens, Ohio

working), and ability to perform activities of daily living are additional pieces of information that should be ascertained.

Information Regarding Education, Learning, and Motivation During the interview, the ability to communicate should be established. The client's primary language, as well as the ability to speak, read, write, and comprehend both that primary language and English, can be determined. Educational level, attention span, long-term and recent memory, and readiness to learn can all be established during either the initial interview or subsequent sessions.

The client's comments about previous prescribed diets, medical treatment, and any issues regarding compliance can and will impact acceptance of newly established goals and interventions. Perceptions of health status and the client's desire to improve health or be involved with health care are crucial determinants of successful nutrition education and interventions.

Tools for Data Collection

Client interview and subsequent data collection can be organized using a number of tools and instruments. Many facilities design their own tools and instruments so that they can collect and organize the health information for easy use. Additionally, there are standardized forms such as the DETERMINE checklist (American Dietetic Association 1991), Subjective Global Assessment (Ottery et al. 1987; Baker et al. 1982), and numerous other instruments that are used for specific populations.

In general, dietary information is assessed either by collecting data **retrospectively** or by summarizing data gathered **prospectively**. All methods have imperfections. The accuracy (or **validity**) of the information and the reliability of the data depend on the experience and skill of the clinician as well as the cooperation and accurate reporting of the client. The ultimate goal is to determine the nutrient

retrospectively—refers to collecting data from events that have already happened

prospectively—refers to collecting data as it occurs or happens

validity—the quality of producing desired results

FIGURE 5.1 DETERMINE: A Nutrition Screening Tool

DETERMINE YOUR NUTRITIONAL HEALTH

The Warning Signs of poor nutritional health are often overlooked. Use this Checklist to find out if you or someone you know is at nutritional risk.

Read the statements below. Circle the number in the "yes" column for those that apply to you someone you know. For each "yes" answer, score the number in the box. Total your nutritional score.

	YES
I have an illness or condition that made me change the kind and/or amount of food I eat.	**2**
I eat fewer than 2 meals per day.	**3**
I eat few fruits or vegetables or milk products.	**2**
I have 3 or more drinks of beer, liquor, or wine almost every day.	**2**
I have tooth or mouth problems that make it hard for me to eat.	**2**
I don't always have enough money to buy the food I need.	**4**
I eat alone most of the time.	**1**
I take 3 or more prescribed or over-the-counter drugs a day.	**1**
Without wanting to, I have lost or gained 10 pounds in the last 6 months.	**2**
I am not always physically able to shop, cook, and/or feed myself.	**2**
TOTAL	

Total Your Nutritional Score. If it's –

0-2 Good! Recheck your nutritional score in 6 months.

3-5 You are at moderate nutritional risk. See what can be done to improve your eating habits and lifestyle. Your office on aging, senior nutrition program, senior citizens center or health department can help. Recheck your nutritional score in 3 months.

6 or more You are at high nutritional risk. Bring this Checklist the next time you see your doctor, dietitian or other qualified health or social service professional. Talk with them about any problems you may have. Ask for help to improve your nutritional health.

Remember that Warning Signs suggest risk, but do not represent a diagnosis of any condition. Turn the page to learn more about the Warning Signs of poor nutritional health.

These materials are developed and distributed by the Nutrition Screening Initiative, a project of:

AMERICAN ACADEMY OF FAMILY PHYSICIANS

THE AMERICAN DIETETIC ASSOCIATION

THE NATIONAL COUNCIL ON THE AGING, INC.

The Nutrition Screening Initiative • 1010 Wisconsin Avenue, NW • Suite 800 • Washington, DC 20007
The Nutrition Screening Initiative is funded in part by a grant from Ross Products Division of Abbott Laboratories, Inc.

(continued on the following page)

FIGURE 5.1 *(continued)*

The Nutrition Checklist is based on the Warning Signs described below.
Use the word DETERMINE to remind you of the Warning Signs.

DISEASE

Any disease, illness or chronic condition which causes you to change the way you eat, or makes it hard for you to eat, puts your nutritional health at risk. Four out of five adults have chronic diseases that are affected by diet. Confusion or memory loss that keeps getting worse is estimated to affect one out of five or more of older adults. This can make it hard to remember what, when or if you've eaten. Feeling sad or depressed, which happens to about one in eight older adults, can cause big changes in appetite, digestion, energy level, weight and well-being.

EAT POORLY

Eating too little and eating too much both lead to poor health. Eating the same foods day after day after day or not eating fruit, vegetables, and milk products daily will also cause poor nutritional health. One in five adults skip meals daily. Only 13% of adults eat the minimum amount of fruit and vegetables needed. One in four older adults drink too much alcohol. Many health problems become worse if you drink more than one or two alcoholic beverages per day.

TOOTH LOSS/MOUTH PAIN

A healthy mouth, teeth, and gums are needed to eat. Missing, loose, or rotten teeth or dentures which don't fit well, or cause mouth sores, make it hard to eat.

ECONOMIC HARDSHIPS

As many as 40% of older Americans have incomes of less than $6,000 per year. Having less — or choosing to spend less — than $25-30 per week for food makes it very hard to get the foods you need to stay healthy.

REDUCED SOCIAL CONTACT

One-third of all older people live alone. Being with people daily has a positive effect on morale, well-being and eating.

MULTIPLE MEDICINES

Many older Americans must take medicines for health problems. Almost half of older Americans take multiple medicines daily. Growing old may change the way we respond to drugs. The more medicines you take, the greater the chance for side effects, such as increased or decreased appetite, change in taste, constipation, weakness, drowsiness, diarrhea, nausea, and others. Vitamins or minerals, when taken in large doses, act like drugs and can cause harm. Alert your doctor to everything you take.

INVOLUNTARY WEIGHT LOSS/GAIN

Losing or gaining a lot of weight when you are not trying to do so is an important warning sign that must not be ignored. Being overweight or underweight also increases your chance of poor health.

NEEDS ASSISTANCE IN SELF CARE

Although most older people are able to eat, one of every five have trouble walking, shopping, buying and cooking food, especially as they get older.

ELDER YEARS ABOVE AGE 80

Most older people lead full and productive lives. But as age increases, risk of frailty and health problems increase. Checking your nutritional health regularly makes good sense.

The Nutrition Screening Initiative • 1010 Wisconsin Avenue, NW • Suite 800 • Washington, DC 20007
The Nutrition Screening Initiative is funded in part by a grant from Ross Products Division of Abbott Laboratories, Inc.

Source: The Nutrition Screening Initiative. http://www.aafp.org/x17367.xml. Reprinted by permission.

FIGURE 5.2 24 Hour Recall form

24 hour recall		Date:		Patient Name:	
Time	Foods and Beverages	Serving Size	How prepared	Where	Comments:

Sample Protocol for Completion of 24-Hour Recall

1. The 24-hour recall consists of obtaining information for food and fluid intake for the 24-hour period preceding the interview. It is assumed that this is a "typical" day. If not, clarify.

2. Patient may not be able to remember all foods eaten. Begin by asking the sequence of events for the previous 24 hours. For example, "Before speaking with me today, when was the last time you ate or drank anything?"; "What was that ____?"; "How much did you eat of ____?" Then proceed backward from that time for the entire 24-hour period.

3. Use food models and food containers to assist patients in clarifying the serving amounts.

4. A checklist may help the interviewer remember to ask or probe all information for each food or beverage.

Components of 24-hour recall:
- Note the time the food or beverage was consumed.
- Record the food or beverage.
- Determine serving size for food or beverage.
- Determine how the food was prepared.
- Determine where the patient had the food or beverage item.
- Include any relevant notes to the food or beverage report.

content of food that is consumed and to then assess the appropriateness of the nutritional intake for that particular individual.

Dietary Assessment Methods

Twenty-Four-Hour Recall

When using a **24-hour recall** as the dietary assessment method, the clinician guides the client through a recall of all food and drink that has been consumed in the previous 24-hour period (see Figure 5.2). Commonly, the clinician asks what food or beverage was consumed most recently prior to the interview and then works backward through the previous 24 hours. The clinician questions the client about activities during the period in order to stimulate the client's memory. At the end of the recall, the clinician reviews the information to verify serving sizes and preparation methods, and to clarify any other uncertainties. The USDA multiple-pass recall, a variation of this method, is a widely accepted and validated method that includes three reviews of information (Conway, Ingwersen, and Moshfegh 2003, 2004).

Advantages of this method include: short administration time, very little cost involved, and little risk for the client. One disadvantage is that a 24-hour recall does not always show typical eating patterns, since day-to-day dietary intake may vary considerably. A second disadvantage is that clients may report information they feel the clinician wants to hear. Research indicates clients may over- or underreport their intake. Additionally, the information obtained may be inaccurate, since this method requires dependence on the client's memory. Accuracy of the method can be strengthened by use of food models, cups, and spoons to improve recall of portion sizes (Conway, Ingwersen, and Moshfegh 2003, 2004; Dwyer 1999; Lee and Nieman 2003; Novotny et al. 2003; Tapsell, Brenninger, and Barnard 2000).

24-hour recall—dietary assessment method in which the clinician interviews the client to obtain a list of all foods/beverages consumed in the previous 24 hours

Food Record/Food Diary

This method has the client document his or her dietary intake as it occurs over a specified period of time. Typically, the record is kept over a three- or five-day period (see Figure 5.3) and should include a sampling of both weekdays and weekends. Clients estimate or weigh their food intake. The advantage of this method is that it is not totally reliant on the client's memory and may be much more representative of the client's actual intake. Problems with validity can occur, however, because underreporting is common, and the client may change food habits for the recording period. Additionally, there is a heavier burden on the client, who must make a commitment to record his or her intake (Dwyer 1999; Lee and Nieman 2003).

FIGURE 5.3 Food Diary

Date/Time	List all foods and drinks	Amount/serving size	Preparation/ cooking method	Seasonings/ Condiments	Where did you eat?	Who were you with?

Directions for Use of Food Diary

READ THE FOLLOWING INSTRUCTIONS CAREFULLY

Record amounts and descriptions of ALL food and drink (including water) for three consecutive days. These days should be "typical" to the way you eat on a normal basis. Please do not try to change your eating habits on the days you are recording. **Please pick two weekdays and one weekend day that are most like your usual daily intake.**

Helpful Hints:

- Record your intake immediately after you have eaten and NOT at the end of the day. This makes it much easier to remember and to record accurately.
- Include all meals and snacks, granola bars, sandwiches, chocolate, sweets, ice cream, fruits—whatever you eat.
- Include all drinks (e.g., water, tea, coffee, beer, sports drinks, and fruit juice).
- Record any additions to food such as mustard, ketchup, mayonnaise, cream or sugar, steak sauce, salsa, dressings, gravy, pickles, honey, or butter.

Describe foods accurately:

- Record cooking methods (e.g., fried, baked, broiled, grilled, frozen, canned, added water, low sodium, and the amount of fat or oil used for cooking).
- Record brand names and the descriptions (e.g., KRAFT, General Mills, Breyers, Campbell's, Del Monte, and whether regular, 2% reduced fat, light, fat free, low carb, or sweetened).
- Name the types of cheese, fish, or meat (e.g., cheddar, American, cod, tilapia, ground, sirloin, shredded).

Describe the amounts as accurately as possible:

- To help with measuring portion size, try to avoid terms such as "one bowl" or "a handful."
- Visualize the following comparisons when figuring portion size:
  - 3 ounces of meat is about the size of a deck of cards or audiotape cassette.
  - A medium-size piece of fruit is about the size of a tennis ball.
  - 1 ounce of cheese is about the size of 4 stacked dice.
  - 1/2 cup of ice cream is about the size of a tennis ball.
  - 1 cup of mashed potatoes or broccoli is about the size of your fist.
  - 1 teaspoon of butter is about the size of the tip of your thumb.
- Use weights marked on packages (e.g., half of a 425-gram can of corn; half of a 16-ounce can; half of a 6-ounce bag of frozen corn).
- Use cups, teaspoons, and tablespoons to record amounts.

Food Frequency

The **food frequency** procedure is a retrospective review of specific food intake. Foods are organized into groups, and the client identifies how often and in what quantities he or she consumes a specific food or food group (see Figure 5.4). The method can be self-administered. Many food frequency instruments have been specialized to identify food group intake for certain disease states such as cardiovascular disease.

Advantages of this methodology are that it is inexpensive and quick to administer. Disadvantages include the fact that response rates tend to be lower since the instrument is self-administered. Also, foods on the pre-prepared list may be inappropriate for the individual who is participating in the food frequency (Dwyer 1999; Lee and Nieman 2003).

food frequency—dietary assessment method in which the client describes the frequency and quantity of his or her consumption of certain foods/food groups

FIGURE 5.4 Food Frequency (MEDFICTS)

In each food category for both Group1 and Group 2 foods check one box from the "Weekly Consumption" column (number of servings eaten per week) and then check one box from the "Serving Size" column. If you check Rarely/Never, do not check a serving size box. See next page for score.

Food Category	Weekly Consumption: Rarely/ never	3 or less	4 or more		Serving Size: Small <5 oz/d 1 pt	Average 5 oz/d 2 pts	Large >5 oz/d 3 pts	Score
Meats								
• Recommended amount per day: ≤5 oz (equal in size to 2 decks of playing cards). • Base your estimate on the food you consume most often. • Beef and lamb selections are trimmed to 1/8" fat.								
Group 1. 10 g or more total fat in 3 oz cooked portion **Beef**—Ground beef, Ribs, Steak (T-bone, Flank, Porterhouse, Tenderloin), Chuck blade roast, Brisket, Meatloaf (w/ground beef), Corned beef **Processed meats**—1/4 lb burger or lg. sandwich, Bacon, Lunch meat, Sausage/knockwurst, Hot dogs, Ham (bone-end), Ground turkey **Other meats, Poultry, Seafood**—Pork chops (center loin), Pork roast (Blade, Boston, Silroin), Pork spareribs, Ground pork, Lamb chops, Lamb (ribs), Organ meats†, Chicken w/skin, Eel, Mackerel, Pompano	□	□ 3 pts	□ 7 pts	×	□ 1 pt	□ 2 pts	□ 3 pts	____
Group 2. Less than 10 g total fat in 3 oz cooked portion **Lean beef**—Round steak (Eye of round, Top round), Sirloin‡, Tip & bottom round‡, Chuck arm pot roast‡, Top Loin‡ **Low-fat processed meats**—Low-fat lunch meat, Canadian bacon, "Lean" fast food sandwich, Boneless ham **Other meats, Poultry, Seafood**—Chicken, Turkey (w/o skin)§, most Seafood†, Lamb leg shank, Pork tenderloin, Sirloin top loin, Veal cutlets, Sirloin, Shoulder, Ground veal, Venison, Veal chops and ribs‡, Lamb (whole leg, fore-shank, sirloin)‡	□	□	□	×	□	□	□ 6 pts	____
Eggs – Weekly consumption is the number of times you eat eggs each week					**Check the number of eggs eaten each time:** ≤1	2	≥3	
Group 1. Whole eggs, Yolks	□	□ 3 pts	□ 7 pts	×	□ 1 pt	□ 2 pts	□ 3 pts	____
Group 2. Egg whites, Egg substitutes (1/2 cups)	□	□	□		□	□	□	____
Dairy								
Milk—Average serving 1 cup **Group 1.** Whole milk, 2% milk, 2% buttermilk, Yogurt (whole milk)	□	□ 3 pts	□ 7 pts	×	□ 1 pt	□ 2 pts	□ 3 pts	____
Group 2. Fat-free milk, 1% milk, Fat-free buttermilk, Yogurt (Fat-free, 1% low fat)	□	□	□		□	□	□	____
Cheese—Average serving 1 oz **Group 1.** Cream cheese, Cheddar, Monterey Jack, Colby, Swiss, American processed, Blue cheese, Regular cottage cheese (1/2 cup), and Ricotta (1/4 cup)	□	□ 3 pts	□ 7 pts	×	□ 1 pt	□ 2 pts	□ 3 pts	____
Group 2. Low-fat & fat-free cheeses, Fat-free milk mozzarella, String cheese, Low-fat, Fat-free milk & Fat-free cottage cheese (1/2 cup) and Ricotta (1/4 cup)	□	□	□		□	□	□	____
Frozen Desserts—Average serving 1/2 cup **Group 1.** Ice cream, Milk shakes	□	□ 3 pts	□ 7 pts	×	□ 1 pt	□ 2 pts	□ 3 pts	____
Group 2. Low-fat ice cream, Frozen yogurt	□	□	□		□	□	□	____

(continued on the following page)

FIGURE 5.4 *(continued)*

Food Category	Weekly Consumption: Rarely/ never	3 or less	4 or more	Serving Size: Small <5 oz/d 1 pt	Average 5 oz/d 2 pts	Large >5 oz/d 3 pts	Score
Frying Foods – Average servings: see below. This section refers to method of preparation for vegetables and meat.							
Group 1. French fries, Fried vegetables (1/2 cup), Fried chicken, fish, meat (3 oz)	□	□ 3 pts	□ 7 pts ×	□ 1 pt	□ 2 pts	□ 3 pts	____
Group 2. Vegetables, not deep fried (1/2 cup), Meat, poultry, or fish—prepared by baking, broiling, grilling, poaching, roasting, stewing: (3 oz)	□	□	□	□	□	□	____
Baked Goods – 1 Average serving							
Group 1. Doughnuts, Biscuits, Butter rolls, Muffins, Croissants, Sweet rolls, Danish, Cakes, Pies, Coffee cakes, Cookies	□	□ 3 pts	□ 7 pts ×	□ 1 pt	□ 2 pts	□ 3 pts	____
Group 2. Fruit bars, Low-fat cookies/cakes/pastries, Angel food cake, Homemade baked goods with vegetable oils, breads, bagels	□	□	□	□	□	□	____
Convenience Foods							
Group 1. Canned, Packaged, or Frozen dinners: e.g., Pizza (1 slice), Macaroni & cheese (1 cup), Pot pie (1), Cream soups (1 cup), Potato, rice & pasta dishes with cream/cheese sauces (1/2 cup)	□	□ 3 pts	□ 7 pts ×	□ 1 pt	□ 2 pts	□ 3 pts	____
Group 2. Diet/Reduced calorie or reduced fat dinners (1), Potato, rice & pasta dishes without cream/cheese sauces (1/2 cup)	□	□	□	□	□	□	____
Table Fats—Average serving: 1 Tbsp							
Group 1. Butter, Stick margarine, Regular salad dressing, Mayonaise, Sour cream (2 Tbsp)	□	□ 3 pts	□ 7 pts ×	□ 1 pt	□ 2 pts	□ 3 pts	____
Group 2. Diet and tub margarine, Low-fat & fat-free salad dressing, Low-fat & fat-free mayonnaise	□	□	□	□	□	□	____
Snacks							
Group 1. Chips (potato, corn, taco), Cheese puffs, Snack mix, Nuts (1 oz), Regular crackers (1/2 oz), Candy (milk chocolate, caramel, coconut) (about 1 1/2 oz), Regular popcorn (3 cups)	□	□ 3 pts	□ 7 pts ×	□ 1 pt	□ 2 pts	□ 3 pts	____
Group 2. Pretzels, Fat-free chips (1 oz), Low-fat crackers (1/2 oz), Fruit, Fruit rolls, Licorice, Hard candy (1 med piece), Bread sticks (1–2 pcs), Air-popped of low-fat popcorn (3 cups)	□	□	□	□	□	□	____

Total from page 1 ____

Total from page 2 ____

Final Score ____

† Organ meats, shrimp, abalone, and squid are low in fat, but high in cholesterol.
‡ Only lean cuts with all visible fat trimmed. If not trimmed of all visible fat, score as if in Group 1.
¥ Score 6 pts if this box is checked.
§ All parts not listed in Group 1 have <10 g total fat.

To Score: For each food category, multiply points in weekly consumption box by points in serving size box and record total in score column. If Group 2 foods checked, no points are scored (except for Group 2 meats, large serving = 6 pts).

Example:

□	□ 3 pts	☑ 7 pts ×	□ 1 pt	□ 2 pts	☑ 3 pts	21 pts

Add score on page 1 and page 2 to get final score.

Key:
≥70 Need to make some dietary changes
40–70 Heart-Healthy Diet
<40 TLC Diet

Source: NCEP, National Heart, Lung and Blood Institute, NIH Reference: NIH Publication no. 02-5215 URL:http://www.nhlbi.nih.gov/guidelines/cholesterol/atp3_rpt.htm.

Observation of Food Intake/ "Calorie Count"

In a clinical setting, actual food intake can be observed and recorded when a kilocalorie (kcal) or kcal-protein count is ordered. Specific procedures for this method vary from institution to institution. If very detailed information is required, as is the case in a research or metabolic study, food may be weighed before and after the meal is served. The patient's food intake is then calculated from differences between the two. Additional food consumed by family members or food brought in from outside the hospital will also need to be recorded.

In most institutions, nursing or dietary staff document what the patient eats from meal trays. The registered dietitian or registered diet technician then calculates nutritional information such as kilocalorie or protein content from this information.

Analysis of Food Intake

In the clinical setting, the requirement for detailed analysis of food intake is not often necessary. Limitations of each assessment method make the assessment of intake an estimation rather than an exact measurement. In general, though, the analysis method will be determined by how the information will be used. For example, in order to determine whether an intervention to change the patient's food choices has improved intake, a direct observation of food intake may be made and then analyzed by simply estimating the energy value and protein content of the food recorded, using an established method such as the diabetic exchange list.

Analysis Based on USDA's MyPyramid

This analysis will quantify food consumed into servings from each of the groups on the food pyramid (published by the United States Department of Agriculture). These data give the clinician an overview of adequacy but do not necessarily quantify macro- or micronutrients. When one simply looks at total energy and protein, there is no way to determine where those nutrients came from. By using food groups, the quality of the diet can also be assessed (see Box 5.2).

Analysis Based on Diabetic Exchanges/Carbohydrate Counting

This method of analysis uses the dietary exchanges established jointly by the American Diabetes Association and the American Dietetic Association (American Diabetes Association and American Dietetic Association 2003). Use of the exchanges provides a quick, rough estimate of protein, kcal, carbohydrate, and fat in the diet. Carbohydrate counting concentrates on estimation of carbohydrate only and is used primarily by individuals with diabetes who are balancing their insulin dosages with dietary intake of carbohydrate. (See Appendix E1.)

Specific Nutrient Analysis

The United States Department of Agriculture (USDA) first published food composition values in 1896. The Nutrient Data Laboratory of the USDA maintains and updates this data banks. Historically, this information has been published in a series of *Agriculture Handbooks*, but now this information is only available online (http://www.ars.usda.gov/main/site_main.htm?modecode=12354500). Even though this database contains information for over 6000 foods and 100 nutrients, it is difficult to keep up with new name-brand foods that are constantly produced.

Other sources of data can be found on nutritional labels, from food manufacturers, and from some restaurants and fast-food establishments. Data on food labels may not be 100% reliable, especially for products from small companies or imported foods.

Computerized Dietary Analysis

Nutrition professionals have access to many sources of computerized data analysis programs. There is significant variation in these programs. Differences include number of food items, nutrients included in the program, sources for the nutrient database, how often the database is updated, cost of the program, and ease of use.

Evaluation and Interpretation of Dietary Analysis Information

After data are collected and analyzed, it is the clinician's job to compare the information to the individual patient's needs. Different levels of dietary recommendations exist—as general as the U.S. Dietary Guidelines (described in this section) or as specific as milligrams of vitamin C that should be consumed—and, like the analysis method, the appropriate level of recommendations is determined by how the information will be used. The Dietary Reference Intake tables are printed on the inside front cover of this book.

USDA's MyPyramid

As stated earlier in this section, comparing the client's diet with the USDA's food pyramid (http://www.mypyramid.gov) provides an overview of the quality of the diet, especially for variety, moderation, and balance. These guidelines are based on decades of nutrition research and reflect the most up-to-date understanding of nutritional requirements.

BOX 5.2 CLINICAL APPLICATIONS

Comparison Assessment of Dietary Intake

Diet 1: 1 egg, 1 slice of toast, coffee with 2 Tbsp half-and-half
2 oz ham sandwich on 2 slices of white bread with 1 Tbsp mayonnaise
1 oz potato chips
4 oz chicken breast fried, 1 roll, iced tea with 2 Tbsp sugar

Diet 2: 1 cup whole grain cereal, 1 banana, 1 c. skim milk
2 oz ham sandwich, ½ cup chopped fresh vegetables on 2 slices of whole grain bread with 1 Tbsp mustard, 1 oz pretzels, 1 medium apple
4 oz chicken breast baked, 1 cup fresh broccoli, asparagus, carrots stir-fried with 1 cup brown rice

Sample 24-hour recall	Analysis for energy and protein [using the dietary exchanges]		Analysis for components of food intake pattern for 1600 kcal [using MyPyramid]	
Diet 1	Nutrient	Total	Grains 5 oz with ≥ 3 oz whole grain	4 – 0 whole grain
	Kcalories	1099.3	Vegetables 2 cups	0
	Pro (g)	53.08	Fruits 1.5 cups	0
	Fat (g)	51.15	Milk 3 cups	0
	Carb (g)	106.2	Meat and Beans 5 oz	6
Diet 2	Nutrient	Total	Grains 5 oz with ≥ 3 oz whole grain	5 – 5 whole grain
	Kcalories	1087.86	Vegetables 2 cups	3
	Pro (g)	70	Fruits 1.5 cups	2
	Fat (g)	24.02	Milk 3 cup	1
	Carb (g)	184.18	Meat and Beans 5 oz	6

They do not, however, allow for quantification of either macro- or micronutrient intake.

U.S. Dietary Guidelines for Americans

The U.S. Dietary Guidelines, published jointly by the USDA and the U.S. Department of Health and Human Services (USDHHS), provide general recommendations for dietary intake that promote health and prevent disease (U.S. Department of Agriculture/U.S. Department of Health and Human Services 2005). These guidelines have been revised five times, most recently in 2005. Although the U.S. Dietary Guidelines are an important tool in nutrition education and planning, they are not the most efficient tool available for evaluating an individual's diet and really only provide a very broad overview. These guidelines are much more useful in setting goals for nutrition and health education or in translating nutrition research into simpler terms for consumers.

Daily Values/Dietary Reference Intakes

One method of evaluating dietary analysis for macro- and micronutrient amounts is using the Daily Values (DV) and the Dietary Reference Intakes (DRI). DRI are standards established by the National Academy of Sciences. These standard reference values allow evaluation of energy, protein, vitamin, and mineral intake for healthy people. There are four different sets of standards within the DRI: Adequate Intakes (AI), Recommended Dietary Allowances (RDA), Tolerable Upper Intake Levels (UL), and Estimated

Average Requirements (EAR). The RDA, AI, and UL can be used to assess diets of individuals (Standing Committee on the Scientific Evaluation of Dietary Reference Intakes 2002, 2004).

It is important when using these standards to understand the context in which the references are established. Values for RDA are determined at approximately two standard deviations above the average (mean) requirement within the healthy population. This margin of safety allows the value of the RDA to meet the needs of most healthy people. Therefore, if the evaluated diet falls below the RDA or AI for a specific nutrient, it does not necessarily mean the client is deficient in this nutrient. Deficiencies of specific nutrients would require additional confirmation using other components of nutrition assessment. Still, the DRI serve as important benchmarks for evaluating the patient's dietary intake, not only from food, but also from dietary supplements.

In the clinical setting, many patients have specific diseases or medical conditions that may have unique nutrient requirements. For example, an individual with a burn injury may require significantly higher doses of vitamin C and zinc in order to ensure appropriate wound healing. Additionally, medications and treatments may alter absorption, utilization, excretion, or storage of specific nutrients. In these situations, patients may need higher or lower levels of these nutrients. The DRI are established for the healthy population and hence may not be appropriate in all clinical situations. Nonetheless, they can always be used as a starting point in dietary evaluation, and as the medical condition and subsequent nutrition therapy are established, adjustments can be made for specific nutrient requirements.

The DV were established by the Food and Drug Administration to assist consumers in interpreting nutrition labeling information. These standards use the RDA and U.S. Dietary Guidelines as their theoretical base and set target goals for fat, saturated fat, cholesterol, total carbohydrate, fiber, sodium, potassium, and protein. These reference standards are expressed with a 2000- and 2500-kcal reference diet. In general, though, the DV are much more useful to the consumer purchasing groceries than the dietitian performing a nutrition assessment. The dietitian will use much more specific, individualized reference data.

Nutritional Physical Examination

In addition to the physical examination performed by medical and nursing staff, the dietitian should perform a nutrition-focused physical examination (Mackle et al. 2003). The purpose of this physical exam is to assess the patient for signs and symptoms consistent with malnutrition or specific nutrient deficiencies. Techniques of inspection, palpation, percussion, and auscultation are used to examine the body for these signs and symptoms. Inspection is a visual assessment of the body conducted in a systematic manner by the clinician in order to note any changes from normal features. Palpation is examination of the body using the sense of touch. Percussion involves the use of sound to identify deviations from standard sounds produced when the body surface is tapped by the fingers of the practitioner. The presence of body organs and cavities will change the resonance and quality of sounds. In the nutrition-focused physical assessment, percussion may be used to assess status of the gastrointestinal tract when assessing feeding route or to identify fluid in the lungs, which may necessitate the need for medical intervention and possibly fluid restriction (Shopbell, Hopkins, and Shronts 2001). Auscultation also uses the sense of hearing to identify deviations from standard sounds. In this technique, a stethoscope is used to evaluate sounds produced by the heart, lungs, and gastrointestinal tract. An example is the identification of bowel sounds.

The methodology and interpretation for both the Subjective Global Assessment and the Nutrition-Focused Physical Assessment are included in Appendices B4 and B5.

Anthropometric/Body Composition Assessment

Anthropometrics

"Anthropometry is the measurement of body size, weight, and proportions" (Lee and Neiman 2003, p. 164). Body composition refers to the distribution of body compartments (e.g., muscle mass and body fat) as part of the total body weight. Evaluating both anthropometric and body composition data allows the clinician to fully assess these compartments.

Because nutrition is a crucial component of normal growth and development, it is accepted practice to measure body compartments in order to evaluate infants and children for appropriate growth. In a normal, healthy individual, the relationships between body storage compartments are relatively stable. But when disease or stress is present, changes in the storage compartments are an important component of determining nutritional status and risk. Results of anthropometric assessment can be used to both identify goals for nutrition intervention and monitor changes that occur as a result of either those interventions or continued effects of disease and stress.

Height/Stature Measurement of supine or standing height is necessary for monitoring growth of infants and children and for interpretation of weight in adults. For children under the age of two years, length is measured recumbently using a length board. This device has a stationary headboard and a movable footboard (Figure 5.5). This measurement requires two clinicians, one of whom holds the child's head touching the headboard while the other extends the leg and bottom of the heel to the footboard. Length is recorded to the nearest 0.1 cm. Over the age of two years, standing height is measured using a tape measure or

FIGURE 5.5 Measuring Infant Length

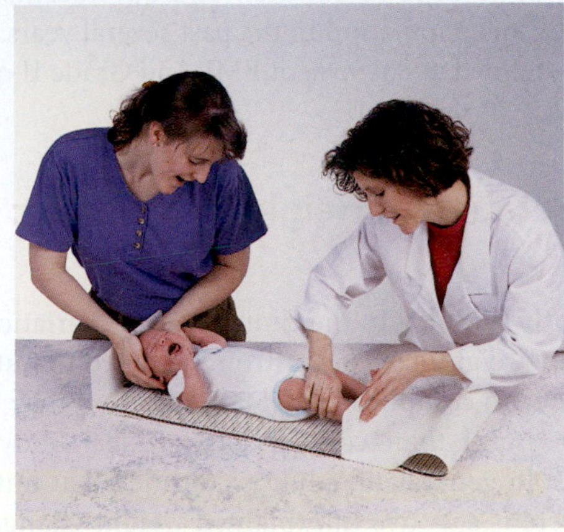

Source: SR Rolfes, K Pinna & E Whitney, *Understanding Normal and Clinical Nutrition 7e*, Fig "How To" box, top photo Page 591.

stadiometer. Procedures for measuring height include having the client stand barefoot and look forward with shoulders, buttocks, and heels touching the vertical surface of either a wall or the stadiometer.

When a client cannot stand for the measurement of height, there are several estimation methods that may be used. Arm span is one method of height estimation for adults. The client extends the arms from the body at a 90-degree angle and distance is measured between the tips of the two middle fingers. A limitation of this method is that it is an estimation of maximum adult height and not actual, current height.

Knee height is another method of height estimation (Figure 5.6). Measurement of knee height, using a knee-height caliper, can be taken when the client is sitting or in a supine position. (Measuring supinely is considered to be more accurate.) The client lies supine with right knee and ankle flexed to 90 degrees. The clinician should place the fixed portion of the caliper under the heel and position the other blade over the anterior portion of the thigh above the knee. The shaft of the caliper is parallel to the tibia. The measurement (repeated two to three times) is recorded to the nearest 0.1 cm. Height is then estimated using the following equations (Chumlea, Guo, and Steinbaugh 1994):

Age 18–60 years:

- White male = 71.85 + (1.88 × knee height)
- Black male = 73.42 + (1.79 × knee height)
- White female = 70.25 + (1.87 × knee height) − (0.06 × age)
- Black female = 68.10 + (1.86 × knee height) − (0.06 × age)

Age 60–80 years:

- White male = 59.01 + (2.08 × knee height)
- Black male = 95.79 + (1.37 × knee height)

FIGURE 5.6 Knee-Height Calipers

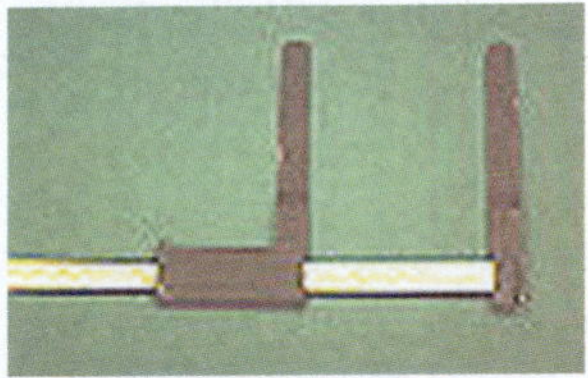

Source: National Resource Center on Nutrition, Physical Activity and Aging (http://nutritionandaging.fiu.edu/) URL: http://www.fiu.edu/~nutreldr/images/knee_caliper.jpg.

- White female = 75.00 + (1.91 × knee height) − (0.17 × age)
- Black female = 58.72 + (1.96 × knee height)

Height has been noted to be one of the most inaccurate measures. In clinical settings, it is often either estimated or recorded from the patient's memory (Kuczmarski, Kuczmarski, and Majjar 2001). Accurate measurement is crucial, because height is used to interpret weight, measure growth for children, calculate energy and protein requirements, and calculate creatinine-height index.

Weight Weight can be measured using a variety of scales, including balance beam and electronic scales. Bathroom scales and those that are moved frequently are not recommended due to problems with calibration. Wheelchair and bed scales are available for nonambulatory patients. In the ideal setting, the client should be weighed with minimal clothing and without shoes, at the same time daily, and after urination.

For those patients with amputation, weight has historically been adjusted using the following factors (Smith, Weiss, and Lehmkuhl 1996):

- Hand: 0.8%
- Forearm and hand: 3.1%
- Entire arm: 6.5%
- Foot: 1.8%
- Lower leg and foot: 7.1%
- Entire leg: 18.6%

For example, for an individual who has had an entire leg amputated and currently weighs 165 lbs., weight would be adjusted by using the following equation:

> Estimated body weight = actual measured weight divided by 100 − %amputation. The whole equation then is multiplied by 100.

stadiometer—a calibrated device used to measure stature

$$\frac{165 \text{ lbs.}}{(100 - 18.6)} \times 100 = 202 \text{ lbs is the estimated body weight.}$$

Equations for estimation have been recently published, but continued validation will be needed (Mozumdar and Roy 2004).

Weight is the most common measure of anthropometrics. Unfortunately, it is a gross measurement of all body compartments and does not distinguish body composition or fluid shifts. Nevertheless, due to its common availability and its relationship to growth, development, and health, it continues to be an important component of nutrition assessment.

Interpretation of Height and Weight: Infants and Children

Growth Charts Weight and height for infants and children are evaluated using growth charts developed by the Centers for Disease Control and Prevention (CDC) and the National Center for Health Statistics (National Center for Health Statistics and National Center for Chronic Disease Prevention & Health Promotion 2000) (see Appendix B1). Determination of height for age and weight for age allows comparison of an infant or child to a reference population. Data for the CDC growth charts are based on the National Health and Nutrition Examination Survey, and were most recently updated in 2000. When infants and children are either < 3rd percentile or > 97th percentile, further assessment should be made to confirm any health problems. There are specific clinical diagnoses, such as genetic and endocrine disorders, that negate use of growth charts or have special growth charts developed for that specialized population.

Weight for height and **percent weight for height** can also be evaluated using CDC growth charts. These measurements allow evaluation to be independent of age and can be used to monitor acute malnutrition (< 5 percentile) or the incidence of obesity (> 95 percentile).

Desirable body weight (DBW) is defined as the 50 percentile for the infant's height or age, or for the child's height and age. From this, a percentage of DBW can be calculated. Weights from 90–109% of DBW are considered to be within normal range.

Body Mass Index Revision of the CDC growth charts in 2000 added the measurement of body mass index (BMI). BMI is: weight (kg)/[height (m)]^2 (National Center for Health Statistics and National Center for Chronic Disease Prevention & Health Promotion 2000). Calculation and interpretation of BMI in children and adolescents has increased for this population in the past several years. Grade I obesity is defined as 85–95% of BMI, and Grade II obesity is defined as > 95% of BMI.

percent weight for height—percentage used to evaluate a child's growth pattern relative to population standards

Interpretation of Height and Weight: Adults

Usual Body Weight In the clinical setting, variations from usual body weight have been strongly linked to nutritional risk and health complications. Such variation may be more clinically useful than comparison to ideal body weight standards. In general, an adult is considered at nutritional risk if there is a > 5% unexplained weight change in less than one month or > 10% in a six-month period (see Table 5.4). The absolute number is also used to describe rapid weight gain as well.

Percent Usual Body Weight and Percent Weight Change

Percent usual body weight is calculated as:

$$\frac{\text{current weight}}{\text{usual body weight}} \times 100$$

Percent weight change is calculated as:

$$\frac{\text{usual body weight} - \text{present weight}}{\text{usual body weight}} \times 100$$

or:

$$100 - \%\text{ usual body weight} = \%\text{ change}$$

Desirable or "Ideal" Body Weight Creation of height-weight tables and estimation of desirable body weight from these data were initially conducted by the life insurance industry. The data have been criticized and their use is discouraged because the original population does not correlate with the current population in the United States. Other criticisms of the development of these standards include the assumption of the weight of an individual's clothes, use of a standard heel height on shoes, and estimation of frame size.

TABLE 5.4

Interpretation of Unintentional Weight Change

Time Frame	Significant Weight Loss	Severe Weight Loss
1 week	1–2 % UBW	> 2 % UBW
1 month	5 % UBW	> 5 % UBW
3 months	7.5 % UBW	> 7.5 % UBW
6 months	10 % UBW	> 10 % UBW

Source: Adapted from American Dietetic Association. *Manual of Clinical Dietetics* 6th ed. Chicago IL: American Dietetic Association. 2000.

In many clinical settings, "ideal" body weight is calculated using the Hamwi equation even though it does not adjust for differences in age, race, or frame size (Hamwi 1964). Calculation of body mass index (see the next two sections) is considered to be the only validated method for estimating desirable or ideal body weight. Though the Hamwi method has been used historically, the validation is questionable.

Men: 106 lbs for 5 foot + 6 lbs per inch over 5 foot or – 6 lbs per inch under 5 foot
Women: 100 lbs for 5 foot + 5 lbs per inch over 5 foot or – 5 lbs per inch under 5 foot

Healthy Body Weight The U.S. Dietary Guidelines established weight guidelines that could be used for general education and population screening. These weight standards for men and women provide a wide range for each height level. These reference standards are not very useful in a clinical setting or for a comprehensive nutrition assessment.

Body Mass Index (BMI)

BMI or Quetelet's Index, as stated previously, is calculated as:

- weight (kg) / [height (m)]^2

or

- weight (lbs) $\times$ 704.5 / [height (inches)]^2.

The use of BMI has been correlated with overall mortality and nutritional risk. It still does not estimate body composition, but it is better at indicating obesity than mere height and weight alone. A client with a BMI $\geq$ 25 is considered to be overweight, and client with a BMI $<$ 18.5 is considered to be underweight. The Expert Panel on Healthy Weight established target healthy weight to be a BMI between 18 and 25 kg/m^2 for adults (NHLBI 1998; National Center for Chronic Disease Prevention & Health Promotion 2004). See Table 5.5 for complete interpretation of BMI.

Frame Size Total body weight varies with gender and age as well as skeletal dimensions. Frame size estimation is necessary for accurate interpretation of weight. Elbow breadth and wrist circumference are typically used to estimate frame size. Elbow breadth is considered superior to wrist circumference because it is less likely to be affected by adipose tissue.

Wrist circumference is measured at the smallest part of the wrist distal to the styloid process of the ulna and radius. Using the following equation, frame size can be determined:

$$r = \frac{\text{height (cm)}}{\text{wrist circumference (cm)}}$$

See Appendix B7 for interpretation of r-value as frame size.

Elbow breadth is measured with elbow flexed at 90 degrees. Calipers are then used to measure the distance between **epicondyles** of the humerus, which is compared to standard values for estimation of frame size. (See Appendix B7.)

TABLE 5.5

Interpretation of BMI in Adults

For adults, classification of weight based on body mass index by the National Institutes of Health is as follows:

BMI	Weight Status	Health Risk
Below 18.5	Underweight	With $<$ 16 suggesting possible eating disorder
18.5–24.9	Normal	Healthy, low health risk
25.0–29.9	Overweight	Associated with increased risk of disease
30.0 and above	Obese	Associated with further increased risk of disease

Source: Adapted from Gropper SS, Smith JL, Groff JL. Advanced Nutrition and Human Metabolism, 4th edition, Belmont, CA: Wadsworth/Thomson Learning, 2005, pg 520.
CDC - http://www.cdc.gov
CDC - http://www.cdc.gov/nccdphp/dnpa/bmi/bmi-means.htm

Waist-Hip Ratio and Waist Circumference Calculation of waist-to-hip ratio has been used to assess distribution of adipose tissue. A ratio of 1.0 or greater in men and a ratio of 0.8 or greater in women indicates android obesity. Waist circumference of $>$ 40 inches for men or $>$ 35 inches for women is considered to be predictive of obesity and chronic disease risk. Fat accumulation, primarily in the abdominal region, has been linked to an increased risk of type 2 diabetes mellitus and other obesity-related diseases.

Body Composition

Height and weight, though crucial components of nutrition assessment, are only gross measurements and cannot distinguish between body compartments. It is not unusual, then, for very muscular individuals to fall into an overweight or obese categorization when simple height and weight are used in their nutrition assessments. Measurements of skinfolds, frame size, waist circumference, and other body composition techniques allow the clinician to make a more thorough and complete nutrition assessment. However, these are not routinely done in most clinical settings where

epicondyle—a rounded projection at the end of a bone, located on or above a condyle, and usually serving as a place of attachment for ligaments and tendons

acute care is the primary goal. In settings where clients are followed for a length of time, these assessments are clinically useful.

Body composition refers to distribution and size of all components contributing to total body weight. In most clinical settings, body composition refers to two major components of total body weight: fat mass and fat-free mass. A more thorough definition would include fat mass, total body water, osseous mineral (bone and teeth), and protein. Nutrition professionals are most concerned about metabolically active tissue and fluid status. Techniques that allow assessment of these compartments are necessary for the design of appropriate nutrition interventions.

Measurement techniques, such as skinfolds, bioelectrical impedance, and near-infrared interactance, use portable equipment and can be easily accomplished in any setting. Other laboratory measurements using hydrodensitometry or imaging techniques require sophisticated and expensive equipment that are not readily available in every clinical setting.

FIGURE 5.7 Triceps Skinfold Measurement

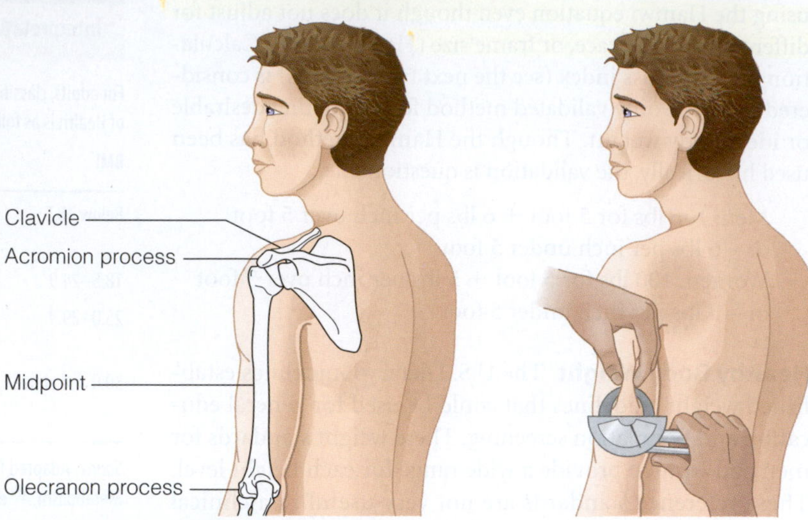

Source: SR Rolfes, K Pinna & E Whitney, *Understanding Normal and Clinical Nutrition 7e*, FigE7, pE8.

BOX 5.3 CLINICAL APPLICATIONS

Arm Muscle Area Assessment

$$\text{UAMA} = \frac{[\text{UAC-(TSF} \times \pi)]^2}{4\pi}$$

Percentile	Category
≤ 5th	Wasted
> 5th but ≤ 15th	Below average
> 15th but ≤ 85th	Average
> 85th but ≤ 95th	Above average
> 95th	High muscle

Percentiles for interpretation of Arm Muscle Area found in Appendix B7.

Source: Frisancho AR. 1990. Anthropometric standards for assessment of growth and nutritional status. Ann Arbor: University of Michigan Press.

Skinfold Measurements Skinfold measurement is used to estimate energy reserves—both fat and somatic protein—in subcutaneous tissue. Skinfold measurement involves measuring a double fold of skin and subcutaneous adipose tissue while excluding muscle tissue (Lee and Neiman 2003). Even though more accurate methods for body composition measurement exist, there are advantages that contribute to the common use of skinfolds, including the ease of measurement and the fact that the measurement is minimally invasive for the client and requires only inexpensive equipment.

Sites for skinfold measurement include the chest, triceps, subscapular, midaxillary, suprailiac, abdomen, thigh, and calf. Using more than one site for measurement may provide a more accurate picture of the individual. The most commonly used site is the triceps. Triceps skinfolds are taken on the nondominant arm, halfway between the olecranon and acromial (see Figure 5.7) processes. Additionally, midarm circumference measurement can be combined with triceps skinfold measurement to indirectly estimate arm muscle area and arm fat area (Box 5.3).

Equipment needed for skinfold measurement includes a tape measure and a skinfold caliper. The Harpendon or Lange caliper is generally recommended because this type was used in development of reference standards and equations.

Interpretation of Skinfold Measurements Skinfold and midarm circumference measurements, as well as the index calculations for mid-upper arm muscle circumference, mid-upper arm muscle area, and mid-upper arm fat area, can be compared to references that are based on NHANES summary data. Comparison data are age, gender, and race specific. They may also be specific to frame size. A recent study indicated that using midarm circumference alone was not a sensitive measure for malnutrition when compared to other measures (Burden et al. 2005). An individual who is below the 5th percentile or greater than the 95th percentile may be at nutritional risk (Table 5.6). See Appendix B7 for these reference standards.

When interpreting these measurements in nutrition assessment, it is important to recognize that these reference

TABLE 5.6

Interpretation of Triceps Skinfold Measurements

Comparing the percentile ranking of a specific individual on the various anthropometric measurements with a classification scheme is the basis for interpreting these values.

Reference data in percentiles for TSF, AMA, and AFA appear in the appendices. Because reference data are those compiled by Frisancho from the NHANES I and NHANES II data, it is appropriate to use classification categories derived statistically from these data. The table below displays percentile categories and their interpretation for arm muscle and arm fat areas as well as total body weight.

Percentile Rank	AMA	AFA	Total body weight
<5	Muscle deficit	Fat deficit	Total body wasting
5.1–15	Below average	Below average	Below average
15.1–85	Average	Average	Average
>85	Above average musculature	Excess fat	Excess total body weight

Source: Reprinted from Contemporary Nutrition Support Practice, L. Matarese and M. Gottschlich; Philadelphia, Pa: WB Saunders Company; 2003, Box 3-1, p. 36, with permission from Elsevier. Ref: Frisancho AF: Anthropometric standards for the assessment of growth and nutritional status, Ann Arbor, Michigan, 1990. The University of Michigan Press, pp. 28, 41-42, 51-51, 54, 60-63.

standards were developed using a healthy population and cannot be used for a patient in a disease state. Furthermore, a one-time skinfold calculation may not really contribute any clinically useful information in the acute care setting. Hydration state and fluid shifts can affect skinfold measurement. When the individual can serve as his or her own control with repeated measures over time, long-term changes in energy stores can be assessed and provide clinically useful information. Accuracy of measurement may be difficult to achieve if more than one clinician is performing the assessment, but error can be minimized when a single well-trained clinician uses the same equipment and method each time the assessment is performed.

Multiple-site skinfold measurements can be used to estimate body density and body fat percentage. Numerous regression equations have been developed. The regression equation originally designed for a population that most closely matches the client should be used.

Bioelectrical Impedance Analysis (BIA) BIA is an increasingly common procedure used to estimate body composition. It is considered to be a precise, rapid, safe, portable, and noninvasive method of assessment. There are numerous types of equipment available to measure BIA, but they are all based on the same scientific principle. A small, low-frequency, alternating electrical current is administered at one extremity of the body. Measurement of impedance of the electrical current can then be used to estimate components of body composition, including body cell mass, fat-free mass, fat mass, and total body water. Tissues that contain low amounts of water such as fat and bone are poor conductors of electricity and therefore have a greater resistance or impedance to flow of the current. Other tissues that have greater water content, such as blood, muscle, and vital organs, are good conductors and therefore have lower impedance. More recent advances in technology provide for segmental measurement, which allows for estimation of regional body fat. Additionally, multiple-frequency BIA may provide a more accurate assessment than a single frequency.

BIA Equipment

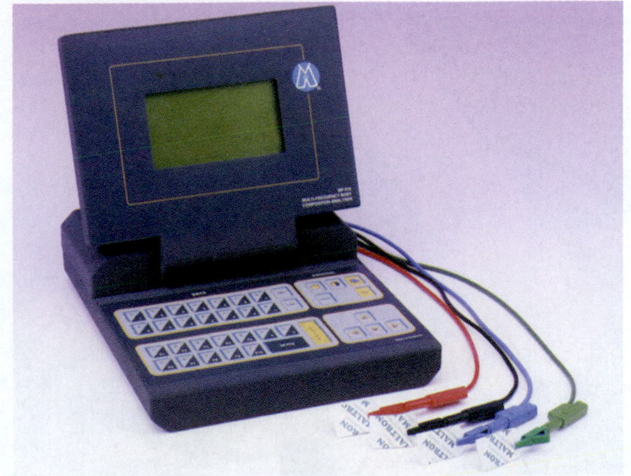

Source: http://www.uky.edu/Education/KHP/Body_comp/EquipmentPhotos.htm

Regression equations include age, gender, weight, height, resistance, and reactance. From this information, components of body composition can be calculated. Additionally, equations have been validated specific to race and physical activity level (Kotler et al. 1996).

The use of BIA has expanded in recent years. **Phase angle,** as a measure of body composition, has been used as an additional measure of prognosis in many chronic conditions. Studies have been published using BIA for HIV, chronic renal failure, cancer, chronic obstructive pulmonary disease, and cirrhosis (Gupta et al., *Am J Clin Nutr* 2004; Gupta et al. *Br J Nutr* 2004; Kotler 2000; Schols et al. 2005; Schwenk et al. 2001). BIA should not be used when there have been major shifts in water balance and distribution. There has previously been a lack of reference data for BIA, but the most recent NHANES III data will include new references for BIA.

phase angle—calculates a mathematical relationship between resistance and reactance; for use with bioelectrical impedance to calculate body composition; higher values for phase angle appear to be consistent with greater body muscle mass and lower risk for morbidity and mortality; values range from 3–12

Hydrodensiometry

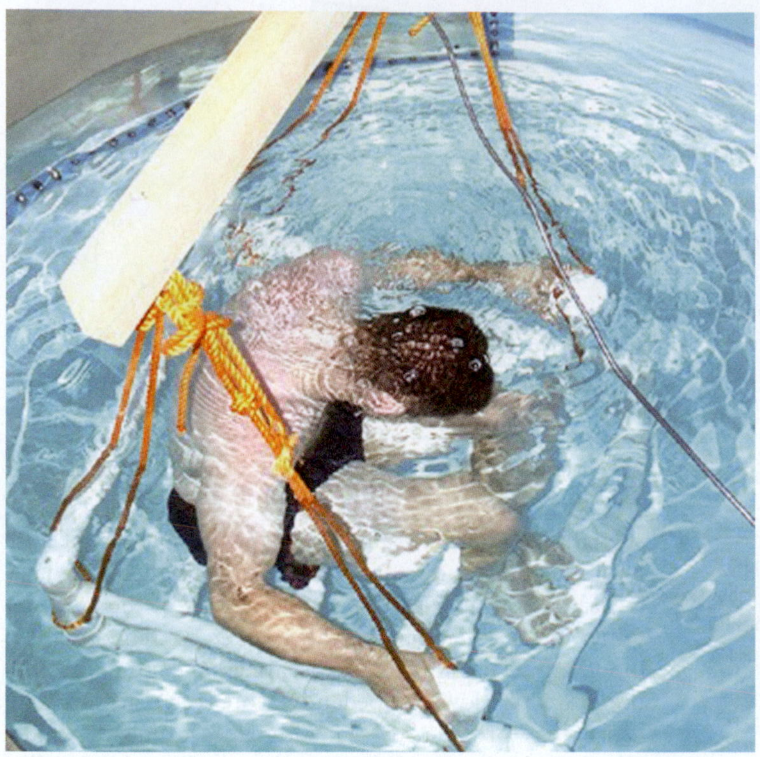

Source: http://kspark.kaist.ac.kr/BodyFat/Body%20Fat%20from%20Underwater%20Weighing.htm.

Dexa

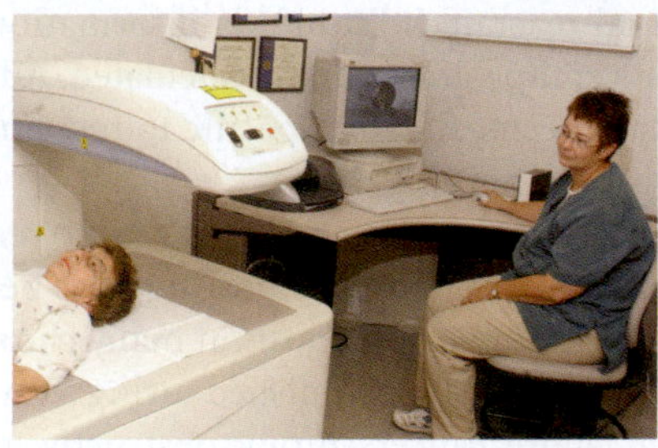

Source: Photo is provided courtesy of Memorial Health System Springfield, Illinois.

Hydrodensitometry—Underwater Weighing Underwater weighing is generally accepted as one of the most accurate methods of measuring body composition. Other body composition methods are often validated against underwater weighing results. Hydrodensitometry does evaluate body composition in the more traditional view, that of the two-compartment model. This method measures body volume (density) and relies on the assumption that the density of fat mass and components of fat-free mass are constant.

Obviously, access to underwater weighing facilities is not readily available, and this method thus has limited clinical use. It is also a difficult procedure for the subject to complete. Additionally, the reference equations were based on Caucasians and may need changes for use with other ethnicities. Other limitations include the possible overestimation of fat mass due to the assumption that all components of fat-free mass are constant. A correction for residual lung volume must also be made, and this volume is difficult to estimate.

Near Infrared Interactance (NIR) New commercial NIR analyzers have gained popularity in fitness and athletic facilities as a method for estimating body fat percentage. This technology was originally developed by the USDA to measure the body composition of livestock and the fat content of various grains (Swatland 1983). It uses the principles of light absorption and reflection to measure body fat. In this method, an electromagnetic signal is sent into the biceps of the nondominant arm, where it is then scattered and reflected back. The amount of energy scattered and reflected can be correlated to the water, protein, and fat composition of the subject. As with BIA, recent research has focused on assessing the accuracy of NIR in various populations. Most studies indicate a variability in results and recommend continued research on its application (Housh et al. 2004; Kalantar-Zadeh et al. 2001; Kamimura et al. 2003; Martinex-Abundis, Gonzalez-Ortiz, and Grover-Paez 2001).

This method's ease of use does not outweigh its limitations. The reference standards used have not been validated in humans. Furthermore, it is based on the assumption that the fat composition of the arm is consistent with the rest of the body, which may not be accurate (Fogelholm and Lichtenbelt 1997).

Dual Energy X-Ray Absorptiometry (DEXA) DEXA was first developed to measure bone mineral content and density. In this method, the body is scanned with photons at two different energy levels. Absorption of the photons is then measured at each of these two levels. The absorption rate of different body tissues (water, fat mass, and osseous material) can allow for calculation of these body compartments (see Figure 5.8). Most practitioners and researchers believe DEXA will allow for the most accurate assessments of body composition (Lee and Neiman 2003).

Biochemical Assessment

Biochemical assessment involves measuring nutritional markers and indicators of organ function, which are found in blood, urine, feces, and tissue samples. Interpretation of biochemical measurements must be made in the context of diagnosis and medical treatment. Disease states, hydration status, and subsequent treatments can have considerable effect on the levels of the biochemical indices. Since

FIGURE 5.8 Sample DEXA Scan Results

DEXA Body Comp FIT

Name: *John Doe*
Age: *48 yrs*
Height: *69.0 in.*
Weight: *163.3 lbs.*
Date: *4/21/03*
Sex: *Male*

Region	% Fat	Total Mass (lbs)	Fat Mass (lbs)	Lean Mass (lbs)	Bone Mass (lbs)
Arms	13.1	18.0	2.2	14.9	0.9
Trunk	17.9	74.5	13.0	59.9	1.6
Legs	25.9	59.6	14.8	42.3	2.5
Total*	19.8	163.3	31.1	125.9	6.4

*The mass values in this Total row include the head. Other mass values do not include the head.

Source: Fitness Institute of Texas, The University of Texas at Austin. http://www.edb.utexas.edu/fit/bodycompsr.htm. Reprinted by permission.

reference values may also vary from lab to lab, data should be interpreted using those values provided by the laboratory that conducted the tests.

Protein Assessment

Protein's own unique function in supporting cellular growth and development elevates its significance in nutrition assessment. Even though there is a large amount of protein stored in muscle and viscera, the body strives to maintain this compartment intact. Under healthy conditions or normal energy deficits, additional energy is drawn from fat and glycogen stores. But when the body is under metabolic stress, it draws protein from the muscles to meet its needs (see Chapter 25). No specific laboratory value can determine the protein status of an individual. Each test has its own limitations. Therefore, a comprehensive nutrition assessment must use a battery of tests in order to delineate the changes that mark the development of a nutrient deficiency.

Historically, assessment has focused on evaluation of two compartments of protein—visceral and somatic. Visceral protein refers to nonmuscular protein making up the organs, structural components, erythrocytes, granulocytes, and lymphocytes, as well as other proteins found in the blood. Somatic protein refers to skeletal muscle.

Somatic Protein Assessment In addition to using anthropometric measures such as midarm muscle area, midarm circumference, and overall body weight and subjective global assessment, biochemical tests such as creatinine height index, 3-methylhistidine, and nitrogen balance can be used to more specifically analyze somatic protein status.

Creatinine Height Index Creatinine is formed at a constant daily rate from muscle creatine phosphate. Creatine phosphate, which is stored in muscle, provides the phosphate group needed to regenerate ATP during high-intensity exercise. Creatinine is not stored in muscle, but is cleared and excreted by the kidney. Daily urine output of creatinine can be correlated with total muscle mass.

For this test, a 24-hour urine collection is performed. Total amount of creatinine excreted in that 24-hour period is compared to either a standard based on height or (as a percentage) to a standard excretion for a particular reference individual of a specific height, gender, and age.

In Table 5.7, expected creatinine excretion is shown for various heights. Creatinine height index (CHI) is usually expressed as a percentage of a standard value.

$$\text{CHI} = \frac{\text{24-hour urine creatinine (mg)}}{\text{expected 24-hour urine creatinine (cm)}} \times 100$$

A value calculated to be 60–80% of the standard suggests mild skeletal muscle depletion, 40–59% suggests moderate skeletal muscle depletion, and <40% is considered to be a severe loss of skeletal muscle.

Limitations to this measurement include that its accuracy depends on a complete collection of urine for 24 hours, which is a common source of error and makes the test more difficult to conduct. Interpretation of this test must also take into account the fact that creatinine excretion can be either higher or lower than the standard depending on certain clinical conditions. Creatinine excretion is increased by meat consumption, emotional stress, sepsis, trauma, fever, and strenuous exercise, and during the second half of the menstrual cycle. Creatinine excretion is decreased with compromised renal function, low urine output, aging, and muscle

TABLE 5.7

Expected 24-Hour Creatinine Excretion in Men and Women of Ideal Weight

Adult Males*		Adult Females †	
Height (cm)	Creatinine (mg)	Height (cm)	Creatinine (mg)
157.5	1288	147.3	830
160.0	1325	149.9	851
162.6	1359	152.4	875
165.1	1386	154.9	900
167.6	1426	157.5	925
170.2	1467	160.0	949
172.7	1513	162.6	977
175.3	1555	165.1	1006
177.8	1596	167.6	1044
180.3	1642	170.2	1076
182.9	1691	172.7	1109
185.4	1739	175.3	1141
188.0	1785	177.8	1174
190.5	1831	180.3	1206
193.0	1891	182.9	1240

*Creatinine coefficient for males = 23 mg/kg of "ideal" body weight

†Creatinine coefficient for females = 18 mg/kg of "ideal" body weight

Source: Reprinted from Blackburn GL, Bistrian BR, Maini BS, Schlamm HT, Smith MF. Nutritional and metabolic assessment of the hospitalized patient. *JPEN J Parenter Enteral Nutr.* 1977;1:15, with permission from the American Society for Parenteral and Enteral Nutrition (A.S.P.E.N.).

atrophy unrelated to malnutrition. Furthermore, standards that are used do not account for creatinine excretion changes with age, disease, physical training, frame size, or weight status (Lee and Neiman 2003).

3-Methylhistidine 3-methylhistidine is an amino acid found in actin and myosin within skeletal muscle. When protein is broken down, 3-methylhistidine is excreted in urine. Healthy adults excrete approximately 0.1–7.8 umol/kg per day. This biochemical measure is rarely used because of its significant limitations (Lee and Neiman 2003). Like creatinine height index, the test relies on an accurate collection of urine for 24 hours. Additionally, there appears to be a significant amount of 3-methylhistidine outside of skeletal muscle, so the amount collected may not accurately reflect skeletal muscle alone. Daily variation is significant and excretion varies with the type of nutritional depletion.

Nitrogen Balance In healthy individuals, nitrogen excretion should be equal to nitrogen intake—thus indicating a state of nitrogen balance. A negative nitrogen balance would occur when nitrogen excretion is greater than nitrogen intake, indicating catabolism or inadequate nitrogen intake, whereas a positive nitrogen balance would occur when nitrogen intake is greater than excretion.

Measuring nitrogen balance assesses overall protein status. Additionally, it can serve as a method of assessing the effectiveness of a nutrition intervention. In order to measure nitrogen balance, the dietary intake of protein for a 24-hour period is estimated while a 24-hour urine collection is gathered to measure total excretion of nitrogen. Nitrogen loss through other routes such as fecal excretion, normal skin breakdown, wound drainage, and nonurea nitrogen is estimated using a constant value of either 3 or 4. The following equation is used for calculation:

$$N_2 \text{ balance} = \frac{\text{dietary protein intake}}{6.25} - \text{urine urea}$$

$$\text{nitrogen} - 4 \text{ (1 g protein} = 6.25 \text{ g nitrogen)}$$

Example: JM consumed approximately 55 grams of protein in the past 24 hours. His 24-hour urine collection indicated a urine urea nitrogen (UUN) of 13 grams. His N_2 balance would be calculated as:

$$\frac{55}{6.25} - 13 - 4 = -8.2 \text{ g}$$

This is interpreted as JM currently being in negative nitrogen balance.

Limitations of measuring nitrogen balance include the inherent error of 24-hour urine collection, failure to account for renal impairment, and inability to measure nitrogen losses from some wounds, burns, diarrhea, and vomiting. Nitrogen intake may also pose difficulties. Oral protein intake may be difficult to measure except when the patient is exclusively on enteral or parenteral nutrition support.

Visceral Protein Assessment As stated previously, visceral protein refers to nonskeletal protein making up the organs, structural components, erythrocytes, granulocytes, and lymphocytes, as well as other proteins found in the blood. Thus, visceral protein assessment indirectly measures these protein stores by assessing proteins made by these organs (primarily the liver) present in blood or lymph fluid. Theoretically, serum protein measurement is affected by a change in the amount of amino acids needed for protein synthesis by the liver. Thus, a change in serum protein levels would be consistent with changes in visceral protein status. However, the synthesis rate can be affected by factors other than protein intake or protein requirements. The sensitivity and specificity of these assessment measures are often described by the term "half-life" of the protein. Half-life, in this clinical situation, means the amount of time it takes before half of the protein is either eliminated or broken down by the body. Therefore, a shorter half-life means that actual changes in these levels will be reflected more quickly than in those proteins with a longer half-life.

TABLE 5.8

Visceral Protein Assessment Overview

Serum Protein	Normal Range	Half-Life	Primary Function	Comments
Albumin	3.5–5.0 mg/dL	17–20 days	Blood transport protein; component of vascular fluid and electrolyte balance	Negative acute phase protein—decreases with illness, infection, trauma, surgery, and metabolic stress; affected by hydration status—decreases with overhydration, increases with underhydration
Transferrin	215–380 mg/dL	8–10 days	Iron transport	Negative acute phase respondent; affected by iron status
Prealbumin/ transthyretin	19–43 mg/dL	2 days	Transport of thyroxine	negative acute phase protein—decreases with illness, infection, trauma, surgery and metabolic stress; decreases with diagnoses of liver disease such as hepatitis or cirrhosis, malabsorption, and hyperthyroidism
Retinol binding protein	2.1–6.4 mg/dL	12 hours	Transport molecule for vitamin A	Negative acute phase respondent; elevated with renal failure; levels are decreased with hyperthyroidism, cystic fibrosis, liver failure, vitamin A deficiency, zinc deficiency, and metabolic stress
Fibronectin	220–400 mg/dL	15 hours	Wound healing and vascular integrity; cell development, regulation of cell growth and differentiation	Affected by coagulation, inflammation, and injury; may be used to assess protein status even during the acute phase process

The other terminology that is used when referring to protein assessment is the measurement of acute-phase proteins. Acute phase proteins are defined as "those whose plasma concentration increases (positive acute-phase proteins) or decreases (negative acute-phase proteins) by at least 25 %" during inflammation, illness, and/or metabolic stress (Gabay and Kushner 1999, p. 448).

An overview of visceral protein assessment is presented in Table 5.8. The specificity, sensitivity, and reliability of each visceral protein measurement are different. An accurate interpretation for nutrition assessment would take these differences into consideration.

Albumin Albumin is probably the most well-known measure of visceral protein status. It is also the most abundant serum protein. Approximately 60% of the albumin in the body is found in extravascular space—in skin, muscle, and organs. Remaining albumin is found within the vascular space. Normal serum levels of albumin are ≥ 3.5 g/dL. Synthesized by the liver, albumin serves many significant functions within the body, most commonly as a transport protein and as a component of vascular fluid and electrolyte balance.

Albumin has been the subject of much nutrition research and thus serves as a good prognostic screening tool, though it is not as reliable as an overall indicator of protein status. Additionally, it is easily measured and has an abundant body pool. Decreased albumin levels have been correlated with increased morbidity, mortality, and length of hospital stay in several clinical populations (Brugler et al. 2002; Charney 2005; Lopez-Hellin et al. 2002; Sullivan, Roberson, and Bopp 2005).

Some of the same factors contribute to its limitations as well. Albumin has a long half-life (approximately 20 days), which decreases its sensitivity to short-term changes in protein status or to short-term interventions to improve protein status. Albumin synthesis is also affected by acute stress and the inflammatory response. Albumin loss occurs with burn injuries, **nephrotic syndrome, protein-losing enteropathy,** and **cirrhosis.** Other medical conditions that may result in hypoalbuminemia include infection, **multiple myeloma,** acute or chronic inflammation, and rheumatoid arthritis. Levels also decrease with aging. On the other hand, albumin

nephrotic syndrome—a clinical state characterized by edema, albuminuria, decreased plasma albumin, usually increased blood cholesterol, and increased permeability of the glomerular capillary basement membranes, often caused by diabetes-induced glomerulosclerosis, systemic lupus erythematosus, renal vein thrombosis, or hypersensitivity to toxic agents

protein-losing enteropathy—increased fecal loss of serum protein, especially albumin, causing hypoproteinemia

cirrhosis—endstage liver disease characterized by damage to hepatic parenchymal cells with modular regeneration and fibrosis, associated with failure of hepatic cell function, interference with hepatic blood flow, frequently jaundice, portal hypertension, ascites, and ultimately hepatic failure

myeloma—a tumor composed of cells derived from hemopoietic tissues of the bone marrow; a plasma cell tumor

levels will be higher with dehydration and when individuals are prescribed anabolic hormones and corticosteroids. Albumin levels must be interpreted carefully. In sick people, they are not a good indicator of protein nutriture, even though they historically have been widely used for that purpose.

Transferrin Synthesized by the liver, transferrin serves as a transporter for iron throughout the body. Due to its shorter half-life (8 to 10 days), it can also serve as an indicator of protein status, because it is sensitive to acute changes in protein intake or requirements. Normal serum levels are 215–380 mg/dL.

Transferrin's primary limitation is that its concentration is directly affected by iron status. When iron stores are decreased, transferrin levels will increase to accommodate the need for increased levels of transport. Other disease states such as hepatic and renal disease, inflammation, and congestive heart failure can also affect transferrin levels. Transferrin can be measured directly or calculated from total iron binding capacity (TIBC).

Prealbumin (Thyroxine Binding Prealbumin or Transthyretin) Prealbumin is another example of an acute phase transport protein synthesized by the liver. Prealbumin is responsible for transport of thyroxine and is associated with retinol binding protein. Research has shown that because of its very short half-life (two days), changes in prealbumin levels respond to short-term modifications in nutritional intake and interventions. Normal serum levels range from 19–43 mg/dL.

Prealbumin is a more expensive test than albumin, but research has indicated that if it were used routinely on admission screening, approximately 44% more hospitalized patients would be identified as being at nutritional risk (Mears 1996). Mears, in a more recent study (2004), confirmed the usefulness of prealbumin in the clinical setting as a means for identifying nutritional risk. Raguso, Dupertuis, and Pichard (2003) found that measurement of prealbumin was indicative of adequate nutrition support during critical illness when markers for inflammation were stable. The clinician evaluating prealbumin levels should recognize that they are increased with renal disease and Hodgkin's disease (a form of lymphoma), and decreased with diagnoses of liver disease such as hepatitis or cirrhosis, malabsorption, and hyperthyroidism.

macrophage—a mononuclear, actively phagocytic cell arising from monocytic stem cells in the bone marrow

fibroblasts—connective tissue cells capable of forming collagen fibers

Retinol Binding Protein (RBP) Retinol binding protein (RBP) is the smallest serum protein and has the smallest body pool and shortest half-life (12 hours). Because of these factors, RBP is considered to be one of the more sensitive indicators of protein status (Winkler, Gerrior, and Pomp 1989). It will reflect short-term changes and responses to nutrition support interventions. RBP is synthesized by the liver, is an acute phase respondent, and serves as the transport molecule for vitamin A (retinol).

RBP levels are elevated with renal failure. Levels are decreased with hyperthyroidism, cystic fibrosis, liver failure, vitamin A deficiency, zinc deficiency, and metabolic stress.

Fibronectin (FN) Fibronectin (FN) is a glycoprotein found as an important component of many cell types, including endothelial cells, **macrophages,** hepatocytes, and **fibroblasts.** This protein appears to have many functions, with its primary function being the regulation of cell growth and differentiation during cell development. FN plays a crucial role in wound healing and vascular integrity. FN can be identified from lymph, amniotic fluid, cerebrospinal fluid, and plasma. Fetal fibronectin (fFN) is a protein produced during pregnancy that functions to attach the fetal sac to the uterine lining. Recent research indicates that fFN may serve as a predictor for preterm labor during pregnancy (Andersen 2000).

Because it is less affected by acute stress than other blood proteins, many believe FN will be increasingly used as a marker of nutritional status as well as a good indicator for the efficacy of nutrition support. FN may be especially pertinent in assessment of patients under acute stress because of its short, 15-hour half-life (Changjiang et al. 1997). At the same time, due to FN's role in wound healing, serum levels may be affected in conditions such as burns, where there is increased deposition at the site of injury and inflammation. Because FN plays a role in thrombosis, a patient's coagulation status may also affect FN levels. Normal serum levels range from 220–400 mg/dL.

Insulin-Like Growth Hormone-1 (IGF-1; Somatomedin C) Insulin-like growth hormone-1 (IGF-1) is a hepatic protein synthesized when stimulated by growth hormone. IGF-1 alterations have been observed in conjunction with several diagnoses, including cancer and Crohn's disease (Huang et al. 2005; Reimund et al. 2005). IGF-1 is often a component of protein status measurements in research settings (Ballard et al. 2005; Delgado et al. 2000; Shiraishi et al. 2005). It is thought that IGF-1 may be more effective than other markers in measuring nutritional status during the acute phase of stress, but at this time it is not routinely used in the clinical setting.

C-Reactive Protein (CRP) C-reactive protein is a positive acute phase protein that is released during periods of

BOX 5.4 CLINICAL APPLICATIONS

CBC and Differential

NAME: Sarah Henley DOB: 10/2
AGE: 31 SEX: F
PHYSICIAN: F. Bowman, MD

**************HEMATOLOGY**************

DAY: 1
DATE: 1/17
TIME:
LOCATION:

	NORMAL		UNITS
WBC	4.3–10	7.2	× 103/mm^3
RBC	4–5 (women) 4.5–5.5 (men)	3.8 L	× 106/mm^3
HGB	12–16 (women) 13.5–17.5 (men)	9.1 L	g/dL
HCT	37–47 (women) 40–54 (men)	33 L	%
MCV	84–96	72 L	μ^3
RETIC	0.8–2.8	0.2 L	%
MCH	27–31	23 L	pg
MCHC	31.5–36	28 L	g/dL
RDW	11.6–16.5	22 L	%
Plt Ct	140–440	282	× 10^3
Diff TYPE			
% GRANS	34.6–79.2	36.2	%
% LYM	19.6–52.7	41.3	%
SEGS	50–62	52	%
BANDS	3–6	4	%
LYMPHS	25–40	31	%
MONOS	3–7	3	%
EOS	0–3	0	%
TIBC	65–165 (women) 75–175 (men)	172 H	μg/dL
Ferritin	18–160 (women) 18–270 (men)	10 L	μg/dL
ZPP	30–80	18 L	μmmol/mol
Vitamin B_{12}	100–700	250	pg/mL
Folate	2–20	2	ng/mL
Total T cells	812–2318		mm^3
T-helper cells	589–1505		mm^3
T-suppressor cells	325–997		mm^3
PT	11–13	12	sec

Source: Reprinted from: Nahikian Nelms M, Long Anderson S. *Medical Nutrition Therapy: A Case Study Approach.* 2nd ed. Belmont, CA: Thomson Wadsworth; 2004. Page 8.

inflammation and infection. The levels of acute phase proteins generally increase as transport protein levels (such as prealbumin or albumin) decrease. Higher CRP levels have been associated with increased nutritional risk during stress, illness, and trauma (Murphy et al. 2000; Slaviero et al. 2003).

Immunocompetence

Historically, evaluation of immunocompetence has been included as a part of any discussion of protein and nutrition assessment. This is logical, since adequate and appropriate immune function is dependent in part on adequate protein status. Protein deficiency routinely results in increased risk of infection as well as altered immune and inflammatory response. In clinical practice, the use of this type of nutrition assessment is complicated by the presence of disease and infection, which of course also affects all components of the immune system. The most valuable use of these measures may be in predicting nutritional risk as well as outcome.

Delayed Cutaneous Hypersensitivity (DCH) Though this test is not commonly used in clinical practice, immune response can be determined by injection of a small amount of antigen **intradermally.** Antigens commonly used are mumps, **candida,** tuberculin, and **trichophyton.** Cell-mediated response to these antigens is measured by degree of redness and swelling at the site of injection. An individual who is unable to mount a sufficient immune response is considered to be anergic.

Total Lymphocyte Count (TLC) When evaluating a complete blood count (CBC) and differential count (Box 5.4), calculation for TLC can be completed as follows:

$$TLC = \frac{WBC \times \%lymphocytes}{100}$$

Total lymphocyte count will be affected by presence of infection, trauma, stress, and presence of disease such as cancer and HIV, as well as medications that influence the immune system (e.g., chemotherapy and corticosteroids). Most recently, a study of 161 elderly subjects indicated that

intradermally—refers to an injection into the corium or substance of the skin

candida—yeastlike fungi found in feces and skin, vaginal, and pharyngeal tissues; GI tract is most important source

trichophyton—pathogenic fungi causing dermatophytosis; attacks the hair, skin, and nails

TLC was not a sensitive marker for malnutrition in the elderly (Kuzuya et al. 2005).

Hematological Assessment

Evaluation of erythrocytes (RBC) can be an important component of nutrition assessment. Hematological assessment is key to diagnosis of all anemia types. A complete blood count includes measurement of total number of blood cells (erythrocytes) in the volume of blood. Many types of anemias exist, including those caused by iron deficiency, folate deficiency, or B_{12} deficiency and anemias arising from chronic disease such as renal failure and congestive heart failure. Anemias are diagnosed by evaluation of the complete blood count and by the microscopic evaluation of the size, shape, and color of erythrocytes. (See Chapter 21 for detailed information on nutritional and other anemias and their effects on blood cells.)

Hemoglobin (Hgb) Hemoglobin is a protein found in erythrocytes that functions to deliver oxygen to cells and to pick up carbon dioxide for expiration by the lungs. Measurement of hemoglobin is common in diagnosis of anemias, particularly iron-deficiency anemia. Additionally, hemoglobin is decreased in some chronic diseases and protein-energy malnutrition. Even though it is commonly measured, it is not the most sensitive or the most specific of hematological assessments of nutritional status. For example, in iron deficiency, iron stores may be depleted before serum hemoglobin levels will be affected.

Hematocrit (Hct) Hematocrit is defined as the percentage of blood that is actually composed of red blood cells. Hematocrit, like hemoglobin, will only be decreased in the final stages of iron deficiency. Hematocrit is affected by other nutrient deficiencies as well as hydration status.

Mean Corpuscular Volume (MCV) Mean corpuscular volume is a measure of the size of red blood cells. Because it reflects the average size of the red blood cell, the value for MCV will be changed in a variety of anemias. MCV is reduced in iron and copper deficiencies and elevated in folic acid and vitamin B_{12} deficiencies.

Mean Corpuscular Hemoglobin (MCH) Mean corpuscular hemoglobin estimates the amount of hemoglobin in each cell. These values can reflect total serum hemoglobin levels. In some situations, MCH can be normal while the number of red blood cells is low, resulting in low Hgb. Abnormalities are generally specific to iron deficiency and other nutritional anemias.

Mean Corpuscular Hemoglobin Concentration (MCHC) Mean corpuscular hemoglobin concentration also estimates the amount of hemoglobin in each red blood cell, but it expresses the value as a percentage.

Ferritin Ferritin is a storage form of iron; therefore, serum ferritin is an estimate of iron stores. Ferritin is a sensitive and specific measure of iron status and will be one of the first indices to change in iron deficiency.

Transferrin Saturation As discussed earlier under "Protein Assessment," transferrin is a serum protein responsible for transport of iron systemically. Each molecule of transferrin can transport two molecules of iron. Under normal conditions, approximately 30% of iron binding sites on the transferrin molecule are saturated. The body's requirement for iron and overall iron status will be reflected by changes in transferrin saturation. When iron status is low, transferrin is less saturated. Transferrin is calculated by using the ratio of serum iron levels to total iron binding capacity (TIBC). TIBC is the test used to measure the saturation ability for transferrin. TIBC is higher during iron deficiency and lower after repletion.

Protoporphyrin When there is inadequate iron available for hemoglobin synthesis, zinc is substituted for iron within hemoglobin. Consequently, zinc protoporphyrin levels rise during iron deficiency and are considered a sensitive measure of iron-deficiency anemia.

Serum Folate Coenzymes associated with folate are necessary factors for amino acid metabolism, including many one-carbon transfer reactions such as the conversion of histidine to glutamate. Folate coenzymes also play a crucial role in the synthesis of purine needed for DNA. Folate deficiency results in a megaloblastic anemia identical to that in vitamin B_{12} deficiency. In order to diagnose folate deficiency, the presence of megaloblastic, macrocytic red blood cells is noted. Serum folate and red cell folate are decreased. Serum B_{12} would be within normal limits. If folate levels are inadequate for conversion of histidine to glutamate, an intermediate product, formiminoglutamate, is formed. Urinary levels of formiminoglutamate (FIGlu) will be elevated and will serve as a diagnostic tool for folate deficiency.

Serum B_{12} Anemia associated with B_{12} deficiency can be diagnosed in several ways. Clinically, it will be similar to folate deficiency but can be distinguished by measuring serum B_{12} levels or by performing the Schillings test. In this test, B_{12} is given as an injection and the amount excreted in urine is measured. This allows distinction between different steps of B_{12} absorption (see Chapter 16). In recent studies, the most sensitive indicator for B_{12} deficiency was homocysteine and methylmalonic acid levels (Oh and Brown 2003).

Vitamin and Mineral Assessment

Laboratory tests are available for the assessment of most vitamins and minerals. These tests vary from high-performance liquid chromatography to radioimmunoassays. Most are cost prohibitive and are not routinely performed in the typical clinical setting (see Table 5.9).

Other Labs with Clinical Significance

Many other biochemical labs are routinely assessed and monitored. These may include measures of lipid status such as LDL cholesterol, HDL-C, or triglycerides. Total cholesterol is often <150 mg/dL in protein-energy malnutrition. Electrolytes, measures of blood urea nitrogen (BUN), creatinine (Cr), and serum glucose are components of routine admission labs. Depending on the patient's diagnosis, hydration status, and medical care, specific labs will be monitored by members of the health care team. See Table 5.10 for a summary of routine admission laboratory measurements.

Functional Assessment

Functional assessment focuses on measurements that assess skeletal muscle function or strength. Additionally, functional assessment could be expanded to include those activities that require adequate strength. For example, in the Subjective Global Assessment (SGA) (see Table 5.11 and Appendix B2), questions that focus on activities and function are included. The SGA identifies the patient's perception of his or her ability to accomplish self-care and the environment where the patient spends the majority of his or her time (e.g., bedridden, in chair, or normal activity). Another method of functional assessment is the identification of specific activities of daily living (ADL) and instrumental activities of daily living (IADL) related to

TABLE 5.9

Laboratory Tests of Vitamin and Mineral Nutrient Status

Test or Test Result	What It Reflects
For Anemia (General)	
Hemoglobin (Hgb)	Total amount of hemoglobin in the red blood cells (RBC)
Hematocrit (Hct)	Percentage of RBC in the total blood volume
Red blood cell (RBC) count	Number of RBC
MCV	Size of the RBC measured in cubic microliters
MCH	Hemoglobin concentration within the average RBC measured in picograms
Mean corpuscular hemoglobin concentration (MCHC)	Hemoglobin concentration within the average RBC measured as a percentage
For Iron-Deficiency Anemia	
(May also be reflected by ↓ or normal Hgb, ↓ or normal Hct, ↓ MCV, ↓ MCH)	
↓ Serum ferritin	Early deficiency state with depleted iron stores
↓ Transferrin saturation	Progressing deficiency state with diminished transport iron
↑ Erythrocyte protoporphyrin	Later deficiency state with limited hemoglobin production
For Folate-Deficiency Anemia	
(May also be reflected by ↓ or normal Hgb, ↓ or normal Hct, ↑ MCV, ↓ or normal MCH, ↓ MCHC)	
↓ Serum folate	Progressing deficiency state
↓ RBC folate	Later deficiency state
For Vitamin B_{12}–Deficiency Anemia	
(May also be reflected by ↓ or normal Hgb, ↓ or normal Hct, ↑ MCV, ↓ or normal MCH, ↓ MCHC)	
↓ Serum vitamin B_{12}	Progressing deficiency state
↑ Serum methylmalonic acid	Progressing deficiency state; inadequate B_{12} available for coenzyme activity

Source: Adapted from: Rolfes RR, Pinna K, Whitney E. Understanding normal and clinical nutrition. 7th ed. Belmont, CA: Thomson Wadsworth; 2006. Table E-2, p. E-11.

TABLE 5.10

Routine Admission Laboratory Measurements

Chem – 7 Panel	
BUN (blood urea nitrogen)	Glucose
Serum chloride	Serum potassium
CO_2 (carbon dioxide)	Serum sodium
Creatinine	
Chem – 20 Panel	
Albumin	Glucose
Alkaline phosphatase	LDH (lactate dehydrogenase)
ALT (alanine transaminase)	Phosphorus - serum
AST (aspartate aminotransferase)	Potassium test
BUN (blood urea nitrogen)	Serum sodium
Calcium - serum	Total bilirubin
Serum chloride	Total cholesterol
CO_2 (carbon dioxide)	Total protein
Creatinine	Uric acid
Direct bilirubin	
Gamma-GT (gamma-glutamyl transpeptidase)	

nutritional status. Different scales have been developed to measure an individual's ability to perform normal daily activities and are scored according to the specific assessment tool. Table 5.12 serves as an example checklist for these activities.

The most common specific assessment of muscle strength is handgrip dynamometry. This is a standardized, simple, and quick means of assessing nutrition in relation to skeletal muscle function. In this assessment, the patient is asked to grip the dynamometer (see photo on p. 130) device as tightly as possible. Handgrip standards are ≥35 kg for males and ≥23 kg for females. Handgrip measure has long been a part of fitness assessment but is now more common in nutrition assessment (Kenjle et al. 2005). This type of test is especially useful for long-term follow-up in outpatient or rehabilitation settings. The test is not valid in the presence of neuromuscular junction, muscle, or joint disease. Motivation to perform the test must be considered as well. Physical assessment can also assist in determining muscle strength and functional status.

Energy and Protein Requirements

The final component of a nutrition assessment is determination of energy and protein requirements for the patient. This is necessary in order to compare intake with needs

TABLE 5.11

Subjective Global Assessment

Scored Patient-Generated Subjective Global Assessment (PG-SGA) History

Patient ID Information

1. Weight:

In summary of my current and recent weight:

I currently weigh about _____ pounds

I am about ______ feet ______ inches tall

One month ago I weighed about ______ pounds

Six months ago I weighted about ______ pounds

During the past two weeks my weight has:

☐ decreased ☐ not changed ☐ increased

☐

2. Food Intake: As compared to my normal, I would rate my food intake during the past month as:

☐ unchanged

☐ more than usual

☐ less than usual

I am now taking:

☐ normal food but less than normal

☐ little solid food

☐ only liquids

☐ only nutritional supplements

☐ very little of anything

☐ only tube feedings or only nutrition by vein

☐

(continued on the following page)

TABLE 5.11 *(continued)*

3. Symptoms: I have had the following problems that have kept me from eating enough during the past two weeks (check all that apply):

- ☐ no problem eating
- ☐ no appetite, just did not feel like eating
- ☐ nausea ☐ vomiting
- ☐ constipation ☐ diarrhea
- ☐ mouth sores ☐ dry mouth
- ☐ things taste funny or have no taste ☐ smells bother me
- ☐ problems swallowing ☐ feel full quickly
- ☐ pain; where? ______________
- ☐ other* ______________________________

* Examples: depression, money, or dental problems

☐

4. Activities and Function: Over the past month, I would generally rate my activity as:

- ☐ normal with no limitations
- ☐ not my normal self, but able to be up and about with fairly normal activities
- ☐ not feeling up to most things, but in bed or chair less than half the day
- ☐ able to do little activity and spend most of the day in bed or chair
- ☐ pretty much bedridden, rarely out of bed

☐

Additive Score of the Boxes 1-4 ☐ A

The remainder of this form will be completed by your doctor, nurse, or therapist. Thank you.

5. Disease and its relation to nutritional requirements

All relevant diagnoses (specify) ______________________________

Primary disease stage (circle if known or appropriate) I II III IV Other ______________

Age ____________

6. Metabolic demand

☐ no stress ☐ low stress ☐ moderate stress ☐ high stress

7. Physical

Numerical score from Box 2 ☐

Numerical score from Box 3 ☐

Numerical score from Box 4 ☐

Global Assessment

- ☐ Well-nourished or anabolic (SGA-A)
- ☐ Moderate or suspected malnutrition (SGA-B)
- ☐ Severely malnourished (SGA-C)

Total numerical score of Boxes A + B + C + D ☐

(See triage recommendations below)

Clinician Signature ______________________________ RD RN PA MD DO Other ______ Date ______________

Nutritional Triage Recommendations: Additive score is used to define specific nutritional interventions including patient and family education, symptom management including pharmacologic intervention, and appropriate nutrient intervention (food, nutritional supplements, enteral or parenteral triage). First line nutrition intervention includes optimal symptom management.

- 0–1 No intervention required at this time. Reassessment on routine and regular basis during treatment.
- 2–3 Patient and family education by dietitian, nurse, or other clinician with pharmacologic intervention as indicated by symptom survey (Box 3) and laboratory values as appropriate.
- 4–8 Requires intervention by dietitian in conjunction with nurse or physician as indicated by symptoms survey (Box 3).
- ≥ 9 Indicates a critical need for improved symptom management and/or nutrient intervention options.

Source: Ottery FD, Kasenic S, DeBolt S, Roger K. Volunteer network accrues >1900 patients in 6 months to validate standardized nutritional triage. Abstract 282. Meeting of the American Society of Clinical Oncology, 1987. Programs/Proceedings of the American Society of Clinical Oncology Volume 17, 1998 © 1998 American Society of Clinical Oncology. Reprinted with permission.

TABLE 5.12

Activities of Daily Living (ADL)

Activities of Daily Living	
Bathing, showering	Personal device care
Bowel and bladder management	Personal hygiene and grooming
Dressing	Personal mobility
Eating	Sexual Activity
Feeding	Sleep/rest
Functional mobility	Toilet hygiene
Instrumental Activities of Daily Living	
Care of others	Health management and maintenance
Care of pets	Home establishment and management
Child rearing	Meal preparation and cleanup
Communication device use	Safety procedures and emergency responses
Community mobility	Shopping
Financial management	

Source: Reprinted from: AJOT (1978) by American Occupational Therapy Assoc. Copyright 2002 by Am Occupational Therapy Assn. Reproduced with permission of Am Occupational Therapy Assn in the format Textbook via Copyright Clearance Center.

Dynamometer

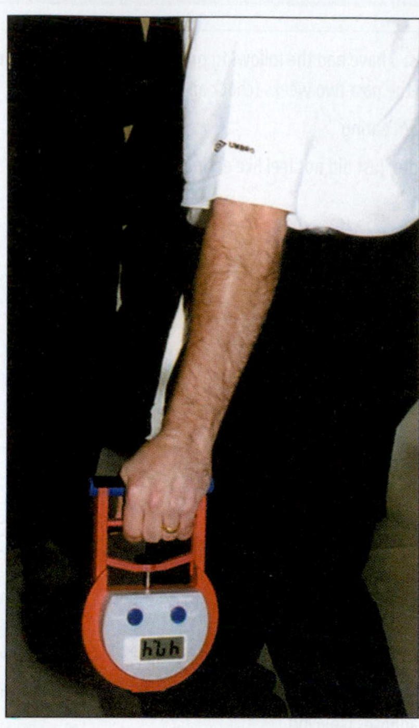

Source: URL for example: http://www.uel.ac.uk/hab/sports/exercise_physiology_facilities.htm

and to establish nutrition goals. In most clinical settings today, protein and energy requirements are estimated using a variety of established equations. In some situations, it may be possible to measure energy needs by using **calorimetry** and protein needs by performing a nitrogen balance study.

The total amount of energy required by an individual consists of three basic components: basal energy expenditure (BEE) or basal metabolic rate (BMR) + energy for physical activity or exercise (PA) + thermic effect of food (TEF) = total energy expenditure (TEE). Basal energy expenditure is defined as energy used for physiological functions to maintain life, such as respiration and heartbeat. Basal energy expenditure accounts for approximately 60% of an individual's energy requirement. When the term basal energy expenditure is used, it refers to a measurement of oxygen consumed by a patient who has gone without food for at least 12 hours and is lying down without movement in a constant temperature environment (Gropper, Smith, and Groff 2005). Due to these strict measurement requirements, actual basal expenditure is in a practical sense theoretical and thus difficult to measure. Therefore, in many discussions regarding energy requirements, the term REE (resting energy expenditure) or RMR (resting metabolic rate) is used. The term "resting" refers to measurement conditions where the individual is resting in a comfortable position without any other restrictions. RMR is usually estimated to be approximately 10% higher than BMR/BEE (Gropper, Smith, and Groff 2005).

Physical activity (PA) is the most variable portion of an individual's energy needs and fluctuates depending on the type, time, and intensity of physical activity. In most individuals, PA accounts for approximately 15–20% of energy requirements. TEF is estimated to be approximately 10% of an individual's caloric intake and represents the energy needed for absorption, transport, and metabolism of nutrient intake.

In hospitalized populations, many patients are hypermetabolic and have additional energy and protein requirements. These will be discussed in general here, but will be covered in more detail for specific disease states in the appropriate chapters within this text.

Measurement of Energy Requirements

The most accurate method of measuring REE/RMR in a clinical setting is to use indirect calorimetry. The equipment (metabolic carts) used for this process have steadily become more sophisticated over the last several decades, but the basic principles for measuring REE/RMR remain the same.

calorimetry—measurement of the flow of heat

Indirect Calorimetry

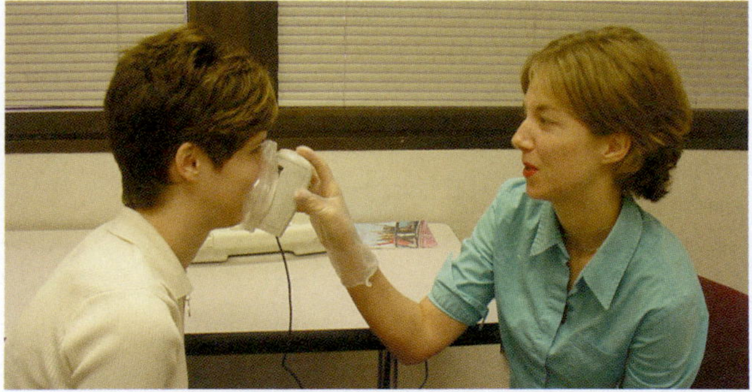

Source: Image used with the permission of the Medical University of South Carolina's Digestive Disease Center (www.ddc.musc.edu)

By measuring the amounts of oxygen and carbon dioxide that are both inspired and expired, VO_2 and VCO_2 can be calculated. These values then are converted to REE/RMR using computer software within the equipment. Most metabolic carts base their calculations on the Weir equation, which necessitates either measurement of nitrogen output or the use of a constant nitrogen output (Forette 2005). From this information, energy substrate utilization and REE can be estimated.

Measurement of Protein Requirements

A nitrogen balance assessment measures urine urea nitrogen and compares it to protein intake. This allows for some degree of measurement for protein requirements. A negative nitrogen balance reflects catabolism and can sometimes be due to inadequate protein intake. A positive nitrogen balance is consistent with anabolism and generally will indicate that the patient is receiving adequate amounts of protein to support current requirements. For those patients who have no renal function, measurement of nitrogen balance is accomplished through a urea kinetic modeling equation. See Chapter 20 for further details.

Estimation of Energy Requirements

As stated previously, it is more common in clinical situations to calculate an estimation of energy requirements. Clinicians use prediction equations to determine an individual's energy requirements. In a recent review, four prediction equations were identified as the most commonly used: Harris-Benedict, Mifflin-St. Jeor, Owen, and World Health Organization/Food and Agriculture Organization/United Nations University (WHO/FAO/UNU) (Frankenfield, Roth-Yousey and Compher 2005; Mifflin et al. 1990). Frankenfield, Roth-Yousey, and Compher's 2005 evidence analysis determined that the Mifflin-St. Jeor was the closest to measured energy expenditure. Other studies have found that the Harris-Benedict was within approximately 14% of measured REE (Frankenfield, Muth, and Rowe 1998; Frankenfield, Roth-Yousey, and Compher 2005).

The Mifflin-St. Jeor equation was established in 1990 and has been validated in over 10 different studies in the past decade (Mifflin et al. 1990). Historically, the most widely used standard for estimation of energy requirements is the Harris-Benedict equation, first published in 1919 (Harris and Benedict 1919). The Food and Agriculture Organization of the United Nations and the World Health Organization have also established equations to estimate basal energy requirements, which are also gender and age specific. All equations are outlined in Table 5.13.

Energy Requirements Based on DRI The Dietary Reference Intake for macronutrients are standards of intake that are age and gender specific and are designed to meet the nutrient requirements of about 98% of the healthy population. The DRI also include Estimated Energy Requirements (EER) that provide guidelines to meet the energy needs of approximately 50% of the healthy population. These values are not recommended for estimating energy requirements in individual patients in a clinical setting because energy requirements vary considerably from individual to individual, and hence the EER are not meant to be goals of nutrient intake for individuals. See the front inside cover of the book for these values.

TABLE 5.13

Estimation of Energy Requirements

Harris Benedict Equation

REE for females = 655.1 + 9.6 W + 1.9 H − 4.7 A

REE for males = 66.5 + 13.8 W + 5.0 H − 6.8 A

[W= weight in kg; H = height in cm; A= age in years]

Mifflin-St. Jeor Equation

Females: 10 W + 6.25 H − 5 Age − 161

Males: 10 W + 6.25 H − 5 Age + 5

[W = weight in kg; H = height in cm; Age = age in years]

FAO/WHO Basal Energy Estimation Equations

	Age	Equation*
Men	18–30 years	Kcal/day = (15.3 × weight) + 679
	30–60 years	Kcal/day = (11.6 × weight) + 879
	>60 years	Kcal/day = (8.8 × weight)+ (1128 × height) − 1071
Women	18–30 years	Kcal/day = (14.7 × weight) + 496
	30–60 years	Kcal/day = (8.7 × weight) + 829
	>60 years	Kcal/day = (9.2 × weight) + (637 × height) − 302

*Height in m; weight in kg

Source: Adapted from: Frankenfield D, Roth-Yousey L, Compher C. Comparison of predictive equations for resting metabolic rate in healthy nonobese and obese adults: a systematic review. J Am Diet Assoc. 2005 May;105(5):775-89. Review.

TABLE 5.14

Calculation of Energy Requirements: Activity and Stress Factors for Hypermetabolic Conditions

To calculate total energy requirements for the hospitalized patient: REE (Resting Energy Expenditure) × Activity Factor × Injury Factor

Activity Factors	*Average Injury Factors*
Out of bed 1.2	Surgery 1.0–1.3
Confined to bed 1.1	Infection 1.0–1.4
	Skeletal trauma 1.2–1.4
	Head injury 1.5

Source: Adapted from: Kudsk K, Sacks G. Ch.91. Nutrition in the care of the patient with surgery, trauma, and sepsis. In Shils M, ed. Modern Nutrition in Health and Disease. Tenth Ed. Philadelphia PA: Lippincott Williams & Wilkins, 2005. Long, CL. The energy and protein requirements of the critically ill patient. In Wright RA, Heymsfield Sb, eds. Nutritional Assessment. Boston: Blackwell Scientific, 1984.

Activity Factor After resting energy expenditure has been determined, energy used in activity also must be estimated in order to estimate total energy requirements. In hospitalized patients, activity is generally estimated using a figure of 1.2 for bedrest and 1.3 for all other patients, which is then multiplied by the REE. Specifically, REE is increased by 20% to account for PA. For example, if REE were 1350, an activity factor of 1.2 would determine the energy requirements to be 1620 kcal (see Table 5.14).

There are many methods used to estimate the amount of energy needed for physical activity, especially in the non-hospitalized population. One total energy requirement formula that was recently developed by the Food and Nutrition Board (Food and Nutrition Board 2002) incorporates a physical activity coefficient. Both the CDC and the American College of Sports Medicine use the exercise metabolic rate, or MET, to estimate the amount of energy used in various physical activities (U.S. Department of Health and Human Services 1999). One MET is estimated to be the energy expenditure for sitting quietly, which for the average adult approximates 3.5 mL of oxygen uptake per kilogram of body weight per minute (1.2 kcal/min for a 70-kg individual) (Ainsworth et al. 1993). An overview of physical activities and the equivalent energy expended is outlined in Table 5.15.

Stress Factors Disease, infection, and trauma all can affect an individual's energy requirements. Hospitalized patients can be hypermetabolic; estimation of their energy needs should take this fact into account. When indirect calorimetry is not available, it is common practice to estimate energy needs using a stress factor in addition to an activity factor if the patient is ambulatory. Unfortunately, these stress factors have not been validated consistently, and thus estimations of energy requirements during stress have been modified considerably over the past decade as understanding of the

TABLE 5.15

Energy Requirements of Common Daily Activities*

Leisure Activities	*METs†*
Mild	
Playing the piano	2.3
Canoeing (leisurely)	2.5
Golf (with cart)	2.5
Walking (2 mph)	2.5
Dancing (ballroom)	2.9
Moderate	
Walking (3 mph)	3.3
Cycling (leisurely)	3.5
Calisthenics (no weight)	4.0
Golf (no cart)	4.4
Swimming (slowly)	4.5
Walking (4 mph)	4.5
Vigorous	
Chopping wood	4.9
Tennis (doubles)	5.0
Ballroom (fast) or square dancing	5.5
Cycling (moderately)	5.7
Skiing (water or downhill)	6.8
Climbing hills (no load)	6.9
Swimming	7.0
Walking (5 mph)	8.0
Jogging (10 min mile)	10.2
Rope skipping	12.0
Squash	12.1
Activities of daily living	
Lying quietly	1.0
Sitting; light activity	1.5
Walking from house to car or bus	2.5
Loading/unloading car	3.0
Taking out trash	3.0
Walking the dog	3.0
Household tasks, moderate effort	3.5
Vacuuming	3.5
Lifting items continuously	4.0
Raking lawn	4.0
Gardening (no lifting)	4.4
Mowing lawn (power mower)	4.5

* These activities can often be done at variable intensities, assuming that the intensity is not excessive and that the courses are flat (no hills) unless so specified. Categories are based on experience or tolerance; if an activity is perceived to be more than indicated, it should be judged accordingly.

† MET indicates metabolic equivalent. One MET is the amount of energy used when sitting quietly.

Source: Reprinted with permission of Lippincott Wiliams & Wilkins. Source: Fletcher GF, Balady GJ, Amsterdam EA, et al. Exercise standards for testing and training: a statement for healthcare professionals from the American Heart Association. Circulation 2001; 104:1694-740. Available from: http://www.hsph.harvard.edu/nutritionsource/Exercise.htm

stress response has deepened. Excessive caloric loads and overfeeding may be much more detrimental than underestimation of energy needs. Table 5.15 estimates stress factors for hypermetabolic conditions.

Additionally, in specific clinical situations certain regression equations have been established but are not commonly used. These include the Ireton-Jones equation used for patients requiring mechanical ventilation (Curreri et al. 1974; Ireton-Jones et al. 1992). The use of these equations will be discussed in appropriate detail within the context of nutrition therapy for a specific diagnosis throughout this text.

Estimation of Protein Requirements

RDA for Protein The Recommended Dietary Allowances provide the best reference for protein requirements in the nonstressed population. For adults, this level is set at 0.8 g protein/kg of body weight. Additional levels are set for infants and children. These can be found on the inside front cover of this book.

Protein Requirements in Metabolic Stress, Trauma, and Disease Protein requirements, like energy requirements, are affected by metabolic stress, trauma, and disease. The type of protein and the need for specific amino acids may additionally be altered within certain diseases. These will be discussed within the context of those specific diagnoses throughout the text. In general, though, if patients are receiving adequate kcal, protein requirements can be met by providing 1.0–1.5 g protein/kg.

Protein-Kilocalorie Ratio Another traditional approach to estimating protein requirements is based on the concept that energy should be provided by lipids and carbohydrates, and protein intake should be reserved for synthesis requirements. Historically, many clinicians have calculated nonprotein kcal using carbohydrates and lipids to provide calculated energy requirements. Protein estimations were considered separately. For healthy individuals, the ratio of 1:200 is recommended. In individuals who have higher protein requirements, ratios range from 1:150 to 1:100.

The subject is controversial and has fueled many professional debates. No standardization of this process currently exists. It is important to realize that biochemistry does not support this practice, since it is known that certain cells use protein for fuel (e.g., enterocyte use of glutamine). Furthermore, overfeeding carbohydrates and lipids can jeopardize medical care by interfering with the function of the lungs, liver, and immune system. This approach is not superior to other systems of estimating protein requirements and does not serve any additional advantage in nutrition assessment.

Interpretation of Assessment Data: Nutrition Diagnosis

In the Nutrition Care Process, after all components of nutrition assessment have been compiled, the clinician determines the nutritional status of the patient and identifies the specific nutrition-related problems. An essential guideline to use for diagnosing the level and type of malnutrition is the International Classification of Disease (World Health Organization 1992), which provides criteria for diagnosing specific types of malnutrition, including kwashiorkor, marasmus, other severe PEM (protein-energy malnutrition), malnutrition of a mild degree, and malnutrition of a moderate degree.

Nutrition diagnoses arise from the nutritional problems identified during the nutrition assessment (ADA 2006; Lacey and Pritchett 2003). The specific criteria for each nutrition diagnosis can be found in Chapter 3. Using the nutrition assessment information, the clinician identifies the nutrition-related problems, determines the probable cause for each of these problems, and substantiates the problem through specific signs and symptoms. The PES (problem, etiology, symptoms) format, as discussed in Chapter 3, is the recommended manner in which to document the nutrition diagnosis.

Conclusion

Health is strongly influenced by nutritional status. It is no surprise that the Joint Commission for Accreditation of Hospital Organizations requires that all patients receive nutrition screening within 48 hours of their hospital admission. No one test exists to assess nutritional status. In this chapter, numerous measures of dietary intake, physical health, and biochemical status have been presented; in practice, these measures allow the clinician to determine not only current nutritional state but also future nutritional risk. Accurate measurement and interpretation of nutritional status can allow a nutritional diagnosis to be made and interventions that improve patient outcomes to be put in place, ultimately decreasing morbidity and mortality.

WEB LINKS

Agricultural Research Service: Nutrient database of 13,000 commonly consumed foods in the United States.

www.ars.usda.gov/foodsearch

Aim for a Healthy Weight: A link from the National Heart, Lung, and Blood Institute provides excellent information for assessment of weight.

http://www.nhlbi.nih.gov/health/public/heart/obesity/lose_wt/index.htm

Dietary Guidelines for Americans: This link is also part of the mypyramid.gov website. Provides all of the background and supporting materials for the 2005 U.S. Dietary Guidelines.

http://www.mypyramid.gov/guidelines/index.html

International Network of Food Data Systems: This website is part of the international Food and Agriculture Organization (FAO). Numerous links for international sources of food composition and nutrition are provided.

http://www.fao.org/infoods/software_en.stm

MyPyramid.gov: United States Department of Agriculture (USDA) website outlining all materials pertaining to the new customizable food pyramid. Provides individual client information as well as information for the professional to use in teaching and counseling.

http://www.mypyramid.gov/professionals/index.html

National Heart, Lung, and Blood Institute: This government website provides patient education material as well as clinical practice guidelines.

http://www.nhlbi.nih.gov/guidelines/index.htm

National Institutes of Health and Office of Dietary Supplements: New database that allows assessment of foods, beverages, and dietary supplements.

www.nal.usda.gov/fnic/foodcomp

The Food Safety Risk Analysis Clearinghouse: The Food Safety Risk Analysis Clearinghouse is the responsibility of the Joint Institute for Food Safety and Applied Nutrition (JIFSAN), a collaboration of the University of Maryland (UM) and the Food and Drug Administration (FDA). This site provides a wealth of nutrition assessment forms and tools as well as links for nutrient analysis.

http://www.foodrisk.org/nutrition_assessment_tools.cfm

The National Guideline Clearinghouse™: The National Guideline Clearinghouse (NGC) is a comprehensive database of evidence-based clinical practice guidelines and related documents. NGC is an initiative of the Agency for Healthcare Research and Quality (AHRQ), U.S. Department of Health and Human Services. NGC was originally created by AHRQ in partnership with the American Medical Association and the American Association of Health Plans (now America's Health Insurance Plans [AHIP]). The NGC mission is (1) to provide physicians, nurses, and other health professionals, health care providers, health plans, integrated delivery systems, purchasers and others an accessible mechanism for obtaining objective, detailed information on clinical practice guidelines and (2) to further their dissemination, implementation, and use.

http://www.guideline.gov/summary/summary.aspx?ss=15&doc_id=3625&nbr=2851

United States Department of Agriculture: This website provides online nutrient analysis that can be used for nutrition assessment.

USDA Agriculture Research Station:

http://www.ars.usda.gov/Services/docs.htm?docid=7783

END-OF-CHAPTER QUESTIONS

1. What is the difference between nutritional status and nutritional risk?
2. How is nutritional screening different from nutritional assessment?
3. Describe the difference between subjective data and objective data that are collected for a nutritional assessment. List three pieces of objective information and three pieces of subjective information that could be collected for nutritional assessment.
4. Name and briefly describe four methods used to collect dietary assessment data. List the advantages and disadvantages of each method.
5. Describe two methods that are used to analyze dietary intake.
6. Which anthropometric measurements are collected for nutritional assessment? Briefly describe each measurement and explain the accuracy of each in determination of body composition and/or health status.
7. List four blood proteins used in nutrition assessment. Describe the effectiveness of each as markers in measuring nutritional status.
8. Describe how energy requirements can be determined or estimated. How is the energy requirement affected by stress?
9. List the hematological measurements collected for nutritional assessment.

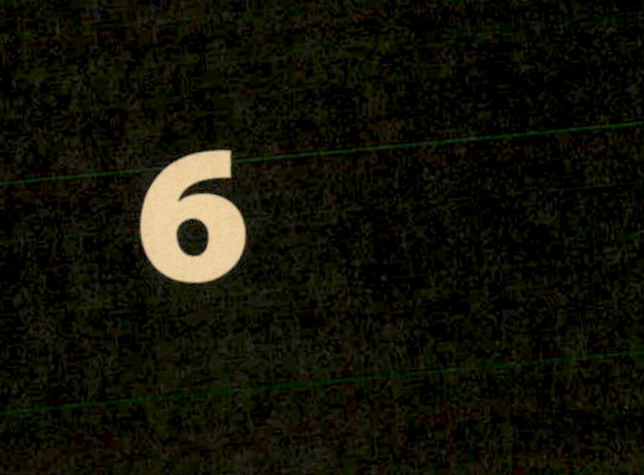

Documentation of Nutrition Care

R. Gerald Nelms, Ph.D.

Associate Professor Rhetoric and Composition, Department of English, Southern Illinois University

CHAPTER OUTLINE

An Overview: Writing in the Profession

Charting: Documentation of the Nutrition Care Process

Standardized Language and Medical Abbreviations • Medical Records • Organization of Nutrition Documentation • Keeping a Personal Medical Notebook • Guidelines for All Charting

Writing for Nonmedical Audiences

Instructional Materials for Patients, their Families, and the Public

Reporting Your Own Research

Conclusions: Your Ethos—Establishing Expertise

An Overview: Writing in the Profession

Students often are surprised to learn how much writing they have to do in clinical settings. As explained in Chapter 3, the nutrition care process (NCP) consists of four interrelated and connected steps: (1) nutrition assessment, (2) nutrition diagnosis, (3) nutrition intervention, and (4) nutrition monitoring and evaluation. Each of these steps includes writing as a vital part of the process (Lacey and Pritchett 2003). The amount of writing and "paperwork" can be extraordinary. In fact, health professionals probably ought to do even more than they do, because keeping personal notes in addition to required documentation of nutrition care provides greater opportunities for insights—and thus better diagnoses and more effective treatment plans.

The Functions, Context, Parts, and Processes of Writing

Writing *functions* in both our personal and professional lives in a variety of different ways:

- To record information, such as agreements among people, documentation of health care in a medical facility, or documentation for legal purposes
- To inform, either as a report of an experience or research findings, or as dissemination of information developed by others
- To persuade
- To entertain

Medicine, including nutrition and dietetics, uses its own language, jargon, and contexts in which practitioners function every day. Sometimes, health professionals may lose sight of the fact that the rest of the "nonmedical" world doesn't participate in, or perhaps even understand, these conversations. Communication is necessary to keep everything moving forward, and thus, writing is a crucial skill that all newcomers to professions must learn.

Three important areas of writing need to be understood: the rhetorical norms—that is, the universal contextual framework within which all communication exists; the levels of discourse—that is, the different levels of writing that one can focus attention on; and writing processes—that is, the cognitive processes and stages of writing.

Rhetorical Norms Each form of writing within the workplace is produced within a context, and every context involves at least four elements:

- *Subject Matter:* What the text is about
- *Purpose:* A reason for writing the text
- *Audience:* A set of readers to whom the text is directed
- *Ethos:* The personality or voice that comes through the text and characterizes the writer for the reader, that is the person the reader assumes the writer to be, based on the ideas, organization, and style of the text.

For example, in medical record documentation, the subject matter is the nutrition care process for a patient. The purposes may include not only the legal documentation of care, but also communication with other health care providers or collection of research data. These purposes will influence the writing style, because the presentation of this information will need to be appropriate for the audience. Your professionalism, clinical knowledge, and expertise will characterize the ethos of the writing. Reflecting on these contextual elements—subject matter, purpose, audience, and ethos—before and during the production of a text can improve its effectiveness.

Levels of Discourse Just as every text exists within a context of rhetorical norms, it also consists of an overall organization of words, sentences, punctuation, paragraphs, and larger passages, all of which communicate the writer's ideas, the content of the writing. These items—rhetorical norms, ideas, organization, and grammar (the sentence structures, punctuation, word choice, and spelling)—are generally referred to as the levels of discourse. Decisions regarding the rhetorical norms at the highest level—that is, decisions about the purpose, audience, subject matter, and writer's ethos—will determine decisions at these other, lower levels of discourse. During the process of writing, successful writers will move—sometimes effortlessly, it seems—between and through these levels of discourse. As they write, successful writers focus on different levels of discourse at different times, as needed. They may be generating ideas at one moment, then reorganizing sections at another. They may consider their choice of particular words and then move on to generating and then revising sentences, and determine the correct punctuation all the way through.

Steps in the Writing Process While texts appear linear, moving from a beginning to a middle to an end, the actual process of writing is virtually never that linear. Successful writers are always stopping and returning to already produced sentences and ideas in order to reorient themselves so that they can go forward again. They may pause to make sure that what they have written will lead the reader logically from one thought to the next. They may pause to reestablish their own connection to the logical flow of their writing, to give themselves direction. A better metaphor for writing than the straight line is the spiral, because writing tends to be "recursive"—that is, writing moves backward in order to then move forward again. Composition scholars have identified the following major steps in the process of writing:

Prewriting or invention, whatever it is that the writer does before actually writing. For longer pieces of writing, successful writers tend to spend time planning a text before beginning to write it. They develop their ideas about the subject matter. For example, in planning a new patient education resource, RDs may read the most current literature supporting nutrition therapy for a particular diagnosis. They identify their purpose and audience. They produce a rough plan (sometimes in writing, sometimes just in their heads) for how the text is to be organized. From that point, an outline may be developed that organizes the most important issues for the patient.

Drafting, or actual sentence generation. As mentioned above, this is not a linear process. It involves setting goals and subgoals and recalibrating one's thinking as more and more text is produced. Successful writers tend to do some revising during drafting, too.

Revision. In addition to revising while drafting, successful writers virtually always make revisions to their text after it has been completely drafted.

Editing. After the text has been drafted and revised, successful writers also spend time making minor corrections at the lowest, sentence and subsentence levels of discourse. This is when proofreading takes place, and these writers make their changes based on errors they identify during that proofing of their manuscripts. Typically, successful writers wait to make editing changes to a text until revision is complete, to avoid wasting time proofing sentences that may later be revised or even eliminated from the text.

As indicated, all writing involves subject matter, a purpose, an audience, and the writer's ethos. But as writers move from being outsiders to being insiders within their communities, they adopt the discourse conventions of their communities—that is, the purposes, audiences, ethos, subject matter, writing processes, textual organizations, and writing styles specific to their communities. These differences in the writing of different communities—different disciplines, different workplaces—coalesce into what are referred to as "genres." Clinical dietitians typically write into *charts* (called charting), but they also write memos, brochures, handouts, and other health information texts as well as research reports. These are all different genres adopted by the dietetics profession.

Charting: Documentation of the Nutrition Care Process

Every medical institution or agency creates a medical record or chart for each individual patient served by that institution or agency. A medical record is a systematic recording of a patient's care, a location in which all data relating to the problem that brought the patient in for medical care are collected. But a medical chart also represents ongoing conversations among the different members of the medical team working on an individual patient's care. The aim of this ongoing conversation is the creation of consensus about the appropriate care and treatment for the patient. The chart is the basis for determining patient care, for documenting communication among health professionals dedicated to that individual patient's care, and for keeping a clear and comprehensive record of all that is done for the patient for legal reasons (see Box 6.1). Several conventional forms for charting have been developed. Different medical institutions and agencies will prefer different forms. When entering a particular workplace, the novice dietitian should make sure to find out what form of charting is used. The information that is documented will be consistent, but the way that information is organized may differ.

The driving forces that impact medical record keeping also include accrediting agencies for health care facilities (including the Joint Commission for Accreditation of Healthcare Organizations/JCAHO), continuous quality improvement programs, and insurance reimbursement for medical care. Decisions regarding how to organize information and what language to use are guided by the need to meet the specific standards of these agencies.

In order for a health care provider to receive payment, the patient's medical record at discharge must contain documentation of the correct DRG (diagnosis-related groups) and indicate that the patient received the appropriate care (see Chapter 1). Clear, concise wording in the medical record, using terminology similar to that used in a prospective payment system, will facilitate reimbursement for services. Trained medical coders are responsible for assigning codes for diagnosis and treatment so that reimbursement occurs.

Medical record charts are used to audit and monitor the health care provided by an institution or a specific group of health care providers. Each state licensing agency, as well as the Joint Commission on Accreditation of Health Care Organizations (JCAHO), requires that all health care facilities monitor, evaluate, and seek ways to improve the quality of care for their patients.

Finally, it is important to remember that the medical record is a legal document. The record serves as a description of exactly what happened during the medical care. Clients frequently request copies of their medical records, and they have the right to read those records. Each institution has policies for controlling the manner in which records are shared. Future clinicians should be aware that institutions and agencies are increasingly moving toward electronic record keeping. Thus, a large percentage of charting is being done directly into the electronic medical record (Dove 2005). This often means that the format for charting is preset, and you have no real choice regarding this organization (see Table 6.1 for a sample template). If this is the case, make sure that you are trained in how to enter charting notes into the system, and ask to read some charts to get an idea of how patient care is charted in your workplace. Familiarity with the various charting formats discussed in this chapter will help you understand the format and conventions of the charting system you are expected to use, whether it is paper-based or electronic.

BOX 6.1 CLINICAL APPLICATIONS

Purposes of Medical Record Charting

- Legal documentation of medical care that the client has received
- Communication between members of the health care team
- Evaluation of medical care for that client
- Funding and resource management
- Continuous quality improvement
- Third-party reimbursement
- Accreditation
- Research

TABLE 6.1

Sample Template for Electronic Medical Record

Blank template—computer fills in requested data.

S: Patient states
Intake:

O: DIET:
AGE: 62 HT: 69 in WT: 192 lb (2/17/2006)
IBW: BMI: 28.4
DIAGNOSIS:
Labs: Glucose: 61 mg/dL L (02/17/06 11:56)
Cholesterol: 180 mg/dl (02/17/06 11:56)
Triglyceride: 83 mg/dl (02/17/06 11:56)
Albumin: 4.7 g/dl (02/17/06 11:56)
TLC: 900mm^3

A: Nutrition status is __________ based upon __________.

P: Care Plans:

Source: Courtesy of Veteran's Administration Medical Center Nutrition Services–Marion, IL

TABLE 6.2

Unacceptable Medical Abbreviations[1]
Note: See Appendix A1 for generally accepted medical abbreviations

Do Not Use	Potential Problem	Use Instead
U (unit)	Mistaken for "0" (zero), the number "4" (four), or "cc"	Write "unit"
IU (International Unit)	Mistaken for IV (intravenous) or the number 10 (ten)	Write "International Unit"
Q.D., QD, q.d., qd (daily)	Mistaken for each other	Write "daily"
Q.O.D., QOD, q.o.d, qod (every other day)	Period after the Q mistaken for "I" and the "O" mistaken for "I"	Write "every other day"
Trailing zero (X.0 mg)*	Decimal point is missed	Write "X mg"
Lack of leading zero (.X mg)		Write "0.X mg"
MS, MSO_4, and $MgSO_4$	Can mean morphine sulfate or magnesium sulfate; mistaken for each other	Write "morphine sulfate" or "magnesium sulfate"

[1] Applies to all orders and all medication-related documentation that is handwritten (including free-text computer entry) or on preprinted forms.

*Exception: A "trailing zero" may be used only where required to demonstrate the level of precision of the value being reported, such as for laboratory results, imaging studies that report size of lesions, or catheter/tube sizes. It may not be used in medication orders or other medication-related documentation.

Source: Copyright © Joint Commission on Accreditation of Healthcare Organizations. Oakbrook Terrace, IL. Reprinted with permission.

Standardized Language and Medical Abbreviations

Steps to assure accuracy of the medical record include the use of standard language and medical abbreviations. Typically, each health care facility designates a list of acceptable abbreviations. The JCAHO recently recommended that certain abbreviations not be used because they are more likely to contribute to patient care errors. See Table 6.2 for that list.

As discussed in Chapter 3, the American Dietetic Association has recently begun to standardize dietetics language through the development of the Nutrition Diagnostic Terminology (Lacey and Pritchett 2003; ADA 2006). A recent study indicated that consistent use of the Nutrition Care Process and standardized language within two midwestern hospitals resulted in improved documentation of nutritional care. The study authors predicted that adopting standardized language terms will improve accountability and reimbursement and will enhance patient care overall (Hakel-Smith, Lewis, and Ethridge 2005).

Medical Records

One common type of medical record is the problem-oriented medical record (POMR). The POMR is divided into five parts: data, problem list, care plan, progress notes, and discharge summary. The data is a collection of subjective and objective information about the patient and is the basis of the problem list. As the problem list is established, a care plan is constructed for each problem. The plan should include: expected outcomes, plans for further data collection, and, if needed, a patient teaching plan. Each health care team member composes progress notes in a narrative format. These notes are used in monitoring a client's care. The frequency of the entry of progress notes is determined by the facility's policies and procedures as well as the individual care plan. Finally, the discharge summary addresses each problem on the problem list and notes whether it was resolved or not. If it was not resolved, a plan is developed to treat the problem after discharge. Such a plan may provide for communication with other facilities, home health agencies, and the patient. The discharge summary often is the only part of the hospital record that other facilities (such as long-term care) receive. Providing accurate nutrition documentation will help ensure that consistent care is given to patients with nutrition problems.

Organization of Nutrition Documentation

Nutrition information within the medical record may be organized using any of several different styles. No matter what style is preferred by the practitioner or the institution, the same data is used to document the Nutrition Care Process. The following sections describe the numerous organization styles used in nutrition documentation.

SOAP SOAP is the oldest and most well known form of medical documentation. It has historically been the format that dietitians have used for daily progress notes. The label "SOAP" refers to the four sections of each entry in the medical chart: subjective data, objective data, assessment, and plan.

Subjective data (see Table 6.3) includes patient information or data collected from the patient or caregiver. This information can be placed into four major categories: diet-related information, lifestyle/psychosocial or emotional information, medical history information, and learning/

TABLE 6.3

Subjective Section of SOAP Note

S – Subjective	
Diet related	
Eating habits and feeding abilities	Food allergies/aversions
Use and fit of dentures	Vitamin, mineral, and nutrient supplement intake
Appetite and digestion problems	Also, complementary/alternative nutrition therapy
Nausea, vomiting, constipation, diarrhea, heartburn, physical problems interfering with adequate oral intake	Nutritional history and family nutritional history
	Adequacy of prior dietary intake
Recent weight change	Method of obtaining foods/nutrients (e.g., Meals on Wheels)
Diet history/previous diet modification or Rx	Previous nutrition education/counseling
Usual pattern of food intake	
Lifestyle/psychosocial/emotional	
Economic situation/income	Smoking
Ability to purchase/prepare/store food	Interaction with/between other family members or caretakers
Living or eating alone	Support systems
Health promotion and exercise practices	Coping mechanisms
Exercise	Occupation
Medically related	
Personal and family medical history	Medications, previous to admission or current PE
Especially, diseases with nutritional implications (e.g., type 2 diabetes)	Prescriptions, OTC (antacids, laxatives, etc.), and any CAM medications
Use of complementary/alternative medical therapies (CAM)	Other physically related problems
Learning and motivation related	
Ability to communicate in English (speaking, comprehending, reading, and writing)	Educational level
Patient's comments about previous prescribed diets/medical tx and compliance issues	Communication patterns
Psychosocial problems, including addiction	Attention span
Perception of health status/reasons for seeking health care	Long-term and recent memory
Desire to improve health or be involved in their own treatment or treatment decisions	Readiness to learn
Learning style/problem-solving abilities	Barriers to change
Intellectual performance	Growth and maturation

motivation information. This section may include symptoms expressed by the patient; descriptions by the patient of her or his pain, discomfort, and/or dysfunction; dietary history; or the presence of symptoms that interfere with the ability to eat.

Objective data (see Table 6.4) includes the empirical information—that is, information drawn from physical tests and medical staff observations that are of consequence to the patient's nutritional status. This information can come from physical examinations, X-ray examinations, other imaging techniques, or biochemical tests that are of nutritional relevance. Examples of objective data would be the patient's age, gender, and anthropometric information. It could include information from the physical examination

TABLE 6.4

Objective Section of SOAP Note

O - Objective	
A	Age, ethnicity, gender, height/weight, BMI; any anthropometric measurement in addition to height and weight.
B	Biochemical lab results that are of nutritional relevance
C	Clinical Diagnosis, medication, treatment orders (including diet orders), any additional clinical findings of nutritional relevance
D	Dietary information, including current intake that has been observed (not subjective) or analysis of diet quality; protein/kcalorie requirements

such as temperature, pulse, blood pressure, or respiratory rate. Medical diagnosis and current medical care are noted here. Objective data can also include sensory information noted by the medical staff member such as smells, how an organ feels during a physical exam, or the visual recording of patient skin coloration.

Assessment (see Table 6.5) is the nutrition diagnosis or interpretation of the patient's nutrition problems. Assessment should include the problem, stated with the supporting data for the nutrition diagnosis. In the assessment section of a SOAP note, conclusions are drawn from the subjective and objective data in order to support the nutrition diagnosis.

Plan (see Table 6.6) will include an outline of interventions necessary to treat each nutrition problem. The plan may include requests for additional information needed to address the patient's nutrition problems. Specific nutrition therapy recommendations are stated here. Finally, goals and objectives may be included with a specific measure and timeline for evaluation of the intervention.

The other health care team members involved in determining and carrying out the individual patient's treatment constitute the audience for SOAP notes. Their purpose is to help create a continuity of appropriate treatment for the patient. The ethos for SOAP notes must be authoritative, knowledgeable, and professional. All other forms of charting are either extensions of SOAP notes or reductions of SOAP notes (see Table 6.7 for a sample SOAP note).

IER Notes IER is a simplified version of SOAP (Klein, Bosworth, and Wiles 1997):

- *Intervention* refers to what has been done for the patient and the patient's response to that treatment.
- *Evaluation* refers to the assessment part of SOAP, the diagnosis and evaluation based on the data gathered. This section often includes a brief summary of the plan and an evaluation of the treatment's effectiveness.
- *Revision of care* refers to any changes recommended or ordered in the patient's treatment. Table 6.8 provides an example of IER charting.

TABLE 6.5

Assessment Section of SOAP Note

A - Assessment (Nutritional Diagnosis)
1. Current nutritional Problems (and medical problems that impact current or long-term nutritional status), Etiology, and Symptoms (PES) 2. Potential nutritional problems (due to prognosis or clinical course of the disease, noncompliance and/or drug nutrient interactions) 3. Prioritization of the nutrition diagnosis

TABLE 6.6

Plan Section of SOAP Note

P – Plan (Nutrition Intervention)	
Gather	Additional information you would need or would like (for current and potential nutritional problems; for instance, whether the patient is lactose intolerant)
Referral	Referral to other health or social professional (examples: psychologist for an eating disorder or depression; social worker if patient is homeless)
Nutrition	Specific nutritional recommendations for the client/patient to address *current* nutritional problem(s) (these may be different than those stated under Assessment; for example, fewer kcalories to help achieve weight loss)
Goals/Education	1. What is/are your short-term goal(s) for this client/patient? 2. For each goal, state the expected outcome(s) of dietary compliance as behavioral objectives for change, or expected outcome of the nutrition support. (Remember that outcomes should be measurable; specify a time frame and criterion [by how much], and encourage client participation, if possible.) • Nutrition support to be recommended (when it is a medical procedure). • Visuals, models, printed material to be given or used, if appropriate. Example: 2. (Goal) Patient will increase dietary fiber consumption. • Patient will eat whole-grain bread instead of white bread and increase consumption by one additional fruit and vegetable every day. • Patient will be given a handout on whole-grain products.
Evaluation—(Nutrition Monitoring and Evaluation)	
When and how will you evaluate the outcome of your Nutrition Plan Goals?	

TABLE 6.7

Sample SOAP Note

Nutrition
12/27/05 11:30 a.m.

S:	Patient's mother relates that Denise's mouth hurts so badly that she can hardly talk. She has had limited oral intake. Patient's mother also describes an "anti-cancer" diet that Denise's aunt and uncle introduced them to. Denise states that she doesn't want to make anyone mad, but those foods on the anti-cancer diet make her mouth hurt worse, and she doesn't know what to believe.
O:	21 yo ♀ Dx: Stage II diffuse large B-cell lymphoma. Admitted with immunosuppression, fungal infection, dehydration R/O pneumonia. s/p first round of chemotherapy/CHOP Ht. 5′6″ Wt. 108# Last adm wt:120# Preillness wt: 130# Labs: Alb 3.0 WBC 1100mm^3 EER: 1700–1800 kcal EPR: 75–80 g protein
A:	Patient has experienced an unintentional 22 lb. wt loss over a three-month period. Albumin indicates a mild to moderate depletion of visceral protein stores. Current PO intake is limited by mucositis 2^0 to fungal infection. Prior to this admission, PO intake had further been limited by adherence to an "anti-cancer" diet that severely limits caloric intake and variety of foods. Pt. also indicates vitamin/mineral supplement intake prior to admission. Nutritional assessment confirms protein-kcalorie malnutrition to a mild degree.
P:	1. Modify PO diet to accommodate soft, easily chewed foods and liquids. Increase nutrient density with use of modular supplements and high kcalorie/high protein liquid supplements as patient tolerates. 6–8 small feedings/day. 2. Initiate calorie count to monitor adequacy of PO intake. 3. Recommend low bacterial/neutropenic precautions. 4. Consult with nursing/MD to assure adequate pain coverage before attempts at oral intake. 5. Provide evidence-based information about all of the components of the diet recommended by her family members for Denise. 6. Monitor daily through patient visitation, calorie counts, daily weights to assess adequacy of current interventions.

Signature:
M. Nahikian-Nelms, Ph.D.,–R.D.,–L.D.

TABLE 6.8

Sample IER Note

12/27/05 11:30 a.m. Nutrition Progress Note

Intervention:

Modify oral intake to 6-8 small feedings; increased nutrient density; addition of high-calorie, high-protein supplement; modification of texture; pain medication prior to meals and supplements.

Evaluation

Per calorie counts and patient visitation, oral intake improved, currently meeting 65% of estimated energy and protein requirements. No further weight loss documented since admission.

Revision

Check prealbumin to monitor visceral protein status.

Focus Notes Focus notes are a blending and reduction of the SOAP and IER formats (Klein, Bosworth, and Wiles 1997):

- *Data* simply collapses SOAP's subjective and objective data sections.
- *Action* refers to the SOAP assessment and IER evaluation sections—that is, the diagnosis and evaluation, based on the data gathered, and the treatment(s) applied.
- *Response* represents the SOAP plan and/or any changes in treatment, the same thing as the IER revision of care. See Table 6.9.

PIE Notes PIE (problem, intervention, evaluation) notes are also a blending of other types of charting:

- *Problem* identifies the specific nutrition problems for the client.
- *Intervention* refers to the nutrition treatments and steps designed to resolve the problems.
- *Evaluation* outlines the progress toward resolution of the nutrition problems (see Table 6.10).

PES, or Problem, Etiology, Signs/Symptoms Statements PES is the organizational structure in which to format the nutrition diagnosis. Chapters 3, 4, and 5 explain the process by which a nutrition diagnosis is determined; the PES is simply the manner in which the nutrition diagnosis is

TABLE 6.9

Sample Focus Note

12/27/05 Nutrition Progress Note

Time	Focus	Data
11:30 a.m.	Inadequate oral intake	**Data:** 22 lb. weight loss over previous 3 months. Albumin 3.0; 24-hour recall indicates <25% of kcal and protein requirements met **Action:** 6–8 small feedings; increased nutrient density; addition of high-calorie, high-protein supplement; modification of texture. **Response:** Caloric intake has increased by 45%. No further weight loss documented.
	Swallowing difficulty	**Data:** Mucositis; dehydration and inadequate caloric intake **Action:** Pain medication prior to meals and supplements; modification of texture of meals and food choices to minimize pain. **Response:** Mucositis resolving; Oral intake improved.

Signature:
M. Nahikian-Nelms, Ph.D.,–R.D.,–L.D.

TABLE 6.10

Sample PIE Note

12/27/05 11:30 a.m. Nutrition Progress Note

Problem

Involuntary weight loss

Intervention

Modify oral intake to 6–8 small feedings; increased nutrient density; addition of high-calorie, high-protein supplement; modification of texture; pain medication prior to meals and supplements.

Evaluation

Per calorie counts and patient visitation, oral intake improved currently, meeting 65% of estimated energy and protein requirements. No further weight loss documented since admission.

Signature:
M. Nahikian-Nelms, PhD,–RD,–LD

documented (Lacey and Pritchett 2003; ADA 2006). A sample PES for a patient with dysphagia might read: "Swallowing difficulty (problem) related to stroke (etiology) as evidenced by coughing following drinking of thin liquids (signs/symptoms)" (ADA 2006, 19).

Assessment, Diagnosis, Intervention, Monitoring/Evaluation (ADIM) The ADIM format is organized to reflect the Nutrition Care Process (Lacey and Pritchett 2003). Relevant data about the patient's condition are recorded in the A (*assessment*) section. These might include, but would not be limited to, referral medical diagnosis; pertinent social, family, and medical history; summary of pertinent data collected; and comparison with standards and/or food-related behaviors. The D (*diagnosis*) section is where the actual PES statements are listed and prioritized. The I (*intervention*) section provides documentation of the specific treatment goals and expected outcomes, interventions, and response of the client. Finally, the M (*monitoring and evaluation*) section records documentation of progress toward goals, factors that are facilitating or hampering progress, any changes in the client's level of understanding or behavior, and future plans for care. (See Table 6.11.)

TABLE 6.11

Sample ADIM Note

Assessment: Patient's mother relates that patient's mouth hurts so badly that she can hardly talk. She has had limited oral intake. Patient's mother also describes an "anti-cancer" diet that patient's aunt and uncle introduced them to.
21 yo ♀ Dx: Stage II diffuse large B-cell lymphoma
Admitted with immunosuppression, fungal infection, dehydration R/O pneumonia. s/p first round of chemotherapy/CHOP
Ht. 5'6" Wt. 108# Last adm wt:120# Preillness wt: 130#

Labs: Alb 3.0 WBC 1100 mm 3
EER: 1700–1800 kcal EPR: 75–80 g protein

Diagnosis: NC–1.1 Swallowing difficulty
Inability to tolerate current oral diet as evidenced by pain from mucositis, admitting dehydration and recent 12 lb. weight change from last hospital admission.
NI-2.1 Inadequate oral food/beverage intake
NC-3.2 Involuntary weight loss

Intervention:

1. Modify PO diet to accommodate soft, easily chewed foods and liquids. Increase nutrient density with use of modular supplements and high-kcalorie/high-protein liquid supplements as patient tolerates. 6–8 small feedings/day.
2. Recommend low bacterial/neutropenic precautions during periods of immunosuppression.
3. Consult with nursing/MD to assure adequate pain coverage before attempts at oral intake.
4. Provide evidence-based information regarding nutritional needs during treatment for lymphoma.
5. Monitoring/Evaluation
6. Patient weight will stabilize as measured by daily weights.
7. Caloric intake will increase to a minimum of 65% of current recommendations as measured by daily calorie count.
8. Patient will express tolerance to current food choices during daily patient visitations.
9. Patient will state understanding of current nutritional needs during chemotherapy and during periods of immunosuppression.

Signature:
M. Nahikian-Nelms, PhD,–RD,–LD

Charting by Exception Charting by Exception (CBE) is an even more abbreviated approach to medical charting that involves recording *only* unusual or out-of-the-ordinary events. The CBE format includes a standardized nutritional care plan. After the initial charting of the care plan, only significant data and/or unanticipated responses to the proposed plan should be included in the record. In most CBE formats, a flowchart is used to document assessments and interventions. An asterisk (*) indicates an abnormal finding on an assessment or an abnormal response to an intervention. The findings are explained in the comments section of the form. The progress notes are used to document revisions in the plan of care and specific interventions.

As you can see, the movement in charting over the last decade or so has tended toward reducing the size of the medical record. All of the formats share similar pieces of information. Regardless of the format or style, there should be a method to document all stages of the Nutrition Care Process.

Novice health professionals do not have enough experience with patient care to know what might be important and what is probably not going to be important. Therefore, it is better to include more information than is needed instead of less information. Medical charting is inherently paradoxical, however. It must be comprehensive, yet it needs to be easy to read; it must be clear, but it also must be concise and to the point. Doctors, nurses, and other health professionals who care for a patient do not have time to read through long chart entries that may seem irrelevant. The goal is to create complete, concise documentation—for legal reasons, and, more important, for medical reasons.

With this last point in mind, health professionals, especially novice health professionals, should consider keeping a personal medical notebook (see the next section) where they can chart everything involved with each patient's treatment, where they can brainstorm problems, and, perhaps most important, where they can include their subjective experiences and express their own emotional and intellectual responses to their experiences.

Keeping a Personal Medical Notebook

The institutional administration and staff typically determine which information and how much of that information goes into a medical record. As indicated earlier, the current trend is toward streamlining the chart; but as records are simplified, potentially important information can be neglected. Also, health professionals face human anxiety, pain, and mortality daily, and they need a way of addressing the emotional and intellectual stress that comes with the job. Keeping a personal medical notebook can help tremendously.

It might mean simply writing in a small notepad or, if charting is done electronically, writing your chart notes into a portable storage device and then cutting and pasting what you feel is most relevant into the official chart. Later, you might take time to review your notes on a patient and write your own personal notes. You might want to follow the SOAP note format, adding in a section at the end for your personal thoughts and musings about the patient's condition and experience.

Another possibly useful format for your personal medical notebook is what is called "the double-entry notebook" (Nahikian-Nelms and Nelms 1994). Devised for journal writing, the double-entry format simply involves dividing your notes into "objective" ones and "subjective" ones. One double-entry format involves drawing a vertical line down the middle of the page, but you also could just write your objective notes first and then respond to them in the next subjective section. "Subjective" here refers to your own subjectivity, not the patient's, although you could certainly include the patient's feelings and statements. You need not be tied to a particular formula, though; organize your personal notes in whatever way works for you.

A crucial point to keep in mind *at all times* is that any notes you take on a patient's condition, care, or treatment must be kept confidential. The official medical chart is protected institutionally, but *you* are responsible for keeping your personal medical notebook safe and confidential.

Guidelines for All Charting

- Make sure to chart whatever you see as significant, even if others may not perceive these changes or events as significant.
- JCAHO standards (2006) require that "all entries in medical records be dated and authenticated, and a method is established to identify the authors of entries." Any medical record entry must include the dietitian's full name and status (e.g., RD, CDE). Students must enter their full names and their status as a student or intern. A student's note will need to be cosigned by the preceptor.
- Always be timely in your charting. Chart just after or shortly after you meet with the patient, receive lab work you ordered, or receive any new data that you find significant.
- Never ask someone else to do your charting, and never do someone else's charting. Legally, an error in the chart could be costly. Medically, an error in the chart could be deadly.
- Never chart a procedure until it is done, and then chart it as soon as possible.
- Remember that medical charts are legal documents as well as medical documents.
- When you chart in handwriting, use only black ink and write legibly. Make sure that your notes will photocopy easily.
- Write clearly. You do not need to use complete sentences, but the meaning of your notes must be obvious and unambiguous to others. Read what you have written and ask yourself, "Will my readers understand what I'm saying?"

- Avoid abbreviations unless you are absolutely certain that anyone needing to read the chart will understand immediately what the abbreviations mean and that they are approved for use within your institution.
- Don't leave "white space." Always begin your notes right after the last note recorded. The chart should follow chronological order, and you should not give someone who handles the chart later the opportunity to add notes that might look like your notes.
- Include too much detail rather than too little. While it is undesirable to write overly long notes, you must make sure that everything significant is noted. Note lengths can differ.
- When writing objective notes, record only what you see, hear, smell, or feel, along with what has been measured. Do not assume or infer anything. Save assumptions, inferences, conclusions, and opinions for the assessment or evaluation and plan sections or for your own subjective notes.
- Bracket your biases. Bracketing simply means you set them aside temporarily. We all meet new experiences with old biases—that is, with our personal preferences, leanings, values, and beliefs—but these biases have no place in objective notes. This does not mean that you have to rid yourself of your values. You will always face patients and situations that run counter to your personal feelings: patients who are drunk, obnoxious, obstinate, and/or abusive; you will always observe doctors and nurses whose treatment of patients is not what you would have ordered. Nonetheless, you can bracket your biases when you are charting (and performing other professional duties) by following these steps:
  (1) Identify your biases—either prior to patient care or as you become aware of them during patient care. Make yourself aware of your feelings, beliefs, and preferences.
  (2) Consciously imagine a box in your mind into which you place your particular bias.
  (3) Imagine how the unbiased health professional would act, and act that way.
  (4) If you have a model of unbiased professionalism, imagine how that person would act in the particular situation you find yourself in, and act that way.
- Use neutral language—that is, avoid emotionally charged words. For example, stating that a patient is receiving 65% of nutritional needs with current nutrition support would be preferable to stating that current nutrition support is inappropriate.
- When referring to a patient, call her or him the "patient" or "client" instead of using his or her name.
- Always keep the medical record intact. Never discard a page from that record for any reason. Such an action could result in legal issues at a later date. Moreover, you can never tell what might prove to be important; diagnoses that are rejected early on may be reconsidered and adopted later. If you spill coffee on the chart and blur some entries, do not discard the page. Simply copy it over and leave both pages in the medical chart. Cross-reference them by writing something like "Recopied from page 2" at the top of the new page.
- If you make a mistake, simply cross the note out with a single horizontal line, write "error" above it, and initial it. Do not scribble through the mistake. It should remain readable. For legal reasons, there should be no question about what was written there.
- Always sign notes after printing your name.
- Also, be sure to include the date and time for each note. Military time is often used in order to prevent confusion between a.m. and p.m.
- Make sure you always follow your institution's or agency's policies and procedures for charting.

Confidentiality All medical record information is confidential. A confidential communication is given by one person to another with the trust and confidence that such information will not be disclosed. The federal U.S. law that assures patients of the confidentiality of their medical information is the Health Insurance Portability and Accountability Act of 1996, or HIPAA (see Box 6.2). HIPAA protects information about clients that is gathered by examination, observation, conversation, or treatment (Gomez 2003). A dietitian cannot discuss a client's status with other clients or staff who are uninvolved in the client's care. A legal suit can be brought against a dietitian who has disclosed information about clients without their consent.

BOX 6.2 HISTORICAL DEVELOPMENTS

Health Insurance Portability and Accountability Act of 1996, Public Law 104-191

Goals of HIPAA are to:

- Create a uniform way for all providers and health plans to send and receive health information electronically.
- Prevent inappropriate use and disclosure of an individual's protected health information.

HIPAA contains three parts relevant to healthcare information:

- Privacy of individually identifiable health information
- Standardization of transaction and code sets
- Security of electronic health information

Dietitians and other health care professionals may have reason to use records for data gathering, research, or continuing education. If this is the case, assurance of patient anonymity is required. This is not a breach of confidentiality as long as the records are used as specified and permission is granted from hospital internal review boards (Willison et al. 2003).

Writing for Nonmedical Audiences

Writing to the community of health professionals within your institution or agency will become easier the more practiced you become in such writing. But writing to audiences of health professionals is not the only kind of writing you may be called on to do. You may also find yourself writing for various nonprofessional audiences, including the public in general. This section describes some of the most common of these forms of writing.

Instructional Materials for Patients, Their Families, and the Public

The purpose of these materials is informative, sometimes with the additional purpose of persuading the reader to take action, if necessary, after reading the material. Examples include informational material outlining nutrition therapies and lifestyle changes. The audience for such writing obviously differs greatly from the professionals that charting addresses. You should avoid most medical abbreviations, because your readers simply will not be familiar with them. You cannot use professional or even academic jargon. Words like "data" will need to be replaced with more generally understood words like "information." In other words, use commonsense language. Remember, the goal here is to instruct. Your reader cannot be instructed if she or he cannot understand what is being said. Establish the appropriate reading level and even have a member of your target audience evaluate your instructional material (Leff 2004).

Tips for Writing Instructional Materials Ask nonprofessionals you know to give you feedback on the instructional materials you write before you prepare them for distribution so that you can revise the text if it is unclear to your test audience. You can also establish a focus group that is representative of your intended audience for the purpose of evaluating and responding to your writing. Their insight can help ensure you that you will meet the audience's needs.

Use numbering and bullets to create easy-to-read lists. As a model, consider the way bullets and numbers are used in this chapter.

Spend time planning the document you want to produce before beginning to actually write it. Put together a rough organizational plan of the information. This prewriting planning will make writing a lot easier.

Leave yourself plenty of time to revise the document once it is drafted. Read through it first to make sure you included all the information (the content) that you planned to convey. Read through it a second time to make sure that the organization makes sense, that information introduced early in the document, for example, is explained immediately rather than much later in the document. Read through the document a third time to check the language and make sure it is understandable to your audience. Make sure you check your spelling, too.

A warning: There is one important similarity between writing to professional peers and writing to nonmedical professionals: the need for clarity. Sometimes, when shifting from technical writing intended for the professional community to writing for nonprofessionals, writers also shift from an ideal of clarity in their writing to an ideal of eloquence. Eloquence is fine for novelists, but it is irrelevant here. Be clear, be concise, and use language appropriate for your audience.

Reporting Your Own Research

When you become a clinical dietitian, you automatically become a member of the professional community of clinical dietitians and other health professionals. You are expected to keep up with the research in your field and to go to conferences and participate in other forms of continuing education. You may also want to contribute to the ongoing deliberations in your field. You may want to do original research and report the findings of that research either in professional journals and books or at professional conferences. Discussion of how to do such research and how to write professional articles, book chapters, and conference papers is beyond the scope of this text. Nevertheless, you need to be aware that this research and professional writing is possible for you. It is considerably more formal than writing instructions for patients, in the sense that there are certain organizational and stylistic conventions that you must adopt when reporting your research. The Web links at the end of the chapter provide information about doing research in dietetics and reporting that research professionally.

Conclusions: Your Ethos—Establishing Expertise

When you write professionally, you establish your professional ethos by making wise recommendations and orders and, most important, by establishing your expertise both in your actions and in your writing. That means continuing your professional education throughout your career, and drawing on that ongoing knowledge when you chart and when you create and update informational and educational material. This is a crucial component that helps to define a profession and is the foundation for maintaining competence.

CASE STUDY

Mr. J, a 52-year-old man, is referred by his physician to the nutrition outpatient clinic for counseling on a weight-reduction diet. While talking with Mr. J, you obtain a quick diet history, which you feel is reasonably accurate. You calculate that his diet contains approximately 2800 kcal per day. Mr. J tells you that he dislikes sweets and fats, rarely eats vegetables, and drinks about 8 cups of fruit juice daily in addition to coffee and tea throughout the day (which has added sugar), and that he is fairly inactive, eats two large meals a day, and never eats breakfast. You measure Mr. J and find that he is 5'7" tall and weighs 195 pounds. You discuss his goals with him and find out that he does want to lose weight because of his family history of diabetes.

One goal that is determined is to reduce fruit juice to 1 cup per day and change to artificial sweetener for coffee and tea. He will keep a food record for one week and return to see you at that time.

- Outline the subjective information for this note.
- List all the objective information.
- Write a PES statement for one nutrition problem.
- Write your assessment.
- Determine your interventions for this nutrition problem.
- Design your evaluation/outcome measures for this problem.

WEB LINKS

American Medical Informatics Association: AMIA develops medical informatics to support patient care, teaching, research, and health care administration.

http://www.amia.org/contact/fcontact.html

Joint Commission for Accreditation for Healthcare Organizations: Website for the national accrediting agency for all health care facilities in the United States.

http://www.jcaho.org

END-OF-CHAPTER QUESTIONS

1. What is the primary purpose of the medical record in the clinical setting? What other functions does it serve?
2. Why should standardized language and abbreviations be used in the medical record?
3. Describe each of the four sections that constitute a SOAP note.
4. How does medical record documentation fit within the Nutrition Care Process?
5. How does charting by exception differ from the other charting methods?
6. Why might a dietitian keep a personal notebook in the clinical setting?
7. Why must information in a chart note be kept confidential?

7

Methods of Nutrition Support

Annalynn Skipper, Ph.D., R.D., F.A.D.A.

Marcia Nahikian Nelms Ph.D., R.D.

Southeast Missouri State University

CHAPTER OUTLINE

Oral Diets

Texture Modifications

Oral Supplements

Appetite Stimulants

Specialized Nutrition Support (SNS)

Enteral Nutrition • Parenteral Nutrition

Introduction and Overview

With an increasing focus on the issues of overweight, obesity, and chronic disease in health care, it is important not to lose sight of the fact that insufficient food intake and resulting protein-energy malnutrition remain prevalent among certain populations. As many as 30 to 50% of hospitalized patients and up to 95% of nursing home patients exhibit some signs of malnutrition (Thomas et al. 2002; Wendland et al. 2003). Malnutrition in hospitals and nursing homes is often the result of chronic illness experienced by patients before admission, but it is exacerbated by pain, anxiety, depression, and unfamiliar foods or meal schedules associated with admission to a health care setting. Insufficient food intake may be related to factors in Figure 7.1.

The consequences of malnutrition include increased risk of infection, delayed wound healing, and delayed return to home, work, or baseline activities of daily living following hospitalization (Stratton et al. 2005). All of these consequences contribute to increased health care costs and diminished quality of life. Health care institutions—and in particular, the registered dietitians on staff—prevent or treat malnutrition and other diseases by providing the patient with appropriate nutritional support. This support can be provided through an oral diet designed to meet the patient's specific nutritional requirements and modified to meet his or her specific physical needs. In many situations, however, modification of the oral diet is not enough. This chapter will cover the continuum of nutritional support: oral diets, supplemental foods, enteral nutrition (or tube feeding), and parenteral nutrition. Nutrition support is a crucial component of the Nutrition Care Process, and the nutrition practitioner will use its principles over and over again. Though interventions will be modified depending on the disease course and the patient's individual needs, each intervention begins with the foundational principles presented here.

Oral Diets

The best prevention and treatment for malnutrition is an adequate supply of acceptable food composing a diet that

FIGURE 7.1 Factors Affecting Nutritional Intake during Illness

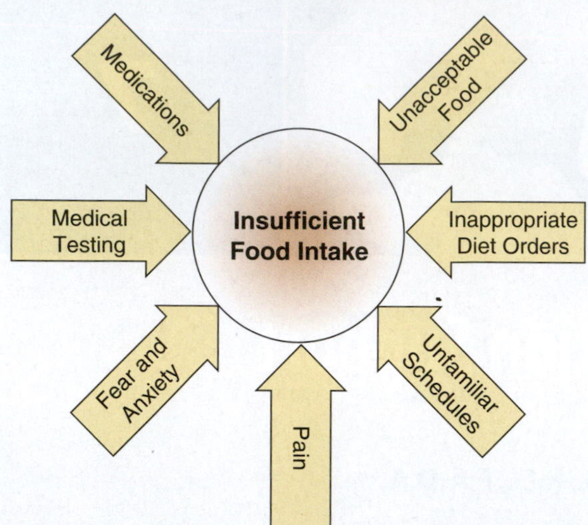

has been individualized to age, height, weight, activity level, and medical condition. The regular or "house" diet served in hospitals and nursing homes is supplied in three meals each day. Regulations govern the timing, frequency, and nutrient content of meals (JCAHO 2006). Menus are written and approved by a qualified dietitian and are designed to provide the Dietary Reference Intake for all nutrients. When patients have the option to select the food items they prefer, as they do in most institutions, one of the simplest yet most helpful interventions may be to assist a patient with menu selection. Offering suggestions and appropriate substitutions can be an efficient method of ensuring that the patient's diet remains adequate and acceptable.

Therapeutic diets have several important functions. They may be used to maintain or restore health an nutritional status. They may be modified to accommodate changes in digestion, absorption, or organ function. Texture and consistency can be adjusted to alleviate mechanical problems. Therapeutic diets can provide the appropriate nutrition therapy to support weight loss or gain, or to assist with treatment of a particular diagnosis through nutrient content changes. Changes from the "house" or regular diet can include: caloric level; consistency; single nutrient manipulation; method of preparation; specific food restriction; number, size, or frequency of meals; and addition of supplements.

dysphagia—difficulty swallowing

NPO—nil per os, which is Latin meaning "nothing per mouth"

clear liquid diet—diet consisting of liquids that contribute minimal residue to the gastrointestinal tract; includes fruit juices without pulp, carbonated sodas, broth, tea, coffee, water, popsicles, fruit ice, Jell-O (gelatin), and liquid nutritional supplements (e.g., Boost Breeze®, Resource Fruit Beverage®)

full liquid diet—diet consisting of all beverages allowed on clear liquid diets with addition of milk, ice cream, yogurt, and liquid nutritional supplements (e.g., Ensure®, Boost®)

osmolality—number of water-attracting particles per weight of water in kilograms (expressed as mOsm/kg)

hyperosmolar—having a higher osmolality than body fluids (>300 mOsm/kg)

Texture Modifications

The regular diet may be modified for patients with impaired chewing ability so that softer foods are served. Soft diets contain foods that are easy to chew and usually avoid raw fruits and vegetables. Individuals with **dysphagia** (difficulty swallowing) may require more specific modifications of texture and consistency. Diets for these individuals are discussed more thoroughly in Chapter 16. For very short periods (two or three meals), liquid diets consisting of broth, juice, cream soups, and milk may be served to patients who are beginning to eat after a long period without food (*nil per os*, or **NPO**). These diets are often referred to as **clear** or **full liquid diets.** A clear liquid diet is intended to provide fluid and energy in a form that requires minimal digestion and contributes to limited residue in the gastrointestinal tract. It may be used during acute gastrointestinal distress, during gastrointestinal medical testing (such as a colonoscopy), or prior to surgery. Clear liquid diets are inadequate in kilocalories (kcal), protein, vitamins, and minerals, so they are used only when necessary. Historically, the clear liquid diet has been used as a progression toward solid food after a surgical procedure, but this may not be warranted. A full liquid diet also has been used as a transitional diet between liquids and solid foods. Because this diet includes milk and milk products, it may present a problem due to the large amounts of lactose. Table 7.1 outlines the basic principles of these liquid diets.

What may be much more important than the content of a clear or full liquid diet is the **osmolality** of the particular liquid that is provided (for more information on osmolality, see the discussion of fluid and nutrient density in the Enteral Nutrition section later in this chapter). **Hyperosmolar** liquids may not be tolerated during these transitional periods or when the gastrointestinal tract has not been stimulated. Table 7.2 provides the osmolality of common liquids that are used in these diets. Choosing those with a lower osmolality may assist in ensuring a successful tolerance for the transition to oral feeding.

Oral diets may be modified to prepare patients for a specific medical test. For example, a high-fat diet (100 g/day) may be administered for two to three days prior to a test for fat malabsorption. Details of the types of diets served are recorded in an institution's diet manual, which lists the types and amounts of foods served on each diet. (See Appendix E for descriptions of various modified diets.)

TABLE 7.1

Principles of Clear and Full Liquid Diets

Diet	Purpose	Foods Acceptable	Limitations
Clear Liquids	Intended to supply fluid and energy in a form that requires minimal digestion and stimulation of the GI tract Clear fluids or foods that are liquid at body temperature and leave minimal residue	Clear fruit juices Bouillon, consommé, clear broth Gelatin, fruit ice, plain hard candy, sugar, honey Commercially prepared low-residue, lactose-free nutritional supplements	Not nutritionally adequate Should be limited to 24 to 48 hours unless supplements are added
Full Liquids	Transition between clear liquids and solid food	Consists of foods or fluids that are or become liquid at body temperature All clear liquids Cream soups Milk Ice cream, pudding, yogurt	May present problem with large amounts of lactose

TABLE 7.2

Osmolality of Soups, Juices, and Beverages

Food/Beverage	Measured mOsm/kg/H_2O
Beef broth	304–543
Chicken broth	293–501
Apple juice	654–734
Apricot juice	973
Grape Juice	1174–1190
Orange Juice	542–710
Peach nectar	915
Prune juice	1174
Cranberry juice cocktail	888–907
Cranapple drink	1087–1212
Cola	591–716
Diet cola	27–29
Ginger ale	515–557
Whole milk	282
Popsicles	665–719
Gelatin	594–847

Source: The table was compiled by the author from: Feldman M, Barnett C. Relationships between the acidity and osmolality of popular beverages and reported postprandial heartburn. Gastroenterology 1995;108:125–131. Wendland BE, Arbus GS. Oral fluid therapy: sodium and potassium content and osmolality of some commercial "clear" soups, juices and beverages. CMAJ 1979;121:564–571.

Given adequate appetite along with sufficient resources to purchase and prepare food, even the most malnourished individual can be rehabilitated with oral diet alone. But maximizing oral intake within the hospital setting is often challenging, because this environment is not always conducive to eating. Add to this environment the stress, fear, pain, and isolation of illness, and it seems a wonder that anyone who is hospitalized can eat adequately. For these individuals, a number of alternatives exist. A primary function of nutrition services in health care institutions is to be the patient's nutrition advocate. When patients present with a suboptimal intake, nutrition services staff members work with the patient and health care team to provide nutritional options.

Oral Supplements

Oral intake may be increased by supplementing a balanced diet with between-meal or evening snacks of nutrient-dense foods acceptable to the individual patient. For these snacks, traditional foods such as fruit, crackers, sandwiches, milk shakes, custard, or pudding may be served. Increasing nutrient density without actually increasing volume is a much more effective tool for an individual who is suffering from decreased appetite. For example, instead of using skim milk, the patient could receive whole milk with 2 T. of dry milk powder to boost both kcal and protein. Adding peanut butter to toast for breakfast is an excellent method to increase both kcal and protein. Table 7.3 provides examples of methods to increase nutrient density using readily available foods.

If extra expense is not a concern, liquid meal replacement formulas may provide a convenient alternative to between-meal snacks. These products typically come in single-portion containers providing 250 to 350 kcal with 7 to 15 grams of protein in 250 mL, and may be available in a variety of flavors. These products are lactose free. Some contain fiber, and others are calorically dense or higher in protein. Examples of these products are Ensure®, Boost® and Nu-Basics Plus®.

Manufacturers have introduced many variations of these products for specific medical conditions, such as wound healing or diabetes. Some of these products may be available in liquid form, as puddings, or as bars. Other products contain single nutrients such as protein or fiber, and may be referred to as "modular" products. Appendix C2 provides information for currently available modular products. A carbohydrate modular supplement, such as Polycose©, provides glucose polymers and generally does not alter the taste of the food to which it is added. Protein modulars, such as ProMod©, can be added to both foods and beverages but will need to be mixed well. There is a slight change in taste and consistency, which is also the case for a lipid modular such as medium-chain triglyceride (MCT) oil. In general, modulars are not as cost-efficient as other types of supplements, and they increase the labor costs as well. Because unopened supplement packages do not require refrigeration, these products may be served at a time convenient to the patient. In nursing homes, they may be administered in place of water with medications as a means of increasing kcal intake. Box 7.1 provides a discussion of the MCT supplement.

Commercial supplements are popular because they are heavily marketed to patients and their caregivers using television and the Internet. However, acceptability and intake are highly individual. Patients receiving oral supplements frequently develop "taste fatigue" a few days after supplements are initiated, and supplement intake then decreases. It is also important to remember that merely providing supplemental feedings will not increase appetite; in fact, many patients complain that extra portions, frequent meals, and supplemental snacks are overwhelming and

TABLE 7.3

Mechanisms to Increase Nutrient Density

Increasing energy content:

- Add butter or margarine to cooked cereals, soups, vegetables, or casseroles.
- Add jam, jelly, or honey to toast or other breads and crackers.
- Use whole milk or cream with soups, casseroles, creamed vegetables, or shakes and smoothies.
- Add sour cream or yogurt to soups, casseroles, creamed vegetables, or shakes and smoothies.
- Add nut butters or cream cheese to raw vegetables, bread, or crackers.

Increasing protein content:

- Add powdered milk to any beverage, soup, or casserole.
- Add liquid egg substitutes to shakes, soups, vegetables, or casseroles.
- Wherever possible, add nuts, nut butters, chopped meats, cooked eggs, cheese, or yogurt to prepared foods.
- Add tofu or soy crumbles to any prepared vegetable, soup, or casserole.

Application:

- Original Breakfast choice: 1 c dry cereal, 1 c skim milk, 1 c orange juice = 305 kcal; 13 g protein
- Increased nutrient density: ½ c cooked oatmeal with 2 T dry milk powder, 1 T brown sugar, 1 T butter, 1 c fruit smoothie made with ½ c fruit yogurt, 1 banana, 2 oz orange juice, 2 T firm tofu = 584 kcal; 36 g protein

Oral Supplements

Source: photo provided courtesy of © Novartis Medical Nutrition

BOX 7.1 CLINICAL APPLICATIONS

A Review of the MCT Supplement

Medium-chain triglycerides (MCT) are eight- and ten-carbon-chain fatty acids, liberated from coconut oil and then re-esterfied to glycerol. Because MCT do not depend on pancreatic lipase or emulsification for digestion, they are used clinically to supply kcal to patients with a variety of pancreatic and gastrointestinal disorders.

MCT are hydrolyzed more readily than LCT (long-chain triglycerides) by lipase, even in the absence of emulsification by bile salts. After transport into the enterocyte, they are not packaged into chylomicrons but instead are absorbed into the portal blood stream and transported bound to albumin.

Medium-chain fatty acids (MCFA) and long-chain fatty acids (LCFA) also differ in their metabolism. In the liver, LCFA must be transformed into acyl carnitine derivatives before they can enter the mitochondria for subsequent beta-oxidation. Carnitine acyl transferase I (CAT I) and carnitine acyl transferase II (CAT II) on the inner mitochondrial membrane are both necessary for the entry of LCFA into the mitochondria. MCFA do not need CAT I or CAT II for entry and their subsequent beta-oxidation.

The oxidation of MCFA in the fed or fasted state will result in increased production of acetoacetate, 3-hydroxybutyrate, and acetone, three molecules known as ketone bodies. LCFA only produce ketones in the fasted state since malonyl CoA, an intermediate of carbohydrate metabolism, inhibits CAT I, thus decreasing the entry of LCFA into the mitochondria in the fed state. Although ketones have a "bad" reputation as a result of high blood concentrations seen during diabetic ketoacidosis, they can be efficiently utilized as oxidative fuels and converted to fatty acids.

reduce appetite. This is why it is essential to include the patient in the decision-making process for changes in the meal plan. The patient needs to understand why these are being offered and how they could improve his or her current medical status. Providing supplement taste tests could be one way to help patients decide which supplement they would prefer to add to their diet. Developing a rotation for snacks or supplements and setting portion goals with the patient may also improve acceptance. Regular follow-up and monitoring are necessary in order to coordinate successful interventions and minimize product waste associated with unused products.

Appetite Stimulants

For patients who are unable to eat amounts sufficient to maintain their weight, it is appropriate to assess other factors that impair intake. Nonfood causes of poor intake range from poorly fitting dentures to lack of interest in unfamiliar foods to depression. If these causes cannot be resolved and poor intake persists, drugs that stimulate appetite are sometimes ordered. These drugs, including prednisone, megestrol acetate, and dronabinol, are available by prescription (McCabe et al. 2003). Like all drugs, appetite stimulants produce significant side effects in some patients.

Prednisone is an inexpensive steroid that is effective over a short period of time and induces an increased sense of well-being and short-term increases in appetite. No studies have shown benefit with this drug in increasing body weight in the long term (Mantovani et al. 2001). Patients receiving prednisone may experience hypokalemia (lowered serum potassium levels), muscle weakness, disordered body fat distribution, hyperglycemia (elevated blood glucose), and impaired immunity.

Megestrol acetate will increase appetite and body weight when administered with exercise and nutrition support, but it can take several weeks to improve intake (Reuben et al. 2005). The drug is also expensive—it can cost several hundred dollars per month—and its side effects can include impotence, vaginal bleeding, and deep vein thrombosis.

Dronabinol is a derivative of marijuana that may improve appetite, but it has not been associated with weight gain. Dronabinol is expensive, and users have experienced nausea, vomiting, and mental status changes, including euphoria and somnolence. As new information becomes available, drug doses may change. Thus, recommendations to use appetite stimulants should be preceded by a thorough review of updated dosing and complication information from reliable sources such as Drug Facts and Comparisons (2005) or the American Hospital Formulary Service's AHFS Drug Information (2005). For more information on appetite stimulants and other antiwasting interventions, see Chapter 24 and Chapter 26.

Specialized Nutrition Support (SNS)

For patients who are unable to maintain their nutritional status using oral diets, supplements, or appetite stimulants, specialized nutrition support is the next alternative. "Specialized nutrition support refers to administration of nutrients with therapeutic intent. This includes (but is not limited to) the provision of total enteral and parenteral nutrition support and the provision of therapeutic nutrients to maintain or restore health" (ASPEN 2002, p. 15). Nutrition support became an important nutrition intervention in the 1970s, as methods were developed to provide adequate feedings by vein. Many technical improvements have been based on extensive experience, the results of research studies, and economic necessity. Nutrition support has developed rapidly, and practice varies widely as clinicians struggle to keep up with changes in the field. See Box 7.2 for a brief history of nutrition support.

Ethical considerations impact decisions related to specialized nutrition support. Patients and their families should be involved in making these decisions. Furthermore, options for specialized nutrition support should be consistent with the level of medical care that the patient is receiving (ASPEN 2002).

BOX 7.2 HISTORICAL DEVELOPMENTS

History of Parenteral Nutrition

Providing solutions intravenously is now standard medical practice. Yet safe infusions of saline and dextrose solution were not available until after World War I, and the technology to adequately nourish patients intravenously was not developed until the 1960s. Successful parenteral nutrition began in 1937, when it was reported that you could "feed" a patient peripherally using IV dextrose and protein hydrolysate solutions (Elman and Weiner 1939). Yet adequate protein and kcal were difficult to provide. As much as 5 liters of the dilute nutritional solutions had to be administered through the peripheral hand and arm veins. It was Dr. Stanley Dudrick, a resident in general surgery, who, building on the work of Drs. Rhoads and Vars, perfected a technique in beagle puppies that allowed hypertonic dextrose and protein solutions to be administered through a central vein (Rhoads and Dudrick 1986).

The adoption of central vein parenteral nutrition and its increasing use in the 1970s led to development of new parenteral solutions, such as crystalline amino acid solutions and IV fat emulsions, which provided essential fatty acids (Rhoads and Dudrick 1986). Parenteral nutrition also led to greater recognition for nutrition support of hospitalized patients, resulting in the increased use of enteral nutrition as well. Today, parenteral nutrition is an accepted mode of treatment for the patient who cannot be adequately nourished via the gastrointestinal tract.

Enteral Nutrition

Enteral nutrition (from the Greek *enteron* or intestine) refers to feeding through the gastrointestinal tract via a tube, catheter, or **stoma** that delivers nutrients distal to (or beyond) the oral cavity (ASPEN 2005). The terms "enteral feeding" and "tube feeding" are used interchangeably in the clinical setting. Medical and nutritional research has increased understanding of the need for gastrointestinal (GI) tract stimulation and the overall health benefits of continuing to provide nutrition via the GI tract, especially in the critically ill. Recognition of these benefits, along with improved formulas and equipment, have supported the increased use of enteral nutrition techniques over the last two decades.

Indications Enteral feeding is indicated and often used for patients who have a functioning gastrointestinal tract but cannot feed themselves adequately. Specifically, enteral nutrition is used for patients with altered mental status, swallowing dysfunction, or disorders of the upper gastrointestinal tract that can be bypassed by inserting a feeding tube below the dysfunction. Enteral feeding is contraindicated if patients have serious medical conditions that affect the gastrointestinal tract, including diffuse peritonitis (inflammation of the peritoneal lining of the abdominal cavity), intestinal obstruction that prevents intestinal contents from passing through the intestine, intractable vomiting not responsive to medical treatment, paralytic ileus that prevents gastrointestinal contents from passing through the gastrointestinal tract, intractable diarrhea that cannot be controlled with medications, and gastrointestinal ischemia (insufficient blood flow to gastrointestinal tissues) (ASPEN 2002; Marian and Charney 2006.)

The relative advantages of enteral nutrition support over intravenous nutrition are emerging as new research becomes available. The use of enteral nutrition support is accompanied by the following advantages (American Dietetic Association Evidence Analysis Library 2006; Braunschweig, Levy, Sheean, and Wang 2001; Byrne et al. 2005; Charney and Malone 2006; Jeejeebhoy 2005:

- Cost-effectiveness
- Reduced rate of infectious complications in critically ill patients
- Improved wound healing
- Reduced surgical interventions
- Maintenance of gastrointestinal function

There are not enough data to support the commonly held belief that enteral nutrition reduces length of hospital stay (American Dietetic Association Evidence Analysis Library 2006). In some settings, in fact, insertion of a feeding tube is associated with increased mortality (Finucane, Christmas, and Travis 1999). Disadvantages of enteral feeding include the difficulty of administration, poor tolerance, and difficulty meeting nutritional requirements of some patients (McClave, Lowen, Kleber, Nicholson, Jimmerson, McConnell, and Jung 1998; McClave et al. 2002). These disadvantages may be minimized by careful patient selection, thorough nutrition-focused physical assessment, and use of standardized protocols (Charney and Malone 2006).

Gastrointestinal Access Once it has been determined that enteral feeding is both feasible and therapeutic, several important decisions must be made in order to design the enteral nutrition prescription. The first is establishing access to the gastrointestinal tract. The access route is often determined by the primary physician according to the patient's diagnosis and the anticipated amount of time the patient will require support (see Figure 7.2). Access is achieved when a feeding tube is placed into the stomach or intestine. Figure 7.3 demonstrates the sites for access.

The type of feeding access is described according to (1) where it enters the body, and (2) where the tip is located. Thus, the tube may extend from the nose (naso-) or mouth (oro-) into the stomach, becoming a **nasogastric or orogastric feeding tube**. Nasogastric (nose to stomach) feeding is the most common, the easiest to achieve, and the easiest to maintain. It is also the least expensive and is acceptable in many circumstances. Small bowel or **nasointestinal feeding tubes** enter the gastrointestinal tract through the nose and reside in the duodenum or jejunum. Nasointestinal access is more difficult to achieve and maintain, but is preferred in some circumstances. For example, nasointestinal access is used to bypass the stomach in cases of gastroparesis (delayed gastric emptying), gastric outlet obstruction, or when previous gastric surgery precludes feeding into the stomach. Nasointestinal feeding may also minimize accidental aspiration (inhalation) of formula into the lung, but data supporting this practice are inconclusive (McClave et al. 2002).

A disadvantage of nasogastric and nasointestinal feeding tubes is discomfort for the patient. Smaller tubes made of pliable material have been developed to improve patient

enteral nutrition (EN)—feeding through the gastrointestinal tract using a tube, catheter, or stoma that delivers nutrients distal to (or beyond) the oral cavity

stoma—an opening

nasogastric feeding tube—a tube that is inserted nasally (through the nose) into the stomach

orogastric feeding tube—a tube that is inserted orally (through the mouth) into the stomach

nasointestinal feeding tube—a tube that is inserted nasally (through the nose) past the stomach into the intestine

FIGURE 7.2 Selecting a Feeding Route

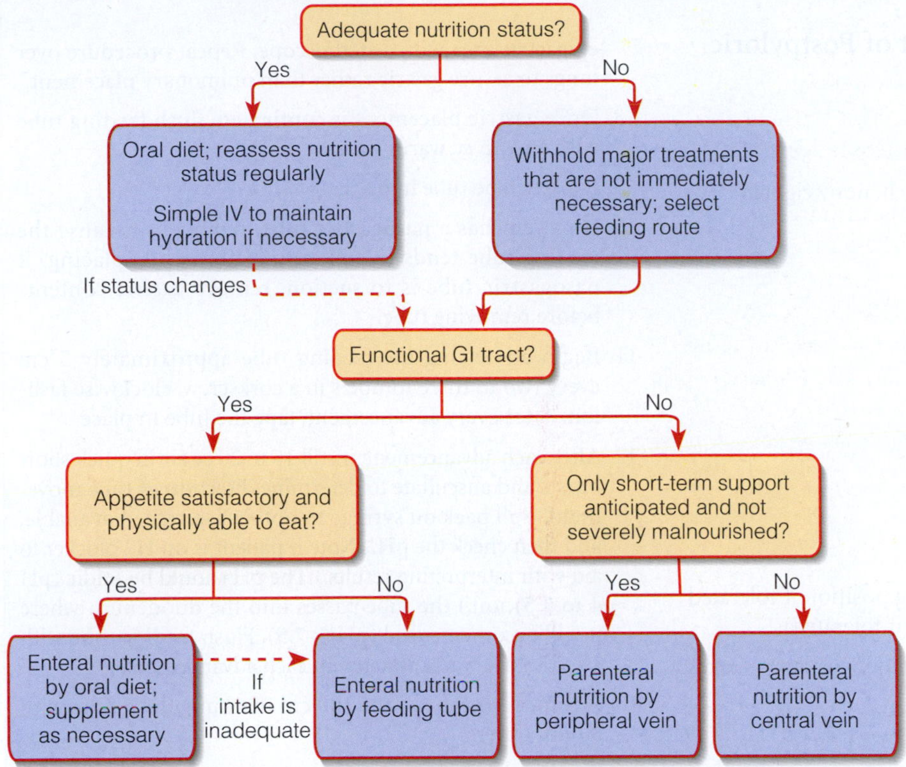

Source: S. Rolfes, K. Pinna and E. Whitney, *Understanding Normal and Clinical Nutrition*, 7e, copyright © 2006, p. 676

FIGURE 7.3 Sites for Enteral Access

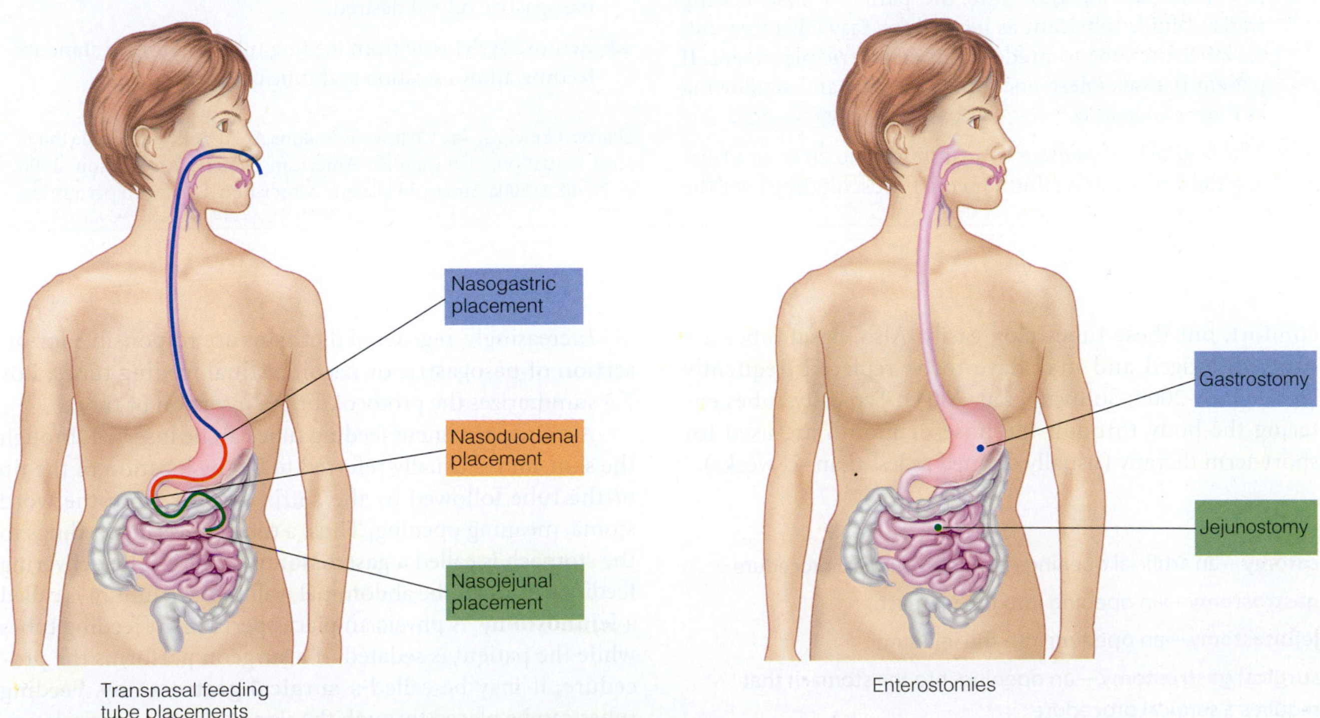

Source: S. Rolfes, K. Pinna and E. Whitney, *Understanding Normal and Clinical Nutrition*, 7e, copyright © 2006, p. 655

BOX 7.3 CLINICAL APPLICATIONS

Protocol for Bedside Placement of Postpyloric Feeding Tubes

Supplies

- 10 Fr feeding tube, with stylet, 43 inch, nonweighted
- 10 mg IV metoclopramide
- water soluble lubricant
- 60 cc syringe with Luer Lock tip
- Silk tape
- pH paper (optional)
- Cup with warm water
- Gloves
- Stethoscope

Procedure

1. Position patient on back or in sitting position if tolerated with head of bed at least 30 degrees (if tolerated).
2. Administer 10 mg IV metoclopramide (Reglan)—Takes about 10 minutes to begin action.
3. Drape towel over patient's chest.
4. Secure stylet into the feeding tube, and then instill 20 to 30 cc warm water through the tube to activate the internal lubricant.
5. Using the tube, measure for gastric placement on the patient (usually 50 to 65 cm).
6. Insert the feeding tube into the patient's nose, adding water-soluble lubricant as needed for easy advancement. Advance the tube to predetermined gastric placement. If patient is awake/alert, encourage relaxing and swallowing as tube is advanced.
7. Check gastric placement by instilling 15 to 20 cc air utilizing the syringe and simultaneously auscultating over the epigastric area with a stethoscope. Repeat procedure over lungs to assure gastric rather than pulmonary placement.
8. Once gastric placement is confirmed, flush feeding tube with 1 to 15 cc warm water.
9. Loosely tape tube in place.
10. If patient has a nasogastric tube, remove it because the feeding tube tends to coil around this while placing. If nasogastric tube is to suction, remove gastric contents before removing tube.
11. Begin advancing the feeding tube approximately 5 cm every two to three minutes in a corkscrew, clockwise fashion. After every advancement, tape the tube in place.
12. After each advancement, instill 15 to 20 cc air in quick short bursts and auscultate to determine direction of tube movement. Pull back on syringe to obtain aspirate, if available, and then check the pH. (Note if patient is on H2 blocker to aid with interpreting results.) The pH should be acidic (pH 4 to 5.5) until the tube passes into the duodenum where it will become neutral (pH 6–7.5). Flush feeding tube with 10 to 15 mL warm water with each advancement.
13. Advance the tube to the 100 cm marking, flush tube with 30 mL water.
14. Tape the feeding tube securely in place.
15. Total procedure takes approximately 15 to 30 minutes.
16. Obtain abdominal radiograph to confirm small bowel placement.
17. Upon radiographic placement confirmation, reinsert nasogastric tube if desired.
18. Remove stylet wire from feeding tube and follow standard feeding administration and tube care procedures.

Source: Cresci, G. In: Charney P, Malone A. ADA Pocket Guide to Enteral Nutrition. Chicago IL: American Dietetic Association, 2006, p. 39–40.

comfort, but these tubes clog easily. Also, nasal tubes are easily dislodged and may have to be replaced frequently (Jeejeehboy 2005; Sullivan et al. 2004). Typically, tubes entering the body through the nose or mouth are used for short-term therapy (usually defined as less than six weeks).

ostomy—an artificial opening created by surgical procedure

gastrostomy—an opening into the stomach

jejunostomy—an opening into the jejunum

surgical gastrostomy—an opening into the stomach that requires a surgical procedure

Increasingly, registered dietitians are responsible for insertion of nasogastric or nasointestinal feeding tubes. Box 7.3 summarizes the protocol for the insertion of tubes.

A more permanent feeding tube can be inserted through the skin, and is usually referred to by the location of the tip of the tube followed by the suffix **ostomy** from the word stoma, meaning opening. Thus, a tube delivering feedings to the stomach is called a **gastrostomy**, while a tube delivering feedings through the abdominal wall to the jejunum is called a **jejunostomy**. A physician places permanent feeding tubes while the patient is sedated. If a surgeon performs the procedure, it may be called a **surgical gastrostomy**. Feeding tubes can be placed through the skin without a surgical incision, which is referred to as percutaneous gastrostomy. If

FIGURE 7.4 PEG Tube Placement

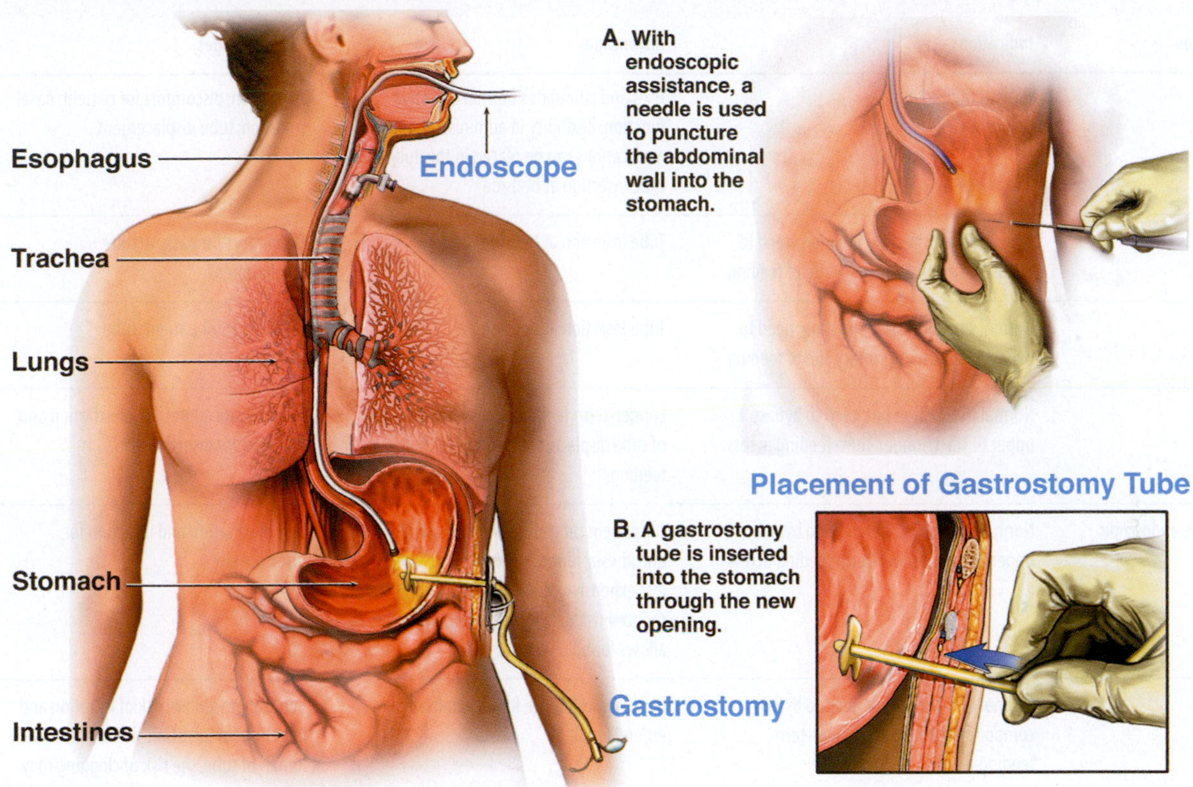

Source: Medical Illustration, Copyright © 2006 Nucleus Medical Art, All rights reserved. www.nucleusinc.com

Gastrostomy Tube

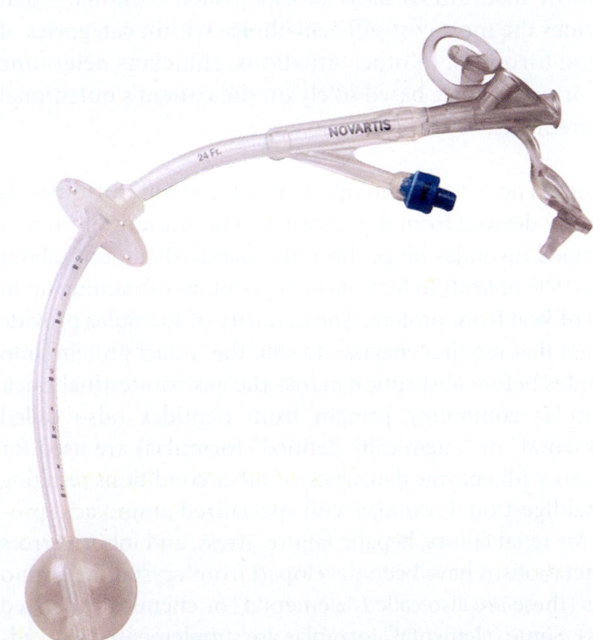

Source: photo provided courtesy of © Novartis Medical Nutrition

this is done using an endoscope, then the procedure is called a **percutaneous endoscopic gastrostomy** or **PEG**. A gastroenterologist, usually in an outpatient GI procedure, inserts these tubes. (See Figure 7.4 and photo of PEG.)

Table 7.4 summarizes the indications, advantages, and disadvantages for each access site.

Formulas The next step in establishing the enteral nutrition prescription is to consider the choice of an enteral formula (see Figure 7.5). Historically, enteral feedings were composed of liquid mixtures such as milk and wine, blenderized foods from a regular diet, or combinations of baby food thinned with milk or juice. Concerns about labor costs, quality control, uncertain formula composition, and sanitation gave rise to commercial formulas. A major consideration in enteral feeding is the selection of the most appropriate formula from among the dozens available on the

percutaneous endoscopic gastrostomy (PEG)—a procedure used by a physician to insert a feeding tube through the skin and into the stomach using an endoscope

TABLE 7.4

Summary of Enteral Access Sites

Enteral Access Sites	Indications	Advantages	Disadvantages
Nasogastric	Normal GI function	Uses and stimulates normal digestive function; flexibility in administration; medications can be placed in this tube; tube insertion at bedside	Aspiration; discomfort for patient; nasal irritation; tube displacement
Nasoduodenal	Normal small intestine function; need to bypass stomach as primary site of feeding	Tube insertion at bedside	Discomfort for patient; tube displacement
Nasojejunal	Normal small intestine function; need to bypass stomach as primary site of feeding	Tube insertion at bedside	Discomfort for patient; tube displacement
Gastrostomy	Normal GI function but need to bypass upper GI tract; longer-term feeding access	Longer-term feeding access; reduced risk of tube displacement; allows for bolus feedings	Surgical procedure; risk of irritation and infection for insertion site
PEG (percutaneous endoscopic gastrostomy)	Normal GI function but need to bypass upper GI tract; longer-term feeding access	Outpatient procedure without risk of anesthesia; longer-term feeding access; less expensive than surgical insertion; reduced risk of tube displacement; allows for bolus feedings	Risk of irritation and infection for insertion site
Jejunostomy	Normal GI function but need to bypass components of GI tract; longer-term feeding access	Increased tolerance for early initiation of enteral feeding	Surgical procedure; risk of irritation and infection for insertion site; with smaller lumen of tube, the risk of clogging may be greater

Gastrostomy

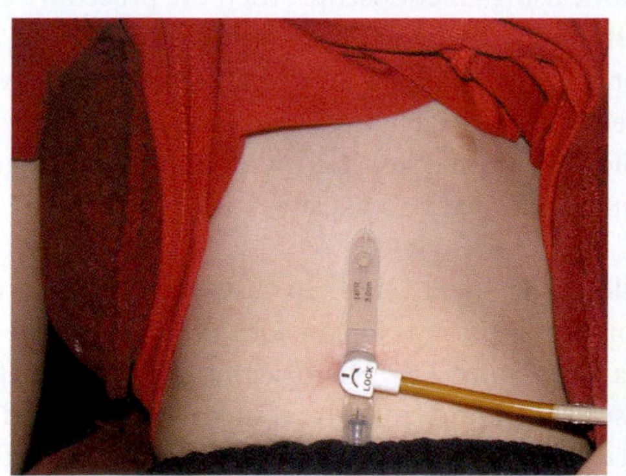

COURTESY OF THE BOBBIE SHONE TRUST

Source: www.fundraising.freeservers.com/images/pc140006.jpg

market. These products are composed of protein, carbohydrate, and fat, with vitamins, minerals, water, and electrolytes in proportions that mimic a balanced diet, or a diet designed for a specific disease or medical condition. Considerations for formula choice will be based on the substrates within the formula, nutrient density, osmolality, and **viscosity.**

Most institutions have an established formulary that provides the most cost-efficient choice within categories of enteral formulas. In other situations, clinicians determine the formula choice based solely on the patient's nutritional requirements.

Protein The protein component of enteral formulas is typically derived from soy or casein. The amount of protein in enteral formulas ranges from the standard amount (about 10%–15% of kcal) to high-protein formulas containing up to 25% of kcal from protein. The majority of formulas provide protein that requires enzymes to split the "intact protein" into peptides before absorption across the gastrointestinal tract. Formulas containing protein from peptides (also called "elemental" or "chemically defined" formulas) are used for patients with enzyme deficiency or other conditions resulting in maldigestion. Formulas with specialized amino acid profiles for renal failure, hepatic failure, stress, and inborn errors of metabolism have been developed from crystalline amino acids (these are also called "elemental" or chemically defined diets). Some "elemental" formulas are supplemented with additional amounts of specific amino acids such as glutamine or arginine. The disadvantages of peptide and crystalline amino acid products include poor patient acceptability due to a

viscosity—thickness of a liquid

FIGURE 7.5 Selecting a Formula

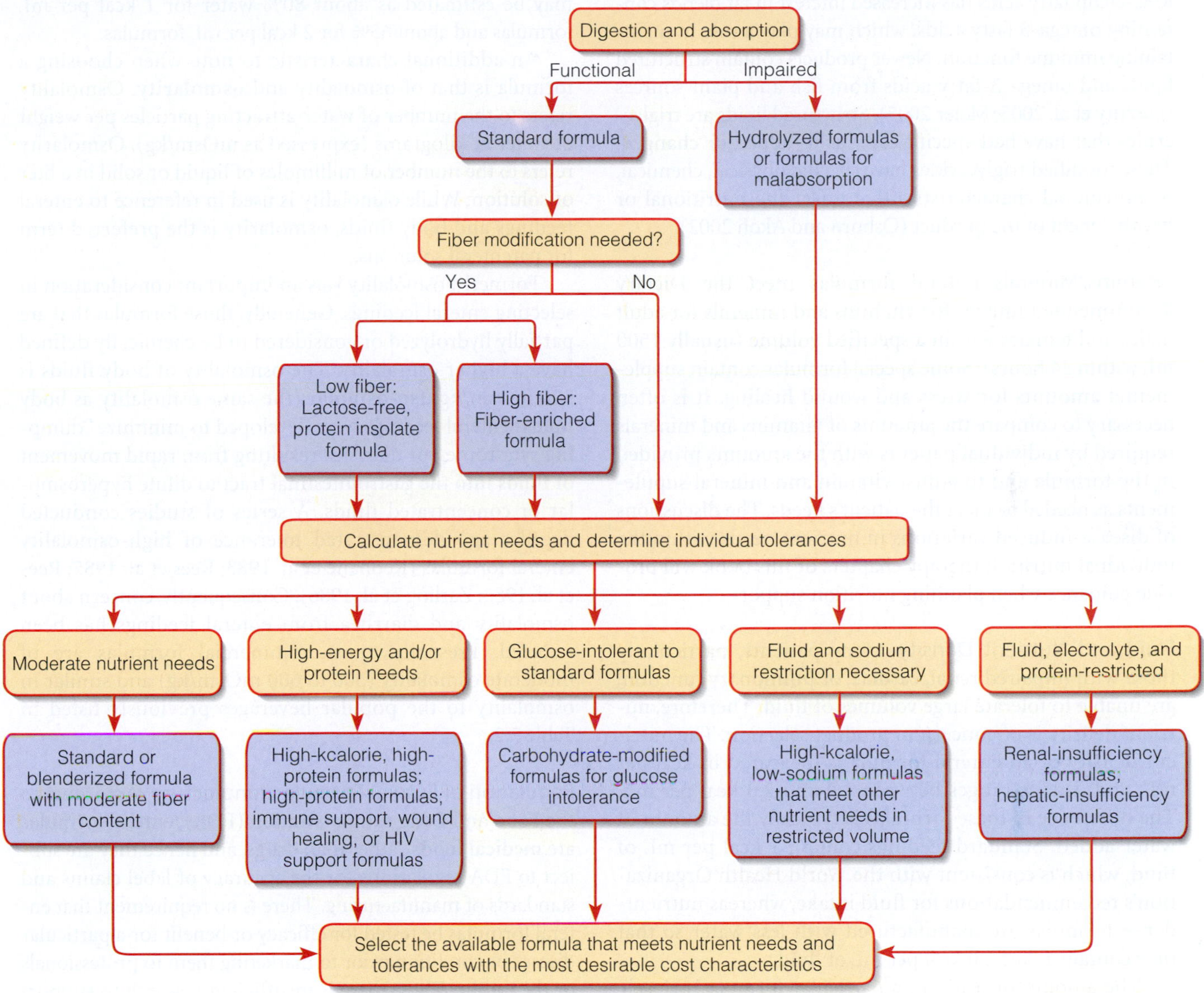

Source: S. Rolfes, K. Pinna and E. Whitney, *Understanding Normal and Clinical Nutrition*, 7e, copyright © 2006, p. 658

disagreeable taste and smell, increased osmolality, high cost, and limited data supporting their efficacy.

Carbohydrate The carbohydrate sources for enteral formulas are large molecules such as monosaccharides, oligosaccharides, dextrins, and maltodextrins. Formulas are lactose free, and sucrose is rarely used. A recent innovation in carbohydrate composition of some enteral products is the addition of fructo-oligosaccharides (FOS), which, like all oligosaccharides, are fermented into short-chain fatty acids. Short-chain fatty acids are used by the **colonocytes** (intestinal cells) as fuel and may play a role in maintaining gastrointestinal integrity (Bengmark 2005).

Originally, formulas were fiber-free and low in residue, but today many products are also available with fiber added. Benefits attributed to fiber, particularly improved bowel function, have more often been associated with soluble fiber. Typically, enteral products contain only small amounts of soluble fiber because it is **hydrophilic** (attracts water). This hydrophilic property causes enteral formulas to thicken and form a gel when fiber is added. Insoluble fibers such as soy polysaccharides are most often found in enteral feedings because they are less hydrophilic, but the advantage of insoluble fiber in enteral formulas is unclear (Malone 2005).

Lipid The fat or lipid sources for enteral formulas include corn and soy oil, which are long- and medium-chain fatty

colonocyte—epithelial cell of the large intestine or colon

hydrophilic—water loving, or attracting water

acids. Concern about the immunosuppressive properties of long-chain fatty acids has increased interest in fat blends containing omega-3 fatty acids, which may play a role in maintaining immune function. Newer products contain structured lipids and omega-3-fatty acids from fish and plant sources (Lasztity et al. 2005; Meier 2005). Structured lipids are triglycerides that have had specific fatty acids added or changed. These modified triglycerides have specific physical, chemical, or nutritional characteristics that affect the nutritional or health benefit of the product (Osborn and Akoh 2002).

Vitamins/Minerals Enteral formulas meet the Dietary Recommended Intakes for vitamins and minerals for adult males and females within a specified volume (usually 1500 mL within 24 hours). Some special formulas contain supplemental amounts for stress and wound healing. It is often necessary to compare the amounts of vitamins and minerals required by individual patients with the amounts provided in the formula and to adjust vitamin and mineral supplements as needed to meet the patient's needs. The discussions of disease-induced variations in nutrient needs within the individual nutrition therapy chapters of this book will provide guidance when planning nutrition support.

Fluid and Nutrient Density Many patients, particularly those with impaired renal, cardiac, or pulmonary function, are unable to tolerate large volumes of fluid. Therefore, nutrient density is of concern in product selection. The nutrient density of an enteral formula is measured in kcal per mL, and usually ranges between 1.0 and 2.0 kcal per mL. The difference in these formulas is typically the amount of water added. Standard feedings contain 1 kcal per mL of fluid, which is consistent with the World Health Organization's recommendations for fluid intake, whereas nutrient-dense formulas are manufactured with less water so that they contain 1.5 or 2.0 kcal per mL of fluid.

The amount of water in a particular formula is closely related to its nutrient density. Enteral formulas are often the sole source of water for patients receiving them; thus, it is important to ensure adequate fluid intake. Patients who receive these products require careful monitoring of their fluid status to ensure that they remain adequately hydrated. The precise water content of formulas may be obtained from the product literature, but in practice the free water content of formulas may be estimated as about 80% water for 1 kcal per mL formulas and about 65% for 2 kcal per mL formulas.

An additional characteristic to note when choosing a formula is that of osmolality and **osmolarity**. Osmolality refers to the number of water-attracting particles per weight of water in kilograms (expressed as mOsm/kg). Osmolarity refers to the number of millimoles of liquid or solid in a liter of solution. While osmolality is used in reference to enteral feedings and body fluids, osmolarity is the preferred term for parenteral solutions.

Formerly, osmolality was an important consideration in selecting enteral feedings. Generally, those formulas that are partially hydrolyzed or considered to be chemically defined have a higher osmolality. The osmolality of body fluids is 300 mOsm/kg. **Iso-osmolar** (the same osmolality as body fluids) enteral feedings were developed to minimize "dumping syndrome," or diarrhea resulting from rapid movement of fluids into the gastrointestinal tract to dilute hyperosmolar or concentrated fluids. A series of studies conducted decades ago demonstrated tolerance of high-osmolality enteral formulas (Keohane et al. 1983; Rees et al. 1985; Rees et al. 1986; Zarling et al. 1986). Consequently, concern about osmolality and diarrhea from enteral feedings has been reduced. Presently, most commercial formulas are of moderate osmolality (300 to 600 mOsm/kg) and similar in osmolality to the popular beverages previously listed in Table 7.2.

Regulation of Enteral Formula Manufacture According to the Food and Drug Administration (FDA), enteral formulas are **medical foods** rather than drugs, and hence they are subject to FDA regulations for the accuracy of label claims and standards of manufacturing. There is no requirement that enteral formulas be tested for efficacy or benefit for a particular disease or condition prior to marketing them to professionals or the public. Thus, there is insufficient research to support use of many specialized enteral formulas on the market (Malone 2005). This point often escapes the attention of the medical community and the public, who think of enteral formulas as thoroughly tested drugs rather than "medical foods."

Cost Cost is an important consideration in selecting enteral products. Traditionally, enteral products have been inexpensive, but newer products and those for specialized indications are increasingly expensive. The cost of products varies a great deal according to volume purchased. As mentioned previously, many institutions implement a formulary system that limits the total number of products, resulting in an overall reduced cost.

Patients who purchase a few cans of formula from a retail grocery store or pharmacy typically pay a much higher price than large institutions that purchase thousands of cases per year. Standard products, purchased in bulk by health care institutions, cost as little as a few dollars a

osmolarity—number of millimoles of liquid or solid in a liter of solution

iso-osmolar—having the same osmolality as body fluids (approximately 300 mOsm/kg)

medical foods—foods administered under the supervision of a physician and intended for the specific dietary management of a disease for which distinctive nutritional requirements are established

case (usually 24 250 mL cans). However, products with specialized amino acid or lipid profiles cost several dollars per 250 mL can. Given the typical requirement of six to eight cans of formula per day, the cost of formula may be several hundred to more than a thousand dollars per month.

Feeding Techniques After the access and formula have been determined, delivery of the enteral feeding should be considered. Several methods to administer enteral feedings have been used successfully (Skipper and Ratz 1998; Thompson 2006). **Bolus feedings** consist of the rapid administration of 250 to 500 mL of formula several times daily. A syringe may be used to inject feedings through the tube (see Figure 7.4). **Intermittent feedings** are also administered several times daily, over 20 to 30 minutes. A pump is typically used to control the flow rate. If pumps are unavailable, formula may flow slowly by gravity into the feeding tube from a container suspended above the patient.

Continuous feedings are administered over 10 to 24 hours daily, using a pump to control the feeding rate. Continuous feedings are typically preferred in hospital or nursing home settings because they are easier and less time consuming to administer than bolus or intermittent feedings. Using a pump to deliver feedings slowly at a continuous rate may improve feeding tolerance, but intermittent feedings are also well tolerated by individual patients. A disadvantage of continuous feeding is the expense of the pump and the disposable equipment that is required. Another disadvantage of continuous feeding is restricted mobility if the pump and other equipment are difficult to move. To overcome this disadvantage, the feeding schedule can be adjusted so that feedings are cycled over 8 to 12 hours rather than 24 continuous hours.

FIGURE 7.6 Bolus Feeding

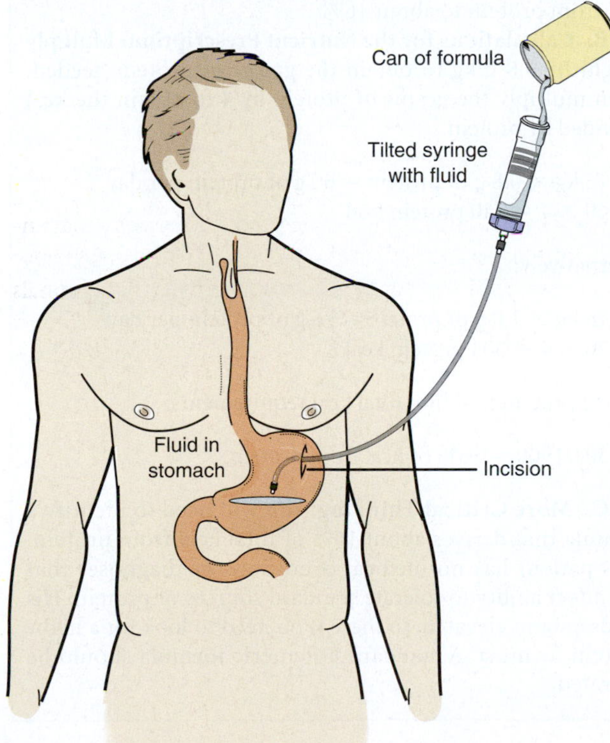

Source: Timby, Website for Fundamental Nursing Skills and Concepts 8e, Copyright © 2005 Lippencott Williams & Wilkins. Instructor's Resource CD-ROM to Accompany Timby's Fundamental Nursing Skills and Concepts 8e, Diana L. Rupert and Geralyn Frandsen.

Equipment Sophisticated feeding tubes, feeding administration sets, and pumps have been developed so that enteral feedings are more comfortable for the patient and easier for patients or caregivers to administer. Feeding tubes have been improved so that they are soft and pliable, resist clogging, and have flexible, weighted tips. Most are made of polyvinyl, silicone, or polyurethane. The outer lumen diameter is described using a measurement called French size (1 Fr = 0.33 mm). Most tubes range from 10 to 14 Fr.

Enteral formulas can be provided in cans or in sealed containers. In years past, pouring formula into bags was a standard procedure for formula delivery (see Photo 7.4). This is no longer necessary.

Enteral feeding is most often delivered using a small pump similar to those used to control intravenous fluids (see photo on page 164). Pumps are available that are small enough for use in ambulatory home care situations and that can be programmed to automatically flush the tube with water. Formula and equipment manufacturers are likely to continue product innovation in order to improve the ease of use and cost-effectiveness of pumps.

Putting It All Together: Determination of the Nutrition Prescription The enteral nutrition prescription will be based on the dietitian's nutrition assessment and recommendations, as described in Box 7.4. The steps in determining the nutrition prescription are the following:

1. Establish dosing weight (the patient's stable weight from which calculations should be made), protein, energy, and fluid requirements.
2. Determine appropriate enteral formula consistent with the patient's nutritional needs and diagnosis.
3. Determine the goal rate (the final rate which will meet the patient's nutritional needs) for the patient. Calculate the total energy needs and divide by the caloric density of the formula (kcal/mL).

bolus feedings—rapid administration of 250 to 500 mL of formula several times daily

intermittent feedings—administration of formula several times daily, over 20 to 30 minutes

continuous feedings—administration of formula for 10 to 24 hours daily, using a pump to control the feeding rate

BOX 7.4 CLINICAL APPLICATIONS

Nutrition Care Process: Determining the Enteral Nutrition Prescription

Nutrition Assessment: You have been consulted concerning a 76-year-old man who has just had a percutaneous endoscopic gastrostomy (PEG) tube placed and will be discharged to a nursing home in a few days. The patient is recovering from a stroke, which left him with severe swallowing difficulty and persistent aspiration. There is a small possibility that he will recover some swallowing function, but for now, he is NPO. The patient weighed 180 pounds on hospital admission, but his weight has declined to 168 pounds at the present time. He is 5'10" tall. The stroke left his right arm paralyzed, but he can walk with assistance and attends physical therapy twice daily for an hour each session. He has been maintained on tube feeding since shortly after hospital admission. Your physical exam reveals that he is appropriately hydrated. His weight has not changed since his medical condition stabilized, and he was transferred from the Neurological Intensive Care Unit several days ago.

Step 1: Determine a "dosing" weight.

A. Critical Thinking: The patient has lost weight but is at his ideal weight for height. In this case, the ideal weight is a good choice for the "dosing" weight.

B. Calculations for the Nutrient Prescription: Example: Convert the weight to kilograms by dividing weight in pounds by 2.2.

168 pounds/2.2 = 76 kg

Step 2: Determine a kcal goal.

A. Critical Thinking: Despite the recent weight loss, it is probably desirable to maintain this patient's current weight, as it is also his desirable weight. Weight gain is probably not desirable as it might further limit his mobility and impede whatever recovery he might have.

To maintain weight, 25 to 30 kcal/kg is often used. As his activity level changes, adjustments in the nutrient prescription may be needed to maintain the desired weight. You could just as easily calculate energy needs using a variety of equations including the Harris-Benedict or Mifflin-St. Jeor.

B. Calculations for the Nutrient Prescription: Example: Multiply the weight by the number of kcal/g selected.

76 kg × 25 kcal = 1900 kcal

Example with Mifflin-St. Jeor:

10 (76) + 6.25 (177.5) − 5 (76) + 5 = 1484 ×
1.2 (activity factor) = 1800 to 1900 kcal*

C. More Critical Thinking: Calculations of kcal goal using two different methods are within 100 kcal of each other. This verifies that you are on the right track.

Step 3: Adjust for activity and injury.

A. Critical Thinking: This patient has limited activity due to his disability, but he may also expend considerable energy during physical therapy. You may wish to read the physical therapy consultation note or ask the therapist directly about activities performed. In this case, the patient will be discharged from physical therapy when he is transferred to the nursing home. There is no reason to add to the maintenance weight based on his current activity level. This patient's medical condition is stable, and there is no reason to add an injury factor at this time.

Step 4: Calculate a protein goal.

A. Critical Thinking: This patient is stable from a nutritional standpoint, with no protein losses. The DRI for protein for a 76-year-old man is 0.8 g/kg of body weight, which would equate to about 10% of kcal from protein. Despite little supporting evidence, it is popular to assume that elderly persons need "more" protein, and many clinicians use a figure of 1.0 g protein/kg to calculate the protein needs of elderly patients, which increases the percentage of kcal from protein to about 16%.

B. Calculations for the Nutrient Prescription: Multiply weight by 0.8 g/kg to obtain the grams of protein needed. Then multiply the grams of protein by 4 to obtain the kcal provided by protein.

76 kg × 0.8 g of protein = 60 g of protein per day
60 × 4 = 240 protein kcal

Alternatively:

76 kg × 1.0 g of protein = 76 g of protein per day
76 × 4 = 304 protein kcal

Divide protein kcal into total kcal requirements:

304/1900 = 16% of kcal from protein.

C. More Critical Thinking: You will need to identify a formula that derives about 16% of total kcal from protein. This patient has no medical conditions or diagnoses that will affect ability to tolerate standard sources of protein. His needs are not elevated, so there is no need to look for a high-protein formula. A standard polymeric formula should be tolerated.

Step 5: Identify an appropriate amount of kcal from lipid.

A. Critical thinking: The minimum amount of lipid for this patient is approximately 10% of kcal (about 190 kcal or 21 grams of fat), while the maximum amount would be about 1.2 grams of fat per kg body weight (810 kcal or about 90 grams of fat). A typical amount would be somewhere near 30% of kcal.

B. Calculation for Nutrient Prescription: Multiply weight by 1.2 to obtain the maximum grams of fat intake. Then multiply by 9 to obtain the kcal provided by the fat.

76 kg × 1.2 g of fat = 91 g of fat per day
91 × 9 = 820 fat kcal
810/1900 = 43% of kcal from fat as the maximum dose

C. More Critical Thinking: This patient has no special needs for types of lipid, so you will need to identify a general formula that is between 10% and 43% fat.

Step 6: Identify an appropriate amount of kcal from carbohydrate.

A. Critical Thinking: The DRI for carbohydrate is 130 grams (520 kcal) or 27% of kcal.

B. Calculation for nutrient prescription: 1900 × .27 = 513 grams of carbohydrate

C. More Critical Thinking: This patient has no medical conditions or diagnoses that will affect ability to tolerate standard sources of carbohydrate. A standard polymeric formula should be tolerated.

Step 7: Consider electrolyte needs.

A. Critical Thinking: This patient has no abnormal electrolyte losses, and at present renal function is normal, so his electrolyte needs should be similar to the DRI. Many formulas provide low electrolyte levels. Try to identify a formula that provides recommended amounts of electrolytes.

Step 8: Consider vitamin and mineral needs.

A. Critical Thinking: Note: It is unlikely that this patient has taken a multiple vitamin supplement each day for years. Example: Consult the DRI tables to identify appropriate vitamin and mineral requirements. Remember to check the vitamin levels in the enteral product so that you will know how much the patient is getting. Try to identify a feeding that provides the DRI for vitamins and minerals within the volume recommended in Step 9.

Step 9: Determine fluid needs.

A. Critical Thinking: Patient is not dehydrated. For this patient, 30 mL/kg/body weight or 1 mL/kcal provided could be used to calculate fluid requirements.

B. Calculation for Nutrient Prescription: Fluid requirements are approximately the same as the energy requirements—1900 to 2000 mL.

Step 10: Establish administration and delivery methods.

A. Critical thinking: This patient has a PEG tube and will be transferred to a nursing home. It will be best to start on a continuous feeding while keeping in mind that he will need to be prescribed a bolus feeding for discharge.

Step 11: Write final enteral nutrition prescription.

A. Critical Thinking: You have determined that this patient needs a standard polymeric formula. From the choices available (see Appendix C2), note that Osmolite® provides 1 kcal/mL and .034 g protein/mL. All other nutrient levels are within the calculations. Since the patient needs 1900 kcal, the total volume of formula will be 1900 mL.

B. Calculation for Nutrient Prescription: For a continuous feeding, take the patient's formula volume and divide by 24 (hours/day) to establish goal rate.

1900/24 = 80 mL/hr

For the bolus feeding, divide the volume of feeding by 4 feedings/day:

1900/4 = 475 mL.

C. More Critical Thinking: Determine initial start rate and recommended progression. As you learned earlier in this text, protocols vary, but a standard isotonic polymeric formula can be initiated at full strength from 25 to 50 mL/hr.

D. Calculation for Nutrient Prescription: Begin Osmolite at 50 mL/hr and increase in 20 mL increments to goal rate of 80 mL/hr every six to eight hours. Prior to discharge, switch patient to bolus regimen of 475 mL/feeding four times daily.

*A note on rounding numbers: Clinicians typically round the results of whole numbers, such as kilocalories, milligrams of sodium, or milliequivalents of potassium to the nearest five or ten. This practice varies widely, however, and attention to local customs is advised.

Many institutions have specific protocols for initiation, advancement, and transition of feedings. Table 7.5 provides examples of these protocols. Most polymeric, isotonic formulas can be initiated at 10–50 mL/hr. The rate is advanced in increments of 10–25 mL/hr every four to eight hours until goal rate is established (ASPEN 2002; Skipper and Ratz 1998; Stroud, Duncan, and Nightingale 2003; Thompson 2006;).

Tube Feeding Container

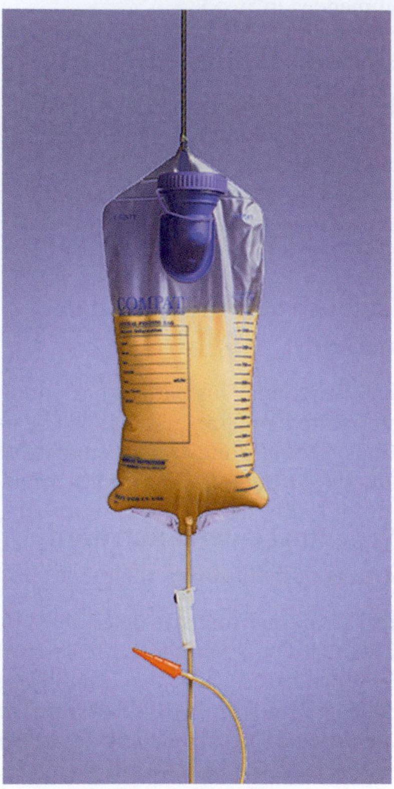

Source: photo provided courtesy of © Novartis Medical Nutrition

Feeding Pump

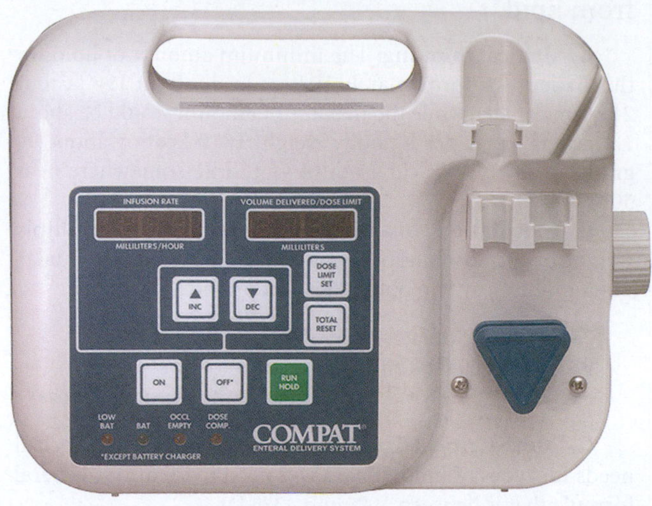

Source: photo provided courtesy of © Novartis Medical Nutrition

Complications Enteral feeding is not a simple procedure. Patients who receive enteral feeding may experience a variety of complications, and some of these—such as aspiration or tube misplacement—may be serious. Complications of enteral feedings may develop at any point during a course of therapy. High-risk patients who have concurrent illnesses require an experienced dietitian to successfully manage enteral feedings.

Mechanical Complications The enteral tube should be placed by a well-trained physician, nurse, or dietitian in order to minimize the risk of tube misplacement. Clogged, twisted, or kinked feeding tubes are common, and may result in reduced or delayed feeding. Clogged tubes most often result from administration of medications through the tube.

To prevent clogged tubes, the feeding tube is flushed with a syringe containing at least 25 mL or more of tap water several times daily. (Note: Some institutions use sterile, distilled, or bottled water for this purpose.) A number of home remedies, usually involving various types of soda or juice, are sometimes recommended to unclog tubes. However, none of these have been shown to be superior to warm water in unclogging tubes (Nicholson 1987). In some institutions, a combination of bicarbonate and pancreatic enzymes is used, and in others a commercial product for unclogging tubes is available. In no circumstances should the **stylet** used to place the tube be reinserted into the tube, because this is potentially painful for the patient and may perforate the feeding tube. Using a small-volume syringe to force liquid into the tube can also rupture the tube, and a large-volume syringe is recommended in order to decrease pressure on the feeding tube wall.

TABLE 7.5

Sample Enteral Protocol

Continuous/Nocturnal Feeding

Initiation: Full strength (all products except 2 kcal/mL) at 50 mL/hour and increase by 25 mL every eight hours to goal rate. A 2.0 cal/mL product is started at 25 mL/hour (as few patients need >50 mL/hour to meet estimated needs). The final goal rate is dependent on the patient's caloric requirements and GI comfort.

Bolus/Intermittent Feeding

Initiation: 125 mL, full strength (regardless of product) every three hours for two feedings; increase by 125 mL every two feedings to final goal volume per feeding during waking hours.

Source: Used with permission from the University of Virginia Health System Nutrition Support Traineeship Syllabus

stylet—wire guide within the enteral tube that assists with insertion

Gastrointestinal Complications The primary gastrointestinal complication of enteral feeding is diarrhea, commonly defined as an abnormal looseness of stool with increased liquidity or decreased consistency and an output of greater than 200 g/day for adults and greater than 20 g/kg for children (Donowitz and Kokke 1995). Diarrhea is often assumed to be caused by enteral feedings, but there are other distinct causes. For example, diarrhea is a well-known side effect of many commonly used antibiotics and is often associated with administration of hyperosmolar medications via the feeding tube. It is important to rule out infectious organisms, another common cause of diarrhea. Manipulation of formula rate, strength, or type is often recommended as a means of reducing diarrhea in tube-fed patients, but data demonstrating the effectiveness of these practices is limited. In practice, antidiarrheal medications are often the only suitable treatment. These medications may include diphenoxylate-atropine (Lomotil©), loperamide (Immodium©), and deodorized tincture of opium, paregoric, and octreotide.

Aspiration **Aspiration** occurs when fluid is inspired into the lungs. Patients who are sedated, who have an endotracheal tube (a tube that allows oxygen into the lungs of patients receiving mechanical ventilation), or who have difficulty swallowing are at risk for aspiration. It is a potentially serious condition that may result in pneumonia or even death. To avoid aspiration, it is important that the patient's head be elevated higher than her or his stomach, or at an angle of about 45 degrees, during feeding. Formerly, blue food coloring was placed in the enteral feedings so that the caregiver could distinguish between the formula and body fluids. This practice has been discontinued due to two associated deaths (Klein 2004). Residual volumes of liquid in the stomach have also been used to determine whether a feeding was emptied from the gastrointestinal tract, but guidelines for this practice are not well established (McClave et al. 2005). In the Consensus Statement presented by the North American Summit on Aspiration in the Critically Ill Patient, it was recommended that enteral feeding be stopped if there is definite regurgitation or aspiration of gastric contents or if a residual greater than 500 mL is measured. In the absence of such a circumstance, careful monitoring and clinical assessment, combined with the residual volume, should be used to make a decision about tolerance of enteral feeding (McClave et al. 2002).

Monitoring for Complications

In order to ensure that complications do not develop, it is important to determine whether the patient is medically stable and how long it has been since the tube feeding started. Patients for whom tube feeding is newly initiated should have more intense monitoring during the time the feeding is being progressed to goal. The frequency of monitoring can be diminished as the patient becomes stable. Table 7.6 outlines recommendations for monitoring the individual patient receiving enteral nutrition.

TABLE 7.6

Suggested Monitoring for Enteral or Parenteral Feedings in Hospitalized Patients

Parameter	Medically or Nutritionally Unstable	Medically or Nutritionally Stable
Sufficiency of nutrient intake: intake/output; weight	Daily	Weekly
Electrolytes, BUN, creatinine	Daily, then 3 x week	3 x week
Magnesium, phosphorus, calcium	Daily, then 3 x week	3 x week
Liver function tests	Weekly	As needed
Triglycerides	Weekly	Every 1–2 weeks
Weight	Daily	Weekly
Hydration/fluid status: physical assessment of skin turgor, presence of edema, temperature; oral cavity for color, texture, moisture/dryness Vital signs: blood pressure, respirations, pulse Intake/output	Daily	3 x week
Bowel function	Daily	As needed
Blood glucose	3 x daily until stable	Every 1–2 weeks
Nitrogen balance	Prn (as necessary)	Prn

Dehydration/Tube Feeding Syndrome Patients with insufficient fluid intake who are receiving tube feedings may develop **"tube feeding syndrome,"** (hyperosmolar-non ketotic dehydration) over a short two to four day period. Tube feeding syndrome may be prevented by providing sufficient fluid (about 1 mL/kcal) with the feeding. Patients receiving less fluid should be monitored with a daily fluid status assessment (see Chapter 8 for more on fluid status and its assessment). Fluid status is easily assessed by testing skin

aspiration—inspiration of foreign matter into the lung

"tube feeding syndrome"—hyperosmolar-nonketotic dehydration, over a short two to four day period

turgor for signs of dehydration or edema, and by estimating the adequacy of fluid intake and output. If the results of the assessment indicate that the patient is taking insufficient fluid, and there is no reason for a fluid restriction, then the amount of fluid administered is increased.

Electrolyte Imbalances In stable patients receiving enteral feedings, the DRI for sodium, potassium, calcium, magnesium, and phosphorus are often used as a guide for electrolyte intake (see Chapter 8 for more on electrolytes). For patients with organ failure, lower electrolyte levels are often appropriate. The electrolyte content of the formula in use may be compared to requirements, and supplemental electrolytes may be provided as needed. Magnesium and potassium administered via the feeding tube may produce a cathartic effect, and gastrointestinal calcium absorption may be poor. Thus, intravenous electrolyte supplementation is sometimes preferred.

It is imperative for the clinician to understand that enteral and parenteral electrolyte requirements differ because of the variable of absorption. Another key distinction is that parenteral electrolytes are measured in mEq or mmol, while oral requirements are stated in milligrams. Details of parenteral and enteral requirements may be found in Table 7.7.

TABLE 7.7

Electrolyte Requirements

		Dietary Reference Intake for Oral/Enteral Feedings	Recommendations for Parenteral Intake
Potassium			
Adults over the age of 14		4700 mg	1 to 2 mEq/kg
Sodium			
14–50 years		1500 mg	1 to 2 mEq/kg
51–70 years		1300 mg	
> 70 years		1200 mg	
Chloride			
14–50 years		2300 mg	To maintain acid-base balance
51–70 years		2000 mg	
> 70 years		1800 mg	
Bicarbonate			
		---	To maintain acid-base balance
Calcium			
14–18 years		1300 mg	10 to 15 mEq
19–50 years		1000 mg	
> 51 years		1200 mg	
Magnesium			
Males	14–18 years	410 mg	8 to 20 mEq
	19–30 years	400 mg	
	> 31 years	420 mg	
Females	14–18 years	360 mg	
	19–30 years	310 mg	
	> 31 years	320 mg	
Phosphorus			
14–18 years		1250 mg	20 to 40 mmol
> 18 years		700 mg	

These are standard intake ranges for generally healthy people with essentially normal organ function who do not have abnormal needs or losses.

Underfeeding or Overfeeding Enteral feeding is based on a "dosing" weight established by the dietitian. Every attempt is made to feed the patient an appropriate amount of nutrients based on this weight. Both underfeeding and overfeeding are thought to be detrimental to the patient. Underfeeding may delay nutritional repletion and wound healing. In some critically ill patients, however, it is thought that "permissive underfeeding" may assist with preventing acute metabolic and respiratory complications (Kudsk and Sacks 2004; Zaloga and Roberts 1994). Overfeeding, on the other hand, may result in hyperglycemia, hypertriglyceridemia, and hepatic steatosis (fatty liver) (Klein, Stanek, and Wiles 1998; Kraft, Btaiche, and Sachs 2005).

There are no clear definitions of either underfeeding or overfeeding. While many clinicians prefer to feed hospitalized patients 25 to 30 kcal per kg, there are others who prefer to use smaller amounts (18 to 20 kcal/kg) for overweight patients. Clinicians commonly complain of overfeeding when caloric intakes approach 35 kcal/kg. However, the accepted maximums of protein (1.8 g/kg/day), carbohydrate (4 mg/kg/min), and fat (1.2 g/kg/day) provide 41 kcal/kg.

Hyperglycemia During periods of physiological stress, such as those caused by severe illness or severe infection (sepsis), hyperglycemia can appear even in patients with no previous history of diabetes. Most recently, the use of intensive insulin therapy to maintain normal blood glucose levels has resulted in a reduction of morbidity and mortality for critically ill patients. It has been proposed that insulin therapy not only controls hyperglycemia seen in metabolic stress but may affect the catabolic state, reduce inflammation, and improve the immune response (Langouche, Vanhorebeek, and Van den Berghe 2005).

The hyperglycemia associated with stress usually resolves as the stress response subsides, and nondiabetic patients do not experience long-term complications.

refeeding syndrome—metabolic alterations that may occur during nutritional repletion of starved patients

Refeeding Syndrome **Refeeding syndrome** is a term used to describe several common metabolic alterations that may occur during nutritional repletion of starved patients (Kraft, Btaiche, and Sachs 2005). This syndrome has been observed in the surviving victims of famine since the beginning of medical history. With the advent of parenteral nutrition, refeeding syndrome gained attention because of its often dramatic and sometimes fatal presentation. With starvation lasting more than a few days, liver gluconeogenesis slows, free fatty acids are used to produce energy in the form of ketones, and basal metabolic rate declines.

The reintroduction of carbohydrate, whether in oral, enteral, or parenteral form, results in a shift from ketones to glucose as the primary energy source. To metabolize glucose into energy, large quantities of phosphorus are required. Magnesium, potassium, and thiamin requirements may also increase to meet anabolic needs. The result is a drop in serum levels of phosphorus, which, if severe, may result in hemolysis, impaired cardiac function, impaired respiratory function, and even death. Hypomagnesemia (low serum magnesium) may result in tremor, muscle twitching, cardiac arrhythmias, and even paralysis (see Chapter 8). Hypokalemia (low serum potassium) is also associated with cardiac abnormalities. Thiamin deficiency has been documented infrequently, but may result in Wernicke's encephalopathy (see Chapter 18).

Patients at risk for refeeding syndrome include those who present with malnutrition, those who have a history of long-term inadequate oral intake, and those who have had minimal intake for several days as a result of physician-ordered NPO status or poor appetite. It is critical to monitor serum levels of phosphorus, magnesium, and potassium, and to provide supplementation as needed until the patient is receiving goal feedings. Clinicians have used the strategy of beginning feedings slowly and avoiding overfeeding in order to prevent refeeding syndrome. As of this date, research that documents the effectiveness of specific protocols to prevent refeeding syndrome has not been conducted.

Parenteral Nutrition

The word "parenteral" means "alongside" or "outside" the gastrointestinal tract, and is now used to describe the administration of drugs or nutrients by vein **(intravenously,** or **IV)**. **Parenteral nutrition (PN)**, developed in the 1960s to sustain the lives of individuals with severe gastrointestinal impairment, may also be called total parenteral nutrition (TPN), central venous nutrition (CVN), or intravenous hyperalimentation (IVH). Generally, parenteral nutrition (PN) is the preferred term. The term "hyperalimentation" originally described the practice of "hyperalimenting" or overfeeding patients. Although deliberately overfeeding or "hyperalimenting" patients is no longer common clinical practice, the term persists in many institutions. The distinguishing feature of PN is administration of concentrated macronutrients, vitamins, minerals, and electrolytes into a large central vein so that the volume of blood flow is sufficient to immediately dilute the concentrated parenteral solutions.

The terms **peripheral parenteral nutrition (PPN)** and peripheral venous nutrition (PVN) refer to the administration of large-volume, dilute solutions of nutrients into a vein in the arm or back of the hand. PVN is irritating to the small veins, and peripheral access is difficult to maintain for more than a few days. PVN also provides insufficient kcal for many patients, and its use is declining.

Indications Parenteral nutrition is indicated in those clinical situations where the patient is unable to meet nutritional needs either by an oral diet or through the use of enteral nutrition. The clinical conditions that may require parenteral nutrition include an inability to digest and absorb nutrients, such as in malabsorption; massive bowel resection or short bowel syndrome; intractable vomiting, as in hyperemesis gravidarum; GI tract obstruction; impaired GI motility; and abdominal trauma, injury, or infection.

Decisions related to parenteral nutrition, like those for enteral nutrition, are based on the Nutrition Care Process. The patient's nutrition assessment, the length of time the patient will require nutrition support, and the patient's diagnosis and current medical condition will assist the clinician in making the decisions that are required to build the parenteral nutrition prescription.

Certification of medical necessity for PN must be established in order to ensure that the patient's care is financially feasible. Reimbursement for nutrition support varies among insurance providers (see Chapter 1).

Venous Access The primary difference between enteral and parenteral feedings is that nutrients are provided via the veins rather than the gastrointestinal tract in PN. The illustration in Figure 7.7 may be of assistance in visualizing the types and location of vascular access used for parenteral nutrition.

Short-Term Venous Access The most common parenteral access is a **central venous catheter (CVC)** or central line inserted percutaneously (through the skin) at the bedside while the patient is under local anesthesia. Central catheters

intravenously (IV)—by vein, in reference to administration of drugs or nutrients

parenteral nutrition (PN)—administration of nutrition directly into the circulatory system (also known as total parenteral nutrition [TPN], central venous nutrition [CVN], or intravenous hyperalimentation [IVH])

peripheral parenteral nutrition (PPN)—administration of nutrition into a vein in the arm or back of the hand (also known as peripheral venous nutrition [PVN])

central venous catheter (CVC)—intravenous access inserted into large veins such as the subclavian, jugular, or femoral veins in the center of the body

FIGURE 7.7 Sites for Parenteral Access

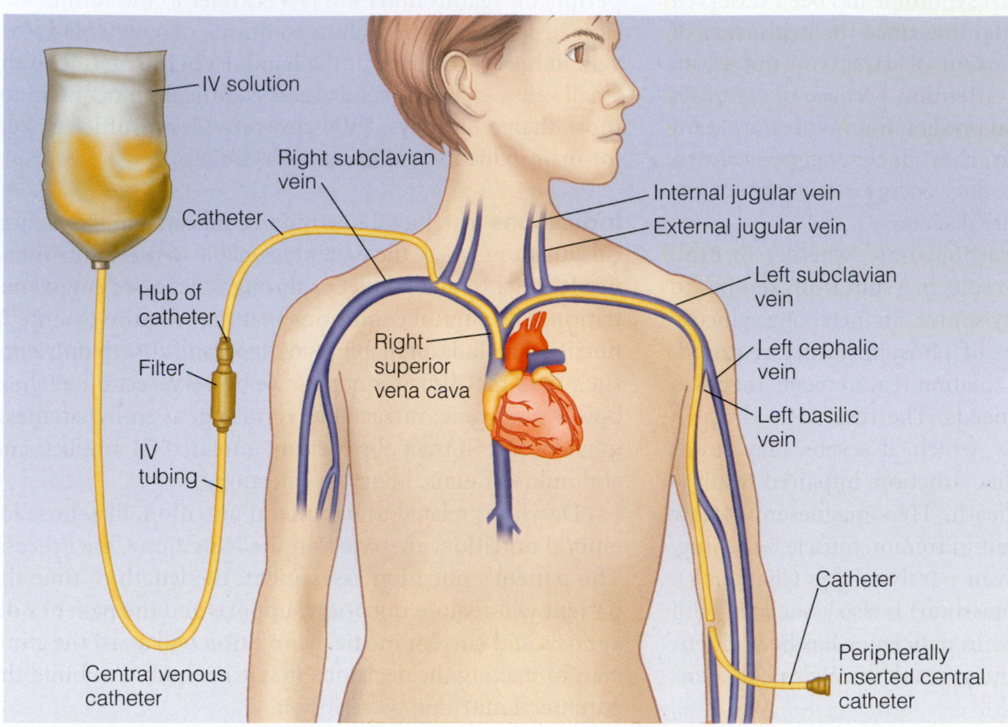

Source: S. Rolfes, K. Pinna and E. Whitney, *Understanding Normal and Clinical Nutrition*, 7e, copyright © 2006, p. 677

are inserted into large veins such as the subclavian, jugular, or femoral veins in the center of the body. Ultimately, these catheters reside in the superior vena cava, or in the inferior vena cava, in the case of femoral placement. These catheters are available in single-, double- or triple-lumen models. The lumen of the catheter refers to the interior of the tube through which the PN solution passes. If a catheter with sufficient lumens is available, a patient may receive medications, fluids, and nutrients at the same time. Catheters are usually changed every few days to help decrease the risk of infection inherent with an opening from the skin into a large, central vein.

peripherally inserted central catheter (PICC)—intravenous access inserted into the arm and threaded into the subclavian vein to the vena cava

tunneled catheter—intravenous access that is placed in the vein on the upper chest wall and exits the body near the xyphoid process, axilla, or abdominal wall

implantable port—intravenous access that is completely under the skin, is placed in the vein on the upper chest wall, and exits the body near the xyphoid process, axilla, or abdominal wall

The **peripherally inserted central catheter (PICC)** is gaining in popularity. Specially trained nurses can insert it, which increases the availability of the procedure and decreases costs, because the use of a central catheter requires a bedside surgical procedure by an MD. PICC lines are inserted into the arm and threaded into the subclavian vein to the vena cava.

Long-Term Venous Access For long-term use, or for home PN, a catheter is tunneled under the skin during a surgical procedure. **Tunneled catheters** most often enter the vein on the upper chest wall and exit the body near the xyphoid process, axilla, or abdominal wall. They are considered permanent, and with proper care can be left in place for several years. If a tunneled catheter contains more than one lumen, it can accommodate infusions of medications, fluids, or blood products in addition to TPN. Thus, it is useful for patients who receive frequent doses of intravenous medications in addition to PN.

Implantable ports are similar to tunneled catheters in that they must be placed in the operating room by a surgeon. They are available with single or double ports, and are suitable for long-term access. Unlike tunneled catheters, they lie completely under the skin, which decreases the risk of infection and makes them more acceptable to patients with body image concerns. Because they are usually placed just below the clavicle on the chest wall, they may be difficult for the patient to access. Nursing intervention may be required to change needles used to gain access to these ports.

Solutions Unlike enteral formulas, which are most often purchased in a form appropriate for patient administration, parenteral formulas are mixed or "compounded" in the hospital pharmacy. The method the pharmacy uses to compound TPN preparations is critical to development of parenteral nutrient prescriptions. In some institutions, a predetermined range of solutions (e.g., standard, high-potassium, high-protein, renal) is available. Since the advent of automated compounding equipment and bulk packaging of concentrated macronutrients, most hospitals provide individualized formulas that are adjusted daily to meet the rapidly changing needs of critically ill patients. An automated compounder is used to combine all nutrients needed for a 24-hour infusion into a single container (see Figure 7.8 for examples of PN solution labels). These automated compounders can be used to manufacture nutrient solutions that combine dextrose and amino acids (two-in-one formulas) or dextrose, amino acids, and lipids (three-in-one formulas). Some hospitals may use both systems, but usually hospital pharmacists prefer one system to the other. Each system has both advantages and disadvantages. For example, when an automated compounder is available, the parenteral prescription may include ingredients in as small as 1 mL increments. If an automated compounder is not available, formula changes and manufacture are more time consuming for the pharmacist.

Most institutions use the two-in-one system. This system provides a greater degree of flexibility in the amounts of dextrose and amino acids that can be given. Because lipids are added separately, they are typically administered based on the available container sizes (100 mL, 250 mL, 500 mL). An advantage of the two-in-one system is that formulas containing only dextrose and amino acids are clear, and any precipitate can be observed. A disadvantage of the two-in-one system is the need for an additional administration set (intravenous tubing and other devices required for the delivery of parenteral nutrition) for the lipids.

The three-in-one system requires a single administration set, which saves nursing time and reduces costs. On the other hand, the addition of lipids with the three-in-one system results in an opaque solution, which obscures precipitate and increases the risk of particulate being infused into the patient. Addition of lipid into the three-in-one solution limits the electrolytes and final concentration of amino acids in solution.

Protein Protein is included in parenteral nutrition in the form of individual amino acids in amounts consistent with the recommendations of the Food and Agriculture Organization and the World Health Organization. Modified products have been developed and marketed for renal failure, hepatic failure, and stress, but they are rarely used today due to the increased expense and limited clinical efficacy. Commercial amino acids are available from various manufacturers in concentrations of 3.5% (35 g/L) to 20% (200 g/L). Lower concentrations (3.5%–5.5%) are used for peripheral administration, while higher ones (8.5%, 10%, 11%, 15%, 20%) are used for central administration. Details for these products are available on websites of the major manufacturers. Parenteral nutrition is typically designed to provide individualized protein requirements, which range from 0.8 g/kg for normal adults to 1.5 to 1.8 g/kg for patients with burns, trauma, or healing wounds (ASPEN 2002; Mirtallo et al. 2002).

Factors that increase protein requirements above the DRI include diagnoses such as trauma, burns, sepsis, wounds, and bone marrow transplant. Protein restrictions are occasionally needed for patients with renal failure who are not receiving dialysis.

Carbohydrate The primary function of parenteral carbohydrate is to serve as an energy source. In the United States, dextrose monohydrate is used as the carbohydrate source for parenteral nutrition. The kcal content of this particular form of carbohydrate is 3.4 kcal/g. The minimum carbohydrate intake was recently specified in the DRI as 130 g/day, and it is known that approximately 100 g of carbohydrate is required daily to allow for protein sparing. The amount of 1 mg/kg/min is often used as the reference for the minimal amount of carbohydrate needed to spare protein. The maximum for glucose oxidation was originally studied in burn patients and found to be 7 mg/kg/min. In practice, lower figures of 3–4 mg/kg/min have been recommended (Mirtallo et al. 2002). Dextrose is commercially available in 5%, 10%, 50%, and 70% concentrations, but other concentrations may also be available.

Excessive carbohydrate may contribute to hyperglycemia, hepatic steatosis, and excessive carbon dioxide production. The standard 100 g of dextrose (appropriate for the reference 70 kg male) would be equivalent to an initial dextrose infusion rate as high as 2.2 mg/kg/min for a small patient—increasing the risk for refeeding syndrome. Elevated carbon dioxide also occurs with overfeeding, and it may jeopardize respiratory status and result in difficulty weaning from mechanical ventilation (see Chapter 23). Insufficient carbohydrate intake may result in protein being catabolized as an energy source.

Lipid The lipid in parenteral solutions is an emulsion of soybean or safflower oil. The lipid substrate provides essential fatty acids, as well as a concentrated source of energy. It also provides an avenue to meet energy needs if the patient is unable to tolerate a higher carbohydrate load. A minimum amount of lipid to prevent essential fatty acid

FIGURE 7.8 (A) Standard PN Label Template for Neonate or Pediatric Patient (B) Standard PN Label Template for Adult Patient

(A)

Institution/Pharmacy Name, Address and Pharmacy Phone Number		
Name	Dosing Weight	Location
Administration Date/ Time	Do Not Use After: Date/time	
Base Formula	Amount/kg/day	Amount/day
Dextrose	g/kg	g
Amino acids[a]	g/kg	g
Electrolytes		
Sodium chloride[b]	mEq/kg	mEq
Sodium acetate[b]	mEq/kg	mEq
Potassium chloride[b]	mEq/kg	mEq
Potassium acetate[b]	mEq/kg	mEq
Potassium phosphate[b]	mmol of P/kg	mmol of P
	(mEq of K)/kg	(mEq of K)
Sodium phosphate[b]	mmol of P/kg	mmol of P
	(mEq of Na)/kg	(mEq of Na)
Calcium gluconate	mEq/kg	mEq
Magnesium sulfate	mEq/kg	mEq
Vitamins, trace elements and medications		
Multiple vitamins[a]	mL/kg	mL
Multiple trace elements[a]	mL/kg	mL
L-cysteine	mg/kg	mg
H_2 antagonists[a]	mg/kg	mg
L-Carnitine	mg/kg	mg
Rate ______ mL/hour	Volume ______ mL	Infuse over 24 hours
Admixture contains _____ mL plus _____ mL overfill		
Central Line Use Only*		

[a]Specify product name.
[b]Since the admixture usually contains multiple sources of sodium, potassium, chloride, acetate, and phosphorus, the amount of each electrolyte/kg provided by the PN admixture is determined by adding the amount of electrolyte provided by each salt.

Source: Reprinted from Mirtallo J, Canada T, Johnson D, Kumpf V, Petersen C, Sacks G, Seres D, Guenter P., Safe Practices for Parenteral Nutrition, JPEN. 2004; 28:S39-S70. Table III, p.S54. Erratum: JPEN J Parenter Enteral Nutr. 2006;30:177, with permission from the American Society for Parenteral and Enteral Nutrition (A.S.P.E.N.). A.S.P.E.N. does not endorse the use of this material in any form other than its entirety.

(B)

Institution/Pharmacy Name, Address and Pharmacy Phone Number		
Name	Dosing Weight	Location
Administration Date/ Time	Do Not Use After: Date/time	
Base Formula	Amount/kg/day	Amount/day
Dextrose	g	(g/L)
Amino acids[a]	g	(g/L)
IVFE[a]	g	(g/L)
Electrolytes		
Sodium chloride	mEq	(mEq/L)
Sodium acetate	mEq	(mEq/L)
Sodium phosphate	mmol of P	(mmol/L)
	(mEq of Na)	(mEq/L)
Potassium chloride	mEq	(mEq/L)
Potassium acetate	mEq	(mEq/L)
Potassium phosphate	mmol of P	(mmol/L)
	(mEq of K)	(mEq/L)
Calcium gluconate	mEq	(mEq/L)
Magnesium sulfate	MEq	(mEq/L)
Vitamins, trace elements and medications		
Multiple vitamins[a]	mL	
Multiple trace elements[a]	mL	
Insulin	Units	(Units/L)
H_2 antagonists[a]	mg	
Rate ______ mL/hour	Volume ______ mL	Infuse over______hours
Formulation contains _____ mL plus _____ mL overfill		
Discard any unused volume after 24 hours		
Central Line Use Only*		

[a]Specify product name.

Source: (B) Fig Standard PN Label Template, Adult Patient Page S50

deficiency is unknown, but common clinical practice is to provide a minimum of 10% of kcal requirements as lipid (Skipper 2002). Much higher amounts of lipid have been administered in the past (up to 50%–60%), but more recent concern about the inflammatory properties of omega-6 fatty acids has reduced the amounts of lipid used in current practice (to 1.0 to 1.2 g/kg).

Electrolytes Using the standards established by the DRI as the beginning benchmark, electrolyte requirements in PN are based on body weight, existing electrolyte deficiencies, ongoing electrolyte losses, and changes in organ function. Because electrolyte requirements are also inextricably linked with the amount of macronutrients provided in the PN, it is impossible to manage PN without a thorough understanding of these complex relationships (see Chapter 8). Recommendations for standard electrolyte intake are found in Table 7.7. Note, however, that in practice electrolytes are individualized according to patient needs and are often considered to be the most difficult component for new registered dietitians working with nutrition support.

Vitamins and Minerals In 1979, the AMA released recommendations for vitamin and mineral additives to PN. These vitamin recommendations were used until 2003, when they were revised to include vitamin K (Helphingstine and Bistrian 2003). Rather than add individual amounts of vitamins to PN, most pharmacies purchase commercial multiple vitamin infusion products that meet the new recommendations. Because the vitamins are administered intravenously, there is no issue with absorption, and the amounts administered may differ from what is recommended for oral intake. The amounts of vitamins in commercial products have been increased over those for well persons based on the assumption that patients receiving PN will have wounds or critical illness. This presents a monitoring challenge for clinicians following patients on long-term PN. Vitamins may be given every other day in situations where excess vitamin intake is of concern. Table 7.8 lists daily adult parenteral vitamin requirements.

Trace Minerals Originally, zinc, copper, chromium, and manganese were added to PN (AMA 1979). Based on reports of deficiencies, newer products have been introduced that contain the original trace minerals plus selenium, iodide, and molybdenum (Skipper 2002). Trace elements are purchased commercially and contain four, five, six, or seven trace elements. In situations where reduced excretion or potential toxicities exist, trace elements are removed from the PN, and individual trace minerals are added according to need. Trace element additions for adult PN are listed in Table 7.9.

Medications In some institutions, PN may be used to deliver medications. In others, this practice is discouraged because limited data document drug compatibility with PN. Nevertheless, it is possible that albumin, aminophylline, cimetidine, famotidine, ranitidine heparin, or regular insulin may be included in PN. Prior to recommending medications be added to PN, the clinician should gain a

TABLE 7.8

Daily Requirements for Adult Parenteral Vitamins

Vitamin	Requirement
Thiamin	6 mg
Riboflavin	3.6 mg
Niacin	40 mg
Folic acid	600 mcg
Pantothenic acid	15 mg
Pyridoxine (B_6)	6 mg
Cyanocobalamin (B_{12})	5 mcg
Biotin	60 mcg
Ascorbic Acid	200 mg
Vitamin A	3300 UI
Vitamin D	200 UI
Vitamin E	10 UI
Vitamin K	150 mcg

Source: Reprinted from J. Mirtallo, T. Canada, D. Johnson, V. Kumpf, C. Petersen, G. Sacks, D. Seres, and P. Guenter. Safe Practices for Parenteral Nutrition. *JPEN* 2004;28(suppl):S54, with permission from the American Society for Parenteral and Enteral Nutrition (A.S.P.E.N.). A.S.P.E.N. does not endorse the use of this material in any form other than its entirety.

TABLE 7.9

Daily Trace Element Additions to Adult PN Formulations*

Trace Element	Standard Intake
Chromium	10–15 mcg
Copper	0.3–0.5 mg
Iron	Not routinely added
Manganese	60–100 mcg†
Selenium	20–60 mcg
Zinc	2.5–5 mg

*Standard intake ranges based on generally healthy people with normal losses.

†The contamination level in various components of the PN formulation can significantly contribute to total intake. Serum concentrations should be monitored with long-term use.

Source: Reprinted from J. Mirtallo, T. Canada, D. Johnson, V. Kumpf, C. Petersen, G. Sacks, D. Seres, and P. Guenter. Safe Practices for Parenteral Nutrition. *JPEN* 2004;28(suppl):S54, with permission from the American Society for Parenteral and Enteral Nutrition (A.S.P.E.N.). A.S.P.E.N. does not endorse the use of this material in any form other than its entirety.

thorough understanding of the practice at his or her institution through observation and consultation with pharmacists and dietitians.

Compounding Parenteral solutions are compounded from as many as 40 different items under the supervision of a licensed pharmacist. In order to maintain sterility, compounding is completed in a "clean room" under a laminar flow hood. Because PN is compounded from amino acid, dextrose, and lipid solutions, solubility is an important consideration affecting both the maximum amount of nutrients and the minimum amount of fluid that can be incorporated. Precipitates may form in PN solutions if greater than recommended maximum amounts of electrolytes and minerals are added, especially when PN is subjected to changes in temperature or pH. Likewise, minimum volumes are impacted by the concentration of amino acids, dextrose, and lipid that are available for compounding.

Putting It All Together: Determination of the Parenteral Nutrition Prescription The parenteral nutrition prescription (see Figure 7.9) will be based on the dietitian's nutrition assessment and the physician's recommendations. Complete detail is provided in Box 7.5. The basic steps include:

1. Establish dosing weight and energy requirements.
2. Calculate a protein goal.
3. Distribute remaining kcal between carbohydrate and lipid.
4. Consider the electrolyte needs for this patient.
5. Consider vitamin and mineral requirements.
6. Establish fluid requirements.
7. Calculate the final parenteral prescription.

Many institutions have specific protocols for initiation, advancement, and transition of feedings. Table 7.10 provides examples of these protocols.

Administration Techniques Parenteral nutrition is administered according to protocol in some institutions, and is guided by tradition in others. Recommendations vary, and there is no "best protocol." One protocol that has worked well is to initiate PN by giving 1 liter the first day, and then increasing to goal volume on day 2. If hyperglycemia (elevated blood sugar, usually defined as 150 to 200 mg/dL since patients receiving PN are not in a fasting state) develops, it is usually treated with insulin. Electrolyte abnormalities are usually treated before or with initiation of PN, but further corrections may also be required on a daily basis until the patient is stable. Initially, most formulas are given continuously over 24 hours. Once it is established that the patient is stable, the length of the infusion may be decreased over a period of two to three days to 12 to 14 hours so that the patient may be free from the pump and other equipment needed to deliver PN.

Patient Monitoring Patients receiving PN can suffer serious, life-threatening consequences including death if the PN is not appropriately monitored and managed. Thus, standard monitoring protocols are in place in many institutions. Monitoring is intense during the first few days, but it decreases as the patient reaches goal feedings and becomes stable.

Intake and output monitoring is usually initiated. Laboratory monitoring includes testing for hyperglycemia three to four times per day and daily measurements of serum electrolytes, BUN and creatinine, magnesium, and phosphorus. At baseline, serum triglycerides are drawn to assess lipid tolerance, and if abnormal, they may be drawn weekly thereafter. The sample protocol for monitoring PN found in Table 7.6 may serve as a guide, although many institutions have protocols in place.

Complications Complications of parenteral feeding may be severe, and they are best prevented through patient monitoring by nutrition support experts. Many of the complications experienced with enteral nutrition occur with parenteral nutrition as well. Patients receiving parenteral feeding may experience electrolyte imbalance, underfeeding and/or overfeeding, hyperglycemia, and refeeding syndrome, just as patients receiving enteral feedings do. These conditions were described in detail earlier, in the Complications section within the Enteral Nutrition section.

Gastrointestinal Gastrointestinal (GI) complications of parenteral feedings have been reported, primarily in those patients whose GI tract is at complete rest. These complications include cholestasis (a condition in which bile accumulates in the gallbladder because it contracts infrequently without enteral stimulation). Increased permeability to bacteria has been noted when atrophic intestinal cells result from lack of enteral stimulation. For this reason, many patients who require PN may receive small amounts of enteral feedings.

If PN is administered continuously for several weeks, transient elevations in liver enzymes may be noted. These usually disappear after PN is discontinued, but may respond to intermittent or cyclic feedings, adjustments in the lipid-to-dextrose ratio, and kcal reduction if overfeeding is operative.

Infectious Patients receiving PN have developed serious infections, and they may have a higher infection rate overall than patients receiving oral or enteral nutrition. Infections may be caused by improperly prepared PN

FIGURE 7.9 Sample Adult PN Order Form

Physician Orders
PARENTERAL NUTRITION (PN) – ADULT

Primary Diagnosis: ______________________ **Ht:** ______ cm **Dosing Wt:** ______ kg

PN Indication: ______________________ **Allergies** ______________________

Instructions: This form must be completed for a new order or continuation of PN and faxed to the Pharmacy by [Insert Time] to receive same day preparation. PN administration begins at [Insert Time]. Contact the Nutrition Support Service at (XXX) XXX-XXXX for additional information.

Administration Route: CVC or PICC *Note: Proper tip placement of the CVC or PICC must be confirmed prior to PN infusion*

Peripheral IV (PIV) (*Final PN Osmolarity ≤ ______ mOsm/L*)

Monitoring: Daily weights, Strict input & output, Bedside glucose monitoring every ______ hours

Na, K, Cl, CO_2, Glucose, BUN, Scr, Mg, PO_4 every ______

T, Bili, Alk Phos, AST, ALT, Albumin, Triglycerides, Calcium every ______

Base Solution: *Select one* — *Parenteral nutrition MUST be administered through a dedicated infusion port and filtered with a 1.2-micron in-line filter at all times. Discard any unused volume after 24 hours.*

PERIPHERAL 2-in-1	CENTRAL 2-in-1	CENTRAL 3-in-1
Dextrose ______ g	Dextrose ______ g	Dextrose ______ g
Amino Acids (*Brand* ______) ______ g	Amino Acids (*Brand* ______) ______ g	Amino Acids (*Brand* ______) ______ g
For patients with PIV and established glucose tolerance; Provides ______ kcal; Maximum Rate not to exceed ______ mL/hour	*For patients with CVC or PICC and established glucose tolerance; Provides ______ kcal; Maximum Rate not to exceed ______ mL/hour*	Fat Emulsion (*Brand* ______) ______ g
		For patients with CVC or PICC and established glucose/fat emulsion tolerance; Provides ______ kcal; Maximum Rate not to exceed ______ mL/hour
		Use of additional fat emulsion not required with 3-in-1 base solution

RATE & VOLUME: ______ mL/hour for ______ hours = ______ mL/day
Must specify

or **CYCLIC INFUSION:** ______ mL/hour for ______ hours, then ______ mL/hour for ______ hours = ______ mL/day

Fat Emulsion ***(Brand*** ______***) – via PIV or CVC with 2-in-1 base solutions*** *(Select caloric density & volume)*

10%	250 mL	Infuse at ______ mL/hour over ______ hours	Frequency ______
20%	500 mL	*(Note: infusions < 4 or > 12 hours not recommended)*	*Discard any unused volume after 12 hours.*

Additives: *(per day)*			***Normal Dosages***
Sodium Chloride	______	mEq	*1-2 mEq Sodium/kg/day*
as Acetate	______	mEq	*pH or CO_2 dependent*
as Phosphate	______	mmol of PO_4	*Consider if hyperkalemic*
Potassium Chloride	______	mEq	*1-2 mEq Potassium/kg/day*
as Acetate	______	mEq	*pH or CO_2 dependent*
as Phosphate	______	mmol of PO_4	*20-40 mmol/day (1 mmol Phos = 1.5 mEq K)*
Calcium **Gluconate**	______	mEq	*5-15 mEq/day*
Magnesium **Sulfate**	______	mEq	*8-24 mEq/day*
Adult **Multivitamins**	______	mL/day	*Contains Vitamin K 150 mcg*
Adult **Trace Elements**	______	mL/day	*Zn ___ mg, Cu ___ mg, Mn ___ mg, Cr ___ mcg, Se ___ mcg (with normal hepatic function)*
H_2 Antagonist ______	______	mg	*___ mg/day with normal renal function*
Other:			

Additives: *(per day)*
Regular Insulin ______ units
Recommend if hyperglycemic, start with 1 unit for every 10 g of dextrose

Pharmacy Use Only: **Ca/PO4 Limit Checked** ______
(Note: Some brands of amino acids contain phosphate)

Physician's Signature: ______________________ **Pager Number:** __________ **Date/time:** __________

Orders transcribed by: __________ **Date/time:** __________ **Orders verified by:** __________ **Date/time:** __________

SEND COMPLETED ORDERS TO PHARMACY

Source: (B) Fig Standard PN Label Template, Adult Patient Page S47

BOX 7.5 CLINICAL APPLICATIONS

Nutrition Care Process: Determining the Parenteral Nutrition Prescription

Nutrition Assessment: You have been consulted concerning a previously healthy woman in the intensive care unit (ICU) who had a major intestinal resection three days ago. Her post-operative course has been complicated by sepsis resulting in respiratory failure. Because she requires mechanical ventilation, she cannot speak, and it is impossible to obtain a nutrition history at this time. Your physical examination is positive for edema. The medical record review reveals a 48-year-old woman with an insignificant medical and surgical history. It is probable that she will need long-term nutritional support, and the surgeon wants to start parenteral nutrition. The patient had emergency surgery, so no height or weight were recorded on the admission form. Her current weight is 168 pounds. Her sister is at the bedside, and tells you that the patient usually weighs 154 to 155 pounds and is 5′6″ tall.

Step 1: Determine a "dosing" weight.

A. Critical Thinking: The hospital bed has a built-in scale, so you can easily weigh the patient. However, this patient has edema, which may add as much as 10 pounds to her "dry" weight. Also, her fluid balance has been positive (intake greater than output) every day since admission. The cumulative positive fluid balance (about 7 liters) confirms your clinical impression that the weight of 168 pounds is accurate, but also reflects her positive fluid balance. For the present, the usual weight reported by the sister is probably a more reasonable choice as the "dosing" weight in this case.

B. Calculations for the Nutrient Prescription: Example: Convert the weight to kilograms by dividing weight in pounds by 2.2.

154 pounds/2.2 = 70 kg

Step 2: Determine a kcal goal.

A. Critical Thinking: In the ICU, 25 kcal/kg is an appropriate initial caloric intake for most patients. If available, indirect calorimetry is helpful if patients do not respond to nutritional therapy. The Mifflin-St. Jeor formula could also be used.

For the present, it is important to select a goal and initiate feeding. The goals for nutrients will change with the patient's condition.

B. Calculations for the Nutrient Prescription: Example: Multiply the weight by the number of kcal/g selected.

70 kg × 25 kcal = 1750 kcal

Example with Mifflin-St. Jeor:

10 (70) + 6.25 (167.4) − 5 (48) + 5 = 1511 kcal

C. More Critical Thinking: Calculations in two different methods are within 200 kcal of each other.

Step 3: Adjust for activity and injury.

A. Critical Thinking: The 25 kcal per kilogram factor is recommended for critically ill patients in intensive care units. The typical intensive care unit patient receives mechanical ventilation, has a major infection or sepsis and has one or more failing organ systems. The 25 kcal/kg figure incorporates the metabolic effect of these changes. Patients with sepsis have an increased resting metabolic rate, but the ventilator reduces basal energy requirements as it does the work of the lungs. Patients on mechanical ventilation are bedfast and have no appreciable activity; therefore, an activity factor is not added here.

Indirect calorimetry may be used to determine energy requirements if available.

Step 4: Calculate a protein goal.

A. Critical Thinking: There is minimal drainage from the surgical wound (100 mL over the last two shifts), which would not result in significant protein loss. To support post-operative wound healing, 1.5 g/kg of protein is appropriate.

B. Calculations for the Nutrient Prescription: Multiply the protein requirement by kcal per gram, and subtract the kcal from protein from the total.

70 kg × 1.5 g of protein = 105 g of protein per day
105 × 4 = 420 protein kcal
1750 kcal − 420 kcal = 1330 kcal

C. More Critical Thinking: The patient will need approximately 105 grams of protein.

Step 5: Distribute remaining kcal between carbohydrate and lipid.

A. Critical thinking: A logical distribution would be to provide about 660 kcal each from carbohydrate and lipid. However, in practice there are wide variations in the amount of carbohydrate and lipid administered. For example, the minimum amount of lipid for this patient would be about 10% of kcal (170 kcal or 17 g), while the maximum amount would be about 1.2 grams of lipid per kg of body weight (84 g or 840 kcal). A typical amount would be somewhere near 30% of kcal (about 600 kcal or 60 g of lipid).

In some hospitals, 20% lipids are packaged in 250 mL containers that provide 50 grams of fat. It is easier for the pharmacy and nursing staff if lipids are administered in multiples of 250 mL. For this patient, a single 250 mL container is appropriate, because it provides 500 kcal and about 28% of kcal as lipid.

B. Calculation for Nutrient Prescription: Multiply the grams of fat by kcal per gram and subtract the kcal from fat from the kcal remaining in Step 3.

50 g fat × 10 kcal = 500 kcal
1330 kcal − 500 kcal = 830 kcal

C. More Critical Thinking: The patient will need 50 grams of lipid. Now what about the carbohydrate?

D. Calculation for Nutrient Prescription: The kcal remaining are divided by 3.4 to obtain grams of parenteral dextrose.

830/3.4 = 245 g of parenteral dextrose

E. More Critical Thinking: Rounding this figure up, you have determined that the patient will need 250 grams of dextrose.

Step 6: Consider electrolyte needs.

A. Critical Thinking: This patient has no abnormal electrolyte losses, and at present renal function is normal. However, she has required ongoing potassium supplementation, suggesting elevated potassium needs. An experienced nutrition support dietitian would negotiate with the attending physician or members of the nutrition support team about incorporating the extra potassium into the PN.

B. Calculation for Nutrient Prescription: Consult Table 7.7 to identify suggestions for electrolyte concentration in the PN.

C. More Critical Thinking: Based on her dosing weight, this patient would need about 70 mEq of sodium (1 to 2 mEq/kg), 70 mEq of potassium (1 mEq/kg), 16 mEq of magnesium, and 10 mEq of calcium with bicarbonate and acetate to balance the solution.

Step 7: Consider vitamin and mineral needs.

A. Critical Thinking: It is unlikely that this patient is deficient in vitamins and minerals, as she was well nourished on admission. She may have elevated needs based on her medical condition, but parenteral vitamin preparations are probably sufficient to account for elevated postoperative needs. For some patients with sepsis, larger doses of antioxidants may be given, but practice varies widely.

B. Calculation for Nutrient Prescription: Consult Tables 7.8 and 7.9 to identify appropriate vitamin and mineral requirements. Remember to check the package insert for the vitamin and mineral preparations so that you will know how much the patient is getting.

Step 8: Determine fluid needs.

A. Critical Thinking: Patients in intensive care settings frequently receive fluids in excess of needs because of the number of intravenous medications they require. For this patient, 30 mL per kilogram of fluid would be appropriate; however, she is in positive fluid balance, with edema. A fluid restriction is likely appropriate, so the PN solution prescribed should be maximally concentrated. The term maximally concentrated simply means that nutrients are provided in their most concentrated form, and no extra fluid is added to the PN. Once the patient is stable, the amount of fluid needed to deliver medications decreases dramatically, and the PN volume is adjusted to meet the patient's changing fluid needs.

B. Calculation for Nutrient Prescription: Provide PN in the smallest volume possible. This patient's minimal fluid requirements using 1 mL/kcal are approximately 1600 mL (halfway between the two calculations in Step 1).

Step 9: Write final parenteral nutrition prescription.

A. Critical Thinking: Parenteral nutrition to provide 105 grams of protein and 250 grams of dextrose with 70 mEq sodium, 70 mEq of potassium, 15 mMol of phosphorus, 16 mEq of magnesium and 10 mEq of calcium with bicarbonate and acetate to balance, with 1 vial of multiple vitamin infusion, and 1 vial of multiple trace element infusion in a volume of 1.6 L daily to run at 65 mL/hour over 24 hours daily.

Administer 50 grams of lipid in a volume of 250 mL to run at 10 mL/hr over 24 hours daily.

TABLE 7.10

Sample Parenteral Protocol

Total Parenteral Nutrition

a. Definitions: IV Nutrition Support using a formulation of amino acids, carbohydrates, lipids, electrolytes, MVI, minerals, and supplemental medications (insulin or H2 blockers).

b. Patient Selection: Inability to use the gut at goal feeds within 5 days.

c. Patient Exclusion: Ability to use the gut at goal feeds or oral intake within 5 days.

d. IV Access: Central Access (TLC, PICC, Hickman, Port-A-Cath).

e. Formula Selection: Based on patient's requirements, critical illness, organ failure, and comorbid disease.

f. Estimating Nutritional needs:

1. Ideal Body Weight (IBW) will be used for nutritional estimates for the majority of patients. Patients greater than 120% of IBW/Ht, registered dietitian will calculate best weight estimate.

- To calculate IBW/Height: Range plus or minus 10%
  - i. Males: 2.3 × (inches over 5′) + 50 kg
  - ii. Females: 2.3 × (inches over 5′) + 45 kg
- Nutritional Calculations: to formulate TPN prescription
  - i. Energy: 25 to 30 kcal/kg (aim for 25 kcals/kg)
  - ii. Protein: 1.0 to 1.8 grams protein/kg (aim for 1.5 gram protein/kg)
  - iii. Lipids: 30 to 70 grams/day (5 to 13 mL/hr), 20 to 30% total
- Monitoring/Management of patient care
  - i. Labs: Basic Metabolic Panel, C- Reactive Protein, Mg and Phos, Day 1, 2 & PRN (LFTs, Pre-albumin, Triglyceride levels check q Thursdays as routine)
  - ii. Metabolic Carts in patients on TPN greater than two weeks

TPN Protocol

A. The Adult Nutrition Support Service should be consulted to assist with prescribing parenteral nutrition (ASPEN guidelines).

B. All TPN is to be ordered or reordered daily, according to the age appropriate order form. Orders must be received by the appropriate time:

1. Adult Medicine and Surgical units by 5 p.m.

C. Monitoring

1. Blood glucose: See intense glucose control Protocol for all ICU patients.

a. For nonunit patients: Blood Glucose testing, adult every 6 hr × 72 hours. Thereafter, renewal is required.

Source: Used with permission: Critical Care Nutrition Vanderbilt University Medical Center. Nashville, TN, 2004.

solutions, and therefore most pharmacies institute rigorous monitoring to minimize this risk. Infection may be introduced into a patient's blood stream while the vascular access device is placed or while a dressing around the line is being changed. Another route for infection is the GI tract, which, according to the indications for PN, should be non-functioning. With disuse, a nonfunctioning GI tract may become permeable to intestinal bacteria, and infection may result.

Conclusion

Many health care providers regard improved nutrition, including the development of parenteral and enteral nutrition, as among the most important medical advances of the twentieth century. Both enteral and parenteral nutrition provide lifesaving therapy to those who cannot eat. Yet enteral and parenteral nutrition can be difficult to manage and may require the specialized expertise of dietitians credentialed in this area of practice. Oral diet is still the preferred method of nutrition, and it is the goal of nutrition intervention.

PRACTITIONER INTERVIEW

Jordan B. Davidson, RD, LD, CNSD, *Clinical Dietitian Specialist*
The Johns Hopkins Bayview Medical Center

How long have you been an RD?

I have only been an RD for five years. When I started my internship, I felt like I didn't know anything! But I also knew that I learned by "doing," so during my internship it all started to make sense—I was actually applying what I had learned in class. One of the more surprising things to me is the realization that the learning and experiencing never stops, even after you become a dietitian. Throughout my work experience, I actually felt the same way I did during my internship, especially when I began to care for pediatric patients. Again there was a learning curve, which also required the hands-on experience to solidify the content.

My comfort level is much better since my internship. With more experience and knowledge you feel more confident explaining the nutrition issues to other members of the health care team. In general I find medical residents and interns more informal, but some departments (like surgery) may be more formal.

Why do you like doing nutrition support?

I like doing nutrition support because it is at a more advanced level of care. I like to understand the pathophysiology and the medical treatment of a disease/disorder plus the nutrition implications and support.

Describe a typical day on the job. Who might your typical patient be?

My typical day is atypical since I do full-time relief and go to whatever service needs me that day. It could be pediatrics, surgery, various ICUs (burn, neonatal, pediatric cardiac), long-term care (with chronic ventilation and wound care), the Medicare PACE program and out-patient clinics (diabetes, weight management). In addition I act as a preceptor for dietetic interns in all areas of clinical nutrition as well as nutrition support. Typically the nutrition support patients we see at my facility are critically ill patients with multiple issues who require aggressive nutrition support. They often have multiple organ failure syndrome (MODS)—commonly renal—they are intubated, have fistulas, have had major bowel surgery, and are hyperglycemic. The most common nutrition diagnosis for patients requiring nutrition support is inability to take nutrition orally because they are intubated or require bowel rest.

How do you keep yourself current in nutrition support topics?

For current information on nutrition support, I attend the ASPEN conferences, read various books (ICU nutrition care and their core curriculum) that can be obtained from the ASPEN website (www.nutritioncare.org), or the NIH PubMed website for journal search and for medical references. I like the paid online service, Uptodate (www.uptodate.com), and I also network with other RDs to discuss topics.

Any advice for students before they start their dietetic internship?

Keep an open mind, adapt and change to different people and scenarios. Ask questions, communicate, admit if you don't understand and you need more help. Review before clinical rotations—classes or notes. Contact the preceptor if there is something else to read or review. Network! Start now, as it is helpful for jobs, information, resources, and so on.

CASE STUDY

Case Study data provided courtesy of Kathy Fitzpatrick, MS, RD, CNSD, OptionCare, Cape Girardeau, Missouri

Fifty-two-year-old female with intractable N & V—unable to control with medications. Referred for home start TPN.

Diagnosis:

Pancreatic cancer currently being aggressively treated with chemotherapy; recent hospitalizations for dehydration/N & V
Ht: 165 cm, wt: 66 kg, usual wt: 80.9 kg

Rx:

Initial TPN formula: 1392 mL, 58 mL/hour continuous 24-hour infusion (21 mL/kg)

Macronutrients:

200 mL 50% dextrose, 800 mL 10% amino acids, 300 mL 10% lipids

Electrolytes:

NaCl—68 mEq, Na Acetate 32 mEq, K Acetate 36 mEq, KPO430 mmol, MgSO4 16 mEq, Ca Gluconate 9.4 mEq, MVI (multivitamin injection) 10 mL, MTE 5 (multiple trace elements) 1 mL

Added medications:

40 mg Pepcid/day

Nutrition Hx:

Taking only clear liquids, small amounts × one week prior to initiation of TPN (very poor prior to that)

Laboratory values:

CHEMISTRY

DAY		Start date for TPN			
DATE		1/8	1/12	1/19	
TIME	NORMAL				UNITS
LOCATION					
Albumin	3.6–5	2.6			g/dL
Total Protein	6–8				g/dL
Prealbumin	19–43	17			mg/dL
Transferrin	200–400				mg/dL
Sodium	135–155	132	130	131	mmol/L
Potassium	3.5–5.5	4.6	4.7	3.9	mmol/L
Chloride	98–108	90	101	98	mmol/l
PO_4	2.5–4.5	3.9	3.4	3.4	mmol/L
Magnesium	1.6–2.6	2.20	2.10	1.70	mmol/L
Osmolality	275–295				mmol/kg H_2O
Total CO_2	24–30	20.8	22.6	23.8	mmol/L
Glucose	70–120	144	122	104	mg/dL
BUN	8–26	16.9	18.9	17.9	mg/dL
Creatinine	0.6–1.3	0.5	0.5	0.5	mg/dL
Uric Acid	2.6–6 (women) 3.5–7.2 (men)				mg/dL
Calcium	8.7–10.2	9.2	9.2	8.7	mg/dL
Bilirubin	0.2–1.3				mg/dL
Ammonia (NH_3)	9–33				µmol/L
SGPT (ALT)	10–60				U/L
SGOT (AST)	5–40				U/L
Alk Phos	98–251				U/L
CPK	26–140 (women) 38–174 (men)				U/L
LDH	313–618				U/L
CHOL	140–199				mg/dL
HDL-C	40–85 (women) 37–70 (men)				mg/dL
VLDL					mg/dL
LDL	< 130				mg/dL
LDL/HDL RATIO	< 3.22 (women) < 3.55 (men)				
Apo A	101–199 (women) 94–178 (men)				mg/dL
Apo B	60–126 (women) 63–133 (men)				mg/dL
TG	35–160				mg/dL
T_4	5.4–11.5				µg/dL
T_3	80–200				ng/dl
Hb A_{1C}	4.8–7.8	7.2			%

Questions:

1. Determine the amount of energy (kcal) and protein provided by the initial TPN solution.
2. Calculate the grams of carbohydrate, protein, and lipid provided by this prescription. How many kcal/kg and grams of protein/kg does it provide? Calculate the patient's nutritional needs. Compare the two.
3. Is this patient at risk for refeeding syndrome? Why? What can be done to prevent it?
4. What clue in the patient's admission history gives support to the patient's low chloride level at the initiation of TPN?
5. On 1/10, the RD recommended that NaCl be increased by 30 mEq, and then by 20 mEq of Na Acetate on 1/12. Why?

WEB LINKS

American Society for Parenteral and Enteral Nutrition (ASPEN) Information on the American Society for Parenteral and Enteral Nutrition may be obtained from this site.

http://www.nutritioncare.org

Dietitians in Nutrition Support Information on Dietitians in Nutrition Support may be obtained from this site.

http://www.dnsdpg.org

National Board of Nutrition Support Certification Inc. Information on the Certified Nutrition Support Dietitian exam.

http://www.nutritioncertify.org

Information on enteral products can be obtained from manufacturer websites:

Nestle Nutrition

http://www.nestle-nutrition.com

Novartis Nutrition

http://www.novartisnutrition.com/us

Ross Products Division of Abbott Laboratories

http://www.ross.com/productHandbook

END-OF-CHAPTER QUESTIONS

1. Describe two ways that the house or regular diet can be modified to accommodate patient needs.
2. What is the difference between clear and full liquid diets? When are they used, and what are their limitations?
3. What are the advantages and disadvantages of enteral and parenteral nutrition support?
4. Describe three ways enteral and two ways parenteral nutrition can be delivered to the patient.
5. List five factors that might influence selection of an enteral formula (e.g., viscosity). Explain why each factor is important when choosing a formula.
6. What are medium-chain triglycerides (MCT), and why are they added to some enteral products? What is the most common source that is currently used?
7. List four complications that might occur when feeding a patient enterally. Describe and provide the rationale for three factors that should be monitored.
8. Calculate the caloric content and protein amount in one liter of parenteral solution composed of 25% dextrose and 4.25% amino acids. If 250 milliliters of a 20% fat emulsion were added, how many more kcal would be provided?

8

Fluid and Electrolyte Balance

Marcia Nahikian Nelms, Ph.D., R.D.

Southeast Missouri State University

CHAPTER OUTLINE

Normal Anatomy and Physiology of Fluids and Electrolytes

Body Solutes

Physiological Regulation of Fluid and Electrolytes

Disorders of Fluid Balance

Alterations in Volume • Alterations in Osmolality • Sodium Imbalances • Potassium Imbalances • Calcium Imbalance • Phosphorus Imbalance • Magnesium Imbalances

Introduction

Humans have long known the importance of water for survival. The geographical distribution of population groups throughout the world has been shaped by the availability of water. In medical care, restoration of normal fluid status is often the first priority in reestablishing homeostasis.

The functions of water in the body include transporting nutrients, transporting and excreting metabolic waste, supporting cell shape and structure, lubricating friction-generating surfaces, and sustaining normal body temperature. A variety of solutes are found in solution with water throughout the body. An important group of solutes found in body fluids is the **electrolytes**. Maintenance of fluid balance is significantly integrated with maintenance of electrolyte balance.

Daily losses of fluid from the body are normal and require replacement. The presence of injury or illness can result in increased fluid losses and increased fluid needs. Detection of fluid and electrolyte imbalance is an essential component of nutrition assessment.

Normal Anatomy and Physiology of Fluids and Electrolytes

Total Body Water

Total body water accounts for approximately 60% of total body weight in the adult male, and for somewhat less, an average of 50%, in the adult female. At birth, total body water accounts for approximately 75% of the infant's weight. Body water content declines throughout the life span and often falls below 50% in the elderly. In general, this is because the proportion of lean body mass to body fat influences the amount of water as a percentage of body weight. Fat tissue has the lowest percentage of water in comparison to all other tissues in the body. As body fat increases, the percentage of body water decreases (see Table 8.1).

electrolytes—those substances that bear an electrical charge (ions)

TABLE 8.1

Total Body Water in Percentage of Body Weight

	% Male	% Female	% Infant
Total Body Water	60	50	70
Extracellular:	20	14	30
Plasma	5	4	4
Interstitial	15	9	25
Intracellular	40	33	40

Adapted from Gropper, Smith, Groff. (2005). *Advanced Nutrition and Human Metabolism*, 4th ed. p. 502.

Fluid Compartments

Membranes separate body fluids into compartments. Approximately two-thirds of body water is found within cells (**intracellular fluid**). The remaining body water is found outside of cells (extracellular fluid). **Extracellular fluid (ECF)** is divided into three compartments: interstitial, intravascular, and transcellular (or transitional). Interstitial fluids surround the cells. Intravascular fluid is found within blood. Transcellular fluids are those fluids found in secretions within organs. These include gastrointestinal secretions, cerebrospinal fluid, and intraocular fluid.

Fluids can accumulate within body cavities in spaces between organs. These are often called the "**third spaces**" and include peritoneal, pericardial, and thoracic cavities as well as the joints and bursae. For the normal healthy individual, these spaces hold insignificant amounts of fluid. However, in illness or injury, fluid accumulation in these spaces may become significant. For example, fluid may accumulate in the peritoneal cavity with liver disease, causing the condition known as **ascites**.

intracellular fluid (ICF)—the fluid within the tissue cells

extracellular fluid (ECF)—the interstitial fluid and the plasma, constituting about 20% of the weight of the body; sometimes used to mean all fluid outside of cells, usually excluding transcellular fluid

"third space" fluid—shift of fluid from the intravascular space to a nonfunctional space

ascites—abnormal accumulation of fluid in the abdominal cavity

osmotic pressure—the pressure that must be applied to a solution to prevent the passage into it of solvent when solution and pure solvent are separated by a membrane permeable only to the solvent

colloid osmotic pressure (oncotic pressure)—the osmotic pressure attributed to proteins and other macromolecules

osmolarity—the number of osmols (standard unit of osmotic pressure) per liter of solution (mOsm/L)

osmolality—the number of osmols per kilogram of solvent (water) (mOsm/Kg)

Movement of Fluid between Blood and Interstitial Spaces Fluid status in the body is in a state of dynamic equilibrium. Water is constantly moving but total volume and concentration remain the same. Fluids move freely between fluid compartments by the processes of osmosis and filtration. In osmosis, only water moves between compartments. But with filtration, water and solutes (except plasma proteins and red blood cells) move. Two types of pressure influence the movement of water and solutes: osmotic and hydrostatic.

Osmosis is the movement of fluid across a semipermeable membrane from an area of low concentration (of solute) to an area of high concentration (of solute). The force that pulls water across the membrane is **osmotic pressure**, which is determined by the number of solute particles in solution. Solutes that do not form a true solution, such as large protein molecules, are called colloids. They also contribute to the osmotic pressure (colloid osmotic pressure). Serum albumin is the protein that exerts the greatest effect on the **colloid osmotic pressure (oncotic pressure)**. The purpose of the movement of fluid is to equalize the concentration of solute, and thus osmotic pressure, on both sides of the membrane.

Hydrostatic pressure is pressure exerted by the fluid on the membrane. For intravascular fluid, hydrostatic pressure (pressure of blood on the arterial walls) is more commonly known as blood pressure. When hydrostatic pressure differs on the two sides of the membrane, fluid is pushed from the area of high pressure to the area of low pressure. The goal of this fluid movement (filtration) is to equalize pressure exerted by the fluids on both sides of the membrane.

Osmotic and hydrostatic pressure work together in favor of moving fluid out of the blood into interstitial areas at the arterial end of the capillary and restoring fluid back into blood at the venous end of the capillary. This phenomenon is called Starling's Law of capillaries. Anatomical differences between capillaries of different organs also affect permeability and therefore affect both osmotic and hydrostatic pressure.

Movement between Extracellular Fluid and Intracellular Fluid Fluid movement between extracellular fluid (ECF) and intracellular fluid (ICF) is directed by osmotic pressure in order to establish osmotic equilibrium. Osmotic pressure can be expressed as either **osmolarity** or **osmolality**. Though technically they have different meanings, they are similar enough that the terms are used interchangeably. Osmolality is the more precise term since the amount of solvent does not vary.

BOX 8.1 CLINICAL APPLICATIONS

Calculation of Osmolality Using Potassium Chloride (KCl)

Milliosmolality (mOsm)
mOsm = atomic wt in mg/particles exerting osmotic pressure.
Example using potassium chloride (KCl):

Step 1: Atomic weight of KCl = 74.5
Step 2: 2 particles in solution: K^+, Cl^-
Step 3: mOsm = 74.5/2 = 32.75 mg
Step 4: 1 mOsm of KCl = 32.75 mg

FIGURE 8.1 Ionic Composition of the Major Body-Fluid Compartments

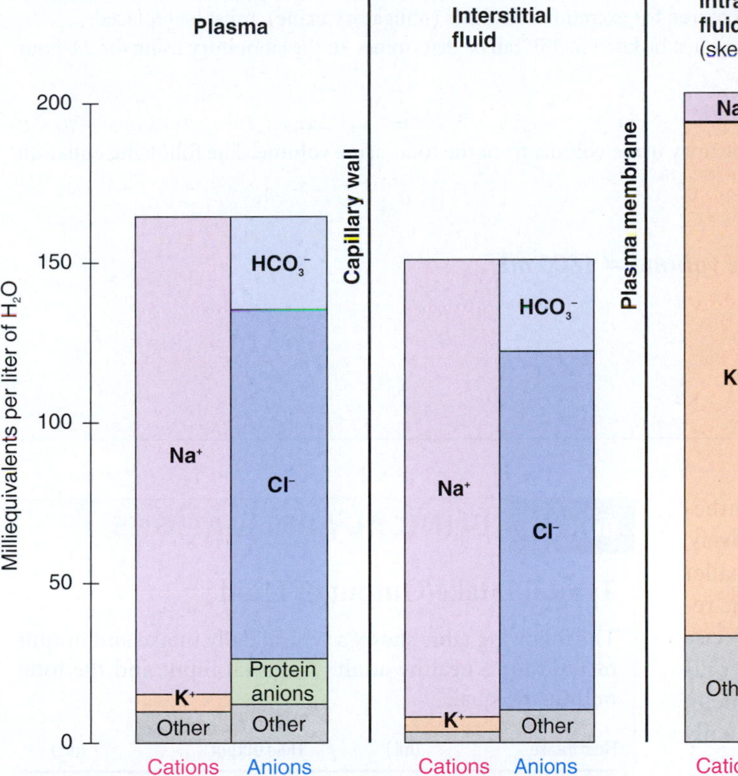

Source: L. Sherwood, *Human Physiology: From Cells to Systems*, 5e, copyright © 2004, p. 562

Osmolality of the blood is used as the normal physiologic range for body fluids. Normal osmolality of the blood is 280 to 320 mOsm/kg H_2O. Estimation of blood osmolality uses the serum concentrations of sodium, potassium, glucose, and urea:

$$\text{mOsm/kg blood} = 2\,(\text{Na}^+\ \text{mEq/L} + \text{K}^+\ \text{mEq/L}) + \frac{\text{glucose mg/dL}}{18} + \frac{\text{BUN mg/dL}}{2.8}$$

In this equation used to calculate osmolality, sodium and potassium concentrations are expressed in mEq/L, and glucose and BUN are expressed in mg/dL. Na^+ and K^+ (cations) are multiplied by 2 to account for the accompanying anions that are needed for electroneutrality (see Box 8.1). Figure 8.1 shows the ionic composition of the major body-fluid compartments.

Fluids that have an osmolality equal to blood are called isotonic. Solutions with an osmolality greater than that of blood are called hypertonic, and those with osmolality less than blood are called hypotonic. In the normal state, osmolalities of the ECF and ICF are assumed to be equal. When cells are exposed to hypertonic solutions, fluid moves out of the cell in an attempt to establish osmotic equilibrium. The result is cellular **dehydration.** Conversely, when a cell is exposed to hypotonic solutions, fluid moves into the cell, resulting in cell swelling.

Total Body Water Balance

The physiological roles of water in excretion of metabolic waste and maintenance of body temperature result in its continual loss from the body. Losses must be replaced in order to maintain equilibrium of fluid in the body. In the clinical setting, this balance is referred to as intake and output, as discussed in Box 8.2.

Fluid Intake Water is taken into the body as part of food and beverages consumed by an individual. Beverages range between 84% and 100% water, with fruit juices being at the lower end of the range and water being 100%. Solid foods range from 0% to 96%, with oils being 0% and cucumbers at 96%. In a clinical setting, anything fluid at room temperature will be calculated as fluid intake. For example, ice cream would be calculated as fluid consumption. Box 8.3 shows a typical daily intake and output of fluid for a healthy adult.

Metabolic reactions often produce water but do not contribute to actual fluid intake in a practical sense. Anabolic reactions such as synthesis of glycogen, triglycerides, or protein release water (condensation reactions). Some catabolic reactions

dehydration—a deficit of water in the body

BOX 8.2 CLINICAL APPLICATIONS

Assessing Fluid Status

Determining Fluid Intake for 24 Hours

Accuracy in estimating fluid intake is dependent upon recording of all oral intake on the I & O (intake and output) sheet. In the hospital setting, the I & O sheet is usually kept at the bedside and recording is done by nursing staff.

Fluid content of solid foods and liquids can be determined using commercial nutrient database programs. If I & O sheets do not include oral intake provided by the nutrition services, a separate record may need to be established to record all solid food and liquid intakes not recorded on the I & O sheet.

Determining Fluid Output for 24 Hours

Fluid lost by feces, lungs, and skin is relatively constant in the normal adult in the absence of extremely hot ambient temperatures and/or extreme exercise. If diarrhea or vomiting is present, fluid losses are higher than normal. Urine output should be collected and measured for the 24-hour period being evaluated.

Determining Obligatory and Facultative Urine

For patients with decreased renal, hepatic, pulmonary or cardiac function, fluid restrictions are often implemented to decrease fluid retention. The goal of a fluid restriction is to eliminate facultative urine production since this represents fluid in excess of requirements and is most likely to be retained.

Fluid losses from feces, skin, and lungs, as well as that required for excretion of solutes (obligatory urine), must be replaced.

To determine obligatory urine, the renal solute load (RSL) must be known. RSL can be determined in the laboratory using the 24-hour urine collection. The following equation can then be used:

Obligatory urine = RSL (mOsm) ÷ 1200 to 1400 mOsm/L

Facultative urine is determined by subtracting the obligatory urine volume from the total urine volume. The following equation can be used:

Facultative urine = Total urine − Obligatory urine

Example: RSL = 950 mOsm; total 24-hour urine volume = 1800 mL

Obligatory urine = 950 mOsm ÷ 1300 mOsm/L
= .731 L (731 mL)

Facultative urine = 1800 mL − 731 mL
= 1069 mL

release water. During aerobic respiration, water is synthesized as a by-product of energy production. Alternatively, catabolic reactions that reduce large molecules to smaller molecules (e.g., proteins → amino acids) are hydrolytic reactions and therefore require the addition of water. Precise determination of **metabolic water** is not possible; it is thus usually estimated by using intakes of carbohydrate, protein, and fat as variables. The role of metabolic water is usually insignificant except in cases of decreased organ function.

Fluid Output Fluid losses from the body are categorized as sensible and insensible. **Sensible losses** are those that are

metabolic water—water that is produced through nutrient metabolism

sensible losses—fluid loss that can be measured (usually refers to fluid lost via urine excretion)

BOX 8.3 CLINICAL APPLICATIONS

Typical Intake/Output of Fluid

The following table shows a typical daily intake and output of fluid for a healthy adult. The total input and the total output are equal.

Fluid Intake	(mL)	Fluid Output	(mL)
Sensible		*Sensible*	
Liquids	1250	Feces	200
Solid food	1000	Urine	1400
		Sweat	100
Insensible		*Insensible*	
Metabolic water	350	Lungs	400
		Nonsweating Skin	500
Total	2600	Total	2600

visible and measurable, such as in urine and feces. **Insensible losses** are usually not seen or measured. These include losses through respiration or through the skin by evaporation. Average daily loss from feces is 200 mL. Insensible losses will range from 700 to 900 mL/day.

Water loss from urine is the most variable and will accommodate changes in dietary fluid intake. Total urine output is the sum of **obligatory urine** and the **facultative urine.** Obligatory urine is the amount that must be excreted in order to remove waste products; it is dependent on the concentrating ability of the kidney. Waste products are referred to as renal solute load (RSL) and include primarily sodium, potassium, chloride, and urea. Obligatory urine is formed even when patients are fasting. The kidneys can concentrate urine to approximately 1200 mOsm. Therefore, minimal RSL of 600 to 700 mOsm/day would require a minimum of 500 mL of obligatory urine even in fasting or starvation. Assessment of the solute concentration in the urine is measured by **specific gravity**. Water has a specific gravity of 1.00 that will rise with each additional solute.

It is assumed the normal adult will have fluid output equal to fluid intake, thus maintaining equilibrium. Facultative urine can vary from negligible amounts when fluid intake is low to large volumes when fluid intake is high (see Box 8.3).

Fluid Requirements Several methods have been developed for estimating fluid requirements. Four methods are listed in Table 8.2, but it should be noted that these methods only estimate fluid needs. Box 8.4 gives an example of patient calculations. Clinical assessment should be used to evaluate whether recommended fluid intake is appropriate. Assessment of hydration status includes evaluation of daily weights, intake and output records, physical evaluation of skin, eyes, lips and oral cavity, respiratory rate and lung sounds, blood pressure and capillary fill, and assessment for peripheral edema. Specific biochemical assessment of blood and urine are discussed later in this chapter.

TABLE 8.2

Calculating Fluid Requirements

Method 1 (based on energy intake): 1 mL of fluid per kcal

Method 2 (based on body weight):

Age/Gender	mL/kg
Infants and Children	
1–10 kg	100–150
11–20 kg	Add 50 mL/kg over 10 kg
≥ 21 kg	Add 25 mL/kg over 20 kg
Adolescents	40–60
Young adult 16–30 yrs	35–40
Average adult	30–35
Adult, 55–65 yrs	30
Adult > 65 yrs	25

Method 3 (based on nitrogen and energy intake): 1 mL/kcal + 100 mL/g N

Method 4 (based on body surface area–BSA): 1,500 mL/m^2

BOX 8.4 CLINICAL APPLICATIONS

Sample Calculations for Estimating Fluid Requirements

Example: 45 y.o. male, 180 lbs, 6′2″, 2200 kcal, 80 g protein

Method 1:

Fluid = 1 ml/kcal × 2200 kcal
= 2200 mL

Method 2:

Fluid = 30–35 ml/kg × 82 kg
= 2460–2870 mL

Method 3:

Fluid = 1 ml/kcal × 2200 kcal + 100 mL/1 g N
× 12.8 g N = 2200 mL + 1280 mL
= 3480 mL

Method 4:

Fluid = 1,500 mL × $(1.87)^2$
= 5250 mL

Body Solutes

Types of Solutes

Solutes in body fluids include both electrolytes and other molecules. Electrolytes (ions) dissociate in fluid to form one or more charged particles. Other molecules, such as glucose, protein, urea, lactate, and other organic acids, remain stable in solution. Major electrolytes in the body are sodium, potassium, calcium, magnesium, chloride, bicarbonate, phosphate, and sulfate. Ions with a positive charge are referred to as cations, and negatively charged ions are called anions.

insensible losses—fluid loss that cannot be easily measured (usually refers to fluid lost via sweat and respirations)

obligatory urine—the amount of fluid necessary for the body to excrete waste products and solutes (approximately 500 mL)

facultative urine—excess water that is excreted through urination

specific gravity—the weight of a solution (e.g., urine) in comparison to an equal amount of distilled water. This is used to measure concentrating ability of the kidney

Distribution of Electrolytes

Some electrolytes are found only in the ECF or the ICF, while others are found in both. Concentrations of electrolytes inside and outside of the cell also differ (see Table 8.3). The key to maintaining normal conditions, however, is ensuring that the amount of cations and anions in the ECF are equal. Likewise, equal amounts of cations and anions must be present in the ICF. The law of thermodynamics has been used to determine that the sum of the cations must be equal to the sum of anions within a given compartment in order to maintain electroneutrality.

In the ECF, the major cation is sodium, and major anions are chloride and bicarbonate. These are only found in small amounts in the ICF. The major cation in the ICF is potassium, and the major anion is phosphate. Potassium and phosphate concentrations are low in the ECF. The distribution of ions in the ECF and ICF are shown in Figure 8.1.

Movement of Solutes

While fluids generally move freely through the semipermeable membranes of the body, cellular membranes can obstruct movement of solutes. Factors influencing movement of solutes include molecular size (smaller molecules move more easily than larger molecules), electrical charge of the molecule, hydrostatic pressure, and method of solute transport. Solutes transported across the membrane by active transport move more easily than those transported by facilitated diffusion or simple diffusion.

TABLE 8.3

Plasma and Intracellular Electrolytes

	Plasma (mEq/L)	Interstitial Fluid (mEq/L H_2O)	Intracellular Water (mEq/L H_2O)
Cations	153	153	195
Na^+	142	145	10
K^+	4	4	156
Ca^{2+}	5	(2–3)	3.2
Mg^{2+}	2	(1–2)	26
Anions	153	153	195
Cl^-	103	116	2
HCO_3^-	28	31	8
Protein	17	—	55
Others	5	(6)	130
Osmolarity (mosm/L)		294.6	294.6
Theoretic osmotic pressure (mm Hg)		5,685.8	5,685.8

Source: Gropper SS, Smith JL, Groff JL. Advanced nutrition and human metabolism. 4th ed. Belmont: Wadsworth, 2005. Table 14.2, p. 508.

Electrolyte Requirements

Adequate Intake (AI) levels for sodium, potassium, and chloride were established in 2005. Table 8.4 summarizes these levels. Electrolyte requirements are generally met by normal dietary intake without difficulty. For example, the AI for sodium is 1500 mg per day and only one teaspoon of table salt has 2300 mg sodium.

Normal serum levels of these electrolytes are listed in Table 8.5. When electrolyte imbalances occur, serum levels are altered to maintain electroneutrality. The kidneys

TABLE 8.4

Adequate Intakes (AI) for Sodium, Chloride, and Potassium

Age (yr)	Sodium AI (mg/day)	Chloride AI (mg/day)	Potassium AI (mg/day)
Infants			
0–0.5	120	180	400
0.5–1	370	570	700
Children			
1–3	1000	1500	3000
4–8	1200	1900	3800
Adults			
9–13	1500	2300	4500
14–50	1500	2300	4700
51–70	1300	2000	4700
>70	1200	1800	4700
Pregnancy	1500	2300	4700
Lactation	1500	2300	5100

Source: Reprinted with permission from *Dietary Reference Intakes for Water, Potassium, Sodium, Chloride, and Sulfate,* Copyright © 2004 by the National Academies of Sciences, courtesy of the National Academies Press, Washington, D.C.

TABLE 8.5

Normal Serum Values for Sodium, Potassium, and Chloride

Electrolyte	Normal Serum Level
Sodium	135 to 142 mEq/L
Potassium	3.8 to 5.0 mEq/L
Chloride	95 to 102 mEq/L

accomplish primary regulation of sodium, potassium, and chloride.

Physiological Regulation of Fluid and Electrolytes

Regulation of fluid and electrolytes is complex and utilizes several integrated mechanisms. The influence of osmotic and hydrostatic pressures has already been discussed. Additional factors necessary for fluid regulation include the hypothalamic thirst mechanism, renal function, and hormonal control.

Thirst Mechanism

Sensors within the interstitial fluid are affected by changes in fluid around them. They trigger the hypothalamus to interpret the signals as thirst and as a result, the individual will be stimulated to increase their fluid intake. This thirst mechanism cannot always be relied on. In the elderly, thirst sensation decreases. In the trained, elite athlete, thirst sensation may not be a valid indication of need for additional fluid.

Renal Function

As mentioned earlier, hydrostatic pressure is pressure exerted by fluid on a membrane. When blood volume increases, hydrostatic pressure also increases. This increase in pressure results in larger amounts of fluid moving from the capillaries into the renal tubules. This fluid is then excreted as urine by the kidney.

Hormonal Influence: Renin-Angiotensin-Aldosterone System (RAAS)

Decreasing hydrostatic pressure is the impetus for the RAAS regulation of fluids and electrolytes. **Baroreceptors** within blood vessels are stimulated by low hydrostatic pressure, which is indicative of a decrease in blood volume. The hormone renin is released from the kidney and stimulates conversion of angiotensinogen to angiotensin I. A second activation converts angiotensin I to angiotensin II. Increasing amounts of angiotensin II stimulate release of aldosterone. Aldosterone is a hormone released from the adrenal cortex. Aldosterone directly influences the kidney to retain Na^+. When Na^+ levels increase, increased osmotic pressure will pull fluid back into the blood; blood volume will thus increase back to its normal range. For individuals with hypertension (high blood pressure), the heart has to work harder to handle a higher blood volume. Dietary intake of sodium is an important component of nutrition therapy for assisting in control of high blood pressure.

Two major factors stimulate the pituitary to release the hormone **arginine vasopressin,** formerly known as antidiuretic hormone (ADH). The first and most important factor is an increasing osmolality of the ECF. The second factor that stimulates release of AVP is detection of a decrease in hydrostatic pressure by baroreceptors in blood vessels. AVP causes fluid to be reabsorbed in the tubules of the kidney. This increases blood volume and lowers blood osmolality (see Figure 8.2).

Electrolyte Regulation

Sodium Controls of electrolytes are interdependent, and are linked to controls of fluid balance, because sodium and water balance are closely related. When AVP and aldosterone act to regulate osmolality and blood volume, sodium is regulated as well. Additionally, atrial natriuretic peptide (ANP) assists in the control of sodium. ANP is released when arterial vessels stretch (as occurs when blood volume increases). ANP is an agonist to the RAAS. This effect results in increased urinary output of sodium and fluid and a decrease in blood volume, and indirectly, an increase in osmolality (see Figure 8.3).

Potassium Aldosterone has an independent effect on potassium levels. High levels of potassium cause the adrenal glands to release aldosterone. Aldosterone secretion results in increased excretion of K^+ by the kidney. Two of the most important components of acid-base balance involve both hydrogen ions and bicarbonate (see Chapter 9 for a detailed discussion of acid-base balance). Since they are both electrolytes, acid-base changes will affect concentrations of other electrolytes in both ECF and ICF. For example, when the body attempts to decrease the concentration of H^+ to restore acid-base balance, K^+ is often exchanged in order to maintain electroneutrality.

Calcium and Phosphorus Serum concentration levels of calcium and phosphorus are dependent on intestinal absorption, exchange between extracellular fluid and bone, and renal excretion of these minerals. These routes for maintenance of serum levels are primarily controlled by hormonal influence. Calcium and phosphorus exist in a reciprocal relationship. This means that when serum calcium levels are high, serum phosphorus levels will be low.

Parathyroid hormone (PTH) is secreted from the parathyroid glands when serum calcium levels are low. PTH works to raise serum calcium levels by pulling calcium from the bone and decreasing excretion of calcium in

baroreceptor—in general, any sensor of pressure changes

arginine vasopressin (AVP)—previously known as antidiuretic hormone

FIGURE 8.2 Influence of RAAS and AVP. The kidneys secrete the hormone renin in response to reduced NaCl, ECF volume, and arterial blood pressure. Renin activates angiotensinogen, a plasma protein produced by the liver, into angiotensin I. Angiotensin I is converted into angiotensin II by angiotensin-converting enzyme (ACE) produced in the lungs. Angiotensin II stimulates the adrenal cortex to secrete the hormone aldosterone, which stimulates Na^+ reabsorption by the kidneys. The resulting retention of Na^+ exerts an osmotic effect that holds more H_2O in the ECF. Together, the conserved Na^+ and H_2O help correct the original stimuli that activated this hormonal system. Angiotensin II also exerts other effects that help rectify the original stimuli.

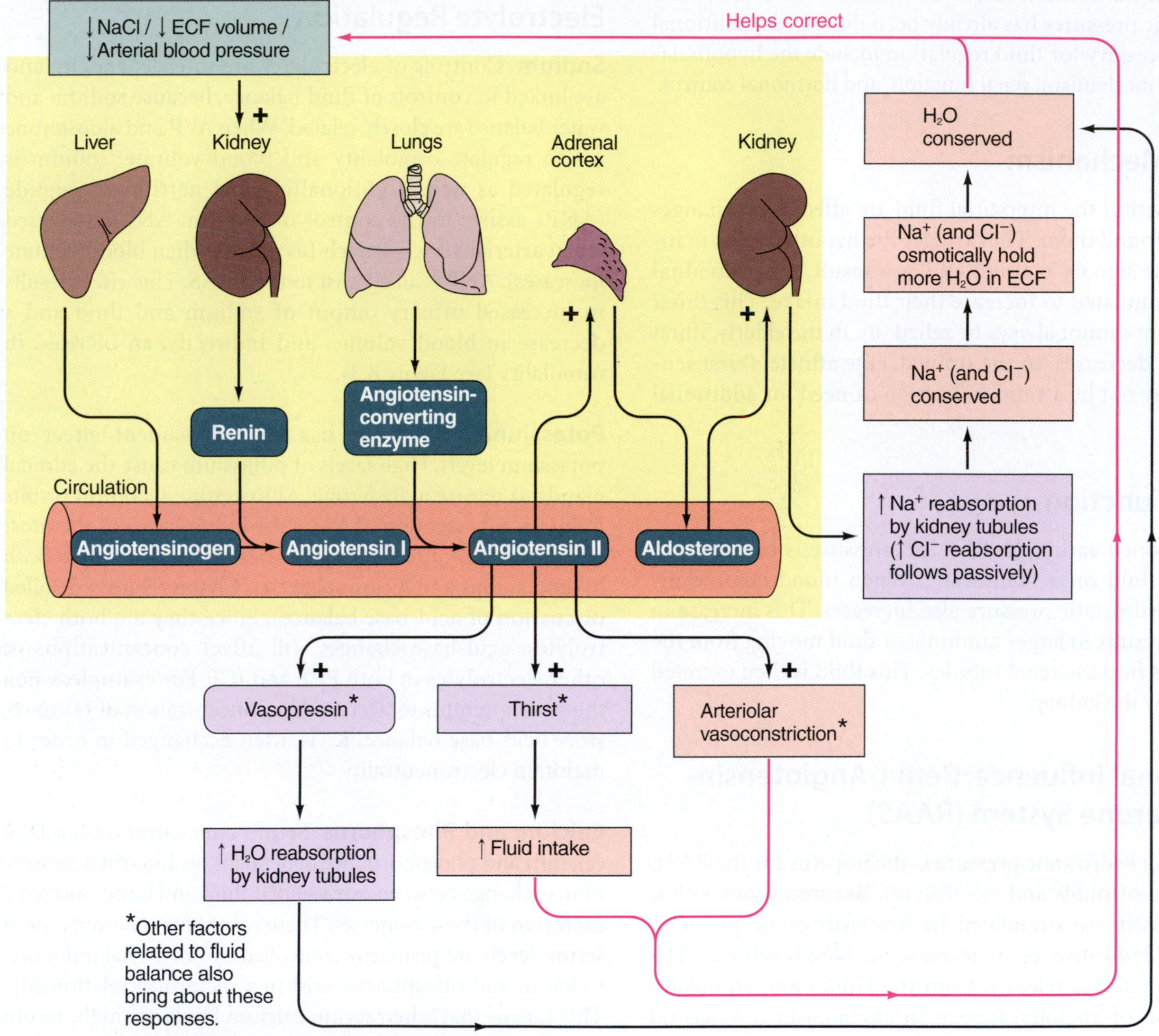

Source: L. Sherwood, *Human Physiology: From Cells to Systems*, 5e, copyright © 2004, p. 529

urine. PTH also stimulates activation of vitamin D. Vitamin D works to maintain serum calcium levels by increasing absorption of calcium in the small intestine. PTH also acts to increase phosphorus excretion when necessary. Calcitonin, another hormone, originates from the thyroid gland. It acts in opposition to PTH by *inhibiting* osteoclasts (cells within the bone that function to break down and resorb bone tissue) and therefore lowering serum calcium levels (see Figure 8.4).

Disorders of Fluid Balance

There are three general categories of alterations that occur in fluid balance. These include:

1. Changes in fluid volume
2. Changes in fluid concentration or osmolality
3. Changes in fluid composition

It is uncommon for these alterations to occur in isolation. Changes in fluid, electrolyte, and acid-base balance

FIGURE 8.3 Dual Effect of a Fall in Arterial Blood Pressure on Renal Handling of Na^+

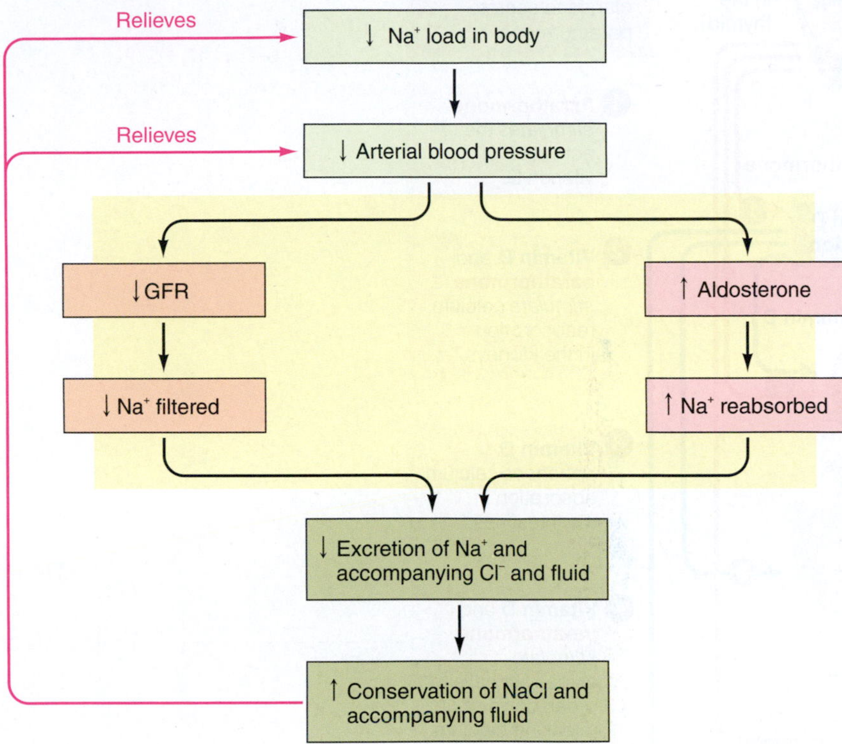

Source: L. Sherwood, *Human Physiology: From Cells to Systems*, 5e, copyright © 2004, p. 565

more commonly occur together, as this typical scenario suggests. Consider, for example, the patient with bacterial gastroenteritis. Excessive loss of fluid volume through vomiting and diarrhea could cause dehydration (change in fluid volume). If untreated, the resulting electrolyte loss of potassium (hypokalemia) results in both a change in osmolality and a change in fluid composition.

Alterations in Volume

Changes in volume primarily affect the ECF compartments. These changes involve relatively equal losses or gains in sodium and water; there is thus very little change in the ICF and no change in the ECF osmolality.

Hypovolemia

Pathophysiology Extracellular fluid deficit or **hypovolemia** is almost always related to renal or extrarenal loss of fluids. This will occur more rapidly when the loss is coupled with decreased oral intake of fluids. Extrarenal losses include any excess loss of fluid outside of renal excretion, including gastrointestinal losses, such as in vomiting or diarrhea. Losses through the skin occur during exposure to heat such as increasing body temperature (fever) or increased environmental heat. An endurance athlete who is involved in physical activity for more than an hour and a half can produce up to 3 liters of sweat per hour.

Excess loss through the skin can occur through burns or draining wounds. A fistula (abnormal opening between the gastrointestinal tract and other organs or peritoneal cavity) can also contribute to ECF losses. These extrarenal losses can be extreme in some clinical situations—as high as 5 to 6 liters in one day.

Other extrarenal losses occur when fluids are trapped in body spaces such as in development of ascites, in congestive heart failure, pulmonary edema, or in burns. This "third spacing" of fluid results in a net loss of ECF.

Renal losses occur in conditions that increase urinary excretion above what the individual has consumed orally. Excessive renal losses may occur as a component of a disease process; for instance, they may occur secondary to **diuresis**, as seen in the recovery phase of acute renal failure. In uncontrolled type 2 diabetes mellitus or **hyperosmolar hyperglycemic nonketotic syndrome**, the body attempts to correct acid-base imbalances and hyperosmolality by increasing urine excretion.

Medications such as diuretics are often prescribed to purposefully decrease ECF. For example, in the treatment of hypertension, diuretics are prescribed to decrease blood volume and therefore reduce blood pressure. Dietary composition can also affect urinary excretion. High-protein diets result in an increase in urine excretion due to the increased renal solute load.

Clinical Manifestations The severity of the signs and symptoms correspond to the severity of the volume deficit. As the ECF volume decreases, the corresponding blood

hypovolemia—decreased blood volume

diuresis—the production of excessive amounts of urine

hyperosmolar hyperglycemic nonketotic syndrome—complication of type 2 diabetes mellitus usually developing after a period of hyperglycemia combined with inadequate fluid intake

FIGURE 8.4 Calcium Balance.

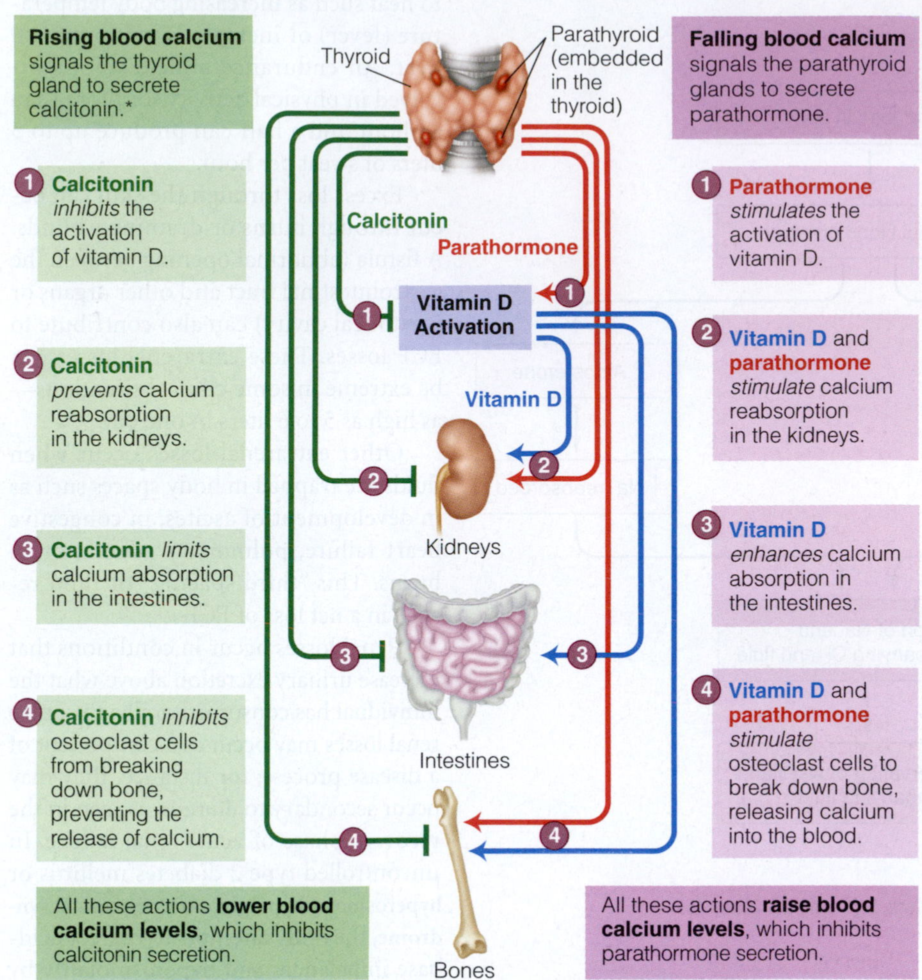
Rising blood calcium signals the thyroid gland to secrete calcitonin.*
Thyroid
Parathyroid (embedded in the thyroid)
Falling blood calcium signals the parathyroid glands to secrete parathormone.
1 Calcitonin inhibits the activation of vitamin D.
Calcitonin
Parathormone
Vitamin D Activation
1 Parathormone stimulates the activation of vitamin D.
2 Calcitonin prevents calcium reabsorption in the kidneys.
Vitamin D
2 Vitamin D and parathormone stimulate calcium reabsorption in the kidneys.
Kidneys
3 Calcitonin limits calcium absorption in the intestines.
3 Vitamin D enhances calcium absorption in the intestines.
Intestines
4 Calcitonin inhibits osteoclast cells from breaking down bone, preventing the release of calcium.
4 Vitamin D and parathormone stimulate osteoclast cells to break down bone, releasing calcium into the blood.
Bones
All these actions lower blood calcium levels, which inhibits calcitonin secretion.
All these actions raise blood calcium levels, which inhibits parathormone secretion.

*Calcitonin plays a major role in defending infants and young children against the dangers of rising blood calcium that can occur when regular feedings of milk deliver large quanities of calcium to a small body. In contrast, calcitonin plays a relatively minor role in adults because their absorption of calcium is less efficient and their bodies are larger, making elevated blood calcium unlikely.

Source: S. Rolfes, K. Pinna and E. Whitney, Understanding Normal and Clinical Nutrition, 7e, copyright © 2006 p. 414

volume will be reduced. The decrease in blood volume will lower blood pressure and decrease cardiac output. In mild cases of volume deficit, compensatory mechanisms will maintain homeostasis and symptoms may not be noticed by the patient. In more severe cases, blood pressure will be low, especially upon changing body position. This is referred to as orthostatic hypotension. The change in cardiac output can also result in tachycardia, weak pulse, and dizziness.

Other physical findings can include poor skin turgor and dry skin and mucous membranes. Rapid weight loss can also be monitored to substantiate the diagnosis of hypovolemia. Actually, any rapid changes in weight should initially considered to be an indication of fluid changes. See Table 8.6 for summary of clinical evaluation of fluid and electrolyte disorders.

Laboratory Findings No single laboratory finding will confirm hypovolemia. Measurements of blood and urine will correspond to the underlying cause of the hypovolemia. Tables 8.7 and 8.8 provide a summary of laboratory assessment.

Hemoconcentration in hypovolemia occurs unless there is also a loss of blood. Hemoconcentration results from the kidneys' compensation by decreasing urinary output. For example, serum sodium and chloride, blood urea nitrogen, hemoglobin, hematocrit, and albumin may be abnormally elevated in dehydration. As urinary output decreases, urinalysis results will reveal concentrated urine with elevated specific gravity, a darker color, and a cloudy appearance.

Treatment Treatment is prescribed according to the underlying cause for the fluid deficit. In mild cases, increasing

TABLE 8.6

Clinical Changes in Fluid and Electrolyte Disorders

Assessment	Evaluation
Daily weights 2% ↑↓: mild fluid volume deficit or excess 5% ↑↓: moderate deficit or excess 8% ↑↓: severe deficit or excess	Rapid changes reflect fluid changes Body weight does not change when fluid shifts to third spaces
Eyes: dry conjunctiva, decreased tearing	Fluid volume deficit
Periorbital edema	Fluid volume excess
Lips and oral cavity: dry, cracked lips; small multifurrowed tongue	Fluid volume deficit
Decreased skin turgor	Fluid volume deficit
Tachycardia	Fluid volume deficit
Slowed pulse, increased BP	Fluid volume excess
Orthostatic BP	Fluid volume deficit
Hand veins	Prolonged filling: volume deficit Prolonged emptying: volume excess
Central venous pressure (CVP)	↓CVP: volume deficit ↑CVP: volume excess
Jugular vein distention (JVD)	Flat neck veins when supine: volume deficit Extended JVD: volume excess
Cardiac dysrhythmias	May indicate deficits or excess of K, Mg, Ca, PO4
Lungs: pulmonary congestion; ↑ respiratory rate, moist rales, rhonchi	Fluid volume excess
Oliguria	Severe fluid volume deficit
Extremities; localized swelling; sacrum: edema present	Fluid volume excess

intake of both sodium and water will allow gradual correction of hypovolemia. In more severe cases, fluid and electrolyte replacement will need to be prescribed through intravenous fluids (see Table 8.9).

Hypervolemia

Pathophysiology In normal, healthy individuals the kidney will excrete excess water, but ECF excess can be a common occurrence in clinical situations. The most common cause of **hypervolemia** is a decrease in urinary output such as seen in acute renal failure. Excess intravenous fluids or the failure of the kidney to accommodate a rapid ingestion of fluids quickly enough may also cause hypervolemia. Excessive secretion of vasopressin can also result in excessive volume retention.

When there is ECF excess, fluid shifts into interstitial spaces so a balance between ECF and ICF is maintained.

TABLE 8.7

Biochemical Evaluation of Fluid and Electrolyte Status

Blood Tests	Normal Value	Discussion: Additional factors that May Affect Levels
Potassium	3.5 to 5.0 mEq/L	↑ in acidosis ↓ in alkalosis
Sodium	135 to 145 mEq/L	Consistent with current osmolality. ↑ Na = hyperosmolar body fluids. Rare that it would indicate high Na level.s ↓ Na = dilutional body fluids in relationship to solute.
Chloride	98 to106 mEq/L	↑ Cl may indicate metabolic acidosis. ↓ Cl often with metabolic alkalosis and hypokalemia.
Calcium	8.7 to 9.2 mg/dL	Evaluate with serum albumin levels –total Ca ↓ when albumin is low; but ionized calcium does not change.
Phosphate	2.5 to 4.5 mg/dL	Elevated in chronic renal failure.
Hematocrit	37% to 47% (women) 40% to 54% (men)	↑ Fluid volume deficit ↓ Fluid volume excess
Glucose	70 to 110 mg/dL	Hyperglycemia causes osmotic diuresis and ↓ blood volume.
BUN	8 to 26 mg/dL	↑ Fluid volume deficit ↓ Fluid volume excess
Osmolality	275 to 295 mOsm/kg	↑ Fluid volume deficit ↓ Fluid volume excess

hypervolemia—increased blood volume

TABLE 8.8

Evaluation of Fluid and Electrolyte Status: Urine Tests

Urine Tests	Normal Value	Comments
Sodium	100 to 260 mEq/24 hr >40 mEq/L in random sample	<10 mEq/24 hr = hyponatremia/edema/volume depletion
Potassium	25 to 100 mEq/24 hr	↑ hyperaldosteronism ↓ adrenal insufficiency
Chloride	110 to 250 mEq/24 hr	<10 mEq/L in metabolic alkalosis secondary to volume deficit >20mEq/L in metabolic alkalosis caused by hyperaldosteronism or ↓ K^+
Color	Pale Yellow	Dark, amber, hazy in dehydration/fluid deficit
Urine osmolality	50 to 1400 mOsm	Reflects concentrating or diluting ability
Specific gravity	1.003 to 1.030	↑ in fluid deficit/dehydration and hyperosmolar urine

Accumulation of fluid in interstitial spaces, as previously discussed, is called **edema**.[1] Overall blood volume will be increased in hypervolemia. This increases blood pressure and overall work of the heart.

Clinical Manifestations Increased blood volume results in elevated blood pressure and jugular venous distention (see Table 8.6). The presence of edema can be noted in peripheral regions such as ankles and feet, or in the face and scrotal areas. When the clinician presses on areas of edema, an indentation will occur. The depth of the "pitting" corresponds to the amount of edema (see photo, "Pitting Edema").

Respiratory symptoms include difficulty breathing (dyspnea) and the presence of **rales**. Rapid weight changes

edema—the accumulation of excess fluid in cells, tissue, or a cavity, resulting in swelling

rales—abnormal respiratory sounds when air flows through liquid present in the airways

hyponatremia—abnormally low concentrations of sodium ions in the circulating blood

Pitting Edema

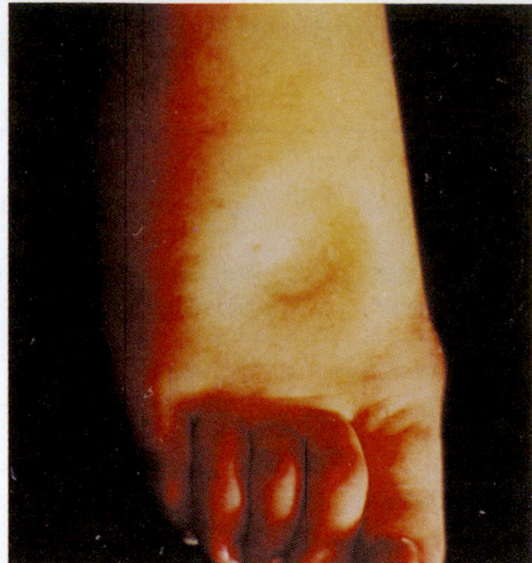

Source: Department of Pathology, Medical College of Virginia

also correlate with hypervolemia. A weight gain of 1 kilogram is equivalent to the retention of approximately 1 liter of fluid.

Hemodilution can also affect laboratory values. Sodium, chloride, hemoglobin, hematocrit, blood urea nitrogen (BUN), and albumin would be abnormally low in this clinical scenario.

Treatment Treatment for hypervolemia will target the underlying cause. Control of other symptoms involves restriction of both fluid and sodium.

Alterations in Osmolality

In general, alterations in osmolality occur when there is a shift of water without a corresponding shift in solute (electrolytes). This most commonly involves changes in the concentration of sodium in the ECF or hyperglycemia associated with uncontrolled diabetes mellitus.

A change in osmolality within the ECF directly influences osmolality of the ICF. This would also be consistent in the opposite situation where change in the osmolality in the ICF directly influences the osmolality of the ECF.

Sodium Imbalances

Maintaining sodium levels involves an intricate balance between water and sodium. The rennin-angiotensin-aldosterone system (RAAS), ECF volume, and overall renal function are most important in maintenance of sodium homeostasis.

Hyponatremia

Pathophysiology Either decreasing amounts of sodium, increasing amounts of water in the ECF, or a combination of both can cause **hyponatremia**. In most cases of hyponatremia,

TABLE 8.9

Commonly Prescribed Intravenous Solutions*

Intravenous Solutions	Content	Osmolality mOsm/L	Use
5% Dextrose	No electrolytes—5 g dextrose/dL; 170 kcal/L	252	Free water; correction of fluid balance and hypernatremia; provides some energy
10% Dextrose	No electrolytes—10 g dextrose/dL; 340 kcal/L	505	
0.45% Saline ("Half Normal Saline")	77 mEq Na^+/L, 77 mEq Cl^+/L	154	No energy provided; correction of fluid balance but doesn't necessarily correct electrolyte imbalances
0.9% Saline ("Normal Saline")	154 mEq Na^+/L, 154 mEq Cl^+/L	308	Na^+ and Cl^+ are greater than in plasma levels; can be administered with blood products
Ringer's Solution	130 mEq Na^+, 4 mEq K^+, 2 mEq $Ca2^+$, 109 mEq Cl^+	309	Similar to plasma composition; does not provide free water or energy
Lactated Ringer's (Hartmann's Solution)	130 mEq Na^+, 4 mEq K^+, 2 mEq $Ca2^+$, 109 mEq Cl^+, 29 g lactate	273	Similar to plasma composition; does not provide free water or energy
Dextrose in Saline: 5% in 0.225%	170 kcal/L, 38.5 mEq Na^+/L, 38.5 mEq Cl^+/L	355	Provides energy, free water, sodium, and chloride
Dextrose in Saline: 5% in 0.455%	170 kcal/L, 77 mEq Na^+/L, 77 mEq Cl^+/L	406	
Dextrose in Saline: 5% in 0.9%	340 kcal/L, 154 mEq Na^+/L, 154 mEq Cl^+/L	560	

*Reference values only. Content may vary slightly between formulations.

Modified from: Whitmire SJ (2001). Fluid and electrolytes. IN: The Science and Practice of Nutrition Support. Dubuque IA: Kendall/Hunt Publishing Company. Heitz UE, Home MM (2001). Pocket Guide to Fluid, Electrolyte, and Acid-Base Balance. 4th Ed. St. Louis MO: Mosby/Elsevier Science.

serum sodium reflects the ratio of water to sodium. It is rare for a deficit in sodium to occur simply from lack of nutritional intake. It could occur through a combination of a sodium restriction used for nutrition therapy and use of diuretics (such as in congestive heart failure or in liver disease) as part of the medical regimen.

The increase in fluid without corresponding increase in sodium can also result in hyponatremia. This might occur, for instance, in a situation in which intravenous fluids are administered without electrolytes. This might also occur in the condition of **water intoxication**, though it is rare. In this situation, usually seen in psychiatric conditions, the patient is unable to control excessive water intake.

The syndrome of inappropriate antidiuretic hormone (SIADH) may also result in hyponatremia. In this condition, total body water increases without subsequent increase in sodium, producing a dilutional effect on serum sodium. It is characterized by high levels of vasopressin (AVP) without the normal stimuli for vasopressin release.

The only situation associated with hyponatremia that does not involve either decreased amounts of sodium or increased amounts of water in the ECF is the increase of solute in the plasma. Hyperglycemia would be the most likely situation that would precipitate this etiology of hyponatremia. The increase in glucose causes a shift of water from the ICF to the ECF in attempts to normalize the osmolality. When this occurs, sodium is diluted and hyponatremia results.

Clinical Manifestations Hyponatremia is often without clinical manifestations until levels of sodium fall below 120 mEq/L. Signs and symptoms that do occur are consistent with changes in osmolality. As serum osmolality falls, water enters the brain cells. Nausea, vomiting, lethargy, confusion, and even seizures can result. All signs and symptoms are more pronounced if hyponatremia occurs rapidly. Laboratory measurements would confirm serum sodium of <135 mEq/L. Plasma osmolality will be <287 mOsm/kg.

water intoxication—uncontrolled, excessive water consumption resulting in dilutional complications

Treatment Treatment focuses on elevating serum sodium and treating underlying causes. This can be accomplished by restricting water intake to less than urinary output. Administration of sodium is reserved for severe cases of hyponatremia. In the case of hyperglycemia, reducing serum glucose is accomplished through the administration of insulin.

Hypernatremia

Pathophysiology **Hypernatremia** can theoretically be caused by either an increase in sodium or decrease in water. But generally, in a normal, healthy person, an increase in sodium results in compensatory increase in renal excretion of sodium, thus maintaining homeostasis.

Hence, situations that change water balance are much more likely to cause hypernatremia than are increases in sodium intake. Insufficient water intake can result in hypernatremia. For example, the elderly have a decreased thirst sensation and may not take in adequate amounts of fluid.

Excessive water losses can also result in hypernatremia. Nonrenal losses might be seen in patients who are febrile, who hyperventilate, or who have open wounds. Conditions such as **diabetes insipidus** or SIADH (discussed earlier) are also potential causes.

Hyperosmolality can also result in hypernatremia. This is seen in conditions with hyperglycemia (such as uncontrolled diabetes mellitus) or in increased urea production (high-protein diets). Although administration of excessive sodium in intravenous solutions would be the most likely culprit in the situation of hypernatremia due to an absolute excess of sodium, this situation is not common.

Clinical Manifestations As in hyponatremia, the central nervous system is most sensitive to changes in fluid and electrolyte balance. Cellular dehydration results in an increasing severity of neurological symptoms as the condition becomes more severe. Possible symptoms range from lethargy and agitation to seizures and coma. Body temperature can be elevated, skin is flushed, and mucous membranes are dry. In cases of hypernatremia, laboratory measurements confirm a serum sodium of $>$145 mEq/L and a plasma osmolality of $>$295 mOsm/kg.

Treatment The goal of treatment is to gradually lower serum sodium to a normal concentration. If possible, the patient should simply increase his or her water intake, which may return the serum Na^+ levels to normal. When levels fall too quickly, cerebral edema can occur, endangering the patient. Intravenous treatment usually involves slow administration of fluids (such as dextrose) without sodium. Diuretics and dialysis may be required in more severe situations.

hypernatremia—abnormally high levels of serum sodium

diabetes insipidus—chronic excretion of very large amounts of pale urine of low specific gravity

hypokalemia—low serum potassium

Potassium Imbalances

The body strives to maintain potassium levels within a very narrow range. Abnormalities in potassium are often life threatening, due primarily to the role of potassium in neurotransmission and muscle contraction.

Small changes in the ECF concentration of potassium may reflect significant abnormalities in the ICF concentration of potassium. Though only a small amount of potassium is found in the ECF, it is essential for neuromuscular function. Distribution of potassium in both ECF and ICF is influenced by acid-base balance and hormonal secretion.

Hypokalemia

Pathophysiology **Hypokalemia** can result from inadequate nutritional intake of potassium, increased renal loss of potassium, or increased loss from the gastrointestinal tract. Hypokalemia also results from a shift of potassium from the ECF to the ICF.

Hypokalemia Secondary to Gastrointestinal Losses Vomiting, nasogastric suction, and diarrhea are common gastrointestinal origins of hypokalemia. Compensatory actions in acid-base balance contribute to hypokalemia. For example, loss of gastric acids through vomiting can lead to metabolic alkalosis. Metabolic alkalosis in turn increases bicarbonate excretion in the distal tubules. Increased amounts of potassium are excreted along with bicarbonate; thus, loss of gastric acids can lead to hypokalemia (see Chapter 9). As a result of gastrointestinal losses during vomiting, there is a decrease in ECF volume. This decrease in volume stimulates the release of aldosterone. While aldosterone causes the retention of sodium (and the subsequent increase in volume), potassium excretion is increased as a secondary consequence.

Hypokalemia Secondary to Renal Losses One of the most common causes of hypokalemia is use of loop diuretics. Drugs such as furosemide (e.g., Lasix™) cause increased urine excretion with accompanying loss of potassium. Nutrition therapy for those individuals receiving loop diuretics will include education for a higher dietary intake of potassium. Notably, natural licorice and chewing tobacco, if swallowed, contain an aldosterone compound that will also stimulate increases in urinary potassium excretion.

Refeeding syndrome (see Chapter 7) may result in hypokalemia as levels of potassium shift from the ECF to ICF to accommodate increased metabolism. When a malnourished individual begins nutrition support, there is potential for inadequate amounts of intracellular electrolytes required to support anabolism. Carbohydrate metabolism also results in shifts of these electrolytes.

Clinical Manifestations Any body system involving muscular action can be affected by hypokalemia. There will generally be muscle weakness and diminished deep tendon reflexes. In advanced hypokalemia, respirations become shallow due to poor lung muscle action. Within the cardiovascular system, rhythm changes can be noted; untreated, these dysrhythmias can lead to cardiac arrest.

Normal serum potassium is 3.5 to 5.0 mEq/L. Hypokalemia is diagnosed when serum levels fall below 3.5 mEq/L. However, it is important to remember if hypokalemia results from ECF-ICF shifts, serum levels may not reflect the true level of electrolyte abnormality. Acid-base imbalance can be confirmed by a serum pH < 7.45.

Treatment Potassium levels in the body can be corrected through dietary sources of potassium, oral supplementation, or intravenous administration of potassium. The underlying cause of potassium loss should be addressed as well.

Hyperkalemia

Pathophysiology The most common cause of **hyperkalemia** is inadequate excretion of potassium. This is commonly found in renal failure. Excessive use of potassium-sparing diuretics, used to treat hypertension, may result in inadequate excretion of potassium.

Shifts in potassium from the ICF to the ECF can result in hyperkalemia. Numerous clinical situations can lead to this shift. Elevated concentrations of potassium in the ECF can occur when there is an increased hemolysis of red blood cells, during **leukocytosis** or **thrombocytosis**. Catabolism and strenuous exercise increase serum potassium. Acidosis also results in hyperkalemia. When hydrogen ions are excreted to correct acidosis, potassium ions are retained. Thus, hyperkalemia is a secondary result (see Chapter 9).

Although it is rare for hyperkalemia to result from excessive ingestion, this can occur with ingestion of potassium-containing salt substitutes, especially if the patient does not have normal renal function. (Salt substitutes are composed of potassium chloride [KCl], while salt is composed of sodium chloride [NaCl].) Hyperkalemia can also occur from blood transfusions or when excessive amounts of potassium are given in intravenous solutions.

Clinical Manifestations Signs and symptoms are a result of the neuromuscular effects of altered potassium levels. A gradual rise in potassium levels, as in chronic renal failure, is better tolerated than the rapid rise in potassium levels that leads to potentially fatal symptoms.

Muscle weakness, paralysis, **paresthesias,** and cardiac dysrhythmias are a consequence of inactivation of electrical transmissions across the cell membrane and the interruption of normal polarization. Severe hyperkalemia, as with hypokalemia, can lead to cardiac arrest. Toxic effects of hyperkalemia are enhanced by the presence of other electrolyte imbalances such as hypocalcemia, hyponatremia, and acidemia.

Serum potassium levels in hyperkalemia are measured at > 5.5 mEq/L. Further diagnosis can be made by electrocardiogram (ECG) changes.

Treatment As with other electrolyte abnormalities, treatment of the underlying cause is crucial in correcting hyperkalemia. In a short-term emergency situation, calcium gluconate can be given intravenously to decrease the abnormalities in cardiac cells that could lead to cardiac arrest. Additionally, both glucose and insulin can be used to shift potassium in the ECF to the ICF. Correction of any acid-base imbalance also results in potassium movement into the ICF. Cation exchange resins such as **Kayexalate** can be given to allow the exchange of sodium for potassium in the large intestine.

For long-term treatment, dialysis and a potassium restricted diet are primary interventions to control hyperkalemia. It is also crucial to prevent malnutrition through adequate nutrition support. Catabolism, as noted above, contributes to the increasing levels of potassium.

Calcium Imbalance

Most calcium found in the body is located within bones and teeth. The remaining 1% found in body fluids is highly regulated due to its function in muscle contraction. Normal serum calcium levels range from 9 to 10.5 mg/dL (4.5 to 5.5 mEq/L). Serum calcium is either protein (albumin) bound (40%), complexed (13%), or ionized (47%). Ionized calcium is essential for cellular processes. Misinterpretation of serum calcium levels is common. Serum calcium is not an indication of bone calcium. Furthermore, total serum calcium must be assessed in relation to serum albumin levels. A decrease in serum albumin by 1 g/dL will decrease serum

hyperkalemia—high serum potassium

leukocytosis—high white blood cell count

thrombocytosis—low number of platelets

paresthesias—symptoms of tingling in fingers and toes; often consistent with electrolyte imbalance

Kayexalate—medication used to reduce high serum potassium; exchanges sodium for potassium in the intestine

calcium by 0.8 mg/dL. Thus, a patient with a low albumin level may falsely appear to have low blood calcium.

$$\text{Total serum Ca}^+ \text{ (mg/dL)} = \text{Measured Ca}^+ \text{ (mg/dL)} + 0.8 \times 4\text{-serum albumin (g/dL)}$$

(Formula is not valid when serum pH is altered.)

Plasma homeostasis is maintained through the interaction of three hormones: parathyroid hormone (PTH), calcitonin, and 1,25-dihydroxycholecalciferol (activated vitamin D). When serum calcium levels fall, PTH is released. PTH acts to decrease bone resorption of calcium, decrease calcium excretion, and increase calcium absorption in the gastrointestinal tract. Activated vitamin D acts to increase absorption of calcium by increasing production of the protein necessary for calcium absorption in the GI tract. Vitamin D also affects uptake of calcium by the bone. Calcitonin, produced by the thyroid gland, inhibits osteoclast activity, and therefore reduces bone resorption.

Hypocalcemia

Pathophysiology **Hypocalcemia** most commonly results from a deficit of PTH or in clinical cases of abnormal vitamin D metabolism such as seen in patients with renal or liver failure. Activated vitamin D requires normal function of both liver and kidneys, and thus abnormalities of organ function are common causes of hypocalcemia. Alkalosis can cause a decrease in ionized calcium in the ECF, and can result in symptoms consistent with hypocalcemia. Finally, because calcium and phosphorus levels are tightly linked, changes in serum phosphorus levels can result in subsequent changes in serum calcium levels.

Clinical Manifestations Hypocalcemia can result in symptoms that will be seen in all body systems. Symptoms are a result of altered nerve transmission and electrical activity of the cell. Neuromuscular symptoms include muscle spasms, tetany, and cardiac dysrhythmias. Untreated hypocalcemia is life threatening. Serum $Ca^+ < 9$ mg/dL or ionized $Ca^+ < 4.5$ mg/dL is consistent with hypocalcemia.

Treatment Treatment of the underlying cause is crucial for long-term control. Intravenous administration of calcium is only given in extreme cases due to the potential side effects.

Hypercalcemia

Pathophysiology The majority of cases of **hypercalcemia** stem from either hyperparathyroidism or from a malignancy. Malignant tumors of the breast, prostate, and cervix commonly metastasize to the bone. Resulting bone resorption can lead to hypercalcemia. Other malignancies produce factors that act in a similar fashion to PTH and cause an increase in bone resorption.

Clinical Manifestations Signs and symptoms will vary according to rapidity of onset and degree of hypercalcemia. Early symptoms may be vague and include fatigue and weakness. Bone pain, confusion, and cardiac dysrhythmias may also be present. This is often seen when a malignancy has spread to bone and causes abnormal release of calcium. Serum calcium levels > 10.5 mg/dL are diagnostic for hypercalcemia.

Treatment The underlying cause of hypercalcemia is the focal point for treatment. Hyperparathyroidism can be treated surgically by removal of the parathyroid gland. Agents can be given to bind calcium in the serum when the treatment is a clinical emergency. Intravenous fluids, diuretics, and dialysis can also be used to dilute serum calcium and increase its excretion.

hypocalcemia—low serum calcium (< 8.7 mg/dL)

hypercalcemia—high serum calcium (> 10.2 mg/dL)

hypophosphatemia—low serum phosphorus (< 1.45 mmol/L)

Phosphorus Imbalance

Phosphate is a crucial anion essential for metabolism of all substrates. It is a crucial component of the cellular energy reservoir, ATP, and an integral part of DNA and RNA. Phosphate also participates in maintenance of acid-base balance and is a structural component of bones, teeth, and phospholipids.

The average diet contains 1 to 1.6 g of phosphorus, which is easily absorbed. Phosphorus imbalances will rarely originate solely from nutritional intake. Serum phosphate exists as inorganic phosphate ions and only about 10% is bound to protein. Calcium and phosphate interact in a reciprocal fashion. Urinary excretion of phosphate increases or decreases in inverse proportions to serum calcium levels.

Hypophosphatemia

Pathophysiology **Hypophosphatemia** can result from vitamin D deficiency or from decreased activation of vitamin D. Hyperparathyroidism can also lead to low serum levels of phosphate. Consumption of aluminum-containing antacids may also bind phosphate.

In respiratory alkalosis, hyperventilation causes a shift of phosphate from the ECF to the ICF resulting in hypophosphatemia. In refeeding syndrome, a rapid shift of phosphate from the ECF to ICF occurs in response to increased metabolism. In hospitalized alcoholic patients, hypophosphatemia can result from withdrawal of alcohol.

Clinical Manifestations When phosphate is unavailable to support ATP and 2,3-diphosphoglycerate in glycolysis, changes are seen in every body system. Respiratory insufficiency and central nervous system abnormalities will eventually lead to encephalopathy and coma.

Low levels of phosphate lead to the mobilization of calcium and phosphorus in the bone, causing osteomalacia and rickets. Metabolic acidosis is often a result of hypophosphatemia, due to decreased hydrogen ion secretion. Hypophosphatemia is defined as a serum level < 2.5mg/dL.

Treatment Treatment should be focused on the underlying cause of the phosphate abnormality. Oral phosphate as food or supplements is the primary route to increase phosphorus levels. Intravenous phosphate is only given in emergency situations due to the risk of precipitation of calcium—the deposition of calcium into soft tissues where it could cause organ damage.

Hyperphosphatemia

Pathophysiology Acute or chronic renal failure is the most common clinical condition associated with **hyperphosphatemia**. As glomerular filtration rate decreases, the ability to excrete phosphorus decreases proportionally. Other causes of hyperphosphatemia involve low levels of PTH and other endocrine disorders.

Phosphate is released when cells break down. This release of phosphate constitutes a significant shift of phosphate from the ICF to the ECF. Drugs or medications that contain phosphorus or a high intake of vitamin D may also result in hyperphosphatemia.

Clinical Manifestations Most signs and symptoms associated with hyperphosphatemia are a result of concurrent hypocalcemia. These symptoms would originate from altered nerve transmission and muscle contraction. Hyperphosphatemia is defined as a serum level > 4.5 mg/dL.

Treatment For treatment of chronic hyperphosphatemia, dietary restriction of phosphorus is necessary. Foods highest in phosphorous include milk, dairy products, and animal protein sources. (See Chapter 20 for complete listing.) Medications that will bind phosphate are also used. In renal failure, where high serum levels of phosphorous are common, calcium supplements are used as a phosphate binder to help control these levels.

Magnesium Imbalances

Magnesium is an abundant mineral in the ICF and is a crucial component of cellular energy metabolism. Its function is closely related to both calcium and potassium, because it assists in maintaining calcium and phosphorus homeostasis.

Serum magnesium levels do not accurately reflect total body magnesium. Most magnesium is located in bone, with the rest being primarily in the ICF. Less than 2% is found in the ECF. Regulation of magnesium is poorly understood. Less than 50% of magnesium consumed in foods is absorbed. Excessive magnesium is excreted in the urine.

Hypomagnesemia

Pathophysiology The most common cause of magnesium imbalance originates from chronic alcoholism and the withdrawal of alcohol. Some medications, such as **cyclosporine**, cause excessive urinary losses of magnesium. In some transplant patient populations, routine magnesium supplementation is necessary due to the excessive excretion in the urine. Magnesium losses in these situations are generally higher than 25 mg/day. Malabsorption syndromes can result in both calcium and magnesium malabsorption, which are common in steatorrhea.

Clinical Manifestations Signs and symptoms of **hypomagnesemia** are difficult to distinguish from those of hypokalemia and hypocalcemia. In a practical sense, the clinician will first rule out other electrolyte deficiencies before attempting to replace magnesium. General symptoms include personality changes, depression, anorexia, nausea and vomiting, and ileus, as well as neuromuscular irritability. Serum magnesium levels < 1.5 mEq/L or < 1.8 mg/dL are considered to be low.

Treatment Imbalances can be treated with diet and oral supplements. The underlying cause of hypomagnesemia should be addressed, and other coexisting electrolyte imbalances should be corrected. Magnesium should be given cautiously intramuscularly or intravenously, because high levels of magnesium will cause cardiac arrest.

Hypermagnesemia

Pathophysiology High serum levels of magnesium are uncommon. If they do exist, it is generally due to declining renal function or from excessive supplementation. Supplementation of magnesium-containing antacids or laxatives (such as Milk of Magnesia) is a common pathway for excessive ingestion. Magnesium is used to treat **preeclampsia**, which is a high-risk condition for both the fetus and the mother.

hyperphosphatemia—high serum phosphorus (>1.45 mmol/L)

cyclosporine—immunosuppressant drug

hypomagnesemia—low serum magnesium levels (<1.1 mmol/L)

preeclampsia—development of hypertension, with symptoms of proteinuria and edema, during pregnancy

Clinical Manifestations Neuromuscular action is impaired and may cause decreased reflexes, muscle weakness, and paralysis. Cardiac function is also impaired, and if uncorrected, may lead to cardiac arrest. Serum levels > 2.5 mEq/L or 3.0 mg/dL is diagnostic for **hypermagnesemia.**

Treatment In mild hypermagnesemia, removal of medications containing magnesium is warranted. Calcium gluconate, given intravenously, can reverse effects of magnesium in more severe cases. Dialysis can also be used to decrease high levels of magnesium.

Summary

Many clinical conditions involve impairment of fluid and electrolyte balance. Nutrition therapy must be coordinated with medical care in the correction of imbalances. Metheny (2000) outlines six important points to address before assessing fluid and electrolyte balance. These include:

- Does the patient have a disease or injury that could affect fluid/electrolyte balance?
- Is there a medication or treatment that could affect fluid/electrolyte balance?
- Is there fluid loss?
- Has nutrition therapy restricted any nutrient that could affect fluid/electrolyte balance?
- Has oral intake for both water and nutrients been adequate?
- Do the intake and output records balance?

The clinician will use this information to rule out any possible fluid and electrolyte disorders. This information will be confirmed by clinical assessment and laboratory values. This is not as easy as it may seem here. Many disorders produce clinical manifestations that are vague and that often overlap each other. Tables 8.6, 8.7, and 8.8 provide a summary of these assessment guidelines. With experience, the clinician will become proficient in assessment of fluid and electrolyte balance.

hypermagnesemia—high serum magnesium levels (>1.1 mmol/L)

PRACTITIONER INTERVIEW

La Paula Sakai, RD, MS, CNSD, *Clinical Nutrition Coordinator, Specialized Nutrition Support Kaiser Permanente Medical Center, Santa Clara, CA*

Do you assess fluids and electrolytes in your practice?

Yes, I give it an 8 on a scale of 1 to 10 with 10 being the most important in patients that require nutrition support.

Has the importance of fluid and electrolyte assessment changed over time?

When I started practicing, RDs didn't really understand the pathophysiology of most disease states and abnormalities of fluid and electrolytes weren't emphasized. Practicing dietetics, I came to realize the importance of fluids and electrolytes, their part in the pathology of the diseases I was seeing, and their relationship to nutritional assessment and support. Over time, I reviewed and researched fluids and electrolytes and furthered my knowledge to better understand their relationships in MNT.

What indicators do you use for assessing fluid and electrolyte status?

I look at the I & Os, degree of edema and ascites (looking at the patient and nursing/MD chart notes, talking to the nurse/MD), abnormal electrolyte labs. I'm careful to try and quantify the output, for example, wound losses (hard to quantify), diarrhea, and so on. Dietetic interns often think that they can look in just one place in the chart to determine the fluid and electrolyte status. For me it's a combination of eyeballing the patient, reading the chart notes, and talking to the appropriate person(s) for more information. Labs that I commonly look at are: sodium, potassium, albumin, and BUN. With TPN I also look at chloride and phosphorus.

What are the most common situations (diagnoses) that present problems with fluid and electrolyte balance?

Common situations are: syndrome of inappropriate ADH (SIADH); failure of certain organs: heart, liver, and kidney; critically ill patient in the ICU d/t fluid shifts; patients receiving large amounts of steroids; chronic diarrhea or fistula losses; emesis losses; excessive wound losses, dehydration of older clients often d/t hot days and illness and the very young (infants); and complication of eating disorders.

How do you interface with the health care team regarding fluid and electrolyte assessment?

I communicate face to face or on the telephone and via chart. I find that many health care personnel don't understand how diet or nutrition support can interface with fluid and electrolyte balance. For example, sometimes a fluid restriction is ordered but I think it would be better for the client's sodium intake to be modified instead, so then I have to be able to communicate that information to the doctor. If my chart note isn't read, then I make the effort to talk with the physician either on the phone or in person.

What would be most important for practitioners just starting to develop their skills in fluid and electrolyte assessment?

You need to start with basic textbook knowledge. If you covered the material in your MNT course, then you need to read it again before seeing patients for the first time. I work in a teaching hospital. I even hear doctors remind medical residents to review fluids and electrolytes because they have forgotten what they learned in medical school. As you gain clinical knowledge and later become an RD, you still need to review and research this topic. The interrelationship between nutrition support and fluids and electrolytes needs is so complex I find myself always going back and reviewing the material again and again.

CASE STUDY

Introduction:

Max Williams is an 18-year-old male who has recently returned from a two-week trip to Quito, Ecuador. He has felt ill for several days, which began the day before he left Ecuador. He describes 5 to 10 episodes of diarrhea each day which have not resolved after taking Kaopectate. Fecal smear indicates gross blood with leukocytes. Admitting diagnosis is moderate dehydration with R/O bacterial versus viral gastroenteritis.

Physical Exam :

General appearance: Lethargic 18-year-old Caucasian male

Vitals:

Temperature 101.5 °F

BP:

80/65

HR:

89 BPM

Respiratory Rate:

22 BPM

Heart:

Moderately elevated pulse

HEENT:

Eyes:

Sunken; Sclera clear without evidence of tears

Ears:

Clear

Nose:

Dry mucous membranes

Throat:

Dry mucous membranes; no inflammation

Genitalia:

Unremarkable

Neurologic:

Alert, oriented x 3. Irritable.

Extremities:

No joint deformity or muscle tenderness. No edema.

Skin:

Warm, dry. Reduced capillary refill (approximately 2 seconds)

Chest/lungs:

Clear to auscultation and percussion

Abdomen:

Tender, nondistended, minimal bowel sounds

Nutrition Assessment:

Ht. 6′2″ Wt. 178 lbs UBW 185 lbs

The Registered Dietitian's interview indicates that prior to this illness, Max had a good appetite with consumption of a wide variety of foods. The patient did remark about consumption of a "lot of new foods" while in Ecuador with some purchased from street vendors.

Labs:

Total Protein 7.2 g/dL, Albumin 4.9 g/dL, Na 154 mmol/L, K^+ 3.2 mmol/L, Cl^- 107, PO_4 4.0 mmol/L, BUN 21 mg/dL, Cr 1.4 mg/dL, Hgb 15.5 g/dL, Hct 41%, WBC 17 × $10^3/mm^3$.

Questions

1. What signs and symptoms in the physical assessment provide evidence for the diagnosis of dehydration?
2. How might Max's laboratory values be affected by his hydration status?
3. What factors should be identified in a urinalysis that may also be consistent with dehydration?
4. Identify at least two nutrition problems revealed by the nutrition assessment and medical history. Next, identify the etiology of each nutrition problem. Finally, identify the signs and symptoms that support the evidence for these nutrition problems.

WEB LINKS

Gatorade Sports Science Institute: The Gatorade Sports Science Institute (GSSI) is a research and educational facility. Provides current information on sports nutrition and exercise science. The materials and services of the Institute are designed as educational tools for sports health professionals.

http://www.gssiweb.com

University of Montana—Sports Nutrition: Information about hydration and electrolytes for athletes.

http://btc.montana.edu/olympics/nutrition/eat02.html

Virtual Chembook—Elmhurst College: Excellent online resource for both fluid and electrolyte balance.

http://www.elmhurst.edu/~chm/vchembook/250fluidbal.html

END-OF-CHAPTER QUESTIONS

1. What are electrolytes? What are anions and cations?
2. List the electrolytes that are primarily found in extracellular fluid and intracellular fluid. What is the normal range of concentration for these electrolytes in the serum?
3. What is the difference between osmolality and osmolarity?
4. Describe three factors that influence the movement of solutes through semipermeable membranes.

5. List three mechanisms by which the body regulates the movement of fluid and solutes to insure homeostasis.

6. Explain how the angiotensin-renin system can affect blood volume.

7. Explain how aldosterone and arginine vasopressin can affect urine volume.

8. Discuss how calcium and phosphate balance are maintained in the body.

9. Physiologically, what does hyper- or hypovolemia describe? What are the common causes of hyper- and hypovolemia?

10. Is there a difference between hypervolemia and hyponatremia? Explain your answer.

11. Do laboratory values of serum $Na^+ > 145$ mEq/L; serum osmolality > 295; and urine osmolality > 800 mOsm/kg indicate hypernatremia or hyponatremia? List three sign/symptoms that accompany this condition.

12. Describe three common conditions that can result in hypokalemia. What are common signs and symptoms of hypokalemia? Hyperkalemia?

13. How do changes in blood pH affect blood potassium levels?

NOTES

[1] An indentation of the skin known as pitting will result when pressure is applied to an area of edema. The degree of pitting is scaled from +1 to +4 as a subjective evaluation. (See photo of +4 edema on page 192.)

9

Acid-Base Balance

Marcia Nahikian Nelms, Ph.D., R.D.

Southeast Missouri State University

CHAPTER OUTLINE

Regulation of Acid-Base Balance

Chemical Buffering • Other Chemical Buffer Systems • Respiratory Regulatory Control • Renal Regulatory Control • Other Renal Regulatory Controls • Effect of Acid and Base on Electrolyte Balance

Assessment of Acid-Base Balance

Acid-Base Disorders

Respiratory Acidosis • Respiratory Alkalosis • Metabolic Acidosis • Metabolic Alkalosis • Mixed Acid-Base Disorders

Introduction

Even minor changes in pH can have significant effects on physiological function. Maintenance of a normal pH, and thus the body's homeostasis, allows for normal cellular function, enzyme activity, and membrane stability. Alterations of pH at the cellular level are manifested in often-dramatic systemic signs and symptoms.

Acid-base imbalances can occur throughout the life span. During infancy, immature kidneys coupled with a high metabolic rate increase risk of acidosis. Changes in respiration result in rapid changes in CO_2 levels in infants due to their small lung capacities. Diseases affecting the kidney and lungs, which are common in the elderly, reduce their ability to maintain homeostasis.

It is crucial for the Registered Dietitian to have a firm understanding of **acid-base balance**, since many nutrition therapies, such as parenteral nutrition, are used to address these metabolic changes. When these concepts are understood, the RD can assist in appropriate clinical interventions to return the body to its normal state. This chapter defines acids, bases, and pH, and describes conditions of acid-base balance.

Acids

Substances that can donate or give up hydrogen ions (H^+) are considered to be acids. In human physiology, two groups of acids are important: volatile and nonvolatile. Volatile acids are those that can be converted to a gaseous form and eliminated by the lungs. Nonvolatile acids include those inorganic acids that occur through metabolism of carbohydrate, protein, and lipid. The lungs cannot eliminate nonvolatile acids.

acid-base balance—maintenance of homeostasis between acidity and alkalinity within body systems

Carbonic acid (H_2CO_3) is the most important volatile acid because it is produced in the largest amount and provides the major source of H^+. The body produces an average of 20,000 mmol of carbonic acid daily. This acid readily dissolves in solution as follows:

$$H_2CO_3 \Leftrightarrow CO_2 + H_2O$$

Because H_2CO_3 does dissolve so readily, it is not possible to measure its exact concentration. Instead, the concentration of CO_2 is used as an indirect measure of acidity. The concentration of CO_2 is expressed as $PaCO_2$, which is a measurement of *partial pressure* exerted by carbon dioxide in blood. It is considered to be partial because additional gases present in the blood such as nitrogen and oxygen also exert their own pressure (Rhoades and Pflanzer 2003).

Nonvolatile or fixed acids are produced as end products of carbohydrate, protein, and lipid metabolism (Remer 2000). Of course, the amount of these nutrients consumed will affect the amount of fixed acids produced, but an average amount of 50 to 100 mmol are produced each day. Fixed acids can be either organic or inorganic. Protein metabolism contributes the most with its addition of inorganic acids such as phosphoric acid and sulfuric acid. Examples of organic acids produced during metabolism include lactic acid and ketoacids such as hydroxybutyric acid. Fixed acids are also produced as a result of starvation or fasting, fever, exercise, and some disease states (see Figure 9.1).

Bases

Bases are substances that can accept or receive a hydrogen ion. The most predominant base involved in human acid-base balance is bicarbonate (HCO_3^-). Other alkaline (basic) substances are added to the body through ingestion of fruits and vegetables. The kidneys provide primary regulation of HCO_3^- concentration by controlling the amount of free hydrogen ions and the amount of bicarbonate that is removed or retained in the body.

Buffers

A buffer is a substance or a group of substances that reacts with either acid or base in order to decrease the effect of acid or base on the pH of a solution. The most important buffer systems will be discussed in detail later in this chapter.

pH

The unit for measuring the relative acidity or alkalinity of a fluid is called pH. Simply stated, pH is the ratio of acids to bases. Hydrogen ion concentration (H^+) is the negative logarithm of hydrogen ions in solution:

$$pH = \log 1/[H^+] = -\log [H^+]$$

Because the scale of (H^+) is logarithmic, in order for the pH to change by one unit (e.g., changing from 3 to 4), there must be a tenfold change in (H^+).

FIGURE 9.1 Overall Schema for Maintenance of Acid-Base Balance. On the usual mixed diet, pH is threatened by production of strong acids (e.g., sulfuric, hydrochloric, and phosphoric), which result mainly from protein metabolism. These strong acids are buffered by chemical buffers in the body. Removal of extra H^+s and the accompanying anions from the body is accomplished by renal excretion. When the kidneys excrete H^+s, they add new bicarbonate to the blood, thereby restoring depleted body buffer bases. The respiratory system eliminates CO_2 produced by metabolism. CO_2 is not a threat to acid-base balance, provided its partial pressure in arterial blood is kept at a normal value.

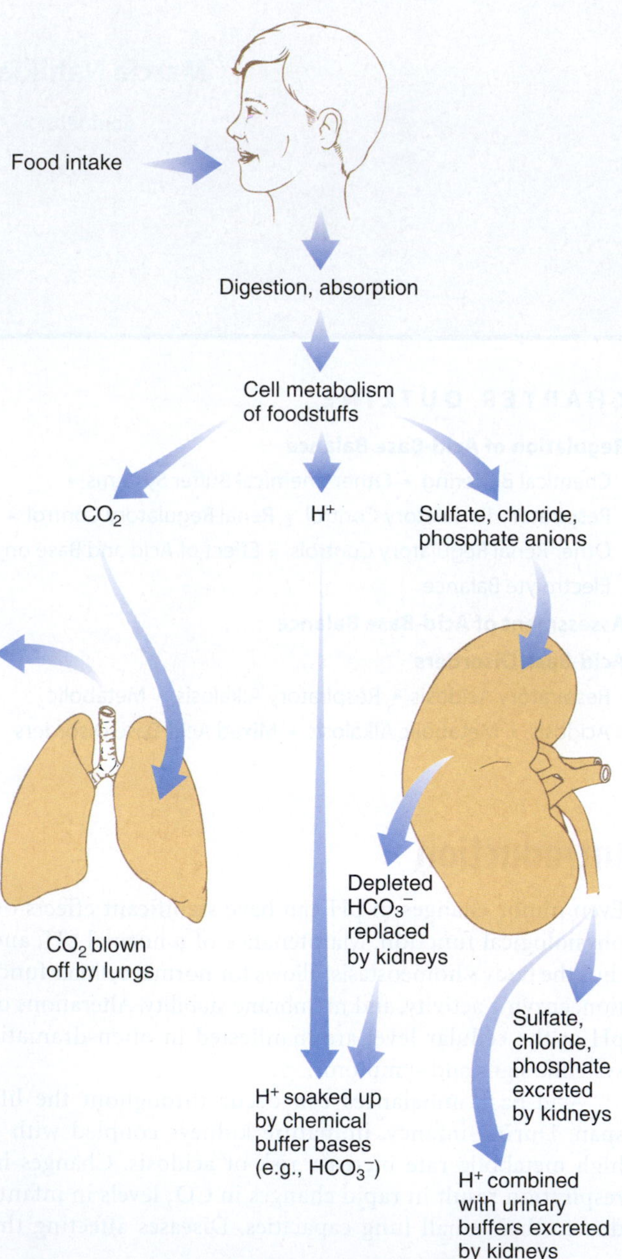

Source: Lauralee Sherwood, Human Physiology: From Cells to Systems, 5e, copyright © 2004, p. 793

The pH of a substance is measured in a range from 1 to 14. A 1 on the pH scale indicates the most acidic, and 14 indicates the most alkaline. Water is considered neutral at 7.0. For humans, a normal serum pH is within the range of 7.35–7.45. The pH of other body fluids varies, with gastric juice being the most acidic.

Terms Describing pH

Acidosis is the process (or processes) that leads to accumulation of acid or loss of base. **Acidemia** is the actual decrease in pH < 7.35. **Alkalosis** is the process (or processes) that leads to accumulation of base or loss of acid. **Alkalemia** is the condition where an actual increase of pH > 7.45 is observed.

Regulation of Acid-Base Balance

The body has to have several lines of defense to accommodate all of the hydrogen ions that are constantly being produced in the body. These include (1) chemical buffers, (2) the respiratory regulation of pH, and (3) the kidney regulation of pH.

FIGURE 9.2 pH of Body Fluids

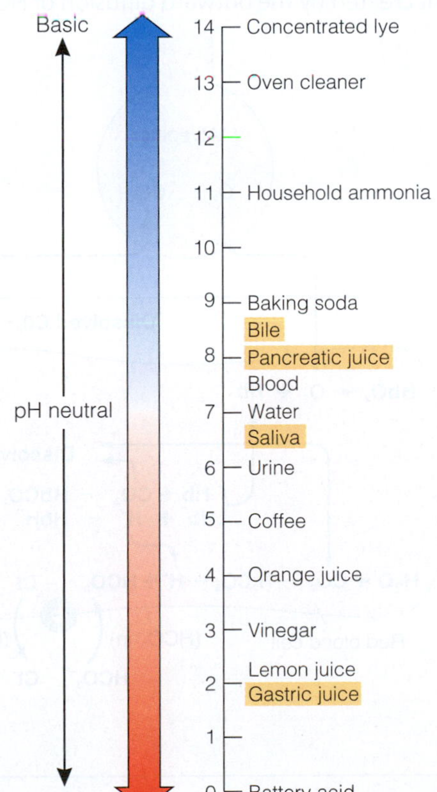

Source: Whitney and S. Rolfes, *Understanding Nutrition*, 10e, Copyright © 2005, p. 81

TABLE 9.1

Chemical Buffers and Their Primary Roles

Buffer System	Major Functions
Carbonic acid: bicarbonate buffer system	Primary ECF buffer against non-carbonic-acid changes
Protein buffer system	Primary ICF buffer; also buffers ECF
Hemoglobin buffer system	Primary buffer against carbonic acid changes
Phosphate buffer system	Important urinary buffer; also buffers ICF

Source: Sherwood L. Human physiology: from cells to systems. 5th ed. Belmont, CA: Brooks/Cole, 2004. Table 15-6, p. 575.

Chemical Buffering

As stated earlier, a buffer reacts with free H^+ in order to maintain acid-base equilibrium. The effectiveness or power of the particular buffer is determined by its association with a cellular salt (**pK**) and by its overall concentration in the fluid compartment. Buffers are present in all body fluids—both extracellularly and intracellularly. Table 9.1 summarizes the chemical buffers.

The primary buffer in the extracellular fluid (ECF) is the bicarbonate-carbonic acid buffer system. This buffer system accommodates more than 80% of the required buffering in the ECF. This buffer system is outlined as follows:

$$H^+ + HCO_3^- \Leftrightarrow H_2CO_3 \Leftrightarrow CO_2 + H_2O$$

As the buffer system reacts with fixed acids, H_2CO_3 is produced. As discussed previously, H_2CO_3 readily dissolves to CO_2 and H_2O. Therefore, the lungs will accommodate the increased load of acids by increasing the rate and depth of breathing and by expiring the CO_2. The kidney helps with this buffer system by either reabsorbing HCO_3^- or regenerating additional HCO_3^- from CO_2 and H_2O. (Rhoades and Pflanzer 2003; Sherwood 2004).

The **Henderson-Hasselbach equation** helps to explain the interrelationships between H_2CO_3, HCO_3^-, and pH. In humans, the pH, or ratio of acids to bases, is 1 part H_2CO_3

acidosis—conditions that produce excess acid in the blood

acidemia—condition of excess acid in the blood consistent with a pH < 7.35

alkalosis—conditions that produce excess base in the blood

alkalemia—condition of excess base in the blood consistent with a pH > 7.45

pK—the constant degree of dissociation (the ability of an acid to release its hydrogen ions) for a given solution; this is a constant amount for any given solution

Henderson-Hasselback equation—
$pH = pK_a + [H_2CO_3] / [HCO_3^-]$

to 20 parts HCO_3^-. In order for pH to remain within the normal range, this ratio has to be maintained. Any change in H_2CO_3 must be accompanied by a proportional change in HCO_3^-. If one part of the equation changes without the other and the ratio is not maintained, pH will move out of the normal range.

Other Chemical Buffer Systems

The body has additional buffer systems in place to compensate for changes that could occur from other sources of acid. An important buffer system within red blood cells and tubules of the kidney is the disodium/monosodium phosphate (Na_2HPO_4) buffer. Excretion of H^+ could potentially make urine so acidic that excretion would be physically damaging to the kidney. Fortunately, phosphate accepts the H^+ and a weaker acid is formed. This is much less harmful to the kidney. This buffer system is outlined as follows:

$$Na_2HPO_4 + H^+ \Leftrightarrow NaH\,_2PO_4 + Na^+$$

Proteins present in the plasma can act as buffers. Their contribution as buffers is most important intracellularly. This buffer system acts in the same fashion as the bicarbonate-carbonic acid buffer system in that protein accepts the H^+. It is important to note, though, that many proteins can also release the H^+ if alkalinity increases. The ability to act in both situations increases the effectiveness of this buffer system.

The action of hemoglobin within the red blood cell is the most important buffer in blood. Carbon dioxide diffuses into the blood as it is produced throughout the body. Most of the CO_2 will combine with water, forming carbonic acid. As stated earlier, carbonic acid will dissociate to bicarbonate and free H^+. Hemoglobin binds the H^+.

The reaction is reversed as blood passes through the lungs and becomes oxygenated. Oxygenated hemoglobin gives up the H^+ to HCO_3^- and thus carbonic acid (H_2CO_3) is generated. As stated earlier, H_2CO_3 dissolves to CO_2 and H_2O. CO_2 is then expired via the lungs. (See Figure 9.3.)

Respiratory Regulatory Control

The next line of defense in maintaining acid-base balance is respiratory control. The lungs have the ability to change respiratory rate and depth of breathing to control either release

FIGURE 9.3 Transport and Exchange of Carbon Dioxide and Oxygen. Carbon dioxide (CO_2) picked up at the tissue level is transported in the blood in three ways: (1) physically dissolved, (2) bound to hemoglobin (Hb), and (3) as bicarbonate ion (HCO_3^-). Hemoglobin is present only in the red blood cells, as is carbonic anhydrase, the enzyme that catalyzes the production of HCO_3^-. The H^+ generated during the production of HCO_3^- also binds to Hb. Bicarbonate moves by facilitated diffusion down its concentration gradient out of the red blood cell into the plasma, and chloride (Cl^-) moves by means of the same passive carrier into the red blood cell down the electrical gradient created by the outward diffusion of HCO_3^-.

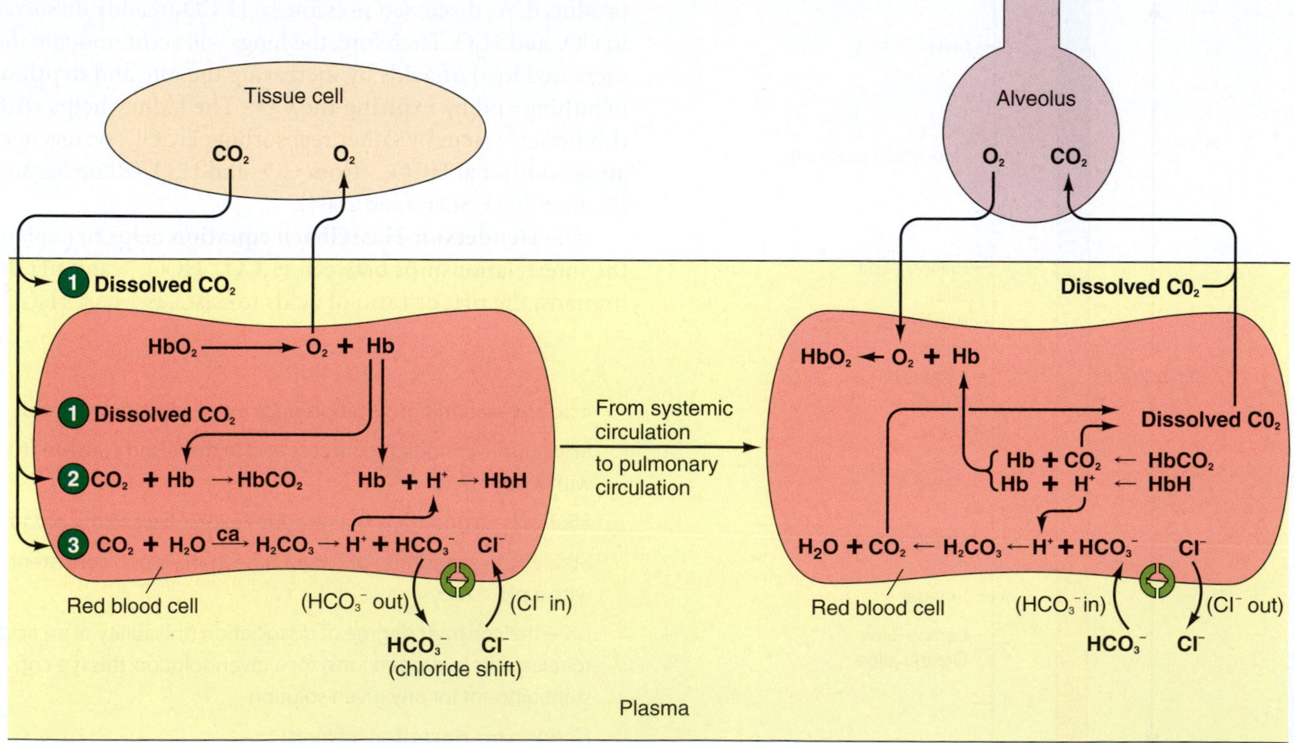

Source: Lauralee Sherwood, *Human Physiology: From Cells to Systems*, 5e, copyright © 2004, p. 493

or retention of CO_2, and hence to assist in management of acid-base balance. This control system is very sensitive and is able to respond spontaneously.

The level of CO_2 in the blood controls the pH of the cerebrospinal fluid, since H^+ and HCO_3^- do not cross the blood-brain barrier. Changes in pH—specifically the level of CO_2—are detected in cerebrospinal fluid by respiratory centers in the brain. When these changes in pH are noted, respiratory rate changes, which will normalize pH.

For example, when acidosis occurs, respiratory rate and depth will increase (hyperventilation). This allows larger amounts of CO_2 to be expired. In the situation when $PaCO_2$[1] has decreased (alkalosis), respirations will slow, CO_2 concentrations will increase and pH will normalize. Any change in anatomy or physiology of the respiratory system, nervous system control of respiration, or muscles that assist in breathing will affect the ability of the respiratory system to respond to changes in pH.

FIGURE 9.4 Control of the Rate of Tubular H^+ Secretion

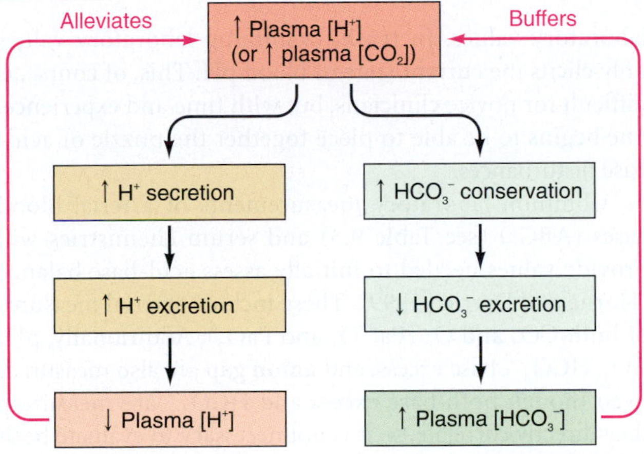

Source: Lauralee Sherwood, *Human Physiology: From Cells to Systems*, 5e, copyright © 2004, p. 579

Renal Regulatory Control

The kidney's role in controlling both H^+ and HCO_3^- is a critical component for the maintenance of pH homeostasis. Respiratory control is ineffectual in dealing with the large amount of nonvolatile (fixed) acids that are constantly produced, since these cannot be expired as gases. To maintain pH, a healthy, normally functioning kidney will reabsorb the majority of all HCO_3^- that is needed. This function requires the kidney to secrete H^+, which combines with HCO_3^-, forming H_2CO_3. H_2CO_3 is dissolved to form CO_2 and H_2O. The H_2O is released and CO_2 is reabsorbed and changed back to HCO_3^-. This allows for constant regeneration of bicarbonate, which is needed to buffer the ongoing supply of fixed acids (see Figure 9.4).

In the situation where alkalosis occurs, the kidney will respond by reducing the amount of HCO_3^- reabsorbed. Likewise, if acidosis occurs, the kidney will reduce secretion of H^+ and increase the amount of HCO_3^- reabsorbed. Renal regulatory control is much slower than respiratory regulation and may take as much as 24 hours to respond to imbalances (Sherwood 2004). See Figure 9.5.

FIGURE 9.5 Hydrogen Ion Secretion Coupled with Bicarbonate Reabsorption. Because the disappearance of a filtered HCO_3^- from the tubular fluid is coupled with the appearance of another HCO_3^- in the plasma, HCO_3^- is considered to have been "reabsorbed."

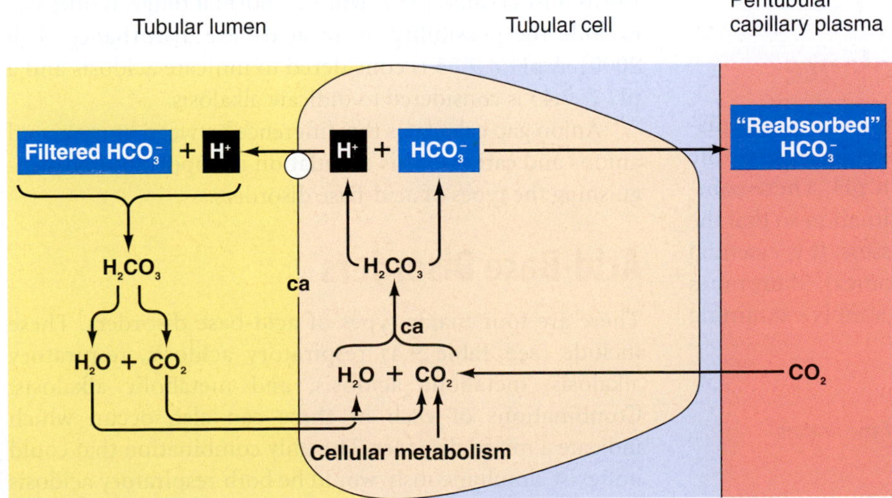

Source: Lauralee Sherwood, *Human Physiology: From Cells to Systems*, 5e, copyright © 2004, p. 580

Secretion of H^+ is a vital component of the renal regulation of acid-base balance. The minimum pH of urine in humans is 4.5. If pH drops below this, the urine's acidity becomes harmful. The kidney cannot use bicarbonate to buffer since it cannot be excreted at the same time as the hydrogen ions. Thus, the kidney uses two other buffers (dibasic phosphate and ammonium) to prevent this damage, as described in the next section.

Other Renal Regulatory Controls

The base NH_3 (ammonia) is formed in renal tubular cells from the amino acid glutamine. Free H^+ combines with NH_3 to form ammonium (NH_4^+). Ammonium cannot cross back across the cell membrane, so H^+ is trapped and is thus excreted in the urine.

TABLE 9.2

Summary of Renal Responses to Acidosis and Alkalosis

Acid-Base Abnormality	H^+ Secretion	H^+ Excretion	HCO_3^- Reabsorption and Addition of New HCO_3^- to Plasma	HCO_3^- Excretion	pH of Urine	Compensatory Change in Plasma pH
Acidosis	↑	↑	↑	Normal (zero; all filtered is reabsorbed)	Acidic	Alkalinization toward normal
Alkalosis	↓	↓	↓	↑	Alkaline	Acidification toward normal

Source: Sherwood L. Human physiology: from cells to systems, 5th ed. Belmont, CA: Brooks/Cole, 2004. Table 15-8, page 581.

Dibasic phosphate and sulfur both function to accept H^+ in order to control acid-base balance. In a situation where a large load of fixed acids is produced, the kidney will respond by increasing formation of acids within this buffer system (Oh 2000). Approximately one-third of the free H^+ is excreted as phosphoric acid (H_2PO_4) and sulfuric acid (H_2SO_4). Table 9.2 summarizes the renal responses to changes in acid-base balance.

Effect of Acid and Base on Electrolyte Balance

Hydrogen ions and bicarbonate are both electrolytes. Acid-base changes will therefore affect concentrations of other electrolytes in both ECF and ICF. For example, the movement of HCO_3^- to the plasma requires the exchange of another negatively charged ion so that **electroneutrality** is maintained. Chloride (Cl^-) is the ion that moves in the opposite direction of the HCO_3^-. Changes in potassium (K^+), chloride (Cl^-), and sodium (Na^+) may accompany acid-base disorders (Milonis et al. 1999; Powers 1999).

Assessment of Acid-Base Balance

It is an understatement to say assessment of acid-base disturbances is difficult. This difficulty arises because of the body's attempt to self-correct changes in pH. These compensatory responses confuse the clinical situation so that the origin of the disturbance is difficult to assess. It is essential to place the assessment within a clinical context. Many times this assessment is more important than simply examining laboratory values. In truth, examining laboratory values only elicits the current state of blood pH. This, of course, is difficult for novice clinicians, but with time and experience, one begins to be able to piece together the puzzle of acid-base disturbances.

Common laboratory measurements of arterial blood gases (ABGs) (see Table 9.3) and serum chemistries will provide values needed to initially assess acid-base balance (Horne and Derrico 1999). These include arterial measures of both CO_2 and O_2 ($PaCO_2$ and PaO_2). Additionally, pH, CO_2, HCO_3^-, base excess, and **anion gap** are also measured. Even though both base excess and HCO_3^- are measured, they directly correlate, so it is not necessary to evaluate both values. See Table 9.3 for an outline of normal values of arterial blood gas parameters and analysis of arterial blood gases.

When evaluating pH, remember that in humans this measures the ratio of acids to bases. If both acid and base increase (or decrease) within the same proportion, pH will remain steady. It is only when one changes out of proportion to the other that a change in pH will be measurable. In other words, just because pH is within a normal range, it does not exclude the possibility of an acid-base disturbance (Fall 2000). A $pH < 7.35$ is considered to indicate acidosis and a $pH > 7.45$ is considered to indicate alkalosis.

Anion gap calculates the difference between unmeasured anions and cations. This calculation is important in distinguishing the types of acid-base disorders.

Acid-Base Disorders

There are four major types of acid-base disorders. These include (see Table 9.4) respiratory acidosis, respiratory alkalosis, metabolic acidosis, and metabolic alkalosis. Combinations of each of these can also occur, which indicate a mixed disorder. The only combination that could not exist simultaneously would be both respiratory acidosis and respiratory alkalosis, since obviously hypoventilation and hyperventilation cannot happen together (Kraut and Madias 2001; Oh 2000).

electroneutrality—the sum of the charges of the anions equals the sum of the charges of the cations

anion gap (AG)—anion gap (AG) = (serum Na^+) − (serum Cl^- + HCO_3^-); normal AG = 12 to 14 mEq/L

TABLE 9.3

Arterial Blood Parameters Used for the Analysis of Acid-Base Status

Parameter	*Normal Value*	*Definition and Implications*
PaO_2	80 to 100 mm Hg	Partial pressure of oxygen in arterial blood (decreases with age) In adults <60 yr: 60 to 80 mm Hg = mild hypoxemia 40 to 60 mm Hg = moderate hypoxemia <40 mm Hg = severe hypoxemia
pH	7.40 (± 0.05 [2 SD]) 7.40 (± 0.02 [1 SD])	Identifies whether there is acidemia or alkalemia; the value using 2 standard deviations (SD) from the mean is the common clinical value pH <7.35 = acidosis; pH >7.45 = alkalosis
$[H^+]$	40 (± 2) nmol/L or nEq/L	The hydrogen ion concentration may be used instead of the pH
$PaCO_2$	40 (± 5.0) mm Hg	Partial pressure of CO_2 in the arterial blood $PaCO_2$ <35 mm Hg = respiratory alkalosis $PaCO_2$ >45 mm Hg = respiratory acidosis
CO_2 content	25.5 (± 4.5) mEq/L	Classic method of estimating $[HCO_3]$; measures HCO_3^- + dissolved CO_2 (latter is generally quite small except in respiratory acidosis)
Standard HCO_3^-	24 (±2) mEq/L	Estimated HCO_3^- concentration after fully oxygenated arterial blood has been equilibrated with CO_2 at a PCO_2 of 40 mm Hg at 38°C; eliminates the influence of respiration on the plasma HCO_3^- concentration
Base excess	0 (± 2) mEq/L	Reflects pure metabolic component Base excess = 1.2 × deviation from 0 Negative in metabolic acidosis Positive in metabolic alkalosis Misleading in respiratory and mixed acid-base disturbances Not essential for interpretation of acid-base disturbances
Anion gap	12 (± 4) mEq/L	Anion gap (or delta) reflects the difference between the unmeasured cations (K^+, Mg^{++}, Ca^{++}) and unmeasured anions (albumin, organic anions, HPO_4^-, SO_4^-) Useful in identifying types of metabolic acidosis Value >16–20 indicates acidosis is caused by retention of organic acids (e.g., diabetic ketoacidosis)

Useful Formulas

Plasma anion gap = $[Na^+] - ([HCO_3^-] + [Cl^-])$

Calculation of third acid-base parameter when two are known: $[H^+] = 24 \times PaCO_2/[HCO_3^-]$

Conversion of pH into [H+] pH of 7.4 = [H+] of 40 mEq/L

For every 0.1 increase in pH above 7.4, multiply 40 × 0.8

For every 0.1 decrease in pH below 7.4, multiply 40 × 1.25

For example, pH of 7.60 = 40 × 0.8 × 0.8 = $[H^+]$ of 26 mEq/L

Source: Reprinted from *Pathophysiology: Clinical Concepts of Disease Processes, 6th ed.* by S.A. Price and L.M. Wilson, Table 22-3, page 298, © 2003, with permission from Elevier.

Respiratory Acidosis

Respiratory acidosis occurs when there is an excess of acid in relationship to base caused by retention of carbon dioxide. This generally occurs when there is an inability of the lungs to expire CO_2. As the level of CO_2 rises, **hypercapnia** occurs, more carbonic acid (H_2CO_3) is formed, and pH is shifted toward acidosis (Sassoon and Arruda 2001).

respiratory acidosis—condition resulting from excess acid in the blood secondary to carbon dioxide retention

hypercapnia—the term used to describe an excess of the blood gas carbon dioxide (CO_2)

TABLE 9.4

Expected Responses to Acid-Base Imbalance in Regard to pH, pCO_2 and HCO_3^-

	Arterial Blood*			
Disturbance	pH	Plasma [HCO_3^-]	PCO2	Compensatory Response
Respiratory acidosis	↓	↑	↑	Kidneys increase H^+ excretion (increase plasma [HCO_3^-])
Respiratory alkalosis	↑	↓	↓	Kidneys increase HCO_3^- excretion (decrease plasma [HCO_3^-])
Metabolic acidosis	↓	↓	↓	Alveolar hyperventilation; normal kidneys increase net acid excretion
Metabolic alkalosis	↑	↑	↑	Alveolar hypoventilation; kidneys increase HCO_3^- excretion

* Heavy arrows indicate the main change.

Source: Rhoades R, Pflanzer, R. Human Physiology, 4th ed. Belmont, CA: Brooks/Cole, 2003. Table 25-2, page 800.

TABLE 9.5

Common Causes of Respiratory Acidosis

Hypoventilation
Chronic obstructive pulmonary disease
Severe pneumonia or asthma
Acute pulmonary edema
Pneumothorax
Drugs: opiate, sedative, anesthetic overdose (acute)
Excessive oxygen treating chronic hypercapnia
Sleep apnea
Neuromuscular disease such as amyotrophic lateral sclerosis (ALS), Guillain-Barré syndrome
Morbid obesity; Pickwickian syndrome
Chest wall injury or skeletal deformity
Aspiration of foreign body or vomitus
Laryngospasm, laryngeal edema, severe bronchospasm
Excessive production of CO_2
Overfeeding, especially with high-carbohydrate components of nutrition support

Etiology Any factor that inhibits the medullary respiratory center can affect ventilation and thus the ability to breathe off CO_2. Medications such as opiates or sedatives can inhibit respiration. Chronic conditions such as sleep apnea or acute events such as cardiac arrest can affect normal ventilation.

Diseases that affect musculature of the respiratory system and chest wall can result in poor ventilation. These may include neurological conditions such as myasthenia gravis or extreme obesity such as seen in Pickwickian syndrome. Additionally, any injury or trauma to the chest wall can potentially result in an inability to expire adequate amounts of carbon dioxide.

Respiratory diseases, such as chronic obstructive pulmonary disease, result in inability to maintain adequate oxygenation or release of carbon dioxide. Other conditions such as pneumonia, acute pulmonary edema, or pneumothorax can result in respiratory acidosis (Kraut 2001; Madias and Adrogué 2003). Common causes are outlined in Table 9.5.

Pathophysiology In respiratory acidosis, the major cellular buffering defense available is ability of the lung to expire CO_2. But since the major reason for respiratory acidosis is respiratory dysfunction, this buffering system is not as efficient. Body stores of HCO_3^- are released in order to maintain the appropriate ratio of CO_2 to HCO_3^-, allowing pH to stay within a normal range. During acute respiratory acidosis, the kidney regulatory systems do not have time to compensate, since these only begin to react within 12 to 24 hours. Chronic respiratory acidosis is less critical because the kidneys have more time to provide for ongoing compensation. Renal compensatory mechanisms work over a longer period and would include increased excretion of H^+ and resorption of HCO_3^-. Other renal buffer systems such as the use of ammonium (NH_4) will also provide compensation.

Clinical Manifestations Laboratory values in acute respiratory failure will indicate a decreased pH and an elevated pCO_2. Bicarbonate levels will be slightly elevated if renal compensation has begun. Compensatory mechanisms allow the pH to remain normal but serum bicarbonate and arterial pCO_2 are elevated. Serum electrolytes will show an increase in serum Ca^+, K^+, and possibly Cl^- due to changes in renal controls (Martinu, Menzies, and Dial 2003).

In both acute and chronic respiratory acidosis, hypoxemia is present. This reduced level of oxygen is responsible for most symptoms associated with the acidosis. In general, the more acute onset will result in an increase in severity of symptoms. Alterations in respiration would include increased respiratory rate (hyperventilation) and an increase in depth of respirations. Other symptoms are a result of the change in oxygenation in the brain and/or a decrease in neurotransmission. These include restlessness, apprehension, lethargy, muscle twitching, tremors, convulsions, and finally, coma (Rhoades and Pflanzer 2003; Sherwood 2004).

Treatment Treatment will focus on correcting the underlying condition causing respiratory changes. The presence of hypoxemia would focus treatment on increasing oxygenation through administration of oxygen or providing mechanical ventilation. In those patients with chronic hypoxemia, it is crucial to realize hypoxemia may be providing the stimulus for ventilation. If oxygen therapy reduces hypoxemia, ventilation may become worse without this stimulus.

Respiratory Alkalosis

Respiratory alkalosis (see Table 9.6) is characterized by a relative excess amount of base (HCO_3^-) as a result of a reduction of CO_2. This acid-base disturbance is generally a result of conditions causing hyperventilation. Rapid breathing results in a decreased $PaCO_2$.

Etiology Hyperventilation is commonly a result of a reduction in serum oxygen levels (hypoxemia). Hypoxemia can be a result of respiratory disease such as pneumonia, asthma, pulmonary embolism, or pulmonary edema. Additionally, high altitudes can result in hypoxemia.

The respiratory center in the brain can also be directly stimulated, causing hyperventilation and resulting loss of CO_2. Disorders of the CNS such as a malignancy or stroke can affect respiratory centers and result in hyperventilation. Hypermetabolic states such as in fever and sepsis can directly stimulate hyperventilation. Drugs, including theophylline, salicylates, progesterone, doxapram, and catchecholamines, can also result in hyperventilation. Even anxiety or other types of emotional distress can result in hyperventilation. Hyperventilation can also occur as an adaptive response to high oxygen demands during strenuous physical activity.

Pathophysiology Acute response to respiratory alkalosis (within the first 24 hours) is a shift of acid from the ICF to ECF with an accompanying movement of bicarbonate into cells in exchange for chloride. Additional H^+ is synthesized by an increase in lactic acid derived from pyruvate within cells. The shifts in H^+ are generally not adequate to handle a continued decrease in $PaCO_2$. For chronic respiratory alkalosis (lasting longer than 24 hours), renal compensation occurs. Kidneys reduce their secretion of H^+ (which also reduces regeneration of HCO_3^-) and increase their excretion of bicarbonate (HCO_3^-).

Clinical Manifestations In acute respiratory alkalosis, pH is > 7.45 and $PaCO_2$ is decreased. In chronic respiratory alkalosis, pH is > 7.45 and plasma HCO_3^- is low. In both situations, alkalosis may be accompanied by electrolyte imbalances. There may be low serum levels of K^+ and Ca^+ as well as high levels of Cl^-.

Other symptoms of respiratory alkalosis are seen in the cardiovascular, central nervous, and respiratory systems. Cardiac arrhythmias can be noted. Symptoms of the respiratory system vary but may include frequent yawning and deeper breaths. Symptoms of the central nervous system are most obvious. "Lightheadedness," mental confusion, anxiety, and seizures can occur. Patients also relate parathesias and cold and clammy extremities.

TABLE 9.6

Common Causes of Respiratory Alkalosis
Hyperventilation
Respiratory infection; pneumonia
Asthma
Change in altitude environment—i.e., high altitude
Drugs that stimulate respirations: theophylline, catecholamines
Anxiety
Cerebrovascular accident
Fever and sepsis

TABLE 9.7

Common Causes of Metabolic Acidosis
Kidney loss of HCO_3^-
Chronic kidney disease
End-stage renal failure
Systemic loss of HCO_3^-
Diarrhea
Fistula drainage
Excessive production of acid
Ketoacidosis secondary to conditions such as diabetes mellitus, alcoholism, or starvation
Lactic acidosis secondary to conditions such as diabetes mellitus, salicylate overdose

Treatment Correction of the underlying cause of respiratory alkalosis is the only significant treatment. Correction of hypoxia by providing oxygen therapy would be a common first step. If the cause is psychological hyperventilation, rebreathing[2] of CO_2 can correct the symptoms.

Metabolic Acidosis

Metabolic acidosis refers to all types of acidosis that are not caused by excessive CO_2. It can result from either excessive loss of base (HCO_3^-) or an excessive gain of fixed (nonvolatile) acids (see Table 9.7). Metabolic acidosis can be seen in both acute and chronic conditions, but due to respiratory compensation, it is most often a chronic condition (see

respiratory alkalosis—condition resulting from excess base in the blood secondary to increased carbon dioxide expiration

metabolic acidosis—condition resulting from either loss of bicarbonate or retention of nonvolatile acid

TABLE 9.8

Respiratory Adjustments to Acidosis and Alkalosis Induced by Nonrespiratory Causes

Respiratory Compensation	Normal (pH 7.4)	Nonrespiratory (metabolic) acidosis (pH 7.1)	Nonrespiratory (metabolic) alkalosis (pH 7.7)
Respiratory rate	Normal	↑	↓
Tidal volume	Normal	↑	↓
Ventilation	Normal	↑	↓
Rate of CO_2 removal	Normal	↑	↓
Rate of carbonic acid formation	Normal	↓	↑
Rate of H^+ generation from CO_2	Normal	↓	↑

Source: Sherwood L. Human physiology: from cells to systems, 5th ed. Belmont, CA: Brooks/Cole, 2004. Table 15-7, page 578.

Table 9.8). Metabolic acidosis is sometimes characterized by using the anion gap calculation to determine origin of the disorder. Although this calculation is an important tool in clinical situations, it is less appropriate for situations of metabolic acidosis, which result from accumulation of both organic and nonorganic acids. Therefore, more specific tests measuring actual levels of acids are being emphasized in practice (Van Biesen and Lameire 2000).

Etiology Conditions that result in an excessive loss of bicarbonate from the gastrointestinal system or from renal excretion of bicarbonate can result in metabolic acidosis. Diarrhea is the most common cause. Additionally, losses from an ileostomy or from pancreatic, biliary, or intestinal fistulas can be another source of excessive HCO_3^- loss (see Figure 9.6) (Fena et al. 2000; Gauthier and Szerlip 2002).

Carbonic anhydrase inhibitors such as the drug acetazolamide can cause excessive loss of base while inhibiting the production of carbonic acid in the kidney. In the condition of renal tubular acidosis (RTA), the ability to reabsorb bicarbonate is decreased. In chronic kidney failure, the ability to restore bicarbonate may fail as well. Other mechanisms to correct acid-base disturbances, such as the production of NH_4^+, may also fail as renal function declines.[3]

Situations that increase the amount of acid can result in metabolic acidosis. These may include administration of ammonium chloride and rapid administration of IV saline. Accidental poisoning with substances such as salicylate (aspirin), ethylene glycol (antifreeze), or formaldehyde can result in metabolic acidosis.

Lactic acidosis (see Figure 9.7) also occurs as a result of increased production of lactate or ketoacids. High levels of lactate may also occur when the kidney or liver fail to convert lactate to pyruvate or bicarbonate. Diabetic ketoacidosis, one of the most common causes, results in metabolic acidosis due to both increased production of betahydroxybuterate and inability to metabolize ketones (see Box 9.1).

In both starvation and chronic alcoholism, synthesis of ketoacids is increased. Starvation, with its reliance on fat stores, can also increase synthesis of ketoacids and result in acidosis (Reddy et al. 2002).

Pathophysiology When H^+ levels increase, the bicarbonate–carbonic acid buffer system is stimulated. This shift of H^+ into ECF reduces serum

FIGURE 9.6 Hydrogen Ion Secretion and Excretion Coupled with the Addition of New HCO_3^- to the Plasma. Secreted H^+ does not combine with filtered HPO_4^{2-} and is not subsequently excreted until all the filtered HCO_3^- has been "reabsorbed," as depicted in Figure 9.5. Once all the filtered HCO_3^- has combined with secreted H^+, further secreted H^+ is excreted in the urine, primarily in association with urinary buffers such as basic phosphate. Excretion of H^+ is coupled with the appearance of new HCO_3^- in the plasma. The "new" HCO_3^- represents a net gain rather than being merely a replacement for filtered HCO_3^-.

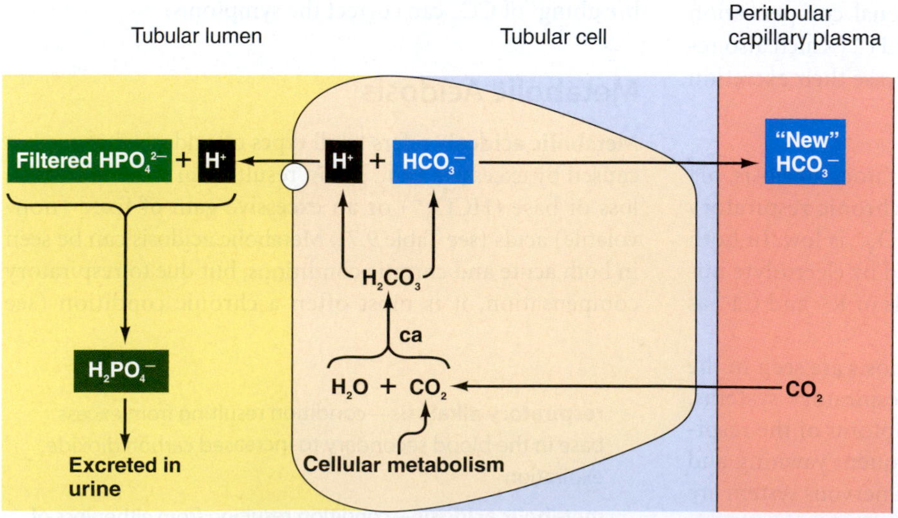

Source: Lauralee Sherwood, *Human Physiology: From Cells to Systems*, 5e, copyright © 2004, p. 581

FIGURE 9.7 Lactic Acid Production. When NAD concentration is decreased compared to NADH + H^+ the scale leans in the direction of lactic acid production and not pyruvic acid. Conditions that result in decreased concentrations of NAD include: decreased oxygenation of the tissues, excessive ketone body production (as in diabetes), and metabolism of ethanol.

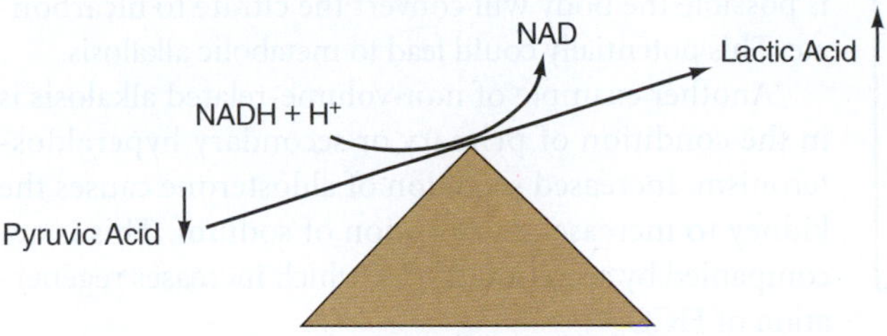

K^+ at the same time in order to maintain equilibrium between the ECF and ICF (see Figure 9.8).

High levels of H^+ in the blood stimulate the respiratory centers in the brain. Lungs respond by increasing rate and depth of breathing. Finally, the kidneys begin their compensatory response by increasing their excretion of H^+ and retaining HCO_3^-. Renal compensation is much slower than respiratory, and if kidney disease is present, effectiveness of the compensation is decreased.

Clinical Manifestations Symptoms of metabolic acidosis are not as clear as those of other acid-base disturbances. Changes in respiration can be noted as **Kussmaul breathing**. The cardiovascular system is affected by decreased contractility and response to catacholamines. Vasodilation may cause hypotension and dysrhythmias. Neurologically, lethargy and stupor with eventual coma are seen as pH falls in cerebrospinal fluid. In chronic renal failure, metabolic acidosis relies on carbonate from bone to handle the acid load. This results in growth failure in children and renal osteodystrophy in adults (Rhoades and Pflanzer 2003).

Treatment Treatment is focused on the underlying cause of the acidosis. Correcting pH too quickly can cause additional complications. The goal is to raise systemic pH to a safe level.

Metabolic Alkalosis

The presence of an excessive amount of base (HCO_3^-) results in **metabolic alkalosis** (see Table 9.9). Generally, this acid-base disturbance is caused by a loss of nonvolatile acids or can be a result of the over-administration of bicarbonate or whole blood.

Etiology Clinical situations resulting in alkalosis can be categorized into those conditions involving fluid imbalance

BOX 9.1 CLINICAL APPLICATIONS

DKA...Diabetic Ketoacidosis

One out of every four emergency room visits for patients with type 1 diabetes mellitus is for diabetic ketoacidosis (DKA). It has been estimated that over one billion dollars is spent each year treating this condition and its complications (Kitabchi 2001).

What is it? How is this condition related to acid-base imbalances that you have learned about in this chapter?

Diabetes mellitus is the disease caused by either the absence or the inefficient use of the hormone insulin (see Chapter 19). Ketoacidosis is one of the most serious acute complications of type 1 diabetes mellitus. Diabetic ketoacidosis typically develops as a result of infections, or because the patient does not take adequate amounts of insulin.

Without adequate insulin, there is an increased dependence on lipids as the primary fuel source. The increased rate of lipolysis results in the production of ketones: acetoacetic acid and hydroxybutyric acid. These ketone bodies are acids that lower serum pH. The kidney reacts by excreting the ketone bodies in the urine (ketonuria). The increased levels of hydrogen ions are buffered by plasma bicarbonate. This combination of events results in metabolic acidosis. High levels of ketoacids lead to an increase in the plasma anion gap.

As explained in this chapter, both the respiratory and renal system serve as compensatory mechanisms in acid-base imbalance. In DKA, as pH lowers, respiratory ventilation changes to accommodate the need to reduce pCO_2. Respirations are deep and labored—Kussmaul's respirations. Secondly, the renal system compensates by conserving bicarbonate (HCO_3^-). Furthermore, there is increased urinary excretion of positively charged cations (Na, K, NH_4^+).

DKA must be treated quickly and accurately to prevent coma and death. Providing adequate insulin, fluids and electrolytes allows the correction of the metabolic acidosis and prevents these complications.

Reference: Kitabchi AE, Umpierrez GE, Murphy MB, Barrett EJ, Kreisberg RA, Malone JI, Wall BM. Management of Hyperglycemic Crises in Patients With Diabetes. Diabetes Care. 2001;24(1):131 and Rhoades T & Pflanzer R. Human Physiology. Fourth Edition. Belmont CA: Brooks-Cole/Thompson Learning.

Kussmaul breathing—rapid, deep, and labored breathing commonly seen in people who have ketoacidosis or who are in a diabetic coma; Kussmaul breathing is named for Adolph Kussmaul, the nineteenth century German doctor who first noted it

metabolic alkalosis—condition resulting from either retention of bicarbonate or loss of nonvolatile acid

FIGURE 9.8 Shift of H^+ into the ECF Reduces Plasma Serum K^+. H^+ moves from an area of high concentration to an area of low concentration. K^+ is exchanged (i.e., moves in the opposite direction) to maintain electroneutrality within the cell.

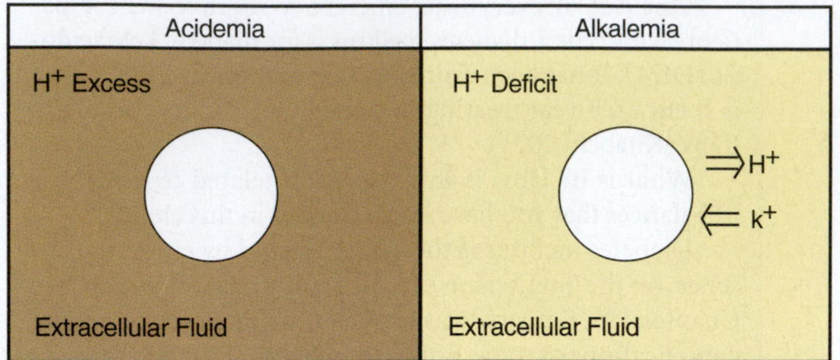

Source: Reprinted from Matarese L., Gottscklich, M.M., Contemporary Nutrition Support. Philadelphia, PA: WB Saunders Company, copyright © 1998 Reprinted with permission from Elsevier.

TABLE 9.9

Common Causes of Metabolic Alkalosis
Loss of acid
Vomiting
Nasogastric suctioning
Hypokalemia
Excessive base
Intravenous therapy
Blood transfusion
Excessive or chronic use of antacids

(alkalosis with volume decrease) or those without fluid imbalances (alkalosis without volume contraction). Conditions that involve fluid imbalance include prolonged vomiting and/or nasogastric suction or use of diuretics.

Conditions leading to alkalosis that do not involve fluid imbalance would include hyperaldosteronism, excessive use of corticosteroids, blood transfusions, chronic use of antacids, and excessive administration of sodium bicarbonate (Fencl et al. 2000).

Pathophysiology Initiation of metabolic alkalosis begins with the underlying event that causes either an excessive loss of acid or accumulation of base. This could be, for example, prolonged vomiting that results in decreased concentration of HCl^-. Normally, the kidney will compensate for the decrease in nonvolatile acid by generation of H^+ and decreased resorption of bicarbonate. In order for metabolic alkalosis to progress, other events need to occur that prevent adequate compensation by the kidney.

In situations where there is also volume depletion, such as in use of diuretics, there is both a fluid loss and a subsequent decrease in K^+ levels. To maintain serum K^+ levels within a safe range, the kidney will excrete H^+ in exchange for K^+. Even though K^+ may increase, the loss of H^+ results in generation of more bicarbonate. This further contributes to alkalosis.

Stored blood contains citrate as a preservative. If an individual receives a large amount of transfused blood, it is possible the body will convert the citrate to bicarbonate. This potentially could lead to metabolic alkalosis.

Another example of non-volume-related alkalosis is in the condition of primary or secondary hyperaldosteronism. Increased secretion of aldosterone causes the kidney to increase reabsorption of sodium. This is accompanied by secretion of H^+, which increases regeneration of HCO_3^-.

Clinical Manifestations Arterial blood gases in metabolic alkalosis will indicate a pH of > 7.45 and elevated levels of $HCO_3^- / > 26$ mEq/L. Accompanying electrolyte imbalances may indicate a $K^+ < 3.5$ mEq/L and $Cl^- < 98$ mEq/L. If compensation by the respiratory system is in place (see Table 9.8), $PaCO_2$ will remain within a normal range or be seen at slightly higher levels.

There are no specific signs and symptoms for metabolic alkalosis. Signs and symptoms are determined by accompanying conditions of volume deficit or electrolyte abnormalities. For example, in the situation where there is hypokalemia and a decrease in ECF, the patient may experience muscle cramping, weakness, and cardiac arrythmias.

Treatment In chloride responsive metabolic alkalosis, correcting volume imbalance with isotonic saline with additional KCl^- will correct alkalosis. Metabolic alkalosis that does not involve a fluid deficit requires treatment of the underlying causes before the alkalosis can be corrected. In severe conditions, the use of a carbonic anhydrase inhibitor will enhance HCO_3^- excretion.

Mixed Acid-Base Disorders

Several acid-base disturbances (see Table 9.10) can coexist in complex medical problems. As stated earlier, the only exception is the combination of respiratory imbalances. When examining ABGs, a mixed disorder should be suspected when $PaCO_2$ and HCO_3^- are not consistent with the measured pH. A mixed disorder may also be present when the compensatory response is exaggerated. For example, metabolic alkalosis and respiratory acidosis may occur when a patient with chronic obstructive pulmonary disease receives diuretics.

Assessment of Acid-Base Disorders

To place all of this into perspective, Table 9.11 summarizes the major components needed to assess all acid-base disorders. As was discussed earlier in this chapter, arterial blood

TABLE 9.10

Common Mixed Acid-Base Disorders

Dual Mixed Disorder	Common Causes
Additive Effect on pH Change	
Metabolic acidosis + respiratory acidosis $PaCO_2$ too high HCO_3^- too low pH very low	Cardiopulmonary arrest Patient with COPD goes into shock Chronic renal failure with fluid volume excess and pulmonary edema Patient with DKA receives potent opiate or barbiturate
Metabolic alkalosis + respiratory alkalosis $PaCO_2$ too low HCO_3^- too high pH very high	Patient with previously compensated respiratory acidosis caused by COPD overventilated on mechanical respirator Hyperventilating patient with CHF or hepatic cirrhosis who is vomiting or is treated with potent diuretics or nasogastric suction Head trauma patient with hyperventilation treated with diuretics
Offsetting Effect on pH Change	
Metabolic acidosis + respiratory alkalosis $PaCO_2$ too low HCO_3^- too low pH near normal	Lactic acidosis complicating septic shock Hepatorenal syndrome Salicylate intoxication
Metabolic alkalosis + respiratory acidosis $PaCO_2$ too high HCO_3^- too high pH near normal	Patient with COPD who is vomiting or who is treated with NG suction or potent diuretics Adult respiratory distress syndrome

CHF, congestive heart failure; COPD, chronic obstructive pulmonary disease; DKA, diabetic ketoacidosis; NG, nasogastric

Source: Reprinted from *Pathophysiology: Clinical Concepts of Disease Processes, 6th ed.* by S.A. Price and L.M. Wilson, Table 22-5, page 306, © 2003, with permission from Elevier.

TABLE 9.11

Summary of CO_2, HCO_3^-, and pH in Acid-Base Abnormalities

Acid-Base Status	pH	$[CO_2]$ (compared to normal)	$[HCO_3^-]$ (compared to normal)	$[HCO_3^-]/[CO_2]$
Normal	Normal	Normal	Normal	20/1
Uncompensated respiratory acidosis	Decreased	Increased	Normal	20/2 (10/1)
Compensated respiratory acidosis	Normal	Increased	Increased	40/2 (20/1)
Uncompensated respiratory alkalosis	Increased	Decreased	Normal	20/0.5 (40/1)
Compensated respiratory alkalosis	Normal	Decreased	Decreased	10/0.5 (20/1)
Uncompensated metabolic acidosis	Decreased	Normal	Decreased	10/1
Compensated metabolic acidosis	Normal	Decreased	Decreased	15/0.75 (20/1)
Uncompensated metabolic alkalosis	Increased	Normal	Increased	40/1
Compensated metabolic alkalosis	Normal	Increased	Increased	25/1.25 (20/1)

Source: Sherwood L. Human physiology: from cells to systems. 5th ed. Belmont, CA: Brooks/Cole, 2004. Table 15-9, p. 586.

gas measurements will provide all needed data for evaluation and begin the steps toward intervention.

Summary

Remembering basic concepts for acid-base balance will keep you on track in evaluating these complex clinical situations.

The scale for measuring acidity or alkalinity of a fluid is the measurement of pH. Simply stated, pH is the ratio of bases to acids. Normal pH in humans is 7.35–7.45. The pH is maintained in a ratio of 20:1 base to acid within body fluids. There will be no change in pH if the ratio remains stable, but changes in either portion of the ratio will result in changes in pH.

The largest source of acid within the body is carbonic acid. We measure concentration of CO_2 as an indirect measure of acidity. Concentration of CO_2 is expressed as $PaCO_2$.

The lungs primarily regulate CO_2 levels.

The largest source of base is HCO_3^-, which is regulated primarily by the kidneys.

Respiratory acidosis is a result of retention of CO_2, whereas respiratory alkalosis is a result of hyperventilation and the decrease in CO_2 levels.

Metabolic acidosis occurs when there is retention of fixed acids or from an excessive loss of bases. Metabolic alkalosis is a result of excessive loss of fixed acids or retention of bases.

CASE STUDY

Mr. N has presented to the physician with complaints of the following: pain, dizziness, and difficulty breathing. Mr. N's daughter also indicates that he has become more and more confused over the previous 24 hours. As the physician proceeds to his physical exam, he notes that blood pressure is out of the normal range, and respiratory rate and heart rate are high. Further tests indicate that Mr. N's hemoglobin is low while his white count is elevated.

Questions:

1. Outline the items from Mr. N's case that fall into each of the following categories:
   a. Signs
   b. Symptoms
   c. Laboratory abnormalities
2. Upon examination of his medical record, you find that Mr. N has the following diagnoses: renal insufficiency, chronic obstructive pulmonary disease, and a history of coronary heart disease. Which of these might interfere with his ability to maintain a normal acid-base balance?
3. The physician has ordered arterial blood gases. The values that you note as abnormal are as follows: pH 7.47; pCO_2 46 mmHg; pO_2 83 mmHg; HCO_3^- 32 mEq/L.
   a. Classify the pH.
   b. Assess pCO_2.
   c. Assess HCO_3^-.
   d. Do you see any indication of compensation? Why or why not?
   e. Identify the primary acid-base disorder.
   f. How do his diagnoses relate to this acid-base imbalance?

WEB LINKS

A. Grogano. Acid-Base Tutorial—Tulane Department of Anesthesiology, Tulane School of Medicine: This tutorial is an excellent method of review for basic concepts of acid-base balance for the health professional.

http://www.acid-base.com/index.php

D. B. Hornick. Iowa School of Medicine: An Approach to the Analysis of Arterial Blood Gases and Acid-Base Disorders: This organizational site for acid-base balance will guide the student through the basic concepts of acid-base homeostasis as well as manifestations of disorders.

http://www.vh.org/adult/provider/internalmedicine/bloodgases

Merck Manual of Medical Information: Acid-Base Imbalance: The Merck Manual historically has provided information for basic medical concepts. The manual describes the components of acid-base balance and how disease may affect this balance.

http://www.merck.com/mmhe/sec12/ch159/ch159a.html

END-OF-CHAPTER QUESTIONS

1. Define the following terms: pH, volatile and nonvolatile (fixed) acid, buffer.
2. What organ controls the level of pCO_2 in the blood? What organ controls HCO_3^- in the blood?
3. What is the basic problem in respiratory acidosis? Respiratory alkalosis?
4. What is the most important difference between metabolic acid-base disorders and those of a respiratory origin?
5. Name some conditions that might result in respiratory acidosis.
6. Name some conditions that might result in metabolic acidosis.
7. What is an anion gap?
8. How can respiratory mechanisms compensate for a metabolic alkalosis? Are there any major limitations to this compensation?

ENDNOTES

[1] Reported blood gas abbreviations. The letter P before CO_2 or O_2 is the partial pressure of the gas but it could be of arterial, venous, or mixed blood. When the letters "Pa" are in front of the gas, it is the partial pressure of the gas in the arterial blood.

[2] Rebreathing into a paper bag: Have you heard of doing this for someone when they are nervous and breathing too fast (e.g., for hyperventilation)? The rationale is that rebreathing into a paper bag will allow the person to replace the carbon dioxide "blown off" while hyperventilating.

[3] When ammonium chloride is administered the chloride ions associate with hydrogen ions to form hydrochloric acid. $NH_4Cl \rightarrow NH_3 + HCl$

10

Cellular and Physiological Response to Injury

Marcia Nahikian Nelms, Ph.D., R.D.

Southeast Missouri State University

CHAPTER OUTLINE

Defining Disease and Pathophysiology

Disease Process: Epidemiology, Etiology, Pathogenesis, Clinical Manifestations, Outcome

Cellular Injury

Mechanisms of Cellular Injury • Cellular Response to Injury

Introduction

Epidemiology, etiology, pathogenesis, clinical manifestations, and disease outcome: this chapter provides the framework and foundation for understanding the disease process. The processes that result in cellular injury and the characteristic response of the cell to injury are similar among various disease states. Thus, an understanding of the basic concepts outlined in this chapter—alterations in cell size, infections, and inflammation—provides the foundation for understanding the complexities of each individual diagnosis and disease process discussed later in this text.

Defining Disease and Pathophysiology

Disease is defined as a process that interferes with or disrupts the body's normal function. The human body strives to maintain a delicate balance among body systems and processes, and it is quite efficient at doing so. But disease or injury leads to a state where that balance is interrupted and homeostasis cannot be maintained. **Pathophysiology** is the study of the disruption of normal physiologic processes. Pathophysiology includes understanding structural changes that occur as a result of disease or injury as well as the clinical course that follows regulatory, metabolic, and structural changes. The clinical course includes impact and duration of the disease, and is monitored throughout diagnoses and conditions. **Pathogenesis** is the clinical course of the disease. Understanding disease mechanisms provides the basis for developing treatment.

Disease Process: Epidemiology, Etiology, Pathogenesis, Clinical Manifestations, Outcome

It is common to organize the study of disease process by analyzing patterns of disease occurrence through epidemiology. **Epidemiology** is the study of the distribution of disease within populations. Epidemiology provides data for

pathophysiology—the study of disease

pathogenesis—the clinical course of disease

epidemiology—the study of the rates of disease within a given population

outcome measures such as **morbidity** and **mortality**, and identifies risk factors associated with disease. Epidemiology commonly provides the first hypotheses for determining disease etiology.

Etiology is the description and identification of the cause of disease. Etiology can be narrowed to a specific causative agent or may be a combination of factors that influence or change disease development. Influential factors could be age, nutritional status, or other coexisting diseases. Etiology of disease can be categorized as *genetic, acquired, multifactorial, idiopathic,* or *iatrogenic.*

Diseases of genetic origin are those that develop from abnormalities in the genetic control of cellular development. These disorders may either be congenital (present at birth) or noncongenital (becoming evident later in life). Examples of genetic diseases include cystic fibrosis, sickle cell anemia, and hemophilia. Acquired disease is one that originates from exposure to environmental agents. Infectious disease may be classified as an acquired disease.

Many disease processes involving nutrition therapy are those considered to be multifactorial. These etiologies include a combination of factors including genetic, environmental, and infectious. Examples include atherosclerosis, osteoporosis, and diabetes mellitus.

An etiology is considered to be idiopathic if the origin is unknown. An iatrogenic disease or complication is any illness or symptom resulting from a medical intervention, treatment, procedure, or error.

Clinical manifestations of disease are evident as cellular injury moves toward systemic changes. You are probably most familiar with this level of discussion of disease as it involves signs, symptoms, and the measurement of laboratory abnormalities. **Signs** are observable phenomena that can be verified. Signs are measurable and include factors such as blood pressure, respiratory rate, weight, or body temperature. Many signs will be noted in the physical examination. **Symptoms** are those factors verbalized by the patient (and/or caregiver) and are frequently noted by the chief complaint in the physician's history and physical. Symptoms are subjective and dependent on the ability of the patient and caregiver to report this information accurately. Laboratory measurements of body fluids and tissue reflect changes occurring from the disease process. Laboratory values falling outside the norm may include blood chemistries, urinalysis, or tissue biopsy. See Box 10.1 for the patient perspective of clinical manifestations.

Outcome of disease is sometimes referred to as the *sequelae* of disease. **Prognosis** is considered to be the expected or usual outcome of the disease. Outcomes include cure, remission, development of chronic disease, or death. Generally, the disease is resolved as a result of treatment or the patient's own defense. Thus, the patient would return to the pre-illness state. Complications could potentially occur as a component of disease outcome. And, as mentioned before, an iatrogenic complication could occur from the treatment of the disease. For example, when a prescribed antibiotic causes diarrhea, nausea, or vomiting, these complications would be considered to be iatrogenic.

BOX 10.1 CLINICAL APPLICATIONS

Clinical Manifestations of Disease

Consider the following patient: A 35-year-old woman was admitted through the emergency room. She was febrile to 105°F and complained of chest pain and dyspnea. Further evaluation of her condition indicated abnormally elevated electrolytes, a chest x-ray that showed areas of infiltrate in the lower left lobe of the lung, and a positive blood culture for legionella pneumonia.

Can you determine which of these descriptions would be signs, symptoms, or laboratory measurements? Symptoms include the patient complaints of difficulty breathing (dyspnea) and chest pains. She could probably describe the symptom of fever. Signs include the measurement of fever and the infiltrate viewed on chest x-ray. The physical examination would also find signs of shortness of breath with changes in respiratory rate or the signs of fever including warm skin, dry mucosal membranes, or changes in heart rate. Laboratory measurements confirmed the etiology of disease with the blood culture and determined the abnormalities in electrolytes that might be consistent with dehydration.

morbidity—the state of being diseased

mortality—the incidence of death in a population

etiology—the cause of disease

clinical manifestations—unique signs and symptoms

signs—observable phenomena such as heart or respiratory rate

symptoms—complaints verbalized by a patient

outcome—the measurable consequence of disease

prognosis—expected outcome; expected response to treatment

Cellular Injury

Practitioners may be much more comfortable discussing clinical manifestations of disease than they are describing the underlying events at the cellular level. But to truly understand pathophysiology, practitioners need to examine causes of cellular injury and cellular response to this injury. The more that is understood about this process, the more efficient and effective treatment can become. Cellular injury may result from physical injury to the cell (trauma), a deficiency of a necessary substance required for cellular function, or interruption of normal cellular processes after exposure to a toxin (Nowak and Handford 2004). The cell's response to injury can include changes in cell growth, inflammation and healing, and/or cell death.

Mechanisms of Cellular Injury

Individual cells can be harmed as a result of hypoxia, (a deficiency of oxygen), nutritional imbalances, or microbiological agents. Processes that interfere with cell function include damage by free radicals, physical agents, immunologic reactions, nutritional imbalances, and genetic defects. Mechanisms that damage the structure of the cell can include physical and microbiological agents, immunologic reactions, genetic defects, and nutritional imbalances.

Hypoxia is a common reason for cell injury. An insufficient oxygen supply (commonly called **ischemia**) will interfere with cellular metabolism. This is easily understood in the process of a myocardial infarction (heart attack) where the oxygen supply to a portion of the heart is reduced and causes death of those cardiac muscle cells.

A free radical is an atom or group of atoms with a single unpaired electron. Free radicals are chemically unstable and are searching for additional electrons. In their search, free radicals can damage cell membranes or alter DNA, resulting in cellular injuries. Therefore, they can either interrupt normal function of the cell or damage the cell structure. An example of a free radical is the reactive oxygen species found in air pollution or produced during the inflammation process.

Physical agents such as ethanol, poisons, or lead and other heavy metals can cause cell damage in all three categories of injury. Some are fast acting, like poisons, or occur over many years, as seen in neurological changes after lead exposure. Other categories of physical injury include effects of burns, radioactive radiation, or actual trauma to the cell such as that seen in wounds and tissue destruction. Intoxication, where toxic by-products accumulate, may inadvertently arise from genetic abnormalities and lead to systemic changes. This process occurs in the metabolic disorder phenylketonuria (PKU). This genetic disorder prevents the normal metabolism of the amino acid phenylalanine.

Cellular Response to Injury

The cell's response to injury depends on the ability of the cell to react, adapt, and repair after exposure to injury. The response may be temporary and completely reversible. But in some situations, the cell is permanently damaged and is no longer functional. Cell responses may include inappropriate accumulation of substances within the cell, changes in cell size, number or shape, and the inflammatory response.

Cellular Accumulations Injury and disease can result in excessive accumulation of water, lipids, proteins, pigments, and minerals within the cell, which disrupts normal metabolism. For example, when a cell is unable to produce adequate ATP to maintain transport of sodium and potassium, fluid will shift, causing a disruption in fluid balance. In some diseases, triglycerides can accumulate in cells of the liver, heart, and pancreas. Alcoholic liver disease, hepatitis, or carbon tetrachloride poisoning can result in abnormal triglyceride deposits (fatty liver). The abnormal fat and protein deposits within arterial walls of the circulatory system characterize the common chronic disease, atherosclerosis.

A common cellular response to injury and disease is deposition of a substance called **hyaline** within and between the cells. Most hyaline deposits are a mixture of different types of proteins such as **collagen**, **fibrin**, and **amyloid**. Hyaline deposits can be found in many different cell types including neurons, hepatocytes, and cells in damaged arteries. The predominant type of protein within the hyaline accumulation can vary depending on the individual disease process. Fibrin masses are found in inflammatory conditions, and collagen predominates in scar tissue. Amyloidosis is a condition where amyloid is deposited in soft tissues, eventually resulting in cell death and organ dysfunction. Amyloid deposits have also been identified in brains of patients who died with Alzheimer's disease.

Pigment accumulation after cell injury can include melanin and derivatives of hemoproteins. Hemosiderin is a yellow-brown pigment produced when hemoglobin is broken down. For example, when you experience a bruise, the skin first appears red-blue, and then breakdown of the red blood cells occurs, causing hemoglobin to be transformed to hemosiderin. Accumulation of this pigment is what causes a bruise to turn the common yellowish green. Bilirubin is another yellow-green pigment found in bile. In some disease conditions, excess bilirubin accumulates in the body causing the symptom of **jaundice**.

ischemia—inadequate supply of oxygen

hyaline—a histological term used to describe tissue injury that has a glassy, pink appearance

collagen—a fibrous protein found in connective tissue

fibrin—a filamentous protein; for blood clotting to occur, fibrinogen must be converted to fibrin

amyloid—a starchlike substance present in diseased tissues

jaundice—a symptom that occurs when excessive bilirubin accumulates in the bloodstream, causing body tissues to become tinted yellow

Cellular Alterations in Size and Number Changes in cell size are both common and a central component of many disease states (see Figure 10.1). Cells respond to variations of hormonal or neurological stimulation by alteration in their size. The cell can increase in number, increase in size, shrink, or have additional functional changes.

Atrophy results from a decrease in cell size. Decreased workload, loss of innervation, diminished blood supply, inadequate nutrition, loss of hormonal stimulation, and aging all contribute to atrophy. A common example of this change is immobility of skeletal muscle. Prolonged bed rest or disuse due to fracture will result in loss of skeletal muscle mass. This can also occur from the lack of neurological stimulation that could be seen in a spinal cord injury.

Hypertrophy is defined as an increase in cell size. This cellular response is prominent in cells that are unable to undergo cell division. Hypertrophy may involve increased synthesis of structural proteins or increased size of organelles. Normal physiologic cellular hypertrophy is demonstrated by increase in skeletal muscle after resistance exercise. Pathologic hypertrophy is seen in valvular heart disease, interstitial lung disease, or even in tonsillitis associated with infectious mononucleosis.

In **hyperplasia**, there is an increase in overall cell number. Various types of anemia result in an increase of the total number of erythrocytes. In **Cushing's syndrome**, there is hyperplasia of the adrenal glands due to excess stimulation of cortisol. After partial resection of the liver (hepatectomy), there is regeneration of hepatocytes to accommodate resulting changes from surgery.

Metaplasia occurs when disease or injury results in displacement of one cell type for another that may be less mature. This is classically seen in vitamin A deficiency, since vitamin A is necessary for cellular differentiation. In deficiency, more mature, functioning cells are replaced with less mature cells. Other examples of metaplasia occur in *Helicobacter pylori* infection or in Barrett's esophagus, a premalignant condition of the esophagus (Slehria and Sharma 2003).

Dysplasia is deranged cellular growth and can result in abnormalities of size, shape, or function. Most often,

atrophy—reduction in size of muscle cells

hypertrophy—increase in cell size

hyperplasia—increased number of cells

Cushing's syndrome—a disorder resulting from prolonged exposure to high levels of glucocorticoid hormones; symptoms include: muscle weakness, thinning of the skin, moon-shaped face, weight gain, and diabetes mellitus

metaplasia—replacement of one cell type with another

dysplasia—abnormal cell growth

FIGURE 10.1A *Atrophy:* Cell Wasting

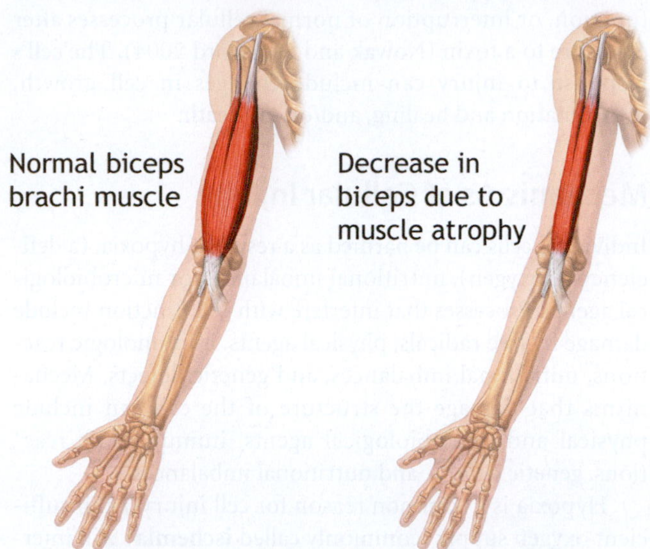

Source: A.D.A.M., Inc., 1600 RiverEdge Parkway, Suite 100, Atlanta, Georgia 30328

FIGURE 10.1B *Hypertrophy:* An Increase in Size of a Cell; Can Be Induced by a Number of Stimuli

Source: copyright © Fabio Cardoso/zefa/Corbis

FIGURE 10.1C *Hyperplasia:* **Increased Cell Production in Normal Tissue or an Organ**

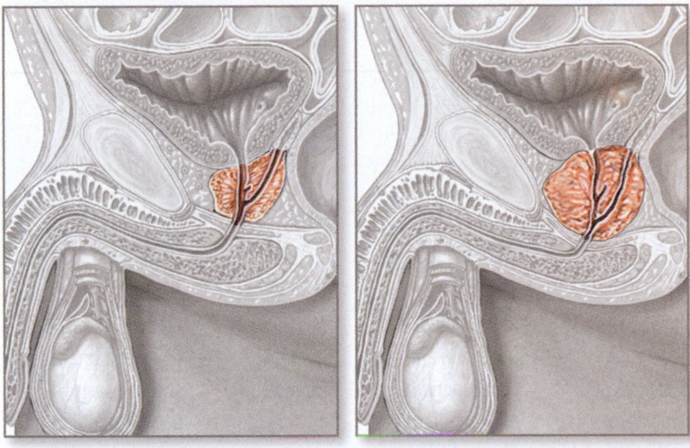

Source: A.D.A.M., Inc., 1600 River Edge Parkway, Suite 100, Atlanta, Georgia 30329

FIGURE 10.1D *Metaplasia:* **Transformation of One Mature Cell Type into Another Mature Cell Type as an Adaptive Response to Some Insult or Injury**

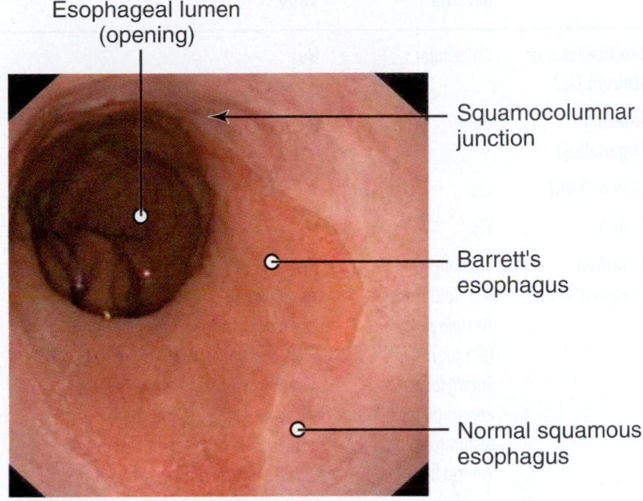

Source: copyright © www.gastrolab.com

FIGURE 10.1E *Dysplasia:* **Abnormal Cell Growth**

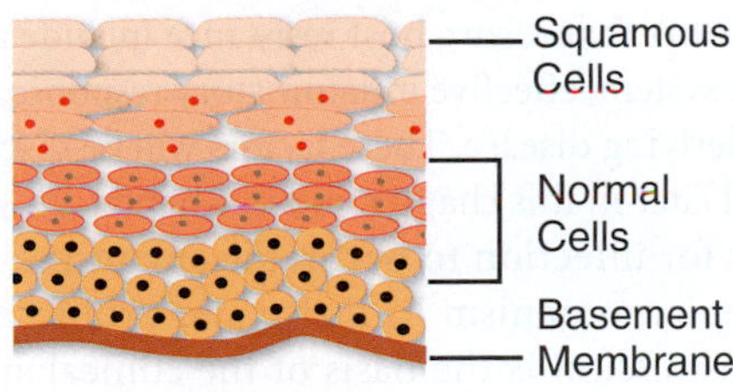

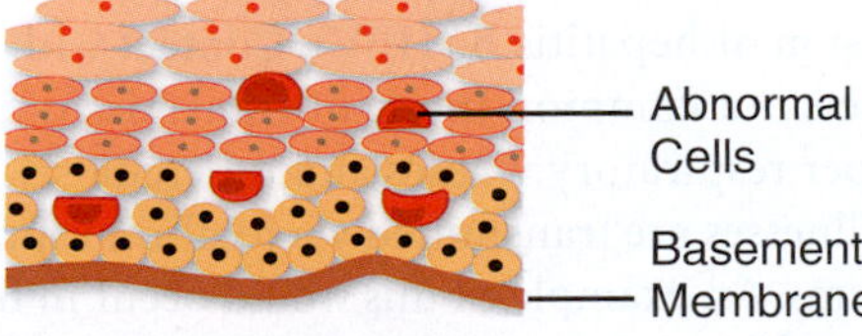

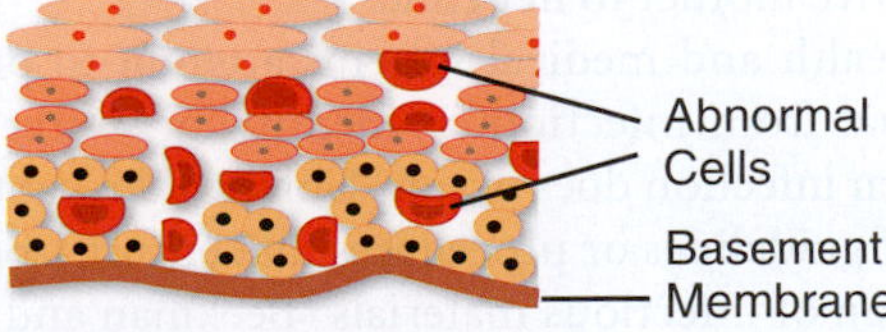

Source: MJ Bovo, M.D. Newport Media Concepts, Inc.

dysplastic changes are considered to be precancerous. For example, dysplastic cells are often noted in an abnormal pap smear (a routine screening test for cervical cancer).

Cellular Injury from Infection Cellular injury can also occur as a result of invasion from microorganisms. This presence of microorganisms is most commonly referred to as infection. In order for an infection to occur, three contributing factors must be in place: (1) the pathogen must be present, (2) the host must be susceptible to the infection, and (3) the environment must be conducive for the pathogen to thrive and proliferate.

Types of Microorganisms Microorganisms involved in human disease include bacteria, fungi, helminth, protozoa, prions, and viruses. Table 10.1 outlines major characteristics of these types of microorganisms and examples of their contributing infections. Microbes are more likely to result in infection if they produce **exotoxins** and **endotoxins**, produce destructive enzymes, produce spores, or develop a bacterial capsule. These characteristics increase pathogenicity of the

exotoxins—toxins produced by bacteria

endotoxins—toxins found in bacteria, often as part of the cell wall, that stimulate an immune response

TABLE 10.1

Major Characteristics of Microorganisms Involved in Human Disease

	Bacteria	Viruses	Fungi	Protozoa	Helminths	Prions
Multicellular or Unicellular?	Unicellular	No	Both	Unicellular	Yes	No
Internal Organelles?	None	No	Yes	Yes	Yes	No
DNA or RNA?	Yes	Yes	Yes	Yes	Yes	No
Living?	Yes	No	Yes	Yes	Yes	No
Common Morphologies?	*Salmonella typhi* (typhoid fever), *Yersinia pestis* (plague), *Staphylococcus aureus* (skin, respiratory, wound infections)	Herpes (chicken pox, cold sores, genital lesions), pox viruses (smallpox), rhinoviruses (common cold), orthomyxovirus (influenza), paramyxoviruses (measles, mumps), retroviruses (gastroenteritis), retroviruses (AIDS, several types of cancer), picornaviruses (polio, SARS), adenoviruses (mild respiratory infections), hepadnaviruses (human hepatitis B)	Ringworm, histoplasmosis, *Candida* yeasts (vaginal yeast infections, thrush)	*Giardia lamblia* (diarrhea) and *Cryptosporidium parvum* (diarrhea), *Plasmodium* (malaria)	*Trichinella spiralis* (roundworm)	Creutzfeldt-Jakob disease (humans), scrapie (sheep), bovine spongiform encephalopathy ("mad cow disease")

References: Madigan, Michael T., Martinko, John M., & Parker, Jack. 2003. Brock Biology of Microorganisms, 10th ed. Prentice Hall, Upper Saddle River, NJ. Microbiology @ Leiceser: Infection & Immunity: Man & microbes. Updated October 21, 2004. Available at URL http://www-micro.msb.le.ac.uk. Accessed February 4, 2005. National Institutes of Health. National Institute of Allergy and Infectious Diseases. Understanding emerging and re-emerging infectious diseases. Available at URL: http://www.science.education.nih.gov. Accessed February 4, 2005.

microorganism. Examples of these microbes can be found in Table 10.2.

Host Resistance In order for infection to occur, the host has to be susceptible to the infection. Susceptibility occurs when there is a break in the host's defense mechanism or in host resistance. The first line of defense against infection is the host's intact skin and mucous membranes. These provide not only a physical barrier against infection and injury but the first chemical response. Skin's epithelial cells provide a physical protection against injury and prevent easy entry into the body. If there is a break in the skin barrier, as might be caused by a splinter in a person's finger, there is a direct route for microorganisms to enter. Skin additionally has a slightly acidic pH that provides a chemical barrier against microorganisms.

Body secretions such as saliva, tears, and gastric juices also protect the host. Saliva produced in the mouth is one of the first protection mechanisms against microorganisms entering the gastrointestinal tract. The presence of lysosomes in saliva provides a nonspecific form of immunity against invading microorganisms. Any process that reduces the presence of these body secretions would decrease normal host resistance.

Other factors that assure host resistance include an effective immune system, effective inflammatory response, and absence of underlying disease. These factors will be discussed in greater detail later in this chapter and in Chapter 12.

In order for infection to occur, the environment must permit the microorganism to thrive and proliferate. Preventing transmission is the basis of the clinical approach for infection control and prevention of disease. Microorganisms can be transmitted from one human to another through contact with blood and body fluids, as seen in the transmission of hepatitis B (HBV). Contact with respiratory droplets is a common mode of transmission, especially for upper respiratory viruses and tuberculosis. Many food-borne illnesses are transmitted via fecal-oral spread of microorganisms. An example of this would occur in hepatitis A (HAV) or from contact with *E. coli*. Finally, microorganisms can be transmitted across the placenta from mother to fetus. This is what occurs in transmission from an HIV-positive mother to her child.

Public health and medical systems attempt to protect the individual from infection by identifying sources and contacts when infection does occur. "Universal Precautions" are the set of guidelines or procedures that ensure isolation and protection of infectious materials (Beekman and Hen-

TABLE 10.2

Microbes with Increased Pathogenicity

Bacterial Pathogenicity	Examples of Microbe
Exotoxins	Gram-negative and gram-negative bacteria
	Clostridium difficile
	Clostridium perfringens
	Staphylococcus aureus
	Streptococcus pyogenes
Endotoxins	Gram-negative bacteria
	E. coli
	Pseudomonas
	Salmonella
	Shigella
Destructive enzymes	*Pseudomonas aeruginosa*
	Streptococcus pneumoniae
	Streptococcus pyogenes
Endospores	Gram-positive bacteria
	Bacillus
	Clostridium
Bacterial capsule	*Pseudomonas aeruginosa*
	Bacillus anthracis

References: Kaiser GE. *Doc Kaiser's Microbiology Website.* 2005. Available at http://www.cat.cc.md.us/%7Egkaiser/goshp.html. Accessed 2/11/05
Todar K. *Todar's Online Textbook of Bacteriology.* 2005. Available at www.textbookofbacteriology.net. Accessed 2/11/05.

derson 2005). Table 10.3 summarizes the basic guidelines for Universal Precautions.

Eliminating a favorable environment includes not only preventing transmission but also interrupting conditions that would allow microorganisms to thrive. This is frequently accomplished through use of disinfectants, antiseptics, and sterilization. **Disinfectants** are chemical and physical agents applied to inanimate objects to kill any microbes. Applying a bleach solution to clean kitchen counters is a typical use of a disinfectant. **Antiseptics** are agents that kill microbes within living tissue. When drawing blood from a patient, the technologist often swabs the skin with either alcohol or **betadine.** This is a common example of antiseptic use. Finally, **sterilization** destroys all microbes. Sterilization is accomplished by use of heat, chemicals, or radiation. For example, microorganisms within foods may be destroyed by exposure to ionizing radiation or exposure to high temperatures during the canning process.

Course of Infection In clinical evaluation of the patient with an infection, particular symptoms and signs correlate

TABLE 10.3

Universal Precautions

Barrier protection	Should be used at all times to prevent skin and mucous membrane contamination with blood, body fluids containing visible blood, or other body fluids (cerebrospinal, synovial, pleural, peritoneal, pericardial, and amniotic fluids, semen, and vaginal secretions). Barrier protection should be used with *all* tissues. The type of barrier protection used should be appropriate for the type of procedures being performed and the type of exposure anticipated. Examples of barrier protection include disposable lab coats, gloves, and eye and face protection.
Gloves	Are to be worn when there is potential for hand or skin contact with blood, other potentially infectious material, or items and surfaces contaminated with these materials.
Face protection	Wear during procedures that are likely to generate droplets of blood or body fluid to prevent exposure to mucous membranes of the mouth, nose, and eyes.
Protective body clothing	Wear (disposable laboratory coats [Tyvek]) when there is a potential for splashing of blood or body fluids.
Wash hands or other skin surfaces	Wash thoroughly and immediately if contaminated with blood, body fluids containing visible blood, or other body fluids to which universal precautions apply. **Wash hands immediately** after gloves are removed.
Avoid accidental injuries	Avoid injuries that can be caused by needles, scalpel blades, laboratory instruments, etc., when performing procedures, cleaning instruments, handling sharp instruments, and disposing of used needles, pipettes, etc.
Sharp items	Place the following in puncture-resistant containers for disposal: used needles, disposable syringes, scalpel blades, pipettes, and other items marked with a biohazard symbol.

Source: Adapted from: National Institute of Environmental Health Sciences (NIEHS), National Institutes of Health (NIH), Department of Health and Human Services (DHHS). Biological Safety. Universal Precautions. Last updated January 29, 2001. Available at http://www.niehs.nih.gov/odhsb/biosafe/univers.htm. Accessed 2/11/05.

disinfectants—agents that kill microbes on inanimate objects or surfaces

antiseptics—agents that kill microbes within living tissue

betadine—a povidone-iodine containing solution that is used topically to destroy microorganisms

sterilization—process that destroys all living organisms

vasomotor—referring to nerves that innervate smooth muscles in the walls of arteries and veins and can cause their constriction or dilation

BOX 10.2 CLINICAL APPLICATIONS

Stages of Infection

In the following example, see if you can identify the correlating stages of infection: Missy has spent the last several days traveling with the golf team on the team bus. Her roommate, who is also on the golf team, has been sick with "tonsillitis." As Missy prepares for the last leg of this tournament, she notices she is more tired than usual and is not playing as well as she did in earlier tournaments. That night she wakes with a fever and severe sore throat. On the second day, her coach sends her to the health center. She tests positive for strep throat. She immediately starts on antibiotics, and by the next morning, she feels much better, has no fever, and is able to return to class. By the next weekend, she feels "back to normal" and shoots an 84 in the next golf tournament.

Missy's travel with her sick roommate represents the incubation period of illness. Her fatigue and decreased performance are consistent with the prodromal stage. The presence of fever and sore throat are in the acute stages of the illness. Finally, after treatment, her fever subsides and within the week she is playing well: recovery and convalescence.

with the stage of infection. Stages of infection include incubation, prodromal, acute, and recovery/convalescence. The incubation period is the time between entry of the microorganism and appearance of any clinical signs or symptoms. The prodromal period is when the individual begins to experience the first, even vague, symptoms of infection. During the acute period of infection, the disease will fully develop. Clinical manifestations of the disease will be noted. Finally, as symptoms begin to resolve and signs subside, recovery and convalescence will proceed. Box 10.2 describes this clinical scenario. Unfortunately, in some situations, acute infection progresses to a chronic infection.

Cellular Response to Injury: Inflammation Inflammation utilizes many of the same cells and chemical messengers that are major components of the immune system (see Chapter 12). However, inflammation differs from the immune response in several important ways. Inflammation is defined as an innate, nonspecific series of interrelated

TABLE 10.4

Inflammation versus Inflammatory Response

Concept	Inflammation	Inflammatory Response
Location	Localized	Systemic
Response time	Seconds to minutes	Hours to days
Responses	Redness, heat, swelling, pain	Fever, neutropenia, anorexia, fatigue

events set into motion as a response to many different types of cellular injury. Examples of initiating events may include foreign invasion—such as infection, chemical exposure, and allergens—physical damage, and trauma. Inflammatory response is localized, but the process of inflammation can result in further systemic response. The inflammatory response occurs within seconds to minutes of injury, whereas immune response occurs much more slowly. In general, inflammatory response is classified as a protective mechanism, even though symptoms may result in systemic signs and symptoms. See Table 10.4.

Stages of Inflammation Inflammation begins with the onset of cellular injury. The body reacts to injury with both a **vasomotor** and a cellular response. Vasomotor response begins with a brief period of vasoconstriction that serves to limit any bleeding to the area of injury. Immediately following this period, capillaries and other blood vessels serving the area of injury experience vasodilation. Increased blood flow to the injury, referred to as **hyperemia**, results in several classic symptoms of inflammation: redness, heat, swelling, pain, and altered function (rubor, calor, tumor, dolor, and functio laesa; see Figure 10.2). Increased blood flow to the injury will supply additional oxygen and nutrients needed for the healing process. It will serve to increase transport of cells needed for repair, healing, and prevention of infection, and will provide a dilutional effect for any toxins present. Finally, the pain caused by increased blood flow will limit movement, which serves as additional protection against further injury.

During the this period of vasodilation, vascular permeability changes. This increased permeability allows proteins and immune cells to pass from blood to tissue spaces where the injury has occurred. Material that accumulates in tissue spaces is called **exudate**. The type of exudate will be consistent with the type and extent of injury. Exudate can contain proteins and general immune cells (serous exudate), blood (hemorrhagic exudate), or microorganisms (purulent exudate).

Chemical messengers or mediators direct vasodilation and change in vascular permeability. These mediators (see

hyperemia—increased blood flow to a body tissue

exudate—fluid produced and released from cells that are inflamed and/or injured

FIGURE 10.2 Cardinal Signs of Inflammation

Inflammation

Five signs of Acute Inflammation:

1. Rubor (Redness)
2. Tumor (Swelling)
3. Calor (Heat)
4. Dolor (Pain)
5. Functio laesa (Loss of Function)

First four by Celsus (30 A.D. Rome)
- Virchow (1858)

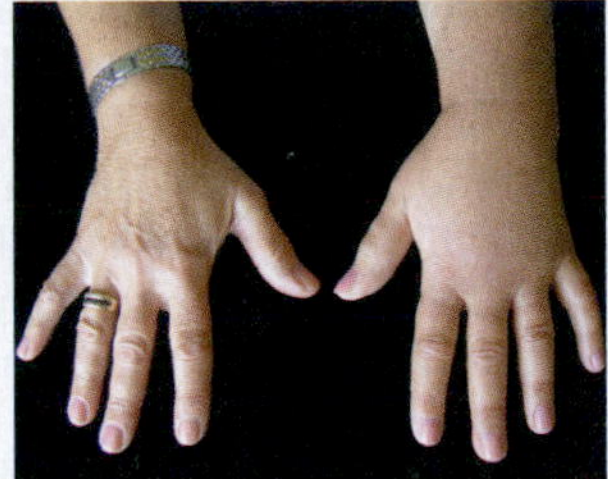

Source: Copyright © Kalab/Custom Medical Stock Photo

Table 10.5) include histamine, nitric oxide, serotonin, prostaglandins, thromboxanes, leukotrienes, complement, and cytokines such as interleukins (IL-1; IL-8). In Figure 10.3, the effects of these chemical messengers are outlined. Further details of these actions are discussed in Chapter 12. These chemical messengers/mediators have particular interest not only for medical and pharmaceutical research but also for nutrition. The precursor for many of these lipid mediators is arachidonic acid (see Figure 10.4). Since many signs and symptoms are controlled by release of these messengers, many current treatments for inflammation center on interrupting their signals. Furthermore, these acute-phase proteins are also used as markers for assessment of the inflammatory process and as an indirect measure of nutritional status. This will be discussed later in this chapter in the Clinical Management of Inflammation section.

Cellular response to inflammation involves the action of particular immunocompetent cells that accomplish both phagocytosis (engulfment of particles or microorganisms) and initiation of the next stages of the immune response, if that is necessary. These cells (neutrophils and macrophages) move into the area and begin destruction of any microorganisms and foreign debris. They also move to clear any dead cells that result from the injury.

Finally, proteins necessary for healing are produced. As noted above, during the period of vascular permeability, proteins escape from the circulatory system into tissue spaces. Fibrinogen (which is activated to fibrin) serves as a major component in wound healing. Fibrin not only works to provide the framework for healing tissue, but also seals off the area of injury and contributes to blood clotting.

Signs and Symptoms of Inflammation The cardinal or classic signs of inflammation, as mentioned previously, are redness, warmth, swelling, and pain resulting from hyperemia and vascular permeability. Table 10.4 compares the clinical features of a localized inflammation to a systemic inflammatory response.

Systemic effects of inflammation may include fatigue, malaise, and fever, an elevation in body temperature. When macrophages begin phagocytosis, they also cause release of prostaglandins, which in turn stimulates the hypothalamus to increase body temperature. Increased body temperature is a protective factor, because most microorganisms do not thrive in higher temperatures. Heat also promotes phagocytosis,

TABLE 10.5

Mediators of Inflammation

Mediator	Action
Chemokines	Chemotaxis
Histamine	Increases blood flow as well as seepage of fluid and proteins from blood
Reactive oxygen species (ROS)	Toxic for microorganisms but also damage tissue
Interleukin 1 (IL-1)	Triggers blood clotting, T cell activation, decrease in blood pressure, fever, and release of prostaglandins
Prostaglandins	Increase vascular permeability and influence platelet aggregation
Leukotrines	Prolong the response; have vasoactive properties

Source: Table created by: Dr. Christina Lee Frazier of Southeast Missouri State University

FIGURE 10.3 Stages of Inflammation

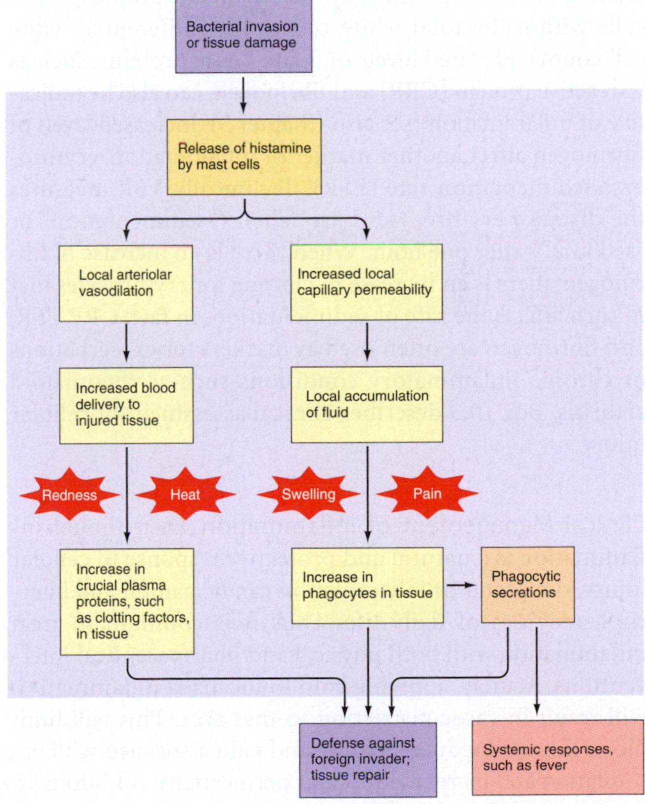

FIGURE 10.4 Arachidonic Acid

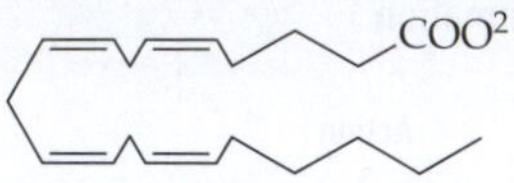

which is a natural step in the inflammatory response. These benefits of higher temperature make the common practice of taking medications to interrupt the fever process problematic. Suppressing a mild fever may do more harm than good. High fevers, of course, can be dangerous, especially in infants and children.

Laboratory markers for inflammation can include increased white cell count (*leukocytosis*). Other components of white cell count may give an indication of the source of infection. This is determined by measuring the proportions of cells within the total white cell count (differential white cell count). Elevated levels of acute phase proteins, such as C-reactive protein (CRP) and fibrinogen, can also be indicative of inflammation (see also Chapter 5). Increased levels of fibrinogen affect another marker of inflammation, erythrocyte sedimentation rate (ESR). Technically, ESR measures the distance erythrocytes have fallen ("sedimentation" or "sed rate") after one hour. When there is an increase in fibrinogen, there is an increase in clotting for erythrocytes that in turn affects the rate of sedimentation. In fact, CRP, ESR, and fibrinogen are often used as markers for exacerbations of chronic inflammatory conditions such as rheumatoid arthritis. Box 10.3 describes clinical assessment of cellular injury.

BOX 10.3 CLINICAL APPLICATIONS

Assessment of Cell Injury

Earlier in this chapter, it was stressed that fully understanding the foundation of pathophysiology allows for application of basic concepts of the disease process to many different diagnoses and conditions. Furthermore, an understanding of basic methods of measuring cellular injury will direct appropriate clinical interventions. When a specific diagnosis or disease process is studied, characteristic responses to cellular injury are apparent.

When cells are injured—whether it is due to the lack of oxygen, nutritional imbalances, or physical injury—cell components may be released into the body, as noted in the section on inflammation. Part of clinical diagnosis and practice is measuring release of these cellular components. For example, many diagnoses depend on measurement of cellular enzymes. In liver disease, enzymes such as alkaline phosphatase (ALP) and aspartate amino transferase (AST) are present in abnormally high levels in the blood. When a myocardial infarction occurs, the contractile protein, troponin, is released from the damaged cell and can be detected in the blood.

Abnormal electrical activity of the cell can also be monitored. Standard medical practice uses the electrocardiogram (ECG, EKG), electroencephalogram (EEG), and electromyogram (EMG) to determine abnormalities in the function of the heart, brain, or muscles. For example, when cardiac tissue is damaged, changes in heart rate or rhythm will be noted on the electrocardiogram.

Finally, actual tissue cells can be examined. This is referred to as a biopsy. Under microscopic examination, abnormalities the cell develops can be detected and clinically evaluated for specific diagnosis and treatment.

Clinical Management of Inflammation Even though inflammation is a natural and protective response to cellular injury, symptoms of inflammation can be painful and interrupt activities of daily life. Thus it is common to treat inflammation with both physical and pharmaceutical interventions. Acutely, applying cold to localized inflammation will result in vasoconstriction to that area. This will limit blood flow and reduce swelling and pain associated with vasodilation and increased vascular permeability. Additionally, pressure and elevation will assist in treatment of these same symptoms. Elevating the injury decreases blood flow as well and will assist in reducing swelling. Constricting the area with pressure may reduce accumulation of exudate.

Several different classes of medications can assist in treating inflammation. Nonsteroidal anti-inflammatory medications (NSAIDs), selective COX-2 inhibitors, and glucocorticoids are all used to treat inflammation (Nowak and Handford 2004; Sherwood 2004). Chemical mediators that signal physiological events for inflammation were discussed earlier. All of these medications block this process in one or more steps. Salicylates (aspirin), an NSAID, inhibit prostaglandin production. Other nonsteroidal anti-inflammatory medications, such as ibuprofen, further prevent synthesis of prostaglandins by blocking cyclooxygenase enzyme 2 (COX-2; see Figure 10.5). Unfortunately, NSAIDs also block COX-1 that has a protective effect on the gastric mucosa (see Figure 10.6). When this enzyme (COX-1) is reduced, side effects of gastritis and upper GI bleeding are possible (Lehman and Beglinger 2005). Selective COX-2 inhibitors only block COX-2 and do not have GI side effects. These medications are generally prescribed for musculoskeletal inflammation (Bertolini, Ottani, and Sandrini 2002).

FIGURE 10.5 NSAIDS Block COX-2

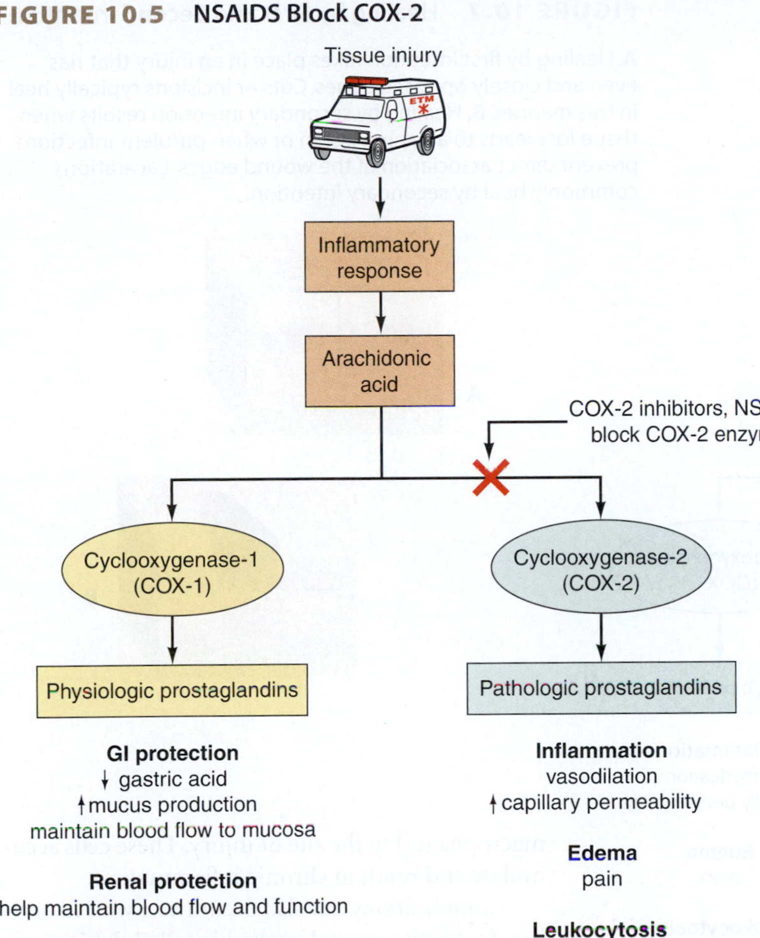

Source: Dr David Gotlieb doc on-line, http://www.arthritis.co.za/nsaids2.htm

Glucocorticoids or steroids can be given not only for musculoskeletal inflammation, but also for systemic inflammatory conditions. Depending on the source of inflammation, glucocorticoids can be given intravenously, orally, topically, or by injection directly to the site of inflammation. Glucocorticoids are synthetic derivatives of endogenously produced cortisone. These medications act on several different aspects of the inflammatory response, affecting vascular permeability or blocking one of the enzymes required for production of pros-taglandins (Perretti and Anluwalia 2000). Long-term use of glucocorticoids is associated with many possible side effects, including hyperglycemia, osteoporosis, and protein loss/muscle wasting.

Cellular Response with Healing Inflammation is the body's natural response to cellular injury. An important component of inflammation is the body's ability to begin the steps toward healing. Healing is defined as repair and restoration of damaged tissue and cells, and is the process by which structure and function are restored after an injury.

Wounds and injuries heal by restoring cells, repairing cells, or replacement of cells. The type of healing will depend on the original injury. If the injured cell is a type able to undergo mitosis (cellular replication), those cells can usually be restored if optimal conditions are maintained. Other cells (specifically cardiac, skeletal, and neurological) are unable to be fully restored after injury. Healing of tissues composed of these cells incurs replacement of original tissue with scar tissue.

Healing by first intention (see Figure 10.7) applies to most wounds that are smaller, where cell loss is minimal and the edges of the wound are close together. In this process, epithelial cells are replaced. Below the surface, granulation tissue is formed with the support of collagen. For deep or large wounds, healing will proceed from the bottom of the wound upward. This is referred to as healing by second intention. The normal progression of wound healing is outlined in Table 10.6.

Nutrition is an important component of successful wound healing, even though clinical research has not provided adequate evidence for verified recommendations (Thompson and Fuhrman 2005). A review of the literature indicates adequate energy, protein, and fluid are the foundation of healing support. Requirements may be increased for arginine and glutamine, but specific levels for these amino acids have not been verified. Specific vitamins and minerals that are involved with wound healing, and that may need supplementation, include vitamin C, vitamin A, vitamin E, vitamin K, and zinc (see Table 10.7). Assuring adequate blood supply and keeping the wound clean and undisturbed during healing additionally support the promotion of healing.

Healing will be delayed if a foreign material is present or if infection develops. Wound healing will be impaired if the wound is exposed to radiation. The elderly are especially at risk for poor wound healing due to changes in their skin and increased risk of poor circulation and poor nutritional status.

FIGURE 10.6 NSAIDS Block COX-1 and COX-2

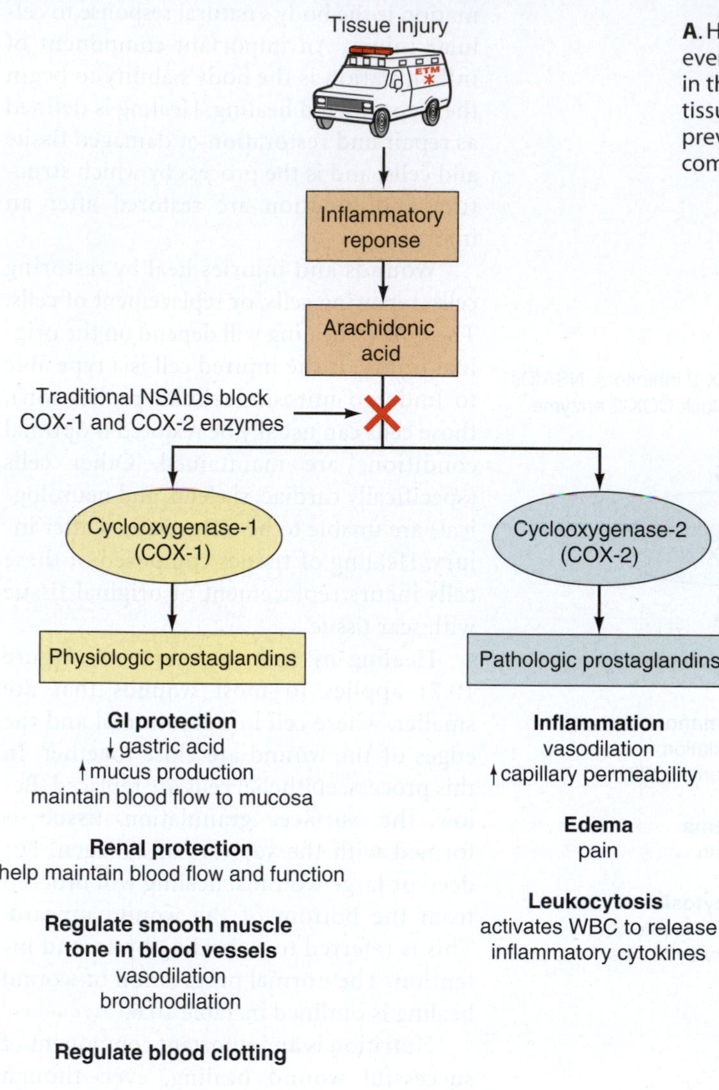

FIGURE 10.7 Healing by First and Second Intention

A. Healing by first intention takes place in an injury that has even and closely opposed edges. Cuts or incisions typically heal in this manner. **B.** Healing by secondary intention results when tissue loss leads to a gaping lesion or when purulent infections prevent direct association of the wound edges. Lacerations commonly heal by secondary intention.

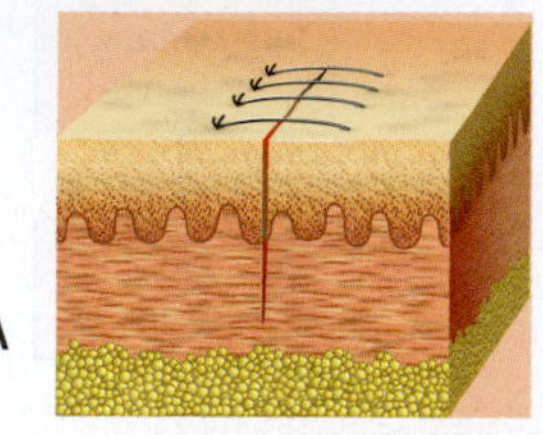

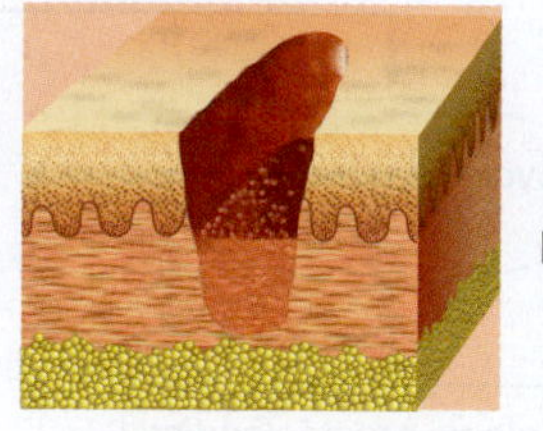

Complications of Inflammation and Wound Healing

Complications of inflammation can occur and may include development of chronic inflammatory conditions. When the injurious agent is not quite strong enough to cause a systemic response but continues to be present, chronic low-level inflammation can occur. This underlying constant inflammation draws cells of the immune system (particularly macrophages) to the site of injury. These cells accumulate and result in chronic inflammation.

Complications can also occur from wound healing. Ineffective wound healing is called **dehiscence**. This simply means that a wound reopens. Dehiscence may result from any of the situations described earlier, such as poor circulation or malnutrition. Obese patients are at higher risk for dehiscence (Thompson and Fuhrman 2005). Other complications include contractures and adhesions. During the healing process, scar tissue can result in a shrinking of the connective tissue. This **contracture** can distort the area around the injury and limit the ability to use that part of the body. Healing of burns often results in development of contractures. **Adhesions** result when two previously unconnected tissues are abnormally joined together. This most often occurs after a surgical procedure and is most common in muscles after abdominal surgery.

dehiscence—separation of wound edges

contracture—shortening of muscle tissue resulting in immobility

adhesion—scar tissue that forms between two body surfaces usually as a result of surgery or injury

Cellular Death If the injurious agent or disease process is severe enough or continued for long enough, the cell will reach a point where adaptation can no longer occur. Compensation is not possible and cellular metabolism ceases. There are noticeable structural changes in an injured cell, but as a cell begins to experience death, distortions become more significant. Increased membrane permeability allows contents of the cell to spill out, cell structures such as mitochondria and endoplasmic reticu-

TABLE 10.6

Progression of Normal Wound Healing

Phase	Typical Duration for Acute Wound Healing	Primary Events
Inflammatory	Begins at the time of injury and continues for about four to six days	Coagulation cascade and fibrin clot formation control bleeding. Vasodilation and increased capillary permeability occur. Neutrophils phagocytize bacteria. Macrophages remove debris and necrotic tissue and secrete growth factors.
Proliferative	Begins about the third to fifth day and continues for two to three weeks	Epithelial cells form a protective covering and framework over the wound. Angiogenesis enables development of granulation tissue. Fibroblasts produce collagen and matrix protein, forming granulation tissue. Collagen deposition and cross-linking begin to strengthen the wound. Myofibroblasts induce wound contraction and the wound begins to close.
Remodeling	Begins about two to three weeks after injury; can continue for up to two years	Collagen maturation and stabilization occur. Fibrous scar tissue matures (decreases in fibroblasts and vascularization) but skin and fascia never regain full strength.

Source: Reprinted from: Thompson C, Fuhrman MP. Nutrients and wound healing: still searching for the magic bullet. *Nutr Clin Pract.* 2005; 20:333, with permission from the American Society for Parenteral and Enteral Nutrition (A.S.P.E.N.). A.S.P.E.N. does not endorse the use of this material in any form other than its entirety.

lum will be grossly altered, cell enzymes (lysosomes) are released to begin cellular destruction, and the nucleus is permanently damaged.

Necrosis refers to the cellular changes that occur during cell death. Different types of necrosis will occur in different organs or tissues and often will indicate the cause of cell death. Coagulation necrosis, caseous necrosis, gangrenous necrosis, and liquefaction necrosis (see Table 10.8) are all specific terms describing the process of cell death.

Apoptosis is a different pattern of cell death. This form of cell death appears to be genetically programmed, which allows for removal of cells in a systematic, orderly fashion. For example, this is the process that organizes removal of cells after inflammation. Likewise, when a woman stops breastfeeding, apoptosis directs actions that clear cells that are no longer needed from the breasts.

Conclusion

In this chapter, we have explored the foundation of pathophysiology. It is crucial that these particular basic concepts of disease are fully understood so they can be applied to specific diagnoses, conditions, and disease processes. This foundation allows the student and clinician to critically evaluate information and develop skills for clinical reasoning. The medical field is constantly changing. If the clinician understands pathophysiology, progress made in treatment is more easily understood and provides the foundation for application.

necrosis—general term referring to cell death

apoptosis—genetically programmed cell death

TABLE 10.7

Nutrition and Wound Healing for Adults Receiving Oral or Enteral Nutrition

Nutrient	Function in Wound Healing	Dietary Reference Intake: RDA, AI, UL	Recommended Intake for Wound Healing	Additional Notes
Kcalories	Energy for hypermetabolic response.		25 to 30 kcal/kg to 30 to 35 kcal/kg depending on type of wound	Monitoring actual kcalories received is essential. Kcalorie goals should be adjusted for ebb and flow phases of the hypermetabolic response, severity of wound, response to healing, comorbidities, and other factors that affect metabolic rate.
Protein	Immune response, nitrogen losses from wounds.	RDA: 0.8 g protein/kg for adults	1.2 to 1.5 g/kg	Unless severe catabolic states or exogenous protein losses are present, > 1.5 to 2.0 g/kg may indicate overfeeding.
Fat n-3/n-6 fatty acids	May play a role in collagen formation.	AI: Men, 19–70+ years: 1.6 g n-3 fatty acids/day Women, 19–70+ years: 1.1 g n-3 fatty acids/day n-3 to n-6 ratio for adults: 10.6:1 (< 51 y); 10:1 (> 51 y)	Not yet determined	One small study found topical use of oils rich in essential fatty acids may prevent skin breakdown (Cardoso 2004).
Fluid	Maintain skin turgor and blood flow to wounded tissue to prevent breakdown of skin.		18–55 yo: 35 mL/kg > 55 yo: > 30 mL/kg or minimum of 1.5 L/d (unless contraindicated)	Provide additional fluid to maintain adequate hydration and compensate for additional losses from fever, draining wounds, etc. Additional fluid needed if fever present.
Vitamin A	Supplementation may assist wound healing that has been retarded by vitamin A deficiency, radiation, chemotherapy, or DM.	RDA Men: 900 RAE/~ 3000 IU RDA Women: 700 RAE/~ 2300 IU UL: 3000 RAE or 10,000 IU	20,000–25,000 IU orally for 10 days if deficiency suspected or confirmed (low serum retinol and/or functional tests such as dark adaptation	Supplementation rarely warranted in ESRD. Patients with fat malabsorption should receive water-soluble form.
Vitamin C	Deficiency can delay wound healing.	RDA Women: 75 mg RDA Men: 90 mg Additional 35 mg/d for smokers UL: 2,000 mg due to GI side effects (nausea, abdominal cramps, or diarrhea)	Up to1 to 2 mg/day for deficiency; during acute illness and injury for up to three months	Evaluation of serum ascorbic acid levels reflects dietary intake and not tissue levels. White blood cell ascorbic acid and ascorbic acid tissue saturation acid and ascorbic acid tissue saturation tests are better indices.
Zinc	Low serum zinc levels associated with impaired healing. Supplementation may enhance wound healing, but only in those who are zinc deficient.	RDA Women: 8 mg RDA Men: 11 mg (Requirements may be up to 50% higher for vegans.) UL: 40 mg	15 to 25 mg elemental zinc/day	Zinc intakes below RDA are commonly seen in the elderly. Increased losses can occur from large skin wounds or urine (after trauma, closed head injury, etc.). Impaired absorption from GI tract can occur with high volume diarrhea or fistulae output.
Iron	Routine supplementation not recommended for wound healing,	RDA premenopausal women: 18 mg RDA postmenopausal women: 8 mg RDA men: 8 mg UL: 45 mg	Not applicable	Low hemoglobin concentration does not seem to impair wound healing, provided adequate tissue perfusion is maintained.

(continued on the following page)

TABLE 10.7 *(continued)*

Vitamin E	No clear role for vitamin E supplementation. Use may adversely affect healing of some types of wounds.	RDA: 15 mg alpha-tocopherol equivalents	Assure 100% of RDA/DRI all nutrients.	Large doses of vitamin E should be avoided before surgery unless deficiency is present. Topical vitamin E is often recommended to reduce scar formation and improve cosmetic appearance of scar tissue, but this has not been documented by research.
Other vitamins, minerals, and trace elements	Thiamin, riboflavin and pantothenic acid play a role in collagen production. Copper and manganese are required for tissue regeneration. Vitamin K is required for prothrombin and synthesis of other clotting factors.		Assure 100% of RDA/DRI all nutrients.	No studies verifying deficiency of these nutrients as having an impact on wound healing could be found.
Arginine	Supplementation may benefit wound healing by increasing collagen deposition, improving nitrogen balance, and enhancing several parameters of immune function. Most, but not all, research demonstrated enhanced nitrogen retention and immune function.	Not applicable	Oral dose of 17.0-18.7 g/day free arginine was used in previous human studies. The recommended dosage, and populations who may benefit most, are unknown at this time.	Supplementation may be contraindicated in patients with severe renal or hepatic failure.
Glutamine	Preferred fuel for enterocytes, lymphocytes, and macrophages. Precursor for nucleotides and glutathione.	Not applicable	Oral doses up to 0.57 mg/kg/d in adults appears safe.	Supplementation may be contraindicated in patients with severe renal or hepatic failure.

Adapted from Thompson, C.W., Nutrition and adult wound healing. Nutrition Week January 18, 2003. Available at http://www.nutritioncare.org/listserv/wound%20healing.pdf

TABLE 10.8

Gangrene

Type of Necrosis	Description	Common Causes	Example
Coagulation	Occurs primarily in kidney, heart, and adrenal glands	Hypoxia caused by nerve ischemia or hypoxia caused by chemical injury	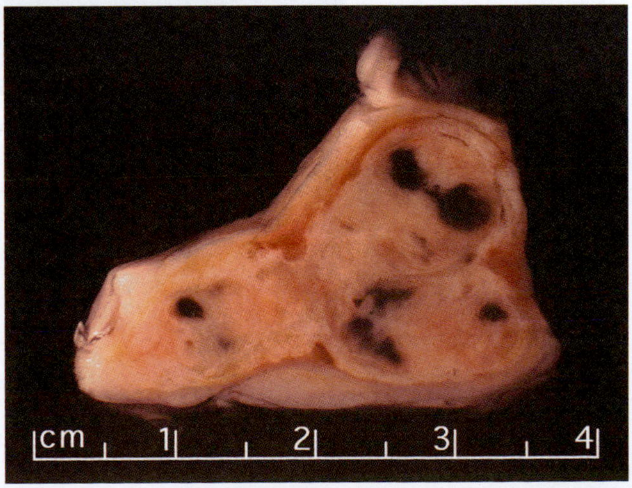 *Source:* From the University of Alabama at Birmingham Department of Pathology PEIR Digital Library © (http://peir.net)

(continued on the following page)

TABLE 10.8 *(continued)*

Gangrene			
Type of Necrosis	**Description**	**Common Causes**	**Example**
Caseous	Combination of coagulation and liquefactive necrosis; tissue appears soft and granular and resembles clumped cheese	TB pulmonary infection, particularly *Mycobacterium tuberculosis*	*Source:* From the University of Alabama at Birmingham Department of Pathology PEIR Digital Library © (http://peir.net)
Gangrenous	Tissue death resulting from severe hypoxic injury, especially in lower leg Dry gangrene—skin color changes to dark brown or black	Usually result of coagulative necrosis	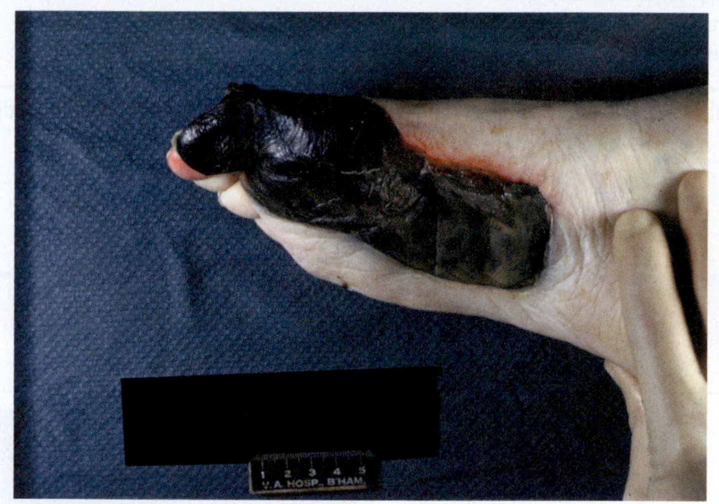 *From the University of Alabama at Birmingham Department of Pathology PEIR Digital Library © (http://peir.net)*

(continued on the following page)

TABLE 10.8 *(continued)*

	Wet gangrene—usually occurs in internal organs, causing site to become cold, swollen, and black; foul odor is present, produced by pus	Develops when neutrophils invade site, causing liquefactive necrosis	 *From the University of Alabama at Birmingham Department of Pathology PEIR Digital Library © (http://peir.net)*
Liquefactive	Brain cells are digested by their own hydrolases; tissue becomes soft, liquefies, and is walled off from healthy tissue, forming cysts	Ischemic injury to neurons and glial cells in brain	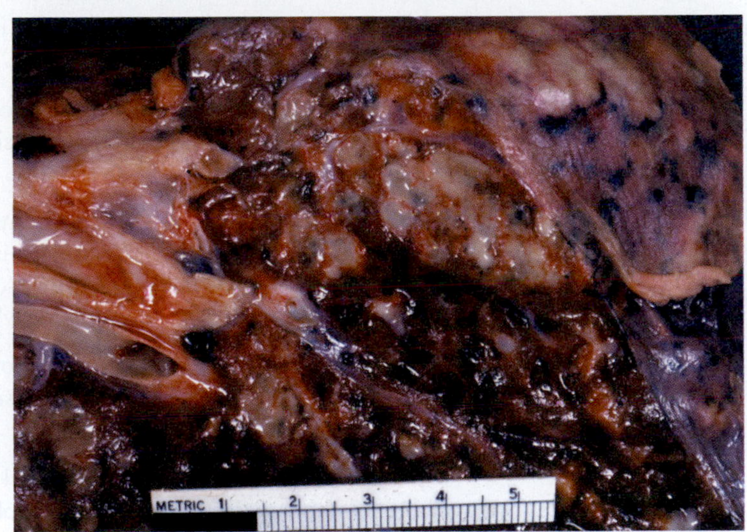 *From the University of Alabama at Birmingham Department of Pathology PEIR Digital Library © (http://peir.net)*

WEB LINKS

Genes and Disease: This website sponsored by the National Center for Biotechnology Information provides an excellent foundation on the etiology of disease, emphasizing the genetic contribution.

http://www.ncbi.nlm.nih.gov/disease/

Centers for Disease Control and Prevention: A division of the U.S. Department of Health and Human Services provides extensive information about every aspect of disease and its prevention throughout the U.S. and the world. This source also provides information on environmental control, including Guidelines for Universal Precautions.

http://www.cdc.gov

Health Information Center at the Cleveland Clinic: This website provides excellent patient education information. This link particularly summarizes information about inflammation and its treatment.

http://www.clevelandclinic.org/health/health-info/docs/0200/0217.asp?index=4857

END-OF-CHAPTER QUESTIONS

1. When researchers study the prevalence of atherosclerosis in developing countries and compare this to the prevalence in industrialized nations, this is an example of:

   A. etiology.
   B. epidemiology.
   C. disease incidence.

2. Infectious disease is an example of which etiological category of disease?

   A. multifactorial
   B. acquired
   C. genetic

   For the other two answers that you did not choose, give an example of that category of disease.

3. Mrs. J is meeting with her physician to discuss her recent diagnosis of breast cancer. As her physician outlines the probable response to therapy and how she expects Mrs. J to respond, the physician is actually discussing:

   A. remission.
   B. prognosis.
   C. cure.

4. Mrs. J's physician also states that the initial goal of her treatment is to find no indication of disease after five years post-chemotherapy. In this discussion, the MD is outlining what we could call the _____ of her disease.

   A. remission
   B. prognosis
   C. cure

5. When Mark twists his ankle in practice, the trainer immediately places cold packs and elevates the ankle. What symptoms do these two actions prevent?

6. How do nonsteroidal anti-inflammatory drugs (NSAID) treat the acute inflammatory process? Give an example of a NSAID.

11

Nutrigenomics

Melissa Hansen-Petrik, Ph.D., R.D., L.D.N.
Research Assistant Professor
Director, Didactic Program in Dietetics
Department of Nutrition, The University of Tennessee–Knoxville

CHAPTER OUTLINE

An Overview of the Structure and Function of Genetic Material
Deoxyribonucleic Acid (DNA) and Genome Structure • Translating the Message from DNA to Protein • Genetic Variation • Epigenetic Regulation • Dietary Regulation and Measurement of Gene Expression

Nutrigenomics in Disease
Cancer • Obesity and Diabetes • Cardiovascular Disease

Nutrigenomics and the Practice of Dietetics
Individual Testing in the Marketplace • Evolving Knowledge and Practice Requirements for Dietitians

Introduction

In 2003, the International Human Genome Sequencing Consortium published the finished version of the human **genome** sequence, thereby marking a historic milestone in science with great implications for the future of health care (see Figure 11.1; Collins et al. 2003). The human genome is the blueprint for approximately 30,000 different proteins (Dennis and Gallagher 2001). In many respects, the human body is a system of proteins. Proteins serve as structural components, hormones, neurotransmitters, and cell-signaling agents that ensure the body is operating smoothly. Production and degradation of each of these proteins is tightly regulated but is also influenced by environmental factors such as nutrition. This interaction between nutrients and **genotype**/gene regulation is known as **nutrigenomics** (Ordovas et al. 2002).

Over the last few decades, various dietary guidelines have been developed for the purpose of optimizing overall health, preventing or treating cardiovascular disease, preventing cancer, treating hypertension, and treating diabetes (U.S. Department of Health and Human Services 2005; National Heart, Lung, and Blood Institute 2005; American Institute for Cancer Research 2006; American Cancer Society 2001; American Diabetes Association 2005). While based upon the best available knowledge of the relationship

genome—the entire set of genes of a given organism

genotype—the specific variants of a gene present in the two alleles in an individual that can result in specific traits or disorders

nutrigenomics—the interaction between nutrients and other food-derived bioactive substances with an individual's genome

FIGURE 11.1 Timeline of Genetics and Genomics from Discovery by Mendel of the Laws of Genetics in 1865 to Completion of the Human Genome Project in 2003

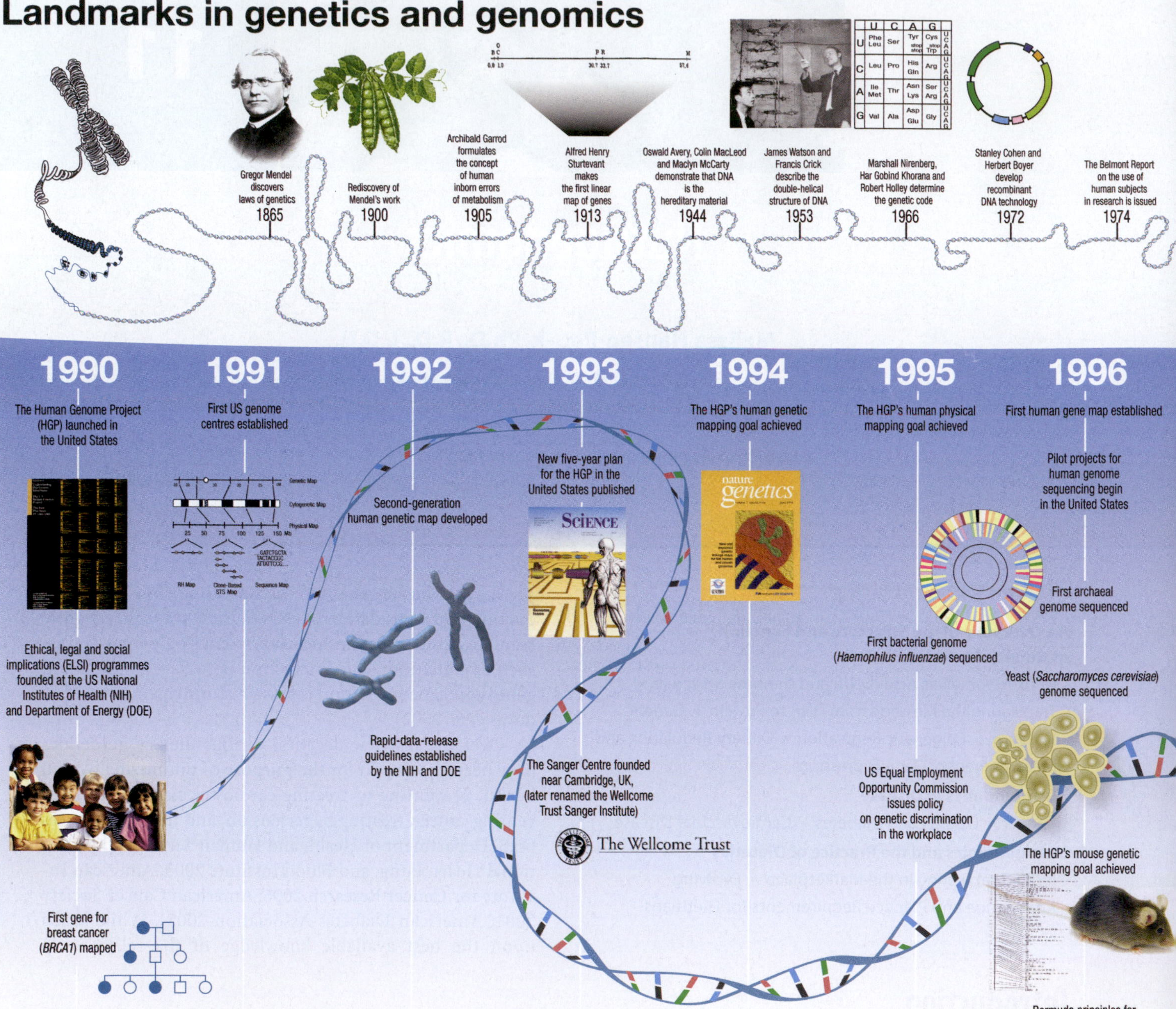

Landmarks in genetics and genomics
Gregor Mendel discovers laws of genetics 1865
Rediscovery of Mendel's work 1900
Archibald Garrod formulates the concept of human inborn errors of metabolism 1905
Alfred Henry Sturtevant makes the first linear map of genes 1913
Oswald Avery, Colin MacLeod and Maclyn McCarty demonstrate that DNA is the hereditary material 1944
James Watson and Francis Crick describe the double-helical structure of DNA 1953
Marshall Nirenberg, Har Gobind Khorana and Robert Holley determine the genetic code 1966
Stanley Cohen and Herbert Boyer develop recombinant DNA technology 1972
The Belmont Report on the use of human subjects in research is issued 1974
1990
The Human Genome Project (HGP) launched in the United States
Ethical, legal and social implications (ELSI) programmes founded at the US National Institutes of Health (NIH) and Department of Energy (DOE)
First gene for breast cancer (BRCA1) mapped
1991
First US genome centres established
1992
Second-generation human genetic map developed
Rapid-data-release guidelines established by the NIH and DOE
1993
New five-year plan for the HGP in the United States published
The Sanger Centre founded near Cambridge, UK, (later renamed the Wellcome Trust Sanger Institute)
The Wellcome Trust
1994
The HGP's human genetic mapping goal achieved
1995
The HGP's human physical mapping goal achieved
First bacterial genome (Haemophilus influenzae) sequenced
US Equal Employment Opportunity Commission issues policy on genetic discrimination in the workplace
1996
First human gene map established
Pilot projects for human genome sequencing begin in the United States
First archaeal genome sequenced
Yeast (Saccharomyces cerevisiae) genome sequenced
The HGP's mouse genetic mapping goal achieved
Bermuda principles for rapid and open data release established
DESIGN BY DARRYL LEJA
PEAS COURTESY J. BLAMIRE, CITY UNIV. NEW YORK; WATSON & CRICK COURTESY A. BARRINGTON BROWN/SPL; SCIENCE COVERS COURTESY AAAS

Source: U.S. Department of Energy Human Genome Program, Oak Ridge National Laboratories, Oak Ridge, Tennessee. http://www.ornl.gov/hgmis

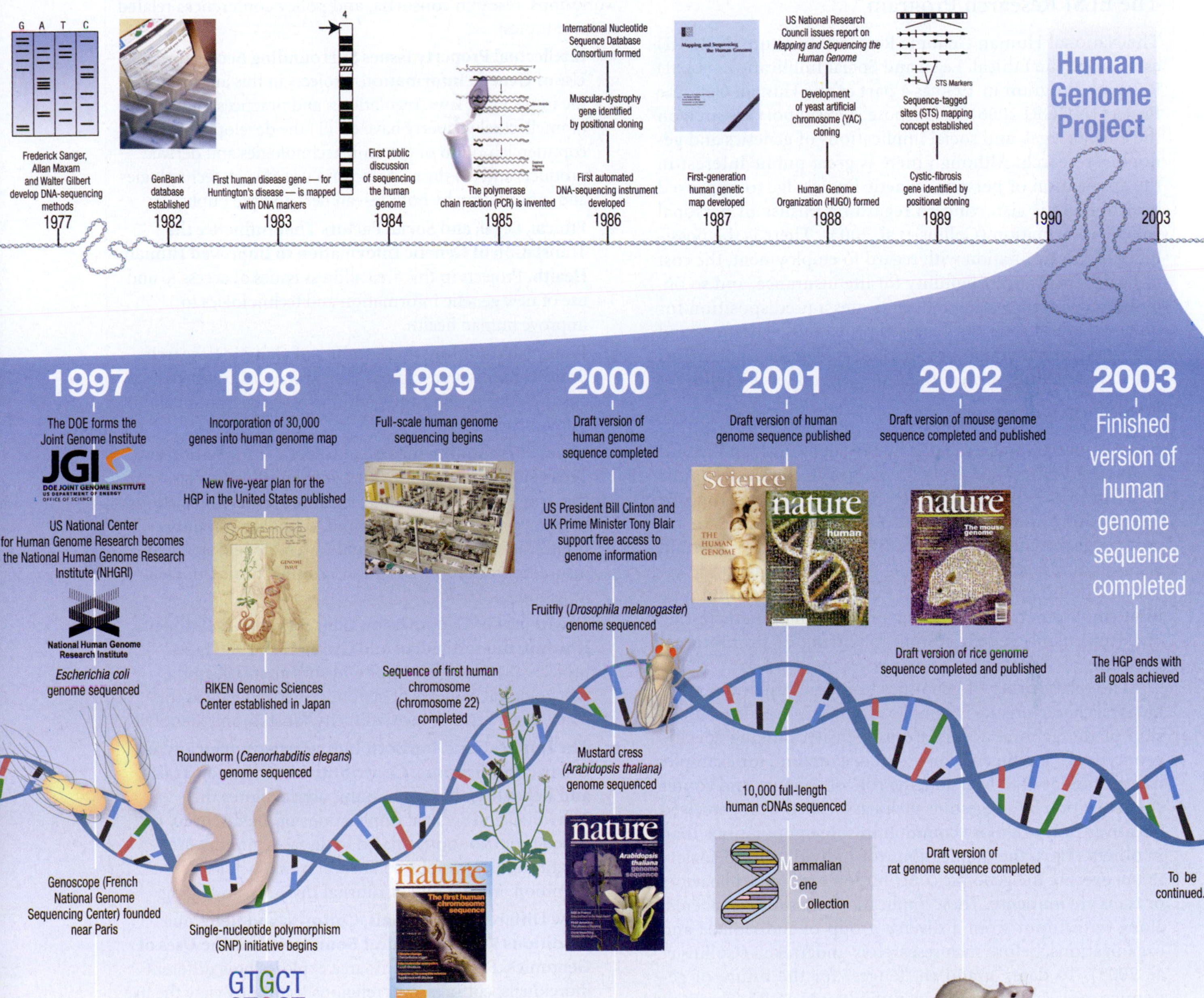
Human Genome Project
G A T C
Frederick Sanger, Allan Maxam and Walter Gilbert develop DNA-sequencing methods
1977
GenBank database established
1982
First human disease gene — for Huntington's disease — is mapped with DNA markers
1983
First public discussion of sequencing the human genome
1984
The polymerase chain reaction (PCR) is invented
1985
International Nucleotide Sequence Database Consortium formed
Muscular-dystrophy gene identified by positional cloning
First automated DNA-sequencing instrument developed
1986
First-generation human genetic map developed
1987
US National Research Council issues report on Mapping and Sequencing the Human Genome
Development of yeast artificial chromosome (YAC) cloning
Human Genome Organization (HUGO) formed
1988
STS Markers
Sequence-tagged sites (STS) mapping concept established
Cystic-fibrosis gene identified by positional cloning
1989
1990
2003
1997
The DOE forms the Joint Genome Institute
JGI
DOE JOINT GENOME INSTITUTE
US National Center for Human Genome Research becomes the National Human Genome Research Institute (NHGRI)
National Human Genome Research Institute
Escherichia coli genome sequenced
Genoscope (French National Genome Sequencing Center) founded near Paris
1998
Incorporation of 30,000 genes into human genome map
New five-year plan for the HGP in the United States published
Science
RIKEN Genomic Sciences Center established in Japan
Roundworm (Caenorhabditis elegans) genome sequenced
Single-nucleotide polymorphism (SNP) initiative begins
GTGCT
GTCCT
Chinese National Human Genome Centers established in Beijing and Shanghai
1999
Full-scale human genome sequencing begins
Sequence of first human chromosome (chromosome 22) completed
nature
The first human chromosome sequence
2000
Draft version of human genome sequence completed
US President Bill Clinton and UK Prime Minister Tony Blair support free access to genome information
Fruitfly (Drosophila melanogaster) genome sequenced
Mustard cress (Arabidopsis thaliana) genome sequenced
nature
Arabidopsis thaliana genome sequence
Executive order bans genetic discrimination in US federal workplace
2001
Draft version of human genome sequence published
Science
THE HUMAN GENOME
nature
human
10,000 full-length human cDNAs sequenced
Mammalian Gene Collection
2002
Draft version of mouse genome sequence completed and published
nature
The mouse genome
Draft version of rice genome sequence completed and published
Draft version of rat genome sequence completed
2003
Finished version of human genome sequence completed
The HGP ends with all goals achieved
To be continued...

BOX 11.1 NEW RESEARCH

The ELSI Research Program

The National Human Genome Research Institute (NHGRI) established the Ethical, Legal and Social Implications (ELSI) Research Program in 1990 as a part of the Human Genome Project (NHGRI 2006). Its purpose is to support research on the ethical, legal, and social implications of genetics and genomics research. Although there is great public interest in the application of personal genetic knowledge to improved health, there is also concern regarding misuse of personal genetic information (Collins et al. 2003). There is the possibility of discrimination with regard to employment, the cost of health insurance, eligibility for life insurance, and so on, based on genetic testing that indicates a predisposition for disease. Several states have passed protective legislation, and U.S. government employees are protected, but a call remains for wide-reaching protection at the federal level (Collins et al. 2003).

In early 2006, the Genetic Information Nondiscrimination Act of 2005 was endorsed by President Bush and passed by the U.S. Senate, but the bill remained in committee in the U.S. House of Representatives (NHGRI 2006). The proposed legislation would prohibit insurers from requesting or requiring genetic testing of an individual or family. It would also prohibit insurers from using genetic information to establish eligibility or premiums. Furthermore, it would prohibit employers from requesting or requiring genetic testing and from using genetic testing for hiring or promotional decisions.

The relationship of genomics to race, ethnicity, and behavioral characteristics is complex, going beyond the relationship of the genome to disease propensity. Linking specific genotypes to intelligence or sexual orientation, for example, has the potential to overstate the role of genetics and confer stigmatization by suggesting alleles associated with perceived negative traits are more common in some populations than in others. Thus, the implications for individuals and society in uncovering the genomic contribution to specific behaviors or traits are immense. These implications must be considered, along with input from a diverse group of individuals and organizations, before such research is undertaken (Collins et al. 2003). To date, "grand challenges" for the future of genomics research have been identified by NHGRI. The ELSI Research Program funds and manages studies and supports workshops, research consortia, and policy conferences related to these topics:

- **Intellectual Property Issues Surrounding Access to and Use of Genetic Information.** Projects in this area examine the impact that laws, regulations, and practices in the area of intellectual property have on (1) the development and commercialization of genomic technologies and derived products and (2) the access to and use of such technologies and information by both researchers and the public.
- **Ethical, Legal, and Social Factors That Influence the Translation of Genetic Information to Improved Human Health.** Projects in this area address issues of access to and use of new genetic information and technologies to improve human health.
- **Issues Surrounding the Conduct of Genetic Research.** Projects in this area explore ethical ways to conduct cutting-edge genetic and genomic research that involves human participants.
- **Issues Surrounding the Use of Genetic Information and Technologies in Non-Health Care Settings.** Projects in this area examine the ethical, legal, and social implications of using genetic information and technologies in non-health care settings, such as in the arenas of employment, insurance, education, adoption, criminal justice, or civil litigation.
- **The Impact of Genomics on Concepts of Race, Ethnicity, Kinship, and Individual and Group Identity.** These projects examine the complex historical, social, and psychological contexts of genomics-derived data as they relate to concepts of race, ethnicity, kinship, and identity.
- **The Implications, for Both Individuals and Society, of Uncovering Genomic Contributions to Human Traits and Behaviors.** Research in this area explores the individual and societal implications of the discovery of genetic contributions related to diseases, nondisease attributes, and various behavioral traits such as cognition, mental illness, diurnal rhythms, and aging.
- **How Different Individuals, Cultures, and Religious Traditions View the Ethical Boundaries for the Uses of Genomics.** Research in this area explores how different individuals, cultures, and religious traditions view the use of genomics.

between diet and disease, these guidelines do not yet take into account the dramatic genetic variation within the population and how that variation can determine individual response to dietary factors and, hence, the propensity to develop disease. Such information has heretofore been unavailable, but recent completion of the Human Genome Project and a heavy research emphasis on identifying the specific interactions between diet and genetic variants are now yielding results (Kaput and Rodriquez 2004). Attention to nutrigenomics and **pharmacogenomics** is likely to escalate in the coming years, because modifying the genes themselves is not yet feasible and also has formidable ethical implications (see Box 11.1).

pharmacogenomics—the interaction between drugs and an individual's genome that can impact drug efficacy and toxicity

FIGURE 11.2 DNA—the Molecule of Life
Genetic material (DNA) is located in the nucleus of each cell in the body except mature red blood cells, which do not contain nuclei. DNA is arranged in chromosomes, and specific sequences of DNA are divided into genes, each of which encodes a protein.

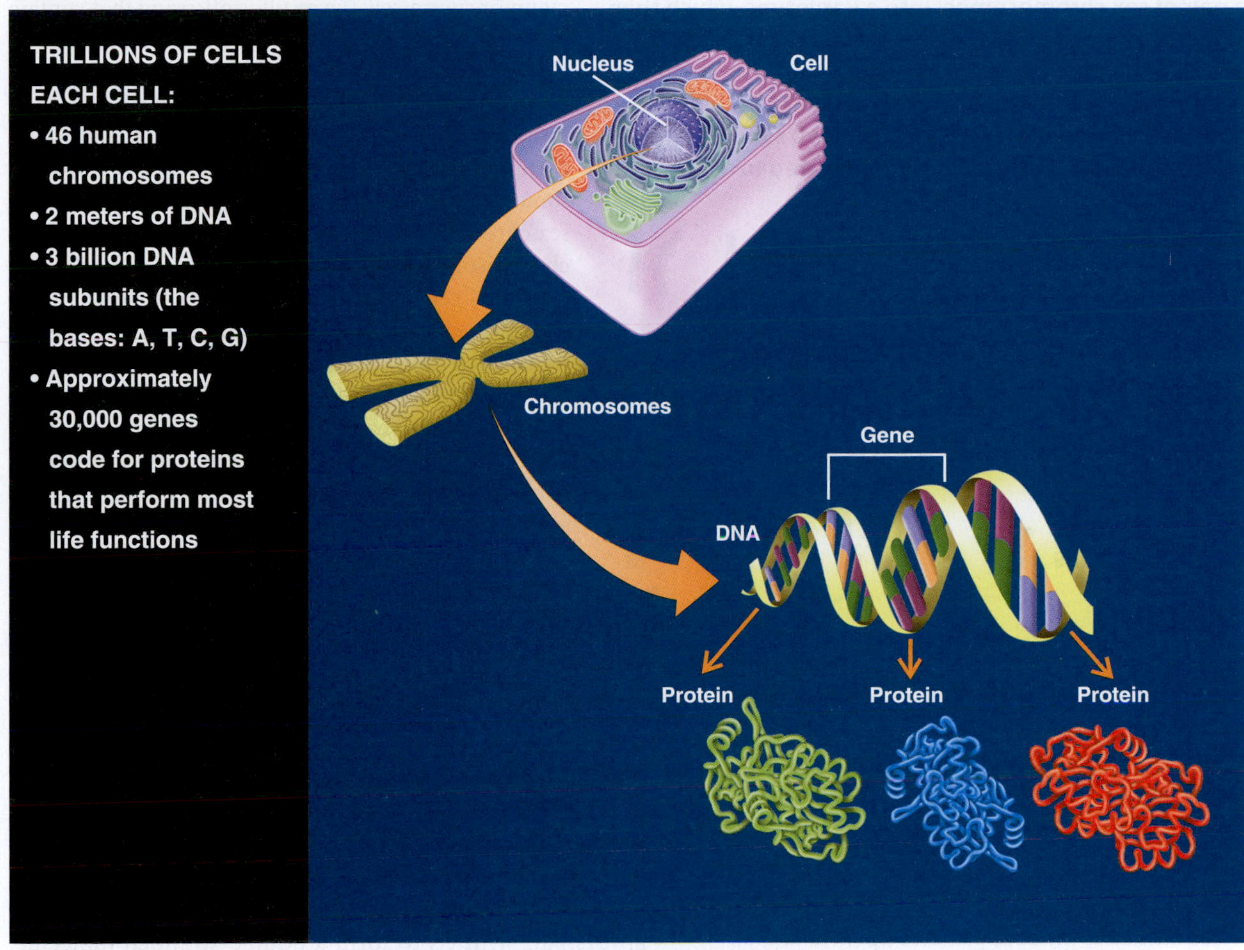

Source: F. Sizer and E. Whitney, *Nutrition: Concepts and Controversies,* 10e, copyright © 2006

An Overview of the Structure and Function of Genetic Material

Deoxyribonucleic Acid (DNA) and Genome Structure

While Mendel, the father of modern-day genetics, discovered the laws of genetics in 1865, deoxyribonucleic acid (DNA) was not itself identified as the blueprint of life until 1944, and its double-helical structure was not discovered until 1953; DeSalle and Yudell 2005). DNA makes up the genome, which does not itself build an organism, but provides the instructions that tell *how* to build an organism. From a human perspective, the genes contained in each person's DNA encode essentially the same proteins, but this code varies from person to person, thus yielding inherited differences in physical characteristics, intellect, and behavioral characteristics as well as the propensity for developing disease (DeSalle and Yudell 2005).

The genetic material or genome lies within each nucleus of each cell in the body (except mature red blood cells, which do not contain nuclei) (see Figure 11.2; DeSalle and Yudell 2005). In humans, the genome is comprised of 23 pairs of **chromosomes**. During **mitosis** (cell division) within an individual, all 23 pairs of chromosomes are copied during the

chromosomes—units of the genome, each consisting of a long molecule of DNA that encodes numerous genes plus histone proteins; there are 22 autosomes and 2 sex chromosomes located within the nucleus of a human cell

mitosis—cell division that produces two cells that are genetically identical to the progenitor cell

FIGURE 11.3 Karyotype: Down Syndrome
Microscopic examination of chromosome size and banding patterns allows medical laboratories to identify and arrange each of the 23 different chromosomes (22 pairs of autosomes and one pair of sex chromosomes) into a karyotype, which then serves as a tool in the diagnosis of genetic diseases. The presence of one X and one Y chromosome indicates this person is male. The extra copy (trisomy) of chromosome 21 in this karyotype identifies this individual as having Down syndrome. The presence of a third chromosome is often referred to as trisomy.

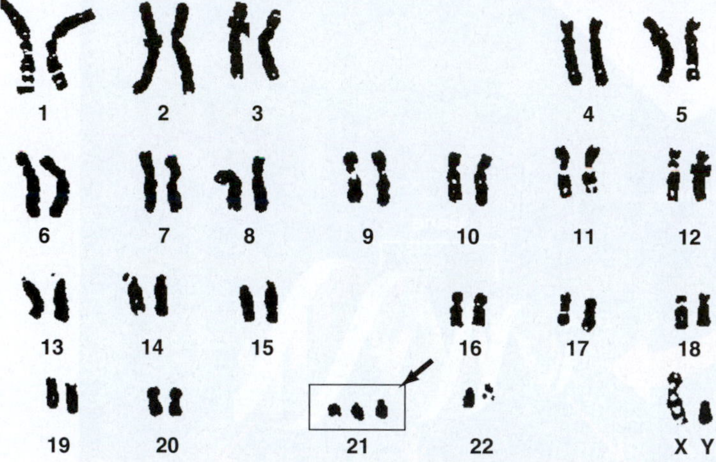

Source: U.S. Department of Energy Human Genome Program, Oak Ridge National Laboratories, Oak Ridge, Tennessee. http://www.ornl.gov/hgmis

creation of a new daughter cell. These chromosomal pairs can be visualized in a **karyotype** (see Figure 11.3). During **meiosis** (reproduction), only one member of each pair of chromosomes is passed on to each ovum or sperm cell; the result is offspring that contain chromosomal pairs created by the donation of one copy of each chromosome from each parent. Of the 23 pairs, 22 are **autosomes** and one pair is comprised of the sex chromosomes. Males have one X and one Y chromosome, while females have two X chromosomes, so the gender of offspring is determined by the sex chromosome passed on by the male parent. Each chromosome consists of DNA containing a linear sequence of genes, each encoding a specific protein. The copy of each gene inherited from the father is the paternal **allele**, and the one inherited from the mother is the maternal allele. Each gene inhabits a particular location on a particular chromosome called its "locus." For example, Figure 11.4 shows a map of chromosome 4 that identifies the location of defects on this chromosome associated with specific disease states.

Each gene is itself a linear sequence of **nucleotides** that are actually responsible for encoding proteins (see Figure 11.5). Nucleotides have three primary components: a purine or pyrimidine nitrogenous base, a ribose (a pentose sugar), and a phosphate group. The backbone of the chain is an alternating strand of the ribose and phosphate residues. The nitrogenous bases project from this backbone and include adenine (A) and guanine (G) (both purines) as well as thymine (T) and cytosine (C), which are both pyrimidines. As DNA, this chain is paired with a complementary strand in which As always pair with Ts and Gs always pair with Cs to form a double-stranded molecule. The DNA is tightly twisted into a double-helical form, which makes each chromosome extremely compact (Lewin 2004).

karyotype—a chart that displays chromosome pairs in order according to size

meiosis—cell division to produce gametes (sperm and ova) that results in the production of cells with half the complement of chromosomes

autosomes—non-sex-determining chromosomes; a human has 22 autosomes

allele—a copy of a specific gene situated in a given locus on a chromosome

nucleotide—the building block of a nucleic acid, consisting of a ribose sugar, a phosphate group, and a nitrogenous base

codon—a series of three nucleotides in mRNA that encodes a specific amino acid

promoter region—regulatory sequence in a gene to which molecules, such as fatty acids, can bind in order to induce expression of that specific gene; molecules can also bind to the promoter region to suppress transcription of a specific gene

intervening sequences—sequences of DNA that lie in between expressed genes and whose function is largely unknown; over 95% of DNA in humans is made up of intervening sequences; sometimes referred to as "junk DNA"

Translating the Message from DNA to Protein

The Genetic Code The code responsible for translation of DNA into proteins was identified as a triplet code in 1961. In other words, a series of three nucleotide bases, called a **codon**, encodes a specific amino acid. Thus, a specific sequence of nucleotides (the genetic code) translates into a specific chain of amino acids (DeSalle and Yudell 2005). This specific chain of amino acids is a protein. Proteins have various functions, including serving as hormones, enzymes, receptors, transporters, cell-signaling agents (transcription factors, etc.), and antibodies. DNA also contains noncoding regulatory sequences called **promoter regions** to which molecules can bind in order to signal unwinding of a specific region of DNA for creation of a needed protein. Furthermore, over 95% of DNA in humans is made up of **intervening sequences**, which lie in between expressed

FIGURE 11.4 Chromosome 4
Sequencing and analysis of human chromosomes have enabled researchers to characterize in detail a number of genes associated with diseases. Identifying the genes on all human chromosomes offers scientists worldwide an invaluable resource for improving human health and combating disease. Knowledge about genes will increase understanding of how genetics influences the development of disease, help researchers find genes associated with particular diseases, and aid in the identification of appropriate dietary interventions and development of new pharmaceuticals. Chromosome 4 (pictured below) contains 203 million bases and is one of the larger human chromosomes. Among the many disease genes it contains is the gene for Huntington's Disease, a rare single-gene disorder.

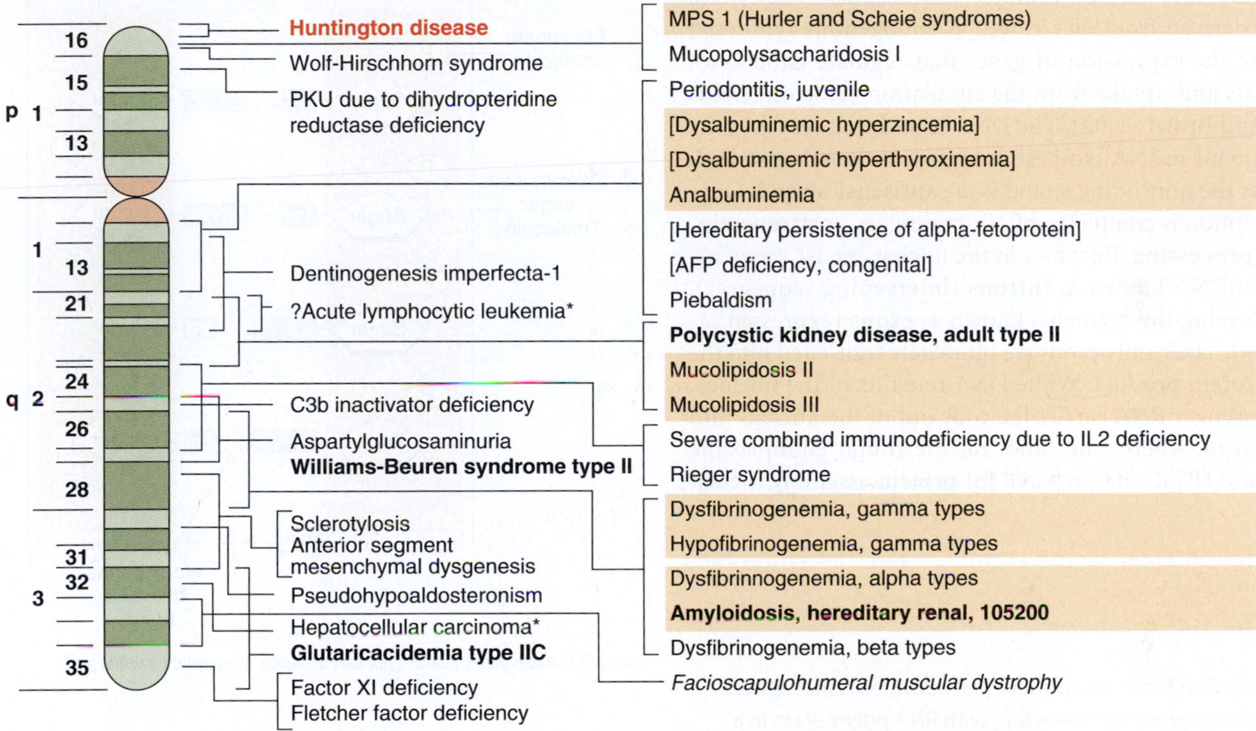

Source: U.S. Department of Energy Human Genome Program, Oak Ridge National Laboratories, Oak Ridge, Tennessee. http://www.ornl.gov/hgmis

genes and whose function is largely unknown (DeSalle and Yudell 2005).

Intervening Sequences Due to their apparent lack of purpose, intervening sequences were initially classified as "junk DNA" (Lewin 2004; DeSalle and Yudell 2005). This view was recently challenged when functions of intervening sequences were initially identified. For example, *SRG1* in yeast is an intervening sequence transcribed into noncoding ribonucleic acid (RNA, an intermediate step in the making of a protein), meaning that the RNA does not undergo translation into a protein. Instead, the *SRG1* RNA itself blocks the adjacent *SER3* gene, thereby preventing synthesis of an enzyme involved in synthesis of the amino acid serine (Martens, Laprade, and Winston 2004). Thus, the sizeable percentage of DNA made up of intervening sequences may play a largely regulatory role. Table 11.1 shows how each of the 64 codons translates into its respective amino acid. Note that in place of the "T" base there is a "U" for uracil, which takes the place of thymine in RNA during the process of creating a protein (Lewin 2004).

Transcription and Translation Progression from code to protein involves two major steps: **transcription** and **translation** (Lewin 2004). In transcription, DNA unwinds in the area encoding the gene of interest. The code is then transcribed (copied) by means of complementary base pairing (see Figure 11.6) into messenger RNA (mRNA), a single-stranded molecule consisting of the bases U, C, A, and G. mRNA is the medium by which the code for a needed protein is carried from the DNA to the cytosol, where the

transcription—the manufacture of RNA from DNA

translation—the assembly of a polypeptide chain based on the sequence of mRNA

new protein is created. Transcription is accomplished by the enzyme RNA polymerase, which first complexes with **transcription factors** in a gene's promoter region before facilitating production of mRNA. Binding of transcription factors can either prevent RNA polymerase from binding, thus repressing transcription of a specific gene, or enhance RNA polymerase binding, thereby increasing transcription of that specific gene. Transcription is very tightly regulated and is dependent in part upon environmental variables such as dietary factors. For example, intracellular cholesterol levels (derived from diet as well as endogenous synthesis) regulate the expression of genes that regulate cholesterol synthesis and uptake from the circulation (Horton, Goldstein, and Brown 2002). The DNA strand that serves as the template for mRNA synthesis is known as the **sense strand**, whereas the noncoding strand is the **antisense strand**. Once transcription is complete, mRNA undergoes **posttranscriptional processing**. Enzymes in the nucleus excise segments of the mRNA known as **introns** (intervening sequences), while leaving the segments known as **exons** (expressed sequences). Thus, only exons are ultimately translated into the final protein product. While DNA remains in the nucleus, the messenger RNA carries the code out of the nucleus into the cytosol, where ribosomes on the rough endoplasmic reticulum (RER) are prepared for protein assembly (Lewin 2004).

FIGURE 11.5 The four nitrogenous bases of DNA are arranged along the sugar-phosphate backbone in a particular order, encoding all genetic instructions for an organism. Adenine (A) pairs with thymine (T), while cytosine (C) pairs with guanine (G). The two DNA strands are held together by weak bonds between the bases.

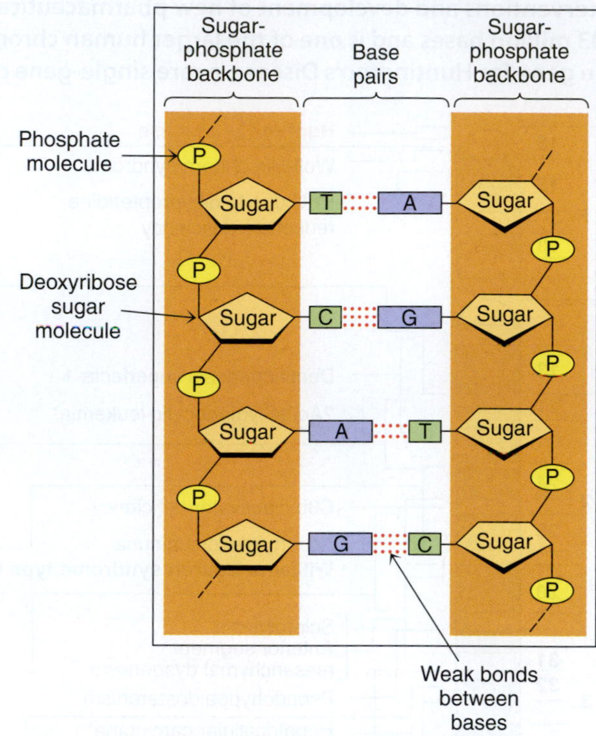

Source: R. Rhoades and R. Pflanzer, *Human Physiology,* 4e, copyright © 2003 p. 57

transcription factor—a protein that activates transcription of a gene or genes by interacting with RNA polymerase in a gene promoter region

sense strand—the coding strand of DNA that is transcribed into RNA

antisense strand—the noncoding strand of DNA

posttranscriptional processing—the processing of newly transcribed RNA to excise introns, thus creating the final mRNA product prior to translation of mRNA into a protein

introns—intervening sequences in mRNA that are enzymatically excised during posttranscriptional processing prior to translation into a protein

exons—expressed sequences in mRNA; sequences that are translated into the final protein product

anticodons—tRNA coding sequences; these sequences are complementary to the codons in mRNA and thus serve as anticodons

stop codon (nonsense codon)—the codon in mRNA that signals completion of translation

posttranslational modification—modification of a newly synthesized protein to its active form through changes such as phosphorylation or cleavage of specific sections

During translation, the triplet codons come into play. As shown in Table 11.1, most amino acids have multiple codons, but each codon only encodes one specific amino acid (Lewin 2004). Small molecules of another form of RNA, transfer RNA (tRNA), serve as **anticodons**. The tRNA molecules each consist of a three-base sequence that is complementary to the codons found in mRNA. After the corresponding amino acids are transferred to the appropriate tRNA, the tRNAs carry each amino acid to the ribosomes, which serve as the protein-making machinery in the cell, and attach to the mRNA via complementary base pairing (A with U and G with C) (see Figure 11.6). After the amino acids are positioned in sequence, peptide bonds are formed between adjacent amino acids and the new protein elongates until a **stop (nonsense) codon** is reached and the newly created protein is released. Additional processing of new proteins is called **posttranslational modification** (Lewin 2004). For example, the insulin polypeptide folds and forms two disulfide bonds, after which it is cut twice in the middle to remove a center section. What remains is two polypeptide chains connected by two disulfide bonds—the active form of insulin.

TABLE 11.1

The Triplet Code

	U		C		A		G	
	UUU	Phe	UCU	Ser	UAU	Tyr	UGU	Cys
U	UUC	Phe	UCC	Ser	UAC	Tyr	UGC	Cys
	UUA	Leu	UCA	Ser	UAA	STOP	UGA	STOP
	UUG	Leu	UCG	Ser	UAG	STOP	UGG	Try
	CUU	Leu	CCU	Pro	CAU	His	CGU	Arg
G	CUC	Leu	CCC	Pro	CAC	His	CGC	Arg
	CUA	Leu	CCA	Pro	CAA	Gln	CGA	Arg
	CUG	Leu	CCG	Pro	CAG	Gln	CGG	Arg
	AUU	Ile	ACU	Thr	AAU	Asn	AGU	Ser
A	AUC	Ile	ACC	Thr	AAC	Asn	AGC	Ser
	AUA	Ile	ACA	Thr	AAA	Lys	AGA	Arg
	AUG	Met	ACG	Thr	AAG	Lys	AGG	Arg
	GUU	Val	GCU	Ala	GAU	Asp	GGU	Gly
G	GUC	Val	GCC	Ala	GAC	Asp	GGC	Gly
	GUA	Val	GCA	Ala	GAA	Glu	GGA	Gly
	GUG	Val	GCG	Ala	GAG	Glu	GGG	Gly

The left-hand column represents the first nucleotide base for each codon in the row, while the row across the top of the table represents the second nucleotide base in each codon. All 64 triplet codons have meaning, with 61 of them encoding amino acids and 3 serving as STOP codons to signal the end of a coding sequence. Methionine (Met) is always the first amino acid in a protein, and its codon, therefore, serves as a START codon. Standard amino acid abbreviations are as follows:

- Ala, alanine
- Asn, asparagine
- Asp, aspartic acid
- Arg, arginine
- Cys, cysteine
- Gln, glutamine
- Glu, glutamic acid
- Gly, glycine
- His, histidine
- Ile, isoleucine
- Leu, leucine
- Lys, lysine
- Met, methionine
- Phe, phenylalanine
- Pro, proline
- Ser, serine
- Thr, threonine
- Try, tryptophan
- Tyr, tyrosine
- Val, valine

Genetic Variation

Polymorphisms (variations) exist within genes throughout the population (DeSalle and Yudell 2005). Most of these variations are not a cause for concern. The outcome of a given variation depends on its nature and location within a given gene. In other words, a specific variation may have no appreciable effect on the production and function of the protein product. However, it is also possible that a single nucleotide change or a more complex alteration in a single gene can have profound effects.

Inheritance Inheritance of specific genes can be classified as **autosomal dominant, autosomal recessive, X-linked dominant, X-linked recessive,** or **Y-linked** (Lewin 2004). Because individuals inherit one copy of each gene from each of their

polymorphisms—DNA sequences of specific genes that vary among individuals

autosomal dominant—an inheritance pattern of a dominant allele on an autosome

autosomal recessive—an inheritance pattern of a recessive allele on an autosome

X-linked dominant—an inheritance pattern of a dominant allele on the X chromosome; such disorders are relatively rare

X-linked recessive—an inheritance pattern of a recessive allele on the X chromosome; related disorders are more common in males, who carry only one X chromosome

Y-linked—inheritance based on the Y chromosome; disorders are extremely rare and occur only in males

FIGURE 11.6 Transcription and Translation
When genes are expressed, the genetic information (base sequence) on DNA is first transcribed (copied) to a molecule of messenger RNA in a process similar to DNA replication. The mRNA molecules then leave the cell nucleus and enter the cytoplasm, where triplets of mRNA bases (codons) forming the genetic code specify the particular amino acids that make up an individual protein. This process, called translation, is accomplished by ribosomes (cellular components composed of proteins and another class of RNA) that read the genetic code from the mRNA, and by transfer RNAs (tRNAs) that transport the corresponding amino acids to the ribosomes for attachment to the growing protein.

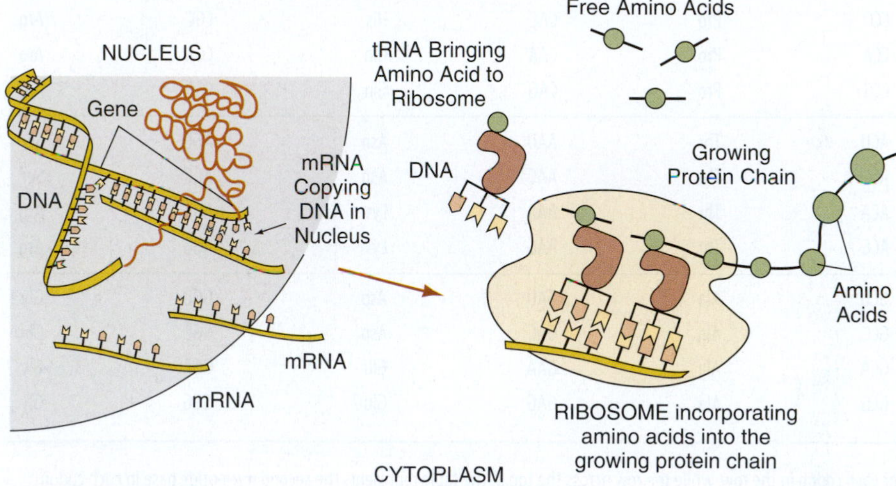

Source: U.S. Department of Energy Human Genome Program, Oak Ridge National Laboratories, Oak Ridge, Tennessee. http://www.ornl.gov/hgmis

Autosomal recessive or dominant traits can be inherited by both males and females. One common example of an autosomal recessive trait with nutritional implications is phenylketonuria, in which affected individuals must inherit one mutated copy of the phenylalanine hydroxylase gene from each parent (homozygous). The resulting inability to convert phenylalanine to tyrosine requires lifelong phenylalanine restriction to prevent mental retardation (see Chapter 28). Cystic fibrosis is another common autosomal recessive disease (see Chapter 18 and Chapter 23). Familial hypercholesterolemia is an autosomal dominant disorder characterized by absence or mutation of LDL receptors leading to severely elevated LDL-cholesterol levels and risk of early myocardial infarction and death (see Chapter 15). Familial hypercholesterolemia homozygotes are rare and have a much more severe manifestation of the disorder than do heterozygotes.

parents, the actual expression of an inherited gene can vary and gene expression is what determines **phenotype**. For example, brown eyes are autosomal dominant whereas blue eyes are autosomal recessive. If an individual inherits the gene for brown eyes from one parent and the gene for blue eyes from the other, his or her eyes will be brown because that is the dominant gene. While the genotype includes genes for both blue and brown eyes, the eye color phenotype is brown. Thus, whether a trait is recessive or dominant determines whether that trait is phenotypically expressed. When the alleles from each parent differ from each other, as in this case, an individual is **heterozygous** for that gene (*hetero* = different). If the alleles from both parents are a match, then that individual is **homozygous** for that gene (*homo* = same).

phenotype—the expressed or physical properties of an organism

heterozygous—having two different alleles or variants of a given gene

homozygous—having two identical alleles or variants of a given gene

The sex chromosomes also contain genes that can result in recessive or dominant disorders (Lewin 2004). Two examples of X-linked recessive disorders are red-green colorblindness and hemophilia. In red-green colorblindness, individuals are unable to distinguish shades of red and green in the color spectrum, whereas in hemophilia, individuals most commonly lack clotting factor VIII, so their blood does not clot normally. Hemophilia requires transfusions to supply the clotting factor and for replacement of blood losses (see Chapter 21). Because these X-linked disorders are recessive disorders, individuals require only one normal copy of the gene for normal function. However, because males have only one X chromosome and, thus, only one copy of this gene, they are much more susceptible to inheriting these disorders. Occurrence in females is rare because they would need to inherit a defective copy of the gene from both the mother *and* father, who would himself have the disorder. Male offspring of affected fathers will not inherit the disorder because only a Y chromosome is inherited from the father. However, female offspring of affected fathers are carriers (heterozygotes), and male children born to them have a 50% chance of having the disorder, depending wholly on which copy of the maternal X chromosome is passed on. X-linked dominant traits are relatively rare. Y-linked disorders are extremely rare, occurring only in males as a result of the inheritance of mutations in the Y chromosome from the father. Y-linked disorders are not considered dominant or

FIGURE 11.7 DNA Sequence Variation in a Gene
Specific codons direct the cell's protein-synthesizing machinery to add specific amino acids. For example, the base sequence ATG codes for the amino acid methionine. Since 3 bases code for 1 amino acid, the protein coded by an average-sized gene (3000 bp) will contain 1000 amino acids. The DNA code is thus a series of codons that specify which amino acids are required to make up specific proteins. Some variations in a person's genetic code will have no effect on the protein that is produced; others can lead to disease or an increased susceptibility to disease.

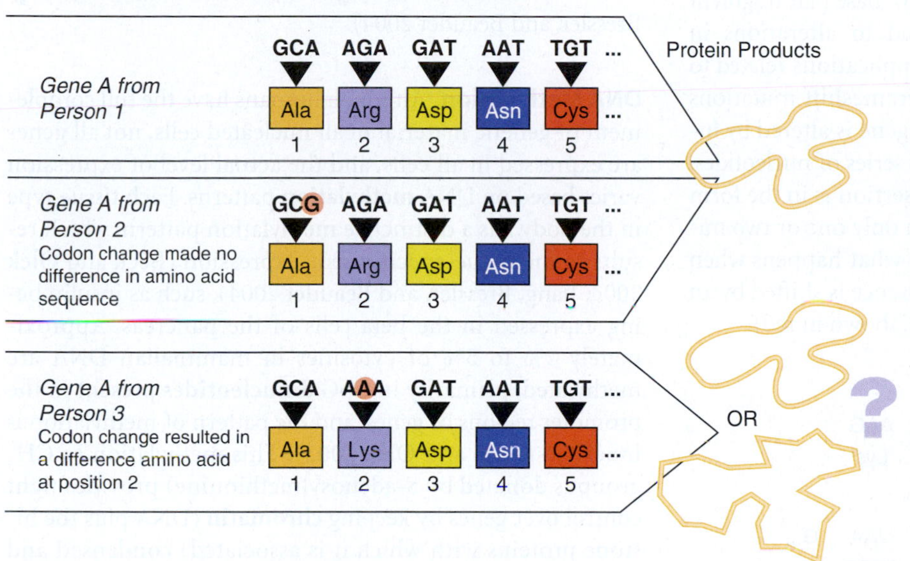

Source: U.S. Department of Energy Human Genome Program, Oak Ridge National Laboratories, Oak Ridge, Tennessee. http://www.ornl.gov/hgmis

recessive because only one copy of the affected chromosome can exist in an individual.

Single Nucleotide Polymorphisms Understanding **monogenic** disorders such as those described above helps lay the groundwork for comprehending the complexities of **polygenic** diseases such as obesity, diabetes, cancer, and cardiovascular disease (Ordovas and Corella 2004). The study of gene-nutrient interactions that are dependent upon gene variance is focused primarily on **single nucleotide polymorphisms** (SNPs, pronounced "snips") (Dennis and Gallagher 2001). SNPs are defined as those genetic variants or polymorphisms in which a single nucleotide has been exchanged for another. For example, the codon UGU is mutated to UGC. Because both encode the amino acid cytosine, this particular SNP results in no alteration in function. However, if UGU is mutated to UGA, that is a potential problem, because UGA is a nonsense or *stop* codon (see example of a SNP in Figure 11.7). Depending on the location of the mutation within a gene, such a mutation could have deleterious effects. If the affected codon is near the end of a coding sequence, it is possible that the final protein product will not be functionally altered. However, if UGU is mutated to UGG, then altering that one amino acid from cytosine to tryptophan has the potential for altering the shape and function of the protein product.

SNPs are generally identified by the gene name, the location of the affected nucleotide within the gene sequence, the common nucleotide in that position, and an arrow indicating that a less common nucleotide is present. For example, *MTHFR* 667C→T (ala→val) indicates that there is a SNP at nucleotide number 667 in the methylene tetrahydrofolate reductase gene characterized by a thymine in place of the more common cytosine. This may also be signified by *MTHFR* 667C>T. As shown in parentheses, this SNP results in an amino acid change in that position from the typical alanine to the less typical valine. This particular SNP has implications for folate metabolism and cancer risk, as discussed later in this chapter. SNPs may also be defined by the amino acid change; for instance, *PPARA*-L162V indicates the 162nd amino acid in the protein sequence for peroxisome proliferator activated receptor-α is a valine (V) when the typical amino acid in this position is a leucine (L).

Identification of SNPs has been a primary focus of genomics research since completion of the Human Genome Project in 2003. In late 2005, it was reported that the human genome has approximately 10 million polymorphisms, meaning that any two unrelated humans have millions of genetic differences (International HapMap Consortium 2005). These polymorphisms are not all independent of each other. Rather, when a specific gene variant is present on a chromosome, it is associated with other particular gene variants on that same chromosome. This group of gene variants that associate together is referred to as a **haplotype**, and these variants may

monogenic—arising from a single gene

polygenic—arising from multiple genes interacting with each other

single nucleotide polymorphisms (SNPs)—situations in which one nucleotide is replaced by another in a gene, potentially leading to altered function

haplotype—a group of gene variants that occur together

work in concert to produce a specific phenotype. The next step in genomics research will be to determine which of these millions of polymorphisms is likely to be functionally important, and to continue to identify how each might relate to environment and health (Goldstein and Cavalleri 2005).

Other Polymorphisms Other types of polymorphisms include insertion or deletion polymorphisms, in which a number of nucleotide base pairs are either added to or deleted from a gene. For example, the angiotensin-converting enzyme (ACE) gene has an insertion/deletion polymorphism characterized by the presence or absence of a 287-base pair fragment in one of its introns, which is linked to alterations in circulating levels of ACE and risk of complications related to type 2 diabetes (Kajantie et al. 2004). Frameshift mutations can occur when the reading frame of a gene is altered by inserting or deleting a single nucleotide or series of nucleotides. These tend to have less impact if the insertion is in the form of a triplet, but can be devastating when only one or two nucleotides are inserted. For example, see what happens when the reading frame for the following sequence is shifted by an insertion of a single nucleotide (adenine, shown in red):

... CUU	AUG	UUA	CGU	AAG ...	
Leu	Met	Leu	Arg	Lys	

... CUU	AAU	GUU	ACG	UAA	G...
Leu	Asn	Val	Thr	STOP	

Other syndromes can occur as a result of inheriting extra copies of chromosomes, as in Down syndrome, or deletions of sections of chromosomes, which is one cause of the neurological disorder Angelman syndrome.

epigenetics—inheritance of information based on gene expression levels rather than gene sequence; regulated by genomic modifications such as DNA methylation and histone acetylation

methylation—the addition of methyl ($-CH_3$) groups; DNA methylation patterns can be inherited and impact patterns of gene expression

dinucleotides—paired nucleotide sequences

chromatin—the entire complement of DNA plus the histone proteins with which it is associated

genetic imprinting—expression of specific genes, which depends on the parent of origin; some genes are expressed only from the maternal allele and others are expressed only from the paternal allele

Epigenetic Regulation

Epigenetics relates not to the genome sequence, but to the pattern of gene expression regulated by modifications to DNA (Gallou-Kabani and Junien 2005, Oommen et al. 2005). Gene expression is regulated in many ways, including DNA methylation, histone methylation, acetylation, or phosphorylation, and transcription factors (Gallou-Kabani and Junien 2005), all of which can be influenced by early programming in response to nutrition in fetal life or infancy as well as throughout the life span. Epigenetic patterns may also be passed from one generation to the next (Jiang, Bressler, and Beaudet 2004).

DNA Methylation Although humans have the full complement of genetic material in all nucleated cells, not all genes are expressed in all cells, and the actual level of expression varies based on DNA **methylation** patterns. Each tissue type in the body has a distinctive methylation pattern, which results in the tissue-specific gene expression (Beck and Olek 2003; Jiang, Bressler, and Beaudet 2004), such as insulin being expressed in the beta cells of the pancreas. Approximately 2% to 5% of cytosines in mammalian DNA are methylated, primarily in CpG **dinucleotides** present in the promoter regions of genes, and the pattern of methylation is inherited (Beck and Olek 2003). This methylation (a CH_3 group is donated by *S*-adenosylmethionine) provides tight control over genes by keeping **chromatin** (DNA plus the histone proteins with which it is associated) condensed and thereby suppressing gene expression or keeping the genes "silenced" (McCabe and Caudill 2005; Oommen et al. 2005). For most genes, both maternal and paternal alleles contribute to production of the protein product, but for others, **genomic imprinting** takes place. In other words, for specific genes, only the maternal or paternal allele is expressed. For example, the gene encoding insulin-like growth factor II is expressed only from the paternal allele, while the maternal allele in mammals is silenced (Beck and Olek 2003). Imprinting errors can result in devastating outcomes in offspring, including the neurological disorders Angelman syndrome and Prader-Willi syndrome (Beck and Olek 2003).

Methyl groups are derived in the diet from sources including folate, choline, methionine, and vitamin B_{12} (McCabe and Caudill 2005). As shown in Figure 11.8, MTHFR catalyzes conversion of 5,10-methylenetetrahydrofolate to 5-methyltetrahydrofolate, which then donates its methyl group to vitamin B_{12}. Vitamin B_{12}, thus activated, then methylates homocysteine in order to form methionine. Alternatively, choline can be converted to betaine, which can also methylate homocysteine to form methionine. Methionine adenosyl transferase then unites methionine with adenosine to form *S*-adenosylmethionine (SAM), which methylates DNA via the action of DNA methyltransferases (McCabe and Caudill 2005). Thus, dietary adequacy plays a role in maintaining appropriate DNA methylation (McCabe and Caudill 2005;

FIGURE 11.8 The Resynthesis of Methionine from Homocysteine, Showing the Roles of Folate and Vitamin B_{12}

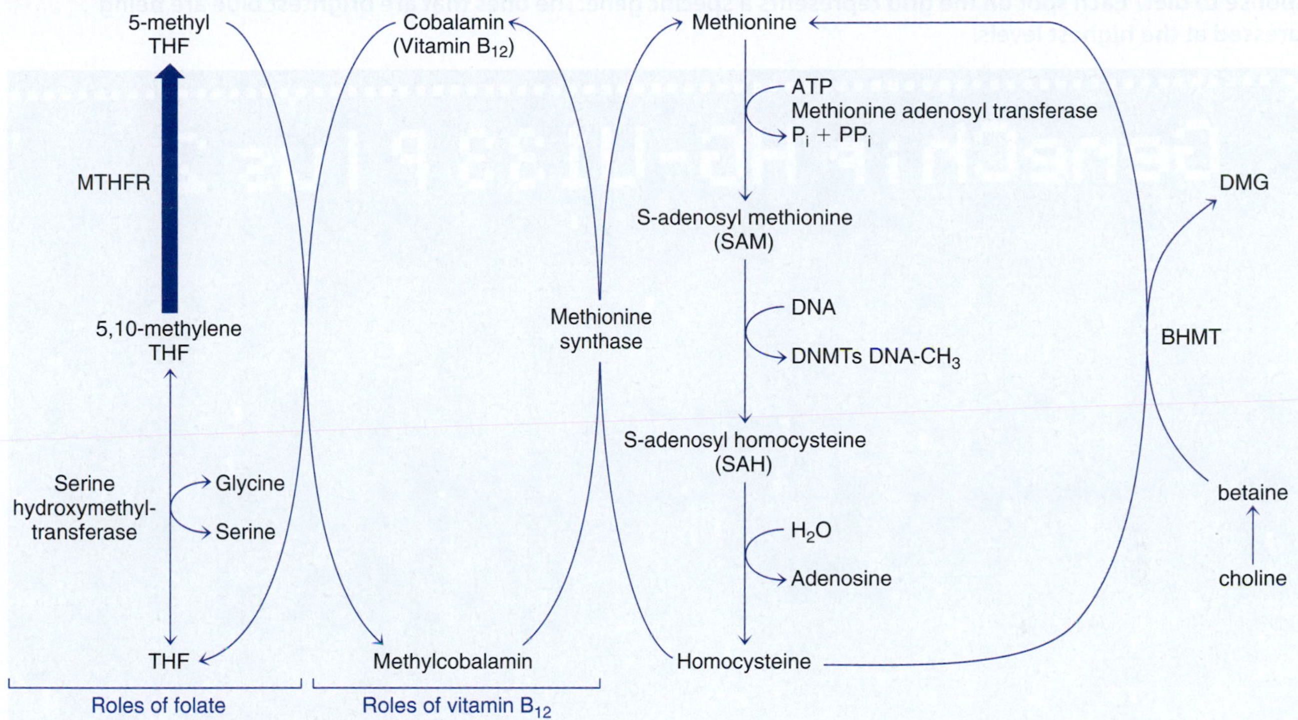

DMG: dimethylglycine
BHMT: betaine homocysteine methyltransferase
MTHFR: methylene tetrahydrofolate reductase
DNMTs: DNA methyltransferases

Source: J. Smith, J. Groff, and S. Gropper, *Advanced Nutrition and Human Metabolism*, 4e, copyright © 2005, p. 305

Oommen et al. 2005). A deficiency of methyl groups related to lack of the above-listed nutrients means that as cells divide, methylation may be reduced, and some of that transcriptional regulation is lost. Impaired methylation of DNA is related strongly to impaired fetal development and cancer (Oommen et al. 2005). For example, hypomethylation of DNA is related to chromosomal instability, including gain or loss of entire chromosomes or increased gene mutation rates during mitosis, both of which can contribute to cancer (McCabe and Caudill 2005).

Histone Modification In addition to DNA methylation, **histone** modification is another form of epigenetic regulation (Beck and Olek 2003). Histones are small proteins around which DNA is wrapped. The histone tail can be modified by methylation, acetylation, phosphorylation, ubiquitination, biotinylation, and so forth, which helps to regulate transcription, DNA repair, apoptosis (programmed cell death), mitosis, and meiosis (McCabe and Caudill 2005; Oommen et al. 2005). Histone modifications work in concert with DNA methylation to determine shape and accessibility of chromatin for transcription. For example, enzymes called histone acetyltransferases attach acetyl groups to histones, and this acetylation is associated with unfolding and accessibility of chromatin for transcription, whereas histone deacetylases, which remove acetyl groups, promote folding of chromatin and block gene transcription (Jiang, Bressler, and Beaudet 2004).

The Epigenotype Because the epigenotype (an individual's unique pattern of DNA methylation and histone modification) displays greater variability than the genotype, it may be more responsive to environmental influences (Jiang, Bressler, and Beaudet 2004). The roles of dietary folic acid, vitamin B_{12}, choline, and methionine are of particular interest, since these are primary sources of methyl groups, and dietary adequacy may influence DNA methylation patterns and thus genomic stability and gene expression. Choline and methionine deficiencies are unlikely, but folic acid and vitamin B_{12} adequacy is of concern (McCabe and Caudill 2005).

histone—a protein around which DNA is wrapped

FIGURE 11.9 A cDNA microarray can be used to determine how the expression of specific genes changes in response to diet. Each spot on the grid represents a specific gene. The ones that are brightest blue are being expressed at the highest levels.

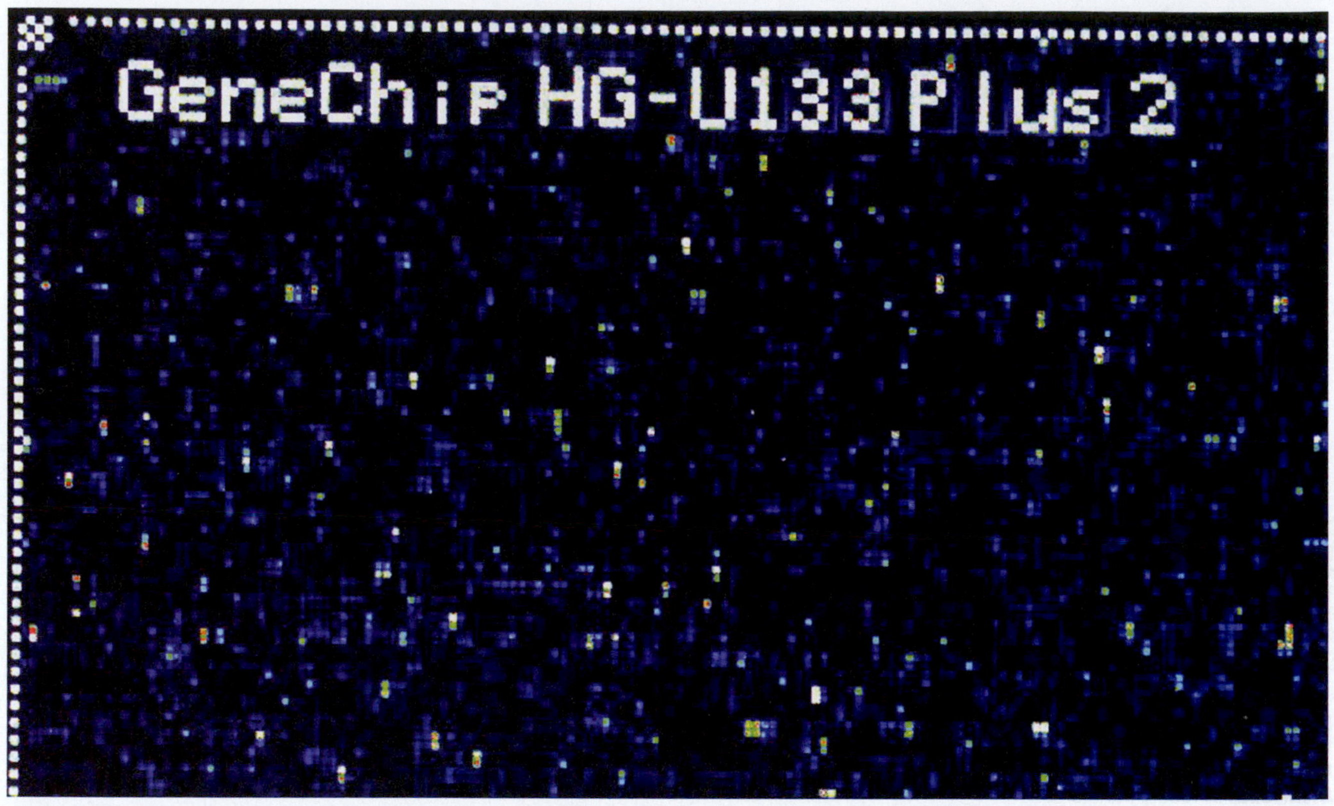

Source: Courtesy of Julia Stair Gouffon, Affymetrix Core Facility, The University of Tennessee–Knoxville.

Dietary Regulation and Measurement of Gene Expression

Individual SNPs may not be the best measure of genotype. Rather, looking at the totality of a gene, including all SNPs in coding, noncoding, and regulatory regions, may be more appropriate as a measure of gene function in combination with epigenetic modifications (Beck and Olek 2003; Syvanen 2005). Additionally, regardless of individual genotypes, environmental factors play a large role in regulating **gene expression**. In other words, diet, activity, smoking, and so forth, can turn specific genes on or off and thus determine the quantity of specific protein products produced as well as the activity of related metabolic pathways. Thus, it is important to understand when and where a gene is expressed as well as the circumstances that influence its expression level.

gene expression—the level of activity of a specific gene in producing mRNA and, subsequently, protein; expression can be regulated by many variables, including diet

xenobiotics—chemicals that are found in an organism but are not produced by it or expected to be there, such as drugs or pollutants

Dietary Components Influence Gene Expression

Examples of dietary components influencing gene expression levels continue to multiply. One of the most illustrative and well-studied examples involves polyunsaturated fatty acids (PUFAs) and their derivatives, which are ligands for (i.e., they bind to and activate) the peroxisome proliferator-activated receptors (PPARs). PPARs (α, γ, and δ) are nuclear receptors, meaning they reside in the nucleus. When activated by binding of a ligand such as a PUFA or PUFA-derived eicosanoid, they respond by altering expression levels of genes involved in lipid metabolism. Effects include adipocyte differentiation, increased fatty acid catabolism and β-oxidation, lower serum triglyceride levels, and improved insulin sensitivity. Recent evidence suggests that soy isoflavones may mediate their effects on improved lipid metabolism, that is, lower total cholesterol, LDL cholesterol, and triglycerides, also by serving as PPAR ligands (Ricketts et al. 2005). Another example of a dietary component that influences gene expression is sulforaphane, a phytochemical found in substantial quantities in broccoli. Human colon cancer cells exposed to sulforaphane *in vitro* demonstrate induction of genes involved in **xenobiotic** metabolism, inhibition of angiogenesis, and inhibition of the cell cycle and, thus, inhibition of cellular proliferation (Traka et al. 2005). All of these actions could contribute to potential anticarcinogenic properties of broccoli simply by means of altering gene expression. It appears this induction may be due in part to induction of a transcription factor by sulforaphane, which then is involved in upregulation of numerous additional genes (Traka et al. 2005).

Measuring Gene Expression Until recently, it was a laborious process to conduct experiments to determine the effects of diet on alterations in expression of a single gene. However, the recent advent of **microarray** technology permits large-scale exploration of the effects of diet on the expression of thousands of genes simultaneously (Davis and Milner 2004). While the entire complement of DNA is pres-ent in all nucleated cells, and genotyping can be done on any sample containing such cells, the sample used must come from the tissue of interest, because not all genes are expressed in all tissues. For example, determining the effects of diet on expression of lipogenic genes (those involved in synthesizing fat in the body, such as fatty acid synthase) would be best accomplished by analyzing liver tissue, because the liver is where lipogenesis occurs. Because tissue biopsy is not realistic in human research, much of the gene expression information is derived from animal and cell research following exposure to various experimental diets. Figure 11.9 shows an example of a microarray gene chip. mRNA is isolated from tissue samples, and the quantity of mRNA for a specific gene provides information about how highly that gene is being expressed at the time the sample was collected. Each blue spot on the microarray gene chip represents a single gene, and the brighter the color, the greater the expression of that particular gene. Animals on different experimental diets often show differing expression levels of multiple genes (Davis and Milner 2004). Further analysis is required to follow up and confirm altered expression of specific genes of interest once identified via microarray (Chuaqui et al. 2002). This new technology is invaluable in determining specific impacts of dietary components or dietary patterns on large-scale gene activity.

Nutrigenomics in Disease

Cancer

From Single Gene Inherited Cancers to Gene-Nutrient Interactions Several well-defined and relatively rare cancers have a clearly established genetic inheritance based on mutations in a single gene. An inherited mutation in the adenomatous polyposis coli (*APC*) tumor suppressor gene, for example, carries a 100% risk of developing the disease familial adenomatous polyposis (FAP) (Ficari et al. 2000). The *APC* mutation causes FAP because it encodes a truncated, or shortened, and therefore dysfunctional, protein product that is unable to act as a tumor suppressor. FAP is characterized by development of thousands of tumors, primarily in the gastrointestinal tract, and requires intensive treatment (Ficari et al. 2000). Remarkably, this well-defined path to intestinal tumorigenesis can be thwarted to some extent by dietary means. For example, mice bearing an inherited *APC* mutation develop 50% fewer tumors when consuming a diet supplemented with the long-chain omega-3 polyunsaturated fats stearidonic acid (SDA, 18:3 n-3) or eicosapentaenoic acid (EPA, 20:5 n-3) (Hansen-Petrik et al. 2000). Fortunately, strictly inherited mutations are rare, although they are devastating to those affected. Heretofore, all other "noninherited" cancers have been attributed primarily to environmental exposures including diet, physical activity, alcohol intake, and tobacco use (Le Marchand 2005). These links between cancer and environmental exposure have formed the basis for public health initiatives and education (e.g., American Institute for Cancer Research, American Cancer Society), although it has not been implicitly clear who will benefit the most from the broad recommendations presented in these initiatives and educational efforts.

Predicting benefit is difficult because people do not all respond in the same way to environmental exposures (Le Marchand 2005). Individuals have consequently been classified as responders or nonresponders to a specific treatment. For example, not all who smoke develop lung cancer. Not all who eat red meat, which has long been associated with higher rates of cancer in epidemiological studies, develop cancer. One important variant has been the lack of knowledge as to each individual's genetic background and how that may interact with nutrients or nonnutritive substances in food (Nowell, Ahn, and Ambrosone 2004). Nutrients have the potential to alter carcinogen metabolism, hormonal status, cell signaling, apoptosis, cell-cycle control, angiogenesis, or a combination thereof. Therefore, current research strives to identify less penetrating polymorphisms or groups of polymorphisms in a single metabolic pathway that may interact with environmental variables such as diet to increase or decrease risk of various disease states, including cancer (Nowell, Ahn, and Ambrosone 2004). While genotyping itself is a straightforward undertaking, linking individual foods, nutrients, or other bioactive components in foods to an interaction with each common genetic variant remains a daunting task that will take time and perseverance to accomplish.

Variations in Xenobiotic Metabolism Influence Risk A study published by Le Marchand et al. (2001) illustrates the complexities of gene-environment interactions. *N*-acetyl

microarray—technology used to measure expression of thousands of genes simultaneously

transferase 2 (NAT2) and cytochrome P450 1a2 (CYP1A2) enzymes in the liver are both involved in biotransformation of incoming xenobiotics into harmless substances for excretion. Individuals exhibiting different phenotypes of these enzymes metabolize xenobiotics at different rates and are thus often classified as slow, intermediate, or rapid acetylators (Le Marchand et al. 2001; Nowell, Ahn, and Ambrosone 2004). This phenotypic variation has potential implications for cancer risk, because such enzymes transform some xenobiotics into genotoxic substances. For example, both NAT2 and CYP1A2 are integral to the biotransformation of heterocyclic amines from cooked meat into genotoxic substances, which by definition have the potential to cause cancer. Furthermore, smoking is known to induce CYP1A2—that is, it increases production of the CYP1A2 enzyme. Epidemiological research has long linked cooked meats to increased risk of colon cancer, and the authors in this study examined how that risk is modified by phenotypic variation in these two enzymes. The only group experiencing a statistically higher risk were those with a rapid NAT2 phenotype combined with an above average CYP1A2 phenotype who were also smokers and consumed their red meat well done. This finding clearly illustrates the oversimplification associated with stating that "eating red meat increases colon cancer risk" when that appears to be true only for a small, well-defined subset of the population and only when the meat is well done. It also illustrates the growing necessity of individualizing dietary recommendations based on specific genomic and environmental variables as research findings begin to more clearly establish such relationships.

***MTHFR* and *ADH* Polymorphisms Interact with Dietary Folate and Alcohol** Low intakes of folate have a long association with cancer risk, including cancer of the colon, and this risk appears to escalate in the presence of high alcohol intake. However, results are not always consistent, suggesting individual effects may vary based on genomic characteristics. Polymorphisms in the *methylenetetrahydrofolate reductase* gene, primarily [MTHFR 667C→T (ala→val)], can reduce MTHFR activity (Beck and Olek 2003). As mentioned previously, MTHFR plays a critical role in metabolizing 5,10-methylenetetrahydrofolate (5,10-methylene THF) to 5-methyl-tetrahydrofolate (5-methyl THF), which is necessary for remethylation of homocysteine to methionine, formation of S-adenosylmethionine, and, therefore, DNA methylation. Because 5,10-methylene THF itself is also necessary for production of thymine, both forms of folate are needed in adequate quantities for genome health. Folate deficiency and reduced activity of MTHFR can thus both contribute to compromised genome integrity and the risk of acquiring genetic damage and cancer (Beck and Olek 2003; McCabe and Caudill 2005).

The interaction between folate status and *MTHFR* polymorphisms in carcinogenesis is illustrated by findings from the Health Professional Follow-Up Study (Giovannucci et al. 2003). That study showed that individuals homozygous for the TT mutation at nucleotide 667 (thus encoding valine from both copies of the gene) tend to accumulate 5,10-methyl THF intracellularly. They appear to be hyperresponders to folate status, meaning they are at low risk for colon cancer if following a low-risk diet (high in folate, low in alcohol), presumably due to accumulation of 5,10-methyl THF and optimal chromosomal stability, but may be at higher risk for developing colon cancer if consuming a low-folate, high-alcohol diet. Folate intakes were relatively high among the study population, and it is possible that more dramatic effects would have been observed with respect to the relationship of folate intake to risk had there been a wider spread in consumption levels. That will likely be difficult to see in U.S.-based studies due to the folate fortification of the food supply in place since 1998.

The interaction of folate status and *MTHFR* genotype with alcohol appears to be a critical point. The same researchers (Giovannucci et al. 2003) also examined the *alcohol dehydrogenase* (*ADH*) genotype of this cohort and found those with a slow metabolizing genotype had significantly higher risk of developing colon cancer with an alcohol intake $\geq$ 20 g/day combined with folate intakes $<$ 338 mcg/day. The low ADH activity could result in slower alcohol metabolism and magnification of alcohol's effects. These effects may include inducing malabsorption of folate, blocking folate release from hepatocytes, and blocking remethylation of 5-methyl THF back to 5,10-dimethyl THF, thereby depleting the latter. In addition, acetylaldehyde, an alcohol metabolite, can cleave and destroy folate. Thus, a slow *ADH* genotype could contribute to colon cancer risk, dependent upon levels of alcohol and folate intake.

Current dietary intake recommendations call for alcohol in moderation (U.S. Department of Health and Human Services 2005). However, even within moderate intake levels, individuals with the slow ADH_3 genotype appear to be at risk (Giovannucci et al. 2003). Another study shows, though, that those with the intermediate ADH_3 genotype benefit from a reduced risk of myocardial infarction with moderate alcohol intake (Hines et al. 2001). Again, this is a clear illustration of the critical relationship between the individual genome and nutrition as well as the complications associated with basing dietary recommendations on a single genotype. Broad dietary recommendations are still likely to be generally accepted, but in the future, as research more clearly elucidates the multitude of gene-gene and gene-nutrient relationships, individualized recommendations will be the key to optimizing health.

Fruits and Vegetables Dietary intake of fruits and vegetables has long been associated with a lower risk of colon cancer (McCullough et al. 2003; Fung et al. 2003), although research results have been contradictory and specific mechanisms have remained somewhat elusive. One complicating factor is that each individual fruit or vegetable is made up of a wide array of nutrients and other bioactive compounds. Many of these have been studied individually, but little research has focused on the potential synergistic effects or interactions of this mix of compounds within a whole vegetable. Furthermore, genomic variation means that individual responses are likely to vary. Van Breda et al. (2005) took the interesting approach of feeding four different whole vegetables (cauliflower, peas, carrots, or onions) to mice and determining how each of these vegetables impacted gene expression in colonic tissue as a clue to their cancer-preventive mechanisms. Expression of several genes known to have either promoting or protective effects with respect to colorectal cancer was altered. For example, mice fed cauliflower or carrots expressed lower levels of the ornithine decarboxylase (ODC) enzyme, which is the rate-limiting step in synthesis of polyamines from the amino acid ornithine (i.e., low ODC levels result in limited polyamine production). Polyamines have a long association with increased risk of colon cancer. ODC is also known to be regulated by the protein product of the *APC* tumor suppressor gene, whose function is lost in many cases of noninherited colon cancer as well as in FAP. Similar effects were observed with high vegetable intake in a parallel human study, along with suppression of genes encoding several cytochrome P450 isozymes (van Breda et al. 2004). Thus, the effects of these vegetables may be protective against colon cancer via interaction with this gene and/or a host of others.

Other evidence of the interaction between vegetables and genes includes the observation that colon cancer risk reduction in humans via intake of cruciferous vegetables is linked to glutathione *S*-transferase genotype (Seow et al. 2002). Isothiocyanates derived from *cruciferae* are known to induce phase II detoxification enzymes, which, like the phase I cytochrome P450 enzymes, are involved in metabolism and removal of potential carcinogens. The glutathione *S*-transferase (GST) family of enzymes is among the most important in this regard. In the Singapore Chinese Health Study (Seow et al. 2002), it was demonstrated that individuals with GST genotypes leading to absence of activity among some of the GST subtypes are the only ones who benefited (lesser risk of colon cancer) from a higher intake of cruciferous vegetables. This is biologically plausible because less active GST would presumably lead to slower clearance of carcinogens and greater cancer risk. A higher intake of cruciferous vegetables would lead to higher isothiocyanate levels and increased activity of remaining GST subtypes to compensate for the loss of others.

Similarly, a study of gene-nutrient interactions in breast cancer causation explored the possible role of antioxidants derived from fruits and vegetables (Ambrosone et al. 1999). The human body produces endogenous antioxidants, including manganese superoxide dismutase (MnSOD), and can also acquire antioxidants exogenously via dietary intake of fruits and vegetables. Since oxidative DNA damage is thought to play a role in carcinogenesis, adequacy of antioxidants is suggested to be anticarcinogenic. Results of a large case-control study found that premenopausal women homozygous for a valine to alanine change at the -9 position (resulting in loss of function) of MnSOD were at increased risk for breast cancer, which was attenuated by high consumption of fruits and vegetables. Presumably, exogenous antioxidant consumption compensated, in part, for the deficiency in endogenous antioxidant function. Although researchers are in the early stages of defining the relationships between genotype, diet, and health, and these are but a few examples, it seems likely that effects will vary among individuals depending on numerous specific genotypes and numerous environmental variables. However, the fact that nutrients and other bioactive compounds in food interact with the genome and thereby impact health and disease, including cancer risk, cannot be disputed.

Obesity and Diabetes

Obesity Obesity has a clear link to genetics demonstrated by numerous studies showing that obesity does persist in families even where food intake and physical activity patterns differ. However, the escalating obesity epidemic in recent decades supports the idea that, while *susceptibility* to obesity is genetically determined, the development of obesity itself is the result of susceptible genetics in the presence of a conducive environment. In other words, placing genetically susceptible individuals in an obesigenic environment—one characterized by plentiful energy-dense, high-fat foods and technological advances requiring little in the way of physical activity—results in obesity. The question then becomes: How do we identify those individuals who are genetically susceptible, and how do we intervene to prevent and/or treat obesity?

The answer is not a simple one. Unlike single-gene inherited traits, obesity susceptibility involves multiple genes and is, therefore, a complex (polygenic) genetic trait. Each gene itself may make only a small contribution to obesity risk, but a multitude of gene variants working together may have a profound effect. These may include genes involved in energy and appetite regulation, metabolism, and storage. In all, over 600 genes, markers, and chromosomal regions have been associated with or linked to human obesity (Perusse et al. 2005). However, at this time information is inadequate to utilize these as markers for screening people or to have an intervention other than current interventions for obesity treatment and prevention. In order for such screening to be useful, the relationship between the gene and obesity must be clearly established and there must be a useful intervention available for those deemed at risk. Research has not yet progressed to that point.

One gene that appears particularly promising as a marker of obesity risk is the gene encoding for the protein perilipin. Perilipins are localized on the surface of fat droplets inside adipocytes and play a regulatory role, primarily by blocking release of stored triglycerides and thereby helping to preserve stored fat (Mottagui-Tabar et al. 2003). Mice lacking perilipin are resistant to diet-induced obesity (Tansey et al. 2001). Furthermore, obese humans have higher perilipin expression (Kern et al. 2004), and variations in the perilipin gene are predictive of obesity risk, particularly among women (Qi et al. 2004). A recent study found that perilipin polymorphism 11482G>A was associated with a lower baseline body weight (234 versus 251 pounds among obese subjects enrolled in the study) and the researchers' findings further suggest that this SNP confers resistance to weight loss while following a reduced-kilocalorie diet for one year (Corella et al. 2005). Such findings have important implications if specific genotypes are definitively proven predictive of weight loss. This was a relatively small study and will require confirmation. Moreover, it is likely there are numerous variables that predict weight loss success, and one must consider the ethics of advising a client to refrain from attempting weight reduction due to a prediction of failure.

Another interesting polymorphism relating to obesity is one occurring in the serotonin (5-hydroxytryptamine or 5-HT) receptor gene promoter. The neurotransmitter serotonin is a key regulator of food intake whose function has been related to obesity and anorexia. A study of 370 children and adolescents ages 10 to 20 suggests that the -1438G>A polymorphism in the 5-HT$_{2A}$ gene promoter does indeed impact food intake (Herbeth et al. 2005). While there was no difference among study subjects in way of age, height, weight, or BMI, there was a significantly higher intake of energy and fat in children with two *G* alleles compared to those with two *A* alleles, and an intermediate effect for those with one *G* and one *A* allele. While the differences in fat and energy intake were not linked to overweight in these children, another study of the same polymorphism in middle-aged men observed significantly higher BMIs and abdominal fat associated with the *GG* genotype (Rosmond, Bouchard, and Bjorntorp 2002). These findings await confirmation, because it is possible these gene polymorphisms coexist with polymorphisms in other genes that predict eating behavior and obesity. It is also possible that 5-HT$_{2A}$ receptor promoter polymorphisms result in mood or personality characteristics that impact food intake (Herbeth et al. 2005). In all, the polygenic nature of obesity paired with its complex environmental interactions will require large studies with thousands of subjects to accurately identify the contribution of various gene polymorphisms.

Developmental Origins of Adult Disease It is important at this point to also discuss the "developmental origins of adult disease" or "thrifty phenotype" paradigm that relates metabolic status and fetal adaptation in the womb to disease risk in later life (Gluckman et al. 2005; Hales and Barker 2001). Beyond maternal and fetal genome sequence and interaction with the immediate environment, nutrient availability during fetal life is predictive of future growth trajectory and disease. For example, nutrient deprivation *in utero* has the outcome of low birth weight but also leads to fetal adaptation to a deprived environment by an increased efficiency in use of nutrients. This has been termed a "predictive adaptive response" in which the fetus predicts the postnatal environment based on fetal nutritional conditions and adapts in order to maximize ability to survive postnatal life (Gluckman et al. 2005). This adaptation is epigenetically regulated (Gallou-Kabani and Junien 2005).

Nutritional deprivation in both the fetal environment and in early childhood has been linked to a proneness to metabolic syndrome, obesity, diabetes, and cardiovascular disease in later life (Barker et al. 2005; Gallou-Kabani and Junien 2005; Gluckman et al. 2005; Hales and Barker 2001; Syddall et al. 2005). This has been particularly true of type 2 diabetes mellitus, in which disease susceptibility is determined by an as-yet undetermined number of genes and a cumulative effect of the environment over a lifetime (McCarthy 2004). Adaptive responses such as this provide an excellent example of phenotypic changes that can occur regardless of genotype (specific gene sequence) and profoundly impact health risk. In fact, it has been hypothesized that the fetal environment is the most critical determining factor in the development of type 2 diabetes (Hales and Barker 2001).

Investigations into interactions between birth weight and genotype have also found that specific genotypes modulate the risk of diabetes. For example, the previously mentioned insertion/deletion polymorphism in the angiotensin-converting enzyme gene (involved in blood pressure regulation) appears to be related to birth weight and propensity for developing type 2 diabetes in later life, such that people with a DD genotype are more likely to be born with a lower birth weight and have an increased risk of diabetes (Kajantie et al. 2004). Thus, genotyping does not tell the entire story in prediction of risk. Fetal environment also plays a role in chronic disease risk; therefore, individual growth history should also be considered in individualized nutrition intervention.

Diabetes Beyond the diabetes risk conferred by the fetal environment, hundreds of genes have been examined for potential roles in the development of type 2 diabetes, including those that may play a role in pancreatic beta cell function, insulin signaling, and so forth (McCarthy 2004). The closely related metabolic syndrome, characterized by insulin resistance, dyslipidemia, abdominal obesity, and hypertension, has garnered much recent attention due to its association with an increased risk of both type 2 diabetes and cardiovascular disease (Roche, Phillips, and Gibney 2005). However, studies of the genetic and environmental contributions to development of metabolic syndrome have not found large single-gene effects (Roche, Phillips, and Gibney 2005). Many of the most telling findings (described in this section) beyond the thrifty phenotype theory have focused on modulation of insulin resistance or prevention of type 2 diabetes in susceptible individuals

(Altshuler et al. 2000; Franks et al. 2004; Laukkanen et al. 2005; Moreno et al. 2005).

The Finnish Diabetes Prevention Study, which aimed to measure the effect of lifestyle intervention in preventing conversion from impaired glucose tolerance (prediabetes) in obese subjects to type 2 diabetes, has also examined the roles of genes (Laukkanen et al. 2005). The researchers observed that polymorphisms in the gene encoding GLUT2 or glucose transporter 2, which helps the pancreatic beta cells detect glucose and secrete insulin accordingly, are related to type 2 diabetes risk. Those in the intervention group had an equally low risk of developing type 2 diabetes regardless of genotype, but those in the control group (continuing their usual lifestyle) were significantly more likely to develop type 2 diabetes if they had a polymorphism in the gene for GLUT2 versus the common allele. Because this subset of the population is at high risk for conversion to type 2 diabetes and benefits from lifestyle intervention, these polymorphisms may serve as a trigger for early and intensive nutritional and physical activity intervention.

One other gene heavily investigated in type 2 diabetes is that for peroxisome proliferator-activated receptor gamma (PPARγ), which is a receptor on the cell nucleus that plays a central role in adipocyte development and function. As previously mentioned, PUFAs are natural ligands for this receptor, but thiazolidinedione drugs (i.e., rosiglitazone) treat diabetes also by interacting with PPARγ to enhance insulin sensitization. Thus, PPARγ activation is associated with greater insulin sensitivity. Because fatty acids with longer chain length and greater desaturation have a higher affinity for PPARγ, diets high in saturated fat are likely to have little effect on PPARγ and have been associated with insulin resistance (Franks et al. 2004). Studies of humans have shown that a relatively common P12A variant, in which alanine (Ala) is substituted for the amino acid proline at the 12 position in one allele, is associated with a protective effect, resulting in a 25% lower risk of type 2 diabetes (Altshuler et al. 2000). It has been suggested that this polymorphism may interact with dietary fatty acid composition and physical activity level to determine diabetes risk by influencing fasting insulin levels. For example, high levels of physical activity and a high dietary polyunsaturated fat to saturated fat (P:S) ratio independently contributed to lowering fasting insulin levels among proline allele homozygotes. In contrast, Ala allele carriers did not benefit at all unless the high P:S ratio and high physical activity level were present simultaneously (Franks et al. 2004). Thus, although the Ala allele carriers may be at lower risk of diabetes, disease development seems to be subject to critical environmental determinants.

While a diet high in saturated fat has been linked to insulin resistance, and diets higher in monounsaturated fats or carbohydrate are linked to improved insulin sensitivity (Riccardi, Giacco, and Rivellese 2004), this does not necessarily hold true in all people. For example, apolipoprotein E or *APOE* genotype has also been linked to insulin resistance in response to dietary fat. Apolipoprotein E plays an important role in lipoprotein metabolism, and specific genotypes have been linked to insulin resistance. In a study of healthy subjects, those with a specific variant in the *APOE* gene promoter did not experience a lowering of glucose and insulin levels when switching from a diet high in saturated fats to diets high in monounsaturated fatty acids (MUFA) or carbohydrates (Moreno et al. 2005). Identification of such polymorphisms can assist in determining which patients will benefit or fail to benefit from specific diet prescriptions to improve insulin sensitivity. However, it must be kept in mind when determining appropriate nutritional intervention that lowering intake of saturated fat in these subjects may have other benefits related to cardiovascular disease risk, even if there is not a direct impact on insulin resistance.

Finally, in addition to predisposition of risk to develop type 2 diabetes, it has also been proposed that the risk of diabetes complications is, in part, genetically determined, and that future identification of those at particular risk will enable more targeted dietary interventions (Kaňková and Šebeková 2005). Beyond the gene polymorphisms discussed here, many other genes involved in glucose regulation and insulin secretion are also under investigation for potential roles in influencing risk of type 2 diabetes and diabetes complications (McCarthy 2004).

Cardiovascular Disease

Individual Variation in Response to Environmental Influences Genetic factors contributing to hyperlipidemia have long been known to have an interplay with environmental factors—diet, tobacco use, physical activity (Corella and Ordovas 2005)—and these environmental influences can impact occurrence, age of onset, and the severity of cardiovascular disease. While dietary guidelines aimed at the public have long been in place, an individual's genomic sequence itself plays a primary role in determining which of these modulations, such as a decrease in saturated fat intake, actually have beneficial effects. It has been known for many years that some individuals are more responsive to dietary intervention than others with respect to hyperlipidemia (Jacobs et al. 1983; Katan et al. 1986). What has not been possible in the past, though, is to determine specifically who will or will not respond to specific dietary measures. In fact, while a low-fat diet is beneficial for many, for others it increases atherogenesis. Like obesity and diabetes, cardiovascular disease is also a complex area of study with numerous gene-gene and gene-diet interactions that have not been completely elucidated. However, more is known about the dyslipidemias, offering perhaps some early opportunities for individualized dietary intervention that are not yet possible for obesity or diabetes. The fact that there is substantial interindividual variation in response to diet provides a clearer basis for individualized rather than generalized population intervention (Corella and Ordovas 2005). Nonetheless, research designs to date have varied widely, and relationships between various genotypes and diet with respect to CVD require much additional study (Masson and McNeill 2005).

Dietary Modification Is Effective in Monogenic Disease Monogenic disorders of lipid metabolism are fairly well understood and provide a basis for examining the more complex polygenic dyslipidemias (Corella and Ordovas 2005). Familial hypercholesterolemia, which results from mutations in the LDL-receptor gene, is perhaps the most readily recognizable example. Over 800 different mutations have been identified as causative, and the variance in mutations results in equally varying phenotypes. Null alleles are mutations resulting in no LDL receptor protein being produced. Other mutations can impair the ability of LDL to bind to the receptor, or impair post-translational processing, and thus function, of the LDL receptor protein. Most people affected by this disorder are heterozygotes, so their functional LDL receptor allele continues to work normally despite being unable to compensate for loss in function of the other. However, clinical presentation continues to vary even among individuals with the same mutation. This is explained by two modulating factors: other genes involved in lipid metabolism can affect the phenotype, and dietary factors can likewise affect the phenotype. In other words, variations in other genes and variations in diet both influence the course of atherosclerosis and life expectancy related to the LDL receptor genotype in familial hypercholesterolemia. This evidence illustrating the efficacy of diet in monogenic disease establishes the likelihood that other gene variations related to lipid metabolism will also be responsive to dietary intervention (Corella and Ordovas 2005).

Dietary Fats Interact with Various Genotypes to Influence Outcomes Fittingly, several examples have already been established. In population studies, PUFA intake has been shown to have a differential effect on HDL concentrations depending on whether the nucleotide base located at the -75 position of the *APOAI* gene promoter is an A or a G (Ordovas et al. 2002). Women in the Framingham Offspring Study with a G/G genotype were observed to have a decrease in HDL levels as PUFA intake increased, whereas HDL levels increased in those with an A/A or G/A genotype. In another example, a variant in the *APOC3* gene promoter region was observed to determine the effectiveness of omega-3 polyunsaturated fatty acids in lowering triglyceride levels (Olivieri et al. 2005). *APOC3* encodes apolipoprotein C-III, which is associated with triglyceride-rich lipoproteins and is a known marker of cardiovascular disease risk. Long-chain omega-3 PUFAs are known to reduce apolipoprotein C-III levels, but specific polymorphisms in the *APOC3* promoter result in a lack of response (Olivieri et al. 2005). As described previously, dietary PUFAs primarily alter lipid metabolism by interacting with transcription factors that regulate genes involved in lipid metabolism. Peroxisome proliferator-activated receptor α (PPARα) is a well-studied nuclear transcription factor. PUFAs are natural ligands for PPARα as well as PPARγ, and polymorphisms in PPARα have also been shown to determine the effect of dietary PUFAs on triglyceride levels and apolipoprotein C-III levels (Tai et al. 2005). Specifically, there was no significant difference in triglyceride or apolipoprotein C-III levels based on level of PUFA in the diet among subjects with the common allele of PPARα. However, a mutation resulting in valine being substituted for leucine at position 162 (*PPARA*-L162V) was associated with significantly lower triglycerides and apolipoprotein C-III levels in response to a high-PUFA diet (Tai et al. 2005).

In another example, an interaction of dietary fats with 5-lipoxygenase genotype determines atherosclerosis risk (Dwyer et al. 2004). Arachidonic acid is an omega-6 PUFA that can be metabolized to inflammatory mediators called leukotrienes via action of the enzyme 5-lipoxygenase. Arterial inflammation plays a role in development of atherosclerosis, so modulation of 5-lipoxygenase could alter cardiovascular disease risk profiles. Researchers compared carotid artery intima-media thickness as an indicator of systemic atherosclerosis and found that those individuals with two variant alleles had significantly greater carotid intima-media thickness. Furthermore, while dietary fat composition had no impact on those with at least one common allele, those with two variant alleles had significantly greater intima-media thickness with higher levels of arachidonic acid or linoleic acid and lower levels of EPA and DHA, which are known to reduce arachidonic acid-derived leukotriene production. There was no relationship with intakes of monounsaturated or saturated fat. All of these examples illustrate how variance in genotype of genes involved in lipid metabolism has the potential to play a role in dictating the most appropriate dietary fat composition to prevent cardiovascular disease on an individual basis.

Nutrigenomics and the Practice of Dietetics

Grasping the intricate interactions between the genome and innumerable dietary factors is critical to the future of nutrition and dietetics practice. As evidenced by the examples in this chapter and the many others that might have been mentioned (but were beyond its scope), this is a complex issue. Appropriate intervention is not simply a matter of genotyping an individual and matching each polymorphism to a specific dietary change. Depending on the outcome sought—decreased risk of colon cancer, breast cancer, or pancreatic cancer; lower triglyceride levels; increased HDL-cholesterol—the interventions may end up contradicting each other. The fact is that much remains unknown about the interactions of genes with each other. Even less is known of nutrient interactions with genotype to define an individualized diet that achieves the best outcome based on the genotypes of 30,000+ genes (Corella and Ordovas 2005). The recommendation of one diet intervention in response to one polymorphism is too simplistic, and as research evolves, practice will in time evolve as well (Corella and Ordovas 2005). In the future, diet counseling is likely to include sequencing of the entire genome to determine disease risk profiling for an individual and planning appropriate lifestyle interventions in accordance with the results of genomic sequencing. In addition to the genome sequence itself, the role of epigenetics and modulation of gene expression by dietary factors must also be considered.

Individual Testing in the Marketplace

Despite the fact that nutrigenomics research is not yet ready for general clinical application, some purveyors of nutrition advice already offer genetic testing and an individualized diet. Direct-to-consumer genetic testing has also recently become available (Sinha 2005). For example, Carolyn Katzin, a Certified Nutrition Specialist in California, has trademarked the "DNA Diet" (Katzin 2006). Clients can mail in a buccal (inner cheek) swab for genotyping of 19 disease-related genes and a personalized diet via telephone or in person. While the advice offered is unlikely to cause harm, it is also unlikely to differ from general dietary recommendations (Sinha 2005). Based on genetic profile, advice may include such recommendations as eating more cruciferous vegetables, legumes, whole grains, and fish—the same advice provided free by the federal government and nonprofit agencies (Sinha 2005). In contrast, the DNA Diet baseline testing costs clients $625 (Katzin 2006). Sciona has also recently begun marketing its genetic assessment kits in U.S. grocery stores, with the comprehensive (19 genes) kit available for $252 (Sciona 2006). While it is not yet clear how various genotypes interact with each other or with diet, to determine risk for polygenic diseases such as cancer, obesity, diabetes, and cardiovascular disease, this is the reality of the marketplace (Sinha 2005). Accordingly, the Evaluation of Genomic Applications in Practice and Prevention (EGAPP) Project was developed by the Office of Genomics and Disease Prevention (OGDP) of the Centers for Disease Control and Prevention (CDC) (Centers for Disease Control and Prevention 2005). The EGAPP Project brings together experts in health care, epidemiology, genomics, public health, laboratory practice, and evidence-based medicine. The goal is to establish a coordinated process for evaluating genetic tests and translating genomic applications that are in transition, such as those predictive for common diseases, from research to clinical practice and health policy. Information regarding the efficacy and cost-effectiveness of testing will ensure that available tests are safe, effective, and used appropriately.

Evolving Knowledge and Practice Requirements for Dietitians

Clearly, registered dietitians must be knowledgeable in genetics and genomics concepts. They must also be able to understand the role of diet in interactions with the genome. As the research base in this area continues to expand, as practice evolves, and as patients more routinely undergo gene sequencing to determine disease risk, registered dietitians with a solid grasp of genomics and diet will perhaps be among the health professionals best equipped to provide genetic counseling to optimize health. Such growth in the field of dietetics will require a substantial expansion of the knowledge base to include pathophysiology of disease at the genomic level as well as at the biochemical, metabolic, and dietary manipulation levels. Practitioners will also need to effectively communicate with consumers not only about diet, but also about the intricacies of genomics and, specifically, how the dietary interventions exert their beneficial effects. It has been proposed that a graduate degree and perhaps a certification exam would be desirable in order to effectively practice in this arena (DeBusk et al. 2005). Nutrigenomics research is the key to closing gaps in the evidence base and strengthening evidence-based nutrition practice (DeBusk et al. 2005)—but practitioners must be up to the task.

Conclusion

Beyond evolution in clinical practice, there will be other changes as well (DeBusk et al. 2005). Research will continue to identify bioactive food components and examine how they interact with specific genes and specific genotypes, and how they influence gene expression to yield changes in health risk. Food scientists will measure bioactive components in foods and develop new functional foods to meet the demand. Clinical trials will examine how functional foods and dietary supplements prevent or slow progression of disease. Dietitians have the opportunity to be involved every step along the way, from development of functional food products to serving as clinical trial coordinators or principal investigators. The new knowledge base will be immense, and dietitians will be called upon to translate new research findings into something consumers can understand and apply. It is also important that dietitians be involved in developing nutrition policies that reflect this new knowledge and find effective ways to communicate to the public dietary recommendations that may contain individualized guidance based on gene polymorphisms. The merit of dietitian involvement is further underscored by the American Dietetic Association's having recently listed "Nutrigenetics and Nutrigenomics" as one of five priority areas in the ADA Strategic Plan (American Dietetic Association 2006). Many opportunities unique to the intersection of nutrition with genomics will arise in the near future and, if the dietetics profession is prepared, dietetics practice will undergo an exciting metamorphosis that will shape the future of health care.

WEB LINKS

National Human Genome Research Institute (NHGRI), National Institutes of Health: The NHGRI home page provides links to a wide array of resources and information for professionals and the public relating to the Human Genome Project. Subjects include grants and research, genomics and health, policy and ethics, educational resources, and careers and training information.

http://www.genome.gov

Centers for Disease Control Genetics and Genomics: The CDC focus is on the relationship of genetics and genomics to public health, including family history and genetic testing. A link to "Six Weeks of Genomic Awareness," a free online training program for public health professionals, is included. A link to the CDC's Office of Genomics and Disease Prevention (OGDP) at http://www.cdc.gov/genomics is also included.

http://www.cdc.gov/node.do/id/0900f3ec8000e2b5

Human Epigenome Project (HEP): HEP aims to identify DNA methylation patterns in the human genome. The website provides basic information about the project.

http://www.epigenome.org

MD Anderson Cancer Center DNA Methylation in Cancer: This site is a resource for professionals interested in the role of DNA methylation in cancer and provides data on specific genes methylated in various cancer types.

http://www.mdanderson.org/departments/methylation

Genomic Imprinting Website resource for students and researchers: This site was established in 1997 to provide information on genomic imprinting for researchers, students, and others. It includes videos of presentations from several international conferences on genomic imprinting.

http://www.geneimprint.com/index.html

National Coalition for Health Professional Education in Genetics: NCHPEG is an organization made up of several health professional organizations dedicated to inclusion of genetics and genomics education in the training of health professionals. The American Dietetic Association is a member of NCHPEG. The site includes their publication of core competencies in genetics for health care professionals.

http://www.nchpeg.org

Genomics and the Future of Public Health: A video presentation of a CDC conference on genomics that took place in 2003.

http://www.cdc.gov/genomics/info/conference/may2003/genomicsday.htm

Obesity Gene Map Database: This database annually updates all markers, genes, and mutations associated with or linked to obesity.

http://obesitygene.pbrc.edu

National Society of Genetics Counselors (NSGC): The official website of NSGC details the profession of genetic counseling.

http://www.nsgc.org

NCMHD Center of Excellence in Nutritional Genomics at the University of California–Davis: This organization is sponsored by the National Center for Minority Health and Health Disparities at the National Institutes of Health. It is dedicated to the promotion of the science of nutritional genomics through news, information, and commentary.

http://nutrigenomics.ucdavis.edu/index.htm

END-OF-CHAPTER QUESTIONS

1. What is a genome? How does knowledge of its content possibly affect dietary recommendations for individuals?
2. What are the differences between genotype, haplotype, epigenotype, and phenotype?
3. Define the following terms: autosomal dominant; autosomal recessive; X-linked dominant; X-linked recessive; Y-linked, heterozygous alleles; and homozygous alleles. Name one autosomal recessive disorder, one autosomal dominant disorder, and an X-linked recessive disorder.
4. What is the difference between a monogenic disorder and a polygenic disorder?
5. Define single nucleotide polymorphisms. How are they identified? Give an example of one and explain what it means.
6. What is meant by epigenetic regulation? How could the nutrients folate, choline, methionine, and vitamin B_{12} affect gene expression?
7. For each of the following disorders, list at least one gene that is linked to its occurrence: obesity, type 2 diabetes, and colon cancer. For each gene listed, describe its possible role in the development of the disorder.
8. Describe an example of "developmental origins of adult disease."

12

Immunology

Christina Lee Frazier, Ph.D.

Professor, Southeast Missouri State University

CHAPTER OUTLINE

Immune System Overview

Cells of the Immune System

Organs of the Immune System

Soluble Mediators

Antigen Recognition Molecules

Immune Response: Attacking Pathogens

Attacking Altered and Foreign Cells: Tumors and Transplants

Immunization

Immunodeficiency

Malnutrition and Immunodeficiency • Inherited Immunodeficiencies • Acquired Immunodeficiencies

Tolerance

Attack on Harmless Antigens: When the Immunological System Causes Harm

Hypersensitivity • Autoimmunity

Introduction to Immunology

Humans are exposed to numerous pathogens as we eat, breathe, and come into contact with environmental objects and other humans. We are protected from disease-causing organisms by natural resistance and the immune system. Natural resistance involves anatomical structures and physiological mechanisms that have other functions in the body. These work predominately by keeping organisms from entering the body and becoming established in the tissues. The immune system includes organs, cells, and soluble factors that respond to pathogens and altered cells that have overcome the natural resistance. In the view of most experts, the immune system developed solely to protect the body from pathogens. Immunity is defined as all those physiological mechanisms that endow the body with the ability to recognize material as foreign and to neutralize, eliminate, and/or metabolize it with or without damage to the body's tissues. Notice that the immune system is not always beneficial. Processes involved in countering organisms that cause infectious disease can damage tissues either as part of the response to a pathogen or when directed at a harmless target, as in allergic reactions or autoimmune disease. Symptoms associated with an infectious disease are often partially or totally caused by the immune response. Plants have elaborate chemical mechanisms that provide resistance to disease, but they do not have an immune system.

Many things influence an individual's susceptibility to infectious disease:

- Gender. Although gender sometimes plays a role, in many cases the underlying mechanism is differential exposure due to occupational and recreational activities.

BOX 12.1 HISTORICAL DEVELOPMENTS

Immunology and the Onion

Immunology is a relatively young field, although immunological phenomena have been reported for years. In his accounts of the plague epidemics that accompanied the Peloponnesian Wars (around 430 BC), Thucydides noted that individuals who had recovered from plague could take care of people with plague without becoming ill. Because immunology is a young field, new discoveries are being made daily, but much is still unknown. In some cases, what is happening is only partially understood or the "what" is understood but not the underlying mechanisms or the "why." In recent years, new treatments have been derived from the growing understanding of some basic immunological processes.

The immune system is composed of a complex set of tissues, organs, cells, fluids, and chemicals that engage in the highly regulated interactions needed to protect the body from diverse pathogens while not attacking healthy tissues. The immune system does not reside in a single organ that can be studied through an anatomical model or dissection. Because the diverse aspects of the immune system are so highly intertwined, it is often necessary to understand concept A in order to understand concept B, while at the same time, understanding elements of concept B is basic to understanding concept A. Due to the newness of the field, established, clear-cut explanations are not always available.

One way to approach this complexity is to imagine peeling an onion. Instead of trying to take a big bite that cuts through all the layers at once to reach the core, you should examine each layer, and wait until you have mastered the concepts for one level before inspecting the next layer of complexity. As deeper layers are scrutinized, new concepts are revealed and old ones revisited at a deeper level of complexity. Like the onion, immunology can bring tears to your eyes in awe and appreciation of its complexity and intricacy, but studying it does not have to provoke tears of frustration if you take your time moving from layer to layer.

- Age. The immune system takes time to develop; therefore, the young do not have the full spectrum of immunological defenses available to the adult. As humans grow older, several immune mechanisms decrease, including secretion of mucous and sebaceous glands and the production of cytokines, including an interferon. However, natural killer cells that attack infected cells and tumor cells increase.
- Nutritional status. Malnutrition is a major cause of **immunodeficiency.**

immunodeficiency—decrease in or lack of an immune response due to absence or defect of one or more components of the immune system

- Hormones. Levels of various hormones play a role in an individual's susceptibility to infectious disease. Individuals with diabetes have an increased risk of fungal and staphylococcal infections, while women with low estrogen have a higher vaginal pH and thus are more susceptible to vaginal infections.
- Stress. Stress activates the fight-or-flight response, resulting in several physiological changes that impact the immune response (Kiecolt-Glaser et al. 2002). Short-term stress boosts the immune system. Increased immune responses were noted in individuals a few hours after surviving the Los Angeles earthquake, performing battle tasks, and completing a math test (Segerstrom 2004). An individual's perception of his/her lack of control of the stressor (Brosschot et al. 1998) and concurrent chronic stress (Pike et al. 1997) have negative immuno-modulating effects. On the other hand, long-term and repetitive stress decreases immune responses. Decreased immune responses have been documented in people who have experienced loss of a loved one or divorce (Kiecolt-Glaser et al. 1987). Corticosteroids from adrenals that have been stimulated by nerves mediate immunosuppression, while innervation of lymphoid tissue and blood vessels influences the movement of cells of the immune system.

Natural Resistance

Several anatomical and chemical barriers contribute to natural resistance (see Figure 12.1). Intact epithelial surfaces such as skin and the lining of the body's tubular structures such as the gastrointestinal, respiratory, and genitourinary tracts, are excellent barriers to most pathogens. *Treponema pallidum*, the causative agent of syphilis, is among the very few organisms that can cross intact skin and mucous membranes. Some organisms, especially fungi, can exploit natural breaks in the skin, such as hair roots, to enter the host.

In addition to providing a physical barrier, skin and mucous membrane components produce chemical barriers, including lysozyme, which is produced by sweat glands. This enzyme damages peptidoglycan, a critical component of bacterial cell walls. Recognition of circumcision as a risk factor for HIV has led to speculation that lysozyme in foreskin might be a factor in host resistance. Sebaceous glands in the skin also secrete lipids that are converted into fatty acids by gram-positive bacteria. The low pH created by fatty acids, which contributes to body odor, inhibits growth of numerous other bacteria.

The surface of mucous membranes can glue or trap microorganisms so they cannot continue their movement into the body. For example, the mucous blanket in the respiratory tract can keep organisms from reaching the lungs. Cilia in respiratory mucosa create the ciliary escalator that helps bring the organisms, which are trapped in mucus, to the surface so they can be coughed out. People who smoke or abuse alcohol

FIGURE 12.1 Components of Natural Resistance

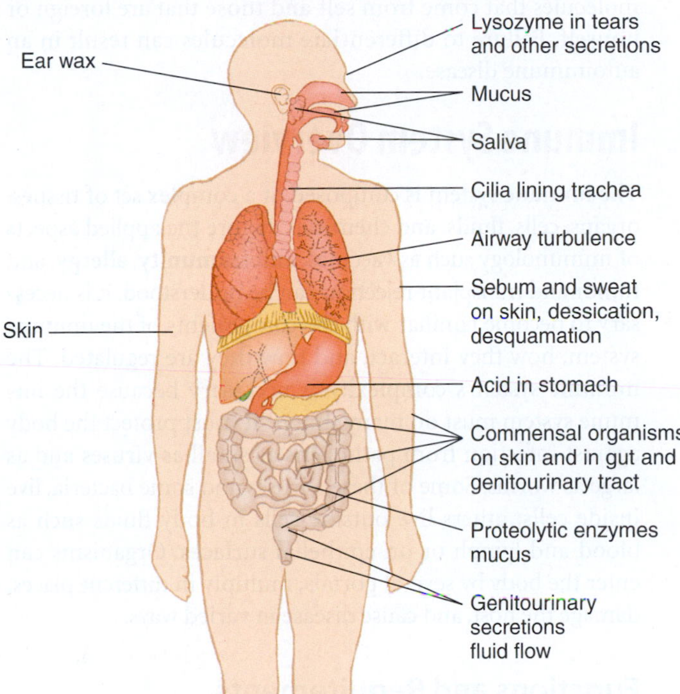

Source: Rhoades/Pflanzer, *Human Physiology*, 4e, copyright © 2003, p.857

damage the respiratory cilia and an increased number of respiratory infections are noted in these individuals (Vander Top et al. 2005). Though mild coughing helps dislodge the mucous blanket and bring trapped organisms up and out of the body, excessive coughing, especially in young children, can lead to a harmful oxygen deficit in the brain.

The body is protected in a number of areas by washing. Tears, urine, and saliva wash organisms out of the eyes, the genitourinary tract, and the mouth. Urine is also acidic, and tears and saliva contain lysozyme and a number of other protective enzymes.

The pH of the stomach protects humans from many of the organisms taken in by mouth, since very few can survive in the acid environment. Foods that act as buffers, such as milk or antacids, can weaken this line of defense, as demonstrated by the association between salmonella infection and gastric acid-lowering medications (Banatvala et al. 1999). Digestive enzymes in the upper GI tract also destroy some microorganisms.

For a microorganism to grow and multiply, environmental temperature must be within its viable range. Therefore, organisms that cannot grow at normal human body temperature do not have the potential to be pathogens in humans. A low level of fever is beneficial, since it enhances the action of the immune system. However, if an individual's temperature rises too high, brain damage can result. Historically, certain infections, such as gonorrhea, were treated by raising individuals' body temperature by infecting them with organisms that cause malaria.

Anaerobic pathogens cannot grow in the presence of oxygen, and microaerophiles require reduced oxygen, so areas in the body where oxygen is found in high concentrations will not provide a good growth environment for these organisms. Anaerobic microenvironments in the highly oxygenated lungs can support the growth of anaerobes including bacteroides, prevotella, and fusobacterium that cause necrotizing pneumonia that can result from abdominal surgery or trauma to the large intestine or bowel.

Antigens: The Key to Recognizing Pathogens and Altered Cells

Antigens and Immunogens

Antigens, small biochemical groups found in and on bacteria, viruses, cells, and larger molecules, are molecules that allow the immune system to recognize potential pathogens and abnormal cells. **Antibodies** are proteins made in response to an antigen. When they bind specifically to an antigen on a pathogen or toxin, they either block the pathogen's action or recruit other cells to destroy it.

The current definition of antigen is a structure that can combine with a cell of the immune system or an antibody but does not necessarily induce activation of the cell or formation of an antibody. An **immunogen** is an antigen that can induce an immune response. The key difference between an antigen and an immunogen is that the immunogen is foreign to the host producing the response. With recognition that the antigen was a small biochemical unit and not an entire organism, the term **antigenic determinant (or epitope)** arose to separate the concept of the whole organism as antigen from the small biochemical molecule.

antigen—a substance that is specifically bound by an antibody or lymphocytes; used by the immune system to recognize pathogens and altered cells; see immunogen

antibody—a protein molecule found is serum and tissues that is secreted by B cells in response to a specific antigen that can bind to that antigen and neutralize or help destroy it. Also called immunoglobulin

immunogen—an antigen capable of inducing an immune response because it is foreign to the host

antigenic determinant—specific part of an immunogen that stimulates a specific immune response and reacts with the resulting antibody or activated T cell; also called epitope

Characteristics of an Antigen

Not all biochemical groupings are antigenic. Although proteins are the best antigens, polysaccharides, lipids, and nucleic acids can be antigenic. In order for a substance to be an antigen, it has to have sufficient size. About 10,000 molecular weights are needed to be antigenic, and the higher the molecular weight, the greater the antigenicity. Molecules too small to be antigenic, **haptens**, can have antigenic activity if coupled to larger molecules. This is a factor in certain autoimmune diseases and allergic responses.

An antigen must have structural stability. Jell-O™ is not a good antigen since its helical structure falls apart when heated, and the helices are not perfectly formed during cooling, resulting in gaps in the helix and a tangled web of polypeptide chains. Trapped water provides the characteristic Jell-O jiggle.

An antigen must be degradable, so it can be processed in order to activate cells of the immune system. Therefore, stainless steel or plastics are not good antigens, and they can be used in prosthetic devices or implanted in the body without causing immunological rejection.

In order for a molecule to be a good antigen, it must be complex. The primary structure of a protein is simply its amino acid sequence. In secondary and tertiary structures, the sequence of amino acids causes the polypeptide to fold. A linear polypeptide of a single type of amino acid is not antigenic, since a variety of amino acids is required to produce folding. Quite often, components that are key to an antigen's structure are dispersed throughout the unfolded molecule and are brought together only with the appropriate folding.

For a molecule to be an immunogenic as well as an antigen, it must be foreign to the organism producing the immune response. Humans and pathogens contain some very similar molecules, and one of the key aspects of the immune system is its ability to differentiate between molecules that come from self and those that are foreign or nonself. Failure to differentiate molecules can result in an autoimmune disease.

hapten—a nonimmunogenic, low-molecular weight molecule that can be recognized by an antibody; it can initiate an immune response if it is conjugated to a "carrier" molecule

autoimmunity—an immune response to one's own tissues

allergy—an inappropriate and harmful immune reaction to a harmless nonpathogenic substance; also called hypersensitivity

humoral immunity—immunity due to soluble factors such as antibodies circulating in the body's fluids, mainly serum and lymph; "humors" is an old term for body fluids

cellular immunity—immune protection provided by the action of immune cells, especially T cells, polymorphonuclear leukocytes, and macrophages

plasma cells—large antibody-producing cells that develop from activated B cells. Also call AFC or antigen forming cells

Immune System Overview

The immune system is composed of a complex set of tissues, organs, cells, fluids, and chemicals. Before the applied aspects of immunology such as vaccines, **autoimmunity**, **allergy**, and tumor and transplant rejection can be understood, it is necessary to become familiar with the components of the immune system, how they interact, and how they are regulated. The immune system's complexity is necessary because the immune system must do many things. It must protect the body against infection from pathogens as small as viruses and as large as worms. Some of these, viruses and some bacteria, live inside cells; others live outside cells in body fluids such as blood and lymph or on epithelial surfaces. Organisms can enter the body by several portals, multiply in different places, damage the host, and cause disease in varied ways.

Functions and Requirements of the Immune System

The immune system has three basic functions. The first is defense. The immune system developed to protect the body from pathogens. However, the immune system is also very active in homeostasis by helping the body to remove damaged and dead cells. It also functions in surveillance by recognizing abnormal cells, such as those infected by viruses and other pathogens. The surveillance function has been adapted to help the immune system identify and attack tumor cells, and it is also the underlying mechanism by which the immune system recognizes transplants and mounts a rejection response.

There are four basic requirements of an effective immune system as it mounts a response: (1) specificity, the ability to react with one and only one antigen, which lowers the chance of a reaction to a pathogen will also harm the person; (2) diversity, so it can respond to many pathogens; (3) adaptivity, the ability to pick the best response to counter the pathogen; and (4) the ability to respond to stimuli not encountered previously.

Immune response consists of two phases: first, the immune system must recognize the pathogen or infected/altered cell; second, it must mount a reaction to it.

Divisions of the Immune System

Humoral and Cellular The immune system is divided into two arms: **humoral** and **cellular immunity**. The humoral arm of the immune system refers to antibodies that appear in serum (clear liquid that separates from the blood after it clots) and B cells that become **plasma cells** that produce antibodies. The cellular part of the immune system refers to T cells, macrophages, monocytes, and polymorphonuclear leukocytes (also known as PMNs, microphages, and granulocytes) that interact with pathogens at the cellular level.

Specific And Nonspecific The immune system is also divided into two branches, both of which contain elements of both the humoral and cellular immune systems. One is referred to as the innate, nonadaptive, or **nonspecific immune system**. The other is called either the acquired, adaptive, or **specific immune system**. Cells of the specific humoral immune system respond to different epitopes on a pathogen than do the cells of the specific cellular immune response.

These two immune systems, although often described separately, are interdependent and are both required for a strong immune response. Cells of the nonspecific immune system (macrophages, monocytes, natural killer cells, and polymorphonuclear leukocytes) react with any antigen; they can thus react immediately. This initial reaction is quite often sufficient for elimination of the pathogen or for reduction of its numbers significantly enough to prevent initiation of the disease process. This response does not increase or improve with repeated exposures.

In the specific immune response, each B and T cell is programmed to attack one specific antigen, but these cells can interact with others that are closely related or very similar. In rheumatic fever, an immune response stimulated by antigens on a Group A *Streptococcus* can attack similar antigens on the heart. The specific immune system takes time to respond initially, but it improves with additional exposures and responds more rapidly on subsequent encounters with the organism. Thus, it normally protects the human from reinfection.

The response to a pathogen normally involves an initial contact with the nonspecific immune system, which often is capable of eliminating the organism by itself. The nonspecific immune system then stimulates the specific immune system to seek out and target remaining pathogens. In some cases, elements of the specific immune system can eliminate the pathogen; in others, they merely tag it or alter it in such a way that it becomes more susceptible to the cells of the nonspecific immune system. The two systems thus work together and are interdependent.

Active and Passive Specific immunity can be described as either **active immunity**, where individuals synthesize their own antibodies or activate immune cells, or **passive immunity**, where they receive antibodies or activated cells produced by another individual. Both active and passive immunity can be described as either natural (occurring without human intervention) or artificial (resulting from human intervention). The four types of immunity are:

- *Active natural immunity:* Mounting an immune response to an infectious organism.
- *Active artificial immunity:* Mounting an immune response to vaccination.
- *Natural passive immunity:* An antibody from the mother goes to the fetus across the placenta. Both regular breast milk and colostrum contain antibodies, but the concentration is higher in colostrum. The antibodies can provide protection from pathogens, but they can also contribute to allergic reactions in the baby.
- *Passive artificial immunity:* Transferring antibodies or immune cells produced in one organism to another organism to prevent the action of a virus or toxin before it does damage. Examples of clinically used antibodies include antirabies or hepatitis globulin, antivenom for snake bites, or antitoxin for tetanus or botulism. Commercially available intravenous immune globulins (Gamimune N, Gammagard, Gammar, Iveegam, Polygam, Sandoglobulin) contain **gamma globulins** from a number of individuals and are used to boost the body's natural defense system against infection in persons with a weakened immune system.

Cells of the Immune System

Origin of Cells of the Immune System

All cells in the immune system are formed in the **bone marrow**, where they mature to varied degrees and are released (see Figure 21.2 in Chapter 21). Like all body cells, cells of the immune system originate from pluripotential **hematopoietic stem cells**. Pluripotential stem cells involved in formation of immune cells are found in the bone marrow, spleen, fetal liver, and fetal yolk sac (where they appear shortly after conception). In a mouse that has had its immune system destroyed through radiation, the addition

nonspecific immunity—all aspects of immunity not directly mediated by antigen-specific lymphocytes

specific immune response—immunity mediated by antigen-specific lymphocytes

active immunity—immunity produced due to exposure to an antigen (e.g., infection or vaccination)

passive immunity—immunity due to the transfer of antibodies or activated T cells produced by another individual

gamma globulins—a group of serum proteins, including most antibody molecules, that migrate fastest toward the cathode during electrophoresis

bone marrow—soft tissue in the cavities of bones where stem cells become red and white blood cells

hematopoietic stem cell—an undifferentiated bone marrow cell that is a precursor for multiple cell types; also called pluripotential stem cells

TABLE 12.1

Clinical Uses of Hemopoietic Inducing Factors

Generic Name	Trade Name	Hemopoietic Inducing Factor	Preparation	Use
Epoetin alfa	Epogen, Procrit	Erythropoietin	Genetically engineered	Treating anemia
Sargramostim	LEUKINE®	Granulocyte-macrophage colony-stimulating factor	Recombinant DNA technology	Restoring neutrophils in individuals undergoing chemotherapy
Pegfilgrastim	Neulasta	Human granulocyte colony-stimulating factor (G-CSF)	Recombinant DNA technology	Increasing the number of neutrophils in individuals undergoing chemotherapy and bone marrow transplantation

of 30 stem cells is all that is required to reconstitute the immune system (Smith et al. 1991). Hemopoietic inducing factors act on pluripotential stem cells to cause them to differentiate, proliferate, and eventually become red blood cells (RBC) or one of the cells of the immune system called leukocytes or white blood cells (WBC). Certain hemopoietic inducing factors are reproduced in the laboratory and used in medical treatment (see Table 12.1).

White blood cells are not white; they are colorless when compared to red blood cells and are found in a number of other tissues in the body besides the blood. Although in a blood smear it appears there are many more RBC than WBC, WBC outnumber RBC three to one in the body. A complete blood count (CBC) determines the percentage of each type of WBC discussed below. Deviations from normal counts can be indicative of certain clinical conditions (see Table 12.2).

macrophage—a large phagocytic antigen-presenting cell derived from the blood monocyte and found in tissues

monocyte—a large mononuclear phagocytic white blood cell that develops into a macrophage when it enters tissue

polymorphonuclear leukocytes (PMN)—leukocytes with a multilobed nucleus and cytoplasmic granules that take up acid and basic dyes; also known as granulocytes, PMNs, and polys

lymphocyte—a small mononuclear cell with a thin rim of cytoplasm that has antigen-specific receptors

T cells—lymphocytes that differentiate in the thymus

B cell—a lymphocyte derived from the bone marrow, which differentiates into a plasma cell that makes an antibody

natural killer cells (NK cells)—large granular lymphocyte cells that attack tumors and virally infected cells but do not exhibit antigenic specificity; also called killer cells (K cells) and null cells

White blood cells are divided into three groups, the **macrophages/monocytes**, microphage/granulocytes/**polymorphonuclear leukocytes**, and **lymphocytes.** The pluripotential stem cell differentiates first into either a myeloid stem cell or a lymphoid stem cell (see Figure 21.2 in Chapter 21). The lymphoid stem cell produces the cells of the specific immune system—**T cells** (T lymphocytes) and **B cells** (B lymphocytes). The myeloid stem cell produces megakaryocytes (bone marrow cells that produce platelets) and cells of the nonspecific immune system, including macrophages/monocytes and polymorphonuclear leukocytes. **Natural killer cells (NK cells or K cells)** are produced from the lymphoid precursor; however, they react in the nonspecific immune system, which demonstrates the continuity between the two divisions.

Cells Derived from the Myeloid Stem Cell

Monocytes and Macrophages As shown in Figure 12.2, the mononuclear phagocyte system includes monocytes, which can differentiate into macrophages. In a stained blood smear, monocytes are larger than most other WBC and have approximately equal amounts of nucleus and cytoplasm. The nucleus may either be roughly circular or horseshoe-shaped. Cytoplasm is grayish in most common stains and appears to contain many little holes due to the presence of the vacuoles. Monocytes circulate in the blood, where they are 1% to 3% of WBC; then they migrate to tissues where they divide and differentiate into macrophages. Differentiation involves morphological, biochemical, and functional changes. These include an increase in the size, number, and complexity of organelles (Golgi bodies, mitochondria); lysosomal enzyme production; and protein synthesis. Also included are changes in surface antigens. Macrophages are divided into two categories, fixed and wandering. Wandering macrophages move around the body so that they can go to areas with infectious organisms. Chemicals produced by pathogens or damaged tissues draw the macrophages to an area. Fixed macrophages (histiocytes) are integrated into the tissues they protect (see Table 12.3) and may live for months or years.

TABLE 12.2

Implications of Abnormal WBC Counts

Cell Type	Normal Value (NLM NIH 2006)	Implication of Abnormal Count Status
Band cells	0% to 3%	Increased ("shift to the left"): • Inflammatory processes, e.g., acute appendicitis or cholecystitis
Monocyte	2% to 8%	Increased: • Viral and parasitic infections • Inflammatory bowel disease • Some cancers
Neutrophil	40% to 60%	Increased: • Obesity • Bacterial infections • Smoking
Eosinophil	1% to 4%	Increased: • Allergic reactions • Worm infection • Myeloproliferative disorders • Malignancies • Autoimmune diseases including rheumatoid arthritis and systemic lupus erythematosus • Eosinophilic gastroenteritis • Addison's disease
Basophils	0.5% to 1%	Increased: • Viral infections • Hemolytic anemia • Inflammatory bowel disease • Hypothyroidism • Increased estrogen levels Decreased: • Stress • Corticosteroid use • Pregnancy • Hyperthyroidism
Lymphocyte	20% to 40%	Increased: • Acute stage of viral infection • Connective tissue disease • Hyperthyroidism • Addison's disease Decreased: • AIDS • Bone marrow suppression • Steroid use • Neurologic disorders, including multiple sclerosis and myasthenia gravis

FIGURE 12.2 Derivation of Cells of the Immune System

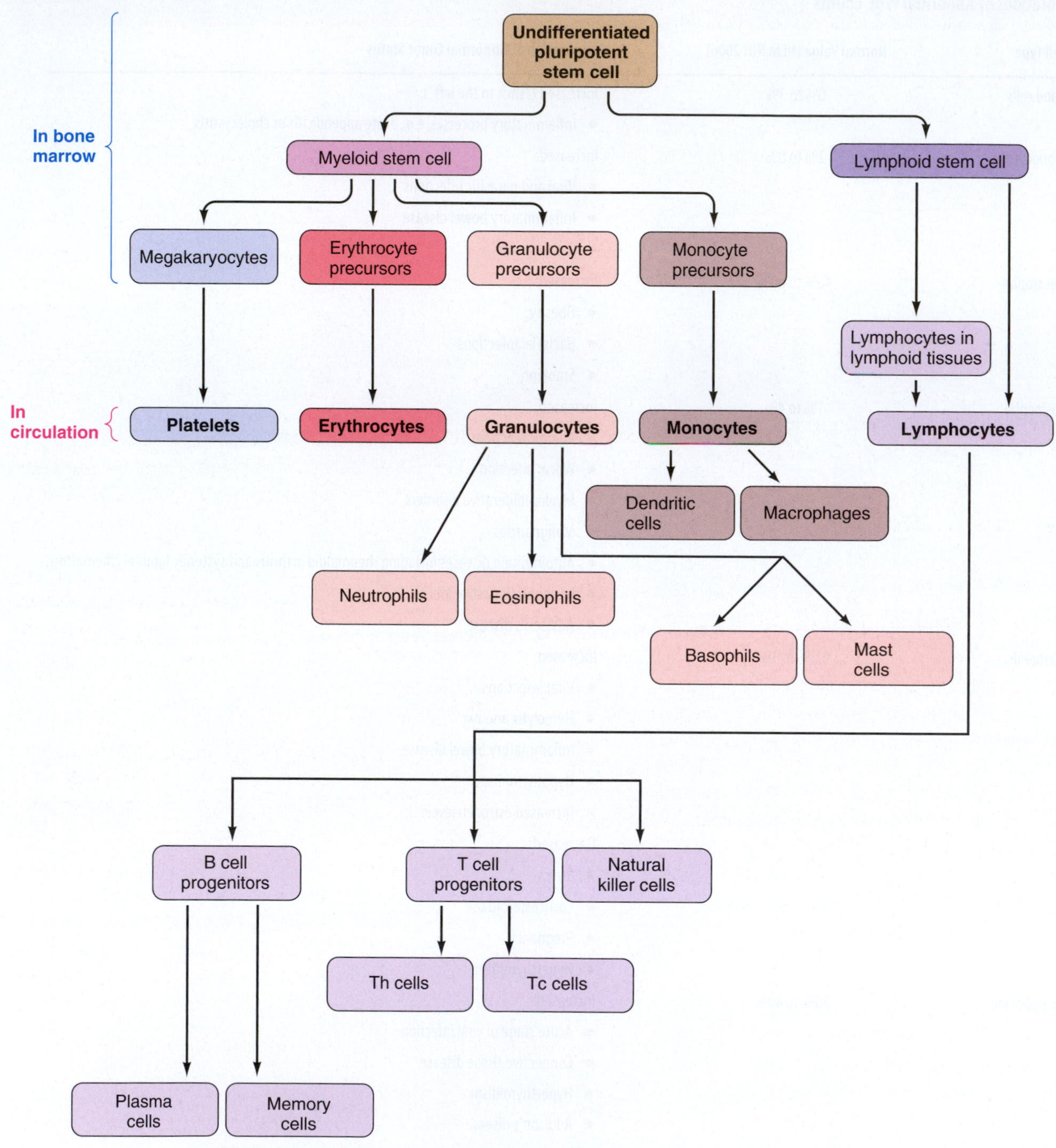

Source: Lauralee Sherwood, *Human Physiology: From Cells to Systems*, 5e, copyright © 2004, p. 400

phagocytosis—the engulfment of a particle or a microorganism by leukocytes such as macrophages and neutrophils, normally followed by destruction of the particle

Monocytes and macrophages are important in removing pathogens—both by themselves and after the pathogen has been targeted by cells of the specific immune system. Monocytes and macrophages are highly specialized; they ingest and destroy particulate matter such as bacteria, aged cells, and neoplastic cells in a process called **phagocytosis**.

TABLE 12.3

Fixed Macrophages

Location	Name
connective tissue	histocyte
spleen	"dust" cells
serous cavity	peritoneal macrophage
bone	osteoclast
brain	microglial cells
lung	alveolar macrophage
liver	Kupffer cells
kidney	mesangial macrophage
joint	synovial A cells

They are major **antigen-presenting cells** (APC) that can break down antigens into small pieces that they "present" on their cell surface. This function is very important for initiating an immune response and will be discussed later in the chapter.

Polymorphonuclear Leukocytes Polymorphonuclear leukocytes (PMNs), also called microphages, granulocytes, or polys, are a second group of cells involved in the nonspecific immune response. *Microphage* was coined to differentiate these cells from the macrophage; they are much smaller. The term *granulocyte* refers to cytoplasmic granules that are visible in commonly used stains. The term *polymorphonuclear leukocyte* refers to nuclei that look very different from cell to cell. In mature cells, the nucleus is segmented; however, in immature cells (also called band cells) the nucleus is in one segment. PMNs move easily between blood and tissues, so their number in blood increases or deceases in infections depending on the type of organism involved. A large pool of PMN is available in bone marrow for rapid response to an infection.

Three types of PMN can be distinguished by the shape of the nucleus and the stains taken up by the granules (see Figure 12.3). The **neutrophil** is the first type and the most common; it comprises about 60% of WBC in blood and about 90% of PMN. Neutrophils are identified by a nucleus that has three or more connected lobes that may appear unattached in stained cells. Their granules have affinity for both acid and basic dyes, so they stain purple. The granules contain lysozyme, with which these cells destroy organisms that they ingest by phagocytosis.

The second type, **eosinophils,** have a bilobed nucleus and are named for their granules, which are bright red when stained with eosin. In the typical adult, they constitute between 1% and 5% of circulating WBC, but the number often increases in a person experiencing an allergic reaction or worm infection. Eosinophils remain in the blood for a short time and then migrate to tissues. Although they are phagocytic, this is not their major role in the immune system since they are adapted to attack larger pathogens, especially worms. Their granules do not contain lysozyme, but eosinophils do produce other chemicals that are secreted and damage pathogens.

The third type of PMN is the **basophil**, so named because the granules, which appear blue-black, take up the basic dye. Basophils are rare, usually less than 1% of the WBC in a typical adult. The nucleus is not always well segmented and is obscured by the granules in some cells. Phagocytic function of these cells is uncertain. They do, however, have receptors for an antibody that is involved in one type of allergic response and immune response to worms. They produce a number of chemicals, most notably **histamine** and serotonin, which are associated with allergic responses in humans.

Other Cells Megakaryocytes divide into platelets, which are pieces that aggregate to help form a blood clot. In addition to their role in blood clotting, platelets are involved in inflammation and are a component of certain types of allergic responses.

Mast cells, which are important in some forms of allergy, share many anatomical and physiological characteristics with basophils, and the relationship between the

antigen-presenting cell (APC)—a cell capable of displaying fragments of antigens from pathogens and altered cells joined to MHC molecules on its surface in a manner that can be recognized by T cells

neutrophil—the most numerous polymorphonuclear leukocytes, with granules that stain with acid and basic dyes; it is phagocytic and enters tissues early in inflammation

eosinophil—a polymorphonuclear leukocyte containing granules that produce substances that damage parasites and decrease inflammation; these granules stain with acid dyes

basophils—polymorphonuclear leukocytes containing granules that stain with basic dyes; they have much in common with mast cells, including the release of histamine and leukotrienes, which contribute to allergic responses and inflammation

histamine—a vasoactive amine that contributes to inflammation and IgE-mediated allergic reaction by causing the dilation of local blood vessels and smooth muscle contraction; histamine release produces some of the symptoms of immediate hypersensitivity reactions

mast cell—a tissue cell found primarily in mucosal and connective tissue that is similar to the basophil (which is found in blood)

FIGURE 12.3 Cell Identification Diagram

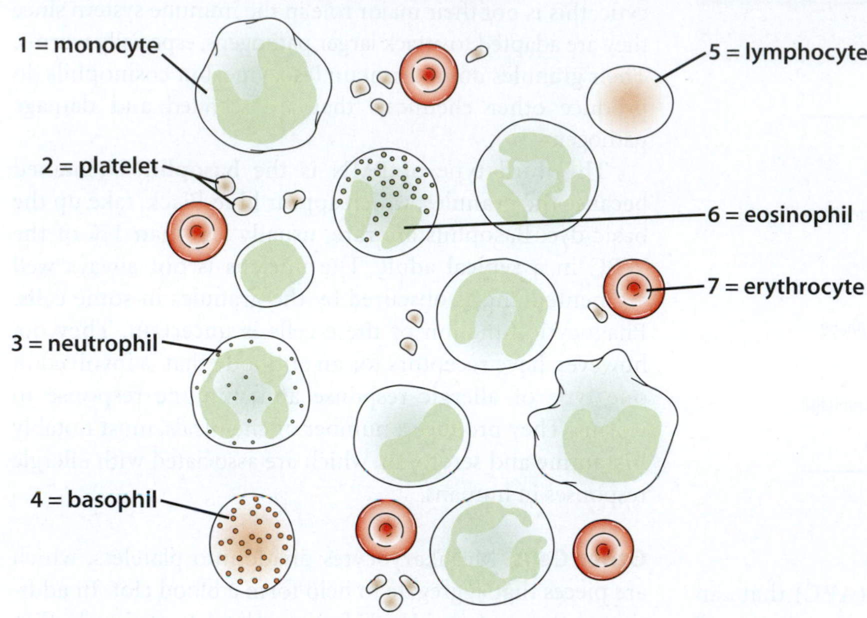

two is controversial. Mucosal mast cells are found in mucosal surfaces, while connective mast cells are found in connective tissues. The myeloid precursor also gives rise to other antigen-presenting cells, including **dendritic cells**, named for their long, nervelike membranous extensions, which are found in **lymph nodes**, the spleen, and blood; **Langerhans cells** in skin; and interdigitating cells in the lymph nodes and thymus.

Cells Derived from the Lymphoid Stem Cell

The lymphoid precursor differentiates into cells of the specific immune system, including B and T cells or lymphocytes (see Figure 12.2). B and T cells share 98% of expressed genes and cannot be distinguished from each other in a stained blood smear, but numerous antigenic markers and biological characteristics can be used to separate them. They are smaller than monocytes and PMN (6–10 μm), they are agranular, and their cytoplasm stains blue. Resting cells are mostly nucleus with a thin ring of cytoplasm, but activated cells have more cytoplasm.

T Cells T cells were named for the thymus, where they differentiate. Most of the lymphocytes in blood, lymph nodes, and lymph are T cells. Most T cells are $\alpha\beta$ T cells since their **T cell receptor** (**TCR**) contains α and β chains. However, some T cells have a TCR made from γ and δ chains. T cells are divided into subcategories based on their role in the immune response. Though the exact number of these categories is debated among immunologists, all agree that helper and cytotoxic T cells are two of the categories. Some immunologists consider the **suppressor T cell** to be a third category. It is unclear whether they are in a specific subcategory of cells or whether they are **regulatory** cytotoxic or helper T cells secreting chemicals that shut down an immune response. The immune system can be powerful, and it is important that there be a shutdown mechanism. If suppression of the immune system does not occur, autoimmune disorders may develop.

Helper T cells (**TH**), also known as T4 cells because of one of the characteristic molecules on their surface, **CD4**, are very important in directing the immune response. They determine to which antigenic determinants on an organism the immune system will respond. They select which cells of the immune system or which chemicals produced by the immune system will be activated, and they interact with other cells of the immune system, causing them to become more immunologically active and to proliferate. There are

dendritic cells—antigen-trapping and antigen-presenting white blood cells with nervelike processes (e.g., Langerhans cells and interdigitating cells)

lymph nodes—small organs of the immune system where mature B and T lymphocytes respond to an antigen; they are distributed widely throughout the body and linked by lymphatic vessels that bring in antigens from surrounding tissue

Langerhans cell—dendritic cell that traps and processes antigens in the epidermal layer of the skin and then migrates through lymphatics to lymph nodes where it presents the antigen to T cells

T cell receptor (TCR)—a two-chain structure on T cells that binds antigen and is associated on the cell with the signal transduction molecules

suppressor T cell—a T lymphocyte that suppresses (turns off) specific immune responses; this may or may not be a separate subclass of T cells

regulatory T cell—a T lymphocyte that turns off specific immune responses

helper T cells (TH)—a subset of T cells that triggers B cells to make antibodies, activates macrophages, and promotes the differentiation of other T cells

CD—"cluster Designation"; an international nomenclature system of leukocyte cell surface molecules (CD number)

CD4—a marker found predominantly on helper T cells that interacts with MHC class II molecules on antigen-presenting cells

two subsets of helper T cells called **Th1** and **Th2**. T1 helper cells activate the cellular immune system, while Th2 cells increase the production of antibodies. **Cytotoxic T cells** (CTL), also called T8 or **CD8** cells, are capable of killing targeted infected, tumor, or transplant cells directly.

B Cells B cells differentiate in bone marrow. When stimulated by antigen and T cells, B cells divide and differentiate into plasma cells and memory B cells. Plasma cells are full of endoplasmic reticulum that facilitates efficient production of protein antibodies. Memory B cells produce a rapid antibody response the next time the person is exposed to the antigen, and they normally block infection and prevent symptoms. B cells can also act as antigen-presenting cells (APC).

Natural Killer Cells Natural killer cells (NK), also called killer cells, arise from the same precursor as B and T cells, but they lack the surface markers that distinguish B and T cells. They act in a nonspecific way, in that they can recognize and attack a tumor or virally infected cells without recognizing specific antigens on the cell.

Organs of the Immune System

Organs of the immune system are classified as either central or peripheral. Central organs, where leucocytes are generated, include bone marrow and the thymus. Peripheral organs, where adaptive immune responses are initiated, include the lymphoid system, spleen, **mucosa-associated lymphatic tissue** (MALT), **bronchial-associated lymphatic tissue** (BALT), and **gut-associated lymphatic tissue** (GALT).

Central Lymphoid Organs

Bone marrow, found in the central core of long bones and in significant amounts in other bones (e.g., the cranium), is a major tissue of the body. Before birth, the fetal liver acts like bone marrow. In order to be protected from diverse pathogens, humans must be capable of generating antibodies that will react with thousands of different antigens. The genetic code for an antibody molecule is divided into seven segments, some of which have many variations. Significant antibody diversity is generated by different combinations of the segments. In the bone marrow, the DNA in B cell progenitors rearranges so that one variation of each segment is expressed, allowing the B cell to make antibodies that will respond to a specific antigen. As a consequence of the need for great diversity, some cells will produce antibodies capable of reacting with a person's own tissue antigens, which could lead to autoimmune diseases. The process of **negative selection** eliminates these self-reactive cells.

The **thymus**, a pouch of epithelial cells filled with lymphocytes, is located below the thyroid in the neck, above the heart (see Figure 12.4). It weighs about 0.5 ounces at birth, grows to about 1 to 1.5 ounces at puberty, and atrophies to about 0.5 ounces by age 40. Stem cells migrate to the thymus due to chemical signals emitted by this organ starting at six to nine weeks of gestation, and then differentiate in the thymus. T cell DNA rearranges, and T cell receptors form so that each cell is capable of responding to a specific antigen. As with B cells, selection occurs in order to retain useful cells and remove potentially self-reactive cells. About 95% of lymphocytes produced die. Some die because they are self-reactive or they did not make a useful receptor, but many die because an excess is produced in order to ensure there are enough of them. Although the thymus is very important early in life, removal of the thymus has minimal impact on the ability to respond to infections once humans reach their late teens. Removal of the thymus can be part of the treatment for the autoimmune disease myasthenia gravis.

Peripheral (Secondary) Lymphoid Organs

The **lymphatic system** (see Figure 12.4) is an extensively branched network of walled vessels with one-way valves that lead to lymph nodes. Interstitial fluid, plasma that leaks out of blood vessels and carries nutrients and WBC, enters lymph vessels to become **lymph**. (Plasma is the fluid

Th1—a subset of the T helper cells that secretes cytokines, which trigger cell-mediated immune responses that promote inflammation and antiviral responses

Th2—helper T cells that predominate in the response to allergens and parasites and that make cytokines that promote antibody responses

cytotoxic T cells (CTL)—T lymphocytes that kill cells infected by viruses or transformed by cancer

CD8—a marker found predominantly on cytotoxic T cells that interacts with MHC class I molecules on target cells

MALT (mucosa-associated lymphatic tissue)—lymphoid tissue found in the surface mucosa of the respiratory, gastrointestinal, and genitourinary tracts

BALT (bronchial-associated lymphatic tissue)—secondary lymphoid organs of the bronchial tree

GALT (gut-associated lymphatic tissue)—lymphoid tissue including Peyer's patches, the appendix, and solitary lymph nodes in the submucosa

negative selection—the process in which B and T cells that react to self molecules are deleted or functionally inactivated during their development

thymus—a primary lymphoid organ, in the chest, where T lymphocytes differentiate, proliferate, and are positively and negatively selected

lymphatic system—a system of vessels through which lymph travels, consisting of lymphatic vessels and lymph nodes at the intersection of vessels

lymph—extracellular fluid containing WBC (mostly lymphocytes) and antibodies that bathe tissues

FIGURE 12.4 The Lymphatic System

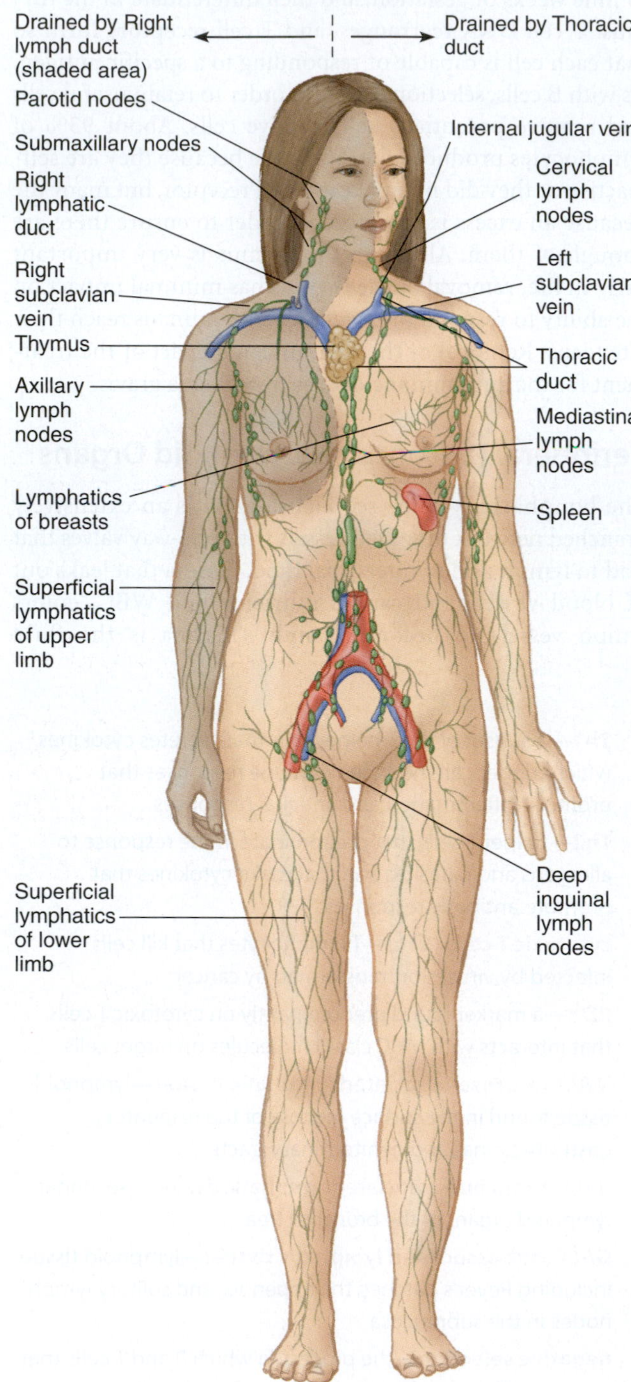

Source: Rhoades/Pflanzer, *Human Physiology*, 4e, copyright © 2003, p.852

spleen—a lymphoid organ in the abdominal cavity that filters blood

Peyer's patches—distinct lymphoid nodules in the intestine that are part of the gut-associated lymphoid tissue (GALT)

component of blood.) Antigens in intracellular spaces are swept into the lymphatic system by lymph and transported (by muscle movement that produces flow) to lymphoid organs, where they encounter cells of the immune system. Lymph nodes are highly organized lymphoid structures containing T cells, B cells, macrophages, and other APC, and are found in areas where lymph from deep and superficial areas of the body is collected. They provide a site where an immune response can be initiated against pathogens picked up in the intracellular spaces.

The major functions of the lymphatic system are to concentrate antigens from all over the body into a few lymphoid organs, to circulate lymphocytes through lymphoid organs in order to allow antigens to interact with antigen-specific cells, and to carry antibodies and effector cells to the bloodstream and tissues. Draining lymph nodes are examined after cancer surgery for cancer cells in order to determine if any of them have spread (metastasized), because some cells leaving the tumor enter the lymph. Lymph nodes become swollen when an immune reaction occurs (some refer to them as swollen glands), but long-term swelling, lymphadenopathy, can be a sign that the immune system is not functioning properly.

The **spleen**, a fist-sized organ behind the stomach, is responsible for destruction of old RBC and is a major site for mounting an immune response. More recirculating lymphocytes pass through the spleen per unit time than through all the lymph nodes combined. The spleen contains red pulp, which filters out old and damaged RBC, and white pulp, where leukocytes react with antigens. In children and young adults, splenectomy, removal of the spleen, can result in overwhelming bacterial infections.

Mucosal surfaces, major sites of pathogen entry, are protected by GALT, BALT, and MALT, which comprise about 50% of lymphoid tissue. Some epithelial cells in mucosal tissue have complex microfolds in their surfaces that contain B and T cells, macrophages, and dendritic cells. These immune cells then migrate to regional lymph nodes after they encounter an antigen. GALT includes (1) **Peyer's patches**, which are large aggregates of lymphoid tissue found in the small intestine; (2) lymphoid aggregates in the appendix and large intestine; (3) lymphoid tissue that accumulates with age in the stomach; and (4) small lymphoid aggregates in the esophagus. BALT is aggregates of lymphocytes that protect the respiratory epithelium. MALT consists of tonsils and adenoids.

Soluble Mediators

Two important soluble mediators (complement and cytokines) assist the immune system's cells in mounting an immune response. The complement proteins are activated by other elements of the immune response and are involved in destroying infected cells and some pathogens. Cytokines mediate communication among the cells of the immune system as well as between the cells of the immune system and other body systems, including the nervous system.

Complement

Complement is a system of plasma proteins that react in tightly regulated cascades activated by antibodies or specific molecules found on pathogens. These cascades result in a final product, the **membrane attack complex**, that lyses cells and produces by-products that trigger inflammation, attract phagocytes to the area, assist in phagocytosis, contribute to activation of naive B lymphocytes, and remove **immune complexes**.

Complement cascades function in both specific and nonspecific modes. Complement is an important part of the body's defenses against infected and tumor cells, but it can also kill beneficial neighboring cells and participates in transplant rejection.

Cytokines

In order for cells in the immune system to work together, they must have mechanisms of communication and regulation. **Cytokines** are proteins produced in cells that, in small amounts, affect behavior of other cells. An inducing stimulus causes a cell to make a cytokine, which then acts on a target cell that has a receptor for it. The result is altered biological activity in the target cell. The target cell may be activated, stimulated to proliferate, or stimulated to differentiate into another cell (see Box 12.2 to learn about the therapeutic uses of cytokines). If the cytokine acts on the cell that produced it, it is called an autocrine; if it acts on a nearby cell, it is called a paracrine; and if acts at a distance it is called an endocrine. Cytokines may be pleiotropic, where the same cytokine will have different effects on different cells, or redundant, where two or more cytokines have the same function. They may work alone or as part of the cascade, where production of one cytokine stimulates production of another cytokine. When more than one cytokine is present, they may be synergetic (work together) or antagonistic (counteract one another).

Initially, cytokines were called **lymphokines** because the first ones discovered were produced by lymphocytes. Later, many of the newly discovered ones were called **interleukins**, because it was thought they only communicated between WBC. Thus, many cytokines are identified by IL and a number (e.g., IL-2). One of the most widely studied is **IL-2**, which is very active in causing cells of the immune system to proliferate and become more immunologically active. Several cytokines are used clinically (Table 12.4), including IL-2, which is given to individuals with compromised immune systems. **Interferon**, abbreviated INF and a major component of the body's defense against viruses, is also a cytokine,

BOX 12.2 CLINICAL APPLICATIONS

Therapeutic Uses of Cytokines and Cytokine Inhibitors

Some cytokines that regulate the immune system have been used to increase selected cell populations damaged by disease or by other therapeutic interventions, or to decrease inflammation. Interferon (Roferon-A, Intron-A, Rebetron, Alferon-N, Peg-Intron, Avonex, Betaseron, Infergen, Actimmune, Pegasys) is used in attempts to block viruses, including those involved in cancer initiation, and to stimulate the immune system. Recombinant Interleukin-10 and Interleukin-4 mimic the action of those anti-inflammatory cytokines. Oprelvekin (Neumega) is a synthetic interleukin-11 (IL-11) used to prevent low platelet counts. Naturally occurring IL-11 is involved in platelet production. Cytokines can be part of the mechanism of some diseases. One new therapeutic strategy is to block cytokines by the use of soluble receptors. Etanercept (Enbrel®), a soluble TNF receptor, is used to treat rheumatoid arthritis. Anakinra (Kineret) is a naturally occurring IL-1 receptor antagonist used to treat rheumatoid arthritis. Infliximab (Remicade) and adalimumab (Humira®) are monoclonal antibodies against TNF used to decrease inflammation in people with rheumatoid arthritis and Crohn's disease. Clinical uses of IL-2 have been hampered since IL-2 has a domain that produces vascular permeability. When administered systemically in clinically effective doses, IL-2 can cause massive leaking of blood, fluids, and serum proteins from the vascular network, which leads to organ failure.

complement—a group of serum proteins activated in a cascade that produces compounds that lyse cells and mediate immune reactions

membrane attack complex—the final product of the complement cascade that forms a pore on the surface of the target cell, which results in lysis of the cell

immune complex—a cluster of antibodies bound to antigens

cytokines—soluble substances secreted by one cell that cause it or other cells to proliferate, differentiate, migrate, or become activated

lymphokine—a soluble molecule used for communication between lymphocytes and other cells

interleukin—now used primarily as a naming convention for cytokines/lymphokines/chemokines/growth factors (IL-number)

IL-2—interleukin-2; a lymphokine required by activated T cells for growth

interferon (INF)—a group of cytokines that regulate the immune system and protect cells from viruses

TABLE 12.4

Cytokines with Clinical Potential

Cytokine	Function in Vivo	Possible Clinical Use
Interleukin-10 (IL-10)	Inhibits IFNg production	Cancer
		Inflammation
		Transplantation
	Down regulates MHC II	Immunodeficiencies
		Parasitic infections
Interleukin-11 (IL-11)	Induces Th1 responses and IFNg production	Breast cancer
	Causes B cell progenitors to differentiate	
Interleukin-12 (IL-12)	Antitumor, possibly by interfering with blood flow to the tumor	Cancer
Interleukin-14 (IL-14)	Induces proliferation of activated B cells	Various lymphomas
	Inhibits antibody synthesis	
IL-1Ra Interleukin-1 receptor antagonist	Anti-inflammatory cytokine; no known agonist function	Anti-inflammation in rheumatoid arthritis

as is **tumor necrosis factor (TNF)**. Three classes of interferons have been identified: (1) Alfa, (2) beta, and (3) gamma. Though their activities overlap, each class has different effects; Alfa is more antiviral than the other classes, and gamma is more active in immune regulation than the other two.

tumor necrosis factor (TNF)—a cytokine that induces programmed cell death, primarily in tumor cells but for any cell with a receptor. Also involved in immunoregulation

BCR—a B-cell receptor made of an antibody molecule and several auxiliary molecules

major histocompatibility complex (MHC)—a cluster of genes encoding polymorphic cell-surface molecules (MHC class I and class II) that help the organism identify pathogens as foreign; they are important in antigen presentation to T cells, play a role in transplantation rejection, and influence the susceptibility to certain autoimmune diseases; MHC antigens are also called HLA antigens

Class I MHC antigen—glycoproteins found on nucleated cells and encoded by the A, B, and C locus of the major histocompatibility complex; they present antigen to cytotoxic (CD8 +) T cells

Class II MHC antigen—glycoproteins found on nucleated cells and encoded by the Dr, Dq, or DP locus of the major histocompatibility complex; they present antigen to helper (CD4 +) T cells

Antigen Recognition Molecules

Antigens must be recognized by T and B cells in order to initiate an immune response. Activation of B cells requires that the antigen react with only the **B cell receptor (BCR)**, but the antigen must be recognized by both the T cell receptor (TCR) on the T cell and the major histocompatibility complex molecule on the antigen-presenting cell in order to activate a T cell.

Major Histocompatibility Complex (MHC)

In the early 1900s, scientists noted that acceptance or rejection of tumors transplanted between mice was influenced by several dominant genes. This was followed by the discovery that skin transplants between identical twins were not rejected. The genes responsible mapped to an area on chromosome 6 called the **major histocompatibility complex (MHC)**, which contains genes for MHC antigens (also called human leukocyte antigens or HLA) and other proteins involved in the immune response. It is highly unlikely that a system as complex as the MHC evolved to regulate Activation of T cells requires that an antigen be "presented" to the T cells attached to the MHC molecules on the surface of the APC. The T cell is not activated unless its receptor recognizes both the antigen and the MHC molecule.

In humans, there are two kinds of MHC antigens, Class I and Class II. The classes differ in structure, function, and cell distribution (see Figure 12.5). The three types of **Class I MHC antigens** are named A, B, and C. MHC Class I molecules contain two proteins, a β2m and an α chain that has three folded regions outside the cell and a transmembrane/cytoplasmic tail.

The genes from both parents are expressed, so two A, two B, and two C antigens (one inherited from each parent) are found on nucleated cells of the body in varying amounts. Antigens from pathogens that infect the cell and pieces of new proteins made in cancerous cells bind to the MHC I protein and are carried to the cell surface, where they can activate TC cells. This allows the immune system to recognize and destroy infected and cancerous cells.

Class II MHC antigens include types DR, DQ, and DP, and are found on cells involved in antigen presentation and activated T cells. They are comprised of two chains, α and β, that are inherited from both parents and can combine in four ways: (1) αDad and βDad, (2) αMom and βMom, (3) αDad and βMom, and (4) αMom and βDad. Thus, each cell has up to 4 different antigens each for DQ, DR, and DP, for a total of 12 possible MHC II antigens per cell. Class II antigens present antigen fragments from molecules or pathogens ingested and digested by the APC to TH cells. This allows the immune system to mount the appropriate immune response to a pathogen that has been phagocytized by the APC (see Figure 12.6).

An individual's MHC haplotype (a combination of closely linked genes on a chromosome inherited as a unit

FIGURE 12.5 MHC Antigens

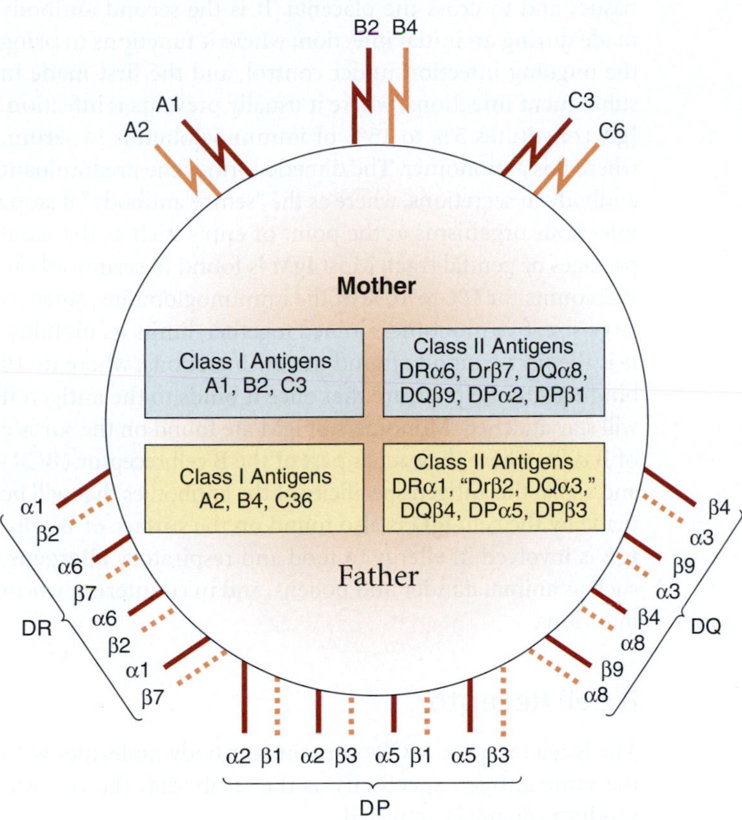

from one parent) is the combination of her MHC Class I and Class II antigens. Multiple alleles and thus multiple antigens exist for A, B, C, DR, DP, and DQ, so there is a nearly infinite number of haplotypes. Thus, few individuals, other than identical twins, have the exact same haplotype. MHC antigens play an important role in transplant rejection, since the presence of a MHC I II antigen on the transplanted organ or tissue that is different from the MHC I II antigens on the recipient's tissues signals the presence of the transplanted tissue and initiates an immune response. The immune system attacks the transplant at MHC I antigens that are different from those found on the recipient's tissues.

MHC antigens play a role in susceptibility to some diseases, including many autoimmune and some infectious diseases, since they select the antigenic fragments from the pathogen that will be presented to the T cells. If a MHC selects a fragment from a pathogen that is similar to an antigen on a human tissue, the resulting immune response might attack the human tissue.

Antibodies

Antibodies, proteins made in response to an antigen that assist in the destruction or neutralization of the antigen, are found primarily in the gamma globulin portion of serum when they are in blood. Thus, they are also called immunoglobulins, which is abbreviated Ig. An antibody molecule must bind specifically to antigens from the pathogen, so part of the molecule must be variable to react specifically with a large variety of antigens. Often the antibody must also recruit other cells or molecules to destroy the pathogen. Thus, part of the antibody must be constant to be recognized by the cells or molecules. The basic antibody monomer is Y shaped and consists of two identical longer protein chains, called the **heavy chains**, and two identical shorter protein chains, known as **light chains** (see Figure 12.7). The COOH end, or **constant region**, of each heavy and light chain has an amino acid sequence that is very similar to the sequence in other antibody molecules. This is the part of the molecule (the **Fc**) that reacts with other cells and chemical mediators (e.g., complement). The NH_3 ends, or **variable region,** of the heavy and light chains are in the two branches of the Y-shaped molecule. They have highly variable amino acid sequences and create the **Fab**, where the antigen binds to the antibody.

The five classes of antibodies are defined by differences in the heavy chain: (1) IgG, (2) IgA, (3) IgM, (4) IgD, and (5) IgE. There are four subclasses of IgG that differ in amino acid sequence in the heavy chain, and two subclasses of IgA, a monomer and a dimer. Each class or subclass differs in structure and function. **IgG** is the most abundant

heavy chain (H chain)—the larger of the two types of immunoglobulin chains

light chain (L chain)—the smaller of the two types of immunoglobulin chains; there are two forms: k or l

constant region (C region)—the carboxyl-terminal portion of an immunoglobulin or TCR molecule that is similar from molecule to molecule

Fc—the part of the antibody without antigen-binding sites made of the C-terminal or constant domains of the immunoglobulin heavy chains

variable region—the part of an antibody or TCR that differs from one antibody or TCR to another and produces a binding site for a specific antigen

Fab—part of the antibody molecule containing one antigen-binding site; contains the variable ends or N terminus of one light and one heavy chain

IgG—the predominant immunoglobulin class produced during secondary immune responses; the most prevalent immunoglobulin in the blood

FIGURE 12.6 Activation of T Cells against a Virally Infected Cell

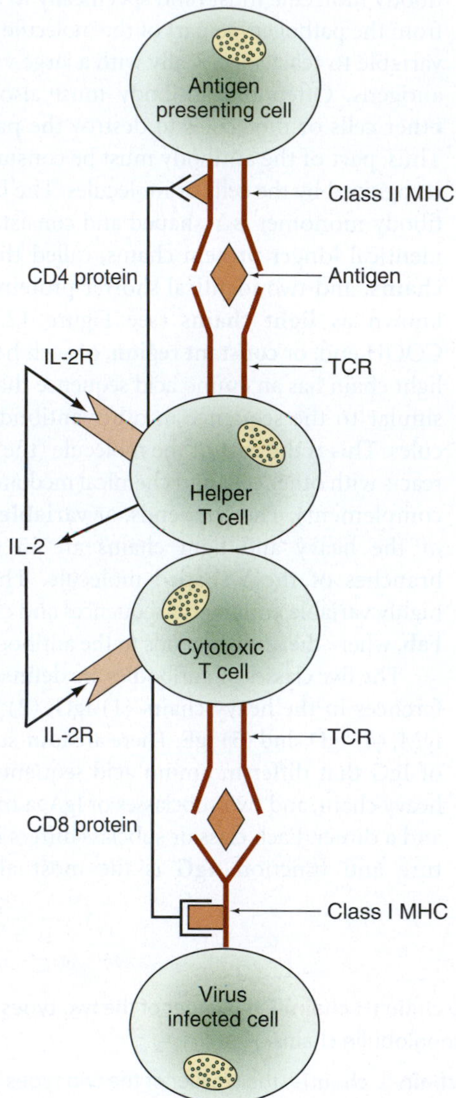

Source: From Sylvia A. Price & Lorraine M. Wilson *Pathophysiology: Clinical Concepts of Disease Process* 6e by Mosby, copyright © 2003, Figure 12.6. Reprinted with permission from Elsevier.

IgA (immunoglobulin A)—the predominant immunoglobulin in secretions

IgM—the predominant immunoglobulin class expressed by virgin B lymphocytes and secreted during primary immune responses

IgD (immunoglobulin D)—an immunoglobulin present in the surfaces of B cells

IgE—the immunoglobulin class that is the predominant mediator of immediate hypersensitivity reactions (allergies)

allergen—an antigen that triggers an allergic response

superantigen—an antigen that activates a large number of T cells by reacting with the TCR and MHC outside of the normal antigen binding sites

(75%). Its size allows it to move easily between serum and tissues and to cross the placenta. It is the second antibody made during an initial infection, where it functions to bring the ongoing infection under control, and the first made in subsequent infections, where it usually prevents reinfection. **IgA** constitutes 5% to 15% of immunoglobulins in serum, where it is a monomer. The dimeric form is the predominant antibody in secretions, where as the "sentry antibody" it stops infectious organisms at the point of entry such as the nasal passages or genital tract. Most **IgM** is found in serum, where it accounts for 5% to 10% of the immunoglobulins, since its large size, five monomers joined together, limits its mobility. It is the first antibody found in new infections, where its 10 binding sites help ensure that once it binds to the antigen it will stay attached. Monomers of IgM are found on the surface of B cells, where they act as part of the B cell receptor (BCR) and signal the antigen specificity of the antibodies that will be made by the cell. **IgD** is also found on the surface of B cells. **IgE** is involved in allergy to food and respiratory **allergens**, such as animal dander and pollens, and in countering worm infections.

B Cell Receptor

The B cell receptor (BCR) contains antibody molecules with the same antigen specificity as the antibodies the cell will produce when it is activated.

T Cell Receptor

The T cell receptor (TCR) has several things in common with the antibody molecule. It is composed of two disulfide bonded chains with a constant transmembrane COOH end and an extra cellular variable amino terminus.

Interaction between the T Cell Receptor and the MHC Antigens

In order to activate a T cell, a TCR must bind with both the antigen and MHC, MHC I for TC and MHC II for TH. This bond is not strong, but co-receptor molecules, CD8 on TC and CD4 on TH, attach to MHC outside of the antigen binding area to provide additional strength to the bond.

Superantigens such as *Staphylococcal enterotoxin* B and toxic shock toxin activate many TH cells by binding to the outside of the MHC II molecules and to the variable region of the TCR. The resulting massive release of cytokines can contribute to the signs and symptoms associated with the infection.

Immune Response: Attacking Pathogens

The immune system must protect against a variety of organisms that enter the body by several mechanisms. Some multiply inside cells, while others reside in blood or lymph

FIGURE 12.7 Antibody Structure

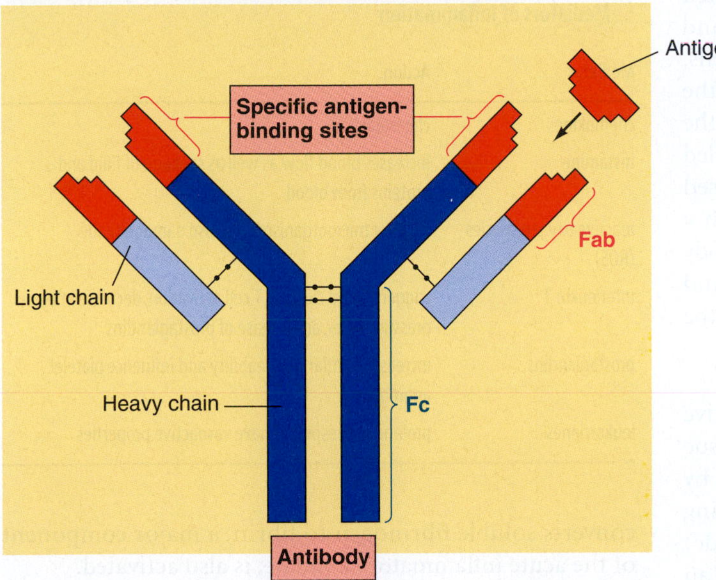

Source: Lauralee Sherwood, *Human Physiology: From Cells to Systems*, 5e, copyright © 2004, p. 426

or on epithelial surfaces. Some cause damage by destroying tissue as they move through the body, while others produce toxins that are responsible for symptoms. Several modes of immune response have developed, and different types of pathogens are attacked by different combinations of these modes of attack.

Modes of Attack

Phagocytosis The immune response begins with phagocytosis of the pathogen. In this process, phagocytic cells—macrophages, monocytes, and neutrophils—are drawn to the pathogen through chemical signals. Soluble factors secreted by bacteria, AG-AB complexes (antibodies bound to antigens), particles produced in complement pathway, and cytokines produced in response to bacterial components (including LPS, peptidoglycan monomers, teichoic acids, mycolic acid, and mannose) attract WBC to the area where they are produced.

Once the phagocytic cell and pathogen are in proximity, the pathogen adheres to the cell and is ingested. The pathogen is brought into a vacuole, the **phagosome**, which fuses with a **lysosome** (containing chemicals to digest the pathogen) and forms the **phago-lysosome**. The pathogen is destroyed by the contents of the phago-lysosome, an increase in superoxide in the cell due to increased oxygen uptake, and changes in pH.

After the pathogen is broken down, most pieces of the pathogen are egested (excreted), but some are retained for antigen presentation. The MHC II molecules of the phagocytic cell dictate which pieces are saved and presented. Once a phagocytic cell displays the antigen via the MHC, it is capable of activating TH cells and thus initiating the specific immune response.

The macrophage is not always successful in breaking down the pathogen. **Granulomas**, such as the tubercle in tuberculosis, form when there is persistent inflammation due to the continued presence of a foreign body or infection.

Cell-Mediated Cytotoxicity Activated CTL and NK cells kill infected or tumor cells using cell-mediated cytotoxicity. CTL react with the target cell when their TCR attach to antigens presented by MHC I molecules on the target cell (see Figure 12.6). Some viruses and cancers cause cells to make fewer MHC I molecules, thus protecting them from attack by TC but enhancing attack by NK cells.

The precise mechanism used by NK cells to recognize target neoplastic or tumor cells is not clear. It is neither antigen specific nor influenced by the specific MHC I antigens found on the cell, but the presence of MHC I antigens inhibits the attack by NK cells.

Cells with an antibody attached to surface antigens can be attacked by NK cells through antibody-dependent cell-mediated cytotoxicity (ADCC). NK cells have receptors for the Fc part of the antibody molecule, so the NK cell and the target cell are brought together by the antibody molecule. Once the cells interact, granules in the TC or NK release **perforin** and enzymes into the space between cells. Perforin molecules make a channel in the target cell membrane, and enzymes and toxins made by the TC or NK cell go through the channels, causing the cell to go into apoptosis (programmed cell death) and die.

phagosome—the cytoplasmic vesicle that encloses an ingested organism during phagocytosis

lysosomes—cytoplasmic granules involved in the digestion of phagocytosed material that contain hydrolytic enzymes

phago-lysosome—intracellular vacuole where the killing and digestion of phagocytosed material occurs; produced by the fusion of a phagosome and a lysosome in a phagocytic cell

granuloma—a mass of macrophages, with some T lymphocytes at the periphery, formed at the site of persisting inflammation due to the continued presence of a foreign body or infection

perforin—molecule produced by cytotoxic T cells and NK cells that forms a pore in the membrane of the target cell to allow chemicals from the T cell or NK cell to enter the target cell and induce apoptosis

Antibodies Antibodies can attack extracellular pathogens and toxins and can contribute to the destruction of infected and tumor cells. Antibodies neutralize bacterial toxins and viruses by blocking their ability to attach to target cells. Antibodies attached to infected or tumor cells trigger the complement cascade, which leads to cell lyses, and target the cell for killing by NK cells through ADCC. Antibodies called opsonins can attach to a cell or pathogen, and the exposed Fc portion of the antibody attaches to Fc receptors on a macrophage. Opsonins are substances, typically an antibody or complement component, that bind to an antigen and enhance its phagocytosis, providing a "handle" by which the phagocytic cell can attach to the antigen.

Inflammation Inflammation originated as a protective response to pathogens, but also occurs in response to tissue injury (see Chapter 10). Pathogens initiate inflammation by activating **alternate complement pathways** or by binding bacterial surface molecules such as LPS (lipopolysaccharide) and peptidoglycan to WBC, especially mast cells. LPS (an endotoxin) found in the outer lipid bilayer of the cell wall of Gram-negative bacteria also triggers complement activation, yielding by-products that contribute to inflammation by causing vasodilation and increasing vascular permeability.

Several mechanisms are activated in addition to increased vascular permeability. These include leukocyte production, chemotaxis and phagocytosis, coagulation, neovascularization, fibrinolysis, fibroplasia, and repair. The cardinal signs of inflammation (increased temperature, redness, swelling, and pain) result from increased blood flow to the area. Interaction with either by-products of the complement cascade or surface molecules from the pathogen causes mast cells to release mediators of inflammation. Some mediators (e.g., histamine) come from granules where they have been stored, while others (e.g., **leukotrienes**) are synthesized. These mediators recruit monocytes (that become macrophages, neutrophils, eosinophils, antigen-presenting dendritic cells, NK cells, and B and T cells) to the site, promote the movement of WBC in and out of the blood through **diapedesis**, and activate cells to produce additional mediators of inflammation (see Table 12.5). The coagulation system, which converts soluble fibrinogen to fibrin, a major component of the acute inflammatory exudate, is also activated.

alternate complement pathway—a complement activation pathway that does not involve activation by an antibody

leukotrienes—metabolic products of arachidonic acid that promote inflammation

diapedesis—part of the of the inflammation response involving movement of blood cells across blood vessel walls into tissues

primary immune response—the immune response when the naive lymphocyte first encounters its antigen

TABLE 12.5

Mediators of Inflammation

Mediator	Action
chemokines	chemotaxis
histamine	increases blood flow as well as seepage of fluid and proteins from blood
reactive oxygen species (ROS)	toxic for microorganisms but also damage tissue
Interleukin 1	triggers blood clotting, T cell activation, decrease in blood pressure, fever, and release of prostaglandins
prostaglandins	increase vascular permeability and influence platelet aggregation
leukotrienes	prolong the response; have vasoactive properties

Inflammation is a two-edged sword. It protects by walling off the damaged area and any pathogens there, bringing effector cells and molecules to the area, and promoting healing. On the other hand, if the response is out of proportion to the threat or becomes chronic, more damage can occur than would have been caused by the pathogen. Inflammation is a component of allergies and some autoimmune diseases including rheumatoid arthritis (RA).

Progression of the Immune Response: Putting the Parts Together

The Primary Response The **primary immune response** (see Figure 12.8) normally begins with phagocytosis, which can be sufficient to block infection if virulence of the pathogens is low and/or the inoculum is small. This process can remove about 90% of an antigen by the time the initial inoculum has circulated through the body once. The phagocytic cells become APC for the helper T cells (TH).

TH activation requires recognition of the MHC II and antigen presented by the APC and interactions between costimulatory molecules on both the T cell and the APC. In addition to providing the MHC-antigen signal, the macrophages secrete cytokines that contribute to activation of TH, B cells, PMN, and NK.

The activated TH cell begins to divide and secretes cytokines that will activate other cells of the immune system. IL-2 made by T cells is the key cytokine involved in inducing proliferation and activating other T cells, B cells, monocytes, and NK. Activated TH cells secrete cytokines that induce proliferation of B cells that can react with antigens on the pathogen and trigger the B cells' differentiation into a plasma cell or antibody-forming cell (AFC). Plasma cells are larger than B cells and packed full of endoplasmic reticulum to produce antibodies at a rate of 30,000 Ig/sec.

FIGURE 12.8 Interactions among Macrophages, B Cells, and Helper T Cells

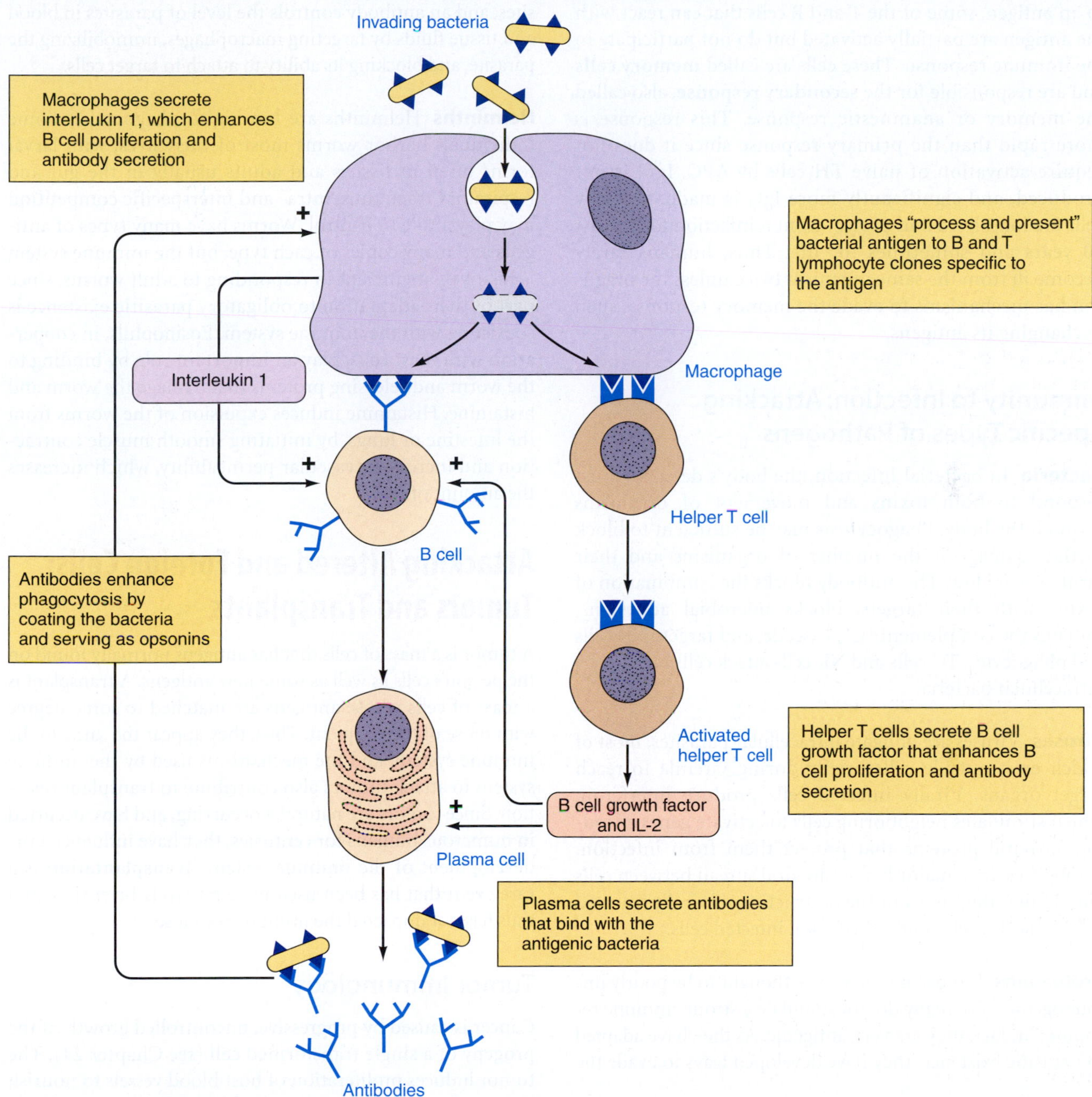
Invading bacteria
Macrophages secrete interleukin 1, which enhances B cell proliferation and antibody secretion
Macrophages "process and present" bacterial antigen to B and T lymphocyte clones specific to the antigen
Interleukin 1
Macrophage
Helper T cell
B cell
Antibodies enhance phagocytosis by coating the bacteria and serving as opsonins
Activated helper T cell
Helper T cells secrete B cell growth factor that enhances B cell proliferation and antibody secretion
B cell growth factor and IL-2
Plasma cell
Plasma cells secrete antibodies that bind with the antigenic bacteria
Antibodies

Source: Lauralee Sherwood, *Human Physiology: From Cells to Systems,* 5e, copyright © 2004, p. 434.

After about five days, the plasma cell will produce IgM for about three days before switching to the production of IgG through class or **isotype switching** (class switching). Some cells will switch to either IgE or IgA instead of IgG due to the influence of cytokines. Since it takes time to produce sufficient antibodies and activated T cells, the response does not prevent disease but terminates the ongoing infection. The longevity of the antibody, which will degrade over time, depends on a number of characteristics of the antigen and host.

isotypes—antibody classes (IgM, IgG, IgD, IgA, and IgE) with specific biological activities due to differences in the heavy chain constant regions

isotype switching—the process by which a plasma cell changes the class (e.g., IgM to IgG) but not the specificity of the antibody it produces; also called class switching

The Secondary Response During the primary response to an antigen, some of the T and B cells that can react with the antigen are partially activated but do not participate in the immune response. These cells are called **memory cells** and are responsible for the **secondary response**, also called the memory or anamnestic response. This response is more rapid than the primary response since it does not require activation of naive TH cells by APC. IgM is not produced, and significantly more IgG is made. Memory cells are long-lived and protect from reinfection for at least 20 years and sometimes for life. Thus, humans rarely become ill from the same organism twice unless the organism has mechanisms to evade the memory response, such as changing its antigens.

Immunity to Infection: Attacking Specific Types of Pathogens

Bacteria In bacterial infection, the body's defenses must respond to both toxins and movement of organisms through the body. Phagocytosis may be sufficient to block further damage if the number of organisms and their virulence are low. The antibody blocks the combination of toxins with their targets, blocks microbial adhesions, initiates the complement (C′) cascade, and targets NK cells and phagocytes. TC cells and NK cells attack cells infected by intracellular bacteria.

Viruses Viruses are obligate intracellular parasites, most of which exist outside cells briefly during viremia to reach target organs. Virally infected cells produce interferon, which stimulates neighboring cells to activate genes encoding antiviral proteins that protect them from infection. Antibodies are a major barrier to viral spread between cells and tissues; they assist in the destruction of virally infected cells while TC cells and NK cells kill infected cells.

Protozoans Protozoans were once thought to be poorly immunogenic since many do not stimulate a strong immune response; in fact, they are very antigenic. As they have adapted to parasitic existence, they have developed ways to evade the immune response. Macrophages can kill extracellular parasites, and an antibody controls the level of parasites in blood and tissue fluids by targeting macrophages, immobilizing the parasite, and blocking its ability to attach to target cells.

Helminths Helminths are too big to phagocytize. Some individuals harbor worms most of their lives, with larval forms often in tissues and adults usually in the gut and respiratory tract; thus, intra- and interspecific competition may play a role in control. Worms have many types of antigens and many copies of each type, but the immune system is relatively inefficient in responding to adult worms, since part of their adaptation to obligatory parasitic existence is coexisting with the immune system. Eosinophils, in cooperation with mast cells, play an important role by binding to the worm and releasing proteins that damage the worm and histamine. Histamine induces expulsion of the worms from the intestine or lungs, by initiating smooth muscle contraction and increasing vascular permeability, which increases the amount of fluid.

Attacking Altered and Foreign Cells: Tumors and Transplants

A tumor is a mass of cells that has antigens normally found on the person's cells as well as some new antigens. A transplant is a mass of cells whose antigens are matched to some degree with those of the recipient. Thus, they appear the same to the immune system, and the mechanisms used by the immune system to attack tumors also contribute to transplant rejection. Since tumors are naturally occurring, and have occurred in numerous humans for centuries, they have influenced the development of the immune system. **Transplantation** is a rare event that has been used for a relatively brief time and thus has not impacted the immune response.

Tumor Immunology

Cancer is caused by progressive, uncontrolled growth of the progeny of a single transformed cell (see Chapter 24). The tumor induces proliferation of host blood vessels to nourish the mass, and out-competes healthy neighboring cells for nutrients and space while producing toxic waste products. There is ample evidence that tumors are attacked by the immune system. Immune cells, including lymphocytes, macrophages, PMN, and dendritic cells, are found in tumors (see Figure 12.9).

Tumor cells have new antigens in addition to those on similar noncancerous cells. When a virus causes a tumor, some of the antigens are specific to the virus, while other antigens, found on tumors of both viral and chemical origin, are unique to the tumor. Cancer cells shed antigens that are picked up by APC and recognized by multiple elements of the immune response as new antigens or an excess of

memory cells—lymphocytes produced on the first encounter with an antigen that produce a rapid, more vigorous response upon subsequent exposures, which often prevents reinfection

secondary immune response—rapid, more vigorous immunologic response by memory lymphocytes after the first encounter with an antigen; produced upon subsequent exposures to the antigen; often prevents reinfection

transplantation—grafting an organ (e.g., kidney or heart) or cells (e.g., bone marrow) from one individual to another

FIGURE 12.9 Immune Surveillance against Cancer

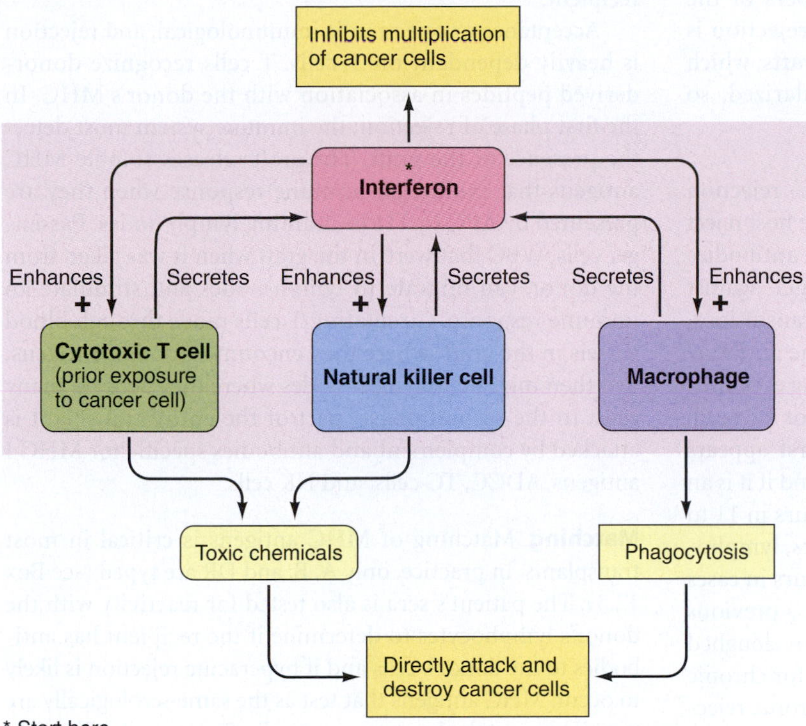

Source: Lauralee Sherwood, *Human Physiology: From Cells to Systems,* 5e, copyright © 2004, p. 444

existing antigens. Solid tumors are attacked by TC, NK cells, complement, ADCC, and macrophages that release antitumor chemicals. Metastatic cells (cells that have broken off a tumor and are traveling to a new site) and single cancer cells are destroyed primarily by antibodies and complement.

The immune system attacks cancer cells, but some escape. Tumor cells are poor APC. Some, such as Hodgkin's disease, suppress the immune system, while others cover themselves in molecules that block lymphocyte attachment (antigen masking). In immunoselection, rare random mutants lose their original surface antigens, which are replaced by new antigens to which the immune attack must be retargeted. Some tumors lose or decrease molecules, including **costimulatory molecules** and MHC I, which prevents attack by cytotoxic T cells (Tc). Immune enhancement occurs when antibodies bind to antigens and block T cell interactions. If the tumor can attain a size where cells lost to immune reactions are rapidly replaced by new cells, the immune response becomes much less effective.

Immunological Approaches to Cancer Therapy Several immunological approaches to therapy have shown various levels of effectiveness against specific cancers. Transfer of live, sensitive lymphocytes from individuals with the same tumor is hampered by **graft-versus-host disease** (see the next section). Use of BCG, a vaccine for tuberculosis that generates many macrophages, has produced some success with melanoma and bladder cancer. Passive antibodies, despite problems with **serum sickness**, have been marginally effective, but **monoclonal antibodies** may produce better results. Use of an antibody to deliver toxins, drugs, and radioisotopes to tumor cells allows reagents that can be harmful to healthy cells to be concentrated on the tumor. Cytokines including IL-2, IL-4, and IL-5, TNF, and INFaα have produced scattered hopeful results. Lymphokine activated killer (LAK) cells are produced from WBC taken from a healthy person and grown in culture with IL-2. Resulting cells, 40% of which are NK cells, are given to the person with cancer, but success has been limited since few cells localize in the tumor. Tumor infiltrate lymphocytes (TIL) are tumor-specific TC cells prepared by growing part of the person's tumor in culture in the presence of IL-2, and are 50 to 100 times more effective than LAK. Designer lymphocytes add genes for cytokines that fight cancer to TIL cells. Efforts to prepare a cancer vaccine are hampered by the lack of common antigens on chemically induced tumors and risks of immune enhancement.

Transplantation Immunology

Transplants can normally occur between one part of a person's body and another (**autograft**) or from an identical

costimulatory molecules—membrane-bound or secreted products required in addition to MHC/TCR interactions for activation of T cells

graft-versus-host disease (GVHD)—a life-threatening reaction in which transplanted immunocompetent cells, usually T cells, attack the tissues of the immunocompromised recipient

serum sickness—a Type III hypersensitivity response following the administration of a passive antibody in foreign serum

monoclonal antibody—an antibody produced by an immortal B cell line that reacts with a single antigenic determinant

autograft—a tissue graft from one area to another on the same individual

twin (**isograft**) without a problematic immune response. Most transplants, however, are between members of the same species (**allografts**), and immunological rejection is an important factor. The vast majority of **xenografts**, which come from a different species, are not vascularized, so immunological rejection is not a concern.

Transplant Rejection Host-versus-graft (HVG) rejection occurs when immunological competent cells in the host reject the graft. Acute rejection is caused by circulating antibodies that are either naturally occurring IgM or IgG against xenografts, the result of multiple pregnancies or transfusions, or a result of improper blood type matching. The antibody reacts with antigens on the surface of cells, blocking establishment of good circulation, and the graft does not succeed. When **first-set rejection** occurs, the graft at first appears healthy, with good vasculature and blood supply, and if it is an organ, starts to function. However, rejection occurs in 11 to 17 days and tissue is infiltrated with macrophages, lymphocytes, and plasma cells. **Second-set rejection** occurs in cases involving a graft with the same MHC antigens as a previous graft. A memory response is mounted so the graft is sloughed off in three to four days. It takes months to years for chronic rejection to occur. In properly matched grafts, chronic rejection is due to antibodies to **minor histocompatibility antigens**, immune complexes, or viruses that stimulate immune responses by placing new antigens on infected cells in the graft. The life span of a kidney transplant is about seven to eight years.

In graft-versus-host (GVH) rejection, which occurs in bone marrow but not heart or kidney transplants, immunocompetent cells (those capable of mounting an immune response) are in the graft, and the host is incompetent due to age, disease, or immunosuppression. Lymphocytes in the graft are sensitized to the recipient's antigens and mount an immunological attack on multiple tissues and organs of the recipient.

Acceptance or rejection is immunological, and rejection is heavily dependant on T cells. T cells recognize donor-derived peptides in association with the donor's MHC. In the first phase of rejection, the immune system must detect the presence of the graft. The graft releases soluble MHC antigens that initiate an immune response when they are presented by APC to TH in draining lymph nodes. Passenger cells, WBC that were in the graft when it was taken from the donor, can migrate to lymph nodes and stimulate an immune response. Circulating T cells move through blood vessels in the graft, where they encounter foreign antigens, and then migrate to lymph nodes where they activate many cells. In the second phase, part of the entire transplant is attacked by complement and antibodies specific for MHC I antigens, ADCC, TC cells, and NK cells.

Matching Matching of MHC antigens is critical in most transplants. In practice, only A, B, and DR are typed (see Box 12.3). The patient's sera is also tested for reactivity with the donor's lymphocytes to determine if the recipient has antibodies to the donor's cells and if hyperacute rejection is likely to occur. MHC antigens that test as the same serologically are not always exactly the same genetically. Since small differences over a long time period can influence graft survival, genetic matching could be beneficial, but such a process makes finding suitable organs for transplant much more difficult. The chance that two random individuals will match on both As and Bs and the beta chain of both the DR and DQ is 1/7,805,952,000 using serology and 1/6,663,259,940,000 with DNA. Family members are often considered as potential donors. Since the genes for MHC A, B, C, DP, DQ, and DR are inherited as a unit (haplotype), each individual will share at least one A, one B, and one C with each parent; hypothetically, with three-quarters of his or her siblings; and possibly with grandparents, aunts, uncles, and cousins.

isograft—tissue transplanted between two genetically identical individuals; also called syngraft

allograft—a tissue/organ graft between two genetically different individuals from the same species

xenograft—tissue transplantation between individuals from different species

first-set reaction—rejection of a foreign tissue graft due to antibodies and activated cells formed in response to the graft; usually occurs one to two weeks after the tissue is transplanted

second-set rejection—accelerated rejection of an allograft due to previous exposure to some of the antigens on the graft

minor histocompatibility antigens—cell surface processed peptides not encoded by the MHC that can contribute to graft rejection

BOX 12.3 CLINICAL APPLICATIONS

Finding the Right Donor: MHC Matching

In a cytotoxic assay, lymphocytes are mixed with antisera to each MHC antigen in the presence of complement, and cell lysis due to antigen antibody reaction is detected by dyes such as trypan blue or Cr51 release. Unless a person inherited the same alleles from both parents, their lymphocytes will react with antisera to two of the A antigens, two of the B antigens, two of the C antigens, and at least two alpha and two beta chains each for DR, DP, and DQ.

In mixed lymphocyte culture, lymphocytes from the donor and recipient are mixed in culture with the donor's lymphocytes inactivated. If they are incompatible, proliferation measured by increased DNA synthesis is initiated. Absence of proliferation is a strong predictor of graft survival.

Immunosuppression Immunosuppression is required both at the time of the transplant and lifelong for the transplant recipient. Physical immune suppression in the form of radiation is fairly nonspecific and relatively ineffective alone, so it is used in combination with other methods around the time of the transplant. Many chemicals used in immunosuppression, especially in the early years, are cell toxins and thus have many side effects. Several have also been used in cancer chemotherapy, because they attack rapidly growing cells. Discovery of Cyclosporine A and its relatives, including FK506 and Rapamycin, advanced chemical immunosuppression significantly, since they selectively inhibit T cells.

Transplantation of Specific Organs and Tissues A major challenge in transplantation is organ shortage. Addressing this issue raises complex ethical and social issues such as organ allocation, whether individuals should be able to opt out of donation, especially to a family member, whether people should be able to donate for payment, and the treatment of brain death/cadaveric donors.

Some sites and tissues are known as privileged and MHC matching is not required. Allograft tissue is protected from rejection in **privileged sites** such as the brain, the anterior chamber of the eye, and the cornea, because they are not vascularized. Trauma at the site can lead to inflammation and then rejection. Privileged tissues such as bone, cartilage, heart valves, and blood vessels are usually not rejected no matter where they are transplanted, since they are more structural than cellular. Most xenografts, such as porcine heart valves in humans, are privileged tissues.

The degree of MHC matching varies significantly according to the tissue, since tissues differ immunologically (e.g., expressing different levels of MHC II antigens). Matching is very important in hemopoietic stem cell transplants (i.e., bone marrow), significant in kidney transplants, and desirable with heart transplants, but it appears to have no net beneficial effect in liver transplants. Often matching only the HLA-DR gives good graft survival. ABO blood type matching is required to avoid acute rejection. Although HLA matching greatly improves graft survival, it does not always prevent rejection, since MHC typing using serological methods does not detect all MHC alleles and minor histocompatibility antigens, which can gradually induce rejection.

Corneal transplants are usually not matched, but if the recipient's cornea has become vascularized, or if previous grafts have been rejected, MHC I matching may be warranted. Skin transplants are usually the patient's own skin, so incompatibility and rejection are not an immunological issue.

Kidneys for transplantation can come from either brain-dead individuals or living donors. MHC matching currently emphasizes three loci—HLA-A, HLA-B, and HLA-DR—but the closer the overall match, the better the success rate. Outcomes for individuals receiving liver transplants are not improved by MHC matching. HLA-DR antigens have a powerful impact in heart transplantation, but heart size and availability often take precedence.

Peripheral blood, bone marrow, and cord blood can all provide cells for hemopoietic stem cell transplants. Such transplants of healthy bone marrow are used to replace non-functioning bone marrow, bone marrow damaged by high levels of chemotherapy or radiation, or bone marrow with genetic defects. With autologous bone marrow transplants, the recipient is also the donor. Stem cells are harvested, stored, and returned to the individual after radiation or chemotherapy, and there is no risk of rejection. Both GVH and HVG rejection can occur in allogeneic bone marrow transplants, but GVH is normally the larger concern because the immunocompromised status of bone marrow recipients limits HVG rejection. For many years, a complete MHC match with an identical twin the ideal donor was the goal. Today, unrelated bone marrow (UBMT) or matched unrelated donor (MUD) transplants using genetically matched marrow or stem cells from donors on national bone marrow registries can be used. In MUD transplants, T cells are selectively removed from the cells to be transplanted in order to minimize the risk of GVH, and the donor and recipient must share some MHC I and II antigens.

Immunization

Passive Immunization

In passive immunization, antibodies are give to either prevent the disease or decrease the severity of the symptoms. Viruses, including rabies, measles, hepatitis A and B, and chicken pox; bacterial toxins such as tetanus, diphtheria, and botulism; and bites from spiders and snakes are common targets of vaccinations. Immunity is fast acting but the antibodies are short lived, and no immunological memory is induced. The major risk is serum sickness (see Type III Allergies).

Active Immunization

In active immunization, the individual is exposed to an antigen in a harmless form, and produces antibodies as well as activated T and B cells and memory cells. The memory immune response, initiated upon exposure to the pathogen or a booster, prevents infection and provides additional antibodies and memory cells. In some cases, such as influenza, the residual antibodies are more important than the memory response due to the short incubation period of the disease, and more frequent vaccinations are often required to maintain the antibody level.

privileged sites—nonvascularized locations in the body where foreign grafts are not rejected

TABLE 12.6

Common Vaccines

Disease	Preparation	Disease	Preparation
Chickenpox (varicella)	Attenuated virus	Anthrax	Extract of attenuated bacteria
Measles	Attenuated virus	*Hemophilus influenzae,* type b (HIB)	Capsular polysaccharide conjugated to protein
Mumps	Attenuated virus	Hepatitis B	HBsAg surface protein
Polio Sabin	Attenuated virus	Influenza	Hemagglutinins
Rubella	Attenuated virus	Meningococcal disease	Polysaccharides
Smallpox	Attenuated virus	Pertussis	Purified components (acellular pertussis = "aP")
Tuberculosis	Attenuated bacteria (BCG)	Pneumococcus	Capsular polysaccharides
Yellow fever	Attenuated virus	Staphylococcus	2 capsular polysaccharides conjugated to protein
Hepatitis A	Inactivated virus	Diphtheria	Toxoid
Polio Salk	Inactivated virus	Tetanus	Toxoid
Rabies	Inactivated virus		

Types of Vaccines

The ability to accomplish acquired immunity has relied on the scientific progress in producing vaccines. Most **vaccines** are whole-organism vaccines (see Table 12.6). Live natural vaccines, such as vaccinia isolated from cows that were used to vaccinate humans for smallpox, immunize with a similar organism. Killed vaccines are inactivated using heat and/or chemicals that may alter the antigens. They are safe but not as effective as **attenuated** vaccines, since they produce primarily IgG and often require more frequent boosters.

Attenuated vaccines are live mutants that have lost their pathogenicity while retaining immunogenicity, and provide more natural protection. Attenuated vaccines are made by passage (serial infections) of the organism in a different species, either using cell culture or living animals, and selecting for mutants. Yellow fever vaccine 17D was made by passing a human isolate in chicken cells. As genes associated with virulence are identified, attenuation may be produced by causing mutations in the virulence genes while retaining the genes needed for immunization. Attenuated vaccines prolong the immune system's exposure to antigens and often can be given by the portal of entry used by the pathogen resulting in mucosal immunity from IgA in addition to IgG in the blood and tissue. The increased immunogenicity can result in activation of TC and a stronger memory response, which results in a need for fewer boosters. The chief drawback of attenuated vaccines is reversion to the pathogenic form. The Sabin vaccine has been a powerful weapon in the control of polio, but the Salk vaccine is now used in the United States, because wild type polio has been eradicated, and all the cases occurring were from vaccine reversions. Sabin vaccine is used where the virus is still endemic and to control epidemics.

Molecular biology techniques are being used to produce vaccines that do not contain the whole organism. Many **subunit vaccines** use surface antigens such as the hemagglutinin in influenza vaccine or a capsular polysaccharide in HIB vaccine. A novel approach incorporating DNA for the antigen in a plasmid and injecting it into a muscle has shown promise, since cellular as well as humoral immunity results (Seder et al. 1999). Researchers are also exploring the possibility of incorporating vaccines into foods such as potatoes (Ariza 2005).

vaccine—a substance made from the whole organism or parts that contain critical antigenic components or genes for those components; it stimulates a primary response that produces antibodies and memory cells that protect against subsequent infection by that organism

attenuated—refers to an antigen rendered less virulent but still capable of eliciting an immune response

subunit vaccine—a vaccine made of a single component of an infectious agent and not the whole organism or toxin

Immunodeficiency

In developing countries, most immunodeficiency is caused by malnutrition, whereas in developed countries most immunodeficiency is genetic. Babies experience transient immunodeficiency of the newborn since the antibodies that come across placenta and in colostrum wane before the infant is able to achieve normal levels. Low levels of "natural" IgM antibodies, derived from neonatal lymphocytes and formed without direct immunization with foreign Ag, are found circulating in the umbilical cord and the neonate. Adult levels of IgM are found at two years of age, but mucosal secretory IgA antibodies do not reach adult levels until age six to eight years. The subclasses of IgG attain adult levels from one to five years of age.

Malnutrition and Immunodeficiency

Malnutrition can result in immunodeficiency. A combination of factors, including insufficient protein, energy, and micronutrients, and not just an insufficient amount of food, are involved. Thus, undernutrition can result from personal dietary choices such as nutritionally unsound fad diets as well as socioeconomic factors. An individual's need for proper nutrition for optimal immune function begins in utero with maternal nutrition, and continues throughout life. Maternal nutritional deficiencies—both large-scale deficiencies due to lack of access to sufficient food and specific nutrient deficiencies due to dietary choice—impair fetal development. Maternal nutrition can impact immune functioning throughout life, not just in the fetal and neonatal stages. For example, adolescents who were prenatally and are currently undernourished produce a significantly lower antibody response to vaccination (McDade et al. 2001). In a series of vicious circles, infants with weakened immune function may benefit less from vaccines and are more susceptible to infections such as diarrhea, which can in turn result in worsened nutrition status. Nutritional deficiencies in the elderly can exacerbate the decline in immune responses associated with aging.

Protein-energy malnutrition in infancy and early childhood has adverse effects on the thymus, including significant reduction in thymic weight, lowered thymic hormone levels, and fewer maturing T cells. It can also cause alterations in the thymic microenvironment and peripheral T-cell function. Lowered helper T cell function will have a negative impact on both the cellular and humoral branches of the immune system. Short-term provision of a high-protein, high-kcalorie diet later can increase levels of serum IgG and IgM and improve the functioning of the cellular immune system, but cell-mediated immune responses diminish within a year of such treatment.

Nutrients critical to the development and effective functioning of the immune system include vitamins A, C, B_6, and E, essential fatty acids, beta-carotene, and the minerals manganese, selenium, zinc, copper, iron, sulfur, magnesium, and germanium (Ashfaq, Zuberi, and Anwar Waqar 2000; Calder and Kew 2002; Chandra 1977; Hughes et al. 1997; Tam et al. 2003). Zinc deficiency promotes apoptosis in B and T lymphocytes (especially helper T cells), hinders the function of the macrophage, alters the production and potency of several cytokines, and is linked to poor thymic development in infants. Low maternal selenium is associated with lowered numbers of cytotoxic and helper T cells, B cells, and NK cells in neonates, while neutrophils and helper T cells are affected by selenium deficits later in life (Dylewski et al. 2002). Vitamin B_6 is a cofactor for many enzymes involved in protein metabolism and is important for cellular growth and maintenance of the thymus, spleen, and lymph nodes. Vitamin A deficiency hinders normal regeneration of mucosal barriers; decreases the function of neutrophils, macrophages, and natural killer cells; negatively impacts the development of helper T cells and B cells; and diminishes antibody-mediated responses. While the essential fatty acids omega-3 and omega-6 are needed for the production and maintenance of immune cells, reduction in total fat intake enhances the immune response by increasing the numbers of monocytes and T and B lymphocytes in the blood. Vitamin C supports phagocyte oxidative burst activity as well as B cell and T cell function (Bowers 2002). While nutritional supplements may be necessary to provide the desirable levels of nutrients, excessive intake of some required nutrients can create adverse reactions.

Inherited Immunodeficiencies

Most congenital/inherited immunodeficiencies are detected in young children because they experience recurrent and/or overwhelming infections, often from opportunists. Males are more apt to have immunodeficiencies, because many such deficiencies involve recessive genes, often on the X chromosome.

Some immunodeficiencies involve just one part of the immune response. In X-linked agammaglobulinemia, individuals have few or no B cells and produce no IgA, IgM, or IgE and small amounts of IgG. They suffer from numerous staphylococcal and streptococcal infections but can be treated with passive antibodies. Some individuals are deficient in a single antibody class, with IgA being the most common. Individuals with deficiencies in phagocytic cells have difficulties in killing intracellular and ingested extracellular bacteria.

Others impact more than one part of the immune response. In DiGeorge syndrome, the thymus epithelium fails to develop, so T cells cannot mature, which affects the production of cell-mediated immunity and T dependant antibodies. In Wiskott-Aldrich syndrome, a defect in a gene on the X chromosome coding for Wiskott-Aldrich syndrome protein affects B and T lymphocytes and platelets, which results in overwhelming pyogenic and opportunistic infections. Bare lymphocyte syndrome, an autosomal recessive condition, is due to a defect in genes that regulate MHC expression; in this condition, there are no MHC II antigens on cells, and thus APC can't stimulate the TH. There are several types of **severe combined immune deficiencies (SCID)**, which are characterized by extreme susceptibility to infection due to the absence of T and B lymphocyte function and often NK cells (see Box 12.4). Some forms can be treated with bone marrow transplants, and gene therapy has been used in others.

severe combined immune deficiency (SCID)—disease due to several mechanisms that produce an early block in differentiation pathways of both B and T lymphocytes, resulting in infants who are born lacking all major immune defenses

BOX 12.4 CLINICAL APPLICATIONS

The Boy in the Bubble: Severe Combined Immune Deficiencies (SCID)

X-linked SCID involves an altered gene on the X-chromosome. About half of SCID cases are due to a mutation in the common γ chain of several cytokines, which results in absence of IL-2, very low T lymphocyte and NK counts, and poor B lymphocyte function. This is the condition that afflicted the "bubble boy," and it is treated with a bone marrow transplant. Other forms of SCID are due to defects in T cell development caused by recessive genes not on the X chromosome that leave the person with few lymphocytes and susceptible to a broad range of infectious agents. About 10% of SCID cases are due to a mutation in the gene for Janus kinase 3 (Jak3), which is necessary for function of the common gamma chain. Thus, infants have the same kinds of T, B, and NK-lymphocyte counts as those with X-linked SCID. A mutation in a gene that encodes the alpha chain of the IL-7 receptor (IL-7Rα), a T lymphocyte growth factor, leaves infants with no T lymphocytes. ADA adenosine deaminase deficiency allows dATP and dGTP, which are toxic to stem cells, to accumulate. Babies with this form of SCID, about 5% of the cases, have the lowest total lymphocyte—T, B, and NK—counts of all. This condition has been treated with gene therapy.

Acquired Immunodeficiencies

Some immunodeficiencies are acquired in later life. The immune system can be suppressed by many cancer drugs; several infectious agents, including HIV; burns where there is a severe loss of Ig through damaged skin; and inflammatory bowel disease with the loss of Ig into the bowel.

Tolerance

Immunological tolerance occurs when an immunocompetent host fails to respond with a specific antigen to an immunogenic challenge (i.e., one that would produce a measurable response in some other, nontolerant host).

immunological tolerance—nonresponsiveness to a particular antigen or group of antigens produced by prior exposure to the antigen under nonimmunizing conditions

clonal deletion—a process by which contact with an antigen, usually self-antigen, early in lymphocyte differentiation leads to cell death by apoptosis

positive selection—the rescue from apoptosis of T cells in the thymus that can recognize self-MHC molecules

Autotolerance

The human immune system must respond to thousands of antigens in order to protect from all pathogens to which humans are exposed. It must also recognize antigens that arise on tumors. As a consequence of the random processes that generate this diversity in response capacity, the immune system will produce B and T cells capable of responding to self-antigens. Autotolerance (tolerance to one's own antigens) mechanisms block these self-reacting cells from reacting with self-antigens, predominantly by eliminating them by **clonal deletion** (in the case of T cells) and/or blocking their development by clonal abortion (in the case of B cells). Failure of autotolerance can result in an autoimmune disease in which the immune system attacks the person's own tissues. Only lymphocytes, cells with Ag-specific receptors, can be tolerized. T cells react faster, remain tolerant longer, and require less antigen to become tolerant than do B cells. Central tolerance is induced during early stages of lymphocyte development, both in utero and in later life, while peripheral tolerance responds to mature lymphocytes.

Induced Tolerance

Tolerance to foreign antigens can be induced. For example, both extremely high and extremely low prolonged doses of antigen can induce tolerance. This can play a role in the failure of the immune system to respond to a tumor or transplant, and must be considered in determining vaccine dosage. Establishment of tolerance is influenced by: (1) stage of maturity of a cell—it is easier in immature cells, (2) affinity of the BCR or TCR for the self-antigen, (3) nature of antigen—large particulate denatured proteins are best, (4) route of exposure, with mucosal and oral most apt to tolerize (this may have evolved to prevent immune response to food), and (5) the concentration, either too high or too low, of antigen.

Central Tolerance

T cells are only activated by antigens presented by APCs with the same MHC antigens, since the T cell receptor must react with both the antigen and the MHC molecule that is holding the antigen to be activated. Helper T cells recognize MHC II antigens, while cytotoxic T cells react with Class I antigens. Thus, only cells with TCRs that react with self-MHC and foreign antigens, but not with self-antigens, will benefit the host. In clonal deletion, cells moving from the cortex or outer part of the thymus through the medulla or inner portion are on a path to programmed cell death (apoptosis). Cells with TCR that react with self-MHC and foreign antigens are rescued during the selection process, while others continue to apoptosis. In **positive selection,** cells with TCR that react with self-MHC are rescued, while reaction with the self-antigen during negative selection causes cells to continue into apoptosis. The removal of self-reactive cells is part of the way the immune system distinguishes self from nonself.

In clonal abortion, self-reactive immature B cells are blocked from further development and eventually die. In a process analogous to negative selection (no analog of positive is required since there is no MHC restriction with B cells), immature B cells that react with self-antigen in bone marrow cease developing. The only antigens present normally are self-antigens. Cross-linking of a cell's BCR as it reacts with a membrane-bound multivalent antigen down-regulates IgM production and halts cell development. Cells are then removed by apoptosis. If the cell encounters soluble self-antigen, it moves to the periphery with IgM.

Peripheral Tolerance

Peripheral tolerance responds to self-reactive cells that evade central tolerance, since not all self-antigens are expressed in the thymus, all self-antigens do not bind with MHC with sufficient strength to induce negative selection, and all TCR do not bind to self-antigens with sufficient strength to signal negative selection. Somatic mutation after selection may create a self-reactive B cell receptor from the nonself receptor that survived selection.

Anergy occurs when a lymphocyte is alive but fails to respond when stimulated through its antigen-specific receptor. In T cells, anergy is usually due to lack of a costimulatory signal (e.g., if APC lacks B7, an accessory surface molecule that is required to activate T lymphocytes to produce IL-2). In that case, activation of other immune cells does not occur and immune response is not initiated. Pathogens do not induce tolerance by anergy, because many have costimulatory molecules (e.g., LPS) and induce B cells to make them. B cells enter into anergy when they are exposed to high concentrations of monomeric antigen, which downregulates surface antibodies so the B cells cannot react with T cells and will not be activated.

Other mechanisms of peripheral tolerance include ignorance, where self-reactive T cells "ignore" antigens, often because they cannot gain access to them; for example, T cells cannot penetrate an endothelial barrier like those found in the testes and brain. B cells can be rendered tolerant due to lack of the T cells needed to stimulate them, which is called functional deletion.

Attack on Harmless Antigens: When the Immunological System Causes Harm

Hypersensitivity

Hypersensitivity (allergy) occurs when extrinsic antigens (allergens) are recognized by presensitized individuals. Allergic responses are identical to responses to pathogens, but the antigen is innocuous, so all pathology is due to immune response. Since the allergic reaction requires presensitization, a response is not seen on first exposure, and reactions can become worse with subsequent exposures due to the memory response. Over 50 million Americans have allergic diseases, making allergies the sixth leading cause of chronic disease (NIAID-NIH 2005) (see Box 12.5 for a possible explanation of this prevalence).

Classifications of Allergic Reactions Gell and Coombs (see Table 12.7) grouped allergic reactions into four classes. Types I, II, and III, which involve antibodies, are considered **immediate hypersensitivities** since initial signs and symptoms can occur within minutes to a few hours after exposure. Type IV, which is driven by T cells, is considered delayed, because it takes one to three days before a reaction is noticed.

Type I or IgE Allergies Type I or IgE allergies involve reactions to respiratory allergens, including pollens, spores, animal dander, and dusts that diffuse across the mucous membrane of nasal passages and activate mucosal mast cells. These cells release mediators that produce sneezing, watery red eyes, runny noses, and respiratory distress. If food containing an allergen is ingested, activation of mucosal mast cells can cause oral inflammation, canker sores, cramps, nausea, diarrhea, gas, **hives** (urticaria), and sometimes respiratory distress. Hives can occur when histamine, released from skin cells due to an allergic reaction, causes blood vessels to dilate, leak fluid, and produce swelling, which in turn irritates nerve endings and results in itching. Mild or **atopic** reactions (translation of *atopic*: strange disease) affect 10% to 20% of the population.

The most severe reaction, **anaphylactic shock** or anaphylaxis, is potentially fatal. Risk of anaphylaxis is greatest when

anergy—antigen-specific nonresponsiveness by a T or B cell in which the cell is present but cannot respond

hypersensitivity—an inappropriate and harmful immune reaction to a harmless, nonpathogenic substance; also called allergy

immediate hypersensitivity—a hypersensitivity reaction that appears within minutes after the exposure to the allergen

hives—an itchy skin condition with raised red lumps, often due to an allergic reaction; also called urticaria

atopic—a milder IgE-mediated allergic response

anaphylactic shock—a life-threatening IgE-mediated allergic reaction; in humans, symptoms include swelling (especially of the lips and face), vomiting, diarrhea, difficulty in breathing, and a sudden drop in blood pressure; also called anaphylaxis

BOX 12.5 NEW RESEARCH

The Hygiene Hypothesis: Have We Become Too "Clean" for Our Own Good?

While children get far fewer serious infections today than they did two centuries ago due to vaccinations, antibiotics, and better sanitation, asthma rates in the U.S. have increased by 75% since 1980, with cases in children mushrooming by 160%. The "hygiene hypothesis," which is both fairly new and controversial, maintains that the rising incidence of allergic and autoimmune disease, especially asthma, inflammatory bowel disease, and multiple sclerosis, may be due at least in part to lifestyle and environmental changes that reduce children's contact with infectious agents and environmental antigens (Bauchner 2002; Carpenter 1999; Renz et al. 2006). Children who are around many children or animals early in life are exposed to more antigens, which causes their immune systems to develop tolerance for the antigens that cause asthma. The underlying idea is that the proper maturation of the immune system requires a stimulus. The children of Asian, Latin American, and African emigrants who have moved to "cleaner" European or North American countries, and have not been exposed to the parasites and early childhood infections that their parents were exposed to, have the same incidence of Crohn's disease, multiple sclerosis, and chronic asthma as the long-time inhabitants (Ullrich 2004).

Experimental evidence suggests that commensal bacteria in the newborn have a strong impact on the maturation of the immune system. Endotoxin is found in the cell wall of Gram-negative bacteria like *E. coli*, many of which are found in the colon. Early childhood exposure to endotoxins influences the occurrence of asthma and allergies in later years. In one study, children were tested for sensitivity to dust mite, cat, dog, cockroach, mouse, milk, egg, and soy, and the endotoxin concentrations of house dust collected from the child's bed, a couch, and floors in the living room, kitchen, and bedroom were measured. The level of endotoxin found in dust samples from bedding was inversely related to the occurrence of hay fever, atopic asthma, and atopic sensitization (Waser et al. 2005). An inverse relationship was observed between the level of endotoxin in the mattresses used by the children and the ability of the children's peripheral blood leukocytes to produce cytokines after innate immune system activation, indicating a marked down regulation of immune responses in the endotoxin-exposed children. Studies also show that children raised on farms when they were very young have a lower incidence of asthma (Elliott, Yeatts, and Loomis 2004).

In Gambia, almost everyone has intestinal worms at some point in their lives, while asthma, Crohn's disease, and multiple sclerosis are extremely rare. Note that genetic and other environmental factors could also explain this observation. When six patients were given worms to treat bowel disease, they converted from chronic illness to complete remission with no diarrhea, no abdominal pain, and no joint problems (Ullrich 2004). Seropositivity for hepatitis A virus, *Toxoplasma gondii*, and herpes simplex virus type 1 have been linked to a decreased risk of hay fever, asthma, and atopic sensitization.

Scientists are not proposing that we return to less sanitary times and risk the resurgence of infectious diseases. The goal is to understand the mechanisms so they can be used to develop preventive strategies.

TABLE 12.7

Characteristics of Type I, II, III, and IV Allergic Responses

	I	II	III	IV
Time	20 to 30 min	5 to 8 hrs	2 to 8 hrs	24 to 72 hrs
Name	Anaphylactic	Cytotoxic	Immune complex	Delayed
Immunoglobulin	IgE	IgG, IgM	IgG, IgM, etc.	None (T cells)
Antigens involved	Heterologous	Autologous or hapten modified	Autologous or heterologous	Autologous or heterologous
Cellular involvement	Mast cells and basophils	RBC, WBC, platelets, etc.	Host tissue cells	Host tissue cells

the allergen is injected directly into circulation so that it activates cells all over the body, as happens with insect stings and IV drugs. About 32 of every 100,000 exposed patients develop an anaphylactic response to penicillin.

IgE allergies have a genetic component, and most humans with allergies are allergic to one or two things. Some MHC types have been linked with specific allergies due to the role of MHC in antigen presentation. Some individuals have genes for high levels of IgE production and often react to numerous antigens. Several other factors influence IgE allergic reactions, including nutrition, level of exposure to the allergen, chronic infections, acute viral infections, being firstborn, and exposure to environmental factors. Environmental factors include sulfur dioxide, nitrous oxide, and diesel fuel, which may increase mucosal permeability and enhance allergen entry.

Antigens involved in Type I allergies are small (15–40,000 mws), highly soluble molecules presented at very low doses that are often inhaled in desiccated particles that diffuse into the mucosa. Cells in mucosa bind allergens and transport

TABLE 12.8

Mediators of Anaphylaxis

Mediator	Source	Smooth muscle contraction	Vasodilation	Increased vasopermeability	Eosinophil attraction	Neutrophil attraction	Increased bronchial mucus secretions	Platelet degranulation	Increased histamine release
Histamine	granule	X	X	X					
Serotonin	granule	X	X	X					
Eosinophilic and neutrophilic chemotactic factors	granule				X	X			
Bradykinin and related kinins	granule	X	X	X					
Platelet activating factor	granule	X						X	
Chymase	granule						X		
Slow-reacting substance of anaphylaxis leukotriene	new	X		X	X				
Prostaglandins	new	X							X

them to lymph nodes. Transmucosal presentation causes TH2 cells to release IL-4, which causes more B cells to make IgE. IgE attaches to basophils and mast cells found throughout the body, and are highly concentrated in connective tissue, the lungs, the uterus, and around blood vessels. Connective tissue mast cells, which are found around blood vessels in most tissues and are found in high numbers in the skin and gut submucosa, have granules with higher histamine concentration and a longer life span than do mucosal mast cells, which are found in the mucosa of the mid-gut and lung and which infiltrate the nasal epithelium in individuals with hay fever during pollen season. The IgE attaches to a specific receptor on mast cells via its Fc, so antigen-binding sites are exposed. Subsequent exposures to the antigen can result in the antigen cross-bridging two IgE antibodies if the concentration of IgE on the cell produces adjacent molecules. Cross-bridging initiates a series of reactions in the cell that results in release of chemicals that cause the signs and symptoms associated with allergy. Other substances, such as lectins, which are found in high amounts in strawberries, can cross-bridge two IgE molecules. Cells can also be activated by other molecules, such codeine and morphine.

In the early phase of the allergic response, granules release preformed chemicals, including histamine, the major mediator of IgE allergy in humans. The result is smooth muscle contraction and an increase in vasopermeability (see Table 12.8). In the late phase, which starts within four to six hours and lasts for one to two days, leukotrienes, derived from arachidonic acid, are released. This release increases vascular permeability; at the same time, mucus secretions contract smooth muscle in the airway, and attract and activate inflammatory cells.

Most treatments for IgE allergic reactions contain antihistamines, which act as competitive inhibitors and block histamine from combining with receptors on nerve endings. This treatment will block further early phase reactions, but corticosteroids are required to block the late phase. In severe reactions, commercial adrenalin, which is actually **epinephrine**, counteracts the actions of histamine.

Allergy Testing Skin scratch testing is relatively inexpensive, safe, and easy to perform, but can produce discomfort in the individual being tested. Commercial inhalant allergens are available for respiratory allergies, but the stability of extracts of food allergens makes testing for food allergies a greater challenge. A small amount of suspected allergens is placed on the skin, pricked into it, or injected under the surface. A reaction, usually swelling and redness, occurs in about 20 minutes at the site of the substance(s) to which the person is allergic. Since "elimination" diets, where suspected food(s) are eliminated from the diet and then gradually reintroduced, can be affected by the person's ideas on what they are allergic to, a double-blind procedure may be used. In such a procedure, suspected foods and placebos are given in a disguised form. Tests employing radioisotopes are used to look for levels of IgE to a specific allergen (RAST) or to any antigen (RIST) in blood.

epinephrine—a chemical made by the adrenal gland that relaxes smooth muscles and constricts blood vessels; when it is used to treat severe allergic reactions, it is sometimes referred to as adrenaline

Allergy Shots Allergy shots contain a regulated dose of the compound(s) to which the person is allergic. Since the antigen does not enter transmucosally, IgG production is stimulated, and little, if any, IgE is produced. The shot series builds and maintains a high level of IgG so that allergens encounter IgG when they enter the body and do not reach the IgE on the mast cells. Since the shot contains the substance(s) to which the person is allergic, there is always a risk that the shot will initiate an allergic reaction. Therefore, recipients are often observed for a period after the shot with epinephrine readily available.

Food Allergies and Intolerances About 1 out of 4 people report they have a food allergy, yet only 2 in 100 adults and about 6 out of 100 children have a clinically documented allergic reaction to food. Many confuse food allergy with food intolerance, an abnormal physiologic response to food, due to similarities in symptoms. A food allergy is an immunologically based abnormal response to a food; in contrast, the immune system plays no role in food intolerance. Several things can lead to intolerance reactions. Histamine, the major mediator in IgE-based allergic reactions, is found in cheese, some wines, and some fish, including tuna and mackerel, and may cause intolerance reactions. Lactase deficiency, which affects about 1 out of 10 people, is the most common intolerance. Deficiency in the lactase enzyme leads to gas formation, bloating, abdominal pain, and diarrhea when dairy products are consumed, because bacteria degrade the lactose that the person cannot. Food additives underlie some intolerance reactions: yellow dye number 5 can cause hives; monosodium glutamate has been linked to flushing, sensations of warmth, headache, facial pressure, chest pain, or feelings of detachment; and sulfites can irritate the lungs and lead to severe bronchospasm in people with **asthma** (Greene 2002).

Epidemiology Ask people if they have a food allergy and one in three will say yes; yet rare double-blind, placebo-controlled food challenge studies indicate that food allergy occurs in only 1% to 2% of the adult population (American Academy of Allergy, Asthma & Immunology 2006). Adults are usually most affected by tree nuts (almonds, Brazil nuts, hazelnuts, and walnuts), fish, shellfish, and peanuts. The prevalence of peanut and tree nut allergy is approximately 1.1% (American Academy of Allergy, Asthma & Immunology 2006), while 0.5% are allergic to shellfish. A small percentage, 0.01% to 0.23%, have adverse reactions to food additives (Food Additives and Ingredients Association 2002). Unlike many conditions related to the immune system, no differences based on gender or race have been detected (James 2004).

Although about one-third of parents believe that food allergies are responsible for a multitude of symptoms in their children, the demonstrated prevalence is only 3% to 7% among young children, and 80% to 90% outgrow their sensitivities by the age of three. IgE-based milk allergy affects about 1% of infants, and soy allergy/soy intolerance affects 1% to 6% of infants but varies with regional diets. The prevalence of egg allergy ranges from 1.6% to 2.6%. Adverse reactions to food additives affect 0.5% to 1% of children. Allergies to egg and cow's milk may disappear, but allergies to nuts, legumes, fish, and shellfish tend to continue to adulthood (Al-Muhsen, Clarke, and Kagan 2003; Fleischer et al. 2005; James 2004).

Pathophysiology and Clinical Manifestations In children, milk, eggs, peanuts, wheat, soy, and tree nuts account for most allergic reactions, while in adults, peanuts, tree nuts, fish, and shellfish are the major causes. Allergies to eggs can be problematic, since egg proteins are found in many food products—which may or may not be easily determined from reading the label. The Food Allergen Labeling and Consumer Protection Act (2004) requires that foods containing milk, eggs, fish, crustacean shellfish, peanuts, tree nuts, wheat, and soy indicate this in plain language on the label.

In most allergic reactions to food, a person with an inherited predisposition produces IgE in response to proteins that cross the gastrointestinal lining. These proteins enter the bloodstream because they are not broken down by cooking, stomach acids, or enzymes. The IgE attaches to mast cells, and subsequent exposures to the food result in the reaction of the allergen with the attached IgE and the release of chemical mediators, especially histamine, by the mast cells.

Symptoms, which appear within minutes to two hours, are influenced by the location of the histamine release. Reactions in the ears, nose, and throat may result in itching in the mouth or trouble breathing or swallowing, whereas interactions in the gastrointestinal tract can lead to abdominal pain, vomiting, or diarrhea. Hives can be a product of histamine release by skin mast cells. Food-initiated anaphylaxis is a severe allergic reaction involving the whole body, with the lungs the major target in humans. Histamine causes constriction of the airways, which causes difficulty in breathing; blood vessel dilation, which lowers blood pressure; and fluid leakage from the bloodstream to tissues, which results in shock, hives, and gastrointestinal symptoms such as abdominal pain, cramps, vomiting, and diarrhea.

Breastfeeding is a way to avoid milk or soy allergies in infants, but if a mother ingests a food to which the child is allergic, allergens can enter the breast milk and cause an allergic reaction in the child.

asthma—a chronic inflammatory lung disease triggered by either an IgE allergic reaction or nonallergic factors that results in inflammation of the airway and reversible airway obstruction

Common Food Allergens Cow's milk protein allergy occurs at higher levels in infants with a family history of allergy. The incidence in older children and adults is much lower. The allergenicity of cow's milk can be reduced by heat treatment and enzymatic digestion of milk proteins.

Peanuts and tree nuts, such as almonds, Brazil nuts, hazelnuts, and walnuts, can cause reactions with minimal exposure through intact skin or by inhalation (see Box 12.6 for a theory regarding the origin of these reactions). These allergies can be present across the life span and in some individuals can result in life-threatening anaphylactic shock.

Fruits, soybeans, eggs, crustaceans (crabs, crayfish, lobster, and shrimp), fish, vegetables, sesame seeds, sunflower seeds, cottonseed, poppy seeds, and mustard seed are common allergens, but the allergenic capacity is often destroyed by cooking or food processing that denatures food proteins.

Diagnosis The first step in diagnosing a food allergy is to obtain a detailed history and perform a complete physical examination to rule out other causes of symptoms. Several tests can be used to test for IgE-mediated food allergy: radioallergosorbent tests (RAST) and the CAP System fluorescent-enzyme immunoassay (FEIA), which use serum and skin prick–puncture tests; elimination diet tests; and food challenge (single- and double-blind) tests that expose the individual to the potential allergen.

Some commercial laboratories offer RAST and FEIA testing with food allergy panels. Unlike the other methods, these do not require that the individual be exposed to potential allergens and thus pose no risk of an adverse allergic reaction during the test. They use a blood sample and provide a convenient method for both patient and physician. Nonetheless, due to lack of consistent quality control from laboratory to laboratory, there are questions about the reliability of the RAST panel. The CAP-FEIA results are as effective as the skin prick tests in predicting food allergy.

In the skin prick test, a drop of food extract is put on the skin and then the top layer of skin is pricked with a small needle, or a pricking device is presoaked in the food extract. Skin prick tests can be used as a preliminary test to narrow the list of potential problem foods, since the negative predictive value is greater than 95%. However, the positive predictive value is about 50%, so the test is not sufficient to establish a positive diagnosis.

In elimination/challenge diet testing, a single food or a combination of suspect foods are not consumed for two weeks. If the symptoms disappear, suspect foods are added back, one at a time, in increasing amounts until normal levels are reached or symptoms occur. Elimination/challenge tests require a high degree of patient motivation and compliance in the elimination phase.

Placebo-controlled food challenges can be either single or double blind, but the double-blind, placebo-controlled food challenge test (DBPCF) is the gold standard because it prevents individuals' beliefs about their allergies from influencing their responses. The suspected foods and placebos (harmless substances) are hidden in another food or in opaque capsules, and neither the person being tested nor the provider know whether it is the suspect food or a placebo being consumed. Due to the risk of reaction, this procedure requires trained personnel and specific facilities: rapid access to emergency medications, including epinephrine, antihistamines, steroids and inhaled beta agonists, and equipment for cardiopulmonary resuscitation.

BOX 12.6 NEW RESEARCH

The Worm and the Peanut: A Basis for IgE Allergies

Hookworms, pinworms, intestinal roundworm, *Schistosoma* worms, and flukes have coexisted with humans for centuries. Although worm infections are not a major health problem in the U.S., in areas of Africa three out of four teenagers are infected with one or more types of worms. Worms are too large for phagocytosis, but trigger an immune response featuring TH2 cells, cytokines (including IL-4, IL-5, IL-9, IL-10, IL-13), IgE, eosinophils, and mast cells. These mediators of immunity have varied importance depending on the worm. Worms release antigens, which move to lymph nodes and activate TH2 cells; these cells release IL-4 and IL-5, leading to the reproduction and activation of eosinophils and the production of IgE. Eosinophils and IgE initiate an inflammatory response in the intestine and lungs to expel the worms. Worm antigens bind to IgE on mast cells, eliciting the release of mediators that may contribute to the expulsion of parasites by inducing the contraction of smooth muscle, the production of excess mucus, and the onset of diarrhea in the gut. They also stimulate coughing and sneezing that might dislodge worms in the respiratory tract.

Scientists have noted the similarities in the mechanisms involved in Type I allergic reactions and the immune response to worms (Yazdanbakhsh, Kremsner, and van Ree 2002). In addition, mast cells are found in the skin, in mucous membranes of the eyes, nose, and throat, and in the lining of the lungs and gut where the body is exposed to the environment, including parasites and allergens. It has been proposed that IgE allergies (which cause an immune response to harmless antigens such as the peanut) are the result of a misdirection of the immune response that evolved to protect against worms. According to evolutionary theory, animals retain traits that are useful for survival and eliminate those that are not useful. The retention of a response that is potentially harmful, such as an allergic reaction to peanuts, violates this theory unless the response also has a beneficial function, such as in the response to worms, that outweighs the negative response.

Asthma People with asthma experience chronic airway inflammation and excessive airway sensitivity to various triggers

BOX 12.7 CLINICAL APPLICATIONS

Triggering Asthma

Asthma can be triggered by many things, including exposure to allergens (e.g., molds, dust, or animal dander), tobacco or wood smoke, polluted air, respiratory irritants (e.g., perfumes, workplace chemicals, or cleaning products), sulfites, and cold, dry weather. Upper respiratory infections (e.g., cold, flu, sinusitis, and bronchitis), excitement or stress, physical exertion or exercise, and reflux of stomach acid (gastroesophageal reflux disease, or GERD) can also trigger an asthmatic response.

(see Box 12.7). There is a genetic component in many, but not all, cases of asthma, and environment and lifestyle can play a role. Individuals with asthma have more mast cells in bronchi, and these cells have a low degranulation threshold.

Although the symptoms may be intermittent, some level of inflammation is constantly present in most people with asthma. When triggered by an allergic reaction, asthma results from activation of submucosal mast cells in lower airways. TH2 is the major T cell found, so when class switching occurs during the production of antibodies, the switch to IgE will be favored. The symptoms of asthma, including wheezing, cough, chest tightness, difficulty breathing, and sputum production, are due to the tightening of muscles around the airways, swelling and thickening of the lining inside the airways, and clogging of the airways with thick mucus.

Type II Allergic Responses In Type II allergic responses, such as transfusion reactions and hemolytic disease of the newborn, IgG or IgM binds to cell surfaces or extracellular matrix molecules, activates C′, and ultimately destroys the cell. Involvement of antigens on the cell and not in serum differentiates Type II from Type III.

Type III Allergies: Immune Complexes Immune complexes (clusters of antibodies bound to antigens) are usually removed by macrophages in the liver and spleen. In Type III allergies, however, these complexes are deposited in blood vessel walls and tissues, especially the synovial membrane of joints and glomerular basement membrane of the kidney. This reaction contributes to the pathology of infectious diseases such as leprosy, malaria, dengue, viral hepatitis; the mechanism of some autoimmune diseases; allergic pneumonitis; and serum sickness.

Immune complexes form due to the continued presence of antigen in blood from low-grade persistent infections in infectious disease, continual inhalation of an antigen in allergic pneumonitis, and passive antibodies in serum sickness. Activation of complement by immune complexes results in release of vasoactive amines (histamines) and chemotaxis for basophils, eosinophils, and neutrophils. Macrophages attach to platelets by Fc receptors and release vasoactive amines that cause endothelial cell retraction, which produces increased vascular permeability. Gaps between cells permit insertion of antibody molecules into blood vessel walls. Failure of neutrophils to engulf immune complexes results in the release of enzymes that produce vessel wall damage. In specific diseases, immune complexes are more apt to be deposited where the charge on the antigen-antibody complex interacts with the charge on the tissue. In allergic pneumonitis, repeated inhalation of antigenic material leads to deposition of immune complexes in alveoli, which produces inflammation and fibrosis.

In serum sickness, immune complexes deposit in blood vessels and tissues, resulting in hives; edema in the face, neck, and joints; joint pain; malaise; and fever that lasts 7 to 10 days. Long-lasting sequelae and fatalities are very rare. Serum sickness develops in 50% of individuals who receive a foreign antibody during passive immune therapy (e.g., tetanus **antitoxin**), due to the presence of the antigen (passive antibody) and the antibody (anti-antibody) in blood at same time.

Type IV Allergic Reaction: Delayed Hypersensitivity Type IV allergic reactions, also called **delayed hypersensitivity**, T cell mediated allergy, or contact dermatitis, include allergies to nickel, rubber accelerators, latex, plant chemicals (poison ivy or poison oak), and the tuberculin reaction used in TB testing. Small, nonprotein antigens or haptens complex with skin proteins when they penetrate the skin, sometimes due to scratching. APC (Langerhans cells) present complexed proteins to TH cells that are activated and produce memory cells that reside in the skin. Upon subsequent exposure, activated memory cells produce cytokines, including IL-17 and IFNγ, that cause skin keratinocytes to secrete IL-1, IL-6, TNF GM-CSF, and chemokines. Chemokines attract monocytes and activate macrophages. These start a generalized attack that results in the characteristic skin irritation. The treatment for delayed hypersensitivity usually involves immunosuppression by chemicals, including hydrocortisone.

Autoimmunity

Autoimmune disease occurs when a specific adaptive immune response is mounted against self, and is a consequence of the open repertoires of B and T cells that allow them to recognize any pathogen. Since many antigens on human cells and pathogens are similar, immune cells targeted at

antitoxin—an antibody to an exotoxin

delayed-type hypersensitivity—a cell-mediated inflammatory allergic reaction in the skin, (e.g., poison ivy) that takes 24 to 48 hours to appear

TABLE 12.9

Major Autoimmune Diseases

Disease	Organ	Mechanism
Hashimoto's thyroiditis	Thyroid	Inflammation is linked to antibodies against thyroglobulin (TG) and thyroid peroxidase (TPO); autoreactive cytotoxic T cells and natural killer cells destroy the thyroid gland.
Graves' disease	Thyroid	The antibody to the thyroid-stimulation hormone receptor on thyroid cells reacts with the receptor and has the same effect as thyroid stimulating hormone, but it is not subject to feedback control, which results in overproduction of thyroid hormone.
Pernicious anemia	Red blood cells	An autoantibody reacts with intrinsic factor produced by parietal cells, resulting in decreased B_{12} absorption in the small intestine.
Addison's disease	Adrenal	Antibodies attack and destroy the adrenal cortex cells that make cortisol and aldosterone.
Premature onset menopause	Ovary	Destruction of ovarian function that is linked to autoimmune responses.
Male infertility	Sperm	Antisperm antibodies bind to the sperm and impair motility, cause them to clump together, and interfere with fertilization of the egg
Type 1 Diabetes mellitus	Pancreas	Insulin-producing ß cells are destroyed by TC cells or antibodies.
Insulin resistant diabetic	Systemic	Insulin-binding antibodies neutralize insulin.
Myasthenia gravis	Muscle	Cells from the immune system cause inflammation in the bowel wall and may also involve antibodies generated in response to an infection that cross-react with cellular antigens.
Goodpasture's syndrome	Kidney, lung	Autoantibodies are deposited in the membranes of the lung and kidneys, causing both inflammation in the kidney and bleeding in the lungs.
AI hemolytic anemia	Red blood cells, platelets	Antibodies bind to cell membrane antigens causing cell lysis.
Ulcerative colitis	Colon	Cells from the immune system cause inflammation; the condition may also involve antibodies generated in response to an infection that cross-reacts with cellular antigens.
Sjögren's syndrome	Secretory glands	The exact trigger and target are unknown, but WBC invade and destroy glands that produce moisture, resulting in dry mouth and dry eyes; problems in other parts of the body also occur in joints, lungs, muscles, kidneys, nerves, thyroid gland, liver, pancreas, stomach, and brain.
Rheumatoid arthritis (RA)	Skin, kidney, joints	The etiology of RA is not fully understood; the presence of rheumatoid factor (an autoantibody, usually IgM, that reacts with IgG), cytokines, and cells of the immune system are indications of an autoimmune link to an acute or chronic inflammation of synovial joints that causes pain, damage, and loss of function.
Systemic lupus erythematosus	Joints, etc.	Immune complexes containing antibodies to DNA, RNA, and nucleoproteins are deposited in the walls of small blood vessels in the kidney and joints.
Rheumatic fever	Heart	Antibodies generated in response to Group A *Streptococcus* cell wall antigens cross-react with cardiac muscle and heart valves, causing damage to the heart.

pathogens can cross-react with human cells. These cannot be eliminated, or there would be a limited response to pathogens. When an **autoantibody** is found in association with disease, the autoimmune response usually produces lesions, but in rare cases, such as the anticardiac antibody found after myocardial infarction, tissue damage simulates an autoantibody.

Autoimmune disease affects 5% to 7% of adults in Europe and North America, with autoantibodies more common in older people. Many clinically normal individuals have low titers of antibodies against some of their own tissues (e.g., against erythrocytes), and these increase with age. Babies can have autoimmune responses for a short time due to maternal antibodies. Most autoimmune diseases are more common in females, but castration of men eliminates the differences. Estrogen and testosterone are thought to play a role, because they activate cells to express different genes. In many autoimmune diseases, including rheumatoid arthritis, disease severity decreases during pregnancy but rapidly rebounds after pregnancy termination.

Autoimmune diseases may be organ specific, with the thyroid, adrenals, stomach, and pancreas common targets, or non-organ specific (see Table 12.9). Systemic lupus erythromatosis (SLE) involves all or almost all tissues in the body. Many autoimmune disorders have spontaneous exacerbations

autoantibody—an antibody to self-antigens

and remissions due to fluctuations between positive and negative regulatory factors. An affected individual can have more than one autoimmune disorder (e.g., the RA cluster), and approximately 15% of all autoimmune patients have two.

Autoimmune diseases have a strong tendency to run in families, with a 40% chance that a family with one affected adult will have another. Genetic factors are clearly involved in autoimmune disease, with combinations of alleles rather than a single predisposing allele the norm. Identical twins both develop a disease 20% to 40% of the time for common autoimmune diseases, so it is highly likely that environmental factors such as diet are also important.

Induction of Autoimmune Disease Autoimmune diseases are induced through evasion of tolerance where tolerance mechanisms remain intact; breakdown of tolerance when mechanisms are defective; and alteration of control of lymphocyte response. Occult/sequestered antigens, which are segregated from circulation so that they are not involved in selection of auto-reactive lymphocytes, are a major mechanism in evasion of tolerance. When they come into contact with the immune system as a result of disease or trauma, they induce the immune response, which attacks them. Anti-sperm antibodies after a vasectomy and anti-heart antibodies after a heart attack are examples. An autoimmune response to the brain or eye lens can occur if damage results in leakage of blood vessels that allows antigens from tissue to enter circulation and encounter responsive lymphocytes. The presence of new epitopes on cells from drugs that act as autocoupling haptens or viral infection can trigger an autoimmune attack on the cell. Molecular mimicry (see Table 12.10) occurs when bacteria or viruses possess determinants similar to cell antigens. These cross-reacting antigens can stimulate TH to activate auto-reactive B cells. Coxsackie virus infection in children is associated with type 1 diabetes, and EBV infection has been linked to Sjögren's syndrome.

A genetic defect in cells involved in establishing tolerance, or damage to those cells by drugs and infectious agents, can lead to a breakdown of tolerance. Several mechanisms can result in alteration of control of lymphocytes, including cytokine imbalance, especially involving expression of IL-2 and IL-2 receptors, and upregulation of INFγ; inappropriate expression of MHC; viral infection; and polyclonal activation of B cells by LPS and EBV. Pancreatic cells of people with type 1 diabetes mellitus have high levels of MHC I and II.

TABLE 12.10

Bacterial and Viral Antigens that Mimic Human Cellular Antigens

Bacterial or Viral Protein	Human Protein	Autoimmune Disease
Polio Vp2	Acetylcholine choline receptor	Myasthenia gravis
Influenza, polyoma, EBV, hepatitis B, measles P3	Myelin basic protein	Multiple sclerosis
Rabies glycoprotein and papilloma E3	Insulin receptor	Diabetes
Streptococcus M protein	Heart valve myosin	Rheumatic fever
Trypanosoma cruzi antigens	Nerve and cardiac tissue	Multiple sclerosis

How Autoimmune Diseases Cause Damage

Damage in autoimmune disease is done by several mechanisms. Antibodies can bind to cell membrane antigens, causing cell lysis (autoimmune hemolytic anemia). They can also bind to receptors, stimulating them (Graves disease). Autoantibodies can also bind to receptors and either block or damage the receptor (myasthenia gravis). Immune complex deposition in walls of small blood vessels in the kidney and joints is a key characteristic of systemic lupus erythematosus (SLE). In Sjögren's syndrome, WBC invade and destroy glands that produce moisture, resulting in dry mouth and dry eyes. Rheumatoid arthritis (RA) is characterized by acute and chronic inflammation of synovial joints causing pain, damage, and loss of function. Etiology of RA is not fully understood, and several etiological factors may cause rheumatoid arthritis even in the same individual. In type 1 diabetes mellitus, insulin-producing β cells are selectively destroyed by TC cells or antibodies. The role of autoimmunity in multiple sclerosis (MS) is a subject of intense study. In MS, immune cells attack and destroy the myelin sheath of neurons in the brain and spinal cord, resulting in a decrease in speed and efficiency when nerve messages are sent.

The category of the mechanism of an autoimmune disease is not always easy to establish. In type 1 diabetes (DM) the beta (islet) cells of the pancreas, which produce insulin, are attacked by the individual's immune system. Islet cell antibodies (ICAs), which react with islet cells in culture, have been identified. The autoantigens initially identified in type 1 DM were a form of glutamic acid decarboxylase (GAD65), a transmembrane protein tyrosine phosphatase-like molecule (IA-2), and insulin (specifically the B chain of human proinsulin or insulin). IA-2ß (phogrin, a protein found in insulin-containing secretory granules in pancreatic beta cells) is 74% identical to IA-2 and also reacts with the ICAs.

In the 1980s, three antigens that are recognized by ICAs were identified. Most newly diagnosed individuals (90%) have autoantibodies to one or more of these antigens. This led to the proposal that attack on the islets by autoantibodies is the cause of the autoimmune damage. However, the autoantibodies precede the development of diabetes by many months or years. Thus, detection of autoantibodies can be an indicator that an otherwise healthy individual is at high risk for type 1 diabetes. Currently it is thought that most, if not all, of the damage to the islets is done by TC cells. Since the onset of type

1 DM often follows a viral infection, many think a virus, especially Coxsackie virus B, may play a role in type 1 DM in some individuals. Other candidate viruses include mumps, rubella, cytomegalovirus, measles, influenza, encephalitis, polio, or Epstein-Barr. Seroepidemiologic studies support this idea (Banatvala et al. 1985). There is also strong evidence of a genetic component. MHC antigens DR3 or DR4, as well as DQA1*0301–B1*0302, are found in many individuals with type 1 DM, while MHC antigens DQA1*0102–B1*0602 show a strong negative association with type 1 diabetes. There is a weak positive association between exposure to cow's milk as a baby and type 1 diabetes, but the role of cow's milk in its causation remains unclear. Studies (Schrezenmeir et al. 2000) indicate the risk may vary with different milk proteins; thus, not all cow's milk would carry the same risk. In one study (Karjalainen et al. 1992) almost all newly diagnosed children had elevated levels of IgG antibodies to a 17-amino acid peptide of whey protein, bovine serum albumin (BSA). However, negative T cell proliferation studies (Atkinson et al. 1993) in response to cow's milk antigen cause some scientists to question the role of cow's milk proteins. Infants fed only cow's milk may also have a higher risk of multiple sclerosis, which complicates interpretation of the data.

Conclusion

The study of both health and disease are dependent on the successful comprehension of the human immune system. Because malnutrition is the leading cause of immunodeficiency, it is especially important for one to understand the interdependent relationship between nutrition and immunity. Equally crucial, however, is the concept that disease response, as the practitioner may see first-hand in many hospitalized patients, is linked to both nutritional status and immunocompetence. Understanding this complex relationship will only further improve the clinician's ability to enhance both.

WEB LINKS

Online lecture notes can be viewed at these sites:

General Immunology

http://www.cehs.siu.edu/fix/medmicro/genimm.htm

Microbiology at Leicester

http://www-micro.msb.le.ac.uk/MBChB/default.html

Microbiology Lecture Guide

http://www.cat.cc.md.us/courses/bio141/lecguide/index.html

Online textbooks are available at these sites:

Immunology

http://users.rcn.com/jkimball.ma.ultranet/BiologyPages/T/TOC.html#Immunology

***Immunobiology* 5th ed.**

http://www.ncbi.nlm.nih.gov/books/bv.fcgi?rid=imm.preface.5

Microbiology and Immunology On-line, University of South Carolina School of Medicine

http://pathmicro.med.sc.edu/book/immunol-sta.htm

The American Academy of Allergy, Asthma & Immunology: Visit the site of the largest professional medical specialty organization in the U.S.

http://www.aaaai.org

American College of Allergy, Asthma & Immunology: You can locate an allergist and find patient education materials at this site.

http://www.acaai.org

Cells of the blood: Photographs of white blood cells can be viewed at this site.

http://www-micro.msb.le.ac.uk/MBChB/bloodmap/Blood.html

Medline Plus: Food Allergy: This site includes descriptions of various conditions and their treatments as well as general overview information on food allergies.

http://www.nlm.nih.gov/medlineplus/foodallergy.html

Medline Plus: Pernicious anemia: This site describes pernicious anemia, its causes, and its treatment.

http://www.nlm.nih.gov/medlineplus/ency/article/000569.htm

National Digestive Diseases Information Clearinghouse: Find information about autoimmune hepatitis and other diseases.

http://digestive.niddk.nih.gov/index.htm

National Institute of Diabetes & Digestive & Kidney Diseases: Find information about type 1 diabetes and other diseases.

http://www.niddk.nih.gov

TransWeb: Links to sites with information on organ transplantation.

http://www.transweb.org

END-OF-CHAPTER QUESTIONS

1. List and describe factors that can influence an individual's susceptibility to infectious disease.
2. Describe an example of natural resistance.
3. What are the differences between antigens, haptens, and immunogens?
4. Describe humoral and cellular immunity, specific and nonspecific immunity, and active and passive immunity.
5. Briefly describe the function of each of the three groups of white blood cells: macrophages/monocytes, microphages/granulocytes/polymorphonuclear leukocytes, and lymphocytes and natural killer cells.
6. How are mast cells involved in the symptoms of allergies?
7. Briefly describe the functions of T helper cells, Th1 and Th2 cells, and cytotoxic T cells. What are CD4 and CD8 cells?
8. Briefly describe the function B cells—plasma cells, memory B cells, and antibody producing cells. What are the immune functions for each of the five antibodies produced by B cells?
9. What is meant by "antigen-presenting cell," and which cells in the body can serve this function?

10. What is meant by the term "negative selection" when self-reacting cells are eliminated? Why is this important?
11. What is the immune function of the lymphatic system?
12. There are several soluble mediators of the immune system. List and briefly describe their function.
13. How do major histocompatibility complexes I and II aid the immune system in distinguishing between self and non-self?
14. How can T helper cells be activated? After activation, what is their response?
15. How are monoclonal antibodies used in cancer treatment? How do they differ from antibodies?
16. Why is it critical to match MHC antigens for tissues used in transplantation? What might happen if they are not matched?
17. What is the difference between active and passive immunization?
18. Describe one way that malnutrition can compromise immunity.
19. List common food allergies.

13

Pharmacology

Marcia Nahikian Nelms, Ph.D., R.D.

Southeast Missouri State University

CHAPTER OUTLINE

Introduction to Pharmacology

Role of Nutrition Therapy in Pharmacotherapy

Drug Mechanisms

Administration of Drugs

Pharmacokinetics: Absorption of Drugs • Pharmacokinetics: Distribution of Drugs • Pharmacokinetics: Metabolism of Drugs • Pharmacokinetics: Excretion of Drugs • Alterations in Drug Pharmacokinetics

How Do Food and Drugs Interact?

Effect of Nutrition on Drug Action • Nutritional Complications Secondary to Pharmacotherapy • At-Risk Populations

Nutrition Therapy

Introduction to Pharmacology

The use of drugs has been a significant component of medical care since ancient times. Historically, drugs were available without a prescription, and alcohol, cocaine, marijuana, and opium were common components of drugs. The Pure Food and Drug Act of 1906, along with the subsequent Food, Drug, and Cosmetic (FD&C) Act, enacted in 1938, began government regulation for drugs in the United States through the Food and Drug Administration (FDA). (See Box 13.1 for the history of the FDA.) As medical care has advanced, so has the development of **pharmacotherapy**. The magnitude of medication use is reflected in its contribution to health care costs. More than two-thirds of all physician visits include a written prescription. Over 2.8 billion outpatient prescriptions were written in 2000, and American spending for prescription medications increased by 14% between 2001 and 2003—more than any other component of health care—costing more than $180 billion dollars each year (Berndt 2002; National Center for Health Statistics 2004; Smith et al. 2005).

Pharmacotherapy is defined as the use of drugs for treatment of disease and health maintenance. A medical drug (or medicine) is defined as a chemical used for the diagnosis, prevention, treatment of symptoms, or cure of diseases. Drugs can be classified by structure or pharmacological action. Many drugs require a physician's prescription, while others are classified as over-the-counter (OTC) medications (not requiring a prescription).

pharmacotherapy—use of drugs for treatment of disease and health maintenance

BOX 13.1 HISTORICAL EVENTS

History of the Food and Drug Administration

The Food and Drug Administration (FDA) has grown from a solitary chemist to over 9,000 employees and an annual budget of almost $1.3 billion. FDA staff includes chemists, pharmacologists, physicians, microbiologists, veterinarians, pharmacists, attorneys, and others in Washington, D.C., and over 150 field offices and laboratories. The following table presents highlights in the FDA's history.

1862	President Abraham Lincoln appoints a chemist to serve in new Department of Agriculture. This was the beginning of the Bureau of Chemistry, predecessor of the Food and Drug Administration.
1902	Biologics Control Act passed to ensure purity and safety of serums, vaccines, and similar products used to prevent or treat diseases in humans. Congress appropriates $5,000 to the Bureau of Chemistry to study chemical preservatives and colors along with their effects on digestion and health.
1906	Original Food and Drugs Act passed by Congress on June 30 and signed by President Theodore Roosevelt to prohibit interstate commerce in misbranded and adulterated foods, drinks, and drugs. Meat Inspection Act is passed the same day.
1907	First Certified Color Regulations, requested by manufacturers and users, list seven colors found suitable for use in foods.
1911	Supreme Court rules 1906 Food and Drugs Act does not prohibit false therapeutic claims, but does prohibit false and misleading statements about ingredients or identity of a drug.
1912	Congress enacts Sherley Amendment to overcome Supreme Court ruling that prohibited labeling medicines with false therapeutic claims intended to defraud the purchaser, a standard difficult to prove.
1913	Gould Amendment requires food package contents to be "plainly and conspicuously marked on the outside of the package in terms of weight, measure, or numerical count."
1914	Supreme Court issues first ruling on food additives: in order for bleached flour with nitrite residues to be banned from foods, the government must show a relationship between the chemical additive and the harm it allegedly caused in humans.
1933	FDA recommends complete revision of obsolete 1906 Food and Drugs Act.
1938	Federal Food, Drug, and Cosmetic (FDC) Act of 1938 is passed by Congress, containing new provisions that: • Extend control to cosmetics and therapeutic devices. • Require new drugs to be shown safe before marketing—starting a new system of drug regulation. • Eliminate Sherley Amendment requirement to prove intent to defraud in drug misbranding cases. • Provide safe tolerances be set for unavoidable poisonous substances. • Authorize standards of identity, quality, and fill-of-container for foods. • Authorize factory inspections. • Add the remedy of court injunctions to previous penalties of seizures and prosecutions. • Under Wheeler-Lea Act, Federal Trade Commission is charged with overseeing advertising associated with products otherwise regulated by FDA, with exception of prescription drugs.
1939	First food standards issued (canned tomatoes, tomato purée, and tomato paste).
1940	FDA transferred from Department of Agriculture to (new) Federal Security Agency.
1941	Insulin Amendment requires FDA to test and certify purity and potency of insulin.
1944	Public Health Service Act is passed to cover a broad spectrum of health concerns, including regulation of biological products and control of communicable diseases.
1950	Court of Appeals rules directions for use on a drug label must include purpose for which the drug is offered. Oleomargarine Act requires prominent labeling of colored oleomargarine to distinguish it from butter. Delaney Committee starts congressional investigation of the safety of chemicals in foods and cosmetics, laying foundation for 1954 Miller Pesticide Amendment, 1958 Food Additives Amendment, and 1960 Color Additive Amendment.
1951	Durham-Humphrey Amendment defines kinds of drugs that cannot be safely used without medical supervision and restricts their sale to prescription by a licensed practitioner.
1953	Agency transferred to Department of Health, Education, and Welfare (HEW).
1954	Miller Pesticide Amendment spells out procedures for setting safety limits for pesticide residues on raw agricultural commodities.
1958	Food Additives Amendment enacted, requiring manufacturers of new food additives to establish safety. Delaney proviso prohibits approval of any food additive shown to induce cancer in humans or animals. FDA publishes in the Federal Register the first list of substances generally recognized as safe (GRAS), which contains nearly 200 substances.
1960	Color Additive Amendment enacted, requiring manufacturers to establish safety of color additives in foods, drugs, and cosmetics. Delaney proviso prohibits approval of any color additive shown to induce cancer in humans or animals. Federal Hazardous Substances Labeling Act requires prominent label warnings on hazardous household chemical products.
1962	Thalidomide, a new sleeping pill, found to have caused birth defects in thousands of babies born in western Europe. Kefauver-Harris Drug Amendments passed to ensure drug efficacy and greater

	drug safety, requiring drug manufacturers to prove to FDA effectiveness of their products before marketing them. The new law also exempts from Delaney proviso animal drugs and animal feed additives shown to induce cancer but which leave no detectable levels of residue in human food supply. Consumer Bill of Rights includes right to safety, right to be informed, right to choose, and the right to be heard.
1966	FDA contracts with National Academy of Sciences/National Research Council to evaluate effectiveness of 4,000 drugs approved on basis of safety alone between 1938 and 1962. Child Protection Act enlarges scope of Federal Hazardous Substances Labeling Act to ban hazardous toys and other articles so hazardous that adequate label warnings could not be written. Fair Packaging and Labeling Act requires all consumer products in interstate commerce to be honestly and informatively labeled.
1968	Reorganization of federal health programs places FDA in Public Health Service. FDA Bureau of Drug Abuse Control and Treasury Department Bureau of Narcotics are transferred to Department of Justice to form the Bureau of Narcotics and Dangerous Drugs (BNDD), consolidating efforts to police traffic in abused drugs. FDA forms Drug Efficacy Study Implementation (DESI) to implement recommendations of National Academy of Sciences investigation of effectiveness of drugs first marketed between 1938 and 1962. Animal Drug Amendments place all regulation of new animal drugs under one section of Food, Drug, and Cosmetic Act—Section 512—making approval of animal drugs and medicated feeds more efficient.
1969	FDA begins administering Sanitation Programs for milk, shellfish, food service, interstate travel facilities, and preventing poisoning and accidents. White House Conference on Food, Nutrition, and Health recommends systematic review of GRAS substances in light of FDA's ban of the artificial sweetener cyclamate.
1970	Court of Appeals upholds enforcement of 1962 Drug Effectiveness Amendments by ruling commercial success alone does not constitute substantial evidence of drug safety and efficacy. FDA requires first patient package insert (oral contraceptives) must contain information for the patient about specific risks and benefits. Comprehensive Drug Abuse Prevention and Control Act replaces previous laws and categorizes drugs based on abuse and addiction potential compared to their therapeutic value. Environmental Protection Agency (EPA) established; takes over FDA program for setting pesticide tolerances.
1971	Artificial sweetener saccharin, included in FDA's original GRAS list, removed from list pending new scientific study.
1972	Over-the-Counter Drug Review begun to enhance safety, effectiveness, and appropriate labeling of drugs sold without prescription. Regulation of Biologics—including serums, vaccines, and blood products—is transferred from NIH to FDA.
1973	U.S. Supreme Court upholds 1962 Drug Effectiveness Law and endorses FDA action to control entire classes of products by regulations rather than rely only on time-consuming litigation. Low-acid food processing regulations issued after botulism outbreaks from canned foods to ensure low-acid packaged foods have adequate heat treatment and are not hazardous. Consumer Product Safety Commission created by Congress takes over programs pioneered by FDA under 1927 Caustic Poison Act, 1960 Federal Hazardous Substances Labeling Act, 1966 Child Protection Act, and PHS accident prevention activities for safety of toys, home appliances, and so on.
1976	Medical Device Amendments passed to ensure safety and effectiveness of medical devices, including diagnostic products, by requiring manufacturers to register with FDA and follow quality control procedures. Vitamins and Minerals Amendments ("Proxmire Amendments") stop FDA from establishing standards limiting potency of vitamins and minerals in food supplements or regulating them as drugs based solely on potency.
1977	Saccharin Study and Labeling Act passed by Congress to stop FDA from banning the chemical sweetener, but requiring a label warning that it has been found to cause cancer in laboratory animals. Introduction of Bioresearch Monitoring Program as an agency-wide initiative ensures quality and integrity of data submitted to FDA and provides for protection of human subjects in clinical trials by focusing on preclinical studies on animals, clinical investigations, and work of institutional review boards.
1980	Infant Formula Act establishes special controls to ensure necessary nutritional content and safety.
1981	FDA and Department of Health and Human Services revise regulations for human subject protections, based on 1979 Belmont Report that had been issued by National Commission for the Protection of Human Subjects of Biomedical and Behavioral Research. Revised rules provide for wider representation on institutional review boards and detail elements of what constitutes informed consent, among other provisions.
1982	Tamper-Resistant Packing Regulations issued by FDA to prevent poisonings such as deaths from cyanide placed in Tylenol capsules. Federal Anti-Tampering Act passed in 1983 makes it a crime to tamper with packaged consumer products.
1983	Orphan Drug Act passed, enabling FDA to promote research and marketing of drugs needed for treating rare diseases.
1984	Drug Price Competition and Patent Term Restoration Act expedites availability of less costly generic drugs by permitting FDA to approve applications to market generic versions of brand-name drugs without repeating research done to prove them safe and effective.
1985	AIDS test for blood approved by FDA in its first major action to protect patients from infected donors.
1988	Food and Drug Administration Act of 1988 officially establishes FDA as an agency of the Department of Health and Human Services with a Commissioner of Food and Drugs appointed by the President with advice and consent of the Senate, and broadly spells out responsibilities of the Secretary and Commissioner for research, enforcement, education, and information. Prescription Drug Marketing Act bans diversion of prescription drugs from legitimate commercial channels.
1989	FDA issues a nationwide recall of all over-the-counter dietary supplements containing 100 milligrams or more of L-tryptophan after a U.S. outbreak of eosinophilia myalgia syndrome (EMS), characterized by fatigue, shortness of breath, and other symptoms.

1990	Congress passes Anabolic Steroid Act of 1990, which identifies anabolic steroids as a class of drugs. Nutrition Labeling and Education Act requires all packaged foods to bear nutrition labeling and all health claims for foods to be consistent with terms defined by the Secretary of Health and Human Services.
1991	Regulations published to accelerate review of drugs for life-threatening diseases. Policy for protection of human subjects in research, promulgated in 1981 by FDA and the Department of Health and Human Services, is adopted by more than 12 federal entities involved in human subject research and becomes known as the Common Rule.
1992	Generic Drug Enforcement Act imposes debarment and other penalties for illegal acts involving abbreviated drug applications. Prescription Drug User Fee Act requires drug and biologics manufacturers to pay fees for product applications and supplements, and other services. Mammography Quality Standards Act requires all mammography facilities in the U.S. to be accredited and federally certified. Nutrition Facts, basic per-serving nutritional information, are required on foods under Nutrition Labeling and Education Act of 1990.
1993	Consolidation of several adverse reaction reporting systems is launched as MedWatch, designed for voluntary reporting of problems associated with medical products to be filed with FDA by health professionals. Revising a policy from 1977 that excluded women of childbearing potential from early drug studies, FDA issues guidelines calling for improved assessments of medication responses as a function of gender.
1994	Dietary Supplement Health and Education Act establishes specific labeling requirements, provides regulatory framework, and authorizes FDA to disseminate good manufacturing practice regulations for dietary supplements. This act defines "dietary supplements" and "dietary ingredients" and classifies them as food.
1995	FDA declares cigarettes to be "drug delivery devices." Restrictions are proposed on marketing and sales to reduce smoking by young people. A series of proposed reforms to reduce regulatory burden on pharmaceutical manufacturers announced.
1996	Federal Tea Tasters Repeal Act repeals Tea Importation Act of 1897 to eliminate Board of Tea Experts and user fees for FDA's testing of all imported tea. Saccharin Notice Repeal Act repeals saccharin notice requirements. Food Quality Protection Act amends Food, Drug, and Cosmetic Act, eliminating application of Delaney proviso to pesticides.
1999	ClinicalTrials.gov is founded to provide public with updated information on enrollment in federally and privately supported clinical research, expanding patient access to studies of promising therapies. A final rule mandates all over-the-counter drug labels must contain data in a standardized format.
2000	U.S. Supreme Court ruled 5–4 FDA does not have authority to regulate tobacco as a drug. Federal agencies required to issue guidelines to maximize quality, objectivity, utility, and integrity of information they generate, and provide a mechanism whereby those affected can secure correction of information that does not meet these guidelines, under the Data Quality Act. Publication of a rule on dietary supplements defines type of statements that can be labeled regarding effect of supplements on structure or function of the body.
2002	Best Pharmaceuticals for Children Act improves safety and efficacy of patented and off-patent medicines for children. In the wake of events of September 11, 2001, Public Health Security and Bioterrorism Preparedness and Response Act of 2002 is designed to improve the country's ability to prevent and respond to public health emergencies, and provisions include a requirement that FDA issue regulations to enhance controls over imported and domestically produced commodities it regulates.
2003	Medicare Prescription Drug Improvement and Modernization Act requires, among other elements, study be made of how current and emerging technologies can be utilized to make essential information about prescription drugs available to the blind and visually impaired. To help consumers choose heart-healthy foods, the Department of Health and Human Services announces FDA will require food labels to include trans fat content, the first substantive change to the nutrition facts panel on foods since the label was changed in 1993. An obesity working group established by Commissioner of Food and Drugs is charged to develop an action plan to deal with the nation's obesity epidemic from perspective of FDA. National Academy of Sciences releases "Scientific Criteria to Ensure Safe Food," a report commissioned by FDA and Department of Agriculture, which buttresses the value of the Hazard Analysis and Critical Control Point (HACCP) approach to food safety already in place at FDA and invokes need for continued efforts to make food safety a vital part of overall public health mission. FDA is given clear authority under the Pediatric Research Equity Act to require sponsors conduct clinical research into pediatric applications for new drugs and biological products.
2004	Project BioShield Act of 2004 authorizes expedited review procedures to enable rapid distribution of treatments as countermeasures to chemical, biological, and nuclear agents that may be used in a terrorist attack against the U.S., among other provisions. Passage of Food Allergy Labeling and Consumer Protection Act requires labeling of any food containing a protein derived from peanuts, soybeans, cow's milk, eggs, fish, crustacean shellfish, tree nuts, and wheat. A ban on over-the-counter steroid precursors increased penalties for making, selling, or possessing illegal steroids precursors, and funds for preventive education to children are features of the Anabolic Steroid Control Act of 2004. FDA issues a public health advisory urging health professionals to limit use of cox-2 selective agents. FDA bans dietary supplements containing ephedrine alkaloids, deeming such products to present unreasonable risk of harm.
2005	Formation of Drug Safety Board to advise FDA on drug safety issues and work with the agency in communicating safety information to health professionals and patients is announced.

Source: FDA Backgrounder, Milestones in U.S. Food and Drug Law History, May 3, 1999, updated August 2005.
Swann JP. History of the FDA. FDA History Office. Available at http://www.fda.gov/oc/history/historyoffda/fulltext.htm. Accessed November 3, 2005.

Pharmacology is the study of drugs, their properties and their effects; **pharmacokinetics** is the study of drug absorption, distribution, metabolism, and excretion. This chapter focuses on the basic principles of pharmacology, with an emphasis on the interaction of medications with nutrition.

All health care practitioners need to understand the basic principles of pharmacology. Such an understanding is especially valuable for registered dietitians (RDs) as they work toward coordination and integration of nutrition therapy with pharmacotherapy. Nutrition therapy (NT) is a "specific nutrition service or procedure used to treat an illness, injury or condition" (ADA 2003). Lifestyle, behavior changes, and alternative and complementary medicines, which include nutrition therapy, are important elements in treatment for many conditions, but use of medications remains a cornerstone of most disease treatment. An understanding of all aspects of medical care, including pharmacotherapy, among practitioners should result in improved patient outcomes, maximized nutritional status, and decreased complications or risks of the prescribed medical care.

The Joint Commission on Accreditation of Hospitals (JCAHO), the organization that accredits medical facilities, requires monitoring, documentation, and patient education for food-drug interactions. Ensuring that this requirement is met necessitates the coordinated efforts of all health care practitioners.

Role of Nutrition Therapy in Pharmacotherapy

Consider the situation of a 52-year-old male currently being treated for hypertension and hyperlipidemia. His physician has prescribed 40 mg Inderal twice daily (BID) to control his blood pressure; 20 mg of Zocor each day; and Niacor 500 mg three times per day (TID) to treat his hyperlipidemia. How does this typical patient situation relate to any nutritional care? Though nutrition's role in pharmacotherapy can be approached from several perspectives, it has traditionally been discussed within the context of the effect of nutrition on the action of the prescribed medication or the effect of the medication on an aspect of nutrition. Drug-nutrient interactions are defined as "an alteration of kinetics or dynamics of a drug or nutritional element, or a compromise in nutritional status as a result of the addition of a drug" (Chan 2002). *The Position of the American Dietetic Association: Integration of Medical Nutrition Therapy and Pharmacotherapy* (ADA 2003) expands this discussion by emphasizing a collaborative model of health care that allows maximum benefit from the use of both pharmacotherapy and nutrition therapy. These classifications will be discussed later in this chapter.

In the patient scenario just presented, the well-trained RD would first recognize that a therapeutically important drug-nutrient interaction could occur between Zocor and grapefruit (Walsky, Gaman, Obach 2005). Grapefruit interferes with absorption of Zocor and could significantly change availability of this medication. Next, it is important to note that first-pass hepatic metabolism of propranolol (Inderal) may be decreased when this medication is taken with food. Drug levels in the body may be increased due to this interaction; to avoid such an event, the patient would be counseled to take this medication on an empty stomach.

Though important, preventing interactions is only one goal of understanding nutrition's contribution to pharmacotherapy. Current recommendations for treatment of hyperlipidemia include The Therapeutic Lifestyle Changes (TLC) from the National Cholesterol Education Program (2001). These recommendations incorporate nutrition therapy as a major component of treatment for cardiac disease, hyperlipidemia, and hypertension (National Cholesterol Education Program 2001). Weight loss, if the patient were overweight, could lower his blood pressure, reduce the required dosage of Inderal, and improve his lipid profile so that his dosage of these medications could be reduced or eliminated. Incorporating principles of the DASH (Dietary Approaches to Stop Hypertension) diet with his weight-loss program may result in a further decrease in his blood pressure (see Chapter 15). Complete counseling for this patient would encompass the following recommendations: avoid grapefruits and grapefruit juice, take Inderal on an empty stomach, and follow a low-kcal, low-saturated fat, and low-sodium diet that is rich in fruits and vegetables and fat-free or low-fat dairy products. Over time, successful nutrition therapy could decrease his yearly prescription costs by $4800. Hence, intervention by the RD provides clinical and economic benefits while also meeting all legal responsibilities (Pronsky and Crowe 2004). This chapter addresses the complex relationship between nutrition therapy and pharmacotherapy in detail, focusing on information the RD will need to successfully integrate the two.

Drug Mechanisms

The most common mechanisms for drug action involve binding of the drug to specific receptors on the cell membrane, which initiates changes in specific enzyme reactions. Drugs react with a cellular receptor site due to their design and shape—as a lock and key might fit together. When this occurs, physiological functions are altered. Most drugs can interact with more than one cell receptor, which may account for various side effects of medication use (see Figure 13.1).

pharmacology—study of drugs, their properties, and their effects

pharmacokinetics—study of drug absorption, distribution, metabolism, and excretion

FIGURE 13.1 Cellular Receptor Site

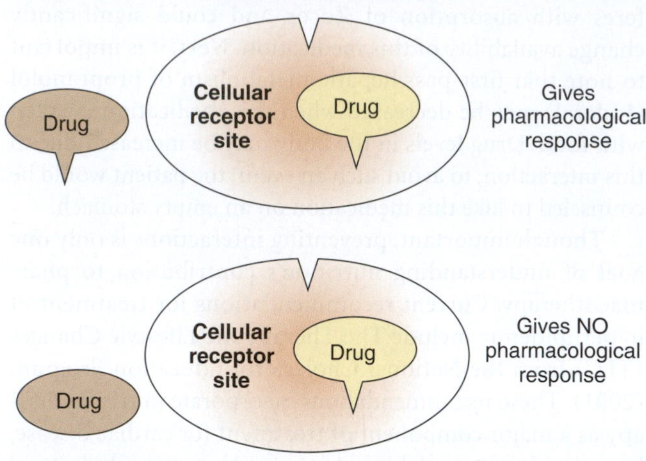

Source: Charles E. Ophardt, Elmhurst College. Reprinted with permission.

Alterations in enzyme systems by medications are caused either by stimulating (induction) or inhibiting an enzyme system (see Pharmacokinetics: Metabolism of Drugs and Figure 13.2). An example of these mechanisms can be found in the class of medications called ACE (angiotensin-converting enzyme) inhibitors. In normal control of blood pressure, angiotensin-converting enzyme stimulates conversion of angiotensin I to angiotensin II. The function of angiotensin II is to constrict blood vessels and cause an increase in blood pressure (see Chapters 8 and 15 for more detail). If this enzyme system is inhibited, blood vessels will vasodilate, causing a decrease in blood pressure. Of course, other physiological functions can change as a result of drug action. These non-specific responses can either be therapeutic or fall into the role of side effects and drug interactions.

FIGURE 13.2 Inhibition of Enzyme System
A. Normal action of ACE.
B. Inhibition of ACE through medication causes blood pressure to drop.

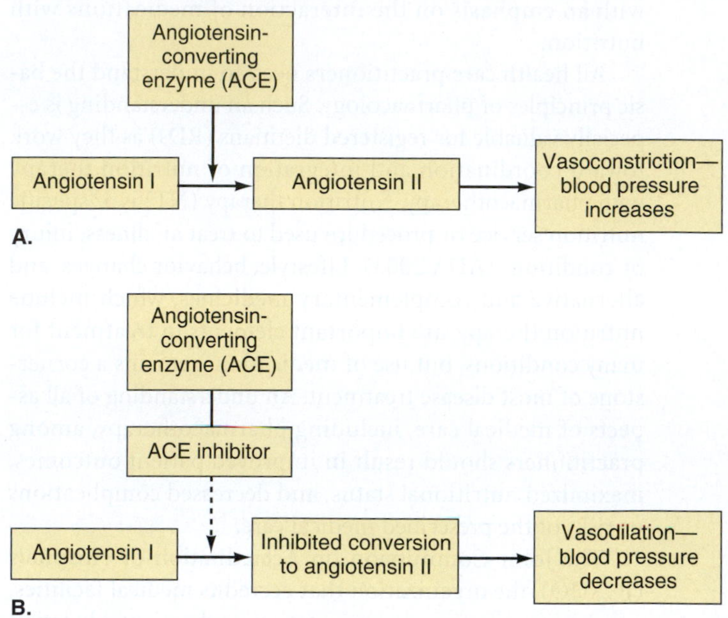

sublingual—refers to placement of a drug under the tongue

buccal—refers to placement of a drug in the cheek

parenteral—refers to injection into the body's circulatory system through a blood vessel

topically—refers to placement of a drug on the skin

inhalation—refers to placement of a drug so that it is breathed into the respiratory system

subcutaneous—refers to injection into the body under the skin

intradermal—refers to injection under the outermost layer of skin

intramuscular—refers to injection into the muscle

intraperitoneal—refers to injection into the body's peritoneal cavity

intravenous—refers to injection directly into a vein

ophthalmic—refers to placement of a drug into the eye

otic—refers to placement of a drug into the ear

epidural—refers to placement of a drug into the spinal fluid

intrathecal—refers to injection of a drug into the membrane surrounding the central nervous system

Administration of Drugs

Drugs can be administered in multiple ways. The administrative route depends on the chemical properties of the drug, the type of effect desired, and, of course, patient characteristics that affect how the medication could be administered. The oral route of administration requires that the patient be able to swallow medication and that the slower rate of absorption of this administration method is acceptable. **Sublingual** or **buccal** administration means the drug is placed under the tongue or in the cheek, respectively. It dissolves there, so it is quickly absorbed across mucous membranes into the circulatory system. When an individual takes nitroglycerin for angina, it is usually via a sublingual route.

Routes of administration can be **parenteral, topical** via skin and mucous membranes, or through **inhalation.**

Parenteral administration requires an injection into the body through routes that are either **subcutaneous (SC), intradermal (ID), intramuscular (IM), intraperitoneal (IP),** or **intravenous (IV).** Topical medications are applied to skin for a direct effect, but can also be absorbed via skin or mucous membranes; for example, Estraderm© is an estrogen patch worn to increase circulatory amounts of estrogen. Drugs that are inhaled have the opportunity to act locally within the respiratory system or to have a systemic effect. When anesthesia is inhaled, systemic effects occur, whereas the medication Combivent© uses two different types of bronchodilators, which act locally to treat asthma and other respiratory conditions. Medications can also be placed directly into target tissue such as the eye (**ophthalmic**), ear (**otic**), or spinal canal (**epidural** or **intrathecal**).

Pharmacokinetics: Absorption of Drugs

Absorption of the drug/medication involves several steps as the substance is transferred from the administrative site (e.g., oral, sublingual, intravenous) to the circulatory or lymphatic system. Absorptive mechanisms for drugs follow the same basic processes as those for nutrients (see Chapters 16 and 17). Collectively, these processes include passive diffusion, facilitated diffusion, and active transport.

The rate and effectiveness of absorption for drugs is dependent on several key factors. First, solubility of the medication determines where in the gastrointestinal tract the medication will dissolve and thus be absorbed. **Dissolution** or dissolving of the medication has to occur before absorption is successful. **Excipients** are those substances added to formulations of medications that affect dissolution. Binders, lubricants, and coating agents decrease dissolution, whereas disintegrants (ingredients that dissolve readily in water) increase dissolution. Coloring and flavoring agents have varying effects on dissolution. Tablet formulation is also a factor; hard, round, and large tablets dissolve more slowly. Dissolution rates of generic equivalents to the original medication may also vary (Epstein et al. 2003).

The amount of time a medication is present in a specific portion of the gastrointestinal (GI) tract, the pH of that portion of the GI tract, and the surface area of the GI tract also affect absorption capability. The largest surface areas for drug absorption are located in the small intestine and lungs. Other factors that affect absorption include the chemical properties of the drug, the integrity of the gastrointestinal tract and other tissues, and the circulation and blood supply (Beers and Berkow 2005). Anatomical regions with the highest blood flow, including the small intestine, lungs, muscle, and buccal and nasal cavities, have efficient rates of absorption and distribution.

The most important chemical properties of medications related to drug absorption include the solubility of the drug in lipid or water and the **ionization** of the medication. Lipid-based drugs will be absorbed across cell membranes quickly, since cell membranes are primarily lipid based. Drugs that are not ionized will also be absorbed much more readily. If the drug is ionized, absorption will be dependent on the pH of the solution where it will be absorbed. For example, if a medication is mildly acidic, absorption will be enhanced in solutions that are also acidic, such as gastric juices (Beers and Berkow 2005). Aspirin is a good example of a medication that is absorbed in the stomach but can also damage the gastric mucosa.

Pharmacokinetics: Distribution of Drugs

After absorption, distribution of the drug occurs. Distribution is defined as the movement of the drug throughout the body to the target sites where it can act. Distribution is variable and is affected by the circulation, the binding of the drug to proteins within the circulation (e.g., albumin, 1-acid glycoprotein) and the binding of the drug to other tissues within the body. Overall, the greater the amount of the drug that binds to another substance, the smaller the amount of active or free drug within circulatory or storage tissues. Physiological or anatomical features also affect distribution of the drug. For example, some drugs cannot cross the placenta or cell membrane into the central nervous system, while others are readily distributed to those sites.

Pharmacokinetics: Metabolism of Drugs

The metabolism of drugs involves **biotransformation** (changing the physical form), which renders the drug inactive so it may be excreted via urine or bile. The liver is the major site for biotransformation, but metabolism occurs within other organs as well. Drug metabolism occurs through the catalysis of enzyme systems including the family of **cytochrome P-450 isoenzymes (CP450).** Approximately 30 enzymes are responsible for the numerous reactions that oxidize drugs within the liver (Wilkinson 2005).

dissolution—dissolving of a medication

excipients—those substances added to formulations of medications, such as color or coating agents

ionization—process of producing negatively or positively charged ions

biotransformation—modification of a drug through metabolism

cytochrome P-450 isoenzymes (CP450)—family of enzyme systems responsible for drug metabolism

FIGURE 13.3 Therapeutic Levels of Drugs

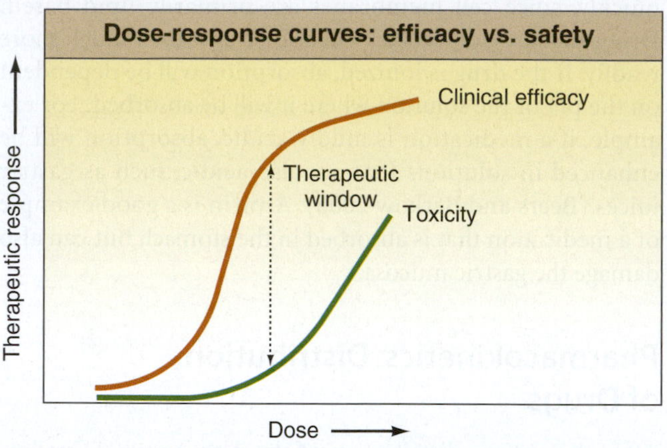

Source: Bottorff M.D., Evans W.E., "Drug concentration monitoring" in Progress in Clinical Biochemistry and Medicine, 1988.

A substance may interact with the CP450 enzymes as either an inhibitor or inducer. An inhibitor reacts with the specific enzyme by competition for the receptor site. An inducer works to stimulate synthesis of the enzymes, increasing action potential. Inhibitors decrease metabolism and generally lead to increased drug effect, whereas inducers will increase metabolism and generally lead to decreased drug effect. Phenobarbital and theophylline are examples of inducers of the CP450 enzymes. Examples of drugs known to be inhibitors include chloramphenicol, cimetidine, valproic acid, allopurinol, and erythromycin. Drug dosages must be adjusted to accommodate metabolism of each medication. Therapeutic levels are determined by measuring blood levels in order to establish the correct effective dose for each person (see Figure 13.3). The dosage range with therapeutic efficacy is referred to as the "therapeutic window." Levels below this window may not be effective, and those above may result in toxicity.

Pharmacokinetics: Excretion of Drugs

Generally, after drugs have been metabolized, the remaining compounds are eliminated from the body. There are exceptions; some drugs can be excreted before they are metabolized. Most drugs are removed by either urinary or biliary excretion, but some can be excreted via the lungs or bowel, depending on the chemical structure of the metabolite. It is important to be aware that some drugs can be excreted in breast milk as well, which means that the nursing infant would be exposed to that drug.

Urinary excretion of drugs can occur in all three stages of urinary filtration and concentration within the nephron, the functional unit of the kidney (see Chapter 20). Each of the over one million nephrons consists of a glomerulus and tubule. Each tubule is divided into several sections, depending on the type of epithelial cells it contains. Sections are referred to as the proximal tubule, Loop of Henle, distal tubule, and the collecting duct. (See Chapter 20, Figure 20.2.) All collecting ducts drain into the ureter and ultimately into the bladder. Most drugs of low molecular weight are filtered out of the blood in the glomerulus unless they are bound to large molecules such as proteins or to erythrocytes. Drugs can be reabsorbed within the tubules. Reabsorption depends on the pH of the urine and the solubility of the drug. Lipid-soluble drugs are more readily reabsorbed. Since the acidity of the urine is quite variable, there is a significant variation in drug reabsorption.

Alterations in Drug Pharmacokinetics

No two people will react in the same way to any given medication. Age; gender; cardiovascular, hepatic, and renal function; presence of disease or infection; diet; and even genetic differences will affect how an individual will respond to a drug dosage. The following sections describe potential alterations in each pharmacokinetic phase.

Altered GI Absorption GI absorption will be altered as health conditions, disease, and treatment modalities interrupt normal absorption processes. Simultaneous consumption of food with medication is one of the more common factors that may change the effectiveness of absorption (Chan 2002). The presence of food stimulates normal digestion and absorption mechanisms, such as changes in rate of gastric emptying and the release of enzymes and hydrochloric acid. All of these normal mechanisms may alter the GI environment so that it is not suitable for absorption of the medication. The presence of food also increases the chance for adhesion of the drug to a food component. Directions for a medication should indicate whether the drug should be taken with or without food.

A classic example of this situation is the effect of different foodstuffs on iron absorption. The absorption rate for iron supplements can vary tremendously depending on the type of food consumed with them. Citrus juices enhance absorption, whereas milk or iced tea would decrease absorption of the iron supplement (Gropper, Smith, and Groff 2005).

Vomiting and diarrhea can influence drug absorption by reducing the time available for solubility and dissolution. Diseases or health conditions that interrupt normal transit time or surface area will decrease the effectiveness of drug absorption. For example, Crohn's disease or other malabsorptive diagnoses will change the ability of the drug to be absorbed across the membrane of the enterocyte. Circulation deficits to and from the GI tract could also reduce the effectiveness of absorption from the small intestine to the rest of the body. The drugs propranolol and dextropropoxyphene increase blood flow to the liver, and thus increase circulation or distribution of other medications.

Drugs, nutrients, and other substances may compete for the carriers needed for active transport. For example, Levodopa, a standard medication for treatment of Parkinson's disease, is transported using the same pathways as neutral amino acids such as leucine and isoleucine. This medication should be taken on an empty stomach so that adequate absorption can be ensured (Howland and Mycek 2006). As mentioned previously, the pH at the absorption site can alter ionization of the drug, which may change the speed and effectiveness of absorption. For example, Ketoconazole, an antifungal agent, must be in an acidic environment for appropriate dissolution and absorption.

Altered Distribution Major factors that change distribution of a drug include variations in circulation, body size and composition, and protein binding of the medication. Factors that could alter circulation include age and disease. Any factor that causes vasodilation would theoretically increase distribution of the drug; for example, physical activity and increased body temperature increase vasodilation and thus distribution of the drug. Body size and body composition can alter drug distribution. The elderly individual may have decreased muscle mass requiring an adjustment for drug dosing (Fulton and Allen 2005). Large amounts of body fat may slow distribution of a medication. Many medications are bound to a protein carrier—most often, albumin. Any situation that could alter albumin concentrations, such as liver or kidney disease or malnutrition, would increase the amount of unbound medication, multiplying the amount of active drug within the body.

Altered Metabolism Age is also a major factor in how drugs are metabolized. Neonates, infants, and young children have vastly different levels of liver function and enzyme systems than adults do, which affects their reactions to different medications (deWildt, Johnson, and Choonara 2002). On the other end of the spectrum, the elderly may also have a decreased ability to metabolize drugs because of the normal physiological changes of aging. For instance, circulation within the liver decreases by approximately 35% with concurrent decreases in liver mass (Wynne 2005). Drug metabolism alterations may appear as decreased effectiveness of some medications or may surface as toxicity symptoms (Fulton and Allen 2005; Kinirons and O'Mahony 2004; Wynne 2005;).

Appropriate metabolism of drugs requires adequate function of organs—especially the liver. When disease and injury interrupt organ functioning, drug metabolism may change as well. The types of drugs or alternative regimens will need to be considered when concurrent drug treatment interferes with metabolism.

Genetic factors may also play a major role. Phenotypic differences are often attributed to differences in genetic coding of metabolic enzyme systems (Okey, Boutros, and Harper 2005). For example, differences in metabolism for **proton pump inhibitors** (such as omeprazole) can affect treatment effectiveness for *H. pylori* infections (Furuta et al. 2004). Gender differences are also apparent in metabolism for some drugs (Cotreau, von Moltke, and Greenblatt 2005).

One of the most common mechanisms for alteration of drug metabolism is concurrent use of other medications, which may interrupt enzyme systems and prevent clearance of metabolites. Numerous drug-drug interactions have been identified that, unless monitored closely, can cause significant adverse symptoms (Roden 2005).

Altered Urinary Excretion Urinary excretion of drugs can change as a result of numerous mechanisms. As stated earlier, the pH of the urine has a direct effect on the type of drugs easily excreted. Nutritionally, different foods can affect the pH of the urine, though these effects are difficult to predict due to variations in digestion and metabolism (Remer and Manz 1995). Excretion can also be changed by the presence of a competitor for active transport across the renal tubule. Finally, urinary excretion can be altered by changes in urinary flow rates or kidney function. This may occur as a result of another medication, as a result of disease or injury, or as a consequence of aging. Changes in **creatinine clearance** significantly alter the effectiveness of medications. If an individual has renal insufficiency from any etiology, drug levels must be adjusted to ensure therapeutic levels. **Digoxin**, **cyclosporine**, and **gentamycin** are examples of medications affected by changes in kidney function. Other medications such as ampicillin or cephalosporins are nephrotoxic and could themselves change kidney function (Loboz and Shenfield 2005).

How Do Food and Drugs Interact?

As stated earlier, drug-nutrient interactions can be organized by examining the effect of nutrition on the action of the prescribed medication, the effect of the medication on nutritional status, or the role of nutrition therapy in maximizing prescribed effect of pharmacotherapy and/or minimizing the side effects (ADA 2003).

proton pump inhibitors—drugs that reduce acid secretion in the stomach

creatinine clearance—rate at which creatinine is filtered through the kidney; often used as a measure of kidney function

digoxin—cardiac glycoside that is prescribed to alter the contractions of the heart

cyclosporine—immunosuppressant medication that is often prescribed after organ transplant

gentamycin—an antibiotic

Effect of Nutrition on Drug Action

This section will discuss the effects of food and nutrition on dissolution, absorption, metabolism, and excretion of medications. Since it is virtually impossible to have a working knowledge of all potential reactions, health professionals in specialty areas become very familiar with medications of their typical patient population. This textbook highlights specific drug-nutrient interactions for each diagnosis. Heightened awareness of potential interactions makes the integration of nutrition and pharmacotherapy a routine component of patient care.

Effect of Nutrition on Drug Dissolution In order for oral drugs to be absorbed, dissolution of the medication is necessary. The pH of the stomach and the gastric emptying rate are two of the most important nutrition-related factors impacting drug dissolution. Medications may require an acidic environment for dissolution. Achlorhydria, which is decreased production of hydrochloric acid, occurs in aging as well as some medical conditions such as HIV and AIDS. Medications that could affect gastric acidity include use of **H_2 blockers** (cimetidine, famotidine), proton-pump inhibitors (omeprazole, lansoprazole), and antacids (TUMS, Rolaids) (see Chapter 16). Gastric emptying rate influences the amount of time in which dissolution can occur; medications that affect gastric emptying time include **prokinetics** such as metoclopramide. The presence of food in the stomach will increase gastric emptying time (i.e., slow emptying rate), especially when a high-fat meal is consumed. This would potentially hinder dissolution.

Any disease, injury, or surgery that affects oral intake or gastric function can affect dissolution of medications. For example, vomiting and diarrhea would certainly decrease dissolution. Gastric surgical resections can dramatically change the rate of gastric emptying as well as the amount of gastric secretions (see Chapter 16 for a discussion of these surgical procedures). Any client who presents with this medical history will need adjustments in the form of the medication to ensure appropriate dissolution. Medications in liquid form are more easily dissolved than those in capsule or tablet form.

H_2 blockers—medications that interrupt the production of acid in the stomach

prokinetics—medications that increase peristalsis

protease inhibitor—a medication that prevents protein replication; a common class of drug that is used to prevent human immunodeficiency virus replication

CYP 3A4—a specific cytochrome enzyme involved in drug metabolism

Effect of Nutrition on Drug Absorption The presence of food, alcohol, or dietary supplements can interact with drugs in several important ways that interfere with drug action. Interactions may increase absorption of medications, hence increasing the amount of available drug. In contrast, if absorption of a medication is decreased by food, therapeutic levels may not be achieved. For instance, the presence of food dramatically reduces absorption of Fosamax©, a medication used to treat osteoporosis. Saquinavir©, a **protease inhibitor** used to treat HIV, is another dramatic example of a drug from which food affects absorption . Taking these medications at the same time as food can reduce absorption considerably. On the other hand, it is recommended that some medications be taken with food in order to decrease the gastric distress associated with them. Examples include Augmentin, ketoconazole, and erythromycin. Chelation is another mechanism that affects absorption. Chelation, the binding of a nutrient or food component with a drug, makes the drug unabsorbable. For example, consumption of calcium with the antibiotic tetracycline causes chelation of the drug, which decreases absorption. Patient education should include specific guidelines for consuming a medication with or without food, if applicable.

Effect of Nutrition on Drug Metabolism Some of the most important food-drug interactions fall into the category of metabolism changes. Research has identified several mechanisms; a summary of these findings is that, in general, some nutrients act either as an inducer or as an inhibitor for metabolic enzyme systems. These actions can change drug effectiveness as well as produce toxic side effects, which increase the potential for morbidity and mortality (Chan 2002; Peng et al. 2004; Pronsky and Crowe 2004; Sorensen 2002). Nutrients can also compete for carrier systems involved in normal drug metabolism.

For example, a recent study found that St. John's wort, an herbal supplement used to treat depression, significantly induced activity of **CYP 3A4.** Long-term use of St. John's wort may result in diminished clinical effectiveness or increased dosage requirements for at least 50% of all marketed medications (Markowitz et al. 2003). These types of interactions appear to pose a much more common and serious risk than was previously recognized. The use of herbal therapies prior to surgical anesthesia, for instance, could prolong the effect of the anesthesia (Norred 2002; Peng et al. 2004).

The potential for nutrient-drug interactions with anticoagulation therapy, a standard component of clinical care in prevention of stroke and heart attack, provides an important illustration of how foods interrupt drug metabolism. Vitamin K improves blood clotting. When foods high in vitamin K or vitamin K supplements are taken during the same time period as warfarin (Coumadin), a vitamin K antagonist, the amount of warfarin needed is increased. Vitamin K intake should therefore be consistent in order to maintain the levels of warfarin within a therapeutic level. Additionally, the dietary supplements feverfew, garlic, gingko biloba, ginger, cayenne, and omega-3 fatty acids can also affect blood coagulation. A change in the dosage of anticoagulation drugs may be required in order to compensate for a patient's dietary intake of these foods and supplements.

A classic example of a drug-nutrient interaction resulting in harmful side effects is the interaction between **pressor agents** in foods (tyramine, dopamine, histamine, phenylethylamine) and **monoamine oxidase (MAO) inhibitors** (e.g., Nardil). This interaction can result in sudden increases in blood pressure with resulting complications. Box 13.2 outlines the specifics for this drug-nutrient interaction.

The interaction of drugs with grapefruit and grapefruit juices has been the subject of recent clinical investigations. Numerous drugs subject to such metabolic interactions, including **statin** medications used to treat hyperlipidemia, several medications used in cardiac care (talinol, nifedipine), and cyclosporines (which are immunosuppressants), have been identified and are a targeted patient education issue for clinicians (Lilja, Neuvonen, and Neuvonen 2004; Odou et al. 2005; Paine, Criss, and Watkins 2005; Schwarz et al. 2005;). See Table 13.1 (tables are grouped at the end of the chapter) for a summary of these interactions.

Effect of Nutrition on Drug Excretion The pH of the urine can vary widely and is one of the most important concerns related to maintenance of consistent drug excretion. Variable urine pH can alter reabsorption of the drug, resulting in fluctuating therapeutic levels. Dietary intake, kidney and respiratory function, acid-base balance, hydration status, and the presence of disease or infection can alter urinary pH and necessitate evaluation of drug dosage. Modification of dietary intake to control urine pH has been applied in the treatment of urolithiasis (kidney stones; see Chapter 20) (Asplin, Coe, and Favus 2005; Remer and Manz 1995). These interventions include increased water intake, limited protein, and overall reduced dietary oxalate.

Nutritional Complications Secondary to Pharmacotherapy

The previous sections of this chapter have focused on the effect of diet and nutrition on drug pharmacokinetics. The other side of drug-nutrient interactions is the effect of drug action on nutritional status. Drugs affect nutrient ingestion, digestion, absorption, and metabolism. The clinical expertise of the registered dietitian is a critical component in the identification, prevention, and correction of these interactions.

Drug Consequence: Effect on Nutrient Ingestion One only has to evaluate the possible side effects of any medication to understand their potential effect on nutrient ingestion. Nausea, vomiting, diarrhea, constipation, increased appetite, and decreased appetite are all common side effects that dramatically affect dietary intake. Further complicating this situation is the fact that many individuals are prescribed numerous medications—one study estimates that most senior citizens take at least five medications each day (Fulton and Allen 2005). Next, consider the additive effect of over-the-counter medications as well as herbal supplements (Fugh-Berman 2000; Peng et al. 2004; Sorensen 2002). Recently, an evaluation of 100 patients with renal disease indicated an average of one to five dietary supplements were used daily (Spanner and Duncan 2005).

Appetite and subsequent food ingestion can be affected by taste, smell, and saliva production. Many medications alter saliva production by either increasing or decreasing saliva, or even by changing its consistency. For example, amitriptyline, a common antidepressant, may cause a decrease in saliva production. Since adequate solution is necessary for taste, many clients on these medications will report difficulty eating, decreased appetite, or anorexia, ultimately due to dry mouth.

Other medications may actually result in a perceived abnormal taste. Patients report experiencing metallic, salty, sweet, and simply foul tastes after taking some medications. Chemotherapy agents, analgesics (pain relievers), antibiotics, and antifungal agents are common groups of medications that result in these patient complaints. For example, methotrexate and cisplatin consistently result in a metallic taste (Pronsky and Crowe 2004).

Increased appetite secondary to medications can result in unplanned weight gain. A common example is treatment with prednisone or other corticosteroids, antiseizure medications, or antidepressants. Zyprexa (olanzapine) and Clozaril (clozapine), used to treat schizophrenia, almost always result in weight gain. These medications appear to block the serotonin receptor associated with satiety, inhibit histamine and dopamine, and increase the hormone prolactin (Hellings et al. 2001). Other antidepressant medications, such as Prozac, can result in the opposite effect—decreased appetite and weight loss. See Table 13.2 for common medications that can result in weight gain and Table 13.3 for those that can result in weight loss.

Drug Consequence: Effect on Nutrient Absorption Any drug that affects gastrointestinal function has the potential to interrupt nutrient absorption. This includes

pressor agents—substances that cause blood pressure to increase

monoamine oxidase (MAO) inhibitors—group of medications that block the enzyme system that inactivates some neurotransmitters

statin—a type of medication that is used to treat hyperlipidemias

BOX 13.2 CLINICAL APPLICATIONS

Monoamine Oxidase Inhibitors (MAOIs) and Nutrient Interactions

Monoamine oxidase (MAO) is an intricate enzyme system distributed predominantly in nervous tissue, liver, and lungs. This enzyme system is responsible for inactivating the neurotransmitters dopamine, norepinephrine, and serotonin once they have played their part in sending messages to the brain. Monoamine oxidase inhibitors (MAOIs) are drugs that block this activity. When the excess neurotransmitters are not destroyed, they accumulate in the brain.

In addition to inactivating these neurotransmitters, MAO breaks down another amine called tyramine. When MAO is blocked by an MAOI, levels of tyramine also rise. Excess tyramine can cause sudden, sometimes fatal increases in blood pressure. To avoid this life-threatening side effect, those taking MAOIs must avoid or limit foods that contain high levels of tyramine.

Tyramine occurs naturally in foods, but it is difficult to quantify the exact amount of tyramine in foods. Tyramine can also vary among different brands of certain foods based on processing, storage, and preparation methods. It is also formed from bacterial breakdown of protein in foods as they age.

MAOIs are most often prescribed for depression, bacterial and protozoal infections, and Hodgkin's disease.

Class	Generic Name	Trade Names
Antidepressants	Isocarboxazid	Marplan
	Phenelzine	Nardil
	Tranylcypromine	Parnate
Antimicrobials	Furazolidone	Furoxone
Antineoplastic	Procarbazine hydrochloride	Matulane

Foods high in tyramine that should be avoided include:

- Aged foods
- Alcoholic beverages (especially chianti, sherry, liqueurs, beer)
- Alcohol-free or reduced-alcohol beer or wine
- Anchovies
- Bologna, pepperoni, salami, pastrami, mortadella, summer sausage, any fermented sausage
- Caviar
- Cheeses (especially strong or aged varieties), except cottage cheese, cream cheese, ricotta, part-skim mozzarella, American
- Chicken livers, smoked or pickled fish, herring
- Fermented foods
- Figs (canned)
- Fruit: raisins, bananas (or any overripe fruit)
- Broad-beans (fava beans), lima beans, bean curd (tofu), eggplant, tomatoes, tomato sauce including ketchup, chili sauce
- Meat prepared with tenderizers; unfresh meat extracts
- Smoked or pickled meat, poultry, or fish
- Soy sauce, teriyaki sauce, soybean paste, fermented bean curd (fermented tofu), miso soup, tamari, natto, shoyu, tempeh

Foods that can be eaten in moderation are:

- Avocados
- Caffeine (including chocolate, coffee, tea, cola)
- Chocolate
- Raspberries
- Sauerkraut
- Soup (canned or powdered)
- Sour cream
- Yogurt

All foods should be very fresh or properly frozen. Meat products should not be refrigerated more than three to four days. Refrigerated cheeses should be eaten within two to three weeks. Combination foods like cheese crackers, submarine sandwiches, and stir-fried dishes containing soy sauce should be avoided. Pizza, lasagna, and other cheese-containing dishes may be eaten only if made with "allowed" cheeses and toppings.

References:

eDrugDigest. Monoamine oxidase inhibitors. Last updated June 2005. Available at http://www.drugdigest.org. Accessed December 2, 2005.

MayoClinic.com. MAOI diet: restrict foods high in tyramine. Available at http://www.mayoclinic.com. Accessed November 6, 2005.

MayoClinic.com. Monoamine oxidase inhibitors (MAOIs). Available at http://www.mayoclinic.com. Accessed November 6, 2005.

University of North Carolina. Verne S. Caviness General Clinical Research Center. Low tyramine diet for use with monoamine oxidase inhibitors. Available at http://gcrc.med.unc.edu. Accessed November 5, 2005.

National Institutes of Health Drug Nutrient Interactions Task Force. Warren Grant Magnuson Clinical Center. Drug-Nutrient Interactions: Monoamine oxidase inhibitor (MAOI) medications. Available at http://www.cc.nih.gov. Accessed December 2, 2005.

medications that cause side effects such as nausea, vomiting, diarrhea, and constipation. Adequate and efficient nutrient absorption requires exposure to enzymes in the appropriate metabolic environment, adequate transit time, sufficient GI tract surface area, and any transporters necessary for absorption. Any medication that speeds gastric emptying or affects the pH of gastric juices could therefore interfere with nutrient absorption. For example, since calcium supplements are absorbed best in an acidic environment, the chronic use of proton pump inhibitors may affect calcium absorption by decreasing stomach acidity (O'Connell et al. 2005). Other examples include **omeprazole** and H_2 blockers, both of which can impair the absorption of vitamin B_{12}. Medications that interfere with lipid metabolism or absorption can interfere with fat-soluble vitamin absorption.

Chronic use of corticosteroids, which are anti-inflammatory and immune-suppressing medications, is a mainstay of several medical conditions, including rheumatoid arthritis, COPD, and others. This class of medications results in decreased absorption of calcium from the GI tract as well as increased urinary loss of calcium. This significant drug consequence places the patient at high risk for bone fracture and osteoporosis (Lindsay and Cosman 2005).

Drug Consequence: Effect on Nutrient Metabolism Drugs can interfere with macronutrient, vitamin, and mineral metabolism. For example, corticosteroids increase the rate of gluconeogenesis, resulting in hyperglycemia and increased nitrogen loss. Numerous medications interfere with vitamin and mineral metabolism. Phenytoin (Dilantin), used for treatment of seizures, inhibits both vitamin D and folate metabolism. Long-term use may result in megaloblastic anemia secondary to folate deficiency. See Table 13.4 for examples of common interactions for nutrient metabolism.

Drug Consequence: Effect on Nutrient Excretion Since most drugs are excreted in urine, any drug that increases urinary output places the patient at risk for accelerated nutrient excretion as well. A classic example is the use of diuretics that are potassium wasting. Use of the diuretic Lasix or any other medications in this class can result in hypokalemia (low serum potassium). Any medication that affects renal function in a significant way—reducing reabsorption of nutrients, for instance—can also cause excessive loss of a nutrient in the urine. An example of a tubular reabsorption deficit involves the use of immunosuppressant medications called cyclosporins (e.g., Neoral, Sandimmune, SangCya). These medications have been associated with large amounts of magnesium loss in the urine. See Table 13.5 for examples of common medications affecting nutrient excretion.

At-Risk Populations

As stated previously, it would be a daunting task to acquire a working knowledge of all potential drug-nutrient interactions. However, a study of the basic principles of pharmacology and categories of interactions reveals that certain situations place individuals at risk. These may include disease state, organ function, or treatment modality. Furthermore, certain groups of individuals are more likely than others, not only to take more medications, but to also have an increased risk of improper or inadequate pharmacokinetics. Knowing these populations are at risk allows the practitioner to target them for monitoring and education.

At-Risk Populations: Drug-Nutrient Interactions in the Elderly The elderly population represents one group with an exceptionally high risk for drug-nutrient interactions (Bergman-Evans 2004; Lindblad et al. 2005; Peng et al. 2004). This risk exists for several reasons. Older individuals generally have the highest rate of chronic disease and are therefore prescribed the largest number of medications; this sheer volume increases risk. Furthermore, the use of over-the-counter and complementary medications compounds the incidence of interactions (Bergman-Evans 2004; Bruno and Ellis 2005). In addition, drug pharmacokinetics are affected by physiological changes that occur with aging. Decreased muscle mass and impaired cardiac, liver, and renal function all are common in the elderly and can change how a drug is absorbed, metabolized, and excreted. For example, the elderly may experience an exacerbation of drug-related confusion if other neurological diseases are present. Finally, compliance with drug regimens can be an important issue for this population. Financial burdens, complex regimens, or lack of proper drug education can lead to inappropriate drug dosing.

Polypharmacy, a term that is often associated with the elder population, is defined as administration of excessive drugs at one time or concurrent use of a large number of drugs, which increases the risk of interactions. Other features of polypharmacy may include the use of medications without a reason; the use of multiple medications for the same condition; the use of medications that interact with one another; the use of inappropriate dosages; the use of additional drugs to treat side effects of medications; and overall improvement when medications are discontinued.

Protocols and clinical guidelines have been developed to prevent adverse drug effects in this population. Beer's Criteria (see Table 13.6) have identified the medications most likely to result in adverse effects (Fick et al. 2003; MacLaughlin et al. 2005). General components of these criteria state that if a patient uses more than five drugs, is noncompliant with medication regimens, and has a history of adverse effects, the risk of continued interactions is high (Chang et al. 2005). Box 13.3 provides guidance for prevention of adverse drug reactions in the elderly.

At-Risk Populations: Drug-Nutrient Interactions in HIV and AIDS Antiretroviral therapy requires concomitant use of multiple medications (see Chapter 26). These medications

omeprazole—a type of proton pump inhibitor used to treat GERD and peptic ulcer disease

BOX 13.3 CLINICAL APPLICATIONS

Prevention of Adverse Drug Reactions in the Elderly

Adverse drug reactions (ADRs) are any harmful, unintentional drug reactions that take place at customarily prescribed doses. These reactions contribute to hospitalizations, disability, morbidity, and mortality, consequently adding billions of dollars to health care expenditures. Elderly patients are considered vulnerable to ADRs as a consequence of adverse physiologic changes that take place as a result of the aging process, a high frequency of comorbid conditions, and the large numbers of medications prescribed to them.

Many ADRs are the result of inescapable patient eccentricities, but many others are believed to be preventable. One way to prevent ADRs is to avoid prescribing inappropriate medications. The Beers criteria (see Table 13.6) are some of the most commonly used methods for assessing appropriateness of prescribing medications for elderly patients, though they are not evidence-based. The most common reasons for ADRs are:

- Decline in physiological functions that naturally occur with aging (May influence disposition of drugs)
- Impaired organ function from prior disease or aging (Alters drug kinetics, organ responses, and homeostatic counter-regulatory drug effects)
- Number of medications prescribed (Probability of toxicity increases with number of medications prescribed)

Noncompliance with medication regimens is another cause for ADRs in elderly patients. Noncompliance may be a result of:

- Inadequate instructions for taking medications
- Switching to alternative medical practices
- Illiteracy
- Poverty
- Misconceptions
- Inability to recall complicated medical regimens

A list of 10 drug interactions frequently identified in long-term care facilities has been developed by the Multidisciplinary Medication Management Project (Brown 2005):

- Warfarin and NSAIDs[1]
- Warfarin and sulfa drugs
- Warfarin and macrolides
- Warfarin and quinolones[2]
- Warfarin and phenytoin
- ACE inhibitors and potassium supplements
- ACE inhibitors and spironolactone
- Digoxin and amiodarone
- Digoxin and verapamil
- Theophylline and quinolones

References:

Beard K. Adverse reactions as a cause of hospital admissions for the aged. Drug Aging 1992; 2: 356–363.

Brown KE. Top ten dangerous drug interactions in long-term care. Multidisciplinary Medication Management Project. Available at http://www.scoup.net/M3Project/topten. Accessed November 6, 2005.

Chang C, Liu PY, Yang YK, Yang Y, Wu C, Lu F. Use of the Beers criteria to predict adverse drug reactions among first-visit elderly outpatients. Pharmacotherapy. 2005; 25(6): 831–8.

Malhotra S, Karan RS, Pandhi P, Jain S. Drug related medical emergencies in the elderly: role of adverse drug reactions and non-compliance. Postgrad Med. J 2001; 77: 703–707.

Montamat SC, Cusack BJ, Verstal RE. Management of drug therapy in the elderly. N Engl J Med. 1989; 321: 303–309.

World Health Organization. International drug monitoring: the role of national centres. WHO technical report series no. 498. Geneva, Switzerland: World Health Organization, 1972.

[1] NSAID class does not include COX-2 inhibitors

[2] Quinolones does not include ciprofloxacin, enoxacin, norfloxacin, and ofloxacin

represent a unique situation that places this population at high risk for drug-nutrient interactions (Panel on Clinical Practices for Treatment of HIV Infection 2005). Many of these medications have specific guidelines for consumption with or without food due to the effect of food on absorption and utilization, and many of them cause significant nutritional side effects such as nausea, vomiting, and diarrhea.

At-Risk Populations: Drug-Nutrient Interactions in Nutrition Support The use of specialized nutrition support (SNS) is another clinical measure that poses a high risk for drug-nutrient interactions (see Chapter 7). Tube feedings have been documented to decrease absorption of some medications (e.g., warfarin, phenytoin, and tetracycline) (AuYeung and Ensom 2000; Chan 2002). Macronutrients present in the tube feeding may cause chelation of some medications. The following guidelines of the American Society for Enteral and Parenteral Nutrition (ASPEN 2002) for medication and tube feedings should be followed closely :

- Medications co-administered with enteral nutrition (EN) should be reviewed periodically for potential incompatibilities with medications.
- When medications are administered via an enteral feeding tube, the tube should be flushed before and after each medication is administered.

- Liquid medication formulations should be used, when available, for administration via enteral feeding tubes.
- EN patients who develop diarrhea should be evaluated for antibiotic-associated causes, including the proliferation of *Clostridium difficile*.
- Co-administration or admixture of medications known to be incompatible with parenteral nutrition (PN) should be prevented.
- In the absence of reliable information concerning the compatibility of a specific drug with an SNS formula, the medication should be administered separately from the SNS.
- Each parenteral nutrition formulation compounded should be inspected for signs of gross particulate contamination, discoloration, particulate formation, and phase separation at the time of compounding and before administration.

Nutrition Therapy

Nutrition Implications

Any use of prescribed drugs, over-the-counter medications, or complementary treatments has the potential to affect nutritional status, interfere with drug pharmacokinetics, and/or alter nutrient metabolism. Additionally, regulatory agencies for health care, including Joint Council on Accreditation of Healthcare Organizations (JCAHO) and Centers for Medicare and Medicaid Services, require an established protocol for identification of, intervention for, and patient education for drug-nutrient interactions.

Nutrition Interventions

Nutrition assessment will focus on factors that could affect absorption, distribution, metabolism, or excretion of drugs (see Chapter 5). First, the clinician should evaluate past and current medical history for any diagnosis affecting kidney, liver, or cardiac function. Baseline laboratory measurements for kidney function (blood urea nitrogen, creatinine), liver function (ALT, AST, bilirubin, alkaline phosphatase, prothrombin time), and glucose should then be evaluated. The medical history should identify any treatment regimens (for example, enteral nutrition or dialysis) that may potentiate drug-nutrient interactions or adverse effects. Overall, nutritional status will need to be quantified to ensure that consistent physiological response to medications is possible. If the patient is malnourished, for example, the amount of protein-bound drug can be reduced due to hypoalbuminemia, increasing the effect of the medication.

Next, all drugs, over-the-counter medications, dietary supplements, and other complementary medical regimens should be identified. Patient interviews and social history should identify any potential barriers to compliance with, understanding of, or access to medical or nutrition therapies. For each prescription drug, over-the-counter medication, and dietary supplement, drug-drug interactions should be identified, along with any nutrition implications

FIGURE 13.4 Nutrition Assessment of Drug-Nutrient Interactions

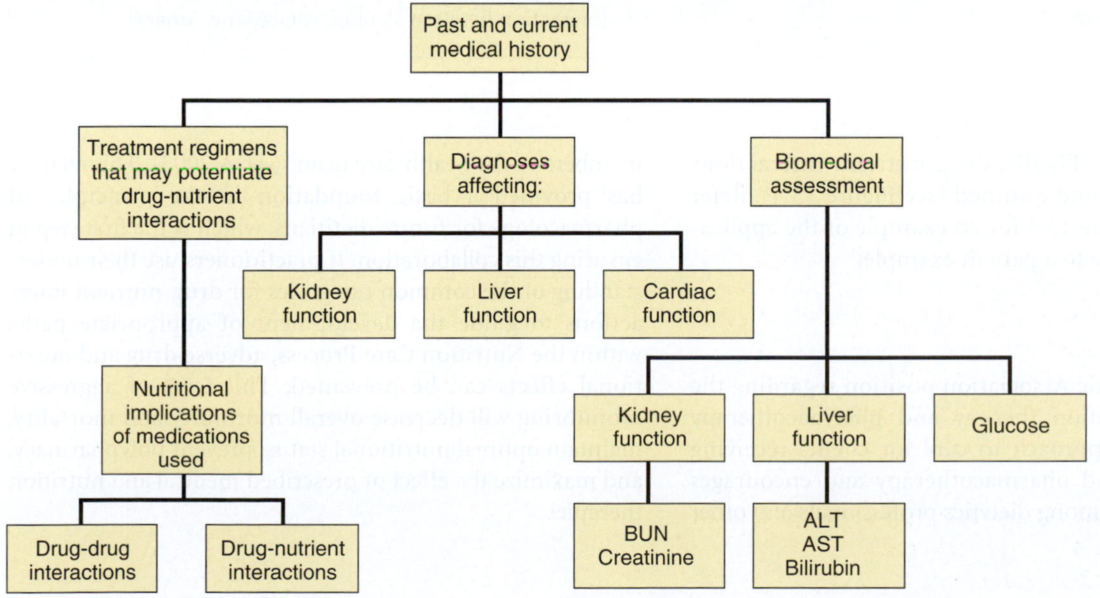

BOX 13.4 CLINICAL APPLICATIONS

Nutrition Assessment of Drug-Nutrient Interactions

Step 1—Past and current medical history: 65-year-old male

Hx of hypertension; myocardial infarction; 4 vessel coronary artery bypass graft; type 2 diabetes mellitus; prostate cancer s/p TURP; long-term use of alcohol

Step 2—Diagnoses affecting:

Cardiac function: hypertension, previous MI, and cardiac surgery

Liver function: probable alcohol abuse

Renal function: type 2 diabetes mellitus; hypertension

Step 3—Treatment regimens that may potentiate drug-nutrient interactions: none at this time

Step 4—Biomedical assessment: Glucose 180 mg/dL; BUN 21 Cr 1.2

Summary: Poor glycemic control; possible renal insufficiency

Step 5—Nutritional implications of medications used:

A. Medication regimen: All once daily: Toprol 50 mg; Plavix 5 mg; Aspirin 325 mg; Altace 5 mg; and Amaryl 2 mg twice daily.

B. Define current drugs:

- *Toprol (metoprolol)*—beta-blocker used to reduce the overall workload of the heart
- *Plavix (clopidogrel)*—inhibits platelet aggregation; used to prevent stroke and myocardial infarction in patients with cardiac history or history of previous stroke
- *Aspirin*—inhibits platelet aggregation; used to prevent stroke and myocardial infarction in patients with cardiac history or history of previous stroke
- *Altace (ramipril)*—ACE inhibitor used to treat hypertension
- *Amaryl (glimepiride)*—oral agent that stimulates insulin release from the beta cells of the pancreas and improves insulin resistance in peripheral tissues in patients with type 2 diabetes mellitus

C. Drug-drug interactions:

- *Toprol and Amaryl—Beta-blockers may increase the risk of hypoglycemia in patients taking Amaryl.*
- *Altace and Amaryl—ACE inhibitors may increase the risk of hypoglycemia in patients taking Amaryl.*
- *Plavix and Aspirin—These drugs together may increase chance of bleeding. Patients should avoid other over-the-counter medications that contain aspirin.*

D. Drug-nutrient interactions:

1. Altace may cause hyperkalemia (high serum potassium). Patients should be instructed to avoid foods high in potassium, especially salt substitutes.
2. Toprol absorption increases with food intake. Daily dosages should be consistently taken with meals so that therapeutic levels can be reached.
3. The patient should avoid all alcohol, because there is an interaction between alcohol, Altace, and Amaryl.

(Sanford et al 2002). Finally, drug-nutrient interactions should be identified and outlined (see Figure 13.4). Refer to Box 13.4 and Figure 13.5 for an example of the application of this procedure to a patient example.

Conclusion

The American Dietetic Association position regarding the integration of nutrition therapy and pharmacotherapy "promotes a team approach to care for clients receiving concurrent MNT and pharmacotherapy and encourages active collaboration among dietetics professionals and other members of the health care team" (ADA 2003). This chapter has provided a basic foundation in the principles of pharmacology for future dietitians, which is the first step in ensuring this collaboration. If practitioners use their understanding of the common categories for drug-nutrient interactions to guide the development of appropriate paths within the Nutrition Care Process, adverse drug and nutritional effects can be prevented. This level of aggressive monitoring will decrease overall morbidity and mortality, maintain optimal nutritional status, prevent polypharmacy, and maximize the effect of prescribed medical and nutrition therapies.

FIGURE 13.5 Nutrition Assessment of Drug-Nutrient Interactions: A Clinical Example

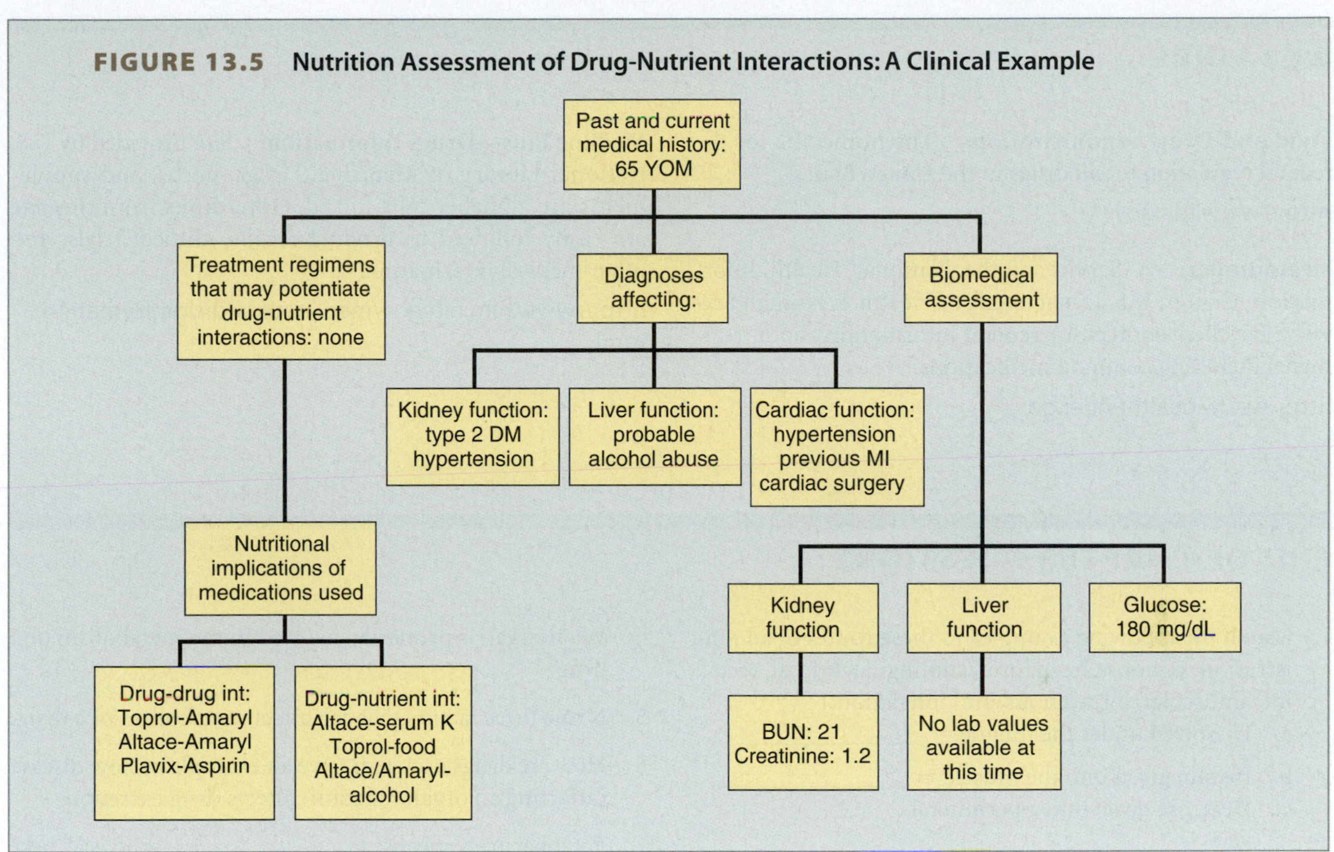

WEB LINKS

Food and Drug Administration: The home site for the federal regulation for all drugs in the United States.

http://www.fda.gov

Healthfinder: A Service of the National Health Information Center, U.S. Department of Health & Human Services. Excellent sources for general information about treatments including common medications.

http://www.healthfinder.gov

Medline Plus—Drugs Information: Site provided by U.S. National Library of Medicine. Drugs, herbs, and supplements are alphabetically linked. Other links from this site are easily followed to product recalls, clinical trials, and other medical information.

http://www.nlm.nih.gov/medlineplus/druginformation.html

END-OF-CHAPTER QUESTIONS

1. Match the following examples to these routes of administration. (Choose from: oral, sublingual, buccal, rectal, intramuscular, intravenous, and inhalation.)
   a. Dissolved under the tongue = _______________
   b. Insulin given into the muscle = _______________
   c. Dextrose given into a peripheral vein = _______________
   d. Asthma medication that is delivered by puffs through a breathing device = _______________
2. What factors could affect the dissolution of a medicine?
3. Distribution of the drug is defined as: _______________. What is the major physiological factor that can affect this?
   a. body temperature
   b. blood flow
   c. presence of food
4. What organ is primarily involved in the metabolism of a drug?
5. Name three factors that can affect metabolism of a drug.
6. How are drugs excreted? Give an example of how disease (affecting an organ function) affects drug excretion.
7. Polypharmacy means: _______________. Who is at risk?
8. Determine how each of the following could be considered a drug-nutrient interaction:
   a. When the use of methotrexate causes a change in taste
   b. When antacids bind phosphorus
   c. When Lasix increases the renal excretion of potassium
   d. When phenobarbital decreases folate metabolism

TABLE 13.1

Grapefruit-Drug Interactions

Confirmed Grapefruit-Drug Interactions

Drug Class	Generic	Trade Name
Antiarrhythmic	amiodarone	Cordarone®
Antihistamines	fexofenadine	Allegra®
	terfenadine	Seldane®
Anti-infective agents	halofantrine (antimalarial)	Halfan®
	indinavir	Crixivan®
Benzodiazepines	diazepam	Valium®
Cholesterol-lowering (HMG-CoA reductase inhibitors) ("-statins")	lovastatin	Mevacor®
	simvastatin	Zocor®
	atorvastatin	Lipitor®
Immunosuppressants	sirolimus	Rapamune®
Psychiatric	buspirone	BuSpar®
	pimozide	Orap®
	ziprasidone	Geodon®
Miscellaneous	cisapride	Prefulside®, Propuslid®
	sildenafil	Viagra®
	cilostazol	Pletal®
	budesonide	Entocort®
	colchicine	None
	eletriptan	Relpax®
	etoposide	Vapesid®
	mifepristone	Mifeprex®
	eplerenone	Inspra®
	itraconazole	Sporanox®
	telithromycin	Ketek®
	propafenone	Rythmol®

References: Elbe D. Grapefruit-Drug Interactions. Available at http://www.powernetdesign.com. Accessed December 2, 2005

Elbe D. Grapefruit-Drug Interactions. In Pronsky ZM. Food-Medications Interactions, 13th ed. Birchrunville (PA): Food-Medication Interactions; 2004.

TABLE 13.2

Medications That May Cause Weight Gain

Drug Class	Generic Name	Trade Name
Antiarthritic	celecoxib	Celebrex®
Antianxiety	alprazolam	Xanax®
	prochlorperazine	Compazine ®
	venlafaxine	Effexor XR®
Anticonvulsants	valproic acid (sodium valproate)	Depakote ®
	chlordiazepoxide	Librium®
	gabapentin	Neurontin®
	topiramate	Topamax ®
Antidepressants	lithium carbonate	Eskalith
		Eskalith CR
		Lithobid®
	nefazodone	Serzone®
	trazodone	Desyrel ®
	phenelzine	Nardil®
	tranylcypromine	Parnate®
	fluoxetine	Prozac®
	sertraline	Sarafem®
		Zoloft®
	paroxetine	Paxil®
	fluvoxamine	Luvox®
	amitriptyline	Elavil®
		Vanatrip®
	doxepin	Sinequan®
	clomipramine HCl	Anafranil ®
	imipramine	Tofranil ®
	nortriptyline	Aventyl®
		Pamelor®
	trimipramine	Surmontil ®
	mirtazapine	Remeron ®
Antihistamines	diphenhydramine	Nytol®
		Benadryl®
	loratadine	Claritin®
		Claritin RediTabs®
Antihypertensives	prazosin	Minipress®
	doxazosin	Cardura
	terazosin	Hytrin®
	propranolol	Inderal®
		Inderal LA®
	metoprolol	Lopressor®
		Toprol XL®
	atenolol	Tenormin®
Anti-osteoporosis	raloxifene	Evista®
Antipsychotics	haloperidol	Haldol®
	loxapine	Loxitane®
	clozapine	Clozaril®
	risperidone	Risperdal®
	olanzapine	Zyprexa®
	chlorpromazine HCl	Thorazine ®
	quetiapine fumarate	Seroquel ®
	thioridazine	Mellaril ®
	thiothixene	Navane®
	ziprasidone	Geodon ®
Appetite stimulant	dronabinol	Marinol®
	megestrol acetate	Megace®
Bronchodilator	albuterol sulfate	Proventil®
		Proventil® Repetabs (SR) ®
		Ventolin®
		Ventolin Repetabs (SR) ®
Corticosteroids	methylprednisolone	Medrol®
	prednisolone	Prelone®
	dexamethasone	Decadron®
	prednisone	Deltasone®
		Orasone®
		Prednicen-M® Liquid Pred®
Hormone	danazol	Danocrine®
	medroxyprogesterone acetate	Cycrin ®
	estrogen	Cenestin®
		Estrace®
		Estradiol oral®
		Ogen®
		Premarin®
		Climara®
		Estraderm ®
	estrogen/progesterone	Activella®
		Femhrt®
		Prempro®
		Premarin®
		CombiPatch ®
Insulin	none	Humalog®
Oral hypoglycemic	glipizide	Glucotrol®
		Glucotrol XL®
	glyburide	Diabeta®
		Micronase®
		Glynase®
	glimepiride	Amaryl ®
	chlorpropamide	Diabinese®
	tolbutamide	Orinase ®
	repaglinide	Prandin ®
	rosiglitazone	Avandia®
	pioglitazone	Actose®

References: Ness-Abramof R, Aprovian CM. Drug-induced weight gain. Drugs Today. 2005 Aug; 41(8): 547-55. Pronsky ZM. Food Medication Interactions, 13th ed. Birchrunville (PA): Food-Medication Interactions; 2004.

TABLE 13.3

Medications That May Cause Weight Loss

Drug Class	Generic Name	Trade Name
Anti-Alzheimer's	donepezil	Aricept®
	rivastigmine	Exelon®
	galantamine	Reminyl®
Antiarrhythmia	digitalis	Digitoxin®
	digoxin	Digoxin, Lanoxin®
	hydroxychloroquine sulfate	Plaquenil ®
Antianxiety	venlafaxine	Effexor®
		Effexor XR®
	alprazolam	Xanax ®
Antibiotic	clindamycin	Cleocin ®
	gentamicin sulfate	Garamycin ®
Anticonvulsant	ethosuximide	Zarontin®
Anticonvulsant/ antiglaucoma	acetazolamide	Diamox ®
Anticonvulsant/ antipanic	clonazepam	Klonopin ®
Antidepressant	bupropion	Wellbutrin®
		Wellbutrin SR®
	fluoxetine	Prozac®
		Prozac Weekly®
		Sarafem®
	fluvoxamine maleate	Luvox ®
	sertraline	Zoloft ®
Anti-ADHD	amphetamines	Adderall®
		Dexedrine®
	methylphenidate	Ritalin ®
Antigout	colchicine	none
Antifungal	amphotericin B	Abelcet®
		AmBisome®
		Amphotec®
		Fungizone®
Antihypertensive	captopril	Capoten ®
	indapamide	Lozol ®
	hydralazine	Apresoline ®
Antihyperlipidemia	cholestyramine	Questran ®
Anti-inflammatory	mesalamine	Asacol®
		Entasa®
		Canasa®
		Rowasa ®
Antineoplastic	cytarabine	Cytosar-U®
	fluorouracil (5-FU)	Adrucil, ®
	tamoxifen citrate	Nolvadex®
		Nolvadex-D®
	anastrozole	Arimidex ®
	cisplatin	Platinol-AQ®
	cyclophosphamide	Cytoxan®
		Cytoxan lyophilized®
	bicalutamide	Casodex ®
	bleomycin sulfate	Blenoxane ®
	mitomycin	Mutamycin ®
	alpha 2a	Roferon-A®
	alpha 2b	Intron-A®
	methotrexate	Methotrexate®
		Rheumatrex ®
	vinblastine sulfate	Velban, ®
		Oncovin®
	vinorelbine tartrate	Navelbine ®
Anti-Parkinson's	levodopa	Depar®
		Larodopa ®
	pramipexole	Mirapex ®
Antipsychotic	loxapine	Loxitane ®
Antiviral	ganciclovir sodium	Cytovene®
Calcium regulator	calcitriol	Rocaltrol®
		Calcijex®
Laxative	bisacodyl	Dulcolax ®
	mineral oil	Agoral plain®
Oral hypoglycemic	metformin	Glucophage ®
Thyroid preparations	Levothyroxine sodium	Synthroid®
		Levoxyl®
		Unithroid ®
Weight control agent	orlistate	Xenical ®
	phentermine	Adipex-P®
		Fastin®
	phentermine resin	Lonamin ®
	sibutramine	Meridian®

References: Pronsky ZM. Food Medication Interactions, 13th ed. Birchrunville (PA): Food-Medication Interactions; 2004.

TABLE 13.4

Medications That Interfere with Nutrient Metabolism

Nutrient	Drug(s)	Effect on Nutrient
Minerals	Diuretics (thiazides), corticosteroids, purgatives	Potassium depletion
	Cortisol, desoxycorticosterone, aldosterone, estrogen-progestogen oral contraceptives, phenylbutazone	Sodium and water retention
	Sulfonylureas, phenylbutazone, cobalt, lithium	Impair uptake or release of iodine
	Oral contraceptives	Lower plasma zinc, elevate copper
	Corticosteroids	Calcium depletion
	Laxatives	Malabsorption of electrolytes and calcium
Vitamins	Phenobarbital and phenytoin (Dilantin)	Increases metabolism of folic acid, vitamins D and K
	Isoniazid (INH), Hydralazine	Pyridoxine and niacin antagonists
	Laxatives	General malabsorption of fat-soluble vitamins
	Pyrimethamine, sulfadoxine, methotrexate	Folate antagonists
Amino acids	Oral contraceptives	Altered tryptophan metabolism

References: Pronsky ZM. *Food Medication Interactions,* 13th ed. Birchrunville (PA): Food-Medication Interactions; 2004.

TABLE 13.5

Medications That Affect Nutrient Excretion

Nutrient	Drug(s)	Effect on Nutrient
Minerals	Loop diuretics	Increase excretion of sodium, potassium, chloride, magnesium, calcium
	Thiazide diuretics	Increase excretion of most electrolytes
	Antifungals	Increase excretion of potassium
	NSAIDs	Increase excretion of potassium
	Caffeine	Increases sodium excretion
	Calcitonin	Increases excretion of phosphorus, magnesium, potassium, chloride, and sodium; may increase or decrease excretion of calcium
	Antihyperlipidemic	Increases excretion of calcium and magnesium
	Antineoplastics	Increase excretion of magnesium, calcium potassium, zinc, copper
	Clonidine	Decreases excretion of sodium and chloride
	Corticosteroids	Decrease excretion of sodium; increase excretion of potassium, calcium, nitrogen, zinc
	Cyclosporine	Increases excretion of magnesium; decreases excretion of potassium
	Digitalis	Increases urinary excretion of magnesium
Vitamins	NSAIDs	Increase excretion of vitamin C
	Corticosteroids	Increase excretion of vitamin C
	Tetracycline	Increases urinary excretion of riboflavin, folacin
Macronutrients	NSAIDs	Increase excretion of protein
	Calcitriol	Increases excretion of albumin
	Antineoplastics	Increase excretion of amino acids

References: Pronsky ZM. Food Medication Interactions, 13th ed. Birchrunville (PA): Food-Medication Interactions; 2004.

TABLE 13.6

Beer's Criteria: Potentially Inappropriate Medications Used with Older Adults with and without Concomitant Diagnoses or Conditions

Key: ↔ Neutral risk; ↑ High Risk; ↓ Low Risk
Medications to Avoid (or Use within Specified Dose/Duration Ranges)

Medications	Problem(s)	Risk
Anti-infectives		
Oral antibiotics	Therapy > four weeks should be avoided except when treating osteomyelitis, prostatitis, tuberculosis, or endocarditis.	↔
Cardiac Agents		
Digoxin (Lanoxin)	Decreased renal clearance, which may increase toxic effects. Doses > 0.125 mg should be avoided except for treatment of atrial arrhythmias.	↑
Disopyramide (Norpace)	May induce heart failure.	↑
EENT Agents		
Antihistamines (alone or in combination, including chlorpheniramine [Clor-Trimeton], diphenhydramine [Benadryl], hydroxyzine [Vistaril and Atarx], cyproheptadine [Periactin], promethazine [Phenergan], and dexchlorpheniramine [Polaramine])	Strong anticholinergic activity.	↓
Decongestants (oxymetazoline [Afrin], phenylephrine [Neo-Synephrine, Vicks Sinex], pseudoephedrine [Sudafed, Suphedrin, Triaminic, Dimetapp])	Avoid daily use for longer than two weeks.	↔
Diphenhydramine (Benadryl)	Should not be used as sleep aid. May cause confusion. Use lowest possible dose for allergies.	↔
Endocrine Agents		
Chlorpropamide (Diabinese)	May cause prolonged and serious hypoglycemia.	↑
Gastrointestinal Agents		
Bisacodyl (Dulcolax), cascara sagrada, and Neoloid except in presence of opiate analgesic use	Long term use of stimulant laxatives may exacerbate bowel dysfunction.	↑
Cimetidine (Tagamet)	Avoid doses > 900 mg/day; do not use >12 weeks.	↔
Dicyclomine (Bentyl)), hyoscyamine (Levsin & Levsinex), propantheline (Pro-Banthine), belladonna alkaloids (Donnatal and others), clidinium-chloridiazepoxide (Librax)	Strong anticholinergic activity. Questionable efficacy as antispasmodic agents. Should be avoided, especially long-term use.	↑
Mineral oil	Potential for aspiration.	↑
Ranitidine (Zantac)	Avoid doses > 300 mg/day; do not use >12 weeks.	↔
Trimethobenzamide (Tigan)	Produces extrapyramidal side effects; one of the least effective antiemetic agents.	↓
Hematopoietic Agents		
Ferrous sulfate iron supplements > 325 mg	Cause constipation; higher doses not effective.	↓

(continued on the following page)

TABLE 13.6 *(continued)*

Beer's Criteria: Potentially Inappropriate Medications Used with Older Adults with and without Concomitant Diagnoses or Conditions

Key: ↔ Neutral risk; ↑ High Risk; ↓ Low Risk

Medications to Avoid (or Use within Specified Dose/Duration Ranges)

Medications	Problem(s)	Risk
Musculoskeletal Agents		
Indomethacin (Indocin and Indocin SR)	Has more CNS side effects than any other NSAID.	↓
Methocarbamol (Robaxin), carisoprodol (Soma), oxybutynin (Ditropan), chlorzoxazone (Paraflex), metaxalone (Skelaxin), cyclobenzaprine (Flexeril)	Anticholinergic side effects, sedation, weakness. Effectiveness of tolerated doses questionable.	↓
Naproxen (Naprosyn, Avaprox, Aleve), oxaprozin (Daypro), piroxicam (Feldane)	Potential to produce GI bleeding, renal failure, high blood pressure & heart failure.	↑
Phenylbutazone (Butazolidin; not on U.S. market)	Serious hematologic side effects.	↓
Psychotropic Agents		
Amitriptyline (Elavil), alone or in combination products (Limbitrol, Triavil)	Anticholinergic and sedating properties.	↑
Barbiturates (other than phenobarbital)	Side effects and addictive properties	↑
Chlordiazepoxide (Librium), alone or in combination; or diazepam (Valium)	Risk of sedation and increased falls.	↑
Doxepin (Sinequan)	Sedating properties and powerful anticholinergic.	↑
Ergot mesylates (Hydergine), cyclandelate isoxsuprine (Cyclospasmol)	Not proven effective.	↓
Flurazepam (Dalmane)	Risk of sedation and increased falls.	↑
Haloperidol (Haldol)	Avoid doses > 3 mg/day.	↔
Lorazepam (Ativan [3 mg]), oxazepam (Serax [60 mg]), alprazolam (Xanax [2 mg]), temazepam (Restoril [15 mg]), zolpidem (Ambien [5 mg]), triazolam (Halcion [0.25 mg])	Avoid higher doses. Avoid single oxazepam dose > 30 mg or > 0.25 mg triazolam.	↓
Meperidine (Demerol)	Not effective orally. More disadvantageous than other narcotic analgesics.	↑
Meprobamate (Miltown and Equanil)	Highly addictive and sedating.	
Pentazocine (Talwin)	Many CNS side effects, including confusion & hallucinations.	↑
Propoxyphene (Darvon)	Few advantages over acetaminophen, produces adverse effects of other narcotic drugs.	↓
Thioridazine (Mellaril)	Avoid doses > 30 mg/day.	↔
Vascular Agents		
Dipyridamole (Persantine)	Causes orthostatic hypotension. Useful only in patients with artificial heart valves.	↓
Hydrochlorothiazide (Hydrodiuril)	Avoid doses > 50 mg/d.	↔
Methyldopa (Aldomet, alone or in combination [Aldoril])	Bradycardia and exacerbates depression.	↑
Propranolol (Inderal)	Better beta-receptor selectivity and less CNS penetration in other beta-blockers.	↔
Reserpine (Harmonyl), alone or in combination	Depression, impotence, sedation, and orthostatic hypotension.	↓

Medications to Avoid with Specific Concomitant Diseases

Disease	Medication	Problem	Risk
Anorexia			
Malnutrition	CNS stimulants: dextroamphetamine (Adderall), methylphenidate (Ritalin), methamphetamine (Desoxyn), pemoline, and fluoxetine (Prozac)	Appetite-suppressing effects.	↑

(continued on the following page)

TABLE 13.6 *(continued)*

Bleeding Disorders			
Blood clotting disorders or receiving anticoagulant therapy	Aspirin, NSAIDs, dipyridamole (Persantin), ticlopidine (Ticlid), and clopidogrel (Plavix)	May prolong clotting time.	↑
Cardiac Disorders			
Arrhythmias	Tricyclic antidepressants (imipramine hydrochloride [Tofranil, Tofranil PM, Janimine], doxepin hydrochloride [Sinequan], amitriptyline hydrochloride [Adepril, Endep, Enovil, Trepiline])	May induce arrhythmias.	↑ if started recently
Heart failure	Disopyramide (Norpace)	May worsen heart failure.	↑
	Drugs with high sodium content (sodium and sodium salts [alginate bicarbonate (Di-Gel, Maalox, Mylanta), biphosphate (Fleet enema), citrate (Bicitra), phosphate (K-Phos), salicylate (Alka-Seltzer, Pepto-Bismol), and sulfate (colyte)])	May lead to fluid retention and worsen heart failure.	↓
Endocrine Disorders			
Diabetes	Beta-blockers (Inderal, Lopressor)	May worsen symptoms in patients treated with insulin or oral hypoglycemic agents.	↓
	Corticosteroids (started recently)	May worsen glycemic control.	↓
Gastrointestinal Disorders			
Constipation	Anticholinergics (Levbid, Anaspaz)	Worsen constipation.	↓
	Calcium channel blockers (Procardia, Cardizem)	Worsen constipation.	↓
	Narcotics	Worsen constipation.	↓
	Tricyclic antidepressants (imipramine hydrochloride [Tofranil], doxepin hydrochloride [Sinequan], and amitriptyline hydrochloride [Limbitrol])	Worsen constipation.	↓
Ulcers	NSAIDs	May exacerbate ulcer disease, gastritis, GERD.	↑
	Aspirin	May exacerbate ulcer disease, gastritis, GERD.	↓
	Potassium supplements	May exacerbate ulcer disease, gastritis, GERD.	↓
Neurologic Disorders			
Cognitive impairment	Barbiturates, anticholinergics/antispasmodics (Levbid, Symax), and muscle relaxants (Paraflex, Remular, Skelaxin), CNS stimulants: dextroamphetamine (Adderall), methylphenidate (Ritalin), methamphetamine (Desoxyn)	CNS-altering effects.	↑
Epilepsy	Clozapine (Clozaril), chlorpromazine (Thorazine), thioridazine (Mellaril), thiothixene (Navane)	Lower seizure threshold.	↓
	Metoclopramide (Reglan)	Lowers seizure threshold	↑
Parkinson's disease	Metoclopramide (Reglan), conventional antipsychotics, and tacrine (Cognex)	Antidopaminergic/cholinergic effects.	↑
Seizure disorder	Bupropion (Wellbutrin)	May lower seizure threshold.	↑

(continued on the following page)

TABLE 13.6 *(continued)*

Beer's Criteria: Potentially Inappropriate Medications Used with Older Adults with and without Concomitant Diagnoses or Conditions

Disease	Medication	Problem	Risk
Psychiatric Disorders			
Depression	Long-term benzodiazepine use	May exacerbate depression.	↑
	Methyldopa (Aldomet), reserpin, & guanethidine (Ismelin)	May exacerbate depression.	↑
Insomnia	Decongestants	May cause or worsen insomnia.	↓
	Theophylline (Theodur)	May cause or worsen insomnia.	↓
	Methylphenidate (Ritalin)	May cause or worsen insomnia.	↓
	Desipramine, SSRIs, MAOIs	May cause or worsen insomnia.	↓
Respiratory Disorders			
Asthma	Beta-blockers	May worsen respiratory function.	↑
COPD	Beta-blockers	May worsen respiratory function	↑
	Sedative-hypnotics	May slow respirations and increase CO_2 retention.	↑
Urologic Disorders			
Benign prostatic hypertrophy	Anticholinergic antihistamines	May cause obstruction.	↑
	Gastrointestinal antispasmodics	May cause obstruction.	↑
	Muscle relaxants	May cause obstruction.	↓
	Narcotic drugs (including propoxyphene)	May cause obstruction.	↓
	Flavoxate, oxybutynin	May cause obstruction.	↓
	Bethanechol	May cause obstruction.	↓
	Anticholinergic antidepressants	May cause obstruction.	↑
Incontinence	Alpha blockers (Doxazosin, Prazosin, Terazosin)	May produce polyuria.	↑
	Anticholinergics	May produce polyuria.	↑
	Tricyclic antidepressants (Imipramine hydrochloride, doxepin hydrochloric, amitriptyline hydrochloride)	May produce polyuria.	↑
	Long-acting benzodiazepines: (chlordiazepoxide [Librium], alone or in combination; or diazepam [Valium])	May produce polyuria.	↑
Vascular Disorders			
Clotting disorders treated with anticoagulants	Aspirin	May cause bleeding.	↑
Hypertension	Amphetamines & other weight control agents	May increase blood pressure.	↑
Peripheral vascular disease	Beta-blockers	Negative chronotropic and inotropic activity.	↓
Syncope	Beta-blockers	Negative chronotropic and inotropic activity.	↓
	Long-acting benzodiazepines	May contribute to falls.	↑
Weight disorders			
Obesity	Olanzapine (Zyprexa)	May stimulate appetite and increase weight gain.	↓

Adapted from: Beers MH, Ouslander JG, Rollingher I, Rueben DB, Brooks J, Beck JC, Explicit criteria for determining inappropriate medication use in nursing home residents. Arch Intern Med. 1991;151:1825-32.

Beers MH. Explicit criteria for determining potentially inappropriate medication use by the elderly: an update. Arch Intern Med. 1997;157:1531-6.

Fick DM, Cooper JW, Wade WF, Waller JL, Maclean R, Beers MH. Updating the Beers criteria for potentially inappropriate medication use in older adults. Results of a US consensus panel of expert. Arch Intern Med. 2003;163:2716-24.

14

Energy Balance and Body Weight

Robert D. Lee, Dr.P.H., R.D.

Central Michigan University

CHAPTER OUTLINE

Energy Balance

Energy Intake • Energy Expenditure • Estimating Energy Requirements • Indirect Calorimetry • Doubly Labeled Water • Direct Calorimetry

Regulation of Energy Balance

Body Composition, Obesity, and Overweight

Epidemiology of Overweight and Obesity

Adverse Health Consequences of Overweight and Obesity

Etiology of Obesity

Treatment of Overweight and Obesity

Eating Disorders

Introduction

Throughout most of recorded history, humans have spent a large proportion of their time and energy obtaining an adequate amount of calories and essential nutrients. The lack of food has historically been a more common condition than one in which there has been a surplus of food. Hunger, nutrient-deficiency diseases, and starvation have been constant threats for most population groups. Tragically, in many **developing nations**, hunger, malnutrition, and starvation continue, resulting in untold suffering, misery, and death. Only within the past century has the mechanization of agriculture produced the abundant harvests that have become commonplace in **developed nations.** Food in developed nations is so readily available and inexpensive that some nutritionists refer to the food situation in these countries as a "toxic food" environment. This has contributed to a marked increase in the preva-

developing nation—a nation that is generally regarded as one with a low standard of living, a low per capita income, a relatively poorly developed infrastructure (e.g., public utilities and systems for transport, public health, and public education), low literacy rates, low life-expectancy, and so on, when compared to the global norm

developed nation—a nation that is generally regarded as one with a high standard of living, a high per capita income, a well-developed infrastructure (e.g., public utilities and systems for transport, public health, and public education), high literacy, long life-expectancy, and so on, when compared to the global average

lence of **overweight** and **obesity** in these countries, and to diseases associated with these conditions, such as type 2 diabetes, hypertension, stroke, coronary heart disease, sleep apnea, gallbladder disease, osteoarthritis, and cancer of the endometrium, breast, prostate, and colon (NHLBI 1998, 2000).

overweight—an excess of body weight in relationship to height; for adults, overweight is generally defined as a body mass index or BMI of 25.0 kg/m^2 to 29.9 kg/m^2; for children and adolescents, overweight can be defined as a BMI-for-age-and-sex at or above the 95th percentile using the CDC growth charts

obesity—an excess of body fat or adipose tissue. Obesity can be defined as a proportion of body weight that is adipose tissue (percent body fat) that is greater than some standard; because it is often impractical in the clinical setting to measure in percent of body fat using body composition analysis, obesity is often defined as a BMI ≥30.0 kg/m^2; the term obesity comes from the Latin *obesus*, meaning, "one who has become plump through eating"

kilojoule (kjoule or kJ)—the SI (Système International d'Unités or International System of Units) unit of measurement for energy; the amount of work required to move 1 kilogram for 1 meter with the force of 1 newton. 1 kcal = 4.2 kJ (to convert kcal to kJ, multiply kcal by 4.2)

kilocalorie (kcalorie or kcal)—the amount of *heat* required to raise 1,000 mL (1 liter) of water 1° Celsius

24-hour energy expenditure—the total amount of energy expended by a human in a 24-hour period, made up of three main components: resting energy expenditure, thermic effect of food, and physical activity energy expenditure

resting energy expenditure—energy expended by the body at rest to keep vital organ systems functioning, including the heart, kidneys, brain, liver, and lungs; it accounts for approximately 60% to 75% of 24-hour energy expenditure and is roughly 1 kcal/kg body weight/hour

thermic effect of food—energy expended by the body to digest, absorb, and metabolize food; it accounts for about 10% of 24-hour energy expenditure

physical activity-related energy expenditure—energy expended in voluntary body movement resulting from the daily activities of life, physical exercise, sports, and play, and nonvoluntary behaviors such as spontaneous muscle contractions, maintenance of posture, and fidgeting; it is the most variable component of 24-hour energy expenditure, depending on how physically active a person is

BOX 14.1 THE MACRONUTRIENTS AND THEIR ENERGY CONTENT IN KILOCALORIES (KCAL) AND KILOJOULES (KJ) PER GRAM

Food Component	kcal/g	kJ/g
Carbohydrate	4	17
Protein	4	17
Fat	9	38
Alcohol	7	29

Energy Balance

Energy Intake

Humans obtain the energy and nutrients that their bodies need from the foods and beverages they consume. The human body derives energy from the oxidation of the macronutrients carbohydrate, protein, and fat, and from alcohol. Internationally, the most commonly used unit of measurement of food energy is the **kilojoule (kJ)**, whereas the **kilocalorie (kcal)** is the unit of measurement of food energy most familiar to those living in the United States. The amounts of energy released by the oxidation of carbohydrate, protein, fat, and alcohol are shown in Box 14.1. These values, rounded for the sake of convenience, initially were derived from experiments in which a small amount of each macronutrient was burned in a device known as a bomb calorimeter, which allowed scientists to accurately measure the amount of heat released from the macronutrient when it was burned. Subsequent experiments have shown that the amounts of energy released by the oxidation of these macronutrients within the human body are similar to the release of energy when burned in the bomb calorimeter (Panel on Macronutrients 2002). Today, the energy content of a food or beverage is generally determined by first measuring the amount of carbohydrate, protein, fat, and alcohol it contains using relatively simple laboratory techniques, and then multiplying the number of grams of carbohydrate, protein, fat, and alcohol in the food or beverage by the energy values for each of the macronutrients shown in Box 14.1. Information on the energy and nutrient content of foods is widely available to consumers and health professionals from a variety of sources, including the Nutrition Facts labels on commercially available food containers, brochures provided by fast-food restaurants, food composition tables and databases, and dietary analysis software.

Energy Expenditure

The total amount of energy expended in one day is referred to as **24-hour energy expenditure** or total energy expenditure, and can be divided into three major compo-

FIGURE 14.1 For most North Americans who are sedentary and rely on labor-saving devices to accomplish most of their work, energy expended in physical activity accounts for less than one quarter of the energy expended in a typical 24-hour period. Surprisingly, resting energy expenditure accounts for about 67% of 24-hour energy expenditure and the thermic effect of food accounts for the remaining 10%

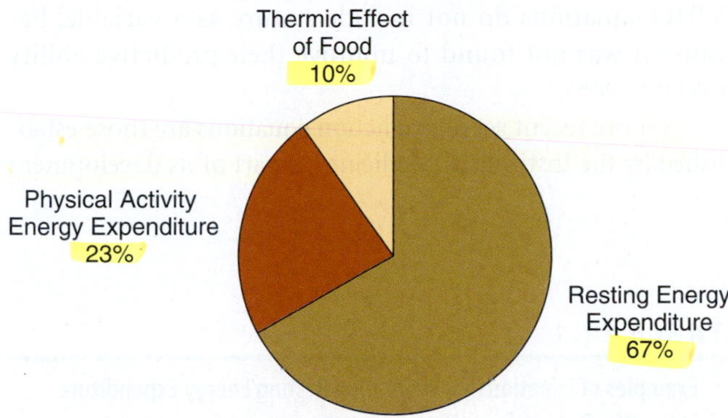

nents: **resting energy expenditure**, the **thermic effect of food**, and **physical activity-related energy expenditure**. Figure 14.1 illustrates the relative proportions of each of these three components for the majority of people living in developed countries, where much of the work is done by labor-saving devices.

Resting Energy Expenditure Resting energy expenditure (REE) is the energy necessary to sustain life and to keep such vital organs as the heart, lungs, brain, liver, and kidneys functioning. For the average North American, REE accounts for approximately 60% to 75% of 24-hour energy expenditure and is roughly 1 kcal/kg body weight/hour. Factors affecting REE are shown in Box 14.2. Of these factors, the most important determinant is lean body mass (or fat-free mass), with REE being greater in persons having a higher lean body mass. **Basal energy expenditure** (BEE) is defined as the lowest rate of energy expenditure of an individual. It is measured in the morning when a subject is in a postabsorptive state (no food consumed during the previous 12 to 14 hours) and is comfortably lying motionless in a supine position (lying on one's back) in a thermally neutral environment (a room temperature that is perceived as neither hot nor cold). These strict conditions often make obtaining a true BEE impractical in the clinical setting. REE, on the other hand, can be measured at any time of day after a subject has quietly rested for the previous 30 minutes. Basal energy expenditure is generally 10% to 20% less than resting energy expenditure (Panel on Macronutrients 2002).

BOX 14.2 FACTORS AFFECTING RESTING ENERGY EXPENDITURE (REE)

Lean Body Mass

Because muscle and other lean tissues are generally more metabolically active than adipose tissue, the greater the lean body mass (also known as the fat-free mass), the greater the REE. This is the primary determinant of REE.

Male Sex

Because males tend to have greater percentage of lean body mass than females, males tend to have a greater REE.

Body Temperature

REE increases in persons who have a fever or an elevated body temperature.

Age

REE decreases about 2% for every decade after age 30 years, even after adjusting for changes in lean body mass.

Energy Restriction

After several weeks of energy restriction, for example to lose weight, resting energy expenditure declines. This is at least part of the reason that, after several weeks of dieting, some people experience a decline in the rate of weight loss or a phenomenon some refer to as a "plateau."

Genetics and the Endocrine System

Depending on genetic influences, some people inherit a predisposition to a higher REE while others are predisposed to a lower REE. Hypothyroidism and hyperthyroidism can dramatically decrease or increase REE, respectively.

Thermic Effect of Food The thermic effect of food (TEF) is a measurable increase in energy expenditure over and above resting energy expenditure that can be measured for several hours following a meal. The thermic effect of food is the energy required to digest, absorb, metabolize, and store the nutrients contained in foods that are consumed and to eliminate the resulting by-products and wastes. Originally referred to as the specific dynamic action of food, it accounts for about 10% of the 24-hour energy expenditure for a person consuming a typical mixed meal

basal energy expenditure—the minimum level of energy expended by the body to sustain life; it is measured in the morning when a subject is in a postabsorptive state, comfortably lying motionless in a supine position, and in a thermally neutral environment

(Panel on Macronutrients 2002). The TEF of a meal is influenced primarily by the amount and macronutrient composition of the food consumed. Large meals have a greater TEF than small meals. Fat has the lowest TEF, while protein has the highest TEF due to the relatively high energy cost of processing the amino acids released from the proteins in food, including the synthesis of urea. TEF peaks at about 60 to 120 minutes following a meal and can last up to four to six hours, depending on the size and composition of the meal.

Physical Activity Energy Expenditure The most highly variable component of 24-hour energy expenditure is that expended in physical activity. For most people in developed nations it accounts for about 20% to 25% of 24-hour energy expenditure. However, in very active individuals, such as heavy laborers and some athletes, the amount of energy expended in physical activity can exceed REE by twofold or more (Panel on Macronutrients 2002). Physical activity energy expenditure is influenced by the person's body weight, the number of muscle groups used in the activity, and the intensity, duration, and frequency of the activity. For any given activity, heavy people expend more energy than lighter-weight people because heavier people have a greater body mass to move. Activities requiring multiple groups of large muscles (e.g., cross-country skiing or handball) expend more energy than those requiring fewer groups of muscles (e.g., walking or golfing).

Estimating Energy Requirements

An individual's energy requirements can either be estimated using a predictive equation or, if a more accurate determination is necessary, using such methods as indirect calorimetry, doubly labeled water, or direct calorimetry. For most patients in the clinical setting, it is usually adequate to estimate energy requirements by means of a predictive equation that uses such variables such as sex, age, weight, stature (height), and physical activity level. However, in critically ill patients, it may be necessary to more accurately determine energy requirements using indirect calorimetry. Doubly labeled water is commonly used in human metabolic research, while use of direct calorimetry is limited by the small number of research facilities having the necessary technology.

Equations In most instances, an individual's energy requirements are estimated using one of several empirically derived equations. Two examples are the Harris-Benedict equations developed in the early 1900s by the researchers J.A. Harris and F.G. Benedict, and those developed in the 1980s by the World Health Organization (WHO), both of which are shown in Table 14.1. Harris and Benedict measured the resting energy expenditures (REE) of 239 healthy young adult males and females using indirect calorimetry (discussed later in this chapter) and then developed a set of regression equations that best predicted REE using the variables sex, weight, stature (height), and age (Roza and Shizgal 1984). Although originally published in 1919, their equations remain in use today. The World Health Organization equations were developed by a group of experts using an approach similar to that used by Harris and Benedict. A key difference between the two sets of equations is that the WHO equations do not include stature as a variable, because it was not found to improve their predictive ability (WHO 1985).

A more recent set of prediction equations are those established by the Institute of Medicine (as part of its development

TABLE 14.1

Examples of Equations for Estimating Resting Energy Expenditure in Healthy Persons[1]

Harris-Benedict			
Females		REE = 655.096 + 9.563 W + 1.850 S – 4.676 A	
Males		REE = 66.473 + 13.752 W + 5.003 S – 6.755 A	
Harris-Benedict (Values Rounded for Simplicity)			
Females		REE = 655.1 + 9.6 W + 1.9 S – 4.7 A	
Males		REE = 66.5 + 13.8 W + 5.0 S – 6.8 A	
World Health Organization (WHO)			**SD[2]**
Females	3–9 years old	22.5 W + 499	± 63
	10–17 years old	12.2 W + 746	± 117
	18–29 years old	14.7 W + 496	± 121
	30–60 years old	8.7 W + 829	± 108
	>60 years old	10.5 W + 596	± 108
Males	3–9 years old	22.7 W + 495	± 62
	10–17 years old	17.5 W + 651	± 100
	18–29 years old	15.3 W + 679	± 151
	30–60 years old	11.6 W + 879	± 164
	>60 years old	13.5 W + 487	± 148

[1] W = weight in kilograms; A = age in years; S = stature in cm.

[2] SD = standard deviation of the differences between actual and computed values—68% of the time actual REE will be within ± 1 standard deviation of the predicted REE.

From Harris JA, Benedict FG. 1919. A biometric study of basal metabolism in man. Publication 279. Washington (DC): Carnegie Institution of Washington; World Health Organization. Energy and protein requirements. Report of a joint FAO/WHO/UNU expert consultation. Technical Report Series 724. Geneva, Switzerland: World Health Organization; 1985.

of the Dietary Reference Intakes) to calculate the **estimated energy requirement (EER)**. The formulas for calculating EER are shown in Table 14.2. The EER is defined as the average dietary energy intake that is predicted to maintain energy balance in a healthy person of a defined age, gender, weight, height, and level of physical activity consistent with good health (Panel on Macronutrients 2002). For infants, children, and adolescents, the EER includes the energy needed for a desirable level of physical activity, as well as the energy needed for optimal growth and development at an age- and gender-appropriate rate that is consistent with good health, including maintenance of a healthy body weight and appropriate body composition. For females who are pregnant or lactating, the EER includes the energy needed for physical activity, maternal and fetal development, and for secretion of milk at a rate consistent with good health.

The EER equations are based on the measurement of 24-hour total energy expenditure using the doubly labeled water technique (discussed later in this chapter) from more than 1,200 healthy-weight subjects of all ages (Panel on Macronutrients 2002). The Dietary Reference Intake (DRI) Committee used these measurements of 24-hour total energy expenditure to develop a series of regression equations that best predicted the energy requirements of healthy-weight individuals using such variables as age, sex, life stage (pregnant or lactating), body weight, stature, and physical activity level. As shown in Table 14.2, EER equations have been developed for infants and young children of both sexes age 0 to 35 months, males and females age 3 to 8 years, males and females age 9 to 18 years, males and females age 19 years and older, and females who are pregnant or lactating.

Except in the case of infants and young children ages 0 to 35 months, a physical activity coefficient (PA) is used in the equations. The PA represents one of four different categories of physical activity level: sedentary, low active, active, and very active. Energy expenditure at each of these levels is as follows:

- *Sedentary:* Includes basal energy expenditure, the thermic effect of food, and physical activities required for independent living.
- *Low active:* Roughly equivalent to the energy expended by a 70 kg (154 lb) adult walking 2.2 miles per day at a rate of 3 to 4 miles per hour (or an equivalent amount of energy expended in other activities) in addition to the activities necessary for independent living.
- *Active:* Roughly equivalent to the energy expended by a 70 kg (154 lb) adult walking 7 miles per day at a rate of 3 to 4 miles per hour in addition to the activities related to independent living.
- *Very active:* Equivalent to walking 17 miles per day in addition to the activities a normal person would ordinarily engage in.

The extra energy needed for growth during infancy, childhood, adolescence, and pregnancy is included in an allowance referred to as "tissue deposition." During pregnancy, the metabolic rate is also increased due to the energy requirements of the uterus and fetus and the increased work of the maternal cardiovascular system. During lactation, extra energy is needed to support milk production, which is somewhat greater in the first six months of breastfeeding than in the second six months. Because most women lose an average of 0.8 kg per month in the first six months postpartum (i.e., after delivery), EER is, on average, 170 kcal per day less (Panel on Macronutrients 2002).

It is important to note that the EER equations apply only to persons having a healthy weight and that EER values have not been established for persons who are overweight or obese (Panel on Macronutrients 2002). Instead, the DRI Committee has developed a separate set of equations for calculating total energy expenditure (TEE) for the maintenance of weight for adults age 19 years and older who are overweight (i.e., have a **body mass index** or BMI between 25.0 kg/m^2 and 29.9 kg/m^2) and/or obese (BMI $\geq$30.0 kg/m^2), and an additional set was developed for children and adolescents age 3 to 18 years who are overweight (a BMI for age and sex $\geq$95th percentile) (Panel on Macronutrients 2002). These are shown in Table 14.3. The DRI Committee adopted the definition of healthy weight for adults (age 19 years and older) used by the *Dietary Guidelines for Americans*, which is a BMI $\geq$18.5 kg/m^2 but

estimated energy requirement (EER)—the average dietary energy intake that is predicted to maintain energy balance in a healthy adult of a defined age, gender, weight, height, and level of physical activity, consistent with good health; in children and pregnant and lactating women, the EER includes the needs associated with the deposition of tissues or the secretion of milk at rates consistent with good health

body mass index (BMI)—weight in kilograms divided by height in meters squared (BMI = kg ÷ m^2); although technically not a body composition assessment technique, it correlates well with estimates of body composition derived from skinfold measurements, and underwater weighing (hydrodensitometry), and can easily be calculated from weight and height; it is also known as **Quetelet's index**, named after its developer, Adolphe Quetelet (1796–1874), a Belgian statistician, astronomer, mathematician, and sociologist; the formula for calculating body mass index is:

$$\text{body mass index} = \frac{\text{weight (kg)}}{\text{height (m)}^2}$$

TABLE 14.2

Equations for Calculating Estimated Energy Requirement (EER) in Kilocalories Per Day[1]

EER for Infants and Young Children

$EER = TEE + \text{Tissue Deposition}$[2]

0–3 months	$(89 \times weight - 100) + 175$
4–6 months	$(89 \times weight - 100) + 56$
7–12 months	$(89 \times weight - 100) + 22$
13–35 months	$(89 \times weight - 100) + 20$

EER for Males 3 through 8 Years

$EER = TEE + \text{Tissue Deposition}$

$EER = 88.5 - 61.9 \times age + PA \times (26.7 \times weight + 903 \times height) + 20$

Where PA is the physical activity coefficient:

$PA = 1.00$ for sedentary

$PA = 1.13$ for low active

$PA = 1.26$ for active

$PA = 1.42$ for very active

EER for Females 3 through 8 Years

$EER = TEE + \text{Tissue Deposition}$

$EER = 135.3 - 30.8 \times age + PA \times (10.0 \times weight + 934 \times height) + 20$

Where PA is the physical activity coefficient:

$PA = 1.00$ for sedentary

$PA = 1.16$ for low active

$PA = 1.31$ for active

$PA = 1.56$ for very active

EER for Males 9 through 18 Years

$EER = TEE + \text{Tissue Deposition}$

$EER = 88.5 - 61.9 \times age + PA \times (26.7 \times weight + 903 \times height) + 25$

Where PA is the physical activity coefficient:

$PA = 1.00$ for sedentary

$PA = 1.13$ for low active

$PA = 1.26$ for active

$PA = 1.42$ for very active

EER for Females 9 through 18 Years

$EER = TEE + \text{Tissue Deposition}$

$EER = 135.3 - 30.8 \times age + PA \times (10.0 \times weight + 934 \times height) + 25$

Where PA is the physical activity coefficient:

$PA = 1.00$ for sedentary

$PA = 1.16$ for low active

$PA = 1.31$ for active

$PA = 1.56$ for very active

EER for Males 19 Years of Age and Older

$EER = TEE$

$EER = 662 – 9.53 \times age + PA \times (15.91 \times weight + 539.6 \times height)$

Where PA is the physical activity coefficient:

$PA = 1.00$ for sedentary

$PA = 1.11$ for low active

$PA = 1.25$ for active

$PA = 1.48$ for very active

EER for Females 19 Years of Age and Older

$EER = TEE$

$EER = 354 - 6.91 \times age + PA \times (9.36 \times weight + 726 \times height)$

Where PA is the physical activity coefficient:

$PA = 1.00$ for sedentary

$PA = 1.12$ for low active

$PA = 1.27$ for active

$PA = 1.45$ for very active

EER for Pregnancy

$EER = EER$ for age + Pregnancy Energy Needs[3] + Tissue Deposition

1st trimester = EER for age + 0

2nd trimester = EER for age + 160 + 180

3rd trimester = EER for age + 272 + 180

EER for Lactation

$EER = EER$ for age + Milk Energy Output[4] − Weight Loss[5]

1st six months = EER for age + 500 − 170

2nd six months = EER for age + 400 − 0

[1] EER = Estimated Energy Requirement; TEE = Total Energy Expenditure; PA = Physical Activity Coefficient; age is in years; height is in meters; weight is in kilograms.

[2] Tissue Deposition represents the energy cost of growth during infancy, childhood, adolescence, and pregnancy as measured in kilocalories.

[3] Pregnancy Energy Needs represents the additional energy required to support the metabolic demands of pregnancy.

[4] Milk Energy Output represents the energy needed to produce the milk during lactation. Milk output is somewhat greater in the first six months than in the second six months of breastfeeding.

[5] Weight Loss represents a average decline in EER of 170 kcal/day that well-nourished lactating women experience during the first six months postpartum, resulting in an average weight loss of 0.8 kg/month.

Source: Adapted from Panel on Macronutrients, Panel on the Definition of Dietary Fiber, Subcommittee on Upper Reference Levels of Nutrients, Subcommittee on Interpretation and Uses of Dietary Reference Intakes, Standing Committee on the Scientific Evaluation of Dietary Reference Intakes. 2002. Dietary reference intakes for energy, carbohydrate, fiber, fat, fatty acids, cholesterol, protein, and amino acids. Washington (DC): National Academy Press.

TABLE 14.3

Equations for Calculating Total Energy Expenditure (TEE) for Weight Maintenance in Kilocalories Per Day for Overweight and Obese Adults and for Overweight Children and Adolescents[1]

TEE for Overweight and Obese Males Aged 19 Years and Older

$TEE = 1086 - 10.1 \times age + PA \times (13.7 \times weight + 416 \times height)$

Where PA is the physical activity coefficient:

PA = 1.00 for sedentary

PA = 1.12 for low active

PA = 1.29 for active

PA = 1.59 for very active

TEE for Overweight and Obese Females Aged 19 Years and Older

$TEE = 448 - 7.95 \times age + PA \times (11.4 \times weight + 619 \times height)$

Where PA is the physical activity coefficient:

PA = 1.00 for sedentary

PA = 1.16 for low active

PA = 1.27 for active

PA = 1.44 for very active

TEE for Overweight Males Aged 3 through 18 Years

$TEE = -114 - 50.9 \times age + PA \times (19.5 \times weight + 1161.4 \times height)$

Where PA is the physical activity coefficient:

PA = 1.00 for sedentary

PA = 1.12 for low active

PA = 1.24 for active

PA = 1.45 for very active

TEE for Overweight Females Aged 3 through 18 Years

$TEE = 389 - 41.2 \times age + PA \times 15.0 \times weight + 701.6 \times height$

Where PA is the physical activity coefficient:

PA = 1.00 for sedentary

PA = 1.18 for low active

PA = 1.35 for active

PA = 1.60 for very active

[1] TEE = Total Energy Expenditure; PA = Physical Activity Coefficient; age is in years; height is in meters; weight is in kilograms. In persons age 19 years and older overweight is defined as a BMI between 25.0 kg/m^2 and 29.9 kg/m^2 and obese is defined as a BMI $\geq$30.0 kg/m^2. In persons age 3 to 18 years, overweight is defined as a BMI for age and sex $\geq$95th percentile.

Adapted from Panel on Macronutrients, Panel on the Definition of Dietary Fiber, Subcommittee on Upper Reference Levels of Nutrients, Subcommittee on Interpretation and Uses of Dietary Reference Intakes, Standing Committee on the Scientific Evaluation of Dietary Reference Intakes. Dietary reference intakes for energy, carbohydrate, fiber, fat, fatty acids, cholesterol, protein, and amino acids. Washington (DC): National Academy Press; 2002.

$\leq$24.9 kg/m^2. Healthy weight for persons age 2 to 18 years is defined as a BMI that is >5th percentile but <85th percentile of BMI for age and sex, as discussed in Chapter 5. The equations developed for overweight or obese persons shown in Table 14.3 allow calculation of the TEE necessary for weight maintenance using the variables gender, age, weight, height, and physical activity level. If weight loss is desired, a recommended approach is reducing energy intake so that it is 500 to 1,000 kcal per day less than that needed for maintenance, and increasing energy expenditure by engaging in moderate physical activity for approximately 60 minutes per day on most days of the week (NHLBI 2000).

Indirect Calorimetry The most commonly used approach for measuring energy requirements in critically ill patients and in human metabolic research is **indirect calorimetry**. It is based on the fact that energy expenditure is proportional to the body's oxygen consumption and carbon dioxide production. Expired air contains less oxygen and more carbon dioxide than inspired air. When the differences in oxygen and carbon dioxide in inspired and expired air are known and the volume of air moving through a subject's lungs is measured, the body's energy expenditure can be calculated.

In the laboratory or clinical settings, indirect calorimetry is accomplished using a portable computerized metabolic monitor that can be brought to the bedside or positioned next to a subject exercising on a treadmill or cycle ergometer. A mask or hood is placed over the subject's face and the amount of air flow through the lungs per minute (known as minute ventilation) is measured by various types of instruments that are built into the mask. Gas analyzers in the monitor measure the oxygen and carbon dioxide content of both inspired and expired air. Technological advances have resulted in the development of lightweight, portable indirect calorimetry units, which can be worn by subjects while working or participating in sports. A typical unit weighs 2.1 lbs (950 g) and allows accurate testing while the subject is engaged in practically any activity at any time or location, without being confined to an artificial laboratory environment. In the clinical setting, indirect calorimetry is useful as a means to accurately determine the energy requirements of critically ill and/or mechanically ventilated patients, and to monitor the adequacy and appropriateness of nutritional support.

indirect calorimetry—an approach to determine energy expenditure by measuring a subject's oxygen consumption, carbon dioxide production, and minute ventilation (the amount of air a subject breathes in one minute)

Doubly Labeled Water The **doubly labeled water** (DLW) technique is a relatively new approach for measuring total energy expenditure in subjects who are engaged in their normal daily routines (i.e., "free-living individuals") over a one- to two-week period, without the use of the instrumentation used in indirect calorimetry (Panel on Macronutrients 2002; Trabulsi et al. 2003). It has been found to be accurate within 1% to 2% when compared to indirect calorimetry, and is considered the "gold standard" for measuring energy expenditure and physical activity in free-living subjects over a one- to two-week period (Hoos et al. 2003; Trabulsi et al. 2003).

DLW involves subjects drinking a known amount of water containing two different stable isotopic forms of water: $H_2{}^{18}O$ and ^2H_2O. Ordinary water is a molecule composed of two atoms of hydrogen, each having an atomic mass of one (^1H), and one atom of oxygen having an atomic mass of 16 (^{16}O). Hydrogen atoms with an atomic mass of two (^2H or deuterium) and oxygen atoms with an atomic mass of 18 (^{18}O) are only naturally present in the environment in extremely minute quantities. Consequently, essentially all of the ^2H and ^{18}O present in the body of test subjects comes from the doubly labeled water. After the subject drinks the two different isotopic forms of water, they mix with the body's water and are gradually eliminated from the body (Panel on Macronutrients 2002; Hoos et al. 2003; Trabulsi et al. 2003). Over the next one to two weeks, the subject provides several urine samples that are used to measure the rate at which the two isotopes disappear from the body. The rate of disappearance is then used to calculate energy expenditure. The method is noninvasive, provides an accurate measurement of energy expenditure over a period of one to two weeks, and, because the two isotopes are stable (nonradioactive), the procedure is considered safe to use even on infants and females who are pregnant or lactating (Panel on Macronutrients 2002).

Direct Calorimetry **Direct calorimetry** involves using a highly sophisticated chamber that is capable of determining a subject's total energy expenditure by measuring the amount of heat given off by the subject's body through evaporation, convection, and radiation (Seale, Rumpler, and Moe 1991; Committee on Metabolic Monitoring 2004). The size of direct calorimeters varies from a chamber just large enough to accommodate a subject lying down to those the size of a small bedroom. Once inside the calorimeter, the subject's activity is monitored, and the subject's response to clinically prepared meals can be studied. If necessary, samples of urine and feces can be collected for analysis. In some instances, subjects may remain inside the calorimeter for up to 24 hours or longer, making it an impractical approach for use with critically ill patients or those afraid of being in an enclosed space for several hours. Because direct calorimetry requires equipment that is bulky, very expensive, and technologically sophisticated, it is rarely used. The U.S. Department of Agriculture has a room calorimeter at its Human Nutrition Research Center in Beltsville, Maryland, which it uses for human research (Seale, Rumpler, and Moe 1991).

Regulation of Energy Balance

The regulation of energy balance and body weight is dependent upon the complex interaction of the nervous system and various hormones (Flier and Maratos-Flier 2005). A decrease in energy intake and loss of body fat mass typically result in **orexigenic** neural and hormonal stimuli that lead to increased appetite and decreased resting energy expenditure. Modest increases in energy intake and increased body fat mass typically result in **anorexigenic** stimuli that lead to decreased appetite and an increase in energy expenditure known as **adaptive thermogenesis**.

Appetite is influenced by a number of signals to the brain that are primarily orchestrated by the hypothalamus region, including neural signals from mouth, stomach, and small intestine during and following eating and the secretion of pancreatic and gastrointestinal hormones such as insulin, glucagon, amylin, cholecystokinin, glucagon-like peptide-1, peptide YY, and ghrelin (Flier and Maratos-Flier 2005; Anderson 2006; Smith 2006). Pleasurable taste sensations within the mouth stimulate appetite and encourage eating. As the stomach fills, it becomes distended, stimulating stretch receptors in the stomach wall that provide neural signals to the hypothalamus that inhibit appetite. Proteins, monosaccharides, and fatty acids in the chyme (semiliquid mass of partially digested food) leaving the stomach stimulate neural and endocrine receptors in the mucosa of the small intestine, resulting in neural signals to the brain that

doubly labeled water—a technique to determine energy expenditure in which subjects drink a known amount of water containing two different stable isotopic forms of water: $H_2{}^{18}O$ and ^2H_2O; the rate that this water disappears from the subject's body is used to calculate the subject's energy expenditure

direct calorimetry—a technique to determine energy expenditure using a highly sophisticated chamber capable of measuring the amount of heat released by a subject's body through evaporation, convection, and radiation

orexigenic—appetite stimulating

anorexigenic—appetite inhibiting

adaptive thermogenesis—energy expenditure above and beyond the thermic effect of food and resting energy expenditure that is seen in response to overfeeding, traumatic injury, changes in hormonal status, and exposure to a cold environment

decrease appetite and food intake, and the release of the hormones cholecystokinin, glucagon-like peptide-1, and peptide YY, which also decrease appetite and food intake (Smith 2006; Anderson 2006).

As plasma glucose level rises following a meal, the β-cells of the pancreas release insulin and amylin, which decrease appetite and food intake. During fasting, the β-cells of the pancreas release glucagon, which also decreases appetite and food intake (Flier and Maratos-Flier 2005; Anderson 2006; Smith 2006). Ghrelin is a peptide hormone that is mainly produced by the stomach and stimulates appetite. Ghrelin levels are normally increased during fasting, but immediately following food intake, ghrelin levels decline. This appears to decrease appetite and food intake (Anderson 2006; Brodsky 2006). However, in patients with Prader-Willi syndrome, a genetic disorder characterized by voracious appetite and massive obesity, ghrelin levels are increased by as much as threefold or fourfold compared to individuals of similar age, sex, and BMI (Paik et al. 2004; Chanione 2005). These numerous and diverse neural and hormonal signals influence the release of various peptides from the hypothalamus, resulting in the final expression of appetite and eating behavior (Flier and Maratos-Flier 2005).

The Adipocyte and Adipose Tissue

The adipocyte (fat cell) is a large, rounded cell primarily filled with a droplet of triglyceride. The cytoplasm containing the nucleus, mitochondria, and other cell organelles is forced to occupy a thin layer immediately beneath the plasma membrane (see Figure 14.2) (Saladin 2004). Throughout the body, adipocytes occur individually or in small groups joined by connective tissue (Pleuss 2005). When found in large aggregations in conjunction with fibrous connective tissue, they form adipose tissue, which serves as the storage site for more than 90% of the body's energy reserves (Flier and Maratos-Flier 2005; Pleuss 2005). In addition, adipose tissue fills body crevices, provides thermal insulation to the body, surrounds and shields internal organs, gives shape and form to the body, and cushions such body areas as the feet, hands, shoulders, and buttocks (Pleuss 2005).

There are two types of adipose tissue: white adipose tissue (WAT) and brown adipose tissue (BAT). The predominant type is WAT, which in reality is a light-yellow in color due to the presence of carotenoids. The cells of WAT store triglycerides derived from dietary fats or those synthesized from carbohydrates and proteins through the process of **lipogenesis**. BAT derives its color from the large number of mitochondria in the adipocytes and from its abundance of blood vessels. BAT is primarily found in fetuses, infants, and young children, and accounts for up to 6% of an infant's body weight. As humans age, the amount of BAT diminishes. In WAT, triglyceride is stored within a single large droplet, whereas in BAT, there are multiple smaller droplets. It appears that the primary function of BAT is maintaining body temperature in human neonates and in hibernating animals by generating heat through a process known as diet-induced thermogenesis or nonshivering thermogenesis. However, because of the small amount of BAT in adult humans, it has a minimal effect on energy expenditure (Saladin 2004; Pleuss 2005).

FIGURE 14.2 An Adipocyte or Fat Cell

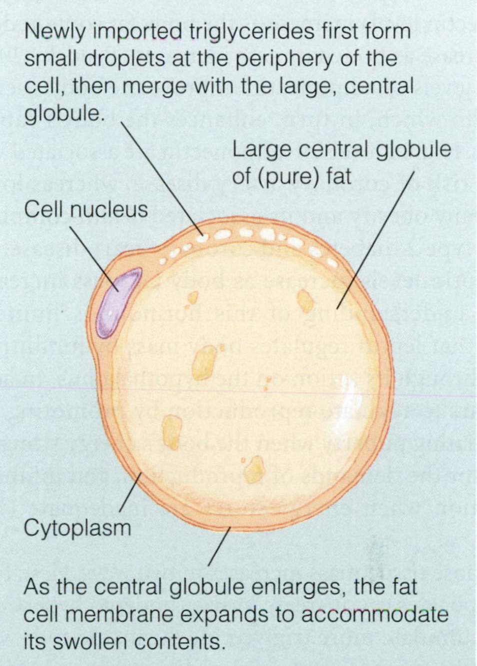
Newly imported triglycerides first form small droplets at the periphery of the cell, then merge with the large, central globule.
Large central globule of (pure) fat
Cell nucleus
Cytoplasm
As the central globule enlarges, the fat cell membrane expands to accommodate its swollen contents.

Source: S. Rolfes, K. Pinna, and E. Whitney, *Understanding Normal and Clinical Nutrition*, 7e, copyright © 2006, p. 157.

Although once thought to be a relatively inert storage site for energy consumed in excess of the body's needs, the adipocyte is metabolically active in the uptake, synthesis, storage, and mobilization of triglycerides. There is a constant turnover of triglycerides in the adipocyte as new triglycerides are synthesized and stored and older triglycerides are hydrolyzed and released from the adipocyte into the circulation (Saladin 2004). Recent research has shown the adipocyte to be an endocrine cell releasing numerous hormones involved in regulating appetite, energy balance, body fat content, and reproduction. In addition, adipocytes produce growth factors and cytokines involved in tissue repair and inflammation (Flier and Maratos-Flier 2005; Brodsky 2006).

Two hormones produced by adipose tissue that are involved in energy balance and fat storage are adiponectin and leptin. Research suggests that adiponectin signals that the

lipogenesis—the synthesis of triglyceride from carbohydrates and proteins

body has the capacity to store fat, while leptin appears to signal that ample fat has been stored (Brodsky 2006). Adiponectin levels increase as the body fat content decreases and decrease as body fat mass increases (Brodsky 2006). Increased levels of adiponectin improve the body's sensitivity to insulin which, in turn, enhances the body's capacity to store fat. Higher levels of adiponectin are associated with decreased risk of coronary artery disease, whereas low levels accompany obesity and its associated health complications such as type 2 diabetes and coronary heart disease. In contrast, leptin levels increase as body fat mass increases. Although understanding of this hormone is limited, it is known that leptin regulates body mass by inhibiting food intake through its action on the hypothalamus. In addition, it appears to regulate reproduction by promoting fertility and initiating puberty when the body's energy stores are adequate for the demands of reproduction, and inhibiting reproduction when energy stores are inadequate (Brodsky 2006).

Adipose tissue mass increases in two ways. First, fully mature adipocytes can increase in size (undergo hypertrophy) as they accumulate more triglyceride during periods when energy intake exceeds energy expenditure. Second, adipocytes can increase in number (undergo hyperplasia) as immature adipocytes divide to produce more cells (Pleuss 2005). Overweight (BMI 25.0 to 29.9 kg/m^2) and moderate obesity (BMI 30.0 to 34.9 kg/m^2) are characterized by hypertrophy (enlargement) of adipocytes, and with weight loss, these enlarged adipocytes become smaller. However, as BMI approaches extreme obesity (BMI $\geq$40.0 kg/m^2), adipocytes reach their maximum size and then experience hyperplasia (an increase in number). As persons with extreme obesity lose weight, the adipocytes become smaller in size but the number of adipocytes does not decrease. The clinical implication of this fact is that overweight and moderately obese people who have experienced only fat cell hypertrophy are more successful at maintaining their weight loss than are extremely obese people who have fat cell hypertrophy and hyperplasia. Achieving and maintaining a healthy body weight is more likely if an increase in fat cell number can be avoided.

During the first year of life, the proportion of fat in the human body typically increases from approximately 15% at birth to about 30% at one year as adipocytes undergo hypertrophy and hyperplasia (Norgan 1998). A higher percentage of body fat is seen in infants who have a high birth weight and infants of diabetic mothers, and these infants are at increased risk of being overweight in later childhood and adolescence (Dietz 2006). Between the ages of 1 and 6 years, the percentage of body fat generally decreases, and then begins to increase at 6 to 8 years of age in a process known as "adiposity rebound" or "BMI rebound" (Norgan 1998; Dietz 2006). Children who experience their adiposity rebound or BMI rebound before age 4 to 6 years are at increased risk of increased BMI in later life (Dietz 2006). It is estimated that 25% to 80% of overweight children remain overweight as adults. This is particularly likely for adolescent females, who have three times the risk of being overweight as adults compared to overweight adolescent males (Dietz 2006).

Body Composition, Obesity, and Overweight

The human body is composed of different types of tissues—adipose tissue or body fat, muscle, bone, blood, cartilage, ligaments, tendons, the brain and nervous tissue, and the viscera located within the thoracic and abdominal cavities. The most common approach to body composition analysis views the body as consisting of two different compartments: fat and fat-free. This is referred to as the "two-compartment model." Using this model, body composition is expressed as a ratio of fat to fat-free mass, or as a percentage of the body composed of adipose tissue or lean tissue.

As discussed in Chapter 5, a variety of methods are available to clinicians for assessing body composition, and the most common of these body composition assessment methods are based on the two-compartment model. They include the following:

- Skinfold measurements
- Underwater weighing or hydrodensitometry
- Bioelectrical impedance analysis
- Air-displacement plethysmography
- Dual-energy x-ray absorptiometry

Technically, obesity is an excess of adipose tissue or body fat. It can be defined as a proportion of body weight composed of adipose tissue (percent body fat) that exceeds a range that is considered healthy. The problem with this definition is that it requires that the body's composition be assessed in order to determine the relative proportions of fat and lean tissue. Most clinicians do not have the time, expertise, or equipment to accurately assess body composition using the techniques just mentioned. In contrast, accurately measuring weight and height is relatively easy and quick, and the necessary equipment is inexpensive and readily available. Consequently, weight and height measurements are often used in place of body composition analysis to determine whether a person is obese. The problem with this approach is that body composition cannot be determined by merely evaluating body weight and height, because measurements of body weight and height alone are incapable of differentiating between the weight of the body's lean tissue and the weight of the body's adipose tissue.

Because it is often impractical to determine body composition in the clinical setting, and because accurate measurements of height and weight can be easily obtained, obesity in adults is often defined as a BMI $\geq$30.0 kg/m^2. Also known as Quetelet's index, BMI is not a direct measure of body fatness. BMI can be considered a proxy for measures of body fatness and is regarded as a convenient and reliable indicator of obesity. It is reasonable to assume that for most people a high BMI (i.e., $\geq$30 kg/m^2) represents an increased amount of adipose tissue in the body rather than unusually well-developed musculature or a large, dense skeleton. Because changes in BMI parallel changes in body composition obtained by direct measures of body fat such as underwater weighing and dual-energy x-ray absorptiometry, it is a convenient and useful approach for tracking improvements in body composition. Throughout North America and Europe, BMI is regarded as the best and most convenient clinical approach to use in evaluating the body weights of patients (NHLBI 2000; IOTF 2002, 2003; Shields 2005; Tjepkema 2005).

Overweight is a body weight in excess of some standard weight, and usually includes a consideration of height. In adults, overweight is generally defined as a BMI of 25.0 kg/m^2 to 29.9 kg/m^2 (NHLBI 2000; USDHHS 2005), "healthy weight" is defined as a BMI of 18.5 kg/m^2 to 24.9 kg/m^2, and underweight is defined as a BMI <18.5 kg/m^2 (USDHHS 2005). Box 14.3 discusses the calculation of BMI and shows these BMI classifications for adults.

For children and adolescents, overweight is defined using the U.S. Centers for Disease Control and Prevention (CDC) growth charts that provide the body mass index-for-age percentiles for males and females 2 to 20 years of age. Using these charts (see Chapter 5 and Appendix B1) "overweight" in persons 2 to 20 years is defined as a BMI-for-age at or above the 95th percentile. Children and adolescents having a BMI-for-age $\geq$85th percentile but <95th percentile are considered at "risk of overweight" (Kuczmarski et al. 2000, 2002).

Clinical judgment must be used in interpreting BMI in situations affecting its accuracy as an indicator of total body fat. Examples of these situations include high muscularity, large skeletal mass, the presence of edema, muscle wasting, and osteoporosis. Some people may have a high body weight but have a low percentage of body fat if they are unusually muscular or have a large skeletal mass. Examples of persons in this category include body builders, gymnasts, weight lifters, and other highly muscular people. In these individuals, BMI overestimates the degree of fatness. On the other hand, a person may have a BMI within the healthy weight range but have a higher than desirable percentage of body fat. An example of this is an elderly person who is frail and has a low muscle mass due to prolonged physical inactivity or has a low skeletal mass due to osteoporosis; in these cases BMI underestimates the degree of fatness. The presence of clinical edema (an excessive accumulation of fluid in the body) will increase body weight without changing the amount of fat in the body. Consequently, in persons with significant edema, BMI overestimates body fatness. In addition, females generally have more body fat for a given BMI than males. Despite these circumstances, in most instances BMI remains a valuable tool for classifying individuals into broad categories of overweight and obesity in order to monitor the weight status of individuals in clinical settings (NHLBI 2000).

BOX 14.3 CALCULATING BMI AND USING BMI TO CLASSIFY ADULTS

Formulas for calculating body mass index or BMI are as follows:

BMI = weight in kilograms ÷ (height in meters)^2

To convert weight in pounds to weight in kilograms: pounds ÷ 2.2 = kilograms

To convert height in inches to height in meters: inches × 0.0254 = meters

For those who have difficulty using the SI units of measurement, the following formula can also be used to calculate BMI using weight in pounds and height in inches:

BMI = (weight in pounds × 703) ÷ (height in inches)^2

The BMI classifications for adults shown here are recommended by the National Institutes of Health in the publication *Clinical Guidelines on the Identification, Evaluation, and Treatment of Overweight* and *Obesity in Adults* (NHLBI 1998) and the World Health Organization in the publication *Obesity: Preventing and Managing the Global Epidemic. Report of a World Health Organization Consultation* (WHO 1999). Similar classification values are used in the latest edition of the *Dietary Guidelines for Americans* (USDHHS 2005) and by a number of highly respected national and international scientific groups (IOTF 2002, 2003; Shields 2005; Tjepkema 2005).

Classification	BMI
Underweight	<18.5 kg/m^2
Healthy weight	18.5 to 24.9 kg/m^2
Overweight	25.0 to 29.9 kg/m^2
Obesity (Class 1)	30.0 to 34.9 kg/m^2
Obesity (Class 2)	35.0 to 39.9 kg/m^2
Extreme obesity (Class 3)	$\geq$40.0 kg/m^2

Body Fat Distribution

The location or distribution of adipose tissue within the body is an important concept when considering the health implications of overweight and obesity (NHLBI 2000; Yusuf et al. 2005; Hill, Catenacci, and Wyatt 2006). Body fat distribution can be divided into two clinically significant categories: (1) abdominal or central body fat distribution, and (2) lower body fat distribution. Abdominal or central fat placement refers to fat located primarily within the abdominal region of the body, both surrounding the organs of the abdomen (intra-abdominal or visceral fat) and located just under the skin around the waist (subcutaneous fat). Abdominal fat distribution is more often seen in males, tends to give the body a shape resembling that of an apple, and is sometimes referred to as *android*, which literally means "manlike." Lower body fat placement refers to fat located primarily in the lower region of the body, particularly within the hips and thighs, and tends to give the body a shape resembling a pear. Lower body fat distribution is more often seen in females and is sometimes referred to as *gynoid* ("womanlike") (Lee and Nieman 2006).

While it is possible to accurately quantify adipose tissue within the regions of the abdomen, hips, and thighs using magnetic resonance imaging (MRI) or computed tomography (CT), routine use of these imaging techniques is not practical in the clinical setting because the instruments are not readily available to most clinicians; they are expensive to acquire, maintain, and operate; and their use is time-consuming. A much more practical approach to quantifying abdominal fat is to measure waist circumference (see Chapter 5). Estimates of abdominal adiposity based on waist circumference measurements compare favorably to abdominal fat measurements using MRI and CT scans (NHLBI 1998, 2000; Lee and Nieman 2006). A practical approach for estimating the amount of adipose tissue in the hips and thighs is to measure the circumference of the hips or buttocks. This measurement is taken at the point yielding the maximum circumference around the hips or buttocks.

Excessive adipose tissue located deep within the abdomen and surrounding the intestines and liver (abdominal obesity) is associated with increased risk of type 2 diabetes, hypertension, dyslipidemia, coronary heart disease, and metabolic syndrome, even when BMI is within the healthy weight range (NHLBI 2000; Pi-Sunyer 2004; Yusuf et al. 2005). In contrast, an increased amount of adipose tissue located within the hips and thighs is not associated with these increased risks; in fact, there is research suggesting that adipose tissue located within the lower body is inversely related to (i.e., lowers) risk of these conditions (Snijder et al. 2003, 2004; Yusuf et al. 2005). Epidemiologic research shows that in adults with a BMI between 25.0 kg/m^2 and 34.9 kg/m^2, risk of type 2 diabetes, hypertension, dyslipidemia, coronary heart disease, and metabolic syndrome increases when the waist circumference exceeds 40 inches in males and 35 inches in females (NHLBI 1998, 2000; USDHHS/USDA 2005). A waist circumference measurement in excess of these values is regarded as "high-risk," as shown in Table 14.4 (NHLBI 1998, 2000; USDHHS/USDA 2005). Waist circumference is particularly useful in assessing the disease risk of patients who are categorized as having a healthy body weight (BMI of 18.5 kg/m^2 to 24.9 kg/m^2), who are considered overweight (BMI of 25.0 kg/m^2 to 29.9 kg/m^2), or who are mildly obese with a BMI of 30.0 kg/m^2 to 34.9 kg/m^2. Waist circumference is a better indicator of disease risk than is BMI for individuals of Asian descent, and it assumes greater value for estimating risk for obesity-related diseases in older persons who often experience a loss of

TABLE 14.4

High-Risk Waist Circumference in Adult Males and Females

Males	>40 in (>102 cm)
Females	>35 in (>88 cm)

Source: National Heart, Lung, and Blood Institute. *The practical guide: identification, evaluation, and treatment of overweight and obesity in adults.* Bethesda (MD): U.S. Department of Health and Human Services, National Institutes of Health; 2000.

muscle mass and an increase in abdominal fat without marked changes in BMI (NHLBI 1998, 2000; Yusuf et al. 2005; Hill, Catenacci, and Wyatt 2006). In persons with a BMI greater than 35.0 kg/m^2, waist circumference is of little value in improving disease risk assessment; therefore, it is not recommended that waist circumference be measured in persons having a BMI >35.0 kg/m^2 (NHLBI 1998, 2000).

An alternative approach to evaluating the impact of body fat distribution on disease risk is the waist-to-hip ratio (WHR), which is calculated by dividing the waist circumference measurement by the hip circumference measurement. Some clinicians prefer using the WHR instead of the waist circumference, citing research suggesting that the WHR is somewhat better at predicting risk of coronary heart disease than is waist circumference alone (Snijder et al. 2003, 2004; Yusuf et al. 2005). Disease risk increases when the WHR is >0.95 in males and >0.8 in females (Yusuf et al. 2005). A WHR >1.0 results when waist circumference is greater than hip circumference and suggests that the amount of abdominal fat is unhealthful. One plausible explanation for the potential superiority of the WHR is that it takes into account the protective effects of larger hip circumferences which may result from increased adipose tissue or from increased muscle mass in the hips and thighs, both of which are associated with a lower risk of type 2 diabetes, hypertension, dyslipidemia, coronary heart disease, and metabolic syndrome (Snijder et al. 2003, 2004; Yusuf et al. 2005). However, both waist circumference and WHR have been shown useful in assessing body fat distribution and evaluating disease risk. The key concept is that fat deep within the abdomen and around the intestines and liver increases disease risk; the technique used to measure it is less critical. Because BMI does not distinguish between lean tissue and adipose tissue or indicate how fat is distributed, it cannot predict disease risk when used alone. This is particularly the case for older persons who, as they age, tend to lose muscle mass and gain fat mass. When evaluating a patient's disease risk in relation to their weight, height, and body fat distribution, BMI and circumferences of the waist and hip should be used. Table 14.5 illustrates how BMI and waist circumference can be

TABLE 14.5

Classification of Overweight and Obesity by BMI, Waist Circumference, and Associated Disease Risk[1]

	BMI (kg/m²)	Obesity Class	Disease Risk[1] (Relative to Normal Weight and Waist Circumference)	
			Men ≤40 in (≤ 102 cm) Women ≤ 35 in (≤88 cm)	Men > 40 in (> 102 cm) Women > 35 in (> 88 cm)
Underweight	< 18.5		—	—
Normal[2]	18.5–24.9		—	—
Overweight	25.0–29.9		Increased	High
Obesity	30.0–34.9	I	High	Very high
	35.0–39.9	II	Very high	Very high
Extreme Obesity	≥ 40.0	III	Extremely high	Extremely high

[1] Disease risk for type 2 diabetes, hypertension, and cardiovascular disease.
[2] Increased waist circumference can also be a marker for increased risk even in persons of normal weight.
Source: National Heart, Lung, and Blood Institute. *The practical guide: identification, evaluation, and treatment of overweight and obesity in adults.* Bethesda (MD): U.S. Department of Health and Human Services, National Institutes of Health; 2000.

used to classify overweight and obesity and to provide an indication of relative disease risk.

Epidemiology of Overweight and Obesity

In nearly every country of the world, the average body weight of children and adults is increasing to such an extent that the World Health Organization (WHO) has coined the term "globesity" to describe what it calls a "global epidemic of obesity" (WHO 2005). While the term "epidemic" is generally used in the context of infectious disease, it can appropriately be applied to any condition or situation having an adverse effect on health, including overweight and obesity (USDHHS 2001). While primarily considered a problem affecting developed nations, overweight and obesity are common in urban areas of many developing nations, where, paradoxically, they coexists with undernutrition occurring in the rural areas of the same country. According to the WHO, the prevalence of obesity ranges from less than 5% of the population of China, Japan, and certain African nations to more than 75% in urban Samoa. Even in a relatively low-prevalence country like China, obesity rates can be as high as 20% in some cities (WHO 2005).

Because of the ease of accurately measuring weight and height, estimates of the prevalence of overweight and obesity are typically based on BMI. Attempting to measure body composition on large numbers of people using such methods as skinfold measurements or underwater weighing is impractical. There are some instances when it is not possible to obtain measured weights and heights on subjects, in which case researchers must rely on self-reported weight and heights. However, estimates based on measured weight and height are more accurate than estimates based on self-reported weight and height, which tend to underestimate the true prevalence of overweight and obesity.

Overweight and Obesity in the United States

In the early 1960s, the National Center for Health Statistics launched a series of surveys examining the health and nutritional status of the U.S. population, in which participants completed questionnaires evaluating diet and lifestyle habits, and underwent diagnostic and laboratory testing as well as extensive anthropometric assessment. In these surveys (initially called the National Health Examination Survey and then renamed the National Health and Nutrition Examination Survey), all anthropometric measurements, including height and weight, were taken by trained health technicians using standardized measuring procedures and equipment yielding highly accurate data for calculating BMI. A key finding from these surveys is that the percentage of people in the United States who are overweight or obese has increased since the early 1960s (see Figures 14.3 and 14.4). When looking at Figure 14.3, note that the percentage of U.S. adults who are *either* overweight *or* obese increased to a much lesser extent than did the percentage of U.S. adults who are obese, as shown in Figure 14.4. While the body weights of Americans have increased in recent decades,

FIGURE 14.3 Prevalence of U.S. Adults Who Are Either Overweight or Obese, 1960 to 2002

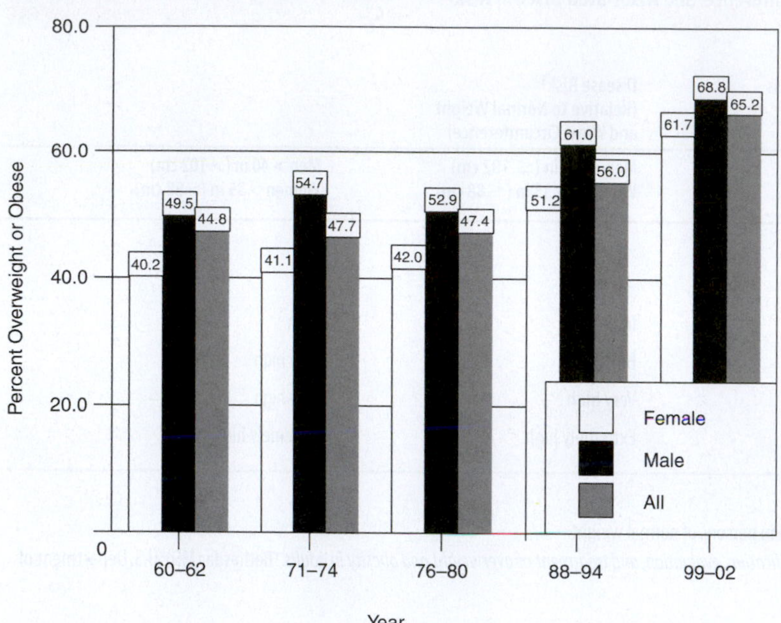

Source: Data from the National Center for Health Statistics.

FIGURE 14.4 Prevalence of Obesity among U.S. Adults, 1960 to 2002

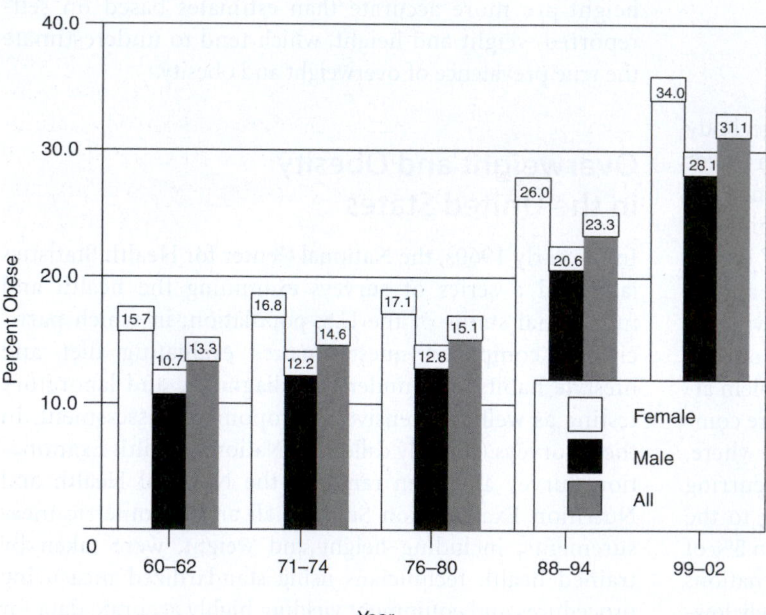

Source: Data from the National Center for Health Statistics.

most of this increase has been due to increases in the obesity category, whereas only minor increases occurred in the prevalence of persons who are overweight but not obese.

Two of the national health objectives for the year 2010 are to increase the proportion of U.S. adults who are at a healthy weight to 60% and to decrease the proportion of U.S. adults who are obese to 15% (USDHHS 2000). Figure 14.5 shows how the percentage of U.S. adult population categorized as having a healthy weight has decreased since the early 1960s to about 33% in the most recent survey period. Figure 14.4 shows that from 1960 to 1980 the prevalence of obesity among U.S. adults was relatively stable, but that between the 1976–80 survey period and the 1999–2004 survey period the prevalence of obesity doubled from 15% to 31%. It is highly unlikely that either of these two national health objectives for the year 2010 will be met, given the upward trend in the body weights of U.S. adults, which appears to be continuing unabated (Hedley et al. 2004).

Figure 14.6 shows how the prevalence of overweight in two age categories of U.S. children and adolescents (6–11 years old and 12–19 years old) has changed since the 1960s. As discussed earlier, in children and adolescents ages 2 to 19 years, overweight is defined as a BMI at or above the 95th percentile for sex and age using the 2000 Centers for Disease Control and Prevention (CDC) BMI-for-age growth charts for the United States. It should be noted that when using the CDC growth charts to evaluate the BMI of children and adolescents, the term "obesity" is intentionally avoided and the word "overweight" is used instead (Kuczmarski et al. 2000, 2002). Figure 14.6 illustrates that from the 1960s to 1980 the prevalence of overweight among U.S. children and adolescents was relatively stable, and that between the 1976–80 survey period and the 1988–94 survey period the prevalence of overweight nearly doubled from approximately 6% to roughly 11%. The prevalence of overweight among children and adolescents in the U.S. is increasing at a faster rate than among U.S. adults (Hill, Catenacci, and Wyatt 2006). One of the national health objectives for the year 2010 is to reduce the proportion of overweight children and adolescents to 5% (USDHHS 2000). However, more recent data show that between the time periods 1988–94 and 1999–2002, the prevalence of overweight increased from about 11% to approximately 16%, representing a 45% increase between the two survey periods. Thus, instead of decreasing or even leveling off, the prevalence of overweight in these two groups is increasing to even higher levels. The data on adolescents are of particular concern in light of the fact that overweight adolescents are at increased risk of becoming overweight adults and

FIGURE 14.5 Prevalence of Healthy Weight among U.S. Adults, 1960 to 2002

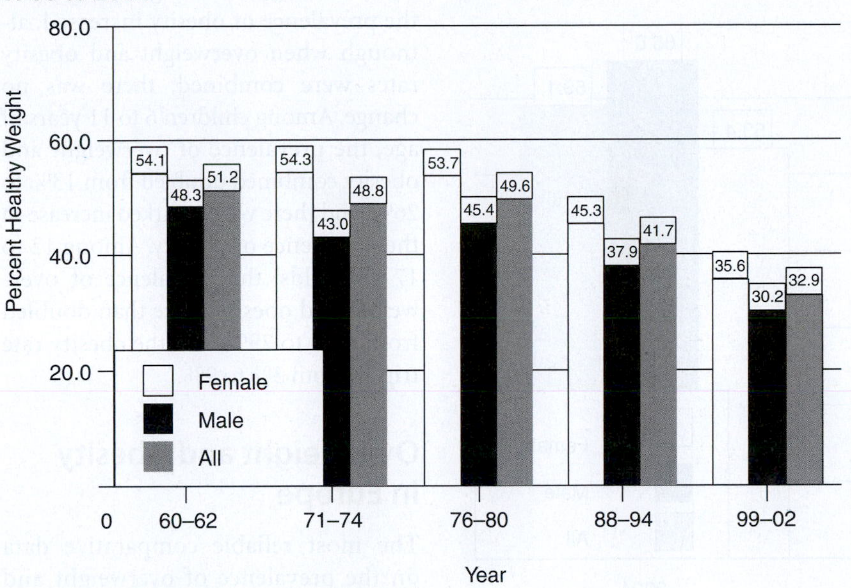

Source: Data from the National Center for Health Statistics.

FIGURE 14.6 Prevalence of Overweight among U.S. Children and Adolescents, 1963 to 2002

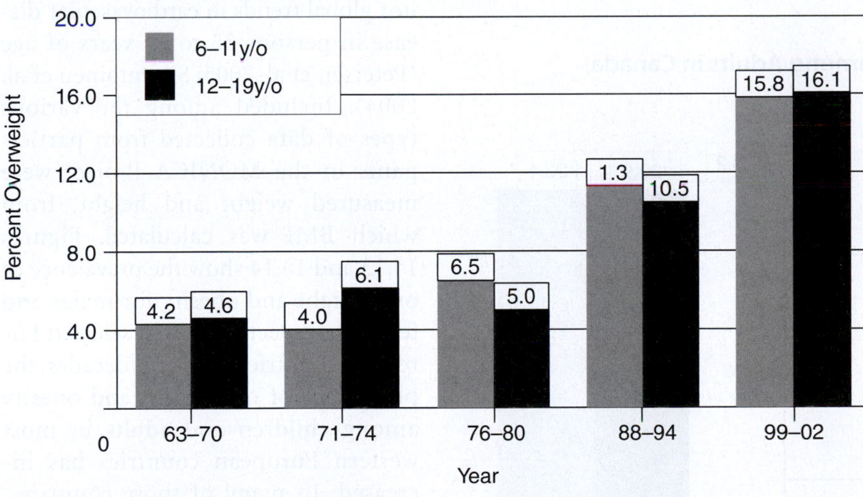

Source: Data from the National Center for Health Statistics.

experiencing the health risks associated with overweight (Hedley et al. 2004).

Overweight and Obesity in Canada

Changes in the prevalence of healthy weight, overweight, and obesity among adult Canadians between 1978–79 and 2004 are shown in Figures 14.7 to 14.10. The data used in these figures come from two different surveys conducted by Statistics Canada: the 1978–79 Canada Health Survey and the 2004 Canadian Community Health Survey, in which the weight and height of subjects were measured by trained health technicians as part of a more comprehensive assessment of nutritional and health status (Tjepkema 2005). In these figures, the definitions of healthy weight, overweight, and obesity are the same as those used in the United States. Between 1978–79 and 2004, the percentage of Canadian adults who were either overweight or obese increased, as shown in Figure 14.7, and most of this increase was due to a marked rise in the prevalence in obesity between the two survey periods, as shown in Figure 14.8. The prevalence of overweight among Canadian adults remained relatively static between the two survey periods, as shown in Figure 14.9. As of 2004, 36% of Canadian adults of both sexes were overweight and 23% were considered obese (Tjepkema 2005). Figure 14.10 shows the decline in the percentage of Canadian adults who had a healthy weight between the two surveys.

When discussing changes occurring in the BMI of children and adolescents in Canada, Statistics Canada takes a somewhat different approach than its American counterpart, the National Center for Health Statistics (NCHS). The terms "overweight" and "obesity" are both used, and these categories are defined using criteria developed by the International Obesity Task Force (IOTF) (Cole et al. 2000). The IOTF definition of overweight and obesity is based on BMI calculated from weight and height measurements obtained from nearly 200,000 children and adolescents ages 2 to 18 years old from Brazil, Great Britain, Hong Kong, the Netherlands, Singapore, and the United States. The IOTF definitions are considered less arbitrary than the NCHS approach and better suited for international comparisons (Cole et al. 2000).

In the 25-year interval between the 1978–79 Canada Health Survey and the 2004 Canadian Community Health Survey, the prevalence of overweight and obesity among Canadian children and adolescents increased by about 70%, and the obesity rate increased 250%, as shown in Figure 14.11 (Shields 2005). There are some notable differences among

FIGURE 14.7 **Prevalence of Canadian Adults Who Are Either Overweight or Obese, 1978–79 to 2004**

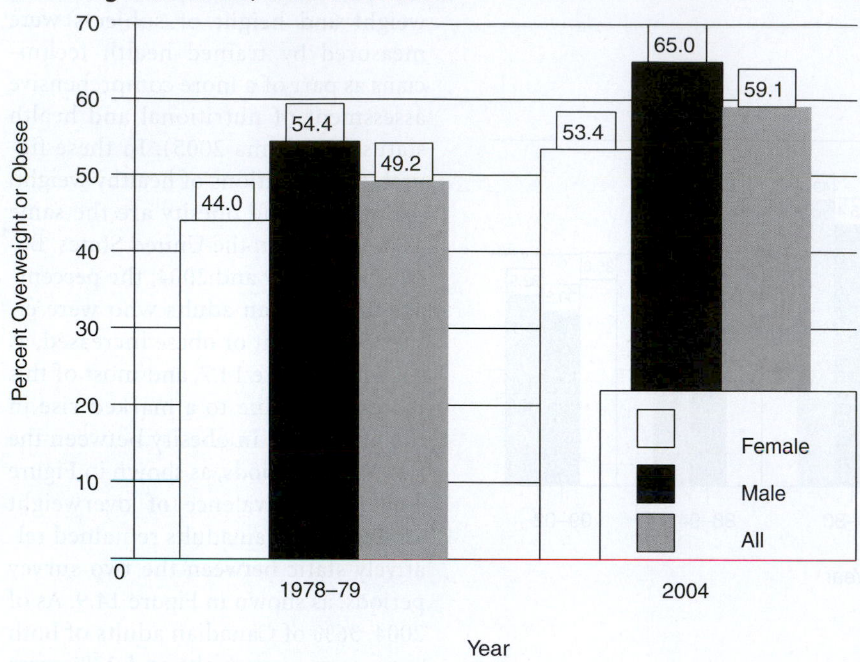

Source: Data from Statistics Canada.

FIGURE 14.8 **Prevalence of Obesity among Adults in Canada, 1978–79 to 2002**

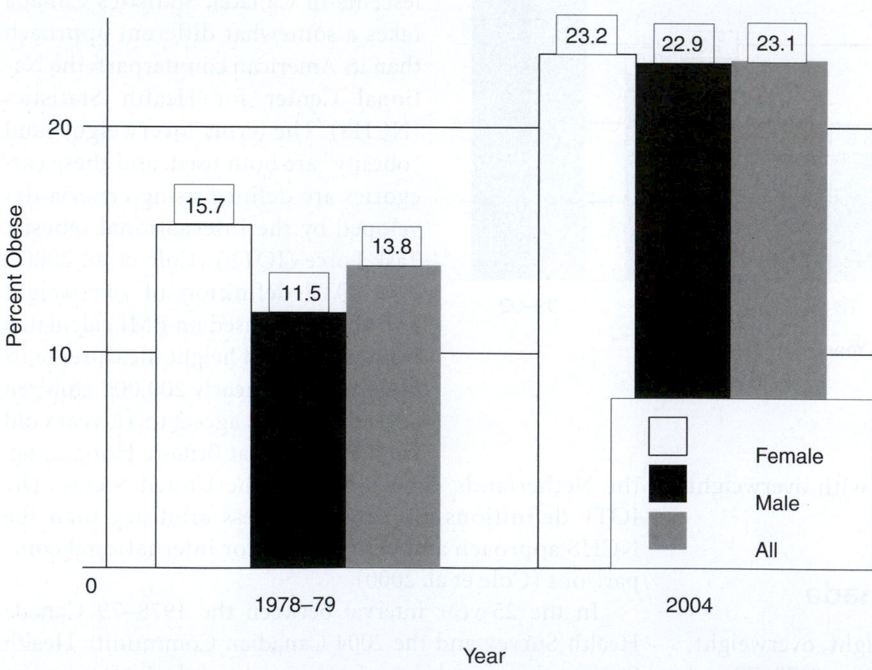

Source: Data from Statistics Canada.

different age groups, as shown in Figure 14.12. Among children ages 2 to 5 years, the prevalence of obesity increased, although when overweight and obesity rates were combined, there was no change. Among children 6 to 11 years of age, the prevalence of overweight and obesity combined doubled from 13% to 26%, and there was a marked increase in the prevalence of obesity. Among 12 to 17 year olds, the prevalence of overweight and obesity more than doubled from 14% to 29%, and the obesity rate tripled from 3% to 9%.

Overweight and Obesity in Europe

The most reliable comparative data on the prevalence of overweight and obesity in Europe come from the World Health Organization's MONICA (Multinational MONItoring of trends and determinants in CArdiovascular disease) Project, an international survey conducted in the 1980s and 1990s to monitor global trends in cardiovascular disease in persons 35 to 64 years of age (Petersen et al. 2005; Silventoinen et al. 2004). Included among the various types of data collected from participants in the MONICA Project were measured weight and height, from which BMI was calculated. Figures 14.13 and 14.14 show the prevalence of overweight and obesity for males and females, respectively, from selected European countries. In recent decades, the prevalence of overweight and obesity among children and adults in most western European countries has increased. In many of those countries, more than half of adults are overweight, and as many as 30% are clinically obese. However, in some central and eastern European countries, such as the Czech Republic, Lithuania, Serbia and Montenegro, and Russia, the average BMI of adults is declining (IOTF 2002, 2003; Silventoinen et al. 2004; Fry and Finley 2005).

Comparing the prevalence of overweight and obesity among children and adolescents in different European coun-

FIGURE 14.9 **Prevalence of Overweight among Adults in Canada, 1978–79 to 2002**

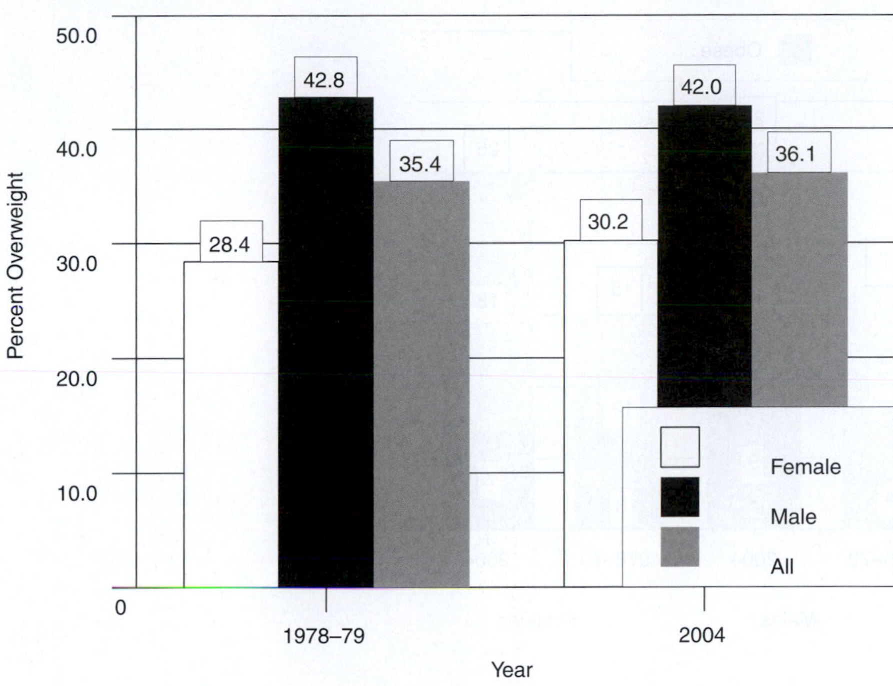

Source: Data from Statistics Canada.

FIGURE 14.10 **Prevalence of Healthy Weight among Canadian Adults, 1978–79 to 2004**

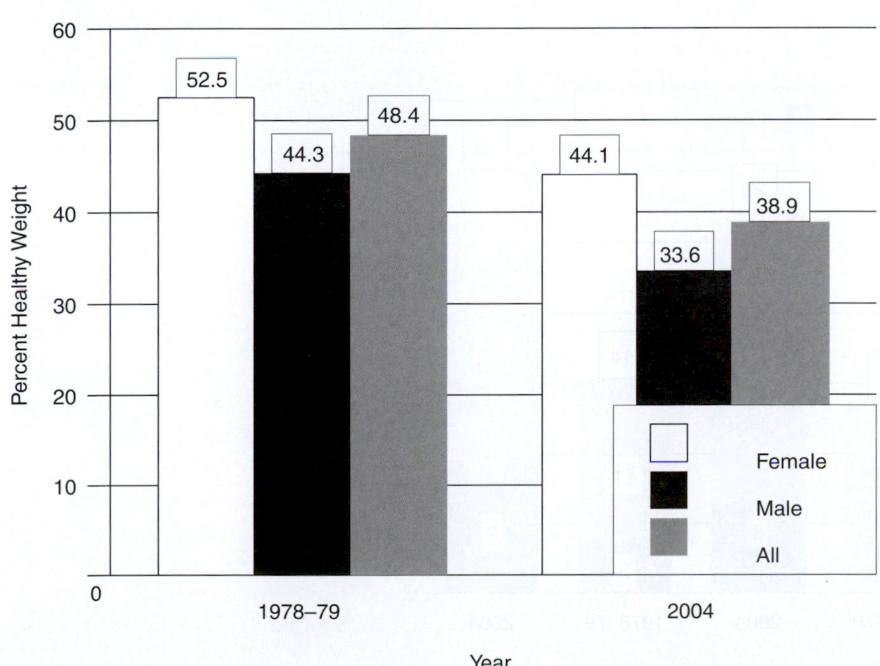

Source: Data from Statistics Canada.

tries is difficult, because the various data sets do not uniformly define overweight and obesity, sometimes rely on self-reported data, and are not representative of the demographic, cultural, and socioeconomic composition of the European population (Livingstone 2000; Lobstein, Baur, and Uauy 2004). Despite these shortcomings, the data indicate that the prevalence of overweight and obesity is increasing throughout most European countries, that the prevalence of obesity in young children is relatively low compared to that of adolescents, and that the highest rates of obesity are observed in eastern and southern European countries, particularly Italy, Greece, and Portugal (Livingstone 2000; Lobstein, Baur, and Uauy 2004).

Effects of Race, Ethnicity, Socioeconomic Status, and Age

According to data from the National Health and Nutrition Examination Survey collected between 1999 and 2002, the prevalence of obesity among adult males in the United States varied little by race or ethnicity, as shown in Figure 14.15. In contrast, there were considerable differences among racial/ethnic groups for adult females. Non-Hispanic black females had the highest prevalence of obesity at 48.8%, non-Hispanic white females had the lowest rate at 30.7%, and Mexican-American females had an intermediate prevalence rate between the two groups. Data from the same survey indicate that higher socioeconomic status is associated with a lower prevalence of obesity.

Between 1999 and 2002, the prevalence of obesity among U.S. adults whose income was below the poverty threshold was 34.7%, while the obesity rate of those whose income was 200% or more above the poverty threshold was 28.7%. As shown in Figure 14.16, the average body weight of U.S. adults increases with age until approximately age 64 years, after which the prevalence

FIGURE 14.11 Prevalence of Overweight and Obesity among Canadian Children and Adolescents Age 2 to 17 years, 1878–79 and 2004

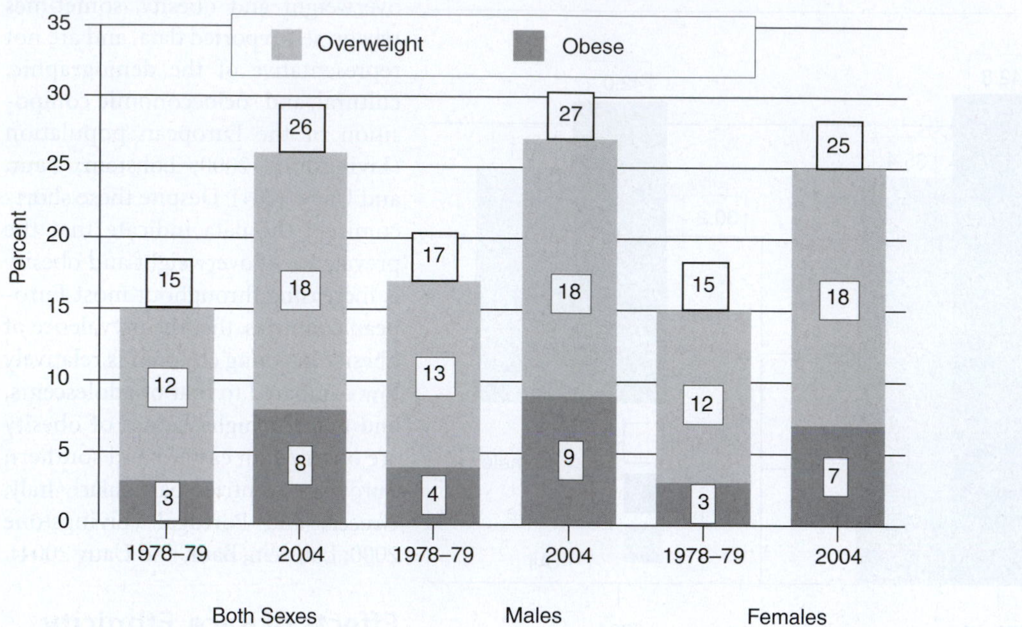

Source: Data from Statistics Canada.

FIGURE 14.12 Prevalence of overweight and obesity among Canadian children and adolescents 2 to 5 years of age, 6 to 11 years of age, and 12 to 17 years of age, 1978–79 to 2004

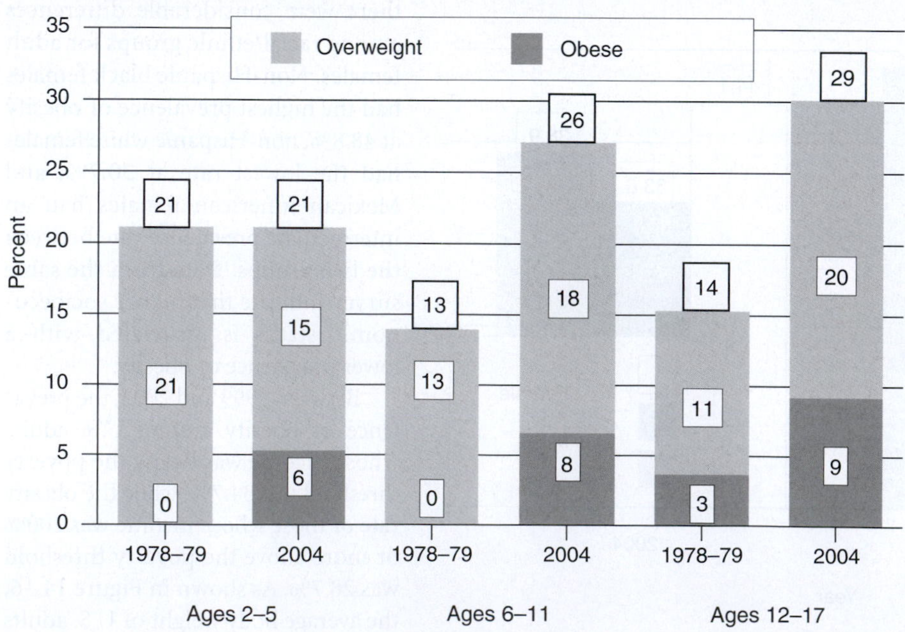

Source: Data from Statistics Canada.

FIGURE 14.13 Prevalence of overweight (BMI 25.0 kg/m² to 29.9 kg/m²) and obesity (BMI ≥30.0 kg/m²) among adult males in select European countries. BMI calculated from measured weight and height

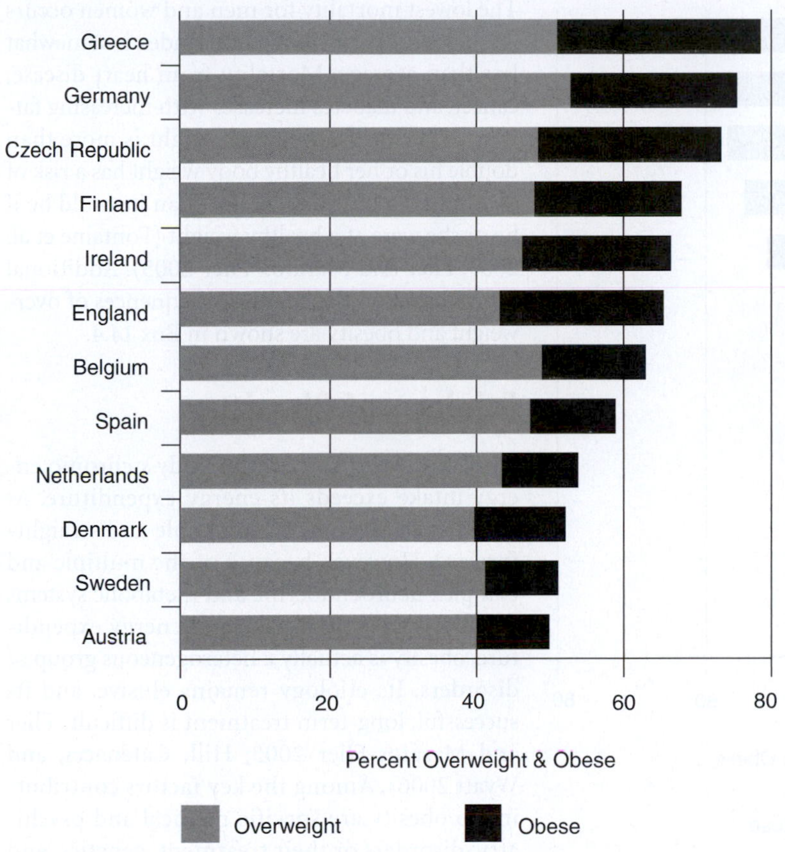

Source: Data from the International Obesity Task Force.

of obesity declines. Some researchers suggest that because obese individuals die at a younger age than those who are not obese, the average BMI of older Americans appears to decline. In addition, many of the chronic conditions commonly seen in the elderly are associated with diminished food intake and weight loss (Hill, Catenacci, and Wyatt 2006).

Adverse Health Consequences of Overweight and Obesity

In North America, the combination of a thin standard of beauty with fat ways of living has resulted in the current era being referred to by some as "the age of caloric anxiety." The media are relentless in promoting the consumption of foods and beverages having a high caloric density while simultaneously advancing an "ideal" body shape that is impossible to attain for practically all females and males. Because of the strong pressures from society to be thin, overweight and obese people often suffer feelings of guilt, depression, anxiety, and low self-worth (Garner et al. 1980; Katzmarzyk and Davis 2001).

Type 2 diabetes is three times as prevalent among the obese as compared with normal-weight persons. Excess body fat, especially when located within the abdominal region, elevates fasting and postprandial levels of plasma free fatty acids. Elevated plasma free fatty acids can stimulate secretion of insulin from the β-cell of the pancreas, cause insulin resistance in peripheral tissues, inhibit cellular uptake of glucose from the blood, reduce glycogen storage, and increase hepatic glucose production, all of which lead to hyperglycemia, hyperinsulinemia, and eventual development of type 2 diabetes (Guven, Kuenzi, and Matfin 2005). Even modest weight loss in persons with type 2 diabetes can result in dramatic improvements in blood glucose control and a reduced need for medications to control blood glucose levels (see Chapter 19) (Mokdad et al. 2003; Manson et al. 2004; Flier and Maratos-Flier 2005). For example, the Diabetes Prevention Program showed that in middle-aged, obese subjects who had impaired glucose tolerance (see Chapter 19), a 7% weight loss, and at least 150 minutes of exercise per week reduced their chance of developing type 2 diabetes by 58% (Diabetes Prevention Program Research Group 2002).

High blood pressure in the obese is three times more common than in normal-weight persons. Even among schoolchildren, increases in obesity are associated with corresponding increases in blood pressure, and weight loss may be an effective treatment for high blood pressure, as it is in adults (Flier and Maratos-Flier 2005). It is thought that the hyperglycemia and hyperinsulinemia associated with obesity increase blood pressure through several mechanisms that are not well understood (Fisher and Williams 2005).

Obese adults are more likely than normal-weight adults to have elevated serum levels of total and low-density lipoprotein (LDL) cholesterol and triglycerides, as well as lower serum levels of high-density lipoprotein (HDL) cholesterol. Elevated serum LDL-cholesterol and low serum HDL-cholesterol are major risk factors for coronary heart disease. Consequently, obesity places individuals at greater risk of coronary heart disease. Obesity results in the overproduction of very-low-density lipoprotein (VLDL) by the liver. Because the body eventually converts VLDL to LDL, increased serum levels of VLDL result in elevations of serum LDL (Grundy 2006). The prevention of the onset of obesity in early life may be important for reducing the risk of coronary heart disease in later life (see Chapter 15) (Wessel et al. 2004).

A number of studies have confirmed that obesity is a significant risk factor for death from cancer generally and from cancer in several specific sites. Obesity in males is associated with increased death from cancer of the esophagus, colon,

FIGURE 14.14 Prevalence of overweight (BMI 25.0 kg/m² to 29.9 kg/m²) and obesity (BMI ≥30.0 kg/m²) among adult females in select European countries. BMI calculated from measured weight and height

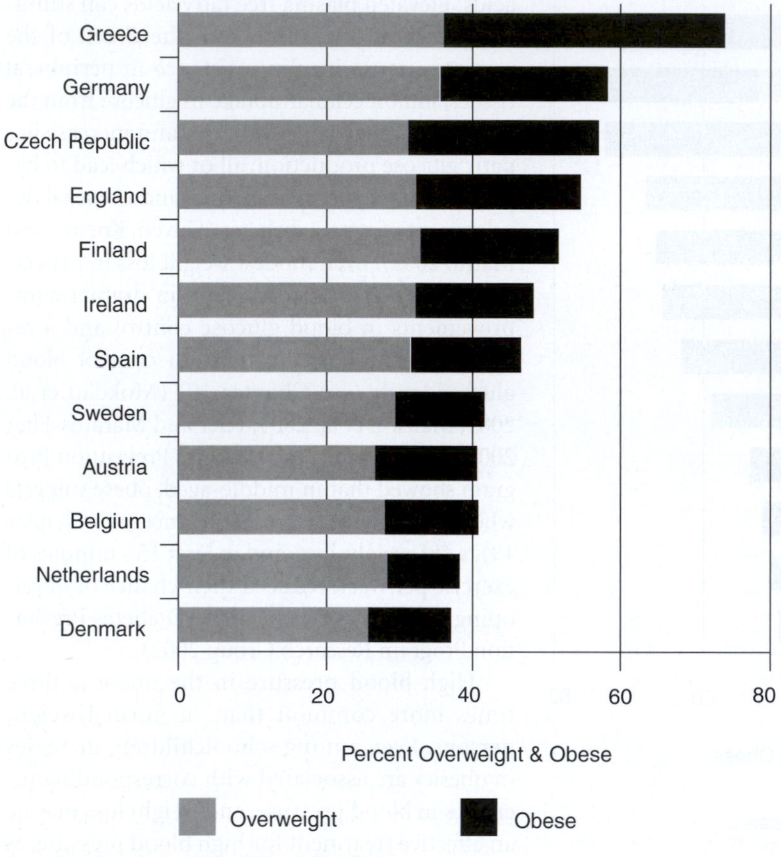

Source: Data from the International Obesity Task Force.

rectum, pancreas, liver, and prostate. In females, obesity increases risk of death from cancer of the gallbladder, bile duct, breast, endometrium, cervix, and ovaries (Flier and Maratos-Flier 2005; Strom et al. 2005). Obesity accounts for 14% and 20% of cancer deaths in U.S. males and females, respectively.(Flier and Maratos-Flier 2005).

Numerous studies by a variety of researchers have consistently shown that, on average, the obese experience earlier death than lean persons. The lowest mortality for men and women occurs among those whose body mass index is somewhat less than average. Mortality from heart disease, cancer, and diabetes increases with increasing fatness. A person whose body weight is more than double his or her healthy body weight has a risk of death that is 12 times greater than it would be if he or she were at a healthy weight (Fontaine et al. 2003; Flier and Maratos-Flier 2005). Additional information on the health consequences of overweight and obesity are shown in Box 14.4.

obesigenic—promoting or encouraging the development of obesity; an obesigenic environment is one that promotes weight gain and the development of obesity by encouraging consumption of energy and discouraging physical activity

iatrogenic—an adverse condition in a patient resulting from treatment, usually by a physician; iatrogenic literally means, "brought forth by a physician"

nonexercise activity thermogenesis (NEAT)—the energy expended through physical activity involved in performing the ordinary activities of daily life; it excludes energy expended in activities to obtain physical exercise or involving sports-like activity

Etiology of Obesity

Obesity develops when the body's chronic energy intake exceeds its energy expenditure. At first glance this may seem simple and straightforward. However, because of the multiple and complex neuroendocrine and metabolic systems influencing energy intake and energy expenditure, obesity is actually a heterogeneous group of disorders. Its etiology remains elusive, and its successful, long-term treatment is difficult (Flier and Maratos-Flier 2005; Hill, Catenacci, and Wyatt 2006). Among the key factors contributing to obesity are specific medical and psychiatric disorders or their treatment, genetics, and an **obesigenic** environment that promotes a high energy intake and discourages physical activity.

Medical Disorders and Medical Treatments

Obesity can result from a specific medical disorder such as Cushing's syndrome, hypothyroidism, or Prader-Willi syndrome, but these are relatively rare. Certain pharmacologic agents (see Table 14.6) are also associated with weight gain (Hill, Catenacci, and Wyatt 2006). When an adverse health condition results from some treatment administered by a physician or other health-care provider, the condition is said to be **iatrogenic** or literally "brought forth by a physician." Weight gain is common when people stop smoking. Compared to the weight gain of males and females who continue to smoke, males who quit smoking gain 9.7 lb (4.4 kg) over 10 years and females who quit smoking gain 11 lb (5.0 kg) over a 10-year period (Flegal et al. 1995).

Two non-normative eating patterns or forms of disordered eating known to contribute to weight gain are night eating syndrome and binge eating (Stunkard and Allison 2003; Tanofsky-Kraff and Yanovski 2004). Night eating syndrome, a common practice among the obese, is defined as consump-

FIGURE 14.15 During 1999–2002, the prevalence of obesity among adult males varied little by racial or ethnic group. Among adult females, non-Hispanic blacks had the highest prevalence of obesity and nonHispanic whites had the lowest. Mexican-American females had a prevalence that was intermediate between the other two groups. Obesity is defined as a BMI ≥30.0 kg/m²

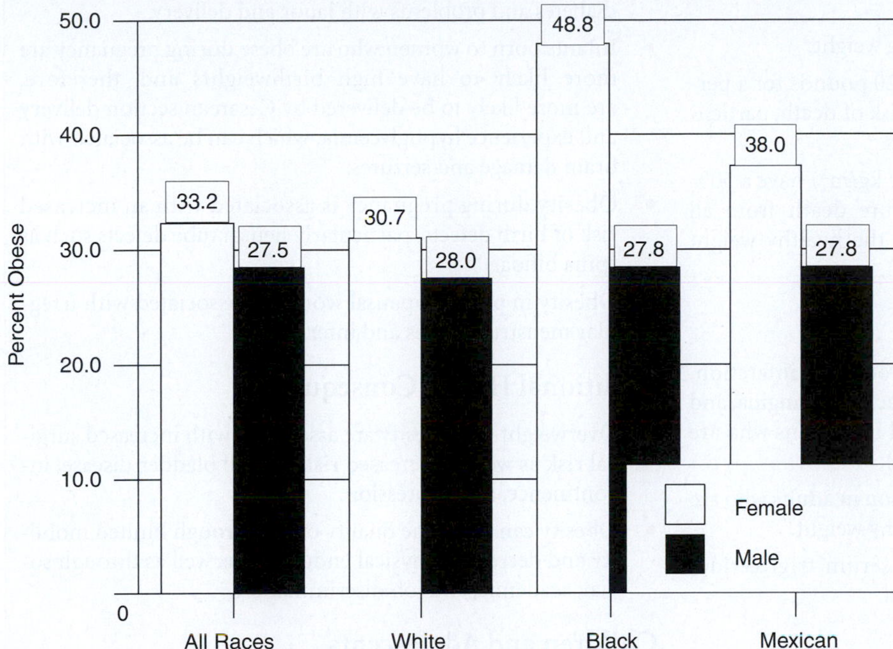

Source: Data from the National Center for Health Statistics.

FIGURE 14.16 As adult Americans age, the prevalence of obesity increases until about age 65 years, at which point the obesity rate declines. Data are from the National Health and Nutrition Examination Survey collected between 1999–2002. Obesity is defined as a BMI ≥30.0 kg/m²

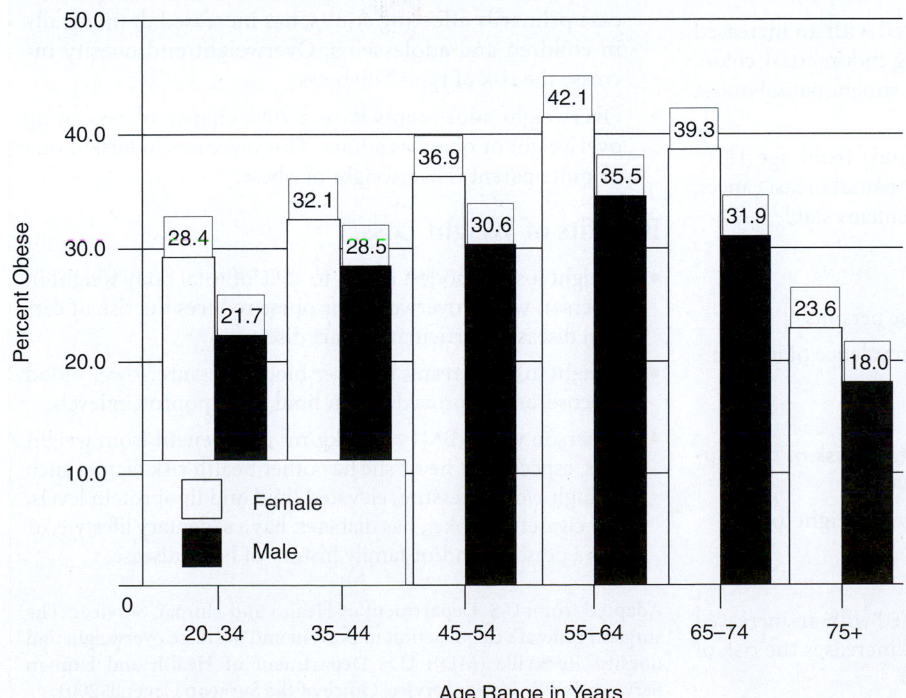

Source: Data from the National Center for Health Statistics.

tion of at least 25% of total energy intake between the evening meal and the next morning. However, some patients with night eating syndrome consume as much as 50% of their total energy intake at night after their evening meal. It is estimated that 10% to 25% of obese persons experience at least occasional episodes of eating large quantities of food in relatively short periods of time, usually in the evening (Stunkard and Allison 2003; Tanofsky-Kraff and Yanovski 2004; Hill, Catenacci, and Wyatt 2006).

Genetics and Body Weight

Genetics affects body weight and body composition by influencing such factors as appetite, taste preferences, energy intake, resting energy expenditure, the thermic effect of food, **nonexercise activity thermogenesis (NEAT)**, and the body's efficiency in storing energy. For example, it has been observed that despite some daily variation in energy intake and energy expenditure, most people maintain their body weight within a fairly narrow range. One explanation for this is the idea that each person's body has a genetically determined metabolic "set-point" that maintains a preferred body weight. While this appears to hold true if the environment remains fairly consistent, significant changes in the past several decades in eating habits and activity levels throughout most of the world have led to a gradual increase in average body weights (Hill, Catenacci, and Wyatt 2006).

Understanding the etiologic role of genetics in obesity is complicated by the fact that obesity is not inherited in families in a predictable manner as are other diseases such as sickle cell anemia, cystic fibrosis, or Huntington's disease. This lack of predictability indicates that multiple genes are involved, with each making a small contribution to body weight and how a person responds to environmental factors like diet, physical activity, and culture (Lyon and Hirschhorn 2005).

BOX 14.4 HEALTH CONSEQUENCES OF OVERWEIGHT AND OBESITY

Premature Death

- An estimated 300,000 deaths per year in the U.S. may be attributable to obesity.
- The risk of death rises with increasing weight.
- Even moderate weight excess (10 to 20 pounds for a person of average height) increases the risk of death, particularly among adults aged 30 to 64 years.
- Individuals who are obese (BMI >30 kg/m^2) have a 50% to 100% increased risk of premature death from all causes, compared to individuals in the healthy weight range (BMI 18.5 kg/m^2 to 24.9 kg/m^2).

Heart Disease

- The incidence of heart disease (myocardial infarction, congestive heart failure, sudden cardiac death, angina, and abnormal heart rhythm) is increased in persons who are overweight or obese (BMI >25 kg/m^2).
- High blood pressure is twice as common in adults who are obese than in those who are at a healthy weight.
- Obesity is associated with elevated serum triglycerides and decreased serum HDL-cholesterol.

Diabetes

- A weight gain of 11 to 18 pounds increases a person's risk of developing type 2 diabetes to twice that of individuals who have not gained weight.
- Over 80% of people with diabetes are overweight or obese.

Cancer

- Overweight and obesity are associated with an increased risk of some types of cancer including endometrial, colon, gall bladder, prostate, kidney, and postmenopausal breast cancer.
- Women gaining more than 20 pounds from age 18 to midlife double their risk of postmenopausal breast cancer, compared to women whose weight remains stable.

Breathing Problems

- Sleep apnea is more common in obese persons.
- Obesity is associated with a higher prevalence of asthma.

Arthritis

- For every 2-pound increase in weight, the risk of developing arthritis is increased by 9% to 13%.
- Symptoms of arthritis can improve with weight loss.

Reproductive Complications

- Obesity during pregnancy is associated with an increased risk of fetal and maternal death and increases the risk of maternal high blood pressure tenfold.
- In addition to many other complications, women who are obese during pregnancy are more likely to have gestational diabetes and problems with labor and delivery.
- Infants born to women who are obese during pregnancy are more likely to have high birthweights and, therefore, are more likely to be delivered by Cesarean section delivery and experience hypoglycemia, which can be associated with brain damage and seizures.
- Obesity during pregnancy is associated with an increased risk of birth defects, particularly neural tube defects such as spina bifida.
- Obesity in premenopausal women is associated with irregular menstrual cycles and infertility.

Additional Health Consequences

- Overweight and obesity are associated with increased surgical risk as well as increased risks of gall bladder disease, incontinence, and depression.
- Obesity can affect the quality of life through limited mobility and decreased physical endurance as well as through social, academic, and job discrimination.

Children and Adolescents

- The most immediate consequence of overweight, as perceived by children themselves, is social discrimination.
- Risk factors for heart disease, such as hyperlipidemia and hypertension, occur more frequently in individuals in the healthy weight range.
- The prevalence of type 2 diabetes, often considered a disease primarily affecting adults, has increased dramatically in children and adolescents. Overweight and obesity increase the risk of type 2 diabetes.
- Overweight adolescents have a 70% chance of becoming overweight or obese as adults. This increases to 80% if one or more parent is overweight or obese.

Benefits of Weight Loss

- Weight loss as modest as 5% to 15% of total body weight in a person who is overweight or obese reduces the risk of certain diseases, particularly heart disease.
- Weight loss can result in lower blood pressure, lower blood glucose, and improved serum lipid and lipoprotein levels.
- A person with a BMI >25.0 kg/m^2 may benefit from weight loss, especially if he or she has other health risk factors such as high blood pressure, elevated lipid and lipoprotein levels, is a cigarette smoker, has diabetes, has a sedentary lifestyle, or has a personal and/or family history of heart disease.

Adapted from: U.S. Department of Health and Human Services. The surgeon general's call to action to prevent and decrease overweight and obesity. Rockville (MD): U.S. Department of Health and Human Services, Public Health Service, Office of the Surgeon General; 2001.

TABLE 14.6

Medical Conditions and Pharmacologic Agents Known to Cause Obesity

Congenital Causes	
• Prader-Willi syndrome	• Alstrom syndrome
• Down syndrome	• Cohen syndrome
• Bardet-Biedel syndrome	• Carpenter syndrome
Neuroendocrine Disorders	
• Cushing syndrome	• Polycystic ovary syndrome
• Hypothalamic disorders	• Growth hormone deficiency
• Hypothyroidism	
Pharmacologic Agents	
Psychiatric Medications	
• Olanzapine, clozapine	• Gabapentin
• Selective serotonin reuptake inhibitors	• Valproate
	• Carbamazepine
• Monoamine oxidase inhibitors	
Steroid Hormones	
• Hormonal contraceptives	• Progestational agents
• Corticosteroids	
Antidiabetic Agents	
• Insulin	• Thiazolidinediones
• Sulfonylureas	
Miscellaneous	
• Antihistamines	• β-adrenergic inhibitors
• α-adrenergic inhibitors	• Protease inhibitors

Source: Hill JO, Catenacci VA, Wyatt HR. Obesity: Etiology. In: Shils ME, Shike M, Ross AC, Cabellero B, Cousins RJ editors. *Modern nutrition in health and disease.* 10th ed. Philadelphia: Lippincott Williams & Wilkins; 2006. 1013–1028.

Furthermore, separating the influence of genetics from the impact of environmental and cultural factors on body weight is difficult. To explore the question of genetics versus environment, investigators have studied individuals within the family unit, pairs of twins, and body weights of adoptees in relation to their biologic and adoptive parents. Having obese family members increases one's risk of obesity, even if the family members do not live together or have similar dietary or physical activity patterns. Studies comparing the body weights of parents and their offspring show that 80% of the offspring of two obese parents eventually become obese, that 40% of offspring of one obese parent eventually become obese, and that when neither parent is obese, the likelihood of obesity in a child is 14% (Mayer 1965).

Studies comparing the BMI and percent body fat/total body fat (as determined by hydrostatic weighing) of identical or monozygotic twins (MZ) and fraternal or dizygotic twins (DZ) have shown that body weights and adiposity of MZ twins tend to be much closer than those of DZ twins. This suggests that genetics plays a role in determining body weight and adiposity. In one study, researchers took 12 pairs of MZ twins who were fed 1,000 kcal/day more than that necessary to maintain their body weight while kept in a sedentary mode of life (Bouchard et al. 1990). This was done for six days a week during a period of 100 days. There was considerable variation in weight gain and change in fat and lean body mass between the 24 individuals. Weight gain ranged from 4 to 13 kg, with mean weight gain being 8.1 kg. However, the variation was not random—there was significant within-pair similarity in weight gain and change in fat and lean body mass in response to the overfeeding. Results of the study suggest that genetics influences the amount of weight gained and the change in fat and lean body mass in response to overfeeding (Stunkard 1991; Sorensen 1995). A study of 540 Danish adoptees gave evidence for a smaller yet still substantial genetic contribution. In this study, the BMI of the adoptees correlated strongly with that of their biologic parents, but not at all with that of their adoptive parents. This finding suggests that in this Danish population, early family environment had apparently little influence in determining the degree of fatness (Stunkard 1991).

The weight of scientific evidence indicates that some people are more prone to obesity than others due to genetic factors, and that 40% to 50% of the variation in BMI is explained by genetic factors (Lyon and Hirschhorn 2005; Hill, Catenacci, and Wyatt 2006). However, environmental factors probably play a greater etiologic role for most people, particularly in light of the fact that famine prevents obesity even in the most obesity-prone individuals. For persons who are genetically predisposed to obesity, it appears that the severity of the disease is largely determined by lifestyle and environmental factors. When the environment changes from one where access to high-energy foods is limited and regular physical activity is required (a "restrictive environment") to one where high-energy foods are easily accessible and the humans are largely sedentary (an "obesigenic environment"), most humans will gain weight. However, those who are genetically predisposed to obesity will gain the most weight while those who are not genetically predisposed to obesity will gain little if any weight (Loos and Rankinin 2005). As important as genetic influences are, persons born with a genetic predisposition to obesity are not necessarily destined to a life of obesity.

Obesigenic Environment

The term "toxic food environment" aptly describes the convenient availability of low-cost, tasty, energy-dense foods, in

BOX 14.5 ENVIRONMENTAL CHANGES OCCURRING IN THE PAST SEVERAL DECADES HAVING AN IMPACT ON THE EATING HABITS OF NORTH AMERICANS

- Growth of the fast food industry—more food is now eaten out of the home and much of this is relatively high in fat.
- Average portion sizes of food consumed in and out of the home have increased.
- Increased availability of foods and beverages, especially those with a high energy density
- Increased numbers of snack and convenience foods.
- Aggressive marketing of foods, particularly to children.
- Decrease in the proportion of disposable income spent on food—average income has increased faster than the increase in the price of food.

TABLE 14.7

Changes in Reported Energy Intakes in the U.S. Adult Population, 1971–74 to 1999–2000

	1971–74	1999–2000	Change
U.S. Adult Females	1,542 kcal	1,877 kcal	+335 kcal (22%)
U.S. Adult Males	2,450 kcal	2,618 kcal	+168 kcal (7%)

Data from the National Center for Health Statistics.

large portion sizes, in North America and the developed world. The toxic food environment, a key component of our obesigenic environment, encourages a high energy intake and has been a major contributing factor in the epidemic of overweight and obesity. This is in sharp contrast to what was the norm throughout most of human history, when considerable energy and time were spent in obtaining food, obesity was rare, and hunger, malnutrition, and starvation were common. In the past, genes favoring the efficient use and storage of energy allowed our ancestors to survive periods of food shortages. Now, these same genes work against maintaining a healthy weight in the present environment where food is plentiful, inexpensive, accessible, and energy-dense (Hill et al. 2003).

Over the past several decades, important changes in the eating habits of North Americans have contributed to the increased prevalence of overweight and obesity. These are outlined in Box 14.5. As shown in Table 14.7, U.S. government surveys attempting to estimate the energy and nutrient intake of Americans suggest that between two surveys conducted during the early 1970s and during the late 1990s, average energy intake of U.S. females and males increased 335 kcal and 168 kcal, respectively. However, it should be noted that between the two surveys, there were changes in the methodology used to quantify dietary intake that could account for some of the differences in reported energy intake between the two surveys.

A growing body of scientific research is demonstrating the value of consuming low-energy-dense foods as an approach to maintaining satiety while controlling energy intake and promoting healthier weights (Ello-Martin, Ledikwe, and Rolls 2005; Rolls, Drewnowski, and Ledikwe 2005). Energy density refers to the energy content of a food relative to its weight (kcal/g). Low-energy-dense foods are relatively low in energy while having a relatively high weight, and include such foods as high-water vegetables and fruits, cooked whole grains, and broth-based soups. In comparison, high-energy-dense foods tend to contain less water and more fat and added sugars. When eating a high-energy-dense diet containing foods high in fat, refined carbohydrates, and added sugars, subjects tend to consume a greater number of kcal than when eating a low-energy-dense diet containing more vegetables, fruits, and broth-based soups (Ello-Martin, Ledikwe, and Rolls 2005; Rolls, Drewnowski, and Ledikwe 2005; Hill, Catenacci, and Wyatt 2006). Food portion size also affects energy intake. When offered larger food portion sizes, subjects tend to consume more energy during mealtimes and while snacking than when eating smaller portion sizes (Ello-Martin, Ledikwe, and Rolls 2005; Rolls, Drewnowski, and Ledikwe 2005). Also, when consuming larger portion sizes, subjects generally do not report an increased or earlier sense of fullness (Ello-Martin, Ledikwe, and Rolls 2005). However, larger portion sizes of low-energy-dense foods have the advantage of maintaining satiety with a lower total energy intake than when high-energy-dense foods are consumed.

A successful strategy for reducing energy intake while maintaining satiety is providing as a first course of a meal satisfying portions of low-energy-dense foods such as vegetable salads or broth-based soups. Greater use of cooked vegetables as side dishes can be an effective way of decreasing the energy density of a meal. An additional strategy to reduce energy density is to prepare the main course of a meal using ingredients that reduce its fat content and increase its water content. Fat can be reduced by using smaller amounts of high-fat ingredients such as meats, dairy products, and oils, or by using leaner cuts of meat and/or reduced-fat dairy products. By using more vegetables in the preparation of a dish such as a pasta salad or casserole, one can increase the water content of that dish while decreasing the energy density (Ello-Martin, Ledikwe, and Rolls 2005; Rolls, Drewnowski, and Ledikwe 2005).

Two barriers to success in promoting a lower-energy-dense diet are cost and convenience. There is an inverse relationship between energy density and cost (Hill, Catenacci, and Wyatt 2006). Refined grains, added sugars, and added fats are among the lowest-cost sources of dietary energy (Drewnowski and Darmon 2005). On a per kcal basis, high-energy-dense foods such as hamburgers and french fries cost considerably less than low-energy dense foods such as fresh fruits and vegetables (Hill, Catenacci, and Wyatt 2006). Heavily processed foods high in added fats and sugars tend to have a longer shelf-life and are generally more convenient to prepare than low-energy-dense foods that require refrigeration, tend to spoil faster, and require time and effort to cook or prepare. Fast-food restaurants are ubiquitous throughout North America and not only offer the consumer convenience, but may be an effective way for families to save money. Is it "elitist" for nutritionists to encourage low- and middle-income families to consume healthier but more costly low-energy-dense foods that they may not be able to afford or have time to prepare? Some experts in the field are examining whether the increasing disparities in income and wealth and the declining value of the minimum wage in North America contribute significantly to an obesigenic diet (Drewnowski and Darmon 2005).

Energy Expenditure

Of the three major components of 24-hour energy expenditure illustrated in Figure 14.1, energy expended through physical activity is the most highly variable and the one humans can most easily control. Physical activity energy expenditure includes movement from the performance of the routine activities of daily life and purposeful exercise as well as energy expended by maintaining posture, fidgeting, and spontaneous muscle contraction. Most studies indicate that obese children and adults are less physically active than their leaner counterparts. However, when obese persons engage in physical activity, they expend more energy than leaner persons performing the same activity. Overall, daily energy expenditure from physical activity by obese persons appears to be no different than that of leaner persons (Hill, Catenacci, and Wyatt 2006). Because obese persons have a greater amount of weight to carry than do lean persons, their lean body mass is greater. Consequently, the obese have a greater resting energy expenditure compared to leaner persons (Hill, Catenacci, and Wyatt 2006).

In the past several decades, a number of environmental changes (outlined in Box 14.6) have impacted the physical activity habits of North Americans by providing inducements to be sedentary and discouraging physical activity. Considerable attention has been paid to the effect of television viewing on body weight in both children and adults. Children in the U.S. spend, on average, as much time watching television in the course of a year as they do attending school (Robinson 2001). Significant associations have been shown between the time spent watching television and obesity in children. In one group of 746 youths aged 10–15 years, those who watched television 5 or more hours per day were 5 times more likely to be overweight than those who watched television 2 hours or less per day (Gortmaker et al. 1996). The association persisted when numerous factors—including being overweight prior to the study, maternal overweight, socioeconomic status, ethnicity, and household structure—were controlled for. Television viewing occupies people for long periods of time in a sedentary activity and exposes them to aggressive marketing of energy-dense foods and beverages, which likely results in increased energy intake. Research supports the idea that reducing time spent watching television may help prevent development of obesity and promote weight loss in young people who are overweight (Robinson 2001). The majority of scientific evidence suggests that the high prevalence of overweight in developed countries is more a function of excessive energy intake than of low activity level. However, increased physical activity is important for the long-term prevention of weight gain and management of healthy body weight.

The choices an individual makes about energy intake and energy expenditure are the most important factors determining his or her body weight. However, an individual's environment is an important factor influencing that person's behavior, either by facilitating or impeding healthy eating and regular physical activity (Hill et al. 2003; Booth, Pinkston, and Poston 2005). The current obesigenic environment in North America and throughout the developed world encourages energy consumption and discourages expenditure of energy, and is widely regarded as a casual factor in the increased prevalence of obesity in North America in the past several decades (Hill et al. 2003; Jeffery

BOX 14.6 ENVIRONMENTAL CHANGES OCCURRING IN THE PAST SEVERAL DECADES HAVING AN IMPACT ON THE PHYSICAL ACTIVITY HABITS OF NORTH AMERICANS

- The variety of electronic media has increased (television, Internet, video games, DVD, wireless communication devices, etc.), which has increased time spent in sedentary activities.
- Physical education programs have been markedly reduced in public schools.
- Many neighborhoods lack sidewalks for safe walking.
- The workplace has become increasingly automated.
- Household chores are assisted by labor-saving machinery.
- Walking or bicycling has been replaced by automobile travel for all but the shortest distances.

and Utter 2003; Booth, Pinkston, and Poston 2005). For example, most areas in the United States and Canada have been designed to be accessed by motor vehicles with little thought, if any, given to the needs of pedestrians or bicyclists. For many people, physical activity is impeded by an environment lacking convenient, pleasant, and safe areas for walking, bicycling, or other forms of recreation. Urban sprawl and lack of public transportation force most to resort to the automobile for commuting to work and school, and for shopping for food and other items. The high density of fast-food restaurants, convenience stores, bars, and vending machines, and the aggressive mass marketing of energy-dense foods, promote consumption of an energy-dense diet. Many areas lack convenient access to supermarkets offering high-quality, low-energy-dense foods such as whole grains, vegetables, and fruits at competitive prices. Compared to wealthier neighborhoods, poorer neighborhoods have one-third as many supermarkets but more convenience stores, fast-food restaurants, and bars (Booth, Pinkston, and Poston 2005).

There is a growing awareness among researchers and public health experts that successfully addressing the problem of overweight and obesity will require identifying feasible ways to cope with and to change the current environment (Hill et al. 2003; Booth, Pinkston, and Poston 2005). A first step in this process would be to give people strategies to better manage within the current environment and to better resist the many factors promoting weight gain. A second, long-term approach would be to build an environment that is more conducive to the adoption and maintenance of healthy dietary and exercise habits (Hill et al. 2003).

Treatment of Overweight and Obesity

The treatment of overweight and obesity is a two-step process: assessment and management (NHLBI 2000). Assessment includes determining the degree of overweight and obesity by calculating BMI, measuring waist circumference, checking for the presence of life-threatening conditions often accompanying obesity, evaluating dietary and exercise habits, and determining the patient's readiness to lose weight. Management includes applying therapies to lose weight and maintain weight loss, and applying measures to control other disease risk factors (NHLBI 1998, 2000). An **algorithm** for the treatment of overweight and obesity developed by the National Institutes of Health is shown in Figure 14.17.

algorithm—a finite set of well-defined instructions for accomplishing a task; given an initial state, an algorithm will terminate in a corresponding recognizable end-point

Assessment

Determining the degree of overweight or obesity is based on BMI calculated from an accurate measurement of the patient's weight and height. Clinical judgment must be used in interpreting the BMI of persons who are very muscular, have lost significant amounts of lean body mass, are short, or who have edema or ascites (NHLBI 2000). Waist circumference is used as an index of abdominal adiposity, and is interpreted using the classifications shown in Table 14.4. In patients with a BMI $\geq$35 kg/m^2, measuring waist circumference is not necessary, because it does not materially contribute to disease risk classification. Table 14.5 incorporates BMI and waist circumference to arrive at a disease risk relative to normal weight and low-risk waist circumference for patients having a BMI <35 kg/m^2. BMI and waist circumference are used to initially assess the degree of obesity and for monitoring the patient's response to treatment (NHLBI 2000; Wadden, Byrne, and Krauthamer-Ewing 2006).

Patients should be evaluated for the presence of diseases that place them at high-risk of morbidity and mortality and that require aggressive treatment. These include established coronary heart disease, the presence of other atherosclerotic diseases (peripheral arterial diseases, abdominal aortic aneurysm, and symptomatic carotid artery disease), type 2 diabetes, impaired glucose tolerance, and sleep apnea (NHLBI 2000). In overweight and obese persons, weight loss lowers elevated blood pressure, elevated blood glucose, elevated serum levels of total cholesterol, LDL-cholesterol, and triglycerides, and raises low levels of HDL-cholesterol. Patients who have three or more of the cardiovascular disease risk factors listed in Box 14.7 are at high risk for obesity-related disorders and will likely require intensive treatment of dyslipidemia and/or management of hypertension, as discussed in Chapter 15. Obese patients should be evaluated for the presence of metabolic syndrome as discussed in Chapter 15.

The dietary and physical activity habits of patients are important considerations. Approaches for evaluating diet are outlined in Chapter 5. Regular physical activity increases energy expenditure, promotes weight reduction, helps maintain weight loss, and reduces risk of hypertension, coronary heart disease, and type 2 diabetes independently of its effect on body weight (NHLBI 2000; Hu et al. 2003; Blair and Church 2004; Weinstein et al. 2004; Flier and Maratos-Flier 2005; Wadden, Byrne, and Krauthamer-Ewing 2006). Assessment of physical activity involves asking patients how much time they spend sleeping, sitting, watching television, driving or riding in motorized vehicles, walking, and standing, and how often they climb stairs, engage in work in and around the home, et cetera (Wadden, Byrne, and Krauthamer-Ewing 2006). Patients should also be queried about their formal exercise habits such as walking, jogging, cycling, swimming, and participating in other types of aerobic activities and weight train-

FIGURE 14.17 An Algorithm for the Treatment of Overweight and Obesity Developed by the National Institutes of Health

Patient Encounter

- Measure weight, height, and waist circumference.
- Calculate body mass index (BMI)

BMI ≥ 25 or high-risk waist circumference*

Yes

Assess Risk Factors (see Box 14.7)

BMI ≥ 30 OR (BMI 25.0 to 29.9 OR high-risk waist circumference*) AND ≥ 2 risk factors

Yes

Clinican and patient devise goals and treatment strategy for weight loss and risk factor control.

No

Does patient want to lose weight?

Yes

No

No

History of BMI ≥ 25?

Yes

No

Brief reinforcement/ educate on weight management

Advise to maintain weight/address other risk factors

Progress being made/goal achieved?

Yes

No

Maintenance counseling:
- Dietary therapy
- Behavior therapy
- Physical activity

Assess reasons for failure to lose weight

Periodic weight, BMI, and waist circumference check

*High-risk waist circumference:
>35 in (88 cm) in females
>40 in (102 cm) in males

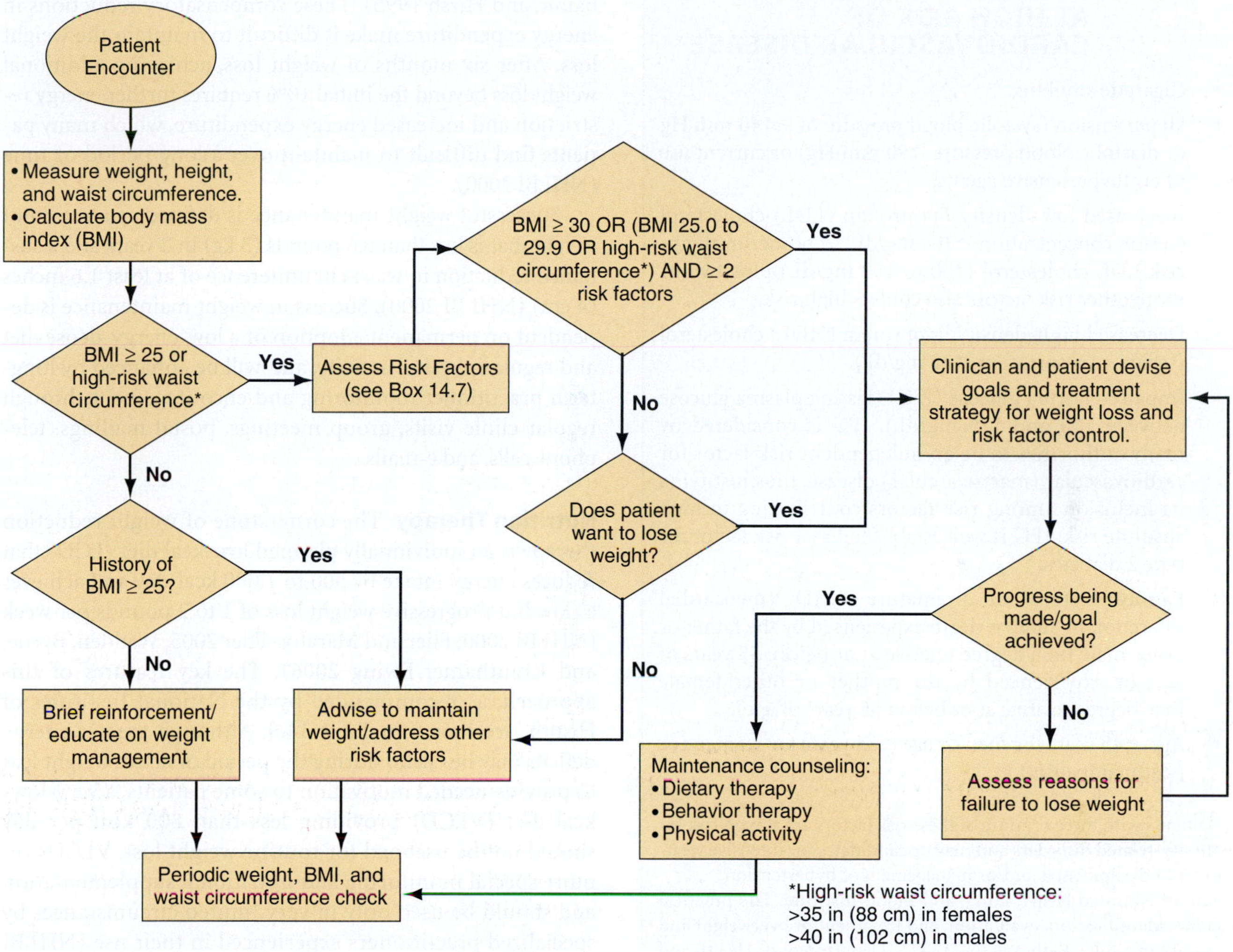

Source: Adapted from National Heart, Lung, and Blood Institute. *The practical guide: identification, evaluation, and treatment of overweight and obesity in adults.* Bethesda (MD): U.S. Department of Health and Human Services, National Institutes of Health; 2000.

ing. Pedometers can be used to count the number of steps walked daily and accelerometers can be used to record the intensity and duration of body movement (Wadden, Byrne, and Krauthamer-Ewing 2006).

Assessing a patient's readiness to lose weight and identifying and addressing potential barriers to that patient's ability to maintain long-term behavior change are important for understanding the patient's needs and achieving successful weight loss (NHLBI 2000; Wadden, Byrne, and Krauthamer-Ewing 2006). Factors associated with successful long-term weight management include a high initial BMI and resting metabolic rate, positive coping skills, and self-efficacy (a patient's belief that he or she can perform the behaviors necessary for weight management). Depression, anxiety, and binge eating tend to be associated with poor success at weight management (NHLBI 2000). A brief behavioral assessment is shown in Box 14.8. Patients who are pregnant, lactating, or have anorexia nervosa, bulimia nervosa, a serious uncontrolled psychiatric illness such as major depression, or active substance abuse should be excluded from weight loss therapy (NHLBI 2000).

Management

Management of overweight and obesity involves the appropriate use of the recommended therapies for initial and long-term successful weight loss, and control of the factors known to increase risk of morbidity and mortality in overweight and obese persons (NHLBI 2000). Recommended therapies for overweight and obesity include diet, physical activity, and behavioral therapy. For some patients, pharmacologic treatment and bariatric surgery are indicated, as shown in Table 14.8.

A minimum goal is to avoid additional weight gain with age once a person reaches his or her healthy, adult weight. Those who are at their normal or healthy weight (BMI 18.5 to 24.9 kg/m^2) should be counseled about effective dietary and

BOX 14.7

CARDIOVASCULAR RISK FACTORS PLACING PATIENTS AT HIGH RISK OF CARDIOVASCULAR DISEASE*

- Cigarette smoking.
- Hypertension (systolic blood pressure of ≥140 mm Hg or diastolic blood pressure ≥90 mm Hg) or current use of antihypertensive agents.
- Inscreased low-density lipoprotein (LDL) cholesterol (serum concentration ≥160 mg/dL). A borderline high-risk LDL-cholesterol (130 to 159 mg/dL) plus two or more other risk factors also confers high risk.
- Decreased high-density lipoprotein (HDL) cholesterol (serum concentration <35 mg/dL).
- Impaired fasting glucose (IFG) (fasting plasma glucose between 110 and 125 mg/dL). IFG is considered by many authorities to be an independent risk factor for cardiovascular (macrovascular) disease, thus justifying its inclusion among risk factors contributing to high absolute risk. IFG is well established as a risk factor for type 2 diabetes.
- Family history of premature CHD (myocardial infarction or sudden death experienced by the father or other male first-degree relative at or before 55 years of age, or experienced by the mother or other female first-degree relative at or before 65 years of age).
- Age ≥45 years for men or age ≥55 years for women (or postmenopausal).

*Patients with three or more of these risk factors are at high risk for obesity-related disorders and may be candidates for intensive treatment of dyslipidemia and/or management of hypertension.

Source: National Heart, Lung, and Blood Institute. The practical guide: identification, evaluation, and treatment of overweight and obesity in adults. Bethesda (MD): U.S. Department of Health and Human Services, National Institutes of Health; 2000.

physical activity habits that can prevent further weight gain. Those who are overweight or obese should set as their goal an initial weight loss of about 10% of their body weight over a 6-month period at a rate of about 1 to 2 pounds lost per week (NHLBI 2000; Wadden, Byrne, and Krauthamer-Ewing 2006). The recommended approach is for patients to reduce their energy intake by 500 to 1,000 kcal/day. Theoretically, this should result in a 26- to 52-pound weight loss after 6 months, but a more typical loss is between 20 and 25 pounds after six months. Continued weight loss after six months is difficult for most patients, in large part because 24-hour energy expenditure declines in response to restricted energy intake and in response to the losses of metabolically active lean body mass that invariably accompany weight loss (NHLBI 2000). Resting energy expenditure begins to decline within days of restricting energy intake, and by 3 to 4 weeks will fall by as much as 25% to 35% below normal in response to total fasting (Hoffer 2006). With loss of body weight, there is loss of both fat and fat-free tissue. A 10% decrease in body weight results in a 15% reduction in 24-hour energy expenditure (Leibel, Rosenbaum, and Hirsh 1995). These compensatory reductions in energy expenditure make it difficult to maintain the weight loss. After six months of weight loss, achieving additional weight loss beyond the initial 10% requires further energy restriction and increased energy expenditure, which many patients find difficult to maintain over a long period of time (NHLBI 2000).

Successful weight maintenance is defined as a regain of weight that is less than 6.6 pounds (3 kg) in 2 years and a sustained reduction in waist circumference of at least 1.6 inches (4 cm) (NHLBI 2000). Success in weight maintenance is dependent on permanent adoption of a low-energy-dense diet and regular physical activity, and will be enhanced by long-term practitioner monitoring and encouragement through regular clinic visits, group meetings, postal mailings, telephone calls, and e-mails.

Nutrition Therapy The cornerstone of weight reduction therapy is an individually planned low-kcal diet (LCD) that reduces energy intake by 500 to 1,000 kcal/day and achieves a slow but progressive weight loss of 1 to 2 pounds per week (NHLBI 2000; Flier and Maratos-Flier 2005; Wadden, Byrne, and Krauthamer-Ewing 2006). The key features of this approach, as recommended by the National Institutes of Health, are shown in Table 14.9. Although greater energy deficits may be useful during the period of active weight loss to provide needed motivation to some patients, a very-low-kcal diet (VLCD) providing less than 800 kcal per day should not be useboxd for routine weight loss. VLCDs require special monitoring and nutritional supplementation and should be used only in very limited circumstances by specialized practitioners experienced in their use (NHLBI 2000). Clinical trials indicate that VLCDs are no more effective in achieving weight loss after 1 year than are LCDs (NHLBI 2000; Wadden, Byrne, and Krauthamer-Ewing 2006). In addition to reducing energy intake, the diet should be modified to minimize CVD risk factors by following the National Cholesterol Education Program's Therapeutic Lifestyle Change diet, which is discussed in Chapter 15 (NCEP 2002). A meal plan providing 1,000 to 1,200 kcal/day is generally recommended for most women. A meal plan providing 1,200 to 1,600 kcal/day is generally recommended for most men and may be suited for women who exercise more or weigh 165 lb or more. A greater reduction in energy intake may be necessary for patients failing to respond to these energy levels, whereas patients complaining of hunger or having difficulty adhering to these recommendations may need a somewhat more liberal intake (NHLBI 2000).

Among the most controversial issues related to dietary therapy for body weight management is whether altering the proportion of energy provided by macronutrients impacts weight loss and weight management. In essence, is a low-carbohydrate diet superior to a more balanced

BOX 14.8 A BRIEF BEHAVIORAL ASSESSMENT

Clinical experience suggests that health care practitioners briefly consider the following questions when assessing an obese individual's readiness for weight loss.

"Has the individual sought weight loss on his or her own initiative?" Weight loss efforts are unlikely to be successful if patients feel that they have been forced into treatment by family members, their employer, or their physician. Before initiating treatment, health care practitioners should determine whether patients recognize the need for and benefits of weight reduction and want to lose weight.

"What events have led the patient to seek weight loss now?" Responses to this question will provide information about the patient's weight loss motivation and goals. In most cases, individuals have been obese for many years. Something has happened to make them seek weight loss. The motivator differs from person to person.

"What are the patient's stress level and mood?" There may not be a perfect time to lose weight, but some times are better than others. Individuals who report higher than usual stress levels with work, family life, or financial problems may not be able to focus on weight control. In such cases, treatment may be delayed until the stressor passes, thus increasing the chances of success. Briefly assess the patient's mood to rule out major depression or other complications. Reports of poor sleep, a low mood, or lack of pleasure in daily activities can be followed up to determine whether intervention is needed; it is usually best to treat the mood disorder before undertaking weight reduction.

"Does the individual have an eating disorder, in addition to obesity?" Approximately 20% to 30% of obese individuals who seek weight reduction at university clinics suffer from binge eating. This involves eating an unusually large amount of food and experiencing loss of control while overeating. Binge eaters are distressed by their overeating, which differentiates them from persons who report that they "just enjoy eating and eat too much." Ask patients which meals they typically eat and the times of consumption. Binge eaters usually do not have a regular meal plan; instead, they snack throughout the day. Although some of these individuals respond well to weight reduction therapy, the greater the patient's distress or depression, or the more chaotic the eating pattern, the more likely the need for psychological or nutritional counseling.

"Does the individual understand the requirements of treatment and believe that he or she can fulfill them?" Practitioner and patient together should select a course of treatment and identify the changes in eating and activity habits that the patient wishes to make. It is important to select activities that patients believe they can perform successfully. Patients should feel that they have the time, desire, and skills to adhere to a program that you have planned together.

"How much weight does the patient expect to lose? What other benefits does he or she anticipate?" Obese individuals typically want to lose 2 to 3 times the 8% to 15% often observed and are disappointed when they do not. Practitioners must help patients understand that modest weight losses frequently improve health complications of obesity. Progress should then be evaluated by achievement of these goals, which may include sleeping better, having more energy, reducing pain, and pursuing new hobbies or rediscovering old ones, particularly when weight loss slows and eventually stops.

Source: National Heart, Lung, and Blood Institute. The practical guide: identification, evaluation, and treatment of overweight and obesity in adults. Bethesda (MD): U.S. Department of Health and Human Services, National Institutes of Health; 2000.

hypocaloric diet such as the one shown in Table 14.9? Although the number of clinical trials addressing this question is small, it appears that low-carbohydrate diets result in greater short-term weight loss (during the first 6 months) than diets providing carbohydrate in the range of 50% to 60% of kcal (Hu et al. 2003; Eckel 2005; Noakes et al. 2005; Wadden, Byrne, and Krauthamer-Ewing 2006). A one-year randomized trial of four popular weight loss programs having widely different macronutrient compositions showed that each of the four diets modestly reduced body weight, waist circumference, and several CVD risk factors (Dansinger et al. 2005). Regardless of the diet followed, about 25% of the subjects sustained a one-year weight loss of more than 5% of their initial body weight, and about 10% of the subjects lost more than 10% of their body weight. In this study, the key determinant of successful weight loss was adherence to the diet, not which one of the four diets was followed (Dansinger et al. 2005).

A growing body of scientific evidence suggests that long-term improvements in body weight, waist circumference, and CVD risk factors appear to be determined more by the total number of kcal consumed and expended than by the proportion of macronutrients in the diet (Dansinger 2005; Eckel 2005; Melanson and Dwyer 2005; Wadden, Byrne, and Krauthamer-Ewing 2006). However, kcal-restricted diets with a modest increase in the proportion of kcal from monounsaturated fats and protein from plant products, poultry, and fish appear to increase satiety, facilitate weight loss, and improve CVD risk factors in some individuals (NCEP 2002; Hu et al. 2003; Eckel 2005; Wadden, Byrne, and Krauthamer-Ewing 2006). Any reduction in the proportion of kcal coming from carbohydrates should be accomplished by reducing intake of foods and beverages containing refined sugars and milled grains, not by sacrificing consumption of whole grains, legumes (dried beans and peas), vegetables, and fruits, which have a low energy density and are

TABLE 14.8

A Guide to Selecting Treatment of Overweight and Obesity

Treatment	BMI Category (kg/m²)				
	25.0–26.7	27.0–29.9	30.0–34.9	35.0–39.9	≥ 40.0
Diet, physical activity, and behavioral therapy	with comorbidites	with comorbidites	+	+	+
Pharmacotherapy		with comorbidites		+	+
Surgery				with comorbidites	

Prevention of weight gain with lifestyle therapy is indicated in any patient with a BMI ≥ 25 kg/m², even without comorbidities, while weight loss is not necessarily recommended for those with a BMI of 25–29.9 kg/m² or a high waist circumference, unless they have two or more comorbidities.

Combined therapy with a low-kcalorie diet (LCD), increased physical activity, and behavior therapy provide the most successful intervention for weight loss and weight maintenance.

Consider pharmacotherapy only if a patient has not lost 1 pound per week after 6 months of combined lifestyle therapy. The + represents the use of indicated treatment regardless of comorbidities.

Source: National Heart, Lung, and Blood Institute. *The practical guide: identification, evaluation, and treatment of overweight and obesity in adults.* Bethesda (MD): U.S. Department of Health and Human Services, National Institutes of Health; 2000.

associated with reduced CVD risk (Schaefer, Gleason, and Dansinger 2005). Adherence to a diet is improved when the patient's food preferences and lifestyle are carefully assessed and modified incrementally. The practitioner and patient must collaboratively establish goals for modifying dietary and physical activity patterns, and the patient must see these modifications as desirable and achievable. In helping the patient be a better informed consumer, particular attention should be given to the topics listed in Box 14.9 (NHLBI 2000).

Physical Activity Although physical activity is less important than an energy-restricted diet in promoting initial weight loss, it is nevertheless considered an important component of weight loss therapy. Moreover, it appears to be crucial for maintaining weight loss (NHLBI 2000; Flier and Maratos-Flier 2005; Wadden, Byrne, and Krauthamer-Ewing 2006). Physical activity has the added benefit of minimizing loss of lean body mass, reducing LDL-cholesterol levels, increasing HDL-cholesterol levels, improving insulin sensitivity, and improving fitness (Wadden, Byrne, and Krauthamer-Ewing 2006). A study of 22,000 men showed that fitness level was a stronger predictor of cardiovascular disease and all-cause mortality than was fatness. Fat but fit men had a significantly lower risk of health complications than did lean men who were unfit (Lee, Blair, and Jackson 1999).

A minimum initial goal for physical activity is 30 to 45 minutes of moderate activity, 3 to 5 days per week (NHLBI 2000). For the sedentary and obese, physical activity should be initiated slowly and then gradually increased in duration and intensity. Physical activity can involve either programmed or lifestyle activities. Programmed or formal activities include regularly scheduled periods of swimming, running, jumping rope, or other aerobic activities performed at a relatively high intensity for a short period of time (30 to 60 minutes). Lifestyle activity involves moving the body more throughout the day in the discharge of the activities of daily life. Examples include walking or bicycling instead of riding in a motor vehicle, climbing stairs instead of using an elevator or escalator, decreasing time spent in sedentary behaviors such as watching television, and increasing time spent performing common chores such as house cleaning and yard work (NHLBI 2000).

Behavior Therapy Behavior therapy provides patients with a set of techniques (self-monitoring, stimulus control, rewards, etc.) to identify and overcome barriers to positive dietary, exercise, and other lifestyle habits (Berkel et al. 2005). The practitioner collaborates with the patient to establish specific, achievable, and measurable goals related to food intake, physical activity, and weight loss. Patients are taught to observe and record their food intake, physical activity, and body weight. Self-monitoring of behavior generally changes behavior in the desired direction and is associated with long-term weight loss (NHLBI 2000; Wadden, Byrne, and Krauthamer-Ewing 2006). Self-monitoring also helps the patient identify social or environmental stimuli that lead to undesirable behaviors or that block the adoption of desirable behaviors. Once these stimuli are identified, steps can be taken to prevent them from occurring or to change one's reaction to them. This is referred to as stimulus control. Rewards are used to encourage attainment of the established goals. Behavioral therapy is a valuable adjunct to diet and physical activity, resulting in marked improvements in weight loss and weight maintenance (Wadden, Byrne, and Krauthamer-Ewing 2006).

Pharmacologic Treatment Drug therapy can be useful as an adjunct to diet, physical activity, and behavior therapy in

TABLE 14.9

Low-Calorie Diet (LCD) Recommended by the National Institutes of Health

Nutrient	Recommended Intake
Calories[1]	Approximately 500 to 1,000 kcal/day reduction from usual intake
Total fat[2]	30% or less of total calories
Saturated fatty acids[3]	8%–10% of total calories
Monounsaturated fatty acids	Up to 15% of total calories
Polyunsaturated fatty acids	Up to 10% of total calories
Cholesterol[3]	<300 mg/day
Protein[4]	Approximately 15% of total calories
Carbohydrate[5]	55% or more of total calories
Sodium chloride	No more than 100 mmol/day (approximately 2.4 g of sodium or approximately 6 g of sodium chloride)
Calcium[6]	1,000 to 1,500 mg/day
Fiber[5]	20 to 30 g/day

[1]A reduction in calories of 500 to 1,000 kcal/day will help achieve a weight loss of 1 to 2 pounds/week. Alcohol provides unneeded calories and displaces more nutritious foods. Alcohol consumption not only increases the number of calories in a diet but has been associated with obesity in epidemiologic studies as well as in experimental studies. The impact of alcohol calories on a person's overall caloric intake needs to be assessed and appropriately controlled.

[2]Fat-modified foods may provide a helpful strategy for lowering total fat intake but will only be effective if they are also low in calories and if there is no compensation by calories from other foods.

[3]Patients with high blood cholesterol levels may need to use the National Cholesterol Education Program's Therapeutic Lifestyle Changes (TLC) diet to achieve further reductions in LDL-cholesterol levels; in the TLC diet, saturated fats are reduced to less than 7% of total calories, and cholesterol levels to less than 200 mg/day.

[4]Protein should be derived from plant sources and lean sources of animal protein.

[5]Complex carbohydrates from different vegetables, fruits, and whole grains are good sources of vitamins, minerals, and fiber. A diet rich in soluble fiber, including oat bran, legumes, barley, and most fruits and vegetables, may be effective in reducing blood cholesterol levels. A diet high in all types of fiber may also aid in weight management by promoting satiety at lower levels of calorie and fat intake. Some authorities recommend 20 to 30 grams of fiber daily, with an upper limit of 35 grams.

[6]During weight loss, attention should be given to maintaining an adequate intake of vitamins and minerals. Maintenance of the recommended calcium intake of 1,000 to 1,500 mg/day is especially important for women who may be at risk of osteoporosis.

Source: Adapted from National Heart, Lung, and Blood Institute. *The practical guide: identification, evaluation, and treatment of overweight and obesity in adults.* Bethesda (MD): U.S. Department of Health and Human Services, National Institutes of Health; 2000.

patients whose BMI is ≥30 kg/m^2 or in patients whose BMI is ≥27 kg/m^2 and who have obesity-related risk factors or diseases (NHLBI 2000). The modest benefits of drug treatment are offset by its cost and side effects, and rebound weight gain following cessation of drug use (Flier and Maratos-Flier 2005). The U.S. Food and Drug Administration has approved two drugs for long-term use for weight loss and the maintenance of weight loss: sibutramine and orlistat. Several have been approved for short-term treatment (6–12 weeks), including mazindol, diethylpropion, benzphetamine, phendimetrazine, and phentermine. Phentermine, the drug most commonly used for short-term treatment, is an amphetamine-like drug with a low addictive potential that acts on the hypothalamus to suppress appetite (Flier and Maratos-Flier 2005).

Sibutramine, marketed under the trade name Meridia, is a serotonin-norepinephrine reuptake inhibitor that acts on receptors in the hypothalamus to suppress appetite. Because it increases heart rate (4–5 beats/minute) and blood pressure (1–2 mm Hg), it should not be used by patients with CVD or uncontrolled hypertension (Wadden, Byrne, and Krauthamer-Ewing 2006). It can result in a 7% reduction in body weight after one year, but when used in conjunction with an intensive program of diet, exercise, and behavior therapy, losses can increase to 10% to 15% of body weight (Wadden, Byrne, and Krauthamer-Ewing 2006). Orlistat, marketed under the trade name Xenical, inhibits the action of gastrointestinal lipase and reduces the digestion of triglyceride by about 30%. Undigested triglyceride is not absorbed, does not provide energy to the body, and is eliminated in the feces resulting in the loss of 150 to 180 kcal/day. It is intended to be taken three times a day with low-fat

BOX 14.9 FOCUS OF EFFORTS IN EDUCATING THE OVERWEIGHT AND OBESE PATIENT

- Energy value of different foods.
- Food composition—fats, carbohydrates (including dietary fiber), and proteins.
- Evaluation of nutrition labels to determine caloric content and food composition.
- New habits of purchasing—give preference to low-kcalorie foods.
- Food preparation—avoid adding high-kcalorie ingredients during cooking (e.g., fats and oils).
- Avoiding overconsumption of high-kcalorie foods (both high-fat and high-carbohydrate foods).
- Adequate water intake
- Reducing portion sizes.
- Limiting alcohol consumption.

Source: National Heart, Lung, and Blood Institute. The practical guide: identification, evaluation, and treatment of overweight and obesity in adults. Bethesda (MD): U.S. Department of Health and Human Services, National Institutes of Health; 2000.

FIGURE 14.18 Surgical Procedures Used in the Treatment of Severe Obesity

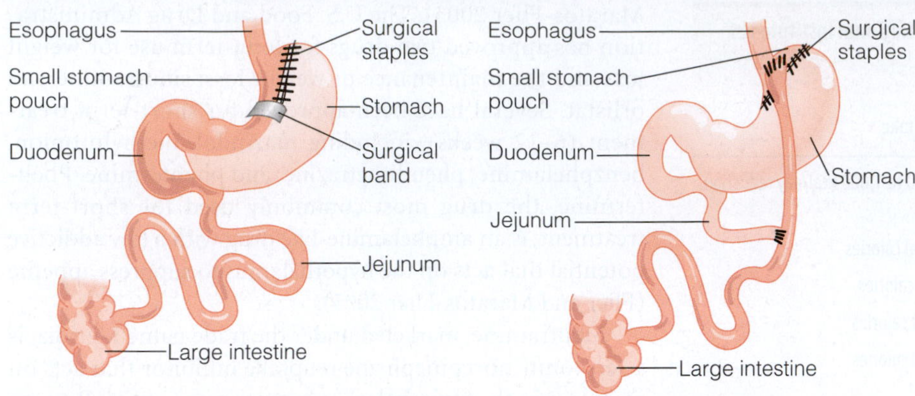

In vertical banded gastroplasty, the surgeon constructs a small stomach pouch and restricts the outlet from the stomach to the intestine.

In gastric bypass, the surgeon constructs a small stomach pouch and creates an outlet directly to the jejunum.

Source: S. Rolfes, K. Pinna, and E. Whitney, *Understanding Normal and Clinical Nutrition*, 7e, copyright © 2006, p. 291.

meals. If taken with meals providing more than 20 grams of fat, there is an increased risk of gastrointestinal side effects such as flatus with discharge, oily stools, and fecal urgency. These adverse events serve as an incentive to consume a low-fat diet, thus further decreasing energy intake. Because Orlistat inhibits the absorption of fat-soluble vitamins, patients are advised to take a multivitamin supplement to offset losses of vitamins A, D, E, and K (Flier and Maratos-Flier 2005; Wadden, Byrne, and Krauthamer-Ewing 2006;). Clinical trials of diet plus Orlistat resulted in a 10% weight loss after 1 year compared to a 6% loss from diet alone. A 4-year trial resulted in a 6.4% weight loss with a significant reduction in the development of type 2 diabetes in persons who had impaired glucose tolerance (Wadden, Byrne, and Krauthamer-Ewing 2006).

Surgery Weight loss or bariatric surgery is reserved for patients who have failed to lose weight by other methods and who have clinically severe obesity: BMI ≥40 kg/m^2 or BMI ≥35 kg/m^2 with obesity-related risk factors or diseases (NHLBI 2000; Wadden, Byrne, and Krauthamer-Ewing 2006). Commonly used bariatric surgical procedures are the Roux-en-Y gastric bypass, the vertical banded gasstroplasty, and the adjustable gastric band (Chapman et al. 2004; Flier 2005; Korenkov, Sauerland, and Junginger 2005; Wadden, Byrne, and Krauthamer-Ewing 2006).

The procedure most commonly used in the U.S. is the Roux-en-Y gastric bypass (see Figure 14.18). The operation creates a small (20–30 mL) pouch at the base of the esophagus that limits food intake by quickly inducing satiety. The remainder of the stomach, the duodenum, and part of the jejunum are bypassed by connecting (or anastomosing) the jejunum to the pouch, decreasing macronutrient absorption. The procedure, which can be performed laparoscopically through small incisions in the abdominal wall, results in a loss of about 30% of initial weight in 1 to 2 years postoperatively and good maintenance of weight loss a decade or more later (Wadden, Byrne, and Krauthamer-Ewing 2006). Other benefits of the surgery include marked improvements in diabetes, sleep apnea, hypertension, and CVD risk factors (Flier and Maratos-Flier 2005; Wadden, Byrne, and Krauthamer-Ewing 2006). Postoperative complications include pulmonary embolism, anastomotic leaks, injury to the spleen, and wound infection (NHLBI 2000; Wadden, Byrne, and Krauthamer-Ewing 2006). Several months following the procedure, patients may experience nausea and vomiting when too much food is consumed at one time and dumping syndrome characterized by nausea, flushing, bloating, and diarrhea after eating refined carbohydrates. Patients should be advised to eat sufficient protein and monitor the status of vitamin B_{12}, iron, calcium, and magnesium (Wadden, Byrne, and Krauthamer-Ewing 2006).

The mortality rate among younger gastric bypass surgery patients without comorbidities and a BMI <50 kg/m^2 is generally less than 1%. The mortality rate among patients with comorbidities and a BMI >60 kg/m^2 ranges between 2% and 4% (NHLBI 2000). Careful preoperative screening and education ensures that candidates for the surgery meet the BMI requirements, are free of major psychopathology, have the physical and emotional stamina to tolerate the procedure, are willing to make the necessary dietary and lifestyle changes, and are committed to long-term follow-up (Flier and Maratos-Flier 2005; Wadden, Byrne, and Krauthamer-Ewing 2006).

In the laparoscopic vertical banded gastroplasty (see Figure 14.18), a small vertical pouch is surgically created at the top of the stomach by placing a line of staples through both walls of the stomach. The bottom of the pouch is secured by a plastic band that controls the volume of the pouch and prevents it from stretching (Flier and Maratos-Flier 2005;

Korenkov, Sauerland, and Junginger 2005). The procedure works by restricting the amount of food a patient can eat at one time. In contrast to gastric bypass, this operation does not interfere with absorption, and digestion proceeds normally. Weight loss can be substantial with good adherence to the recommended dietary plan. However, some patients fail to meet their weight-loss goal by consuming excessive amounts of energy-dense foods and beverages—a practice some clinicians refer to as "eating out the pouch." In terms of operative safety and postoperative recovery, the procedure is comparable to the laparoscopic Roux-en-Y gastric bypass, but the amount of weight lost is generally less with the vertical banded gastroplasty (Olbers et al. 2005).

In laparoscopic adjustable gastric banding, an inflatable silicone ring or band is secured around the upper part of the stomach to create a small pouch and a narrow opening or stoma at the bottom of the pouch through which food passes into the rest of the stomach. The pouch restricts the amount of food that can be consumed at one time. The band is connected by a tube to an access port that is placed under the skin of the abdomen and is inflated when saline is injected into the access port using a syringe. The inflation of the band can be adjusted by the amount of saline injected into or removed from the access port. As the band is inflated, the stoma becomes more narrow and the emptying of the pouch is delayed, giving the patient a greater sense of fullness and further restricting the patient's food intake. If it is necessary to permit greater food intake, the band can be deflated to enlarge the stoma and allow the pouch to empty faster (Vella and Galloway 2003; Flier and Maratos-Flier 2005; Korenkov, Sauerland, and Junginger 2005; Provost 2005). Placement of the adjustable gastric band is considerably less invasive than the Roux-en-Y gastric bypass or the vertical banded gastroplasty because the procedure does not require any cutting or stapling in the stomach. Consequently, hospitalization and postoperative recovery are shorter. The band can be adjusted to suit the patient's needs and the procedure is fully reversible (Vella and Galloway 2003; Al-Momen, El-Mogy, and Ibrahim 2005; Provost 2005). Patients receiving the adjustable gastric band lose weight at a slower rate compared to patients receiving the Roux-en-Y gastric bypass, but over time the total amount of weight lost is comparable (Provost 2005).

Between 1998 and 2002, the number of Americans seeking bariatric surgery more than quadrupled and the inpatient death rate from the surgeries declined 64% (Encinosa et al. 2005). The future demand for bariatric surgery will likely increase as the number of eligible patients rises and as the safety of the procedure continues to improve. The average cost of the procedure is approximately $20,000.

Eating Disorders

Eating disorders are psychiatric conditions characterized by severe disturbances in eating behavior, resulting in significant physiologic impairment and, in some instances, death (APA, Diagnostic and Statistical Manual, 2000; ADA 2001; Coughlin and Guarda 2006). The American Psychiatric Association recognizes three categories of eating disorders: anorexia nervosa, bulimia nervosa, and eating disorder not otherwise specified (APA, Diagnostic and Statistical Manual, 2000). The diagnostic criteria for these are listed in Box 14.10. The most prominent clinical characteristic of anorexia nervosa (AN) is a refusal or inability to maintain a minimally normal body weight, leading to a body weight that is less than 85% of what is expected for age and height (APA, Diagnostic and Statistical Manual, 2000; Mitchell, Cook-Myers, and Wonderlich 2005; Walsh 2005). Bulimia nervosa (BN) is characterized by repeated episodes of binge eating followed by abnormal compensatory weight-loss behaviors such as self-induced vomiting, fasting, excessive exercise, and misuse of laxatives and diuretics (APA, Diagnostic and Statistical Manual, 2000; Walsh 2005). Unlike AN, patients with BN generally maintain a body weight within normal limits. The category eating disorder not otherwise specified (EDNOS) encompasses behaviors that fail to meet the specific diagnostic criteria for either AN or BN. EDNOS also includes binge eating disorder (BED), in which the patient engages in binging behavior without the compensatory weight-loss behaviors characteristic of BN.

While in theory these three conditions exist as distinct categories, in practice they share common features: all patients with eating disorders have a disturbed body image that leads them to overestimate their body shape and weight, perceive themselves a being obese even though they may have a very low body weight, have an intense fear of weight gain and obesity, and have a relentless drive to lose weight (APA, Diagnostic and Statistical Manual, 2000; ADA 2001; Fairburn and Harrison 2003; Walsh 2005; Coughlin and Guarda 2006). This body image disturbance drives a set of abnormal behaviors. While most people evaluate themselves on the basis of their perceived performance in such areas as relationships, work, parenting, athletics, and accumulation of possessions and wealth, persons with eating disorders base their self-worth primarily on the ability to control their shape and body weight (Fairburn and Harrison 2003). Further blurring the distinctions among the three categories is the fact that, over time, patients with eating disorders tend to migrate back and forth between the diagnostic categories of AN, BN, and EDNOS, as illustrated in Figure 14.19 (ADA 2001; Fairburn and Harrison 2003). Many BN patients have a past history of AN, and many AN patients have engaged in binge eating and compensatory weight-loss behaviors characteristic of BN. Body weight is a critical distinction between the two disorders: patients with AN are significantly underweight while those with BN generally have a body weight within normal limits (Fairburn and Harrison 2003; Walsh 2005).

Common characteristics of AN and BN are shown in Table 14.10. Although males are diagnosed with eating

BOX 14.10 **AMERICAN PSYCHIATRIC ASSOCIATION'S DIAGNOSTIC CRITERIA FOR ANOREXIA NERVOSA, BULIMIA NERVOSA, EATING DISORDERS NOT OTHERWISE SPECIFIED, AND RESEARCH CRITERIA FOR BINGE-EATING DISORDER**

Diagnostic Criteria for Anorexia Nervosa

A. Refusal to maintain body weight at or above a minimally normal weight for age and height (e.g., weight loss leading to maintenance of body weight less than 85% of that expected).

B. Intense fear of gaining weight or becoming fat, even though underweight.

C. Disturbance in the way in which one's body weight or shape is experienced, undue influence of body weight or shape on self-evaluation, or denial of the seriousness of the current low body weight.

D. In postmenarchal females, amenorrhea, i.e., the absence of at least three consecutive menstrual cycles. (A woman is considered to have amenorrhea if her periods occur only following hormone, e.g., estrogen administration.)

Subtypes of Anorexia Nervosa:

Restricting Type:

During the current episode of anorexia nervosa, the person has not regularly engaged in binge-eating or purging behavior (i.e., self-induced vomiting or the misuse of laxatives, diuretics, or enemas)

Binge-Eating/Purging Type:

During the current episode of anorexia nervosa, the person has regularly engaged in binge-eating or purging behavior (i.e., self-induced vomiting or the misuse of laxatives, diuretics, or enemas)

Diagnostic Criteria for Bulimia Nervosa

A. Recurrent episodes of binge eating. An episode of binge eating is characterized by both of the following.

1. Eating, in a discrete period of time (e.g., within any 2-hour period), an amount of food that is definitely larger than most people would eat during a similar period of time and under similar circumstances
2. A sense of lack of control over eating during the episode (e.g., a feeling that one cannot stop eating or control what or how much one is eating)

B. Recurrent inappropriate compensatory behavior in order to prevent weight gain, such as self-induced vomiting; misuse of laxatives, diuretics, enemas, or other medications; fasting; or excessive exercise.

C. The binge eating and inappropriate compensatory behaviors both occur, on average, at least twice a week for three months.

D. Self-evaluation is unduly influenced by body shape and weight.

E. The disturbance does not occur exclusively during episodes of anorexia nervosa.

Subtypes of Bulimia Nervosa:

Purging Type:

During the current episode of bulimia nervosa, the person has regularly engaged in self-induced vomiting or the misuse of laxatives, diuretics, or enemas

(continued on the following page)

disorders, approximately 90% of cases of AN and BN are seen in white females living in Western societies where food is plentiful and where being thin is associated with attractiveness (Fairburn and Harrison 2003; Walsh 2005). AN is more often seen among adolescents with the peak age of onset being 14 to 19 years, although it is diagnosed in prepubertal children and, much less frequently, in middle-aged and older adults (APA, Diagnostic and Statistical Manual, 2000; Bulik et al. 2005). On the other hand, BN is generally diagnosed in persons in their mid-to-late twenties with a history of binge eating and purging for as long as 5 to 10 years (Fairburn and Harrison 2003; Walsh 2005). Although there is disagreement among experts regarding whether eating disorders are now being diagnosed with greater frequency than in past decades, it appears that BN is more prevalent than in past decades, while the observed increase in AN is more likely due to a greater number of AN patients seeking help and better detection of cases by clinicians (Fairburn and Harrison 2003; Hoek and van Hoeken 2003; Bulik et al. 2005).

Mortality from AN is one of the highest of any psychiatric disorder, with 5% of AN patients dying every decade. In contrast, mortality from BN is much less likely, but is greater than in similar women in the general population who do not have an eating disorder (Mehler, Bulimia Nervosa, 2003). Patients with AN often deny that they have a problem and seldom seek medical treatment unless urged by concerned family or friends. Compared with patients with AN, those with BN are painfully aware that they have a problem and are more likely to seek treatment because of their inability to control their chaotic eating behavior. When interviewed in a supportive, nonjudgmental environment, patients with BN are able to provide details about eating behaviors and discuss their condition (Walsh 2005).

BOX 14.10 *(continued)*

Nonpurging Type:

During the current episode of bulimia nervosa, the person has used other inappropriate compensatory behaviors, such as fasting or excessive exercise, but has not regularly engaged in self-induced vomiting or the misuse of laxatives, diuretics, or enemas

Eating Disorder Not Otherwise Specified

The eating disorder not otherwise specified category is for disorders of eating that do not meet the criteria for any specific eating disorder. Examples include:

1. For females, all of the criteria for anorexia nervosa are met except that the individual has regular menses.
2. All of the criteria for anorexia nervosa are met except that, despite significant weight loss, the individual's current weight is in the normal range.
3. All of the criteria for bulimia nervosa are met except that the binge-eating inappropriate compensatory mechanisms occur at a frequency of less than twice week or for a duration of less than 3 months.
4. The regular use of inappropriate compensatory behavior by an individual of normal body weight after eating small amounts of food (e.g., self-induced vomiting after the consumption of two cookies).
5. Repeatedly chewing and spitting out, but not swallowing, large amounts of food.
6. Binge-eating disorder; recurrent episodes of binge eating in the absence of the regular use of inappropriate compensatory behaviors characteristic of bulimia nervosa.

Research Criteria for Binge-Eating Disorder:

A. Recurrent episodes of binge eating. An episode of binge eating is characterized by both of the following:
   1. Eating, in a discrete period of time (e.g., within any 2-hour period), an amount of food that is definitely larger than most people would eat in a similar period of time under similar circumstances
   2. A sense of lack of control over eating during the episode (e.g., a feeling that one cannot stop eating or control what or how much one is eating)

B. The binge-eating episodes are associated with three (or more) of the following:
   1. Eating much more rapidly than normal
   2. Eating until feeling uncomfortably full
   3. Eating large amounts of food when not feeling physically hungry
   4. Eating alone because of being embarrassed by how much one is eating
   5. Feeling disgusted with oneself, depressed, or very guilty after overeating

C. Marked distress regarding binge eating is present

D. The binge eating occurs, on average, at least 2 days a week for 6 months

E. The binge eating is not associated with the regular use of inappropriate compensatory behaviors (e.g., purging, fasting, excessive exercise) and does not occur exclusively during the course of anorexia nervosa or bulimia nervosa.

Etiology of Eating Disorders

Eating disorders often begin when an individual starts dieting, initially in a manner similar to that followed by many adolescent and young adult females (Fairburn and Harrison 2003; Walsh 2005; Coughlin and Guarda 2006). As weight loss progresses, persons predisposed to eating disorders experience an intense fear of gaining weight, diet more strictly, and begin developing the characteristic psychological, behavioral, and medical problems associated with eating disorders. For those developing AN, there is a sustained and obsessive pursuit of self-starvation. These patients view weight loss as a desired accomplishment rather than an affliction, and they have little motivation to change their behavior. In patients developing BN, similar attempts to lose weight and control body size and shape are thwarted by regular episodes of uncontrolled overeating followed by abnormal compensatory behaviors to lose weight. Persons with bulimia sometime describe themselves as "failed anorexics" (Fairburn and Harrison 2003).

Although the etiology of eating disorders is unknown, it is clear that multiple factors are associated with an increased risk of being diagnosed with an eating disorder, including environmental factors, certain character traits, and genetics (Fairburn and Harrison 2003; Walsh 2005; Coughlin and Guarda 2006). Environmental risk factors include a family history of mood disturbances, childhood physical or sexual abuse, the perception that there is a low degree of social support from family members, and societal pressures on females to attain a degree of thinness that is not only unhealthy but impossible for most females to attain. Of particular relevance to BN are childhood and parental obesity, early menarche, and parental alcoholism (Fairburn and Harrison 2003; Walsh 2005). Character traits associated with both AN and BN include low self-esteem and elevated harm avoidance, while traits more often seen in patients with AN

FIGURE 14.19 Over time patients with eating disorders tend to migrate back and forth between the various categories of eating disorders. An arrow's size represents the likelihood of patients migrating from category to another in the indicated direction. Recovery is represented by an arrow pointing outside the circle

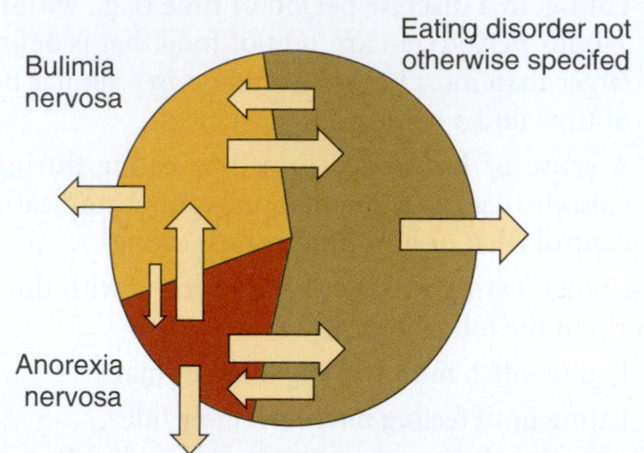

Source: Fairburn CG, Harrison PJ. *Eating disorders.* Lancet. 2003;361:407–416.

include perfectionism, conscientiousness, persistence, and obsessiveness. Traits more often associated with BN include impulsiveness, novelty seeking, negative emotionality, and stress reactivity (Coughlin and Guarda 2006). Although no specific gene for AN, BN, or EDNOS has yet been conclusively identified, the observation that an individual's risk of developing an eating disorder is greater when another first-degree family member is affected suggests that genetics plays an etiologic role. The role of genetics in AN has been more clearly elucidated than in it has in BN (APA, Diagnostic and Statistical Manual, 2000; Fairburn and Harrison 2003; Walsh 2005). A useful approach for examining the etiologic role of genetics is to study twins. For example, if one identical twin has AN, the other identical twin has an approximately 55% chance of having AN, whereas if one fraternal twin has AN, the other fraternal twin has an approximately 5% chance of having AN (Walsh 2005; Bulik et al. 2006).

Anorexia Nervosa

As outlined in Box 14.10, the diagnostic criteria for AN are a refusal to maintain a minimally normal body weight for age and height, an intense fear of gaining weight, a disturbed perception of body shape and/or size, and amenorrhea in postmenarchal females (APA, Diagnostic and Statistical Manual, 2000). In the *Diagnostic and Statistical Manual of Mental Disorders*, 4th ed. (DSM-IV, 2000), the American Psychiatric Association recognizes two mutually exclusive subtypes of AN: restricting subtype and binge eating/purging subtype. Those with the restricting subtype accomplish their weight loss through dieting, fasting, or excessive exercise and do not regularly engage in binge eating or any of the compensatory weight loss behaviors such as purging. Those with the binge eating/purging subtype engage in regular binge eating and/or purging.

Patients with AN experience a severe and selective restriction of food intake, particularly of foods perceived as fattening, resulting in a marked weight loss. The term "anorexia," which literally means "without appetite," is a misnomer considering the fact that most AN patients do not lose their desire for food and experience profound hunger (APA, Diagnostic and Statistical Manual, 2000; Fairburn and Harrison 2003). AN patients are generally obsessed with exercising and dieting, and preoccupied with thoughts of food to the extent that some AN patients collect cookbooks and recipes and are drawn to food-related occupations. As weight loss progresses, AN patients typically become irritable, moody, socially withdrawn, and isolated, and they experience loss of libido (Fairburn and Harrison 2003; Walsh 2005). Although amenorrhea in postmenarcheal females is a diagnostic criterion for AN, some females continue to menstruate while meeting every other criterion for the disease (Walsh 2005).

In applying the criterion that AN patients refuse to maintain body weight at or above a minimally normal body weight for age and height, the DSM-IV suggests that a body weight less than 85% of that expected for age and height be used as a guideline. Clinical judgment is necessary in applying this criterion, and consideration must be given to the patient's body build, weight history, and developmental stage, if the patient is an adolescent (ADA 2001; Mitchell, Cook-Myers, and Wonderlich 2005). For patients 20 years of age and older, a BMI $\leq$18.5 kg/m^2 has been suggested by some as meeting this guideline (Walsh 2005), while others suggest using a BMI $\leq$17.5 kg/m^2 as the cutpoint (Becker et al. 1999; ADA 2001). Another approach for evaluating the body weight of an adult is the Hamwi equation (see Chapter 5). For patients less than 20 years of age, underweight is defined as an BMI for age and sex that is $\leq$5th percentile using the CDC growth charts (see Chapter 5) (ADA 2001; Kuczmarski et al. 2000, 2002). Because body weight can be manipulated by such means as excessive water or fluid intake or severe restriction of water or fluid intake, a body weight that has unexpectedly changed or is inconsistent with other clinical findings should be interpreted with caution (ADA 2001).

For AN patients with the binge eating/purging subtype, compensatory weight-loss behaviors include self-induced vomiting, fasting, excessive exercise, and misuse of laxatives, diuretics, and enemas (APA, Diagnostic and Statistical Manual, 2000; Walsh 2005). The most common compensatory weight-loss behavior is purging through self-induced vomiting.

Health Complications of Anorexia Nervosa Numerous health complications and abnormalities, some of which can be life-threatening, are seen in patients with AN, as outlined in Table 14.11 and illustrated in Figure 14.20. Most of these are the direct result of malnutrition due to the self-imposed state of starvation and are reversed as healthy eating habits are

TABLE 14.10

Common Characteristics of Anorexia Nervosa (AN) and Bulimia Nervosa (BN)

	Anorexia Nervosa	Bulimia Nervosa
Global and ethnic distribution	Predominantly seen in Western societies among white people	Predominantly seen in Western societies among white people
Sex	90% female 10% male	Female to male proportion ranges from 10:1 to 20:1
Age of onset	Mid-adolescence	Late adolescence, early adulthood
Socioeconomic status	Appears evenly distributed across all social classes	Appears evenly distributed across all social classes
Prevalence in females	0.5 to 0.7% of adolescent females	1 to 3% in females 16 to 35 years of age
Incidence (per 100,000 per year)	19 in females, 2 in males	29 in females, 1 in males
Weight status	Markedly decreased	Typically within normal limits
Menstruation	Absent	Usually normal
Mortality	Approximately 5% per decade	Lower than in AN but higher than similar women in the general population who do not have an eating disorder.
Prognosis	Less favorable than BN. Full recovery seen in 25%–50% of patients with treatment.	More favorable than AN. Full recovery occurs in 50% of patients within 10 years.

Source: Adapted from Fairburn CG, Harrison PJ. Eating disorders. Lancet. 2003;361:407–16; Bulik CM, Reba L, Siega-Riz AM, Reichborn-Kjennerud T. Anorexia nervosa: definition, epidemiology, and cycle of risk. Int J Eat Disord. 2005;37:S2–S9; Mehler PS. Bulimia nervosa. N Eng J Med. 2003;349:875–81; Walsh BT. Eating disorders. In Kasper DL, Braunwald E, Fauci AS, Hauser SL, Longo DL, Jameson JL editors. Harrison's Principles of Internal Medicine. 16th ed. New York: McGraw-Hill; 2005. p. 430–33.

restored, body weight returns to a more normal level, and nutritional status improves, especially if AN is diagnosed early in the course of the illness and is treated by a skilled interdisciplinary team (Walsh 2005; Mehler 2001). One notable exception to this pattern of recovery is bone mineral density, which may not reach the level expected for the patient's sex and age, particularly if AN occurs during adolescence, a critical time period for bone development because the rate of bone mineralization is at its peak (see Chapter 27) (Walsh 2005; Mehler, Osteoporosis, 2003; Mehler 2001).

Common physical findings of AN include cold intolerance, reduced gastric emptying, constipation, and the presence on the skin of fine, downy-like hair known as lanugo. In some instances patients experience alopecia (hair loss). Despite profound weight loss, the face may have a fullness due to salivary gland enlargement resulting from both starvation and frequent vomiting. The fingers and toes may take on a bluish tint (known as acrocyanosis), while patients consuming large quantities of vegetables rich in carotenoids may experience hypercarotenemia, which can result in the skin having a slightly yellow or orange color to it. There may be abnormalities in vital signs, including bradycardia (heart rate <60 beats/minute), hypotension (systolic blood pressure <90 mm Hg), orthostatic hypotension (low blood pressure upon standing), and hypothermia (APA, Diagnostic and Statistical Manual, 2000; Walsh 2005; Coughlin and Guarda 2006).

Abnormal laboratory test values can include anemia, leukopenia (abnormally low white blood cell count), low plasma glucose, elevated serum total cholesterol, and decreased serum triiodothyronine (T_3), and low-normal values for the thyroid hormones serum thyroxine (T_4). Dehydration can result in slight increases in blood urea nitrogen and serum creatinine. Self-induced vomiting can result in hypokalemia (low serum potassium), hypochloremia (low serum chloride), and metabolic alkalosis (elevated serum bicarbonate). Excessive water consumption, a common tactic to increase body weight, can result in hyponatremia (low serum sodium). AN has a marked impact on the endocrine system, particularly on the female reproductive system, resulting in amenorrhea, which is the absence of menses when they would normally be expected to occur. AN results in decreased secretion of luteinizing hormone and follicle-stimulating hormone from the anterior pituitary; this results in decreased estrogen, which causes the amenorrhea (APA, Diagnostic and Statistical Manual, 2000; Saladin 2004; Walsh 2005; Coughlin and Guarda 2006).

Reduced bone mineral density, a common feature of AN, results from multiple nutritional deficiencies and amenorrhea (Mehler 2001; Walsh 2005; Coughlin and Guarda 2006). Depending on the severity and length of the disease, AN during adolescence can result in premature cessation of linear bone growth, failure to achieve expected adult height, and reduced bone mineral density. Even after several years of recovery from AN, these patients may never achieve their peak bone mineral density, and as they enter middle and late adulthood are at increased risk of painful fractures, disfiguring kyphosis, loss of height, and increased

TABLE 14.11

Physical and Diagnostic Findings in Patients with Anorexia Nervosa

Skin & Extremities	
• Cold hands & feet	• Alopecia
• Dry skin	• Acrocyanosis
• Lanugo	• Dependent edema
Cardiovascular	
• Bradycardia	• Cardiac arrhythmias
• Hypotension	• Electrocardiographic abnormalities
• Orthostatic hypotension	
Gastrointestinal	
• Salivary gland enlargement	• Constipation
• Delayed gastric emptying	
Plasma/Serum Values	
• Elevated BUN & creatinine	• Low T_3
• Hyponatremia	• Low-normal T_4
• Hypokalemia	• Low luteinizing hormone
• Hypercholesterolemia	• Low follicle-stimulating hormone
• Hypoglycemia	• Hypophosphatemia (during refeeding)
Bone	
• Decreased bone mineral density	

Sources: American Psychiatric Association. Practice guideline for the treatment of patients with eating disorders. *Am J Psychiatry.* 2000;157(Suppl):1–39. Fairburn CG, Harrison PJ. Eating disorders. *Lancet.* 2003;361:407–416. Walsh BT. Eating disorders. In Kasper DL, Braunwald E, Fauci AS, Hauser SL, Longo DL, Jameson JL, editors. *Harrison's Principles of Internal Medicine.* 16th ed. New York: McGraw-Hill; 2005. pp. 430–433.

risk of death (Mehler, Osteoporosis, 2003). On average, more than 50% of females with AN eventually develop osteoporosis and more than 50% of males with AN eventually develop a marked reduction in the mineral density of the femoral neck and the lumbar vertebrae. The importance of early diagnosis and treatment of AN is underscored by the observation that clinically significant bone mineral loss does not usually occur in the first 12 months of the illness (Mehler, Osteoporosis, 2003).

Bulimia Nervosa

As outlined in Box 14.10, the American Psychiatric Association's diagnostic criteria for BN are recurrent episodes of binge eating and use of inappropriate compensatory behaviors to prevent weight gain that occur, on average, at least twice a week for 3 months and that do not occur exclusively during episodes of AN. In addition, BN patients are preoccupied with their body shape and weight and their self-esteem is primarily based on their perceptions of body shape and weight, which are often distorted (APA, Diagnostic and Statistical Manual, 2000; Mehler, Bulimia Nervosa, 2003; Walsh 2005). The diagnostic criteria specify that the episode of binge eating must be characterized by eating an amount of food during a relatively short period of time (e.g., within a 2-hour time period) that is definitely larger than what most people would ordinarily eat in a similar situation. In addition, the episode of binge eating must be characterized by the subject sensing a lack of control (e.g., the subject feels he or she cannot stop or control what or how much he or she is eating) (APA, Diagnostic and Statistical Manual, 2000). When defining a binge, it is important to consider the context in which the food is consumed. For example, normal food consumption during a holiday or celebration might be considered excessive during a typical meal. Foods consumed during a binge are often those that are sweet and kcal-rich, such as ice cream or cake.

The American Psychiatric Association recognizes two mutually exclusive subtypes of BN: purging and nonpurging. In the purging subtype, the patient regularly engages in some type of purging behavior such as self-induced vomiting or the misuse of laxatives, diuretics, or enemas. The most common form of purging is self-induced vomiting, which is used by 80% to 90% of persons treated for BN. Initially, these subjects may use their fingers or some object to stimulate the gag reflex. Eventually, however, many patients develop the ability to initiate vomiting at will. Although rarely used to induce vomiting, syrup of ipecac can be toxic to the myocardium (the heart muscle) if regularly used (Mehler, Bulimia Nervosa, 2003). The next most common purging method is laxative abuse, which is used by about one-third of patients with BN (APA, Diagnostic and Statistical Manual, 2000). In the nonpurging subtype, the patient does not resort to purging, but does use some other abnormal compensatory weight loss behavior such as fasting or excessive exercise (APA, Diagnostic and Statistical Manual, 2000). Exercise is considered excessive if it significantly interferes with important activities, it occurs at inappropriate times or in inappropriate settings, or if the subject continues to exercise despite an injury or medical complication. Subjects with type 1 diabetes may resort to weight loss by omitting or reducing insulin doses (APA, Diagnostic and Statistical Manual, 2000).

Patients with BN are ashamed of their chaotic eating habits, binge eat in secret, and keep their disorder hidden from friends and family members (APA, Diagnostic and Statistical Manual, 2000; Fairburn and Harrison 2003; Walsh 2005). Although the diagnostic criteria for BN focus on the binge/purge cycles, much of the time patients with BN are restricting their eating in an attempt to control body shape and weight. Triggers for binge eating include hunger resulting from restricting food intake, anxiety,

FIGURE 14.20 Physical Changes and Health Complications Seen in Patients with Anorexia Nervosa

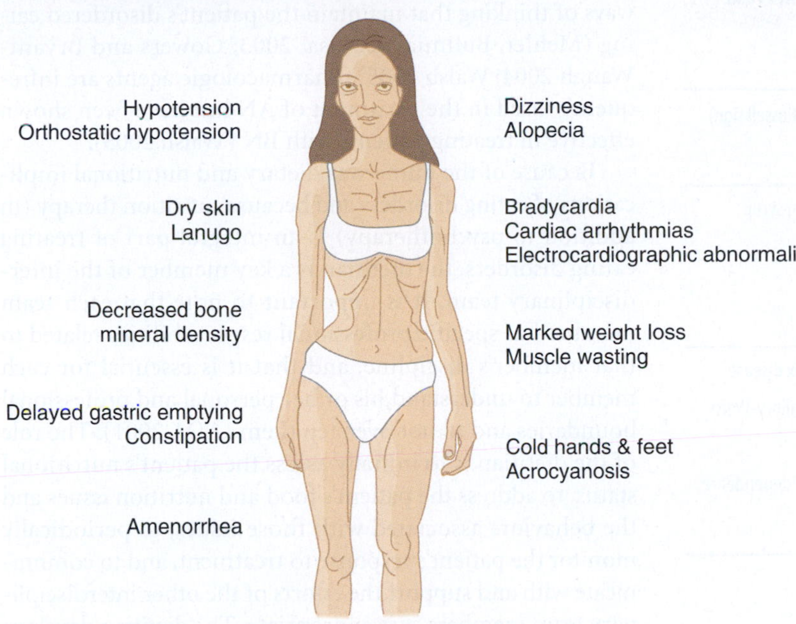

Source: http://www.pdrhealth.com/content/nutrition_health/chapters/fgnt09.shtml.

depression, and low self-esteem. The amount of food consumed in a binge is variable, with energy intake per binge ranging between 1,000 and 2,000 kcal (Fairburn and Harrison 2003). The binge typically results in shame and an unpleasant feeling of fullness which then triggers the purging or other compensatory weight loss behavior; this behavior initially provides some degree of relief, but this is often soon followed by guilt and shame (ADA 2001). Although the diagnostic criteria in Box 14.10 indicate a minimum frequency of twice a week for three months, the frequency of purging can vary considerably, with some patients reporting purging as often as 5 to 10 times per day (Mehler, Bulimia Nervosa, 2003). Despite purging and other abnormal compensatory behaviors, body weights of BN patients tend to be within normal limits, indicating that these attempts to control body weight are not as effective as some patients believe.

Health Complications of Bulimia Nervosa The health complications of BN are usually the result of regular purging or the other compensatory weight loss behaviors and are generally not life-threatening (APA, Diagnostic and Statistical Manual, 2000; Fairburn and Harrison 2003; Mehler, Bulimia Nervosa, 2003; Walsh 2005). These are outlined in Table 14.12 and illustrated in Figure 14.21. Frequent self-induced vomiting using the fingers to stimulate the gag reflex can result in the teeth leaving a scar or callus on the back of the hand, referred to as the Russell sign (Coughlin and Guarda 2006). Painless, bilateral enlargement of the salivary glands may give the face a full or puffy appearance. Frequent vomiting can permanently erode the enamel of the teeth, especially the lingual surface of the front teeth, giving them a ragged or "moth-eaten" appearance. Frequent vomiting can increase the risk of dental caries and cause esophagitis, gastroesophageal reflux disease (GERD), and, less frequently, tearing of the esophagus (Mallory-Weiss tears) and rupture of the stomach. Both vomiting and laxative abuse can cause fluid and electrolyte imbalances including hypokalemia, hyponatremia, hypochloremia, and alkalosis. Emetine, the active ingredient in syrup of ipecac, is cardiotoxic, has a long half-life, and it tends to accumulate in cardiac muscle (Coughlin and Guarda 2006). Frequent use of syrup of ipecac to induce vomiting can cause cardiomyopathy, electrocardiographic changes, and cardiac arrhythmias, potentially leading to congestive heart failure and death (APA, Diagnostic and Statistical Manual, 2000; ADA 2001; Fairburn and Harrison 2003; Mehler, Bulimia Nervosa, 2003; Walsh 2005). Laxative abuse can result in laxative dependence, constipation, dehydration, and potentially renal damage (Coughlin and Guarda 2006). Patients with BN generally have normal bone density (Mehler, Bulimia Nervosa, 2003).

Eating Disorders Not Otherwise Specified

Patients who have atypical eating disorders or with disordered eating behaviors that fail to meet all the criteria for either AN or BN are categorized as having an eating disorder not otherwise specified (EDNOS), as outlined in Box 14.10 (APA, Diagnostic and Statistical Manual, 2000; Coughlin and Guarda 2006). A common example of EDNOS is a female who meets all the criteria for AN but who still has regular menses or whose body weight has not fallen below 85% of that expected for age and height. Another example of EDNOS is the patient meeting all the diagnostic criteria for BN except that the binge eating and purging or other compensatory weight loss behaviors occur at a frequency of less than twice a week or for a duration of less than three months. Some individuals may engage in repeatedly chewing and then spitting out food rather than swallowing it (APA, Diagnostic and Statistical Manual, 2000; ADA 2001).

The diagnostic criteria for binge-eating disorder (BED), which is classified as an EDNOS, are outlined in Box 14.10 (APA, Diagnostic and Statistical Manual, 2000). BED is characterized by a person binge eating at least twice a week for six or more months and sensing a lack of control over eating during the binge. Patients with BED do not engage in purging or the other compensatory weight loss behaviors seen in patients

TABLE 14.12

Physical and Diagnostic Findings in Patients with Bulimia Nervosa	
Skin & Extremities	
• Callus on back of hand from stimulating gag reflex to induce vomiting (Russell sign)	
Cardiovascular	
• Cardiomyopathy from syrup of ipecac	• Cardiac arrythmias from syrup of ipecac
• Electrocardiographic changes from syrup of ipecac	
Gastrointestinal	
• Loss of dental enamel	• Gastroesophageal reflux disease
• Dental caries	• Esophageal tearing (Mallory-Weiss tears)
• Salivary gland enlargement	
• Esophagitis	• Constipation & laxative dependence
Plasma/Serum Values	
• Alkalosis	• Hypokalemia
• Hypochloremia	• Hyponatremia

Sources: American Psychiatric Association. Practice guideline for the treatment of patients with eating disorders. *Am J Psychiatry.* 2000;157(Suppl):1–39. Fairburn CG, Harrison PJ. Eating disorders. *Lancet.* 2003;361:407–416. Walsh BT. Eating disorders. In Kasper DL, Braunwald E, Fauci AS, Hauser SL, Longo DL, Jameson JL, editors. *Harrison's Principles of Internal Medicine.* 16th ed. New York: McGraw-Hill; 2005. p. 430–433.

with BN (APA, Diagnostic and Statistical Manual, 2000). The estimated prevalence of BED is 1% to 2% of the population (ADA 2001). Most BED patients are obese, seek treatment because of their obesity, and are subject to the same medical problems as obese patients (ADA 2001).

Nutrition Therapy for Eating Disorders

Because eating disorders are psychiatric illnesses with major medical complications, their treatment requires an interdisciplinary team of health care professionals whose primary focus is on psychiatric management (ADA 2001; Walsh 2005). For patients with AN, there are a range of psychiatric treatment options and various treatment settings, including outpatient care, intensive outpatient care, partial hospitalization as a day patient, residential treatment, and inpatient hospitalization (APA, Practice Guideline, 2000; Vandereycken 2003; Gowers and Bryant-Waugh 2004). Very briefly, patients with AN require considerable emotional support to overcome their strenuous resistance to gaining weight and extensive counseling to help them find healthier ways of developing self-esteem than achieving an inappropriately low body weight (Walsh 2005). The psychiatric model shown to be most effective for treating patients with BN is known as cognitive behavioral therapy, and is an approach focusing on modifying the specific behaviors and ways of thinking that maintain the patient's disordered eating (Mehler, Bulimia Nervosa, 2003; Gowers and Bryant-Waugh 2004; Walsh 2005). Pharmacologic agents are infrequently used in the treatment of AN but have been shown effective in treating patients with BN (Walsh 2005).

Because of the numerous dietary and nutritional implications of eating disorders and because nutrition therapy (in addition to psychotherapy) is an integral part of treating eating disorders, the dietitian is a key member of the interdisciplinary team. It is important to note that each team member has specific professional responsibilities related to that member's discipline, and that it is essential for each member to understand his or her personal and professional boundaries and to not overstep them (ADA 2001). The role of the dietitian is to initially assess the patient's nutritional status, to address the patient's food and nutrition issues and the behaviors associated with those issues, to periodically monitor the patient's response to treatment, and to communicate with and support the efforts of the other interdisciplinary team members as is appropriate. The dietitian develops the nutrition component of the treatment plan, takes a leading role in implementing the nutritional component of the treatment plan, and provides ongoing support to the patient in accomplishing the goals set out in the treatment plan (ADA 2001).

Nutrition Therapy for Anorexia Nervosa In theory, the primary care physician's decision as to where and how to treat the patient with AN should be primarily determined by the subject's current weight, the rapidity of recent weight loss, the severity of medical and psychological complications, and the necessity of removing the patient from an unhealthful environment (APA, Practice Guideline, 2000; Vandereycken 2003). In practice, however, the treatment setting is more often determined by the availability of care and its cost. For example, a suitable residential or inpatient treatment program may be several hours away from where the patient lives, or the cost of intensive treatment may be prohibitive, depending on the patient's insurance coverage or whether the patient is insured (Vandereycken 2003). Medical conditions warranting residential or inpatient treatment include severe electrolyte imbalances or a body weight <75% of expected regardless of the patient's electrolytes or other laboratory values. In less severe cases of AN, the day patient or outpatient treatment can be less expensive and somewhat more convenient to the patient and family members. The primary treatment goal in AN is restoring the patient's weight to at least 90% of the expected weight, and in most cases this degree of nutritional repletion can be accomplished by normal oral feedings without resorting to administering nutrients enterally (via nasogastric tube) or parenterally (ADA 2001; Walsh 2005). Additional goals are cessation of weight loss behaviors, improvement in eating

FIGURE 14.21 Physical Changes and Health Complications Seen in Patients with Bulimia Nervosa

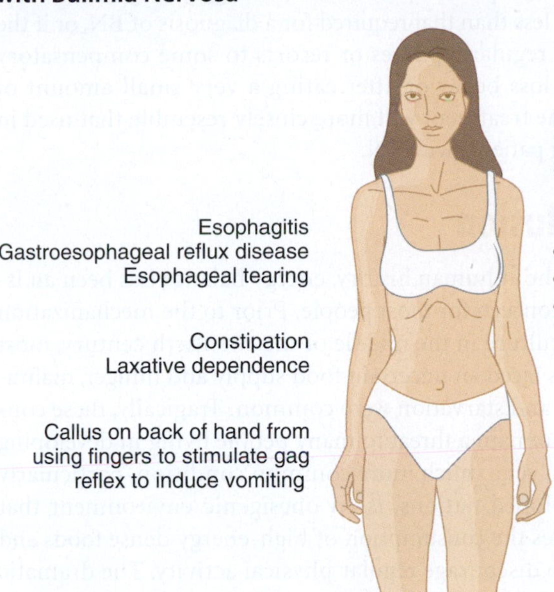

Source: http://www.pdrhealth.com/content/nutrition_health/chapters/fgnt09.shtml.

behaviors, and improvement in emotional and psychological health. Regardless of the treatment setting, the goals remain the same; the only difference is the intensity of treatment. In the outpatient setting, the treatment is somewhat less intense than in the inpatient setting (ADA 2001).

Key to the success of residential treatment or inpatient hospitalization is a duration of treatment sufficient to allow adequate weight gain and weight stabilization, and sufficient to provide the therapy necessary to allow the patient to adjust emotionally to the healthier weight. A low body weight at the beginning of treatment and inadequate weight gain during inpatient treatment are both associated with poor treatment outcome and inpatient readmission (Vandereycken 2003). The recommended weight gain is 2 to 3 lb/week for inpatient treatment and 0.5 to 1 lb per week for outpatient treatment (APA, Practice Guideline, 2000). Initially, the energy intake should be 30 to 40 kcal/kg of body weight per day, which can then be advanced as tolerated by the patient. During the phase of active weight gain, the energy intake may need to be as high as 70 to 100 kcal/kg of body weight while an intake of 40 to 60 kcal/kg of body weight may be sufficient for weight maintenance and to support adequate growth and development in children and adolescents (APA, Practice Guideline, 2000). For some patients, vitamin and mineral supplements may be helpful, and it is important to ensure adequate intake of vitamin D (400 IU/d) and calcium (1500 mg/d) to minimize bone losses (Walsh 2005). Meals should be supervised by staff members who firmly stress the importance of adequate food consumption, are empathetic about the patient's challenges, and provide encouragement and reassurance about the patient's eventual recovery (Walsh 2005).

The patient's response to nutrition therapy can be assessed by monitoring the patient's vital signs, food intake, fluid intake and output, and changes in weight, height, body mass index, body composition, and laboratory test values. An unexpected increase in body weight may indicate fluid retention during refeeding or excessive water or fluid intake by the patient to artificially increase body weight. Treatment programs generally have specific protocols for weighing patients that address when patients are weighed, who weighs the patients, and whether patients are informed of their weights. In some instances, a patient may be weighed with his or her back to the scale and not immediately informed of the value (ADA 2001). Children's and adolescents' height and BMI can be assessed using the CDC growth charts. Changes in body composition can be monitored by measuring skinfold thicknesses (see Chapter 5). Patients should also be observed for signs of congestive heart failure and gastrointestinal problems such as constipation and bloating. Cardiac monitoring may be warranted for severely malnourished patients whose weight is <70% of expected (ADA 2001; APA, Practice Guideline, 2000).

During nutritional repletion, serum electrolytes should be closely monitored for signs of refeeding syndrome (see Chapter 7), which is characterized by serum electrolyte depletion, fluid shifts, cardiac arrhythmias, and glucose derangements occurring in severely malnourished patients when they receive nutritional repletion either orally, enterally, or parenterally (APA, Practice Guideline, 2000; ADA 2001; Marinella 2003). Common electrolyte disturbances seen in refeeding syndrome include hypokalemia, hypomagnesemia, and, most notably, hypophosphatemia. Adverse effects of hypophosphatemia include cardiac failure, muscle weakness, immune dysfunction, and possibly death (Marinella 2003).

Nutrition Therapy for Bulimia Nervosa BN can often be treated on an outpatient basis, and compared to AN there is a greater likelihood of recovery. The primary treatment goal in BN is to reduce the chaotic cycle of binging and purging and to normalize the patient's eating habits (APA, Practice Guideline, 2000; ADA 2001). Because the weight of most patients with BN is within the normal range, weight restoration is not the focus of therapy as it is with patients who have AN. Some BN patients may initially be underweight and could benefit emotionally and physiologically from weight gain. Others, because of their disturbed body image, overestimation of their body shape and weight, and intense fear of obesity and weight gain, may insist on losing weight. Any efforts to change the patient's body weight should be postponed until after eating habits are normalized (APA, Practice Guideline, 2000; ADA 2001).

The role of the dietitian is to work with the other members of the interdisciplinary team to develop an eating plan to normalize the patient's eating habits. A common recommendation for patients with BN is to consume three meals per day with one to three snacks per day in a structured manner, providing order to eating and breaking the chaotic eating pattern of binging and purging. Initially, food intake should be sufficient to prevent hunger, which is a common trigger of binge eating. Nutrition counseling should focus on helping the patient expand the diet to include the patient's self-imposed "forbidden" or "feared" foods. As eating habits normalize, patients may experience fluid retention and will need education and support to deal with this temporary and disturbing phenomenon. The laxative-dependent patient will benefit from information on prevention of bowel obstruction resulting from laxative withdrawal through consumption of foods rich in dietary fiber and adequate amounts of fluid (APA, Practice Guideline, 2000; ADA 2001).

Nutrition Therapy for Eating Disorders Not Otherwise Specified Medical nutrition therapy for EDNOS will depend on the patient's specific abnormal eating behaviors. For example, if the EDNOS patient presents with many, but not all, of the criteria for AN, the nutrition therapy will more closely follow that recommended for AN. Likewise, if the EDNOS patient presents with binge eating but at a frequency less than that required for a diagnosis of BN, or if the patient regularly purges or resorts to some compensatory weight loss behavior after eating a very small amount of food, the treatment will more closely resemble that used in treating patients with BN.

Conclusion

Throughout human history, energy balance has been an issue of concern for most people. Prior to the mechanization of agriculture in the middle of the twentieth century, most humans faced an uncertain food supply and hunger, malnutrition, and starvation were common. Tragically, these conditions remain a threat to many people living in developing nations; yet a much more common condition, particularly in developed nations, is an obesigenic environment that promotes the consumption of high-energy dense foods and tends to discourage regular physical activity. The dramatic rise in the prevalence of overweight and obesity in recent decades has been aptly described as an epidemic, and the term "globesity" has been coined to represent the global nature of the obesity epidemic. The complex etiology of obesity requires a multifactorial approach to its prevention and treatment. This will involve better understanding of the biological basis of body weight regulation and the control of hunger and appetite, more informed personal choices about food intake and physical activity, and modifying the environment to promote the adoption and maintenance of healthy dietary and physical activity behaviors. At the same time, eating disorders remain a threat to health and demand attention, particularly considering the high mortality associated with anorexia nervosa.

PRACTITIONER INTERVIEW

Betty Kovacs, M.S., R.D. *Co-Director & Director of Nutrition, New York Obesity Research Center Weight Loss Program – New York Obesity Research Center St. Luke's-Roosevelt Hospital, New York, NY*

How long have you been an RD? How long have you worked in weight management?

I have been an RD for nine years and worked in weight management for eight years.

Describe a typical client that you see at the clinic.

The average age of the clients in our program is 44 years old, 80% are women, and they tend to be professionals who have tried all of the diets and programs available.

In your practice, what do you find is key information for you to obtain in order to assess a client? Are there common nutritional problems associated with these patients? If so, what are they?

Before beginning the program, each client sees me for an hour and a half. The point of this appointment is to assess if the client is appropriate for the program and if the program is appropriate for him or her. Just because someone comes to us to lose weight does not mean that it's the best way for them to do it. It's important to discuss what they have tried to do to lose weight and what did and did not work for them. It can give you insight into what to do with them next. For example, if structure helped them stay focused, then a plan with structure would be the starting point. If having to count calories "makes them crazy" I would never begin with that. The most important thing that I do is to adjust their plan based on their likes and dislikes. The majority of the people that I see come in dreading having to keep food records. Like most dietitians, I used to discuss the importance of this and require that they do it. Now I know that they can lose weight initially without food records . . . so why add to their stress by requiring them? Instead, we agree that when they are struggling with their weight loss, they will keep records and hand them in. Their relief is clear and they realize that they do have options. I spend a great deal of the appointment going over what their eating style and food preferences are. I ask for when they eat, what kinds of foods they eat, who they eat with, and where their food comes from and end by asking what they think is contributing to their weight gain or inability to lose weight. It helps get them ready for me to begin designing a plan with them.

Their nutritional problems tend to be related to their health conditions and medications that they are taking. The most common health conditions are diabetes, hypertension, and high cholesterol. Along with the medications for these conditions, psychotropic drugs are very common and may be contributing to their weight problem. It's a very difficult situation when the medication is helping them psychologically but I know that it's affecting their weight. I try not to say it to the client; instead I speak with the doctor or recommend that they speak with him or her. The decision to change the medication has to be made by the prescribing physician, not me. One last nutritional concern is deficiencies. A lot of people assume that if you are overweight you could never be deficient in any nutrients. Once I discuss what they are eating and evaluate their labs, it's clear what vitamins and minerals they are not getting enough of. There are some cases when their intake of a macronutrient is too low as well. You can't use their weight as an indication of their nutritional status.

Do you have specific goal or approach for counseling clients?

My first goal is to make them feel comfortable opening up to me, so I try to make light of what it's like to see a dietitian. During their orientation to find out about the program, I ask how many people are actually looking forward to seeing me. They tend to be very honest and most do not raise their hand. I then ask them what they imagine it will be like, and the common answer is that I am going to judge what they are doing and take away everything that they like to eat. I use that time to assure them that they are not coming to see me to hear how I think that they should eat. The purpose of seeing me is to figure out how to take what they like to eat and adjust it so that they can lose weight. I acknowledge that we all know that fruits and vegetables are good for us, so I won't spend their appointment preaching that to them. I don't need to take them from where they are with their eating to the "perfect" diet. All that you need to do to lose weight is cut 500 to 1,000 kcal per day from their maintenance needs. If I can find a way to keep their favorite foods in their plan and help them lose weight, they will be much more likely to stick with this for the long term.

What is the biggest challenge in working with patients who are overweight?

This would be undoing the misconceptions about nutrition and weight loss. I spend more time telling people about why other diets don't work and explaining what the latest study in the news is really about. I am fortunate that I get a year or more with my clients, so there is time for them to learn to trust what I say. If I only had a couple of appointments, it would be difficult to get them to believe me over the media. The public doesn't really understand how a registered dietitian is the nutrition expert. They hear physicians and that is who they listen to.

Has the treatment of obesity changed much in the last ten years? How do you stay abreast of changes in this field?

There have been a lot of changes in our treatment options and understanding of the causes and consequences of being overweight and obese. The National Heart, Lung, and Blood Institute's Obesity Education Initiative convened a panel of experts to develop The Clinical Guidelines on the Identification, Evaluation, and Treatment of Overweight and Obesity in Adults.

Objectives of the Guidelines:

- To identify, evaluate, and summarize published information about the assessment and treatment of overweight and obesity
- To provide evidence-based guidelines for physicians, other health care practitioners, and health care organizations for the evaluation and treatment of overweight and obesity in adults
- To identify areas for future research

In order to keep abreast of the latest research, it's imperative that I read the appropriate journals, am a member of the professional associations related to this field, attend meetings, and read any news and books that my clients may be exposed to. The most important membership to have in this field is with the North American Association for the Study of Obesity, The Obesity Society. Their journal, *Obesity Research*, and their annual meeting keep us updated on all of the advances happening in our field.

Any advice for dietetic students about counseling clients with weight issues?

My advice is to help the clients improve on what they are doing; don't try to get them to eat a "perfect" diet. This is a chronic problem and the progress is slow and difficult at times. You need to be patient and understand that there is a lifetime of habits behind the way that they eat and think about food. When I was a recent graduate, I can remember thinking that I could teach people about the food guide pyramid and serving sizes and that would be enough to help them lose weight. I didn't agree with using meal replacements, medication, or surgery. I now know that all of these tools exist because they are needed and that my job is to know how and when to use them. I have been doing this for eight years, and I am still learning. The more that I have learned, the more I realize how much there is to still learn. My training in school was the foundation, and it's my responsibility to become an expert in this and to learn all that I can to be able to do whatever possible to help them. As long as my client is willing to continue trying, I have to be willing to keep looking for ways to help. Obesity research clearly indicates that weight loss requires three interventions: behavioral, nutrition, and exercise. Ideally, you want to work with experts in the other two areas. When that's not possible, it's my responsibility to know when to refer my clients to experts in the other fields. A client can tell that I genuinely want to do what is best for them and that I believe that they can be successful. Society is constantly judging them for their weight, so they need to be able to seek help from people who will not do the same.

CASE STUDY

CC:

Patient admitted through ER after complaining of dizziness at the end of a training run.

Patient History:

16-year-old premenarchal junior in high school who competes on high school cross country and track teams. She is an endurance athlete competing at regional and state events. After a training run today, she complained of dizziness. Mother states that patient has lost a lot of weight over the past 6 months. Her mother does not know exact amount but seven months ago patient began training for her first marathon. Patient states she has not lost enough weight and her recent performance has been hampered by weight. Mother states she trains constantly, even going out for additional runs after homework in the evening.

Labs:

Alb 3.8 mg/dL, Prealbumin 22 mg/dL; Na^+ 137 mmol/L, K^+ 3.2 mmol/L, Cl 104 mmol/L, Hgb 14 mg/dL, Hct 37%, Ferritin 159 pg/mL, Glu 48 mg/dL

Nutrition Assessment:

Ht. 5'4" Wt. 91 lbs. UBW 105 lbs. (nine months previous)

Diet History:

AM: 8 oz. skim milk, bagel with 1 tsp peanut butter; lunch: 1 c. ice milk, 1 banana, pretzels—about 20, water; PM: meat (usually chicken)—skinless, baked 3 oz., 1 c. vegetable, 2 c. salad without dressing, skim milk—8 oz. Mother says that patient rarely eats at the hours of the rest of the family and often will not eat evening meal until after 8 p.m. Meals, particularly breakfast and lunch, are missed at least twice per week. Patient states she eats no red meat and no cheese and never adds sugar or salt to food.

Exercise History:

Trains 2.5 hours/day seven days per week. Includes a minimum of 2 hours of running at 8 miles/hour with weight lifting at least 4 times per week for 30–45 minutes.

Physical Assessment:

% body fat – 10.6%, triceps skinfold – 11.75 mm, upper arm circumference 8.75 inches.

Questions:

1. Identify signs and symptoms that may be important to consider when assessing this patient's nutritional risk.
2. Calculate this patient's BMI and evaluate her weight change. Assess her anthropometric data. Does this place her at nutritional risk?
3. Evaluate her biochemical indices. Do they indicate that she may be at nutritional risk?
4. Using her diet history, estimate her daily energy intake. Compare this with her estimated energy and protein requirements.
5. Why is this patient at age 16 still premenarchal? Does this place her at any medical risk?
6. Using the diagnostic criteria for eating disorders, compare your nutritional assessment to identify her risk for an eating disorder. What are your conclusions?

NUTRITION CARE PROCESS FOR OVERWEIGHT AND OBESITY

Step One: Nutrition Assessment

Medical/Social History

Diagnose/determine comorbidities that may indicate risk of metabolic syndrome: hypertension, diabetes mellitus/impaired glucose tolerance, dyslipidemia

Past medical history/previous methods used for weight loss if applicable

Medications (especially medications that might cause weight gain: antidepressants, lithium, beta-blockers, corticosteroids)

Socioeconomic status/food security

Support systems

Education – primary language

Dietary Assessment

Ability to chew; use and fit of dentures

Problems swallowing

Nausea, vomiting

Constipation, diarrhea

Heartburn

Any other symptoms interfering with ability to ingest normal diet

Ability to consistently purchase adequate amounts of food on a daily basis

Ability to feed self

Ability to cook and prepare meals

Food allergies, preferences, or intolerances

Previous food restrictions

Ethnic, cultural, and religious influences

Use of alcohol, vitamin, mineral, herbal or other type of supplements

Previous nutrition education or nutrition therapy

Eating pattern: 24-hour recall, diet history, food frequency; focus on portion sizes, meals eaten away from home, food preparation methods, sources of high energy density (fat, concentrated sugar content)

Anthropometric

Height

Current weight

Weight history: highest adult weight; usual body weight

Reference weight/BMI

Waist and hip circumference

Biochemical Assessment/Physical Assessment

Laboratory measures for comorbidities/assessment of metabolic syndrome: serum glucose, HgBA1C, total cholesterol, LDL, HDL, triglycerides

Blood pressure

Visceral protein assessment: standard

Hematological assessment: standard

Step Two: Common Diagnostic Labels

- Excessive fat intake
- Food and nutrition-related knowledge deficit
- Disordered eating pattern
- Undesirable food choices
- Overweight/Obesity
- Involuntary weight gain
- Physical inactivity

Sample PES: Obesity related to undesirable food choices and food and nutrition-related knowledge deficit as confirmed by current BMI of 36 kg/m^2 and diet history indicating average energy intake 175% of estimated requirements.

Step Three: Sample Intervention

1. Reduce restaurant meals to twice per week. Provide education on appropriate restaurant choices to reduce both fat and energy intake.
2. Substitute kcal-free beverages for current soft drink choices.
3. Add 2 servings of either fruit or vegetable daily. Provide education on food choices, variety, and preparation methods.

Step Four: Monitoring and Evaluation

1. Assess food diary information. Provide phone call and follow-up with patient at a minimum of one time weekly.
2. Monitor body weight one time weekly.
3. Evaluate ability to maintain initial food/behavior/lifestyle choices via interview and food diary.

WEB LINKS

National Heart, Lung, and Blood Institute (NHLBI): The NHLBI, a division of the National Institutes of Health, supports research into the causes, treatment, and prevention of conditions affecting risk of cardiovascular diseases, including obesity. It is a valuable source of authoritative and scientifically accurate information on obesity.

http://www.nhlbi.nih.gov

Centers for Disease Control and Prevention, Division of Nutrition and Physical Activity: The website of the Centers for Disease Control and Prevention's Division of Nutrition and Physical Activity is a source of current scientific information on nutrition, physical activity, overweight, and obesity. It provides links to resources on topics relevant to obesity that are available from a variety of public and private organizations.

http://www.cdc.gov/nccdphp/dnpa

National Institute of Diabetes and Digestive and Kidney Diseases (NIDDK) Weight-control Information Network: The NIDDK's Weight-control Information Network provides the general public, health professionals, the media, and Congress with up-to-date, science-based information on weight control, obesity, physical activity, and related nutritional issues.

http://win.niddk.nih.gov

National Center for Health Statistics (NCHS): The NCHS is a source of information on the health of the U.S. population, including the most recent statistics on the prevalence of overweight and obesity. The NCHS publication, "Health: United States," is a highly recommended collection of statistics on the health determinants of the U.S. population and is updated annually.

http://www.cdc.gov/nchs

Statistics Canada: Statistics Canada is a source of data of various types relating to Canada, including the most recent prevalence estimates of overweight and obesity.

http://www.statcan.ca

National Eating Disorder Information Centre (NEDIC): The NEDIC is a Canadian not-for-profit organization established in 1985 to provide information and resources on eating disorders and weight preoccupation. Its goal is to promote healthy lifestyles that allow people to be fully engaged in their lives.

http://www.nedic.ca

END-OF-CHAPTER QUESTIONS

1. How is the energy content of food determined today? What is a kilojoule? How does it differ from a kilocalorie?

2. Describe the three main components of energy expenditure. What is the difference between basal energy expenditure and resting energy expenditure?

3. Describe three methods that can be used to estimate or determine a person's energy requirement.

4. List five substances produced in the body that can affect appetite. Pick two of them and describe them in more detail (source, function, and effect).

5. What is the clinical implication of excess adipose tissue in specific body locations? How should it be measured—BMI, waist circumference, or waist-to-hip ratio (WHR)? Why?

6. List at least five factors that have contributed to growing problem of obesity. Pick one factor and explain how you would address this concern with a client who needs to lose weight.

7. What are the key elements of nutrition therapy for weight loss? What additional benefits are associated with increasing physical activity?

8. What medications are currently approved for weight loss? Pick one, describe its mechanism of action, and list possible side effects. Describe one surgery that is performed for weight reduction. List possible complications associated with the surgery.

15

Diseases of the Cardiovascular System

Thomas J. Pujol, Ed.D., F.A.C.S.M.

Southeast Missouri State University

Joshua E. Tucker, M.S.

Michigan State University College of Osteopathic Medicine

Anatomy and Physiology of the Cardiovascular System
Hypertension
Atherosclerosis
Ischemic Heart Disease
Peripheral Arterial Disease
Heart Failure

Introduction

Diseases of the cardiovascular system are the leading causes of death in the United States, accounting for 37.3% of all deaths, or 1 in every 2.7 deaths (see Figure 15.1). Cardiovascular disease was an underlying or contributing cause of death in over 1.4 million cases in 2002. Over 70 million American adults live with one or more cardiovascular diseases, and 27.4 million of these individuals are over the age of 65. Projections for 2006 indicated that the direct and indirect costs of cardiovascular disease will exceed $400 billion (American Heart Association 2006). In the past 30 years, however, the age-adjusted cardiovascular disease death rate has dropped by almost 40% as a result of developments in treatment and prevention of these diseases (Gordon, Leighton, and Mooss 2005).

In this chapter, all major forms of cardiovascular disease—hypertension, atherosclerosis, ischemic heart disease, peripheral vascular disease, and heart failure—will be discussed. These conditions are interrelated and more often than not will coexist. For example, atherosclerosis is often synonymous with ischemic heart disease and peripheral vascular disease, while hypertension is a risk factor for all other cardiovascular diseases. As with all chronic disease, the risk of developing a cardiovascular disease is determined by a combination of hereditary, environmental, and lifestyle factors. The lifestyle modifications described in this chapter aid in the prevention and treatment of these conditions.

Anatomy and Physiology of the Cardiovascular System

The role of the cardiovascular system is to regulate blood flow to the tissues in order to deliver oxygenated blood and nutrients as well as to retrieve waste products from cellular metabolism. Other major functions include thermoregulation, hormone transport, and gas exchange. The cardiovascular system forms a closed loop of blood vessels for which the heart acts as two pumps.

The Heart

The heart is a hollow, muscular organ whose walls are composed of three layers. The outer layer is the epicardium, and the inner layer is the endocardium. The middle layer, the

FIGURE 15.1 Rates of Death Due to Diseases of the Heart, 2001*

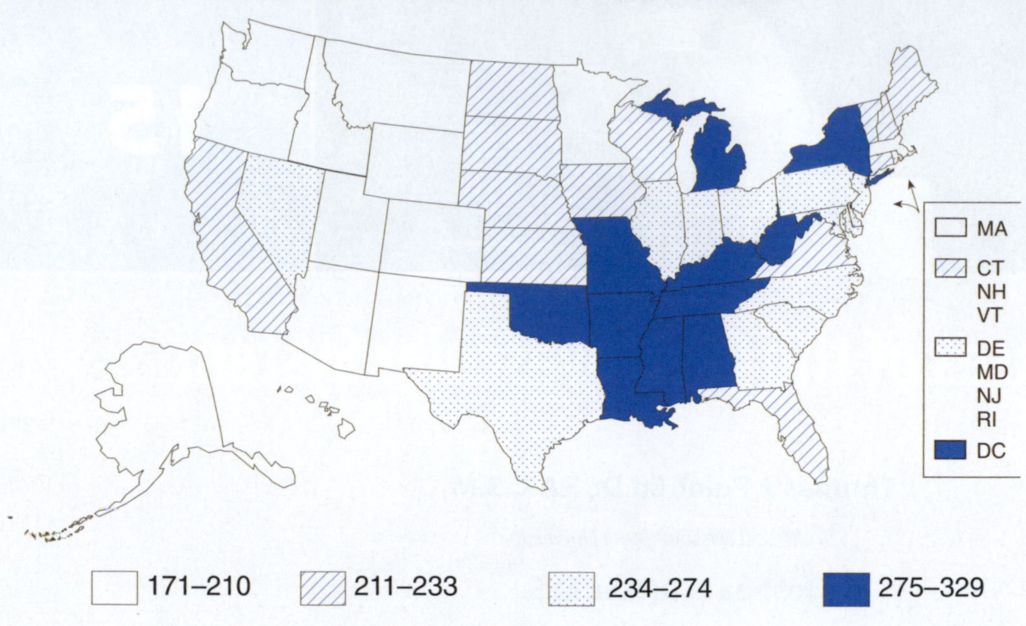

*Deaths per 100,000, age adjusted to 2000 total U.S. population.

Source: Center for Disease Control and Prevention.

myocardium, is responsible for the muscle contraction which moves the blood from the heart. The heart is divided into four chambers (see Figure 15.2). The upper two chambers are the left and right atria, and the lower two chambers are the left and right ventricles. The right atrium and right ventricle make up the right heart and pump blood through the pulmonary circulation for oxygenation. The left atrium and left ventricle make up the left heart and pump oxygenated blood through the systemic circulation.

Blood moves from the right atrium into the right ventricle, where it is pumped through the pulmonary trunk into the pulmonary circulation. As blood flows through the pulmonary circulation, it enters the lungs, where carbon dioxide is removed and oxygen is added. The blood returns to the heart via the pulmonary veins and enters the left atrium. Then it moves from the left atrium to the left ventricle (LV), from which it is pumped into the systemic circulation through the aorta, a large artery. Adequate function of the LV is crucial for maintaining adequate cardiovascular function; this becomes obvious during the conditions of hypertension, congestive heart failure, and **left ventricular hypertrophy (LVH)**. The aorta branches into smaller arteries, which direct blood to the organs of the body. As an artery enters into an organ, it branches into progressively smaller arteries and eventually into arterioles. The arterioles then branch into the body's smallest vessels, the capillaries, where gas exchange occurs.

The blood is then returned to the heart through progressively larger vessels. The capillaries unite to form venules, and then the venules unite to form veins. The veins from the upper part of the body drain into and form the superior vena cava, and those from the lower part of the body drain into and form the inferior vena cava. The superior and inferior vena cava unite to return the blood to the right atrium.

Electrical Activity of the Heart Heart muscle cells are connected to one another by membranes called intercalated discs, which allow electrical impulses to pass from one cell to the next. Though many **myocardial cells** have the capability of

left ventricular hypertrophy (LV hypertrophy)—enlargement of the left ventricle; most commonly related to hypertension and/or congestive heart failure

myocardial cells—cells found in the myocardium

FIGURE 15.2 Blood Flow Through and Pump Action of the Heart

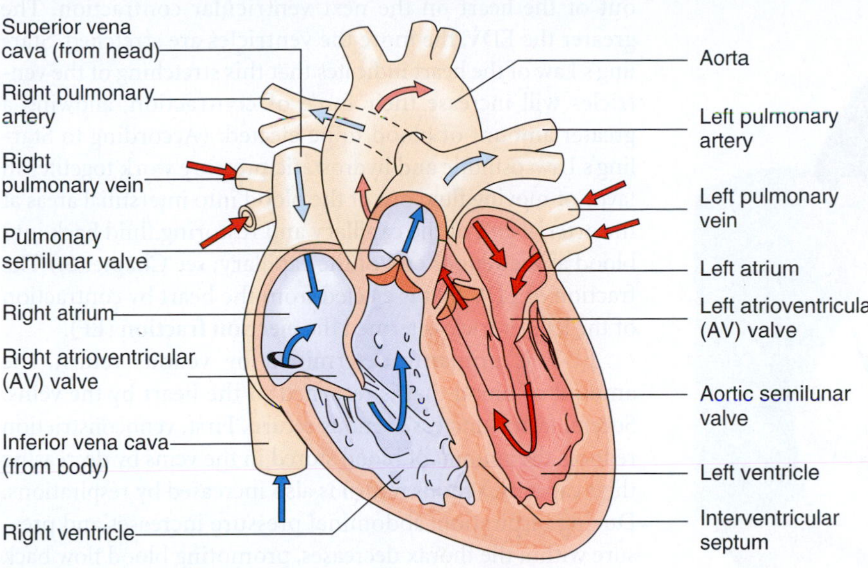

Source: L. Sherwood, *Human Physiology: From Cells to Systems*, 5e, copyright © 2004, p. 306.

FIGURE 15.3 Specialized Conduction System of the Heart

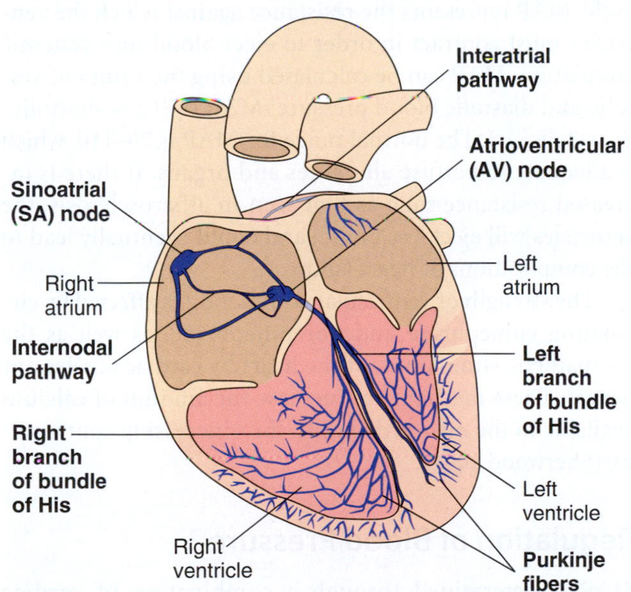

Source: L. Sherwood, *Human Physiology: From Cells to Systems*, 5e, copyright © 2004, p. 311.

generating spontaneous electrical activity, under normal conditions electrical activity is initiated in the heart at the sinoatrial (SA) node (see Figure 15.3). The change in electrical membrane potential (depolarization) in the SA node causes the contraction of the atria. The depolarization is carried from the atria to the ventricles by way of the atrioventricular (AV) node located in the base of the right atrium. The depolarization of the AV node is carried into the ventricles by the atrioventricular bundle (bundle of His), which splits into the right and left bundle branches. The depolarization is carried down the bundle branches, and then spread throughout the ventricles by the Purkinje fibers. As resting membrane potential decreases, the SA node will reach its depolarization threshold, and the process will repeat. The electrical activity of the heart can be measured through an electrocardiogram (ECG) (see Box 15.8 later in the chapter).

Cardiac Cycle The repeating contraction and relaxation of the heart is termed the cardiac cycle. The heart alternates between the phase of contraction (the **systole**) and the phase of relaxation (the **diastole**). Since the atria and ventricles depolarize and thus contract separately, there is an atrial systole and diastole as well as a ventricular systole and diastole. The force exerted by the blood on the walls of blood vessels during the contraction of the ventricles is termed **systolic blood pressure**; the force exerted during relaxation of the ventricles is **diastolic blood pressure** (see Figure 15.4).

Cardiac Function

The volume of blood ejected with each contraction of the left ventricle is termed **stroke volume**. It is regulated by

systole—contraction phase of the cardiac cycle; during this phase blood is ejected from the ventricles into the aorta and pulmonary artery

diastole—relaxation phase of the cardiac cycle; during this phase, ventricles empty and blood fills the atria

systolic blood pressure—pressure exerted when ejected from the ventricles (systole phase of the cardiac cycle)

diastolic blood pressure—pressure that occurs as ventricles relax (diastole phase of the cardiac cycle)

stroke volume—the volume of blood that is ejected from the left ventricle with each systolic phase; defined mathematically as LVEDV-LVESV

FIGURE 15.4 Measurement of Blood Pressure using a Sphygmomanometer

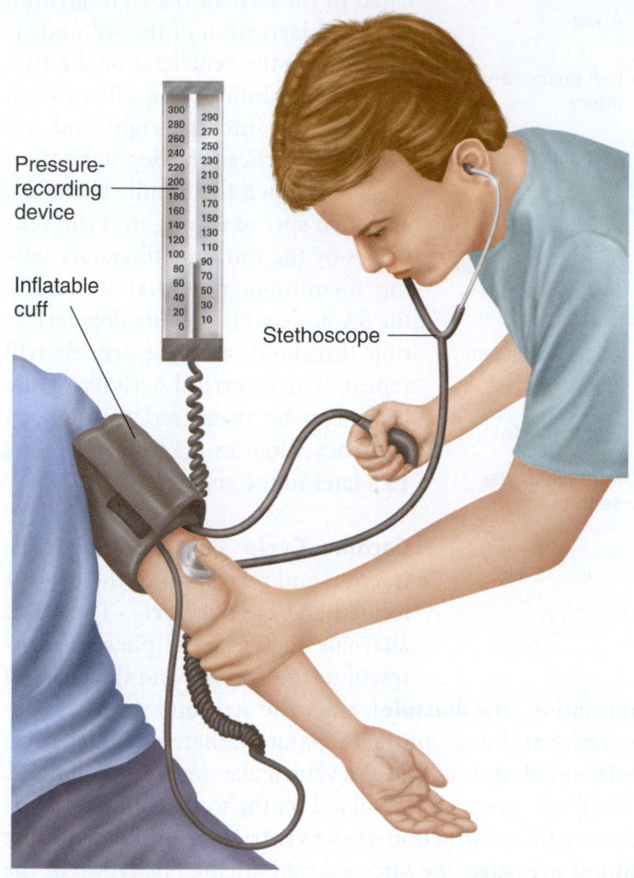

Source: L. Sherwood, *Human Physiology: From Cells to Systems*, 5e, copyright © 2004, p. 350.

end-diastolic volume (EDV), mean aortic blood pressure (mean arterial pressure/MAP), and strength of ventricular contraction (Sherwood 2004).

EDV refers to the amount of blood in the ventricles at the end of diastole. Often EDV is referred to as preload, since it is indicative of the possible amount of blood that can be forced out of the heart on the next ventricular contraction. The greater the EDV, the more the ventricles are stretched. Starling's Law of the heart indicates that this stretching of the ventricles will increase their force of contraction, allowing a greater amount of blood to be ejected. (According to Starling's Law, osmotic and hydrostatic pressure work together in favor of moving fluid out of the blood into interstitial areas at the arterial end of the capillary and restoring fluid back into blood at the venous end of the capillary; see Chapter 8.) The fraction of EDV that is ejected from the heart by contraction of the left ventricle is termed the **ejection fraction (EF)**.

EDV is primarily determined by venous return, the amount of blood that is returned to the heart by the veins. Several factors increase venous return. First, venoconstriction reduces the amount of blood stored in the veins by decreasing their capacity. Venous return is also increased by respirations. During inspiration, abdominal pressure increases and pressure within the thorax decreases, promoting blood flow back to the heart. Rhythmic skeletal muscle contractions, termed a skeletal muscle pump, also affect venous return. As the muscles contract, blood is pushed toward the heart because the muscles compress the veins. Valves within the veins prevent blood from flowing away from the heart (Sherwood 2004).

The mean arterial pressure (MAP) also affects stroke volume. MAP (also termed afterload) is the average force exerted by blood against the walls of the arteries over a cardiac cycle. MAP represents the resistance against which the ventricles must contract in order to eject blood into systemic circulation. MAP can be calculated using the values of systolic and diastolic blood pressure: MAP = [(2 × diastolic) + systolic] / 3. The normal range for MAP is 70–110, which is adequate to perfuse all tissues and organs. If there is increased resistance such as that seen in atherosclerosis, the ventricles will eject less blood, and could eventually lead to the complications of heart failure.

The strength of ventricular contraction is affected by circulating epinephrine and norepinephrine, as well as the sympathetic stimulation of the heart by cardiac accelerator nerves. These mechanisms increase the amount of calcium available to the myocardial cells, thus increasing contractility (Sherwood 2004).

left ventricular end-diastolic volume (LVEDV)—the amount of blood in the left ventricle at the end of the diastolic phase and immediately prior to systolic ejection of blood

left ventricular end-systolic volume (LVESV)—the amount of blood that remains in the left ventricle at the conclusion of the systolic phase

ejection fraction—the percentage of the LVEDV that is ejected in the systolic phase; in normal, apparently healthy adults, the typical ejection fraction is 50% to 60%; defined mathematically as stroke volume ÷ LVEDV

cardiac output—the volume of blood ejected from the left ventricle each minute; mathematically defined as heart rate × stroke volume

Regulation of Blood Pressure

MAP is determined through a combination of **cardiac output** and total peripheral resistance. The MAP must be regulated so that it is high enough to force blood through systemic circulation without being so high as to cause vascular damage.

The regulation of MAP (see Figure 15.5) involves the sympathetic nervous system, the renin-angiotensin system, and renal function. All three affect cardiac output and thus blood pressure (BP). First, cardiac output is equal to heart rate multiplied by stroke volume. Heart rate is dependent

FIGURE 15.5 Factors Influencing Arterial Blood Pressure

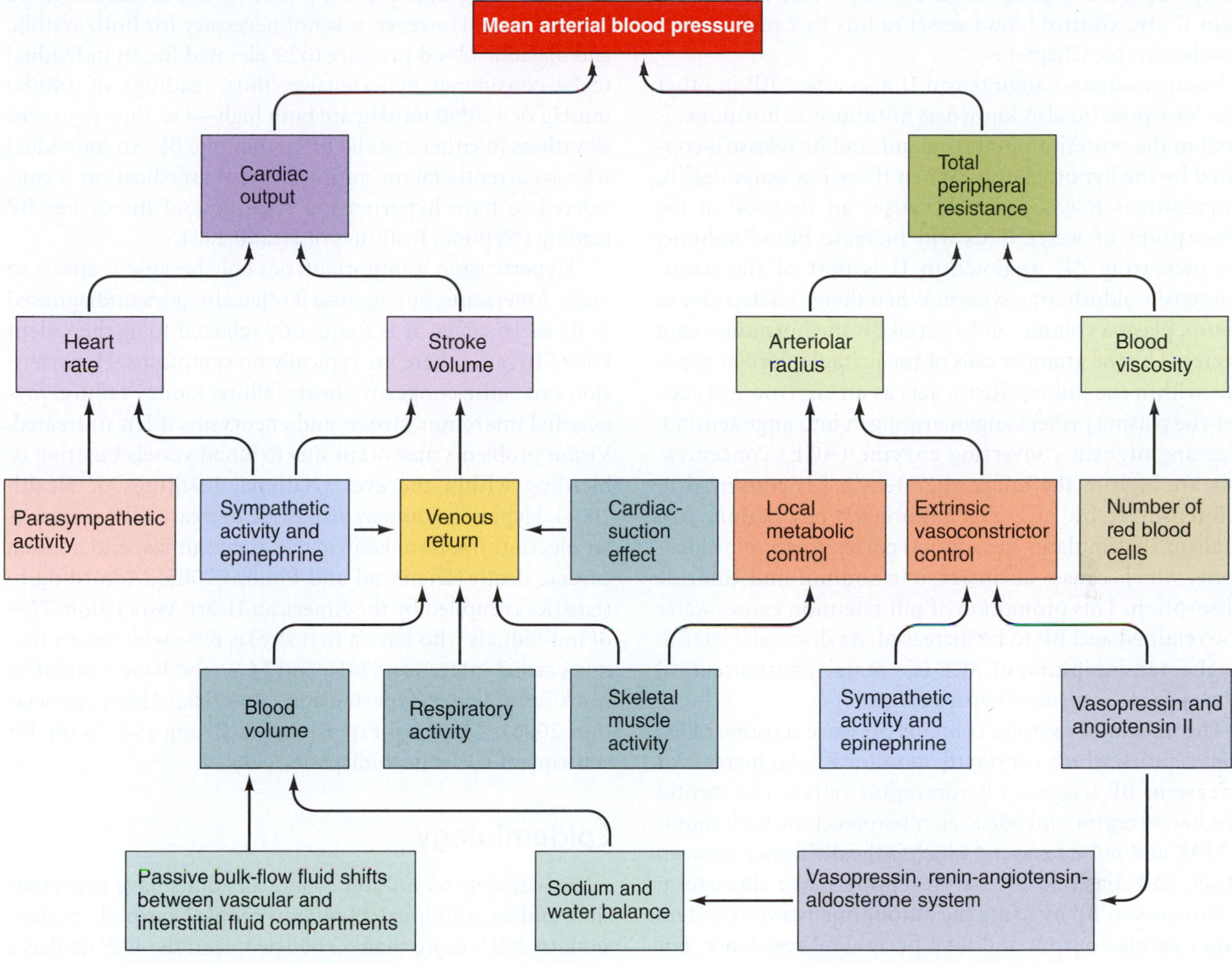

Source: L. Sherwood, *Human Physiology: From Cells to Systems*, 5e, copyright © 2004, p. 376.

upon the balance between parasympathetic activity, which decreases heart rate, and sympathetic activity, which increases heart rate. The parasympathetic nervous system acts to decrease heart rate through part of the tenth cranial nerve (the vagus nerve), which stimulates both the SA and AV nodes. When stimulated, the fibers release acetylcholine, which causes a decrease in heart rate. The sympathetic fibers, which are part of the cardiac accelerator nerves, stimulate the SA node and ventricles. When stimulated, these fibers release norepinephrine, which causes an increase in heart rate (Sherwood 2004).

Blood flow is directly proportional to the change in pressure and inversely proportional to resistance to flow.

$$\text{Flow} = \Delta \text{ pressure / resistance}$$

$$\text{Resistance} = \frac{(\text{length of vessel} \times \text{viscosity of the blood})}{(\text{radius})^4}$$

Resistance is dependent upon the radius of all arterioles, length of the vessel, and the blood viscosity. Arteriolar radius is the more important factor in determining peripheral resistance. Resistance to flow is inversely proportional to the fourth power of the radius of the vessel. Thus, a small reduction in the radius of a vessel would cause a great increase in resistance (Rhoades and Pflanzer 2003). Blood viscosity is determined by the number of formed elements (hematocrit). If the hematocrit is higher than normal, this will result in greater viscosity and greater resistance to flow. Thus, an athlete who is supplementing with erythropoietin could potentially have a greater blood viscosity than normal.

The radius is controlled by several factors. Local metabolic controls in skeletal muscles within a particular region of the body may cause vasodilation and increase blood flow to those muscles in order to match metabolic needs. The vasodilation would decrease resistance by increasing the radius of the vessel. Vasoconstriction will decrease vessel radius and increase resistance. This is caused by sympathetic activity and epinephrine. The hormones vasopressin and angiotensin II also control blood vessel radius by causing vasoconstriction (see Chapter 8).

Vasopressin and angiotensin II also affect BP in other ways. Vasopressin, also known as antidiuretic hormone, is stored in the posterior pituitary gland, and its release is controlled by the hypothalamus. When there is a water deficit, vasopressin is released, which causes an increase in the reabsorption of water. This will increase blood volume, thus increasing BP. Angiotensin II is part of the renin-angiotensin-aldosterone system. When there is a decrease in sodium, plasma volume, and arterial BP, the hormone renin is secreted by the granular cells of the juxtaglomerular apparatus within the kidney. Renin acts as an enzyme and activates the plasma protein angiotensinogen into angiotensin I. Since angiotensin converting enzyme (ACE) concentrations are high in the lungs, angiotensin I is converted to angiotensin II by ACE via pulmonary circulation. Angiotensin II stimulates the adrenal cortex to secrete aldosterone, which causes an increase in sodium and chloride reabsorption. This promotion of salt retention causes water to be retained and BP to be increased. As discussed later in this chapter, inhibition of ACE is a major pharmaceutical pathway for treatment of hypertension.

The circulatory system contains pressure sensors, called baroreceptors, which constantly monitor BP. An increase or decrease in BP triggers a baroreceptor reflex. The carotid sinus baroreceptor and aortic arch baroreceptor both monitor MAP and pulse pressure, which is the difference between systolic and diastolic BP. Baroreceptors make short-term adjustments to BP by using the autonomic nervous system to alter cardiac output and total peripheral resistance. For long-term adjustments, urine output and thirst are regulated in order to restore normal sodium, chloride, and water balance. If either cardiac output or peripheral resistance increases without a compensatory decrease in the other, BP will increase (Sherwood 2004).

Hypertension

Hypertension refers to a chronic elevation in BP. The Seventh Report of the Joint National Committee on Prevention, Detection, Evaluation and Treatment of High Blood Pressure (JNC-7) classifies hypertension according to the criteria shown in Figure 15.6.

A measurement of blood pressure is expressed using the reading for systolic pressure as the first (higher) number and the reading for diastolic pressure as the second (lower) number. A reading above 140/90 mmHg is considered to be hypertensive. However, it is not necessary for both systolic and diastolic blood pressure to be elevated for an individual to be considered hypertensive; thus, readings of 140/80 mmHg or 120/90 mmHg are both high—i.e., they represent elevations in either systolic BP or diastolic BP. An individual who is currently taking antihypertensive medication is considered to have hypertension regardless of his or her BP reading (National Institutes of Health 2004).

Hypertension is important, not only because it affects so many Americans, but because it often also goes undiagnosed in its early stages. It is frequently referred to as the "silent killer," because there are typically no symptoms. Hypertension can cause congestive heart failure, kidney failure, myocardial infarction, stroke, and aneurysms if left untreated. Vision problems may occur due to blood vessels bursting or bleeding within the eyes (National Institutes of Health 2004). Hypertension may also cause decreased left ventricular ejection fraction, ventricular arrhythmias, and sudden cardiac death (Diamond and Phillips 2005). According to statistics compiled by the American Heart Association, 77% of individuals who have a first stroke, 69% who have a first myocardial infarction (MI), and 74% who have congestive heart failure have hypertension (American Heart Association 2006). Thus, hypertension is a strong risk factor for subsequent CVD morbidity/mortality.

hypertension—condition of chronically elevated blood pressure

Epidemiology

Approximately 65 million American adults have hypertension, and an additional 59 million people have prehypertension. In 2003, more than 52,600 people in the U.S. died as a direct result of hypertension. Hypertension occurs in one out of every three adults. In addition, hypertension was listed as a primary or contributing cause of more than 277,000 deaths in the U.S. that year (American Heart Association 2006).

The rates of hypertension vary by gender and ethnic group. Hispanics/Latinos have the lowest prevalence among ethnic groups, at 19.0%. Mexican-Americans have higher rates, at 27.8 and 28.7% for males and females, respectively. White males, at 30.6%, have a lower prevalence of hypertension than white females, at 31.0%. The prevalence of hypertension is highest among blacks, at 41.8% for males and 45.4% for females (American Heart Association 2006).

Etiology

There are two types of hypertension. Primary or essential hypertension is idiopathic, which means there is no known cause, and accounts for about 90% of all cases. Hence, the overwhelming majority of hypertension is idiopathic (Gordon, Leighton, and Mooss 2006). Secondary hypertension

FIGURE 15.6 Seventh Report of the Joint National Committee on Prevention, Detection, Evaluation, and Treatment of High Blood Pressure (JNC-7)

EVALUATION

Classification of Blood Pressure (BP)*

Category	SBP mmHg		DBP mmHg
Normal	<120	and	<80
Prehypertension	120–139	or	80–89
Hypertension, Stage 1	140–159	or	90–99
Hypertension, Stage 2	≥160	or	≥100

* See *Blood Pressure Measurement Techniques* (reverse side)
Key: SBP = systolic blood pressure DBP = diastolic blood pressure

Diagnostic Workup of Hypertension

- Assess risk factors and comorbidities.
- Reveal identifiable causes of hypertension.
- Assess presence of target organ damage.
- Conduct history and physical examination.
- Obtain laboratory tests: urinalysis, blood glucose, hematocrit and lipid panel, serum potassium, creatinine, and calcium. Optional: urinary albumin/creatinine ratio.
- Obtain electrocardiogram.

Assess for Major Cardiovascular Disease (CVD) Risk Factors

- Hypertension
- Obesity (body mass index ≥30 kg/m²)
- Dyslipidemia
- Diabetes mellitus
- Cigarette smoking
- Physical inactivity
- Microalbuminuria, estimated glomerular filtration rate <60 mL/min
- Age (>55 for men, >65 for women)
- Family history of premature CVD (men age <55, women age <65)

Assess for Identifiable Causes of Hypertension

- Sleep apnea
- Drug induced/related
- Chronic kidney disease
- Primary aldosteronism
- Renovascular disease
- Cushing's syndrome or steroid therapy
- Pheochromocytoma
- Coarctation of aorta
- Thyroid/parathyroid disease

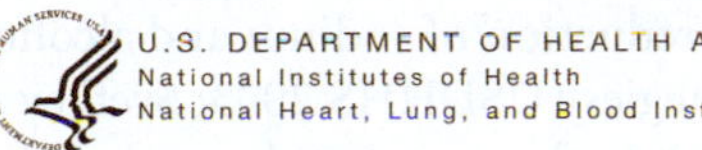

U.S. DEPARTMENT OF HEALTH AND HUMAN SERVICES
National Institutes of Health
National Heart, Lung, and Blood Institute

TREATMENT

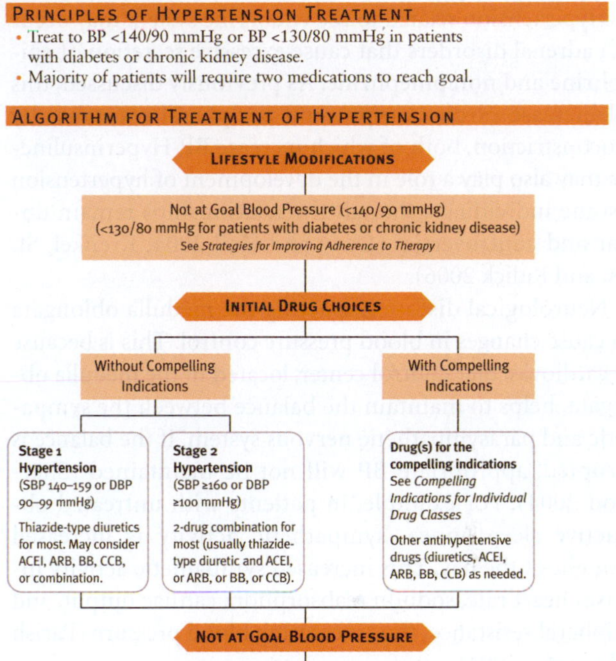

Optimize dosages or add additional drugs until goal blood pressure is achieved. Consider consultation with hypertension specialist.

See *Strategies for Improving Adherence to Therapy*

Source: National Heart Lung Blood Institute.

occurs as a result of another primary problem, such as renal disease, other cardiovascular disease, endocrine disorders, or neurogenic disorders, such as might occur in the compression of cranial nerves (Sherwood 2004).

Though its cause is unknown, primary hypertension may be a result of a variety of factors. Lifestyle factors such as diet (including excessive sodium intake), exercise, smoking, stress, and obesity all contribute to the development of primary hypertension. Poor lifestyle choices may exacerbate the problem, since it appears to have a strong genetic component. Numerous genes have been found that contribute to the management of sodium balance and most probably play a role in the development of hypertension (Cambien 2005; Diamond and Phillips 2005; Meneton et al. 2005). The development and progression of hypertension may also be due to inflammatory responses and individual differences within the renin-angiotensin-aldosterone control of blood pressure. (Savoia and Schiffrin 2006). Dietary factors also play a role in the development of hypertension (Svetkey et al. 2004; Appel et al. 2006). This is discussed in detail under Nutrition Therapy later in this chapter.

Pathophysiology

Vasopressin and angiotensin II, as previously mentioned, cause vasoconstriction and fluid retention. Both will increase BP. Often hypertensive individuals have excessive secretion of vasopressin from the hypothalamus. Hypertensive individuals may also have a variation in the gene that produces angiotensinogen. An increased production of angiotensinogen may increase production of angiotensin II, thus increasing BP.

Though the mechanisms are not fully understood, smoking is known to be a risk factor for the development of hypertension. Smoking causes acute and chronic elevations in blood pressure. The relationship may be partially explained by the fact that cigarette smoking interferes with the action of nitrous oxide, thus impairing endothelial relaxation and vasodilation (Marks 2001).

In renal disease, blood flow is reduced through the kidney, because of either atherosclerosis within the lumen of a renal artery or compression of a vessel by a tumor. In order to improve blood flow, angiotensin II is released. This causes vasoconstriction and promotes sodium, chloride, and water retention, which increase blood volume. The increase in blood volume and vasoconstriction both act to increase arterial pressure (Appel et al. 2006).

Hypertension related to the endocrine system may occur with adrenal disorders that cause excessive secretion of epinephrine and norepinephrine. As previously discussed, this will increase cardiac output and peripheral resistance by vasoconstriction, both of which increase BP. Hyperinsulinemia may also play a role in the development of hypertension in some individuals, though the relationships remain unclear and controversial (Davy and Hall 2004; Krenkel, St. Joer, and Kulick 2006).

Neurological disease impacting the medulla oblongata can cause changes in blood pressure control. This is because the cardiovascular control center, located in the medulla oblongata, helps to maintain the balance between the sympathetic and parasympathetic nervous system. If the balance is disrupted, appropriate BP will not be maintained (Sherwood 2004). For example, in patients with untreated obstructive sleep apnea, sympathetic activity is increased throughout the day. This increased sympathetic activity increases heart rate, sodium reabsorption, cardiac output, and peripheral resistance, thus increasing blood pressure (Parish and Somers 2004).

Treatment

The goals of treatment for hypertension are (1) reduction in the risk of cardiovascular and renal disease, and (2) reduction of BP to <140/80 mmHg (or to <130/80 mmHg in those individuals with diabetes or chronic renal disease). This is achieved through a comprehensive plan involving weight reduction, physical activity, nutrition therapy, and pharmacological interventions. Lifestyle modifications, nutrition therapy, and physical activity will be discussed later in this section and within this chapter.

In order to change BP, either cardiac output or peripheral resistance must be altered. Pharmacological interventions include several major classes of medications that use one or both of these mechanisms (see Table 15.1). Major groups of diuretics include "loop" diuretics, thiazides, carbonic anhydrase inhibitors, and potassium-sparing diuretics. Loop diuretics (furosemide, bumetanide, torsemide) act by inhibiting sodium, chloride, and potassium reabsorption in the loop of Henle of the kidney (see Chapter 20 for the anatomy/physiology of the kidneys). Loop diuretics also increase prostaglandins, resulting in vasodilation. Thiazides (hydrochlorothiazide) also inhibit the reabsorption of sodium, chloride, and potassium, but primarily act in the distal tubule and the ascending loop of Henle. Carbonic anhydrase inhibitors (acetazolamide, methazolamide), prevent the exchange of hydrogen ions with sodium and water by blocking the enzyme carbonic anhydrase. Potassium-sparing diuretics (spironolactone, amiloride) act within the collecting and convoluted tubules, where they prevent sodium-potassium exchange and reduce aldosterone stimulation.

Medications called ACE-inhibitors (Captopril) competitively block the c enzyme that converts angiotensin I into angiotensin II. This results in vasodilation, a decrease in vasopressin release, and a resulting decrease in BP.

Beta-adrenergic blocking agents (propanolol, atenolol, acebutolol) block β-receptors in the heart to decrease rate and cardiac output. Alpha-receptor antagonists (Cardura, Minipress, Hytrin) block the vascular muscle action that normally responds to sympathetic stimulation. This reduces stroke volume and thus BP. Calcium channel blocking agents (Verapamil) affect the movement of calcium, which causes the blood vessels to relax and therefore reduces vasoconstriction. The final class of medications that can be used in treatment for hypertension are the aldosterone antagonists (spironolactone and eplerenone), which suppress the actions of aldosterone. The type of medication regimen is determined by the classification of hypertension and other risk factors (see Figure 15.6).

Nutrition Therapy

Nutrition Implications Nutritional treatment of hypertension includes both lifestyle modifications and nutrition therapy. Increased physical activity, smoking cessation, and weight loss, as well as reduction of sodium and alcohol intake, are primary strategies (USDHHS 2003; Svetkey et al. 2004; Appel et al. 2006). In the past decade, several clinical trials—including the landmark Dietary Approaches to Stop Hypertension (DASH) and the PREMIER trials—have revealed that nutrition interventions that include decreasing sodium, saturated fat, and alcohol while increasing calcium, potassium, and fiber have demonstrated significant effects for lowering blood pressure (see Table 15.2) (Svetkey et al 2003; McGuire et al. 2004; Brunner et al. 2005; Carey et al. 2005; Geleijnse, Grobbee and Kok 2005; NHLBI 2005; Steffen et al. 2005; Appel et al. 2006).

Nutrition Interventions A comprehensive approach that addresses multiple lifestyle factors has the most significant effect on blood pressure control for hypertensive individuals (see Table 15.3) (Chobanian et al. 2003; Thompson et al. 2003; McGuire et al. 2004; Khan et al. 2005; Appel et al. 2006). The sections that follow address each component of the current recommendations for blood pressure control.

Weight Loss Weight reduction is a standard component of nutrition therapy for treatment of hypertension (He et al. 2000; Thompson et al. 2003; Avenell et al. 2004). Meta-analyses of studies conducted between 1966 and 2002 continue to support this recommendation. Within the 26 studies recently reviewed, weight loss of greater than 5 kg reduced both diastolic and systolic BP (Neter et al. 2003). An

TABLE 15.1

Selected Drugs Used in Cardiac Care

Classification	Mechanism	Generic	Brand Name	Possible Food–Drug Interactions/Side-Effects
Diuretics	Decrease blood volume by increasing urinary output; inhibit renal sodium and water reabsorption	Furosemide Hydrochlorothiazide	Lasix HydroDIURIL	Hypokalemia, hyperlipidemia, hypertriglyceridemia hypercholesterolemia, glucose intolerance; N/V, anorexia, dry mouth, diarrhea, constipation; potassium supplements may be necessary; contraindications: effect antagonized by by NSAIDS, avoid natural licorice
Calcium Channel Blockers	Affect the movement of calcium, cause blood vessels to relax; therefore, reduce vasoconstriction	Nisoldipine Nifedipine Nicardipine Bepridil Diltiazem Verapamil	Sular Adalat/Procardia Cardene Vascor Cardizem Calan/Isoptin	Edema, nausea, heartburn; contraindications: heart failure or greater than 1st degree heart block, avoid natural licorice, limit caffeine, avoid or limit alcohol
ACE Inhibitors	Vasodilators that reduce BP by decreasing peripheral vascular resistance by interfering with the production of angiotensin II from angiotensin I and inhibiting degradation of bradykinin	Captopril Benazepril Enalapril Lisinopril Ramipril	Capoten Lotensin Vasotec Prinivil/Zestril Altace	Hypotension, esp. in elderly patients; can worsen renal function, hyperkalemia, dysgeusia; causes dry, nonproductive cough, hyperkalemia; contraindications: pregnancy, avoid natural licorice, avoid salt substitutes
Angiotensin II Receptor Blockers	Interferes with renin-angiotensin system without inhibiting degradation of bradykinin	Candesartan Eprosartan Irbesartan Telmisartan Valsartan	Atacand Teveten Avapro Micardis Cozaar	May increase serum potassium; avoid salt substitutes; nausea, dysgeusia
Alpha-Adrenergic Blockers	Blocks the vascular muscle response to sympathetic stimulation; reduces stroke volume	Alfuzosin Terazosin Tamsulosin Prazosin		Avoid natural licorice; nausea/vomiting, diarrhea, mouth dryness
Nitrate	Vasodilation	Nitroglycerin	Nitro-Bid Nitro-Dur Nitrostat Transderm-Nitro Minitran Deponit Nitrol	Nausea, vomiting, abdominal pain, dryness of mouth
Beta-1-Blocker	Blocks β-receptors in heart to decrease heart rate and cardiac output	Metoprolol Atenolol Acebutolol	Lopressor Tenormin Sectral	Nausea, diarrhea; calcium may interfere with absorption; upset stomach, dry mouth, stomach pain, gas or bloating, heartburn
Aldosterone Antagonists	Interrupt aldosterone, which increases sodium and water excretion	Spironolactone	Aldactone	May increase serum potassium; avoid salt substitutes; dysgeusia, upset stomach, vomiting, diarrhea, stomach pain
Digitalis	Increases strength of heart contractions; slows the electrical conduction between the atria and ventricles	Digoxin	Lanoxin	Diarrhea, loss of appetite, lower stomach pain, nausea, and/or vomiting
Fibrinolytic Therapy	Interrupts prothrombin which reduces ability of blood to clot	Heparin Alteplase Reteplase Streptokinase	Heparin Activase Retavase Streptase	Abdominal pain, nausea, constipation
Positive Inotropic Drugs	Stimulate heart rate; increase heart contractions	Dopamine HCL Milrinone	Dopamine Primacor	May decrease serum potassium; proteinuria; nausea, vomiting

TABLE 15.2

Effects of Dietary Factors and Dietary Patterns on BP: A Summary of the Evidence

	Hypothesized Effect	Evidence
Weight	Direct	++
Sodium Chloride (salt)	Direct	++
Potassium	Inverse	++
Magnesium	Inverse	+/−
Calcium	Inverse	+/−
Alcohol	Direct	++
Fat: Saturated	Direct	+/−
Omega-3 Polyunsaturated Fat	Inverse	++
Omega-6 Polyunsaturated Fat	Inverse	+/−
Monounsaturated Fat	Inverse	+
Protein: Total	Uncertain	+
Protein: Vegetable	Inverse	+
Protein: Animal	Uncertain	+/−
Fiber	Inverse	+
Cholesterol	Direct	+/−
Dietary Patterns: Vegetarian Diets	Inverse	++
DASH Type Dietary Patterns	Inverse	++

Key: +/− indicates limited or equivocal evidence; + suggestive evidence, typically from observational studies and some clinical trials; ++ persuasive evidence, typically from clinical trials.

Source: Table 2, p. 305: Appel LJ, Brands MW, Daniels SR, Karania N, Elmer PJ, Sacks FM; American Heart Association. Dietary approaches to prevent and treat hypertension: a scientific statement from the American Heart Association. *Hypertension.* 2006; Feb;47(2):296–308.

TABLE 15.3

Effects of Lifestyle Modification to Manage Hypertension

Recommendation	Average Systolic BP Reduction
Weight reduction to maintain a BMI 18.5–24.9	5–20 mmHg/10 kg
Diet rich in fruits, vegetables, & low-fat dairy products with reduced saturated & total fat − the DASH Eating Plan	8–14 mmHg
Intake of not >100 mEq/day (2.4 g sodium or 6 g sodium chloride)	2–8 mmHg
Aerobic activity, such as brisk walking for 30 min/day, most days of the week	4–9 mmHg
Most men: Not >2 drinks/day Women & lighter weight men: Not >1 drink/day	2–4 mmHg

*DASH—Dietary Approaches to Stop Hypertension.

University of Washington Department of Medicine.

Sources: Appel LJ, et al. A clinical trial of the effects of dietary patterns on blood pressure. (*N Engl J Med* 1997;336:1117–1124). Sacks FM, et al. Effects on blood pressure of reduced dietary sodium and the dietary approaches to stop hypertension (DASH) diet. (*N Engl J Med* 2001;344:3–10). JNC 7 USDHHS, 2003.

approximate 20 lb weight loss will result in lowered systolic BP (Appel et al. 1997; Sacks et al. 2001), and even less than 10% weight loss has a sustained effect on BP (Stevens et al. 2001).

Though waist circumference (see Figure 15.7) is related to body weight, it is an independent predictor of hypertension risk. For those patients who fall within a normal or overweight BMI, waist circumference should be measured. It is not necessary to measure waist circumference for those patients with a BMI >35, because it adds no additional predictive power (NHLBI 2006).

salt-sensitive—describes an individual who experiences an increase in blood pressure as a result of salt intake

salt-resistant—describes an individual whose body presents resistance to change in blood pressure as a result of salt intake

Sodium Although the use of sodium restriction to manage BP has been a controversial issue, consistent evidence has supported the efficacy of a reduction of sodium for controlling BP (Conlin et al. 2000; Stamler et al. 2000; Vollmer et al. 2001; He and MacGregor 2002; Luepker et al. 2003). Large population studies, such as the INTERSALT Study, have confirmed that urinary sodium excretion has a significant and direct relationship with systolic blood pressure (Stamler 1997; Freedman and Petitti 2001; Beevers 2002). It has been estimated that sodium modifications may reduce incidence of hypertension by as much as 17% (Beevers 2002). The DASH trials have further supported the role of sodium reduction in treatment for hypertension (Vollmer et al. 2001; Svetkey et al. 2004). BP control through sodium restriction could reduce the incidence of cardiovascular disease, renal disease, and stroke (Beevers 2002; Geleijnse, Kok and Grobbee 2005).

Individual response to sodium restriction can vary (Obarzanek et al. 2003). Nonetheless, Appel et al. (2006) argue that, despite descriptions of people as **"salt sensitive"** or **"salt resistant,"** changes in BP as a result of sodium restriction are seen across all ages and all ethnicities. The DASH-Sodium trial tested the response to three different sodium levels and provided further evidence of a reduction in blood pressure with sodium restriction. The most signifi-

FIGURE 15.7 Waist Circumference

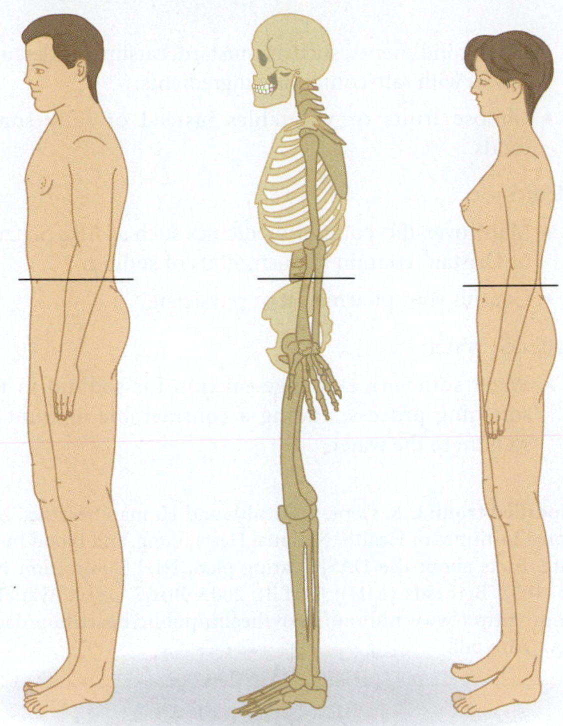

Source: National Heart, Lung, and Blood Institute.

cant reductions were seen in blacks, older individuals, and individuals with comorbidities such as diabetes mellitus (Sacks et al. 2001; Appel et al. 2006).

Americans consume high amounts of sodium, in part because of the amounts used in processed foods (Appel et al. 2006). Average sodium intake for Americans ranges between 3,000 and 4,500 mg/day (130–195 mEq Na or 8–10 g of sodium chloride). The *Dietary Guidelines for Americans* recommend an intake of less than 2,300 mg of sodium, the equivalent of 6 g of sodium chloride (table salt), each day (USFDA 2005). This goal is supported by the most recent statement from the American Heart Association (Appel et al. 2006).

Because only small amounts of sodium occur naturally in food, effective reduction of sodium intake requires limiting the intake of highly processed foods, avoiding those foods that are cured using salt, and omitting salt during the cooking and preparation of foods. The practitioner should teach the client strategies for limiting intake to 2,400 mg/day (104 mEq) and provide information on the sodium content of foods (see Table 15.4). The DASH diet in Appendix E10 gives specific guidelines for comprehensive nutrition therapy. Boxes 15.1 and 15.2 list practical steps for controlling sodium intake.

TABLE 15.4

Sodium Content of Foods

Sodium content varies by processing. The AI for Sodium is 1,500 mg.

Food Groups	Sodium (mg)
Grains and grain products	
Cooked cereal, rice, pasta, unsalted, ½ cup	0–5
Ready-to-eat cereal, 1 cup	100–360
Bread, 1 slice	110–175
Vegetables	
Fresh or frozen, cooked without salt, ½ cup	1–70
Canned or frozen with sauce, ½ cup	140–460
Tomato juice, canned ¾ cup	820
Fruit	
Fresh, frozen, canned, ½ cup	0–5
Low-fat or fat free dairy foods	
Milk, 1 cup	120
Yogurt, 8 oz	160
Natural cheeses, 1 ½ oz	110–450
Processed cheeses, 1 ½ oz	600
Nuts, seeds, and dry beans	
Peanuts, salted, 1/3 cup	120
Peanuts, unsalted, 1/3 cup	0–5
Beans, cooked from dried, or frozen, without salt, ½ cup	0–5
Beans, canned, ½ cup	400
Meats, fish, and poultry	
Fresh meat, fish, poultry, 3 oz	30–90
Tuna canned, water pack, no salt added, 3 oz	35–45
Tuna canned, water pack, 3 oz	250–350
Ham, lean, roasted, 3 oz	1,020

Reprinted from: U.S. Dept. of Health and Human Services, National Institutes of Health, National Heart, Lung, and Blood Institute. Facts about the DASH eating plan. NIH Publication No. 03–4082. Bethesda (MD): NHLBI; 2003. Box 5, p. 7. Available from: http://www.nhlbi.nih.gov/health/public/heart/hbp/dash/new_dash.pdf.

Alcohol As alcohol intake increases above 2 drinks per day for men (and 1 drink/day for women), the risk of hypertension increases accordingly, in a dose-dependent relationship (Xin et al. 2001; Appel et al. 2006). The U.S. *Dietary Guidelines* and JNC 7 recommend limiting alcohol intake to ≤2 drinks per day for men and ≤1 drink per day for women. One drink is defined as 12 oz. of beer, 5 oz. of wine, or 1.5 oz. of 80 proof distilled spirits (USDHHS 2003; USFDA 2005).

BOX 15.1 CLINICAL APPLICATIONS

Other Methods to Reduce Sodium Intake

Important Tips:

- One teaspoon of table salt is equivalent to 2,300 mg sodium.
- If you wish to stay within a 4 gram Na diet, each food item should have no more than 300 mg of sodium per serving. To stay within a 2 gram Na diet, each food item should have no more than 200 mg of sodium per serving.

Eating Out:

- Avoid fast-food restaurants, because these food choices are prepared with large amounts of salt.
- Ask how foods are prepared. Request that they be prepared without added salt, monosodium glutamate (MSG), or salt-containing ingredients. Most restaurants are willing to accommodate requests.
- Know the terms that indicate high sodium content: pickled, cured, soy sauce, broth.
- Do not use any added salt in food preparation or at the table.
- Limit condiments, such as mustard, catsup, pickles, and sauces with salt-containing ingredients.
- Choose fruits or vegetables instead of salty snack foods.

Medicines:

- Many over-the-counter medicines such as Alka Seltzer® or Dristan® contain high amounts of sodium.
- Consult your pharmacist or physician.

Softened Water:

- Water softeners exchange calcium for sodium in the softening process, adding a considerable amount of sodium to the water.

Modified from: U.S. Dept. of Health and Human Services, National Institutes of Health, National Heart, Lung, and Blood Institute. Facts about the DASH eating plan. NIH Publication No. 03–4082. Bethesda (MD): NHLBI; 2003. Box 7, p. 10. Available from: http://www.nhlbi.nih.gov/health/public/heart/hbp/dash/new_dash.pdf.

Potassium, Calcium, and Magnesium Potassium, calcium, and magnesium have all been positively correlated with reduction of BP and treatment of hypertension. The role of these minerals as part of the nutrition therapy for hypertension is highlighted by the results of the DASH studies. All three minerals appear to have an inverse relationship to hypertension—suggesting that as dietary intakes increase, BP decreases (Appel et al. 2006; Svetkey et al. 2004; Jee et al. 2002; Vollmer et al. 2001).

The relationship between potassium and BP is a strong inverse relationship (Geleijnse, Kok, and Grobbee 2005). The diet used in the DASH trials provided an average of 4–6 g of potassium/day from fruits and vegetables (see Chapter 20 for food sources of potassium). These intakes were associated with reduced blood pressures. As Appel et al. (2006) emphasize, an increased potassium intake does not pose a health risk in healthy individuals. In those clients who may have an impaired urinary excretion of potassium, however, these recommendations may need to be modified.

The relationship between calcium and hypertension has been studied for over 25 years. The most dramatic relationship between calcium and blood pressure reduction was seen in the DASH trials. The DASH diets provided the equivalent of 3 cups of dairy products as their major source of calcium. In a recent trial, intakes of lower-fat milk and milk products were correlated with lower rates of hypertension, but this relationship was not sustained for whole-milk products (Alonso et al. 2005). At present, more specific recommendations for calcium intake in hypertension have not been established beyond the recognized DRI levels (Appel et al. 2006).

It is important to remember that the nutritional effects demonstrated by the DASH study—and in particular, the relationship between K, Ca, Mg, and blood pressure reduction—were a result of a dietary *pattern* rich in these nutrients rather than mineral intake from *supplements*.

DASH—Dietary Approaches to Stop Hypertension As discussed earlier in this section, the concept of approaching nutrition therapy for hypertension with a comprehensive dietary method was brought to the forefront with the Dietary Approaches to Stop Hypertension (DASH) in the late 1990s (see Box 15.3) (Appel et al. 1997). These clinical trials focused on the use of a variety of foods that not only reduced sodium intake but increased potassium, magnesium, calcium, and fiber intakes within a moderate energy intake. At 2000 kcal a day, the DASH-Sodium Diet provides approximately 4,700 mg (120 mEq) potassium, 500 mg magnesium, 1,240 mg calcium, 90 g protein, 30 g fiber, and 2,400 mg (100 mEq) sodium (NHLBI 2005). The Canadian Hypertension Education Program Evidence-Based Recommendations Task Force has further supported this multi-faceted plan for nutrition therapy and lifestyle change (Khan et al. 2005). It has been proposed that the blood pressure reductions seen in both the DASH and OMNI HEART trials are most likely a synergistic effect of increasing potassium, magnesium, calcium, and fiber while reducing sodium and saturated fat (Appel et al. 2006; Appel et al. 2005; Carey et al. 2005). Initiating these broad dietary changes and then continuing them lifelong is a realistic ap-

BOX 15.2 CLINICAL APPLICATIONS

Using the Food Label to Identify Foods Lower in Sodium and Fat

1. **Check the ingredient list.** Ingredients that may indicate higher sodium content include baking soda, brine, monosodium glutamate (MSG), baking powder, disodium phosphate, or sodium benzoate.
2. **Assess the label language:**

Phrase	What It Means
Sodium	
Sodium free or salt free	Less than 5 mg per serving
Very low sodium	35 mg or less of sodium per serving
Low sodium	140 mg or less of sodium per serving
Low sodium meal	140 mg or less of sodium per 3½ oz. (100 g)
Reduced or less sodium	At least 25 percent less sodium than the regular version
Light in sodium	50 percent less sodium than the regular version
Unsalted or no salt added	No salt added to the product during processing
Fat	
Fat free	Less than 0.5 g per serving
Low saturated fat	1 g or less per serving
Low-fat	3 g or less per serving
Reduced fat	At least 25 percent less fat than the regular version
Light in fat	Half the fat compared to the regular version

Modified from: U.S. Dept. of Health and Human Services, National Institutes of Health, National Heart, Lung, and Blood Institute. Facts about the DASH eating plan. NIH Publication No. 03–4082. Bethesda (MD): NHLBI; 2003. Box 9, p. 10. Available from: http://www.nhlbi.nih.gov/health/public/heart/hbp/dash/new_dash.pdf.

proach to both the treatment of hypertension and the prevention other diseases (Brunner et al. 2005).

Physical Activity According to the JNC 7, physical activity of 30 minutes per day does decrease blood pressure. Moreover, increasing physical activity decreases the relative workload on the heart for all forms of activity, a benefit important for all forms of cardiovascular disease. For instance, mowing the lawn requires a certain percentage of one's maximal functional capacity. If a person starts a walking program and improves their cardiorespiratory fitness, then mowing the lawn will require a lower percentage of their functional capacity. Since the relative strain on the cardiovascular system will be reduced, the BP response to the activity will be reduced as well. Furthermore, increasing physical activity will facilitate weight management.

Smoking Cessation When an individual quits smoking they realize health benefits almost immediately. Smoking cessation may be the most important change any individual can make to reduce their risk of hypertension and all forms of cardiovascular disease (Gordon, Leighton, and Mooss 2006). All smoking cessation plans are not equal, and each individual should seek out a program that suits his or her needs. In order to achieve success, the smoker should also be able to identify his or her reasons for quitting. The American Lung Association has developed a three-step Quit Smoking Action Plan (2006) that can be accessed online at http://www.lungusa.org.

Developing the Nutrition Therapy Prescription Evidence-based guidelines should be utilized to provide nutrition therapy to patients/clients. Education is a key component of providing nutrition therapy. A recent Cochrane data analysis of 23 clinical trials confirmed that nutrition education increased fiber, fruit, and vegetable intake; lowered total dietary fat intake; and reduced blood pressure, LDL-cholesterol, and total serum cholesterol (Brunner et al. 2005).

Nutrition therapy is guided by the patient's hypertension history, other medical risk factors, and current medical treatment (see Box 15.4). First, the clinician should evaluate the need for weight reduction in order to move toward the goal of reaching a BMI of 18.5–24.9 (USDHHS 2003). Since the DASH diet is the foundation of nutrition therapy for hypertension, the clinician's next step should be to assess dietary intake for alcohol, sodium, potassium, calcium, and fiber. The practitioner and client should then work together to prioritize the methods to meet the DASH dietary goals. It is just as important to assess the client's physical activity levels and then tailor exercise goals to the individual, in order to meet the current recommendations for 30–60 min of aerobic exercise on a minimum of four days per week (Thompson et al. 2003; Khan et al. 2005; Appel et al. 2006).

Atherosclerosis

Definition

The term **atherosclerosis (AS)** comes from the Greek *athero,* meaning gruel, and *sclerosis,* meaning hardening. The terms

atherosclerosis (AS)—thickening of the blood vessel walls specifically caused by the presence of plaque

BOX 15.3 HISTORICAL DEVELOPMENTS

The History of DASH

Scientists from Brigham and Women's Hospital, (Boston, MA), Duke University Medical Center (Durham, NC), Johns Hopkins University (Baltimore, MD), and Pennington Biomedical Research Center, Louisiana State University (Baton Rouge, LA), who were supported by the National Heart, Lung, and Blood Institute (NHLBI), conducted two key studies in the 1990's. The first was called "DASH," and it tested the effects of nutrients, as they occur together in food, on BP. This study found that blood pressures were reduced with an eating plan that is low in saturated fat, cholesterol, and total fat, and that emphasizes 8–9 servings of fruits and vegetables, and three servings of low-fat dairy foods. This eating plan—known as the DASH eating plan—also includes whole-grain products, fish, poultry, and nuts. It is reduced in red meat, sweets, and sugar-containing beverages. It is rich in magnesium, potassium, and calcium, as well as protein and fiber.

The DASH study involved 459 adults with systolic blood pressures of less than 160 mmHg and diastolic pressures of 80–95 mmHg. About 27% of the participants had hypertension. About 50% were women and 60% were African-Americans. DASH compared three eating plans: a plan similar in nutrients to what many Americans consume; a plan similar to what Americans consume but higher in fruits and vegetables; and the DASH eating plan. All three plans included about 3,000 milligrams of sodium daily. None of the plans was vegetarian or used specialty foods. Results were dramatic: both the fruits and vegetables plan and the DASH eating plan reduced BP. But the DASH eating plan had the greatest effect, especially for those with high BP. Furthermore, the BP reductions came fast—within 2 weeks of starting the plan.

The second study was called "DASH-Sodium," and it looked at the effect on blood pressure of a reduced dietary sodium intake as participants followed either the DASH eating plan or an eating plan typical of what many Americans consume. DASH-Sodium involved 412 participants. Their systolic blood pressures were 120–159 mmHg and their diastolic blood pressures were 80–95 mmHg. About 41% of them had high blood pressure.

About 57% were women and about 57% were African-Americans. Participants were randomly assigned to one of the two eating plans and then followed for a month at each of three sodium levels. The three sodium levels were: a higher intake of about 3,300 milligrams per day (the level consumed by many Americans); an intermediate intake of about 2,400 milligrams per day; and a lower intake of about 1,500 milligrams per day. Results showed that reducing dietary sodium lowered BP for both eating plans. At each sodium level, BP was lower on the DASH eating plan than on the other eating plan. The biggest BP reductions were for the DASH eating plan at the sodium intake of 1,500 milligrams per day. Those with hypertension saw the biggest reductions, but those without it also had large decreases. These reductions occurred even when body weight remained stable. The magnitude of BP reduction with this dietary pattern was similar to the reduction noted with BP-lowering medications.

Modified from: U.S. Dept. of Health and Human Services, National Institutes of Health, National Heart, Lung, and Blood Institute. Facts about the DASH eating plan. NIH Publication No. 03–4082. Bethesda (MD): NHLBI; 2003. pp. 3–4. Available from: http://www.nhlbi.nih.gov/health/public/heart/hbp/dash/new_dash.pdf.

arteriosclerosis—a general term for thickening of the walls of the blood vessels with a resulting loss of vascular elasticity and narrowed lumen

infarct—cellular necrosis as a result of lack of oxygen

myocardial infarction (MI)—necrosis of the myocardial cells as a result of oxygen deprivation.

coronary artery disease (CAD)—general term for all causes of heart disease characterized by narrowing of vessels supplying blood to the heart

peripheral vascular disease (PVD)—atherosclerotic heart disease of all vessels except specific coronary vessels

congestive heart failure (CHF) or heart failure (HF)—impairment of the ventricles' capacity to eject blood from the heart or to fill with blood

AS and arteriosclerosis are often used interchangeably. **Arteriosclerosis** is a general term defined as a thickening of the walls of the vessels and a loss of vascular elasticity. Arteriosclerosis may be caused by an atherosclerotic plaque.

AS is the development of plaque in the vascular wall that will occlude the lumen of the vessel and create ischemic conditions (see Figure 15.8). The plaque begins as a fatty and fibrous growth, and over time may calcify. The development of an atherosclerotic plaque can result in a restriction of blood flow severe enough to cause an **infarct** resulting in a **myocardial infarction (MI)** or in a cerebrovascular accident (stroke). Therefore, AS is the root cause of two of the three leading causes of death in the United States, **coronary artery disease (CAD)** and stroke. In addition, an atherosclerotic plaque in the leg can result in **peripheral vascular disease (PVD)** that may result in the tissue death associated with gangrene and loss of a limb. Severe CAD may impair cardiac function to the point that **congestive heart failure (CHF)** results.

BOX 15.4 CLINICAL APPLICATIONS

Brief Nutrition Counseling for Hypertension: A Model for Physicians

Steps in Behavioral Counseling

Due to limited time, your approach will be more directive, with less opportunity for patient input.

Assess

- Food intake and diet habits in the context of health risks
- Current physical activity
- Readiness to change behavior
- From WAVE Nutrition Counseling Tool (http://bms.brown.edu/nutrition/acrobat/wave.pdf)
  - W=Weight: Review BMI, blood pressure, lipids, blood sugar to screen for metabolic syndrome.
  - A=Activity: Conduct physical activity assessment. Ask about:
    - Moderate physical activity? Goal—30 minutes/day or more
  - V=Variety and E=Excess: Based on DASH-Sodium Diet:
    - Number of low-fat dairy foods? Goal—2 to 3 servings/day
    - Number of fruits and vegetables? Goal—8 to 10 servings/day
    - Salt added to food? Goal—no salt added at the table, only half the usual amount in cooking
    - Use of frozen, canned, or dried processed foods (soup, spaghetti sauce, frozen dinners, helper mixes)? Goal—reduced sodium from processed foods
    - Alcohol intake? Goal—no more than one drink/day for women and two drinks/day for men

Advise

- Give clear, specific, and personalized behavior change advice. You might say:
  - "Diet changes, exercise, and weight loss can reduce your blood pressure as much as medicine."
  - For patients taking medication for diabetes, lipids, or hypertension: "Diet choices are important even if you are taking medication, since eating carefully helps the medicine do a better job. You may be able to save money by cutting down on the amount of medicine you take."
  - For patients NOT ready to change behavior, add: "I'd like to help you when you are ready to make changes in your diet and be more active."

Agree

- Collaborate with patient to select treatment goals and methods.
- Base goals on readiness to change behavior.
- For patient NOT ready to change behavior: "Is it okay if I ask you again at our next visit?"
- Possible goals for patient ready to change:
  - Return for further discussion in 2–4 weeks.
- Keep food and exercise records to increase awareness, if patient is willing.
  - Refer for registered dietitian visit.

Assist

- Help patient acquire knowledge, skills, and support for behavior.
- Provide hand-outs and Web resources, based on patient interest and need.
- Provide lists and recommendations for community resources (exercise and diet programs, health clubs, etc.).

Arrange

- Schedule follow-up appointments.

Source: Medical Nutrition Handbook, Department of Medicine, University of Wisconsin School of Medicine and Public Health (http://www.medicine.wisc.edu/mainweb/includes/viewfile.php?fileid=899&viewtype=inline§ion=naa).

Epidemiology

CVD is the leading cause of death in the United States and throughout the world, and greater than 50% of all diagnoses related to CVD result from atherosclerosis (American Heart Association 2006). Most of what we know about the epidemiology of AS and heart disease has come from large epidemiological studies that have provided an endless flow of data for over three decades. These studies—particularly the Framingham Study, National Health and Nutrition Examination Survey (NHANES), National Cholesterol Education Program (NCEP), and Heritage Study—continue to provide us with valuable information about factors related to the development of heart disease. The American Heart Association (2006) statistics indicate that over 12 million individuals are affected by AS, and that it results in more than one half million deaths each year.

FIGURE 15.8 Stages of Plaque Progression

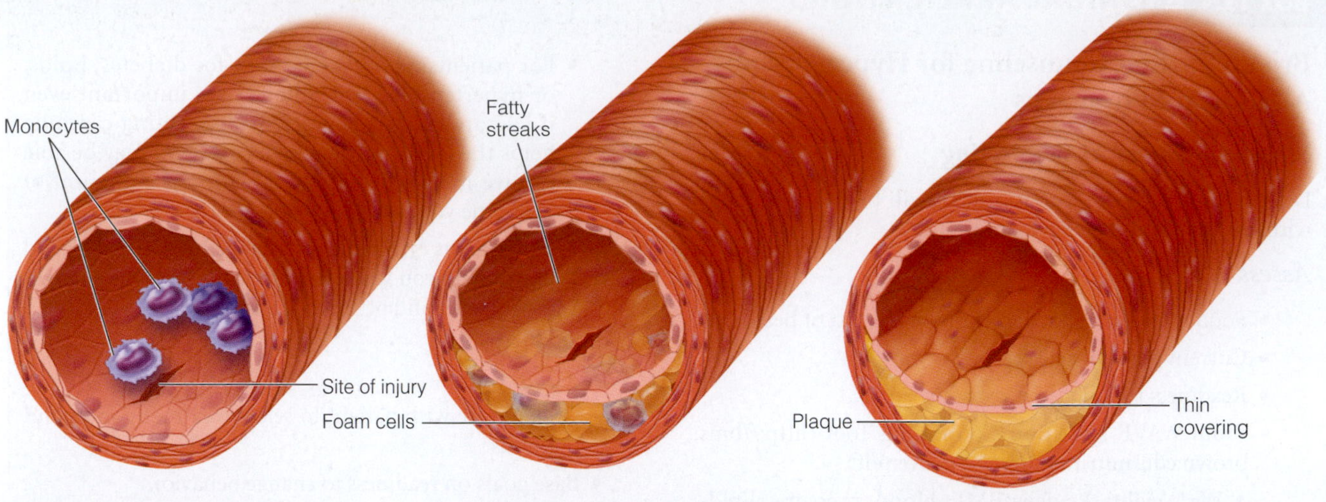

Monocytes—phagocytic white blood cells—circulate in the bloodstream and respond to injury on the artery wall.

Monocytes slip under blood vessel cells and engulf LDL cholesterol, becoming foam cells. The thin layers of foam cells that develop on artery walls are known as *fatty streaks.*

A fatty streak thickens and forms plaque as it accumulates additional lipids, smooth muscle cells, connective tissue, and cellular debris.

The artery may expand to accommodate plaque. When this occurs, the plaque that develops often contains a large lipid core with a thin fibrous covering and is vulnerable to rupture and thrombosis.

Source: S. Rolfes, K. Pinna, and E. Whitney, *Understanding Normal and Clinical Nutrition,* 7e, copyright © 2006, p. 821.

Etiology

These long-term studies have allowed researchers to identify risk factors for CAD, PVD, and stroke. These risk factors include: family history, age, sex, obesity, dyslipidemia, hypertension, diabetes, physical inactivity, and cigarette smoking. The risk factors are additive in their predictive power; thus, the more risk factors one demonstrates, the greater the risk of development of AS.

Typically, the risk factors are divided into categories based on whether they are alterable or unalterable. Family history, age, and sex are considered unalterable risk factors. Alterable risk factors include obesity, dyslipidemia, hypertension, physical inactivity, atherogenic diet, and cigarette smoking. Many sets of risk factors include a category termed *impaired fasting glucose*, which may be alterable; diabetes itself is an unalterable risk factor because, although one can be in metabolic control, this does not change that individual's risk status (see Chapter 19).

Family History There is certainly a genetic component to the development of AS. This genetic component may be related to endothelial function or cholesterol metabolism. CAD death rates are higher in individuals with these disorders of cholesterol metabolism than in the population as a whole. In some cases, familial hypercholesteremia that is caused by genetic abnormalities in lipoprotein clearance and lipid metabolism results in early death from disease (see Chapter 11). In many other cases, hypercholesterolemia is the result of environmental influences rather than a genetic trait. Dietary and physical activity habits are learned from the social network in which one is reared.

Age and Sex Because AS is a disease that develops over a span of years, a greater age allows for a greater period of time for the disease to develop. Most heart disease is seen in people over the age of 65; however, the number of sudden cardiac deaths among people 15–34 has increased in recent years (Centers for Disease Control and Prevention 2004). As one ages there are associated changes in endothelial control of vascular relaxation and in the elasticity of the arteries. Males tend to develop AS at a faster rate than females, though the differences between the sexes tend to decline after the woman reaches menopause. The purported reason for the differences has to do with a protective effect of estrogen. Thus, as women age past the point of menopause, their risk of development of clinically significant AS increases greatly in the absence of estrogen therapy. The lifetime risk of development of ischemic heart disease (discussed later in this chapter) after the age of 40 is reported to be one in two for men and one in three for women (Barnard 2005).

TABLE 15.5

Classification of Overweight and Obesity by BMI, Waist Circumference, and Associated Disease Risk*

	BMI (kg/m²)	Obesity Class	Disease Risk* Relative to Normal Weight and Waist Circumference: Men ≤102 cm (≤40 in) Women ≤88 (≤35 in)	Men >102 cm (>40 in.) Women >88 cm (>35 in.)
Underweight	< 18.5		——	——
Normal⁺	18.5–24.9		——	——
Overweight	25.0–29.9		Increases	High
Obesity	30.0–34.9	I	High	Very high
	35.0–39.9	II	Very high	Very high
Extreme Obesity	≥40	III	Extremely high	Extremely high

* Disease risk for type 2 diabetes, hypertension, and CVD.
⁺ Increased waist circumference can also be a marker for increased risk even in persons of normal weight.
National Heart, Lung, and Blood Institute. Determination of relative risk status based on overweight and obesity parameters [monograph on the Internet]. Table IV-2. Bethesda (MD): National Heart, Lung, and Blood Institute; 1998. Available from: http://www.nhlbi.nih.gov/guidelines/obesity/e_txtbk/txgd/4121.htm.

Obesity Only in the last decade has obesity been listed as a separate and independent risk factor for CAD. Obesity can be defined in several different ways when used for risk assessment, but is most commonly identified as a body mass index (BMI, measured in kg/m^2) of 30 or greater (see Chapter 14). A BMI of 30 or greater is associated with a proportionally higher all-cause mortality rate than is a BMI of 25–29.9 (see Table 15.5). Alternatively, obesity may be defined by waist girth or waist-to-hip ratio (WHR). The NCEP Adult Treatment Panel III (ATP III) (2001) report identified waist girth alone as a suitable predictor of risk. A waist girth of >102 cm for men or >88 cm for women is used as the criterion for increased risk (USDHHS 2001). Waist girth and WHR are related to the way in which humans store fat. Android obesity is the form in which more mass is stored in the upper body; gynoid obesity is the form in which more mass is stored below the waist. Those who store more fat below the waist are at lower risk than those who store more fat above their waist. This relationship between weight distribution and AS risk exists for two reasons: one is a purported relationship between abdominal fat and insulin resistance, and the second is that men store more fat above their waist. Men, who do not enjoy the protective benefit of estrogen, are more likely to develop AS at an earlier age.

Obesity is reaching epidemic proportions in the United States. Using a classification of overweight of a BMI of 25–29.9 kg/m^2, NHANES 1999–2002 data indicate that 65% of adults age 20–74 years are overweight or obese. This age-adjusted figure is up from 56% in the 1988–1994 data. The age-adjusted 1999–2002 NHANES data show that 31% are obese, an increase from 23% from 1988–1994 data and 15% from 1976–1980 data (National Center for Health Statistics 2004).

Obesity is positively associated with dyslipidemia, hypertension, physical inactivity, and diabetes, all of which are also associated with AS (Poirier et al. 2006). These associations make it impossible to determine just how many deaths are associated with obesity alone, but obesity is estimated to be the cause of over 300,000 deaths annually (see Chapter 14).

Hypothyroidism (see Chapter 19) leading to obesity has also been identified as a factor leading to increased risk of coronary AS. This relationship is due at least in part to the altered lipid metabolism in this population. There is a decrease in the activity of the lipogenic enzyme that down-regulates LDL receptors. Studies of patients with poorly managed hypothyroidism have revealed evidence of greater progression of coronary atherosclerotic lesions than in patients who were well treated (Nichols 2005).

Dyslipidemia Lipids are transported via lipoproteins comprised of a lipid interior and protein shell. There are several different types of lipoproteins. These lipoproteins vary in their protein makeup, in their lipid-to-protein ratio, and in the proportion of lipid components they contain. The protein portion of the lipoprotein is called the **apolipoprotein**. The apolipoprotein provides structural integrity and allows for receptors to recognize the lipoprotein particle. The lipid-to-protein ratio and lipid composition of a lipoprotein affect the density of the structure and allow for classification. Table 15.6 identifies the lipid composition of different

apolipoprotein—protein portion of the lipoprotein; provides cellular stability and allows for cellular recognition and binding

TABLE 15.6

Chemical and Physical Properties of Plasma Lipoproteins in Humans

Property	Chylomicrons	VLDL	IDL	LDL	HDL
Density (g/mL)	<1.006	<1.006	1.006–1.019	1.019–1.063	1.063–1.21
Diameter (nm)	80–500	40–80	24.5	20	7.5–12
Lipids (% by wt.)	98	92	85	79	50
Cholesterol	9	22	35	47	19
Triglyceride	82	52	20	9	3
Phospholipid	7	18	20	23	28
Apolipoproteins (%)	2	8	15	21	50
Major	A-1, A-2	B-100	B-100	B-100	A-1, A-2
	B-48	C-1,2,3	C-1,2,3		C-1,2,3
	C-1,2,3	E	E		E
	E				

Adapted from table 6–3 Gropper SS, Smith JL, Groff JL. Advanced Nutrition and Human Metabolism. 2005, p.140.

classifications of lipoproteins and the apolipoproteins associated with each type.

Chylomicrons transport dietary lipids after intestinal absorption. The other lipoproteins transport endogenous lipid from the liver to the rest of the body. Note from Table 15.6 that very-low-density lipoproteins (VLDL) are similar to chylomicrons in density but are smaller in size. The triglyceride content is much higher in VLDL than in low-density lipoproteins (LDL), which are much higher in cholesterol content. VLDL is produced in the liver, and as it travels through the body, triglycerides are removed until the particle density and size are reduced. It becomes a transient intermediate-density lipoprotein (IDL), then finally an LDL. Thus, the cholesterol-rich LDL may be viewed as the end point of the forward transport of lipids to tissues. Ultimately, LDL particles are removed from circulation by tissues in need of cholesterol for structural purposes. Cells have LDL-receptors that are activated when the need arises. The LDL particle is bound to the receptor and then engulfed by the cell. LDL receptors are found on liver cells and cells in a number of other tissues.

Dyslipidemia refers to a lipid profile that increases the risk of atherosclerotic development. Typically, dyslipidemia is a condition in which LDL levels are elevated and high-density lipoprotein (HDL) levels are low. (A variety of other dyslipidemic conditions can also exist, such as the combination of normal LDL and high triglyceride levels.) HDL particles are involved in reverse cholesterol transport, in that they transport cholesterol from tissues and other lipoproteins to the liver. ATP III indicated that serum LDL levels are the single strongest indicator of CVD risk. Among the lipoproteins, LDL are most heavily involved in the atherosclerotic process. Oxidation of the LDL causes this lipoprotein to be altered and can initiate damage, starting the atherosclerotic process. Additionally, oxidized LDL is more likely to be taken up into the atherosclerotic plaque (see the Pathophysiology section). Thus, the higher the serum LDL levels, the greater the risk of the initiation of an atherosclerotic plaque and the greater the risk of continual development of an existing atherosclerotic plaque. In the ATP III, recommendations included maintaining an LDL level below 100 mg/dL (see Table 15.7) (USDHHS 2001).

HDL offers a protective effect against AS. Since HDL removes cholesterol from tissues and returns that cholesterol to the liver, it reduces cholesterol in plaques. ATP III recommends that an HDL <40 mg/dL be considered low, indicating a greater level of risk. An HDL of >60 mg/dL is considered high and will in turn reduce risk for AS (USDHHS 2001).

In assessing risk for AS, one would use LDL as the primary marker; if LDL is unavailable, then HDL should be used as the indicator of risk. Total cholesterol should be used as an indicator of risk only if it is the only measure available. Total cholesterol levels should be maintained below 200 mg/dL.

Hypertension Hypertension is both a cardiovascular condition and a risk factor for other forms of cardiovascular disease. An increase in BP increases the forces applied to the endothelium and can cause the initiation of an atherosclerotic lesion (see the Pathophysiology section). Changes in pressure may also cause established plaques to rupture, which not only can initiate an event such as an infarct but can also cause a proliferation of existing plaques. From 40 to

TABLE 15.7

Interpretation of Laboratory Values

Total Cholesterol Levels	
Less than 200 mg/dL	"Desirable" level that puts you at lower risk for heart disease. A cholesterol level of 200 mg/dL or greater increases your risk
200 to 239 mg/dL	"Borderline-high"
240 mg/dL and above	"High" blood cholesterol. A person with this level has more than twice the risk of heart disease compared to someone whose cholesterol is below 200 mg/dL
HDL-Cholesterol Levels	
Less than 40 mg/dL	A major risk factor for heart disease
40 to 59 mg/dL	The higher your HDL, the better
60 mg/dL and above	An HDL of 60 mg/dL and above is considered protective against heart disease
LDL-Cholesterol Levels	
Less than 100 mg/dL	Optimal
100 to 129 mg/dL	Near optimal/Above optimal
130 to 159 mg/dL	Borderline high
160 to 189 mg/dL	High
190 mg/dL and above	Very high
Triglyceride Levels	
Less than 150 mg/dL	Normal
150 to 199 mg/dL	Borderline-high
200–499 mg/dL	High
500 mg/dL or above	Very high

Source: Grundy SM, Cleeman JI, Bairey Merz CN, Brewer HB, Clark LT, Hunninghake DB, Pasternak RC, Smith SC, Stone NJ, for the Coordinating Committee of the National Cholesterol Education Program. Implications of Recent Clinical Trials for the National Cholesterol Education Program Adult Treatment Panel III Guidelines Circulation. Adult Treatment Panel III Guidelines 2004; 110:227–239. http://www.americanheart.org/presenter.jhtml?identifier=11206.

70 years of age, an increase of systolic BP by 20 mmHg or diastolic BP by 10 mmHg increases risk of cardiovascular disease twofold. One estimate is that a reduction of 12 mmHg in systolic blood pressure of hypertensives will prevent one death for every 11 patients treated (Gordon, Leighton, and Mooss 2006). The sheer forces of blood against the arterial wall alone can cause endothelial damage. AS that occurs at bifurcations and trifurcations (branch points) of blood vessels, where these forces are amplified, demonstrates the importance of BP in the initiation of AS. Obstructive AS in the epicardial coronary arteries occurs most frequently in the first 5 cm where the forces associated with the branching of the arteries are greatest.

Physical Inactivity Many professionals think of AS and heart disease in general as hypercaloric diseases. In other words, they think that the increase in the patient's mass due to chronic positive caloric balance, along with the associated dyslipidemia, hypertension, and insulin resistance, results in AS. The exact mechanism by which physical inactivity increases AS risk has not been identified. However, increasing physical activity is known to impact several factors related to AS by lowering blood pressure and triglycerides, increasing HDL, improving endothelial function, and decreasing platelet aggregation (Lichtenstein and Deckelbaum 2001; Marshall et al. 2005; Altena, Michaelson, Ball, Guilford, and Thomas 2006). Increases in physical activity also aid in weight maintenance and reduce the relative workload of any activity of daily living on the cardiovascular system. For example, if vacuuming a room required a person to work at 70% of his or her functional capacity (or maximal workload) prior to beginning an exercise program, then after beginning the exercise program, the same activity may only require 65% of that individual's functional capacity. This would mean that the relative strain on the cardiovascular system would be reduced.

Atherogenic Diet Naturally, diet plays a role in obesity, which is directly and indirectly associated with AS. This is discussed in much more detail under Nutrition Therapy later in this section.

Several studies of CVD over the years have used the term "Westernized Diet" to describe a diet high in saturated fat and low in fiber. When compared to populations that eat diets high in saturated fat, populations that consume diets higher in fruits, vegetables, whole grains, and unsaturated fats have lower rates of atherosclerotic disease than can be explained by other risk factors (Gordon et al. 2006). The National Cholesterol Education Program (NCEP) Adult Treatment Plan (ATP III) includes an LDL-lowering diet (Therapeutic Lifestyle Changes diet) that limits saturated fat intake to $<7\%$ of total kcal or less than 16 g for an individual on a 2,000 kcal/day diet (Grundy et al. 2004).

Diabetes Mellitus CAD is the most common cause of death among diabetics (see Chapter 19). The risk of death from cardiovascular disease among type 1 and type 2 diabetics is 2–4 times greater than in non-diabetics (American Heart Association 2006). In 2003, 3.5 million diabetics age 35 or older self-reported as having CAD (National Center for Health Statistics 2004). While the risk of death from cardiovascular disease has decreased greatly among adults in the U.S. since 1970, it has decreased less in diabetics (Centers for Disease Control and Prevention 11/7/2003).

Impaired Fasting Glucose and Metabolic Syndrome

Impaired fasting glucose and diabetes are closely associated with the risk of cardiac death. An impaired fasting glucose is defined as blood glucose of 100 mg/dL or higher (blood glucose of >126 mg/dL is diagnostic for diabetes). Although impaired fasting glucose is listed by many organizations as a separate and independent risk factor, it does have a close association with other factors of the *metabolic syndrome*. The metabolic syndrome is a constellation of metabolic risk factors, including abdominal obesity, insulin resistance, dyslipidemia, hypertension, and prothrombotic state (a state in which the formation of blood clots is facilitated). This close relationship causes many to question the independent predictive power of impaired fasting glucose alone (Gordon et al. 2006).

It has been estimated that approximately 25% of the U.S. population has metabolic syndrome (Vitarius 2005). According to the NCEP guidelines, displaying three of the five risk factors classifies an individual as having metabolic syndrome (see Table 15.8). The NCEP criterion predicts both all-cause and cardiovascular mortality, while the World Health Organization criteria predict cardiovascular mortality but not all-cause mortality (Vitarius 2005). The risk of metabolic syndrome increases with BMI; thus, with the changes in prevalence of overweight and obesity since the 1988–1994 NHANES data were collected, the number of individuals with metabolic syndrome has likely increased significantly (National Center for Health Statistics 2004).

Given that the metabolic syndrome consists of multiple CAD risk factors, it is a potent predictor of risk for AS. Risk is increased with the number of metabolic syndrome factors present in an individual. Persons with metabolic syndrome are more likely to exhibit what may be termed a fibrinolytic profile (i.e., an impaired ability to dissolve blood clots), which in turn would translate to atherosclerotic development and arterial stiffness (Bodary et al. 2003; Gordon et al. 2006). The evidence of this lies in the C-reactive protein

TABLE 15.8

Diagnosis of Metabolic Syndrome

Risk Factor	NCEP ATP III Criteria[1]	IDF Criteria[2]
Abdominal Obesity*	Men: waist circumference** >102 cm (>40 in) Women: waist circumference >88 cm (>35 in)	Europoid, Sub-Saharan, Eastern Mediterranean and Middle East (Arab) men: waist circumference ≥94 cm for men Europoid, Sub-Saharan, Eastern Mediterranean and Middle East (Arab) women: waist circumference ≥80 cm South Asian, Chinese, Ethnic South and Central American men: waist circumference ≥90 cm South Asian, Chinese, Ethnic South and Central American women: waist circumference ≥80 cm Japanese men: waist circumference ≥85 cm Japanese women: waist circumference ≥90 cm
Triglycerides	≥150 mg/dL (1.7 mmol/L)	>150 mg/dL (1.7 mmol/L) or treatment for this lipid abnormality
HDL-Cholesterol	Men: <40 mg/dL (0.9 mmol/L) Women: <50 mg/dL (1.1 mmol/L)	Men: <40 mg/dL (0.9 mmol/L) Women: <50 mg/dL (1.1 mmol/L) or specific treatment for this lipid abnormality
Blood Pressure	≥130/≥85 mmHg	≥130/≥85 mmHg or treatment of previously diagnosed hypertension
Insulin Resistance	Fasting plasma glucose ≥100 mg/dL	Fasting plasma glucose ≥100 mg/dL (5.6 mmol/L) or previously diagnosed T2DM

[1] Overweight and obesity are associated with insulin resistance and the metabolic syndrome. However, the presence of abdominal obesity is more highly correlated with the metabolic risk factors than is an elevated body mass index (BMI). Therefore, the simple measure of waist circumference is recommended to identify the body weight component of the metabolic syndrome.

[2] Some male patients can develop multiple metabolic risk factors when the waist circumference is only marginally increased, e.g., 94–102 cm (37–39 in). Such patients may have a strong genetic contribution to insulin resistance. They should benefit from changes in life habits, similarly to men with categorical increases in waist circumference.

References: Expert Panel on the Detection, Evaluation, and Treatment of High Blood Cholesterol in Adults. Executive summary of the Third Report of the National Cholesterol Education Program (NCEP) Expert Panel on the Detection, Evaluation, and Treatment of High Blood Cholesterol in Adults (Adult Treatment Panel III). *JAMA.* 2001;385:2486–2497. International Diabetes Federation: The IDF consensus worldwide definition of the metabolic syndrome. Available from http://www.idf.org/webdata/docs/Metabolic_syndrome_definition.pdf. Accessed March 18, 2006.

(CRP) levels and plasminogen activator inhibitor type 1 (PAI-1) levels. These levels are typically higher in persons with metabolic syndrome. CRP and PAI-1 are likely related to the excess of adipose tissue. CRP is a powerful predictor of coronary events associated with AS. The full importance of CRP to the atherosclerotic process, hypertension, and acute coronary events is just beginning to be recognized (Savoia and Schiffrin 2006).

Cigarette Smoke Cigarette smoking remains a major health problem. Over 44 million adult Americans smoke, and between 1990 and 2000, the largest increase in smokers was in the group aged 18–24 years and in Hispanic females. At least in part because of smoking and obesity, younger white males and females ages 18–44 years are at higher risk for noncommunicable diseases than in previous years. Over 437,000 Americans die as the result of smoking-related illnesses, 35% of which are cardiovascular-disease related (American Heart Association 2006). Use of low-tar cigarettes increases risk of cardiovascular disease and MI compared to that of non-smokers. Even passive smoke exposure is associated with an increase in cardiovascular disease (Ambrose and Barua 2004). Compared to non-smokers, smokers have significantly higher levels of serum cholesterol, triglycerides, and LDL cholesterol, as well as lower HDL cholesterol levels. Exercise may attenuate the effect of smoking on lipid profile.

Cigarette smoking as a strong risk factor for AS and causative factor in CAD mortality has been supported by a number of studies. Environmental smoke exposure or passive smoking has also been associated with a significant increase in CAD risk. In addition to the relationship noted for coronary AS, aortic and peripheral AS are associated with smoking as well (Ambrose and Barua 2004). Cigarette smoking is associated with one of every 5 deaths annually (Centers for Disease Control and Prevention 2004) and from 1997–2001 it was estimated that cigarette smoking caused 438,000 premature deaths (Centers for Disease Control and Prevention 2005). Cigarette smokers are 2–4 times more likely to develop heart disease than nonsmokers (Centers for Disease Control and Prevention 2005). Annual productivity losses related to deaths from smoking are estimated at $92 billion. The total economic costs of smoking are more than $167 billion, including an additional $75.5 billion in smoking-related medical expenditures (Centers for Disease Control and Prevention 2005).

Endothelial dysfunction, inflammation, and modification of lipids that initiate and progress atherosclerotic development are affected by cigarette smoke. Endothelial relaxation is impaired by cigarette smoking. Nitric oxide (NO), which is primarily responsible for vasodilation of the endothelium, is decreased in endothelial cells exposed to components of cigarette smoke such as nicotine. Cells exposed to blood from cigarette smokers demonstrate a decrease in the activity of the endothelial NO synthase enzyme (Ambrose and Barua 2004).

Inflammatory markers such as CRP are increased in response to cigarette smoke. The increased leukocyte count and proinflammatory cytokines that occur in response to cigarette smoking increase endothelium-leukocyte interaction (Ambrose and Barua 2004). Cigarette smokers also have significantly higher total cholesterol and LDL and lower HDL than non-smokers. Moreover, cigarette smoking increases oxidative modification of LDL, which is a major step in the development of atherosclerosis. LDL exposed to cigarette smoke extract in culture was taken up by macrophages after modification of the lipoprotein. The clearance of LDL by macrophages is an integral part of atherosclerotic plaque formation. Increased LDL oxidation is likely the result of a decrease in the activity of protective enzymes in the plasma (Ambrose and Barua 2004).

Pathophysiology

The prevailing theory for the pathophysiology of AS is that the onset of disease begins as a response to injury that results in an inflammatory process. The injury is damage to the endothelial lining of the arterial wall. This injury may be caused by pressure on the wall exerted by the blood (as a result of hypertension) or by vasospasm. Chemical irritants from tobacco, oxidized LDL, glycated substances resulting from diabetic metabolism, and homocysteine are also possible culprits that are receiving a great deal of attention from researchers (Squires 2006).

Much of the most recent research into the initiation of the atherosclerotic lesion has focused on normal vasodilatatory control and the impairment of that control mechanism as a point of disease initiation (Sanders et al. 2001; Ambrose and Barua 2004; Poirier et al. 2006; Savoia and Schiffrin, 2006). NO is a substance naturally produced by endothelial cells and a number of other cells. NO produced in the endothelial cells controls the normal relaxation of smooth muscle in the arteries and arterioles. NO also helps regulate other mechanisms important to the atherosclerotic process, including leukocyte adhesion, platelet adhesion, and thrombosis. Research has indicated that factors that have been implicated as causative in AS decrease NO production (Ambrose and Barua 2004; Poirier et al. 2006; Savoia and Schiffrin 2006).

At the onset of lesion formation, the damage to the endothelial layer attracts platelets to the area (see Chapter 21). These platelets attach to the endothelium and form a small clot termed a mural thrombi. The platelets secrete platelet-derived growth factor (PDGF), which attracts monocytes and promotes smooth muscle cell mitosis from the media. Growth factors such as PDGF have a chemotactic or attracting effect, drawing smooth muscle cells, fibroblasts, and other cells to the injured area. The net result is an increase in collagen and a harder, more fibrous growth. Platelets also secrete thromboxane A2, which causes vasoconstriction and promotes additional vascular injury. Greater amounts of CRP and PAI-1 promote a fibrinolytic state, impairing the ability to dissolve thrombi. Monocytes adhere to the damaged section, and migrate between endothelial cells and into subendothelial spaces where they convert to macrophages. The macrophages will express receptors for oxidized LDL, and PDGF stimulates an increase in LDL receptors. Oxidized LDL enters the cell at a much more rapid rate than does non-oxidized LDL. Smooth muscle cells, which migrate to the intima and macrophages, take up LDL until they are transformed into **foam cells**. Foam cells are filled with cholesterol and are released into the extracellular spaces where they form fatty streaks (Pradka 2000; Squires 2006). The fatty streak, which may occur as early as age 5, is the earliest visible sign of AS (Poirier et al. 2006). Since clot formation is integral to the AS process, it is understood that AS thrives on a prothrombotic state—one in which the formation of clots is facilitated. In addition to the factors already mentioned, catecholamine secretion as the result of stress can also introduce a prothrombotic state (Squires 2006).

The atherosclerotic lesion progresses with continued migration of cells into the area, proliferation of the plaque, and growth of tissue. Over time, this plaque develops into a fibromuscular complex with a fibrous cap. Smooth muscle cells migrate into the intima from the media by mitosis, and platelets secrete collagen, adding to the fibrous makeup of the plaque's exterior surface. Inside the complex is a mix of connective tissue, lipids, macrophages, smooth muscle cells, **thrombus**, and calcium. As this plaque grows, the artery compensates by expanding outward and leaving the lumen size basically unchanged. As the plaque continues to grow, it eventually decreases the size of the lumen (Pradka 2000; Squires 2006).

The rate at which a plaque progresses varies. Some plaques may grow rapidly and then stabilize; some may slowly and steadily progress, while still others may grow at a rapid rate. The structure and composition affect the likelihood for rupture. Rupture may mean a portion of the fibrous cap is lost as the plaque ruptures, exposing the underlying tissue to the blood. The process begins anew at that location; thus, a new thrombus is formed. This scenario may play out repeatedly, with layers of thrombi being continually incorporated into the lesion (Pradka 2000; Squires 2006). Cigarette smoke increases the risk of plaque rupture (Ambrose and Barua 2004).

foam cells—macrophage cells containing lipid; found within the fatty streaks in the development of atherosclerosis

thrombus—blood clot

Clinical Manifestations

AS is asymptomatic except when the patient begins to experience the symptoms of ischemic heart disease (see the Ischemic Heart Disease section later in this chapter). Signs of AS are assessed by a lipid profile and determination of other specific risk factors. These are summarized in the ATP-III Treatment Guidelines (see Table 15.9).

Treatment

Current treatment for and prevention of AS are guided by the Adult Treatment Panel III Guidelines developed by the National Cholesterol Education Program (NCEP) and approved by the National Heart, Lung, and Blood Institute, the American College of Cardiology, and the American Heart Association (see Table 15.8) (Grundy et al. 2004; Fletcher et al. 2005). These practice guidelines are a synthesis of existing research and are the currently accepted practice. The most recent change in treatment occurred with the release of the ATP III guidelines, when the focus of treatment shifted from total serum cholesterol levels to LDL and triglyceride levels.

Table 15.10 outlines current classifications of medications used to reduce the risk of AS (Fletcher et al. 2005). Severe AS involving obstruction of blood flow often requires surgical intervention. Current procedures include the percutaneous transluminal coronary angioplasty (PTCA), laser angioplasty, or coronary artery bypass graft (CABG). Each of these procedures is described in Box 15.5.

Nutrition Therapy

Nutrition Implications Poor nutrition has historically been considered a risk factor for the development of AS (Pollack 1953; Keys, Arvanis, and Blackburn 1980; Keys et al. 1984). In general, it is believed that nutrition therapy affects AS by interfering with plaque formation and/or by inhibiting the inflammatory response that causes the physiological changes within the blood vessels (De Caterina et al. 2006).

But specific nutritional risk factors and recommendations have changed as scientific understanding of the disease process has deepened. The use of single nutrition interventions to reduce cardiovascular risk reflects a rather simplistic view of the cardiovascular disease process. Certainly, diet is a modifiable risk factor, but to impact the disease process with interventions based on one dietary component is improbable. The clinician should focus on the cumulative effect of the entire diet as well as other lifestyle factors when planning dietary changes (Koertge et al. 2003). The currently accepted nutrition therapy for the prevention of atherosclerosis is the Therapeutic Lifestyle Changes (TLC) plan developed as a component of the ATP III guidelines (U.S. Department of Health and Human Services 2001; Grundy et al. 2004). Table 15.11 summarizes the TLC diet (see Appendix E9 for the complete guidelines).

TABLE 15.9

Summary of ATP III Guidelines

STEP 1: Determine lipoprotein levels—obtain complete lipoprotein profile after 9- to 12-hour fast.

STEP 2: Identify presence of clinical atherosclerotic disease that confers high risk for coronary heart disease (CHD) events (CHD risk equivalent):

- Clinical CHD
- Symptomatic carotid artery disease
- Peripheral arterial disease
- Abdominal aortic aneurysm

STEP 3: Determine presence of major risk factors (other than LDL):

Major Risk Factors (Exclusive of LDL Cholesterol) that Modify LDL Goals

- Cigarette smoking
- Hypertension (BP ≥140/90 mmHg or on antihypertensive medication)
- Low HDL cholesterol (<40 mg/dL)*
- Family history of premature CHD (CHD in male first degree relative <55 years; CHD in female first degree relative <65 years)
- Age (men ≥45 years; women ≥55 years)

* HDL cholesterol ≥60 mg/dL counts as a "negative" risk factor; its presence removes one risk factor from the total count.

Note: in ATP III, diabetes is regarded as a CHD risk equivalent.

STEP 4: If 2+ risk factors (other than LDL) are present without CHD or CHD risk equivalent, assess 10-year (short-term) CHD risk.

Three levels of 10-year risk:

- >20%—CHD risk equivalent
- 10%–20%
- <10%

STEP 5: Determine risk category:

- Establish LDL goal of therapy
- Determine need for therapeutic lifestyle changes (TLC)
- Determine level for drug consideration

LDL Cholesterol Goals and Cutpoints for Therapeutic Lifestyle Changes (TLC) and Drug Therapy in Different Risk Categories

Risk Category	LDL Goal	LDL Level at Which to Initiate Therapeutic Lifestyle Changes (TLC)	LDL Level at Which to Consider Drug Therapy
CHD or CHD Risk Equivalents (10-year risk >20%)	<100 mg/dL	≥100 mg/dL	≥130 mg/dL (100–129 mg/dL: drug optional)*
2+ Risk Factors (10-year risk ≤20%)	<130 mg/dL	≥130 mg/dL	10-year risk 10%–20%: ≥130 mg/dL 10-year risk <10%: ≥160 mg/dL
0–1 Risk Factor**	<160 mg/dL	≥160 mg/dL	≥190 mg/dL (160–189 mg/dL: LDL-lowering drug optional)

* Some authorities recommend use of LDL-lowering drugs in this category if an LDL cholesterol <100 mg/dL cannot be achieved by therapeutic lifestyle changes. Others prefer use of drugs that primarily modify triglycerides and HDL, e.g., nicotinic acid or fibrate. Clinical judgment also may call for deferring drug therapy in this subcategory.

** Almost all people with 0–1 risk factor have a 10-year risk <10%, thus 10-year risk assessment in people with 0–1 risk factor is not necessary.

STEP 6: Initiate therapeutic lifestyle changes (TLC) if LDL is above goal.

TLC Features

- TLC Diet:
  - Saturated fat <7% of kcal, cholesterol <200 mg/day
  - Consider increased viscous (soluble) fiber (10–25 g/day) and plant stanols/sterols (2 g/day) as therapeutic options to enhance LDL lowering
- Weight management
- Increased physical activity

STEP 7: Consider adding drug therapy if LDL exceeds levels shown in Step 5 table:

- Consider drug simultaneously with TLC for CHD and CHD equivalents.
- Consider adding drug to TLC after 3 months for other risk categories.

STEP 8: Identify metabolic syndrome and treat, if present, after 3 months of TLC.

Clinical Identification of the Metabolic Syndrome - Any 3 of the risk factors defined in Table 15.A (see following table).

Treatment of the metabolic syndrome

- Treat underlying causes (overweight/obesity and physical inactivity):
  - Intensify weight management
  - Increase physical activity
- Treat lipid and non-lipid risk factors if they persist despite these lifestyle therapies:
  - Treat hypertension
  - Use aspirin for CHD patients to reduce prothrombotic state
  - Treat elevated triglycerides and/or low HDL (as shown in Step 9 below)

(continued on the following page)

TABLE 15.9 *(continued)*

STEP 9: Treat elevated triglycerides.

ATP III Classification of Serum Triglycerides (mg/dL)

< 150	Normal
150–199	Borderline high
200–499	High
≥500	Very high

Treatment of elevated triglycerides (≥150 mg/dL)

- Primary aim of therapy is to reach LDL goal.
- Intensify weight management.
- Increase physical activity.
- If triglycerides are ≥200 mg/dL after LDL goal is reached, set secondary goal for non-HDL cholesterol (total - HDL) 30 mg/dL higher than LDL goal.
- Comparison of LDL Cholesterol and Non-HDL Cholesterol Goals for Three Risk Categories

Risk Category	LDL Goal (mg/dL)	Non-HDL Goal (mg/dL)
CHD and CHD Risk Equivalent (10-year risk for CHD >20%)	<100	<130
Multiple (2+) Risk Factors and 10-year risk ≤20%	<130	<160
0–1 Risk Factor	<160	<190

If triglycerides 200–499 mg/dL after LDL goal is reached, consider adding drug if needed to reach non-HDL goal:

- intensify therapy with LDL-lowering drug, or
- add nicotinic acid or fibrate to further lower VLDL.

If triglycerides ≥500 mg/dL, first lower triglycerides to prevent pancreatitis:

- very-low-fat diet (≤15% of calories from fat)
- weight management and physical activity
- fibrate or nicotinic acid
- when triglycerides <500 mg/dL, turn to LDL-lowering therapy

Treatment of low HDL cholesterol (<40 mg/dL)

- First reach LDL goal, then:
- Intensify weight management and increase physical activity
- If triglycerides 200–499 mg/dL, achieve non-HDL goal
- If triglycerides <200 mg/dL (isolated low HDL) in CHD or CHD equivalent, consider nicotinic acid or fibrate

Source: U.S. Department of Health and Human Services, Public Health Service, National Institutes of Health, National Heart, Lung, and Blood Institute, NIH Publication No. 01–3305 May 2001.

Nutrition Interventions The nutrition recommendations in the TLC diet can be summarized using the current approved health claim for food labeling: "Diets low in saturated fat and cholesterol and rich in fruits, vegetables, and grain products that contain some types of dietary fiber, particularly soluble fiber, may reduce the risk of heart disease" (Hung et al. 2004; Rosner, Spiegelman and Willett 2004; USFDA 2005). Long-term clinical trials also support the following major components for dietary intervention in reducing cardiac risk: modification of dietary fat, saturated fat, and cholesterol, with increased amounts of fruits, vegetables, and fiber (Lauer 2000; Hu and Willett 2002; Appel et al. 2005; Jenkins et al. 2005; De Caterina et al. 2006).

Weight Loss Achieving a BMI within the normal range of 18.5–24.9 kg/m^2 is thought to decrease cardiovascular risk and complications associated with cardiovascular disease. Obesity negatively affects many of the known risk factors for atherosclerosis including dyslipidemia, high BP, and insulin resistance (Poirier et al. 2006).

Achieving weight reduction along with reduced waist circumference and visceral/abdominal obesity should be a priority goal in development of nutrition therapy interventions for treatment and prevention of AS and cardiovascular disease. Chapter 14 provides an in-depth discussion of the risks of overweight as well as treatment options.

Physical Activity Patients who experience angina or who are post-MI and/or post operative should participate in medically supervised exercise programs, commonly referred to as cardiac rehabilitation. Cardiac rehabilitation is designed to improve functional capacity while controlling risk factors in order to improve outcomes. Data indicate that cardiac rehabilitation is underused after an MI, particularly in women and older patients. Of those 70 years of age and older, only 32% participate in cardiac rehabilitation compared to 66% of 60–69 year olds and 81% of the patients under 60 years (American Heart Association 2006).

TABLE 15.10

Drug Therapy for Primary Prevention

Drug Class	Agents and Daily Doses	Lipid/Lipoprotein Effects	Side Effects	Contraindications
HMG CoA Reductase Inhibitors (statins)	Lovastatin (20–80 mg), Pravastatin (20–40 mg), Simvastatin (20–80 mg), Fluvastatin (20–80 mg), Atorvastatin (10–80 mg), Cerivastatin (0.4–0.8 mg)	LDL-C ↓ 18%–55% HDL-C ↑ 5–15% TG ↓ 7–30%	Myopathy Increased liver enzymes	Absolute: Active or chronic liver disease Relative: Concomitant use of certain drugs*
Bile Acid Sequestrants	Cholestyramine (4–16 g), Colestipol (5–20 g), Colesevelam (2.6–3.8 g)	LDL-C ↓ 15%–30% HDL-C ↑ 3–5% TG No change or increase	Gastrointestinal distress Constipation Decreased absorption of other drugs	Absolute: dysbeta-lipoproteinemia TG > 400 mg/dL Relative: TG > 200 mg/dL
Nicotinic Acid	Immediate release (crystalline) nicotinic acid (1.5–3 mg), extended release nicotinic acid (Niaspan®) (1–2 g), sustained release nicotinic acid (1–2 g)	LDL-C ↓ 5–25% HDL-C ↑ 15%–35% TG ↓ 20%–50%	Flushing Hyperglycemia Hyperuricemia (or gout) Upper GI distress Hepatotoxicity	Absolute: Chronic liver disease Severe gout Relative: Diabetes Hyperuricemia Peptic ulcer disease
Fibric Acids	Gemfibrozil (600 mg BID) Fenofibrate (200 mg) Clofibrate (1,000 mg BID)	LDL-C ↓ 5–20% (may be increased in patients with high TG) HDL-C ↑ 10%–20% TG ↓ 20%–50%	Dyspepsia Gallstones Myopathy	Absolute: Severe renal disease Severe hepatic disease

* Cyclosporine, macrolide antibiotics, various anti-fungal agents, and cytochrome P-450 inhibitors (fibrates and niacin should be used with appropriate caution).

Source: U.S. Department of Health and Human Services, Public Health Service, National Institutes of Health, National Heart, Lung, and Blood Institute, NIH Publication No. 01–3305 May 2001.

Total Dietary Fat Current recommendations by the NCEP ATP III guidelines include maintenance of dietary fat intake within 25%–35% of total caloric intake. Fat that is consumed as a part of the normal diet is a mixture of **saturated**, **polyunsaturated**, and **monounsaturated** fatty acids. For example, olive oil is composed primarily of monounsaturated fat but also contains lesser amounts of saturated and polyunsaturated fat. The specific type of fat that is consumed has been the emphasis for research in nutrition and cardiovascular disease for the previous five decades (U.S. Department of Health and Human Services 2001; Grundy et al. 2004). Current guidelines have emphasized reducing specific types of saturated fat rather than a strict adherence to a reduced-fat diet.

The benefit from restricting total dietary fat remains controversial. A recent controlled trial with over 48,000 women in the Women's Health Initiative study indicated that a lower-fat diet that was higher in fruits and vegetables affected cardiovascular risk only slightly (Howard et al. 2006). Critics of this study, however, point out that the reduction of total dietary fat was only minimal, and suggest that a larger reduction is needed to achieve significant results. Adherence plays a significant role in any dietary intervention. A comparison of popular diets for cardiovascular disease indicated that patients who lost weight were able to reduce cardiovascular risk as measured by LDL-to-HDL ratios (Dansinger et al. 2005).

saturated fats—sources of fat that have a predominant amount of fatty acids that contain all single bonds within their chemical structures

polyunsaturated fats—sources of fat that have a predominant amount of fatty acids that contain more than one double bond in their chemical structures

monounsaturated fats—sources of fat that have a predominant amount of fatty acids with one carbon-carbon double bond within their chemical structures

BOX 15.5 CLINICAL APPLICATIONS

Surgical Treatment Procedures for Atherosclerosis and Ischemic Heart Disease

Percutaneous Transluminal Coronary Angioplasty (PTCA)

- Slender balloon-tipped tube—a catheter—from an artery in the groin to a trouble spot in an artery of the heart. The balloon is then inflated, compressing the plaque and dilating (widening) the narrowed coronary artery so that blood can flow more easily. This is often accompanied by inserting an expandable metal stent. Stents are wire mesh tubes used to prop open arteries after PTCA.

Atherectomy

- Procedure similar to PTCA, but plaque is removed by high-speed drill.

Laser Angioplasty

- Procedure similar to PTCA, but the catheter uses a laser that is able to remove enough plaque to permit a balloon to be inflated in order to dilate the stenosis.

Percutaneous transmyocardial revascularization (PTMR)

- Catheter and laser are directed through artery in the leg to the heart. Small holes within the blocked vessel that are created by the laser increase blood flow to heart.

Coronary Artery Bypass Graft

- Surgical procedure that uses the saphenous vein or internal mammary artery to "bypass" the blocked vessel.

TABLE 15.11

Nutrient Composition of the TLC Diet

Nutrient	Recommended Intake
Saturated fat	Less than 7% of total kcal
Polyunsaturated fat	Up to 10% of total kcal
Monounsaturated fat	Up to 20% of total kcal
Total fat	25%–35% of total kcal
Cholesterol	<200 mg/day
Carbohydrate	50%–60% of total kcal
Fiber	20–30 g/day
Protein	Approximately 15% of total kcal
Sodium	<2,400 mg/day
Stanol esters	3–4 grams

Source: U.S. Department of Health and Human Services, Public Health Service, National Institutes of Health, National Heart, Lung, and Blood Institute, NIH Publication No. 01–3305 May 2001.

Very-low-fat diets combined with other lifestyle modifications such as increased physical activity and smoking cessation appear to have the most dramatic results in reducing cardiovascular risk factors (Aldana et al. 2004; Marshall et al. 2006; Mohanka et al. 2006).

Saturated Fat A saturated fatty acid is a fatty acid that has only single bonds between carbons in its chemical structure. Saturated fats are primarily found in animal sources, though there are highly saturated plant sources, such as palm and coconut oil as well. Research over the past five decades has led to the recommendation that no more than 7% of total kcal should be from saturated fat sources (Expert Panel on Detection, Evaluation, and Treatment of High Blood Cholesterol in Adults 2001, AHA 2006). But controversy remains, because not all saturated fatty acids appear to affect serum lipids in the same manner (German and Dillard 2004; Knopp and Retzlaff 2004). For example, **stearic acid**, found in beef, has neither a positive nor a negative effect on cholesterol and LDL levels. While it is difficult to identify genetic differences in responses to dietary fat modification, it is these differences that may ultimately, in the future, allow more individualized nutrition therapy based on genetic profile to evolve. In the interim, current practice recommends moderation in saturated fat intake.

Trans Fatty Acids In the **hydrogenation** of fatty acids, monounsaturated and polyunsaturated fats are made into solid fats so that they can be used as margarine or shortening. When hydrogen is introduced, a double bond is formed. If the carbons are on the opposite side of the carbon-carbon double bond (which occurs during hydrogenation), this fat is designated a *trans* fatty acid (see Figure 15.9). In nature, carbons are generally found on the same side of the carbon-carbon double bond, in what is designated as a *cis* configuration (Gropper, Smith, and Groff 2005).

A *trans* fat is used in food products to increase the shelf life. Unsaturated products have a lower melting point and

stearic acid—an 18-carbon saturated fatty acid

FIGURE 15.9 Fatty Acids

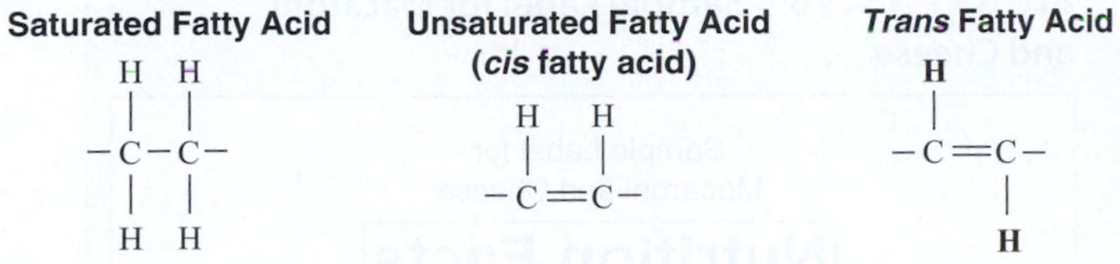

Saturated Fatty Acid	Unsaturated Fatty Acid (*cis* fatty acid)	*Trans* Fatty Acid
H H / –C–C– / H H	H H / –C=C–	H / –C=C– / H

can reach rancidity faster than saturated fatty acids. In the diet, *trans* fatty acids appear to behave similarly to saturated fatty acids in that they increase total cholesterol and LDL levels and perhaps lower HDL levels (de Roos, Bots, and Katan 2001; Wijendran and Hayes 2004). Additional research indicates that an increased amount of *trans* fatty acids are associated with an increased risk of myocardial infarction (Baylin et al. 2003; ADA Evidence Library 2006). Furthermore, *trans* fatty acids appear to be associated with an increased inflammatory response that may contribute to the atherogenic process (Mozaffarian et al. 2004). This has led to the most recent change in food labeling (which took effect in January 2006) that requires the listing of *trans* fatty acids on the Nutrition Facts panel (USFDA 2003). The American Hearth Association (2006) recommends to consume <1% of energy from trans fat (see Box 15.6).

Monounsaturated Fat A monounsaturated fatty acid is a fatty acid with one carbon-carbon double bond within its chemical structure. Intakes of monounsaturated fats are related to cardiovascular disease, since they affect serum lipid levels positively (Keys et al. 1980, 1984, 1986; Kris-Etherton et al. 1999). Specifically, monounsaturated fatty acid intake appears to lower LDL while having no affect on HDL levels. More recent evidence indicates that mo1nounsaturated fatty acids tend to lower both LDL and apolipoprotein AII-lipoprotein levels (U.S. Department of Health and Human Services 2001; Rodenas et al. 2005). Though olive and canola oils are rich sources of monounsaturated fatty acids, a recent review of longitudinal data indicates that the majority of monounsaturated fat intake in the U.S. is from oleic acid. In the U.S., unfortunately, the most common food sources of oleic acid include french fries, whole milk, peanut butter, and pizza (Nicklas et al. 2004).

The benefits of monounsaturated fat intake have been linked to the dietary habits of populations living within the Mediterranean region of the world. This connection is based on the epidemiological evidence that, despite higher dietary fat intakes, individuals in this area have lower cardiovascular disease rates than people from other regions (Trichopoulou et al. 2003). The Mediterranean diets that have been examined do indeed have greater amounts of monounsaturated fat, but additionally include more seafood—which provides increased amounts of omega-3- fatty acids—as well as more whole grains, fruits, and vegetables than the typical U.S. diet. Finally, overall amounts of saturated fat from animal sources are much lower than in typical diets in the U.S. and other western countries. It is obvious that more than one dietary factor may be involved in the positive effects seen in these population studies.

Ω-3 (Omega-3) Fatty Acids: Linolenic Acid Linolenic acid, an omega-3 fatty acid, is considered an essential fatty acid because humans lack the enzymes Δ^{12} and Δ^{15} desaturases that are necessary to add double bonds to this 18-carbon fatty acid (Gropper, Smith, and Groff 2005). Two additional fatty acids—eicosapentaenoic acid (EPA) and docosahexenoic acid (DHA)—are considered conditionally essential fatty acids because their synthesis is dependent on adequate amounts of linolenic acid. EPA is a 20-carbon fatty acid that is a precursor of the important eicosanoids. Eicosanoids include families of substances called thromboxanes, prostaglandins, and leukotrienes, as were discussed in Chapters 10 and 12. EPA, therefore, is important in cellular processes, vasoconstriction, vasodilation, platelet function, immune system response, and inflammatory response, and has been implicated as a mediator in asthma and allergic reactions.

Cold-water fishes and fish oils are particularly rich sources of linolenic acid. Flaxseed and flaxseed oil are also significant sources of alpha-linolenic acid (ALA). Several large clinical trials have examined these sources of linolenic acid and cardiovascular disease outcomes. This research has demonstrated reduced mortality with increased intakes of these specific types of lipids, though other studies have had mixed results (De Caterina and Massaro 2005; Harper and Jacobson 2005; De Caterina et al. 2006). AHA (2006) recommends that patients with CHD to consume 1 g of EPA and DHA daily.

Polyunsaturated Fatty Acids Polyunsaturated fat sources have a predominant amount of fatty acids that contain more than one double bond in their chemical structures. In the typical American diet, these fatty acids are primarily n-6 linoleic acid, which is an essential fatty acid. Polyunsaturated fat sources are oils from vegetables such as corn, cottonseed, soybean, safflower, and sunflower oils. When substituted for saturated fatty acids, polyunsaturated fatty acids have been linked to a reduction of LDL and are associated with decreased cardiovascular disease risk (U.S. Department of Health and Human Services 2001).

Cholesterol Humans have a consistent requirement for cholesterol, as it is a precursor for hormones such as estrogen and testosterone and for the vitamin D provitamin (dehydrocholesterol). Cholesterol is also a major component of cell

BOX 15.6 CLINICAL APPLICATIONS

Practical Tips for Consumers (Use with sample food package label)

- Check the Nutrition Facts panel to compare foods, because the serving sizes are generally consistent in similar types of foods. Choose foods lower in saturated fat, *trans* fat, and cholesterol. For saturated fat and cholesterol, use the Quick Guide to %DV: 5%DV or less is low and 20%DV or more is high. (Remember, there is no %DV for *trans* fat.)
- Choose Alternative Fats. Replace saturated and *trans* fats in your diet with monounsaturated and polyunsaturated fats. These fats do not raise LDL (or "bad") cholesterol levels and have health benefits when eaten in moderation.
- Sources of monounsaturated fats include olive and canola oils.
- Sources of polyunsaturated fats include soybean oil, corn oil, sunflower oil and foods like nuts and fish.
- Choose vegetable oils (except coconut and palm kernel oils) and soft margarines (liquid, tub, or spray) more often because the amounts of saturated fat, *trans* fat, and cholesterol are lower than the amounts of these substances in solid shortenings, hard margarines, and animal fats, including butter.
- Consider Fish. Most fish are lower in saturated fat than meat is. Some fish, such as mackerel, sardines, and salmon, contain omega-3 fatty acids that are being studied to determine if they offer protection against heart disease.
- Choose Lean Meats. These include poultry (without skin and not fried), lean beef, and pork (visible fat trimmed and not fried).
- Ask Before You Order When Eating Out. A good tip to remember is to ask which fats are being used in the preparation of your food when eating or ordering out.
- Watch Calories. Don't be fooled! Fats are high in calories. All sources of fat contain 9 calories per gram, making fat the most concentrated source of calories. By comparison, carbohydrates and protein have only 4 calories per gram.
- Here are two actions consumers can take to keep their intake of saturated fat, *trans* fat, and cholesterol "low":
  - Look at the Nutrition Facts panel when comparing products. Choose foods low in the combined amount of saturated fat and *trans* fat and low in cholesterol as part of a nutritionally adequate diet.
  - When possible, substitute alternative fats that are higher in monounsaturated and polyunsaturated fats like olive oil, canola oil, soybean oil, sunflower oil, and corn oil.

FIGURE 15.10 Sample Label for Macaroni and Cheese

Sample Label for Macaroni and Cheese

Start Here

Nutrition Facts

Serving Size 1 cup (228g)
Servings Per Container 2

Amount Per Serving

Calories 250 Calories from Fat 110

Limit these Nutrients

	% Daily Value*
Total Fat 12g	18%
Saturated Fat 3g	15%
Trans Fat 1.5g	
Cholesterol 30mg	10%
Sodium 470mg	20%
Total Carbohydrate 31g	10%
Dietary Fiber 0g	0%
Sugars 5g	
Protein 5g	
Vitamin A	4%
Vitamin C	2%
Calcium	20%
Iron	4%

Get Enough of these Nutrients

Quick Guide to % DV

5% or less is low
20% or more is high

Footnote

* Percent Daily Values are based on a 2,000 calorie diet. Your Daily Values may be higher or lower depending on your calorie needs.

	Calories	2,000	2,500
Total Fat	Less than	65g	50g
Sat Fat	Less than	20g	25g
Cholesterol	Less than	300mg	300 mg
Sodium	Less than	2,400mg	2,400mg
Total Carbohydrate		300g	375g
Dietary Fiber		25g	30g

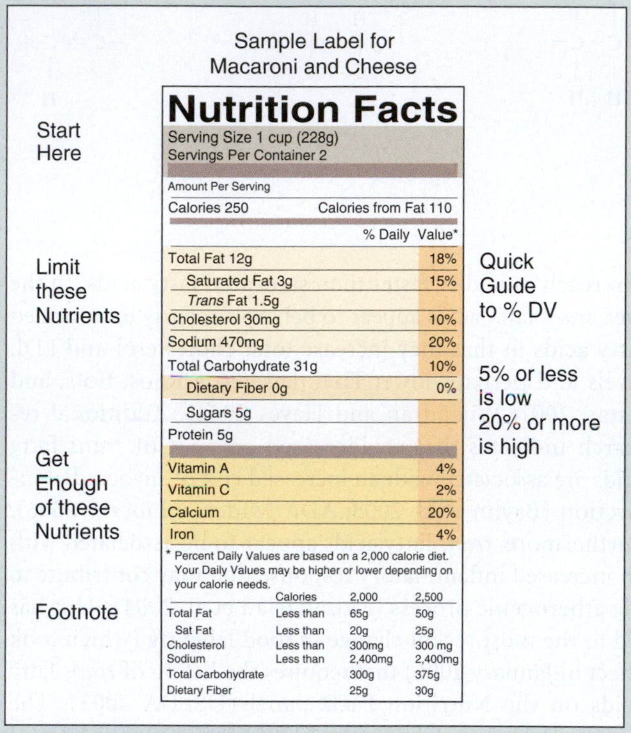

Source: Center for Disease Control and Prevention.

Reprinted from: Food and Drug Administration, Center for Food Safety and Applied Nutrition. *Trans* Fat Now Listed With Saturated Fat and Cholesterol on the Nutrition Facts Label [monograph on the Internet]. College Park (MD): FDA/CFSAN; 2006 [cited 2006 March 5]. Available from: http://www.cfsan.fda.gov/~dms/transfat.html.

membranes and other cellular structures. Total daily cholesterol absorption and synthesis is approximately 1,000 mg/day. The multi-step process for cholesterol synthesis, occurring primarily in the liver, is a negative feedback reaction: as the body pool of cholesterol increases, synthesis rate will decrease (Gropper, Smith, and Groff 2005). The rate-limiting step in synthesis involves the enzyme HMG CoA reductase, which has been a target for pharmaceutical intervention in treating hypercholesterolemia through the use of statin medications.

Historically, dietary cholesterol intake was a major point of nutrition therapy for treatment of heart disease. It is now understood that the primary concern for dietary cholesterol intake centers around its effect in raising LDL levels, not on its effect on serum cholesterol levels (Connelly 2005; Jenkins et al. 2005). In the U.S., dietary intake of cholesterol has steadily decreased over the past two decades. The National Cholesterol Education Program currently recommends intake of less than 200 mg/day (U.S. Department of Health and Human Services 2001).

Fiber Current recommendations for dietary intake of soluble fiber in cardiovascular disease are based on its ability to reduce LDL and total serum cholesterol levels. Meta-analyses of clinical trials have supported the hypothesis that diets high in soluble fiber decrease total and LDL cholesterol (Brown et al. 1999; Pereira et al. 2004).

Fiber is a type of polysaccharide—the classification of complex carbohydrates that include starches, dietary fiber, and functional fibers (Gropper, Smith and Groff 2005). Dietary fiber is defined as "nondigestible carbohydrates and lignin that are intact and intrinsic in plants" (Gropper, Smith, and Groff 2005, p. 109; Food and Nutrition Board 2002). Functional fiber is defined as "Nondigestible carbohydrates that have been isolated, extracted or manufactured and have a demonstrated human benefit" (Gropper, Smith, and Groff 2005, p. 109; Food and Nutrition Board 2002). Though fiber has several characteristics that impact human health, fiber's ability to bind molecules is the most important for its impact on cardiovascular disease. Soluble viscous fiber may reduce serum cholesterol and LDL levels in several ways. First, soluble fiber may decrease overall absorption of lipids. Secondly, soluble fiber is thought to bind bile acids and increase their excretion rather than allow bile to enter enterohepatic circulation. This excretion decreases the overall body pool of cholesterol. Finally, in response to the decreasing amounts of cholesterol, the body transfers low-density lipoproteins (LDL) and the cholesterol they contain to the liver to support the increased synthesis of bile (Marlett 2001; American Dietetic Association 2002; Gropper, Smith, and Groff 2005).

Since people eat a mixture of fibers, distinguishing the physical effects from individual types of fiber is difficult. The U.S. *Dietary Guidelines* recommend consuming 14 g of fiber for every 1,000 kcal, while the Food and Nutrition Board recommends that men under 50 years of age consume 38 g of fiber/day and women in the same age group consume 25 g of fiber/day (USFDA 2005). The TLC dietary recommendations additionally support the use of 20–30 grams of fiber/day. The best sources of soluble fiber with the binding ability that can lower serum cholesterol and LDL levels include gums, beta glucans, psyllium, resistant starches, and pectin. Food sources that are recommended include fruits, vegetables, and oat and soy products.

Plant/Stanol Esters Plants do not contain cholesterol but they do have similar sterol components. There are over 60 different types of plant sterols, but the most common is sitosterol. Humans do not synthesize these sterols as they do cholesterol, nor are they well absorbed. Research has demonstrated that when these plant sterols are esterified to a common fatty acid (stanol ester or sterol ester), they can assist in lowering serum cholesterol and LDL levels (Hallakainen, Sarkkinen, and Uusitupa 2000; Law 2000; Lichtenstein and Deckelbaum 2001). The exact mechanism for the lipid-lowering effect is not completely understood, but plant sterols may inhibit endogenous cholesterol synthesis or they may interrupt lipid absorption at the micelle. Research has not indicated significant risk from the use of stanol esters, though interference with fat-soluble vitamin absorption has been discussed (USFDA 2003). No safe levels have been established for pregnant women or children. The FDA approved the following health claim for plant stanol/sterol esters and reduced risk of heart disease: "Diets low in saturated fat and cholesterol that include at least 1.3 grams of plant sterol esters or 3.4 grams of plant stanol esters, consumed in 2 meals with other foods, may reduce the risk of heart disease" (USFDA 2000; 2003). Even though typical daily diets include small amounts of plant stanols, supplementation with products such as Benecol® or Take Control® is necessary in order to consume the amounts demonstrated to have a therapeutic effect. Exceeding this amount does not appear to have additional benefit.

Developing the Nutrition Therapy Prescription As with any diagnosis, the first step in the nutrition care process is completion of a thorough nutrition assessment. Specifically, dietary assessment should focus on those components, such as dietary fat intake and saturated fat intake, that will assist the clinician to identify specific nutrition problems and develop individualized nutrition therapy. Assessment tools that help target these specific nutrients include the MEDFICTS assessment tool (see Chapter 5), the Dietary CAGE questions (see Table 15.12) (Expert Panel on Detection, Evaluation, and Treatment of High Blood Cholesterol in Adults 2001), and the REAP (see Table 15.13) (Institute for Community Health Promotion, Brown University 2005).

During the nutrition assessment, the patient's target weight is calculated. Any of the established methods for estimating energy requirements can be used to determine

TABLE 15.12

Dietary CAGE Questions for Assessment of Intakes of Saturated Fat and Cholesterol

C: Cheese	(and other sources of dairy fats – whole milk, 2% milk, ice cream, cream, whole fat yogurt)
A: Animal fats	(hamburger, ground meat, frankfurters, bologna, salami, sausage, fried foods, fatty cuts of meat)
G: Got it away from home	(high fat meals either purchased and brought home or eaten in restaurant)
E: Eat (extra) high-fat commercial products	candy, pastries, pies, doughnuts, cookies

Table V 2–4 Adopting Healthful Habits to Lower LDL Cholesterol and Reduce CHD Risk. ATP Guidelines III, 2001, p. V-5.

TABLE 15.13

Rapid Eating Assessment for Patients (REAP)

Client is asked to respond to each of the following questions with "usually/often," "sometimes," "rarely/never," or "does not apply to me."

Topic	In an average week, how often do you:
Meals	1. Skip breakfast?
	2. Eat 4 or more meals from sit-down or take out restaurants?
Grains	3. Eat less than 3 servings of whole-grain products a day? *Note:* Serving = 1 slice of 100% whole-grain bread; 1 cup whole-grain cereal like Shredded Wheat, Wheaties, Grape Nuts, high-fiber cereals, oatmeal, 3–4 whole-grain crackers, ½ cup brown rice or whole-wheat pasta
Fruits and Vegetables	4. Eat less than 2–3 servings of fruit a day? *Note:* Serving = ½ cup or 1 med. fruit or 4 oz. 100% fruit juice
	5. Eat less than 3–4 servings of vegetables/potatoes a day? *Note:* Serving = ½ cup vegetables/potatoes, or 1 cup leafy raw vegetables
Dairy	6. Eat or drink less than 2–3 servings of milk, yogurt, or cheese a day? *Note:* Serving = 1 cup milk or yogurt; 1 ½–2 ounces cheese
	7. Use 2% (reduced fat) or whole milk instead of skim (non-fat) or 1% (low-fat) milk?
	8. Use regular cheese (like American, cheddar, Swiss, Monterey jack) instead of low-fat or part-skim cheeses as a snack, on sandwiches, pizza, etc.?
Meats/Chicken/ Turkey	9. Eat beef, pork, or dark meat chicken more than 2 times a week?
	10. Eat more than 6 ounces (see sizes below) of meat, chicken, turkey, or fish per day? *Note:* 3 ounces of meat or chicken is the size of a deck of cards or ONE of the following: 1 regular hamburger, 1 chicken breast or leg (thigh & drumstick), or 1 pork chop.
	11. Choose higher fat red meats like prime rib, T-bone steak, hamburger, ribs, etc. instead of lean red meats?
	12. Eat the skin on chicken and turkey or the fat on meat?
	13. Use regular processed meats (like bologna, salami, corned beef, hot dogs, sausage, or bacon) instead of low-fat processed meats (like roast beef, turkey, lean ham; low-fat cold cuts/hot dogs)?
Fried Foods	14. Eat fried foods such as fried chicken, fried fish, or french fries?
Snacks	15. Eat regular potato chips, nacho chips, corn chips, crackers, regular popcorn, nuts instead of pretzels, low-fat chips or low-fat crackers, air-popped popcorn?
Fats and Oils	16. Use regular salad dressing and mayonnaise instead of low-fat or fat-free salad dressing and mayonnaise?
	17. Add butter, margarine, or oil to bread, potatoes, rice or vegetables at the table?
	18. Cook with oil, butter, or margarine instead of using non-stick sprays like Pam or cooking without fat?
Sweets	19. Eat regular sweets like cake, cookies, pastries, donuts, muffins, and chocolate instead of low-fat or fat-free sweets?
	20. Eat regular ice cream instead of sherbet, sorbet, low-fat or fat-free ice cream, frozen yogurt, etc.?
	21. Eat sweets like cake, cookies, pastries, donuts, muffins, chocolate, and candies more than 2 times per day?
Soft Drinks	22. Drink 16 ounces or more of non-diet soda, fruit drink/punch or Kool-Aid a day? *Note:* 1 can of soda = 12 ounces
Sodium	23. Eat high-sodium processed foods like canned soup or pasta, frozen/packaged meals (TV dinners, etc.), chips?
	24. Add salt to foods during cooking or at the table?

(continued on the following page)

TABLE 15.13 *(continued)*

Alcohol	25. Drink more than 1–2 alcoholic drinks a day? *Note:* One drink = 12 oz. beer, 5 oz. wine, one shot of hard liquor or mixed drink with 1 shot.
Activity	26. Do less than 30 total minutes of physical activity 3 days a week or more? *Examples:* Walking briskly, gardening, golf, jogging, swimming, biking, dancing, etc. 27. Watch more than 2 hours of television or videos a day?

Client is asked to respond to each of the following questions with "yes" or "no."
Do you...

28. Usually shop and prepare your own food?
29. Ever have trouble being able to shop or cook?
30. Follow a special diet, eat or limit certain foods for health or other reasons?

Client is asked to circle the number that best describes how he/she feels on a scale of 5 to 1, with 5 = "Very willing" and 1 = "Not at all willing."

31. How willing are you to make changes in what, how or how much you eat in order to eat healthier?

weight maintenance energy requirements and/or energy requirements to facilitate weight loss. Optimally, the modifications that are priorities for the TLC (reducing fat, increasing physical activity, and increasing fruits, vegetables, and fiber) result in subsequent kcal reduction and weight loss.

Next, using the assessment of the patient's dietary history, nutrition problems are prioritized in order to determine nutrition diagnoses. For example, suppose the dietary assessment indicates that the patient's overall fat intake is >45% of total kcal and that most of the dietary fat comes from large servings of animal protein several times a day as well as high-fat dairy products. The first nutrition problem could be the intake of excessive fat and perhaps kcal, which would be labeled as nutrition diagnosis NI-51.1: Excessive fat intake or NI 1.5: Excessive energy intake. Identifying ways to assist the patient to reduce serving sizes and choose substitutions for each of the high-fat foods is the first step toward accomplishing several of the target TLC goals (see Box 15.7) and is incorporated into the third step of the nutrition care process—nutrition intervention. For many individuals, it is overwhelming to make these dietary changes all at one time. Dietary intake and physical activity plans should be used for a minimum of six weeks, and if results are not achieved, it is recommended that pharmaceutical intervention be considered by the primary physician as a means to assist with reducing LDL and total cholesterol levels (U. S. Department of Health and Human Services 2001).

The American Dietetic Association has made the following recommendations: "Referral to a registered dietitian for Medical Nutrition Therapy (MNT) is recommended whenever an individual has an abnormal lipid profile, based on ATP III Risk category and LDL-C goals, or has CHD. A planned initial visit lasting from 45–90 minutes and at least two to six planned follow-up visits (30–60 minutes each, with an RD) can lead to improved dietary pattern; improved lipid profile; reduced plasma total cholesterol, LDL-C, and triglycerides; and improved weight status. The number and duration of visits in the course of Medical Nutrition Therapy will need to be greater if the client is in a higher risk category, if there is a large number of Therapeutic Lifestyle Changes (TLC) that need to be made, and if the individual is not motivated to make TLC changes. Increasing the number of visits and length of time spent with a dietitian can improve serum lipid levels and CVD risk" (American Dietetic Association, Disorders of Lipid Metabolism, Evidence Analysis Library 2006).

Ischemic Heart Disease

Definition

A sedentary individual may develop an atherosclerotic plaque that occludes up to 50% of the lumen of a coronary artery, and remain completely asymptomatic. However, if the individual becomes active they may experience a symptom, called **angina**, which is directly associated with reduced blood flow to parts of the heart. When the coronary arteries are occluded to the point that the blood flow to portions distal to the blockage is compromised, the individual is said to have myocardial ischemia. The term **ischemic heart disease** (IHD) is often used interchangeably with the term CAD.

angina—chest pain caused by oxygen deficit to the heart
ischemic heart disease (IHD)—heart disease characterized by inadequate blood supply to the heart

BOX 15.7 CLINICAL APPLICATIONS

Brief Nutrition Counseling for Hyperlipidemia: A Model for Physicians

Steps in Behavioral Counseling

Due to limited time, your approach will be more directive, with less opportunity for patient input.

Assess

- Food intake and diet habits in the context of health risks
- Current physical activity
- Readiness to change behavior
- From WAVE Nutrition Counseling Tool (http://bms.brown.edu/nutrition/acrobat/wave.pdf)
  - W=Weight: Review BMI, blood pressure, blood sugar, lipids to screen for metabolic syndrome.
  - A=Activity: Conduct physical activity assessment. Ask about:
    - Moderate physical activity? Goal—30 minutes/day or more
  - V=Variety and E=Excess: Conduct brief diet assessment. Ask about:
    - High-saturated-fat foods like cheese, ice cream, butter, fatty meats? Goal—low-fat dairy, lean meat, vegetable oils
    - High-fiber foods like oats, barley, or legumes? Goal—daily or several times/week
    - Number of fruits and vegetables? Goal—at least five/day
    - Number of meals and snacks? Goal—at least three meals/day
    - Use of sweetened beverages? Goal—if triglycerides high, eliminate or reduce significantly

Advise

- Give clear, specific, and personalized behavior change advice. You might say:
  - "Changes in your diet and exercise habits can lead to significant improvement in your blood fats and reduce your risk of heart disease."
  - For patients taking medication for lipids, blood pressure, diabetes: "Diet choices are important even if you are taking medication, since eating carefully helps the medicine do a better job. You may be able to save money by cutting down on the amount of medicine you take."
  - For patients NOT ready to change behavior, add: "I'd like to help you when you are ready to make changes in your diet and be more active."

Agree

- Collaborate with patient to select treatment goals and methods.
- Base goals on readiness to change behavior.
- For patient NOT ready to change behavior: "Is it okay if I ask you again at our next visit?"
- Possible goals for patient ready to change:
  - Return for further discussion in 2–4 weeks.
  - Keep food and exercise records to increase awareness, if patient willing.
  - Refer for registered dietitian visit.

Assist

- Help patient acquire knowledge, skills, and support for behavior change.
- Provide hand-outs and Web resources, based on patient interest and need.
- Provide lists and recommendations for community resources (exercise and diet programs, health clubs, etc.).

Arrange

- Schedule follow-up appointments.

References:

Medical Nutrition Handbook, Department of Medicine, University of Wisconsin School of Medicine and Public Health (http://www.medicine.wisc.edu/mainweb/includes/viewfile.php?fileid=893&viewtype=inline§ion=naa).

Severe and prolonged myocardial ischemia can precipitate a myocardial infarction (MI), during which necrosis of heart tissue occurs due to the lack of oxygen. Depending upon the site of the infarct, the result may be necrosis of a small area of myocardium, cardiac rhythm abnormalities due to damage to neural pathways such as the AV node or bundle branches, or sudden cardiac death.

Epidemiology

IHD is the single largest killer of Americans, accounting for 20% of all deaths in 2003 (American Heart Association 2006). It has also been estimated that by the year 2020, IHD will be the leading cause of death and disability worldwide (Pasternak et al. 2004). About 40% of individuals who suffer a coronary event in a given year will die as a result. The estimated average number of years of life lost as a result of an MI is 14.2. It is estimated that in 2006 1.2 million Americans will have a new or recurrent coronary attack (American Heart Association 2006).

As large as some of these numbers are, it is important also to note that deaths from IHD have declined dramatically over the last 60 years. In the last half of the twentieth century, IHD death rates declined by 59%, and from 1999–2003 the death rate decreased by 30.2%. This suggests that progress is being made. However, this condition still impacts the duration and the quality of life of millions of Americans.

Ischemia and MI can lead to heart failure and rhythmic abnormalities, which result in death. In individuals over the age of 35, 80% of all sudden cardiac death is related to IHD. Contrast this to causes of sudden cardiac death before the age of 35 where only 10% result from IHD and almost 50% result from **hypertrophic cardiomyopathy** (Saffitz 2005). Even though there have been declines in IHD and deaths from MI, there have been increases in the prevalence of risk factors. From 1991 to 2001, the prevalence of hypertension, dyslipidemia, diabetes, and obesity increased, while the prevalence of smoking remained stable. The result is a decreased prevalence of individuals who have no risk factors for IHD. From 1991 to 2001, the prevalence of persons with one or more risk factors increased from 58.2% to 64%, and the prevalence of persons with no known risk factors decreased from 41.8% to 36% (Morbidity and Mortality Weekly Report, 1/16/2004). Given that risk factors account for 90% of the risk of an initial MI, this leads one to predict that IHD and cardiovascular diseases as a whole will increase (Morbidity and Mortality Weekly Report, 1/16/2004; American Heart Association 2006).

Etiology

Acute coronary syndrome is a term used to describe the condition of persons who present with either an acute MI or **unstable angina**. An estimated 879,000 persons were discharged from hospitals in 2003 with acute coronary syndrome (American Heart Association 2006). The causes of an acute MI or unstable angina are plaque erosion, rupture of a plaque resulting in formation of a thrombus, and vasoconstriction. The type of acute coronary syndrome depends on the duration of the occlusion (Squires 2006). Unstable angina is likely to be caused by a transient occlusion of the artery due to vasoconstriction or a thrombus that dissolves rapidly. Longer-term occlusion would result in MI.

Traditional risk factors for AS apply to IHD. A prospective study of lipid and non-lipid risk factors among healthy middle-aged females indicated that the addition of C-reactive protein (CRP) improved prediction of increased risk for MI (Smith et al. 2004).

Individuals who survive an MI have a chance of illness and death that is 1.5 to 15 times that of the general population. Within six years of having an MI (American Heart Association 2006):

- 18% of men and 35% of women will have another one. The higher rate for women may well be associated with the lower percentage of women participating in cardiac rehabilitation programs.
- 7% of men and 6% of women will experience sudden cardiac death.
- 22% of men and 46% of women will be disabled with heart failure.
- 8% of men and 11% of women will suffer a stroke.

Pathophysiology

Any of the following four mechanisms can initiate an MI or angina in an individual with IHD:

- Sudden blockage of a coronary artery
- Hemorrhage into an atherosclerotic plaque
- Arterial spasm
- Increase in myocardial oxygen demand

All of the aforementioned mechanisms have one factor in common: an atherosclerotic plaque that is contributing to the occlusion of the lumen of the artery. Recall that resistance to flow is inversely proportional to diameter of the vessel. A very minor occlusion results in a great increase in resistance; thus, after the lesion has caused remodeling of the artery and has progressed beyond that point, resistance to flow will increase with every small increase in plaque size. While all plaques have similar features, they are not all the same and some plaques are more likely to rupture and form thrombi.

Large, hardened plaques contain more smooth muscle cells that have migrated into the plaque than smaller, softer ones. These types of plaques are less likely to rupture, but over time will gradually create occlusion of more than 70% of the lumen. These plaques may cause angina, but because of collateral circulation that can develop over time, will occlude the artery without infarct. Thus, the large hardened plaques are not typically the cause of acute events (Pradka 2000).

hypertrophic cardiomyopathy—a genetic disorder causing abnormal thickening of the left ventricular wall

acute coronary syndrome—condition characterized by an episode of acute unstable angina

unstable angina—chest pain that occurs at rest

Soft, lipid-rich plaques are more likely to cause acute MIs. These plaques are less likely to occlude large areas within the lumen, but are more prone to rupture. These plaques have a fibrous cap covering the lipid-rich core, which contains a great number of macrophages. Inflammation from within the plaque weakens the bond between the cap and the interior. Macrophages within the core secrete matrix metalloproteinases and other substances that break down the cap. As the cap's collagen is dismantled, the bonding plaque becomes more unstable (Faxon et al. 2004). This weakening, in combination with the physical forces of blood flow and sometimes augmented by vasoconstriction, along with nicotine and immune complexes, causes a rupture. The rupture or tearing of the cap exposes the plaque to the flowing blood. Blood pushes into the fissure, where there is hemorrhaging into the plaque, and clotting ensues. The result may be abrupt thrombotic occlusion of the artery (Pradka 2000; Squires 2006). In some cases, occlusion will occur, because blood seeping into the plaque causes it to enlarge and eventually obstruct the coronary artery (Crowley 2001, p. 300).

The rupture may be predicated by a change in blood pressure or flow dynamics in the area of the plaque. If a vasospasm occurs in the area of the plaque, it can certainly cause angina because of temporarily obstructed blood flow. Spasms do occur in the area of atherosclerotic plaques, possibly the result of endothelial dysfunction and impaired NO production. The spasm would change flow dynamics and could initiate a plaque rupture which would lead to occlusion.

It is not unusual for individuals to experience episodes of angina in response to an increase in cardiac workload. An increase in physical activity will cause the heart rate to increase with a concomitant increase in BP. As these variables increase the workload on the heart, the oxygen requirements of the myocardium are also increased. As the oxygen demand increases, the ischemic artery becomes unable to supply adequate flow to satisfy this demand. In such a case, the individual will experience anginal pain. If the activity is stopped, the pain will likely fade because myocardial oxygen needs will be met when the workload is removed. However, it is also possible that the increasing pressure in the coronary arteries will cause rupture of a plaque, occlusion, and a myocardial infarction. Severe ischemia in the coronary arteries can even cause an abnormal rhythm that results in sudden cardiac death.

ventricular tachycardia—rapid heartbeat originating from the ventricle

ventricular fibrillation—uncontrolled contractions of the ventricle; often associated with myocardial infarction

These rupture-prone lesions are usually fairly small, occluding less than 50% of the lumen, and therefore most individuals who experience an acute MI have had no symptoms previously. Occlusion typically has to be much more significant to cause angina. Even though the smaller lesions are more likely to rupture, angiographically significant blockage does provide an indication of the extent of the disease (Squires 2006). The plaques that are more likely to rupture are those with a larger lipid core relative to total plaque area. A lipid-rich plaque is one with more than 50% lipid by volume, and while only approximately 15% of plaques fall into this category, they account for an estimated 80% of all acute MI and episodes of unstable angina (Pradka 2000). In men, cigarette smoke increases the risk of lipid-rich plaque rupture and resultant sudden cardiac death (Ambrose and Barua 2004).

Though efforts to lower lipids tend to have only a minimal effect on the harder, more fibrous plaques, they do result in regression of the lipid-rich plaques. Lipid-lowering efforts retard further plaque formation and reduce the chance for plaque rupture (Pradka 2000).

The same type of atherothrombotic occlusion that was described earlier can initiate **ventricular tachycardia** or **ventricular fibrillation** and sudden cardiac death. In sudden cardiac death, there is an abrupt loss of heart function and death occurs immediately or within an hour of the appearance of symptoms (Squires 2006). There are other potential causes of sudden cardiac death, but not all are triggered by ischemia and/or infarct. Hypertrophic cardiomyopathy, left ventricular hypertrophy, and valvular disease are all causes of sudden cardiac death. In adults 35 years of age and under, hypertrophic cardiomyopathy is the cause of almost half of all sudden cardiac death, particularly in competitive athletes. In combination with left ventricular hypertrophy and congenital coronary abnormalities, cardiomyopathy accounts for over 75% of all sudden cardiac deaths in this population (Saffitz 2005).

When an MI occurs, necrosis of myocardial cells will occur because of the ischemia caused by prolonged arterial occlusion. If the ischemia is severe enough to cause an infarct, then the damage to the myocardium will be irreversible. This means that the membrane of the myocardial cell is disrupted and the contents escape the cell. These include certain biological markers of an MI: creatine kinase and cardiac troponin T and are used to clinically diagnose an MI (see Table 15.14). The necrotic cardiac cells do not regenerate and are replaced instead by scar tissue. How long scar tissue formation takes is dependent upon the size of the area impacted by the infarct. The size of the tissue damage is determined by the location of the occlusion and by the amount of collateral circulation to the area affected. As blood flow to tissues may be compromised for a number of years, a secondary pathway for blood flow may be developed from smaller vessels; this is collateral circulation. An infarct may only affect part of the cardiac wall; this is known as a subendocardial infarct. If the entire width of the cardiac wall is damaged by the infarct, it is termed a transmural infarct.

TABLE 15.14

Cardiac Biomarkers

Biomarker	Normal Levels (Standard reference range dependent on individual patient factors and laboratory diagnostic methods)	Time to Initial Elevation	Time to Peak Elevation	Time to Return to Normal
CK-MB	Total CPK Male 55–170 U/L Female 30–135 U/L CPK-MB – 0%	4–8 hours	12–24 hours	72–96 hours
CK-MB Isoforms	0%	2–6 hours	18 hours	<24 hours
Myoglobin	<90 μ/L	2–4 hours	8–10 hours	24 hours
LD-I	313–618 U/L	10–12 hours	48–72 hours	7–10 days
cTnI	<0.5 ng/dl	4–6 hours	12 hours	3–10 days
cTnT	<0.5 ng/dl	4–6 hours	12–48 hours	7–10 days

CK-MB, MB isoenzyme of creatine kinase; LD-I, lactate dehydrogenase isoenzyme; cTnI, cardiac troponin I; cTnT, cardiac troponin T. http://www.abbottdiagnostics.com/Your_Health/Heart_Disease/troponin-physicians-brochure.cfm.

Source: Abbott Laboratories. Troponin—Physician's Brochure [monograph on the Internet]. Table 2. Abbott Park (IL): Abbott Laboratories; 2006. Available from: http://www.abbottdiagnostics.com/Your_Health/Heart_Disease/troponin-physicians-brochure.cfm.

In some cases the infarct may result in adverse changes to the left ventricle. These changes include an expansion of the left ventricular chamber and a thinning of the left ventricular wall, both of which severely impact ventricular contractility and result in heart failure (Squires 2006).

The left ventricle and the septum of the heart contain most of the muscle mass of the heart and therefore most of the blood flow. The oxygen requirements of the left ventricle are much greater than those of the right ventricle. To visualize this, consider that the left ventricle must move blood against an average systolic BP of around 120 mmHg, whereas the right ventricle moves against a force that is around 1/6 of that. The workloads of the right ventricle are low enough that collateral circulation can often meet the needs of the tissue if flow is disturbed. Since most of the blood flow and oxygen demands are made by the myocardium of the left ventricle and the septum, these areas are most susceptible to disturbances in oxygen supply. Infarcts involve these areas of the heart almost exclusively (Crowley 2001).

Persons who suffer an MI are subject to complications, which may be classified as:

- Disturbances of cardiac rhythm
- Heart failure
- Intracardial thrombi
- Pericarditis
- Cardiac rupture
- Papillary muscle dysfunction
- Ventricular aneurysm

Cardiac muscle tissue adjacent to an infarcted area will become irritable and can cause arrhythmias. The most serious of these, which has been mentioned previously, is ventricular fibrillation. Ventricular fibrillation causes circulation to come to a halt. In some cases the nervous tissue that transmits impulses from the atria to the ventricles will become impaired, causing what is termed heart block. In heart block the atria and ventricles may be contracting on separate rhythms, depending on the degree to which the heart block is occurring. Over time the rhythmic disturbances resulting from the infarct may subside (Crowley 2001).

In cases where the infarct affects the endocardium, a thrombus may form on the interior surface of the ventricular wall. This thrombus will cover the damaged area, forming what is known as a mural thrombus. Some parts of this thrombus may break loose and travel as **emboli**. These are particularly dangerous and may cause an infarct in other tissues, such as the brain, with catastrophic results.

Infarcts involving the epicardium may cause fluid accumulation in the pericardial sac. The inflammation associated with the damage to the myocardial cells will be the source of the fluid. The fluid accumulation in the pericardial sac is known as pericarditis.

embolus—blood clot that breaks from the cellular surface and freely moves through the circulation

Much more serious is cardiac rupture. In the case of a transmural infarct, it is possible that a leak may develop in the cardiac wall through the necrotic tissue. If this occurs, the pericardial sac will fill with blood, eventually placing enough pressure on the heart so that it cannot expand to accept blood during diastole, a condition known as cardiac tamponade. Circulation will cease because the heart can no longer move blood. In transmural infarcts involving the septum, this same scenario can occur between the right and left ventricles. The perforation of the septum causes blood to move from the left to the right ventricle during systole. This will compromise cardiac output, resulting in heart failure.

Papillary muscles contract during systole to hold the leaflets of the valves closed, restricting blood from moving back into the atria. Infarcts damage papillary muscles and can result in their inability to keep valve leaflets in place. Mitral valve insufficiency causes blood to leak back into the left ventricle during ventricular systole.

A ventricular **aneurysm** is an outward bulging of the healing infarct during ventricular systole. The aneurysm fills with blood during systole and therefore impacts the total distribution of blood. This can result in heart failure. The damaged tissue does not contract and overall cardiac efficiency is reduced.

Clinical Manifestations

Practitioners should be aware that the majority of men and women who die from heart disease have reported no previous symptom. In fact, only 20% of coronary attacks are preceded by angina, the primary symptom of ischemic heart disease. **Stable angina** is the substernal pain experienced when the workload on the heart is increased due to physical or emotional stress. While angina is typically referred to as a dull ache in the substernal area radiating to the arm and neck, it should be noted that individuals experience angina pain differently. Ischemic heart disease is often misdiagnosed or undiagnosed in females because they tend to experience angina pain differently than men, reporting more intense levels of pain (Barsky, Peekna, and Borus 2001). Diabetics may present differently as well (Smitherman and Reis 1997). Unstable angina is angina pain that is not associated with increases in workload, and may even occur in a person at rest. The discomfort associated with unstable angina may be more severe and prolonged (American Heart Association 2006). Other symptoms of angina may include indigestion, nausea, vomiting, sweating, shortness of breath, weakness, and fatigue.

There are other symptoms associated with heart disease that may precede a significant coronary event. These include a number of signs and symptoms that are most commonly associated with rhythmic abnormalities and may not necessarily be caused by coronary ischemia. This is particularly true in individuals who have no history of previous ischemic disease. These may include sudden-onset bradycardia, palpitations, and syncope and/or dizziness.

aneurysm—a weakened portion of the blood vessel wall

stable angina—chest pain associated with increased oxygen demand such as occurs with physical exertion

Diagnosis

Diagnostic procedures for ischemic heart disease may include a variety of noninvasive tests, as previously discussed in the section for AS (see Box 15.8). These include chest x-ray, electrocardiogram, or exercise stress tests. Imaging tests include radionucleotide imaging, PET/CT scan, echocardiogram, and cardiac catherization.

The World Health Organization criteria for diagnosis of myocardial infarction state that an individual must meet two out of the following three criteria: clinical history of ischemic type chest pain, changes in serial ECG readings, and a rise and fall of serum cardiac enzymes (Ryan et al. 1999; Beers and Berkow 2006). These diagnostic criteria focus on measurement of a series of enzymes or proteins that are released from damaged or dying cells. During necrosis, cellular contents, such as enzymes and proteins, are released (see Chapter 10). If the patient's blood contains enzymes/proteins normally found in large amounts within myocardial cells, this would be indicative of myocardial injury. These include MB isoenzyme of creatine kinase (CK-MB), lactate dehydrogenase isoenzyme (LD-I), cardiac troponin I (cTnI,), cardiac troponin T (cTnT), and myoglobin (see Table 15.14).

Treatment

The goals of immediate medical treatment after myocardial infarction are to reduce pain, reduce the work of the heart, stabilize cardiac function, and prevent or limit complications. Medical interventions include the use of oxygen, aspirin (for antithrombolytic effect), and morphine (for pain). A variety of medications may be used to provide fibrinolytic therapy, reduce the overall work of the heart, and treat other cardiac dysfunctions. See Table 15.1 for a description of these classes of medications.

Myocardial infarction treatment protocols provide a structure for initiation of activity. Usually bed rest is only recommended for the initial 24–48 hours after the cardiac event. Stages of physical activity slowly increase and the patient is usually discharged within 5–7 days (Beers and Berkow 2006).

BOX 15.8 CLINICAL APPLICATIONS

Cardiac Diagnostic Procedures

Noninvasive

- Chest x-ray: Assesses anatomy of the heart; pulmonary congestion.
- Electrocardiogram (ECG/EKG): Graphic recording of the heart's electrical activity.
- Holter Monitor: Portable ECG.
- Exercise Stress Test: Heart rate, blood pressure, and ECG are measured after a session of prescribed exercise on a treadmill or bicycle.

Imaging Tests

- Radionucleotide Imaging: Involves the intravenous injection of small quantities of radioactive isotopes into a peripheral vein. The distribution of the radioactive tracers can be detected by gamma cameras from the radiation emitted as the radionuclide decays. Depending on the radioactive isotope used, RI is used to measure myocardial perfusion and detect ischemia, perform infarct imaging, evaluate ventricular function, and to detect and evaluate coronary artery disease.
- MUGA (Multiple Gated Acquisition scan): Radionucleotides attached to red blood cells allow for visualization of the heart while beating. Especially useful to measure ejection fraction of the ventricles.
- Thallium Stress Test: After performing an exercise stress test, the patient is injected with thallium, which allows for visualization of blood flow to the heart after stress. As the patient rests, continued visualization allows for documentation of blood flow during rest.
- PET/CT (Positron Emission Tomography/Computer Tomography): The combined use of PET and CT allows for anatomical visualization as well as metabolic assessment of tissues and organs.
- SPECT (Single photon emission computed tomography): Allows for 3-D visualization of the heart as well as blood flow.
- Echocardiogram: Uses sound waves to document anatomy and function of the heart.
- Coronary angiography/cardiac catherization: Catheter is inserted from appropriate artery or vein into the ventricle. Allows visualization of heart anatomy, function of the valves. When contrast dye is injected, blood flow can be documented in order to identify any obstructions in circulation.

Physical Assessment and Laboratory Tests

- Pulse: Measurement of heart rate.
- Blood Pressure: Measurement of cardiac output and peripheral resistance.
- Doppler Studies: Measurement of audible blood flow within the peripheral vessels.
- Auscultation: Listening to heart sounds with stethoscope.
- Cardiac Enzymes: Enzymes and isoenzymes are released as cells die from oxygen deprivation. These are not necessarily specific to cardiac tissue. These include lactic dehydrogenase (LDH-1), aspartate aminotransferase (AST), creatinine phosphokinase (CK-MB or CPK-2).
- Cardiac Troponin I: Protein released from myocardial cells—elevated after cardiac injury.
- Myoglobin: Equivalent to hemoglobin that is present in skeletal muscle. Will be released when tissue is damaged.
- Lipid Profile: Serum cholesterol, HDL-C, LDL-C and TC:HDL-C ratio; serum triglyceride. Others may include apolipoprotein-B.

Nutrition Therapy

Nutrition Implications Immediate medical care after myocardial infarction strives to reduce pain, stabilize cardiac function, and, when appropriate, begin the rehabilitation post-MI. Nutrition therapy after MI will be consistent with these medical goals.

Nutrition Interventions During the immediate post-MI period, oral intake may be decreased due to pain, anxiety, fatigue, and shortness of breath. Many institutions' treatment protocols limit initial oral intake to clear liquids without caffeine in order to prevent arrythmias and to decrease risk of vomiting or aspiration (Escott-Stump 2002). Oral diets usually progress from liquids to soft, easily chewed foods with smaller, more frequent meals. As the patient stabilizes, the goals of nutrition therapy will be individualized according to the patient's risk factors and should follow the Therapeutic Lifestyle Changes Dietary Recommendations (U.S. Department of Health and Human Services 2001; Carson et al. 2004).

Peripheral Arterial Disease

Definition

Peripheral arterial disease (PAD) is a term used to describe occlusion of blood flow in non-coronary arteries, and for the purpose of this section of the text is limited to the lower extremities. Most PAD occurs in the pelvis and legs. Peripheral vascular disease is a term that includes diseases of the veins (Chant 2004). PAD has been used by some to describe all non-coronary atherosclerotic disease including involvement in the carotid arteries (Pasternak et al. 2004).

Epidemiology

PAD affects about 8 million Americans. The prevalence of the condition increases with age and disproportionately affects blacks. Hispanics have a slightly higher risk than whites (American Heart Association 2006; Pasternak et al. 2004). One study found a prevalence of PAD of 2% to 3% by age 50 and approximately 20% by age 75. This study found that 10% of the individuals with PAD had **claudication**, 50% had atypical leg pain, and 40% had no leg pain associated with physical activity (Pasternak et al. 2004).

There is a strong association between vascular disease and damage to the coronary, cerebral, and carotid arteries; thus, the 5-year mortality rate is relatively high. The long-term prognosis for symptomatic PAD patients is poorer than that of asymptomatic patients. Those with severe symptoms have a much worse prognosis compared to those who have mild symptoms (Pasternak et al. 2004).

Risk factors for the development of AS are important to the development of PAD and are targets for treatment. The impact of the individual risk factors on atherosclerotic disease development and progression in the periphery are not the same as in the coronary arteries. The most influential independent risk factors for PAD development and progression are cigarette smoking and diabetes mellitus (American College of Cardiology/American Heart Association Task Force/American Diabetes Association 2003; Faxon et al. 2004; Pasternak et al. 2004). In smokers, PAD development is increased 2 to 5 times. Smokers also are 8–10 times more likely to develop intermittent claudication. Cessation of smoking is associated with decreased amputation rates and increased longevity (Faxon et al. 2004).

peripheral arterial disease (PAD)—atherosclerotic heart disease of all vessels except specific coronary vessels; term used interchangeably with peripheral vascular disease

claudication—pain in arms and legs due to inadequate blood flow to those muscles

The single strongest risk factor for PAD is diabetes. Data from the Framingham study indicate that 20% of symptomatic PAD patients were diabetics (American Diabetes Association 2004; Pasternak et al. 2004). Diabetic men have a higher rate of claudication than any other group. Diabetes in women eliminates the protective effect of estrogen so that their risk is elevated to that of men with similar risk profiles. Diabetes is the leading cause of non-traumatic amputation in the United States (see Chapter 19) (Faxon 2004).

While dyslipidemia and hypertension are important and do increase risk of PAD, they do not appear to be as influential as diabetes and cigarette smoking. Elevated LDL, low HDL, and elevated serum triglycerides do impact risk of PAD. It has been estimated that each 10 mg/dL rise in total cholesterol increases the relative risk of PAD by approximately 1.1 times. Thus, the risk is increased by 10% over adults with the same risk profiles and optimal total cholesterol levels (Faxon et al. 2004). Studies of lipid and non-lipid risk factors for PAD have shown that the total cholesterol-to-HDL ratio is the strongest lipid predictor. When CRP was added to lipid screening, the predictive value was greatly enhanced (Smith et al. 2004).

The effect of hypertension on PAD appears to be much more subdued than the effect in the cerebral or coronary arteries. Because the data from several studies appears to provide mixed results, the degree of influence hypertension exerts on atherosclerotic development in the peripheral arteries is unclear (Faxon et al. 2004).

Pathophysiology

In PAD, the occlusion of an artery, typically in the pelvis or lower leg, restricts blood flow to tissues. The pathophysiology of PAD is quite similar to that of AS and IHD. In these conditions, an inflammatory response precedes the plaque rupture with subsequent embolus formation. The presence of PAD is also an indicator of ischemic disease in other vascular beds; thus, the risk of MI, stroke, unstable angina, and sudden cardiac death is increased in PAD patients. There are some small differences between the pathophysiology of PAD and that of IHD. The focus in this section will be on those differences.

While thrombosis is known to play a critical role in all acute ischemic incidents, it has been postulated that it has an even more important role in acute ischemic events in those with PAD. Fibrinolytic therapy in PAD patients is more effective in reducing cardiovascular events than the same therapy in those with IHD. Reducing platelet adhesion lowers the risk of emboli and thrombosis, the formation of which is key in PAD and other ischemic events, such as stroke and transient ischemic attack (Faxon et al. 2004).

In previous sections, the pathophysiology of thrombus formation and plaque rupture was described as occurring in a prothrombotic state. That prothrombotic state included high levels of PAI-1 and CRP. The inflammatory process is a part of both the development of the plaque and its rupture. An elevated level of CRP is a serum marker of inflammation and is commonly found not only in persons suffering acute coronary events such as MI and unstable angina, but also in persons with PAD. The importance of CRP to the progression of PAD is unknown at this time, but it is suspected that the severity of acute events and symptoms are associated with increased CRP levels (Faxon et al. 2004).

There is a poor correlation between variations in BP across the leg and ischemic symptoms. The variables that contribute to claudication and their function in eliciting this symptom are somewhat complex. What is known is that in response to ischemia collateral vessels will be developed to allow some blood flow. The extent to which these vessels are formed and their contribution to function seem to be determined to a great degree by the region in which the ischemia occurs. In some cases the new vasculature may be sufficient to cause regression of symptoms (Faxon et al. 2004).

Patients with PAD will eventually suffer from denervation of affected muscle tissue. The ischemia-reperfusion of the tissue over time will cause this damage, which is thought to be mediated by oxidative damage. The damage of denervation and alterations in muscle fiber type will reduce muscle function. There is additional evidence of abnormal muscle cell metabolism in affected tissues (Brass et al. 2004; Faxon et al. 2004).

There are several changes in muscle tissue as a result of ischemia in PAD. Mitochondrial expression increases in patients with PAD. While this is not uncommon to conditions that impair mitochondrial function, the implication is that activities inside the muscle lead to its dysfunction. Accumulation of metabolic intermediates and products of incomplete metabolism lend credence to the idea that there is a state of metabolic disorder inside the muscle. Lactate levels at rest and during light workloads are elevated beyond that which might be attributable to low blood/oxygen supply. Thus, the muscle may have made an adaptation to shunt pyruvate away from complete oxidation. This may also be a function of the relative activity of pyruvate dehydrogense and lactate dehydrogenase. Pyruvate dehydrogenase has been found to be altered in patients with PAD (Brass et al. 2004). Acylcarnitines, a class of intermediates of substrate oxidation, accumulate in skeletal muscle and blood of PAD patients. The degree of accumulation of acylcarnitines at rest is strongly correlated with functional impairment of the individual during exercise (Brass et al. 2004; Faxon et al. 2004).

The electron transport chain is not only a major source of reactive oxygen species formation, but also suffers extensive oxidative damage as a result of PAD. Muscle tissue from PAD patients shows specific defects in the enzymes of the electron transport chain. This damage can led to increased reactive oxygen species formation and may further metabolic injury. In addition, the increased free radical production can contribute to further endothelial injury (Brass et al. 2004).

PAD can cause ulcerations of the lower leg. **Ulceration** is defined as a nonhealing break in the skin. Inadequate perfusion to the tissues (i.e., lack of oxygen) is the primary etiology for skin breakdown. The ischemic conditions cause a breakdown of tissue, forming the ulcer. The arterial ulcer associated with PAD commonly occurs on the foot or toes. This type of ulcer is said to be more painful than other types and has a pitted or punched out appearance (Chant 2004). This ulceration occurs more often in those with an **Ankle Brachial Index (ABI)** <0.4 (Pasternak et al. 2004).

Clinical Manifestations and Diagnosis

The symptom associated with the ischemic conditions of PAD is intermittent claudication (Brass, Hiatt, and Green 2004; Chant 2004). Intermittent claudication is a cramp-like pain that is associated with activity and then subsides with rest. This symptom is most common in the calf but may occur in the thigh, buttocks, or feet. Intermittent claudication has an earlier onset and is more intense if the activity is more strenuous, such as walking up a slope or stairs. Therefore, individuals with PAD may unconsciously alter their physical activity patterns.

Intermittent claudication may be used as a means of diagnosing PAD (Jude and Gibbons 2005). An individual who walks a certain distance and senses claudication pain, then rests until symptoms subside, then walks the same distance and experiences the same level of pain is more likely to have PAD. This is particularly true if the pulses in the foot and ankle are not present (Jude and Gibbons 2005).

Though the presence of PAD may be identified by claudication with activity and absence of a peripheral pulse, a more accurate indicator is the Ankle Brachial Index (ABI). This test is more sensitive and specific for PAD than observation of symptoms (American Diabetes Association 2003; Pasternak et al. 2004). The ABI is the ratio of the systolic blood pressures of the upper and lower extremities measured with Doppler recordings, which use low-intensity ultrasound to detect blood flow in arteries or veins. The ABI is generally a reliable indicator of PAD, but possesses one flaw: poorly compressible vessels in the elderly and diabetics in particular yield falsely high readings (greater than 1.3). Poorly compressible vessels have a great degree of calcifica-

ulceration—nonhealing break in skin or tissue surface

ankle brachial index (ABI)—ratio of Doppler-recorded systolic blood pressures between upper and lower extremities; a measure of peripheral vascular disease

tion compared to normal vessels. The use of ABI is less reliable in these cases (American Diabetes Association 2003).

In some cases, a treadmill test may be needed for diagnostic purposes. Individuals who experience claudication will typically have a decrease in ankle BP of 20 mmHg after exercise (American Diabetes Association 2003). In cases of mild obstruction the post-exercise ABI will drop to >0.5, and in moderate obstruction, to >0.2; below 0.2 indicates severe obstruction (Squires 2006).

Because the ischemic conditions of PAD can lead to gangrene of the distal tissues, PAD is a major risk factor for lower extremity amputation. Even in cases where the individual is asymptomatic PAD would be indicative of systemic vascular disease. If AS has progressed to the point that it can cause occlusion in the vasculature of the leg, then this same progression has probably occurred in the coronary, carotid, and cerebral arteries, where it increases the risk of death from other vascular diseases (American Diabetes Association 2004; American Heart Association 2006).

Heart Failure

Definition

Heart failure is an impairment of the ventricles' capacity to eject blood from the heart or to fill with blood. The underlying cause of this disorder can be either structural or functional in nature. Heart failure represents the end stage of all forms of cardiovascular disease. Many heart failure patients will have well-preserved left ventricular function, while others may display signs of significant left-ventricular impairment (Barnard 2005; Hunt et al. 2005).

Epidemiology

Data for 2003 indicate that 5 million Americans were diagnosed with heart failure and that it was a contributing cause of death in 286,700 individuals. Between 1999 and 2003, the overall death rate declined 2%, while the death rate from heart failure increased 20.5%. At least part of this change in death rate can be explained by improvements in treatment and management of MI and other conditions, which extend life expectancy. Death rates from heart failure are higher for black males and females compared to their white counterparts (American Heart Association 2006).

Data from the Framingham Heart Study indicate that (American Heart Association 2006):

- By age 65, the incidence of heart failure is slightly below 10 per 1,000 in the population.
- 75% of persons diagnosed with heart failure have hypertension; lifetime risk of heart failure doubles for those with a resting blood pressure of 160/90 mmHg compared to those with a blood pressure below 140/90 mmHg.
- Among MI survivors, 22% of males and 46% of females will be disabled with heart failure within 6 years.
- At age 40, lifetime risk of developing heart failure is one in five; for individuals who have not suffered an MI, the risk drops to one in nine for males and one in six for women.

Heart failure is closely associated with aging. Women have a higher relative risk of heart failure primarily because women comprise over 60% of the population over 65 years of age and 75% of the population over the age of 85 years. (For more on heart failure among women, see Box 15.9.)

The prevalence, incidence, and mortality for heart failure in diabetics is very high. For every 100 individuals with diabetes who are free of heart failure at the start of a year, twelve will develop heart failure and six will die within that year. The prevalence and incidence of heart failure in diabetics is associated with age and comorbidities such as nephropathy, IHD, and PAD (Bertoni et al. 2004; Centers for Disease Control and Prevention 2003).

Etiology

Heart failure may result from disorders of the pericardium, myocardium, endocardium, or vessels, but the majority is due to impaired left ventricular myocardial function. Heart failure is a broad term, and may be used to describe conditions in which left ventricular size and ejection fraction are maintained as well as those in which the left ventricle is dilated and ejection fraction is severely reduced (Hunt et al. 2005). Classifications for heart failure are outlined in Table 15.15. There are common differences between men and women: women tend to have preserved left ventricular systolic function, whereas men tend to have greater impairment in systolic function (Barnard 2005).

The primary causes of heart failure are IHD, hypertension, and dilated cardiomyopathy. In women, hypertension is the most common cause, while in men IHD is the most common cause. Of those with dilated cardiomyopathy, approximately 30% may have a genetic cause. Valvular disease is an additional common cause of heart failure (Barnard 2005; Hunt et al. 2005).

Pathophysiology

Heart failure is a process beginning with an injury to the heart (as described earlier in this chapter) or with left ventricle hypertrophy that impairs overall function of the heart. To compensate for the impairment in function, the renin-angiotensin-aldosterone system initiates changes in BP that exacerbate the dysfunction. As a result, the heart becomes weakened and dilated, myocardial fibrosis limits the ability of the walls to respond to stresses, oxidative damage further impairs contractility, and the overall structure of the heart is

BOX 15.9 NEW RESEARCH

Women and Heart Failure

Even though other diseases that are prevalent among women, such as breast cancer, seem to get more media attention, heart disease is the number one cause of death among women. Preliminary data in 2003 indicate that over 483,800 women died from heart disease. Heart failure was the primary cause of death in 34,900 of those cases, compared to 22,300 heart failure-related deaths among men during the same period (American Heart Association 2006).

At age 40, the lifetime risk of developing heart failure is equal between men and women, whereas the lifetime risk of developing IHD is 1 in 2 for men and 1 in 3 for women (Barnard 2005). The incidence of heart failure among women surpasses that of men after the age of 75. This may be partly explained by the fact that women make up a much larger percentage of the population over the age of 65 (American Heart Association 2006).

Why is heart failure more common in women than in men? It appears that the risk factors for heart failure do not contribute equally to the development of the disease when comparing the sexes (Barnard 2005).

- Hypertension is a stronger risk factor for heart failure among women.
- Left ventricular hypertrophy that is associated with a history of hypertension and diabetes is a stronger predictor of heart failure among women. Prevalence of physician-diagnosed diabetes between men and women is roughly equal.
- Women with heart failure are more likely than men to have diabetes. Compared to non-diabetics, the risk of heart failure among diabetic women is 8 times greater, compared to 4 times greater for men.
- Even though men are more likely to have an MI, women are more likely to develop heart failure after an MI or coronary artery bypass graft surgery.
- Other risk factors such as metabolic syndrome, physical inactivity, obesity, valvular disease, and renal insufficiency are more common in women than in men.

Sex-based differences in normal cardiac physiology and in the remodeling associated with heart failure can explain why women suffer from heart failure with preserved left ventricular systolic function. The cardiac remodeling that occurs as a result of IHD differs between men and women. Heart weight tends to increase to a greater degree in men, with greater cellular hypertrophy. Women with heart failure tend to demonstrate a decrease in diastolic compliance, which results in a decreased left ventricular end diastolic volume (Barnard 2005).

Studies in animal models have identified a potential role for estrogen deficiency in the development of left ventricular hypertrophy. This has led researchers to examine the effects of estrogen on the renin-angiotensin-aldosterone system. Oral administration of estrogen in a rat model up-regulates angiotensinogen production, while down-regulating rennin, angiotensin converting enzyme, and angiotensin-1 receptors. Human females with heart failure do exhibit higher circulating levels of angiotensinogen than do men. These differences may explain at least part of the sex differences in heart failure (Barnard 2005).

For females, addressing the two most prevalent risk factors, hypertension and diabetes, should be the priority in the prevention of heart failure. All cardiovascular risk factors need to be targeted, but specific efforts to control these two risk factors are imperative. Interventions to address metabolic syndrome, smoking, dyslipidemia, weight management, and physical activity should help to reduce the development of heart failure. Angiotensin converting enzyme inhibitor therapy has also been demonstrated to improve outcomes to a greater extent in women than in men by decreasing cardiovascular mortality and hospitalizations (Barnard 2005).

TABLE 15.15

Stages of Heart Failure

Stage	Definition
A	Patients who are at high risk for developing heart failure but have no structural abnormalities
B	Patients who have structural heart disease but demonstrate no symptoms of heart failure
C	Patients with past or current symptoms of heart failure who have underlying structural heart disease
D	Patients with end-stage disease requiring specialized treatment, such as mechanical circulatory support, procedures to facilitate fluid removal

Source: Hunt et al. 2005.

changed in such a way that it cannot function properly (see Figure 15.11).

Often the pathophysiology of heart failure can be traced to damage resulting from IHD. The most common scenario is that damaged sections of tissue resulting from an MI impair contractile function of the left ventricle and reduce ejection fraction. Heart failure is described as a progressive disorder because even if there is no additional injury to the myocardium, there will be a continued deterioration in function. Post-MI, the left ventricle will undergo a change in structure and geometry. The change, referred to as cardiac remodeling, will result in a hypertrophied and/or dilated left ventricular chamber. The change in structure impairs performance of the heart and may even cause some regurgitation

FIGURE 15.11 Effects of Congestive Heart Failure

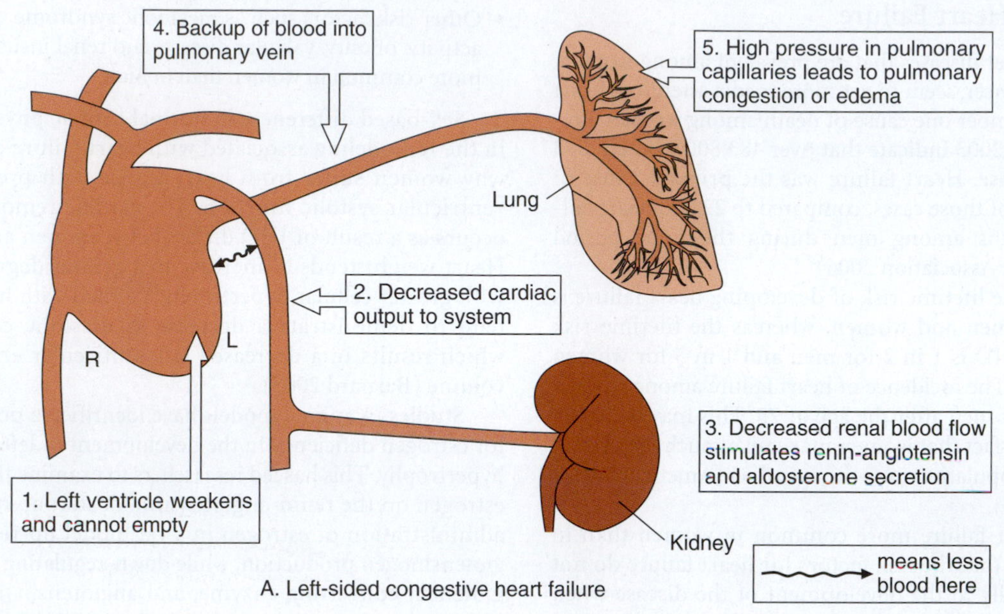
4. Backup of blood into pulmonary vein
5. High pressure in pulmonary capillaries leads to pulmonary congestion or edema
Lung
2. Decreased cardiac output to system
R
L
3. Decreased renal blood flow stimulates renin-angiotensin and aldosterone secretion
1. Left ventricle weakens and cannot empty
Kidney
means less blood here
A. Left-sided congestive heart failure

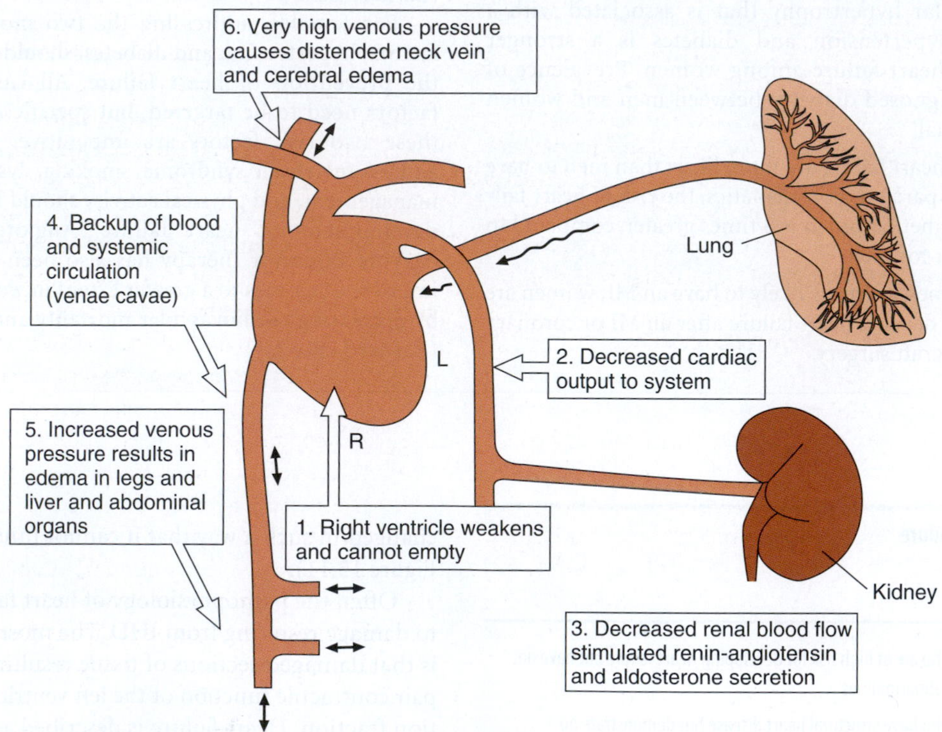
6. Very high venous pressure causes distended neck vein and cerebral edema
4. Backup of blood and systemic circulation (venae cavae)
Lung
L
2. Decreased cardiac output to system
5. Increased venous pressure results in edema in legs and liver and abdominal organs
R
1. Right ventricle weakens and cannot empty
Kidney
3. Decreased renal blood flow stimulated renin-angiotensin and aldosterone secretion
B. Right-sided congestive heart failure

Source: Pathophysiology for the Health Professions Third Edition. Philadelphia, PA: Saunders Elsevier. Figure 18-22, p. 333.

through the mitral valve back into the left atria (Hunt et al. 2005).

Cardiac remodeling precedes symptoms and will continue after symptoms appear. The progression may result in a worsening of symptoms even when treatment is ongoing. Progression of heart failure may be accelerated by progression of IHD, diabetes, hypertension, or the onset of atrial fibrillation (Hunt et al. 2005).

The progression of heart failure and cardiac remodeling is mediated to some extent by neurohormonal systems. Heart failure patients typically have elevated blood and tissue levels of norepinephrine, angiotensin II, aldosterone, endothelin, vasopressin, and cytokines. All of these substances, either alone or in tandem, can have adverse effects on cardiac structure. Sodium retention and peripheral vasoconstriction result in increases in arterial BP, thereby increasing myocardial workload. Other substances mediate oxidative stress, causing myocardial cell damage and myocardial fibrosis that further alter the structure of the heart and reduce its function (Barnard 2005; Diamond and Phillips 2005; Hunt et al. 2005).

Left ventricular hypertrophy as a result of extended periods of hypertension can initiate heart failure by reducing extensibility of the left ventricular wall and contractility. A characteristic of hypertension-related cardiac hypertrophy is a phenotype change of cardiac fibroblasts. When stimulated, myofibroblasts proliferate and increase production of fibrous substances such as fibronectin and collagens. Under normal circumstances, matrix metalloproteinases increase the degradation of fibrillar collagen, but an inhibitor will prevent their activation in the presence of soluble collagen. The balance between the matrix metalloproteinases and their inhibitors is disrupted in heart failure. The metalloproteinases destroy normal collagens and leave poorly crosslinked collagens intact. This weakens the structure of the cardiac wall and results in dilatation (Diamond and Phillips 2005).

The dilatation of the left ventricle impairs contractility, resulting in a decreased cardiac output and ejection fraction. A decrease in cardiac output will also cause decreases in renal blood flow and glomerular filtration (see Chapter 20). The kidneys respond by activating the rennin-angiotensin-aldosterone system to raise blood pressure and restore blood flow to the filtration units. This attempt to maintain homeostasis increases circulating levels of angiotensin II and aldosterone. Subsequently, afterload increases, edema develops, and heart failure progresses. The typical heart failure patient has serum levels of aldosterone 20 times higher than what is considered normal. These increased aldosterone levels are also associated with myocardial fibrosis. Administration of aldosterone antagonists and angiotensin converting enzyme inhibitors to MI patients after revascularization surgery decreases fibrosis and increases left ventricular ejection fraction (Jennings et al. 2005). Increased aldosterone concentrations have been implicated in a number of other mechanisms associated with heart failure. These include endothelial dysfunction, reduced variability in heart rate, reduced cardiac norepinephrine uptake, and increased risk of cardiac arrhythmias (Jennings et al. 2005).

Reductions in renal blood flow and glomerular filtration will reduce the rate of solute and water delivery to the distal diluting segment of the nephron. The end results are an inability to excrete a dilute urine and hyponatremia. This is more common in severe heart failure as compared to mild-to-moderate heart failure. Only about 5% of heart failure patients suffer from hyponatremia; however, the hormonal abnormalities that cause this imbalance are present in most heart failure patients (Sica 2005).

Conditions such as aortic regurgitation and mitral regurgitation can result in impaired left ventricular function. In these conditions, structural abnormalities, infection, and/or damage to tissues can result in hypertrophy of the left ventricle as a result of volume overload. While mitral regurgitation may occur as a result of structural changes to the heart in cases of severe heart failure, it often occurs as a genetic disorder or secondary to damage to the chordae tendinea or valve leaflets (Squires 2006).

Clinical Manifestations

Signs and symptoms of heart failure are a result of the basic pathophysiology of the disease. They will vary depending on the predominance of the disorder—either left- or right-sided failure. In general, decreased blood flow and oxygen supply lead to dyspnea, fatigue, weakness, exercise intolerance, and poor adaptation to cold temperatures. Dyspnea is an unusual shortness of breath not appropriate to the workload, and this symptom in conjunction with fatigue may limit the individual's functional capacity. When left-sided failure is predominant, dyspnea is more predominant and also includes **orthopnea**.

Right-sided failure is characterized by the signs and symptoms caused by systemic backup of the circulatory system. This fluid retention can cause pulmonary congestion and edema in the periphery. Eventually, edema affects the gastrointestinal tract and results in **hepatomegaly**, **splenomegaly**, and ascites (fluid retention within the abdominal cavity). These may further impair respirations by pressing up on the diaphragm and limiting pulmonary function. Other signs of right-sided failure include distended neck veins, headache, and a flushed face. Signs of compensation include tachycardia, pallor, polycythemia, and oliguria. Dyspnea, fatigue, and

orthopnea—shortness of breath associated with lying in the supine position

hepatomegaly—enlargement of the liver

splenomegaly—enlargement of the spleen

edema affect quality of life and limit functional capacity (Hunt et al. 2005).

Treatment

The goals of medical treatment for heart failure are to treat the underlying cause of the cardiac disorder, control the symptoms associated with heart failure, and prevent continued damage to the heart (see Table 15.1) (Hunt et al. 2005; Little and Brucks 2005). Many of the same medications used to treat other cardiovascular disorders are also used to treat heart failure. Diuretics such as furosemide, bumetanide, and torsemide are crucial for control of edema and fluid retention. Some diuretics, such as Spironolactone, are not as effective in this stage of heart disease but may be used in conjunction with other diuretics. Control of BP is essential and clinical trials indicate an improved outcome for those individuals treated with ACE inhibitors. Other medications associated with improved outcome include beta-adrenergic blockers (Veterans Health Administration 2003; Hunt et al. 2005). Medications used to improve heart function by increasing myocardial contraction include digitalis, dopamine, and dobutamine (Veterans Health Administration 2003). The newest classes of medications for treatment—levosimendan, nesiritide, and L-NAME—may prove to be alternatives that can improve long-term outcomes (Rauch, Motch, and Bottiger 2006). Other components of care include prevention of respiratory infections, exercise, and nutrition therapy.

Nutrition Therapy

Nutrition Implications It has been estimated that as many as 50% of patients with heart failure are malnourished (Tangalos 2002; Schwengel, Gottlieb, and Fisher 1994; Akahsi, Springer, and Anker 2005). Nutritional care during CHF is difficult. Nutrition therapy that restricts both sodium and fluid is crucial to control acute symptoms and may assist with reducing the overall work of the heart. But at the same time, individuals with heart failure have difficulty eating and many experience a syndrome of malnutrition called **cardiac cachexia**. Cardiac cachexia is a form of malnutrition similar to the wasting syndrome seen in AIDS and cancer and characterized by extreme skeletal muscle wasting, fatigue, and anorexia (Akahsi, Springer, and Anker 2005; Filippatos, Anker, and Kremastinos 2005). The etiology is not completely understood, but it is assumed that it is multifactorial and involves both metabolic and hormonal abnormalities (Filippatos, Anker and Kremastinos 2005; Azhar and Wei 2006). Additional contributing mechanisms for cachexia in heart failure include myocardial nutrient deficiencies of L-carnitine, coenzyme Q10, creatine, thiamin, and taurine (Brady, Rock, and Horneffer 1995; Sole and Jeejeebhoy 2000, 2002; Witte, Clark, and Cleland 2001).

Further complications from heart failure that contribute to nutrition problems include (1) decreased blood flow to the gastrointestinal tract causing slowed peristalsis and early satiety, (2) decreased blood flow to the gastrointestinal tract which may impair nutrient absorption, and (3) side effects from drugs such as nausea, vomiting, and anorexia, which are common with the use of ACE inhibitors, beta blockers, cardiac glycosides, and digoxin (Sharma and Anker 2002; Tangalos 2002; Berger and Mustafa 2003; Anker, Steinborn, and Strassburg 2004; Carson et al. 2004). Nutrient deficiencies are also a common side effect from the use of diuretics and other medications.

cardiac cachexia—CVD-associated malnutrition/wasting syndrome characterized by extreme skeletal muscle wasting, fatigue, and anorexia

Nutrition Interventions Nutrition counseling for individuals with CHF is a priority. In a study by Kuehneman, Saulsbury, Splett, and Chapman (2002), readmissions and cost of hospital stays were found to have a direct relationship to excessive sodium intake. When a registered dietitian provides specific nutrition education for a patient, it can lead to fewer readmissions and an overall improved response to medical treatment (Colin et al. 2004; Arcand et al. 2005).

Nutrition therapy for CHF focuses on the control of signs and symptoms associated with the diagnosis and on the promotion of overall nutritional rehabilitation (see Table 15.16). Components of nutrition therapy include sodium and fluid restriction, correction of nutrient deficiencies, and nutrition education for increasing nutrient density and making food choices that enhance oral intake (Kuehneman et al. 2002; Neily et al. 2002; Berger and Mustafa 2003; Carson et al. 2004).

Sodium A 2,000 mg sodium diet is a standard initial recommendation for individuals with CHF (Carson et al. 2004; Hunt et al. 2005). Adjustments to levels of 2,000 mg, 1,000 mg, or 500 mg may be prescribed depending on the patient's individual medical condition—specifically, fluid and volume states as well as overall oral intake. Guidelines for these diets are outlined in Tables 15.17, 15.18, and 15.19. Because it is a challenge to manage this level of restriction outside of a hospitalized setting, it is crucial to critically evaluate the patient's actual PO intake to determine the level of sodium the patient is consuming prior to putting any further modifications into place. Anorexia, fatigue, and shortness of breath lead to such poor oral intake that many patients consume much less than 2,000 mg.

TABLE 15.16

Goals for Nutritional Care in CHF
• Stabilization/improvement in cardiac function • Stabilization/improvement in body weight • Prevention of/improvement in diet-related diseases or conditions associated with the development of CHF • Prevention of/improvement in adverse health outcomes associated with CHF • Prevention/minimization of drug/nutrient interaction

Source: Tangalos, 2002.

Fluid Fluid requirements are typically calculated at 1 mL/kcal or 35 mL/kg (see Chapter 7). To treat fluid overload in CHF, a fluid limitation of 1,500 mL/day is the standard recommendation, with an upper level of 2,000 mL. Again, adjustments will need to be made based on renal and cardiac status in order to prevent volume overload. Weighing the patient daily will allow the practitioner to monitor fluid status.

Fluid restriction is one of the most difficult diet orders for patients to tolerate. When providing nutrition education on fluid restriction, the clinician should make sure the patient understands the specific volume that is allowed, what items are considered to be fluids, and the suggestions to aid with controlling thirst. Visually demonstrating the amount of fluid the patient is allowed may support the patient's understanding and compliance. All beverages and foods such as soups, Popsicles®, sherbet, ice cream, yogurt, custard, and gelatin should be counted within the fluid allowance. Finally, good mouth care, rinsing the mouth frequently, and using cold or frozen foods can help control thirst. (See Table 20.10 in Chapter 20 for additional tips for controlling fluid intake.)

Drug-Nutrient Interactions The use of multiple diuretics in the medical treatment of CHF may lead to losses of multiple water-soluble nutrients, including potassium, magnesium, and thiamin (Zenuk et al. 2003; Hanninen et al. 2006). Nutrition education for increasing these nutrients within the diet is the first level of intervention. However, since meeting the patient's increased needs may be difficult because of the patient's overall decreased oral intake, supplementation may be warranted. Levels of supplementation will start with providing the DRI for each and then adjusting dosages after monitoring biochemical indices. Serum levels of magnesium (normal = 1.6–2.6 mmol/L) and potassium (normal = 3.5–5.5 mg/dL) would be used as benchmark levels for comparison. Thiamin levels are assessed by measuring the activity of erythrocyte thiamin pyrophosphate (thiamin pyrophosphate effect). Adequate or normal levels are evaluated at 0% to 15%; mild deficiency is indicated at >15% to 25%; and severe deficiency is indicated at greater than 25% stimulation.

Some treatment protocols for thiamin supplementation include thiamin prescribed at 200 mg/day orally for six weeks, while others recommend an initial parenteral dose of 100 mg followed by daily supplementation (Seligmann et al. 1991; Shimon et al. 1995; McCabe-Sellers, Sharkey, and Browne 2005).

Other Nutrients of Concern Additional conditionally essential nutrients that have been examined to determine their possible role in treatment for heart failure include arginine, carnitine, and taurine (as mentioned earlier, they have been linked to cardiac cachexia as well). Hawthorn, an herbal supplement, has also been studied as a complementary treatment for heart failure but is without significant demonstrated benefit (see Appendix F, Table F9) (Pittler, Schmidt, and Ernst 2003). The rationale for use of arginine supplementation in heart failure is to increase the production of nitric oxide. As discussed earlier, nitric oxide plays a significant role in intiating vasodilation in the vascular endothelium. Initial studies have indicated a possible role for L-arginine supplementation in heart failure, although more research is necessary to confirm this benefit (Bednarz et al. 2004; Tousoulis, Charakida, and Stefanadis 2005; American Dietetic Association, Disorders of Lipid Metabolism, Evidence Analysis Library 2006). Carnitine is responsible for carrying fatty acids intracellularly into the mitochondria for oxidation. Patients with heart failure have been shown to have lower levels of carnitine, and when supplemented with carnitine have demonstrated postive outcomes, though most have been in small clinical studies (Ferrari et al. 2004; American Dietetic Association, Disorders of Lipid Metabolism, Evidence Analysis Library 2006). It will be important to monitor future research in larger clinical trials for substantiated evidence to support supplementation.

Conclusion

This chapter has illustrated the impact of nutrition as a controllable risk factor, as a means to prevent disease, and as a critical component of medical treatment. Management of hypertension, dyslipidemia, and diabetes, weight management, physical activity, and smoking cessation are common targets. Lifestyle modifications begin with management of these controllable risk factors.

The initial strategy should be weight management and increased physical activity. A well-established relationship exists between increased body weight, hypertension, dyslipidemia, and a pro-thrombotic state. In addition to its relationship with high triglycerides and low HDL cholesterol, visceral adiposity is associated with thrombotic and inflammatory markers such as CRP, homocysteine, and fibrinogen, and type 2 diabetes mellitus (Mora et al. 2006). Since visceral obesity is associated with dyslipidemia, pro-thrombotic state, glucose

TABLE 15.17

Guidelines for Food Selection for 2,000 mg Sodium Diet

Food Category	Allowed	Excluded or Limited
Beverages	Milk (limit to 16 oz or 480 mL daily), buttermilk (limit to 1 cup or 240 mL per week); eggnog; all fruit juices; low-sodium, salt-free vegetable juices; low-sodium carbonated beverages	Malted milk, milkshake, chocolate milk; regular vegetable or tomato juices; commercially softened water used for drinking or cooking
Breads and Cereals	Enriched white, wheat, rye, and pumpernickel bread, hard rolls, and dinner rolls; muffins, cornbread, and waffles; most dry cereals, cooked cereal without added salt; unsalted crackers and breadsticks; low-sodium or homemade bread crumbs	Breads, rolls, and crackers with salted tops; quick breads; instant hot cereals; pancakes; commercial bread stuffing; self-rising flour and biscuit mixes; commercial bread crumbs or cracker crumbs
Desserts and Sweets	All; desserts and sweets made with milk should be within allowance	Instant pudding mixes and cake mixes
Fats	Butter or margarine; vegetable oils; unsalted salad dressings limited to 1 tbsp (15 mL); light, sour, and heavy cream	Regular salad dressings containing bacon fat, bacon bits, and salt pork; snack dips made with instant soup mixes or processed cheese
Fruits	Most fresh, frozen, and canned fruits	Fruits processed with salt or sodium-containing compounds (i.e., some dried fruits)
Meats and Meat Substitutes	Any fresh or frozen beef, lamb, pork, poultry, fish, and shrimp; canned tuna or salmon, rinsed; eggs and egg substitutes; low-sodium cheese including low-sodium ricotta and cream cheese; low-sodium cottage cheese; regular yogurt; low-sodium peanut butter; dried peas and beans; frozen dinners (<500 mg or 22 mmol sodium/serving)	Any smoked, cured, salted, koshered, or canned meat, fish, or poultry including bacon, chipped beef, cold cuts, ham, hot dogs, sausage, sardines, anchovies, crab, lobster, imitation seafood, marinated herring, and picked meats; frozen breaded meats; pickled eggs; regular hard and processed cheese, cheese spreads, and sauces; salted nuts
Potatoes and Potato Substitutes	White or sweet potatoes; squash; enriched rice, barley, noodles, spaghetti, macaroni, and other pastas cooked without salt; homemade bread stuffing	Commercially prepared potato, rice, or pasta mixes; commercial bread stuffing
Soups	Low-sodium commercially canned and dehydrated soups, broths, and bouillons; homemade broth and soups without added salt and made with allowed vegetables; cream soups within milk allowance	Regular canned or dehydrated soups, broths, or bouillon
Vegetables	Fresh, frozen vegetables and low-sodium canned vegetables	Regular canned vegetables, sauerkraut, pickled vegetables, and others prepared in brine; frozen vegetables in sauces; vegetables seasoned with ham, bacon, or salt pork
Miscellaneous	Salt substitute with physician's approval; pepper, herbs, spices; vinegar, lemon, or lime juice; hot pepper sauce; low-sodium soy sauce (1 tsp or 5 mL); low-sodium condiments (catsup, chili sauce, mustard); fresh ground horseradish; unsalted tortilla chips, pretzels, potato chips, popcorn, salsa (2 tbsp or 30 mL)	Any seasoning made with salt including garlic salt, celery salt, onion slat, and seasoned salt; sea salt, rock salt, kosher salt; meat tenderizers; monosodium glutamate; regular soy sauce, barbecue sauce, teriyaki sauce, steak sauce, Worcestershire sauce, and most flavored vinegars; canned gravy and mixes; regular condiments; salted snack foods; olives

Source: American Dietetic Association. *Manual of Clinical Dietetics.* 6th edition. Table 68.3, pp. 773–774.

intolerance, and hypertension, it is strongly implicated as the root cause of metabolic syndrome. Over 47 million (age adjusted prevalence, 23.7%) Americans have metabolic syndrome (American Heart Association 2006).

Close to 100 million American adults (49.8% of the adult population) have a total serum cholesterol level greater than 200 mg/dL, and 17.3% have a total serum cholesterol greater than 240 mg/dL; 39.5% of adults have an LDL cholesterol level greater than 130 mg/dL, and 22.6% have an HDL cholesterol level lower than 40 mg/dL (American Heart Association 2006). Using the research and principles of the DASH diet, the Therapeutic Lifestyle Changes diet, and extensive skills for individualization of patient education and behavior modification, the registered dietitian is uniquely situated to impact the extent of cardiovascular disease within the population and to impact an individual's quality of life with overall improvement of health and well-being.

TABLE 15.18

Guidelines for Food Selection for 1,000 mg Sodium Diet

Food Category	Allowed	Excluded or Limited
Beverages	Milk (limit to 16 oz. or 480 mL daily), eggnog; all fruit juices; low-sodium, salt-free vegetable juices; low-sodium carbonated beverages	Malted milk, milkshake, buttermilk, chocolate milk; regular vegetable or tomato juices; commercially softened water used for drinking or cooking
Breads and Cereals	Enriched white, wheat, rye, and pumpernickel bread, hard rolls, and dinner rolls (2 servings/day); low-sodium bread, crackers, matzo, and melba toast; muffins, cornbread, pancakes, and waffles made with low-sodium baking powder; cooked cereal without added salt; low-sodium dry cereals including puffedrice, puffed wheat, and shredded wheat; unsalted crackers and breadsticks; low-sodium or homemade bread crumbs and cracker crumbs	Breads, rolls, and crackers with salted tops or made with regular baking powder or baking soda; graham crackers; quick breads; instant hot cereals; pancakes; commercial bread stuffing; self-rising flour and biscuit mixes; regular bread crumbs or cracker crumbs
Desserts and Sweets	Ice cream, pudding, and custard made with milk should be within allowance; fruit ice; unsalted bakery goods, homemade or commercial; sherbet and flavored gelatin (not to exceed ½ cup or 120 mL/d); low-sodium baking powder	All candies made with sweet chocolate, nuts, or coconut; desserts made with rennin or rennin tablets; instant pudding mixes; commercial cakes; cookie and brownie mixes
Fats	Unsalted butter or margarine; vegetable oils; unsalted salad dressings; low-sodium mayonnaise; nondairy cream (up to 1 oz. or 28 g daily)	Salted butter and margarine; regular salad dressings containing bacon bits and salt pork; snack dips made with instant soup mixes or processed cheese
Fruits	Most fresh, frozen, and canned fruits	Fruits processed with salt or sodium-containing compounds
Meats and Meat Substitutes	Any fresh or frozen beef, lamb, pork, poultry, fish; low-sodium canned tuna or salmon; eggs; low-sodium cheese, cottage cheese, ricotta, and cream cheese; regular yogurt; low-sodium peanut butter; dried peas and beans; frozen dinners (<150 mg or 6.5 mmol sodium/serving)	Any smoked, cured, salted, koshered, or canned meat, fish, or poultry including bacon, chipped beef, cold cuts, ham, hot dogs, sausage, sardines, anchovies, marinated herring, and picked meats; all shellfish; frozen breaded meats; pickled eggs; egg substitutes; regular hard and processed cheese, cheese spreads, and sauces; salted nuts
Potatoes and Potato Substitutes	White or sweet potatoes; squash; unsalted enriched rice, barley, noodles, spaghetti, macaroni, and other pastas cooked without salt; homemade bread stuffing	Commercially prepared potato, rice, or pasta mixes; commercial bread stuffing
Soups	Low-sodium commercially canned and dehydrated soups, broths, and bouillons; homemade broth, soups without added salt and made with allowed vegetables; low-sodium cream soups within milk allowance	Regular canned or dehydrated soups, broths, or bouillon
Vegetables	Fresh, frozen vegetables and low-sodium canned vegetables	Regular canned vegetables, sauerkraut, pickled vegetables, and others prepared in brine; frozen peas, lima beans, and mixed vegetables; all frozen vegetables in sauces; vegetables seasoned with ham, bacon, or salt pork
Miscellaneous	Salt substitute with physician's approval; pepper, herbs, spices; vinegar, lemon, or lime juice; hot pepper sauce; low-sodium condiments (catsup, chili sauce, mustard); fresh ground horseradish; unsalted tortilla chips, pretzels, potato chips, popcorn	Salt and any seasoning made with salt including garlic salt, celery salt, onion slat, and seasoned salt; sea salt, rock salt, kosher salt; meat tenderizers; monosodium glutamate; regular and low-sodium soy sauce (check label), barbecue sauce, teriyaki sauce, steak sauce, Worcestershire sauce, and cooking wine or sherry; canned gravy and mixes; regular condiments including olives, horseradish, pickles, relish, catsup, mustard, and commercial salsa

Source: American Dietetic Association. *Manual of Clinical Dietetics.* 6th edition. Table 68.4, pp. 775–776.

TABLE 15.19

Guidelines for Food Selection for 500 mg Sodium Diet

Use the 1,000 mg sodium diet guidelines with the following modifications:

Use low-sodium bread only.

Omit sherbet and flavored gelatin.

Limit meat to 5 oz. (140 g) per day. One egg may be used per day in place of 1 oz. (28 g) meat.

Omit the following vegetables: beets, beet greens, carrots, kale, spinach, celery, white turnips, rutabagas, mustard greens, chard, frozen peas, and dandelion greens.

Use distilled water (depending on natural sodium content of water supply).

Limit milk and milk products to 16 oz. (480 mL) daily.

Source: American Dietetic Association. *Manual of Clinical Dietetics*. 6th edition. Table 68.5, p. 777.

PRACTITIONER INTERVIEW

Eileen MacKusick, M.S., R.D. *Chief Clinical Dietitian, Watsonville Community Hospital, Watsonville California*

I have been a clinical dietitian for over 11 years. In my facility, roughly 25% of the patients I see have congestive heart failure (CHF). Only one of the attending doctors routinely classifies the CHF patient but the most common etiology is cardiomyopathy.

How do you nutritionally assess these patients, and is there anything unique that would be useful for students to aware of?

Physical: Obtaining a dry weight can be challenging because of edema, especially if the patient is obese. Whenever a dry weight can be established, it is helpful. It is also important to do a visual assessment, because patients don't appear to be underweight based on their weight but they are actually cachectic with edema. It seems that patients with CHF are either morbidly obese with poor somatic protein stores or cachectic. I recently had a patient that weighted 484 pounds. We took 80 lbs off of him over a 1½ month period and it was still difficult to determine the water weight vs. the actual wt. Our goals for him were weight loss while increasing his visceral protein, which we achieved.

Diet Assessment: This generally focuses on salt/sodium and fluid intake. Most of our physicians are using a no added salt diet vs. a low-sodium diet these days. Trying to maintain a 2 g sodium diet was resulting in undernutrition and exacerbating fluid shifts and edema. The fluid limits are individualized by the physicians. We monitor the patients for dehydration due to over diuresis.

Biochemical: I always look at albumin/prealbumin with C Reactive Protein (CRP) to distinguish acute phase reaction. Visceral protein determination is difficult in this population because of the fluid shifts. B-Type Natriuretic Peptide (BNP) provides the degree or severity of the CHF. When the BNP is extremely high, you may see a more severe fluid/sodium restriction.

What are the common nutritional problems associated with CHF?

There are several nutrition problems associated with CHF. Patients find it difficult complying with dietary fluid and sodium restrictions while consuming adequate nutrition to increase or maintain adequate blood protein levels to prevent third spacing of extracellular fluid. Adequate nutrition intake is also compromised by shortness of breath (SOB), decreased appetite, and lethargy; many patients have additional complications as the result of polypharmacy with drug nutrient interactions.

What changes have you seen with CHF nutrition therapy in the last ten years?

The use of the BNP is relatively new, and sodium restrictions have been liberalized over the last 10 years. When I first started practicing, you would see 500–1,000 mg sodium diets ordered. A two gram sodium was the standard diet. Now we barely use two grams and never use 500–1,000 mg diets. Journal articles and CEU programs are primarily how I have stayed abreast of changes for this disorder.

Any advice for dietetic students about counseling clients with CHF?

Pick your battles. Ideally your diet instruction would focus on a heart-healthy diet. However, just getting people to stop using salt can be a victory. You really need to assess your patient and customize a plan that works with their life and their home situation. Being overly aggressive often results in noncompliance. I also find it very important to give a really basic description of sodium, fluid retention, and the pathophysiology of CHF. I find that patients are much more willing to comply when they understand the role of sodium with congestive heart failure.

CASE STUDY 15

CC:

"I really want to control this blood pressure—my parents both died of complications from blood pressure and heart disease."

General:

49-year-old African-American male; diagnosed with Stage 2 essential HTN × six months. Current treatment includes lower salt diet, smoking cessation, and a sporadic walking program.

Labs:

BP 160/100; TChol 300 mg/dL; HDL-C 35 mg/dL; LDL-C 135 mg/dL; TG 250 mg/dL.

Rx:

To begin thiazide diuretic and ACE inhibitor.

Past Med Hx:

Noncontributory but positive family history for cardiovascular disease, stroke, and hypertension.

Nutrition Hx:

Ht. 6'2: Wt. 245# UBW/Highest adult weight 245–250#

Usual intake:

(AM) oatmeal or cold cereal, 2% milk, coffee; (Snack) coffee, sweet roll, or doughnut; (LUNCH) sandwich (ham or salami with cheese) or soup from home; chips; diet Coke; (Dinner) 6–8 oz. meat, baked, broiled, or grilled; salad; potato, pasta, or rice; roll or bread; diet Coke; (Bedtime snack) peanut butter and crackers, or popcorn.

1. What are the criteria for diagnosis with Stage 2 essential HTN? What factors allow for that diagnosis for this patient?
2. Evaluate the patient's weight history. Is this a risk factor? What other possible risk factors does this patient present with?
3. What would be the complications of untreated hypertension?
4. What are the mechanisms that thiazide diuretics, and ACE inhibitors use to treat high blood pressure?
5. What is the recommended nutrition therapy for hypertension? Evaluate this patient's usual dietary intake; are there specific nutrition problems that can be identified?

NUTRITION CARE PROCESS FOR DISEASES OF THE CARDIOVASCULAR SYSTEM

Step One: Nutrition Assessment

Medical/Social History

Diagnoses/date of diagnosis

Comorbidities

Medications

Previous medical conditions or surgeries

Socioeconomic status/food security

Support systems

Education level—primary language

Physical Assessment

Blood pressure

Anthropometric

Height/length

Current weight

Weight history if adult: highest adult weight; usual body weight

Body mass index

Waist circumference

Biochemical Assessment

Erythrocyte thiamin pyrophosphate effect

Glucose, BUN, Cr

Electrolytes

Lipid Assessment

Triglyceride, total cholesterol, HDL, LDL, LDL:HDL ratio

Visceral Protein Assessment

Albumin

Prealbumin

Dietary Assessment (adapted from U of Wisconsin School of Medicine)

Meals/ snacks—patterns, frequency

Portion sizes

Saturated and trans fat from dairy products and fatty meats, commercial snack foods and pastries, fried foods, and added fats and oils

Refined carbohydrates from baked products, desserts, cookies, and other sweets

Sweetened beverages (juice drinks, soda) and alcohol

Major sources of sodium from processed foods, eating out, and added salt

Frequency of restaurant meals, fast food, take-out food

Step Two: Nutrition Diagnosis

Excessive energy intake

Excessive fat intake

Inadequate vitamin intake

Inadequate mineral intake

Food medication interaction

Food and nutrition-related knowledge deficit

Limited adherence to nutrition-related recommendations

Sample PES: Inadequate thiamin intake related to furosemide prescription as evidenced by presence of angular cheilitis and deficient erythrocyte thiamin pyrophosphate levels.

Step Three: Intervention

1. Recommend 100 mg IV thiamin twice daily for seven days. Follow with daily oral thiamin supplementation of 100 mg/day.
2. Recommend multivitamin supplementation daily to meet DRI for all vitamins and minerals.

3. Increase nutrient density of food choices with intake goal of 1,500 kcal/day.

Step Four: Monitoring and Evaluation

1. Reassess erythrocyte thiamin pyrophosphate levels.
2. Monitor physical lesions for improvement.

WEB LINKS

National Heart Lung Blood Institute: Provides all guidelines and patient teaching materials for the DASH diet.

http://www.nhlbi.nih.gov/health/public/heart/hbp/dash/index.htm

National Heart Lung Blood Institute/National Cholesterol Education Program: Provides clinical guidelines for treatment of dyslipidemia.

http://www.nhlbi.nih.gov/about/ncep

American Heart Association: Provides education materials and clinical information for all aspects of cardiovascular disease.

http://www.americanheart.org

JustMove Physical Activity Program: Information regarding exercise and nutrition designed by the American Heart Association.

http://www.justmove.org/home.cfm

American Heart Association Journal: Provides access to all professional journals published by American Heart Association.

http://www.ahajournals.org

National Institutes of Health: Compilation of other web sources for information about all aspects of diagnosis and treatment for cardiovascular disease.

http://www.nlm.nih.gov/medlineplus/heartdiseases.html

University of Wisconsin—Department of Medicine. Nutrition Academic Award Program. Grant-developed educational tools and guidelines for hypertension, lipid disorders, obesity, and diabetes.

http://www.medicine.wisc.edu/mainweb/DOMPages.php?section=naa&page=medicalnutritionhandbook

The Framingham Heart Study: This website chronicles this important longitudinal study which began in 1948.

http://www.framingham.com/heart

END-OF-CHAPTER QUESTIONS FOR CHAPTER 15

1. Define the following terms: systolic, diastolic, stroke volume, and cardiac output.
2. Describe the factors that will influence stroke volume and mean arterial pressure (MAP).
3. What is the definition of hypertension? Explain the pathophysiology for the lifestyle factors known to contribute to the development of hypertension.
4. List the major classifications of medications used to treat hypertension. Describe their mechanism of effect. Describe the DASH diet.
5. List the risk factors associated with development of atherosclerosis. Which risk factors are alterable? List the dietary changes in the ATP III guidelines.
6. What are the four mechanisms that can initiate an myocardial infarction (MI)? How does the rupture of an atheromatous plaque result in an MI?
7. Compare peripheral arterial disease (PAD) to atherosclerosis—how are they similar and how do they differ? Describe several complications associated with PAD.
8. What are the primary causes of heart failure? Describe the clinical progression of heart failure. What is meant by "cardiac remodeling" and what is its etiology?
9. What are the nutritional implications associated with heart failure and cardiac cachexia?

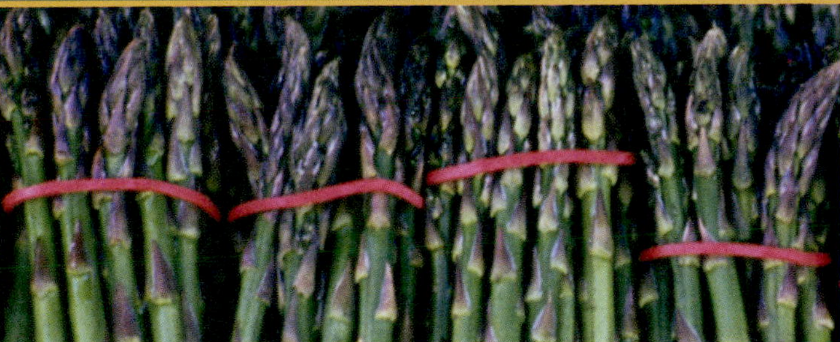

16

Diseases of the Upper Gastrointestinal Tract

Marcia Nahikian Nelms, Ph.D., R.D.

Southeast Missouri State University

CHAPTER OUTLINE

Normal Anatomy and Physiology of the Upper Gastrointestinal Tract
Motility, Secretion, Digestion, and Absorption • Anatomy and Physiology of the Oral Cavity • Normal Anatomy and Physiology of the Esophagus • Normal Anatomy and Physiology of the Stomach
Pathophysiology of the Upper Gastrointestinal Tract
Pathophysiology of the Oral Cavity
Nutrition Therapy for Pathophysiology of the Oral Cavity
Nutritional Implications • Nutrition Intervention
Pathophysiology of the Esophagus
Gastroesophageal Reflux Disease (GERD) • Barrett's Esophagus—A Complication of GERD • Nutrition Therapy—Gastroesophageal Reflux Disease • Nutrition Therapy for Dysphagia • Achalasia • Hiatal Hernia
Pathophysiology of the Stomach
Indigestion • Nausea and Vomiting • Nutrition Therapy for Nausea and Vomiting • Gastritis • Peptic Ulcer Disease
Nutrition Therapy for PUD
Nutrition Therapy for Gastric Surgery • Other Conditions of Gastric Pathophysiology

Introduction

No other system of the human body is so intimately involved in preservation of an optimal nutritional status than the gastrointestinal (GI) tract. Any pathology involving the GI tract can have a significant effect on maintenance of nutritional status. A wide array of diagnoses affects normal digestion and absorption. Nausea, vomiting, diarrhea, constipation, and malabsorption, all common symptoms in GI disease, can potentially jeopardize the individual's nutritional status. In some diseases, such as celiac disease, nutrition therapy is the only treatment.

GI disease is estimated to affect more than 70 million people in the United States each year. More than 50 million physician office visits and 13% of all hospitalizations are related to GI disease. Health care costs for these individuals are staggering, at over $107 billion dollars every year (Everhart 1994; National Center for Health Statistics 2005).

Normal Anatomy and Physiology of the Upper Gastrointestinal Tract

A good working knowledge of normal anatomy and physiology of the GI tract is essential to understanding pathophysiology, and subsequent medical and nutritional care of GI disease. The GI tract is often described as a long tube approximately 15 feet in length. This description, though technically accurate, minimizes the complexity of this body system. Cell and organ function and intricacy of control factors throughout the GI tract are highly differentiated. The upper GI tract is composed of the mouth, pharynx, esophagus, and stomach, while the lower GI tract in-

FIGURE 16.1 Anatomy of the Gastrointestinal System

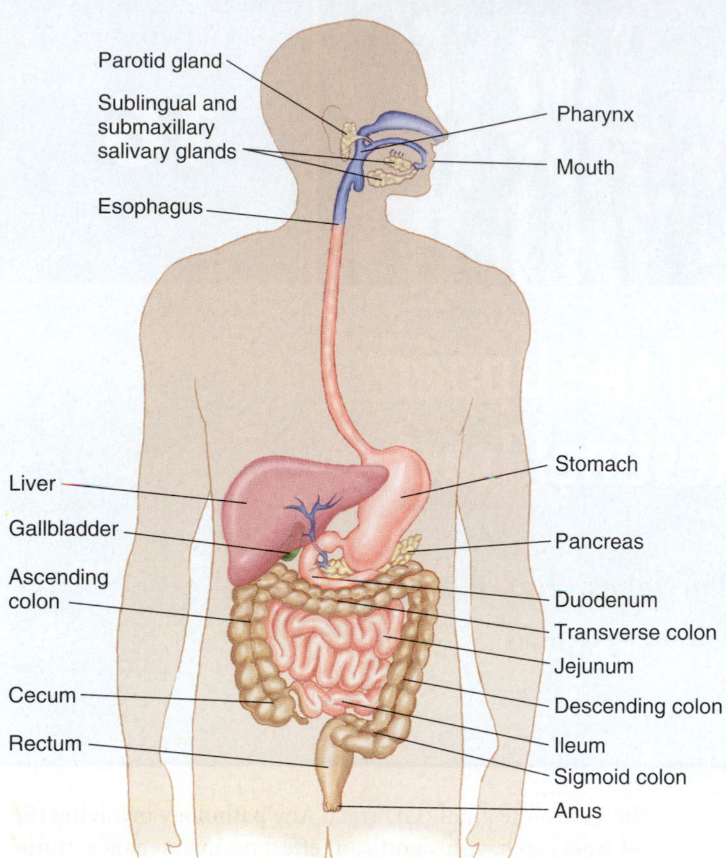

Source: R. Rhoades and R. Pflanzer, *Human Physiology,* 4e, copyright © 2003 p. 686

cludes the small and large intestine (see Figure 16.1). These continuous organs differ from one another in both anatomical structure and specialized function. Accessory or ancillary organs that contribute to the function of the GI tract include the liver, biliary system, and pancreas.

This section will focus on each organ of the upper GI tract and discuss its normal function in the context of the four basic functions of the GI tract: motility, secretion, digestion, and absorption. GI disease can affect any one or all of these functions.

Motility, Secretion, Digestion, and Absorption

Motility is the movement of the food consumed along the GI tract. Both propulsive contractions and mixing movements serve not only to move foodstuffs toward sites of digestion and absorption, but to mix foods with digestive secretions and maximize potential absorption. Secretions of the GI tract include water, electrolytes, enzymes, bile salts, and mucus. Through the process of digestion, complex molecules are converted to their simplest form. Carbohydrates are digested from their most complex form of polysaccharides to the monosaccharides: glucose, fructose, and galactose. Proteins are converted from polypeptides to single amino acids, di- and tri-peptides. Lipids are digested to their simplest forms: free fatty acids, monoglycerides, glycerol, phospholipids, and cholesterol. After digestion, these basic molecules are absorbed along with water, electrolytes, vitamins, and minerals to provide essential nutrients to every cell.

Contraction and motility are regulated through the GI tract's own specialized pacemaker cells called interstitial cells of Cajal. The autonomic nervous system relays input for the smooth muscle of the GI tract through a specialized enteric nervous system. Two types of neurons regulate contraction of smooth muscle, motility, and secretory functions of the GI tract. Primary neurotransmitters involved in transmission of impulses include acetylcholine, substance P, vasoactive intestinal polypeptide (VIP), and nitric oxide. Other neurotransmitters and modulators involved in neuroregulation for the GI tract include norepinephrine, serotonin, neuropeptide Y, and γ-aminobutyric acid (GABA). See Table 16.1.

Major innervation for the enteric nervous system is supplied by parasympathetic and sympathetic fibers of the autonomic nervous system. Most sympathetic

TABLE 16.1

Neurotransmitters of the Enteric Nervous System and Their Effects on Gastrointestinal Motility

Neurotransmitter	Effect on Motor Activity
Nonapeptides	
Acetylcholine	Usually excitatory
Serotonin	Excitatory
ATP	Inhibitory
Dopamine	Inhibitory
Nitric oxide	Inhibitory
Peptides	
Cholecystokinin	Excitatory
Enkephalins	Excitatory
Gastrin-releasing peptide	Excitatory
Neuropeptide Y	Excitatory
Substance P	Excitatory
Somatostatin	Inhibitory
Vasoactive intestinal peptide	Inhibitory

Source: Rhoades R, Pflanzer, R., *Human Physiology,* 4th ed. Belmont: Brooks/Cole; 2003. Table 22-1, p. 690.

impulses are carried along the splanchnic nerves. Parasympathetic impulses are carried by the **vagus nerve**.

Anatomy and Physiology of the Oral Cavity

The oral cavity or mouth serves as entry into the digestive tract. The mouth consists of the lips, teeth, tongue, and palate. The lips assist in directing food into the oral cavity. By separating the mouth and nasal cavity, the palate (roof of the mouth) allows for chewing, swallowing, and breathing to occur all at the same time. The tongue serves its primary role in moving food from the front of the mouth to the pharynx in preparation for swallowing. Major taste buds are also located on the tongue. The tongue and lips also play a primary role in speech.

Oral Cavity Motility After food is voluntarily placed in the oral cavity, teeth begin their work of mastication (chewing). The purposes of mastication are: (1) to break food into smaller pieces, (2) to mix food with saliva, and (3) to stimulate taste buds. The tongue assists by moving food in place for chewing. When the jaw is closed, upper and lower teeth fit together to mash, grind, and tear food.

Oral Cavity Secretions The primary secretion in the oral cavity is saliva. Saliva is produced in the mouth by three pairs of salivary glands. These include the parotid, submandibular, and sublingual. Saliva is made of water (99.5%), electrolytes, and protein. Electrolytes include sodium chloride, bicarbonate, and potassium. Proteins include enzymes, mucus, and lysozyme. Lysozyme functions as part of the first level of defense for the immune system and are capable of destroying bacteria in the mouth.

One to two liters of saliva are produced each day. Both autonomic and acquired reflexes can stimulate the amount that is produced. An example of an autonomic reflex is production of saliva in response to the presence of food in the mouth. An acquired reflex is learned; for example, the "mouth waters" at the sight or smell of food (saliva is produced without oral stimulation).

Functions of saliva include: (1) moistening and lubricating food to facilitate swallowing; (2) initiating digestion of carbohydrate; (3) providing antibacterial protection with lysozyme and by rinsing away food from the oral cavity; (4) enhancing taste by providing a solution that can interact with taste buds; (5) serving as a buffer—the pH of saliva is approximately 6.8, which neutralizes acids and protects the teeth from dental caries; (6) promoting oral hygiene by dissolving food, dead cells, and foreign substances; and (7) assisting speech by allowing free movement of the lips and tongue.

After food is chewed and mixed with saliva, it is shaped into a sticky ball called a bolus. The bolus is moved from the front of the oral cavity to the back where swallowing is then initiated. Pressure of the bolus on the pharynx stimulates nerve impulses to the swallowing center. The swallowing center, located in the medulla, stimulates the sequence of muscle actions that coordinate the swallow. Thus, the initiation of swallowing is voluntary, but thereafter swallowing is under autonomic control.

Normal Anatomy and Physiology of the Esophagus

The esophagus is a straight, hollow tube approximately 25 cm long and 2 cm in diameter. The esophagus has **sphincter** muscles at either end.

Walls of the esophagus consist of four layers of tissue. The inner layer is the mucosa, which is made of stratified squamous epithelial cells. The next layer is the submucosa, which contains secretory cells that produce mucus to facilitate movement of the bolus during swallowing. The muscle layer consists of both longitudinal and circular muscles that coordinate movement of the food bolus by alternately contracting. This "squeezing" contraction easily moves the bolus down the esophagus. The outer layer (or adventitia) of tissue for the esophagus is connective tissue and has no additional outer covering.

The chief function of the esophagus is motility. Transporting the bolus of food from the oral cavity to the stomach is its primary task. Though it sounds simplistic, several disorders of the upper GI tract actually involve derangements in this task.

An individual unconsciously swallows over 600 times each day (Baker 1993). Each swallow is composed of four phases. The first phase, described previously, is the **oral preparatory phase** where food is chewed and mixed with saliva. The second phase or **oral phase** was also described previously and consists of voluntary movement of the bolus of food from the front of the oral cavity to the back. The

vagus nerve—tenth cranial nerve; one of its major functions is to coordinate the autonomic nervous system communication between organs of digestion

sphincter—a circular muscle that prevents movement or passage through the circle when contracted; sphincter muscles are located throughout the GI tract and are crucial control factors for peristalsis

oral preparatory phase—tongue, teeth, and mandible involved in chewing of food and preparation of bolus; food is mixed with saliva, pressed against the hard palate, and formed into a bolus

oral transit phase of swallowing—tongue moves bolus to back of throat

FIGURE 16.2 Peristalsis in the Esophagus

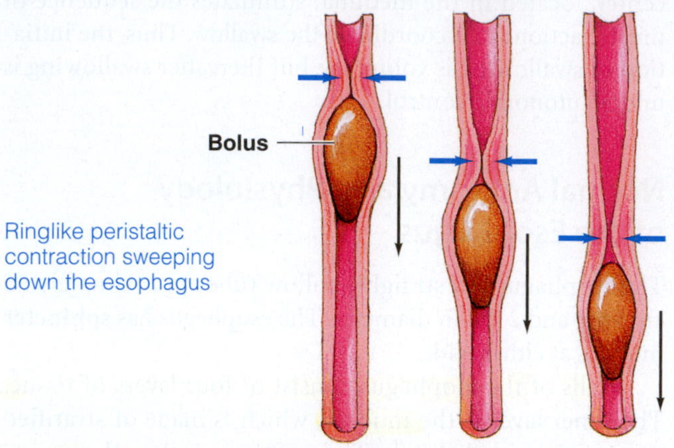

Source: L. Sherwood, *Human Physiology: From Cells to Systems*, 5e, copyright © 2004, p. 603

third phase of swallowing is known as the **pharyngeal phase.** The most important part of this phase is to assure the bolus is directed into the esophagus and is prevented from entering the trachea. This is initially accomplished when the uvula seals off the nasal passage so food does not leak into the nose. Next, laryngeal muscles contract and seal off the glottis (entrance to the larynx). The epiglottis also tilts upward to assist in preventing food from entering the larynx.

The final phase of swallowing is the **esophageal phase.** The cricopharyngeal sphincter or pharyngoesophageal sphincter is located at the top of the esophagus. This sphincter, when open, allows the bolus to enter the esophagus. When the sphincter is closed, it prevents air from entering the GI tract during breathing. After the bolus of food moves through the cricopharyngeal sphincter into the esophagus, the sphincter closes and normal breathing will resume.

When the esophageal phase of swallowing begins, autonomic control initiates the peristaltic wave that moves the bolus of food down the esophagus into the stomach (see Figure 16.2). At the end of the esophagus, another sphincter muscle—the **lower esophageal sphincter** (LES)—controls release of the bolus from the esophagus into the stomach. This sphincter is closed except during swallowing. It serves as a barrier to protect the esophageal mucosa from stomach contents. Atmospheric pressure is greater in the esophagus than in the stomach under normal conditions. This positive pressure of approximately 30 cm H_2O assists in preventing stomach contents from refluxing back into the esophagus. (Gastric pressure is approximately 10 cm H_2O.) Two major neurotransmitters are responsible for allowing the LES to relax and oppose the stimulatory action of acetylcholine. Nitric oxide and VIP inhibit closure of the LES, allowing it to relax so that the movement of the bolus slides through the esophagus into the stomach. The swallow is complete when the bolus of food moves through the LES. Pharyngeal and esophageal phases of swallowing take only 6 to10 seconds under normal conditions. If for some reason all food is not cleared from the esophagus, secondary peristaltic waves are initiated. This might occur when a sticky substance is eaten that does not move as readily down the esophagus. The individual, in this case, would usually be unaware of these secondary peristaltic waves.

pharyngeal phase of swallowing—the involuntary swallowing reflux begins, and the bolus is carried through the pharynx to the top of the esophagus; the entrance to the trachea (larynx) closes, and the soft palate lifts and closes off entrance to the nose

esophageal phase of swallowing—esophageal peristalsis carries the bolus through the esophagus and LES and into the stomach

lower esophageal sphincter (LES)—the junction between the esophagus and the stomach

Normal Anatomy and Physiology of the Stomach

The final portion of the upper GI tract is the stomach (see Figure 16.3). This organ lies from left to right across the upper abdomen directly under the diaphragm. Portions of the stomach (fundus, corpus, antrum, and pylorus) differ by anatomy and function. Sphincters at both ends of the stomach regulate flow of foodstuffs from the esophagus to the small intestine. Major functions of the stomach include all four digestive processes: motility, secretion, digestion, and absorption.

Gastric Motility Gastric motility includes filling of the stomach, storage of foodstuffs, mixing with gastric juices, and finally emptying into the small intestine. When empty, the stomach's volume is only about 50 mL, but it can stretch to hold more than 1000 mL. Storage occurs primarily in the body (corpus) of the stomach. Mixing occurs in the antrum where muscle is much thicker and can accommodate the strong peristaltic waves. Each peristaltic wave moves foodstuffs toward the pyloric sphincter at the bottom of the stomach. Rate of gastric emptying through the pyloric sphincter into the upper portion of the small intestine is controlled by the anatomical structure of the stomach, nutrient content of the foodstuffs, the nervous system, and by influence of specific hormones.

Gastric Secretions The stomach secretes approximately 1 to 3 liters of gastric juices each day. Gastric juice is composed of water, mucus, hydrochloric acid, enzymes, and electrolytes.

The mucosa, which lines the fundus and the body of the stomach, contains gastric glands. Several different types of cells are located within the gastric glands. The mucous cells secrete mucus, which protects the lining of the stomach

FIGURE 16.3 Parts of the Stomach

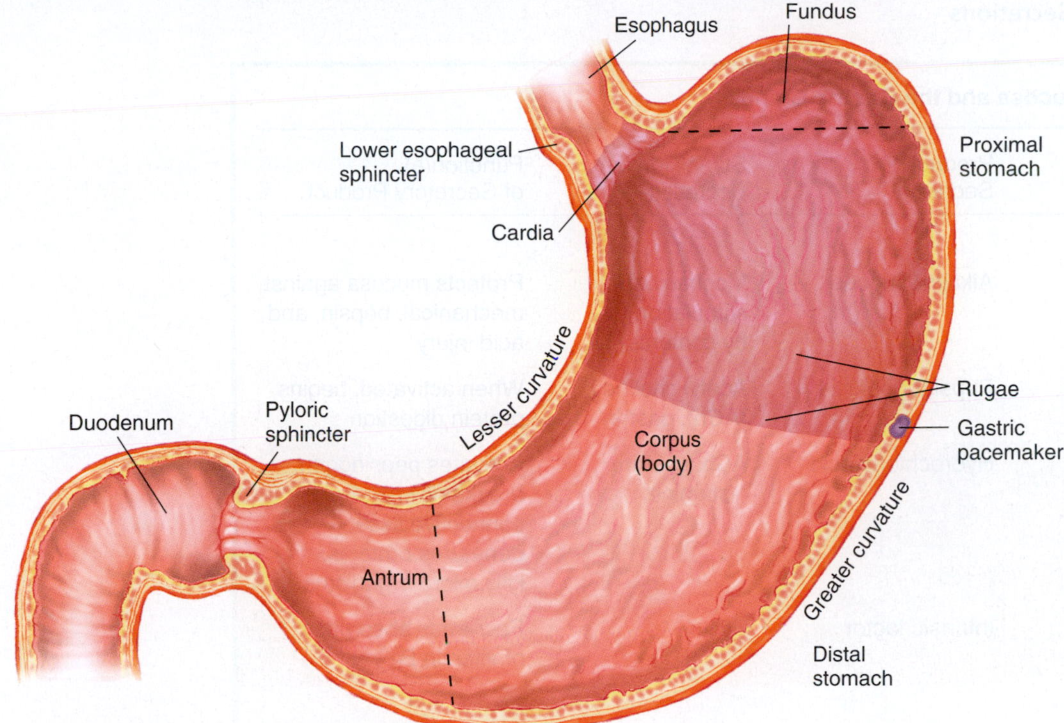

Source: R. Rhoades and R. Pflanzer, *Human Physiology,* 4e, copyright © 2003 p. 696

from mechanical or acid insult.[1] Chief cells secrete zymogen, pepsinogen, and the enzyme gastric lipase. Zymogen is an inactive enzyme. Pepsinogen, when activated, will begin protein digestion. Gastric lipase provides some preliminary digestion for lipids. **Parietal cells** secrete hydrochloric acid and intrinsic factor. Hydrochloric acid serves to activate pepsinogen, kill microorganisms, and denature proteins. Intrinsic factor is a protein necessary for the absorption of vitamin B_{12}.

In the pylorus, enterochromaffin (ECL) cells secrete histamine. G cells secrete gastrin, and D cells secrete somatostatin. All three of these substances assist in overall control and production of gastric juices.

Control of Gastric Secretions Control of gastric secretions is accomplished through complementary actions of nervous and endocrine systems, and involves four major chemical messengers: acetylcholine, histamine, and gastrin, which stimulate gastric secretions, and somatostatin, which inhibits gastric secretions (see Tables 16.2, 16.3, and 16.4). Acetylcholine is a neurotransmitter that stimulates parietal, chief, and ECL cells. Histamine, a **paracrine**, acts on parietal cells to increase hydrochloric acid (HCl) release. Gastrin, a hormone, stimulates chief and parietal cells as well as ECL cells to release histamine. **Somatostatin** works as an inhibitory paracrine by providing negative feedback to the stimulatory pathways. When gastric pH falls (becomes more acidic), somatostatin acts on each of the stimulatory mechanisms to slowly decrease gastric secretions.

All stimulatory pathways for production of HCl work in similar ways. The enzyme H^+, K^+-ATPase drives production of hydrogen ions (H^+). Gastrin, acetylcholine, and histamine act to increase the amount of H^+ available by the parietal cell to form HCl. Many medications designed to decrease gastric acidity work at this cellular level to either prevent production of stimulatory factors or block transport of H^+ needed for production of HCl. This will be discussed in more detail in the sections on gastroesophageal reflux disease and peptic ulcer disease.

Release of Gastric Secretions Gastric secretions are produced even before food enters the stomach and serve to

parietal cell—one of the gastric gland cells that lies on the basement membrane covered by chief cells, and secretes hydrochloric acid

paracrine—a name for a neurotransmitter that is released from a cell that is close to the target cell

somatostatin—a hormone and neurotransmitter that inhibits release of peptide hormones in several tissues

TABLE 16.2

Control of Gastric Secretions

The Stomach Mucosa and the Gastric Glands

Type of Secretory Cell	Product Secreted	Stimuli for Secretion	Function(s) of Secretory Product
Exocrine cells			
Mucous cells	Alkaline mucus	Mechanical stimulation by contents	Protects mucosa against mechanical, pepsin, and acid injury
Chief cells	Pepsinogen	Acetyl choline (ACh), gastrin	When activated, begins protein digestion
Parietal cells	Hydrochloric acid	ACh, gastrin, histamine	Activates pepsinogen, breaks down connective tissue, denatures proteins, kills micro-organisms
	Intrinsic factor		Facilitates absorption of vitamin B_{12}
Endocrine/ paracrine cells			
Enterochromaffi-like (ECL) cells	Histamine	ACh, gastrin	Stimulates parietal cells
G cells	Gastrin	Protein products, ACh	Stimulates parietal, chief, and ECL cells
D cells	Somatostatin	Acid	Inhibits parietal, G, and ECL cells

Source: L. Sherwood, *Human Physiology: From Cells to Systems*, 5e, copyright © 2004, p. 609

TABLE 16.3

Stimulation of Gastric Secretions

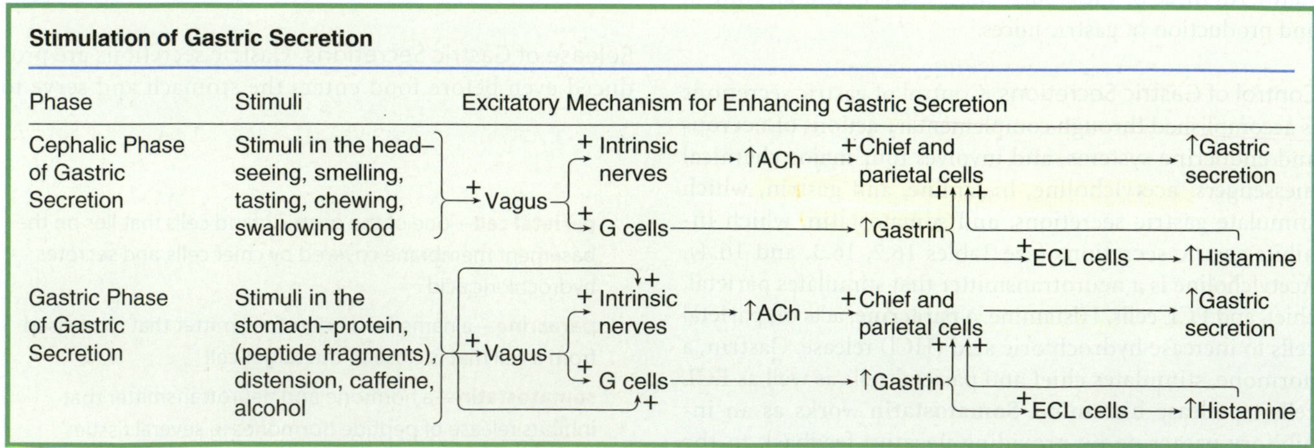

Source: L. Sherwood, *Human Physiology: From Cells to Systems*, 5e, copyright © 2004, p. 612

TABLE 16.4

Inhibition of Gastric Secretions

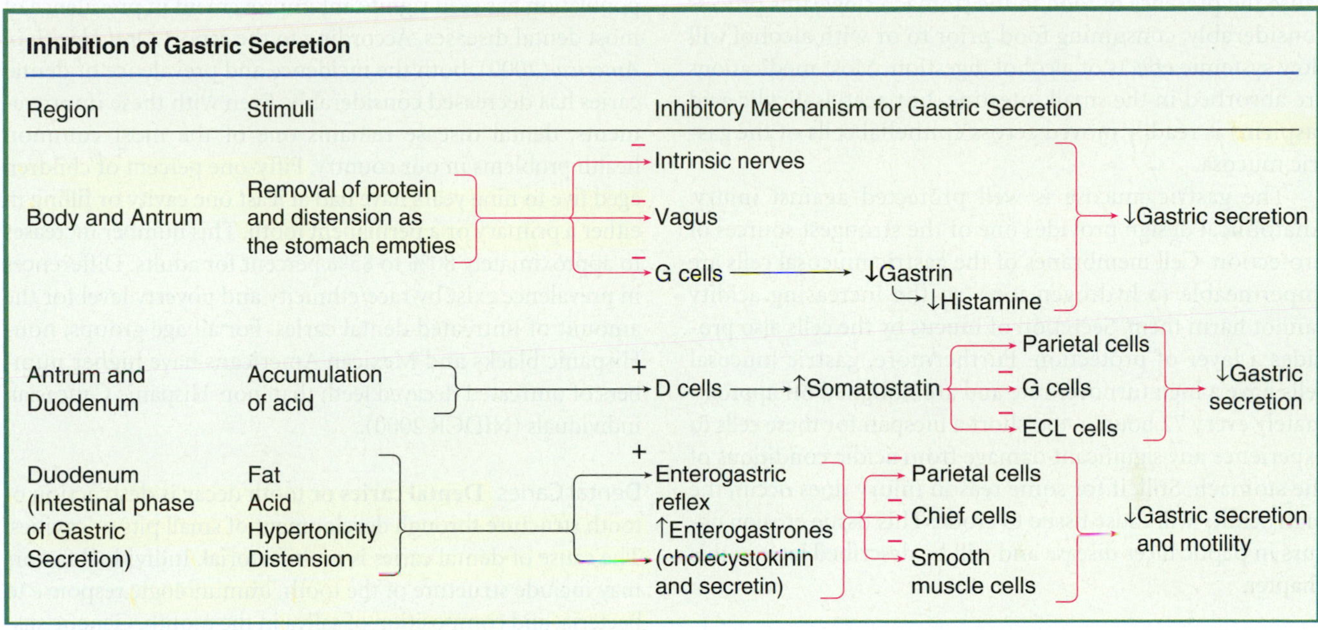

Inhibition of Gastric Secretion
Region
Stimuli
Inhibitory Mechanism for Gastric Secretion
Body and Antrum
Removal of protein and distension as the stomach empties
–
Intrinsic nerves
–
Vagus
–
G cells
↓Gastrin
↓Histamine
↓Gastric secretion
Antrum and Duodenum
Accumulation of acid
+
D cells
↑Somatostatin
–
Parietal cells
–
G cells
–
ECL cells
↓Gastric secretion
Duodenum (Intestinal phase of Gastric Secretion)
Fat
Acid
Hypertonicity
Distension
+
Enterogastric reflex
↑Enterogastrones (cholecystokinin and secretin)
–
Parietal cells
–
Chief cells
–
Smooth muscle cells
↓Gastric secretion and motility

Source: L. Sherwood, *Human Physiology: From Cells to Systems*, 5e, copyright © 2004, p. 613

prepare the stomach for its eventual role in digestion. Thus, release of gastric secretions in response to a meal is divided into three phases: cephalic, gastric, and intestinal phases. The cephalic ("head") phase refers to release of HCl and pepsinogen when stimulated by tasting, smelling, or even seeing food. The gastric phase begins when food enters the stomach. Both cephalic and gastric phases are stimulatory phases, meaning they result in production of gastric secretions. Several factors contribute to stimulation of gastric juices, including presence of protein, distention of the stomach, caffeine, and alcohol. Stimulation of HCl by alcohol and caffeine may be damaging if food is not present in the stomach. Thus, alcohol and caffeine are restricted for persons with peptic ulcer disease (PUD) or gastroesophageal reflux disease (GERD). (See Nutrition Therapy—Gastroesophageal Reflux Disease for further discussion.)

In contrast to cephalic and gastric phases that stimulate gastric juices, the intestinal phase is inhibitory in that it slows gastric secretions and prepares the small intestine for receipt of the acidic **chyme.** Distention of the stomach, accumulation of acid, presence of fat, and increasing osmolality of partially digested food result in release of two major hormones: cholecystokinin and secretin. These hormones act on the smooth muscle of the antrum to slow gastric motility. Additionally, somatostatin is released from the antrum of the stomach. Together these chemical messengers inhibit action of the parietal and chief cells, which reduces the amount of gastric juices.

Gastric Digestion and Absorption

Digestion in the Stomach Digestion in the stomach is both mechanical and chemical in nature. Mechanical digestion occurs as contractions of the stomach shear the foodstuffs and mix the bolus of food with gastric juices. Pepsin, the primary enzyme responsible for chemical digestion in the stomach, works to cleave amino acids, di- and tripeptides from inner portions of the protein, and results in shorter chains of amino acids called peptones. Of the three macronutrients, proteins are subjected to the most active chemical digestion in the stomach. Hydrochloric acid causes proteins to unravel or denature. Additionally, HCl converts pepsinogen to the active enzyme, pepsin. Carbohydrate and lipid digestion are fairly limited in the stomach. Because lipids remain insoluble in gastric juices, the action of gastric lipase is minimal. Carbohydrate digestion that began in the oral cavity is slowed down because salivary amylase is inactivated by HCl. A small amount of carbohydrate digestion may continue in the inner portions of the food bolus, which are not exposed to the acid.

Absorption in the Stomach Absorption in the stomach is limited; no food and only a small amount of water are

chyme—partially digested food in a semifluid state

absorbed. Exceptions include alcohol and some medications. Alcohol can be absorbed in the gastric mucosa and enters the bloodstream through capillaries of the stomach. Because the presence of food in the stomach slows this process considerably, consuming food prior to or with alcohol will slow systemic effects of alcohol ingestion. Most medications are absorbed in the small intestine, but acetylsalicylic acid (aspirin) is readily moved across epithelial cells of the gastric mucosa.

The gastric mucosa is well protected against injury. Anatomical design provides one of the strongest sources of protection. Cell membranes of the gastric mucosal cells are impermeable to hydrogen ions, so the increasing acidity cannot harm them. Secretion of mucus by the cells also provides a layer of protection. Furthermore, gastric mucosal cells have a high turnover rate and are sloughed off approximately every 72 hours—too short a lifespan for these cells to experience any significant damage from acidic conditions of the stomach. Still, if for some reason injury does occur, the high acidity will cause tissue to erode. This tissue erosion occurs in peptic ulcer disease and will be described later in this chapter.

Pathophysiology of the Upper Gastrointestinal Tract

Pathophysiology of the Oral Cavity

In this section, general problems associated with mouth, teeth, and gums will be discussed. These are of nutritional concern because most of these conditions involve problems that interfere with adequate oral intake and can lead to nutritional deficits.

periodontal disease— a bacterial infection that destroys the attachment fibers and supporting bone that hold the teeth in the mouth

dental caries—decay of the teeth that begins when acid dissolves the enamel that covers the tooth

enamel—Hard outer layer of teeth consisting of hydroxyapatite; this mineral is composed of calcium, phosphorous, fluoride, chloride, sodium, and magnesium

dentin—the hard tissue of the tooth surrounding the central core of nerves and blood vessels

plaque—the noncalcified accumulation of oral microorganisms and their by-products that adhere to the teeth

calculus—calcified deposits that have formed around the teeth

Oral Disease Conditions affecting oral health include the most common dental diseases: dental caries, gingivitis, and **periodontal disease**. Over the past several decades, the U.S. population has seen significant improvement in prevalence of most dental diseases. According to the report *Oral Health in America* (2000), both the incidence and prevalence of dental caries has decreased considerably. Even with these improvements, dental disease remains one of the most common health problems in our country. Fifty-one percent of children aged five to nine years have had at least one cavity or filling in either a primary or a permanent tooth. This number increases to approximately 80% to 85% percent for adults. Differences in prevalence exist by race/ethnicity and poverty level for the amount of untreated dental caries. For all age groups, non-Hispanic blacks and Mexican Americans have higher numbers of untreated decayed teeth than non-Hispanic Caucasian individuals (NIDCR 2000).

Dental Caries **Dental caries** or tooth decay is destruction of tooth structure through development of small pits or cavities. The cause of dental caries is multifactorial. Individual factors may include structure of the tooth, immunologic response to bacteria, and composition of saliva in the mouth. Genetic susceptibility to dental caries has not been well established, though it is frequently mentioned in the literature (Shuler 2001). It is currently understood that several different types of bacteria commonly found in the mouth, including streptococcus mutans, are involved in development of dental caries. Steps in caries development can be described as follows: after dietary carbohydrates are hydrolyzed by salivary amylase, bacteria preferentially will use those simple sugars as fuel and produce acids as a waste product. As a result, pH falls. The lowered pH causes demineralization of the tooth and ultimately results in decay of tooth **enamel** and **dentin** (Touger-Decker and van Loveran 2003). Foods that are sticky adhere to the surface of the tooth and are more likely to cause caries. Frequent snacking increases the time acids are in contact with the tooth and additionally increase risk of caries.

The combination of acids, bacteria, food, and saliva make up biofilm, a sticky substance commonly called **plaque**. Plaque adheres to the tooth and, if it is not removed, will mineralize into tartar or **calculus.** Factors that contribute to development of caries and plaque include frequency of eating, stickiness of the food, composition of saliva, presence of buffers, and overall oral hygiene (Touger-Decker and van Loveran 2003).

Cavities are usually painless until they are large enough to destroy internal structures of the tooth. Dental caries can also lead to death of the nerve and blood vessels in the tooth.

Individuals at Risk for Dental Disease It is estimated that after age 50, most Americans have lost at least 12 permanent teeth. Dental caries, tooth loss, and ill-fitting dentures may contribute to inadequate or improper intake of nutrients. At the same time, malnutrition and weight loss may contribute to the poor fit for dentures and loss of teeth. This may be a significant issue for the elderly, as it is estimated that one-third of those persons over 65 years are **edentulous.** Individuals who have difficulty chewing may rely on soft foods with limited variety, resulting in an inadequate nutrient intake.

Other medical conditions may be associated with increased dental caries. Eating disorders are frequently linked with dental disease. In bulimia nervosa, for example, the exposure to gastric juices during repetitive vomiting contributes to the demineralization of the teeth (Studen-Pavolvich and Elliott 2001).

Infants and children are considered to be at high risk for dental disease. "Baby bottle tooth decay" can occur when an infant falls asleep with a bottle containing a high-sugar beverage such as fruit juice or infant formula. During sleep, the beverage may pool in the mouth, contributing to widespread caries. Cavities usually form on upper front teeth and back molars. It is recommended that bottles be removed from a child's mouth during sleep. Other preventions include brushing a child's teeth as soon as they erupt, or wiping them with a wet cloth. A child's first dental examination should occur by the end of his or her first year (American Academy of Pediatric Dentistry 2004; American Dietetic Association 2003).

Prevention of Dental Disease Prevention of dental caries focuses on both nutrition and public health interventions. Fluoridation of water supplies, use of topical fluoride treatments, and use of dental sealants are primary methods of preventing dental caries. Fluoride ingested when teeth are developing is incorporated into the structure of the enamel and protects it against the action of acids. Topical fluoride also provides protection for the outer covering of the tooth (Navia 1994). Nutritional choices can play a role in the prevention of dental disease. Recent research in nutrition and dental health has indicated both milk and cheese provide an excellent buffer that prevents decrease in pH and resulting demineralization (Kashket and DePaola 2002; Stephenson 2002). The sugar alcohol xylitol is often used as an artificial sweetener in products such as chewing gum, toothpaste, and mouthwashes. This carbohydrate is not metabolized by bacteria in the mouth, and current research indicates it may have an additional protective effect against dental caries (Soderling 2001). Additional research continues to focus on other dietary factors, such as black tea, that may assist in reducing the risk of dental caries (American Dental Association 2001). According to the current U.S. Dietary Guidelines, the following recommendations can assist in the prevention of dental caries:

- Choose fiber-rich fruits, vegetables, and whole grains often.
- Choose and prepare foods and beverages with little added sugars or caloric sweeteners, such as amounts suggested by the USDA Food Guide and the DASH Eating Plan.
- Reduce the incidence of dental caries by practicing good oral hygiene and consuming sugar- and starch-containing foods and beverages less frequently (United States Department of Agriculture 2005).

Inflammatory Conditions of the Oral Cavity It is not uncommon for patients to suffer from inflammatory conditions of the mouth. When the mouth is inflamed or infected, maintaining oral intake is very difficult. Inflammatory conditions may result from poor dental hygiene and lack of dental care, and can also be seen in persons who are immunosuppressed or who have undergone chemotherapy or radiation therapy. Gingivitis is an inflammation of the **gingiva** (gums). Gums appear red and swollen and often bleed. The tissue is very tender and painful. Other symptoms include fever, loss of appetite, foul breath, and a bad taste in the mouth.

Stomatitis or mucositis is inflammation of the oral mucosa and is often associated with fungal infections such as *Candida albicans* or with herpeslike viruses. It is common in stomatitis to have open ulcerations on the oral mucosa, gingiva, and palate. Nutritional strategies for individuals with stomatitis are presented in Table 16.5.

Glossitis and cheilosis are inflammatory symptoms of the oral cavity classically associated with vitamin deficiencies. Glossitis involves increased redness, swelling, and pain of the tongue and lips. Cheilosis is fissuring and scaling at the corners or angles of the mouth. Both cheilosis and glossitis may be a result of riboflavin, niacin, or pyridoxine deficiency.

Conditions Resulting in Altered Salivary Gland Function The importance of saliva production in assuring normal chewing and swallowing is emphasized above (see Oral Cavity Secretions). In many clinical conditions, saliva production may be altered, resulting in decreased oral intake and ultimately placing the patient at nutritional risk.

Xerostomia may occur as a result of a disease process, but can also be a result of medical treatment (see Table 16.6). Infection or damage to salivary glands through surgical resection or radiation therapy can interrupt normal saliva production. Blockage of the salivary ducts may also occur as a result of a tumor or other medical condition. Systemic changes such as seen in **Sjögren's syndrome** or with

edentulous—without any teeth

gingiva—the gums

stomatitis—inflammation of the membrane in the mouth

xerostomia—decreased saliva production and dry mouth

Sjögren's syndrome—a chronic systematic inflammatory disorder, etiology unknown, characterized by dryness of mucous membranes

Stomatitis

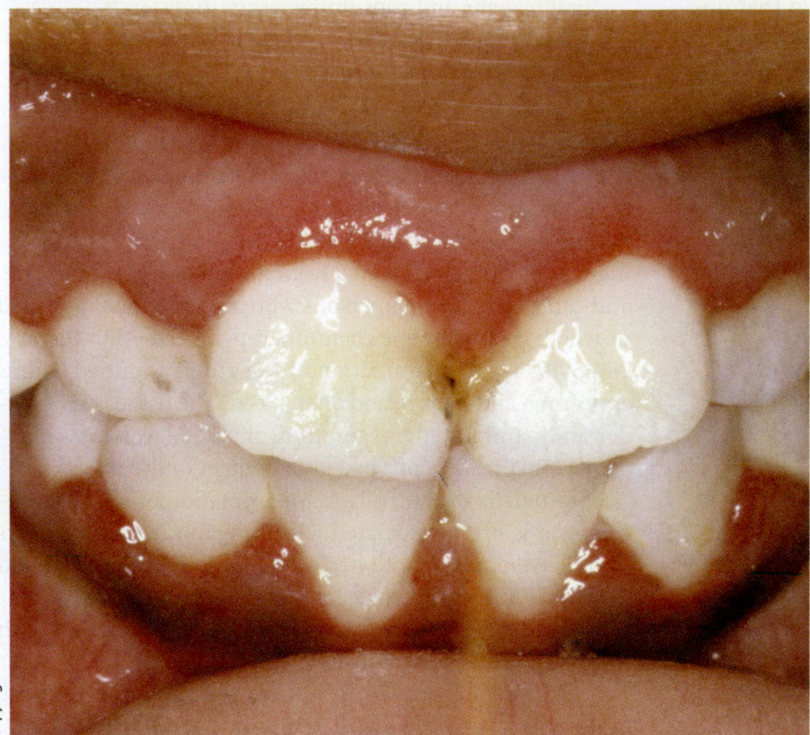

TABLE 16.5

Nutrition Strategies: Stomatitis

Prevention is key—good oral hygiene with frequent mouth rinses is important (avoid alcohol-based products).

Treat oral lesions pharmacologically as appropriate (antifungal medications if needed).

Consider using oral topical agents and anesthetics, such as viscous lidocaine and institution-specific mouth rinses, that are combinations of nystatin, Maalox® (Novartis, Parsippany, NJ), diphenhydramine, hydrocortisone, and viscous lidocaine.

Adjust texture and temperature as tolerated. Extremes of very hot or cold are often not tolerated.

Avoid carbonated beverages.

Avoid caffeine, alcohol, and tobacco products.

Avoid other irritants (e.g., acidic, spicy foods).

Try oral glutamine supplementation—optimal dose is 10 g tid.

Consider enteral nutrition support if unable to maintain nutritional status orally.

Note. From "Appendix C. Nutrition Impact Symptoms and Interventions" (pp. 348–349), in V.J. Kogut and S.K. Luthringer (Eds.), *Nutritional Issues in Cancer Care,* 2005, Pittsburgh, PA: Oncology Nursing Society. Copyright 2005 by the Oncology Nursing Society. Reprinted with permission

dehydration can also change saliva production. Many medications cause a reduction of saliva by affecting the parasympathetic nervous system. For instance, medications called anticholinergics, which act to block the effect of acetylcholine, reduce the amount of saliva as a major side effect. Groups of medications that have anticholinergic effects include antihistamines and antidepressants. Table 16.7 outlines interventions to assist with xerostomia.

Excessive saliva production may also be a concern, but in general does not pose the nutritional problems that xerostomia can. Nervous system diseases such as Parkinson's disease may interfere with autonomic control of saliva production. Other situations that increase saliva production, such as seeing, tasting, or smelling food or the consumption of sour-tasting food, involve the autonomic reflex.

Surgical Procedures for the Oral Cavity Surgical resections of the tongue, palate, or pharynx occur with head/neck malignancies. The mandible is a frequent site for fractures due to accidents. These procedures require significant nutrition interventions to assure maintenance of nutritional status (American Association of Oral and Maxillofacial Surgeons 1999).

Surgical procedures involving the jaw require immobilization of the jaw. The most common procedure is the maxillomandibular fixation (MMF), which is commonly known as "wiring the jaw." Nutritional intake is limited to those liquids and blenderized foods that can be put through a straw or syringe (see Table 16.8 for blenderized diets).

Impaired Taste: Dysgeusia/Ageusia Dysgeusia is the condition of altered or impaired sense of taste. Ageusia is the inability to taste or "mouth blindness." Many clinical conditions affect the ability to taste. Cells of the oral mucosa have a high turnover rate and thus are affected by treatments such as chemotherapy or radiation therapy (see Chapter 24). Diseases of the tongue and palate can interfere with normal function of taste buds. Nervous system diseases can also affect transmission of sensory information. Medications can also change the ability to taste.

Nutrition Therapy for Pathophysiology of the Oral Cavity

Nutritional Implications

The primary nutrition problem in diseases involving the oral cavity is the impaired ability to maintain adequate oral intake. Foods are difficult to swallow due to lack of saliva, thickened saliva, or the pain of an inflammatory condition or dental disease. When saliva is impaired, there is an additional risk of infection and dental caries. If intervention

Cheilosis

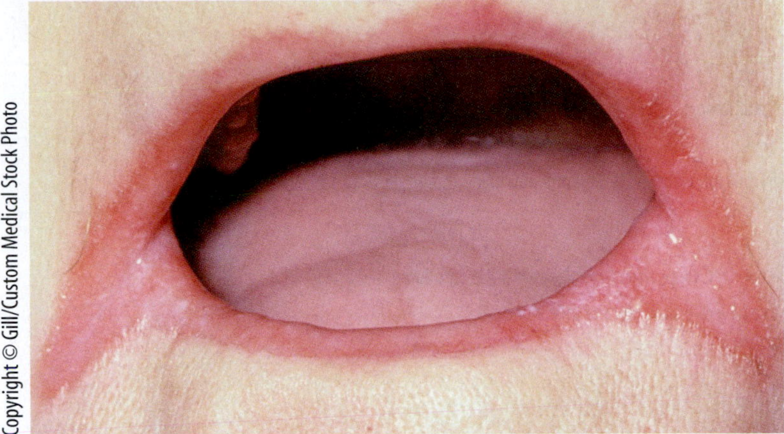

TABLE 16.6

Possible Causes of Xerostomia	
Side effects of some medicines	More than 400 medicines can cause the salivary glands to make less saliva. Medicines for high blood pressure and depression often cause dry mouth.
Disease	Some diseases affect the salivary glands, for example, Sjögren's Syndrome, HIV/AIDS, diabetes, and Parkinson's disease.
Radiation therapy	The salivary glands can be damaged if they are exposed to radiation during cancer treatment.
Chemotherapy	Drugs used to treat cancer can make saliva thicker, causing the mouth to feel dry.
Nerve damage	Injury to the head or neck can damage the nerves to the salivary gland.

TABLE 16.7

Nutrition Strategies: Xerostomia
Try tart foods to stimulate saliva.
Sip on liquids or suck on ice chips throughout the day.
Avoid caffeine, alcohol, and tobacco products.
Try using a cool mist humidifier at bedtime.
Try drinking through a straw.
Rinse mouth frequently with mild saline solution.
Add extra sauces and gravies to foods.

Note. From "Appendix C. Nutrition Impact Symptoms and Interventions" (pp. 348–349), in V.J. Kogut and S.K. Luthringer (Eds.), *Nutritional Issues in Cancer Care*, 2005, Pittsburgh, PA: Oncology Nursing Society. Copyright 2005 by the Oncology Nursing Society. Reprinted with permission

and treatment of the underlying condition do not occur, and without subsequent improved dietary intake, the patient will be at risk for malnutrition.

Nutrition Intervention

Texture modification of the current diet will be necessary for most of these clinical conditions. Changing the diet to include soft, moist foods, liquids, or blenderized foods will allow for increased intake. Use of gravies, sauces, and soft casseroles should be encouraged, while foods that are dry, crunchy, or have sharp edges should be avoided. Liquids should be consumed with meals. Table 16.9 outlines nutrition interventions that provide texture modifications.

Kcalorie and protein density can be increased by using modular components (such as dry skim milk powder) and high-kcalorie, high-protein liquid supplements (such as Ensure®, Boost®, or milk shakes), and by increasing fat intake (serving cream soups or adding margarine to mashed potatoes). Increasing frequency of meals may also allow for overall improved intake. Many patients may need to be prescribed a general multivitamin supplement. If specific deficiencies are confirmed, then appropriate supplementation can proceed.

Extreme temperatures and spices in foods may increase pain and intolerance to oral intake. Foods at room temperature and foods that are bland in flavor such as custards, yogurt, or pudding are generally well tolerated.

Liberally using fluids will increase moisture in the mouth and assist with increasing solid food intake. Assure adequate hydration at all times. Spraying the mouth with water or sucking on ice chips may be helpful. Sugar-free beverages containing citric acid (lemonade, etc.) may stimulate saliva production. Though usually a temporary intervention, using sugar-free gum or mints can also help. Cold and frozen foods are sometimes preferred. In extreme conditions, artificial saliva can be used.

Oral hygiene is an important part of nutritional care. Frequently rinsing the mouth to remove food particles can help prevent a bad taste in the mouth and an increase in dental caries. Alcohol-containing mouthwashes tend to dry the mouth, so other choices such as lemon-glycerine solutions or warm water with baking soda can be used. Using ¼ tsp of baking soda in 1 cup of water is recommended.

If pain is severe, coordinating pain medication with oral intake is important. Oral agents that provide localized numbing can be used, but in severe inflammatory conditions, systemic pain medications are frequently prescribed.

Evaluation and Monitoring Oral intake can be monitored by observation, kcalorie count, or food diary.

TABLE 16.8

Blenderized Diets

General recommendations:

You will need a blender or food processor to prepare foods to the appropriate consistency.

Cut into small pieces and add liquid to make blending easier.

Commercially prepared baby food is available to use (do not use "junior" foods), but you will need to experiment with herbs and spices to add flavor.

Adding extra protein:

Dry milk powder—this can be added to thin cooked cereal, cocoa, eggnog, blenderized soups and casseroles, milk shakes, pudding, and gravy.

Double strength milk—add 1 cup dry milk powder to 1 cup whole milk. Use as a beverage or in place of milk in recipes.

Liquid egg substitutes—Add EggBeaters or other pasteurized egg substitutes.

Blenderized meats.

Cottage cheese.

Yogurt, smooth blended.

Commercial liquid nutritional supplements such as Ensure, Boost, or Carnation Instant Breakfast.

To add extra kcalories:

Use whole milk, half and half, or evaporated milk in place of water as the fluid when blenderizing.

Add extra butter or margarine to *blenderized* vegetables, soups, or thin cereals. Add gravy to *blenderized* meats.

Add sugar or honey to *blenderized* fruits, squash, pumpkin, or carrots.

Food Groups	Foods That Blend Well	Foods That Do Not Blend Well
Beverages	Milk and milk beverages; yogurt drinks; juices	Milk or yogurt products with nuts or seeds
Breads and cereals	Breads and crackers without nuts, seeds, or dried fruits; cooked or ready-to-eat cereals; pancakes, waffles, French toast, and other baked products made with "foods that blend well"	Coarse, whole-grain breads or breads with nuts; seeds or dried fruit; granola and other coarse, whole-grain cereals
Dessert	Smooth custards and puddings; sherbet, shakes; gelatin; baked goods made with "foods that blend well"	Desserts or baked goods made with nuts or seeds; coconut; chocolate; butterscotch; peanut butter; chips
Fats	Butter or margarine; cream and cream substitutes; cream cheese; cooking fats and oils; smooth sauces or gravy; whipped topping	None
Fruits	All cooked or canned fruits without skins or seeds; fresh peeled apples, apricots, bananas, melons, peaches, and pears; fruit juices; nectars	Fruits with seeds, membranes, or tough skins (e.g., strawberries, raspberries, watermelon, tomato, pineapple, orange and grapefruit sections, cherries, grapes); dried fruits
Meat and meat substitutes	Cooked, tender meat, fish, poultry; infant-strained meats, commercially prepared pureed meats; cooked legumes; tofu; cottage cheese; smooth peanut butter; eggs; cheese sauce; casseroles made with "foods that blend well"	Fried meats, sausages, or wieners with tough skins; poultry skin; fish with bones; anchovies; fried eggs; most hard cheeses; nuts; crunchy peanut butter
Potato/Grains	Cooked, peeled potatoes; cooked rice, pasta, and noodles	Fried rice; fried noodles; potato skins
Soups	Broth, bouillon; all blended, strained stock or cream soups	None
Sweets	Jelly; honey; sugar; sugar substitute; chocolate syrup; maple syrup	Marmalade
Vegetables	Vegetable juices; well-cooked or canned vegetables except those that "do not blend well"	Raw or fried vegetables; vegetables with seeds, membranes, or tough skins; corn; celery
Miscellaneous	Ground seasonings and spices; tomato paste, mustard, ketchup, and other smooth condiments	Nuts; coconut; seeds; popcorn; relishes

Source: Manual of Clinical Dietetics, 6th Edition, 2000, pp. 659-661. American Dietetic Association. Original source: Manual of Nutritional Care 4e, Vancouver: BC: British Columbia Dietitians and Nutritionists Assoc.; 1992.

TABLE 16.9

Nutrition Interventions: Texture Modifications for Problems in Chewing and Swallowing

Food Groups	Foods Recommended	Foods Not Recommended
Beverages	All	None
Breads and Cereals	Oatmeal, Cream of Wheat, grits, and other hot, cooked cereals; dry ready-to-eat cereals softened with milk or other liquid; pasta; rice; other cooked grains	Initially not recommended to consume high-fiber cereals and bran; tough or chewy breads (i.e., bagels or French bread); breads with tough crusts; crackers, chips with hard edges; popcorn; anything containing coconut, seeds, or dried fruits
Fats	All	None unless not tolerated
Fruits	All cooked fruits without skins or seeds; soft fresh fruits without skins or seeds; fruit juices without pulp	Tough, crunchy raw fruits with hard edges; coconut; dried fruits
Meats and Protein Sources	Tender cooked meats without skin or bones; eggs; soy tofu; cheese, cottage cheese; dried beans and peas	Tough, stringy, high-fat, or fried meats; cooked cheeses; crunchy peanut butter
Vegetables	All cooked soft without skins or seeds; vegetable juices and sauces	Raw, stringy, or crunchy vegetables; cooked vegetables with skins or seeds

BOX 16.1 CLINICAL APPLICATIONS

Nutrition Assessment of the Upper GI Tract

Anthropometric Assessment

Height, weight, and usual body weight should be determined and assessed for any weight loss that may have already occurred. If other areas of the nutrition assessment indicate that the patient may be a nutritional risk, other anthropometric measures would be assessed to substantiate the presence of malnutrition.

Biochemical Assessment

Laboratory measures of protein status are evaluated to establish any visceral protein deficit. Other laboratory indices would be assessed depending on the underlying medical condition, such as those that might be seen in dehydration and/or anemia.

Clinical Assessment

Physical assessment of the head, neck, and oral cavity is a standard component of physical assessment. It is a crucial component of nutrition assessment for conditions of the oral cavity. Review results of swallowing evaluations. Assess for clinical symptoms of malnutrition; perform physical assessment for symptoms of dehydration.

Dietary

Dietary assessment will focus on determining the changes in oral intake that have occurred due to the disease condition. Methods to obtain this information can be a 24-hour recall, diet history, or direct observation of the patient's intake. Evaluate tolerance to different textures and consistencies. Evaluate for supplement intake.

Adequacy of intake is compared to estimated kcalorie and protein requirements. Every step should be taken to meet the patient's food preferences and tolerances. Success of nutrition interventions will be measured by weight gain or weight maintenance and the evaluation of biochemical parameters.

Pathophysiology of the Esophagus

This section will discuss the most common diseases or conditions involving the esophagus, including gastroesophageal reflux disease (GERD), Barrett's esophagus, achalasia, hiatal hernia, and dysphagia.

Gastroesophageal Reflux Disease (GERD)

Each year, more than 20 million Americans suffer daily symptoms of gastroesophageal reflux, while more than 100 million suffer occasional symptoms (National Center for Health Statistics 2005). **Gastroesophageal reflux disease (GERD)** occurs as a result of reflux of gastric contents into the esophagus. The etiology of the reflux is multifactorial and can include both physical and lifestyle factors. The most common factor involves incompetence of the lower esophageal sphincter (LES) that normally serves as a bar-

gastroesophageal reflux disease (GERD)—chronic or recurrent gastric pain due to reflux of gastric secretions into the lower esophagus

TABLE 16.10

Conditions and Substances Associated with Esophageal Reflux
Conditions That Increase the Likelihood of Reflux
Ascites (accumulation of fluid in the abdomen)
Delayed gastric emptying
Eating large meals
Lying flat after eating
Obesity
Pregnancy
Wearing clothes that fit tightly across the waist or abdomen
Substances That Decrease Lower Esophageal Sphincter Pressure
Alcohol
Medications such as anticholinergic agents, calcium channel blockers, Meperidine, Diazepam, Theophylline
Caffeine
Chocolate
Cigarette smoking
Garlic
High-fat foods
Onions
Peppermint and spearmint oils
Progesterone

Source: Rolfes SR, Pinna K, Whitney E. Understanding normal and clinical nutrition. 7th ed. Belmont: Wadsworth; 2006. Table 23-3, p. 721.

rier between the esophagus and stomach. As discussed earlier, atmospheric pressure is greater in the esophagus than in the stomach under normal conditions. The pressure differential prevents reflux of gastric contents. Many factors that can lower LES pressure and thus contribute to LES incompetence and GERD have been identified (DeVault and Castell 2005). These include (1) increased secretion of the hormones gastrin, estrogen, and progesterone; (2) presence of other medical conditions, such as hiatal hernia or scleroderma; (3) cigarette smoking; (4) use of medications, including dopamine, morphine, and theophylline; and (5) specific foods. Foods high in fat, chocolate, spearmint, peppermint, alcohol, and caffeine all may decrease LES pressure. Table 16.10 outlines factors that lower LES pressure.

Symptoms of GERD may include dysphagia (difficulty swallowing), heartburn, increased salivation, and belching. In some situations, pain is severe and may radiate to the back, neck, or jaw. In fact, pain from GERD can be confused with pain that is cardiac in origin because of the diffuse spread of pain into these other areas. For many patients, pain is worse at night when they are lying down. Complications of untreated or unresponsive GERD may include impaired swallowing, aspiration of gastric contents into the lungs, ulceration, and perforation or stricture of the esophagus. Barrett's esophagus is also considered to be a complication of GERD (see Barrett's Esophagus—A Complication of GERD).

Treatment for GERD includes three major goals: (1) increasing LES competence; (2) decreasing gastric acidity, and thus decreasing the severity of symptoms; and (3) improving clearance of contents from the esophagus. Surgical intervention may be warranted if the disease is unresponsive to medical management or if the patient experiences complications (DeVault and Castell 2005; Society of American Gastrointestinal Endoscopic Surgeons 2001).

Lifestyle factors that compromise LES competence such as cigarette smoking, use of medications, and nutritional history should be modified. Patient education can focus on ways to improve clearance in the esophagus; for example, patients may be instructed to remain upright after eating, to lose weight, to wear loose-fitting clothing, and to raise the head of their bed for sleeping.

Decreasing gastric acidity involves use of medication and nutrition therapy. Medications fall into five major categories: (1) antacids or buffering agents, (2) histamine blocking agents, (3) prokinetic agents, (4) proton pump inhibitors, and (5). mucosal protectants. Gastric secretions are controlled in several different ways (see Control of Gastric Secretions) and medications used to treat GERD interfere with control of gastric secretions by blocking several of those control pathways. See Table 16.11 for a summary of these medications.

The surgical procedure used for GERD is **fundoplication** (see Figure 16.4). This procedure takes the fundus of the stomach and wraps it around the lower esophagus. This provides additional strength to the LES and assists in preventing the reflux. This procedure can be done **laparoscopically**, which avoids an abdominal incision and considerably reduces the recovery time (DeVault and Castell 2005; Patti and Fisichella 2004; Society of American Gastrointestinal Endoscopic Surgeons 2001). The most recent advancement in treatment is the Stretta procedure. In this procedure, radiofrequency energy is delivered to the lower esophageal sphincter and gastric cardia. The procedure has been shown

fundoplication—a surgical technique used to suture the fundus of the stomach around the esophagus to prevent reflux

laparoscopically—the process of using laparoscopic procedure through which an instrument is used to see structures within the abdomen and pelvis; in this way, a number of surgical procedures can be performed without the need for a large surgical incision

TABLE 16.11

Medications for Treatment of GERD

Classification of Medication	Generic and/or Trade Names	Precautions
Antacids	Alka-Seltzer, Maalox, Mylanta, Pepto-Bismol, Rolaids, and Riopan, are usually the first drugs recommended to relieve heartburn and other mild GERD symptoms. Many brands on the market use different combinations of three basic salts—magnesium, calcium, and aluminum—with hydroxide or bicarbonate ions to neutralize the acid in your stomach.	Antacids, however, have side effects. Magnesium salt can lead to diarrhea, and aluminum salts can cause constipation. Aluminum and magnesium salts are often combined in a single product to balance these effects. Calcium carbonate antacids, such as Tums, Titralac, and Alka-2, can also be a supplemental source of calcium. They can cause constipation as well.
Foaming agents	Work by covering your stomach contents with foam to prevent reflux. These drugs may help those who have no damage to the esophagus (Gaviscon is one type of agent).	
H_2 blockers	Impede acid production—medications available include: cimetidine (Tagamet HB), famotidine (Pepcid AC), nizatidine (Axid AR), and ranitidine (Zantac 75). They are available in prescription strength and over the counter.	These drugs provide short-term relief, but over-the-counter H_2 blockers should not be used for more than a few weeks at a time. They are effective for about half of those who have GERD symptoms. Many people benefit from taking H_2 blockers at bedtime in combination with a proton pump inhibitor.
Proton pump inhibitors	Proton pump inhibitors are more effective than H_2 blockers and can relieve symptoms in almost everyone who has GERD and include omeprazole (Prilosec), lansoprazole (Prevacid), pantoprazole (Protonix), rabeprazole (Aciphex), and esomeprazole (Nexium), which are all available by prescription.	
Prokinetics	Helps strengthen the sphincter and makes the stomach empty faster. This group includes bethanechol (Urecholine) and metoclopramide (Reglan). Metoclopramide also improves muscle action in the digestive tract.	These drugs have frequent side effects that limit their usefulness.

Reference: National Digestive Diseases Information Clearinghouse [homepage on the Internet]. Bethesda: National Digestive Diseases Information Clearinghouse; June 2003 [cited 2005 Oct 13]. Heartburn, Hiatal Hernia, and Gastroesophageal Reflux Disease (GERD); NIH Publication No. 03-0882. Available from: http://digestive.niddk.nih.gov/ddiseases/pubs/gerd/index.htm#4

to improve the function of the LES in a minimally invasive and less expensive manner (Richards et al. 2003).

Barrett's Esophagus—A Complication of GERD

Barrett's esophagus, or Barrett's metaplasia, involves a change in the epithelial cells of esophageal mucosa and is usually considered a complication of GERD. Barrett's esophagus is detected in approximately 10% of patients undergoing **endoscopy** for GERD. Furthermore, it has been established that patients with refractory GERD (unresponsive to treatment) are much more likely to develop Barrett's esophagus. For example, patients with Barrett's esophagus have been shown to have persistent abnormal pH monitoring results even on maximum pharmacological treatment (Slehria and Sharma 2003).

In this condition, the normal squamous cell epithelium of the esophagus changes to metaplastic columnar cell epithelium. This dysplastic cellular change is considered to be a precursor to a malignancy. Those patients with Barrett's esophagus are at higher risk for adenocarcinoma of the esophagus (Spechler 2002). Current research is focusing on determining better ways to identify high-risk patients. Such research might establish biomarkers and would allow for early detection and treatment (Slehria and Sharma 2003).

Patients do not experience any specific symptoms with this condition. It is generally undetected unless the patient has a biopsy done as part of an upper GI diagnostic work-up. This is usually conducted as a result of GERD. There are no specific nutritional concerns unless the patient is diagnosed with esophageal cancer. In the case of a malignancy,

Barrett's esophagus—a complication of severe chronic GERD involving changes in the cells of the tissue that line the bottom of the esophagus; these esophageal cells become irritated when the contents of the stomach back up, and there is a small but definite increased risk of cancer of the esophagus

endoscopy—examination of the interior of a canal by means of an endoscope

TABLE 16.12

Nutrition Interventions for GERD

Foods Not Recommended

I. Foods that relax the lower esophageal sphincter

Peppermint or spearmint
Chocolate
Fried foods or those with high amounts of added fat
Alcohol
Coffee (decaffeinated and caffeinated)

II. Foods that may increase gastric acid secretion

Coffee (decaffeinated and caffeinated)
Alcohol
Pepper

Food Group	Foods to Avoid (if symptomatic)
Beverages	Cola, coffee, tea, cocoa, alcohol
Milk and milk products	2% milk, whole milk, cream, high-fat yogurts, chocolate milk
Eggs	Fried or scrambled using high-fat cooking methods
Cereals/grains	None
Meat and protein sources	Fried meats, bacon, sausage, pepperoni, salami, bologna, frankfurters/hot dogs
Vegetables	Only those that aggravate individual symptoms
Fruits	Only those that aggravate individual symptoms
Fat	As tolerated within current recommendations of U.S. Dietary Guidelines
Desserts	Those considered high fat or those that are fried
Miscellaneous	Pepper

Source: Adapted from: Nahikian-Nelms M. Gastrointestinal Disease. Nutrition Care Manual. Chicago, IL: American Dietetic Association. © 2005 American Dietetic Association. Adapted with permission.

nutritional issues are addressed as the patient begins treatment for that diagnosis.

Nutrition Therapy—Gastroesophageal Reflux Disease

Nutritional Implications Most patients identify foods they feel make their symptoms worse and thus decrease intake of those foods. In these situations, restriction of food groups may result in weight loss or nutritional deficiency. Nutritional therapy may assist not only by addressing these nutritional problems but also by decreasing the symptoms that the patient is experiencing.

aspiration—the accidental inhalation of food particles or fluids into the lungs

Assessment For the patient with GERD, 24-hour recall, diet history, or food diary should be used to focus on consumption of foods that lower LES pressure, increase gastric acidity, or cannot be tolerated by the patient. Additionally, lifestyle factors such as smoking and physical activity patterns are important for their contribution to LES incompetence.

Nutrition Interventions The goals of nutrition therapy are consistent with the goals of medical care discussed earlier. These goals include reducing gastric acidity and restricting foods that lower LES pressure. To reduce gastric acidity, black and red pepper, coffee (both caffeinated and decaffeinated), and alcohol should be avoided, because all have been identified as stimulants for gastric acid production. Likewise, meals of larger quantity tend to produce more acid, delay gastric emptying, and increase the risk of reflux. Thus, smaller, more frequent meals may be indicated. Foods that lower LES pressure should also be restricted. These include chocolate, mint, and foods with a high fat content. Furthermore, any food the client identifies as irritating should be avoided. If the patient is obese, weight reduction should be a component of the plan for nutrition therapy. Nutrition interventions for GERD are outlined in Table 16.12.

Dysphagia

Dysphagia, or difficulty swallowing, is not generally considered a diagnosis but a symptom caused by a variety of disorders. Since many conditions of the esophagus involve dysphagia, it is important to understand this "symptom" and the importance of nutrition therapy in its treatment. There are numerous medical conditions and treatments that can ultimately affect one or more of the four phases of swallowing: oral preparatory, oral, pharyngeal, and esophageal. See Table 16.13 for a full list of potential causes of dysphagia.

Symptoms of dysphagia that a patient will experience depend on the phase of swallowing impaired. For example, if the problem originates in the oral preparation phase, food may be pocketed in the buccal mucosa (cheek area) because the patient cannot propel the bolus of food effectively from the front of the oral cavity to the pharyngeal area. Other general symptoms may include drooling, coughing, and choking. Many patients will experience weight loss and generalized malnutrition due to inadequate nutritional intake (Finestone and Greene-Finestone 2003; Preshaw 2004). **Aspiration** or inhalation of oropharyngeal contents is a primary complication. This can lead to aspiration pneumonia with the accompanying infections. Table 16.14 provides a review of dysphagia symptoms with nutritional considerations.

Diagnosis and treatment of dysphagia involve many different members of the health care team. Many institutions have a dysphagia team consisting of physicians, nurses, speech language therapist, dietitian, physical therapist, and occupational therapist. Diagnosis of dysphagia begins with a bedside swallowing assessment usually performed by the speech language therapist. Conclusive evaluation uses a videofluoroscopy swallowing study or fiberoptic endoscopic evaluation of swallowing (see Figure 16.5 and Box 16.2). In these diagnostic procedures, barium is added to a variety of textures of foods and liquids. The patient is then monitored to determine his or her ability to swallow each of these foods. From these evaluations, a specific site of the dysphagia can be determined and a care plan can be developed.

FIGURE 16.4 Fundoplication

Nissen fundoplication

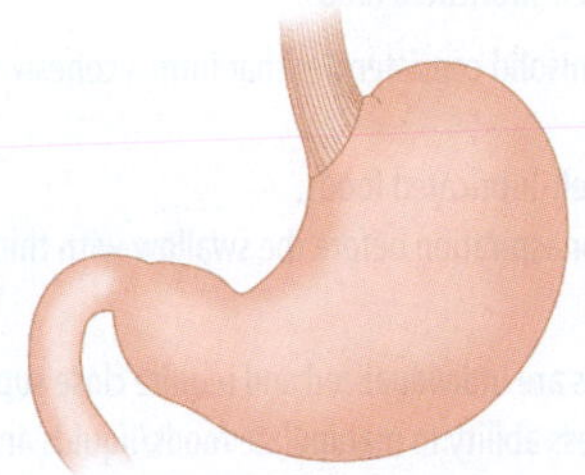

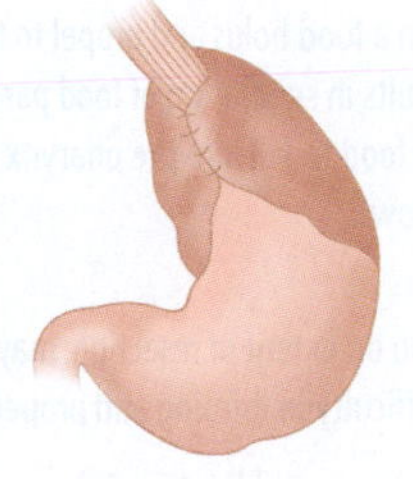

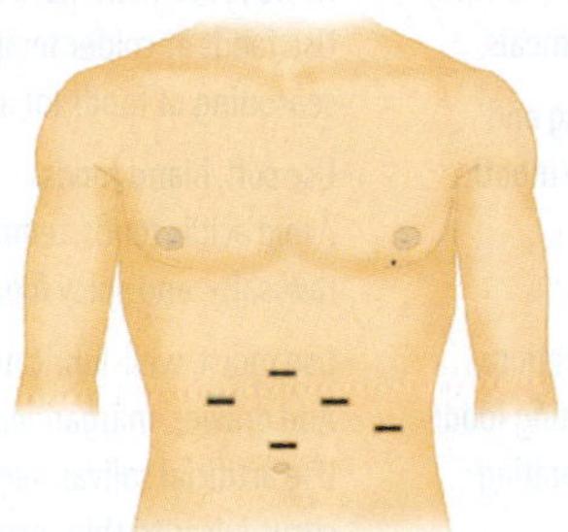

Source: www.mayoclinic.org/gerd/refluxsurgery.html

Nutrition Therapy for Dysphagia

Nutrition Implications The primary nutrition implication is weight loss and subsequent development of nutritional deficiencies that can occur due to an inadequate dietary intake.

Nutrition Interventions After reviewing results of swallowing diagnostic tests, the health care team will be able to determine how the patient handles various textures of foods and liquids. The registered dietitian can then use acceptable textures for the development of an adequate menu. Previously, standards for dysphagia diets were confusing, because health team members and institutions used conflicting terms and definitions. In 2002, however, a United States task force developed standard definitions for foods and liquids as well as levels of nutrition intervention. The levels of diet intervention are the *National Dysphagia Diets 1, 2, and 3*. Table 16.15 outlines the foods allowed on each level of the National Dysphagia Diet (McCallum 2003; National Dysphagia Diet Task Force 2002).

TABLE 16.13

Diseases and Conditions Associated with Dysphagia

Neurogenic	Polymyositis	**Structural Lesions**	**Iatrogenic Conditions**
Cerebrovascular accidents	Dermatomyositis	Diverticula	Medication-induced injury
Strokes	Sarcoidosis	Oropharyngeal tumors	Postoperative, postradiation, chemotherapy
Parkinson's disease	Myotonic dystrophy	Thyromegaly	Pharmacological effects
Multiple sclerosis	Oculopharyngeal dystrophy	Abscess	**Psychiatric**
Neoplasms of the brainstem, meninges, etc.	Cerebral palsy	Webs and rings	Dementia (not only psychiatric but also neurological)
Muscular dystrophy	Poliomyelitis	Cervical osteoarthritis	Depression
Amyotrophic lateral sclerosis	Spinocerebellar degeneration	Vascular lesions (dysphagia aortica)	**Other**
Huntington's disease	Progressive supranuclear palsy	Peptic stricture	Tracheostomy patients
Myasthenia gravis	Alzheimer's disease	**Motility Disorders**	Prolonged intubation
Myopathy	Head injuries	Achalasia	
		Esophageal spasm, scleroderma	

Source: Manual of Clinical Dietetics, 6th Edition, 2000, p. 669. American Dietetic Association.

TABLE 16.14

Dysphagic Symptoms that Have Nutritional Implications

Problem/Condition	Signs and Symptoms	Dietary Considerations
Oral Preparation Phase		
Reduced buccal (cheek) and/or lip tone, decreased lip closure, poor tongue control for lateralization and/or anterior to posterior movement (tipping)	Food falls into the lateral sulcus during chewing and is difficult to retrieve (pocketing); poor bolus formation; food (before and after swallow) or liquid leaks from the mouth (drooling); slow oral transit time with solids; increased risk of aspiration of thin liquids (before the swallow).	Maintain semisolid consistencies that form a cohesive bolus. Provide nutrient-dense meals secondary to poor feeding efficiency. Assess the risk of aspiration before and during the swallow with thin liquids.
Poor rotary jaw movement secondary to increased tone or apraxia; reduced tongue lateralization or reduced range of motion of the jaw	Difficulty chewing food may result in aspiration or choking during and after swallow. Particles may remain lodged in pharynx, causing aspiration after swallow.	Maintain semisolid or chopped/minced consistencies. Adjust texture to severity of jaw dysfunction. Use moist, well-lubricated foods.
Reduced tongue movement; partial glossectomy, neurological insult, etc.	Limited ability to form a food bolus and propel to the back of the throat results in separation of food particles and increased risk for food to fall into the pharynx before initiation of the swallow.	Maintain semisolid consistencies that form a cohesive bolus. Use moist, well-lubricated foods. Assess risk for aspiration before the swallow with thin liquids.
Total glossectomy; floor of mouth resection; palate resection	Effects vary depending on extent of resection; may result in varying levels of difficulty in forming and propelling the food bolus.	Modifications are individualized and require close supervision to assess ability to manipulate foods/liquids and swallow safely.
Reduced oral sensation or awareness	Food lodges or becomes pocketed in areas of reduced sensitivity. Particles may fall over base of tongue, causing aspiration before the swallow, drooling during meals.	Position food in the most sensitive area. Avoid foods with more than one texture. Use foods at colder temperatures; increase texture and seasoning of foods for increased sensation/awareness.
Mucositis	Severe mouth soreness associated with chewing and swallowing; difficulty manipulating food in the mouth.	Use soft, bland foods. Avoid acidic foods, temperature extremes, and rough, raw, salty, and spicy foods.
Xerostomia, associated with cancer treatment radiation and Sjögren's disease; dry mouth, associated with pharmaceutical side effects	Excessive or thick saliva; unable to manage secretions, resulting in difficulty lubricating and manipulating food; thick, ropy saliva; gagging and spitting/expectorating spittle during meals.	Use moist, well-lubricated foods. Add gravies, margarine, and sauces. Use artificial salivas, sugarless lemon drops, papain, or citrus juices to thin secretions. Avoid dry, crumbly foods. Ensure adequate fluid intake. Some patients may benefit from avoidance of or dilution of fresh dairy products if secretions become thick or unmanageable with milk products. Dairy products do not stimulate increased production of saliva.
Oral Transit Phase		
Delayed or absent swallow reflex and/or reduced coordination associated with progressive neuromuscular disease; resections of hard/soft palate and/or tongue.	Aspiration before the swallow is initiated due to pooling (into valleculae, pyriform sinus, etc.) and overflow of food and liquid into the airway.	Use cohesive foods. Density depends on the level of oral sensation and swallowing dysfunction. Temperature extremes and highly seasoned foods may help to excite nerves. Use thickened liquids. Avoid sticky or bulky foods. Assess ability to control liquids.
Esophageal Transit Phase		
Weakened or lazy cricopharyngeus	Food material returns from the esophagus into the pharynx and may spill into the airway, resulting in aspiration after the swallow.	Use semisolid, moist foods that maintain a cohesive bolus. GERD precautions.
Reduced esophageal peristalsis	Food bolus remains in the esophagus.	Avoid sticky and dry foods. Try dense foods followed by liquids. Medical follow-up.
Esophageal obstruction from fistulas, soft bone growth, or tissue growth	Narrowing of esophageal passage.	Use thin liquids and pureed or soft solids; avoid sticky, dry foods. Medical follow-up.

(continued on the following page)

TABLE 16.14 *(continued)*

Pharyngeal Transit Phase		
Reduced laryngeal closure, associated with supraglottic laryngectomy	Airway protection is nonexistent or incomplete, resulting in risk for aspiration before the swallow.	Use cohesive foods that do not fall apart and thickened liquids.
Reduced or slowed movement of bolus through the pharynx	Food residue remains at the base of the tongue and high in the throat, resulting in aspiration after the swallow if particles fall into the airway.	Use moist, well-lubricated foods that maintain a cohesive bolus.
Decreased laryngeal elevation	Food remains on top of the larynx and may result in aspiration after the swallow when the larynx opens to restore breathing.	Use soft solids and thick to spoon-thick liquids. Avoid sticky and bulky foods that tend to fall apart.
Dysfunctional/hypertonic cricopharyngeus	Food may collect in the pyriform sinuses and overflow into the airway, resulting in risk for aspiration after the swallow.	Use thickened liquids and pureed foods. Medical follow-up.

Note: Coughing and/or wet/"gurgly" vocal quality are potential symptoms associated with pharyngeal phase disorders.
Source: Manual of Clinical Dietetics, 6th Edition, 2000, p. 672–4. American Dietetic Association. © 2005 American Dietetic Association. Used with permission.

TABLE 16.15

National Dysphagia Diet 1, 2, and 3

National Dysphagia Diet 1 (NDD-1) "Dysphagia Pureed"

Food Allowed	Food Not Allowed	Sample Menu
Includes foods of "pudding-like" consistency that are smooth or pureed with no lumps.	Gelatin desserts, fruited yogurt, peanut butter, unblenderized cottage cheese, scrambled, fried, or hard cooked eggs.	Pureed chicken, mashed potatoes with gravy, pureed carrots, applesauce, and chocolate pudding.

National Dysphagia Diet 2 (NDD-2) "Dysphagia Mechanically Altered"

Food Allowed	Food Not Allowed	Sample Menu
Foods that are moist and soft textured such as tender ground or finely diced meats, soft cooked vegetables, soft ripe or canned fruit, and some moistened cereals.	Bread, dry cake, rice, cheese cubes, corn, and peas.	Scrambled egg, pancake with syrup, flaked cold cereal with milk, banana, orange juice (beverages thickened as appropriate).

Dysphagia Mixed is a term used by some institutions to designate a customized puree (NDD-1) diet that also allows one mechanically altered (NDD-2) item. Sample meal: orange juice, vanilla yogurt, cream of wheat cereal, scrambled egg.

Mechanical Soft is another alternative diet that allows bread, cake, and rice in addition to the NDD-2 mechanically altered diet. Sample menu: diced chicken with gravy, steamed rice, Harvard beets, pound cake, and fresh strawberries.

National Dysphagia Diet 3 (NDD-3) "Dysphagia Advanced"

Food Allowed	Food Not Allowed	Sample Menu
Includes most regular foods except very hard, sticky, or crunchy items. Bread, rice, cake, shredded lettuce, and tender, moist meats are allowed.	Not allowed are hard fruit and vegetables, corn skins, nuts, and seeds.	Vegetable soup, shredded lettuce salad with dressing, turkey sandwich with mayonnaise, fresh ripe melon, and chocolate chip cookie with no nuts.

Liquids

The following terminology is recommended by the National Dysphagia Diet Task Force to describe the viscosity of beverages and other liquids on the dysphagia diet:

Spoon-thick

Honey-like

Nectar-like

Thin liquids: Allows all liquids, including water, ice, milk, milk shakes, juice, coffee, tea, carbonated beverages, frozen desserts, and gelatin.

Source: Reprinted from *Journal of the American Dietetic Association,* 103(3), SL McCallum, The National Dysphagia Diet: implementation at a regional rehabilitation center and hospital system. pages 381–384, Copyright 2003, with permission from American Dietetic Association.

BOX 16.2 CLINICAL APPLICATIONS

Diagnostic Procedures: Upper GI Pathology

A variety of standard procedures are available to assist in the diagnosis of pathology involving the upper GI system, and most specifically for the esophagus and stomach. These procedures can be divided into those tests that involve visualization of the GI tract, and those that involve the measurement of secretions by these organs.

Endoscopy

During esophagogastroduodenoscopy (EGD), a fiberoptic endoscope (see Figure 16.5) is introduced through the oral pharynx and is moved through the esophagus and stomach and into the duodenum. The endoscope is an optical instrument that includes a lens viewer, a long flexible tube, and a light source. The lens viewer allows the physician to visually inspect the mucosa of the organs and determine any abnormalities. The patient may have clear liquids from midnight the night before and then should be NPO for the last 6 hours prior to the procedure. The patient will also receive a variety of premedications to assist in the procedure. These may include Versed, a medication that is given as an agent to induce sedation and/or amnesia prior to and during the procedure. Other medications include pain medications and antibiotic prophylaxis. A needle biopsy can be passed through the endoscope that allows for determination of the presence of abnormalities in the tissue or the presence of infection.

Barium Radiology Studies

In these procedures, the patient is given a contrast medium to drink. The most common media is barium sulfate, which is a chalky, white radiopaque substance that the patient drinks like a milk shake. The barium can be visualized by fluoroscopy or by x-ray. This study allows the physician to monitor swallowing and the movement through the stomach into the duodenum. It also can distinguish many other abnormalities such as ulcers, tumors, or inflammation.

Esophageal Manometry

Motor function of the esophagus and the lower esophageal sphincter is evaluated in this procedure. A tube is passed through the oral pharynx into the stomach. An instrument called a transducer is attached to the outer end of the tube, which records the barometric pressure as the tube is pulled back from the stomach through the LES. The patient should be NPO for 8 hours prior to the procedure, and is premedicated with both antianxiety and sedative drugs. A normal manometry will indicate that the LES pressure is normal and relaxes during a swallow. It will also indicate that the pattern of muscle contractions is systematic and coordinated during a swallow.

24-Hour pH Monitoring with Intraesophageal pH Electrode and Recorder

This test involves placing a pH probe into the distal esophagus for a 12- to 24-hour period in order to generate a graph depicting continuous pH readings. Information is obtained regarding quantity and pattern of gastroesophageal (GE) reflux events, the correlation with symptoms, and the efficiency of esophageal acid clearance.

Bernstein Test (Acid Perfusion Test)

The Bernstein test may be used to differentiate between chest pain that is cardiac in origin and pain caused by acid reflux. A nasogastric tube is inserted through the nasopharynx into the esophagus. The physician attempts to replicate the symptoms of acid reflux by alternately injecting either a mild hydrochloric acid solution or a saline solution through the tube. The patient reports the differences in symptoms with each solution.

Electrogastrography (EGG)

This procedure is used to diagnose and study stomach rhythm. Disturbances in the stomach's pacemaker can produce nausea and vomiting. This procedure can assist in determining the etiology of nausea and vomiting.

Antroduodenal Manometry

The purpose of 24-hour antroduodenal manometry is to measure the pressure in the small intestine before and after a meal and during sleep and waking hours. The measurements provide documentation of gastric emptying and peristalsis of the small intestine.

Gastric Analysis (Basal Acid Output)

This procedure measures the amount of hydrochloric acid produced by the stomach under baseline conditions. This test is often used to diagnose pernicious anemia, achlorhydria, and Zollinger-Ellison syndrome or to evaluate the effectiveness of surgical or pharmacological interventions.

FIGURE 16.5 A. Endoscopy and B. Barium Swallow

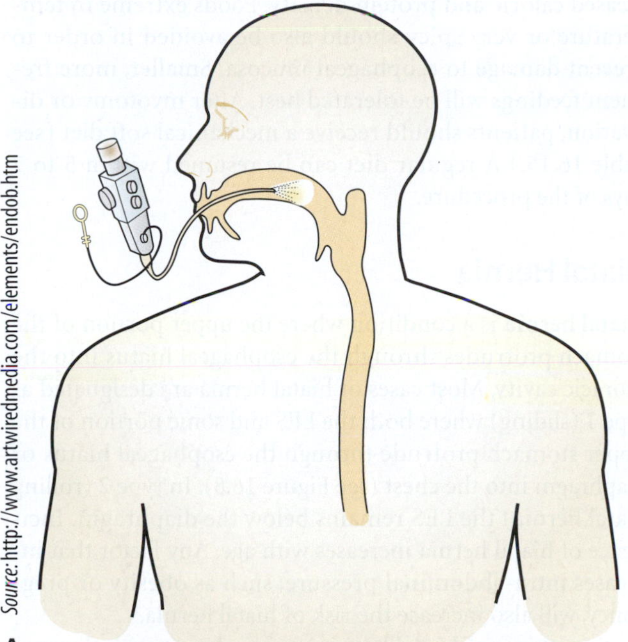

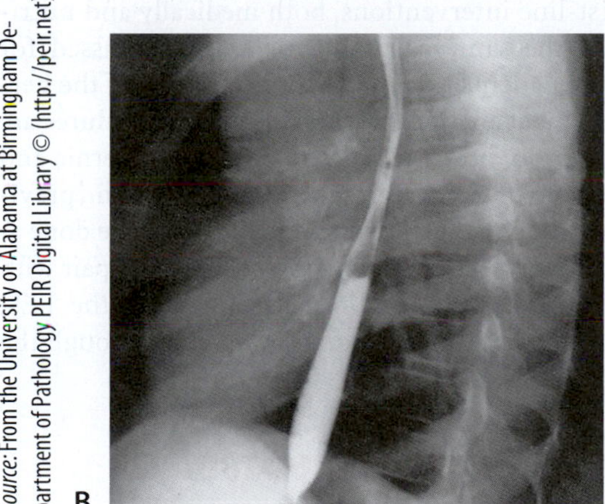

Source: From the University of Alabama at Birmingham Department of Pathology PEIR Digital Library © (http://peir.net)

Source: http://www.artwiredmedia.com/elements/endob.htm

FIGURE 16.6 Dysphagia Products. Can you tell that the foods in this photo are pureed foods shaped with commercial thickeners?

Source: S. Rolfes, K. Pinna and E. Whitney, *Understanding Normal and Clinical Nutrition,* 7e, copyright © 2006 p. 719

TABLE 16.16

Thickening Agents and Specialty Food Products Used to Treat Dysphagia

Product	Manufacturer
Thickening Agents for Liquids	
Thicken up	Novartis
Frutex	Crescent Foods
Thick n Easy	Hormel Health Labs
Thick-it	Precision Foods, Inc.
Thick Set	Bernard Fine Foods, Inc.
Thixx	Bernard Fine Foods, Inc.
Specialty Food Products	
Shape&Serve™	Hormel Health Labs
Puree Appeal	Novartis
NutraBalance	Ross Labs
Pureed meats	Tavis Meats, Inc.
Pureed foods	Zartc Foods

There are many specialty products available to assist in development of foods for dysphagia diets. They include a variety of thickening agents and specialized products (see Figure 16.6) pre-prepared to meet a specific consistency for the diets. See Table 16.16 for thickening agents and specialty food products.

Monitoring and Evaluation Depending on the origin of the patient's dysphagia, tolerance of an oral diet may improve with treatment. The registered dietitian, with the other health care team members, will reevaluate the ability of the patient to progress in use of the prescribed diet. If problems arise, the diet may also need to be further restricted or changed in texture or consistency. Patients' weight, nutritional parameters, and hydration should be monitored closely to assure adequacy of nutritional intake.

Achalasia

Achalasia is a motility disorder in which there is an absence of peristalsis or a weakened peristalsis within the esophagus. Additionally, there is often elevated LES pressure and impaired relaxation of the LES. This condition is relatively uncommon. It is estimated in that there are approximately eight cases per one million individuals in the U.S. and throughout the world (Vaezi and Richter1999).

The etiology of achalasia is unknown at this time, but current research has focused on the role of infectious disease and on autoimmune origins (Vaezi and Richter 1999). There are two distinguishing types of achalasia, primary and secondary. Primary achalasia is the most common type and results from loss of ganglion cells in the myenteric plexus of the lower portion of the esophagus. Secondary achalasia is due to other disease states such as diabetes mellitus, **Chagas disease**, and certain malignancies.

In primary achalasia, damaged ganglion cells result in a subsequent loss of the inhibitory neurotransmitters nitric oxide and VIP. The autonomic swallow is coordinated by autonomic nervous system control with nitric oxide and VIP acting as the primary inhibitory neurotransmitters. Without these neurotransmitters, the LES will not relax and appropriate swallowing cannot occur.

Treatments for achalasia include medications and invasive procedures. Primary medications focus on relaxation of smooth muscle. Calcium channel blockers such as nifedipine or nitrates can provide temporary relief. Most recently, botulinum toxin has been used. This toxin blocks acetylcholine, resulting in prolonged release of nitric oxide and relaxation of the LES (Meier 2001; Vaezi, Richter, and Wilcox 1999). Unfortunately, most patients experience a recurrence within 1 year of treatment.

Pneumatic dilatation involves mechanical dilation of the LES. Esophageal myotomy is a surgical procedure that divides the muscle fiber of the LES. Most recently, this procedure has been done via laparoscopy, which allows for reduced complications and shorter hospital stays. The most common complication is gastroesophageal reflux, which occurs in 7% to 20% of patients (Patti and Fisichella 2004). Figure 16.7 outlines the treatment options for achalasia.

Nutrition Therapy for Achalasia

Nutrition Implications In achalasia, patients experience dysphagia, vomiting, and substernal pain upon swallowing. Foods and fluids accumulate in the lower esophagus, causing the body of the esophagus to lose its muscle tone and become dilated or stretched. These symptoms result in poor oral intake with subsequent weight loss.

Chagas disease—a parasitic disease caused by *Trypanosoma cruzi*

hiatal hernia—protrusion of part of the stomach through the diaphragm into the space normally occupied by the esophagus, heart, and lungs

dyspepsia—vague upper abdominal symptoms that may include upper abdominal pain, bloating, early satiety, nausea, or belching

Nutrition Interventions Prior to treatment, patients with achalasia will need to have texture modification with increased caloric and protein density. Foods extreme in temperature or very spicy should also be avoided in order to prevent damage to esophageal mucosa. Smaller, more frequent feedings will be tolerated best. After myotomy or dilatation, patients should receive a mechanical soft diet (see Table 16.15.) A regular diet can be resumed within 5 to 7 days of the procedure.

Hiatal Hernia

Hiatal hernia is a condition where the upper portion of the stomach protrudes through the esophageal hiatus into the thoracic cavity. Most cases of hiatal hernia are designated as type 1 (sliding) where both the LES and some portion of the upper stomach protrude through the esophageal hiatus or diaphragm into the chest (see Figure 16.8). In type 2 (rolling hiatal hernia) the LES remains below the diaphragm. Incidence of hiatal hernia increases with age. Any factor that increases intra-abdominal pressure, such as obesity or pregnancy, will also increase the risk of hiatal hernia.

Symptoms of hiatal hernia are consistent with those of GERD. First-line interventions, both medically and nutritionally, are the same as those previously discussed for GERD. Some patients do require surgical repair of the hernia. Conventional and laparoscopic surgical procedures are used. In this procedure, the surgeon retracts the hernia and repairs the hole in the diaphragm. Fundoplication (previously described in the section on GERD) can also be done at this time, if needed. The combination of surgical repair with fundoplication provides additional support of the LES, which prevents the stomach from sliding back through the diaphragm (Richards 2003).

Pathophysiology of the Stomach

Disease and clinical disorders that affect the stomach can certainly influence normal nutritional status. Some disorders, such as indigestion, are mild and temporary conditions that resolve easily. Others, such as peptic ulcer disease, are chronic and require aggressive medical intervention.

Indigestion

Indigestion or **dyspepsia** is not considered to be a specific condition. Most people use the term "indigestion" to refer to a wide range of symptoms that may include abdominal pain, abdominal fullness, gas, bloating, belching, nausea, or even gastroesophageal reflux.

Nausea and Vomiting

Nausea is the unpleasant sensation that there is a need to vomit; vomiting is the expulsion of gastric contents. Even

FIGURE 16.7 Treatment Options for Achalasia

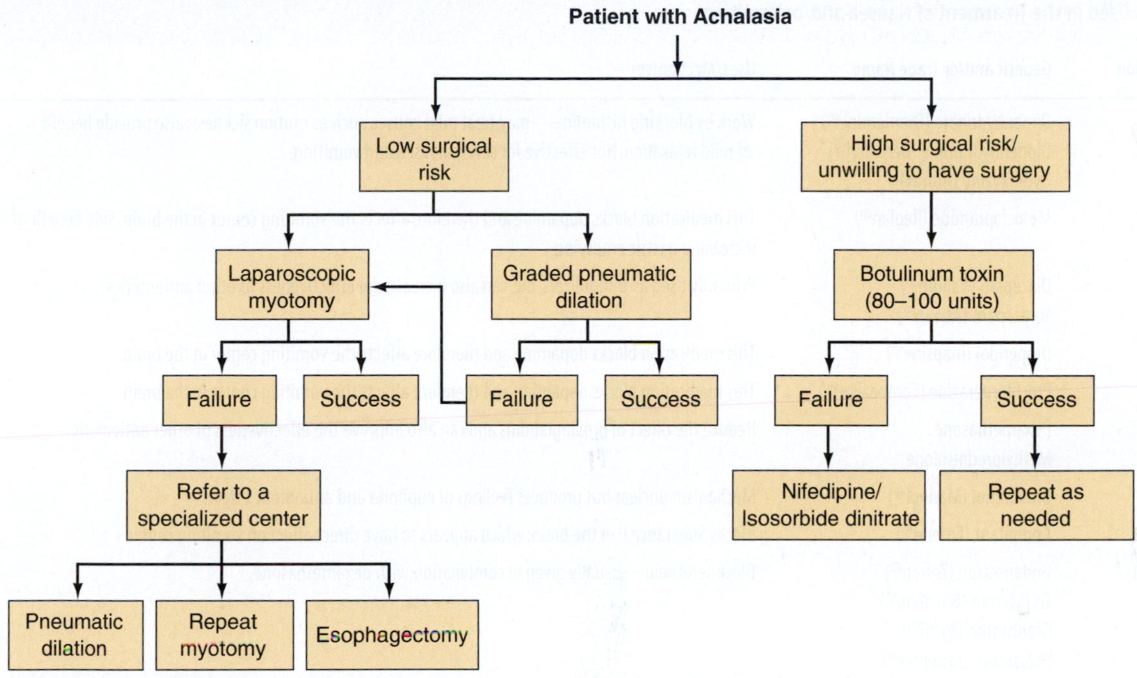

Source: M. Vaezi & J. Richter, Diagnosis and Management of Achalasia. J of Gastroenterology. 94: 3406-3412, 1999.

though nausea does not always lead to vomiting, they are often considered together because they are controlled through the same neural pathways. Neural signals are sent to the vomiting center located in the medulla. As a result of these stimuli, the steps of vomiting or emesis occur. In this sequence of events, gastric contents are pushed upward by the constriction of the respiratory muscles, the esophageal sphincter opens, the glottis closes (to prevent aspiration), and gastric contents are expelled through the mouth. Additionally, chemoreceptor zones in the medullary nucleus can also trigger the vomiting center. Drugs, toxins, metabolic conditions (such as renal failure or acid-base imbalances), and motion affect chemoreceptor zones, which can lead to nausea and emesis. Vomiting may also occur as a result of stress or extreme emotions.

FIGURE 16.8 Hiatal Hernia

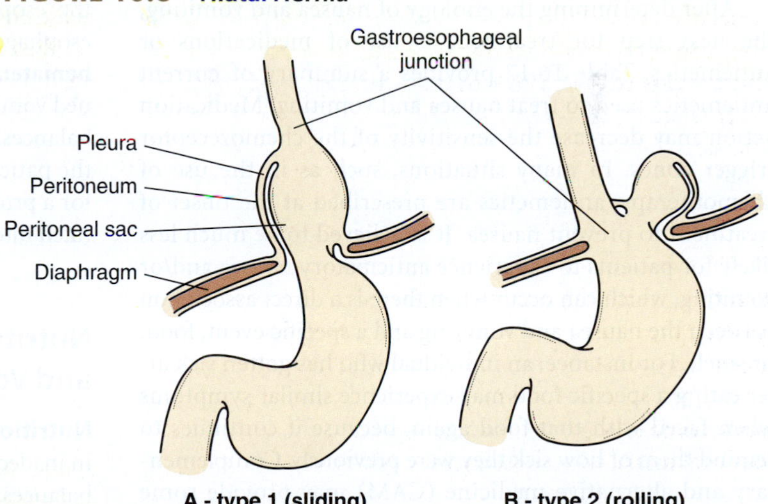

Reprinted from Price and Wilson: *Pathophysiology: Clinical Concepts of Disease Processes,* 6e © 2006 Mosby with permission from Elsevier.

Nausea and vomiting occur with many different medical conditions. These may include infection, pain, pregnancy, **syncope,** headache, metabolic disorders, motion sickness, kidney failure, myocardial infarction, and a host of other possibilities. Therefore, treatment of the underlying cause is the most important step in treating nausea and vomiting. The patient's history and physical examination will assist in determining the cause of nausea and vomiting. The etiology may be further clarified after assessment of the symptoms the patient experiences prior to and after vomiting. For example, if vomiting occurs within a very short time after eating, it may be indicative of an obstruction. Abdominal pain is symptomatic of an inflammatory process. Simple regurgitation of food occurs when gastric contents move easily from stomach to the mouth and is not a forceful movement, which is seen in vomiting.

syncope—temporary loss of consciousness; fainting

TABLE 16.17

Antiemetic Agents Used in the Treatment of Nausea and/or Vomiting

Classification of Medication	Generic and/or Trade Name	Uses/Mechanism
H_1 Antihistamines	Dimenhydrinate (Dramamine®) Diphenhydramine (Benadryl®) Hydroxyzine (Atarax) ®	Work by blocking histamine—may treat mild nausea such as motion sickness; also provide benefit of mild relaxation. Not effective for severe nausea and vomiting.
Benzamides	Metoclopramide (Reglan®)	This medication blocks dopamine and therefore affects the vomiting center in the brain. Side benefit of increasing gastric emptying.
Benzodiazepines	Diazepam (Valium®) Lorazepam (Ativan®)	Primarily used as tranquilizers, but can also increase the effectiveness of other antiemetics.
Butyrophenones	Droperidol (Inapsine®)	This medication blocks dopamine and therefore affects the vomiting center in the brain.
Phenothiazines	Prochlorperazine (Compazine®)	This medication blocks dopamine and therefore affects the vomiting center in the brain.
Corticosteroids	Dexamethasone Methylprednisolone	Reduce the effect of prostaglandins and can also improve the effectiveness of other antiemetics.
Cannabinoids	Dronabinol (Marinol®)	Mechanism unclear but produces feelings of euphoria and antiemetic effect.
NK1-receptor antagonists	Aprepitant (Emend®)	Blocks Substance P in the brain, which appears to have direct affect on vomiting center.
5-HT_3 receptor antagonists/ serotonin antagonists	Ondansetron (Zofran®) Tropisetron (Navoban®) Granisetron (Kytril®) Dolasetron (Anzemet®)	Block serotonin—usually given in combination with dexamethasone.

After determining the etiology of nausea and vomiting, the next step for treatment is use of medications or antiemetics. Table 16.17 provides a summary of current antiemetics used to treat nausea and vomiting. Medication action may decrease the sensitivity of the chemoreceptor trigger zones. In many situations, such as in the use of chemotherapy, antiemetics are prescribed at the onset of treatment to prevent nausea. It is believed to be much less likely for patients to experience anticipatory nausea and/or vomiting, which can occur when there is a direct association between the nausea and vomiting and a specific event, food, or smell. For instance, an individual who has gotten sick after eating a specific food may experience similar symptoms when faced with that food again, because it continues to remind them of how sick they were previously. Complementary and alternative medicine (CAM) may provide some additional avenues for controlling and treating symptoms experienced with nausea as well as other symptoms experienced with diagnoses involving the upper gastrointestinal tract (see Appendix F.1).

Prolonged nausea and vomiting can have significant clinical consequences. Forceful vomiting can rupture either the esophagus (Boerhaave's syndrome) or tear the lower esophageal sphincter (Mallory-Weiss tear). Bleeding or **hematemesis** is a serious outcome of these injuries. Continued vomiting also can result in dehydration and acid-base imbalances. Malnutrition can be a long-term consequence for the patient if he or she is not able to ingest an adequate diet for a prolonged amount of time. If gastric contents are aspirated into the lungs, aspiration pneumonia is a likely result.

hematemesis—the vomiting of blood

learned food aversion—avoidance of certain foods due to association with unpleasant GI symptoms

Nutrition Therapy for Nausea and Vomiting

Nutritional Implications Nausea and vomiting can result in inadequate nutrient intake, dehydration, and acid-base imbalances, and over time can lead to **learned food aversions**. This is similar to anticipatory nausea and vomiting. When a negative consequence is linked to a particular food, most people want to avoid eating that food.

Nutritional Interventions Nutrition therapy does not necessarily treat nausea and vomiting but can minimize symptoms and discomfort. If patients can manage oral intake, foods that are cold and have minimal smell usually are best tolerated. Table 16.18 outlines suggestions for foods and food-related activities that may reduce nausea and vomiting. Close monitoring of hydration status and length of time that the patient is without adequate oral intake will be crucial in preventing long-term nutritional consequences.

TABLE 16.18

Food/Nutritional Suggestions to Reduce Nausea and Vomiting
Provide optimal antiemetic medication for planned therapy.
Use relaxation techniques for anticipatory vomiting.
Evaluate other factors possibly contributing to vomiting, such as constipation, brain metastasis, and other medications.
Eat small, frequent, low-fat meals with minimal odors.
Try dry, starchy, and/or salty foods, such as pretzels, saltines, potatoes, noodles, and cereals.
Try foods cold or at room temperature.
Sip on ginger ale, tea, or candied dried ginger.
Avoid favorite foods until symptoms resolve.
Consume clear liquids such as broth, gelatin, or juice drinks on chemotherapy days, as these may be better tolerated than solid food.

Note. From "Appendix C. Nutrition Impact Symptoms and Interventions" (pp. 348–349), in V.J. Kogut and S.K. Luthringer (Eds.), *Nutritional Issues in Cancer Care,* 2005, Pittsburgh, PA: Oncology Nursing Society. Copyright 2005 by the Oncology Nursing Society. Reprinted with permission

Gastritis

Gastritis is inflammation of the gastric mucosa. This condition is not a single disorder and may be a result of numerous conditions. Acute gastritis is due to local irritation of the gastric mucosa. This irritation can be from infections, such as with *Helicobacter pylori (H. pylori)*, food poisoning, alcohol ingestion, or from medications such as nonsteroidal anti-inflammatory drugs (NSAIDs). The form of gastritis caused by NSAIDs is usually short lived and causes no long-term problems. Symptoms of gastritis can include belching, anorexia, abdominal pain, vomiting, and, in the more severe cases, bleeding and hematemesis.

Chronic gastritis is usually classified by either the etiology or the region of the stomach involved. Type A chronic gastritis involves the fundus and is often associated with an autoimmune process, which results in the formation of antibodies against the parietal cells. Type A chronic gastritis also occurs with pernicious anemia. Type B chronic gastritis results in atrophy of the gastric mucosa and is most frequently associated with infection from *H. pylori* (Gelfand and Ott 1999). Incidence of chronic gastritis increases with age and is often seen with **achlorhydria.** Treatment for gastritis includes identifying and treating the cause of the gastritis; for example, an antibiotic and medication regimen may be used to treat infections caused by *H. pylori* that are responsible for gastritis.

Peptic Ulcer Disease

Peptic ulcer disease (PUD) involves ulcerations of the gastric mucosa that penetrate the submucosa, usually in the antrum of the stomach or in the first few centimeters of the duodenum. Erosion may proceed to other levels of tissue and can eventually perforate. Breakdown in the tissue allows for continued insult by the highly acidic environment of the stomach as well as damage from other secretions of the stomach, such as pepsin. One out of every ten Americans develops peptic ulcer disease and this condition accounts for over 150,000 hospitalizations each year.

Peptic Ulcer Disease

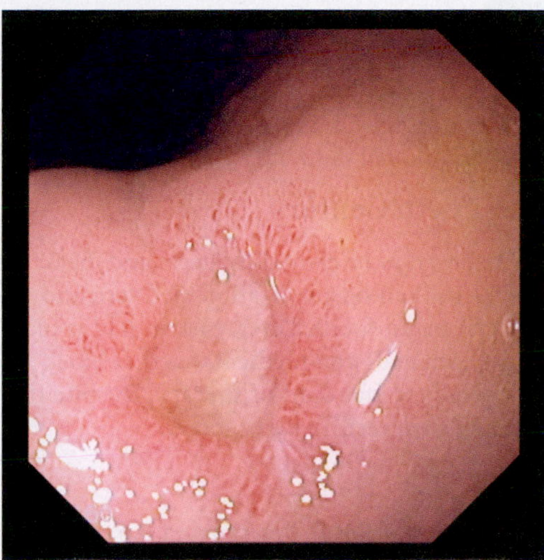

Source: Atlas of Gastrointestinal Endoscopy www.EndoAtlas.com

Peptic ulcer disease has been redefined over the last decade because *H. pylori* is now recognized as a pivotal factor in development of gastric and duodenal ulcers. It is estimated that 92% of duodenal ulcers and 70% of gastric ulcers are caused by *H. pylori.* Recent research indicates that there is also an increased risk of gastric cancer associated with *H. Pylori* infection (Take et al. 2005). Nonetheless, even with this progress in research on PUD, there are still quite a large number of individuals who suffer from ulcer disease and are not infected with *H. pylori.* This section will describe the role of *H. pylori* and other factors that have been correlated to the development of PUD, (see Box 16.3.)

Helicobacter pylori is a spiral-shaped, flagellated, Gram-negative rod that lives under the mucous layer of the stomach and attaches to mucus-secreting cells lining the stomach. These organisms break down urea to produce ammonia, which helps neutralize acid in the immediate vicinity of these bacteria and enhances their survival. The *H. pylori* organisms subsequently produce various proteins that damage mucosal cells, attracting lymphocytes and causing persistent

gastritis—inflammation of the gastric mucosa

achlorhydria—lack of gastric hydrochloric acid secretions

peptic ulcer disease—ulceration or perforation in the lining of the stomach, duodenum, or esophagus

BOX 16.3 HISTORY OF ULCER DIAGNOSIS AND TREATMENT

The road to a cure for ulcers has been a long and bumpy one. Research over the previous decade that indicates that ulcers are caused by a bacterium and can be cured with antibiotics has changed traditional thinking.

Early Twentieth Century

Ulcers are believed to be caused by stress and dietary factors. Treatment focuses on hospitalization, bed rest, and prescription of special bland foods. Later, gastric acid is blamed for ulcer disease. Antacids and medications that block acid production become the standard of therapy. Despite this treatment, there is a high recurrence of ulcers.

1982

Australian physicians Robin Warren and Barry Marshall first identify the link between *Helicobacter pylori* (*H. pylori*) and ulcers, concluding that the bacterium, not stress or diet, causes ulcers. The medical community is slow to accept their findings.

1994

A National Institutes of Health Consensus Development Conference concludes that there is a strong association between *H. pylori* and ulcer disease and recommends that ulcer patients with *H. pylori* infection be treated with antibiotics.

1995

Data show that about 75% of ulcer patients are still treated primarily with antisecretory medications, and only 5% receive antibiotic therapy. Consumer research by the American Digestive Health Foundation finds that nearly 90% of ulcer sufferers are unaware that *H. pylori* causes ulcers. In fact, nearly 90% of those with ulcers blame their ulcers on stress or worry, and 60% point to diet.

1996

The Food and Drug Administration approves the first antibiotic for treatment of ulcer disease.

1997

The Centers for Disease Control and Prevention (CDC), with other government agencies, academic institutions, and industry, launches a national education campaign to inform health care providers and consumers about the link between *H. pylori* and ulcers. This campaign reinforces the news that ulcers are a curable infection and the fact that health can be greatly improved and money saved by disseminating information about *H. pylori*. Medical researchers sequence the *H. pylori* genome. This discovery can help scientists better understand the bacterium and design more effective drugs to fight it.

2005

Nobel Prize for physiology or medicine awarded to Drs. Barry J. Marshall and Robin Warren for proving that bacteria and not stress was the main cause of painful ulcers of the stomach and intestine.

1. Munnangi S. and Sonnenberg A. Time Trends of Physician Visits and Treatment Patterns of Peptic Ulcer Disease in the United States. *Arch Intern Med.* vol. 175, July 14, 1997. pp 1489–1494.

2. *Helicobacter pylori* in Peptic Ulcer Disease, National Institutes of Health Consensus Development Panel on *Helicobacter pylori* in Peptic Ulcer Disease, Journal of the American Medical Association. Volume 272, no. 1. July 6, 1994, pp 65–69.

Source: Centers for Disease Control and Prevention [homepage on the Internet]. Atlanta: Centers for Disease Control and Prevention; 2001 Feb 2 [cited 2005 Oct 13]. History of Ulcer Disease and Treatment. Available from: http://www.cdc.gov/ulcer/history.htm.

inflammation (Baron 2000; Graham et al.1999; Qureshi and Graham 1999). By-products released by the organism result in damage to the epithelium and impair the mucous barrier within the stomach.

The etiology of PUD also involves factors that may decrease mucosal integrity, such as use of NSAIDs (e.g., ibuprofen) or alcohol, excessive glucocorticoid secretion or medication, and factors that decrease the blood supply, such as smoking, stress, or shock. Factors that increase acid secretions, including certain foods, rapid gastric emptying, or increased gastrin secretions, also contribute to the development of PUD. The genetic link to PUD has also been explored; ulcers are approximately three times more common in first-degree relatives than in the general population. This may be related to an increased susceptibility to infection from *H. pylori* (Del Valle et al. 2005).

The most common symptom related to PUD is **epigastric** pain, but this pattern is not consistent. In general, patients will complain of abdominal pain and a burning sensation, which may be precipitated by certain types of foods or accentuated by food intake. For others, epigastric pain may be relieved by food intake due to its ability to dilute any irritants. For a duodenal ulcer, pain characteristically occurs from 90 minutes to 3 hours after eating, and is usually relieved within minutes either by eating or by use of antacids. Unfortunately, partial neutralization of gastric acid is followed by a rebound of gastrin release, causing additional stimulation of HCl and probably more pain (Chan and Leung 2002; Harbison and Dempsey 2005).

epigastric—referring to the upper abdominal region

The presence of blood in stool or vomit may be indicative of active bleeding from the ulcer. Changes in hematological indices such as hemoglobin or hematocrit will also be indicative of active bleeding. If there is an active infection, changes in white blood cell count will be consistent with the inflammatory process.

The same diagnostic procedures that have been discussed earlier in this chapter will allow for a definitive diagnosis. Endoscopy coupled with a tissue biopsy will allow for visualization of the ulcer and confirmation of *H. pylori* infection.

Treatment of peptic ulcer disease associated with *H. pylori* infections includes regimens of three to four medications (triple/quadruple therapy). The most common therapy at present involves a 7- to 14-day course of two antibiotics with bismuth and one of the acid-pump inhibitors. Eradication rates associated with triple/quadruple therapy range from 86% to 98% if patients comply with triple/quadruple therapy treatment regimens (see Table 16.19). However, frequently occurring adverse effects such as nausea, vomiting, and abdominal pain associated with these regimens significantly hinder patient compliance. Approximately 10% of patients discontinue treatment early due to unwanted effects (Chan and Leung 2002; Kiyota et al. 1999; Qureshi and Graham 1999; Soll 1996).

TABLE 16.19

FDA-Approved Treatment Options for Eradication of *H. pylori* Infection

Omeprazole 40 mg QD + clarithromycin 500 mg TID × 2 wks, then omeprazole 20 mg QD × 2 wks

-OR-

Ranitidine bismuth citrate (RBC) 400 mg BID + clarithromycin 500 mg TID × 2 wks, then RBC 400 mg BID × 2 wks

-OR-

Bismuth subsalicylate (Pepto Bismol®) 525 mg QID + metronidazole 250 mg QID + tetracycline 500 mg QID* × 2 wks + H_2 receptor antagonist therapy as directed × 4 wks

-OR-

Lansoprazole 30 mg BID + amoxicillin 1 g BID + clarithromycin 500 mg TID × 10 days

-OR-

Lansoprazole 30 mg TID + amoxicillin 1 g TID × 2 wks**

-OR-

Ranitidine bismuth citrate 400 mg BID + clarithromycin 500 mg BID × 2 wks, then RBC 400 mg BID × 2 wks

-OR-

Omeprazole 20 mg BID + clarithromycin 500 mg BID + amoxicillin 1 g BID × 10 days

-OR-

Lansoprazole 30 mg BID + clarithromycin 500 mg BID + amoxicillin 1 g BID × 10 days

*Although not FDA approved, amoxicillin has been substituted for tetracycline for patients for whom tetracycline is not recommended.

**This dual therapy regimen has restrictive labeling. It is indicated for patients who are either allergic or intolerant to clarithromycin or for infections with known or suspected resistance to clarithromycin.

Source: Centers for Disease Control and Prevention [homepage on the Internet]. Atlanta: Centers for Disease Control and Prevention; 2001 Feb 2 [cited 2005 Oct 13]. *Helicobacter pylori* and Peptic Ulcer Disease. Available from: http://www.cdc.gov/ ulcer/md.htm

TABLE 16.20

Drugs Used in Treatment of Peptic Ulcer Disease

Antibiotics	Metronidazole; tetracycline; clarithromycin; amoxicillin
Antacids	Mylanta, Maalox; Tums; Gaviscon
H_2 blockers	Cimetidine; ranitidine; famotidine; nizatidine
Proton pump inhibitors	Lansoprazole; rabeprazole; esomeprazole; omeprazole; pantoprozole
Stomach-lining protector	Sucralfate; prostaglandin analogue (Misoprostol); bismuth subsalicylate

Other treatment for PUD focuses on use of medications to suppress acid secretion, which will ultimately promote healing of the ulceration (see Table 16.20). These medications include antacids, proton pump inhibitors, histamine blocking agents, prokinetic agents, and mucosal protectants. Because salicylates (aspirin) and NSAIDs are linked to increased gastric irritation, these medications should never be taken by someone with PUD.

For those patients who are refractory to treatment or who suffer from complications such as hemorrhage, perforation, or gastric outlet obstruction, surgical resection may be warranted (this will be discussed later in this chapter).

Nutrition Therapy for PUD

Nutritional Implications For patients with PUD, symptomatic abdominal pain can impair oral intake and result in weight loss and/or nutrient imbalances.

Nutrition Interventions For several decades, dietary factors have gained and lost favor as a significant component in both the cause and treatment of peptic ulcers. Currently, goals for nutrition therapy include supporting medical treatment, maintaining or improving nutritional status, and providing a diet that minimizes symptoms of PUD.

Current nutrition therapy for PUD restricts only foods known to increase acid secretion or cause direct irritation to gastric mucosa. These foods include black and red pepper, caffeine, coffee (including decaffeinated), and alcohol. Additionally, it is recommended that patients avoid any foods they do not individually tolerate. Historically, milk and cream were used to treat PUD, but it is now known that their consumption increases both gastrin and pepsin secretion. Furthermore, pH of a food prior to its consumption has little effect after it is consumed. Restricting acidic juices or other foods is not warranted unless the patient identifies intolerance to them.

Other components of NT will include timing and size of meals. Patients should not lie down after eating and avoid eating large meals close to bedtime. Smaller, more frequent meals may be better tolerated, but there is some controversy regarding whether this might increase the overall amounts of acid that are secreted.

Evaluation Follow-up for the patient with PUD will focus on adequacy of the patient's nutritional intake and tolerance to the oral diet. Normal nutrition assessment indices will be monitored to ensure maintenance of nutritional status.

Gastric Surgery

When peptic ulcer disease does not respond adequately to medical treatment, or when the patient experiences a complication of PUD, surgery is often the next step. Complications from PUD may include **hemorrhage, perforation,** or **obstruction** of the pyloric sphincter. The surgical procedure chosen is based on the patient's current medical status and prior surgical history.

Vagotomy The purpose of the **vagotomy** is to eliminate the **cholinergic** stimulation to the stomach. Selective vagotomy eliminates innervations from the vagus nerve to parietal cells, resulting in decreased acid production and a decreased response to gastrin. Other functions of the vagus nerve would remain intact, and the normal pathway for gastric emptying and peristalsis would continue. In many patients, total vagotomy with **pyloroplasty** is chosen. In this procedure, innervations to parietal cells are severed, and the portion of the vagus nerve controlling gastric emptying is also eliminated. Pyloroplasty enlarges the pyloric sphincter. Gastric resection is also an option depending on location of the ulcer and extent of the stomach that requires removal. Reconstruction after pyloroplasty or gastric resection will generally use one of three procedures: a gastroduodenostomy (Billroth I), gastrojejunostomy (Billroth II), or Roux-en-Y procedure (see Figure 16.9).

Gastroduodenostomy (Billroth I); Gastrojejunostomy (Billroth II); Roux-en-Y Procedure In the procedure gastroduodenostomy, or Billroth I, a partial **gastrectomy** or pyloroplasty is performed with a reconstruction that consists of an **anastomosis** of the proximal end of the duodenum to the distal end of the stomach. A gastrojejunostomy, or Billroth II, is a partial gastrectomy with a reconstruction that consists of an anastomosis of the proximal end of the jejunum to the distal end of the stomach. In this surgical procedure, a blind loop of the duodenum is created. The Roux-en-Y procedure accomplishes the same thing as the Billroth II but creates a very small pouch after the gastric resection and connects the jejunum to the upper portion of the stomach. Although the Roux-en-Y procedure (or gastric bypass) has most recently featured prominently as a treatment for morbid obesity, it originated as a treatment for PUD and other gastric diseases.

Nutrition Therapy for Gastric Surgery

Nutritional Implications Nutritional risk is due to reduced capacity of the stomach and potential change in gastric emptying and transit time when the normal pathway for digestion and absorption is interrupted. Additionally, when portions of the stomach are resected, valuable components of digestion may be altered or lost. These issues combine to place the patient at significant nutritional risk due to decreased oral intake, maldigestion, and/or malabsorption.

Dumping Syndrome One of the most common complications after gastric surgery is **dumping syndrome.** Dumping syndrome occurs when an increased osmolar load enters the small intestine too quickly from the stomach. Severity of the symptoms varies depending on the extent of gastric surgery and the overall change in gastric emptying. When the stomach is removed or partially resected, important steps in digestion are missed. As discussed earlier, food may remain in the stomach anywhere from 1 to 3 hours as it becomes liquefied and partially digested. It then enters the duodenum via the pyloric sphincter slowly, so the acidic chyme is neutralized by pancreatic bicarbonate. When the pyloric portion of the stomach is removed, bypassed, or destroyed, the rate of gastric emptying is increased. Additionally, when the duodenum

hemorrhage—bleeding

perforation—a break in the integrity of the tissue

obstruction—blockage

vagotomy—severing of the vagus nerve; often a component of gastric surgery

cholinergic—resembling acetylcholine; stimulated by or releasing acetylcholine or a related compound

pyloroplasty—enlarging the pyloric sphincter

gastrectomy—surgery to resect a portion of or the entire stomach

anastomosis—the surgical connection of body parts, especially hollow tubular parts like those of the GI tract

dumping syndrome—a group of symptoms that occurs with rapid passage of large amounts of food into the small intestine; symptoms include dizziness, sweating, decreased blood pressure, and diarrhea

FIGURE 16.9 Gastric Surgeries. A. Roux-en-Y gastric bypass; B. Billroth I; C. Billroth II

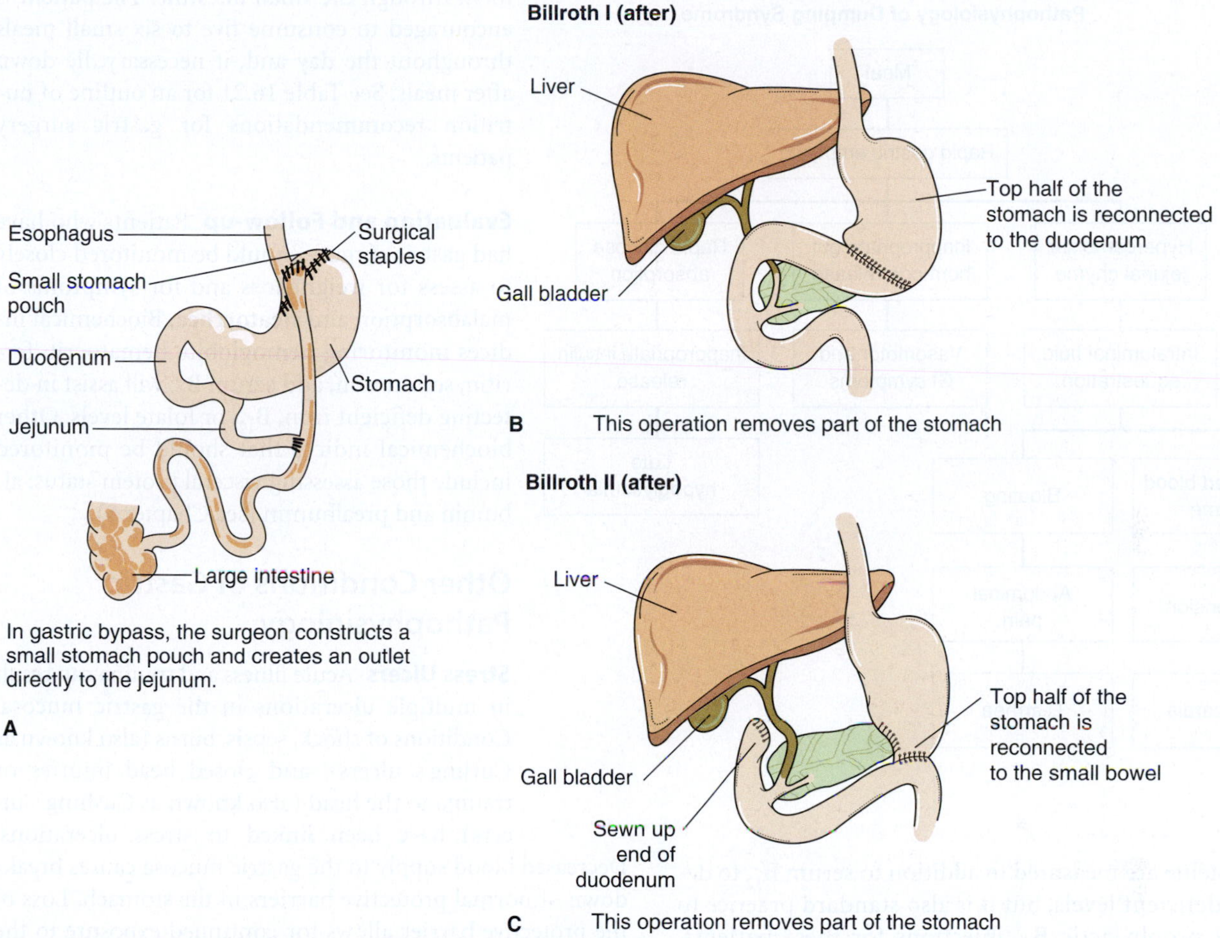

Source: (a) John E. Pandolfino, Brintha Krishnamoorthy, Thomas J. Lee. Gastrointestinal Complications of Obesity Surgery, Medscape General Medicine 6(2), 2004. http://www.medscape.com/viewarticle/471952; (b, c) Diagrams reproduced with permission from CancerHelp UK, a free information service about cancer and cancer care for people with cancer and their families. It is brought to you by Cancer Research UK. www.cancerhelp.org.uk

is bypassed, feedback inhibition is lost. Furthermore, surgery will affect the release of hormones, enzymes, and other secretions. If this process is altered (as it is in gastric resections), food "dumps" into the small intestine (see Figure 16.10). Because the chyme is hyperosmolar, fluid is drawn into the small intestine from the vascular compartment in an attempt to dilute intestinal contents. These processes result in cramping, abdominal pain, hypermotility, and diarrhea. Furthermore, fluid changes in the vascular compartment result in dizziness, weakness, and tachycardia. These symptoms constitute what is generally referred to as "early" dumping syndrome, which actually occurs within 10 to 20 minutes after eating. "Intermediate" dumping syndrome occurs approximately 20 to 30 minutes after eating. As foodstuffs enter the colon, fermentation and action of microflora cause the production of gas, abdominal pain, cramping, and diarrhea. "Late" dumping syndrome can occur anywhere from 1 to 3 hours after eating, and is especially common after consuming simple carbohydrates. In this situation, rapid absorption in the small intestine stimulates insulin release. After quick movement and absorption of food through the small intestine, there is no longer any substrate for the insulin to act upon. This results in **hypoglycemia** and its symptoms of shakiness, sweating, confusion, and weakness.

Additional nutrition concerns include the potential for vitamin and mineral deficiencies. With changes in gastric anatomy, there may be a lack of intrinsic factor secreted. This would prevent normal B_{12} absorption and lead to a subsequent deficiency. Research has confirmed that patients who have had gastric surgery have a high prevalence of vitamin B_{12} deficiency. Treatment of the deficiency can prevent cardiovascular, hematologic, and neurologic abnormalities seen with B_{12} deficiency or pernicious anemia (Sumner et al. 1996). Levels of methylmalonic acid and

hypoglycemia—a low serum glucose; generally considered to be <70 mg/dL

FIGURE 16.10 Dumping Syndrome

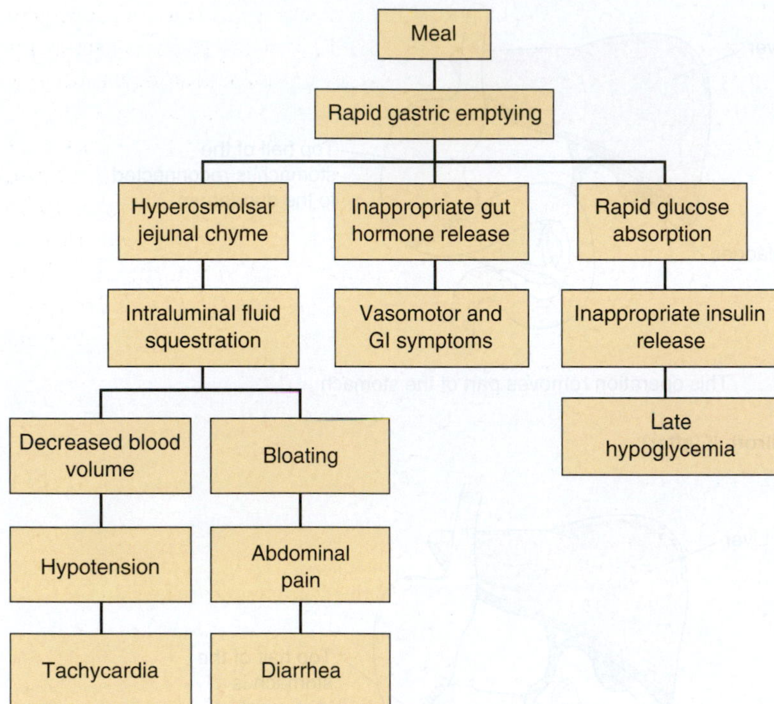

homocysteine are measured in addition to serum B_{12} to determine deficient levels, but it is also standard practice to prescribe prophylactic B_{12} injections for these patients. Iron deficiency is also common. The cause is multifactorial, including a decrease in HCl, decreased dietary intake, and possible malabsorption. Risk of osteoporosis is also increased due to decreased absorption of calcium. It is recommended that both calcium and vitamin D supplements be prescribed for these patients.

Nutrition Interventions Nutrition therapy can prevent and treat many complications of gastric surgery. The post-gastrectomy or "anti-dumping" diet encourages a well-balanced diet slightly higher in protein and fat than what is recommended by the U.S. Dietary Guidelines. Simple sugars are avoided in order to prevent some hyperosmolality and hypoglycemia associated with the dumping syndrome. Lactose is often not tolerated. If the patient is lactose intolerant, commercial products that provide lactase or are lactose-free can be recommended. This is an additional reason to recommend calcium and vitamin D supplements. Liquids should be consumed between meals to prevent their contribution to dumping syndrome, because they facilitate quick movement through the small intestine. The patient is encouraged to consume five to six small meals throughout the day and, if necessary, lie down after meals. See Table 16.21 for an outline of nutrition recommendations for gastric surgery patients.

Evaluation and Follow-up Patients who have had gastric surgery should be monitored closely to assess for weight loss and for symptoms of malabsorption and **steatorrhea**. Biochemical indices monitoring hemoglobin, hematocrit, ferritin, serum iron, and serum B_{12} will assist in detecting deficient iron, B_{12}, or folate levels. Other biochemical indices that should be monitored include those assessing visceral protein status: albumin and prealbumin (see Chapter 5).

steatorrhea—excessive fat in the feces

Other Conditions of Gastric Pathophysiology

Stress Ulcers Acute illness and trauma can result in multiple ulcerations in the gastric mucosa. Conditions of shock, sepsis, burns (also known as Curling's ulcers), and closed head injuries or trauma to the head (also known as Cushing's ulcers) have been linked to stress ulcerations. Decreased blood supply to the gastric mucosa causes breakdown of normal protective barriers in the stomach. Loss of the protective barrier allows for continued exposure to the highly acidic gastric juices. Ulcers generally develop very early in the trauma or shock period.

The most common symptoms are acute hemorrhage or perforation. It is standard practice in the clinical treatment of trauma and sepsis to provide preventive care with the use

TABLE 16.21

Nutrition Interventions after Gastric Surgery
Initially avoid all simple sugars. Do not start clear liquids as first oral feeding.
The first meals should consist of protein, fat, and complex carbohydrate, but with only one or two food items at a time. Patients may be initially lactose intolerant.
Slowly progress to five or six small meals each day with each containing a protein source such as eggs, meat, poultry, fish, milk, yogurt, cottage cheese, cheese, peanut butter, dried beans, lentils, or tofu.
Consume liquids 30 minutes to 1 hour after consuming solid food.
Lie down after eating.
Consider addition of functional fibers to delay gastric emptying and assist with treatment of diarrhea.

Source: Nahikian-Nelms, M. Gastrointestinal Disease. Nutrition Care Manual. Chicago IL: American Dietetic Association, 2005. © 2005 American Dietetic Association. Adapted with permission.

of a continuous infusion of H_2 blockers or liquid antacids every 2 to 3 hours.

Zollinger-Ellison Syndrome Zollinger-Ellison (ZE) syndrome is a condition of gastric acid hypersecretion. Symptomatically, ZE is initially similar to PUD, but is typically unresponsive to standard therapy. Hypersecretion is caused by presence of a non-B-cell endocrine tumor, or what is commonly called a gastrinoma. Zollinger-Ellison syndrome can be distinguished by measuring serum gastrin levels. In ZE patients, gastrin levels are >150 to 200 pg/mL compared to the normal levels of <150 pg/mL. Treatment consists of the use of proton pump inhibitors or surgical resection for those unresponsive to medical intervention (Chan and Leung 2002; Jensen and Fraker 1994).

Conclusion

A thorough understanding of the pathophysiology involving the upper gastrointestinal tract is a fundamental component of nutrition therapy. This chapter has covered numerous physical signs and symptoms—such as dysphagia, nausea, and vomiting—that are a concern for the hospitalized patient with any number of conditions. Additionally, this chapter has covered multiple diagnoses, conditions, and symptoms that contribute to significant nutritional risk and cause frequent nutrition problems. For many, nutrition therapy is the major form of medical treatment. The role of the registered dietitian in planning, implementing, and evaluating nutrition therapy for the upper gastrointestinal tract should be of primary importance.

PRACTITIONER INTERVIEW: PRACTITIONER'S PERSPECTIVE ON BARIATRIC SURGERY

Valerie Simler, M.S., R.D., C.D.E., *Bariatric Surgery Program Coordinator*
Valley Care Medical Center, Pleasanton, California

Our facility predominantly performs Roux-en-Y (RNY) gastric bypass surgeries, but also performs laparoscopic gastric banding. Common nutritional problems after surgery can be divided into early (0–3 months), post op, and long term (3 months and beyond). In the early post-op period, patients are put on an all-liquid diet. They are unable to tolerate solid food well and have an increased risk of obstruction while there is still swelling from the surgery. The protocol that I recommend is clear, then full, liquids over the first month, then soft diet for the second month, and finally a return to solid foods in the third month. The most common problem immediately after RNY is an inability to consume adequate fluids. It takes considerable effort on the part of the patient to drink enough to stay well hydrated, when all liquids must be swallowed very slowly to prevent nausea and vomiting. Often getting enough protein is challenging in the first several months. Patients are instructed to use liquid protein supplements until they are able to consume enough protein in their meals. For many, they will always need to include a protein supplement as the meal capacity remains too small to enable them to eat enough protein as well as a variety of foods from other food groups. Some patients experience chronic nausea for a few weeks, but that is not really common. Vomiting is always a sign that something is wrong, either medically or eating/drinking too fast or too much or the wrong texture. Some will have problems with early constipation until the body adapts to the new physiology and fiber is reintroduced. After about 3 months, most patients have really good food tolerance and are more easily able to meet their protein and fluid needs. It is always challenging to get protein in most, if not all, meals. If they take the vitamin and mineral supplements as recommended, only a few patients will still experience deficiencies that require more substantial supplementation (iron, B_{12}, calcium). Patients may have varying intolerances to certain foods (beef, lactose, and so on) but this is not usually a significant problem.

The standard dietary recommendations after food tolerance has been achieved are:

1. Eat three meals each day. No snacking except protein supplements. Meal size will start at 2 to 3 oz, but will enlarge over the first 6 to 8 months.
2. Eat slowly (20 to 30 min/meal), and stop at the first sign of fullness. Extra bites are likely to cause nausea and/or vomiting.
3. Choose healthful foods (same general recommendations as for everyone), but make sure each meal has a significant protein source.
4. Don't drink anything with meals. Wait 60 minutes or longer before drinking after finishing the meal. (Only wait 30 minutes during the second month, while on soft diet.) Choose only low-sugar beverages.
5. Read labels for protein, sugar, and fat. Choose foods and supplements that will contribute substantially to your protein goal. Limit saturated fat as is appropriate for everyone. Limit sugar intake to <5 g per meal or supplement to avoid dumping syndrome.
6. Take your multivitamin, calcium, and B_{12} supplements daily. Have your labs checked regularly to assess for nutritional adequacy.
7. Exercise daily for at least 30 minutes.

Would I recommend the surgery? Of course there are surgical risks, which should be investigated thoroughly and not taken lightly. For many obese individuals, it seems like the surgical risks are worth taking, given their medical risks living with obesity. Even for healthy obese people, quality of life is an important consideration. If your health is good and you are not unhappy being obese, why take the risk? But if you are unhappy, self-conscious, unable to do what you would really like to be doing in your life because of your weight, then I think it is worth it. Almost every single patient I have worked with—even those who have some complication, or find that surgery did not cure all their problems—still says it has been worth all the hard work and sacrifices. Even though there are significant challenges to keeping the weight off permanently, most people say that is better than what they were dealing with before the surgery. Struggling to keep off 20 to 30 lbs puts them on the same playing field with most other Americans, and that is someplace they say they want to be.

This is an amazing, satisfying career choice. People who come to me for help in achieving success with weight loss surgery are usually very motivated individuals. Not only that, they are also very hopeful and optimistic about the outcome of their decision. This gives me a unique opportunity to provide education at a time when it is likely to be well received. I try to give them a very realistic perspective of their choice of surgery to achieve weight loss: what the role of the surgery is *and* what their personal role is in maximizing their weight loss and keeping the weight off permanently.

One of the real challenges working in the field of bariatric surgery is that there is not an abundance of "evidence-based scientific research" on the nutrition component of this procedure. While we fight for more studies to be conducted, we use what literature is available, as well as our own experience and that of other dietitians specializing in bariatric surgery. There is a real move right now for creating some standardized nutrition protocols that hopefully will be endorsed by both the American Dietetic Association and the American Society of Bariatric Surgery. This is likely to take awhile, so in the meantime, we continue to do the best we can to help our patients of today.

Just recently, I have been promoted to the role of bariatric surgery program coordinator. This job has traditionally been held by an RN, and this is still the case at most institutions. I feel that a dietitian can do this job equally well as long as a nurse is involved in the multidisciplinary team. My role is changing by this promotion, in that I am involved in more administrative functions of the program now. I have other dietitians to do some of the patient care, but I will continue to teach all of the classes and facilitate the support groups. This job enables me to help patients move through our program more effectively—completing their screening appointments, meeting their prerequisites for surgery, getting insurance authorization, and so on. I will be conducting in-service presentations for hospital staff that have a role in caring for our surgery patients. I will also take a more active role in the marketing of our program within the community (advertisements as well as community presentations). I will act as a liaison between the bariatric program and the individual surgeon's offices, as well as the administrative staff of the hospital.

CASE STUDY

Introduction:

Sara Flores is a 45-year-old Hispanic female who has a 5-year history of gastroesophageal reflux disease. Mrs. Flores presented to her physician with complaints of chronic indigestion and increased abdominal pain. She is now s/p endoscopy, which revealed a 2 cm. duodenal ulcer and generalized gastritis. Biopsy positive for *Helicobacter pylori.* She was prescribed a 14-day course of Bismuth subsalicylate—525 mg QID, metronidazole—250 mg QID, tetracycline—500 mg QID. Omeprazole 20 mg BID × 28 days.

Nutrition Assessment:

Ht 5′2″ Wt 110# UBW 145#

Labs: Total Protein 5.9 g/dL, Albumin 3.4 g/dL, Prealbumin 22 mg/dL, Hgb 11.5 g/dL, Hct 36%.

The registered dietitian's interview indicates that the patient describes appetite as poor. States that she is afraid to eat because it makes the pain worse. Specific food intolerances include anything fried or "spicy," coffee, and chocolate. Patient relates her usual weight to be about 145 pounds. The last time she weighed was 6 weeks ago. Her admission weight is 110 pounds.

Usual dietary intake (prior to current illness):

AM: coffee, dry toast. On weekends, cooks large breakfasts for family, which includes omelets, rice or grits, or pancakes, waffles, fruit. Lunch: sandwich from home, fruit, cookies. Dinner: rice, some type of meat, fresh vegetables, coffee. Has previously consumed 8 to 10 cups coffee daily. Drinks one to two sodas each day.

Questions:

1. Mrs. Flores' endoscopy indicated that her biopsy was positive for *Helicobacter pylori.* What is this and how is it related to her duodenal ulcer?
2. What admission laboratory values are abnormal? Interpret their significance in relationship to both her diagnosis and nutritional status.
3. This patient was prescribed four different medications. How do each of these work? What are the current recommendations for treatment of *Helicobacter pylori* infection? Are there any drug-nutrient interactions that need to be addressed?
4. Identify at least two nutrition problems that can be found as a result of the nutrition assessment and medical history. Next, identify the etiology of each nutrition problem. Finally, identify the signs and symptoms that support the evidence for these nutrition problems.

NUTRITION CARE PROCESS: UPPER GASTROINTESTINAL TRACT

Step One: Assessment

Medical/Social History

Diagnoses

Medications

Previous medical conditions or surgeries

Socioeconomic status/food security

Support systems

Education level—primary language

Anthropometric

Height

Current weight

Weight history: highest adult weight; usual body weight

Reference weight (BMI)

Biochemical Assessment
Visceral(Transport) Protein Assessment:

Albumin

Prealbumin

Hematological Assessment:

Hemoglobin

Hematocrit

Lipid Assessment

Total Cholesterol

HDL

LDL

Triglyceride

Dietary Assessment

Ability to chew; use and fit of dentures

Problems swallowing

Nausea, vomiting

Constipation, diarrhea

Heartburn

Any other symptoms interfering with ability to ingest normal diet

Ability to feed self

Ability to cook and prepare meals

Food allergies, preferences, or intolerances: spicy foods, high-fat foods, pepper, caffeine, coffee, tea, alcohol, spearmint, peppermint, chocolate

Previous food restrictions

Ethnic, cultural, and religious influences

Use of alcohol, vitamin, mineral, herbal, or other type of supplements

Previous nutrition education or nutrition therapy

Eating Pattern: 24-hour recall, food history, food frequency

Step Two: Common Diagnostic Labels

Inadequate oral food/beverage intake

Inadequate fluid intake

Increased nutrient needs

Swallowing difficulty

Chewing difficulty

Altered GI function

Involuntary weight loss

Food and nutrition-related knowledge deficit

Sample: Chewing difficulty (NC-1.2)

PES: Chewing difficulty related to recent jaw fracture and postoperative status for maxillomandibular fixation resulting in inability to consume food of normal texture and consistency.

Step Three: Sample Intervention

1. Modify texture, viscosity, and consistency of all foods to assure passage of blenderized diet through opening between teeth left after surgical repair.
2. Provide nutrition education regarding selection, preparation and storage of foods and supplements, and selection and use of feeding equipment.
3. Recommend multivitamin daily.
4. Maintain adequate hydration by consuming a minimum of 3000 mL of fluid daily.

Step Four: Monitoring and Evaluation

1. Monitor weight and nutrition assessment indices.
2. Evaluate 24-hour recall for nutrition adequacy.

WEB LINKS

National Digestive Diseases Information Clearinghouse: A service of the National Institute of Diabetes and Digestive Diseases. This government resource provides basic information and statistics on diagnoses of all digestive tract diseases.

http://digestive.niddk.nih.gov

Directory of Digestive Disease Organizations for Patients: Contact information for patient advocate, nonprofit organizations serving those individuals with diagnoses of digestive disease.

http://digestive.niddk.nih.gov/resources/patient.htm

Digestive Center for Excellence, University of Virginia Health System: This site provides excellent summaries of current research and patient education materials.

http://www.healthsystem.virginia.edu/internet/digestive-health

International Foundation for Functional Gastrointestinal Disorders (IFFGD) Inc.: Nonprofit education and research organization publishing several quarterly newsletters and patient education pamphlets.

http://www.iffgd.org

Pediatric/Adolescent Gastroesophageal Reflux Association Inc. (PAGER): Provides information on pediatric gastroesophageal reflux and related disorders.

http://www.reflux.org

END-OF-CHAPTER QUESTIONS

1. Define and describe the four basic functions of the GI tract as discussed in this chapter.
2. Considering the basic functions of saliva, what are the possible consequences of xerostomia?
3. An imbalance of pressure at the lower esophageal sphincter (LES) may result in the symptoms associated with gastroesophageal reflux disease. What factors may affect LES pressure?
4. Explain the potential nutritional and metabolic consequences of prolonged vomiting.
5. Peptic ulcer disease (PUD), in many cases, is linked to an infection. What is the origin of this infection, and how is it treated?
6. Identify three major goals for nutrition interventions to assist in the control of symptoms associated with PUD.
7. Complications of peptic ulcer disease may result in surgical resection. What are the physiological consequences of gastric resection?
8. What are the potential nutritional complications of gastric resection?
9. Explain the symptomatic and etiological differences between early and late dumping syndrome.

NOTE

[1] "Mucus" is a noun; "mucous" is an adjective.

17

Diseases of the Lower Gastrointestinal Tract

Marcia Nahikian Nelms, Ph.D., R.D.

Southeast Missouri State University

CHAPTER OUTLINE

Normal Anatomy and Physiology of the Lower Gastrointestinal Tract

Small Intestine Anatomy • Small Intestine Motility • Small Intestine Secretions • Small Intestine Digestion • Small Intestine Absorption • Large Intestine Anatomy • Large Intestine Motility • Large Intestine Secretions • Large Intestine Digestion and Absorption

Pathophysiology of the Lower Gastrointestinal Tract

Diarrhea • Constipation • Malabsorption • Celiac Disease • Irritable Bowel Syndrome • Inflammatory Bowel Disease • Diverticulosis/Diverticulitis • Common Surgical Interventions for the Lower GI Tract • Short Bowel Syndrome • Bacterial Overgrowth

Introduction

An exploration of pathophysiology affecting the small and large intestine quickly reveals that nutrition therapy is the foundation of treatment for many of these diagnoses. For some of these, such as celiac disease, nutrition therapy is the only treatment. Additionally, many gastrointestinal symptoms such as diarrhea, constipation, or malabsorption place an individual at significant nutritional risk by impairing adequate or appropriate utilization of nutrients.

The discussion of the upper gastrointestinal tract (mouth, esophagus, and stomach) in Chapter 16 focused on four basic functions: motility, secretion, digestion, and absorption. These four functions are also of primary importance in both the small and large intestine, because more than 98% of all digestion and absorption occurs in the lower GI tract.

Normal Anatomy and Physiology of the Lower Gastrointestinal Tract

The small intestine is composed of three distinct parts: duodenum, jejunum, and ileum. These are not separate compartments, but each does differ in anatomy, motility, secretion, digestion, and absorption.

Small Intestine Anatomy

The anatomy of the small intestine is both unique and highly functional. This anatomy is organized to provide maximum surface area and allow for complete digestion and absorption of most foodstuffs. First, the tissue of the small intestine is circularly folded into what are referred to as the folds of Kerckring. Rising from the mucosal surface are fingerlike projections called villi. On the surface of villi are fine hairlike projections called microvilli. This area is often referred to as the "brush border." The combined features of these anatomical structures increase the surface area of the small intestine to such an extent that it is 600 times greater

FIGURE 17.1 Sphincters of the Gastrointestinal Tract

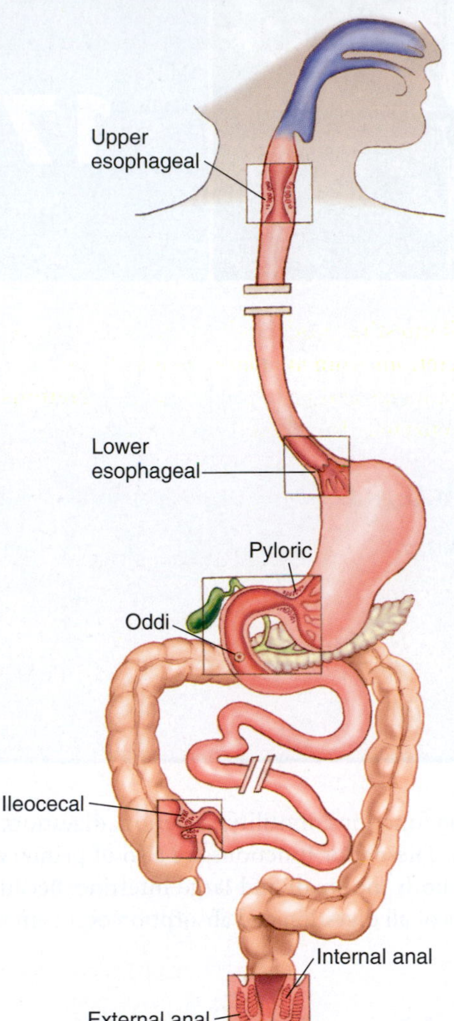

Source: R. Rhoades and R. Pflanzer, *Human Physiology,* 4e, copyright © 2003 p. 693

FIGURE 17.2 Segmentation. Segmentation consists of ringlike contractions along the length of the small intestine. Within a matter of seconds, the contracted segments relax and the previously relaxed areas contract. The oscillating contractions thoroughly mix chyme within the small intestine lumen.

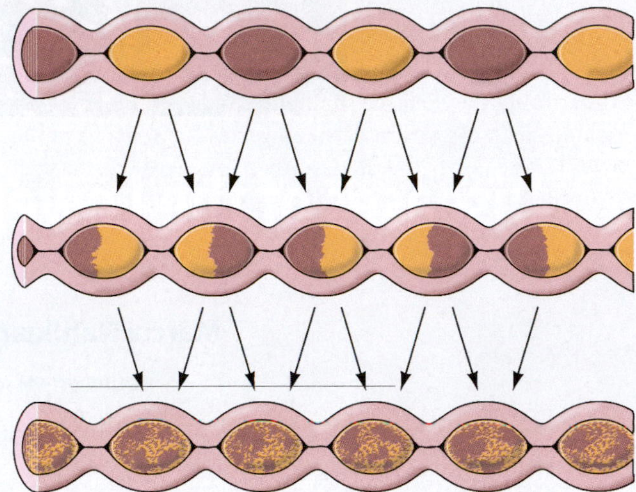

Source: L. Sherwood, *Human Physiology: From Cells to Systems,* 5e, copyright © 2004, p. 623

than it would be if the intestine were a straight, flat tube. Areas between villi are called crypts. These crypts are the location of **stem cells** from which specialized epithelial cells (enterocytes) for the small intestine develop. They migrate up the villi where, after serving their particular physiological function, they are sloughed off and replaced with newly generated enterocytes.

The functional anatomy of the small intestine is strongly influenced by nutritional status and disease. Enterocytes of the small intestine have a high turnover rate and therefore have a high nutrient need. Malnutrition and disease affect the ability of these cells to regenerate and result in decreased villous height. Ultimately, they reduce the ability of the small intestine to perform digestion and absorption. Maintaining nutritional health of the small intestine has been the focus of much nutrition research over the past decade (Reeds 2001).

The small intestine is very adaptive and can adjust its function rather efficiently. More than 50% of the small intestine has to be removed before any significant reduction in its capability is observed. The duodenum and jejunum can perform each other's role in both digestion and absorption. The ileum can also adapt in this way—up to a certain point. The ileocecal sphincter (valve) protects the small intestine from bacteria translocation from the large intestine by remaining closed except during the digestive process (see Figure 17.1). This sphincter also maintains an appropriate transit time to ensure adequacy of both digestion and absorption. This sphincter relaxes when stimulated by either increasing pressure, such as seen in the presence of fluid, or by chemical irritation. At that time, relaxation allows slow movement of remaining digestive contents into the upper portion of the large intestine.

Small Intestine Motility

Motility of the small intestine is controlled by the enteric nervous system and influenced by hormones. This peristaltic reflex was discovered in the early twentieth century. An understanding of the reflex has led not only to an understanding of digestion and absorption but also to researchers' deeper understanding of conditions where motility may be disturbed (Smout 2004).

When food enters the duodenum from the stomach, major movement of foodstuffs (or chyme) evolves from seg-

stem cells—nondifferentiated, primitive cells that have the ability both to multiply and to differentiate into more specialized cells that display unique functions

mental contractions (see Figure 17.2) of the small intestine. This segmentation motility allows mixing of chyme with digestive secretions in the small intestine. In the duodenum, contractions occur approximately every 9 minutes, with a slower rate further down the small intestine. The hormone gastrin initiates the action of segmentation when chyme first enters the duodenum from the stomach. It may take as long as 3 to 5 hours to complete the process of movement through the small intestine (Sherwood 2004).

Additionally, when the small intestine is empty, motility continues with the action of the migrating motility complex (MMC). The MMC, first described in 1969, consists of much weaker contractions that occur approximately every 100 to 150 minutes and serves the purpose of cleaning out the small intestine of any leftover bacteria or waste (Smout 2004). Motilin, a hormone secreted by the small intestine, assists in the control of the MMC. Animal studies are now providing evidence of the role of additional hormones in the regulation of small intestine motility, including cholecystokinin, orexin, and leptin (Ehrstom et al. 2003; Sherwood 2004; Wu et al. 2002).

Motility of the small and large intestine is of much interest due to its possible role in several diseases and its importance in enteral nutrition support (Smout 2004). For example, irritable bowel syndrome, constipation, and diarrhea, which are discussed later in this chapter, are common disorders and may originate from abnormal motility.

Small Intestine Secretions

The small intestine produces its own secretions and also receives secretions from ancillary organs of digestion, including the pancreas and gallbladder. These secretions include digestive enzymes, bicarbonate, and bile.

As chyme moves from the stomach into the duodenum, the hormones cholecystokinin, gastrin, and secretin stimulate release of pancreatic and gallbladder secretions, (see Table 17.1.) Bicarbonate from the pancreas allows neutralization of

TABLE 17.1

Gastrointestinal Hormones

Hormone	Source	Stimuli for Release	Function/Action
Gastrin	G cells in gastric atrium and proximal duodenum	Gastrin-releasing peptide Ingestion of protein, amino acids, peptides, coffee, alcohol, calcium; gastric distension, vagal stimulation, HCl in contact with gastric mucosa	Stimulates: acid secretion, pancreatic HCO_3 secretion, pancreatic enzyme secretion, gallbladder contraction, gastric motility, intestinal motility, insulin release, gastric oxyntic gland mucosa growth, pancreatic growth Relaxes ileocecal sphincter Inhibits gastric emptying
Secretin	S cells in duodenal mucosa	Acid in duodenal lumen	Stimulates: pancreatic HCO_3 secretion, pancreatic enzyme secretion, gallbladder contraction, insulin release, pancreatic growth Inhibits: acid secretion, gastric emptying and gastric motility, intestinal motility, mucosal growth
Cholecystokinin (CCK)	I cells of proximal duodenal mucosa	Nutrients in duodenal lumen, especially fat and to a lesser extent protein	Stimulates: acid secretion, pancreatic HCO_3 secretion, pancreatic enzyme secretion, gallbladder contraction, intestinal motility, insulin release, mucosal growth, pancreatic growth Inhibits: gastric emptying, gastric motility
Glucose-dependent insulinotropic peptide (GIP)	K cells of duodenum and jejunum mucosa	Glucose, amino acids, fatty acids Interdigestive state	Stimulates: insulin release Inhibits: gastric emptying, gastric motility
Motilin	M cells of duodenum and jejunum mucosa	Interdigestive state	Stimulates gastric motility between meals

References: Gropper SS, Smith JL, and Groff JL: *Advanced Nutrition and Human Metabolism,* 4th ed., Belmont, CA: Thomson Wadsworth, 2005; Rhoades R and Pflanzer R: *Human Physiology,* 4th ed., Belmont, CA: Thomson Learning, Inc, 2003; Sherwood L: *Human Physiology: From Cells to Systems,* 5th ed., Belmont, CA: Thomson Learning, Inc, 2004.

the very acidic chyme as it enters from the stomach. Neutralization protects the duodenum from acidity and allows for a more favorable environment for both digestion and absorption. Bile from the gallbladder supplies emulsification needed for adequate lipid digestion.

Other secretions of the small intestine include approximately 1.5 liters of intestinal "juices" or *succus entericus*. These secretions, which are primarily water and mucus, provide the appropriate water-soluble environment for digestion and provide protection to the mucosa of the small intestine. Important digestive enzymes are secreted at the brush border of the small intestine and will be discussed in the next section.

Small Intestine Digestion

Pancreatic juices provide the primary digestive enzymes in the small intestine. These enzymes include trypsinogen, chymotrypsinogen, procarboxypeptidases, and elastase, which are all involved in protein digestion. Pancreatic amylase is the primary enzyme involved in starch digestion. Pancreatic lipase and colipase accomplish the largest proportion of lipid digestion.

Brush border enzymes in the small intestine include lactase, alpha-dextrinase, sucrase, maltase, and glucosidase, which provide final digestion of all carbohydrates (see Figure 17.3). Other brush border enzymes include enterokinase, which activates the pancreatic enzyme trypsinogen. Trypsin then activates other trypsinogen molecules and pancreatic proenzymes. Together, these enzymes degrade protein into smaller units (oligopeptides of two to six amino acids and free amino acids). Peptidases, located in the brush border, are responsible for digestion of oligopeptides into free amino acids, dipeptides, and tripeptides that can then be absorbed (see Figure 17.4).

Small Intestine Absorption

The anatomy of the small intestine, as discussed earlier, is uniquely constructed to accomplish maximal digestion and absorption. Each villus contains access to the circulatory and lymphatic systems via capillaries and lymphatic vessels, and villi thus provide necessary routes for absorbed nutrients.

Absorption for end products of digestion occurs primarily through active transport and may utilize a **Na^+/K^+ pump** system at the brush border. Glucose, galactose, and amino acids utilize this type of absorption mechanism. Fructose uses facilitated diffusion as its absorptive mechanism (see Figure 17.5). Recent research indicates some nutrients, including glucose, may be absorbed in part through a paracellular route. This means that small amounts of nutrients may leak between epithelial cells. Other recent research has indicated the amount of glucose absorbed may be directly related to motility of the duodenum (Schwartz et al. 2002).

Lipid absorption is much more difficult due to its insolubility in water. For lipids to be successfully absorbed, they must undergo several steps and utilize several protein carriers. Fatty acids and other lipid components must be first incorporated into micelles in the gut lumen before absorption into the enterocytes. Then, in order for them to be absorbed in the enterocytes, they must be incorporated into chylomicrons (a type of lipoprotein). Chylomicrons then enter the lymphatic system via passive absorption (see Figure 17.6). Due to this complex absorption mechanism, many diseases of the small intestine can interrupt normal fat digestion and absorption. Understanding steps of fat digestion and absorption can assist in differentiation of diseases affecting the small intestine or ancillary organs of digestion. For example, pancreatic disease may reduce the amount of pancreatic lipase required for adequate digestion. Crohn's disease may decrease transit time and reduce the ability of the small intestine to accomplish all the steps required for digestion and absorption. **Steatorrhea** is the condition that exists when lipid is not digested or absorbed correctly. The result is an abnormal amount of fat in the stool (see Fat Malabsorption for further discussion).

As mentioned earlier, portions of the small intestine can adapt to absorb most nutrients. However, in a normal, healthy individual, most nutrients will be absorbed in the duodenum and jejunum (see Figure 17.7 for sites of absorption). The ileum can also accommodate absorption of many nutrients if foodstuffs remain there long enough (Gropper, Smith, and Groff 2004). One exception to this is the absorption of B_{12}, which can only occur at specific receptor sites in the ileum.

The ileum is also the primary site for reabsorption of bile acids. This process is referred to as the enterohepatic circulation of bile acids. Some bile acids may also be reabsorbed in the jejunum and colon. Since bile acids are exclusively produced in the body, these substances need to be recirculated back to the liver in order to maintain an adequate body pool of approximately a total of 4 grams. Bile acids recirculate from the small and large intestine back to the liver approximately six to eight times per day. When disease interrupts enterohepatic circulation, fat malabsorption can occur. Furthermore, in hypercholesterolemia, the medication cholestyramine is used to bind bile acids in order to decrease the body pool and reduce serum levels of cholesterol.

Large Intestine Anatomy

The anatomy of the large intestine has both significant differences from and important similarities to the anatomy of

Na^+/K^+ pump—the enzyme-based mechanism that moves potassium ions into and sodium ions out of a cell by active transport

steatorrhea—excess fat in the stool (> 6 g/24 hrs)

FIGURE 17.3 Carbohydrate Digestion and Absorption

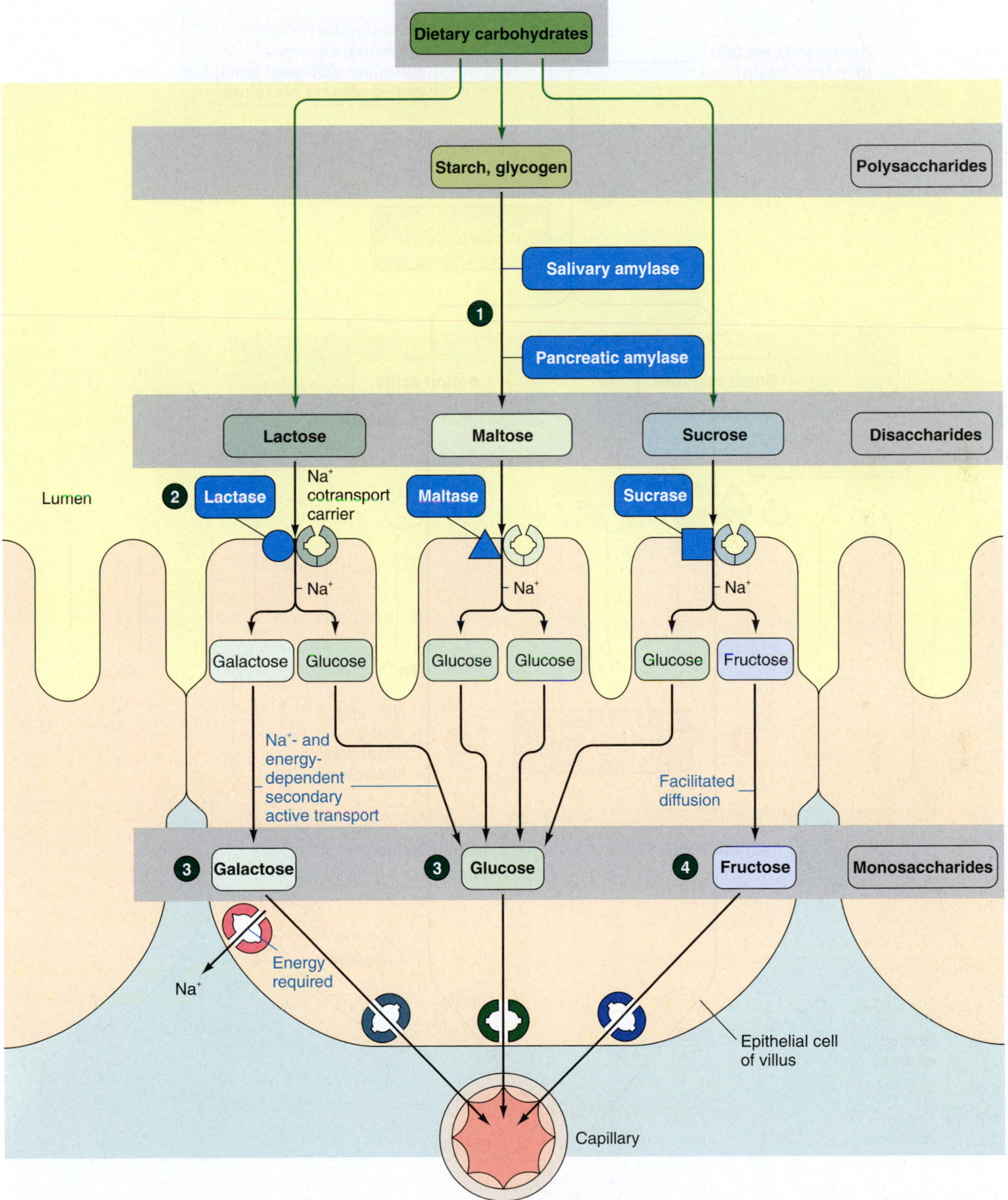

1. The dietary polysaccharides starch and glycogen are converted into the disaccharide maltose through the action of salivary and pancreatic amylase.
2. Maltose and the dietary disaccharides lactose and sucrose are converted to their respective monosaccharides by the disaccharidases (maltase, lactase, and sucrase) located in the brush borders of the small-intestine epithelial cells.
3. The monosaccharides glucose and galactose are absorbed into the interior of the cell and eventually enter the blood by means of Na^+- and energy-dependent secondary active transport.
4. The monosaccharide fructose is absorbed into the blood by passive facilitated diffusion.

Source: L. Sherwood, *Human Physiology: From Cells to Systems,* 5e, copyright © 2004, p. 630

FIGURE 17.4 Protein Digestion and Absorption

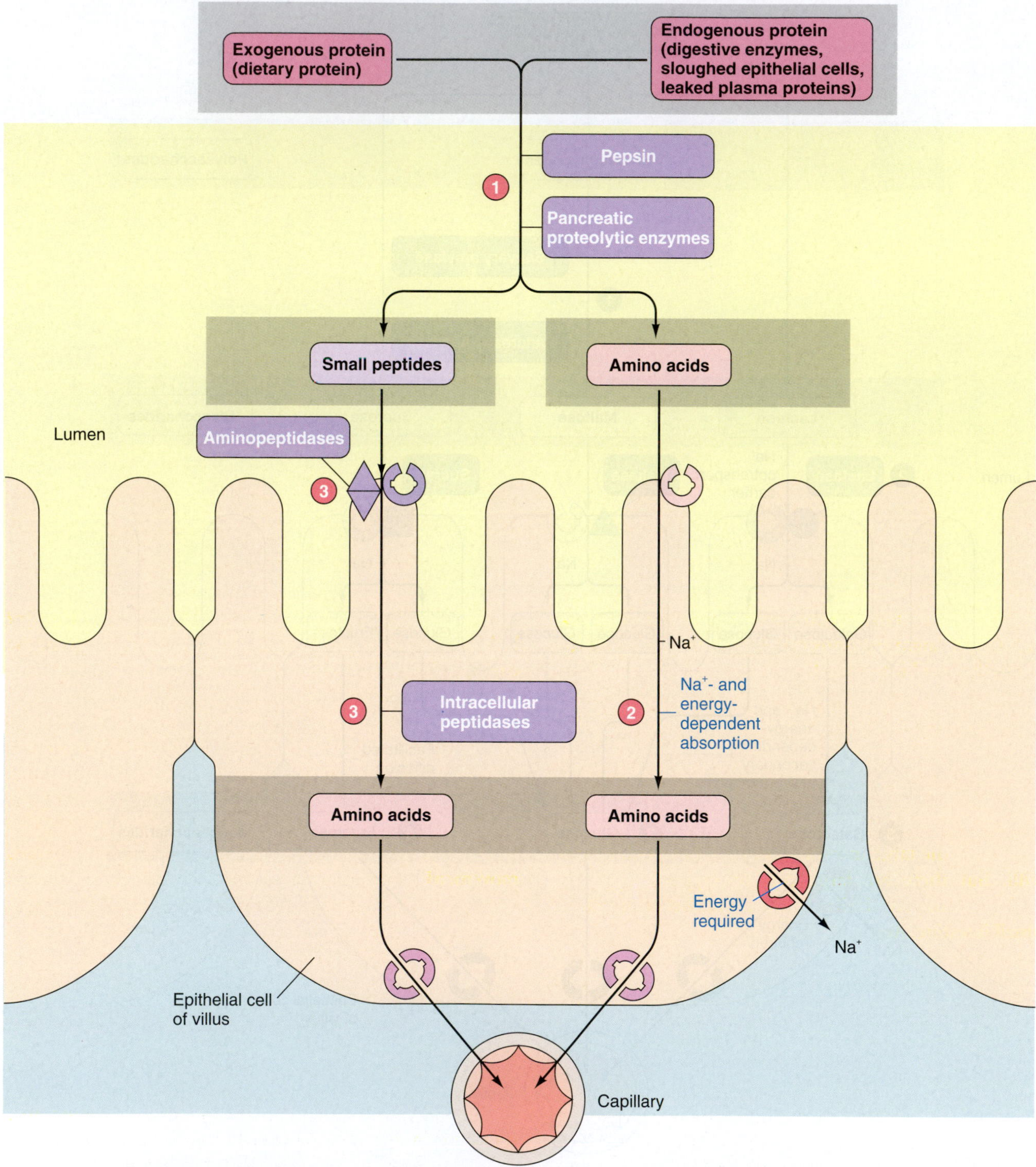

1. Dietary and endogenous proteins are hydrolyzed to their constituent amino acids and a few small peptide fragments by gastric pepsin and the pancreatic proteolytic enzymes.
2. Amino acids are absorbed into the small-intestine epithelial cells and eventually enter the blood by means of Na^+- and energy-dependent secondary active transport. Various amino acids are transported by carriers specific for them.
3. The small peptides, which are absorbed by a different type of carrier, are broken down into their amino acids by aminopeptidases in the epithelial cells' brush borders or by intracellular peptidases.

Source: L. Sherwood, *Human Physiology: From Cells to Systems*, 5e, copyright © 2004, p. 631

FIGURE 17.5 Absorption of Nutrients

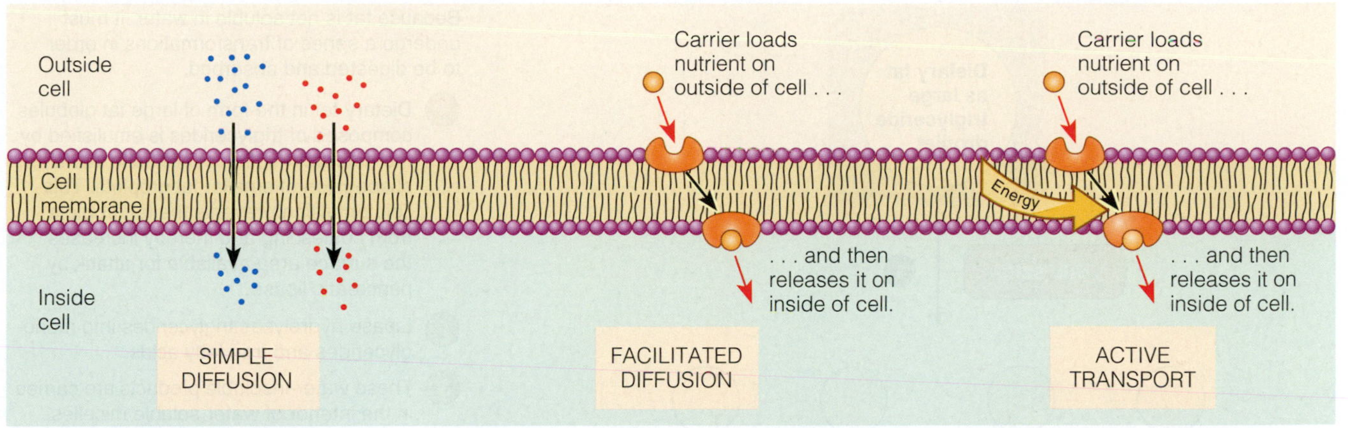

Some nutrients (such as water and small lipids) are absorbed by simple diffusion. They cross into intestinal cells freely.

Some nutrients (such as the water-soluble vitamins) are absorbed by facilitated diffusion. They need a specific carrier to transport them from one side of the cell membrane to the other. (Alternatively, facilitated diffusion may occur when the carrier changes the cell membrane in such a way that the nutrients can pass through.)

Some nutrients (such as glucose and amino acids) must be absorbed actively. These nutrients move against a concentration gradient, which requires energy.

Source: E. Whitney and S. Rolfes, *Understanding Nutrition,* 10e, Copyright © 2005, p. 84

the small intestine. First, mucosa of the large intestine form three nearly straight portions rather than the circular folds found in the small intestine. The colon's major portions are referred to as the ascending colon, the transverse colon and the descending colon (see Figure 17.8). The final section of the colon is referred to as the sigmoid colon due to its "S" shape. The sigmoid colon ends in the rectum where another sphincter (the anal sphincter) controls voluntary release of intestinal contents.

Secondly, the large intestine does not have villi or microvilli. But there are large pits or crypts (Crypts of Lieberkuhn) that are similar to the crypts between villi in the small intestine. Again, similar to the small intestine, cells are generated within these crypts and, after migration, differentiate into specialized epithelial cells such as goblet cells, which produce mucus.

Large Intestine Motility

Differences in musculature of the large intestine provide the basic structure that supports motility of the large intestine. The large intestine has repeating bands of longitudinal skeletal muscle (called taeniae coli) that follow the length of the colon and circular smooth muscle that covers the entire organ.

Motility of the large intestine can be categorized into several distinct types. In the small intestine, motility includes segmentation, which allows for mixing of intestinal contents. Likewise, the large intestine also uses a type of segmentation, called haustration. Haustration occurs when circular muscle forms small sacs called haustra. Haustra hold amounts of chyme as it is mixed with secretions of the colon. Haustra can form and then disappear when intestinal contents are moved through the colon (see Figure 17.8). Other types of movement that accomplish motility include propulsion, mass movements, and defecation. Propulsion is accomplished by alternating waves of relaxation and contraction of smooth muscle lasting for several minutes. Intestinal contents can move in both directions within the colon, allowing for adequate absorption of fluid and electrolytes. Mass movements occur when there is a significant contraction of a large portion of the colon. This generally occurs several times a day and will accomplish moving a large portion of intestinal contents along the colon. Finally, defecation occurs when distention of the rectum relaxes the anal sphincter. This final movement is ultimately (usually) under voluntary control.

Large Intestine Secretions

Compared to the small intestine, the large intestine produces relatively few secretions. As mentioned earlier, goblet cells produce mucus that serves to protect the epithelium and assists in formation of feces. Potassium and bicarbonate are both released in the large intestine, and they play a role in the electrolyte and fluid absorption that occurs there.

Large Intestine Digestion and Absorption

No enzymatic digestion occurs in the large intestine. In normal, healthy individuals, digestion has already been

FIGURE 17.6 Lipid Digestion and Absorption

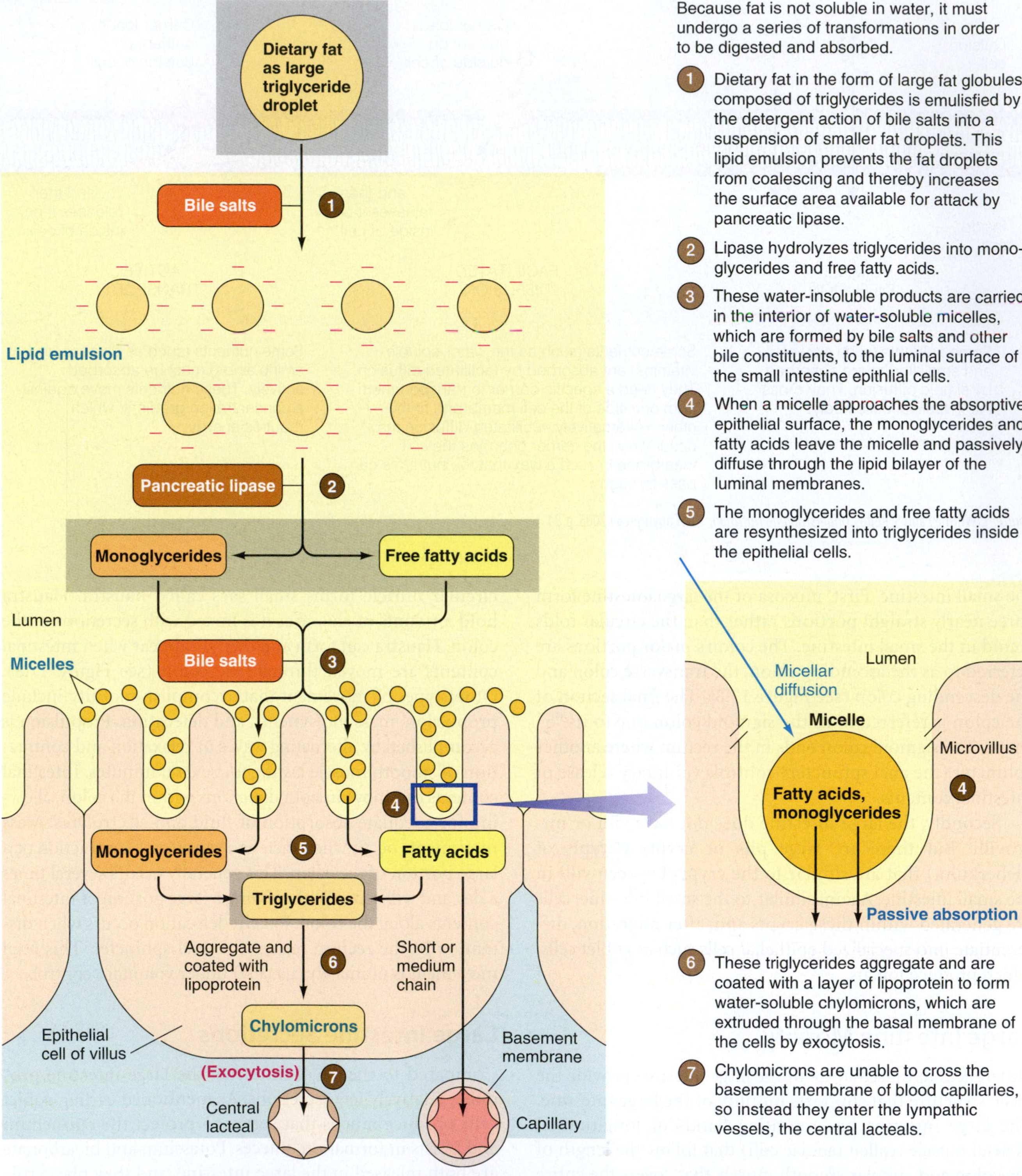

Because fat is not soluble in water, it must undergo a series of transformations in order to be digested and absorbed.

1. Dietary fat in the form of large fat globules composed of triglycerides is emulsified by the detergent action of bile salts into a suspension of smaller fat droplets. This lipid emulsion prevents the fat droplets from coalescing and thereby increases the surface area available for attack by pancreatic lipase.
2. Lipase hydrolyzes triglycerides into monoglycerides and free fatty acids.
3. These water-insoluble products are carried in the interior of water-soluble micelles, which are formed by bile salts and other bile constituents, to the luminal surface of the small intestine epithlial cells.
4. When a micelle approaches the absorptive epithelial surface, the monoglycerides and fatty acids leave the micelle and passively diffuse through the lipid bilayer of the luminal membranes.
5. The monoglycerides and free fatty acids are resynthesized into triglycerides inside the epithelial cells.
6. These triglycerides aggregate and are coated with a layer of lipoprotein to form water-soluble chylomicrons, which are extruded through the basal membrane of the cells by exocytosis.
7. Chylomicrons are unable to cross the basement membrane of blood capillaries, so instead they enter the lymphatic vessels, the central lacteals.

Source: L. Sherwood, *Human Physiology: From Cells to Systems*, 5e, copyright © 2004, p. 632

FIGURE 17.7 Sites of Nutrient Absorption

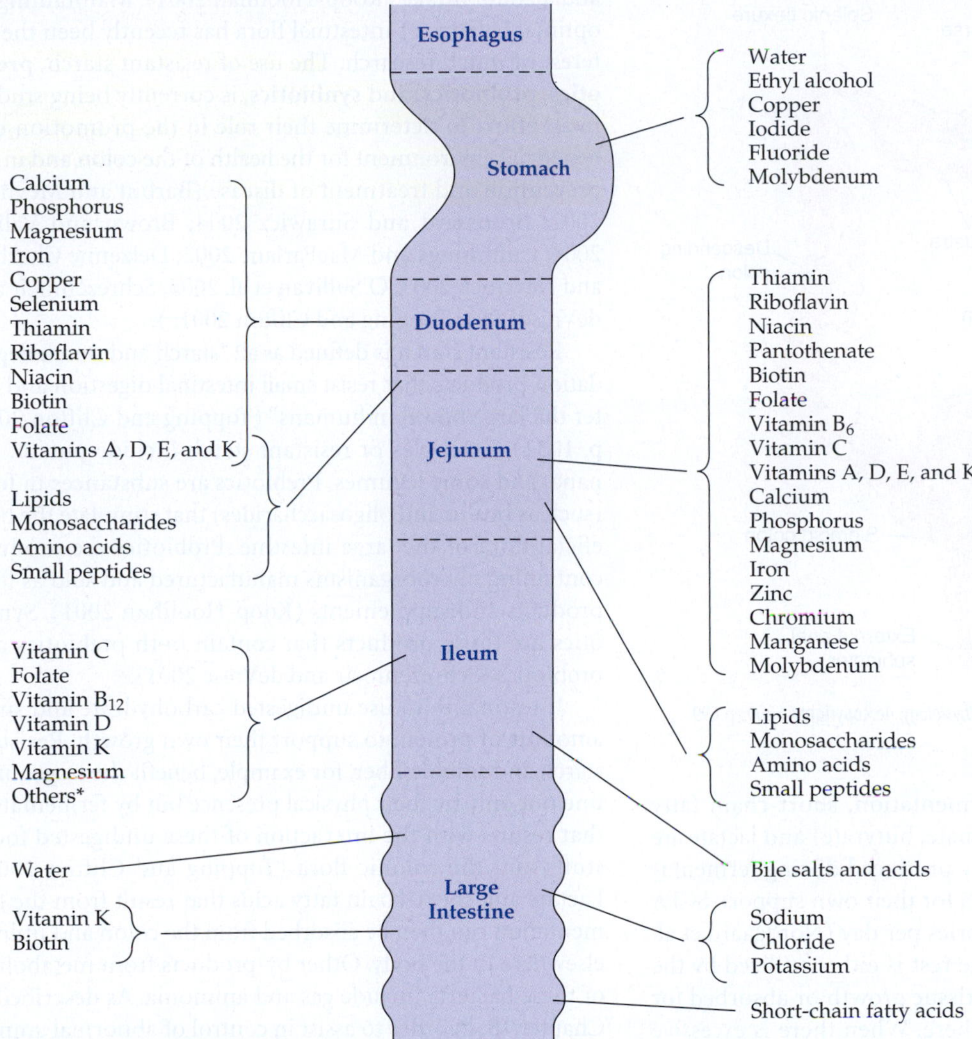

*Many additional nutrients may be absorbed from the ileum depending on transit time.

Source: J. Smith, J. Groff, and S. Gropper, *Advanced Nutrition and Human Metabolism*, 4e, copyright © 2005, p. 47

accomplished by the time chyme exits the small intestine. The primary function of the large intestine is to provide a site for reabsorption of water, electrolytes, and some vitamins. The colon can increase its absorption significantly—as much as 3 to 5 times more than normal (Rees-Parrish 2005). The colon's role in absorption is even more important when disease affects the small intestine. In conditions where digestion and absorption have not successfully occurred in the small intestine, these nutrients are lost in the feces (human waste) unless the substrate can be fermented to short chain fatty acids. The second major function of the large intestine is to serve as the site for formation and storage of feces.

When chyme enters the large intestine from the ileum, it is primarily liquid. During movement along the colon, water is reabsorbed, resulting in a drier mass of fecal matter. Sodium, potassium, and other electrolytes are absorbed along with water. Feces contain undigested foodstuffs—primarily insoluble fiber, **bilirubin**, and bacteria. This entire process may take anywhere from 12 to 72 hours (Groff and Gropper 2004).

As many as 400 different species of bacteria—including bifidobacteria, coliforms, bacteroides, peptococci, clostridia, lactobacteria, and methanogens—live within the colon. They provide fermentation of fiber and sugar alcohols. As

bilirubin—the breakdown product of hemoglobin molecules; it is normally excreted from the body via bile secretions

FIGURE 17.8 Large Intestine Anatomy

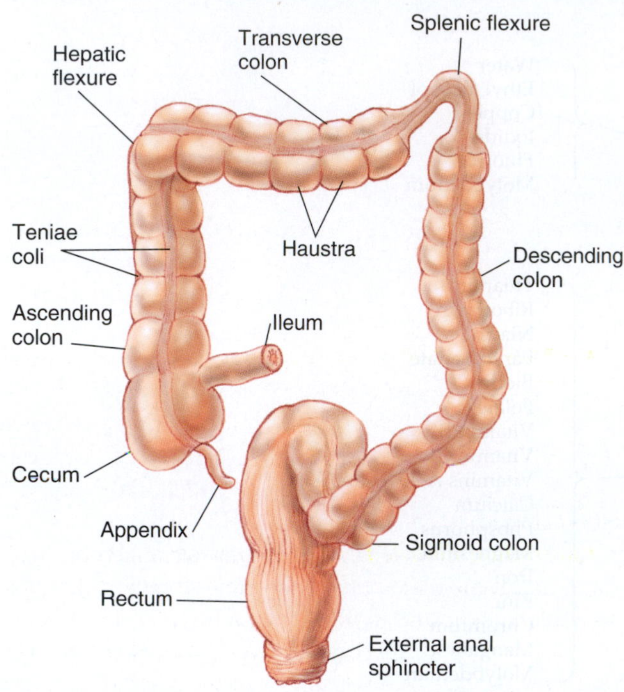

Source: R. Rhoades and R. Pflanzer, *Human Physiology*, 4e, copyright © 2003 p. 699

these substrates undergo fermentation, short-chain fatty acids (SCFA) (acetate, propionate, butyrate) and lactate are produced. Some of the energy produced during fermentation is used directly by bacteria for their own support. SCFA can provide 500 to 1200 kcalories per day (Nordgaard et al. 1996; Rees-Parrish 2005). The rest is either utilized by the colon for support of its own tissue growth or absorbed for utilization by the body elsewhere. When there is excessive substrate in the colon—such as undigested carbohydrates—gas and **flatulence** may result. This will be discussed in greater detail later in this chapter.

Many factors affect not only the amount but also the types of bacteria in the colon. These factors include age and health status, composition of the diet, transit time, stress, and alcohol intake (Koop-Hoolihan 2001). Maintaining an optimal balance of intestinal flora has recently been the interest of much research. The use of resistant starch, **prebiotics**, **probiotics**, and **synbiotics**, is currently being studied in an effort to determine their role in the promotion of a beneficial environment for the health of the colon and in the prevention and treatment of disease (Barbut and Meynard 2002; Broussard and Surawicz 2004; Brown and Valiere 2004; Cummings and MacFarlane 2002; Delzenne Cherbut, and Nevrinck 2003; O'Sullivan et al. 2005; Schrezenmeir and deVrese 2001; Topping and Clifton 2001;).

Resistant starch is defined as all "starch and starch degradation products that resist small intestinal digestion and enter the large bowel in humans" (Topping and Clifton 2001, p. 1031). Examples of resistant starch include potato, banana, and some legumes. Prebiotics are substances in food (such as **inulin** and oligosaccharides) that stimulate the beneficial flora of the large intestine. Probiotics are products containing microorganisms manufactured and sold as food products and supplements (Koop-Hoolihan 2001). Synbiotics are those products that contain both prebiotics and probiotics (Schrezenmeir and deVrese 2001).

Intestinal flora use undigested carbohydrate and small amounts of protein to support their own growth. Resistant starch and soluble fiber, for example, benefit the large intestine not only by their physical presence but by fermentation that results with the interaction of these undigested foodstuffs and the colonic flora (Topping and Clifton 2001). Lactate and short-chain fatty acids that result from the fermentation can then be absorbed from the colon and utilized elsewhere in the body. Other by-products from metabolism of these bacteria include gas and ammonia. As described in Chapter 18, in order to assist in control of abnormal ammonia levels in persons with liver disease, medications are given to induce diarrhea, which decreases the ability of the colon to reabsorb this ammonia.

Vitamin K and biotin are two endogenously produced vitamins. Biotin is produced by normal intestinal flora in the colon and is absorbed via passive diffusion (Said 1999). *E. coli* and *Bacteroides fragilis* in the colon synthesize vitamin K. The absorption route for this endogenous vitamin K is not clear at this time. Truly, for both of these vitamins, it is difficult to estimate the contribution of endogenous synthesis (Suttie 2000). Diseases of the lower GI tract, the use of antibiotics, and the presence of prebiotics and probiotics may all potentially interfere with or promote endogenous synthesis.

flatulence—perceived excess gas in the intestinal tract

prebiotics—substances in food that stimulate the beneficial flora of the large intestine

probiotics—products containing microorganisms manufactured and sold as food products and supplements

synbiotics—products that contain both prebiotics and probiotics

inulin—a fructooligosaccharide derived from chicoy; intravenous inulin is used as a diagnostic test for kidney function since it is not utilized by the body and is excreted in the urine

Pathophysiology of the Lower Gastrointestinal Tract

This section will begin with a focus on several common conditions that may occur by themselves but may also be symptoms associated with certain diseases. These conditions, which include diarrhea, constipation, and malabsorption, often require medical and nutritional intervention. It is important to remember that their etiology is multifactorial and may be a symptom of an underlying disorder. In the latter sections of this chapter, the discussion of pathophysiology will focus on specific disorders and diagnoses.

Diarrhea

Definition Diarrhea is defined as an increase in frequency of bowel movements and/or an increase in water content of stools that affects either the consistency or the volume of fecal output. Other definitions describe abnormality in stool production as >200 g/day for adults and >20 g/kg for children (Donowitz, Kokke and Saidi 1995).

Epidemiology Incidence of infectious diarrhea is estimated to be approximately 99 million new cases each year, resulting in 3,100 deaths each year in the United States. Food-borne illnesses are a major cause of these cases. Hedburg and colleagues estimate there are approximately 2.9 cases of *E. coli* infection per 100,000 persons (Hedberg et al. 1997). Over 2 million children throughout the world will die from dehydration secondary to diarrhea each year; therefore, impacting the incidence of diarrhea worldwide is the focus of many world health agencies.

Etiology Diarrhea can be classified in several different ways. First, diarrhea can be either acute or chronic in origin. Diarrhea can also be classified as either osmotic or secretory. The etiology of the diarrhea can also serve as a framework for discussion and understanding of diarrheal conditions. Acute diarrhea is short-term, whereas diarrhea lasting several weeks is considered chronic and is usually associated with a greater number of health concerns, such as electrolyte imbalances, malabsorption, dehydration, and malnutrition.

Osmolality is a measurement of concentration of particles in solution. Normal osmolality of the gastrointestinal tract is approximately 300 mOsm/L. When there is an increase in osmotically active particles in the intestine, the body reacts by pulling water into the lumen in an attempt to normalize osmolality. When this occurs, increased water efflux results in what we refer to as osmotic diarrhea. Osmotic diarrhea can be caused by maldigestion of nutrients, excessive **sorbitol** or **fructose** intake, enteral feeding, and some laxatives. In general, when the causative agent is removed, osmotic diarrhea will cease. An example would be a patient whose diarrhea resolves when he is made NPO. This is one of the major differentiations between osmotic and secretory diarrhea.

Secretory diarrhea also results from excessive fluid and electrolyte secretions into the intestine. The difference here is that the underlying disease is what causes excessive secretions. Furthermore, secretory diarrhea does not resolve when the patient is made NPO. Bacterial infections often produce enterotoxins that result in this type of diarrhea. Protozoa, viruses, and other infections can also cause secretory diarrhea (Goodman and Segreti 1999). (See Table 17.2 for summary of potential infectious diarrheal agents.) Traveler's diarrhea is a common health problem affecting those who travel to other countries. The major infectious agents resulting in traveler's diarrhea are enterotoxigenic *Escherichia coli*, enteroaggregative *E. coli*, and *Shigella* spp, *Salmonella, Campylobacter, Yersinia, Aeromonas,* and *Plesiomonas* spp (Vila et al. 2003). Other factors that could potentially cause secretory diarrhea include medications, hormone-producing tumors, prostaglandins, and excessive amounts of bile acids or unabsorbed fatty acids in the colon.

Antibiotics and other medications may cause diarrhea as a side effect. These medications generally cause diarrhea either by increasing GI motility or by disturbing the normal flora of the colon (Bartlett 2002; Kyne, Farrell and Kelly 2001). See Table 17.3 for a list of frequently used antibiotics and other medications that can cause diarrhea.

Clostridium difficile is a Gram-positive anaerobic bacterium that is the major cause of antibiotic-related diarrhea and is generally found in hospital and long-term care environments. When antibiotics are prescribed, their use can disturb the balance of normal flora of the colon. When this occurs, *C. difficile* has the potential to proliferate. Its endotoxins result in injury and inflammation to the colon. The most common antibiotics associated with *C. difficile* infection are *ampicillin, amoxicillin, cephalosporins,* and *clindamycin*. Symptoms can range from mild diarrhea to severe colitis. In mild cases, stopping the prescribed antibiotic will be enough to stop the infection. In more severe cases, antibiotics, such as metronidazole or vancomycin, are generally used to treat *C. difficile* infections (Joyce and Burns 2002; Kyne, Farrell, and Kelly 2001;). Antibiotics were identified as the primary reason for diarrhea in a recent study of hospitalized patients (McErlean et al. 2005).

Many gastrointestinal diseases have diarrhea as a common symptom. If the underlying disease disrupts normal digestion and absorptive capabilities, diarrhea will most often result. Examples of these conditions that will be covered in this chapter include Crohn's disease, ulcerative colitis, and

diarrhea—frequent or unusually liquid bowel movements

sorbitol—a sugar alcohol; it is used as a sugar substitute

fructose—a monosaccharide absorbed by facilitated transport mechanism but not against a concentration gradient; when the concentration of fructose in the small intestine is greater than that of glucose, its rate of absorption slows and the unabsorbed fructose is fermented in the colon, causing diarrhea; osmotic diarrhea has been reported in persons who have overconsumed sodas sweetened with high-fructose corn syrup or fruit juices

TABLE 17.2

Infectious Organisms Associated with Diarrhea

Etiology	Organism	Likely Source of Contamination	Symptoms	Onset of Symptoms	Duration of Illness	Treatment
Bacteria	*Bacillus anthracis*	Undercooked, contaminated meats	Bloody diarrhea, nausea, vomiting, malaise, acute abdominal pain	2 days to weeks	Weeks	Penicillin is first drug of choice; ciprofloxacin is second
	Bacillus cereus (diarrheal toxin)	Meats, stews, gravies, vanilla sauce	Sudden onset of severe nausea and vomiting; diarrhea may be present	10–16 hours	24–48 hours	Supportive care
	Brucella abortus, B. melitensis and *B. suis*	Raw milk, goat cheese made from unpasteurized milk, contaminated water	Diarrhea, fever, chills, sweating, weakness, headache, muscle and joint pain; bloody stools during acute phase	7–21 days	Weeks	Acute: rifampin and doxycycline daily for > 6 weeks; infections with complications require combination therapy with rifampin, tetracycline, and an aminoglycoside
	Campylobacter jejuni	Raw or undercooked poultry, unpasteurized milk, contaminated water	Diarrhea, cramps, fever, and vomiting; diarrhea may be bloody	2–5 days	2–10 days	Supportive care; for severe cases erythromycin may be prescribed
	Clostridium botulinum—children and adult (preformed toxin)	Home-canned foods with low acid content, improperly canned commercial foods, home-canned or fermented fish, herb-infused oils, baked potatoes in aluminum foil, cheese sauce, bottled garlic, foods held warm for extended periods of time (e.g., in a warm oven)	Diarrhea, vomiting, blurred vision, diplopia, dysphasia, and descending muscle weakness	12–72 hours	Variable (from days to months)	Supportive care
	Clostridium perfringens	Contaminated meats, poultry, or gravy cooked < 50°C; or precooked foods, time- and/or temperature-abused foods	Watery diarrhea, nausea, crampy abdominal pain w/out vomiting	8–16 hours	24–48 hour	Supportive care
	Clostridium difficile	Endogenous bacterial flora is altered by an antibiotic (ampicillin, clindamycin, and cephalosporins most commonly implicated)	Watery diarrhea and crampy diarrhea, crampy lower abdominal pain; tenesmus, nausea, vomiting; and anorexia may also be present	1–10 days after cessation of antibiotic therapy	10–12 days once antibiotic is discontinued	Discontinuance of offending antibiotic; oral vancomycin ororal metronidazole required for more severe cases; antidiarrheal agents should be avoided
	Enterohemorrhagic (EHEC) *(E. coli* 0157:H7) and other Shiga toxin producing *E. coli* (STEC)	Undercooked ground beef, especially hamburger, unpasteurized milk and juice, raw fruits and vegetables (e.g., sprouts), salami (rarely), and contaminated water	Severe diarrhea that is often bloody, abdominal pain, and vomiting, usually little or no fever; more common in children < 4 years	1–8 days	5–10 days	Supportive care, monitor renal function, hemoglobin and platelets closely; *E. coli* 0157:H7 infection is also associated with hemolytic uremic syndrome (HUS), which causes lifelong complications; antibiotics may promote development of HUS

(continued on the following page)

TABLE 17.2 *(continued)*

	Enterotoxic *E. coli* (ETEC)	Watery or food contaminated with human feces	Watery diarrhea, abdominal cramps, some vomiting	1–3 days	3 to < 7 days	Supportive care
	Listeria monocytogenes	Fresh soft cheeses, unpasteurized milk, inadequately pasteurized milk, ready-to-eat deli meats, hot dogs	Fever, muscle aches, and nausea or diarrhea	0–48 hours for gastrointestinal symptoms, 2–6 weeks for invasive disease	Variable	Supportive care and antibiotics; IV ampicillin, penicillin, or TMP-SMX recommended for invasive disease
	Salmonella spp.	Contaminated water, eggs, poultry, dairy products (fecal-oral contamination); pets (turtles, snakes, ducklings, hedgehogs)	Sudden onset of headaches, chills, abdominal pain, nausea, vomiting, diarrhea	12–36 hours	1–4 days	Supportive care; antibiotic therapy contraindicated
	Shigella spp.	Contaminated food, water, swimming pools, participation in day care centers (fecal-oral route)	Crampy abdominal pain, rectal burning, fever, multiple small-volume bloody mucoid bowel movements; complications include intestinal perforation, severe protein loss, respiratory symptoms, meningismus, seizures, hemolytic uremic syndrome, arthritis, rashes	36–72 hours after exposure	Usually < 5 days	Supportive care
	Staphylococcus aureus	Foods high in salt, protein, or sugar (such as ham, potato salad, or similar foods), meats and dairy products	Nausea, vomiting, profuse diarrhea	1–6 hours	24 hours	Supportive care
	Vibrio cholerae	Fecally contaminated water or food	Profuse, watery diarrhea without fever or abdominal cramps	1–3 days	3–7 days; causes life-threatening dehydration	Restore fluids and electrolyte balance and maintain intravascular volume
	Vibrio parahaemolyticus	Undercooked or raw seafood, fish, or shellfish	Explosive, watery diarrhea, abdominal cramps, nausea, vomiting	2–48 hours	2–5 days	Supportive care
	Vibrio vulnificus	Undercooked or raw shellfish, especially oysters, other contaminated seafood, and open wounds	Diarrhea, vomiting, abdominal pain, bacteremia, and wound infections; more common in the immunocompromised	1–7 days	2–8 days	Supportive care and antibiotics (tetracycline, doxycycline, and ceftazidime)
	Yersinia enterocolitica and *Y. pseudotuberculosis*	Undercooked pork, unpasteurized milk, tofu, contaminated water; infection has occurred in infants whose caregivers handle chitterlings	Appendicitis-like symptoms (diarrhea and vomiting, fever and abdominal pain) occur primarily in older children and young adults	24–48 hours	1–3 weeks, usually self-limiting	Supportive care; antibiotics (doxycycline and ciprofloxacin also effective) recommended if septicemia or other invasive disease occurs
Viruses	Hepatitis A	Shellfish harvested from contaminated waters, raw produce, contaminated drinking water, uncooked foods and cooked foods that are not reheated after contact with infected food handler	Diarrhea, dark urine, jaundice, and flu-like symptoms, i.e., fever, headache, nausea, and abdominal pain	28 days average (15–50 days)	Variable: 2 weeks–3 months	Supportive care; prevention with immunization

(continued on the following page)

TABLE 17.2 *(continued)*

Infectious Organisms Associated with Diarrhea

Etiology	Organism	Likely Source of Contamination	Symptoms	Onset of Symptoms	Duration of Illness	Treatment
	Noroviruses (& other calicivirures)	Shellfish, fecally contaminated foods, ready-to-eat foods touched by infected food workers (salads, sandwiches, ice, cookies, fruit)	Diarrhea, nausea, vomiting, abdominal cramping, fever, myalgia, and some headache; diarrhea is more prevalent in adults and vomiting is more prevalent in children	12–48 hours	12–60 hours	Supportive care such as rehydration
	Reovirus (rotavirus)	Invade mucosal epithelial cells	More severe than that caused by Norwalk virus	1–3 days	4–8 days	Supportive care; severe diarrhea may require fluid and electrolyte replacement
	Other viral agents (astroviruses, adenoviruses, parvovirus)	Fecally contaminated foods, ready-to-eat foods touched by infected food workers, some shellfish.	Variable combination of fever, anorexia, nausea, vomiting, myalgia, abdominal pain and diarrhea	10–70 hours	2–9 days	Supportive care; usually mild and self-limiting
	Enteric adenovirus (serotypes 40 and 41)	Person-to-person via fecal-oral route	Diarrhea, fever, respiratory symptoms, and sometimes vomitings	10–70 hours after contaminated food or water is consumed	2–9 days	
Parasitic	*Cryptosporidium parvum*	Any undercooked food or food contaminated by an ill food handler after cooking; drinking water	Watery diarrhea, stomach cramps, upset stomach, slight fever	2–10 days	May be remitting and relapsing over weeks	Supportive care, usually self-limiting
	Cyclospora cayetanensis	Various types of fresh produce (imported berries, lettuce)	Diarrhea (usually watery), loss of appetite, substantial loss of weight, stomach cramps, nausea	1–14 days; usually at least 1 week	May be remitting and relapsing over weeks	TMP-SMX for 7 days
	Entamoeba histolytica	Any uncooked food or food contaminated by an ill food handler after cooking; drinking water	Diarrhea (often bloody), frequent bowel movements, lower abdominal pain	2–3 days to 1–4 weeks	May be protracted (several weeks to several months)	Metronidazole and a luminal agent (iodoquinol or paromomycin)
	Giardia lamblia	Any uncooked food or food contaminated by an ill food handler after cooking; drinking water	Diarrhea, stomach cramps, gas	1–2 weeks after infection	Days to weeks	Metronidazole
	Trichinella spiralis	Raw or undercooked contaminated meat, usually pork or wild game meat (e.g., bear or moose)	Acute nausea, diarrhea, vomiting, fatigue, fever, abdominal discomfort followed by muscle soreness, weakness, and occasional cardiac and neurologic complications	1–2 days for initial symptoms; others begin 2–8 weeks after infection	Months	Supportive care plus mebendazole and albendazole

References: Centers for Disease Control. *MMWR Recommendations and Reports.* Diagnosis and management of foodborne illnesses. A primer for physicians and other health care professionals. April 16, 2004. Available at URL: http://www.cdc.gov/mmwr/preview/mmwrhtml/rr5304a1.htm. Accessed September 11, 2004.
Guerrant RL et al. Practice guidelines for the management of infectious diarrhea: IDSA Guidelines. *CID* 2001:32 (1 February) 331–351.
Reisdorff EJ, Pflug VJ. Infectious diarrhea. EMR Textbook. Available at URL: http://www.thrombosis-consult.com/articles/Textbook/81_infecdiarr.htm. Accessed August 22, 2004.
Thomson ABR, Shaffer EA (eds) First Principles of Gastroenterology, 3rd ed. Canada: AstraZeneca, 2000. Available at URL: http://www.gastroresource.com/GITextbook/En/Chapter7/7-10.htm. Accessed August 27, 2004. *USDA Center for Food Safety and Applied Nutrition. Foodborne pathogenic microorganisms and natural toxins handbook.* Available at URL: http://vm.cfsan.fda.gov/~mow/chap22.html. Accessed September 11, 2004.

TABLE 17.3

Medications That May Cause Diarrhea

Classification	Medication: Generic Name	Medication: Trade Name
Antacids		
Magnesium-containing antacids		Milk of Magnesia
H_2 receptor antagonists	Ranitidine	Zantac
	Cimetidine	Tagamet
	Famotidine	Pepcid
	Nizatidine	Axid
Proton pump inhibitors (PPI therapy)	Omeprazole,	Prilosec
	Lansoprazole	Prevacid
	Pantoprazole	Protonix
	Rabeprazole	AcipHex
	Esomeprazole	Nexium
Antibiotics	Clindamycin	Cleocin
	Ampicillin	Penicillin, Unasyn
	Cephalosporins	Ancef, Cefizox, Cefobid, Cefotan, Ceptaz, Claforan, Fortaz, Kefzol, Mandol, Maxipime, Mefoxin, Monocid, Rocephin, Tazicef, Tazidime, Zefazone
	erythromycin	E-Mycin
	etracycline	Achromycin-V, Sumycin, Tetracycline
	sulfonamides	Gantanol, Gantrisin, Novo-Soxazole
	Any broad-spectrum antibiotic	
Anti-inflammatory medications		
NSAIDs	Acetylsalicylic acid	Aspirin, Ascriptin, Bufferin, Ecotrin
	Ibuprofen	Advil, Motrin
	Naproxen	Aleve, Anaprox
Antigout	Colchicine	Colchicine
Cardiac Drugs		
Sodium channel blockers	quinidine	Quinaglute
	procainamide	Pronestyl
	disopyramide	Norpace
	phenytoin	Dilantin
	bretylium	Bretylol
Beta blockers	esmolol hydrochloride	Brevibloc
	carvedilol	Coreg
	timolol	Blocadren
	propranolol	Inderal
	metoprolol	Lopressor
	atenolol	Tenormin
	nadolol	Corgard
ACE inhibitors	Captopril	Capoten
	Enalapril	Vasotec
	Lisinopril	Prinivil, Zestril
	Fosinopril	Monopril
	Benazepril	Lotensin
	Moexipril	Univasc
	Perindopril	Aceon
	Quinapril	Accupril
	Ramipril	Altace
	Trandopril	Mavik
Antiarrhythmic	Digitalis	Digitek, Digitoxin, Digoxin, Lamoxin
Antihypertensives	Reserpine	Serpalan, Serpasil
	Guanethidine	Ismelin
	Methyldopa	Aldomet, Aldoril
	Guanabenz	Wytensin
	Guanadrel	Hylorel
	Hydralazine	Apresoline
Cholinergics	Bethanechol	Duvoid, Reglan, Urabeth,
	Metoclopramide	Urecholine
	Neostigmine	
Hypolipidemic agents	Clofibrate	Abitrate, Atromid-S, Novofibrate
	Gemfibrozil	Lopid
	HMG-CoA reductase Inhibitors	(this is only a partial listing)
Laxatives	Lovastatin, fluvastatin, pravastatin)	Castor oil, Citrucel, Colace, Correctol, Dulcolax, Ex-Lax, Fiberall, Fleet laxative, herbal laxatives (senna, etc.), Metamucil, Milk of Magnesia, Purge, Senokot,
Neuropsychiatric drugs	Lithium	Eskalith CR (SR), Lithobid (SR), Lithotabs
	Fluoxetine	Prozac, Prozac Weekly, Sarafem
	Alprazolam	Xanax
	Valproic acid	
	Ethosuximide	Zarontin
Miscellaneous agents	Theophylline (bronchodilator)	Elixophyllin, Slo-Phyllin, Theo-24, Theobid, Theo-Dur, Theolair, Uniphyl
	Thyroid hormones	Synthroid, Levoxyl, Unithroid
	Misoprostol	Cytotec, PGEI gel
	some chemotherapeutic agents	Methotrexate

References: Horne Js, Swanson LN. Diarrhea. U.S. Pharmacist. Available at URL: http://www.uspharmacist.com/oldformat.asp?url=newlook/files/Feat/ACF2EF8.cfmandpub. Accessed August 27, 2004. Family Practice Notebook.com. Diarrhea secondary to medications: drug-induced diarrhea. Available at URL: http://www.fpnotebook.com/GI180.htm. Accessed August 27, 2004. Medline Plus. Available at URL: http://www.nlm.nih.gov/medlineplus/medlineplus.html. Accessed August 28, 2004.

celiac disease. These diagnoses can also result in malabsorption of lipids and other nutrients, which further contribute to the diarrhea.

Other diseases that do not originate in the gastrointestinal tract can also present with symptoms of diarrhea. These may include, but are not limited to, AIDS enteropathy with HIV infection, thyroid dysfunction, and some malignancies.

Clinical Manifestations Diarrhea presents as a change from the normal bowel function. This is generally a watery stool that is increased in frequency. Other characteristics of stool output will vary depending on the etiology of the diarrhea. For example, foul-smelling, frothy stools are associated with steatorrhea, which means fat in the stool. This would occur with fat malabsorption.

Composition and volume of stool is also consistent with the etiology of the diarrhea. Blood may be present in the stool, which is characterized by being either "frank" blood, occult blood, or as melena. Frank blood is bright red blood on the surface of the stool, and represents contamination of blood from the rectum or anus. Occult blood is detected by testing the stool and usually has occurred from bleeding in the lower gastrointestinal tract. Melena is a dark stool and is caused by contamination of blood in the upper GI tract. Hemoglobin from blood contributes to the dark color. Mucus in the stool may also be indicative of secretory diarrhea and may be infectious in origin. High amounts of electrolytes are also consistent with secretory diarrhea (Guerrant et al. 2001). The presence of leukocytes in the stool indicates an inflammatory process such as inflammatory bowel disease, which will be discussed later in this chapter.

Other clinical manifestations that may occur with diarrhea are abdominal pain and cramping. When defecation relieves cramping, diarrhea is generally from the distal colon. If abdominal pain and cramping continue after defecation, the origin is generally from the small bowel. Other symptoms such as dehydration, weight loss, and electrolyte and acid-base imbalances are dependent on volumes of stool lost and represent one of the most serious consequences of diarrhea.

Diagnosis Diagnosis of the underlying etiology is the most important step in determining treatment of diarrhea. Considerations that will direct diagnostic procedures are the age of the patient, hydration status, the presence of blood in the stool, and whether the patient is immunocompromised. Other important symptoms to note are any recurring characteristics of diarrheal episodes, including time of day or any relationship to food intake.

Typical diagnostic work-up will begin with stool cultures that will be examined for microorganisms, ova and parasites, leukocytes, **lactoferrin**, and for the presence of blood. Further invasive procedures such as upper endoscopy, flexible sigmoidoscopy, or colonoscopy may assist in diagnoses not determined with initial stool cultures.

Osmolality and electrolyte content of the stool can also be determined, and will assist in differentiation between osmotic and secretory diarrhea. Other clinical tests will measure complete blood count, electrolytes, albumin, and thyroid stimulating hormone. Specific diagnostic tests for *C. difficile* include tissue culture using the cytotoxicity assay or the enzyme immunoassay for Toxins A and B (Guerrant et al. 2001; Joyce and Burns 2002).

Treatment Treatment of the underlying disorder is the most important component of therapy. If the diarrhea is infectious in nature, antibiotics will be the first line of treatment. Restoring normal fluid, electrolyte, and acid-base balance is crucial. This is accomplished through either intravenous therapy or through use of rehydration solutions (discussed in the Nutrition Therapy section) (Bergogne-Berezin 2000; Guerrant et al. 2001).

Other medications can be used to treat the symptoms of diarrhea. These agents work either to decrease motility or to thicken the consistency of the stool. These include medications such as LoMotil®, Immodium, Tincture of Opium, paregoric, Kaopectate®, or bismuth subsalicylate. It is important to note any medication side effects for these drugs. See Table 17.12 for a list of antidiarrheal medications and possible side effects/drug-nutrient interactions.

Prevention of diarrhea should be a major focus of any discussion regarding this condition. Recommendations for the prevention of diarrhea worldwide include strategies such as (Bateman and McGahey 2001):

- Improving access to clean water and safe sanitation
- Promoting hygiene education
- Exclusive breast-feeding
- Improving weaning practices
- Immunizing all children, especially against measles
- Using latrines
- Keeping food and water clean
- Washing hands with soap (the baby's as well) before touching food
- Sanitary disposal of stools

Nutrition Therapy

Nutrition Implications Nutrition implications of diarrhea are initially dependent on the volume of gastrointestinal

lactoferrin—a protein in plasma and secretions (milk, mucus, bile), secreted by leukocytes, that can bind iron; it helps prevent infection by depriving bacteria of the iron necessary for their growth

losses and then on the length of the disease course. Large volume losses can quickly lead to dehydration, and electrolyte and acid-base imbalances. Hyponatremia and hypokalemia are both common with diarrhea. Metabolic acidosis may occur due to excessive loss of bicarbonate ions in stool output (see Chapters 8 and 9). Infants and elderly are at particular risk because their systems are much more sensitive to rapid shifts in both fluids and electrolytes. Maintaining homeostasis is much more difficult for both of these populations, in part due to the inability of their renal systems to act quickly enough for adequate compensation. Chronic diarrhea can cause fluid and electrolyte complications and can result in malnutrition and specific nutrient deficiencies. Diarrhea can affect appetite and thus impair adequate ingestion. Diarrhea also results in decreased transit time, which interferes with the ability of the gastrointestinal tract to perform adequate digestion and absorption.

Nutrition Interventions Historically, nutrition therapy for diarrhea has included making the patient NPO or prescribing clear liquids. Current research indicates it is important to stimulate the gastrointestinal tract by feeding the patient. This speeds recovery of damaged cells. In addition, clear liquids are typically high in simple carbohydrates, which increase osmolality of the gastrointestinal tract. This actually can make diarrhea worse due to hyperosmolality.

Oral rehydration solutions are designed to both restore fluid and electrolyte balance and enhance absorption in the intestinal tract (Khan et al. 2005). There are several commercially prepared rehydration solutions such as Pedialyte®, Resol®, Ricelyte®, and Rehydralyte®. The World Health Organization also has a standard "recipe" for an oral rehydration solution (see Table 17.4).

Infants with diarrhea are of special concern. Their risks of dehydration and electrolyte and/or acid-base imbalances are high. It is recommended that infants who breast-feed continue to do so. Formula-fed infants can be fed half-strength formula. Banana flakes, apple powder, or other pectin sources can be added to formula in an attempt to thicken the stool. If the infant has begun solid foods, strained bananas, applesauce, and rice cereal are the best initial food choices.

In adults and older children, introducing solid foods should begin with a **low-residue diet** (see Appendix E13, Low-Residue Diet and Box 17.1 for the definition of residue). Beginning with starches, and then slowly adding foods as they are tolerated, is the recommended scenario. Use of products with pectin such as banana flakes can also assist increasing the consistency of the stool.

Another step in treating diarrhea is using foods with probiotics or prebiotics (Barbut and Meynard 2002; Broussard and Surawicz 2004; O'Sullivan 2005). These foods and supplements support growth of healthy flora and/or repopulate the intestinal tract with these healthy bacteria (see Box 17.2 for suggestions of food sources of probiotics). As mentioned above, undigested substrates are fermented to short-chain fatty acids. Probiotics and prebiotics increase the amount of short-chain fatty acids (SCFA) produced. Recent research indicates SCFA promote water and electrolyte absorption in the colon, which reduces the incidence of diarrhea (Cummings and McFarlane 2002). This may play a significant role in reducing diarrhea associated with enteral feeding (Delzenne, Cherbut, and Nevrinck 2003). Other research has studied the effect of using probiotics as part of the treatment for radiation-induced diarrhea, diarrhea secondary to rotavirus, and traveler's diarrhea (de Roos and Katan 2000; Unger et al. 2001; Urbancsek et al. 2001). Dosages for probiotics and prebiotics are still being established. For example, dosages of acidophilus

TABLE 17.4

WHO Rehydration Solution

Reduced Osmolarity Oral Rehydration Solution (ORS)	grams/liter
Sodium chloride	2.6
Glucose, anhydrous	13.5
Potassium chloride	1.5
Trisodium citrate, dehydrate	2.9
Total weight	20.5
Reduced Osmolarity Oral Rehydration Solution (ORS)	**mmol/liter**
Sodium	75
Chloride	65
Glucose, anhydrous	75
Potassium	20
Citrate	10
Total weight	245

Source: WHO Dept. of Child and Adolescent Health and Development. The treatment of diarrhoea. A manual for physicians and other senior health workers. 2005 ISBN 52-4-159318.0.

low-residue diet—a diet low in fiber and other food constituents that may contribute to bulk in the large intestine

BOX 17.1 CLINICAL APPLICATIONS

Is It Fiber . . . or Is It Residue? What's the Difference?

You're on grand rounds and the attending physician asks you to explain the difference between fiber and residue to the residents. What do you tell them?

Residue is defined as any solid contents that end up in the large intestine after digestion. Residue is primarily dietary fiber plus bacteria and any remaining gastrointestinal secretions.

One of the residents asks the difference between a low-residue diet and low-fiber diet. Can you enlighten this curious resident?

A low-residue diet is a low-fiber diet with some additional restrictions: milk is limited to 2 cups per day and prune juice is excluded. Short term, a low-residue diet can be nutritionally sound, but long-term use may result in vitamin C or folic acid deficiencies.

The first patient seen on rounds is Joseph Blough. His diet is being advanced from liquid to low-residue diet. Mr. Blough tells you he'd really like some oatmeal with his breakfast and a baked potato with his lunch. Can he have oatmeal and a baked potato on a low-residue diet?

Even though oatmeal appears "soft," it is actually high in fiber. A better choice would be corn flakes (or any low-fiber dry cereal) or white toast. Fiber in potatoes is in the skin; thus taking off the skin removes that fiber. All refined grains are acceptable. So Mr. Blough could have a boiled potato without skin, white rice, or a white dinner roll.

Mr. Blough wonders if he can have any fresh fruit. What would you tell him?

Fruits like fresh melons and bananas are allowed since they contain very little fiber. Typically, most fiber is found in the skin on fruit; therefore, some fruits are allowed if skin is removed (examples: peaches, apricots, nectarines). Fresh or canned pears are high in fiber.

One of the residents comments that since bean burritos are soft, they should be allowed on a low-residue diet. This is a teachable moment . . . go for it!

Beans, like oatmeal, appear soft, but they are very high in fiber. All beans and peas should be avoided on a low-residue diet. More appropriate choices would be chicken rice soup with saltine crackers, reduced-fat cottage cheese, or creamy peanut butter and jelly sandwich on white bread.

Mr. Blough is asking about desserts . . . He states his wife makes excellent brownies (with walnuts) and German chocolate cake. Additionally, he really likes to have pecan pie for his birthday, which is coming up in a week or so. Another teachable moment has presented itself.

Pecan pie and brownies contain nuts. German chocolate cake is made with coconut. Nuts and coconuts are high in fiber. Low-fat frozen yogurt and low-fat ice cream are low residue as long as they do not contain fruit chunks or nuts.

References:

Beyer P. Medical nutrition therapy for lower gastrointestinal tract disorders. In Mahan K and Escott-Stump S (eds) *Krause's Food, Nutrition, and Diet Therapy,* 11th ed. Philadelphia: Saunders, 2004.

Nutrition Services. Self-Test: Low Fiber, Low Residue Diet. Northwestern Memorial Hospital. 2000. Available at URL: https:// www .nmh.org/nmh/patientinformation/selftestlowiberlowresiduediet .htm. Accessed September 6, 2004.

LeWine H. Ask the Expert: What is a low residue diet? Intelihealth .com. Available at URL: http://www.intelihealth.com/IH/ihtIH/ WSIHW000/8270/8438/386755.html. Accessed September 6, 2004.

are expressed not in grams or milligrams, but in billions of organisms. It has been estimated that a typical daily dose should supply about 3 to 5 billion live organisms. Other probiotic bacteria are used similarly. "The typical dose of *S. boulardii* yeast is 500 mg twice daily (standardized to provide 3×10^{10} colony-forming units per gram), to be taken while traveling, or at the start of using antibiotics and continuing for a few days after antibiotics are stopped" (Bergogne-Berezin 2000, p. 524). The amount of probiotic bacteria in a product is left up to the manufacturer. For example, culture manufacturers recommend formulation of yogurt or milk products with added probiotics at 10^6 probiotic bacteria per gram or mL (California Dairy Research Association 2005). The actual count may change as products age.

constipation—a decrease in frequency of bowel movements with straining with defecation and/or hard stools

Constipation

Definition There are many subjective definitions for **constipation.** Surveys have attempted to interpret the individual's complaints of gastrointestinal distress. Individuals will commonly describe constipation as a decrease in frequency of bowel movements. In order to provide some general framework for this diagnosis, The Rome Consensus Criteria were established. The Rome Consensus Criteria define constipation as a condition where at least two of the following symptoms have occurred in the previous year for at least 12 nonconsecutive weeks (Locke, Pemberton, and Phillips 2000; Thompson et al. 2000):

- "Straining" with > ¼ of defecations
- Hard stools in > ¼ defecations
- Sensation of incomplete evacuation or anorectal obstruction in > ¼ defecations
- Manual maneuvers to facilitate > ¼ of defecations
- < 3 defecations each week

BOX 17.2 CLINICAL APPLICATIONS

Food Sources of Probiotics

Consumer Guidelines for Choosing Probiotics

Look for the "Live Active Culture" seal. This means that, if refrigerated, yogurt contains 108 viable lactic acid bacteria per gram at the time of manufacture; or, if frozen, contains 107 viable lactic acid bacteria per gram. Remember that this does not specifically identify probiotic bacteria from starter culture bacteria.

Find the genus, species, and strain designation of the bacteria available in the product.

Look for the numbers of each probiotic strain at the end of the shelf-life.

Find suggested serving size that will deliver the effective dose of probiotics.

Find the proper storage conditions for the product. Does it need to be refrigerated or frozen?

Identify the corporate contact details for consumer information.

Sources	Description/Information
Yogurt	Most commonly added strains include: *Lactobacillus acidophilus, L. casei, L. reuteri,* and *Bifidobacterium bifidum.* Label should read: "Live, active cultures."
Sour cream, buttermilk, milk	Contain acidophilus, lactobacilli, and bifidobacteria.
Tempeh	Fermented soy that can be added to casseroles, soups, stews, and stir-fry.
Miso	A thick paste made from fermented and processed soy beans. Red miso is a combination of barley and soy beans, and yellow miso is a combination of rice and soy beans. Contains variety of lactic acid bacteria.
Kim chi	Mixture of fermented vegetables. Contains variety of lactic acid bacteria.
Kefir	Kefir® is a fermented milk product that contains a variety of probiotic strains. Manufactured by Lifeway Foods, Inc.

Sources:

Dixon S. http://www.cancer.med.umich.edu/news/pro09spr02.htm
www.ific.org/foodinsight/2003/ma/friendlybugsfi203.cfm

Joint FAO/WHO Working Group Report on Drafting Guidelines for the Evaluation of Probiotics in Food London, Ontario, Canada, April 30 and May 1, 2002. Available from: http://www.who.int/foodsafety/fs_management/en/probiotic_guidelines.pdf.

Regulatory Issues. Available from: http://www.usprobiotics.org/.

Epidemiology As previously stated, constipation is one the most common gastrointestinal complaints. Studies have estimated that as much as 25% of the U.S. population experiences symptoms of constipation. Estimates differ depending on criteria that are used to define constipation. These symptoms result in almost 2.5 million physician visits each year (Sonnenberg and Koch 1989).

Etiology Constipation can be a result of several different distinct causes (Müller-Lissner et al. 2005). Slowed colonic transit can result in constipation. Constipation can be due to rectal outlet obstruction or other sources of obstruction such as fecal impaction, adhesions, or even the presence of a tumor. **Pelvic floor dysfunction** (University of Southern California 2003) results not only in slowed colonic transit but also storage of fecal contents in the rectum for long periods of time. Constipation can be a major component of irritable bowel syndrome, which will be discussed later in this chapter. Constipation can also be a component of other medical conditions, including scleroderma, amyloidosis, and neurological diseases such as multiple sclerosis (MS) or Parkinson's disease. Finally, constipation can be a side effect of many different classes of medications. These include very common prescription drugs such as calcium channel blockers, antidepressants such as amitriptyline, pain medications such as morphine, diuretics, and antihistamines. Other over-the-counter medications that often cause constipation include iron, calcium, and other vitamin supplements, and, for some individuals, even nonsteroidal anti-inflammatory drugs can result in constipation.

Clinical Manifestations Symptoms of constipation include decreased frequency of bowel movements. Bowel movements are often hard and pelletlike. Abdominal pain, bloating, and gas are common accompanying symptoms.

Diagnosis Correct diagnosis will include a complete history and physical. Laboratory screening tests will include a complete blood count, thyroid-stimulating hormone, and serum glucose. Additional tests can include a colonoscopy or flexible sigmoidoscopy if further evaluation is needed.

Treatment Treatment of the underlying etiology will direct medical care for constipation. Common interventions include bowel retraining and use of enemas or cathartic and laxative medications. Other medications involve bulking

pelvic floor—refers to the pelvic diaphragm, the sphincter mechanism of the lower urinary tract, the upper and lower vaginal supports, and the internal and external anal sphincters; it is a network of muscles, ligaments, and other tissues that hold up the pelvic organs (vagina, rectum, uterus and bladder)

pelvic floor dysfunction—weakening of the pelvic floor that can cause the organs to shift, bulge, and push outward against each other, resulting in urinary or fecal incontinence or obstruction, vaginal prolapse or pain, sexual dysfunction, and other problems.

BOX 17.3 CLINICAL APPLICATIONS

Medications Used in Treatment of Constipation

Type	Mechanism	Drug/Agent	Onset of Action	Side Effects (Selected)	Comments
Bulk	Increases stool bulk, decreases colonic transit time, increases GI motility	Fiber (bran), psyllium (e.g., Metamucil), methylcellulose (e.g., Citrucel), calcium polycarbophil (e.g., Fibercon)	Days	Bloating, flatulence (iron and calcium malabsorption may occur with fiber)	Safest for promoting elimination; only laxative acceptable for long-term use
Emollient laxative (stool softener)	Cyclic AMP stimulates secretion of water, sodium, and chloride into GI lumen	Docusate sodium (Colace), glycerin, mineral oil	12–72 hours	Often used to prevent constipation, ineffective for severe constipation	Nausea
Osmotic laxative	Poorly absorbed polyvalent ions (e.g., Mg, phosphate, sulfate) or disaccharides (Lactulose, Sorbitol) remain in colon, increasing intraluminal osmotic pressure and drawing water into intestine; increased volume stimulates peristalsis	Poorly absorbed disaccharides (Lactulose, Sorbitol), magnesium laxatives (Milk of Magnesia), sodium salts (Fleets enema), polyethylene glycol lavage solution (GoLytely)	Within 3 hours	Cramps, flatulence (Lactulose, Sorbitol)	Better for treating constipation than preventing it
Stimulant laxative	Irritate intestinal mucosa or directly stimulate submucosal and myenteric plexus	Anthraquinone (cascara extract [Casanthranol] and senna extract [Senokot]), diphenylmethane (bisacodyl [Dulcolax] and phenolphthalein [Correctol, Ex-Lax])	6–8 hours	Abdominal pain (cramps), prolonged use can damage large intestine	Other laxative types are preferred over these; risk of laxative abuse; not used if there is possibility of intestinal obstruction
Enemas and suppositories	Evacuation induced by distended colon, mechanical lavage	Tap water (500 ml rectally); soapsuds (1500 ml rectally); glycerine suppository	5–15 minutes	Rectal irritation, mechanical trauma	Never use hot water

References: Laxative. Family Practice Notebook.com. Available at URL: http://www.fpnotebook.com/GI175.htm. Accessed September 19, 2004. Chapter 27 Diarrhea and Constipation. The Merck Manual. Available at URL: http://www.merck.com. Accessed September 19, 2004.

agents and stool softeners (Locke, Pemberton, and Phillips 2000). See Box 17.3 for a listing of medications used in the treatment of constipation.

Nutrition Therapy

Nutrition Implications Research regarding nutrition and constipation has concentrated on the role of adequate fiber and fluid intake. The National Health and Nutrition Examination Surveys (NHANES) indicated most Americans consume an average of 14 to 15 grams of fiber each day (Alaimo et al. 1998). When compared to recommendations, most individuals are consuming only 40% to 50% of the recommended intake. Adequate intake of whole grains, fruits, and vegetables has been one of the primary focuses of both the Dietary Guidelines for Americans and the Nutrition Recommendations for Canadians (Department of Health and Human Services 2005; Health and Welfare Canada 2005). Inadequate fiber intake has long been associated with many gastrointestinal conditions, including constipation, diverticular disease, and hemorrhoids.

Nutrition Interventions Twenty to thirty-five grams of dietary fiber are recommended for adults each day. Based on caloric intake, this would be approximately 10 to 13 g of dietary fiber per 1000 kcal. For children over the age of 2 years, fiber intake is recommended to be the amount equal to their age plus 5 grams/day (American Dietetic Association 2000, 2002). More recent recommendations suggest adults under the age of 50 should consume 38 grams of fiber per day (Food and Nutrition Board and Institute of Medicine 2002).

Ensuring adequate fiber intake is crucial for treating constipation. After a thorough diet history and nutrition assessment, the clinician should make recommendations for slowly increasing fiber intake. This can be accomplished by adding one to two high-fiber foods each day. Foods have a mixture of different kinds of fiber, but in general, the recommendations are for a 3:1 ratio of insoluble to soluble fiber. See Box 17.4 for ways to increase fiber and a sample menu to meet 25 to 35 grams of fiber each day. Due to individual tolerance, some individuals—especially the elderly—may not be able to achieve levels in the highest ranges without using fiber supplementation. Table 17.12 lists several bulking agents that can be used to supplement fiber intake.

At the same time fiber intake is increased, the clinician should also emphasize adequate water intake. This should be at a minimum of 2000 mL/day (approximately 8 cups/day).

Use of probiotics and prebiotics has also been recommended for treatment of constipation. For example, consumption of fructooligosaccharides has been shown to soften feces and to assist in relieving constipation (Broussard and Surawicz 2004; Brown and Valiere 2004; Garleb et al. 2002; O'Sullivan 2005).

Malabsorption

Definition Malabsorption is a general term referring to malabsorption of fat, carbohydrate, or protein as a result of maldigestion or from damage to the anatomy and physiology of the small intestine.

Etiology Damage to the anatomy and physiology of the small intestine due to disease is the most common cause of malabsorption. Conditions such as celiac disease, Crohn's disease, and even protein-calorie malnutrition result in decreased villous height, decreased enzyme production, and resulting malabsorption and/or maldigestion. Dysfunction of the accessory organs of digestion (liver, pancreas, and gallbladder) may also serve as the origin of the maldigestion.

Decreased transit time, as seen in diarrhea or from surgical changes in the anatomy, also can result in either/or maldigestion or malabsorption. For example, after a gastrectomy, dumping syndrome causes a rapid transit through the small intestine, which prevents adequate exposure to enzymes and adequate time for the absorptive mechanisms. See Table 17.5 for a list of potential causes of malabsorption.

Pathophysiology Nutrient digestion and absorption are dependent on normal anatomy; normal physiology; adequate production of enzymes, hormones, and other secretions such as bile; and appropriate motility. Malabsorption may be examined in the context of each of the main macronutrients: lipid, carbohydrate, and protein.

Fat Malabsorption The digestion and absorption process for lipid or fat is the most complex, and therefore the easiest to disrupt. Fat malabsorption is called steatorrhea—literally meaning fat in the stool. Digestion and absorption of fat requires adequate colipase and pancreatic lipase, adequate emulsifier—bile—from the liver and gallbladder, and adequate secretion through the common bile duct and pancreatic ducts. Motility needs to be normal due to the lengthy process lipid has to undergo from micelle to chylomicron for absorption. When any of these processes is disrupted, fat remains in the stool and travels undigested and unabsorbed to the large intestine. Fat-soluble vitamins are malabsorbed as well. An additional concern for fat malabsorption is the potential presence of excess oxalate. Under normal conditions, calcium within the GI tract binds with oxalate and allows for its excretion or metabolism. In fat malabsorption, calcium often binds with the malabsorbed fat. This allows oxalate to be absorbed and then excreted through the kidney. Excessive amounts of oxalate have been linked to development of urothiasis or kidney stones. Hyperoxaluria (excessive oxalate in the urine) is responsible for about 30% of kidney stones and is considered to be the most common cause.

Persons with fat malabsorption will experience abdominal pain, cramping, and diarrhea. Stools produced will be frothy, foul-smelling, and greasy in appearance.

Specific laboratory tests can diagnose steatorrhea and then assist in determination of the etiology of the malabsorption. Diagnostic tests include the 72-hour quantitative fecal fat test that involves collection of stool output for three days after ingesting 100 grams fat/day. If more than 6 grams of fat are present in the stool after 24 hours, the diagnosis of steatorrhea can be made. This establishes malabsorption but does not determine the specific cause. This test is costly and unpleasant for the patient. It often takes up to 10 days to receive results (Rees-Parrish 2005).

The D-xylose absorption test assists in distinguishing between pancreatic dysfunction and small bowel malabsorption. D-xylose is easily absorbed in the small intestine and is not metabolized. Its absorption does not require pancreatic or biliary function. In this test, after the patient drinks 25 grams of D-xylose, urine and blood samples are collected. Normal findings would show that blood levels of D-xylose are 25 to 40 mg/dL after 2 hours. Excretion in the urine should be 80% to 95% after 5 hours. Normal values point to malabsorption originating from pancreatic or biliary dysfunction. In patients with intestinal malabsorption, blood and urine levels would be diminished.

BOX 17.4 CLINICAL APPLICATIONS

Bulking Up: Ways to Increase Patients' Fiber Intake

Fiber is carbohydrate that enters the large intestine undigested. While it is found only in plants, all fiber is not the same. One way to categorize fiber is by its solubility in water. Soluble fiber dissolves in water, forming a gel-like material. Insoluble fiber does not dissolve in water and increases GI transit time.

Sources of Fiber	Soluble	Insoluble
Grains	Oats (oatmeal, oat bran), rice bran, barley	Whole grain breads and cereals, wheat bran, barley, couscous, brown (whole grain) rice, bulgur, rye
Legumes	Dried peas, beans, lentils	
Nuts/seeds	Nuts and seeds	Seeds
Fruits	Apples, citrus fruits, pears, strawberries, blueberries, tomatoes	
Vegetables		Cabbage, carrots, cucumbers, zucchini, celery

It is important to maintain a high fluid intake when consuming foods high in fiber. Six to eight glasses of liquid a day is a minimum. Furthermore, many people become aware of bloating, cramping, or gas when increasing fiber intake. Small, incremental changes can alleviate this. Advise patients to start with one of the changes listed below, then wait 5 to 7 days before making further changes.

Getting More Fiber

Use the USDA's Food Guide and MyPyramid to guide food choices to increase fiber intake.

Cereals and grains	Begin the day with high-fiber (5 grams of fiber per serving) breakfast cereals.
	Or add unprocessed wheat bran to cereal.
	Add bran cereal or unprocessed wheat bran to baked products (meatloaf, breads, muffins, casseroles, cookies, etc.).
	Add bran as a crunchy topping for casseroles, salads, or cooked vegetables.
	Use whole-grain breads (whole wheat or another whole grain flour is listed as the first ingredient).
Vegetables	Snack on raw vegetables instead of chips or snack crackers.
	Try to eat raw vegetables as much as possible; cooking may reduce fiber content.
	Frozen vegetables have the same fiber content as raw vegetables.
Fruits	Eat fruit at every meal.
	Eat whole fruits instead of drinking fruit juices.
	Do not peel fruits (fiber is found in the skins).
	Eat fresh and dried fruit for snacks.
Other high fiber foods	Eat cooked beans at least once a week (substitute cooked beans for meat to lower saturated fat).
	Add beans to soups, stews, and salads.
	Low-fat popcorn and whole-grain crackers are good snack choices.

Sample Menu to Meet 25 to 35 g Fiber Daily

Breakfast	
1 cup oatmeal w/fresh sliced strawberries	4 g
1 egg scrambled in 1 tsp butter	
1 cup orange juice	<2 g
1 cup low fat milk	
Lunch	
Sliced avocado, sprouts, provolone cheese, and sliced tomato on 2 slices whole-grain bread	3 g
1 cup baby carrots	4 g
1 medium apple	5+ g
Unsweetened iced tea	
Dinner	
2 cups black beans and rice	10+ g
2 corn bread muffins w/1 tsp butter	<2 g
Ice water	
Unsweetened iced tea	
Snack	
3 cups popped popcorn	5+ g
12 oz cola soft drink	

For detailed information about the dietary fiber content of foods, the USDA website has information at: http://www.nal.usda.gov/fnic/foodcomp.

References:

Roughing it: Fitting more fiber into your diet (July 28, 2003). MayoClinic.com. Available at URL: http://www.mayoclinic.com/invoke.cfm?id=NU00033. Accessed September 19, 2004.

American Academy of Family Physicians. Fiber: How to increase the amount in your diet. familydoctor.org. Accessed at URL: http://familydoctor.org/x1718.aml?printxml. Accessed September 19, 2004.

American Dietetic Association. Manual of Clinical Dietetics, 6th ed. Chicago: American Dietetic Association, 2000.

Harvard School of Public Health. Fiber. Start roughing it! Nutrition Source. Available at URL: http://www.hsph.harvard.edu/nutritionsource/fiber.html. Accessed September 19, 2004.

Hwant MY. Are you getting enough fiber? JAMA and Archives Journals Medical Library. Available at URL: http://www.medem.com/medlb/article_detaillb.cfm?article_id=zzz5naczmachandsub_cat=377. Accessed September 19, 2004.

TABLE 17.5

Possible Causes of Malabsorption

Category	Mechanism	Examples
Inadequate digestion	Postgastrectomy*	
	Deficiency or inactivation of pancreatic lipase	
	Exocrine pancreatic insufficiency	Chronic pancreatitis Pancreatic carcinoma Cystic fibrosis Pancreatic insufficiency-congenital or acquired
	Gastrinoma-acid inactivation of lipase*	
	Drugs	Orlistat
Reduced intraduodenal bile acid concentration/impaired micelle formation	Liver disease	Parenchymal liver disease Cholestatic liver disease
Bacterial overgrowth in small intestine:	Anatomic stasis	Afferent loop stasis/blind loop/strictures/fistulae
Functional stasis	Diabetes* Scleroderma* Intestinal pseudoobstruction	
Interrupted enterohepatic circulation of bile salts	Ileal resection Crohn's disease* Drugs (bind or precipitate bile salts): neomycin, cholestyramine, calcium carbonate	
Impaired mucosal absorption/mucosal loss or defect	Intestinal resection or bypass	
	Inflammation, infiltration, or infection	Crohn's disease* Amyloidosis Scleroderma* Lymphoma* Eosinophilic enteritis Mastocytosis Tropical sprue Celiac disease Collagenous sprue Whipple's disease Radiation enteritis Folate and vitamin B_{12} deficiency Infections: salmonellosis, giardiasis Graft-vs.-host disease
	Genetic disorders	Disaccharidase deficiency Agammaglobulinemia Abetalipoproteinemia Hartnup disease Cystinuria
Impaired nutrient delivery to and/or from intestine:	Lymphatic obstruction	Lymphoma* Lymphangiectasia
	Circulatory disorders	Congestive heart failure Constrictive pericarditis Mesenteric artery atherosclerosis Vasculitis
Endocrine and metabolic disorders	Diabetes* Hypoparathyroidism Adrenal insufficiency Hyperthyroidism Carcinoid syndrome	

*Malabsorption caused by more than one mechanism.

Finally, a small bowel x-ray with contrast dye can indicate delays of motility such as an obstruction or **ileus**. A faster motility might also support the diagnosis of malabsorption.

Carbohydrate Malabsorption The most common example of carbohydrate malabsorption is lactose malabsorption. When there is inadequate lactase available for digestion, or if anatomy or motility does not allow adequate exposure to lactase, lactose will travel to the large intestine undigested and unabsorbed. Bacteria in the large intestine will cause the lactose to undergo fermentation, which creates increased gas and abdominal cramping. Undigested lactose also pulls additional water into the large intestine, contributing to abdominal cramping and resulting diarrhea.

Lactose malabsorption can be diagnosed using either a lactose tolerance test or lactose breath hydrogen test. In the lactose breath hydrogen test, baseline breath hydrogen concentration is measured. Then the patient consumes 25 to 50 grams of lactose. Breath hydrogen concentration is remeasured in 3 to 8 hours. An increase >20 ppm suggests lactose malabsorption. This test is preferred over the lactose tolerance test and has approximately 90% sensitivity.

ileus—decreased or absent motility of the bowel and forward movement of bowel contents

TABLE 17.6

Malabsorption Guide and Laboratory Tests Used to Identify Deficiencies

Diagnosis Exacerbated or Caused by Malabsorption	Nutrients at Risk for Deficiency	Symptoms
Celiac disease	Fat, protein, carbohydrate, vitamin K, folate, vitamin B_{12}, calcium, iron	Diarrhea, steatorrhea, weight loss, bone loss
Crohn's disease	Protein, fat, fat-soluble vitamins, vitamin B_{12}, iron, calcium, copper, zinc, selenium, folic acid, vitamin C	Diarrhea, steatorrhea, weight loss, bone loss, anemia
Cystic fibrosis	Fat, fat-soluble vitamins, carbohydrates, protein, sodium, chloride	Steatorrhea, pancreatic juices, hormones and enzymes secretions impaired, insulin secretion impaired growth to cystic fibrosis symptoms
HIV virus	Sodium, chloride, potassium, fat-soluble vitamins, fat, protein, water-soluble vitamins	Anorexia, diarrhea, infections, weight loss
Liver disease	Folate, vitamin B_6, riboflavin, thiamin, fat-soluble vitamins, calcium, zinc, protein stores	Steatorrhea, anorexia, esophageal varices
Short gut syndrome	Fat, fat-soluble vitamins, protein, carbohydrate, calcium, magnesium, vitamin B_{12}	Bone loss, anemia, weight loss, muscle wasting, diarrhea
Ulcerative colitis	Protein, fat, fat-soluble vitamins, vitamin B_{12}, iron, calcium, copper, zinc, selenium, folic acid, vitamin C	Diarrhea, weight loss, bone loss, anemia

Laboratory Tests to Assess Deficiencies	
Biotin	Serum biotin, urinary biotin
Folate	Erythrocyte folate, free folate, (FIGLU) urinary formiminoglutamic acid
Niacin	Urinary N-methyl nicotinamide
Riboflavin	Urinary riboflavin, erythrocyte glutathione reductase
Thiamin	Blood pyruvate and lactate, urinary thiamin excretion, erythrocyte transketolase, apoenzyme levels
Vitamin A	Serum carotene, retinol binding protein
Vitamin B_6	Whole blood level of pyridoxal phosphate
Vitamin B_{12}	Schilling test, erythrocyte B_{12}, DUMP test, serum B_{12}
Vitamin C	Plasma vitamin C, leukocyte vitamin C, urinary vitamin C
Vitamin D	Serum alkaline phosphatase
Vitamin E	Serum tocopherol, erythrocyte hemolysis
Vitamin K	Prothrombin time

Reprinted with the permission of *Today's Dietitian* © Great Valley Publishing, Co. Source: Clairmont MA. Malabsorption assessment guide. Today's Dietitian. 2000 April.

Protein Malabsorption Protein malabsorption is most commonly referred to as protein-losing enteropathy. This is not a specific disease, but it occurs, as do most all malabsorption disorders, as a result of other diseases. Excessive protein loss in the gastrointestinal tract has been noted in over 65 different diagnoses (Binder 2005). Excessive protein is lost in the stool, and the patient will experience reduced serum levels of proteins and an increasing amount of peripheral edema due to the reduced **oncotic pressure**.

oncotic pressure—pressure exerted by large protein molecules in blood plasma, which usually do not cross the capillaries; these molecules decrease the fluid that can leak out of the capillaries into the tissue

Treatment Appropriate treatment for malabsorption will depend on the nutrient that is malabsorbed and the underlying disease causing malabsorption. These will be discussed in much more depth as specific diseases are covered in this chapter.

Nutrition Therapy

Nutrition Implications Nutritional implications for malabsorption include weight loss, vitamin and mineral deficiencies, and chronic protein-calorie malnutrition. See Table 17.6 for a malabsorption guide as well as laboratory tests used to test for deficiencies (Clairmont 2000).

BOX 17.5 CLINICAL APPLICATIONS

Suggestions for Adding MCT Oil to the Diet

Made from coconut oil, MCT oil can be added to foods in small amounts throughout the day, or in the form of MCT-containing liquid supplements. Some consider the nutritional supplements to be more palatable forms of MCT. To prevent osmotic diarrhea and decreased absorption of LCT, it is important not to exceed the threshold dose of 50 g/d (8 tablespoons).

MCT oil may be added to such foods as fruit juices, salads, fat-free salad dressings, and vegetables. It can also be incorporated into sauces. Because of its low smoking point, MCT can only be used in baking or cooking where temperature does not exceed 66° to 75°C.

Products Containing MCT

Manufacturer	*Product*	*Description*
Mead Johnson http://www.meadjohnson.com (800) 361-6323	M.C.T.® Oil	Oil
	Portagen	Milk-based powder
Novartis Medical Health, Inc http://www.novartisnutrition.com	Lipisorb®	Nutritionally complete liquid supplement
Nestle http://www.nestleclincalnutrition.com (800) 776-5446	Nutren® 1.5	Nutritionally complete liquid supplement
	Nutren® 2.0	Nutritionally complete liquid supplement
	Peptamen®	Nutritionally complete liquid supplement
Ross http://www.ross.com (800) 258-7677	Promote®	Nutritionally complete liquid supplement
Universal Nutrition http://www.universalnutrition.com (800) 872-0101	MCT Oil	Oil

References:

Food and Agriculture Organization of the United Nations and the World Health Organization. Fats and oils in human nutrition. Report of a joint expert consultation. *FAO 1994.*

American Diabetes Association. Manual of Clinical Dietetics, 6th ed. Chicago: American Dietetic Association, 2000.

Nutrition Interventions The purpose of nutrition therapy is to provide structure for elimination of the malabsorbed nutrient, but at the same time provide appropriate substitutions in order to ensure maintenance of nutritional status.

Nutrition Therapy for Fat Malabsorption Restriction of fat to 25 to 50 grams per day is a standard first step in reducing the symptoms of fat malabsorption. Appendix E8 outlines the dietary interventions for a fat-restricted diet. Additionally, **medium-chain triglyceride** (**MCT**) supplements can be used to increase caloric intake. Triglycerides that contain fatty acids considered medium chain are those with 6 to 12 carbons. Most MCT oil products contain primarily caprylic (C8) and capric (C10) fatty acids. One half-ounce (15 mL) is 115 kcal or 8.3 kcal/g. MCT is absorbed directly into the circulatory system from the small intestine and does not require normal lipid digestion and absorption routes that long-chain fatty acids require. See Box 17.5 for suggestions for adding MCT Oil to the diet.

If the etiology for steatorrhea originates from pancreatic dysfunction, use of pancreatic enzymes is a primary mode of treatment. An example is the product Pancrease® that consists of pancreatic lipase, amylase, and protease. Dosages are individualized and taken with each meal or snack to ensure adequate digestion.

Nutrition Therapy for Lactose Malabsorption Lactose is the simple carbohydrate found in milk and dairy products. Lactose is also found as an ingredient in many other food products in which it is often used as filler. Milk provides approximately 11 grams of lactose per cup. Other dairy products have varying amounts, with ice cream having approximately 9 grams per cup and cheese having 1 to 2 grams per ounce. Restriction of all milk and dairy products is the major step to treat lactose malabsorption—individuals do vary on amounts of lactose they can tolerate. Products such as Lact-Aid® provide the lactase enzyme and can be used when milk and dairy foods are ingested. See Appendix E14 for lactose sources.

Celiac Disease

Definition **Celiac disease** (**CD**), previously referred to as gluten-sensitive enteropathy, gluten intolerance, or non-tropical sprue, is a complex disease whose etiology originates from both genetic and autoimmune factors. In this disease, exposure to gluten results in damage to the intestinal mucosa. Other diseases or conditions associated with CD are type 1 diabetes mellitus and thyroid dysfunction as well as dermatitis and muscle and joint pain. Persons with CD may also be at higher risk for lymphoma and other malignancies.

Epidemiology Recent epidemiological studies indicate CD is much more common than previously thought, especially in Caucasians and other individuals of European ancestry (Kamin and Furuta 2004). A recent study indicates the prevalence of CD in the United States was 1:22 in first-degree relatives, 1:39 in second-degree relatives, and

medium-chain triglycerides (MCTs)—triglycerides composed of fatty acids with 8 carbons (octanoic and decanoic fatty acids)

celiac disease (CD)—inflammation of the small intestine caused by gluten found in various grains, including wheat

FIGURE 17.9 Small Intestine Villi in Celiac Disease

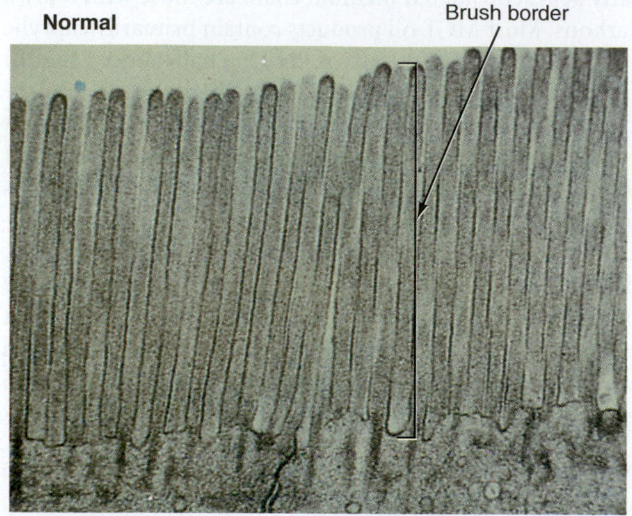

(a)

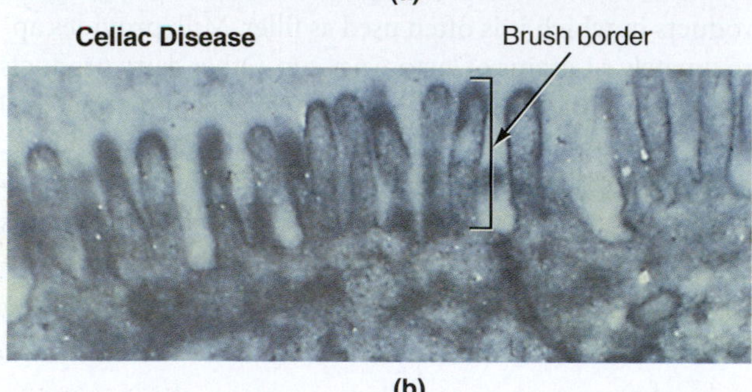

(b)

Source: L. Sherwood, *Human Physiology: From Cells to Systems,* 5e, copyright © 2004, p. 627

1:56 in symptomatic patients. Overall, prevalence of CD in not-at-risk groups was 1:133 (Fasano et al. 2003). Other countries, including Argentina, Italy, Germany, Denmark, and Finland, have published similar studies indicating a much higher prevalence in these countries as well.

Etiology It is well understood that damage to the intestinal mucosa, which is observed in CD, occurs when the small intestine is exposed to the prolamin fraction—α-gliadin and other protein components of gluten. Gluten is found in wheat, rye, malt, barley, and, in smaller amounts, in oats.

More recent research indicates damage to the intestinal mucosa is accompanied by an infiltration of white blood cells into the mucosa. This inflammatory response is also reflected in production of IgA antigliadin and antiendomysial antibodies. These antibodies, which now serve as components of the diagnostic procedures for CD, reflect the autoimmune nature of this disease.

Genetic markers for CD have not been completely established, but as previously stated, evidence is significantly convincing that a genetic component exists. Current research indicates CD is strongly associated with a group of genes on chromosome 6. These genes for HLA class II antigens are involved in regulation of the body's immune response to gluten protein fractions (Jennings and Howdle 2003; Robins and Howdle 2004).

Pathophysiology When the small intestinal mucosa is exposed to certain sequences of amino acids found in the prolamin fraction of wheat (gliadin), rye (secalin), and barley (hordein), there appears to be both a toxic and inflammatory response (Thompson 2003). This response damages villi; height is reduced, and they are flattened in appearance (see Figure 17.9). Lack of surface area and reduction of enzyme production cause both malabsorption and maldigestion. Celiac disease is often accompanied by other systemic autoimmune disorders, including dermatitis herpetiformis, type 1 diabetes mellitus, thyroid disease, systemic lupus erythematous, primary biliary cirrhosis, rheumatoid arthritis, and Sjögren's syndrome. Persons with CD are considered to be at higher risk for lymphoma and osteoporosis as well as the complications of nutrient deficiencies and malnutrition.

Clinical Manifestations Classic clinical symptoms of CD include diarrhea, abdominal pain and cramping, bloating, and gas production. Other symptoms that can occur in the absence of GI problems include bone and joint pain, muscle cramping, fatigue, peripheral neuropathy, seizures, skin rash, and mouth ulcerations.

Diagnosis Previously, diagnosis for CD was confirmed by biopsy of the small intestinal mucosa and subsequent indication of villous atrophy, crypt hyperplasia, and lymphocytic and plasma cell infiltrate in the lamina propria. Reversal of symptoms after restriction of gluten generally provided the final evidence. Even though biopsy remains the gold standard for diagnosis, it is common now to diagnose CD after identifying antibodies to gliadin (AGA), endomysium (EMA), reticulin (ARA), and transglutaminase (tTG). Theses simple blood tests are easily accomplished. Of the four, EMA is considered to be close to 100% for both specificity and sensitivity (Lagerqvist et al. 2001). Because symptoms vary widely between individuals and prevention of complications is of paramount importance, the use of antibody screening and diagnosis is logical.

Prognosis and Treatment The only treatment for CD is nutrition therapy consisting of a gluten-free diet. After avoidance of all gluten, villous height returns to normal. As the anatomy returns to normal, maldigestion and malabsorption resolve. **Refractory CD** or nonresponsive CD is

refractory celiac disease—initial or subsequent failure of a strict gluten-free diet to restore normal intestinal architecture and function in patients who have celiac-like enteropathy

BOX 17.6 CLINICAL APPLICATIONS

Nutritional Management for Diarrhea

Diarrhea may possibly lead to weight loss, dehydration, and electrolyte imbalance that may compromise nutritional status. Preventing dehydration and malnutrition is the cornerstone of nutrition management for diarrhea.

Replace fluids and electrolytes	Drink liquids at room temperature throughout the day Drink fluids that provide water, electrolytes, and calories: Diluted fruit juices Soups and broths Gatorade™ and Powerade™ Oral rehydration formulas (Pedialyte™, Ceralyte™, Infalyt™)
Include	Foods high in soluble fiber: Oatmeal Ripe bananas (also a good source of potassium) Applesauce Fruit sections without membranes (orange [good source of potassium] and grapefruit) Easy to tolerate foods: Mashed potatoes Breads made with refined white flour White, refined rice, pasta, macaroni, or noodles Cream of Wheat™ or farina White toast, saltine crackers Ready-to-eat cereals (Cheerios™, Rice Krispies™)
Limit	Fats Fried or greasy foods Foods high in insoluble fiber Raw vegetables Fruit seeds, skins and stringy fibers, dried fruits Whole kernel corn, nuts, seeds Whole, unrefined grains or granolas, high-fiber cereals (bran cereals)
Avoid caffeine	Avoid chocolate, coffee, tea, soft drinks Choose decaf (herbal) teas and decaffeinated beverages
Lactose may not be well tolerated	Avoid milk and milk products for a few days Yogurt with live cultures is usually well tolerated

References: National Digestive Diseases Information Clearinghouse (NDDIC). Digestive Diseases: Diarrhea. NIH Publication #04-2749. National Institute of Diabetes and Digestive and Kidney Diseases, National Institutes of Health. October 2003. Available at http://digestive.niddk.nih.gov. Accessed November 20, 2004.

defined as "initial or subsequent failure of a strict gluten-free diet to restore normal intestinal architecture and function in patients who have celiac-like enteropathy" (Abdulkarim et al. 2002, 2006). Abdulkarim and colleagues (2002) found the most common reasons for nonresponsive CD were unknown gluten contamination and the presence of coexisting diseases such as pancreatic insufficiency and irritable bowel syndrome as well as the presence of malignancies.

Nutrition Therapy

Nutrition Implications Nutritional consequences of CD are dependent on the extent of malabsorption present. Severe malabsorption will result in significant weight loss, vitamin and mineral deficiencies, and, ultimately, protein-energy malnutrition.

Nutrition Intervention Nutrition therapy will be consistent with the level of damage to the intestinal mucosa and the degree of malabsorption. Most often, the individual diagnosed with CD will need to initially begin on nutrition therapy using a low-residue, low-fat, lactose-free, gluten-free diet. A low-residue dietary modification will allow the symptoms of diarrhea to be minimized (see Box 17.6, Nutritional Management for Diarrhea). A low-fat diet of approximately 45 to 50 grams/day can assist in minimizing symptoms of steatorrhea. Lactase deficiency will be common in this disorder due to damaged villi and enzyme secretion. As villi are regenerated and absorptive capability returns, these nutrients (fiber, lactose, and fat) can be added back to the diet slowly and usually do not require a lifelong restriction.

On the other hand, gluten does require a lifelong restriction. The patient will need to avoid all foods and other products that contain wheat, rye, barley, and malt. Restriction of oats is still controversial, but most recently research has indicated some individuals may tolerate oats (Thompson 2003). The major controversy regarding use of oats is contamination within oat products by wheat, barley, or rye. Thompson recommends that oats should be limited to 1/2 cup per day, and that steps should be taken to reduce the chance of contamination. This would include, in part, contacting manufacturers regarding the methods of production and avoiding products sold in bulk bins (Thompson 2003).

Specific modifications that allow for avoidance of all gluten are outlined in Table 17.7. It is important that patients understand that fillers used in over-the-counter medications, toothpastes, and mouthwashes may contain gluten. The manufacturers of such products should be contacted to confirm any gluten content. Labeling for gluten-free products in the U.S. states a product does not contain any more than 200 ppm of prolamin, but the most recent studies indicate 20% of products labeled wheat-free actually contain some wheat protein. Current labeling does not require sources of ingredients be designated on the label; this would be an important improvement to ensure adherence to a gluten-free diet. Many specialty products are now available that provide alternatives for wheat, rye, and barley. Food products using rice, corn, and soy allow for greater variety today than was previously available. National organizations, including the Celiac Disease Foundation, Canadian Celiac Association, and the Gluten Intolerance Group of North America, for example, provide excellent resources for persons diagnosed with CD.

TABLE 17.7

Specific Modifications That Allow for Avoidance of Gluten

Gluten-Free	Contain Gluten	Questionable Products[1]	Products That May Contain Gluten	Contamination
Amaranth	Ale	Brown rice (can be made with barley)	Breading	Oats do not naturally contain gluten, but they may become contaminated during harvesting, transporting, milling, and processing. Research has shown most adults with celiac disease can safely consume moderate amounts of contaminated oats without problem.
Arrowroot	Barley	Caramel color	Broth	Gluten-free foods can become contaminated if foods are prepared on common surfaces or with utensils not thoroughly cleaned after preparing gluten-containing foods (i.e., toasters, flour sifters, deep fat fryer).
Bean	Beer	Dextrin (check to see if made with corn or wheat)	Coating mixes	
Buckwheat	Farina	Flour or cereal products	Communion wafers	
Corn	Kamut	Hydrolyzed vegetable protein (HVP), vegetable protein, hydrolyzed plant protein (HPP), or textured vegetable protein (TVP)	Croutons	
Distilled alcohol	Malt vinegar	Malt or malt flavorings (check to see if made from barley or corn)	Imitation bacon and seafood	
Distilled vinegars	Rye	Modified food starch or modified starch	Marinades	
Miller	Spelt	Natural and artificial flavors	Pastas	
Nut flours	Triticale	Soy sauce or soy sauce solids (may contain wheat)	Processed meats	
Potato	Wheat (durum, semolina)		Roux	
Quinoa			Sauces	
Rice			Self-basting poultry	
Sorghum			Soup base	
Soy			Stuffing	
Tapioca			Thickeners	
Tef				

[1] verify absence of gluten-containing grains

References: Celiac Disease Foundation/Gluten Intolerance Group. *Quick Start Diet Guide for Celiac Disease.* March 2004. Celiac Disease Foundation/Gluten Intolerance Group. *Gluten Free Diet.* Available at http://www.gluten.net, Accessed November 22, 2004. Thompson T. Gluten contamination of commercial oats in the United States. New England Journal of Medicine. 2004, 351:2021–2022, Nov 4, 2004.

BOX 17.7 CLINICAL APPLICATIONS

Rome II Criteria to Define IBS

In the prior 12 months, >12 weeks of continuous or recurrent symptoms:

Abdominal pain or discomfort along with >2 of the following:

- Relieved with defecation *and/or*
- Associated with change in frequency of stool *and/or*
- Associated with change in form (appearance) of stool

Reproduced with permission from the BMJ Publishing Group. *Source:* Drossman DA, Corazziari E, Talley NJ, Thompson WG, Whitehead WE. Rome II: the functional gastrointestinal disorders: a multinational consensus. Gut 1999; 45(suppl 2):1–81.

irritable bowel syndrome (IBS)—a bowel disorder characterized by abdominal pain with diarrhea and/or constipation

Irritable Bowel Syndrome

Definition As early as 1849, symptoms of this disorder were described (Cummings 1849). Throughout the last century, varying names such as spastic colon, irritable colon syndrome, and neurogenic mucous colitis have been given to this syndrome. Currently, **irritable bowel syndrome (IBS)** is defined using the Rome II criteria (see Box 17.7) that for at least 12 weeks (they need not be consecutive weeks) and in the past 12 months, the individual has experienced abdominal pain that has at least two of the following three criteria: (1) pain relieved with defecation, (2) onset associated with change in frequency of stool, (3) onset associated with change in form of stool. Subtypes of IBS include conditions that primarily involve diarrhea, conditions that primarily involve constipation, and conditions where diarrhea and constipation alternate (Adeniji, Barnett and DiPalma 2004; Thompson et al. 2000; Holten, Wetherington, and Bankston 2003).

Certain "red-flag" symptoms should be eliminated prior to diagnosis of IBS. These symptoms may actually signal other conditions. They include age at onset over 50, progressively severe symptoms, symptoms at night that wake the patient, persistent diarrhea, bleeding, anemia, weight loss, vomiting, fever, or a family history of colon cancer.

Epidemiology Irritable bowel syndrome is the most common gastrointestinal complaint in the United States and Canada. It is estimated to affect as many as 20% of the population. This condition affects women more than men and the prevalence of IBS varies minimally with age (Saito, Schoenfeld and Locke 2002). The cost of medical care for IBS is estimated to be as much as $8 million each year.

Etiology IBS historically has been designated as a "functional" disorder. This means a diagnosis is made after ruling out all other causes of the patient's symptoms. Contrary to widespread belief among the public and in the medical community, IBS is a real condition. It is not a psychosomatic disorder, although its symptoms can be aggravated by stress, anxiety, depression, or emotional trauma. The specific cause of IBS is unknown, but etiological factors may include increased levels of serotonin, an elevated inflammatory response to infection, and an increased sensitivity of the enteric nervous system that causes abnormal motility and pain. There are often other conditions associated with IBS, which include anxiety, panic, mood, and **somatization** disorders.

Pathophysiology The pathophysiology of IBS is complex and, as previously stated, not completely understood. Proposed etiological factors may be examined in context of what we know about the normal physiology of the gastrointestinal tract. In IBS, abnormal motility is considered to be one of the major factors involved in symptoms of abdominal pain and altered bowel habits (Lyford et al. 2002). As discussed in Chapter 16, contraction and motility are regulated through the gastrointestinal tract's own specialized pacemaker cells called the interstitial cells of Cajal. The autonomic nervous system relays input for the smooth muscle of the GI tract through a specialized enteric nervous system. Two types of neurons regulate the contraction of smooth muscle, motility, and secretory functions of the GI tract. Primary neurotransmitters involved in transmission of impulses include acetylcholine, substance P, vasoactive intestinal polypeptide (VIP), and nitric oxide. Other neurotransmitters and modulators involved in neuroregulation for the GI tract include norepinephrine, serotonin, neuropeptide Y, and Γ-aminobutyric acid (GABA). Altered serotonin or 5-HT_4 (5-hydroxytryptamine) levels have been documented in IBS and may lead to abnormal motor and secretory function (American College of Gastroenterology 2002; Cash 2004; Crowell 2004; Elsenbruch 2004). The migrating motility complex (MMC), discussed earlier in the section Small Intestine Motility, provides an additional type of motility whose weak contractions serve to constantly sweep through the small intestine and remove leftover waste. Patients with IBS appear to have abnormal periods of MMC contractions when compared to controls (Evans 1997).

Individuals with IBS have been found to have an increased sensitivity to stimulation of the gastrointestinal tract. This means the same stimuli in normal patients do not result in symptoms that patients with IBS experience: abdominal pain, urgency, diarrhea, or constipation. When IBS patients were evaluated using balloon-distention, they experienced abdominal pain and gastrointestinal symptoms at much lower levels of distention than controls (American College of Gastroenterology Functional Gastrointestinal Disorders Task Force 2002; Cash 2004). An infectious and inflammatory component of IBS has also been proposed. Increases in inflammatory cells within the colon have been noted (Talley 2001). Other studies have correlated development of IBS after infectious enteritis. Specific organisms that have been documented include blastocystis hominis, campylobacter, salmonella, and parasites such as trichinella spiralis (Gomez-Escudero 2003). Abnormal cellular immune responses to certain nutrients have also been documented (Elsenbruch et al. 2004).

Though correlations with certain psychiatric conditions have been documented frequently in patients with IBS, no specific connections have been made. Some researchers feel previous physical and sexual abuse may increase sensitivity to GI stimulation and thus may contribute to symptoms. Stress is known to worsen symptoms of IBS.

Clinical Manifestations Abdominal pain, alteration in bowel habits, gas, and flatulence as well as some upper GI symptoms (reflux and noncardiac chest pain) are major symptoms for IBS. Abdominal pain can be acute and relieved by defecation. At the same time, some patients with IBS experience constant, chronic abdominal pain.

Alterations in bowel habits are seen in both major types of IBS—constipation and diarrhea. In some patients, both constipation and diarrhea are experienced.

Increased levels of gas and flatulence are also experienced. Gas is produced when food passes into the large intestine and is only partially digested. Intestinal bacteria act on these foodstuffs, and by-products of their metabolism result in gas production. In people with IBS, there appears to be an increased sensitivity to certain foods such as lactose, wheat, or high-fiber foods, which results in an exaggerated response to these nutrients.

Other concurrent diagnoses that occur with IBS include **fibromyalgia**, chronic fatigue syndrome, **temporomandibular joint (TMJ) syndrome**, and food allergies.

Treatment Since the etiology of IBS is currently unknown, treatment for IBS is guided by the patient's symptoms (Brandt

somatization—the physical manifestation of stress

fibromyalgia—a condition characterized by chronic pain (in muscle and soft tissues surrounding joints) and fatigue

temporomandibular joint (TMJ) syndrome—a condition of facial pain in the joints of the lower jaw

2002). For those patients with diarrhea-predominant IBS, antidiarrheal agents can be used. These medications assist by decreasing motility and increasing consistency of the stool. (See previous discussion regarding diarrhea.) These medications include diphenoxylate (Lomotil®), loperamide, atropine, and cholestyramine. Loperamide is the only medication that has been used in a controlled study of IBS, and its effectiveness appears limited (American College of Gastroenterology Functional Gastrointestinal Disorders Task Force 2002; Brandt 2002). Unfortunately, antidiarrheal agents do not necessarily affect the abdominal pain. Most recently, clonidine (Captapress®) has been used in treating IBS in hopes that both diarrhea and abdominal pain could be treated. Clonidine is an alpha-adrenergic stimulating agent that acts to inhibit norepinephrine activity. It is more commonly used to treat hypertension but has been used in clinical trials to treat IBS (Talley 2001).

Tricyclic antidepressants and selective serotonin reuptake inhibitors (SSRIs) are also used to treat IBS, and can commonly assist in the control of chronic pain. These medications include amitriptyline, desipramine, doxepin, or trazodone and fluoxetine, paroxetine, or sertraline, respectively. A summary of the research indicates that these medications can be used effectively in treating the pain of IBS. Antispasmodics, including dicyclomine and hyoscyamine, are also commonly prescribed, but the American College of Gastroenterology Functional Gastrointestinal Disorders Task Force suggests these are no more effective than a placebo in controlling symptoms of altered motility and abdominal pain.

Medications used to treat constipation-predominant IBS should include bulking agents and osmotic laxatives. Bulking agents are supplements or medications that add psyllium, bran, or other sources of fiber to the diet. Studies have indicated bulking agents may help with constipation, but are not necessarily effective for IBS. Osmotic laxatives include Milk of Magnesia®, polyethylene glycol, and sorbitol.

The newest medications for IBS are drugs that work as either agonists or antagonists for the 5-HT_4 receptors. Tegaserod® is the only 5-HT_4 receptor agonist and works to treat constipation-predominant IBS. Controlled clinical trials have shown the effectiveness of this medication (American College of Gastroenterology Functional Gastrointestinal Disorders Task Force 2002; Beradi 2004; Brandt et al. 2002). Alosetron®, which is a 5-HT_3 receptor antagonist, has been tested in diarrhea-predominant IBS. This medication is prescribed in the United States under a strict limited marketing program. Severe constipation and ischemic colitis have been reported as potential side effects of the drug.

Other treatments for IBS include behavioral therapies (hypnosis, relaxation techniques, guided imagery), antibiotics, probiotics, and nutrition therapy (Beradi 2004; Crowell 2004;). It is recommended by the American Dietetic Association that patients with IBS see a registered dietitian (RD) for a minimum of three visits upon diagnosis (American Dietetic Association 1996, 1997).

TABLE 17.8

Gas-Producing Foods

Vegetables	Legumes
Beets	Black-eyed peas
Broccoli	Bog beans
Brussels sprouts	Broad beans
Cabbage	Chickpeas
Carrots	Field beans
Cauliflower	Lentils
Corn	Lima beans
Cucumbers	Mung beans
Leeks	Peanuts
Lettuce	Peas
Onions	Pinto beans
Parsley	Red kidney beans
Peppers, sweet	Soybeans
Grains/Cereals/Seeds/Nuts	**Others**
Barley	Bagels
Breakfast cereals	Baked beans
Granola	Bean salads
Oat bran	Chili
Oat flour	Lentil soup
Pistachios	Pasta
Rice bran	Peanut butter
Rye	Soy milk
Sesame flour	Split-pea soup
Sorghum, grain	Stir-fried vegetables
Sunflower flour	Stuffed cabbage
Wheat bran	Tofu
Whole-wheat flour	Whole-grain breads

Source: http://www.beanogas.com/foodlist.asp

Nutrition Therapy

Nutrition Implications Symptoms of IBS can lead to changes in oral intake that lead to nutrient deficiencies, potential underweight, and malnutrition.

Nutrition Interventions Nutrition therapy goals for IBS will focus on decreasing anxiety; normalizing dietary patterns; assuring adequate nutritional intake, including sufficient fiber; and taking the necessary steps to reduce gas production. Nutrition therapy should be initiated after a careful diet history. Food diaries or diet history should focus on dietary components that the patient has associated with any increase in gastrointestinal symptoms. Common food intolerances include high-fat foods, lactose, caffeine, and sorbitol. Specific questions for fruits, vegetables, legumes,

TABLE 17.9

Steps to Decrease Gas Production

Avoid swallowing air	Foods and beverages should be consumed slowly. Foods should be chewed slowly before swallowing. Liquids should not be drunk through straws. Chewing gum or hard candy should not be used. Don't smoke or use tobacco products. If dentures are worn, they should be checked by a dentist for proper fit.
Avoid gas-forming foods	See Table 17.8. Identify and avoid problem foods. Check for lactose intolerance. Soak dry beans in water overnight; then discard water and cook soaked beans in new water.
Eating tips	Cut back on fried and fatty foods. Temporarily cut back on high-fiber foods; add back gradually. Eat smaller meals throughout the day. Avoid eating while anxious, upset, or on the run. Avoid carbonated beverages.
Personal habits	Avoid constipation (can cause bloating). Exercise regularly.
Medications	Avoid laxatives. Nonprescription antigas products (e.g., Beano™) help break down sugars found in vegetables and grains. Lactase enzyme supplements (such as Dairy Ease™ and Lact-Aid™) can be taken with dairy products to help break down lactose in foods. Try acidophilus capsules or liquid if gas is result of loss of beneficial bacteria due to antibiotic use. Peppermint tea contains menthol which may have an antispasmodic effect on the digestive tract. (Caution: peppermint tea may contribute to heartburn or acid reflux.)

References: American Gastroenterological Association. *Gas in the Digestive Tract.* Available at http://www.gastro.org. Accessed November 23, 2004. Mayo Clinic. *Diseases and Conditions: Gas and gas pains.* MayoClinic.com. Available at http://www.mayoclinic.com. Accessed November 23. 2004. WebMDHealth. *Gas, Bloating, and Burping Prevention.* WebMD.com. Available at http://www.webmd.com. Accessed November 23, 2004.

whole grains, and other significant sources of fiber should be noted. Other concerns or foods the patient routinely avoids should be identified.

Once a baseline nutritional history has been established, the RD and patient can begin to identify any needed changes in diet. Overall nutritional adequacy should be addressed first. Many patients with IBS tend to eat erratically due to their gastrointestinal symptoms, and often eating is associated with a high level of anxiety and stress. Establishing a regular eating pattern that does not exacerbate symptoms is a crucial initial step.

The next goal is to focus on increasing fiber intake to approximately 25 grams/day. Higher fiber intakes may not be initially tolerated. Patients may be wary of increasing fiber but should be encouraged to increase slowly by adding one high-fiber food at a time. Fiber supplements may be used if necessary. Of course, adequate fluid is also necessary as fiber intake is increased (Nobaek et al. 2000). (See Appendix E11, High-Fiber Diet.)

Both prebiotics and probiotics have received attention for their potential use in IBS. Adding these foods and supplements may be beneficial in the overall MNT plan (see Box 17.2).

Due to problems with gas and flatulence, providing recommendations to relieve these symptoms will also be beneficial. Simple carbohydrates that cause gas are raffinose, lactose, fructose, and sorbitol (see Table 17.8 for examples of gas-producing foods). Avoiding foods that produce gas and taking steps to decrease swallowed air will decrease gas production. Products such as Bean-O® or Bean-zyme® provide alpha-galactosidase, which potentially decreases the presence of undigested carbohydrate entering the large intestine and thus decreases the amount of gas produced. Table 17.9 outlines steps to decrease gas production.

BOX 17.8 CLINICAL APPLICATIONS

Ulcerative Colitis versus. Crohn's Disease

Ulcerative Colitis	Characteristic	Crohn's disease (regional enteritis)
Unknown, but hereditary factors seem to play a role (family history is the most significant risk factor)	Etiology	Unknown, but cigarette smoking contributes to development or exacerbation
Both sexes affected equally; higher prevalence in Ashkenazi Jews; approximately 10% of those with UC have a first-degree relative with disease; peak onset 20 to 30 years, secondary peak in middle age	Epidemiology	White males and females in temperate regions of North America, South Africa, Australia; north American and Western European Jews (Ashkenazi Jews) have highest incidence; no age group exempt, but peak age of onset is teens to twenties.
GI tract unable to distinguish foreign from self-antigens; characterized by chronic inflammation of colonic mucosa and submucosa, atrophy and possible dysplasia limited to colon; extent of disease varies and may involve only the rectum (ulcerative proctitis) left side of colon to splenic flexure, or entire colon (pancolitis).	Pathology	Localized inflammation in bowel mucosa progressing through bowel wall; tends to be localized in terminal ileum and right colon fistulizing
Bloody diarrhea with mucus Abdominal and/or rectal pain Fever Weight loss Possibly constipation and rectal spasm Arthritis Dermatological changes Ocular manifestations	Signs and symptoms	Chronic diarrhea Abdominal pain and cramping Blood and/or mucus in stool Anorexia Weight loss Fever Abdominal tenderness Delayed growth in prepubescent patients Perianal fistula
Severe bleeding Toxic colitis Toxic megacolon Strictures Impending perforation Intolerance to immunosuppression Colonic strictures Dysplasia Carcinoma Growth failure in children	Complications	Malabsorption Abdominal fistulas and abscesses Intestinal obstruction Bacterial overgrowth (blind loop syndrome) Gallstones Kidney stones Urinary tract infections Thromboembolic complications Perianal disease Neoplasia
Clinical presentation Computed tomography (CT) Endoscopy Biopsy Possibly barium enema	Diagnosis	Barium contrast x-rays CT scans Sigmoidoscopy and colonoscopy with biopsies
Infectious colitis	Differential	Appendicitis

(continued on the following page)

inflammatory bowel disease (IBD)—an autoimmune, chronic inflammatory condition of the gastrointestinal tract; IBD is actually the term designating a syndrome consisting of two diagnoses: **ulcerative colitis** and **Crohn's disease**

ulcerative colitis (UC)—a chronic inflammatory bowel disease (IBD) primarily located in the colon and rectum

Crohn's disease—a chronic inflammatory bowel disease (IBD) that can affect the entire gastrointestinal tract but most commonly affects the ileum and colon

Inflammatory Bowel Disease

Definition Inflammatory bowel disease (IBD) is characterized as an autoimmune, chronic inflammatory condition of the gastrointestinal tract. IBD is actually the term designating a syndrome consisting of two diagnoses: **ulcerative colitis** (UC) and **Crohn's disease**[2]. These diagnoses are very similar but also have very distinct differences. See Box 17.8 for a comparison of the characteristics of UC and Crohn's disease.

Epidemiology Prevalence of IBD is higher in countries within the Northern Hemisphere—North America and

BOX 17.8 *(continued)*

Antibiotic-associated colitis Amyloidosis Solitary rectal ulcer syndrome AIDS-related diarrhea	Diagnosis	Bacterial overgrowth Bowel tuberculosis Small-bowel cancer Nonsteroidal enteropathy Diverticulitis Celiac sprue Lymphoma Postsurgical adhesions Behçet's disease Ischemic colitis Radiation enteropathy
Chronic with repeated exacerbations and remissions; nearly 30% of those with extensive ulcerative colitis require surgery; patients with localized UC have best prognosis, surgery rarely required, and life expectancy normal.	Prognosis	Rarely cured, but characterized by intermittent exacerbations. Disease never extends into new areas of small bowel beyond initial distribution at first diagnosis. Approx 70% require surgery. Pts prone to reactive depression and potential abuse of pain medications.
Reduce acute and chronic inflammation eventually resulting in remission Drugs: Anti-inflammatory drugs Adrenocorticosteroids 5-aminosalicylic acid (5-ASA) Antidiarrheals Steroids Antibiotics Diet: Avoid raw fruits and vegetables, caffeine, alcohol, and pepper for those with frequent bowel movements; milk-free diet for those with lactose intolerance; psyllium or bran may be helpful for those with proctitis and constipation Surgery: Colectomy Total proctocolectomy with Brooke ileostomy Intra-abdominal Koch pouch Restorative proctocolectomy with ileal pouch–anal anastomosis Psychological support necessary for any form of colon resection	Treatment	Based on severity of disease. Drugs: Anti-inflammatory drugs Antibiotics Steroids Immunomodulatory drugs Biologic therapies Diet: Elemental TPN Surgery: Surgical removal of affected areas

References:

Bayless T, Talamini M, Kaufman H, Norwitz L and Kalloo AN. Crohn's disease. The Johns Hopkins Medical Institutions Gastroenterology and Hepatology Resource Center. Available at URL http://hopkins-gi.nts.jhu.edu. Accessed January 1, 2005.

Harris M, Kaufman H, Talamini M, Dassapoulus T, Norwitz L, and Kalloo AN. Ulcerative colitis. The Johns Hopkins Medical Institutions Gastroenterology and Hepatology Resource Center. Available at URL http://hopkins-gi.nts.jhu.edu. Accessed January 1, 2005.

Lashner BA. Inflammatory bowel disease. The Cleveland Clinic Disease Management Project. Published May 29, 2002, revised June 17, 2004. Available at URL http://www.clevelandclinicmeded.com. Accessed Jan. 2, 2005.

The Merck Manual, Sec. 3, Ch. 31, Inflammatory Bowel Diseases. Available at URL http://www.merck.com. Accessed December 31, 2004.

Northern European countries. Respectively, prevalence is much lower in the Southern Hemisphere, in countries of Southern Europe, and in Australia. There is very low prevalence in Asia and South America (Lashner 2003; Loftus et al. 2000). Incidence of IBD in the United States ranges from 5 to 15 per 100,000 persons. Prevalence is estimated to be approximately 90 per 100,000 for Crohn's disease and 200 per 100,000 for UC. Prevalence is fairly equal in both males and females, but it is higher in those populations with Caucasian and Ashkenazi Jewish descent. The estimated cost of medical care for these patients is approximately $1.6 billion per year in the United States.

Etiology The complete etiology for both Crohn's disease and UC is unknown at this time. It is understood though that multiple factors play a role in this condition (Thompson-Chagoyán 2005). These may include environmental factors such as smoking, infectious agents, intestinal flora, and physiological changes in the small intestine from which an abnormal inflammatory response is triggered (see Figure 17.10). There is a strong genetic association for IBD. There is a positive family history in approximately 5% to 15% of patients with IBD. In identical twins, the incidence of IBD is 44% versus only 3.8% in fraternal twins. Susceptible genes have been identified on chromosomes 2 and 6 (Hanauer 2003; Ogura et

FIGURE 17.10 Disease Pathogenesis of IBD: Comparison of colonic mucosa in normal, Crohn's, and ulcerative colitis patients; (top), gross; (center), histological; (bottom), endoscopic appearance.

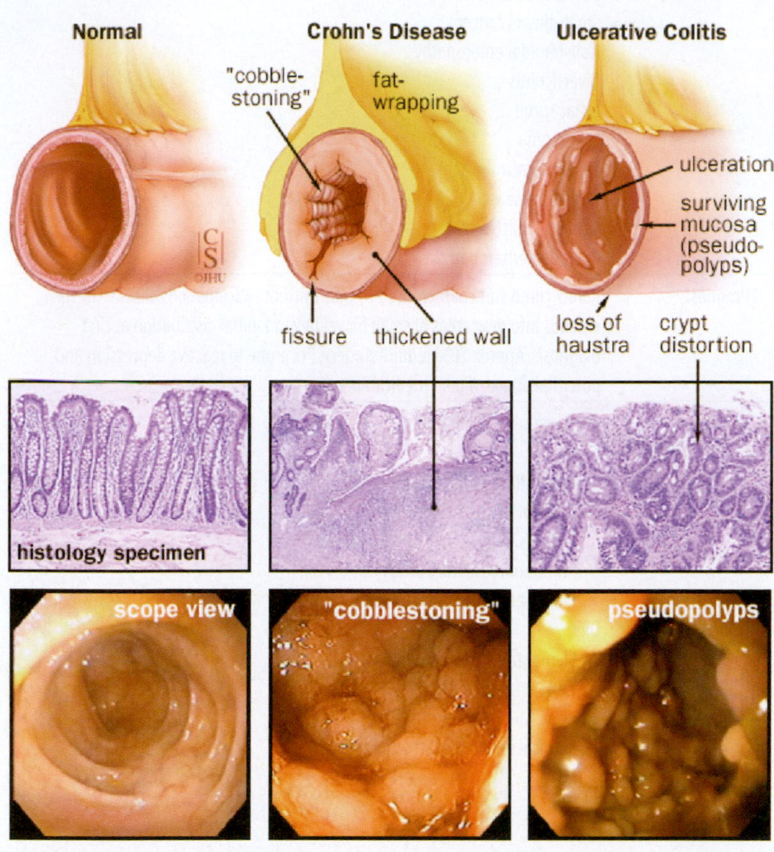

Source: Artwork is reproduced, with permission, from the Johns Hopkins Gastroenterology and Hepatology Resource Center, www.hopkins-gi.org, copyright 2006, Johns Hopkins University, all rights reserved.

al. 2001;). One specific gene, NOD2/CARD15, has been identified in Crohn's disease (Nayar and Rhodes 2004).

Pathophysiology Even though not all aspects of IBD are understood, it is theorized that those individuals who are genetically susceptible and who are exposed to certain triggers experience an abnormal immune response. This immune response results in release of cytokines that direct an excessive inflammatory reaction. It is this inflammatory reaction that destroys the intestinal mucosa. UC and Crohn's are also characterized by exacerbations of the disease process interspersed with periods of remission.

toxic megacolon—a very inflated colon with abdominal distention, and sometimes fever, abdominal pain, or shock

fistula—an abnormal opening or passage between two internal organs or from an internal organ to the surface of the body

UC disease is primarily located in the colon and rectum. Approximately 50% of patients with UC have disease only involving the rectum. Damage to intestinal mucosa in UC also usually only involves the first two layers of tissue (mucosa and superficial submucosa). But with chronic disease, the intestinal wall can become so thin the mucosa is ulcerated. This is referred to as **toxic megacolon**. UC disease usually affects one section of the gastrointestinal tract at a time whereas Crohn's disease often presents with a "skipping" pattern affecting multiple portions of the gastrointestinal tract.

Crohn's disease can affect any portion of the gastrointestinal tract from mouth to anus even though it most commonly affects the ileum and colon. Crohn's disease can damage all layers of gastrointestinal mucosa. Its inflammatory process is characterized by the development of **fistulas** that when healed are replaced by fibrotic tissue. This fibrosis can result in recurrent strictures and bowel obstructions.

Clinical Manifestations Patients with UC present with signs and symptoms including abdominal pain, bloody diarrhea, and tenesmus (urgency for defecation). Patients with severe disease often are febrile, tachycardic, and have diarrhea that contains pus and mucous. Disease activity is rated using the Truelove and Witts Criteria (Truelove and Witts 1955). (See Table 17.10.) Radiology testing (lower GI series with barium enema) often will show severe inflammation of the large bowel with thickened walls and superficial ulcerations, and over time, the haustra become edematous and thickened. During attacks of the disease, acute phase reactants such as C-reactive protein and the erythrocyte sedimentation rate (ESR) are elevated. Most recently, elevated lactoferrin levels in stool have been found indicative of exacerbations for UC (Kornbluth and Sachar 2004). White blood count can also be elevated, and leukocytes in stool confirm the inflammatory process. Biochemical indices for anemia are generally depressed, and in severe disease often confirm significant anemia.

Patients with Crohn's disease experience abdominal pain, diarrhea, and tenesmus. They are much less likely to have blood in their stool but usually experience more abdominal pain and cramping than patients with UC. Weight loss is very common due to both increased requirements and decreased oral intake. Crohn's disease can also be insidious, presenting with only mild symptoms or only those that are extraintestinal in nature. In research and clinical trials, Crohn's disease is described using the CDAI (Crohn's Disease Activity Index) that suggests that patients who score

TABLE 17.10

True love and Witts Criteria for Assessing Disease Activity in Ulcerative Colitis

	Mild Activity	Severe Activity
Daily bowel movements (no.)	≤5	>5
Hematochezia	Small amounts	Large amounts
Temperature	<37.5°C	≥37.5°C
Pulse	<90/min	≥90/min
Erythrocyte sedimentation rate	<30 mm/h	≥30 mm/h
Hemoglobin	>10 g/dl	≤10 g/dl

Patients with fewer than all 6 of the below criteria for severe activity have moderately active disease.

Reproduced with permission from BMJ Publishing Group: Truelove SC, Witts LJ. Cortisone in ulcerative colitis: final report on a therapeutic trial. *Br Med J.* 1955;2:1041–1048.

over 150 are experiencing a flare-up of the disease, and those who score over 300 are experiencing severe exacerbation of the disease (Best et al. 1976). Factors such as diarrhea, abdominal pain, abdominal mass, decreased sense of well-being, extraintestinal manifestations, weight loss, and laboratory features are evaluated in this index. Box 17.9 outlines the CDAI. The most current AGA practice guidelines describe Crohn's with four definitions: mild-moderate disease; moderate-severe disease; severe-fulminant disease; and remission (Kornbluth and Sachar 2004). See Table 17.11 for definitions of stages of Crohn's disease. As in UC, the ESR and C-reactive protein are generally elevated in active disease. In severe exacerbations, low albumin levels and elevated WBC are common. Radiology testing (barium enema, small bowel follow-through) can demonstrate deep ulcerations that often skip over portions of the GI tract. Fistulas and tracts between ulcerations can be observed in severe disease (see Box 17.10).

BOX 17.9 **CLINICAL APPLICATIONS**

Crohn's Disease Activity Index

The Crohn's Disease Activity Index was developed using data from the prospective National Cooperative Crohn's Disease Study group. Eight selected variables are used to estimate disease progress or lack of progress. Index values of < 150 are associated with quiescent disease. Values >150 indicate active disease, and values > 450 are seen in those with extremely severe disease.

Variable	Rating	Multiplication Factor
Number of liquid/very soft stools	Total for 1 week	2
Daily abdominal pain	0 = none, 1 = mild, 2 = moderate, 3 = severe	5
Daily ratings of general well-being for 1 week	0 = generally well, 1 = slightly under par, 2 = poor, 3 = very poor, 4 = terrible	7
Symptoms/complaints presumed related to Crohn's disease (select each set that applies)	Arthritis or arthralgia Iritis or uveitis Erythema nodosum, pyoderma gangrenosum, aphthous stomatitis Anal fissure, fistula or perirectal abscess Other bowel-related fistula Febrile episode > 100 degrees during past week	20
Taking Lomotil or opiates for diarrhea	0 = no, 1 = yes	30
Abdominal mass	0 = none, 2 = questionable, 5 = definite	10
Hematocrit (normal average): Male = 47 Female = 42	Skip this section if typical and/or current are unknown. Enter typical value AND current value	6
Weight factor % below predicted body weight 100 × [(standard weight − actual weight) / standard weight]	Recommendation: Skip this section unless weight changes related to Crohn's are known. The purpose is to check for weight change = change in conditions. (Example: wt loss caused by dehydration)	1

References:

Best WR, Becktel JM, Singleton JW and Kern F. Development of a Crohn's disease activity index: National Cooperative Crohn's Disease Study. Gastroenterology 1976; 70(3):438–444.

Ball C and Wotton C. Crohn's disease: the Crohn's disease activity index may identify patients in remission. Available at URL: http://www.eboncall.org. Accessed Jan. 3, 2005.

TABLE 17.11

Definitions of Stages of Crohn's Disease

Stage	Definitions
Mild-Moderate Disease	Ambulatory individuals able to tolerate oral alimentation without development of dehydration, toxicity (high fevers, rigors, prostration), abdominal tenderness, painful mass, obstruction, or >10% weight loss
Moderate-Severe Disease	Individuals who have failed to respond to treatment for mild-moderate disease or those with more major symptoms of fevers, significant weight loss, abdominal pain or tenderness, intermittent nausea or vomiting (without obstructive findings), or significant anemia
Severe-Fulminant Disease	Individuals with persisting symptoms in spite of introduction of steroids as outpatients, or those presenting with high fever, persistent vomiting, evidence of intestinal obstruction, rebound tenderness, cachexia, or evidence of an abscess
Remission	Asymptomatic individuals or those without inflammatory sequelae and includes those who have responded to acute medical intervention or have undergone surgical resection without gross evidence of residual disease

Note: Individuals requiring steroids to maintain well-being are considered to be "steroid-dependent" and are usually not considered to be "in remission."
Adapted from Hanauer ST, Sandborn W, and The Practice Parameters Committee of the American College of Gastroenterology. Management of Crohn's Disease in Adults. Practice guidelines. The American Journal of Gastroenterology. 2001;96(3):635–643.

BOX 17.10 CLINICAL APPLICATIONS

Diagnostic Testing for Lower Gastrointestinal Tract Disorders

In addition to a comprehensive medical history and physical examination, diagnostic evaluation including laboratory tests, imaging tests, and/or endoscopic procedures may be used to diagnose lower gastrointestinal disorders.

Category	Procedure	Can Be Used to Diagnose
Laboratory tests	Fecal occult blood	Irritable bowel syndrome
	Stool culture	Antibiotic-associated diarrhea and colitis, Irritable bowel syndrome
	Enzyme-linked immunoassay	Antibiotic-associated diarrhea and colitis
	CBC, electrolytes, ESR, urinalysis	Irritable bowel syndrome
Imaging tests	CT scan	Diverticulitis, Crohn's disease
	Barium (enema)	Diverticulitis, ulcerative colitis
	X-ray	Crohn's disease
	Ultrasound	Diverticulitis, Crohn's disease
Endoscopic procedures	Colonoscopy	Diverticulitis, ulcerative colitis, Crohn's disease
	Sigmoidoscopy	Ulcerative colitis, antibiotic-associated diarrhea and colitis
Other procedures	Mucosal biopsies	Ulcerative colitis, Crohn's disease

References:
Digestive disorders. Diagnostic procedures. Available at URL: http://www.rushcopley.com. Accessed January 4, 2005.

The Merck Manual. Chapter 107. Lower gastrointestinal tract disorders. Available at URL: http://www.merck.com. Accessed January 4, 2005.

Antibody testing has been used to distinguish between UC and Crohn's disease. Identification of ASCA (antisacchromyces antibodies) is now thought to be consistent with Crohn's. The presence of ANCA (antineutrophil cytoplasmic antibodies) is most consistent for ulcerative colitis (Dubinsky et al. 2003).

Patients with IBD can experience disease manifestations (referred to as extraintestinal) outside the GI tract. These include osteopenia and osteoporosis, dermatitis, rheumatological conditions such as ankylosing spondylitis, ocular symptoms, and hepatobiliary complications (Hanauer et al. 2002; Kornbluth and Sachar 2004).

Treatment Treatments for both UC and Crohn's disease include antibiotics, immunosuppressive medications, immunomodulators, and biologic therapies as well as surgical intervention (Hanauer and Sandborn 2001). Medical treatment for ulcerative colitis historically has used combinations of both antibacterial coverage with sulfapyridine and anti-inflammatory 5-aminosalicylic acid (5-ASA) therapy (see Table 17.12 for information about this and other medications used for diseases of the lower GI tract). The most commonly used medications in this category today include olsalazine, balsalazide, Asacol, Claversal, and Pentasa. Immunomodulators work to inhibit inflammatory cell proliferation by interrupting cellular RNA and by inhibiting the overall immune response. These medications include azathioprine (AZA) and 6-mercaptopurine (6-MP). Corticosteroids work to inhibit the overall inflammatory response and are commonly used to treat UC. Antibiotics are used in UC only when there is an acute infection.

Treatment for Crohn's disease can utilize all categories of medical treatments listed previously. Aminosalicylate medications are typically used in Crohn's disease that has ileal and colon involvement. These include mesalamine and sulfasalazine. As in UC, the immunomodulators, azathioprine (AZA) and 6-mercaptopurine (6-MP) are used. Corticosteroids, such as prednisone or budenoside, are often used in acute exacerbations, especially in severe-fulminant disease, but patients are at risk for becoming steroid dependent. Corticosteroids have been shown not to be successful when used to maintain remissions of the disease. Antibiotics

TABLE 17.12

Selected Medications for Lower Gastrointestinal Diseases

Classification	Mechanism	Generic	Brand Names	Possible Food-Drug Interactions
Antidiarrheal: Adsorbents	Provide protective coating for intestinal walls; adsorb toxins, virus, or bacteria	Kaolin, pectin, methylcellulose, activated attpulgite, magnesium aluminum silicate	Kaopectate Advanced Formula, Donnagel, Diasorb, Rheaban Maximum Strength, Equalactin	Constipation, bloating
		Polycarbophil	FiberCon, Fiberall, Mitrolan, Equalactin	Constipation, bloating
Antidiarrheal: Anticholinergics	Decrease intestinal muscle tone and peristalsis of GI tract to slow movement of fecal material through GI tract	Belladonna		Dry mouth/constipation
		Alkaloids		Dry mouth/constipation
		Atropine		Dry mouth/constipation
		Hyoscyamine	Anaspaz, Cystospaz, Cystospaz-M, Levsin, Neoquess	Dry mouth/constipation
Aminosalicylates: Anti-inflammatory	Works as an anti-inflammatory agent in the colon and may also act as an immune-suppressant	Sulfasalazine Olsalazine Mesalamine	Asulfidine Dipentum Asacol, Pentasa, Canasa, Balsalazide	Take with 8 oz water after meals or with food, take Fe or folate separately, folate supplement (1 mg/d), ↑ dietary folate, adequate hydration, anorexia, N/V, GI distress, diarrhea
Antibiotics	Enters bacteria and destroys DNA	Metronidazole	Flagyl	Anorexia, dry mouth, metallic taste, N/V, epigastric distress, diarrhea, constipation; avoid alcohol during use and 3 days after
Antidiarrheal: Bismuth compounds	Decrease fluid secretions; reduce stool output	Bismuth subsalicylate	Pepto Bismol	Dry mouth/constipation
Antidiarrheal: Intestinal flora modifiers	Cultures of *lactobacillus* organisms supply missing bacteria in GI tract and suppress growth of diarrhea-causing bacteria	*L. acidophilus*	Lactinex	None
Antidiarrheal: Opiates	Inhibit acetylcholine and decrease peristalsis	Loperamide Diphenoxylate Paregoric	Immodium Lomotil, Logen Camphorated Tincture of Opium	Dry mouth, N/V, abdominal pain, bloating, constipation
Glucocorticoids Anti-inflammatory	Mimics the action of cortisol; redistribution of white blood cells—reduction of lymphocytes; increase neutrophils; decreased production of prostaglandins	Prednisone Hydrocortisone Prednisolone Dexamethasone	Deltasone Depo-Medrol, Solu-Medrol, Medrol Decodron, Hexadrol, Dexameth, Dexone	Caution with DM; ↑ glucose, highly protein bound, may need ↑ K, PO_4, Ca, and ↑ vits A, C, D, ↑ protein, and ↓ dietary Na, avoid alcohol
Immunosuppressive	Antagonizes purine metabolism and may inhibit synthesis of DNA, RNA, and proteins; it may also interfere with cellular metabolism and inhibit mitosis	Azathioprine	Imuran	Anorexia, N/V, diarrhea, steatorrhea

(continued on the following page)

TABLE 17.12 *(continued)*

Selected Medications for Lower Gastrointestinal Diseases

Classification	Mechanism	Generic	Brand Names	Possible Food-Drug Interactions
Laxative: Bulking agent	Soluble fiber forms gel in the colon; retains water and increases peristalsis	Psyllium	Fiberall (powder and wafers) Metamucil (powder, wafers, sugar-free formula) Perdiem Fiber	Diarrhea, cramping, malabsorption of nutrients
		Calcium polycarbophil	FiberCon (tablets)	Diarrhea, cramping, malabsorption of nutrients
		Methylcellulose	Citrucel (powder and sugar-free formula)	Diarrhea, cramping, malabsorption of nutrients
Laxative: Stimulants	Stimulates peristalsis in the colon	Castor oil bisacodyl		Diarrhea, cramping, malabsorption of nutrients
Laxative: Stool softener	Emulsifier that softens stool	Glycerin Mineral oil Docusate sodium		Diarrhea, cramping, malabsorption of nutrients
Monoclonal antibodies	Antibody that blocks the effects of tumor necrosis factor alpha (TNF alpha)	Infliximab	Remicade	N/V, abdominal pain

Reference: Horne JS Swanson LN. Diarrhea. U.S. Pharmacist. Available at URL: http://www.uspharmacist.com/oldformat.asp?url=newlook/files/Feat/ACF2EF8.cfmandpub. Accessed 3 March 2006.
Marks JW, Lee D. Diarrhea. MedicineNet.com. Available at URL: http//www.medicinenet.com Accessed 3 March 2006. MedicineNet. Available at: http://www.medicinenet.com/infliximab/article.htm. Accessed 3 March 2006. Pronsky, Zaneta. *Powers and Moore's Food Medication Interactions*, 13th edition. Food-Medication Interactions, Birchrunville, PA, 2004. Shapiro W. Inflammatory bowel disease (last updated 6/04). Emedicine. Available at URL: http://www.emedicine.com. Accessed Jan. 4, 2005. U.S. National Library and National Institutes of Health. MedlinePlus. Available at URL: http://www.nlm.nih.gov/medlineplus/druginformation.html. Accessed 3 March 2006.

used include metronidazole and ciprofloxacin (Hanauer et al. 2002; Lichenstein and MacDermott 2002). The newest class of medications is biologic therapies. Infliximab is a monoclonal antibody that works to interrupt tumor necrosis factor-alpha (TNF-alpha) and thus cytokine-directed inflammatory activity seen in Crohn's disease. Clinical trials with infliximab have shown improvement in over 80% of patients treated (Hanauer et al. 2002).

Surgical intervention is required in both UC and Crohn's disease in over 60% of patients. The most common procedure in UC is a total colectomy, and in Crohn's disease, the ileostomy. Surgery is performed due to nonresponsive disease and due to acute complications such as perforation, obstruction, or abscess. These surgical procedures will be described in greater detail later in this chapter in the section Common Surgical Interventions for the Lower GI Tract.

Nutrition Therapy

Nutrition Implications Patients with IBD are at significant nutritional risk. It is estimated that from 60% to 75% of patients with Crohn's disease experience malnutrition (see Box 17.11) (Krok and Lichenstein 2003). Both Crohn's disease and UC have dramatic effects on nutritional status and often require nutritional support during periods of exacerbation. These diagnoses affect normal digestion and absorption; may increase caloric, protein, and micronutrient requirements; can result in protein-energy malnutrition; and additionally may require nutrition therapy to minimize symptoms. Nutrition therapy may also be implicated in treatment of the disease process.

Symptoms of both UC and Crohn's usually involve diarrhea and abdominal pain. Because increased motility decreases the success of digestion and absorption, severe diarrhea can result in malabsorption of all nutrients. These symptoms can also result in decreased oral intake. Many patients, during acute exacerbations of the disease, electively restrict eating in order to minimize symptoms. Pain commonly causes generalized anorexia, which further decreases dietary intake.

When infection is present or when the patient is febrile, energy needs are increased. Protein needs are increased, in some cases up to 150% of normal requirements. This is due, in part, to increased protein losses in inflammatory **exudate.** Micronutrients, especially iron, zinc, magnesium, and electrolytes, are at risk for deficiency due to their losses

exudate—fluid and cellular debris that seeps from blood vessels, usually as a result of inflammation

BOX 17.11 CLINICAL APPLICATIONS

Common Nutrient Deficiencies Seen with Crohn's Disease

Malnutrition can be present even when Crohn's disease is in remission. Protein-calorie malnutrition and other nutrient deficiencies can be caused by decreased nutrient intake, malabsorption, drug-nutrient interactions, anorexia, and protein-losing enteropathy. These, in turn, can lead to growth retardation (in children), anemia, osteoporosis, poor wound healing, and a compromised immune system.

Nutrient Deficiency	Probable Cause
Calories	Insufficient intake Anorexia Fear of abdominal pain and diarrhea after eating
Protein	Increased protein needs (losses from GI tract caused by inflammation) Catabolism (when infection or abscesses present) Healing from surgery
Fluid and electrolytes	Short bowel syndrome
Iron	Blood loss
Magnesium, zinc	Intestinal losses, especially from short bowel syndrome
Calcium and Vitamin D	Long-term steroid use Decreased intake of dairy foods as result of lactose-restricted diets
B_{12}	Surgical resections of stomach (loss of intrinsic factor) and/or terminal ileum (site of absorption)
Folate	Medications used to treat IBD

References:

Eden KA. Nutritional considerations in inflammatory bowel disease. *Practical Gastroenterology* 2003; May:33–54.

Jeejeebhoy KN. Clinical nutrition: 6. Management of nutritional problems of patients with Crohn's disease. *CMAJ* 2002; 166(7): 913–918.

Krok KL and Lichtenstein GR. Nutrition in Crohn (sic) disease. *Curr Opin Gastroenterol* 2003; 19(2):148–153. Available at URL: http://www.medscapte,com. Accessed Jan. 4, 2005.

in blood and diarrhea. IBD is common in both children and young adults. Meeting nutritional needs of the growing child or adolescent poses its own challenge. It is crucial for nutrition therapy to be designed to ensure adequate nutrients to support growth and development.

Since the mainstay of treatment for IBD involves multiple medications and often surgery, these nutritional risks compound those of the disease process. For example, use of corticosteroids can result in hyperglycemia, nitrogen wasting, and increased risk of osteoporosis. Another example is the use of sulfasalazine, which interferes with folate metabolism (see Table 17.12 for specific drug-nutrient interactions seen in IBD). Surgery increases calorie and protein requirements and additional nutrients are needed to support wound healing. Depending on the extent and type of surgery, normal absorption and digestive pathways may be interrupted. Specific nutrition therapy may be required if the patient has either an ileostomy or colostomy, as discussed in the section Common Surgical Interventions for the Lower GI Tract.

Nutrition Interventions

Nutrition Therapy during Exacerbation of Disease During acute exacerbations of both UC and Crohn's disease, the extent of diarrheal output and bleeding will direct the level of nutritional intervention. For **fulminant** disease, parenteral nutrition support or enteral nutrition with a chemically defined formula will probably be necessary. Most research indicates parenteral nutrition is not necessarily advantageous, and the gastrointestinal tract can benefit from exposure to enteral nutrition (Dominioni 2003). Glutamine and arginine supplementation may assist with modifying inflammatory response in the disease process (Akisu et al. 2003; Kanauchi et al. 2003; Panigrahi et al. 1997; Van der Hulst et al. 1993).

Energy needs for adults can be estimated using the Harris-Benedict or Mifflin-St. Jeor equation with appropriate stress factor (1.3–1.5). The amount of prior weight loss and the presence of infection will support the need for higher energy provision. To meet growth needs of infants, children, and adolescents, specific attention to their unique requirements is important. As much as 120 kcal/kg for infants and 80 kcal/kg for adolescents may be required (see Chapter 5). If available, indirect calorimetry provides the most reliable indicator of energy needs in the hospitalized patient.

Estimation of protein requirements will be based on the presence of any lean body mass wasting and biochemical parameters measuring protein status such as prealbumin and albumin. Protein needs may be as high as 1.5 to 1.75 g protein/kg for adults and 2.0 to 2.5 g/kg for infants, children, and adolescents.

If oral intake can be initiated, a low-residue, lactose-free diet with small, frequent meals is best tolerated. If steatorrhea is present, then fat should be reduced with added MCT or an MCT-containing supplement to assist with meeting energy requirements. As the patient responds to medical therapy, adding small amounts of fiber and then lactose as the patient can tolerate will advance the diet. Other foods that may need to be initially restricted may be gas-producing foods, spicy or fried foods, caffeinated beverages, or any other food the individual patient identifies as problematic. The addition and advancement of an oral diet will need to be highly individualized. Some research has indicated that there is no improvement in complication rates when comparing a low-residue diet to a regular diet in patients with Crohn's (Levenstein et al. 1985).

All patients should receive a multivitamin that meets the RDA or AI for all nutrients. Patients with IBD are at higher risk for deficiencies of vitamin B_{12} and iron. In a normal small intestine, the ileum has specific receptor sites that allow for B_{12} absorption. Therefore, disease affecting the ileum specifically can potentially result in B_{12} deficiency. Supplementation of B_{12} to prevent pernicious anemia can be accomplished using nasal gel or oral tablets, or by intramuscular injection (Little 1999; Eiden 2003). Micronutrient requirements are additionally increased during exacerbations of disease. It is recommended that additional supplementation should include zinc (12 to 15 mg/liter of stool output); calcium (10 to 25 mEq/day); magnesium (15 to 30 mEq/day); and copper (0.5 to 1.5 mg/day) (Eiden 2003; Jeejeehboy 2002).

Research has shown patients with Crohn's disease have lower serum levels of antioxidants (vitamin E, vitamin C, and beta-carotene) (Krok and Lichenstein 2003). It is thought that this might contribute to higher levels of **oxidative stress** in this disease. Higher levels of antioxidants may be warranted, but specific levels have not been established at this time; nor is it clear that supplement forms produce the same effect as foods. The most convincing evidence for the relationship between antioxidants and disease prevention has been in epidemiologic studies where strong associations have been demonstrated between dietary sources of fruits and vegetables and disease risk (McDermott 2000). Patients with IBD may avoid fruits and vegetables due to their disease symptoms and perceived intolerance to these foods.

Nutrition Therapy for Rehabilitation during Periods of Remission Maximizing energy and protein intake to facilitate rehabilitation should be the initial goal. Gaining weight within a normal healthy range combined with physical activity will ensure rebuilding of protein stores and muscle mass. Depending on the extent of disease and the response to treatment, specific dietary modifications will need to be individualized. It is always a goal to normalize dietary patterns and encourage a variety of all foods as the patient is able to tolerate them.

Consumption of foods high in antioxidants (for example, carotenoids, vitamin E, vitamin C, and selenium) and omega-3 fatty acids has been associated with protection against inflammation. These would include fruits, vegetables, vegetable oils, nuts, and fishes such as tuna and salmon. Although some reports have indicated that glutamine, short-chain fatty acids, antioxidants, and immunonutrition with omega-3 fatty acids are an important therapeutic alternative in the management of inflammatory bowel diseases, the reported beneficial effects have yet to be translated into clinical practice. The real efficacy of these nutrients still needs further evaluation through prospective and randomized trials (Campos et al. 2003).

Foods high in oxalate may increase risk for urolithiasis or kidney stones, which can occur in IBD. These foods include, for example, cocoa, tea, wheat germ, strawberries, nuts, spinach, beets and baked beans, peanut butter, tofu, and high doses of vitamin C supplements (>2 g/day).

As has been previously discussed in this chapter, use of probiotics and prebiotics enhance the normal flora of the GI tract. Several recent studies have indicated consumption of foods and supplements with probiotics and prebiotics have been associated with decreased symptoms for patients with IBD (Galvez, Rodriguez-Cabezas, and Zarzuelo 2005; Gassull 2005; Guarner 2005; Shanahan 2004; Schultz and Sartor 2000). See the previous discussion regarding prebiotics and probiotics (in the section Large Intestine Digestion and Absorption) and Box 17.2 for foods and supplements that provide probiotics and prebiotics.

oxidative stress—a disturbance in the pro-oxidant–antioxidant balance in favor of the former, leading to potential damage; indicators include damaged DNA bases, protein oxidation products, and lipid peroxidation products

Diverticulosis/Diverticulitis

Definition **Diverticulosis** is defined as the abnormal presence of outpockets or pouches on the surface of the small intestine or colon. Meckel's diverticulum is a type of diverticulosis present at birth. Meckel's diverticula are usually found near the ileocecal valve and may cause gastrointestinal bleeding or obstruction for the newborn.

Epidemiology Estimations of prevalence or incidence are difficult because diverticulosis, in most people, is asymptomatic. Best estimates, however, indicate diverticulosis is most common in Western and industrialized countries where it is thought that approximately 5% to 10% of the population will have diverticula by age 50. Incidence will increase with age, with some estimates as high as 65% to 70% in persons over the age of 85 (Beitz 2004; Kang, Melville and Maxwell 2004; Society for Surgery of the Alimentary Tract 2003).

Etiology Evidence suggests development of diverticulosis is related to low fiber intake, history of constipation, and the resulting long-term increased colonic pressure. Recently it has been proposed that low fiber intake also contributes to an increased risk for an inflammatory response in the colon, which further contributes to the incidence of diverticulosis (Floch and Bina 2004; Ye, Losada and West 2005). Factors that may increase risk for development of diverticulosis include obesity, decreased physical activity, steroids, alcohol and caffeine intake, and cigarette smoking (Aldoori et al. 1995, 1998).

Pathophysiology Diverticula do occur in the small intestine but are most common in the colon. Factors that affect integrity of the mucosa of the colon appear to contribute to development of the diverticula. The aging process and differences within parts of the colon may

FIGURE 17.11 Diverticula

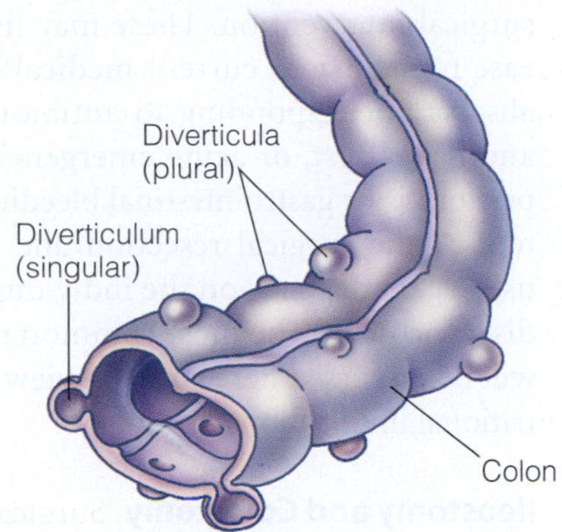

Source: E. Whitney, C. Cataldo, and L. DeBruyne, *Nutrition & Diet Therapy*, 6e, copyright © 2003 p. 469

account for the pattern of development that has been observed. Specific pathophysiology indicates that within the colon, two or more of the muscular bands (taeniae coli) contract at the same time. This hinders motility of the colon, and thus its ability to move waste products. Fecal matter becomes trapped and exerts excessive pressure against the wall of the colon. This pressure causes development of small pouches on the wall of the colon, which are referred to as diverticula (Beitz 2004; West and Losada 2004). (See Figure 17.11.) Constipation increases colonic pressure by excessive straining involved in bowel movements. This further increases the probability for development of diverticula (Stollman and Raskin 1999).

Diverticulitis is acute inflammation of the diverticula. Foodstuffs and bacteria can collect in diverticula and can become infected. Further complications can include development of bleeding, abscess, obstruction, fistula, or perforation (American Society of Colon and Rectal Surgeons 2000).

Clinical Manifestations Diverticulosis is asymptomatic for most individuals. Diverticula are usually only diagnosed when other tests such as a colonoscopy identify them. In approximately 20% of individuals, complications from diverticula may develop that include development of diverticulitis. Signs and symptoms of diverticulitis can include fever, abdominal pain, gastrointestinal bleeding, and elevated white blood cell count. Radiology testing (ultrasound and CT scan) may be used to diagnoses diverticulitis. Test results can demonstrate thickened walls of the colon, abscess, or inflammation (American Society of Colon and Rectal Surgeons 2000).

Treatment Treatment for diverticulosis involves only nutrition therapy, with a specific focus on fiber intake, and use of probiotic and prebiotic supplementation. Treatment for acute diverticulitis begins with making the patient NPO with complete bowel rest until symptoms subside. Antibiotics are used to treat any infection. The most common antibiotics involve treatment for Gram-negative rods and anaerobes (American Society of Colon and Rectal Surgeons 2000; Stollman and Raskin 1999). For those patients with complications (such as abscess or sepsis), surgical resections may be necessary.

Nutrition Therapy

Nutrition Implications Research indicates dietary habits may be strongly linked to the etiology of diverticulosis. Nutrition therapy should then focus on those nutrition interventions that could impact disease course. The patient with diverticulosis is not at any more risk for malnutrition than any other individual. The presence of diverticulitis with infection and inflammation does impact nutritional requirements if this condition is prolonged or if other complications, such as sepsis, occur.

Nutrition Interventions As mentioned earlier in the discussion of constipation, many Americans consume only limited amounts of fiber (Cordain et al. 2005). Nutrition therapy to treat and prevent diverticulosis will include a **high-fiber diet** of 6 to 10 grams above and beyond the recommendations of 20 to 35 grams/day. It is common practice to avoid nuts, seeds, and hulls, which are sharp enough, hard enough, or large enough to irritate or get caught in diverticula. This will prevent opportunity for these foodstuffs to lodge in diverticula and potentially result in diverticulitis. Foods to omit include caraway seeds, nuts, popcorn hulls, and sunflower, pumpkin, and sesame seeds. A food with small seeds such as a tomato, squash, cucumber, strawberry, or raspberry is usually tolerated. Many patients, especially the elderly, will need to use a fiber supplement if they are unable to consume adequate fiber from foods. Fiberall® and Metamucil® are bulk-forming agents made from psyllium, a source of insoluble fiber. Benefiber® is soluble dietary fiber extracted from guar gum. They may be used to normalize GI function. Other sources of fiber supplementation such as methylcellulose have been used successfully. Certainly the preferred method to increase fiber intake is through foods, but, if needed, these supplements are available.

The patient with acute diverticulitis will be progressed from being at bowel rest to clear liquids. The patient can then move toward a low-residue diet, avoiding nuts, seeds, and fibrous vegetables until inflammation and bleeding are no longer a risk.

diverticulosis—an abnormal presence of outpockets or pouches (diverticula) on the surface of the small intestine or colon

diverticulitis—an acute inflammation of the diverticula

high-fiber diet—a diet high in fiber (6 to 10 g above the usual recommendation of 20 to 35 g/day)

FIGURE 17.12 Surgical options for treatment of ulcerative colitis; A. proctocolectomy; B. Brooke ileostomy; C. Koch pouch ileostomy; D. restorative proctocolectomy.

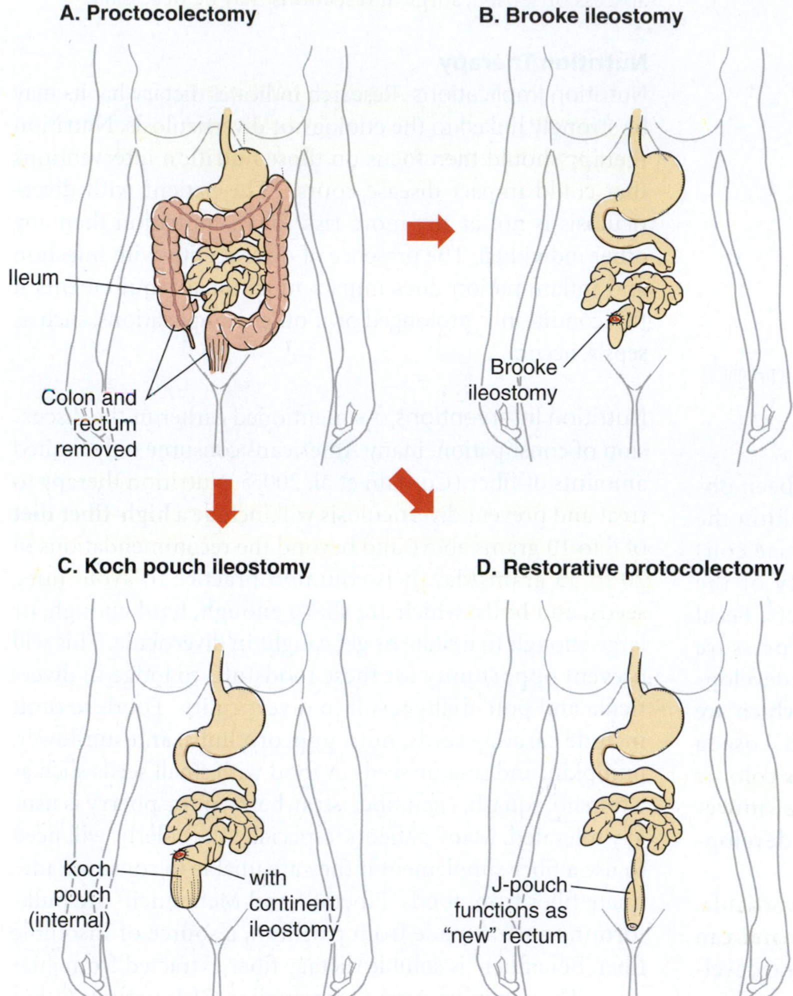

Source: The Johns Hopkins Medical Institutions Gastroenterology & Hepatology Resource Center. Digestive Disease Library – Colon & Rectum. Ulcerative Colitis, 2004

Common Surgical Interventions for the Lower GI Tract

Surgical resections may be warranted for many diagnoses discussed in this chapter, including Crohn's disease, UC, and diverticulitis. Specific details of each individual's disease course will determine need for surgical intervention. These may include disease refractory to current medical treatment, abscess not responding to antibiotic therapy and bowel rest, or acute emergencies such as peritonitis or gastrointestinal bleeding. The extent of the surgical resection and procedure used again depends on the individual patient's disease course. The most common procedures will be discussed here with a review of the nutritional implications.

stoma—a surgically created artificial opening into the abdomen

ileostomy—a procedure in which the colon and rectum are surgically removed, and the end of the ileum is attached to the stoma

colostomy—a procedure in which the rectum only is surgically removed, and the end of the colon is attached to the stoma

Ileostomy and Colostomy Surgical resection of the colon and rectum requires development of a new path for feces to be excreted from the body. Any of these surgeries create a stoma from which waste products can be excreted. A **stoma** is a surgically created artificial opening into the abdomen. An **ileostomy** is when the colon and rectum are removed. The end of the ileum is surgically attached to the stoma. The individual then uses an appliance (pouch) where feces and other waste products are collected. A **colostomy** exists when the rectum only is removed and the end of the colon is surgically attached to the stoma. Again, the individual utilizes a pouch appliance to collect waste products.

Other alternatives for these procedures exist where the outside surgical appliance can be avoided. These procedures create internal pouches where waste products can collect. These procedures include ileoanal reservoir surgery, ileal pouch-anal anastomosis, or continent ileostomy (see Figure 17.12).

Nutrition Therapy for Ileostomy and Colostomy

Nutrition Implications The surgical implications of intestinal resection may be analyzed in the context of anatomy and physiology of the gastrointestinal tract. When a certain part of the intestinal tract is removed, normal physiology and function of that portion is lost to the individual. This loss of function will produce change in motility, change in absorption, and change in how waste products are handled—all of which potentially can impact nutritional status. Larger resection will result in the most nutritional complications. Resections of the terminal ileum and loss of the ileocecal valve tend to result in significant fluid, electrolyte, vitamin, and mineral deficiencies. The ileocecal valve controls the rate of movement from the small intestine to the large; hence, when it is absent, motility is much faster, which interrupts normal absorption.

The location of the stoma on the GI tract will also determine the type of fecal matter produced. As explained earlier, the function of the colon is to reabsorb water and electrolytes.

Fecal matter further along the colon will produce firmer, less watery stool. Output in an ileostomy, then, will be much more liquid, while output in a colostomy, depending upon where it is located along the colon, will result in firmer, more normal stool (Beyer 2001).

Nutrition Interventions Goals for nutrition therapy include: decrease risk of obstruction, maintain normal fluid and electrolyte balance, reduce excessive fecal output, and minimize gas and flatulence (to reduce odor and inflation of the appliance) (American Dietetic Association 2000). After surgery, the patient will be transitioned to an oral diet. This begins with clear liquids and progresses as tolerated to a low-residue diet with four to six small feedings each day. Foods that may not be completely digested and that can cause stoma obstruction should be avoided for the first 6 to 8 weeks after surgery. These include tough fibrous meats; vegetables such as spinach, corn, and peas; dried fruits such as raisins; fruit skins and seeds; and popcorn. The patient will need to be instructed to eat slowly, chew thoroughly, and drink adequate fluids. (See Diet for Ostomies in Appendix E19.) Generally, oral intake should resemble the regular diet, meeting all nutritional needs by the eighth week postoperatively.

If the patient experiences excessive or watery fecal output, the amount of insoluble fiber should be reduced while increasing the amount of soluble fiber. Applesauce, bananas, tapioca, potatoes, oatmeal, oat bran, rice, and pasta may help decrease diarrhea. Foods that cause gas and flatulence should be avoided (see Table 17.8). It is these same foods that can cause difficulty for the patient with an ostomy. Use of yogurt, parsley, and buttermilk may decrease gas and odor.

Recent research has focused on absorption of vitamins and minerals in the diets for patients with ileostomy. It was found that most vitamins, minerals, and phytochemicals appear to have adequate absorption in this population. (Chen et al. 2004; Faulks et al. 2004). Livny and colleagues did find beta-carotene was best absorbed from cooked carrots rather than raw carrots in these patients (Livny 2003). However, since most patients' intake and tolerance varies widely, a general multivitamin is recommended. Vitamin B_{12} supplementation may also be required, as previously discussed.

Short Bowel Syndrome

Definition **Short bowel syndrome (SBS)** (also known as short gut syndrome) results from a large resection of the small intestine. Specific definitions vary, but Buchman (2004) describes the most important concerns of this condition: patients who are at the greatest nutritional and dehydration risk generally have less than 115 cm of residual small intestine in the absence of colon in continuity or less than 60 cm of residual small intestine with colon in continuity (Buchman 2004). Patients with less than 100 cm of residual jejunum often have a net secretory response to food and may actually secrete more fluid than they ingest. The American Gastroenterological Association (2003) states, in its clinical guidelines, that "short bowel syndrome occurs when, after surgery or congenitally, a patient is left with less than 200 cm of functional small intestine" (American Gastroenterological Association 2003, p. 1105). Those patients who also experience resections of the large intestine will have additional symptoms that contribute to complications of their SBS.

TABLE 17.13

Etiology of Short Bowel Syndrome in Children and Adults

Children	*Adults*
Necrotizing enterocolitis	Massive surgical resection
Intestinal atresia (volvulus, hernia, intussusception)	Crohn's disease
	Malignancy
Congential short bowel syndrome	Radiation enteritis
Trauma	Trauma
Gastroschisis	Vascular catastrophes (embolus/thrombus)
Apple peel anomaly	
Crohn's disease	Volvulus
Abdominal tumors	Strangulated hernias
Radiation enteritis	SB fistulas
Hirschsprung's disease	Surgical bypass
	Surgical error or obesity treatment
	Chronic intestinal pseudo-obstruction

Source: Ree Parrish, C. The Clinician's Guide to Short Bowel Syndrome. Practical Gastroenterology. September 2005, p.70.

Epidemiology Incidence of SBS is estimated to be approximately two to three cases per million individuals per year. Prevalence is approximately four cases per million individuals per year (Buchman 2004). The most common causes of SBS are malignancy, damage from radiation therapy (radiation enteritis), Crohn's disease with resulting multiple resections, vascular accident, trauma, or **volvulus** (Beyer 2001; Buchman 2004).

Etiology Surgical resections of the small intestine and colon due to disease and trauma can result in extensive loss of surface area of the small intestine and colon. Without normal anatomy and physiology, malabsorption of nutrients, fluids, and electrolytes will result (see Table 17.13).

Pathophysiology Several factors will determine the prognosis of this condition: extent of remaining small intestine,

short bowel syndrome (SBS)—decreased digestion and absorption that result from a large resection of the small intestine

volvulus—the twisting of the bowel causing obstruction

presence of the colon, presence of the ileocecal valve, health of the remaining gastrointestinal tract, and any comorbid conditions the individual may have. Though each case is highly individualized, most research agrees that a resection of more than 70% of the GI tract will result in severe nutritional and metabolic complications (American Gastroenterological Association 2003; Buchman 2004; Lykins and Stockwell 1998).

The postoperative period for SBS generally follows three distinct phases. The first period ranges anywhere from 7 to 10 days and is characterized by extensive fluid and electrolyte losses within large volumes of diarrhea. During this phase, patients are dependent on total parenteral nutrition, which provides not only required nutrients but manages fluid and electrolyte balance.

The second postoperative phase may last for several months and is characterized by reduction in diarrhea volumes with the initial stages of adaptation of the remaining bowel. It is during this phase that enteral nutrition can be introduced with a gradual transition to an oral diet (Rees-Parrish 2005).

During the third phase, there is continued adaptation of the remaining bowel. There is some evidence the intestinal tract increases in both length and diameter with additional increase in villous height. This time frame varies, but may range from 1 to 2 years (Buchman 2004).

As mentioned earlier, the amount of remaining bowel determines the extent of this condition. Loss of the ileum prevents B_{12} absorption and reabsorption of bile salts. Reduction in bile salts further contributes to fat malabsorption. No other part of the intestinal tract can compensate for these losses. The ileocecal valve not only controls intestinal motility but also prevents translocation of bacteria from the colon to the small intestine. When this control is lost, nutritional and metabolic complications are much more prominent.

Vitamin and mineral losses are major issues in SBS. When there is fat malabsorption, there is an inability to absorb adequate amounts of vitamins A, D, E, and K. These will need to be supplemented appropriately, and levels within the body will need to be evaluated. Other nutrients often deficient include sodium, magnesium, iron, zinc, selenium, and calcium, because they are often lost in the large volumes of diarrhea (Rees-Parrish 2005).

Treatment Initially, medical treatment will focus on managing fluid and electrolyte balance. This is generally managed by total parenteral nutrition and intravenous support initially, and then as the patient is able, by oral rehydration solutions. Motility is controlled by medications used to treat symptoms of diarrhea. These agents work to either decrease motility or to thicken consistency of the stool. These include medications such as LoMotil® (diphenoxylate), Immodium (loperamide), paregoric, codeine, Tincture of Opium, Kaopectate, or bismuth subsalicylate. Octreotide, which is given intravenously, is a somatostatin analog that reduces the levels of growth

TABLE 17.14

Osmolality of Selected Liquids

Beverage	*(mOsm/kg)*	*Beverage*	*(mOsm/kg)*
Milk	275	Prune juice	1265
Malted milk	940	Grape juice	863
Ice cream	1905	Apple juice	683
Eggnog	695	Orange juice	614
Fruit yogurt	871	Tomato juice	595
Sherbet	125	Punch with sugar	448
Popsicles	720	Sugar-free punch	29
Ensure/Boost	590/640	Mineral water	74
Ensure Plus/Boost Plus	680/720	Broth	445
Boost Breeze	920	Polycose	900
Enlive!	840	Flavored gelatin	735
Resource fruit beverage	750	D_{10} (10% dextrose in water)	505
Enteral formulas	250–710		

Source: Rees Parrish C. The clinician's guide to short bowel syndrome. Nutrition issues in gastroenterology, series #31. Practical Gastroenterology. Sept. 2005, pp. 88–89.

hormone and has been used to treat diarrhea in short bowel syndrome (Gomez-Herrera, Farias-Llamas, and Gutierrez-de la Rosa 2004; Rees-Parrish 2005). Gastric hypersecretion, common after extensive bowel resection, is treated by use of proton pump inhibitors or H_2 antagonists. In general, though, oral medications are not consistently absorbed and will need to be monitored closely.

New classes of drugs have been used to enhance cell proliferation in the remaining portions of the intestinal tract. These include growth hormone and GLP-II (glucagon-like peptides) (Jeppesen et al. 2001; Seguy et al. 2003).

Nutrition Therapy

Nutrition Implications Maintenance of nutritional and hydration status is critical for individuals with SBS. Aggressive nutrition support and careful progression to an oral diet require careful attention by the health care team.

Nutrition Interventions Immediately postoperatively, patients will receive total parenteral nutrition. Prescription for this therapy will be based on energy, protein, and micronutrient requirements (see Chapter 7). As diarrhea begins to decrease (anywhere from 2 to 6 weeks postoperatively), oral diets can begin (Sundaram, Koutkia, and Apovian 2002). Many patients require combinations of parenteral, enteral, and oral nutrition support in order to accommodate degrees of malabsorption and patient requirements for nutrition and fluids.

Sugar-free isotonic clear liquids should be the first items offered. Table 17.14 outlines examples of isotonic beverage choices. Diet is then progressed slowly to a low-residue, low-fat, lactose-free, low-oxalate diet. Caffeine and alcohol should

TABLE 17.15

Diet Guidelines for Short Bowel Syndrome

General Guidelines

Patients with jejunostomies/ileostomies (higher fat): approximately 20–30% CHO, 20–30% protein, 50–60% fat.

Patients with intact colon (higher CHO): approximately 50–60% CHO, 20–30% protein, 20–30% fat.

Avoid concentrated sweets and fluids.

Chew foods well.

Add salty meals and snacks if no colon.

Eat smaller meals, more often.

Decrease total nutrient load over the day and space out over time.

Trial of oral rehydration solutions.

Limit fluids with meals; drink isotonic beverages.

Separate solids and liquids at meals as much as possible (solids before liquids).

Solids slow emptying.

Too much liquid creates a column effect (imagine the swelling of a stream when it rains and the increased flow generated).

Use MCT containing beverages if necessary vs MCT oil (45).

Lactose restriction if necessary (may try Lactaid).

Avoid high oxalate foods in those patients with kidney stones.

Liquid or chewable vitamin/mineral supplements if necessary.

Limit or avoid enteral stimulants such as alcohol and caffeine.

Good Choices	***Avoid***
Starches/breads	
Breads, pita bread, rolls	Donuts, sweet rolls, pastries, Pop-Tarts
Bagels, English muffins	
Plain waffles or pancakes	
Corn bread, plain muffins	
Banana or zucchini bread	
Tortillas—whole wheat or white flour, corn—toasted	
Pasta, macaroni, noodles	
Rice, brown rice, wild rice	
Cereals	
Unsweetened cereals (wet or eaten dry as a snack)	Sugary cereals, high fiber cereals (>1–2 grams fiber/serving), bran cereals
Cheerios, cornflakes, Rice Krispies, Rice Chex, Spoonfuls, Special K, Kix, puffed rice or wheat	Flavored hot cereals
Hot cereals: cream of rice or wheat, grits, oatmeal	
Vegetables	
Canned or cooked vegetables	Creamed vegetables, legumes such as lima, kidney, pinto beans, etc.
Potatoes, sweet potatoes, yams	
Small amounts of lettuce (1/2 cup)	
Fruits	
Bananas, melons, unsweetened canned fruits (applesauce, pears, peaches, mandarin oranges, apricots, cherries, plums, etc.)	Dried fruits, fruit canned in syrup, fruit juice, fruit drinks, watch out for high fructose corn syrup in drinks (e.g., CAPRI SUN)

(continued on the following page)

TABLE 17.15 *(continued)*

Diet Guidelines for Short Bowel Syndrome

Good Choices	*Avoid*
Meats/fish/poultry	
Meats, fish, shellfish, poultry, tuna fish, ham	Heavily fried meats, fish, poultry
Dairy/Soy	
Cheese, cottage cheese, plain yogurt or yogurt sweetened with artificial sweeteners, cream cheese Plain soy milk	Highly sweetened yogurts or kefir, chocolate or other flavored milks, cream, half and half, Go-GURT, flavored soy milks
Eggs	
Poached, hard or soft cooked, omelet, scrambled	Eggs prepared with ingredients not allowed
Nut butters	
Peanut, almond, cashew	Nutella, peanut butter with jam/jelly mixed in it
Beverages	
Oral Rehydration solution Soups, broth—4 oz per day Lactaid milk	>4 oz coffee, tea, ice tea, flavored coffees or teas, hot cocoa, Ovaltine, Quick, fruit juices or fruit drinks (watch out for high fructose corn syrup in drinks), Kool-Aid, Tang, regular sodas (all kinds), alcohol, water, sugar-free beverages, supplements such as Boost or Ensure
Snacks	
Crackers—saltines, soda, and so on Pretzels, matzo Corn or potato chips Bagel snack crackers	
Desserts	
Animal crackers, graham crackers, angel food cake, vanilla wafers, shortbread, plain pound cake, cake donuts—no icing, marshmallows	Iced cakes, cookies, Little Debbie Cakes, pie, ice cream, sherbet, candies, donuts, sweetened gelatin
Miscellaneous	
Salt, pepper, herbs, spices, dill pickles, Splenda, Equal, Sweet 'n Low	Sugar, sorbitol-containing sweets, maple or other syrups, jams, jellies, chocolate syrup, honey, molasses

Source: Rees Parrish C. The clinician's guide to short bowel syndrome. Nutrition issues in gastroenterology, series #31. Practical Gastroenterology. Sept. 2005, pp. 88–89. http://www.healthsystem.virginia.edu/internet/digestive-health/nutritionarticles/September2005.pdf (Table 16). Used with permission from the University of Virginia Health System Nutrition Support Traineeship Syllabus

not be initially consumed. Alcohol sugars such as xylitol, mannitol, and sorbitol are usually not tolerated. Insoluble fiber is generally not tolerated initially, but sources of soluble fiber may actually assist in promoting mucosal health. Soluble fiber, like other sources of prebiotics, assists in production of short-chain fatty acids that are a primary fuel for the colon. Overall, it is crucial to remember that the diet truly needs to be designed for the individual patient and that significant differences between patients will be likely. As Rees-Parrish states, "Fat is an important calorie source. Maximize medication delivery before imposing strict dietary guidelines —no diet is a good diet if not eaten" (Rees-Parrish 2005, p. 86). Table 17.15 summarizes nutrition therapy for short bowel syndrome.

One item at a time is added to the diet to ensure tolerance. If GI symptoms are exacerbated, the food added should be removed from the diet. It may be added again at a later date, depending on the patient's adaptation after surgery. Lykins and Stockwell suggest it may be best to retry categories of restricted foods even as long as 6 months, since bowel adaptation can take as long as 1 to 2 years (Lykins and Stockwell 1998).

Many patients with SBS are discharged on home parenteral nutrition (PN) or home enteral nutrition support (EN) in addition to the limited oral diet. PN or EN is usually cycled over 10 to 12 hours at home, which will allow a patient to resume normal activity.

Bacterial Overgrowth

Definition **Bacterial overgrowth syndrome** results from cross-contamination of bacteria from the colon to the small intestine. This may be a result of surgery, disease, or trauma to the GI tract.

Pathophysiology In this condition, motility of the gastrointestinal tract is delayed due to disease, surgery, or trauma, and stasis develops. There is a high risk for development of small bowel bacterial overgrowth for those individuals with short bowel syndrome. Bacteria numbers increase and begin to compete with the host for nutrients. Malabsorption, maldigestion, and malnutrition can result (Parisi et al. 2003; Singh and Toskes 2003).

Clinical Manifestations Signs and symptoms are similar to all conditions of malabsorption. Diarrhea, steatorrhea, anemia, and weight loss all may be present in this condition. Hydrogen breath tests (see previous section, Carbohydrate Malabsorption) can assist in diagnosis, but according to Rees-Parrish (2005), most clinicians initiate treatment with antibiotics due to the cost and complications of administering the hydrogen breath test.

Treatment Bacterial overgrowth syndrome is treated by both correcting the underlying cause and by the use of broad-spectrum antibiotics (Singh and Toskes 2003).

Nutrition Therapy

Nutrition Interventions Nutrition therapy will be consistent with the level of malabsorption that is present. Nutrients most commonly malabsorbed (fat and lactose) should be eliminated from the diet initially until the underlying condition is treated. Methods to increase nutrient density will accompany any restrictions, and steps should be made to maximize caloric and protein intake to replenish nutrient stores.

Conclusion

Nutrition is intimately involved in treatment for all diseases of the gastrointestinal tract. This chapter has discussed digestion, absorption, and transport of nutrients, which is crucial when evaluating the effect of disease on the gastrointestinal tract. Any diagnosis involving maldigestion and malabsorption has the potential to alter the nutritional status of the patient. The genetics of disease and the use of nutrition supplementation to treat disease (such as the use of fructooligosaccharides, glutamine supplementation, and probiotics) are current topics that have been addressed and will be of utmost interest to any student. Finally, nutrition therapy must be based on a thorough understanding of the effect of malnutrition on the integrity and function of the intestinal tract. This foundational knowledge allows the clinician not only to understand the disease process but also to plan nutritional rehabilitation for the patient. This chapter has discussed nutrition interventions for the lower GI tract, including specific diet modifications to assist in the treatment of disease.

bacterial overgrowth syndrome—malabsorption and malnutrition that result from cross contamination of bacteria from the colon to the small intestine

PRACTITIONER INTERVIEW

Shelly Case, B.Sc., R.D. *Nutrition Counselor and Author of the Book* Gluten-Free Diet

Working with patients with celiac disease is very rewarding. Nutrition therapy is the only treatment for this disorder. I see many patients or family members who don't feel well, and by educating them about what they can and can't eat, they start to feel better, often within 2 weeks. They think of me as a miracle worker. What could be better than that?

We now know that celiac disease is more common than previously thought, and more individuals are being diagnosed. As a result, I am seeing more clients. When I first meet with the client I try to focus on the positive; for example: "This is a 'good' autoimmune disorder with a known treatment; the diet will make you feel better and improve your quality of life; and the diet is healthful and may help prevent other chronic diseases." But at the same time I am up front with them since the diet is complex and challenging, and explain that I will give them the tools they need so they know which food to eat or not eat.

First, I work with them so they can identify the foods they should avoid and then plan meal menus. I use the grocery store as an education tool. I give them a list of the cereals that are allowed and those to be avoided, and have them walk the perimeter of the store to identify foods they can eat. Less processed food is located in these aisles. I then have them move into the inside aisles and again identify foods that they can eat that are more processed. Obviously, learning to read the ingredient label is very important. Lastly, I go over the special gluten-free products. The breakfast and lunch menus are the hardest, since they are often wheat based. Kids are more willing to try nontraditional items, like fruit smoothies for breakfast. I encourage dinner leftovers for lunch—meat, rice, cheese, fruit, and so on. For dietitians, food is our profession; but for patients newly diagnosed with celiac disease, we forget that they now have to think about every bite they take, and it's overwhelming. I try to hook them up with a support group, so they can network with others, and it helps them cope with the diet and the disease.

Although working with celiac disease clients is very rewarding, I have frustrating moments as well. Some clients who have been diagnosed with the disease have no symptoms, and they are often noncompliant, even though one of the long-term complications of this disorder, if untreated, is an increased incidence of several types of cancer. I also have to convince doctors to not tell patients to try the diet before they have the diagnostic serological tests and biopsy as these tests may come back negative if the patient is avoiding gluten.

When I started out, I wasn't an expert in celiac disease, but I had clients who needed help, so I did my homework. Remember—you are not taught everything you need to know in school or your internship, but you are taught how to look for information. My search for information resulted in a career just by developing resources for clients. I networked with celiac patients, which led to being the local advisor of a celiac chapter, which led to being on the national board. I was asked to be a speaker, which led to more referrals and then a book. I am now an expert RD on celiac disease and have been on the *Today Show* and a National Institutes of Health panel. My advice to young dietitians is to be on the lookout for a niche that you are interested in, do your homework, and be a bit of a risk taker.

CASE STUDY

Introduction:

KM is a 36-year-old female whose small bowel biopsy indicates: flat mucosa with villus atrophy and hyperplastic crypts—inflammatory infiltrate in lamina propria. A 72-hour fecal fat test (11.5 g) indicates steatorrhea and malabsorption. Positive AGA, EMA antibodies. Her gastroenterologist has informed her of a positive diagnosis of celiac disease with secondary malabsorption and anemia.

Nutrition Assessment:

Ht. 5′3″ Wt. 92 lbs UBW 112 lbs

Labs: Total Protein 5.5 g/dL, Albumin 2.9 g/dL, Prealbumin 14 mg/dL Hgb 10.5 g/dL, Hct 35%, Ferritin 12 μg/dL, Vitamin B_{12} 82 ng/mL.

The registered dietitian's interview with the patient reveals that she is hungry all the time. "I do eat, but it seems that every time I eat in any large amount that I almost immediately have diarrhea. I do not have nausea or vomiting." Foods that are fried and meat—especially beef—tend to make the diarrhea worse. Relates that she has been relying on chicken noodle soup, crackers, and Sprite for the last several days. Patient relates that her highest weight was prior to her last pregnancy where she weighed 112 lbs. She gained 11 lbs with her pregnancy and her full-term son weighed 5 lbs 2 oz.

Questions:

1. What is a 72-hour fecal fat test? Interpret her result of 11.5 g/24 hours.
2. What do the results of her small bowel biopsy tell you about the change in the anatomy of the small intestine? How is celiac disease related to this change?
3. What is the etiology of Celiac disease? How do AGA and EMA antibodies assist in this diagnosis?
4. Identify at least two nutrition problems that can be found as a result of the nutrition assessment and medical history. Next, identify the etiology of each nutrition problem. Finally, identify the signs and symptoms that support the evidence for these nutrition problems.

NUTRITION CARE PROCESS FOR THE LOWER GASTROINTESTINAL TRACT

Step 1 : Nutrition Assessment

Medical/Social History

Diagnoses:

Previous medical conditions or surgeries

Medications:

Socioeconomic status/food security

Support systems

Education—primary language

Dietary Assessment

Ability to chew; use and fit of dentures

Problems swallowing

Nausea, vomiting

Constipation, diarrhea

Heartburn

Any other symptoms interfering with ability to ingest normal diet

Ability to consistently purchase adequate amounts of food on a daily basis

Ability to feed self

Ability to cook and prepare meals

Food allergies, preferences, or intolerances

Previous food restrictions

Ethnic, cultural and religious influences

Use of alcohol, vitamin, mineral, herbal or other type of supplements

Previous nutrition education or medical nutrition therapy

Eating Pattern: 24-hour recall, diet history, food frequency

Anthropometric

Height

Current weight

Weight history: highest adult weight; usual body weight

Reference weight (BMI)

Biochemical Assessment

Visceral Protein Assessment:

Albumin

Prealbumin

Transferrrin

Retinol Binding Protein

Fibronectin

3-Methyl Histidine

C Reactive Protein

Immunocompetence:

Delayed cutaneous hypersensitivity

Total Lymphocyte Count

Hematological Assessment:

Hemoglobin

Hematocrit

MCV

MCHC

MCH

TIBC

Lipid Assessment

Total cholesterol

HDL

LDL

Triglyceride

Lower GI: Specific Labs

Folate	Erythrocyte folate, free folate, (FIGLU) urinary formiminoglutamic acid
Vitamin B_{12}	Schilling test, erythrocyte B_{12}, DUMP test, serum B_{12}
Vitamin C	Plasma vitamin C, leukocyte vitamin C, urinary vitamin C
Vitamin D	Serum alkaline phosphatase

Vitamin K	Prothrombin time
Vitamin A	Serum carotene, retinal binding protein
Vitamin E	Serum tocopherol, erythrocyte hemolysis
Biotin	Serum biotin, urinary biotin
Niacin	Urinary N-methyl nicotinamide
Riboflavin	Urinary riboflavin, erythrocyte glutathione reductase
Vitamin B_6	Whole blood level of pyridoxal phosphate
Thiamin	Blood pyruvate and lactate, urinary thiamin excretion, erythrocyte transketolase, apoenzyme levels

Step Two: Common Diagnostic Labels

Altered GI function

Inadequate oral food/beverage intake

Inadequate energy intake

Impaired nutrient utilization

Altered nutrition-related laboratory values

Food-medication interaction

Involuntary weight loss

Underweight

Sample: Altered GI function

PES: Altered GI function related to decreased functional length of GI tract as evidenced by postoperative status >65% of small bowel resected, diarrhea, and dehydration.

Step Three: Sample Intervention

1. Initiate parenteral nutrition: 200 mL 50% dextrose, 800 mL 10% amino acids, 300 mL 10% lipids in total volume of 1392 mL, 58 mL/hour continuous 24-hour infusion (21 mL/kg).
2. Increase rate 80 mL/hr as patient tolerates.
3. Adjust electrolytes per daily laboratory values.

Step Four: Monitoring and Evaluation

1. Patient will meet goal rate of 80 mL/hr to meet 95% of estimated energy and protein needs within 48 hours.

WEB LINKS

National Institutes of Health—National Institute of Diabetes, Digestive, and Kidney Disease: This site provides information about diagnoses and treatment for diseases of the gastrointestinal tract, including excellent information about anatomy and physiology. There are links to current research and clinical trials.

http://www.niddk.nih.gov

National Guideline Clearinghouse: Resource for evidence-based clinical practice guidelines which is a collaboration between the Agency for Healthcare Research and Quality, U.S. Department of Health and Human Services, American Medical Association and American Association of Health Plans.

http://www.guideline.gov

Crohn's Colitis Foundation: This organization provides information about these diseases and current treatments for patients. Additionally, this organization funds research, provides and sponsors educational workshops and symposia, and publishes the journal *Inflammatory Bowel Diseases.*

http://www.ccfa.org

Celiac Foundation: This organization provides support, information, and assistance to people affected by celiac disease/dermatitis herpetiformis (CD/DH). Links to current research and gluten-free products are prominently included at this site.

http://www.celiac.org

U.S. Probiotics Organization: Background research about probiotics along with consumer information about products.

http://www.usprobiotics.org

END-OF-CHAPTER QUESTIONS

1. What are pre- and probiotics? How do they affect the health of the GI tract?
2. Describe the types of diarrhea and compare/contrast their possible etiologies. Are there nutritional consequences of diarrhea? Describe dietary measures that are commonly recommended for diarrhea.
3. Describe the pathophysiology of irritable bowel syndrome (IBS) and its recommended medical treatment. What is the role of NT in the treatment of IBS?
4. Describe the pathophysiology of inflammatory bowel disease (IBD) by comparing Crohn's disease and ulcerative colitis. Medically, what is recommended for the treatment of IBD? What are the potential nutritional consequences of IBD? Describe common NT recommendations for IBD.
5. How can diet help prevent and treat diverticulosis? Describe the pathophysiology of diverticulitis.
6. What is short bowel syndrome, and what factors increase its incidence after surgery? Describe the role of NT in the treatment of short bowel syndrome.
7. Describe the primary nutrition-related concerns of people who have undergone colostomies and ileostomies.

NOTES

[1] A "BRAT diet" traditionally used to treat diarrhea in children consists of ripe bananas, rice, applesauce, and toast.

[2] Although not the first physician to suspect the disorder, Dr. Burrill Crohn, working at the Mount Sinai Hospital in New York, was the first to report on several cases in 1932, which he referred to as regional ileitis. Later, this inflammatory disorder was named after Dr. Crohn.

18

Diseases of the Hepatobiliary: Liver, Gallbladder, Exocrine Pancreas

Mildred Mattfeldt-Beman, Ph.D., R.D., L.D.

Department of Nutrition and Dietetics, Saint Louis University

CHAPTER OUTLINE

Normal Anatomy and Physiology of the Liver
Anatomy • Functions of the Liver • Bile
Jaundice
Pertinent Laboratory Values and Procedures
Use of Tests
Pathophysiology of the Liver
Alcoholism and Malnutrition • Hepatitis • Cystic Fibrosis-Associated Liver Disease (CFALD)
The Gallbladder
Normal Function of the Gallbladder • Cholelithiasis (Gallstones)
The Pancreas
Normal Anatomy and Physiology of the Pancreas • Pancreatitis

Introduction

The liver is one of the largest organs in the body, weighing about three pounds in an adult. It influences nutritional status through the synthesis of bile salts and metabolism of protein, carbohydrate, fat, and vitamins. It is the first stop for nutrient-rich blood from the intestines and protects the body through modification of toxic substances. It is central to hemopoiesis and blood clotting, both synthesizing and storing of needed compounds for these processes. The liver, pancreas, and gallbladder are accessory organs to the gastrointestinal (GI) tract (see Figure 18.1). All three organs are important in digestion, absorption, and/or metabolism of nutrients from food, and constitute the hepatobiliary processes. (The Greek word for liver is "hepatos," hence "hepatic.") An indication of the importance of the liver is its ability to regenerate itself: Even if up to 70% of a healthy liver is removed, it will regenerate to its original size (Fausto 2000).

Normal Anatomy and Physiology of the Liver

Anatomy

The liver is located in the upper right quadrant of the abdomen, following the curve of the diaphragm. The healthy human liver is brownish-red in color; when highly infiltrated with fat, it becomes a dark, yellowish-brown mustard color. There are four anatomical lobes of the liver: the right lobe (largest), quadrate lobe, caudate lobe, and left lobe. Based on blood supply and biliary drainage, there are two functional lobes: right and left. The right lobe receives blood from the right hepatic artery, and the left lobe (including the caudate and quadrate lobes) receives blood from the left hepatic and middle hepatic arteries. The right and left hepatic ducts drain each lobe, respectively.

The basic functional unit of the liver is the lobule, which is a cylindrical structure several millimeters in length and 0.8 to 2 mm in diameter (see Figure 18.2). Each lobe is made up

FIGURE 18.1 Gross Anatomy of the Hepatobiliary System

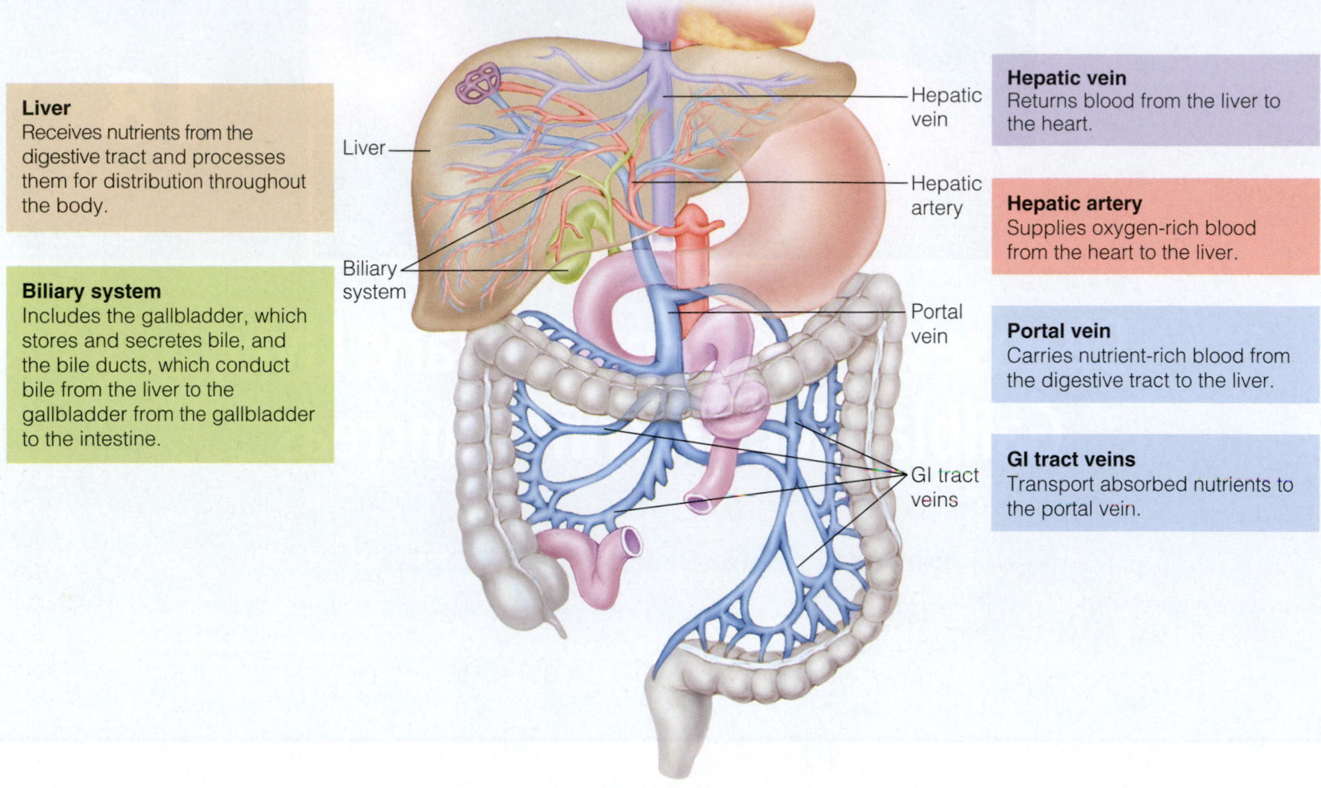

Liver
Receives nutrients from the digestive tract and processes them for distribution throughout the body.
Biliary system
Includes the gallbladder, which stores and secretes bile, and the bile ducts, which conduct bile from the liver to the gallbladder from the gallbladder to the intestine.
Liver
Biliary system
Hepatic vein
Hepatic artery
Portal vein
GI tract veins
Hepatic vein
Returns blood from the liver to the heart.
Hepatic artery
Supplies oxygen-rich blood from the heart to the liver.
Portal vein
Carries nutrient-rich blood from the digestive tract to the liver.
GI tract veins
Transport absorbed nutrients to the portal vein.

Source: S. Rolfes, K. Pinna, and E. Whitney, *Understanding Normal and Clinical Nutrition,* 7e, copyright © 2006, p. 770.

FIGURE 18.2 Anatomy of the Lobule (a) Hepatic lobule (b) Wedge of a hepatic lobule

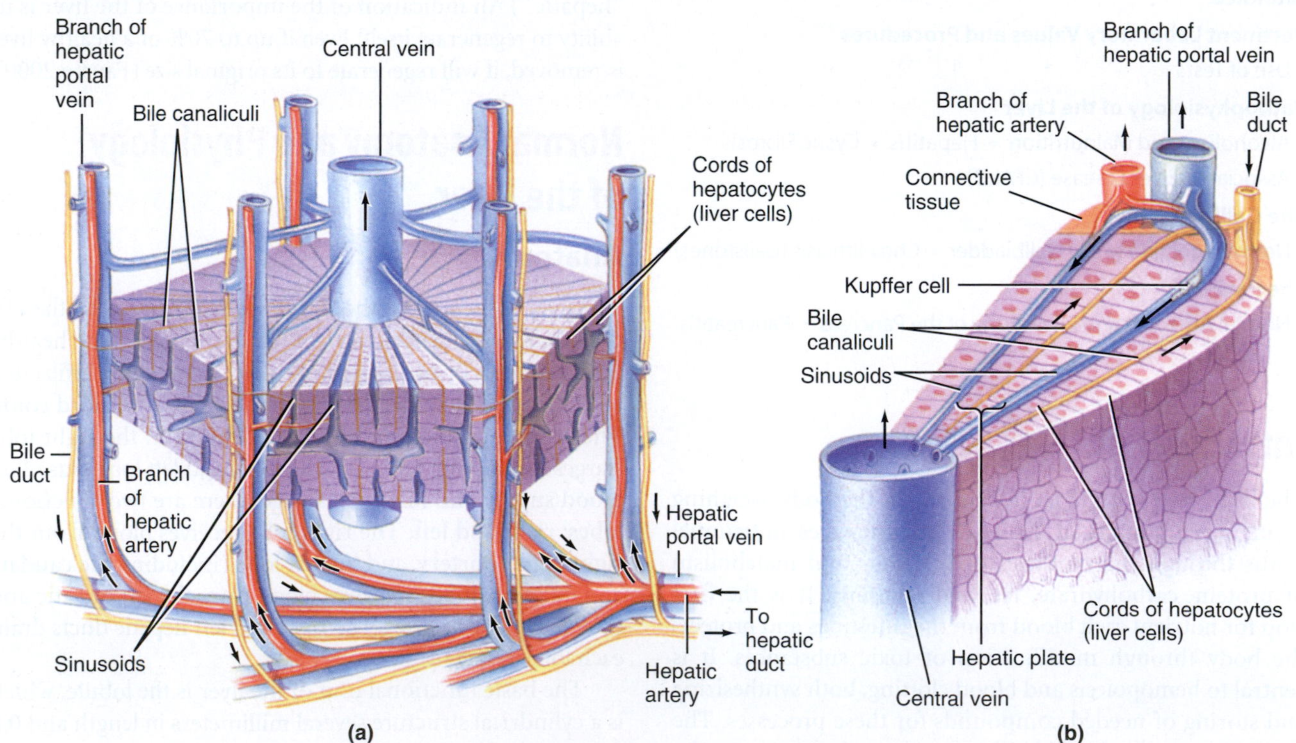

Source: L. Sherwood, *Human Physiology: From Cells to Systems,* 5e, copyright © 2004, p. 619.

of thousands of lobules (50,000–100,000 individual lobules) (Mayer 1997). The liver lobule is constructed around a central vein that empties into the hepatic vein and then into the vena cava. The lobule itself is composed of many hepatic cellular plates that radiate from the central vein like spokes of a wheel.

FIGURE 18.3 Circulation of Blood to and from the Liver

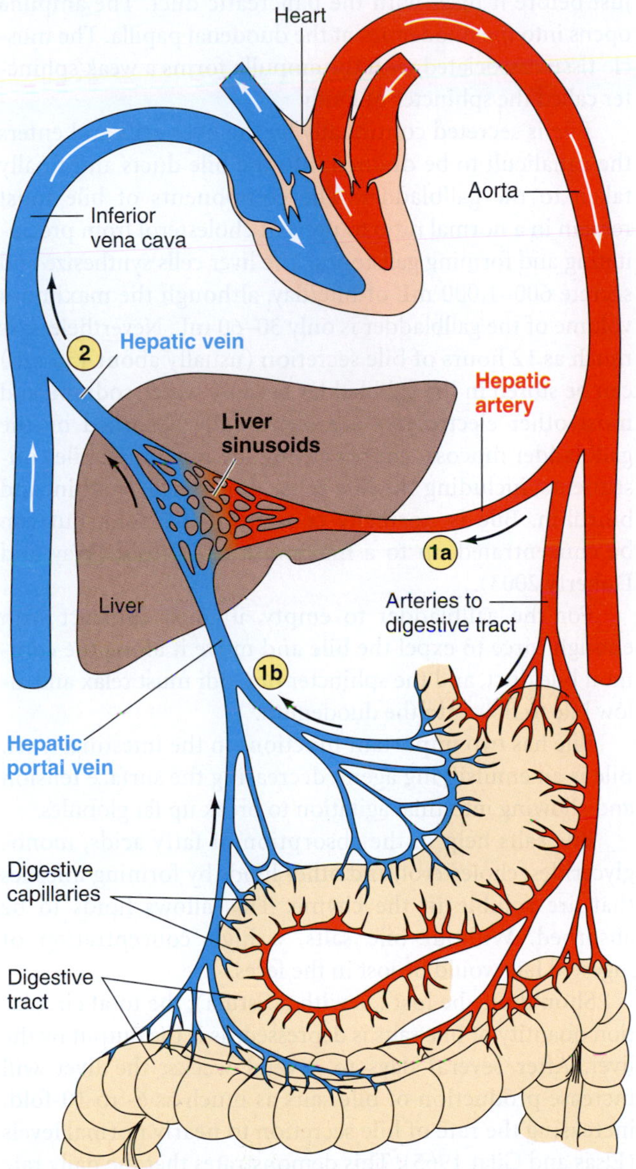

1 The liver receives blood from two sources:

a Arterial blood, which provides the liver's O_2 supply and carries blood-borne metabolites for hepatic processing, is delivered by the **hepatic artery**.

b Venous blood draining the digestive tract is carried by the **hepatic portal** vein to the liver for processing and storage of newly absorbed nutrients.

2 Blood leaves the liver via the **hepatic vein**.

Source: L. Sherwood, *Human Physiology: From Cells to Systems,* 5e, copyright © 2004, p. 618.

Each plate is about two cells thick and between the adjacent cells are bile canaliculi, which collect bile produced in the cells and carry it to the hepatic terminal bile duct.

The liver's structure allows for the many functions of this organ. Understanding the vascular supply, both serum and lymph, is essential. The main blood and lymph vessels, bile ducts, and nerves enter and leave the liver through the hilus. The liver receives blood from two sources. As you look at the schematic diagram of the lobule, you will see a portal venule that comes from the portal vein. The portal vein drains the intestine, spleen, and pancreas, bringing nutrients to the liver. The portal venules flow into the flat, branched venous sinusoids that lie between the hepatic cell plates and into the central hepatic vein. Three-quarters of the blood flow to the liver comes from the central hepatic vein. The hepatic arteries carry oxygenated blood from the abdominal aorta to the hepatic cellular plates (see Figure 18.3). In the liver sinusoids, there is a mixing of the oxygenated blood coming from the hepatic artery with the venous blood (rich in nutrients) coming from the portal vein. The size of the venous sinusoids is a function of the amount of the blood within them. Smaller blood vessels within the sinusoids (capillaries) differ from other capillaries in that they have a greater permeability to macromolecules, especially protein.[1]

The venous sinusoids are lined with at least four types of cells: typical endothelial cells, large **Kupffer cells**, perisinusoidal fat (and vitamin A) storing cells, and pit cells (least common cells with natural killer cell functions) (Worobetz et al. 2005). The endothelial lining of the venous sinusoids has very large pores, allowing hepatocytes to have ready access to nutrients in the plasma. A narrow space is present between the surface of the hepatocyte and the surface of the endothelial cell. This is called the space of Disse; it is filled with numerous microvilli from the hepatocytes, allowing an increase in the surface area coming in contact with the blood and facilitating the exchange of molecules between hepatocytes and the blood. Substances of the plasma move freely into the space. There are also many terminal lymphatics to remove excess fluid from these spaces.

A summary of blood flow through the liver can be traced as follows:

- The portal vein from the gastrointestinal tract divides into very fine branches that discharge portal blood into the venous sinusoids.
- Each branch of the portal vein is accompanied by networking branches of the hepatic artery.

Kupffer cells—specialized phagocytic cells of the reticuloendothelial system found on the luminal surface of the hepatic sinusoids; they filter bacteria and small foreign proteins out of the blood and dispose of worn-out red blood cells

- The blood eventually flows into the sinusoids. Kupffer cells lining the sinusoids system remove bilirubin, dyes, bacteria, damaged red blood cells, and other debris from the plasma through phagocytosis. These cells also modulate the immune response through the release of cytotoxins.
- The content of the sinusoids enters the central vein and finally reaches the vena cava.
- A network of lymphatic vessels in the portal canals drain the space of Disse.

The liver is one of the primary blood reservoirs in the body, and can store 200–400 mL of blood (Shier, Butler, and Lewis 2002). If an individual hemorrhages, and large amounts of blood are lost from the circulatory system, much of the blood normally in the liver sinusoids drains into the general circulation to help replace blood loss. In disease states such as **cirrhosis**, in which the sinusoids are obstructed, blood cannot flow through the liver, and fluid exits through the liver surface into the abdominal cavity, causing **ascites**.

In cardiac failure, as the damaged heart muscle decreases cardiac output, venous pressure rises, causing the liver to fill with additional blood. The continual stretching of liver sinusoids leads to necrosis of hepatic cells and results in cardiac failure with accompanying symptoms of liver disease.

Functions of the Liver

The liver has over 500 known functions. Most of the liver's activities depend upon its unique structure as well as its location in the body. The construction of the liver lobule meets the requirements of a "secreting gland." The lobule contains cells that form secretions (i.e., bile), blood vessels to provide raw materials, and a system of ducts to carry the secretions away. The versatile liver cells also play an important role in monitoring all the nutrient-rich blood delivered by the portal vein (Table 18.1).

Bile

Bile is a complex aqueous solution secreted by the liver. Ultimately, all bile drains into one large duct from each lobe of the liver. Two main trunks, one from the right lobe and one from the left, unite to form the common hepatic duct. The hepatic duct descends to the right for a few inches and is then joined by the cystic duct from the gallbladder to form the common bile duct. The common bile duct joins the pancreatic duct, forming a single tube called the ampulla of Vater. There is a strong sphincter of Boyden in the bile duct just before it fuses with the pancreatic duct. The ampulla opens into the duodenum at the duodenal papilla. The muscle tissue associated with the ampulla forms a weak sphincter called the sphincter of Oddi.

Bile is secreted continually by the liver cells and enters the canaliculi to be drained into the bile ducts and finally taken to the gallbladder. The components of bile must remain in a normal ratio to prevent cholesterol from precipitating and forming gallstones. The liver cells synthesize and secrete 600–1,000 mL of bile/day, although the maximum volume of the gallbladder is only 30–60 mL. Nevertheless, as much as 12 hours of bile secretion (usually about 450 mL) can be stored in the gallbladder because water, sodium, and most other electrolytes are continually absorbed by the gallbladder mucosa, concentrating the remaining bile constituents, including the bile salts, cholesterol, lecithin, and bilirubin. Bile is normally concentrated 5-fold, but can be concentrated up to a maximum of 20-fold (Way and Doherty 2003).

For the gallbladder to empty, it must contract with enough force to expel the bile and move it along the common bile duct, and the sphincter of Oddi must relax and allow bile to flow into the duodenum.

Bile has two important functions in the intestinal tract. Bile is an emulsifying agent, decreasing the surface tension and allowing intestinal agitation to break up fat globules.

Bile salts help in the absorption of fatty acids, monoglycerides, cholesterol, and other lipids by forming micelles that are soluble in the chyme. This allows lipids to be absorbed. Without bile salts, a high concentration of ingested fats would be lost in the feces.

Should bile be lost (as with a fistula), the total circulation quantity of bile salts is depressed, as is the output by the liver. After several days to several weeks, the liver will increase production of bile salts as much as 6- to 10-fold, increasing the rate of bile secretion to nearly normal levels (Elsas and Gilat 1965). This demonstrates that the daily rate of bile salt secretion is actively controlled in the enterohepatic circulation.

Enterohepatic Circulation About 95% of the bile salts are reabsorbed into the blood from the small intestine, about half by diffusion through the mucosa in the upper small intestine and half by active transport though the intestinal mucosa in the distal ileum. Once absorbed, the bile salts enter the portal blood and are returned to the liver. Once in the liver, the venous sinusoids absorb these salts

cirrhosis—any pathological condition where fibrous connective tissue invades any organ, usually as a consequence of inflammation or other injury

ascites—accumulation or retention of free fluid within the peritoneal cavity

bile—an emulsifying agent produced in the liver and eventually secreted into the duodenum

TABLE 18.1

Summary of Liver Functions

Metabolic Functions	
Carbohydrate Metabolism	Glycogenesis, gluconeogenesis, oxidation via TCA cycle, glycogenolysis, glycolysis
Lipid Metabolism	Lipogenesis, lipolysis, saturation/desaturation, ketogenesis, esterfication of fatty acids, fatty acid oxidation, uptake/formation/breakdown of phosphotides, synthesis/degradation/esterification/excretion of cholesterol, formation of lipoproteins
Protein Metabolism	Synthesis of serum proteins, synthesis of prothrombin, globin of hemoglobin, apoferritin, nucleoproteins and serum mucoprotein, degradation of some proteins to peptides and amino acids, synthesis of urea
Enzyme Metabolism	Synthesis of alkaline phosphatase, mono-amine oxidases (MAOs), acetylcholine esterase, oxidases, cholesterol esterase, dehydrogenases, beta glucuronidase, glutamic oxalacetic transaminase (SGOT-AST), and glutamic pyruvic transaminase (SPGT-ALT)
Vitamin Metabolism	Formation of acetyl CoA from pantothenic acid, hydroxylation of vitamin D to 25-OH D3, formation of 5-methyl tetrahydrofolic acid (THFA), methylation of niacinamide, phosphorylation of pyridoxine, de-phosphorylation of thiamin, formation of coenzyme B_{12}
Bile Acid Metabolism	Transformation of cholesterol to 7-hydroxycholesterol to cholic acid and chenodeoxycholic acid
Heme Metabolism	Heme is oxidized to biliverdin, which is then reduced to bilirubin; bilirubin is transported to the liver where it is converted to bilirubin diglucuronide to be excreted with the bile pigments
Storage	Storage of glycogen, fats, fatty acids, and fat-soluble vitamins
Other Functions	Conjugation, detoxification and degradation, Reticuloendothelial System (RES) activity, water movement regulation, fetal hematopoiesis, excretion

almost entirely into the hepatic cells and then resecrete them into the bile. About 95% of all the bile salts are recirculated into the bile; this happens, on average, at least two to three times per meal, and 18 times before the salts are excreted in the feces (Pauli-Magnus et al. 2005). The small quantities of bile salts lost into the feces are replaced by new amounts formed continually by the liver cells. This recirculation of bile salts is called the enterohepatic circulation (see Figure 18.4).

Jaundice

The word **jaundice** means a yellowish tint to the body tissues, including the yellowness of the skin and the deep tissues. Large quantities of bilirubin in the extracellular fluids, either unconjugated or conjugated bilirubin, is the usual cause of jaundice. The normal plasma concentration of bilirubin, including both the unconjugated and conjugated forms, averages less than 1.1 mg per dL of plasma (Tintinalli et al. 2004). In certain abnormal conditions, this can rise to as high as 5 mg/dL, but does not exceed this level because the bone marrow cannot increase production of red blood cells beyond eightfold. The skin begins to appear jaundiced when serum bilirubin exceeds 2.4 to 3.0 mg/dL (LaBrecque and Moody 1991).

Jaundice is a symptom generally classified based on one of two causes:

- Increased destruction of red blood cells with rapid release of bilirubin into the blood—hemolytic jaundice.
- Obstruction of the bile ducts or damage to the liver cells so that even the usual amounts of bilirubin cannot be excreted into the gastrointestinal tract—obstructive jaundice.

Pertinent Laboratory Values and Procedures

Laboratory tests, often referred to as liver function tests (LFTs), are useful in the evaluation and management of patients with hepatic dysfunction. First, they provide a sensitive, noninvasive method of screening for the presence of liver dysfunction. This is particularly important in patients

jaundice—a clinical manifestation of hyperbilirubinemia, consisting of deposition of bile pigments in the skin, resulting in a yellowish staining of the skin and mucous membranes

FIGURE 18.4 Enterohepatic Circulation

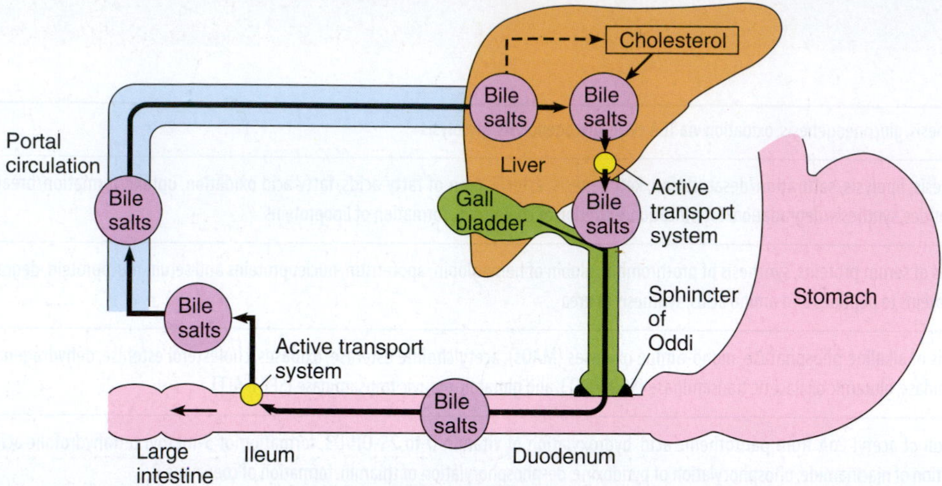

Source: R. Rhoades and R. Pflanzer, *Human Physiology*, 4e, copyright © 2003, p. 712.

without jaundice who may have unsuspected disorders such as viral hepatitis, chronic active hepatitis, cirrhosis, or partial bile duct obstruction. Second, once the presence of hepatic dysfunction is recognized, the pattern of laboratory test abnormalities may allow practitioners to recognize the general type of liver disorder. Third, laboratory tests allow practitioners to assess the severity of liver dysfunction and, occasionally, to predict outcome early in the course of the disease. Finally, LFTs allow the physician to follow the course of liver disease, to accurately evaluate the response to treatment, and to adjust treatment when necessary. Table 18.2 lists the laboratory methods commonly used in the clinical management of hepatic disease.

Use of Tests

Pratt and Kaplan (2003) recommend performing the following group of tests during the initial encounter with a patient with jaundice or suspected liver disease:

- Serum bilirubin: direct and total
- Urine bilirubin
- Aminotransferases: ALT and AST
- Alkaline phosphatase
- Total protein with albumin and globulin
- Prothrombin time

alcoholic liver disease (ALD)—liver diseases associated with alcoholism; it usually refers to the coexistence of two or more subentities, that is, alcoholic fatty liver, alcoholic hepatitis, and alcoholic liver cirrhosis, but may be the general entity when subentities are not specified

Pathophysiology of the Liver

Alcoholism and Malnutrition

Diagnosis and Epidemiology of Chronic Alcoholism Alcoholism represents one of the largest health problems in the United States. It is a costly disease with serious physical, psychosocial, and nutritional implications. Alcohol abuse is generally defined as chronic consumption of more than 80 g of ethanol per day. This means about 250 mL of 80-proof liquor (about six 1.5-oz. shots), 870 mL of table wine (about six 5-oz. glasses), or 2,218 mL beer (about six 12-oz cans) (U.S. Department of Agriculture, Agricultural Research Service 1998).[2] Alcohol abuse is associated with many health problems and causes 85,000 deaths annually, making it the third leading actual cause of death in the United States (Mokdad et al. 2004). Alcohol abuse is also associated with an annual economic loss of $184.6 billion in the United States (National Institute on Alcohol Abuse and Alcoholism 2000). This estimate cannot begin to account for the social consequences of spousal abuse, child abuse and neglect, disruption of family life, crime, accidents, and job absenteeism associated with alcohol abuse.

Americans consume approximately 4.5% of total kcalories as alcohol, with adult drinkers consuming over 10% of their kcalories from alcohol. A large part of the morbidity and mortality of alcohol abuse relates to its pervasive effects on all the major organ systems, with **alcoholic liver disease (ALD)** the primary cause of chronic medical illness and death. It is the twelfth leading cause of death in the United States (Kochanek et al. 2004). For Hispanic males and females, alcoholic liver disease and cirrhosis is the seventh and tenth leading cause of death, respectively, and for American Indian/Alaska Native males, it is the fifth leading cause of death (National Center for Health Statistics 2004). Despite these facts, the importance of alcoholism and related liver diseases is underestimated

TABLE 18.2

Laboratory Methods Used in Clinical Management of Hepatic Disease

Test	Normal Value	Clinical Implications
Blood		
Ammonia	19–60 μg/dL	Increased in cirrhosis, liver failure, and with portacaval shunting of the blood
Cholesterol Ester	60%–75% of cholesterol	Elevated in biliary obstruction; decreased in parenchymal liver disease
Dye Clearances		
BSP (sulfobromophthalein Na)	4% retention	Normal clearance depends on hepatic blood flow, functioning liver cell mass, & lack of obstruction; retention associated with hepatic damage
Indocyanine Green	0% retention	Similar sensitivity in detecting liver dysfunction compared with BSP
Protein Studies		
Albumin	3.5–4.5 g/dL	Decreased value associated with hepatic disease; generally parallels the functional status of parenchymal cells, but normal with considerable cellular damage
Globulin	1.5–3.8 g/dL	Often increased as it reflects inflammation
Total Protein	6.5–8.3 g/dL	
Prothrombin Time	9–11 seconds, 100% return	Prolonged with hepatic disease
Urine		
Bilirubin	0	Presence in the urine is indicative of biliary obstruction or RBC hemolysis
Urobilinogen	0–4 mg/24 hrs	Same as bilirubin
Stool		
Color	Brown	Alterations in color occur due to a decrease or absence of urobilinogen
Urobilinogen	75–400 mg/24 hrs	Stools are clay colored with biliary obstruction and light brown with hepatocellular damage
Enzymes		
Alkaline Phosphatase	30–95 U/L	Increased activity occurs in hepatic disease & malignancy & in chronic obstruction of the biliary duct, but is non-specific; increased also in bone diseases, bone trauma & bone growth
GGT	30 U/L or less	Elevation highly indicative of hepatocellular injury secondary to ethanol abuse
AST or SGOT	30–40 U/L or less	Less specific enzyme to detect hepatic disease secondary to cellular necrosis; also elevated with severe cardiac and muscle damage
ALT or SGPT	7–40 U/L or less	Most sensitive test to detect hepatocellular injury secondary to exacerbation of infectious hepatitis; high levels of 300 observed in acute hepatocellular damage
SGOT/SGPT Ratio	1	Ratio is useful in differential diagnosis; SGOT/SGPT >2—most pts have alcoholic liver disease; increased ratio is due primarily to the low activity of SGPT in the liver of alcoholic pts
LDH	280 IU/L or less	Iso-enzyme used to differentiate hepatitis from mononucleosis
Pigment Studies		
Serum Bilirubin		Reflects the ability of the liver to conjugate and excrete bilirubin; increased in liver and biliary disease causing jaundice clinically
Direct	0.1–0.3 mg	Indicates biliary tree obstruction
Indirect	0.1–0.5 mg	(Protein bound); indicate RBC hemolysis or liver damage
Total	0.2–0.9 mg	

From: Pratt DS, Kaplan MM. *Evaluation of the liver: laboratory tests.* In: Schiff, ER, Sorrell, MF, Maddrey WC, editors. *Schiff's diseases of the liver.* 9th ed. Baltimore: Lippincott Williams and Wilkins; 2003. Friedman LS, Martin P, Muñoz SJ. *Laboratory Evaluation of the Patient with Liver Disease.* In: Zakim D, Boyer TD, editors. *Hepatology: a textbook of liver disease.* 4th ed. Philadelphia: Saunders; 2003. Bishop ML, Fody EP, Schoeff L, editors. *Clinical chemistry: principles, procedures, correlations.* Baltimore: Lippincott Williams and Wilkins; 2005.

by both the general public and health professionals, and the variable susceptibility to alcoholic liver disease is not often appreciated.

Gender, Age, and Racial Factors Studies on gender differences in alcohol use disorders have found that, compared to men, women become intoxicated after drinking half as much, metabolize alcohol differently, develop hepatitis and cirrhosis of the liver over a shorter period of time, and have a greater risk of dying from alcohol-related accidents (Greenfield 2002). The proposed reason for the increased susceptibility to ethanol activity in women is the increased concentration of ethanol due to the lower body water content and weight compared with men. Women have also been found to have decreased gastric alcohol dehydrogenase (ADH) activity, which may further increase blood alcohol levels.

Drinking patterns appear to stabilize in middle age, though about one-third of older adults who are alcoholics became alcoholics later in life (Rigler 2000). Age does not seem to affect the rate of absorption and elimination of alcohol. However, between the ages of 25 and 60 the proportion of body fat almost doubles in men, and increases by 50% in women (Dufour, Archer, and Gordis 1992), while the volume of total body water decreases. Therefore, since alcohol is a water-soluble compound, an alcohol dose given to an older individual produces a higher blood alcohol concentration than it would for a similarly sized younger individual of the same gender.

Racial susceptibility to the toxic effects of ethanol is poorly understood. However, there are increased rates of alcoholic liver disease and alcohol-related traffic accidents among some minority groups, even when the quantity of alcohol consumed is equivalent. These rates suggest differential vulnerability to alcohol problems. Biological, socioeconomic, psychological-cultural, and environmental factors may contribute to variations in vulnerability.

Alcohol Withdrawal Syndrome The habitual, compulsive use of alcohol causes psychologic and/or physical dependency on the drug. Psychological dependency means that the patient requires alcohol to achieve an adequate level of functioning or well-being. Physical dependency means a physiologic adaptation to chronic alcohol use. This adaptation may be evident as both tolerance (the need for increasing amounts of alcohol to achieve the same effect) and withdrawal (the appearance of symptoms when alcohol is discontinued). The alcoholic patient must be observed for evidence of alcohol withdrawal syndrome (see Table 18.3) during the first three to five days of detoxification (Blondell 2005).

Metabolism of Alcohol Knowledge of ethanol metabolism is the key to understanding the biochemical, clinical, and pathological processes in alcoholism. Ethanol is rapidly and completely absorbed from the GI tract, even in malabsorptive states. It cannot be stored in the body and is metabolized by both oxidative and non-oxidative mechanisms, with oxidation as the primary means of metabolism (Matsumoto and Fakui 2002). Only 2% to 10% of ethanol is eliminated through the kidneys and lungs; the rest is principally oxidized by the liver.

TABLE 18.3

Characteristics of Alcohol Withdrawal Syndrome

Symptoms	Time of Appearance after Cessation of Alcohol Use
Minor withdrawal symptoms: insomnia, tremulousness, mild anxiety, gastrointestinal upset, headache, diaphoresis, palpitations, anorexia	6 to 12 hours
Alcoholic hallucinosis: visual, auditory, or tactile hallucinations	12 to 24 hours*
Withdrawal seizures: generalized tonic-clonic seizures	24 to 48 hours†
Alcohol withdrawal delirium (delirium tremens): hallucinations (predominately visual), disorientation, tachycardia, hypertension, low-grade fever, agitation, diaphoresis	48 to 72 hours‡

*Symptoms generally resolve within 48 hours.
†Symptoms reported as early as two hours after cessation.
‡Symptoms peak at five days.

Reproduced with permission from "Alcohol Withdrawal Syndrome," March 15, 2004, *American Family Physician*.

Studies in humans and rats have shown that a fraction of a small dose of ethanol is metabolized before reaching the peripheral circulation (Julkunen, Di Padova, and Lieber 1985; Frezza et al. 1990). Controversy exists, however, as to whether this first-pass metabolism by alcohol dehydrogenase (ADH) occurs primarily in the gastric mucosa or in the liver (Matsumoto and Fakui 2002). The liver has much greater ADH activity than does the stomach (Boleda, Moreno, and Pares 1989), but since three forms of gastric ADH exist in the stomach, it has been claimed that the gastric mucosa is responsible for nearly all first-pass metabolism (Matsumoto and Fakui 2002).

The hepatocytes contain three pathways for ethanol oxidative metabolism, each located in a different subcellular compartment: the ADH pathway in the cytosol, the microsomal ethanol oxidizing system (MEOS) in the endoplasmic reticulum, and catalase in peroxisomes (responsible for metabolizing less than 10% of ethanol). Each of these pathways produces acetaldehyde, a highly toxic metabolite that is converted to acetate. Figure 18.5 shows these pathways.

FIGURE 18.5 Acetaldehyde Pathways

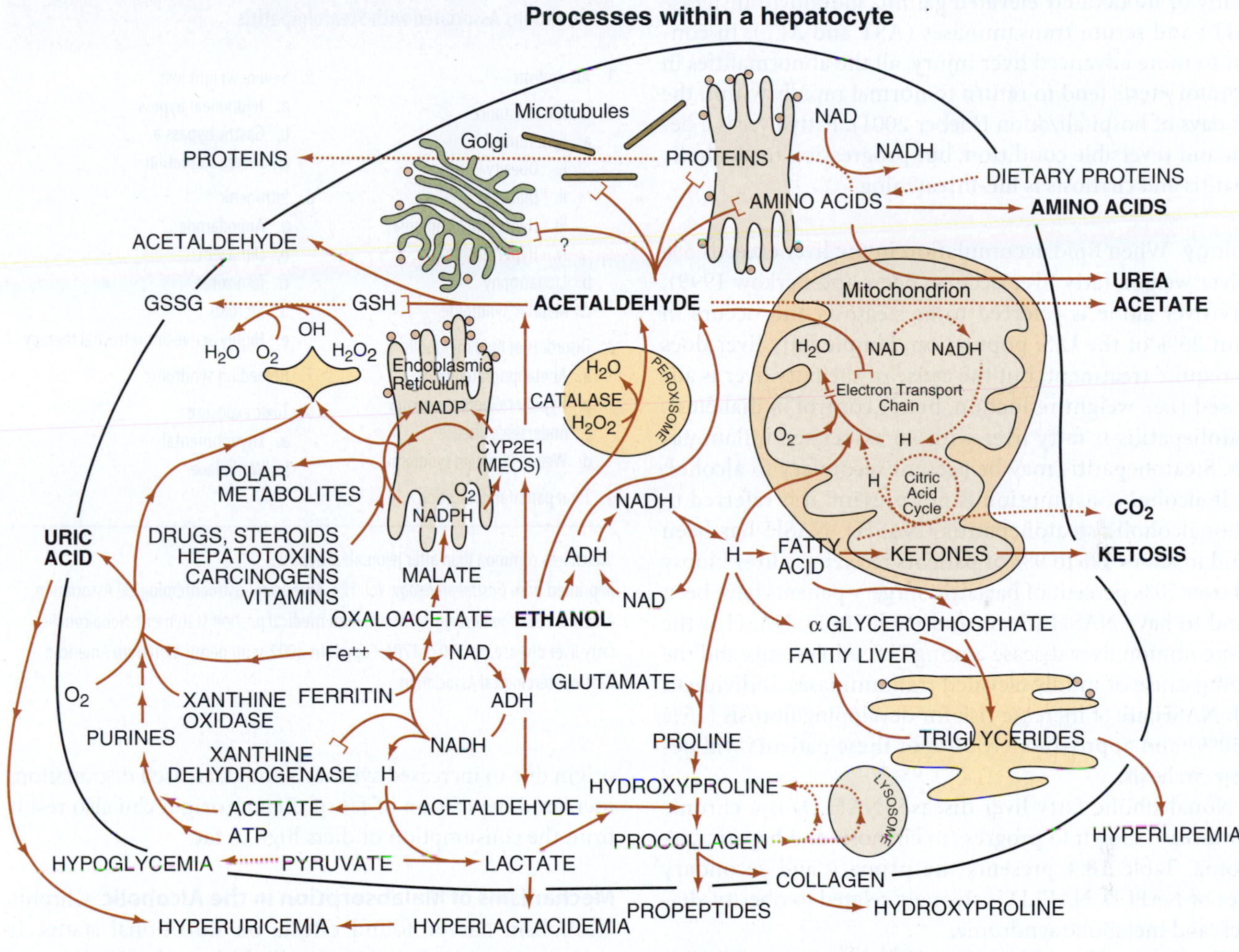

Source: CS Lieber, "Hepatic and Metabolic Effects of Ethanol: Pathogenesis and Prevention," *Annals of Medicine*, 26, 325–330, 1994.

In the ADH pathway, ADH catalyzes the conversion of ethanol to acetaldehyde, coupled with the reduction of NAD+ to NADH. This acetaldehyde, in turn, is converted to acetate and reducing equivalents by ALDH. The acetate may be oxidized further to carbon dioxide and water or enter the TCA cycle. Metabolism of ethanol via the ADH/ALDH pathway produces a marked increase in the NADH/NAD+ ratio, which produces an alteration in the cell's redox potential (Svensson et al. 1999).[3] Damage to the mitochondria occurs, inhibiting reoxidation of NADPH and causing lactic acidosis, hypoglycemia, and hyperuricemia. Another important consequence of this redox change is decreased fatty acid oxidation related to decreased TCA cycle activity. The increased fatty acids form triglyceride by combining with alpha-glycerophosphate, causing an increase in plasma triglycerides.

With chronic alcoholism, an additional pathway contributes to the oxidation of ethanol, namely, the microsomal ethanol oxidizing system (MEOS) (Lieber 2004). The MEOS can be used for both ethanol and drug metabolism. The MEOS system differs from the ADH/ALDH system in that it is triggered by exposure to ethanol, utilizes oxygen and NADPH (using up energy rather than generating it), and has a lower optimum pH.

Fatty Liver

Clinical Manifestations **Fatty liver** is present in 90% of chronic alcohol abusers (Sherman and Williams 1994), but most patients with fatty liver are virtually asymptomatic. Hepatomegaly is the most common clinical sign. Severe forms of fatty liver may present a clinical picture mimicking extrahepatic obstructive jaundice with dark urine and

fatty liver—yellow discoloration of the liver due to fatty degeneration of liver parenchymal cells

alcoholic stools. Typical abnormalities in laboratory tests are slightly or moderately elevated gamma glutamyltransferase (GGT) and serum transaminases (AST and ALT). In contrast to more advanced liver injury, all the abnormalities in laboratory tests tend to return to normal rapidly within the first days of hospitalization (Lieber 2001). Fatty liver is a benign and reversible condition, but progression to alcoholic hepatitis and cirrhosis is life-threatening.

Etiology When lipid accumulation in the liver exceeds 5% of liver weight, fatty liver occurs (Beers and Berkow 1999). Fatty liver alone is referred to as steatosis and occurs in about 25% of the U.S. population. Simple fatty liver does not require treatment, but the cause of the fatty liver is addressed (i.e., weight reduction, better control in diabetes). Steatohepatitis is fatty liver with an associated inflammation. Steatohepatitis may be present secondary to alcohol, but if alcohol consumption is not present, it is referred to as nonalcoholic steatohepatitis (NASH). NASH has been found in about 1% to 9% of patients receiving a liver biopsy and over 50% percent of bariatric surgery patients have been found to have NASH (Sears and Patel 2005). NASH is the most common liver disease among U.S. adolescents and the leading cause of mildly elevated transaminases. Individuals with NASH are at increase risk for developing fibrosis (15% to 50%) and approximately 30% of these patients will develop cirrhosis.

Nonalcoholic fatty liver disease (NAFLD) is a chronic liver disease shown to progress to cirrhosis and hepatic carcinoma. Table 18.4 presents the primary and secondary types of NAFLD. NAFLD is strongly related to obesity, diabetes, and metabolic syndrome.

The cause of fatty liver after alcohol ingestion is the increased availability of fatty acids in the liver. The sources of fatty acids are adipose tissue, lipids synthesized by the liver itself, and dietary lipids. The source of fatty acids depends on the fat content of the diet and whether alcohol is ingested acutely or chronically, since fatty acids originate from adipose tissue after the acute ingestion of a large dose of alcohol. The release of fatty acids under these conditions is similar to that observed in stress mediated by epinephrine release (Mezey 2000). Fatty liver develops in most people who abuse alcohol for a period of days (Menon et al. 2001).

Increased synthesis and decreased degradation of fatty acids in the liver occur during the chronic ingestion of alcohol. The latter effects of alcohol are principally related to the increase in the NADH/NAD+ ratio which occurs during the metabolism of alcohol. Synthesis of fatty acids is stimulated by increases in NADPH produced when reducing equivalents from NADH are transferred to NADP+, while the oxidation of fatty acids is reduced by the depressant effect of the increased NADH/NAD+ ratio on the TCA cycle (see Figure 18.6).

Finally, although the fatty acids that accumulate during chronic alcohol ingestion are primarily of endogenous origin due to increased synthesis and decreased degradation, increased deposition of fat of dietary origin can also result from the consumption of diets high in fat.

TABLE 18.4

Conditions Associated with Steatohepatitis

1. Alcoholism
2. Insulin resistance
   a. Syndrome X
      i. Obesity
      ii. Diabetes
      iii. Hypertriglyceridemia
      iv. Hypertension
   b. Lipoatrophy
   c. Mauriac syndrome
3. Disorders of lipid metabolism
   a. Abetalipoproteinemia
   b. Hypobetalipoproteinemia
   c. Andersen's disease
   d. Weber-Christian syndrome
4. Total parenteral nutrition
5. Severe weight loss
   a. Jejunoileal bypass
   b. Gastric bypass a
   c. Severe starvation
6. Iatrogenic
   a. Amiodarone
   b. Diltiazem
   c. Tamoxifen
   d. Steroids
   e. Highly active antiretroviral therapy
7. Refeeding syndrome
8. Toxic exposure
   a. Environmental
   b. Workplace

[a]Much less common than after jejunoileal bypass.

Reprinted from *Gastroenterology,* Vol. 123, American Gastroenterological Association, American Gastroenterological Association medical postion statement: Nonalcoholic fatty liver disease, pp. 1702–1704, Copyright 2002, with permission from American Gastroenterological Association.

Mechanisms of Malabsorption in the Alcoholic Chronic alcoholism has profound effects on nutritional status. It causes primary malnutrition by displacing other nutrients in the diet, either because of the high energy content of the alcoholic beverages or because of associated medical disorders. Secondary malnutrition may result from either maldigestion or malabsorption of nutrients caused by GI complications associated with alcoholism.

Esophagus Alcohol intake is commonly considered to exacerbate heartburn. Disordered motility has been noted in patients with alcoholic neuropathy, and lower esophageal sphincter pressure is reduced following alcohol intake. Changes in saliva and esophageal motility and the direct effect of ethanol may be important causes of esophagitis and stricture, commonly found in alcoholics and known to interfere with food intake. Patients who drink alcohol habitually may vomit, causing Mallory-Weiss tears of the lower esophageal mucosa; however, alcohol per se is not the direct cause of tears (Kortas et al. 2001). Esophageal cancer has been clearly linked with alcohol intake (Lieber 1993).

Stomach Alcohol ingestion is a cause of gastritis and duodenitis. There is clear evidence that alcohol intake causes atrophy of the gastric mucosal barrier due to a decrease in

FIGURE 18.6 Effects of the Increased NADH/NAD+ RATIO

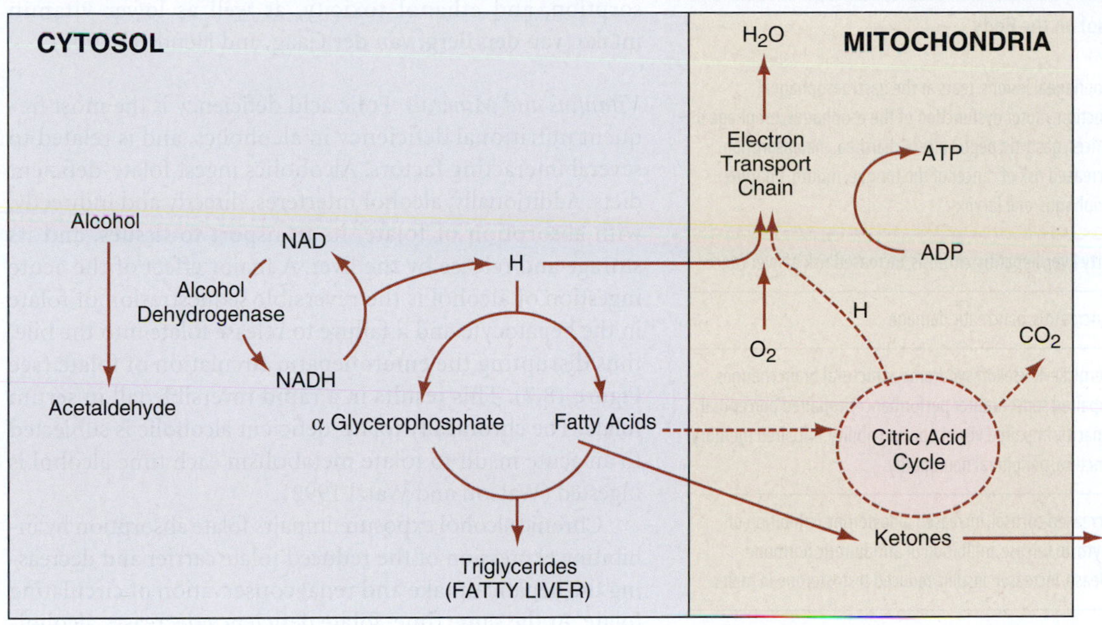

Source: E Mezey, "Alcoholic Liver Disease: Roles of Alcohol and Malnutrition," *The American J. of Clinical Nutrition*, 33, 1980.

hydrochloric acid production. Aspirin and alcohol appear to have additive effects in impairing mucosal barrier integrity. Alcohol intake may increase shedding of plasma protein into the stomach, and hemorrhage is commonly observed in alcohol abuse (Bode and Bode 1997). The incidence of peptic ulcer is much higher in cirrhotic patients (38.5% vs. 2% to 4% in the general population) but the etiology of this phenomenon is unclear and is most likely multifactorial (Kitano and Dolgor 2000). Increases in vitamin B_{12} deficiency and pernicious anemia (due to loss of intrinsic factor production by the stomach) and stomach cancer have been reported. The effects of alcohol on gastric emptying of meals are dose-dependent. Higher concentrations cause more consistent delays in passage of solid contents while enhancing movement of liquid contents.

Intestine Alcohol is readily absorbed from the stomach and upper small intestine. Alcohol dehydrogenase (ADH) found in the stomach and upper small intestine is responsible for the "first pass" metabolism of ethanol. The abnormal intestinal function produced by ethanol may be related to structural and morphological changes seen within the intestinal mucosa (see Table 18.5). Alcohol has been shown to produce small hemorrhagic lesions in the small intestine villi tips. Acute alcohol ingestion increases duodenal, jejunal, and ileal motility, which can contribute to diarrhea and malabsorption in the alcoholic patient (Bode and Bode 1997). In contrast, chronic alcohol abuse increases digestion time, possibly due to toxic effects of alcohol on the smooth muscle of the intestinal tract (Addolorato et al. 1997). In the case of alcoholism, chronic alcohol abuse damages the small intestine mucosa, leading to malabsorption of nutrients and contributing to nutrient deficiencies seen in alcoholics. In addition, bacterial overgrowth is commonly seen (Bode and Bode 2000). Table 18.5 summarizes the effects of alcohol on various systems.

Nutrition Implications of Alcoholism

Alcohol in the Diet The significant caloric contribution of excessive ethanol consumption may lead to obesity. Dessert or cocktail wines, liqueurs, cordials, beer, and mixers provide additional kilocalories in the form of carbohydrate. Conversely, in chronic alcoholics, irregular eating habits and decreased appetite during alcoholic binges may account for weight loss.

The number of kilocalories derived from ethanol can be calculated by using the following formula (see Table 18.6 for examples):

$$0.8 \times \text{proof} \times \text{ounces} = \text{kilocalories}$$

Malnutrition and Alcohol Protein-energy malnutrition is found in the majority of patients with alcoholic hepatitis and cirrhosis as well as aggravated liver injury. Multiple factors are involved, including poor dietary intake, malabsorption, the hypercatabolic state of chronic liver disease, altered energy storage, biochemical changes due to ethanol metabolism, and occasionally GI protein loss (Sherman and Williams 1994). Malnutrition is a major cause of liver damage and resulting dysfunction due to the replacement of normal liver substrates with alcohol (Lieber 2001).

TABLE 18.5

Adverse Effects of Alcohol on the Body	
GI Tract	Esophageal lesions, tears at the gastroesophageal junction, motor dysfunction of the esophagus, esophageal varices, gastritis, peptic ulcers, diarrhea, malabsorption, increased risk of cancer of the tongue, mouth, pharynx, esophagus, and larynx
Liver	Fatty liver, hepatitis, cirrhosis, increased risk of liver cancer
Pancreas	Pancreatitis, pancreatic damage
Nervous System	Wernicke-Korsakoff syndrome, structural brain changes, impaired sensorimotor performance, impaired perceptual capacity, impaired visual-spacing ability, impaired memory function, peripheral neuropathy
Endocrine System	Increased cortisol, increased aldosterone, inhibition of oxytocin release, inhibition of antidiuretic hormone release, increased insulin, reduced testosterone in males
Muscles	Muscle weakness and breakdown
Blood	Anemia, abnormal platelet production, decreased WBCs
Cardiovascular System	High blood pressure, ECG abnormalities, enlarged heart, biventricular heart failure
Lungs	Tuberculosis, airway obstruction
Reproductive System	Impotency, early postmenopausal amenorrhea, fetal alcohol syndrome
Other	Interferences with drug metabolism, malnutrition

TABLE 18.6

Derivation of Kilocalories from Alcohol

Alcohol Consumed Daily	Calculation	Kilocalories from Ethanol
6 12-oz cans of 4% beer	0.8 × 8 proof × 72 oz	460.8
3 4-oz glasses of 12% wine	0.8 × 24 proof × 12 oz	230.4
1/2 pint of 40% vodka	0.8 × 80 proof × 8 oz	512.0

Reprinted from *Journal of the American Dietetic Association*, Vol. 83, BJ Visocan, "Nutritional management of alcoholism," pp. 693–696, Copyright 1983, with permission from the American Dietetic Association.

Alcohol also interferes with nutrient activation, resulting in changes in nutritional requirements (Lieber 2000), and promotes nutrient degradation or impaired activation. This primary and secondary malnutrition can affect virtually all nutrients. Excessive chronic alcohol intake is generally associated with vitamin deficiency (especially deficiencies of folate, thiamin, and vitamin B_6) due to malnutrition, malabsorption, and ethanol toxicity, as well as lower vitamin intake (van den Berg, van der Gaag, and Hendriks 2001).

Vitamins and Minerals Folic acid deficiency is the most frequent nutritional deficiency in alcoholics, and is related to several interacting factors. Alcoholics ingest folate-deficient diets. Additionally, alcohol interferes, directly and indirectly, with absorption of folate, its transport to tissues, and its storage and release by the liver. A major effect of the acute ingestion of alcohol is the reversible sequestration of folate in the hepatocyte and a failure to release folate into the bile, thus disrupting the enterohepatic circulation of folate (see Figure 18.7). This results in a rapid reversible fall in serum folate. The chronically folate-deficient alcoholic is subjected to an acute insult to folate metabolism each time alcohol is ingested (Watson and Watzl 1992).

Chronic alcohol exposure impairs folate absorption by inhibiting expression of the reduced folate carrier and decreasing the hepatic uptake and renal conservation of circulating folate. At the same time, folate deficiency decreases alcohol-induced changes in hepatic methionine metabolism, promoting enhanced oxidative liver injury and the histopathology of alcoholic liver disease (Halsted et al. 2002).

Folate deficiency can cause morphological changes in the intestine, including villus shortening, decreased mitosis, macrocytosis, and enlargement of epithelial cell nuclei. However, folate deficiency can alter intestinal function prior to the production of these morphological changes. Since folate is required for many metabolic functions relating to nucleoprotein synthesis and cell turnover, the most common sign of folate deficiency is megaloblastic anemia. Structural and functional changes found in the intestinal mucosa are less obvious (Davidson and Townley 1977; Boyer, Wright, and Manns 2006). Studies have also demonstrated a synergistic effect between alcohol and folic acid deficiency (Halsted 2004).

Alcoholics rarely show deficiency of vitamin B_{12}, usually retaining normal serum levels even when folate is deficient. There is controversy among researchers regarding whether the binding of intrinsic factor-vitamin B_{12} complex to ileal sites is abnormal due to alcohol use (Lieber 2000).

Thiamin (B_1) is required for all tissues and is found in high concentrations in skeletal muscle, heart, liver, kidneys, and brain. Severe depletion can be seen in patients on a strict thiamin-deficient diet within 18 days (Singleton and Martin 2001). Thiamin diphosphate, the active form of thiamin, is a cofactor for enzymes primarily involved in carbohydrate metabolism. These enzymes are important in the biosynthesis of neurotransmitters, the production of reducing equivalents used in oxidant stress defenses, and for synthesis of the nucleic acid precursors pentoses.

The most common cause of thiamin deficiency in the United States is alcoholism. Alcohol affects thiamin uptake and other aspects of thiamin utilization, and these effects may contribute to the prevalence of thiamin deficiency in alcoholics. The major manifestations of thiamin deficiency in

FIGURE 18.7 Folate homeostasis and enzyme and binding proteins potentially affected by chronic alcoholism. J: jejunum, major site for folate absorption; insert shows events at mucosal surface. L: liver, major site of folate storage (7 to 10 mg) and metabolism; insert shows events in hepatocyte. K: kidneys; insert shows events at renal tubule. EHFC: enterohepatic folate circulation, approximately 10% of total folate pool; <0.1% excreted in feces per day. SFC: systemic folate circulation from liver to other organs; <1% excreted in urine per day. (1) Jejunal BBFH, required for initial digestion of dietary polyglutamyl folates ($PteGlu_n$). (2) FBP and transport mechanism, probably required at interstinal brush border, hepatocyte plasma membrane, and renal tubular brush border for transport of monoglutamyl folate (PteGlu). (3) Folate synthetase, required for synthesis of intracellular $PteGlu_n$ from PteGlu. (4) Intracellular folate hydrolase, required for conversion of $PteGlu_n$ to PteGlu prior to efflux from the cell. See text for further details.

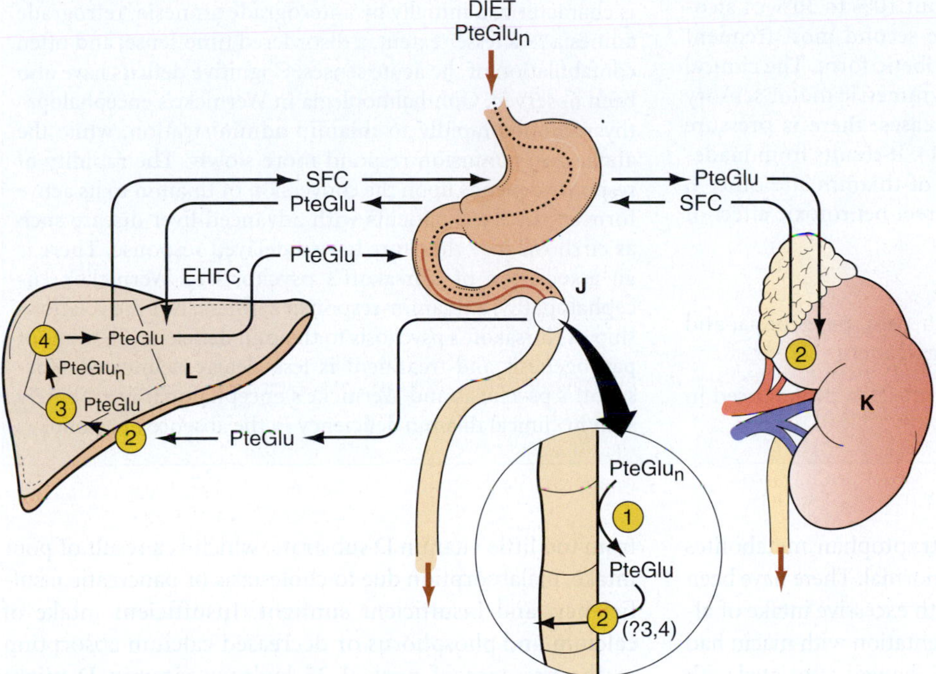

Source: Watson and Walzl (Eds.), *Nutrition and Alcohol,* 1992, Fig. 1, p. 263.

humans involve the cardiovascular (wet beriberi) and nervous (dry beriberi, and **Wernicke-Korsakoff syndrome**) systems (see Box 18.1).

The nervous tissue is highly vulnerable to thiamin deficiency. In the nervous system, thiamin participates in a number of enzymatic reactions, primarily those concerned with energy metabolism. In particular, thiamin pyrophosphate functions as a cofactor for three enzymes critically involved in the pentose-phosphate pathway and the TCA cycle. A number of mechanisms may be involved in the pathogenesis of thiamin deficiency in the alcoholic population, though this subject remains controversial. Among the mechanisms proposed are an inadequate thiamin intake, impairment of thiamin absorption, and decreased conversion to the biologically active form of thiamin (thiamin pyrophosphate).

Low plasma levels of pyridoxine have been reported in over 50% of alcoholics without hematologic findings or abnormal liver function tests (Gloria et al. 1997). Inadequate intake may explain some of it, but increased destruction and reduced formation may play a role. Some studies have shown a displacement of pyridoxal-5′-phosphate (PLP) by acetaldehyde in rats, though high levels of alcohol were used in these studies. Clinical management involves the provision of pyridoxine in the usual multivitamin dosage. Large doses (as little as 200 mg) should be avoided, because there is a danger of B_6 toxicity causing ataxia due to sensory neuropathy.

Evidence of niacin deficiency is difficult to find in alcoholics. Nevertheless, the concurrent administration of

Wernicke-Korsakoff syndrome—a syndrome consisting of Wernicke's encephalopathy in the acute phase and followed by Korsakoff's syndrome, usually associated with severe alcoholism

BOX 18.1 CLINICAL APPLICATIONS

Nervous System Involvement

The relative contribution of alcohol and nutrient deficiency to the deterioration of the nervous system in alcoholism is controversial (Manzo et al., 1994), though many neurological disorders are associated with alcohol abuse. A large body of evidence does exist that suggests ethanol is neurotoxic. The two most common disorders will be discussed here.

Motility and mental function disorders are among the complications of chronic alcoholism. They have been known for more than two centuries as "alcoholic paralysis," and are caused by alcoholic neuropathy (Kucera et al. 2002). Alcoholic polyneuropathy occurs in about 10% to 30% of alcoholics (Neundorfer 2001); it is the second most frequent type of polyneuropathy after the diabetic form. The clinical pattern is a symmetric sensory or symmetric motor sensory manifestation type. In almost all cases, there is pressure pain in the calves (Neundorfer 2001). It results from inadequate nutrition, mainly deficiency of thiamin and other B vitamins. Additionally, there is a direct neurotoxic effect of alcohol.

Signs and symptoms are:

1. Distal sensory disturbances with pain, paresthesia, and numbness in a glove and stockings-pattern
2. Weakness and atrophy of distal muscles, pronounced in the lower limbs
3. Loss of tendon jerks
4. Decreased response in autonomic fibers

Therapy consists of absolute alcohol abstinence, high-caloric nutrition, parenteral thiamin, and other vitamins. The prognosis of alcoholic polyneuropathy is favorable, with alcohol abstinence, within several months up to a few years (Schuchardt 2000).

The Wernicke-Korsakoff syndrome is the most spectacular CNS-related neurologic problem in alcoholism. Wernicke's encephalopathy is characterized by weakness of eye movements, gait disturbance, and confusion. Korsakoff's psychosis is characterized initially by anterograde amnesia, retrograde amnesia to a lesser extent, a disordered time sense, and often confabulation of the acute senses. Cognitive deficits have also been observed. Ophthalmoplegia in Wernicke's encephalopathy responds rapidly to thiamin administration, while the ataxia and confusion respond more slowly. The rapidity of response depends upon the conversion of thiamin to its active form in the liver; patients with advanced liver disease such as cirrhosis may therefore have a delayed response. There is an association of Korsakoff's psychosis to Wernicke's encephalopathy, a thiamin-responsive illness, but the relationship of Korsakoff's psychosis to thiamin deficiency in terms of pathogenesis and treatment is less clearly delineated. Korsakoff's psychosis and Wernicke's encephalopathy are rarely seen in clinical thiamin deficiency in the absence of alcohol.

vitamin B_6 and niacin returns the tryptophan metabolites excreted in the urine of alcoholics to normal. There have been case reports of pellagra associated with excessive intake of alcohol. In one reported case, supplementation with niacin had a prompt effect on correcting the skin changes associated with pellagra (Lorentzen, Fugleholm, and Weismann 2000).

Vitamin C is deficient in alcoholics with and without liver disease, and levels correlate with dietary intake. Low leukocyte concentrations of ascorbic acid (a measure of tissue stores) are found in patients with alcoholic cirrhosis. Daily supplementation with 175–500 mg of ascorbic acid may be necessary for weeks or months to restore plasma and urinary ascorbate to normal.

Ethanol impairs osteoblastic activity, which results in reduced bone formation and mineralization. Affecting the osteoblasts' function, alcohol is also able to induce bone resorption. Despite the direct toxic effect of ethanol on bone tissue, the indirect influence of ethanol on metabolism of hormones participating in bone homeostasis has been revealed. In chronic alcoholics, the deficiency of active metabolites of vitamin D is often observed (Medras and Jankowska 2000).

Alcoholics have been found to have low, normal, and increased levels of 25-hydroxy vitamin D. In patients with alcoholic liver disease, vitamin D deficiency probably derives from too little vitamin D substrate, which is a result of poor intake, malabsorption due to cholestasis or pancreatic insufficiency, and insufficient sunlight. Insufficient intake of calcium and phosphorus or decreased calcium absorption in the presence of normal 25-hydroxy vitamin D might accelerate bone loss in alcoholics. Bone disease in those with liver disease should be treated by increasing intake of vitamin D, ultraviolet light therapy, and correction of fat malabsorption to keep plasma calcium and phosphorus normal, along with abstinence from alcohol (Lieber 2000).

Vitamin K deficiency may be present in alcoholics when there is fat malabsorption due to pancreatic insufficiency, biliary obstruction, or intestinal mucosal abnormality secondary to folate deficiency. It is unlikely to be due to low intake, since the microflora of the gut are a reliable source of the vitamin. Any prolongation of the prothrombin time is probably related to the associated hepatic disease causing failure of prothrombin synthesis. Vitamin K may be given intramuscularly to clinically test whether hepatocellular dysfunction or lack of availability of vitamin K to the liver is responsible for low levels of vitamin K-dependent clotting factors in the blood (Lieber 2000).

The interaction of alcoholism and vitamin A involves the intake (and possibly the absorption) of the vitamin and its metabolism. Alcoholics may suffer from vitamin A deficiency.

An important clinical consequence of low tissue vitamin A is night blindness. Abnormal dark adaptation occurs in alcoholics with and without cirrhosis and is more common among those with cirrhosis, those with elevated bilirubin, and those at an older age (Hussaini et al. 1998). Several factors make vitamin A therapy complicated when alcoholism is present: assessment of tissue stores of vitamin A is difficult; vitamin A in high doses is toxic, and even the usual doses of vitamin A are potentially toxic with continued intake of alcohol; and monitoring vitamin A hepatotoxicity is difficult in the presence of continued alcohol intake. Replacement of vitamin A via supplementation should only be considered for patients who are confirmed as deficient and who assuredly practice abstinence from alcohol. Low serum levels and night-blindness can be considered evidence of a deficiency.

There are clear interactions between zinc and vitamin A, with zinc participating in the absorption, mobilization, transport, and metabolism of vitamin A. Similarly, vitamin A affects zinc absorption and use. Therefore, changes in the status of either could affect the metabolism of the other. Vitamin A and zinc metabolism are affected both by ethanol and by hepatic cirrhosis. Ethanol causes abnormal dark adaptation by acting as a competitive inhibitor with retinol alcohol dehydrogenase in the eye. Vitamin A malnutrition in cirrhotics may be caused by poor diet, malabsorption, decreased hepatic vitamin A uptake, and decreased hepatic storage capacity for vitamin A. In some cirrhotic patients, zinc deficiency and/or protein deficiency may limit the ability to respond to vitamin A. In animals, oral ethanol intake results in increased losses of zinc by the urinary and fecal routes. Combined vitamin A and zinc deficiencies are common in cirrhotics and either may result in abnormal dark adaptation or impaired taste or smell. A low serum zinc level can be treated with doses of 600 µg $ZnSO_4$ per day; considering the interrelationship of vitamin A and zinc metabolism, zinc therapy might be tried when vitamin A therapy fails. However, recent literature suggests that patients with ALD who have hypozincemia responded poorly to oral zinc supplementation (McClain et al. 2002). Some clinical trials suggest that vitamin A should be parenterally replaced when there is documented fat malabsorption.

Chronic alcohol abuse is associated with both an altered response to infection and deranged iron homeostasis (Potter and Wang 2002). Iron metabolism is important in alcoholism because there may be a deficiency or an excess of iron in the body. The question of the metabolism of iron is particularly relevant because of the association of hepatic injury with excess iron. It is not clear whether an increase in hepatic iron results from alcohol increasing intestinal absorption or hepatic uptake of iron from serum in es-

TABLE 18.7

The Anemias of Liver Disease

Type of Anemia	RBC MCV	Serum Fe	TIBC	Transferritin Percentage of SAT	Ferritin	Folate	B_{12}	Type of Liver Disease
	Tests							
Megaloblastic Normochromic	>110	N	N	N	N	↓	N	Alcoholism with injury
		N	N	N	N/↑/	↓	↑	Acute injury
		↓	↓	↓	N	N	N	Chronic injury
Microcytic Normochromic	<80	↓	↑/N	↓	↓	N	N	Chronic Disease with blood loss
Hemolytic	<95	N	N	N	N/↑	N	↑/N	Autoimmune
		N	N/↓	N	N	N	↑/N	Zieve's Syndrome
								Wilson Disease
Normochromic Normocytic	80–95	↓	↓	↓	N	N	N	Chronic Disease
Pancytopenis	80–95	N	N	N	N	N	↑	Acute Viral Hepatitis, Drug toxicity, Hypersplenism
Megaloblastic Hyperchomic	>95	↑	↑	↑	↑	N	N	Hemochypochromic
		↑	↑	↑	N	↓	N	Thalassemia

tablished alcoholic liver disease. Alcoholics may receive excessive dietary iron from the beverages they drink, such as certain wines, or through inadvertent treatment with iron-containing vitamin preparations. In addition, anemias unrelated to iron deficiency may be incorrectly treated with iron (see Table 18.7). Pancreatic insufficiency, folate deficiency, portosystemic shunting, and cirrhosis may increase iron absorption. Of greatest potential significance is the contribution hepatic iron may make to liver damage via its role in lipid peroxidation and its possible role in promoting fibrogenesis (Lieber 2000).

Alcoholics may be iron-deficient as a result of GI lesions that may bleed (esophagitis, esophageal varices, gastritis, duodenitis). The usual laboratory tests (serum iron, serum binding capacity) are helpful. Iron supplements should be restricted to clearly diagnosed cases of deficiency.

Plasma ferritin concentration has been positively associated with alcohol use among men. In a study that was published in Australia, it was found that inquiries could usefully be made into the alcohol consumption of men with a plasma ferritin concentration >652 μg/L, because approximately one in three would admit to drinking hazardously (Peach and Bath 1999).

Acute and chronic exposure to ethanol influences intracellular calcium homeostasis (Bondy et al. 1998). The results of a study done in Hungary suggest that alcohol dependence can develop at the cellular level and that changes in calcium homeostasis, likely due to the effects of ethanol on ion channels, may play a role in processes leading to adaptation of cells to alcohol (Nagy 2000). Alcoholics have illnesses related to abnormalities of calcium homeostasis. They have decreases in bone density and bone mass, increased susceptibility to fractures, increased bone cell death (osteonecrosis), and an earlier occurrence of osteoporosis. The direct toxic affects of alcohol on bone and calcium absorption are not fully understood.

Hypokalemia is common in patients with alcoholic liver disease. Even if the serum potassium level is normal, body stores of potassium are likely to be low due to poor dietary intake, vomiting, and particularly diarrhea. In patients with ascites, secondary aldosteronism and the administration of diuretics contribute to hypokalemia (Halsted 2004).

Vitamin supplementation is commonly provided in the treatment of alcoholism, whether deficiencies are present or not. The administration of thiamin is justified, because when it is given in large amounts, passive absorption overcomes deficiencies resulting from active absorption of thiamin at low concentrations. It is generally recommended that a multiple water-soluble vitamin be given at twice the RDA. During "drying out" or recovery, the alcoholic also requires increased amounts of protein and vitamins.

hepatitis—inflammation of the liver and liver disease involving degenerative or necrotic alterations of hepatocytes

TABLE 18.8

Nutrition Therapy for Chronic Alcoholism

Nutrient	Nutrition Therapy
Energy	30–35 kcal/kg to support protein sparing
Protein	.8 g–1.5 g/kg
Carbohydrate	Minimum of 300 grams (55–65% of total kcal) with avoidance of simple sugars
Lipid	25–30% of total kcal unless steatorrhea is present
Vitamins	Supplementation with folic acid, thiamin, and a general multivitamin (unless contraindicated)
Minerals	May require supplementation for calcium, iron, zinc

Summary of the Nutritional Effects of Alcoholism The chronic alcoholic with liver disease is usually also malnourished, which can aggravate the liver disease (see Table 18.8). Halsted (2004) lists the following reasons for malnutrition in the alcoholic:

- Imbalanced diet and/or anorexia. Alcohol often replaces food in the diet. Though it is possible to obtain maintenance energy needs from the kcal in the alcohol consumed, malnourishment results because of an inadequate intake of nutrients; though high in kcal, alcohol contains no protein, vitamins, or minerals. Alcohol is also preferentially metabolized over whatever nutrients are obtained through food. Cytokines, such as tumor necrosis factor, and leptin release from the cells also contribute to malnutrition.
- Intestinal maldigestion and malabsorption. Alcohol causes inflammation of the stomach, pancreas, and intestine, interfering with the normal processes of digestion and absorption and resulting in secondary malnutrition. The subsequent decrease in pancreatic enzyme secretion, intestinal transporters, and micelle formation (due to inadequate bile salt secretion) results in altered absorption of fat-soluble vitamins, thiamin, and folic acid. In addition, alcohol and acetaldehyde have a hepatotoxic effect that interferes with metabolism and activation of vitamins by liver cells.
- Increased catabolism of visceral protein and skeletal muscle. This occurs due to cytokine release in response to inflammation.
- Increased excretion of selected vitamins. The metabolism of alcohol increases the need for certain nutrients, particularly the B-vitamins and magnesium. Alcohol

causes increased urinary excretion of folate and pyridoxine (B_6), as well as increased retinoid metabolism.

Hepatitis

Definition and Epidemiology **Hepatitis** is an inflammation of the liver caused by a virus, bacteria, toxins, obstruction, parasites, or drugs (chloroform, carbon tetrachloride). Viral hepatitis is caused by five identified viruses (types A, B, C, D, E), with at least 10 other viruses under study. In the United States, apart from sexually transmitted diseases, viral hepatitis is the fourth most frequently reported infectious disease (National Center for Health Statistics 2004). About half of the world's reported viral hepatitis cases are hepatitis A (HAV, formerly infectious hepatitis). HAV is generally transmitted via the oral-fecal route. Sources of contamination include drinking water, food (many times seafood), and sewage. NHANES III found that slightly more than a third of the U.S. population had serological evidence of having had HAV infection (CDC 1996).

Serum hepatitis, or type B hepatitis (HBV), is transmitted through transfusions of blood or blood-derived fluids, or through improperly sterilized medical instruments, dental drills, tattooing needles, or other skin-puncturing instruments that have come in contact with contaminated blood. It can also be transmitted by other than the parenteral route, which makes its specific cause more obscure. Some people may be carriers of the hepatitis B antigen (HBsAG) while remaining asymptomatic. The prevalence of chronic HBV in the United States is low (5%) and about 30% of these individuals have no signs or symptoms (McQuillan et al. 1999). Hepatitis vaccinations are now required and are considered to be the first anti-cancer vaccine.

Hepatitis C (HCV) occurs when an individual is exposed to blood or body fluids from an infected person—sharing needles (most are from this cause) or recipient of clotting factors before 1987. Hemodialysis patients and infants born to infected mothers are also at increased risk. Unlike HAV and HBV, there is no vaccine to prevent HCV. While the number of new infections per year has declined, about 85% will become chronically infected, and 20% of these individuals will develop cirrhosis. HCV is now the most common diagnosis treated with liver transplantation. Individuals with HCV who develop cirrhosis are also at an increased risk of developing hepatocellular carcinoma. It is unknown how to predict which patient will have a benign course and which will develop cirrhosis or carcinoma.[4]

Clinical Manifestations The clinical manifestations are similar for all types. Common clinical symptoms include jaundice, dark urine, anorexia, fatigue, headache, nausea, vomiting, and fever. The liver becomes enlarged (hepatomegaly) and, in some cases, the spleen enlarges (splenomegaly). Bilirubin, alkaline phosphatase, and serum AST are generally elevated.

Nutrition Therapy Nutritional care is the same for all types of hepatitis. The primary objective is to spare the liver and provide it with the nutrients needed for regeneration. Treatment includes adequate rest, fluids, good nutrition, and avoidance of further damage to the liver—in particular, avoidance of alcohol to prevent further liver cell damage. Patients suffering from anorexia are frequently unable to consume an adequate diet, and the first effort should be to increase dietary intake through the use of foods. Frequent small feedings are generally better tolerated. Considering the protein-sparing role of adequate kilocalories to prevent the body from using body tissue or protein for energy, 30–35 kcal/kg body weight are recommended, with a target of 3,000 kcal or more. If achieving this goal with food is not possible, concentrated liquid formulas can be given until the person is able to consume an adequate diet. In cases of severe vomiting or diarrhea, a 5% to 10% solution of glucose is administered intravenously. If prolonged parenteral feeding is indicated, protein hydrolysates or amino acids are added (Matarese and Gottschlich 1998).

Adequate protein seems to protect the liver against chemical poisons. Fatty changes in the liver occur in the peripheral cells around the portal system and may be due to a failure in the synthesis of lipoprotein carriers. A protein level of 1–1.2 g/kg is recommended. In a few patients with hepatitis or ESLD, hepatic coma may result, and protein intake must be decreased during acute phases. This should only be done during the acute phase, and protein should be increased as quickly as feasible.

Normally, a diet of 30% to 40% of kcalories from fat is well tolerated during convalescence, making the diet more palatable and allowing the transport of fat-soluble vitamins. There is no evidence that the course of hepatitis is influenced by the fat content of the diet. However, when liver damage interferes with bile production, fat and fatty foods may not be well tolerated. Some patients have anorexia, vomiting, and steatorrhea on high-fat diets.

The remaining kcal will come from carbohydrate. Carbohydrates provide a good source of efficient kcal as well as vitamins, minerals, and a source of liver glycogen. Recommendations are that the diet should provide 50% to 55% of kcal from carbohydrate (Escott-Stump 2002).

Supplemental vitamin K, either parenteral or water-soluble oral, may be necessary due to reduced plasma prothrombin levels, usually the result of disturbed liver function. The damaged liver is unable to synthesize prothrombin and is therefore unable to use vitamin K. However, intrahepatic biliary obstruction does occur in infectious hepatitis, and occasionally the exclusion of bile from the gut will interfere with absorption of vitamin K. Under these conditions, parenteral or oral, water-soluble vitamin K will effectively raise prothrombin levels. Potassium and sodium may be needed if excessive vomiting or diarrhea occurs.

Alcoholic Hepatitis

Alcoholic hepatitis is a form of toxic liver injury associated with chronic ethanol consumption. Patients with alcoholic hepatitis frequently have increased susceptibility to infections—pneumonia, spontaneous bacterial peritonitis, cellulitis, and even septicemia. Antibiotics are often required, and occasionally corticosteroids are recommended in severe alcoholic hepatitis with cholestasis. The catabolic effect of these medications adds to the complexity of the nutritional management. Symptoms include fatigue, weakness, anorexia, fever, and hepatomegaly.

Treatment and Nutrition Therapy First steps in treatment include abstention from alcohol and the treatment of delirium tremors, acute episodes of delirium caused by alcohol withdrawal and characterized by anxiety, confusion, hallucinations, shaking, and sweating (The American Heritage Stedman's Medical Dictionary 2002). Apart from these, the major aspect of the treatment of alcoholic hepatitis is the correction of nutritional deficiencies. A multivitamin preparation that includes B_{12}, folate, thiamin, pyridoxine, vitamin A, vitamin D, and a mineral supplement that includes zinc, magnesium, calcium, and phosphorus should be provided. Adequate kcal and protein are needed, in supplement form if necessary.

Cirrhosis

Definition Cirrhosis is a chronic liver disease in which healthy tissue is replaced by scar tissue, blocking the flow of blood through the organ and resulting in the loss of liver function. Cirrhosis is the ninth leading cause of death in the U.S. The most common causes of cirrhosis are chronic alcoholism and HCV. The accumulation of fat in the liver (**steatosis**) is generally considered the first stage of cirrhosis. Malnutrition is thought to be a major factor in the development of steatosis, but a causal relationship between malnutrition and cirrhosis has not been established. The etiology of the injury to the liver in the alcoholic patient has been attributed to alcohol toxicity, genetic predisposition, malnutrition, or a combination of all these factors.

While a large percentage of the patients with cirrhosis are alcoholics, it should not be assumed that all patients with cirrhosis are alcoholics. Not all heavy drinkers develop liver cirrhosis. Genetic factors can increase susceptibility. It has been proven that the predisposition for many different diseases is associated with specific histocompatibility antigens (HLA) (Bohinjec 2005).

steatosis—accumulation of fat in the interstitial tissue of an organ

Even when irreversible liver complications are present, therapeutic intervention, primarily nutritional treatment, can alleviate major complications of cirrhosis, such as encephalopathy (which responds to protein restriction) and manifestations of portal hypertension, such as ascites (which responds favorably to salt restriction) (see Table 18.9). At present, a major task is to avoid the development of these serious complications at an early stage and to arrest the disease process prior to the medical or social disintegration of the individual.

Etiology The most common form of cirrhosis is Laennec's cirrhosis. It is generally associated with alcoholism, although not all alcoholics will develop Laennec's cirrhosis and it may be seen in nonalcoholics. A variety of metabolic activities are impaired in Laennec's cirrhosis. Scar tissue forms in the liver, and the conversion of fat to lipoproteins is impaired; hence the accumulation of fat in the liver. Portal hypertension may develop and blood flow becomes obstructed, with esophageal varices (varicose veins in the esophagus) being the end result. Esophageal varices are a serious complication, since the danger of rupture with ensuing hemorrhage is an imminent possibility.

Clinical Manifestations The cirrhotic patient has an enlarged liver as a result of fat accumulation and necrosis of the liver cells. Ascites (accumulation of fluid in the abdomen) and edema of the extremities may be present since hepatic filtering is impaired and serum protein levels are low. SGOT is elevated and BSP (sulfobromophthalein) clearing time is

TABLE 18.9

Nutritional Support for the Cirrhotic Patient without Encephalopathy

Kcal	40–50 kcal/kg dry body weight or BEE × 1.5 to 1.75 (High kcalories to minimize endogenous protein catabolism)
Protein	1.0–1.5 g/kg—may tolerate dairy and vegetable proteins better (Based on dry weight, protein tolerance, and degree of malnutrition)
Fat	40–50% of nonprotein kcalories (Decrease LCTs with steatorrhea, MCT oil may be used; fat may be restricted if evidence of jaundice)
Carbohydrate	300–400 g to spare protein
Vitamins/ Minerals	Vitamin B complex, vitamins C and K, zinc, magnesium, phosphorus, monitor needs for vitamins A and D
Meal Frequency	Smaller meals (4 to 6 per day) may be helpful to encourage intake

Adapted from Zeman, 1991; Escott-Stump S. *Nutrition and diagnosis-related care.* 5th ed. Baltimore: Lippincott, Williams, & Wilkins; 2002.

reduced, resulting in elevated levels. Vitamin deficiencies and depressed hematocrit and hemoglobin values are commonly seen, and may be due to malnutrition, gastrointestinal bleeding, or both. Patients appear jaundiced, may lack appetite, and may have delirium tremors. Other signs and symptoms are fever, gallstones, ulcers, gastroesophageal reflux, gastritis, and diarrhea (Escott-Stump 2002).

Complications Regardless of the etiology of cirrhosis, with progressive loss of liver function the body is threatened by three major complications:

1. Portal hypertension with its accompanying variceal hemorrhage and splenomegaly
2. Ascites
3. Hepatic encephalopathy

Portal Hypertension **Portal hypertension** is always present in patients with ascites secondary to cirrhosis of the liver. Portal hypertension results from a decrease in hepatic vascular bed due to:

- Fibrosis and destruction of liver parenchymal cells
- Increased postsinusoidal resistance secondary to pressure caused by regenerating nodules
- Arteriovenous anastomoses (an interconnection between an artery and a vein) between tributaries of the hepatic artery and the portal vein due to pressure changes

Portal systemic shunts, passages connecting the portal flow with other anatomical channels, occur when blood flow from the portal vein through the liver sinusoids into the hepatic vein is obstructed by cirrhosis of the liver. Collateral veins drain from the portal system in response to a high pressure gradient at the sites of natural anastomoses between portal and systemic veins in the esophagus and lower rectum, in falciform ligament, in the retroperitoneal space, and in adhesions between the visceral and parietal peritoneum. A common symptom in cirrhosis is esophageal varices. The transjugular intrahepatic portosystemic shunt (TIPS) procedure (see Figure 18.8) reroutes blood flow in the liver and reduces pressure in all abnormal veins, not only in the stomach and esophagus, but also in the bowel and the liver.

Ascites Ascites is the accumulation of fluid in the peritoneal cavity, and is the most common complication of cirrhosis. The three principal factors responsible for ascites in chronic liver disease are: (1) portal hypertension and lymphatic obstruction secondary to hepatic fibrosis, (2) reduced osmotic pressure of plasma due to failure of the liver to synthesize albumin, (3) increased retention of sodium.

Reduced oncotic pressure is due to a failure of the diseased liver to synthesize serum proteins, particularly albumin. In addition, when water leaves the liver sinusoids, protein goes with it. The ascitic fluid in the abdominal cavity can amount to as much as 15.5 liters and contains protein concentrations of 1–2 g/100 mL (Wang et al. 1996). This fluid can be removed by **paracentesis** (tapping the abdomen), but this results in protein loss as well. For instance, if 4 L of ascitic fluid are removed from a patient, he or she loses 40–80 g of protein via this route.

The excessive renal absorption of sodium is a result of increased aldosterone production and decreased inactivation of aldosterone. There is also lowered renal blood flow and decreased filtration rate, contributing to sodium retention. The etiology of ascites in liver disease is illustrated in Figure 18.9.

Nutrition Therapy The standard treatment of ascites due to cirrhosis consists of:

- Encouraging oral proteins in order to replenish the serum albumin and raise the colloidal osmotic pressure of the serum
- Restricting the daily salt intake to less than 2 g/day
- Restricting daily fluid intake to 1500 cc/day if serum Na drops below 120 mEq/L
- Adequate kcal to meet energy requirements
- Diuretics (spironolactone or amiloride)

Often these factors are contradictory. A high-protein diet contains abundant endogenous salt, and it is difficult to combine a high-protein diet and low salt intake. Utilization of a modular protein diet is one approach to achieving desired goals. Powdered protein products, such as Propac® or Promod®, supply 4 g protein/tablespoon with only 7 mg of salt. If the patient with cirrhosis and ascites also manifests encephalopathy, then the daily protein intake may need to be reduced accordingly.

The salt substitute potassium chloride may be hazardous in patients with ascites secondary to the frequent use of aldosterone antagonist diuretics (Spironolactone, "aldactone"). These diuretics may conserve potassium to dangerous toxic levels, which can lead to cardiac arrhythmias and death.

Restriction of sodium has been one of the most useful procedures in the treatment of ascites and edema. Hidden sources of sodium in intravenously administered plasma and whole blood (sodium citrate) transfusions, in the diet—such as is contained in bread or artificially softened drinking water—or in some oral and parenteral drugs may cause an unexpected reaccumulation of fluid (Runyon 1994).

portal hypertension—abnormally increased pressure in the portal venous system; frequently seen in cirrhosis of the liver and in other conditions that cause obstruction of the portal vein

paracentesis—a procedure in which fluid is withdrawn from a body cavity via a trocar and cannula, needle, or other hollow instrument

FIGURE 18.8 Diagrams of Portal Hypertension Blood Flow and TIPS Procedure

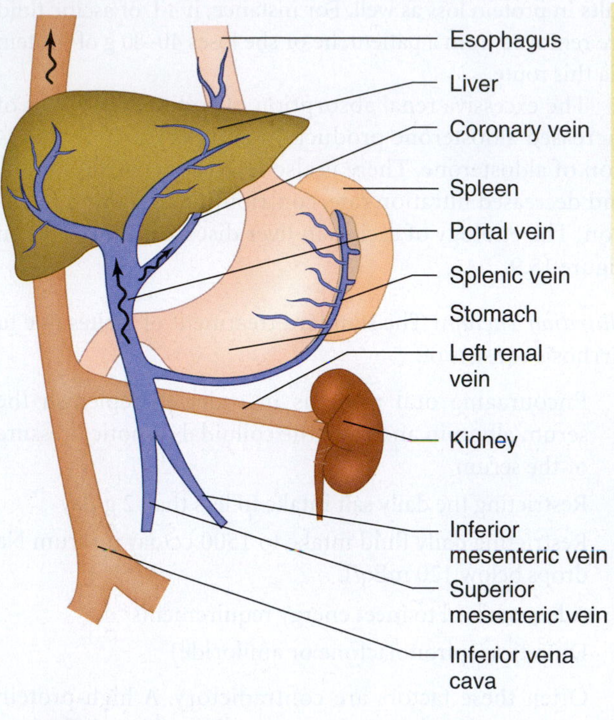

Portal Hypertension before the TIPS procedure is performed.

Portal hypertension causes blood flow to be forced backward, causing veins to enlarge and varices to develop across the esophagus and stomach from the pressure in the portal vein. The backup of pressure also causes the spleen to become enlarged.

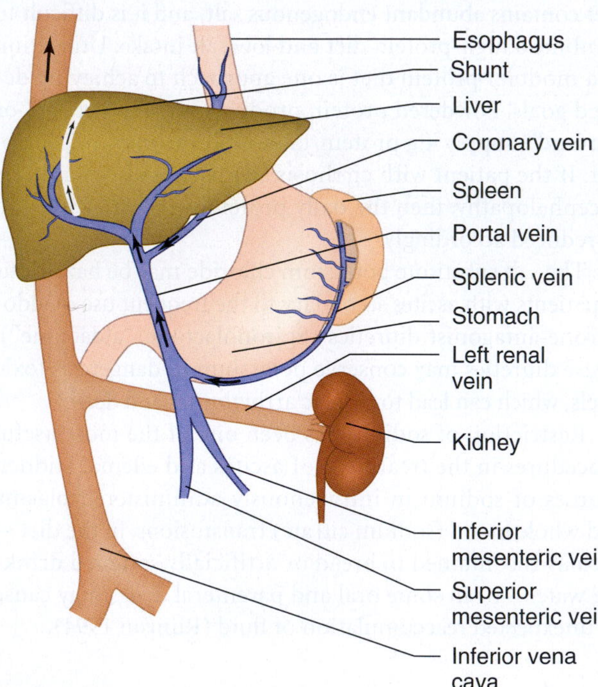

After the TIPS procedure is performed.

A radiologist makes a tunnel through the liver with a needle, connecting the portal vein to one of the hepatic veins. A metal stent is placed in this tunnel to keep the track open.

The shunt allows the blood to flow normally through the liver to the hepatic vein. This reduces portal hypertension, and allows the veins to shrink to normal size, helping to stop variceal bleeding.

Source: The Cleveland Clinic Foundation.

FIGURE 18.9 Theories of Etiology of Ascites Formation in Liver Disease

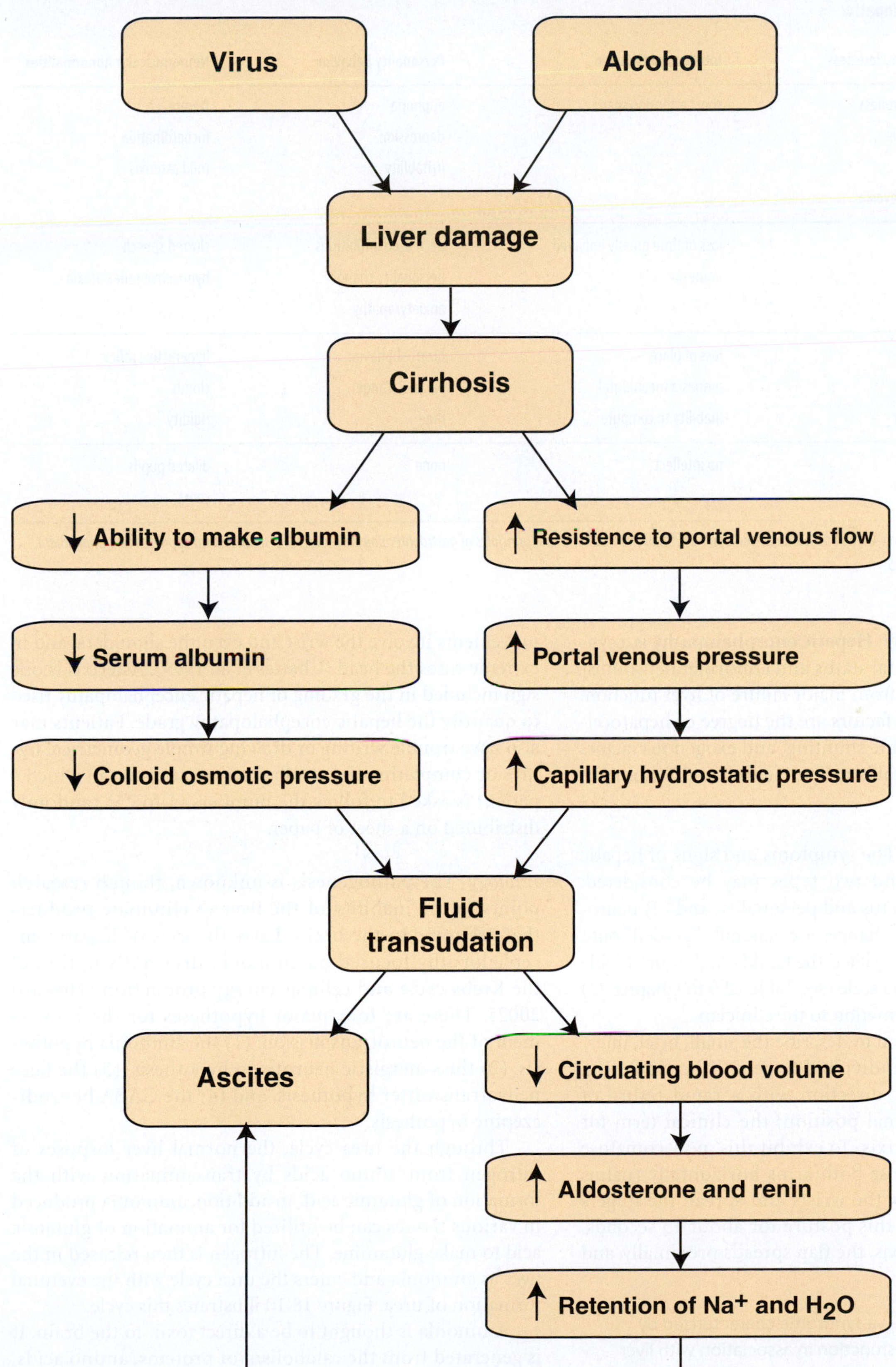

TABLE 18.10

Stages of Hepatic Encephalopathy

Grade	Level of Consciousness	Intellectual Function	Personality Behavior	Neuromuscular Abnormalities
1	lack of awareness hypersomnia insomnia day/night reversal	short attention span	euphoria depression irritability	tremor incoordination mild asterixis
2	lethargic	loss of time grossly impaired amnesia	decreased inhibitions personality change anxiety/apathy	slurred speech hypoactive reflex ataxia
3	somnolence confusion semi-stupor	loss of place amnesia meaningful inability to compute	bizarre behavior paranoia/anger rage	hyperactive reflex clonus rigidity
4	unrousable	no intellect	none	dilated pupils coma

Reprinted with permission of Janssen-Ortho and editors.Thomson ABR, Shaffer EA, editors. *First principles of gastroenterology: the basis of disease and an approach to management.* 5th ed. Toronto: Janssen-Ortho; 2005.

Hepatic Encephalopathy **Hepatic encephalopathy** is a syndrome of impaired mental status and abnormal neuromuscular function resulting from major failure of liver function. Important contributing factors are the degree of hepatocellular failure, portosystemic shunting, and exogenous factors such as sepsis and variceal bleeding (Gerber and Schomerus 2000).

Clinical Manifestations The symptoms and signs of hepatic encephalopathy vary, and two types may be considered: (1) changes in mental status and personality, and (2) neuromuscular changes. These changes are generally "graded" onto one of four clinical stages, called the Child-Pugh score (Table 18.10). The Glasgow coma scale (see Table 22.9 in Chapter 22) provides additional information to the clinician.

The term "flap" is used to describe the small, brief, intermittent movements of individual fingers, either in flexion or laterally in an ulnar direction with a rapid return of the fingers to the original position; the clinical term for this clinical sign is asterixis. To exhibit this, non-comatose patients are asked to raise both arms horizontally (palms downward), to dorsiflex the wrists and spread the fingers wide apart and to hold this posture for about 15 seconds. With more severe asterixis, the flap spreads proximally, and movements involve the wrist and even the shoulders, and in extreme cases the head (Charles et al. 1999). Asterixis is one sign included in the grading of hepatic encephalopathy used to quantify the hepatic encephalopathy grade. Patients may also have trouble writing or drawing simple geometrical figures or completing the number connection test in which a patient is asked to follow the numbers (1 to 25) randomly distributed on a sheet of paper.

Etiology The pathogenesis is unknown, though research points to the inability of the liver to eliminate products that are toxic to the brain. Early theories of hepatic encephalopathy focused on ammonia-driven disruption of the Krebs cycle and cellular energy production (Howard 2002). There are four major hypotheses for the impairment of the neurotransmission: (1) the ammonia hypothesis, (2) the synergistic neurotoxin hypothesis, (3) the false neurotransmitter hypothesis, and (4) the GABA benzodiazepine hypothesis.

Through the urea cycle, the normal liver disposes of nitrogen from amino acids by transamination with the formation of glutamic acid. In addition, ammonia produced in various tissues can be utilized for amination of glutamic acid to make glutamine. The nitrogen is then released in the liver as ammonia and enters the urea cycle with the eventual formation of urea. Figure 18.10 illustrates this cycle.

Ammonia is thought to be a direct toxin to the brain. It is generated from the catabolism of proteins, amino acids, purines, and pyrimidines. Ammonia synthesized in the gut is absorbed and transported in the intestinal venous blood

hepatic encephalopathy—a syndrome characterized by central nervous system dysfunction in association with liver failure

FIGURE 18.10 Normal Utilization of Ammonia and Formation of Urea

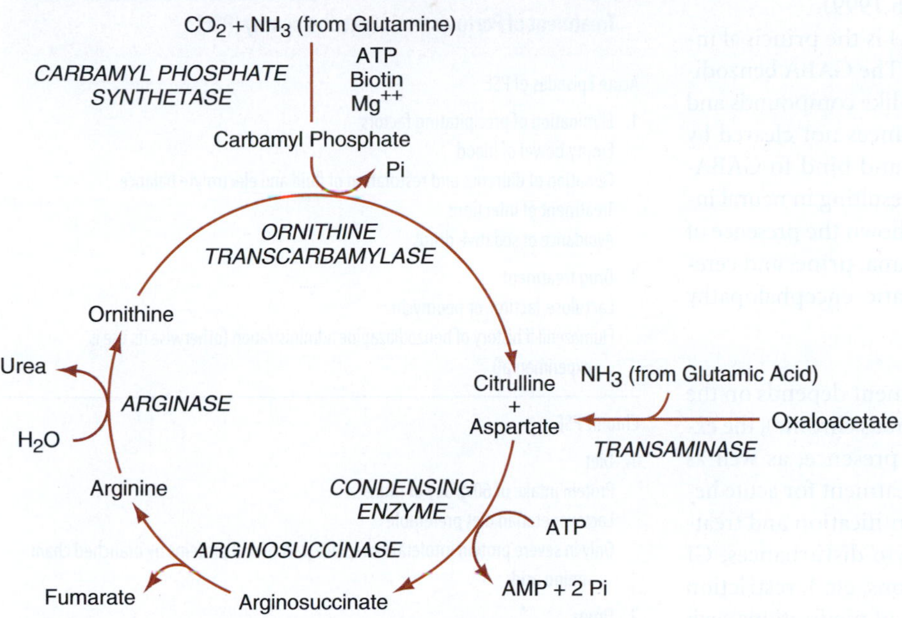

Source: Robert E. Hodges (ed.), *Nutrition-Metabolic and Clinical Applications,* New York: Plenum Press; 1979, p. 147.

to the liver where it is metabolized to urea. Liver disease interferes with this detoxification process and shifts ammonia metabolism to skeletal muscle, where it is used in the conversion of glutamate to glutamine. Muscle wasting reduces the capacity of patients with liver failure to detoxify ammonia. Ammonia therefore accumulates in arterial blood and leads to elevated levels in the brain. The brain lacks significant activity of urea cycle enzymes and cannot synthesize urea from ammonia. The ammonia in the brain is detoxified via the conversion of glutamate to glutamine. According to the ammonia hypothesis, the depletion of glutamate and accumulation of glutamine may contribute to hepatic encephalopathy.

The following evidence indicates that altered ammonia metabolism may be responsible for the symptoms of hepatic encephalopathy:

- There is a good correlation between blood ammonia levels and the abnormalities in mental state, though 10% of patients with hepatic encephalopathy have normal blood ammonia levels.
- Hepatic encephalopathy can be induced in animals by giving toxic doses of ammonium salts.
- Hepatic encephalopathy can be precipitated in the cirrhotic patient by feeding a high-protein diet or by increased endogenous production of ammonia.
- Cirrhotic patients often have abnormal ammonium tolerance tests.

The symptoms of hepatic coma are often corrected by decreasing endogenous ammonia production (Cordoba and Blei 1997).

Mercaptans, ammonia, indoles, scatoles, short-chain fatty acids, and phenols also accumulate in liver failure. The synergistic neurotoxin hypothesis proposes that these neurotoxins, many of which are produced by intestinal bacteria, are involved in hepatic encephalopathy. While there were initial studies in rats supporting this hypothesis, more recent tests in rabbits and rats refute these findings. These compounds may cause cerebral edema, but it is doubtful that they could cause hepatic encephalopathy.

Hyperammonemia is thought to stimulate glucagon secretion and enhance gluconeogenesis from amino acids. The resulting hyperglycemia stimulates hyperinsulinemia, enhancing the uptake of **branched-chain amino acids (BCAA)** by muscle and lowering the plasma levels of BCAAs relative to **aromatic amino acids (AAA)**. The two types of amino acids compete for the same carrier transport system across the blood-brain barrier. This leads to an accumulation of AAAs in the brain. The subsequently raised tryptophan level leads to formation of the inhibitory neurotransmitter serotonin, and the raised phenylalanine level leads to the inhibition of dihydroxyphenylalanine (DOPA) production and formation of so called "false neurotransmitters." The false neurotransmitter hypothesis suggests that these false transmitters may displace catecholamines from their receptors.

Aromatic amino acids, such as tryptophan, tyrosine, and phenylalanine, are elevated in the brains of patients with **fulminant hepatic failure** and hepatic encephalopathy. Pronounced decrease in BCAA to AAA ratio has been described in chronic liver disease but correlates poorly with the presence or absence of hepatic encephalopathy. In theory, administration of BCAA metabolized by skeletal muscle rather than the liver would restore this imbalance between BCAA and AAA, and would decrease production of the toxic false neurotransmitters. This hypothesis has provided the

branched-chain amino acids (BCAA)—one of the amino acids that has a branch chain, namely, leucine, isoleucine, and valine

aromatic amino acids (AAA)—amino acids containing an aromatic side chain (phenylalanine, tyrosine, tryptophan)

fulminant hepatic failure—the severe impairment of hepatic functions in the absence of pre-existing liver disease

rationale for the use of BCAAs in the treatment of hepatic encephalopathy (Hazell and Butterworth 1999).

Gamma-amino butyric acid (GABA) is the principal inhibitory neurotransmitter in mammals. The GABA benzodiazepine hypothesis proposes that GABA-like compounds and endogenous benzodiazepine-like substances not cleared by the liver cross the blood-brain barrier and bind to GABA-benzodiazepine complexes in the brain, resulting in neural inhibition and coma. Recent research has shown the presence of benzodiazepine-like compounds in plasma, urine, and cerebral spinal fluid of patients with hepatic encephalopathy (Quevedo et al. 1999).

Treatment and Nutrition Therapy Treatment depends on the type of hepatic encephalopathy (acute versus chronic), the extent of neurological disorder, and the presence, as well as the cause of, precipitating factors. The treatment for acute hepatic encephalopathy consists in the identification and treatment of precipitating factors (electrolyte disturbances, GI bleeding, constipation, bacterial infections, etc.), restriction of protein ingestion, and administration of medications such as neomycin, lactulose, or lactitol. When acute hepatic encephalopathy is resolved, it is essential to increase protein to maximum tolerance and to maintain treatment with lactulose or lactitol to avoid constipation and obtain at least two bowel movements per day. A summary of hepatic encephalopathy treatment is presented in Table 18.11 and drug-nutrient interactions in Table 18.15.

Several potential toxins are thought to arise from the action of intestinal bacteria on gut protein. Older studies have shown that dietary protein precipitated deterioration, and that encephalopathy improved with dietary protein restriction, making this the standard course of treatment. Protein breakdown rates in cirrhosis, with or without encephalopathy, are significantly higher than normal and are associated with lower rates of protein synthesis. The minimum protein requirement of such patients is about 50 g/day. Thus, although protein restriction has been a cornerstone of treatment of portosystemic encephalopathy, failure to provide adequate protein may lead to progressive depletion of body protein, a situation which is associated with reduced host defense to infection. Thus, attention has been focused on alternative ways of providing adequate protein without precipitating encephalopathy. More recent studies show that allowing normal protein levels (1.2 g/kg) does not impede treatment of encephalopathy and decreases protein breakdown (Cordoba et al. 2004).

Early studies have suggested that the type of dietary protein may be as important as the quantity in terms of its ability to produce hepatic encephalopathy. Meat proteins have the highest concentration of AAA and are thus not the best protein option for patients with encephalopathy. Greater amounts of protein from vegetable sources have been recommended to patients with mild encephalopathy. The mechanism whereby vegetable protein is better tolerated is unclear, but may be related to the increased fiber content, with positive effects on gut transit (Jones 1993). However, vegetable diets are very bulky and may not be well tolerated because of abdominal bloating and gaseous distension. Thus, while a diet of only vegetarian proteins may not be practical, there is evidence that a combination of vegetable and dairy protein may be beneficial (Amodio et al. 2001). Milk protein has the lowest AAA concentration and increased BCAA. Milk and cheese have been found to be better tolerated than an equivalent amount of protein from meat, and dairy protein may have lower ammonia content than meat protein.

Branched-chain amino acids (BCAAs) may be given as a single amino acid combination of valine, leucine, and isoleucine, or in combination with other amino acids such as arginine or ornithine. However, such preparations are expensive and may not benefit all patients with hepatic encephalopathy, though beneficial effects have been shown in both acute and chronic encephalopathy patients. Branched chain amino acid supplementation is best reserved for patients who cannot meet their protein needs through diet alone, in whom this has been shown to be the only effective treatment (Bianchi et al. 2005). Commercially available enteral solutions high in BCAAs are recommended for patients with hepatic coma; parenteral solutions high in BCAAs and low in AAAs, if needed, are also best reserved

TABLE 18.11

Treatment of Portosystemic Encephalopathy (PSE)

Acute Episodes of PSE

1. Elimination of precipitating factors
   Empty bowel of blood
   Cessation of diuretics and restoration of fluid and electrolyte balance
   Treatment of infections
   Avoidance of sedative drugs
2. Drug treatment
   Lactulose, lactitol, or neomycin
   Flumazenil if history of benzodiazapine administration (otherwise its use is experimental)

Chronic PSE

1. Diet
   Protein intake of 60 g/day or more
   Lactovegetarian diet preferable
   Only in severe protein intolerance replace part of oral proteins by branched chain amino acids
2. Drugs
   Nonabsorbable disaccharide in doses sufficient to produce two soft bowel movements/day
   Diuretics reduced to a minimum
   Sedatives to be avoided, (exception: small doses of oxazepam if absolutely needed)

With kind permission of Springer Science and Business Media: Rodes J. *Clinical Manifestations and therapy of hepatic encephalopath, Cirrhosis, hyperammonemia, and hepatic encephalopathy.* New York: Plenum Press; 1994. p. 43, Table 4.

for patients with hepatic coma and abnormal BCAA to AAA ratios (Escott-Stump 2002).

A low serum potassium level may precipitate encephalopathy in a cirrhotic patient, especially if diuretics have been administered chronically or if repeated diarrhea has occurred. Correction of the potassium deficiency often results in a dramatic resolution of the encephalopathic status. The serum potassium is an inaccurate assessment of the body's potassium status; ideally, total body potassium should be measured, but this is complex and impractical. A concurrent urinary potassium level is often valuable; if it is low with marginal serum potassium, then body potassium is probably depleted and renal mechanisms to conserve potassium are operative. Attention should also be directed to correcting hypoglycemia as well as thiamin, magnesium, folate, and vitamin K deficiencies, which are prevalent in cirrhosis and may cause symptoms. In the presence of steatorrhea, water-soluble forms of vitamins A, D, and E may be necessary.

Liver Transplant

Transplantation is considered in cases where the effects of liver disease have the potential to cause mortality greater than the short- and long-term mortality from liver transplant. To be placed on the waiting list for a liver transplant requires a Child-Turcotte-Pugh score of at least 7 (this score indicates the severity of liver disease), but patients with conditions that have low short-term survival rates, such as ascites and encephalopathy, or with advanced cirrhosis, also qualify. If liver disease was caused by alcohol or substance abuse, most centers require at least six months of abstinence from these substances before transplant is considered. Candidates also go through a series of other evaluations, such as psychosocial and nutritional, in order to determine whether transplantation is a viable option (Everson and Trotter 2003). In those who receive transplants, one year survival rates range from 85% to 88% (American Liver Foundation, Facts on Liver Transplantation 2005).

Nutrition Therapy Dietary management before and after transplantation is individualized. However, a general goal before transplant is to lessen the effects of malnutrition and complications of liver disease, such as ascites, if present. Typical kcal and protein goals are 35–45 kcals/kg and 1.0–1.5 g/kg, respectively, depending on the presence of encephalopathy. The major goal of pre-transplant nutrition is to normalize, to the extent possible, macronutrient and micronutrient metabolism—normalizing blood sugar, nitrogen balance, and other relevant lab values while avoiding further damage to the liver. After transplant, the patient is usually on a regular diet, with adequate kcal and protein—slightly lower than pre-transplant goals, at 30–35 kcals/kg and 1.0–1.2 g/kg, respectively. Other nutrients are individualized, based on the immunosuppressant drug regimen used to prevent rejection and the functioning of the new liver. All immunosuppressant drugs may cause hyperglycemia, which is managed by decreasing simple sugars, with carbohydrates providing 50% to 60% of total kcal. Corticosteroids, such as prednisone, often cause sodium retention, managed by a 2–4 g sodium restriction. Cyclosporine and tacrolimus may both require potassium restrictions (See Table 18.15). Other minerals and electrolytes are monitored. Providing the DRI of vitamins remains important before and after transplant in order to encourage proper wound healing and maintain overall health status (Escott-Stump 2002).

Cystic Fibrosis-Associated Liver Disease (CFALD)

Cystic fibrosis (CF) is an inherited disorder of epithelial transport—one of the most common lethal inherited disorders among Caucasians, affecting 1 in every 3,500 live births (Cystic Fibrosis Association, "About Cystic Fibrosis" 2005).

Epidemiology Liver disease is the second leading cause of death among patients with cystic fibrosis (Cystic Fibrosis Foundation, "Patient Registry" 2005); hence, cystic fibrosis-associated liver disease (CFALD) is an important predictor of outcome in patients with cystic fibrosis. It occurs mainly in the first decade of life, with prevalence reported between 9% (Scott-Jupp, Lama, and Tanner 1991) and 41% of patients at 12 years of age (Lamireau et al. 2004). The mean age of presentation of liver disease in CF patients is between 7 and 9 years of age, with prevalence increasing with age, peaking at about 16–20 years of age (Scott-Jupp, Lama, and Tanner 1991). Males are significantly more likely to develop liver disease.

Etiology The mutated gene in CF codes for a defective protein, namely, cystic fibrosis transmembrane conductance regulator (CFTR). There have been greater than 1,000 mutations of the responsible gene identified (Cutting 2005). CFTR is a chloride (Cl) transporter found in the membrane of epithelial cells. In normal cells, the membrane allows the release of Cl from the cell, creating an electrolyte imbalance that draws water out of the cell through osmosis. Water keeps mucus moist and prevents infection. If CFTR does not function, Cl is prevented from leaving the cell and water cannot exit. Without water, mucus thickens, cilia (hairlike structures on cells) cannot function properly, and bacteria can collect on the cells, which can lead to infections.

Clinical Manifestations and Treatment Typical manifestations of the CFTR defect(s) include lung infection (see Chapter 23); pancreatic insufficiency, typically requiring enzyme replacement; and obstruction of the common bile duct. In the normal liver, the CFTR gene is expressed in the epithelia of the intrahepatic and extrahepatic bile ducts and

TABLE 18.12

Hepatobiliary Manifestations of Cystic Fibrosis

Condition	Approximate Frequency
Asymptomatic elevation of liver blood tests	10%–35%
Neonatal cholestasis	>2%
Hepatic steatosis and steatohepatitis	20%–60%
Focal biliary cirrhosis	11%–70%
Multilobular cirrhosis	5%–15%
Cholelithiasis and cholecystitis	1%–10%
Microgallbladder	30%
Sclerosing cholangitis	<1%
Common bile duct stenosis	>2%
Cholangiocarcinoma	rare

Reprinted with permission of Lippincott Williams & Wilkins from: Sokol, RJ, Durie PR (for the Cystic Fibrosis Foundation Hepatobiliary Disease Concensus Group). Recommendations for Management of Liver and Biliary Tract Disease in Cystic Fibrosis. *J Pediatr Gastroenterol Nutr.* 1999;28(1):S1–S13. Table I, p. S3.

the gallbladder (Grubman et al. 1995). Since CFTR is not expressed in hepatocytes or other cells of the liver, it is likely that CFTR affects Cl and water secretion into the bile at the ductal level. If this is the case, CFTR mutations can lead to abnormalities in the composition, consistency, alkalinity and flow of bile, contributing to the pathogenesis of liver lesions and bile duct damage (Lenaerts undated). (Table 18.12 lists hepatobiliary manifestations of CF.)

Hepatic stenosis is the most common hepatic lesion in CF, and may be related to malnutrition, essential fatty acid deficiency, or to the effect of elevated levels of cytokines from the genetic defect. Pancreatic-sufficient patients appear to have a lower incidence of liver disease (Wilchanski et al. 1999). Malnutrition early in life may increase the risk for development of CFALD, emphasizing the importance of diagnosing and intervening as early as possible. Children diagnosed with CF later seem to be at greater risk for the development of CFALD (Corbett et al. 2004). Height faltering happens early in CF children with CFALD, before CFALD, and may help to identify those at increased risk.

Most CF patients with cirrhosis are asymptomatic, with the most common clinical presentation being hepatomegaly or splenomegaly. The usual clinical signs of liver disease, such as jaundice, ascites, and encephalopathy, are rarely present or occur very late. The most deleterious complication of CFALD is portal hypertension. Complications of portal hypertension need to be monitored, because variceal bleeding occurs in one-third of CF patients with associated liver cirrhosis. Fortunately, only a small proportion progress to advanced liver disease. Liver transplantation is an option, and has been successfully performed in CF patients with end-stage liver disease (Curry and Hegarty 2005). There is evidence that liver transplant does have beneficial effects on nutritional status (Columbo et al. 2005). However, this may not be true for patients with variceal bleeding as the only indication of liver disease, and transplantation is generally not recommended for these patients (Gooding et al. 2005).

Ursodeoxycholic acid (UDCA), an exogenous, hypophilic, non-toxic bile acid with choleretic, hepatoprotective, and immunomodulatory properties, is increasingly used to treat the hepatobiliary complications of CF (Kowdley 2000). The mechanisms of action are not well understood; however, it most likely works by reducing bile viscosity and preventing plugging of the bile ducts. Early studies have indicated that functional improvement is substantially lower in CF patients with advanced liver disease, suggesting that treatment should be started as early as possible in the course of liver disease. The long-term risks or benefits of taking this medication are not known, though currently under evaluation.

Nutrition Therapy The emphasis in the management of CF patients with CFALD is maintenance of a normal nutrition status and prevention of deficiencies (Sokol and Durie 1999). Counseling provided to patients with CFALD should include the risks associated with intake of ethanol and encouragement to avoid herbal therapies without consultation with their physicians. (Table F3 in Appendix F lists herbal and other complementary/alternative remedies used for various hepatobiliary disorders.)

Kilocalories CF patients may have energy needs that exceed normal recommendations by 20% to 40%. This is secondary to malabsorption and increased oxygen consumption associated with cholestasis and cirrhosis.

Fat Patients with significant cholestasis may need medium-chain triglycerides to promote absorption of dietary lipids.

Protein Protein intake should not be restricted unless the patient is exhibiting encephalopathy secondary to decompensated hepatic failure.

Fat-Soluble Vitamins Status of fat-soluble vitamins should be assessed every 6–2 months in CF patients with CFALD (Sokol 1994). All supplements should be given with a meal and with pancreatic enzyme supplements. If a vitamin dosage is changed, testing should reoccur 1–2 months following the change (Sokol and Durie 1999).

Vitamin A In general, CF patients are at risk for vitamin A deficiency. They are liable to develop night blindness and conjunctival xerosis, particularly in the presence of liver disease or failure to take daily vitamin supplements. CF patients with low serum retinal (15-20 μg/dL) should receive two to four times the RDI for age (Sokol and Durie 1999). However, CF patients can have a 3.5-fold greater hepatic vitamin reserve compared to normal individuals; thus, care must be taken to avoid hypervitaminosis (Wood, Gibson, and Manohar 2005). Rayner et al. (1989) found that CF patients taking supplements (of 5,000 IU vitamin A and 100–200 mg vitamin E per day) had low plasma levels of vitamin A and decreased retinol binding protein (RBP) in the presence of elevated concentrations of vitamin A in the liver. It was suggested that zinc status be evaluated, because this mineral is needed for the release of vitamin A and RBP from the liver.

Vitamin E Vitamin E status tends to worsen with age (Back et al. 2004), and deficiency often persists despite supplementation. Vitamin E may play a vital role in protecting the lungs from oxidative stress. Supplementation of vitamin E (D-[alpha]-tocopheryl polyethylene glycol-1000 succinate) at a dose of 15–25 IU per day is suggested to prevent or correct vitamin E deficiency (Sokol and Durie 1999).

Vitamin D Vitamin D deficiency is rarely seen in pediatric CF patients, but it becomes more common in older, sedentary patients and those with advanced cholestatic liver disease (Wood, Gibson, and Manohar 2005). Large doses of vitamin D (800–1600 IU/day of vitamin D_2 or D_3 or 2 to 4 μg/dL per day 25-hydroxy vitamin D calcidiol) may be needed by patients with CFALD to normalize serum 25-hydroxyvitamin D concentrations (Sokol and Durie 1999).

Vitamin K Prothrombin time (PTT) can be used to indirectly monitor vitamin K status. With prolonged PTT treatment, supplementation with 2.5 mg (infants) to 10 mg (adolescents) doses of vitamin K, administered daily to twice per week, depending on the response, is recommended (Sokol and Durie 1999).

Essential Fatty Acids (EFA) Normal levels of essential fatty acids are difficult to maintain in CF patients, with 85% of CF patients having EFA deficiency (van Egmond et al. 1996). Supplementation with polyunsaturated fatty acids is recommended to restore EFA status (Sokol 1994).

The Gallbladder

Normal Function of the Gallbladder

The gallbladder is on the underside of the liver and the right side of the abdomen. The gallbladder stores bile produced in the liver (as discussed in the section anatomy and physiology of the liver above), concentrates it and secretes it into the intestines in response to the presence of food (particularly fat). Bile is composed of cholesterol, bile and bilirubin, and is secreted into the intestines to help digest fats. The functions of the gallbladder are the following:

- Removal of water and some inorganic electrolytes from the bile, thus increasing the concentration of larger organic solutes.
- Storage of bile.
- Control of the delivery of bile salts to the duodenum. Following the eating of a meal, the enzyme cholecystokinin pancreozymin (CCK-PZ) is released from the mucosa of the small intestine by the hydrolytic products of digestion. CCK-PZ stimulates the gallbladder to contract and sends bile into the proximal small intestine. Bile salts are essential for the solubilization of fat before lipids can be absorbed.

Cholelithiasis (Gallstones)

Definition **Cholelithiasis** (formation of stones [calculi] within the gallbladder or biliary duct system) is one of the most common medical problems leading to surgery. There are basically three types of stones: cholesterol (80%, with only ~10% being purely cholesterol), pigment, and mixed stones (Ahmed et al. 2000). Cholesterol, a major component of bile, is normally kept in solution by bile acids, lecithin, and phospholipids. However, when bile is supersaturated with cholesterol, it crystallizes and gallstones are formed.

Epidemiology There are approximately 700,000 cholecystectomies performed each year (Schaffer 2005). Nearly two-thirds of patients with gall stones are asymptomatic (Gupta and Shukla 2004). The exact cause of this disorder is not known. However, the presence of obesity, diabetes, and inflammatory bowel disease increases the risk for developing gallstones, as does rapid weight loss and cholesterol-lowering medications. Prolonged parenteral nutrition and short bowel can also cause biliary stasis and therefore increased risk of stones (American Liver Foundation, Gallstones, 2005).

It has been estimated that 20% of adults over 40 years of age and 30% of those over 70 years of age have gallstones. The prevalence of gallstones is higher in individuals of northern European and Hispanic descent. However, Pima Indians have the highest prevalence (up to 75% in the elderly) with other Native American groups also having increased risk. Premenopausal women are 4 times more likely to have this disorder, but, as women age, their risk nears equality with that of males. It has been known for some time that variations in estrogen cause increased secretion of cholesterol and progesterone, leading to bile stasis. While oral contraceptives do not seem to pose a greater risk for gallbladder disease, the

cholelithiasis—the presence or formation of gallstones

Women's Health Initiative postmenopausal hormone trial found that hormone replacement therapy is causally related to gallbladder disease (Cirillo et al. 2005) (see Table 18.13).

Etiology The etiology of the formation of gallstones has been under investigation for years. In the process of secreting bile salts, approximately one-tenth as much free cholesterol is also secreted into the bile. The cholesterol is thought to be a by-product of bile formation, since it has no known function. Cholesterol is insoluble in water, but the bile salts and lecithin keep it in solution in the form of micelles. Because bile salts and lecithin are concentrated with cholesterol in the gallbladder, cholesterol is kept in solution. In abnormal conditions, cholesterol precipitates as gallstones. The amount of cholesterol in bile is determined partly by the quantity of fat eaten, because the hepatic cells synthesize cholesterol as one of the products of fat metabolism in the body. For this reason, persons on high-fat diets over a period of many years are prone to the development of gallstones.

Inflammation of the gallbladder epithelium often results from low grade chronic infection; this changes the absorptive characteristics of the gallbladder mucosa, sometimes allowing excessive absorption of water, bile salts, or other substances that are necessary to keep the cholesterol in solution. As a result, cholesterol begins to precipitate, usually forming many small crystals of cholesterol on the surface of the inflamed mucosa. These, in turn, act as nidi (cluster) for further precipitation of cholesterol, and the crystals grow larger and larger. Occasionally, tremendous numbers of sand-like stones develop, but much more frequently they coalesce to form a few large gallstones, or even a single stone that fills the entire gallbladder. Four conditions that may cause the precipitation of gallstones are illustrated in Figure 18.11.

Complications

Biliary Obstruction When a gallstone passes from the gallbladder through the cystic duct and lodges in the common duct, or in the head of the pancreas, it is called **choledocholithiasis**. The bile is no longer carried to the duodenum and the excretion of bile pigments in the urine gives the urine a dark color. The feces are no longer colored with bile pigments and become grayish (clay colored). Additionally, there is marked disturbance in digestion and absorption of fat. Patients usually experience severe right upper quadrant pain. If uncorrected, back up of bile can result in jaundice and liver damage (**secondary biliary cirrhosis**). Obstruction of the distal common bile duct can lead to pancreatitis if the pancreatic duct is blocked (Sugiyama and Atomi 2004).

choledocholithiasis—gallstones that are present in the common bile duct but are usually formed in the gallbladder

biliary cirrhosis—liver cirrhosis in which there is interference with intrahepatic bile flow

cholecystitis—inflammation of the gallbladder

FIGURE 18.11 Formation of Gallstones

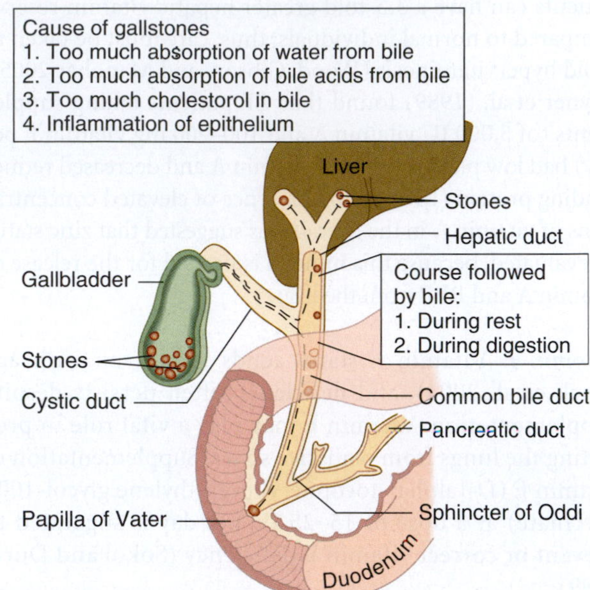

Source: AA Guyton, *A Textbook of Medical Physiology*, W.B. Saunders Co.; 2000. Fig. 64-12.

Cholecystitis Inflammation of the gallbladder, **cholecystitis**, generally develops secondary to obstruction, infection, and ischemia of the gall bladder. This condition can be chronic or acute. Obstruction of the cystic duct by gallstones is the most common cause of gall bladder inflammation, which can lead to infection and necrosis.

Cholangitis Cholangitis is an inflammation of the biliary ducts, usually secondary to obstruction of the common bile duct leading to infection. The infection can ascend into the hepatic ducts, then into the biliary canaliculi, hepatic veins, and perihepatic lymphatics, leading to bacteremia (25% to 40%) (Santen 2006). It is a life-threatening complication of biliary obstruction, particularly in the elderly. Initial therapy is generally with antibiotics, fluid resuscitation, and correction of blood clotting (Bornman, van Beljon, and Krige 2003).

Roentgenography of the Biliary System Cholecystography, the radiologic examination of the gallbladder, is performed to detect calculi and to estimate the ability of the gallbladder to fill, concentrate bile, contract, and empty in a normal fashion.

Since stones are not usually sufficiently radiopaque, an iodide-containing contrast medium that is excreted into the bile by the liver and concentrated in the gallbladder is administered. It is given by mouth or by intravenous injection. Drugs are given 10 to 12 hours before the x-ray study. Injections are given approximately 10 minutes before the x-ray.

When the gallbladder has filled and concentrated the dye it is seen as a pear shaped shadow 2–3 inches long. When stones are present, there are mottled densities within this shadow corresponding to their outline. X-ray examination is repeated at intervals until the gallbladder has expelled all of the dye. If the gallbladder fills and empties normally and there are no stones, it is concluded that no disease is present.

Endoscopic Retrograde Cholangiopancreatography (ERCP) is an endoscopic procedure in which radiographs are taken to view the pancreas and the biliary tree.

Treatment The primary treatment of cholelithiasis is surgery (**cholecystectomy**). However, if surgery is not suitable or the patient refuses, bile acid therapy is a safe option for some, but it is unsuitable for patients with recurrent symptoms or large stones. If the stones are radiolucent (permit the penetration and passage of x-rays or other forms of radiation), good results have been found with ursodeoxycholic acid (750 mg daily) treatment (Bateson 1999).

Nutrition Therapy Diet has long been suspected to be a risk factor for gallstones. However, whether or not altering dietary intake can affect an individual's risk for gallbladder disease has not been established (Tseng, Everhart, and Sandler 1999). It is interesting to note that none of the prospective studies in Table 18.13 indicate an association between fat intake and gallbladder disease. Likewise, research does not support high-cholesterol diets as a cause of gallbladder disease. There is strong evidence for an inverse association between alcohol intake and gallbladder disease. This may be due to alcohol's association with reduced biliary cholesterol saturation and higher serum high-density lipoprotein (HDL) (Thornton, Symes, and Heaton 1983).

Consumption of a highly refined carbohydrate diet, particularly when combined with a low fiber intake, may increase risk for gallstones. Consumption of refined carbohydrates increases bile cholesterol saturation, alters gut micro flora, and leads to hyperinsulinemia, factors that can lead to gallstone formation. Insoluble fiber may provide some protection by decreasing intestinal transit time, and thereby reducing the production of secondary bile acids, such as deoxycholate, which stimulate the liver to increase output of bile (Marcus and Wheaton 1986). The evidence that vegetables, vegetable protein, and vegetable fiber have a protective effect is also strong. Clearly, part of this effect is secondary to the fiber content, but the impact and role of the antioxidants from these foods has not been fully studied. Using NHANES III data, one study found an inverse relationship between serum ascorbic acid levels and gallbladder disease among women (Simon and Hudes 2000). This may have been secondary to ascorbic acid's role in the catabolism of cholesterol to bile acids or to the inhibition of oxidative changes within the gallbladder that decrease mucoprotein production and gallstone formation.

Like many other diseases, gallbladder disease may be caused by combined effects of dietary factors as well as other lifestyle factors that relate to weight gain and obesity, such as exercise. Physical activity is inversely and independently related to gallbladder disease (Leitzmann et al. 1998). The current recommended guidelines for a healthy diet, including minimizing consumption of sweets, increasing intake of fruits, vegetables, and whole grains, and drinking alcohol in moderation, are also the diet recommendations to reduce the risk of gallbladder disease.

The necessity of low-fat or fat-free diets for asymptomatic patients has been questioned. Once gallstones are discovered, a fat-free or low-fat diet is often prescribed, even though there is little evidence as to its therapeutic effect. The gallbladder contracts after administration of fat, since contraction is correlated with CCK release, but little is known of gallbladder dynamics after nonfatty meals.

However, because fat stimulates gallbladder secretion and sphincter of Oddi action, it should be avoided by the patient with symptomatic gallbladder disease. The symptomatic patient will learn through experience what is more comfortable. Plain, simple foods are best tolerated; rich pastries, nuts, chocolate, and fatty, fried, and gas-forming foods are associated with increased discomfort. Condiments and highly seasoned foods may cause distention and increase peristalsis, which ultimately results in irritation of the gallbladder. However, the disturbance varies by individual and dietary management should be individualized.

Acute Attack An acute attack almost always occurs in connection with an obstruction. When it does occur, the gallbladder should be kept as inactive as possible, which is achieved through an NPO diet and complete bowel rest until symptoms lessen, with nutrition administered parenterally as needed (Kalloo and Kantsevoy 2001). The diet is advanced as tolerated to liquids, though only low-fat liquids are typically used. In such a diet, protein may be supplied by skim milk, and carbohydrate is obtained from sugary liquids, such as fruit juices. As tolerated, limited amounts of fat and solid food are added (Escott-Stump 2002), and diet progresses to that prescribed for those with chronic gallstones.

Chronic Condition A diet low in fat (about 25% of total kcalories) is desirable for the dietary treatment of patients with chronic cholecystitis. Many patients with cholecystitis or gallstones are overweight, and attention should be given to weight reduction through a fat- and kcal-controlled diet in order to decrease incidence of attacks. Because rapid

cholecystectomy—surgical removal of the gallbladder

TABLE 18.13

Prospective Cohort Studies on Energy, Macronutrients and GBD

Study Population (reference)	No. of Cases[1]	No. in Study	Period of Follow-Up	Dietary Assessment	Energy	Total Fat	Cholesterol	Carbohydrates	Simple Sugars	Fiber	Protein
Framingham, MA (Friedman et al. 1966)	226	5,209	1957–59	Questions on intake at fifth biennial exam, 8 years after initial exam		N	N				N
U.S.[2] (Sichieri et al. 1991)	216	4,730	1971–75 to 1982–84	24-hour recall	–[3]	N[4]	N[4]	N[4]		–	
Nurses, U.S.[2] (Maclure et al. 1989, 1990)	612	88,837	1980–84	61-item food frequency questionnaire given in 1980	+	N[4]	N[4]	N[4]	N[4]	–[4]	N[4]
Zutphen, the Netherlands[5] (Moerman et al. 1994)	54	860	1960–85	Diet history taken in 1960	N	N	–	N	+[4]	–	N
Japanese Americans, Oahu, HI[5] (Kato et al. 1992)	471	7,831	1965–68 to 1990	24-hour recall and questions on usual monthly intake of alcohol	–	N		N			N

N, no association; +, positive association; –, inverse association; blank indicates the factor was not examined.
[1] Cases were identified as patients diagnosed with gallstones in hospital or clinical setting.
[2] Women only.
[3] Individuals under the age of 50 years.
[4] Adjusted for energy intake.
[5] Men only.
Reprinted with permission of the Nutrition Society from: Tseng M, Everhart JE, Sandler RS. Dietary intake and gallbladder disease: a review. *Public Health Nutrition.* 1999:2(2);161–172, Table 3.

weight loss also has the potential to cause gallstones, however, gradual weight loss via a moderate-fat diet is recommended (Erlinger 2000).

Normal dietary protein levels are maintained, and the carbohydrate allowance is adjusted as needed to maintain the patient's weight at the desired level. Increasing the amount of carbohydrate may also serve as a therapeutic measure in cases complicated by jaundice, because adequate fiber serves to bind excess bile acids. Due to poor absorption of fat, a water-soluble form of vitamins A, D, E, and K may be necessary.

Because the presence of fats in the intestinal lumen stimulates the release of bile by the gallbladder, fat intake is frequently restricted (30–45 grams per day) in patients with cholecystitis. In patients with chronic gallbladder problems or suspected fat malabsorption, administration of water-soluble forms of fat-soluble vitamins may be appropriate. When the gallbladder is removed, the large common duct connecting the liver to the small intestine adapts by stretching to store bile. Diet is generally resumed once bowel sounds are present and advanced as tolerated to a regular diet.

Postoperative Cholecystectomy Diet Following surgical removal of the gallbladder, oral feedings are usually resumed once bowel sounds return and the patient can tolerate nasogastric drainage tube removal. The diet can be advanced as tolerated to a regular diet. In the absence of the gallbladder, the liver secretes bile directly into the intestine, which can cause diarrhea. This diarrhea may be managed through increased fiber intake to increase fecal bulk, and patient avoidance of foods that are known to cause diarrhea (The Mayo Clinic 2005). However, symptoms may lessen over time, since the bile duct dilates to allow bile to be held in a manner similar to the original gallbladder.

Dietetic professionals should be concerned that nearly 50% of cholecystectomy patients complain of digestive symptoms one year after surgery (Bateson 1999). The mechanism by which food provokes pain is unknown. The most common nonpain symptoms are in the area of digestion, fatty food intolerance, heartburn, and nausea. This suggests that these patients might benefit from nutrition therapy to minimize these digestive symptoms. An individual approach needs to be taken with each patient, examining their tolerances and adjusting their diet accordingly. Research on the effectiveness of dietary intervention is needed.

The Pancreas

Normal Anatomy and Physiology of the Pancreas

The pancreas (see Figure 18.12) is a large accessory digestive gland located behind the stomach and opposite both the duodenum and the spleen. The pancreas is divided into different portions: the head, which is the expanded portion of the pancreas, followed by the neck, the body, and the tail.

The pancreas has two major functions:

- Exocrine function: Produces the enzymes necessary for digestion (see Table 18.14).
- Endocrine function: Produces hormones to regulate the use of body fuels, mainly glucose (see Chapter 19).

The pancreas is the only organ of the body that has both exocrine and endocrine functions.

The pancreas is composed of two major types of tissues: the acini, ducted exocrine tissues that are responsible for secreting digestive juices into the duodenum; and the islets of Langerhans, ductless endocrine tissues (meaning they have no means of secreting externally) that secrete the hormones insulin and glucagon directly into the blood. The pancreas has almost a million islets of Langerhans, organized around small capillaries (making it very well vascularized) into which its cells secrete their hormones. There are 3 major types of endocrine cells: alpha—which compose 25% of the cells and secrete glucagons; beta—which make up 60% of the cells and are responsible for the synthesis and secretion of insulin; and delta—which make up the remaining 10% of cells and secrete somatostatin. Endocrine secretions are carried to the digestive tract via the pancreatic duct, which then joins with the common bile duct and enters the duodenum.

Pancreatitis

Definition and Clinical Manifestations **Pancreatitis**, an inflammation of the pancreas, is characterized by edema, cellular exudate, and fat necrosis. The disease is also characterized by autodigestion, necrosis, and hemorrhage of pancreatic tissue and can be either acute or chronic. While some cases are asymptomatic, symptoms may include upper abdominal pain radiating to the back, generally worsening with ingestion of food. Clinical presentation may also include nausea, vomiting, abdominal distention, and steatorrhea. Severe cases are complicated by hypotension and dehydration.

Etiology and Pathogenesis The most common causes of pancreatitis are alcoholism and cholelithiasis (70% of cases) (National Digestive Diseases Information Clearinghouse 2005). However, the exact mechanisms that lead to pancreatic injury are not fully understood. One theory involves blockage or reflux of the ductal contents into the pancreatic duct (Calleja and Barkin 1993). A common characteristic seems to be premature activation of trypsin

pancreatitis—inflammation of the pancreas

FIGURE 18.12 Anatomy of the Pancreas

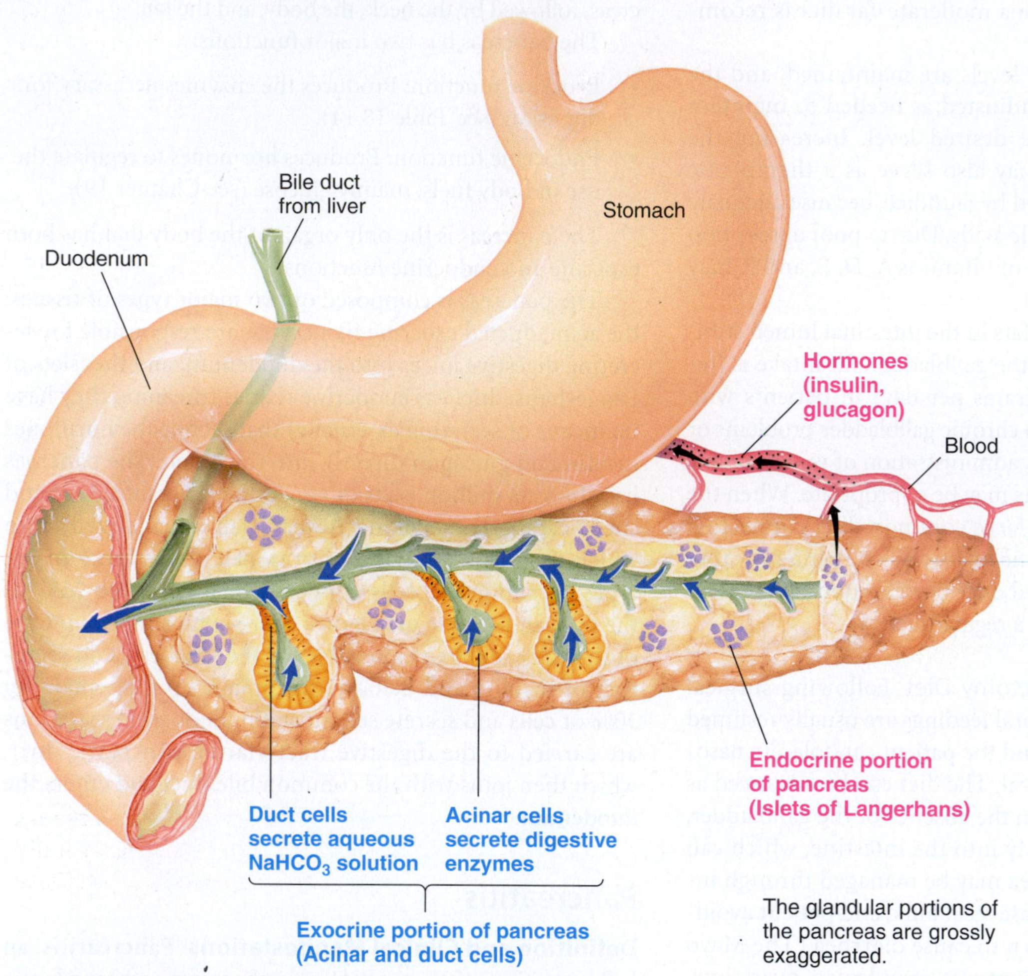

Source: L. Sherwood, *Human Physiology: From Cells to Systems*, 5e, copyright © 2004, p. 616.

within the pancreas, resulting in autodigestion of the pancreatic cells. The enzymes released by destroyed pancreatic cells eventually reach the bloodstream, causing elevated serum amylase and lipase levels. In chronic pancreatitis, if the islet cells have been damaged, diabetes may also develop.

Nutrition Therapy Pancreatitis frequently results in malnutrition due to poor intake, abdominal pain, nausea/vomiting, fistula drainage, protein losses, infections, lengthy disease course, and increased nutritional needs of those with chronic alcohol abuse. The pain associated with pancreatitis is partially related to the secretory mechanisms of pancreatic enzymes and bile. One goal of nutrition intervention is to provide minimal stimulation of these systems. During severe, acute attacks, all oral feedings are withheld. In less severe attacks, a clear liquid diet may be given in a few days. The diet should be progressed as tolerated to easily digested foods with a low fat content and then advanced as tolerated. Foods may be better tolerated if they are divided into six small meals to compensate for the diminished exocrine function of the pancreas (Schneider and Singer 2005).

Nutrition Support in Acute Pancreatitis Early, aggressive nutrition support is suggested in severe, acute pancreatitis, when a prolonged NPO status is anticipated to allow for pancreatic rest. Nutrition support has not been shown to benefit those with mild to moderate pancreatitis. In patients with acute pancreatitis, protein catabolism increases by 80% and caloric needs by 20%; nutrition support in these patients may help prevent nutrient deficits while preserving lean body mass. The use of nutrition support to resolve negative nitrogen balance in acute pancreatitis is associated with improved outcomes (Abou-Assi and O'Keefe 2002). Goals, therefore, are to provide adequate kcal and protein, minimize nitrogen losses, and manage imbalances. These include imbalances in blood glucose, hypertriglyceridemia, and micronutrients such as calcium, magnesium, and vitamins. At the same time, pancreatic stimulation should be minimized.

While in the past parenteral feedings were the norm, enteral feedings have now become the preferred method.

TABLE 18.14

Pancreatic Exocrine Secretions

Enzyme	Substrate	Action & Product of Action	Absorption
Trypsin	Protein and polypeptides	Hydrolysis of interior peptide bonds to form polypeptides	
Chymotrypsin	Protein and polypeptides	Hydrolysis of interior peptide bonds to form polypeptides	Pinocytosis of small peptides
Carboxypolypeptide	Polypeptides	Hydrolysis of terminal peptide bonds to form amino acids	Amino acids absorbed into blood
Ribonuclease Deoxyribonculease	Ribonucleic acids Deoxyribonucleic acids	Hydrolysis to form mononucleotides	
Elastrase	Fibrous protein	Hydrolysis to form peptides and amino acids	
Lipase	Fat	Hydrolysis to form simple glycerides, fatty acids, and glycerol	Micelles → mucosal cells → chylomicrons → lymph
Cholesterol Esterase	Cholesterol	Hydrolysis to form cholesterol and fatty acids	
α-Amylase	Starch and dextrin	Hydrolysis to form dextrins and maltose	

Recent studies suggest that early enteral feedings do not increase hospital stay, time to normalization of amylase, or time to advancement to an oral diet (McClave et al. 1997). They also reduce septic and metabolic complications and are much less costly to administer (Abou-Assi and O'Keefe 2002). In addition, they maintain gut integrity. A prospective study of pancreatitis patients comparing TPN and gut disuse to enteral feeding found that enteral feeding maintained near normal villi, while TPN and gut disuse showed significant villous atrophy (Groos, Hunefeld, and Luciano 1996). This is important, since there is evidence of leaky gut in pancreatitis patients with loss of gut integrity, which can increase the stress response and disease severity (McClave 2005).

Enteral Support In order to minimize pancreatic stimulation, feeding occurs below the ligament of Treitz, usually via insertion of a nasoenteric tube. Recommended initiation of feeding is 25 mL/hour with advancement to a goal kcal level of 25 kcal/kg over the first 24–48 hours (McClave 2005). However, even hypocaloric jejunal feedings have been shown to be superior to TPN (Abou-Assi, Craig, and O'Keefe 2002). If the tube is low enough, any formula may be used, but nearly fat-free elemental formulas result in least stimulation of the pancreas; small peptide formulas with 70% of the fat as medium chain triglycerides stimulate slightly more, but are better absorbed. Some small trials have indicated that this population may benefit from immune-enhanced formulas. Advancement to an oral diet may occur when amylase and lipase begin decreasing towards normal levels and the patient has been pain free for at least 24 hours (McClave 2005).

Parenteral Support TPN should only be considered in patients in whom enteral access is not possible or who cannot tolerate enteral feeding. It is suggested that initiation of TPN be delayed about five days, until after the peak inflammatory response has occurred. Mixed fuel should be used, and the volume should be increased slowly to a goal kcal level of 25 kcal/kg. Intralipid should be less than 15% to 30% of kcal, and protein should be individualized to meet patient needs. Electrolytes, especially calcium, should be monitored closely, as should triglyceride and blood sugar levels, which should be maintained as near to normal as possible (McClave 2005).

Nutrition in Chronic Pancreatic Insufficiency The object of therapy in such patients is to prevent further damage to the pancreas, forestall further attacks of acute inflammation, alleviate pain, treat steatorrhea, and correct malnutrition, preventing weight loss and promoting weight gain as appropriate. The frequency of attacks may be reduced by frequent small meals of a moderate- to low-fat diet. Pancreatic enzymes, such as Viokase®, Pancrease®, or Cotazym®, are prescribed to be taken at each meal or snack to improve digestion and absorption. The amounts of these enzymes necessary with each meal may vary depending on the fat content of the food consumed. Alcohol and other gastric stimulants such as coffee, tea, spices, and condiments should be avoided (Schneider and Singer 2005).

Individuals with chronic pancreatitis and scarring of the tissue often lose weight, even when appetite and eating habits are normal. This weight loss is secondary to insufficient secretion of pancreatic enzymes, causing malabsorption, which leads to loss of fat, protein, and sugar into the stool. To promote weight gain, the level of fat in the diet should be the maximum a patient can tolerate without increased steatorrhea or pain. Medium-chain triglycerides may be added to the diet, since they do not require lipase for digestion (Escott-Stump 2002).

TABLE 18.15

Drug–Nutrient Interactions for Hepatobiliary Medications

Classification	Mechanism	Generic	Brand Names	Possible Food–Drug Interactions
K sparing diuretic	Aldosterone receptor antagonist	Spironolactone	Aldactone, Aldactazide	Avoid excess K intake & K supplementation, avoid salt substitutes, ↓ kcal and Na may be recommended, avoid natural licorice, anorexia, ↑ thirst, N/V, diarrhea, gastritis
K sparing diuretic	Aldosterone receptor antagonist	Amiloride	Midamor	N/V, dry mouth, diarrhea, avoid salt substitutes
Antibiotic	Interferes with bacterial protein synthesis	Neomycin	Mycifradin, Neo-Fradin, Neo-Tab	Neomycin impairs absorption (and may also increase excretion) of a broad variety of nutrients including carbohydrates, fats, calcium, iron, magnesium, nitrogen, potassium, sodium, folic acid, and vitamins A, B_{12}, D, and K
Laxative, antihyperammonemic		Lactulose, lactitol	Cephulac, Chronulac, Duphalac, Kristalose	High fiber with 1,500–2,000 mL fluid/d to prevent constipation, N/V, belching, cramps, borborygmi, diarrhea, flatulence
Benzodiazepine receptor antagonist	Benzodiazepine receptor antagonist	Flumazenil	Romazicon	N/V
Glucocorticoids Anti-inflammatory, immunosuppressant, hormone	Mimics the action of cortisol	Prednisone	Deltasone	Caution with DM - ↑ glucose, highly protein bound, may need ↑ K, PO_4, Ca, and ↑ vitamins A, C, D, ↑ protein, and ↓ dietary Na, avoid alcohol
Immunosuppressant	Attacks the white blood cells	Cyclosporin, tacrolimus	Neoral, Sandimmune	No K supplement or salt sub, caution with grapefruit, anorexia, N/V, diarrhea, ↑ glucose
Antacid, antiflatulent		Aluminum hydroxide, magnesium hydroxide	Mylanta	Take 1 hr after meals, take Fe or folate supplement separately by 2 hr, Take separately from citrus fruit/juice or Ca citrate by 3 hr (juice ↑ Al abs), ↓ absorption folate, PO_4, Fe, diarrhea
Antacid		Aluminum hydroxide, magnesium hydroxide	Maalox	Take 1 hr after meals, take Fe or folate supplement separately by 2 hr, Take separately from citrus fruit/juice or Ca citrate by 3 hr (juice ↑ Al abs), ↓ absorption folate, PO_4, Fe, diarrhea
Antiulcer, antiGERD, antisecretory	H_2 receptor antagonist	Famotidine	Pepcide AC, Pepcid, Pepcid IV	Bland diet may be recommended, take >1 hour after Fe supplement, take Mg supplement or Al/Mg antacids separately by >2 hr; limit caffeine; may ↓ absorption of Fe and B_{12}
Antiulcer, antiGERD, antisecretory	H_2 receptor antagonists	Gaviscon cimetidine	Tagamet HB Zantac	Bland diet may be recommended, take at least 2 hr after Fe supplement, take Mg supplement or Al/Mg antacids separate by at least 2 hr, limit caffeine, ↓ Fe and vitamin B_{12} absorption, Mg or Al/Mg antacids ↓ drug abs, liquid cimetidine precipitates tube feeding

(continued on the following page)

TABLE 18.15 *(continued)*

Antiulcer, antiGERD, antisecretory	H_2 receptor antagonists	Nizatidine	Axid AR, Axid	Bland diet may be rec., take at least 2 hr after Fe supplement, take Mg supplement or Al/Mg antacids sep by at least 2 hr, limit caffeine, ↓ Fe and vitamin B_{12} abs, Mg or Al/Mg antacids ↓ drug abs, liquid cimetidine precipitates tube feeding
Antiulcer, antiGERD, antisecretory	H_2 receptor antagonists	Ranitidine	Zantac 75	Bland diet may be rec., take at least 2 hr after Fe supplement, take Mg supplement or Al/Mg antacids separate by at least 2 hr, limit caffeine, ↓ Fe and vitamin B_{12} abs, Mg or Al/Mg antacids ↓ drug absorption, liquid cimetidine precipitates tube feeding
Antiulcer, antiGERD, antisecretory	Proton pump inhibitors Block production of acid by the stomach	Esomeprazole	Nexium	May ↓ absorption of Fe and B_{12}
Antiulcer, antiGERD, antisecretory	Proton pump inhibitor	Lansoprazole	Prevacid	Take 30–60 min before meals, may ↓ Fe and vitamin B_{12} absorption, diarrhea
Antiulcer, antiGERD, antisecretory	Proton pump inhibitor	Omeprazole	Prilosec	Take 30–60 min before meals, may ↓ Fe and vitamin B_{12} absorption, diarrhea
AntigGERD	Proton pump inhibitor	Pantoprozole	Protonix	Take 30–60 min before meals, may ↓ Fe and vitamin B_{12} absorption, diarrhea
AntiGERD	Proton pump inhibitor	Rabeprazole	Aciphex	Take 30–60 min before meals, may ↓ Fe and vitamin B_{12} absorption, diarrhea

Steatorrhea is a common occurrence and results in malabsorption of fat-soluble vitamins. Also, deficiency of pancreatic protease, necessary to cleave vitamin B_{12} from its carrier protein, could potentially lead to vitamin B_{12} deficiency. With appropriate supplemental enzyme therapy at meals, vitamin absorption should be improved; however, the patient should be monitored periodically for vitamin deficiencies. Water-soluble forms of the fat-soluble vitamins or parenteral administration of vitamin B_{12} may be necessary.

Because pancreatic bicarbonate secretion is frequently defective, medical management may also include maintenance of an optimal intestinal pH to facilitate enzyme activation. Antacids, H2-receptor antagonists, or proton pump inhibitors that reduce gastric acid secretion may be used to achieve this effect (Draganov and Toskes 2004).

Efforts should be made to cater to the patient's tolerances and preferences for nutrition management. In chronic cases with extensive pancreatic destruction, the insulin-secreting capacity of the pancreas decreases and glucose intolerance develops. Treatment with insulin and nutrition care similar to that used for a patient with diabetes mellitus is then required.

Conclusion

This chapter has presented evidence-based practices for maintaining nutritional status and providing nutrition therapy for patients with hepatobiliary disease. Diseases of the hepatobiliary system have a significant impact on the nutritional status of the patient. The clinical manifestations—jaundice, anorexia, fatigue, abdominal pain, steatorrhea, and malabsorption—all impact nutritional status. Furthermore, the disease processes have the potential to interrupt normal metabolism, placing the patient at significant nutritional risk. Hence, nutrition therapy is a vital component of medical treatment.

PRACTITIONER INTERVIEW

Mary Ellen Beindorff, RD, LD *Abdominal Organ Transplant Nutrition Specialist, Barnes-Jewish Hospital, St. Louis, MO*

Background

I have been a dietitian for 25 years, and currently my primary responsibility is patients with abdominal organ transplant (liver, kidney, pancreas) and hepatobiliary surgery.

How many liver patients do you see in a day, and what are the primary disorders?

I see four to ten patients with liver disease everyday. This is significantly higher than the number a dietitian would see in a community hospital, which would be more likely to be a handful a month. I see liver patients with a wide variety of diagnoses, but the most common liver diseases are hepatitis C (HCV) and alcoholic liver disease. However, I also see patients with primary biliary cirrhosis (PBC), primary sclerosing cholangitis (PSC), non-alcoholic steatohepatitis (NASH), hepatitis B (HBV), autoimmune hepatitis (AIH), and other genetic liver disorders (Wilson's disease, hemochromatosis, biliary atresia, etc.). I see less hepatitis B (HBV) patients because the vaccine has eliminated the virus in patients under the age of 40. Now most of our HBV patients are foreign-born.

How has MNT practice for liver disease changed over the years?

When I first started practice, all liver patients were put on low-protein diets. Now we know that only in cases of refractory encephalopathy (in 5%–10% of patients) should protein be restricted. We also used to severely restrict fluid in liver patients; now we restrict only when hyponatremia occurs (usually with serum Na levels less than 125 mEq/L). So both protein and fluid restrictions may occur, but only temporarily. Unfortunately, I find that many MDs who do not see liver patients on a regular basis still recommend these restrictions and many RDs (not knowing any better) educate the patient on these restrictions.

How do you nutritionally assess a liver patient?

The key information needed for nutritional assessment is a detailed weight history, thorough diet history, functional capacity, physical assessment, past medical history, current medications (including herbals), and labs. By functional capacity I mean, are they still working? Are they independent with sufficient activities of daily living (ADL). Do they need help for some things? Do they mostly stay in bed or on the couch all day? Most labs are not helpful in assessing liver patients, but I do look at electrolytes, to correct if necessary. I also look at bilirubin and international normalized ratio (INR), which measures the speed of prothrombin time for coagulation, to determine the extent of liver disease.

What is a common nutritional diagnosis?

The "typical" diagnosis for liver disease patients is some degree of malnutrition. In the past years, I have seen many more severely obese liver patients that are malnourished. To assess this, I rely on a thorough diet history and physical assessment. For diet history, I look to see if adequate protein and other nutrients are ingested, and on the physical exam, I look for muscle wasting and to what degree in addition to noting volume of ascites and/or edema. Physicians often assume that because patients are obese they are well nourished, but this is not necessarily the case.

What are the challenges working with this population?

I see the greatest challenges in working with the moderately to severely malnourished patients who, because of symptoms of their liver disease, have a difficult time getting adequate dietary intake. They usually have ascites, which often causes early satiety and reflux. If they're getting tapped often, they're losing protein there too. They often are also taking lactulose and/or neomycin to control encephalopathy which causes malabsorption and diarrhea.

What resources to you use to stay current?

I currently rely on many resources for information on liver disease and nutrition support. Both the Dietitian in Nutrition Support (DNS) practice group, ADA, and ASPEN have numerous resources available on liver disease and nutrition support. One of our hepatologists just completed an article in *Nutrition in Clinical Practice*, ASPEN's publication, and another had an article in DNS's *Support Line* a few years ago. In addition, I scan the medical journals available each month in the hepatologists' and liver surgeons' offices. Lastly, I find it invaluable to "network" with my colleagues at national meetings.

Any advice for the nutrition students?

In my opinion, many interns are not confident of their clinical skills at first. Jump right in and try to do it. (We all learn from our mistakes.) Be self-confident; we are the nutrition experts, and most of the time, we know much more than our audience.

CASE STUDY

CC:

Acute abdominal pain extending into the lower back; severe nausea and vomiting. DOB: 1/12/1953

General:

Middle-aged Caucasian male who is employed as a sales representative for a Midwestern trucking firm. He is married with 2 children. Ht: 5′10″ Wt: 192 lbs. Usual body weight: 200–210 lbs.

Labs:

TP 3.6 g/dL; Alb 3.5 mg/dL; Lipase 521 U/L; Amylase 925 U/L

Rx:

Captopril 25 mg; Hydrochlorothiazide 50 mg
NKDA, NKFA

Past Med Hx:

Hypertension × 10 years. Pt. describes epigastric pain off and on for previous 3 months. Most recently the pain has started to radiate to his back and lasts from hours to several days.

Nutrition Hx:

Normally consumes 3 meals/day with breakfast and evening meal at home with family. Rarely eats out. Has no food intolerances except that recently his stomach becomes "upset" when he eats any food that is fried. Smokes 2 ppd. Describes alcohol intake as 6–8 beers/night. Denies regular "hard liquor" intake but does consume occasionally. Denies any drug use except prescribed medicines.

Questions

1. This patient was diagnosed with acute pancreatitis. What is this disorder?
2. What signs and symptoms support the diagnosis?
3. What factors in this client's history are consistent with risk for pancreatitis?
4. As he is admitted to the hospital, what will be the most likely steps taken for treatment?
5. What will be his primary nutrition diagnosis?
6. What is the nutrition therapy for acute pancreatitis?

NUTRITION CARE PROCESS FOR DISEASES OF THE HEPATOBILIARY SYSTEM

Step One: Nutrition Assessment

Medical/Social History

Standard

Dietary Assessment

Standard, but close attention to: use of alcohol, vitamin, mineral, herbal, or other type of supplements

Previous nutrition education, or medical nutrition therapy

Eating Pattern: 24-hour recall, diet history, food frequency

Anthropometric

Standard

Biochemical Assessment

Visceral and Acute Phase Protein Assessment: (All may be affected by hepatobiliary disease).

Total protein

Albumin

Prealbumin

Transferrin

Retinol Binding Protein

Fibronectin

3-Methyl Histidine

C-Reactive Protein

Liver function:

Bilirubin: conjugated, unconjugate

Alkaline phophastase

ALT

AST

Prothrombin time

Hematological Assessment:

Hemoglobin

Hematocrit

MCV

MCHC

MCH

TIBC

Lipid Assessment:

Total cholesterol

HDL

LDL

Triglyceride

Vitamin/Mineral Status Specific Labs	*Hepatobiliary*
Folate	Erythrocyte folate, free folate, (FIGLU) urinary formiminoglutamic acid
Vitamin B_{12}	Schilling test, erythrocyte B_{12}, DUMP test, serum B_{12}
Vitamin C	Plasma vitamin C, leukocyte vitamin C, urinary vitamin C
Vitamin D	Serum alkaline phosphatase provides indirect measure
Vitamin K	Prothrombin time
Vitamin A	Serum carotene, retinal binding protein
Vitamin E	Serum tocopherol, erythrocyte hemolysis
Niacin	Urinary N-methyl nicotinamide
Riboflavin	Urinary riboflavin, erythrocyte glutathione reductase
Vitamin B_6	Whole blood level of pyridoxal phosphate
Thiamin	Blood pyruvate and lactate, urinary thiamin excretion, erythrocyte transketolase, apoenzyme levels
Zinc	Serum zinc

Step Two: Nutrition Diagnosis

Excessive alcohol intake

Evident protein-energy malnutrition

Impaired nutrient utilization

Altered nutrition-related laboratory values

Involuntary weight loss

Altered GI function

Sample PES: Impaired nutrient utilization as evidenced by history of chronic pancreatitis and presence of steatorrhea after meals.

Step Three: Intervention

1. Reinforce use of pancreatic enzymes before each meal and snack.
2. Provide nutrition education on lower-fat dietary choices while limiting oils and fats added as condiments to 3 tsp/day.

Step Four: Monitoring and Evaluation

1. The patient will consume adequate pancreatic enzymes as prescribed before each meal or snack.
2. The patient will limit avoid adding fat as a condiment or as a cooking medium to 3 tsp/day.

WEB LINKS

Columbia University Division of Liver Diseases: Cute picture, good information on different types of liver disease, links to other websites, but not recently updated.

http://cpmcnet.columbia.edu/dept/gi/disliv.html

American Liver Foundation: Good liver health information, links to recent clinical trials.

http://www.liverfoundation.org

American Association for the Study of Liver Diseases: Good patient resources, especially the "Fast Facts."

https://www.aasld.org/eweb/StartPage.aspx

Colorado State University: Nice review of anatomy and physiology of liver and gallbladder. Good pictures and explanations of systems.

http://arbl.cvmbs.colostate.edu/hbooks/pathphys/digestion/liver/anatomy.html

Hardin MD: Liver Disease: Links to lots of places with good pictures.

http://www.lib.uiowa.edu/hardin/md/liverdisease.html

Virtual Hospital by IA State: Has areas for patients and providers.

http://www.vh.org/index.html

American Gastroenterological Association: Patient center web site.

http://www.gastro.org/wmspage.cfm?parm1=478

National Digestive Diseases Information: Clearinghouse (NIH)—stats and information about each disease.

http://digestive.niddk.nih.gov

Hepatitis information network: Provides links to articles.

http://www.hepnet.com

National Center for Complementary and Alternative Medicine: Visitor can search disease states for related CAM therapies.

http://nccam.nih.gov/health

Up to Date Patient Information: Patient information of lots of disease states.

http://patients.uptodate.com

END-OF-CHAPTER QUESTIONS

1. List the major functions of the liver, pancreas, and gallbladder.
2. What is jaundice? What is the difference between conjugated and unconjugated bilirubin? What disorders could cause elevated unconjugated jaundice, and what disorders could cause elevated conjugated jaundice?
3. What is portal hypertension (include causes, signs, and symptoms)? Why would portal hypertension cause ascites and esophageal varices? What is the medical and nutrition treatment for portal hypertension?
4. List the types of viral hepatitis and their modes of transmission. How is alcoholic hepatitis different from viral hepatitis?
5. What is cirrhosis? List some of the causes of this liver disorder. What are the common complications of cirrhosis? How does cirrhosis cause hypoglycemia and hyperglycemia? What parameters can you use to nutritionally assess a patient with cirrhosis?
6. What are the possible biochemical causes of hepatic encephalopathy? What is the amount and type of protein used in the medical nutrition therapy for hepatic encephalopathy?
7. What is recommended for a post operative cholecystectomy diet?
8. Describe the difference between acute and chronic pancreatitis. List the pertinent labs for each type of pancreatitis. What are common nutritional problems associated with chronic pancreatitis?

ENDNOTES

[1] Since the pores of the hepatic sinusoids are large enough to allow easy passage of protein from the plasma into the lymph system, hepatic lymph has the highest protein content (nearly equal to blood plasma). This permeability of the liver sinusoids allows large quantities of lymph to form. The lymph coming from the liver is estimated to comprise one third to one half of all the lymph formed in the body.

[2] The National Institute on Alcohol Abuse and Alcoholism (NIAAA) defines 1 drink as 12 oz of beer, 5 oz of wine, or 1.5 oz distilled spirits, all of which contain approximately 0.5 oz of ethanol.

[3] A very high NADH/NAD+ greatly affects glucose metabolism, resulting in inhibition of gluconeogenesis and a drop in blood glucose; this is the main reason that diabetics must avoid ethanol.

[4] The NIH has published a consensus statement on the treatment and management of HCV (http://www.hivandhepatitis.com/2002conf/nih/2.html).

19

Diseases of the Endocrine System

Sara Long, Ph.D., R.D.

Southern Illinois University

CHAPTER OUTLINE

Normal Anatomy and Physiology of the Endocrine System
Classification of Hormones • Endocrine Function • Endocrine Control of Energy Metabolism

Endocrine Disorders
Thyroid Disorders

Diabetes Mellitus
Type 1 Diabetes Mellitus • Type 2 Diabetes Mellitus • Gestational Diabetes Mellitus (GDM) • Hypoglycemia

Introduction

Endocrine glands that make up the endocrine system are not attached anatomically, but scattered all through the body. All the same, these glands make up a system in a functional sense. Functions are carried out by secreting **hormones** (chemical messengers) into the blood, and numerous interactions occur between the various glands (Sherwood 2004).

The endocrine system is more of a complex functional system than an anatomical system (Copstead and Banasik 2005). A single endocrine gland may generate several hormones: the pituitary gland secretes six hormones that have distinct functions and are under different control mechanisms. A single hormone may be secreted by more than one endocrine gland; for example, somatostatin is secreted by both the hypothalamus and the pancreas. A single hormone can have more than one type of target cell and thus can generate more than one type of effect. Vasopressin acts as a vasoconstrictor throughout the body in addition to promoting H_2O reabsorption by kidney tubules. Some single hormones have assorted target-cell types and are capable of coordinating and integrating activities of various tissues toward a common end. Such is the effect of insulin on muscle, liver, and fat in the storage of nutrients after absorption.

Secretion rates of specific hormones fluctuate in a cyclic pattern over the course of time, providing chronological synchronization of function. Reproductive cycles, such as the menstrual cycle, are managed by endocrine hormones. Single target cells may be influenced by more than one hormone. Some cells contain an assortment of receptors for reacting in different ways to different hormones. Insulin promotes conversion of glucose into glycogen in liver cells by stimulating one particular hepatic

hormones—blood-borne chemical messengers that act on target cells located a long distance from the endocrine gland that produces them

enzyme, whereas glucagon activates another hepatic enzyme to enhance degradation of glycogen into glucose in liver cells. Some chemical messengers may function as a hormone or neurotransmitter depending on the sources and mode of delivery to the target cell. Norepinephrine is secreted as a hormone by the adrenal medulla and released as a neurotransmitter from sympathetic postganglionic nerve fibers. Some endocrine organs, like the anterior pituitary, exclusively secrete hormones, while other endocrine organs perform additional functions. The testes, for example, both produce sperm and secrete testosterone (Sherwood 2004). Specific functions of major hormones are listed in Table 19.1.

Normal Anatomy and Physiology of the Endocrine System

Many functional interactions take place among the various ductless glands scattered throughout the body (see Figure 19.1) that make up the endocrine system (Sherwood 2004). This chapter will focus only on endocrine glands and disorders related to nutrition and nutritional status.

Classification of Hormones

Hormones released from endocrine glands regulate activities throughout the body. In a healthy state, hormones are released when their actions are required and inhibited when effects are achieved. Endocrine diseases manifest through either hyperfunction (exceptionally high blood concentrations of a hormone), hypofunction (depressed levels of hormones in the blood) (Copstead and Banasik 2005), or abnormal target-cell responsiveness (Sherwood 2004).

Hormones travel through the blood to target cells. Their functions may be grouped into four categories (Kaplan and Conway 2004):

- Reproduction and sexual differentiation
- Growth and development
- Homeostasis
- Regulation of metabolism and nutrient supply

There are three chemical classes of hormones: (1) peptides and proteins, (2) amines, and (3) steroids (see Table 19.2) (Sherwood 2004). The majority of hormones fall into

FIGURE 19.1 The Endocrine System

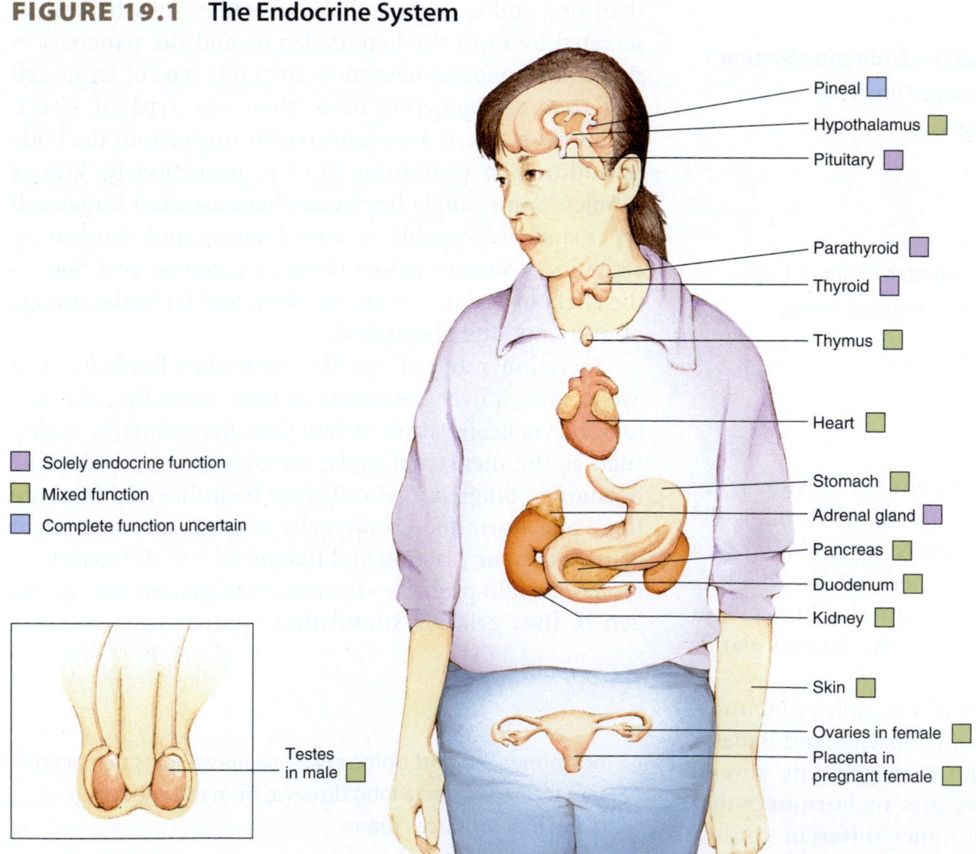

Source: L. Sherwood, *Human Physiology: From Cells to Systems*, 5e, copyright © 2004, p. 668.

TABLE 19.1

Summary of the Major Hormones

Endocrine Gland	Hormones	Target Cells	Major Functions of Hormones
Hypothalamus	Releasing and inhibiting hormones (TRH, CRH, GnRH, GHRH, GHIH, PRH, PIH)	Anterior pituitary	Controls release of anterior pituitary hormones
Posterior Pituitary (hormones stored in)	Vasopressin (antidiuretic hormone)	Kidney tubules	Increases H_2O reabsorption
		Arterioles	Produces vasoconstriction
	Oxytocin	Uterus	Increases contractility
		Mammary glands (breasts)	Causes milk ejection
Anterior Pituitary	Thyroid-stimulating hormone (TSH)	Thyroid follicular cells	Stimulates T_3 and T_4 secretion
	Adrenocorticotropic hormone (ACTH)	Zona fasciculata and zona reticularis of adrenal cortex	Stimulates cortisol secretion
	Growth hormone	Bone; soft tissues	Essential but not solely responsible for growth; stimulates growth of bones and soft tissues; metabolic effects include protein anabolism, fat mobilization, and glucose conservation
		Liver	Stimulates somatomedin secretion
	Follicle-stimulating hormone (FSH)	*Females:* ovarian follicles	Promotes follicular growth and development; stimulates estrogen secretion
		Males: seminiferous tubules in testes	Stimulates sperm production
	Luteinizing hormone (LH) (interstitial cell-stimulating hormone—ICSH)	*Females:* ovarian follicle and corpus luteum	Stimulates ovulation, corpus luteum development, and estrogen and progesterone secretion
		Males: interstitial cells of Leydig in testes	Stimulates testosterone secretion
	Prolactin	*Females:* mammary glands	Promotes breast development; stimulates milk secretion
		Males	Uncertain
Thyroid Gland Follicular Cells	Tetraiodothyronine (T_4 or thyroxine); triiodothyronine (T_3)	Most cells	Increases the metabolic rate; essential for normal growth and nerve development
Thyroid Gland C Cells	Calcitonin	Bone	Decreases plasma calcium concentration
Adrenal Cortex *Zona glomerulosa*	Aldosterone (mineralocorticoid)	Kidney tubules	Increases Na^+ reabsorption and K^+ secretion
Zona fasciculata and zona reticularis	Cortisol (glucocorticoid)	Most cells	Increases blood glucose at the expense of protein and fat stores; contributes to stress adaption
	Androgens (dehydroepiandrosterone)	*Females:* bone and brain	Responsible for the pubertal growth spurt and sex drive in females
Adrenal Medulla	Epinephrine and norepinephrine	Sympathetic receptor sites throughout the body	Reinforces the sympathetic nervous system; contributes to stress adaption and blood pressure regulation
Endocrine Pancreas (Islets of Langerhans)	Insulin (β cells)	Most cells	Promotes cellular uptake, use, and storage of absorbed nutrients
	Glucagon (α cells)	Most cells	Important for maintaining nutrient levels in blood during postabsorptive state
	Somatostatin (D cells)	Digestive system	Inhibits digestion and absorption of nutrients
		Pancreatic islet cells	Inhibits secretion of all pancreatic hormones

(continued on the following page)

TABLE 19.1 *(continued)*

Summary of the Major Hormones

Endocrine Gland	Hormones	Target Cells	Major Functions of Hormones
Parathyroid Gland	Parathyroid hormone (PTH)	Bone, kidneys, intestine	Increases plasma calcium concentration; decreases plasma phosphate concentration; stimulates vitamin D activation
Gonads			
Female: ovaries	Estrogen (estradiol)	Female sex organs; body as a whole	Promotes follicular development; governs development of secondary sexual characteristics: stimulates uterine and breast growth
		Bone	Promotes closure of the epiphyseal plate
	Progesterone	Uterus	Prepares for pregnancy
Male: testes	Testosterone	Male sex organs; body as a whole	Stimulates sperm production; governs development of secondary sexual characteristics; promotes sex drive
		Bone	Enhances pubertal growth spurt; promotes closure of the epiphyseal plate
Testes and ovaries	Inhibin	Anterior pituitary	Inhibits secretion of follicle-stimulating hormone
Pineal Gland	Melatonin	Brain; anterior pituitary; reproductive organs; immune system; possibly others	Entrains body's biological rhythm with external cues; believed to inhibit gonadotropins; initiation of puberty possibly caused by a reduction in melatonin secretion; acts as an antioxidant; enhances immunity
Placenta	Estrogen (estradiol); progesterone	Female sex organs	Help maintain pregnancy; prepare breasts for lactation
	Chorionic gonadotropin	Ovarian corpus luteum	Maintains corpus luteum of pregnancy
Kidneys	Renin (→ angiotensin)	Zona glomerulosa of adrenal cortex (acted on by angiotensin, which is activated by renin)	Stimulates aldosterone secretion
	Erythropoietin	Bone marrow	Stimulates erythrocyte production
Stomach	Gastrin	Digestive-tract exocrine glands and smooth muscles; pancreas; liver; gallbladder	Control of motility and secretion to facilitate digestive and absorptive processes
Duodenum	Secretin; cholecystokinin		
	Glucose-dependent insulinotropic peptide	Endocrine pancreas	Stimulates insulin secretion
Liver	Somatomedins	Bone; soft tissues	Promotes growth
	Thrombopoietin	Bone marrow	Stimulates platelet production
Skin	Vitamin D	Intestine	Increases absorption of ingested calcium and phosphate
Thymus	Thymosin	T lymphocytes	Enhances T lymphocyte proliferation and function
Heart	Atrial natriuretic peptide	Kidney tubules	Inhibits Na^+ reabsorption

Reprinted from: Sherwood L. *Human physiology: from cells to systems.* 5th ed. Belmont (CA): Thomson-Brooks/Cole; 2004. Table 18–1 p. 666–671.

the category of peptides and proteins, which are amino acid derivatives. Amines are derivatives of the amino acid tyrosine. Steroid hormones are derived from cholesterol.

Endocrine Function

Pituitary Gland The pituitary gland is located in the bony cavity at the base of the brain just below the hypothalamus (see Figure 19.2). It is connected to the hypothalamus by a thin connecting stalk (Sherwood 2004) called the pituitary stalk (Rhoades and Pflanzer 2003).

The pituitary gland actually consists of two anatomically and functionally distinct glands: the anterior pituitary and the posterior pituitary. Location is the only thing they have in common. The anterior pituitary secretes six hormones (see Figure 19.3) that control secretion of various other hormones. None of the hormones is secreted at a constant rate, but secretion is regulated by hypothalamic hormones and feedback from target gland hormones. The posterior pituitary releases hormones synthesized by the hypothalamus, **vasopressin** and **oxytocin** (Sherwood 2004). Box 19.1 provides a summary of pituitary disorders.

Thyroid Gland The thyroid gland, which is responsible for controlling metabolic rate, lies over the trachea just below the larynx, and consists of two lobes connected by a thin strip called the isthmus (see Figure 19.4) (Sherwood 2004).

TABLE 19.2

Classifications of Endocrine Hormones and Their Endocrine Gland of Origin

Hormone Classification	Secreted by
Peptide Hormones	Hypothalamus Anterior pituitary Posterior pituitary Pineal gland Pancreas Parathyroid gland Gastrointestinal tract Kidneys Liver Thyroid C cells Heart Thymus
Amine Hormones	Thyroid gland Adrenal medulla
Steroid Hormones	Adrenal cortex Gonads Placenta

Source: Sherwood L. *Human Physiology: From Cells to Systems.* Belmont (CA): Thomson; 2004.

The thyroid hormones are two iodine-containing hormones derived from the amino acid tyrosine: thyroxine (T_4 or tetraiodothyronine) and triiodothyrone (T_3). The prefixes tetra- and tri- and subscripts 4 and 3 denote the number of iodine atoms incorporated into each of these hormones. T_4 is the major hormone secreted by the thyroid, but T_3 is more active. The conversion of T_4 to T_3 within the anterior pituitary, liver, and kidney accounts for approximately two-thirds of T_3 production (Kaplan and Conway 2004; Sherwood 2004).

Response from increased secretion of thyroid hormone (which may affect several different organs and processes in the body) takes several hours to become apparent. Maximal response does not become apparent for several days. Because thyroid hormone is not rapidly degraded, the response to increased secretion continues to be expressed over a period of days or even weeks after plasma thyroid hormone concentrations return to normal (Sherwood 2004).

Adrenal Glands The two adrenal glands are embedded above each kidney (see Figure 19.5) and encapsulated in fat. Each adrenal gland is composed of two endocrine organs (Sherwood 2004). The inner portion, the adrenal medulla, forms part of the sympathetic nervous system and secretes **epinephrine**, **norepinephrine**, and **catecholamines** (Kaplan and Conway 2004; Sherwood 2004).

The outer layers, known as the adrenal cortex, compose 80% to 90% of the adrenal gland and produce over 50 known **adrenocortical hormones**. Structural variations in these hormones confer different functional capabilities and allow them to perform different primary actions (see Table 19.3) (Kaplan and Conway 2004; Sherwood 2004). Box 19.2 summarizes adrenal cortex disorders.

vasopressin—the primary endocrine factor that regulates urinary H_2O loss and overall H_2O balance; regulates blood pressure via this hormone's pressor effects on blood vessels; also known as antidiuretic hormone (ADH)

oxytocin—a hormone that stimulates contraction of the uterus during childbirth, and promotes ejection of milk from mammary glands during breast-feeding

epinephrine—a hormone that is secreted from the adrenal medulla; regulates arterial blood pressure and prepares body for "fight or flight" responses; formerly referred to as adrenaline

norepinephrine— a neurotransmitter released form sympathetic postganglionic fibers; formerly referred to as noradrenaline

catecholamines—the chemical classification of adrenodedullary hormones

adrenocortical hormones—steroids derived from the precursor cholesterol

FIGURE 19.2 Anatomy of the Pituitary Gland
(a) Relation of the pituitary gland to hypothalamus and rest of the brain.
(b) Schematic enlargement of pituitary gland in connection to hypothalamus.

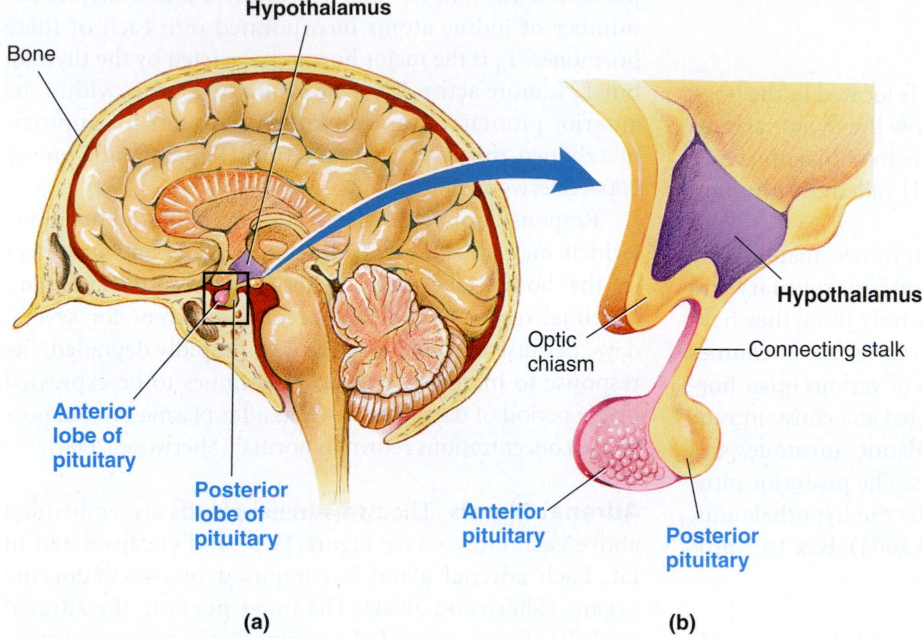

Source: L. Sherwood, *Human Physiology: From Cells to Systems,* 5e, copyright © 2004, p. 683.

TABLE 19.3

Functions of Adrenal Cortex Hormones

Category	Primary Hormone	Function
Mineralocoticoids	Aldosterone	Long-term regulation of blood pressure: promotes Na+ retention and enhances K+ elimination during formation of urine.
Glucocorticoids	Cortisol	Metabolic effects: stimulates hepatic gluconeogenesis; inhibits glucose uptake and use by tissues (except brain); stimulates protein degradation (especially in muscle) for use in gluconeogenesis or protein synthesis; facilitates lipolysis. Permissive actions: permits catecholamines to induce vasoconstriction. Adaptation to stress: secreted in response to physical (trauma, surgery, intense heat or cold), chemical (reduced O2 supply), physiologic (heavy exercise, hemorrhagic shock, pain), psychological or emotional (anxiety, fear sorrow), and social (personal conflict, change in lifestyle) stresses. Anti-inflammatory and immunosuppressive effects: administration of glucocorticoid inhibits almost every step of the inflammatory process, making them effective drugs in treating conditions like rheumatoid arthritis.
Sex hormones	Dehydroepiandrosterone (a male "sex" hormone)	Identical or similar to those produced by gonads.

Source: Sherwood L. *Human Physiology: From Cells to Systems.* Belmont (CA): Thomson; 2004.

exocrine pancreas—part of the pancreas that secretes digestive enzymes and bicarbonate into the duodenal lumen

Endocrine Pancreas The pancreas is located in the abdominal cavity adjacent to the upper part of the small intestine (see Figure 19.1). Different groups of cells within the pancreas carry out different functions. Cells making up the **exocrine pancreas** are responsible for secretion of fluid and various digestive enzymes that are secreted via the

FIGURE 19.3 Functions of Anterior Pituitary Hormones

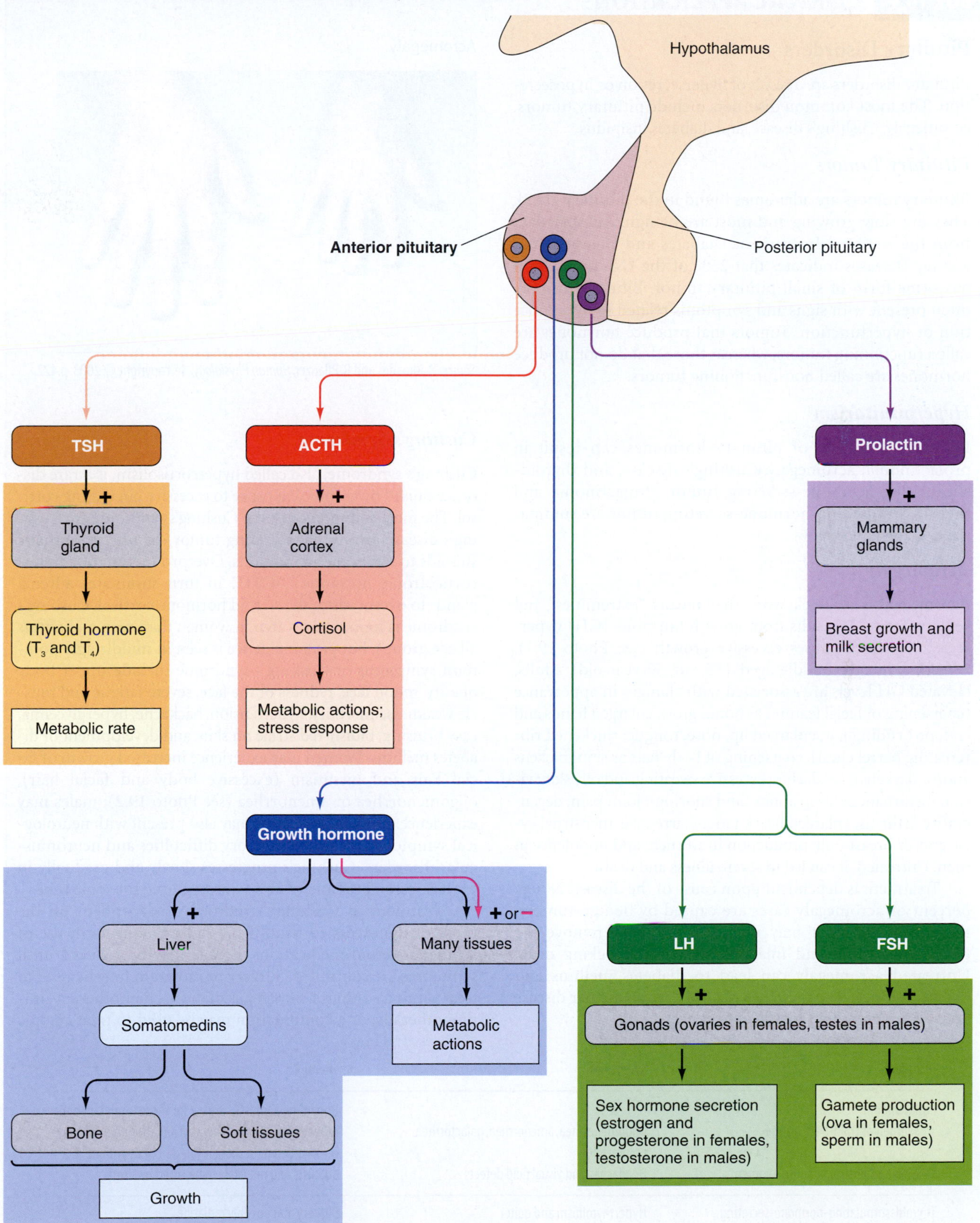

Source: L. Sherwood, *Human Physiology: From Cells to Systems,* 5e, copyright © 2004, p. 686.

BOX 19.1 CLINICAL APPLICATIONS

Pituitary Disorders

Pituitary disorders are a result of hypersecretion or hyposecretion. The most common disorders include pituitary tumors, acromegaly, Cushing's disease, and diabetes insipidus.

Pituitary Tumors

Pituitary tumors are adenomas found in the pituitary gland. They are slow growing and most are benign. Autopsy data from the National Institute of Diabetes and Digestive and Kidney Diseases indicates that 25% of the U.S. population has some form of small pituitary tumor. Pituitary tumors often present with signs and symptoms related to hypofunction or hyperfunction. Tumors that produce hormones are called functioning tumors, whereas those that do not produce hormones are called non-functioning tumors.

Hyperpituitarism

Increased secretion of pituitary hormones can result in prolactinoma, acromegaly, Cushing's disease, and thyroid-stimulating-hormone-secreting tumor. Prolactinoma and thyroid-stimulating-hormone-secreting tumor are summarized in the table below.

Acromegaly

Acromegaly is a Greek word that means "extremities" and "enlargement." It results from growth hormone (GH) hypersecretion, which causes excessive growth (see Photo 19.1). Affecting mostly middle-aged (35- to 50-year-old) adults, elevated GH levels are associated with changes in appearance (coarsening of facial features as bones grow, enlarged hands and feet, protruding jaw, enlarged lip, nose, tongue; thickened ribs [creating barrel chest], coarsening of body hair as skin thickens and/or darkens); headaches; excessive sweating and oily skin; vision disturbances; sleep apnea (and snoring); joint pain; degenerative arthritis; enlarged heart; fatigue; irregular menstrual cycle and/or breast milk production in women; and impotence in men. Untreated, it can led to severe illness and death.

Treatment is dependent upon cause of the disease. Ninety percent of acromegaly cases are caused by benign tumors; therefore, treatment may include surgery to remove the tumor, radiation, and injection of a GH-blocking drug. Untreated acromegaly can lead to diabetes mellitus and hypertension. It also increases risk for cardiovascular disease and colon polyps that may lead to cancer.

Acromegaly

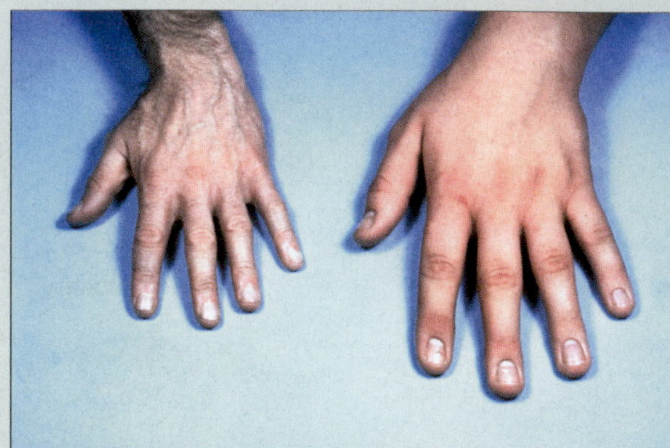

Source: R. Rhoades and R. Pflanzer, *Human Physiology,* 4e, copyright © 2003, p.422.

Cushing's Syndrome

Cushing's syndrome, also called hypercortisolism, is a rare disorder caused by chronic exposure to excessive circulating cortisol. The most common causes of Cushing's syndrome are Cushing's disease (an ACTH-secreting tumor) or use of synthetic steroids to treat other conditions. Overproduction of adrenocorticotropic hormone (ACTH) in turn stimulates adrenal glands to overproduce the steroid hormone cortisol. Cushing's syndrome is found more often in women than men and affects all age groups, but peak incidence is seen in middle age. Common symptoms of Cushing's syndrome include upper body obesity, moon face, redness of the face, severe fatigue and muscle weakness, infertility, hypertension, backache, hyperglycemia, easy bruising, bluish-red striae on skin, and development of diabetes mellitus. Women may experience increased growth of facial hair and hirsutism (excessive body and facial hair), oligomenorrhea or amenorrhea (see Photo 19.2); males may experience impotence. Patients may also present with neurological symptoms, including memory difficulties and neuromuscular disorders. Cushing's progresses slowly and gradually in most cases, and can therefore go unrecognized for some time.

Treatment of Cushing's syndrome is dependent on the cause of the excess cortisol. If the cause is long-term use of synthetic steroid medications, dosage may be reduced until symptoms are controlled. Surgery or radiation may be used to treat pituitary tumors. Surgery, radiation, chemotherapy, immunotherapy, or a combination may be used to treat ectopic

Disorder	Symptoms	Therapy
Prolactinoma		Dopamine agonists, surgery for those intolerant of or refractory to medical therapy; radiation therapy may be considered for those intolerant of dopamine agonists and not likely to be cured by surgery
Women of child-bearing age	Oligomenorrhea, amenorrhea, galactorrhea, or infertility	
Men and postmenopausal women	Headaches and visual field defects	
Thyroid-stimulating-hormone-secreting (TSH) tumor	Hyperthyroidism and goiter	Surgery and possibly radiation

(continued on the following page)

BOX 19.1 *(continued)*

Cushing's syndrome

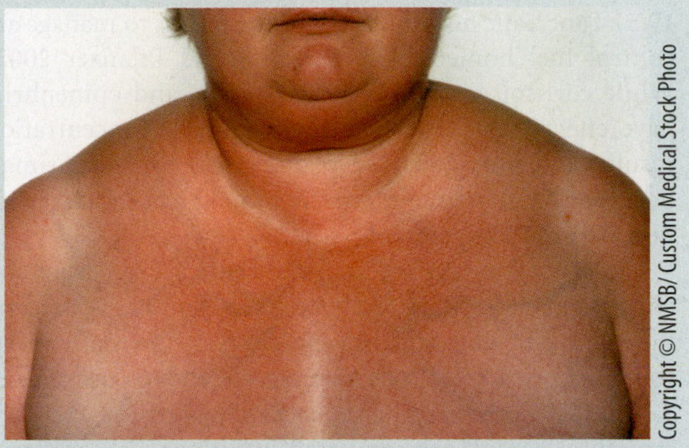

ACTH syndrome. Prognosis is also dependent upon cause of the disease. Most cases can be cured, although recovery may be complicated by various aspects of the causative illness.

Hypopituitarism

Pituitary tumors are the most common cause of hypopituitarism, which results in deficiencies of GH, gonadotropin, adrenocorticotropic hormone (ACTH), and thyrotropin (TSH). Manifestations of these disorders are shown in the following table.

Disorder	Symptoms	Therapy
Growth Hormone Deficiency	Decreased muscle strength, exercise intolerance, reduced sense of well-being, increased body fat (particularly intra-abdominally), decreased lean body mass	GH replacement
Gonadotropin Deficiency		
Women	Infertility, oligomenorrhea or amenorrhea, lack of libido, hot flashes, dyspareunia	Estrogen replacement to prevent osteoporosis and treat hot flashes, decreased libido, and vaginal dryness
Men	Decreased libido, impotence	IM (intramuscular) testosterone injections or transdermal testosterone (patch or gel)
Adrenocorticotropic Hormone Deficiency (ACTH)	Chronic malaise, fatigue, anorexia, hyponatremia	Replacement regimen of hydrocortisone
Thyrotropin (TSH) Deficiency	Malaise, leg cramps, fatigue, dry skin, cold intolerance	Levothyroxine replacement therapy

Diabetes Insipidus

Diabetes insipidus (DI), not to be confused with diabetes mellitus, results from insufficient production of antidiuretic hormone by the hypothalamus (the portion of the brain that stimulates the pituitary gland). Antidiuretic hormone is produced by the hypothalamus, but stored and released into the bloodstream by the pituitary gland. Normally, antidiuretic hormone controls the kidneys' output of urine. DI causes polyuria (>3 L/24 hr) and polydipsia resulting from excessive loss of fluid.

Treatment of DI depends upon its cause. Treatment of the cause usually resolves DI. Common causes of DI include:

- Malfunctioning hypothalamus
- Malfunctioning pituitary gland
- Brain injury
- Tumor
- Tuberculosis
- Blockage of cerebral arteries
- Encephalitis
- Meningitis
- Sarcoidosis

Sources:

Hamrahian A. Pituitary disorders. The Cleveland Clinic. Available from: http://www.clevelandclinicmeded.com. Accessed January 6, 2006.

National Institute of Neurological Disorders and Strokes. NINDS Pituitary tumors information page. Available from: http://www.nids.nih.gov. Accessed January 6, 2006.

National Institute of Neurological Disorders and Strokes. NINDS Cushing's syndrome information page. Available from: http://www.nids.nih.gov. Accessed January 6, 2006.

The Pituitary Foundation. Pituitary Foundation fact sheet: acromegaly. Available from: http://www.pituitary.org.uk. Accessed January 6, 2006.

The Pituitary Foundation. Pituitary Foundation fact sheet: Cushing's disease. Available from: http://www.pituitary.org.uk. Accessed January 6, 2006.

The Pituitary Foundation. Cushing's disease. Available from: http://www.pituitary.org.uk. Accessed January 6, 2006.

The Pituitary Foundation. Diabetes inspidus. Available from: http://www.pituitary.org.uk. Accessed January 6, 2006.

The Pituitary Foundation. Pituitary Foundation fact sheet: diabetes insipidus. Available from: http://www.pituitary.org.uk. Accessed January 6, 2006.

University of Maryland Medical Center. Endocrinology health guide: acromegaly. Available from: http://www.umm.edu. Accessed January 6, 2006.

University of Maryland Medical Center. Endocrinology health guide: pituitary tumor University of Maryland Medical Center. Endocrinology health guide: diabetes insipidus. Available from: http://www.umm.edu. Accessed January 6, 2006.

FIGURE 19.4 Anatomy of the Thyroid Gland

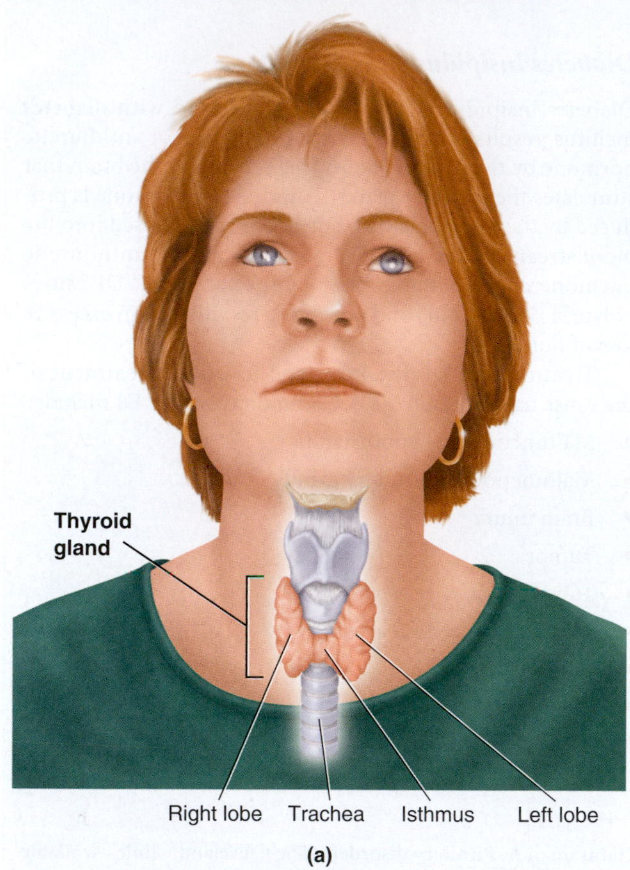

Source: L. Sherwood, *Human Physiology: From Cells to Systems*, 5e, copyright © 2004, p. 702.

pancreatic duct into the duodenum (see Chapter 18). Endocrine pancreas cells are an anatomically small portion of the pancreas. They secrete hormones that regulate energy metabolism and fuel homeostasis.

Islets of Langerhans The groups of cells that make up the endocrine pancreas, called the islets of Langerhans, are embedded in the exocrine portion of the gland (see Figure 19.6). The average human pancreas contains approximately 1 million islet cells which make up only 1% to 2% of total pancreatic mass (Copstead and Banasik 2005). Islets are composed of four major types of cells, each of which synthesizes and secretes a different hormone. The types of islet cells and the hormones they produce are listed in Table 19.4.

anabolic—refers to building up or synthesis of larger organic molecules from smaller organic molecular subunits

GLUT-4—glucose transporter that transports glucose between blood and cells; it is the only glucose transporter responsive to insulin

Effects of Pancreatic Hormones on Metabolism Energy use in the body is constant, but ingestion of energy-yielding nutrients is sporadic. This means that excess energy taken in meals must be stored for later use between meals (see Table 19.5). Pancreatic hormones afford the means to manage and control fuel homeostasis (Rhoades and Pflanzer 2003). While cortisol, growth hormone (GH), and epinephrine have effects that influence blood glucose concentration, insulin and glucagon (see Figure 19.7) are the primary hormones that maintain normal blood glucose concentration (70 to 110 mg/100 mL).

Following meals, ingested nutrients are absorbed and enter the bloodstream (fed state) (see Figure 19.8). During this period of time, glucose functions as the main energy source, because most cells have a preference to use glucose. Additional nutrients not immediately used for energy or structural repairs are converted into their storage forms: glycogen or triglycerides (Sherwood 2004).

It takes approximately four hours for a typical meal to be absorbed. Afterwards, during the time period when no nutrients are in the gastrointestinal tract (fasting state), endogenous energy stores are mobilized for energy (see Figure 19.9). Synthesis of protein and fat is abbreviated, and stored forms of these nutrients are catabolized for glucose formation and energy production, respectively. Through mechanisms of gluconeogenesis and glucose sparing, the blood glucose level is sustained to nourish the brain (see Tables 19.6 and 19.7) (Sherwood 2004).

As an **anabolic** hormone, insulin controls the metabolic fate of carbohydrate, protein, and fat. When intakes of these nutrients are high, insulin provides the signal that directs storage of excesses while suppressing mobilization of preexisting stores (Rhoades and Pflanzer 2003; Sherwood 2004).

Action of Insulin on Carbohydrate Metabolism The most commonly known function of insulin is maintenance of blood glucose homeostasis. Most tissues in the body depend on insulin for transportation of glucose from the bloodstream into cells to be used for energy (see Table 19.8 and Figure 19.10). There are three exceptions: cells of the brain, liver, and working muscles are readily permeable to glucose even in the absence of insulin (Rhoades and Pflanzer 2003; Sherwood 2004).

Action of Insulin on Fat Metabolism Insulin promotes storage of excess energy as triglycerides via **GLUT-4** recruitment, and stimulates fatty acid synthesis in liver and adipose tissue. The net effect of insulin is to promote triglyceride storage and inhibit lipolysis (Rhoades and Pflanzer 2003; Sherwood 2004).

Action of Insulin on Protein Metabolism The most pronounced effect of insulin on protein metabolism is seen in

FIGURE 19.5 Anatomy of Adrenal Glands

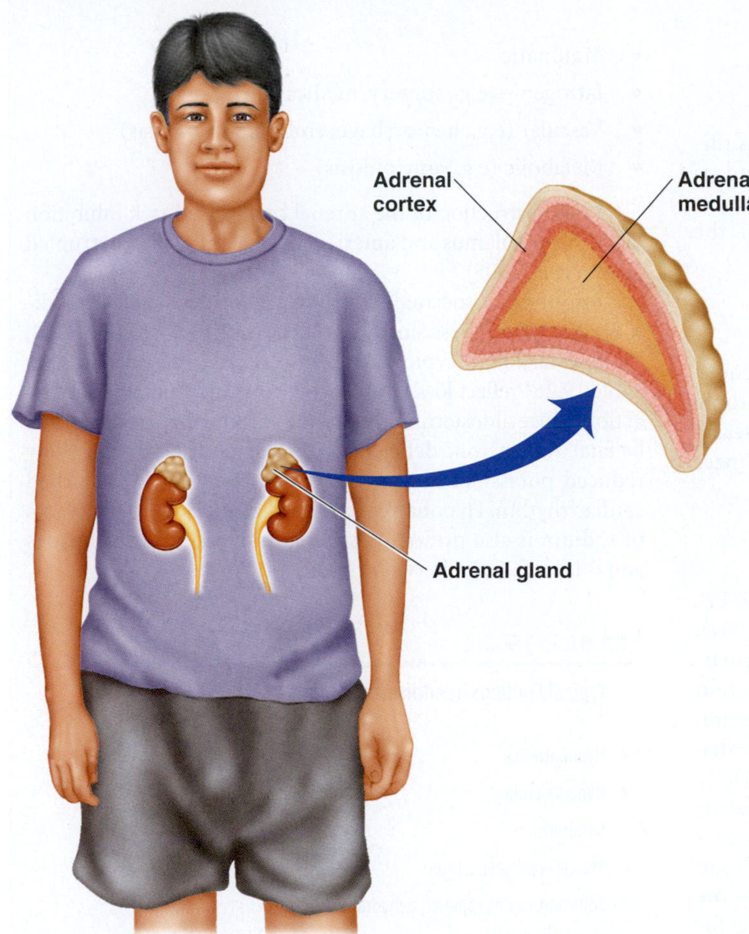

Source: L. Sherwood, *Human Physiology: From Cells to Systems,* 5e, copyright © 2004, p. 708

skeletal muscle and the liver. It promotes active transport of amino acids from the blood into muscle and other tissues, thus promoting protein synthesis (see Figure 19.11) within cells. This anabolic effect of insulin on protein metabolism produces a **positive nitrogen balance**. When insulin is deficient, there is net loss of protein, or **negative nitrogen balance**. These effects demonstrate the importance of insulin in tissue growth (Rhoades and Pflanzer 2003; Sherwood 2004).

Endocrine Control of Energy Metabolism

The previous sections discussed the effect of the pancreatic hormone insulin on energy metabolism. Table 19.9 summarizes other hormonal factors governing normal energy metabolism. When macronutrients are converted to their absorbable forms, they pass through the intestinal wall into the bloodstream, where they are distributed throughout the body (see Figure 19.12).

Endocrine Disorder

Endocrine disorders are the result of hyposecretion or hypersecretion of hormones (Sherwood 2004), or of **hyporesponsiveness** of target organs (Copstead and Banasik 2005). Table 19.10 outlines the most common causes of endocrine dysfunction.

Primary hyposecretion occurs when an endocrine organ releases an inadequate amount of hormone to meet physiological needs. Secondary hyposecretion occurs when secretion of a **tropic hormone** is inadequate to cause an endocrine organ to secrete adequate amounts of a hormone. For instance, if the thyroid gland produces inadequate amounts of thyroid hormone, this would be considered primary hyposecretion, whereas inadequate production of thyroid hormone that is caused by insufficient secretion of a tropic hormone such as thyroid stimulating hormone (TSH) is secondary hyposecretion. The interrelationship of the various hormones makes diagnosis of hormone deficiency quite complex. Evaluation of multiple lab values may assist in differentiating between primary and secondary deficiencies. In a primary thyroid hormone deficiency, for example, thyroid hormone would be low, but TSH levels would be high; in secondary thyroid hormone deficiency, both thyroid hormone and TSH levels would be abnormally low (Copstead and Banasik 2005).

Hypersecretion disorders can also be primary or secondary. When an endocrine gland is secreting abnormally high amounts of a hormone due to a primary disorder, the tropic hormone will be at unusually low levels. When hypersecretion is secondary (to elevated tropic hormone levels), plasma concentrations of both hormones will be elevated (Copstead and Banasik 2005).

Hyporesponsiveness of the target organ will cause the same symptoms as hyposecretion, but hormone levels will be normal or high instead of low. Most cases of hyporesponsiveness are caused by a lack or deficiency of hormone receptors on the target cells (Copstead and Banasik 2005).

positive nitrogen balance—net accumulation of protein in the body

negative nitrogen balance—net loss of protein in the body

hyporesponsiveness—hormone resistance

tropic hormone—a hormone that regulates secretion of another hormone

BOX 19.2 CLINICAL APPLICATIONS

Abnormalities of Adrenal Cortex Function: Altered Physiological Regulation

A number of common deficiencies result from either insufficient or excess secretion of adrenal cortex hormones. And, as with other endocrine disorders, symptoms are the result of either the absence or magnification of effects of the hormones involved (Rhoades and Pflanzer 2003).

Excess Secretion of Glucocorticoids

Prolonged exposure to high levels of endogenous or exogenous glucocorticoids results in the condition Cushing's syndrome (Adler and Lawrence 2001; Endocrine and Metabolic Diseases Information Service, Cushing's, 2002; Rhoades and Pflanzer 2003), which is discussed in Box 19.1.

Insufficient Secretion of Adrenal Cortex Steroids

Both adrenal glands must be nonfunctional (or removed) before adrenocortical insufficiency can occur (Sherwood 2004). As a result of either occurrence, both glucocorticoid (cortisol) and mineralcorticoid (aldosterone) hormone production is lacking (Endocrine and Metabolic Diseases Information Service, Cushing's, 2002). Death may result from untreated adrenocortical insufficiency (Rhoades and Pflanzer 2003).

Primary adrenal insufficiency is uncommon, but iatrogenic (caused by medical treatment) adrenal insufficiency is more frequent, although exact incidence is unknown (Wilson and Speiser 2003). Autoimmune Addison's disease is the more common form of adrenal insufficiency, responsible for about 75% of cases (Endocrine and Metabolic Diseases Information Service, Addison's, 2004; Kaplan and Conway 2004). Addison's can occur at any age, but is most common in people aged 30–50 years of age (Liotta et al. 2005), and afflicts men and women equally (Endocrine and Metabolic Diseases Information Service, Addison's, 2004).

Adrenal insufficiency can be classified as either primary or secondary (Wilson and Speiser 2003). Primary adrenal insufficiency (Addison's disease) is caused by a dysfunctional adrenal cortex that impairs both glucocorticoid and mineralcorticoid production. Secondary adrenal insufficiency results from inadequate ACTH production by the anterior pituitary, resulting primarily in deficient glucocorticoid secretion (Wilson and Speiser 2003; Sherwood 2004). Adrenal insufficiency can further be classified as congenital or acquired (Wilson and Speiser 2003).

Primary adrenal insufficiency results from destruction of the adrenal cortex. Aldosterone is produced by the medulla of the adrenal gland; cortisol is produced in the adrenal cortex. Clinical findings manifest after 90% of the adrenal cortex has been destroyed. Causes of this destruction are as follows (Liotta et al. 2005):

- Autoimmune
- Infectious (e.g., mycobacterial, fungal)
- Neoplastic (e.g., primary, metastatic)
- Traumatic
- Iatrogenic (e.g., surgery, medication)
- Vascular (e.g., hemorrhage, emboli, thrombosis)
- Metabolic (e.g., amyloidosis)

With destruction of the adrenal cortex, feedback inhibition of the hypothalamus and anterior pituitary gland is interrupted (Liotta et al. 2005).

Symptoms associated with aldosterone deficiency in Addison's disease progress slowly and insidiously (Sherwood 2004; Liotta et al. 2005). Typical symptoms of Addison's disease (see Table 19.26) reflect loss of glucocorticoid and mineralcorticoid action. Since aldosterone is essential for life, the condition can be fatal. Aldosterone deficiency causes hyperkalemia caused by reduced potassium loss in the urine, resulting in disturbed cardiac rhythm. Hyponatremia caused by excessive urinary loss of sodium is also present, resulting in hypotension (Rhoades and Pflanzer 2003; Sherwood 2004).

TABLE 19.26

Typical Findings in Addison's Disease

- Hyponatremia
- Hyperkalemia
- Hypotension
- Muscle weakness, fatigue
- Vomiting, loss of appetite, dehydration
- Hypoglycemia
- Excess pigmentation of skin in some people

Source: Rhoades R, Pflanzer R. *Human Physiology,* 4th ed. Pacific Grove (CA): Thomson Learning, 2003.

Cortisol deficiency results in poor response to stress, hypoglycemia (resulting from reduced gluconeogenesis), and hyperpigmentation (from excessive secretion of ACTH) (Sherwood 2004). Addison's disease may coexist with other autoimmune disorders, especially thyroid disease, premature ovarian failure, and type 1 diabetes mellitus (Kaplan and Conway 2004).

Treatment of Addison's disease involves replacing or substituting hormones not being produced by the adrenal gland (Liotta et al. 2005; Endocrine and Metabolic Diseases Information Service, Addison's, 2004). Medications used to treat Addison's disease, as well as potential drug-nutrient interactions, are outlined in Table 19.13.

Patients with Addison's disease should not restrict salt in their diets. Patients with concurrent primary hypertension may restrict salt intake rather than discontinue mineralcorticoid replacement. Patients living in warm climates should increase salt intake due to increased loss of salt through perspiration (Endocrine and Metabolic Diseases Information Service, Addison's, 2004; Liotta et al. 2005).

FIGURE 19.6 Release of Hormones from the Pancreas

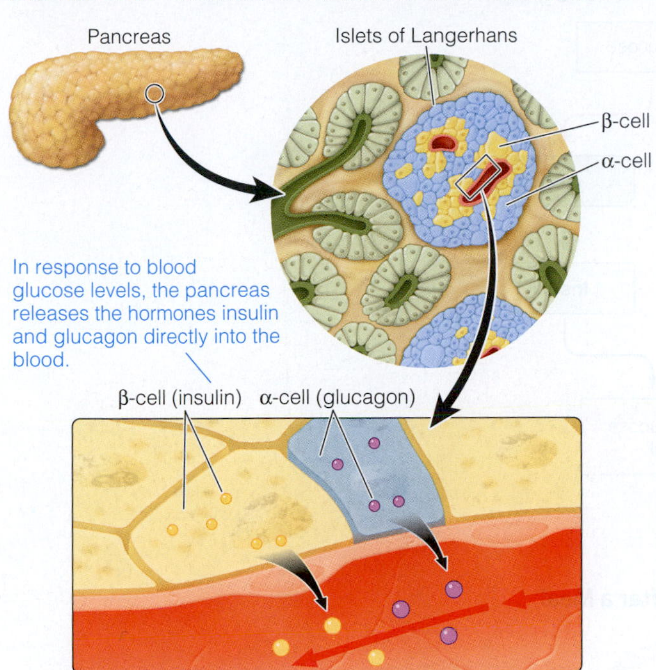
Pancreas
Islets of Langerhans
β-cell
α-cell
In response to blood glucose levels, the pancreas releases the hormones insulin and glucagon directly into the blood.
β-cell (insulin)
α-cell (glucagon)

Source: M. McGuire and K. Beerman, *Nutritional Sciences,* 1e, copyright © 2007, p. 144.

TABLE 19.4

Major Cell Types of Islets of Langerhans and the Hormones They Produce

Cell Type	Hormone Produced	Function
Alpha cells	Glucagon	Regulation of blood glucose.
Beta cells	Insulin	Regulation of blood glucose.
Delta cells	Somatostatin	Identical to somatostatin produced in hypothalamus. Controls secretion of growth hormone from anterior pituitary gland.
F cell	Pancreatic polypeptide	Not yet known.

Source: Rhoades R, Pflanzer R. *Human Physiology,* 4th ed. Pacific Grove (CA): Thomson Learning; 2003.

TABLE 19.5

Stored Fuels in the Body

Metabolic Fuel	Circulating Form	Storage Form	Major Storage Site	Percentage of Total Body Energy Content (and kcal)	Reservoir Capacity	Role
Carbohydrate	Glucose	Glycogen	Liver, muscle	1% (1,500 kcal)	Less than a day's worth of energy	First energy source; essential for the brain
Fat	Free fatty acids	Triglycerides	Adipose tissue	77% (143,000 kcal)	About two months' worth of energy	Primary energy reservoir; energy source during a fast
Protein	Amino acids	Body proteins	Muscle	22% (41,000 kcal)	Death results long before capacity is fully used because of structural and functional impairment	Source of glucose for the brain during a fast; last resort to meet other energy needs

Reprinted from: L. Sherwood, *Human Physiology: From Cells to Systems.* 5th ed. Belmont (CA): Thomson-Brooks/Cole; 2004. Table 19–4, page 721.

FIGURE 19.7 Complementary Interactions of Insulin and Glucagons

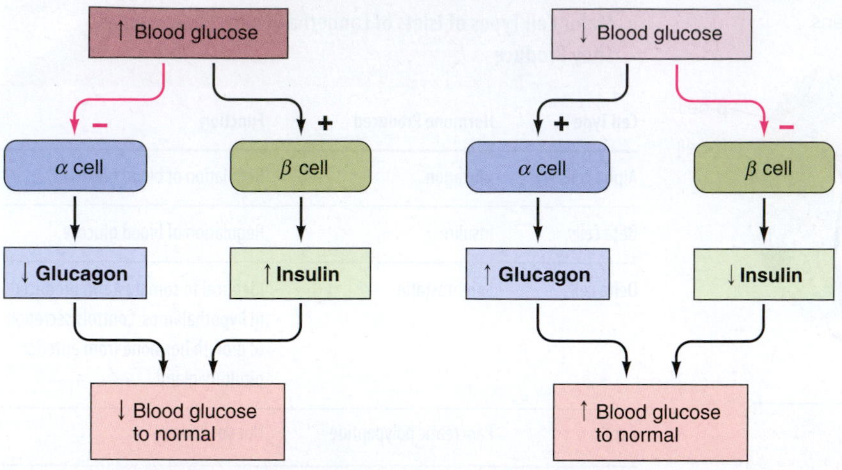

Source: L. Sherwood, *Human Physiology: From Cells to Systems*, 5e, copyright © 2004, p. 731.

FIGURE 19.8 Summary of Nutrient Flow Immediately after a Meal

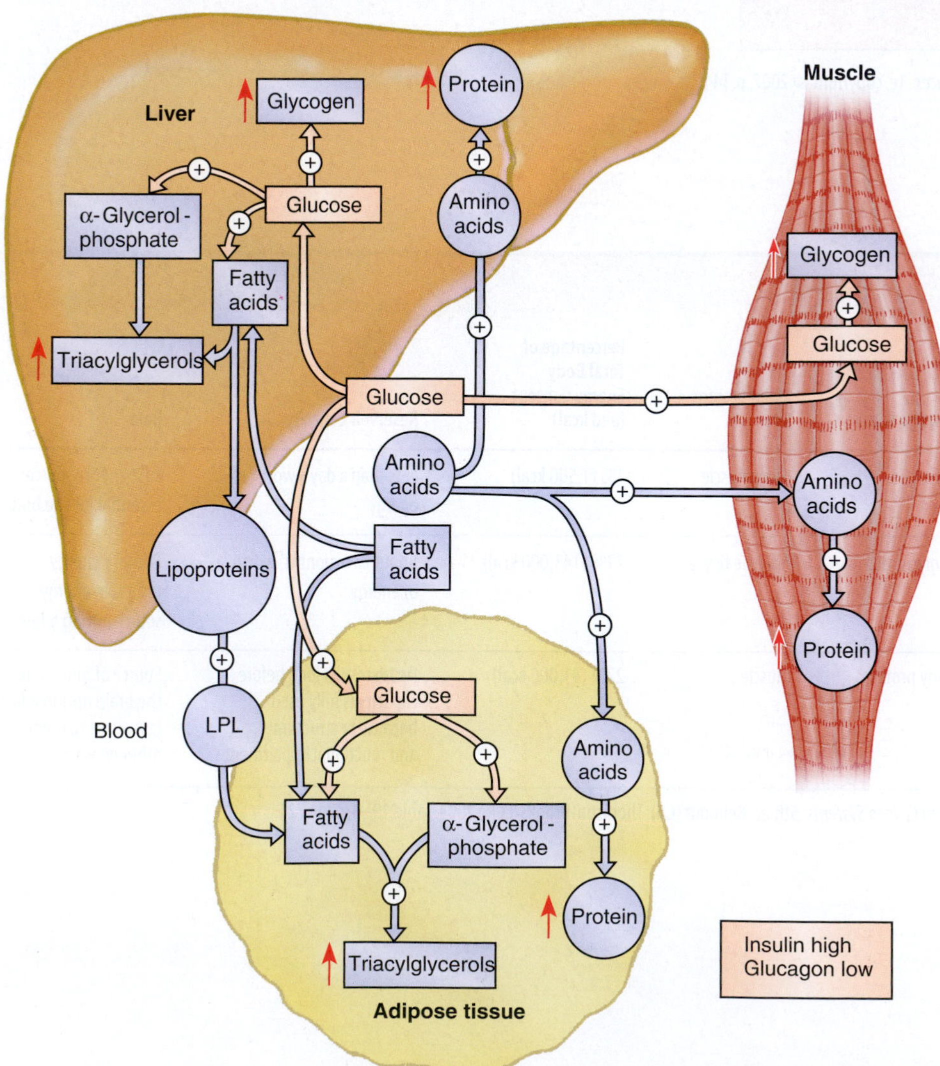

Source: R. Rhoades and R. Pflanzer, *Human Physiology*, 4e, copyright © 2003, p. 466.

FIGURE 19.9 Summary of Nutrient Flow During Fasting

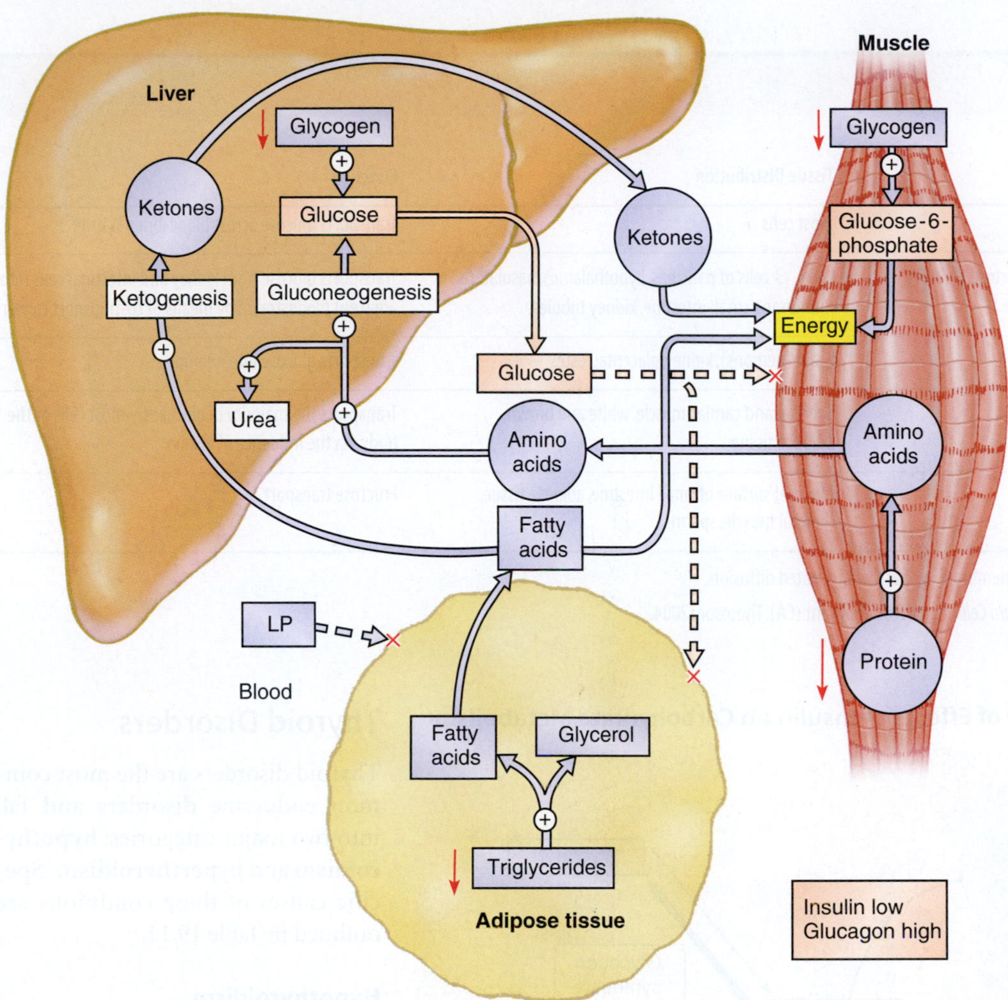

Source: R. Rhoades and R. Pflanzer, *Human Physiology*, 4e, copyright © 2003, p.467.

TABLE 19.6

Comparison of Fed and Fasting States

Nutrient	Fed State	Fasting State
Carbohydrate	Glucose = major energy source	Glycogen degraded and depleted
	Glycogen synthesized and stored	Glucose sparing conserves glucose for the brain
	Excess converted and stored as triglyceride (fat)	New glucose produced via gluconeogenesis
Fat	Triglyceride synthesized and stored	Triglyceride catabolized
		Fatty acids provide major energy source for non-glucose dependent tissues
Protein	Protein synthesized	Protein catabolized
	Excess converted and stored as triglyceride (fat)	Amino acids used for gluconeogenesis

Source: Sherwood L. *Human Physiology: From Cells to Systems.* Belmont (CA): Thomson; 2004.

TABLE 19.7

Roles of Key Tissues in Metabolic States

Tissue	Fed State	Fasting State
Liver	Stores glycogen when excess glucose is available	Releases glucose into blood when needed
		Principle site for metabolic interconversions (such as gluconeogenesis)
Adipose tissue	Primary energy storage site	Regulates serum fatty acid levels
Muscle	Primary site of amino acid storage	
	Major energy user	
Brain	Normally uses only glucose for energy	
	Relies on maintenance of serum glucose levels	

Source: Sherwood L. *Human Physiology: From Cells to Systems.* Belmont (CA): Thomson; 2004.

TABLE 19.8

Glucose Transportation

Glucose Transporter (GLUT)[1]	Substrate	Tissue Distribution	Function
GLUT-1	Glucose	Most cells	Transports glucose across blood-brain barrier
GLUT-2	Glucose, galactose, fructose	Liver, β cells of pancreas, hypothalamus, basolateral membrane small intestine, kidney tubules	Transports glucose from kidney and intestinal cells into adjacent bloodstream by means of co-transport carriers
GLUT-3	Glucose	Brain (neurons), kidney, placenta, testes	Transports glucose into neurons
GLUT-4	Glucose	Skeletal and cardiac muscle, white and brown adipose tissue	Transports the majority of glucose by most cells of the body via the influence of insulin
GLUT-5	Fructose	Mucosal surface of small intestine, adipose tissue, skeletal muscle, sperm	Fructose transport

[1]All glucose is transported across plasma membrane by passive facilitated diffusion.

Source: Sherwood L. *Human Physiology: From Cells to Systems.* Belmont (CA): Thomson; 2004.

FIGURE 19.10 Summary of Effects of Insulin on Carbohydrate Metabolism

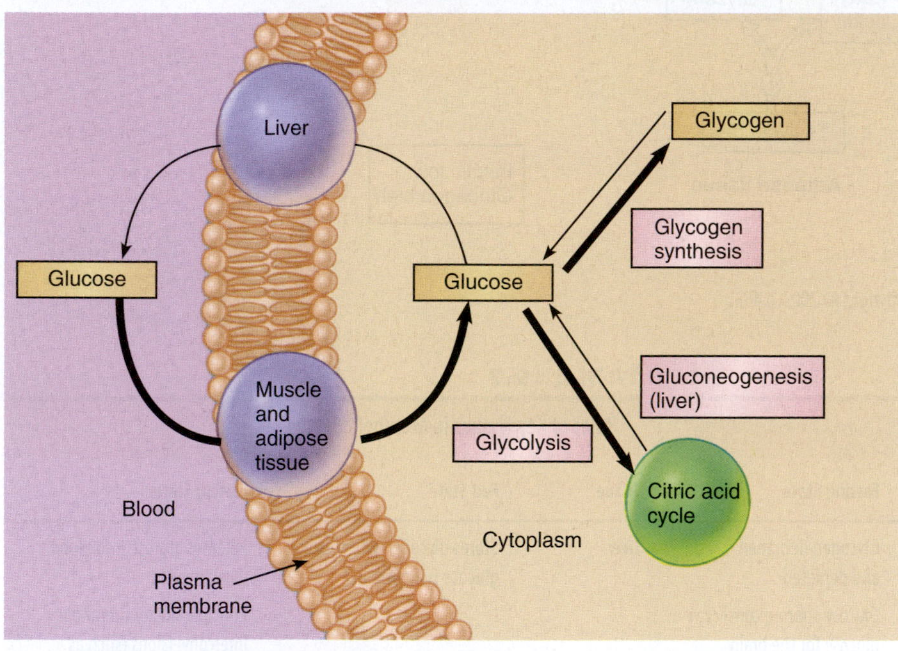

Source: R. Rhoades and R. Pflanzer, *Human Physiology,* 4e, copyright © 2003, p. 461.

Thyroid Disorders

Thyroid disorders are the most common endocrine disorders and fall into two major categories: **hypothyroidism** and **hyperthyroidism**. Specific causes of these conditions are outlined in Table 19.11.

Hypothyroidism

Definition Hypothyroidism is the clinical state resulting from decreased production and secretion of thyroid hormones (Kaplan and Conway 2004) and is the most common pathologic hormone deficiency (Orlander, Woodhouse, and Davis 2005). Many medical conditions may directly or indirectly affect the thyroid gland. Thyroid hormone influences growth, development, and many cellular processes; therefore, insufficient thyroid hormone has extensive consequences throughout the body.

Cretinism refers to congenital hypothyroidism, which affects 1 in 4,000 newborns (Orlander, Woodhouse, and Davis 2005). Because adequate levels of thyroid hormone are essential for normal growth and central nervous system (CNS) development, cretinism is characterized by dwarfism and mental retardation in addition to other general symptoms of thyroid deficiency (discussed later) (Sherwood 2004). This discussion will focus specifically on hypothyroidism in adults.

hypothyroidism—deficient thyroid secretion
hyperthyroidism—excess thyroid secretion

FIGURE 19.11 Summary of Effects of Insulin on Muscle Protein Metabolism

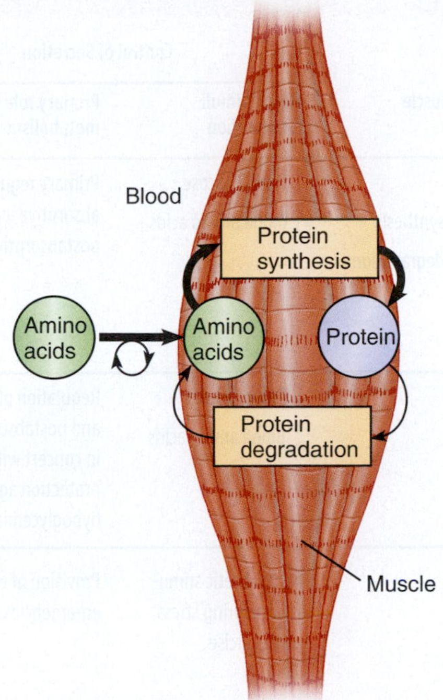

Source: R. Rhoades and R. Pflanzer, *Human Physiology,* 4e, copyright © 2003, p.462.

Epidemiology In the U.S., hypothyroidism is found in 4.6% of the population and is most prevalent in the elderly. Generally, thyroid disease is more common in females, increases with age, and is more common in whites and Mexican-Americans than in blacks (Hollowell et al. 2002). Death from hypothyroidism is uncommon (Orlander, Woodhouse, and Davis 2005).

Etiology Worldwide, the most frequent cause of hypothyroidism is iodine deficiency. In the U.S. and other areas where iodine intake is adequate, autoimmune thyroid disease and previous treatment for hyperthyroidism are the most common causes (Orlander, Woodhouse, and Davis 2005). Table 19.12 outlines common causes of hypothyroidism.

Pathophysiology Decreased thyroid hormone production resulting from localized disease of the thyroid gland is the most common cause of hypothyroidism. The thyroid gland normally releases 100–125 μg of T_4 (thyroxine) and only small amounts of T_3 daily. The half-life of T_4 is approximately 7–10 days. As mentioned previously, T_4 is converted to T_3 in peripheral tissues. Early in the disease process, compensatory mechanisms maintain T_3 levels. Decreased production of T_4 causes increased secretion of TSH by the pituitary gland. TSH stimulates hypertrophy and hyperplasia of the thyroid gland, causing the thyroid to release more T_3 (Orlander, Woodhouse, and Davis 2005).

Thyroid hormone deficiency causes a variety of system-wide effects throughout the body. **Myxedematous** infiltration of heart tissue results in decreased contractility, cardiac enlargement, pericardial effusion, decreased pulse rate, and decreased cardiac output. Infiltration of the GI tract can cause **achlorhydria** and decreased intestinal transit time with gastric stasis. Other common occurrences are delayed puberty, **anovulation**, menstrual irregularities, and infertility. Additionally, hypothyroidism can cause increased levels of total cholesterol and LDL-cholesterol, and increased **insulin resistance** (Orlander, Woodhouse, and Davis 2005).

Clinical Manifestations Symptoms of hypothyroidism are generally subtle. They are not specific (meaning they can mimic symptoms of many other conditions) and are often attributed to aging. Patients with mild hypothyroidism may have no signs or symptoms, but symptoms usually become more noticeable as the condition worsens and are, for the most part, correlated with reduction in overall metabolic activity. Patients with hypothyroidism may present with the following symptoms (Sherwood 2004; Orlander, Woodhouse, and Davis 2005):

- Reduced basal metabolic rate
- Intolerance of cold
- Weight gain
- Easily fatigued
- Bradycardia
- Slow reflexes and movement
- Slow mental responsiveness (diminished alertness, slow speech, poor memory)
- Pitting edema of lower extremities
- Periorbital puffiness
- Myxedema
- Goiter (see Figure 19.13)
- Loss of scalp hair, axillary hair, pubic hair, or a combination
- Abdominal distension

myxedematous—non-pitting edema

achlorhydria—absence of hydrochloric acid from the gastric juice

anovulation—lack of ovulation during the menstrual cycle

insulin resistance—resistance of body cells to the action of insulin

TABLE 19.9

Summary of Hormonal Control of Energy Metabolism

	Major Metabolic Effects				Control of Secretion	
Hormone	Effect on blood glucose	Effect on blood fatty acids	Effect on blood amino acids	Effect on muscle protein	Major stimuli for secretion	Primary role in metabolism
Insulin	↓ + Glucose uptake + Glycogenesis − Glycogenolysis − Gluconeogenesis	↓ + Triglyceride synthesis − Lipolysis	↓ + Amino acid uptake	↑ + Protein synthesis − Protein degradation	↑ Blood glucose ↑ Blood amino acids	Primary regulator of absorptive and postabsorptive cycles
Glucagon	↑ + Glycogenolysis + Gluconeogenesis − Glycogenesis	↑ + Lipolysis − Triglyceride synthesis	No effect	No effect	↓ Blood glucose ↑ Blood amino acids	Regulation of absorptive and postabsorptive cycles in concert with insulin; protection against hypoglycemia
Epinephrine	↑ + Glycogenolysis + Gluconeogenesis − Insulin secretion + Glucagon secretion	↑ + Lipolysis	No effect	No effect	Sympathetic stimulation during stress and exercise	Provision of energy for emergencies and exercise
Cortisol	↑ + Gluconeogenesis − Glucose uptake by tissues other than brain; glucose sparing	↑ + Lipolysis	↑ + Protein degradation	↓ + Protein degradation	Stress	Mobilization of metabolic fuels and building blocks during adaptation to stress
Growth Hormone	↑ − Glucose uptake by muscles; glucose sparing	↑ + Lipolysis	↓ + Amino acid uptake	↑ + Protein synthesis − Protein degradation + Synthesis of DNA and RNA	Deep sleep Stress Exercise Hypoglycemia	Promotion of growth; normally little role in metabolism; mobilization of fuels plus glucose sparing in extenuating circumstances

Up arrows (↑) and plus signs (+) indicate increases; down arrows (↓) and minus signs (−) indicate decreases.

Reprinted from: Sherwood L. *Human Physiology: From Cells to Systems.* 5th Ed. Belmont (CA): Thomson-Brooks/Cole; 2004. Table 19–6, p. 733.

Treatment Treatment of hypothyroidism consists of administration of exogenous thyroid hormone to supplement or replace endogenous thyroid hormone production, with one exception. If hypothyroidism is caused by iodine deficiency, it can be remedied through adequate intake of dietary iodine (Sherwood 2004; Orlander, Woodhouse, and Davis 2005).

Clinical benefits of either treatment begin in 3–5 days and level off after 4–6 weeks. Patients should be monitored for signs and symptoms of over treatment (tachycardia, palpitations, nervousness, tiredness, headache, increased excitability, sleeplessness, tremors, possible angina) (Orlander, Woodhouse, and Davis 2005).

Nutrition Therapy No special nutrition therapy is required for patients with hypothyroidism other than correcting iodine deficiency where it exists (Orlander, Woodhouse, and Davis 2005). However, there are nutrition implications for patients receiving hypothyroidism medications. Drug-nutrient interactions for these medications (and others used to treat endocrine disorders) are outlined in Table 19.13.

Hyperthyroidism

Definition Hyperthyroidism is characterized by excessive secretion of thyroid hormones (Kaplan and Conway 2004). It is a relatively rare condition in children (Gold and Sadeghi-Nejad 2004).

FIGURE 19.12 Summary of Major Pathways Involving Nutrient Absorption and Metabolism

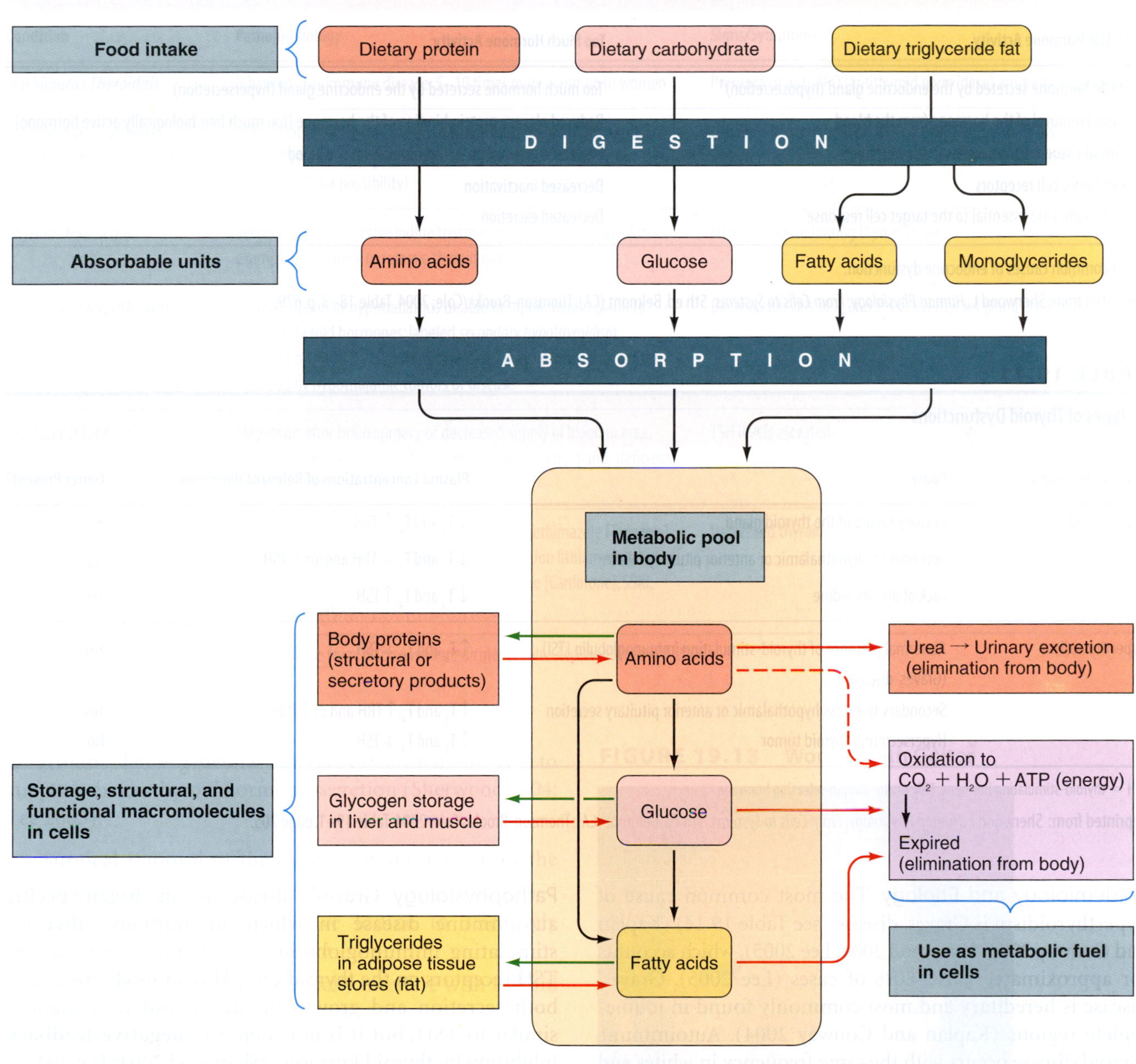

Source: L. Sherwood, *Human Physiology: From Cells to Systems*, 5e, copyright © 2004, p. 720.

Nutrition Therapy While no special diet must be followed (Lee 2005) by individuals with hyperthyroidism, they should be monitored for drug-nutrient interactions related to any medications they receive (see Table 19.13).

Diabetes Mellitus

Over 20 million individuals in the U.S., or 7% of the population, have **diabetes mellitus.** An estimated 14.6 million have been diagnosed, but 6.2 million (nearly one third) are unaware they have the disease (American Diabetes Association 2005). In 2002, diabetes cost the U.S. an estimated $132 billion in medical expenses and lost productivity (American Diabetes Association, Economics, 2003). In 2005, 1.5 million new cases of diabetes were diagnosed in people aged 20 years and older. It is the sixth leading cause of death in the U.S. and is likely to be underreported. Risk for death among people with diabetes is about twice that of those without diabetes (NDIC 2005). Table 19.15 outlines complications of diabetes in the U.S.

FIGURE 19.14 Exophthalmos: Abnormal Fluid Retention behind Eyeballs Causes Them to Bulge Forward

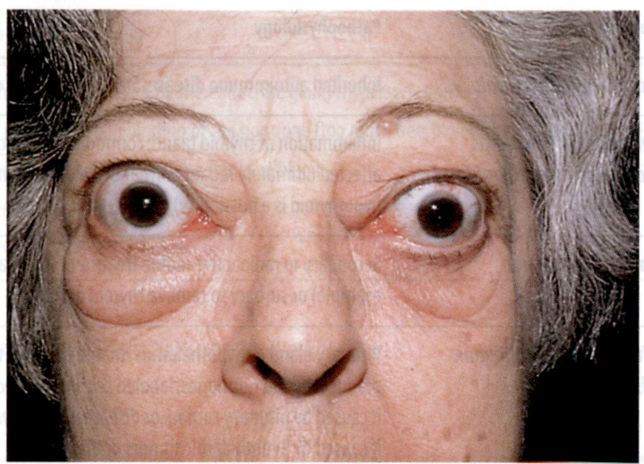

Source: L. Sherwood, Human Physiology: From Cells to Systems, 5e, copyright © 2004, p. 706.

Definition

By far the most common of all endocrine disorders, and a worldwide health problem, diabetes mellitus is not a single disease but a diverse group of disorders that differ in origin and severity (World Health Organization 2002; Rhoades and Pflanzer 2003; Kaplan and Conway 2004; Sherwood 2004; American Diabetes Association, Fact Sheet, 2006). Yet all forms of diabetes mellitus share one common characteristic: hyperglycemia resulting from defects in insulin production, insulin action, or both (Rhoades and Pflanzer 2003; Kaplan and Conway 2004; American Diabetes Association, Diagnosis, 2006; American Diabetes Association, Fact Sheet, 2006). Chronic hyperglycemia is correlated with long-term damage, dysfunction, and failure of numerous organs, particularly the eyes, kidneys, nerves, heart, and blood vessels (see Table 19.15) (American Diabetes Association, Fact Sheet, 2006).

Insulin deficiency (see Figure 19.15) is generally due to either insufficient insulin secretion by beta (β) cells or comparative deficient response by target tissue cells to insulin (Rhoades and Pflanzer 2003). Whatever the cause of insulin deficiency, it results in glucose intolerance. Various conditions associated with glucose intolerance were used by the Expert Committee on the Diagnosis and Classification of Diabetes Mellitus in 2006 to diagnose and classify diabetes (see Box 19.3) (American Diabetes Association, Diagnosis, 2006).

diabetes mellitus—a diverse group of disorders that share the primary symptom of hyperglycemia resulting from defective insulin production, insulin action, or both

Management

Medical care of diabetes mellitus should be the coordinated effort of a team with expertise and special interest in diabetes. The team should be comprised of (but not limited to) the individual with diabetes and the following care providers (American Diabetes Association, Standards, 2006):

- Physicians
- Registered dietitians or dietetic technicians, registered
- Certified Diabetes Educators (CDE)
- Nurse practitioners
- Nurses
- Physician's assistants
- Pharmacists
- Mental health professionals

The management plan should be individualized for the patient and family. Diabetes self-management education (DSME) is a vital element of care. Development of the management strategy should integrate the patient's age, school/work schedule and circumstances, eating patterns, physical activity, social environment and personality, cultural issues, other medical conditions, and presence of complications. It is important that each aspect of the management plan is understood and agreed upon by the patient and care providers, and that the goals and treatment plan are reasonable (American Diabetes Association, Standards, 2006).

Type 1 Diabetes Mellitus

Over the years, several terms have been used to classify the different types of diabetes. In order to prevent individuals from

TABLE 19.13

Drug-Nutrient Interactions for Medications Used to Treat Non-Diabetic Endocrine Disorders

Classification	Mechanism	Generic	Brand Names	Possible Food-Drug Interactions
Thyroid Hormone Replacement	Influences growth and maturation of tissues; involved in normal growth, metabolism and development; produces stable levels of T_3 and T_4.	Levothyroxine	Synthroid Levoxyl Levothroid Unithroid	Take iron, calcium, or magnesium supplements separately from drug by ≥ 4 hr (may decrease absorption), decreased absorption also reported with soy, soy milk, soy infant formula, walnuts, cottonseed meal, and high fiber foods; appetite changes may result in weight loss.
Antithyroid Medications	Blocks oxidation of iodine in thyroid gland, inhibiting T_4 to T_3 conversion.	Propylthiourcil	Propylthiour	Antivitamin K activity.
	Inhibits thyroid hormone by blocking iodine oxidation (not known to inhibit peripheral conversion of thyroid hormone).	Methimazole	Tapazole	Inhibits vitamin K activity.
Beta-Adrenergic Receptor Blockers	Used to reduce symptoms of tachycardia, tremor, and anxiety in hyperthyroidism.	Propranolol	Inderal, Betachron E-R	Low-sodium, low-kcal diet may be recommended; avoid natural licorice; avoid alcohol.
Corticosteroids	Used to restore corticosteroid levels.	Cortisone	Cortone	Low sodium, high calcium, high vitamin D, high protein; may need high potassium, vitamin A, vitamin C, phosphorus (or supplements); calcium-vitamin D supplement recommended with long-term use; caution with grapefruit/grapefruit juice (with methylprednisolone); increased appetite, increased weight; avoid alcohol; negative N balance due to protein catabolism; calcium wasting with long term use; chromium deficiency may increase risk of steroid-induced diabetes.
	Used for partial replacement therapy in adrenocortical insufficiency.	Fludrocortisone	Florinef	Decrease dietary sodium unless increased sodium is used to manage hypotension, increase calcium, vitamin D. May need increased potassium (or supplement), calcium-vitamin D supplement recommended with long term use; avoid alcohol.

Sources: Lee SL. Hyperthyroidism. Last updated 7/20/05. eMedicine.com, Inc. Available from: http://www.emedicine.com. Accessed January 13, 2005. Liotta EA, Brough A, Erickson QL, Elston DM. Addison disease. Last updated 12/2/05. eMedicine.com, Inc. Available from: http://www.emedicine.com. Accessed January 15, 2005. Orlander PR, Woodhouse WR, Davis AB. Hypothyroidism. Last updated 9/23/05. eMedicine.com, Inc. Available from: http://www.emedicine.com. Accessed January 7, 2005. Pronsky ZM. *Food Medication Interactions,* 13th ed. Birchrunville, PA; 2004.

TABLE 19.14

Common Causes of Hyperthyroidism

Graves' Disease	Autoimmune thyroid disease distinguished by overactive thyroid gland
Iodine Induced	Usually seen in those who already have underlying abnormal thyroid gland; a number of medications (such as amiodarone [Cordarone] used to treat heart problems) contain large amounts of iodine and may be associated with thyroid function irregularity
Excessive Intake of Thyroid Hormones	Arises often due to lack of follow-up of patients taking thyroid medications
Toxic Multi-modular Goiter	Thyroid gland become lumpier as people age; the lumps do not produce thyroid hormones, but occasionally a nodule may become "autonomous" and not respond to pituitary regulation via TSH and produces thyroid hormones independently
Abnormal Secretion of TSH	Pituitary gland tumor may produce abnormally high secretion of TSH which excessively signals thyroid gland to produce thyroid hormones
Inflammation of Thyroid (thyroiditis)	May come about after viral illness; associated with a fever and sore throat that is often painful upon swallowing; thyroid gland tender to touch

Source: Axford J, O'Callaghan C (eds). *Medicine,* 2nd ed. Oxford, UK: Blackwell Science Ltd.; 2004 and Mathur R, Shiel WC. Hyperthyroidism. Last editorial review 11/09/05. MedicineNet.com. Available from: http://www.medicinenet.com. Accessed January 8, 2006.

TABLE 19.15

Complications of Diabetes in the United States

Complication	Risk for Individuals with Diabetes
Heart Disease and Stroke	Approximately 65% of all deaths in individuals with DM Death rates 2–4 times higher 2–4 times risk for stroke
Hypertension	Present in approximately 73% of adults with diabetes
Blindness	Leading cause of new cases in adults over 20 years 12,000 to 24,000 new cases each year due to diabetic retinopathy
Kidney Disease	Accounts for 44% of new cases
Nervous System Disorders	60%–70% have mild to severe forms resulting in: Impaired sensation/pain in feet or hands Gastroparesis Carpal tunnel syndrome Other nerve problems Approximately 30% of those over age 40 have impaired sensation in feet
Amputations	Cause of more than 60% of nontraumatic lower-limb amputations
Dental Disease	Periodontal disease more common Risk doubled in young adults
Complications of Pregnancy	Can cause major birth defects in 5 to 10% of women with poorly controlled DM before conception and during first trimester Can result in spontaneous abortions in 15%–20% of pregnancies
Other Complications	Often leads to biochemical imbalances that can cause acute life-threatening events: Ketoacidosis Hyperosmolar (nonketotic) coma More susceptible to many other illnesses More likely to die with pneumonia or influenza

Sources: American Diabetes Association. National Diabetes Fact Sheet, 2005. Available from: http://www.diabetes.org/diabetes-statistics.jsp. Accessed February 10, 2006. Centers for Disease Control and Prevention. National diabetes fact sheet: general information and national estimates on diabetes in the United States, 2005. Atlanta (GA): U.S. Department of Health and Human Services. Centers for Disease Control and Prevention; 2005. National Diabetes Information Clearinghouse (NDIC). National diabetes statistics. NIH Publication #06–3892, November 2005. Available from: http://diabetes.niddk.nih.gov. Accessed February 11, 2006.

being classified by treatment modality rather than disease characteristics, the terms insulin-dependent diabetes mellitus (IDDM), juvenile diabetes, brittle diabetes, non-insulin-dependent diabetes mellitus (NIDDM), or adult-onset diabetes should not be used (Kaplan and Conway 2004).

Epidemiology Type 1 diabetes mellitus (T1DM)[2] accounts for 5% to 10% of all diagnosed cases of diabetes. Approximately 400 to 600 children and adolescents have T1DM (Centers for Disease Control and Prevention 2005; NDIC 2005; American Diabetes Association 2006). While this form of diabetes develops most frequently in children and adolescents, it is being increasingly noted later in life (World Health Organization 2002), even in individuals in their 80s and 90s (American Diabetes Association, Fact Sheet, 2006). Gender distribution of T1DM is equal (Kaplan and Conway 2004).

autoantibodies—self-antibodies; in the case of autoimmunity affecting the pancreas, these include islet cell autoantibodies, autoantibodies to insulin, and autoantibodies to glutamic acid decarboxylase (GAD_{65})

Etiology Immune-mediated type 1 diabetes mellitus results from a cellular-mediated autoimmune destruction of β-cells of the pancreas. One or more **autoantibodies** are present in 85% to 89% of individuals diagnosed with T1DM (Notkins and Lernmark 2001). Rate of β-cell destruction is variable, fast in certain individuals (primarily infants and children) and slow in others (primarily adults). The first sign of T1DM

FIGURE 19.15 Acute Effects of Insulin Deficiency
Acute consequences of insulin deficiency can be grouped according to effects on carbohydrate, protein, and fat metabolism. These effects ultimately cause death through a variety of pathways.

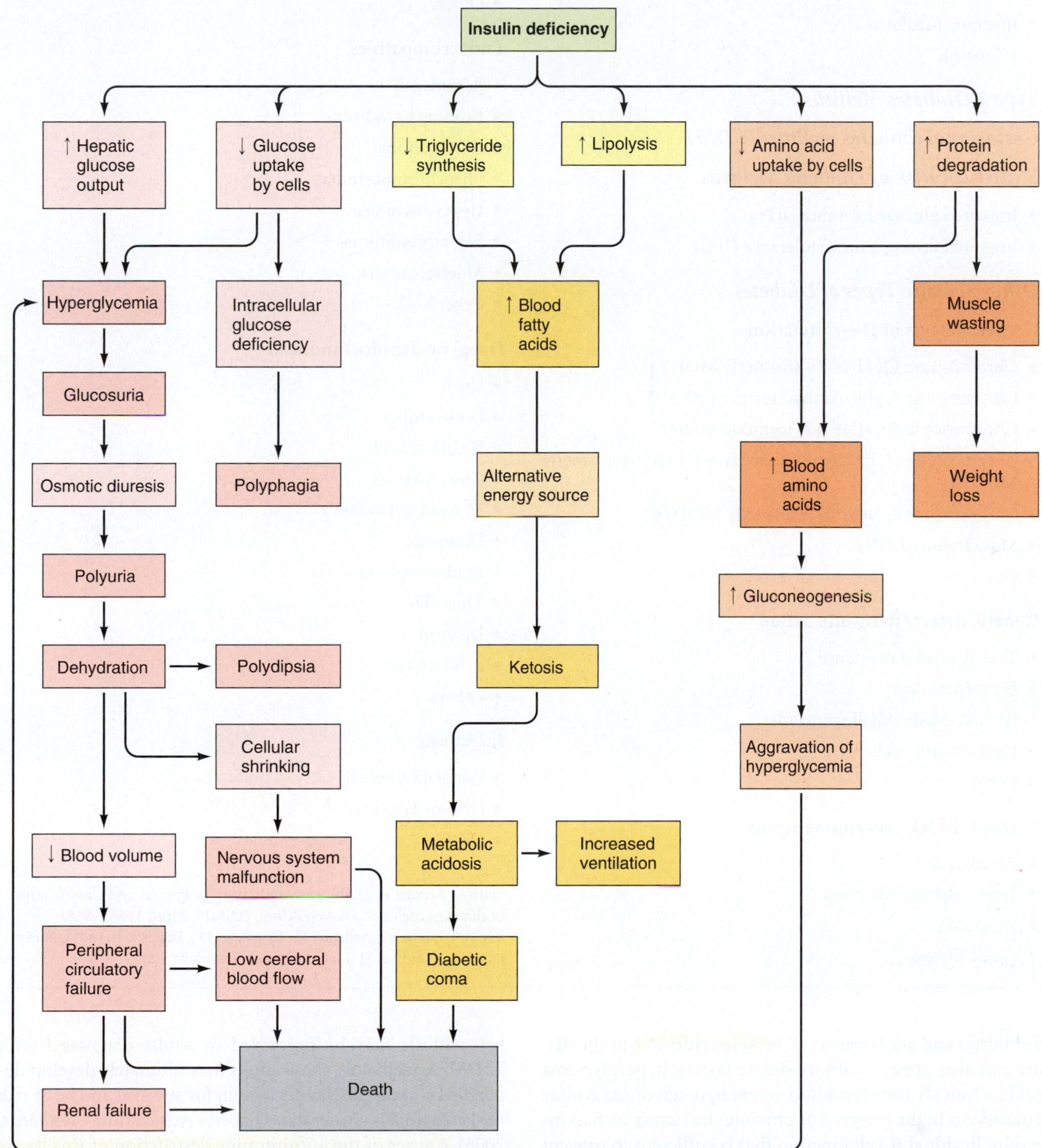

Source: L. Sherwood, *Human Physiology: From Cells to Systems*, 5e, copyright © 2004, p. 727.

BOX 19.3 CLINICAL APPLICATIONS

Etiologic Classifications of Diabetes Mellitus

Type 1 Diabetes Mellitus

- Immune mediated
- Idiopathic

Type 2 Diabetes Mellitus

Gestational diabetes mellitus (GDM)

Statistical Risk of Diabetes Mellitus

- Impaired glucose tolerance (IGT)
- Impaired fasting glucose tolerance (IFG)

Other Specific Types of Diabetes

Genetic defects of β-cell function

- Chromosome 12, HNF-1α (formerly MODY3)
- Chromosome 7, glucokinase (formerly MODY2)
- Chromosome 20, HNF-4α (formerly MODY1)
- Chromosome 13, insulin promoter factor-1 (IPF-1, formerly MODY4)
- Chromosome 2, NeuroD1 (formerly MODY6)
- Mitochondrial DNA
- Others

Genetic defects in insulin action

- Type A insulin resistance
- Leprechaunism
- Rabson-Mednenhall syndrome
- Lipoatrophic diabetes
- Others

Diseases of the exocrine pancreas

- Pancreatitis
- Trauma/pancreatectomy
- Neoplasia
- Cystic fibrosis
- Hemochromatosis
- Fibrocalculous pancreatopathy
- Others

Endocrinopathies

- Acromegaly
- Cushing's syndrome
- Glucagonoma
- Pheochromocytoma
- Hyperthyroidism
- Somatostatinoma
- Aldosteronoma
- Others

Drug- or chemical-induced

- Vacor
- Pentamidine
- Nicotinic acid
- Glucocortoids
- Thyroid hormone
- Diazoxide
- β-adrenergic agonists
- Thiazides
- Dilantin
- α-Interferon
- Others

Infections

- Congenital rubella
- Cytomegalovirus
- Others

Sources: American Diabetes Association. Diagnosis and classification of diabetes mellitus. *Diabetes Care.* 2006;29(Suppl 1):S43-S48.
Asp AA. Diabetes mellitus. In: Copstead LC, Banasik JL (eds). *Pathophysiology.* 3rd ed. St. Louis: Elsevier Saunders; 2005.

in children and adolescents can be **ketoacidosis,** but the disease can also present with moderate fasting hyperglycemia that can quickly transform into severe hyperglycemia and/or ketoacidosis in the presence of physiological stress such as infection. Residual β-cell function that is sufficient to prevent ketoacidosis may be preserved in adults diagnosed with T1DM. Nonetheless, these adults may ultimately develop dependence upon exogenous insulin for survival and be at risk for ketoacidosis (American Diabetes Association, Standards, 2006). Causes of the autoimmune destruction of β-cells are not clearly understood, but multiple genetic predispositions and unidentified environmental factors appear to contribute to T1DM (Rhoades and Pflanzer 2003; American Diabetes Association, Diagnosis, 2006). Though the causal environmental agent is currently unknown, viruses are suspected to be probable candidates (Rhoades and Pflanzer 2003).

ketoacidosis—an acid-base imbalance caused by an increase in concentration of ketones in the blood

Some forms of T1DM have no known cause and are referred to as idiopathic diabetes. Individuals with idiopathic diabetes produce no insulin and are prone to ketoacidosis, but have no evidence of autoimmunity. Individuals with T1DM who fall into this category represent a very small minority, and most are of African or Asian ancestry.

Pathophysiology and Clinical Manifestations T1DM is characterized by an absolute deficiency of insulin due to destruction of pancreatic β-cells, resulting in the inability of cells to use glucose for energy (Notkins and Lernmark 2001; Copstead and Banasik 2005). By the time clinical symptoms occur, 60% to 80% of β-cells have been destroyed. Cells that produce glucagon, somatostatin, and pancreatic polypeptide are typically conserved but may be redistributed within the islets (Notkins and Lernmark 2001).

When glucose cannot enter cells, two things happen: plasma glucose levels rise (hyperglycemia) and cells starve. To compensate for the hyperglycemia, excess glucose is lost in the urine because the kidneys can filter only so much glucose from the blood. As a result, **glycosuria** and frequent urination (**polyuria**) occur. Loss of fluid stimulates the thirst mechanism and leads to **polydipsia**. Cells dependent on glucose for energy have none available. In turn, the body responds to this emergency by promoting hunger (polyphagia) (Copstead and Banasik 2005).

As the insulin deficiency persists, production of additional hormones (catecholamines, cortisol, glucagon, and growth hormone) increases, leading to lipolysis. As the body breaks down fat stored in adipose tissue, the resulting fatty acids are transformed into keto acids in the liver. In the non-diabetes state, keto acids can be used for energy by muscle and brain cells. As increased production of keto acids occurs, pH falls (7.3 to 6.8), and ketone bodies are secreted in the urine. Metabolic acidosis develops as bicarbonate concentration is reduced, and ketoacidosis results (Copstead and Banasik 2005).

As total body water decreases, potassium, sodium, magnesium, and phosphorus are also lost. Serum levels of these ions may be normal or elevated due to decreased fluid volume in the body (hypovolemia). Hypovolemia also accounts for increased hematocrit, hemoglobin, protein, white blood cell count, creatinine, and **serum osmolality**. Hypovolemia and muscle catabolism are the cause for considerable, imminent weight loss in persons with ketoacidosis, and often present at diagnosis of T1DM. Hypovolemic shock can lead to death if left untreated. The body tries to offset metabolic acidosis through deep, labored respirations (Kussmall respirations) (Copstead and Banasik 2005). This process is summarized in Figure 19.15.

Diabetic Ketoacidosis (DKA) Diabetic ketoacidosis (DKA), a severe form of hyperglycemia, is a life-threatening situation that commands prompt medical attention (Umpierrez, Murphy, and Kitabchi 2002; Arnold 2005). DKA occurs more often in T1DM, but is also a risk for individuals with type 2 diabetes mellitus (T2DM) during acute illness and/or when they have become insulin deficient (American Diabetes Association, Hyperglycemia, 2004; Arnold 2005). When adequate insulin is not available, glucose production is stimulated by counter-regulatory hormones via gluconeogenesis and lipolysis in an effort to avoid starvation. One of the by-products of lipolysis is the generation of ketones. As glucose and ketones accumulate in the bloodstream, osmotic diuresis occurs, resulting in dehydration and electrolyte imbalances. As fluid is lost, the blood becomes concentrated, bringing about hyperglycemia (Umpierrez, Murphy, and Kitabchi 2002; American Diabetes Association, Hyperglycemia, 2004; Arnold 2005).

Risk for DKA intensifies during illness, infection, and emotional stress. Omission of insulin is also a frequent cause. Individuals may not take their insulin (a primary treatment for T1DM) when they feel too sick to eat, or because they are afraid of developing hypoglycemia (Umpierrez, Murphy, and Kitabchi 2002; American Diabetes Association, Hyperglycemia, 2004; Arnold 2005). Symptoms of DKA include (American Diabetes Association, Hyperglycemia, 2004; Arnold 2005):

- Nausea and/or vomiting
- Stomach pain
- Fruity or acetone breath
- Kussmaul respirations
- Mental status changes

Treatment typically involves hospitalization for assessment and/or administration of IV fluids, insulin, and electrolytes. Supplemental doses of insulin are administered until metabolic stability returns (American Diabetes Association, Hyperglycemia, 2004; Arnold 2005). While the vast majority of DKA cases are resolved, 2% to 5% of cases are fatal (Umpierrez, Murphy, and Kitabchi 2002; Arnold 2005).

Long-Term Complications of Hyperglycemia Diabetes is a complicated chronic metabolic disorder that requires attention to issues beyond glycemic control (American Diabetes Association, Standards, 2006). Long-term hyperglycemia from either type 1 or type 2 DM results in microvascular and macrovascular complications that substantially increase morbidity and mortality associated with the disorder and reduce quality of life (American Diabetes Association, Implications, 2002). Occurrence and rate of development of chronic complications of diabetes can be reduced, however, as demonstrated by two groundbreaking studies (Wheeler

glycosuria—the presence of glucose in the urine

polyuria—frequent urination

polydipsia—excessive thirst

serum osmolality—a measure of the concentration of solute molecules in the blood

2005). The Diabetes Control and Complications Trial (DCCT) demonstrated that intensive treatment (see the section Insulin Regimens) of individuals with T1DM resulted in significantly lower A1C and reductions in incidence and rate of progression of retinopathy, nephropathy, and neuropathy (Diabetes Control and Complications Trial Research Group 1993). The United Kingdom Prospective Diabetes Study (UKPDS) demonstrated that individuals with newly diagnosed T2DM who received intensive treatment (insulin, sulfonylureas, or metformin) experienced relatively improved A1C (glycated hemoglobin) results, significant reductions in all microvascular complications, and reductions in cardiovascular disease outcomes (U.K. Prospective Diabetes Study Group 1998).

Macrovascular Complications: Cardiovascular Disease (CVD) About 65% of deaths among individuals with diabetes are due to heart disease or stroke. Adults with diabetes have heart disease-related death rates about 2 to 4 times higher than those of adults without diabetes (American Diabetes Association, Fact Sheet, 2005; Centers for Disease Control 2005; NDIC 2005). Diabetes is an independent risk factor for macrovascular disease in addition to the common coexisting risk factors of hypertension and dyslipidemia (see Chapter 15). Hypertension is not only a major risk factor for CVD, but also a complication for microvascular complications of DM such as retinopathy and nephropathy. Hypertension is often the consequence of underlying nephropathy in individuals with T1DM. In T2DM, hypertension may manifest as part of the metabolic syndrome (see Chapter 15), which is accompanied by high rates of CVD. Hypertension can be improved by increasing physical activity and consumption of fruits, vegetables, and low-fat dairy products, by decreasing sodium intake and body weight (when indicated), and by avoiding excessive alcohol intake. Dyslipidemia can be improved by a reduction of saturated fat and cholesterol intake, weight loss if indicated, and increased physical activity (American Diabetes Association, Standards, 2006).

Microvascular Complications

Nephropathy

Nephropathy occurs in 20% to 40% of individuals with diabetes, and is the single leading cause of chronic kidney disease (CKD)—kidney failure that must be treated with dialysis or transplantation (see Chapter 20) (American Diabetes Association, Standards, 2006). Diabetes is the cause of 44% of new cases of CKD (American Diabetes Association, Fact Sheet, 2005; Centers for Disease Control 2005; NDIC 2005). The earliest stage of nephropathy in T1DM, which is also a marker for development of nephropathy in T2DM, is persistent albuminuria in the range of 30–299 mg/24 hours (microalbuminuria). Microalbuminuria is also a risk factor for increased risk of CVD. When individuals with microalbuminuria progress to macroalbuminuria (≥300 mg/24 h), they are apt to develop CKD in the coming years. Onset of microalbuminuria and progression to macroalbuminuria in individuals with diabetes can be delayed by intensive diabetes management (defined as achieving near normoglycemia). Protein restriction and use of ACE inhibitors may also slow progression of albuminuria, glomerular filtration rate decline, and occurrence of CKD (American Diabetes Association, Standards, 2006).

Retinopathy

Retinopathy is the most frequent cause of new cases of blindness in adults (American Diabetes Association, Fact Sheet, 2005; Centers for Disease Control 2005; NDIC 2005), and prevalence of retinopathy is strongly associated with duration of diabetes (American Diabetes Association, Standards, 2006). In addition, other eye ailments, including glaucoma and cataracts, occur earlier in individuals with diabetes (American Diabetes Association, Standards, 2006). Glycemic control is not the only risk factor for retinopathy. If nephropathy is present, it is likely that retinopathy is also present. Hypertension is an established risk factor for development of macular edema, and is associated with the development of retinopathy (American Diabetes Association, Standards, 2006). Progression of retinopathy can be decreased by glycemic control and lowering of blood pressure (U.K. Prospective Diabetes Study Group 1998).

Nervous System Diseases Approximately 60% to 70% of individuals with diabetes have some form of nervous system damage that causes impaired sensation or pain in the feet or hands, slowed digestion of food in the stomach, carpal tunnel syndrome, and other nerve problems (American Diabetes Association, Fact Sheet, 2005; Centers for Disease Control 2005; NDIC 2005).

Autonomic Neuropathy

The significance of autonomic neuropathy, a serious and common complication of diabetes, is often underappreciated (Vinik et al. 2003). It can involve one or more mechanisms of the autonomic nervous system (Wheeler 2005). Autonomic neuropathy commonly coexists with other peripheral neuropathies and other complications of diabetes. It may affect many organ systems throughout the body, including the gastrointestinal tract, genitourinary tract, and cardiovascular system. Gastrointestinal (GI) disturbances are common and can occur along any section of the GI tract (Vinik et al. 2003). Gastroparesis, delayed gastric emptying, results from damage to the vagus nerve, which controls peristalsis. It can cause anorexia, nausea, vomiting, early satiety, postprandial bloating, and erratic glycemic control (Wheeler 2005). See Box 19.4 for nutrition therapy for treatment of gastroparesis. Constipation is the most frequent lower GI symptom, but can alternate with episodes of diarrhea. Bladder and/or sexual dysfunction are common genitourinary tract disturbances

BOX 19.4 CLINICAL APPLICATIONS

Nutrition Therapy for Treatment of Gastroparesis

No controlled trial of nutrition therapy for management of symptoms of gastroparesis has been reported. The following recommendations have been established using professional judgment, clinical practice, and interpretation of gastric physiology.

- **Small, frequent meals** may decrease the feeling of bloating and early satiety, thus decreasing the possibility of impaired nutritional status.
- **Reduced fat intake** may shorten the time for gastric emptying.
- **Physical activity, such as walking, after meals** may increase gastric emptying rates.
- **Foods with soft or liquid consistency** may be more easily digested, although hypertonic enteral formulas should be avoided because they further delay gastric emptying.
- **Adjustment of insulin doses and timing** to better match delayed nutrient absorption and postprandial rise in glucose levels should be considered. An example would be administering regular insulin after meals instead of before meals.

Source: Wheeler ML. Diabetic gastropathy and medical nutrition therapy. In: Franz MJ, Bantle JP, editors. *American Diabetes Association guide to medical nutrition therapy for diabetes.* Alexandria (VA): American Diabetes Association; 1999.

associated with autonomic neuropathy (Vinik et al. 2003) and may manifest as recurrent urinary tract infections, pyelonephritis, or incontinence. Males and females may suffer sexual dysfunction (American Diabetes Association, Standards, 2006), including retrograde ejaculation in males (Vinik et al. 2003). Cardiovascular autonomic neuropathy (CAN) is considered the most clinically important form of autonomic neuropathy. It may manifest through resting tachycardia (>100 bmp), orthostatic hypotension (a fall in systolic blood pressure >20 mmHg upon standing), or increased risk of silent heart disease (Vinik et al. 2003; American Diabetes Association, Standards, 2006).

Diagnosis General criteria for diagnosis of diabetes are outlined in Box 19.5. There are three ways to diagnose diabetes. If diagnosis based on one of these three methods is made in the absence of hyperglycemia, then that diagnosis must be confirmed on a subsequent day by any one of the three methods. Use of glycosylated hemoglobin (A1C) is not recommended for diagnosis of diabetes (American Diabetes Association, Diagnosis, 2006).

Diagnosis of T1DM can be made on the basis of a casual plasma glucose ≥200 mg/dL (≥11.1 mmol/L) in addition to certain symptoms (unexplained weight loss, polydipsia, polyuria), or fasting plasma glucose ≥126 mg/dL (≥7.0 mmol/L).

Laboratory Measurements

Oral Glucose Tolerance Test (OGTT) **Oral glucose tolerance tests (OGTTs)** are rarely needed to diagnose T1DM (Mulcahy and Lumber 2004) due to the sudden onset of symptoms accompanied by hyperglycemia. In fact, OGTT is contraindicated in infants and young children (Mulcahy and Lumber 2004), but it is commonly used to diagnose gestational diabetes, impaired glucose tolerance (IGT), and impaired fasting glucose (IFG) (American Diabetes Association, Diagnosis, 2006). An OGTT is administered after at least 3 days of an unrestricted diet providing at least 150 grams of carbohydrate daily and normal physical activity. The test is preceded by an overnight fast of 8 to 14 hours, during which water may be drunk. Smoking is not permitted during the test. After collection of a fasting blood glucose sample, a drink containing 75 grams of anhydrous glucose (100 grams for pregnant women) in 250 to 300 mL of water should be consumed over a 5 minute period. Timing of the test begins at the beginning of the drink. Blood samples are collected 2 hours after the test load (World Health Organization 1999). In a person without diabetes, blood glucose levels rise, then fall quickly to normal. In individuals with diabetes, blood glucose levels rise higher than normal, then fall slowly back to normal. IFG is diagnosed when fasting plasma glucose is found to be 100 to 125 mg/dL; IGT is diagnosed when 2-hour postprandial load is found to be between 140 and 199 mg/dL. Details of diagnostic criteria for diabetes are found in Box 19.5.

Diabetes-Related Autoantibodies Autoantibody testing is not compulsory to diagnose T1DM, but it can be valuable in screening individuals at high risk for developing diabetes (e.g., siblings of individuals with T1DM and offspring of parents with T1DM) up to 7 years before clinical onset (American Association for Clinical Chemistry 2006; Biomerica 2004). Diabetes-related autoantibody testing is largely performed to differentiate between autoimmune T1DM and diabetes resulting from obesity and/or insulin resistance. These autoantibodies do not trigger T1DM,

oral glucose tolerance test (OGTT)—timed glucose challenge to examine efficiency of the body in metabolism of glucose

BOX 19.5 CLINICAL APPLICATIONS

Criteria for Diagnosis of Diabetes Mellitus[1]

Symptoms of diabetes[2] casual[3] plasma glucose concentration ≥200 mg/dL (11.1 mmol/L)	OR	Fasting plasma glucose[4] ≥126 mg/dL (7.0 mmol/L)	OR	2-hour post prandial glucose ≥200 mg/dL (11.1 mmol/L) during an oral glucose tolerance test (OGTT)[5]

Source: Adapted from: American Diabetes Association. Diagnosis and classification of diabetes mellitus. *Diabetes Care.* 29(Suppl 1):S43–48, 2006. American Diabetes Association. Standards of medical care in diabetes. *Diabetes Care.* 2006;29(Suppl 1):S4.

[1] In absence of unequivocal hyperglycemia, these criteria should be confirmed by repeat testing on a differrent day.

[2] Polyuria, polydipsia, and unexplained weight loss.

[3] Casual is defined as any time of day without regard to time since last meal.

[4] Defined as no caloric intake for at least 8 hours.

[5] OGTT should be performed as described by WHO, using a glucose load containing the equivalent of 75 g anhydrous glucose dissolved in water. OGTT is not recommended for routine clinical use.

but serve as indicators of the body's destructive immune response against its own β-cells (American Association for Clinical Chemistry 2006). Tests used to measure diabetes-related autoantibodies include islet cell cytoplasmic autoantibodies (ICA), insulin autoantibodies (IAA), glutamic acid decarboxylase autoantibodies (GADA), and insulinoma-associated-2 autoantibodies (IA-2A).

Glutamic Acid Decarboxylase Autoantibodies (GADA) Tests for glutamic acid decarboyxlase autoantibodies (GADA) measure specific islet cell antigens. GADA have been found in 70% to 90% of individuals with T1DM, and have been shown to be the most sensitive marker for identifying persons at risk for developing T1DM. They are generally more prevalent in older children and individuals with **latent autoimmune diabetes of adulthood (LADA)** (Biomerica 2004).

Islet Cell Autoantibodies (ICA) The ICA test measures a group of islet cell autoantibodies. ICA have been found in 70% to 80% of individuals younger than 30 years of age with newly diagnosed diabetes. Among individuals with T1DM, prevalence of ICA decreases the longer an individual has diabetes. Existence of ICA in relatives without diabetes has been shown to be a sign of increased risk for the disease (Biomerica 2004). There is an increased risk for T1DM in individuals without diabetes who test positive for one or more islet autoantibodies. The more islet autoantibodies present, the greater the individual's risk for developing T1DM (American Association for Clinical Chemistry 2006).

Insulin Autoantibodies (IAA) The presence of IAA is evidence of ongoing destruction of β-cells (Biomerica 2004). IAA testing must be performed before insulin therapy is initiated since the test does not determine whether the body's immune system is making autoantibodies against endogenous or exogenous insulin (American Association for Clinical Chemistry 2006). IAA are found primarily, though not exclusively, in young children developing T1DM as an early predictive marker. They are rarely established in adults with T1DM (Biomerica 2004).

Measures of Glycemic Control **Glycemic control** is fundamental to the management of diabetes. Prospective randomized clinical trials (Diabetes Control and Complications Trial Research 1993; UK Prospective Diabetes Study 1998) have demonstrated that improved glycemic control is correlated with sustained reduced rates of retinopathy, nephropathy, and neuropathy. Efficacy of glycemic control and the management plan can be evaluated by health providers and the individual with diabetes through several methods (see Table 19.16) (Diabetes Control and Complications Trial 1993; American Diabetes Association, Standards, 2006). Recommended glycemic goals are outlined in Table 19.17. The combination of self-monitoring of blood

latent autoimmune diabetes of adulthood (LADA)—sometimes called T1.5DM, a slowly progressive form of T1DM; individuals are often diagnosed as T2DM, but have positive pancreatic islet antibodies, especially to glutamic acid decarboxylase (GADA)

glycemic control—control of blood glucose

TABLE 19.16

Techniques Used to Assess Glycemic Control

Technique	Benefit	Recommendations for T1DM	Recommendations for T2DM	Comments
Self-Monitoring of Blood Glucose (SMBG)	Allows patients to individualize response to therapy Useful in preventing hypoglycemia Useful in adjusting medications, nutrition therapy, and physical activity	≥ 3 times daily	Nutrition therapy and exercise alone: 1–2 times daily at alternating times throughout day (e.g., Monday before dinner and bedtime, Tuesday before breakfast and lunch) Oral glucose-lowering medication: 1–2 times daily, rotating test times each day Insulin: ≥ 3 times daily	Accuracy instrument- and user-dependent Evaluate patients' monitoring techniques initially and subsequently at regular intervals Regularly evaluate patients' ability to use SMBG data to adjust food intake, exercise, or pharmacological therapy to achieve specific glycemic goals
A1C	Allows measurement of average glycemia over preceding 2–3 months	Every 3 months	Patients whose therapy has changed or who are not meeting glycemic goals: test every 3 months Patients with stable glycemic control: test (at least) twice yearly	Regular A1C testing allows detection of departures from target management goals

Sources: American Diabetes Association. Standards of medical care in diabetes. *Diabetes Care.* 2006;29(Suppl 1):S4. Sacks DB, Bruns DE, Goldstein DE, Maclaren NK, McDonald JM, Parrott M. Guidelines and recommendations for laboratory analysis in the diagnosis and management of diabetes mellitus. *Clin Chem.* 2002;48:436–472.

TABLE 19.17

Recommendations for Glycemic Control

Parameter	Recommendations
A1C[1,2]	< 7.0%
A1C for pregnant patients	< 6%
Preprandial capillary plasma glucose	90–130 mg/dL (5.0–7.2 mmol/L)
Peak postprandial capillary plasma glucose[3]	< 180 mg/dL (<10.0 mmol/L)

[1] Primary target for glycemic goals

[2] goals of < 6% may further reduce complications at increased risk or hypoglycemia (particularly in T1DM)

[3] Measured 1–2 hours after the beginning of a meal

Source: Adapted from American Diabetes Association. Standards of medical care in diabetes. *Diabetes Care.* 2006;29(Suppl 1):S4.

TABLE 19.18

Correlations Between A1C Level and Mean Plasma Glucose Levels[1]

A1C (%)	Mean Plasma Glucose mg/dL
4	60
5	90
6	120
7	150
8	180
9	210
10	240
11	270
12	300
13	330

[1] Based on a normal A1C of 6.

Source: Rickheim P, Flader J, Reynolds K, Radosevich G, Carstensen K. *Insulin Basics.* Minneapolis, MN: International Diabetes Center; 2001.

glucose (SMBG) and A1C is the best indicator of glycemic control. Table 19.18 shows the correlation between A1C levels and mean plasma glucose levels.

Glycated Hemoglobin Assays (A1C) Glycated hemoglobin assays (hemoglobin A1C or A1C) measure the amount of glucose bound to hemoglobin protein. The higher the glucose concentration in the blood, the more hemoglobin is glycated (addition of a glucose molecule to amino acid side-chains), thus making it a valid test to measure degree of hyperglycemia. Because red blood cells have a lifespan of 120 days, A1C can measure the average glucose concentration for the previous 2–3 months. Because A1C values cannot be significantly affected by manipulating diet or treatment in the week before a clinic appointment, they are valuable tools to assess glycemic control for a period of time. A1C is a monitoring tool for individuals with known diabetes and is not specific or sensitive enough to be used for diagnosis of diabetes (Baynes and Betteridge 2004). In addition, since it measures hemoglobin-bound glucose, A1C is inappropriate as a gauge of glycemic control for individuals with blood disorders such as anemia (see Chapter 21).

A1C should be tested at least twice yearly in individuals who are meeting treatment goals and who have stable glycemic control. For individuals not meeting glycemic goals, or whose therapy has changed, A1C testing should be performed quarterly (American Diabetes Association, Standards, 2006).

Self-Monitoring of Blood Glucose (SMBG) Daily home glucose monitoring indicates what an individual's glucose level is at the very moment the measurement is taken. Information provided by SMBG can assist in adjusting daily eating patterns and medications as necessary to maintain glycemic control. SMBG is also useful in identifying patterns and the ways in which food, exercise, or other factors affect glycemic control. Adjustments to an individual's treatment program can be made immediately in order to prevent hyperglycemia, hypoglycemia, and long-term complications of diabetes (American Diabetes Association, Standards, 2006).

A typical SMBG test includes a drop of blood obtained via a finger prick that is applied to a chemically treated reagent strip. Home monitors are generally used to determine results. Frequency and timing of SMBG should be determined by the specific needs and goals of the individual with diabetes and the health care team. As a means to monitor for asymptomatic hypoglycemia and hyperglycemia, daily SMBG is particularly valuable for all individuals with diabetes. SMBG is recommended three or more times daily for most individuals with T1DM and diabetic pregnant women. Individuals with T2DM taking insulin usually need to perform SMBG more often than those not using insulin. Individuals with T1DM or T2DM ought to test more often than usual when therapy is modified (American Diabetes Association, Standards, 2006).

Accuracy of SMBG is instrument- and user-dependent. For this reason, the patient's monitoring techniques should be assessed at the onset and at regular intervals thereafter. Patients should also be taught how to use SMBG data to modify food intake, exercise, and pharmacological therapy in order to realize individual glycemic goals (American Diabetes Association, Standards, 2006).

Fructosamine Test Fructosamine assays have nothing to do with fructose; instead, the name refers to the chemical structure formed when glucose attaches to a molecule of protein, which resembles a fructose molecule. The fructosamine test is another index of time-averaged plasma glucose. This assay is used to monitor glycemic control over a 1–3-week period and is neither influenced by variant hemoglobin nor attached to the lifespan of a red blood cell. This assay is not reliable in individuals with renal failure or liver disease (American College of Physicians 1998).

Urine Testing for Glucose Glycosuria occurs when **renal threshold** for glucose is exceeded. Renal threshold varies between individuals, but usually occurs when blood glucose levels are >250 mg/dL. Hypoglycemia cannot be detected by urine testing (Baynes and Betteridge 2004). SMBG is a much more accurate method of monitoring glycemic control.

Urine Testing for Ketones Urine tests are used to detect the presence of ketones in the urine. These tests should always be performed regularly during periods of illness or stressful situations when glucose levels are likely to be elevated. In individuals with T1DM, urine ketones should be tested when blood glucose is consistently over 300 mg/dL (American Diabetes Association, Standards, 2006).

Other Testing In addition to monitoring glycemic control, other parameters that should be monitored include lipids and blood pressure. Total cholesterol, low-density lipoproteins (LDL-cholesterol), high-density lipoprotein (HDL-cholesterol), and triglycerides should be monitored annually or more frequently as needed. The Accuracy of these tests is dependent upon an overnight fast. Goal values are as follows (American Diabetes Association, Standards, 2006):

- Total cholesterol: <200 mg/dL
- LDL-cholesterol: <70mg/dL
- HDL-cholesterol: >40 mg/dL (men); >50 mg/dL (women)
- Triglycerides: <150 mg/dL

Treatment Treatment goals include avoiding hyperglycemia and retarding development of complications within an acceptable level of treatment side effects. The closer to the normal range blood glucose can be maintained over the long term, the lower the risk of microvascular complications (Diabetes Control and Complications Trial 1993). Box 19.6 describes risk factors and treatments for short- and long-term complications of diabetes.

In individuals without diabetes, endogenous insulin is secreted by the β-cells of the pancreas in response to changes in blood levels of glucose and other nutrients. When the pancreas is stimulated, proinsulin is cleaved into insulin and c-peptide which that are secreted into the bloodstream in equal amounts. A healthy, nonpregnant, nonobese adult normally secretes 0.5 to 0.7 units of insulin per kg of body weight per day (Rystrom 2005).

To survive, individuals with T1DM must depend on daily administration of exogenous insulin (WHO 2002; National Diabetes Information Clearinghouse 2005) in conjunction with nutrition therapy and physical activity

renal threshold—a concentration level of glucose in the blood above which the kidneys pass it through into the urine

(Rystrom 2005) to mimic the insulin secretion in an individual without diabetes. The variety of available insulin options permits development of suitable insulin regimens that correspond to an individual's preferential meal routine, food choices, and lifestyle (Kulkarni and Franz 1999; Franz et al. 2002; American Diabetes Association 2004). A clear understanding of insulin pharmacokinetics and the ability to recognize trends in blood glucose data are essential in order to monitor patients' meal and insulin plans and determine appropriate insulin-to-carbohydrate ratios (Rystrom 2005). The best possible insulin management can only be achieved by evaluating blood glucose monitoring records, adjusting food and exercise activities, and proposing insulin adjustments (Kulkarni and Franz 1999).

Types of Insulin Types of insulin are frequently combined in relation to timing of activity of the insulin in an effort to mimic normal physiological action of insulin. To accomplish this, insulin is classified based on expected onset of action, peak time of action, and duration of action (American Diabetes Association 2000). It is important for health care professionals to thoroughly understand these features before developing a care plan and meal plan (Rystrom 2005). Table 19.19 summarizes the features of available insulin types.

Determining Insulin Doses The correct insulin dosage is often determined by using algorithms based on body weight. These algorithms (see Box 19.7) can be used as a general guideline, but insulin dosage is adjusted based on blood glucose levels (Rystrom 2005).

Insulin Regimens A single dose of insulin is rarely capable of providing optimal glycemic control in T1DM (Kulkarni and Franz 1999). There are three basic types of insulin administration regimens: fixed (conventional or standard therapy), flexible (intensive insulin therapy), and continuous subcutaneous insulin infusion (CSII) (see Box 19.8).

Conventional or standard insulin therapy consists of a constant dose of basal (or background) insulin combined with short- or rapid-acting (or bolus) insulin. This is referred to as a mixed dose and the individual may mix the insulins him/herself or use premixed insulins (for example, 30 units of 70/30 insulin). If more than one injection is used, such as 15 units lispro plus 25 units NPH before breakfast and 10 units lispro and 20 units NPH before the evening meal, this is referred to as a split (mixed) dose. Individuals using conventional therapy must synchronize administration of their insulin and food intake to avoid hypoglycemia. A good understanding of onset, peak, and duration of their insulin dose in relation to their meals and snacks in addition to consistency of food intake is also important (Kulkarni and Franz 1999; Rystrom 2005). Nutrition goals are based on overall diabetes management goals: target glycemic goals and nutrition-related behaviors that affect these goals (Kulkarni and Franz 1999).

Flexible or intensive insulin therapy requires multiple daily injections (MDIs) of bolus insulin before meals in addition to basal insulin once or twice daily. Insulin can be adjusted to correspond to food intake, therefore replicating endogenous insulin secretion in a person without diabetes. This also allows for adjustment of insulin dose in response to hyperglycemia, variable carbohydrate intake, or alteration in usual physical activity (Kulkarni and Franz 1999; Rystrom 2005). Results of the Diabetes Control and Complications Trial indicate intensive insulin therapy (when compared to conventional therapy) delays onset and slows the progression of retinopathy, nephropathy, and neuropathy in patients with T1DM (Diabetes Control and Complications Trial Research Group 1993).

It is still important to integrate the insulin regimen with the patient's lifestyle (American Diabetes Association, Implications, 2002; Kulkarni and Franz 1999). Individuals employing intensive therapy are required to know their basic doses for background and bolus (meal time) insulins. This permits them to fine-tune the short- and rapid-acting insulin dose when they digress from customary meal plans and/or exercise programs (Kulkarni and Franz 1999). Before initiating intensive insulin therapy, the individual needs (Kulkarni and Franz 1999):

- A consistent eating pattern or food plan that has been individualized to his or her lifestyle and kcal requirements
- A basic insulin regimen that covers his or her customary meal plan
- The aptitude to recognize portions and macronutrient content of usual food items and/or the ability to read labels to ascertain carbohydrate content
- The motivation to test blood glucose levels 4–5 times daily with occasional 3 a.m. testing (usually only in T1DM) to determine success of adjustments

This type of therapy may not be appropriate for everyone. Patients should be reminded that out-of-target blood glucose results can be brought about by circumstances (stress, illness, unpredictable insulin absorption, changes in exercise) other than food intake (Kulkarni and Franz 1999).

Continuous subcutaneous insulin infusion (CSII) (or pump therapy) is a form of intensive therapy. Basal rapid- or short-acting insulin is pumped continuously in micro-amounts through a subcutaneous catheter and is received 24 hours a day. Boluses of rapid- or short-acting insulins are given before meals (Kulkarni and Franz 1999).

Syringes and Pens The two conventional methods of insulin administration are by means of syringes and pens (see Figure 19.16). Insulin syringes are disposable (should only be used once), have short, fine beveled needles, and are designed for U-100 insulin. To make the injection easier and reduce tissue damage, needles are lubricated (Rystrom 2005).

BOX 19.6 CLINICAL APPLICATIONS

Risk Factors for and Treatment of Complications of Diabetes Mellitus

Short-Term Complications

Complication	Symptoms[1]	Causes	Treatment
Hyperglycemia	Polyuria, polydipsia, blurred vision, polyphagia, weight loss, fatigue, low energy, delayed healing, irritability	Excess food and/or CHO; large meals or excess snacking	Eat less food and/or CHO; distribute food/CHO appropriately
		Physical inactivity	Gradually increase activity
		Lack of blood glucose monitoring	Regular self-monitoring of blood glucose
		Inadequate diabetes medication	Add, adjust, or change medication(s)
		Inappropriate timing of medications	Coordinate timing of medication(s) and food
		Over-treatment of hypoglycemia	Use appropriate amounts of CHO sources
		Adverse effect of nondiabetes medications medications	Seek information about effect of medication on glucose
		Illness	Know how to manage diabetes during illness
		Variability in insulin absorption	Proper site rotation; proper insulin storage; check expiration date and discard 30 days after opening
		Variability in rates of digestion/absorption of food	Address issues of gastroparesis (delayed stomach emptying)
		Stress	Practice relaxation techniques
Ketoacidosis	Nausea and/or vomiting; stomach pain, fruity (acetone) breath, heavy (or Kussmal) breathing, mental status change	Lack of blood glucose self monitoring	Regular self monitoring; test for ketones if glucose > 250 mg/dL
		Severe illness or infection	Closely monitor effects of illness on blood glucose; increase frequency of glucose measurement; treat illness if indicated; take DM medications even when eating less; maintain hydration; plan for sick-day management
		Insulin omitted	Investigate rationale
		Increased insulin needs with growth spurts	Frequent blood glucose monitoring
		Inappropriately stored insulin	Discard expired insulin; protect insulin from excessive heat or cold
Hyperglycemic Hyperosmolar Syndrome	Polyuria, polydipsia, polyphagia, weight loss; symptoms persist and worsen over several days or hydration status worsens	Dehydration	Monitor fluid intake; establish plan to take fluids regularly
		Excessive fluid losses	Monitor fluid status; address causal factors, replace fluids
		Prolonged hyperglycemia	Monitor blood glucose regularly; treat mild hyperglycemia
Mild Hypoglycemia	Trembling, nervousness, trouble concentrating, anxiety, blurred vision, sweating, irritability, rapid heart rate, inability to think clearly, tingling in extremities, dizziness, hunger, nausea, fatigue, weakness, headache	Excess medication, or inappropriate timing of medications	Adjust amount and/or type of medication; coordinate timing of medications with food and activity
		Overcorrection of hyperglycemia with insulin	Use appropriate amount of insulin for correction
		Too little food and/or CHO	Consume appropriate amount of food and/or CHO
		Missed or delayed meal	Eat meals on time, or eat snack if meal will be late
		Increased activity	Increase food intake or reduce insulin
		Side effects from non-diabetes medication	Seek information about effect of medication on glucose
		Variability in insulin absorption	Proper site rotation; proper insulin storage; check insulin expiration date
		Variability in rates of digestion/absorption of food	Address issues of gastroparesis
		Alcohol consumption	Consume food when drinking alcohol; limit amount of alcohol consumed

(continued on the following page)

BOX 19.6 *(continued)*

Severe hypoglycemia	Mental confusion, argumentative, combative, lethargy, seizures, unconsciousness	Glucose level 51–70 mg/dL Glucose $\leq$ 50 mg/dL	Consume 15 g CHO[2], repeat if glucose does not return to normal range after 15 minutes Consume 20–30 g CHO, repeat if glucose does not return to normal range after 15 minutes

Long-term complications

Complication	Risk Factors	Treatment
Macrovascular Complications		
Cardiovascular disease	T2DM	Reduce LDL-cholesterol to 110 mg/dL ($\downarrow$ foods high in saturated or trans fats) Lower triglycerides to $<$ 150 mg/dL (wt loss, $\uparrow$ consumption of fish and n-3 vegetable sources, fish oil supplements) Increase HDL-cholesterol to $>$ 40–50 mg/dL (weight loss, increased physical activity, smoking cessation)
Macrovascular Complications		
Nephropathy	Hypertension Hyperglycemia Native American, Hispanic American, or African American descent	Optimize glycemic control (preprandial plasma glucose 90–130 mg/dL; postprandial plasma glucose $<$ 180 mg/dL) Aggressive BP control ($<$ 130/80 mm Hg) $\downarrow$ dietary protein
Retinopathy	Duration of DM Hyperglycemia Hypertension	Optimal glycemic control Optimal BP control
Nervous System Disease		
Peripheral neuropathy	DM $\geq$ 10 years Poor glucose control Other DM-related complications	Optimal glycemic control Daily foot care, walking, gentle stretching, relaxation exercises Medication for pain relief (topical capsaicin, antidepressants, and anticonvulsants)
Autonomic neuropathy Cardiovascular (postural hypotension and "silent" heart disease) Genitourinary (sexual dysfunction, bladder emptying problems) Gastroparesis (delayed emptying of stomach)	Longstanding DM	Optimized glycemic control Dietary modifications Therapeutic lifestyle changes small, frequent meals, reduce fat intake, reduce fiber intake, use foods with soft consistency, exercise after meals, adjust insulin doses and timing) Pharmacologic options (cholinergic drugs, dopamine antagonists, motilin-receptor agonists) Surgical treatment (jejunostomy, gastrectomy)

Source: Adapted from: Arnold MS. Hypoglycemia and hyperglycemia. In: Ross TA, Boucher JL, O'Connell BS, editors. *American Dietetic Association guide to diabetes: medical nutrition therapy and education.* Chicago: American Dietetic Association, 2005; Wheeler ML. Long-term complications. In: Ross TA, Boucher JL, O'Connell BS, editors. *American Dietetic Association guide to diabetes: medical nutrition therapy and education.* Chicago: American Dietetic Association, 2005; American Diabetes Association. Standards of medical care in diabetes. *Diabetes Care.* 2006;29(Suppl 1):S4.

[1] All symptoms may not be experienced by all individuals

[2] ½ cup fruit juice or regular (nondiet) soft drink, 3–4 glucose tablets, 3–5 hard candies.

TABLE 19.19

Types of Insulin and Pramlintide

Insulin Type	Brand Name (Manufacturer)	Onset of Action	Peak of Action (hours)	Duration of Action (hours)	Comments
Rapid-Acting Analog (clear)					
Lispro Aspart Glulisine	Humalog (Eli Lilly) NovoLog (Novo Nordisk) Apidra (Aventis)	10–20 min.	1–3	3–5	Can be used in pump therapy
Inhalation Powder	Exubera (Pfizer)	30 min	2–3	5–6	Should not be used by smokers or individuals with lung disease
Short-Acting (clear)					
Regular	Humulin R (Eli Lilly) Novolin R (Novo Nordisk)	30–60 min.	2–4 2.5–5	5–8	Can be mixed with longer-acting insulin
Intermediate-Acting (cloudy)					
NPH	Humulin N (Eli Lilly) Novolin N (Novo Nordisk)	1–3 hours	8	20	Usually given in 2 daily doses
Lente	Novolin L (Novo Nordisk)	1–2.5 hours	7–15	18–24	
Extended Long-Acting Analogue (clear)					
Insulin glargine Insulin detemir	Lantus (Aventis) Levemir (Novo Nordisk)	1 hour	None	24	Cannot be mixed with other insulins
Premixed (cloudy)					
70/30	Mixtard (Novo Nordisk) Humulin 70/30 (Eli Lilly)	30–60 min	Dual	10–16	70% NPH, 30% regular
50/50	Humulin 50/50 (Eli Lilly)	30–60 min	Dual	10–16	50% NPH, 50% regular
60/40	Mixtard 40 (Novo Nordisk)	30 min	2–8	24	60% NPH, 40% regular
Antihyperglycemic Drug (clear)					
Pramlintide (synthetic analog amylin)	Symlin (Amylin Pharmaceuticals)	Slows transit of digesting food through intestine; given at mealtimes to increase efficacy of insulin; should not be mixed with insulin			

Sources: Eli Lilly and Company. http://www.lilly.com; Novo Nordisk. http://www.novonordisk.com; Aventis Pharmaceuticals Inc. (Sanofi Aventis). https://www.lantus.com/consumer/index.do; Pfizer Inc. http://www.pfizer.com/pfizer/main.jsp; Amylin Pharmaceuticals, Inc., http://www.symlin.com/Ind; Rystrom JK. Insulin therapy. In: Ross TA, Boucher JL, O'Connell BS, editors. *American Dietetic Association guide to diabetes: medical nutrition therapy and education.* Chicago: American Dietetic Association; 2005.

BOX 19.7 CLINICAL APPLICATIONS

Algorithms Used to Determine Insulin Dose

Diabetes Type	Daily Insulin Dose
T1DM	0.6 units/kg actual body weight
T1DM with trace to small ketones or T2DM with BMI ≤ 27	0.3–0.5 units/kg actual body weight
T1DM with moderate to large ketones or T2DM with BMI ≥ 27	0.5–0.7 units/kg actual body weight
T2DM with oral hypoglycemic medications	0.1–0.3 units/kg ideal body weight

Source: Adapted from Rystrom JK. Insulin therapy. In: Ross TA, Boucher JL, O'Connell BS, editors. *American Dietetic Association guide to diabetes: medical nutrition therapy and education.* Chicago: American Dietetic Association; 2005.

Insulin pens look a lot like large marking pens. Reusable (with prefilled cartridges) or disposable pens are available and come filled with either 150 units or 300 units of insulin. Cartridges and prefilled pens are available for rapid-acting, regular, and extended long-acting insulins, some premixed insulins, and glargine. Needles in pens are used once, discarded, and then replaced for each injection (Rystrom 2005).

Insulin Pumps Insulin pumps (see Figure 19.16) are approximately the size of pagers, and are powered by batteries. Regular or rapid-acting (aspart, lispro, glulisine, and apridra) insulin is delivered through flexible tubing and is attached to the individual via an infusion set. Continuous subcutaneous insulin infusion (CSII) allows creation of variable and adjustable insulin dosing to meet specific, individual insulin needs (Rystrom 2005). Pump therapy duplicates endogenous insulin secretion more closely than other methods of insulin delivery. Detailed instructions and training are necessary, and mastering this method requires time and effort.

Inhaled Insulin The first non-injectable insulin option in the U.S. since the introduction of insulin over 80 years ago was approved for use by the FDA in January 2006. Exubera® (insulin human [rDNA origin]) Inhalation Powder is a rapid-acting, dry powder insulin inhaled through the mouth into the lungs at mealtimes (10 minutes before meals). It is approved for use in adults with T1DM and T2DM. Individuals with T2DM may use Exubera® alone, or in combination with oral hypoglycemic medications or longer-acting insulins (Pfizer 2006). Exubera® is as effective as short-acting insulin, significantly improves glycemic control when added to oral hypoglycemic medications (Hollander et al.

BOX 19.8 CLINICAL APPLICATIONS

Insulin Regimens

Conventional therapies:	Short- or rapid-acting insulin mixed with intermediate-acting insulins[1] given before breakfast and before evening meal.
or	
	Combination of short- and intermediate-acting insulins before breakfast, short-acting insulin before evening meals, and intermediate-acting insulin at bedtime. Used to control **dawn phenomenon**.
Intensive insulin therapy (multiple daily injections [MDIs]):	Intermediate insulin given once or twice daily and rapid- or short-acting insulin is given prior to meals. Allows more flexibility in type and timing of meals. Amount of rapid- or short-acting insulin can be adjusted based on meal composition and/or its carbohydrate content.
Continuous subcutaneous insulin infusion (CSII):	Provides basal rapid- or short-acting insulin pumped continuously in micro-amounts through a subcutaneous catheter and is monitored 24 hours a day. Boluses of rapid- or short-acting insulins are given before meals.

Source: Adapted from: Kulkarni K, Franz MJ. Nutrition therapy for type 1 diabetes. In: Franz MJ, Bantle JP, editors. *American Diabetes Association guide to medical nutrition therapy for diabetes.* Alexandria (VA): American Diabetes Association; 1999.

[1]See Table 19.20 for descriptions of insulin types.

2004), and is preferred over subcutaneous insulin (Rosenstock et al. 2004). Inhaled insulin is not recommended for smokers.

Side Effects and Complications of Insulin Therapy Although individuals with T1DM would not survive without exogenous insulin treatment, it is not without risks. The appropriate insulin dose and regimen necessary to avoid hyperglycemia and development of resulting complications may produce side effects. This is especially true of intensive

dawn phenomenon—an increase in blood glucose in the early morning, most likely due to increased glucose production in the liver after an overnight fast

FIGURE 19.16 Insulin Syringes, Insulin Pens, and Insulin Pumps

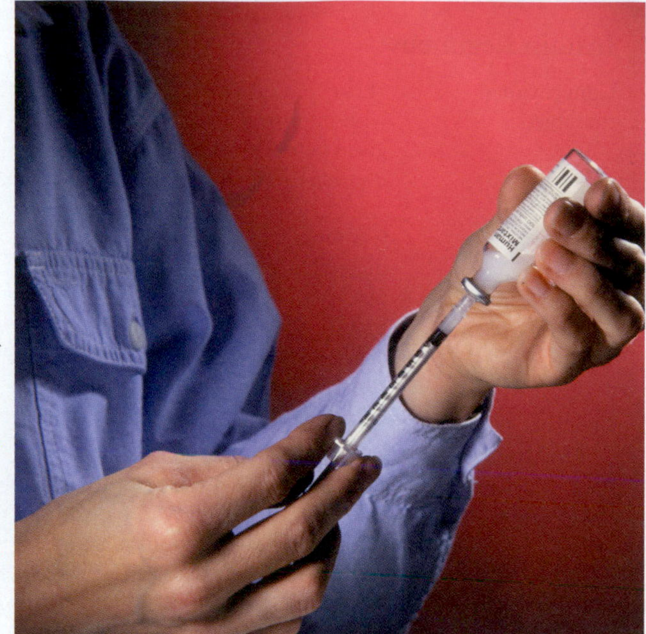

(a) insulin syringe

Source: Saturn Stills/ Photo Researchers, Inc.

(b) disposable insulin pen

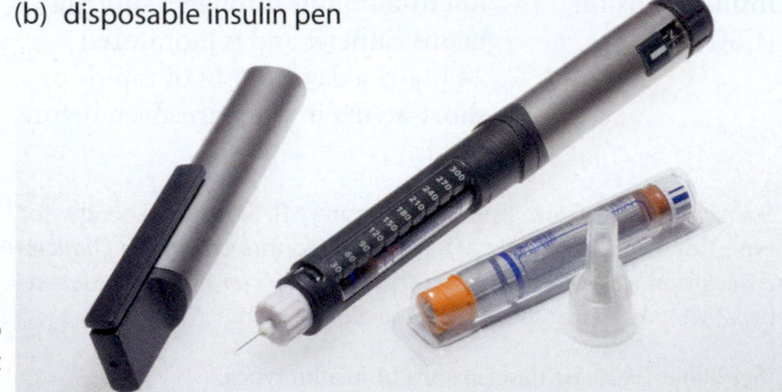

Copyright © Creatas/Fotosearch

(c) refillable insulin pen

Copyright © Creatas/Fotosearch

(d) insulin pump

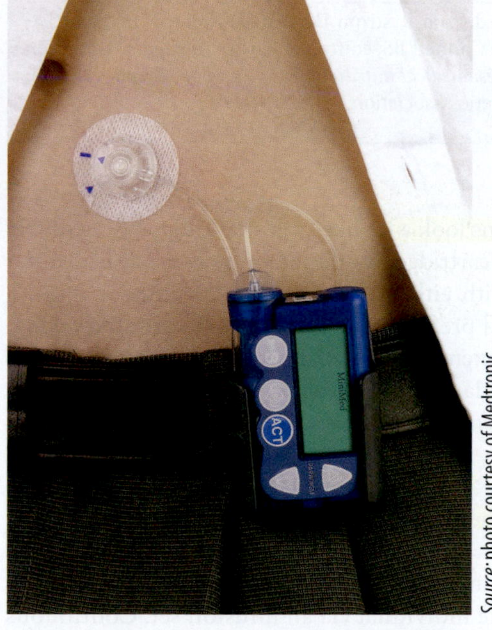

Source: photo courtesy of Medtronic

insulin therapy (Diabetes Control and Complications Trial Research Group 2003).

The most universal side effect of insulin is hypoglycemia (blood glucose level <70 mg/dL). Thus, it is imperative that individuals taking insulin be educated on the subject of signs, symptoms, and remedies for hypoglycemia. Each individual may experience hypoglycemia at varying blood glucose levels, and while symptoms are individual, the most common are weakness, shakiness, perspiration, hunger, and rapid heart beat (Arnold 2005).

Mild hypoglycemia can be self-treated by following several steps (Arnold 2005):

1. If blood glucose level is <70 mg/dL, the individual should consume 10 to 15 grams of any carbohydrate that contains fast acting carbohydrate (for example: ½ cup fruit juice, ½ cup regular soft drink, 3–4 glucose tablets). If blood glucose is <50 mg/dL, 20 to 30 grams of carbohydrates should be consumed.
2. Fifteen minutes after treatment, blood glucose levels should be rechecked to ascertain whether blood glucose levels have been restored to the normal range. If not, the process outlined in #1 should be repeated.
3. The individual should determine whether additional snacks are required. If blood glucose normalizes, but the individual will not eat in less than an hour, has recently exercised, or is going to bed, an additional snack may be needed. Blood glucose should be tested and treated as appropriate.

Severe hypoglycemia is defined as hypoglycemia that cannot be self-treated (Cryer et al. 2003). With severe hypoglycemia, an individual can lose consciousness or become so confused that he or she is unable to think clearly. If conscious, an individual may seem lethargic or belligerent. Conscious or unconscious, the individual may not be grateful for help, but he or she still needs it. A trained family member, friend, or emergency medical professional must inject glucagon or glucose to restore blood glucose levels to normal. All persons at risk for severe hypoglycemia should have a glucagon emergency kit available at all times (Arnold 2005).

Weight gain from improved glycemic control is another common side effect. Instead of losing glucose through urine, the body actually utilizes glucose or stores it as glycogen or fat. Injection of insulin itself may produce complications. Subcutaneous tissue where insulin is injected may **lipoatrophy** and/or suffer **lipohypertrophy** (Rystrom 2005) if the injection site is not rotated. In addition to the side effects of insulin itself, there are several drugs that alter the effect of insulin. These drugs are listed in Table 19.20.

Physical Activity For most individuals with diabetes, benefits of physical activity far exceed the risks (Hayes 2005). These benefits include (American Diabetes Association, Physical Activity, 2004; Hayes 2005):

- Improved glycemic control (A1C)
- Improved blood lipids and blood pressure, with subsequent lower cardiovascular risks and overall mortality
- Positive impact on metabolic abnormalities characteristic of T2DM
- Prevention or delay of onset of T2DM for individuals at high risk for developing diabetes or with **pre-diabetes mellitus**
- Reduced risk of development of cardiovascular disease, since physical inactivity and diabetes are independent risk factors for it
- Improved coping and stress management and reduced feelings of depression
- Improved physical fitness and functional capacity
- Enhanced quality of life

Both hypoglycemia and hyperglycemia are acute risks of exercise. Hypoglycemia can occur during exercise that lasts longer than 1 hour, and for up to 24 hours after unusually strenuous, prolonged, and/or sporadic exercise. Blood glucose levels should be monitored, and carbohydrates should be increased and/or insulin adjustments should be made (Kulkarni and Franz 1999).

In those individuals whose diabetes is poorly controlled (underinsulinized), exercise can cause hyperglycemia. When insulin is deficient, the rise in counter-regulatory hormones that takes place during exercise causes an increase in hepatic glucose production and free fatty acids. Cellular uptake of glucose is minimal, resulting in both hyperglycemia and increased production of ketones (Kulkarni and Franz 1999; American Diabetes Association, Standards, 2006).

Blood glucose levels should be monitored both before and after exercise to understand how diabetes affects glycemic control and to determine appropriate insulin and carbohydrate adjustments. For moderate exercise lasting less than 30 minutes, additional carbohydrate or insulin adjustment is rarely necessary. On the other hand, if blood glucose levels are <80 mg/dL before exercise, a small snack is needed. As a rule, an additional 15 grams of carbohydrate should be adequate for one hour of moderate physical activity. For more strenuous exercise, 30 grams of carbohydrate per hour may be required. For exercise before breakfast or later in the afternoon, extra carbohydrate should be consumed before the exercise. For exercise after meals, the additional carbohydrate can be taken after exercise (Kulkarni and Franz 1999).

It may be necessary to adjust insulin dosage before exercise to avoid hypoglycemia, especially if exercise lasts for 45 to 60 minutes. For most, a modest decrease (~20%) in the insulin component corresponding to the period of exercise is a good place to start. More prolonged or vigorous exercise may necessitate a larger reduction in initial insulin dosage (by as much as one-third to one-half) to avoid hypoglycemia. Regular exercise (at least every other day) usually does not require adjustments to insulin dosage. Since the bodies of patients who exercise regularly will have adjusted to this activity level, the total insulin doses prescribed will already be lower (Kulkarni and Franz 1999).

Nutrition Therapy There is no one "diabetic diet" or "ADA diet." Even though the term "ADA diet" has never been clearly defined, in the past it usually meant a physician-determined kcal level with explicit percentages of carbohydrate, protein, and fat based on the exchange lists. The American Dietetic Association (ADA) recommends the term "ADA diet" not be used since the ADA no longer sanctions any single meal plan or specified percentages of nutrients (American Dietetic Association, Translation, 2002).

Nutrition therapy is an essential element of glycemic control and diabetes self-management education (DSME).

lipoatrophy—an immune response related to source and purity of insulin resulting in thinning of subcutaneous fat at the injection site, which causes concaving or pitting of fatty tissue

lipohypertrophy—thickening of subcutaneous fat at an insulin injection sit

pre-diabetes mellitus—blood glucose levels that are higher than normal but not yet high enough to be diagnosed as diabetes

TABLE 19.20

Drugs that Alter the Effect of Insulin

Drugs that *decrease* hypoglycemic effect and increase blood glucose	Drugs that *increase* hypoglycemic effect and decrease blood glucose
Acetazolamide	ACE inhibitors
AIDS antivirals	Alcohol
Albuterol	Anabolic steroids
Asparaginase	β-blockers[1]
Calcitonin	Calcium
Corticosteroids	Chloroquine
Cyclophosphamide	Clofibrate
Danazol	Clonidine
Dextrothyroxine	Disopyramide
Diazoxide	Fluoxetine
Diltiazem	Guanethidine
Diuretics	Lithium carbonate
Dobutamine	MAO inhibitors
Epinephrine	Mebendazole
Estrogens	Oral antidiabetic product
Ethacrynic acid	Pentamidine[2]
Isoniazid	Phenylbutazone
Lithium carbonate	Propoxyphene
Morphine sulfate	Pyridoxine
Niacin	Salicylates
Nicotine	Somatostatin analog (e.g., octreotide)
Oral contraceptives	Sulfinpyrazone
Phenothiazines	Sulfonamides
Phenothiazines	Tetracyclines
Phenytoin	
Somatropin	
Terbutaline	
Thiazide diuretics	
Thyroid hormones	

[1] May delay recovery from hypoglycemia or mask signs and symptoms.

[2] May sometimes be followed by hyperglycemia.

Source: Adapted from: Mulcahy K, Lumber T. *The diabetes ready reference for health professionals.* 2nd ed. Alexandria (VA): American Diabetes Association; 2004. Table II, p. 24.

Individualized nutrition therapy is required to achieve treatment goals (American Diabetes Association, Standards, 2006). A comprehensive nutrition assessment, a self-care treatment plan, and the client's health status, learning ability, readiness to change, and current lifestyle should be the basis for nutrition therapy and DSME. Every meal planning method has advantages and drawbacks. Consequently, tailoring the meal planning approach to each individual's needs is key (Pastors, Waslaski, and Gunderson 2005). Individuals receiving conventional insulin therapy must be consistent with timing of their meals and amounts of food consumed. Those receiving intensive insulin therapy have more flexibility in when and what they eat (Kulkarni and Franz 1999).

Nutrition recommendations for total fat, saturated fat, cholesterol, fiber, vitamins, and minerals are the same for individuals with diabetes as for the general population. Protein intake can range from 15% to 20% of daily kcalories from animal and vegetable protein sources. If the patient has nephropathy, lower intakes of protein (about 10% of daily energy intake) may be warranted. Carbohydrate recommendations are individualized based on the individual's eating habits, blood glucose goals, and lipid goals, but at least 130 grams/day are recommended. Blood glucose control is not impaired by use of sucrose in the meal plan, but sucrose-containing foods should be substituted for other carbohydrates and foods, and should not be eaten in addition to a meal plan. Blood glucose levels are not affected by moderate alcohol use if diabetes is well controlled. Alcohol kcal should be considered an addition to regular food or meals, and no food should be omitted (Franz et al. 2002).

Goals of Nutrition Therapy Four main goals of nutrition therapy are as follows (Franz et al. 2002; American Diabetes Association, Nutrition, 2004; American Diabetes Association, Standards, 2006):

- Attain and maintain optimal metabolic outcomes, including:
  - Glucose level in normal range, or as close to normal range as is safely possible, to prevent or reduce risk of complications
  - Lipid or lipoprotein profile that reduces risk for macrovascular disease
  - Blood pressure levels that reduce risk for vascular disease
- Prevent and treat chronic complications. Modify nutrient intake and lifestyle as appropriate for prevention and treatment of obesity, dyslipidemia, cardiovascular disease, hypertension, and nephropathy.
- Enhance health using healthy food choices and physical activity.
- Address individual nutritional needs with regard to personal and cultural preferences and lifestyles while respecting the individual's wishes and willingness to change.

Meal Planning No one meal planning method has been scientifically validated or shown to be superior to any other (Kulkarni and Franz 1999). Regardless of which meal planning method is used, it should be individualized based on customary food intake and the patient's contribution, which will permit diet modification to be put into practice with less difficulty. Meals and snacks should be distributed in a manner that is consistent with the individual's way of life, activity patterns, and diabetes medications (Pastors, Waslaski, and Gunderson 2005). Meal planning approaches range from simple guidelines to more complex counting methods (Nutrition Care Manual 2006). The approach selected should be the one that best fits the individual patient's situation. Carbohydrate counting and the exchange lists are described in the next section; other meal planning approaches are outlined in Appendix E4.

Carbohydrate Counting Of the four meal planning approaches used in the Diabetes Control and Complications Trial (DCCT), the most successful was carbohydrate counting (Anderson et al. 1993). The basic concept of carbohydrate counting is that the carbohydrate found in foods is the major macronutrient influencing postprandial glucose variations (Nuttall 1993), and that it influences pre-meal insulin requirements more than the protein and fat content of the meal (Gillespie, Kulkarni, and Daly 1998). The total amount of daily carbohydrate intake, not its source, is the focus of this meal planning approach (Kulkarni and Franz 1999).

Emphasis on eating consistent amounts of carbohydrate at meals and snacks can make carbohydrate counting a simpler method of meal planning (Kulkarni and Franz 1999). Food carbohydrate sources are starches, fruits, milk/yogurt, and sweets. (Non-starchy vegetables do not need to be counted unless eaten in servings containing >15 g of carbohydrates.) Carbohydrates can be counted in one of the following two ways (Kulkarni and Franz 1999):

- The amount of food containing 15 g carbohydrate counts as one carbohydrate choice.
- Total grams of carbohydrate in a meal or snack can be counted.

This is not to say that meats and fats can be totally ignored. The kcal and fat content of meats and fats can contribute to weight gain and/or lipid abnormalities (Kulkarni and Franz 1999). By allowing individuals to make appropriate adjustments in their diabetes management, carbohydrate counting empowers them to learn relationships between food, insulin, physical activity, and blood glucose levels (Gillespie, Kulkarni, and Daly 1998). Box 19.9 outlines the three levels of carbohydrate counting (including matching carbohydrate content to insulin doses).

Exchange System Since 1950, a widely used method for planning food intake has been the Exchange List for Meal Planning (Appendix E1) (American Diabetes Association, Exchange, 2003). It provides uniformity in meal planning and allows a wide variety of foods to be included in the diet. This method uses the concept of "exchange" or substitution of different foods within each of three groups: carbohydrate (starch, fruit, milk, non-starchy vegetables, and other carbohydrates), meat and meat substitutes (very lean meats, lean meats, medium-fat meats, high-fat meats), and fats (monounsaturated, polyunsaturated, saturated) (Kulkarni and Franz 1999; Pastors, Waslaski, and Gunderson 2005). Each food portion on a particular list can be exchanged with any other food portion on the same list (Nutrition Care Manual 2006).

The following guidelines can be used to calculate a meal plan using exchanges (American Dietetic Association 2000):

1. Assess current food intake and eating pattern using diet history or food records.
2. Categorize usual food intake into exchange amounts based on portions and foods consumed at each meal and snack. Calculate total grams of carbohydrate, protein, and fat and translate into energy. Refer to Chapter 5 for guidelines for calculation and Appendix E1 for the Exchange Lists.
3. Determine appropriate energy prescription. Subtract energy if weight loss is desired; add energy if weight gain is desired. Generally 250 to 500 kcal per day can be subtracted/added for a ½ to 1 pound per week weight loss/gain.
4. Translate energy prescription into exchanges, staying as close to current pattern of intake (from diet history) as possible. Calculate grams of carbohydrate, protein, and fat from exchanges and determine percentages of energy contributed by each macronutrient:
    - % carbohydrate = grams carbohydrate × 4 / total kcal
    - % protein = grams protein × 4 / total kcal
    - % fat = grams fat × 9 / total kcal
5. Adjust exchanges as needed to reach goal percentages for each macronutrient.
6. Compare usual intake to energy prescription and mutually determine how to distribute exchange groups among meals and snacks.

While there are many advantages to using the exchange lists, they are often not the most appropriate meal planning system for individuals with diabetes. The exchange lists are written at a ninth to tenth grade reading level; therefore, individuals using them must be able to read at this level and understand the concept of "exchanging" foods. It may require several educational sessions and practice for them to be used effectively (Nutrition Care Manual 2006).

Short-Term Illness Everyday maladies like colds, fever, nausea, vomiting, and diarrhea can cause havoc with glycemic control for individuals with diabetes. Left untreated, diabetic ketoacidosis (DKA) or HHNS can develop. Treatment includes supplemental insulin; replacement fluids, electrolytes, and glucose; blood glucose monitoring; and urine testing for ketones. Sometimes, medical intervention is necessary (Nutrition Care Manual 2006).

BOX 19.9 CLINICAL APPLICATIONS

Three Levels of Carbohydrate Counting

Level 1: Basic Carbohydrate Counting Skills

- Knowing.
  - Carbohydrate sources.
  - How to count grams of carbohydrate in foods.
  - Relationship between portion size and carbohydrate content.
- Usual carbohydrate intake recorded.
  - Shared with RD.
  - Target amounts of carbohydrates for meals and snacks determined.

Level 2: Intermediate Carbohydrate Counting Skills

- Pattern management.
  - Identify blood glucose patterns impacted by food, insulin, and physical activity
  - Identify and interpret patterns to make adjustments in diabetes regimens
- Rapid- or short-acting insulins matched to carbohydrate content of usual meals.
- Insulin doses adjusted when deviations from usual carbohydrate content are made.
  - For every 15–20 g CHO added or subtracted from a meal, 1–2 units rapid- or short-acting insulin suggested.
  - Each person's requirements should be individualized.

Level 3: Advanced Carbohydrate Counting Skills

- Used by individuals on intensive insulin therapy.
- Insulin adjusted on basis of ratio of grams of carbohydrate intake to doses of rapid- or short-acting insulin.
- RD calculates carbohydrate-to-insulin ratio for each meal.
  - Uses food, insulin, blood glucose monitoring records.
  - Ratios may vary from meal to meal, from workdays to weekend days, from exercise days to non-exercise days, and they may change over time.
  - Periodic reevaluation is required.
- Calculation of carbohydrate-to-insulin ratios
  - Grams of CHO eaten at a meal divided by number of units of rapid- or short-acting insulin necessary to meet blood glucose goals. For example:
    - 45 g CHO (3 CHO choices) at a meal and requires 5 units insulin.
    - Ratio of 1 U insulin to 9 g CHO or 2 U insulin for every 1 carbohydrate choice.
  - There may be a need for food-specific insulin doses.
  - Large amounts of meat and/or fat at a meal may require adjustment of insulin administration after the meal instead of before the meal.
  - Grams of fiber may be subtracted from total carbohydrate content of a food if it contains ≥5 g fiber per serving, since fiber is not considered an available source of glucose.

Source: Adapted from: Kulkarni K, Franz MJ. Nutrition therapy for type 1 diabetes. In: Franz MJ, Bantle JP, editors. *American Diabetes Association guide to medical nutrition therapy for diabetes.* Alexandria (VA): American Diabetes Association; 1999; Ahren JA, Gancomb PM, Held NA, Pettit WA, Tamborlane WV. Exaggerated hyperglycemia after a pizza meal in well-controlled diabetes. *Diabetes Care.* 1993;16:578–580; Vlachokosta FV, Poper CM, Gleason R, Kinzel L, Kahn CR. Dietary carbohydrate, a Big Mac, and insulin requirement in type 1 diabetes. *Diabetes Care.* 1988;11:330–336.

Individuals with diabetes should be provided with a list of carbohydrate-containing foods that are tolerated during acute illness and are easy to digest. Furthermore, for illnesses lasting less than 24 hours, the following guidelines are recommended (Nutrition Care Manual 2006):

- Take usual insulin doses during acute illness. Insulin is still necessary, and insulin needs may even increase, because fever, infection, or stress can trigger release of counter-regulatory hormones.
- Monitor blood glucose and test for ketones at least 4 times a day: before each meal and at bedtime. Additional insulin is needed if blood glucose >240 mg/dL and moderate to large ketones are present.
- Drink a large glass of liquid (e.g., water, tea, broth) every hour. Small sips of 1 to 2 tablespoons every 15 to 30 minutes should be consumed if nausea or vomiting is present. A primary care provider should be notified if vomiting persists longer than 6 hours.
- If regular foods are not tolerated, replace meals with small amounts of liquid or soft carbohydrate-containing foods eaten every 3 to 4 hours. Consume 3 to 4 carbohydrate servings (45–60 grams of carbohydrate) of foods such as:
  - Regular soft drinks (do not use sugar-free soft drinks)
  - Soup
  - Juices
  - Jell-O®
  - Ice cream
- Contact a primary care provider if illness continues >24 hours or if unable to eat regular foods for >24 hours.

- Call a physician if any of these symptoms develop, especially in children. They are signs of developing DKA:
  - Moderate to large ketones in urine with elevated blood glucose levels
  - Fruity-smelling breath
  - Severe nausea
  - Vomiting
  - Diarrhea
  - Abdominal pain
  - Rapid breathing

Type 2 Diabetes Mellitus

Type 2 diabetes was previously called non-insulin diabetes mellitus (NIDDM) or adult-onset diabetes, but these terms perform a disservice to individuals with diabetes, because they classify them by treatment modality rather than disease characteristics (Baynes and Betteridge 2004).

Epidemiology In the U.S. and worldwide, about 90% to 95% of all diagnosed cases of diabetes are T2DM. T2DM occurs most frequently in adults, but is being diagnosed with increasing frequency in children and adolescents as well (World Health Organization 2002; Centers for Disease Control 2005; NDIC 2005; American Diabetes Association, Diagnosis, 2006). Box 19.10 describes traits associated with increased risk for T2DM. Gender distribution of T2DM is equal, but prevalence increases with age (Kaplan and Conway 2004).

T2DM is not an equal-opportunity disease. The elderly and persons of color are disproportionately affected. Prevalence of T2DM in minority populations is shown in Figure 19.17. Prevalence of diabetes in non-Hispanic whites 20 years or older is 8.7% of this population. The prevalence of diabetes for other groups of individuals of similar age is as follows (Centers for Disease Control 2005; NDIC 2005; American Diabetes Association 2006):

- American Indians and Alaska Natives are 2.2 times as likely as non-Hispanic whites to have diabetes.
- Prevalence among Alaska Natives is 8.1%.
- 26.7% among American Indians in the southern United States
- 27.6% in southern Arizona Indians.
- Native Hawaiians and other Pacific Islanders are 2 times as likely to have diabetes.
- Non-Hispanic blacks are 1.8 times as likely to have diabetes.
- Mexican-Americans are 1.7 times as likely to have diabetes.

Etiology In some, heredity may be a factor in development of T2DM. A clear autosomal dominant pattern of inheritance (maturity onset diabetes of the young [MODY]) has been found in those who develop T2DM before the age of 25 years. (Box 19.11 outlines criteria for testing for T2DM in children.) Identifiable gene defects have been found in most families with this pattern of diabetes. Obesity—body fat distribution in particular—also appears to play a role in development of T2DM. Central body adiposity seems to increase the degree of insulin resistance, but the mechanism is unclear (Kaplan and Conway 2004).

Physical inactivity increases risk of T2DM unrelated to body weight. Exercise seems to reduce risk of T2DM by enhancing whole-body insulin sensitivity. High birth weight also appears to increase the risk for T2DM during adulthood. Low-birth weight infants tend to develop T2DM later in life when compared to higher-birth weight babies. Poor placental growth (or food insecurity for the mother during pregnancy) can bring about poor fetal nutrition, producing defective pancreatic organogenesis in utero. Later in life, obesity-related insulin resistance results in increased stress on the pancreas, causing essential insulin deficiency which ultimately leads to diabetes (Kaplan and Conway 2004).

BOX 19.10 CLINICAL APPLICATIONS

Risk Characteristics for Type 2 Diabetes

- Older age
- Obesity
- Family history of diabetes
- History of gestational diabetes
- Impaired glucose metabolism
- Physical inactivity
- Race/ethnicity such as African-Americans, Hispanic/Latino Americans, American Indians, some Asian Americans and Native Hawaiians, or other Pacific Islanders

Sources: American Diabetes Association. National Diabetes Fact Sheet, 2005. Available from: http://www.diabetes.org/diabetes-statistics.jsp. Accessed February 10, 2006.
Centers for Disease Control and Prevention. National diabetes fact sheet: general information and national estimates on diabetes in the United States, 2005. Atlanta, GA: U.S. Department of Health and Human Services. Centers for Disease Control and Prevention, 2005; National Diabetes Information Clearinghouse (NDIC). National diabetes statistics. NIH Publication #06–3892, November 2005. Available from: http://diabetes.niddk.nih.gov. Accessed February 11, 2006.

Pathophysiology Whereas T1DM results from lack of insulin caused by destruction of β-cells, individuals with T2DM produce insulin, but their tissues are insulin resistant. This causes increased need for insulin, so the pancreas increases production. Eventually the pancreas loses its ability to produce insulin (Kaplan and Conway 2004; Centers for

FIGURE 19.17 Prevalence of Diabetes by Race/Ethnicity

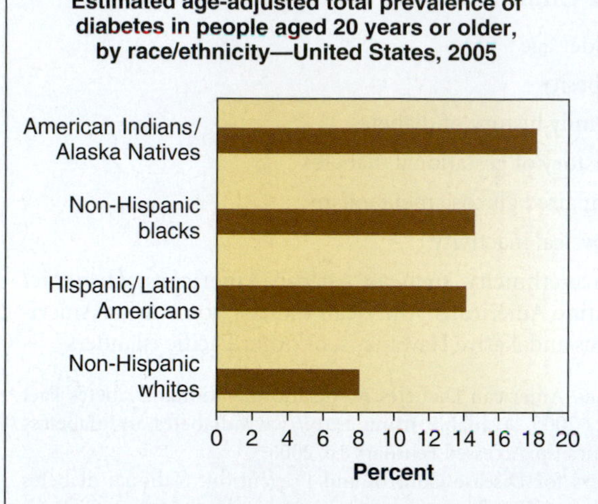

Source: For American Indians/Alaska Natives, the estimate of total prevalence was calculated using the estimate of diagnosed diabetes from the 2003 outpatient database of the Indian Health Service and the estimate of undiagnosed diabetes from the 1999–2002 National Health and Nutrition Examination Survey. For the other groups, 1999–2002 NHANES estimates of total prevalence (both diagnosed and undiagnosed) were projected to year 2005.

BOX 19.11 CLINICAL APPLICATIONS

Testing for T2DM in Children

Criteria:	Overweight (BMI >85th percentile for age and gender, or weight for height >85th percentile, or weight >120% of ideal for height)
Plus any 2 of the following risk factors:	Family history of T2DM in first- or second-degree relative Race/ethnicity (Native American, African-American, Latino, Asian American, Pacific Islander) Signs of insulin resistance or conditions associated with insulin resistance (**acanthosis nigricans**, hypertension, dyslipidemia, or polycystic ovarian syndrome
Age of initiation:	10 years or at onset of puberty, if puberty occurs at a younger age
Frequency:	Every 2 years
Test preferred:	Fasting plasma glucose

Source: Adapted from: American Diabetes Association. Standards of medical care in diabetes. *Diabetes Care.* 2006;29(Suppl 1):S4.

Disease Control 2005; Copstead and Banasik 2005; NDIC 2005; American Dietetic Association, Diagnosis, 2006). Consequently, two metabolic defects are observed in individuals with T2DM: insulin resistance and relative insulin deficiency. Although insulin resistance develops many years before onset of diabetes in individuals with predisposition to T2DM, clinical onset is correlated with the diminishing pancreatic release of insulin (Kaplan and Conway 2004).

T2DM is typified by peripheral insulin resistance with an insulin secretory defect that varies in severity. Insulin resistance is caused by a cell-receptor defect resulting in the body's inability to use insulin. As a result, glucose is unable to be absorbed into cells for fuel. Defective insulin secretory response results in excess production of glucose from the liver (Mulcahy and Lumber 2004). For T2DM to manifest, both defects must be present. Every overweight individual has insulin resistance, but diabetes develops only in those lacking the capacity to step up β-cell production of insulin. At first, postprandial glucose levels rise; subsequently, hepatic gluconeogenesis steps up, resulting in fasting hyperglycemia (Votey and Peters 2005).

acanthrosis nigricans—diffuse, velvety-thickening hyperpigmentation of the skin; it may be present at the nape of the neck, axillae, area beneath the breasts, intertriginous areas, and exposed areas (elbows, knuckles); thought to be the result of insulin resistance

Risk of each of the microvascular and macrovascular complications (Table 19.15) of T2DM, including mortality is strongly associated with hyperglycemia. The UK Prospective Diabetes Study (UKPDS) demonstrated a 37% decreased risk for macrovascular complications and a 21% decreased risk of death related to diabetes for each 1% reduction in hemoglobin A1C. The lowest risk appears to be associated with A1C levels in the normal range of <6% (Stratton et al. 2000).

Metabolic Syndrome Another condition related to insulin resistance is metabolic syndrome, which shares some characteristics of T2DM (O'Connell 2005). Pathogenesis of this syndrome and each of its components is complicated and not well understood. Central obesity and insulin resistance are significant contributing features (International Diabetes Federation 2005) along with atherosclerotic risk factors including dyslipidemia and hypertension (Scott 2003). Metabolic syndrome places individuals at increased risk for coronary artery disease. Treatment of metabolic syndrome is multifaceted and includes diet, exercise, and pharmacologic therapy including statins, fibrates, angiotensin-converting enzyme (ACE) inhibitors, and thiazolidinediones (Scott 2003).

Prevalence of the metabolic syndrome in the U.S. varies depending upon the criteria used to define it; prevalence is approximately 35% based on the National Cholesterol Education Program (NCEP) definition. Using a newer definition

BOX 19.12 CLINICAL APPLICATIONS

Criteria for Testing for Diabetes in Asymptomatic Adults

	BMI	Other Criteria
Individuals ≥ 45 years of age	≥ 25[1] kg/m^2	If normal, test every 3 years
Individuals ≤ 45 years of age	≥ 25[2] kg/m^2	Habitually physically inactive
		Have first-degree relative w/ diabetes
		Are members of high-risk ethnic populations (Figure 19.17)
		Delivered a baby weighing >9 lbs or been diagnosed with GDM[3]
		Hypertensive (≥140/90 mmHg)
		Have HDL-cholesterol , <35 mg/dL (0.90 mmol/L) and/or triglyceride >250 mg/dL (2.82 mmol/L)
		Have polycystic ovarian syndrome (see Box 19.13)
		Had IGT or IFG on previous testing (see section on impaired glucose tolerance and impaired fasting glucose)
		Have other clinical conditions associated with insulin resistance (acanthrosis nigricans)
		History of vascular disease

Source: Adapted from: American Diabetes Association. Standards of medical care in diabetes. *Diabetes Care.* 2006;29(Suppl 1):S4.

[1]May not be correct for all ethnic groups.

[2]May not be correct for all ethnic groups.

[3]Gestation diabetes mellitus.

from the International Diabetes Federation (IDF), almost 40% of the U.S. population would be classified as having the metabolic syndrome (Ford 2005). Diagnostic criteria are outlined in Chapter 15.

Clinical Manifestations While onset of T1DM is sudden, onset of T2DM is insidious. Many individuals will be asymptomatic for as long as 6–10 years but present with complications associated with diabetes (World Health Organization 2002; Kaplan and Conway 2004). For example, an optician may detect retinopathy during a visit provoked by blurred vision (Kaplan and Conway 2004). The estimate that as many as one-third of all individuals with T2DM are undiagnosed reinforces the need for screening of individuals at high risk (American Diabetes Association 2006). Criteria for testing and screening for diabetes in asymptomatic, undiagnosed adults are listed in Box 19.12, and one of these criteria—polycystic ovarian syndrome—is discussed further in Box 19.13.

Hyperglycemic Hyperosmolar Nonketotic Syndrome (HHNS) HHNS is characterized by blood glucose levels >600 mg/dL, serum osmolality >320 mOsm/kg of water, and absence of significant ketoacidosis. Infection and dehydration are precipitating factors of HHNS and it occurs most often in individuals with T2DM (Umpierrez, Murphy, and Kitabchi 2002; American Diabetes Association, Hyperglycemia, 2004; Arnold 2005).

Symptoms of HHNS are comparable to moderate hyperglycemia: polyuria, polydipsia, polyphagia, and weight loss. These symptoms develop gradually, and for that reason are less conspicuous and more easily overlooked than DKA. Without judicious monitoring, elderly individuals with T2DM who are unable or unwilling to self-hydrate may slide into a spiraling process of gradual but steady fluid losses and rising blood glucose levels leading to severe dehydration (Umpierrez, Murphy, and Kitabchi 2002; American Diabetes Association, Hyperglycemia, 2004; Arnold 2005).

Treatment entails hospitalization for slow rehydration as well as treatment for complications and underlying medical problems (e.g., infection). Insulin may or may not be required to adequately reduce hyperglycemia (Umpierrez, Murphy, and Kitabchi 2002; Arnold 2005). Mortality rate for HHNS is approximately 15%, much higher than for DKA (American Diabetes Association, Hyperglycemia, 2004; Arnold 2005). Box 19.14 compares and contrasts HHNS with DKA.

Treatment Three factors contribute to glycemic control: hepatic glucose production, glucose uptake by the periphery, and absorption of glucose from food. Medical management of T2DM entails a combination of nutrition therapy, physical activity, and medication when required to counteract abnormalities of glycemic control. Successful management requires awareness of the potential of each therapy, the synergistic connection between therapies, and maximal use of each therapy (Beebe 1999).

BOX 19.13 CLINICAL APPLICATIONS

Polycystic Ovary Syndrome: More Than Infertility

What is Polycystic Ovarian Syndrome (PCOS)?

PCOS is a health problem that can affect a woman's menstrual cycle, fertility, hormones, insulin production, heart, blood vessels, and appearance. Women with PCOS have these characteristics:

- High levels of male hormones, also called androgens.
- An irregular or no menstrual cycle.
- May or may not have many small cysts in their ovaries. Cysts are fluid-filled sacs.

PCOS is the most common hormonal reproductive problem in women of childbearing age.

How many women have PCOS?

An estimated 5% to 10% of women of childbearing age have PCOS.

What causes PCOS?

No one knows the exact cause of PCOS. Women with PCOS frequently have a mother or sister with PCOS. But there is not yet enough evidence to say there is a genetic link for this disorder. Many women with PCOS have a weight problem, so researchers are looking at the relationship between PCOS and the body's ability to make insulin. Since some women with PCOS make too much insulin, it's possible the ovaries react by making too many male hormones, called androgens. This can lead to acne, excessive hair growth, weight gain, and ovulation problems.

Why do women with PCOS have trouble with their menstrual cycle?

In women with PCOS, the ovary doesn't make all of the hormones it needs for any of the eggs to fully mature. They may start to grow and accumulate fluid. But no one egg becomes large enough. Instead, some may remain as cysts (see Figure 19.18). Since no egg matures or is released, ovulation does not occur and the hormone progesterone is not made. Without progesterone, a woman's menstrual cycle is irregular or absent. Also, the cysts produce male hormones, which continue to prevent ovulation.

What are the symptoms of PCOS?

These are some of the symptoms of PCOS:

- Infrequent menstrual periods, no menstrual periods, and/or irregular bleeding
- Infertility or inability to get pregnant because of not ovulating
- Increased growth of hair on the face, chest, stomach, back, thumbs, or toes
- Acne, oily skin, or dandruff
- Pelvic pain

FIGURE 19.18 Polycystic Ovarian Syndrome

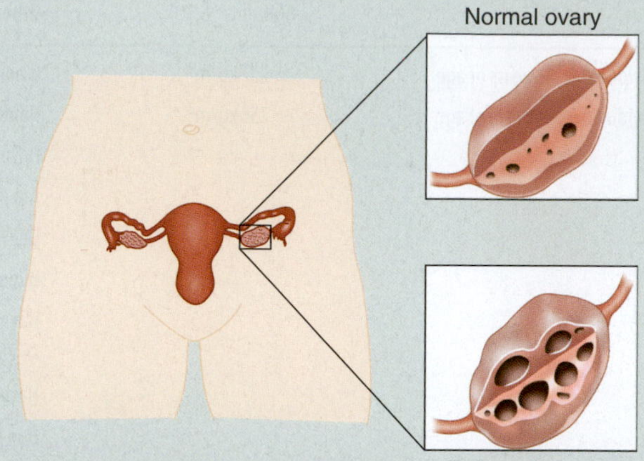

Source: http://www.4woman.gov/faq/pcos.htm

- Weight gain or obesity, usually carrying extra weight around the waist
- Type 2 diabetes
- High cholesterol
- High blood pressure
- Male-pattern baldness or thinning hair
- Patches of thickened and dark brown or black skin on the neck, arms, breasts, or thighs
- Skin tags, or tiny excess flaps of skin in the armpits or neck area
- Sleep apnea excessive snoring and breathing stops at times while asleep

What tests are used to diagnose PCOS?

There is no single test to diagnose PCOS. Physicians will take a medical history, perform a physical exam—possibly including an ultrasound, check hormone levels, and measure glucose in the blood. At the physical exam, the doctor will want to evaluate the areas of increased hair growth. During a pelvic exam, the ovaries may be enlarged or swollen by the increased number of small cysts. This can be seen more easily by vaginal ultrasound, or screening, to examine the ovaries for cysts and the endometrium. The endometrium is the lining of the uterus. The uterine lining may become thicker if there has not been a regular period.

How is PCOS treated?

Because there is no cure for PCOS, it needs to be managed to prevent problems. Treatments are based on the symptoms each patient is having and whether she wants to conceive or needs contraception. Below are descriptions of treatments used for PCOS.

Birth control pills. For women who don't want to become pregnant, birth control pills can regulate menstrual cycles, reduce male hormone levels, and help to clear acne. However,

(continued on the following page)

BOX 19.13 *(continued)*

the birth control pill does not cure PCOS. The menstrual cycle will become abnormal again if the pill is stopped. Women may also think about taking a pill that only has progesterone, like Provera, to regulate the menstrual cycle and prevent endometrial problems. But progesterone alone does not help reduce acne and hair growth.

Diabetes Medications. The medicine, Metformin, also called Glucophage, which is used to treat T2DM also helps with PCOS symptoms. Metformin affects the way insulin regulates glucose and decreases the testosterone production. Abnormal hair growth will slow down and ovulation may return after a few months of use.

Fertility Medications. The main fertility problem for women with PCOS is the lack of ovulation. Even so, her husband's sperm count should be checked and her tubes checked to make sure they are open before fertility medications are used. Clomiphene (pills) and Gonadotropins (shots) can be used to stimulate the ovary to ovulate. PCOS patients are at increased risk for multiple births when using these medications. In vitro fertilization (IVF) is sometimes recommended to control the chance of having triplets or more. Metformin can be taken with fertility medications and helps to make PCOS women ovulate on lower doses of medication.

Medicine for increased hair growth or extra male hormones. If a woman is not trying to get pregnant there are some other medicines that may reduce hair growth. Spironolactone is a blood pressure medicine that has been shown to decrease the male hormone's effect on hair. Propecia, a medicine taken by men for hair loss, is another medication that blocks this effect. Both of these medicines can affect the development of a male fetus and should not be taken if pregnancy is possible. Other nonmedical treatments such as electrolysis or laser hair removal are effective at getting rid of hair. A woman with PCOS can also take hormonal treatment to keep new hair from growing.

Surgery. Although it is not recommended as the first course of treatment, surgery called ovarian drilling is available to induce ovulation. The physician makes a very small incision above or below the navel, and inserts a laparoscope. The ovary is then punctured with a small needle carrying an electric current to destroy a small portion of the ovary. This procedure carries a risk of developing scar tissue on the ovary. This surgery can lower male hormone levels and help with ovulation. But these effects may only last a few months. This treatment does not help with increased hair growth and loss of scalp hair.

Healthy weight. Maintaining a healthy weight is another way women can help manage PCOS. Since obesity is common with PCOS, a healthy diet and physical activity help maintain a healthy weight, which will help the body lower glucose levels, use insulin more efficiently, and may help restore a normal period. Even loss of 10% of her body weight can help make a woman's cycle more regular.

How does PCOS affect a woman while pregnant?

There appears to be a higher rate of miscarriage, gestational diabetes, pregnancy-induced high blood pressure, and premature delivery in women with PCOS. Researchers are studying how the medicine Metformin prevents or reduces the chances of having these problems while pregnant, in addition to looking at how the drug lowers male hormone levels and limits weight gain in women who are obese when they get pregnant. No one yet knows if Metformin is safe for pregnant women. Because the drug crosses the placenta, doctors are concerned that the baby could be affected by the drug. Research is ongoing.

Does PCOS put women at risk for other conditions?

Women with PCOS can be at an increased risk for developing several other conditions. Irregular menstrual periods and the absence of ovulation cause women to produce the hormone estrogen, but not the hormone progesterone. Without progesterone, which causes the endometrium to shed each month as a menstrual period, the endometrium becomes thick, which can cause heavy bleeding or irregular bleeding. Eventually, this can lead to endometrial hyperplasia or cancer. Women with PCOS are also at higher risk for diabetes, high cholesterol, high blood pressure, and heart disease. Getting the symptoms under control at an earlier age may help to reduce this risk.

Does PCOS change at menopause?

Researchers are looking at how male hormone levels change as women with PCOS grow older. They think that as women reach menopause, ovarian function changes and the menstrual cycle may become more normal. But even with falling male hormone levels, excessive hair growth continues, and male pattern baldness or thinning hair gets worse after menopause.

For More Information...

You can find out more about PCOS by contacting the National Women's Health Information Center (NWHIC) at (800) 994–WOMAN (9662) or the following organizations:

National Institute of Child Health and Human Development (NICHD), NIH, HHS
Phone: (800) 370–2943
Internet Address: http://www.nichd.nih.gov/womenshealth

American Association of Clinical Endocrinologists (AACE)
Phone: (904) 353–7878
Internet Address: http://www.aace.com

American Society for Reproductive Medicine (ASRM)
Phone: (205) 978–5000
Internet Address: http://www.asrm.org

Center for Applied Reproductive Science (CARS)
Phone: (423) 461–8880
Internet Address: http://www.ivf-et.com

InterNational Council on Infertility Information Dissemination, Inc. (INCIID)
Phone: (703) 379–9178
Internet Address: http://www.inciid.org

PolyCystic Ovarian Syndrome Association, Inc. (PCOSA)
Phone: (877) 775–7267
Internet Address: http://www.pcosupport.org

The Hormone Foundation
Phone: (800) 467–6663
Internet Address: http://www.hormone.org

Source: Adapted from: The National Women's Health Information Center. Polycystic Ovarian Syndrome (PCOS) [monograph on the Internet]. Washington (DC): U.S. Department of Health and Human Services Office on Women's Health; 2004. Available from: http://www.4woman.gov/faq/pcos.htm.

The optimal method for achieving glycemic control is targeted blood glucose control, whereby individuals know and attempt to achieve their blood glucose goals for various times of the day. This is accomplished using feedback from daily SMBG and routine laboratory evaluations (A1C) (Beebe 1999; Mulcahy and Lumber 2004). This has been found to be effective in individuals with T1DM, those with T2DM using insulin, and those with T2DM who are not using insulin (Welschen et al. 2005). For example, an individual with T2DM exercises every morning for 30 minutes and takes 18 units glargine at bedtime, 6 units lispro before breakfast, and 5 units lispro before lunch and dinner. In reviewing his or her blood glucose log, it becomes evident that glucose levels are consistently below glucose target range before lunch. In consultation with the registered dietitian, he or she can discuss possible solutions, such as adding a mid-morning snack or reducing pre-breakfast lispro insulin (Kulkarni 2005).

Frequency of monitoring is influenced by selected tightness of control, capacity to execute tests unaided, affordability, and motivation to test. It is suggested that individuals with T2DM test as often as needed to achieve glycemic goals (see Table 19.21), before and after physical activity, and to ascertain existence of hypoglycemia and reaction to treatment. When ill, testing every 4–6 hours is suggested (Mulcahy and Lumber 2004). Glucose should also be checked before driving.

Medications for Type 2 Diabetes As T2DM progresses, use of glucose-lowering medications is indicated if glycemic control cannot be achieved with nutrition therapy and regular physical activity alone. There are seven classes of medications (Freeman 2005; Votey and Peters 2005) used to treat T2DM, as shown in Table 19.22:

- **Alpha-glucosidase** inhibitors (AGIs)
- **Amylin** analogs
- Biguanides
- Incretin mimetics
- Meglitinides
- Sulfonylurea agents
- Thiazolidinediones

Drug-nutrient interactions for these medications are listed in Table 19.23.

alpha-glucosidase—a digestive enzyme found in the brush border cells of the small intestine that cleaves more complex carbohydrates into sugars

amylin—a hormone synthesized by pancreatic β-cells that contributes to glucose control during the postprandial period

Physical Activity Because the benefits of physical activity for glycemic control are well documented, exercise (along with nutrition therapy) is generally prescribed for all individuals with T2DM. Physical activity improves blood glucose levels by enhancing muscle blood glucose uptake during or shortly after activity and by improving insulin sensitivity. Furthermore, it enhances weight loss efforts, which in turn improve insulin sensitivity and glycemic control (Beebe 1999).

Thirty to forty-five minutes of moderate-intensity physical activity 3–5 days a week is recommended to improve glycemic control, assist with weight maintenance, and reduce risk of CVD. No more than two consecutive days should go by without physical activity. As long as there are no contraindications, resistance exercise targeting major muscle groups should be performed three times a week. If the client begins an exercise program more vigorous than brisk walking, however, conditions that could be related to increased likelihood of CVD or cause injury to the individual should be assessed. Age and previous physical activity should also be considered (American Diabetes Association, Standards, 2006).

In those taking insulin and/or insulin secretagogues (sulfonylurea or meglitinides), physical activity can cause hypoglycemia if medication dose or carbohydrate consumption is not changed. Hypoglycemia rarely occurs in individuals with T2DM not being treated with insulin or insulin secretagogues. If pre-exercise glucose levels are <100 mg/dL, additional carbohydrate should be ingested. Supplementary carbohydrate is generally not necessary for individuals treated with diet alone, metformin, α-glucosidase inhibitors, and/or TZDs without insulin or a secretagogue (American Diabetes Association, Physical, 2004; American Diabetes Association, Standards, 2006).

Nutrition Therapy The impact of dietary modification on overall health, metabolic control, and treatment for acute and chronic complications is substantial (Beebe 1999; American Diabetes Association, Standards, 2006). Nutrition therapy is an integral component of diabetes management and self-management (American Diabetes Association, Standards, 2006).

T2DM is a complex disorder that is heterogeneous in nature, in that it affects individuals of different ages, lifestyles, and cultural backgrounds. Therefore, nutrition therapy should be implemented by prioritizing metabolic problems and instituting a plan based on those priorities (Beebe 1999). Individuals with T2DM are often overweight and insulin resistant (Holzmeister and Geil 2005). Treatment goals and lifestyle changes that the patient is willing and able to make, rather than predetermined energy levels and percentages of carbohydrate, protein, and fat, should be paramount in determining the nutrition prescription. The aim of nutrition intervention is to support and facilitate lifestyle and behavior modifications that will result in improved metabolic control (Franz et al. 2002).

Weight Management Overweight and obesity are strongly associated with development of T2DM. Moreover, obesity is an independent risk factor for hypertension, dyslipidemia, and CVD, the major cause of death in those with diabetes. Moderate weight loss improves glycemic control and reduces CVD risk; therefore, weight loss is recommended for individuals with BMIs >25.0 kg/m^2. Therapeutic lifestyle changes (see Chapters 14 and 15) that include a reduction in energy intake and an increase in physical activity are recommended (American Diabetes Association, Standards, 2006).

Carbohydrates Because the total amount of dietary carbohydrate is a strong predictor of glycemic response, monitoring total grams of carbohydrate by either the use of exchanges or carbohydrate counting is strategic in achieving glycemic control. Low-carbohydrate diets are not suggested, because carbohydrates are significant sources of energy, water-soluble vitamins and minerals, and fiber. Less than 130 grams of carbohydrate per day is not recommended because the brain and central nervous system have an absolute requirement for glucose as an energy source (American Diabetes Association, Standards, 2006).

Protein Intake of dietary protein exceeding 20% of energy intake may be a risk factor for development of nephropathy. Protein intake for individuals with diabetes and nephropathy should not exceed 0.8 g/kg or ~10% of total kcal (American Diabetes Association, Standards, 2006).

Fat The goal for dietary fat intake (amount and type) for individuals with diabetes is the same as for those without diabetes who have a history of CVD. Total fat intake should not exceed 25% to 35% of total kcal, and saturated fat intake should not exceed 7%. Intake of trans fat should be minimal (American Diabetes Association, Standards, 2006).

Fiber A variety of fiber-containing foods such as legumes and fiber-rich cereals (≥5 g fiber/serving), as well as fruits, vegetables, and whole-grain products, are recommended (American Diabetes Association, Standards, 2006). Foods contain a mixture of fibers, but those foods that have high amounts of gums, beta glucans, psyllium, resistant starches, and pectin appear to have the biggest effect on serum glucose levels by slowing the absorption of glucose from the small intestine. The foods highest in these types of fiber include legumes, fruits, vegetables, and oats. The U.S. Dietary Guidelines (U.S. Dept. of Health and Human Services and U.S. Dept. of Agriculture 2005) recommend consuming 14 grams of fiber for every 1,000 kcal, while the Food and Nutrition Board recommends that men under 50 consume 38 grams of fiber/day and women consume 25 grams of fiber/day (Institute of Medicine 2002).

Gestational Diabetes Mellitus (GDM)

Definition Gestational diabetes (GDM) is a form of glucose intolerance first diagnosed during pregnancy (American Diabetes Association, Gestational, 2004; American Diabetes Association, Fact Sheet, 2005; Centers for Disease Control 2005; NDIC 2005).

Epidemiology Approximately 7% of all pregnancies are complicated by GDM, and, women who have had GDM have a 20% to 50% chance of developing diabetes in the next 5 to 10 years. Women at risk for GDM have the following characteristics (American Diabetes Association, Gestational, 2004; American Diabetes Association, Fact Sheet, 2005; Centers for Disease Control 2005; NDIC 2005; Thomas, Classification, 2005):

- Obesity (BMI >30.0)
- Personal history of GDM
- Glycosuria
- Strong family history of diabetes (1st degree relative)
- Prior poor obstetrical outcome (stillbirth, birth defects, or baby >9 lbs)
- Member of a high-risk ethnic group (Hispanic, African American, Native American, South or East Asian, Pacific Islander)

Etiology During the second or third trimesters of pregnancy, metabolic alterations occur to meet maternal and fetal demands for energy and nutrients. In addition to alterations in insulin secretion, these alterations affect glucose, amino acid, and lipid metabolism (Thomas, Pathophysiology, 2005). Although most women with GDM revert to normal glucose tolerance postpartum, there is increased likelihood of developing GDM in subsequent pregnancies and T2DM later in life. Increasing physical activity and reducing postpartum weight gain can reduce risk of subsequent diabetes (American Diabetes Association, Nutrition, 2004).

Pathophysiology GDM is pathophysiologically similar to T2DM. Islet cell function abnormalities or peripheral insulin resistance are thought to decrease insulin secretory response and insulin sensitivity. Inability of the β-cells to meet increased insulin needs during pregnancy results in higher levels of circulating glucose (Thomas, Pathophysiology, 2005).

GDM also affects the fetus. When maternal blood glucose levels are elevated, the fetus is constantly exposed to these levels as well, and fetal insulin production is increased. It appears maternal hyperglycemia induces fetal hyperglycemia leading to fetal hyperinsulinemia and **macrosomia** (World Health Organization 2002; Thomas, Pathophysiology, 2005).

Clinical Manifestations Maternal complications associated with GDM include hypertension (preeclampsia), **polyhydramnios**, difficult birth, preterm delivery (before 38 weeks

macrosomia—refers to the condition of abnormally large infants whose mothers have diabetes

polyhydramnios—excessive accumulation of amniotic fluid

BOX 19.14 CLINICAL APPLICATIONS

Hyperglycemic Hyperosmolar Nonketotic Syndrome and Diabetic Ketoacidosis

	Hyperglycemic Hyperosmolar Nonketotic Syndrome (HHNS)	Diabetic Ketoacidosis (DKA)
Characteristics	Adequate insulin to prevent lipolysis and ketogenesis but inadequate to maintain no rmoglycemia Occurs most often in T1DM	Hyperglycemia Metabolic acidosis Ketogenesis Occurs most often in T2DM Those with T2DM at risk during acute illness
Causes	Dehydration from inadequate fluid intake or excess fluid losses Prolonged hyperglycemia	Infections Acute illnesses (CVA, alcohol/drug abuse, pancreatitis, pulmonary embolism, MI, trauma) Psychological stress Lack of SMBG Insulin omitted Increased insulin needs with growth spurts
Symptoms	Undiagnosed diabetes Between 55 and 70 years old Frequently residents of long-term care Progresses slowly (over days and weeks) Polyuria Polydipsia Progressive decline in level of consciousness Fever (due to underlying infection) Volume depletion	Develops rapidly Polyuria Polydipsia Weight loss Vomiting Abdominal pain Dehydration (loss of skin turgor, dry mucous membranes, tachycardia, hypotension) Acetone breath Kussmaul respirations

(continued on the following page)

gestation), and a higher rate of cesarean sections. Fetal and neonatal complications include macrosomia, hypoglycemia, respiratory distress syndrome, hypocalcemia, hyperbilirubinemia, and polycythemia (American Diabetes Association, Gestational, 2004; Thomas and Gutierrez 2005).

Diagnosis In the U.S., a two-step approach is generally used to diagnose GDM (American Diabetes Association, Gestational, 2004). Table 19.24 outlines the diagnosis of GDM using the glucose challenge test (GCT; 50 g) and 100 g OGTT. Use of this two-step approach identifies approximately 80% of women with GDM (American Diabetes Association, Gestational, 2004; Thomas, Pathophysiology, 2005).

Treatment The American Diabetes Association recommends all women with GDM receive nutrition counseling and therapy by a registered dietitian. Nutrition therapy should be individualized and based on maternal weight and height. Energy and nutrients adequate to meet the needs of pregnancy should be incorporated with established maternal blood glucose goals (Thomas, Pathophysiology, 2005). When nutrition therapy alone fails to maintain SMBG at the following levels, insulin therapy is added to reduce fetal mortality (American Diabetes Association, Gestational, 2004):

Fasting plasma glucose ≤105 mg/dL (5.8 mmol/L)

or

1 hour postprandial plasma glucose ≤155 mg/dL (8.6 mmol/L)

or

2 hour postprandial plasma glucose ≤130 mg/dL (7.2 mmol/L)

Use of oral diabetes medications is generally not recommended during pregnancy (American Diabetes Association, Gestational, 2004).

Monitoring Maternal hyperglycemia increases fetal risk. SMBG, not A1C, is considered to be the best method to detect maternal hyperglycemia. Postprandial monitoring is

BOX 19.14 *(continued)*

	Hyperglycemic Hyperosmolar Nonketotic Syndrome (HHNS)	Diabetic Ketoacidosis (DKA)
Laboratory Findings		
Plasma glucose	> 600 mg/dL	> 250 mg/dL
Arterial pH	> 7.3	< 7.0 to 7.30
Serum bicarbonate	> 15 mEq/L	< 10 to 18 mEq/L
Urine ketones	Small	Positive
Serum ketones	Small	Positive
Serum osmolality	> 320 mOsm/kg	Variable
Treatment	Hospitalization for slow rehydration Treatment for underlying medical problems Insulin may or may not be required	Hospitalization for administration of: IV fluids Insulin Assessment of serum electrolytes
Prevention	Routine hydration Adequate monitoring	Identify cause(s) to determine approach to prevention. Can include: Regular self-monitoring Test for ketones if BG > 250 mg/dL Monitor effects of illness on BG closely Take medications even when eating less Sick day management plan Probe rationale for omitting insulin

Source: Adapted from: Arnold MS. Hypoglycemia and hyperglycemia. In: Ross TA, Boucher JL, O'Connell BS, editors. *American Dietetic Association guide to diabetes: medical nutrition therapy and education.* Chicago: American Dietetic Association; 2005, and Umpierez GE, Murphy MB, Kitabchi AE. Diabetic ketoacidosis and hyperglycemic hyperosmolar syndrome. *Diabetes Spectrum.* 15(1):28–36, 2002.

TABLE 19.21

Target Plasma Glucose Levels

Glycemic Indicator	Normal	Goal[1]	Goal in Pregnancy
Preprandial glucose	< 100 mg/dL (< 6.7 mmol/L)	90–130 mg/dL (5.0–7.2 mmol/L)	60–90 mg/dL (3.3–5.0 mmol/L)
Postprandial glucose	< 130 mg/dL (< 7.2 mmol/L)	< 180 mg/dL (< 10.0 mmol/L)	< 120 mg/dL (< 6.7 mmol/L)
A1C	< 6%	< 7%	< 7%

[1] For nonpregnant adults. Different treatment goals may be warranted by individuals with comorbid diseases, the very young, older adults, and others with unusual conditions or circumstances.

Source: Adapted from: Mulcahy K, Lumber T. *The diabetes ready reference for health professionals.* 2nd ed. Alexandria (VA): American Diabetes Association; 2004. Table IV, page 12.

TABLE 19.22

Types of Diabetes Medications

Class	Generic	Trade Name	Action	Susceptibility to Hypoglycemia	Disadvantages	Advantages
α-Glucosidase Inhibitors (AGIs)	Acarbose Miglitol	Precose Glyset	Delays intestinal absorption of glucose	No	Flatulence, diarrhea, less efficacy frequent dosing Contraindicated in individuals with intestinal diseases, must take with meals 3 times/day	Safety, postprandial effect
Amylin Analogs (Injectable medication)	Pramlintide acetate	Symlin	Delays gastric emptying, decreases postprandial glucagon release, suppresses appetite	Increases risk of insulin-induced hypoglycemia (can be used to treat T1DM or T2DM taking insulin)	GI complaints, must be used in syringe separate from insulin, hypersensitivity to pramlintide	Improves long-term control (A1c) compared to insulin alone; lowers insulin use and body weight
Biguanides	Metformin	Glucophage	Decreases hepatic glucose production, increases insulin uptake in muscles	No	Transient diarrhea, nausea, bloating, anorexia, flatulence, lactic acidosis (rare); contraindicated in individuals with renal insufficiency, liver failure, or treated CHF	Weight control, no hypoglycemia with monotherapy, may be CV benefits
Incretin Mimetics (Injectable medication)	Exenatide	Byetta	Mimics glucose-dependent insulin secretion, suppresses elevated glucagon secretion, delays gastric emptying	Can cause hypoglycemia when used with sulfonylureas	May decrease absorption of orally administered drugs (drugs requiring rapid absorption such as oral contraceptives, antibiotics)	Better glycemic control
Meglitinides	Repaglinide Nateglinide	Prandin Starlix	Stimulates insulin secretion in presence of glucose, short-acting	Yes	Hypoglycemia, frequent dosing, expensive	Short action with less hypoglycemia at night or with missed meal, glucose- dependent effect on insulin, postprandial effect
Sulfonylurea Agents			Stimulates insulin secretion	Yes	Hypoglycemia (more with glyburide); contraindicated in individuals with renal insufficiency, weight gain	Inexpensive, long history of effectiveness, only needed once daily for most patients
First Generation	Acetohexamide Chlorpropamide Tolazamide Tolbutamide	Dymelor Diabinese Tolinase Orinase				
Second Generation	Glipizide Glipizide-XL Glyburide Glimepiride	Glucotrol Glucotrol XL DiaBeta Micronase PresTab Glynase Amaryl				

(continued on the following page)

TABLE 19.22 *(continued)*

Class	Generic	Trade Name	Action	Susceptibility to Hypoglycemia	Disadvantages	Advantages
Thiazolidinediones	Pioglitazone Rosiglitazone	Actos Avandia	Decreases insulin resistance	No	Weight gain, edema, worsened CHF, most expensive, slow onset of action; contraindicated in individuals with CHF	Very effective in highly insulin-resistant individuals, okay with renal insufficiency, potential CV benefit, usually needed only once daily
Insulin (not an oral medication, but often used to treat T2DM)	Rapid, short, intermediate, long	See Table 19.20	Replaces endogenous insulin	Yes	See Table 19.20	See Table 19.20

Sources: Ahmann AJ, Riddle MC. Current oral agents for type 2 diabetes; many options, but which to choose when? *Postgrad Med.* 2002;111:32. Beebe CA. Nutrition therapy of type 2 diabetes. In: Franz MJ, Bantle JP, editors. *American Diabetes Association guide to medical nutrition therapy for diabetes.* Alexandria (VA): American Diabetes Association; 1999. Inzucchi SE. Oral antihyperglycemic therapy for type 2 diabetes. *JAMA.* 2002;287(3):360–372. Luna B, Feinglos MN. Oral agents in the management of type 2 diabetes mellitus. *American Family Physician.* 2001;63(9):1747–1756. Votey SR, Peters AL. Diabetes mellitus, type 2 – a review. Last updated 7/14/05. eMedicine.com, Inc. Available from: http://www.emedicine.com. Accessed March 4, 2006.

superior to preprandial monitoring when insulin therapy is used. Postprandial blood glucose levels are directly related to rates of macrosomia, neonatal hypoglycemia, and cesarean delivery. Additionally, maternal blood pressure and urine protein should be monitored to detect hypertensive disorders (American Diabetes Association, Gestational, 2004; Gutierrez and Reader 2005).

Nutrition Therapy Nutrition therapy goals for GDM include a goal shared by all pregnant women: to promote nutrition for maternal and fetal health while providing adequate energy for appropriate gestational weight gain. Furthermore, achievement and maintenance of normoglycemia and absence of ketones are goals specific to treatment of GDM (American Diabetes Association, Gestational, 2004; Gutierrez and Reader 2005).

Adequate energy is necessary for desirable weight gain during pregnancy (Franz et al. 2002; American Diabetes Association, Nutrition, 2004). Energy needs are evaluated indirectly by monitoring the woman's physical activity, appetite, food intake, blood glucose levels, ketone records, and weight change (American Diabetes Association, Nutrition, 2004; Gutierrez and Reader 2005). It is not necessary to calculate energy needs unless problems with excessive weight loss or gain are experienced (Gutierrez and Reader 2005). Formulas useful for estimating energy requirements for women with GDM are outlined in Box 19.15.

Protein requirements increase during the second and third trimesters of pregnancy to 25 grams per day or 1.1 g protein per kg desirable body weight (Institute of Medicine 2002). Two factors should be contemplated concerning dietary fat intake: impact on the woman's body weight and plasma lipoprotein profiles. Reduced fat intake may be necessary if total energy intake should be decreased, and saturated fat, trans fat, and cholesterol intake should be curtailed (Gutierrez and Reader 2005).

Consequences of folate deficiency in pregnancy (i.e., neural tube defects) have been well documented. All women of reproductive age capable of becoming pregnant should take 400 μg additional folate from food or supplements (Institute of Medicine 2000). Whereas about 10% of the iron is absorbed from food in the nonpregnant state, iron absorption increases to 25% at the beginning of the second trimester (Gutierrez and Reader 2005). Supplementation of 30 mg ferrous iron in the second and third trimester is recommended (Institute of Medicine 1990). Nutrition recommendations for GDM are outlined in Table 19.25.

Hypoglycemia

Definition Hypoglycemia is an abnormally low blood glucose level. It occurs when glucose is utilized too rapidly, glucose release rate falls behind tissue demands, or excess insulin enters the bloodstream. Spontaneous[3] hypoglycemia in adults is either fasting or postprandial (Brabson et al. 2001; Masharani and Karam 2003).

Etiology Reactive hypoglycemia may occur in individuals with diabetes due to administration of too much insulin or oral diabetes medications. In those without diabetes, reactive hypoglycemia may occur due to a sharp increase in insulin release after a meal. It usually disappears when the individual eats something (Brabson et al. 2001).

Fasting hypoglycemia usually results from excess insulin or insulin-like substance from external factors such as alcohol or drug ingestion (Brabson et al. 2001).

TABLE 19.23

Drug-Nutrient Interactions for Medications Used to Treat Diabetes Mellitus

Classification	Mechanism	Brand Names	Possible Food-Drug Interactions
Insulin	Replaces endogenous insulin.	See Table 19.20	Increased weight; use alcohol with caution.
α-Glucosidase Inhibitors (AGIs)	Delays intestinal absorption of glucose.	Precose Glyset	Take with first bite of meal, limit alcohol.
Amylin Analogs	Delays gastric emptying, decreases postprandial glucagon release, suppresses appetite.	Symlin	Caution with alcohol.
Biguanides	Decrease hepatic glucose production, increases insulin uptake in muscles.	Glucophage	Decreases folate and vitamin B_{12} absorption, avoid alcohol, take with meals to decrease GI distress.
Incretin Mimetics	Mimics glucose-dependent insulin secretion, suppresses elevated glucagon secretion, delays gastric emptying.	Byetta	Caution with alcohol, may cause GI disturbances.
Meglitinides	Stimulates insulin secretion in presence of glucose.	Prandin Starlix	Limit alcohol.
Sulfonylurea Agents (first generation)	Stimulates insulin secretion.	Dymelor Diabinese Tolinase Orinase	Avoid alcohol.
Sulfonylurea Agents (second generation)	Stimulates insulin secretion.	Glucotrol Glucotrol XL DiaBeta Micronase PresTab Glynase	Avoid alcohol.
Thiazolidinediones	Decreases insulin resistance. Rosiglitazone	Actos Avandia	None.

Sources: Pronsky ZM. *Food medication interactions,* 13th ed. Birchrunville (PA): Food-Medication Interactions; 2004. Ahmann AJ, Riddle MC. Current oral agents for type 2 diabetes; many options, but which to chose when? *Postgrad Med.* 2002;111:32. Beebe CA. Nutrition therapy of type 2 diabetes. In: Franz MJ, Bantle JP, editors. *American Diabetes Association guide to medical nutrition therapy for diabetes.* Alexandria (VA): American Diabetes Association; 1999. Inzucchi SE. Oral antihyperglycemic therapy for type 2 diabetes. *JAMA.* 2002;287(3):360–372. Luna B, Feinglos MN. Oral agents in the management of type 2 diabetes mellitus. *American Family Physician.* 2001;63(9):1747–1756. Votey SR, Peters AL. Diabetes mellitus, type 2 – a review. Last updated 7/14/05. eMedicine.com, Inc. Available from: http://www.emedicine.com. Accessed March 4, 2006.

Pathophysiology When blood glucose levels fall too low, glucagon releases stored hepatic glucose to raise blood glucose levels. Epinephrine is also released, causing the symptoms of weakness, fatigue, sweating, and tachycardia (American Dietetic Association 2000).

Fasting Hypoglycemia Fasting hypoglycemia can be a primary or secondary manifestation. Primary causes of fasting hypoglycemia include hyperinsulinism due to pancreatic β-cell tumors or surreptitious administration of insulin or oral diabetes medications, and non-insulin producing extrapancreatic tumors. Secondary fasting hypoglycemia may be caused by certain endocrine disorders such as hypopituitarism, Addison's disease, or myxedema; liver disorders such as acute alcoholism or liver failure; and renal failure, especially in individuals undergoing dialysis (Masharani and Karam 2003).

Postprandial (Reactive) Hypoglycemia Postprandial hypoglycemia is classified as either early (within 2 to 3 hours after eating) or late (3 to 5 hours after eating). Early hypoglycemia occurs after rapid discharge of ingested foods from the stomach into the small intestine. This is followed by rapid glucose absorption and hyperinsulinemia. This is often observed after gastrointestinal surgery, particularly with dumping syndrome after gastrectomy (Masharani and Karam 2003).

TABLE 19.24

Diagnosis of Gestational Diabetes Mellitus

Time	50 g glucose challenge test (GCT)[1]	100 g oral glucose tolerance test[2]
Fasting		95 mg/dL (5.3 mmol/L)
1 hour	≥ 140 mg/dL (7.8 mmol/L)	180 mg/dL (10.0 mmol/L)
2 hour		155 mg/dL (8.6 mmol/L)
3 hour		140 mg/dL (7.8 mmol/L)

[1] Administered regardless of time of day or if food was consumed. If results exceed 140 mg/dl threshold, the 3-hour 100 g OGTT is administered

[2] Two or more venous plasma concentrations must be met or exceeded for a positive diagnosis. Test should be done in the morning after an overnight fast of 8 to 14 hours and after at least 3 days of unrestricted diet (≥ 150 g carbohydrate per day) and unlimited physical activity.

Source: Adapted from: American Diabetes Association. Gestational diabetes mellitus. *Diabetes Care.* 27(Suppl 1):S88, 2004, and Thomas AM. Classification, screening, and diagnosis. In: Thomas AM, Gutierrez YM, editors. *American Dietetic Association guide to gestational diabetes mellitus.* Chicago: American Dietetic Association; 2005.

BOX 19.15 CLINICAL APPLICATIONS

Estimating Energy Requirements in GDM

To calculate a woman's (19 years and older) energy needs during pregnancy, estimated energy requirements (EER) must first be calculated:

$$EER = 354 - (6.9 \times A) + PA \times (9.36 \times W + 726 \times H)$$

Where A = age in years; PA = physical activity coefficient [1.0 (sedentary), 1.12 (low active), 1.27 (active), 1.45 (very active)], W = weight in kg; H = height in meters.

To estimate energy requirements for pregnant women who have normal weight:

1st trimester = Adult EER + 0
2nd trimester = Adult EER + 160 kcal[1] + 180 kcal
3rd trimester = Adult EER + 272 kcal[2]

There is no formula supported by research to determine energy needs in overweight or obese pregnant women. For these women, weight gain should be monitored regularly to maintain an approximate weight gain of 0.5 lb/week.

Sources: Gutierrez YM, Reader DM. Medical nutrition therapy. In: Thomas AM, Gutierrez YM, editors. *American Dietetic Association guide to gestational diabetes mellitus.* Chicago: American Dietetic Association; 2005.
Institute of Medicine. Dietary reference intake for energy, carbohydrates, fiber, fat, fatty acids, cholesterol, protein, and amino acids (macronutrients). Washington (DC): National Academy Press; 2002.
Institute of Medicine. Nutrition during pregnancy: Part I: Weight gain, Part II: Nutrient supplements. Washington (DC): National Academy Press; 1990.

[1] 8 kcal/wk × 20 wk.
[2] 8 kcal/wk × 34 wk.

Clinical Manifestations Symptoms of hypoglycemia occur when plasma glucose levels reach 70 mg/dL. Impairment of brain function occurs at approximately 50 mg/dL. Fasting hypoglycemia usually presents with neuroglycopenia, while reactive hypoglycemia manifests with symptoms of sweating, palpitations, anxiety, and tremulousness (Masharani and Karam 2003).

Regardless of cause, characteristics of hypoglycemia consist of (Masharani and Karam 2003):

- History of hypoglycemic symptoms
- Fasting blood glucose ≤40 mg/dL
- Immediate recovery upon administration of glucose

Hypoglycemic symptoms often develop in the early morning or after missing a meal, and may occasionally occur after exercise. Symptoms can include blurred vision, headache, feelings of detachment, slurred speech, and weakness. Personality and mental changes vary from anxiety to psychotic behavior. Sweating and palpitations may not occur (Masharani and Karam 2003).

Treatment Anticholinergic agents may be used in treatment of reactive hypoglycemia in order to slow gastric emptying and intestinal motility and inhibit vagal stimulation of insulin release. Surgery and drug therapy are generally necessary for fasting hypoglycemia. Tumor removal is the treatment of choice for patients with insulinoma (insulin-producing tumor). Medications may include nondiuretic thiazides, such as diazoxide to inhibit insulin secretions; streptozocin; and hormones, such as glucagon or glucocorticoids (Brabson et al. 2001).

Nutrition Therapy Effective treatment of reactive hypoglycemia requires dietary modification to help delay glucose absorption and gastric emptying.

Small, frequent meals of complex carbohydrates, fiber, and a protein source are used to treat reactive hypoglycemia (Brabson et al. 2001; American Dietetic Association 2000).

Simple carbohydrates (such as candy, sugar, jam, jelly, syrup, honey, and soft drinks) and alcohol should be avoided (American Dietetic Association 2000).

It may be beneficial to restrict caffeine, which may reduce cerebral blood flow and consequently glucose supply to the brain (American Dietetic Association 2000).

Use of carbohydrate counting may be helpful in regulating total carbohydrate intake (American Dietetic Association 2000).

TABLE 19.25

Nutrition Recommendations for GDM

Nutrient or Food Type	Recommendation	Meal-Planning Tips
Energy	Intake should be sufficient to promote adequate, but not excessive, weight gain and to avoid ketonuria.	Include 3 small- to moderate-sized meals and 2–4 snacks. Space snacks and meals at least 2 h apart. A bedtime snack (or even a snack in the middle of the night) is recommended, to diminish the number of hours fasting.
Carbohydrate	Recommendations are based on effect of intake on blood glucose levels. Intake should be distributed throughout the day. Frequent feedings, smaller portions, with intake sufficient to avoid ketonuria.	Common carbohydrate guidelines: 2 carbohydrate choices (15–30 g) at breakfast, 3–4 choices (45–60 g) for lunch and evening meal, and 1–2 choices (15 to 30 g) for snacks. Recommendations should be modified based on individual assessment and blood glucose self-monitoring test results.
High-Sucrose/ High-Energy Foods	Inclusion should be based on individual's ability to maintain blood glucose goals, nutritional adequacy of diet, and contribution of these foods to total meal plan.	Eliminate foods containing large amounts of carbohydrates, such as sweets and sweetened drinks.
Protein	RDA for adult women (0.8 g/kg DBW) + 25 g/day, or 1.1 g/kg DBW.	Protein foods do not raise post-meal blood glucose levels. Add protein to meals and snacks, to help provide enough calories and to satisfy appetite.
Fat	Limit saturated fat.	Fat intake may be increased because of increased protein intake; focus on leaner protein choices.
Sodium	Not routinely restricted.	
Fiber	For relief of constipation, gradually increase intake and increase fluids.	Use whole grains and raw fruits and vegetables. Activity and fluids help relieve constipation.
Nonnutritive Sweeteners	Generally safe in pregnancy. Use in moderation.	Saccharin crosses the placenta but has not been shown to be harmful.
Vitamins and Minerals	Preconception folate. Assess for specific individual needs: multivitamin throughout pregnancy, iron at 12 weeks, and calcium especially in last trimester and while lactating.	Take prenatal vitamin. If it causes nausea, try taking at bedtime.
Caffeine	Limit to < 300 mg/day.	
Alcohol	Avoid.	

DBW = desired body weight; RDA = Recommended Dietary Allowance

Source: © 2005 American Dietetic Association. Reprinted with permission. Ross TA, Boucher JL, O'Connell BS, editors. *American Dietetic Association guide to diabetes: medical nutrition therapy and education*. Chicago: American Dietetic Association, 2005, Chapter 17, Table 17.1, p. 191.

Conclusion

This chapter has described the physiology behind endocrine function, endocrine control of metabolism, thyroid disorders, diabetes mellitus, and hypoglycemia, and their clinical consequences for the patient and practitioner. Nutrition therapy must address not only endocrine disorders but the effect of the medications used to treat these disorders on nutrition and nutrition status. Therapeutic nutrition interventions have been discussed, with particular emphasis on specific modifications to assist in the treatment of disease. Nutrition therapy for endocrine disorders, especially DM, pose a challenge for the RD but also offer the opportunity to greatly improve clinical outcomes and quality of life for clients and patients.

PRACTITIONER INTERVIEW

Jill Weisenberger, M.S., R.D., C.D.E. *Research Dietitian, Hampton Roads Center for Clinical Research Norfolk, VA and Health & Nutrition Writer*

How did you become a Certified Diabetes Educator?

I haven't been a Certified Diabetes Educator for very long—only since 2002. In fact, dietetics was a second career for me—I didn't become an RD till 1991. The opportunity to become a CDE fell into my lap and I jumped at the chance. I am a consulting dietitian and felt the CDE would make me more marketable by widening my scope of practice.

How has dietetics changed since you have begun your practice?

During my career, I have seen substantial changes in how we treat diabetes. Today the medications are better, the diet is much less restrictive, and we focus on the clients being able to manage themselves. When I started out, I mostly just educated the client on what to eat via the exchange system and to not eat sugar. Now I counsel people. I help them to use information to solve their own problems and recognize their barriers. This change didn't happen overnight but rather with experience and after recognizing what worked and didn't work.

What is your typical patient like? Tell us about how you accomplish nutrition therapy with your clients.

Almost all my clients have type 2 diabetes but many of them are on insulin. Some are as young as 20, yet most (75%) are over 45 years old. About half my clients are African Americans. One of my clients was a 45-year-old gentleman who came in with HbA_1C of 13.7%, overweight, and on oral medications. We worked on: (1) monitoring blood sugar, (2) eating three meals a day, 60 g of carbohydrate per meal, (3) eating 0–20 g carbohydrate for a snack if hungry, and (4) identifying foods in the diet high in saturated fat. Over three visits (3 months), his HbA_1C dropped to 6.5%. He only lost a small amount of weight but was able to go off medication.

During a client's first visit I have a basic plan—chart the outcomes for specific behavioral changes. For example, the goal may be to normalize blood sugar by spreading the consumption of carbohydrates throughout the day. The behavioral outcome would be to consume three meals a day (60 g of carbohydrate per meal) and if hungry a snack of 0–20 g carbohydrates. On the second visit, I find out what worked and what didn't work for the client and if they were happy with the plan. Then, I find out if they anticipate any issues coming up (wedding, vacation, work) that might interfere with the plan. Lastly, I look at blood glucose records but they don't always bring them.

When I work with people I only suggest small changes in their diet each time I see them. Changing someone's diet is very hard. They're busy, tired, have kids, and often have few cooking skills. But I do insist that they monitor their blood sugars. If they refuse, we negotiate. They learn so much from knowing their blood sugar and it allows them to manage their own disease by seeing first hand the impact of food. Although another goal is to also help prevent heart disease, it is hard to get clients to focus on heart disease if they are dealing with blood sugar, and they may still eat high-saturated fat and foods high in cholesterol that don't have carbohydrates.

My advice to new dietitians is to have compassion for each patient. You will do a better job and will enjoy what you do. Wherever you work, surround yourself with smart, capable people who work hard, because that is how you become successful. Keep up to date on all aspects of nutrition and feel confident to discuss your patient with their physician.

CASE STUDY

General:

71-year-old female who arrives at the emergency room presents with non-healing wound on right foot between the 2nd and 3rd digits. Patient has a history of frequent bladder infections, slight tingling and numbness in her feet, and today, serum blood glucose of 325 mg/dL. She is admitted for antibiotics, probable surgical debridement of a wound and stabilization and treatment for T2DM.

Labs:

BUN 26 mg/dL; Cr 1.2 mg/dL; Chol 300 mg/dL; HDL 35 mg/dL; LDL 140 mg/dL; Glucose 325 mg/dL; HbA1C 8.5%

Rx:

None at this time

Past Med Hx:

HTN treated with 50 mg Captopril two times daily.

Nutrition Hx:

Lives with sister who has T2DM; prepares own meals—rarely eats at restaurants. Likes all foods but avoids "foods with sugar."

Ht. 5′0″, Wt. 155 lbs., Usual adult body weight—145–165 lbs.

Treatment Plan:

Admit for surgical debridement of wound and antibiotics; normalize blood glucose levels with sliding scale insulin, and then initiate discharge planning with comprehensive diabetic education and Rx of oral hypoglycemic agent.

Questions:

1. What is the difference between type 1 and type 2 DM? Why is it assumed that this patient has type 2?
2. What risk factors does this patient present with? What symptoms may indicate that she has complications of type 2 DM?
3. Evaluate each of the available labs and relate them to the disease process.
4. Assess this patient's current weight. Determine the initial nutrition therapy prescription.
5. What would be the possible options to provide nutrition education for this patient?

NUTRITION CARE PROCESS FOR DISEASES OF THE ENDOCRINE SYSTEM

Step One: Assessment

Medical/Social History

Diagnoses

Family medical history

Medications

Previous medical conditions or surgeries

Socioeconomic status/food security

Support systems

Education—primary language

Anthropometric

Height

Current weight

Weight history (if available): highest adult weight; usual body weight

Children and adolescents: birthweight, growth chart results

Reference weight (BMI)

Waist/hip ratio; waist circumference

Biochemical Assessment

Visceral (Transport) Protein Assessment:

Albumin

Prealbumin

Hematological Assessment:

Hemoglobin, Hematocrit

Lipid Assessment

Total Cholesterol

HDL

LDL

Triglycerides

Other Laboratory Indices:

Electrolytes, BUN, Cr, Osmolality

Glucose, HbA1C, results of self-monitoring of blood glucose if available

Dietary Assessment

Ability to chew; use and fit of dentures

Problems swallowing

Nausea, vomiting

Constipation, diarrhea

Any other physical symptoms interfering with ability to ingest normal diet

Ability to consistently purchase food on a regular basis

Ability to feed self

Ability to cook and prepare meals

Food allergies, preferences, or intolerances

Previous food restrictions

Ethnic, cultural and religious influences on food choices, preparation

Use of alcohol, vitamin, mineral, herbal, or other type of supplements

Previous nutrition education or nutrition therapy

Eating pattern: 24-hour recall, diet history, food frequency with particular focus on carbohydrate intake

Physical activity: determine activity types and frequency

Determine limitations that may hinder exercise; assess willingness and ability to become more physically active

Step Two: Diagnosis

Impaired nutrient utilization

Altered nutrition-related laboratory values

Impaired intake of food fats

Excessive carbohydrate intake

Inadequate carbohydrate intake

Inconsistent carbohydrate intake

Inappropriate intake of types of carbohydrate

Overweight/obesity

Food and nutrition-related knowledge deficit

Sample PES: NI-53.4 Inconsistent carbohydrate intake

Inconsistent carbohydrate intake related to inability to calculate carbohydrate content of current diet as evidenced by inadequate knowledge of carbohydrate counting and wide ranges of carbohydrate content per meal. Inconsistency further evidenced by wide swings in blood glucose level.

Step Three: Intervention

1. Adjust carbohydrate prescription to meet patient's nutritional needs and pre/post prandial glucose goals.
2. Provide nutrition education for identification of carbohydrate content using food labels, food models, and carbohydrate counting methods.

Step Four: Monitoring and Evaluation

1. Patient will be able to state carbohydrate goals with appropriate insulin coverage.
2. Patient will be able to correctly identify carbohydrate content in a variety of regularly consumed foods.

WEB LINKS

National Diabetes Education Program (NDEP) (1–800–860–8747): NDEP is a federally sponsored initiative that involves public and private partnerships to improve treatment and outcomes for people with diabetes. Single copies of most materials are available free of charge or can be downloaded from the NDEP website.

http://ndep.nih.gov

American Diabetes Association (ADA) (1–800–DIABETES): The American Diabetes Association publishes many health professional and client materials in addition to its scientific journals. In an annual supplement to *Diabetes Care*, ADA reissues practice guidelines that address a wide array of clinical issues including nutrition. These guidelines can be downloaded from the ADA website free of charge.

http://diabetes.org

Diabetes Care and Education (DCE) Dietetic Practice Group of the American Dietetic Association: DCE publishes On the Cutting Edge, a theme-centered newsletter on timely topics related to nutrition and diabetes. Educational materials developed by the DCEP are available for purchase from the ADA. The DCE website provides information about Medicare MNT benefits for persons with diabetes, Health Care Financing Administration (HFCA) rules for ADA Education Recognition Programs reimbursement, and CPT codes for MNT.

http://eatright.org or http://www.dce.org

Division of Diabetes Translation of the Centers for Disease Control and Prevention (1–877–CDC–DIAB): The website has information about the Diabetes Control Program in each state and diabetes-related statistics such as the rise in the prevalence and incidence of diabetes. The *National Diabetes Fact Sheet CDC Information* is available as a downloadable file.

http://www.cdc.gov/diabetes

National Institutes of Diabetes & Digestive & Kidney Diseases (NIDDK): The NIDDK website provides information about its diabetes-related clinical trials and other research programs, a directory of diabetes organizations, health education programs, and diabetes-related topics. Downloadable files include Diabetes Dateline, an NIDDK newsletter; client education materials; and information for health professionals.

http://www.niddk.nih.gov/health/diabetes/diabetes.htm

American Association of Diabetes Educators (AADE): As a multidisciplinary organization of health professionals who teach about diabetes, the AADE and its website provide information about the scope of practice and standards related to diabetes education. The website also has information and AADE publications.

http://aadenet.org

END-OF-CHAPTER QUESTIONS

1. List the three chemical classes of endocrine hormones. For each class, pick one hormone, name its production site, and briefly describe its function.
2. Describe the action of insulin on carbohydrate, lipid, and protein metabolism.
3. What is the definition of diabetes mellitus (DM)? List the classifications for DM and briefly explain similarities and differences for their epidemiology, etiology, pathophysiology, and clinical manifestations. Describe three ways diabetes can be diagnosed.

4. What is meant by glycemic control, and why is it important? Describe the physiological consequences of poor glycemic control. Which laboratory measurements are an indicator of short- and long-term glycemic control? How often are they checked?

5. List and describe the types of insulin that are now available. How is insulin dosage determined, and how can insulin be administered? What are the differences in nutrition therapy recommendations for a person with diabetes who is using insulin on the conventional plan versus a person using intensive insulin therapy?

6. Briefly describe several meal planning approaches that are used with individuals with diabetes mellitus. Select a meal plan you would use for a 70-year-old man with an 8th grade education (type 2 DM), a 13-year-old teenage female athlete (type 2 DM), and a 32-year-old pregnant women (gestational DM), and justify your answers.

7. List the 7 classes of diabetes medications. Briefly describe their effects and mechanisms of action.

8. For individuals with type 2 diabetes, why is weight management often included as a component of nutrition therapy? Why is it important for the treatment of type 2 diabetes?

ENDNOTES

[1]Only endocrine disorders with nutritional implications will be discussed.

[2]The abbreviations T1DM and T2DM used in this chapter are not standardized abbreviations supported by the ADA, but are used for the sake of brevity.

[3]As opposed to hypoglycemia related to diabetes.

20

Diseases of the Renal System

Maria Karalis, M.B.A., R.D., L.D.N., Commercial Development Manager

Abbott Renal Care, Abbott Park, IL

Jessie M. Pavlinac, M.S., R.D., C.S.R., L.D., Clinical Nutrition Manager

Oregon Health & Science University, Portland, OR

Jordi Goldstein-Fuchs, D.Sc., R.D., Kidney Nutrition Specialist

Division of Nephrology, University of California, San Francisco, Sparks Dialysis, Sparks, NV

CHAPTER OUTLINE

The Kidneys
Normal Anatomy • Normal Physiology
Nephrotic Syndrome
Chronic Kidney Disease
Acute Renal Failure
Nephrolithiasis

Introduction

Kidney and urological diseases affect approximately 20 million Americans each year, and the medical interventions for these diseases constitute some of the most expensive areas of medical treatment. In 2003, medical care for individuals with kidney disease cost more than $20 billion dollars (CDC 2006). The primary causes of kidney disease are directly related to diabetes mellitus and hypertension. As the prevalence of these two diagnoses rise, the incidence of kidney disease will also rise.

The Kidneys

Normal Anatomy

The kidneys are two **retroperitoneal** organs the size of a fist. The right kidney is usually found to be slightly lower than the left kidney (see Figure 20.1). Each kidney is 11–12 cm long, 5–7.5 cm wide, and 2.5–3 cm thick. The average weight of a kidney in adults is 125–170 grams in men and 115–155 grams in women (Madsen and Verlander 2005). The kidney is made up of a complex capillary network and an array of **tubules** to perform regulatory and metabolic functions that

retroperitoneal—lying behind the peritoneum (lining of the abdominal cavity)

tubules—component of the nephron responsible for reabsorption and secretion; designated as the proximal convoluted tubule, the loop of Henle, and the distal convoluted tubule

FIGURE 20.1 Anatomy of Urinary Tract and Kidney
(a) The urine formed in the kidneys flows into the renal pelvis. It is carried by the ureters to the urinary bladder, which is drained by the urethra. (b) The human kidney.

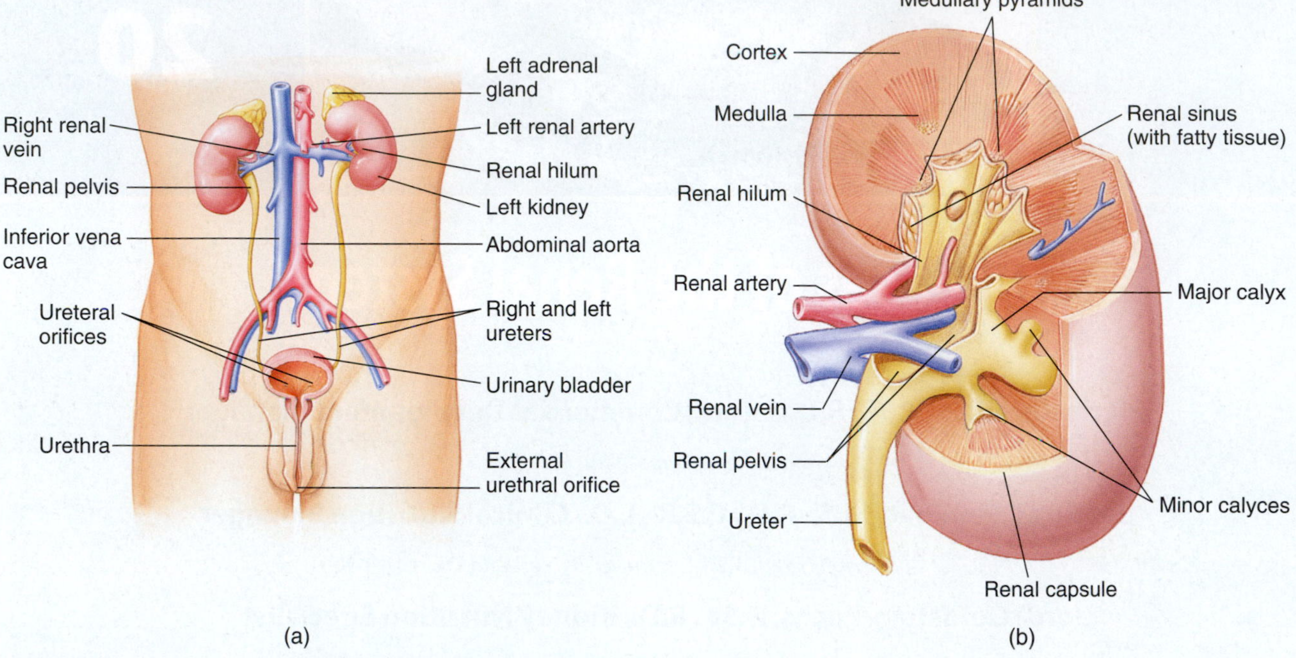

Source: R. Rhoades and R. Pflanzer, *Human Physiology*, 4e, copyright © 2003, p. 731.

are vital to life (Madsen and Verlander 2005; Rose 1987; Valtin and Schafer 1995).

The functioning unit of the kidney is called the **nephron**. Each kidney consists of approximately 1.2 million nephrons. Each nephron is made up of a **glomerulus**, which is a capillary tuft located between two arterioles (the **afferent** and the **efferent**), and a network of tubules lined by epithelial cells (see Figure 20.1) (Rose 1987). The afferent arteriole carries blood to the glomerulus, and the efferent arteriole carries blood from the glomerulus. The nephron extends through three sections of the kidney called the cortex, outer medulla, and inner medulla (see Figures 20.2 and 20.3) (Madsen and Verlander 2005). The tubular portion is divided into three subdivisions: the proximal convoluted tubule, the loop of Henle, and the distal convoluted tubule (Valtin and Schafer 1995). The cortex contains the glomeruli and the proximal and distal convoluted tubules. The medulla consists of the collecting ducts, loops of Henle, and vasa recta (Rose 1987; Madsen and Verlander 2005).

As blood passes through the kidney, an **ultrafiltrate** similar in composition to the blood is formed by the glomerulus, which filters large proteins and blood cells. This filtrate is then modified as it passes through the network of tubules by either reabsorption of amino acids, glucose, selective minerals, and water or by secretion of solutes and water. The cells located throughout the nephron vary in terms of structure and function. The proximal tubule contains cells with a complex brush border, tight intercellular junctions, and large numbers of mitochondria (Rose 1987; Valtin and Schafer 1995; Madsen and Verlander 2005). These cells are suited for active transport, and 65% of filtered sodium and water are reabsorbed here. The cells of the loop of Henle, unlike the structure of the cells in the proximal tubules, are suited for passive diffusion and do not perform active transport.

nephron—basic functioning unit of the normal kidney; each nephron has two main parts: the glomerulus and the tubule

glomerulus—a network of thin-walled capillaries closely surrounded by a pear-shaped epithelial membrane called the Bowman's capsule

afferent—carrying blood to the designated site; for example, the afferent arteriole carries blood to the glomerulus

efferent—carrying blood away from the designated site; for example, the efferent arteriole carries blood away from the glomerulus

ultrafiltrate—referring to the initial filtration of metabolic by-products from the filtered blood within the tubule

FIGURE 20.2 Component Parts of the Nephron

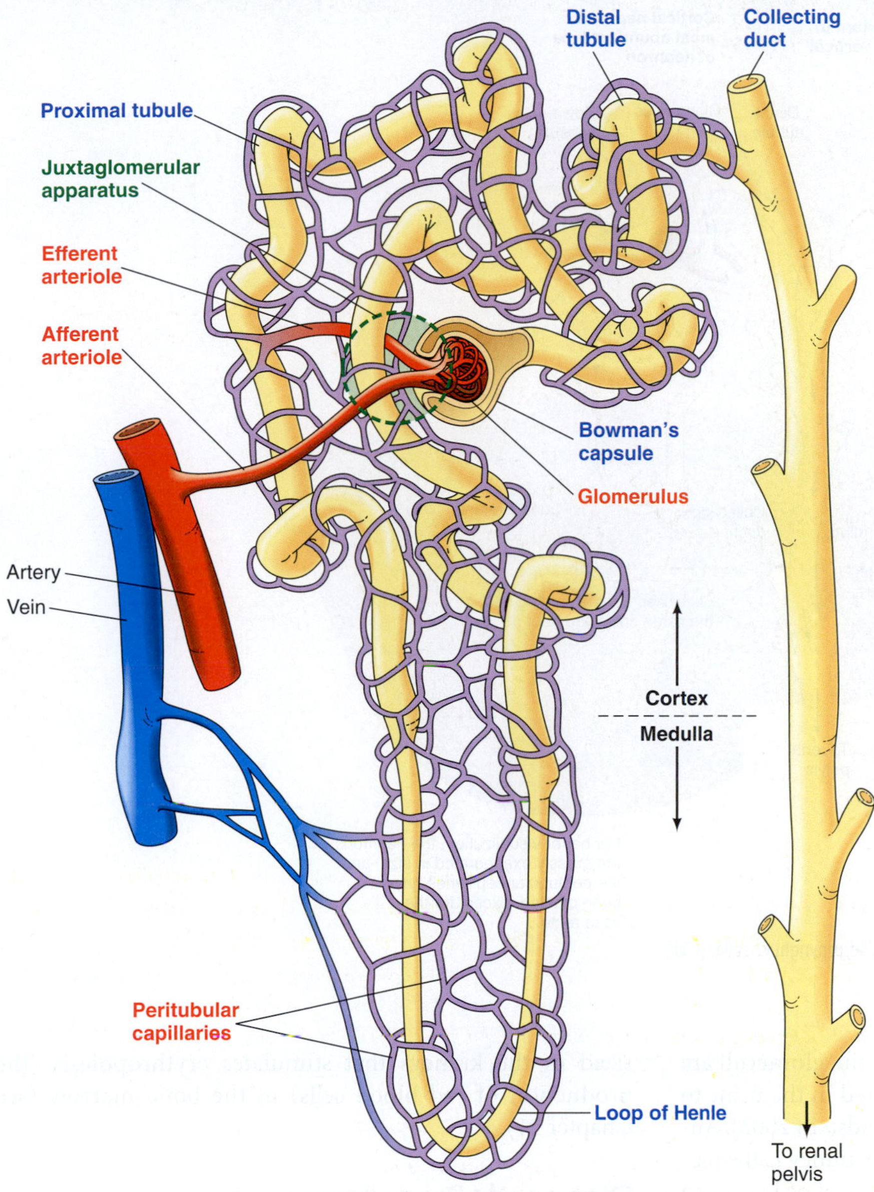

Overview of Functions of Parts of a Nephron

Vascular component

- Afferent arteriole—carries blood to the glomerulus
- Glomerulus—a tuft of capillaries that filters a protein-free plasma into the tubular component
- Efferent arteriole—carries blood from the glomerulus
- Peritubular capillaries—supply the renal tissue; involved in exchanges with the fluid in the tubular lumen

Combined vascular/tubular component

- Juxtaglomerular apparatus—produces substances involved in the control of kidney function

Tubular component

- Bowman's capsule—collects the glomerular filtrate
- Proximal tubule—uncontrolled reabsorption and secretion of selected substances occur here
- Loop of Henle—establishes an osmotic gradient in the renal medulla that is important in the kidney's ability to produce urine of varying concentration
- Distal tubule and collecting duct—variable, controlled reabsorption of Na^+ and H_2O and secretion of K^+ and H^+ occur here; fluid leaving the collecting duct is urine, which enters the renal pelvis

Source: L. Sherwood, *Human Physiology: From Cells to Systems*, 5e, copyright © 2004, p. 514.

The nephron completes several vital functions. Nephrons maintain the extracellular environment that is required for cell function (Rose 1987). This is completed by excretion of waste products of metabolism (i.e., uric acid, drugs, creatinine—a by-product of muscle breakdown—and urea converted from ammonia from protein metabolism in the liver) and by adjustment of the urinary excretion of water and electrolytes to maintain fluid, electrolyte, and acid-base balance. For example, the tubules reabsorb or secrete individual constituents such as potassium and water. The distal tubule, the primary site for potassium secretion, feeds into the collecting duct, which allows the urine to travel to the renal pelvis through the ureters, into the bladder, and finally through the urethra for excretion. Vasopressin, secreted by the pituitary gland, works at the collecting duct level in response to blood volume by either increasing or decreasing absorption of water to maintain fluid balance. Nephrons also secrete hormones that modulate renal hemodynamics, red blood cell production via production of erythropoietin (discussed in Chapter 21), and bone metabolism by activation of 1,25-dihydroxycholecalciferol (active vitamin D).

Normal Physiology

The kidney has three types of functions: excretory, metabolic, and endocrine. Compromise of these functions by renal disease requires special nutrition and medical management. Excretion and regulation of body water, minerals, and organic compounds are the most important functions of the kidney. Without excretory function, a patient rarely survives longer than 4 or 5 weeks, especially if the patient is hypercatabolic (Kopple 1994).

The kidneys remove nonessential solutes from the blood and conserve those essential to the body. Solutes

FIGURE 20.3 Comparison of Juxtamedullary and Cortical Nephrons

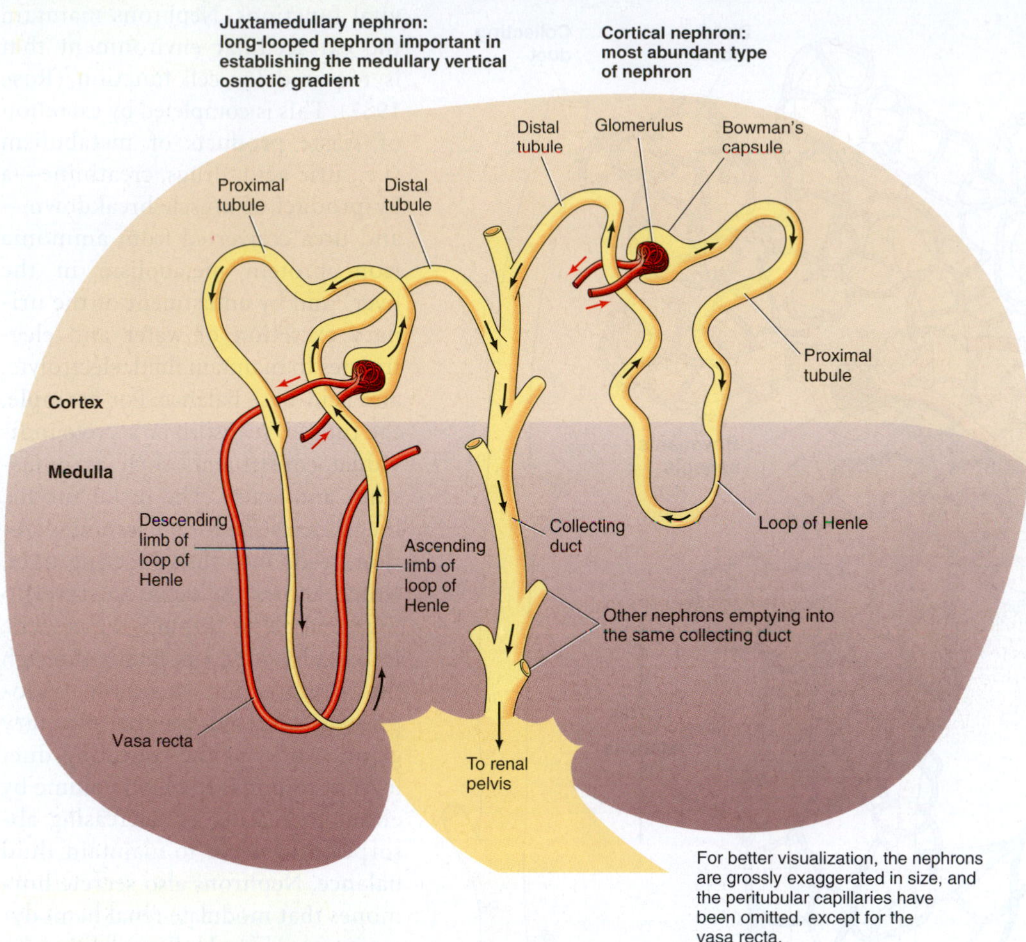

Source: L. Sherwood, *Human Physiology: From Cells to Systems*, 5e, copyright © 2004, p. 515.

that are freely filtered from the blood by the glomeruli are either reabsorbed by the tubules or excreted in the urine to maintain fluid and electrolyte balance (Goldstein 2002). Another important metabolic function of the kidney is the regulation of acid-base balance through removal of hydrogen ions from extracellular fluid in the tubules (see Chapter 8).

In addition to maintaining electrolyte and fluid balance, the kidney produces important hormones, including 1,25-dihydroxycholecalciferol and erythropoietin. The active form of vitamin D (1,25-dihydroxycholecalciferol, 1,25-$(OH)_2D_3$) is synthesized in the kidney after the inactive direct precursor (25-hydroxycholecalciferol, 25$(OH)D_3$) is hydroxylated in the liver. Erythropoietin (EPO) is a glycoprotein synthesized in the kidneys that stimulates erythropoiesis (the production of red blood cells) in the bone marrow (see Chapter 21).

glomerular filtration rate (GFR)—the filtration ability of the glomerulus; used as an index of kidney function; normal value is approximately 125 mL/min

Diagnostic Procedures

Diagnosis of kidney disease and overall assessment of kidney function require biochemical tests and morphological evaluation of the organ function. Biochemical tests involve estimation of the kidneys' ability to perform their normal physiological function. Most commonly, kidney function is measured by the **glomerular filtration rate (GFR)**, which is reflected in clearance tests that measure the rate at which substances are cleared from the plasma by the glomeruli. The normal GFR is 135–180 liters per day. Of this large volume, 98% to 99% of the filtrate is reabsorbed with urine output, usually averaging 1–2 liters per day. The GFR is used to evaluate kidney health, estimate the severity of diagnosed disease, and monitor kidney disease progression (Rose 1987).

Clearance, defined as the volume of plasma cleared of a particular solute in a given time, is expressed in moles, or weight of the substance per volume per time (Barri 2001; Zawada 2001). The mean GFR, expressed in milliliters per minute per 1.73 m^2, can be calculated as follows (Mitch and Walser 1996; Barri 2001; Goldstein 2002):

GFR = 122.49 – 0.37 (age) for adults younger than 45 years

GFR = 153.9 – 1.07 (age) for those 45 and older

In the clinical setting, endogenous creatinine clearance was once the "gold standard" used to approximate the actual GFR. It is now thought that approximation of GFR through calculations is the method of choice to approximate an individual's kidney function. The National Kidney Foundation (NKF) K/DOQI (Kidney Disease Outcomes Quality Initiative) Guideline recommends using equations based on serum creatinine but adjusted for ethnicity, gender, and age (National Kidney Foundation, K/DOQI Evaluation, Classification, and Stratification, 2002). The two equations most frequently cited are the Modification of Diet in Renal Disease (MDRD) Study equation and the Cockcroft-Gault equation (see Box 20.1). (See National Kidney Foundation, K/DOQI Evaluation, Classification, and Stratification, Clinical practice guidelines . . . Evaluation, 2002). Web-based GFR calculators, such as the one on the NKF website (http://www.kidney.org), are available.

Until commercial laboratories start calculating GFR, the formula most often used to calculate creatinine clearance will be (Zawada 2001) the following:

$$(140 - \text{Age}) \times \frac{(\text{Weight})}{72} \times \text{PCr}$$

where age is stated in years, body weight is in kilograms, and P-Cr is plasma creatinine concentration in milligrams per deciliter (Zawada 2001; National Kidney Foundation 2002).

This formula applies to white males. For women and African-American males, the result should be multiplied by 0.85 and 1.12, respectively. The formula overestimates GFR for persons who have edema or are obese (Kasiske and Keane 1996).

Plasma creatinine concentration varies inversely with GFR. The normal range of serum creatinine is 0.8 to 1.2 mg/dL for males and 0.6 to 1.0 for females. Many different laboratory techniques are available for measuring serum creatinine, and the upper limit of normal varies significantly. Although the serum creatinine value cannot be used as a measure of GFR, levels start to increase as kidney function decreases. The clinical definition of CRF includes a long-term reduction in GFR, decreased creatinine clearance, and a corresponding increase in serum creatinine concentration (Shaver 2001).

BOX 20.1 CLINICAL APPLICATIONS

Estimating Glomerular Filtration Rate (GFR)

The most widely used method for estimating GFR is the Cockcroft-Gault equation. This equation considers the effects of age, sex, and body weight on creatinine generation, thereby adjusting serum creatinine values to accurately reflect creatinine clearance:

$$\text{GFR} = [(140 - \text{age}) \times \text{body weight (kg)} \times 0.85 \text{ if female}] \div [72 \times \text{serum creatinine (mg/dL)}]$$

More recently, the Modification of Diet in Renal Disease (MDRD) modified GFR equation has come to be considered to be the "gold standard" of measurement (in addition to incorporating the influence of age and gender, and the effects of race, three biochemical measures are included):

$$\text{GFR} = 170 \times \text{serum creatinine}^{-0.999} \times \text{age}^{-0.176} \times \text{female}^{0.762} \times (1.18 \times \text{black race}) \times \text{SUN}^{-0.17} \times \text{serum albumin}^{0.318}$$

Source: National Kidney Foundation: Kidney Disease Outcomes Quality Initiative Clinical Practice Guidelines for Nutrition in Chronic Renal Failure. Adult Guidelines. *American Journal of Kidney Diseases,* 2000;35 (suppl 2):s17–s104.

Other biochemical assessments may involve tubular function tests. These assessments include evaluation of concentration and dilution, urine acidification, and sodium conservation. Morphological evaluation of the kidney includes microscopic evaluation of the urine (see Table 8.8), radiological evaluation, and biopsy of the organ. Radiological procedures include **intravenous pyelogram (IVP)**, renal ultrasonography, renal radionuclide imaging, computing tomography, MRI, and renal arteriogram.

Nephrotic Syndrome

Definition

Nephrotic syndrome (NS) is an abnormal condition that is marked by deficiency of albumin in the blood and its excretion in the urine due to altered permeability of the glomerular basement membranes. Nephrotic syndrome is

intravenous pyelogram (IVP)—radiographic imaging of the kidneys, ureter, and bladder using x-ray and contrast dye that is injected intravenously

nephrotic syndrome—a clinical condition consisting of losses of protein in the urine exceeding 3.5 g/day, hyperlipidemia, and low albumin levels (<3.5 g/dL) with edema

FIGURE 20.5 Example of a Hemodialysis System

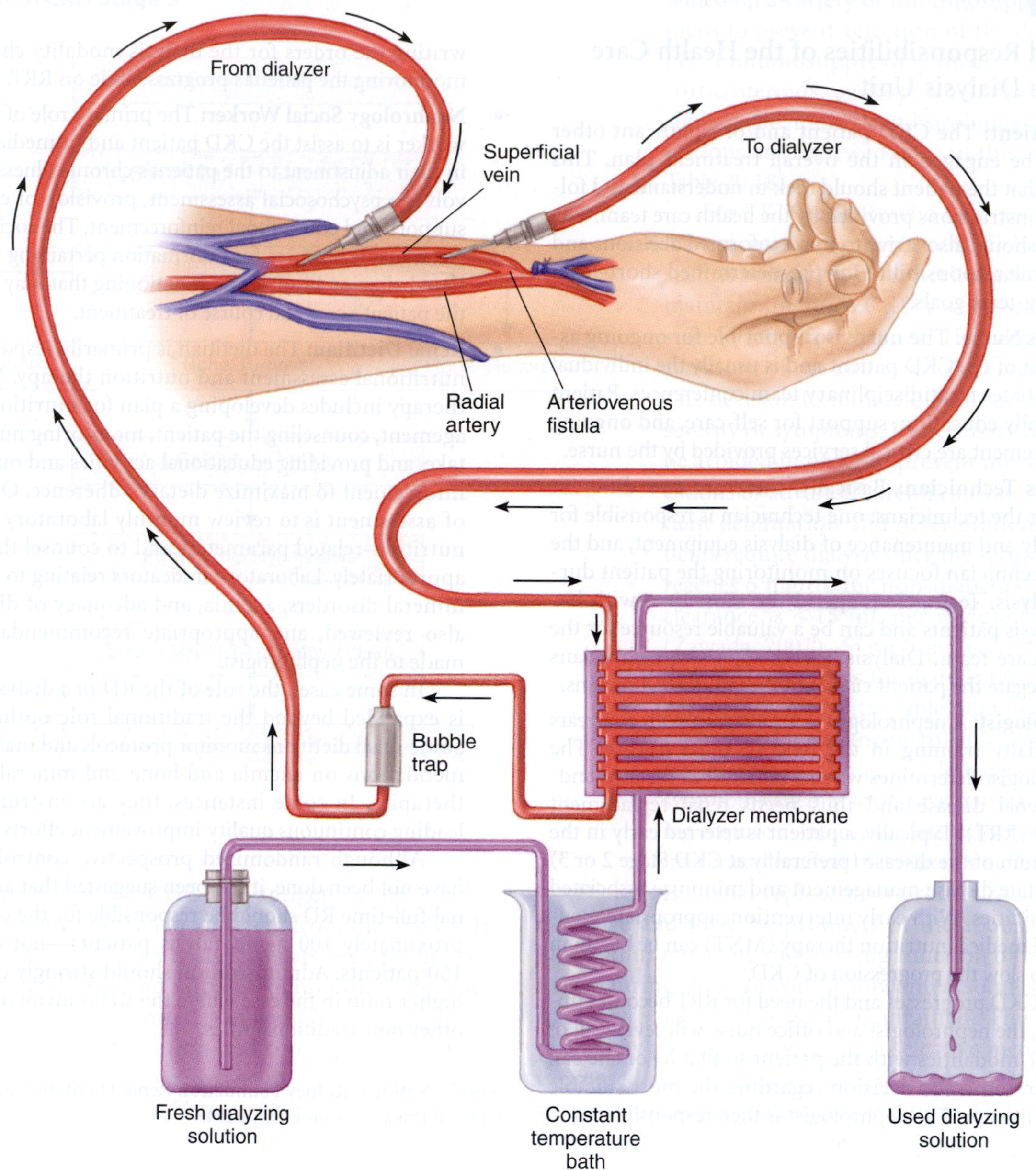

Source: R. Rhoades, and R. Pflanzer, *Human Physiology*, 4e, copyright© 2003, p. 782.

Vascular Access, 2001). If the patient's veins are not adequate for this procedure, an **arteriovenous graft (AVG)** can be created with polytetrafluoroethylene (Teflon) (see Figure 20.7) (National Kidney Foundation, K/DOQI Vascular Access, 2001). The AV fistula requires 4–6 weeks to become fully functional. The subclavian route may be used temporarily if HD is required before the AV fistula is ready for use.

arteriovenous graft (AVG)—a connection of an artery and vein to provide circulatory access for hemodialysis

The grafts can be punctured repeatedly by the arterial and venous needlesticks required for each dialysis treatment. Blood travels through a needle placed into the arterial side of the graft. The needle is attached to tubing that leads to the hollow fibers of the dialyzer, or between the sheets of membranes in the parallel plate design (Daugirdas, Van Stone, and Boag 2001). While blood passes through the dialyzer, dialysate simultaneously passes around the artificial membrane. Pressure gradients applied to the dialysate affect fluid and solute removal (Daugirdas, Van Stone, and Boag 2001). The blood then returns to the patient through the venous side (see Figure 20.5).

FIGURE 20.6 Peritoneal Dialysis

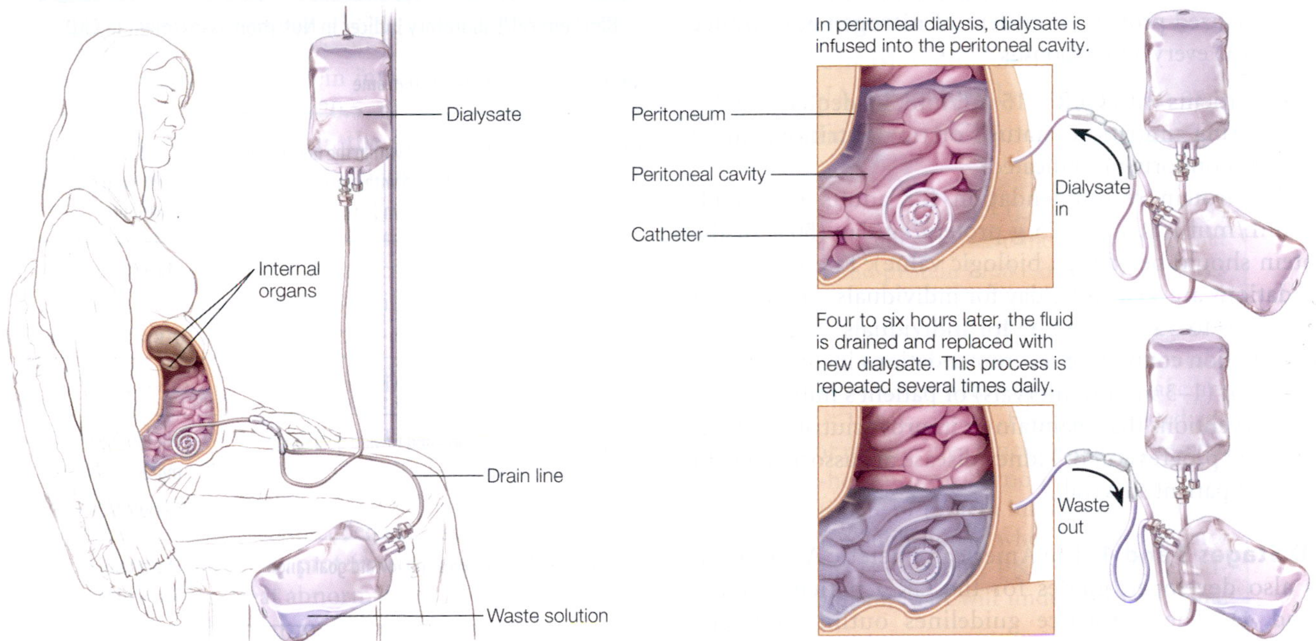
Dialysate
Internal organs
Drain line
Waste solution
In peritoneal dialysis, dialysate is infused into the peritoneal cavity.
Peritoneum
Peritoneal cavity
Catheter
Dialysate in
Four to six hours later, the fluid is drained and replaced with new dialysate. This process is repeated several times daily.
Waste out

Source: S. Rolfes, K. Pinna, and E. Whitney, *Understanding Normal and Clinical Nutrition,* 7e, copyright© 2006, p. 876.

FIGURE 20.7 Diagram of a Graft

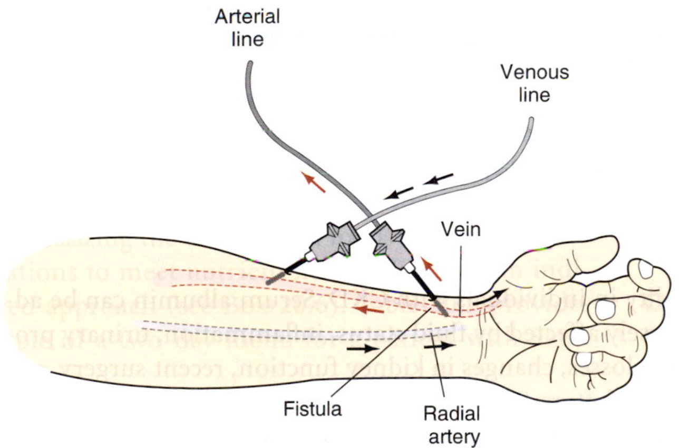
Arterial line
Venous line
Vein
Fistula
Radial artery

Source: Handbook of Dialysis, 3e, 2001. Lippincott, Williams & Wilkins.

In peritoneal dialysis, access to the patient's blood supply is gained via a catheter of silicone rubber or polyurethane, placed surgically into the peritoneal cavity (Blake and Daugirdas 2001; Sorkin and Blake 2001). In this procedure, dialysate is introduced into the peritoneum through the peritoneal catheter. Solutes from the plasma circulating in the vessels and capillaries perfusing the peritoneal wall pass across the peritoneal membrane into the dialysate, which is subsequently removed and discarded (Figure 20.6) (Blake and Daugirdas 2001; Sorkin and Blake 2001).

The dialysate for PD is available with a range of dextrose concentrations that alter the osmolality of the dialysate and assist in fluid removal (Blake and Daugirdas 2001; Sorkin and Blake 2001). In addition, the dwell time (i.e., how long the dialysate remains in the peritoneum) and the number of exchanges (i.e., how many bags of dialysate and the total volume of each used in 24 hours) also affect the amount of fluid and solute removal. Since PD dialysate is high in dextrose, weight gain may result.

Nutrition Therapy

CKD Stages 1 and 2 Specific nutrition goals for Stages 1 and 2 have not been identified, and nutrition therapy should focus on the comorbid conditions—diabetes, hypertension, and hyperlipidemia—and on slowing the progression of potential cardiovascular disease. Glucose control, decreasing blood pressure, and lipid management all have well-documented nutrition guidelines (National Kidney Foundation, K/DOQI Evaluation, Classification, and Stratification, 2002).

The panel of experts that comprised the K/DOQI Nutrition Work Group outlined three clinical practice guidelines specifically for adults with CKD who are not on dialysis (Stage 4). These guidelines (National Kidney Foundation, K/DOQI Evaluation, Classification, and Stratification, 2002) outline recommended nutrition measures, protein intake, energy requirements, and nutrition counseling. Nutrition measures identified for individuals with GFR <20 mL/min include:

- Serum albumin and actual or percent standard body weight and/or subjective global assessment (SGA) every 1 to 3 months

TABLE 20.18

Commonly Used Immunosuppressants and Potential Adverse Effects

Generic	Brand Name(s)	Possible Adverse Effects	Nutrition Implications
Cyclosporine A	Neoral, Sandimmune	N/V, diarrhea, hyperkalemia, hyperglycemia, gingival hyperplasia, hypertension, hypomagnesemia, GI distress	No K supplement or salt substitute; caution with grapefruit; anorexia
Azathioprine	Imuran	N/V, mouth ulcers, GI distress, esophagitis	Anorexia, steatorrhea
Corticosteroids	Prednisone, Deltasone	Cushingoid appearance, sodium retention, enhanced appetite, osteoporosis, protein catabolism, hyperglycemia	Caution with DM patients; highly protein bound; may need increased K, PO_4, Ca, and vitamins A, C, and D, increased protein, decreased dietary Na; avoid alcohol
Tacrolimus	Prograf	Hypertension, hyperglycemia, hyperkalemia, hypomagnesemia, GI distress	No K supplement or salt substitute; caution with grapefruit; anorexia
Mycophenolate Mofetil (MMF)	CellCept	GI distress, GI bleeding, hypophosphatemia, hyperglycemia, hypercholesterolemia	Take on an empty stomach; take Mg supplements separately
Sirolimus	Rapamune	Hyperlipidemia, hypokalemia, delayed wound healing	Avoid grapefruit and grapefruit juice
Cyclophosphamide	Cytoxan	N/V, abdominal pain, dry mouth, diarrhea	Increased fluid needs

Source: Manual of Clinical Dietetics, Sixth Edition, American Dietetic Association; 2000.

TABLE 20.19

Drug-Nutrient Interactions for Medications Used for Diseases of the Renal System

Classification	Mechanism	Generic	Brand Names	Possible Food–Drug Interactions
K Sparing Diuretic	Aldosterone receptor antagonist	Spironolactone	Aldactone, Aldactazide	Avoid excess K intake & K supplementation, avoid salt substitutes, ↓ kcal and Na may be recommended, avoid natural licorice, anorexia, ↑ thirst, N/V, diarrhea, gastritis
K Sparing Diuretic	Aldosterone receptor antagonist	Spironolactone	Aldactone, Aldactazide	Avoid excess K intake & K supplementation, avoid salt substitutes, ↓ kcal and Na may be recommended, avoid natural licorice, anorexia, ↑ thirst, N/V, diarrhea, gastritis
K Sparing Diuretic	Aldosterone receptor antagonist	Amiloride	Midamor	N/V, dry mouth, diarrhea, avoid salt substitutes
Antibiotic	Interferes with bacterial protein synthesis	Neomycin	Mycifradin, Neo-Fradin, Neo-Tab	Neomycin impairs absorption (and may also increase excretion) of a broad variety of nutrients including carbohydrates, fats, calcium, iron, magnesium, nitrogen, potassium, sodium, folic acid, and vitamins A, B_{12}, D, and K
Recombinant Human Erythropoietin	Stimulates RBC production	Epoetin alfa	Epogen, Procrit	May need Fe, B_{12}, and/or folate supplementation, N/V, diarrhea

Even in the absence of hyperparathyroidism, hypophosphatemia occurs in as many as 50% of post-transplant patients (Massari 1997). This is common, due to renal tubular phosphate wasting and increased urinary phosphate loss from the effects of immunosuppressive medications. Most post-transplant patients will need to increase their phosphorus. Phosphorus supplementation is often needed in the early post-transplant period. Serum potassium levels should be monitored closely, because some phosphorus supplements (Neutraphos-K) contain large amounts of potassium.

Rejection Because doses of corticosteroids are increased during periods of acute rejection, protein and kcal requirements are increased due to increasing catabolism. The same guidelines used for calculating protein and kcal requirements in the acute post-transplant phase apply. In the presence of chronic graft rejection, several studies have alluded to reduced protein intakes (0.55 g per kg body weight) as providing a beneficial effect in protecting the graft (Kasiske et al. 2000; Martinez-Castelao et al. 2002). However, the long-term efficacy of lower protein intakes needs to be further studied.

Acute Renal Failure

Definition

Acute renal failure (ARF) is a disorder in which the kidneys suddenly stop functioning, characterized by abrupt cessation or reduction in GFR and accumulation of nitrogenous wastes.

Epidemiology and Etiology

The prevalence of ARF is estimated as 1% for all hospitalized patients, 3% to 5% for general medical-surgical patients, 5% to 25% for those in intensive care units, 5% to 20% for open heart surgery patients, 10% to 30% for those receiving aminoglycoside therapy (a group of antibiotics used to treat Gram-negative bacteria), 20% to 60% for those with severe burns, 20% to 30% for those with rhabdomyolysis (destruction of muscle tissue accompanied by the release of myoglobin into the bloodstream resulting in sometimes acute renal failure), and 15% to 25% for those treated with cisplatinum, bleomycin, and vinblastine (chemotherapeutic agents) (Anderson and Schrier 1993; Thadhani, Pascual, and Bonventre 1996; Albright 2001).

Data indicate that the mortality rate from ARF has remained around 50% for the past 25 years (Leblanc, Tapolyai, and Paganini 1995; Shah 2001). The high mortality rate is attributed to the fact that many patients who present with ARF are older and often have a complicated medical or surgical course. In addition, patients with ARF who do not survive often die from extrarenal (occurring outside of the kidney) disease rather than from ARF itself.

It has been estimated that death occurs in 40% of nonsurgical patients with severe ARF, in as many as 80% of surgical patients, and in 20% of those with noncatabolic conditions. This high morbidity is associated with the degree of hypercatabolism and infection (Druml 1995; Shah 2001). No method has yet proved to reduce the catabolism observed in this patient population. ARF is associated with many clinical situations that result in a stress or injury-induced hypercatabolic state.

Pathophysiology

It is common for the nutritional status of patients with acute renal failure to decline within a short period of time, owing to nitrogen losses (up to 30 g per day), which leads to loss of lean body mass; toxicity-related symptoms (anorexia, nausea, vomiting, bleeding); loss of essential and nonessential amino acids and plasma proteins during intervention dialysis therapy; and intermediary metabolic disturbances (impaired glucose utilization and protein synthesis) from uremia. Energy and protein malnutrition often result, although one may predominate.

Clinical Manifestations

Normal urine output is 1 to 1.5 L per day. Oliguria is urine output of less than 400 mL per day; anuria, less than 100 mL per day; and polyuria, more than 3 L per day (Wolk and Swartz 1986; Mehta 1994). ARF patients are likely to develop fluid and electrolyte disorders, azotemia, and wasting, particularly if they are both oliguric and hypercatabolic (common complications of ARF).

Electrolytes Serum levels of potassium, magnesium, and phosphorus are generally elevated in patients with ARF because of decreased renal clearance and marked net protein breakdown. On the other hand, decreased levels of serum potassium, magnesium, and phosphorus may also occur as a result of intracellular shifts associated with carbohydrate delivery and anabolism. In addition, serum phosphorus may be decreased secondary to severe respiratory alkalosis as a result of increased clearance across the dialysis membrane with **continuous renal replacement therapy** (CRRT) or because of intracellular shifts. Hypophosphatemia also occurs in the refeeding syndrome, malnutrition, and diuretic therapy. Serum levels of potassium, magnesium, and phosphorus should therefore be monitored frequently to assess the need for additional supplementation. Delivery of potassium, magnesium, and phosphorus should be individualized according to serum levels (Mehta 1994).

Blood Urea Nitrogen and Creatinine Blood urea nitrogen (BUN) and creatinine are elevated in ARF, although the ratio of BUN to creatinine may be normal (10:1 or higher). Insufficient dietary kcal and protein and altered blood levels of proteases contribute to high levels of protein catabolism. Dialysis may be required to remove metabolic wastes and excess water. When recovery of renal function is expected to take several weeks, or when wasting is severe, aggressive dialysis is often recommended. Medical and nutrition management typically aim to maintain BUN in the range of 80 to 100 mg/dL (Mehta 1994).

continuous renal replacement therapy (CRRT)—type of renal replacement therapy used to treat patients in acute renal failure, particularly those with multiple organ failure; the types of patients treated tend to be hemodynamically unstable, have poor cardiac output, and be unable to tolerate hemodialysis

FIGURE 21.5 Blood Cells in Iron-Deficiency Anemia
A normal red blood cell smear is shown on the top, using Wright's stain under a microscope. The cell in the center is a neutrophil or white blood cell type. The red blood cell smear shown on the bottom is of severe iron deficiency anemia. There is a lymphocyte slightly left of center. Notice the cells are microcytic (small sized) and hypochromic (pale colored). There is a good deal of variation in the size of the cells and in their shapes (anisocytosis, poikilocytosis). The red cell is normally rounded and all are about the same size.

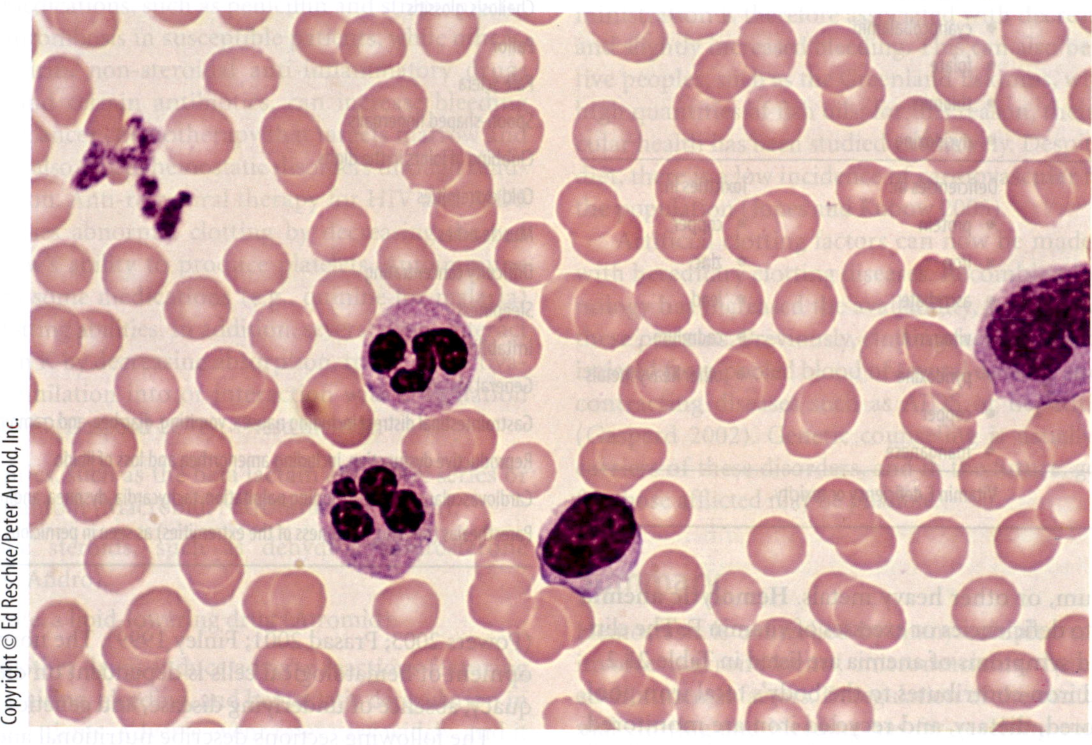

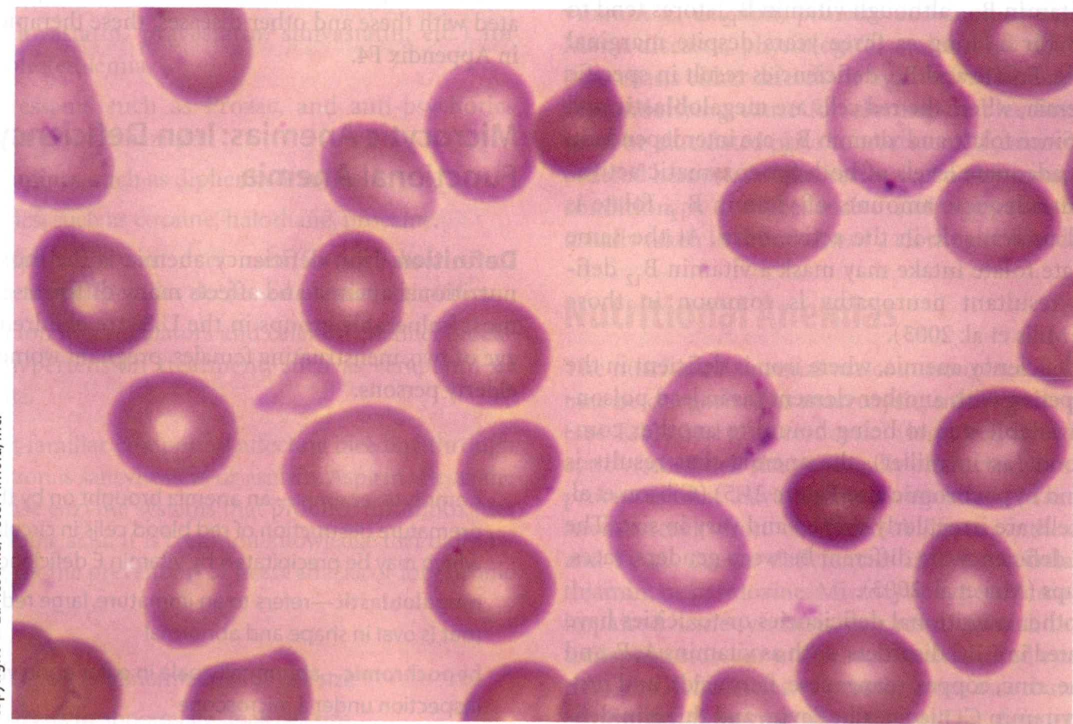

Iron-deficiency anemia is a condition where there is a decrease in the number of normal circulating RBC per cubic millimeter of blood, decreased levels of hemoglobin, or decreased volume of packed RBC per deciliter of blood as a result of greater demand on stored iron than can be supplied. The diagnosis of anemia is based on visual inspection of RBC as well as laboratory indices. Iron-deficiency anemia is a microcytic anemia, and is staged as follows:

- Subclinical with no overt symptoms
- Clinical with laboratory value alterations and some observable signs and symptoms
- Overt clinical iron-deficiency anemia with alterations in laboratory values and observable signs and symptoms upon physical examination of the patient

These stages correspond to negative iron balance, iron-deficient erythropoiesis, and iron-deficiency anemia (see Figure 21.6).

Epidemiology The epidemiology of iron-deficiency anemia worldwide varies by socioeconomic status, with citizens of poor nations and poor persons within wealthy nations the most susceptible. Because minority groups occupy lower socioeconomic strata, the incidence and prevalence of iron deficiency with overt clinical features is higher in African-American and Hispanic women than in white women (Anonymous 2002).

Prevalence of iron deficiency in the United States is estimated to be greatest among females aged 12–49 (12%) and children aged 1–2 (7%) (CDC 2002). The rates of iron deficiency in the U.S. have increased as a percentage of the population from 1994 to 2000. Predictions are that these numbers will continue to increase, in light of the "graying of America" or the aging of the U.S. population, which will leave greater numbers of older, poor women and minority women at greater risk for iron-deficiency anemia (Guralnik et al. 2004).

Etiology The etiology of iron-deficiency anemia varies greatly. It can result from blood loss, as in the event of gastric ulceration or dysmenorrhea (abnormal menses). Blood losses require homeostatic restoration of blood volume (Ashorn 2004). If volume is restored, but new RBC have not been produced at the same rate, the number of viable RBC in a given volume of blood is decreased; this results in a functional anemia, a situation where oxygen is insufficient

FIGURE 21.6 Sequential Changes in Iron Status

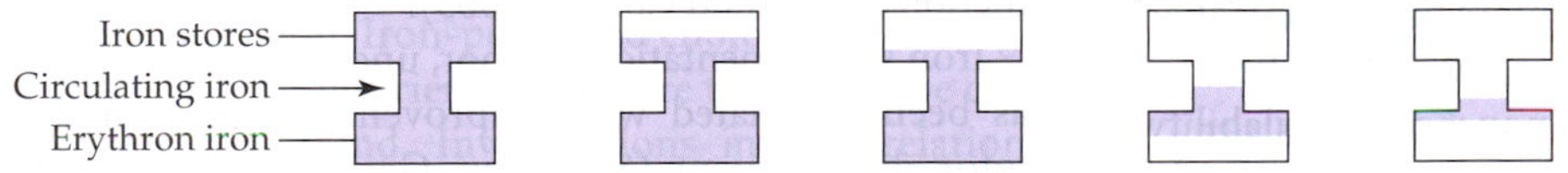

Iron stores — Circulating iron — Erythron iron

	Normal	Early Negative Iron Balance	Iron Depletion	Iron-Deficient Erythropoiesis	Iron Deficiency Anemia
Reticuloendothelial marrow iron	2–3+	1+	0–1+	0	0
Transferrin iron-binding capacity (μg/dL)	330±30	330–360	360	390	410
Plasma ferritin (μg/L)	100±60	<25	20	10	<10
Iron absorption (%)	5–10	10–15	10–15	10–20	10–20
Plasma iron (μg/dL)	115±50	<120	115	<60	<40
Transferrin saturation (%)	35±15	30	30	<15	<15
Sideroblasts (%)	40–60	40–60	40–60	<10	<10
Erythrocyte protoporphyrin (μg/dL)	30	30	30	100	200
Erythrocytes	Normal	Normal	Normal	Normal	Microcytic Hypochromic
Serum transferrin receptors	Normal	Normal–high	High	Very high	Very high
Ferritin iron	Normal	Normal–low	Low	Very low	Very low

Source: Adapted with permission by The American Journal of Clinical Nutrition, Copyright © American Journal of Clinical Nutrition, American Society for Nutrition.

animal origin and vitamin C sources between meals. Iron-fortified meal replacements and energy drinks should not be used unless specified by a physician. If the patient is being treated with binders, the diet should be modified accordingly, and more liberalization of the diet may occur. More iron sources and iron–vitamin C combinations are permitted. Continued monitoring of blood levels should dictate dietary constraints and patient education strategies for dietary adherence.

Future Research The genetic mutations for hemochromatosis have been mapped. Isolating the genetic sequence responsible for enhanced iron absorption in hemochromatosis may lead to discoveries of compounds that would enhance iron absorption from nutritionally poor substrates. Such a discovery could decrease the incidence of iron-deficiency anemia worldwide. Iron-deficiency anemia is prevalent and causes many deaths secondary to decreased immunity. This application could improve global health.

Hemoglobinopathies: Non-Nutritional Anemias

The five major classes of hemoglobinopathy are structural (e.g., sickle cell anemia), thalassemias, thalassemic hemoglobin variants, hereditary persistence of fetal hemoglobin, and acquired hemoglobinopathy (e.g., anemia secondary to exposure to toxins or a disease state such as cancer). Table 21.10 summarizes the etiology, manifestations, and medical treatment for selected hemoglobinopathies. Pertinent nutrition interventions, which are typically designed to support medical treatments, are described in the following sections.

Sickle Cell Anemia

Sickle cell anemia is the most common structural hemoglobinopathy, defined as homozygous abnormal hemoglobin polymerization resulting in a sickling of cells in symptomatic individuals. The crescent-shaped cell morphology is obvious when stained and magnified. Combinations of sickle cell and thalassemia occur as a result of inheritance of the traits from each parent and result in variant syndromes not strictly categorized as sickle cell anemia (Gaspard 2002).

Nutrition Therapy Treatment for an acute sickle cell crisis should include increased macronutrients, because there is an increased level of energy expenditure. Oral glutamine may be beneficial for the patient, because glutamine is an important amino acid for rapidly dividing cells (Williams 2004). Antioxidant vitamin and mineral status should be evaluated and maintained at recommended levels. Optimal folate, vitamin B_{12}, and pyridoxine levels should be maintained through supplementation. Fluids and hydration must also be closely monitored in these patients (Fey 2003).

Thalassemia

The thalassemias are inherited disorders of abnormal alpha or beta globin synthesis. Reduction in globin availability results in decreased hemoglobin synthesis. The severity of the resulting anemia is dependent upon the degree to which synthesis is impaired. There can be several inherited abnormal genes, which worsens the clinical phenotype. The RBC are hypochromic, elliptical, and irregular. The bone marrow becomes hyperplastic in severe cases, with increased anemia stimulating excessive erythropoietin production with no concomitant increase in functional hemoglobin synthesis (Benz 2005).

Nutrition Therapy Transfusions are the standard treatment for hemoglobinopathies. Iron overload is often treated with an iron chelator, deferoxamine. Patients often undergo periods where they become unresponsive to deferoxamine treatment. Ascorbic acid administration enhances the efficacy of the chelation regimen (Forget and Cohen 2005).

Polycythemia

Defined as an increase in circulating RBC, the syndrome can be spurious or real. Spurious polycythemia relates to decreased plasma volumes showing an increase in RBC above normal within a deciliter of blood; real polycythemia is a result of dysregulated feedback and increased production of red cells by marrow, usually detected by abnormal epoietin levels. Very low erythropoietin levels usually indicate polycythemia vera (**chronic myeloproliferative disease**). Very high erythropoietin levels indicate either polycythemia due to abnormal production of RBC (primary) or polycythemia as a physiological response to hypoxia (secondary) (Hoffman, Baker, and Prchal 2005).

Nutrition Therapy Routine phlebotomy induces iron-deficiency anemia. Patients should be educated regarding increasing their dietary iron (refer to Table 21.7) and avoiding very high-dose supplementation, which causes complications from oxidative stress and the increased competition among minerals for absorption (Hoffman, Baker, and Prchal 2005).

sickle cell anemia—a hereditary disease of genetically altered red blood cells that have a sickled shape, carry abnormally formed hemoglobin, and have abnormal transport capabilities for oxygen; the disease is thought to confer protection against malaria

chronic myeloproliferative disease—long-term hyperplasia of hematological tissues, with concomitant overproduction of abnormal cells, growth factors, chemokines, cytokines, and hormones involved in hematopoiesis

TABLE 21.10

Characteristics of Non-Nutritional Anemias (Hemoglobinopathies)

Anemia	Epidemiology	Etiology	Pathophysiology	Clinical Manifestations	Treatment
Sickle Cell Anemia	Found in areas where malaria is endemic. 1 in 400 African-Americans in the U.S. are homozygous for sickle cell (Benz 2005).	The substitution shaped cell glutamic acid in the hemoglobin causes it to turn into a gel. There is occlusion of blood vessels due to increased stickiness	The crescent of valine for decreases oxygen carrying capacity, makes the cell die off prematurely and causes cell stickiness (Tiosano and Hochberg 2001).	Sickle cell crisis involves severe pain and fatigue. Blood vessel occlusion can occur anywhere in the body. Infections, jaundice, hyperbilirubinemia are common as a result of premature lysis of RBC (Gaspard 2002; Stettler et al. 2001; Terlouw et al. 2004).	Transfusions Bone marrow transplants Chronic antibiotic and pain medication (Benz 2005; Forget and Cohen 2005)
Thalassemia	Present in 15% of African-Americans and persons of Mediterranean heritage (Forget and Cohen 2005).	Abnormal globin proteins are produced, and severity varies with hetero- or homozygous genetics.	Abnormal proteins cause premature RBC death and inability to carry oxygen (Benz 2005).	Stunting, birth defects, organ damage, and severe hypoxia (Forget and Cohen 2005).	Transfusion Bone marrow transplant
Polycythemia	Prevalence is estimated at 2 per 100,000 persons (Hoffman, Baker, and Prchal, The Polycythemias, 2005).	Hyperproliferation of stem cells in bone marrow, with high levels of circulating blood cells.	Increased cell numbers result in increased blood viscosity.	Impaired circulation, hypertension, brain ischemia, and organ damage (Hoffman, Baker, and Prchal, The Polycythemias, 2005).	Phlebotomy Radiation and chemotherapy to reduce cell numbers Continuous airway pressure pumps to deliver oxygen Anticoagulants (Hoffman, Baker, and Prchal, The Polycythemias, 2005)
Hemolytic Anemia	Prevalence of autoimmune hemolytic anemia is 1 per 100,000 with boys 2.5x more likely to be affected. Prevalence of Rh alloimmunization is estimated at 11 cases per 100,000 live births (Gaspard 2002).	Anemia results in insufficient oxygen carrying capacity due to fewer numbers of RBC, trauma (Cunningham and Silberstein 2005; Schrier and Reid 2005).	Immature RBC are produced as a result of negative feedback on hematopoiesis by premature RBC destruction. Hypoxia and tissue damage occur (Rutigliano et al. 2002).	Build-up of the byproducts of hemoglobin metabolism result in fatigue, bleeding, infection, jaundice, and presence of breakdown products in the urine (hemoglobinuria) (Gaspard 2002; Schrier and Reid 2005; Cunningham and Silberstein 2005).	Transfusion Immunosuppressive treatment
Anemia of Prematurity	Prematurity occurs in approximately 12% of live births in the U.S. 84% of these births exhibit anemia (Kramer and Cohen 2005).	Decreased production of erythropoietin by the premature kidney (Rowe and Avivi 2005).	Bone marrow is understimulated without epoetin; there is a decreased iron absorption by the infant and insufficient oxygenation of tissues results.	Cyanosis (bluish appearance with lack of oxygen and build-up of carbon dioxide in blood); congestive heart failure; hemorrhage; multiple organ system failure (Rowe and Avivi 2005).	Recombinant human erythropoietin administration Transfusion Iron supplementation (Whitehall, Patole, and Campbell 1999; Rowe and Avivi 2005)
Aplastic Anemia	Prevalence is estimated at 1 in 100,000 live births (Fanconi's Autosomal Recessive Aplastic Anemia) (Young and Maciejewski 2005).	Bone marrow failure to produce cells. This can be due to heredity or secondary to toxic infection or exposure to chemicals (Young and Maciejewski 2005).	Marrow failure to make cells reduces clotting ability, immunity, and ability to carry oxygen (Young and Maciejewski 2005).	Sepsis (systemic infection); hemorrhage; failure to thrive in infants (Young and Maciejewski 2005).	Transfusion Bone marrow transplant
Blackfan-Diamond Anemia	Prevalence is estimated at 7 per 1 million live births.	Inherited bone marrow failure.	Anemia, hypoxia resulting from abnormal stem cells in the marrow failing to produce viable blood constituents (Gunasekaran et al. 2000; Cipolli et al. 1999).	Tissue and organ damage is widespread.	Bone marrow transplant Transfusions Il-3 and corticosteroid administration (Freedman 2005)
Schwachmann's Sndrome	Prevalence is estimated at 100 per 1 million live births.	Inherited bone marrow failure.	Anemia, hypoxia due to bone marrow stem cell failure (Gunasekaran et al. 2000; Cipolli et al. 1999).	Tissue and organ damage is widespread.	Bone marrow transplant Transfusions Il-3 and corticosteroid administration (Freedman 2005)

PRACTITIONER INTERVIEW

Mary Ellen Beindorff, RD LD, Specialty—*Liver Transplant*, *Barnes-Jewish Hospital, St. Louis, MO*

In your practice, which patients are typically anemic?

From my experience, I expect to see anemia in certain populations—such as chronic kidney disease, cancer, liver disorders—usually resulting from the disease and/or its treatment. Anemia is common in chronic inflammatory diseases and disorders that cause blood loss. It is also not unusual to see anemia in the very young, pregnant women, or seniors. In the very young and during pregnancy, iron deficiency would typically be the issue, but with seniors it is more likely macrocytic pernicious anemia related to decreased intake of vitamin B_{12} and/or impaired absorption resulting from decreased production of intrinsic factor.

How has the treatment of anemia changed?

Since I became a dietitian, the medical treatment of non-nutritional anemias has dramatically improved because of the medication erythropoietin. Yet it too has nutritional implications. The patient must have adequate protein, energy, and iron for the medication to be effective.

Advice to dietetic students?

Nutritional assessment should include looking at the patient's hematological lab values. I find that new dietetic interns frequently overlook theses labs, mainly because they are so overwhelmed with everything else that needs to be taken into consideration for nutritional assessments. My advice to interns is to look at the "big picture." As with all signs and symptoms, the dietitian needs to distinguish between nutritional or non-nutritional causes for altered labs. Understanding the pathophysiology of the disease is essential to determine these distinctions.

CASE STUDY

CC:

Mother relates that child has poor appetite; appears more tired than usual; having some difficulty in kindergarten—when asked, describes behavior issues and "acting out."
DOB: 5/4/2000

General:

5-year-old African-American male; birthweight 5 lbs. 2 oz. at 38 weeks gestation.
Ht. 40 inches (5 percentile stature for age) Wt. 32 lbs. (5–10 percentile for weight for age)
BMI: 14 (5 percentile)

Labs:

MCV 65; MCHC 27; Hgb Hematocrit

Rx:

None

Past Med Hx:

Frequent upper respiratory infections; multiple ear infections.

Nutrition Hx:

Consumes 24–48 oz. milk per day; does not like a lot of meat, and mother relates that they really don't have it very often. Favorite foods include peanut butter and jelly sandwiches, french fries, and hot dogs. Family receives WIC supplemental foods for younger sibling and $200.00 in food stamps each month. Mother is employed, but with four children, appears to have inadequate access to food.

1. Patient was diagnosed with hypochromic microcytic anemia, most likely secondary to iron deficiency. Define the terms hypochromic and microcytic. How are they related to his diagnosis?
2. Define each of his laboratory values.
3. What are signs and symptoms in the patient's history that are consistent with this diagnosis?
4. Evaluate his anthropometric information. Are there any particular concerns?
5. What additional dietary history information will be important for the RD to determine?
6. What will be his primary nutrition diagnosis?
7. How will his iron deficiency anemia be treated?

NUTRITION CARE PROCESS FOR DISEASES OF THE HEMATOLOGICAL SYSTEM

Step One: Nutrition Assessment

Medical/Social History

Standard

Dietary Assessment

Standard but close attention to major food sources of iron, folate, zinc, and B_{12}

Previous nutrition education or medical nutrition therapy

Eating pattern: 24 hour recall, diet history, food frequency

Anthropometric

Standard

Biochemical Assessment
Visceral Protein Assessment:

Total Protein

Albumin

Prealbumin

Transferrin

Retinol Binding Protein

Fibronectin

3-Methyl Histidine

C-Reactive Protein

Hematological Assessment:

Hemoglobin

Hematocrit

Serum Iron

Total Iron Binding Capacity

Iron binding capacity saturation

Ferritin

MCV

MCHC

MCH

Methemoglobin

Protoporphyrin, free erythrocytic

Transferrin receptor

Transferrin

Sickle Cell Test

Lipid Assessment

Total Cholesterol

HDL

LDL

Triglyceride

Vitamin/Mineral Status Specific Labs	*Hepatobiliary*
Folate	Erythrocyte folate, free folate, (FIGLU) urinary formiminoglutamic acid
Vitamin B_{12}	Schilling test, erythrocyte B_{12}, DUMP test, serum B_{12}
Vitamin B_6	Whole blood level of pyridoxal phosphate
Zinc	Serum zinc

Step Two: Nutrition Diagnosis

Inadequate mineral intake

Altered nutrition-related laboratory values

Food-medication interaction

Food and nutrition-related knowledge deficit

Sample: Inadequate iron intake – NI-55.1

PES: Inadequate dietary iron as evidenced by average daily consumption of less than 6 g of iron—<50% of DRI for age and gender.

Step Three: Intervention

1. Increase iron through consumption with a source of vitamin C.
2. Increase iron through consumption of foods with higher iron bioavailability.
3. Increase the use of high-iron foods in the diet.
4. Increase iron intake through the use of iron-fortified foods.
5. Increase iron absorption through proper supplementation.

Step Four: Monitoring and Evaluation

1. Monitor iron supplementation to assure compliance with instructions for optimal absorption.
2. Evaluate 24-hour recall for intake of foods with higher iron bioavailability, foods with higher iron content, and iron-fortified foods.

WEB LINKS

National Heart Lung and Blood Institute: This site is the official website of the branch of the National Institute of Health and the Department of Health and Human Services that deals with diseases of the hematopoietic system as well as with other acute and chronic diseases. The site contains scientifically based evidence and links to authoritative sources for further information.

http://www.nhlbi.nih.gov

FIGURE 22.4 The Spinal Cord

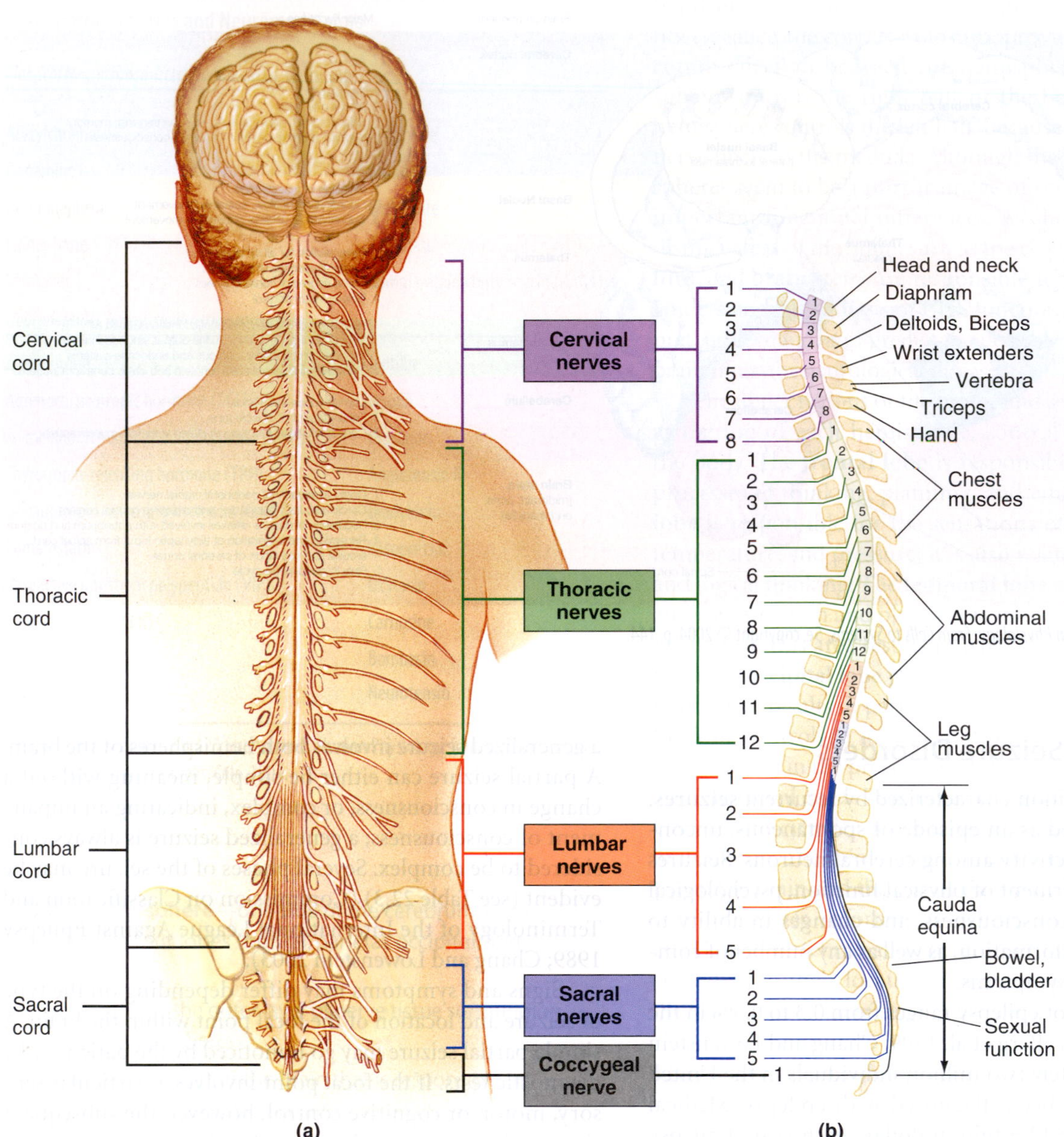

Source: L. Sherwood, *Human Physiology: From Cells to Systems*, 5e, copyright © 2004 p. 173.

Electroencephalograms (EEG) are used to record the electrical activity of the brain. This diagnostic tool provides information about abnormalities and can assist in locating the focal point of the seizure.

Epilepsy medications have been in use since 1912. Traditional medications are phenytoin, valproic acid, carbamazepine, and ethosuximide (Kwan and Brodie 2004; Ross et al. 2004). These medications may have significant drug-nutrient interactions (see Table 22.4 for a summary). It is important to note specifically the interaction of phenytoin with folate metabolism; this will be discussed in detail later in this section.

Over the last decade, advancements in technology and research have made newer anti-epileptic medications available for the treatment of partial and generalized seizures. These medications include levetiracetam, zonisamide, and oxcarbazepine (French et al. 2004; Sankar and Holmes 2004). For those patients who are not responsive to medications, newer experimental treatments include vagus nerve and deep brain stimulation. Surgical options can include temporal lobectomy and amygdalo hippocampectomy (Zimmerman and Sirven 2003; Ross et al. 2004). Some individuals with epilepsy or other neurological disorders may pursue complementary and alternative medicine (CAM) remedies along with or instead

TABLE 22.3

Classification of Seizures

Partial Seizures

Simple partial

Complex partial

Partial seizures with secondary generalization

Primary Generalized Seizures

Absence (petit mal)

Tonic-clonic (grand mal)

Tonic

Atonic

Myoclonic

Unclassified Seizures

Neonatal seizures

Infantile spasms

Source: Commission on Classification and Terminology of the International League Against Epilepsy. Proposal for revised classification of epilepsies and epileptic syndromes. Epilepsia 1989; 30: 389–399.

of conventional medicine; practitioners should be aware of the possible medical and nutritional implications of CAM for these patients (see Appendix Table F5).

Historically, and now more recently, a ketogenic diet has been used to treat those individuals with refractory seizures (Wilder 1921; Kossoff, McGrogan, and Freeman 2004; Stafstrom and Bough 2003; Liu et al. 2003; Levy and Cooper 2003; Mandel et al. 2002). A retrospective chart review of 34 children who were treated with a ketogenic diet revealed a reduction in seizures and seizure medication for those children who were maintained in the highest level of ketosis (Peterson et al. 2005). A previous review of the literature indicated that there has been reported success in observational studies, but no randomized controlled trials to verify the overall outcomes (Levy and Cooper 2003). Still, others report significant cost savings with utilization of this diet as well as anecdotal benefit (Mackay et al. 2005; Liu et al. 2003).

This nutrition therapy treatment induces a state of ketosis by using a diet composition that provides the majority of energy from fat (70%–90%) and the remaining kcalories from protein and carbohydrate. The exact mechanism is not fully understood. It has been proposed that the increased

TABLE 22.4

Selected Medications Used in Epilepsy

Anti-Epileptic Drug	Generic	Brand Names	Possible Food–Drug Interactions
Carbamazepine	Carbamazepine	Tegretol Carbatrol	Do not take with grapefruit juice; consistently take either with food or without food to ensure consistent absorption; may cause hyponatremia
Felbamate	None available	Felbatol	Aplastic anemia, hepatic failure
Gabapentin	Gabapentin	Neurontin	Do not take with antacids; sedation, pedal edema
Lamotrigine	Lamotrigine	Lamictal	Consistently take either with food or without food to ensure consistent absorption
Gamma-Amino Butyric Acid (GABA)	Pregabalin	Lyrica	May cause weight gain if on Avandia or other medications for type 2 diabetes; sedation, dizziness; contraindicated with alcohol
Levetiracetam	Levetiracetam	Keppra	Sedation, cognitive effects, dizziness, contraindicated with alcohol
Oxcarbazepine	Oxcarbazepine	Trileptal	Hyponatremia
Phenobarbital	Phenobarbital	Luminal	Sedation, cognitive effects
Phenytoin	Phenytoin	Dilantin; Phenytek	Ataxia, cognitive disturbance
Topiramate	Topiramate	Topamax	Nephrolithiasis
Valproic Acid	Valproic acid	Depakote	Weight gain, tremor
Zonisamide	None available	Zonegran	Weight loss, nephrolithiasis

Source: Adapted from: American Academy of Neurologists (2005): http://aan.com/professionals/practice/pdfs/patient_ep_onset_c.pdf

BOX 22.1 CLINICAL APPLICATIONS

Outline for Ketogenic Nutrition Therapy Used in Seizure Control

- Nutrition therapy should not be initiated except under strict clinical supervision.
- Hospital-based fasting for initial 24–72 hours until 4+ ketonuria is achieved.
- Establish energy requirements and protein requirements for individual patient.
- Grams of fat calculated at a 4:1 ratio (4 grams of fat to each gram of protein and carbohydrate—approximately 75% of total energy intake).
- Fluid 65 mL/kg/day with maximum of 2000 mL/day.
- Sugar-free multivitamin.
- Calcium supplement to meet AI for age.
- Other supplements as needed.
- Monitor urine for maintenance of ketosis.

Adapted from: Kinsman et al. 1992; Carroll and Koenigsberger 1998; MacCracken and Scalisi 1999; Mandel et al. 2002; Dahlin et al. 2005.

levels of ketones change neuron metabolism and that the ketone body acts to change the balance of neurotransmitters, resulting in an anticonvulsant effect (Hemingway et al. 2001; Dahlin et al. 2005).

Nutrition Implications Individuals with epilepsy and seizure disorders can be at nutritional risk. For infants, children, and adolescents, impaired ability to consume adequate nutrients, limited food choices (if on a ketogenic diet), and drug-nutrient interactions may interfere with the ability to achieve optimal growth and development (Peterson et al. 2005). Ensuring adequate energy, protein, vitamin, and mineral intake is the major component of nutrition therapy for these populations (Hemingway et al. 2001; Liu et al. 2003; Stafstrom and Bough 2003). Box 22.1 outlines some general principles of the ketogenic diet; nonetheless, it is crucial that each diet be carefully calculated and monitored closely in order to both control seizures and ensure nutrient needs are met.

ischemic stroke—stroke caused by an interruption of blood flow to the tissue

hemorrhagic stroke—stroke caused by rupture of a blood vessel (e.g., aneurysm)

aneurysm—weakened area of a wall of a blood vessel

Since pharmacotherapy is the mainstay of seizure treatment, identifying potential interactions for other medications as well as nutrients is a critical component of nutritional care. The most common issues include weight gain with Valproate (Depakote), carbamazepine (Tegretol, Carbatrol), gabapentin (Neurontin), and Felbamate (Felbatol); and potential weight loss with topiramate (Topamax) and zonisamide (Zonegran).

Phenytoin (Dilantin) inhibits both vitamin D and folate metabolism. It is estimated that more than 50% of patients on phenytoin have decreased serum levels of folate (Berg et al. 1995). Long-term use may result in megaloblastic anemia secondary to folate deficiency (see Chapter 13). When folate is supplemented, the effectiveness of phenytoin (i.e., seizure control) should be monitored closely so that a steady state is achieved.

Stroke and Aneurysm

Stroke (cerebrovascular accident) is defined as an interruption of brain function due to blockage or interruption of blood flow to the brain. When the arteries supplying blood to the brain are obstructed, the stroke is classified as an **ischemic stroke.** A **hemorrhagic stroke** occurs when a vessel bursts and releases blood into the brain tissue. Transient ischemic attacks (TIA) are defined as an episode of ischemia where blood flow is quickly restored, but if symptoms last more than 24 hours, the event is considered a stroke. An **aneurysm** is the dilation of smooth muscle usually found at the points where cerebral arteries divide or split (bifurcation). There is risk that this weakened portion of the vessel will burst (Adams et al. 2000).

According to the National Institute of Neurological Diseases and Stroke (2005) and the American Heart Association (Thom et al. 2006), stroke is the third leading cause of death in the U.S., the fourth leading cause in Canada, and a significant health issue throughout the world. In the U.S., women and African-Americans are at highest risk of stroke.

Ruptured aneurysms are often the cause of a stroke (Bederson et al. 2000). Incidence of aneurysm is estimated to be approximately 6 per 100,000 individuals in the U.S.

Multiple factors are thought to place an individual at risk for stroke (Sacco 2001; Adams et al. 2005). As with many conditions, risk factors can be grouped as modifiable and non-modifiable. Non-modifiable risk factors include age, gender, ethnicity, and genetics. Age has the strongest association with stroke risk, with risk doubling for each decade after age 55 (Thom et al. 2006).

Risk factors that can be modified include the presence of hypertension, cardiovascular disease, diabetes mellitus, hyperlipidemia, asymptomatic carotid stenosis, cigarette smoking, alcohol use, and illicit drug use. Lifestyle factors include dietary intake and the use of oral contraceptives. Smoking, for instance, almost doubles the risk of ischemic stroke where it acts synergistically with other risk factors (Kurth et al. 2003). Data from the Northern Manhattan Stroke Study provides new insights into these stroke risk factors (White et al. 2005).

In this study, African-Americans and Hispanics had a greater incidence of stroke, with an almost twofold increase in risk for stroke compared with Caucasians. The protective effect of physical activity and moderate alcohol consumption was confirmed and further established these behaviors as modifiable risk factors. The independent effects of lipids, apolipoproteins, and lipoproteins were also clarified. High-density lipoprotein was shown to be protective against ischemic stroke (particularly atherosclerotic stroke subtypes). The ratio of apolipoprotein b to apolipoprotein a-1 was shown to be associated with carotid atheroma. In addition, newer risk factors, including homocysteine and chronic infection (*Chlamydia pneumoniae* and periodontal disease), are being studied as predictors of ischemic stroke.

Risk factors for aneurysm include cigarette smoking and alcohol ingestion. Other conditions that place an individual at risk for aneurysm include familial intracranial aneurysm syndrome and polycystic kidney disease.

In Chapter 10, hypoxia was introduced as a common mechanism for cell injury. An insufficient oxygen supply (or ischemia) will disrupt cellular metabolism. When circulation to a particular region of the brain ceases, as occurs during a stroke, the cells within that region die. Without a blood supply, necrosis (cell death) occurs within 4–10 minutes due to a lack of oxygen and glucose.

Signs and symptoms of stroke vary depending on the area of the brain that is involved, and include loss of vision or speech, and paralysis or muscle weakness. Other signs and symptoms include a change in mental status. This change can be as dramatic as the onset of coma, but may present more subtly as symptoms of confusion or changes in memory.

Diagnostic criteria used in acute evaluation for stroke include the National Institutes of Health Stroke Scale, which evaluates the following patient characteristics: level of consciousness, speech, and ability to follow commands (Bushnell, Johnston, and Goldstein 2001). According to the Guidelines for the Early Management of Patients with Ischemic Stroke (Adams et al. 2005), imaging of the brain is ultimately required for diagnosis of stroke, even though acute care may have to occur prior to these tests. Imaging allows isolation of damage as well as confirmation of diagnosis and the classification of type of stroke. Typical diagnostic testing for stroke includes computed tomography (CT) and magnetic resonance imaging (MRI). (See Figure 22.5.) Possible diagnoses using CT include identification of infarction, hemorrhage, calcification, embolism, and other anatomical changes such as aneurysm. MRI results allow estimation of the time since injury—a stroke may be acute, subacute (one week), or old (several weeks to years) (Latchaw 2004). Other imaging techniques that may be used include positive emission tomography (PET), ultrasound, and angiography.

Treatment of stroke involves both acute and chronic care. The type of stroke, extent of damage, and individual patient characteristics will guide these decisions. Guidelines for emergency treatment of stroke have been firmly established by the National Institute of Neurological Disorders and Stroke (NINDS) as well as others (Adams et al. 2005). Treatment begins after initial stabilization of the patient and confirmation of diagnosis. Types of treatment include medical support (e.g., oxygen therapy), thrombolysis (e.g., intravenous [IV] tissue plasminogen activator [tPA]), anticoagulation (e.g., heparin), antiplatelet therapy (e.g., aspirin), and neuroprotection (Albers, Amarenco, Easton, Sacco, and Teal 2004). Significant evidence supports these modes of therapy. The NINDS recombinant tPA (rtPA) Stroke Study showed a clear benefit for intravenous rtPA in selected patients with acute stroke. The International Stroke Trial and the Chinese Acute Stroke Trial found that use of aspirin within 48 hours of the acute stroke event resulted in improved outcomes (Chen et al. 2003).

FIGURE 22.5 CT Scan: Diagnosis of Stroke

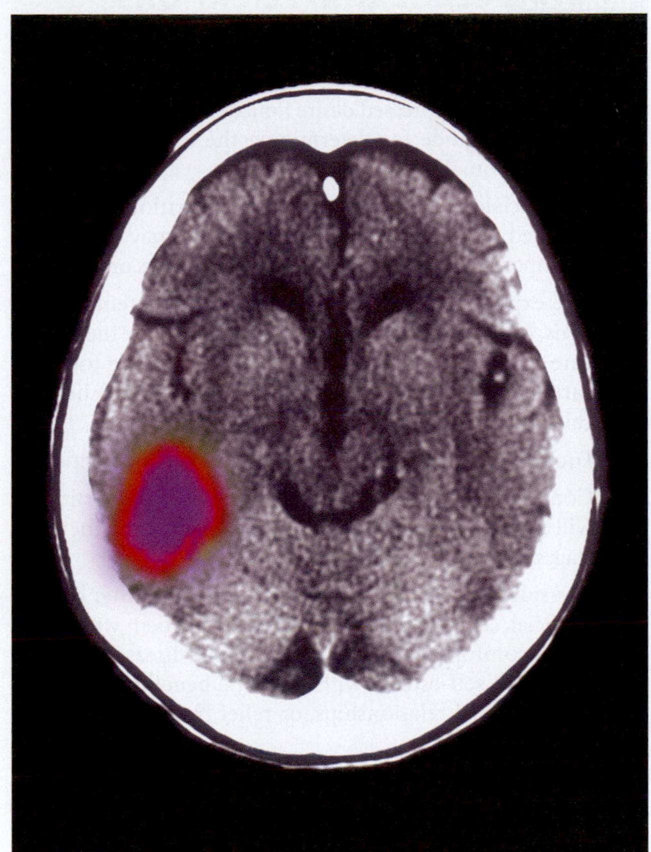

Source: http://www.strokecenter.org/pat/diagnosis/ct.htm

Chronic therapy focuses on prevention of further stroke and rehabilitation for functional recovery. Rehabilitation goals set by the American Heart Association practice guidelines are to prevent complications, minimize impairments, and maximize function. Comprehensive rehabilitation can occur in several different settings, such as an in-patient facility, nursing facility, and outpatient or in-home care setting. The program of care should utilize a multidisciplinary team involving physical therapy, occupational therapy, speech and language pathology, nutrition therapy, kinesiotherapy, and physical medicine (Duncan et al. 2005).

BOX 22.2 CLINICAL APPLICATIONS

Ethical Decisions in Nutrition Support

1. The patient's expressed desire for extent of medical care is a primary guide for determining the level of nutrition intervention.
2. The decision to forgo hydration or nutrition should be weighed carefully because such a decision may be difficult or impossible to reverse with a period of days or weeks.
3. The expected benefits, in contrast to the potential burdens, of nonoral feeding must be evaluated by the health care team and discussed with the patient. The focus of care should include the patient's physical and psychological comfort.
4. Food and hydration are considered medical interventions.
5. Consider whether or not nutrition, either oral or artificial, will improve the patient's quality of life during the final stages of life.
6. Consider whether or not nutrient support, either oral or artificial, can be expected to provide the patient with emotional comfort, decreased anxiety about disease cachexia, improved self-esteem with cosmetic benefits, improved interpersonal relationships, or relief from fear of abandonment.
7. If death is imminent and feeding will not alter condition consider whether or not nutrient support will be burdensome.
8. When oral intake is appropriate:
   a. Oral feeding should be advocated whenever possible. Food and control of food intake may give comfort, pleasure and sense of autonomy and dignity. The most important priority is to provide food according to the individual patient's wishes.
   b. Efforts should be made to enhance the patient's physical and emotional enjoyment of food by encouraging staff and family assistance in feeding the patient.
   c. Nutrition supplements, including commercial products and other alternatives, should be used to encourage intake and ameliorate symptoms associated with hunger, thirst, or malnutrition.
   d. The therapeutic rationale of previous diet prescriptions for an individual patient should be reevaluated. Many dietary restrictions can be liberalized. Coordination of medication or medication schedules with the diet should be discussed with the physician, with the objective of maximizing food choice and intake by the patient.
   e. The patient's right to self-determination must be considered in determining whether to allow the patient to consume foods that are not generally permitted within the diet prescription.
   f. Suboptimal oral feedings may be more appropriate than burdensome tube or parenteral feeding.
9. When tube feeding or parenteral feeding is being considered:
   a. The patient's informed preference for the level of nutrition intervention is primary. The patient or substitute decision maker should be advised on how to accomplish whatever feeding the patient desires.

(continued on the following page)

The goal of treatment for aneurysm depends on the size and symptoms of the aneurysm as well as individual patient characteristics. Options include surgical treatment or conservative intervention intended to prevent further hemorrhage.

Individuals who have experienced stroke are at varying degrees of nutritional risk, with the severity of nutritional risk and the resulting nutritional interventions depending on the area of the brain that has been affected by the stroke. For example, in the acute period after stroke, an individual may be comatose, cognitively impaired, or unable to swallow successfully. Nutrition intervention will be individualized to maximize nutritional support, whether it is administered orally, enterally, parenterally, or through a combination of routes. Additionally, the nutritional status of patients at the onset of the injury is an important consideration for a successful recovery. The FOOD (Feed or Ordinary Diet) randomized clinical trial for 3,012 stroke patients indicated that poor nutritional status was related to complications and negative outcomes post-injury (Dennis, Lewis, and Warlow 2005).

Nutrition Implications As stated in the previous section, acute nutritional problems for stroke or aneurysm typically involve impairment of the ability to chew, swallow, or self-feed. Enteral nutrition support will be necessary if an oral diet cannot meet nutritional needs. Evidence supports the early initiation of nutritional support to prevent complications, reduce hospital stay, and promote rehabilitation (see Chapter 7). Ongoing changes in the nutritional plan will certainly be necessary as rehabilitation progresses.

Dysphagia (see Chapter 16) is a common condition that accompanies stroke. The dysphagia symptoms that a patient will experience are dependent on the phase of swallowing that is impaired. For example, if the problem originates in the oral preparation phase, food may be pocketed in the buccal mucosa (cheek area) because the patient cannot propel the bolus of food effectively from the front of the oral cavity to the pharyngeal area. Other general symptoms may include drooling, coughing, and choking. Many patients will experience weight loss and generalized malnutrition due to inadequate nutritional intake. Aspiration or inhalation of oropharyngeal

BOX 22.2 *(continued)*

b. When palliative care is the agreed goal, nutritional support must be part of the palliative plan. A palliative care plan does not automatically preclude aggressive nutrition support. The decision to forgo "heroic" medical treatment does not preclude baseline nutrition support. All options for nutritional support can be considered.

c. Feeding may not be desirable if death is expected within hours or a few days and the effects of partial dehydration or the withdrawal of nutrition support will not adversely alter patient comfort.

d. Facilities should provide and distribute written protocols for the provision of and termination of tube feedings and parenteral feedings. The protocols should be reviewed periodically, and revised if necessary, by the health care team. Legal and ethical counsel should be routinely sought during the development and interpretation of the guidelines. The institution's ethics committee, if available, should assist in establishing and implementing defined, written guidelines for nutrition support protocol. The registered dietitian should be a contributing member of or consultant to such a committee.

e. Conflict within the family or among stakeholders can be resolved by referring to an ethics committee or consultant if available within the institution.

f. The potential benefits vs. burdens of tube feeding or parenteral feeding should be weighed on the basis of specific facts concerning the patient's medical and mental status, as well as on the facility's options and limitations.

g. Facility options and limitations—one should consider the following:

(1) lack of staffing—no one to manage or monitor feeding

(2) too costly without financial help

(3) if a feeding strategy is started in one site it will have to be stopped when the patient is transferred to another site, which can lead to a sense of abandonment.

10. Either short- or long-term parenteral nutrition should be considered only when other routes are impossible or inadequate to meet the comfort needs of the patient.

11. The physician's written diet order in the medical chart documents the decision to administer or forgo nutrition support.

a. The registered dietitian should participate in the decision.

b. If a decision is made that the registered dietitian does not agree with, appeal to the facility's ethics mechanism (committee or consultant) is appropriate.

c. If the court has ordered feeding or no feeding and you do not agree with the court's decision, appeal to the facility's ethics mechanism is appropriate.

Reprinted from *Journal of the American Dietetic Association*, V102(5), O'Sullivan Maillet J, Potter RL, Heller L, Position of the American Dietetic Association: Ethical and legal issues in nutrition, hydration, and feeding, pp. 716–726, Copyright 2002, with permission from the American Dietetic Association.

contents into the lungs is a primary complication. This can lead to aspiration pneumonia with accompanying infection (see Chapter 23). Table 16.15 (in Chapter 16) provides a review of dysphagia symptoms with nutritional considerations.

Diagnosis and treatment of dysphagia involve different members of the health care team. Many institutions have a dysphagia team consisting of physicians, nurses, a speech language pathologist, a registered dietitian, a physical therapist, and an occupational therapist, just as they may have a stroke rehabilitation team. Diagnosis of dysphagia begins with a bedside swallowing assessment, usually performed by the speech language pathologist. Conclusive evaluation uses a videofluoroscopy swallowing study or fiberoptic endoscopic evaluation of swallowing. From these evaluations, the specific site of the swallowing disorder can be determined, a diagnosis can be established, and a care plan can be developed. The nutrition care plan may include modifying the consistency of food or liquids, positioning of the patient, or swallowing exercises.

Modifiable risk factors for patients who have had or are at risk for a stroke include the secondary diagnoses of hypertension, atherosclerosis, and diabetes. Each of these diagnoses involves nutrition therapy as an important component of medical care, as discussed in the appropriate chapter of this text, and should be foremost in the practitioner's plan of care.

Progressive Neurological Disorders

The medical and social management of progressive neurological (neurodegenerative) disorders is challenging. The physiological changes accompanying progression of the disease affect both nutritional needs and nutritional status. As symptoms increase in severity, the ability to meet nutritional needs by an oral route diminishes. Nutrition therapy and individual patient goals will address the required changes for procurement, ingestion, and digestion of food that are necessitated by the neurological impairment. As in the acute period after a stroke, enteral nutrition support often becomes necessary at some point.

Treatment decisions for these conditions often involve complex ethical dilemmas, such as stem cell transplant or the possibility of long-term nutrition support. Box 22.2 explores ethical decisions related to nutrition support.

distinct forms of the disease exist: bulbar and spinal. Initial presentation is due to the differences in where the neuron deterioration begins.

Diagnosis of ALS involves a complete neurological examination that allows the exclusion of other diseases, because no single diagnostic test exists. Tests utilized may include an electromyography (EMG), or a nerve conduction velocity test (NCV), and other imaging studies such as magnetic resonance imaging (MRI).

Clinical Manifestations Classic signs and symptoms of ALS include **asymmetric muscle weakness** and atrophy, **hyperreflexia** (hyper-stiffening of the muscles), and **fasciculations** (uncontrolled twitching of the muscles). The way these signs and symptoms manifest in the patient again depends on the area of neuron destruction. There may be difficulty with gross motor actions such as walking, or in fine motor actions such as grasping an object. As the disease progresses, the extensive muscle atrophy leads to increasing paralysis, usually requiring mechanical ventilation and nutrition support.

Treatment There is only one current treatment for ALS: Riluzole, which affects the release of the neurotransmitter glutamate. Most current research indicates that muscle weakness and complication rate improved in patients who were prescribed Riluzole, though no long-term effects have been demonstrated (Nirmalananthan and Greensmith 2005).

Many supportive therapies exist that assist with control of symptoms. Different classes of medications can control excessive saliva or secretions, muscle spasms, and pain. Physical therapy, massage, and heat can treat muscle pain and contractures (Miller et al. 1999).

Nutrition Therapy Nutrition interventions will be designed to meet the specific nutritional needs of the patient as the disease progresses. For example, oral intake may decrease, because the patient may fatigue easily. Increasing nutrient density maximizes kcal and protein in the food that is consumed. The patient may require alterations in texture and consistency or alternate routes of nutrition support in order to maintain adequate intake. (See Chapter 16 for a discussion of texture and consistency modification, and Chapter 7 for nutrition support guidelines.)

asymmetric muscle weakness—muscle weakness occurring unequally in different parts or sides of the body

hyperreflexia—overresponse or exaggeration of response to a neural stimulus (e.g., twitching)

fasciculations—involuntary twitching or movement of muscle

Guillain-Barré—an acute peripheral nervous system disease characterized by progressive paralysis

plasmapheresis—treatment that removes blood from the body, separates out certain cells from the plasma, and then returns the blood back to the body

Guillain-Barré

Definition **Guillain-Barré** (GB) is defined as an acute peripheral nervous system disease characterized by progressive paralysis.

Epidemiology Incidence in the U.S. is approximately 3.0 cases per 100,000 in the general population. Disease rates are similar throughout the world (Hughes and Rees 1997).

Etiology GB appears to be an autoimmune response to an infectious trigger. Most patients report a history of infection in the period prior to the onset of GB. The most common infections that have been associated with GB include *Campylobacter jejuni* and *Cytomegalovirus* (van der Meche and Schmitz 1992; van der Meche et al. 1997).

Pathophysiology Several variants of the syndrome exist, including acute inflammatory demyelinating polyneuropathy (AIDP), acute motor axonal neuropathy (AMAN), acute motor sensory axonal neuropathy (AMSAN), and Miller Fisher syndrome. In all forms, it appears that autoantibodies attack specific cells of the nervous system after infectious exposure. This results in damage to the myelin sheath, axons, sensory nerves, and roots, depending on the variant of the disease (Newswanger and Warren 2004; Hauser and Asbury 2005).

Clinical Manifestations Symptoms include a rapidly progressive paralysis that can involve all limbs. Cranial nerve involvement results in severe dysphagia and respiratory failure. Approximately 30% of patients with GB require mechanical ventilation (Hauser and Asbury 2005).

Treatment Treatment includes the use of high-dose intravenous immunoglobulin, an antibody produced by the B-lymphocytes. **Plasmapheresis**, which is a procedure that removes the antibodies from the plasma, is also used effectively. Depending on the component of the nerve cell that is affected, full recovery can occur within several months. Damage to the axons and nerve roots, on the other hand, may result in a prolonged recovery. Complications may arise from respiratory failure or infection.

Nutrition Therapy As in other neurological disorders, nutrition problems will stem from the disease process. For Guillain-Barré, these will depend on the specific cranial nerve(s) involved. Difficulty swallowing and chewing as well as modifications related to mechanical ventilation are issues that may arise.

Myasthenia Gravis

Definition **Myasthenia gravis** is a progressive neuromuscular disorder that affects the skeletal muscles.

Epidemiology Myasthenia gravis is one of the more common neuromuscular disorders, with an incidence of approximately 1 in 7,500 individuals (Drachman and Brodsky 2005).

Etiology In myasthenia gravis, an autoimmune reaction damages or destroys the cellular receptors for **acetylcholine**. Because the thymus gland is abnormal in many individuals with this disorder, it is thought that this gland may be responsible for the autoantibody production (Onodera 2005). The cause of the autoimmune response is not known.

Pathophysiology As discussed under anatomy and physiology of the nervous system, acetylcholine is an excitatory neurotransmitter. Normally, acetylcholine receptors allow for stimulation of the muscle for contraction by acetylcholine. When there are reduced numbers of these receptors, the muscle will tire easily and the individual will experience muscle weakness that may improve with rest. The disease may have periods of remission and exacerbation which will vary between patients (Lindstrom 2000; Drachman and Brodsky 2005; Romi, Gilhus, and Aarli 2005).

Clinical Manifestations Signs and symptoms include skeletal muscle weakness that is exacerbated by physical activity. The most common muscles that are affected include muscles of the face, eyes, arms, and legs. Some individuals have drooping eyelids and double vision.

Diagnostic tests include the Tensilon test, repetitive nerve stimulation test, and assessment for anti-acetylcholine receptor (anti-AChR) antibodies. The Tensilon test requires injection of edrophonium (Tensilon), a drug that inhibits the enzyme anticholinesterase and allows acetylcholine to accumulate in the muscles, resulting in improved muscle functioning. This rapid-acting test can be given at bedside. Repetitive nerve stimulation tests measure the response of a single nerve to a series of electrical stimulations. The muscle's potential action is measured by electrodes and is documented. The third test measures antibodies against acetylcholine receptors in the blood. A positive test for anti-AChR antibodies is almost always diagnostic for myasthenia gravis, but a negative result does not rule out the disease (Meriggioli 2005).

Treatment Currently, there is no cure for myasthenia gravis, but effective treatment is available. Medications include anticholinesterase drugs such as Neostigmine and Pyridostigmine as well as immunosuppressive drugs such as glucocorticoids, azathioprine, or intravenous immunoglobulin. Other treatments, such as stem cell transplant, have been used to treat autoimmune progressive diseases (Drachman and Brodsky 2005). As mentioned previously, the thymus gland is often abnormal in those with myasthenia gravis, and the removal of the gland (thymectomy) has had positive results in these patients. Finally, plasmapheresis can be used to remove antibodies from the plasma for symptomatic relief (Fergusson et al. 2005; Romi, Gilhus, and Aarli 2005; Sieb 2005; Schneider-Gold et al. 2005).

Nutrition Therapy Specific nutrition problems for individuals with myasthenia gravis may originate from decreased oral intake or difficulty with meal preparation. These problems are linked to fatigue, muscle weakness, visual problems, and pain. Nutrition interventions will focus on increasing nutrient density and modifying the timing of meals as well as food textures and consistency.

Multiple Sclerosis

Definition Sclerosis means hardening of tissue, and **multiple sclerosis** (MS) was first defined as the presence of many scars within the brain. MS is currently defined as a disorder characterized by demyelination of cells within the CNS, inflammation, and development of scar tissue. There are several distinct types of MS, which are classified as: relapsing remitting MS (RRMS), secondary progressive stage MS (SPMS), and primary progressive MS (PPMS).

Epidemiology The highest prevalence rates for MS are in Scotland, Northern Europe, the northern U.S., and Canada. Rates as high as 250 individuals per 100,000 have been observed (Hauser and Goodin 2005). Low rates of this disease are found throughout the rest of the world, especially in Asia and Africa.

Etiology The exact cause of MS has not been established. There are currently three major areas of study regarding etiology: genetics, autoimmune processes, and infectious factors. Genetic etiology has been examined through epidemiology within various ethnic groups as well as twin and family studies (Sospedra and Martin 2005). There are consistent factors among these groups that point to a genetic tendency, but this certainly has not been firmly established at this time.

Experimental evidence and results from patients with MS indicate that autoantibodies and inflammatory cells that damage the myelin and axonal portions of the nerve cell are involved (Hafler et al. 2005). Cells that are key to the

myasthenia gravis—a progressive neuromuscular disorder that affects the skeletal muscles and causes muscle weakness, particularly of the face, eyes, arms, and legs

acetylcholine—an excitatory neurotransmitter

multiple sclerosis—a disorder characterized by demyelination of cells within the CNS, inflammation, and development of scar tissue, causing numbness, tingling, incoordination, weakness, and varying degrees of blindness

FIGURE 23.1 Anatomy of the Pulmonary System

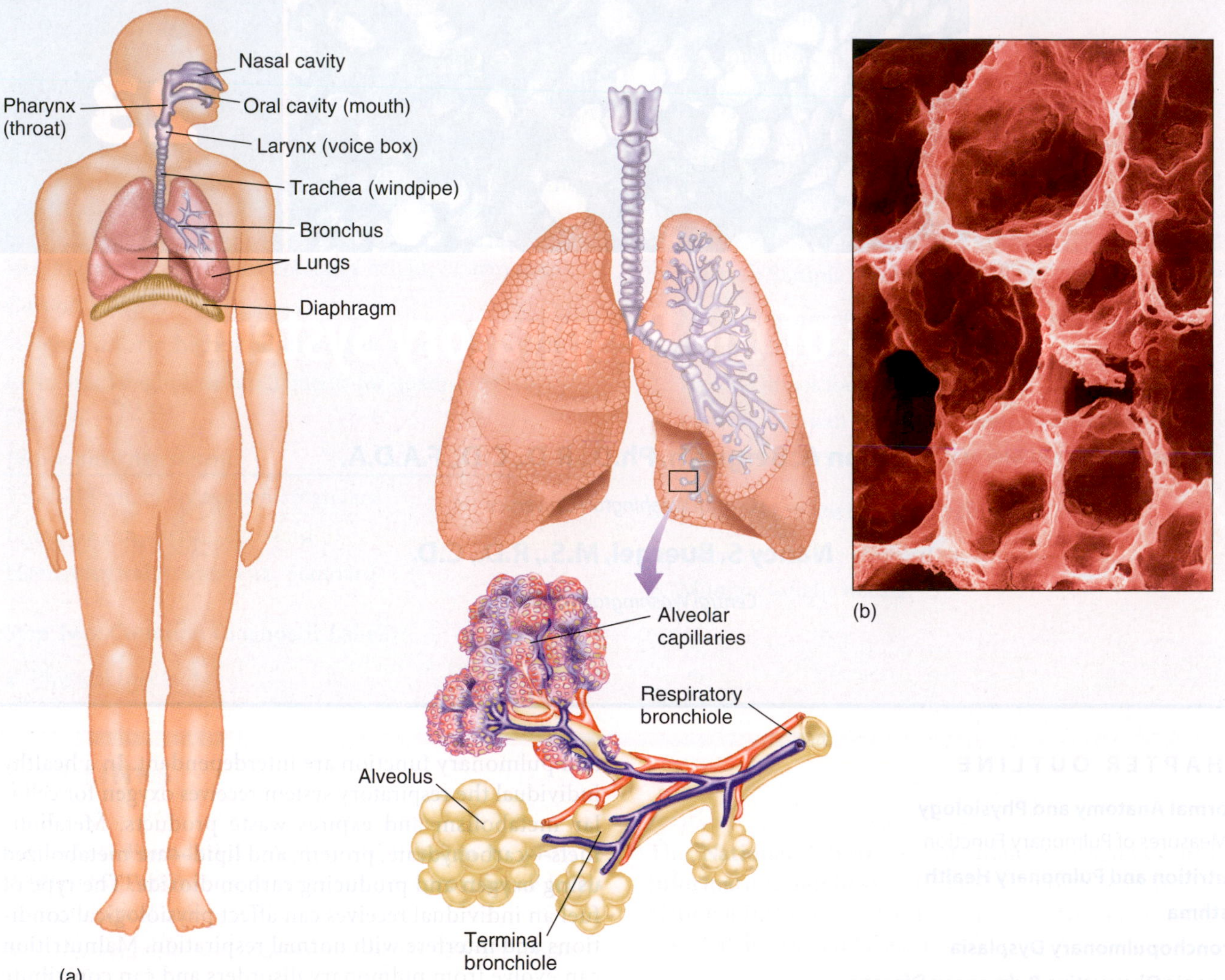

Source: R. Rhoades and R. Pflanzer, *Human Physiology*, 4e, copyright © 2003, p. 633.

bronchi, supplying the right and left lung (see Figure 23.1). The bronchi further divide again and again into smaller and smaller bronchioles. The bronchioles end in small air sacs called alveoli which are paper thin. Each alveoli is imbedded with millions of capillaries that are responsible for the exchange of oxygen and carbon dioxide. The skeleton and muscles surrounding the lungs, particularly the intercostal and diaphragmatic muscles, support the respiratory system.

The lungs undergo a period of growth and maturation during the first two decades of life. By 10–12 years of age the maximal number of alveoli is attained. Full maturation of the respiratory system is achieved by age 20 for females and 25 for males. Aging is associated with a progressive decrease in lung function; however, unless affected by disease, the lungs are capable of providing adequate gas exchange during the entire lifespan (Janssens, Pache, and Nicod 1999).

Gas exchange occurs in the alveolar-capillary unit which, in the adult lung, covers approximately the area of a tennis court and contains more than 100 million capillaries (see Figure 23.2). The alveolar-capillary unit consists of the capillary endothelium and its basement membrane, the interstitial space, and the alveolar epithelium and its basement membrane. The interstitial space consists of the tissue layers between the lungs' air sacs (alveoli), and is very thin, which allows for efficient exchange of gas if ventilation is adequate. The alveolar epithelium consists of two types of cells: type I and type II cells. Type I cells form the structure of the alveolar wall. Type II cells are responsible for producing **surfactant**, a fluid secreted by the cells of the alveoli that reduces the

surfactant—substance secreted by the alveolar cells of the lung that serves to maintain the stability of pulmonary tissue by reducing the surface tension of fluids that coat the lung

FIGURE 23.2 Gas Exchange between Alveoli and Capillaries

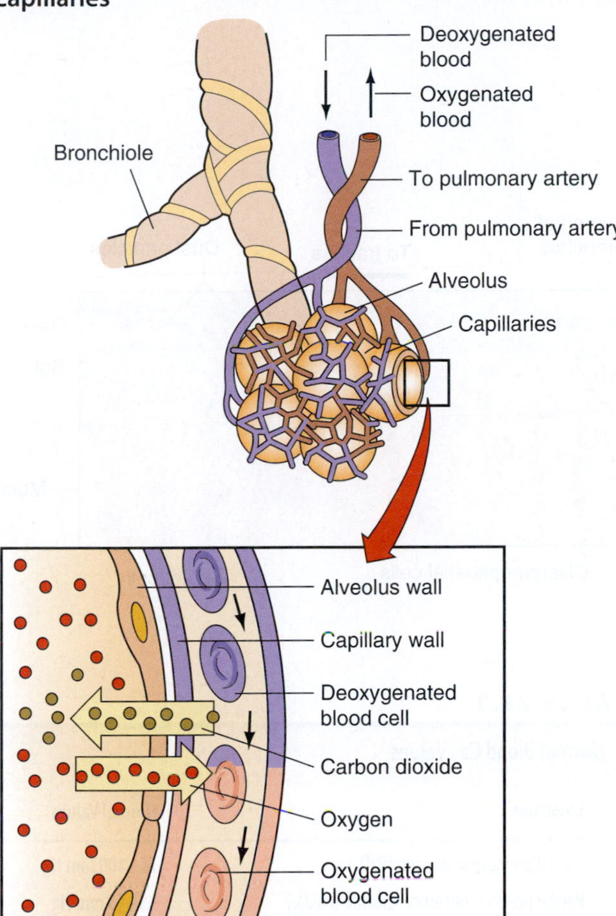

Source: From the Merck Manual of Medical Information, Second Home Edition, p. 246, edited by Mark H. Beers. Copyright © 2003 by Merck & Co., Inc., Whitehouse Station, NJ. Reprinted by permission.

surface tension of pulmonary fluids and contributes to the elastic properties of pulmonary tissue. At rest, the alveolar-ventilation is approximately equal to the cardiac output.

The lungs also play an important role in protecting the body against infection and harmful environmental toxins. Inhaled particles such as smoke, bacteria, and viruses pass through the nose and are trapped in the lungs by a sticky mucus substance, which serves to keep the airway moist. The cells that line the trachea, bronchi, and bronchioles contain tiny hair-like cells called cilia, which beat with a rhythm fast and forceful enough to propel the mucus and unwanted cells upward toward the pharynx where they can be coughed out or swallowed (see Figure 23.3). Additionally, the epithelial surface of the alveoli contains macrophages or scavenger cells that engulf and destroy the inhaled bacteria.

Measures of Pulmonary Function

The initial evaluation of pulmonary function is generally accomplished with the physical examination and the tools of **percussion** and **auscultation**. Using these techniques for evaluating sounds, abnormalities in breathing and underlying organs may be detected. For example, dull or low-pitched sounds may suggest **pulmonary consolidation** or **pleural effusion**, which may occur with damaged cells during a disease process such as pneumonia. **Rales** are bubbly sounds created by a change in air flow that could occur when inflammation is present or when excessive mucus changes air flow.

Pulmonary function tests are used to detect lung diseases or to monitor the progression of a particular disease. In pulse oximetry, which is often used at bedside or in any outpatient setting, light waves measure the oxygenation of arterial blood. The pulse oximeter is able to detect the percentage of oxygen within the hemoglobin molecule based on the color of the blood. The finger or ear lobe is the most common site for using the pulse oximeter. The most common pulmonary function test is done with a machine called a spirometer (Petty 1999). During spirometry, the patient breathes into a tube attached to the machine, which calculates the amount of air the lungs can hold and the rate the air can be inhaled and exhaled. The results of the test are compared with those of healthy individuals of similar height and age, and of the same sex and race.

Common spirometry measurements include FVC (Forced Vital Capacity), the total volume of air expired after a full inspiration, and FEV1 (Forced Expiratory Volume in One Second), the volume of air exhaled in the first second after a deep inhalation. In addition to spirometry, another test used to evaluate lung function is gas diffusion. Gas diffusion measures how well oxygen and other gases pass through the lung's air sacs and are absorbed by the blood. A reduced diffusing capacity could indicate pulmonary disease.

Evaluation of arterial blood gases (ABGs) determines the pH (acidity), oxygen content, and carbon dioxide content of the blood and can also be used to measure pulmonary function (see Table 23.1). Changes in pulmonary

percussion—a technique used during physical examination in which the hands are used to strike the body's surface, and the sounds that are transmitted from the underlying tissues and organs are evaluated

auscultation—a technique used during physical examination in which a stethoscope is used to evaluate the sounds created in body organs

pulmonary consolidation—changes in tissue structure of the lungs; often visualized as opaque components on a chest x-ray

pleural effusion—accumulation of fluid between the two outer membranes surrounding the lungs

rales—bubbly sounds heard via stethoscope that may indicate pulmonary pathology

FIGURE 23.3 Mucociliary Transport System

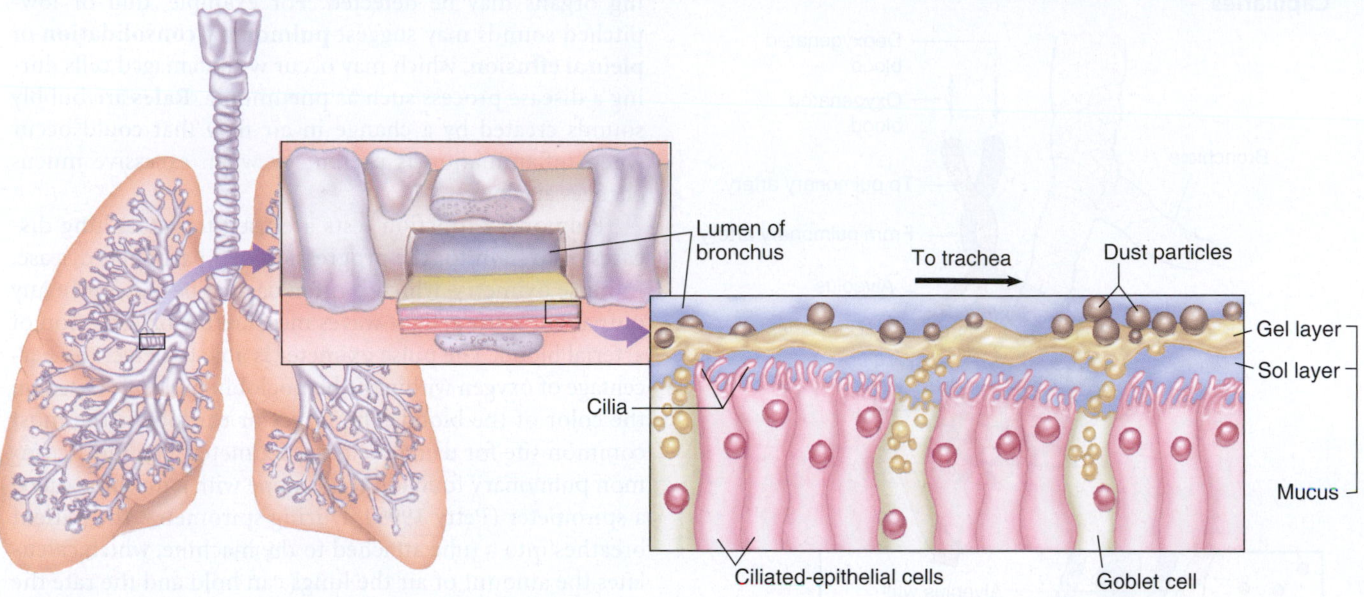

Source: R. Rhoades and R. Pflanzer, *Human Physiology,* 4e, copyright © 2003, p. 635.

Pulse Oximeter

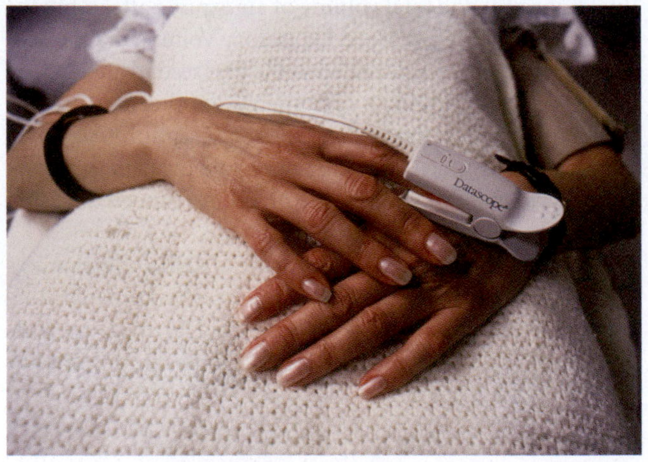

Source: Josh Sher/ Photo Researchers, Inc.

TABLE 23.1

Normal Blood Gas Values

Blood Gas	Normal Values
Partial pressure of oxygen (PaO_2)	75–100 mm Hg
Partial pressure of carbon dioxide ($PaCO_2$)	35–45 mm Hg
pH	7.35–7.45
Oxygen saturation (O_2Sat)	94%–100%
Bicarbonate—(HCO_3)	22–26 mEq/liter

function are reflected by changes in the partial pressure of dissolved carbon dioxide ($PaCO_2$), and by changes of partial pressure of dissolved oxygen (PaO_2). Changes in the $PaCO_2$ measure how well carbon dioxide is able to move out of the blood into the airspaces of the lung, and out in the exhaled air. Changes in PaO_2 measure how well oxygen is able to move from the air into the lungs. Oxygen saturation is the measure of the amount of oxygen carried by the red blood cells and can be calculated using the partial pressure of dissolved oxygen (PaO_2). In patients with pulmonary disease, fewer red blood cells carry the usual load of oxygen, and oxygen saturation is decreased.

minute ventilation—the volume of air per unit time moved into or out of the lungs; measured by collecting expired volume for a fixed time

In addition to the exchange of oxygen and carbon dioxide, the lungs play a major role in regulating acid-base balance. The pH is a measure of hydrogen ion concentration (H^+) in blood, which indicates its acid or base (alkaline) nature; a pH of less than 7 is acidic, and a pH greater than 7 is alkaline. Respiratory acidosis, caused by decreased ventilation, results in carbon dioxide retention, while respiratory alkalosis, caused by increased ventilation, results in loss of carbon dioxide. Respiratory changes in **minute ventilation** occur quickly in response to acid-base disturbances by quickly altering blood pH and regulating the retention or excretion of carbon dioxide. Carbon dioxide dissolves more readily in the blood than oxygen and forms bicarbonate and smaller amounts of carbonic acid. When present in normal amounts, the ratio of carbonic acid to bicarbonate helps to keep the body pH normal. In situations where acidosis occurs, respiratory activity is increased and the lungs quickly compensate to excrete excess CO_2. If alkalosis is present, respiratory activity automatically decreases and CO_2 is retained, producing a compensatory respiratory acidosis (see Chapter 9).

Nutrition and Pulmonary Health

One of the difficulties in studying the relationship between nutrition and diet and respiratory diseases involves the scarcity of evidenced-based research, particularly with specific respiratory diseases. Without evidenced-based research, valid practice recommendations are not possible (see Box 23.1). Many studies looking particularly at food and nutrient intake related to the etiology of respiratory disease are based on retrospective population studies, rather than intervention trials. Unfortunately, population studies are often contradictory due to a variety of factors, including sampling errors, selection bias, and low statistical power. Even intervention studies sometimes show conflicting outcomes, resulting from differences in study design.

Malnutrition has been shown to have an adverse affect on clinical outcomes. The impact of protein-energy malnutrition on lung function has been examined in both clinical and animal studies. Also, the effects of weight loss on pulmonary function in individuals without lung disease have been described (Keys et al. 1951; Aora and Rochester 1982; Sahebjami 2003). Malnutrition associated with poor intake appears to have an impact on the strength and endurance of respiratory muscles, particularly the diaphragm, and may also cause reductions in lung parenchyma. With continued malnutrition, increased incidence of pulmonary infection may also occur, as a result of depressed immune function (Pingleton 1998).

There is increased evidence correlating the role of dietary antioxidants such as vitamin C, vitamin E, β-carotene, and selenium with healthy lung functions. A variety of antioxidants are present in the extra cellular fluid (ECF) and appear to play an important role in protecting the lungs from oxidant injury as the result of the inflammatory process caused by the inhalation of cigarette smoke and other pollutants (Young, Roxborough, and Woodside 1999). Results of a 3-year population-based study looking at dietary intake of antioxidants and lung function found a positive association between vitamin C, vitamin E, and carotenoids intake with pulmonary function (Schunemann et al. 2001, 2002). Of all the carotenoids studied, lutein/zeaxanthin had the strongest relationship to pulmonary function as measured by FEV and FEV1. The Children's Health Study, a 10-year longitudinal study of respiratory health in school-age children, found that low intakes of antioxidant vitamins were associated with decreased pulmonary function in boys and girls (Gilland et al. 2003). Low intakes of both vitamin C and A were positively associated with deficits in FEV and FEV1.

Cigarette smoking is associated with reduced levels of antioxidants in various body fluids. Smokers have shown depleted levels of serum ascorbate, α-tocopherol, β-carotene, and selenium (Young, Roxborough, and Woodside 1999). A meta-analysis examining the relationship between cigarette smoking and nutrient intake showed that smokers had higher intakes of energy, total fat, saturated fat, cholesterol, and alcohol and lower intakes of antioxidants and fiber compared to non-smokers (Dallongeville et al. 1998). The metabolic turnover for vitamin C is about 35 mg/day greater for smokers than non-smokers. Because of this, it is recommended that people who smoke consume an additional 35 mg/day of vitamin C beyond the DRI (Food and Nutrition Board 2000).

BOX 23.1 NEW RESEARCH

Evidence-Based Guidelines

It is important that the dietitian use the most up-to-date information when providing care for patients with pulmonary disease. This information must be based on evidence supported by well-controlled research studies and clinical practice. Evidence-based recommendations or guidelines are scientifically developed to assist health care professionals in making appropriate decisions about patient care. (See Tables 23.11 and 23.13 for examples.)

The following are sources of evidence-based research designed to help health care professionals choose the best clinical approach to patient care:

1. American Dietetic Association. Evidence Analysis Library. Available at: http://www.adaevidencelibrary.com. The Library has been created to summarize the best available research in dietetics and nutrition.
2. National Guideline Clearinghouse (NSG). Available at: http://www.guideline.gov. The NSG is a resource for evidence-based clinical practice guidelines for physicians, nurses, and other health care professionals, including dietitians.
3. Cochrane Library. Available at http://www.cochrane.org. The Cochrane Library publishes Cochrane Reviews, which are based on the best available information about health care interventions. The reviews explore the evidence for and against the effectiveness and the appropriateness of various types of medical treatment.
4. Agency for Health Care Research and Quality (AHRQ). Available at: http://www.ahrq.gov. AHRQ is the research arm of the U.S. Department of Health and Human Services (HHS). It examines how people access health care, its cost, and the results of this care. The main goals of AHRQ are to identify the most effective ways to organize, manage, finance, and deliver high-quality health care.

Respiratory disease often includes a variety of symptoms that may affect dietary intake, including early satiety, anorexia, weight loss, cough, and **dyspnea** during eating. As the disease progresses, these symptoms may have a marked impact on nutritional status. For this reason, it is important that a nutritional assessment be initiated that includes an evaluation of weight history, nutrient intake, medication use, biochemical markers (albumin and prealbumin), and functional status.

Asthma

Definition

Asthma is a chronic inflammatory disorder of the airway involving many cells and cellular elements, such as mast cells, eosinophils, T lymphocytes, macrophages, neutrophils, and epithelial cells. Inflammation is the primary problem in asthma and is thought to be primarily immunoglobulin E (IgE) mediated (see Chapter 12). In susceptible individuals, this inflammation causes recurrent episodes of wheezing, breathlessness, chest tightness, and coughing, particularly at night or in the early morning. These episodes are usually associated with airflow obstruction that is often reversible either spontaneously or with treatment.

Epidemiology

The prevalence of asthma in western countries has increased since the early 1980s across all age, sex, and racial groups (Asthma and Allergy Foundation of America 2006). Asthma is the leading cause of hospitalization and chronic illness among children under 15 years of age (American Lung Association, Asthma & Children, 2005).

Etiology

Asthma is usually divided into two types: allergic and non-allergic asthma (Asthma and Allergy Foundation of America 2006). Of the two, allergic asthma is the most common and is triggered by inhaled allergens such as dust mite allergen, pet dander, pollen, and mold. Non-allergic asthma is caused by other factors, such as anxiety, stress, exercise, cold air, dry air, hyperventilation, smoke, viruses, or other irritants. Patients with asthma are at greater risk for life-threatening allergic reactions to foods (Spergel and Fiedler 2004). Persistent asthma has been associated with elevated IgE to egg and wheat, though food allergies are rarely a cause of asthma.

dyspnea—shortness of breath or difficulty breathing

asthma—a chronic inflammatory disorder of the airway involving many cells and cellular elements such as mast cells, eosinophils, t lymphocytes, macrophages, neutrophils, and epithelial cells

BOX 23.2 NEW RESEARCH

Serum Cholesterol and Pulmonary Health

The relationship between high serum cholesterol levels and the incidence of cardiovascular disease is well established. However, additional research has demonstrated an association between low serum cholesterol levels (<160 mg/dL) and increased risk for non-cardiovascular mortality, including respiratory disease. Low serum cholesterol is thought to be an indicator of poor health and may be lowered in the wasting syndrome associated with an underlying disease process (Jacobs, Blackburn, and Higgins 1992; Volpato et al. 2001; Brescianini et al. 2003). Low serum cholesterol has also been used as one risk factor to identify poor health and nutrition in the elderly (American Academy of Physicians, American Dietetic Association, and National Council on the Aging 2006).

The relationship between low cholesterol levels and poor respiratory health is not clearly understood. Since cholesterol is a component of pulmonary surfactant, it has been suggested that low circulating cholesterol levels could result in impaired production of pulmonary surfactant (Volpato et al. 2001). Using data from the Third National Health and Examination Survey, Cirillo et al. (2002) investigated the relationship between serum lipids and pulmonary health (Iribarren et al. 1997). Among young and middle-aged persons with low cholesterol levels, higher LDL components were associated with worse lung function, while higher HDL components were associated with better lung function, suggesting possible association between the lipid sub-fractions and pulmonary health. However, in older persons with low cholesterol levels, lower LDL components were associated with a greater risk of respiratory disease. No association with HDL levels and lung function was seen in this age group. The authors concluded that the lipoprotein components may be markers for some aspect of diet or lifestyle that may affect lung function. Further research is needed to identify how cholesterol levels affect pulmonary health. Of specific interest is whether the use of drug treatment to lower cholesterol levels may affect pulmonary health in the elderly.

Pathophysiology

When asthma occurs, bronchi and bronchioles respond to stimuli by contraction of smooth muscle (bronchoconstriction). The mucosa is inflamed and edematous, with the increased production of mucus. This results in a partial or totally obstructed airway.

Clinical Manifestations

The initial symptoms the patient may experience include cough, dyspnea, and a tight feeling in the chest. Signs may include wheezing, increased respiratory rate, and labored breathing. Increased heart rate (tachycardia) and hypoxia may also be observed. Longer, prolonged episodes of asthma may result in respiratory alkalosis that can proceed to respiratory acidosis as well.

Treatment

An acute episode of asthma requires immediate attention to dilate airways and improve oxygenation. These interventions would include the use of a bronchodilator with a β-adrenergic agent such as ipratropium and theophylline (see Table 23.4). Chronic and long-term control of asthma is achieved using a variety of agents, including steroids and leukotriene receptor antagonists. Other treatments will include the use of environmental control of potential allergens and the development of controlled breathing techniques.

Nutrition Therapy

Nutrition Implications It is thought that diet and nutrition play a role in the development and treatment of asthma. Increases in asthma, particularly in developed countries, have paralleled increases in obesity. Data from NHANES I showed that adults with asthma had a 46% higher prevalence of obesity than those without asthma; this ratio remained relatively constant from NHANES 1 (1971–1975) through NHANES III (1988–1994) (Ford and Mannino 2005). This relationship has been consistently observed in women (Camargo et al. 1999; Mishra 2004; Hancox et al. 2005). The strongest association of BMI with asthma severity in women has been associated with early menarche (Varraso et al. 2005). In children, increased risk of new-onset asthma has been associated with increased BMI (Oddy et al., The relation of breastfeeding, 2004).

Reasons for the relationship between obesity and asthma are not clear. Some possible reasons include the direct effects of obesity on the mechanical functioning of the lungs, changes in the immune system or an inflammatory response related to obesity, hormonal influences, and the interrelationship between the genes responsible for asthma and obesity (Weiss 2005).

A number of studies have demonstrated a protective effect of breast-feeding against the development and severity of asthma in children, while other studies have not shown this effect (Rust et al. 2001; Peat et al. 2003; Sears, Taylor, and Poulton 2003; Kemp and Kakakious 2004). Differences in study design and sampling errors have been cited as factors that may contribute to these conflicting results. Using specific criteria to assess study design, Oddy et al. evaluated 29 previous studies, looking at the relationship between breast-feeding and the development of asthma and allergic diseases (Oddy and Peat 2003). The criteria included standards for assessing breast-feeding status and duration; the use of strict diagnostic criteria for asthma; controlling for other confounding factors such as the mother's education, mother's diet, socioeconomic status, smoking status, and housing type and allergen exposure; and the appropriate use of the statistics, including sufficient statistical power. Of the 15 studies that met these criteria, all demonstrated a protective effect of breast-feeding; of the remaining 12 studies that did not meet the criteria, 4 showed a positive effect of breast-feeding, 3 showed a negative effect, and 5 showed no effect. Based on this review, these authors concluded that there is clear evidence to demonstrate that breast-feeding does protect against asthma and allergy in childhood. However, the protective effect of breast-feeding on asthma and allergy in adolescence and adulthood still needs to be confirmed.

Leukotrienes are chemical mediators produced by the body that contribute to the development of asthma. Leukotrienes produce tissue edema, mucus secretion, smooth-muscle proliferation, and powerful bronchoconstriction. Leukotrienes are synthesized from arachidonic acid by a specific synthesis pathway. Two types of leukotriene-based medications have been developed to combat asthma: leukotriene inhibitors that interfere with the actual synthesis of leukotrienes (Zyflo™), and leukotriene antagonists that block the action of leukotrienes at the receptor level (Accolate™ and Singulair™).

One possible approach to preventing the synthesis of leukotrienes is through dietary modification. Normally, human inflammatory cells contain high amounts of the omega-6 fatty acid, arachidonic acid, and low amounts of omega-3 fatty acids (Simopoulos 2002). Because both omega-6 and omega-3 fatty acids are metabolized by a common pathway, an excess of omega-3 fatty acids has been shown to interfere with the metabolism of the omega-6 fatty acids and reduce their incorporation into tissue lipids. Although there is some evidence that omega-3 fatty acid supplementation can decrease the production of inflammatory agents, primarily leukotrienes, in asthmatic patients, evidence that omega-3 fatty acid supplementation decreases the clinical severity of asthma in controlled trials has been inconsistent (Dry and Vincent 1991; Hodge et al. 1998; Oddy et al., Ratio of omega-6, 2004; Wong 2005; Woods et al. 2005). A Cochrane review of nine randomized controlled

leukotrienes—powerful inflammatory mediators produced by the body that are important in inflammation and allergic reactions because of their ability to constrict blood vessels and attract a variety of types of immune cells

studies of omega-3 fatty acid supplementation from fish or fish products in asthmatic patients found no consistent effects on clinical outcome measures of asthma, including pulmonary function tests, asthmatic symptoms, medication use, and **bronchial hyperreactivity** (Thien et al. 2002).

Interest has also been focused on the antioxidant vitamins—vitamins A, C, and E—and the carotenoids. Studies examining the protective effects of antioxidants and the consumption of fruits and vegetables on the development and treatment of asthma have not been conclusive (Green 1999; Shaheen et al. 2001; Harik-Khan, Muller, and Wise 2004; Kalantar-Zadeh, Lee, and Block 2004; Ram, Rowe, and Kaur 2004). Of all the antioxidant vitamins, the data linking vitamin C with asthma appears the strongest (Weiss 1997). As an antioxidant, vitamin C may modify oxidative insults from inhaled or infectious agents and reduce cellular inflammation. However, a Cochrane review of eight randomized, controlled trials investigating the treatment of asthma using vitamin C supplementation concluded that there is insufficient data to either refute or confirm the role of vitamin C in the management of patients with asthma (Ram, Rowe, and Kaur 2004).

Nutrition Interventions Treatment of asthma includes removing any items from the patient's environment that are known to be asthma triggers. When these measures do not work, there are a variety of medications used to control symptoms. Generally, medications are divided into those that provide quick relief and those designed to provide long-term control. Quick relief medicines, usually bronchodilators, are used to ease the wheezing, coughing, and tightness of the chest that occur during asthma episodes. Long-term medications include anti-inflammatory agents, such as oral steroids, that are designed to make the airways less sensitive and prevent them from reacting as easily to triggers.

Medications prescribed for treatment of asthma may have a number of nutritionally relevant side effects, including dry mouth, throat irritation, nausea, vomiting, and diarrhea. Long-term use of corticosteroids has been associated with increased serum glucose levels and sodium retention, as well as changes in bone mineral density. The short-term use of inhaled corticosteroids at conventional or usual doses (for two to three years) has not been associated with loss of bone mineral density (BMD) or fractures in adult patients with asthma or mild chronic obstructive pulmonary disease (Jones et al. 2002). Higher doses of inhaled steroids have been associated with biochemical markers of increased bone turnover, but data on BMD and fractures are not available. There is a need for further randomized studies that look at the effects of long-term use of conventional and higher doses of inhaled corticosteroids on BMD.

Until the relationship between diet and the etiology of asthma is confirmed, it is most important to recommend a nutritionally adequate diet for individuals suffering from asthma. The dietitian can play a significant role in assessing nutrition status and making appropriate recommendations based on specific needs. Because of the potential benefits of breastfeeding, mothers should be encouraged to breast-feed whenever possible. In light of the positive relationship between BMI and the development asthma, assisting individuals and families with weight control is also important.

bronchial hyperreactivity—tendency of the smooth muscle of the tracheobronchial tree to narrow in response to a stimulus; present in virtually all symptomatic patients with asthma

bronchopulmonary dysplasia (BPD)—a chronic lung disorder that may affect infants who have been exposed to high levels of oxygen therapy and ventilator support

Bronchopulmonary Dysplasia

Definition

Bronchopulmonary dysplasia (BPD), also called chronic lung disease of prematurity, is characterized by pulmonary inflammation and impaired growth and development of the alveoli. The common criteria to diagnose BPD are the requirement for supplemental oxygen beyond 28 days of life and signs of chronic respiratory changes on x-ray (American Thoracic Society Documents 2003). Factors associated with BPD include extreme prematurity (birth weight <1500 g), perinatal infection, and the presence of patent ductus arteriosus (PDA) (American Thoracic Society Documents 2003; Carlson 2004).

Etiology

The etiology of BPD is complex and multifactorial. Infants born before 28 weeks have immature lung development and require supplemental oxygen on a ventilator for extended periods of time. Prolonged exposure to high oxygen concentrations has been identified as one cause for BPD. The premature infant has a poorly developed antioxidant system and therefore is at risk of oxygen free radical damage.

Poor vitamin A status is another possible factor in the development of BPD. Vitamin A is important in normal alveolar development and surfactant production, and supports the integrity and regeneration of respiratory epithelial cells. Vitamin A deficiency in animals results in epithelial lesions similar to those seen in BPD (Zachman 1995). Most premature infants are born with low serum vitamin A levels and lower levels of vitamin A transport carrier protein, a retinol-binding protein.

Infant Receiving Supplemental Oxygen via a Nasal Cannula

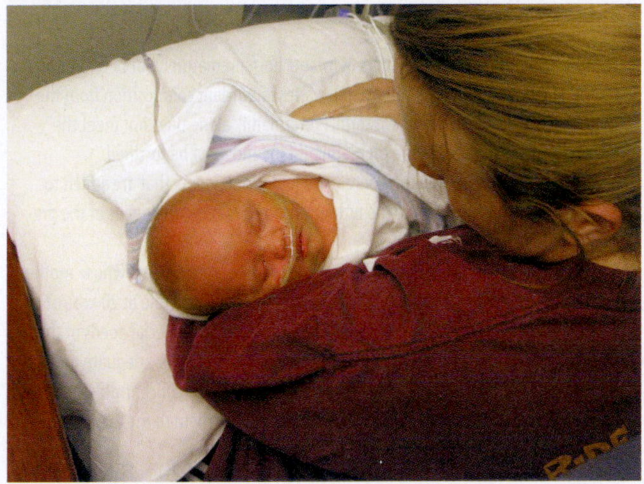

Source: http://www.jimmcintosh.com.

Treatment

Infants with BPD often require prolonged, intensive hospitalization accompanied by **mechanical ventilation**. Parenteral or enteral nutrition support or both are often required. The best method for preventing BPD is good prenatal care for the pregnant woman, including good nutritional status, which helps ensure that infants are born full term.

Nutrition Therapy

Nutrition Implications The causes of growth failure include concomitant organ dysfunction resulting in congestive heart failure and renal insufficiency in some infants, and decreased nutrient intake, **hypoxemia** and increased energy requirements in others. Poor nutrient intake is often related to a variety of factors, including poor swallowing function, oral aversion, reflux esophagitis, fatigue during feeding, and the need for fluid restriction resulting from fluid retention (Carlson 2004). Increased oxygen consumption reflected by increased work of breathing is thought to be one factor related to the increased energy requirement seen in these infants. However, the resting metabolic rate is also higher and contributes to their increased energy and nutrient needs (Kurzner et al. 1988). Infants with BPD are also at risk for delayed skeletal mineralization and osteopenia.

Nutrition Interventions A complete nutrition assessment to evaluate specific nutrient needs is essential. Infants with BPD have difficulty maintaining weight gain and achieving development similar to normal healthy infants. Since infants with BPD are at nutrition risk, nutrition screening should be done to identify specific nutrition needs. Anthropometric measurements of length, weight, and head-circumference should be taken routinely, using growth charts for very low birth weight children (Sherry et al. 2003). Because of immature swallowing function and oral aversion related to endotracheal and suctioning stimuli, nutrition support initially often includes nasogastric tube feedings. Nutrient requirements are often complicated by the need for fluid and sodium restrictions and the use of diuretics and other medications, such as steroids, that may cause nutrient losses and catabolism. When there is no longer risk of aspiration, and swallowing functions have matured, specially designed infant formulas to meet kcal, protein, calcium, phosphorus, vitamin, and mineral needs are used.

Energy and Macronutrient Needs The energy needs of infants with BPD are generally 15% to 25% higher than those of healthy, normal infants (Denne 2001). Estimated energy needs of 120 to 130 kcal/kg are often required to provide for appropriate growth (Specific guidelines for disease—pediatrics 2002). Energy intake and weight/length gains should be monitored closely, and energy goals should be adjusted to maintain proper growth. In some instances, energy intakes of 130 to 160 kcal/kg may be required due to increased metabolic demand. The metabolism of carbohydrate versus fat may increase the production of carbon dioxide (CO_2). Although high-carbohydrate feedings have been shown to increase CO_2 production in infants with BPD, they have not been associated with worsening of pulmonary function (Pereira et al. 1994; Chessex et al. 1995).

Protein catabolism induced by corticosteroid medications can be significant. Protein intakes of 3–4 g/kg/day have been recommended to meet needs (American Thoracic Society Documents 2003; Carlson 2004). Intakes of greater than 4 g/kg/day should be avoided because of risk of acidosis seen in preterm infants with immature kidneys. This occurs because the premature infant has a reduced ability to excrete both acid and the renal solutes (urea nitrogen and electrolytes). Elevations in blood urea nitrogen (BUN) and serum ammonia are used to monitor protein tolerance.

mechanical ventilation—artificial ventilation using a ventilator or respirator; performed with a piece of equipment designed to intermittently or continuously assist or control pulmonary ventilation

hypoxemia—condition in which there is an inadequate supply of oxygen in the blood

Vitamins and Minerals Lung injury in the preterm infant appears to occur within a short time after delivery, and oxidation appears to be a major contributor to this process. Unfortunately, attempts to deliver antioxidants directly to the lung have not been successful. As discussed earlier, preterm infants are born with reduced stores of vitamin A. Spears et al. found that low plasma retinol concentrations during the first month of life were significantly associated with increased risk for developing BPD in very low birth weight infants (<1250 g) (Spears, Cheney, and Zerzan 2004). It has been determined that low plasma retinol concentration up to six months of age can produce a similar effect. The need for long-term oxygen support was significantly greater for infants with low vitamin A status versus those with higher vitamin A status. Supplementation of 1,500 to 2,800 IU/kg/day or 450 to 840 μg/kg/day of vitamin A appears to be safe and has led to decreased incidence of BPD, decreased days on mechanical ventilation and supplemental oxygen, and decreased number of days in intensive care (American Thoracic Society Documents 2003).

Electrolyte balance, particularly of sodium and chloride, is essential to maintain growth. The use of diuretics to treat pulmonary edema and fluid overload is common. Diuretics increase urinary losses of sodium, potassium, chloride, and calcium. Preterm infants receiving diuretic therapy often exhibit signs of electrolyte depletion. Close monitoring of electrolyte balance and correction of electrolyte abnormalities is necessary.

Infants are at significant risk for delayed skeletal maturation and osteopenia. Low body calcium stores are aggravated by the use of diuretics. The use of corticosteroids is also likely to reduce bone mineral deposition. Infant formulas specifically designed for the premature infant provide sufficient minerals to support bone growth (see Table 23.2).

Feeding Practices Early, aggressive nutrition support may improve the nutritional status of infants with BPD and may provide protection from the development of BPD. Once BPD has developed, nutrition appears to play an important role in preventing further damage and promoting healing. Initially very low birth weight infants may be placed on parenteral nutrition support. Gradual transition from parenteral to enteral feedings should be initiated as soon as possible; however, this transition may take weeks or even months in some infants (Carlson 2004). Enteral feeding advancement may be delayed due to poor peristalsis and medical and respiratory instability. Once initiated, continuous nasogastric feedings are often used, with transition to bolus feedings as respiratory status improves. Stimulation of oral-motor skills should also occur to prepare the infant for oral feeding.

Because these infants often have higher energy needs, they are often placed on high-kcal enteral formulas to meet these needs. These increased nutrient requirements can complicate the need for fluid restriction. Breast milk is the preferred feeding because it reduces sepsis and the **necrotizing enterocolitis (NEC)** often seen in the premature infant. However, breast milk alone cannot meet the high energy needs of the premature infant. Several human milk fortifiers are available to increase the nutrient content of breast milk. If human milk is not available, formulas for premature infants may be substituted. Premature infant formulas are available in 20 kcal/oz and 24 kcal/oz concentrations. (See Table 23.2.)

When it is necessary to increase the caloric density of infant formula beyond 24 kcal/oz, it is preferable that the formula be prepared with the addition of either powdered or concentrated preterm infant formula rather than carbohydrate or fat additives. While carbohydrate or fat additives can be used to increase the caloric density of breast milk or formula, they also dilute the protein and mineral concentrations of the feeding. The use of formulas fortified with carbohydrate and fat may contain inadequate amounts of calcium and phosphorus; supplementation of these minerals may be needed. Infants fed high-energy formulas supplemented with carbohydrate or fat have demonstrated increased weight gain and fat mass, but not lean body mass (Romera et al. 2004).

TABLE 23.2

Preterm Infant Formulas

- **Human milk** produced by mothers of infants born prematurely (or before term) is referred to as *preterm milk*. Preterm milk varies in nutrient composition from the milk of mothers of infants born at term. Preterm milk alone does not meet the increased nutrient needs of the premature infant and must be fortified.
- **Human milk fortifiers** are powdered or liquid supplements that are added to breast milk in order to increase the nutrient content to meet the needs of the premature infant.
- **Preterm infant formulas** are also available for premature infants whose mothers cannot provide human milk. Preterm infant formulas are available in 20 kcal/oz and 24 kcal/oz concentrations and contain nutrients to meet the special needs of the premature infant. Most premature infants require 24/kcal/oz for adequate growth. Infant formulas are available ready to feed, or as powders or concentrates; powders and concentrates require the addition of water to dilute them to the appropriate caloric concentration.

necrotizing enterocolitis (NEC)—a condition that occurs primarily in premature infants or sick newborns, in which intestinal tissue dies. The cause for this disorder is unknown, but it is thought to be due to decreased blood flow to the bowel, which keeps it from producing the normal protective mucus. If an infant is suspected of having necrotizing enterocolitis, feedings are stopped to allow the bowel to rest

BOX 23.3 CLINICAL APPLICATIONS

Common Feeding Problems for Low Birth Weight Infants

"My baby falls asleep when I feed her. She doesn't seem to have the energy for feeding."

This is a fairly common problem in the first weeks after the baby comes home. The solution depends on the reason for the problem:

- The infant isn't able to maintain oxygen status during feeding and might benefit from supplemental oxygen during feedings, or from "pacing" the feeding to allow for feeding breaks to maintain adequate oxygen status.
- The infant is overwhelmed by too much adjacent activity while feeding. Decrease the exposure to light, noise, and movement while feeding.
- Small babies have limited stomach capacity and are not able to take much at each feeding. Provide smaller, more frequent feedings or concentrate the formula or breast milk to increase the caloric density.
- Some babies easily become fatigued and do not get enough total formula or breast milk in a 24-hour period. Provide smaller, more frequent feedings; concentrate the formula to increase kcal; or provide supplemental tube feedings.

"My baby gets really upset when I try to feed her. Feeding doesn't seem to bring her pleasure."

Some possible reasons for this include:

- Some infants may have gastroesophageal reflux with or without aspiration; this should be explored and treated immediately.
- Some infants associate feeding with unpleasant things that happened to and around their mouths early in life. They should never be force-fed. A feeding therapist can help develop a more structured approach to feeding that will gradually desensitize the infant to oral feeding.
- The flow of milk from the bottle or the breast may be too rapid or too slow. Changing the bottle nipple type or size, or the hole in the nipple, may help. Some breast-fed infants do better if the breast milk "let down" is established before they are put to the breast; the initial volume of milk which follows "let down" overwhelms some babies.

"My baby coughs and gags when I feed her. Sometimes this even leads to spitting up."

It is important to find the cause for the coughing and gagging. Coughing or gagging that results in apnea or color changes is serous and should be evaluated promptly. Other causes might be:

- Infants have trouble coordinating suck-swallow-breathe. A feeding therapist can help establish a program to pace the feedings until the infant has developed the neurological maturity to overcome this problem.
- Infants might tire at the end of the feeding and lose ability to coordinate suck-swallow-breathe. Pacing the feedings so the infant doesn't get too tired (as above) or providing smaller, more frequent feedings may help.
- Milk may flow too rapidly from bottle or breast. See discussion above for possible solutions.

Adapted with permission from: Some Common Feeding Problems for Low Birth Weight Infants. Gaining and Growing: Ensuring Nutrition Care of Preterm Infants. At: http://depts.washington.edu/growing/Feed/Oralprob.htm.

Infants often continue to have a number of feeding problems after hospital discharge (see Box 23.3). In a study designed to identify the nutritional risk factors seen in infants after hospital discharge, Johnson et al. found that parents expressed a number of concerns about feeding, including getting the infant to take enough food; the frequency of feeding; resistance to feeding; the occurrence of gagging, vomiting, and choking; and the need for knowledge about feeding techniques (Johnson, Cheney, and Monsen 1998). Infants with feeding problems often require several hours a day to feed. Dietitians can play an important role in helping families and caregivers manage the nutritional care of these infants. Comprehensive nutrition counseling and follow-up after discharge are necessary. Providing education and support to caregivers helps to ensure that these infants receive adequate nutrition to support appropriate growth.

Chronic Obstructive Pulmonary Disease

Definition

Chronic obstructive pulmonary disease (COPD) is a progressive disease which limits airflow through either inflammation of the lining of the bronchial tubes

chronic obstructive pulmonary disease (COPD)—a disease that limits airflow through either inflammation of the lining of the bronchial tubes or destruction of alveoli (**bronchitis**) or destruction of alveoli (**emphysema**). Frequently, both conditions coexist as part of this disorder.

Epidemiology

COPD is the fourth leading cause of death in America and the number of deaths of women from COPD exceeded those of men for the first time in 2002 (American Lung Association, Chronic Obstructive, 2005). The primary risk factor for the development of COPD is smoking. Other risk factors include air pollution, second-hand smoke, history of childhood infections, and occupational exposure to certain industrial pollutants. Even though normal lung function gradually declines with age, individuals who are smokers have a more rapid decline—twice the rate of non smokers (Rennard 1998). Low body weight has also been shown to be a risk factor for the development of COPD, even after adjusting for other potential risk factors including smoking and age (Harik-Khan, Fleg, and Wise 2002).

Etiology

Chronic bronchitis is one of the principal manifestations of COPD and is characterized by inflammation and eventual scarring of the lining of the bronchial tubes accompanied by restricted airflow, excessive mucus production, and a persistent cough. As stated previously, cigarette smoking is the primary cause; the longer and more heavily a person smokes, the more likely it is that he or she will develop bronchitis. Secondhand smoke may also cause bronchitis. Chronic bronchitis is seen in people of all ages but is more common in individuals over the age of 45 (American Lung Association, Chronic Obstructive, 2005). Females are more than twice as likely to be diagnosed with chronic bronchitis as males.

Emphysema develops gradually over years, usually as a result of the exposure to cigarette smoke. Approximately 95% of Americans diagnosed with emphysema are 45 years of age or older (American Lung Association, Chronic Obstructive, 2005). In the past, more males than females suffered from emphysema; however, in the past few years the incidence in women has significantly increased so that the difference in the prevalence rates between the sexes has become statistically insignificant.

In rare cases emphysema is caused by the deficiency of a protein called alpha 1-antitrypsin (ATT) or alpha 1-protease inhibitor. ATT is produced by the liver and is released into the bloodstream, where it travels to the lungs to protect them from the destructive actions of common illnesses and exposures, particularly tobacco smoke. Only about 5% of emphysema in the U.S. is caused by ATT deficiency (American Lung Association, Chronic Obstructive, 2005). Unlike the common form of emphysema seen in otherwise healthy individuals who have smoked for many years, this ATT deficiency form of emphysema may occur at a younger age (between 32 and 41 years) and after minimal exposure to tobacco smoke.

Pathophysiology: Chronic Bronchitis

In chronic bronchitis, repeated exposure to cigarette smoke and other pollutants results in a generalized inflammatory response. This includes decreased cilia function, increased phagocytosis, and suppressed amounts of immunoglobulin A (IgA). Chronic inflammation causes hyperplasia of the mucus-secreting cells, resulting in edema of the bronchioles. The walls of the airways thicken and mucus glands become hyperplastic. The damaged cilia are unable to clean mucus from the airways, and the patient is unable to increase the work of breathing enough to overcome the signs and symptoms of the disease.

Clinical Manifestations: Chronic Bronchitis

Chronic bronchitis is characterized by decreased air flow rates (↓FEV), dyspnea, hypoxemia, and hypercapnia. Signs of chronic hypoxemia include **cyanosis**, **clubbing**, and **secondary polycythemia**.

Cyanosis

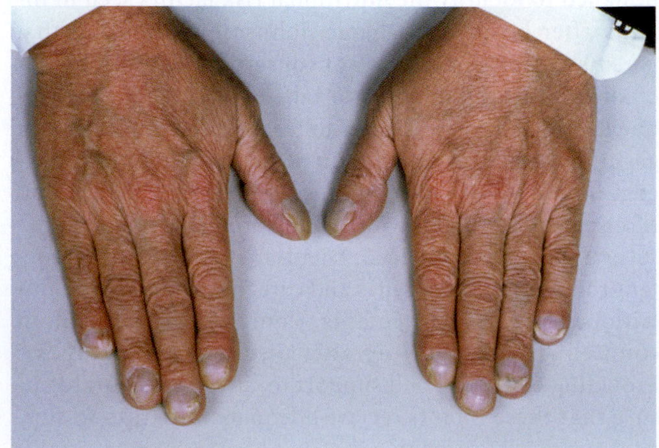

Source: John Radcliffe Hospital/ Photo Researchers, Inc.

bronchitis—a condition characterized by inflammation and eventual scarring of the lining of the bronchial tubes accompanied by restricted airflow, excessive mucus production, and a persistent cough

emphysema—a condition characterized by thinning and destruction of the alveoli, resulting in decreased oxygen transfer into the blood stream and shortness of breath

cyanosis—blue-tinged mucous membranes and skin due to inadequate oxygen supply

clubbing—changes in fingers and toes due to hypoxemia; fingers and toes show a curve at a tip of the nail with flattening surface

secondary polycythemia—condition in which an excessive number of red blood cells are produced; occurs in response to compensation for chronic hypoxemia

TABLE 23.3

Classifications of COPD by Severity

Stage 0 (at risk)

- Normal spirometry
- Chronic symptoms (cough, sputum production)

Stage 1 (mild COPD)

- FEV1/FVC <70%
- FEV1 > 80% predicted
- With or without chronic symptoms

Stage II (moderate COPD)

- FEV1/FVC <70%
- FEV1 < 80% predicted
- With or without chronic symptoms

Stage III (severe COPD)

- FEV1/FVC <70%
- FEV1 < 50% predicted
- With or without chronic symptoms

Stage IV (very severe COPD)

- FEV1/FVC <70%
- FEV1 < 30% predicted or
- FEV1 < 50% predicted with symptoms of chronic respiratory failure
- FVC = forced vital capacity, FEV1 = forced expiratory volume in one second

Source: National Guideline Clearinghouse. Global strategy for the diagnosis, management, and prevention of chronic obstructive pulmonary disease. [cited: 2006 January 15] Available from: http://www.guideline.gov/summary/summary.aspz?doc_id=8128&nbr=004530&string=ex.

The quality of life for persons suffering from COPD diminishes as the disease progresses, resulting in an inability to work and possibly limiting normal day-to-day physical exertion (see Table 23.3). Often, individuals with COPD eventually require supplemental oxygen and may have to rely on mechanical respiratory assistance. **Cor pulmonale** may also occur late in the development of the disease, further complicating the individual's health status.

Pathophysiology: Emphysema

The destruction of actual lung tissue differentiates emphysema from chronic bronchitis, even though emphysema often develops as a late complication of chronic bronchitis. The loss of connective tissue results in a loss of surface area and decreased amounts of surfactant. Since the bronchioles lose their elasticity, they collapse during exhalation and trap air in the lungs.

Clinical Manifestations: Emphysema

Emphysema results in a decreased forced expiratory volume (FEV). Though inspiration is not impaired, expiration is, because air is trapped within the lungs. This inability to expire results in dyspnea and **orthopnea**, and causes hypercapnia and respiratory acidosis. The increased use of accessory muscles for expiration causes the development of a "barrel chest." Patients with emphysema do not experience hypoxemia until the last stages of the disease, when extreme fatigue and physical exhaustion prevent adequate oxygen intake.

Treatment

Medical treatment for individuals with COPD involves lifestyle changes, including smoking cessation, avoiding smoke and other air pollutants, exercising as tolerated, and good nutrition (American Lung Association, Chronic Obstructive, 2005). Treatment strategies are based on assessment of disease severity and response to various therapies. Pharmacologic treatment is used to prevent and control symptoms, improve health status, and improve exercise tolerance. Medications include bronchodilators, β-agonists to relax smooth muscle, and anticholinergics to decrease airway contraction and mucus production. Steroids are used to decrease swelling, and mucolytic agents work to decrease the viscosity of secretions (see Table 23.4).

Pulmonary rehabilitation programs provide a comprehensive, multidisciplinary approach to treatment that combines education with therapeutic exercise. Team members may include nurses, occupational therapists, registered dietitians, respiratory therapists, physical therapists, and physicians. These programs help the patient understand and cope with these chronic conditions, with the ultimate goal of improving overall quality of life.

Nutrition Therapy

Malnutrition occurs in 24% to 35% of patients with moderate to severe COPD, with average weight loss of 5% to 10% of initial body weight (Congleton 1999). The incidence of malnutrition depends on the severity of the disease. Malnutrition due to COPD can weaken respiratory muscles, resulting in altered ventilation and impaired immune function (Engelen 2003). Weight loss occurs frequently, particularly in individuals with emphysema, and is associated with increased resting energy expenditure secondary to the work of breathing, reduced nutrient intake, and inefficient fuel metabolism. In contrast, individuals with bronchitis frequently have normal or above normal BMI. Losses of lean body mass (LBM), however, have been seen in both conditions. In a study looking at body composition of individuals

cor pulmonale—an increase in size of the right ventricle of the heart caused by resistance to the passage of blood through the lungs; can lead to heart failure

orthopnea—difficulty breathing while lying down

TABLE 23.4

Medications Used in Diseases of the Respiratory System

Type of Medication	Brand Names	Action	Common Side-Effects
Bronchodilators *Types of bronchodilators:* β2-agonist Anticholinergics Theophyllines	*β2-Agonists:* Albuterol® Pirbuterol®, Salbutamol® Terbutaline® *Anticholinergics:* Ipatropium® Tiotropium®	Used to open or relax the bronchial tubes and relieve shortness of breath. May be taken as an inhaler or pill.	Fast heartbeat, shakiness, and cramping of hands, legs, and feet; dry mouth particularly with the anticholinergics. Severe nausea and vomiting with theophyllines.
Steroids (*corticosteroids*)	Predinisone®; Prednisolone® Solu-Medrol® Solu-Cortef®	Used to reduce inflammation in the bronchial tubes. May be used as an inhaler or taken orally.	Side effects depend on the dose, length of use, and whether taken orally or inhaled. For inhaled steroids, most common side effects include sore mouth, hoarse voice, and infections in throat and cough. Orally in high doses or low doses for a long period of time, the side effects include: altered fluid/electrolyte balance, hypertension, mood swings, increased appetite, weight gain, hyperglycemia, osteoporosis, hyperlipidemia, poor wound healing, growth retardation in children.
Antibiotics Doxycycline Hyclate Amoxicillin Macrolides Fluoroquinolones	 Doryx®, Vibramycin® Augmentin®	Treat respiratory infection.	Nausea, vomiting, diarrhea.
Leukotriene Inhibitors Zafirlukast Aileuton Montelukast	 Accolate® Zyflo® Singulair®	Mediates inflammation of COPD.	Headache, nausea, diarrhea, infection. Singulair®—caution with grapefruit/grapefruit juice.
Mucolytics Acetylcysteine	 Mucomyst®	Makes mucus in the lungs thinner and less sticky.	Nausea, vomiting, runny nose, drowsiness, clammy skin.
Immune System Modifier Xolair	 Omalizumab®	Inhibits the binding of IgE and allergic response	Cold- or flu-like symptoms. Muscle or joint pain. Itching; fever.
Pancreatic Enzymes	Creon®, Pancrease®, Pancrease MT® Pancrecarb®, Ultrase®, Viokase®, Ultrase MT®	Replace deficient pancreatic enzymes for patients with cystic fibrosis.	Take immediately before meals; do not take on empty stomach.

with COPD, Engelen et al. found depletion in LBM in 37% of individuals with emphysema and 12% of those with chronic bronchitis. Even in individuals with normal body weights, depletion of LBM was found in 16% of those with emphysema and 8% of those with chronic bronchitis (Engelen et al. 1999).

Energy and Macronutrients Needs A complete nutrition assessment is necessary to identify patients who are at nutrition risk. This assessment should include an evaluation of weight history, nutrient intake, functional status, and medication usage. Based on this assessment, nutrition goals for the individual patient can then be established.

Low dietary intake and weight loss occur in individuals with moderate to severe COPD because of symptoms of dyspnea, fatigue, and early satiety. Taste perceptions may be altered with chronic mouth breathing and appetite may be further reduced as a result of depression. Even individuals with adequate dietary intake may lose weight, as elevations in both the resting energy expenditure (REE) and total energy expenditure independent of the REE are seen in individuals with COPD (Shols et al. 1991; Congleton 1999; Gosker et al. 2000).

In COPD, the respiratory muscles need to generate a large force to expand the thoracic rib cage. This increase in the energy cost of breathing is often cited as a major cause of the increased REE. However, this does not appear to be the only cause of the hypermetabolism. Increased systemic inflammation has also been cited as a cause of hypermetabolism evidenced by elevations in **tumor necrosis factor (TNF-α)** in these patients (Shols 2001). In addition, the thermogenic effects of medications used to treat COPD, including the bronchodilating drugs, may also play a role.

Maintaining optimal energy balance in the individual with COPD is essential in order to preserve body weight, lean body mass, and general well being. Respiratory muscle function is severely affected by declining nutrition status and is closely linked to body weight and lean body mass. It is essential that the individual with COPD receive sufficient kcal and protein to maintain body weight, lean body mass, and adequate nutrition status. Thorsdottir and Gunnarsdottir found that energy intakes of 125% to 156% (average 140%) above basal energy expenditure and protein intakes of 1.2 to 1.7 grams/kg body weight (average 1.2 g/kg) were adequate to avoid protein losses in patients admitted to the hospital with exacerbation of their COPD (Thorsdottier and Gunnarsdottir 2002). Malnourished patients need additional energy and protein to provide for repletion. Because the predicted REE using the Harris-Benedict formula significantly underestimates the measured REE by 10% to 15% for patients with COPD, indirect calorimetry is the best method to assess kcal needs without overfeeding or underfeeding (Mallampalli 2004). When indirect calorimetry is not available, providing 25–30 kcal/kg of body weight appears appropriate with approximately 20% of total kcal from protein (1.2–1.7 grams of protein/kg body weight), depending on the patient's individual needs (Felbinger et al. 2000; Shols 2001; Thorsdottir and Gunnarsdottir 2002). In the hospitalized patient with COPD, particularly patients who have compromised pulmonary function, it may be necessary to provide ventilatory support using mechanical ventilation. In these patients, overfeeding is of primary concern because it is associated with increased CO_2 production, which can further complicate ventilation. Although glucose and protein have been shown to stimulate ventilatory drive, excess glucose administration (>5 mg/kg/minute) increases CO_2 production and makes it difficult to wean or remove patients from mechanical ventilation (Specific guidelines for disease—adults, Pulmonary, 2002). In spite of this, when total kcal are provided in moderate amounts (approximately 30% above basal needs), the macronutrient composition of the feeding has little effect on CO_2 production (Talpers et al. 1992). The production of excess CO_2 occurs when patients are overfed (>1.5 X REE).

Commercial enteral formulas that have been specifically designed for individuals with respiratory disease contain a lower carbohydrate content (30%) and higher lipid content (50%). Compared to the other macronutrients, and fat in particular, metabolized carbohydrate yields the greatest amount of CO_2. Controlled clinical trials using these modified formulas have demonstrated decreased CO_2 production when compared to standard formulas equal in kcal but higher in carbohydrate (al-Saddy, Blackmore, and Bennett 1989; Kuo, Shiao, and Lee 1993). However, improvement in clinical outcomes with the use of these formulas has not been consistently demonstrated (Akrabawi et al. 1996; Pingleton 1998; Specific guidelines for disease—adults, Pulmonary, 2002). One potential negative side effect of higher-fat meals or supplements is delayed gastric emptying, which may result in abdominal discomfort, bloating, or early satiety (Akrabawi et al. 1996).

Vitamins and Minerals As discussed earlier, a number of dietary factors, particularly the antioxidants, influence respiratory health. Numerous studies have shown that smokers have lower intakes of antioxidant vitamins, specifically vitamins C, A, and E, and β-carotene (Dallongeville et al. 1998; Ma, Hampl, and Betts 2000; Watson et al. 2002). Individuals with COPD undergo oxidative damage during both exacerbations of the disease and stable periods. During periods of exacerbations, serum concentration of vitamins A and E have been shown to decrease (Tug, Karatas, and Terzi 2004). Further investigation is needed to determine whether

tumor necrosis factor (TNF-α)—one type of cytokine which has been found to possess a wide range of proinflammatory actions

supplementation with these nutrients could prevent or reverse this damage.

Phosphate is essential for the synthesis of adenosine triphosphate (ATP) and **2,3-diphosphoglycerate (DPG)**, both of which are critical for pulmonary function (Specific guidelines for disease—adults, Pulmonary, 2002). Respiratory and peripheral muscle stores of phosphate have been shown to be depleted in individuals with COPD (Fiaccadori et al. 1994). Medical treatment with drugs commonly used for COPD, including corticosteroids, diuretics, and bronchodilators, is associated with hypophosphatemia and likely contributes to the depleted phosphate stores (Fiaccadori et al. 1990). Serum phosphate levels need to be closely monitored in patients with pulmonary disease or respiratory failure to ensure adequate levels.

Osteoporosis, with resulting bone fractures, has been shown to be a significant problem in patients with advanced COPD (Biskobing 2002; Ionescu and Schoon 2002; Katsura and Kida 2002). A number of risk factors have been related to the pathophysiology of osteoporosis in these individuals. These include smoking, suppression of estrogen or testosterone levels, vitamin D deficiency, low BMI, and decreased mobility (Shols 2003). Smoking has been shown to be an independent risk factor for osteoporosis in both men and women. Kanis et al. found that a smoking history was associated with a significant increased risk for fractures, particularly for hip fractures, compared to a nonsmoking history (Kanis et al. 2005).

Body weight is closely related to bone mineral density (BMD). Weight loss and malnutrition are likely involved in the pathogenesis of low BMD in individuals with COPD. Low serum 25-hydroxyvitamin D levels have also been documented in individuals with COPD, suggesting that vitamin D deficiency due to poor intake and decreased sun exposure may also play a role in the bone disease. In addition, the use of glucocorticosteroids in the treatment of COPD has been shown to increase the incidence of osteoporosis (Biskobing 2002). Glucocorticosteroids decrease the intestinal absorption of calcium and increase urinary excretion, resulting in an increase in parathyroid hormone levels and bone resorption. Bone mineral density should be measured in individuals who have COPD, particularly in those receiving long-term glucocorticoid treatment (>7.5 mg prednisone/day), with follow-up testing every two years (Biskobing 2002). The intake of calcium and vitamin D should be assessed, particularly in individuals with a reduced intake. According to the current nutritional recommendations, 1,200 to 1,500 mg/day of calcium and at least 400 IU of vitamin D should be provided (Biskobing 2002; Ionescu and Schoon 2002).

Feeding Strategies Weight loss and low BMI have been associated with increased mortality in patients with COPD, regardless of disease severity (Specific guidelines for disease—adults, Pulmonary, 2002). Providing adequate nutrition support is essential to helping maintain body weight and skeletal muscle mass.

Designing a nutrition care plan requires identification of the possible causes of reduced or inadequate intake. Oral intake in individuals with COPD is often inadequate because of a number of factors, including anorexia, early satiety, dyspnea, bloating, and fatigue (Donahoe 1998; Martin-Harris 2000). A common complaint is that the basic activities of daily living, including eating, require effort. Individuals with COPD often complain that they tire easily when eating or experience dyspnea during eating and drinking. Fatigue resulting from dyspnea may also interfere with eating. Chewing and swallowing may be impaired since both activities change breathing patterns and reduce oxygen uptake. Chronic mouth breathing or certain medications may also cause changes in taste perceptions and/or xerostomia (dry mouth).

Patients with COPD often suffer from **hyperinflation of the lungs** with accompanying flattening of the diaphragm and reduced abdominal volume, leading to unnecessary fullness and bloating at mealtime (Shols 2003). If the individual eats too much at a meal, there is increased positive pressure applied to the diaphragm which results in breathing difficulty. **Aerophagia** is often seen in COPD and may also cause bloating. Additional factors that may contribute to poor nutrition include depression and difficulty in shopping and preparing foods.

The dietitian can play an important role by completing a thorough nutrition assessment of the individual, including a complete nutrition history and identification of specific problems that may interfere with dietary intake (see Table 23.5). An evaluation of medication usage is also important in order to determine any resultant impacts on appetite and nutrition status. Particular attention needs to be paid to consuming foods that are not only good sources of both kcal and protein but are also nutrient dense. Instructing individuals to rest before meals to avoid fatigue may be helpful. Eating smaller, more frequent meals rather than three larger ones may help to alleviate the feeling of fullness and bloating. The use of nutrition supplements to provide additional kcal and protein has shown mixed results, suggesting that nutritional supplementation alone is

2,3-diphosphoglycerate (DPG)—an important regulator for the affinity of hemoglobin for oxygen. The synthesis of 2,3-biphosphoglycerate in red blood cells (rbc) is critical for controlling hemoglobin affinity for oxygen

hyperinflation of the lungs—results from loss of elasticity of the alveoli, causing air to be trapped; often seen in emphysema

aerophagia—the swallowing of too much air resulting in gas and bloating

TABLE 23.5

Symptoms and Nutrition Counseling Strategies for Individuals with COPD

Complaint	Recommendations
Anorexia	• Eat high-kcalorie foods first. • Have favorite foods available. • Try more frequent meals and snack throughout the day. • Add margarine, butter, sauces, and gravies to add kcalories.
Early Satiety	• Eat high-kcalorie foods first. • Limit liquids with meals; drink fluids an hour after meals.
Dyspnea	• Rest before meals. • Use bronchodilators before meals. • Eat more slowly. • Have pre-prepared meals available for periods of increased shortness of breath.
Fatigue	• Rest before meals. • Have ready prepared meals available for periods of increased shortness of breath.
Bloating	• Eat smaller, more frequent meals. • Avoid rushed meals. • Avoid gas-forming foods.
Constipation	• Incorporate exercise as tolerated. • Eat high-fiber foods with adequate fluids. • Refer for medical evaluation for stool softener as appropriate.
Xerostomia	• Avoid foods that are dry; add gravies, sauces to food to add moisture. • Limit dry salty foods. • Use artificial saliva before meals.

Reprinted from: *Respiratory Care Clinics of North America*, Vol. 4(1), Donahoe M, Nutritional aspects of lung disease, pages 85–112, Copyright 1998, with permission from Elsevier.

not sufficient to improve physical and nutrition status (Ferreira et al. 2005).

Physical exercise as part of the nutritional rehabilitation of patients with COPD is important. Nutrition support combined with exercise as part of a pulmonary rehabilitation program has been shown to have the best overall effect on increasing body weight, fat-free mass, and respiratory muscle strength in stable patients with COPD (Puhan et al. 2005). The type of exercise prescribed depends on the severity of COPD. As the disease progresses, patients often experience dyspnea, particularly on exertion, which limits their ability to do strenuous exercise. An evaluation of randomized controlled trials comparing different exercise protocols for patients with COPD concluded that strength/resistance training (versus endurance training) was associated with the greatest improvements in quality of life (Puhan et al. 2005; National Guideline Clearinghouse 2006). Skeletal muscle dysfunction has been recognized as an indicator of the advanced stages of COPD; strength training may help to improve skeletal muscle function and overall well being.

Excessive weight gain, particularly excessive body fat, may be deleterious by increasing the workload of an already compromised respiratory system (Donahoe 1998). Individuals who are morbidly obese have difficulty breathing caused by restrictions on the chest wall due to the accumulation of fat in and around the thoracic cage, diaphragm, and abdomen. This results in reduced lung volume accompanied by poor oxygen and carbon dioxide exchange.

Patients who are more than 40% above IBW should be evaluated individually to determine the most appropriate intervention that will provide long-term benefits (Donahoe 1998). The primary goal should be to prevent further weight gain and promote moderate weight loss, if appropriate. For patients whose health status is borderline, particularly patients who have a history of weight or appetite fluctuations, weight loss during exacerbations of the disease, or weight gain associated with prolonged steroid use, weight reduction is contraindicated. Weight reduction may exacerbate an existing risk for weight loss associated with the disease and lead to reduced pulmonary function.

Cystic Fibrosis

Definition

Cystic fibrosis (CF) is a disease characterized by abnormally thick mucus secretions from the epithelial surfaces of various organ systems, including the respiratory tract, the gastrointestinal tract, the liver, the genitourinary system and the sweat glands.

Epidemiology

CF is the most common **autosomal recessive** disease in the United States, affecting approximately 30,000 children and adults (Cystic Fibrosis Foundation 2005). One in 31 Americans are carriers of the defective gene which causes CF. To have CF, an individual must inherit two defective genes, one from each parent. Each time two carriers conceive, there is a 25% chance that their child will have CF; a 50% chance that

cystic fibrosis (CF)—disease characterized by abnormally thick mucus secretions from the epithelial surfaces of various organ systems, including the respiratory tract, the gastrointestinal tract, the liver, the genitourinary system, and the sweat glands

autosomal recessive—method of hereditary disease transmission in which the patient receives two chromosomes bearing the gene anomaly, one from each parent

the child will be a carrier of the CF gene; and a 25% chance that the child will be a non-carrier. Most individuals with CF are diagnosed by age three; however, nearly 10% of new cases are diagnosed at age 18 or older. The median survival age for individuals with CF is now more than 30 years.

Etiology

CF is caused by an abnormal mutation of the cystic fibrosis transmembrane conductance regulator (CFTR), which is a type of protein classified as an ATP-binding cassette (ABC) transporter (Human Genome Project Information 2004). Mutations of this protein prevent CFTR from functioning normally.

Pathophysiology

The CFTR proteins are responsible for the transport of sugars, peptides, inorganic phosphate, chloride, and metal cations across the cellular membrane. CFTR transports chloride ions (Cl^-) across the membranes of cells in the lungs, liver, pancreas, digestive tract, reproductive tract, and skin. In individuals with CF, CFTR's failure to function properly results in thick viscous secretions that eventually lead to obstruction of the glands and ducts in the affected organs.

Approximately 50% of individuals with CF have pulmonary symptoms that include a chronic cough and wheezing. Respiratory symptoms may begin during the first month of life. Pulmonary insufficiency leading to pulmonary failure is the major cause of death in individuals with CF. Because of this, early effective respiratory treatment is essential. A variety of aerosol therapies are used to increase airflow, reduce the thick mucus accumulation, and reduce infection. They include **inhaled bronchodilators**, **inhaled anti-inflammatory agents**, **inhaled mucolytics**, and **inhaled antibiotics**. **Chest physiotherapy**, designed to reduce airway obstruction by improving the clearance of secretions, is also an integral part of pulmonary care (see Table 23.4). Although CF affects many organ systems, the lungs are the most seriously involved, and respiratory failure and death result for more than 90% of patients (Venuta et al. 2001). Lung transplantation has become an important treatment option for CF patients with end-stage respiratory failure and may involve either a bilateral lung transplant or a heart lung transplant. (See discussion of lung transplantation later in this chapter.)

Pancreatic insufficiency occurs in 85% to 90% of CF patients (Beers and Berkow 2005). Symptoms of pancreatic involvement include frequent passage of bulky, foul-smelling, oily stools; abdominal distension; and poor growth pattern with decreased subcutaneous tissue and muscle mass despite a normal or voracious appetite. Insulin-dependent diabetes develops in approximately 10% of adults, and 4% to 5% of adolescents and adults develop cirrhosis of the liver with portal hypertension and varices (see Chapter 18 for a discussion of these problems). Children and adults with CF have an increased amount of sodium and chloride in their sweat. Because of this, excessive sweating in hot weather or fever may lead to dehydration and circulatory failure.

In addition to a complete medical history and physical examination, the diagnosis of CF is made using one or more of the following tests: a **sweat chloride test**, a blood test to confirm mutations of the CFTR gene, sputum cultures to test for infections typical in CF, **pancreatic function tests**, and pulmonary function tests. Fasting blood levels of carotenoids, vitamins A and E, essential fatty acids, and cholesterol are reduced in patients with steatorrhea (dietary fat malabsorption). Approximately 40% of older individuals have abnormal glucose tolerance due to delayed insulin secretion (Beers and Berkow 2005).

In May 2005, the FDA approved the first DNA-based test to detect CF. The test directly analyzes human DNA to find genetic variations indicative of the disease and will be used to diagnose both children with the disease and adults who are carriers of the CF gene (U.S. Food and Drug Administration, FDA News, 2005).

Nutrition Therapy

Nutrition Implications The relationship between nutrition status and long-term survival of individuals with CF is well documented. Pancreatic insufficiency results in

inhaled bronchodilators—medications used to maximize airway size and improve clearance of mucus

inhaled anti-inflammatory agents—class of medications that often includes inhaled corticosteroids

inhaled mucolytics—class of medications, including pulmozyme™, which hydrolyzes the DNA in sputum of cystic fibrosis patients and reduces sputum viscosity

inhaled antibiotics—medications designed to reduce airway infection, particularly *pseudomonas aeruginosa*, commonly seen in CF

chest physiotherapy—physical therapy that includes a variety of techniques designed to reduce or prevent infection by clearing pooled secretions and/or infected materials from the lungs

sweat chloride test—a test to measure the amount of chloride in the sweat by stimulating the skin to produce a large amount of sweat that is then absorbed by a special filter paper and analyzed for chloride content

pancreatic function tests—tests to measure pancreatic function, including serum amylase or lipase, a test for the amount of fat in the stool, and an x-ray of the anatomical features of the pancreas and common bile duct

poor digestion, poor absorption of fat and fat-soluble vitamins, and loss of bile and bile salts. Chronic pulmonary infections and deteriorating pulmonary function may lead to anorexia, increased energy requirements, and malnutrition. Children with CF have abnormal growth at all ages. Twenty percent of children in the 1993 National CF Patient Registry were found to be below the 5th percentile for both height- and weight-for-age (Lai et al. 1998). Lower-than-average height and weight are particularly pronounced in infants, adolescents, and individuals newly diagnosed with CF. Individuals with CF are also at risk for osteopenia and osteoporosis because of pancreatic insufficiency, the malabsorption that occurs (calcium, phosphorus, magnesium, vitamins D and K), and the chronic use of corticosteroid medications.

Nutrition Interventions A multidisciplinary approach to treatment of the individual with CF is essential in order to assist individuals and their families in meeting their complex medical and nutritional needs. The dietitian is responsible for assessing nutrition status, including determining energy requirements, providing nutrition counseling, and helping to plan intervention strategies when individuals are at risk for under nutrition or are diagnosed with nutritional failure. Early identification of CF is important in ensuring proper nutrition. The Cystic Fibrosis Foundation has developed ongoing consensus guidelines for treatment of CF, including nutrition treatment. These helpful guidelines provide up-to-date information about nutrition management. The most recent nutrition consensus report was published in 2002 (Borowitz, Baker, and Stalling 2002).

Early detection of poor growth allows for appropriate intervention and treatment. There are three periods when special attention needs to be focused on growth and nutritional status: during the first 12 months after diagnosis of CF, from birth until 12 months of age for infants diagnosed at birth, and during the peripubertal growth period (9 to 16 years for girls and 12 to 18 years for boys) (Borowitz, Baker, and Stalling 2002). Growth charts from the National Center for Health Statistics/Center for Disease Control (NCHD/CDC) which plot weight, head circumference, length, and height should be used to monitor growth status (available at http://www.cdc.gov/growthcharts). Children are considered to be at nutrition risk if they are between the 10th and 25th weight-for-length percentile and are considered to have nutritional failure if they are at less than the 10th weight-for-length percentile (Borowitz, Baker, and Stalling 2002). Mid-arm circumference and triceps fatfold measurement provide information about adequacy of subcutaneous fat stores (energy) and lean body mass (muscle). Table 23.6 outlines the timeline for nutrition evaluation, including growth status, for individuals with CF.

TABLE 23.6

Assessment of Nutrition Status for Individuals with Cystic Fibrosis

Nutrition Parameter	Frequency of Assessment
Anthropometric Measurements:	
Weight (to 0.1 kg)	Every 3 months
Height (length $<$2 years) (to 0.1 cm)	Every 3 months
Head circumference (to 0.1 cm)	Every 3 months
Mid-arm circumference (to 1.0 mm)	Annually
Triceps skinfold	Annually
Mid-arm muscle area, mm^2	Annually
Mid-arm fat area, mm^2	Annually
Nutritional Assessment:	
Dietary intake (24-hour recall)	Annually
Nutritional supplement intake	Annually
Anticipatory dietary and feeding behavior guidance	As indicated
Laboratory Status:	
β-carotene	At physician's discretion
Vitamins, A, D, E, K	Annually
Essential fatty acids	Check in infants and those who are failure to thrive (FTT)
Calcium/bone status	$>$age 8 years if risk factors are present
Iron	Annually
Zinc	Consider 6-month supplementation trial and follow growth
Sodium	If dehydration is suspected or exposed to heat stress
Albumin	Annually

Adapted from: Borowitz D, Baker RD, Stalling V. Consensus report on nutrition for pediatric patients with cystic fibrosis. *J Pediatr Gastroenterol.* 2002; 35:246–259.

Pancreatic Enzyme Therapy Because a significant number of individuals with CF have pancreatic insufficiency, malabsorption of dietary fat, protein, fat-soluble vitamins, and other nutrients is often seen. Pancreatic insufficiency has a strong influence on nutrition status and is a predictor of long-term outcome. The thickened secretions obstruct the pancreatic ducts and prevent the secretions of lipase, amylase, proteases, and bicarbonate. When pancreatic insufficiency is present, individuals are treated with pancreatic enzyme extracts. All enzyme products contain the various enzymes synthesized by the pancreas, including amylase, proteases, and lipase in varying amounts. The potency of the enzymes is usually based on the content of lipase in each capsule, because lipase is required to treat malabsorption of fat, which is the macronutrient most often malabsorbed in CF. Commercial enzyme products vary in lipase activity from 4,000 to 25,000 U lipase/capsule. They are available in powder form as tablets that are acid labile, or as enteric-coated microspheres—the enteric coating is designed to protect the enzyme from destruction by the acid environment of the stomach.

Pancreatic enzymes are always given when food or beverages are consumed. The dosage for enzymes is individualized based on the patient's diet, nutritional status, degree of pancreatic insufficiency, intestinal pH, and GI anatomy and physiology (Borowitz, Grand, and Durie 1995). Infants may be given 2,000 to 4,000 lipase units per 120 mL of formula or breast-feeding. The recommended enzyme dose for children begins with 1,000 lipase units/kg per meal for children under 4 years and 500 lipase units/kg per meal for children over 4 years of age. As children grow older, they require less lipase units per kilogram body weight because they tend to ingest less fat per kilogram of body weight. Usually, one half the standard dosage is given with snacks.

Because of inconsistencies in enzyme formulations, the FDA has issued a new rule requiring manufacturers of pancreatic enzyme supplements to obtain approval for their products (U.S. Food and Drug Administration, Questions, 2005). Prior to obtaining approval, manufacturers will need to test the enzymes in clinical trials and demonstrate that they are safe and effective. This rule means that the FDA now requires pancreatic enzymes to meet the same standards of testing as any other new drug.

Energy and Macronutrient Needs Adequate kcal to support normal growth and development are essential, especially in the presence of pancreatic insufficiency. Energy intake should be based on the patterns of weight gain and growth in children. Energy needs for children with CF without respiratory infection are comparable to that of healthy children (100% to 110% of Recommended Dietary Allowances [RDA]) (Consensus conference 2003; Marin et al. 2004). However, if an individual has significant lung disease or malabsorption, energy requirements may be significantly increased (120% to 150% of the RDA) (Creveling et al. 1997). The 2002 Nutrition Consensus Report states that there is no perfect method to estimate the kcal needs of a person with CF; instead, a steady rate of weight gain in growing children should be the goal (Borowitz, Baker, and Stalling 2002). For adults, the desired outcome is to maintain an acceptable weight in relation to height with optimal fat and muscle stores.

To obtain adequate kcal and compensate for any fat malabsorption, individuals with CF often require a greater fat intake (35%–40% of total kcal) than what is normally recommended for the general population (25% to 35% of total kcal) (Borowitz, Baker, and Stalling 2002). Fat restriction is not recommended, because fat is an important energy source, and pancreatic enzyme replacement therapy is used to aid its absorption. Medium-chain triglycerides (MCT) require less lipase activity than long-chain fatty acids and may be utilized as a better source of fat kcal. MCT have a fatty acid chain length between 6 and 12 carbons, making them short enough to be water soluble. They require less bile salt for solubilization and can be transported as free fatty acids through the portal system. Adequate protein intake (approximately 15% to 20% of kcal) is essential to meet the needs of the growing child and maintain protein stores. Good nutrition also plays an important role in preparing the individual with cystic fibrosis for potential transplant later in life.

Patients with CF do have biochemical essential fatty acid deficiency, but clinical signs and symptoms of the deficiency are rare (Borowitz, Baker, and Stalling 2002). There are no current recommendations for supplementation of essential fatty acids, but this is an area of current research. Vegetable oils such as canola, soy, and flaxseed are good sources of both energy and linolenic acids and should be encouraged in the diets of patients with CF.

Glucose intolerance and cystic fibrosis-related diabetes (CFRD) occur in approximately 10% to 15% of CF patients over the age of 20 (Moran et al. 1999; Wilson et al. 2000; Borowitz, Baker, and Stalling 2002). CFRD is rarely found in young children and occurs more frequently between 18 and 21 years (Moran et al. 1999). The exact cause of diabetes mellitus in CF is not clearly understood but is thought to be related to the accumulation of fibrous tissues in the pancreas that interfere with normal insulin production. Individuals with CFRD have an increased morbidity and mortality. They are often underweight and have more advanced pulmonary involvement than those without CFRD. Malnutrition and low BMI are warning signs of the development of CFRD. Wilson et al. (2000) found that, at diagnosis of CFRD, 57% of teenagers and adults attending the Toronto Cystic Fibrosis Clinics had body mass index (BMI) below the 10th percentile for age.

Like type 2 diabetes (see Chapter 19), CFRD may be present for years before diagnosis. Symptoms of CFRD include polyuria (excessive urination) and polydipsia (excessive thirst), failure to gain or maintain weight despite aggressive nutrition intervention, poor growth velocity, unexpected decline in pulmonary function, and a failure to progress normally through puberty (Moran et al. 1999; Borowitz, Baker, and Stalling 2002). Diagnosis of CFRD is made using an oral glucose tolerance test (OGTT), a test of the body's ability to metabolize carbohydrate (see Chapter 19). Since a majority of patients do not have fasting hyperglycemia, using hemoglobin A1c (which measures the amount of glucose that has adhered to the hemoglobin cell) as a screening tool for CFRD is not appropriate; results of this test are normal in a majority of cases.

Dietary management is critical for the health and survival of patients with CFRD (see Table 23.7). Since a majority of these patients have difficulty maintaining weight, kcal restriction is never appropriate. For patients on insulin, carbohydrate counting offers a great degree of flexibility. Patients should be able to eat as they choose with appropriate insulin coverage (Moran 2000). Although carbohydrate is not restricted, patients should be taught to distribute carbohydrate kcal throughout the day and to avoid concentrated carbohydrate loads.

TABLE 23.7

Dietary Management of Cystic Fibrosis Related Diabetes

- Combine management of diet of both CF and diabetes.
- Aim for >100% RDA for energy.
- Provide 3 meals and 3 snacks per day.
- Fat should account for 35–45% of total calories.
- No restrictions should be placed on total carbohydrate intake; carbohydrate calories should be distributed throughout the day.
- If insulin for diabetes management, carbohydrate counting should be taught.
- Allow flexibility in meal planning.

Sources: Moran A. Cystic fibrosis-related diabetes: an approach to diagnosis and management. Pediatr Diabetes. 2000;1:41–8. Reprinted from *Clinical Nutrition*, Vol. 19(2), Wilson, DC, Kalnins D, Stewart C, Hamilton N, Hanna K, Durie PR, et al, Challenges in the dietary treatment of cystic fibrosis related diabetes, pages 87–93, Copyright 2000, with permission from Elsevier.

TABLE 23.8

Recommendations for Daily Vitamin Supplementation for Children and Adolescents with Cystic Fibrosis

	Vitamin A (IU)	Vitamin E (IU)	Vitamin D (IU)	Vitamin K (IU)
0–12 months	1,500	40–50	400	0.3–0.5*
1–3 years	5,000	80–150	400–800	0.3–0.5*
4–8 years	5,000–10,000	100–200	400–800	0.3–0.5*
>8 years	10,000	200–400	400–800	0.3–0.5*

*Currently commercially available products do not have ideal doses for supplementation.

Note: These fat-soluble vitamins are given in addition to an age appropriate dose of non-fat-soluble vitamins.

Adapted from: Borowitz D, Baker RD, Stalling V. Consensus report on nutrition for pediatric patients with cystic fibrosis. *J Pediatr Gastroenterol.* 2002;35:246–259.

Vitamins and Minerals Individuals with CF who are adequately treated with pancreatic enzymes may continue to have malabsorption of fat-soluble vitamins (A, D, E, and K) (Feranchak et al. 1999; Borowitz, Baker, and Stalling 2002). Patients with liver disease and accompanying disturbances of enterohepatic bile circulation are at significant risk for malabsorption of fat-soluble vitamins (Borowitz, Baker, and Stalling 2002). The combination of oral pancreatic enzyme therapy and supplementation of the fat-soluble vitamins will improve vitamin status. Fat-soluble vitamin supplementation needs to be individually adjusted based on age, dietary intake, and disease progression (see Table 23.8). These vitamins can be given in liquid, chewable, or pill form, depending on the age of the individual. Laboratory monitoring of each of these vitamins should be done at the time of diagnosis and at least yearly to ensure that the patient is receiving adequate amounts (see Table 23.6).

Vitamin A is important for vision, the integrity and proliferation of epithelial cells, and for normal immunity. Studies examining the vitamin status in individuals with CF indicate that deficiency of vitamin A may be common (Lancellotti et al. 1996; Huet et al. 1997; Borowitz, Baker, and Stalling 2002; Mrugacz, Tobolvzyk, and Minarowska 2005). Since plasma vitamin A levels can be decreased in infection, levels measured during acute illness can be misleading. Beta-carotene, a precursor of vitamin A, also functions as an antioxidant. Serum β-carotene levels in patients with CF have been shown to be low; however, data demonstrating deficiencies of β-carotene are lacking (Rust et al. 1998; Borowitz, Baker, and Stalling 2002).

The major function of vitamin D is to increase calcium absorption. Numerous studies have documented low serum 25-hydroxy vitamin D concentrations despite daily supplementation of vitamin D (Hahn et al. 1979; Aris et al. 1999; Elkin et al. 2001; Chavasse et al. 2004). This is of particular importance because of the increased prevalence of osteoporosis and bone fractures seen in patients with CF (Hahn et al. 1979; Elkin et al. 2001; Flohr et al. 2002). Serum vitamin D levels need to be carefully monitored to ensure that the patient is receiving an appropriate dosage of the vitamin (see Table 23.8).

Vitamin K is necessary for the biosynthesis of normal clotting factors and plays an important regulatory role in bone formation and mineralization. Data show that vitamin K deficiency may be common in CF (Von Horn et al. 2003; Conway et al. 2005). Since measurement of serum vitamin K levels is not practical, **plasma prothrombin concentrations** are used to measure vitamin K status instead. Because colonic bacteria are also a source of vitamin K, and periods of antibiotic therapy can disrupt vitamin K synthesis in the colon, vitamin K status during prolonged antibiotic treatment needs to be evaluated.

Vitamin E is a powerful antioxidant. Deficiencies lead to hemolytic anemia, neuromuscular degeneration, and retinal and cognitive changes. Low vitamin E levels as well as symptomatic deficiency states have been reported in patients with CF, even those taking pancreatic enzymes and multivitamins (Lancellotti et al. 1996; Borowitz, Baker, and Stalling 2002). Current recommendations for supplementation of vitamin E may not be adequate, and require further study (see Table 23.8).

As described earlier, individuals with CF have an increased incidence of osteopenia and osteoporosis, and an increased risk of fractures. Contributing factors include deficiencies of vitamin D, vitamin K, and calcium; use of corticosteroids; disease severity; inactivity and low body mass;

plasma prothrombin concentrations—a measure of blood clotting ability

hypogonadism; and malnutrition (Borowitz, Baker, and Stalling 2002). Children age 8 and older who have risk factors for bone disease should have their bone mass evaluated using dual-energy x-ray absorptiometry (DEXA). Additionally, children at risk for poor bone health should have serum calcium, phosphorus, parathyroid hormone, and 25-hydroxyvitamin D levels measured. To prevent bone disease, emphasis should be placed on adequate nutrition, appropriate weight-bearing exercise, and minimizing malabsorption. Calcium intake equal to the DRI for age should be encouraged.

Infants and children with CF are at risk for developing hyponatremia (low serum sodium levels) because of salt loss through the skin. When sodium intake is inadequate, lethargy, vomiting, and dehydration may occur. Individuals with CF should consume a high-salt diet, particularly during the summer months or if they live in hot climates. Breast-fed infants or individuals participating in sports activities are particularly susceptible to sodium depletion. For infants and small children, sodium chloride solutions available through pharmacies provide the most accurate method for supplying additional sodium. Adequate sodium intake for adolescents and adults is usually not a problem because of the amount of sodium available in the food supply, particularly in processed foods.

Children and adolescents with CF have also been shown to have iron and zinc deficiency (Borowitz, Baker, and Stalling 2002). Iron status should be monitored yearly by checking hemoglobin and hematocrit levels. Zinc deficiency may be present even though plasma zinc levels are normal (Easley et al. 1998; Krebs et al. 2000). Zinc supplementation (~1 mg elemental zinc/kg/day) has been recommended for children who are failing to thrive or who have short stature.

Feeding Strategies Breast-feeding is recommended for infants under the first year of age. Proprietary infant formulas may also be used. Caloric density greater than 20 kcal/ounce may be needed to support growth. This can be achieved by fortifying breast milk or by concentrating formula. Solid foods should be added at the appropriate age (see Table 23.9). If infants taking solid food are not achieving appropriate growth, additional kcal can be added to infant formula in the form of carbohydrate polymers (e.g., Polycose® from Ross Products Division, Abbott Laboratories) and/or fats such as vegetable oil or MCT oil. Families need to be counseled that when table food is introduced, the diet must be moderately high in fat and protein. When breast milk is discontinued, whole milk should be added. The diet should be supplemented with both fat-soluble and water-soluble vitamins in age-appropriate doses (see Table 23.8). Supplemental iron and fluoride may need to be given if the dietary intake is inadequate. Supplementation of sodium chloride may be necessary for infants, particularly during the summer months. At all ages, appropriate doses of pancreatic enzymes must be given prior to meals and snacks to ensure nutrient absorption (Borowitz, Baker, and Stalling 2002).

TABLE 23.9

Developmental Approaches to Nutrition Counseling for Individuals with Cystic Fibrosis

Infants

Breast-feeding is recommended for most infants as the primary source of nutrition during the first year of life. Human milk fortifiers may be used to increase the nutrient density of breast milk if needed.

Iron fortified infant formula may be used. The caloric density of the standard infant formula (20 kcal/ounce) may need to be increased by concentrating the formula or using fat and/or carbohydrate additives, as appropriate.

Solid foods should be added at 4–6 months developmental age, according to recommendations of the American Academy of Pediatrics.

Infant cereal should be prepared with breast milk or infant formula, not water or juice; additional fat or carbohydrate kcal may be added to infant cereal if needed to achieve expected rate of growth.

Pancreatic enzymes should be given prior to each feeding.

Vitamin supplements with additional fluoride and iron should be given.

Toddlers or Preschool Age (1–4 Years)

Whole milk should be encouraged for the child with CF.

Adding kcal to table food may help with maintenance of growth at this stage; avoid giving low-fat or low-kcal foods.

Regular mealtimes and snack times should be encouraged.

Dietitians should inquire about feeding behaviors to promote positive interactions and prevent negative behaviors; grazing behavior should be discouraged.

Pancreatic enzymes and vitamins are continued.

School Age (5–10 Years)

A normal, healthy diet with a variety of food should be the basis of the diet.

This may be a high-risk period for decreased growth rate in children with CF; identify factors that may interfere with meeting nutritional needs, such as activities that may lead to limited time for snacks and enzyme adherence and the progression of the disease.

Pancreatic enzymes and vitamins are continued.

Adolescence (11–18 Years)

Associated with high nutrient requirements due to accelerated growth, pubertal development and high levels of physical activity.

Nutritional counseling will be more effective if directed toward the adolescent, not the parent.

Adolescents may be more receptive to efforts to improve muscular strength and body image rather than stressing weight gain and improved disease status.

Continue pancreatic enzymes and vitamins.

Adapted from: Borowitz D, Baker RD, Stalling V. Consensus report on nutrition for pediatric patients with cystic fibrosis. J Pediatr Gastroenterol. 2002;35:246–259.

Children need to be monitored for appropriate growth. Additional kcal need to be added to the diet to support growth in the form of in-between meal snacks and nutrient-dense foods, particularly foods high in fat. Low-fat and low-kcalorie foods should be avoided. Appropriate feeding behaviors should be encouraged at each age. School-aged children (5–10 years) are at higher risk for decreased growth rate. Participation in activities increases energy expenditure and may also lead to limited time for consumption of snacks and taking of enzymes. Adolescence (11–18 years) is a time associated with accelerated growth and pubertal development; nutritional counseling directed specifically toward the adolescent may be necessary. Female adolescents are at greatest risk for poor nutrition because of their higher energy and nutrient requirements and poor eating habits (Lai et al. 1998). At this age, pulmonary infections, CFRD, or liver disease may develop, making nutrition management more complex.

The use of nutritional supplements may be helpful in an effort to add additional kcal, protein, and nutrients to the diet. Homemade foods high in kcal and protein, such as fortified beverages or puddings, may be beneficial. The addition of supplemental enteral feedings may be needed if adequate kcal intake cannot be achieved or growth is compromised. Nocturnal tube feedings are encouraged to promote normal eating behaviors during the day. Standard formulas (1.5–2.0 kcal/cc) containing complete protein and long-chain fatty acids are usually well tolerated (Borowitz, Baker, and Stalling 2002). Pancreatic enzyme should be taken before these feedings are given.

Pneumonia

Definition

Pneumonia is defined as an inflammation of the lungs, usually caused by bacteria, viruses, or fungi. Once the offending agent enters the lungs, it usually settles in the alveoli where it can grow rapidly. The infection causes deterioration of lung function resulting in fluid accumulation and breathing difficulty.

Epidemiology

Prior to 1936, pneumonia was the leading cause of death in the U.S., but with the use of antibiotics, the incidence of pneumonia has been substantially reduced. In 2000, pneumonia and influenza collectively ranked as the 7th leading cause of death (American Lung Association, What is pneumonia?, 2005).

Etiology

Although the microorganisms responsible for pneumonia are present in the environment and are inhaled into the lungs all the time, the cilia and microphages present in the lungs help prevent them from entering the alveoli. However, in certain populations the normal defense mechanisms may be compromised. These include the elderly, infants and young children, and individuals with health problems including those with COPD, diabetes mellitus, asthma, alcoholism, congestive heart failure, and sickle cell anemia. Individuals living with HIV/AIDS and those undergoing chemotherapy or organ transplants who are immunocompromised are also at substantial risk for developing pneumonia.

There are two major categories of pneumonia based on the cause: that which is community acquired and that which is hospital acquired. Community-acquired pneumonia occurs when infected persons cough or sneeze and spread the bacteria, primarily *streptococcus pneumoniae* or *pneumococcal pneumonia*, to those around them (American Lung Association, Pneumonia Fact Sheet, 2005). Hospital-acquired pneumonia, also known as nosocomial pneumonia, often affects patients who are in the intensive care unit (ICU) or are on a mechanical ventilator. Hospital-acquired pneumonia occurs fairly frequently and has high rate of mortality. The incidence of hospital-acquired pneumonia is highest among the elderly, the very young, and those who are already debilitated by other diseases (Beers, Hospital-acquired, 2006).

Nutrition Implications

Nutrition screening can help to identify those individuals in the population who are more susceptible to the development of pneumonia. In a study of patients admitted to the hospital for community-acquired pneumonia, the most important risk factor associated with increased mortality was a low serum albumin (≤3.0 g/dl) (Hedlund, Hansson, and Ortqvist, Hypoalbuminemia, 1995). This depressed albumin was associated with an inflammatory response rather than malnutrition as such. Other indices of poor nutritional status associated with death during a six-month follow-up of the study population included a low triceps skinfold (TSF) measurement and low body mass index (BMI) (Hedlund, Hansson, and Ortqvist, Short- and long-term, 1995).

Aspiration Pneumonia

Another common cause for the development of pneumonia is aspiration of inhaled materials (saliva, nasal secretions, bacteria, liquids, food, or gastric contents) into the airway below the level of the vocal cords. **Aspiration pneumonia** occurs when the aspirated material causes an inflammatory response in the lung.

There are a number of normal defense mechanisms which help to prevent aspiration from occurring (McClave et al. 2002). During the swallowing process the epiglottis, a thin cartilage structure located at the base of the tongue,

pneumonia—inflammation of the lungs, usually caused by bacteria, viruses, or fungi

aspiration pneumonia—aspiration of inhaled materials (saliva, nasal secretions, bacteria, liquids, food, or gastric contents) into the airway below the level of the vocal cords that results in an inflammatory response in the lung

TABLE 23.10

Risk Factors Associated with Aspiration

- Decreased level of consciousness
- Neuromuscular disease and structural abnormalities of the upper gastrointestinal tract
- Endotracheal intubation or mechanical ventilation
- Vomiting
- Delayed gastric emptying (seen in diabetes, hyperglycemia, electrolyte abnormalities, and with drugs known to reduce gastric emptying)
- Advanced age (60+ years)
- Poor oral care
- High gastric residual volumes
- Prolonged supine position (head of the bed flat)
- Presence of a nasogastric tube
- Non-continuous or intermittent tube feeding
- Large diameter of feeding tube and/or malpositioned feeding tube

Source: McClave N. Risk factors for aspiration. *JPEN.* 2002;226:S26–S33.

folds over the top of the larynx to prevent food and liquid from entering the trachea (see Chapter 16). The lower esophageal sphincter (LES) also prevents the upward movement of gastric contents into the esophagus. In addition, food particles and fluids that may be aspirated into the lungs are entrapped in the mucus layer of the respiratory epithelium. The mechanical beating of cilia on the respiratory epithelium advances the mucus and entrapped particles upward so it can be cleared by the cough. Cough is an important mechanism that allows clearance of foreign material and secretions from the airway.

Aspiration is the leading cause of pneumonia in the ICU and contributes to the morbidity and mortality of patients. A number of risk factors may contribute to aspiration (see Table 23.10) (Metheney 2002). Patients with head injuries often have delayed gastric emptying, decreased gag and cough reflex, and are at high risk for aspiration. Individuals who have neurological impairments such as stroke and Parkinson's disease are a high risk for dysphagia (difficulty in swallowing) and subsequent aspiration (see Chapter 16). Hyperglycemia can result in disordered motility throughout the gastrointestinal tract. Patients with diabetes who have gastroparesis (delayed gastric emptying, vomiting, nausea, or bloating caused by stomach nerve or muscle damage) may have aspiration, particularly if they receive gastric tube feedings. In addition, patients with abnormalities of the gastrointestinal tract including **esophageal stricture** and gastroesophageal reflux disease (GERD), and those who require mechanical ventilation, are also at high risk for developing aspiration.

esophageal stricture—a significant narrowing of the esophagus that may significantly interfere with swallowing

Aspiration is also a serious side effect of enteral tube feeding, particularly gastric feeding (see Chapter 7) (Gomes et al. 2003). Patients fed using both a nasogastric feeding tube and a gastrostomy tube are at high risk for aspiration. The source of aspiration is likely due to the reflux of gastric contents from the stomach into the pharynx, where it is aspirated into the lungs. In addition, the presence of a nasogastric feeding tube may interfere with the effectiveness of the lower esophageal sphincter, resulting in gastrointestinal reflux.

A number of clinical monitors have been used to identify the presence of aspiration in patients who are tube fed (McClave et al. 2002). For a number of years, blue dye (Blue No. 1) was added directly to the tube feeding product to visually detect the presence of formula in the pulmonary aspirate. However, studies examining the effectiveness of this practice have demonstrated that it is not a reliable indicator of aspiration (Pots et al. 1993; McClave et al. 2002; Gomes et al. 2003). Additionally, in 2003 the FDA issued a public health bulletin concerning the toxic absorption of blue dye resulting in both discolored skin and patient deaths after its use (U.S. Food and Drug Administration, FDA Public Health, 2005). For these reasons, the practice of using blue dye to detect the presence of pulmonary aspiration in tube-fed patients is no longer recommended.

Glucose oxidase reagent strips, which detect the presence of glucose-containing formula in the pulmonary aspirate, are sometimes used to monitor for the presence of aspiration (McClave et al. 2002). Glucose reagent strips are commonly used to monitor blood glucose levels. The test strips contain the enzyme glucose oxidase and color indicators. When blood is placed on the strip, the glucose present is oxidized, resulting in a color change. However, the use of these strips can lead to false-positive results when used to test for aspiration, particularly if blood is present in tracheal secretions.

Monitoring gastric residual volume is another method used to reduce the risk of aspiration in tube-fed patients. Gastric residuals represent the volume of stomach contents present at any one time. Fluids that commonly accumulate in the stomach of a patient receiving tube feedings include the enteral formula, swallowed saliva, gastric secretions, and regurgitated secretions from the small bowel. High gastric residual volumes may increase the risk for regurgitation of stomach contents into the esophagus, resulting in aspiration.

Gastric residuals are monitored by removing and measuring them periodically using a syringe connected to the feeding tube. Because these gastric contents include both tube feeding product and digestive juices, they are returned to the stomach after they are measured. Unfortunately, the practice of measuring gastric residual volumes is poorly standardized. Gastric residual volumes do not correlate well with the gastric emptying, the volume of stomach contents, changes in tube feeding infusion rates, and the incidence of regurgitation or aspiration in tube fed patients (McClave et al. 2002, 2005). In addition, there is lack of agreement on the volume of gastric residuals that should be considered an aspiration risk during tube feeding. Practice guidelines issued

TABLE 23.11

Evidence-Based Guidelines for Prevention of Aspiration Associated with Enteral Feeding

- In the absence of medical contraindication(s), elevate the head of the bed at an angle of 30–45 degrees. (II)*
- Routinely verify appropriate placement of the tube feeding tube. (IB)*
- No recommendation can be made for the preferential use of small-bore tubes for enteral feeding. (Unresolved issue)*
- No recommendation can be made for preferentially administering enteral feedings continuously or intermittently. (Unresolved issue)*
- No recommendation can be made for preferentially placing the feeding tubes distal to the pylorus (e.g., jejunal tubes). (Unresolved issue)*

* Refers to strength of evidence to support the recommendations based on the following criteria: IA = strong recommendations for implementation supported by well designed research; IB = strong recommendations for implementation supported by certain clinical and epidemiological studies and strong theoretical rationale; IC = required for implementation, as mandated by federal or state regulation or standard; II = suggested for implementation and supported by suggestive clinical or epidemiological studies or strong theoretical rationale; Unresolved issue = insufficient evidence to recommend practice.

Source: Centers for Disease Control and Prevention, Guidelines for preventing health-care-associated pneumonia, 2003: recommendations of CDC ad Healthcare Infection Control practices Advisory Committee. *MMWR.* 2004;53(No. RR-3): [inclusive page numbers].

by the Society for Enteral and Parenteral Nutrition recommend checking gastric residuals frequently when a tube feeding is being initiated and withholding the tube feeding if residual volumes exceed 200 mL on two successive assessments (see Chapter 7) (Access for administration of nutrition support 2002).

Since the methods of detecting aspiration in patients receiving tube feeding are unconfirmed, preventing its occurrence is very important. The dietitian plays in important role in monitoring patients who are at risk for the development of aspiration and recommending appropriate plans to prevent its occurrence. In 2003, evidence-based guidelines for preventing health-care-associated pneumonia were published by the Department of Health and Human Services (HHS) and the Centers for Disease Control (Centers for Disease Control and Prevention 2004). These guidelines include recommendations for prevention of aspiration, including aspiration associated with tube feeding (see Table 23.11).

Two evidence-based recommendations for prevention of aspiration associated with tube feeding include (1) elevating the head of the patient's bed 30° to 45° to prevent gastroesophageal reflux, and (2) routinely verifying the placement of the feeding tube to ensure that it has been properly placed and does not move upward beyond the pyloric sphincter. Although there is insufficient evidence to definitively recommend their implementation, the following practices have also been proposed to prevent aspiration: Use small bowel feedings, rather than gastric feedings to prevent reflux of gastric contents, deliver the tube feeding product using a continuous method versus an intermittent or bolus method to reduce the amount delivered at any one time, and the use of **prokinetic agents**, which have been shown to reduce gastrointestinal reflux and increase peristalsis (McClave et al. 2002). Additionally, if the patient has diabetes, maintaining tight blood glucose control may be helpful.

Patients with Tracheostomies

A **tracheostomy** is a surgical opening made in the trachea to assist breathing. A tracheostomy tube is inserted through the surgical opening (stoma), as shown in Figure 23.4. A tracheostomy is usually done for one of the following reasons: (1) to bypass an obstruction in the trachea, (2) to clean and remove secretions from the trachea and prevent them from going into the lungs, or (3) to more easily and safely deliver oxygen to the lungs when the patient is not able to breathe without assistance. Sometimes children or adults require permanent tracheostomies to breathe. Patients who are unable to breathe on their own usually also require a mechanical ventilator in addition to the tracheostomy.

There are many potential complications related to the presence of a tracheostomy tube, including the inability to speak or swallow normally (Murray and Brzozowski 1998). Patients with tracheostomy tubes who are on mechanical ventilation are often at high risk for aspiration. Frequently these patients require tube feeding for nutritional support. Once they are weaned (removed) from the ventilator, the tracheostomy tube may remain in place to remove secretions from the trachea.

When it is safe for these patients to eat orally, the viscosity of the food often makes a difference (Murray and Brzozowski 1998). The dietitian often works closely with the speech pathologist to determine whether the patient can safely swallow food without aspiration and the consistency of food that is appropriate to be fed. The speech pathologist is specially trained to evaluate swallowing and make recommendations about food consistencies. Because respiration is momentarily halted during swallowing, for a patient to eat orally, the act of swallowing needs to be combined successfully with respiration. Normally, exhalation of air both precedes and follows the swallow, ensuring that remnants of the food bolus are not aspirated. Successful feeding also requires the presence of an effective cough and intact upper airway reflexes (Hughes 2003). Liquids that are thin in consistency are more frequently aspirated than those with more viscosity, because they move quickly into the pharynx, which requires rapid closure of the

prokinetic agents—medications that cause the lower esophageal sphincter to close tightly, preventing gastric reflux. They also act to increase transit time (peristalsis) of stomach contents

tracheostomy—a surgical opening placed in the trachea to assist breathing

FIGURE 23.4 Tracheostomy Tube

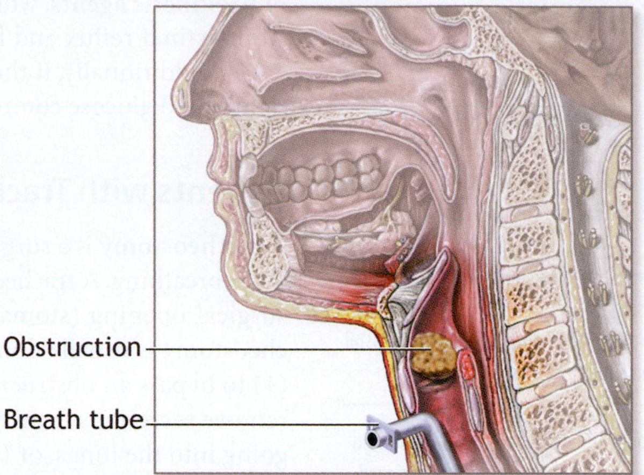

Source: Copyright © 2006 A.D.A.M., Inc.

larynx (Murray and Brzozowski 1998) (see Chapter 16). Also, solid foods that generate thin liquids during chewing (from oral secretions) may also present a problem, because the liquid can leak into the airway before the solid food bolus is swallowed. Thicker liquids with a honey-like texture can often be swallowed more efficiently, bypassing the incomplete laryngeal closure. Foods that require little or no chewing, such as pureed food, may be better tolerated by patients who aspirate thin liquids. Aspiration risk may also be increased if a patient becomes fatigued during mealtime. Smaller, more frequent meals may be beneficial by helping to reduce fatigue.

Tracheostomy tubes come in many varieties, including both cuffed and uncuffed tubes. A cuff is a soft balloon around the distal (far) end of the tube that can be inflated to prevent oral secretions from entering the lungs. The cuffs are inflated with air, foam, or sterile water. When the cuff is deflated, the tube allows air around it for vocalization. It is not completely clear whether the cuff should be inflated or deflated when feeding patients orally. A small study of 12 patients with tracheostomy tubes found that there was nearly a threefold increase in the aspiration rate when the cuff was inflated, versus deflated (Davis et al. 2002).

Respiratory Failure

Definition

Respiratory failure (RF) occurs when the respiratory system is no longer able to perform its normal functions. It can result from long-standing chronic lung disease like COPD or

acute respiratory distress syndrome (ARDS)—respiratory failure (RF) resulting from an acute insult to the lungs that occurs when the respiratory system is no longer able to perform its normal functions

TABLE 23.12

Conditions Associated with the Development of Acute Respiratory Distress Syndrome

Direct Injury	Indirect Injury
Pneumonia	Sepsis
Aspiration of gastric contents	Severe trauma
Inhalation injury	Acute pancreatitis
Near drowning	Cardiopulmonary bypass
Pulmonary contusion	Massive transfusions
Fat embolism	Drug overdose
Pulmonary edema	
Post-lung translation	

Adapted from Borowitch D, Baker RD, Stalling V. Consensus Report on Nutrition for Pediatric Patients With Cystic Fibrosis. Journal of Pediatric Gastroenterology © 2002; 35:246-259.

cystic fibrosis, or as a result of an acute insult to the lung such as **acute respiratory distress syndrome (ARDS)**. The conditions that lead to ARDS can be categorized as either (1) those that directly cause injury to the lung, such as pneumonia, aspiration, or an inhalation injury, or (2) those that result in indirect injury precipitated by events outside the lung, such as sepsis, trauma, or pancreatitis (see Table 23.12) (Levy et al. 1995). Aspiration of gastric contents, sepsis, and trauma account for approximately 70% to 80% of all ARDS cases (Levy et al. 1995). ARDS usually occurs within 24–72 hours of the predisposing factors (Ubodi and Childs 2003). Although the actual mechanism for the lung injury seen in ARDS is not totally understood, it occurs in response to a variety of cytokines (small, hormone-like proteins) released from inflammatory cells and is characterized by dyspnea, severe hypoxemia, decreased lung compliance, loss of surfactant, and leakage of a protein-rich fluid into the interstitium and alveolar lumen (Zwischenberger, Alpard, and Bidani 1999; Ware and Matthay 2000; Michaels 2004).

FIGURE 23.5 Mechanical Ventilation

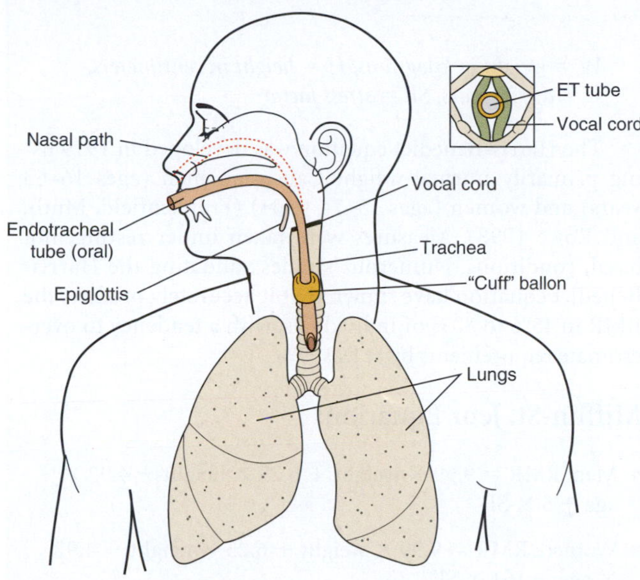

TABLE 23.13

Nutrition Practice Guidelines for Patients with Respiratory Failure
• Patients should undergo nutrition screening to identify those who require nutrition assessment and development of a nutrition care plan. (B)
• Energy intake should be kept at or below estimated needs in patients with demonstrated carbon dioxide retention. (B)
• Modified enteral formulas consisting of n-3 fatty acids may be beneficial in patients with early ARDS. (B)
• A fluid-restricted nutrient formulation should be used in patients with ARDS whose hemodynamic status necessitates fluid restriction. (B)
• Serum phosphate levels should be monitored closely. (B)

Note: (B) refers to strength of evidence to support the recommendations based on the following criteria: A = good research based evidence supported by prospective, randomized trials; B = fair evidence based on well designed studies without randomization; C = based on expert opinion.
Adapted from: Specific guidelines for disease—adults. *JPEN.* 2002;26(1):SA63–SA64.

Although patients with RF, particularly those with ARDS, may be initially managed with supplemental oxygen, progressive hypoxemia often requires intubation and mechanical ventilation (see Figure 23.5). Patients who require ventilator support are not able to consume nutrition by the normal oral route. If ventilator support is required for more than 48 hours, an alternative form of nutritional support is required. The route of nutritional support will be determined by the patient's underlying illness and gastrointestinal function; nonetheless, enteral nutrition is the preferred method of support due to its role in maintaining gastrointestinal function, the reduced risk of sepsis, and its lower cost. In cases where enteral nutrition support is not possible, parenteral nutrition is used.

Nutrition Therapy

Nutrition Implications The goals for nutrition care for patients with RF are to meet their nutrition needs; preserve and restore lean body mass, particularly respiratory muscle mass; maintain fluid balance; and facilitate weaning from mechanical ventilation (see Table 23.13). Nutritional needs vary widely depending on the age of the patient, the underlying disease process, and the patient's prior nutritional status. Often patients with RF are poorly nourished, particularly if they have a long-standing history of COPD or CF. A complete nutrition assessment to evaluate the patient's individual nutrition needs, including anthropometric and laboratory status, should be completed. Because these patients often have fluid imbalances, careful interpretation of laboratory data is essential.

Nutrition Interventions Energy requirements for patients with RF vary considerably based on the underlying disease state (often hypermetabolic). One problem is that these patients often do not have the pulmonary reserve needed to clear excess carbon dioxide. Because overfeeding is associated with increased CO_2 production, it is important that a careful assessment of the patient's nutrient requirements be made, particularly if the patient requires mechanical ventilation. The increased ventilatory demand associated with overfeeding is related to both excess glucose administration (>5 mg/kg per minute) and excess energy intake (Specific guidelines for disease—adults, Pulmonary, 2002). Increases in carbon dioxide production from overfeeding may result in difficulty in weaning or removing patients from mechanical ventilation, which further complicates their medical care (see Chapter 7 for a discussion of overfeeding).

Total caloric requirements can either be estimated using a predictive equation or directly measured using indirect calorimetry. Although indirect calorimetry is the preferred method for assessing kcalorie needs, a number of predictive equations are also available (see Box 23.4). The provision of 25 kilocalories per kilogram of usual body weight (or 130% of basal energy expenditure) appears to be adequate for most patients (Cerra et al. 1997; Specific guidelines for disease—adults, Pulmonary, 2002). Once the kcalorie requirements have been estimated, the patient's pulmonary status, body weight, and fluid balance must be closely monitored to ensure that overfeeding does not occur. Patients with ARDS do have increased protein requirements resulting from hypermetabolism. Providing 1.2 to 1.5 g/kg/day of protein appears appropriate and should be adjusted to promote nitrogen retention, without being excessive (Cerra et al. 1997). An evaluation of urine urea nitrogen is one method for assessing adequacy of protein intake, though it does have its limitations (see Chapter 5).

Several enteral products, high in fat and low in carbohydrate, have been developed specifically for patients with RF and ARDS and have demonstrated decreases in both $PaCO_2$ and time on mechanical ventilation (Specific guidelines for

BOX 23.4 CLINICAL APPLICATIONS

Methods for Predicting the Caloric Needs of Patients with Respiratory Disease

Although the use of indirect calorimetry to determine the kcal needs of patients with respiratory disease is preferred, the expense of calorimeters and the need for trained personnel to run these tests have limited their widespread use, particularly outside the critical care area. For this reason, the use of predictive equations to determine kcal needs has become common. Most of these methods are derived from regressive equations developed from direct or indirect calorimetry measures (Reeves 2003; da Rocha et al. 2005). However, the use of these methods does have some limitations. Predictive regressive equations tend to work best with groups of people (Frankenfield, Roth-Yousey, and Compher 2005; ADA Evidence Analysis Library 2006). When they are applied to individuals, significant errors can occur because the individual may not share important characteristics with the group of people from whom the equation was developed, including age, sex, body composition, ethnicity, and specific disease traits (ADA Evidence Analysis Library 2006). For this reason, the dietitian must recognize that using these equations may result in an overestimate or underestimate of the patient's actual kcal needs.

There are two approaches used to calculate the kcal needs of critically ill patients using predictive equations. The first involves calculating the resting metabolic rate (RMR) and then multiplying it by a stress factor reflecting the patient's diagnosis (e.g., sepsis, trauma, surgery).

Two equations to determine resting metabolic rate are commonly used:

Harris-Benedict Equation:

- Men: RMR $= 66 + 13.75(W) + 5(H) - 6.8(A) \times$ SF
- Women: RMR $= 655 + 9.6(W) + 1.8(H) - 4.7(A) \times$ SF

W = weight in kilograms, H = height in centimeters, A = age in years, SF = stress factor

The Harris-Benedict equation was developed in 1919 using primarily normal-weight Caucasian men (ages 16–63 years) and women (ages 15–74 years) (Frankenfield, Muth, and Rowe 1998). Measures were taken under resting, not basal, conditions. Numerous studies validating the Harris-Benedict equation have shown that it accurately predicts the RMR in 45% to 80% of individuals, with a tendency to overestimate requirements by at least 5%.

Mifflin-St. Jeor Equation:

- Men: RMR $= 9.99 \times$ weight $+ 6.25 \times$ height $- 4.92 \times$ age $+ 5 \times$ SF
- Women: RMR $= 9.99 \times$ weight $+ 6.25 \times$ height $- 4.92 \times$ age $- 161 \times$ SF

W = weight in kilograms, H = height in centimeters, A = age in years, SF = stress factor

The Mifflin-St. Jeor equation was derived from a sample of 498 normal-weight, obese, and severely obese individuals (ages 19–78 years) (Mifflin et al. 1990). The racial composition of the sample is not specified and the representation of older adults (ages 75–84 years) is limited. An expert panel convened by the American Dietetic Association (ADA) to determine the most accurate predictive equation for RMR in healthy non-obese and obese adults determined that the Mifflin-St. Jeor equation was more likely than the other equations tested to estimate the RMR within 10% of that measured using calorimetry.

(continued on the following page)

disease—adults, Pulmonary, 2002). Because ARDS is also associated with the development of pulmonary edema, the use of fluid restricted enteral formulations (1.5–2 kcal/cc) may be helpful in patients whose hemodynamic status requires fluid restriction (Specific guidelines for disease—adults, Pulmonary, 2002).

ARDS is also associated with the production of oxygen free radicals and inflammatory mediators derived from arachidonic acid. These mediators and free radicals have been shown to cause lung inflammation, edema, alveolar damage, and lung collapse. Recent research has shown that the dietary fatty acids, particularly eicosapentaenoic acid (EPA), found in fish oil, and γ-linolenic acid (GLA), found in borage oil, can reduce the severity of the inflammatory injury by altering the availability of arachidonic acid in tissue phospholipids (Gadek et al. 1999). Patients with ARDS receiving a high-fat enteral product supplemented with n-3 fatty acids (fish oil and borage oil) and antioxidants (Oxepa™, Ross Products Division, Abbott Laboratories) spent less time on mechanical ventilation, had briefer stays in the ICU, and had a decreased incidence of organ failure (Gadek et al. 1999).

The antioxidant status of patients with ARDS is often severely depressed, and markers of lipid peroxidation are increased (Metniz et al. 1999; Nelson et al. 2003). Supplementation with α-tocopherol, β-carotene and vitamin C at levels higher than the DRI has been associated with substantial increases in serum α-tocopherol and β-carotene, and appears to prevent further oxidative damage (Nelson et al. 2003).

Phosphate is essential for optimal pulmonary function and normal contractibility of the diaphragm. The length of hospital stay and dependence on mechanical ventilation have been shown to be increased in critically ill patients who have hypophosphatemia (Marik and Bedigian 1996). The phosphate balance of patients in respiratory failure needs to be closely monitored, particularly for those patients

BOX 23.4 *(continued)*

When the dietitian chooses to use one these two equations to predict the kcal needs of patients in the clinical setting, he or she should consider several factors. First, both of these formulas were originally developed with normal, healthy individuals, not individuals with a specific disease process. Second, these formulas have not been validated for use in older individuals (60–82 years), non-white racial populations, underweight adults, or children (da Rocha et al. 2005). Lastly, energy requirements and metabolism are often altered when a disease process is present. To account for this, injury or stress factors have been used to account for the differences between the RMR and the energy cost of the disease. However, data supporting the validation of these stress factors is limited. When a stress factor is applied to these formulas, this requires clinical judgment that in many cases is an "educated guess" and may further increase its inaccuracy.

The second approach used to predict the kcal needs of critically ill patients is to calculate the metabolic rate from a regressive equation that incorporates healthy resting metabolism as well as clinical markers of illness such as degree of trauma, or body temperature and minute ventilation. Although there are a number of these equations that have been developed, few of them have been adequately validated. Two equations that have been studied in the clinical setting include:

Ireton-Jones Equation:

- Spontaneously breathing men or women: EEE = 629 − 11(A) +25(W) − 609(O)
- Ventilated men or women: EEE = 1784 − 11(A) + 5(W) + 244(S) + 239(T) + 804(B)

EEE = estimated energy expenditure, including stress and activity; W = weight in kg.; A = age; O = 1 for obese patients over 130% of ideal body weight and 0 if not obese; S = sex (male = 1, female = 0); T = 1 if trauma present and 0 if no trauma present; B = 1 if burn present and 0 if no burn present.

These two equations were developed by Ireton-Jones et al. in 1992 using indirect calorimetry measurements on 200 spontaneously breathing or ventilator-dependent patients. Corrections to the equations were made in 1997 by Ireton-Jones (Ireton-Jones et al. 1992; Ireton-Jones 1997).

Penn State Equation:

- RMR = Harris Benedict (1.1) + V_E (32) + T_{Max} (140) − 5340

V = minute ventilation in liters/minute read from ventilator, and T = maximum body temperature (degrees Centigrade in previous 24 hours).

The Penn State Equation was developed in 1998 from a retrospective analysis of indirect calorimetry measurements on 169 mechanically ventilated critical care patients (Frankenfield, Smith, and Cooney 2004). It was revised in 2003 using indirect calorimetry measurements on 47 mechanically ventilated, critically ill patients. The equation uses an adjusted Harris-Benedict equation to calculate the RMR plus minute ventilation in liters read directly from the ventilator and the patient's maximum 24-hour body temperature (degrees Centigrade).

Both of these equations have been developed for use in critically ill patients, specifically patients who are on mechanical ventilation. For this reason, they are appropriate for that population only. An expert panel supported by ADA is currently looking at evidence-based data to support the use of specific disease-based predictive equations in the clinical setting.

receiving parenteral nutrition support. Phosphate supplementation should be initiated whenever hypophosphatemia is present.

Transplantation

Definition and Epidemiology

Lung transplantation is the surgical procedure to replace one or both lungs with healthy organs from a human donor. Lung transplantation is an accepted option for patients with end-stage lung disease. In 2004, approximately 4,000 people in the U.S. were waiting for a lung transplant, but only 25% received a transplant (United Network for Organ Sharing 2006). In some cases where the heart has also been weakened, both the heart and lungs will be replaced. Until 1989, combined heart-lung transplants were the most common form of lung transplantation, but currently single and double lung transplants have become more common (United Network for Organ Sharing 2006).

Pathophysiology

Transplant recipients are at high risk for rejection of the transplanted lung. The body's immune system considers the transplanted organ as an invader (similar to infection) and may attack it. Because of the rejection risk, patients must take immunosuppressive (anti-rejection) medications, which suppress the body's immune response (see Chapter 12). Rejection occurs most often during the first three months after transplantation, but the immunosuppressive medication may need to be taken indefinitely (Neuringer, Chalermskulrat, and Aris 2005). Common immunosuppressive drugs used are: cyclosporine, tacrolimus, mycophenolate mofetil, azathioprine, and prednisone. The nutritional side effects of these medications may include GI distress (nausea, vomiting, diarrhea, and/or constipation), increases in blood pressure, edema, and alterations in blood sugar levels. They also lower the body's immunity and increase the risk of infection.

Patients with lung disease may be underweight, normal weight, or overweight. Overweight patients with lung disease, however, are often very sedentary and have significant increases in body fat mass rather than lean body mass (LBM). Lean body mass depletion has been associated with a higher rate of mortality in patients awaiting transplantation. Poor nutritional status and low LBM have a significant impact on the length of time post-operative transplant patients spend on ventilation, and can double the length of stay in the ICU (Schwebel et al. 2000). Patients with a pretransplant BMI <17 kg/m^2 or >25 kg/m^2 have an increased risk of dying within 90 days post-transplant (Madill et al. 2001). This risk is significantly higher in patients with a BMI >27 kg/m^2.

Nutrition Therapy

A comprehensive nutrition assessment of transplant recipients should include a physical assessment, dietary history, anthropometric measures, and laboratory values. Specific assessment issues are included in the NCP feature. Direct measurement of body composition using DEXA or dilution techniques may be more helpful than weight or BMI to assess changes in LBM. Although bioelectrical impedance analysis (BIA) has become widely available for assessing body composition, single tests using BIA may not be valid because of the fluid shifts that are often seen in these patients.

The physical and nutrition status of patients awaiting transplant may decline. Therefore, nutrition support is very important during the pretransplant period. The goal is to optimize nutrition status as much as possible by increasing kcal and protein. Most patients waiting for an organ transplant are able to eat (Hasse 2001). Small, frequent meals composed of nutrient-dense foods and supplements should be encouraged. Tube feedings may be indicated if the patient is unable to eat adequate amounts.

The acute posttransplant period may be complicated by rejection, infection, and surgical complications (Hasse 2001). The nutrition goals during this time are to provide adequate nutrients to promote wound healing, treat changes in electrolyte balance, and achieve optimal blood glucose control (Hasse 2001; Specific guidelines for disease—adults, Solid organ, 2002). Most transplant patients are allowed to eat 3 to 5 days after transplantation (Levy et al. 1995). Nutrition support should be considered when the patient is not able to eat or if oral feeding is delayed. When nutrition support is required, the use of enteral nutrition is preferred. Patients who are malnourished or at risk for extended NPO (nothing by mouth) status may benefit from immediate posttransplant tube feeding. When enteral nutrition support is not possible, parenteral nutrition is advocated. Parenteral nutrition should be administered cautiously because of postoperative hyperglycemia due to metabolic stress, infection, and the use of corticosteroids. Hyperglycemia has been shown to impair wound healing and increase the risk of infection (Specific guidelines for disease—adults, Solid organ, 2002). When the oral diet is initiated, simple carbohydrates should be limited.

Protein and energy requirements are affected by the stress of surgery, postoperative complications, episodes of rejection, and the use of immunosuppressant drugs, particularly corticosteroids. The use of indirect calorimetry to assess kcal needs is indicated when the patient's medical condition is complicated by posttransplant complications. When indirect calorimetry is not available, 130% to 150% of BEE (or 35 kcal/kg body weight) is usually adequate (Levy et al. 1995; Ubodi and Childs 2003). Adequate amounts of protein are required for wound healing and to prevent infection. Protein needs may be increased due to surgical stress and the use of corticosteroids. Nitrogen balance studies suggest protein requirements range from 1.5 to 2.0 g/kg per day (Hasse 2001; Specific guidelines for disease—adults, Solid organ, 2002). These requirements may decrease to 1 g/kg as the dose of corticosteroids is reduced to maintenance levels (Specific guidelines for disease—adults, Solid organ, 2002).

Upper Respiratory Infection

Definition and Epidemiology

Upper respiratory infection (URI), generally known as the common cold, is a nonspecific term used to describe acute infections which involve the nose, sinuses, pharynx, larynx, trachea and bronchi (Beers, Respiratory tract, 2006). URIs occur more frequently during the winter months. Adults develop an average of 2–4 colds per year; while children develop an average of 6 viral respiratory tract infections each year (Beers, Respiratory tract, 2006; Mossad 2006).

Pathophysiology

Transmission of the organisms which cause URIs occurs primarily by aerosol, droplet (sneezing or coughing), or direct hand-to-hand contact with infected secretions. Onset of symptoms occurs 1–3 days after exposure to the infectious agent (Beers, Respiratory tract, 2006). Runny nose, nasal

upper respiratory infection (URI)—a nonspecific term used to describe acute infections involving the nose, sinuses, pharynx, larynx, trachea, and bronchi; often referred to as the common cold

congestion, and sneezing are common symptoms of URIs, often accompanied by sore throat, cough, and headache. Treatment for URIs is usually directed towards minimizing symptoms. Rest and increased fluid intake are recommended. Hundreds of over-the-counter (OTC) medications are available; however, none have proven to be suitable in controlling the course of the infection, only effective in improving symptoms (Beers, Respiratory tract, 2006).

Clinical Manifestations

The signs and symptoms associated with an upper respiratory infection are usually mild and brief in duration. These include the cardinal symptoms of inflammation (rubor, calor, dolor, and functio laesa; see Chapter 10). Fluid and mucus released in response to the inflammation obstruct upper airways. If the inflammation includes the larynx, trachea, and upper bronchi, the patient will experience labored breathing. These symptoms may be more difficult for infants and children due to their immature respiratory system and accessory muscles of the neck.

Nutrition Implications

Both vitamin C and zinc have been studied in preventing or decreasing the cold symptoms. Twenty years ago, Linus Pauling first stimulated public interest in the use of large doses of vitamin C (>1 gram/day) to prevent infections associated with the common cold (Douglas et al. 2004). Subsequent research has concluded that large doses of vitamin C do not have a significant effect on the incidence of the cold. However, a few studies have demonstrated that taking supplemental vitamin C (~2 grams) may decrease the duration and severity of the cold, an effect likely related to the antihistamine effects that occur with large vitamin C doses (Johnston, Martin, and Cai 1992; Douglas et al. 2004). These doses do result in intakes above the Tolerable Upper Intake Level (UL) for vitamin C and may result in gastrointestinal disturbances.

The use of zinc lozenges has been promoted to reduce the duration of the common cold. However, numerous controlled trials have found conflicting results regarding the effectiveness of zinc lozenges in reducing both its symptoms and duration (Marshall 1999; Jackson, Lesho, and Peterson 2000). Taking zinc lozenges at recommended levels (12.8 mg of zinc/lozenge) for every 2–3 hours while awake may result in a dietary intake of zinc above the UL of 40 mg/day. Short-term use of zing lozenges (~5 days) has not resulted in serious side effects; however, long-term use (6–8 weeks) may result in copper deficiency.

Conclusion

This chapter has reviewed the anatomy and physiology of the respiratory tract along with the detailed changes that occur during the disease process. Many individuals with respiratory disease are at high nutritional risk and need specialized nutrition support. Nutritional care during respiratory disease functions both to support the normal function of the respiratory tract and to provide important interventions that serve as crucial components of medical care.

Use of alcohol, vitamin, mineral, herbal or other type of supplements

Previous nutrition education or nutrition therapy

Eating pattern: 24-hour recall, diet history, food frequency

Anthropometric

Height (measured, recumbent, knee height, or arm span)

Current Weight

Weight History: highest adult weight; usual body weight

Reference Weight (BMI)

Physical assessment: temporal wasting; presence of edema

Biochemical Assessment

Visceral Protein Assessment: Standard

Immunocompetence: Standard

Hematological Assessment: Standard

Other specific labs: Electrolytes, pH, glucose, arterial blood gases

Vitamin/Mineral Status Specific Labs	*Respiratory*
Vitamin D	Serum alkaline phosphatase provides indirect measure
Vitamin K	Prothrombin time
Vitamin A	Serum carotene, retinal binding protein
Vitamin E	Serum tocopherol, erythrocyte hemolysis
Zinc	Serum zinc

Step Two: Common Diagnostic Labels

- Inadequate oral food/beverage intake
- Hypermetabolism
- Increased energy expenditure
- Inadequate energy intake
- Food medication interaction
- Underweight
- Involuntary weight loss
- Swallowing difficulty
- Chewing difficulty
- Physical inactivity
- Poor nutrition quality of life

Sample: Involuntary weight loss NC-3.2

PES: Patient has experienced involuntary weight loss as evidenced by unplanned 18% weight loss over the previous eight months secondary to fatigue, shortness of breath, and increased caloric requirements to support increased work of breathing.

Step Three: Sample Intervention

1. Patient will consume a minimum of 2,200 kcal/day.
2. Increase nutrient density of foods consumed—provide education for meal preparation and food choices.
3. Provide education on appropriate physical activity regimen per pulmonary rehabilitation team.
4. Recommend utilizing supplemental oxygen during all meals and snacks.

Step Four: Monitoring and Evaluation

1. Monitor weight and nutrition assessment indices.
2. Patient will document intake per daily food diary.
3. Meet with patient weekly during pulmonary rehabilitation visits.

WEB LINKS

American Academy of Allergy, Asthma, and Immunology: This website provides many resources for patients and for health care professionals regarding allergies, asthma, and other associated diseases.

http://www.aaaai.org

American Lung Association: A website for the American Lung Association which contains excellent information about numerous respiratory-related disorders.

http://www.lungusa.org

American Thoracic Society: The American Thoracic Society publishes many papers and guidelines for health professionals related to various respiratory diseases. They also publish three online respiratory-related journals (see Relevant Online Journals).

http://www.thoracic.org

Centers for Disease Control and Prevention (CDC): A governmental agency whose mission is to promote health and quality of life by preventing and controlling disease, injury, and disability. The CDC maintains national health statistics, conducts research, and provides services concerning prevention of illness and injury.

www.cdc.gov

Cystic Fibrosis Foundation: A nonprofit organization dedicated to improving the care of patients with cystic fibrosis and helping to find a cure for cystic fibrosis. The Cystic Fibrosis Foundation also supports research related to the treatment of cystic fibrosis. This website contains excellent information for professionals and patients.

http://www.cff.org

National Heart, Lung and Blood Institute: This website contains excellent information for health care professionals and patients about various diseases related to the heart, blood vessels, and lungs.

http://www.nhlbi.nih.gov

National Institute of Allergy and Infections Disease: Conducts and supports basic and applied research to better understand, treat, and ultimately prevent infectious, immunologic, and allergic diseases. The website contains good information about numerous infectious diseases.

http://www.niaid.nih.gov/default.htm

United Network for Organ Sharing: A nonprofit organization that collects and managesdata related to organ transplants. They also assist to facilitate organ transplants.

http://www.unos.org

Relevant Online Journals

The following journals are online publications for health care professionals who treat patients with respiratory diseases:

American Journal of Respiratory and Critical Care Medicine: An online journal published by the American Thoracic Society.

http://ajrccm.atsjournals.org

The Journal of Respiratory Care Practitioners: On line journal for health care professionals who treat respiratory related disorders. Contains articles on nutrition management of various pulmonary diseases.

http://www.rtmagazine.com

END-OF-CHAPTER QUESTIONS

1. Describe the three major functions of the respiratory system in human health. What methods are used to measure pulmonary function?
2. Describe the role of nutrition in pulmonary health. Which nutrients have been associated with normal pulmonary function? How does smoking affect vitamin C requirements?
3. Based on supportive evidence, what are the important nutrition factors to keep in mind when treating patients of various age groups with asthma?
4. Define bronchopulmonary dysplasia (BPD). Why does it occur? How is vitamin A related to BPD?
5. Define cystic fibrosis (CF). What organ systems are involved in the disease? How does this organ involvement affect nutrition status? You receive a nutrition referral from a physician for a 9-year-old male with CF. He is below the 10th percentile weight for height. Outline an appropriate nutrition protocol for someone his age.
6. Describe aspiration pneumonia. As a dietitian, what procedures or methods would you recommend to help to prevent the occurrence of aspiration pneumonia in a patient receiving tube feeding?
7. Define respiratory failure (RF). What are the goals of nutrition therapy for RF? Outline a nutrition protocol for a patient with RF. Which nutrients are of specific concern?

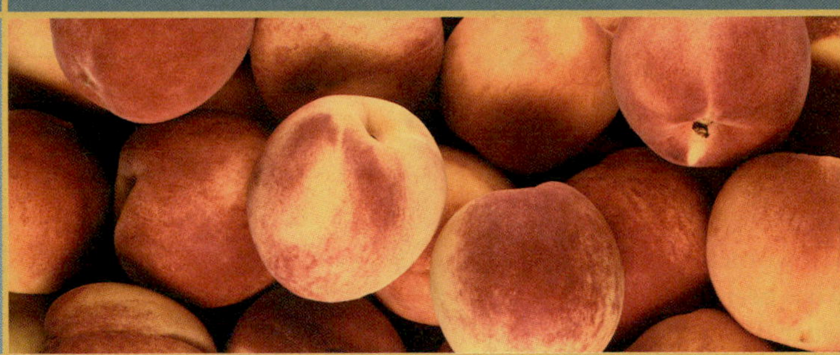

24

Neoplastic Disease

Deborah A. Cohen, M.M.Sc., R.D.

Clinical Faculty Associate, Southeast Missouri State University

CHAPTER OUTLINE

Definition

Epidemiology

Etiology

Cancer Screening and Prevention

Pathophysiology

Diagnosis

Clinical Manifestations

Treatment

Surgery • Chemotherapy • Radiation • Other Therapies

Nutrition Therapy

Nutrition Implications

Overview

When people speak about cancer, they may fail to understand that this term encompasses over one hundred different disease types. Each disease type has its own unique characteristics, and all types share some common characteristics. Therefore, this chapter will focus on the universal features of the disease process, consistent modes of therapy, and most importantly, the relationship to nutrition. Nutrition can be discussed as a factor in prevention, and compromised nutrition can be considered a complication of the disease or the treatment process. Some cancers, particularly lung and head and neck cancers, affect an individual's nutritional status even before the cancer is diagnosed. In addition, treatment of cancer, including surgery, chemotherapy, and radiation, can have a significant impact on an individual's nutritional status. Weight loss, **anorexia**, alterations in metabolism, and lean body mass wasting that often occur in cancer patients can have a profound impact on morbidity, mortality, the ability to withstand or tolerate treatment, and quality of life.

Definition

Cancer is a disorder of cell growth and regulation. These abnormal cells know no limits for cellular replication and produce cells that serve no purpose.

anorexia—lack of appetite

cancer—a class of diseases characterized by uncontrolled cell division and the ability of these cells to invade other tissues, either by direct growth into adjacent tissue (invasion) or by migration of cells to distant sites (metastasis)

Epidemiology

Cancer is a major cause of mortality in the United States (U.S.), second only to cardiovascular disease (see Figure 24.1). In 2004, 1.36 million new cases of cancer (699,560 men, 668,470 women) were diagnosed and 563,700 persons (290,890 men, 272,810 women) died from cancer. Cancer incidence has been declining by about 2% each year since 1992. Most cases of cancer occur in older individuals. Statistics indicate that two-thirds of all cases were in those over age 65 (American Cancer Society 2005). As of 2005, cancer of the lung and bronchus remains the number one killer in both men and women. Prostate and colorectal cancers are second and third leading causes of cancer cases and mortality in men, while breast and colorectal cancers are the second and third leading causes of cancer cases and mortality in women (see Figures 24.2 and 24.3).

Cancer rates do vary by ethnicity. For all cancer sites combined, African-American men have a 25% higher cancer incidence rate and a 43% higher cancer mortality rate than white men. African-American women have lower incidence rates than white females for all cancer sites combined, and yet they have a 20% higher mortality rate (Zoorob et al. 2001). Possible explanations include risk factors (hepatitis C for liver cancer), access to regular screening (breast, cervical, and colorectal cancers), and timely, high-quality treatments (many cancers) (Jemal et al. 2004).

FIGURE 24.1 Change in the U.S. Death Rates* by Cause, 1950 & 2002

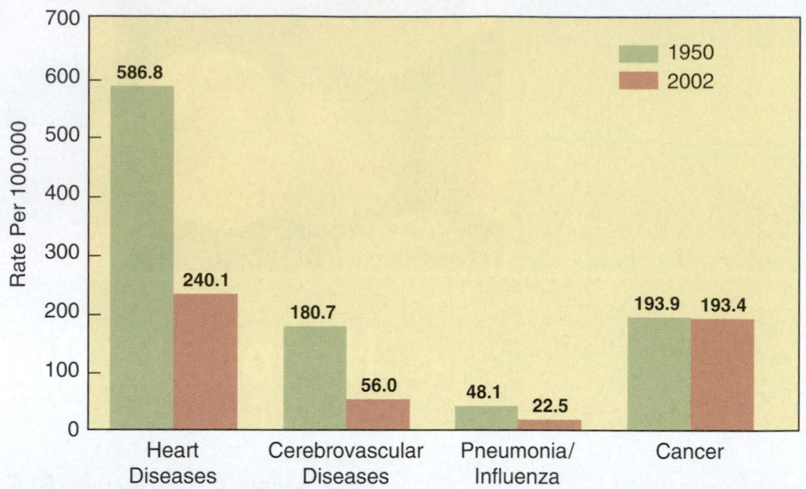

*Age-adjusted to 2000 U.S. standard population.

Source: American Cancer Society.

FIGURE 24.2 2005 Estimated U.S. Cancer Deaths

Men 295,280	
Lung and bronchus	31%
Prostate	10%
Colon and rectum	10%
Pancreas	5%
Leukemia	4%
Esophagus	4%
Liver and intrahepatic bile duct	3%
Non-Hodgkin Lymphoma	3%
Urinary bladder	3%
Kidney	3%
All other sites	24%

Women 275,000	
27%	Lung and bronchus
15%	Breast
10%	Colon and rectum
6%	Ovary
6%	Pancreas
4%	Leukemia
3%	Non-Hodgkin Lymphoma
3%	Uterine corpus
	Multiple myeloma
2%	Brain/ONS
22%	All other sites

ONS = other nervous system.

Source: American Cancer Society.

Etiology

Carcinogenesis is a multistep process in which normal cells are transformed into cancer cells. Many factors play a role in carcinogenesis, including exposure to **carcinogens** such as chemicals, physical agents, radiation, and infectious microorganisms. Genetics and nutritional factors can also play a role. Though only a small percentage of cancers are actually considered hereditary, all cancers involve genetics to a certain degree. Genetic research has explained the mechanisms of cancer development. Damage to a gene may occur as a result of exposure to chemicals, physical agents (ionizing radiation, ultraviolet radiation, asbestos), viral agents (Epstein-Barr virus, Human papilloma virus), and bacterial agents (*Helicobacter pylori*). Genes may also be affected by nutritional components such as antioxidants, soy protein, fat, kcalories, and alcohol. Nutritional genomics (nutrigenomics), the study of genetic variations that cause different phenotypic responses to diet among humans, is a

carcinogenesis—the multistep process (initiation, promotion, and progression) through which normal cells are transformed into cancer cells

carcinogen—substance that causes cancer

FIGURE 24.3 2005 Estimated U.S. Cancer Cases*

Men 710,040		Women 662,870	
Prostate	33%	32%	Breast
Lung and bronchus	13%	12%	Lung and bronchus
Colon and rectum	10%	11%	Colon and rectum
Urinary bladder	7%	6%	Uterine corpus
Melanoma of skin	5%	4%	Non-Hodgkin Lymphoma
Non-Hodgkin Lymphoma	4%	4%	Melanoma of skin
Kidney	3%	3%	Ovary
Leukemia	3%	3%	Thyroid
Oral Cavity	3%	2%	Urinary bladder
Pancreas	2%	2%	Pancreas
All Other Sites	17%	21%	All other sites

***Excludes basal and squamous cell skin cancers and in situ carcinomas except urinary bladder**

Source: American Cancer Society, 2005.

relatively new field that is likely to spawn exciting opportunities for research in the field of nutrition and may alter the way we apply nutrition therapy and cancer prevention strategies (see Chapter 11).

Evidence suggests that one third of the more than 500,000 cancer deaths that occur in the U.S. each year can be attributed to diet and physical activity habits, with another third due to cigarette smoking (Byers et al. 2002). Epidemiologic studies and laboratory experiments have provided substantial evidence for a relationship between environmental factors and carcinogenesis relating to cigarettes and lung cancer, and ultraviolet radiation and skin cancer. Establishing a strong link between diet and cancer has proved to be much more difficult due to the complexity of metabolism and variation in nutrient intake.

Biomarkers, which are distinctive biological or biologically derived indicators (such as a biochemical metabolite in the body), may be used to identify nutrient exposure and help improve the precision of epidemiologic studies. In studies of dietary interventions, nutritional biomarkers can be used as a measure of internal dose, which is an indication of the amount of nutrient available to the tissues after absorption and metabolism. The marker can also be used as a measure of dietary change or compliance with a new dietary regimen (LeMarchand et al. 1994). Biomarkers for many nutrients are not always reliable, sensitive, or specific. Many factors can affect the use of nutritional biomarkers to validate nutrient intake, including physiology, absorption, nutrient-nutrient interactions in the body, cooking methods, and tissue and renal saturation levels. For example, vitamin C, which is often studied for its antioxidant properties in cancer research, is stored largely in white blood cells, and thus the amount of vitamin C in these cells does not correlate with vitamin C intake. Nonetheless, biomarkers have a significant potential for helping to establish the cause-and-effect relationship between diet and cancer in the future.

Information obtained from animal models is useful and does provide important evidence to substantiate certain relationships that are alluded to in epidemiologic studies. However, data from animals often cannot be directly applied to humans. Genetic technology will soon be useful in applying animal/laboratory research to human biology. The role of genetic research and nutrigenomics in determining the mechanisms and the role of nutrients in cancer formation will have a significant impact on detection, screening, and prevention of cancer.

Epidemiologic studies compare patterns of intake of nutrients between certain population groups having high and low incidence of a particular type of cancer. Most individuals consume thousands of compounds each day. Relating a nutrient to cancer has been exceedingly difficult due to compounding factors that include environmental exposures (cigarette smoking, for example), age, sex, socioeconomic status, genetics, and numerous lifestyle and occupational hazards. The critical limiting feature of most human studies is the imprecision of quantifying nutrient intake. A number of tools are utilized, including food records/diaries, diet histories, food frequencies, and 24-hour recalls. All methodologies have inherent strengths and weaknesses. The method used must be appropriate for the population and nutrient(s) being studied.

Despite the limitations of the various research designs, epidemiologic studies currently have provided strong, consistent data to support an inverse relationship between cancer risk and consumption of fruits and vegetables. This means that as the intake of fruits and vegetables increases, the risk of cancer decreases. Research also supports inverse relationships between cancer risk and whole grains, fiber, certain micronutrients, and certain types of fat (e.g., omega-3 fatty acids and omega-3/omega-6 fatty acid ratio), as well as physical activity. Research supports direct relationships between

biomarker—a biological molecule used as a marker to measure or indicate the effects or progress of a disease or condition

cancer risk and intakes of total fat/certain types of fat (e.g., saturated fat) and alcohol; obesity (as measured by a high body mass index [BMI]); and certain food preparation methods such as smoking, salting, and pickling foods, and high-temperature cooking of meats (Greenwald, Clifford, and Milner 2001).

In a study of more than 900,000 U.S. adults, death rates from all cancers combined for individuals with a BMI of >40 were 52% higher for men and 62% higher for women than the death rates in men and women of normal weight. In both men and women, BMI of at least 40 was also significantly associated with higher rates of death due to cancer of the esophagus, colon and rectum, liver, gallbladder, pancreas, and kidney. On the basis of associations observed in that study, the authors estimated that current patterns of overweight and obesity in the U.S. could account for 14% of all deaths from cancer in men and 20% of those in women (Calle et al. 2003).

Cancer Screening and Prevention

Prevention of cancer can be addressed on two levels: primary and secondary. In primary prevention, specific factors are identified as part of the cancer process, and these factors are acted upon to decrease their potential activity as a carcinogen. For example, smoking cessation would be a primary prevention of cancer. Primary prevention of cancer refers to personal and community-wide efforts, whereas secondary prevention of cancer consists of measures for early detection and intervention. Screening for cancer risk is considered to be a secondary level of prevention. The American Cancer Society (ACS) publishes guidelines on nutrition and cancer prevention every five years. These guidelines, developed by a national panel of experts in cancer research, prevention, epidemiology, public health, and policy, represent the most current scientific evidence relating to dietary/physical activity patterns and cancer risk (Byers et al. 2002). The ACS guidelines (see Table 24.1) are consistent with both guidelines from the American Heart Association for the prevention of coronary heart disease and recommendations for general health promotion, as defined by the Department of Health and Human Services' 2005 *Dietary Guidelines for Americans* (U.S Department of Health and Human Services 2005).

The ACS report summarizes the rationale for the guidelines and presents recent data and issues pertaining to early cancer detection. It also includes the most recent data on adult cancer screening rates (Zoorob et al. 2001). Screening is recommended by medical organizations for breast (clinical breast examination and mammography for women aged 50–70 every one to two years), cervical (Papanicoloaou test and pelvic examination at least every 3 years in women aged 20–65), and colorectal cancer (annual fecal occult blood test along with flexible sigmoidoscopy at 5–10 year intervals in persons over the age of 50). Screening for prostate cancer (serum prostate-specific antigen, or PSA, for men over the age of 50) is currently under debate (Zoorob et al. 2001). There are no general screening guidelines for lung, oral, endometrial, or ovarian cancers, but there are warning signs and symptoms that may aid with detection of all cancers. See Table 24.2 for a summary of these warning signs.

Another level of cancer prevention includes chemoprevention. Research in this area uses specific agents to "reverse, suppress, or prevent carcinogenesis before the development of invasive malignancy" (Brawley and Kramer 2005). For example, the use of Tamoxifen to block hormonally driven breast cancers has been the target of several large clinical trials (Fabian and Kimler 2005). Nutritional factors and vaccines are currently being studied as potential chemopreventive agents. These include alpha-tocopherol (vitamin E), beta-carotene, vitamin C, and other antioxidants (Tamimi et al. 2002).

Pathophysiology

In order to understand the pathophysiology of cancer cells, it is important to review the basic principles of normal cell growth. All cells reproduce during the embryonic phase, but only some cells continue to reproduce after the first few months following birth. Cells that do reproduce, such as those of the liver, bone marrow, skin, and gastrointestinal tract, copy their DNA exactly and then split into two new daughter cells, which allows these types of cells to constantly regenerate. Cells that reproduce do so at an innate rate—the rate at which they are genetically programmed to reproduce—and this rate may be decreased or increased depending on genetic factors. In general, the cells of the bone marrow and the gastrointestinal tract have the fastest rate of replication in a normal environment.

Cells are classified as cycling cells, non-dividing cells, and resting cells. Cycling cells divide continuously; the epithelial cells that line the gastrointestinal tract are an example. Non-dividing cells divide before they differentiate (specialize), and then they do not divide again. Resting cells remain dormant initially, but certain conditions can stimulate their replication and growth.

Genetic controls for cellular division and growth include two basic sets of genes called oncogenes and tumor-suppressor genes. Oncogenes stimulate growth, and suppressor genes, as their name implies, suppress cellular growth. Examples of suppressor genes include the RB gene, which codes for the master "brake" of the cell cycle, and the P53 gene, which codes for the protein monitoring of cell health and the reliability of the cellular DNA. It is thought that cellular growth is also controlled by a counting system based on **telomeres.** Telomeres are end pieces of chromosomes that be-

telomere—the end section of a human chromosome

TABLE 24.1

Cancer Prevention Guidelines

Eat a healthful diet.

Eat a variety of healthful foods, with an emphasis on plant sources.

Eat five or more servings of a variety of vegetables and fruits each day.

- Include vegetables and fruits at every meal and for snacks.
- Eat a variety of vegetables and fruits.
- Limit French fries, snack chips, and other fried vegetable products.
- Choose 100% juice if you drink fruit or vegetable juices.

Choose whole grains in preference to processed (refined) grains and sugars.

- Choose whole grain rice, bread, pasta, and cereals.
- Limit consumption of refined carbohydrates, including pastries, sweetened cereals, soft drinks, and sugars.

Limit consumption of red meats, especially those high in fat and processed.

- Choose fish, poultry, or beans as an alternative to beef, pork, and lamb.
- When you eat meat, select lean cuts and smaller portions.
- Prepare meat by baking, broiling, or poaching, rather than by frying or charbroiling.

Choose foods that help maintain a healthful weight.

- When you eat away from home, choose food low in fat, calories, and sugar, and avoid large portions.
- Eat smaller portions of high-calorie foods. Be aware that "low fat" or "fat free" does not mean "low calorie" and that low-fat cakes, cookies, and similar foods are often high in calories.
- Substitute vegetables, fruits, and other low-calorie foods for calorie-dense foods such as French fries, cheeseburgers, pizza, ice cream, doughnuts, and other sweets.

Adopt a physically active lifestyle.

Adults: Engage in at least moderate activity for 30 minutes or more on 5 or more days of the week; 45 minutes or more of moderate to vigorous activity on 5 or more days per week may further reduce the risk of breast and colon cancer.

Children and adolescents: Engage in at least 60 minutes per day of moderate-to-vigorous physical activity for at least 5 days per week.

Helpful Ways to Be More Active

- Use stairs rather than an elevator.
- If you can, walk or bike to your destination.
- Exercise at lunch with your workmates, family, or friends.
- Take a 10-minute exercise break at work to stretch or take a quick walk.
- Walk to visit co-workers instead of sending an e-mail.
- Go dancing with your spouse or friends.
- Plan active vacations rather than only driving trips.
- Wear a pedometer every day and watch your daily steps increase.
- Join a sports team.
- Use a stationary bicycle while watching TV.
- Plan your exercise routine to gradually increase the days per week and minutes per session.

Maintain a healthful weight throughout life.

Balance caloric intake with physical activity.

Lose weight if currently overweight or obese.

Being overweight or obese is associated with an increased risk of developing several types of cancer:

- Breast (among postmenopausal women)
- Colon
- Endometrium
- Esophagus
- Gallbladder
- Pancreas
- Kidney

If you drink alcoholic beverages, limit consumption.

People who drink alcohol should limit their intake to no more than 2 drinks per day for men and 1 drink a day for women. The recommended limit is lower for women because of their smaller body size and slower metabolism of alcohol. A drink is defined as 12 ounces of beer, 5 ounces of wine, or 1.5 ounces of 80 proof distilled spirits.

- Alcohol is an established cause of cancers of the:
- Mouth
- Pharynx (throat)
- Larynx (voice box)
- Esophagus
- Liver
- Breast

Alcohol may also increase the risk of colon cancer.

Reprinted from: American Cancer Society [homepage on the Internet]. Philadelphia: American Cancer Society [cited 2005 Oct 31]. The Complete Guide—Nutrition and Physical Activity. Available from: http://www.cancer.org/docroot/PED/content/PED_3_2X_Diet_and_Activity_Factors_That_Affect_Risks.asp?sitearea=PED.

come shorter after each cell division. When the telomere shortens to a specific length, the cell will stop dividing.

Normal cellular reproduction is controlled by a combination of factors: genetic controls, hormones, and growth substances secreted by distant cells; local growth factors; and chemical cues from neighboring cells. Examples of hormones and systemic growth factors include epidermal growth factor, fibroblast growth factor, erythropoietin, insulin-like growth factors, and platelet-derived growth factor. Local growth factors include interleukins and cytokines. Cells also receive messages from neighboring cells that provide information about cellular type and the physical space available for cellular growth.

Unlike a normal cell, whose growth is closely regulated, a cancer cell reproduces at an uncontrolled rate. The cancer cell becomes autonomous from the normal growth signals and genetic control, and may even secrete its own growth factor. In a cancer cell, an enzyme is secreted that destroys

TABLE 24.2

Major Warning Signs of Cancer
Change in bowel habits or bladder function
Sores that do not heal
Unusual bleeding or discharge
Thickening or lump in breast or other parts of the body
Indigestion or difficulty swallowing
Recent change in a wart or mole
Nagging cough or hoarseness

Reprinted from: American Cancer Society [homepage on the Internet]. Philadelphia: American Cancer Society [cited 2005 Oct 31]. Specific Cancer Signs and Symptoms. Available from: http://www.cancer.org/docroot/CRI/content/CRI_2_4_3X_What_are_the_signs_and_symptoms_of_cancer.asp_cancer.asp.

the telomere, leading to the loss of the cell's internal clock or counting mechanism—which controls cellular replication. The process of cell differentiation may change, and a specific cell type may take on other traits. The physical characteristics of the cancer cell are altered: the nucleus and cytoplasm may be enlarged or misshapen, the mitosis rate is usually higher, and there may be derangements in the chromosome sequence.

The change from a normal cell to a cancer cell theoretically involves several steps. These include **initiation, promotion**, and **progression.** Figure 24.4 outlines this process. It is difficult in some situations to distinguish between initiation and promotion, but in general, initiation occurs as a result of exposure to an initiating agent, such as tobacco (Longo 2005). An initiating agent predisposes the cell to genetic mutation. Factors that promote the cell's movement through the carcinogenic changes include some hormones such as estrogen or testosterone. These promoters require an activation of the carcinogen as well as a failure of natural immunity and cellular repair mechanisms. Conditions must be conducive for the **neoplasm** (tumor) to continue to grow. Tumor growth rate is dependent on characteristics of the host such as age, sex, overall health, nutritional status, and immune function.

Characteristics of the tumor also affect the cell's ability to grow and the growth rate. The original cell type (and its natural rate of proliferation), as well as the availability of an adequate blood supply for the cancer cells, are crucial factors that determine how quickly the cancer cells will grow. Cancer cells may grow locally at the original (primary) site of cell transformation or spread to distant sites. This distant spread is called **metastasis.** Specific cancer types have typical routes for metastasis that include the lymphatic system, circulatory system, or nearby body cavities. For example, breast cancer typically metastasizes to brain and lung tissue through both the circulatory and lymphatic systems.

Diagnosis

When cancer is diagnosed, a series of blood and physical tests, **cytologic** tests, imaging, and biochemical tests is performed. Biochemical analysis of blood, serum, urine, and other body fluids can detect tumor biomarkers and also help to determine if the cancer has metastasized. Tumor markers are also used to determine if a patient is responding to treatment. For example, the **carcinoembryonic antigen (CEA)** is often measured to monitor colon cancer while **CA-125** is useful in monitoring treatment for ovarian cancer.

Diagnostic procedures provide useful information regarding tumor size, localization of the tumor for biopsy or resection, and assessment of the anatomical extent of the disease. Tumor imaging techniques are valuable for visualization of the tumor in relation to internal organs. Imaging techniques include magnetic resonance imaging (MRI), computerized axial tomography (CT), x-rays, ultrasound, positron emission tomography (PET), mammograms, bone scans, and endoscopy. Invasive diagnostic techniques allow for direct visualization of the tumor and may include needle or excisional biopsy (see Figure 24.5), cytologic aspiration, and laparoscopy. All diagnostic procedures provide information that is clinically useful in determining the tissue type of the tumor, the primary site of the malignancy, the extent of disease in the body, and the tumor's potential to recur. This

initiation—the first phase in cancer cell development; the exposure of cells to an appropriate dose of a carcinogen (initiator)

promotion—the second phase in cancer cell development; process induced in a normal cell that has been exposed to a carcinogen to transform into a cancer cell (promoters are not necessarily carcinogenic)

progression—the third phase in cancer cell development; the orderly transformation of a preneoplastic lesion to a tumor and, ultimately, invasive cancer

neoplasm—literally means "new growth"; an abnormal mass of tissue, the growth of which exceeds and is uncoordinated with that of normal tissue

metastasis—spread of cancer from the primary site to nearby or distant areas through the blood or lymph

cytologic—refers to tests that help to determine the morphologic features of a cell

carcinoembryonic antigen (CEA)—a glycoprotein present in fetal gastrointestinal tissue and in the cells or serum of adults having certain types of cancers. It is used clinically to monitor the effectiveness of a treatment, such as for colorectal cancer

CA-125—a protein that is secreted into the blood by ovarian cells and is used to monitor progress in the treatment of ovarian cancer

FIGURE 24.4 Cancer Progression

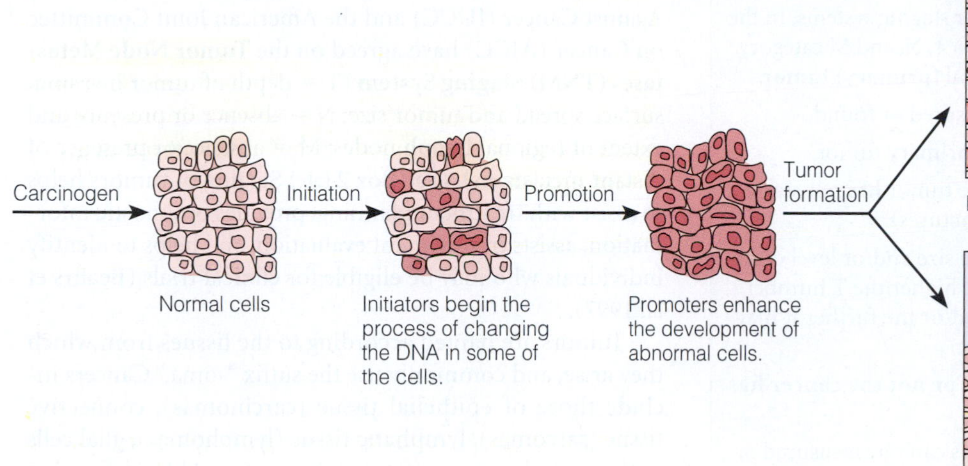

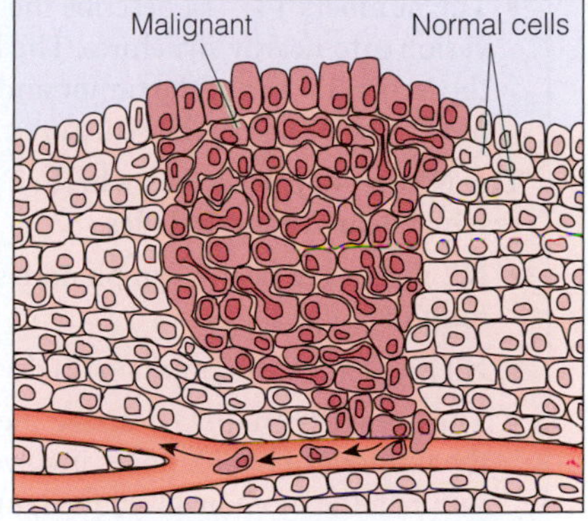

Source: S. Rolfes, K. Pinna, and E. Whitney, *Understanding Normal and Clinical Nutrition*, 7e, copyright © 2006 p. 880.

FIGURE 24.5 Biopsy Procedure

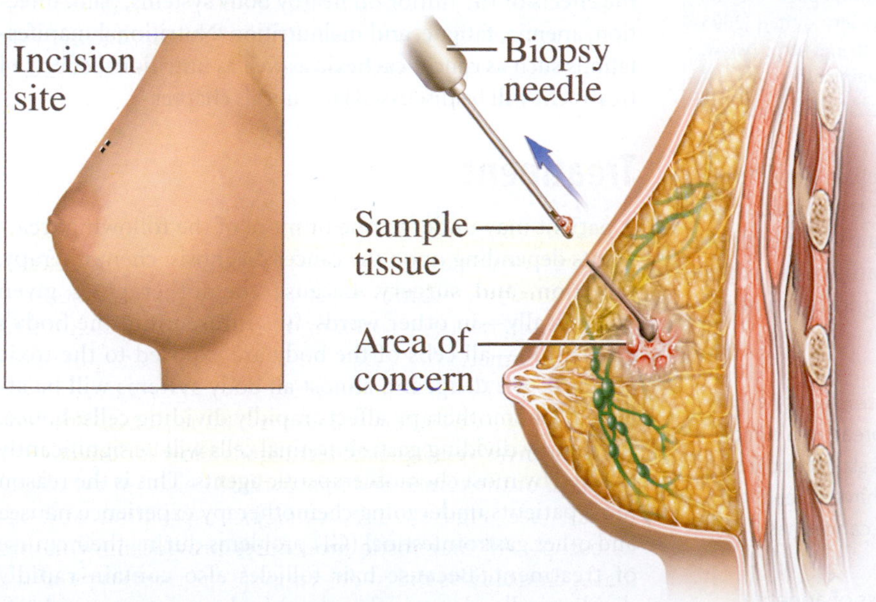

Source: Medical Illustration Copyright © 2006 Nucleus Medical Art, All rights reserved. www.nucleusinc.com.

BOX 24.1 CLINICAL APPLICATIONS

Tumor Node Metastases (TNM) Staging System

The American Joint Committee on Cancer (AJCC) developed the TNM classification system as a tool for doctors to stage different types of cancer based on certain standard criteria. It has replaced many of the older staging systems. In the TNM system, each cancer is assigned a T, N, and M category.

The T category describes the original (primary) tumor.

- TX means the tumor can't be measured or found.
- T0 means there is no evidence of primary tumor.
- Tis means the cancer is in situ (the tumor has not started growing into the surrounding structures).
- The numbers T1 – T4 describe the size and/or level of invasion into nearby structures. The higher the T number, the larger the size of the tumor and/or the further it may have grown into nearby structures.

The N category describes whether or not the cancer has reached nearby lymph nodes.

- NX means the nearby lymph nodes can't be measured or found.
- N0 means nearby lymph nodes do not contain cancer.
- The numbers N1 – N3 describe the size, location, and/or the number of lymph nodes involved. The higher the N number, the more involved the lymph nodes are.

The M category tells whether there are distant metastases (spread of cancer to other parts of body).

- MX means metastasis can't be measured or found.
- M0 means there are no known distant metastases.
- M1 means that distant metastases are present.

Reprinted from: American Cancer Society [homepage on the Internet]. Philadelphia: American Cancer Society [cited 2005 Oct 31]. The TNM Staging System. Available from: http://www.cancer.org/docroot/ETO/content/ETO_1_2X_Staging.asp.

information comprises the critical first step in developing a treatment plan. Precisely which diagnostic technique is used depends on a number of factors, including the patient's presenting signs and symptoms, the clinical status of the patient, the anticipated goal of treatment when diagnosis is made, biologic characteristics of the suspected malignancy, diagnostic equipment available in the community, and insurance approval of diagnostic procedures (McCance and Roberts 2002).

The tumor is then classified and assigned a "stage" on the basis of cell type, tissue of origin, whether it is benign or malignant, degree of differentiation, anatomic site, and function. The TNM Committee of the International Union Against Cancer (IUCC) and the American Joint Committee on Cancer (AJCC) have agreed on the **Tumor Node Metastases (TNM) Staging System** (T = depth of tumor invasion, surface spread and tumor size; N = absence or presence and extent of regional lymph nodes; M = absence or presence of distant metastases). (See Box 24.1.) Staging of tumors helps to assist with treatment planning, provides prognostic information, assists in treatment evaluation, and helps to identify individuals who may be eligible for clinical trials (Beahrs et al. 1997).

Tumors are named according to the tissues from which they arise, and commonly use the suffix "-oma." Cancers include those of epithelial tissue (carcinomas), connective tissue (sarcomas), lymphatic tissue (lymphomas), glial cells of the central nervous system (gliomas), and blood-forming organs, primarily the bone marrow (leukemias) (McCance and Roberts 2002).

Clinical Manifestations

As mentioned at the beginning of this chapter, cancer encompasses over one hundred different disease types. Each disease type has its own unique characteristics; therefore, specific signs and symptoms will correlate with the specific diagnosis. Yet there are common signs and symptoms that an individual may experience. These signs and symptoms may result from the effects of the tumor on nearby body systems, pain, infection, anemia, fatigue, and malnutrition. Nutritional manifestations such as cancer **cachexia** as well as nutritional effects of treatment will be discussed later in this chapter.

Treatment

A patient may undergo one or more of the following treatments depending upon the cancer diagnosis: chemotherapy, radiation, and surgery. Because chemotherapy is given systemically—in other words, it is infused into the body's circulation—all cells of the body are exposed to the toxic effects of the drug, and almost all body systems will be affected. Chemotherapy affects rapidly dividing cells; hence, the rapidly dividing gastrointestinal cells will be significantly affected by most chemotherapeutic agents. This is the reason most patients undergoing chemotherapy experience nausea and other gastrointestinal (GI) problems during their course of treatment. Because hair follicles also contain rapidly dividing cells, alopecia (hair loss) is also a common side effect of chemotherapy. The effects of surgical treatment and radiation therapy depend on the particular diagnosis.

Tumor Node Metastases (TNM) Staging System—a systematic way of describing the size, location, and spread of a tumor; T describes the primary tumor according to its size, N applies to the lymph nodes and whether cancer cells have spread to them, M refers to metastases and whether the cancer has spread to distant sites

cachexia—weight loss, wasting of muscle, loss of appetite, and general debility that can occur during a chronic disease

The type of treatment an individual may undergo depends on a number of factors, including location of the tumor, size of the tumor, and the health of the individual. In general, large bulky tumors are not curable with drugs. Drugs may not be able to penetrate into a solid tumor in amounts sufficient to kill the cells. Also, most cells in a bulky tumor may not be replicating at the time of treatment and thus survive to reestablish the tumor mass. The longer a tumor has been present, the greater the likelihood that it has already metastasized, resulting in ineffective drug therapy. Understanding the treatment pathways for cancer is crucial in planning nutrition therapy for the cancer patient. The extent and type of treatment plan will determine the potential adverse effects on the patient's nutritional status.

Surgery

Surgical treatment can be considered **primary**, **adjuvant**, **combination**, **salvage**, and **palliative**. Primary (definitive) treatment indicates that surgery will be the only therapy an individual will receive, for example, surgical removal of a small tumor. Adjuvant therapy includes chemoreductive therapy (debulking) or chemotherapy to reduce the size/bulk of a tumor in addition to surgical removal of the mass. Salvage therapy involves use of extensive surgery to treat local recurrence after a less extensive primary approach has been implemented, for example, mastectomy after lumpectomy and radiation (Pfeifer 2001). Palliative surgery is used to ameliorate disease and/or treatment-related symptoms without attempting to cure the cancer. Surgical removal of a tumor that is causing a spinal cord compression is an example of palliative therapy. Surgical removal of a tumor may be undertaken if it has been established that the tumor can be excised without damaging too much of the surrounding area, if the tumor is blocking an important pathway (GI tract, esophagus, or trachea), or if the tumor is small enough to be successfully excised. Chemotherapy or radiation is often employed before surgical excision in order to shrink the tumor, making it easier to remove. Because surgical removal of any tumor can cause cancer cells to leak or metastasize, surgery may not be an option for all cancers. If a tumor has invaded a major organ, successful surgical resection may be impossible.

If a tumor is localized or has limited local-regional spread, the goal of surgery may be to cure the disease. This is particularly applicable to cancer of the bladder, breast, cervix, colon, endometrium, larynx, head and neck, kidney, lung, ovaries, and testes. Cure is achieved by the mechanical removal of all cancer cells. In circumstances in which tumor resection cannot be performed, multiple therapies with radiation, chemotherapy, or a combination of chemotherapy and radiation may reduce the size of the cancer, making it amenable to surgical resection for cure (Merck 2005). However, cancer cells are frequently found in operative wound washings or in the drainage from postoperative wounds. Because many of these patients never develop recurrent cancer, it is thought that host immune defenses are effective in destroying any tumor cells missed at the time of resection (Pfeifer 2001).

Second-look procedures involve follow-up surgery after the original surgery or adjuvant treatment to check for the presence or absence of disease, especially in cancers that tend to recur locally. This type of surgery is not used as often today as it was in the past because of the development and use of less invasive diagnostic procedures.

Cancer Diagnoses Requiring Surgery for Treatment

This section reviews the more common diagnoses requiring surgery for treatment. Nutritional concerns are of high importance for those surgical procedures that have impact on nutritional intake, digestion, and/or absorption. Chapters 16 and 17 also address common gastrointestinal surgical procedures and their resulting nutritional concerns.

Significant and long-term nutritional side effects due to cancer surgery are seen most often for head and neck surgery and gastric surgery (partial and total gastrectomy). Surgical procedures may potentially alter an individual's physical appearance (mastectomy), alter organ function (colostomy), or both (radical neck dissection), which can contribute to depression and result in reduced food intake and weight loss. Surgical treatment that involves resection of the stomach or small intestine or removal of the esophagus or tongue will have the greatest impact on nutritional status.

Cancers of the Head and Neck Surgery that involves the head and neck area may result in difficulty in chewing and swallowing, **dysgeusia**, **xerostomia**, alterations in smell and difficulty with speaking. In addition, patients with head and neck cancers typically have a history of alcohol, tobacco, and/or substance abuse and significant weight loss prior to diagnosis, and therefore often present with malnutrition. It is estimated that 60% of patients with head

primary cancer—the location or organ/cells from which the cancer originated

adjuvant—usually, treatment "in addition to" initial treatment. For example, one or more anticancer drugs may be used in combination with surgical therapy as part of the cancer treatment regimen

combination—the use of two or more therapeutic agents/processes for the treatment of a neoplasm

salvage—additional treatment, used in hope of a cure or to prolong life, in a patient with recurrence of a malignancy following initial treatment

palliative—a non-curative treatment which reduces symptoms such as pain

dysgeusia—altered taste

xerostomia—dry mouth, often the result of damage to the salivary glands

and neck cancers will present with malnutrition at the time of diagnosis (Dudrick, Brown, and Biggs 1999). Surgical procedures for head and neck cancers can significantly alter both appearance and function, depending on the surgical and reconstructive procedures employed. In addition to the primary surgery, the patient may also undergo lymph node dissection. The head and neck have approximately 300 lymph nodes, all of which have the potential to contain metastatic cancer cells that may need to be removed. A radical neck dissection (RND) involves removal of all lymph nodes on one side of the neck, internal and external veins, external carotid artery, the sternocleidomastoid muscle (one of the muscles that functions to flex the head), internal jugular (neck) vein, submandibular gland (one of the salivary glands), the hypoglossal nerve, portions of the vagus nerve, tail of the parotid gland, and the spinal accessory nerve (a nerve that helps control speech, swallowing, and certain movements of the head and neck). While there is a good probability that most, if not all, of the cancer has been removed, a RND can cause severe nutritional deficits. In anticipation of the long rehabilitation and healing that is required before oral intake can be resumed, placement of a gastrostomy or jejunostomy at the time of surgery is prudent. The patient will also require speech therapy soon after surgery.

A modified RND involves removal of all lymph nodes on both sides of the neck with preservation of the spinal accessory nerve, internal jugular vein, and/or sternocleidomastoid muscle. Physical deformity is less severe with a modified RND. Selective neck dissection involves selective removal of one or more lymph node groups with preservation of the spinal accessory nerve.

Esophageal Cancer Esophageal cancer is relatively uncommon in the U.S. In 2002, 13,100 cases and 12,600 deaths were reported (Jemal et al. 2003), representing 1% of all cancer cases. However, the **prognosis** for esophageal cancer remains poor due to the advanced stage in which most patients present. In 2002, the 5-year survival rate for esophageal cancer was 14%. Risk factors for esophageal cancer include smoking and alcohol abuse, especially in combination (alcohol and tobacco appear to have a synergistic effect on carcinogenesis), **Barrett's esophagus** (a condition caused by long-term gastroesophageal reflux disease), and a diet low in fruits and vegetables. Males have a three times higher risk of developing esophageal cancer than females, and African-American males under the age of 55 are twice as likely to develop this disease compared to Caucasian males (Jemal et al. 2003).

Surgery and radiation therapy remain the mainstays of treatment for esophageal cancer; nonetheless, the overall results are disappointing. Many patients with esophageal cancer develop cachexia at some point in the progression of their disease (Burt and Brennan 1984). Trans-hiatal and transthoracic esophagectomy, while controversial, are the most common procedures for treatment of esophageal cancer. In most cases, the stomach is used for reconstruction of the esophagus. If the disease involves the stomach and a gastric resection is necessary, the esophagus can be reconstructed from part of the small or large intestine (Vaporciyan and Swisher 1999). Surgical procedures for esophageal cancer can substantially delay recovery of oral intake, and often patients require placement of a jejunal feeding tube during surgery.

Gastric Cancer Despite a universal decrease in the incidence and mortality of gastric cancer, it remains the second most common cause of cancer-related death in the world (Chan, Wong, and Lam 2001). The overall 5-year survival rate of gastric cancer is less than 25%. Prognosis for patients with gastric cancer in the western world is poor, with recent 5-year survival rates of only 22% overall and 59% for localized disease in the U.S. (Jemal et al. 2003). One of the important predisposing factors for the development of gastric cancer is repeated infection with *H. pylori*. Surgical intervention is the only potentially curative therapy for the treatment of gastric cancer. Partial (subtotal) or total gastrectomy increases a patient's risk for vitamin B_{12} deficiency, due to removal of parietal cells and loss of intrinsic factor, which is necessary for the transport and absorption of B_{12}. Calcium and iron absorption will also be reduced due to a reduction in the secretion of hydrochloric acid, which is also produced by the parietal cells in the gastric mucosa; this may lead to deficiencies of these minerals. **Dumping syndrome**, delayed gastric emptying, early satiety, nausea, and vomiting may also occur as a result of a partial or total gastrectomy.

Intestinal Cancers Cancers of the small intestine are relatively uncommon. Surgical excision of parts of the small bowel can have significant effects on the ability to digest and absorb nutrients. Malabsorption of nutrients (especially vitamin B_{12}) and **steatorrhea** are common following small bowel resection. The extent of the side effects depends in large part on the amount of small bowel that is resected, the ability of the remaining portion to adapt, and whether the ileocecal valve could be spared during the surgery.

Colorectal cancer remains the third most common cancer and the third leading cause of cancer deaths in men and women (Jemal et al. 2003). Cancers of the colon and rectum are treated primarily with surgery and chemotherapy. Risk factors for colorectal cancer include family history, diet (see

prognosis—a prediction of the probable course and outcome of a disease

Barrett's esophagus—pre-malignant condition that is considered a risk factor for esophageal adenocarcinoma; a complication of severe chronic GERD involving changes in the cells of the tissue that line the bottom of the esophagus

dumping syndrome—condition in which food moves too quickly from the stomach to the small intestine

steatorrhea—fat malabsorption resulting in severe diarrhea

BOX 24.2 CLINICAL APPLICATIONS

Dietary Risk Factors for Colorectal Cancer

The risk of colorectal cancer is higher for those with a family history of colorectal cancer. In addition to diet and physical activity, several other factors are linked to this cancer. Risk is increased by tobacco use and may be decreased by use of aspirin or other nonsteroidal anti-inflammatory drugs (NSAIDs) and, possibly, by hormone replacement therapy. Currently, however, neither aspirin-like drugs nor postmenopausal hormones are recommended to prevent colorectal cancer because of their potential side effects.

Some studies show a lower risk of colon cancer among those who are moderately active on a regular basis, and more vigorous activity may even further reduce the risk of colon cancer. Being inactive is linked more to an increased risk of cancer of the colon than cancer of the rectum. Diets high in vegetables and fruit may lower the risk, and diets high in red meat may increase the risk of colon cancer. Some evidence shows that folic acid supplements may reduce the risk of colon cancer.

The best advice to reduce the risk of colon cancer is to:

- Increase your physical activity
- Eat more vegetables and fruit
- Limit intake of red meats
- Avoid obesity
- Avoid excess alcohol

In addition, it is very important to follow the American Cancer Society guidelines for regular colorectal screening because finding and removing polyps in the colon can prevent colorectal cancer.

Reprinted from: American Cancer Society [homepage on the Internet]. Philadelphia: American Cancer Society [cited 2005 Oct 31]. Diet and Physical Activity Factors that Affect Risks for the Most Common Cancers: Colorectal Cancer. Available from: http://www.cancer.org/docroot/PED/content/PED_3_2X_Diet_and_Activity_Factors_That_Affect_Risks.asp?sitearea=PED.

Box 24.2), lack of physical activity, obesity, smoking, and history of inflammatory bowel disease. Surgeries performed for colorectal cancer are colon resection with reanastomosis, colostomy (temporary or permanent), and abdominal perianal resection (Murphy 2001). A colostomy is created when a portion of the colon is resected, the distal part of the colon is brought through the abdominal wall, and an artificial opening is created through which waste material passes out of the body from the bowel (see Figure 24.6). When the descending colon is resected, the effects on nutrition are usually minimal because most water and electrolytes have already been reabsorbed. However, when the ascending or transcending portion of the colon is resected, there is a higher risk for electrolyte and fluid loss resulting in electrolyte abnormalities, especially potassium, and dehydration. In general, most colostomies result in a stool that is more liquid in consistency.

FIGURE 24.6 Colostomy and Ileostomy

Colostomy

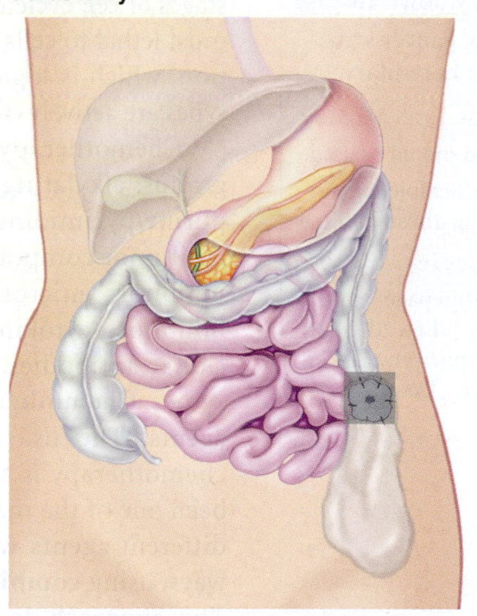

In a colostomy, the rectum and anus are removed, and the stoma is formed from the remaining colon.

Ileostomy

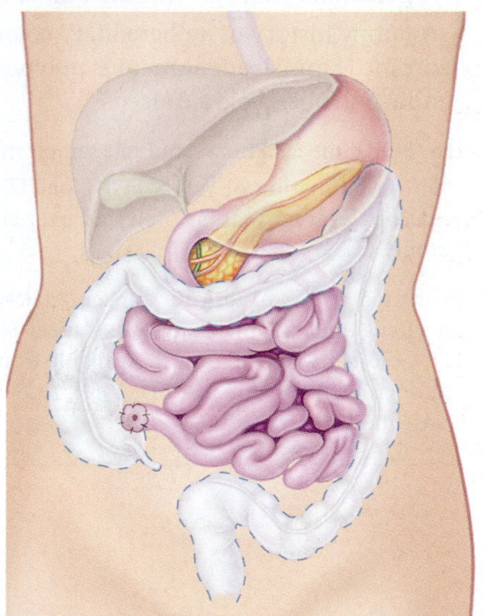

In an ileostomy, the entire colon, rectum, and anus are removed, and the stoma is formed from the ileum.

Source: S. Rolfes, K. Pinna, and E. Whitney, *Understanding Normal and Clinical Nutrition,* 7e, copyright © 2006, p. 761.

Colon cancer, if detected and treated at an early stage, before it has metastasized, may be cured by surgical therapy. See Box 24.3 for colorectal cancer screening guidelines.

Pancreatic Cancer Pancreatic cancer is the fifth leading cause of cancer-related death for both men and women, and is responsible for 5% of all cancer-related deaths (Jemal et al. 2003). Because of the frequent inability to diagnose pancreatic cancer when it is still localized and surgically resectable, and the lack of effective systemic therapies, the incidence rates (30,300 in 2002) are virtually the same as the mortality rates (Wolff, Abbruzzese, and Evans 2005). Pancreatic cancer is frequently diagnosed late due to the vague and nonspecific symptoms that accompany the disease; however, weight loss

BOX 24.3 CLINICAL APPLICATIONS

Colon and Rectal Cancer Screening Guidelines

Beginning at age 50, both men and women at average risk for developing colorectal cancer should follow one of these five testing schedules:

- Yearly fecal occult blood test (FOBT)* or fecal immunochemical test (FIT)
- Flexible sigmoidoscopy every 5 years
- Yearly FOBT* or FIT plus flexible sigmoidoscopy every 5 years**
- Double-contrast barium enema every 5 years
- Colonoscopy every 10 years

All positive tests should be followed up with colonoscopy.

People should begin colorectal cancer screening earlier and/or undergo screening more often if they have any of the following colorectal cancer risk factors.

- A personal history of colorectal cancer or adenomatous polyps
- A strong family history of colorectal cancer or polyps (cancer or polyps in a first-degree relative younger than 60 or in two first-degree relatives of any age)
  Note: a first degree relative is defined as a parent, sibling, or child.
- A personal history of chronic inflammatory bowel disease
- A family history of an hereditary colorectal cancer syndrome (familial adenomatous polyposis or hereditary non-polyposis colon cancer)

*For FOBT, the take-home multiple sample method should be used.

**The combination of yearly FOBT or FIT plus flexible sigmoidoscopy every 5 years is preferred over either of these options alone.

Reprinted from: American Cancer Society [homepage on the Internet]. Philadelphia: American Cancer Society [cited 2005 Oct 31]. Colon and Rectal Cancer. Available from: http://www.cancer.org/docroot/PED/content/PED_2_3X_ACS_Cancer_Detection_Guidelines_36.asp?sitearea=PED

and anorexia are common symptoms at diagnosis. Ninety percent of patients present with weight loss, 75% with malnutrition, and 60% with anorexia at the time of diagnosis (Spitz et al. 1999). Pancreatic exocrine insufficiency due to obstruction of the pancreatic duct commonly results in malabsorption and steatorrhea. Most patients with pancreatic cancer also have hyperglycemia.

The standard surgical procedure for neoplasms of the pancreatic head and periampullary region is pancreaticoduodenectomy, also known as the Whipple procedure, which involves removal of the pancreatic head, duodenum, gallbladder, and bile duct with or without the gastric antrum (Wolff, Abbruzzese, and Evans 2005). Delayed gastric emptying is common after pancreaticoduodenectomy. The nutrition consequences of delayed gastric emptying are most significant in those patients with some degree of nutrition depletion preoperatively and in older patients with significant medical comorbid conditions (Wolff, Abbruzzese, and Evans 2005). There are numerous potential nutrition complications for this procedure and thus, the placement of a jejunostomy feeding tube is common practice.

Chemotherapy

Chemotherapy includes medications that interrupt different stages of cell cycle replication. Chemotherapeutic agents are most lethal to cells that are undergoing continual proliferation, which is logical since cells of many common tumor types are actively dividing.

Chemotherapy agents are classified into the following groups: alkylating, anti-metabolites (folate antagonists), purine/pyrimidine antagonists, anthracyclines, platinum antitumor compounds, antibiotics, nitrosureas, mitototic inhibitors, microtubule targeting agents, topoisomerase inhibitors, cytokines, biologic response modifiers, monoclonal antibodies, immunotherapy, hormones, and enzymes. Chemotherapeutic agents are rarely used as single agents. The use of combinations of drugs in cancer chemotherapy is a commonly employed practice and has been one of the major advances made in this field. Because different agents work to kill the cancer cell in different ways, using **combination chemotherapy** is advantageous. This approach decreases incidence of drug resistance, allows for an additive or synergistic effect of the drugs, and decreases the potential overall toxicity or at least the toxicity to any one organ system. **Adjuvant chemotherapy** refers to chemotherapy after surgery. This chemotherapy has the theoretical advantage of eliminating any residual or metastatic cells, thus improving patient survival. **Neoadjuvant chemotherapy** refers to chemotherapy that is administered before surgery; this chemotherapy is indicated when the tumor size is too large for an effective resection.

Because most of the normal cells in the body are in a resting stage, they are somewhat protected from the lethal effects of most chemotherapeutic agents. Normal, healthy

combination chemotherapy—the use of two or more antineoplastic agents to achieve maximum kill of malignant cells

adjuvant chemotherapy—the use of drugs as additional treatment for patients with cancers that are thought to have spread outside their original sites

neoadjuvant chemotherapy—refers to chemotherapy used prior to primary treatment, which is typically surgery

cells in the human body that are susceptible to the effects of these lethal agents include those that are frequently dividing, such as the cells of the bone marrow (red blood cells, white blood cells, and platelets), the epithelial lining of the gastrointestinal tract, and the cells of the hair follicles. The most common side effects, therefore, are due to toxicity to these cells, and include **neutropenia**, thrombocytopenia, anemia, diarrhea, **mucositis** (discussed later in this chapter), and alopecia. Some chemotherapeutic agents are also known to cause cardiotoxicity, neurotoxicity, and nephrotoxicity. (See Table 24.3, which presents the side effects of chemotherapeutic agents.) Specific nutritional side effects of chemotherapy will also be discussed later in this chapter.

Radiation

Radiation therapy (RT) alone is the most common treatment for certain types of head and neck cancers, such as cancer of the nasopharynx, larynx, and oropharynx (Hoffman et al. 1998). Radiation may be used to cure the cancer, as in Hodgkin's disease, testicular seminomas, thyroid carcinomas, localized cancers of the head and neck, and cancers of the uterine cervix. RT may also be used to control malignant disease when a tumor cannot be removed surgically or when local nodal metastasis is present, or it can be used prophylactically to prevent leukemic infiltration of the brain or spinal cord.

RT is delivered with electromagnetic rays (gamma rays and x-rays) and charged particles (electrons). RT destroys cancer cells by altering cellular and nuclear material, especially DNA. The most harmful tissue disruption is due to the alteration of the DNA molecule within the cells of the tissue. Ionizing radiation breaks the strands of the DNA helix, leading to cell death. As with chemotherapy, those cells that are continually proliferating are the cells most sensitive to the effects of radiation. As discussed earlier, these include epithelial cells, bone marrow cells, lymph tissue, and hair cells. But, unlike chemotherapy's toxic effects, toxicity of RT is localized to the region being irradiated. Toxicity may be exacerbated when concomitant chemotherapy is administered.

The goal of treatment planning is to uniformly irradiate a specified target while minimizing the dose to surrounding normal tissues (Goitein et al. 1991). Normal cells as well as the targeted cancer cells are susceptible to the toxicity of RT; therefore, custom-made lead blocks are developed for each patient to protect vital organs that may be in the RT field and which may be damaged during RT.

RT may be administered externally or internally, either alone or in combination. External beam RT is administered by linear accelerators on a daily basis (in an outpatient setting), usually five days per week, for a period of time (approximately 5–6 weeks) depending on the cancer being treated. **Brachytherapy** involves the placement of radioactive sources either within an existing body cavity (e.g., the vagina) in close proximity to the tumor (intracavitary), or directly within a tumor (interstitial) (Mundt et al. 2003). Brachytherapy often involves a hospital stay, and is used to treat cancers of the prostate and cervix, pituitary adenomas, and acoustic neuromas. Stereotactic radiosurgery (SRS) is a technique used to precisely deliver a single, high-dose fraction of external beam radiation to a small, intracranial volume (Schell et al. 1995). A common use of RT is in combination with surgery and/or chemotherapy. When combined with surgery, RT may be given prior to (preoperative), following (postoperative), or during (intraoperative) surgery. Delayed wound healing is common when RT is used in combination with surgery, especially when RT is administered postoperatively. Chemotherapy may be administered prior to (neoadjuvant), during (concomitant), or following RT (maintenance) (Chu and DeVita 2005). RT may also be used as a palliative form of therapy; for instance, it can be used to control intracranial swelling due to metastatic disease in order to help alleviate pain. Whole brain RT for metastatic disease is not used with the intention of curing the disease, but rather to control symptoms such as headaches and visual problems. Palliative therapy is often delivered at higher doses for shorter periods of time.

Common adverse effects of RT to the head and neck area include fatigue, mucositis, dysgeusia, xerostomia secondary to salivary gland destruction, **dysphagia**, **odynophagia**, and severe esophagitis. Symptoms will manifest by approximately day ten of RT and continue until about two to three weeks after RT has ended. Mucositis, xerostomia, and odynophagia due to esophagitis may be severe and debilitating enough to warrant a temporary discontinuation of RT. The individual is at high risk of dehydration due to inadequate fluid intake, and may need intravenous fluids for hydration and electrolyte correction. Esophageal tissue may become extremely irritated and friable to the extent that oral intake is impossible. A surgically placed feeding tube may be indicated to provide nutritional support for these patients. Nasoenterally placed feeding tubes may be contraindicated due to the fragile nature of the esophagus and increased risk of bleeding during the placement. In addition, the esophagus may be totally or partially obstructed by the tumor mass, which prevents placement of a feeding tube in this manner.

neutropenia—low white blood cell count

mucositis—inflammation of a mucous membrane (e.g., mouth sores)

brachytherapy—a type of radiation therapy in which radioactive materials are placed in direct contact with the tissue being treated

dysphagia—difficulty swallowing

odynophagia—painful swallowing

TABLE 24.3

Side Effects of Antineoplastic Agents

Drug Class and Examples	Mechanism of Action	Cell Cycle Specificity	Common Side Effects
Alkylating Agents			
busulfan, carboplatin, chlorambucil, cisplatin, cyclophosphamide, dacarbazine, hexamethyl melamine, ifosfamide, melphalan, nitrogen mustard, thiotepa	Alters DNA structure by misreading DNA code, initiating breaks in the DNA molecule, cross-linking DNA strands	Cell cycle-nonspecific	Bone marrow suppression, nausea, vomiting, cystitis (cyclophosphamide, ifosfamide), stomatitis, alopecia, gonadal suppression, renal toxicity (cisplatin)
Nitrosureas			
carmustine (BCNU), lomustine (CCNU), semustine (methyl CCNU), streptozocin	Similar to the alkylating agents; cross the blood-brain barrier	Cell cycle-nonspecific	Delayed and cumulative myelosuppression, especially thrombocytopenia; nausea, vomiting
Topoisomerase I Inhibitors			
irinotecan, topotecan	Induce breaks in the DNA strand by binding to enzyme topoisomerase I, preventing cells from dividing	Cell cycle-specific	Bone marrow suppression, diarrhea, nausea, vomiting, hepatotoxicity
Antimetabolites			
5-azacytidine, cytarabine, edatrexate fludarabine, 5-fluorouracil (5-FU), FUDR, gemcitabine, hydroxyurea, Leustatin, 6-mercaptopurine, methotrexate, pentostatin, 6-thioguanine	Interfere with the biosynthesis of metabolites or nucleic acids necessary for RNA and DNA synthesis	Cell cycle-specific (S phase)	Nausea, vomiting, diarrhea, bone marrow suppression, proctitis, stomatitis, renal toxicity (methotrexate), hepatotoxicity
Antitumor Antibiotics			
bleomycin, dactinomycin, daunorubicin, doxorubicin (Adriamycin), idarubicin, mitomycin, mitoxantrone, plicamycin	Interfere with DNA synthesis by binding DNA; prevent RNA synthesis	Cell cycle-nonspecific	Bone marrow suppression, nausea, vomiting, alopecia, anorexia, cardiac toxicity (daunorubicin, doxorubicin)
Mitotic Spindle Poisons			
Plant alkaloids: etoposide, teniposide, vinblastine, vincristine (VCR), vindesine, vinorelbine.	Arrest metaphase by inhibiting mitotic tubular formation (spindle); inhibit DNA and protein synthesis	Cell cycle-specific (M phase)	Bone marrow suppression (mild with VCR), neuropathies (VCR), stomatitis
Taxanes: paclitaxel, docetaxel	Arrest metaphase by inhibiting tubulin depolymerization	Cell cycle-specific (M phase)	Bradycardia, hypersensitivity reactions, bone marrow suppression, alopecia, neuropathies
Hormonal Agents			
androgens and antiandrogens, estrogens and antiestrogens, progestins and antiprogestins, aromatase inhibitors, luteinizing hormone-releasing hormone analogs, steroids	Bind to hormone receptor sites that alter cellular growth; block binding of estrogens to receptor sites (antiestrogens); inhibit RNA synthesis; suppress aromatase of P450 system, which decreases estrogen level	Cell cycle-nonspecific	Hypercalcemia, jaundice, increased appetite, masculinization, feminization, sodium, and fluid retention, nausea, vomiting, hot flashes, vaginal dryness
Miscellaneous Agents			
asparaginase, procarbazine	Unknown or too complex to categorize	Varies	Anorexia, nausea, vomiting, bone marrow suppression, hepatotoxicity, anaphylaxis, hypotension, altered glucose metabolism

RT to the abdominal and pelvic area can result in radiation enteritis, a severe, often debilitating disease that can take up to ten years after RT to manifest. Radiation enteritis is one of the most feared complications of abdominal and pelvic radiation. Bowel injuries that result in fistulas, strictures, and chronic malabsorption are potentially life-threatening complications, and have a significant impact on quality of life (Mann 1991). Symptoms include severe diarrhea and malabsorption. Medical therapies are primarily supportive. Total parenteral nutrition may be necessary to provide nutrition to prevent weight loss and correct electrolyte abnormalities during periods when bowel rest is necessary.

Other Therapies

Bone Marrow Transplantation Bone marrow transplantation is a therapeutic option for those with hematalogic and some non-hematologic malignancies (see Chapter 21). Most transplants are now performed with peripheral blood stem cells as opposed to bone marrow. Bone marrow or hematopoietic stem cell transplantation (HSCT) is a potentially curative treatment for hematologic malignancies, including chronic leukemias, acute leukemias, Hodgkin's disease, non-Hodgkin's lymphoma, multiple myeloma, and myelodysplastic syndromes. In addition, HSCT is used in the treatment of some solid tumors (breast cancer, ovarian cancer, testicular cancer, and lung cancer), bone marrow failure (aplastic anemia), and it is used experimentally for autoimmune disorders (multiple sclerosis, systemic lupus erythematous, rheumatoid arthritis, systemic sclerosis) (Lenssen and Aker 2002).

Stem cells used in HSCT may be obtained from one of three sources: a donor (related or unrelated), a genetically identical twin, or from the individual who will be undergoing HSCT. Donor transplantation is referred to as allogeneic HSCT. If the donor is a genetically identical twin, it is referred to as syngeneic HSCT. When patients donate their own stem cells, the procedure is referred to as autologous HSCT. Those patients undergoing an allogeneic HSCT are at highest risk for developing transplant-related morbidity and mortality due to complications secondary to graft-versus-host disease.

The stem cells are harvested (either surgically, if bone marrow cells are used, or by a procedure called apheresis if peripheral stem cells are used) from the donor or the patient. After a certain period of time, the patient undergoes high-dose chemotherapy and/or total body irradiation (TBI), which is referred to as the preparative or conditioning regimen. Stem cells are infused once the conditioning regimen is complete, usually after several days, and the day of stem cell infusion is referred to as day 0.

The conditioning regimen serves two purposes: (1) to provide sufficient immunosuppression to prevent rejection and allow engraftment of donor stem cells, and (2) to eradicate malignant cells. Major toxicities associated with HSCT are those related to the conditioning regimen, toxicities that occur as a result of graft rejection, known as graft-versus-host disease (GVHD), and infectious complications associated with immunosuppression. GVHD only occurs in those individuals receiving an allogeneic HSCT. The most common toxicities—nausea, vomiting, mucositis, pancytopenia (including neutropenia), fevers, and infections—which are associated with the high doses of chemotherapy (and TBI) are typically seen immediately after the conditioning regimen is delivered (see Chapter 21). Severity of the toxicities depends on the intensity and type of agent used in the conditioning regimen. If oral intake is inadequate, parenteral nutrition support is initiated to meet nutritional needs. Tube feeding has not been successful in these patients due to the severity of nausea, vomiting, mucositis, delayed gastric emptying, and diarrhea. In addition, low platelet counts associated with the conditioning regimen significantly increase risk of bleeding during placement of a feeding tube.

Late complications are more commonly observed after an allogenic HSCT, and occur as a consequence of the effects of long-term damage to normal tissues, either from the conditioning regimen or from immunosuppressive agents and GVHD. Long-term complications include: growth retardation, infertility, endocrine failure, avascular joint necrosis, osteopenia, cataracts, renal insufficiency, restrictive pulmonary defects, neurocognitive defects, and secondary malignancies (Childs 2005). Delayed gastric emptying may persist for months following a bone marrow transplant.

GVHD, which occurs in a high percentage of allogeneic HSCT recipients, is associated with serious long-term medical problems, including immunosuppression, organ dysfunction, and infections. Acute GVHD is graft rejection that occurs in the first 100 days after HSCT infusion, while chronic GVHD is graft rejection that occurs after day 100. GVHD is thought to be mediated by donor T-lymphocytes that induce tissue damage in the skin, liver, and/or GI tract. Skin manifestations may include a mild to severe erythematous papular rash. Liver manifestations may include bile duct damage with cholestatic liver test abnormalities. The GI tract can be affected anywhere from the mouth to the anus. GI manifestations range from dry mouth and oral ulcerations to small intestinal GI mucosal damage leading to severe diarrhea (>1 liter per day) and malabsorption.

Anorexia, weight loss (especially loss of lean body mass), and nutritional deficiencies are common and often severe. Patients with moderate to severe GI GVHD usually require total parenteral nutrition (TPN). These patients have elevated energy needs. Protein needs increase as well, to support replacement GI losses (and dermal losses if skin GVHD is also present) and to meet amounts needed for protein synthesis (see Chapter 21). Protein needs are also elevated because these patients are given high doses of glucocorticoids for immunosuppression. These drugs increase nitrogen losses and cause lean body mass/skeletal muscle loss. Enteral nutrition is usually contraindicated in patients with severe GI GVHD due to severe delayed gastric emptying and severely damaged GI mucosa, which causes malabsorption.

Primary treatment of either acute or chronic GVHD includes the administration of high doses of immunosuppressant medications (including cyclosporine, tacrolimus, methotrexate, and prednisone). Corticosteroids, such as prednisone, are administered at such high doses that the side effects of these medications, which have significant nutritional implications, can be seen within days after the drug is first given. Nutrition-related side effects include hyperglycemia (which is usually resistant to diet manipulation and requires insulin), nitrogen catabolism, hypertension, and sodium retention. Long-term complications of corticosteroids include diabetes, osteoporosis, Cushing's syndrome, weight gain, muscle wasting (noted especially in the extremities), and poor wound healing.

All patients undergoing autologous, syngeneic, or allogeneic HSCT should be considered at high nutritional risk and should undergo a complete nutritional assessment by a clinical dietitian prior to beginning the conditioning regimen. In addition, those undergoing allogeneic HSCT need to be monitored closely throughout their transplant course and in the outpatient setting after discharge for nutritional complications due to immunosuppression or infections, late complications due to the conditioning regimen, and acute and/or chronic GVHD.

Biological Therapies Biological therapies are a relatively new form of treatment used to combat cancer. Biological therapies, which are also called immunotherapy or biological response modifiers, use the body's own immune system, either directly or indirectly, to eradicate cancer cells. Biological therapies use interferons, interleukins, monoclonal antibodies, colony-stimulating factors, gene therapy, and nonspecific immunomodulating agents.

Biologic response modifiers (BRMs) are used to stop, control, or suppress processes that permit cancer growth; make cancer cells more recognizable for destruction by the body's own immune system; enhance the killing power of the body's own immune system; enhance the body's ability to repair or replace normal cells damaged or destroyed by other forms of cancer treatment; and prevent cancer cells from spreading to other parts of the body (National Cancer Institute). While some BRMs are part of standard treatment, some are only available to patients enrolled in clinical trials.

Interferons and interleukins are cytokines that are produced by the human body and can be synthesized by drug companies. They are used to fight certain types of cancer, including metastatic renal cancer and metastatic melanoma. Colony-stimulating factors (CSFs) are given to increase bone marrow cell production, including red blood cells, white blood cells, and platelets. They are used to decrease the period of neutropenia or thrombocytopenia in those patients receiving chemotherapeutic agents that cause a reduction of these cells. The patient receiving CSFs is less prone to develop infections or anemia and less likely to bleed excessively. G-CSF (Filgrastim), GM-CSF (SARGRAMOSTIM), and erythropoietin (Epogen) are commonly prescribed CSFs used in cancer therapy. Monoclonal antibodies react with certain types of cancers to make the body's immune response more effective in fighting the cancer. Sometimes monoclonal antibodies are "attached" to a chemotherapeutic agent to make destruction of the cancer cells more effective. Rituxan® (rituximab) and Herceptin® (trastuzumab) are two monoclonal antibodies recently approved by the Food and Drug Administration (FDA). Side effects of these drugs include mild to moderate flu-like symptoms.

Gene therapy is currently only available in experimental clinical trials. Gene therapy works by inserting a gene into a person's immune cell to enhance its ability to eradicate cancer cells. Nonspecific immunomodulating agents work to directly or indirectly stimulate the body's own immune system, especially immunoglobulins and cytokines, to destroy malignant cells.

Common side effects of BRM's include bone pain, fatigue, fever, anorexia, rashes at the site of injection, and flu-like symptoms (nausea, vomiting, chills, fever, and anorexia).

Nutrition Therapy

The primary goal of nutrition therapy for the cancer patient is to prevent malnutrition, because reversing it may prove to be very difficult. The metabolic alterations discussed in the following section can cause severe anorexia and cachexia that may not be amenable to common nutrition interventions. Nutritional treatment relies heavily on screening for those patients who are at high risk for developing malnutrition, which may result from either the cancer itself or the medical treatment, including surgery, chemotherapy, and/or radiation. Clinical dietitians need to be aware of those cancer diagnoses and treatments that are most likely to cause malnutrition in this population. Pharmacologic agents, such as appetite stimulants, **prokinetics**, **anti-emetics**, and anabolic agents, may be useful for control of symptomatic treatment in combination with nutrition therapy.

Nutrition Implications

Cachexia Cachexia is one of the most common causes of death among patients with cancer, and is present in 80% at death (Nelson 2000). The word "cachexia" is derived from the Greek words *kakos*, meaning "bad," and *hexis*, meaning

prokinetic—a pharmacologic agent that promotes gastric emptying
anti-emetic—a pharmacologic agent that reduces nausea

"condition" (Bruera 1997). Cachexia is characterized by involuntary weight loss, tissue wasting (particularly lean body mass and adipose tissue), inability to perform daily activities, and metabolic alterations. These alterations in glucose, amino acid/protein, and lipid metabolism can have an impact on the patient's nutritional and medical status with a subsequent impact on quality of life, morbidity, and mortality.

The pathophysiology of cancer cachexia is not completely understood. However, it seems to be attributable, at least in part, to metabolic alterations that lead to intermediary metabolites (e.g., lactate, ketones, oligonucleotides) that accumulate along an abnormal pathway; to other substances released by the tumor itself; or to substances released by normal cells in response to the tumor (Bruera 1992). It has long been suspected that circulating chemical mediators released by tumors increase metabolic rates and thus induce a hypermetabolic catabolic state. Chemical mediators involved in cachexia include cytokines, hormones, neurotransmitters, serotonin, interleukins, interferons, prostaglandins, tumor necrosis factor, neuropeptide Y, substance P, bradykinins, and glutamate (Sanchez 2004). These "cachectic factors" are presumably tumor-specific since, for example, lung and gastrointestinal tumors including pancreatic cancer are well known for causing cachexia with much higher incidence than breast and hematopoietic tumors (Capra, Ferguson, and Ried 2001). Changes in taste and smell perception, psychologic factors, uncontrolled pain, and therapy-induced side effects also play an important role in the severity of cachexia, but vary from one patient to another.

Standard methods of nutritional therapy, including enteral and parenteral nutrition support, may not be effective in improving the outcomes of cancer patients due to alterations in metabolism. In addition, cancer chemotherapy, radiation, and surgery can exacerbate an already abnormal metabolic milieu.

There are no standard criteria with which to diagnose cachexia. Diagnosis usually stems from the presenting signs and symptoms. These include weight loss, anorexia, muscle wasting, fatigue, and early satiety. Treatment of the anorexia-cachexia syndrome (discussed in the section Nutrition Interventions) may include nutrition therapy strategies.

Abnormalities in Carbohydrate, Protein, and Lipid Metabolism The normal physiologic conservation mechanisms seen during periods of acute starvation do not occur in the presence of a malignant tumor (Tayek 1992). Normally, during periods of simple starvation, free fatty acids from adipose tissue supply energy to the liver and muscle. The free fatty acids are converted to ketone bodies that can then be utilized by most tissues in the body as a source of energy. Ketone bodies inhibit glucose utilization and protein degradation from lean body mass. Therefore, protein is not used as a primary energy source. Serum insulin levels decline with increasing ketone body formation.

In malignancy, several biochemical changes occur. The most important carbohydrate abnormalities are insulin resistance, increased glucose synthesis, gluconeogenesis, increased Cori cycle activity, and decreased glucose tolerance and turnover (Mantovani et al. 2001). Most solid tumors produce large amounts of lactate, which is converted back to glucose in the liver. This cyclic metabolic pathway, in which glucose is converted back to lactate by glycolysis and then reconverted to glucose in the liver, is known as the Cori cycle. Abnormal elevations in Cori cycle activity, which have been noted in malnourished cancer patients, are reported to account for up to 300 kcal/d loss of energy (Eden et al. 1984). Gluconeogenesis, or the production of glucose in the liver from non-glucose sources such as lactate, uses ATP molecules and is very inefficient for the host. This is known as futile cycling and may be responsible, at least in part, for the increased energy expenditure seen in many cancer patients (Inui 2002). A 40% increase in hepatic glucose production has been reported in weight-losing cancer patients, in contrast to the reduced production seen in patients with anorexia nervosa. Thus, changes in carbohydrate metabolism in cancer patients probably arise as a consequence of meeting the metabolic demands of the tumor, and may contribute to the development of the cachectic state (Tisdale 2000).

In cancer cachexia, amino acids are not spared as they are during simple starvation, and depletion of lean body mass occurs. Muscle wasting may be due to increased protein catabolism (hypercatabolism) or decreased protein synthesis; the simultaneous presence of both results in the most intense muscular atrophy. Simple anorexia alone cannot fully explain wasting of lean body mass and increased protein breakdown observed in cancer cachexia. In some animal models, cachexia develops in the complete absence of anorexia. It seems likely that cancer cachexia is provoked by chemical mediators originating from the host and/or the tumor, including insulin, insulin-like growth factor, growth hormone, glucagons, glucocorticosteroids, ketone bodies, arginine, β-adrenergic agonists, prostaglandins, interferon, interleukins, tumor necrosis factor, and roteolysis-inducing glycoprotein (Baracos 2000).

Alterations in lipid metabolism also occur in the presence of malignancy. Fat is the body's primary fuel source, both under normal physiologic conditions and during simple starvation. Abnormalities that occur in the presence of cancer include increased lipid metabolism, decreased lipogenesis, and decreased activity of lipoprotein lipase (LPL), the enzyme responsible for triglyceride clearance from the plasma (Inui 2002). Mobilization of fatty acids from adipose tissue may occur before weight loss, suggesting the presence of a lipid-mobilizing factor (LMF) produced either by the tumor or host tissues (Tisdale 2000) (see Figure 24.7).

Nutritional Implications of Cancer Treatment Patients who are receiving treatment for cancer may experience nausea, vomiting, early satiety, dysgeusia, diarrhea, mucositis,

FIGURE 24.7 Factors Influencing Development of Anorexia and Cachexia

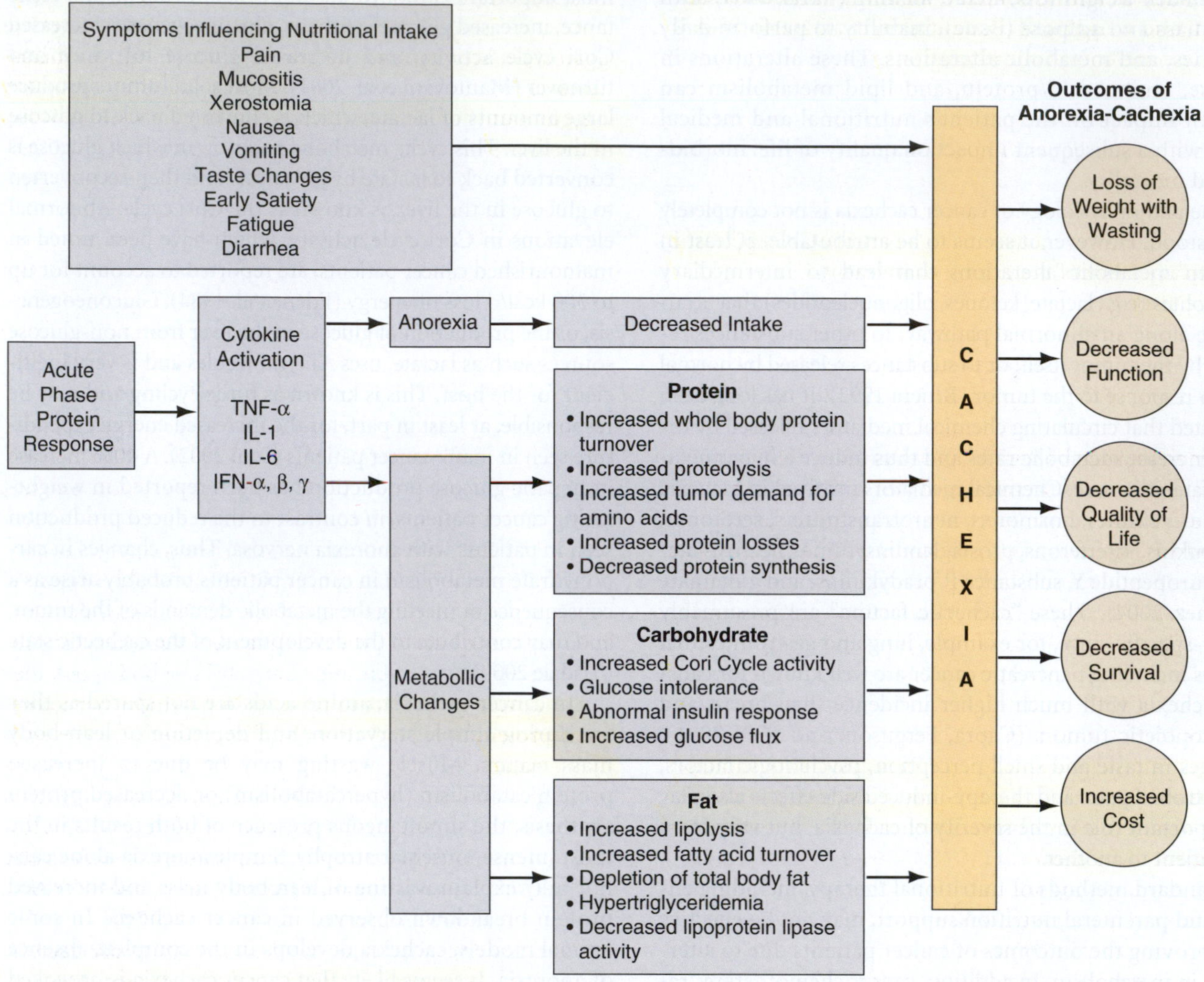

Source: C. Yarbro, M. Frogge, and M. Goodman, *Cancer Symptom Management,* 3e, copyright © 2004, Jones and Bartlett.

xerostomia, constipation, weight loss, and anemia. All of these side effects have the potential to place the individual at nutrition risk, and, if not successfully treated, may result in malnutrition. Each of these symptoms will be covered in more detail in the sections that follow.

Nutrition Interventions

As discussed in Chapters 3 and 5, the nutrition care process identifies patients who are potentially at risk, allows for the early identification of nutrition-related problems, and allows for a nutrition diagnosis to be made from which intervention and nutrition therapy can take place. Aggressive identification and treatment of nutrition-related side effects can stabilize or reverse weight loss in 50% to 88% of oncology patients (Ottery et al. 1998). Malnutrition is a common cause of morbidity and mortality in cancer patients, and is more prevalent in those who have been sick for longer periods of time, those who have had multiple treatments, and those with certain types of cancer including lung, pancreatic, GI cancers, head and neck cancers, and ovarian cancer. Data indicate that malnutrition reduces responsiveness to chemotherapy and RT, increases perioperative morbidity in cancer patients, worsens quality of life, and diminishes the likelihood of survival (Andreyev et al. 1998; Van Bokhorst-de van der Schuer et al. 1999; Bosaeus, Daneryd, and Lundholm 2002).

Nutrition Assessment A recent study showed subjective global assessment (SGA) to be as reliable as other methods of nutrition assessment. The study demonstrated a positive correlation between the use of SGA and objective assessments such as anthropometry, albumin, total cholesterol, weight, and weight loss (Sungurtekin et al. 2004). SGA is a validated method of nutritional assessment based on the features of a medical history (weight change, dietary intake change, and gastrointestinal symptoms that have persisted for more than two weeks; changes in functional capacity) and physical examination (loss of subcutaneous fat, muscle wasting, ankle/sacral edema, and ascites). Biochemical tests or skinfold measurements are not a part of SGA. Application of clinical judgment is a critical component.

The Patient Generated SGA (PG-SGA) and the Scored PG-SGA are recent modifications to the original SGA. The health care professional completes the SGA, while the patient completes the PG-SGA. Both are applicable in the inpatient and outpatient oncology settings. The Scored PG-SGA is a further adaptation of the PG-SGA and allows for triaging of specific nutrition interventions, as well as facilitating quantitative outcomes data collection (McCallum 2000). (See Appendix B2 for SGA, PG-SGA, Scored PG-SGA.) The Scored PG-SGA has been shown to be accurate at distinguishing well-nourished patients from malnourished patients, and has a high sensitivity and specificity. It is a quick, valid, and reliable nutrition assessment tool that enables malnourished hospital patients with cancer to be identified and triaged for nutrition support (Bauer, Capra, and Ferguson 2002).

Accurate height and weight measurements should be performed as part of an initial nutrition assessment. A detailed weight history obtained either from the patient or a significant other allows the clinical dietitian to determine if weight has changed significantly in the past six months. Usual or pre-illness weight should be obtained. If weight loss has occurred, a determination is made as to whether weight loss was voluntary or involuntary. If weight loss occurred involuntarily, the cause of the weight loss should be investigated. In the cancer patient, weight change may occur due to any one or a combination of the following: anorexia, depression, surgery, anxiety, nausea, vomiting, taste changes, xerostomia, diarrhea, or constipation. Once a weight history has been established, current weight needs to be documented. Weighing a patient can be achieved using a variety of standard scales available in the facility. A chair scale can be employed if a cancer patient is physically weak and lacks the strength to stand on a scale. Many outpatient clinics, as well as inpatient facilities, have chair scales available. If a patient has edema (sacral, pedal, ascites), this should be documented and taken into account when assessing current weight.

All weight changes should be considered within the context of time. The Scored PG-SGA includes criteria with which to evaluate weight changes. Weight loss as well as weight gain (which could signify edema) can have important nutritional implications. Lack of weight loss, however, may not necessarily indicate that lean body mass has not been lost. In patients with ovarian cancer, for example, a significant amount of lean body mass may have been lost due to anorexia and poor food intake due to early satiety (usually as a result of the growing tumor which pushes up against the stomach wall). However, ovarian cancer metastases to the liver may cause significant ascites, and thus weight gain due to fluid retention. The gain from ascites may negate the loss from lean body mass; therefore, it is possible that neither the patient nor the health care professional will detect a weight change.

Anthropometric measurements that are also useful include skinfold measurements (to measure subcutaneous fat) and mid-arm muscle circumference (to assess lean body mass). Serial measurements are useful when monitoring weight to determine if fat and/or lean body mass is being lost or accrued. These measurements, while providing useful information, need to be assessed cautiously in cancer patients, because the "norms" on which they are based represent healthy individuals and hence may not have direct applications to the cancer population.

Serum hepatic proteins, such as albumin, prealbumin, and transferrin, have been, historically, the most commonly monitored and assessed biochemical markers utilized in the nutrition assessment process. Because serum hepatic proteins have a long half-life and relatively large body pool, they are relatively insensitive to changes in nutrition. Serum hepatic protein levels can assist the clinical dietitian in identifying patients who are the most ill and thus at risk for developing serious nutritional deficits. A patient with a decreased albumin, prealbumin, or transferrin level is less likely to meet energy and nutrient requirements volitionally and therefore will probably require aggressive nutrition therapies (Fuhrman, Charney, and Mueller 2004).

Serum albumin levels are affected by many factors. These include changes in plasma volume (for example, intravenous fluids given in preparation for chemotherapy—dehydration secondary to nausea, vomiting, diarrhea, mucositis), GI bleeding, severe diarrhea, renal and liver disease, burns, massive trauma, blood losses (such as those that occur during surgery or with trauma), and chemotherapy. Because serum albumin is affected by so many different factors, some or all of which may be present in the cancer patient, it may not be the best biochemical tool with which to assess nutritional status in these patients.

Nonetheless, serum albumin has been shown to be an excellent predictor of survival in patients with cancer. Several studies have shown that a serum albumin level below normal can be used to predict disease outcomes in many groups of patients, including those with Hodgkin's disease and lung cancer (Tayek 1999). Low serum hepatic protein concentration is not necessarily indicative of nutrient deficit, but is more likely a response to inflammatory processes.

C-reactive protein (CRP) is the most sensitive indicator of inflammation because it increases in serum concentration as much as 1,000-fold (see Chapter 5 for a full discussion of CRP). Testing for the serum concentration of CRP at baseline may identify a subset of patients for whom a decline in nutritional status is linked to the presence of an active inflammatory response, a recognized precursor of cachexia (Slaviero et al. 2003).

Delayed cutaneous hypersensitivity (DCH) has been used to measure immunological competence and to indirectly measure nutritional status. Because DCH is significantly affected by a compromised immune system, and many cancer patients have depressed immune function secondary to the disease process, to chemotherapy, or to radiation, use of DCH as part of the nutrition assessment process is not appropriate for oncology patients.

A thorough assessment of the patient's clinical signs and symptoms, as well as previous therapies, is imperative before determining a nutrition care plan. A complete physical assessment should include observation for signs of edema, ascites, temporal lobe and muscle wasting (especially in all four extremities). Gastrointestinal tract assessment includes determining whether there is any history of anorexia, change in appetite, nausea and vomiting, diarrhea, constipation, abdominal pain, early satiety, mouth sores (mucositis), taste changes, or dysphagia. A complete oral assessment should also be completed to determine the individual's ability to chew and to assess the condition of the teeth, gums, and tongue.

Upon initial assessment, a review of an individual's most recent dietary intake is important to determine what kinds of foods can be tolerated, and what, if any, special diets he or she is on, including the use of alternative diets, herbal therapy, vitamin supplements, or nutritional supplements. Many individuals seek out complementary and alternative medicine (CAM) therapies (see Chapter 4) after a cancer diagnosis and may or may not report this to their health care providers. It is important to carefully obtain detailed information regarding alternative therapies and special supplements, because some therapies may be nutritionally inadequate, interfere with conventional therapy, and be harmful to the individual. By developing a strong rapport with the patient, the healthcare professional can elicit the proper information without making the patient feel as if he or she is "doing something wrong or dangerous." Patients are more likely to divulge information if they feel that they can trust the health care professional. The primary priority for the clinical dietitian is to assess whether the alternative therapy has the potential to cause harm. See Appendix F, Table F6 for a list of the numerous contemporary and alternative remedies associated with neoplastic disease, and Box 24.3 outlines the potentially harmful alternative treatments most commonly used by cancer patients. Finally, it is important to determine if the individual has tried any of the several liquid nutritional supplements available on the market and, if so, which one(s) they preferred and/or disliked.

Determining Nutrient Requirements After the patient is screened and assessed, a nutrition care plan should be devised that is individualized to that particular patient's identified nutrition diagnoses and that takes into account the medical diagnosis and treatment(s) that may have nutritional implications. See the nutrition care process feature at the end of this chapter for examples of nutrition diagnoses that may be commonly identified in cancer patients.

Provision of adequate kcalories is essential to maintain current weight and/or prevent treatment- or disease-associated loss. The following energy needs equations may be used (in addition to the Harris-Benedict or Mifflin-St. Jeor equations) to determine kcalorie needs of cancer patients (Martin 2000):

- Obese patients: 21–25 kcalories/kg
- Non-ambulatory or sedentary adults: 25–30 kcalories/kg
- Slightly hypermetabolic patients or those patients who need to gain weight, or are anabolic: 30–35 kcalories/kg
- Hypermetabolic or severely stressed patients or those with malabsorption: 35 kcalories/kg or greater as needed

Meeting protein needs is important to prevent or reduce negative nitrogen balance and to meet protein synthesis needs. Protein needs are especially elevated in those patients with severe diarrhea and/or malabsorption. Protein needs may be calculated based on body weight (kg) using the following guidelines (Martin 2000):

- Normal or maintenance protein needs: 0.8–1.0 g/kg
- Non-stressed cancer patients: 1.0–1.5 g/kg
- Bone marrow transplant or HSCT patients: 1.5 g/kg
- Increased protein needs (protein-losing enteropathy, hypermetabolism, extreme wasting): 1.5–2.5 g/kg
- Hepatic or renal compromise including BUN approaching 100 mg/dL or elevated ammonia: 0.5–0.8 g/kg

Many cancer patients, especially those undergoing chemotherapy and/or radiation, can become dehydrated easily. Those patients receiving chemotherapeutic agents that damage the GI mucosa and cause diarrhea are at particularly high risk for developing dehydration. Patients undergoing radiation to the head and neck area are also at high risk for dehydration due to their inability to take adequate oral fluids secondary to pain and inflammation of the mouth, throat, and esophagus. High-risk patients need to be assessed frequently for signs and symptoms of dehydration (dark, concentrated urine, decreased urine output, dry mouth, acute weight loss). Fluid needs can be calculated using the same formulas used for most other patients without renal disease (30–35 mL/kg).

Vitamins and minerals act as cofactors for essential processes both in health and illness. Deficiencies of vitamins (especially folate, vitamin C, and retinol) and minerals (magnesium, zinc, copper, and iron) can occur in cancer patients due to the direct effects of the tumor, effects of cytokines, infectious processes, chemotherapy, radiation, or inadequate food intake. Micronutrient requirements have not been established for those individuals diagnosed with cancer, because the precise needs of these patients have not been well documented. Use of a daily multivitamin and mineral supplement that contains <150% of the DRI may be beneficial for most patients undergoing chemotherapy and/or radiation therapies.

Nausea and Vomiting Nausea/vomiting is one the most common side effects that occurs as a result of oncologic therapies and can be debilitating.

Nausea/vomiting has multifactorial etiologies (see Chapter 16). Causes of nausea and vomiting in cancer patients include chemotherapy, radiation, narcotic analgesics, odors (including food odors, perfumes), and delayed gastric emptying. Nausea and vomiting associated with chemotherapy can be classified as acute, delayed, or anticipatory (Schnell 2003). Acute nausea and vomiting occur within 24 hours of administration of chemotherapy. The most **emetogenic** chemotherapeutic agents include cisplatin, methotrexate, doxorubicin, and cyclophosphamide (see Table 24.4). Delayed nausea and vomiting usually begin 24 hours after the chemotherapy has been administered and may last up to a week. Delayed nausea and vomiting are most commonly seen after the administration of cisplatin, carboplatin, cyclophosphamide, or doxorubicin (Gralla et al. 1999). Anticipatory nausea and vomiting most commonly occur before the initiation of chemotherapy, but may also occur during or after the initiation of chemotherapy. This type of nausea and vomiting often results from inadequate prevention and/or poorly controlled nausea and vomiting during the first chemotherapy and is more commonly seen in pediatric patients. Nausea and vomiting related to RT are dependent on the field being irradiated. Almost 100% of patients undergoing total body irradiation (TBI) during bone marrow transplantation experience emesis, while radiation of the cranium only is considered low risk (about 10% to 30% of patients experience emesis) (Gralla et al. 1999). Upper- and mid-abdominal RT can also result in nausea and vomiting starting 1–2 hours after treatment and persisting for several hours (Harding, Young, and Anno 1993).

A thorough assessment of the causes of nausea and vomiting will help with treatment. Patients who are experiencing nausea and vomiting due to certain odors are encouraged to take precautions in avoiding noxious odors. Nausea from cooking odors can be minimized by using a microwave oven, opening windows when cooking, taking a walk when meals are being cooked, and avoiding frying of foods, which emits more odors than most other forms of cooking. Patients should ask friends and family members to avoid perfumes when they are visiting. In addition, a patient's medication list should be reviewed for potential causes of nausea and vomiting. A common cause of nausea and vomiting is the use of narcotic analgesics (morphine, codeine, fentanyl), which are prescribed for many cancer patients for chronic pain. Usually the nausea and vomiting that result from these agents occurs acutely at the beginning of therapy and resolves with chronic use. Other medications known to cause nausea and vomiting include antibiotics, digoxin, and anticholinergic agents. Assessing the patient for signs and symptoms of early satiety is also important. Delayed gastric emptying can result in nausea and vomiting. Small, frequent meals may be helpful, as well as the administration of prokinetics.

By far, the most common cause of nausea and vomiting in cancer patients is chemotherapy; this is referred to as chemotherapy-induced nausea and vomiting (CINV). A thorough review of the patient's chemotherapy regimen will assist the dietitian in determining the severity of the nausea and vomiting that can be anticipated. The patient should be advised to eat only a small, low-fat meal the morning of the first treatment and to avoid fried, greasy, and favorite foods for several days following the treatment. A clear liquid diet for the first few days after therapy may be indicated. To provide kcalories and maintain hydration, consumption of electrolyte-fortified beverages such as Gatorade; nutritional fruit beverages such as Resource (Novartis), Enlive (Ross Laboratories), and NuBasics (Nestlé); and non-acidic fruit drinks (apple and grape juice, nectars) should be encouraged. It is important for patients to avoid favorite foods at any time the chance for emesis is high, since once a favorite food has been vomited, the likelihood of its subsequent consumption is low. The same principle applies to the use of "creamy" liquid nutritional drinks. A patient who has vomited a nutritional beverage will associate vomiting with that beverage, even if told that their vomiting was probably caused by chemotherapy. The likelihood that the clinical dietitian will be able to encourage the use of nutritional drinks later in therapy for weight gain is unlikely.

Whenever possible, the clinical dietitian should play an active role in assisting the patient to avoid the onset or at least minimize the occurrence of nausea and vomiting. One important intervention is to encourage patients to take their anti-emetics as instructed by their physician. To encourage adequate intake and maximal control of nausea and vomiting, anti-emetics should be taken at least 30–45 minutes before a meal is consumed. Patients should be encouraged to take their anti-emetics even if they do not feel nauseated at the time, especially while actively receiving treatment.

emetogenic—an agent that causes nausea and/or vomiting

BOX 24.4 CLINICAL APPLICATIONS

Popular Alternative Therapies used by Cancer Patients

	Chemical Components	Clinical Efficacy as Cancer Therapy	Potential for Adverse Reactions	Caution
Pau D'Arco	Quinine compounds	No published clinical evidence	High—includes mild to moderate nausea, vomiting + anticoagulant effects	People taking anticoagulants and patients with thrombocytopenia
DHEA	Steroid precursor (pregnenolone)	May stimulate cancer growth in patients with hormone sensitive cancers including prostate, breast and endometrial	High—aggressiveness, fatigue, headache, insomnia, elevated liver function tests, decreased Hgb and RBC	Patients with cancer sensitive hormones should avoid; decreases HDL levels and may increase the risk for heart disease
Goldenseal	Alkaloids	No published clinical evidence	High—can produce significant changes in blood pressure + anticoagulant effects	Patients on antihypertensive medications (beta-blockers, calcium channel blockers, digoxin) should avoid
Mistletoe (Iscador)	Amines, including tyramine, histamine, acetylcholine, beta-phentolamine	Mistletoe does demonstrate immune system modulation and antineoplastic activity—studies too small and short duration to demonstrate evidence	High—may increase hypotensive effects of antihypertensive medications; CNS depressant	Mistletoe plant is toxic and should not be used as a home remedy
Kombucha Tea	Made by incubating the Kombucha mushroom in black tea	No published clinical evidence	Several deaths have been reported after ingestion	Mushrooms may be contaminated with potentially pathogenic bacteria
Astragalus	Betaine, beta-sitosterol, choline, glycosides, plant acids	Demonstrates immune enhancing properties, however, large clinical trials have not been conducted	Low—may increase hypotensive effects of antihypertensive medications	Those patients who are taking immunosuppressants and those with autoimmune disorders should avoid
714X	Nitrogen, camphor, ammonia, ethanol	No published clinical evidence	Moderate—flu-like symptoms, inflammation at injection site	FDA has placed an import ban on this compound
PC-SPES	Combination of chrysanthemum, isatis, licorice, *Ganoderma lucidum*, *Panax pseudoginseng*, *Rabdosia rubescens*, saw palmetto, and skullcap	No published clinical evidence	High—similar to hormonal drugs: gynecomastia, dyspepsia, nausea, fatigue, leg cramps, diarrhea, angina, hot flashes, and thromboembolic effects	Recalled and removed from the market by the FDA after several batches were found to be contaminated with prescription drugs

(continued on the following page)

Effective and consistent use of anti-emetic agents by the patient with nausea and vomiting is important for the control of both delayed and anticipatory CINV (Schnell 2003). In addition, control of nausea and vomiting may assist in maintaining a patient's nutritional intake. Anti-emetics are classified as 5-HT$_3$ receptor antagonists, dopamine receptor antagonists, corticosteroids, cannabinoids, and benzodiazepines. The 5-HT$_3$ receptor antagonists, developed in the mid-1980s, have become the accepted gold standard for the control of CINV. 5-HT$_3$ receptor antagonists specifically block the binding of serotonin to the receptors on the vagal nerve that trigger the emetic response (Doherty 1999). They include Granisetron, Ondansetron, Dolasetron, and Palonosetron. Corticosteroids, such as prednisone and dexamethasone, are occasionally used in combination with other anti-emetics for short-term control of nausea and vomiting; however, their long-term use is contraindicated due to their significant adverse effects, including immunosuppression, hyperglycemia, osteoporosis, skeletal muscle wasting, GI irritation, and mood changes. Dronabinol (Marinol®) is a cannabinoid that contains delta-9-tetrahydrocannabinol and is FDA approved for the treatment of CINV. Cannabinoids, however, are only used in selected patients due to the cultural and societal constraints as well as their relatively low therapeutic index (Walsh, Nelson, and Mahmoud 2003).

BOX 24.4 *(continued)*

	Chemical Components	Clinical Efficacy as Cancer Therapy	Potential for Adverse Reactions	Caution
Essiac	Usually prepared as a tea; contains burdock root, rhubarb root, sheep sorrel, and slippery elm bark	No published clinical evidence	Moderate—nausea, vomiting, increased urination, flu-like symptoms	
Laetrile	Amygdalin, cyanide	No controlled clinical trials have ever been conducted	High—attributed to cyanide poisoning and include nausea, vomiting, headache, dizziness, low blood pressure, bluish discoloration of the skin due to lack of oxygen in the blood, liver damage, fever, mental confusion, coma, death	Not approved by the FDA as a cancer treatment in the US, preparations have been found to be contaminated by bacteria
Cartilage (bovine, shark)	Collagen, glycosaminoglycans	Inhibitors of angiogenesis have been found in cartilage; ongoing clinical trials in patients with advanced cancer, however, the cumulative data is inconclusive regarding its effectiveness as a cancer treatment	Mild to moderate Injectable form: inflammation at injection site, dysgeusia, fatigue,nausea, dyspepsia, and dizziness. Powdered shark cartilage form: nausea, vomiting, abdominal cramping/bloating, constipation, hypotension, hyperglycemia, generalized weakness, and hypercalcemia	More published clinical evidence needed
Milk Thistle	Silymarin (a flavonoid mixture)	No published clinical trials in patients with cancer	Mild to moderate—mild laxative effect, allergic reaction	Decreases the activity of the cytochrome P450 enzyme system and may affect the clearance of certain chemotherapeutic agents
Hydrazine sulfate	Hydrazine sulfate	No evidence of anticancer activity in randomized clinical trials	Mild to moderate—nausea, vomiting, dizziness, sensory and motor neuropathies; highly toxic when taken with alcohol or barbiturates	Increases the incidence of lung, liver, and breast tumors in animals
Antineoplaston Therapy	Substances (amino acid derivatives, peptides, and essential amino acids) isolated from normal human blood and urine	No published clinical data	Mild to moderate—nausea, vomiting, flatulence, chills, rashes, fever, joint pain, changes in blood pressure, and body odor during therapy	Research needed

Non-pharmacologic alternative methods that have been used with varying success to prevent or treat nausea and vomiting include acupressure, acupuncture, hypnosis, and guided imagery (see Table 24.4). The website for the National Center for Complementary and Alternative Medicine (http://nccam.nih.gov) is useful for finding complementary and alternative therapies in the treatment of nausea and vomiting, as well as for finding general information about cancer.

Early Satiety A common complaint expressed by cancer patients is "I just can't eat as much as I used to" or "I get full right after I start eating." This describes the symptom of early satiety, which is caused primarily by delayed gastric emptying. It is important for the patient with early satiety to eat small, frequent meals that are nutrient dense. Beverages should also contain nutrients and should be consumed between meals rather than with meals so as not to add to the feeling of fullness. Consumption of raw vegetables, such as salads, and other high-fiber foods should be avoided. Prokinetics, medications that increase gastric emptying, may be useful. Metoclopramide, for example, is a motility agent that selectively stimulates gastric emptying and may be useful for the patient with early satiety. A potential side effect of metoclopramide is diarrhea; therefore, it should not be used by patients that already have loose stools.

TABLE 24.4

Relative Emetogenic Potential of Antineoplastic Drugs

Emetogenic Potential (% of patients)	Agent	Dosage	Onset/Duration of Response (hours)
High (>90%)	Cisplatin	>50 mg/m^2	1.5–56
	Cyclophosphamide	>1,000 mg/m^2	9–28
	Dacarbazine		4–24
	Mechlorethamine		0.5–24
Moderately high (60%–90%)	Cisplatin	<50 mg/m^2	1.5–56
	Cyclophosphamide	750–1,000 mg/m^2	9–28
	Methotrexate	>1,000 mg/m^2	4–12
	Carboplatin		6–46
	Doxorubicin	>60 mg/m^2	3.5–34
Moderate (30%–60%)	Cyclophosphamide	<750 mg/m^2	9–28
	Methotrexate	250–1,000 mg/m^2	4–12
	Doxorubicin	<60 mg/m^2	3.5–34
Moderately low (10%–30%)	Methotrexate	<250 mg/m^2	4–12
	Fluorouracil		3–10
	Etoposide		3.5–34
Low (<10%)	Hydroxyurea		8–48
	Vinblastine		3.5–34
	Bleomycin		3.5–24
	Tamoxifen		12–36
	Chlorambucil		48–56

Reprinted from *Clin J Oncol Nurs* 1999;3:113–119.

Mucositis Mucositis, also known as stomatitis, is irritation and inflammation of the epithelial cells of the mucosal membranes lining the gastrointestinal tract that can occur at any point in the GI tract from the mouth to the anus. Its manifestations may range from generalized swelling and inflammation to obvious ulceration and hemorrhage (Kwong 2004). Mucositis-associated pain is the main source of cancer treatment-related pain, which afflicts from 40% to 70% of patients receiving chemotherapy or radiotherapy (Berger et al. 1995). It has been proposed that mucositis is related to direct and indirect cytotoxicity, local tissue cytokine and immune activity, and bacterial colonization of the ulcerative lesion (Sonis 1998). Refer to Tables 24.3 and 24.4 for a listing of the most emetogenic chemotherapeutic agents. The growth factor filgrastim may also cause mucositis. Other causes of mucositis include viral, bacterial and fungal infections, radiation, stem cell transplant therapy, and graft-versus-host disease.

The patient with oral mucositis should have a thorough and systematic assessment of the mouth. Alterations in the oral mucosa may include color changes of the tongue, lips and gingiva, changes in moisture, and changes in integrity, including cracks, fissures, ulcers, blisters and lesions. The presence of white plaques is generally indicative of fungal infections such as candidiasis. The disruption of the mucosal barrier in the oral cavity increases the risk of infections. Chemotherapy-induced mucositis commonly occurs five to seven days after chemotherapy is initiated and may continue until the patient recovers from the immunosuppression (referred to as the **nadir**) or until the **absolute neutrophil count** is >500. Symptoms will include pain and burning with

nadir—the lowest point, usually in reference to the white blood cell count

absolute neutrophil count (ANC)—a measure of the number of neutrophil granulocytes (also known as polymorphonuclear cells, PMNs, polys, granulocytes, segmented neutrophils, or segs) present in the blood. Neutrophils are a type of white blood cell that fights against infection

BOX 24.5 CLINICAL APPLICATIONS

Guidelines for Preventing and Treating Mouth Sores

Check mouth twice a day using a small flashlight and padded popsicle stick. If you wear dentures, remove them before inspecting your mouth. Report any changes in appearance, taste, or feeling to your doctor or nurse.

Follow this plan for mouth care 30 minutes after eating and every four hours while awake:

- Brush your teeth using a soft nylon bristle toothbrush. To soften the bristles even more, soak the brush in hot water before brushing and rinse brush with hot water during brushing. If the toothbrush hurts, use a popsicle stick with gauze wrapped around it or a cotton swab instead.
- Rinse toothbrush well after use and store in a cool, dry place.
- Use a nonabrasive toothpaste or a baking soda solution.
- Remove and clean dentures between meals on a regular time schedule. If sores are under dentures, leave dentures out between meals and at night. Clean dentures well between uses.
- Gently rinse mouth before and after meals and at bed time with one of the following solutions (swish solution around mouth and gently gargle, then spit out):

  ½ teaspoon baking soda

  2 cups water

  or

  ½ teaspoon salt

  1 teaspoon baking soda

  1 quart water
- Avoid commercial mouthwashes, which often contain alcohol or other irritants.
- Keep lips moist with petroleum jelly, mild lip balm, or cocoa butter.
- Drink at least two to three quarts of fluids daily with approval from your doctor.
- If mouth pain is severe or interferes with eating, ask your doctor about medicine that can be swished 15–20 minutes before meals or painted on each painful sore with a cotton swab before meals. Hold the solution in your mouth for several minutes before swallowing or spitting it out.
- To promote healing ask your doctor about: Maalox or Milk of Magnesia. (Allow this to settle and separate, pour the liquid off the top of the solution, and swab the pasty part onto the sore area with a cotton swab. Rinse with water after 15–20 minutes.)
- Sip warm tea slowly.
- Rinse mouth with 1 teaspoon of Kaopectate, then spit out.
- Eat chilled foods and fluids (e.g., popsicles, ice cubes, frozen yogurt, sherbet, ice cream).
- Eat soft foods that are moist and easy to swallow.
- Eat small, frequent meals of bland, nonspicy foods. Avoid raw vegetable and fruits, and other hard or crusty foods such as chips or pretzels.
- Avoid acid fruits and juices, such as tomato, orange, grapefruit, lime, or lemon.
- Avoid carbonated drinks.
- Create a pleasant mealtime atmosphere.

Reprinted from: American Cancer Society [homepage on the Internet]. Philadelphia: American Cancer Society [cited 2005 Oct 31]. Mouth Sores: What the Patient Can Do. Available from: http://www.cancer.org/docroot/MBC/content/MBC_2_3x_Mouth_Sores.asp.

chewing and swallowing. Mucositis may be severe enough to cause the patient to completely forgo any food or fluids, which can lead to dehydration and acute weight loss. Good oral hygiene is important for the patient with oral mucositis in order to prevent infection. Box 24.5 outlines the ACS recommendations for preventing and treating mouths sores. Oral glutamine has been utilized for the prevention and treatment of oral mucositis; however, at the present time there is insufficient evidence to make recommendations for patients.

Narcotic analgesics may be required for pain. Topical therapies including agents such as sucralfate, nystatin, and clotrimazole troches also help treat pain and infections, but may cause taste changes. Patients with oral mucositis may need nutrition education to provide guidelines for eating until the mucositis resolves. The patient should be encouraged to eat only soft, non-fibrous, non-acidic foods. Hot foods should be avoided as they can burn the already tender, fragile mucosa. Liquids should be encouraged to prevent dehydration; non-acidic juices such as nectars may be helpful. High-kcalorie, high-protein milkshakes or nutritional supplements may be beneficial at this time (see Table 24.5).

Diarrhea Because antineoplastic agents target those cells that have the highest replication rate, they often cause diarrhea. In the GI tract, antineoplastic agents, especially antimetabolites (e.g., 5-fluorouracil), inhibit mitosis in rapidly proliferating crypt cells, leading to a disproportionate increase in the number of immature crypt cells (Ippoliti 1998).

When mucositis is present in the oral mucosa, it can be assumed that it may also be present in the stomach and in the small and large intestine, resulting in diarrhea, which may at times become severe. Dehydration can occur rapidly. The patient with diarrhea should be encouraged to drink small amounts of fluid frequently throughout the day. Large amounts of fruit juices should be avoided as excessive fructose can exacerbate diarrhea. Gatorade®, Pedialyte®, clear liquid

TABLE 24.5

High-Kilocalorie, High-Protein Nutritional Beverages

	Manufacturer	Kcalories (per 240 mL)*	Protein (g/240 mL)*
Ensure HP	Ross	230	12
Ensure Plus	Ross	360	13
Ensure Plus HN	Ross	355	15
Prosure Shake	Ross	300	17
Boost	Novartis	240	10
Boost High Protein	Novartis	240	15
Boost Plus	Novartis	360	14
Resource Standard	Novartis	250	9
Resource Plus	Novartis	360	13
Resource 2.0	Novartis	475	21
Resource Fruit Beverage	Novartis	250	9
Resource Nutritious Juice Drink 6 oz	Novartis	210	6
Resource Healthshake			
4 oz	Novartis	200	6
6 oz		300	9
Resource Shake Plus	Novartis	480	15
Scandishake (made with 8 oz whole milk)	Scandipharm	600	12
Carnation Instant Breakfast drink (made with 8 oz whole milk)	Nestlé	300	12
Carnation Instant Breakfast Plus	Nestlé	375	12
Carnation Instant Breakfast VHC	Nestlé	560	22.5
Replete	Nestlé	250	15.6
Probalance	Nestlé	300	13.5
Nutren 1.5	Nestlé	375	15
Nutren 2.0	Nestlé	500	20

* Unless otherwise noted.

nutritional beverages, and other oral rehydration fluids are recommended. Patients should be encouraged to use the antidiarrheal medications as prescribed by their physicians. Antidiarrheals include loperamide (Imodium®) and diphenoxylate (Lomotil®). Instructing the patient to increase their intake of foods high in soluble fiber may help with the treatment of diarrhea; however, often these patients have a poor appetite and may have a difficult time increasing their intake of foods in general. (For further details on nutrition therapy for diarrhea, see Chapter 17 and Appendices E12,13.)

Dysgeusia "Meats have a bitter taste." "My morning coffee just doesn't taste the same." These common complaints may be related to another typical nutritional problem in cancer patients—dysgeusia. Dysgeusia, or alterations in taste, can have a profound effect on a patient's ability to ingest an adequate amount of nutrition. The presence of certain tumors can elicit taste changes even before a diagnosis is made. Many chemotherapeutic agents, specifically cisplatin, and radiation to the head and neck area, cause dysgeusia. Taste changes that occur include a metallic taste (usually due to the chemotherapeutic agent cisplatin), no taste sensation (aguesia), a heightening of certain tastes (especially sweets), or aversions to foods the patient liked to eat in the past.

Patients who experience a metallic taste in their mouth should be advised to avoid metal utensils and instead use plastic utensils. If nutritional supplements are consumed, they should be poured into a glass first, as often the metal container may also be offensive. Meats are often not tolerated. To ensure an adequate protein intake, the patient should be encouraged to incorporate other high-protein foods into the diet, including peanut butter, cottage cheese, cheese, poultry, and soy meat substitutes. Patients with aguesia should be encouraged to use more highly spiced and flavorful foods, such as marinated foods. Sweet foods often taste too sweet to individuals undergoing cancer therapy. Many homemade drinks and nutritional beverages may be too sweet for these patients. Alternative options may be to have the patient try a non-sweet supplement such as Osmolite® (Ross Laboratories) or one that is juice or yogurt based.

Xerostomia Xerostomia, a sensation of dry mouth, is a common side effect of head and neck radiation and chemotherapy (methotrexate, 5-fluorouracil, paclitaxel, carboplatin, cisplatin). Other causes of xerostomia include dehydration, chronic graft-versus-host disease of the GI tract, medications (narcotic analgesics, antianxiety agents, antihistamines, beta blockers, antidepressants, diuretics), Sjögren's syndrome, and aging. The severity of xerostomia is correlated with the severity of oral discomfort, dysgeusia, dysphagia, and **dysphonia.** Drugs used to treat cancer can make saliva thicker, causing the mouth to feel dry (Kwong 2004).

Treatment of xerostomia may include use of artificial saliva (saliva substitutes) and/or mouth moisturizers. There are several artificial salivas available on the market; however, patient compliance may be a problem due to their consistency, taste, and cost (Porter, Scully, and Hegarty 2004). In addition, the duration of action is short as they are quickly removed from the mouth with swallowing. Mouth moisturizing lubricants come in the form of gels, lozenges, and mouthwashes. Sugar-free gum and sour-flavored sugar-free hard candies may help increase the flow of saliva in the mouth and are less expensive than artificial saliva. One study

dysphonia—difficulty speaking

found chewing gum to be more effective than artificial saliva for the treatment of radiation-induced xerostomia (Davies 2000). Denture wearers may not be able to chew gum for the treatment of xerostomia.

Anorexia Lack of appetite, or anorexia, is a challenging problem for both patients and clinical dietitians. The prevalence of anorexia in cancer patients is estimated at approximately 50% of patients upon diagnosis. While the exact prevalence is unknown, it is generally acknowledged that anorexia and reduced food intake are frequent occurrences in cancer patients (Laviano, Meguid, and Rossi-Fanelli 2003). Anorexia has multiple etiologies in the cancer patient, including circulating cytokines, hormones, depression, therapy (surgery, radiation, chemotherapy), learned food aversions, fatigue, and certain medications. Chronic anorexia and reduced energy intake can lead to weight loss and exacerbate the development of cancer cachexia. In general, patients with hematologic malignancies and breast cancer seldom exhibit weight loss and cachexia; however, most other solid tumors, including those of the GI tract, head and neck cancer, and lung cancer, are associated with anorexia and cachexia (Inui 2002). While nutrition therapy for the treatment of anorexia (see Table 24.6) may be helpful for some patients, for the majority, manipulation of the diet does little to help improve a poor appetite. Exercise may help to increase appetite, but many patients may be unable to increase their physical activity for a variety of reasons, including profound fatigue, severe thrombocytopenia (platelets $<20,000\ \mu^3$), severe immunosuppression, and side effects from therapy, such as nausea, vomiting, or diarrhea. Exercise, on the other hand, may actually relieve fatigue, prevent muscle wasting, and improve the ability to perform activities of daily living by improving endurance levels.

Pharmacologic interventions have been found to be relatively useful for stimulating appetite in cancer patients. To date, two agents have been recognized as useful for appetite stimulation: megestrol acetate and corticosteroids agents. Agents that have been found to be less effective or unproven include dronabinol, cyproheptadine, hydrazine, and metoclopramide.

Megestrol acetate is available in tablet form and as a liquid suspension. Patient compliance will be significantly enhanced when the liquid suspension is taken as opposed to tablets. The optimal dose is 800 mg/day, which can be provided in 20 mL liquid suspension, whereas the tablets contain only 20 mg each, meaning the patient must take 40 tablets to achieve a full appetite stimulating dose. Appetite may improve in as little as 24 hours after initial administration. Adverse effects of these drugs include hyperglycemia, peripheral edema, increased risk of thromboembolic events, breakthrough uterine bleeding, hypertension, and Cushing's syndrome (Jatoi et al. 2002).

Corticosteroids (dexamethasone, prednisone, methylprednisolone) are known appetite stimulants. Although several randomized, placebo-controlled studies demonstrated that corticosteroids induce a usually temporary (limited to a few weeks) effect on indicators such as appetite, food intake, sensation of well-being, and performance status, none of the studies showed a beneficial effect on body weight (Moertel et al. 1974; Willox et al. 1984, Bruera et al. 1985; Robustelli Della Cuna, Pellagrini, and Piazzi 1989; Mantovani et al. 2001). Corticosteroids have multiple adverse side effects, including sodium retention, fluid retention, hyperglycemia, lean body mass wasting, hypertension, Cushing's syndrome, osteoporosis, immunosuppression, and delayed wound healing. They should thus be used cautiously in cancer patients.

TABLE 24.6

Nutrition Therapy for the Treatment of Anorexia
Eat small, frequent meals.
Eat at times when appetite is most normal.
Limit fluid with meals to avoid feeling of fullness.
Keep favorite foods readily available at all times.
Mild exercise, as tolerated (check with physician).
Eat meals in a pleasant environment.
A glass of wine before a meal may help to stimulate the appetite (check with the physician first).
Avoid noxious odors; ventilate eating area.
Find a nutritional supplement that is appealing and drink only 2–4 ounces at a time (to avoid a feeling of fullness); keep unopened beverage in the refrigerator.
Try relaxation exercises before mealtimes.
Consider pharmacologic agents/appetite stimulants.

Pharmacologic agents used in the treatment of weight loss in cancer patients include growth hormone, insulin-like growth factor-1 (IGF-1), testosterone, dihydrotestosterone, and the testosterone analogues oxandrolone and nandrolone decanoate. Oxandrolone is the only one of these approved for use by the FDA for weight gain following disease-related weight loss. It has been used clinically for the treatment of cancer-induced wasting. Oxandrolone is a synthetic agent with a chemical structure similar to that of testosterone and is currently a schedule III controlled substance. It is administered orally and, in general, appears to be safe and well tolerated. Oxandrolone is associated with gains in weight and lean body mass as well as improvements in nitrogen balance and functional status in many patients. It is indicated as an adjuvant therapy to promote weight gain after weight loss following extensive surgery, chronic infection, or severe trauma, and in some patients who, without pathophysiologic reasons, fail to gain or maintain normal weight (Langer, Hoffman, and Ottery 2001). Oxandrolone has been studied in patients with human immunodeficiency virus (HIV), acquired immunodeficiency syndrome (AIDS), chronic obstructive pulmonary disease, and severe muscle wasting due to paraplegia or hemiplegia.

Physical activity is a key factor influencing the maintenance of lean body mass. Whenever appropriate, cancer patients should be advised to engage in some form of exercise. Prolonged inactivity due to bed rest, certain medications (corticosteroids), aging, and weight loss are associated with the loss of skeletal muscle tissue. Many cancer patients, however, may be unable to participate in regular physical activity, either due to their medical condition, fatigue, or advanced age. Physical therapists will be helpful in determining the appropriate activity for individual patients.

In cancer patients, feeding studies have shown generally disappointing results in altering the course of catabolic changes that occur in cancer cachexia (Baracos 2001). Protein anabolism mechanisms are impaired; anabolic hormone levels such as insulin and IGF-1 are decreased, while levels of cortisol, a catabolic hormone, are elevated. In addition, the ability to utilize nutrients effectively appears to be altered in cancer patients.

Several nutrients are being studied to determine if they can play a role in the reversal of cancer cachexia. The use of an oral supplement with a combination of β-hydroxy-β-methylbutyrate (HMB), arginine, and glutamine was effective in increasing fat-free mass of advanced stage cancer patients. The increase in fat-free mass contributed to an overall increase in total body weight of the subjects. Omega-3-fatty acids, specifically eicosapentaenoic acid (EPA), have been shown to have antitumor and anticachectic effects (Beck, Smith, and Tisdale 1991). Administration of omega-3-fatty acids or high purity EPA capsules has been associated with weight stabilization in weight-losing patients with advanced pancreatic cancer (Wigmore et al. 1996, 2000). On the other hand, in a study on 200 patients with unresectable pancreatic cancer, those who were provided with an oral liquid supplement enriched with omega-3-fatty acids and antioxidants failed to show a therapeutic advantage compared with subjects provided with an identical supplement without the omega-3-fatty acids or antioxidants; both supplements were equally effective at arresting weight loss (Fearon et al. 2003). Further studies are needed to determine if omega-3-fatty acids are beneficial for the treatment of cancer cachexia. Patient compliance with nutritional supplements enriched with omega-3-fatty acids may be problematic due to the taste and cost of the product.

Pharmacologic agents are being investigated as potential therapy for the treatment of cancer cachexia. These include anticytokine therapies (pentoxifylline, thalidomide), nonsteroidal anti-inflammatory drugs, melatonin, β_2 agonists, and anabolic agents (growth hormone, insulin-like growth factor, testosterone, dihydrotestosterone, and testosterone analogues such as oxandrolone, discussed previously).

terminal—a condition or disease in which there is no cure

Nutrition Support The use of specialized nutritional support (enteral and parenteral nutrition) has been a controversial area in the field of oncology nutrition. A thorough nutrition assessment is critical when determining whether or not a patient needs nutrition support. The principle "when the gut works, use it" applies to cancer patients. When oral intake continues to decline despite aggressive efforts at dietary and pharmacological interventions, or when the GI tract becomes non-functional, the RD and the health care team managing the patient should consider nutrition support options. While there is no evidence to support the routine use of enteral or parenteral nutrition in well-nourished cancer patients undergoing surgery, chemotherapy, or radiation, enteral nutrition is indicated for any malnourished cancer patient whose GI tract is functional, who is undergoing anticancer therapy, and who has a reasonable prognosis. Nutrition support is considered an aggressive form of therapy and should be utilized only when other aggressive medical approaches (i.e., chemotherapy, surgery, radiation) are also being used to treat the cancer. Nutrition support is inappropriate for most **terminal** cancer patients or for patients with poor prognoses in whom all medical anti-cancer therapies have been exhausted.

The practice guidelines for nutrition support of adults with cancer of the American Society for Parenteral and Enteral Nutrition (ASPEN) include:

- Specialized nutrition support (SNS) should not be used routinely in patients undergoing major cancer operations.
- Preoperative SNS may be beneficial in moderately or severely malnourished patients if administered for 7 to 14 days preoperatively, but the potential benefits of nutrition support must be weighed against the potential risks of the SNS itself and of delaying the operation.
- SNS should not be used routinely as an adjunct to chemotherapy.
- SNS should not be used routinely in patients undergoing head and neck, abdominal, or pelvic irradiation.
- SNS is appropriate in patients receiving active anti-cancer treatment who are malnourished and who are anticipated to be unable to ingest and/or absorb adequate nutrients for a prolonged period of time.
- The palliative use of SNS in terminally ill cancer patients is rarely indicated (American Society for Parenteral and Enteral Nutrition 2002).

Enteral nutrition support may be beneficial in malnourished patients undergoing RT for head and neck cancers. Enteral nutrition may allow patients to complete their entire course of RT without interruption while minimizing weight loss. Nonetheless, there are few clinical trials investigating the routine use of SNS as an adjunct to RT in cancer patients (Klein et al. 1997). Enteral nutrition may also be indicated in patients with head and neck cancer undergoing extensive surgery, patients with esophageal cancer (many of whom suffer from dysphagia) following esophagectomy, and gastric cancer patients following total or subtotal gastrectomy.

Providing parenteral nutrition support to cancer patients with advanced disease has significant ethical implications, increases the risk of metabolic and infectious complications, and is expensive.

Home Nutrition Support Adult cancer patients account for a high percentage of those patients on home nutrition support, despite the lack of evidence that parenteral nutrition support can improve length of survival. In a subsample of 37 home nutrition support programs that have consistently reported their data to the North American Home Parenteral and Enteral Nutrition Patient Registry since 1985, more than 90% of their program growth was accounted for by new patients with active cancer. This is now the largest single diagnosis of patients starting home parenteral and enteral nutrition (Howard 1993).

Some existing data suggest that quality of life may be maintained temporarily at an acceptable level in some patients with advanced cancer on home parenteral nutrition (HPN), but quality of life is a subjective assessment and difficult to apply to all patients. Adult patients with active cancer on home parenteral and enteral nutrition do not have a greater incidence of therapy-related readmissions than other patient groups; however, their overall rehospitalization rate is much higher (Howard 1993).

A patient's lack of appetite and food intake is often a greater source of concern to health care providers, family members, and caretakers than it is to the patient. Caretakers feel a sense of frustration and powerlessness, and feeding the patient gives them some sense of control over their loved one's care and well-being. Fear of death by starvation is common. Orrevall et al. (2004) studied patients with advanced cancer who had received HPN and found that the desperate and chaotic situation in the family influenced a patient's willingness to accept HPN. It is unknown whether HPN can improve a patient's quality of life. Quality of life is very difficult to measure, and few studies have documented quality of life in advanced cancer patients who are on HPN. Bozzetti et al. (2002) studied patients with advanced cancer who received HPN with respect to nutritional status, length of survival, and quality of life. Their study suggests that in malnourished, chronically obstructed patients with advanced cancer resistant to conventional curative therapy, HPN may help to prolong survival past 7 months (in approximately one third of patients), improve their quality of life (in 20% to 40% of them), or at least maintain it until 2 months prior to death.

Before discharging a patient from the hospital, insurance (including Medicare) guidelines regarding HPN reimbursement need to be clarified. Medicare will not approve the use of HPN unless it is documented in the medical record that HPN will be required for at least three months and that enteral nutrition is not a viable means of feeding the patient (Ireton-Jones 2002).

Conclusion

Research has consistently demonstrated that a healthy, nutritious diet that is low in fat, low in saturated fat, and high in fruits, vegetables, and whole grains is important for the prevention of many types of cancers. Physical activity also appears to play a role in cancer prevention. Paying close attention to diet and nutrition is important during the treatment of cancer, because many treatment options can significantly affect a person's nutritional status. Obtaining adequate kcalories, protein, fluids, vitamins, and minerals, and preventing weight loss, which occurs so frequently in cancer patients, may help to improve quality of life for many of those undergoing surgical and/or pharmacologic treatment of cancer. Research in the areas of nutrigenomics and the biologic basis of anorexia-cachexia promises to offer new insight into cancer prevention as well as new treatment options.

Calculation of upper arm muscle area—will need midarm circumference and triceps skinfold

Bioelectrical impedance

Biochemical Assessment

Visceral protein assessment: albumin, prealbumin, retinol binding protein

Hematological assessment: hemoglobin, hematocrit, ferritin, MCV, MCHC, MCH, TIBC, platelet count

White blood cell count, absolute neutrophil count

Step Two: Common Diagnostic Labels

- Inadequate oral food/beverage intake
- Inadequate fluid intake
- Inadequate bioactive substance intake
- Inadequate vitamin intake
- Hypermetabolism
- Increased nutrient needs
- Swallowing difficulty
- Chewing difficulty
- Altered GI function
- Altered nutrition-related laboratory values
- Food-medication interaction
- Involuntary weight loss
- Food, nutrition, nutrition-related knowledge deficit

Sample PES: NI-2.1 Inadequate oral food/beverage intake secondary to mucositis post-radiation as evidenced by dietary history, presenting dehydration and 7% weight change in last two weeks.

Step Three: Sample Intervention

1. Modify texture and consistency of meals avoiding extremes in temperature.
2. Increase nutrient density of foods offered and initiate oral high-kcalorie, high-protein beverage of choice.
3. Support initiation of pain medications prior to eating and adequate, appropriate mouth care.

Step Four: Monitoring and Evaluation

1. The patient will consume 50% of estimated energy and protein needs within 48 hours of initiating interventions for mucositis.
2. Patient will be able to meet basic fluid requirements within 24 hours of initiating interventions for mucositis.

WEB LINKS

American Institute for Cancer Research: Nonprofit organization that supports both research on diet and cancer prevention and public health education. Numerous sources for both consumers and professionals.

http://www.AICR.org

American Cancer Society: This site provides specific information for patients, survivors, and professionals. Information for diagnosis and treatment—including nutrition—is available.

http://www.cancer.org

Cancer Facts: This online resource for cancer patients, their families, and caregivers focuses on providing background and treatment-related information about specific diagnoses.

http://www.cancerfacts.com

National Cancer Institute: This government-sponsored site provides links and information about specific diagnoses and current research. Links to nutrition education for therapy regimens as well as cancer prevention.

http://www.nci.nih.gov/cancerinfo

END-OF-CHAPTER QUESTIONS

1. Carcinogenesis is the process by which normal cells are transformed into cancer cells. How could a nutritional factor act as a carcinogen?
2. In general, how does the growth of a cancer cell differ from that of a normal, healthy cell?
3. Explain the following diagnosis in terms of how a malignancy is classified: Stage II diffuse large B-cell lymphoma.
4. A patient diagnosed with cancer may be treated with one or a combination of the following treatments:

chemotherapy, radiation, and/or surgery. Describe the basic mechanism or rationale behind each method used to treat a malignancy.

5. How do radiation and chemotherapy affect healthy cells within the body? What are the cells that are primarily affected? How is this related to the side effects that are often experienced with treatment?

6. Another category of treatment includes biological response modifiers or immunotherapy. What is the general mechanism for these treatments?

7. What is the Subjective Global Assessment? What are the primary nutrition assessment factors that this tool identifies?

8. Many cancer patients present with some nutritional symptoms such as weight loss or taste changes. How might a malignancy affect nutritional status even before it is diagnosed?

9. Name three common nutritional problems that a cancer patient might experience. Identify interventions for each.

10. Identify one complementary and alternative therapy that is commonly used by cancer patients. Explain the indications for this therapy and/or its risks.

25

Metabolic Stress

Marcia Nahikian Nelms, Ph.D., R.D.

Southeast Missouri State University

CHAPTER OUTLINE

Physiological Response to Starvation

Physiological Response to Stress

Burns

Surgery

Sepsis, Systemic Inflammatory Response Syndrome (SIRS), and Multi-Organ Distress Syndrome (MODS)

Introduction

The nutrition care process identifies those patients who are at significant nutritional risk or are suffering from malnutrition and determines the interventions necessary for recovery.

Chapter 10 discussed the body's response to injury at the cellular level. The type of cellular response depends on the ability of the cell to react, adapt, and repair itself after exposure to injury. This response may be temporary and completely reversible, but in some situations, the cell is permanently damaged and is no longer functional. Cell responses may include inappropriate accumulation of substances within the cell; changes in size, number, or shape; and the inflammatory response.

In this chapter, the body's unique responses to specific types of stress and injury are explored. The physiological effects of metabolic stress place the patient at the highest level of nutritional risk. The system is initially overwhelmed and unable to respond appropriately, allowing *protein-energy malnutrition* (PEM) to develop. In no other situation can nutritional deficits result in such dramatic and often severe consequences for a patient.

Physiological Response to Starvation

Malnutrition occurs when there is an inadequate nutrient supply—that is, during starvation. But the body can also experience malnutrition when it is unable to utilize nutrients appropriately or when its nutritional needs are so high that current intake cannot meet those demands—that is, during metabolic stress. The body's reaction to these two situations is quite different.

The most important difference between the physiological response to starvation and the response to metabolic stress is the adaptation that occurs during starvation. It was not until completion of the hallmark studies of Dr. Ancel Keys and his subsequent publication of *The Biology of Human Starvation* (1950) that the manner in which the body responds to starvation through physiological adaptations was understood (see Box 25.1). Keys and his research team followed the effects of starvation in 36 conscientious objectors during World War II (Kalm 2005).

BOX 25.1 HISTORICAL DEVELOPMENTS

Ancel Keys: Historical Perspective on Human Starvation

By Taylor Nelms, The Ohio State University

In 1939, as the war in Europe intensified and the United States prepared to enter the conflict, Ancel Keys, a new member of the faculty of the University of Minnesota, was asked by the War Department to develop a food ration for paratroopers. Keys had just founded the Laboratory of Physiological Hygiene beneath the bleachers of Memorial Stadium. The solution Keys and his collaborators concocted in the lab—the infamous K-ration—was so successful that the U.S. military assigned it to *all* their troops. Partly because of this success, Keys received support from the U.S. government as the war came to a close to conduct a large-scale comprehensive study on the physiological and psychological effects of starvation.

Keys put 36 conscientious objectors through a diet and exercise program designed to recreate the living conditions of occupied Europe. His human subjects ate simple, starchy foods and root vegetables, and were required to walk at least 22 miles per week. After three months of this semi-starvation, the young men were re-fed and rehabilitated. The resulting two-volume publication, *The Biology of Human Starvation*, was an instant classic, not merely for its detailed investigation of the physical consequences of protein-energy malnutrition, but also for its description of the condition's psychological effects (Keys et al. 1950). Deprived of food, the men became depressed and lost their motivation; later, they became obsessed with food, licking their plates, hoarding food, and even cheating (Kalm and Semba 2005).

Ancel Keys was one of the twentieth century's most important physiologists and nutritionists. Besides inventing the K-ration and studying starvation, he was the first to uncover a relationship between a high intake of saturated fat, the level of cholesterol, and the development of cardiovascular disease. He was one of the first medical scientists to utilize mathematical regression in the study of human health (VanItallie 2005). Finally, and perhaps most importantly, Keys demonstrated the importance of cultural and economic factors in the health of human populations. Throughout his career in physiology, nutrition, and public health, Keys emphasized the mutability of the body and our ability to prevent disease through simple modifications in lifestyle.

One of the distinctions between starvation and metabolic stress is the difference in energy requirements. During starvation, the body responds to a reduction in food intake by reducing its overall energy needs; the basal metabolic rate is reduced so that fewer kcal are needed. In contrast, energy requirements are increased during metabolic stress and injury. The next major difference between starvation and metabolic stress is the source of fuel that is used to meet energy requirements. Under normal circumstances, the body uses a mixture of fuel (primarily carbohydrate and lipid) to meet energy requirements. But since humans have a limited ability to store carbohydrate, the primary source of fuel during periods of starvation shifts as glucose availability decreases. Lipolysis becomes preferential and the accumulated lipid stores serve as the primary energy source. This adaptation for the use of lipid as the primary fuel and the subsequent metabolism of ketones allows for preservation of muscle mass and prevents the complications of protein deficiency (infection and decreased transport protein synthesis) (Gropper, Smith, and Groff, Protein, 2005). Table 25.1 outlines both normal nutrient metabolism and the key adaptations that occur during starvation. Figure 25.1 summarizes the changes in metabolism during starvation.

Unfortunately, when the body is faced with an injury or infection causing metabolic stress, the normal adaptations that should occur do not occur.

TABLE 25.1

Comparison of Metabolism during Normal Nutritional States versus Starvation

Normal Nutritional State	Starvation
Metabolic rate matches current physical activity requirements and body composition.	Decrease in metabolic rate to ensure conservation of energy.
Carbohydrate and lipid are efficient metabolized sources of energy providing 55%–85% of energy requirements.	Decreased need for glucose utilization.
Protein is used for maintenance of protein structures and to meet ongoing protein synthesis requirements.	Utilization of the lipid as main source of energy.
	Preservation of lean mass, minimizing protein loss.

Physiological Response to Stress

Definition

Metabolic stress is the hypermetabolic, catabolic response to acute injury or disease. Diagnoses that may lead to metabolic stress include trauma as seen in a gunshot wound or motor vehicle accident (MVA); closed head injury (see Chapter 22); burns; severe inflammation such as

FIGURE 25.1 Changes in Metabolism during Starvation

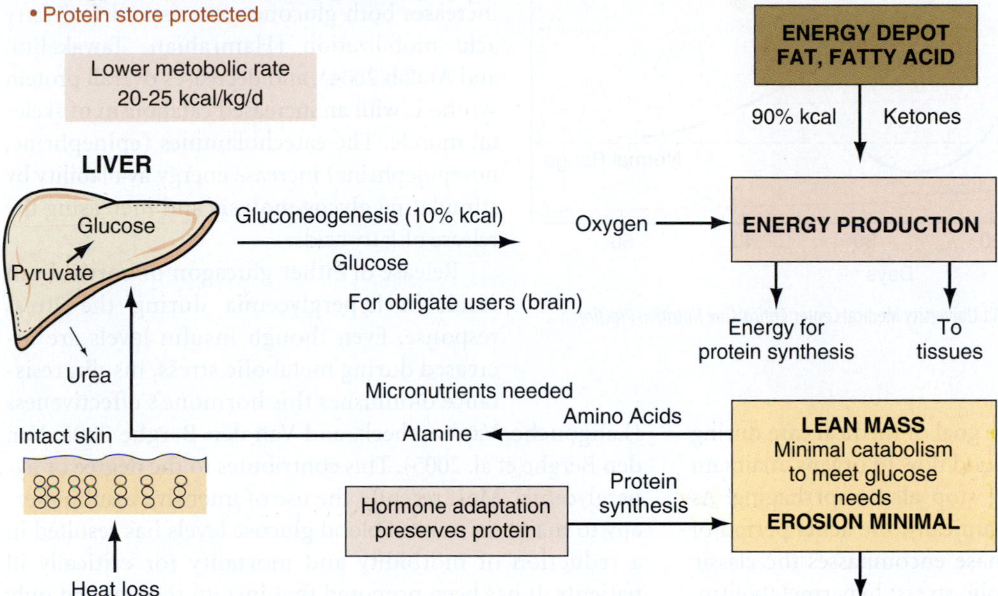

Source: Robert H. Demling, MD, Leslie De Santi, RN, *Effect of a Catabolic State With Involuntary Weight Loss on Acute and Chronic Respiratory Disease.*

in pancreatitis; cancer; **sepsis**; **hypoxic injury** as seen in acute renal failure; and necrosis of tissue such as in **gangrene** or after major surgery. The degree of metabolic stress generally correlates with the seriousness of the injury (Kudsk and Sacks 2004; Wooley and Btaiche 2005). In critical care medicine, rankings for severity of illness use scoring systems such as the Glasgow Coma Scale (see Table 22.9 in Chapter 22), the Acute Physiology and Chronic Health Evaluation (APACHE), the Injury Severity Score (ISS), or the Abdominal Trauma Index (ATI).

Epidemiology

Injuries caused by physical force (falls, gunshot wounds, stabbing, drowning, and other accidents) are classified as trauma. Trauma is a leading cause of death for young people, accounting for more than 70% of all deaths for those aged 15–24 years (National Center for Health Statistics, Deaths, 2006).

Etiology

The metabolic consequences of injury and stress are a result of numerous factors including hormone release, acute-phase protein synthesis, hypermetabolism, increased reliance on gluconeogenesis and its subsequent production of glucose, and shifts in fluid balance and decreased urine output (Gabay and Kushner 1999; Demling and DeSanti 2000; Cresci and Martindale 2001; Kudsk and Sacks 2004).

Clinical Manifestations

The stress response has been described as a progression through three phases: the ebb phase, the flow phase, and finally the recovery or resolution phase (Cuthbertson 1979). The ebb phase encompasses the immediate period after injury (2–48 hours). This period is characterized by shock resulting in hypovolemia and decreased oxygen availability to tissues. The decrease in blood volume results in decreased cardiac

sepsis—systemic inflammatory response and immunosuppressive process that prevents an adequate response to infection or trauma; may result in organ dysfunction or hypoperfusion abnormalities

hypoxic injury—cellular injury as a result of oxygen deprivation

gangrene—tissue death due to lack of blood flow and oxygen

FIGURE 25.2 Increased Nitrogen Loss during Metabolic Stress

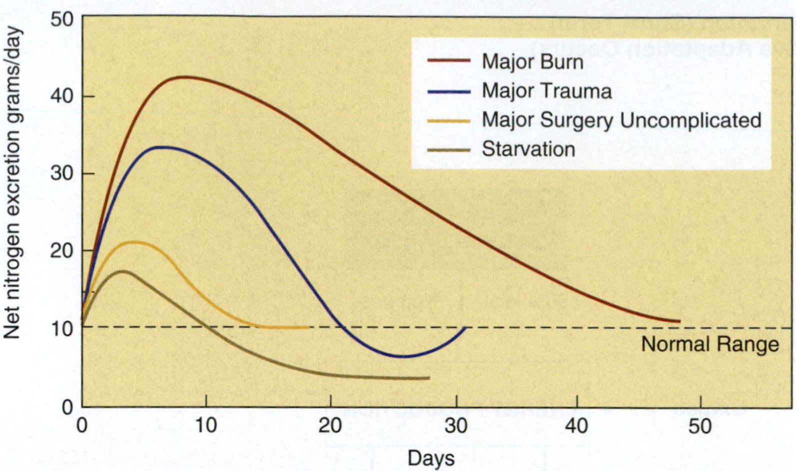

Source: JJ Diaz, R Pousman, G Jensen, Vanderbilt University Medical Center, *Critical Care Nutrition Practice Management Guidelines*.

output and urinary output. The goal of medical care during this acute period is to restore blood flow to organs, maintain oxygenation to all tissues, and stop all hemorrhaging. As the patient stabilizes hemodynamically, the acute period of the flow phase begins. This phase encompasses the classic signs and symptoms of metabolic stress: hypermetabolism, catabolism, and altered immune and hormonal response. The final adaptation phase or recovery phase indicates a resolution of the stress with a return to anabolism and normal metabolic rate.

Pathophysiology

Hormones, acute-phase proteins, the immune system, and altered cellular metabolism direct the physiological changes that characterize metabolic stress. Stress and injury activate the hormones that direct a "flight or fight" response, including glucagon, cortisol, epinephrine, and norepinephrine. Their primary purpose is to mobilize nutrient stores to meet the immediate energy demand. Increased levels of glucagon serve to increase glucose production from amino acids (**gluconeogenesis**). Cortisol increases both gluconeogenesis and free fatty acid mobilization (Hamrahian, Tawakalitu, and Arafah 2004), and decreases overall protein synthesis with an increased catabolism of skeletal muscle. The catecholamines (epinephrine, norepinephrine) increase energy availability by stimulating **glycogenolysis** and increasing the release of fatty acids.

Release of either glucagon or cortisol can result in hyperglycemia during the stress response. Even though insulin levels are increased during metabolic stress, insulin resistance diminishes this hormone's effectiveness (Langouche, Vanhorebeek, and Van den Berghe 2005; Van den Berghe et al. 2005). This contributes to the degree of hyperglycemia. Most recently, the use of intensive insulin therapy to maintain normal blood glucose levels has resulted in a reduction of morbidity and mortality for critically ill patients. It has been proposed that insulin therapy not only controls hyperglycemia seen in metabolic stress but may reduce catabolism and inflammation, which improves the immune response (Langouche, Vanhorebeek, and Van den Berghe 2005).

The increased rate of gluconeogenesis creates reliance on protein as a source of glucose (Wolfe 2005). The need for the amino acids alanine and glutamine is particularly increased. Since alanine is the primary substrate required for gluconeogenesis, there is an increased catabolism of skeletal muscle to make it available. Glutamine is a nonessential amino acid that has been found to be significant in both metabolic and immunologic pathways. For example, glutamine is the primary fuel for enterocytes within the gastrointestinal tract and for T-lymphocytes (Kudsk and Sacks 2004). In injury and stress, the synthesis rate may be unable to accommodate the increased need. Negative nitrogen balance is a consistent sign during metabolic stress. Figure 25.2 demonstrates the substantial nitrogen loss and complications that accompany skeletal muscle catabolism. (Figure 25.3 and Table 25.2 provide a summary of metabolic changes in stress and trauma.) Complications of catabolism may include: immunosuppression, increased infection rates, delayed or impaired wound healing and increased mortality (Demling & Desanti 2001).

Positive acute-phase proteins are often used as markers of the stress response (see Chapter 5). These include **fibronectin**, **C-reactive protein**, **ceruloplasmin**, and **serum amyloid A**. An acute-phase protein is defined as "one whose

gluconeogenesis—metabolic pathway from which glucose is formed from noncarbohydrate sources

glycogenolysis—metabolic pathway from which glycogen is converted to glucose

fibronectin—acute-phase glycoprotein involved in the regulation of cell growth and differentiation, wound healing, and vascular integrity

C-reactive protein—protein released as a response to inflammation

ceruloplasmin—protein used in copper transport

serum amyloid A—family of apolipoproteins associated with high-density lipoprotein (HDL) in plasma; considered to be an acute-phase protein released in response to inflammation

FIGURE 25.3 Summary of Metabolic Changes in Stress and Trauma

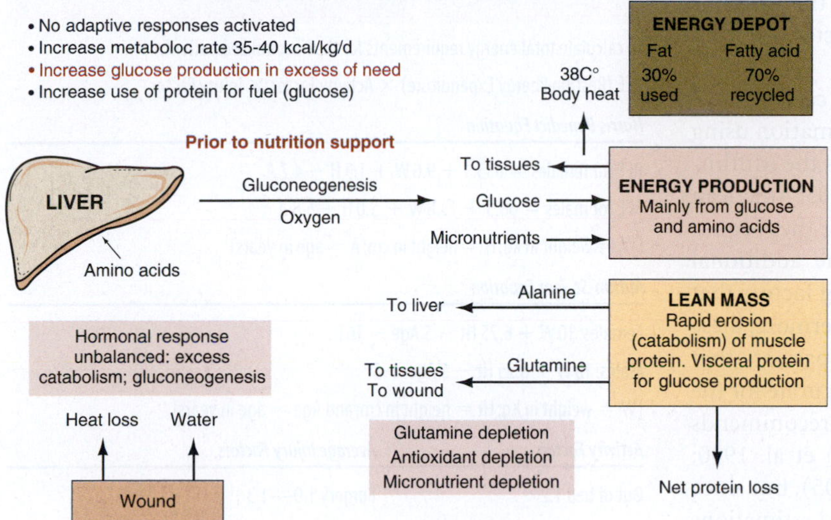

Source: JJ Diaz, R Pousman, G Jensen, Vanderbilt University Medical Center, *Critical Care Nutrition Practice Management Guidelines.*

TABLE 25.2

Summary of Metabolic Abnormalities Observed in Stress Response

- Increased levels of glucagon, cortisol, epinephrine, norepinephrine
- Hyperglycemia and insulin resistance
- Increased basal metabolic rate
- Increased rate of gluconeogenesis
- Catabolism of skeletal muscle
- Increased urinary nitrogen excretion—negative nitrogen balance
- Increased synthesis of positive acute-phase proteins—CRP, fibronectin, ceruloplasmin
- Decreased synthesis of negative acute-phase proteins

TABLE 25.3

Proinflammatory Cytokines and Their Metabolic Effects

Cytokines	Metabolic Effect
Tumor necrosis factor (TNF)	Altered metabolism: catabolism, hypermetabolism
Interleukin-1 (IL-1)	Increased body temperature
Interleukin-6 (IL-6)	Activation and release of cellular communication/ mediators

plasma concentration increases (positive acute-phase proteins) or decreases (negative acute-phase proteins) by at least 25% during inflammatory disorders" (Gabay and Kushner 1999, p. 448). The release of the acute-phase proteins is regulated by a variety of cytokines and other communication molecules within the immune system. Cytokines include interleukins (interleukin-1/IL-1, interleukin-6/IL-6), leukotrienes, tumor necrosis factor and interferons (see Chapter 12). IL-6 directly affects protein metabolism by decreasing acute-phase proteins such as albumin and prealbumin and increasing other acute-phase proteins such as C-reactive protein (Kudsk and Sacks 2004). Table 25.3 summarizes the metabolic effects of pro-inflammatory cytokines.

As discussed previously, cytokines are proteins that, in small amounts, affect behavior of other cells (see Chapters 10 and 12). The injury or stress induces cytokine production, and cytokines then act on target cells whose behavior can result in anorexia, fever, inflammation, and metabolic abnormalities such as hyperglycemia and catabolism.

Nutrition Therapy

Nutrition therapy for metabolic stress must take into consideration the critical illness the patient is experiencing in addition to accommodation of the metabolic changes that have occurred. There is a delicate balance between prevention of PEM and prevention of the possible complications of nutrition support. As stated previously, critically ill patients are at significant nutritional risk. Nutritional status prior to the current illness is an important predictor of morbidity and mortality. The level of injury will determine the potential level of metabolic stress (Demling and DeSanti 2000; Biolo, Grimble, and Preiser 2003). As already mentioned, the level of injury and the subsequent risks are often quantified using scoring systems; the Glasgow Coma Scale or the APACHE II score are common in critical care medicine. Use of a scoring system may give the clinician an idea of the level of metabolic stress that might be expected in the individual patient.

Nutrition Assessment Many of the standard measures for nutritional assessment are neither valid nor reliable in the critically ill population. Information regarding prior dietary history or weight status is not always available. Patients with this level of injury or disease are often on mechanical ventilation or are unable to communicate. Family members will be an important source of data to consider.

Measured weight may not be reflective of actual weight due to changes in fluid balance. Biochemical indices to assess visceral protein status (transport proteins) may be more reflective of the level of stress than of the patient's actual protein status, and may also be affected by fluid balance.

In this population, energy requirements can be measured by using indirect calorimetry or by estimation using a variety of equations (see Chapter 5) such as the Mifflin-St. Jeor or Harris-Benedict equations (EAST 2003; American Association for Respiratory Care 2004). While these equations estimate basal energy expenditure, the additional needs for stress and injury are estimated using factors that are actually based on much less research. Furthermore, these populations include such a wide variety of patients that generic stress and injury factors may not be accurate for the individual patient. A recent evidence analysis recommends use of the Mifflin-St. Jeor equation (Mifflin et al. 1990; Frankenfield, Roth-Yousey, and Compher 2005), but since there are limitations to both measurements and estimations in the intensive care environment, it is essential to use clinical judgment for each case (Kudsk and Sacks 2004; Reid, Campbell, and Little 2004; Vasquez Martinez et al. 2004; Campbell, Zander, and Thorland 2005). (See Table 25.4 for these equations). It is most important to avoid overfeeding, because the associated complications can easily occur in these critically ill populations (Jeejeebhoy 2004). Even less research has focused on establishing protein requirements in the critical care environment (Reid 2004; Campbell, Zander, and Thorland 2005). Most initial caloric goals can be met at 25–35 kcal/kg. Depending on the individual patient diagnosis and nutrition assessment, protein requirements can be estimated at 1.2–1.5 g protein/kg (Kudsk and Sacks 2004; Reid, Campbell, and Little 2004; DeSouza and Greene 2005).

Since it is crucial to avoid overfeeding, determining energy requirements for the morbidly obese patient poses a specific challenge for the nutrition practitioner. No validated approach to determining a goal weight to be used in calculations has been established. It is felt that "permissive underfeeding" may assist with preventing acute metabolic and respiratory complications. Hypocaloric recommendations aim at providing approximately 14 kcal/kg and 1.2 g protein/kg (Choban and Dickerson 2005; Dickerson 2005).

Nutrition Interventions Oral nutrition is the preferred route for meeting nutritional needs. For critically ill individuals, however, nutritional needs can rarely be met in this manner, and alternate feeding routes, including both enteral and parenteral nutrition, must be considered. There is evidence to support early initiation of nutrition support with specific metabolically stressed diagnoses such as acute pancreatitis, closed head injury, and burns (ASPEN 2002; Braga et al. 2002; Dominioni et al. 2003; EAST 2003; DiFronzo et al. 2003; Malhotra, Mathur, and Gupta 2004). See Figure 25.4.

TABLE 25.4

Calculation of Energy and Protein Requirements: Activity and Stress Factors for Hypermetabolic Conditions

To calculate total energy requirements for the hospitalized patient:

REE (Resting Energy Expenditure) × Activity Factor × Injury Factor

Harris Benedict Equation

REE for females = 655.1 + 9.6 W + 1.9 H − 4.7 A

REE for males = 66.5 + 13.8 W + 5.0 H − 6.8 A

[W = weight in kg; H = height in cm; A = age in years]

Mifflin-St. Jeor Equation

Females: 10 W + 6.25 Ht − 5 Age − 161

Males: 10 W + 6.25 Ht − 5 Age + 5

[W = weight in Kg; Ht = height in cm; and Age = age in years]

Activity Factors	*Average Injury Factors*
Out of bed 1.2	Surgery 1.0—1.3
Confined to bed 1.1	Infection 1.0–1.4
	Skeletal trauma 1.2–1.4
	Head Injury 1.5

Protein Requirements

RDA 0.8 g protein/kg

Minor surgery 1–1.1 g protein/kg

Major surgery 1.2–1.5 g protein/kg

Burn 1.5–2.0 g protein/kg

Enteral nutrition (EN) support should be considered first (see Figure 25.5). When compared to parenteral nutrition (PN), EN is more cost-effective and is associated with reduced infectious complications, fewer surgical interventions, and, in some studies, fewer hospital days (ASPEN 2002; Heyland et al. 2002; EAST 2003; Marik 2004; Binnekade 2005; Farber, Moses, and Korn 2005; Jeejeebhoy 2005). The first step in this process, once access to the GI tract has been established, is choosing the appropriate enteral formula. Decisions will be influenced by the ability to meet energy and protein needs within the fluid volume that can be best tolerated by the patient.

In addition to energy and protein needs, formula selection should address the specific types of nutrients that may have increased requirements during metabolic stress. Specialized formulas (see Table 25.5) with immunosupport and synbiotics have been developed that may be helpful in the metabolically stressed patient (Cynober 2003; Grimm and Kraus 2003; Calder 2004; DeSouza and Greene 2005; Grimble 2005). (Synbiotics contain both prebiotics—substances in food that stimulate the beneficial flora of the large intestine—and probiotics—products containing

FIGURE 25.4 Nutrition Support Protocols Overview

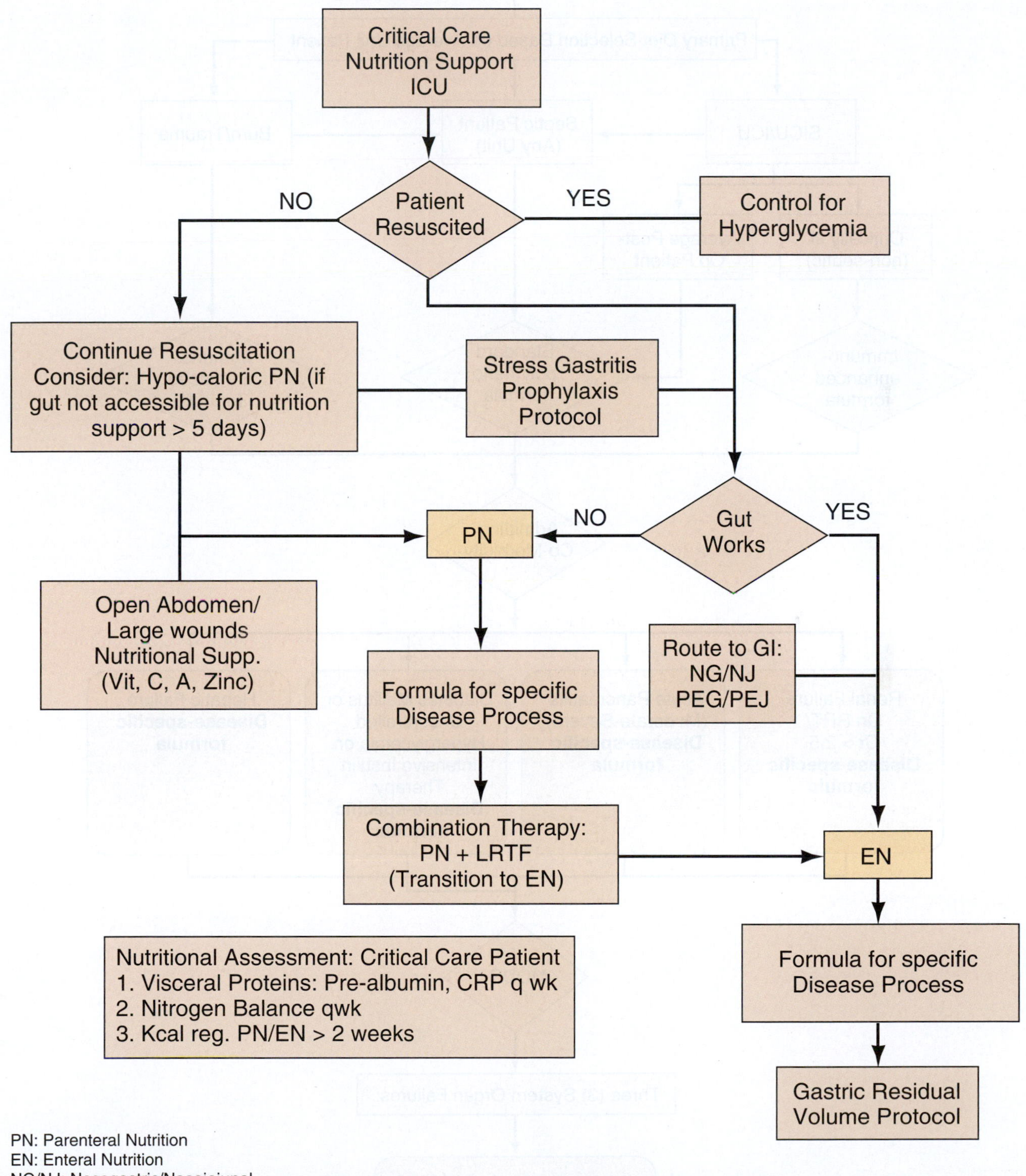

PN: Parenteral Nutrition
EN: Enteral Nutrition
NG/NJ: Nasogastric/Nasojejunal
PEG/PEJ: Percutaneous Endoscopic Gastrostomy/ Percutaneous Endoscopic Jejunostomy

Source: JJ Diaz, R Pousman, G Jensen, Vanderbilt University Medical Center, *Critical Care Nutrition Practice Management Guidelines*.

FIGURE 25.5 Enteral Nutrition Protocol

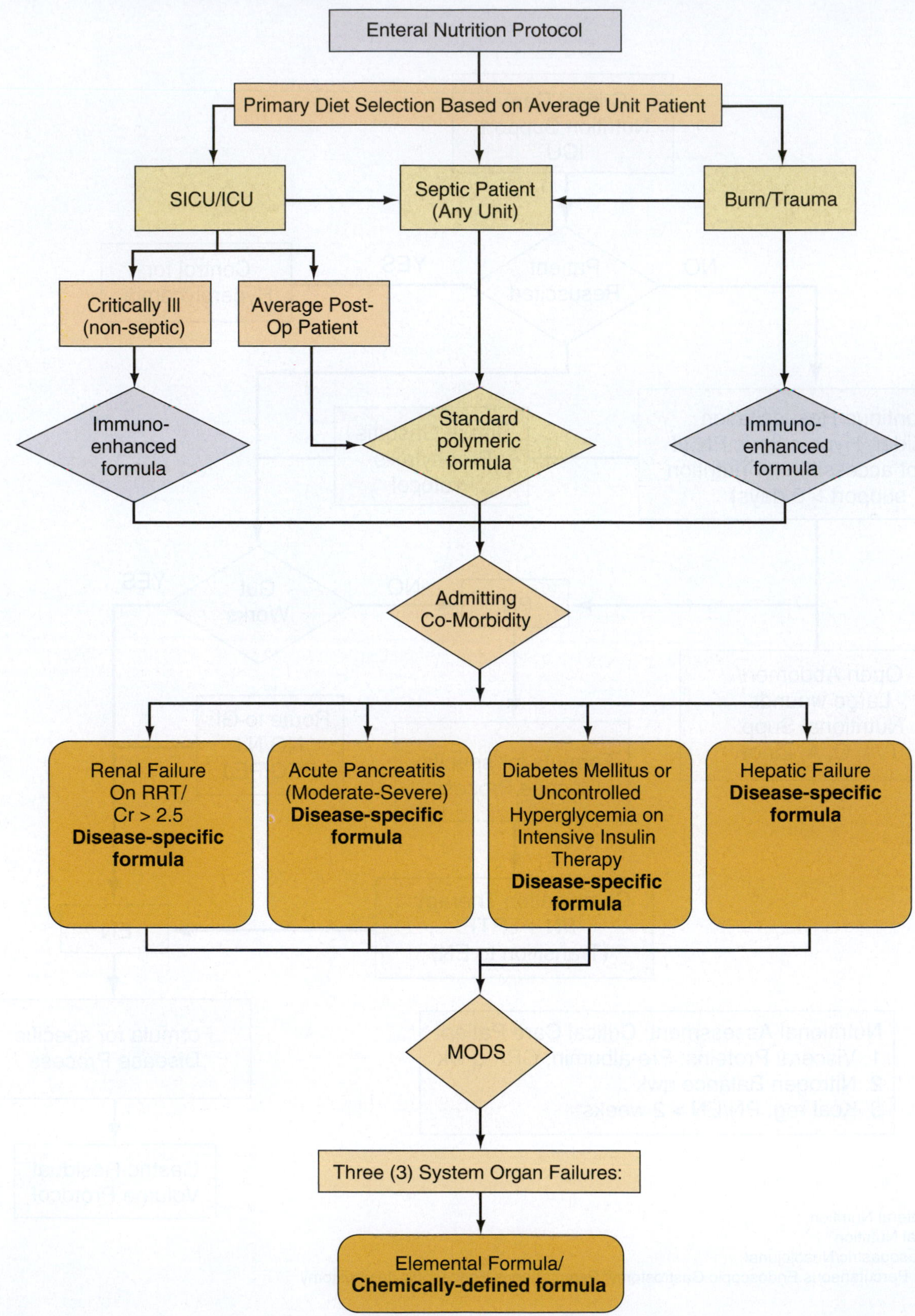

Source: JJ Diaz, R Pousman, G Jensen, Vanderbilt University Medical Center, *Critical Care Nutrition Practice Management Guidelines*.

TABLE 25.5

Enteral Formulas in Metabolic Stress

Formula/Manufacturer/ Basic Nutrition Information	Carbohydrate Source	Protein Source	Lipid Source	Rationale for Metabolic Stress
Pivot© (Ross) **Caloric Density:** 1.5 kcal/mL **Osmolality:** 595 mOsm/kg water	46% (corn syrup solids)	25% partially hydrolyzed sodium caseinate, whey protein hydrolysate	Structured lipid (interesterified sardine oil and medium chain triglycerides), soy oil, canola oil	Arginine—13 g/L; glutamine (inherent): 6.5 g/L; omega-3 fatty acids (EPA, 2.6 g/L; DHA: 1.3 g/L); fructooligosaccharides (FOS) and increased antioxidants
Crucial© (Nestle) **Caloric Density:** 1.5 kcal/mL **Osmolality:** 375 mOsm/kg water	36% (maltodextrin)	25% (hydrolyzed casein with added amino acid fortification)	Marine oil, MCT, and soybean oil	n6:n3 ratio of 1.5:1; fortified with arginine, vitamin C, A, zinc and beta-carotene
Impact© (Novartis) **Caloric Density:** 1.0 kcal/mL **Osmolality:** 375 mOsm/kg water	53% (hydrolyzed cornstarch)	22% (sodium and calcium caseinates, L-arginine 12.5 g/L)	25% (palm kernel oil, sunflower oil, menhaden oil)	EPA/DHA: 1.7 g/L; n-6 : n-3 ratio: 1.4:1.0; fortified with arginine—also available with added fiber
Impact Glutamine© (Novartis) **Caloric Density:** 1.3 kcal/mL **Osmolality:** 630 mOsm/kg water	46% (maltodextrin)	24% (wheat protein hydrolysate, free amino acids, sodium caseinate)	30% (palm kernel oil, menhaden oil, sunflower oil)	Glutamine: 15 g/L; L-arginine: 16.3 g/L; dietary nucleotides: 1.6 g/L; n-6: n-3 ratio: 1.4:1.0; added probiotic with soy and hydrolyzed guar gum
Periative© (Ross) **Caloric Density:** 1.3 kcal/mL **Osmolality:** 460 mOsm/kg water	55% (corn maltodextrin)	20.5% (partially hydrolyzed sodium caseinate, hydrolyzed lactalbumin)	Canola oil, MCT	.6 g of FOS/8 fl oz (6.5 g/L and 9.8 g/ 1,500 mL); fortification with arginine

Information obtained directly from manufacturer information: http://www.ross.com/productHandbook/adultNut.asp; http://www.novartisnutrition.com/us/productList; http://www.nestle-nutrition.com.

beneficial microorganisms.) The amino acids glutamine and arginine have been the most thoroughly studied in respect to critical illness. Both of these amino acids are nonessential for humans, meaning that under normal circumstances endogenous synthesis is adequate. But during the stress of critical illness, the synthesis rate cannot meet the increased needs. The use of arginine is controversial, in that even though research has suggested that arginine may assist with inhibiting the immunosuppression that occurs during the stress response as well as improving nitrogen balance (Pan 2004; Zaloga et al. 2004), additional evidence demonstrates negative outcomes for use of arginine in patients with sepsis (Bower et al. 1995; Bertolini et al. 2003).

Glutamine is a primary fuel for hyperplastic cells such as enterocytes or immunocompetent cells. Again, during critical illness and metabolic stress, endogenous synthesis rate may not be able to match the increased requirements for this amino acid. Supplementation with glutamine may decrease infection rate and prevent translocation of bacteria from the GI tract (Kelly and Wischmeyer 2003; DeSouza and Greene 2005; Jeejeebhoy 2005). Glutamine is added to enteral formulations and is available for oral supplementation, but it is not approved for intravenous use.

Branched-chain amino acids include isoleucine, leucine, and valine. Interest in their role during starvation and metabolic stress is linked to their particular metabolism within the skeletal muscle (Laviano et al. 2005). For example, leucine is completely oxidized for energy within the muscle and provides more ATP than glucose. The increased use of branched-chain amino acid metabolism within the skeletal muscle may spare other substrates that are required to meet the metabolic demands of stress (Gropper, Smith, and Groff, Metabolism, 2005).

In Chapter 7, the common substrate sources for protein, carbohydrate, and lipid were discussed. In metabolic stress, the type of protein, carbohydrate, and lipid are of concern. Many recent studies have focused on how these may either exacerbate the stress response or in some situations favorably affect the outcome of the stress or injury. For example, higher percentages of kcal from fat have been proposed as a

branched-chain amino acid—amino acid that has a branched side chain; includes isoleucine, leucine, and valine

cause of immunosuppression, which may exacerbate the stress response. On the other hand, specific types of fats such as omega-3 fatty acids play an important role in improving the body's ability to respond to stress (Calder 2004; Roy, Bouthillier, and Seidman 2004; Babcock, Dekoj, and Espat 2005; Jeejeebhoy 2005; Calder 2006), though some researchers stress the importance of continued study for establishing recommended levels (American Association of Clinical Endocrinologists 2003). The use of immunonutrition in the patient who is septic is contraindicated. For enteral formulas, **menhaden oil**, marine oil, and structured lipids can be used to modify the type of lipid in nutrition support products (see Chapter 7).

Sources of fiber added to enteral feedings provide substrates that assist in maintenance of beneficial bacteria in the gastrointestinal tract (probiotics, prebiotics, and synbiotics) and may assist in preventing diarrhea (Spapen et al. 2001). As discussed in Chapter 17, the substrates such as inulin, guar gum, and other soluble fibers are fermented to short-chain fatty acids and lactate. Enteral nutrition formulas with added probiotics, prebiotics, and synbiotics are important options to consider for the prevention of infectious complications associated with critical illness (Bengmark 2005).

Complications of enteral feeding may include metabolic complications (hyperglycemia and electrolyte imbalances), aspiration, or mechanical complications (clogging or misplacement of the tube).

Total parenteral nutrition (TPN) should be reserved for those cases of prolonged nothing by mouth (NPO) status, when enteral access cannot be obtained or when enteral nutrition support cannot meet the patient's needs or is not tolerated. It is important to understand that, in this population, even when the gastrointestinal tract is functional, nutritional needs may not be met with EN alone. This may be because the patient cannot tolerate the volume of feeding necessary to meet the increased nutritional needs (ASPEN 2002). Chapter 7 provides details regarding the design of PN prescriptions. The most important concern for PN in metabolic stress may be its contribution to hyperglycemia. As stated previously, recent research has established the importance of maintaining normal blood glucose ranges in the critically ill patient. Doing so has resulted in improved overall outcomes (Langouche, Vanhorebeek, and Van den Berghe 2005 ; Van den Berghe et al. 2005). Other complications of PN may include catheter occlusion, catheter-related infection, hypertriglyceridemia, intestinal atrophy (usually only from long-term TPN without any oral intake), electrolyte disturbances, and refeeding syndrome in previously malnourished individuals.

menhaden oil—hydrogenated and partially hydrogenated oils from the menhaden fish—a small plankton-feeding fish

Burns

Definition

Burns are a result of tissue injury caused by exposure to heat, chemicals, radiation, or electricity. They may result from injury to the skin, but the damage may extend into muscle and bone. The depth of the wound and the percent of the body surface area that is affected are used to classify burn injury. The wounds are described as superficial, superficial partial thickness, deep partial thickness, or full thickness.

Epidemiology

Medical treatment is required for more than 1.1 million burn victims each year, with approximately 45,000 hospitalizations (National Institute of Medical Sciences, Trauma, Shock, Burn, 2006). The mortality rate from burns has declined significantly over the previous several decades due to major advances in medical care.

Etiology

Burns may be a result of thermal exposure, which includes direct contact with a heat source such as hot water or flames. These are the most common type of burns and occur both in the home and the workplace. Burns can also be caused by chemical or electrical exposure. Damage to the body occurs when the electrical current moves through tissues or bone. The severity of the burn will be correlated with the amount of voltage, the location where it enters the body, and the amount of time that the exposure continues. Chemical burns occur when the body is exposed to an acid or alkali such as battery acid, drain cleaner, or other environmental sources.

Clinical Manifestations

Signs and symptoms that an individual experiences will be determined by the extent of the burn injury—by both depth and body surface area affected. Age, nutritional status, and other comorbidities will have an impact on the physiological response to the injury, treatment, and recovery.

Table 25.6 summarizes the characteristics of burn classifications. Superficial burns involve the top layer of epidermis, which will characteristically redden. An example of this type of burn may result from sunburn. Superficial burns are painful but typically heal easily and quickly. Superficial partial thickness burns produce open, weeping wounds that are very painful. Partial thickness deep burns involve destruction of the epidermis and dermis. Full thickness burns destroy all layers of the skin and can involve underlying muscle, organs, and bone (Morgan, Bledsoe, and Barker 2000; Gould 2004).

TABLE 25.6

Characteristics of Burns Based on Depth

Classification	Cause	Characteristics			
		Appearance	Sensation	Healing Time	Scarring
Superficial	Ultraviolet light, very short flash (flame exposure)	Dry and red; blanches with pressure	Painful	3 to 6 days	None
Superficial partial-thickness	Scald (spill or splash), short flash	Blisters; moist, red and weeping; blanches with pressure	Painful to air and temperature	7 to 20 days	Unusual; potential pigmentary changes
Deep partial-thickness	Scald (spill), flame, oil, grease	Blisters (easily unroofed); wet or waxy dry; variable color (patchy to cheesy white to red); does not blanch with pressure	Perceptive of pressure only	More than 21 days	Severe (hypertrophic) risk of contracture
Full-thickness burn	Scald (immersion), flame, steam, oil, grease, chemical, high-voltage electricity	Waxy white to leathery gray to charred and black; dry and inelastic; does not blanch with pressure	Deep pressure only	Never (if the burn affects more than 2% of the total surface area of the body)	Very severe risk of contracture

Adapted with permission of Robert Hugh Demling, M.D. (Editor, BurnSurgery.org) from: http://www.burnsurgery.org/Modules/BurnWound/rationale/burn_injury/depth_assessment.htm.

The Rule of "Nines" is one method that is used to make a rapid estimation of body surface area (BSA) that has been burned. In this method, the body is divided into portions with a value or derivative of nine. For example, an entire leg and foot is estimated to be approximately 9% of the total body surface area (see Figure 25.6). Estimation of the affected body surface area assists in assessment of the extent of the injury, and helps provide the basis for prescribing fluid and medications.

Pathophysiology

While superficial burns and some superficial partial thickness burns are generally treated on an outpatient basis, other levels of burn require treatment in specialized burn units. The initial burn shock is a result of the extensive inflammatory process and involves rapid fluid shifts and accumulation and, therefore, fluid loss from the wound. Initial treatment of a burn will focus on fluid resuscitation and stabilization of all organ systems (Monafo 1996; Morgan, Bledsoe, and Barker 2000; Gould 2004).

The physiological response to burn injury progresses through the same phases of metabolic stress as described earlier in this chapter: hypermetabolism, catabolism, and altered immune and hormonal response. Respiratory complications are multifactorial. They may originate from inhalation of smoke and other toxic substances or occur as complications secondary to fluid resuscitation, pain, inflammation, and infection.

Treatment

Management of the burn wound involves application of topical agents such as **silver sulfadiazine cream** and **silver nitrate** to prevent infections. Other treatment involves the complex procedures used to clean, **debride,** and dress the wounds. Full thickness burns require skin grafting or the use of skin substitutes for acceptable closure of the wounds.

Nutrition Therapy

Nutrition Implications The patient with a burn injury is at significant nutritional risk due to the hypermetabolic, catabolic response that occurs after the injury. It is estimated that as much as 20% of body protein can be lost within the

silver sulfadiazine cream—sulfa medicine used to prevent and treat bacterial or fungal infections

silver nitrate—colloidal silver used as an antibacterial treatment in burns

debride—to remove dead or injured tissue

FIGURE 25.6 Rule of Nines to Estimate Body Surface Area

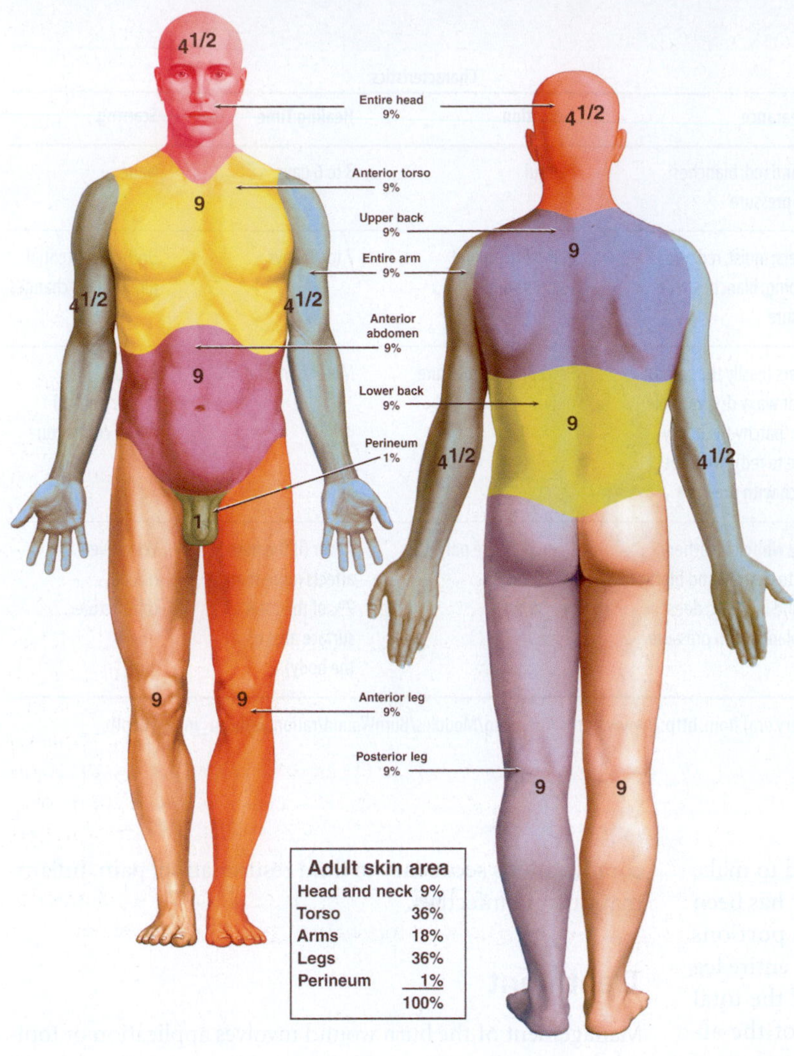

Source: Medical Illustration Copyright © 2006 Nucleus Medical Art, All rights reserved. www.nucleusinc.com.

first two weeks of burn injury (Lee, Benjamin, and Herndon 2005). Fluid imbalance, pain, and immobility make it difficult for the patient with extensive burns to maintain his or her nutritional status. The patient also requires optimal nutrition therapy to support wound healing during the treatment and healing process.

Nutrition Assessment Standard nutritional parameters are affected by both the nature of the burn injury and the metabolic stress response. Weight may fluctuate considerably due to fluid shifts and resuscitation. High protein losses from wounds and the overall acute inflammatory response can affect accurate interpretation of the protein markers albumin, prealbumin, or C-reactive protein in relationship to nutritional status.

Energy requirements are measured by indirect calorimetry or estimated using standard equations (Mifflin et al. 1990; Campbell, Zander, and Thorland 2005). A specific equation that incorporates and accounts for the extent of the burn injury, called the Curreri equation, was developed for use in burn patients (Curreri 1979). Research and practice, however, have indicated that this equation best estimates energy requirements at the peak of burn injury and that calculations do not necessarily accommodate the changes that occur from day to day in the burn injury. While the degree of hypermetabolism and catabolism correlate with the patient's level of injury, researchers agree that the level of hypermetabolism does not increase beyond that reached for a 50% to 60% total body surface area burn (Mayes and Gottschlich 2001; Thompson and Fuhrman 2005). Other factors that contribute to increased energy needs include fever, infection, or the development of sepsis (Gore et al. 2003; Jeschke and Barrow 2004). If indirect calorimetry is unavailable, current practice suggests using the Mifflin-St. Jeor equation to estimate resting energy expenditure and then using injury factors of 1.3–1.5 to set initial kcal goals for the patient. This initial estimate falls between 25 and 35 kcal/kg.

Protein requirements can be based on 1.5–2 g protein/kg. It is unlikely that the negative nitrogen balance that occurs during the catabolic phase of burn injury can be totally prevented, but it is crucial to set a goal of minimizing losses and promoting wound healing. Monitoring daily calorie counts, wound closure, and acceptance of engraftment provides practical measurements of adequate nutritional support.

Nutrition Interventions Most severe burn injuries require nutrition support in addition to or as a substitute for an oral diet. As previously discussed, early enteral feeding is beneficial for some diagnoses. In burn patients, EN has been associated with prevention of infections (in particular bacterial translocation), the prevention of **Curling's ulcer**, and the reduction of protein catabolism (Hart 2003; Flynn 2004; Lee, Benjamin, and Herndon 2005; Magnotti and Deitch 2005). Curling's ulcer is also treated prophylactically with medications such as H2 antagonists.

Curling's ulcer—ulceration of gastric or duodenal tissue as a result of burn or trauma

The enteral feeding prescription for a burn patient is developed by following the same steps used with other diagnoses, but will need to accommodate the special metabolic requirements of this stressed state. In severe burn patients, ileus (general paralysis of the GI tract) is common during the burn shock period, but enteral feeding is generally tolerated when delivered to the small bowel. When choosing a formula, the clinician should focus on those with higher amounts of protein (20% to 25% of kcal) and consider those formulas with supplemental arginine, glutamine, and omega-3 fatty acids (Jeschke et al. 2001; DeSouza and Greene 2005; Peng et al. 2005). More research is needed to determine the exact amounts or duration of therapy for glutamine, arginine, or omega-3 fatty acids for promotion of wound healing (Thompson 2005).

When enteral feeding cannot meet nutritional needs, PN should be prescribed. This may be used in combination with enteral or oral feedings, or can provide all of the patient's nutritional needs. In a recent study examining nutrition support for patients with **necrotizing fasciitis**, 94% of patients required either EN or PN for an average of 24 days (range 1–68 days) (Graves et al. 2005).

Careful attention will be necessary to avoid overfeeding and control hyperglycemia. There is a tenuous line between providing the amount of energy needed to meet metabolic requirements and contributing to further metabolic complications.

Oxandrolone, an anabolic steroid, is often used to promote protein synthesis in burns affecting >15% BSA (Demling and DeSanti 2001). Additional vitamins, minerals, and trace elements are often supplemented for wound healing (see Chapter 10, Table 10.5). In burn patients, higher amounts of these nutrients are prescribed to replace the large amounts lost via the wound exudates, but also to ensure adequate support for engraftment and overall wound healing. Supplementation with vitamin C, vitamin A, vitamin E, and zinc are routinely included for nutrition protocols in burn intensive care units.

As recovery proceeds and the patient is able, oral feedings can be initiated. Weaning from nutrition support is recommended when the patient is able to meet at least 60% of nutritional needs orally (see Figure 25.7). Nutritional requirements will need to be adjusted as the patient heals and the focus of therapy is rehabilitation.

Surgery

Definition

Surgery is defined as an operative procedure used to diagnose, repair, or treat an organ or tissue. Surgery can be further classified by the seriousness of the procedure (major or minor), the necessity (elective or emergency), or the specific purpose of the procedure (diagnostic, excision, palliative, reconstructive, or transplant).

Epidemiology

In 2003, over 43.9 million surgical procedures were performed in the United States (National Center for Health Statistics, Surgery, 2006). These included a wide range of procedures such as cardiac catherizations, mobilization of fracture, and hysterectomy.

Etiology

Many surgical procedures do not pose nutritional risk. Nonetheless, if the patient enters surgery malnourished or overnourished, or if the surgical procedure will interrupt normal nutrition processes, the individual will be at nutritional risk postoperatively. Age and coexisting diagnoses will have an impact on the outcome and recovery from the surgical procedure. Malnutrition increases the risk of the most common complications postoperatively, including wound dehiscence (improper wound opening after suture closure) and infections (Kudsk and Sacks 2004).

Screening and prognostic tools have been developed to identify those patients most likely to be at nutritional risk (see Chapter 5). Preoperative changes in weight, albumin, and C-reactive protein have been quantified to predict outcome in a number of research studies. One of these, The National VA Surgical Risk Study, evaluated the relationship of numerous characteristics to complications and mortality rate in over 50,000 surgical patients (Gibbs et al. 1999). These researchers found that preoperative albumin was a better predictor of complications and mortality than any of the other characteristics such as age, smoking, and other laboratory values.

Clinical Manifestations

The signs and symptoms experienced with surgery depend on the type of procedure. Patients are required to refrain from eating or drinking at least 12 hours before surgery. The patient will receive **general**, **epidural**, or **local anesthesia**. Postoperatively, the patient may also have a nasogastric tube in place to remove gastric secretions and/or a urinary catheter in

necrotizing fasciitis—inflammation of the connective tissue leading to necrosis of the tissue; may be caused by infection, injury, or an autoimmune reaction

general anesthesia—total loss of sensation and consciousness as a result of anesthesia drug

epidural anesthesia—anesthetic drug placed into the epidural space of the lumbar or sacral region of the spine, causing loss of sensation from the abdomen and pelvis to the lower limbs

local anesthesia—loss of sensation only in the area where an anesthetic drug is placed

FIGURE 25.7 Combination Feeding Protocol

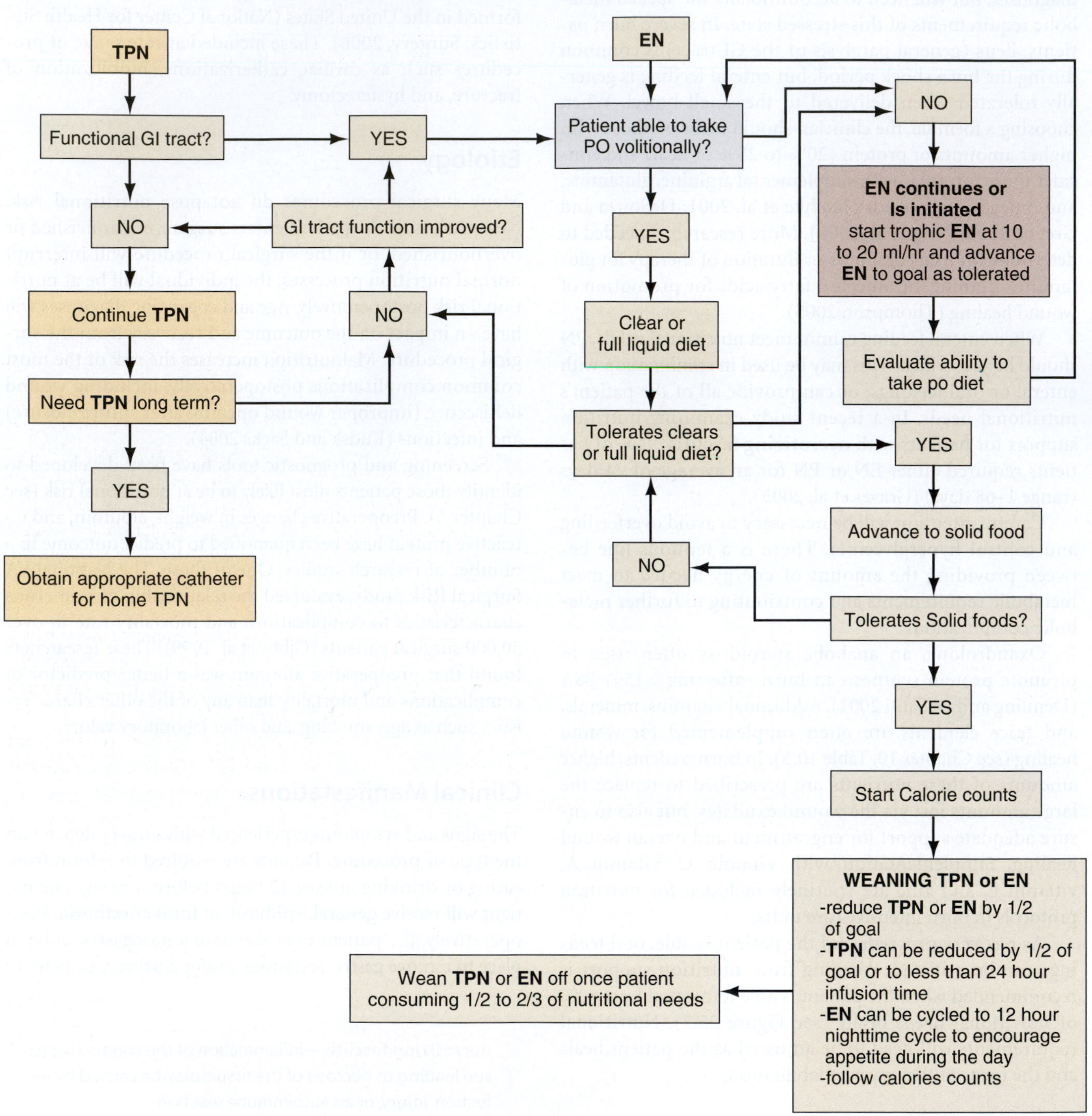

Source: JJ Diaz, R Pousman, G Jensen, Vanderbilt University Medical Center, *Critical Care Nutrition Practice Management Guidelines.*

place to remove urine until normal sensations for urination have returned. Other general concerns for surgical patients include maintenance of respiratory function, circulation, prevention of infection, wound healing, and pain control.

General anesthesia may result in a postoperative ileus (lack of motility), a general paralysis of the gastrointestinal tract. Resolution of the ileus generally occurs within 24–48 hours, depending on the type of procedure. Traditionally, the patient was prevented from eating or drinking until the ileus was resolved, and the production of gas or bowel movement was a sign of resolution of ileus. Because it is difficult to ascertain when GI function has returned, however, the determination of when to actually begin postoperative feeding has been a topic of recent debate (Schulman and Sawyer 2005). In a study by Fettes et al. (2002), 34% of 200 patients evaluated lost more than 5% of their weight during the postoperative period. Many patients cannot withstand additional weight loss if they have entered surgery with the presence of nutritional deficits. Further weight loss may increase the chances of complications and lengthen hospital stay. Allowing a patient to eat as soon after surgery as it is possible and safe is recommended.

Nutrition Therapy

Nutrition Implications Major surgical procedures, including gastrointestinal procedures such as **esophagectomy** or pancreatic resections, result in metabolic stress postoperatively (Kudsk and Sacks 2004). These procedures result in the cycle of metabolic stress described throughout this chapter. Furthermore, the ability to meet nutritional needs by an oral diet may be prevented by the postoperative status of the GI tract, pain, and other systemic effects of a major surgery. Nutritional needs are increased in order to accommodate repair and healing postoperatively. Recovery of functional status requires adequate nutritional support.

Nutrition Interventions The surgeon and registered dietitian will recommend the progression for postoperative feeding on an individual basis. It is suggested, though, that the patient should be progressed from nil per os (NPO) to solid food as quickly as possible (Schulman and Sawyer 2005). In a randomized trial of 96 patients undergoing major abdominal surgery, patients tolerated advancement to solid food after an initial trial of 500 mL of clear liquids (Steed et al. 2002).

Energy and protein requirements are established individually and should support postoperative healing. Resting energy expenditure (REE) with appropriate activity and injury factors are typically used to set target energy requirements, but as stated earlier, are not without limitations. Protein requirements are elevated above the RDA of 0.8 g/kg but vary depending on the type of surgery and preoperative nutritional status. Table 25.4 provides guidelines for injury factors to consider when estimating energy and protein requirements for surgical patients.

For those patients who experience metabolic stress postoperatively, nutritional needs will increase to meet the demands of hypermetabolism and catabolism. Enteral and parenteral feeding may be initiated immediately after surgery if prolonged NPO status is anticipated. Examples include major bowel or pancreatic resections (see Chapters 7 and 16).

Sepsis, Systemic Inflammatory Response Syndrome (SIRS), and Multi-Organ Distress Syndrome (MODS)

Definition

Sepsis has historically been defined as an uncontrolled inflammatory response to infection or trauma. It is now thought that sepsis is actually an immunosuppressive process that prevents an adequate response to infection (Hotchkiss and Karl 2003). The Society for Critical Care Medicine, along with eleven other international organizations, defined severe sepsis as "infection-induced organ dysfunction or hypoperfusion abnormalities." Septic shock is defined as "hypotension not reversed with fluid resuscitation and associated with organ dysfunction or hypoperfusion abnormalities" (Dellinger et al. 2004).

Systemic Inflammatory Response Syndrome (SIRS) is an additional classification of this condition not necessarily caused by an infectious process. SIRS may occur after major surgery or trauma, or with other conditions such as myocardial infarction (Brun-Buisson C 2000).

Multi-Organ Distress Syndrome (MODS) results from the complications of sepsis and SIRS. MODS is defined as "the presence of the altered function of 2 or more organs in an acutely ill patient, such that homeostasis cannot be maintained without intervention" (Dellinger et al. 2004).

Epidemiology

Over 10 million cases of sepsis were identified in the 750 million hospitalizations analyzed from 1974–2000 (Martin et al. 2003). In 2001, sepsis was rated as the 11th leading cause of death with approximately 25,000 cases (National Center for Health Statistics, Data for Injury, 2006).

Etiology

As stated previously, sepsis was originally thought to be a result of an overwhelming systemic infection. Researchers are now beginning to realize the complexity of the cascade of

esophagectomy—surgical procedure resecting or removing the esophagus

TABLE 25.7

Diagnostic Criteria for SIRS

SIRS is present with any 2 of the following conditions:

- Temperature > 38.0°C or < 36.0°C
- Heart rate > 90 beats per minute
- Respiratory rate > 20 breaths per minute
- Partial pressure of carbon dioxide (PCO_2) < 32 mm Hg
- Leukocytosis (white blood cell [WBC] count > 12,000 μL^{-1})
- Leukopenia (WBC count < 4,000 μL^{-1})
- Normal WBC count with > 10% immature forms

Reference: Dellinger RP, Carlet JM, Masur H, Gerlach H, Calandra T, Cohen J, Gea-Banacloche J, Keh D, Marshall JC, Parker MM, Ramsay G, Zimmerman JL, Vincent JL, Levy MM. Surviving sepsis campaign guidelines for management of severe sepsis and septic shock. *Crit Care Med.* 2004 Mar;32(3):858–873.

TABLE 25.8

Diagnostic Criteria for MODS

In MODS, organ dysfunction is identified by:

- Arterial hypoxemia (PaO_2/fraction of inspired oxygen [FiO_2] ratio of < 300 torr)
- Acute oliguria (urine output < .5 mL · kg^{-1} · $hour^{-1}$ or 45 mmol/L for at least 2 hours)
- Creatinine > 2.0 mg/dL
- Coagulation abnormalities (international normalized ratio > 1.5 or activated partial thromboplastin time > 60 seconds)
- Thrombocytopenia (platelet count < 100,000 μL^{-1})
- Hyperbilirubinemia (plasma total bilirubin > 2.0 mg/dL or 35 mmol/L)
- Tissue-perfusion variable: hyperlactatemia (> 2 mmol/L)
- Hemodynamic variables:
  arterial hypotension (systolic blood pressure [SBP] < 90 mm Hg)
  mean arterial pressure [MAP] < 70 mm Hg,
  SBP decrease > 40 mm Hg

Reference: Dellinger RP, Carlet JM, Masur H, Gerlach H, Calandra T, Cohen J, Gea-Banacloche J, Keh D, Marshall JC, Parker MM, Ramsay G, Zimmerman JL, Vincent JL, Levy MM. Surviving sepsis campaign guidelines for management of severe sepsis and septic shock. *Crit Care Med.* 2004 Mar;32(3):858–873.

FIGURE 25.8 Pathophysiology of Sepsis, SIRS, and MODS

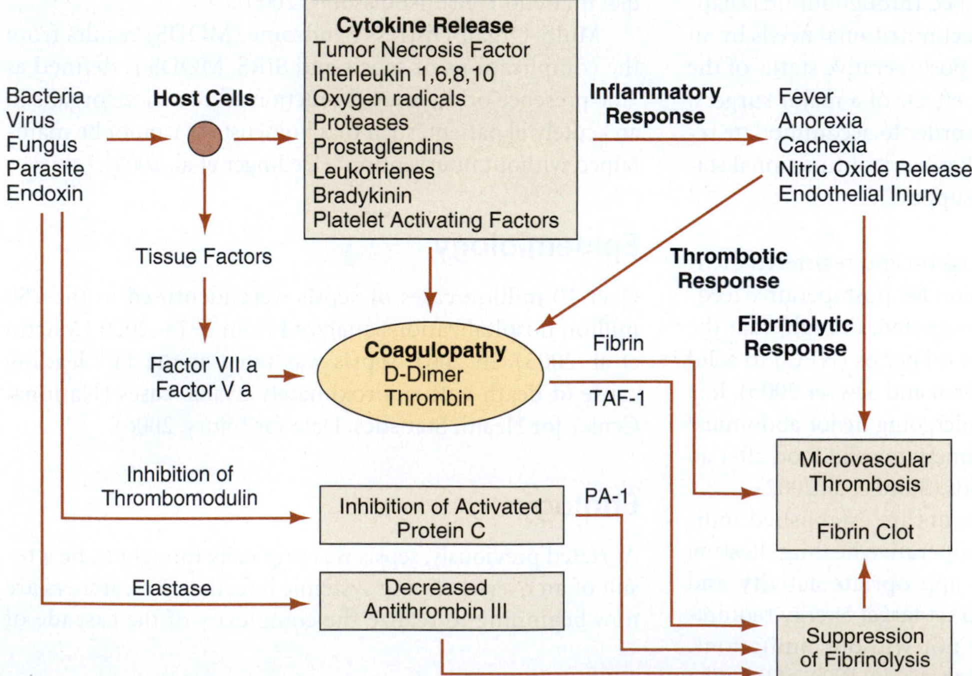

Source: Sat Sharma, MD, FRCPC, FCCP, Anand Kumar, MD, FRCPC, FCCP, FCCM, *Current Opinion in Pulmonary Medicine* 9(3):199–209, 2003.

events associated with sepsis and that the similar syndrome of SIRS can occur without infection. A combination of proinflammatory cytokine release, altered cellular metabolism, **hypoperfusion**, and **hypotension** direct the physiological changes that occur with sepsis, SIRS, and MODS (Sharma and Fumar 2003; Dellinger et al. 2004).

Clinical Manifestations

The initial major signs of sepsis include increased white blood cell count (>12,000 mm^3), increased heart rate (>90 beats per minute) and respirations (>20 breaths/minute), and fever (>38°C) or hypothermia (<36°C) (Dellinger et al. 2004; Munford 2005). C-reactive protein, fibrinogen, complement proteins, and other acute-phase proteins are elevated (Gabay and Kushner 1999). Other laboratory values may include increased serum lactate and serum glucose.

The diagnostic criteria for SIRS and MODS as defined by Dellinger and colleagues (2004) are presented in Table 25.7 and Table 25.8, respectively.

Pathophysiology

This complex multifactorial condition is initiated by an originating source of infection and/or trauma. The systemic response causes the release of inflammatory mediators including TNF, interferon, and interleukins (see Chapter 12). As sepsis continues, there is a shift from an inflammatory response to an anti-inflammatory response with resulting anergy (inability to mount an immune response). Deaths of immunocompetent cells contribute to the inability of the body to respond appropriately to the sepsis (Hotchkiss and Karl 2003). The increased rate of gluconeogenesis results in significant catabolism of skeletal muscle mass. These metabolic abnormalities result in hyperglycemia and increased serum lactate (see Figure 25.8).

Treatment

Treatment of sepsis will center on removal of the source of infection or trauma with hemodynamic, renal, respiratory, and metabolic support (Rivers et al. 2001; Hotchkiss and Karl 2003; Dellinger et al. 2004; Schrier and Wang 2004). The use of intensive insulin therapy, antimicrobial agents and coagulation-modulating drugs (such as drotrecogin alfa activated), and nutrition support are all crucial components of the treatment protocols for sepsis (Sharma and Fumar 2003; Dellinger et al. 2004).

Nutrition Therapy

Nutrition Implications Nutrition support is a critical step in meeting the challenges of treatment and resolution of sepsis. Challenges to meet the needs of these critically ill patients include abnormalities of metabolism, difficulty estimating and/or measuring nutritional requirements, fluid/volume restrictions, and multi-system organ dysfunctions. Nutrition support plans will follow the standard procedures outlined previously in this chapter.

Summary

Metabolic stress represents the most challenging and demanding environment for nutritional care. In no other situation are the complications of inadequate nutritional support more evident. Research continues to provide practical application of nutrition support to optimize outcomes for the critically ill.

hypoperfusion—reduced blood flow

hypotension—low blood pressure

(Mann 1989). In the late 1970s and early 1980s, there were conditions later attributed to HIV infection called "Slim Disease" in Zaire (now the Democratic Republic of Congo), Uganda, and Tanzania. HIV may have affected at least five continents, including North and South America, Europe, Africa, and Australia, by the early 1980s (AVERT 2006). There may have been between 100,000 to 300,000 persons infected by the time the first discussions that led to the description of AIDS were documented in 1981 (Hymes, Greene, and Marcus 1981; Osmond 2003). Rare cases of *Pneumocystis carinii* pneumonia (abbreviated PCP, though now referred to as *Pneumocystis jiroveci* in the form that is infectious to humans) were documented and on the rise in both California and New York when the Centers for Disease Control and Prevention (CDC) took notice in April of 1981 and published a report noting an unidentified cause for five cases of PCP in the Los Angeles area.

In 1982, the disease syndrome was dubbed with several names, including the gay-related immune deficiency (GRID) and the community-acquired immune dysfunction. By July 1982, there were a total of 452 cases from 23 states reported to CDC, including cases in Haitians and patients with hemophilia. The disease was dubbed the Acquired Immune Deficiency Syndrome at that time, suggesting the immune deficiency was acquired and the manifestation was multifactorial. When a child who had received multiple blood transfusions was diagnosed and the first cases of possible mother-to-child transmission were reported in December of 1982, it became apparent that the disease was caused by an infectious agent. By mid-1983, a virus was isolated as an etiologic agent for AIDS and was named the lymphadenopathy-associated virus (LAV). Later that year, the first globally oriented meeting to discuss an "AIDS epidemic" was convened, and surveillance was initiated. At that time, there were 3,064 reported cases and 1,292 reported deaths attributed to AIDS in the United States. In 1984, isolation of the human T-lymphotropic virus-type 3 (HTLV-III) was announced, along with the hope that a vaccine for testing would be developed within two years. By 1986, the name human immunodeficiency virus (HIV) was adopted, and the director of the WHO suggested that there may have been as many as 10 million people with HIV infection worldwide. By the end of that year, 38,401 cases had been reported, 31,741 of which were in the Americas. During that time, the first anti-HIV treatment, azidothymidine (AZT), was tested. Studies were discontinued early due to a significant difference in survival noted in the first six months.

Many more developments in care and treatment were achieved over the subsequent decade, including medication combinations effective in reducing viral burden and progression of the disease. However, most of these advantages that blunted the deadliness of the infection were not realized by people living in developing countries. By the end of 2005, there were an estimated 40.3 million people with HIV infection in the world, most of whom resided in southern and eastern African countries (UNAIDS 2005). Estimates of new infections in 2005 were 4.9 million people, including approximately 700,000 children. More than 3 million people may have died from complications related to HIV infection in 2005. In the United States, there were an estimated 850,000 to 950,000 people living with HIV infection and AIDS (**PLHA**), with an estimated 40,000 new infections each year (Centers for Disease Control and Prevention 2005). The majority of new infections are in minority populations, women, and youth who may have less than optimal access to health care.

HIV and AIDS present a wide variety of challenges for nutrition status maintenance. Changes in nutritional status can result from HIV infection, disease complications and co-infections, and disease treatments. The interactions with nutritional status include social, economic, and clinical issues. Populations already at risk for nutritional compromise, because of lack of health care access, lifestyle choices (such as smoking, alcohol, and drug abuse), **food and nutrition insecurity** (the lack of consistent access to an adequate and appropriate food supply), and comorbidities such as hepatitis, diabetes, or other conditions, may find their health status worsened with HIV infection and AIDS. Alterations in nutrient intake, absorption, metabolism, and excretion have been documented throughout the disease spectrum. While these interactions are complex and multifactorial in nature, this chapter will concentrate on clinical aspects of the disease and potential interventions.

PLHA—people living with HIV and AIDS; other acronyms include PLWHA (people living with HIV/AIDS) and PWA (people with AIDS)

food insecurity—lack of adequate access by all people, at all times, to sufficient food for an active and healthy life, including at a minimum a readily available supply of nutritionally adequate and safe foods and an assured ability to acquire acceptable foods in a socially acceptable way

nutrition insecurity—the provision of an environment that encourages and motivates society to make food choices consistent with short- and long-term good health

Normal Anatomy and Physiology of the Immune System

This section provides a brief review of immune functions targeted by HIV and the interactions between nutrients and nutritional status that may be helpful to keep in mind when working with persons infected with HIV. Immunity can be classified as: (1) primary or secondary, according to whether it is an initial or subsequent contact with the antigen, (2) humoral or cellular, depending on the activities of B cells or T cells, respectively, and (3) active or passive, depending on whether it is acquired through contact with an antigen or provided by a transfer of pre-sensitized or activated immune cells (see Chapter 12). B cells can neutralize and destroy invaders that are not incorporated into host cells, and T cells will kill a host cell based on the expression of a foreign antigen on its surface. T cells are a primary target of HIV. Helper T cells can secrete interleukins, which stimulate proliferation of T cells. Cytotoxic/suppressor T cells kill infected host cells and modulate B and T cell responses.

Activated **CD4 cells** and several other types of cells, including macrophages (white blood cells that engulf foreign material), are infected and rendered dysfunctional by HIV. CD4 cells are particularly targeted by HIV infection. These CD4 cells are manufactured in the bone marrow and mature in the thymus. CD4+ cells (matured CD4 cells) are also called "T helper" cells, which help to mediate cell recognition events that direct some immune activities to target foreign antigens, promote the differentiation of B cells and cytotoxic T cells, and activate macrophages. The impairment of CD4+ cell activity associated with HIV infection results in defects in other immune cell activities and leaves the body open to many types of infection and malignancy.

Though HIV infection is widespread and has a tremendous impact on the immune function of the HIV-infected person, malnutrition remains the most prevalent cause of immune dysfunction in the world today. While any malnutrition that disturbs the normal synthesis and function of proteins in the body will impair immunity, general protein-energy malnutrition is the most commonly noted cause of immune dysfunction (Serog 1990). However, even changes in single nutrients, whether related to altered nutrient intake, disease, or aging, have the potential to impair immune defenses (Amati et al. 2003). Some of the micronutrients closely tied to immune function include iron, zinc, copper, selenium, and vitamins A, D, thiamine, riboflavin, pyridoxine, pantothenic acid, cobalamin, biotin, folic acid, and ascorbic acid (Cunningham-Rundles, McNeeley, and Moon 2005).

Etiology

HIV is a **retrovirus** and thus contains RNA that is transcribed to dual strand DNA for incorporation into the host cell DNA (see Chapter 11 for more on DNA and RNA). HIV is approximately 0.1 microns in diameter, about 1/70th of the diameter of the CD4 immune cell that it particularly targets. It contains nine genes, six of which are essential to penetrate and infect the target cells and produce copies of the virus. The other three genes are used to provide the necessary information to produce new viral particles in the host cell. HIV is most typically transmitted via blood through sexual contact (generally penile-anal or penile-vaginal intercourse), blood transfusion, intravenous needle sharing, and perinatally (from mother to child) through blood or breast milk. HIV transmission, like other disease transmission, depends on the inoculation dose and number of exposures (Quinn et al. 2000).

This retrovirus targets many cells in the body, including gastrointestinal cells, organ cells, and immune cells, among others. The resulting immunodeficiency syndrome is most closely related to the infection of activated CD4 T helper cells, which become a viral factory, as shown in Figure 26.1. The virus identifies the target CD4 cell and fuses to the surface. HIV injects RNA, enzymes, and other substances that assist in viral integration and replication. Using its own kit of injected substances, HIV RNA is transcribed to DNA particles using reverse transcriptase enzymes. The DNA is carried to the nucleus and integrated into the host DNA using its own integrase enzymes. At this point, the integrated viral materials can remain dormant until they are activated, at which time they command the cell to become a viral factory that manufactures viral components. Protease enzymes cleave the viral proteins for assembly into viral cores. Once fully assembled, the virus is ready to bud out of the infected host cell. As the host CD4 cell manufactures, assembles, and releases viruses, it is incapacitated and destroyed. In addition, macrophages harboring HIV are rendered dysfunctional. It is through this process that the immune system is compromised and HIV disease progresses.

Pathophysiology

Primary HIV infection is often accompanied by flu-like symptoms and a reduction in CD4 cell counts. As CD4 and other cells are damaged and rendered dysfunctional, the body's defenses against infection and malignancy may decline. There is a strong relationship between CD4+ cell counts, **viral load** (in copies per milliliter) and progression to a diagnosis of AIDS (Mellors et al. 1997). The higher the viral burden of HIV, the more CD4 cells are infected, rendered dysfunctional, and destroyed, leaving the body open to opportunistic infections and cancers. Lower viral

CD4 cell—immune cell that is one of the primary targets of HIV for infection

retrovirus—a virus that carries RNA rather than DNA; RNA must be transcribed prior to integrating into the host cell DNA to reproduce

primary HIV infection—the time of the initial seroconversion to HIV infection; usually involves a spike in the level of the virus and sometimes is accompanied by a flu-like syndrome

viral load—the level of virus or viral markers measured in the blood

TABLE 26.2

WHO Clinical and Immune Cell Categories of HIV Infection

Categories*	Sample Criteria
Primary HIV Infection	Acute retroviral syndrome, but no complicating opportunistic infection or immune dysfunction
Clinical Stage 1	Primarily asymptomatic as above, possible persistent generalized lymphadenopathy
Clinical Stage 2	Weight losses that are <10% of body weight, minor mucocutaneous manifestations, recurrent bacterial upper respiratory tract infections, fungal infections of fingers
Clinical Stage 3	Weight loss of >10% of body weight, persistent constitutional symptoms (fever, diarrhea), oral thrush or hairy leukoplakia, pulmonary tuberculosis, severe bacterial infections, unexplained anemia, neutropenia, and/or thrombocytopenia for more than a month (confirmatory testing is required for anemias)
Clinical Stage 4	HIV wasting syndrome (>10% weight loss with chronic diarrhea, weakness, fever), opportunistic events as described in Clinical Category 3

Source: World Health Organization. Interim WHO Clinical Staging of HIV/AIDS and HIV/AIDS Case Definitions for Surveillance (2005).

Group for the Study of the WHO Staging System, 1993). This system attempts to categorize patients into four clinical stages according to the severity of disease manifestations (see Table 26.2). The current set of staging guidelines has separated primary infection from three clinical stages, similar to the CDC classifications (World Health Organization, Interim WHO Clinical Staging, 2005). Additional diagnoses that may be recognized in the WHO set of guidelines include fungal nail infection and dermatitis for Clinical Stage 2 and unexplained but confirmed anemias for Clinical Stage 3 (World Health Organization, HIV/AIDS Clinical Staging, 2005).

Clinical Manifestations

Chronic HIV infection and related complications or cofactors affect the individual's presentation. HIV infection and related immune dysfunction can lead to disturbances and damage to a number of body systems. A summary of infections and other complications along with selected nutritional implications is shown in Table 26.3.

As noted in the Pathophysiology section, HIV infection can lead to immune dysfunction, leaving the body open to a number of opportunistic conditions. In addition, the initiation of inflammatory processes can lead to altered nutrient metabolism, including a subclinical wasting effect on the body's protein stores.

Neurologic disorders, such as neuropathy and dementia, are common in HIV infection and treatment. Neuropathy is often peripheral and associated with pain, numbness, tingling, and burning sensations. This type of neuropathy may start with the feet and can spread. Peripheral neuropathy has been related to a number of the antiretroviral therapies. Dementia and altered cognitive functions can be related to HIV infection, other infections, and nutrient deficiencies. The loss of ability to adequately provide self-care and maintain nutritional status can include inability to self-feed and engage in routine physical activity.

Pulmonary disorders may occur both related to and unrelated to HIV infection. The level of CD4+ cell count is related to the risk for the development of several types of pulmonary diseases. With some of the many symptoms associated with pulmonary problems, including persistent coughing, chest pain, fatigue, fever, and shortness of breath, it may become more difficult for the patient to maintain adequate food intake. Pulmonary hypertension has been demonstrated in patients with HIV infection without AIDS-defining diagnoses. The use of **antiretroviral therapy (ART)** has significantly diminished the effect of pulmonary hypertension in patients with HIV infection.

Cardiac manifestations may include infections and inflammation, cardiomyopathy, pulmonary hypertension, and coronary artery disease. Cardiomyopathy may be related to infectious agents, inflammation processes, and medication therapies. For instance, the **antiretroviral (ARV)** medication zidovudine has been associated with skeletal muscle myopathy. However, overall it appears that there may be a protective effect of ART on the mortality associated with myocardial disease (Bijl et al. 2001).

antiretroviral therapy (ART)—refers to the combination of medications that are typically used for controlling and reducing viral load

antiretroviral (ARV)—refers to medications targeted to interrupt the retrovirus life cycle

TABLE 26.3

Selected Clinical Manifestations of HIV Infection and Treatment

System	Examples	Nutritional Implications
Cardiac	Pericarditis/endocarditis Pulmonary hypertension Coronary artery disease	Risk for congestive heart failure and myocardial infarction may benefit from dietary modulation
Central and Peripheral Nervous System	HIV can cross the blood-brain barrier and lead to cognitive disorders; HIV is also associated with neuropathy	HIV-related dementia can interfere with activities of daily living, including self-care and feeding; neuropathy can limit the ability to exercise and maintain nutritional status
Gastrointestinal Tract and Symptoms	Rapid intestinal cell turnover; infection of intestinal immune cells Nausea/vomiting Abdominal pain Diarrhea	Immature enterocytes, reduction in intestinal enzyme production, and blockage of intestinal surface can lead to malabsorption with or without diarrhea. Often the side effect of medication therapies, nausea, vomiting, abdominal pain, and diarrhea may lead to inadequate nutrient intake or nutrient losses and require both dietary and medication interventions
Hematologic	Anemias	Anemias and related fatigue can impair physical capacity and the ability to maintain body composition
Hepatic	Hepatitis Biliary tract disorders	Common co-infections and disorders with HIV may lead to dietary restrictions consistent with hepatic disease; intolerance to fat and deficiency of pancreatic hormones can result in malabsorption
Immune System	Infiltration and destruction of immune cell function and numbers; increased threat of opportunistic events	The body's response to infection is dependent on the severity of the infection or trauma; repeated opportunistic events can lead to progressive wasting, particularly of body cell mass, and greater risk for morbidities and mortality
Musculoskeletal	Myopathy	Fatigue and muscle weakness limit the capacity to exercise and maintain body composition
Neurologic	Neuropathy Dementia/altered mental status	Pain, weakness, and hypersensitivity to touch may impair capacity to exercise and appetite; impairment in cognitive function can interfere with self-care, including food-related activities
Oral	Candidiasis Herpes Periodontal disease Salivary gland disease	Oral pain, taste changes, and dry mouth can lead to a reduction in food intake and increased risk for weight loss
Pulmonary	Bacterial pneumonia Kaposi's sarcoma Fungal pneumonia	Fever, fatigue, difficulty breathing, and persistent coughing may impair adequate food intake; pulmonary involvement may reduce exercise capacity and limit the ability to maintain nutritional status through exercis
Renal	HIV-associated nephropathy (HIVAN)	Progression to end-stage renal disease is less common with ART, but can lead to significant dietary restrictions
Systemic	Inflammatory process as the body fights HIV infection	Catabolism of muscle stores and fluid shifts lead to subclinical wasting with or without weight losses; hormonal changes that occur with chronic inflammation can lead to alterations in nutrient metabolism

Chronic inflammation, immune dysfunction, concomitant conditions (such as insulin resistance, diabetes, renal involvement, and hypertension) and opportunistic conditions place a patient infected with HIV at a higher risk for cardiac disease. Much of the coronary artery disease risk is associated with medication interactions that increase blood lipid levels. However, in addition to HIV infection, there are many other cofactors to consider such as smoking, drinking, and other risk factors. Myocardial infarctions have been reported at higher rates in antiretroviral-treated patients in one long-term study (Jutte et al. 1999), but that finding was refuted by a review of records that found a decrease in hospital admissions for cardiac-related events with the use of ARV treatments (Bozzette et al. 2003).

Hepatic diseases include disorders related to opportunistic infections or cancers, toxicity related to anti-HIV treatments, and alcohol or substance abuse. Common co-infections and disorders with HIV may lead to dietary restrictions consistent with hepatic disease, including a deficiency of pancreatic hormones resulting in malabsorption.

Anemias are a common occurrence in chronic HIV infection, with higher prevalence in more symptomatic phases of the disease. Anemia can be related to chronic disease, hormonal alterations, infections, and medications. Nutrient-related anemias can be caused by inadequate intake of protein and micronutrients and can be exacerbated by altered metabolism seen in chronic inflammation of HIV and other infections.

Renal failure in the form of HIV-associated nephropathy (HIVAN) is associated with risk factors such as infection, certain medications, male gender, and black African-descent race. Tubular necrosis is associated with several medications commonly used in HIV-infected patients, such as acyclovir, tenofovir, adefovir, and cidofovir, among other nephrotoxic medications. Nephrolithiasis (formation of kidney stones) has been associated with indinavir, and adequate hydration is a key recommendation for its use. Volume depletion can occur in patients experiencing diarrhea related to infections or medication intolerance, which increases the risk for renal involvement and hyponatremia, hypokalemia, and other imbalances.

Oral lesions associated with immune suppression can have a strong impact on normal food intake. Oral manifestations can be caused by fungal infection, viral infection, bacterial conditions, neoplastic problems, salivary gland disease and others. Oral lesions can lead to mouth itching, pain, a burning sensation (especially when eating spicy or acidic foods), and taste changes. Salivary gland disease is sometimes associated with dry mouth and xerostomia, which can cause discomfort in eating foods that are dry or crisp. Gastrointestinal problems include malabsorption and constitutional symptoms related to HIV and other gut pathogens, side effects of medications, and gastrointestinal damage.

Treatment

Mortality rates to date are nearly 60% for adults and children worldwide. The advent of effective combinations of antiretroviral drugs has reduced the reported cases of its symptomatic manifestation as AIDS by nearly one-third, and reported deaths by nearly half, since 1995 (Centers for Disease Control and Prevention 2002). Prior to the common use of combination medication treatments (which became widespread in 1996), HIV infection was considered a progressive disease with little chance for a return to health because of recurrent infections and other illnesses that led inevitably to death. Wasting and nutritional decline were commonly reported, with the anticipation that though approximately 20% of the initial AIDS diagnoses were made based on AIDS-related wasting, nearly all patients would experience this type of decline prior to their death. In 1989 it was reported that wasting was a strong predictor of the timing of death (Kotler et al. 1989). The progression of HIV infection to AIDS and the rate of mortality have dropped significantly since the introduction of combinations of therapies that effectively reduce viral burden and destruction of immune cells in order to allow immune reconstitution. Treatment for chronic HIV infection includes antiretroviral medications, prevention and treatment for opportunistic events, modulation of altered hormonal milieu, and maintenance and restoration of nutritional status.

Anti-HIV Therapies

Antiretroviral medications are used to lower viral load, and the goal is to achieve and maintain an undetectable level of less than 50 copies/mL in serial tests. There are currently five classes of antiretroviral medications, including **fusion inhibitors**, nucleoside/nucleotide **reverse transcriptase inhibitors** and non-nucleoside **reverse transcriptase inhibitors**, **integrase inhibitors**, and **protease inhibitors**. A summary of ARV treatments is shown in Table 26.4.

fusion inhibitors—medications that interrupt the viral replication cycle by inhibiting fusion of the HIV virus to the target cell

reverse transcriptase inhibitors—medications that interrupt the viral replication cycle by inhibiting reverse transcriptase enzymes that allow the viral RNA to be transcribed to DNA before being integrated into the host cell DNA

integrase inhibitors—medications that interrupt the viral replication cycle by inhibiting integrase enzymes that allow the transcribed viral DNA to integrate into the host cell DNA

protease inhibitors—medications that interrupt the viral replication cycle by inhibiting protease enzymes that allow the viral proteins to be cleaved for reassembly into viral cores

TABLE 26.4

Food- and Nutrition-Related Medication Interactions for Selected Antiretroviral Therapies

Drug/ Abbreviation	Class	Brand Name	Diet Requirements	Diarrhea	Nausea/ Vomiting	Appetite Loss	Abdominal Pain	Taste Change	Lipid Alterations*	Glucose Intolerance	Lipo-dystrophy
Abacavir/ABC	PI	Ziagen	No food restrictions	X	**X**	X	X				
Amprenavir/APV	PI	Agenerase	Avoid taking with high-fat meal	**X**	**X**		X	**X**		**X**	
Atazanavir/ATV	PI	Reyataz	Take with food	X	X			X		X	
Darunavir/TMC114/r	PI	Prezista	Take with food	X	X		X		X	X	X
Delavirdine/DLV	NNRTI	Rescriptor	No food restrictions; take with acidic beverage	X	X	X					
Didanosine/ddI	NRTI	Videx	Take without food	**X**	**X**	**X**	X	X	X		
Efavirenz/EFV	NNRTI	Sustiva	Avoid taking with high-fat meal	**X**	X	X	X		**X**		
Emtricitabine/FTC	NRTI	Emtriva	No food restrictions	**X**	X	X	X				
Enfuvirtide/T-20	FI	Fuzeon	No food restrictions			X		X	X		
Fosamprenavir/FPC	PI	Lexiva	No food restrictions	X	X		X				
Lamivudine/3TC	NRTI	Epivir	No food restrictions	X	**X**	X	X			X	
Lopinavir/LPV/r	PI	Kaletra	Take with food (no food restrictions)	**X**	**X**		**X**		X	X	X
Nelfinavir/NFV	PI	Viracept	Take with food	**X**	X	X	X			X	
Nevirapine/NVP	NNRTI	Viramune	No food restrictions		**X**		X				
Ritonavir/RTV	PI	Norvir	Take with food	**X**	**X**	X	X	X	X	X	X
Saquinavir/SQV	PI	Fortovase or Invirase	Take with food	**X**	X		X	X	X		X
Stavudine/d4T	NRTI	Zerit	No food restrictions	**X**	**X**	X	X		X		X
Tenofovir/TDF	NNRTI	Viread	No food restrictions (high fat meal increases bioavailability)	X	**X**	X	X				
Tipranavir/TPV	PI	Aptivus	Take with a full meal, preferably high fat	**X**	X	X			X	X	X
Zidovudine/ AZT or ZVD	NNRTI	Retrovir	No food restrictions		**X**	**X**					

Bolded X suggests increased incidence or severity.

* Protease inhibitors are often associated with elevated lipids; complete information on the effect of darunavir on lipids was unavailable, however elevated total cholesterol was reported by Grinsztejn B et al. 2005.

Source: Drug Facts and Comparisons, 2006.

Combinations of ARV medications that inhibit the various segments of the life cycle of HIV infection effectively have been referred to as "highly active antiretroviral therapy" or **HAART**, and generally include the use of three or more medications. Recommendations for the use of ARV therapies in adults and children have been developed by the CDC and are generally followed by most practitioners (Centers for Disease Control and Prevention March 2004, January 2004).

The categories of HIV infection used in guidelines to determine treatment strategies include the degree of immunosuppression as well as opportunistic events (Department of Health and Human Services 2005; Health Resources and Services Administration 2005). Anti-HIV treatments are aimed at interrupting the viral life cycle at one or more points. However, because the virus can reproduce rapidly, generating between a billion and a trillion virons per day, the potential for alterations in genetic structure (or "mutation"), and hence drug resistance, is high. Several strains may exist in a single person, and depending on the suppression of the particular strains, some strains may still replicate, leading to immune destruction despite treatment. Combination medications that attack different parts of the viral life cycle or that can be used with strains that are resistant to some of the drugs in a class can be used to prevent disease progression and further health decline.

The transmission of a mutated virus from one person to another can also limit the number and types of medications that can successfully suppress viral burden and prevent disease progression. A virus that has not undergone mutations due to the use of ART is referred to as a "Wild Type" virus and is generally susceptible to most ARVs. If the presence of ARVs causes the virus to mutate, then drug resistance can follow. Once mutations occur, ART can "fail" as viral load increases and CD4+ cell destruction accelerates. ARVs may be introduced and discontinued based on observation of these effects or through direct and quicker genotype and phenotype testing for specific types of mutations (see the Diagnosis section).

Adherence to therapy regimens is a major determinant of ART success. Successful ART requires nearly perfect adherence at 95% or better. However, actual adherence rates are estimated at 20% to 50%, largely because of fear and experience of side effects and symptoms of these potent chemotherapy combinations (World Health Organization 2003). The most common challenges to adherence include pill burden, complexity of regimen, understanding of appropriate use of the medications (including diet interactions), and potential or experienced adverse effects. With adequate preparation, screening for the most appropriate therapy regimens, and careful management of adverse effects of therapy, adherence can be improved. Treatment is only one aspect of the medical management of HIV infection. Interestingly, even in so-called asymptomatic states where viral load is fully suppressed to undetectable levels, the effects of HIV infection continue to exist with evidence of continuing **inflammatory responses** that can alter body functions and maintenance.

HAART—highly active antiretroviral therapy; a combination of ARVs that is able to fully suppress the virus

inflammatory response—the body's response to infection or injury that is cortisol-driven and allows for the breakdown of labile body protein to increase the amino acid pool for the purpose of synthesizing protective and healing proteins

Prevention and Treatment of Opportunistic Disease

Opportunistic infections and other diseases that occur as a result of immune dysfunction are associated with further nutritional and health decline. Opportunistic infections may initiate a reduction in food intake and lead to episodic wasting (Macallan et al. 1993) as well as activate CD4+ cells, making them active viral factories and a target for further spread of the infection in the body. Prophylaxis and early treatment for opportunistic events can help to prevent nutritional decline and disease progression. Guidelines for the prevention and treatment of opportunistic infections and other events in PLHA have been published by the CDC (Benson et al. 2004). Prevention efforts with medications are recommended for varying CD4+ levels according to the risk or history of infection. Table 26.5 shows a summary of selected treatments for opportunistic events and nutritional implications.

Nutrition Therapy

Support for the maintenance and restoration of nutritional status is essential to the management of chronic HIV infection. Poor clinical outcomes have been documented in patients who are malnourished compared to those who are not. Intervention with macronutrient and micronutrient therapies to support immune function could play an important role in the course of HIV disease (Macallan 1999).

HIV infection itself can present a formidable challenge to the maintenance of nutritional status and health. Because there is no cure for HIV infection, treatment strategies are aimed at preventing and slowing immune compromise and containment of complications of immune suppression, opportunistic disease, and nutritional compromise.

TABLE 26.5

Nutritional Implications of Selected Treatments for Opportunistic Events

OI	Medications	Nutritional Implications
Candidiasis	Fluconazole	Take with/without food; may cause nausea, vomiting, abdominal pain, taste changes
Cytomegalovirus	Valganciclovir Ganciclovir Foscarnet	Valganciclovir/ganciclovir: take with food; may cause nausea, diarrhea, anemia Foscarnet: may cause anorexia, nausea, vomiting, abdominal pain, diarrhea, taste changes, anemia, renal abnormalities
Hepatitis C	Peg interferon	Ensure hydration; may cause anorexia, nausea, vomiting, abdominal pain, diarrhea, dry mouth and taste changes, anemia
Herpes Simplex	Acyclovir	Ensure hydration; may cause nausea, vomiting, abdominal pain, diarrhea, renal abnormalities
Mycobacterium Avium Complex (MAC)	Azithromycin or clarithromycin	Azithromycin: take without food; may cause nausea, diarrhea Clarithromycin: take with/without food; may cause dyspepsia, abnormal taste, abdominal pain, diarrhea
P. jeroveci (PCP)	Trimethoprim-sulfamethoxazole (TMP-SMX)	Take with food, ensure hydration, long-term use may require supplemental folic acid; may cause nausea, vomiting, anorexia

Sources: Bartlett JG, *Pocket Guide to Adult HIV/AIDS Treatment;* January 2005. Pronsky ZM, *HIV Medications-Food Interactions;* 2001.

Nutrition Implications of Disease and Treatment

After exposure, seroconversion to an HIV-positive status starts the lifelong inflammatory process. This process is similar to other infections in that it is the severity of infection that determines the level of inflammatory response. A simplified overview from a nutritional response standpoint includes a continued cortisol response to HIV infection, leading to breakdown of labile body protein stores to feed the inflammatory response. This response includes alterations in nutrient intake and utilization along with changes in hormone sensitivity and levels that regulate nutritional status (Fraker 1994; Semba 1998). Even during "asymptomatic" phases of HIV infection, body composition testing shows alterations reflecting the inflammatory process and suggesting that subclinical wasting can continuously and progressively occur (Ott et al., Early Changes, 1993).

The **AIDS-related wasting syndrome (AWS)** is an AIDS-defining diagnosis that was added into the CDC definition in 1987 (Centers for Disease Control and Prevention 1987). The definition states that weight loss of 10% without any known cause accompanied by fever or diarrhea for more than a month is AWS and qualifies for the diagnosis of AIDS. The limitations of this definition are many. Significant weight loss often occurs during opportunistic infection or other events, the definition does not suggest a time frame for weight loss that would help to qualify the severity of compromise, weight loss may not be accompanied by fever or diarrhea for more than 30 days, and the data may not be captured as an AIDS-defining event if an opportunistic infection is concurrently identified. With the advent of combination antiretroviral therapies that can achieve low and undetectable viral loads, wasting should be less prevalent. Yet in studies on the effect of antiretrovirals on nutritional status, the results continue to be mixed, with AWS continuing to define nearly 20% of AIDS cases in the United States (Carbonnel et al. 1998; Moore and Chaisson 1999). Ongoing cohort studies suggest that the lack of changes in prevalence of wasting may reflect nutritional insecurity more than the disease process when compared to the pre-HAART era (Wanke 2004).

Changes in body composition during weight loss in HIV infection appear to fit the profile of starvation or marasmus (Forrester et al. 2001). This suggests that, though weight loss episodes are generally matched to an opportunistic event, a combination of increased nutrient needs, decreased nutrient intake, and malabsorption may play a role in precipitating

AIDS-related wasting syndrome (AWS)—defined by the Centers for Disease Control and Prevention (CDC) as a 10% weight loss without an identifiable cause that is accompanied by fever or diarrhea for 30 days or more

weight loss episodes. Patients with advanced HIV infection may even consume more kilocalories than their asymptomatic and HIV-negative counterparts, while still losing ground (Grunfeld et al. 1992). Patients who lose weight during opportunistic infection may not fully recover their weight or normalize their body composition (Macallan et al. 1993). Conflicting evidence suggests that wasting may be more complicated than simple starvation, and weight loss might not always be significant while a wasting process occurs. A longitudinal evaluation of 172 HIV-infected men revealed a loss of lean tissues without significant changes in body weight. This suggests that body composition changes may be a common occurrence in chronic HIV infection, and be driven by cytokine mediators rather than inadequate dietary intake or by altered androgen levels (Roubenoff et al. 2002).

In addition, other factors influence the wasting process. Some medications may have the effect of reducing muscular protein synthesis, making it difficult to recover lost protein stores as HIV viral burden and associated inflammation are decreased (Hong-Brown, Brown, and Lang 2005). Treatment for wasting and for events that can trigger a wasting process, such as opportunistic infections, are both important to the restoration of patient health and quality of life.

In addition to weight loss and wasting, there are many other nutrition implications of HIV disease (Ambrus and Ambrus 2004). Because of the inflammatory nature of chronic HIV infection, evident even when viral load is fully suppressed, nutrient metabolism can be expected to change. Positive and negative **acute phase proteins** are continually produced, and exert an effect on both macronutrient and micronutrient status. Documented micronutrient changes include lower serum levels of selenium, zinc, magnesium, calcium, iron, manganese, copper, carotene, choline, glutathione, and vitamins A, B_6, B_{12}, and E (Skurnick et al. 1996; Bogden et al. 1990; Semba and Tang 1999). Elevated levels of folate, niacin, and carnitine have been documented as well.

Of specific interest are the associations between micronutrient levels and complications of immune impairment, HIV infection process and progress, and treatment interactions (Tang and Smit 1998). Low levels of vitamin B_{12} are associated with neurologic changes, bone marrow toxicity in patients taking zidovudine, and accelerated progression of HIV disease. **Oxidative stress** conditions have been associated with lower vitamin E and C levels (Evans and Halliwell 2001). While there are clearly associations between micronutrient status and disease management outcomes, the ability to reverse these complications remains unclear (Lanzillotti and Tang 2005).

acute phase proteins—proteins that increase or decrease in the blood during an inflammatory process, such as C-reactive protein and fibrinogen

oxidative stress—the imbalance of pro-oxidant production and the body's antioxidant supplies that yields cell damage

body cell mass (BCM)—kcalorie-using protein stores in the body; primarily muscle and organ tissues

Nutrition Assessment

With long-term survival and chronic polypharmacy, intermittent changes in nutritional status are likely. Nutritional risk screening should be conducted on all patients with HIV infection, with follow-up assessment for identified risk factors. Table 26.6 illustrates various aspects of nutrition screening and assessment. Initial screening may include items from a self-screener that can identify general risk factors, such as access to food, consumption of food groups, symptoms, and others. Full assessment can then be completed on patients who are at risk, or if a baseline of measurements is desired. Assessment factors include physical, biochemical, and nutrition-related behaviors and social factors. Physical evaluation should include at least the minimum data set of height, weight, and body mass index (BMI). In addition, physical measures such as anthropometry, body composition, and examination for clinical signs of overnutrition and undernutrition can help to provide an overview of the patient's nutritional status and identify issues to address. Biochemical measures will include disease-related measures, nutrition-related measures, and selected measures of concomitant disease factors such as diabetes, cardiovascular disease, and others. Patient food access and food choices should be explored with consideration to psychosocial and economic factors. Table 26.6 outlines assessment factors.

Physical Assessment Physical changes, such as weight or body composition changes, are commonly seen throughout the spectrum of HIV disease and with or without antiretroviral treatment. Weight loss continues to be a complication of HIV infection, despite HAART use (Tang 2003). A history of weight changes will help to establish the level of nutritional risk. Weight losses of 10% are strong predictors of a fourfold to sixfold increase in mortality in HIV infection (Tang et al. 2002). Smaller weight losses of 3% or 5% between six-month intervals may also be predictive of mortality. BMI is a strong predictor of survival and may also be a reasonable surrogate marker for changes in CD4 counts for the initiation of ART in resource-limited settings (van der Sande et al. 2004; Zachariah et al. 2006).

The causes of weight loss episodes should be considered, because the health risk may vary according to the type of weight lost. Starvation, including dieting efforts and malabsorption, can yield lower losses of **body cell mass (BCM)** than the kcal imbalances that result from infections. Weight

TABLE 26.6

Nutritional Screening and Assessment Factors in HIV Infection

Factor	Description	Comments
Dietary Evaluation	Access to food, food consumption	Self screeners can contain questions on food group intake, meals per day, and food access Assessments may include food resources and intake analysis Food intake can be compared to estimated requirements to determine counseling needs
Physical Assessment	Weight, body mass index, physical examination for clinical signs of deficiency and excess of macronutrients and micronutrients, anthropometry to characterize body composition and patterning (including body shape changes associated with lipodystrophy); body composition analysis	Body mass index of <20, weight loss of 5%–10%, or body cell mass loss of 5% over any time period is associated with risk for morbidity and mortality Three-compartment body composition analysis using equations validated in HIV can be used to evaluate wasting and inflammatory responses associated with infection
Biochemical Assessment	Immunologic profile, hematologic profile, lipid profile, liver function, renal function, electrolytes, glucose and insulin levels; inflammatory markers	Biochemical testing is based on the need for routine evaluation for medication and disease cofactors Additional testing can be done if problems are expected or anticipated according to an individual patient's medical history
Medical History	Past and current information on diagnoses and symptoms; family history of diabetes, cardiovascular diseases, cancers, and renal disease; history of smoking, alcohol, and drug use; medication history and current profile (including herbal, supplement, and other complementary therapies)	In conjunction with other assessment criteria, risk for nutrition-related problems can be anticipated

losses that are related to inadequate food intake and malabsorption are likely to be mostly fat. Weight losses related to infections and injuries can be mostly body cell mass, a more detrimental condition. Patterns of weight loss in patients with HIV infection suggest that full reconstitution of weight and body composition is not always achieved when weight loss is associated with opportunistic infection (Macallan et al. 1993; Sheehan and Macallan 2000). Careful monitoring to catch the problem of weight loss early may help to prevent this gradual decline due to a series of weight loss events.

Alternate definitions of wasting have been offered to include both weight losses and changes in body composition that increase risk for functional impairment (Polsky, Kotler, and Steinhart 2001). Weight loss and wasting can accompany opportunistic infection. Evidence suggests that 10% weight loss for any reason and over any period of time is a strong predictor of death in HIV-infected patients, that even less than 5% weight loss may be a risk factor for mortality, and that there may be an increase at the 5% level in the post-HAART era (Wheeler 1999; Tang et al. 2002; Tang et al. 2005). A body mass index (BMI) of less than 20 suggests an increased risk for mortality in patients with HIV infection (Wanke et al. 2003). BMIs of less than 18 are markers of mild to moderate malnutrition and are associated with a significantly reduced survival (van der Sande et al. 2004). Patients with higher BMI levels, on the other hand, have a decreased risk for HIV disease progression (Jones et al. 2003).

TABLE 26.7

Suggested Criteria for the Diagnosis of Wasting in HIV Disease

Parameter	Criteria
Weight Loss	10% loss over 12 months or 7.5% loss over 6 months
Body Mass Index (BMI)	<20
Body Cell Mass	5% loss over 6 months or <35% of weight if BMI is less than 27 in men <23% of weight if BMI is less than 27 in women

Source: Polsky et al., 2001.

In addition to weight loss, body composition changes without weight loss can compromise clinical and nutritional status. Additional wasting criteria may include a loss of 5% BCM over a period of six months and a BMI of <20 (see Table 26.7).

BOX 26.2 CLINICAL APPLICATIONS

Bioelectrical Impedance Analysis (BIA) in Adults with HIV Infection

Bioelectrical Impedance Analysis (BIA) testing has been used to estimate fat compartments in primarily healthy people. The linear equations developed to estimate fat volume were typically based on the gold standard of underwater weighing. In populations with fluid shifts, usually seen in injury and infection, such equations were not considered valid, limiting the use of BIA in the clinical setting. More recent investigation has provided the clinician with nonlinear equations to estimate fat-free mass (FFM) and the FFM subcompartment of body cell mass (BCM). These equations have been validated in healthy populations, obese populations, and people living with HIV infection (Kotler 1996). This difference in the use of BIA technology can be used to anticipate wasting and monitor the severity and resolution of injury and infection according to the recommended criteria shown in Table 26.6. Thus, BIA utilization in chronic inflammatory diseases (such as chronic HIV infection) can be used to monitor the status of infections and effectiveness of treatments that affect nutritional status.

The body can be compartmentalized according to body functions. A two-compartment model may be FFM and fat mass. Using the nonlinear equations validated for use in HIV infection, a three-compartment model helps to differentiate two functional masses within fat-free mass: BCM and extracellular mass (ECM). Descriptions of these tissues are shown here:

- BCM: the fat-free mass body compartment that contains muscle and organ tissues, the most volatile of which is muscle; responsible for nearly all kcal use in the adult body; related to health maintenance and survival.
- ECM: the fat-free mass body compartment that contains bone, collagen, and extracellular fluids, the most volatile of which is extracellular fluids; provides structure and transport in the body.
- Fat: all tissue weight that remains after the calculations of fat-free mass; fat mass by BIA includes both stored and essential fat. It should be noted that because the estimates of fat are not directly estimated, but rather calculated by subtracting FFM from total body weight, the accuracy of this compartment estimate may be reduced. In addition, linear equations developed with the gold standard of underwater weight may not consider the essential fat compartment as a component of fat as it is in the case of the HIV-validated nonlinear equations.

Normal levels of each compartment are important to maintain in order to preserve normal body processes. Expected levels of BCM vary according to sex and height. Expected levels of ECM and fat vary according sex and weight. For purposes of this discussion, the BIA predictive equations that are validated for HIV infection will be the foundation for expected values and results, and expected values may vary if other equations are used. For men, the functional level of BCM starts at approximately 40% of ideal body weight by height (using Hamwi equations). For women, the functional level of BCM begins at approximately 30% of ideal weight. Expected ranges of ECM are approximately 40% to 45% of current weight for men and 37% to 45% of current weight for women. Fat ranges are functional at between 11% and 22% of current weight for men and between 20% and 32% of current weight for women.

Movement of compartments can give clues about changes in body function and metabolic processes such as infection. Multiple types of and the potential reasons for nutritional compromise can be identified using the nonlinear equations. BIA results can be used to determine the

(continued on the following page)

Both total volumes of body compartments and patterning of fat tissues will be useful in monitoring nutritional and clinical status in HIV-infected patients, as well as evaluating this risk factor for mortality (Ott et al. 1995). Body composition evaluation can be achieved in a number of ways, including DEXA, MRI, CT, total body potassium counting, deuterium hydroxide, underwater weighing, and others (see Chapter 5). In the clinical setting, bioelectrical impedance analysis (BIA) is a convenient, inexpensive, and relatively common method for evaluating body composition (see Box 26.2). Tetrapolar BIA provides resistance and reactance readings that can be used in equations that have been validated for use in fluid-shifted populations, such as people living with chronic HIV infection (Kotler et al. 1996). Such evaluation has allowed the observation that during the wasting of protein stores, muscle tissues are likely reduced, while the organ tissue component of body cell mass is relatively preserved, as demonstrated by increased resting energy expenditure per kilogram of body cell mass without indications of hypermetabolism at a cellular level (Schwenk et al. 1996).

While anthropometric evaluations, such as abdominal and peripheral (mid-upper arm, thigh, and calf) circumferences and fatfolds, are helpful in estimating body composition, it is likely that their most appropriate use in chronic HIV infection will be to monitor fat patterns and changes over time. Changes seen in fat patterns include losses of subcutaneous fat stores (lipoatrophy) and gains in central fat stores. Fat losses are most apparent in the peripheral limbs and in the facial area. Fat gains tend to be centralized, concentrated around the dorsocervical area, in

BOX 26.2 *(continued)*

level and source of a problem that affects nutritional status. Serial measures can assist in monitoring both the effects of nutritional therapies and the nutritional response to acute events. The following table gives an overview of how body compartments move and the interpretation of that movement in fluid-shifted patients with chronic inflammatory conditions.

Body Compartment and Change	Interpretation in HIV-Infected Patients
Decreased BCM	Rapid and more severe losses are usually associated with acute phases of infectious disease or injury. Slower and less severe losses are associated with a starvation process (including inadequate intake, malabsorption, and increased losses of nutrients) or reduced physical activity.
Increased BCM	Indicates a resolution of starvation and/or acute events, exercise/activity, and anabolic medications. The inability to improve BCM after the apparent resolution of an acute injury or infection suggests a non-responsive condition or anabolic block that may occur with chronic HIV infection.
Decreased ECM	A drop below expected values indicates dehydration. A return to expected levels indicates the resolution of an acute event that caused extracellular fluid increases (fluid shifts). Note that increases in the fat compartment (which is approximately 11% fluid compared to fat-free mass, which is more than 75% fluid) may reduce the percentage of weight that is ECM.
Increased ECM	An increase to expected levels indicates rehydration. Increases to levels above expected values indicates fluid shifts related to acute events, such as injury or infection. Note that a decrease in fat volume can increase relative ECM levels. Note that BIA values for ECM should be carefully interpreted with additional information on hydration and infection or injury; this is because only total ECM volume is reflected in the estimated levels by BIA. Mixed events can include both the intravascular dehydration at the same time as extravascular fluid shifts (edema). The shifts in each compartment (BCM and ECT) can add up to a total volume within the expected range.
Decreased fat	A decrease in fat volume indicates a loss of fat and imbalance in kcal related to starvation-related problems or exercise/activity. Fat volume can also indicate metabolic alterations (such as lipoatrophy). The evaluation of the fat compartment should include additional indicators to differentiate types of fat losses.
Increased fat	An increase after a previous loss indicates a resolution of starvation or caloric imbalance. Increases in fat without concurrent increases in BCM or weight can indicate metabolic alterations such as fat accumulation and non-response to nutrient-based therapies or anabolic block.

upper back and breast areas, and in the belly region as visceral fat (see photo).

In addition to anthropometry, measures using dual energy x-ray absorptiometry (DEXA or DXA) and computed tomography (CT) scans or magnetic resonance imaging (MRI) scans have been used to identify and differentiate changes in subcutaneous and visceral fat stores (Schwenk 2002). Additional anthropometric measures that are specific to fat changes seen in chronic HIV infection include facial fat measures, dimension measures of breast and dorsocervical fat pad areas, and abdominal measures to differentiate subcutaneous fat gain from potential visceral fat gains (Fields-Gardner 2001). Losses in subcutaneous fat and/or gains in abdominal fat stores can both result from and lead to problems of insulin resistance, which can be addressed (at least in part) through dietary and exercise modulation.

FIGURE 26.2 Dorsocervical Fatpad

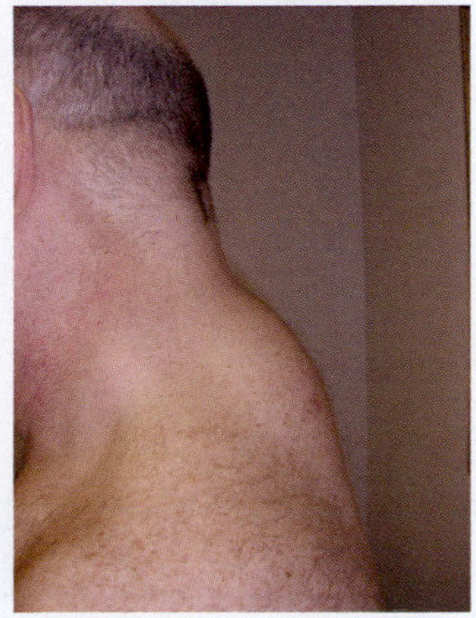

Source: © 2007 Cade Fields-Gardner.

In addition to the more quickly lethal problems of weight loss and body composition changes, weight gain and obesity have become important aspects of nutrition assessment and therapy. Weight gain and obesity should be categorized according to the type of additional weight that is seen. With efforts to restore body cell mass, weight gain to a high body mass index may be a result of muscle hypertrophy. However, fat gain can be differentiated as normal, subcutaneous fat deposition or abnormal, visceral fat accumulation. A description of selected lipodystrophy-specific anthropometry is shown in Box 26.3.

Biochemical Assessment Biochemical measures that assist in nutrition-related assessment include measures of disease progression (e.g., viral load, CD4 count), current inflammatory process (e.g., C-reactive protein), and general nutritional status, with inflammatory processes in mind (e.g., albumin, transferrin, others). Selected measures are shown in Table 26.8. Because viral load and CD4 counts have been associated with changes in weight, a high viral load or a low CD4 count should be considered a nutritional risk factor (Batterham, Garsia, and Greenop 2002; van der Sande et al. 2004). Other traditional biochemical measures of nutritional status such as

BOX 26.3 CLINICAL APPLICATIONS

Anthropometry for Fat Patterning in Lipodystrophy

In addition to standard anthropometric measures, there are some additional circumferences and skinfolds that can help to characterize fat patterning and changes associated with lipodystrophy. This box describes the anthropometric measures and methods that may be used to evaluate abdominal fat deposition in order to differentiate between normal gains and deep-tissue gains of fat, to track changes in dorsocervical fat deposition, and to track peripheral fat losses in the facial areas.

Abdominal fat evaluation requires a circumference at the abdominal level and four skinfold measures at the same level, including abdominal, right side, left side, and back. Methods for circumferences are the same as for standardized methods. The abdominal circumference should be measured at the level of the navel so that it can be repeated in the same place over time (see Figure 26.2A).

Abdominal skinfolds are taken from the front approximately one inch to the right of the navel; on the right and left sides at the mid-axillary line, and on the back about one inch from the spinal column (see Figures 26.2B, C, and D).

The abdominal circumference and skinfolds are then entered into the following equation in order to monitor differences over time and estimate changes in both subcutaneous and deep-tissue fat accumulation. It should be noted

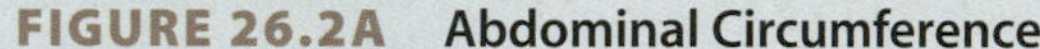

FIGURE 26.2A Abdominal Circumference

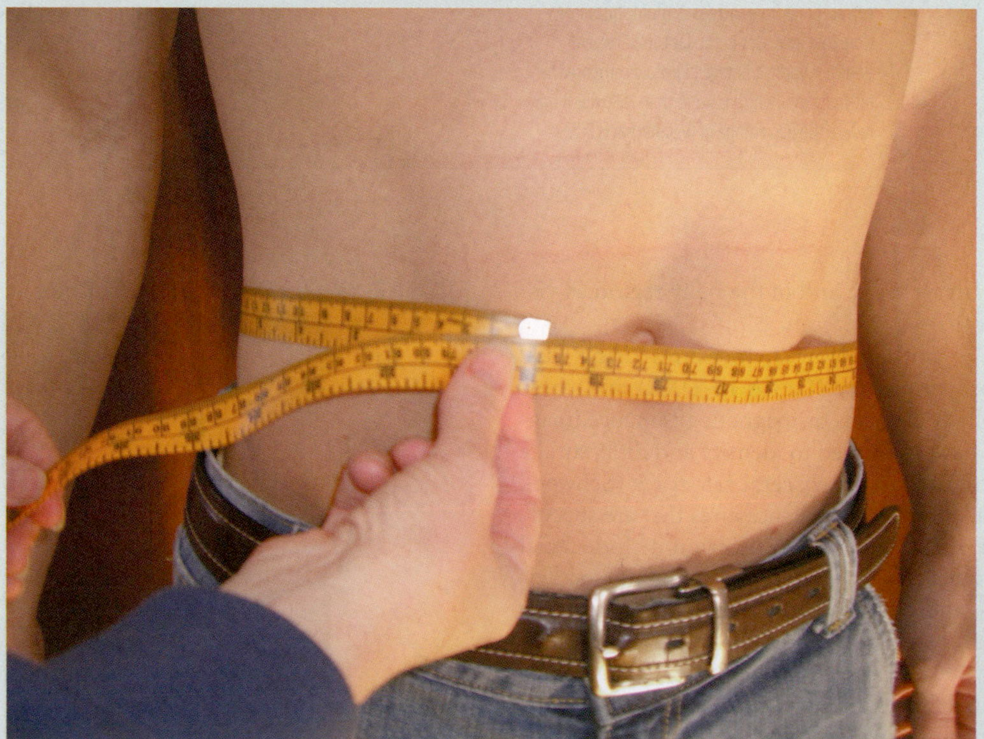

Source: © 2007 Cade Fields-Gardner.

(continued on the following page)

BOX 26.3 *(continued)*

Figure 26.2B Abdominal Skinfold

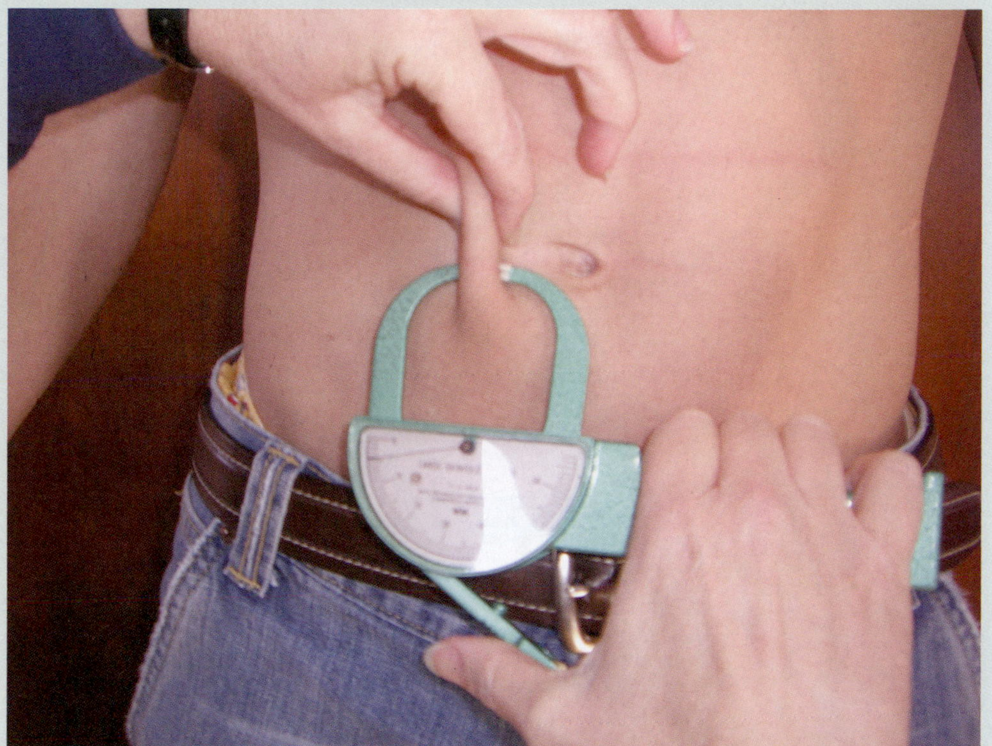

Source: © 2007 Cade Fields-Gardner.

FIGURE 26.2C Right Side Skinfold

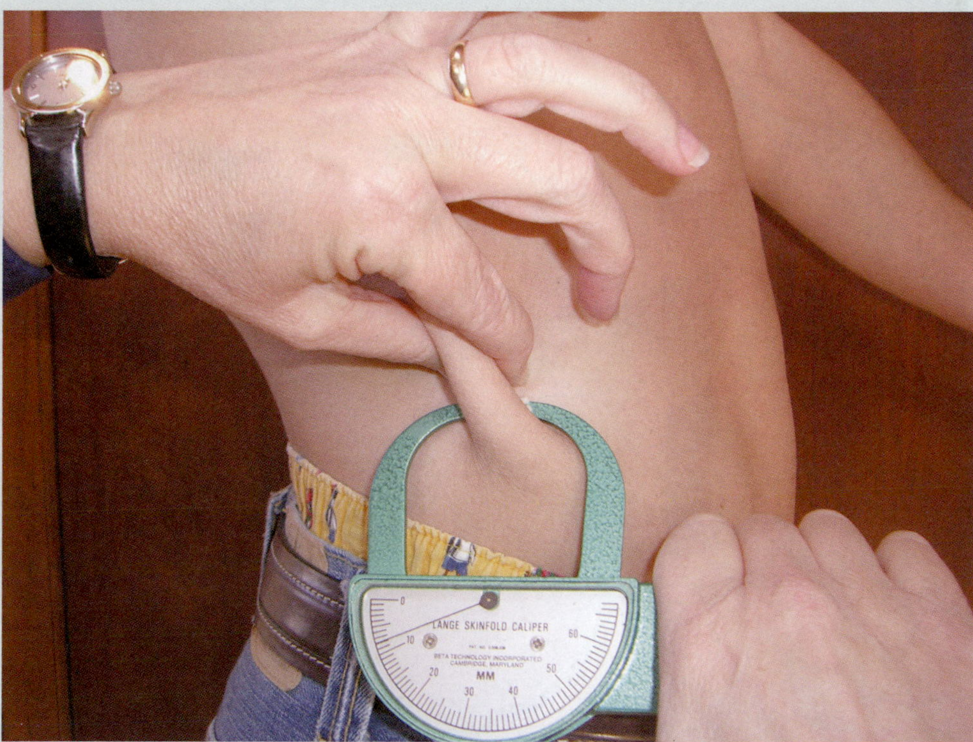

Source: © 2007 Cade Fields-Gardner.

(continued on the following page)

BOX 26.3 *(continued)*

Figure 26.2D Back Skinfold

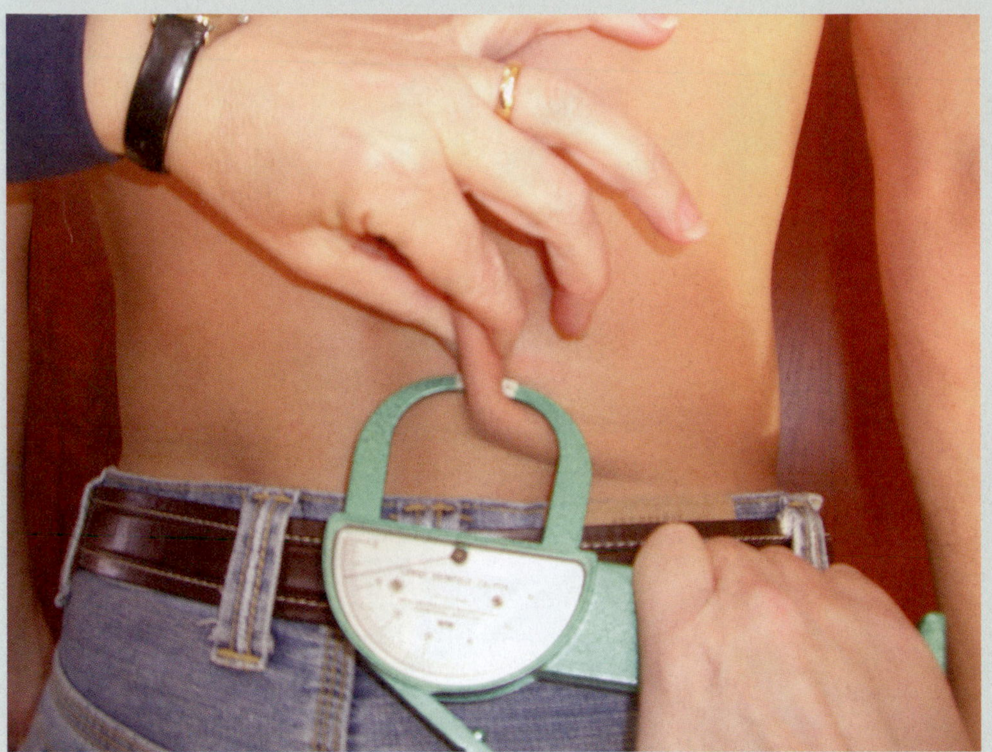

Source: © 2007 Cade Fields-Gardner.

FIGURE 26.2E Infraorbital Skinfold

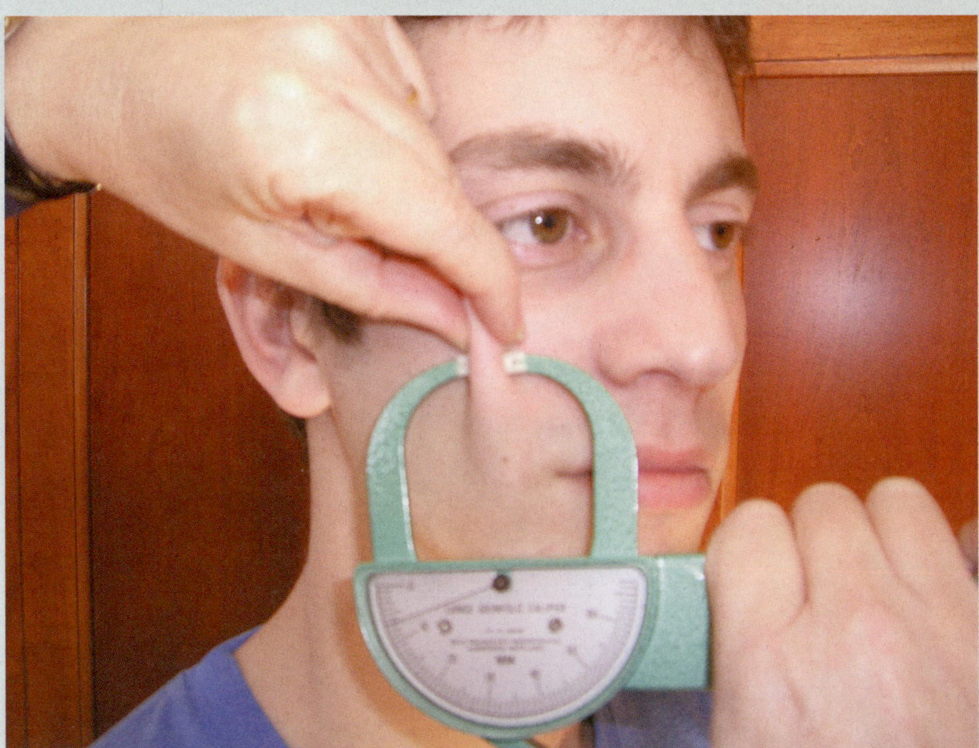

Source: © 2007 Cade Fields-Gardner.

(continued on the following page)

BOX 26.3 *(continued)*

Figure 26.2F Buccal Skinfold

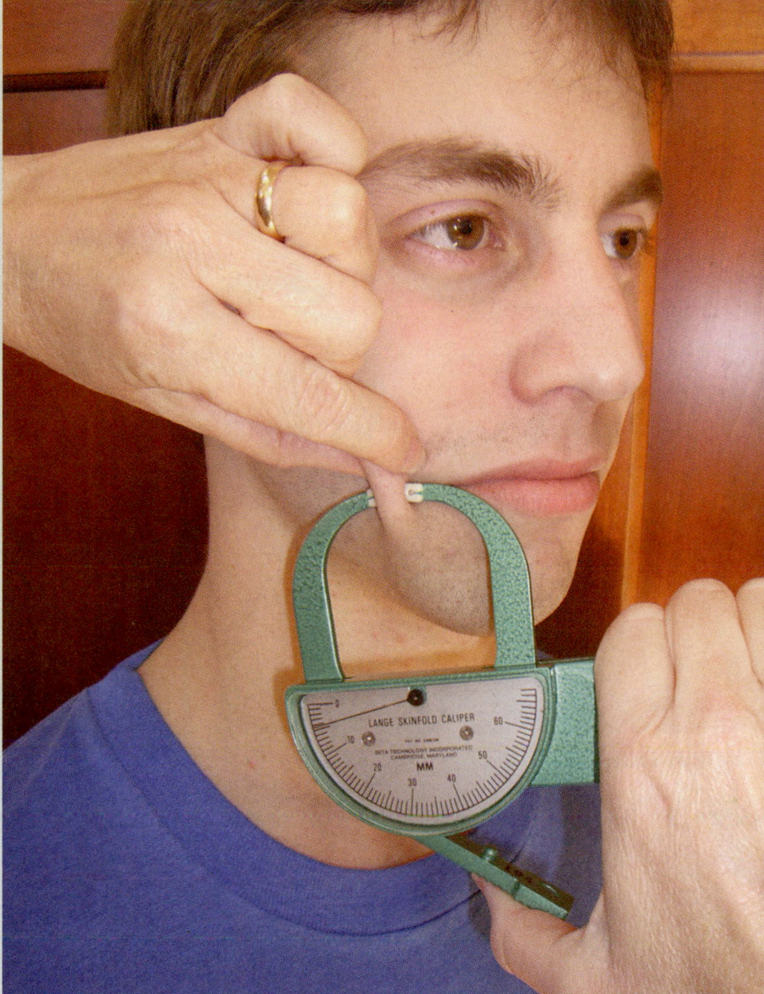

Source: © 2007 Cade Fields-Gardner.

that deep or visceral fat accumulation should be differentiated from ascites in patients at risk for ascites.

Total abdominal area (TAA) $cm^2 = \pi \times ((\text{abdominal circumference in cm}/2) \times \pi)^2$

Total visceral area (TVA) $cm^2 = \pi \times ((((\text{abdominal circumference in cm}/2) \times \pi)^2) - (((\text{abdominal skinfold} + \text{right side skinfold} + \text{back skinfold} + \text{left side skinfold})/8) \times 10))^2$

When the abdominal circumference increases and the ratio of total visceral area (TVA) to total abdominal area (TAA) increases, deep fat accumulation may be occurring. Conversely, when increased abdominal circumference shows a decreased TVA:TAA, fat may be accumulated in the subcutaneous area and would be considered a normal fat gain. As abdominal circumferences decrease, noting which compartment decreases will help to differentiate between the loss of fat in the subcutaneous area and deep fat losses (see Figure 26.3).

To effectively monitor peripheral lipoatrophy, it is best to have a baseline measure to compare with follow-up measures. Changes over time in both circumferences and skinfolds can help to differentiate muscle from fat wasting in the arm, thigh, and calf. Facial fat changes are often the most distressing to patients and can be tracked through the facial skinfold measures shown in Figures 26.2E, F, and G.

The infraorbital skinfold can be measured on the zygomatic process under the eye (see Figure 26.2E). This measure should be done very carefully because it may be somewhat painful to some who have lost fat padding in this area. The buccal skinfold can be measured just to the right of the corner of right side of the mouth (see Figure 26.2F). The sub-mandible skinfold is measured at the midpoint between the bottom-middle of the chin and the back curve point of the mandible. The thumb and index finger hold the skinfold at the edge of the bone (see Figure 26.2G).

While there are currently no "normal levels" in the literature for each of these measures, comparison of serial measures will give clues about both the process and treatment efforts toward reversing facial lipoatrophy.

(continued on the following page)

BOX 26.3 *(continued)*

FIGURE 26.2G Sub-Mandible Skinfold

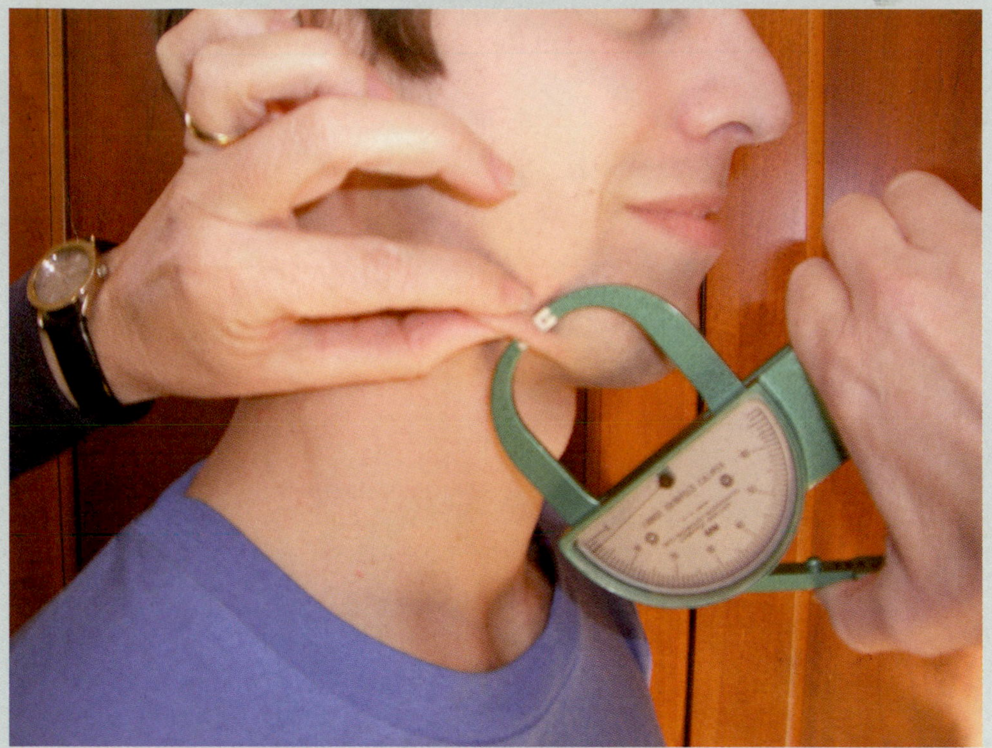

Source: © 2007 Cade Fields-Gardner.

FIGURE 26.3 Ratio of Total Visceral Area to Total Abdominal Area (TVA:TAA)

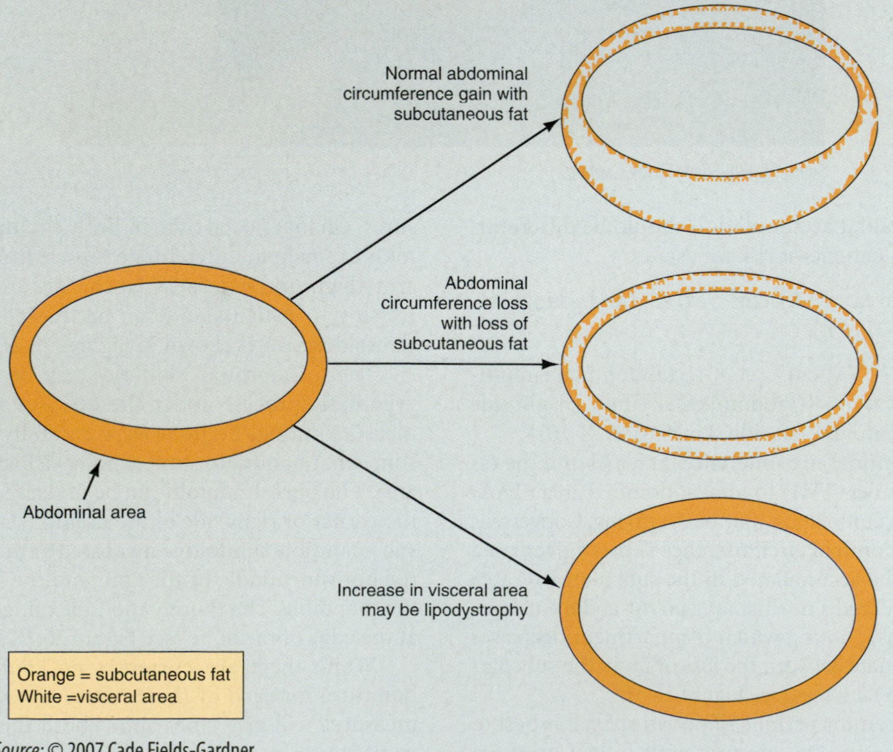

Source: © 2007 Cade Fields-Gardner.

TABLE 26.8

Selected Biochemical Measures in HIV Disease

Measure	Criteria or Expected Values	Evaluation in HIV Infection
Immunologic		
CD4 cell count	398–1535/µL	<200/µL defines AIDS; decreased levels are prognostic for opportunistic disease and often associated with body mass index levels
Viral load (PCR)	Undetectable	Elevated levels are prognostic for immune deficits
Hematologic		
Hemoglobin	F: 12.1–15.6 g/dL; M: 14.6–17.5 g/dL	Decreased in anemia; elevated in dehydration, chronic testosterone replacement
Hematocrit	F: 34–45%; M: 41–51%	Decreased in anemia; elevated in dehydration, chronic testosterone replacement
Mean corpuscular volume	78–93 cubic microns/RBC	Increased in folate or vitamin B_{12} deficiency anemia, associated with zidovudine; decreased in iron-deficiency anemia
Ferritin	F: 12–150 ng/mL; M: 30–320 ng/mL	Elevated in inflammation; decreased in iron-deficiency anemia
Transferrin	212–360 mg/dL	Elevated in iron deficiency; decreased in malnutrition
Albumin	3.5–5.0 mg/dL	Decreased in malnutrition; rapid decrease with acute inflammation
Prealbumin (Transthyretin)	18–38 mg/dL	Decreased in acute catabolism, inflammation, malnutrition
Organ Function		
AST	M: 10–37 U/L; F: 10–31 U/L	Elevated in hepatitis or due to medication interactions
ALT	M: 4–40 U/L; F: 4–31 U/L	
BUN	8–23 mg/dL	Elevated in diabetes; low in malnutrition
Creatinine	Adult: 0.4–1.2 mg/dL	Elevated in renal disease, wasting
Endocrine		
Glucose	Fasting: 70–99 mg/dL	Elevated in diabetes, pancreatitis, chronic malnutrition
Insulin	Fasting: 427 uIU/mL	Elevated in metabolic syndrome, type 2 diabetes
Glycated hemoglobin A1c	4–6%	Elevated in diabetes, iron deficiency
Testosterone	350–1,080 ng/dL	Decreased in hypogonadism, AIDS wasting
Cardiovascular		
Total cholesterol	120–199 mg/dL	Elevated in hyperlipidemia, diabetes, obesity, infection
HDL	40–60 mg/dL	Decreased in starvation, obesity, diabetes, smoking, liver disease, AIDS
LDL	<100 mg/dL	Elevated in hyperlipidemia, lower in advanced AIDS
Triglycerides	Fasting: <150 mg/dL	Elevated in hyperlipidemia, AIDS
C-reactive protein (CRP)	Regular: <0.8 mg/dL	High sensitivity CRP will provide risk as low (<1 mg/L), average (1–3 mg/L), or high (>3 mg/L)
Electrolytes		
Sodium	136–144 mEq/L	Decreased in diarrhea, vomiting, AIDS
Potassium	3.5–5.5 mEq/L	Decreased in diarrhea, vomiting, chronic stress/fever

albumin can be used, but the clinician should take into account the potential for changes in interpretation of results in the presence of chronic inflammatory processes.

Evaluation of the severity of co-diagnoses can assist in determining limitations for diet and nutrition-related therapies. These tests may include liver function tests, renal function tests, insulin and blood sugar testing, testosterone and other hormone levels, and others. Anemias are relatively common in chronic HIV infection, and should be routinely monitored for patients at risk and carefully differentiated in order to match nutrition-related problems with appropriate treatment strategies. While **lactic acidosis** is a serious finding often associated with reverse transcriptase inhibitor ARVs and metformin, there does not appear to be adequate evidence to support the routine measurement of lactic acid levels unless symptoms or other findings suggest a problem (Imhof et al. 2005). A finding of altered levels of micronutrients in HIV infection is not universal. There should be a reason to suspect deficiency before routinely checking micronutrient biochemical status markers (Henderson et al. 1997).

Medical History Assessment In addition to biochemical and physical assessment, current and past medical conditions and therapies should be included in a full nutrition assessment. History of nutrition-related problems, such as past episodes of wasting or bouts of diarrhea, can also help to establish reasons for nutritional status changes and appropriate choices for interventions. Current uses and histories of medications, including antiretrovirals, antibiotics, and other medications, will assist in establishing potential for nutritional risk and the need for educational interventions (see Tables 26.4 and 26.5). Current and past history of acute or chronic infection, disease, or injury will also assist to determine the causes of and risks for altered nutritional status. Concomitant diseases, such as diabetes, insulin resistance, hepatitis, renal dysfunction, pancreatic dysfunction, cardiovascular disease, osteoporosis, cancers, lactic acidosis, and others, should be prioritized along with HIV infection for nutrition-related intervention.

Over-the-counter medications and herbal or other nonnutrient supplements should be included in the medical history. Table 26.9 describes some potential effects with the use of selected herbal therapies. Such supplements should be considered as pharmaceutical interventions with a potential for interactions with prescribed therapies (Power et al. 2002). For instance, garlic has been recommended as a natural antifungal agent and large doses have been recommended to assist in treatment of infections. However, there is the potential for garlic supplements to interact with ARVs or other therapies because of its effects on the liver's enzyme processing (Mills et al., Natural Health, 2005). In some cases, therapies can increase the serum level of the medications, causing a risk for toxicity, and others can reduce the medication levels, causing a reduction in effectiveness of the drug (Mills et al., Impact of African, 2005). If the latter occurs with ARV medications, viral resistance can result, leading to the ineffectiveness of other drugs in the same medication classification.

Potential medication interactions have been suggested for some of the supplements shown in Table 26.9. There are a few herbal medications that have known toxicities that are particularly important to avoid in diseases such as chronic HIV infection. Liver toxicities are associated with borage, coltsfoot, and germander. Renal toxicity is associated with calamus. Hypertension is associated with ephedra and glycerin. Vaso-occlusive disease is associated with comfrey and life root. Complete information regarding contents and potency may not be available for such supplements, and they may not be regulated as traditional medications are. While some of the uses of supplemental herbal and nutrient preparations have the potential to be beneficial (Collins et al. 1997), it is especially important to assess the use of such supplements and to discuss and anticipate any adverse effects or interactions wherever possible. Additional studies are required to determine safety and efficacy of complementary and alternative medicines when used in conjunction with ARVs and other medications.

Dietary Evaluation Food behaviors and food/nutrition insecurity should be included in full nutrition assessments. An assessment of food intake can be accomplished in many ways, including a food recall, food records, and food frequencies. Dietary intake can be compared with estimated needs for fluids, kcal, protein, and micronutrients to determine the most appropriate diet-related interventions. The need for food and nutrient supplements should be evaluated on an individual basis, with consideration given to medication profiles and the potential for interactions or toxicities.

Nutrition Interventions

In order to provide integrated nutritional care, the clinician should have a good working knowledge of the disease, common complications and co-diseases, treatments, and potential for interactions and adverse events (Gerbert et al. 2004; Heslin et al. 2005; Wilson et al. 2005). Nutrition therapy can include the restorative or modulating effects of macronutrients, micronutrients, or other therapies aimed at improving nutritional status. Because chronic HIV infection results in a heterogeneous set of complications, it is likely that each HIV-infected person will require a different set of dietary and other recommendations according to their assessment (Coyne-Meyers and Trombley 2004). While there

lactic acidosis—an accumulation of lactic acid in the body characterized by abdominal pain, vomiting, and rapid breathing; this condition occurs in diabetes and as a potential side effect of medications

TABLE 26.9

Effects of Selected Herbal and Other Complementary Therapies in HIV Disease (see Chapter 4 for detailed descriptions and cautionary information)

Therapy (Category)	Description, Claims, and Comments
Aloe vera; kumari, sabila (Ayr, Folk, Nat)	Used to inhibit HIV; may reduce absorption of some medications
Alpha-lipoic acid (Nat)	Antioxidant; possible improvement of neuropathy; slowed in vitro replication of HIV
Anemarrhena, zhi mu (TCM)	Used for potential anti-HIV and anti-cancer properties; may increase insulin stimulation
Ascorbic acid (vitamin C) (Meg)	Improve antioxidant capacity of body and immune system; increases iron absorption from non-heme sources; upper limit is 2000 in adults and less in children; increases urinary losses of oxalate and calcium
Ashwagandha; winter cherry, Indian ginseng (Ayr, Nat)	Used for antidementia, antibacterial, anticancer, and antioxidant; reduces gastric acidity; possible increase in serum T4 (thyroid); can potentiate barbiturate drugs
Astragalus, milk vetch; huang qi (TCM, Nat)	In combination with other herbal medications for anti-HIV (in vitro information) as immune enhancer
Atractylodes; bai zhu; cang zhu (TCM)	Anti-HIV activity effect on enzymes; potential for increased insulin and hypoglycemia or loss of blood glucose control
Burdock; niu bang zi (TCM, Folk, Nat)	Immune modulator; increases insulin and may cause hypoglycemia or loss of blood glucose control
Cat's claw; gambir; gao teng (Ayr, TCM, Folk, Nat)	Used as antifungal, anti-herpes; may inhibit platelet aggregation (anticoagulant)
Chromium (Meg)	Used to improve insulin sensitivity and improve lipid profile; speculated as potential treatment for lipodystrophy/metabolic syndrome in HIV; in vitro and animal studies suggest cellular DNA damage associated with long-term intake; picolinate form may have increased absorption rate; suggested competition for iron binding to proteins
Cobalamin (vitamin B_{12}) (Meg)	Used to reverse low levels of B_{12}; in vitro inhibition of HIV-1; suggested to reduce effects of fatigue, anemia, neuropathy, cognitive impairment
Coenzyme Q10 (Nat)	Decreased blood levels are seen in HIV infection; suggested to improve immunity, energy; some gastric distress is associated with supplementation
Dehydro-epiandrosterone (DHEA) (Nat)	Adrenal hormone supplement used to prevent and treat muscle wasting, improve resistance exercise performance; suggested to increase testosterone levels; possible small improvement in immune function; improved mood; caution is suggested for cancer risk
Dong quai	Anticoagulant; may change effectiveness of estrogen replacement; may work synergistically with calcium channel blockers
Echinacea, purple coneflower (Folk, Nat)	Anticancer; immunostimulatory; may inhibit metabolism of drugs using the cytochrome P-450 enzyme pathway
Garlic; ajo; lasuna (Folk, Nat)	Antifungal and antimicrobial properties; hypoglycemic agent should be used with caution in patients with altered blood glucose control; may work synergistically with anticoagulants, antihypertensives, lipid-lowering agents, competes with liver processing of ARVs; inhibits uptake of iodine by thyroid gland
Ginko biloba (Nat)	Antioxidant; memory enhancer; anticoagulant; may interact adversely with ARVs
Glutamine (Nat, Meg)	Immune enhancing; antiwasting; treatment in diarrhea
Goldenseal; yellow root; yellow puccoon; yellow Indian paint; ground raspberry (Folk, Nat)	Antibiotic properties; long-term use can decrease absorption and utilization of B vitamins; may exacerbate hypertension; may reduce effectiveness of some medications (tetracycline, vibramycin, doxycycline); hypoglycemic and may reduce blood glucose control
Gotukola; brahmi (Ayr, Nat)	Immunomodulatory; possible anticancer; suggested to have antioxidant effects; hypoglycemic and may reduce blood glucose control
Grape seed extract (Nat)	Antioxidant, anticancer, anti-inflammatory; anti-HIV effects hypothesized

(continued on the following page)

TABLE 26.9 *(continued)*

Effects of Selected Herbal and Other Complementary Therapies in HIV Disease (see Chapter 4 for detailed descriptions and cautionary information)

Therapy (Category)	Description, Claims, and Comments
L-carnitine (Nat, Meg)	Large doses required; suggested for antioxidant (decrease TNF-alpha cytokine), immunomodulation (increase CD4 cells), cognitive enhancement, and other problems; treatment for vomiting and lactic acidosis associated with NRTI ARVs; IV form used for neuropathy treatment related to NRTI ARVs, oral form shows promise in some neuropathy improvement; speculated as potential treatment for lipoatrophy; some nausea, gastritis, cramps, and diarrhea associated with oral doses; diet high in carbohydrate and low in fat is recommended as adjunctive treatment for carnitine deficiency; slight improvements in lipid profile may be seen
Licorice root; glycerrhizin; mulethi; gan cao (Ayr, TCM, Folk, Nat)	Antiviral, antibiotic; suggested to reduce HIV infection by increasing cell fluidity; suggested to reduce liver damage related to infection and medications; interacts with potassium-losing diuretics to increase risk of hypokalemia; may cause hyperglycemia, electrolyte imbalance, and increase risk for side effects with MAOI medications; oral contraceptives may have reduced effectiveness; may increase interferon and isoniazid drugs
N-acetyl cysteine, NAC (Nat, Meg)	Glutathione precursor; antioxidant; anti-inflammatory; decreased viral load and increased CD4 cell count
Selenium (Meg)	Immune enhancement (lymphocyte function), antioxidant cofactor; high doses are associated with nausea, diarrhea, peripheral neuropathy, fatigue, and immune dysfunction
St. John's Wort; hypericum (Nat)	Antidepressant, antianxiety, anti-HIV; contraindicated with the use of medications processed by the CYP3A4 and P-glycoprotein pathways, including protease inhibitors and NNRTIs; reduced effectiveness of oral contraceptives; antagonistic to antihypertensive medications
Zinc (Meg)	Immune enhancement; antifungal; competes with copper and long-term high doses are associated with anemias and immunotoxicity; adverse effect on LDL:HDL; high dose can cause nausea, vomiting, taste changes

AYR-Ayurvedic Medicine
TCM-Traditional Chinese Medicine
FOLK-Folk Remedy
NAT-Natural Product
MEG-Vitamin/Mineral Megadose

are occasional reports of toxicities and potential for adverse interactions with medication therapies, food and supplement levels of nutrients can be compared to recommended intakes as well as to upper limits that have been documented in the literature. Some components of foods or supplements may also have the potential for interactions, so careful monitoring should be established to identify any adverse effects (Kupferschmitdt et al. 1998; Piscitelli et al. 2002).

Macronutrient Therapy Alterations in macronutrients for therapeutic purposes can address disease and treatment complications, such as symptom management, weight maintenance (including gain and loss according to needs), and management of concomitant disease processes. A summary of nutrient interventions is shown in Table 26.10 and an overview of dietary recommendations for symptom relief is shown in Table 26.11.

Fluid recommendations are based on needs for hydration maintenance with any limitations or enhanced needs, such as renal dysfunction or fever and sweating. Kcal recommendations are based on the balance needed to maintain a healthy weight. This may include the need for additional kcal to gain weight, or for fewer kcal in order to lose weight. Although recovery of fat before recovery of lean tissues is a normal process during weight regain, emphasis on lean tissue gain can be achieved with aggressive efforts to introduce kcal, protein, and muscle supporting activities. In some cases, weight loss may be appropriate to improve health. When body composition has been altered, however, weight loss through dieting may be risky, especially if there is no buffer in the volume of body cell mass to lose when lean tissue is lost during dieting. Kcal balance and other nutrient-based therapies may be best used as one of several therapies aimed at improving overall health.

Protein recommendations usually include extra protein to buffer the catabolism of protein stores that accompanies the inflammatory response to viral and other infections. This additional protein helps to improve immune and other inflammation-mediated responses. Diets are generally high in protein in developed countries, allowing this recommendation to be made without any particular diet change. However, in cases where baseline levels of protein intake are low or food insecurity limits the ability to obtain adequate high-quality protein, enhancing protein intake may be required. For patients with renal dysfunction, bone mineral losses, and long-term diabetes, the quality of protein and the best sources may be more specifically tailored. In such instances, a balance of animal- and plant-based protein sources will be

TABLE 26.10

Nutrient-Based Therapy Recommendations

Nutrient	Recommendations
Fluids	Hydration maintenance is the goal, and standard fluid intake recommendations can apply; additional fluids are recommended in cases of dehydration, fluid losses through diarrhea or sweating; restrictions are recommended in cases renal insufficiency
Kcal	Weight maintenance is the goal with additional kcal typically recommended if weight gain is desired and a mild restriction in kcal to achieve desired weight losses; additional energy may be required during bouts of opportunistic conditions that increase metabolic rate; increased kcal requirements during pregnancy and lactation should be incorporated into recommendations
Carbohydrate	The amount and types of carbohydrate recommended are based on both energy needs and carbohydrate tolerance; insulin resistance and diabetes may require dietary modification to modulate glucose and insulin levels
Protein	The amount and types of protein recommended are based on the need for protein stores maintenance; additional protein is likely to be needed in cases of inflammation, fever, and during pregnancy; any protein losses should be restored with increased protein intake and activity to promote protein stores maintenance; renal disease or other conditions may require protein restriction or other changes in protein recommendations
Fat	The amount and types of fat recommended are based on energy needs, cardiovascular risk, and inflammatory conditions; weight maintenance may require increases or decreases in fat kcals; cardiovascular risk may require lower fat intake and a higher ratio of unsaturated fats; omega-3 fatty acid sources may be recommended to help reduce inflammation effects as well as improve lipid profiles
Vitamins and minerals	Recommendations are based on individual needs; for instance, during bouts of diarrhea, the replacement of electrolytes and any potential losses of vitamins and minerals (such as fat-soluble vitamins during steatorrhea and zinc during larger volume losses of fluids) are essential to balance; upper limits of toxicity, the potential for nutrient interactions with medications and disease, and balance in micronutrient intake should be considered in recommendations (for instance, iron supplementation is controversial due to the potential for increasing the risk of opportunistic infection without overcoming the inflammatory-mediated drop in iron availability); specific conditions that alter micronutrient requirements, such as pregnancy and lactation or child growth and development, should be considered in recommendations
Fiber	Fiber recommendations are similar to those in healthy individuals and fiber has been suggested to improve glucose tolerance, affect glycemic response to foods, and reduce the potential for cardiovascular risk and altered fat deposition seen in lipodystrophy

especially important. Overdoses of protein pose additional burden and risk in patients with potential for renal or bone problems. Special care should be taken to tailor protein recommendations to individual needs and limitations.

Carbohydrate and fat kcal may be modulated according to assessed risk for insulin resistance, cardiovascular disease, and other conditions where treatment is related to dietary interventions. Because HIV infection is associated with chronic inflammatory response, it may be prudent to provide education on balancing daily dietary intake to minimize problems and reduce associated risks.

Nutritional supplements have been recommended for patients who require a boost in macronutrient intake. Patients may use recipes to create their own nutritional supplements and/or obtain ready-to-use commercial supplements in the form of beverages and bars. While some food-based kcal may be displaced by such therapy, kcal-containing supplements have been shown to yield an increase in energy intake and may assist in adherence to a weight gain regimen (Schwenk, Steuck, and Kremer 1999).

Micronutrient Therapy Altered micronutrient levels have been documented in both pediatric and adult HIV disease, in research that suggests a worsening associated with disease progression. Alterations in micronutrients may exacerbate immune compromise, disease progression, damage due to oxidative stress, morbidity, and even risk for HIV transmission independent of CD4 cell counts (Evans and Halliwell 2001; Mehendale et al. 2001). While there have been many speculations about the benefit of reversing a low level of selected micronutrients, evidence of the impact on HIV disease to date has been equivocal (Hegde, Woodman, and Sankaran 1999; Lanzillotti and Tang 2005; Patrick 1999, 2000). Table 26.12 provides information on selected micronutrients that have been targeted for supplementation in PLHA.

Reversal of many true nutrient deficiencies is possible with supplementation (Coodley 1995). Micronutrient supplementation remains controversial in most areas, but is generally recommended when nutrient intake may be inadequate (Buys et al. 2002). The use of multivitamin/mineral supplementation may have a supportive role in slowing disease progression, particularly in patients without access to antiretroviral therapies.

Additional micronutrient therapies may include those targeted to treat anemias, diarrhea, and potential dietary deficiencies. Special care should be taken in each instance to

TABLE 26.11

Dietary Recommendations for Selected Symptom Management

Symptom	Education and Counseling Recommendations	Symptom	Education and Counseling Recommendations
Appetite Loss	Determine the cause of anorexia and pursue treatment Offer tips: Eat favorite foods often in relaxed settings Add flavors and have foods of various colors for more interest Keep a supply of snacks handy when appetite increases Fix and eat foods that don't take as much energy or enjoy take-out food If significant weight loss occurs, consider appetite stimulants Refer to resources for food if food insecurity is a problem Monitor nutrient intake, weight, and body composition	**Nausea/ Vomiting**	Determine the cause of nausea and vomiting and pursue treatment Offer tips: Replace any lost fluids and electrolytes Try bland, non-odorous foods Drink beverages between meals and not with meals Eat smaller, more frequent meals Keep upper body elevated for at least an hour after meals Reduce fatty foods if early satiety is a problem If vomiting leads to chronic reduced food intake, evaluate for antiemetic medications Monitor food and food intake, fluid status, weight
Diarrhea	Determine the cause of diarrhea and pursue treatment Offer tips: Replace fluids and electrolytes with juices, sports drinks, gelatin, broths Eat bland foods that are lower in fiber and residue Avoid fatty and gassy foods If lactose is a problem, find substitutes for lactose-containing products Anti-diarrheal medications should be considered for significant acute and chronic diarrhea Monitor fluid intake, hydration, weight, and body composition	**Oral Lesions**	Determine and pursue appropriate treatment for oral lesions Offer tips: Eat moist, soft foods and finely dice foods and keep mouth moist between meals Avoid irritating spicy or acid-containing foods Eat foods at room temperature or cooler temperatures Avoid tough foods or foods that require chewing Consider topical medications to ease pain in eating and advise on good oral hygiene Monitor food intake, weight
Heartburn/ Reflux	Determine any cause of heartburn and reflux and pursue treatment Offer tips: Eat small amounts of food more often throughout the day Keep the upper body elevated for at least an hour after eating Avoid alcohol, caffeinated beverages, very spicy, and fatty foods Stimulate saliva production by chewing sugarless gum If overweight, lose weight and avoid tight-fitting clothes If reflux is chronic, consider medications to control the symptoms Monitor food intake, weight		

consider the effect of chronic inflammation in HIV infection. While there is some hope for health benefits and reversing nutrient deficits or otherwise low serum levels, in the case of diarrhea and wasting, a randomized placebo-controlled trial showed no significant benefit for short-term oral supplementation of vitamins A, C, and E, along with zinc and selenium, in reducing the risk for mortality or reversing hematologic parameters (Kelly et al. 1999).

Children who are perinatally infected with HIV often show multiple nutritional problems, including micronutrient deficiencies, which hinder normal growth and development as well as immune function (Cunningham-Rundles et al. 2002).

In general, studies are equivocal and have not provided an adequate basis to make generalized recommendations. Individualized assessment, care plans, and counseling are likely to be most appropriate for micronutrient supplementation in addition to diet-related strategies.

Pregnancy Outcomes and Infant Feeding Much research has concentrated on the role of nutrients in pregnancy outcome and prevention of mother-to-child transmission of HIV infection. The supplementation of multivitamins/minerals shows some promise in improving pregnancy outcomes in women who are malnourished. Improvements have been seen in birth weight and a reduced number of preterm births (Fawzi et al. 1998). The mother's nutritional status does not always have bearing on transmission rates of HIV. For instance, a low level of vitamin A has been associated with greater transmission risk between mother and child, yet supplementation with vitamin A has shown disappointing results in curbing this problem (Dreyfuss and Fawzi 2002). In fact, single nutrient supplementation of vitamin A may be related to an increase in mother-to-child transmission of HIV, whereas multiple vitamin supplementation showed promising results in the reduction of transmission through breast-feeding and reduction of child

TABLE 26.12

Selected Micronutrients in HIV Infection

Nutrient	DRI/UL*	Comments
Vitamin A	700–900 μg/ 3,000μg	Vitamin A and beta carotene have been associated with immune function, pregnancy outcomes, and growth and development of children (especially with diarrhea); supplementation may not have an impact beyond normalizing serum values; maternal supplementation of vitamin A is associated with greater risk of mother-to-child transmission of HIV infection; however, supplementation along with zinc in undernourished children may assist in reducing diarrhea and opportunistic infections
Riboflavin	1.1–1.3 mg/ -	Supplementation has been suggested along with thiamin in cases of lactic acidosis
Folic Acid	400 mg/ 1,000 mg	Folate supplementation is generally recommended in cases of pregnancy
Pyridoxine	1.5–1.7 mg/100 mg	Has been touted to reduce peripheral neuropathy; however, excesses can contribute to peripheral neuropathy
Cyanocobalamin	2.4 mcg/ -	Vitamin B_{12} may be malabsorbed due to changes in stomach pH; it is related to cognitive function and has been commonly supplemented as a part of oral multivitamin/mineral supplements, through sublingual, nasal, and intravenous doses
Ascorbic Acid	75–90 mg/ 2,000 mg	Supplemented to improve antioxidant status and reduce oxidative stress; high doses, but as little as 1000 mg per day, can interact with ARVs
Alpha Tocopherol	15 mg/ 1,000 mg	Supplementation has the potential for immunostimulation; should be avoided when taking the ARV amprenavir
Iron	8–18 mg/ 350 mg	Red blood cell iron is affected by inflammatory processes and the shunting to storage forms; iron supplementation is not generally recommended unless the threat of iron-deficiency anemia-related mortality is imminent because of the potential to increase the risk of opportunistic infection and progression of disease
Selenium	55 μg/ 400 μg	Low values are an independent predictor of survival; low selenium values may be related to inflammatory process selenium is a cofactor in antioxidant protection, and low values are related to immune dysfunction; excess selenium is associated with immune dysfunction
Zinc	8–11 mg/ 40 mg	Low zinc values are seen in HIV infection and associated with inflammatory processes and an increased mortality; zinc supplementation has been associated with fewer opportunistic infections and slower disease progression; high zinc intake has been associated with disease progression; zinc restoration is associated with reversal of weight losses and amelioration of diarrhea; effective ART does not appear to reverse low levels of zinc

*DRI/UL-Dietary Reference Intake/Upper Level

Sources: Abrams B, Duncan D, Hertz-Piddiotto. A prospective study of dietary intake and acquired immune deficiency syndrome in HIV-seropositive homosexual men. *J Acquir Immune Defic Syndr.* 1993;6(8):949–958.

Baum MK. Role of micronutrients in HIV-infected intravenous drug users. *J Acquir Immune Defic Syndr.* 2000;25 suppl 1: S49–S52.

Baum MK, Campa A, Lai S, Lai H, Page JB. Zinc status in human immunodeficiency virus type 1 infection and illicit drug use. *Clin Infect Dis.* 2003;37 suppl 2:S117–S123.

Baum MK, Shor-Posner G. Micronutrient status in relationship to mortality in HIV-1 disease. *Nutr Rev.* 1998;56:S135–S139.

Bowers JM, Bert-Moreno A. Treatment of HAART-induced lactic acidosis with B vitamin supplements. *Nutr Clin Pract.* 2004;19(4):375–378.

Butensky E, Kennedy CM, Lee MM, Harmatz P, Miaskowski C. Potential mechanisms for altered iron metabolism in human immunodeficiency virus disease. *J Assoc Nurses AIDS Care.* 2004;15(6):31–45.

Duggan C, Fawzi W. Micronutrients and child health: studies in international nutrition and HIV infection. *Nutr Rev.* 2001;59:358–369.

Fawzi W. Nutritional factors and vertical transmission of HIV-1. Epidemiology and potential mechanisms. *Ann NY Acad Sci.* 2000;918:99–114.

Slain D, Amsden JR, Khakoo RA, Fisher MA, Lalka D, Hobbs GR. Effect of high-dose vitamin C on the steady-state pharmacokinetics of the protease inhibitor indinavir in healthy volunteers. *Pharmacotherapy.* 2005;25(2):165–170.

Wellinghausen N, Kern WV, Jochle W, Kern P. Zinc serum level in human immunodeficiency virus-infected patients in relation to immunological status. *Biol Trace Elem Res.* 2000;73:139–149.

mortality (Fawzi et al. 2002). Maternal micronutrient status may have less effect on child mortality than the child's nutritional status and should be considered regardless of breast-feeding (Fawzi 2000).

General recommendations for feeding newborns of HIV-infected mothers in developed countries include the preferred use of breast milk substitutes and formulas that will meet nutritional needs of the infant and eliminate the exposure to breast milk virus (Jackson et al. 2003). In developing countries, this recommendation may be less feasible, affordable, and sustainable. For these instances, the WHO has outlined breast-feeding recommendations and the need to provide an unbiased representation of risks and benefits so that mothers can make educated and personalized decisions about breast-feeding their infants (Gupta, Mathur, and Sobti 2002; Habicht 2004).

Non-Nutrient Therapy for Nutritional Status Maintenance Several types of non-nutrient therapies target the improvement of nutritional status. Many of these therapies are used in combination with nutrient-based therapies or with each other to improve the effect to improve nutritional status, quality of life, and overall health status.

Education and Counseling Education on relevant nutrition-related topics and counseling on nutritional and other therapies is a key feature of all interventions. Baseline education on nutrition principles, including balanced diet and food safety, should be provided to all patients (see Box 26.4). Additional education on topics such as drug-nutrient interactions or symptom management can be provided in individual or group sessions. Individualized counseling should include all facets of the nutrition care plan, and how other aspects of the overall health care plan may affect nutritional status and well-being. In addition, recommendations for food intake can be customized to a prescription level in counseling sessions (McDermott et al. 2003). Referrals should be made for other aspects of care that affect nutritional status and health, including smoking cessation, alcohol or drug rehabilitation, HIV and other infectious disease transmission, and psychosocial-economic factors that can affect food access, choices, and metabolism (Normen et al. 2005). Counseling activities may be essential to the successful use of nutritional supplements to achieve weight gain objectives (Rabeneck et al. 1998). Medication interactions with food and nutrients are an important feature of education and counseling for patients taking ARVs and other therapies (see Tables 26.4 and 26.5).

Treatments for HIV and Opportunistic Infection Suppression of viral load by antiretrovirals has shown varying effects on nutritional status. Weight gains are commonly seen in patients on successful ARV therapy (Carbonnel et al. 1998). When changes in body shape and composition were noted after the introduction of HAART (Silva et al. 1998), the effects of these medications were explored (Lee, Rao, and Grunfeld 2005). Varying speculations suggested that alterations in body shape and metabolic indices are multifactorial and may be related to many risk factors, including ART, other medications, HIV as a chronic inflammation, and lifestyle behaviors, such as smoking, drug and alcohol abuse, and diet and exercise.

In one study, the type of medication regimen did not appear to affect triglyceride or insulin responses (Thomas-Geevarghese et al. 2005), while others showed a specific impact on these, hormonal balances, or other indicators of

BOX 26.4 CLINICAL APPLICATIONS

Counseling Concepts in HIV Infection

Clinicians should include nutrition-related counseling in their patient care plans. Basic education on nutrition-related concepts and principles will help the patient to make diet and health-related decisions. Tailored counseling can address specific problems and conditions that may benefit from nutrients or nutrition-related care.

Nutrition Education Basics Checklist

- Nutrition priorities: water and fluids, kilocalories, protein, and micronutrients
- Weight maintenance and kilocalorie balance
- Role of exercise and lifestyle choices, including alcohol, drug abuse, and smoking
- Food and water safety
  - Water safety
  - Food shopping and storage
  - Food preparation and dining out

Nutrition Counseling Checklist

- Food interactions with medication regimens
- Symptom management
  - Nausea and vomiting
  - Diarrhea and constipation
  - Appetite loss
  - Heartburn and bloating
- Weight losses
- Weight gains: overweight and obesity
- Additional conditions
  - Lipodystrophy: fat gain, fat loss
  - Insulin resistance and diabetes
  - Hyperlipidemias/altered blood fats
  - Hepatic conditions
  - Gastrointestinal conditions
  - Renal conditions

nutritional risk (Schutt et al. 2004; Hong-Brown et al. 2005). However, it appears to be a combination of successful ARV therapy, pre-therapy viral load, nadir CD4 count, and improvement of CD4 counts that are predictive of body composition and other changes. For instance, baseline CD4 count, change in CD4 count, higher baseline viral load, and zidovudine use were associated with changes in fat deposition. Protease inhibitor and zidovudine use were predictive of bone mineral losses. Interestingly, this study suggested that HAART was not associated with fat mass changes, despite the popular opinion that there is a connection (McDermott et al. 2005). It has been speculated that a reduction in inflammatory processes can be accomplished with effective ARV therapy, improving markers of health risk that are commonly associated with inflammation (Young et al. 2004). However, contradictory studies have suggested that the potential side effects of ARV therapy may overcome the positive effects of weight improvement (Schwenk et al. 1999).

In addition to ARV therapy, the prevention and early treatment of opportunistic infection is an important strategy to maintain and restore nutritional status. Weight loss episodes and health decline are triggered by both acute and chronic co-infection. Medications, vaccinations, and other treatment strategies are employed in conjunction with nutrition-related therapies to preserve normal body composition and functions as much as possible.

Exercise Exercise has been recommended to balance activity with kcal intake, improve muscle volume and function, and normalize lipid and energy metabolism (O'Brien et al. 2004). Exercise has the potential to mitigate the loss of muscle mass in wasting conditions and to improve the recovery from bouts of protein wasting (Zinna and Yarasheski 2003). The combination of diet interventions and exercise is considered first line therapy for cardiovascular disease and altered fat metabolism and deposition in chronic HIV infection (Thoni et al. 2002; Scevola et al. 2003). General recommendations include a tailored, routine exercise program that includes both aerobic and resistance components at least three times per week. As with any exercise program, adherence is an important factor in realizing benefits, and may present a challenge to the health care team designing and tailoring such interventions. Exercise and diet are both cost-effective therapies to improve nutritional status, and may be combined with medications or other therapeutic strategies to improve effectiveness (Shevitz et al. 2005).

For metabolic alterations, including fat accumulation, exercise may prove to be an essential and effective therapy. Aerobic training, in particular, may improve lipid metabolism, lower blood lipids, and reduce visceral fat accumulation (Jones et al. 2001; Thoni et al. 2002). A summary of exercise effects is shown in Table 26.13.

TABLE 26.13

Effects of Exercise

Type	Effects	Uses in HIV Infection
Progressive Resistance	Preserves, improves muscle mass; improves strength, improves effect of anabolic therapies	Restoration of BCM during recovery from weight loss and wasting, prevention of BCM volume and function loss, improvement of muscle function
Aerobic	Improves insulin sensitivity, increases HDL-C, improved endurance, improves effect of anabolic therapies	Restoration of normal activity level after illness or debilitation, reduced effect of insulin resistance on body shape and fat changes, prevention and treatment adjunct for cardiovascular health efforts

Anti-Wasting Medications Medication interventions targeted to reduce risk and reverse wasting and weight loss include appetite stimulants, anti-catabolic and anabolic medications, and hormone replacement therapies. A summary of selected medications is shown in Table 26.14.

Two of the primary medications used for treatment of anorexia in HIV infection include megestrol acetate (Megace) and dronabinol (Marinol). Megestrol acetate is an antineoplastic and progestational agent that has been used to successfully improve appetite and weight gain in cancer-related and HIV-related wasting. Recent pharmacokinetic information on megestrol acetate suggests that the original formulation (Megace) had a diminished absorption when taken with food. However, a new formulation utilizing nanocrystal technology overcomes this effect and decreases the required dose volume in the product Megace ES as compared to Megace (Facts and Comparisons 2005). Dronabinol (Marinol), a cannabinoid, has been used in cases of anorexia, nausea, and pain. Some concern has been expressed about the potential for abuse and immunosuppression of cannabinoid drugs; however, research has explored and refuted these issues as serious concerns (Kraft and Kress 2004). There has also been some exploration of "medical marijuana" to improve appetite in chronic HIV infection. While both oral and smoked forms of the drug appear to have some effect on appetite, the smoked form may show localized immunosuppressive effects in the lungs (Haney 2002). In addition, there is always a concern about the interaction between a medication and the primary treatments for HIV. One study examined the potential for smoked and oral cannabinoids to interact with nelfinavir and indinavir. While there appeared to be an effect in decreasing concentrations of the antiretrovirals, the authors stated that it is unlikely that there will be concern for short-term clinical effects of the reduction of medication levels (Kosel et al. 2002). Even so, it may be of concern that cannabinoids delivered through smoking may lead to some destruction of lung immunity and slowed ability to heal lung tissues.

TABLE 26.14

Selected Non-Nutrient Therapies for Wasting

Therapy	Description
Androgens	*Nandrolone:* injection, occasional liver function and alkaline phosphatase increase, enhanced effect on improved body cell mass with exercise regimen *Oxandrolone:* occasional nausea and vomiting, potential for hepatotoxicity, enhanced improvement of body cell mass when used with exercise; monitor liver function and blood lipids *Oxymetholone:* possible edema and potential for hepatoxicity, glucose intolerance, enhanced improvement of body cell mass when used with exercise; monitor liver function and blood lipids *Testosterone:* injection, potential for hepatoxicity, elevated blood lipids, enhanced improvement of body cell mass when used with exercise; monitor liver function and blood lipids
Appetite Stimulants	*Dronabinol:* some CNS effects; modest weight gains *Megestrol acetate:* hyperglycemia, hypogonadism (with long-term use), adrenal insufficiency (especially in children); megestrol acetate ES (nano-crystal form) eliminates food effect; may be used with testosterone therapy
Exercise	Both resistance and aerobic exercises are important to maintain body composition and normalize metabolism, especially in conditions such as wasting, insulin resistance, cardiovascular risk, and lipodystrophy
Growth Hormone	Used for both wasting and lipodystrophy (fat accumulation) in lower doses, may cause muscle discomfort, potential for fluid retention, hypertension, lipoatrophy, insulin resistance
Insulin Sensitizing Agents	*Metformin:* reduces hepatic insulin resistance and may be used to reduce the effect of fat accumulation; caution with renal or hepatic dysfunction; may cause nausea, vomiting, fullness, diarrhea, flatulence; monitor for lactic acidosis *Glitazones:* reduces peripheral insulin resistance and may be used to reduce the effect of fat atrophy; caution with liver dysfunction; may cause increases in blood lipids

Several medications have been used to specifically support and restore lean tissues, normalize hormonal balances and function, and improve health and quality of life. Low sex hormone levels have been treated with hormone replacement therapy (Crum et al. 2005). Both men and women experiencing hypogonadism may lose lean and fat tissues, including bone mineral density, and experience related anemias (Clay and Lam 2003; Behler et al. 2005; Miller and Mulligan 2005). Testosterone replacement is available in injection, patch, and gel forms, and acceptability and efficacy may vary between patients (Clay 2004). Physiologic dosing of androgens have been used in women less commonly, but may provide some benefit, including support for the maintenance of lean tissues (Mylonakis, Koutkia, and Grinspoon 2001; Mazer and Shifren 2003). Improved muscle function may be enhanced with the combination use of hormone replacement, anabolic therapies, exercise, and adequate diet (Strawford et al. 1999). Oral agents include oxandrolone and oxymetholone. Injection and patch agents are testosterone-based and the injectable nandrolone decanoate is a synthetic hormone. Caution should be taken for patients with active liver disease, and liver function should be carefully monitored at higher doses (Orr and Fiatarone Singh 2004). Growth hormone and related treatments have demonstrated effects in both improving body cell mass and reducing central fat accumulation in **lipodystrophy syndrome** (Falutz et al. 2005; Haugaard et al. 2005). The dose used for reversal of wasting is substantially higher than the dose recommended for use in reducing central fat accumulation.

lipodystrophy syndrome—loss or absence of fat, or the abnormal distribution of fat in the body, in HIV infection, these changes are likely hormonally mediated; subcutaneous fat loss is most apparent in peripheral limbs and facial areas; fat deposits are most commonly central, located in the dorsocervical area, breast area, and abdominal region

Symptom Management Side effects of medications and disease-related symptoms require adequate management to support medication adherence, prevent additional disease progression, and improve quality of life (Spirig et al. 2005). Common symptoms related to the complications of disease and treatments include those that can interfere with nutritional maintenance and the apparent acceleration of chronic diseases such as cardiovascular disease, diabetes, osteoporosis, anemias, and others. The most commonly reported symptoms include fatigue and diarrhea.

Symptom management often includes a nutrition component, but may also include non-nutrient therapies. For instance, diarrhea can result from opportunistic infection, medications, and organ dysfunction. If adequately evaluated, the appropriate therapy may include treatment of opportunistic infections identified, alteration of medication or doses, additional medication therapy to reduce diarrhea, and dietary management. Some of the medications used to manage diarrhea may include anti-diarrheals, pancreatic and/or lactase enzymes, or supplementation of nutrients used as pharmaceutical agents (such as L-glutamine). Table 26.15 shows selected medication strategies for symptom management.

Nutrition Care Process

The nutrition care process should be well integrated into the medical care process. Interventions for chronic HIV infection span from education and counseling to social and economic support, to medication therapies for HIV and comorbidities. The care plan must prioritize items according

TABLE 26.15

Selected Medication Therapies for Symptom Management

Symptom	Medication Therapies: Generic (Brand Name)
Anorexia	Dronabinol (Marinol): synthetic THC/cannabinoid, may cause dry mouth, avoid alcohol Megestrol acetate (Megace, Megace ES): progestational agent, may cause edema, potential for hypogonadism; take with food if needed to prevent dyspepsia.
Constipation	Cisapride (Propulsid) Note: This drug can interact unfavorably with protease inhibitors and non-nucleoside reverse transcriptase inhibitors, increasing risk of potentially fatal changes in heart rhythms.
Diarrhea	Loperimide (Immodium), diphenolxylate/atropine (Lomotil), Kaopectate, tincture of opium, Pepto-Bismol, psyllium (Metamucil), pancreatic enzymes (Ultrase, Pancrease)
Gastroesophageal Reflux Disease (GERD)	*Proton Pump Inhibitors:* rabeprazole (AcipHex), esomeprazole (Nexium), lansoprazole (Prevacid), omeprazole (Prilosec), pantoprazole (Protonix) Note: These drugs reduce acid and may affect ARV absorption, particularly atazanavir; in some cases, ARVs may speed the metabolism of these medications, rendering them less effective. *H2 Blockers:* nizatidine (Axid), famotidine (Pepcid), cimetidine (Tagamet), ranitidine (Zantac) Note: These drugs reduce acid and may affect ARV absorption; some interactions with ARVs have the potential to increase toxicity (e.g., cimetidine increases protease inhibitor levels).
Heartburn	Antacids (Alka-Selzer, Bromo-Selzer, Maalox, Mylanta, Rolaids, Tums) Note: Caution should be taken when administered with ARVs or other medications that require an acid environment for absorption, such as agenerase, atazanavir, tipranavir, delavirdine; Alka-Selzer contains aspirin and Bromo-Selzer contains acetaminophen.
Nausea and/or Vomiting	Procholorperazine (Compazine), cola syrup (Emetrol), metoclopramide (Reglan), chlorpromazine (Thorazine), thiethylperzaine (Roecan), ondansetron (Zofran) Note: Many of these medications can interact with ARVs, causing a need for dose adjustments.

to the most urgent issues. For instance, if a patient is diagnosed with diabetes, nutrition-related intervention should be targeted to improve blood sugar control and insulin sensitivity as well as restore general nutrition well-being. The plan should be created in collaboration with the patient and other health care team members, and should be tailored to support adherence. Any plan for nutritional care should also carefully consider the priority of HIV disease management and potential interactions of therapies and well-being.

Intervention Plans are implemented according to the setting and wishes of the patient and his or her care provider. The health care team, including the patient and/or appropriate care providers, should agree to and be capable of implementing the nutrition care plan. Goals for the nutrition care plan may include prevention of adverse events related to therapies, restoration of adequate nutritional status, management of co-conditions (e.g., diabetes, liver disease, renal dysfunction), and others. Implementation may include the coordination of services and products to target maintenance and improvement of nutritional status as well as direct patient care. All aspects of patient care should be considered in order to reduce the possibility of adverse events and effects of therapy, and to enhance the efficacy of and adherence to recommended therapies.

With the advent of HAART, there are many more opportunities to implement plans in the outpatient setting. Routine appointments should be scheduled according to the nutrition care plan and as often as necessary to monitor and adjust the plan. Because the coordination with the overall health care plan is an important feature of the nutrition care plan in chronic HIV infection, communication and documentation should be shared between key members of the health care team.

Monitoring/Evaluation Chronic HIV infection is a multifactorial disease that presents in a variety of ways that may differ between patients. Problems, associated interventions, and monitoring and evaluation of appropriate outcomes are important features of the nutrition care plan. Evaluations should be conducted routinely according to the nutrition care plan. Interventions that should yield results should be monitored when those achievements are expected. For instance, if the goal is to reduce weight and maintain a buffer of body cell mass, measurable changes may be achieved in three months and an appointment for follow-up should be made to confirm that progress is being made and to adjust the plan appropriately to support and improve outcomes. If a short-term plan is put into place in a hospital setting, follow-up should be scheduled prior to discharge or should be a part of the discharge plan and communication with outpatient service providers.

Optimal outcomes should be matched to the problem list based on assessment, as with any other disease state. Prevention of nutritional compromise, restoration of nutritional well-being, and support for the overall health care plan are important general features of desired outcomes. Improved knowledge, therapy adherence, management of disease and co-diseases, and support for physical and psychosocial/economic well-being related to nutrition and food are more specific goals that can be tailored to each patient's needs. Clear documentation is necessary to make the case for the supportive role of nutrition therapy in HIV infection and disease management.

In addition to individualized outcomes, monitoring and research on the benefits of providing support to prevent malnutrition and improve nutritional status will be important to demonstrate the benefits that nutrition therapy can have in patients living with chronic HIV infection (Young 1997).

Conclusion

HIV infection creates a number of challenges to the maintenance of nutritional status. While weight loss and wasting are not an inevitable part of the natural history of HIV infection (Kotler et al. 1989), chronic inflammation due to HIV infections, co-conditions, and complications assault nutritional status continuously.

With long-term survival, it will be an important practice to prioritize health issues, including conditions directly and indirectly related to nutritional status. Treatments for HIV infection can both support and inhibit nutritional status maintenance and improvement. There is an ongoing need for education, counseling, and a variety of other interventions to mitigate nutritional decline and disease progression. Successful nutrition-related therapies will help to stabilize nutritional and other types of clinical status markers. Patient-centered care through inclusion of the patient in the health care team that specializes in HIV-related care may be best suited for providing ongoing care to patients living with this complex and life-long disease (American Dietetic Association 2004).

PRACTITIONER INTERVIEW

Margaret Davis, M.B.A., R.D. *Clinical Nutrition Consultant, San Francisco Bay Area*

I have worked as a registered dietitian for 29 years and for 20 years in direct HIV/AIDS patient care. In San Francisco, I would estimate 75% of the clients I see have been infected with HIV for over 10 years.

I nutritionally assess my patients not only for HIV/AIDS nutritional markers but also for chronic comorbid diseases. I look for family medical history; the patient's weight history (highest and lowest weight since diagnosis); history of opportunistic infections; current medications; appetite; symptoms of the entire gastrointestinal tract; allergies and/or food intolerances; dietary supplements; any other comorbidities such as type 2 diabetes mellitus; hyperlipidemias, hepatitis C, tuberculosis; and current lab values—HIV viral load, CD4 count, electrolytes panel, renal function tests (BUN and creatinine), liver function tests, lipid panel, and glucose. In the past, I would have been more concerned about protein (muscle) wasting, but now with the newer categories of medication for HIV, I need to be more concerned with chronic long-term effects of HIV, the medications they are taking, and the conditions associated with aging.

The most common nutrition problems I see in my clients are hyperlipidemia, dyslipidemia, metabolic syndrome, type 2 diabetes, and acute weight loss related to opportunistic infections. Many clients who have lived with HIV for 15–20 years or longer still think that they are going to die from AIDS and thus may continue to smoke and/or practice other poor health habits. In reality, though, they should be concerned with developing chronic diseases that will more likely cause their death than HIV.

Some of my biggest nutrition challenges working with patients with HIV/AIDS are keeping up with all the medications they take and their side effects, treating nutrition problems associated with chronic conditions, and treatments such as those associated with hepatitis C, which often causes acute weight loss and depression. There have been tremendous changes in the treatment of HIV/AIDS in the last 15 years. The varieties of medications that are available have now made my job a bit more like a specialist in geriatric nutrition, since clients are older and developing chronic diseases. The upside of working with HIV/AIDS in the San Francisco Bay Area is that it is a hub for community-based HIV research. Working with other health professions involved in research helps keep me abreast of the field, and they know that nutrition is a critical part of HIV treatment—I am appreciated and considered part of the HIV health care team.

My advice to dietetic students is to not get overwhelmed by all the numbers; listen and learn from your clients. If you ask them about their HIV history, they will tell you their stories, and from both a medical and physiological standpoint, it is one of the most fascinating areas of medicine to work in. Working in the HIV/AIDS community, you are able to utilize almost all of your clinical skills and MNT knowledge, and because it has become a more chronic disease, I have developed many long-term relationships with clients, which are not common in the acute-care setting.

CASE STUDY 26

Introduction:

RE is a 49-year-old Caucasian male who has been HIV positive for approximately 14 years. He presented to his physician with complaints of abdominal discomfort and heartburn related to his enlarged abdomen. His doctor previously diagnosed Mr. E with lipodystrophy syndrome based on physical examination and insulin resistance. He was referred to the dietitian for recommendations for nutrition-related interventions.

Nutrition Assessment:

Ht. 5'11" Wt. 205 lbs. UBW 175 lbs.

Labs:

Viral load: undetectable (below 50 copies per mL), CD4+ count: 660 per mL

Medications:

Currently taking efavirenz and combovir (lamivudine and zidovudine), previously taking lopinavir/ritonavir and combovir, just prescribed rosiglitazone/metformin combination

Additional Anthropometry:

Abdominal circumference: 104 cm, abdominal fatfold: 3 mm

The registered dietitian's interview indicates that the patient describes gaining weight quickly over the last six months, and most appears to be in the abdomen. Heartburn began in the last few weeks and he is concerned about his body shape. He stopped exercising regularly about a year ago after he moved to the suburbs and away from his usual gym.

Dietary Intake:

AM: scrambled eggs with sausage and toast, orange juice, coffee; Mid-day: meatloaf sandwich with cheese or gravy-dipped Italian beef with cola; PM: casserole with bread or rolls and water. Snacks throughout the day are typically potato chips or crackers. Drinks 3 large mugs of coffee and 1–2 colas throughout the day. Partner cooks large meals for family and friends on weekends.

Questions:

1. What role might Mr. E's insulin resistance play in his current concern for fat accumulation and heartburn?
2. What potential effect does each of his previous and current medications have on lipodystrophy and insulin resistance?
3. What parts of his diet history appear to put him at risk for or exacerbate lipodystrophy and heartburn?
4. Identify and prioritize nutrition-related problems.
5. What-diet related recommendations would you discuss with this patient? Include diet recommendations to support medications.
6. What non-nutrient recommendations would you suggest that this patient, his physician, and his health care team explore in order to improve lipodystrophy-related discomfort and heartburn?
7. What criteria would you monitor in this patient for improvement?

NUTRITION CARE PROCESS FOR HIV/AIDS

Step One: Nutrition Assessment

Medical/Social History

Diagnoses:
HIV disease status
Medications: Past and present
Previous medical conditions or surgeries; previous opportunistic infections
Socioeconomic status/Food security
Support systems
Education level – primary language

Physical Assessment

Signs and symptoms of nutrient deficiencies or toxicities
Appearance of muscle wasting

Anthropometric

Height
Current weight
Weight history: highest adult weight; usual body weight
Reference weight (BMI)
Baseline BIA
Abdominal girth—appearance of lipodystrophy

Biochemical Assessment
HIV Laboratory Values:

CD4+
CD8+
Viral load

Inflammatory/Acute Phase Proteins:

C-reactive protein
Fibronectin

Visceral Protein Assessment:

Albumin
Prealbumin

Hematological Assessment:

Hemoglobin
Hematocrit

Lipid Assessment:

Total cholesterol
HDL
LDL
Triglyceride

Dietary Assessment

Ability to chew; use and fit of dentures
Problems swallowing
Nausea, vomiting
Constipation, diarrhea
Heartburn
Any other symptoms interfering with ability to ingest normal diet
Ability to feed self
Ability to cook and prepare meals
Food allergies, preferences, or intolerances: spicy foods, high fat foods, pepper, caffeine, coffee, tea, alcohol, spearmint, peppermint, chocolate
Previous food restrictions
Ethnic, cultural, and religious influences
Use of alcohol, vitamin, mineral, herbal, or other type of supplements
Previous nutrition education or nutrition therapy
Eating pattern: 24-hour recall, food history, food frequency

Step Two: Nutrition Diagnosis

Inadequate oral food/beverage intake
Involuntary weight loss
Food–medication interaction
Food-and nutrition-related knowledge deficit

Sample: Involuntary weight loss NC-3.1
PES: Involuntary weight loss as evidenced by 9% weight change over previous six months.

Step Three: Intervention

1. Provide nutrition education regarding methods to increase nutrient density for all foods.
2. Consume smaller, more frequent meals.
3. Add high-kcal, high-protein beverage between meals.
4. Recommend multivitamin daily.

Step Four: Monitoring and Evaluation
Monitor weight and nutrition assessment indices.
Evaluate 24-hour recall for nutrition adequacy.

WEB LINKS

Research and Reviews: Adult AIDS Clinical Trials Group: Includes information and summaries; research links.

http://www.aactg.org

Medscape AIDS Page: Covers conference and journal reviews and provides online access to AIDS Reader articles.

http://www.medscape.com/hiv-home

HIV InSite: Clinician and Patient Materials and Information: Online Textbook on HIV/AIDS through University of California, San Francisco School of Medicine.

http://hivinsite.ucsf.edu

Health Resources and Services Administration (HRSA) Guide to Nutrition in HIV/AIDS: Includes clinician guidelines/algorithms and patient handouts.

http://www.aidsetc.org/pdf/p02-et/et-30-20-01/nutr_guide_0602.pdf

Johns Hopkins AIDS Service:

http://hopkins-aids.edu/publications/pocketguide/pocketgd0105.pdf

Pocket Guide for ARVs: Literature reviews.

http://www.hopkins-hivguide.org/literature_review/index.html?categoryId=9351&siteId=7151

The AIDS InfoNet: Fact sheets and guides for patients and health care providers.

http://www.aidsinfonet.org

National Institutes of Health AIDS Information: Patient and Provider Fact Sheets on topics related to HIV/AIDS.

http://www.aidsinfo.nih.gov/other/factsheet.aspx

Guidelines and Other Resource Links:

AIDS Resource List: Links for information resources.

http://www.specialweb.com/aids

Food and Nutrition Technical Assistance/Academy of Educational Development: Funded by United States Agency for International Development, FANTA provides services and monographs appropriate for use in resource-limited settings and developing countries.

http://www.fantaproject.org

AVERT: Provides information on HIV/AIDS-related care in resource-limited settings.

http://www.avert.org/hivcare.htm

HIV/AIDS Dietetic Practice Group of the American Dietetic Association: Nutrition and HIV/AIDS information, including continuing education.

http://www.hivaidsdpg.org

Project Inform: Information on treatment, community programs, and research.

http://www.projinf.org

END-OF-CHAPTER QUESTIONS

1. What kind of virus is the human immunodeficiency virus (HIV)? Which types of cells does it target?
2. Describe the inflammatory process that is seen in early and asymptomatic HIV infection. How can it trigger a wasting process?
3. List two factors that are predictive of disease progression. List one nutrition-related factor that is predictive of survival.
4. List two examples of how HIV or opportunistic infection compromises gastrointestinal function.
5. List two cofactor diseases or conditions that are related to HIV infection and its treatment and may require nutritional therapies.
6. Define AIDS-Related Wasting Syndrome (AWS).
7. At which levels of weight loss and body cell mass loss is there an increased risk for morbidity and mortality in HIV infection?
8. Which antiretroviral (ARV) medications should be taken on an empty stomach? Which should be taken with food?
9. Describe potential side effects of two medications that alter nutritional status.
10. Describe how HIV infection and its treatment may contribute to insulin resistance and diabetes.
11. Describe three types of dietary interventions that may be required in treated HIV infection.
12. List four types of non-nutrient interventions that are used to improve nutritional status.
13. Describe the characteristics of lipodystrophy syndrome and the potential uses of nutrition-related therapies to reduce its effects.

27

Diseases of the Musculoskeletal System

Robert D. Lee, Dr. P.H., R.D.

Central Michigan University

CHAPTER OUTLINE

Normal Anatomy and Physiology of the Skeletal System

Cartilage • Bone • Hormonal Control of Bone Metabolism

Osteoporosis

Prevention • Medical Management • Pharmacologic Prevention and Treatment

Paget Disease

Rickets and Osteomalacia

Arthritic Conditions • Osteoarthritis • Rheumatoid Arthritis • Gout

Fibromyalgia

Introduction

The musculoskeletal system comprises the bones and cartilage of the skeleton, as well as the muscles, tendons, and ligaments attached to the bones. The skeleton is divided into the axial and appendicular skeleton. The axial skeleton forms the axis of the body and includes the bones of the skull, vertebral column, and thorax. The appendicular skeleton consists of the bones of the upper extremities, the lower extremities, the shoulder, and the hip (Porth 2005). The musculoskeletal system makes movement in the external environment possible, gives shape and stability to the body, and protects and maintains the position of soft tissues. The bones also serve as a reservoir for calcium and phosphorus, so that the levels of these important minerals in bodily fluids can be maintained within certain physiologically acceptable ranges (Heaney 1999). Consequently, diseases of this system can severely limit mobility and the performance of the activities of daily life, can cause considerable pain and disability, and have the potential to deform the body, which, in turn, can adversely affect respiration, digestion, nervous system function, and nutritional status.

This chapter addresses the more common diseases of the musculoskeletal system that have the potential to be prevented or managed by nutrition therapy. These include osteoporosis, Paget disease, rickets, osteomalacia, osteoarthritis, rheumatoid arthritis, and gout.

Normal Anatomy and Physiology of the Skeletal System

Composed of cartilage, ligaments, tendons, and bones, the skeletal system forms a strong, tightly bound but flexible framework for the body. Far from being an inert scaffold for the body, the skeletal system is composed of numerous metabolically active cells and tissues that interact physiologically with other organ systems of the body and are in a continual state of change throughout the life cycle.

Cartilage

Cartilage is a flexible yet firm connective tissue consisting of cells and collagen fibers surrounded by an amorphous gel-like matrix composed primarily of water and protein-carbohydrate complexes known as proteoglycans. Cartilage is formed by cells called **chondroblasts** that actively secrete and surround themselves with a gel-like matrix until they become trapped in little cavities known as lacunae. Once the chondroblasts are enclosed in lacunae, they are called **chondrocytes.** The matrix is composed of collagen fibers, proteoglycans, and water (Saladin 2004; Porth 2005). Collagen is a fibrous protein that serves as a major constituent of cartilage, tendons, ligaments, bone, and skin, and is the most common protein found in the body. Collagen fibers are tough and flexible, resist stretching, and give cartilage its form and tensile strength. Proteoglycans are large macromolecules shaped like a test-tube brush, with a central protein core to which are attached numerous polysaccharides that project out much like bristles do. The most common polysaccharide found in the proteoglycans of cartilage is **chondroitin sulfate** (Saladin 2004). The proteoglycans attract and hold water, which gives cartilage its elasticity, stiffness, and ability to resist compression. It is estimated that 65% to 80% of the wet weight of cartilage is water (Porth 2005).

Cartilage gives shape to such structures as the external ear, the tip of the nose, and the larynx. It is the precursor of the axial and appendicular skeleton during embryonic development, the growth zone of the long bones of children, and an important component of various types of joints in the mature skeleton (Saladin 2004). The costal cartilage of the rib cage firmly binds the ribs and clavicles to the sternum, while the fibrocartilage of the intervertebral discs join the bodies of the vertebral bones and allow limited movement between adjacent vertebrae. The articular cartilage of the synovial joints allows free movement of the bones of the hands, feet, arms, legs, and so on. The articular cartilage covering the tips of the bones at the joints provides a remarkably smooth surface, resulting in extremely low friction between the two bones during movement of the joint. It increases the surface area between the two bones, helps to dissipate mechanical stresses applied to bones, and aids in transmitting the load down the bone (Saladin 2004; Brandt 2005).

Cartilage is free of blood vessels except for the period when it is changing into bone during the skeletal development of childhood. Consequently, gasses, nutrients, and wastes must travel by solute diffusion between the blood vessels outside the cartilage and the chondrocytes embedded within the matrix of the cartilage. Although this diffusion is facilitated by the high water content of cartilage, it is a relatively slow process. As a result, chondrocytes have low rates of metabolism and cell division, and when injured, cartilage heals slowly (Saladin 2004; Porth 2005). Cartilage undergoes constant turnover as its worn-out matrix components are degraded by enzymes produced by the chondrocytes, which then secrete new matrix to replace the old. Thus, the integrity of joints is dependent upon healthy cartilage, which is maintained by properly functioning chondrocytes (Hightower and Gunta 2005). However, in the latter decades of life, the secretion of new matrix fails to keep-up with losses, and the articular cartilage gradually thins, or in the case of osteoarthritis, is absent (Saladin 2004).

chondroblasts—cells that are actively forming cartilage

chondrocytes—cells surrounded by cartilage and located inside small cavities known as lacunae

chondroitin sulfate—the most common polysaccharide found in the proteoglycan molecules of cartilage

osseous tissue—the group of cells and cell products that collectively form bone; bone tissue

hydroxyapatite—a crystallized calcium phosphate salt that gives bones their stiffness

Bone

There are two ways in which the term bone is used in this chapter. It is used to refer to specific body structures such as the pelvis, femur, and mandible, which are organs composed of multiple tissue types including nerves, blood vessels, adipose tissue, bone marrow, cartilage, and fibrous tissue. The term bone is also used to refer to bone tissue or **osseous tissue**, which is the major component of bone and makes up most of the mass of individual bones (Saladin 2004). Osseous tissue is a connective tissue having an *organic* component that is mineralized or calcified by being impregnated with an *inorganic* component. The organic or protein component includes bone cells embedded in a matrix of collagen fibers and proteoglycans. The protein collagen comprises 90% of this organic matrix (Heaney 1999). The organic component of osseous tissue contributes approximately one-third of the dry weight of bones and gives bone a degree of flexibility. Without this component, bones would be brittle and likely to shatter when stressed. The inorganic or mineral component consists of approximately 85% **hydroxyapatite** (a crystallized calcium phosphate salt), about 10% calcium carbonate, and smaller amounts of magnesium, sodium, potassium, fluoride, sulfate, and carbonate (Heaney 1999; Saladin 2004; Porth 2005). The inorganic component, responsible for about two-thirds of the dry weight of bones, gives stiffness to bones and allows them to easily support the weight of the body without bending. When bones lack this mineral component, they are soft and can easily bend, as is the case with the childhood disease rickets and the condition known as osteomalacia, which is seen in adults (Saladin 2004; Hightower and Gunta 2005).

In addition to serving as the body's framework, bones are a ready source of calcium and phosphorus for maintaining physiologic concentrations of these minerals in the extracellular fluids. The calcium content of the adult human body ranges from 1,100 to 2,000 g, with 99% of this in the bones. The bones provide two calcium reserves: a massive (99% of total available calcium) but less readily accessible one in the form of hydroxyapatite (crystallized calcium phosphate salt) and a much smaller pool (1% of total available calcium) that can be quickly released into the extracellular fluid. The average adult body contains about 500 to 800 g of phosphorus and 85% to 90% of this is in the bones (Heaney 1999; Saladin 2004; Bringhurst et al. 2005). In health, the body maintains tight control of these minerals in the serum. The normal serum concentrations of calcium and phosphorus are 9.0 to 10.5 mg/dL and 3.0 to 4.5 mg/dL, respectively (Bringhurst et al. 2005). Of the two minerals, abnormalities of serum calcium concentrations are the most critical. Hypocalcemia, an abnormally low serum calcium concentration, can cause excessive excitability of the nervous system and lead to facial grimacing, muscle spasms, and tetany (an inability of the muscle to relax). **Carpopedal** spasm results from tetany occurring in the hands and feet and can be a sign of hypocalcemia. In extreme hypocalcemia, laryngeal spasm, respiratory arrest, and convulsions can occur. Hypercalcemia, an abnormally high serum calcium concentration, can result in fatigue, depression, mental confusion, anorexia, nausea, vomiting, and constipation.

The Cells of Osseous Tissue Four principal types of cells are found in osseous tissue:

1. **Osteogenic cells** are stem cells capable of differentiating (developing a more specialized form or function) into *osteoblasts*, and are the only source of new osteoblasts. Osteogenic cells are active during normal growth of the skeletal system during childhood and adolescence, and in adulthood are activated in response to bone injury such as a fracture and in response to the stress placed on bones during weight-bearing exercise.
2. **Osteoblasts**, or bone-building cells, synthesize, deposit, and then orient the fibrous proteins of the organic matter of the bone matrix (collagen, proteoglycans, and other proteins) and then participate in the calcification or mineralization of the bone matrix, in a process known as bone formation or mineral deposition (Heaney 1999; Saladin 2004). Active osteoblasts are found on the surface of newly forming bone. They remove ions of calcium, phosphate, and other minerals from the blood plasma and deposit them within the bone matrix, thus hardening it. In response to bone fracture or the stress of weight-bearing exercise, the osteogenic cells multiply more rapidly and then differentiate to become osteoblasts (Saladin 2004; Porth 2005).
3. **Osteocytes** are mature osteoblasts surrounded and entrapped by the matrix they have synthesized and calcified, and represent the vast majority of cells in bone (Saladin 2004; Porth 2005). Osteocytes reside in the tiny, fluid-filled cavities within the calcified matrix called lacunae, which are interconnected by narrow channels called canaliculi. Delicate cytoplasmic processes extend from the osteocytes and pass through the canaliculi. These processes connect with those of other osteocytes, thus allowing the osteocytes to chemically signal each other. The fluid-filled canaliculi also serve as channels for the passage of nutrients and metabolites between the osteocytes and nearby blood vessels. Although the osteocytes neither deposit nor remove bone, they are actively involved in maintaining the bony matrix by monitoring the amount of strain (bending) a bone experiences when it is mechanically loaded (for example, by weight-bearing exercise) and then communicating this information to osteoblasts on the bone surface (Heaney 1999; Saladin 2004; Bringhurst et al. 2005). The osteoblasts can then build up and strengthen the bone where needed in response to the stress.
4. **Osteoclasts** are bone-removing cells that secrete hydrochloric acid (pH of about 4) to dissolve the mineral component of bone matrix and an enzyme called acid phosphatase that digests the collagen and other protein components of the bone matrix. This process is known as bone resorption or mineral resorption (Saladin 2004). The dissolved minerals are released into the blood and made available for other uses.

Skeletal Growth and Development Throughout life, the bones of the skeleton are in a continual state of change. Linear and circumferential growth of the long bones are the most prominent observable features of skeletal development

carpopedal—referring or pertaining to the hand and foot

osteogenic cells—stem cells capable of developing into osteoblasts

osteoblasts—cells that synthesize, deposit, and then orient the fibrous proteins of the organic matter of the bone matrix

osteocytes—mature osteoblasts surrounded and entrapped by the matrix they have synthesized and then calcified

osteoclasts—bone-removing cells that dissolve the mineral component of the bone matrix, playing a major role in bone resorption during childhood and adolescence. In females, mature height is reached at about age 16 to 18 years, and in males, at about age 18 to 20 years. In addition to linear and circumferential skeletal growth during childhood and adolescence, bones are continually changing their size and shape in response to changes in forces applied to the skeleton in a process known as remodeling (Lindsay and Cosman 2005). For example, as children begin to walk and then become increasingly physically active, bones develop ridges, spines, and bumps on their surfaces in response to stresses placed on bones by the exercising muscles. Remodeling also involves the repair of microscopic bone damage resulting from excessive or accumulated stresses and the maintenance of serum calcium levels by moving calcium from the bones into the blood (Saladin 2004; Lindsay and Cosman 2005). Compared to sedentary people, the bones of those regularly engaged in physical activity or heavy manual labor tend to be stronger and have a greater density and mass. Even within the same person, the bones of the dominant arm (e.g., the right arm for a person who is right-handed) are stronger than the bones in the non-dominant arm (Heaney 1999). In remodeling, the osteoclasts remove bone from low-stress areas where it is not needed, while the osteoblasts lay down new bone in high-stress areas where it is needed (Saladin 2004; Porth 2005).

Cortical and Trabecular Bone The bones of the skeleton are made from two different types of osseous tissue: **cortical bone** and **trabecular bone**. Cortical bone (also known as compact bone) is dense and has no open spaces visible to the naked eye. It forms the external surfaces of all bones, the shafts of the long bones (e.g., the femur and humerus), and a shell that caps the ends of the long bones. Trabecular bone (also known as cancellous bone), on the other hand, is loosely organized with a sponge-like appearance. Trabecular bone is found at the ends (or "heads") of the long bones. The vertebrae, pelvis, sternum, and scapulae are primarily trabecular bone with a thin covering of cortical bone (Heaney 1999; Saladin 2004). Figure 27.1, which shows the proximal end of the femur cut longitudinally, illustrates some of the differences between cortical and trabecular bone. The shaft or middle part of the bone has a dense, thick wall composed of cortical bone enclosing a space inside the shaft called the medullary cavity, which contains bone marrow. The end of the bone has a spongy appearance due to the trabecular bone that predominates there. The trabecular bone at the end is capped with a thin layer of cortical bone (Saladin 2004).

FIGURE 27.1 In this photo of the proximal end of a human femur cut longitudinally, cortical bone forms a dense outer shell around the sponge-like trabecular bone. At the top or head of the bone, trabecular bone predominates and the shell of cortical bone is thin, unlike the shaft where thick cortical bone provides strength and rigidity.

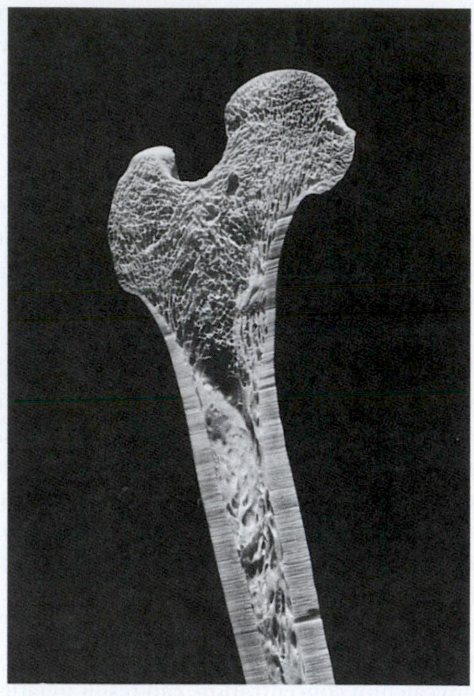

Source: S. Rolfes, K. Pinna, and E. Whitney, *Understanding Normal and Clinical Nutrition*, 7e, copyright © 2006, p. 429.

Cortical and trabecular bone are considered lamellar bone because the osseous tissue of both is organized in layers known as lamellae. However, the arrangement of the lamellae differs between the two types of bone, as shown in Figure 27.2, which represents the microscopic appearance of a biopsy specimen removed from the shaft of the long bone. In the cortical bone, the lamellae are arranged concentrically, in onion-like layers, around a central canal that runs parallel to the long axis of the bone. Each concentric lamella is connected to its adjacent lamellae by canaliculi. A central canal and its surrounding lamellae constitute an osteon, which is the basic structural unit of cortical bone. The central canals are joined by traverse or diagonal canals. These canals provide passages through which blood vessels and nerves pass (Saladin 2004; Porth 2005).

Trabecular bone consists of a framework of crossed and interconnected plates, rods, and spicules called trabeculae that form a lattice-like pattern (Heaney 1999; Saladin 2004). The space between the rods and plates is filled with bone marrow. The matrix of the rods and plates is arranged in

cortical bone—dense bone that forms the external surfaces of all bones, the shafts of the long bones, and a shell that caps the ends of the long bones

trabecular bone—loosely organized bone having a sponge-like appearance and found at the ends of long bones

FIGURE 27.2 Organization of a long bone
(a) A long bone cut longitudinally shows the dense cortical bone covering the sponge-like trabecular bone. Enclosed within the shaft is the medullary cavity, which contains bone marrow. (b) A small piece removed from the long bone and enlarged illustrates the organization of cortical and trabecular bone. In cortical bone, the lamellae are arranged concentrically around a central canal running parallel to the long axis of the bone. A central canal and its surrounding lamellae constitute an osteon. (c) A magnified spicule of trabecular bone shows that the lamellae do not surround a central canal because the osteocytes have a nearby blood supply. Consequently, in trabecular bone no osteons are formed. (d) A magnification of lamellae from cortical bone shows the osteocytes and each concentric lamella connected to its adjacent lamellae by canaliculi.

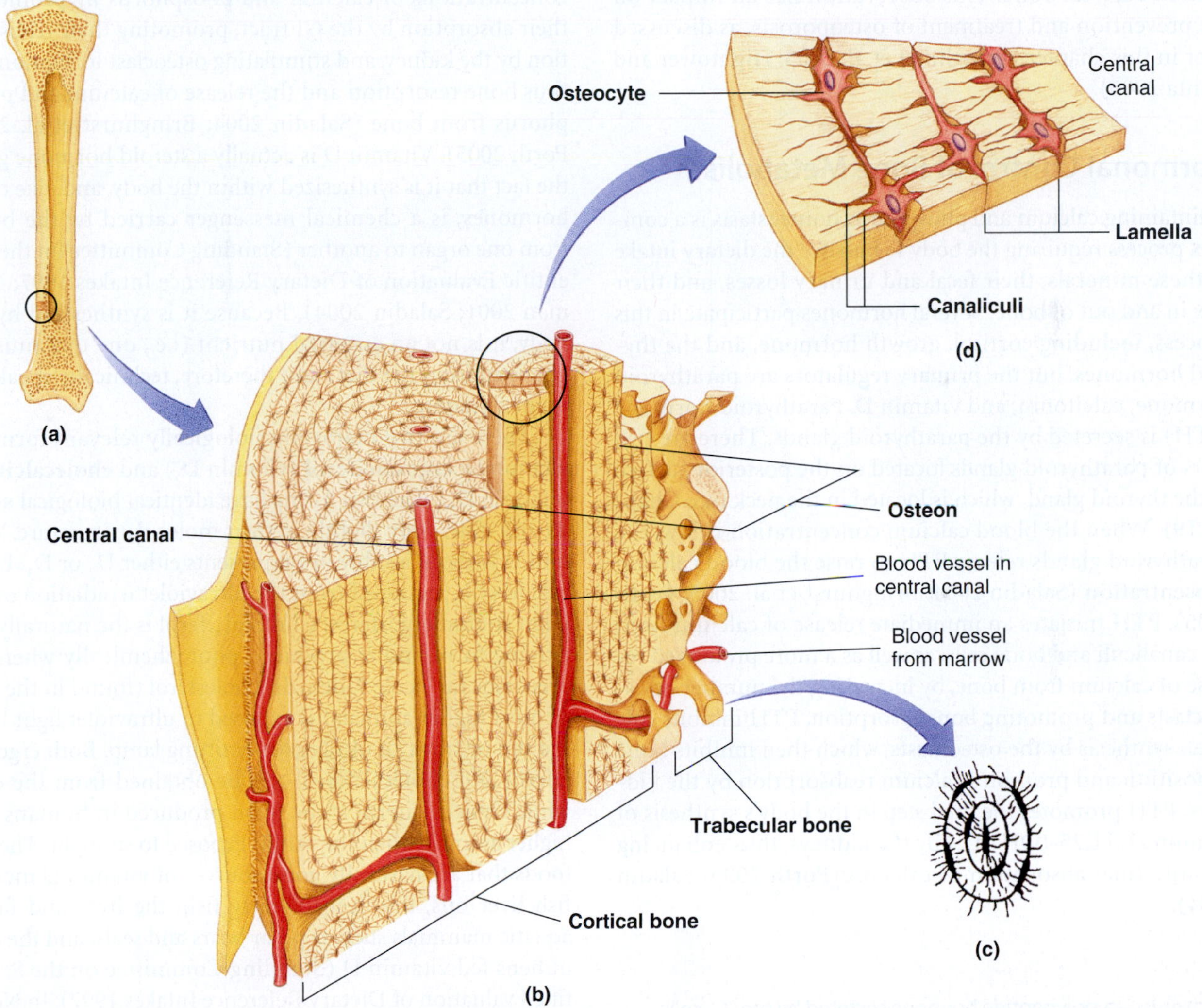

Source: L. Sherwood, *Human Physiology: From Cells to Systems*, 5e, copyright © 2004, p. 738.

lamellae but not in layers surrounding a central canal. The central canal is not needed because no osteocyte is very far from a source of blood. Consequently, no osteons are formed in trabecular bone. The arrangement of the rods and plates occurs in response to the stresses placed on the bone and is not random. As mentioned earlier, the osteocytes act as strain sensors and signal the osteoblasts to form bone where it is needed in response to the mechanical stresses bones receive, for example, from weight-bearing exercise (Saladin 2004; Bringhurst et al. 2005). The unique structure of trabecular bone imparts considerable tensile strength and weight-bearing properties while simultaneously being relatively light. The densely packed calcified matrix of cortical bone makes it much more rigid than trabecular bone (Saladin 2004; Porth 2005).

Although roughly 75% of the weight of the skeleton is composed of cortical bone and about 25% is composed of trabecular bone, the relative quantity of cortical and trabecular bone varies considerably in different types of bones and even within the same bone depending on the need for lightness and strength (Saladin 2004; Bringhurst et al. 2005; Porth 2005). Through the process of remodeling, bones are able to adapt and respond to changing physical demands and stresses imposed on them. Studies indicate that each year as much as 18% of the total mineral content of the skeleton is deposited and removed and that the rate of turnover in trabecular bone is much greater than in cortical bone. This observation has an impact on the prevention and treatment of osteoporosis, as discussed later in this chapter (Bringhurst et al. 2005; Hightower and Gunta 2005).

Hormonal Control of Bone Metabolism

Maintaining calcium and phosphorus homeostasis is a complex process requiring the body to balance the dietary intake of these minerals, their fecal and urinary losses, and their flux in and out of bone. Several hormones participate in this process, including cortisol, growth hormone, and the thyroid hormones, but the primary regulators are parathyroid hormone, calcitonin, and vitamin D. Parathyroid hormone (PTH) is secreted by the parathyroid glands. There are two pairs of parathyroid glands located on the posterior surface of the thyroid gland, which is located in the neck (see Chapter 19). When the blood calcium concentration is low, the parathyroid glands release PTH to raise the blood calcium concentration (Saladin 2004; Bringhurst et al. 2005; Porth 2005). PTH initiates an immediate release of calcium from the canaliculi and bone cells, as well as a more prolonged release of calcium from bone, by increasing the number of osteoclasts and promoting bone resorption. PTH inhibits collagen synthesis by the osteoblasts, which then inhibits bone deposition and promotes calcium reabsorption by the kidneys. PTH promotes the final step in the body's synthesis of vitamin D_3 [1,25-$(OH)_2D_3$] by the kidneys, thus enhancing the intestinal absorption of calcium (Porth 2005; Saladin 2004).

calcitonin—a polypeptide hormone secreted by the "C" cells of the thyroid gland when the blood calcium concentration is high. It lowers blood calcium by inhibiting bone resorption, promoting bone formation, and reducing renal reabsorption of calcium and phosphorus

ergocalciferol—a form of vitamin d produced by exposing the plant steroid ergosterol to ultraviolet irradiation. Also known as vitamin D_2

cholecalciferol—a naturally occurring form of vitamin D produced in humans when the precursor molecule 7-dehydrocholesterol present in the skin is exposed to sunlight or to ultraviolet radiation. Also known as vitamin D_3

Calcitonin (also known as thyrocalcitonin) is secreted by the parafollicular or "C" cells of the thyroid gland when the blood calcium concentration is abnormally high. Its secretion lowers blood calcium concentration by inhibiting the activity of osteoclasts (and thus bone resorption), stimulating the activity of osteoblasts (and thus bone formation), and reducing the renal reabsorption of calcium and phosphate (Saladin 2004; Bringhurst et al. 2005; Porth 2005).

The major function of vitamin D is to increase blood concentrations of calcium and phosphorus by promoting their absorption by the GI tract, promoting their reabsorption by the kidney, and stimulating osteoclast formation and thus bone resorption and the release of calcium and phosphorus from bone (Saladin 2004; Bringhurst et al. 2005; Porth 2005). Vitamin D is actually a steroid hormone given the fact that it is synthesized within the body, and, like other hormones, is a chemical messenger carried by the blood from one organ to another (Standing Committee on the Scientific Evaluation of Dietary Reference Intakes 1997; Norman 2001; Saladin 2004). Because it is synthesized by the body, it is not an essential nutrient (i.e., one that must be obtained from the diet) and therefore, technically speaking, is not a vitamin.

There are two major physiologically relevant forms of vitamin D: **ergocalciferol** (vitamin D_2) and **cholecalciferol** (vitamin D_3). The two forms have identical biological activity and differ only slightly in their molecular structure. Vitamin D without a subscript represents either D_2 or D_3. Ergocalciferol is produced from the ultraviolet irradiation of the plant steroid ergosterol. Cholecalciferol is the naturally occurring form and is produced photochemically when the precursor molecule 7-dehydrocholesterol (found in the skin of most higher animals) is exposed to ultraviolet light from the sun or from an ultraviolet-emitting lamp. Both ergocalciferol and cholecalciferol can be obtained from the diet, while cholecalciferol is the form produced in humans and higher animals when the skin is exposed to sunlight. The few foods that are naturally good sources of vitamin D include fish liver oils, the flesh of fatty fish, the liver and fat of aquatic mammals such as polar bears and seals, and the eggs of hens fed vitamin D (Standing Committee on the Scientific Evaluation of Dietary Reference Intakes 1997). In North America, most dietary vitamin D is obtained from fortified milk and milk products and other fortified foods such as breakfast cereals, soy milk, and margarine. In Canada and the United States, all commercially available milk, regardless of its fat content, must be fortified with 385 IU/liter or 400 IU/quart. However, several studies have shown that the actual vitamin D content of fortified milk varies considerably and that as much as 70% of milk sampled did not contain the required amount of added vitamin D (Standing Committee on the Scientific Evaluation of Dietary Reference Intakes 1997). Because it is fat soluble, vitamin D absorption is dependent on many of the same mechanisms involved in the digestion and absorption of fat, and any intestinal, biliary, or lymphatic condition impeding fat digestion and/or absorption has the potential to impede vitamin D absorption.

Cholecalciferol (vitamin D_3) is synthesized in the human body when the skin is exposed to ultraviolet radiation from sunlight or an artificial source, resulting in the conversion of 7-dehydrocholesterol into vitamin D_3 (see Figure 27.3). A circulating vitamin D-binding protein moves the vitamin D_3 from the skin into the blood circulation. When sun exposure is sufficient, vitamin D production by the skin is adequate to meet the body's physiologic requirements, and in most of the world, this is the major source of the vitamin for humans. However, several factors can limit vitamin D synthesis in the skin, including the topical application of a sunscreen, clothing that covers the skin and limits sun exposure, increased melanin pigmentation in the skin, limited time outside in the sun due to being housebound or institutionalized, and advanced age. Cutaneous vitamin D synthesis is decreased by as much as fourfold in persons aged 65 years compared to those aged 20 to 30 years. The season of the year, time of day, and latitude can also impact cutaneous vitamin D production. Persons living above 40° north latitude or above 40° south latitude may have no cutaneous vitamin D production during three to four months of winter. Persons living in far northern or southern latitudes may not experience vitamin D production for up to six months (Standing Committee on the Scientific Evaluation of Dietary Reference Intakes 1997; Calvo and Whiting 2003). As shown in Figure 27.4, much of the United States and all of Canada lie above 40° north latitude. Research suggests that a large proportion of North Americans and Europeans, for example, have suboptimal sun exposure and, thus, inadequate vitamin D synthesis. Consequently, for these individuals, adequate dietary intake of vitamin D is necessary to meet their requirements, and in this situation vitamin D becomes a true vitamin (Norman 2001).

Whether obtained from the diet or synthesized in the skin, ergocalciferol and cholecalciferol are both biologically inactive until they are modified by the liver and then by the kidney to form the biologically active form of vitamin D: 1,25-dihydroxyvitamin D (1,25-$(OH)_2$D) (Standing Committee on the Scientific Evaluation of Dietary Reference Intakes 1997; Norman 2001). The first step in this activation process occurs in the liver, where a hydroxyl group (OH) is added to the molecule to form 25-hydroxyvitamin D. From the liver, 25-hydroxyvitamin D is transported to the kidney where a second hydroxyl group is added to form the biologically active 1,25-dihydroxyvitamin D. Because kidney disease can interfere with this final step, patients with renal failure will be vitamin D-deficient unless they receive supplements of the vitamin in its biologically active form. Calcitriol is a commercially available form of 1,25-dihydroxyvitamin D available by prescription that can be administered orally or by intravenous injection. Over-the-counter vitamin D supplements contain either cholecalciferol or ergocalciferol and are not suitable for renal failure patients because both forms require activation by the liver and kidney before becoming biologically active.

FIGURE 27.3 Vitamin D can be obtained in the diet or synthesized when skin is exposed to ultraviolet light. Cholecalciferol and ergocalciferol must then be changed into their biologically active forms by having added to them two hydroxyl (OR) groups, in a process known as hydroxylation. The first hydroxylation occurs in the liver and the second occurs in the kidneys, resulting in 1,25-dihydroxycholecalciferol, the active form of vitamin D.

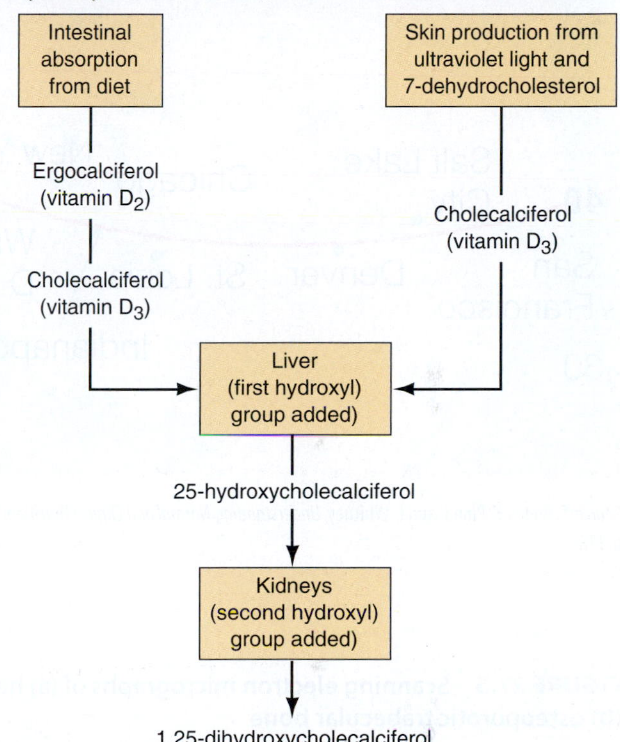

Source: CM Porth, *Pathophysiology: Concepts of Altered Health Status*, 7e fig. 56-4 pg. 1362.

Osteoporosis

Osteoporosis, the most common bone disease in humans, is characterized by the loss of bone mass and deterioration of bone microarchitecture, compromised bone strength, and an increased susceptibility to fracture and painful morbidity (National Osteoporosis Foundation 2003; Kirk and Fish 2004; Lindsay and Cosman 2005). The scanning electron micrographs shown in Figure 27.5 dramatically illustrate the

osteoporosis—a disease resulting from a decreased amount of bone mineral and organic matrix which weakens bones, making them more susceptible to fracture

FIGURE 27.4 Much of the United States and all of Canada lie above 40° north latitude. Research indicates that people residing in this area have insufficient duration and intensity of sunlight exposure for optimal synthesis of vitamin D. Unless vitamin D is obtained from fortified foods or from supplements, many North Americans are at risk of deficiency of this important vitamin.

Source: S. Rolfes, K. Pinna, and E. Whitney, *Understanding Normal and Clinical Nutrition*, 7e, copyright © 2006, p. 378.

differences between normal and osteoporotic trabecular bone in terms of bone mass and microarchitecture. The U.S. National Institutes of Health (NIH) defines osteoporosis as "a skeletal disorder characterized by compromised bone strength predisposing a person to an increased risk of fracture. Bone strength reflects the integration of two main features: bone density and bone quality" (National Institutes of Health 2000). Bone quality is influenced by factors such as bone architecture, bone turnover, mineralization, and the accumulation of damage to the bone (e.g., microfractures). Because bone quality is difficult to quantify, clinicians rely on a diagnosis or history of a fragility fracture as the only clinically applicable index of bone quality (Brown and Josse 2002). A fragility fracture is one that occurs in the absence of trauma or following minimal trauma (National Osteoporosis Foundation 2003; Lindsay and Cosman 2005). For example, it is not uncommon for a fracture of an osteoporotic hip to *precede* and result in a fall (note that the fracture occurs *before* the fall) or for a fracture to occur after bumping into a table or kitchen counter top.

FIGURE 27.5 Scanning electron micrographs of (a) healthy trabecular bone, and (b) osteoporotic trabecular bone

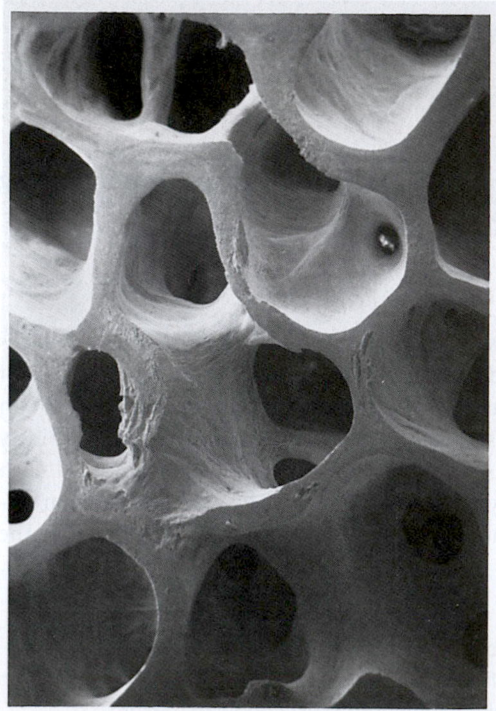

A

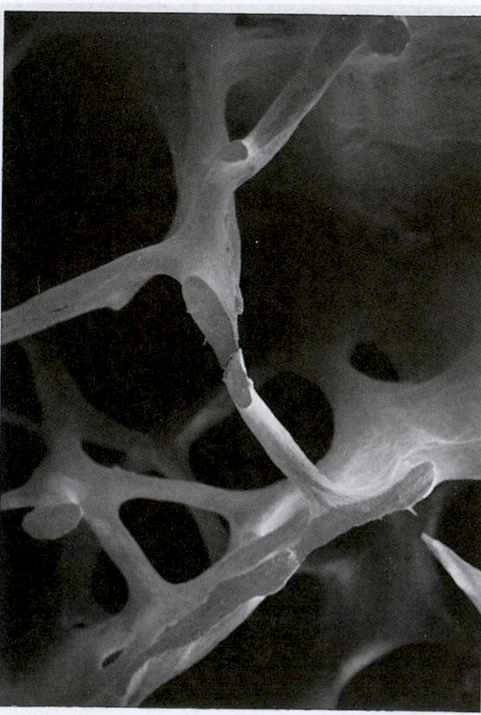

B

Source: S. Rolfes, K. Pinna, and E. Whitney, *Understanding Normal and Clinical Nutrition*, 7e, copyright © 2006, p. 429.

FIGURE 27.6 A DXA scan of the left hip

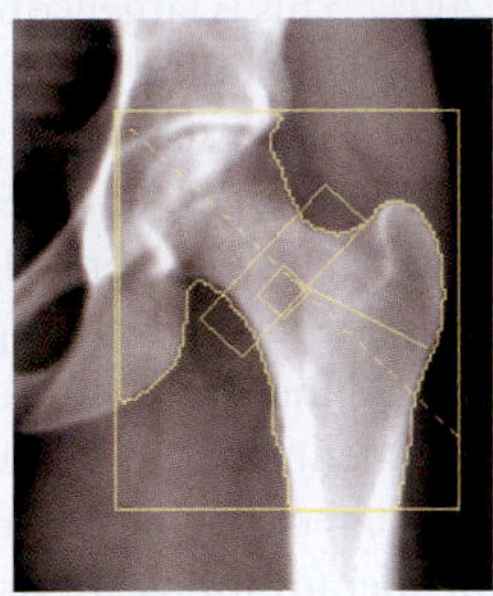

Source: Photo courtesy of Hologic, Inc.

Diagnosis

The most widely used method for diagnosing osteoporosis is measurement of bone mineral density (Writing Group for the ISCD Position Development Conference, Diagnosis, 2004; Writing Group for the ISCD Position Development Conference, Nomenclature, 2004). Bone mineral density (BMD) accounts for approximately 70% of bone strength, and because there is currently no accurate measure of overall bone strength, BMD is frequently used as a surrogate measure of bone strength (National Institutes of Health 2000). BMD is expressed in grams of mineral per area (g/cm^2) and is best measured using dual-energy x-ray absorptiometry (DXA) (Theodorou, Theodorou, and Sartoris 2003). In DXA, x-ray beams of two different energy levels are projected through the body and received by a detector opposite the x-ray source (see Chapter 5). As the x-ray beam passes through the body, some of its energy is absorbed by the body's tissues, particularly dense tissue such as bone; this reduction in energy is known as attenuation. Because bone and soft tissue have different densities, they attenuate or absorb x-ray energy differently, and the differences in attenuation between soft tissues and bone at the two different energy levels are used to calculate BMD (Krall and Dawson-Hughes 1999; Richmond 2003). Figure 27.6 shows a DXA scan of the hip (see Chapter 5 for a photo of a DXA instrument).

Osteoporosis is generally diagnosed by comparing a patient's BMD (determined by DXA) with the mean normal BMD in a population of healthy young adults of the same sex and race by using what is referred to as a "T-score" (Brown and Rosen 2003; Richmond 2003; Writing Group, Nomenclature 2004; Lindsay and Cosman 2005). The **T-score** compares a patient's BMD with the mean in a healthy young reference population, which is considered the standard for peak bone mass. The T-score is the number of standard deviations above or below the mean BMD for normal young adults of the same sex and race as the patient. The most widely accepted diagnostic criterion for osteoporosis is that developed by the World Health Organization (WHO), which defines osteoporosis as a T-score at or below −2.5. For example, a diagnosis of osteoporosis can be made in the case of a 60-year-old white female if her BMD is 2.5 standard deviations or more below the mean BMD of young adult white females. When the T-score is between 1.0 and 2.5 standard deviations below the mean BMD of the young adult mean value, a condition known as **osteopenia** is present. Osteopenia is not considered a diagnosis but rather a term used to describe a bone mineral density that is somewhat low but not so low as to warrant a diagnosis of osteoporosis. Risk of fracture is increased as well, but not to the extent seen in osteoporosis (Hightower and Gunta 2005; National Osteoporosis Foundation 2003; Krall and Dawson-Hughes 1999; U.S. Department of Health and Human Services, Bone Health, 2004).

Measurements of BMD at any skeletal site have value in predicting fracture risk, and the lower the BMD, the greater the fracture risk (National Osteoporosis Foundation 2003). It is estimated that fracture risk increases 1.5 to 3.0 times for each standard deviation decrease in bone density (Krall and Dawson-Hughes 1999; Prentice 2004). Clinical determinations of BMD are usually made at the lumbar spine and hip, but hip BMD is the best predictor of risk of hip fracture and is useful for predicting fractures at other sites (National Osteoporosis Foundation 2003; Lindsay and Cosman 2005). Table 27.1 summarizes the WHO criteria for evaluating BMD.

TABLE 27.1

World Health Organization T-Score Criteria for Classifying Normal Bone Mineral Density (BMD), Osteopenia, and Osteoporosis

Classification	T-Score
Normal BMD	−1.0 or greater
Osteopenia	Between −1.0 and −2.5
Osteoporosis	−2.5 or less
Severe Osteoporosis	−2.5 or less and fragility fracture

Adapted from the World Health Organization and the International Society for Clinical Densitometry.

T-score—the number of standard deviations that the patient's BMD is either above or below the mean BMD for healthy young adults of the same sex and race; measure that compares a patient's bone mineral density (BMD) to a standard, healthy BMD, which is set at the mean BMD of healthy young adults of the same sex and race as the patient

osteopenia—a term used to describe a bone mineral density that is low but not low enough to meet the diagnostic criterion for osteoporosis

FIGURE 27.7 This compact and lightweight instrument measures the transmission of high-frequency sound waves (ultrasound) through the calcaneus (heel bone). Quantitative ultrasonography (QUS) is a clinically useful and cost-effective screening approach for identifying persons at risk for osteoporosis who would benefit from more definitive evaluation of bone mineral density using DXA.

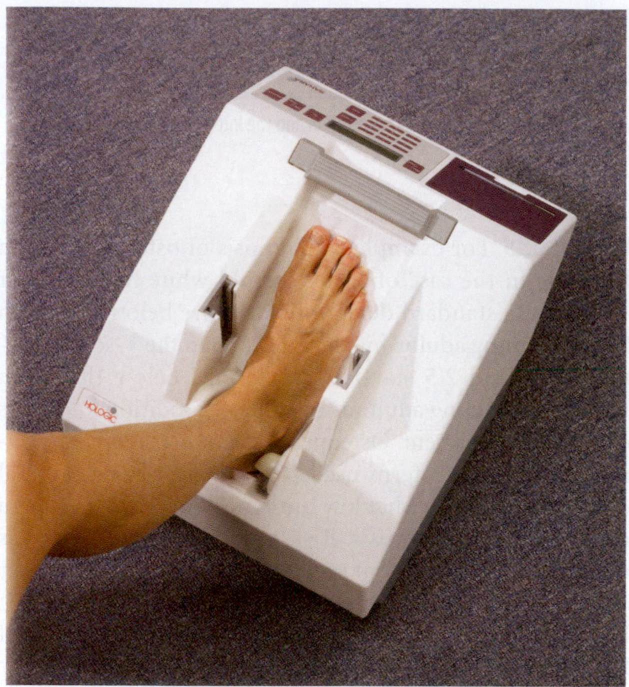

Source: Photo courtesy of Hologic, Inc.

Quantitative ultrasound of the calcaneus (the bone of the heel) is gaining recognition as a useful screening method for identifying persons at high risk of osteoporosis who might then be referred for additional evaluation of BMD using DXA (Dargent-Molina, Piault, and Breart 2003). Quantitative ultrasound is radiation free, relatively simple to use, portable, and inexpensive, and measurements take one or two minutes to perform (Theodorou, Theodorou, and Sartoris 2003; Rothenberg, Boyd, and Holcomb 2004). When used in conjunction with clinical risk assessment, quantitative ultrasound is particularly useful at identifying people at high risk of fracture who warrant additional assessment of BMD with DXA (Dargent-Molina, Piault, and Breart 2003). A commercially available quantitative ultrasonography instrument is shown in Figure 27.7.

Many North Americans are unaware that their bone health is in jeopardy because osteoporosis is frequently not diagnosed (Mazanec 2004). Data from the third National Health and Nutrition Examination Survey show that only 25% of men aged 65 years and older who have osteoporosis are aware that they have the condition and that about 42% of women aged 65 years and older who have osteoporosis are aware of the fact (U.S. Department of Health and Human Services, Bone Health, 2004). This is largely due to physicians and other health care providers failing to discuss osteoporosis with patients or to recommend screening for persons at risk. In one study of 1,500 women aged 40 to 69 years who participated in a primary care health plan, only 49% reported that a health care provider had discussed the topic of osteoporosis with them (Gallagher, Geline, and Comite 2002). In another study of hip fracture patients in four Midwestern health care systems, fewer than 25% received bone mineral density testing. Prescription drug therapy for osteoporosis was prescribed in only 3% to 26% of these hip fracture patients, depending on the site of hospitalization (Harrington et al. 2002). Despite the relative ease of screening for osteoporosis and availability of effective measures for preventing and treating the condition, it all too often is under-recognized and under-treated (Mazanec 2004).

Epidemiology

Osteoporosis is a major global public health problem, particularly among postmenopausal white women, but its prevalence is increasing across all groups, including males and nonwhite persons, as life expectancy increases and the world population expands. In the United States and Canada, approximately 25% of postmenopausal white women have osteoporosis and an additional 50% of postmenopausal white women have low BMD, and thus are at increased risk of fracture and future development of osteoporosis (Brown and Josse 2002; National Osteoporosis Foundation 2003). One out of every two white women in North America will experience an osteoporotic fracture as some time in her lifetime (National Osteoporosis Foundation 2003). Although osteoporosis is often considered a disease of women, it is a major health care problem in men. According to World Health Organization (WHO) estimates, the risk of suffering an osteoporotic fracture over the course of life is about 40% for women and about 13% for men (Brown and Josse 2002).

Bone mineral density rapidly increases during the growth spurt of adolescence (11 to 14 years of age in females and 13 to 17 years of age in males), after which it continues to increase at a much slower rate until maximum bone mineral density is reached in the late 20s or 30s (Krall and Dawson-Hughes 1999). As early as adolescence, differences in BMD between the sexes begin to appear, with females having a lower BMD than males. Adult females begin losing bone mineral earlier and at a faster rate than adult males. The rate of bone mineral loss in females averages about 1% per year as menopause approaches, increases to 2% to 6% per year for one to five years after the onset of menopause, and then by the tenth year after the onset of menopause returns to about 1% per year and remains at that level throughout the remainder of life (Krall and Dawson-Hughes 1999). On average, U.S. females live seven to eight years longer than U.S. males, which contributes to a female's greater lifetime loss of bone mineral and increased incidence of osteoporosis and risk of fracture. By age 70 years, the average white woman has lost 30% of her bone mass while some white women can lose as much as 50%. Between the age of 50 and 60 years, bone mineral loss in males averages about 1% per year, and continues at this rate throughout the remainder of life, with total losses rarely exceeding 25% (Krall and Dawson-Hughes 1999). Other factors responsible for the lower prevalence of osteoporosis among males compared to females is their greater body mass, greater bone size, absence of a decrease in endogenous sex hormone production analogous to menopause, and shorter average lifespan (Olszynski et al. 2004).

Race and ethnicity influence BMD and risk of fracture. Beginning in early childhood and on through the remainder of the life cycle, blacks tend to have higher BMD values than whites and Asians (Krall and Dawson-Hughes 1999). Hip fractures are much more frequent among whites than nonwhites, with U.S. white women reporting three times as many hip fractures as African-American women (Brown and Rosen 2003). Risk of osteoporosis and fracture increases with age, and as the number of older people increases worldwide, osteoporosis and fracture will become even greater concerns. The current global estimate of 330 million persons aged 65 years or over is expected to increase nearly fivefold to 1.6 billion persons by the year 2050, in large part due to increased life expectancy in Asia and Latin America. The WHO estimates that the number of hip fractures will rise from 1.7 million in 1990 to as many as 6.3 million by 2050. In 1990, approximately 50% of hip fractures in the elderly occurred in North America and Europe, but by 2050 it is estimated that this number will drop to 25% and that more than half of all hip fractures will occur in Asia (Black et al. 2001; Brown and Rosen 2003).

Health and Economic Impact of Fractures

The most common fracture sites are the vertebrae (spine), proximal femur (hip), and distal forearm (wrist) (National Osteoporosis Foundation 2003). Fractures of the spine occur two to three times as frequently as hip fractures, but in contrast to hip fractures, are not as easily diagnosed and are often asymptomatic (without symptoms) (Black et al. 2001; Cheung et al. 2004). Spine fractures are a major source of chronic back pain, disfigurement, low self-esteem, and depression, and result in a slight increase in mortality. In the elderly, multiple compression fractures of the vertebral bodies can cause a loss in stature of several inches and an unnatural curvature of the back known as kyphosis. Postural and height changes associated with kyphosis can restrict the activities of daily life such as bending and reaching. Severe kyphosis is associated with restrictive lung disease and such abdominal symptoms as pain, distension, loss of appetite, early satiety, gastroesophageal reflex disease, and constipation (National Osteoporosis Foundation 2003; Lindsay and Cosman 2005). Fractures of the wrist are the least debilitating and only 20% of these require hospitalization (U.S. Department of Health and Human Services, Bone Health, 2004).

Hip fractures are the most severe in terms of impact on morbidity and mortality, and each year account for nearly 300,000 hospitalizations in the U.S. and more than 30,000 in Canada (U.S. Department of Health and Human Services, Bone Health, 2004; Cheung et al. 2004). Of these hip fracture patients, 20% die within the first year of a fracture, 20% end up in a nursing home within a year, and many become isolated, depressed, or afraid to leave home because of fear of another fall. So devastating are the consequences of a hip fracture that in a study of women 75 years and older, 80% preferred death to a severe hip fracture resulting in nursing home placement (U.S. Department of Health and Human Services, Bone Health, 2004). Although the overall prevalence of hip fractures is greater among females, the likelihood of death from hip fractures is considerably greater among males. In some studies, males hospitalized for hip fracture were nearly twice as likely to die than were hospitalized females (Olszynski et al. 2004). Despite the serious nature of fractures, including those of the hip, few people die as a direct result of fractures. Death is generally the indirect result of complications from the fracture that either trigger or hasten a downward spiral in health. This is particularly the case among frail, elderly persons with underlying health conditions or those who are already living in a nursing home and who might have died even in the absence of a hip fracture (U.S. Department of Health and Human Services, Bone Health, 2004).

Morbidity resulting from osteoporosis is expensive in terms of both indirect and direct costs. Indirect costs include reduced productivity due to disability and premature death and reduced earnings because of workdays lost by patients and caregivers. Indirect costs are difficult to estimate but are considered sizeable (U.S. Department of Health and Human Services, Bone Health, 2004). Direct costs include those related to inpatient and outpatient health care, nursing home care, home health care, durable medical equipment, and pharmaceuticals. The annual direct costs of osteoporotic fractures in the United States in 2002 were estimated at $12 to $18 billion dollars. A majority of these costs are due to the inpatient treatment of fractures and their sequelae (diseases or illnesses occurring as a consequence of another condition or event) (National Institutes of Health 2000; Brown and Josse 2002; U.S. Department of Health and Human Services, Bone Health, 2004). Fractures of the hip are the most expensive to treat. The initial cost of hospitalization following a hip fracture can vary from $30,000 to $44,000, with an additional $15,000 in costs for follow-up and outpatient care during the first year following the fracture, and over a lifetime costs can exceed $81,000 (Braithwaite, Col, and Wong 2003; U.S. Department of Health and Human Services, Bone Health, 2004). The annual costs related to osteoporosis could more than triple by the year 2040 due to the rising cost of health care and the increasing number of persons likely to experience osteoporosis and osteoporosis-related fractures (National Osteoporosis Foundation 2003).

Etiology

Osteoporosis can be categorized as either primary or secondary. In primary osteoporosis, no specific cause of the condition can be identified, whereas secondary osteoporosis results from or is "secondary to" a specifically identifiable cause, such as a disease or the use of certain drugs. Primary osteoporosis is more common than secondary and is generally a disease of the elderly, resulting from the cumulative impact of bone mineral loss and deterioration of bone structure that occur with aging (U.S. Department of Health and Human Services, Bone Health, 2004). Primary osteoporosis is sometimes referred to as "age-related osteoporosis" or as "postmenopausal osteoporosis," because it is often diagnosed in elderly, postmenopausal women. Table 27.2 outlines factors associated with increased risk of osteoporosis fracture. Risk factors which are modifiable are considered prime targets for interventions used in osteoporosis prevention or nonpharmacologic treatment. Secondary osteoporosis is more commonly seen in premenopausal women than in men. Diseases associated with increased risk of osteoporosis are listed in Table 27.3, and drugs associated with increased risk of osteoporosis are listed in Table 27.4. Secondary osteoporosis accounts for approximately 10% to 30% of cases of osteoporosis in postmenopausal women and as many as two-thirds of cases in men (Kirk and Fish 2004; U.S. Department of Health and Human Services, Bone Health, 2004).

Risk of osteoporosis (regardless of whether it is primary or secondary) and osteoporosis-related fractures is influenced by a number of factors, as outlined in Table 27.2. Risk factors having a particularly strong bearing on the development of peak BMD from childhood through early adulthood include genetic susceptibility and family history, female sex, Caucasian race, premenopausal amenorrhea, physical inactivity, and low lifetime calcium and vitamin D intake. Low lifetime intake of fluoride, magnesium, and zinc may also have a bearing (Feskanich, Willett, and Colditz 2002; Prentice 2004). Genetics is a particularly important determinant of peak bone mass and of subsequent fracture

TABLE 27.2

Risk Factors for Osteoporosis Fracture

Major Risk Factors
- Low bone mineral density
- Personal history of fracture as an adult
- History of fracture in a parent or sibling
- Female sex
- Age 65 years or older
- Caucasian race
- Menopause before age 45 years
- Premenopausal amenorrhea for 1 year or more
- Glucocorticoid therapy for $>$3 months
- Recurrent falls

Minor Risk Factors
- Impaired vision
- Dementia
- Alcoholism
- Low lifelong calcium and vitamin D intake
- Physical inactivity
- Poor health/frailty
- Current cigarette smoking
- Body weight $<$127 lb ($<$58 kg)

Adapted from: National Osteoporosis Foundation. *Physician's guide to prevention and treatment of osteoporosis.* Washington (DC): National Osteoporosis Foundation, 2003; Lindsay R, Cosman F. Osteoporosis. In: Kasper DL, Braunwald E, Fauci AS, Hauser SL, Longo DL, Jameson JL, editors. *Harrison's Principles of Internal Medicine,* 16th ed. New York: McGraw-Hill; 2005. pp. 2268–2278; Brown JP, Josse RG. 2002 clinical practice guidelines for the diagnosis and management of osteoporosis in Canada. *CMAJ.* 2002;167(10 suppl):S1–S34.

TABLE 27.3

Diseases Associated with Increased Risk of Osteoporosis

Nutritional Disorders
- Biliary cirrhosis
- Celiac disease
- Gastrectomy
- Inflammatory bowel disease
- Malabsorption syndromes
- Malnutrition
- Parenteral nutrition

Genetic Disorders
- Ehlers-Danlos syndrome
- Glycogen storage diseases
- Hemochromatosis
- Homocystinuria
- Marfan syndrome
- Menkes' syndrome
- Osteogenesis imperfecta

Endocrine Disorders
- Acromegaly
- Cushing's syndrome
- Hyperparathyroidism
- Thyrotoxicosis
- Type 1 diabetes mellitus

Hypogonadal States
- Amenorrhea
- Anorexia nervosa
- Hyperprolactinemia
- Klinefelter syndrome
- Turner syndrome

Miscellaneous
- Alcoholism
- Ankylosing spondylitis
- Immobilization
- Organ transplantation
- Renal failure
- Rheumatoid arthritis

Adapted from: Lindsay R, Cosman F. Osteoporosis. In Kasper DL, Braunwald E, Fauci AS, Hauser SL, Longo DL, Jameson JL, editors. *Harrison's Principles of Internal Medicine,* 16th ed. New York: McGraw-Hill; 2005. pp. 2268–2278; U.S. Department of Health and Human Services. Bone Health and Osteoporosis: A Report of the Surgeon General. Rockville (MD): U.S. Department of Health and Human Services, Office of the Surgeon General, 2004; Kirk D, Fish SA. Medical management of osteoporosis. *Am J Manag Care.* 2004;10:445–455.

TABLE 27.4

Drugs Associated with Increased Risk of Osteoporosis

- Glucocorticoids
- Cyclosporine A and Tacrolimus
- Cytotoxic drugs
- Anticonvulsants
- Excessive alcohol
- Excessive thyroxine
- Gonadotropin-releasing hormone agonists
- Heparin
- Lithium

Adapted from: Lindsay R, Cosman F. Osteoporosis. In: Kasper DL, Braunwald E, Fauci AS, Hauser SL, Longo DL, Jameson JL, editors. *Harrison's Principles of Internal Medicine,* 16th ed. New York: McGraw-Hill; 2005. pp. 2268–2278; U.S. Department of Health and Human Services. Bone Health and Osteoporosis: A Report of the Surgeon General. Rockville (MD): U.S. Department of Health and Human Services, Office of the Surgeon General, 2004; Kirk D, Fish SA. Medical management of osteoporosis. *Am J Manag Care.* 2004;10:445–455.

risk (Feskanich, Willett, and Colditz 2002; Lindsay and Cosman 2005).

Factors increasing the rate and/or degree of bone demineralization once peak BMD has been achieved include female sex, premenopausal amenorrhea, menopause before age 45 years, advanced age, glucocorticoid therapy, cigarette smoking, physical inactivity, and low intakes of calcium and vitamin D. Low intakes of fluoride, magnesium, zinc, and vitamin K may also have a bearing on bone demineralization.

It is important to note that low bone mass is only one of several risk factors for osteoporotic fracture, and that consideration should be given to any factor increasing the likelihood of falling. Among these are impaired vision, dementia, alcoholism, physical inactivity, poor health, and frailty. In addition, the use of any medication that might impair balance or alertness should be considered, especially among the elderly who often simultaneously take several prescription and over-the-counter drugs (a practice known as polypharmacy), some of which may affect balance and alertness when used in combination. Unsafe conditions in the physical environment that might increase the chance of falls should also be considered, including uneven floors, slick floors, poorly illuminated halls and stairways, extension cords stretched across walkways, the upturned edges or corners of rugs or carpets, and inadequate railings or hand-holds in stairways and bathrooms.

Prevention

Preventing osteoporosis involves strategies to encourage individuals to change certain behaviors during childhood, adolescence, and early adulthood in order to reduce their risk of developing osteoporosis in the latter decades of life. The goal is to promote an individual's genetically determined peak BMD, slow the rate of bone mineral loss once demineralization begins in middle age, and reduce overall fracture incidence. Preventive strategies are generally recommended for the entire population and must be shown to be safe and reasonably effective. It should be noted, however, that the results of research on the efficacy of some preventive strategies are sometimes inconclusive and perhaps even contradictory. This is largely due to the difficulty of establishing cause–effect relationships between disease outcome and diet and other lifestyle factors, particularly when those relationships are weak or are influenced by factors not included in the experimental design. Despite the sometimes inconclusive nature of the research upon which preventive strategies are based, the scientific community has reached a consensus on a set of key recommendations for reducing the risk of osteoporosis and fractures. The key recommendations include adequate calcium and vitamin D intake, weight-bearing and muscle-strengthening exercise, fall prevention, smoking cessation, and avoiding excessive alcohol intake.

Calcium Calcium is a primary bone-forming mineral required in adequate amounts throughout the entire life cycle to achieve peak bone mass, maintain bone mass, minimize bone mineral loss, and reduce the incidence of osteoporosis-related fracture. Inadequate calcium intake during growth will result in suboptimal development of peak bone mass and increased risk of osteoporosis in the latter decades of life (Heaney 1999; Krall and Dawson-Hughes 1999). During adulthood, a calcium intake insufficient to maintain serum calcium levels leads to increased secretion of parathyroid hormone which, in turn, stimulates the resorption of bone in order to raise serum calcium levels (Lindsay and Cosman 2005). The Adequate Intakes (AI) for calcium and vitamin D for various life stage and gender groups as established by the Institute of Medicine of the National Academy of Sciences are shown in Table 27.5. These recommended intake levels are based on the best available scientific data and appear sufficient to achieve peak bone mass during growth and to minimize bone mineral loss in adulthood (Standing Committee on the Scientific Evaluation of Dietary Reference Intakes 1997). The mean (average) estimated daily calcium intakes for American females and males of different age groups are shown in Table 27.6. These data, from the 1999–2000 National Health and Nutrition Examination Survey (NHANES 1999–2000), indicate that the average calcium intakes of Americans are, in most instances, well below the recommended levels.

There is convincing evidence that for women and men living in a Western-style environment where the risk of osteoporosis tends to be high (e.g., North America and Europe), calcium intakes of less than 400 mg/day have an adverse impact on bone mineral density and are associated with increased risk of osteoporosis and fracture (Prentice 2004; Lindsay and Cosman 2005). However, there is less certainty about osteoporosis and fracture risk among these

TABLE 27.5

Adequate Intakes (AI) for Calcium and Vitamin D by Life Stage and Gender Group

Life Stage/Gender Group	AI for Calcium	AI for Vitamin D*
Children, 1–3 years old	500 mg/d	5 μg/d
Children, 4–8 years old	800 mg/d	5 μg/d
Adolescents, 9–18 years old	1,300 mg/d	5 μg/d
Females & males, 19–50 years old	1,000 mg/d	5 μg/d
Females & males, 51–70 years old	1,200 mg/d	10 μg/d
Females & males, >70 years old	1,200 mg/d	15 μg/d
Pregnancy and Lactation		
≤18 years old	1,300 mg/d	5 μg/d
19–50 years old	1,000 mg/d	5 μg/d

*As cholecalciferol. 1 μg of cholecalciferol = 40 international units (IU) of vitamin D.

Adapted from: Standing Committee on the Scientific Evaluation of Dietary Reference Intakes, Food and Nutrition Board, Institute of Medicine. Dietary Reference Intakes for Calcium, Phosphorus, Magnesium, Vitamin D, and Fluoride. Washington (DC): National Academy Press, 1997.

TABLE 27.6

Mean Estimated Daily Calcium Intakes of Americans by Age and Gender*

Age	Females	Males
All ages	765 mg/d	966 mg/d
Under 6 years	785 mg/d	916 mg/d
6–11 years	860 mg/d	915 mg/d
12–19 years	793 mg/d	1,081 mg/d
20–39 years	797 mg/d	1,025 mg/d
40–59 years	744 mg/d	969 mg/d
60 years and over	660 mg/d	797 mg/d

*Data are from the 1999–2000 National Health and Nutrition Examination Survey (NHANES 1999–2000).

Source: U.S. Department of Health and Human Services, Centers for Disease Control and Prevention, National Center for Health Statistics.

women and men when their calcium intakes are in the range of 600 to 800 mg/day, intakes that are typical of women living in North America and Europe. Calcium requirements vary considerably among humans, depending on each individual person's biology, diet, lifestyle, and environment. Furthermore, risk of osteoporosis is influenced by numerous factors other than calcium intake. Paradoxically, in many developing countries where average calcium intake tends to be lower than in developed nations, hip fracture incidence also tends to be lower than in developed nations, whereas in highly developed Western nations that have the highest rates of hip fracture, average calcium intake is greater than in developing nations. Although research to explain this paradox is lacking, it is assumed that the explanation lies with certain dietary and lifestyle practices of people living in less-industrialized countries, where calcium intakes are low. Among these practices are lower intakes of animal proteins, sodium, and caffeine; increased consumption of fruits, vegetables, legumes, and whole grain products; and higher levels of physical activity and sun exposure, all of which have the potential to reduce calcium losses (and thus calcium requirements) and to enhance bone mineral density (Standing Committee on the Scientific Evaluation of Dietary Reference Intakes 1997; Prentice 2004).

Because calcium requirements vary among individuals, calcium intakes less than those recommended by the AI may be sufficient to maintain calcium nutriture in some persons. However, given the impracticality of determining one's calcium requirement and the relative ease of achieving an adequate calcium intake, North Americans are advised to maintain calcium intakes at the levels suggested by the AI (shown in Table 27.5), preferably by consuming a variety of calcium-rich foods. These include milk (liquid and powdered), milk products (e.g., cheese, yogurt, and kefir), dark green vegetables (mustard and turnip greens, kale, and broccoli), some nuts (e.g., almonds), some seeds (e.g., sesame), tofu (manufactured using calcium sulfate as opposed to magnesium chloride), corn tortillas, and a variety of calcium-fortified foods such as citrus juice and soy beverages. Persons who are lactose intolerant typically avoid dairy products because they have insufficient production of lactase, the gastrointestinal enzyme necessary to digest lactose, the disaccharide present in milk. Approximately 75% of the world's population and about 20% of North Americans are lactose intolerant. This condition is more common in persons who are of African, Hispanic, Asian, and Native North American ancestry. Persons with lactose intolerance should be advised to consume a variety of calcium-rich, lactose-free foods. Lactose-free milk and tablets containing the enzyme lactase are available for those who wish to consume dairy products. Examples of dairy and non-dairy food sources of calcium are shown in Tables 27.7 and 27.8, respectively.

Calcium supplements may be required for persons having difficulty achieving the recommended intake of calcium from foods alone (Dawson-Hughes 2006; Heaney 2006). The most common and least expensive calcium supplements are those containing calcium carbonate, which should be taken with meals because calcium carbonate requires acid to make the calcium more soluble and absorbable. Supplements containing calcium citrate may be taken anytime. Calcium carbonate interferes with the absorption of iron and should not be taken at the same time as an iron supplement; however, calcium citrate does not

TABLE 27.7

Dairy Food Sources of Calcium [1]

Food, Standard Amount	Calcium (mg)	Kcalories
Plain yogurt, non-fat (13 g protein/8 oz), 8-oz container	452	127
Romano cheese, 1.5 oz	452	165
Pasteurized process Swiss cheese, 2 oz	438	190
Plain yogurt, low-fat (12 g protein/8 oz), 8-oz container	415	143
Fruit yogurt, low-fat (10 g protein/8 oz), 8-oz container	345	232
Swiss cheese, 1.5 oz	336	162
Ricotta cheese, part skim, ½ cup	335	170
Pasteurized process American cheese food, 2 oz	323	188
Provolone cheese, 1.5 oz	321	150
Mozzarella cheese, part-skim, 1.5 oz	311	129
Cheddar cheese, 1.5 oz	307	171
Fat-free (skim) milk, 1 cup	306	83
Muenster cheese, 1.5 oz	305	156
1% low-fat milk, 1 cup	290	102
Low-fat chocolate milk (1%), 1 cup	288	158
2% reduced fat milk, 1 cup	285	122
Reduced fat chocolate milk (2%), 1 cup	285	180
Buttermilk, low-fat, 1 cup	284	98
Chocolate milk, 1 cup	280	208
Whole milk, 1 cup	276	146
Ricotta cheese, whole milk, ½ cup	255	214
Blue cheese, 1.5 oz	225	150
Mozzarella cheese, whole milk, 1.5 oz	215	128
Feta cheese, 1.5 oz	210	113

[1] Foods are ranked by milligrams of calcium per standard amount. The number of kcalories are per standard amount.
Source: Agricultural Research Service (ARS) Nutrient Database for Standard Reference, Release 17.

TABLE 27.8

Non-Dairy Food Sources of Calcium [1]

Food, Standard Amount	Calcium (mg)	Kcalories
Fortified ready-to-eat cereals (various), 1 oz	236–1,043	88–106
Soy beverage, calcium fortified, 1 cup	368	98
Sardines, Atlantic, in oil, drained, 3 oz	325	177
Tofu, firm, prepared with calcium sulfate, ½ cup	253	88
Pink salmon, canned, with bone, 3 oz	181	118
Collards, cooked from frozen, ½ cup	178	31
Molasses, blackstrap, 1 Tbsp	172	47
Spinach, cooked from frozen, ½ cup	146	30
Soybeans, green, cooked, ½ cup	130	127
Turnip greens, cooked from frozen, ½ cup	124	24
Ocean perch, Atlantic, cooked, 3 oz	116	103
Oatmeal, instant, fortified, 1 packet prepared	99–110	97–157
Cowpeas, cooked, ½ cup	106	80
White beans, canned, ½ cup	96	153
Kale, cooked from frozen, ½ cup	90	20
Okra, cooked from frozen, ½ cup	88	26
Soybeans, mature, cooked, ½ cup	88	149
Blue crab, canned, 3 oz	86	84
Beet greens, cooked from fresh, ½ cup	82	19
Pak-choi, Chinese cabbage, cooked from fresh, ½ cup	79	10
Clams, canned, 3 oz	78	126
Dandelion greens, cooked from fresh, ½ cup	74	17
Rainbow trout, farmed, cooked, 3 oz	73	144

[1] Foods are ranked by milligrams of calcium per standard amount. The number of kcalories are per standard amount. The bioavailability of calcium may vary. Both calcium content and bioavailability should be considered when selecting dietary sources of calcium. Some plant foods have calcium that is well absorbed, but the large quantity of plant foods that would be needed to provide as much calcium as in a glass of milk may be unachievable for many. Many other calcium-fortified foods are available, but the percentage of calcium that can be absorbed is unavailable for many of them.
Source: Agricultural Research Service (ARS) Nutrient Database for Standard Reference, Release 17.

interfere with the absorption of iron from supplements. Persons taking prescription or over-the-counter medications should check for any incompatibility with calcium supplements prior to their use. The amount of elemental calcium provided by a supplement is listed on the Supplement Facts label, and should be noted. Calcium absorption is enhanced when smaller doses (<500 mg) are taken two or more times a day. The Tolerable Upper Intake Level (UL) for calcium for all persons 1 year of age and older, including pregnant or lactating females, is 2,500 mg/day, and total calcium intake (calcium obtained from food plus any taken in supplemental form) should not routinely exceed this level. There is no evidence that calcium intake in excess of the AI is beneficial. A patient's usual calcium intake should be evaluated and, if low, the patient should be encouraged to increase the consumption of calcium-rich foods to achieve the recommended calcium intake. A simple approach for arriving at a rough estimate of a patient's calcium intake is shown in the Box 27.1. If necessary, a commercially available calcium supplement with vitamin D should be recommended at a dosage sufficient to meet the AI without routinely exceeding the UL. Calcium supplements containing dolomite and bonemeal should not be used because they may be contaminated with lead (Dawson-Hughes 2006; Heaney 2006).

As long as total calcium intake is kept at or below the UL, calcium supplementation is not harmful unless contraindicated by some medical condition or by a potential adverse

BOX 27.1 **CLINICAL APPLICATIONS**

A Simple Approach for Estimating Dietary Calcium Intake

Step 1: For each of the following dairy or calcium-fortified foods, enter the number of servings consumed per day. Multiply the number of servings for each food by the calcium content per serving to arrive at the total calcium from the food per day.

Food	*Servings of food/day*	*Calcium/serving of food*	*Total calcium from food*
Milk, fat-free or 1% low-fat, 1 cup	_____	× 300 mg	= ________ mg
Soy beverage, calcium-fortified, 1 cup	_____	× 350 mg	= ________ mg
Yogurt, nonfat or low-fat, 8-oz container	_____	× 400 mg	= ________ mg
Cheese, 1.5 oz	_____	× 300 mg	= ________ mg
Orange juice, calcium-fortified, 1 cup	_____	× 350 mg	= ________ mg
Breakfast cereals, calcium-fortified, 1 oz	_____	× 200–1,000 mg[a]	= ________ mg

Step 2: Add together the amounts of calcium from the above dairy and calcium-fortified foods. ________ mg

Step 3: Add an additional 300 mg to account for calcium obtained from food sources other than the above dairy and calcium-fortified foods to arrive at a rough estimate of total calcium intake per day from all foods. ________ mg

[a] The calcium content of fortified ready-to-eat breakfast cereals will vary. Check the Nutrition Facts label for calcium content.

Adapted from: *Physician's Guide to Prevention and Treatment of Osteoporosis.* Washington (DC): National Osteoporosis Foundation, 2003.

drug or nutrient interaction. To avoid **hypercalciuria**, patients with a history of kidney stones should have a 24-hour urine calcium determination before beginning calcium supplementation (Lindsay and Cosman 2005). Most authorities support supplementation if dietary sources of calcium do not provide sufficient calcium to reach the AI on most days of the week. The best sources of calcium are foods which, in addition to calcium, provide other key nutrients needed for optimal bone development and overall good health. Supplemental calcium will not protect a person against bone loss caused by poor diet, estrogen deficiency, physical inactivity, smoking, alcohol abuse, or various medical disorders or treatments (Prentice 2004; Lindsay and Cosman 2005).

Vitamin D As previously mentioned, the major function of vitamin D is to increase blood concentrations of calcium and phosphorus by (1) promoting their absorption by the GI tract, (2) promoting their reabsorption by the kidney, and (3) stimulating osteoclast formation, and thus bone resorption and the release of calcium and phosphorus from bone (Saladin 2004; Bringhurst et al. 2005; Porth 2005). An overt deficiency of vitamin D is linked to rickets in children and osteomalacia in adults, both of which are discussed in detail later in this chapter. Although the fortification of milk with vitamin D (introduced in the 1930s) has nearly eliminated rickets in North America, vitamin D insufficiency continues to be common throughout North America, particularly among those living above 40° North latitude who have inadequate sun exposure during the winter months (Standing Committee on the Scientific Evaluation of Dietary Reference Intakes 1997; Calvo and Whiting 2003). The best indicator of vitamin D status is the concentration of serum 25-hydroxyvitamin D, because it represents both dietary vitamin D intake and vitamin D synthesized by the body. Vitamin D status is considered adequate when the serum 25-hydroxyvitamin D concentration is consistently >50 μmol/L (20 ng/mL) (Calvo and Whiting 2003; Lindsay and Cosman 2005). Research shows that the prevalence of vitamin D insufficiency is particularly high in persons living year-round in northern latitudes who have dark skin and who are older. The high prevalence of vitamin D insufficiency observed in these groups indicates that vitamin D fortification of milk is not an effective strategy for ensuring adequate vitamin D nutriture (Calvo and Whiting 2003). Because of its central role in maintaining calcium homeostasis and promoting bone health, vitamin D supplementation should be considered for people who have limited sunlight exposure for cultural or medical reasons, people who are dark skinned and live outside the tropics, and older individuals (Prentice 2004).

Vitamin D insufficiency in the elderly is linked to age-related bone loss and increased risk of fracture, and

hypercalciuria—excessive calcium in the urine

vitamin D supplementation is an effective strategy for preventing fracture in the frail elderly (Krall and Dawson-Hughes 1999). Research has shown that supplementation with calcium and vitamin D slows bone loss in elderly males and females, and reduces fracture risk, including fractures of the hip, by as much as 20% to 30% (Krall and Dawson-Hughes 1999; Olszynski et al. 2004; Prentice 2004; Lindsay and Cosman 2005). Providing an adequate intake of vitamin D and calcium is a safe, effective, and inexpensive way to reduce risk of fracture (National Osteoporosis Foundation 2003). Vitamin D and calcium supplementation is now advocated as the basic minimum for treating osteoporosis and reducing risk of fracture in older females and males (Olszynski et al. 2004; Prentice 2004). The potential risk of vitamin D toxicity is generally overstated. Although the tolerable upper intake level (UL) for vitamin D for children and adults has been set at 2,000 IU/day, scientific evidence indicates that toxicity rarely occurs until intakes exceed 10,000 IU/day (Holick 2006).

Physical Activity Bone mineral density increases in response to the stress bones receive from weight-bearing or impact-type physical activities such as physically demanding occupations, walking, jogging, jumping rope, climbing stairs, high-impact aerobics, weight/resistance training, and a variety of sports such as tennis, soccer, basketball, and gymnastics. Persons living in rural communities and in countries where physical activity is maintained into adult life have a lower fracture risk compared to those living in more urban, sedentary societies (Prentice 2004). Research has shown that athletes have a higher BMD than those in the general population and that significant bone mineral loss can result from prolonged bed rest, paralysis, and periods of weightlessness (Brown and Josse 2002; Lindsay and Cosman 2005). Although improvements in BMD are most notable when impact exercise begins during growth and before the age of puberty, benefits have been observed in adult males and premenopausal and postmenopausal females. Research into the effects of physical activity on BMD and fracture risk in adults are complicated by small sample sizes, the short-term nature of most studies (less than two years), poor subject compliance, and high dropout rates. The results show that regular, moderate-intensity exercise initiated during adult life increases BMD by 1% to 2% and that impact-type activities result in greater improvements in BMD than do non-impact activities (e.g., stretching, yoga, weight lifting, and resistance training). Subjects demonstrating a higher degree of compliance with recommendations to engage in regular, moderate physical activity experienced a greater degree of improvement in BMD. Very high levels of physical activity can be detrimental to bone health, particularly in premenopausal women who experience **oligomenorrhea** or **amenorrhea**. In adult male runners, the greatest improvements in BMD are seen in those averaging 15 to 20 miles per week, whereas longer weekly distances result in little additional benefit or, in some instances, an actual reduction in BMD (Brown and Josse 2002).

In postmenopausal females, impact activities such as walking, dancing, and jumping have been shown to slow or prevent bone loss and reduce risk of hip fracture (Krall and Dawson-Hughes 1999; Brown and Josse 2002; Kirk and Fish 2004). In a prospective cohort study of more than 61,000 postmenopausal women, those who walked for at least 4 hours per week had a 41% lower risk of hip fracture compared to women who walked less than 1 hour per week (Feskanich, Willett, and Colditz 2002). Although the benefits of non-impact activities on BMD in postmenopausal females are inconsistent, activities leading to improvements in flexibility, balance, agility, and muscle strength are associated with fewer falls and a significant reduction in fracture risk. This is particularly the case when coupled with an assessment and modification of fall hazards in the home, reviewing prescription medications for side effects that may affect stability and balance, and checking and correcting hearing and vision (Brown and Josse 2002; National Osteoporosis Foundation 2003).

Cigarette Smoking Cigarette smoking has been shown to be causally related to lower bone density, increased bone mineral loss, and increased risk of fracture in males and females (Krall and Dawson-Hughes 1999; Kirk and Fish 2004; Olszynski et al. 2004; U.S. Department of Health and Human Services, Health Consequences, 2004; Lindsay and Cosman 2005). There are several mechanisms that place smokers at increased risk of poor bone health. Nicotine and cadmium in tobacco smoke are toxic to osteoblasts. Tobacco smoke appears to reduce intestinal calcium absorption, and smokers generally have lower intakes of vitamin D and lower serum levels of 25-hydroxyvitamin D compared with nonsmokers (U.S. Department of Health and Human Services, Health Consequences, 2004). Smokers are more likely to consume excessive amounts of alcohol than are nonsmokers, which can adversely impact bone health. Smokers tend to have lower body weights, are less physically active, are more frail, and experience higher rates of chronic disease. Some of the medications used to treat these chronic diseases (e.g., glucocorticoids used to treat lung disease) can indirectly impact bone health and increase fracture risk (U.S. Department of Health and Human Services, Health Consequences, 2004; Lindsay and Cosman 2005). Smoking accelerates the metabolism of estrogen and may reduce the benefits of hormone replacement therapy in females. On average, female cigarette smokers reach menopause one to two years earlier than nonsmokers, which extends the

oligomenorrhea—abnormally infrequent menstrual cycles

amenorrhea—the absence of menstrual cycles when they would be expected to occur

postmenopausal period during which the rate of bone loss is accelerated (Kirk and Fish 2004; U.S. Department of Health and Human Services, Health Consequences, 2004).

Tobacco smoking is the leading cause of preventable death in the United States and in most developed nations. It has been shown to have negative health impacts on people at all stages of life and to harm nearly every organ of the body, causing many diseases and damaging the overall health of smokers. Quitting smoking has immediate and long-term benefits, reducing risks for diseases caused by smoking and improving health (U.S. Department of Health and Human Services, Health Consequences, 2004). Members of the health care team have a professional obligation to model good health behaviors to their patients by not smoking, to encourage their patients not to smoke, and to support the efforts of their patients who wish to quit smoking.

Alcohol There is no consistent evidence that moderate alcohol consumption has an adverse impact on bone health and fracture risk (Krall and Dawson-Hughes 1999; Prentice 2004). The *Dietary Guidelines for Americans* (U.S. Department of Health and Human Services and U.S. Department of Agriculture 2005) define moderate alcohol consumption as no more than one drink per day for women and no more than two drinks per day for men. Some research suggests that moderate alcohol consumption may increase BMD and reduce bone mineral loss, although this beneficial effect may only be seen in women (Krall and Dawson-Hughes 1999; Prentice 2004; U.S. Department of Health and Human Services, Health Consequences, 2004). However, heavy alcohol use is associated with decreased BMD, reduced bone formation, and increased risk of fracture. Chronic alcoholism is considered a major risk factor for osteoporosis. Heavy alcohol use increases calcium and magnesium losses from the body and adversely impacts vitamin D and overall nutritional status (U.S. Department of Health and Human Services, Health Consequences, 2004). Even moderate alcohol consumption increases the risk of falling and other types of skeletal trauma, thus increasing the risk of fracture (U.S. Department of Health and Human Services, Health Consequences, 2004).

Other Nutrients and Food Components The role of several other nutrients and food components in the prevention of osteoporosis has been studied. Among these are phosphorus, protein, fruits and vegetables, sodium, caffeine, fluoride, and trace minerals. Phosphorus is an essential bone-forming mineral, and an adequate supply is necessary for optimal bone health throughout life. According to data from NHANES 1999–2000, mean intakes of phosphorus are well above the RDAs for both sexes at all age levels except for males 6 to 11 years of age and females 9 to 18 years of age. In these two groups, mean intakes are slightly less than the RDA. Apart from these two age categories, it appears that the phosphorus intakes of Americans are adequate except for instances of malnutrition, intestinal malabsorption, and excessive use of phosphorus-binding antacids (Krall and Dawson-Hughes 1999; Prentice 2004). Concerns have been raised about diets high in phosphorus and low in calcium resulting in a high phosphorus:calcium ratio, particularly in relation to the consumption of carbonated soft-drinks. The ratio of phosphorus to calcium in the diet can vary widely with no detectable effect on the absorption and retention of either mineral or on BMD. Phosphorus-rich carbonated soft-drinks appear to only have a negligible effect on calcium excretion (Krall and Dawson-Hughes 1999; Prentice 2004).

The observation that hip fractures are common in countries where meat and dairy food consumption is high has led to interest in the relationship between high-protein diets and risk of osteoporosis. As dietary protein intake increases, so does urinary calcium excretion. This has led some to believe that excess protein intake in industrialized nations, particularly from meat and dairy products, is causally linked to increased risk of osteoporosis despite the relatively high calcium intakes typically seen in these Western nations. However, the association between protein intake (both total and animal protein) and risk of osteoporosis and fracture is not clear. Dietary protein is an essential component of the organic matrix of bone and is necessary to maintain production of hormones and growth factors that modulate bone synthesis (Atkinson and Ward 2001). A low protein intake in the elderly appears to increase risk of osteoporotic fracture, and elderly patients who have protein-energy malnutrition are more likely to be frail and to fall. Hip-fracture patients receiving dietary protein supplements during hospitalization experience lower rates of complications and mortality following surgery and have shorter hospitalizations than similar patients who do not receive protein supplements (Krall and Dawson-Hughes 1999; Prentice 2004; U.S. Department of Health and Human Services, Bone Health, 2004).

Because animal proteins are rich in sulphur-containing amino acids, they contribute to an acidic environment within the body (see Chapter 9), which places a demand on bone as a source of skeletal salts to neutralize the acid generated from a high-meat diet (Tucker et al. 1999; Atkinson and Ward 2001; Prentice 2004). It is thought that calcium is removed from bone in order to neutralize the acid generated by high-meat diets. However, diets providing ample fruits and vegetables and vegetable proteins result in a more alkaline environment within the body that does not necessitate the removal of calcium from bone to maintain an appropriate acid-base homeostasis. Research shows that diets providing potassium, magnesium, fruits, vegetables, and vegetable proteins are associated with higher BMD (Tucker et al. 1999; Prentice 2004). Patients should be encouraged to maintain good nutritional status and adequate protein intakes, and to consume plant foods from various groups consistent with the recommendations of the *Dietary Guidelines for Americans* (U.S. Department of Health and Human Services and U.S. Department of Agriculture 2005).

As dietary sodium intake increases, so does urinary calcium excretion. High-sodium diets are associated with increased excretion of calcium in the urine (Krall and

Dawson-Hughes 1999; Atkinson and Ward 2001; U.S. Department of Health and Human Services and U.S. Department of Agriculture 2005). Although there is convincing evidence that reducing sodium intake results in a lowering of urinary calcium excretion, research has not yet demonstrated whether a low-sodium diet is an effective approach for preventing osteoporosis and reducing risk of fracture. However, considering the beneficial effects on blood pressure of diets providing ample amounts of vegetables and fruits, adequate intakes of non-fat dairy products, and moderate amounts of lean meats, there is a strong rationale for patients to follow such a dietary pattern (Prentice 2004; U.S. Department of Health and Human Services and U.S. Department of Agriculture 2005).

Medical Management

Once a patient is diagnosed with osteoporosis or found to have low BMD, the patient should be evaluated to determine whether this condition is primary or is secondary to some other disease (see Table 27.3) or to drug use (see Table 27.4). This evaluation should begin with a thorough history and physical examination as well as several basic laboratory tests (see Table 27.9), which will rule out the most common causes of secondary osteoporosis. In patients with secondary osteoporosis, treatment of the underlying causes should be addressed before initiating osteoporosis-specific treatment. Patients with primary osteoporosis or those with secondary osteoporosis in which the underlying cause is being addressed should then undergo risk factor modification, dietary treatment, and drug therapy as needed. Risk factor modification includes efforts to encourage smoking cessation, control of excessive alcohol use, and moderate physical activity, while dietary treatment would include adequate intakes of calcium and vitamin D and promotion of overall good nutrition. If necessary, pharmacologic agents can be used to treat the osteoporosis or, in patients with low BMD, to prevent further loss of BMD in order to prevent osteoporosis.

TABLE 27.9

Basic Laboratory Tests to Determine Secondary Osteoporosis

Laboratory Test	Normal Range for Adults
Albumin, serum	3.5–5.0 g/dL (35–50 g/L)
Alkaline phosphatase, serum	30–120 U/L (0.5–2.0 µKat/L)
Calcium, total, serum	9.0–10.5 mg/dL (2.25–2.75 mmol/L)
Creatinine, serum	
Female	0.5–1.1 mg/dL (44–97 µmol/L)
Male	0.6–1.2 mg/dL (53–106 µmol/L)
Free testosterone, serum	
Female	<1 ng/mL (<3.5 nmol/L)
Male	3–10 ng/mL (10–35 nmol/L)
Parathyroid hormone, serum	10–65 pg/mL (10–65 ng/L)
Phosphorus, serum	3.0–4.5 mg/dL (0.97–1.45 mmol/L)
Thyroid-stimulating hormone, serum	2–10 µU/mL (2–10 mU/L)
Total protein, serum	6.4–8.3 g/dL (64–83 g/L)

Pharmacologic Prevention and Treatment

Several drugs have been approved by the U.S. Food and Drug Administration (FDA) and Health Canada's Therapeutic Products Directorate (TPD) for the prevention and treatment of osteoporosis (Brown and Josse 2002; National Osteoporosis Foundation 2003; Lindsay and Cosman 2005). Included among these are estrogens, selective estrogen receptor modulators, bisphosphonates, and teriparatide, a synthetic form of parathyroid hormone. Even when drugs are used in the prevention and treatment of osteoporosis, it is essential to ensure adequate intakes of calcium and vitamin D, as well as to advise patients to refrain from smoking, engage in regular weight-bearing exercise, and use alcohol in moderation, if at all. All drugs have the potential for adverse side effects or drug-nutrient interactions (see Table 27.12) and in many instances they are expensive. Consequently, the therapeutic benefits of drug treatment must be weighed against their potential adverse effects and their cost, which is sometimes high.

Estrogen is a generic term for a group of female sex hormones that promote fertility in female mammals. They include estradiol, estriol, and estrone, and are marketed under a variety of names and forms. Estrogens are approved by the U.S. FDA and Health Canada's TPD for the prevention and treatment of osteoporosis in postmenopausal women. Estrogens reduce bone turnover, prevent bone loss, increase BMD by 5% to 10% in the hip, spine, and total body, significantly reduce fracture risk, and are effective in relieving the hot flashes and night sweats that often accompany menopause (Reginster 2004; Lindsay and Cosman 2005). Although several randomized controlled trials have shown that use of estrogens to treat women with postmenopausal osteoporosis reduces vertebral and hip fractures by about 33% and osteoporotic fractures at other sites by about 23%, there are potential serious side effects to such treatment, including irregular vaginal bleeding, breast tenderness, increased risk of breast cancer, and increased risk of thrombus (a stationery blood clot attached to the wall of a blood vessel that obstructs blood flow) and embolus (a blood clot carried by blood flow from the site of formation to a smaller blood vessel). In one study, women receiving combined estrogen-progestin treatment experienced an increased risk of breast cancer (26% increased risk), coronary heart disease (29%), stroke (41%), and thromboembolism (obstruction of a blood vessel by a thrombus) (111%) compared to women receiving a placebo (Rossouw et al. 2002). For the past several decades, hormone therapy has been the primary pharmacologic approach for preventing and treating postmenopausal osteoporosis, but the potential risks of hormone treatment and the availability of other effective drugs is casting doubt on the use of hormones as the preferred therapeutic option.

Selective estrogen receptor modulators (SERMs) are non-hormonal agents that have tissue-specific effects in estrogen-responsive target tissues. In the estrogen-responsive tissues of the skeletal and cardiovascular system, the effects of SERMs are similar to those of estrogen. In breast and uterine tissue, however, SERMs have no estrogen-like effects (Brown and Rosen 2003; Reginster 2004; Favus and Vokes 2005). Raloxifene is a SERM approved by the FDA and TPD for the prevention and treatment of osteoporosis in postmenopausal women. When raloxifene binds to estrogen receptors in bone, it has an estrogen-like effect that increases BMD and reduces spine fracture risk by 30% (National Osteoporosis Foundation 2003; Reginster 2004). In contrast, raloxifene has no estrogen-like effects when it binds to estrogen receptors in breast and uterine tissue, and consequently it does not cause breast tenderness or vaginal bleeding, and it does not increase the risk of breast or uterine cancer. Thus, SERMs have several advantages over the use of estrogens for the prevention and treatment of postmenopausal osteoporosis. However, raloxifene use is associated with increased risk of thromboembolism and it is not effective in treating perimenopausal symptoms (e.g., hot flashes and night sweats), and may actually worsen hot flashes (Brown and Rosen 2003; Reginster 2004).

Bisphosphonates are potent non-hormonal drugs that reduce bone resorption. They act exclusively in bone by binding to hydroxyapatite (the primary mineral component of bone) at sites of bone resorption and impairing osteoclast function, reducing osteoclast number, and promoting the death of osteoclasts (Brown and Josse 2002; Page et al. 2002; Brown and Rosen 2003; Olszynski et al. 2004; Lindsay and Cosman 2005). Bisphosphonates approved by the FDA and TPD for preventing and treating postmenopausal osteoporosis, osteoporosis in men, and glucocorticoid-induced osteoporosis in both sexes are alendronate, risedronate, and ibandronate. Bisphosphonates are considered the first choice for the pharmacological treatment of osteoporosis in men (Olszynski et al. 2004). Bisphosphonates have been shown to decrease bone turnover and increase bone mass in the lumbar spine and hip by as much as 8% and 6%, respectively, and to reduce spine and hip fracture risk by as much as 50% (Olszynski et al. 2004; Reginster 2004; Lindsay and Cosman 2005). The bisphosphonates differ from each other in terms of their potency, ability to inhibit bone resorption, toxicity, and dosing regimen.

Bisphosphonates are poorly absorbed by the gastrointestinal tract and should only be taken first thing in the morning on an empty stomach with at least one cup of plain water. Taking bisphosphonates with food or any beverage other than plain water markedly reduces their absorption. Although generally well tolerated by patients, they may cause esophageal and gastric irritation. Consequently, after taking bisphosphonates, patients should remain in an upright position (sitting or standing) and avoid bending over for at least 30 to 60 minutes.

Bisphosphonates are as effective at treating osteoporosis when taken once weekly as they are when taken once daily, and the once-weekly dosing schedule has the advantage of improved patient convenience and adherence, and reduced risk of gastrointestinal complications. There is ongoing research on the safety and efficacy of once-monthly oral and once-yearly intravenous administration of bisphosphonates (Lindsay and Cosman 2005). They are poorly absorbed even when taken on an empty stomach with plain water, and their absorption is markedly reduced if taken with food or with any beverage other than water (Brown and Josse 2002; Reginster 2004; Lindsay and Cosman 2005).

Parathyroid hormone (PTH) is a peptide hormone largely responsible for calcium homeostasis within the body (Saladin 2004; Lindsay and Cosman 2005). Teriparatide is a synthetic version of PTH produced using recombinant DNA technology. Structurally similar to human PTH and having similar effects within the human body, teriparatide acts on the osteoblasts (the bone-building cells) to stimulate new bone growth and increase BMD; thus, it has an anabolic (or tissue building) effect on the skeleton. In this respect, teriparatide's mode of action is different than the previously mentioned drugs, which primarily reduce osteoclast activity and thus are considered antiresorptive agents. Teriparatide has been approved by the FDA and the TPD for treatment of osteoporosis in both females and males who have a high risk of fracture (Brown and Josse 2002; Lindsay and Cosman 2005). Because it is a peptide hormone, it must be administered by injection, typically by daily subcutaneous injections. Research has shown that teriparatide stimulates new bone formation and lowers risk of spinal fractures by 65% and risk of fracture at other sites (wrist, ribs, hip, ankle, and foot) by 53%.

Paget Disease

Paget disease is a localized, progressive, often crippling disorder of bone remodeling resulting from overactive osteoclasts that cause rapid bone resorption followed by rapid formation of new bone by osteoblasts. The structure of the new bone is haphazard, disorganized, and structurally inferior, leaving the diseased bone more subject to bowing, deformity, fracture, and poor healing following a fracture (U.S. Department of Health and Human Services, Bone Health, 2004; Favus and Vokes 2005). The bones most often affected by Paget disease are the upper femur, pelvis, vertebral bodies, skull, and tibia (Favus and Vokes 2005). While pain (including headaches) is the most common presenting symptom, approximately 70% of patients with Paget disease are asymptomatic. Other clinical manifestations include bowing of the long bones with resulting gait abnormalities, nerve paralysis, hearing loss, facial deformity, tooth loss, and cardiovascular disease (Favus and Vokes 2005; Hightower and Gunta 2005).

After osteoporosis, Paget disease is the second most common bone disorder and is diagnosed in about 3% of persons over age 40 years. The prevalence is greater in males and increases with age, and there are wide geographic variations in its frequency. The disease typically begins insidiously, and then slowly progresses over many years. Although the etiology of Paget disease is unknown, there is evidence suggesting that genetic and viral factors play a role. Diagnosis is based on x-rays showing characteristic bone changes and deformities. Biochemical markers of bone formation and resorption, such as serum alkaline phosphatase and urinary hydroxyproline, are used to confirm the diagnosis, evaluate the severity of the disease, and determine the patient's response to treatment (Favus and Vokes 2005; Hightower and Gunta 2005). Treatment of Paget disease involves use of nonsteroidal anti-inflammatory agents for pain and use of bisphosphonates to suppress osteoclast activity and decrease bone resorption and formation (Favus and Vokes 2005). Persons with Paget disease should also maintain an adequate intake of calcium and vitamin D.

Rickets and Osteomalacia

Rickets and **osteomalacia** are related diseases in which there is insufficient mineralization of the organic matrix of bone, in most cases due to vitamin D deficiency. As discussed earlier in this chapter, approximately two-thirds of bone is mineral and the remaining one-third is an organic matrix. In rickets and osteomalacia, the organic matrix is present but it is not sufficiently mineralized, unlike in osteoporosis, where there is a loss of both the organic matrix and bone mineral content. Rickets is a disease of childhood and osteomalacia is seen in adults.

Rickets

Epidemiology, Etiology, and Clinical Manifestations

Rickets is characterized by inadequate maturation and mineralization of the cartilaginous growth plate and inadequate mineralization of the organic matrix within the bones of children (Heaney 1999; Hightower and Gunta 2005). The most common cause of rickets is vitamin D deficiency due to inadequate vitamin D intake and/or inadequate sunlight exposure, which leads to decreased intestinal calcium absorption, inadequate bone mineralization, and a low serum calcium concentration. Risk factors for rickets are outlined in Table 27.10. A low serum calcium concentration stimulates secretion of parathyroid hormone (PTH), which in turn mobilizes calcium and phosphorus from bone in order to maintain an acceptable blood calcium concentration. Other causes of rickets include calcium deficiency, disorders of vitamin D metabolism, and hypophosphatemia (low serum phosphate concentration) (Pettifor 2004; Bringhurst et al. 2005). The symptoms of rickets are generally seen between 6 and 36 months of age and include lethargy, weakness, growth stunting, enlargement of the ends of the long bones and ribs, an abnormally shaped thorax, and bowing of the legs (Hightower and Gunta 2005). Figure 27.8 illustrates the bowing of legs due to rickets. Although rickets is primarily seen in developing countries and among immigrants in developed countries, cases are reported in the United States and Canada (Weisberg et al. 2000). Inadequate sunlight exposure to provide optimum vitamin D synthesis is likely to be seen in those living in temperate climates above the 40th parallel, and in persons living closer to the equator who have limited sun exposure because of social or religious customs.

TABLE 27.10

Risk Factors for Rickets

- Maternal vitamin D deficiency
- Prolonged breast-feeding without vitamin D supplementation
- Living in a temperate climate
- Lack of sunlight exposure
- Dark skin pigmentation
- Calcium deficiency
- Intake of phytates from diets high in unrefined grains

Prevention At birth, most infants have adequate stores of vitamin D to cover their needs for the first few months of life. The American Academy of Pediatrics and the Canadian Paediatric Society regard breast-feeding as the optimal method for feeding infants (Canadian Paediatric Society 1998; Gartner and Eidelman 2005). Despite the nutritional and immunological superiority of breast-feeding compared to feeding commercially available infant formulas or cow's milk, the vitamin D content of human milk is normally low and is insufficient to meet the infant's needs for vitamin D. Commercially available infant formulas sold in the United States and Canada are required to contain vitamin D at levels sufficient to meet the needs of infants (Gartner and Greer 2003). Vitamin D-fortified cow's milk is also a suitable source of vitamin D and can be recommended for children when they reach an appropriate age, which is at about one year. Commercially available soy beverages (except soy-based infant formulas), rice beverages, or other vegetarian beverages, regardless of whether they are fortified, are inappropriate alternatives to breast milk, infant formulas, or pasteurized whole cow's milk in the first two years of life (Canadian Paediatric Society 1998).

rickets—a condition characterized by inadequate mineralization of the organic matrix in the bones of children usually caused by a deficiency of vitamin D and resulting in bowing of the legs and skeletal deformity of the rib cage

osteomalacia—a condition in which the organic matrix of the bones of adults is inadequately mineralized, resulting in muscular weakness, bone pain, and, in advanced cases, deformities of the ribs, pelvis, and bones of the legs

FIGURE 27.8 Rickets often results in bowing of the legs. A deficiency of vitamin D and/or calcium leads to inadequate mineralization of the bones, which then lack the strength to support the weight of the upper body once a child begins to stand and walk.

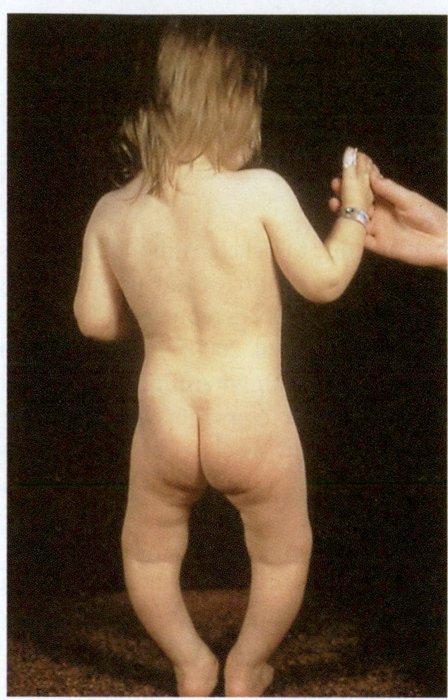

Source: S.Rolfes, K.Pinna, and E.Whitney, *Understanding Normal and Clinical Nutrition*, 7e, copyright © 2006, p.377.

Although an infant's vitamin D requirements can be met through sunlight exposure, determining what is adequate sunlight exposure for any given infant or child is difficult. Sunlight exposure varies considerably due to such factors as the season of the year, cloud cover, pollution, time spent in the shade, and the amount of body surface area covered by clothing when outdoors. Furthermore, vitamin D synthesis in response to sunlight exposure will be less for individuals who have darker skin pigmentation and for those who use sunscreens to limit exposure to ultraviolet light in order to reduce their risk of skin cancer (Gartner and Greer 2003). Consequently, the American Academy of Pediatrics recommends that beginning at two months of age, exclusively breast-fed infants be given a multivitamin supplement designed for infants containing 5 μg (200 IU) of vitamin D. Infants receiving less than 500 mL per day of infant formula should also be given a multivitamin supplement designed for infants. Children and adolescents who do not receive regular sunlight exposure and who do not consume at least 500 mL (2 cups) per day of vitamin D-fortified milk should also take an age-appropriate multivitamin supplement containing at least 5 μg (200 IU) of vitamin D (Gartner and Eidelman 2005).

Treatment Rickets is treated by a balanced diet that is age appropriate and that provides adequate intake of vitamin D, calcium, and phosphorus. Skeletal deformities can be prevented by maintaining good posture, body positioning, and bracing. After the disease is controlled, skeletal deformities may require surgical treatment (Bringhurst et al. 2005; Hightower and Gunta 2005).

Osteomalacia

Etiology and Clinical Manifestations Osteomalacia is a generalized bone condition affecting adults in which the organic matrix of bone is inadequately mineralized. Sometimes regarded as the adult form of rickets, the primary clinical manifestations of osteomalacia are muscular weakness, bone pain, and in advanced cases, deformities of the ribs, pelvis, and bones of the legs (Hightower and Gunta 2005). The most common causes of osteomalacia are vitamin D deficiency or impaired vitamin D action, calcium deficiency, and hypophosphatemia. Vitamin D deficiency can result from inadequate sun exposure, low dietary intake, or malabsorption of vitamin D due to biliary tract or intestinal diseases that impair the absorption of fat and fat-soluble vitamins (Hightower and Gunta 2005). Drugs such as phenytoin, phenobarbital, and rifampin stimulate hepatic breakdown of vitamin D and accelerate its loss from the body (Bringhurst et al. 2005). Diseases of the liver and kidney can impair the body's ability to convert vitamin D absorbed from the diet or synthesized in the skin into its biologically active form. A major consequence of long-standing vitamin D deficiency is reduced intestinal calcium absorption leading to hypocalcemia which, in turn, results in increased secretion of PTH. Increased secretion of PTH stimulates the removal of mineral from the organic matrix of bone in order to raise the serum calcium concentration but also results in a weakening of bone. Hypophosphatemia can result from excessive renal phosphate losses seen in renal tubular acidosis or from inadequate absorption due to long-term use of antacids that bind dietary phosphate in the GI tract and prevent its absorption (Hightower and Gunta 2005).

Treatment Elderly persons are at increased risk of osteomalacia due to diets low in calcium and vitamin D, a lack of sun exposure, decreased efficiency in synthesizing vitamin D when exposed to sunlight, and increased incidence of intestinal malabsorption problems that accompany aging (Hightower and Gunta 2005). The treatment of osteomalacia should address the underlying causes. Effective treatment of vitamin D deficiency may require a multivitamin providing as much as 20 μg of vitamin D, which is somewhat greater than the AI shown in Table 27.5 but still well below the Tolerable Upper Intake Level of 50 μg. Calcium intake should be adequate and may require use of an appropriate supplement providing as much as 1,500 to 2,000 mg of elemental calcium per day (Bringhurst et al. 2005). Patients with severe, long-standing vitamin D deficiency may initially require pharmacologic doses of up to 1,250 μg per week for 3 to 12 weeks, followed by maintenance therapy of 20 μg per day. Patients taking drugs that accelerate hepatic breakdown of vitamin D (e.g., phenytoin, phenobarbital, and rifampin) and those with intestinal malabsorption problems will require vitamin D in doses much greater than the AI (Bringhurst et al. 2005). Patients with liver or kidney diseases that prevent the steps necessary to activate vitamin D into its biologically active form (1,25-dihydroxyvitamin D) will require a form of vitamin D that is already biologically active, such as calcitriol (Bringhurst et al. 2005; Hightower and Gunta 2005).

Arthritic Conditions

Definition and Epidemiology

Arthritic conditions encompass more than 100 different diseases and conditions affecting the joints, the tissues surrounding the joints, and the connective tissues. Often referred to as "arthritis," arthritic conditions are among the most common diseases in the world and include osteoarthritis, rheumatoid arthritis, and gout, which are addressed in this section (Rizzo 2005). Arthritic conditions affect nearly one in six North Americans and are the leading cause of disability among Americans 18 years of age and older (U.S. Department of Health and Human Services, Prevalence of disabilities, 2001). Among Americans 65 years of age and older, arthritic symptoms are second only to hypertension as the most commonly reported chronic condition. It is estimated that by the year 2020, approximately 60 million Americans will be affected by arthritic conditions.

Common misconceptions about arthritic conditions are that they only affect older persons, that they are an inevitable consequence of aging, and that there are limited options for managing the symptoms of arthritis. Although arthritic conditions affect one out of every two people age 65 years and older, they are diagnosed in people of all ages, including children and teens. In fact, most people with arthritic conditions are younger than 65 years. Juvenile rheumatoid arthritis affects 70,000 to 100,000 children in the U.S. and is one of the most common chronic conditions of childhood (Rizzo 2005).

There are several factors known to increase the risk of arthritic conditions, three of which are modifiable: overweight, joint injuries, and infections. Overweight and obesity increase the risk of several arthritic conditions, particularly osteoarthritis of the knee in females and gout in males. Precautions should be taken in the workplace to avoid repetitive joint use which increases risk of certain arthritic conditions. Sport-related injuries to joints and connective tissues can be prevented by following injury prevention strategies such as using protective equipment, exercising to strengthen muscles, and warming up before exercising and cooling down afterwards. Because infectious diseases (e.g., Lyme disease) can be associated with arthritic conditions, their prevention and treatment in a timely and effective manner are important strategies. Nonmodifiable risk factors include female sex, age, and family history. Compared with males, arthritic conditions are diagnosed more frequently in females, who account for 60% of cases in persons aged 15 years and older. Risk increases with age and in persons with a family history (Brandt 2005; Rizzo 2005).

Osteoarthritis

Epidemiology, Etiology, and Clinical Manifestations

Osteoarthritis (OA) is the most common arthritic condition and is a leading cause of physical disability, increased health care costs, and impaired quality of life (Towheed and Anastassiades 2000; Rizzo 2005). It is estimated that 12% of the U.S. population 25 years of age and older have clinical signs and symptoms of OA (Towheed and Anastassiades 2000). OA is not a specific entity but a disease process involving all the structures of the joint, including the articular cartilage, the **subchondral** bone, the **synovial fluid** and membranes, the ligaments, and the nerves and muscles supporting the joint. However, the most striking changes seen in OA are those within the load-bearing articular cartilage (Brandt 2005). The progressive loss of articular cartilage and the inflammation of the other tissues composing the joint result in joint pain, stiffness, limited joint movement, wasting of **periarticular muscles**, and in some instances joint instability and deformity (Rizzo 2005). The joints of the fingers, feet, lumbar and cervical vertebrae, hips, and knees are those most commonly affected by OA (Saladin 2004; Rizzo 2005).

Figure 27.9 illustrates the basic structures of the normal joint and a joint with OA. In the early stages of OA, the articular cartilage thickens, but as the condition progresses the cartilage thins and softens. Surface cracks develop in the articular cartilage and it loses its smooth surface. Eventually these cracks extend completely through the articular cartilage down to the bone, portions of the articular cartilage become completely eroded, and the exposed surface of the subchondral bone becomes thickened and polished. Cysts form within the bone as synovial fluid leaks through the cracked and eroded articular cartilage. Dislodged fragments of cartilage and bone may float freely within the joint cavity. As the OA progresses, growth of cartilage and bone at the joint margins leads to the formation of abnormal bony outgrowths called **osteophytes** or bone spurs (Brandt 2005; Rizzo 2005).

osteoarthritis—a condition involving progressive loss of articular cartilage and inflammation of the tissues composing the joint, resulting in joint pain, stiffness, and limited joint movement

subchondral bone—bone located beneath the articular cartilage of a joint

synovial fluid—a protein-rich, slippery fluid contained inside a fibrous capsule that lubricates and nourishes the cartilage covering the ends of bones at their joints

periarticular muscles—those muscles located near a joint

osteophyte—a bony outgrowth near the joint affected by osteoarthritis. also referred to as a bone spur

Figure 27.9 A normal joint (left) and a joint affected by osteoarthritis (right) are compared. Early osteoarthritic changes shown on the left side of the affected joint include thickening of the articular cartilage, narrowing of the joint space, and development of surface cracks in the cartilage. Late changes shown on the right side of the affected joint include erosion of cartilage, development of bone cysts as synovial fluid comes into contact with bone, and formation of osteophytes or bone spurs.

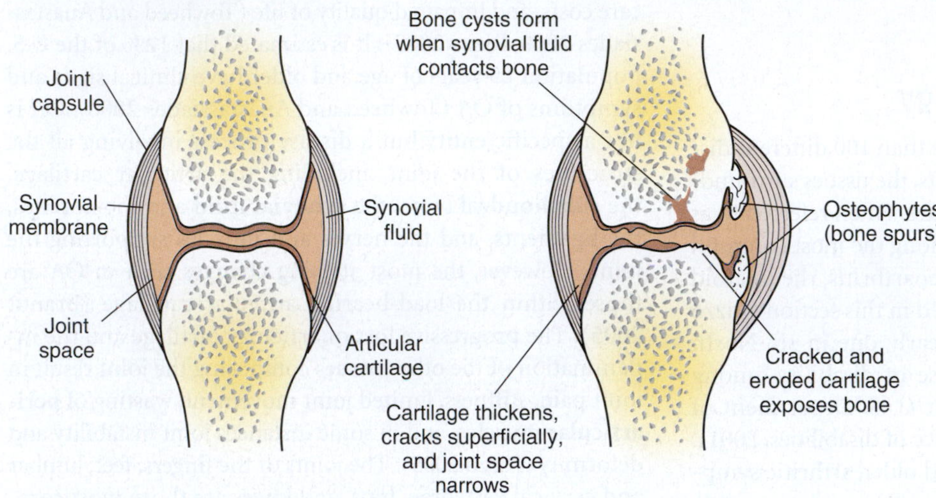

*Source: CM Porth, **Pathophysiology: Concepts of Altered Health Status** 7e.*

Major risk factors for OA include age (considered the most powerful risk factor), female sex, family history, major trauma to a joint or to soft tissues surrounding a joint, repetitive joint stress related to occupation, and obesity. Although OA is common in the joints typically stressed by ballet dancers, baseball pitchers, and prizefighters, there is no evidence that long-distance running or jogging increases risk of OA in the joints of the lower extremities. In addition to the added stress on the joints of the lower extremities resulting from obesity, excess body fat may also have a direct metabolic effect on the cartilage of joints (Rizzo 2005). Weight loss has great potential for reducing risk of OA of the knee. A loss of only 5 kg (11 lb) can reduce the risk of symptomatic knee OA by 50% (Brandt 2005).

Treatment Treatment of OA is focused on reducing joint inflammation, reducing pain, maintaining joint mobility, and minimizing disability. Nonpharmacologic treatments include improving body posture, proper footwear, weight reduction as indicated, periodic rest of the affected joint (but rarely is complete immobilization advised), and application of heat to the affected joint. In patients with OA, physical activity has been shown to decrease joint pain and disability while improving function, quality of life, cardiovascular fitness, and muscular strength (Brandt 2005). Periarticular muscle strength is a major factor protecting the articular cartilage from stress and injury. Therapeutic exercises to strengthen the periarticular muscles of the knee can decrease pain to an extent similar to that achieved by the use of non-prescription analgesics. Therapeutic exercise also can reduce disability, anxiety, and depression, and improve functional status.

Because no pharmacologic agent has been shown to prevent OA, slow its progression, or reverse it, the aim of drug therapy is pain relief as an adjunct to nonpharmacologic treatment (Brandt 2005). The mainstays of drug therapy are the nonsteroidal anti-inflammatory drugs (NSAIDs) which include aspirin, acetaminophen, indomethacin, ibuprofen, naproxen, and the cyclooxygenase-2 (COX-2) inhibitors celecoxib and valdecoxib. NSAIDs provide effective relief from mild to moderate pain associated with OA and other inflammatory conditions such as rheumatoid arthritis, gout, and toothache (Page et al. 2002). Except for the COX-2 inhibitors, all the above NSAIDs are available at low cost without a prescription. Notwithstanding consumer familiarity with NSAIDs and their non-prescription status, they are associated with gastrointestinal (GI) blood loss and ulceration, reduced platelet aggregation, prolonged bleeding time, and greater risk of death, particularly among the elderly and those with a history of peptic ulcer disease or upper GI bleeding. Use of NSAIDs has been linked to as many as 30% of all hospitalizations and deaths related to peptic ulcer disease in persons aged 65 years and older (Brandt 2005). Although the COX-2 inhibitors are much more expensive and no more effective than the other NSAIDs in reducing inflammation and pain, they have been aggressively marketed as being associated with a lower incidence of peptic ulcer disease. However, use of

COX-2 inhibitors significantly increases risk of stroke, heart attack, and death from coronary heart disease, and there is uncertainty whether celecoxib and valdecoxib protect against GI complications (Topol 2005). Despite concerns about physical and psychological dependence associated with their use, opioids such as codeine can be useful for some OA patients with chronic pain. Side effects associated with opioids include nausea, vomiting, constipation, urinary retention, mental confusion, drowsiness, and respiratory depression (Page et al. 2002; Brandt 2005).

In recent years, glucosamine and chondroitin have attracted considerable attention in the public and medical literature as dietary supplements that are potentially useful in treating OA (McAlindon et al. 2000; Towheed and Anastassiades 2000; Biggee and McAlindon, Glucosamine for osteoarthritis: part I, 2004; Biggee and McAlindon, Glucosamin for osteoarthritis: part II, 2004). **Glucosamine** and **chondroitin** have been used for decades in Europe as dietary supplements for treating OA. In the United States and Canada, they are aggressively marketed for treatment of OA and widely sold in supermarkets, pharmacies, and health food stores. Glucosamine and chondroitin are naturally occurring components of articular cartilage. Glucosamine is an amino sugar and a raw material for synthesizing glycosaminoglycans and proteoglycans, which are important constituents of articular cartilage. Chondroitin is a specific glycosaminoglycan found in the proteoglycans of articular cartilage. They are extracted from animal tissue; glucosamine comes from the shells of crab, lobster, and shrimp, and chondroitin comes from the cartilage of calves, cows, steers, whales, and sharks. Between the years 1997 and 2000, glucosamine supplements were ranked as the third best-selling nutritional product, with more than $300 million in sales in 2000 (Biggee and McAlindon, Glucosamine for osteoarthritis: part I, 2004).

Research on cell cultures and animals shows that both compounds have anti-inflammatory activity, favorably affect cartilage metabolism, and have anti-arthritic effects (Towheed and Anastassiades 2000). Preliminary human research indicates that glucosamine and chondroitin can favorably modify the progression of OA and that both compounds may be effective therapies in the symptomatic treatment of OA (McAlindon et al. 2000; Towheed and Anastassiades 2000; Biggee and McAlindon, Glucosamine for osteoarthritis: part I, 2004). However, the only long-term randomized controlled trial on the safety and efficacy of glucosamine and chondroitin is the Glucosamine/Chondroitin Arthritis Intervention Trial (GAIT), conducted by National Center for Complementary and Alternative Medicine in collaboration with the National Institute of Arthritis and Musculoskeletal and Skin Diseases. The GAIT studied the effectiveness of glucosamine alone, chondroitin alone, and the combination of glucosamine and chondroitin in treating knee pain from OA compared to a placebo. The GAIT is discussed in Box 27.2. Additional complementary and alternative therapies used for musculoskeletal disorders are listed in Table F1 in Appendix F.

Rheumatoid Arthritis

Epidemiology, Etiology, and Clinical Manifestations

Rheumatoid arthritis (**RA**) is a chronic inflammatory disease in which the synovial membrane of the joint becomes inflamed, resulting in swelling, stiffness, pain, limited range of motion, joint deformity, and disability (Lipsky 2005; Rizzo 2005). RA affects approximately 0.8% of the population (range 0.3% to 2.1%), is more common in older persons, and affects females three times more often than males, although this difference between the sexes diminishes in older groups (Pattison, Symmons, and Young 2004; Lipsky 2005). Although RA primarily affects the joints, it also can affect other tissues, resulting in anorexia, weight loss, fatigue, and generalized aching and stiffness.

The clinical course of RA is variable and characterized by periods of exacerbation and remission. In some patients the joint inflammation results in mild to moderate pain and stiffness lasting for short periods of time, while in others it can progress to debilitating, irreversible joint deformity and destruction. In most instances, RA begins insidiously with patients experiencing fatigue, anorexia, generalized weakness, and vague musculoskeletal symptoms for weeks or months, during which time diagnosis is difficult. Eventually, inflammation of the **synovial membrane** in the joints of the hands, wrists, knees, and feet results in warmth, redness, swelling, stiffness, and pain around these joints. As the inflammation progresses, rapid division and growth of cells in the synovial membrane cause an abnormal thickening of the synovial membrane known as **pannus**, which can eventually fill the synovial cavity and invade the joint margin, as illustrated in

glucosamine—a nutritional supplement used by some as a treatment for osteoarthritis; it is an amino sugar and a raw material for synthesizing glycosaminoglycans and proteoglycans, which are important constituents of articular cartilage

chondroitin—a nutritional supplement used by some to treat osteoarthritis; it is a specific glycosaminoglycan found in the proteoglycans of articular cartilage

rheumatoid arthritis—a chronic inflammatory disease in which the synovial membrane of the joint becomes inflamed, resulting in swelling, stiffness, pain, limited range of motion, joint deformity, and disability

synovial membrane—a membrane lining the capsule that encloses synovial joints and secretes synovial fluid, which lubricates and nourishes the cartilage at the end of bones

pannus—an abnormal destructive tissue that develops on the synovial membrane of patients with advanced rheumatoid arthritis. Inflammatory cells in pannus secrete enzymes that are destructive to articular cartilage and subchondral bone

BOX 27.2 NEW RESEARCH

The Glucosamine/chondroitin Arthritis Intervention Trial (GAIT)

Glucosamine and chondroitin sulfate are the two most widely used dietary supplements used to treat osteoarthritis (OA), the most common form of arthritis in North America (Clegg et al. 2006; Hochberg 2006; Towheed and Anastassiades 2000). The pain, stiffness, and limited mobility associated with OA of the knees are among the most common reasons that people seek alternative medical treatments. In 2001, more than 5 million Americans had used either glucosamine or chondroitin, and in 2004, Americans spent an estimated $734 million on these supplements, making them among the most widely used dietary supplements sold in the U.S. (Kolata 2006). Despite the popularity of these supplements, there have been unanswered questions about the safety and efficacy of glucosamine and chondroitin as a treatment of knee pain for OA patients. A meta-analysis of 17 double-blind, randomized placebo-controlled clinical trials indicated that glucosamine and chondroitin are likely beneficial in treating the symptoms of OA, but the degree of benefit was unclear due to methodological flaws in some of the trials (McAlindon et al. 2000). Those trials that were publicly funded tended to show little, if any, benefit from the supplements, while those that were funded by the supplement makers showed positive effects (Kolata 2006).

To provide a more definitive answer to this question, the National Center for Complementary and Alternative Medicine and the National Institute of Arthritis and Musculoskeletal and Skin Diseases funded the Glucosamine/chondroitin Arthritis Intervention Trial (GAIT), a double-blind, randomized, placebo-controlled trial in which 1,583 patients with symptomatic knee OA were randomly assigned to one of five different groups. Each group received one of the following treatments: 1,500 mg per day of glucosamine hydrochloride, 1,200 mg per day of chondroitin sulfate, both glucosamine hydrochloride and chondroitin sulfate, 200 mg of celecoxib per day, or a placebo (Clegg et al. 2006). The primary outcome measure was a 20% reduction in knee pain between the beginning and ending of the 24-week study. The results of the GAIT were that the glucosamine hydrochloride and chondroitin sulfate alone or in combination were not significantly better than the placebo in reducing knee pain by 20%. There was some evidence suggesting that, in OA patients with moderate to severe knee pain, the combination of glucosamine and chondroitin sulfate is effective in reducing pain. However, because GAIT was not specifically designed to address this patient subgroup, further research is needed to conclusively establish a benefit in these patients.

Figure 27.10. With further progression, inflamed cells in the pannus release enzymes that digest the adjacent bone and cartilage, causing joint deformity, severe pain, and consolidation and immobility of the joint known as ankylosis. This leads to muscle atrophy from disuse, stretching of ligaments, and changes in tendons. Although treatment can slow the progression of the disease, many of these destructive changes are permanent once they occur (Lipsky 2005; Rizzo 2005).

The body's immune system plays an important role in RA and it is thought that the disease results from an aberrant immune response in which a person's immune system mistakenly regards the body's own healthy joint tissues as foreign and mounts an attack against those healthy tissues. Conditions such as RA in which a person's immune system attacks his or her own healthy tissues are known as autoimmune diseases. It is not known what causes RA or even whether RA is a single disease or several different diseases having common features. Research suggests that the autoimmune response is triggered by an infectious agent in a genetically susceptible person (Lipsky 2005; Rizzo 2005).

Treatment The goals of RA treatment are to reduce pain and inflammation, protect the joint from destruction, maintain the function of the joint and surrounding structures, and control any systemic manifestations (Lipsky 2005). Treatment is multidisciplinary and involves physicians, physical therapists, occupational therapists, clinical psychologists, and dietitians. A number of pharmacologic agents are used such as NSAIDs (discussed earlier in this chapter), glucocorticoids, immunosupressive drugs, and what are referred to as disease-modifying antirheumatic drugs (DMARDs). The glucocorticoid drug prednisone is commonly used to treat RA, but its use is associated with side effects such as increased risk of osteoporosis, peptic ulcer disease, esophagitis, and glucose intolerance, increased appetite, weight gain, and retention of sodium and fluids (Page et al. 2002). When prednisone is used systemically at relatively low doses or injected directly into a joint, these side effects can be minimized. DMARDs are a chemically and pharmacologically disparate group of drugs that modify the course of RA but are also toxic and have potentially serious side effects. The most commonly used DMARD is the folic acid antagonist methotrexate, which can cause gastrointestinal upset, mouth ulcers, and liver toxicity. The simultaneous administration of folic acid can reduce the severity of side effects without diminishing treatment effectiveness (Page et al. 2002; Lipsky 2005).

Diet and Rheumatoid Arthritis The role of diet and nutritional supplements in the etiology and treatment of RA has been a topic of considerable interest and speculation for decades. However, studying the relationship between diet and RA is problematic given the multiple risk factors for RA, the episodic nature of RA symptoms, the expense of conducting long-term randomized controlled trials, the difficulty of measuring diet, and the difficulty of assessing subjects' compliance to dietary change (Martin 1998).

There is evidence that lower intakes of vegetables, fruits, and dietary vitamin C are associated with increased risk of developing RA. It has been observed that there is a lower risk of developing RA in countries such as Italy and Greece,

FIGURE 27.10 A normal joint (left) and a joint affected by rheumatoid arthritis (right) are compared. Early changes shown on the left side of the affected joint include inflammation and thickening of the synovial membrane. Late changes shown on the right side of the affected joint include development of pannus, erosion of articular cartilage and bone, and filling of the joint space by pannus.

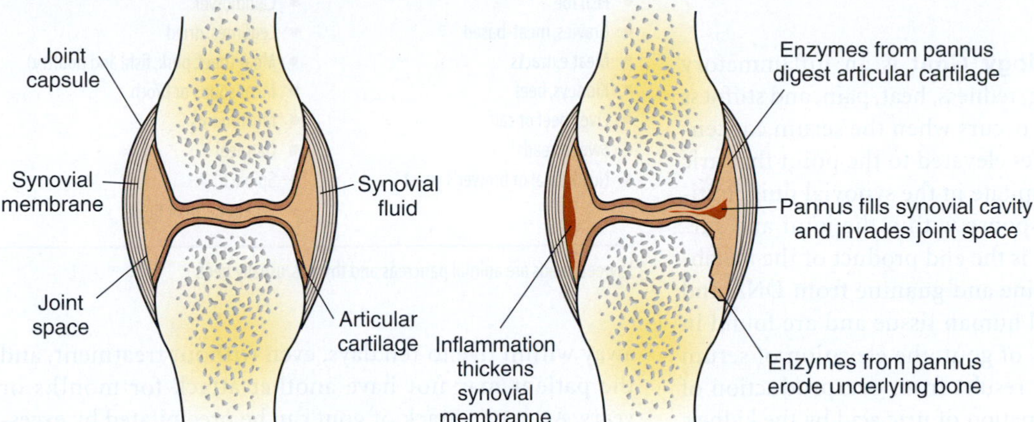

Source: CM Porth, *Pathophysiology: Concepts of Altered Health Status 7e.*

where oil-rich fish, olive oil, vegetables, and fruits are consumed in greater amounts compared to other countries that have a higher risk of developing RA (Rennie, Hughes, and Jebb 2003; Pattison, Symmons, and Young 2004). Several randomized, double-blind, placebo-controlled trials have shown that dietary fish oil supplementation is effective in reducing the symptoms associated with chronic RA (Cleland, James, and Proudman 2003; Rennie, Hughes, and Jebb 2003; Lipsky 2005). Fish oils contain two important omega-3 long chain polyunsaturated fatty acids: eicosapentaenoic acid (EPA) and docosahexaenoic acid (DHA). Consumption of EPA and DHA reduces the synthesis of chemicals known to stimulate joint inflammation and cartilage degradation, and RA patients taking EPA and DHA report a reduction of morning stiffness, decreased joint tenderness, and less need to take NSAIDs to relieve pain (Volker et al. 2000; Cleland, James, and Proudman 2003). Consumption of omega-6 polyunsaturated fatty acids, on the other hand, tends to increase the synthesis of proinflammatory chemicals (Adam et al. 2003; Rennie, Hughes, and Jebb 2003). When used in conjunction with drugs to treat RA, EPA and DHA can potentially reduce the severity of side effects of NSAIDs, glucocorticoids, immunosupressive drugs, and DMARDs (Cleland, James, and Proudman 2003). Current research suggests that the effective dose of fish oils is between 2.6 g/d and 7.1 g/d, but it generally takes two to three months before the anti-inflammatory effects of fish oils are noticed. Because most commercially available fish oil preparations provide approximately 300 mg of EPA or DHA per capsule, more than eight capsules per day would be necessary to achieve a dose of 2.6 g/d. Because of the high cost of fish oil capsules, a more economical approach for some patients might be to purchase bottled fish oils (Cleland, James, and Proudman 2003). Considering the observation that high doses of fish oil can suppress immunity in animals, further research is needed to study the long-term safety of fish oil supplementation in humans (Rennie, Hughes, and Jebb 2003).

There is considerable speculation about the benefits of excluding foods from the diet that are thought to aggravate the symptoms of RA, and many patients report an improvement in symptoms after excluding such foods as red meats, dairy products, cereals, and wheat gluten (Martin 1998; Rennie, Hughes, and Jebb 2003). Some of this improvement is likely due to the episodic nature of RA: the fact that it is characterized by periods of remission and exacerbation. Although controlled studies in which these foods have been eliminated have yielded inconsistent results, a better experimental design is to give patients capsules containing the alleged problem food antigen without the patient knowing what is in the capsule. This technique is referred to as a blind challenge test, because the patient does not know what he or she is taking and is therefore "blinded" to the treatment. Most patients who report an improvement in symptoms during elimination diets when aware of what they are eating notice no change in symptoms during blind challenge tests (when unaware of what they are eating). Because a small number of patients do report changes in symptoms in response to the elimination or addition of certain foods, further evaluation and testing to confirm a food allergy may be warranted. However, the prevalence of immune sensitivity to specific foods by patients with RA is similar to that found in the general population (Martin 1998; Rennie, Hughes, and Jebb 2003). In some RA patients, symptoms improve after eliminating meat from the diet. However, it is not known whether the improvement is due to eliminating meat or a consequence of consuming more foods rich in antioxidants such as vegetables and fruits (Martin 1998; Rennie, Hughes, and Jebb 2003). Because poor nutritional status is common in patients with RA, emphasis must be given to maintaining optimal nutritional status.

Gout

Epidemiology and Etiology Gout is an inflammatory disease resulting in swelling, redness, heat, pain, and stiffness in the affected joint. Gout occurs when the serum concentration of uric acid becomes elevated to the point that uric acid crystals begin to precipitate in the synovial fluid, initiating an inflammatory response within the joint and surrounding tissues. Uric acid is the end product of the metabolism of the purines adenine and guanine from DNA and RNA, which are part of all human tissue and are found in many foods. In most cases of gout, the elevation of serum uric acid (hyperuricemia) results from overproduction of uric acid, inadequate elimination of uric acid by the kidney, or a combination of both (Kim et al. 2003; Rizzo 2005).

Gout is one of the most painful arthritic conditions, is much more common in males than in females, and is the most common cause of inflammatory arthritis in males >40 years of age (Kim et al. 2003). Risk factors associated with gout include genetics, male sex, older age, overweight, excessive alcohol consumption (3 or more drinks per day), eating foods rich in purines (see Table 27.11), exposure to lead, and use of certain drugs such as aspirin, diuretics, nicotinic acid, cyclosporine, and levodopa. It rarely occurs in children, young adults, or premenopausal females.

Clinical Manifestations Compared with plasma, synovial fluid is a poor solvent for uric acid, particularly at temperatures less than 37° C, which helps explain why the joint most commonly affected by gout is the great toe. Other sites commonly affected by gout are joints in the periphery of the body such as the instep, ankles, heels, knees, wrists, fingers, and elbows (Kim et al. 2003; Reginato 2005; Rizzo 2005). The symptoms of gout usually occur rapidly, sometimes overnight, resulting in sudden, severe joint pain and swelling, shiny, red skin around the joint, and extreme tenderness around the joint. These symptoms typically go away within five to ten days, even without treatment, and the patient may not have another attack for months or years. An acute attack of gout can be precipitated by excessive exercise, certain medications, purine-rich foods, excessive alcohol consumption, or crash dieting (Galperin, German, and Gershwin 1999; Rizzo 2005). Over a period of years, chronic hyperuricemia can result in the formation of uric acid crystals in the joints and the soft tissues surrounding the joints, which form large deposits called tophi that resemble lumps just below the skin. If left untreated, recurrent attacks of gout will result in persistent swelling, stiffness, mild to moderate joint pain, and eventually permanent joint damage and disability (Kim et al. 2003; Reginato 2005; Rizzo 2005).

TABLE 27.11 Foods That Are Very High or Moderately High in Purines

Very-High-Purine Foods	Moderately-High-Purine Foods
• Anchovies and Sardines	• Asparagus
• Fish roe	• Cauliflower
• Gravies, meat-based	• Legumes, dried
• Meat extracts	• Meat (beef, pork, fish, and poultry)
• Kidneys, beef	• Meat soups or broth
• Liver, beef or calf	• Mushrooms
• Sweetbreads[1]	• Oatmeal
• Nutritional or brewer's yeast	• Spinach
	• Wheat germ or bran

[1] Sweetbreads are animal pancreas and thymus, usually beef.

Treatment Initial treatment of gout involves use of nonsteroidal anti-inflammatory drugs (NSAIDs) to relieve pain and reduce inflammation. Glucocorticoids and colchicine are also used to reduce inflammation. Once the pain and inflammation of the acute attack are controlled, the hyperuricemia is treated using the drug allopurinol or one of several other drugs that increase urinary excretion of uric acid. Lifestyle modifications helpful in managing gout include moderate alcohol consumption, avoiding purine-rich foods, and achieving and maintaining a healthy body weight. Crash dieting resulting in overly rapid weight loss will reduce urinary excretion of uric acid and thus possibly precipitate an attack of gout (Galperin, German, and Gershwin 1999). As shown in Table 27.11, purine-rich foods include sardines, fish roe, anchovies, meat-based gravies, meat extracts, food yeast and several animal organs used for food including the liver, kidneys, the pancreas, and the thymus, the latter two euphemistically known as "sweetbreads."

gout—swelling, redness, heat, pain, and stiffness in a joint due to the formation of uric acid crystals in the synovial fluid, resulting in inflammation within the joint and in the surrounding tissues

fibromyalgia—a chronic musculoskeletal disorder characterized by widespread muscle pain, joint stiffness, disturbed sleep, fatigue, headache, cognitive and memory problems, paresthesias, and numerous tender points. The word comes from the Latin term for fibrous tissue (fibro) and the Greek terms for muscle (myo) and pain (algia)

Fibromyalgia

Definition and Epidemiology

Fibromyalgia is a chronic musculoskeletal disorder characterized by widespread muscle pain, joint stiffness, disturbed sleep, fatigue, headache, cognitive and memory problems (sometimes referred to as "fibro fog"), **paresthesias**, and numerous tender points, which are specific muscle-tendon sites throughout the body that are painful when pressed (National Institute of Arthritis and Musculoskeletal and Skin Diseases 2004; Gilliland 2005). Fibromyalgia—also referred to as fibromyalgia syndrome or FMS—is the second most common musculoskeletal condition encountered by **rheumatologists**, affecting about 3 to 6 million Americans, or about 1 in 50 (National Institute of Arthritis and Musculoskeletal and Skin Diseases 2004). In the United States, the prevalence is estimated to be about 2% overall, with 3.4% of females and 0.5% of males affected (Goldenberg, Burckhardt, and Crofford 2004). Approximately 80% to 90% of cases of fibromyalgia are diagnosed in females, and the condition is more common in middle-aged and older persons, with the prevalence increasing with age (Gilliland 2005). The condition is found in all types of climates, among most ethnic groups, and in most countries (Gilliland 2005). Unlike osteoarthritis or rheumatoid arthritis, fibromyalgia is not crippling, deforming, or disabling (Gilliland 2005).

Etiology

The etiology of fibromyalgia is unknown, but it can be associated with a traumatic or emotionally stressful event such as a motor vehicle crash or with injuries due to repetitive motion. Fibromyalgia is more often seen in persons diagnosed with autoimmune diseases such as rheumatoid arthritis or systemic lupus erythematosis. Other factors linked to an increased risk of a fibromyalgia diagnosis include sleep disturbances, low levels of serotonin in the brain, reduced levels of growth hormone, and such psychological abnormalities as depression, anxiety, somatoform disorders (physical ailments stemming from psychological problems), and **hypochondriasis** (Gilliland 2005).

Diagnosis and Clinical Manifestations

Establishing a diagnosis of fibromyalgia is complicated by the fact that there are no objective diagnostic tests specific for the condition, and the symptoms of fibromyalgia are common to many other illnesses. Consequently, clinicians generally rule out other potential causes of symptoms before diagnosing fibromyalgia, and patients typically see several physicians before ultimately being diagnosed. The American College of Rheumatology (ACR) has established the following diagnostic criteria for fibromyalgia: (1) a history of pain that is widespread (i.e., on both sides of the body as well as above and below the waist) which has lasted for at least 3 months and (2) excessive tenderness or pain when pressure is applied to at least 11 of 18 different tender points that the ACR has identified in the neck, shoulders, arms, back, and legs (Goldenberg, Burcckhardt, and Crofford 2004; National Institute of Arthritis and Musculoskeltal and Skin Diseases 2004; Gilliland 2005). The 18 different tender points designated by the ACR are shown in Figure 27.11.

Treatment

Treating fibromyalgia involves a multidisciplinary approach aimed at improving the quality of sleep, treating depression, anxiety, and pain, increasing physical activity, and using various approaches to help patients better cope with stressful events and improve their ability to relax (Goldenberg, Burckhardt, and Crofford 2004; National Institute of Arthritis and Musculoskeletal and Skin Diseases 2004; Lemstra and Olszynski 2005). The initial aim of therapy is improving the quality of sleep through the use of antidepressant drugs shortly before bedtime and educating patients about lifestyle practices that can improve the quality of sleep. Depression and anxiety are treated using psychiatric counseling and, if necessary, pharmacologic agents. Some fibromyalgia patients may periodically require analgesics for pain control (Goldenberg, Burckhardt, and Crofford 2004; Gilliland 2005). Regular physical activity involving aerobic exercise, muscle strengthening, and flexibility training have all been shown to improve aerobic fitness, overall well-being, and sleep quality, and to reduce pain, fatigue, and depression (Goldenberg, Burckhardt, and Crofford 2004). Low-impact activities such as water aerobics, bicycling, walking, and aerobic dance are particularly helpful. Initially, patients should begin exercising at a low level, and then gradually increase the duration and intensity of activity until they are exercising 20 to 30 minutes 3 to 4 days per week (Goldenberg, Burckhardt, and Crofford 2004; Gilliland 2005).

Psychological counseling and cognitive behavioral therapy have been shown to decrease the severity of pain and fatigue, assist in alleviating depression and anxiety, and improve the ability to cope with stressful life events. Fibromyalgia patients have reported improvement in symptoms from instruction in meditation and relaxation techniques and participation in stress management workshops (Goldenberg,

paresthesia—an abnormal sensation in the skin that may be described as burning, pricking, or like ants crawling on the skin

rheumatologist—a medical doctor specializing in diseases of the muscles and joints that are classified as rheumatic diseases

hypochondriasis—a somatoform disorder (i.e., a physical ailment stemming from a psychological problem) characterized by an unfounded belief that one is suffering from a serious illness

Figure 27.11 **Location of the 18 different tender points designated by the American College of Rheumatology in its criteria for diagnosing fibromyalgia**

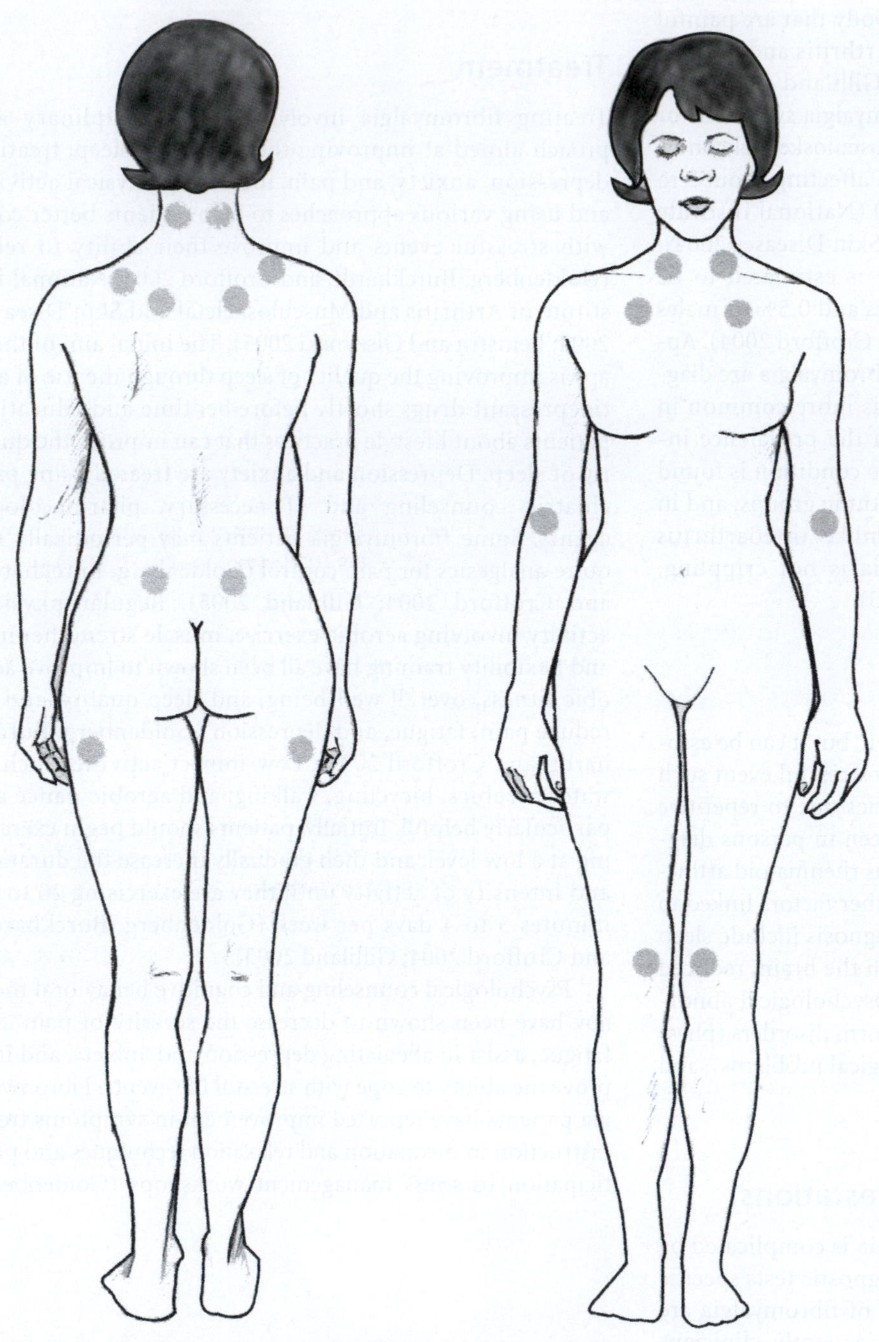

Source: Courtesy of National Institute of Arthritis and Musculoskeletal and Skin Diseases. Questions and Answers about Fibromyalgia. Bethesda, MD: National Institutes of Health, 2004.

Burckhardt, and Crofford 2004; Gilliland 2005). Some patients report improvement of symptoms from acupuncture, chiropractic spinal manipulation, massage, hypnotherapy, biofeedback, ultrasound treatments, and **balneotherapy** (Holdcraft, Assefi, and Buchwald 2003; Goldenberg, Burckhardt, and Crofford 2004).

Intensive patient education about fibromyalgia and the various available treatment approaches has been shown effective at improving the quality of sleep, increasing adherence to a regular exercise program, reducing pain, reducing depression and anxiety, and improving coping skills and the overall quality of life (Nicassio et al. 1997; Alamo, Moral, and Pérula de Torres 2002; Goldenberg, Burckhardt, and Crofford 2004). An effective approach to patient education has been a group format involving lectures, distribution of educational literature, group discussions, and demonstrations (Goldenberg, Burckhardt, and Crofford 2004). Participation in a multidisciplinary treatment program involving group exercise sessions, group pain and stress management lectures, dietary education lectures, and massage therapy sessions resulted in improvements in pain, depression, and self-perceived health status compared to fibromyalgia patients who did not participate in the program (Lemstra and Olszynski 2005).

Diet and Fibromyalgia

Although some fibromyalgia patients report symptom improvement after eating or avoiding certain foods, there is no consensus within the scientific community that the condition is significantly improved by any specific dietary practice, in large part due to the absence of rigorous randomized controlled trials on the effect of diet on fibromyalgia (National Institute of Arthritis and Musculoskeletal and Skin Diseases 2004). In one relatively small, short-term observational study, 18 fibromyalgia patients who followed a low-sodium, uncooked vegan diet for three months reported an improvement in pain, sleep quality, joint stiffness, and self-perceived health status compared to a group of 15 other fibromyalgia patients who followed their conventional diet (Kaartinen et al. 2000). Uncooked vegan diets include liberal quantities of berries, fruits, vegetables, nuts, germinated grains, and sprouts, all of which are sources of numerous vitamins, minerals, dietary fibers, and phytochemicals (Hanninen et al. 2000). Because this diet has a very low caloric density, it promotes weight loss in overweight persons, which could explain some of the reported improvements in symptoms (Kaartinen et al. 2000).

Studying the relationship between diet and fibromyalgia and other musculoskeletal conditions is complicated by the cyclic nature of the symptoms patients experience; in other words, patients experience periods of remission followed by periods of exacerbation. When an improvement in symptoms coincides with a dietary change, one must ask whether the change in symptoms is due to the dietary change or to the patient experiencing a transient remission. The psychosomatic nature of fibromyalgia and the placebo effect must be considered as well. Often, patients will report an improvement in symptoms because they believe or hope that a dietary change will be helpful. Many of the prevalent ideas about the influence of diet on fibromyalgia and other musculoskeletal conditions are based on anecdotal evidence instead of rigorously designed, randomized controlled trials (Holdcraft, Assefi, and Buchwald 2003). The common notion that monosodium glutamate or MSG in the diet can exacerbate pain in persons with fibromyalgia is based on anecdotal evidence and is not supported by sound scientific research (Geenen et al. 2004).

Our current knowledge of the role of diet in treating fibromyalgia and other musculoskeletal conditions is limited, and there is much that is yet to be learned in this area. It is important to keep an open mind and to actively research these questions. However, the dietitian should avoid giving a patient hope of relief from symptoms or hope of a cure by recommending some dietary regimen if there is no sound scientific evidence supporting both the safety and efficacy of that dietary approach.

Conclusion

Historically, the role of nutrition has been long recognized in the development of both rickets and osteomalacia. Today, there is increasing concern regarding the impact of osteoporosis on both health care expenditures and overall quality of life within the population of the United States and Canada. This chapter has focused on the role of nutrition therapy in both treatment and prevention of this most common bone disease. New technology and treatment for osteoporosis furthers provides opportunities for the registered dietitian in impacting the effects of this disease. As research continues, additional roles for nutrition may be identified to assist in both treatment and prevention of other musculoskeletal diseases.

balneotherapy—treating disease by bathing. It can involve hot water, steam, or application of hot packs to the body; cold water; contrasting hot and cold baths or showers; a steam bath followed by immersion in cold water; or immersion in mineral springs or a hot bath to which various medicinal herbs or minerals have been added

TABLE 27.12

Drug–Nutrient Interactions for Medications Used to Treat Diseases of the Musculoskeletal System

Classification	Mechanism	Generic	Brand Names	Possible Food–Drug Interactions
Anti-gout	Xanthine oxidase Inhibitor	Allopurinol	Aloprim, Zyloprim	Drink 2.5–3 L fluids/day; avoid large doses of vit C; N/V, gastritis, abdominal pain, diarrhea; limit alcohol
Anti-gout, anti-inflammatory	Inhibits mitosis	Colchicine	Colchicine	↓ purine diet during acute attack; ↓ kcal if wt loss needed; anorexia, ↓ wt.; may ↓ absorption of B_{12}; N/V, diarrhea, abdominal pain; avoid alcohol
Disease-Modifying Antirheumatic Drugs (DMARDs)				
Antineoplastic, antipsoriatic, antiarthritic (rheumatoid)	Dihydrofolate Inductase Inhibitor	Methotrexate	Methotrexate, Rheumatrex	Encourage ↑ fluid intake to ↑ urine output; food delays absorption and ↓ peak concentration & bioavailability; ↓ folate absorption; anorexia, ↓ wt., dehydration, altered taste, N/V, diarrhea; avoid alcohol
Glucocorticoids				
Anti-inflammatory, immunosuppressant, hormone	Mimics the action of cortisol	Prednisone	Deltasone	Caution with DM—↑ glucose; highly protein bound; may need ↑ K, PO_4, Ca, and ↑ vits A, C, D, ↑ protein, and ↓ dietary Na; avoid alcohol
Anti-osteoporosis	Selective estrogen receptor modulator	Raloxifene	Evista	Adequate Ca & vit D intake essential; ↑ wt.; limit alcohol
Parathyroid hormone	PTH acts on bone-building cells called osteoblasts to stimulate new bone growth and improve bone density	Teriparatide	Forteo	Nausea, constipation
Hormone replacement therapy	Increases calcium abs, decreases osteoclast activity in bone	Estrogen, Estrogen/progesterone	Estrogen: Cenestin, Estrace, Estradiol, Ogen, Premarin, Climara, Estraderm Estrogen/Progesterone: Activella, Femhrt, Prempro, Premphase, Combipatch	↑ foods high in Ca, vit D, Mg, folate, B_6; may need Ca suppl, vit C suppl >1 g/d; appetite changes, ↑ Ca absorption, N/V, diarrhea; limit alcohol; ↓ glucose tolerance
Bisphosphonates	Bind permanently to the surfaces of the bones and slow down the osteoclasts	Alendronate, etidronate risedronate	Fosamax, Didrocal, Actonel	Nausea, abdominal pain, loose bowel movements, no Ca supplement or MVI for 2 hrs before or after
Opioids	Bind to specific opioid receptors in the central nervous system and in other tissues	Morphine, codeine, oxycodone, hydrocodone, loperamide, heroin	Oxycontin, Roxicodone, Imodium, Roxinal	N/V, constipation, dry mouth

(continued on the following page)

TABLE 27.12 *(continued)*

Steroids	Mimics the action of cortisol	Prednisone	Deltasone	Caution with DM—↑ glucose; highly protein bound; may need ↑ K, PO_4, Ca, and ↑ vits A, C, D, ↑ protein, and ↓ dietary Na; avoid alcohol
Immunosuppressant, antineoplastic	Attacks the white blood cells	Cylclophosphamide	Cytoxan	↑ fluid needs; N/V, abdominal pain, dry mouth, diarrhea
Immunosuppressant	Attacks the white blood cells	Mycophenolate mofetil (MMF)	CellCept	Take on an empty stomach; take Mg supplement separately; N/V, diarrhea
Immunosuppressant	Attacks the white blood cells	Azathioprine	Imuran	Diarrhea, vomiting, anorexia, steatorrhea
Antineoplastic	Blocks the metabolism of cells	Methotrexate	Rheumatrex, Trexall	Mouth ulcers, diarrhea; ↑ fluid needs; food delays absorption and ↓ bioavailability; folate or MVI may ↓ response of drug; may ↓ absorption of fat, vit B_{12}, Ca, and folate; anorexia, ↓ wt.; avoid alcohol
Immunosuppressant	Blocks the production of folic acid	Cotrimoxazole	Bactrim, Septra	May need folate supplement; anorexia, N/V, diarrhea; adequate fluid
Immunosuppressant	Attacks the white blood cells	Ciclosporin, tacrolimus	Neoral, Sandimmune	No K supplement or salt substitute, caution with grapefruit; anorexia, N/V, diarrhea; increases glucose

PRACTITIONER INTERVIEW

Nancy Duhaime, M.S., R.D., L.D. *Clinical Dietitian Bartlett Regional Hospital (BRH), Juneau, Alaska, and Wildflower Court, a long-term care hospital that contracts with BRH for 20 hours/week*

How long have you been an RD? How long have you worked in long-term care?
I've been an RD for 20 years but have only worked in long-term care consistently for the past two and a half years. Previously, I ran an elderly nutrition program in Connecticut for three years, and I was program director for an AmeriCorps program (in Vermont) that helped match community volunteers with elders in their own homes. Although this was not a "nutrition" job, the volunteers played a vital role in getting meals and groceries delivered, cleaning out refrigerators, and even helping with some food preparation. It was a very rewarding job.

Roughly what percentage of your residents have arthritis or osteoporosis? How does this affect their nutritional status?
Approximately 35% of the residents have some type of degenerative joint disease, including gout and osteoarthritis. Many have other comorbidities that may impact on their nutritional status as well. I pay attention to their ability to hold a utensil and feed themselves. To gain insight into how arthritis, especially in the hands, can impact nutrition, "buddy" tape your thumb to the index finger so you cannot use it. Now try eating a meal using utensils, especially soup! We had a resident who had arthritis along with COPD. When he first came to live at the home, he would come to the common dining area, but he wouldn't eat the food put in front of him. He complained of the smell or look. As the RD, I was called in to talk with him regarding meal preferences. As I spoke with him he disclosed that he didn't like people to see him eat because he had a difficult time holding the utensils and made a mess all over his clothing. We put together a care plan that allowed him to eat in his room, with assistance as needed, as well as eat in the common dining room when "finger foods" were on the menu, for example a sandwich (hamburger) that could be cut up and held in his hand. His food intake improved and he still had times when he came out for some of the social aspects of eating.

Are there common nutritional problems associated with arthritis?
The effects of medications on the stomach and appetite are the biggest common problems I have seen. NSAIDS for pain relief are hard on the GI tract. Also, some of the treatments for osteoporosis are hard on the gut. Fosamax, for example, requires one to sit upright for at least 30 minutes after taking it and to take it 30 minutes before eating (not to mention the risk for esophageal erosion if not done properly). It's quite a "dance" between nurse's aides who get folks up and out of bed in the morning and medication nurses who are on a time schedule to provide prescribed medications; and then the cooks have a scheduled time for breakfast—if Mrs. Smith doesn't get up early so that the medication nurse can give her the Fosamax in a timely manner, she gets a cold breakfast when everyone else gets a hot one. Also, if posture is compromised, as it usually is with osteoporosis, the slow compression of the spine leads to more pressure on the internal organs, especially the stomach, causing early satiety. Additionally, it can impact on regular bowel movements, which can also decrease food intake.

So many people are involved in one person's care; it takes a great deal of teamwork for it to all fall into place. I always try to find something positive to say to the nurse's aides, nurses, or food service staff to let them know they are all part of a very important team. If they all work together, it's fewer nutritional problems for the resident, and in the long run, less work for them.

What is the biggest challenge working with patients with arthritis?
Because they move slowly and are in pain, they usually need some assistance walking or getting to the toilet. It's imperative that all staff working with these folks have patience and understanding. If you have to ask someone to take you to the bathroom every time you feel the urge and that person is busy and speaks in harsh tones while telling you "Just a minute!" you don't want to eat or drink because it's too uncomfortable to get help to use the toilet. As an RD, I have a lot of things to do, and I have to keep reminding myself that I must be available when the resident is available mentally and physically.

Any advice for dietetic students about counseling clients with arthritis?
Taking time to build rapport is so important. Constant pain is not pleasant, and many of these residents have been in constant pain for years. They learn to live with it, but usually upon the first meeting, they can be grouchy, mean, insulting, and not much fun to visit. However, I have come to understand that because of their constant pain, they strike out but don't intend to be mean or insulting. Taking a gentle, caring approach and validate that they are in pain. Ask if it would be better to come back at another time. Doing that often softens them to a point where their "kind" self shines through. We all have bad days and we very likely snap at someone ourselves sometimes. I give the resident a chance to know that they have snapped because of their pain. Then I offer them an opportunity to start over again. It's worked every time.

Any advice for dietetic students?
Well, it's been 20 years since my internship, but I remember feeling overwhelmed—yet I've felt that more recently as

well. I was out of the field of dietetics for over 12 years, and two and a half years ago I took the plunge to return to clinical dietetics. I was very scared. How well would I remember all of the information I had learned in school some 20 years ago—and what would I do about the fact that some things had changed since that time? Well, I'm happy to say I survived, just like the students will survive their first experiences. What I found helpful was being able to communicate with people in a friendly way, making sure they knew that I was interested in what they were saying, and when I didn't have an answer to their questions, I would reassure them that I would get back to them. (I can't tell you how many times I had to run to the nearest diet manual to look things up just to be sure I was remembering them correctly.)

I was even scared of the doctors, and my worst fears came true one day (but my good training came through for me!). One day, in front of the patient, a doctor asked me a very detailed question about the patient's insulin dose and carbohydrates intake. I knew it was something I should have had an immediate answer for, but I didn't. I had to admit to that I did not know the answer, but quickly followed with the statement, "I need to check a reference to be sure I'm accurate." They were fine with this, and I was able to look up the information and provide the answer some time later. I keep in mind that if I were the patient in the bed, I'd feel better knowing that someone was checking their own accuracy rather than trying to look like they know it all.

CASE STUDY

CC: Presents to emergency room after fall on ice—diagnosed with fracture of her right wrist. DOB: 9/15/1950

General: Sixth grade teacher, married with three college-aged children. Ht: 5′6″ Wt: 172 lbs. Usual body weight: 160–180 lbs. for past five years

Labs: All WNL

Rx: Multivitamin

Past Med Hx: Has not had menstrual cycle for past year; no other contributory history.

Family History: Grandmother and 2 great aunts have osteoporosis. Her mother died at an early age.

Nutrition Hx: Reports good appetite; eats a variety of foods but states that she has routinely tried to lose weight which means she has restricted kcal off and on for previous five years. When asked, calcium intake is from yogurt 3–4 times per week and the amount of milk on her breakfast cereal. She states that she has purposefully stopped eating cheese in order to help her lose weight.

This patient was referred for a DEXA (Dual Energy x-Ray Absorptiometry) scan to evaluate her bone mineral density by the orthopedic surgeon who treated her wrist fracture. She was diagnosed with osteopenia as her BMD is 2.0 standard deviations below the mean.

Questions

1. What is osteopenia?
2. What risk factors does this patient present with? What additional questions would be important for the medical team to establish in evaluating her diagnosis?
3. Explain why calcium is vital to bone health.
4. What level of calcium supplementation should be initiated for this patient?
5. Discuss the importance of vitamin D to calcium absorption. Should she begin vitamin D supplementation also?
6. What other medications might her physician consider to prevent osteoporosis in this patient?

NUTRITION CARE PROCESS FOR MUSCULOSKELETAL DISEASES

Step One: Assessment

Medical/Social History

Standard

Medications: NSAIDS, steroids

Dietary Assessment

Standard but close attention to dietary intake of calcium, vitamin D, protein, and sodium. Use of alcohol, vitamin, mineral, herbal, or other type of supplements.

Evaluate physical activity patterns

Eating pattern: 24-hour recall, diet history, food frequency

Anthropometric

Standard

Biochemical Assessment

Visceral Protein Assessment: Standard

Hematological Assessment:

Hemoglobin

Hematocrit
MCV
MCHC
MCH
TIBC

Lipid Assessment

Total Cholesterol
HDL
LDL
Triglyceride

Specific Labs

Alkaline phosphatase, serum
Calcium, total, serum
Creatinine, serum
Free testosterone, serum
Parathyroid hormone, serum
Phosphorus, serum
Thyroid-stimulating hormone, serum

Step Two: Nutrition Diagnosis

Food medication interaction
Underweight
Inadequate protein-energy intake
Evident protein-energy malnutrition
Inadequate bioactive substance intake
Inadequate mineral intake
Physical inactivity

Sample PES: Evident Protein Energy Malnutrition (NI-5.2):

Protein energy malnutrition as evidenced by temporal wasting, atrophy of lower extremity musculature, 12% weight loss from stable adult weight, and diet history indicating intake of approximately 50% of EPR and 75% of EER.

Step Three: Intervention

1. Increase caloric and protein density of food choices through the use of modular components.
2. Increase the number of meals to 5 small feedings.
3. Substitute nutrient-dense beverage for coffee.

Step Four: Monitoring and Evaluation

1. Patient will maintain current weight.
2. Patient will increase consumption of both energy and protein to 90% of estimated requirements (2,000–2,200 kcal/80–90 g protein) within two weeks.
3. Patient will decrease coffee intake to 2 c. per day.
4. Patient will add juice, milk, or milkshake to each meal.

WEB LINKS

National Osteoporosis Foundation (NOF): The NOF is a voluntary health organization based in Washington, D.C., that promotes education, awareness, training, and research on osteoporosis and bone health. It is a resource for information on osteoporosis for the general public and health professionals.

http://www.nof.org

National Institute of Arthritis and Musculoskeletal and Skin Diseases (NIAMS): The NIAMS, a division of the National Institutes of Health, supports research into the causes, treatment, and prevention of arthritis and musculoskeletal and skin diseases. It is a valuable source of authoritative and scientifically accurate information on various musculoskeletal diseases.

http://www.niams.nih.gov

Osteoporosis Canada: Osteoporosis Canada is a not-for-profit foundation serving people who are at risk of or who have osteoporosis and working with individuals and communities in the prevention and treatment of osteoporosis. It is a source of scientifically sound information on osteoporosis for the general public and for health professionals.

http://www.osteoporosis.ca

International Society for Clinical Densitometry (ISCD): The ISCD is a multidisciplinary, not-for-profit organization that provides a central resource for a number of scientific disciplines with an interest in bone mass measurement. It is a source of scientifically sound information on bone mass measurement.

http://www.iscd.org

International Osteoporosis Foundation (IOF): The IOF is an international non-governmental organization based in Berne, Switzerland, whose mission is to advance the understanding of osteoporosis and to promote prevention, diagnosis, and treatment of the disease worldwide. It is a source of scientifically sound information on osteoporosis for the general public and for health professionals, presented from an international perspective.

http://www.osteofound.org

END-OF-CHAPTER QUESTIONS

1. Describe osseous tissue and its composition.
2. List the four principal types of cells found in osseous tissue and describe their functions.
3. Briefly describe the hormonal control of bone metabolism.
4. What is osteoporosis? List risk factors for this disorder. Describe the diagnostic measurement of bone mineral density (BMD). Is it an appropriate measure of osteoporosis?
5. Describe the dietary and pharmacologic prevention and treatment of osteoporosis.
6. Describe Paget disease and its treatment.
7. What are the similarities and differences between osteoarthritis, rheumatoid arthritis, and gout? Describe the dietary and pharmacologic treatments for these disorders.

28

Metabolic Disorders

Elaina Jurecki, M.S., R.D.

Regional Metabolic Nutrition Coordinator

Kaiser Permanente Medical Center, Northern California

Joyce Wong, M.S., R.D.

Regional Metabolic Nutrition Coordinator

Kaiser Permanente Medical Center, Northern California

CHAPTER OUTLINE

Epidemiology and Inheritance
Pathophysiology of Impaired Metabolism
Diagnosis/Newborn Screening
Clinical Manifestations of Inborn Errors of Metabolism
Approaches to Treatment
Amino Acid Disorders
Urea Cycle Disorders
Mitochondrial Disorders
Disorders Related to Vitamin Metabolism
Disorders of Carbohydrate Metabolism
Galactosemia • Hereditary Fructose Intolerance • Glycogen Storage Diseases
Disorders of Fat Metabolism

Introduction and Definition

Inborn errors of metabolism are a group of diseases that affect a wide variety of metabolic processes. Disease results when there is defective processing or transport of small molecules such as amino acids, fatty acids, metals, or sugars in the body. For some of these small molecules, cells in the body possess enzymes that can catalyze synthesis of a given substrate when it is needed. The classical inborn error of metabolism is caused by a defect in the activity of one of these enzymes. The resulting disorder may lead not only to the accumulation of compounds with harmful effects, but also to deficiencies of substances that are essential for normal growth and development (Collins and Leonard 1985).

History

Sir Archibald Garrod first postulated that metabolic disorders were inherited biochemical blocks in normal metabolic pathways in 1908. Garrod used the term "inborn errors of metabolism" to describe these disorders of lifetime duration. He hypothesized that specific enzymes were synthesized under the direction of genes. This one-gene-one-enzyme relationship was later proven in 1940 by La Du et al. when they were able to identify the defective or missing enzyme resulting in an abnormal metabolic process in four different disorders. This led to the concept that genes controlled metabolism and that disease states were created by blocks in this metabolic flow that yielded accumulated precursors and deficient products. Garrod's observations were the beginning of human biochemical genetics, and he defined the principle of genetically determined biochemical individuality, noting that "no two individuals of a species are absolutely identical in bodily structure, neither are their chemical processes carried out on exactly the same line" (Garrod 1908). Consequently, there is considerable human variation in the structure and activity of these proteins in their role of synthesizing and breaking down compounds, but only a few individuals are so impaired that ingestion of the recommended daily allowances of nutrients will create severe disease (Packman 1986; Schmidt 1989).

Epidemiology and Inheritance

Over 200 genetic disorders have been reported in which there is toxicity, deficiency, or overproduction of normally occurring **substrates** and/or products of metabolic flow. Most inborn errors of metabolism are inherited through an autosomal recessive transmission of the affected gene, meaning that both parents, who usually have no clinical symptoms, must each pass the defective gene to the offspring for the disease to be manifested (see Chapter 11). With **autosomal recessive inheritance**, carrier parents have a 25% chance of having an affected child with each pregnancy. A few disorders are X-linked, meaning that the defective enzyme is produced from an abnormal gene found on the X chromosome. In these disorders, males are symptomatic, but females are rarely affected because they have a functioning gene on their other X chromosome.

The defect or deficiency of a needed substance at any stage of a metabolic reaction causes a partial or total block. Two individuals can be affected with the same metabolic disorder, but with different degrees of severity and time of onset of the disease. The mutant or absent gene (the allele, one of two or more forms of a gene at the same site in a chromosome) responsible for the metabolic defect is present from fertilization, but the appearance of its effects may not be seen immediately. Mutations—permanent, transmissible changes in the genetic material, usually in a single gene—can affect an enzyme and cause differences in degree of stability and activity of an enzyme. A less severe enzyme alteration may allow an individual to function adequately for years until a stressor such as infection, dehydration, increased protein intake, or a growth spurt results in an accumulation of toxic levels of the substance. Thus, severity of the clinical manifestations of the inborn error of metabolism can be described by time of onset. Examples of more severe onset of disease include rapid onset, progressive, and severe neonatal course, whereas milder forms include subacute juvenile onset and progression of a mild adult form. The classical form of the disorder is considered the most severe form of the disease. Intermediate, intermittent, variant, malignant, and benign are all additional qualifiers of this classification system relating to biochemical variability. This biochemical variability in presentation is called the heterogeneity of a genetic disorder (Packman 1986; Schmidt 1989).

In the United States, approximately 4% of individuals are born each year with a genetic or partial genetic disorder. Inherited metabolic diseases contribute significantly to this total. Approximately one out of every 1,000 children born will have an inborn error of metabolism, a relatively high cumulative frequency given the rarity of each individual disorder. Hence, a large tertiary care nursery may expect to deal with one or more such diseases yearly. With current diagnostic methods and mass screening programs, practitioners are finding many of these disorders to be significantly more common than previously believed. There are also many children and adults afflicted with disorders in metabolism yet to be identified (Wallen and Packman 1985; Cederbaum 1989; Elsas and Acosta 1999).

The diagnosis of an inborn error of metabolism is devastating to a family, for it represents not only an inherited disorder ("something I gave to my child"), but also a chronic disorder, often associated with a shortened life span. Despite major diagnostic achievements, patients and their families are often disappointed by the absence of specific therapies and by the investigational nature of many of the treatment options. Modifications of the diet can alleviate the manifestations of many of these disorders. In a large number of patients, however, irreversible damage has already occurred by the time symptoms appear. Optimal management of these disorders depends on identifying affected subjects while they are presymptomatic or before irreversible disease has occurred (Elsas and Acosta 1999).

substrates—any substance that an enzyme acts on to make a product

autosomal recessive inheritance—inheritance of a trait as the result of inheriting a recessive gene for a particular trait from each parent

binding site—active site of the enzyme that binds to and acts on a particular substrate

feedback inhibition—regulatory mechanism to limit production of certain substrates, which when present in significant amounts will limit the enzyme involved in making it to decrease the amount of this substrate being produced

Pathophysiology of Impaired Metabolism

Impaired metabolism of nutrients can occur as a direct result of deficient or absent enzyme activity. It can also be caused by a defective gene that results in a change to the **binding site** of cofactors that are needed for a specific enzyme to function properly. Figure 28.1 shows that as a result of the defective enzyme 3, the precursors, both immediate (B) and distant (A), are unable to be metabolized and can build up. The precursors can accumulate as a direct result of the block or as a result of **impaired feedback inhibition** related to the inability to produce the end substrate. Toxic metabolites can be produced from the precursors that are built up as a result of the blocked pathway (enzyme 3).

FIGURE 28.1 Schematic of Enzymatic Pathway

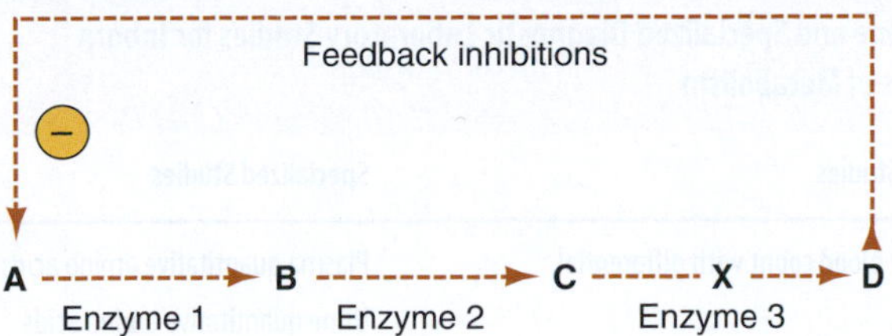

In other disorders, the enzymatic block can result in a deficiency of much-needed end products. In Figure 28.1, the absence of enzyme 3 leads to the inability to produce substrate D. The inability to produce vital nutrients such as glucose, essential amino acids, or ketone bodies can lead to serious consequences such as hypoglycemia, coma, growth failure, kidney dysfunction, and even death.

Dietary interventions for treating many of these disorders consist of rigid restrictions of protein, fat, or carbohydrates that would in turn limit intake of many micronutrients. Therefore, a secondary consequence of these metabolic defects can be nutritional deficiencies. Nutrition therapy for these disorders is aimed at not only treating the underlying disorder, but at also providing sufficient nutrients to promote adequate growth and development. Frequent assessment of anthropometric, biochemical, and clinical indices is required to ensure optimal care of these patients (Elsas and Acosta 1999).

Diagnosis/Newborn Screening

Newborn babies are required to perform myriad complex physiologic tasks. They must assume independent control of body temperature, blood glucose levels, toxic metabolites, intake and assimilation of nutrients, and many other physical processes. Any significant illness in a newborn impedes these adaptive processes. The metabolic diseases greatly complicate the clinical picture for ill neonates, because symptoms of one disease may closely resemble those of others that have quite different etiologic and pathogenic mechanisms. Waiting until sepsis and other more common causes of illness are ruled out before initiating a specific diagnostic evaluation is inadvisable, as is indiscriminate study of all ill newborns for metabolic disorders (Cederbaum 1989).

The detection of inborn errors of metabolism is complicated by the fact that the symptoms are nonspecific and can be similar from one inborn error of metabolism to the next. Some of the symptoms of these disorders include failure to feed, vomiting, hypotonia, hypertonia, seizures, and lethargy progressing to coma. These symptoms can also be seen in neonates suffering from infection, cardiopulmonary dysfunction, intracranial hemorrhage, congenital structural abnormalities of the brain, or trauma. Most neonates with an inborn error of metabolism look normal at birth and have no gross physical anomalies. Clinical detection of an inborn error of metabolism is elusive. Genetic screening for these metabolic disorders will test for gene products or resulting metabolites with the aim of identifying disorders caused by mutant genes. Nonselective neonatal screening, or newborn screening, is associated with the screening of all newborns for a limited number of the more common inborn errors. The benefit of testing all infants in a population is to identify a neonate suspected of having an inborn error of metabolism so that treatment can start before the infant has signs of disease. Selective neonatal screening is the testing of an individual known to be at increased risk for a genetic disorder. Testing of siblings in a family with a child diagnosed with an inborn error of metabolism or confirmatory testing of a positive neonatal screening test are examples of selective screening (Cederbaum 1989; Burton 1998).

In the United States, all states have screening programs that test newborns for various metabolic disorders at 24 to 48 hours of age. All states screen for phenylketonuria, a disorder in amino acid metabolism. Beyond that, there is great variability between states in the types and number of disorders screened. A new methodology, **tandem mass spectroscopy**, allows clinicians to screen for over 30 disorders by analyzing metabolites in a blood spot collected on a filter paper from the newborn. This technique is relatively inexpensive and provides the potential to screen for a large number of inborn errors of metabolism pre-symptomatically. Many states have already included this procedure in their newborn screening program, and many more states are anticipated to do so in the near future (Cederbaum 1989; Schmidt 1989).

Clinical Manifestations of Inborn Errors of Metabolism

Major and less common clinical manifestations of inborn errors of metabolism are summarized in Table 28.1. These symptoms usually appear 24 hours or more after birth and may be attributed to ingestion of the precursor substrate of the defective enzyme. It is more likely, however, that the primary source of the accumulated precursor is endogenous, resulting from the normal catabolism that accompanies transfer from the intrauterine to the extrauterine environment. This response can also occur later in life as a result of episodes causing endogenous protein catabolism such as infection, fasting, or fever.

Any neonate or infant who presents with a history of postnatal onset of neurologic dysfunction after an interval

tandem mass spectroscopy—the methodology used to detect a large number of organic acid compounds on a filter paper blood spot for diagnosing an inborn error of metabolism

TABLE 28.1

Major Clinical Manifestations of Inborn Errors of Metabolism

Common Signs and Symptoms

Neurologic Signs:	*Gastrointestinal Signs:*
• Poor suck	• Poor feeding
• Lethargy (progressing to coma)	• Vomiting
• Abnormalities of tone (hypertonia and hypotonia)	• Diarrhea
• Loss of reflexes	• Reflux
• Seizures	
Respiratory Signs:	***Organ Dysfunction:***
• Hyperpnea	• Hepatomegaly
• Respiratory failure	• Hepatic dysfunction
• Tachypnea	• Cardiomegaly
• Apnea	• Cardiomyopathy
Rarer Findings:	
• Abnormal smell	• Renal stones
• Abnormal hair	• Ectopia lentis
• Self-mutilation	• Corneal clouding
• Blood in urine	• Arthritis
• Ataxia	• Dystonia
• Myopathy	• Cirrhosis
• Dysmorphic features	

Adapted from: Burton BK. Inborn errors of metabolism in infancy: a guide to diagnosis. 1998. *Pediatr.* 102(6):1–15, and Cederbaum SD. Diagnosing metabolic disease in the neonate. 1989. *Metabolic Currents.* 2(1):1–7.

of good health should raise the suspicion of a metabolic disorder. Neonates with a metabolic disorder generally show acute central nervous system symptoms, including generalized or partial seizures. Infants and older children with metabolic disorders present with neurologic symptoms that are characteristically intermittent in their severity. The child may show poor growth or failure to thrive, significant developmental delay, and/or specific neurologic deficits such as ataxia or visual dysfunction (Burton 1998).

Although the symptoms of inborn errors of metabolism are nonspecific and sometimes subtle, a few of these disorders will have blatant signs. There can be a striking, unusual odor present in the diaper such as burnt sugar, sweaty feet, or a musty odor. These odors can also present in the ear wax and the sweat of the older infant and child. Even without other indications, a mother's observation that she thinks the infant "smells funny" should be a clue to further investigate an inborn error of metabolism (Schmidt 1989).

ketosis—an abnormally elevated concentration of ketone bodies in the body tissues and fluids

skin fibroblasts—connective tissue cells found in the skin

hyperpnea—rapid breathing

ectopia lentis—displaced lenses in the eye

TABLE 28.2

Routine and Specialized Diagnostic Laboratory Studies for Inborn Errors of Metabolism

Routine Studies	Specialized Studies
Complete blood count with differential	Plasma quantitative amino acids
Urinalysis	Urine quantitative amino acids
Blood gases	Plasma carnitine profile
Serum electrolytes and blood pH	Blood acylcarnitine profile
Blood glucose	Urine organic acids
Plasma ammonia	
Urine reducing substances	
Urine ketones	
Blood lactate and pyruvate	

Adapted from: Burton BK. Inborn errors of metabolism in infancy: a guide to diagnosis. 1998. *Pediatr.* 102(6):1–15, and Cederbaum SD. Diagnosing metabolic disease in the neonate. 1989. *Metabolic Currents.* 2(1):1–7.

The lack of specificity in the clinical manifestations of inborn errors of metabolism puts greater weight on the laboratory evaluation in making the correct diagnosis. Laboratory studies, as identified in Table 28.2, fall into two categories: routine studies available in many clinical laboratories, and specialized studies only done at laboratories set up to do these tests. The routine studies can identify the child with hypoglycemia, a disturbance of acid-base balance, hyperammonemia, or **ketosis**. A sequence of selected screening tests, performed in consultation with a biochemical geneticist, follows the finding of abnormal routine chemistries.

The more directed analysis (for amino acids and organic acids) can be done on one or more specimens of serum, plasma, and/or urine. A urine organic acid analysis, a plasma and urine amino acid screen, a plasma carnitine profile, and an acylcarnitine profile are sent to laboratories specializing in biochemical genetic disorders. The enzyme defect itself can often be identified in **cultured skin fibroblasts** (grown from a skin biopsy) or peripheral blood leukocytes. If the death of the child is imminent, the geneticist and pathologist should be consulted even prior to death, to prepare for the retrieval of critical tissues for diagnostic purposes, with parental consent. Because an inborn error of metabolism may be the cause, and such a diagnosis is of major importance for genetic counseling, efforts to make a diagnosis post-mortem should be made (Burton 1998; Sauderbray et al. 2002).

Approaches to Treatment

Acute Therapy

Some metabolic disorders may appear to be benign, though many others cause a variety of severe problems beginning early in life unless an effective treatment is quickly initiated. Any metabolic stress that results in endogenous catabolism, such as an infection or surgery, can lead to the accumulation of toxic metabolites; these metabolites will lead to rapid illness and require acute management. The correction of acid-base status and hydration are essential and of immediate importance. Maintenance of an adequate caloric intake is necessary for prevention of tissue catabolism and must be achieved by parenteral or oral routes of administration. Offending metabolites, such as protein for urea cycle defects and aminoacidopathies, should be restricted. Long-term management of many inborn errors of metabolism is quite distinctive from the acute management (Packman 1986; Ogier de Bawny 2002).

Chronic Therapy

Approaches to the chronic management of metabolic disorders is based on one or more of the therapy options (Elsas and Acosta 1999) described in this section.

Restriction of Precursors This treatment is aimed at limiting the substrate or substrates prior to the block. In Figure 28.1, the block in enzyme 3 requires the restriction of the precursors A and B. This usually results in restriction of one of the macronutrients. In the case of amino acid disorders, this treatment involves the restriction of dietary protein intake, since proteins found in natural foods contain varying amounts of each amino acid. In order to limit the individual offending amino acid(s), a dietary protein restriction is typically implemented. In disorders of fat metabolism, total fat intake is generally restricted, because all fats will eventually be oxidized to the point where the enzymatic block occurs in the body. In certain disorders of carbohydrate metabolism, sources of foods containing particular monosaccharides, such as fructose, must be restricted.

Replacement of the End Products In Figure 28.1, with the block of enzyme 3, substrate D needs to be replaced. Substrates that should have been formed from the enzymatic reactions are still needed for energy production, protein turnover, feedback inhibition of metabolic pathways, and/or production of other substrates. In the case of metabolic blocks in these enzymatic reactions, alternative means of achieving the end goal of the reaction must be provided. In glycogen storage disease type I, for example, the enzymatic block results in an inability to produce glucose. Consequently, providing a continuous source of glucose through frequent feedings is the therapy for this disorder. Another example is seen in the amino acid disorder, homocystinuria, where the amino acid, homocysteine, is unable to be converted to cysteine and consequently cysteine needs to be supplemented in the diet.

Providing Alternate Substrates for Metabolism In the case of long-chain fatty acid oxidation defects, medium chain triglycerides (MCT)—fatty acids that are 6 to 8 carbons in chain length—are used as a supplemental energy source as they are able to bypass the metabolic defect and be metabolized for energy in place of the longer-chain fats found in the food supply. In glycogen storage disease type III, protein is used as an alternative source of glucose production, via the **alanine shunt**, since the use of carbohydrates is compromised due to the enzymatic block.

Use of Scavenger Drugs to Remove Toxic By-Products Pharmaceutical products such as sodium benzoate and sodium phenylbutyrate are needed to remove the excess nitrogen that is produced as a result of a block in the urea cycle. **Carnitine** is used in many organic acid disorders to remove and detoxify waste products that accumulate in disorders affecting the metabolism of proteins, carbohydrates, and fats. (See Table 28.16 at the end of this chapter for drug-nutrient interaction information for these and other medications/supplements used in the treatment of metabolic disorders.)

Supplementation of Vitamins or Other Cofactors Use of certain vitamins, at pharmacologic doses, can in certain instances increase the enzymatic activity or facilitate biochemical processes in various inborn errors of metabolism. For example, the five respiratory chain complexes in the mitochondria use many different vitamins, not only as **cofactors** but also as electron receptors. Methylmalonic acidemia and maple syrup urine disease (MSUD) are both examples of conditions that have a "vitamin responsive" form of that disorder where the vitamin administered can increase residual enzymatic activity. These disorders are discussed further in subsequent sections.

alanine shunt—process that allows for the production of glucose from protein sources by the conversion of the amino acid alanine to pyruvate to glucose in the liver; the glucose produced can then be used in the muscle to generate additional alanine

carnitine—substrate needed for the normal metabolism of fat for energy. It is mostly found in beef and lamb but can be synthesized endogenously from the amino acids l-lysine and l-methionine

cofactors—vitamins or other nutrients needed for the proper function of certain enzymes

Several examples of more common inborn errors of metabolism will be discussed in the remainder of the chapter; references are available with more detailed listings of less common metabolic disorders (Scriver et al. 2001). Examples of disorders to be discussed are amino acid disorders; urea cycle disorders; mitochondrial disorders; disorders related to vitamin metabolism; disorders of carbohydrate metabolism, including galactosemia; hereditary fructose intolerance; glycogen storage diseases; and disorders of fat metabolism.

Amino Acid Disorders

Epidemiology, Etiology, and Clinical Manifestations

Amino acid disorders include conditions affecting the metabolism of a single amino acid, such as phenylketonuria (PKU) or isovaleric acidemia (IVA) as well as disorders involving multiple amino acids, such as maple syrup urine disease (MSUD) and propionic acidemia (PPA). Table 28.3 lists some of the more common disorders of amino acid metabolism and their affected amino acids. Discovery of most of these conditions is relatively recent, with most disorders discovered less than 50 years ago. Incidence of these disorders can vary significantly, with the most common amino acid disorder, PKU, having an incidence of 1 in 10,000 births. The rarer disorder, methylmalonic academia, however, has an incidence of only 1 in 50,000 births. Incidence of these disorders can vary based on ethnicity. Those populations that tend to be more homogenous have been found to have higher rates of certain disorders. For example, MSUD has been detected at an incidence of 1 in 290,000 births in state screening programs. In the Mennonite Community, a homogeneous group in Pennsylvania, there is a much higher prevalence for this condition at 1 in every 790 births (Nyhan and Pinar 1998).

TABLE 28.3

Amino Acid Disorders and Affected Amino Acid(s)

Metabolic Condition	Affected Amino Acid(s)
Glutaric acidemia	Lysine, tryptophan
Homocystinuria	Methionine
3-hydroxyisobutyric aciduria	Valine
Isovaleric acidemia	Leucine
Maple syrup urine disease	Leucine, isoleucine, valine
3-methylcrotonyl glycinuria	Leucine
Methylmalonic acidemia	Valine, isoleucine, methionine, threonine
Phenylketonuria	Phenylalanine
Propionic acidemia	Valine, isoleucine, methionine, threonine
Tyrosinemia	Tyrosine, phenylalanine

Treatment recommendations have changed over time and can still vary quite a bit between different parts of the world. This is attributed to the rarity of these conditions, which limits patient populations and makes it difficult to develop consensus treatment recommendations. With screening programs, treatments are now able to be instituted earlier, and subsequent outcomes have been better. Many of these patients can now live into adulthood, and it is even possible for some affected women to have successful pregnancies. However, those patients with late-treated or untreated amino acid disorders can have serious complications, including mental retardation, coma, and even death. In general, amino acid disorders are usually characterized by infants who are initially poor feeders and hypotonic. Some of these patients may require long-term gastrostomy feedings due to their poor feeding skills. Many of the milder forms of these disorders may not present in individuals until after the introduction of infant formula or milk, both of which have higher protein content compared to breast milk.

Phenylketonuria Phenylketonuria (PKU) is the most common amino acid disorder. The deficiency or absence of the phenylalanine hydroxylase enzyme (as shown in Figure 28.2) leads to the inability to convert the amino acid phenylalanine into the amino acid tyrosine. Consequently, tyrosine becomes a conditionally essential amino acid for individuals with PKU. As a result of the buildup of phenylalanine, phenylacetic and phenylpyruvic acid levels also increase.

An individual with PKU who goes untreated will most likely be mentally retarded with severe behavioral problems that result in the need for institutionalization for the majority of her or his adult life. Untreated PKU can also lead to neurologic abnormalities, seizures, and eczema. Affected individuals are typically found to have a musty or mousy odor because of the accumulation of phenylketones in their urine. Decreased pigmentation has been related to the inhibition of the tyrosinase enzyme by phenylalanine. Hyperphenylalaninemia has been found to be toxic to the brain, and demyelination of white matter has been seen on MRI (magnetic resonance imaging) results from affected individuals.

Accumulation of phenylketones, including phenylpyruvate, phenyllactic acid, phenylacetic acid, and phenylacetylglutamine, create an abnormal chemical milieu that is believed to cause the manifestations of this disease. Decreased production of serotonin, epinephrine, norepinephrine, and dopamine are presumably caused by inhibition in their synthesis as a result of the accumulation of phenylketones. The metabolites that accumulate in PKU also inhibit the production of 4-aminobutyric acid (GABA) in the brain. The exact mechanism of how the elevation of these phenylketones, the deficiency of tyrosine, and the additional compounds just mentioned result in these complications is not known at this time.

Previously, health care providers allowed for the discontinuation of dietary treatment when a child reached school age, since the brain was believed to have been fully devel-

FIGURE 28.2 Enzymatic Defect in Phenylketonuria

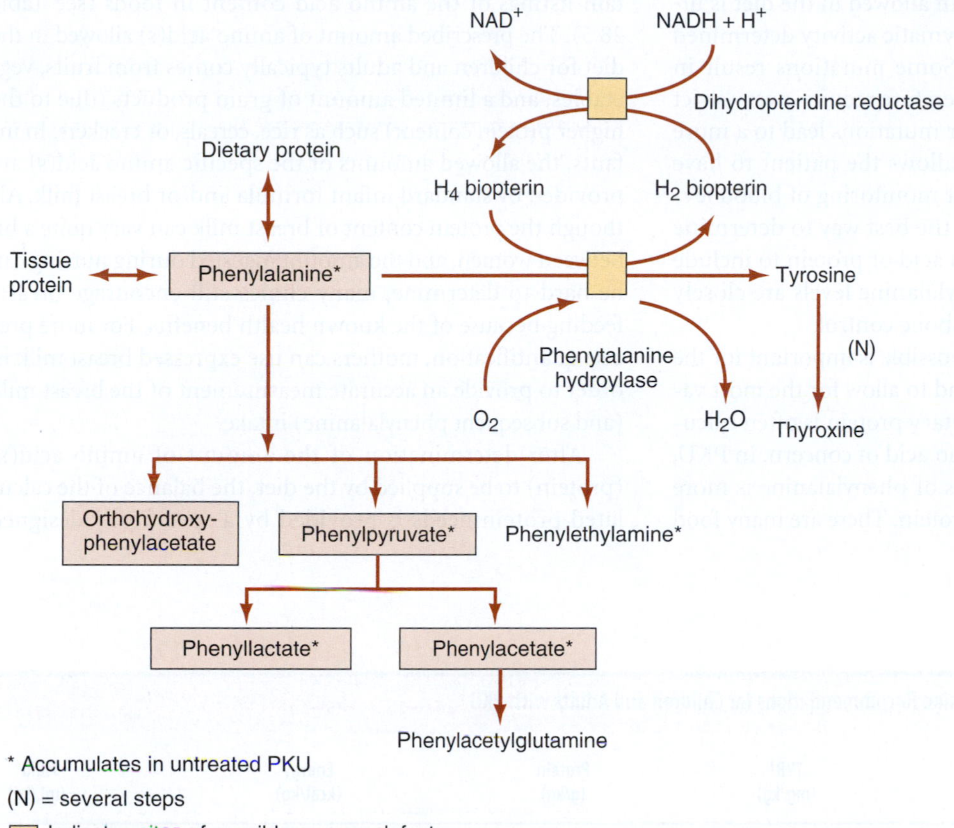

Source: Modification of Figure A, p. 1 of Acosta PA, Yannicellis S, *Nutrition Support Protocols,* 4e. Ross Metabolic Formula System; © 2001.

oped, and hence not impacted by the accumulation of these phenylketones. Subsequent research revealed that problems such as phobias and depression developed in those individuals who were off the PKU diet for extended periods of time (Ishimaru et al. 1993; Fisch et al. 1995). Today, most clinics recommend dietary treatment for life in order to optimize the cognitive and psychosocial abilities of these patients and to improve their overall quality of life.

Whereas the classic form of PKU is tied specifically to the activity of the phenylalanine hydroxylase enzyme, some of the more milder, variant forms of PKU are thought to be related to the inability of the binding site of the enzyme to recognize and utilize the coenzyme **tetrahydrobiopterin** effectively. More research looking at this substrate's ability to enhance the phenylalanine hydroxylase enzyme activity may provide a new therapy option in the future (Schuett 2003). Individuals with PKU who are well treated can expect to lead normal and complete lives without a shortened life span.

Nutrition Interventions

Treatment of amino acid disorders is based on the restriction of dietary protein, which is the source of the offending amino acids. Consequently, protein needs must be met by using a synthetic formula containing all the essential amino acids except for the specific offending amino acid(s). The main principle of dietary treatment is to allow the maximum amount of dietary protein in the diet, in order to promote growth, while still maintaining laboratory levels of the specific amino acid(s) in the desired range. Target blood phenylalanine levels for PKU can vary at different medical centers, but the National Institute of Health consensus guidelines from the year 2000 recommend blood levels between 2–6 mg/dL for children until 12 years of age and a more relaxed recommendation of 2–15 mg/dL for those over 12 years of age (National Institute of Child Health and Development 2000).

A diet prescription is developed by initially assessing the protein and kcal needs of the patient. Although there are

tetrahydrobiopterin—the cofactor needed to stabilize the enzyme phenylalanine hydroxylase

guidelines available for protein requirements by age (see Table 28.4), the amount of protein allowed in the diet is ultimately based on the level of enzymatic activity determined by the specific gene mutation. Some mutations result in almost no enzyme activity and therefore require a very strict protein restriction, whereas other mutations lead to a more mild form of the disorder that allows the patient to have a greater protein intake. Frequent monitoring of blood levels after the diagnosis is made is the best way to determine the amount of the specific amino acid or protein to include in the diet. In PKU, blood phenylalanine levels are closely monitored to help maintain metabolic control.

Allowing as much protein as possible is important for the promotion of adequate growth and to allow for the most variety in a very restrictive diet. Dietary protein is often calculated in terms of the specific amino acid of concern. In PKU, for example, counting milligrams of phenylalanine is more precise than counting grams of protein. There are many food tables and nutrition software applications available that contain listings of the amino acid content in foods (see Table 28.5). The prescribed amount of amino acid(s) allowed in the diet for children and adults typically comes from fruits, vegetables, and a limited amount of grain products (due to the higher protein content) such as rice, cereals, or crackers. In infants, the allowed amounts of the specific amino acid(s) are provided by standard infant formula and/or breast milk. Although the protein content of breast milk can vary quite a bit between women, and the amount ingested during nursing can be hard to determine, many clinics still encourage breastfeeding because of the known health benefits. For more precise quantification, mothers can use expressed breast milk in order to provide an accurate measurement of the breast milk (and subsequent phenylalanine) intake.

After determination of the amount of amino acid(s) (protein) to be supplied by the diet, the balance of the calculated protein needs is provided by a specifically designed

TABLE 28.4

Protein, Kcal, Phenylalanine, and Tyrosine Recommendations for Children and Adults with PKU

Age	PHE[1,2] (mg/kg)	TYR[1] (mg/kg)	Protein (g/kg)	Energy (kcal/kg)	Fluid[3] (mL/kg)
Infants					
0 to < 3 mo	25–70	300–350	3.50–3.00	120 (145–95)	160–135
3 to < 6 mo	20–45	300–350	3.50–3.00	120 (145–95)	160–130
6 to < 9 mo	15–35	250–300	3.00–2.50	110 (135–80)	145–125
9 to < 12 mo	10–35	250–300	3.00–2.50	105 (135–80)	135–120
Girls and Boys	(mg/day)	(g/day)	(g/day)	(kcal/day)	(mL/day)
1 to < 4 yr	200–400	1.72–3.00	≥ 30	1,300 (900–1,800)	900–1,800
4 to < 7 yr	210–450	2.25–3.50	≥ 35	1,700 (1,300–2,300)	1,300–2,300
7 to < 11 yr	220–500	2.55–4.00	≥ 40	2,400 (1,650–3,300)	1,650–3,300
Women					
11 to < 15 yr	250–750	3.45–5.00	≥ 50	2,200 (1,500–3,000)	1,500–3,000
15 to < 19 yr	230–700	3.45–5.00	≥ 55	2,100 (1,200–3,000)	1,200–3,000
≥ 19 yr	220–700	3.75–5.00	≥ 60	2,100 (1,400–2,500)	2,100–2,500
Men					
11 to < 15 yr	225–900	3.38–5.50	≥ 55	2,700 (2,000–3,700)	2,000–3,700
15 to < 19 yr	295–1,100	4.42–6.50	≥ 65	2,800 (2,100–3,900)	2,100–3,900
≥ 19 yr	290–1,200	4.35–6.50	≥ 70	2,900 (2,000–3,300)	2,000–3,300

[1] Modify prescription based on frequency obtained blood and/or plasma values and growth in infants and children and frequently obtained plasma values and weight maintenance in adults.

[2] PHE requirements of premature infants may be greater than highest value noted.

[3] Under normal circumstances, offer minimum of 1.5 mL fluid to neonates and 1.0 mL to children and adults for each kcal ingested.

Reprinted with permission from: Table 1-1, p. 12. Acosta PB, Yannicelli S. *Nutrition support protocols.* 4th ed. Columbus (OH): Ross Products Division of Abbott Laboratories; 2001.

metabolic formula (e.g., Phenex, Phenylfree). The metabolic formula is usually a powder that contains all of the amino acids except for the one(s) that are not able to be metabolized appropriately. The powder formula can contain a carbohydrate source, a fat source that includes essential fatty acids, and vitamins and minerals. Many of the metabolic formulas also contain carnitine, utilized to scavenge toxic metabolites in certain amino acid disorders.

Formulas containing little carbohydrate or fat have recently been developed in an attempt to minimize the caloric contribution of the formulas. It is important to determine whether extra kcal are needed to promote satiety and to decrease the intake of foods containing the offending amino acid(s). On the other hand, as more individuals are successfully treated and reaching adulthood, lower caloric needs are indicated due to decreasing rates of growth. Higher-kcal formulas can result in excessive weight gain in these individuals. Newer formulas that contain little or no vitamins and minerals have also been developed in order to improve the palatability of the product. These products need to be used in conjunction with a vitamin/mineral supplement. Additional calcium may be needed if using an unfortified product, given the restricted use of dairy products and inadequate amount of calcium in a general multiple vitamin/mineral supplement. Formula bars and tablets are now available for certain disorders in order to meet the demands of the more active, older patients who do not want to prepare a liquid formula each day.

Intake of small amounts of metabolic formula throughout the day provides for the best utilization of the synthetic amino acids. Studies have shown increased nitrogen loss when the formula is consumed 1 to 2 times per day versus 4 to 5 times per day (MacDonald et al. 2003). Although an essential part of the treatment, the coverage of metabolic formulas by various health insurance plans can vary considerably between states.

Calculated protein needs are determined by weight in a growing child and are higher than the standard recommended dietary intakes (RDI) for age due to the synthetic nature of the protein in the formula (Kindt and Halvorsen 1980). Although guidelines are available to help estimate protein and amino acid requirements, needs are best determined

TABLE 28.5

Phenylalanine Content of Selected Foods

Food	Weight (g)	Approximate Measure	PHE (mg)	TYR (mg)	Protein (g)	Energy (kcal)
Cereals, Cooked, measure after cooking						
Corn grits						
instant						
cheese flavor	36	¼ packet	36	29	0.7	27
plain	34	¼ packet	27	22	0.5	20
regular, quick (plain)	45	3 Tbsp	33	28	0.6	28
Cream of rice	81	⅓ cup	30	40	0.7	42
Cream of Wheat						
instant	30	2 Tbsp	30	18	0.6	19
Mix'n Eat						
flavored	38	¼ packet	33	20	0.6	33
plain	36	¼ packet	37	22	0.7	26
quick	30	2 Tbsp	25	14	0.4	16
regular	30	2 Tbsp	30	18	0.5	19
Oats, regular, quick, and instant	20	1 Tbsp + 1 tsp	28	18	0.5	12
Cereals, Ready to Eat						
All Bran®	5	1 Tbsp	29	22	0.8	13
Apple Jacks®	9	⅓ cup	28	18	0.5	36
Cap'n Crunch®	12	⅓ cup	33	25	0.6	51
Cheerios®	4	3 Tbsp	35	23	0.6	17
Cinnamon Toast Crunch®	19	½ cup	30	18	0.7	80

Reprinted with permission from: Table 1-3, p. 16. Acosta PB, Yannicelli S. *Nutrition support protocols.* 4th ed. Columbus (OH): Ross Products Division of Abbott Laboratories; 2001.

by frequent monitoring of growth and laboratory indices (Ney et al. 1985; Parsons et al. 1990). Protein and amino acid needs taper per unit body weight as the child ages, due to decreased growth velocity. The kcal supplied by the metabolic formula and the allowed amount of dietary protein are subtracted from the calculated caloric needs of the individual. Caloric needs vary by disorder and are sometimes estimated higher to minimize catabolism. Calculated caloric requirements may also be lower than the DRI for age if the disease results in hypotonia and inactivity, leading to decreased energy use (Feillet et al. 2000). The balance of the caloric needs are provided by a protein-free kcal supplement and/or by the inclusion of specially designed low-protein foods. There are many companies that can supply these foods, but as with the metabolic formulas, coverage of these products by health insurance plans and state agencies varies considerably by state.

During illness or metabolic crisis, the main goals of therapy are to (1) decrease intake of offending amino acid(s) by eliminating or reducing the intake of natural proteins, (2) provide sufficient kcal to prevent catabolism, and (3) provide enough fluids to flush out any toxic metabolites. Aggressive measures such as temporary nasogastric feedings, parenteral nutrition, and/or intravenous hydration may be necessary to achieve these goals. It is advisable to start some metabolic formula feedings (not containing the offending amino acids) after 2 days of illness to prevent further catabolism and the creation of essential amino acid deficiencies. There are several companies throughout the United States that specialize in the preparation of custom parenteral amino acid solutions individualized for the various aminoacidopathies. Treatment for any associated illness that leads to **metabolic decompensation** must be started immediately. Many patients are provided with emergency care protocols to help them to receive prompt treatment and prevent further metabolic decompensation.

Nutritional Concerns

Given the restricted nature of the diet, patients with amino acid disorders are at risk for nutritional deficiencies despite the fact that the majority of the metabolic formulas used are fortified with adequate amounts of vitamins and minerals. Most patients who are drinking the prescribed amount of metabolic formula should have their nutritional needs met. However, since the majority of their diet comes from synthetic sources, there can still be problems associated with the absorption and utilization of these nutrients. For example, there have been several reports of selenium deficiencies in PKU patients on their prescribed diet (Greeves et al. 1990).

Nutritional deficiencies become a concern for those patients taking a suboptimal amount of metabolic product. They are not only receiving insufficient protein for growth, but may also be showing signs of other nutritional deficiencies such as anemia, neuropathy (peripheral nervous system damage), or osteopenia (decreased bone mass) associated with such nutrients as iron, vitamin B_{12}, and calcium. Growth retardation, associated with the diet restriction and metabolic imbalances, has been noted in patients with PKU and propionic acidemia (Wolf et al. 1981; Dobbelaere et al. 2003). The development of newer, more palatable formulas that are incomplete in nutrient composition necessitates frequent monitoring to ensure that patients are receiving adequate intakes of kcal, protein, essential fats, vitamins and minerals. Patients should have their laboratory studies evaluated periodically for nutritional indices such as albumin, total protein, complete blood count, transthyretin, essential fatty acids, calcium, magnesium, and zinc (see Box 28.1).

Bone status can also be a concern in certain amino acid disorders. Results from several studies have shown decreased bone mineral density in PKU patients despite adequate intakes of calcium and magnesium (Allen et al. 1994; Al-Qadresh et al. 1998). The explanation for this observation is currently unclear. Studies have indicated that compliance could play a part in the decreased bone mineralization, because it was noted that decreased densities were more significant in patients greater than 10 years of age with compromised metabolic control (Al-Qadresh et al. 1998; Yanicelli and Medeiros 2002).

Another nutritional concern is the possibility of amino acid deficiencies. It is important to provide enough essential amino acids in the diet for protein synthesis and turnover. The diet prescription aims to provide just enough amino acids to meet this need. Overrestriction of the diet may occur because of the parents' desire to achieve metabolic stability, resulting in low blood levels rather than levels in the recommended range. This overrestriction can lead to anorexia and/or increased catabolism, both of which can compromise metabolic control. Overrestriction of amino acids may also occur secondary to insufficient advancement of the diet to account for growth. This may happen if patients fail to keep appointments with the clinic or get regular blood testing.

In some amino acid disorders, low levels of certain offending amino acids can develop as a result of attempts to achieve the desired level of one particular amino acid. In MSUD, for instance, there is a metabolic block affecting the metabolism of the essential amino acids leucine, isoleucine, and valine. When leucine is restricted to maintain normal plasma levels, then intakes of the other two amino acids (valine and isoleucine) will also be limited and plasma levels will be subsequently low. **Periorificial acrodermatitis**

decompensation—inability to maintain metabolic balance, leading to derangements in biochemical and clinical parameters

periorificial acrodermatitis—disease of the skin surrounding the mouth area

BOX 28.1 CLINICAL APPLICATIONS

Case Example for Phenylketonuria

Mary is a 4-year-old girl with PKU. Weight = 17 kg, Height = 100 cm. She is currently taking 150 grams of Phenex1, a metabolic formula powder, which is mixed with water to make 32 oz. of liquid formula. This provides her with 22.5 g of (phenylalanine-free) protein and 720 kcal. The remainder of her protein and kcal are provided by her diet. Mom reports that Mary is often not hungry because her formula fills her up. She also reports difficulty getting Mary to eat the recommended amount of phenylalanine.

Flow Sheet

Month	Phe Level	Caloric Intake	Protein Intake	Phe Intake
January	4.1 mg/dL	1,400	30 g	325 mg
February	3.8 mg/dL	1,450	29.5 g	315 mg
March	5.3 mg/dL	1,350	29.5 g	315 mg
April	7.2 mg/dL	1,400	29.5 g	320 mg

Estimated needs (based on Table 28.4):

- 30–35 g protein
- 1,400–1,700 kcalories
- 310–320 mg phenylalanine

Discussion:

Mary's phenylalanine intake has been generally within the prescribed range, but blood phenylalanine levels have increased over the last few months. Mary's weight is between the 50–75th percentile; her height is at the 25–50th percentile growth curves. Kcal and protein intakes are on the lower end of recommended range.

The increase in blood levels could be related to increasing protein needs due to growth. Mary would benefit from an increased protein intake, but if she consumed more protein from her diet, she would also receive more phenylalanine. Consuming more formula would likely further decrease her appetite and daily food intake. This could lead to compromised metabolic control since phenylalanine is an essential amino acid, and a lower than required intake could lead to catabolism and cause an increase in her blood level.

Plan:

Change Mary's metabolic formula to one that is more concentrated in protein and provides fewer kcal. This type of formula is designed for older children and adults and contains greater concentrations of protein along with more vitamins and minerals to better meet her needs. The plan would be to transition Mary to the more concentrated formula to provide her more protein in less volume. The new formula Phenex 2 will contain 350 kcal and 25 grams of protein in 85 grams of powder. Mixed with water, this amount of powder will make 16 oz. of formula. If Mary keeps her dietary protein intake restricted at 7 grams, she will now be receiving 32 grams of protein each day. A decrease in her phenylalanine intake to 300–310 mg each day will also promote improved blood levels. The decreased volume and kcal from formula will help to stimulate Mary's appetite, but since she has to also slightly decrease her dietary phenylalanine intake, she will have to include more special low-protein foods in her diet. These specially designed foods often are fairly high in kcal and will help ensure she meets her caloric needs while not exceeding her phenylalanine prescription. See the adjusted meal plan shown in the table.

Original Meal Plan			Adjusted Meal Plan		
Food/beverage	Phe (mg)	kcal	Food/beverage	Phe (mg)	kcal
Breakfast			*Breakfast*		
¾ cup Froot Loops	56	82	½ cup Froot Loops	37	55
1 banana	43	105	½ cup low-protein cereal loops	1	52
8 oz. formula	0	180	1 banana	43	105
			4 oz. formula	0	88
Lunch			*Lunch*		
1 slice low-protein bread	15	100	2 slices low-protein bread	30	200
1 slice low-protein cheese	30	60	1 slice low-protein cheese	30	60
20 goldfish crackers	36	52	20 low-protein pretzels	34	112
½ cup canned peaches	17	29	½ cup canned peaches	17	29
1 tsp. mayonnaise	2	40	1 tsp. mayonnaise	2	40
8 oz. formula	0	180	4 oz. formula	0	88
			Juice box	0	100
Dinner			*Dinner*		
1 cup low-protein pasta	8	150	1 cup low-protein pasta	8	150
¼ cup spaghetti sauce	27	25	¼ cup spaghetti sauce	27	25
¾ cup broccoli	49	16	½ cup broccoli	33	11
8 oz. formula	0	180	4 oz. applesauce	6	97
			4 oz. formula	0	88
Snack			*Snack*		
8 oz. formula	0	180	4 oz. formula	0	88
1 cup popcorn	35	35	1 cup popcorn	35	35
Total	318	1414	Total	303	1418

can be seen as a result of specific amino acid deficiencies (Giacoia and Berry 1993; De Raeve et al. 1994). Supplementation with specific amino acids is usually required to normalize levels of the compromised amino acids and needs to be continued along with diet therapy for life.

As previously noted, successful pregnancies are now possible for some women with these amino acid disorders. Pregnancy adds additional nutrition concerns, in terms of ensuring that intakes of protein, kcal, vitamins, and minerals are adequate to promote the growth and development of the fetus, yet restricted enough to maintain metabolic balance to optimize the health of the mother. It is important to maintain good metabolic control because elevations of certain metabolites are known to be teratogenic to the developing fetus. This is the case in maternal PKU, where elevated blood phenylalanine levels in the mother with PKU are associated with microcephaly, cardiac defects, and developmental delay in the affected child. The International Maternal PKU Study has developed guidelines for the management of PKU during pregnancy (Matalon et al. 1998). There are additional case reports of successfully managed pregnancies in other amino acid disorders. It is usually recommended that blood levels be within the goal range prior to conception. Tolerance to dietary protein is much lower during the first trimester, and will increase in the second and third trimester due to increasing needs of the fetus. This allows for a greater dietary protein intake for the mother. These women have high-risk pregnancies, and thus need to be followed closely in the metabolic clinics as well as by their obstetrician. This will help ensure the most optimal outcome for mother and baby.

Adjunct Therapies

There are many additional pharmacologic and dietary treatments used to manage amino acid disorders. Certain antibiotics have been used to decrease intestinal bacteria production of toxic metabolites, such as propionate and ammonia (Bain et al. 1998). Other medications have been utilized to detoxify the accumulated metabolites. Carnitine is used to bind a variety of organic acids, and sodium benzoate and sodium phenylbutyrate can bind with excess nitrogen to decrease ammonia production (Walter 2003). The use of these therapies in conjunction with protein restriction is needed to optimize the outcomes of these patients.

Urea Cycle Disorders

Epidemiology, Etiology, and Clinical Manifestations

The overall incidence of urea cycle disorders has been reported to be approximately 1 in every 30,000 births; however, there are no population studies to support this frequency. The lack of newborn screening programs targeted at these disorders prevents researchers from obtaining prevalence data.

Urea cycle disorders result in an impaired capacity of the body to excrete nitrogen in the form of urea. Ammonia is normally converted to urea in the liver by means of several biochemical steps and eventually excreted in the urine (see Figure 28.3). The urea cycle resides primarily in the hepatocytes (cells of the liver), but this process can also take place to a lesser extent in the kidneys and small intestine. It is an essential biochemical pathway for excretion of waste nitrogen extracted from the amino acids in the body. A disruption in any one of the eight biochemical pathways involved results in disorders of the urea cycle. These pathways are governed by enzymes (see Table 28.6) that undergo a cascade of enzymatic transformations to convert the toxic ammonia molecule to nontoxic, water-soluble urea, which contains 2 nitrogen groups and is eliminated in the urine. A block in the urea cycle can result from an enzyme deficiency (carbamyl phosphate synthetase I, ornithine transcarbamylase, argininosuccinic acid synthetase, argininosuccinic acid lyase, arginase) in the urea cycle pathway. Alternatively, in other disorders a transport defect in the intestine and/or the kidneys (hyperornithinemia-hyperammonemia-homocitrullinuria syndrome, and lysinuric protein intolerance) can result in the depletion of an amino acid essential to the normal function of the cycle.

With the exception of ornithine transcarbamylase (OTC) deficiency, all have an autosomal recessive mode of inheritance. OTC deficiency is inherited as an X-linked dominant trait that is usually lethal in males. Urea cycle disorders are a common cause of inherited hyperammonemia, but because of the severe consequences of these disorders for the patient, they should be distinguished from other inborn errors of metabolism with secondary hyperammonemia, such as fatty acid oxidation disorders and organic acidemias. In general, the earlier in the process of converting nitrogen atoms into urea that the defect in the biochemical pathway occurs, the more severe and resistant to treatment the hyperammonemia will be (e.g., CPS and OTC are most severe). There is considerable heterogeneity in the magnitude of hyperammonemia and in the age of initial presentation. This is based on both the position of the block within the urea cycle and the degree of the enzyme deficiency. The most severe cases have no enzyme activity and present with hyperammonemic coma in the first week of life, with a very poor rate of survival. Patients with the milder forms have some residual enzyme activity and their clinical presentation occurs later in life (ranging from infancy to adulthood) with recurrent episodes of hyperammonemia (Summar and Tuckman 2001). If diagnosed early and aggressively treated, these individuals can expect to live to adulthood with minimal neurologic damage.

FIGURE 28.3 Complete Urea Cycle Pathway with Enzymes Identified in Individuals with Urea Cycle Disorders, and Pharmacologic Pathways Used for Nitrogen Removal

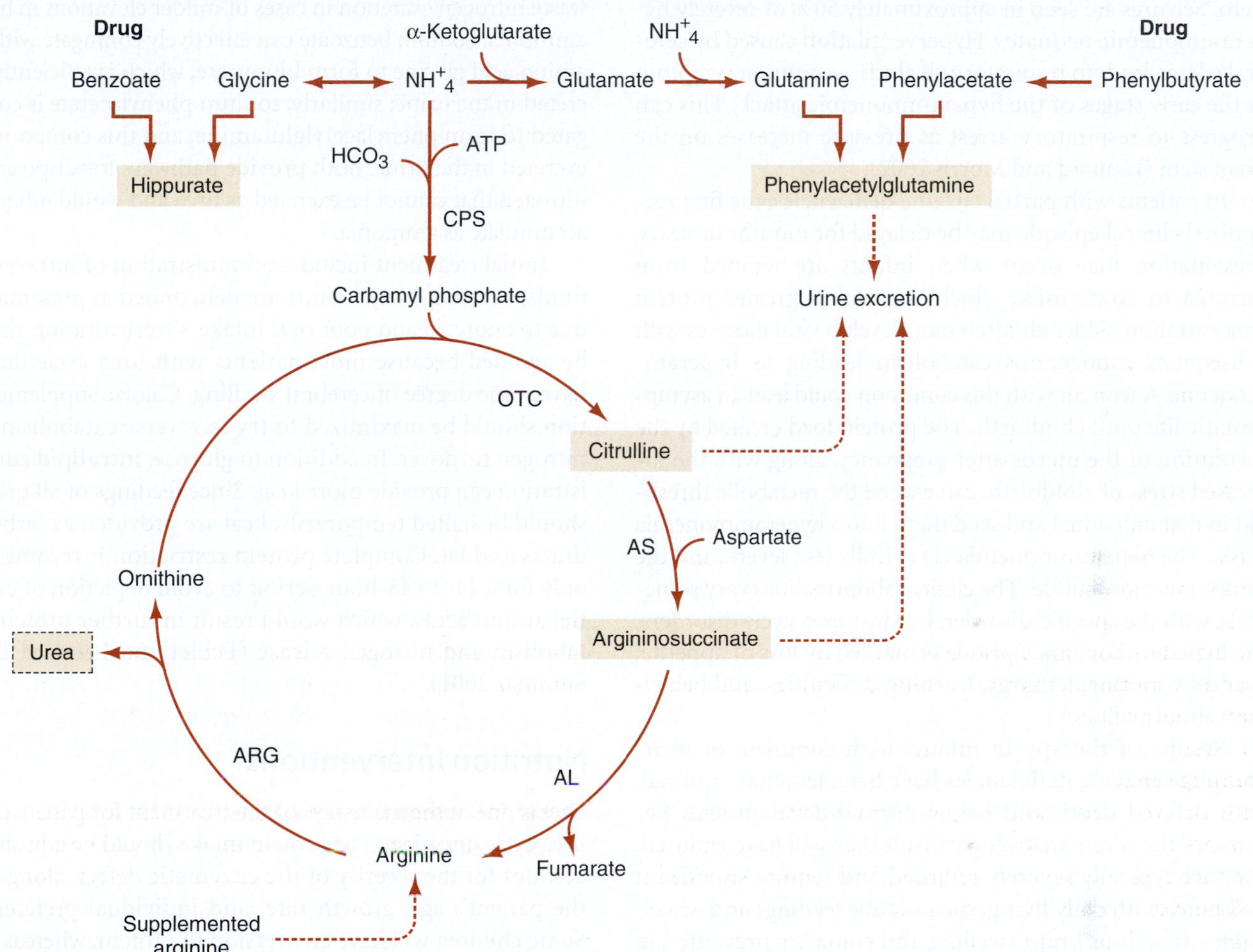

Source: Modification of Figure 2, p. 57 of Summas M, Tuckman M, "Proceedings of a consensus conference for the management of patients with urea cycle disorders," *J Pediatrics.* vol. 138:57, 2001.

TABLE 28.6

Listing of Enzymes in the Urea Cycle
Enzyme Defects in the Urea Cycle Pathway:
CPS – carbamyl phosphate synthetase I OTC – ornithine transcarbamylase AS – argininosuccinate synthetase AL – argininosuccinate lyase ARG – arginase
Enzyme Defects Outside of the Urea Cycle Pathway Leading to Deficiencies of Amino Acids Needed for the Urea Cycle Pathway:
HHH – hyperornithinemia, hyperammonemia, homocitrullinemia LPI – lysinuric protein intolerance NAGS – N-acetylglutamate synthase

Infants with complete enzyme deficiencies are usually born at term with no prenatal complications, because the maternal circulation detoxifies the accumulating ammonia. The typical initial symptoms of a child with hyperammonemia include poor feeding and somnolence, which may not be recognized by new parents. As a result, advice and care are sought later when the child's illness has progressed to become more severe. When the parents finally seek medical care for their infant, his or her plasma ammonia levels are elevated to 100 to 200 μmol/L (normal level is less than or equal to about 35 μmol/L). These levels are usually associated with clinical symptoms of lethargy, confusion, and vomiting, and higher levels usually result in coma because of the cerebral edema secondary to the elevation of this compound. There is a loss of thermoregulation, with a low core temperature and feeding disruption that correlates with the somnolence. Abnormal posturing and **encephalopathy** are often related to the degree of central nervous system swelling and pressure on the brain stem. Seizures are seen in approximately 50% of severely hyperammonemic neonates. Hyperventilation caused by cerebral edema leads to respiratory alkalosis, a common symptom in the early stages of the hyperammonemic attack. This can progress to respiratory arrest as pressure increases on the brain stem (Leonard and Morris 2002).

In patients with partial enzyme deficiencies, the first recognized clinical episode may be delayed for months or years. Presentation may occur when infants are weaned from formula to cow's milk, which contains a greater protein concentration. Older children may develop viral illnesses with subsequent endogenous catabolism leading to hyperammonemia. A woman with this condition could lead an asymptomatic life until childbirth. The protein load created by the involution of the uterus after pregnancy, along with the increased stress of childbirth, can exceed the metabolic threshold in that individual and send them into a hyperammonemia crisis. The hyperammonemia is typically less severe, and the symptoms more subtle. The clinical abnormalities vary somewhat with the specific disorder. In most urea cycle disorders, the hyperammonemic episode is marked by loss of appetite, cyclical vomiting, lethargy, learning difficulties, and behavioral abnormalities.

Results of therapy in infants with complete or near-complete enzyme deficiencies have been less than optimal, with delayed death and below-normal development. Because of the severe neurologic insult they will have endured, they are typically severely retarded and require significant assistance with daily living, such as tube feedings and wheelchairs. If serious brain swelling and coma are prevented in the neonatal period, or if onset of the hyperammonemic crisis is delayed, physical growth and mental development are more nearly normal with nutrition and pharmacological support. If diagnosis is anticipated and treatment is begun during the neonatal period in affected siblings, a relatively normal outcome is observed, even with severe enzyme defects (Brusilow and Horwich 2001).

Acute Treatment

The most effective treatment of the acute hyperammonemic crisis that occurs in the newborn period is hemodialysis. Initially, these patients will be dialyzed in order to detoxify them by rapidly removing the ammonia. Once the dialysis phase is complete, the drugs sodium benzoate and sodium phenylacetate are used to scavenge excess ammonia from the bloodstream. The pharmacologic approach (see Figure 28.4), rather than dialysis, can be used as an alternative method of waste nitrogen excretion in cases of milder elevations in blood ammonia. Sodium benzoate can effectively conjugate with the amino acid glycine to form hippurate, which is efficiently excreted in the urine; similarly, sodium phenylacetate is conjugated to form phenylacetylglutamine, and this compound is excreted in the urine. Both provide pathways for disposing of nitrogen that cannot be excreted as urea and would otherwise accumulate as ammonia.

Initial treatment includes administration of intravenous fluids, because many infants are dehydrated at presentation due to anorexia and poor oral intake. Overhydration should be avoided because most patients with urea cycle defects have some degree of cerebral swelling. Caloric supplementation should be maximized to try to reverse catabolism and nitrogen turnover. In addition to glucose, **intralipid** administration can provide more kcal. Since feedings of all protein should be halted temporarily, kcal are provided as carbohydrates and fat. Complete protein restriction is recommend only for a 24- to 48-hour period to avoid depletion of essential amino acids, which would result in further protein catabolism and nitrogen release (Feillet and Leonard 1998; Summar 2001).

Nutrition Interventions

Diet is one of the mainstays of the treatment for patients with urea cycle disorders. The protein intake should be adjusted to account for the severity of the enzymatic defect, along with the patient's age, growth rate, and individual preferences. Some children will have an aversion to protein, whereas others do not. Most patients, except for those with arginase deficiency, will need supplemental arginine. In normal children and adults, arginine is not an essential amino acid, because it can be synthesized by the urea cycle. However, this amino acid will become essential because of the enzymatic block and hence needs to be supplemented back into the diet.

Ideally, protein should be given in the exact amount needed for growth and maintenance without any excess. Table 28.7 shows the difference in protein recommendations for age for the general population vs. those affected with a urea cycle disorder (Leonard 2001). In reality, however, this has proven to be impossible. Tissue protein is constantly being synthesized and degraded. During growth, a net protein synthesis occurs, but during fasting or illnesses, with an increased rate of degradation exceeding the protein synthesis rate, the outcome is a negative nitrogen balance. Negative nitrogen balance results in an increased flux through the urea cycle and ultimately an elevated ammonia level. During adulthood, protein needs are decreased, because there is less need for growth and a decreased rate of tissue turnover. Protein-restricted diets need to be maintained throughout life.

For patients with the more severe variants of the disorder, some of the natural protein that is provided from foods,

encephalopathy—degenerative disease of the brain

intralipid—an intravenous fat emulsion used to prevent or correct deficiency of essential fatty acids and to provide kcal

TABLE 28.7

Comparison of Daily Protein Recommendations for the General Population and Those Affected with a Urea Cycle Disorder

Age	FAO/WHO/UNU Safe Protein Recommendations (mean + 2 SD) (1985)	Protein Recommendations for Individuals with UCDs
0–3 months	2.25 g/kg	1.25–2.20 g/kg
3–6 months	1.86 g/kg	1.8–2.0 g/kg
6–9 months	1.65 g/kg	1.6–1.8 g/kg
9–12 months	1.48 g/kg	1.4–1.6 g/kg
1–4 years	1.09–1.26 g/kg	8–12 g/day
4–7 years	1.01–1.06 g/kg	12–15 g/day
7–11 years	1.0 g/kg	14–17 g/day
11- to 15-year-old girls	0.9–1.0 g/kg	20–23 g/day
11- to 15-year-old boys	0.96–0.98 g/kg	20–23 g/day
15- to 19-year-old girls	0.8–0.9 g/kg	20–23 g/day
15- to 19-year-old boys	0.86–0.92 g/kg	21–24 g/day

Sources: Leonard JV. The nutritional management of urea cycle disorders. *J Pediatr.* 2001;138:S40–S45. Acosta PB, Yannicelli S. *Nutrition support protocols.* 4th ed. Columbus (OH): Ross Products Division of Abbott Laboratories; 2001.

which includes a mixture of essential and nonessential amino acids, may be replaced with an essential amino acid mixture to ensure that the individual receives a sufficient amount of these proteins. When one of these mixtures is ingested, the body's surplus nitrogen is used to synthesize the nonessential amino acids, thereby reducing the load on the urea cycle. In general, it is suggested that about 25% to 50% of total protein intake be provided as essential amino acids. There are special metabolic formulas designed in the treatment of urea cycle disorders that provide essential amino acids and that may or may not also include kcal along with vitamins and minerals (Leonard 2001).

Nutritional Concerns

As with any low-protein diet, care must be taken to ensure the diet is nutritionally complete. The amino acid intake needs to be balanced, and the essential amino acid intake adequate. Risk of micronutrient deficiency is equally as great as risk of an essential amino acid deficiency, particularly for iron and zinc; hence, these nutrients need to be supplemented. It is also important to ensure an adequate energy intake. Many patients may be anorexic secondary to elevated ammonia levels, and may not only have a low protein intake but an energy deficit as well. Enteral nutrition support, such as tube feedings, may need to be considered to ensure a sufficient intake.

Clinical, biochemical, and nutritional monitoring should be continuous. The lower the protein intake is, the more meticulous the monitoring must be. Any changes to the diet are made in an incremental manner, by no more than 10% at any one time. Clinically, the patient's general health and well-being should be taken into account as well as his or her growth rate. Various other parameters are monitored to confirm that the protein intake is adequate. These include clinical features such as the growth of the patient and the appearance of the hair, skin, and nails. Biochemical tests include the plasma concentrations of ammonia and essential amino acids. Other tests to monitor include hemoglobin, hematocrit, albumin, prealbumin, transferrin, and total protein. A detailed flow sheet can help document trends that relate dietary protein intake to certain laboratory indices. Table 28.8 shows an example of a clinical flow sheet. Medications used in the management of these conditions should also be noted, since many are prescribed on a per kilogram body weight basis and need to be adjusted to account for growth in relation to laboratory results. Including all of these parameters on the flow sheet helps the clinician to make the necessary adjustments to treatments promptly and accurately.

When patients, even those with severe defects, are clinically stable, there is some flexibility in giving additional protein without causing an increase in the blood ammonia level. However, for patients who are doing poorly, there is no flexibility, because their metabolic status may deteriorate rapidly. If a patient is not on an adequate diet, and the plasma ammonia concentration is satisfactory, the dietary protein may be increased. However, if the ammonia levels are high, possibly because of insufficient protein or kcal, the dietitian should be able to analyze the diet and determine the likely cause. The diet should then be adjusted accordingly, not only to maintain normal ammonia levels but to also provide a nutritionally adequate intake (Acosta and Yannicelli 2001; Berry and Steiner 2001).

All patients are at risk of decompensation with metabolic stress, including fasting and intercurrent illness, where a rapid increase in blood ammonia can occur. It is important to change the diet to increase energy content and reduce protein intake orally while it is still tolerated. If a child is unable to tolerate the diet orally or is decompensating, then the parents need to take their child to the hospital for administration of intravenous medications, fluids, and kcal. An emergency care protocol should be developed for the patient so that when he or she arrives at the hospital, appropriate and urgent care can be administered.

Adjunct Therapies

Because developmental disabilities or even death can result from a hyperammonemic crisis, some patients with the more severe forms of this disorder have been treated with liver transplantation. However, until liver transplantation becomes more widely accessible or gene therapy technically feasible, alternative pathway therapy combined with protein restriction will remain the mainstay of therapy for urea cycle disorders (Berry and Steiner 2001).

TABLE 28.8

Sample Flow Sheet Used for the Monitoring of Patients with Urea Cycle Defects

Patient Name: **Date of Birth:**
Patient Medical Record #: **Diagnosis:**

Date	Age	Wt/Ht	OFC	Medications	Plasma Citrulline	Plasma Glutamine	NH_3/alb	H/H	Formula	Protein Intake	Kcal

Mitochondrial Disorders

Etiology and Clinical Manifestations

Mitochondria are essential for the production of energy in all types of tissues. They are found in all cells in the body and are considered the "power house" or energy producers of the cell. Mitochondrial disorders result either from defects of the respiratory chain that produces energy or from defects affecting the overall number and function of the mitochondria. A defect in function of the mitochondrial system results in diseases affecting multiple organs with a wide range of symptoms, including cardiomyopathy, blindness, deafness, endocrine problems, and muscle disorders. Individuals presenting with a mitochondrial disorder can range from infants, as in the case of Leigh's disease, to adults with progressive neurologic or muscle disorders. Table 28.9 lists the spectrum of conditions associated with mitochondrial mutations leading to disruption in function. Mitochondrial disorders are renowned for their variability in clinical features and genetic causes, making it difficult to determine their true prevalence (Wallace 1997; Smeitink 2003).

Diagnosis of some of the most common mitochondrial disorders such as MELAS (mitochondrial encephalomyopathy with lactic acidosis and stroke like episodes) or NARP (neurogenic muscular weakness, ataxia, and retinitis pigmentosa) can be made through **DNA mutation testing**. Skin and muscle tissue may also be needed for further **histological** and biochemical analyses in order to make a diagnosis. In a healthy person, energy is produced through several metabolic processes involving the use of fatty acids and carbohydrates within the mitochondria. Disorders of the mitochondria include fatty acid transport disorders, fatty acid oxidation defects, **pyruvate complex disorders**, and respiratory chain defects. Because the respiratory chain defects are the predominant types of mitochondrial disorders, the remainder of this section will focus on these defects. Disorders of fatty acid transport and oxidation will be discussed in a subsequent section.

The respiratory chain (see Figure 28.4) is made up of five complexes that undergo changes in their oxidative state to produce ATP (adenosine triphosphate). Carbohydrates are eventually metabolized to pyruvate, which will then enter into the Krebs cycle. Electrons generated from the Krebs cycle and from beta oxidation of fatty acids are used in the production of energy via the complexes of the respiratory chain. Defects affecting several of the individual complexes

TABLE 28.9

Spectrum of Conditions Related to Mitochondrial Dysfunction

- Ataxia
- Autism
- Blindness
- Cardiomyopathy
- Deafness
- Dementia
- Developmental delay
- Diabetes
- Epilepsy
- Gastrointestinal motility problems (diarrhea, dysmotility)
- Hypoglycemia
- Hypotonia
- Language delays
- Liver failure
- Neuropathy
- Pancreatitis
- Renal failure
- Seizures
- Strokes
- Weakness

DNA mutation testing—blood test that screens for known mutation(s) that have been shown to result in a particular disease

histological—pertaining to the minute structure, composition, and function of the tissues

pyruvate complex disorders—dysfunction in the metabolism of pyruvate, the end product of glycolysis, via either the Krebs cycle or gluconeogenesis, resulting in the production of lactic acid

FIGURE 28.4 Respiratory Chain Pathway

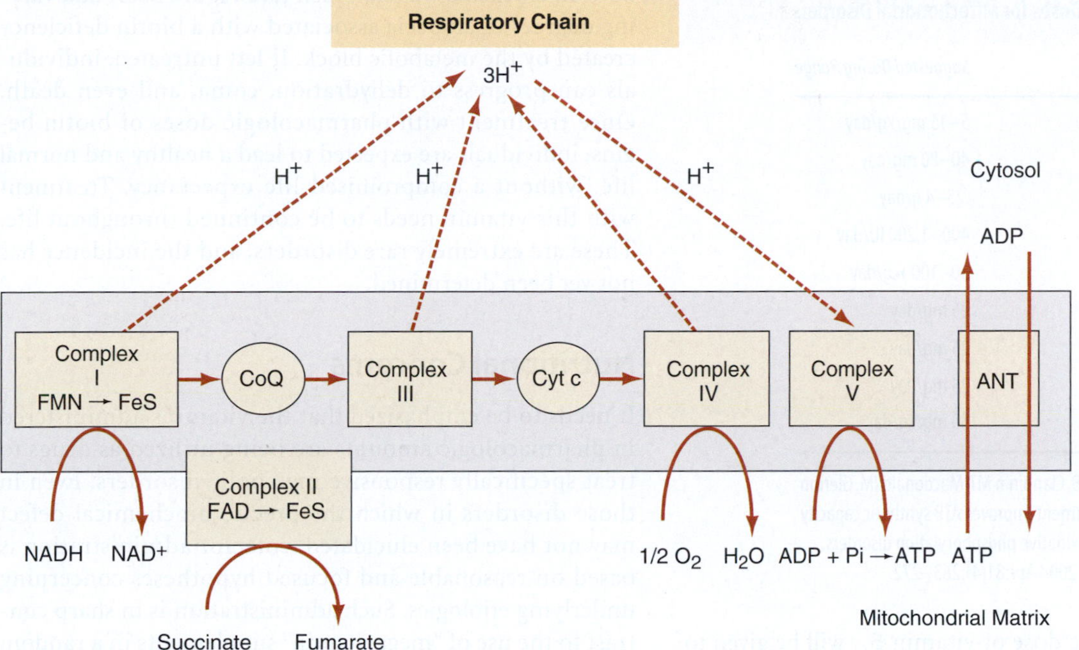

Source: From slide presentation by Marriage B, "Cofactor treatment in oxidative phosphorylation disorders: is it effective?" Presented at 9th Ross Metabolic Conference: Advances in Management of Inherited Metabolic Disorders, April 4–6, 2003.

have been identified. These defects lead to decreased energy production in various tissues, and subsequently to clinical symptoms, such as hypotonia, developmental delay, and failure to thrive. Attempts to facilitate the function of the respiratory chain through the administration of pharmacological doses of several vitamins and nutrients have resulted in limited success thus far (Smeitink 2003).

Nutrition Intervention

There is no definite treatment for mitochondrial disorders. The use of dietary intervention may help alleviate some of the symptoms and/or delay progression of the disease, but will not prevent the debilitating effects of the disorder. Therapy for defects of the respiratory chain entails the use of vitamin cofactors in pharmacological amounts equal to approximately 100 to 1,000 times the DRI for age to enhance the activity of the various complexes. Riboflavin and thiamin serve as cofactors, while vitamin E and lipoic acid are used to protect against free radicals that are produced from ongoing aberrant reactions (see Chapter 15 for a discussion of free radicals). Vitamins C and K and coenzyme Q10 are used as artificial electron receptors and transporters (Przyrembel 1987). Prescription doses of these cofactors vary widely between clinics, but Table 28.10 shows the ranges prescribed (Acosta and Yanicelli 2001). The reported efficacy of vitamin therapy can vary significantly, but given the benign nature of most of the vitamins prescribed, many clinics will recommend a trial of one or more of these supplements. Because fasting produces stress on the respiratory chain, frequent feedings are recommended.

Adjunct Therapies

Carnitine and glycine have been used in the treatment of several of these disorders. They act by conjugating with toxic metabolites, removing them from the body (Przyrembel 1987).

Disorders Related to Vitamin Metabolism

Etiology and Clinical Manifestations

As noted earlier, vitamins are needed for various functions. They can be used as cofactors for various enzymatic reactions, antioxidants, or electron receptors. Sometimes a pharmacologic dose of a vitamin is all that is necessary to maintain normal enzymatic function and eliminates the need for other treatment modalities, such as a restrictive diet. In general, the vitamin-responsive form of the disorder tends to be less common than the non-responsive form of the same condition.

Nutrition Interventions

There are some amino acid disorders, including methylmalonic acidemia and MSUD, that have a vitamin-responsive

TABLE 28.10

Recommended Cofactor Doses for Mitochondrial Disorders

Cofactor	*Suggested Dosing Range*
Co-Enzyme Q_{10}	5–15 mg/kg/day
Vitamin K	40–80 mg/day
Vitamin C	.25–4 g/day
Vitamin E	400–1,200 IU/day
Selenium	50–100 μg/day
Thiamine	25 mg/day
Riboflavin	25 mg/day
Pantothenate	25 mg/day
Carnitine	50 mg/kg/day

Source: Marriage B, Clandinin MT, Macdonald IM, Glerum DM. Cofactor treatment improves ATP synthetic capacity in patients with oxidative phosphorylation disorders. *Mol Genet Metab.* 2004 Apr:81(4):263–272.

form. A pharmacologic dose of vitamin B_{12} will be given to a newly diagnosed methylmalonic acidemia patient to determine if the enzymatic defect is vitamin responsive (Wolf et al. 1981). Only a small percentage of patients with methylmalonic acidemia have been found to be responsive to vitamin B_{12}, but those who are responsive do not require additional treatment and need not adhere to a restrictive diet. Vitamin B_{12} is a cofactor for the methylmalonyl-CoA mutase enzyme (see Figure 28.5), because this enzyme needs the adeno form of this vitamin in order to form the product, which is the amino acid cysteine. In order for this enzyme to function properly, it requires the presence of vitamin B_{12}. Consequently, there have been reports of patients diagnosed with methylmalonic acidemia with a malfunctioning enzyme due to a vitamin B_{12} deficiency, and not because of a genetic mutation. These individuals will have the classical clinical presentation of methylmalonic acidemia, including failure to thrive, developmental delay, and hypotonia. This can be caused by following a strict vegan diet, or may even develop in an infant who is breast-feeding from a vitamin B_{12}-deficient mother (Schrock-Kelley et al. 1998). Supplementation with B_{12} leads to normalization of methylmalonic acid levels.

Administration of a vitamin can be the primary treatment modality in some metabolic conditions. In holocarboxylase synthetase deficiency and biotinidase deficiency, biotin in doses from 100 to 1,000 times the DRI for age is all that is required to treat these disorders and prevent serious complications. These disorders result from an inability to regenerate biotin from endogenous sources. Individuals with one of these conditions typically present in infancy with feeding and respiratory difficulties, vomiting, seizures, lethargy, and hypotonia related to the lactic acidosis, ketosis, and hyperammonemia created by the accumulation of the intermediate compounds created by the metabolic block. Frequently infants with these conditions have an erythematous skin rash (see Figure 28.6) and varying degrees of alopecia associated with a biotin deficiency created by the metabolic block. If left untreated, individuals can progress to dehydration, coma, and even death. Once treatment with pharmacologic doses of biotin begins, individuals are expected to lead a healthy and normal life, without a compromised life expectancy. Treatment with this vitamin needs to be continued throughout life. These are extremely rare disorders, and the incidence has not yet been determined.

Nutritional Concerns

It needs to be emphasized that the vitamins administered in pharmacologic amounts are being utilized as drugs to treat specifically responsive metabolic disorders. Even in those disorders in which the precise biochemical defect may not have been elucidated, cofactor administration is based on reasonable and focused hypotheses concerning underlying etiologies. Such administration is in sharp contrast to the use of "megavitamin" supplements in a random fashion in conditions for which a response has not been documented, or in patients with ill-defined dysfunctions presenting no clinically valid justifications for therapeutic trials (Packman 1986).

Most of the vitamins used in the treatment of metabolic disorders are water soluble, so issues of toxicity are minimized. When fat-soluble vitamins (E and K) are used, such as in the treatment of mitochondrial disorders, toxicity can be of concern. Because these vitamins can affect clotting factors, laboratory studies assessing clotting times should be assessed periodically. Compliance with vitamin supplementation can be a concern, especially when a large amount and number of vitamins and/or cofactors are prescribed, as in the case of mitochondrial disorders. The cost associated with purchasing these supplements is of concern, as many health insurance companies will not cover them. Further research documenting the effectiveness of these nutrients in the treatment of metabolic disorders is needed in order to justify advocacy of insurance coverage for these products that is similar to current coverage for medications.

Disorders of Carbohydrate Metabolism

Disorders of carbohydrate metabolism include problems with processing the simple sugars galactose and fructose in the body. There is also a group of disorders called glycogen storage diseases that result in various defects in synthesizing and releasing glucose. A summary of the carbohydrate disorders to be discussed here, along with the main clinical symptoms and treatments used, appears in Table 28.11. Prompt diagnosis and dietary treatment significantly help to improve the overall outcome for patients affected with one of these types of conditions.

FIGURE 28.5 Vitamin B_{12} (adenosyl cobalamin) and Methylmalonic Acidemia (MMA): Adenosyl cobalamin (AdoCbl) is necessary for the conversion of L-methylmalonyl CoA to succinyl CoA by the methylmalonyl CoA mutase enzyme

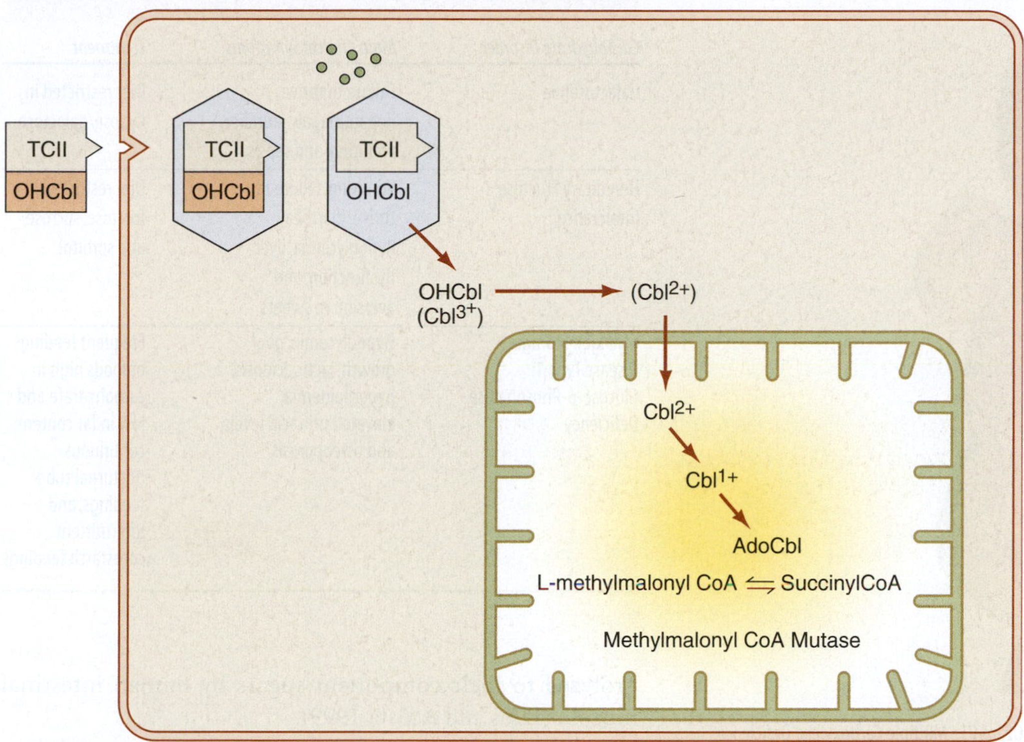

TCII = transcobalamin II
OHCbl = hydroxycobalamin
Source: Modification of Figure 1, p. 620 of Shevell MI, Matiaszuk N, Ledley F, Rosenblatt DS, "Varying neurological phenotypes among muto & mut- patients with methylmalonyl CoA mutase deficiency," *Am J Med Genetics;* 1993.

Galactosemia

Epidemiology, Etiology, and Clinical Manifestations

Classical galactosemia is an enzyme defect in galactose metabolism leading to failure to thrive, hepatomegaly (enlarged liver), and life-threatening sepsis in the newborn period. Vomiting and jaundice may develop as early as a few days after milk feedings are begun. Anorexia, failure to gain weight or to grow, or even weight loss ensues. In the absence of treatment, **parenchymal** damage to the liver leads to the development of cirrhosis. Patients may also have edema, ascites, bleeding problems, and an enlarged spleen. If milk feedings are continued, the disease may be rapidly fatal. Other complications associated with the continued ingestion of milk include cataract formation, mental retardation, and renal tubular dysfunction. Many states now screen for galactosemia in the newborn period before symptoms are present or complications of the disease have taken place (Holton et al. 2001). Galactosemia has been reported to occur in approximately one in every 60,000 to 80,000 births in the United States.

In patients with galactosemia, the enzyme defect is in the uridyltransferase, which normally converts galactose, in the form of galactose-1-phosphate, to glucose-1-phosphate (see Figure 28.7). Glucose-1-phosphate can then produce free glucose or further be utilized to produce energy via **glycolysis**. The other enzymes involved in galactose metabolism are normal in these patients. Galactose-1-phosphate accumulates in tissues and may be responsible for many of the clinical manifestations of the disease. The abnormalities in the lens of the eye, for example, are produced by **galactitol**, a by-product of galactose accumulation. Therapeutic measures that result in reduction of intracellular concentrations of galactose-1-phosphate lead to the prevention or disappearance of symptoms. When diagnosis is early and compliance with therapy is good, levels of patient IQ have been normal, allowing them to lead full and normal lives (Nyhan and Pinar 1998).

parenchymal—referring to the essential elements of an organ

glycolysis—the anaerobic enzymatic conversion of glucose to lactate or pyruvate, which results in the production of energy in the form of ATP

galactitol—sugar alcohol produced from galactose

FIGURE 28.6 Infant with Erythematous Skin Rash Associated with Biotin Deficiency Secondary to Holocarboxylase Synthetase Deficiency

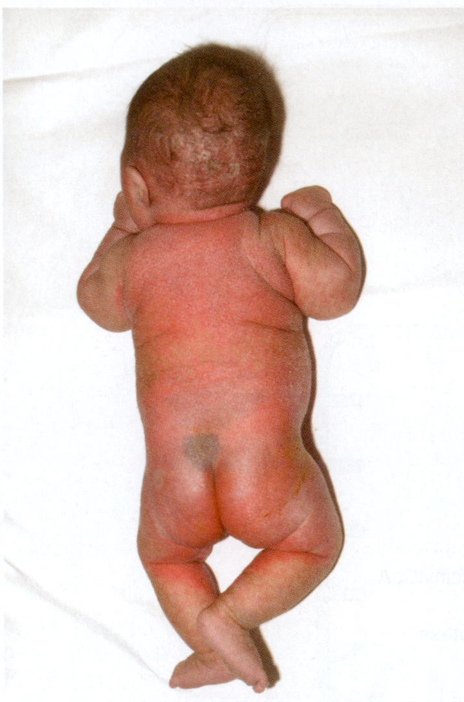

Source: Figure 4.3, p. 28 of Nyhan WL and Ozard PT, "Multiple Carboxylase Deficiency," *Atlas of Metabolic Diseases* 1st ed. Chapman and Hall Medical; 1998.

TABLE 28.11

A Summary of Carbohydrate Disorders, Associated Symptoms, and Treatments Used

Carbohydrate Disorder	*Main Clinical Symptoms*	*Treatment*
Galactosemia	Failure to thrive, hepatomegaly jaundice, vomiting, and sepsis	Diet restricted in lactose/galactose
Hereditary Fructose Intolerance	Vomiting, failure to thrive, diarrhea, hypoglycemia, liver dysfunction, and aversion to sweets	Diet restricted in fructose, sucrose, and sorbitol
Glycogen Storage Disease Type 1: Glucose-6-Phosphatase Deficiency	Hypoglycemia, poor growth, lactic acidosis, hyperlipidemia, elevated uric acid levels, and osteoporosis	Frequent feedings of foods high in carbohydrate and low in fat content, continuous nocturnal tube feedings, and intermittent cornstarch feedings

Nutrition Interventions The treatment for galactosemia is exclusion of galactose from the diet. Treatment with a lactose/galactose-free diet leads to immediate reversal of symptoms. Lactose, the principal sugar of mammalian milks, is the predominant dietary source of galactose. It is a disaccharide in which glucose and galactose are linked together. The mainstay of the diet for an infant is the substitution of **casein hydrolysate-containing** formula for a milk-based formula. The casein hydrolysate may contain small amounts of lactose, since it is prepared from milk, but this appears not to affect the therapeutic efficacy of the preparation. Infant soy formulas have also been used, though this practice has been questioned because of the presence of sugars containing galactose, such as raffinose and stachyose. However, it is believed that these **galactose oligosaccharides** are not hydrolyzed to their component sugars by human intestinal mucosa (Elsas and Acosta 1999).

casein hydrolysate—product made from the breakdown of casein, a milk-based protein, to smaller components, which are then easier to digest

oligosaccharide—a carbohydrate that through hydrolysis yields a small number of monosaccharides

As the children grow, it is important for them and their families to be aware of sources of galactose in foods other than milk. Education of the parents, and of the child, as he or she grows older, on the galactose content of foods is important. A listing of foods containing galactose should be provided, along with the identification of potential galactose-containing ingredients. Besides the obvious dairy products, there are a number of other foods that these families should be made aware of. Table 28.12 lists potential galactose-containing ingredients. Many processed foods, such as cakes, cookies, muffins, and French fries, contain milk products, and consequently galactose. Several fruits and vegetables also contain galactose, such as persimmons, dried figs, and tomatoes. Some of the galactose in fruits and vegetables may be found as the galactose oligosaccharides, raffinose and stachyose, and hence is unavailable for absorption. Drugs should also be checked because they could contain galactose when lactose is used as a sweetener. The determination of the galactose-1-phosphate content in the red blood cells is helpful in monitoring adherence to the diet. There is no good evidence that at a prescribed age the diet can be relaxed (Acosta and Yannicelli 2001).

Nutrition Concerns Maintaining a galactose-restricted diet is not harmful as long as alternative sources of those nutrients abundant in dairy foods, such as calcium and vitamin D, are provided. Calcium supplements are often required by children and adults with galactosemia. Kcal, protein, vitamin, and

FIGURE 28.7 Enzymatic Pathway for Galactose Metabolism and Site of Enzymatic Defect, Galactose-1-Phosphate Uridyltransferase, in Galactosemia

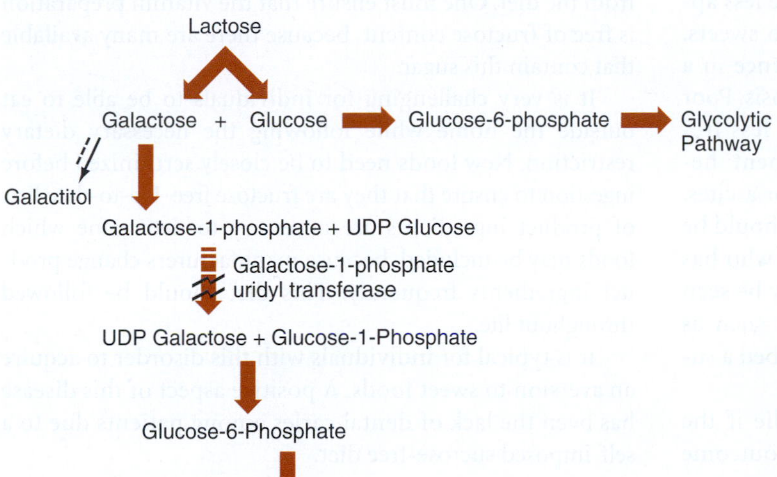

TABLE 28.12

Ingredients that Contain Galactose	
Butter	Whey and whey solids
Calcium caseinate	Buttermilk and solids
Nonfat milk	Casein
Dry milk and milk protein	Cream
Hydrolyzed protein made from casein or whey (avoid if unspecified)	Lactose
Lactalbumin	Milk chocolate
Milk and milk solids	Cheese
Organ meats	Sour cream
Sodium caseinate	Yogurt

Source: Parents of Galactosemic Children, Inc. [homepage on the Internet]. Gauiter (MS): Parents of Galactosemic Children, Inc.; 2006. Available from: http://www.galactosemia.org/.

mineral needs are similar to those of other individuals unaffected by this disease.

Despite treatment, however, some individuals with galactosemia develop long-term complications, including mild to moderate developmental delay, especially in the areas of speech, growth failure, and ovarian failure in females. Inadequate dietary restriction of plant sources containing galactoproteins and other galactose complexes has been suggested as a possible cause of these complications. More recent analyses of food composition have revealed a more extensive listing of fruits and vegetables containing galactose. Further research is needed to address actual galactose ingestion versus clinical outcome.

Several other theories have been proposed to explain the poor outcomes in patients with galactosemia. These theories include fetal damage in utero when exposed to galactose via maternal circulation, damage before nutrition intervention, overrestriction of dietary galactose, de novo galactose synthesis, and a deficiency of uridine-diphosphate-galactose. More questions than answers regarding optimal treatment to alleviate the clinical manifestations associated with galactosemia still abound. Further research may lead to modifications in nutrition support (Waggoner et al. 1990; Gross and Acosta 1991).

Adjunct Therapies Oral uridine therapy has been proposed in an attempt to improve clinical outcomes. Because of the enzymatic block, not only is there an accumulation of galactose, in the form of galactose-1-phosphate, but there is also a decreased production of uridine-diphosphate-galactose. Supplementation of this nutrient has resulted in normalized red blood cell levels of this compound. Unfortunately, trials providing uridine supplementation in individuals with galactosemia have failed to show any beneficial impact on clinical status (Kaufman et al. 1989). More research is needed in order to improve the treatment of galactosemia.

Hereditary Fructose Intolerance

Etiology and Clinical Manifestations Fructose is predominantly metabolized in the liver, kidney, and small intestine in a specialized pathway composed of three enzymes—fructokinase, aldolase type B, and triokinase—that convert fructose into intermediates of the glycolytic-gluconeogenic pathway. Hereditary fructose intolerance is caused by a deficiency in activity of fructose-1-phosphate aldolase (aldolase B). A deficiency of the fructoaldolase B enzyme results in accumulation of fructose-1-phosphate in tissues that possess fructokinase, which causes depletion of **inorganic phosphate** and ATP. Liver and renal impairment is thought to be due to either the direct toxic effect of fructose-1-phosphate, or due to a deficiency in tissue ATP content. Fructose-induced hypoglycemia results from inhibition of both **gluconeogenesis** and

inorganic phosphate—phosphate compound not derived from an organic origin

gluconeogenesis—the process of synthesizing glucose from fatty acids or glycerol

glycogenolysis. Symptoms appear only when fructose, sucrose, or sorbitol is introduced into the diet, and they are not specific, so diagnosis may be overlooked. Symptoms may occur early in life and be severe; later they are less apparent, because the child develops an aversion to sweets. Vomiting is such a constant finding that its absence in a subject ingesting fructose argues against the diagnosis. Poor feeding, diarrhea, and later failure to thrive are less frequent. Some manifestations reflect liver impairment: hepatomegaly, bleeding tendency, jaundice, edema, or ascites. The diagnosis of hereditary fructose intolerance should be strongly suspected in an infant with liver failure who has had intake of sucrose or fructose. This can readily be seen in an infant ingesting a sucrose-based formula such as Isomil, or who has been given fruit juice or prescribed a sucrose-containing medication.

Some infants with severe liver failure may die if the diagnosis is overlooked. In the majority of cases, outcome is excellent. With a fructose-free diet, vomiting disappears immediately, as does the bleeding tendency, in less than 24 hours. All the clinical and laboratory findings are normal within 1 to 2 weeks, except for hepatomegaly, which persists for many years. Normal growth in length and weight is reached within 2 to 3 years. Hepatomegaly and **steatosis** disappear with the fructose-free diet between the ages of 5 and 10 years (Morrow 1997; Steinmann et al. 2001), but diet recommendations are continued for life.

Nutrition Interventions Fructose is a widely distributed natural compound. As the free monosaccharide, it is found in honey and in numerous vegetables and fruits, where it can account for up to 40% of the dry weight. As the disaccharide sucrose, which consists of one molecule of fructose attached to a molecule of glucose, it is found in many more nutrients and constitutes an important source of dietary carbohydrate. Listings of the fructose content of various foodstuffs are available. Fructose is extensively used in the diet as a sweetening additive in foods, medications, and even infant formulas. Average intake of this sugar has steadily increased since the beginning of this century to amounts exceeding 100 grams per day in adults. Successful treatment requires strict avoidance of all dietary fructose and sucrose. Kcal, protein, vitamin, and mineral requirements are similar to those of normal individuals (Acosta and Yannicelli 2001).

glycogenolysis—the process of breaking down glycogen to produce glucose

steatosis—excessive amounts of fat found in the stool

Nutritional Concerns Sometimes a vitamin supplement may be indicated in order to compensate for the nutrients limited by the omission of most fruits and some vegetables from the diet. One must ensure that the vitamin preparation is free of fructose content, because there are many available that contain this sugar.

It is very challenging for individuals to be able to eat outside the home while following the necessary dietary restriction. New foods need to be closely scrutinized before ingestion to ensure that they are fructose free. Up-to-date lists of product ingredients are necessary to determine which foods may be included, because manufacturers change product ingredients frequently. This diet should be followed throughout life.

It is typical for individuals with this disorder to acquire an aversion to sweet foods. A positive aspect of this disease has been the lack of dental caries among patients due to a self-imposed sucrose-free diet.

Adjunct Therapies Currently, there are no therapeutic interventions other than the avoidance of ingestion of fructose, sucrose, and sorbitol.

Glycogen Storage Diseases

Epidemiology, Etiology, and Clinical Manifestations

Glycogen storage diseases are caused by deficiencies of enzymes that regulate the synthesis or degradation of glycogen. Glycogen, a polysaccharide composed of glucose units, is the main carbohydrate reserve substance in the body. This polysaccharide is assembled through a process of chain elongation by the sequential addition of glucose units. The process of chain elongation and branching is achieved by glycogen synthetase and branching enzymes. Catabolism of the polysaccharide is achieved by the opposite process, through phosphorylase and debranching enzymes (see Figure 28.8). Glycogen is abundant in the liver, where it is used predominantly to form blood glucose, and in muscle, where it is a fuel for muscle contraction. The brain, on the other hand, although utilizing glucose preferentially for its metabolic needs, does not store glycogen to any significant extent. Hence, the brain is dependent on receiving glucose from the blood supply (Ryman 1974).

The syndromes of glycogen storage disease can be divided into at least eight different types with respect to clinical and chemical manifestations and according to the enzymatic deficiency. Most of the various types affect the liver and are a consequence of deficient activity of one or the other of the enzymes directly involved in degradation of glycogen to glucose-6-phosphate, or rarely, the synthesis of glycogen from glucose-6-phosphate. The enzymes noted by an asterisk in Figure 28.8 are those whose absences are known to lead to the accumulation of an abnormal amount of glycogen. The common clinical manifestation is abnor-

FIGURE 28.8 Summary of the Enzymes Involved in the Synthesis and Degradation of Glycogen in Tissue. Those enzymes carrying an asterisk are associated with a specific type of glycogen storage disease.

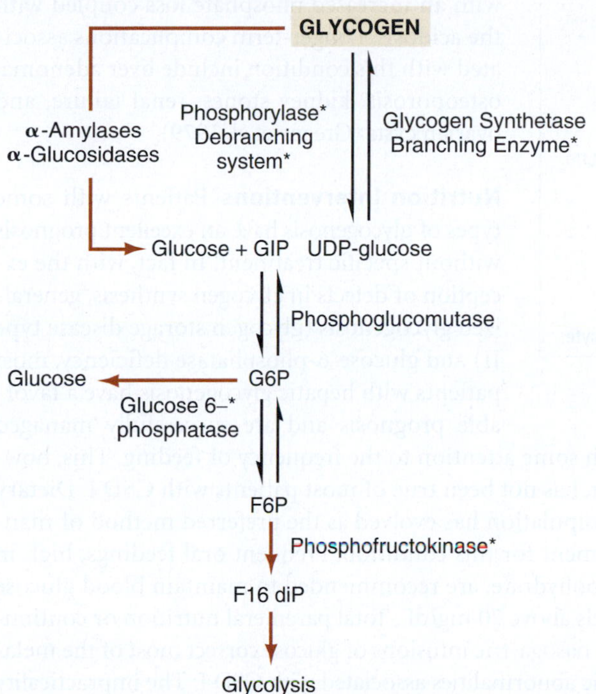

Source: Figure 3, p. 107 of Ryman BE, "The glycogen storage diseases," *J Clin Pathology,* vol. 27, suppl. 8, 1974.

mal glycogen deposition, primarily in liver, muscle, or both. In some of these defects, the accumulated glycogen is of normal structure. On the other hand, rare defects, that is, type IV, result in glycogen of an abnormal structure, more resembling **amylopectin**, which is the component of starch that is the storage product of the plant world. It is probably treated as a foreign substance despite the fact that the body produces it, and such a hostile reaction may be the basis of the early cirrhosis that is characteristic of this disease. Table 28.13 summarizes the types of glycogen storage disease that are now recognized and the main tissue affected. The main clinical symptoms for each disorder are also listed in the table (Chen and Burchell 1995). There is a wide range of incidence in the occurrence of glycogen storage diseases. The more common glycogen storage disease type I occurs about one in every 100,000 births, while frequency of the less common forms such as type II has not yet been determined.

TABLE 28.13

Listing of the More Common Glycogen Storage Diseases, Tissue Affected, and Clinical Symptoms

Glycogen Storage Disease and Enzyme Defect	*Tissue Affected*	*Symptoms*
Type I		
Glucose-6-phosphatase deficiency	Liver, kidney, small intestines	Hypoglycemia, hepatomegaly, lactic acidosis, hyperlipidemia, hyperuricemia, poor growth, and osteoporosis
Type II		
Lysosomal acid alpha glucosidase deficiency	Most tissues	Cardiomegaly in infants, hypotonia
Type III		
Glycogen debranching enzyme	Liver and muscle	Hepatomegaly, variable hypoglycemia, similar to Type I
Type IV		
Branching enzyme	Liver, leukocytes, and fibroblasts	Progressive cirrhosis with hepatosplenomegaly, failure to thrive, and ascites
Type V		
Muscle phosphorylase deficiency	Muscle	Easy fatigability, painful cramps after strenuous exercise, myoglobinuria
Type VI		
Liver phosphorylase defect	Liver, leukocytes	Mild hepatomegaly, mild hypoglycemia and hyperlipidemia, growth retardati on
Type VII		
Muscle phosphofructokinase deficiency	Muscle	Easy fatigability, painful cramps after strenuous exercise, myoglobinuria Degenerative brain disease
Type IX		
Defect in 1 of the 4 subunits of phosphorylase kinase	Liver and muscle	Mild hepatomegaly, mild hypoglycemia and hyperlipidemia, growth retardation

Glycogen Storage Disease Type I Glycogen storage disease type I (GSD I), the most commonly diagnosed type of glycogen storage disease, actually occurs despite correct functioning in the enzymes required for both the synthesis of glycogen and its degradation to glucose-6-phosphate. As a consequence, in this enzymatic defect, many of the clinical and chemical features seen in this disorder are unique among the other types of glycogen storage disease. GSD I is characterized by a deficiency of the enzyme glucose-6-phosphatase. The in-

amylopectin—the insoluble component of starch

FIGURE 28.9 Metabolic Pathway Affected in Glycogen Storage Disease Type I

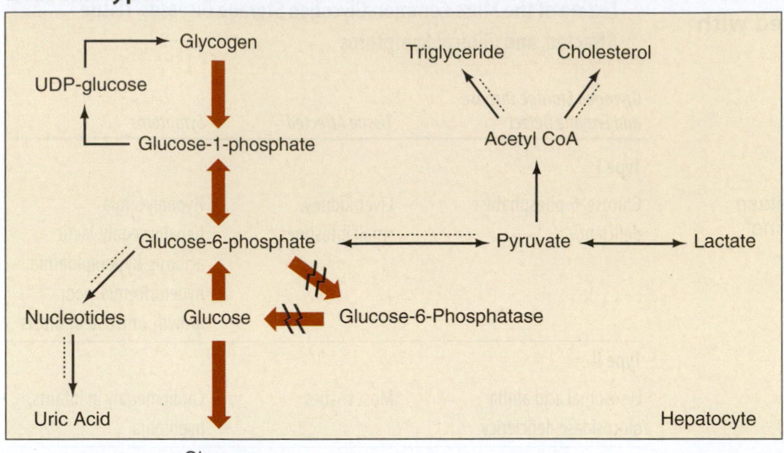

ability to **dephosphorylate** glucose-6-phosphate results in hypoglycemia and its metabolic consequences. Figure 28.9 illustrates the metabolic pathway involved in GSD I. Since patients with GSD I are at risk of death or hypoglycemic damage to the brain in early infancy, prompt diagnosis, the avoidance of fasting, and the provision of free glucose are important in getting the patient through this critical period (Greene et al. 1979).

The most constant and life threatening feature of this disease is the low blood glucose levels which result from relatively short periods of fasting. Fasting for as little as 2 to 4 hours is almost always associated with a decrease in blood glucose to less than 70 mg/dL, and it is not uncommon to observe 6- to 8-hour fasting levels of 5 to 10 mg/dL. In normal individuals, blood glucose levels are maintained within a relatively narrow range by hormones excreted by the liver, such as glucagon, which releases glucose from stored glycogen, as well as through gluconeogenesis, the process of synthesizing glucose from amino acids. In patients with GSD I, it is possible to degrade glycogen to glucose-6-phosphate, but, in the absence of glucose-6-phosphatase, no glucose is released and blood glucose levels continue to fall. The liver is further stimulated, which leads to continued glycogen degradation resulting in secondary manifestation of GSD I that causes the characteristic elevations in blood lipids, lactate, and uric acid. Hepatomegaly is often present at birth and can progresses to huge enlargement of the liver if the disease is poorly controlled. This enlargement is due to the accumulation of glycogen as the metabolic block compromises the ability to release free glucose. The chronic lactic acidosis with elevated glucagon and low insulin levels seems to be related to the poor growth seen in children with this condition. Bones may be osteoporotic, and some patients show delayed bone age associated with an increased phosphate loss coupled with the acidosis. Longer-term complications associated with this condition include liver adenoma, osteoporosis, kidney stones, renal failure, and ovarian cysts (Greene et al. 1979).

Nutrition Interventions Patients with some types of glycogenosis have an excellent prognosis without specific treatment. In fact, with the exception of defects in glycogen synthesis, generalized glycogenosis (glycogen storage disease type II) and glucose-6-phosphatase deficiency, most patients with hepatic glycogenosis have a favorable prognosis and are successfully managed with some attention to the frequency of feeding. This, however, has not been true of most patients with GSD I. Dietary manipulation has evolved as the preferred method of management for this condition. Frequent oral feedings, high in carbohydrate, are recommended to maintain blood glucose levels above 70 mg/dL. Total parenteral nutrition or continuous nasogastric infusions of glucose correct most of the metabolic abnormalities associated with GSD I. The impracticality of these forms of management, however, led to the development of a dietary treatment plan that includes frequent daytime meals followed by continuous drip nocturnal enteral feedings. Hypoglycemia and death have been reported following malfunctions of the pump or dislodging of the tube (Daeschel et al. 1983; Folk and Greene 1984). Dietary treatment, consisting of frequent glucose feedings, needs to be maintained for life.

The use of uncooked cornstarch has been introduced due to its ability to maintain blood glucose levels over an extended period of time. The cornstarch is slowly digested by the body, allowing for the gradual release of glucose. This helps the individual to maintain a fed state for an extended period of time, from about 3 to 6 hours. The typical dose of cornstarch is in amounts of 1 to 2 grams cornstarch per kg body weight per dose, administered every 3 to 6 hours. Some individuals have even been able to forgo nocturnal feedings by ingesting cornstarch at bedtime and once in the middle of the night. Cornstarch should not be mixed in beverages containing citric acid, nor should it be heated, because that will result in a breakdown in the starch molecules that allows more rapid digestion and interferes with the release of glucose over an extended time period (Chen et al. 1993).

The goals of nutritional therapy for GSD I are to prevent hypoglycemia, correct metabolic derangement, and provide optimal nutrition to support growth and development. Carbohydrates should be provided from complex sources and should provide about 60% to 70% of the total caloric intake.

dephosphorylate—to remove a phosphate group from an organic molecule

A portion of the carbohydrate may be provided by the uncooked cornstarch. Protein should come from lean sources, providing 10% to 15% of total kcal. Fat should provide less than 30% of total kcal, and dietary saturated fat should be limited. Two-thirds of the kcal will be given during daytime feedings to allow for carbohydrates to be distributed frequently and evenly over the 24-hour period. One-third of kcal is administered overnight, primarily from a carbohydrate source via cornstarch or tube feedings. Meals high in starch, without free glucose, tend to provide a more sustained increase in blood glucose levels and, except during symptoms suggestive of hypoglycemia, glucose solutions alone are not recommended (Daeschel et al. 1983; Folk and Greene 1984).

Nutritional Concerns Compliance with dietary treatment reduces the risk of developing hypoglycemia. Daytime schedules must be flexible enough to include ingestions of feedings as needed. School-age children need to carry high-carbohydrate snacks with them at all times. Infants and toddlers are given a formula concentrated in glucose polymers for symptomatic hypoglycemia. Refusal by toddlers of breakfast can cause potential problems. In such cases, mothers are encouraged to use a nasogastric tube for overnight feedings and to keep it in place until after the first meal in the morning. Illness, particularly vomiting and diarrhea, can be life threatening to GSD I patients. Sick-day guidelines should be provided. If sick children are not eating in adequate amounts, the tube should be left in place to permit continuous infusion of carbohydrate. If this is not tolerated, then the child must be hospitalized to receive intravenous glucose. (See Box 28.2 for a clinician's perspective on management of GSD I in children.)

Another significant issue with dietary management is the patient's adjustment to a decreased oral intake, since a significant amount of their caloric needs must be administered during overnight feedings. The omission of fructose and galactose (fruits, juice, milk, dairy products, and sweets that contain sucrose) from the GSD I patient's diet further limits the already restricted diet. This omission is necessary, because these sugars cannot be converted into glucose and are subsequently metabolized via alternative pathways leading to an increased production of lactic acid and lipids. Consequently, infants are recommended to use a formula that contains glucose polymers as the carbohydrate source.

A multivitamin/mineral supplement is usually indicated, since a high percentage of the kcal may be coming from "empty" kcal sources, such as cornstarch, and because of the limitation of food groups, including fruits and dairy, in an attempt to limit ingestion of fructose and galactose. Calcium supplementation is usually needed due to a limited intake of dairy products. Additional iron supplementation is frequently warranted for patients relying on cornstarch for night feedings, because the cornstarch can chelate and inhibit the absorption of iron. Dietary compliance is necessary to achieve the full growth potential of GSD I children and for the prevention of secondary complications of the disease (Daeschel et al. 1983; Acosta and Yannicelli 2001).

Adjunct Therapies Prognosis for patients with GSD I is still unclear. Prognosis during infancy has dramatically improved with the administration of dietary management. Biochemical and clinical aberrations can be substantially improved by treatment aimed at decreasing the hepatic stimulus for glycogenolysis by maintenance of blood glucose concentration between 70 and 150 mg/dL. This treatment has been very effective in providing the necessary milieu for normal growth and development. The fact that a number of patients have reached adulthood suggests that this type of dietary management is sufficient for some patients. There are many long-term complications reported in individuals with this disease, including nephropathy, hepatic adenomas, hepatocellular carcinoma, osteoporosis, anemia, pulmonary hypertension, acute pancreatitis, and polycystic ovaries. Transplantation of the liver provides a definitive cure of this otherwise life-time disease; however, the magnitude of the procedure would indicate its reservation for a small number of patients with refractory disease or hepatic malignancy (Moses 2002).

Disorders of Fat Metabolism

Etiology and Clinical Manifestations

Fatty acids are transported into the mitochondria for oxidation via a complex system (Pollitt 1995; Bennett 1998). The fatty acids must be taken into the mitochondria using the carnitine transport system (see Figure 28.10). Carnitine palmitoyltransferase I (CPTI) is an enzyme that allows a carnitine molecule to bind to a fatty acid molecule outside of the mitochondria while releasing the coenzyme A segment. This acylcarnitine compound is then taken into the mitochondria via the carnitine-acylcarnitine translocase enzyme. In the mitochondria, another enzyme, carnitine palmitoyltransferase II (CPT II), attaches the acylated fatty acid released from the carnitine with the coenzyme A segment. The carnitine is then released to allow for the transport of more fatty acids into the mitochondria.

Once they are inside the mitochondria, the process of beta oxidation is used to convert the fatty acids to ketone bodies, which can be used as an energy substrate in various tissues. Each time a fatty acid enters the beta oxidation pathway, a two-carbon segment of the fatty acid is cleaved off; this segment can subsequently be used to produce ATP via the Krebs cycle. Electrons produced in these reactions are transferred to the electron transport chain. The resulting shorter fatty acid then reenters the pathway to produce additional two-carbon segments. The process of cleaving off a two-carbon segment requires four separate enzymes that are

BOX 28.2 CLINICAL APPLICATIONS

Case Report for Glycogen Storage Disease Type 1

Case Presentation

Johnny is a 2-year-old boy who was diagnosed with glycogen storage disease type 1 (GSD 1) by 4 months of age, after presenting with hypoglycemia due to an extended fasting period of 6 hours. A gastrostomy tube was placed during the initial hospitalization to accommodate overnight feedings. Mother also administers tube feedings when Johnny refuses to eat meals during the day. Since then, he has been maintained on a regimen that consists of daytime feedings administered every 3 hours and continuous overnight feedings. Daytime meals consist of high-complex carbohydrate, low-fat foods with limited intakes of dairy and fruit. When Johnny refuses to eat, he is administered the formula that is used for overnight feedings. This formula is low in fat (containing only 10% of total kcal from fat) with added modulars including Polycose formula to provide glucose polymers as the carbohydrate source. It contains ample amounts of protein, providing up to 20% of total kcal, provided from the protein modular, Provimin, which also contains vitamins and minerals. The carbohydrate is administered at a rate of 6 to 8 mg glucose per kg body weight per minute. Mother checks Johnny's blood glucose level whenever he seems cranky or sleepy, and if it is less than 70 mg/dL, she will administer a mixture of glucose polymers and water via the gastrostomy tube followed by some crackers or cereal taken orally. Johnny has maintained fairly good metabolic control with recent labs of lactate of 2.7 mmol/L (normal is less than 2.2), triglyceride of 200 (normal is less than 150), and uric acid of 6.0 mg% (normal is less than 8). He has been growing appropriately, following the 25–50th percentile growth curve for height and 50–75th percentile for weight. Biochemical parameters used to monitor nutritional status, which include complete blood cell count, albumin, total protein, and transthyretin, have been within normal limits. Intake of total kcal and protein adequately meet estimated nutrient needs based on age and weight. He takes an additional vitamin/mineral supplement to ensure that his DRI for age are met.

During a recent clinic visit to the Metabolic Center, mother indicated that Johnny was refusing more daytime feedings and, consequently, she needed to use the tube to administer over one-third of his meals. There was a tendency to overfeed Johnny via gastrostomy tube feedings as mother was attempting to compensate for a decreased oral intake. His weight velocity has also increased, from the 50th to the 75th percentile growth curve.

Discussion

A decreased oral intake is a common problem encountered by children with this disease. This is because parents, desiring to maintain euglycemia, tend to overcompensate with increased frequency of feedings. The ability for the child to experience the sensation of hunger is significantly blunted. There is also a tendency to overfeed the child leading to excessive weight gain. Cornstarch is considered since it can help to reduce frequency of feedings. Since it is a concentrated source of carbohydrate, which is free of fat, it can help reduce the amount of kcal administered.

Plan

Cornstarch has been added to Johnny's daytime feedings to allow for more extended periods between meals. Johnny now receives 2 tbsp. cornstarch, providing approximately 2 g cornstarch/kg body weight/dose, at the completion of nocturnal tube feedings at 9 a.m. He is then allowed to go for a 4- to 5-hour period without eating or receiving a gastrostomy tube feeding to allow him an opportunity to become hungry and eat an afternoon meal. Johnny receives another bolus of cornstarch at 2 p.m. and again at 7 p.m. in between meals. Since the institution of cornstarch, Johnny has been eating more meals and has actually been expressing hunger prior to mealtime. Blood glucose control has remained fairly stable, with levels remaining between 70 and 140 mg/dL. He did experience a moderately low level one day, around 5 p.m., after having a very active day. Mother was encouraged to ensure that Johnny is eating more snacks or glucose-containing beverages, such as sports drinks, during these times. Mother is happy with the new feeding regimen as Johnny has started eating a greater variety of foods and he has become less dependent on the tube for daytime feedings.

specific to the fatty acid, based on its carbon chain length. There are enzymes specific for short-, medium- and long-chain fatty acids. Given the complexity of the many enzymes involved in the metabolism of fat, a number of disorders are known to affect this macronutrient's metabolism (see Table 28.14) (Przyrembel et al. 1991; Hale and Bennett 1992; Pollitt 1995; Bennett 1998; Tein 2003).

The presentation of these disorders in patients can vary considerably. An affected individual can present in the neonatal period when breast milk or an infant formula, both of which are high in long-chain fatty acids, is introduced. In the case of medium-chain acyl-CoA dehydrogenase deficiency (MCADD), presentation does not usually occur until the infant or child is sleeping through the night. The fasting period during the night leads to the release of fatty acids that cannot be properly utilized for energy and results in hypoglycemia, hyperammonemia, and/or death. MCADD is the most common of the fatty acid oxidation disorders and is reported to occur as frequently as PKU occurs, one in every 15,000 births. Many individuals, however, can go undiagnosed for years until a metabolic stressor such as a viral illness or prolonged fast occurs. Disorders of carnitine me-

FIGURE 28.10 Fatty Acid Transport System: The enzyme CPT 1 is needed to bring the fatty acyl CoA across the outer mitochondria. Inside the mitochondria, CPT 2 is needed to release the fatty acyl CoA from the carnitine so that it can be oxidized.

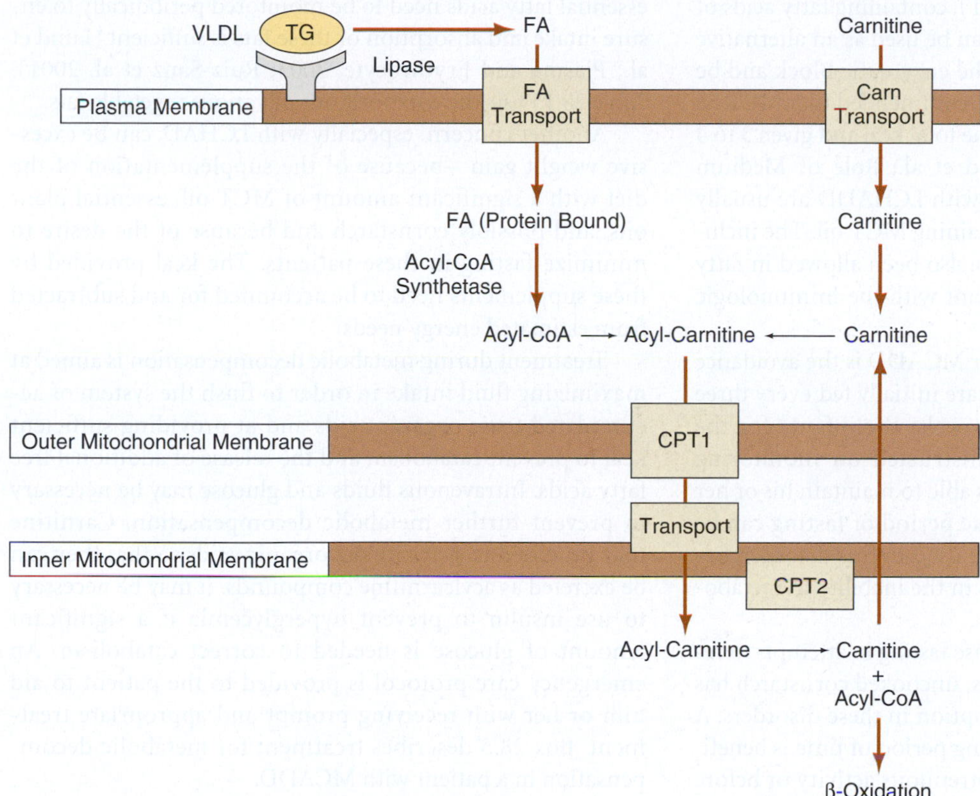

Source: Figure 1, p. 1 of Bennett MJ, Ross. *Metabolic Currents*, Vol 9, #1.

tabolism that can present at various ages with symptoms associated with the deficiency of this compound have also been found. Secondary carnitine deficiency, associated with these conditions, can result in cardiac hypertrophy, muscle weakness, liver enlargement, or **rhabdomyolysis**. Many states are now screening for these disorders at birth.

TABLE 28.14

Disorders of Fat Metabolism

- Carnitine acylcarnitine translocase deficiency
- Carnitine palmotyl transferase (CPT) I deficiency
- Carnitine palmotyl transferase (CPT) II deficiency
- Very long chain acyl-CoA dehydrogenase (VLCAD) deficiency
- Tri-functional protein (TFP) deficiency
- Long-chain 3 hydroxy acyl-CoA dehydrogenase (LCHAD) deficiency
- Medium-chain acyl-CoA dehydrogenase (MCAD) deficiency
- Short-chain 3 hydroxy acyl-CoA dehydrogenase (SCHAD) deficiency
- Short-chain acyl-CoA dehydrogenase (SCAD) deficiency

Nutrition Interventions

The main goals of therapy involve prevention of fasting, limiting intake of fatty acids, and/or providing an alternate substrate for metabolism (Hale and Bennett 1992). Dietary intake of fatty acids may be mildly restricted in some disorders, because the individual fatty acids are not transported appropriately in the mitochondria (in carnitine transporter disorders) and can accumulate. When the fatty acids are not metabolized correctly, there is an increased production of acylated fatty acids, which can build up within the mitochondria. A moderate fat restriction is recommended, at a level of no more than 30% of total kcal consumed. Other general dietary guidelines include consuming foods that are high in complex carbohydrates and low in simple sugars in order to maintain **euglycemia** and prevent the release of free fatty acids.

rhabdomyolysis—disintegration of muscle; associated with excretion of myoglobin in the urine

euglycemia—maintenance of normal blood sugar levels

Intake of long-chain fats in long chain 3-hydroxy acyl-CoA dehydrogenase deficiency (LCHADD) is restricted to no more than 15% of total kcal consumed. In this disorder, medium-chain triglycerides (MCT), containing fatty acids of 6 to 10 carbon chains in length, can be used as an alternative substrate since they can bypass the enzymatic block and be used for energy production. It is recommended that MCT be supplemented at 15% to 20% of the total kcal and given 3 to 4 times throughout the day (Lund et al., Role of Medium Chain, 2003). Infants diagnosed with LCHADD are usually started on an infant formula containing MCT oil. The inclusion of expressed breast milk has also been allowed in fatty acid disorders to provide the infant with the immunologic benefits of maternal breast milk.

The mainstay of treatment for MCADD is the avoidance of fasting. Infants with MCADD are initially fed every three hours. Parents are instructed to wake the infants for the feeding. The parents are also instructed on monitoring blood glucose levels. If the baby is able to maintain his or her blood sugar consistently, then the period of fasting can be extended. MCT oil should not be used in this disorder, because the enzymatic block results in the inability to metabolize these fats into energy.

Because of the desire to decrease fasting in attempts to reduce the release of free fatty acids, uncooked cornstarch has been suggested as a therapeutic option in these disorders. A steady release of glucose over a long period of time is beneficial during extended periods of strenuous activity or before bedtime. Despite good metabolic control, **retinopathies** often develop in LCHAD. Research has been conducted to determine if supplementation of the essential fatty acid, **docosahexaenoic acid** (**DHA**) can help to prevent the progression of retinal degeneration, but results thus far have shown little benefit (Lund et al., Plasma and Erythrocyte, 2003).

Nutritional Concerns

Due to the fat restriction and the current availability of fat-free food products, overrestriction of fat intake is a concern. Fat is essential for fat-soluble vitamin absorption and is needed to supply the body with essential fatty acids and energy. Essential fatty acid deficiency can occur as a result of the dietary restriction; therefore, the prescribed diet should contain at least 3% of the allowed long-chain fats as a source of linoleic acid and 2% as linolenic acid. Red blood cell levels of essential fatty acids need to be monitored periodically to ensure intake and absorption of these fats is sufficient (Lund et al., Plasma and Erythrocyte, 2003; Ruiz-Sanz et al. 2001). Table 28.15 lists the compositions of various vegetable fats.

Another concern, especially with LCHAD, can be excessive weight gain—because of the supplementation of the diet with a significant amount of MCT oil, essential plant oils, and possibly cornstarch and because of the desire to minimize fasting in these patients. The kcal provided by these supplements need to be accounted for and subtracted from estimated energy needs.

Treatment during metabolic decompensation is aimed at maximizing fluid intake in order to flush the system of accumulated toxic organic acids and at providing sufficient kcal to prevent catabolism and the release of additional free fatty acids. Intravenous fluids and glucose may be necessary to prevent further metabolic decompensation. Carnitine may be added to detoxify the organic acids so that they can be excreted as acylcarnitine compounds. It may be necessary to use insulin to prevent hyperglycemia if a significant amount of glucose is needed to correct catabolism. An emergency care protocol is provided to the patient to aid him or her with receiving prompt and appropriate treatment. Box 28.3 describes treatment for metabolic decompensation in a patient with MCADD.

Adjunct Therapies

The use of carnitine in these disorders is controversial: some believe that it can be harmful because it can promote a greater influx of fatty acids into the mitochondria, disrupting their function, while others believe it is helpful because it acts as a scavenger, removing the acylated fatty acids. Carnitine is usually supplemented if the amount of free carnitine is low and/or if the ratio of esterified to free carnitine is high. Overall benefit of carnitine supplementation in these disorders is still to be determined.

retinopathies—non-inflammatory diseases in the retina of the eye

docosahexanoic acid (DHA)—a fatty acid that can be produced from the essential omega-3 fatty acid linolenic acid; this fatty acid can also be found in many types of fish. DHA has been found to be beneficial for proper brain and vision development as well as in the management of hypertriglyceridemia

TABLE 28.15

Fatty Acid Composition of Various Fats (Per 1 tsp./5 cc)

Source	*Linoleic*	*Linolenic*
Canola	990	495
Corn	2,600	Trace
Flaxseed	575	2,425
Peanut	1,440	Trace
Safflower	3,350	Trace
Walnut	2,380	470

BOX 28.3 CLINICAL APPLICATIONS

Case Report of Medium Chain Acyl-CoA Dehydrogenase Deficiency

Case Presentation

James is a 6 month old with medium chain acyl-CoA dehydrogenase deficiency (MCADD). He is receiving 4–6 oz. of formula every 4 hours, and has been demonstrating appropriate growth. His early morning blood sugars have been in the 70–90 mg/dL range in the past month. James just started on rice cereal and some jarred baby foods, and typically eats 2–3 tbsp. of these foods twice a day. James is also starting to teethe, and his mom noted a low grade fever before she put him to bed. He had only half of his 4 a.m. feeding, and his blood sugar at 7:30 in the morning was 70 mg/dL. He only had half a bottle at 8 a.m. At 9 a.m., his mother tried to feed him some cereal, but he only ate 1 tbsp. and vomited a half hour later. At 11 a.m., his mother tried to give James another bottle due to his poor early morning intake, but he only took 1 oz. of formula and fell asleep.

Discussion

Given James's low-normal blood glucose level in the morning, his decreased intake, vomiting, and history of a low grade fever, James is likely catabolic and at risk for metabolic decompensation. Falling asleep after his morning feeding could be suggestive of lethargy associated with metabolic derangement and need for immediate attention.

Plan

James's mother contacted the Metabolic Clinic and was advised to take him with his emergency care protocol to the urgent care clinic for IV hydration with glucose. The glucose will provide kcal to stop the catabolic process, while the IV fluids will help to flush out any toxic metabolites that have accumulated. He will also be given carnitine through the IV, as it will be used to conjugate the toxic metabolites, detoxifying them so that they are excreted via urine output. James will receive an antipyretic to reduce his fever, in order to prevent further catabolism and help promote oral intake. The health care team will evaluate James to ensure that he does not have an infection or other condition necessitating further treatment. Once he is able to take liquids and foods orally, without difficulty, he will be discharged home.

Conclusion

Major advances have been made in identifying the specific enzymatic defects in more than 200 disorders caused by inborn errors of metabolism, allowing for accurate diagnosis. However, despite these major diagnostic achievements, patients and their families have been disappointed by the absence of specific therapies for some of the more debilitating disorders. Therapeutic endeavors have primarily involved attempts to alter the disease course by manipulations at the level of the metabolic or biochemical defects. Restriction of the accumulated precursor by dietary management, chelation (combining with a metal) or administration of appropriate metabolic inhibitors, and supplementation of the deficient metabolic products are the main treatment strategies. The uses of alternative metabolic pathways for excretion of toxic metabolites, and the use of cofactors to stimulate residual enzymatic function, have been employed in various disorders. Several drugs and hormones have been used in attempts to correct for the metabolic imbalance. Growth hormone, in its function of stimulating anabolism, has helped to stabilize patients and ultimately has led to improvement in metabolic control. Transplantation of organs capable of producing the normal enzyme has been utilized when other treatment options result in a less than optimal outcome. The applicability of this approach is restricted by the significant surgical risk as well as the limited number of organs available for transplantation.

Research has been directed at enzyme replacement as a potential means of treating selected inborn errors of metabolism, particularly with the lysosomal storage diseases. Work has been under way to synthesize large quantities of stable, non-immunogenic, sterile enzymes with high specific activities that can be used in appropriately designed human trials. Limitations in this methodology include the short circulating and intracellular half-lives of the enzyme, the inability to target enzymes to specific tissue, the inability to closely follow the fate of administered enzymes, and immunologic complications. Theoretically, the ideal cure for inborn errors of metabolism would be the insertion of a normal segment of DNA coding for the synthesis of the normal gene product. This in turn would produce a functioning enzyme that would correct for the metabolic derangement. Gene therapy is precluded by our limited biochemical and cellular technology to insert a gene that will be under the proper genetic regulation for normal expression. Much research is under way in this area, though it will be quite some time before the benefits of this treatment are achieved (Desnick and Grabowski 1981).

Management of patients with inborn errors of metabolism in the clinical setting has also been changing. The main focus of the metabolic genetics team has been on making the correct diagnosis and implementing a specific treatment to optimize patient outcome. Because improved treatment options mean that these patients are now surviving longer, the emphasis has turned to more completely understanding the potential medical and psychiatric complications associated with these disorders in adolescence and adulthood. It is essential that the family fully understands the necessity for the diet and for medical management; failure to do so usually leads to poor compliance. Every effort should be made to enable the child to grow up normally. It is important to continuously reinforce the principles of the diet and promote self-management skills so that each patient can both cope with the dietary and medical therapeutic regimens required by adolescence or early adulthood and lead as independent a life as possible.

TABLE 28.16

Drug–Nutrient Interactions for Medications/Supplements Used in Metabolic Disorders

Classification	*Mechanism*	*Generic*	*Brand Names*	*Possible Food–Drug Interactions*
Nutrient Supplement	Anti-atherogenic, antioxidant and immunomodulatory actions. It may also have wound-repair activity	Arginine	None	Nausea, abdominal cramps, and diarrhea
Nutrient Supplement	Replaces biotin in metabolic conditions resulting in increased excretion	Biotin	None	None
Nutrient Supplement	Provides alternate metabolic pathway by correcting deficiency of urea cycle intermediates	Citrulline	None	None
Antihomocystinuric	Binds with homocysteine to form the amino acid, methionine	Cystadane	Betaine	Diarrhea, nausea, stomach upset
Bowel Disease, Inflammatory, suppressant	Decreases bacterial synthesis of propionates in the intestine to aid with maintaining metabolic control	Metronidazole	Flagyl	Diarrhea; loss of appetite; nausea or vomiting; stomach pain or cramps; change in taste sensation; dryness of mouth; unpleasant or sharp metallic taste
Nutrient Supplement	Carnitine binds with toxic organic acids to allow excretion from the body	Levocarnitine	Carnitor	Abdominal or stomach cramps; diarrhea; headache; nausea or vomiting
Nutrient Supplement	Provides alternate metabolic pathway in urea cycle disorders and in disorders of leucine catabolism	Glycine	None	None
Nutrient Supplement	Used for cofactor therapy to stimulate residual enzymatic activity	Riboflavin	None	None
Antihyperammonemic	Decreases ammonia production by providing alternative pathway for nitrogen excretion	Sodium Benzoate	Ammonul	Dry mouth; increased thirst; loss of appetite; N and V; unusual tiredness or weakness; increased hunger
Antihyperammonemic	Decreases ammonia production by providing alternative pathway for nitrogen excretion	Sodium phenylbutyrate	Buphenyl	Nausea or vomiting; stomach pain; unpleasant taste; changes in taste; decreased appetite; unusual tiredness or weakness
Nutrient Supplement	Used for cofactor therapy to stimulate residual enzymatic activity	Thiamin (B_1)	None	↑ kcal intake requires ↑ thiamin supplementation
Supplement	Antioxidant and cofactor for respiratory chain	Alpha-Lipoic Acid	None	None reported

Practitioner Interview

Kathleen Huntington, M.S., R.D., L.D. *Metabolic Clinic, Child Development & Rehabilitation Center (CDRC) Oregon Health & Science University, Portland*

How long have you been an RD?

I've been an RD for about 20 years and I've worked in the Metabolic Clinic at CDRC since March 1987. The most common metabolic disorder we see is phenylketonuria (PKU), but we see all of them. For PKU our center nutritionally assesses for growth velocity, essential fatty acids, trace elements, urine methylmalonic acid (MMA) levels (to check vitamin B_{12} level), CBC, ferritin, and phenylalanine/tyrosine (PHE/TYR) levels when they come to the clinic, and we review annual historical PHE levels and average the lab results. In addition we use a database that reports out an ordering profile for medical foods for the clients. This database is unique to our program and is not available in other metabolic centers. This ordering profile tells us immediately whether there could be a potential problem, because if the care providers aren't ordering the prescribed medical protein, it means the clients are at risk nutritionally.

What are the common nutrition problems that you see in your clients?

Common nutrition problems for PKU clients are inadequate protein intake because of their protein-restricted diets, and there seems to be a bit of a problem with the nutrient deficiencies of zinc, essential fatty acids, sometimes iron. Working with this population, my biggest challenges are: 1) parents who are unable to organize themselves to meet the challenge of day-to-day planning and implementing medical nutritional therapy in the home, plus 2) health plan providers who deny coverage for medical foods and dietitian services, which creates more costs for the parents to shoulder.

How has the science of dietetics changed for you in your practice? What do you do to make sure that you correlate your practice with the latest science? What advice can you give to newly practicing dietitians and students?

The biggest change I have seen in the last ten years is in the expanded choices and variety of medical foods, which is a good thing. To keep up with changes in the field, I read journal articles and go to conferences. My advice to students is: to focus on biochemistry and metabolism; take as many courses as you can that have to do with motivation, counseling behavior change, and family support; and enhance your skills in desk top publishing for development of patient education materials.

CASE STUDY

General:

9-day-old female who had a positive newborn screen for PKU at 2 days of age.

Labs:

Day 2: phenylalanine 15 mg/dL; Day 8: phenylalanine 34.4 mg/dL and tyrosine 1.2 mg/dL.

Treatment:

None at this time

Past Medical History:

40 weeks gestation −5.5 lb birthweight; 18.5 inches length. Physical exam indicates normal infant reflexes and all other assessments WNL.

Nutrition History:

Soy formula—approximately 4 oz. every four hours

Treatment Plan:

Initial nutrition therapy goal to reduce phenylalanine levels to treatment range of 2–6 mg/dL. Prescription included 52 grams of Phenex-1 and 15 grams of Similac powder with iron providing 25 mg/kg of phenylalanine, 316 mg/kg of tyrosine, 3.5 g protein/kg, and 122 kcal/kg. 16 oz. of formula yields 20 kcal/oz.

Questions:

1. What is the screening procedure for PKU?
2. Why is it important to accomplish screening as soon after birth as possible?
3. What is the goal of the PKU diet?
4. What is Phenex? How is it different from Similac or other standard infant formulas?
5. What would be the possible long-term complications of untreated PKU?
6. How long will the infant have to remain on a special diet?
7. In general, how will her diet change as she grows?

NUTRITION CARE PROCESS FOR METABOLIC DISORDERS

Step One: Assessment

Medical/Social History

Diagnoses/date of diagnosis

Family history/genetic history

Birth/prenatal history

Medications

Previous medical conditions or surgeries

Socioeconomic status/food security

Support systems

Education level – primary language

Physical Assessment

Signs and symptoms of nutrient deficiencies or toxicities

Tone

Physical activity level

Signs and symptoms associated with metabolic condition

Anthropometric (will vary depending on age of patient)

Height/length

Current weight/birth weight

Weight history if adult: highest adult weight; usual body weight

Body mass index

Biochemical Assessment
Visceral Protein Assessment (for adult client)

Albumin

Prealbumin

Complete blood count

General chemistries (i.e., electrolytes, LFTs, BUN, creatinine, etc.)

Newborn screening results

Labs pertinent to metabolic condition (i.e., phenylalanine level, plasma amino acids, ammonia, carnitine profile)

Dietary Assessment

Ability to suck in infant or chewing in older child

Problems swallowing, feeding aversions

Nausea, vomiting, reflux

Behavioral management

Compliance to dietary restrictions

Constipation, diarrhea

Feeding pattern: 24-hour recall and/or 3-day diet record

Nutrition support (i.e., tube feedings, TPN)

Step Two: Common Diagnostic Labels

Inadequate bioactive substance intake

Excessive bioactive substance intake

Impaired nutrient utilization

Underweight

Food-medication interaction

Food- and nutrition-related knowledge deficit

Limited adherence to nutrition-related recommendations

Limited resources to obtain special dietary products

Sample: Excessive bioactive substance intake NI-4.2

PES: Excessive bioactive substance intake as evidenced by increasing blood levels of phenylalanine and dietary record indicating an average daily intake of 345 mg Phenylalanine (125% of recommended intake).

Step Three: Sample Intervention

1. Change current formula to Phenex at 16 oz./day to replace protein not provided from diet.
2. Provide suggestions for substitutions of current food choices with lower-protein choices.

Step Four: Monitoring and Evaluation

1. Monitor serum phenylalanine levels; weight and nutrition assessment indices.
2. Evaluate food diary for nutrition adequacy and consistent phenylalanine levels within recommended amounts of 276 mg/day.

WEB LINKS

United Mitochondrial Disease Foundation: Information about vitamin therapy for mitochondrial diseases.

http://www.umdf.org

Parents of Galactosemic Children: Support for parents regarding dietary management of galactosemia.

http://www.galactosemia.org

PKU News: Latest research on PKU and information on new products.

http://www.pkunews.org

National Organization for Rare Disorders: Information about many rare disorders.

http://www.rarediseases.org

Fatty Acid Oxidation Disorders: Parent support organization with latest research on fatty acid oxidation disorders, family stories, and new products.

http://www.fodsupport.org

Organic Acidemia Association: Parent support organization with latest research on organic acid disorders, family stories, and new products.

http://www.oaanews.org

MSUD Newsletter: Parent support organization with latest research on maple syrup urine disease, family stories, and new products.

http://www.msud-support.org

National Coalition for PKU & Allied Disorders: General information related to supporting parents of children with a variety of metabolic conditions.

http://www.pku-allieddisorders.org

Genetic Metabolic Dietitians, International: Professional organization for dietitians and other health practitioners working in the field of metabolic disorders.

http://www.gmdi.org

Society for Inherited Metabolic Diseases: Professional organization for dietitians and other health practitioners working in the field of metabolic disorders.

http://www.simd.org

National Urea Cycle Disorders Foundation: Parent support organization with latest research on urea cycle disorders, family stories, and new products.

http://www.nucdf.org

Online Mendelian Inheritance in Man (OMIM): Detailed description of inherited metabolic disorders.

http://www3.ncbi.nlm.nih.gov

END-OF-CHAPTER REVIEW QUESTIONS

1. What is the difference between selective and non-selective screening for metabolic disorders?
2. What amino acids are restricted in maple syrup urine disease?
3. What amino acid becomes essential secondary to the enzymatic block in PKU?
4. What is periorfacial acrodermatitis, and when does it occur?
5. What is the role of carnitine?
6. How does fasting affect disorders of fat metabolism?
7. Which vitamins are used as cofactors in mitochondrial disorders?
8. What are some clinical signs/symptoms that would indicate the need for a further metabolic work-up?
9. What vitamin deficiency can present as methylmalonic acidemia?
10. What vitamins are used for their antioxidant properties in the treatment of mitochondrial disorders?
11. What sugars are restricted in GSD I?
12. What percentage of protein intake should come from essential amino acids in urea cycle disorders?
13. Which food groups are restricted in galactosemia, and why?
14. What is the mainstay of dietary management for GSD I?
15. What becomes significantly elevated in the blood of a person with a urea cycle disorder prior to treatment?
16. How is the diet prescription different in an individual with a more severe enzymatic defect vs. a less severe defect?

17. What are the initial symptoms of a child with a UCD?
18. How much protein should be given for an individual with a UCD?
19. What parameters should be monitored in an individual with a UCD?
20. In GSD, there are deficiencies of enzymes involved in synthesizing and degrading which polysaccharide?
21. Describe the diet composition on an individual with GSD I.
22. What typically happens when a person with HFI ingests fructose?
23. What are some of the complications that an individual with HFI will encounter after ingestion of fructose?
24. Describe successful dietary treatment of HFI.
25. What is monitored in the blood to measure dietary compliance in galactosemia?
26. What are some possible theories on why some patients with galactosemia experience poor outcomes?
27. What types of formula are used to treat infants with galactosemia?
28. Why is MCT oil used in disorders of long-chain fat metabolism? Why can't it be used in MCADD?
29. In what conditions can you breast-feed a child with a metabolic disorder? When would it be contraindicated?
30. What are the three main goals for the acute treatment of an amino acid disorder?

References

2001–2002 Alzheimer's disease progress report. National Institutes of Health publication number 03–5333. 2003 Jul; p. 2.

2006 Comprehensive accreditation manual for hospitals. Oak Brook (IL): Joint Commission on Accreditation of Healthcare Organizations; 2006.

A.S.P.E.N. Board of Directors and Standards Committee. Definition of terms, style, and conventions used in A.S.P.E.N. Guidelines and standards. Nutr Clin Prac. 2005;20:281–5.

A.S.P.E.N. Board of Directors and the Clinical Guidelines Task Force. Guidelines for the use of parenteral and enteral nutrition in adult and pediatric patients. JPEN J Parenter Enteral Nutr. 2002;26(1 Suppl):1SA–138SA.

A.S.P.E.N. Drug-nutrient interactions. JPEN J Parenter Enteral Nutr. 2002;26(1 Suppl):42SA–44SA.

Abdulkarim AS, Burgart LJ, See J, Murray JA. Etiology of nonresponsive celiac disease: results of a systematic approach. Am J Gastroenterol. 2002;97:2016–8.

Abou-Assi S, O'Keefe SJ. Nutrition support during acute pancreatitis. Nutrition. 2002;18:938–43.

Abou-Assi S, Craig K, O'Keefe SJ. Hypocaloric jejunal feeding is better than total parenteral nutrition in acute pancreatitis: results of a randomized comparative study. Am J Gastroenterol. 2002;97:2255–62.

Abrams B, Duncan D, Hertz-Piddiotto. A prospective study of dietary intake and acquired immune deficiency syndrome in HIV-seropositive homosexual men. J Acquir Immune Defic Syndr. 1993;6:949–58.

Abrams SA. Using stable isotopes to assess mineral absorption and utilization by children. Am J Clin Nutr. 1999;70: 955–64.

Abramson RG, Harrington CA, Missmar R, Li SP, Mendelson DN. Generic drug cost containment in Medicaid: Lessons from five state MAC programs. Health Care Financing Review 2004;25:25–34.

Access for Administration of nutrition support. JPEN J Parenter Enteral Nutr. 2002;26:SA33–SA41.

Achaya KT. Indian food: a historical companion. Delhi, India: Oxford University Press; 1994.

Acosta PA, Yannicelli S. Nutrition support protocols. 4th ed. [Need CITY]: Ross Metabolic Formula System; 2001.

Adam O, Beringer C, Kless T, Lemmen C, Adam A, Wiseman M, et al. Anti-inflammatory effects of a low arachidonic acid diet and fish oil in patients with rheumatoid arthritis. Rheumatol Int. 2003;23:27–36.

Adams H, Adams R, Del Zoppo G, Goldstein LB. Stroke Council of the American Heart Association; American Stroke Association. Guidelines for the early management of patients with ischemic stroke: 2005 guidelines update a scientific statement from the Stroke Council of the American Heart Association/American Stroke Association. Stroke. 2005;36:916–23.

Adams KE, Cohen MH, Eisenberg D, Jonsen AR. Ethical considerations of complementary and alternative medical therapies in conventional medical settings. Ann Intern Med. 2002;137:660–4.

Adamson J. Iron deficiency and other hypoproliferative anemias. In: Kasper DL, Braunwald E, Fauci AS, Hausner SL, Longo DI, Jameson JL, editors. Harrison's principles of internal medicine. 16th ed. New York: McGraw Hill Medical Publishing; 2005. p. 586–93.

Adamson JW, Longo DL. Hematological alterations. In: Kasper DL, Braunwald E, Fauci AS, Hausner SL, Longo DI, Jameson JL, editors. Harrison's principles of internal medicine. 16th ed. New York: McGraw Hill Medical Publishing; 2005. p. 329–37.

Addolorato G, Montalto M, Capristo E, Certo M, Fedeli G, Gentiloni N, et al. Influence of alcohol on gastrointestinal motility: lactulose breath hydrogen testing in orocecal transit time in chronic alcoholics, social drinkers and teetotaler subjects. Hepatogastroenterology. 1997;44:1076–81.

Adelson PD, Bratton SL, Carney NA, Chesnut RM, du Coudray HE, Goldstein B, et al. American Association for Surgery of Trauma; Child Neurology Society; International Society for Pediatric Neurosurgery; International Trauma Anesthesia and Critical Care Society; Society of Critical Care Medicine; World Federation of Pediatric Intensive and Critical Care Societies. Guidelines for acute medical management of severe traumatic brain injury in infants, children and adolescents. Pediatr Crit Care Med. 2003;4(3 Supp):S1–S491.

Adeniji OA, Barnett CB, DiPalma JA. Durability of the diagnosis of irritable bowel syndrome based on clinical criteria. Dig Dis Sci. 2004;49:572–4.

Adler G, Lawrence J. Cushing syndrome. New York: eMedicine.com, Inc.; 2001 [updated 2001 Sep 26; cited 2005 Jan 15]. Available from: http://www.emedicine.com.

Adler S, Fairley K. The patient with hematuria, proteinuria, or both, and abnormal findings on urinary microscopy. In: Schrier R, editor. Manual of nephrology. 6th ed. Philadelphia: Lippincott Williams & Wilkins; 2005. p. 116–33.

Agarwal DP. Genetic polymorphisms of alcohol metabolizing enzymes. Pathol Biol. 2001;49:703–9.

Agarwal KN. Iron and the brain: neurotransmitter receptors and magnetic resonance spectroscopy. Br J Nutr. 2001;85 Suppl 2:S147–50.

AHFS Drug Information. Bethesda (MD): American Hospital Formulary Service; 2005.

Ahluwalia JS. Health care in the United States: our dynamic jigsaw puzzle. Arch Int Med. 1990;150:256–8.

Ahluwalia N. Diagnostic utility of serum transferrin receptors measurement in assessing iron status. Nutr Rev. 1998; 56(Pt 1):133–41.

Ahmann AJ, Riddle MC. Current oral agents for type 2 diabetes; many options, but which to choose when? Postgrad Med. 2002;111:32.

Ahmed A, Cheung RC, Keeffe EB. Management of gallstones and their complications. Am Fam Physician. 2000;61: 1673–80, 1687–8.

Ainsworth BE, Haskell WL, Leon AS, et al. Compendium of physical activities:classification of energy costs of human physical activities. Med Sci Sports Exerc. 1993;25: 71–80.

Akahsi YJ, Springer J, Anker SD. Cachexia in chronic heart failure: prognostic implications and novel therapeutic approaches. Curr Heart Fail Rep. 2005;2:198–203.

Akisu M, Baka M, Huseyinov A, Kultursay N. The role of dietary supplementation with L-glutamine in inflammatory mediator release and intestinal injury in hypoxia/reoxygenation-induced experimental necrotizing enterocolitis. Ann Nutr Metab. 2003;47:62–6.

Akrabawi SS, Mobarhan S, Stolz RR, Ferguson PW. Gastric emptying, pulmonary function, gas exchange, and respiratory quotient after feeding a moderate versus high fat enteral formula meal in chronic obstructive pulmonary disease. Nutrition. 1996;12:260–5.

Alaimo K, Briefel RR, Frongillo EA Jr, Olson CM. Food insufficiency exists in the United States: results from the third National Health and Nutrition Examination Survey (NHANES III). Am J Public Health. 1998;88:419–26.

Alaimo K, McDowell MA, Briefel RR, Bischof AM, Caughman CR, Loria CM, Johnson CL. Dietary intake of vitamins, minerals, and fiber of persons ages 2 months and over in the United States: Third National Health and Nutrition Examination Survey, Phase 1, 1988–1991. Advance Data from Vital and Health Statistics, No. 258, Hyattsville, MD, National Center for Health Statistics.

Alamo MM, Moral RR, Pérula de Torres LA. Evaluation of a patient-centered approach in generalized musculoskeletal chronic pain/fibromyalgia patients in primary care. Patient Educ Couns. 2002;48:23–31.

Albers GW, Amarenco P, Easton JD, Sacco RL, Teal P. Antithrombotic and thrombolytic therapy for ischemic stroke: the Seventh ACCP Conference on Antithrombotic and Thrombolytic Therapy. Chest. 2004;126(3 Suppl): 483S–512S.

Albonico M, Saviolo L. Hookworm infection and disease: advances for control. Ann Ist Super Sanita. 1997;33:567–79.

Albright RC. Acute renal failure: a practical update. Mayo Clin Proc. 2001;76:67–74.

Aldana SG, Greenlaw R, Thomas D, Salberg A, DeMordaunt T, Fellingham GW, Avins AL. The influence of an intense cardiovascular disease risk factor modification program. Prev Cardiol. 2004;7:19–25.

Aldoori WH, Giovannucci EL, Rimm EB, et al. Prospective study of physical activity and the risk of symptomatic diverticular disease in men. Gut. 1995;36:276–82.

Aldoori WH, Giovannucci EL, Rimm EB, Wing AL, Trichopoulos DV, Willett WC. A prospective study of alcohol, smoking, caffeine, and the risk of symptomatic diverticular disease in men. Ann Epidemiol. 1995;5:221–8.

Aldoori WH, Giovannucci EL, Rockett HR, Sampson L, Rimm EB, Willett WC. A prospective study of dietary fiber types and symptomatic diverticular disease in men. J Nutr. 1998;128:714–9.

Alexopoulos, GS, Katz IR, Reynolds CF, Ross RW. Depression in older adults. J Psych Prac. 2001;7:441–6.

Alfonzo-Gonzalez G, Doucet E, Almeras N, Bouchard C, Tremblay A. Estimation of daily energy needs with the FAO/WHO/UNU 1985 procedures in adults: comparison to whole-body indirect calorimetry measurements. Eur J Clin Nutr. 2004;58:1125–31.

Al-Hilaly N, Kwiatkowski D, Gould S, Mitchell C, Sullivan PB. Gastric leiomyosarcoma presenting as severe iron-deficiency anemia. J Pediatr Gastroenterol Nutr. 1999:29: 354–7.

Allen LH. Advantages and limitations of iron amino acid chelates as iron fortificants. Nutr Rev. 2002;60 (Pt 2): S18–21.

Allen, JR, Humphries IR, Waters DL, Roberts DC, Lipson AH, Howman-Giles RG, Gaskin KJ. Decreased bone mineral density in children with phenylketonuria. Am J Clin Nutr. 1994;59:419–22.

Al-Momen A, El-Mogy I, Ibrahim A. Initial experience with Swedish adjustable gastric band at Saad Specialist Hospital, Al-Khobar, Saudi Arabia. Obes Surg. 2005;15:506–9.

Al-Muhsen S, Clarke AE, Kagan RS. Peanut allergy: an overview. CMAJ. 2003;168:1279–85.

Alonso A, Beunza JJ, Delgado-Rodriguez M, Martinez JA, Martinez-Gonzalez MA. Low-fat dairy consumption and reduced risk of hypertension: the Seguimiento Universidad de Navarra (SUN) cohort. Am J Clin Nutr. 2005;82: 972–9.

Alpert JS, Thygesen K. Myocardial infarction redefined—a consensus document of the Joint European Society of Cardiology/American College of Cardiology Committee for the Redefinition of Myocardial Infarction. Eur Heart J. 2000;21:1502–13.

Al-Qadresh A, Schulpis KH, Athanasopoulou H, Mengreli C, Skarpalezou A, Voskaki I. Bone mineral status in children with phenylketonuria under treatment. Acta Paediatr. 1998;87:1162–6.

al-Saddy NM, Blackmore CM, Bennett ED. High fat, low carbohydrate, enteral feeding lowers PaCO2 and reduces the period of ventilation in artificially ventilated patients. Intensive Care Med. 1989;15:290–5.

Altena TS, Michaelson JL, Ball SD, Guilford BL, Thomas TR. Lipoprotein subfraction changes after continuous or intermittent exercise training. Med Sci Sports Exerc. 2006; 38:367–72.

Alvarez-Leite JI. Nutrient deficiencies secondary to Bariatric surgery. Curr Opin Clin Nutr Metab Care. 2004;7:569–75.

Alvear J, Andreani S, Cortes F. Fetal alcohol syndrome and fetal alcohol effects: importance of early diagnosis and nutritional treatment. Rev Med Chil. 1998;126:407–12.

Amati L, Cirimele D, Pugliese V, Covelli V, Resta F, Jirillo E. Nutrition and immunity: laboratory and clinical aspects. Curr Pharm Des. 2003;9:1924–31.

Ambrose JA, Barua RS. The pathophysiology of cigarette smoking and cardiovascular disease. J Am Coll Cardiol. 2004;43:1731–7.

Ambrus JL Sr, Ambrus JL Jr. Nutrition and infectious diseases in developing countries and problems of acquired immunodeficiency syndrome. Exp Biol Med. 2004;229: 464–72.

American Academy of Allergy, Asthma & Immunology. Patient/public education: fast facts: food allergy [monograph on the Internet]. Milwaukee: American Academy of Allergy, Asthma & Immunology; 2006 [cited 2006 Mar 5]. Available from: http://www.aaaai.org/patients/contact.stm.

American Academy of Physicians, American Dietetic Association, National Council on the Aging, Inc. Nutrition Intervention Manual for Professionals Caring for Older Americans. Executive Summary [cited 2006 January 14]. Available from: http://www.eatright.org/ada/files/NSI OlderAmericansComplete1.pdf.

American Association for Clinical Chemistry. Diabetes-related autoantibodies [updated 2006 Jan 30; cited 2006 Mar 18]. Washington (DC): Lab Tests Online. Available from: http://www.labtestsonline.org.

American Association for Respiratory Care. Metabolic measurement using indirect calorimetry during mechanical ventilation—2004 revision & update. Respir Care. 2004;49:1073–9.

American Association of Clinical Endocrinologists medical guidelines for the clinical use of dietary supplements and nutraceuticals. Endocr Pract. 2003;9:417–70.

American Association of Colleges of Pharmacy. 2006. Available from: http://www.aacp.org.

American Cancer Society. Statistics for 2005. Retrieved July 2005. Available from: http://www.cancer.org/docroot/ STT/stt_0.asp.

American College of Cardiology/American Heart Association Task Force. American Diabetes Association. Peripheral arterial disease in people with diabetes. Diab Care. 2003;26:3333–41.

American College of Gastroenterology Functional Gastrointestinal Disorders Task Force. Evidence-based position statement on the management of irritable bowel syndrome in North America. Am J Gastro. 2002; 97:S1–S5.

American College of Physicians. Access to health care. Ann Int Med. 1990;112:641–61.

American College of Physicians. Diabetic testing and monitoring. Philadelphia: Focus On; 1998, Issue 6 [cited 2006 Mar 17]. Available from: http://www.acponline.org/mle/ diabetic_test.htm.

American Diabetes Association. Diagnosis and classification of diabetes mellitus. Diabetes Care. 2006;29(Suppl 1): S43–8.

American Diabetes Association. Economic costs of diabetes in the U.S. in 2002. Diabetes Care. 2003;26:917–32.

American Diabetes Association. Gestational diabetes mellitus. Diabetes Care. 2004;27(Suppl 1):S88–90.

American Diabetes Association. Hyperglycemic crisis in diabetes. Diabetes Care. 2004;27(Suppl 1):S94–102.

American Diabetes Association. Implications of the United Kingdom prospective diabetes study. Diabetes Care. 2002; 25(Suppl 1):S28–32.

American Diabetes Association. National Diabetes Fact Sheet, 2005. Alexandria (VA): American Diabetes Association; 2005 [cited 2006 Feb 10]. Available from: http://www .diabetes.org/diabetes-statistics,jsp.

American Diabetes Association. Nutrition principles and recommendations in diabetes. Diabetes Care. 2004;27(Suppl 1):S36–46.

American Diabetes Association. Physical activity/exercise and diabetes. Diabetes Care. 2004;27(Suppl 1):S58–62.

American Diabetes Association. Standards of medical care in diabetes—2006. Diabetes Care. 2006;29(Suppl 1): S4–42.

American Diabetes Association. The Diabetes Ready-Reference Guide for Health Care Professionals. Alexandria (VA): American Diabetes Association; 2000.

American Diabetes Association. Translation of the diabetes nutrition recommendations for health care institutions. Diabetes Care. 2002;35(Suppl 1):S61–3.

American Diabetes Association and American College of Cardiology. Peripheral arterial disease in diabetes. Diab Cardiovasc Dis Rev. 2004;6:1–8.

American Diabetes Association and American Dietetic Association. Exchange Lists for Meal Planning. Alexandria (VA) and Chicago (IL): American Diabetes Association and American Dietetic Association; 2003.

American Diatetic Association Evidence Analysis Library [homepage on the Internet]. Chicago: American Dietetic Association; 2006 [cited 2006 Jan 12]. Determining resting metabolic rate. Evidence analysis. Estimating RMR with prediction Equations: What does the evidence tell us? Available from: http://www.adaevidencelibrary.com/topic .cfm?format_tables=0&cat=2694.

American Diatetic Association Evidence Analysis Library. Disorders of lipid metabolism. Available by subscription at: http://www.adaevidencelibrary.com. Accessed: 6 April 2006.

American Diatetic Association Evidence Analysis Library. Heart failure. Available by subscription at: http://www

.adaevidencelibrary.com/topic.cfm?cat=1398. Accessed: 15 April 2006.

American Dietetic Association Evidence Analysis Library [homepage on the Internet]. Chicago (IL): American Dietetic Association; 2006 [cited 2006 Jan]. Available from: http://www.adaevidencelibrary.com.

American Dietetic Association Evidence Analysis Library. Calorie and protein needs during acute and rehabilitative phases. Available from: http://www.adaevidencelibrary.com/topic.cfm?cat=1443. Accessed: 18 November 2005.

American Dietetic Association Evidence Analysis Library. Disorders of lipid metabolism. Available from: http://www.ebg.adaevidencelibrary.com/template.cfm?template=guide_summaryandkey=229. Accessed: 2 June 2006.

American Dietetic Association Evidence Analysis Library. Effects of enteral versus parenteral nutrition. Available from: http://www.adaevidencelibrary.com/topic.cfm?cat=1032. Accessed: 4 February 2006.

American Dietetic Association Evidence Analysis Library. Nutritional care for prevention of pressure ulcers in SCI. Available from: http://www.adaevidencelibrary.com/evidence.cfm?evidence_summary_id=250072. Accessed: 18 November 2005.

American Dietetic Association. 2005 [cited 2005 June 29]. Available from: http://www.eatright.org/Public/Continuing Education/index_dpg10.cfm.

American Dietetic Association. Code of ethics for the profession of dietetics. J Am Diet Assoc. 1999;99:109–10.

American Dietetic Association. Health implications of fiber. J Am Diet Assoc. 2002;102:993–1000.

American Dietetic Association. High fiber diet. In: Manual of clinical dietetics. 6th ed. Chicago (IL): American Dietetic Association; 2000.

American Dietetic Association. Legislative highlights: Finn offers testimony on nutrition and the elderly. J Am Diet Assoc. 1992;92:1064–965.

American Dietetic Association. Manual of clinical dietetics. 6th ed. Chicago (IL):American Dietetic Association; 2000.

American Dietetic Association. Medical nutrition therapy across the continuum of care. Chicago (IL): American Dietetic Association; 1996.

American Dietetic Association. Nephrotic syndrome. In: Manual of clinical dietetics. 6th ed. Chicago (IL): American Dietetic Association; 2000. p. 455–82.

American Dietetic Association. Nutrition diagnosis: a critical step in the nutrition care process. Chicago (IL): American Dietetic Association; 2006.

American Dietetic Association. Nutrition screening initiative, a project of the American Academy of Family Physicians, American Dietetic Association and the National Council on the Aging, Inc. 1991.

American Dietetic Association. Ostomy. In: Manual of clinical dietetics. Chicago (IL): American Dietetic Association; 2000.

American Dietetic Association. Position of the American Dietetic Association and Dietitians of Canada: nutrition intervention in the care of persons with human immunodeficiency virus infection. J Am Diet Assoc. 2004;104:1425–41.

American Dietetic Association. Position of the American Dietetic Association: integration of medical nutrition therapy and pharmacotherapy. J Am Diet Assoc. 2003;103:1363–70.

American Dietetic Association. Position of the American Dietetic Association: nutrition intervention in the treatment of anorexia nervosa, bulimia nervosa, and eating disorders not otherwise specified (EDNOS). J Am Diet Assoc. 2001;101:810–9.

American Dietetic Association. Position of the American Dietetic Association: cost-effectiveness of medical nutrition therapy. J Am Diet Assoc. 1995;95:88–91.

American Dietetic Association. Position of the American Dietetic Association: nutrition services in managed care. J Am Diet Assoc. 1996;96:391–5.

American Dietetic Association. Reimbursement and insurance coverage for nutrition services. Chicago (IL): American Dietetic Association; 1991.

American Dietetic Association. Supplement to Medical nutrition therapy across the continuum of care. Chicago (IL): American Dietetic Association; 1997.

American Dietetic Association. The American Dietetic Association standards of professional practice for dietetic professionals. J Am Diet Assoc. 1998;98:84–5.

American Dietetic Association. Urolithiasis. In: Manual of clinical dietetics. 6th ed. Chicago (IL): American Dietetic Association; 2000. p. 483–6.

American Gastroenterological Association medical position statement: short bowel syndrome and intestinal transplantation. Gastroenterology. 2003;124:1105–10.

American Heart Association. Heart disease and stroke statistics—2006 update. Dallas (TX): American Heart Association; 2006.

American Heritage Stedman's Medical Dictionary. Houghton Mifflin Company; 2002.

American Holistic Medical Association. 2004. Available from: http://www.holisticmedicine.org/about/about.shtml.

American Liver Foundation. 2005. Facts on liver transplantation. http://www.liverfoundation.org/db/articles/1085.

American Liver Foundation. 2005. Gallstones: a national health problem. http://www.liverfoundation.org/db/articles/1047.

American Lung Association [homepage on the Internet]. New York: American Lung Association; 2005 [cited 2005 February 16]. Asthma & children fact sheet. Available from: http://www.lungusa.org/site/pp.asp?c=dvLUK9O0E&b=44352.

American Lung Association [homepage on the Internet]. New York: American Lung Association, 2005 [cited 2005

May 10]. Chronic Obstructive Pulmonary Disease Fact Sheet. Available from: http://www.lungusa.org/site/pp.asp?c=dvLUK9O0E&b=35020.

American Lung Association [homepage on the Internet]. New York: American Lung Association; 2005 [cited 2005 February 7]. How our lungs work. Available from: http://www.lungusa.org/site/pp.asp?c=dvLUK9O0E&b=22551.

American Lung Association [homepage on the Internet]. New York: American Lung Association; 2005 [cited 2005 November 9]. Pneumonia fact sheet. Available from: http://www.lungusa.org/site/pp.aspx?c=dvLUK9O0E&b=35692.

American Lung Association [homepage on the Internet]. New York: American Lung Association; 2005 [cited 2005 November 18]. What is pneumonia? Available from: http://www.lungusa.org/site/pp.aspx?c=dvLUK9O0E&b=35691.

American Lung Association. New York: American Lung Association [updated 2006 April 14; cited 2006 April 14]. Quit smoking action plan. Available from: http://www.lungusa.org.

American Medical Association Department of Foods and Nutrition. Multivitamin preparations for parenteral use; a statement by the Nutrition Advisory Group. JPEN J Parenter Enteral Nutr. 1979;3:258–62.

American Medical Association. 2006. Available from: http://www.ama-assn.org.

American Psychiatric Association. Diagnostic and statistical manual of mental disorders. 4th ed. Text revision (DSM-IV-TR). Washington (DC): American Psychiatric Association; 2000.

American Psychiatric Association. Practice guideline for the treatment of patients with eating disorders. Am J Psychiatry. 2000;157(Suppl):1–39.

American Society for Parenteral and Enteral Nutrition. Guidelines for the use of parenteral and enteral nutrition in adult and pediatric patients. JPEN J Parenter Enteral Nutr. 2002;26(Suppl 1):82S–83S.

American Society of Colon and Rectal Surgeons. Practice parameters for the treatment of sigmoid diverticulitis, March 2000. Available from: http://www.fascrs.org/displaycommon.cfm?an=1&subarticlenbr=149.

American Thoracic Society Documents. Statement on the care of the child with chronic lung disease of infancy and childhood. Am J Crit Care Med. 2003;168:356–96.

Ames BN. The metabolic tune-up: metabolic harmony and disease prevention. J Nutr. 2003;133:1544S–1548S.

Ames BN, Elson-Schwab I, Silver EA. High-dose vitamin therapy stimulates variant enzymes with decreased coenzyme binding affinity (increased K(m)): relevance to genetic disease and polymorphisms. Am J Clin Nutr. 2002;75:616–58.

Amodio P, Caregaro L, Patteno E, Marcon M, Del Piccolo F, Gatta A. Vegetarian diets in hepatic encephalopathy: facts or fantasies? Dig Liver Dis. 2001;33:492–500.

Amylin Pharmaceuticals. Symlin® (pramlintide acetate injection). 812004-CC. 2005.

Amylin Pharmaceuticals and Eli Lilly Company. Byetta® (exenatide injection). 02-05-1136-A; EX-35924. 2005.

Andersen HF. Use of fetal fibronectin in women at risk for preterm delivery. Clin Obstet Gynecol. 2000;43:746–8.

Andersen HS, McArdle HJ. How are genes measured? Examples from studies on iron metabolism in pregnancy. Proc Nutr Soc. 2004;63:481–90.

Anderson EJ, Delahanty L, Richardson M, Castle G, Cercone S, Lyon R, Mueller D, Snetselaar L. Nutrition interventions for intensive therapy in the diabetes control and complications trial. J Am Diet Assoc. 1993;93:768–72.

Anderson EN. Why is humoral medicine so popular? Soc Sci Med. 1987;25:331–7.

Anderson JW. Diabetes mellitus: medical nutrition therapy. In: Shils ME, Shike M, Ross AC, Cabellero B, Cousins RJ, editors. Modern nutrition in health and disease. 10th ed. Philadelphia: Lippincott Williams & Wilkins; 2006. p. 1043–66.

Anderson PM, Ramsay NK, Shu XO, et al. Effect of low-dose oral glutamine on painful stomatitis during bone marrow transplantation. Bone Marrow Transplant. 1998;22:339–44.

Anderson R, Schrier R. Acute tubular necrosis. In: Schrier R, Gottschalk CW, editors. Diseases of the Kidney. Boston: Little, Brown; 1993. p. 1287–318.

Andrès E, Noel E, Kaltenbach G. Comment: treatment of vitamin B(12) deficiency anemia: oral versus parenteral therapy. Ann Pharmacother. 2002;36:1809–10.

Andres T. Hepatic encephalopathy. In: Bircher J, Benhamou P, McIntyre N, Rizzeto M, Rodes J, editors. Oxford textbook of clinical hepatology. 2nd ed. Oxford: Oxford University Press; 1999. p. 765–83.

Andreyev H, Norman A, Oates J, Cunningham D. Why do patients with weight loss have a worse outcome when undergoing chemotherapy for gastrointestinal malignancies? Eur J Cancer. 1998;34;503–9.

Angell M. The American Health Care System revisited. N Engl J Med. 1999;340:48.

Anker SD, Steinborn W, Strassburg S. Cardiac cachexia. Ann Med. 2004;36:518–29.

Anonymous. August 7, 2002a. Canadian Institute for Health Information. Review of Reviewed Item.

Anonymous. August 7, 2002b. Canadian Institute for Health Information.

Anonymous. August 7, 2003c. Japanese Health Insurance System.

Anonymous. From the centers for disease control and prevention. Iron deficiency—United States, 1999–2000. JAMA. 2002;288:2114–6.

Anonymous. Parenteral nutrition in patients receiving cancer chemotherapy. American College of Physicians. Ann Intern Med. 1989;110:734–6.

Anonymous. World Health Report. World Health Organization; 2005.

Aora NS, Rochester DF. Respiratory muscle strength and maximal voluntary ventilation in undernourished patients. Am Rev Respir Dis. 1982;126:5–8.

Appel LJ, Brands MW, Daniels SR, Karania N, Elmer PJ, Sacks FM. American Heart Association. Dietary approaches to prevent and treat hypertension: a scientific statement from the American Heart Association. Hypertension. 2006;47: 296–308.

Appel LJ, Moore TJ, Obarzanek E, Vollmer WM, Svetkey LP, Sacks FM, et al. A clinical trial of the effects of dietary patterns on blood pressure. DASH Collaborative Research Group. N Engl J Med. 1997;336:1117–24.

Appel LJ, Sacks FM, Carey VJ, Obarzanek E, Swain JF, Miller ER 3rd, et al. OmniHeart Collaborative Research Group. Effects of protein, monounsaturated fat, and carbohydrate intake on blood pressure and serum lipids: results of the OmniHeart randomized trial. JAMA. 2005;294: 2455–64.

Appelbaum FR. Hematopoietic cell transplantation. In: Kasper DL, Braunwald E, Fauci AS, Hausner SL, Longo DI, Jameson JL, editors. Harrison's principles of internal medicine. 16th ed. New York: McGraw Hill Medical Publishing; 2005. p. 668–73.

Aquino VM, Harvey AR, Garvin JH, Godder KT, Nieder ML, Adams RH, et al. A double-blind randomized placebo-controlled study of oral glutamine in the prevention of mucositis in children undergoing hematopoietic stem cell transplantation: a pediatric blood and marrow transplant consortium study. Bone Marrow Transplant. 2005;36: 611–6.

Arcand JA, Brazel S, Joliffe C, Choleva M, Berkoff F, Allard JP, Newton GE. Education by a dietitian in patients with heart failure results in improved adherence with a sodium-restricted diet: a randomized trial. Am Heart J. 2005;150:716.

Arfons LM, Lazarus HM. Total parenteral nutrition and hematopoietic stem cell transplantation: an expensive placebo? Bone Marrow Transplant. 2005;36:281–8.

Argilés JM, Busquets S, FJ López-Soriano. Cytokines in the pathogenesis of cancer cachexia. Curr Opin Clin Nutr Metab Care. 2003;6:401–6.

Arinzon Z, Fidelman Z, Peisakh A, Adunsky A. Folate status and folate related anemia: a comparative cross-sectional study of long-term care and post-acute care psycho-geriatric patients. Arch Gerontol Geriatr. 2004;2039:133–42.

Aris RM, Lester GE, Dingman S, Ontjes DA. Altered calcium homeostasis in adults with cystic fibrosis. Osteoporos Int. 1999;10:102–8.

Ariza LM. Defensive eating food vaccines show promise—now forget about them [monograph on the Internet]. New York: Scientific American, Inc.; 2005 [cited 2006 26 Jan]. Available from: http://www.sciam.com/article.cfm?articleID=0007FFAD-7626-1264-B1DB83414B7F0000.

Armitage JO, Longo DL. Malignancies of lymphoid cells. In: Kasper DL, Braunwald E, Fauci AS, Hausner SL , Longo DI, Jameson JL, editors. Harrison's principles of internal medicine. 16th ed. New York: McGraw Hill Medical Publishing; 2005. p. 641–56.

Armstrong SA, Golub TR. Genomic approaches to the study of hematologic science. In: Hoffman R, editor. Hematology, basic principles and practice. 4th ed. Philadelphia (PA): Elsevier Medical Publishing; 2005. p. 17–27.

Arnadottir M, Berg A. Treatment of hyperlipidemia in renal transplant recipients. Transplantation. 1997;63:339–45.

Arnold GL, Kirby R, Preston C, Blakely E. Iron and protein sufficiency and red cell indices in phenylketonuria. J Am Coll Nutr. 2001;20:65–70.

Aronoff S, Rosenblatt S, Braithwaite S, Egan JW, Mathisen AL, Scheider RL. The Proglitazone 001 Study Group. Pioglitazone hydrochloride monotherapy improves glycemic control in the treatment of patients with type 2 diabetes. Diabetes Care. 2000;23:1605–11.

Ashfaq MK, Zuberi HS, Anwar Waqar M. Vitamin E and beta-carotene affect natural killer cell function. Int J Food Sci Nutr. 2000;51 Suppl:S13–20.

Ashford JW. APOE genotype effects on Alzheimer's disease onset and epidemiology. J Mol Neurosci. 2004;23:157–65.

Ashorn M. Acid and iron-disturbances related to helicobacter pylori infection. J Pediatr Gastroenterol Nutr. 2004;38: 137–9.

Asplin JR, Coe FL, Favus MJ. Nephrolithiasis. In: Kasper DL, Braunwald E, Fauci AS, Hauser SL, Longo DL, Jameson JL, Isselbacher KL, editors. Harrison's principles of internal medicine. 16th ed. 2005.

Asthma and Allergy Foundation of America [homepage on the Internet]. Washington (DC): Asthma and Allergy Foundation of America; 2006 [cited 2005 February 16]. Asthma overview. Available from: http://www.aafa.org/display.cfm?id=8.

Astin JA. 1998. Why patients use alternative medicine: results of a national study. JAMA. 279:1548–53.

Astin JR, Shapiro SL, Eisenberg DM, Forys KL. Mind-body medicine: state of the science, implications for practice. J Am Board Fam Pract. 2003;16:131–47.

Astor BC, Muntner P, Levin A, Eustace JA, Coresh J. Association of kidney function with anemia: The third national health and nutrition examination survey (1988–1994). Arch Intern Med. 2002;162:1401–8.

Atanasova B, Mudway IS, Laftah AH, et al. Duodenal ascorbate levels are changed in mice with altered iron metabolism. J Nutr. 2004;134:501–5.

Atkinson MA, Bowman MA, Kao K, Campbell L, Dush PJ, Shah SC, et al. Lack of immune responsiveness to bovine serum albumin in insulin-dependent diabetes. N Engl J Med. 1993;329:1853–8.

Atkinson SA, Ward WE. Clinical nutrition 2: the role of nutrition in the prevention and treatment of adult osteoporosis. CMAJ. 2001;165:1511–4.

Atwood KC. The ongoing problem with the National Center for Complementary and Alternative Medicine. Skeptical Inquirer. 2003;27:23–29.

August DA. Nutritional care of cancer patients. In: Norto JA, editor. Surgery: scientific basis and current practice. New York: Springer Verlag; 2001.

AuYeung SC, Ensom, MH. Phenytoin and enteral feedings: does evidence support an interaction? Ann Pharmaco. 2000; 34:896–905.

Avenell A, Brown TJ, McGee MA, Campbell MK, Grant AM, Broom J, et al. What are the long-term benefits of weight reducing diets in adults? A systematic review of randomized controlled trials. J Hum Nutr Diet. 2004;17:317–35.

AVERT. The History of AIDS 1981–1986 [monograph on the Internet]. West Sussex (UK): AVERT; 2006. Available from: http://www.avert.org/his81_86.htm.

Avram MM. Improving prognosis for kidney disorders in the 21st century: hypertension, anemia, nutrition, and lipids: Introduction. Am J Kidney Dis. 2001;38:1334–6.

Azhar G, Wei JY. Nutrition and cardiac cachexia. Curr Opin Clin Nutr Metab Care. 2006;9:18–23.

Babcock TA, Dekoj T, Espat NJ. Experimental studies defining omega-3 fatty acid antiinflammatory mechanisms and abrogation of tumor-related syndromes. Nutr Clin Pract. 2005;20:62–74.

Babior BM, Bunn HF. Megaloblastic anemia. In: Kasper DL, Braunwald E, Fauci AS, Hausner SL, Longo DI, Jameson JL, editors. Harrison's principles of internal medicine. 16th ed. New York: McGraw Hill Medical Publishing; 2005. p. 601–7.

Back EI, Frindt C, Nohr D, Frank J, Ziebach R, Stern M, Ranke M, Bieslski HK. Antioxidant deficiency in cystic fibrosis: when is the right time to take action? Am J Clin Nutr. 2004;80:374–84.

Baer, Hans. Biomedicine and alternative healing systems in America. Madison (WI): University of Wisconsin Press; 2001.

Baik HW, Russell RM. Vitamin B12 deficiency in the elderly. Ann Rev Nutr. 1999;19:357–77.

Baile GR, Johnson CA, Mason NA. Parenteral iron use in the management of anemia in end-stage renal disease patients. Am J Kid Dis. 2000;35:1–12.

Bain MD, Borriello SP, Tracey BM, Jones M, Reed PJ, Chalmers RA, et al. Contribution of gut bacterial metabolism to human metabolic disease. Lancet. 1998;1078–9.

Baker DM. Assessment and management of impairments in swallowing. Nurs Clin North Am. 1993; 28:793–805.

Baker JP, Detsky AS, Wesson DE, Wolman SL, Stewart S, Whitewell J, et al. A comparison of clinical judgment and objective measurements. N Engl J Med. 1982;306: 969–72.

Baliga MS, Jagetia GC, Ulloor JN, Baliga MP, Venkatesh P, Reddy R, et al. The evaluation of the acute toxicity and long term safety of hydroalcoholic extract of Sapthaparna (Alstonia scholaris) in mice and rats. Toxicol Lett. 2004; 15:317–26.

Ballard TL, Clapper JA, Specker BL, Binkley TL, Vukovich MD. Effect of protein supplementation during a 6-mo strength and conditioning program on insulin-like growth factor I and markers of bone turnover in young adults. Am J Clin Nutr. 2005;81:1442–8.

Banatvala JE, Cramp A, Jones IR, Feldman RA. Salmonellosis in North Thames (East), UK: associated risk factors. Epidemiol Infect. 1999;122:201–7.

Banatvala JE, Bryant J, Schernthaner G, Borkenstein M, Schober E, Brown D, et al. Coxsackie B, mumps, rubella, and cytomegalovirus specific igm responses in patients with juvenile-onset insulin-dependent diabetes mellitus in Britain, Austria, and Australia. Lancet. 1985;1(8443): 1409–12.

Bannister DK, Acchiardo SR, Moore LW, et al. Nutritional effect of peritonitis in continuous ambulatory peritoneal dialysis. J Am Diet Assoc. 1987;87:53–6.

Barabino A, Dufour C, Marino CE, Claudiani F, De Alessandri A. Unexplained refractory iron-deficiency anemia associated with helicobacter pylori gastric infection in children: further clinical evidence. J Pediatr Gastroenterol Nutr. 1999;28:116–9.

Baracos VE. Management of muscle wasting in cancer-associated cachexia: understanding gained from experimental studies. Cancer. 2001;92:1669–77.

Baracos VE. Regulation of skeletal-muscle-protein turnover in cancer-associated cachexia. Nutrition, 2000;16:1015–8.

Barbut F, Meynard JL. Managing antibiotic associated diarrhoea. BMJ. 2002;324:1361–4, 1345–6.

Barnard DD. Heart failure in women. Curr Cardiol Rep. 2005;7:159–65.

Barnes J, Abbott C. Articles on complementary medicine in the mainstream medical literature: an investigation of MEDLINE, 1966–1996. Arch Intern Med. 1999;159: 1721–5.

Barnes PM, Powell-Griner E, McFann K, Nathin RL. Complementary and alternative medicine use among adults: United States, 2002. Advance Data from Vital and Health Statistics, 343. Hyattsville (MD): National Center for Health Statistics; 2004.

Baron JH. Peptic ulcer. Mt Sinai J Med. 2000;67:58–62.

Barrett B, Marchand L, Scheder J, Appelbaum D, Plane MB, Blustein J, et al. What complementary and alternative medicine practitioners say about health and health care. Annals of Family Medicine. 2004;2:253–9.

Barri Y. Vascular disorders of the kidney. In: Carpenter G, Griggs R, Loscalzo J, editors. Cecil essentials of medicine. Philadelphia: WB Saunders; 2001. p. 278–82.

Barsky, AJ., Peekna HM, Borus, JF. Somatic symptom reporting in women and men. J Gen Intern Med. 2001;16: 266–75.

Bartlett JG. Antibiotic associated diarrhea. N Engl J Med. 2002;346:334.

Bartlett JG. Pocket guide to adult HIV/AIDS treatment. January 2005. Available from: http://hopkins-aids.edu/publications/pocketguide/pocketgd0106.pdf.

Bateman M, McGahey C. A framework for action: child diarrhea prevention. Global Health Link; 2001. Available from: http://www.ehproject.org/Pubs/GlobalHealth/GHCArticle.htm. Accessed: 3 March 2006.

Bateson MC. Gallbladder disease. BMJ. 1999;318:1745–8.

Batmanglij NK. New food of life: ancient Persian and modern Iranian cooking and ceremonies. Washington (DC): Mage; 2000.

Batterham MJ, Garsia R, Greenop P. Prevalence and predictors of HIV-associated weight loss in the era of highly active antiretroviral therapy. Int J STD AIDS. 2002;13:744–7.

Bauchner H. The hygiene hypothesis: is it good to be clean? Journal Watch Pediatrics and Adolescent Medicine [serial on the Internet]. 2002 Jul 29. Available from: http://pediatrics.jwatch.org/cgi/content/full/2002/729/1.

Bauer J, Capra S, Ferguson M. Use of the scored patient-generated subjective global assessment (PG-SGA) as a nutrition assessment tool in patients with cancer. Eur J Clin Nutr. 2002;56:779–85.

Bauer KA. Hypercoagulable states. In: Hoffman R, editor. Hematology, basic principles and practice. 4th ed. Philadelphia, PA: Elsevier Medical Publishing; 2005. p. 2197–224.

Baum MK. Role of micronutrients in HIV-infected intravenous drug users. J Acquir Immune Defic Syndr. 2000;25 suppl 1:S49–S52.

Baum MK, Shor-Posner G. Micronutrient status in relationship to mortality in HIV-1 disease. Nutr Rev. 1998;56: S135–S139.

Baum MK, Campa A, Lai S, Lai H, Page JB. Zinc status in human immunodeficiency virus type 1 infection and illicit drug use. Clin Infect Dis. 2003;37 suppl 2:S117–S123.

Bausell RB, Lee WL, Berman BM. Demographic and health-related correlates to visits to complementary and alternative medical providers. Med Care. 2001;39:190–6.

Bayard M, McIntyre J, Hill KR, Woodside J. Alcohol withdrawal syndrome. Am Fam Physician. 2004;69:1443–50.

Baylin A, Kabagambe EK, Ascherio A, Spiegelman D, Campos H. High 18:2 trans-fatty acids in adipose tissue are associated with increased risk of nonfatal acute myocardial infarction in Costa Rican adults. J Nutr. 2003;133:1186–91.

Baynes K, Betteridge DJ. Diabetes mellitus, lipoprotein disorders and other metabolic diseases. In: Axford J, O'Callaghan C, editors. Medicine. 2nd ed. Oxford (UK): Blackwell Science Ltd.; 2004.

Beahrs OH, Henson DE, Hutter DVP, et al. American Joint Committee on Cancer: Manual for staging of cancer (5th ed.). Philadelphia: Lippincott; 1997.

Beard JL. Iron biology in immune function, muscle metabolism and neuronal functioning. J Nutr. 2001;131: 568S–579S; discussion 580S.

Beard JL, Connor JR. Iron status and neural functioning. Annu Rev Nutr. 2003;23:41–58.

Beard K. Adverse reactions as a cause of hospital admissions for the aged. Drug Aging 1992;2:356–63.

Beck SA, Smith KL, Tisdale MJ. Anticachectic and antitumor effect of eicosapentaenoic acid and its effect on protein turnover. Cancer Res, 1991;51:6089–93.

Becker AE, Grinspoon SK, Klibanski A, Herzog DB. Eating disorders. N Engl J Med. 1999;340:1092–8.

Bederson JB, Awad IA, Wiebers DO, Piepgras D, Haley EC Jr, Brott T, et al. Recommendations for the management of patients with unruptured intracranial aneurysms: a statement for healthcare professionals from the Stroke Council of the American Heart Association. Stroke. 2000;31: 2742–50.

Bednarz B, Jaxa-Chamiec T, Gebalska J, Herbaczynska-Cedro K, Ceremuzynski L. L-arginine supplementation prolongs exercise capacity in congestive heart failure. Kardiol Pol. 2004;60:348–53.

Beebe CA. Nutrition therapy of type 2 diabetes. In: Franz MJ, Bantle JP, editors. American Diabetes Association guide to medical nutrition therapy for diabetes. Alexandria (VA): American Diabetes Association; 1999.

Beekman SE, Henderson DK. Protection of healthcare workers from bloodborne pathogens. Curr Opin Infect Dis. 2005;18:331–6.

Beers MH, editor. The Merck Manual Medical Library. Hospital-acquired and institutional-acquired pneumonia. Whitehouse Station (NJ): Merck & Co., Inc.; 1995–2006 [cited 2006 January 30]. Available from: http://www.merck.com/mmhe/print/sec04/ch042/ch042c.html.

Beers MH, editor. The Merck Manual Medical Library. Respiratory tract infections. Whitehouse Station (NJ): Merck & Co., Inc.; 1995–2006 [cited 2006 January 13]. Available from: http://www.merck.com/mmhe/print/sec23/ch273/ch273i.html.

Beers MH. Explicit criteria for determining potentially inappropriate medication use by the elderly: an update. Arch Intern Med. 1997;157:1531–6.

Beers MH, Berkow R, editors. Merck manual diagnosis and therapy. 17th ed. Sect. 19, Pediatrics, Chapter 267, Cystic fibrosis. [cited 2005 May 16]. Available from: http://www.merck.com/mrkshared/mmanual/section19/chapter267/267a.jsp.

Beers MH, Berkow R, editors. Merck manual of diagnosis and therapy. 17th ed. Whitehouse Station (NJ): Merck & Co.; 1999.

Beers MH, Berkow R, editors. Merck manual of diagnosis and therapy 17th ed. Ch. 202 Coronary Heart Disease. [Accessed: 2 April 2006]. Available from: http://www.merck.com/mrkshared/mmanual/sections.jsp.

Beers MH, Berkow R, editors. Clinical pharmacology. In: Merck manual of diagnosis and therapy. Seventeenth ed. Whitehouse Station (NJ): Merck & Co., Inc.; 2005.

Beers MH, Berkow R, editors. Merck manual of diagnosis and therapy. 17th ed. Whitehouse Station (NJ): Merck & Co.; 2005 [cited 2006 Feb 19]. Available from: http://www.merck.com/mrkshared/mmanual/home.jsp.

Beeuwkes AM. The prevalence of scurvy among voyageurs to America 1493-1600. J Am Diet Assoc. 1948;24:300–3.

Beevers DG. The epidemiology of salt and hypertension. Clin Auton Res. 2002;12:353–7.

Behler C, Shade S, Gregory K, Abrams D, Volberding P. Anemia and HIV in the antiretroviral era: potential significance of testosterone. AIDS Res Hum Retroviruses. 2005; 21:200–6.

Beitz JM. Diverticulosis and diverticulitis spectrum of a modern malady. J Wound Ostomy Continence Nurs. 2004 Mar–Apr 31;(2):75–82.

Bell A, Dorsch KD, McCreary DR, Hovey R. A look at nutritional supplement use in adolescents. J Adolesc Health. 2004;34:508–16.

Bellinghieri G, Santoro D, Calvani M, Savica V. Role of carnitine in modulating acute-phase protein synthesis in hemodialysis patients. J Ren Nutr. 2005;15:13–7.

Bender DA. Megaloblastic anemia in vitamin B12 deficiency. Br J Nutr. 2003;89:439–41.

Bengmark S Bio-ecological control of acute pancreatitis: the role of enteral nutrition, pro and synbiotics. Curr Opin Clin Nutr Metab Care. 2005;8:557–61.

Bennett MJ. Mitochondrial fatty acid oxidation defects: biochemistry, diagnosis and therapy. Metabolic Currents (Ross). 1998;9:1–6.

Benson CA, Kaplan JE, Masur H, Pau A, Holmes KK. Treating opportunistic infections among HIV-infected adults and adolescents: recommendations from CDC, the National Institutes of Health, and the HIV Medicine Association/Infectious Diseases Society of America. MMWR Morb Mortal Wkly Rep. 2004;53 (RR15):1–112.

Benz E. Hemoglobinopathies. In: Kasper DL, Braunwald E, Fauci AS, Hausner SL, Longo DI, Jameson JL, editors. Harrison's principles of internal medicine. 16th ed. New York: McGraw Hill Medical Publishing; 2005. p. 593–601.

Beradi RR. Safety and tolerability of Tegaserod in irritable bowel syndrome management. J Am Pharm Assoc. 2004; 44:41–51.

Berg AT, Shinnar S, Levy SR, Testa FM. Newly-diagnosed epilepsy in children: presentation at diagnosis. Epilepsia. 1999;40:445–52.

Berg MJ, Stumbo PJ, Chenard CA, Fincham RW, Schnieder PJ, Schottelius D. Folic acid improves phenytoin pharmacokinetics. J Am Diet Assoc. 1995:95:352–6.

Berger A, Henderson M, Nadoolman W, et al. (1995). Oral capsaicin provides temporary relief for oral mucositis pain secondary to chemotherapy/radiation therapy. J Pain Symptom Manage, 10(3), 243–248.

Berger MM, Mustafa I. Metabolic and nutritional support in acute cardiac failure. Curr Opin Clin Nutr Metab Care. 2003;6:195–201.

Bergman-Evans B. Improving medication management for older adult clients. Iowa City (IA): University of Iowa Gerontological Nursing Interventions Research Center, Research Dissemination Core; 2004.

Bergogne-Berezin E. Treatment and prevention of antibiotic associated diarrhea. Int J Antimicrob Agents. 2000 Dec 16;(4):521–6.

Bergström S. Infection-related morbidities in the mother, fetus and neonate. J. Nutr. 2003;133 Suppl 2: 1656S–1660S.

Berkel LA, Poston WSC, Reeves RS, Foreyt JP. Behavioral interventions for obesity. J Am Diet Assoc. 2005;105: S35–S43.

Berman BM, Lao L, Langenberg P, Lee WL, Gilpin AMK, Hochberg MC. Effectiveness of acupuncture as adjunctive therapy in osteoarthritis of the knee: a randomized, controlled trial. Ann Intern Med. 2004;14:901–10.

Berndt ER. Pharmaceuticals in US Health Care: determinants of quantity and price. J Econ Persp. 2002;16:45–66.

Berrios GE. Alzheimer's disease: a conceptual history. Int J Geriatr Psychiatry. 1990;5:355–65.

Berry GT, Steiner RD. Long-term management of patients with urea cycle disorders. J Pediatr. 2001;138:S56–S61.

Berseth CL, Van Aerde JE, Gross S, Stolz SI, Harris CL, Hansen JW. Growth, efficacy, and safety of feeding an iron-fortified human milk fortifier. Pediatrics. 2004;114:e699–706.

Bertolini A, Ottani A, Sandrini M. Selective COX-2 inhibitors and dual acting anti-inflammator drugs: critical remarks. Curr Med Chem. 2002;9:1033–43.

Bertolini G, Iapichino G, Radrizzani D, Facchini R, Simini B. Early enteral immunonutrition in patients with severe sepsis: results of an interim analysis of a randomised multicentre clinical trial. Intensive Care Med. 2003;29:834–40.

Bertoni AG, Hundley WG, Massing MW, Bonds DE, Burke GL, Goff DC. Diab Care. 2004;27:699–703.

Besarab A, Raja R. Vascular access for hemodialysis. In: Daugirdas J, Blake P, Ing T, editors. Handbook of dialysis. 3rd ed. Philadelphia: Lippincott Williams & Wilkins; 2001. p. 67–101.

Bessey PQ. Metabolic response to critical illness. ACS Surgery; 2004. [Posted 01/09/2004]. Available from: http://www.medscape.com/viewarticle/466714.

Best WR, Becktel JM, Singleton JW, Kern F. Development of a Crohn's disease activity index. National Cooperative Crohn's Disease Study. Gastroenterology. 1976;70:439–44.

Beutler E. Iron absorption in carriers of the C282Y hemochromatosis mutation. Am J Clin Nutr. 2004;80: 799–800.

Beyer PL. Short bowel syndrome. In: Coulston AM, Rock CL, Monson ER, editors. Nutrition in the prevention and treatment of disease. San Diego (CA): Academic Press; 2001.

Bharshankar JR, Bharshankar RN, Deshpande VN, Kaore SB, Gosavi GB. Effect of yoga on cardiovascular system in subjects above 40 years. Indian J Physiol Pharmacol. 2003; 47:202–6.

Bianchi G, Marzocchi R, Agostini F, Marchesini G. Update on branched-chain amino acid supplementation in liver diseases. Curr Opin Gastroenterol. 2005;21:197–200.

Bielory L. Complementary and alternative interventions in asthma, allergy, and immunology. Ann Allergy Asthma Immunol. 2004;93:S45–S54.

Biggee BA, McAlindon T. Glucosamine for osteoarthritis: part I, review of the clinical evidence. Med Health R I. 2004;87:176–9.

Biggee BA, McAlindon T. Glucosamine for osteoarthritis: part II, biological and metabolic controversies. Med Health R I. 2004;87:180–1.

Biglan KM, Holloway RG, Shoulson I, Parkinson Study Group. An item analysis of the mental unified Parkinson's disease rating scale (UPDRS) in individuals with early Parkinson's disease (PD) treated initially with pramipexole vs. levodopa: a sub analysis of the 4-year CALM-PD trial. Mov Disord. 2002;17:1107.

Bijl M, Dieleman JP, Simoons M, van der Ende ME. Low prevalence of cardiac abnormalities in an HIV-seropositive population on antiretroviral combination therapy. J Acquir Immune Defic Syndr. 2001;27:318–20.

Binder HJ. Chapter 286: Disorders of absorption. In: Eugene Braunwald, Anthony S. Fauci, Kurt J. Isselbacher, Dennis L. Kasper, Stephen L. Hauser, Dan L. Longo, J. Larry Jameson, editors. Harrison's principles of internal medicine. 15th ed. New York: McGraw Hill, Inc; 2005. Online edition available from: http://harrisons.accessmedicine.com. Accessed 3 March 2006.

Binnekade JM. Review: enteral nutrition reduces infections, need for surgical intervention, and length of hospital stay more than parenteral nutrition in acute. Evid Based Nurs. 2005;8:19.

Biolo G, Grimble G, Preiser JC. Position paper of the ESICM Working Group on Nutrition and Metabolism. Metabolic basis of nutrition in intensive care unit patients: ten critical questions. Intensive Care Med. 2003;28:1512–20.

Biomerica [homepage on the Internet]. Newport Beach (CA): Biomerica; 2004. Available from: http://www.biomerica.com/.

Birks J, Flicker L. Selegiline for Alzheimer's disease. Cochrane Database Syst Rev. 2003;CD000442.

Bishop ML, Fody EP, Schoeff L, editors. Clinical chemistry: principles, procedures, correlations. Baltimore: Lippincott Williams & Wilkins; 2005.

Biskobing D. COPD and osteoporosis. Chest. 2002;121: 609–20.

Bissoli L, Mazzali G, Gambina S, Residori L, Pagliari P, Guariento S, et al. Energy balance in Alzheimer's disease. J Nutr Health Aging. 2002;6:247–53.

Black DM, Steinbuch M, Palermo L, Dargent-Molina P, et al. An assessment tool for predicting fracture risk in postmenopausal women. Osteoporos Int. 2001;12:519–28.

Blair SN, Church TS. The fitness, obesity, and health equation: is physical activity the common denominator? JAMA. 2004; 292:1232–4.

Blake P, Daugirdas J. Physiology of peritoneal dialysis. In: Daugirdas J, Blake P, Ing T, editors. Handbook of dialysis. 3rd ed. Philadelphia: Lippincott Williams & Wilkins; 2001. p. 281–96.

Block GA, Hulbert-Shearon TE, Levin NW, Port FK. Association of serum phosphorus and calcium x phosphorus product with mortality risk in chronic hemodialysis patients: a national study. Am J Kidney Dis. 1998;31:607–17.

Block GA, Klassen PS, Lazarus JM, Ofsthun N, Lowrie EG, Chertow GM. Mineral metabolism, mortality and morbidity in maintenance HD. J Am Soc Nephrol. 2004;15: 2208–18.

Blondell RD. Ambulatory detoxification of patients with alcohol dependence. Amer Fam Phys. 2005;71:495–502.

Blumenkrantz MJ, Gahl GM, Kopple JD, Kamdar AV, Jones MR, Kessel M, et al. Protein losses during peritoneal dialysis. Kidney Int. 1981;19:593–602.

Blumenkrantz MJ, Kopple JD, Moran JK, Coburn JW. Metabolic balance studies and dietary protein requirements in patients undergoing continuous ambulatory peritoneal dialysis. Kidney Int. 1982; 21:849–61

Blumenthal M, Klein J, editors. The Complete German Commission E Monographs: Therapeutic guide to herbal medicines. Austin (TX): American Botanical Council; 1998.

Boccio JR, Iyengar V. Iron deficiency: causes, consequences, and strategies to overcome this nutritional problem. Biol Trace Elem Res. 2003;94:1–32.

Bodary PF, Yasuda N, Watson DD, Brown AS, Davis JM, Pate RR. Effects of short-term exercise training on Plasminogen Activator Inhibitor (PAI-1). Med Sci Sport Exer. 2003; 35:1853–8.

Bode C, Bode JC. Alcohol's role in gastrointestinal tract disorders. Alcohol Health Res World. 1997;21:76–83.

Bode JC, Bode C. Alcohol, the gastrointestinal tract and pancreas. Ther Umsch. 2000;57(4):212–9.

Bodnar DM, Busch S, Fuchs J, Piedmonte M, Schreiber M. Estimating glucose absorption in peritoneal dialysis using peritoneal equilibration tests. Adv Perit Dial. 1993;9: 114–8.

Boeing KL, Holben DH. Self-identified food security knowledge and practices of licensed dietitians in Ohio: implications for dietetics and clinical nutrition practice. Top Clin Nutr. 2003;18:185–91.

Bogden JD, Baker H, Frank O, Perez G, Kemp F, Bruening K, Louria D. Micronutrient status and human immunodeficiency virus (HIV) infection. Ann NY Acad Sci. 1990;587: 189–95.

Bohinjec M. Clinical immunogenetics and cell therapy. Transpl Immunol. 2005;14:171–4.

Boleda MP, Julia P, Moreno A, Pares X. Role of extrahepatic alcohol dehydrogenase in rat ethanol metabolism. Arch Biochem Biophys. 1989;274:74–81.

Bondy B, Engel RR, de Jonge S, Schutz CG, Soyka M. Phytohemagglutinin-stimulated calcium signed in lymphocytes of alcoholics before, during, and after detoxification. Psychiatry Res. 1998;81:157–62.

Bonuck KA, Kahn R. Prolonged bottle use and its association with iron-deficiency anemia and overweight: a preliminary study. Clin. Pediatr. 2002;1:603–7.

Boon H, Stewart M, Kennard MA, Guimond J. Visiting family physicians and naturopathic practitioners. Comparing patient-practitioner interactions. Can Fam Physician. 2003;49:1481–2, 1435–7.

Booth KM, Pinkston MM, Poston WSC. Obesity and the built environment. J Am Diet Assoc. 2005;105:S110–S117.

Bordreaux JP, McHugh L, Canfax DM Ascher N, Sutherland DE, Payne W, Simmons RL, Najarian JS, Fryd DS. The impact of cyclosporine and combination immunosuppression on the incidence of posttransplant diabetes in renal allograft recipients. Transplantation. 1987;44:371–81.

Bornman PC, van Beljon JI, Krige JE. Management of cholangitis. J Hepatobiliary Pancreat Surg. 2003;10:406–14.

Borowitz D, Baker R, Stalling V. Consensus report on nutrition for pediatric patients with cystic fibrosis. J Pediatr Gastroenterol. 2002;35:246–59.

Borowitz DS, Grand RJ, Durie PR. Use of pancreatic enzyme supplements for patients with cystic fibrosis in the context of fibrosing colonpathy. Consensus committee. J Pediatr. 1995;127:681–4.

Bosaeus I, Daneryd P, Lundholm K. Dietary intake, resting energy expendture, weight loss, and survival in cancer patients. J Nutr. 2002;132(Supp 11):3465S–3466S.

Bouchard C, Tremblay A, Despres JP, Nadeau A, Lupien PJ, et al. The response to long-term overfeeding in identical twins. N Eng J Med. 1990;322:1477–82.

Bower RH, Cerra FB, Bershadsky B, Licari JJ, Hoyt DB, Jensen GL, Van Buren CT, Rothkopf MM, Daly JM, Adelsberg BR. Early enteral administration of a formula (Impact) supplemented with arginine, nucleotides, and fish oil in intensive care unit patients: results of a multicenter, prospective, randomized, clinical trial. Crit Care Med. 1995;23:436–49.

Bowers JM. Nutrition and immunity: you are what you eat [monograph on the Internet]. The Body; 2002. Available from: http://www.thebody.com/cria/spring02/nutrition_immunity.html.

Bowers JM, Bert-Moreno A. Treatment of HAART-induced lactic acidosis with B vitamin supplements. Nutr Clin Pract. 2004;19:375–8.

Boyer TD, Wright TL, Manns MP. Zakim and Boyer's hepatology. 5th ed. Philadelphia: WB Saunders; 2006.

Boyer Z. Hepatology: a textbook of liver disease. 3rd ed. Philadelphia: WB Saunders; 2006.

Bozzette SA, Ake CF, Tam HK, Chang SW, Louis TA. Cardiovascular and cerebrovascular events in patients treated for human immunodeficiency virus infection. N Engl J Med. 2003;348:702–10.

Bozzetti F, Cozzagliio L, Biganzloi E, Chiavenna G, De Cicco M, Donati D, Gilli G, Percolla S, Pironi. Quality of life and length of survival in advanced cancer patients on home parenteral nutrition. Clin Nutr, 2002;21:281–8.

Brabson TA, Chussil JT, Daack-Hirsch S, Dixon D, Falk KM, Ferrelra BF, Gruener RC, et al. Professional guide to diseases. 7th ed. Springhouse (PA): Springhouse Corporation; 2001.

Brady JA, Rock CL, Horneffer MR. Thiamin status, diuretic medications, and the management of congestive heart failure. J Am Diet Assoc. 1995;95:541–5.

Braga M, Gianotti L, Gentilini O, Liotta S, Di Carlo V. Feeding the gut early after digestive surgery: results of a nine-year experience. Clin Nutr. 2002;21:59–65.

Brain Trauma Foundation, American Association of Neurological Surgeons. Clinical guidelines: management and prognosis of severe traumatic brain injury. 2000. Available from: http://www2.braintrauma.org/guidelines/downloads/btf_guidelines_management.pdf.

Braithwaite RS, Col NF, Wong JB. Estimating hip fracture morbidity, mortality, and costs. J Am Geriatr Soc. 2003; 51:364–70.

Brandt KD. Osteoarthritis. In: Kasper DL, Braunwald E, Fauci AS, Hausner SL, Longo DL, Jameson JL, editors. Harrison's principles of internal medicine. 16th ed. New York: McGraw-Hill; 2005. p. 2036–45.

Brandt LJ, Bjorkman D, Fennerty B, Locke GR, Older K, Peterson W, et al. Systematic review on the management of irritable bowel syndrome in North America. Am J Gastro. 2002;97:S7–10.

Brannan T, Martinez-Tica J, Yahr MD. Effect of dietary protein on striatal dopamine formation following L-dopa administration: an in vivo study. Neuropharmacology. 1991;30:1125–7.

Brass EP, Hiatt WR, Green S. Skeletal muscle metabolic changes in peripheral arterial disease contribute to exercise intolerance: a point-counterpoint discussion. Vasc Med. 2004;9:293–301.

Braunschweig CL, Levy P, Sheean PM, Wang X. Enteral compared with parenteral nutrition: a meta-analysis. Am J Clin Nutr. 2001;74:534–42.

Brawley OW, Kramer BS. Prevention and early detection of cancer. In: Kasper DL, Braunwald E, Fauci AS, Hauser SL, Longo DL, Jameson L, and Isselbacher KJ, editors. Harrison's principles of internal medicine. 16th ed. Boston: McGraw Hill; 2005.

Brescianini S, Maggi S, Farchi G, Mariotti S, Carlo A, Baldereschi M, et al. Low cholesterol and increased risk of

Macro-Nutrient Intake Heart trial to prevent heart disease (OMNI-Heart). Clin Trials. 2005;2:529–37.

Carlson SJ. Current nutrition management of infants with chronic lung disease. Nutr Clin Pract. 2004;19:581–6.

Carlston M. Classical homeopathy. New York: Churchill Livingston; 2003.

Carmel R. Cobalamin (vitamin B12). In: Shils ME, Shike M, Ross CA, Caballero BC, Cousins RJ, editors. Modern nutrition in health and disease. 10th ed. Philadelphia (PA): Lippincott Williams & Wilkins; 2005. p. 482–97.

Carmel R. Cobalamin, the stomach, and aging. Am J Clin Nutr. 1997;66:750–9.

Carmel R. Folic acid. In: Shils ME, Shike M, Ross CA, Caballero BC, Cousins RJ, editors. Modern nutrition in health and disease. 10th ed. Philadelphia (PA): Lippincott Williams & Wilkins; 2005. p. 470–81.

da Rocha EE, Alves VG, Silva MH, Chiesa CA, da Fonseca RB. Can measured resting energy expenditure be estimated by formulae in daily clinical practice? Curr Opin Clin Nutr Metab Care. 2005;8:319–28.

Daeschel IE, Janick LS, Kramish MJ, Coleman RA. Diet and growth of children with glycogen storage dieases types I and III. J Am Dietet Assoc. 1983;83:135–41.

Dahlin M, Elfving A, Understedt U, Amark P. The ketogenic diet influences the excitatory and inhibitory amino acids in the CSF in children with refractory epilepsy. Epilepsy Res. 2005;64:115–25.

Dallongeville J, Marecaux N, Fruchart J, Amouyel P. Cigarette smoking is associated with unhealthy patterns of nutrient intake: a meta-analysis. J Nutr. 1998;128:1450–7.

Dansinger ML, Gleason JA, Griffith JL, Selker HP, Schaefer EJ. Comparison of the Atkins, Ornish, Weight Watchers, and Zone diets for weight loss and heart disease risk reduction. JAMA. 2005;293:43–53.

Dargent-Molina P, Piault S, Breart G. A comparison of different screening strategies to identify elderly women at high risk of hip fracture: results of the EPIDOS prospective study. Osteoporos Int. 2003;14:969–77.

Darlow BA, Graham PJ. Vitamin A supplementation for preventing morbidity and mortality in very low birthweight infants. The Cochrane Database of Systematic Reviews 2002, Issue 4. Art. No.: CD000501.

Das BS, Devi U, Mohan Rao C, Srivastava VK, Rath PK, Das BS. Effect of iron supplementation on mild to moderate anaemia in pulmonary tuberculosis. Br J Nutr. 2003;90: 541–50.

Davidsson L. Approaches to improve iron bioavailability from complementary foods. J Nutr. 2003;133 Suppl 1:1560S–2S.

Davis S. Clinical sequelae affecting quality of life in the HIV-infected patient. J Assoc Nurses AIDS Care. 2004;15 Suppl:28S–33S.

Deitcher SR. Antiplatelet, anticoagulant and fibrinolytic therapy. In: Kasper DL, Braunwald E, Fauci AS, Hausner SL, Longo DI, Jameson JL, editors. Harrison's principles of internal medicine. 16th ed. New York: McGraw Hill Medical Publishing; 2005. p. 687–94.

Daugirdas J, Van Stone J, Boag J. Hemodialysis apparatus. In: Daugirdas J, Blake P, Ing T, editors. Handbook of dialysis. 3rd ed. Philadelphia (PA): Lippincott Williams & Wilkins; 2001. p. 46–66.

Davidson GP, Townley RR. Structural and functional abnormalities of the small intestine due to nutritional folic acid deficiency in infancy. J Pediatr. 1977;90:590–4.

Davies AN. A comparison of artificial saliva and chewing gum in the management of xerostomia in patients with advanced cancer. Palliat Med. 2000;14;197–203.

Davis BC, Kris-Etherton PM. Achieving optimal essential fatty acid status in vegetarians: current knowledge and practical implications. Am J Clin Nutr. 2003;78:640S–646S.

Davis, DG, Bears S, Barone JE, Corvo PR, Tucker JB. Swallowing with a traceostomy tube in place: does cuff inflation matter? J Inten Care Med. 2002;17:132–5.

Davy KP, Hall JE. Obesity and hypertension: two epidemics or one? Am J Physiol Regul Integr Comp Physiol. 2004; 286:R803–13.

Dawson-Hughes B. Osteoporosis. In: Shils ME, Shike M, Ross AC, Cabellero B, Cousins RJ, editors. Modern nutrition in health and disease. 10th ed. Philadelphia (PA): Lippincott Williams & Wilkins; 2006. p. 1339–52.

DCCT Research Group. Nutrition interventions for intensive therapy in the Diabetes Control and Complications Trial. J Am Diet Assoc. 1993;93:768–72.

De Caterina R, Massaro M. Omega-3 fatty acids and the regulation of expression of endothelial pro-atherogenic and pro-inflammatory genes. J Membr Biol. 2005;206:103–16.

De Caterina R, Zampolli A, Del Turco S, Madonna R, Massaro M. Nutritional mechanisms that influence cardiovascular disease. Am J Clin Nutr. 2006;83:421S–426S.

De Raeve L, De Meirleir L, Ramet J, Vandenplas Y, Gerlo E. Acrodermatitis enteropathica-like cutaneous lesions in organic aciduria. J Peds. 1994;124:416–20.

de Roos NM, Bots ML, Katan MB. Replacement of dietary saturated fatty acids by trans fatty acids lowers serum HDL cholesterol and impairs endothelial function in healthy men and women. Arterioscler Thromb Vasc Biol. 2001;21:1233–7.

de Roos NM, Katan MB. Effects of probiotic bacteria on diarrhea, lipid metabolism, and carcinogenesis: a review of papers published between 1988 and 1998. Am J Clin Nutr. 2000;71:405–11.

de Wildt SN, Johnson TN, Choonara I. The effect of age on drug metabolism. Paediatric and Perinatal Drug Therapy. 2002;5:101–6.

Debusk RM. A practical guide to herbal supplements for nutrition practitioners. Topics in Clinical Nutrition. 2001;16:53.

Del Valle J. Peptic ulcer disease and related disorders. In: Harrison's pinciples of internal medicine. 15th ed. 2006 [cited 2005 Oct 11]. Available from: http://harrisons.accessmedicine.com.

Del Valle L, White MK, Enam S, Oviedo SP, Bronner MQ, Thomas RM, Parkman HP, Khalili K. Detection of JC virus DNA sequences and expression of viral T antigen and agnoprotein in esophageal carcinoma. Cancer. 2005;103:516–27.

Del Vecchio L, Pozzoni P, Andrulli S, Locatelli F. Inflammation and resistance to treatment with recombinant human erythropoietin. J Ren Nutr. 2005;15:137–41.

Delgado AF, Kimura HM, Cardoso AL, Uehara D, Carrazza FR. Nutritional follow-up of critically ill infants receiving short term parenteral nutrition. Rev Hosp Clin Fac Med Sao Paulo. 2000;55:3–8.

Dellinger RP, Carlet JM, Masur H, Gerlach H, Calandra T, Cohen J, et al. Surviving sepsis campaign guidelines for management of severe sepsis and septic shock. Crit Care Med. 2004;32:858–73.

DeLong MR, Juncos JL. Parkinson's disease and other movement disorders In: Kasper DL, Braunwald E, Fauci AS, Hausner SL, Longo DL, Jameson JL, editors. Harrison's principles of internal medicine. 16th ed. New York: McGraw-Hill; 2006.

Delphi Report. Critical thinking: a statement of expert consensus for purposes of educational assessment and instruction (Vol. ERIC ED 315-423). Millbrae (CA): California Academic Press; 1990.

Delzenne N, Cherbut C, Nevrinck A. Prebiotics: actual and potential effects in inflammatory and malignant colonic diseases. Curr Opin Clin Nutr Metab Care. 2003;6:581–6.

Demling RH, DeSanti L. The rate of restoration of body weight after burn injury, using the anabolic agent oxandrolone, is not age dependent. Burns. 2001;27:46–51.

Demling RH, DeSanti L. The stress response to injury and infection: role of nutritional support. Wounds. 2000;12:3–14.

Denne SC. Energy expenditure in infants with pulmonary insufficiency: is there evidence for increased energy needs? J Nutr. 2001;131:935S–37S.

Dennis MS, Lewis SC, Warlow C. FOOD Trial Collaboration. Routine oral nutritional supplementation for stroke patients in hospital (FOOD): a multicentre randomised controlled trial. Lancet. 2005;365(9461):755–63.

Department of Health and Human Services (DHHS). Guidelines for the use of antiretroviral agents in HIV-1-infected adults and adolescents. October 6, 2005. Available from: http://aidsinfo.nih.gov/ContentFiles/AdultandAdolescentGL.pdf.

Department of Health and Human Services (DHHS) and the Department of Agriculture (USDA). Dietary guidelines for Americans. 2005.

Department of Health and Human Services. Final MNT regulations. CMS-1169-FC. 1 November 2001. Available from: http://cms.hhs.gov/physicians/pfs/cms1169fc.asp.

Desnick RJ, Grabowski GA. Advances in the treatment of inherited metabolic diseases. Adv Hum Genet. 1981;11:281–369.

DeSouza DA, Greene LJ. Intestinal permeability and systemic infection. Crit Care Med. 2005;233(5): 1125–35.

DeVault KR, Castell DO, The Practice Parameters Committee of the American College of Gastroenterology. Updated guidelines for the diagnosis and treatment of gastroesophageal reflux disease. Am J Gastroenterol. 2005;100:190–200.

Devine W. Review of nutritional status on diet in dialysis and transplant patients. Dial Transplant. 1994;23:38–41, 47–8.

deVise P. Misuses and misplaced hospitals and doctors: a locational analysis of the urban health care crisis. Resource Paper no. 2. Commission on College Geography; 1973.

Dhingra D, Parle M, Kulkarni SK. Memory enhancing activity of Glycyrrhiza glabra in mice. J Ethnopharmacol. 2004;91:361–5.

Diabetes Control and Complications Trial Research Group. The effect of intensive treatment of diabetes on the development and progression of long-term complications in insulin-dependent diabetes mellitus. N Engl J Med. 1993;329:977–86.

Diabetes Education Society, Inc. Byetta™ (Exenatide injection) Lifeskills Teaching Guide; 2006.

Diabetes Prevention Program Research Group. Reduction in the incidence of type 2 diabetes with lifestyle intervention or metformin. N Engl J Med. 2002;346:393–403.

Diamond JA, Phillips RA. Hypertensive heart disease. Hypertens Res. 2005;28:191–202.

Diaz JJ, Pousman R, Jensen G. Critical care nutrition practice management guidelines. Critical care nutrition support flow diagram. Nashville (TN): Vanderbilt University Medical Center; 2004.

Diaz JJ, Pousman R, Jensen G. Critical care nutrition practice management guidelines. Total enteral nutrition protocol flow diagram. Nashville (TN): Vanderbilt University Medical Center; 2004.

Díaz JR, de las Cagigas A, Rodríguez R. Micronutrient deficiencies in developing and affluent countries. Eur J Clin Nutr. 2003;57 Suppl 1:S70–2.

Dietcher SR. Antiplatelet, anticoagulant, and fibrinolytic therapy. In: Kasper DL, Braunwald E, Fauci AS, Hausner SL, Longo DI, Jameson JL, editors. Harrison's principles of internal medicine, 16th ed. New York, NY: McGraw Hill Medical Publishing; 2005. p. 687–94.

Dickerson RN. Hypocaloric feeding of obese patients in the intensive care unit. Curr Opin Clin Nutr Metabol Care. 2005;8:189–96.

Dierkes J, Domrose U, Ambrosch A, Bosselman HP, Neumann KH, Luley C. Response of hyperhomocysteinemia to folic acid supplementation in patients with end stage renal disease. Clin Nephrol. 1999;51:108–15.

Dietary reference intakes for calcium, phosphorus, magnesium, vitamin D, and fluoride. Standing Committee on the Scientific Evaluation of Dietary Reference Intakes, Food and Nutrition Board, Institute of Medicine; 1997.

Dietary reference intakes for energy, carbohydrate, fiber, fat, fatty acids, cholesterol, protein, and amino acids (macronutrients). A report of the Panel on Macronutrients, Subcommittees on Upper Reference Levels of Nutrients and Interpretation and Uses of Dietary Reference Intakes, and the Standing Committee on the Scientific Evaluation of Dietary Reference Intakes, Food and Nutrition Board, Institute of Medicine; 2002.

Dietary reference intakes for thiamin, riboflavin, niacin, vitamin B6, folate, vitamin B12, pantothenic acid, biotin, and choline. A Report of the Standing Committee on the Scientific Evaluation of Dietary Reference Intakes and its Panel on Folate, Other B Vitamins, and Choline and Subcommittee on Upper Reference Levels of Nutrients, Food and Nutrition Board, Institute of Medicine; 1998.

Dietary reference intakes for vitamin A, vitamin K, arsenic, boron, chromium, copper, iodine, iron, manganese, molybdenum, nickel, silicon, vanadium, and zinc. Panel on Micronutrients, Subcommittees on Upper Reference Levels of Nutrients and of Interpretation and Use of Dietary Reference Intakes, and the Standing Committee on the Scientific Evaluation of Dietary Reference Intakes, Food and Nutrition Board, Institute of Medicine; 2000.

Dietary reference intakes for vitamin C, vitamin E, selenium, and carotenoids. Panel on Dietary Antioxidants and Related Compounds, Subcommittees on Upper Reference Levels of Nutrients and Interpretation and Uses of DRIs, Standing Committee on the Scientific Evaluation of Dietary Reference Intakes, Food and Nutrition Board, Institute of Medicine; 2000.

Dietary reference intakes: proposed definition of dietary fiber. Panel on the Definition of Dietary Fiber, Standing Committee on the Scientific Evaluation of Dietary Reference Intakes, Food and Nutrition Board, Institute of Medicine; 2001.

Dietitians in Nutrition Support. Study guide: nutrition focused physical assessment skills for dietitians. Chicago (IL): American Dietetic Association; 1998.

Dietz WH. Childhood obesity. In: Shils ME, Shike M, Ross AC, Cabellero B, Cousins RJ, editors. Modern nutrition in health and disease. 10th ed. Philadelphia (PA): Lippincott Williams & Wilkins; 2006. p. 979–90.

DiFronzo LA, Yamin N, Patel K, O'Connell TX. Benefits of early feeding and early hospital discharge in elderly patients undergoing open colon resection. J Am Coll Surg. 2003;197:747–52.

Digh EW, Dowdy RP. A survey of management tasks, completed by clinical dietitians in the practice setting. J Am Diet Assoc. 1994;94:1381–4.

Dimeny E, Fellstrom B. Metabolic abnormalities in renal transplant recipients. Risk factors and predictors of chronic graft dysfunction. Nephrol Dial Transplant. 1997;12:21–4.

Division of Government Affairs. Health care reform. In: Legislative Newsletter. Division of Government Affairs; 1991.

Dobbelaere D, Michaud L, Debrabander A, Vanderbecken S, Gottrand F, Turck D, et al. Evaluation of nutritional status and pathophysiology of growth retardation in patients with phenylketonuria. J Inherit Metab Dis. 2003;26:1–11.

Doherty KM. Closing the gap in prophylactic antiemetic therapy: patient factors in calculating the emetogenic potential of chemotherapy. Clin J Oncol Nurs, 1999;3:113–9.

Dominioni L, Rovera F, Pericelli A, Imperatori A. The rationale of early enteral nutrition. Acta bio-medica de L'Ateneo parmense: organo della Societa di medicina e scienze naturali di Parma. 2003;74 Suppl 2:41–4.

Don B, Kaysen G. Nutritional and nonnutritional management of the nephrotic syndrome. In: Kopple JD, Massry SG, editors. Nutritional management of renal disease. 2nd ed. Philadelphia (PA): Lippincott Williams & Wilkins; 2004. p. 415–32.

Donahoe M. Nutritional aspects of lung disease. Respir Care Clin N Am. 1998;4;85–112.

Donowitz M, Kokke FT, Saidi R. Evaluation of patients with chronic diarrhea. N Engl J Med. 1995;332:725–9.

Doody RS. Refining treatment guidelines in Alzheimer's disease. Geriatrics. 2005;Suppl:14–20.

Dorland WAN. Dorland's illustrated medical dictionary. 30th ed. Elsevier; 2003.

Douglas RM, Hemila H, Chalker E, D'Souza RRD, Treacy B. Vitamin C for preventing and treating the common cold. The Cochrane Database of Systematic Reviews 2004, Issue 4. Art. No.: CD000980.

Dove JT. The electronic health record—the time is now. Am Heart Hosp J. 2005;3:193–200.

Drachman DB, Brodsky RA. High-dose therapy for autoimmune neurologic diseases. Curr Opin Oncol. 2005;17:83–8.

Draganov P, Toskes PP. Chronic pancreatitis: controversies in etiology, diagnosis and treatment. Rev Esp Enferm Dig. 2004;96:649–54, 654–9.

Drewnowski A, Darmon N. The economics of obesity: dietary energy density and energy cost. Am J Clin Nutr. 2005;82(suppl):265S–273S.

Dreyfuss ML, Fawzi WW. Micronutrients and vertical transmission of HIV-1. Am J Clin Nutr. 2002;75:959–70.

Dreyfuss ML, Stoltzfus RJ, Shrestha JB, et al. Hookworms, malaria and vitamin A deficiency contribute to anemia and iron deficiency among pregnant women in the plains of Nepal. J Nutr. 2000;130:2527–36.

Drug facts and comparisons. St. Louis (MO): Walters Klewer; 2005.

Druml W. Nutritional management of acute renal failure. In: Jacobson H, Striker G, Skahr S, editors. The principles and practice of nephrology. St. Louis (MO): CV Mosby; 1995. p. 745–53.

Druss BG, Rosenbeck RA. Association between use of unconventional therapies and conventional medical services. JAMA. 1999;282:651–6.

Dry J, Vincent D. Effect of fish oil diet on asthma: results of a 1-year double-blind study. Int Arch Allergy Appl Immunol. 1991;95(2–3):156–7.

Dubinsky MC, Lee-Uy N, Lin Y-C, et al. Synergism of Nod2 and ASCA contribute to disease behavior in pediatric Crohn's patients. Gastroenterology. 2003;124:A-2. [Abstract #27]

Dudrick SJ, Brown W, Biggs CL. Nutritional management of patients with head and neck tumors. In: Thawley SE, Panje WR, Batsakis JG, Lindberg RD, editors. Comprehensive management of head and neck tumors. Philadelphia (PA): Saunders; 1999.

Dufour MC, Archer L, Gordis E. Alcohol and the elderly. Health Promotion and Disease Prevention. 1992;8: 127–40.

Duggan C, Fawzi W. Micronutrients and child health: studies in international nutrition and HIV infection. Nutr Rev. 2001;59:358–69.

Duncan PW, Zorowitz R, Bates B, Choi JY, Glasberg JJ, Graham GD, et al. Management of adult stroke rehabilitation care: a clinical practice guideline. Stroke. 2005;36:e100.

Dwyer JL. Dietary assessment in modern nutrition in health and disease. 9th ed. Philadephia (PA): Lippincott Williams & Wilkins; 1999.

Dylewski ML, Mastro AM, Picciano MF. Maternal selenium nutrition and neonatal immune system development. Biol Neonate. 2002;2:122–7.

Dzieczkowski JS, Anderson KC. Transfusion biology and therapy. In: Kasper DL, Braunwald E, Fauci AS, Hausner SL, Longo DI, Jameson JL, editors. Harrison's principles of internal medicine. 16th ed. New York: McGraw Hill Medical Publishing; 2005. p. 662–8.

Easley D, Krebs N, Jefferson M, Miller L, Erskine J, Accurso F, et. al. Effect of pancreatic enzymes on zinc absorption in cystic fibrosis. JPEN J Parenter Enteral Nutr. 1998;26: 136–9.

EAST Practice Management Guidelines Work Group. Practice management guidelines for nutritional support of the trauma patient. Allentown (PA): Eastern Association for the Surgery of Trauma (EAST); 2003.

Eckardt VF. BoTox or NoTox—which therapy is best for patients with achalasia? Med Gen Med. 2001;3(1) [formerly published in Medscape Gastroenterology eJournal 2001;3(1)]. Available from: http://www.medscape.com/viewarticle/407968.

Eckel RH. The dietary approach to obesity: is it diet or the disorder. Am J Clin Nutr. 2005;293:96–7.

Eden D, Edstrom S, Bennegard K, Schersten T, Lundholm K. Glucose flux in relation to energy expenditure with and without cancer during periods of fasting and feeding. Cancer Res. 1984;44:1718–24.

eDrugDigest. Monoamine oxidase inhibitors. Last updated June 2005. Available from: http://www.drugdigest.org. Accessed December 2, 2005.

Ehrstom M, Naslund E, Ma J, et al. Physiological regulation and NO-dependent inhibition of migrating myoelectric complex in the rat small bowel by orexin A. Am J Physio Gastrointest Liver Physiol. 2003; 285:688–95.

Eiden KA. Nutritional considerations in inflammatory bowel disease. Practical Gastroenterology. 2003 May:33–54.

Eisenberg DM, Davis RB, Ettner SL, Appel S, Wilkey S, Van Rompay M, et al. Trends in alternative medicine use in the United States, 1990–1997: results of a follow-up national survey. JAMA. 1998;280:1569–75.

Eisenberg DM, Kessler RC, Foster C, Norlock FE, Calkins DR, Delbanco TL. Unconventional medicine in the United States. N Engl J Med. 1993;328:246–52.

Eisenberg L. Complementary and alternative medicine: what is its role? Harv Rev Psychiatry. 2002;10:221–30.

Elbe D. Grapefruit-Drug Interactions. Available at http://www.powernetdesign.com. Accessed December 2, 2005

Elbe D. Grapefruit-drug interactions. In: Pronsky ZM. Food-medication interactions. 13th ed. Birchrunville (PA): Food-Medication Interactions; 2004.

Elkin SL, Fairney A, Burnett S, Kemp M, Kyd P, Burgess J, et al. Verebral deformities and low bone mineral density in adults with cystic fibrosis: a cross-sectional study. Osteoporos Int. 2001;12:366–72.

Elliott L, Yeatts K, Loomis D. Ecological associations between asthma prevalence and potential exposure to farming. Eur Respir J. 2004;24:938–41.

Ello-Martin JA, Ledikwe JH, Rolls BJ. The influence of food portion size and energy density on energy intake: implications for weight management. Am J Clin Nutr. 2005; 82(Suppl):236S–241S.

Elman, R, Weiner, D. Intravenous alimentation with special reference to protein (amino acid) metabolism. JAMA. 1939;112:796–802.

Elsas LJ, Acosta PB: Nutrition support of inherited metabolic disease. In: Shils ME, Olson, JA, Shike M, Ross AC, editors. Modern nutrition in health and disease. [CITY]: Williams & Wilkins, A Waverly Company; 1999. p. 1003–56.

Elsas LJ, Gilat T. Cholecystocolonic fistula with malabsorption. Ann Intern Med. 1965;63:481–6.

Elsenbruch S, Holtmann G, Oezcan D, Lysson, Janssen O, Goebel MU, Schedlowski M. Are there alterations of neuroendocrine and cellular immune responses to nutrients in women with irritable bowel syndrome? Am J Gastroenterol. 2004;99:703–10.

Elstein AS, Shulman LS, Sprafka SA. Medical problem solving: an analysis of clinical reasoning. Cambridge (MA): Harvard University Press; 1978.

Encinosa WE, Bernard DM, Steiner CA, Chen CC. Use and costs of bariatric surgery and prescription weight-loss medication. Health Affairs. 2005;24:1039–46.

Endocrine and Metabolic Diseases Information Service. Addison's disease. Washington (DC): NIH Publication #04-3054. 2004 June [cited 2006 Jan 15]. Available from: http://www.niddk.nih.gov.

Endocrine and Metabolic Diseases Information Service. Cushing's syndrome. Washington (DC): NIH Publication #02-3007. 2002 June [cited 2006 Jan 15]. Available from: http://www.niddk.nih.gov.

Engelen MP. Protein metabolism in chronic respiratory disease. Eur Respir Mon. 2003;24:25–33.

Engelen MP, Schols AM, Lamers RJ, Wouters EF. Different patterns of chronic tissue wasting among patients with chronic pulmonary disease. Clin Nutr. 1999;18:275–80.

Epilepsy Foundation [homepage on the Internet]. Available from: http://www.efa.org. Accessed December 30, 2005.

Epstein S, Cryer B, Ragi S, Zanchetta JR , Walliser J, Chow J, et al. Disintegration/dissolution profiles of copies of Fosamax (alendronate). Curr Med Res Opin. 2003;19:781–9.

Erikson KM, Jones BC, Beard JL. Iron deficiency alters dopamine transporter functioning in rat striatum. J Nutr. 2000;130:2831–7.

Erlinger S. Gallstones in obesity and weight loss. Eur J Gastroenterol Hepatol. 2000;12:1347–52.

Ernst RL, Hay JW. The U.S. economic and social costs of Alzheimer's disease revisited. Am J Public Health. 1994; 84:1261–4.

Escott-Stump S. Myocardial infarction. In: Nutrition and diagnosis related care. Philadelphia (PA): Lippincott Williams & Wilkins; 2002.

Escott-Stump S. Nutrition and diagnosis-related care. 5th ed. Baltimore: Lippincott Williams & Wilkins; 2002.

Essama-Tjani JC, Guilland JC, Potier de Courcy G, Fuchs F, Richard D. Folate status worsens in recently institutionalized elderly people without evidence of functional deterioration. J Am Coll Nutr. 2000;19:392–404.

Evans P, Halliwell B. Micronutrients: oxidant/antioxidant status. Br J Nutr. 2001;85 suppl 2:S67–S74.

Evans PR, Bak YT, Shuter B, Hoschl R, Kellow JE. Gastroparesis and small bowel dysmotility in irritable bowel syndrome. Dig Dis Sci. 1997;42:2087–93.

Everhart, JE, editor. Digestive diseases in the United States: epidemiology and impact (NIH Publication No. 94-1447). U.S. Department of Health and Human Services, National Institutes of Health, National Institute of Diabetes and Digestive and Kidney Diseases. Washington (DC): U.S. Government Printing Office; 1994.

Everson, GT, Trotter, JF. Transplantation of the liver. In: Schiff ER, Sorrell MF, Maddrey WC, editors. Schiff's diseases of the liver. 9th ed. Baltimore: Lippincott Williams & Wilkins; 2003.

Expert Panel on Detection, Evaluation, and Treatment of High Blood Cholesterol in Adults. Executive summary of the Third Report of the National Cholesterol Education Program (NCEP) Expert Panel on Detection, Evaluation, and Treatment of High Blood Cholesterol in Adults (Adult Treatment Panel III). JAMA. 2001;285:2486–97.

Ethical implications of differences in the availability of health services. In: President's Commission for the Study of Ethical Problems in Medical and Biomedical and Behavioral Research Report. Washington (DC): U.S. Government Printing Office; 1983.

Fabian CJ, Kimler BF. Selective estrogen-receptor modulators for primary prevention of breast cancer. J Clin Oncol. 2005;23:1644–55.

Facts and Comparisons. Par announces approval of Megace ES. 2005. Available from: http://www.factsandcomparisons.com/News/ArticlePage.aspx?cat=update&id=6784. Accessed August 15, 2005.

Fairburn CG, Harrison PJ. Eating disorders. Lancet. 2003; 361:407–16.

Fall PJ. A stepwise approach to acid-base disorders. Practical patient evaluation for metabolic acidosis and other conditions. Postgrad Med. 2000;107:249–50, 253–4, 257–8.

Falutz J, Allas S, Kotler D, Thompson M, Koutkia P, Albu J, et al. A placebo-controlled, dose-ranging study of a growth hormone releasing factor in HIV infected patients with abdominal fat accumulation. AIDS. 2005;19:1279–87.

Farber MS, Moses J, Korn M. Reducing costs and patient morbidity in the enterally fed intensive care unit patient. JPEN J Parenter Enteral Nutr. 2005;20(1 Suppl):S62–S69.

Fasano A, Berti I, Gerarduzzi T, Not T, Colletti RB, Drago S, et al. Prevalence of celiac disease in at-risk and not-at-risk groups in the United States: a large multicenter study. Arch Intern Med. 2003;163:286–92.

Faulks RM, Hart DJ, Brett GM, Dainty JR, Southon S. Kinetics of gastro-intestinal transit and carotenoid absorption and disposal in ileostomy volunteers fed spinach meals. European Journal of Nutrition. 2004;43:15–22.

Fausto N. Liver regeneration. J Hepatol. 2000;32(1 Suppl): 19–31.

Fava M. Weight gain and antidepressants. J Clin Psychiatry. 2000;61(Suppl1):37–41.

Favus MJ, Vokes TJ. Paget disease and other dysplasias of bone. In: Kasper DL, Braunwald E, Fauci AS, Hausner SL, Longo DL, Jameson JL, editors. Harrison's principles of internal medicine. 16th ed. New York: McGraw-Hill; 2005. p. 2279–86.

Fawzi W. Nutritional factors and vertical transmission of HIV-1. Epidemiology and potential mechanisms. Ann NY Acad Sci. 2000;918:99–114.

Fawzi WW, Mbise R, Spiegelman D, Fataki M, Hertzmark E, Ndossi G. Vitamin A supplements and diarrheal and respiratory tract infections among children in Dar es Salaam, Tanzania. J Pediatr. 2000;137:660–7.

Fawzi WW, Msamanga GI, Hunter D, Renjifo B, Antelman G, Bang H, et al. Randomized trial of vitamin supplements in relation to transmission of HIV-1 through breast-feeding and early child mortality. AIDS. 2002;16; 1935–44.

Fawzi WW, Msamanga GI, Spiegelman D, Urassa EJ, McGrath N, Mwakagile D, et al. Randomised trial of effects of vitamin supplements on pregnancy outcomes and T cell counts in HIV-1-infected women in Tanzania. Lancet. 1998;351:1477–82.

Faxon DP, Fuster V, Libby P, Beckman, JA, Hiatt WR, Thompson RW, et al. Atherosclerotic vascular disease conference: writing group III: pathophysiology. Circulation. 2004;109:2617–25.

Fearon KCH, von Meyenfeldt MF, Moses AGW, et al. Effect of a protein and energy dense n-3 fatty acid enriched oral supplement on loss of weight and lean tissue in cancer cachexia: a randomized double blind trial. Gut. 2003;52: 1479–86.

Feillet F, Leonard JV. Alternative pathway therapy for urea cycle disorders. J Inherit Metab Dis. 1998;21:101–11.

Feillet F, Bodamer AF, Dixon, MA, Sequeira S, Leonard JV. Resting energy expenditure in disorders of propionate metabolism. J Peds. 2000;136:659–63.

Felbinger, T, Suchner U, Peter K, Askanazi J. Nutrition support in respiratory disease. In: Payne-James J, Grimble G, Silk D, editors. Artificial nutrition support in clinical practice. Cambridge University Press; 2000. p. 537–52.

Fencl V, Jabor A, Kazda A, Figge J. Diagnosis of metabolic acid-base disturbances in critically ill patients. Am J Respir Crit Care Med. 2000;162:2246–51.

Feranchak AP, Sontag MK, Wagener JS, Hammond KB, Accurso FJ, Sokol RJ. Prospective, long-term study of fat soluble vitamin status in children with cystic fibrosis identified by newborn screen. J Pediatr. 1999;135:601–10.

Fergusson D, Hutton B, Sharma M, Tinmouth A, Wilson K, Cameron DW, et al. Use of intravenous immunoglobulin for treatment of neurologic conditions: a systematic review. Transfusion. 2005;45:1640–57.

Ferrari R, Merli E, Cicchitelli G, Mele D, Fucili A, Ceconi C. Therapeutic effects of L-carnitine and propionyl-L-carnitine on cardiovascular diseases: a review. Ann N Y Acad Sci. 2004;1033:79–91.

Ferreira IM, Brooks, D, Lacasse Y, Goldstein RS, White J. Nutritional supplementation for stable obstructive pulmonary disease. The Cochrane Database of Systematic Reviews 2005, No.: CD000998.pub2.

Feskanich D, Willett W, Colditz G. Walking and leisure-time activity and risk of hip fracture in postmenopausal women. JAMA. 2002;288:2300–6.

Fettes SB, Davidson HIM, Richardson RA, Pennington CR. Nutritional status of elective gastrointestinal surgery patients pre- and post-operatively. Clin Nutr. 2002;21:249–54.

Fey MF. ESMO minimum clinical recommendations for the diagnosis, treatment and follow up of AML in patients. 2003;14:1161–2.

Fiaccadori E, Coffrini E, Fraccia C, Rampulla C, Montaga T, Borghetti A. Hypophosphatemia and phosphorus depletion in respiratory and peripheral muscles of patients with respiratory failure due to COPD. Chest. 1994;105: 1392–8.

Fiaccadori E, Coffrini E, Ronda N, Vezzani A, Cacciani G, Fracchia C, et al. Hypophosphatemia in course of chronic obstructive pulmonary disease. Prevalence, mechanisms, and relationships with skeletal muscle phosphorus content. Chest. 1990;97(4):857–68.

Fick DM, Cooper JW, Wade WE, Waller JL, Maclean JR, Beers MH. Updating the Beers criteria for potentially inappropriate medication use in older adults: results of a US consensus panel of experts. Arch Intern Med. 2003;163: 2716–24.

Fields-Gardner C. Anthropometry Measures [accessed 23 Nov 2005]. Cary (IL): Hi-R-Ed; 2001. Available from: http://www.Hi-R-Ed.org.

Filippatos GS, Anker SD, Kremastinos DT. Cardiac cachexia. Curr Opin Clin Nutr Metab Care. 2005;8:249–54.

Finestone HM, Greene-Finestone LS. Rehabilitation medicine: 2. Diagnosis of dysphagia and its nutritional management for stroke patients. CMAJ. 2003;169.

Finley JW. Manganese absorption and retention by young women is associated with serum ferritin concentration. Am J Clin Nutr. 1999;70:37–43.

Finucane TE, Christmas C, Travis K. Tube feeding in patients with advanced dementia. JAMA. 1999;282:1365–70.

Fisch RO, Chang PN, Weisberg S, Guldberg P, Guttler F, Tsai MY. Phenylketonuric patients decades after diet. J Inherit Metab Dis. 1995;18:347–53.

Fishbane S, Frei GL, Maesaka J. Reduction of recombinant human erythropoietin doses by use of chronic intravenous iron supplementation. Am J Kidney Dis. 1995;26: 41–6.

Fishbane S, Mittal SK, Maesaka JK. Beneficial effects of iron therapy in renal failure patients on hemodialysis. Kidney Int. 1999;55:S67–S70.

Fisher NDL, Williams GH. Hypertensive vascular disease. In: Kasper DL, Braunwald E, Fauci AS, Hauser SL, Longo DL, Jameson JL, editors. Harrison's principles of internal medicine. 16th ed. New York: McGraw-Hill; 2005. p. 1463–81.

Fitzgerald JF, Troncone R, Rutigliano V, et al. Clinical quiz. Complications of small bowel resection. J Pediatr Gastroenterol Nutr. 2002;35:205, 219.

Flaws B, Sionneau P. The treatment of modern Western medical diseases with Chinese medicine: a textbook and clinical manual. Boulder (CO): Blue Poppy Press; 2001.

Flegal KM, Caroll D, Ogden CL, Johnson CCL. Prevalence and trends in obesity among US adults, 1999–2000. JAMA. 2002;288:1723–7.

Flegal KM, Troiano RP, Pamuk ER, Kuczmarski RJ, Campbell SM. The influence of smoking cesssation on the prevalence of overweight in the United States. N Engl J Med. 1995;333:1165–70.

Fleischer DM, Conover-Walker MK, Matsui EC, Wood RA. The natural history of tree nut allergy. J Allergy Clin Immunol. 2005;116:1087–93.

Fleming DJ, Jacques PF, Tucker KL, et al. Iron status of the free-living, elderly Framingham heart study cohort: an iron-replete population with a high prevalence of elevated iron stores. Am J Clin Nutr. 2001;73:638–46.

Fletcher B, Berra K, Ades P, Braun LT, Burke LE, Durstine JL, et al. Managing abnormal blood lipids: a collaborative approach. Circulation. 2005;112:3184–209.

Flier JS, Maratos-Flier E. Obesity. In: Kasper DL, Braunwald E, Fauci AS, Hauser SL, Longo DL, Jameson JL, editors. Harrison's principles of internal medicine. 16th ed. New York: McGraw-Hill; 2005. p. 422–9.

Floch MH, Bina I. The natural history of diverticulitis: fact and theory. J Clin Gastroenterol. 2004;38(5 Suppl):S2–7.

Flohr F, Lutz A, App EM, Matthys H, Reincke M. Bone mineral density and quantitative ultrasound in adults with cystic fibrosis. Eur J Endo. 2002;146:531–6.

Flynn MB. Nutritional support for the burn-injured patient. Crit Care Nurs Clin North Am. 2004;16:139–44.

Fogelholm M, van Marken Lichtenbelt W. Comparison of body composition methods:a literature analysis. Eur J Clin Nutr. 1997;51:495–503.

Foley RN, Parfrey PS, Sarnak MJ. Clinical epidemiology of cardiovascular disease in chronic renal disease. Am J Kidney Dis 1998;32: S112–S119.

Folk CC. Greene HL. Dietary management of type I glycogen storage disease. J Am Diet Assoc. 1984;84:293–301.

Folstein, MF, Folstein, SE, McHugh PR. Mini-mental state: a practical method for grading the cognitive state of patients for the clinician. J Psych Res.1975;12:189–98.

Fontaine KR, Redden DT, Wang C, Westfall AO, Allison DB. Years of life lost due to obesity. JAMA. 2003;289:187–93.

Food Additives and Ingredients Association. Food intolerance and allergy [monograph on the Internet]. Maidstone (UK): Food Additives and Ingredients Association; 2002 [cited 2006 Mar 5]. Available from: http://www.faia.org.uk/foodallergy.php.

Food Allergen Labeling and Consumer Protection Act of 2004, Title II of Pub. L. No. 108–282 [cited 2006 Apr 11]. Available from: http://www.cfsan.fda.gov/~dms/alrgact.html.

Food and Drug Administration. Exenatide (marketed as Byetta™) patient information sheet. Washington (DC): Food and Drug Administration [updated 2005 May; cited 2006 Mar 4]. Available from: http://www.fda.gov/cder/drug/InfoSheets/patient/exenatidePIS.htm.

Food and Nutrition Board and Institute of Medicine. Dietary reference intakes for energy, carbohydrate, fiber, fat, fatty acids, cholesterol, protein, and amino acids (Macronutrients). Washington (DC): National Academy Press; 2002.

Food and Nutrition Board and Institute of Medicine. Dietary reference intakes for vitamin C, vitamin E, selenium, and carotenoids. Washington (DC): National Academy Press; 2000. p. 152–3.

Food and Nutrition Board. Dietary reference intakes for water, potassium, sodium, chloride, and sulfate. Panel on Dietary Reference Intakes for Electrolytes and Water, Standing Committee on the Scientific Evaluation of Dietary Reference Intakes; 2004.

Food Assistance and Nutrition Research Report No. (FANRR35). 58 pp. Economic Research Service, USDA; October 2003.

Ford ES. Prevalence of the metabolic syndrome defined by the International Diabetes Federation among adults in the U.S. Diabetes Care. 2005;8:2745–9.

Ford ES, Mannino DM. Time trends in obesity among adults with asthma in the United States: findings from three national surveys. J Asthma. 2005;42:91–5.

Forette TL. Indirect calorimetry; principles and applications for managing critically ill patients. Available from: http://www.medscape.com/viewprogram/4704_pnt. Accessed December 30, 2005.

Forget BG, Cohen AR. Thalassemia syndromes. In: Hoffman R. Hematology, basic principles and practice. 4th ed. Philadelphia (PA): Elsevier Medical Publishing; 2005. p. 557–90.

Forrester JE, Spiegelman D, Woods M, Knox TA, Fauntleroy JM, Gorbach SL. Weight and body composition in a cohort of HIV-positive men and women. Public Health Nutr. 2001;4:743–7.

Fote-Ardah CE. The meaning of complementary and alternative medicine practices among people with HIV in the United States: strategies for managing everyday life. Sociology of Health & Illness. 2003;25:481–500.

Fouque D, Laville M, Boissel JP. Low protein diets for chronic kidney disease in non diabetic adults. Cochrane Database Syst Rev. 2006 Apr 19;(2):CD001892.

Fouque D, Laville M, Boissel JP, et al. Controlled low protein diets in chronic renal insufficiency: meta-analysis. BMJ. 1992;304:216–20.

Fox MK. 2000. Mr. Galley, Editorial, Boston Globe, September 1991, as quoted by M. K. Fox, Reimbursement practices and trends, presented at the American Dietetic Association's annual meeting in Dallas.

Fox MK. Adapted from: Overview of third-party reimbursement. Chicago (IL): American Dietetic Association; 1991.

Fraker P. Nutritional immunology: methodological considerations. J Nutr Immunol. 1994;2:87–92.

Frank C. Approach to skin ulcers in older patients. Can Fam Physician. 2004;50:1653–9.

Frankenfield D, Muth E, Rowe W. The Harris-Benedict studies of human basal metabolism: history and limitations. J Am Diet Assoc. 1998;98:439–45.

Frankenfield D, Roth-Yousey L, Compher C. Comparison of predictive equations for resting metabolic rate in healthy nonobese and obese adults: a systematic review. J Am Diet Assoc. 2005;105:775–89.

Frankenfield D, Smith S, Cooney R. Validation of 2 approaches to predicting resting metabolic rate in critically ill patients. JPEN J Parenter Enteral Nutr. 2004;28:259–64.

Frankenfield DC, Muth ER, Rowe WA. The Harris-Benedict studies of human basal metabolism: history and limitations. J Am Diet Assoc. 1998;98:439–45.

Franz MJ, Bantle JP, Beebe CA, Brunzell JD, Chiasson J, Garg A, et al. Evidence-based nutrition principles and recommendations for the treatment and prevention of diabetes and related complications. Diabetes Care. 2002;25: 148–98.

Freedman DA, Petitti DB. Salt and blood pressure. Conventional wisdom reconsidered. Eval Rev. 2001;25:267–87.

Freedman MH. Inherited forms of bone marrow failure. In: Hoffman R. Hematology, basic principles and practice. 4th ed. Philadelphia (PA): Elsevier Medical Publishing; 2005. p. 339–81.

Freeman J. Oral diabetes medications. In: Ross TA, Boucher JL, O'Connell BS, editors. American Dietetic Association guide to diabetes: medical nutrition therapy and education. Chicago (IL): American Dietetic Association; 2005.

Freidenberg J, Mulvihill M, Caraballo LR. From ethnology to survey: some methodological issues in research on health seeking in east Harlem. Human Organization. 1993;52:151–61.

French JA, Kanner AM, Bautista J, Abou-Khali B, Browne T, Harden CL, et al. Efficacy and tolerability of the new antiepileptic drugs I: treatment of new onset epilepsy: report of the Therapeutics and Technology Assessment Subcommittee and Quality Standards Subcommittee of the American Academy of Neurology and the American Epilepsy Society. Neurology. 2004;62:1252–60.

Frezza M, di Podova C, Pozzato G, Terpin M, Baraona E, Lieber CS. High blood alcohol levels in women. The role of decreased gastric alcohol dehydrogenase activity and first-pass metabolism. N Engl J Med. 1990;322:95–9.

Friedman EA, Shyh T, Beyer MM, Maris T, Butt KM. Post transplant diabetes in kidney transplant recipients. Am J Nephrol. 1985;5:196–202.

Friedman LS, Martin P, Muñoz SJ. Laboratory evaluation of the patient with liver disease. In: Zakim D, Boyer TD, editors. Hepatology: a textbook of liver disease. 4th ed. Philadelphia (PA): Saunders; 2003.

Fry J, Finley W. The prevalence and costs of obesity in the EU. Proc Nutr Soc. 2005;64:359–62.

Fuchs M, Bashshur R. Use of traditional Indian medicine among urban Native Americans. Medical Care. 1975;13: 915–27.

Fugate SE, Church CO. Nonestrogen treatment modalities for vasomotor symptoms associated with menopause. Ann Pharmacother. 2004;38:1482–99.

Fugh-Berman A. Herb-drug interactions. Lancet. 2000;355: 134–8.

Fuhrman MP, Charney P, Mueller CM. Hepatic proteins and nutritional assessment. J Am Diet Assoc. 2004;104:1258–64.

Fuhrman MP, Herrmann V, Masidonski P, Eby C. Pancytopenia after removal of copper from total parenteral nutrition. JPEN Parenter Enteral Nutr. 2000;24:361–6.

Fulton MM, Allen ER. Polypharmacy in the elderly: a literature review. J Am Acad Nurse Pract. 2005;17:123–32.

Furuta T, Shirai N, Sugimoto M, Ohashi K, Ishizaki T. Pharmacogenomics of proton pump inhibitors. Pharmacogenomics. 2004;5:181–202.

Gabay C, Kushner I. Acute phase proteins and other systemic responses to inflammation. N Engl J Med. 1999; 340:448–54.

Gabel JG, Claxton I, Gil J, Pickreign H, Whitmore E, Holve B, et al. Health benefits in 2004: four years of double-digit premium increases take their toll on coverage. Health Affairs. 2004;23:200–9.

Gabriel DA, Shea T, Olajida O, Serody JS, Comeau T. The effect of oral mucositis on morbidity and mortality in bone marrow transplant. Semin Oncol. 2003;30 Suppl 18: 76–83.

Gadducci A, Cosio S, Fanucchi A, Genazzani AR. Malnutrition and cachexia in ovarian cancer patients: pathophysiology and management. Anticancer Res. 2001;21:2941–7.

Gadek J, DeMichele SJ, Karlstad MD, Pacht ER, Donahoe M, Albertson TE, et al. Effect of enteral feeding with eicosapentaenoic acid, _-linolenic acid, and antioxidants in patients with acute respiratory distress syndrome. Crit Care Med. 1999;27:1409–20.

Gallagher PG, Jarolim P. Red blood cell membrane disorders. In: Hoffman R. Hematology, basic principles and practice. 4th ed. Philadelphia (PA): Elsevier Medical Publishing; 2005. p. 669–93.

Gallagher TC, Geline O, Comite F. Missed opportunities for prevention of osteoporotic fracture. Arch Intern Med. 2002;162:450–6.

Gallop, CNN, and USToday. 2005. Health care: people's chief concerns. Survey results available from: http://www.publicagenda.org/issues/pcc_detail.cfm?issue_type=healthcare&list=2.

Galperin C, German BJ, Gershwin ME. Nutrition and diet in rheumatic diseases. In: Shils ME, Olson JA, Shike M, Ross AC, editors. Modern nutrition in health and disease. 9th ed. Baltimore (MD): Williams & Wilkins; 1999. p. 1339–51.

Galvez J, Rodriguez-Cabezas ME, Zarzuelo A. Effects of dietary fiber on inflammatory bowel disease. Mol Nutr Food Res. 2005;49:601–8.

Ganesh SK, Stack AG, Levin NW, Hulbert-Shearon T, Port FK. Association of elevated serum phosphorus, calcium x phosphorus product and parathyroid hormone with cardiac mortality risk in chronic hemodialysis patients. J Am Soc Nephrol. 2001;12:2131–8.

Gantt EL. Medical management of ascites in clinician-2, liver disease. New York: Medcon, Searle & Co.; 1971. p. 61.

García-Casal MN, Leets I, Layrisse M. Beta-carotene and inhibitors of iron absorption modify iron uptake by caco-2 cells. J Nutr. 2000;130:5–9.

Garleb KA, et al. Application of fructooligosaccharides to medical foods and fermentable dietary fiber. Bioscience Microflora. 2002;21:43–54.

Garner DM, Garfinkel PE, Schwartz D, Thompson M. Cultural expectations of thinness in women. Psychol Rep. 1980;47:483–91.

Garrard J, Harms S, Eberly LE, Matiak A. Variations in product choices of frequently purchased herbs: caveat emptor. Arch Intern Med. 2003;163:2290–5.

Garrod AE. Lancet. 1908;2:142.

Gartner LM, Eidelman AI. American Academy of Pediatrics policy statement: breast-feeding and the use of human milk. Pediatr. 2005;115:496–506.

Gartner LM, Greer FR. Prevention of rickets and vitamin D deficiency: new guidelines for vitamin D intake. Pediatr. 2003;111:908–10.

Gaspard KJ. The hematopoietic system. In: Porth CM. Pathophysiology: concepts of altered health states. 7th ed. Philadelphia (PA): Lippincott Williams & Wilkins; 2002. p. 249–306.

Gassull MA. Macronutrients and bioactive molecules: is there a specific role in the management of inflammatory bowel disease? JPEN J Parenter Enteral Nutr. 2005;29 Suppl:S179–83.

Gaudet G, Laplante J. Soft drink abuse, malnutrition, and folic acid deficiency. Am J Hematol. 1999;60:311–2.

Gauthier PM, Szerlip HM. Metabolic acidosis in the intensive care unit. Crit Care Clin. 2002;18:289–308, vi.

Geenen R, Janssens EL, Jacobs JW, van Staveren W. Hypothesis: dietary glutamate will not affect pain in fibromyalgia. J Rheumatol. 2004;31:785–7.

Geleijnse JM, Grobbee DE, Kok FJ. Impact of dietary and lifestyle factors on the prevalence of hypertension in Western populations. 2005;19 Suppl 3:S1.

Geleijnse JM, Kok FJ, Grobbee DE. Blood pressure response to changes in sodium and potassium intake: a metaregression analysis of randomised trials. J Hum Hypertens. 2005;19 Suppl 3:S1–4.

Gelfand DW, Ott DJ, Chen MYM. Radiologic evaluation of gastritis and duodenitis. AJR Am J Roentgenol. 1999;173:357–61.

Gerber T, Schomerus H. Hepatic encephalopathy in liver cirrhosis: pathogenesis, diagnosis, and management. Drugs. 2000;60:1353–70.

Gerbert B, Caspers N, Moe J, Clanon K, Abercrombie P, Herzig K. The mysteries and demands of HIV care: qualitative analyses of HIV specialists' views on their expertise. AIDS Care. 2004;16:363–76.

German JB, Dillard CJ. Saturated fats: what dietary intake? Am J Clin Nutr. 2004;80:550–9.

Gerster H. High-dose vitamin C: a risk for persons with high iron stores? Int J Vitam Nutr Res. 1999;69:67–82.

Giacoia GP, Berry G. Acrodermatitis enteropathica-like syndrome secondary to isoleucine deficiency during treatment of maple syrup urine disease. Am J Dis Child. 1993;147:954–6.

Gibbs J, Cull W, Henderson W, Daley J, Hur K, Khuri SF. Preoperative serum albumin level as a predictor of operative mortality and morbidity: results from the National VA Surgical Risk Study. Arch Surg. 1999;134:36–42.

Gibson RS, Yeudall F, Drost N, Mtitimuni BM, Cullinan TR. Experiences of a community-based dietary intervention to enhance micronutrient adequacy of diets low in animal source foods and high in phytate: a case study in rural Malawian children. J Nutr. 2003;133 Suppl 2:3992S–3999S.

Gilden DH. Infectious causes of multiple sclerosis. Lancet Neurol. 2005;4:195–202.

Gilgun-Sherki Y, Melamed E, Offen D. Oxidative stress induced neurodegenerative diseases: the need for antioxidants that penetrate the blood brain barrier. Neuropharmacology. 2001;40:959–75.

Gilland FD, Berhane KJ, Li Y, Gauderman WJ, McConnell R, Peters J. Children's lung function and antioxidant vitamin, fruit, juice, and vegetable intake. Am J Epidemiol. 2003;158:576–84.

Gillespie S, KulkarniK, Daly A. Using carbohydrate counting in diabetes clinical practice. J Am Diet Assoc. 1998;98: 897–9.

Gilliland BC. Fibromyalgia, arthritis associated with systemic disease, and other arthritides. In: Kasper DL, Braunwald E, Fauci AS, Hausner SL, Longo DL, Jameson JL, editors. Harrison's principles of internal medicine. 16th ed. New York: McGraw-Hill; 2005. p. 2055–64.

Giuliano B, Leiserowitz M, Shamir E, et al. Nutritional and metabolic implications of acute renal failure. In: Renal nutrition. Report of eleventh Ross roundtable on medical issues. Columbus (OH): Ross Laboratories; 1991. p. 58–64.

Gloria L, Cravo M, Camilo ME, Resende M, Cardoso JN, Oliveira AG, et al. Nutritional deficiencies in chronic alcoholics: relation to dietary intake and alcohol consumption. Am J Gastroenterol. 1997;92:485–9.

Goitein M, Laughlin J, Purdy JA, et al. State-of-the-art of external photon beam treatment planning. Int J Radiat Oncol Biol Phys. 1991;21:9–24.

Gold J, Sadeghi-Nejad A. Hyperthyroidism. New York: eMedicine.com, Inc. 2004 [updated 2004 May 26, cited 2005 Jan 11]. Available from: http://www.emedicine.com.

Goldberg B. Alternative medicine: the definitive guide. 2nd ed. Berkeley (CA): Celestial Arts; 2002.

Goldenberg DL, Burckhardt C, Crofford L. Management of fibromyalgia syndrome. JAMA. 2004;292:2388–95.

Goldin R. The pathogenesis of alcholic liver disease. Int J Exp Path. 1994;75:71–8.

Goldstein DJ. Assessment of nutritional status in renal disease. In: Mitch W, Klahr S, editors. Handbook of nutrition and the kidney. 4th ed. Philadelphia (PA): Lippincott-Raven; 2002.

Goldstein DJ. Assessment of nutritional status in renal disease. In: Mitch W, Klahr S, editors. Handbook of nutrition and the kidney. 5th ed. Philadelphia: Lippincott-Raven; 2005.

Gomes GF, Pisani JC, Macedo ED, Campos AC. The nasogastric feeding tube as a risk factor for aspiration and aspiration pneumonia. Curr Opin Clin Nutr Metab Care. 2003;6:327–33.

Gomez E. Health Insurance Portability and Accountability Act protects privacy of medical records. Oncology Nursing Society News. 2003;18(1):13.

Gomez RS, Carneiro MA, Souza LN, et al. Oral recurrent human herpes virus infection and bone marrow transplantation survival. Oral Surg Oral Med Oral Pathol Oral Radiol Endod. 2000;91:552–6.

Gomez-Escudero O, Schmulson-Wasserman MJ, Valdovinos-Diaz MA. Post-infectious irritable bowel syndrome. A review based on current evidence. Rev Gastroenterol Mex. 2003;68:55–61.

Gomez-Herrera E, Farias-Llamas OA, Gutierrez-de la Rosa JL, Hermosillo-Sandoval JM. The role of long-acting release (LAR) depot octreotide as adjuvant management of short bowel disease. Cir Cir. 2004;72:379–86.

Gooding I, Dondos V, Gyi KM, Hodson M, Westaby D. Variceal hemorrhage and cystic fibrosis: outcomes and implications for liver transplantation. Liver Transpl. 2005; 11:1522–6.

Goodman L, Segreti J. Infectious diarrhea. Dis Mon. 1999:45:268–99.

Gordon NF, Leighton RF, Mooss A. Factors associated with increased risk of coronary heart disease. In: Kaminsky LA, editor. American College of Sports Medicine's resource manual for guidelines for exercise testing and prescription. 5th ed. Baltimore (MD): Lippincott Williams & Wilkins; 2005. p. 95–114.

Gore DC, Chinkes D, Sanford A, Hart DW, Wolf SE, Herndon DN. Influence of fever on the hypermetabolic response in burn-injured children. Arch Surg. 2003;138:169–74.

Gore-Felton C, Vosvick M, Power R, Koopman C, Ashton E, Bachmann MH, et al. Alternative therapies: a common practice among men and women living with HIV. J Assoc Nurses AIDS Care. 2003;14:17–27.

Goringe AP, Brown S, O'Callaghan U, et al. Glutamine and vitamin E in the treatment of hepatic veno-occlusive disease following high-dose chemotherapy. Bone Marrow Transplant. 1998;21:829–32.

Gortmaker SL, Must A, Sobol AM, Peterson K, Colditz GA, Dietz WH. Television viewing as a cause of increasing obesity among children in the United States, 1986–1990. Arch Pediatr Adolesc Med. 1996;150:356–62.

Gosker HR, Wouter EFM, van der Vusse GJ, Schols AM. Skeletal muscle dysfunction in chronic obstructive pulmonary disease and chronic heart failure: underlying mechanisms and therapy perspectives. Am J Clin Nutr. 2000;71:1033–47.

Gottschlich MM, Baumer T, Jenkins M, Khoury J, Warden GD.The prognostic value of nutritional and inflammatory indices of patients with burns. J Burn Care Rehabil. 1992;13:10513.

Gould B. Inflammation and healing. In: Pathophysiology for the health professions. Philadelphia (PA): WB Saunders; 2004.

Gould R. The next rung on the ladder: achieving and expanding reimbursement for nutrition services. J Am Diet Assoc. 1991;1:1383–4.

Gowers S, Bryant-Waugh R. Management of child and adolescent eating disorders: the current evidence base and future directions. J Child Psychol Psychiatry. 2004;45:63–83.

Graham DY, et al. Recognizing peptic ulcer disease. Keys to clinical and laboratory diagnosis. Postgrad Med. 1999; 105:113–6, 121–3, 127–8.

Gralla RJ, Osoba D, Kris MG, et al. Recommendations for the use of antiemetics: evidence based, clinical practice guidelines. American Society of Clinical Oncology. J Clin Oncol. 1999;17:2971–94.

Graves C, Saffle J, Morris S, Stauffer T, Edelman L. Caloric requirements in patients with necrotizing fascitis. Burns. 2005;31:55–9.

Gray GE, Gray LK. Evidence-based medicine: applications in dietetic practice. J Am Diet Assoc. 2002;102:1263–72.

Green LS. Asthma, oxidant stress, and diet. Nutrition. 1999;15:899–907.

Green R, Miller JW. Folate deficiency beyond megaloblastic anemia: hyperhomocysteinemia and other manifestations of dysfunctional folate status. Semin Hematol. 1999;36: 47–64.

Greene A, editor. Food allergies vs. food intolerances [monograph on the Internet]. Danville (CA): DrGreene.com; 2002 [cited 2006 Mar 5]. Available from: http://www.drgreene.com/21_1259.html.

Greene HL, Slonim AE, Burr IM. Type I glycogen storage disease: a metabolic basis for advanced in treatment. Adv Pediatr. 1979;26:63–92.

Greenwald P, Clifford CK, Milner JA. Diet and cancer prevention. Eur J Cancer. 2001;37:948–65.

Greeves LG, Carson DJ, Craig, BG, McMaster D. Potentially life-threatening cardiac dysrhythmia in a child with selenium deficiency and phenylketonuria. Acta Paediatr Scand. 1990;79:1259–62.

Greger JL. Nutrition versus toxicology of manganese in humans: evaluation of potential biomarkers. Neurotoxicology. 1999;20:205–12.

Grimble RF. Immunonutrition. Curr Opin Gastroenterol. 2005;21:216–22.

Grimm H; Kraus A. Immunonutrition—supplementary amino acids and fatty acids ameliorate immune deficiency in critically ill patients. Langenbeck's Archives of Surgery. 2001;386:369–76.

Groff JL, Gropper SS. Advanced nutrition and human metabolism. 3rd ed. Wadsworth-Thomson Learning; 2000.

Hasse JM. Nutrition assessment and support of organ transplant recipients. JPEN J Parenter Enteral Nutr. 2001;25: 120–31.

Hasse JM. Recovery after organ transplantation in adults: the role of postoperative nutrition therapy. Top Clin Nutr. 1998;13:15–26.

Hathcock JN. Vitamins and minerals: efficacy and safety. Am J Clin Nutr. 1997;67:351–3.

Haugaard SB, Andersen O, Flyvbjerg A, Orskov H, Madsvad S, Iversen J. Growth factors, glucose and insulin kinetics after low dose growth hormone therapy in HIV-lipodystrophy. J Infect. 2005. in press.

Hauser SL, Asbury AK. Chapter 365. Guillain-Barré syndrome and other immune-mediated neuropathies. In: Kasper DL, Braunwald E, Fauci AS, Hausner SL, Longo DL, Jameson JL, editors. Harrison's principles of internal medicine. 16th ed. New York: McGraw-Hill; 2005.

Hauser SL, Goodin DS. Chapter 359. Multiple sclerosis and other demyelinating diseases. In: Kasper DL, Braunwald E, Fauci AS, Hausner SL, Longo DL, Jameson JL, editors. Harrison's principles of internal medicine. 16th ed. New York: McGraw-Hill; 2005.

Hauser SL, Ropper AH. Chapter 356. Diseases of the spinal cord. In: Kasper DL, Braunwald E, Fauci AS, Hausner SL, Longo DL, Jameson JL, editors. Harrison's principles of internal medicine. 16th ed. New York: McGraw-Hill; 2005.

Hawk C, Long CR, Boulanger KT. Prevalence of nonmusculoskeletal complaints in chiropractic practice: report from a practice-based research program. J Manipulative Physiol Ther. 2001;24:157–69.

Hawk C, Long CR, Perillo M, Boulanger KT. A survey of US chiropractors on clinical preventive services. J Manipulative Physiol Ther. 2004;27:287–98.

Hayes C. Physical activity and exercise. In: Ross TA, Boucher JL, O'Connell BS, editors. American Dietetic Association guide to diabetes: medical nutrition therapy and education. Chicago (IL): American Dietetic Association; 2005.

Hazell AS, Butterworth RF. Hepatic encephalopathy: an update of pathophysiologic mechanisms. Proceedings of the Society for Experimental Biology and Medicine. 1999; 222:99–112.

He FJ, MacGregor GA. Effect of modest salt reduction on blood pressure: a meta-analysis of randomized trials: implications for public health. J Hum Hypertens. 2002;16: 761–70.

He J, Whelton K, Appel LJ, Charleston J, Klag MJ. Long-term effects of weight loss and dietary sodium reduction on incidence of hypertension. Hypertension. 2000;35:544–9.

Health and Welfare Canada. Canada's food guide to healthy eating. Minister of Supply and Services Canada; 2005.

Health Resources and Services Administration (HRSA). Guidelines for the use of antiretroviral agents in pediatric HIV infection. November 3, 2005. Available from: http://aidsinfo.nih.gov/ContentFiles/PediatricGuidelines_PDA.pdf.

Heaney RP. Bone biology in health and disease. In: Shils ME, Shike M, Ross AC, Cabellero B, Cousins RJ, editors. Modern nutrition in health and disease. 10th ed. Philadelphia (PA): Lippincott Williams & Wilkins; 2006. p. 1314–25.

Heath AL, Skeaff CM, Gibson RS. The relative validity of a computerized food frequency questionnaire for estimating intake of dietary iron and its absorption modifiers. Eur J Clin Nutr. 2000;54:592–9.

Hebert LE, Beckett LA, Scherr PA, Evans DA. Annual incidence of Alzheimer disease in the United states projected to the years 2000 through 2050. Alzheimer Dis Assoc Disord. 2001;15:169–73.

Hebert LE, Scherr PA, Bennett DA, Evans DA. Alzheimer Disease in the U.S. Population: Prevalence Estimates Using the 2000 Census. Arch Neuro. 2003;60:1119–22.

Hedberg C, Angulo F, Townes J, Hadler J, Vugia D, Farley M, and the CDC/USDA/FDA Foodborne Diseases Active Surveillance Network. Differences in Escherichia coli O157:H7 annual incidence among FoodNet active surveillance sites. 5th International VTEC Producing Escherichia coli Meeting. Baltimore, MD, July 1997.

Hedley AA, Ogden CL, Johnson CL, Carroll MD, Curtin LR, Flegal KM. Overweight and obesity among US children, adolescents, and adults, 1999–2002. JAMA. 2004;291: 2847–50.

Hedlund J, Hansson LO, Ortqvist A. Short- and long-term prognosis for middle-aged and elderly patients hospitalized with community-acquired pneumonia: impact of nutritional and inflammatory facotors. Scand J Infect Dis. 1995;27:32–7.

Hedlund JU, Hansson LO, Ortqvist AB. Hypoalbuminemia in hospitalized patients with community acquired pneumonia. Arch intern Med. 1995;155:1438–42.

Hegde HR, Woodman RC, Sankaran K. Nutrients as modulators of anergy in acquired immune deficiency syndrome. J Assoc Physicians India. 1999;47:318–25.

Heimburger DC, McLaren DS, Shils ME. Clinical manifestations of nutrient deficiencies and toxicities: a resume. In: Shils ME, Shike M, Ross CA, Caballero BC, Cousins RJ, editors. Modern Nutrition in Health and Disease. 10th ed. Philadelphia (PA): Lippincott Williams & Wilkins; 2005. p. 595–614.

Heimburger O, Stevinkel P, Lindholm B. Chronic peritoneal dialysis. In: Kopple J, Massry S, editors. Nutritional management of renal disease. 2nd ed. Philadelphia (PA): Lippincott Williams & Wilkins; 2004. p. 477–511.

Hellings HA, Zarcone JR, Crandall K, Wallace D, Schroder SR. Weight gain in a controlled study of Risperidone in children, adolescents and adults with mental retardation and autism. J Child Adolesc Psychopharmacol. 2001;11:229–38.

Helman CG. Culture, health and illness: an introduction for health professionals. 2nd ed. London: Wright PSG Pub.; 1990.

Helphingstine C, Bistrian B. New Food and Drug Administration requirements for inclusion of vitamin K in adult parenteral multivitamins. JPEN J Parenter Enteral Nutr. 2003;27:220–4.

Helzer JE, Canino GJ. Alcoholism in North America, Europe, and Asia. Oxford: Oxford University Press; 1992.

Hemingway C, Freeman JM, Pillas DJ, Pyzik PL. The ketogenic diet: a 3- to 6-year follow-up of 150 children enrolled prospectively. Pediatrics. 2001;108:898–905.

Henderson RA, Talusan K, Hutton N, Yolken RH, Caballero B. Serum and plasma markers of nutritional status in children infected with the human immunodeficiency virus. J Am Diet Assoc. 1997;97:1377–81.

Henry PH, Longo DL. Enlargement of lymph nodes and spleen. In: Kasper DL, Braunwald E, Fauci AS, Hausner SL, Longo DI, Jameson JL, editors. Harrison's principles of internal medicine. 16th ed. New York: McGraw Hill Medical Publishing; 2005. p. 343–9.

Hernandez DG, Paisán-Ruíz C, McInerney-Leo A, Jain S, Meyer-Lindenberg A, Evans EW, et al. Clinical and positron emission tomography of Parkinson's disease caused by LRRK2. Ann Neurol. 2005;57:453–6.

Hertrampf E, Olivares M. Iron amino acid chelates. Int J Vitam Nutr Res. 74:435–43.

Heslin KC, Andersen RM, Ettner SL, Kominski GF, Belen TR, Margenstern H, et al. Do specialist self-referral insurance policies improve access to HIV-experienced physicians as a regular source of care? Med Care Res Rev. 2005;62:583–600.

Hess SY, Zimmermann MB, Arnold M, Langhans W, Hurrell RF. Iron-deficiency anemia reduces thyroid peroxidase activity in rats. J Nutr. 2004;132:1951–5.

Heyland DK, Drover JW, Dhaliwal R, Greenwood J. Optimizing the benefits and minimizing the risks of enteral nutrition in the critically ill: role of small bowel feeding. JPEN J Parenter Enteral Nutr. 2002;26(6 Suppl): S51–55; discussion S56–57.

HHS. Heath and Human Services fact sheet: the State Children's Health Insurance Program (SCHIP) Washington (DC): U.S. Government Printing Office; 2002.

HHS. The facts about upcoming new benefits in Medicare. Publication No. CMS-11054. February 17, 2004. Edited by U.S. Department of Health and Human Services. Washington (DC): U.S. Government Printing Office; 2004.

HHS. The initiative to eliminate racial and ethnic disparities in health. Edited by U.S. Department of Health and Human Services. Washington (DC): U.S. Government Printing Office; 1998.

HHS. U.S. Department of Health and Human Services, Healthy People 2010. Washington (DC): U.S. Government Printing Office; 2000.

Hightower MK, Gunta KE. Disorders of skeletal function: developmental and metabolic disorders. In: Porth CM, editor. Pathophysiology: concepts of altered health status. 7th ed. Philadelphia (PA): Lippincott Williams & Wilkins; 2005. p. 1393–415.

Hill JO, Catenacci VA, Wyatt HR. Obesity: etiology. In: Shils ME, Shike M, Ross AC, Cabellero B, Cousins RJ, editors. Modern nutrition in health and disease. 10th ed. Philadelphia (PA): Lippincott Williams & Wilkins; 2006. p. 1013–28.

Hill JO, Wyatt HR, Reed GW, Peters JC. Obesity and the environment: where do we go from here? Science. 2003;299: 853–5.

Hixon AL, Chapman RW. Healthy People 2010: The role of family physicians in addressing health disparities. Am Fam Phys. 2000;62(9):1971–5.

Ho C, Kauwell GP, Bailey LB. Practitioners' guide to meeting the vitamin B-12 recommended dietary allowance for people aged 51 years and older. J Am Diet Assoc. 1999; 99:725–7.

Hochberg MC. Nutritional supplements for knee osteoarthritis—still no resolution. N Engl J Med. 2006;354: 858–60.

Hodge L, Salome CM, Hughes JM, Liu-Brennan D, Rimmer J, Allman M, et al. Effect of dietary intake of omega-3 and omega-6 fatty acids on severity of asthma in children. Eur Respir J. 1998;11:361–5.

Hoehn MM, Yahr MD. Parkinsonism: onset, progression and mortality. Neurology. 1967;17:427–42.

Hoek HW, van Hoeken D. Review of the prevalence and incidence of eating disorders. Int J Eat Disord. 2003;34: 383–96.

Hoffbrand AV, Herbert V. Nutritional anemias. Semin Hematol. 1999;36 Suppl 7:13–23.

Hoffer LJ. Metabolic Consequences of Starvation. In: Shils ME, Shike M, Ross AC, Cabellero B, Cousins RJ, editors. Modern nutrition in health and disease. 10th ed. Philadelphia (PA): Lippincott Williams & Wilkins; 2006. p. 730–48.

Hoffman HT, Karnell LH, Funk GF, Robinson RA, Menck HR. The national cancer data base report on cancer of the head and neck. Arch Otolaryngol Head Neck Surg. 1998;124:951–62.

Hoffman R, Ravandi F-Kashani. Idiopathic myelofibrosis. In: Hoffman R. Hematology, basic principles and practice. 4th ed. Philadelphia (PA): Elsevier Medical Publishing; 2005. p. 1255–76.

Hoffman R, Baker KR, Prchal JT. The polycythemias. In: Hoffman R. Hematology, basic principles and practice. 4th ed. Philadelphia (PA): Elsevier Medical Publishing; 2005. p. 1209–45.

Holben DH. Position of the American Dietetic Association: food insecurity and hunger in the United States. J Am Diet Assoc. 2006;106: in press.

Holben DH, Myles W. Food insecurity in the United States: how it affects our patients. Am Fam Physician. 2004;69: 1058–63.

Holdaas, H Fellstrom B, Jardine A, et al. Prevention of cardiac death and non-fatal coronary events with fluvastatin in renal transplant patients: a multicentre randomized placebo controlled trial. Lancet. 2003;361:2024–31.

Holdcraft LC, Assefi N, Buchwald D. Complementary and alternative medicine in fibromyalgia and related syndromes. Best Pract Res Clin Rheumatol. 2003;17:667–83.

Holden K. Eat well, stay well with Parkinson's disease. Winfield (KS): Five Star Living; 2004.

Holden K, Remig VM. Parkinson's disease: assessing and managing unique nutrition needs. Chicago (IL): American Dietetic Association; 1999.

Holick MF. Vitamin D. In: Shils ME, Shike M, Ross AC, Cabellero B, Cousins RJ, editors. Modern nutrition in health and disease. 10th ed. Philadelphia (PA): Lippincott Williams & Wilkins; 2006. p. 376–95.

Holland SM, Gallin JI. Cases of granulocytes and monocytes. In: Kasper DL, Braunwald E, Fauci AS, Hausner SL, Longo DI, Jameson JL, editors. Harrison's principles of internal medicine. 16th ed. New York: McGraw Hill Medical Publishing; 2005. p. 349–57.

Hollander PA, Blonde L, Rowe R, Mehta AE, Milburn JL, Hershon KS, Chiasson J, Levin SR for the Exubera Phase III Study Group. Efficacy and safety of inhaled insulin (Exubera) compared with subcutaneous insulin therapy in patients with type 2 diabetes. Diabetes Care. 2004;27: 2356–62.

Hollowell JG, Staehling NW, Flanders WD, Hannon WH, Gunter EW, Spencer CE, et al. Serum TSH, T4, and thyroid antibodies in the United States population (1988 to 1994): National Health and Nutrition Examination Survey (NHANES III). J Clin Endocrinol Metab. 2002;87: 489–99.

Holman H. Chronic disease—the need for a new clinical education. JAMA. 2004;92:1057–9.

Holten KB, Wetherington A, Bankston L. Diagnosing the patient with abdominal pain and altered bowel habits: is it irritable bowel syndrome? Am Fam Physician. 2003;67: 2157–62.

Holton JB, Walker JH, Tyfield LA. Galactosemia. In: Scriver C, Beaudet A, Sly W, Valle D, editors. The metabolic basis of inherited diseases. Vol. 1. New York: McGraw-Hill; 2001. p. 1553–87.

Holzmeister LA, Geil P. Evidence-based nutrition care and recommendations. In: Ross TA, Boucher JL, O'Connell BS, editors. American Dietetic Association guide to diabetes: medical nutrition therapy and education. Chicago (IL): American Dietetic Association; 2005.

Hong-Brown LQ, Brown DR, Lang CH. HIV antiretroviral agents inhibit protein synthesis and decrease ribosomal protein S6 and 4EBP1 phosphorylation in C2C12 myocytes. AIDS Res Hum Retroviruses. 2005;21:854–62.

Hong-Brown LQ, Pruznak AM, Frost RA, Vary TC, Lang CH. Indinavir alters regulators of protein anabolism and catabolism in skeletal muscle. Am J Physiol Endocrinol Metab. 2005;289:E382–E390.

Hoos MB, Plasqui G, Gerver WJM, Westerterp KR. Physical activity level measured by doubly labeled water and accelerometry in children. Eur J Appl Physiol. 2003;89:624–6.

Horne C, Derrico D. Mastering ABGs. The art of arterial blood gas measurement. Am J Nurs. 1999;99:26–32; quiz 33.

Hornick CA, Myers A, Sadowska-Krowicka H, Anthony CT, Woltering EA. Inhibition of angiogenic initiation and disruption of newly established human vascular networks by juice from Morinda citrifolia (noni). Angiogenesis. 2003; 6:143–9.

Horowski R, Horowski L, Vogel S, Poewe W, Kielhorn FW. An essay on Wilhelm von Humboldt and the shaking palsy: first comprehensive description of Parkinson's disease by a patient. Neurology. 1995;45(3 Pt 1):565–8.

Horwich TB, Fonarow GC, Hamilton MA, MacLellan WR, Borenstein J. Anemia is associated with worse symptoms, greater impairment in functional capacity and a significant increase in mortality in patients with advanced heart failure. J Am Coll Cardiol. 2002;39:1780–6.

Hostetler JA. Folk medicine and sympathy healing among the Amish. In: Hand WD, editor. American Folk Medicine. Berkeley (CA): University of California Press; 1976.

Hotchkiss RS, Karl IE. The pathophysiology and treatment of sepsis. N Engl J Med. 2003;348:138–50.

Housh TJ, Johnson GO, Housh DJ, Cramer JT, Eckerson JM, Stout JR, et al. Accuracy of near-infrared interactance instruments and population-specific equations for estimating body composition in young wrestlers. J Strength Cond Res. 2004;18:556–60.

Howard BV, Van Horn L, Hsia J, Manson JE, Stefanick ML, Wassertheil-Smoller S, et al. Low-fat dietary pattern and risk of cardiovascular disease: the Women's Health Initiative Randomized Controlled Dietary Modification Trial. JAMA. 2006;295:655–66.

Howard JJ. Branched chain amino acids in hepatic encephalopathy. Am J Surg. 2002;183:424–9.

Howard L. Home parenteral and enteral nutrition in cancer patients. Cancer. 1993;72:3531–41.

Howland RD, Mycek MJ, editors. Drugs used in Parkinson disease. In: Pharmacology. 3rd ed. Baltimore (MD): Lippincott Williams & Wilkins; 2006.

Hsu CY, McCulloch CE, Curhan GC. Epidemiology of anemia associated with chronic renal insufficiency among adults in the United States: results from the third national health and nutrition examination survey. J Am Soc Nephrol. 2002;13:504–10.

Hu FB, LI TY, Colditz GA, Willett WC, Manson JE. Television watching and other sedentary behaviors in relation to risk of obesity and type 2 diabetes mellitus in women. JAMA. 2003;289:1785–91.

Huang Q, NaiYJ, Jiang ZW, Li JS. Change of the growth hormone-insulin-like growth factor-I axis in patients with gastrointestinal cancer: related to tumour type and nutritional status. Am J Clin Nutr. 2005;81:1163–7.

Huet F, Semama K, Maingueneu C, Charavel A, Nivelon JL. Vitamin A deficiency and nocturnal vision in teenager with cystic fibrosis. Eur J Pediatr. 1997;156:949–51.

Hughes DA, Wright AJ, Finglas PM, Peerless AC, Bailey AL, Astley SB, et al. The effect of beta-carotene supplementation on the immune function of blood monocytes from healthy male nonsmokers. J Lab Clin Med. 1997;129: 309–17.

Hughes RA, Rees JH. Clinical and epidemiologic features of Guillain-Barre syndrome. J Infect Dis. 1997;176 Suppl 2:S92–8

Hughes T. Neurology of swallowing and oral feeding disorders: assessment and management. J Neurol Neurosurg Pyschiatry. 2003;70(Suppl III):48–52.

Human Genome Project Information [homepage on the Internet]. Washington (DC): Human Genome Program; 2004 [cited 2005 May 6]. CFTR: the gene associated with cystic fibrosis. Available from: http://www.ornl.gov/sci/techresources/Human_Genome/posters/chromosome/cftr.shtml.

Hung HC, Joshipura KJ, Jiang R, Hu FB, Hunter D, Smith-Warner SA, et al. Fruit and vegetable intake and risk of major chronic disease. J Natl Cancer Inst. 2004;96:1577–84.

Hunt JR, Zeng, H. Iron absorption by heterozygous carriers of the HFE C282Y mutation associated with hemochromatosis. Am J Clin Nutr. 2004;80:924–31.

Hunt SA, Abraham WT, Chin MH, Feldman AM, Francis GS, Ganiats TG, et al; American College of Cardiology; American Heart Association Task Force on Practice Guidelines; American College of Chest Physicians; International Society for Heart and Lung Transplantation; Heart Rhythm Society. ACC/AHA 2005 guideline update for the diagnosis and management of chronic heart failure in the adult—summary article: a report of the American College of Cardiology/American Heart Association task force on practice guidelines (Writing committee to update 2001 guidelines for the evaluation and management of heart failure). Circulation. 2005;112:1825–52.

Hurrell R. How to ensure adequate iron absorption from iron-fortified food. Nutr. Rev. 2002;60 Pt 2:S7–15.

Hussaini SH, Henderson T, Morrell AJ, Losowsky MS. Dark adaptation in early primary biliary cirrhosis. Eye. 1998;12(Pt 3a):419–26.

Hyder SM, Persson LA, Chowdhury AM, EC Ekström. Do side-effects reduce compliance to iron supplementation? A study of daily- and weekly-dose regimens in pregnancy. J Health Popul Nutr. 2002;20:175–9.

Hymes KB, Greene JB, Marcus A. Kaposi's sarcoma in homosexual men: a report of eight cases. Lancet. 1981;2: 598–600.

Igic PG, Lee E, Harper W, Foach KW. Toxic effects associated with consumption of zinc. Mayo Clinic Proceedings. 2002;77:713–6.

Iglehart JK. The American health care system. N Eng J Med. 1999;340:70–6.

Iglesias-Gutiérrez E, García-Rovés PM, Rodríguez C, Braga S, García-Zapico P, Patterson AM. Food habits and nutritional status assessment of adolescent soccer players. A necessary and accurate approach. Can J Appl Physiol. 2005;30:18–32.

Ikeda R, Uehara M, Takasaki M, et al. Dose-responsive alteration in hepatic lipid peroxidation and retinol metabolism with increasing dietary beta-carotene in iron deficient rats. Int J Vitam Nutr Res. 2002;72:321–8.

Ikizler TA, Flakoll PJ, Parker RA, Hakim, RM. Amino acid and albumin losses during hemodialysis. Kidney Int. 1994;46:830–7.

Imhof A, Ledergerber B, Gunthard HF, Haupts S, Weber R; Swiss HIV Cohort Study. Risk factors for and outcome of hyperlactatemia in HIV-infected persons: is there a need for routine lactate monitoring? Clin Infect Dis. 2005;41:721–28.

Inborn errors of metabolism. 2nd ed. London: Oxford University Press; 1923. p. 43.

Inman-Felton A, Smith KG, Johnson EQ, eds. Medical Nutrition Therapy across the Continuum of Care. Chicago (IL): American Dietetic Association, 1997.

Institute for Community Health Promotion, Brown University. Rapid eating assessment for patients. 2005. Available from: http://bms.brown.edu/nutrition/acrobat/REAP%206.pdf. Accessed June 17, 2006.

Institute of Medicine. America's health care safety net: intact but endangered. Washington (DC): National Academy Press; 2000.

Institute of Medicine. Committee on Quality of Health Care in America. Crossing the chasm: a new health system for the 21st century. Washington (DC): National Academy Press; 2001.

Institute of Medicine. Dietary reference intake for energy, carbohydrates, fiber, fat, fatty acids, cholesterol, protein, and amino acids (macronutrients). Washington (DC): National Academy Press; 2002.

Institute of Medicine. Dietary reference intake for thiamin, riboflavin, niacin, vitamin B6, folate, vitamin B12, pantothenic acid, biotin, and choline. Washington (DC): National Academy Press; 2000.

Institute of Medicine. Nutrition during pregnancy: part I: weight gain, part II: nutrient suppls. Washington (DC): National Academy Press; 1990.

Institute of Medicine. Unequal treatment: confronting ethnic and racial disparities in health care. Washington (DC): National Academy Press; 2002.

Institute of Medicine of the National Academies. Complementary and alternative medicine in the United States. Washington (DC): National Academies Press; 2005.

International Food Information Council. Functional foods: attitudinal research. Washington (DC): author; 1999.

International Obesity Task Force, European Association for the Study of Obesity. Obesity in Europe. London: International Obesity Task Force; 2002. Available from: http://www.iotf.org/media/euobesity.pdf.

International Obesity Task Force, European Association for the Study of Obesity. Obesity in Europe 2: waiting for a green light for health. London: International Obesity Task Force; 2003. Available from: http://www.iotf.org/media/euobesity2.pdf.

Inui A. Cancer anorexia-cachexia syndrome: current issues in research and management. CA Cancer J Clin. 2002; 52:72–91.

Inzucchi SE. Oral antihyperglycemic therapy for type 2 diabetes. JAMA. 2002;287:360–72.

Ionescu AA, Schoon E. Osteoporosis in chronic obstructive pulmonary disease. Eur Respir J. 2002;22(Suppl 46): 64s–75s.

Ioannou GN, Rockey DC, Bryson CL, Weiss NS. Iron deficiency and gastrointestinal malignancy: a population-based cohort study. Am J Med. 2002;113:276–80.

Ioannou GN, Dominitz JA, Weiss NS, Heagerty PJ, Kowdley KV. The effect of alcohol consumption on the prevalence of iron overload, iron deficiency, and iron-deficiency anemia. Gastroenterology. 2004;126:1293–301.

Ippoliti C. Antidiarrheal agents for the management of treatment-related diarrhea in cancer patients. Am J Health Syst Pharm. 1998;55:1573–80.

Ireton Jones CS. Why use predictive equations for energy expenditure assessment? J Amer Diet Assoc. 1997;97:A-44.

Ireton-Jones C. (2002). Home enteral nutrition from the provider's perspective. JPEN J Parenter Enteral Nutr. 2002;26;S8–S9.

Ireton-Jones CS, Turner WW, Liepa GW, Baxter CR. Equations for estimating energy expenditure in burn patients with special reference to ventilatory status. J Burn Care Rehabil. 1992;13:330–3.

Iribarren C, Jacobs D, Sidney S, Claxton AJ, Gross M, Sadler M, et al. Serum total cholesterol and risk of hospitalization and death from respiratory disease. Int J Epidemiol. 1997;26:1191–202.

Ishimaru K, Tamasawa N, Baba M, Matsunaga M, Takeg K. Phenylketonuria with adult-onset neurological manifestation. Clinical Neurol. 1993;33:961–5.

Isolauri E. Probiotics in human disease. Am J Clin Nutr. 2001;73:1142S–1146S.

Jensen RT, Fraker DL. Zollinger-Ellison syndrome: advances in treatment of gastric hypersecretion and the gastrinoma. JAMA. 1994;271:1429–36.

Jackson DJ, Chopra M, Witten C, Sengwana MJ. HIV and infant feeding: issues in developed and developing countries. J Obstet Gynecol Neonatal Nurs. 2003;32:117–27.

Jackson JJ. Urban black Americans. In: Harwood A, editor. Ethnicity and medical care. Cambridge (MA): Harvard University Press; 1981.

Jackson JL, Lesho E, Peterson C. Zinc and the common cold: a meta-analysis revisited. J Nutr. 2000;1512S–15S.

Jacobs D, Blackburn H, Higgins M. Report of the Conference on Low Cholesterol: mortality associations. Circulation. 1992;86:1046–60.

Jacobs, EA, Copperman SM, Jeffe A, Kulig J. Fetal alcohol syndrome and alcohol related neurodevelopmental disorders. Pediatrics. 2000;106:358–61.

Jagtap AG, Shirke SS, Phadke AS. Effect of polyherbal formulation on experimental models of inflammatory bowel diseases. J Ethnopharmacol. 2004;90:195–204.

Jahoor F, Abramson S, Heird WC. The protein metabolic response to HIV infection in young children. Am J Clin Nutr. 2003;78:182–9.

James JM. Food allergies [monograph on the Internet]. Omaha (NE): eMedicine; 2004 [cited 2006 Mar 5]. Available from: http://www.emedicine.com/med/topic806.htm.

Janssen JC, Beck JA, Campbell TA, Dickinson A, Fox NC, Harvey RJ, et al. Early onset familial Alzheimer's disease: mutation frequency in 31 families. Neurology. 2003;60: 235–9.

Janssens JP, Pache JC, Nicod JP. Physiological changes in respiratory function associated with aging. Eur Respir J. 1999;13:197–205.

Jatoi A, Windschitl HE, Loprinzi CL, Sloan JA, Dakhil SR, Mailliard JA, et al. Dronabinol versus megestrol acetate versus combination therapy for cancer-associated anorexia: a North Central Cancer Treatment Group study. Journal of Clin Oncol. 2002;20:567–73.

JCAHO. 2006 Comprehensive Accreditation Manual for Hospitals. Oak Brook (IL): Joint Commission on Accreditation of Healthcare Organizations; 2006. p. 425.

JCAHO. 2006 Comprehensive Accreditation Manual for Hospitals. Oak Brook (IL): Joint Commission on Accreditation of Healthcare Organizations; 2006. p. 750.

Jee SH, Miller ER 3rd, Guallar E, Singh VK, Appel LJ, Klag MJ. The effect of magnesium supplementation on blood pressure: a meta-analysis of randomized clinical trials. Am J Hypertens. 2002;15:691–6.

Jeejeebhoy KN. Enteral feeding. Curr Opin Gastroenterol. 2005;21:187–91

Jeejeebhoy KN. Permissive underfeeding. Nutr Clin Pract. 2004;19:477–80.

Jeejeehboy KN. Clinical Nutrition 6. Management of nutritional problems of patients with Crohn's disease. CMAJ. 2002;166:913–8.

Jeffery RW, Utter J. The changing environment and population obesity in the United States. Obes Res. 2003; 11(suppl):12S–22S.

Jeffery, KM, Harkins, B, Cresci GA, Martindale RG. The clear liquid diet is no longer a necessity in the routine postoperative management of surgical patients. American Surgery. 1996;63:167–70.

Jemal A, Murray T, Samuels A, Ghafoor A, Ward E, Thun MJ. Cancer Statistics 2003. CA Cancer J Clin. 2003;53:5–26.

Jemal A, Tiwari RC, Murray T, Ghafoor A, Samuels A, Ward E, et al. Cancer statistics. CA Cancer J Clin. 2004;54:8–29.

Jenkins DJ, Kendall CW, Marchie A, Faulkner DA, Wong JM, de Souza R, et al. Aldosterone receptor antagonism in heart failure. Pharmacotherapy. 2005;25:1126–33.

Jennings JSR, Howdle PD. New developments in celiac disease. Curr Opin Gasroenterol. 2003;19:118–29.

Jeppesen PB, Hartmann B, Thulesen J, et al. Glucagon-like peptide 2 improves nutrient absorption and nutritional status in short-bowel patients with no colon. Gastroenterology. 2001;120:806–815.

Jeschke MG, Barrow RE, Herndon DN. Extended hypermetabolic response of the liver in severely burned pediatric patients. Arch Surg. 2004;139:641–747.

Jeschke MG, Herndon DN, Ebener C, Barrow RE, Jauch KW. Nutritional intervention high in vitamins, protein, amino acids, and omega-3 fatty acids improves protein metabolism during the hypermetabolic state after thermal injury. Arch Surg. 2001;136:1301–6.

JL Jr, Jones DW, Materson BJ, Oparil S, Wright JT Jr, Roccella EJ, for Jude E and Gibbons J. Identifying and treating intermittent claudication in people with diabetes. The Diab Foot. 2005;8:84–92.

Johnson DB, Cheney C, Monsen ER. Nutrition and feeding in infants with bronchopulmonary dysplasia after initial hospital discharge: risk factors for growth failure. J Am Diet Assoc. 1998;98:649–56.

Johnson MA, Hawthorne NA, Brackett WR, et al. Hyperhomocysteinemia and vitamin B-12 deficiency in elderly using title IIIc nutrition services. Am J Clin Nutr. 2003;77:211–20.

Johnson RK, Coulston AM. Medicare: reimbursement rules, impediments, and opportunities for dietitians. J Am Diet Assoc. 1995;95:1378–80.

Johnson R. The Lewin Group—what does it tell us and why does it matter? J Am Diet Assoc. 1999;99:426–7.

Johnston CS, Martin LJ, Cai X. Antihistamine effect of supplemental ascorbic acid and neutrophil chemotaxix. J Am Coll Nutr. 1992;11:172–6.

Joint Commission on Accreditation of Healthcare Organizations. 2004 comprehensive accreditation manual for hospitals: the official handbook (CAMH). Chicago (IL).

Jones A, Fay Jk, Burr M, Stone M, Hood K, Roberts G. Inhaled corticosteroid effects on bone metabolism in asthma and mild chronic obstructive pulmonary disease. The Cochrane Database of Systematic Reviews 2002, Issue 1. Art No.: CD003537.DOI: 10.1002/14651858.CD003537.

Jones CY, Hogan JW, Snyder B, Klein RS, Rompalo A, Schuman P, Carpenter CC, HIV Epidemiology Research Study Group. Overweight and human immunodeficiency virus (HIV) progression in women: association HIV disease progression and changes in body mass index in women in the HIV epidemiology research study cohort. Clin Infect Dis. 2003;37(Suppl 2):S69–S80.

Jones DB. Hepatic encephalopthy. J Gastroenterol Hep. 1993;8:363–9.

Jones PJ, Kubow S. In: Shils ME, Shike M, Ross CA, Caballero BC, Cousins RJ, editors. Modern nutrition in health and disease. 10th ed. Philadelphia (PA): Lippincott Williams & Wilkins; 2005. p. 92–122.

Jones SP, Doran DA, Leatt PB, Maher B, Pirmohamed M. Short-term exercise training improves body composition and hyperlipidaemia in HIV-positive individuals with lipodystrophy. AIDS. 2001;15:2049–51.

Jones WO, Nidus BD. Biotin and hiccups in chronic dialysis patients. J Ren Nut. 1991;2:80–3.

Joyce AM, Burns DL. Recurrent Clostridium difficile colitis. Postgrad Med. 2002;112:53–65. Available from: http://www.postgradmed.com/issues/2002/11_02/joyce3.htm.

Julkunen R.J, Di Padova C, Lieber CS. First-pass metabolism of ethanol—a gastrointestinal barrier against the systematic toxicity of ethanol. Life Sci. 1985;37:567–73.

Jumisko E, Lexell J, Soderberg S. The meaning of living with traumatic brain injury in people with moderate or severe traumatic brain injury. J Neurosci Nurs. 2005;37:42–50.

Jutte A, Schwenk A, Franzen C, Romer K, Diet F, Diehl V, Fatkenheuer G, Salzberger B. Increasing morbidity from myocardial infarction during HIV protease inhibitor treatment? AIDS. 1999;13:1796–7.

Kaartinen K, Lammi K, Hypen M, Nenonen M, Hanninen O, Rauma AL. Vegan diet alleviates fibromyalgia symptoms. Scand J Rheumatol. 2000;29:308–13.

Kaduszkiewicz H, Zimmerman T, Beck-Bornholdt HP, van den Busssche H. Cholinesterase inhibitors for patients with Alzheimer's disease: systematic review of randomized clinical trials. BMJ. 2005;331(7512):321–7.

Kaiser LL, Townsend MS. Food insecurity among US children: implications for nutrition and health. Top Clin Nutr. 2005;20:313–20.

Kalantar-Zadeh K, Lee GH, Block G. Relationship between dietary antioxidants and childhood asthma: more epidemiological studies are needed. Med Hypotheses. 2004;62:280–90.

Kalantar-Zadeh K, Block G, Kelly MP, Schroepfer C, Rodriguez RA, Humphreys MH. Near infra-red interactance for longitudinal assessment of nutrition in dialysis patients. J Ren Nutr. 2001;11:23–31.

Kalantar-Zadeh K, Kleiner M, Dunne E, et al. Total iron-binding capacity-estimated transferrin correlates with the nutritional subjective global assessment in hemodialysis patients. Am J Kidney Dis. 1998;31:263–72.

Kalarchian MA, Marcus MD, Levine MD, Haas GL, Greeno CG, Weissfeld LA, et al. Behavioral treatment of obesity in patients taking antipsychotic medications. J Clin Psychiatry. 2005;66:1058–63.

Kallet RH, Liu K, Tang J. Management of acidosis during lung-protective ventilation in acute respiratory distress syndrome. Respir Care Clin N Am. 2003;9:437–56.

Kalloo AN, Kantsevoy SV. Gallstones and biliary disease. Prim Care. 2001;28:591–606.

Kalm LM, Semba RD. They starved so that others are better fed: remembering Ancel Keys and the Minnesota experiment. J Nutr. 2005;135:1347–52.

Kamalaporn P, Sobhonslidsuk A, Jatchavala J, Atisook K, Rattanasiri S, Pramoolsinsap C. Factors predisposing to peptic ulcer disease in asymptomatic cirrhotic patients. Aliment Pharmacol Ther. 2005;21:1459.

Kamboh MI. Molecular genetics of late-onset Alzheimer's disease. Ann Hum Genet. 2004;68(Pt 4):381–404.

Kamimura MA, Jose Dos Santos NS, Avesani CM, Fernandes Canziani ME, Draibe SA, Cuppari L. Comparison of three methods for the determination of body fat in patients on long-term hemodialysis therapy. J Am Diet Assoc. 2003; 103:195–9.

Kamin DS, Furuta GT. The iceberg cometh: establishing the prevalence of celiac disease in the United States and Finland. Gastroenterology. 2004;126:359–61.

Kan MN, Chang HH, Sheu Wf, Cheng CH, Lee BJ, Huang YC. Estimation of energy requirements for mechanically ventilated, critically ill patients using nutritional status. Crit Care. 2003;7:R108–15.

Kanauchi O, Mitsuyama K, Araki Y, Andoh A. Modification of intestinal flora in the treatment of inflammatory bowel disease. Current Pharmaceutical Design. 2003;9: 333–46.

Kang JY, Melville D, Maxwell JD. Epidemiology and management of diverticular disease of the colon. Drugs Aging. 2004; 21:211–28.

Kanis JA, Johnell O, Oden A, Johansson H, De Laet C, Eisman JA, et al. Smoking and fractures: a meta-analysis. Osteoporos Int. 2005;16:155–62.

Kao GD, Devine P. Use of complementary health practices by prostate carcinoma patients undergoing radiation therapy. Cancer. 2000;88:615–9.

Kaplan F, Conway G. Endocrine disease. In: Axford J, O'-Callaghan C, editors. Medicine. 2nd ed. Oxford (UK): Blackwell Science Ltd.; 2004.

Kaptchuk TJ, Millar FG. Viewpoint: what is the best and most ethical model for the relationship between mainstream and alternative medicine: opposition, integration, or pluralism? Acad Med. 2005;80:286–90.

Kapur D. Effectiveness of nutrition education, iron supplementation or both on iron status children. Indian Pediatr. 2003;40:1131–44.

Karjalainen J, Martin JM, Knip M, Ilonen J, Robinson BH, Savilahti E, et al. A bovine albumin peptide as a possible trigger of insulin-dependent diabetes mellitus. N Engl J Med. 1992;327:302–7

Kashket S, DePaola DP. Cheese consumption and the development and progression of dental caries. Nutr Rev. 2002;60:97–104.

Kasiske B, Magdalena AA. Nutritional management of renal transplantation. In: Kopple J, Massry S, editors. Nutritional management of renal disease. 2nd ed. Philadelphia (PA): Lippincott Williams & Wilkins; 2004. p. 513–25.

Kasiske BL, Keane WF. Laboratory assessment of renal disease: clearance, urinalysis, and renal biopsy. In: Brenner BM, Rector FC, editors. The kidney. 5th ed. Philadelphia (PA): WB Saunders; 1996. p. 1137–74.

Kasiske BL, Bazquez MA, Harmon WE, et al. Recommendations for the outpaitnet surveillance of renal transplant recipients. J Am Soc Nephrol. 2000:11(Suppl 1).

Kataoka-Yahiro ,M, Saylor A. critical thinking model for nursing judgment. J Nurs Educ. 1994;33:351.

Katsura H, Kida K. A comparison of bone mineral density in elderly female patients with COPD and bronchial asthma. Chest. 2002;122:1949–55.

Katzmarzyk PT, Davis C. Thinness and body shape of Playboy centerfolds from 1978 to 1998. Int J Obes Relat Disord. 2001;25:590–2.

Kaufman FR, Ng WG, Xu Yk, et al. Normalization of uridine in patients with classical galactosemia. Clin Res. 1989;37: 184A.

Kaysen GA, Yeun JY. Nephrotic syndrome: nutritional consequences and dietary management. In: Mitch WE, Klahr S, editors. Nutrition and the kidney. 5th ed. Philadelphia (PA): Lippincott Williams & Wilkins; 2005. p. 160–75.

Keenan NL, Mark S, Fugh-Berman A, Browne D, Kaczmarczyk J, Hunter C. Severity of menopausal symptoms and use of both conventional and complementary/alternative therapies. Menopause. 2003;10:507–15.

Kelly D, Wischmeyer PE. Role of L-glutamine in critical illness: new insights. Current opinion in clinical nutrition and metabolic care. 2003;6:217–22.

Kelly P, Musonda R, Kafwembe E, Kaetano L, Keane E, Farthing M. Micronutrient supplementation in the AIDS diarrhea-wasting syndrome in Zambia: a randomized controlled trial. AIDS. 1999;13:495–500.

Kemp A, Kakakious A. Asthma prevention: breast is best? J Paediatr Child Health. 2004;40:337–9.

Kenjle K, Limaye S, Ghurgre PS, Udipi SA. Grip strength as an index for assessment of nutritional status of children aged 6–10 years. J Nutr Sci Vitaminol. 2005;51:87–92.

Kennedy E, Meyers L. Dietary reference intakes: development and uses for assessment of micronutrient status of women—a global perspective. Am J Clin Nutr. 2005;81: 1194S–1197S.

Keohane P, Attrill H, Love M, Frost P, Silk DB. Relation between osmolality of diet and gastrointestinal side effects in enteral nutrition. Br Med J. 1983;288:678–80.

Keys A, Arvanis C, Blackburn H. Seven countries: a multivariate analysis of death and coronary heart disease. Cambridge (MA): Harvard University Press; 1980. p. 381.

Keys A, Brozek J, Henschel A, Mickelsen O, Taylor HL. The biology of human starvation. Minneapolis: University of Minnesota Press; 1950.

Keys A. Food items, specific nutrients, and "dietary" risk. Am J Clin Nutr. 1986;43:477–9.

Keys A, Brozek J, Henschel A, Mickelson O, Taylor H. The biology of human starvation. Volume I. Minneapolis: University of Minnesota Press; 1951. p. 601–6.

Keys A, Menotti A, Aravanis C, Blackburn H, Djordjevic BS, Buzina R, et al. The seven countries study: 2,289 deaths in 15 years. Prev Med. 1984;13:141–54.

Khan AM, Sarker SA, Alam NH, Hossain MS, Fuchs GJ, Salam MA. Low osmolar oral rehydration salts solution in the treatment of acute watery diarrhoea in neonates and young infants: a randomized, controlled clinical trial. J Health Popul Nutr. 2005;23:52.

Khan NA, McAlister FA, Lewanczuk RZ, Touyz RM, Padwal R, Rabkin SW, et al. Canadian Hypertension Education Program. The 2005 Canadian Hypertension Education Program recommendations for the management of hypertension: part II - therapy. Can J Cardiol. 2005;21: 657–72.

Khaodhiar L, Keane-Ellison M, Tawa NE, Thibault A, Burke PA, Bistrian BR. Iron-deficiency anemia in patients receiving home total parenteral nutrition. JPEN J Parenter Enteral Nutr. 2002;26:114–9.

Kiecolt-Glaser, JK, McGuire L, Robles TF, Glaser R. Psychoneuroimmunology and Psychosom Med: back to the future. Psychosom Med. 2002;64:15–28

Kiecolt-Glaser JK, Fisher LD, Ogrocki P, Stout JC, Speicher CE, Glaser R. Marital quality, marital disruption, and immune function. Psychosom Med. 1987;49:13–34.

Kim H, Lee S, Lee G, Hwangbo Y, Ahn K, Lee B. The protective effect of aminolevulinic acid dehydratase 1-2 and 2-2 isozymes against blood lead with higher hematologic parameters. Environ Health Perspect. 2004;112:538–41.

Kim J, Chan MM. Factors influencing preferences for alternative medicine by Korean Americans. Am J Chin Med. 2004;32:321–9.

Kim KY, Schumacher HR, Hunsche E, Wertheimer AI, Kong SX. A literature review of the epidemiology and treatment of acute gout. Clin Ther. 2003;25:1593–617.

Kindt E, Halvorsen S. The need of essential amino acids in children: an evaluation based on the intake of phenylalanine, tyrosine, leucine, isoleucine and valine in children with phenylketonuria, tyrosine amino transferase defect and maple syrup urine disease. Am J Clin Nutr. 1980;33: 279–86.

King LE, Fraker PJ. Zinc deficiency in mice alters myelopoiesis and hematopoiesis. J Nutr. 2002;132:3301–7.

Kinirons MT, O'Mahony MS. Drug metabolism and aging. Br J Clin Pharmacol. 2004;57:540–4.

Kinsman SL, Vining EP, Quaskey SA, Mellits D, Freeman JM. Efficacy of the ketogenic diet for intractable seizure disorders: review of 58 cases. Epilepsia. 1992;33:1132–6.

Kirk D, Fish SA. Medical management of osteoporosis. Am J Manag Care. 2004;10:445–55.

Kirksey KM, Goodroad BK, Kemppainen JK, Holzemer WL, Bunch EH, Corless IB, et al. Complementary therapy use in persons with HIV/AIDS. Journal of Holistic Nursing. 2002;20:264–78.

Kitano S, Dolgor B. Does portal hypertension contribute to the pathogenesis of gastric ulcer associated with liver cirrhosis? Gastroenterol. 2000;35:79–86.

Kittler PG, Sucher KP. Food and culture. 4th Ed. Belmont (CA): Wadsworth/Thomson Learning; 2004.

Kiyota K, Habu Y, Sugano Y, Inokuchi H, Mizuno S, Kimoto K, Kawai K. Comparison of 1-week and 2-week triple therapy with omeprazole amoxicillin, and clarithromycin in peptic ulcer patients with Helicobacter pylori infection: results of a randomized trial. J Gastroenterol. 1999;34 Suppl 11:76–9.

Klahr S. Effects of renal insufficiency on nutrient metabolism and endocrine function. In: Mitch WE, Klahr S, editors. Handbook of nutrition and the kidney. Lippincott-Raven; 1998. p. 25–44.

Klahr S, Levey AS, Beck GJ, Caggiula AW, Hunsicker L, Kusek JW, et al. The effects of dietary protein restriction and blood-pressure control on the progression of chronic renal disease. Modification of diet in renal disease study group. N Engl J Med. 1994;330:877–84.

Klee GG. Cobalamin and folate evaluation: measurement of methylmalonic acid and homocysteine vs vitamin B(12) and folate. Clin Chem. 2000;46 Pt 2:1277–83.

Kleignen J, Knipschield P, ter Riet G. Clinical trials of homeopathy: a meta analysis. BMJ. 1991;302:316–23.

Klein CJ, Bosworth JB, Wiles CE. Physicians prefer goal-oriented note format more than three to one over other outcome-focused documentation. J Am Diet Assoc. 1997;97:1306–10.

Klein CJ, Stanek GS, Wiles CE. Overfeeding macronutrients to critically ill adults: metabolic complications. J Am Diet Assoc. 1998;98:795–806.

Klein L. Is blue dye safe as a method of detection for pulmonary aspiration? J Am Diet Assoc. 2004;104:1651–2.

Klein S, Kinney J, Jeejeebhoy K, et al. (1997). Nutrition support in clinical practice: review of the published data and recommendations for future research directions. JPEN J Parenter Enteral Nutr. 1997;21:133–56.

Knopman DS, DeKosky ST, Cummings JL, Chui H, Corey-Bloom J, Relkin N, et al. Practice parameter: diagnosis of dementia (an evidence-based review): report of the quality standards subcommittee of the American Academy of Neurology. Neurology. 2001;56:1143–53.

Knopp RH, Retzlaff BM. Saturated fat prevents coronary artery disease? An American paradox. Am J Clin Nutr. 2004;80:1102–3.

Knudtson ML, Wyse DG, Galbraith PD, Brant R, Hildebrand K, Paterson D, et al. Chelation therapy for ischemic heart disease: a randomized controlled study. JAMA. 2002;287: 307–9.

Ko RJ. Adulterants in Asian patent medicine. N Engl J Med. 1998;339:847.

Kochanek, KD, Murphy, SL, Anderson, RN, Scott, C. Deaths: final data for 2002. National vital statistics reports. 2004; 53(5).

Koertge J, Weidner G, Elliott-Eller M, Scherwitz L, Merritt-Worden TA, Marlin R, et al. Improvement in medical risk factors and quality of life in women and men with coronary artery disease in the Multicenter Lifestyle Demonstration Project. Am J Cardiol. 2003;91:1316–22.

Kolata G. Supplements fail to stop arthritis pain, study says. New York Times, February 23, 2006, p. A23.

Konefal J. The challenge of educating physicians about complementary and alternative medicine. Academic Medicine. 2002;77:847–50.

Koo LC. The use of food to treat and prevent disease in Chinese culture. Soc Sci Med. 1984;18:757–66.

Kopp-Hoolihan L. Prophylactic and therapeutic uses of probiotics: a review. J Am Diet Assoc. 2001;101:229–41.

Kopple J. Nutrition, diet, and the kidney. In: Shils M, Olson J, Shike M, editors. Modern nutrition in health and disease. Philadelphia (PA): Lea & Febiger; 1994. p. 1102–46.

Kopple JD. Nutrition, diet, and the kidney. In: Shils ME, Shike M, Ross CA, Caballero BC, Cousins RJ, editors. Modern nutrition in health and disease. 10th ed. Philadelphia (PA): Lippincott Williams & Wilkins; 2005. p. 1475–511.

Korenkov M, Sauerland S, Junginger T. Surgery for obesity. Curr Opin Gastroenterol. 2005;21:679–83.

Kornblut AE, Wilson D. How one pill escaped place on steroid list. New York Times, April 17, 2005.

Kornbluth A, Sachar DB. Ulcerative colitis practice guidelines in adults (update). American College of Gastroenterology, Practice Parameters Committee. Am J Gastroenterol. 2004;99:1371–85.

Kortas DY, Haas LS, Simpson WG, Nickl NJ 3rd, Gates LK. Mallory-Weiss tear: predisposing factors and predictors of a complicated course. Am J Gastroenterol. 2001;96: 2863–5.

Kosel BW, Aweeka FT, Benowitz NL, Shade SB, Hilton JF, Lizak PS, Abrams DI. The effects of cannabinoids on the pharmacokinetics of indinavir and nelfinavir. AIDS. 2002;16:543–50.

Kossoff EH, McGrogan JR, Freeman JM. Benefits of an all-liquid ketogenic diet. Epilepsia. 2004;45:1163.

Kotler DP. Body composition studies in HIV-infected individuals. Ann NY Acad Sci. 2000;904:546–52.

Kotler DP. Human immunodeficiency virus-related wasting: malabsorption syndromes. Semin Oncol. 1998;25 Suppl 6:70–5.

Kotler DP, Burastero S, Wang J, Pierson RN Jr. Prediction of body cell mass, fat-free mass, and total body water with bioelectrical impedance analysis: effects of race, sex, and disease. Am J Clin Nutr. 1996;64(3 Suppl):489S–497S.

Kotler DP, Tierney AR, Wang J, Pierson RN Jr. Magnitude of body-cell-mass depletion and the timing of death from wasting in AIDS. Am J Clin Nutr. 1989;50:444–7.

Koury MJ, Ponka P. New insights into erythropoiesis: the roles of folate, vitamin B12, and iron. Annu Rev Nutr. 2004;24:105–31.

Kowdley KV. Ursodeoxycholic acid therapy in hepatobiliary disease. Am J Med. 2000;108:481.

Kraft B, Kress HG. Cannabinoids and the immune system. Of men, mice and cells. Schmerz. 2004;18:203–10.

Kraft MD, Btaiche IF, Sachs GS. Review of Refeeding Syndrome. Nutr Clin Prac. 2005;20:625–33.

Krall EA, Dawson-Hughes B. Osteoporosis. In: Shils ME, Olson JA, Shike M, Ross AC, editors. Modern nutrition in health and disease. 9th ed. Baltimore (BD): Williams & Wilkins, 1999. p. 1353–64.

Kramer K, Cohen HJ. Antenatal diagnosis of hematologic disorders. In: Hoffman R. Hematology, basic principles and practice. 4th ed. Philadelphia (PA): Elsevier Medical Publishing; 2005. p. 2697–710.

Krauss RM, Eckel RH, Howard B, et al. American Heart Association dietary guidelines: revision 2000: a statement for healthcare professionals from the Nutrition Committee of the American Heart Association. Circulation 2000;102:2284–99.

Kraut JA, Madias NE. Approach to patients with acid-base disorders. Respir Care. 2001;46:392–403.

Krebs NF, Westcott JE, Arnold TD, Kluger DM, Accurso FJ, Miller LV, et al. Abnormalities in zinc homeostasis in young infants with cystic fibrosis. Pediatr Res. 2000;48: 256–61.

Krenkel J, St. Joer S, Kulick D. Relationship of nutrition to chronic diseases. In: Kaminsky LA, editor. American College of Sports Medicine's resource manual for guidelines for exercise testing and prescription. 5th ed. Baltimore: Lippincott Williams & Wilkins; 2006. p. 146–64.

Krevsky B, Godley J. Nutritional support in advanced liver disease. Nutritional Support Service. 1985;5:9–14.

Kris-Etherton PM. AHA Science Advisory. Monounsaturated fatty acids and risk of cardiovascular disease. American Heart Association. Nutrition Committee. Circulation. 1999;100:1253–8.

Kris-Etherton PM, Pearson TA, Wan Y, Hargrove RL, Moriarty K, Fishell V, et al. High-monounsaturated fatty acid diets lower both plasma cholesterol and triacylglycerol concentrations. Am J Clin Nutr. 1999;70:1009–15.

Krok KL, Lichenstein GR. Nutrition in Crohn disease. Curr Opin Gastroenterol. 2003;19:148–53.

Krucoff MW, Crater SW, Gallup D, Blankenship JC, Cuffe M, Guarneri M, Krieger RA, et al. Music, imagery, touch, and prayer as adjuncts to interventional cardiac care: the Monitoring and Actualisation of Noetic Trainings (MANTRA) II randomized study. Lancet. 2005;366:211–7.

Kucera P, Balaz M, Varsik P, Kurca E. Pathogenesis of alcoholic neuropathy. Bratish Lek Listy. 2002;103:26–9.

Kuczmarski MF, Kuczmarski RJ, Najjar M. Effects of age on validity of self-reported height, weight, and body mass index: findings from the Third National Health and Nutrition Examination Survey, 1988–1994. J Am Diet Assoc. 2001;101:28–34.

Kuczmarski RJ, Ogden CL, Grummer-Strawn LM, et al. CDC growth charts: United States. Advance data from vital and health statistics; no. 314. Hyattsville (MD): National Center for Health Statistics; 2000.

Kuczmarski RJ, Ogden CL, Guo SS, et al. 2000 CDC growth charts for the United States: methods and development. National Center for Health Statistics. Vital Health Stat 11(246); 2002.

Kudsk KA, Sacks GS. Nutrition in the care of the patient with surgery, trauma and sepsis. In: Modern nutrition in health and disease. 10th ed. Philadelphia (PA): Lippincott Williams & Wilkins; 2004.

Kuehneman T, Saulsbury D, Splett P, Chapman DB. Demonstrating the impact of nutrition intervention in a heart failure program. J Am Diet Assoc. 2002;102:1790–4.

Kulkarni K, Franz MJ. Nutrition therapy for type 1 diabetes. In: Franz MJ, Bantle JP, editors. American Diabetes Association guide to medical nutrition therapy for Diabetes. Alexandria (VA): American Diabetes Association; 1999.

Kulkarni K. Pattern management. In: Ross TA, Boucher JL, O'Connell BS, editors. American Dietetic Association guide to diabetes: medical nutrition therapy and education. Chicago (IL): American Dietetic Association; 2005.

Kuo CD, Shiao GM, Lee JD. The effect of high-fat and high-carbohydrate diet loads on gas exchange and ventilation in COPD patients and normal subjects. Chest. 1993;104: 189–96.

Kupfer DJ. The pharmacological management of depression. J Clin Psychiatry. 2005;66:1058–63.

Kupferschmitdt HH, Fattinger KE, Ha HR, Follath F, Krahenbuhl S. Grapefruit juice enhances the bioavailability of the HIV protease inhibitor saquinavir in man. Br J Clin Pharmacol. 1998;45:355–9.

Kurth T, Kase CS, Berger K, Gaziano JM, Cook NR, Buring JE. Smoking and risk of hemorrhagic stroke in women. Stroke. 2003;34:2792–5.

Kurzner ST, Garg M, Bautista DB, Bader D, Merritt RJ, Warburton D, et al. Growth failure in infants with bronchopulmonary dysplasia: nutrition and elevated resting metabolic expenditure. Pediatrics. 1988;81:379–84.

Kuzuya M, Kanda S, Koike T, Suzuki Y, Iguchi A. Lack of correlation between total lymphocyte count and nutritional status in the elderly. Clin Nutr. 2005;24:427–32.

Kwan P, Brodie MJ. Phenobarbital for the treatment of epilepsy in the 21st century: a critical review. Epilepsia. 2004;45:1141–9.

Kwong K. Prevention and treatment of oropharyngeal mucositis following cancer therapy: are there new approaches? Cancer Nurs, 2004;27:183–205.

Kyne L, Farrell RG, Kelly CP. Clostridium difficile. Gastroenterol Clin North Am. 2001;30:753–77, ix–x.

Labbé RF, Vreman HJ, Stevenson DK. Zinc protoporphyrin: a metabolite with a mission. Clin Chem. 1999;45:2060–72.

LaBrecque DR, Moody FG. Diseases of the liver and biliary tract. St. Louis: Mosley Year Book; 1991.

Lacey JM, Houser RA. Time for dietetics and mental health alliance. J Am Diet Assoc. 2001;101:744.

Lacey K, Pritchett E. Nutrition care process and model: ADA adopts road map to quality care and outcomes management. J Amer Diet Assoc. 2003;103:1061–72.

Ladefoged S, Pedersen S, Skielboe M, et al. Renal functional reserve after an acute intravenous lipid load. J Renal Nutr. 1993; 4:186–90.

Lagerqvist C, Ivarsson A, Juto P, Persson LA, Hernell O. Screening for adult coeliac disease: which serological marker(s) to use? J Intern Med. 2001;250:241–8.

Lai HC, Kosorok MR, Sondel SA, Chen ST, FitzSimmons SC, Green CG, et. al. Growth status in children with cystic fibrosis based on the national cystic fibrosis patient registry data: evaluation of various criteria used to identify malnutrition. J Pediatr. 1998;132:478–85.

Lake AM. Food-induced eosinophilic proctocolitis. J Pediatr Gastroenterol Nutr. 2000;30 Suppl:S58–60.

Lamireau T, Monnereau S, Martin S, Marcotte JE, Winnock M, Alvaez F. Epidemiology of liver disease in cystic fibrosis: a longitudinal study. J Hepatol. 2004;41:920–5.

Lancellotti L, D'Orazio C, Mastella G, Mazzi G, Lippi U. Deficiency of Vitamins E and A in cystic fibrosis is independent of pancreatic function and current enzyme and vitamin supplementation. Eur J Pediatr. 1996;155:281–5.

Langer CJ, Hoffman JP, Ottery FD. Clinical significance of weight loss in cancer patients: rationale for the use of anabolic agents in the treatment of cancer-related cachexia. Nutrition. 2001;17(suppl 1):S1–S18.

Langouche L, Vanhorebeek I, Van den Berghe G. The role of insulin therapy in critically ill patients. Treat Endocrinol. 2005;4:353–60.

Lanzillotti JS, Tang AM. Micronutrients and HIV disease: a review pre- and post-HAART. Nutr Clin Care. 2005;8:16–23.

Lashner BA. Clinical Advances in ulcerative colitis. 2003. Available from: http://www.medscape.com/viewarticle/463425.

Lasztity N, Hamvas J, Biro L, Nemeth E, Marosvolgyi T, Decsi T, et al. Effect of enterally administered n-3 polyunsaturated fatty acids in acute pancreatitis—a prospective randomized clinical trial. Clin Nutr. 2005;24:198–205.

Latchaw RF. Cerebral perfusion imaging in acute stroke. J Vasc Interv Radiol. 2004;15(1 Pt 2):S29–46.

Lauer RM, Obarzanek E, Hunsberger SA, Van Horn L, Hartmuller VW, Barton BA, et al. Efficacy and safety of lowering dietary intake of total fat, saturated fat, and cholesterol in children with elevated LDL cholesterol: the Dietary Intervention Study in Children. Am J Clin Nutr. 2000;72(5 Suppl):1332S–1342S.

Laviano A, Meguid M, Rossi-Fanelli F. Improving food intake in anorectic cancer patients. Curr Opin Clin Nutr Metab Care. 2003;6:421–6.

Laviano A, Muscaritoli M, Cascino A, Preziosa I, Inui A, Mantovani G, et al. Branched chain amino acids: the best compromise to achieve anabolism? Curr Opin Clin Nutr Metab Care. 2005;8:408–14.

Law M. Plant sterol and stanol margarines and health. Br Med J. 2000;320:861–4.

Lazier JN, Kessler CM. Clinical aspects and therapy of hemophilia. In: Hoffman R. Hematology, basic principles and practice. 4th ed. Philadelphia (PA): Elsevier Medical Publishing; 2005. p. 2047–70.

Le Marchand L, Hankin JH, Carter FS, Essling C, Luffey D, Franke AA, et al. A pilot study on the use of plasma carotenoids and ascorbic acid as markers of compliance to a high fruit and vegetable dietary intervention. Cancer Epidemio Biomarkers Prev. 1994;3:245–51.

Leblanc M, Tapolyai M, Paganini E. What dialysis dose should be provided in acute renal failure? Adv Ren Replace Ther. 1995;3:255–64.

Lee AN, Werth VP. Activation of autoimmunity following use of immunostimulatory herbal supplements. Arch Dermatol. 2004;140:723–7.

Lee CD, Blair SN, Jackson AS. Cardiorespiratory fitness, body composition, and all-cause and cardiovascular disease mortality in men. Am J Clin Nutr. 1999;69:373–80.

Lee GA, Rao MN, Grunfeld C. The effect of HIV protease inhibitors on carbohydrate and lipid metabolism. Curr HIV/AIDS Rep. 2005;2:39–50.

Lee JO, Benjamin D, Herndon DN. Nutrition support strategies for severely burned patients. Nutr Clin Pract. 2005; 20:325–30.

Lee MM, Lin SS, Wrensch MR, Adler SR, Eisenberg D. Alternative therapies used by women with breast cancer in four ethnic populations. J Natl Cancer Inst. 2000;92:42–7.

Lee RD, Neiman DC. Nutrition assessment. St. Louis Missouri: McGraw-Hill Companies, Inc.; 1993.

Lee RD, Nieman DC. Nutritional Assessment. 4th ed. Boston: McGraw-Hill; 2006.

Lee RD, Nieman DC. Nutritional assessment. St. Louis (MO): McGraw-Hill Companies, Inc.; 2003.

Lee SL. Hyperthyroidism. New York: eMedicine.com, Inc. [updated 2005 Jul 20; cited 2005 Jan 13]. Available from: http://www.emedicine.com.

Leff D. How to write for the public. J Am Diet Assoc. 2004;104:730–2.

Lehman FS, Beglinger C. Impact of COX-2 inhibitors in common clinical practice a gastroenterologist's perspective. Curr Top Med Chem. 2005;5:449–64.

Leibel RL, Rosenbaum M, Hirsch J. Changes in energy expenditure resulting from altered body weight. N Engl J Med. 1995;332:621–8.

Leitzmann C. Vegetarian diets: what are the advantages? Forum Nutr. 2005;57: 147–56

Leitzmann MF, Giovannucci E, Rimm EB, Stampfer MJ, Willett WC. Physical activity and the risk of gallstone disease in men. Ann Intern Med. 1998;128:415–25.

Lemstra M, Olszynski WP. The effectiveness of multidisciplinary rehabilitation in the treatment of firbomyalgia: a randomized controlled trial. Clin J Pain. 2005;21:166–74.

Lenaerts C. Cystic Fibrosis-associated liver disease [monograph on the Internet]. Amiens, France: Gastroenterology and Hepatology Unit, Department of Pediatrics, Centre Hospitalier Universitaire [cited 2006 Jan 9]. Available from: http://www.md.ucl.ac.be/pedihepa/CysticFibrosis.htm.

Lenssen P, Aker S. In: Hasse JM, Blue LS, editors. Comprehensive guide to transplant nutrition. Chicago (IL): American Dietetic Association; 2002.

Leo MA, Lieber CS. Alcohol, vitamin A, and beta-carotene: adverse interactions, including hepatotoxicity and carcinogenicity. Am J Clin Nutrition. 1999;69:1071–85.

Leonard JV. The nutritional management of urea cycle disorders. J Pediatr. 2001;138:S40–S45.

Leonard JV, Morris AA. Urea cycle disorders. Semin Neonatal. 2002;7:27–35.

Leong WI, B Lönnerdal. Hepcidin, the recently identified peptide that appears to regulate iron absorption. J Nutr. 2004;134:1–4.

Levenstein S, Prantera C, Luzi C, D'Ubaldi A. Low residue or normal diet in Crohn's disease: a prospective controlled study in Italian patients. Gut. 1985;26:989–93.

Levin LI, Munger KL, Rubertone MV, Peck CA, Lennette ET, Spiegelman D, et al. Temporal relationship between elevation of Epstein-Barr virus antibody titers and initial onset of neurological symptoms in multiple sclerosis. JAMA. 2005;293:2496–500.

Levy PC, Utell MJ, Sickel JZ, Apostolakos MJ. The acute respiratory distress syndrome: current trends in pathogenesis and management. Compr Ther. 1995;21:438–44.

Levy R, Cooper P. Ketogenic diet for epilepsy. The Cochrane Database of Systematic Reviews 2003, Issue 3. Art. No.: CD001903. DOI: 10.1002/14651858.CD001903.

Lewin Group Inc., editor. The cost of covering medical nutrition therapy services inder TRICARE: benefits costs, cost avoidance and savings prepared for the Department of Defense Health Affairs. Chicago (IL): American Dietetic Assocition, Medical Nutrition Therapy Works; 2001.

Li FX, Verhoef MJ, Best A, Otley A, Hilsden RJ. Why patients with inflammatory bowel disease use of do not use complementary and alternative medicine: a Canadian national survey. Can J Gastroenterol. 2005;19:567–73.

Lichenstein GR, MacDermott RP. Crohn's disease: advances in treatment. 2002. Available from: http://www.medscape.com/viewprogram/2118_pnt.

Lichtenstein AH, Deckelbaum RJ. American Heart Association Nutrition Committee. Stanol/Sterol Ester–containing foods and blood cholesterol levels. A statement for healthcare professionals from the nutrition committee of the council on nutrition, physical activity, and metabolism of the American Heart Association. Circulation. 2001;103:1177–9.

Lieber CS. Alcohol: its metabolism and interaction with nutrients. Annu Rev Nutr. 2000;20:395–430.

Lieber CS. Alcoholic liver injury: pathogenesis and therapy in 2001. Pathol Biol. 49:738–52.

Lieber CS. Hepatic and metabolic effects of ethanol: pathogenesis and prevention. Ann Med. 1994;26:325–30.

Lieber CS. Herman award lecture: a personal perspective on alcohol, nutrition and the liver. Am J Clin Nutr. 1993;58: 430–42.

Lieber CS. Mechanisms of ethanol-drug-nutrition interactions. Clin Toxicol. 1994;32: 631–81.

Lieber CS. The discovery of the microsomal ethanol oxidizing system and its physiologic and pathologic role. Drug Metab Rev. 2004;36:511–29.

Lilia JJ, Neuvonen M, Neuvonen PJ. Effects of regular consumption of grapefruit juice on the pharmacokinetics of simvastatin. Br J Clin Pharmacol. 2004;58:56–60.

Lin PH, Proschan MA, Bray GA, Fernandez, Hoben K, Most-Windhauser M, et al, DASH Collaborative Research Group. Estimation of energy requirements in a controlled feeding trial. Am J Clin Nutr. 2003;77:639–45.

Lin YC, Bioteau AB, Ferrari LR, Berde CB. The use of herbs and complementary and alternative medicine in pediatric preoperative patients. J Clin Anesth. 2004;16:4–6.

Lindblad CI, Artz MB, Pieper CF, Sloane RJ, Hajjar ER, Ruby CM, et al. Potential drug–disease interactions in frail, hospitalized elderly veterans. Ann Pharmacother. 2005;39:412–7.

Lindsay R. Daily hemodialysis: the time has come. Am J Kidney Dis. 2005;45:793–7.

Lindsay R, Cosman F. Chapter 333. Osteoporosis. In: Kasper DL, Braunwald E, Fauci AS, Hauser SL, Longo DL, Jameson JL, Isselbacher KJ, editors. Harrison's principles of internal medicine. 16th ed. New York: McGraw-Hill; 2005. p. 2268–78.

Lindstrom JM. Acetylcholine receptors and myasthenia. Muscle Nerve. 2000;23:453.

Liotta EA, Brough A, Erickson QL, Elston DM. Addison disease. New York: eMedicine.com, Inc. [updated 2005 Dec 2; cited 2005 Jan 15]. Available from: http://www.emedicine .com.

Lipsky PE. Rheumatoid arthritis. In: Kasper DL, Braunwald E, Fauci AS, Hausner SL, Longo DL, Jameson JL, editors. Harrison's principles of internal medicine. 16th ed. New York: McGraw-Hill; 2005. p. 1968–77.

Lipson JG, Meleis AI. Issues in health care of Middle Eastern patients. West J Med. 1983;139:854–61.

Little DR. Ambulatory management of common forms of anemia. Am Fam Physician. 1999;59:1598–604.

Little WC, Brucks S. Therapy for diastolic heart failure. Prog Cardiovasc Dis. 2005;47:380–8.

Liu YM, Williams S, Basualdo-Hammond C, Stephens D, Curtis R. A prospective study: growth and nutritional status of children treated with the ketogenic diet. J Am Diet Assoc. 2003;103:707–12.

Livingstone B. Epidemiology of childhood obesity in Europe. Eur J Pediatr. 2000;159(Suppl 1):S14–S34.

Livny O, Reifen R, Levy I, Madar Z, Faulks R, Southon S, Schwartz B. Beta-carotene bioavailability from differently processed carrot meals in human ileostomy volunteers. Eur J Nutr. 2003;42:338–45.

Llach F. Chronic renal failure. In: Wilcox C, Tisher CC, editors. Handbook of nephrology & hypertension. 5th ed. Philadelphia (PA): Lippincott Williams & Wilkins; 2005. p. 267–74.

Lo HC, Tu ST, Lin KC, Lin SC. The anti-hyperglycemic activity of the fruiting body of Cordyceps in diabetic rats induced by nicotinamide and streptozotocin. Life Sci. 2004;74:2897–908.

Loboz KK, Shenfield GM. Drug combinations and impaired renal function—the 'triple whammy'. Br J Clin Pharmacol. 2005;59:239–43.

Lobstein T, Baur L, Uauy R. Obesity in children and young people: a crisis in public health. Obes Rev. 2004;5:4–85.

Locke GR III, Pemberton JH, Phillips SF. AGA technical review on constipation. Gastroenterology. 2000;119: 1861–78.

Loftus EV, Silverstein MD, Sandborn WJ, et al. Ulcerative colitis in Olmsted County, Minnesota. 1940–1993: incidence, prevalence and survival. Gut. 2000;46:335–43.

Longo DI, Anderson KC. Plasma cell disorders. In: Kasper DL, Braunwald E, Fauci AS, Hausner SL, Longo DI, Jameson JL, editors. Harrison's principles of internal medicine. 16th ed. New York: McGraw Hill Medical Publishing; 2005. p. 656–62.

Longo DL. Approach to the patient with cancer. In Kasper DL, Braunwald E, Fauci AS, Hauser SL, Longo DL, Jameson L, Isselbacher KJ, editors. Harrison's principles of internal medicine. 16th ed. Boston: McGraw Hill; 2005.

Loos RJF, Rankinin T. Gene-diet interactions on body weight changes. J Am Diet Assoc. 2005;105:S29–S34.

Lopez JA, Thiagarajan P. Acquired disorders of platelet function. In: Hoffman R. Hematology, basic principles and practice. 4th ed. Philadelphia (PA): Elsevier Medical Publishing; 2005. p. 2347–68.

Lopez-Hellin J,Baena-Fuestegueras JA, Schwartz-Riera S, Garcia-Arumi E. Usefulness of short-lived proteins as nutritional indicators in surgical patients. Clin Nutr. 2002;21:119–25.

Lorefalt B, Ganowiak W, Palhagen S, Toss G, Unosson M, Granerus AK. Factors of importance for weight loss in elderly patients with Parkinson's disease. Acta Neurol Scand. 2004;110:180–7.

Lorentzen HF, Fugleholm AM, Weismann K. Zinc deficiency and pellagra in alcohol abuse. Ugeskr Laeger. 2000;162: 6854–6.

Ludman EK, Newman JM. Yin and yang in the health-related practices of three Chinese groups. J Nutrition Education. 1984;16:3–5.

Luepker RV, Apple FS, Christenson RH, Crow RS, Fortmann SP, Goff D, et al; AHA Council on Epidemiology and

Prevention; AHA Statistics Committee; World Heart Federation Council on Epidemiology and Prevention; European Society of Cardiology Working Group on Epidemiology and Prevention; Centers for Disease Control and Prevention; National Heart, Lung, and Blood Institute. Case definitions for acute coronary heart disease in epidemiology and clinical research studies: a statement from the AHA Council on Epidemiology and Prevention; AHA Statistics Committee; World Heart Federation Council on Epidemiology and Prevention; the European Society of Cardiology Working Group on Epidemiology and Prevention; Centers for Disease Control and Prevention; and the National Heart, Lung, and Blood Institute. Circulation. 2003;108:2543–9.

Lukaski HC. Vitamin and mineral status: effects on physical performance. Nutrition. 2004;20:632–44.

Luna B, Feinglos MN. Oral agents in the management of type 2 diabetes mellitus. Am Fam Physician. 2001;63: 1747–56.

Lund AM, Dixon MA, Vreken P, Leonard JV, Morris AAM. Plasma and erythrocyte fatty acid concentrations in long-chain 3-hydroxyacyl-CoA dehydrogenase deficiency. J Inherit Metab Dis. 2003;26:410–2.

Lund AM, Dixon MA, Vreken P, Leonard JV, Morris AAM. What is the role of medium-chain triglycerides in the management of long-chain 3-hydroxyacyl-CoA dehydrogenase deficiency? J Inherit Metab Dis. 2003;26:353–60.

Lundberg G. The American Healthcare "System" in 2005—part 1; 2005. [cited 2005 July 1]. Available from: http://www.medscape.com/viewarticle/496865.

Lundholm K, Daneryd P, Bosaeus I, U Körner, Lindholm E. Palliative nutritional intervention in addition to cyclooxygenase and erythropoietin treatment for patients with malignant disease: effects on survival, metabolism, and function. Cancer. 2004;100:1967–77.

Lyford GL, He CL, Soffer E, et al. Pan-colonic decrease in interstitial cells of Cajal in patients with slow transit constipation. Gut. 2002;51:496–501.

Lykins TC, Stockwell J. Comprehensive modified diet simplifies nutrition management of adults with short-bowel syndrome. J Am Diet Assoc. 1998;98:309–15.

Lyon HN, Hirschhorn JN. Genetics of common forms of obesity: a brief overwiew. Am J Clin Nutr. 2005;82(suppl): 215S–217S.

Ma J, Hampl JS, Betts NM. Antioxidant intakes and smoking status: data from the continuing survey of food intakes by individuals 1994–1996. Am J Clin Nutr. 2000;71:774–80.

Macallan DC. Nutrition and immune function in human immunodeficiency virus infection. Proc Nutr Soc. 1999; 58:743–8.

Macallan DC, Noble C, Baldwin C, Foskett M, McManus T, Griffin GE. Prospective analysis of patterns of weight change in stage IV human immunodeficiency virus infection. Am J Clin Nutr. 1993;58:417–24.

MacArthur RD, Perez G, Walmsley S, Baxter JD, Mullin CM, Neaton JD; Terry Beirn Community Programs for Clinical Research on AIDS (CPCRA) 042/045; Canadian HIV Trials Network (CTN) 102 Protocol Teams. Comparison of prognostic importance of latest CD4+ cell count and HIV RNA levels in patients with advanced HIV infection on highly active antiretroviral therapy. HIV Clin Trials. 2005;6:127–35.

MacBurney M, Young LS, Ziegler TR, Wilmore DW. A cost-evaluation of glutamine-supplemented parenteral nutrition in adult bone marrow transplant patients. J Am Diet Assoc. 1994;94:1263–6.

MacCracken KA, Scalisi JC. Development and evaluation of a ketogenic diet program. J Am Diet Assoc. 1999;99:1554–8.

MacDonald A, Rylance G, Davies P, Asplin D, Hall SK, Booth IW. Administration of protein substitute and quality of control in phenylketonuria: a randomized study. J Inherit Metab Dis. 2003;26:319–26.

Mackay MT, Bicknell-Royle J, Nation J, Humphrey M, Harvey AS. The ketogenic diet in refractory childhood epilepsy. J Paediatr Child Health. 2005;41:353–7.

Mackle TJ, Touger-Decker R, Maillet JO, Holland BK. Registered dietitians' use of physical assessment parameters in professional practice. J Am Diet Assoc. 2003;103:1632–8.

MacLaughlin EJ, Raehl CL, Treadway AK, Sterling TL, Zoller DP, Bond CA. Assessing medication adherence the elderly: which tools to use in clinical practice? Drugs Aging. 2005;22:231–55.

MacLean DB, Luo LG. Increased ATP content/production in the hypothalamus may be a signal for energy-sensing of satiety: studies of the anorectic mechanism of a plant steroidal glycoside. Brain Research. 2004;10:1–11.

Madhavan Nair K, Bhaskaram P, Balakrishna N, Ravinder P, Sesikeran B. Response of hemoglobin, serum ferritin, and serum transferrin receptor during iron supplementation in pregnancy: a prospective study. Nutrition. 2004;20: 896–9.

Madias NE, Adrogué HJ. Cross-talk between two organs: how the kidney responds to disruption of acid-base balance by the lung [electronic resource]. Nephron Physiol. 2003;93:61–6.

Madill J, Gutierrez C, Grossman J, Allard J, Chan C, Hutcheon M, et al. Nutritional assessment of the lung transplant patient: body mass index as a predictor of 90-day mortality following transplantation. J Heart Lung Transplant. 2001;20:288–96.

Madsen K, Verlander J. Renal structure in relation to function. In: Wilcox C, Tisher CC, editors. Handbook of nephrology & hypertension. 5th ed. Philadelphia (PA): Lippincott Williams & Wilkins; 2005. p. 3–13.

Maduro R. Curanderismo and Latin view of disease and curing. West J Med. 1983;139:868–74.

Magnotti LJ, Deitch EA. Burns, bacterial translocation, gut barrier function, and failure. J Burn Care Rehabil. 2005; 26:383–91.

Mahan LK, Escott-Stump S. 2000. Krause's food, nutrition and diet therapy. 10th ed. Philadelphia (PA): WB Saunders Company; 2000.

Maillot F, Farad S, Lamisse F. Alcohol and nutrition. Pathologie et Biologie. 2001;49:683–8.

Malhotra A, Mathur AK, Gupta. Early enteral nutrition after surgical treatment of gut perforations: a prospective randomised study. J Postgrad Med. 2004;50:102–6.

Malhotra S, Karan RS, Pandhi P, Jain S. Drug related medical emergencies in the elderly: role of adverse drug reactions and non-compliance. Postgrad Med J. 2001;77:703–7.

Malhotra V, Singh S, Tandon OP, Madhu SV, Prasad A, Sharma SB. Effect of yoga asanas on nerve conduction in type 2 diabetes. Indian J Physiol Pharmacol. 2002;46: 298–306.

Mallampalli A. Nutritional management of the patient with chronic obstructive pulmonary disease. Nutr Clin Prac. 2004;19:550–6.

Malone A. Enteral formula selection: a review of selected product categories. Practical Gastroenterology. 2005: 44–74.

Malone DL, Genuit T, Tracy JK, Gannon C, Napolitano LM. Surgical site infections: reanalysis of risk factors. J Surg Res. 2002;103:89–95.

Mandel A, Ballew M, Pina-Garza JE, Stalmasek V, Clemens LH. Medical costs are reduced when children with intractable epilepsy are successfully treated with the ketogenic diet. J Am Diet Assoc. 2002;102:396–8.

Mann J. AIDS: a worldwide pandemic in Current Topics in AIDS, volume 2. Gottlieb MS, Jeffries DJ, Mildvan D, Pinching AJ, Quinn TC, editors. John Wiley & Sons; 1989.

Mann SK, Kaur S, Bains K. Iron and energy supplementation improves the physical work capacity of female college students. Food Nutr Bull. 2002;23:57–64.

Mann WJ. Surgical management of radiation enteropathy. Surg Clin North Am. 1991;71:977–90.

Manson JE, Skerrett PJ, Greenland P, VanItallie TB. The escalating pandemics of obesity and sedentary lifestyle: a call to action for clinicians. Arch Intern Med. 2004;164: 249–58.

Mantovani G, Maccio A, Massa E, Madeddu C. Managing cancer related anorexia/cachexia. Drugs. 2001;61:499–514.

Manzo L, Locatelli C, Candura SM, Costa LG. Nutrition and alcohol neurotoxicity. Neuro Toxicol. 1994;15:555–66.

Marcus SN, Wheaton KW. Effects of a new, concentrated wheat fiber preparation on intestinal transit, deoxycholic acid metabolism and the omposition of bile. Gut. 1986;27:893–900.

Marian M, Charney P. Patient selection and indications for enteral feedings. In: Charney P, Malone A, editors. ADA pocket guide to enteral nutrition. Chicago (IL): American Dietetic Association; 2006.

Marik PE, Bedigian MK. Refeeding hypophosphatemia in critically ill patients in an intensive care unit. Arch Surg. 1996;131:1043–7.

Marik PE, Raghavan M. Stress-hyperglycemia, insulin and immunomodulation in sepsis. Intensive Care Med. 2004; 30:748–56.

Marik PE, Zaloga GP. Early enteral nutrition in acutely ill patients: a systematic review. Crit Care Med. 2001;29: 2264–70.

Marik PE, Zaloga GP. Meta-analysis of parenteral nutrition versus enteral nutrition in patients with acute pancreatitis. BMJ. 2004;328(7453):1407.

Marin VB, Velandia S, Hunter B, Gattas V, Fielbauum O, Herrera O, et al. Energy expenditure, nutrition status, and body composition in children with cystic fibrosis. Nutrition. 2004;20:181–6.

Marinella MA. The refeeding syndrome and hypophosphatemia. Nutr Rev. 2003;61:320–3.

Markell MS, Armenti V, Danovitch G, et al. Hyperlipidemia and glucose intolerance in the renal transplant patient. J Am Soc Nephrol. 1994;27:117–23.

Markowitz JS, Donovan JL, DeVane CL, Taylor RM, Ying, Ruan Y, et al. Effect of St. John's Wort on drug metabolism by induction of Cytochrome P450 3A4 Enzyme. JAMA. 2003;290:1500–4.

Marks B. Tobacco exposure and chronic illness. In: Roitman JL, editor. American College of Sports Medicine's resource manual for guidelines for exercise testing and prescription. 4th ed. Baltimore (MD): Lippincott Williams & Wilkins; 2001. p. 41–6.

Marks G, Solis J, Richardson JL, Collins LM, Birba L, Hisserich JC. Health behavior of elderly Hispanic women: does cultural assimilation make a difference? Am J Public Health. 1987;77:1315–9.

Marlett JA. Dietary fiber and cardiovascular disease. In: Cho SS, Dreher ML, eds. Handbook of dietary fiber. New York: Marcel Dekker, Inc; 2001. p. 17–30.

Maroni BJ, Hirschberg R. The importance of nutritional status on outcomes from acute renal failure. Contemp Dialysis Nephrol. 1996;17:22–5.

Maroni BJ, Staffeld C, Young VR, et al. Mechanisms permitting nephritic syndrome patients to achieve nitrogen equilibrium with a protein restricted diet. J Clin Invest. 1997;99:2749–87

Marriage B, Clandinin MT, Macdonald IM, Glerum DM. Cofactor treatment improves ATP synthetic capacity in patients with oxidative phosphorylation disorders. Mol Genet Metab. 2004;81:263–72.

Marshall DA, Vernalis MN, Remaley AT, Walizer EM, Scally JP, Taylor AJ. The role of exercise in modulating the impact of an ultralow-fat diet on serum lipids and apolipoproteins in patients with or at risk for coronary artery disease. Am Heart J. 2006;151:484–91.

Marshall I. Zinc for the common cold. The Cochrane Database of Systematic Reviews 1999, Issue 2. Art. No.: CD001364.

Martin C. Calorie, protein, fluid, and micronutrient requirements. In: MacCallum PD, Polisena CG, editors. The clinical guide to oncology nutrition. Chicago (IL): American Dietetic Association; 2000.

Martin CM DG, Heyland DK, Morrison T, Sibbald WJ. Multicentere, clusterrandomized clinical trial of algorithms for critical-care enteral and parenteral therapy (ACCEPT). CMAJ. 2004;170:197–204.

Martin GS, Mannino DM, Eaton S, Moss M. The epidemiology of sepsis in the United States from 1979 through 2000. N Engl J Med. 2003;348:1546–54.

Martin RH. The role of nutrition and diet in rheumatoid arthritis. Proc Nutr Soc. 1998;57:231–4.

Martinez-Abundis E, Gonzalez-Ortiz M, Grover-Paez F. Association of adiposity assessed by means of near-infrared interactance with the beta-cell function, insulin resistance and leptin concentrations in non-obese subjects. Exploratory study. J Diabetes Complications. 2001;15:181–4.

Martinez-Castelao A, Grinyo JM, Gil-Vernet S, et al. Lipid lowering long term effects of six different statins in hypercholesterolemic renal transplant patients under cyclosporine immunosuppression. Transplant Proc. 2002; 34:398–400.

Martin-Harris B. Optimal patterns of care in patients with chronic obstructive pulmonary disease. Semin Speech Lang 2000;21:311–9.

Martinu T, Menzies D, Dial S. Re-evaluation of acid-base prediction rules in patients with chronic respiratory acidosis. Can Respir J. 2003;10:311–5.

Masharani U, Karam JH. Diabetes mellitus and hypoglycemia. In: Tierney LM, McPhee SJ, Papadakis MA, editors. Current medical diagnosis and treatment 2003. New York: Lange Medical Books/McGraw-Hill; 2003.

Mason J, Bailes A, Beda-Andourou M, et al. Recent trends in malnutrition in developing regions: vitamin A deficiency, anemia, iodine deficiency, and child underweight. Food Nutr Bull. 2005;26:59–108.

Massari PU. Disorders of bone and mineral metabolism after renal transplantation. Kidney Int. 1997;52:1412–21.

Matalon K, Acosta PB, Castiglioni L, Austin V, Rohr F, Wenz E, et al. Protocol for nutrition support of maternal PKU. Maternal PKU Collaborative Study funded by the National Institute of Child Health and Human Development; 1998.

Matarese LE, Gottschlich MM. Contemporary nutrition support practice: a clinical guide. Philadelphia (PA): WB Saunders Company; 1998.

Mathieu J, Foust M, Ouellette P. Implementing nutrition diagnosis, step two in the nutrition care process and model: challenges and lessons learned in two health care facilities. J Am Diet Assoc. 2005;105:1636–40.

Matsuhashi Y. Thinness: drives and results. J Adolesc Health. 2000;27:149–50.

Matsumoto H, Fakui Y. Pharmokinetics of ethanol: a review of methodology. Addition Biology. 2002;7:5–14.

Mattson MP. Pathways towards and away from Alzheimer's disease. Nature. 2004;430(7000):631–9.

Maxer SE, West JR. Drinking patterns and alcohol related birth defects. Alcohol Res Health. 2001;25:168–74.

May PE, Barber A, D'Olimpio JT, Hourihane ANP, Abumrad NN. Reversal of cancer-related wasting using oral supplementation with a combination of [beta]-hydroxy-[beta]-methlybutyrate, arginine, and glutamine. Am J Surg. 2002:183:471–9.

Mayer J. Genetic factors in human obesity. Ann NY Acad Sci. 1965;131:412–21.

Mayer JJ. Exploring alcohol's effect on liver function. Alcohol Res Health. 1997;21(1).

Mayes T, Gottschlich MM. Burns and wound healing. In: The science and practice of nutrition support. Dubuque (IA): Kendall Hunt; 2001. p. 391–420.

Mayo Clinic. 2005; Diarrhea: a concern after gallbladder removal? Available from: http://www.mayoclinic.com/invoke.cfm?objectid=2F5BC8D1-F09D-495F-B0D4A305788B7E81.

MayoClinic.com. MAOI diet: restrict foods high in tyramine. Available from: http://www.mayoclinic.com. Accessed November 6, 2005.

MayoClinic.com. Monoamine oxidase inhibitors (MAOIs). Available from: http://www.mayoclinic.com. Accessed November 6, 2005.

Mazanec D. Osteoporosis screening: time to take responsibility. Arch Intern Med. 2004;164:1047–8.

Mazer NA, Shifren JL. Transdermal testosterone for women: a new physiological approach for androgen therapy. Obstet Gynecol Surv. 2003;58:489–500.

McAlindon TE, LaValley MP, Gulin JP, Felson DT. Glucosamine and chondroitin for treatment of osteoarhritis: a systematic quality assessment and meta-analysis. JAMA. 2000;283:1469–75.

McCabe BJ, Frankel EH, Wolfe JJ, editors. Handbook of food and drug interactions. New York: CRC Press/Taylor & Francis; 2003.

McCabe-Sellers BJ, Sharkey JR, Browne BA. Diuretic medication therapy use and low thiamin intake in homebound older adults. J Nutr Elder. 2005;24:57–71.

McCallum PD. Patient-generated subjective global assessment. In: The clinical guide to oncology nutrition. Chicago (IL): American Dietetic Association; 2000.

McCallum SL. The national dysphagia diet: implementation at a regional rehabilitation center and hospital system. J Am Diet Assoc. 2003;103:381–4.

McCance KL, Roberts LK. (2002). Biology of cancer. In: McCance KL, Huether SE, editors. Pathophysiology: the biologic basis for disease in adult and children. 4th ed. Philadelphia (PA): Mosby; 2002.

McClain CJ, Hill DB, Song Z, Deaciuc I, Barve S. Monocyte activation in ALD. Alcohol. 2002;27:53–61.

McClave SA. Nutrition support in acute pancreatitis. Missouri Dietetic Association Annual Meeting Program; 2005.

McClave SA, DeMeo MT, DeLegge MH, DiSario JA, Heyland DK, Maloney JP, et al. North American Summit on

Aspiration in the Critically Ill Patient: consensus statement. JPEN J Parenter Enteral Nutr. 2002;26:S80–5.

McClave SA, Greene LM, Snider HL, Makk LJ, Cheadle WG, Owens NA, et al. Comparison of the safety of early enteral vs parenteral nutrition in mild acute pancreatitis. JPEN J Parenter Enteral Nutr. 1997;21:14–20.

McClave SA, Lowen CC, Kleber MJ, Nicholson JF, Jimmerson SC, McConnell JW, et al. Are patients fed appropriately according to their caloric requirements? JPEN J Parenter Enteral Nutr. 1998;22:375–81.

McClave SA, Lukan JK, Stefater JA, Lowen CC, Looney Sw, Matheson PJ, et al. Poor validity of residual volumes as a marker for risk of aspiration in critically ill patients. Crit Care Med. 2005;3:324–30.

McClave SA, Lukan JK. Stefater JA, Lowen CC, Looney SW, Matheson PJ, et al. Poor validity of residual volumes as a marker for risk of aspiration in critically ill patients. Crit Care Med. 2005;33:324–30.

McDade, TW, Beck MA, Kuzawa C, Adair LS. Prenatal undernutrition, postnatal environments, and antibody response to vaccination in adolescence. Am J Clin Nutr. 2001;74:543–8.

McDermott AY, Shevitz A, Must A, Harris S, Roubenoff R, Gorbach S. Nutrition treatment for HIV wasting: a prescription for food as medicine. Nutr Clin Pract. 2003; 18:86–94.

McDermott AY, Terrin N, Wanke C, Skinner S, Tchetgen E, Shevitz AH. CD4+ cell count, viral load, and highly active antiretroviral therapy use are independent predictors of body composition alterations in HIV-infected adults: a longitudinal study. Clin Infect Dis. 2005;41:1662-70.

McDermott JH. Antioxidant nutrients: current dietary recommendations and research update. J Am Pharm. 2000; 40:785–99.

McErlean A, Kelly O, Bergin S, Patchett SE, Murray FE. The importance of microbiological investigations, medications and artificial feeding in diarrhoea evaluation. Ir J Med Sci. 2005;174:21–5.

McGonigle RSR, Wallin JD, Shadduck RK, Fisher JW. Erythropoietin deficiency and erythropoiesis in renal insufficiency. Kidney Int. 1984;25:437–44.

McGuire HL, Svetkey LP, Harsha DW, Elmer PJ, Appel LJ, Ard JD. Comprehensive lifestyle modification and blood pressure control: a review of the PREMIER trial. J Clin Hypertens. 2004;6:383–90.

McKenna DJ, Jones K, Hughes K. Botanical medicines: the desk reference of major herbal supplements. New York: The Haworth Herbal Press; 2002.

McQuillan GM, et al. Prevalence of hepatitis B virus infection in the United States: the National Health and Nutrition and Examination Surveys, 1976 through 1994. Am J Public Health. 1999;89:14–8.

Mears E. Outcomes of continuous process improvement of a nutritional care program incorporating serum prealbumin measurements. Nutrition. 1996;12:479–84.

Mears E. Outcomes of continuous process improvement of a nutritional care program incorporating TTR measurement. Clin Chem Lab Med. 2002;40:1355–9.

Medras M, Jankowska EA. The effect of alcohol on bone density in men. Przeglad Lekarski. 2000;57:743–6.

Medstat PULSE Survey 2000. Medstat: alternative medicine. Available from: http://www.medstat.com/healthcare/althernative.asp; 2003.

Medstat PULSE Survey. HealthLeaders, October. 2002. Available from: http://www.healthleaders.com/magazine/print.php?contentid=39103§ion=factfile.

Mehendale SM, Shepherd ME, Brookmeyer RS, Semba RD, Divekar AD, Gangakhedkar RR, et al. Low caratenoid concentration and the risk of HIV seroconversion in Pune, India. J Acquir Immune Defic Syndr. 2001;26: 352–9.

Mehler PS. Bulimia nervosa. N Engl J Med. 2003;349:875–81.

Mehler PS. Diagnosis and care of patients with anorexia nervosa in primary care settings. Ann Intern Med. 2001;134: 1048–59.

Mehler PS. Osteoporosis in anorexia nervosa: prevention and treatment. Int J Eat Disorder. 2003;33:113–26.

Mehta R. Therapeutic alternatives to renal replacement for critically ill patients in acute renal failure. Semin Nephrol. 1994;14:64–82.

Mei Z, Parvanta I, Cogswell ME, Gunter EW, Grummer-Strawn LM. Erythrocyte protoporphyrin or hemoglobin: which is a better screening test for iron deficiency in children and women? Am J Clin Nutr. 2003;77:1229–33.

Meier PN. Management of achalasia [monograph on the Internet]. New York: Medscape Portals Inc.; 2001 [cited 2004 Jan 6]. Available from: http://www.medscape.com/viewarticle/420090.

Meier R. Enteral fish oil in acute pancreatitis. Clin Nutr. 2005;24:169–71.

Melanson K, Dwyer J. Popular diets for treatment of overweight and obesity. In: Wadden TA, Mitchell JE, Cook-Myers T, Wonderlich SA. Diagnostic criteria for anorexia nervosa: looking ahead to DSM-V. Int J Eat Disord. 2005;37:S95–S97.

Melchart D, Streng A, Hoppe A, Brinkhaus B, Witt C, Wagenpfeil S, et al. Acupuncture in patients with tension-type headaches: a randomized controlled study. BMJ. 2005;331:376–82.

Mellors JW, Munoz A, Giorgi JV, Margolick JB, Tassoni CJ, Gupta P, et al. Plasma viral load and CD4+ lymphocytes as prognostic markers of HIV-1 infection. Ann Intern Med. 1997;126:946–54.

Meneton P, Jeunemaitre X, De Wardener HE, Macgregor GA. Diseases handling, blood pressure, and cardiovascular links between dietary salt intake, renal salt. Physiol Rev. 2005;85:679–715.

Menon KV, Gores GJ, Shah VH. Pathogenesis, diagnosis, and treatment of alcoholic liver disease. Mayo Clinic Proceedings. 2001;76:1021–9.

Merck. http://www.merck.com/mrkshared/mmanual/section11/chapter144/144b.jsp. Accessed August 1, 2005.

Meriggioli MN. Use of immunoassays in neurological diagnosis and research. Neurol Res. 2005;27:734–40.

Messa P, Sindici C, Cannella G, et al. Persistant SHPT after renal transplantation. Kidney Int. 1998;54:1704–13.

Metheney NA. Risk factors for aspiration. JPEN J Parenter Enteral Nutr. 2002;26:S26–S33.

Metheny NM. Fluid and electrolyte balance: nursing considerations. Philadelphia (PA): Lippincott Williams & Wilkins; 2000.

Metniz PGH, Bartens C, Fischer M, Fridrich P, Steltzer H, Druml W. Antioxidant status in patients with acute respiratory distress syndrome. Intensive Care Med. 1999;25: 180–5.

Mezey E. Alcoholic liver disease: roles of alcohol and malnutrition. Am J Clin Nutr. 1980;33:2709–18.

Mezey E. Influence of sex hormones on alcohol metabolism. Alcohol Clin Exp Res. 2000;24:421.

Mezey E, Jow E, Slavin RE, et al. Pancreatic function and intestinal absorption in chronic alcholism. Gastroenterology. 1970;59:657.

Michael P. Impact and components of the Medicare MNT benefit. J Am Diet Assoc. 2001;101:1140–1.

Michaels AJ. Management of post traumatic respiratory failure. Crit Care Clin. 2004;20:83–99.

Mifflin MD, St Joer ST, Hill LA, Scott BJ, Daugherty SA, Young OK. A new predictive equation for resting energy expenditure in healthy individuals. Am J Clin Nutr. 1990:51:241–7.

Mikhail N, Wali S, Ziment I. Use of alternative medicine among Hispanics. J Altern Complement Med. 2004;10: 851–9.

Milionis HJ, Bourantas CL, Siamopoulos KC, Elisaf MS. Acid-base and electrolyte abnormalities in patients with acute leukemia. Am J Hematol. 1999;62:201–7.

Millen AE, Dodd KW, Subar AF. Use of vitamin, mineral, nonvitamin, and nonmineral supplements in the United States: the 1987, 1992, and 2000 National Health Interview Survey results. J Am Diet Assoc. 2004;104:942–50.

Miller GW. Principles of alcohol detoxification. Am Fam Phys. 1984;30:147.

Miller JW, Selhub J, Nadeau MR, Thomas CA, Feldman RG, Wolf PA. Effect of L-dopa on plasma homocysteine in PD patients: relationship to B-vitamin status. Neurology. 2003;60:1125–9.

Miller MG, Mulligan T. Human immunodeficiency virus and hypogonadal bone disease. Pharmacotherapy. 2005; 25:632–4.

Miller RG, Rosenberg JA, Gelinas DF, Mitsumoto H, Newman D, Sufit R, et al. Practice parameter: the care of the patient with amyotrophic lateral sclerosis (an evidence-based review): report of the Quality Standards Subcommittee of the American Academy of Neurology: ALS Practice Parameters Task Force. Neurology 1999;52:1311–23.

Mills E, Foster BC, van Heeswijk R, Phillips E, Wilson K, Leonard B, et al. Impact of African herbal medicines on antiretroviral metabolism. AIDS. 2005;19:95–7.

Mills E, Montori V, Perri D, Phillips E, Koren G. Natural health product-HIV drug interactions: a systematic review. Int J STD AIDS. 2005;16:181–6.

Mills JL, Von Kohorn I, Conley MR, et al. Low vitamin B-12 concentrations in patients without anemia: the effect of folic acid fortification of grain. Am J Clin Nutr. 2003;77: 1474–7.

Mirtallo J, Canada T, Johnson D, Kumpf V, Petersen C, Sacks G, et al; Task Force for the Revision of Safe Practices for Parenteral Nutrition. Safe practices for parenteral nutrition. JPEN J Parenter Enteral Nutr. 2004;28:S39–S70.

Mishra V. Effect of obesity on asthma among adult Indian women. Int J Obes. 2004;28:1048–58.

Mitch WE, Walser M. Nutritional therapy for the uremic patient. In: Brenner BM, Rector FC, editors. The kidney. 5th ed. Philadelphia (PA): WB Saunders; 1996. p. 2382–423.

Mitchell JE, Cook-Myers T, Wonderlich SA. Diagnostic criteria for anorexia nervosa: looking ahead to DSM-V. Int J Eat Disord. 2005;37:S95–S97.

Mitrache C, Passweg JR, Libura J, et al. Anemia: an indicator for malnutrition in the elderly. Ann Hematol. 2001;80: 295–8.

Mock V, Olsen M. Current management of fatigue and anemia in patients with cancer. Semin Oncol Nurs. 2003;19 Suppl 2:36–41.

Moertel C, Scutt AG, Reiteneier RJ, et al. Corticosteroid therapy of pre-terminal gastrointestinal cancer. Cancer. 1974;33:1607–9.

Mohanka M, Irwin M, Heckbert SR, Yasui Y, Sorensen B, Chubak J, et al. Serum lipoproteins in overweight/obese postmenopausal women: a one-year exercise trial. Med Sci Sports Exerc. 2006;38:231–9.

Mokdad A, Marks J, Stroup D, Gerberding J. Actual cause of death in the United States. JAMA. 2004;291:1238–45.

Mokdad AH, Ford ES, Bowman BA, Dietz WH, Vinicor F, Bales VS, Marks JS. Prevalence of obesity, diabetes, and obesity-related health risk factors, 2001. JAMA. 2003;289: 76–9.

Molony D. The American Association of Oriental Medicine's complete guide to chinese herbal medicine: how to treat illness and maintain wellness with chinese herbs. New York: Berkley Books; 1998.

Molteno C, Smit I, Mills J, Huskisson J. Nutritional status of patients in a long-stay hospital for people with mental handicap. S Afr Med J. 2000;90:1135–40.

Monafo WW. Initial management of burns. N Engl J Med. 1996;335:1581–5.

Monheit AC, Vistnes VP. Race/ethnicity and health insurance status: 1987 and 1996. Med Care Res Rev. 2000; 57(suppl 1):11–35.

Monroe JG, Turka LA. Regulation of activation of B and T lymphocytes. In: Hoffman R. Hematology, basic principles and practice. 4th ed. Philadelphia (PA): Elsevier Medical Publishing; 2005. p. 157–77.

Monsen ER. From the environment to MNT: dietitians face key issues. J Am Diet Assoc. 1997;97:360.

Monson P, Mehta R. Nutrition in acute renal failure: a reappraisal for the 1990s. J Renal Nutr. 1994;2:5–B77.

Montamat SC, Cusack BJ, Verstal RE. Management of drug therapy in the elderly. N Engl J Med. 1989;321:303–9.

Moore RD, Chaisson RE. Natural history of HIV infection in the era of combination antiretroviral therapy. AIDS. 1999;13:1933–42.

Mora A, Lee IM, Buring JE, Ridker PM. Association of physical activity and body mass index with novel and traditional cardiovascular biomarkers in women. JAMA. 2006; 295:1412–9.

Moran A. Cystic fibrosis-related diabetes: an approach to diagnosis and management. Pediatr Diabetes. 2000;1:41–8.

Moran A, Hardin D, Rodman D, Allen HF, Beall RJ, Borowitz C, et al. Diagnosis, screening and management of cystic fibrosis related diabetes mellitus: a consensus conference report. Diabetes Res Clin Pract. 1999;45:61–73.

Morgan BW, Kori S, Thomas JD. Adverse effects in 5 patients receiving EDTA at an outpatient clinic. Vet Hum Toxicol. 2002;44:274–6.

Morgan ED, Bledsoe SC, Barker J. Ambulatory management of burns. Am Fam Phys. 2000;62:2015–26.

Moriarty-Craige SE, Ramakrishnan U, Neufeld L, Rivera J, Martorell R. Multivitamin-mineral supplementation is not as efficacious as is iron supplementation in improving hemoglobin concentrations in nonpregnant anemic women living in Mexico. Am J Clin Nutr. 2004;80: 1308–11.

Morris SS, Ruel MT, Cohen RJ, Dewey KG, de la Brière B, Hassan MN. Precision, accuracy, and reliability of hemoglobin assessment with use of capillary blood. Am J Clin Nutr. 1999;69:1243–8.

Morrow, G. Diagnosis and discussion. Hereditary fructose intolerance. Arch Pediatr Adolesc Med. 1997;151:1166.

Moses SW. Historical highlights and unsolved problems in glycogen storage disease type 1. Eur J Pediatr. 2002; 161(Suppl 1):s2–9.

Mossad SB. Upper respiratory tract infections [monograph on the Internet]. Cleveland (OH): The Cleveland Clinic Disease Management Project; 2006 [cited 2006 January 13]. Available from: http://www.clevelandclinicmeded.com/diseasemanagement/infectiousdisease/urti/urti.htm.

Moura VL, Warber SL, James SA. CAM providers' messages to conventional medicine: a qualitative study. Am J Med Qual. 2002;17:10–4.

Mozaffarian D, Pischon T, Hankinson SE, Rifai N, Joshipura K, Willett WC, et al. Dietary intake of trans fatty acids and systemic inflammation in women. Am J Clin Nutr. 2004; 79:606–12.

Mozumdar A, Roy SK. Method for estimating body weight in persons with lower-limb amputation and its implication for their nutritional assessment. Am J Clin Nutr. 2004;80:868–75.

Mrugacz M, Tobolvzyk J, Minarowska A. Retional binding protein status in relation to ocular surface changes in patients with cystic fibrosis treated with daily vitamin A supplements. Eur J Pediatr. 2005;164:202–6.

Mulcahy K, Lumber T. The diabetes ready reference for health professionals. 2nd ed. Alexandria (VA): American Diabetes Association; 2004.

Mullally AM, Vogelsang GB, Moliterno AR. Wasted sheep and premature infants: the role of trace metals in hematopoiesis. Blood Rev. 2004;18:227–34.

Muller-Felber W. Therapy of polyneuropathies, casual and symptomatic. MMW Fortschr Med. 2001;143 Suppl 2:54–9.

Müller-Lissner SA, Kamm, MA, ScarpignatoC, Wald A. Myths and Misconceptions About Chronic Constipation. Am J Gastroenterol. 2005;100:124–9.

Mundt AJ, Roeske JC, Chung TD, Weichselbaum RR. Principles of radiation oncology. In: Kufe DW, Pollock RE, Weichselbaum RR, Bast RC, Gansler TS, Holland JF, Frei E, editors. Holland-Frei: Cancer Medicine. 6th ed. Hamilton: BC Decker Inc.; 2003.

Munford RS. Chapter 254. Severe sepsis and septic shock. In: Kasper DL, Braunwald E, Fauci AS, Hausner SL, Longo DL, Jameson JL, editors. Harrison's principles of internal medicine. 16th ed. New York: McGraw-Hill; 2005.

Murphy MC, Brooks CN, New SA, Lumbers ML. The use of the Mini-Nutritional Assessment (MNA) tool in elderly orthopaedic patients. Eur J Clin Nutr. 2000;54:555–62.

Murphy ME. Colorectal cancers. In: SE Otto, editor. Oncology nursing. 4th ed. Philadelphia (PA): Mosby; 2001.

Murphy SP, Allen LH. Nutritional importance of animal source foods. J Nutr. 2003;133 Suppl 2:3932S–3935S.

Murray KA, Brzozowski LA. Swallowing in patients with tracheotomies. AACN Clinical Issues: Advanced Practice in Acute and Critical Care 1998; volume 9, no. 3. [cited 2006 January 10]. Available from: www.aacn.org/AACN/jrnlci.nsf/GetArticle/ArticleTen93?openDocument=.

Murray RH, Robel AJ. Physicians and healers—unwitting partners in health care. N Engl J Med. 1992;326:61–4.

Murray SM, Pindoria S. Nutrition support for bone marrow transplant patients. Cochrane Database Syst. Rev. 2: CD002920; 2002.

Mylonakis E, Koutkia P, Grinspoon S. Diagnosis and treatment of androgen deficiency in human immunodeficiency virus-infected men and women. Clin Infect Dis. 2001;33:857–64.

Nadeau RG, Groner W. The role of a new noninvasive imaging technology in the diagnosis of anemia. J Nutr. 2001;131:1610S–4S.

Nagano T, Toyoda T, Tanabe H, et al. Clinical features of hematological disorders caused by copper deficiency during long-term enteral nutrition. Internal Med. 2005;44: 554–9.

Nagel G, Hoyer H, Katenkamp D. Use of complementary and alternative medicine by patients with breast cancer: observations from a health-care survey. Supportive Care in Cancer; 2004.

Nagy J. Stud. Alcohol dependence at the cellular level: effects of ethanol on Ca homeostasis of IM-9 human lymphoblast cells. Alcohol. 2000;61:225–31.

Nahikian-Nelms ML, Nelms, RG. Use of double-entry journals in a required dietetic course: encouraging critical thinking skills. J Nutrition Ed. 1994;26:93–6.

Nair KM, Bhaskaram P, Balakrishna N, Ravinder P, Sesikeran B. 2004. Response of hemoglobin, serum ferritin, and serum transferring receptor during iron supplementation in pregnancy: a prospective study. Nutrition. 20:896–9.

Nakai I, Omoni Y, Aikawa I, Yasumura T, Suzuki S, Yoshimura N, et al. Effect of cyclosporine on glucose metabolism in kidney transplant recipients. Transplant Proc. 1988;20;969–78.

National Cancer Institute. Cancer facts. Available from: http://cis.nci.nih.gov/fact/7_2.htm. Accessed on August 3, 2005.

National Center for Chronic Disease Prevention and Health Promotion. Defining overweight and obesity, nutrition & physical activity, 2004. Available from: http://www.cdc.gov/nccdphp/dnpa/obesity/defining.htm#References.

National Center for Complementary and Alternative Medicine. NCCAM Publication No. D158; 2002. Available from: http://www.nih.gov/about/aboutnccam/index.htm.

National Center for Health Statistics [homepage on the Internet]. Centers for Disease Control and Prevention. Hyattsville (MD): Centers for Disease Control and Prevention, United States Department of Health and Human Services [cited 2005 Oct 11]. Available from: http://www.cdc.gov/nchs/.

National Center for Health Statistics. Data for injury. Hyattsville (MD): Center for Disease Control. Available from: http://www.cdc.gov/nchs/data/factsheets/injury.pdf. Accessed January 28, 2006.

National Center for Health Statistics. Data for surgery. Hyattsville (MD): Center for Disease Control. Available from: http://www.cdc.gov/nchs/fastats/insurg.htm. Accessed January 28, 2006.

National Center for Health Statistics. Health, United States, 2004, with Chartbook on Trends in the Health of Americans. 2004.

National Center for Health Statistics. Health, United States. Hyattsville (MD): Center for Disease Control, 2003.

National Center for Health Statistics. Hyattsville (MD): Centers for Disease Control and Prevention; c2004 [updates 2004 December 16;cited 2006 March 20]. Prevalence of overweight and obesity among adults: United States, 1999–2002. Available from: http://www.cdc.gov/nchs/products/pubs/pubd/hestats/obese/obse99.htm

National Center for Health Statistics. Hyattsville (MD): Centers for Disease Control and Prevention; c2004 [updates 2005 February 15;cited 2006 March 20]. Obesity still a major problem, new data show. Available from: http://www.cdc.gov/nchs/pressroom/04facts/obesity.htm.

National Center for Health Statistics. National Vital Statistics Reports. Table E: deaths and percentage of total deaths for the ten leading causes of death, by race: United States, 2001. National Center for Health Statistics. 2003;52:9. Available from: http://www.cdc.gov/nchs/data/dvs/nvsr52_09p9.pdf Accessed January 28, 2006.

National Center for Health Statistics. NCHS data on injury. Bethesda (MD): Centers for Disease Control, 2005. Available from: http://www.cdc.gov/nchs/data/factsheets/injury.pdf. Accessed February 24, 2006.

National Center for Health Statistics. NCHS data on Parkinson's disease. Bethesda (MD): Centers for Disease Control; 2005. Available from: http://www.cdc.gov/nchs/data/factsheets/Parkinsons.pdf. Accessed February 24, 2006.

National Center for Health Statistics. Trends in the health of Americans. Hyattsville (MD): National Center for Health Statistics; 2004.

National Center for Health Statistics, in collaboration with the National Center for Chronic Disease Prevention and Health Promotion (2000). Available from: http://www.cdc.gov/growthcharts.

National Cholesterol Education Program. Third Report of the National Cholesterol Education Program Expert Panel on Detection, Evaluation, and Treatment of High Blood Cholesterol in Adults (Adult Treatment Panel III). Bethesda (MD): National Institutes of Health, National Heart, Lung, and Blood Institute; 2002.

National Diabetes Information Clearinghouse (NDIC). National diabetes statistics. Washington (DC): NIH Publication #06–3892, 2005. [cited 2006 Feb 11]. Available at http://diabetes.niddk.nih.gov.

National Digestive Diseases Information Clearinghouse. (2005). Pancreatitis. Available from: http://digestive.niddk.nih.gov/ddiseases/pubs/pancreatitis.

National Dysphagia Diet Task Force. National Dysphagia Diet: standardization for optimal care. Chicago (IL): American Dietetic Association; 2002.

National Guideline Clearinghouse [homepage on the Internet]. Rockville (MD): National Guideline Clearinghouse; 1998–2006 [cited 2006 January 15]. Physical activity in the prevention, treatment and rehabilitation of diseases. Available from: http://www.guideline.gov/summary/summary.aspx?doc_id=6536&nbr=004102&string=ex.

National Guidelines Clearinghouse (NGC). Resource for evidence-based clinical practice guidelines. NGC is an initiative of the Agency for Healthcare Research and Quality (AHRQ), U.S. Department of Health and Human Services. http://www.guidelines.gov. Accessed February 2006.

National Heart, Lung, and Blood Institute. Clinical guidelines on the identification, evaluation, and treatment of overweight and obesity in adults. Bethesda (MD): U.S. Department of Health and Human Services, National Institutes of Health; 1998.

National Heart, Lung, and Blood Institute. The practical guide: identification, evaluation, and treatment of overweight and obesity in adults. Bethesda (MD): U.S. Department of Health and Human Services, National Institutes of Health; 2000.

National Institute of Arthritis and Musculoskeletal and Skin Diseases. Questions and answers about fibromyalgia. Bethesda (MD): National Institutes of Health; 2004. Available from: http://www.niams.nih.gov/hi/topics/fibromyalgia/fibrofs.htm.

National Institute of Child Health and Development. Phenylketonuria: screening and management. National Institutes of Health, Consensus Development Conference Statement, National Institute of Child Health and Development. 2000 Oct 16–18;17(3):1–27.

National Institute of Dental and Craniofacial Research. Oral health in America: a report of the Surgeon General 2000. Executive Summary. Rockville (MD): U.S. Department of Health and Human Services, National Institutes of Health; 2000. [cited 2006 February 17]. Available from: http://www.nidcr.nih.gov/AboutNIDCR/SurgeonGeneral/ExecutiveSummary.htm.

National Institute of General Medical Sciences. Trauma, shock, burn, and injury: facts and figures. Bethesda (MD): National Institute of General Medical Sciences, National Institutes of Health. Available from: http://publications.nigms.nih.gov/factsheets/trauma_burn_facts.html. Accessed January 27, 2006.

National Institute of Health. Nurse practitioner in Medline Plus Encyclopedia. 2005. Available from: http://www.nlm.nih.gov/medlineplus/encyclopedia.html.

National Institute of Mental Health. Depression [monograph on the Internet]. Bethesda (MD): National Institute of Mental Health; 2000 [accessed 28 Nov 2005]. Available from: http://www.nimh.nih.gov/publicat/depression.cfm.

National Institute of Neurological Diseases and Stroke. Available from: http://www.ninds.nih.gov/disorders/tbi/detail_tbi.htm.

National Institute on Alcohol Abuse and Alcoholism. 2000. Updating estimates of the economic costs of alcohol abuse in the United States.

National Institutes of Health Drug Nutrient Interactions Task Force. Warren Grant Magnuson Clinical Center. Drug-nutrient interactions: monoamine oxidase inhibitor (MAOI) medications. Available at http://www.cc.nih.gov. Accessed December 2, 2005.

National Institute of Health. Bethesda (MD): U.S. Department of Health and Human Services, National Institutes of Health; c2004 [updates 2004 August; cited 2006 April 18]. What is high blood pressure?; [about 2 screens]. Available from: http://www.nhlbi.nih.gov/health/dci/Diseases/Hbp/HBP_WhatIs.html.

National Institute of Health. Osteoporosis prevention, diagnosis, and therapy. NIH consensus statements. 2000; 17(1):1152. Available from: http://consensus.nih.gov/cons/111/111_statement.pdf.

National Kidney Foundation. K/DOQI Clinical practice guidelines for bone metabolism and disease in chronic kidney disease. Am J Kidney Dis. 2003;42(suppl 3):S1–S202.

National Kidney Foundation. K/DOQI Clinical practice guidelines for cardiovascular disease in dialysis patients. Am J Kidney Dis. 2005;45(suppl 3):S1–S154.

National Kidney Foundation. K/DOQI Clinical practice guidelines for chronic kidney disease: evaluation, classification, and stratification. Am J Kidney Dis. 2002;39(suppl 1):S1–S200.

National Kidney Foundation. K/DOQI Clinical practice guidelines for managing dyslipidemias in chronic kidney disease. Am J Kidney Dis. 2003;41(suppl 3):S1–S91.

National Kidney Foundation. K/DOQI Clinical practice guidelines for nutrition in chronic renal failure adult guidelines. Am J Kidney Dis. 2000;35(suppl 2):S17–S104.

National Kidney Foundation. K/DOQI Clinical practice guidelines for the treatment of anemia of chronic renal failure. Am J Kidney Dis. 2001;37(suppl 1):S182–S235.

National Kidney Foundation. K/DOQI Vascular access, access. Am J Kidney Dis. 2001;37(suppl 1):S137–S173.

National Kidney Foundation. Kidney Early Evaluation Program 2004. Am J Kidney Dis. 2005;45(suppl 2):S54–S70.

National Kidney Foundation. Pocket guide to nutrition assessment of the patient with chronic kidney disease. 3rd ed. New York: National Kidney Foundation; 2002.

National Osteoporosis Foundation. Physician's guide to prevention and treatment of osteoporosis. Washington (DC): National Osteoporosis Foundation; 2003.

Navia JM. Carbohydrates and dental health. Am J Clin Nutr. 1994;59:719S.

Nayar M, Rhodes JM. Management of inflammatory bowel disease. Postgrad Med J. 2004;80:206–13.

Nead KG, Halterman JS, Kaczorowski JM, Auinger P, Weitzman M. Overweight children and adolescents: a risk group for iron deficiency. Pediatrics. 2004;114:104–8.

Neel J. Healthcare: U.S. looks to German model. Nature. 1992;351:433.

Neily JB, Toto KH, Gardner EB, Rame JE, Yancy CW, Sheffield MA, et al. Potential contributing factors to noncompliance with dietary sodium restriction in patients with heart failure. Am Heart J. 2002;143:29–33.

Nelson J, DeMichele SJ, Pacht ER, Wennberg AK. Effect of enteral feeding with eicosapentaenoic acid, _-linolenic acid, and antioxidants on antioxidant status in patients with acute respiratory distress syndrome. JPEN J Parenter Enteral Nutr. 2003;27:98–104.

Nelson KA. The cancer anorexia-cachexia syndrome. Semin Oncol. 2000;27:64–8.

Nelson M, Poulter J. Impact of tea drinking on iron status in the UK: a review. J Hum Nutr Diet. 2004;17:43–54.

Nelson MC, Zemel BS, Kawchak DA, et al. Vitamin B6 status of children with sickle cell disease. J Pediatr Hematol Oncol. 2002;24:463–9.

Neri A, Sabah G, Samra Z. Bacterial vaginosis in pregnancy treated with yoghurt. Acta Obstet Gynecol Scand. 1993;72:17–9.

Ness-Abramof R, Aprovian CM. Drug-induced weight gain. Drugs Today. 2005 Aug;41(8):547–55.

Neter JE, Stam BE, Kok FJ, Grobbee DE, Geleijnse JM. Influence of weight reduction on blood pressure: a meta–analysis of randomized controlled trials. Hypertension. 2003;42:878–84.

Neundorfer B. Alcohol polyneuropathy. Firtschr Neurol Psychiatr. 2001;69:341–5.

Neuringer I, Chalermskulrat WJ, Aris RM. Special problems in long-term survivors of lung transplantation [monograph on the Internet]. Northbrook (IL): American College of Chest Physicians; 1999–2005 [cited 2005 Jun 30]. Available from: http://www.chestnet.org/education/online/pccu/vol19/lessons11_12/index.php.

Newman CR, Downes NJ, Tseng RY, McProud LM, Newman LK. Nutrition-related backgrounds and counseling practices of doctors of chiropractic. J Am Diet Assoc. 1989; 89:939–43.

Newswanger DL, Warren CR. Guillain-Barre syndrome. Am Fam Physician. 2004;69:2405–10.

Newton-Sanchez OA, Basurto-Celaya G, Richardson V, Belkind-Gerson J. Hemorrhagic disease of the newborn, a resurgent disease. Implications for prevention. Salud Publica Mex. 2002;44:57–9.

Ney D, Bay C, Saudubray JM, Kelts DG, Kulovich S, Sweetman L, and Nyhan WL. An evaluation of protein requirements in methylmalonic academia. J Inherit Metab Dis. 1985;8:132–42.

NHLBI. National Heart, Lung, and Blood Institute. The DASH eating plan. 2005. Available at: http://www.nhlbi.nih.gov/health/public/heart/hbp/dash/. Accessed March 17, 2006.

NIAID-NIH. Allergy statistics [monograph on the Internet]. Bethesda (MD): National Institute of Allergy and Infectious Diseases and National Institutes of Health; 2005 [cited 2006 Jan 26]. Available from: http://www.niaid.nih.gov/factsheets/allergystat.htm.

Nicassio PM, Radojevic V, Weisman MH, Schuman C, Kim J, Schoenfeld-Smith K, et al. A comparison of behavioral and educational interventions for fibromyalgia. J Rheumatol. 1997;24:2000–7.

Nichols N. Hypothyroidism and cardiovascular disease. Can J Cardiovasc Nurs. 2005;15:68–73.

Nicholson L. Declogging small–bore feeding tubes. JPEN J Parenter Enteral. 1987;11:594–7.

Nicklas TA, Hampl JS, Taylor CA, Thompson VJ, Heird WC. Monounsaturated fatty acid intake by children and adults: temporal trends and demographic differences. Nutr Rev. 2004;62:132–41.

Ning B, Nowell S, Sweeney C, Ambrosone CB, Williams S, Miao X, et al. Pharmacogenetics and Genomics. 2005;15: 465–73.

Nirmalananthan N, Greensmith L. Amyotrophic lateral sclerosis: recent advances and future therapies. Curr Opin Neurol. 2005;18:712–9.

Nishiyama S, Irisa K, Matsubasa T, Higashi A, Matsuda I. Zinc status relates to hematological deficits in middle-aged women. J Am Coll Nutr. 1998;17:291–5.

Nitenberg G, Raynard B. Nutritional support of the cancer patient: issues and dilemmas. Crit Rev Oncol Hematol. 2000;34:137–68.

NLM-NIH. Medical encyclopedia: blood differential [monograph on the Internet]. Bethesda (MD): National Library of Medicine and National Institutes of Health; 2006 [cited 2006 Jan 26]. Available from: http://www.nlm.nih.gov/medlineplus/ency/article/003657.htm#Normal%20Values.

Noakes M, Keogh JB, Foster PR, Clifton PM. Effect of an energy-restricted, high-protein, low-fat diet relative to a conventional high-carbohydrate, low-fat diet on weight loss, body composition, nutritional status, and markers of cardiovascular health in obese women. Am J Clin Nutr. 2005;81:1298–306.

Nobaek S, Johansson ML, Molin G, Ahrn'e S, Jeppsson B. Alteration of intestinal microflora is associated with reduction in abdominal bloating and pain in patients with irritable bowel syndrome. Am J Gastroenterol. 2000;95: 1231–8.

Nolan S. Traumatic brain injury: a review. Crit Care Nurs Q. 2005;28:188–94.

Nord N, Andrews M, Carlson S. Household food security in the United States; 2002.

Nordgaard I, Hansen BS, Mortensen PB. Importance of colonic support for energy absorption as small bowel failure proceeds. Am J Clin Nutr. 1996:64:222–31.

Norgan NG. Body composition. In: Ulijaszek SJ, Johnston FE, Preece MA, editors. The Cambridge encyclopedia of human growth and development. London: Cambridge University Press; 1998. p. 212–5.

Norman AW. Vitamin D. In: Bowman BA, Russell RM, editors. Present knowledge in nutrition. 8th ed. Washington (DC): International Life Sciences Institute Press; 2001. p. 146–55.

Normen L, Chan K, Braitstein P, Anema A, Bondy G, Montaner JS, et al. Food insecurity and hunger are prevalent among HIV-positive individuals in British Columbia, Canada. J Nutr. 2005;135:820–5.

Norred CL. Complementary and alternative medicine use by surgical patients. AORN J. 2002;76:1013–21.

Notkins AL, Lernmark A. Autoimmune type 1 diabetes: resolved and unresolved issues. J Clin Invest. 2001;108: 1247–52.

Novotny JA, Rumpler WV, Riddick H, Herbert JR, Rhodes D, Judd JT, et al. Personality characteristics as predictors of underreporting of energy intake on 24-hour dietary recall interviews. J Am Diet Assoc. 2003;103:1146–51.

Nowak TJ, Handford AG. Chapter 1. Cell injury. In: Pathophysiology: concepts and applications for health care professionals. New York: McGraw-Hill; 2004.

Nussbaum RL, Polymeropoulos MH. Genetics of Parkinson's disease. Hum Mol Genet. 1997;6:1687–91.

Nutrition Care Manual [database on the Internet]. Chicago (IL): American Dietetic Association; 2006 [cited 2006 Mar 18]. Available by subscription from: http://nutritioncare manual.org.

Nutrition Screening Initiative. Managing nutrition care in health plans. Washington, (DC): Greer, Margolis, Mitchell, Burns, & Associates; 1996.

Nutt DJ, Peters TJ. Alcohol: the drug. Br Med Bull. 1994;50: 5–17.

Nuttall FQ. Carbohydrate and dietary management of clients with insulin-requiring diabetes. Diabetes Care. 1993;16:1039–42.

Nyhan WL, Pinar TO. Atlas of metabolic diseases. London: Chapman & Hall Medical; 1998.

Nyvad O, Danielsen H, Madsen S. Intravenous iron-sucrose complex to reduce epoetin demand in dialysis patients. Lancet. 1994;344;1305-6.

The NPD Group/NPD Foodworld. The NPD's new dieting monitor tracks America's dieting habits. 2004. Available from: http://www.npd.com/press/releases/press_040503a .htm.

O'Brien K, Nixon S, Glazier RH, Tynan AM. Progressive resistive exercise interventions for adults living with HIV/AIDS. Cochrane Database Syst Rev. 2004;4: CD004248.

O'Connell BS. Diabetes classification, pathophysiology, and diagnosis. In: Ross TA, Boucher JL, O'Connell BS, editors. American Dietetic Association guide to diabetes: medical nutrition therapy and education. Chicago (IL): American Dietetic Association; 2005.

O'Connell MB, Madden DM, Murray AM, Heaney RP, Kerzner LJ. Effects of proton pump inhibitors on calcium carbonate absorption in women: a randomized crossover trial. Am J Med. 2005;118:778–81.

O'Connor RG, Schottenfeld RS. Patients with alcohol problems. N Engl J Med. 1998;338:592–601.

O'Hare NE. Treatment of cardiovascular disease. In: Kaminsky LA, editor. American College of Sports Medicine's resource manual for guidelines for exercise testing and prescription. 5th ed. Baltimore (MD): Lippincott Williams & Wilkins; 2006. p. 427–38.

O'Sullivan Maillet J, Potter RL, Heller L. Position of the American Dietetic Association: ethical and legal issues in nutrition, hydration, and feeding. J Am Diet Assoc. 2002; 102:716–26.

Obzaranek E, Proschan MA, Vollmer WM, Moore TJ, Sacks FM, Appel LJ, et al. Individual blood pressure responses to changes in salt intake: results from the DASH-Sodium trial. Hypertension. 2003;42:459–67.

Ochs M. New medicare changes to expand MNT access in 2002. J Am Diet Assoc. 2002;102:30.

O'Connor DL, Latulippe ME, Campos C, Merlos C, Villalpando S, Picciano MF. Folate deficiency does not alter the usefulness of the serum transferrin receptor concentration as an index for the detection of iron deficiency in Mexican women during early lactation. J Nutr. 2005;135: 144–9.

O'Connor JA, Cogley C, Burton M, Lancaster-Weiss K, Cordle RA. Posttransplantation lymphoproliferative disorder: endoscopic findings. J Pediatr. Gastroenterol Nutr. 2000; 31:458–61.

Oddy WH, Peat JK. Breastfeeding, asthma, and atopic disease: an epidemiological review of the literature. J Hum Lact. 2003;19:250–61.

Oddy WH, de Klerk NH, Kendall GE, Mihrshahi S, Peat JK. Ratio of omega-6 to omega-3 fatty acids and childhood asthma. J Asthma. 2004;41:319–26.

Oddy WH, Sherriff JL, de Klerk NH, Kendall GE, Sly P, Beilin LJ, et al. The relation of breastfeeding and body mass index to asthma and atopy in children: a prospective cohort study to age 6 years. Am J Public Health. 2004;94: 1531–7.

Odou P, Ferrari N, Barthelemy C, Brique S, Lhermitte M, Vincent A, et al. Grapefruit juice-nifedipine interaction: possible involvement of several mechanisms. J Clin Pharmacol Ther. 2005;30:153–8.

Ogier de Bawny H. Management and emergency treatments of neonates with a suspicion of inborn errors of metabolism. Semin Neonatal. 2002;7:17–26.

Ogura Y, Bonen DK, Inohara N, et al. A frameshift mutation in NOD2 associated with susceptibility to Crohn's disease. Nature. 2001;411:603–6.

Oh MS. New perspectives on acid-base balance. Semin Dial. 2000;13:212–9.

Oh RC, Brown DL. Vitamin B12 Deficiency. Am Fam Phys. 2003;67:979–93.

Okey AB, Boutros PC, Harper PA. Polymorphisms of human nuclear receptors that control expression of drug-metabolizing enzymes. Pharmacogenet Genomics. 2005; 5:371–9.

Okunade AL, Hufford CD, Richardson MD, Peterson JR, Clark AM. Antimicrobial properties of alkaloids from Xanthorhiza simplicissima. J Pharm Sci. 1994;83:404–6.

Olbers T, Fagevik-Olsen M, Maleckas A, Lonroth H. Randomized clinical trial of laparoscopic Roux-en-Y gastric bypass versus laparoscopic vertical banded gastroplasty for obesity. Br J Surg. 2005;92:557–62.

Olszynski WP, Davison KS, Adachi JD, Brown JP, Cummings SR, et al. Osteoporosis in men: epidemiology, diagnosis, prevention and treatment. Clin Ther. 2004;26:15–28.

Olt GJ. Fatigue and gynecologic cancer. Curr Womens Health Rep. 2003;3:14–8.

Omenn GS. Challenges facing public health policy. J Am Diet Assoc. 1993;93:643.

Onodera H. The role of the thymus in the pathogenesis of myasthenia gravis Tohoku J Exp Med. 2005;207:87–98.

Oppenheimer SJ. Iron and its relation to immunity and infectious disease. J Nutr. 2001;131:616S–633S; discussion 633S–635S.

Orlander PR, Woodhouse WR, Davis AB. Hypothyroidism. New York: eMedicine.com, Inc. [updated 2005 Sep 23; cited 2006 Jan 7]. Available from: http://www.emedicine.com.

Orque MS. Nursing care of Filipino American patients. In: Orque MS, Bloch B, Monrroy LSA, editors. Ethnic nursing care: a multicultural approach. St. Louis (MO): Mosby; 1983.

Orr R, Fiatarone Singh M. The anabolic androgenic steroid oxandrolone in the treatment of wasting and catabolic disorders: review of efficacy and safety. Drugs. 2004;64: 725–50.

Orrevall Y, Tishelman C, Herrington MK. The path form oral nutrition to home parenteral nutrition: a qualitative interview study of the experiences of advanced cancer patients and their families. Clin Nutr. 2004;23:1280–7.

Osborn HT, Akoh CC. Structured lipids-novel fats with medical, nutraceutical, and food applications. Comprehensive Reviews in Food Science and Food Safety. 2002;1:93–103.

Osmond DH. Epidemiology of HIV/AIDS in the United States. HIV InSite Knowledge Base. 2003. Available from: http://hivinsite.ucsf.edu?InSite?pages=kb-01-03#S1X.

O'Sullivan GC, Kelly P, O'Halloran S, Collins C, Collins JK, Dunne C, Shanahan F. Probiotics: an emerging therapy. Curr Pharm Des. 2005;11:3–10.

Ots T. The angry liver, the anxious heart, and the melancholy spleen: the phenomenology of perceptions in Chinese culture. Culture, Medicine, and Psychiatry. 1990;14: 21–58.

Ott M, Wegner A, Caspary WF, Lembcke B. Intestinal absorption and malnutrition in patients with the acquired immunodeficiency syndrome. Z Gastroenterol. 1993;31: 661–5.

Ott M, Fischer H, Polat H, Helm EB, Frenz M, Caspary WF, et al. Bioelectrical impedance analysis as a predictor of survival in patients with human immunodeficiency virus infection. J Acquir Immune Defic Syndr Hum Retrovirol. 1995;9:20–5.

Ott M, Lembcke B, Fischer H, Jager R, Polat H, Geier H, et al. Early changes of body composition in human immunodeficiency virus-infected patients: tetrapolar body impedance analysis indicates significant malnutrition. Am J Clin Nutr. 1993;57:15–9.

Ottery FD, Kasenic S, DeBolt S, Rodgers K. Volunteer network accrues >1900 patients in 6 months to validate standardized nutritional triage. Proceedings of ASCO. 1998; 17: abstract 282.

Packman, S. Nutritional therapy in inborn errors of metabolism. Perinatol Neonatol. 1986;10:33–45.

Pagano G, Korkina LG. Prospects for nutritional interventions in the clinical management of Fanconi anemia. Cancer Causes and Control. 2000;11:881–9.

Page C, Curtis M, Sutter M, Walker M, Hoffman B. Integrated pharmacology. 2nd ed. Edinburgh, Scotland: Mosby International; 2002.

Paik KH, Jin DK, Song SY, Lee JE, Ko SH, Song SM, et al. Correlation between fasting plasma ghrelin levels and age, body mass index (BMI), BMI percentiles, and 24-hour plasma ghrelin profiles in Prader-Willi syndrome. J Clin Endocrinol Metab. 2004;89:3885–9.

Paine MG, Criss AB, Watkins PB. Two major grapefruit components differ in time to onset of intestinal CYP3A4 inhibition. J Pharmacol Exp Ther. 2005;312:1151–60.

Pak E, Esrason KT, Wu VH. Hepatotoxicity of herbal remedies: an emerging dilemma. Progress in Transplantation. 2004;14:91–6.

Palhagen S, Lorefalt B, Carlsson M, Ganowiak W, Toss G, Unosson M, et al. Does L-dopa treatment contribute to reduction in body weight in elderly patients with Parkinson's disease? Acta Neurol Scand. 2005;111:12–20.

Pan M. Arginine transport in catabolic disease states. J Nutr. 2004;134:2826S–2829S, discussion 2853S.

Panel on Clinical Practices for Treatment of HIV Infection. Guidelines for the use of antiretroviral agents in HIV-1-infected adults and adolescents. Bethesda (MD): Department of Health and Human Services (DHHS); 2005 Apr 7.

Panel on Macronutrients, Panel on the Definition of Dietary Fiber, Subcommittee on Upper Reference Levels of Nutrients, Subcommittee on Interpretation and Uses of Dietary Reference Intakes, Standing Committee on the Scientific Evaluation of Dietary Reference Intakes. Dietary reference intakes for energy, carbohydrate, fiber, fat, fatty acids, cholesterol, protein, and amino acids. Washington (DC): National Academy Press; 2002.

Pang KYC. The practice of traditional Korean medicine in Washington DC. Soc Sci Med. 1994;28:875–84.

Panigrahi P, Gewolb I, Bamford P, et al. Role of glutamine in bacterial transcytosis and epithelial cell injury. JPEN J Parenter Enteral Nutr. 1997;21:75–80.

Papayannopoulou T, D'Andrea AD, Abkowitz JL, Migliaccio AN. Biology of erythropoiesis, erythroid differentiation, and maturation. In: Hoffman R. Hematology, basic principles and practice. 4th ed. Philadelphia (PA): Elsevier Medical Publishing; 2005. p. 267–83.

Pare S, Barr SI, Ross SE. Effect of daytime protein restriction on nutrient intakes of free-living Parkinson's disease patients. Am J Clin Nutr. 1992;55:701–7.

Parents of Galactosemia Children [homepage on the Internet]. [updated 2006 Feb 14, cited 2006 Feb 27]. Available from: http://www.galactosemia.org.

Pari L, Saravanan R. Antidiabetic effect of diasulin, a herbal drug, on blood glucose, plasma insulin and hepatic enzymes of glucose metabolism in hyperglycaemic rats. Diabetes Obes Metab. 2004;6:286–92.

Parish JM, Somers VK. Obstructive sleep apnea and cardiovascular disease. Mayo Clin Proc. 2004;79:1036–46.

Parisi G, Leandro G, Bottona E, Carrara M, Cardin F, Faedo A, et al. Small intestinal bacterial overgrowth and irritable bowel syndrome. Am J Gastroenterol. 2003;98:2572; author reply 2573–4.

Parkinson Study Group. Levodopa and the progression of Parkinson's disease. N Engl J Med.2004;351:2498–508.

Parsons HG, Carter RJ, Unrath M, Snyder FF. Evaluation of branched-chain amino acid intake in children with maples syrup urine disease and methylmalonic aciduria. J Inherit Metab Dis. 1990;13:125–36.

Pasternak RC, Criqui MH, Benjamin EJ, Fowkes FGR, Isselbacher EM, McCullough PA, et al. Atherosclerotic vascular disease conference: Writing group I: Epidemiology. Circulation. 2004;109:2605–12.

Pastors JG, Franz MJ, Warshaw H, Daly A, Arnold MS. How effective is medical nutrition therapy in diabetes care? J Am Diet Assoc. 2003;103:827–31.

Pastors JG, Waslaski J, Gunderson H. Diabetes meal-planning strategies. In: Ross TA, Boucher JL, O'Connell BS, editors. American Dietetic Association guide to diabetes: medical nutrition therapy and education. Chicago (IL): American Dietetic Association; 2005.

Patel MG. The effect of dietary intervention on weight gain after renal transplantation. J Renal Nutr. 1998;8:137–41.

Pathak A, Roth P, Piscitelli J, Johnson L. Effects of vitamin E supplementation during erythropoietin treatment of the anaemia of prematurity. Arch Dis Child Fetal Neonatal Ed. 2002;88:F324–8.

Paton NI, Ng YM, Chee CB, Persaud C, Jackson AA. Effects of tuberculosis and HIV infection on whole-body protein metabolism during feeding, measured by the [15N]glycine method. Am J Clin Nutr. 2003;78:319–25.

Patrick L. Nutrients and HIV: part one – beta carotene and selenium. Altern Med Rev. 1999;4:403–13.

Patrick L. Nutrients and HIV: part two – vitamins A and E, zinc, B-vitamins, and magnesium. Altern Med Rev. 2000;5:39–51.

Patti MG, Fisichella PM. Minimally invasive esophageal procedures [monograph on the Internet]. ACS Surgery: Principles and Practice; 2004 [cited 2004 Jan 6]. Available from: http://www.medscape.com/viewarticle/451426.

Pattison DJ, Symmons DPM, Young A. Does diet have a role in the aetiology of rheumatoid arthritis? Proc Nutr Soc. 2004;63:137–43.

Paul R. The art of redesigning instruction. In: Willsen J, Blinker A, editors. Critical thinking: how to prepare students for a rapidly changing world. Santa Rosa (CA): Foundation for Critical Thinking; 1993.

Pauli-Magnus C, Stieger B, Meier Y, Kullak-Ublick GA, Meier PJ. Enterohepatic transport of bile salts and genetics of cholestasis. J Hepatol. 2005;43:342–57.

Payne A. Nutrition and diet in the clinical management of multiple sclerosis. J Hum Nutr Dietet. 2001;14:349–57.

PDR Health. Latest thinking on diet and immunity [monograph on the Internet]. Stamford (CT): Thomson Scientific and Healthcare; 2004 [accessed 28 Nov 2005]. Available from: http://www.pdrhealth.com/content/nutrition_health/chapters/fgnt27.shtml.

Peach HG, Bath NE. Post test probability that men in the community with raised plasma ferritin concentrations are hazardous drinkers. J Clin Pathol. 1999;52:853–5.

Pearl ML, Frandina M, Mahler L, Valea FA, DiSilvestro PA, Chalas EA. Randomized controlled trial of a regular diet as the first meal in gynecologic oncology patients undergoing intraabdominal surgery. Obstet Gynecol. 2002; 100(s):230–4.

Peat J, Allen J, Oddy W, Webb K. Breastfeeding and asthma: appraising the controversy. Pediatr Pulmon. 2003;35: 331–4.

Pedrini MT, Levey AS, Lau J, Chalmers TC, Wang PH. The effect of dietary protein restricition on the progression of diabetic and nondiabetic renal disease: a meta-analysis. Ann Intern Med. 1996;124:627–32.

Pegram AA, Kennedy LD. Prevention and treatment of veno-occlusive disease. Ann Pharmacother. 2001;35: 935–42.

Peng CC, Glassman PA, Trilli LE, et al. Incidence and severity of potential drug-dietary supplement interactions in primary care patients. Arch Intern Med. 2004;164:630–6.

Peng X, Yan H, You Z, Wang P, Wang S. Clinical and protein metabolic efficacy of glutamine granules-supplemented enteral nutrition in severely burned patients. Burns. 2005;31:342–6.

Pentieva K, McNulty H, Reichert R, et al. The short-term bioavailabilities of [6S]-5-methyltetrahydrofolate and folic acid are equivalent in men. J Nutr. 2004;134:580–5.

Pereira GR, Baumgart S, Bennett MJ, Stalling VA, Georgieff MK, Hamosh M, et al. Use of high-fat formula for premature infants with bronchopulmonary dysplasia;metabolic, pulmonary, and nutritional needs. J Pediatr. 1994;124: 605–11.

Pereira MA, O'Reilly E, Augustsson K, Fraser GE, Goldbourt U, Heitmann BL, et al. Dietary fiber and risk of coronary heart disease: a pooled analysis of cohort studies. Arch Intern Med. 2004;164:370–6.

Perkin JE, Wilson WJ, Schuster K, Rodriguez J, Allen-Chabot A. Prevalence of nonvitamin, nonmineral supplement usage among university students. J Am Diet Assoc. 2002; 102:412–4.

Perretti M, Anluwalia A. The microcirculation and inflammation: site of action for glucocorticoids. Microcirculation. 2000;7:147.

Petersen S, Peto V, Rayner M, Leal J, Luengo-Fernandez R, Gray A. European cardiovascular disease statistics. London: British Heart Foundation; 2005.

Peterson SJ, Tangney CC, Pimental-Zablah EM, Hjelmgren B, Booth G, Berry-Kravis E. Changes in growth and seizure reduction in children on the ketogenic diet as a treatment for intractable epilepsy. J Am Diet Assoc. 2005;105:718–24.

Pettifor JM. Nutritional rickets: deficiency of vitamin D, calcium, or both? Am J Clin Nutr. 2004;80(suppl): 1725S–1729S.

Petty TL. Spirometry made simple. National Lung Health Education Program [homepage on the Internet]. Irving (TX): American Association for Respiratory Care; 1999 [cited 2005 February 9]. Available from: http://www.nlhep.org/resources/SpirometryMadeSimple.htm.

Pfeifer K. Surgery. In Otto SE, editor. Oncology nursing. 4th ed. Philadelphia (PA): Mosby; 2001.

Pfeiffer RF. Gastrointestinal dysfunction in Parkinson's disease. Lancet Neurol. 2003;2(2):107–16.

Pfizer Corporation. Lipid Intervention Program [1999 December; cited 2005 October 25]. Available from: http://healthproject.stanford.edu/koop/pfizer99/documentation.html.

Pfizer Inc. Facts about Exubera. Project: NN261789P. 2006.

Piccirillo N, De Matteis S, Sora F, et al. Glutamine parenteral supplementation in stem cell transplant. Bone Marrow Transplant. 2004;33:455.

Pierson CE. Phytoestrogens in botanical dietary supplements: implications for cancer. Integrative Cancer Therapies. 2003;2:120–38.

Pike, JL, Smith TL, Hauger RL, Nicassio PM, Patterson TL, McClintick J, et al. Chronic life stress alters sympathetic, neuroendocrine, and immune responsivity to an acute psychological stressor in humans. Psychosom Med. 1997; 59:447–57.

Pingleton SK. Enteral nutrition in patients with respiratory disease. Eur Respir J. 1998;9:364–70.

Pinhas-Hamiel O, Newfield RS, Koren I, Agmon A, Lilos P, Phillip M. Greater prevalence of iron deficiency in overweight and obese children and adolescents. Int J Obes Relat Metab. 2003;27:416–8.

Pugh MB, Werner B, editors. Stedman's medical dictionary. 27th ed. Baltimore (MD): Lippincott Williams & Wilkins; 2000.

Powers HJ. Riboflavin (vitamin B-2) and health. Am J Clin Nutr. 2003;77:1352–60.

Prasad AS. Recognition of zinc-deficiency syndrome. Nutrition 2001;17:67–9.

Pronsky Z, editor. Food Medication Interactions. 13th ed. FMI Publications. Birchrunville (PA); 2004.

Pirsch JD, Armbrust MD, Knechtle SJ, D'Allessandro AM, Sollinger HW, Heisey DM, et al. Obesity as a risk factor following renal transplantation. Transplantation. 1995;59: 631–63.

Piscitelli SC, Berstein AH, Welden N, Gallicano KD, Falloon J. The effect of garlic supplements on the pharmacokinetics of saquinavir. Clin Infect Dis. 2002;34:234–8.

Pi-Sunyer FX. The epidemiology of central fat distribution in relation to disease. Nutr Rev. 2004;62:S120–6.

Pitetti R, Singh S, Hornyak D, Garcia SE, Herr S. Complementary and alternative medicine use in children. Pediatr Emerg Med. 2001;17:165–9.

Pittler MH, Schmidt K, Ernst E. Hawthorn extract for treating chronic heart failure: meta-analysis of randomized trials. Am J Med. 2003;114:665–74.

Planta M, Gunderson B, Petitt JC. Prevalence of the use of herbal products in a low-income population. Family Medicine. 2000;32:252–7.

Pleuss J. Alterations in nutritional status. In: Porth CM, editor. Pathophysiology: concepts of altered health status. 7th ed. Philadelphia (PA): Lippincott Williams & Wilkins; 2005. p. 217–238.

Poirier P, Giles TD, Bray GA, Hong Y, Stern JS, Pi-Sunyer FX, et al. American Heart Association; Obesity Committee of the Council on Nutrition, Physical Activity, and Metabolism. Obesity and cardiovascular disease: pathophysiology, evaluation, and effect of weight loss: an update of the 1997 American Heart Association Scientific Statement on Obesity and Heart Disease from the Obesity Committee of the Council on Nutrition, Physical Activity, and Metabolism. Circulation. 2006;113:898–918.

Pollak OJ. Reduction of blood cholesterol in man. Circulation. 1953;2:702–6.

Pollitt RJ. Disorders of mitochondrial long-chain fatty acid oxidation. J Inherit Metab Dis. 1995;18:473–90.

Polsky B, Kotler DP, Steinhart C. HIV-associated wasting in the HAART era: guidelines for assessment, diagnosis, and treatment. AIDS Patient CARE STDS. 2001;15:411–23.

Porter SR, Scully C, Hegarty AM. (2004). An update of the etiology and management of xerostomia. Oral Surg Oral Med Oral Pathol Oral Radiol Endod 2004;97:28–46.

Porth CM. Structure and function of the musculoskeletal system. In: Porth CM, editor. Pathophysiology: concepts of altered health status. 7th ed. Philadelphia (PA): Lippincott Williams & Wilkins; 2005. p. 1357–66.

Potter BJ, Wang F. Molecular regulation of iron homeostasis and resistance to infection in alcoholics. Front Bioscience. 2002;7:1396–409.

Potter P, Perry A. Fundamentals of Nursing, Concepts, Process and Practice (Fourth ed.). St. Louis: Mosby; 1997.

Potter PA, Perry AG. Fluid, electrolyte, and acid base balances. In: Fundamentals of nursing. St. Louis (MO): Mosby; 1999.

Potts RG, Zaroukian MH, Guerrero PA, Baker CD. Comparison of blue dye visualization and glucose oxidase test strip methods for detecting pulmonary aspiration of enteral feedings in intubated adults. Chest. 1993;103: 117–21.

Power R, Gore-Felton C, Voxvick M, Israelski Dm, Spiegel D. HIV: effectiveness of complementary and alternative medicine. Prim Care. 2002;29:361–78.

Powers F. The role of chloride in acid-base balance. J Intraven Nurs. 1999;22:286–91.

Pradka LR. Lipids and their role in coronary heart disease: what they do and how to manage them. Nurs Clin North Am. 2000;35:981–91.

Pratt DS, Kaplan, MM. Evaluation of the liver: laboratory tests. In: Schiff ER, Sorrell MF, Maddrey WC, editors. Schiff's diseases of the liver. 9th ed. Baltimore (MD): Lippincott Williams & Wilkins; 2003.

Prentice A. Diet, nutrition and the prevention of osteoporosis. Public Health Nutr. 2004;7:227–43.

Preshaw R. Management of dysphagia. Can Med Assoc J. 2004;170:1079.

Price SA, Wilson LM. Fluid and electrolyte disorders. In: Pathophysiology: clinical concepts of disease process. 6th ed. St. Louis (MO): Mosby; 2003.

Pronsky ZM. Food Medication Interactions. 13th ed. Birchrunville (PA): Food-Medication Interactions; 2004.

Pronsky ZM, Crowe JP. Food-drug interactions. In: Krause's food, nutrition, and diet therapy. 11th ed. Philadelphia (PA): Saunders; 2004. p. 455–74.

Pronsky ZM, Meyer SA, Fields-Gardner C. HIV medications food interactions. 2nd Edition. 2001.

Pronsky ZM, Crowe JP. Mechanisms of food-medication interaction. 13th ed. Birchrunville (PA): Food-Medication Interactions; 2004.

Provost DA. Laparoscopic adjustable gastric banding: an attractive option. Surg Clin North Am. 2005;85:789–805.

Przyrembel H. Therapy of mitochondrial disorders. J Inherit Metab Dis. 1987;10:129–46.

Przyrembel H, Jakobs C, Ijlst L, De Klerk JBC, Wanders RJA. Long-chain 3-hydroxyacyl-CoA dehydrogenase deficiency. J Inherit Metab Dis. 1991;14:674–80.

Ptachcinski RJ, Burckart GJ, Venkataramanan R. Cyclosporine. Drug-Intell-Clin Phar. 1985;19:90–100.

Puhan MA, Schunemann HJ, Frey M, Scharplatz M, Bachmann LM. How should COPD patients exercise during respiratory rehabilitation? Comparison of exercise modalities and intensities to treat skeletal muscle dysfunction. Thorax. 2005;60:367–75.

Purnell LD, Paulanka BJ. Transcultural health care: a culturally competent approach. 2nd ed. Philadelphia (PA): FA Davis Co.; 2002.

Pytlik R, Benes P, Patorkova M, Chocenska E, Gregora E, Prochazka B, et al. Standardized parenteral alanyl-glutamine dipeptide supplementation is not beneficial in autologous transplant patients: a randomized, double-blind, placebo controlled study. Bone Marrow Transplant. 2002;30:953–61.

Qian H, Nihorimbere V. Antioxidant power of phytochemicals from Psidium guajava leaf. Journal of the Zhejiang University: Science. 2004;5:676–83.

Quevedo J, Amaral O, Walz R, Kapczinski C. Pathogenesis of hepatic encephalopathy—a role for the benzodiazepine receptor? Medicina. 1999;32:82–96.

Quinn TC, Wawer MJ, Sewankambo N, Serwadda D, Li C, Wabwire-Mangen F, et al. Viral load and heterosexual transmission of human immunodeficiency virus type 1. N Engl J Med. 2000;342:921–9.

Qureshi WA, Graham DY. Diagnosis and management of Helicobacter pylori infection. Clin Cornerstone. 1999;1: 18–28.

Rabeneck L, Palmer A, Knowles JB, Seidehamel RJ, Harris CL, Merkel KL, et al. A randomized controlled trial evaluating nutrition counseling with or without oral supplementation in malnourished HIV-infected patients. J Am Diet Assoc. 1998;98:434–8.

Raggi P. Cardiac calcification in adult hemodialysis patients. J Am Coll Cardiol. 2002:39:695–701.

Raguso CA, Dupertuis YM, Pichard C. The role of visceral proteins in the nutritional assessment of intensive care unit patients. Curr Opin Clin Nutr Met Care. 2003;6: 211–6.

Ram FSF, Rowe BH, Kaur B. Vitamin C supplementation for asthma. The Cochrane Database of Systematic Reviews. 2004, Issue 3. Art. No.: CD000993. DOI: 10.1002/14651858.CD000993.pub2.

Ramakrishna J, Weiss MG. Health, illness, and immigration. East Indians in the United States. West J Med. 1992;157: 265–70.

Rasmu HS, Aurup P, Goldstein K, et al. Influence of magnesium substitution therapy on blood lipid composition in patients with ischemic heart disease: a double blind placebo controlled study. Arch Int Med. 1989;149:1050–3.

Rauch H, Motsch J, Bottiger BW. Newer approaches to the pharmacological management of heart failure. Curr Opin Anaesthesiol. 2006;19:75–81.

Rayner RJ, Tyrrell JC, Hiller EJ, Marenah C, Neugebauer MA, Vernon SA, et al. Night blindness and conjunctival xerosis caused by vitamin A deficiency in patients with cystic fibrosis. Arch Dis Child 1989;64:1151–6.

Reddy ST, Wang CY, Sakhaee K, Brinkley L, Pak CY. Effect of low-carbohydrate high-protein diets on acid-base balance, stone-forming propensity, and calcium metabolism. Am J Kidney Dis 2002;40:265–74.

Reeds PJ, Burrin DG. Glutamine and the bowel. J Nutr. 2001;131:2505S–2508S.

Rees Parrish C, Yoshida CM. Nutrition intervention for the patient with gastroparesis; an update. Practical Gastroenterology. 2005 Aug;29–66.

Rees R, Keohane P, Grimble G, Frost P, Attrill H. Elemental diet administered nasogastrically without starter regimens to patients with inflammatory bowel disease. JPEN J Parenter Enteral Nutr. 1986;10:258–62.

Rees R, Keohane P, Grimble G, Frost P, Attrill H, Silk D. Tolerance of elemental diet administered without starter regimen. BMJ. 1985;290:1869–70.

Rees-Parrish C. The clinician's guide to short bowel syndrome. Practical Gastroenterology. 2005 Sept: 67–106.

Rutigliano V, De Venuto D, Brindicci D, Prete F, Tota V. Clinical quiz. J Pediatr Gastroenterol Nutr. 2002;35:205–219.

Ryan TJ, Antman EM, Brooks NH, et al. 1999 update: ACC/AHA guidelines for the management of patients with acute myocardial infarction. A report of the American College of Cardiology/American Heart Association Task Force on Practice Guidelines (Committee on Management of Acute Myocardial Infarction). J Am Coll Cardiol. 1999;34:890–911 and Circulation. 1999;100:1016–30.

Ryman BE. The glycogen storage diseases. J Clin Path. 1974;(Suppl 8):106–21.

Rystrom JK. Insulin therapy. In: Ross TA, Boucher JL, O'-Connell BS, editors. American Dietetic Association guide to diabetes: medical nutrition therapy and education. Chicago (IL): American Dietetic Association; 2005.

Saadeh S, Davis GL. Management of ascites in patients with end-stage liver disease. Rev Gastroenterol Disord. 2004;4: 175–85.

Saavedra JM, Abi-Hanna A, Moore N, Yolken RH. Long-term consumption of infant formulas containing live probiotic bacteria: tolerance and safety. Am J Clin Nutr. 2004;79:261–7.

Sacco RL. Newer risk factors for stroke. Neurology. 2001;57(5 Suppl 2):S31–4. Review.

Sachedev H, Gera T, Nestel P. Effect of iron supplementation on mental and motor development in children: systematic review of randomised controlled trials. Public Health Nutr. 2005;8:117–32.

Sacks FM, Svetkey LP, Vollmer WM, Appel LJ, Bray GA, Harsha D, et al; DASH-Sodium Collaborative Research Group. Effects on blood pressure of reduced dietary sodium and the Dietary Approaches to Stop Hypertension (DASH) diet. DASH-Sodium Collaborative Research Group. N Engl J Med. 2001;344:3–10.

Saffitz JE. The pathology of sudden cardiac death in patients with ischemic heart disease—arrhythmology for anatomic pathologists. Cardiovasc Pathol. 2005;14:195–203.

Sage WM. The forgotten third: liability insurance and the medical malpractice crisis. Health Affairs. 2004;23:10–21.

Sahebjami H. Effects of nutritional depletion on lung parenchyma. Eur Respir J. 2003;24:113–24.

Said HM. Cellular uptake of biotin: mechanisms and regulation. J Nutr. 1999;129(2S Suppl):490S–493S.

Saito YA, Schoenfeld P, Locke GR. The epidemiology of irritable bowel syndrome in North America: a systematic review. Am J Gastroenterol. 2002;97:1910–5.

Saladin KS. Anatomy and physiology: the unity of form and function. 3rd ed. Boston: McGraw-Hill; 2004.

Sanchez O. Insights into novel biological mediators of clinical manifestations in cancer. AACN Clin Issues Adv Pract Acute Crit Care. 2004;15:112–8.

Sanders DB, Hunter K, Wu Y, Jablonowski, C, Bahl JJ, Larson DF. Modulation of the inflammatory response in the cardiomyocyte and macrophage. J Extra-Corporeal Tech. 2001;33:167–74.

Sanford MG, Ryan C, Cummings AD, Hunt A, Hackes B. Protocols for identifying drug-nutrient interactions in patients: the role of the dietitian. J Am Diet Assoc. 2002;102:729–30; discussion 730–1.

Sankar R, Holmes GL. Mechanisms of action for the commonly used antiepileptic drugs: relevance to antiepileptic drug-associated neurobehavioral adverse effects. J Child Neurol. 2004;19 Suppl 1:S6–14.

Santen S. Cholangitis. http://www.emedicine.com/emerg/topic96.htm. Accessed January 2006.

Sari R, Yildirim B, Sevinc A, Buyukberber S. Gluten-free diet improves iron-deficiency anaemia in patients with coeliac disease. J Health Popul Nutr. 2000;18:54–6.

Sassoon CS, Arruda JA. Acid-base disturbance. Respir Care. 2001;46:327.

Satcher D. Press Release: HHS report finds health improves for most racial, ethnic groups but disparities remain in some areas. [2002 January 24; cited 2005 October 25]. Available from: http://www.cdc.gov/nchs/pressroom/02news/healthimpr.htm.

Saudebray JM, Nassogne MC, deLanley P, Towati G. Clinical approach to inherited metabolic disorders in neonates: an overview. Semin Neonatal. 2002;7:17–26.

Savage D, Lindenbaum J. Anemia in alcoholics. Medicine. 1986;65:322–38.

Savilahti E. Food-induced malabsorption syndromes. J Pediatr Gastroenterol Nutr. 2000;30 Suppl:S61–6.

Savoia C, Schiffrin EL. Inflammation in hypertension. Curr Opin Nephrol Hypertens. 2006;15:152–8.

Sawyers JE, Eaton L. Gastric cancer in Korean-Americans: cultural implications. Oncology Nursing Forum. 1992;19: 619–23.

Saxon DW, Tunnicliff G, Browkaw JJ, Raess Bu. Status of complementary and alternative medicine in the osteopathic medical school curriculum. J Am Osteopath Assoc. 2004;104:121–6.

Scevola D, DiMatteo A, Lanzarini P, Uberti F, Scevola S, Bernini V, et al. Effect of exercise and strength training on cardiovascular status in HIV-infected patients receiving highly active antiretroviral therapy. AIDS. 2003;17(Suppl 1):S123–S129.

Schaefer EJ, Gleason JA, Dansinger ML. The effects of low-fat, high-carbohydrate diets on plasma lipoproteins, weight loss, and heart disease risk reduction. Curr Atheroscler Rep. 2005;7:421–7.

Schaffer DM, Gordon NP, Jensen CD, Avins AL. Nonvitamin, nonmineral supplement use over a 12–month period by adult members of a large health maintenance organization. J Am Diet Assoc. 2003;103:1500–5.

Schaffer EA. Epidemilogy and risk factors for gallstone disease: has the paradigm changed in the 21st century? Curr Gastroenterol Rep. 2005;7:132–40.

Schell MC, Bova FJ, Larson V, et al. AAPM Report 54: Stereotactic radiosurgery. Am Assoc Physicists Med. 1995.

Schiff ER, Sorrell MF, Maddrey WC, editors. Schiff's diseases of the liver. 9th ed. Baltimore (MD): Lippincott Williams & Wilkins; 2003.

Schiff L, Schiff ER. Diseases of the liver. 7th ed. Philadelphia (PA): JB Saunders Co.; 1993.

Schmidt K. A primer to the inborn errors of metabolism for perinatal and neonatal nurses. J Perinat Neonatal Nurs. 1989;2:60–71.

Schneider A, Singer MV. Conservative treatment of chronic pancreatitis. Schweiz Rundsch Med Prax. 2005;18:94:831–8.

Schneider T, Ullrich R, Zeitz M. The immunologic aspects of human immunodeficiency virus infection in the gastrointestinal tract. Semin Gastrointest Dis. 1996;7:19–29.

Schneider-Gold C, Gajdos P, Toyka KV, Hohlfeld RR. Corticosteroids for myasthenia gravis. Cochrane Database Syst Rev. 2005 Apr 18;(2):CD002828.

Schnell FM. Chemotherapy-induced nausea and vomiting: the importance of acute antiemetic control. Oncologist. 2003;8:187–98.

Schnuchardt V. Alcohol and the peripheral nervous system. Ther Umsch. 2000;57:169–9.

Scholl TO. Iron status during pregnancy: setting the stage for mother and infant. Am J Clin Nutr. 2005;81: 1218S–1222S.

Schölmerich J. Postgastrectomy syndromes—diagnosis and treatment. Best Pract Res Clin Gastroenterol. 2004;18: 917–33.

Schols Am, Broekhuizen R, Weling-Schepers CA, Wouters EF. Body composition and mortality in chronic obstructive pulmonary disease. Am J Clin Nutr. 2005;82:53–9.

Schrezenmeir J, de Vrese M. Probiotics, prebiotics and synbiotics—approaching a definition. American Journal of Clinical Nutrition. 2001;73:361S–364S.

Schrezenmeir J, Jagla A. Milk and diabetes. J Am Coll Nutr. 2000;19:176S–190S.

Schrier RW, Wang W. Acute renal failure and sepsis. N Engl J Med. 2004;351:159–69.

Schrier SL, Reid EG. Extrinsic non-immune hemolytic anemia. In: Hoffman R. Hematology, basic principles and practice. 4th ed. Philadelphia (PA): Elsevier Medical Publishing; 2005. p. 709–19.

Schrock-Kelley S, Abbott M, Jurecki E, Packman S. Acquired methylmalonic acidemia in a breast fed infant: issues for the genetic counselor. National Society of Genetic Counselors, 17th Annual Education Conference; 1998.

Schroeder SA. Prospects for expanding health insurance coverage. N Eng J Med. 2001;344:847–52.

Schuett V. Will Tetrahydrobiopterin have a role in PKU treatment? National PKU News. Spring/Summer 2003;15:1–3.

Schulman As, Sawyer RG. Have you passed gas yet? Time for a new approach to feeding patients postoperatively. Practical Gastroenterology. 2005:10:82–88.

Schultz M, Sartor RB. Probiotics and inflammatory bowel disease. Am J Gastroenterol. 2000; 95:19S–21S.

Schunemann HJ, Bryden JB, Grant B, Freudenheim, J, Muti P, Browne R, et al. The relation of serum levels of antioxidants vitamins C and E, retinol and carotenoids with pulmonary function in the general population. An J Respir Crit Care Med. 2001;163:1246–55.

Schunemann HJ, McCann S, Grant B, Trevisan M, Muti P, Freudenheim J. Lung function in relation to intake of carotenoid and other antioxidant vitamins in a population-based study. Am J Epidemiol. 2002;155:463–71.

Schutt M, Zhou J, Meier M, Klein HH. Long-term effects of HIV-1 protease inhibitors on insulin secretion and insulin signaling in INS-1 beta cells. J Endocrinol. 2004;183: 445–54.

Schwartz MP, Samsom M, Renooij W, et al. Small bowel motility affects glucose absorption in a healthy man. Diabetes Care. 2002;25:1857–61.

Schwarz S, Leweling H. Multiple Sclerosis and nutrition. Multiple Sclerosis. 2005;11:24–32.

Schwarz UI, Seemann D, Oertel R, Miehlke S, Kuhlisch E, Fromm MF, et al. Grapefruit ingestion significantly reduces talinolol bioavailablity. J Clin Pharmacol Ther. 2005;77:291–301.

Schwebel C, Pin I, Barnoud D, Devouassoux G, Brichon PY, Chaffanjon PH, et al. Prevalence and consequences of nutritional depletion in lung transplant candidates. Eur Respir J. 2000;18:1050–5.

Schwengel RH, Gottlieb SS, Fisher ML. Protein-energy malnutrition in patients with ischemic and nonischemic dilated cardiomyopathy and congestive heart failure. Am J Cardiol. 1994;73:908–10.

Schwenk A, Breuer P, Kremer G, Ward L. Clinical assessment of HIV-associated lipodystrophy syndrome:bioelectrical impedance analysis, anthropometry and clinical scores. Clin Nutr. 2001;20:243–9.

Schwenk A, Steuck H, Kremer G. Oral supplements as adjunctive treatment to nutritional counseling in malnourished HIV-infected patients: randomized controlled trial. Clin Nutr. 1999;18:371–4.

Schwenk A, Hoffer-Belitz E, Jung B, Kremer G, Burger B, Salzberger B, et al. Resting energy expenditure, weight loss, and altered body composition in HIV infection. Nutrition. 1996;12:595–601.

Schwenk A, Kremer G, Cornely O, Diehl V, Fatkenheuer G, Salzberger B. Body weight changes with protease inhibitor treatment in undernourished HIV-infected patients. Nutrition. 1999;15:453–7.

Schwenk A. Methods of assessing body shape and composition in HIV-associated lipodystrophy. Curr Opin Infect Dis. 2002;15:9–16.

Scott CL. Diagnosis, prevention, and intervention for the metabolic syndrome. Am J Cardiol. 2003;82(suppl): 35i–42i.

Scott-Jupp R, Lama M, Tanner MS. Prevalence of liver disease in cystic fibrosis. Arch Dis Child. 1991;66:698–701.

Scrimshaw NS, SanGiovanni JP. Synergism of nutrition, infection, and immunity: an overview. Am J Clin Nutr. 1997;66(suppl):464S–477S.

Scriver C, Beaudet A, Sly W, Valle D, editors. The metabolic basis of inherited diseases. 8th ed. New York: McGraw-Hill; 2001.

Seale JL, Rumpler WV, Moe PW. Description of a direct-indirect room-sized calorimeter. Am J Physiol 1991;260: E306–E320.

Seaman S. Considerations for the global assessment and treatment of patients with recalcitrant wounds. Ostomy Wound Manage. 2000;46 Suppl:10S–29S.

Sears D, Patel T. Fatty liver [monograph on the Internet]. Omaha: eMedicine/WebMD; 2005 [cited 2006 Mar 24]. Available from: http://www.emedicine.com/MED/topic 775.htm.

Sears M, Taylor DB, Poulton R. Breastfeeding and asthma: appraising the controversy—a rebuttal. Pediatr Pulmon. 2003;36:366–8.

Seder RA, Gurunathan S. DNA vaccines designer vaccines for the 21st century. N Engl J Med. 1999;341:277–8.

Segerstrom SC, Miller GE. Psychological stress and the human immune system: a meta-analytic study of 30 years of inquiry. Psychol Bull. 2004;130:601–30.

Seguy D, Vahedi K, Kapel N, et al. Low dose growth hormone in adult home parenteral nutrition-dependent short bowel syndrome patients: a positive study. Gastroenterology. 2003;124:293–302.

Seligmann H, Hallkin H, Rauchfleisch S, Kaufmann N, Motro M, Vered Z, et al. Thiamine deficiency in patients with congestive heart failure receiving long-term furosemide therapy: a pilot study. Am J Med. 1991;91:151–5.

Semba RD, Bloem MW. The anemia of vitamin A deficiency: epidemiology and pathogenesis. Eur J Clin Nutr. 2002; 56:271–81.

Semba RD, Tang AM. Micronutrients and the pathogenesis of human immunodeficiency virus infection. Br J Nutr. 1999;81:181–9.

Semba RD. Micronutrients and the pathogenesis of human immunodeficiency virus infection. In: Fitzpatrick DW, Anderson JE, L'Abbe ML, eds. Proceedings of the 16th International Congress of Nutrition. Ottawa: Canadian Federation of Biological Societies. 1998:349–51.

Sempos CT. Iron and colorectal cancer. Nutr Rev. 2001;59: 344–6.

Serog P. Clinical status of nutritional origin involving immune deficiency. Food Addit Contam. 1990;7 Suppl 1:S87–S93.

Shad JA, Chinn CG, Brann OS. Acute hepatitis after ingestion of herbs. South Med J. 1999;92:1095–7.

Shah S. Acute renal failure. In: Carpenter G, Griggs R, Loscalzo J, editors. Cecil essentials of medicine. Philadelphia (PA): WB Saunders; 2001. p. 283–90.

Shaheen M, Broxmeyer HE. The humoral regulation of hematopoiesis. In: Hoffman R. Hematology, basic principles and practice. 4th ed. Philadelphia (PA): Elsevier Medical Publishing; 2005. p. 233–65.

Shaheen, S, Sterne JAC, Thompson RL, Songhurst CE, Margetts BM, Burney PGJ. Dietary antioxidants and asthma in adults. Population-based case-control study. Am J Respir Crit Care Med. 2001;164:1823–8.

Shanahan F. Probiotics in inflammatory bowel disease—therapeutic rationale and role. Adv Drug Deliv Rev. 2004;56:809–18.

Sharma N, Trope B, Lipman TO. Vitamin supplementation: what the gastroenterologist needs to know. J Clin Gastroenterol. 2004;38:844–54.

Sharma R, Anker SD. From tissue wasting to cachexia: changes in peripheral blood flow and skeletal musculature. Eur Heart J Supp. 2002;4:D12–D17.

Sharma S, Fumar A. Septic shock, multiple organ failure, and acute respiratory distress syndrome. Curr Opin Pulm Med. 2003;9:199–209.

Shaver MJ. Chronic renal failure. In: Carpenter G, Griggs R, Loscalzo J, editors. Cecil essentials of medicine. Philadelphia (PA): WB Saunders; 2001. p. 291–300.

Shaw S, Lieber CS. Nutrition and alcoholism. In: Goodhard RS, Shils ME, editors. Modern nutrition in health and disease. Philadelphia (PA): Lea and Febiger; 1980.

Sheashaa H, El-Husseini A, Sabry A, et al. Parenteral iron therapy in treatment of anemia in end-stage renal disease patients: a comparative study between iron saccharate and gluconate. Nephron Clin Pract. 2005;99:c97–101.

Sheehan LA, Macallan DC. Determinants of energy intake and energy expenditure in HIV and AIDS. Nutrition. 2000;16:101–6.

Sheikh AA, Sheikh KS, editors. Eastern and Western approaches to healing: ancient wisdom and modern knowledge. New York: Wiley; 1989.

Sheng HP. Body fluids and water balance. In: Stipanuk M, editor. Biochemical and physiological aspects of human nutrition. Philadelphia (PA): WB Saunders; 2000. p. 843–66.

Sheng HP. Sodium, chloride and potassium. In: Stipanuk M, editor. Biochemical and physiological aspects of human nutrition. Philadelphia (PA): WB Saunders; 2000. p. 686–70.

Sherlock S. Nutrition and the alcoholic. Lancet. 1984; 1(8374):436–9.

Sherman DIN, Williams R. Liver damage: mechanisms and management. Br Med Bull. 1994;50:124–38.

Sherry B, Mei Z, Grummer-Strawn L, Dietz WH. Evaluation of and recommendations for growth references for very low birth weight (< or = 1500 grams) infants in the United States. Pediatrics. 2003;111:750–8.

Sherwood L. Fluid and acid-base balance. In: Human physiology: from cells to systems. 5th ed. Belmont (CA): Brooks/Cole; 2004. p. 559–589.

Sherwood L. Human physiology: from cells to systems. 5th ed. Belmont (CA): Brooks/Cole; 2004.

Sherwood L. The blood vessels and blood pressure. In: Human physiology: from cells to systems. 5th ed. Belmont (CA): Brooks/Cole; 2004.

Sherwood L. The body defenses. In: Human physiology: from cells to systems. 5th ed. Belmont (CA): Brooks/Cole; 2004.

Sherwood L. The urinary system. In: Human physiology: from cells to systems. 5th ed. Belmont (CA): Brooks/Cole; 2004. p. 511–57.

Sherwood, L. The central nervous system. In: Human physiology: from cells to systems. 5th ed. Belmont (CA): Brooks/Cole; 2004.

Shevitz AH, Wilson IB, McDermott AY, Spiegelman D, Skinner SC, Antonsson K, et al. A comparison of the clinical and cost-effectiveness of 3 intervention strategies for AIDS wasting. J Acquir Immune Defic Syndr. 2005;38: 399–406.

Shi L. The convergence of vulnerable characteristics and health insurance in the U.S. Soc Sci Med. 2001;53:519.

Shi L, Singh DA. Delivering health care in America. Gaithersburg (MD): Aspen Publishers; 2001.

Shields M. Overweight Canadian children and adolescents. Ottawa: Statistics Canada; 2005.

Shier D, Butler J, Lewis R. Hole's human anatomy and physiology. 9th ed. St. Louis: McGraw Hill; 2002.

Shimon I, Almog S, Vered Z, Seligmann H, Shefi M, Peleg E, et al. Improved left ventricular function after thiamine supplementation in patients with congestive heart failure receiving long-term furosemide therapy. Am J Med. 1995;98:485–90.

Shiraishi T, Kawahara K, Yamamoto S, Maekawa T, Shirakusa T. Postoperative mangagement after esophagectomy: is TPN the standard of nutritional care? Int Surg. 2005;90:30–5.

Shols AM. Nutrition and respiratory disease. Clin Nutr. 2001;20(Suppl. 1):173–9.

Shols AM. Nutritional modulation as part of the integrated management of chronic obstructive pulmonary disease. Proc Nutr Soc. 2003;62:781–91.

Shols AM, Fredric EW, Soeters PB, Westerterp KB, Wouter EF. Resting energy expenditure in patients with chronic obstructive pulmonary disease. Am J Clin Nutr. 1991;54: 983–7.

Shopbell JM, Hopkins B, Shronts EP. Nutrition screening and assessment. In: Gottschlich MM, ed. The science and practice of nutrition support. Dubuque (IA): Kendall-Hunt; 2001.

Shuler CF. Inherited risks for susceptibility to dental caries. J Dent Educ. 2001;65:1038–45.

Sica DA. Hyponatremia and heart failure—pathophysiology and implications. Congest Heart Fail. 2005;11:274–7.

Sieb JP. Myasthenia gravis: emerging new therapy options. Curr Opin Pharmacol. 2005;5:303–7.

Sikland G. Medical nutrition therapy lowers serum cholesterol and saves medication costs in Medicare population with hypercholsterolemia. J Am Diet Assoc. 1998;98: 889–94.

Silva M, Skolnik PR, Gorbach SL, Spiegelman D, Wilson IB, Fernandez-DiFranco MG, et al. The effect of protease inhibitors on weight and body composition in HIV-infected patients. AIDS. 1998;12:1645–51.

Silventoinen K, Sans S, Tolonen H, Monterde D, Kuulasmaa K, Kesteloot K, et al. Trends in obesity and energy supply in the WHO MONICA Project. Int J Obes. 2004;28:710–8.

Simko MD, Conklin MT. Focusing on the effectiveness side of the cost effectiveness equation. J Am Diet Assoc. 1989; 89:485–7.

Simon JA, Hudes ES. Serum ascorbic acid and gallbladder disease prevalence among US adults: the third national health and nutrition examination survey (NHANES III). Arch Intern Med. 2000;160:931–6.

Simopoulos A. Omega-3 fatty acids in inflammation and autoimmune diseases. J Amer Coll Nutr. 2002;21:495–505.

Singh PN, Sabate J, Fraser GE. Does low meat consumption increase life expectancy in humans? Am J Clin Nutr. 2003; 78:526S–532S.

Singh VV, Toskes PP. Small bowel bacterial overgrowth: presentation, diagnosis, and treatment. Curr Gastroenterol Rep. 2003;5:365–72.

Singhi S, Ravishanker R, Singhi P, Nath R. Low plasma zinc and iron in pica. Indian J Pediatr. 2003;70:139–43.

Singleton CK, Martin PR. Molecular mechanisims of Thiamin utilization. Curr Mol Med. 2001;1:197–207.

Skipper A, Ratz N. Enteral nutrition. In: Skipper A, editor. Dietitian's handbook of enteral and parenteral nutrition. 2nd ed. Rockville (MD): Aspen Systems, Inc.; 1998.

Skipper A. Parenteral nutrition. In: Matarese L, Gottschlich M, editors. Contemporary nutrition support practice. Philadelphia (PA): WB Saunders; 2002. p. 714.

Skroubis G, Sakellaropoulos G, Pouggouras K, Mead N, Nikiforidis G, Kalfarentzos F. Comparison of nutritional deficiencies after roux-en-Y gastric bypass and after biliopancreatic diversion with roux-en-Y gastric bypass. Obes Surg. 2002;12:551–8.

Skurnick JH, Bogden JD, Baker H, Kemp FW, Sheffet A, Quattrone G, et al. Micronutrient profiles in HIV-1-infected heterosexual adults. J Acquir Immune Defic Syndr Hum Retrovirol. 1996;12:75–83.

Slain D, Amsden JR, Khakoo RA, Fisher MA, Lalka D, Hobbs GR. Effect of high-dose vitamin C on the steady-state pharmacokinetics of the protease inhibitor indinavir in healthy volunteers. Pharmacotherapy. 2005;25:165–70.

Slaviero KA, Read JA, Clarke SJ, Rivory LP. Baseline nutritional assessment in advanced cancer patients receiving palliative chemotherapy. Nutr Cancer. 2003;46:148–57.

Slehria S, Sharma P. Barrett's esophagus. Curr Opin Gastroenterol. 2003;19:387–93.

Sleivert G, Burke V, Palmer C, Walmsley A, Gerrard D, Haines S, et al. The effects of deer antler velvet extract or powder supplementation on aerobic power, erythropoiesis, and muscular strength and endurance characteristics. Int J Sport Nutr Exerc Metab. 2003;13:251–65.

Slone DS. Nutritional support of the critically ill and injured patient. Crit Care Clin. 2004;20:35–57.

Smedley BD, Stith AY, Nelson AR, eds. Unequal treatment: confronting racial and ethnic disparities in health care. Washington (DC): National Academy Press; 2002.

Smeitink JAM. Mitochondrial disorders: clinical presentation and diagnostic dilemmas. J Inherit Metab Dis. 2003; 26:199–207.

Smith C, Cowan C, Sensenig A, Catlin A. Health spending growth slows in 2003. Health Aff. 2005;24:185–94.

Smith GP. Controls of food intake. In: Shils ME, Shike M, Ross AC, Cabellero B, Cousins RJ, editors. Modern nutrition in health and disease. 10th ed. Philadelphia (PA): Lippincott Williams & Wilkins; 2006. p. 707–19.

Smith LG, Weissman IL, Heimfeld S. Clonal analysis of hematopoietic stem-cell differentiation in vivo. Proc Natl Acad Sci U S A. 1991;88:2788–92.

Smith LK, Weiss EL, Lehmkuhl LD. Brunnstrom's clinical kinesiology. 5th ed. Philadelphia (PA): FA Davis Company; 1996.

Smith PD, Mai UE. Immunopathophysiology of gastrointestinal disease in HIV infection. Gastroenterol Clin North Am. 1992;21:331–5.

Smith PD, Meng G, Salazar-Gonzalez JF, Shaw GM. Macrophage HIV-1 infection and the gastrointestinal reservoir. J Leukoc Biol. 2003;74:642–9.

Smith RE, Patrick S, Michael P, Hager H. Medical nutrition therapy: the core of ADA's advocacy efforts. J Am Diet Assoc. 2005;105:987–96.

Smith SC, Milani RV, Arnett DK, Crouse JR, McDermott MM, Ridker PM, et al. Atherosclerotic vascular disease conference: writing group II: risk factors. Circulation. 2004;109:2613–6.

Smith SM. Red blood cell and iron metabolism during space flight. Nutrition. 2002;18:864–6.

Smitherman TC, Reis, SE. Heart disease in women with diabetes. Diabetes Spectrum. 1997;10:207–15.

Smout AJP. Small intestinal motility. Curr Opin Gastroenterol. 2004;20:77–81.

Snijder MB, Dekker JM, Visser M, Yudkin JS, Stehouwer CD, et al. Larger thigh and hip circumferences are associated with better glucose tolerance: the Hoorn Study. Obes Res. 2003;11:104–11.

Snijder MB, Zimmet PZ, Visser M, Dekker JM, Seidell JC, et al. Independent and opposite associations of waist and hip circumferences with diabetes, hypertension and dyslipidemia: the AusDiab Study. Int J Obes Relat Metab Disord. 2004;28:402–9.

Snively CS, Gutierrez C. Chronic kidney disease: prevention and treatment of common complications. Am Fam Physician. 2004;70:1921–8.

Society of American Gastrointestinal Endoscopic Surgeons (SAGES). Guidelines for surgical treatment of gastroesophageal reflux disease (GERD) [monograph on the Internet]. Santa Monica (CA): Society of American Gastrointestinal Endoscopic Surgeons (SAGES); 2001 [cited 2004 Jan 4]. Available from: http://www.guideline.gov.

Soderling E. Nutrition, diet, and oral health in the 21st century. Int Dent J. 2001; 51(Suppl):389–91.

Sokol RJ. Fat-soluble vitamins and their importance in patients who have cholestatic liver disease. Gastroenterol Clin North Am. 1994;23:673–705.

Sokol RJ, Durie PR (for the Cystic Fibrosis Foundation Hepatobiliary Disease Concensus Group). Recommendations for management of liver and biliary tract disease in cystic fibrosis. J Pediatr Gastroenterol Nutr. 1999;28:S1–S13.

Sokol RJ, et al. Alcoholism. Clin Exp Research. 1980;4:135.

Sole MJ, Jeejeebhoy KN. Conditioned nutritional requirements and the pathogenesis and treatment of myocardial failure. Curr Opin Clin Nutr Metab Care. 2000;3:417–24.

Sole MJ, Jeejeebhoy KN. Conditioned nutritional requirements: therapeutic relevance to heart failure. Herz. 2002;27:174–8.

Solis JM, Marks G, Garcia M, Shelton D. Acculturation, access to care, and use of preventative services by Hispanics: findings from the HHANES 1982-84. Am J Public Health. 1990;80:11–9.

Soll AH. Medical treatment of peptic ulcer disease: practice guidelines. JAMA. 1996;275:622–9.

Sonis ST. Mucositis as a biological process: a new hypothesis for the development of chemotherapy-induced stomatotoxicity. Oral Oncol. 1998;34:39–43.

Sonnenberg A, Koch TR. Physician visits in the United States for constipation: 1958 to 1986. Dig Dis Sci. 1989;34: 606–11.

Sorensen JM. Herb-drug, food-drug, nutrient-drug, and drug-drug interactions: mechanisms involved and their medical implications. J Altern Complement Med. 2002;8: 293–308.

Sorensen TI. The genetics of obesity. Metabolism. 1995; 44(Suppl 3):4–6.

Sorkin M, Blake P. Apparatus for peritoneal dialysis. In: Daugirdas J, Blake P, Ing T, editors. Handbook of dialysis. 3rd ed. Philadelphia (PA): Lippincott Williams & Wilkins; 2001. p. 297–308.

Sospedra M, Martin R. Immunology of multiple sclerosis. Annu Rev Immuno. 2005;23:683–747.

Soucie JM, Thun MJ, Coates RJ, et al. Demographic and geographic variability of kidney stones in the US. Kidney Int. 1994;46:893–9.

Spanner ED, Duncan AM. Prevalence of dietary supplement use in adults with chronic renal insufficiency. J Ren Nutr. 2005;15:204–10.

Spapen H, Diltoer M, Van Malderen C, Opdenacker G, Suys E, Huyghens L. Soluble fiber reduces the incidence of diarrhea in septic patients receiving total enteral nutrition: a prospective, double-blind, randomized and controlled trial. Clin Nutr. 2001;20:301–5.

Spears K, Cheney C, Zerzan J. Low plasma retinol concentrations increase risk of developing bronchopulmonary dysplasia and long-term respiratory disability in very-low-birth-weight infants. Am J Clin Nutr. 2004;80:1589–94.

Spechler SJ. Clinical practice: Barrett's esophagus. N Engl J Med. 2002;346:836–43.

Specific guidelines for disease – adults. Pulmonary disease. JPEN J Parenter Enteral Nutr. 2002;26:SA63–SA64.

Specific guidelines for disease – adults. Solid organ transplantation. JPEN J Parenter Enteral Nutr. 2002;26:SA74–SA75.

Specific guidelines for disease – pediatrics. Pulmonary: bronchopulmonary dysplasia. JPEN J Parenter Enteral Nutr. 2002;26:118SA–19SA.

Spector RE. Cultural diversity in health and illness. 6th ed. New York: Prentice-Hall; 2003.

Spergel JM, Fiedler J. Food allergy and additives: triggers in asthma. Immunol Allergy Clin N Am. 2004;25:149–67.

Spinal Cord Injury Information Network. Facts and figures at a glance. 2005. Available from: http://www.spinalcord.uab.edu. Accessed November 18, 2005.

Spirig R, Moody K, Battegay M, DeGeest S. Symptom management in HIV/AIDS. Advancing the conceptualization. Adv Nurs Sci. 2005;28:333–4.

Spitz FR, Bouvet M, Fuhrman GM, Berger DH. Pancreatic adenocarcinoma. In Feig BW, Berger DH, Furhman GM, editors. The MD Anderson surgical oncology handbook. Philadelphia (PA): Lippincott Williams & Wilkins; 1999.

Spivak JI. Polycythemia vera and other myeloproliferative diseases. In: Kasper DL, Braunwald E, Fauci AS, Hausner SL, Longo DI, Jameson JL, editors. Harrison's principles of internal medicine. 16th ed. New York: McGraw Hill Medical Publishing; 2005. p. 626–31.

Splett P. Effectiveness and cost effectiveness of nutrition care: a critical analysis with recommendations. J Am Diet Assoc. 1991;91:S1–S50.

Squires RW. Pathophysiology and clinical features of cardiovascular diseases. In: Kaminsky LA, editor. American College of Sports Medicine's resource manual for guidelines for exercise testing and prescription. 5th ed. Baltimore (MD): Lippincott Williams & Wilkins; 2006. p.411–26

Sreedhar B. Conflicting evidence of iron and zinc interactions in humans: Does iron affect zinc absorption? Am J Clin Nutr. 2003;78:1226; author reply 1226–7.

Srigiridhar K, Nair KM. Supplementation with alpha-tocopherol or a combination of alpha-tocopherol and ascorbic acid protects the gastrointestinal tract of iron-deficient rats against iron-induced oxidative damage during iron repletion. Br J Nutr. 2000;84:165–73.

Stabler SP, Allen RH. Vitamin B12 deficiency as a worldwide problem. Annu Rev Nutr. 2004;24:299–326.

Stacy M. Pharmacotherapy for advanced Parkinson's disease. Pharmacotherapy. 2000;20:8S–16S.

Stafstrom CE, Bough KJ. The ketogenic diet for the treatment of epilepsy: a challenge for nutritional neuroscientists. Nutr Neurosci. 2003;6:67–79.

Stamler J, Appel L, Cooper R, Denton D, Dyer AR, Elliott P, et al. Dietary sodium chloride (salt), other dietary components and blood pressure: paradigm expansion, not paradigm shift. Acta Cardiol. 2000;55:73–8.

Stamler J. The INTERSALT Study: background, methods, findings, and implications. Am J Clin Nutr. 1997;65: 626S–642S.

Standing Committee on the Scientific Evaluation of Dietary Reference Intakes, Food and Nutrition Board, Institute of Medicine. Dietary reference intakes for calcium, phosphorus, magnesium, vitamin D, and fluoride. Washington (DC): National Academy Press; 1997.

Staveteig S, Wigton A. Racial and ethnic disparities: key findings from the National Survey of America's Families. The Urban Institute New Federalism Series B. 2000 Feb. No. B-5.

Steed HL, Capstick V, Flood C, Schepansky A, Schulz J, Mayes DC. A randomized controlled trial of early versus "traditional" postoperative oral intake after major abdominal gynecologic surgery. Am J Obstet Gynecol. 2002;186:861–5.

Steffen LM, Kroenke CH, Yu X, Periera MA, Slattery ML, Van Horn L, et al. Associations of plant food, dairy product, and meat intakes with 15-y incidence of elevated blood pressure in young black and white adults: the Coronary Artery Risk Development in Young Adults (CARDIA) Study. Am J Clin Nutr. 2005;82:1169–77.

Steinmann B, Gitzelmann R, Van den Berghe G. Disorders in fructose metabolism. In: Scriver C, Beaudet A, Sly W, Valle D, editors. The metabolic basis of inherited diseases. Vol. 1. New York: McGraw-Hill; 2001. p. 1489–520.

Steketee RW. Pregnancy, nutrition and parasitic diseases. J Nutr. 2003;133 Suppl 2:1661S–1667S.

Stephenson J. Combating cavities. JAMA. 2002;287:2937.

Stern RS, Juhn PI, Gertler PJ, Epstein AM. A comparison of length of stay and costs for health maintenance organization and fee-for-service patients. Arch Int Med. 1989;149: 1185–8.

Stettler N, Zemel BS, Kawchak DA, Ohene-Frempong K, Stallings VA. Iron status of children with sickle cell disease. JPEN J Parenter Enteral Nutr. 2001;25:36-8.

Stevens VJ, Obarzanek E, Cook NR, Lee IM, Appel LJ, Smith West D, et al. Trials for the Hypertension Prevention Research Group. Long-term weight loss and changes in blood pressure: results of the Trials of Hypertension Prevention, phase II. Ann Intern Med. 2001;134:1–11.

Stollman L. Nutrition Entrepreneur's Guide to Reimbursement Success. Chicago IL: American Dietetic Association; 1995.

Stollman NH, Raskin JB. Diagnosis and management of diverticular disease of the colon in adults. Ad Hoc Practice Parameters Committee of the American College of Gastroenterology. Am J Gastroenterol. 1999;94:3110–21.

Stone EG, Morton SC, Hulscher ME, Maglione MA, Roth EA, Grimshaw JM, et al. Interventions that increase use of adult immunization and cancer screening services: a meta-analysis Ann Int Med. 2002;36:641–51.

Stone J, Doube A, Dudson D, Wallace J. Inadequate calcium, folic acid, vitamin E, zinc, and selenium intake in rheumatoid arthritis patients: results of a dietary survey. Semin Arthritis Rheum. 1997;27:180–5.

Stratton IM, Adler AI, Neil AW, Matthews DR, Manley SF, Cull CA, et al. Association of glycaemia with macrovascular and microvascular complications of type 2 diabetes (UKPDS 35): prospective observational study. BMJ. 2000;321:405–12.

Stratton R, Bircher G, Fouque D, Stenvinkel P, DeMutssert R, Engfer M, et al. Multinutrient oral supplements and tube feeding in maintenance dialysis: a systematic review and meta-analysis. Am J Kid Dis. 2005;46:387–405.

Strawford A, Barbieri T, Van Loan M, Parks E, Catlin D, Barton N, et al. Resistance exercise and supraphysiologic androgen therapy in eugonadal men with HIV-related weight loss: a randomized controlled trial. JAMA. 1999;281:1282–90.

Streiff MB, Mehta S, Thomas DL. Peripheral blood count abnormalities among patients with hepatitis C in the United States. Hepatology. 2002;35:947–52.

Strom SS, Wang X, Pettaway CA, Logothetis CJ, Yamamura Y, Do KA, et al. Obesity, weight gain, and risk of biochemical failure among prostate cancer patients following prostatectomy. Clin Cancer Res. 2005;11:6889–94.

Stroud M, Duncan H, Nightingale J. Guidelines for enteral feeding in adult hospital patients. Gut. 2003;52(suppl 7):vii1–vii12.

Strube YN, Beard JL, Ross AC. Iron deficiency and marginal vitamin A deficiency affect growth, hematological indices and the regulation of iron metabolism genes in rats. J Nutr. 2002;132:3607–15.

Strunk BC, Cunningham PJ. Trends in Americans' access to needed medical care, 2001–2003. Track Report 10. 2004;(August):1–4.

Studen-Pavolvich D, Elliott MA. Eating disorders in women's oral health. Dent Clin North Am. 2001;45:491–511.

Stunkard AJ. Genetic contributions to human obesity. Res Publ Assoc Res Nerv Ment Dis. 1991;69:205–18.

Stunkard AJ, Allison KC. Two forms of disordered eating in obesity: binge eating and night eating. Int J Obes Relat Metab Disord. 2003;27:1–12.

Sugiyama M, Atomi Y. Risk factors for acute biliary pancreatitis. Gastrointest Endosc. 2004;60:210–2.

Sullivan DH, Nelson CL, Klimberg VS, Bopp MM. Nightly enteral nutrition support of elderly hip fracture patients: a pilot study. J Am Coll Nutr. 2004;23:683–91.

Sullivan DH, Roberson PK, Bopp MM. Hypoalbuminemia 3 months after hospital discharge:significance for long-term survival. J Am Geriatr Soc. 2005;53:1222–6.

Sullivan SS, Anderson EJ, Best S, et al. The effect of diet on hypercholesterolemia in renal transplant recipients. J Renal Nutr. 1996;6:141–51.

Summar M. Current strategies for the management of neonatal urea cycle disorders. J Pediatr. 2001;138:S30–S39.

Summar M, Tuckman M. Proceedings of a consensus conference for the management of patients with urea cycle disorders. J Pediatr. 2001;138:S6–S10.

Sumner AE, Chin MM, Abrahm JL, Berry GT, Gracely EJ, Allen RH, et al. Homocysteine levels show high prevalence of Vitamin B 12 deficiency after gastric surgery. Ann Surg. 1996;124:469–76.

Sundaram A, Koutkia P, Apovian CM. Nutritional management of short bowels syndrome in adults. J Clin Gastroenterol. 2002;34:207–20.

Sunder-Plassmann G, Hurl WH. Erythropoietin and iron. Clin Nephrol. 1997;47:141–57.

Sungurtekin H, Sungurtekin U, Hanci V, Erdem E. Comparison of two nutrition assessment techniques in hospitalized patients. Nutrition. 2004;20:428–32.

Suttie JW. Vitamin K. In: Shils ME, Shike M, Ross CA, Caballero BC, Cousins RJ, editors. Modern nutrition in health and disease. 10th ed. Philadelphia (PA): Lippincott Williams and Wilkins; 2005. p. 412–25.

Suttie JW. Vitamin K. In: Stipanauk MH, editor. Biochemical and physiological aspects of human nutrition. Philadelphia (PA): WB Saunders; 2000.

Svensson S, Some M, Lundsjo A, Helander A, Cronholm T, Hoog JO. Activities of human alcohol dehydrogenases. In: The metabolic pathways of ethanol and serotonin. Eur J Biochem. 1999;262:324–9.

Svetkey LP, Harsha DW, Vollmer WM, Stevens VJ, Obarzanke E, Elmer PJ, et al. Premier: a clinical trial of comprehensive lifestyle modification for blood pressure control: rationale, design and baseline characteristics. Ann Epidemiol. 2003;13:462–71.

Svetkey LP, Simons-Morton DG, Proschan MA, Sacks FM, Conlin PR, Harsha D, et al. Effect of the dietary approaches to stop hypertension diet and reduced sodium intake on blood pressure control. J Clin Hypertens. 2004; 6:373–81.

Swank RL, Dugan BB. Effect of low saturated fat diet in early and late cases of multiple sclerosis. Lancet. 1990;336:37–9.

Swann IL, Kendra JR. Severe iron-deficiency anemia and stroke. Clin Lab Haematol. 2000;22:221–3.

Swatland HG. Infrared fiber optics spectrophotometry of meat. J Anim Sci. 1983;56:1329.

Tabet N, Birks J, Grimley Evans J, Orrel M, Spector A. Vitamin E for Alzheimer's disease. Cochrane Database Syst Rev; 2003: CD002854.

Tahsin M. Trace elements and vitamins in renal disease. In: Mitch W, Klahr S. Handbook of nutrition and the kidney. 4th ed. Philadelphia (PA): Lippincott Williams & Wilkins; 2002. p. 233–52.

Take S, Mizuno M, Ishiki K, Nagahara Y, Yoshida T, Yokota K, Oguma K, Okada H, Shiratori Y. The effect of eradicating helicobacter pylori on the development of gastric cancer in patients with peptic ulcer disease. Am J Gastroenterol. 2005;100:1037–42.

Talley NJ. New therapeutic insights into irritable bowel syndrome. 2001. Available from: http://www.medscape.com/viewarticle/418546.

Talpers SS, Romberger DJ, Bunce SB, Pingleton SK. Nutritionally associated increased carbon dioxide production. Excess total calories vs high proportion of carbohydrate calories. Chest. 1992;102:551–5.

Tam M, Gomez S, Gonzalez-Gross M, Marcos A. Possible roles of magnesium on the immune system. Eur J Clin Nutr. 2003;57:1193–7.

Tamimi RM, Lagiou P, Adami HO, Trichopoulous D. Prospects for chemoprevention of cancer. J Int Med. 2002;251:286–300.

Tan EK, Cheah SY, Fook-Chong S, Yew K, Chandran VR, Lum SY, et al. Functional COMT variant predicts response to high dose pyridoxine in Parkinson's disease. Am J Med Genet B Neuropsychiatr Genet. 2005;137B(1):1–4.

Tang AM, Forrester J, Spiegelman D, Knox TA, Tchetgen E, Gorbach SL. Weight loss and survival in HIV-positive patients in the era of highly active antiretroviral therapy. J Acquir Immune Defic Syndr. 2002;31:230–6.

Tang AM, Jacobson DL, Spiegelman D, Knox TA, Wanke C. Increasing risk of 5% or greater unintentional weight loss in a cohort of HIV-infected patients, 1995–2003. J Acquir Immune Defic Syndr. 2005;40:70–6.

Tang AM, Smit E. Selected vitamins in HIV infection: a review. AIDS Patient Care STDS. 1998;12:249–50.

Tang AM. Weight loss, wasting and survival in HIV-positive patients: current strategies. AIDS Read. 2003;13(12 Suppl):S23–S27.

Tangalos EG. Congestive heart failure. Nutrition management for older adults. Washington (DC): Nutrition Screening Initiative (NSI); 2002.

Tanner EM, Finn-Stevenson M. Nutrition and brain development social policy implications. Am J Orthopsychiatry. 2002;72:182–93.

Tanofsky-Kraff M, Yanovski SZ. Eating disorder or disordered eating? Non-normative eating patterns in obese individuals. Obes Res. 2004;12:1361–6.

Tapsell LC, Brenninger V, Barnard J. Applying conversation analysis to foster accurate reporting in the diet history interview. J Am Diet Assoc. 2000;100:818–24.

Tarng D, Huang T, Chen TW, Yang W. Erythropoietin hyporesponsiveness: from iron deficiency to iron overload. Kidney Int. 1999;55:S107–S118.

Tayek JA. A review of cancer cachexia and abnormal glucose metabolism in humans with cancer. J Am Coll Nutr. 1992; 11:445–56.

Tayek JA. Nutritional and biochemical aspects of the cancer patient. In: Heber D, Blackburn GL, editors. Nutritional oncology. San Diego: Academic Press; 1999.

Teasdale G, Jennett B. Assessment and prognosis of coma after head injury. Acta Neurochir. 1976;34:45–55.

Teasdale G, Jennett B. Assessment of coma and impaired consciousness. Lancet. 1974;2(7872):81–4.

Tein I. Carnitine transport: pathophysiology and metabolism of known molecular defects. J Inherit Metab Dis. 2003;26:147–69.

Temme EH, Van Hoydonck PG. Tea consumption and iron status. Eur J Clin Nutr. 2002;56:379–86.

Terlouw DJ, Desai MR, Wannemuehler KA, et al. Relation between the response to iron supplementation and sickle cell hemoglobin phenotype in preschool children in western Kenya. Am J Clin Nutr. 2004;79:466–72.

Tesch BJ. Herbs commonly used by women: an evidence-based review. Am J Obstet Gynecol. 2003;188:S44–S55.

Thadhani R, Pascual M, Bonventre J. Acute renal failure. N Engl J Med. 1996;334:1448–60.

Theodorou SJ, Theodorou DJ, Sartoris DJ. Evaluation of osteoporosis in orthopedic practice: a review of current diagnostic modalities. Am J Orthop. 2003;32:178–88.

Thien FK, Woods, R, De Luca S, Abramson MJ. Dietary marine fatty acids (fish oil) for asthma in adults and children. The Cochrane Database of Systematic Reviews 2002, Issue 2, Art. No.: CD001283.

Thom T, Haase N, Rosamond W, Howard VJ, Rumsfeld JR, Manolio T, et al. Heart disease and stroke statistics—2006 update; A report from the American Heart Association Statistics Committee and Stroke Statistics Subcommittee. Circulation. Published online Jan 11, 2006. Available from: http://www.circ.ahajournals.org; Accessed February 24, 2006.

Thomas AM, Gutierrez YM. Maternal and fetal complications associated with gestational diabetes mellitus. In: Thomas AM, Gutierrez YM, editors. American Dietetic Association guide to gestational diabetes mellitus. Chicago (IL): American Dietetic Association; 2005.

Thomas AM. Classification, screening, and diagnosis. In: Thomas AM, Gutierrez YM, editors. American Dietetic Association guide to gestational diabetes mellitus. Chicago: American Dietetic Association; 2005.

Thomas AM. Pathophysiology of gestational diabetes mellitus. In: Thomas AM, Gutierrez YM, editors. American Dietetic Association guide to gestational diabetes mellitus. Chicago (IL): American Dietetic Association; 2005.

Thomas DR, Ashmen W, Morley JE, Evans WJ. Nutritional management in long-term care: development of a clinical guideline. Council for Nutritional Strategies in Long-Term Care. J Gerontol A Biol Sci Med Sci. 2000;55:M725–34.

Thomas DR, Zdrowski CD, Wilson M-M, Conright KC, Lewis C, Tariq S, et al. Malnutrition in subacute care. Am J Clin Nutr. 2002;75:308–13.

Thomas-Geevarghese A, Raghavan S, Minolfo R, Holleran S, Ramakrishnan R, Ormsby B, et al. Postprandial response to a physiologic caloric load in HIV-positive patients receiving protease inhibitor-based or nonnucleoside reverse transcriptase inhibitor-based antiretroviral therapy. Am J Clin Nutr. 2005;82:146–54.

Thompson C, Fuhrman MP. Nutrients and wound healing: still searching for the magic bullet. Nutr Clin Prac. 2005; 331–47.

Thompson C. Initiation, advancement and transition of enteral feedings. In: Charney P, Malone A, editors. ADA pocket guide to enteral nutrition. Chicago (IL): American Dietetic Association; 2006.

Thompson PD, Buchner D, Pina IL, Balady GJ, Williams MA, Marcus BH, et al. American Heart Association Council on Clinical Cardiology Subcommittee on Exercise, Rehabilitation, and Prevention; American Heart Association Council on Nutrition, Physical Activity, and Metabolism Subcommittee on Physical Activity. Exercise and physical activity in the prevention and treatment of atherosclerotic cardiovascular disease: a statement from the Council on Clinical Cardiology (Subcommittee on Exercise, Rehabilitation, and Prevention) and the Council on Nutrition, Physical Activity, and Metabolism (Subcommittee on Physical Activity). Circulation. 2003;107:3109–16.

Thompson T. Oats and the gluten free diet. J Am Diet Assoc. 2003;103:376–9.

Thompson WG, Longstreth G, Drossman DA, Heaton K, Irvine EJ, Muller-Lissner S. Functional bowel disorders. In: Drossman DA, Corazziari E, Talley NJ, Thompson WG, Whitehead WE, editors. Rome II. The functional gastrointestinal disorders. Diagnosis, pathophysiology and treatment: a multinational consensus. 2nd ed. McLean, VA: Degnon Associates; 2000. p. 382–91.

Thompson-Chagoyán OC, Maldonado J, Gil A. Aetiology of inflammatory bowel disease (IBD): role of intestinal microbiota and gut-associated lymphoid tissue immune response. Clin Nutr. 2005;24:339–52.

Thoni GJ, Fedou C, Brun JF, Fabre J, Renard E, Reynes J, et al. Reduction of fat accumulation and lipid disorders by individualized light aerobic training in human immunodeficiency virus infected patients with lipodystrophy and/or dyslipidemia. Diabetes Metab. 2002;28:397–404.

Thornton J, Symes C, Heaton K. Moderate alcohol intake reduces bile cholesterol saturation and raises HDL cholesterol. Lancet 1983;2:819–822.

Thorsdottir I, Gunnarsdottir I. Energy intake must be increased among recently hospitalized patients with chronic obstructive pulmonary disease to improve nutritional status. J Am Diet Assoc. 2002;102:247–9.

Tintinalli JE, Kelen GD, Stapczynski JS, Ma OJ, Cline DM. Tintinalli's emergency medicine: a comprehensive study guide. 6th ed. St. Louis: McGraw Hill; 2004.

Tiosano D, Hochberg Z. Endocrine complications of thalassemia. J Endocrinol Invest. 2001;24:716–23.

Tirtha SS. The Ayurveda encyclopedia: natural secrets to healing, prevention, and longevity. Bayville (NY): The Ayurveda Holistic Center Press; 1998.

Tisdale MJ. Metabolic abnormalities in cachexia and anorexia. Nutrition. 2000;16, 1013–4.

Tjepkema M. Adult obesity in Canada, measured height and weight. Ottawa: Statistics Canada; 2005.

Tondeur MC, Schauer CS, Christofides AL, et al. Determination of iron absorption from intrinsically labeled microencapsulated ferrous fumarate (sprinkles) in infants with different iron and hematological status by using a dual-stable-isotope method. Am J Clin Nutr. 2004;80: 1436–44.

Topaloglu AK, Hallioglu O, Canim A, Duzovali O, Yilgor E. Lack of association between plasma leptin levels and appetite in children with iron deficiency. Nutrition. 2001;17: 657–9.

Treem WR. Emerging concepts in celiac disease. Curr Opin Pediatr. 2004;16:552–9.

Topol EJ. Arthritis medicines and cardiovascular events—house of coxibs. JAMA. 2005;293:366–8.

Topping DL, Clifton PM. Short-chain fatty acids and human colonic function: roles of resistant starch and nonstarch polysaccharides. Physiol Rev. 2001;81:1031–64.

Toth MJ, Fishman PS, Poehlman ET. Free-living daily energy expenditure in patients with Parkinson's disease. Neurology. 1997;48:88–91.

Touger-Decker R, van Loveran C. Sugars and dental caries. Am J Clin Nutr. 2003;78:S881.

Tousoulis D, Charakida M, Stefanadis C. Inflammation and endothelial dysfunction as therapeutic targets in patients with heart failure. Int J Cardiol. 2005;100:347–53.

Towheed TE, Anastassiades TP. Glucosamine and chondroitin for treating symptoms of osteoarthritis, evidence is widely touted by incomplete. JAMA. 2000;283:1483–4.

Trabulsi J, Troiano RP, Subar AF, Sharbaugh C, Kipnis V, Schatzkin A, et al. Precision of the doubly labeled water method in a large-scale application: evaluation of a streamlining-dosing protocol in the Observing Protein and Energy Nutrition (OPEN) study. Eur J Clin Nutr. 2003;57:1370–7.

Travers P, Walport M, Capra D, Janeway C. Immunobiology, the immune system in health and disease. New York: Garland Publishing Co.; 2004.

Trichopoulou A, Costacou T, Bamia C, Trichopoulos D. Adherence to a Mediterranean diet and survival in a Greek population. N Engl J Med. 2003;348:2599–608.

Truelove SC, Witts LJ. Cortisone in ulcerative colitis: final report on a therapeutic trial. Br Med J. 1955; 2:1041–8.

Tsalis G, Nikolaidis MG, Mougios V. Effects of iron intake through food or supplement on iron status and performance of healthy adolescent swimmers during a training season. Int J Sports Med. 2004;25:306–13.

Tseng M, Everhart JE, Sandler RS. Dietary intake and gallbladder disease: a review. Public Health Nutr. 1999;2: 161–72.

Tucker JM, Townsend DM. Alpha-tocopherol: roles in prevention and therapy of human disease. Biomed Pharmacother. 2005;59:380–7.

Tucker KL, Hannan MT, Chen H, Cupples LA, Wilson PWF, Kiel DP. Potassium, magnesium, and fruit and vegetable intakes are associated with greater bone mineral density in elderly men and women. Am J Clin Nutr. 1999;69: 727–36.

Tucker KL, Olson B, Bakun P, Dallal GE, Selhub J, Rosenberg IH. Breakfast cereal fortified with folic acid, vitamin B-6, and vitamin B-12 increases vitamin concentrations and reduces homocysteine concentrations: a randomized trial. Am J Clin Nutr. 2004;79:805–11.

Tug T, Karatas F, Terzi SM. Antioxidant vitamins (A, C and E) and malondialdehyde levels in acute exacerbation and stable periods of patients with chronic obstructive pulmonary disease. Clin Invest Med. 2004;27:23–8.

Tung J, Hadzic N, Layton M, et al. Bone marrow failure in children with acute liver failure. J Pediatr Gastroenterol Nutr. 2000;31:557–61.

Tungtrongchitr R, Pongpaew P, Soonthornruengyot M, et al. Relationship of tobacco smoking with serum vitamin B12, folic acid and hematological indices in healthy adults. Public Health Nutr. 2003;6:675–81.

Turner RB, Bauer R, Woelkart K, Hulsey TC, Gangemi JD. An evaluation of Echinacea angustifolia in experimental rhinovirus infections. N Engl J Med. 2005;353:341–8.

U.S. Census. 2003. Children with Health Insurance: 2001 edited by U. S. C. Bureau: U.S. Department of Commerce.

U.S. Census. 2004 Health Insurance (Table HI05). Washington (DC): U.S. Census Bureau; 2004.

U.S. Census. 2005. Historical health insurance tables. Edited by U.S.C. Bureau. Available from: http://www.census.gov/hhes/hlthins/historic/index.html.

U.S. Census. Children with health insurance, edited by U.S.C. Bureau. U.S. Department of Commerce; 2001.

U.S. Department of Agriculture and U.S. Department of Health and Human Services. Nutrition and your health: dietary guidelines for Americans. 7th ed. Hyattsville (MD): U.S. Department of Agriculture. Center for Nutrition Policy and Promotion; 2005.

U.S. Department of Agriculture. Dietary Guidelines for Americans 2005: Chapter 7 carbohydrates [monograph on the Internet]. United States Department of Agriculture; 2005 [cited 2005 Oct 11]. Available from: http://www.health.gov/dietaryguidelines/dga2005/document/html/chapter7.htm

U.S. Department of Agriculture. Food composition and nutrient database. Available from: http://www.nal.usda.gov/fnic/foodcomp.

U.S. Department of Agriculture. USDA Nutrient Database for Standard Reference, Release 12. Washington (DC): U.S. Department of Agriculture, Agricultural Research Service; 1998.

U.S. Department of Health and Human Services and U.S. Department of Agriculture. Dietary guidelines for Americans, 2005. 6th ed. Washington (DC): U.S. Government Printing Office; 2005.

U.S. Department of Health and Human Services and U.S. Department of Agriculture. Dietary guidelines for Americans 2005 [monograph on the Internet]. Washington (DC): U.S. Department of Health and Human Services/U.S. Department of Agriculture; 2005 [cited 2006 March 20]. Available from: http://www.health.gov/dietaryguidelines/dga2005/document/.

U.S. Department of Health and Human Services, National Institutes of Health, National Heart Lung Blood Institute. Seventh Report of the Joint National Committee on Prevention, Detection, and Treatment of High Blood Pressure. (JNC 7). NIH Publication 03–5231; 2003. Available from: http://www.nhlbi.nih.gov/guidelines/hypertension/jncintro.htm. Accessed 1 June 2006.

U.S. Department of Health and Human Services, Public Health Service, Centers for Disease Control and Prevention, National Center for Chronic Disease Prevention and Health Promotion, Division of Nutrition and Physical Activity. Promoting physical activity:a guide for community action. Champaign (IL): Human Kinetics; 1999.

U.S. Department of Health and Human Services. Bone health and osteoporosis: a report of the Surgeon General. Rockville (MD): U.S. Department of Health and Human Services, Office of the Surgeon General; 2004.

U.S. Department of Health and Human Services. Healthy People 2010. 2nd ed. Washington (DC): U.S. Government Printing Office; 2000.

U.S. Department of Health and Human Services. Healthy People 2010 (2 vols). Washington (DC): U.S. Dept. of Health and Human Services; 2000.

U.S. Department of Health and Human Services. Prevalence of disabilities and associated health conditions among adults—United States, 1999. MMWR Morb Mortal Wkly Rep. 2001;50:120–5.

U.S. Department of Health and Human Services. Public Health Service National Institutes of Health National Heart, Lung, and Blood Institute Third Report on the National Cholesterol Education Program Adult Treatment Panel III Guidelines NIH Publication No. 01–3305; 2001.

U.S. Department of Health and Human Services. The health consequences of smoking: a report of the Surgeon General. Rockville (MD): U.S. Department of Health and Human Services, Office of the Surgeon General; 2004.

U.S. Department of Health and Human Services. The Surgeon General's call to action to prevent and decrease overweight and obesity. Rockville (MD): U.S. Department of Health and Human Services, Public Health Service, Office of the Surgeon General; 2001.

U.S. Department of Health and Human Services. USDA dietary guidelines for Americans. 6th ed. 2005. Available from: http://www.health.gov/dietaryguidelines. Accessed July 27, 2005.

U.S. Department of Labor, Bureau of Labor Statistics, Occupational Outlook Handbook. http://www.bls.gov/oco; 2004–2005.

U.S. Food and Drug Administration [homepage on the Internet]. Rockville (MD): U.S. Food and Drug Administration; 2005 [cited 2005 May 19]. FDA News. FDA approves first DNA-based test to detect cystic fibrosis. Available from: http://www.fda.gov/bbs/topics/NEWS/2005/NEW 01178.html.

U.S. Food and Drug Administration [homepage on the Internet]. Rockville (MD): U.S. Food and Drug Administration; 2005 [cited 2006 Jan 26]. FDA Public Health Advisory. Reports of blue discoloration and death in patients receiving enteral feedings tinted with the dye, FD&C Blue NO. 1. U.S. Food and Drug Administration/Center for Food Safety and Applied Nutrition. Available from: http://www.cfsan.fda.gov/~dms/col ltr2.html.

U.S. Food and Drug Administration [homepage on the Internet]. Rockville (MD): U.S. Food and Drug Administration; 2005 [cited 2006 January 23]. Questions and answers on exocrine pancreatic insufficiency drug products. Available from: http://www.fda.gov/cder/drug/infopage/pancreatic_drugs/pancreatic_QA.htm.

U.S. Food and Drug Administration. 1997. Food labeling: health claims; oats and coronary heart disease. Fed. Register 62:3584–3601.

U.S. Food and Drug Administration. 1998. Food labeling: health claims; soluble fiber from certain foods and coronary heart disease. Fed. Register 63:8103–21.

U.S. Food and Drug Administration. 1999. Food labeling: health claims; soy protein, and coronary heart disease. Fed. Register 64:57700–57733.

U.S. Food and Drug Administration. 2000. Food labeling: health claims; plant sterol/stanol esters and coronary heart disease. Interim final rule. Fed. Register 65:54686–54739.

U.S. Food and Drug Administration. Center for Food Safety and Applied Nutrition. A Food Labeling Guide, 2005. Available from: http://www.cfsan.fda.gov/~dms/flg 6c.html. Accessed 1 March 2006.

U.S. Food and Drug Administration. Dietary supplement enforcement Report. 2002. Available from: http://www .fda.gov/oc/nutritioninitiative/report.html.

U.S. Food and Drug Administration. FDA authorizes new coronary heart disease health claim for plant sterol and plant stanol esters. FDA Talk Paper. September 5, 2000. Available at: http://www.cfsan.fda.gov/~lrd/tpsterol.html .Accessed March1, 2006.

U.S. Food and Drug Administration. Trans fatty acids in nutrition labeling; consumer research to consider nutrient content and health claims and possible footnote or disclosure statements; final rule and proposed rule. July 11, 2003. Available from: http://www.cfsan.fda.gov/~lrd/fr03711a.html. Accessed 1 March 2006.

U.S. Food and Drug Administration. United States Department of Agriculture and U.S. Department of Health and Human Services, Dietary Guidelines for Americans 2005. 6th ed. Home and Garden Bulletin no. 232. Washington (DC); 2005). Available at: http://www.usda.gov/cnpp/DG2005/index.html. Accessed March 1, 2006.

U.S. National MS Society. 2005. Comparing the disease-modifying drugs. Available from: http://www.national mssociety.org/Brochures-Comparing.asp. Accessed November 11, 2005.

U.S. Preventive Services Task Force. Screening for presence of deficiency, toxicity, and disease. Nutr Clin Care. 2003; 6:120–2.

U.S. Renal Data System. USRDS 2003 annual data report: atlas of end-stage renal disease in the United States. Bethesda (MD): National Institutes of Health, National Institute of Diabetes and Digestive and Kidney Diseases; 2003.

Uauy R, Hertrampf E, Dangour AD. Food-based dietary guidelines for healthier populations: international considerations. In: Shils ME, Shike M, Ross CA, Caballero BC, Cousins RJ, editors. Modern nutrition in health and disease. 10th ed. Philadelphia (PA): Lippincott Williams & Wilkins; 2005. p. 1701–15.

Ubodi K, Childs E. Acute respiratory distress syndrome. Am Fam Physician. 2003;67:315–22.

UK Prospective Diabetes Study (UKPDS) Group. Intensive blood-glucose control with sulphonylureas or insulin compared with conventional treatment and risk of complications in patients with type 2 diabetes (UKPDS 33). Lancet 1998;352:837–53.

Ullrich D. Hygiene hypothesis: are we too "clean" for our own good? [monograph on the Internet]. Milwaukee: Medical College of Wisconsin; 2004 [cited 2006 Jan 26]. Available from: http://healthlink.mcw.edu/article/1031002421.html.

Umpierrez GE, Murphy MB, Kitabchi AE. Diabetic ketoacidosis and hyperglycemic hyperosmolar syndrome. Diabetes Spectrum. 2002;15:28–36.

UNAIDS. AIDS Epidemic Update December 2005. Available from: http://www.unaids.org/Epi2005/doc/report.html. Accessed November 21, 2005.

Unger C, Haring B, Kruse A, et al. Double-blind randomised placebocontrolled phase III study of an E. coli extract plus 5-fluorouracil versus 5-fluorouracil in patients with advanced colorectal cancer. Arzneimittelforschung. 2001;51: 332–8.

United Network for Organ Sharing [homepage on the Internet]. Richmond (VA): United Network for Organ Sharing; 2006 [cited 2005 Jun 30]. Organ Procurement and Transplantation Network. Available from: http://www .unos.org/data/default.asp?displayType=usData.

United Network of Organ Sharing. Available from: http://www.unos.org/. Accessed June 6, 2006.

University of North Carolina. Verne S. Caviness General Clinical Research Center. Low tyramine diet for use with monoamine oxidase inhibitors. Available from: http://gcrc.med.unc.edu. Accessed November 5, 2005.

University Southern California, University Hospital; 2003. Available from: http://www.uscuh.com/CWSContent/uscuh/ourServices/medicalServices/USCUniversityHospitalPelvicFloorDisordersProgram/Patient+Education.htm.

Urbancsek H, Kazar T, Mezes I, et al. Results of a double-blind, randomized study to evaluate the efficacy and safety of Antibiophilus in patients with radiation-induced diarrhoea. Eur J Gastroenterol Hepatol. 2001;13:391–6.

Vaezi MF, Richter JE. Diagnosis and management of achalasia. J Gastroenterol. 1999;94:3406–12.

Vaezi MF, Richter JE, Wilcox CM, et al. Botulinum toxin versus pneumatic dilatation in the treatment of achalasia: a randomised trial. Gut. 1999;44:231–9.

Valtin H, Schafer J. Components of renal function. In: Valtin H, Schafer J, editors. Renal function. 3rd ed. Boston: Little, Brown; 1995.

Van Biesen W, Lameire N. General aspects of physiology and pathophysiology of metabolic acidosis in the critically ill. Acta Clin Belg. 2000;55:133–40.

Van Bokhorst-de van der Schuer, Van Leeuwen PA, Kulk DJ, Klop WM, Sauerwein HP, Snow GB, et al. The impact of nutritional status on the prognoses of patients with advanced head and neck cancer. Cancer 1999:86:519–27.

Van Camp G, Flamez A, Cosyns B, Weytiens C, Mutldermans L, Van Zandijcke M, et al. Treatment of Parkinson's disease with pergolide and relation to restrictive valvular heart disease. Lancet. 2004;363(9416):1179–83.

Van Den Berg H, Van Der Gaag M, Hendricks H. Influence of lifestyle on vitamin bioavailability. Int J Vitamin Nutrition Resource. 2001;72:53–9.

Van den Berghe G, Wouters P, Weekers F, Verwaest M, Brucynincick F, Schetz M, et al. Intensive insulin therapy protects the endothelium of critically ill patients. J Clin Invest. 2005;115:2277–86.

van der Dijs FP, Fokkema MR, Dijck-Brouwer DA, et al. Optimization of folic acid, vitamin B(12), and vitamin B(6) supplements in pediatric patients with sickle cell disease. Am J Hematol. 2002;69:239–46.

Van der Hulst RR, van Kreel BJ, von Meyenfeldt MF, et al. Glutamine and the preservation of gut integrity. Lancet. 1993;341:1363–5.

van der Meche FG, Schmitz PI. A randomized trial comparing intravenous immune globulin and plasma exchange in Guillain-Barre syndrome. Dutch Guillain-Barre Study Group. N Engl J Med. 1992;326:1123–9.

van der Meche FG, Visser LH, Jacobs BC, Endtz HP, Meulstee J, van Doorn PA. Guillain-Barre syndrome: multifactorial mechanisms versus defined subgroups. J Infect Dis. 1997;176 Suppl 2:S99–102.

van der Sande MA, Schim van der Loeff MF, Aveika AA, Sabally S, Togun T, Sarge-Nije R, et al. Body mass index at time of HIV diagnosis: a strong and independent predictor of survival. J Acquir Immune Defic Sydr. 2004;37: 1288–94.

van Egmond AW, Kosorod MR, Koscik R, Laxova A, Farrell PM. Effect of linoleic acid intake on growth of infants with cystic fibrosis. Am J Clin Nutr. 1996;63:746–52.

Van Hasselt P, Gashe BA, Ahmad J. Colloidal silver as an antimicrobial agent: fact or fiction? J Wound Care. 2004;13: 154–5.

Vander Top EA, Wyatt TA, Gentry-Nielsen MJ. Smoke exposure exacerbates an ethanol-induced defect in mucociliary clearance of streptococcus pneumoniae. Alcohol Clin Exp Res. 2005;29:882–7.

Vandereycken W. The place of inpatient care in the treatment of anorexia nervosa: questions to be answered. Int J Eat Disord. 2003;34:409–22.

Vanherweghem JL. A new form of nephropathy secondary to the absorption of Chinese herbs. Bulletin et Memoires de l'Academie Royale de Medicine de Belgique. 1994;149: 128–40.

VanItallie TB. Ancel Keys: a tribute. Nutr Metab. 2005;2:4.

Vaporciyan AA, Swisher SG. Esophageal cancer. In: Feig BW, Berger DH, Furhman GM, editors. The MD Anderson surgical oncology handbook. Philadelphia (PA): Lippincott Williams & Wilkins; 1999.

Varela-Moreiras G. Nutritional regulation of homocysteine: effects of drugs. Biomedicine and Pharmacotherapy. 2001;55:448–53.

Varraso, R, Siroux V, Maccario J, Pin I, Kauffmann F. Asthma severity is associated with body mass index and early menarche in women. Am J Resp Crit Care Med. 2005;171: 334–9.

Vasquez Martinez JL, Martinez-Romillo PD, Diez Sebastian J, Ruza Tarrio F. Predicted vs. measured energy expenditure by continuous, online indirect calorimetry in ventilated, critically ill children during the postinjury period. Pediatr Crit Care Med. 2004;5:96–7.

Vella M, Galloway DJ. Laparoscopic adjustable gastric banding for severe obesity. Obes Surg. 2003;13:642–8.

Venuta F, Quattrucci S, Rendina EA, De Giacomo T, Mercadante E, Moretti M, et al. Improved results with lung transplantation for cystic fibrosis: a 6-year experience. Interactive Cardiovascular and Thoracic Surgery 2001 3:21–24. [cited 2006 January 29]. Available from: http://www.icvts.org.

Veterans Health Administration, Department of Veterans Affairs. The pharmacologic management of chronic heart failure. Washington (DC): Veterans Health Administration, Department of Veterans Affairs; 2003.

Vila J, Ruiz J, Gallardo F, Vargas M, Soler L, Figueras MJ, et al. Aeromonas spp. and traveler's diarrhea: clinical features and antimicrobial resistance. Emerg Infect Dis [serial online]. 2003 May. Available from: http://www.cdc.gov/ncidod/EID/vol9no5/02-0451.htm.

Vinik AI, Maser RE, Mitchell BD, Freeman R. Diabetic autonomic neuropathy (technical review). Diabetes Care. 2003;26:1552–79.

Virdi J, Sivakami S, Shahani S, Suthar AC, Banavalikar MM, Biyani MK. Antihyperglycemic effects of three extracts from Momordica charantia. Journal of Ethnopharmacology. 2003;88:107–11.

Visocan, BJ. Nutritional management of alcoholism. J Am Diet Assoc. 1983;83:693–6.

Vitarius JA. The metabolic syndrome and cardiovascular disease. Mount Sinai J Med. 2005;72:257–62.

Volker D, Fitzgerald P, Major G, Garg M. Efficacy of fish oil concentrate in the treatment of rheumatoid arthritis. J Rheumatol. 2000;27:2343–6.

Vollmer WM, Sacks FM, Ard J, Appel LJ, Bray GA, Simons-Morton DG, et al. DASH-Sodium Trial Collaborative Research Group. Effects of diet and sodium intake on blood pressure: subgroup analysis of the DASH-sodium trial. Ann Intern Med. 2001;135:1019–28.

Volpato S, Zuliani G, Guralnik J, Palmieri E, Fellin R. The inverse association between age and cholesterol level amount older patients: the role of poor health status. Gerontology. 2001;47:36–45.

Von Horn, JHL, Hendriks, JJE, Vermeer C, Forget P. Vitamin K supplementation in cystic fibrosis. Arch Dis Child. 2003;88:974–5.

Votey SR, Peters AL. Diabetes mellitus, type 2—a review. New York: eMedicine.com, Inc. [updated 2005 Jul 14; cited 2006 Mar 4]. Available from: http://www.emedicine.com.

Wadden TA, Byrne KJ, Krauthamer-Ewing S. Obesity: management. In: Shils ME, Shike M, Ross AC, Cabellero B, Cousins RJ, editors. Modern nutrition in health and disease. 10th ed. Philadelphia (PA): Lippincott Williams & Wilkins; 2006. p. 1029–42.

Waggoner DD, Buist NR, Donnell GN. Long term prognosis in galactosemia: results of a survey of 350 cases. J Inherit Metab Dis. 1990;7:365–9.

Wagner EH. The role of patient care teams in chronic disease management. BMJ. 2000;320:569–72.

Waldmann A, Koschizke JW, Leitzmann C, Hahn A. Dietary intakes and lifestyle factors of a vegan population in Germany: results from the German Vegan Study. Eur J Clin Nutr. 2003;57:947–55.

Waldmann A, Koschizke JW, Leitzmann C, Hahn A. Dietary iron intake and iron status of German female vegans: results of the German vegan study. Ann Nutr Metab. 2004;48:103–8.

Walji M, Sagaram S, Meric-Bernstam F, Johnson C, Bernstam E. Cancer-related complementary and alternative medicine online: factors affecting information retrieval. Medinfo. 2004;1318–22.

Wallace DC. Common causes of complex disorders. Exceptional Parent. 1997 June:39–52.

Wallen MA, Packman S. Nutrition and inborn errors of metabolism. Nutrition Updates. 1985;2:71–89.

Walling A. Amyotrophic lateral sclerosis: Lou Gehrig's disease. Am Fam Physician. 1999;59:1489–96.

Walsh BT. Eating disorders. In: Kasper DL, Braunwald E, Fauci AS, Hauser SL, Longo DL, Jameson JL, editors. Harrison's principles of internal medicine. 16th ed. New York: McGraw-Hill; 2005. p. 430–3.

Walsh D, Nelson KA, Mahmoud FA. Established and potential therapeutic applications of cannabinoids in oncology. Support Care Cancer. 2003;11:137–43.

Walsky RL, Gaman EA, Obach RS. Examination of 209 drugs for inhibition of cytochrome P450 2C8. J Clin Pharmacol. 2005;45:68–78.

Walter JH. L-Carnitine in inborn errors of metabolism: what is the evidence? J Inherit Metab Dis. 2003;26:181–8.

Wang SS, Chen CC, Chao Y, Wu SL, Lee FY, Lin HC, et al. Sequential hemodynamic changes for large volume paracentesis in post-hepatitic cirrhotic patients with massive ascites. Proc Natl Sci Counc Repub China B. 1996;20:117–22.

Wang SY. Weight loss and metabolic changes in dementia. J Nutr Health Aging. 2002;6:201–5.

Wanke C. Pathogenesis and consequences of HIV-associated wasting. J Acquir Immune Defic Syndr. 2004;27(Suppl4):S277–S279.

Wanke CA, Silva M, Ganda A, Fauntleroy J, Spiegelman D, Knox TA, et al. Role of acquired immune deficiency syndrome-defining conditions in human immunodeficiency virus-associated wasting. Clin Infect Dis. 2003;37(Suppl 2):S81–S84.

Ware LB, Matthay MA. The acute respiratory distress syndrome. New Eng J Med. 2000;342:1334–40.

Waser M, von Mutius E, Riedler J, Nowak D, Maisch S, Carr D, et al; The ALEX Study team. Exposure to pets, and the association with hay fever, asthma, and atopic sensitization in rural children. Allergy. 2005;60:177–84.

Watson L, Margetts B, Hawarth P, Dorward M, Thompson R, Little P. The association between diet and chronic obstructive pulmonary disease in subjects selected from general practice. Eur Respir J. 2002;20:313–8.

Watson, Watzl. Nutrition and Alcohol. Florida: CRC Press, Inc.; 1992.

Way LW, Doherty GM. Current surgical diagnosis and treatment. 11th ed. St. Louis (MO): McGraw Hill; 2003.

Weiner Feldman R. Nutrition in acute renal failure. J Renal Nutr. 1994;4:97–9.

Weinstein AR, Sesso HD, Lee IM, Cook NR, Manson JE, Buring JE, et al. Relationship of physical activity vs. body mass index with type 2 diabetes in women. JAMA. 2004;292:1188–94.

Weisberg P, Scanlon KS, Li R, Cogswell ME. Nutritional rickets among children in the United States: review of cases reported between 1986 and 2003. Am J Clin Nutr. 2000; 80(suppl):1697S–1705S.

Weiss S. Diet as a risk factor for asthma. Ciba Found Symp. 1997;206:244–57.

Weiss ST. Obesity: insight into the origins of asthma. Nature Immunol. 2005;6:537–9.

Wellinghausen N, Kern WV, Jochle W, Kern P. Zinc serum level in human immunodeficiency virus-infected patients in relation to immunological status. Biol Trace Elem Res. 2000;73:139–49.

Welschen LMC, Bloemendal E, Nupels G, Dekker JM, Heine RJ, Stalman WAB, et al. Self-monitoring of blood glucose in patients with type 2 diabetes who are not using insulin. Diabetes Care. 2005;28:1510–7.

Wendland BE, Greenwood CE, Weinberg I, Young KW. Malnutrition in institutionalized seniors: the iatrogenic component. J Am Ger Soc. 2003;51:85–90.

Wenger SL. Spontaneous chromosome breakage in pernicious anemia. Clin Nutr. 2000;19:467–8.

Wessell TR, Arant CB, Olson MB, Johnson BD, Reis SE, Sharaf BL, et al. Relationship of physical fitness vs. body mass index with coronary artery disease and cardiovascular events in women. JAMA. 2004;292:1179–87.

West AB, Losada M. The pathology of diverticulosis coli. J Clin Gastroenterol. 2004;38(5 Suppl):S11–6.

Wethers DL. Sickle cell disease in childhood: part I. Laboratory diagnosis, pathophysiology and health maintenance. Am Fam Physician. 2000;62:1013–20, 1027–8.

Wetzler M, Byrd JC, Bloomfield CD. Acute and chronic myeloid leukemia. In: Kasper DL, Braunwald E, Fauci AS, Hausner SL, Longo DI, Jameson JL, editors. Harrison's principles of internal medicine. 16th Ed. New York: McGraw Hill Medical Publishing; 2005. p. 631–41.

Wheeler DA. Weight loss and disease progression in HIV infection. AIDS Read. 1999;9:347–53.

Wheeler ML. Long-term complications. In: Ross TA, Boucher JL, O'Connell BS, editors. American Dietetic Association guide to diabetes: medical nutrition therapy and education. Chicago (IL): American Dietetic Association; 2005.

White H, Boden-Albala B, Wang C, Elkind MS, Rundek T, Wright CB, et al. Ischemic stroke subtype incidence among whites, blacks, and Hispanics: the Northern Manhattan Stroke Study. Circulation 2005;111:1327–31.

White JR. The pharmacologic management of patients with type II diabetes mellitus in the era of new oral agents and insulin analogs. Diabetes Spectrum. 1996;9:227–34.

White JR, Davis SN, Cooppan R, Davidson MB, Mulcahy K, Mank GA, Nelinson D, Diabetes Consortium Medical Advisory Board. Clarifying the role of insulin in type 2 diabetes management. Clinical Diabetes. 2003;21:14–21.

White KC. Anemia is a poor predictor of iron deficiency among toddlers in the United States: for heme the bell tolls. Pediatrics. 2005;115:315–20.

White LJ, McCoy SC, Castellano B, Gutierrez G, Stevens JE, Walter GA, et al. Resistance training improves strength and functional capacity in persons with multiple sclerosis. Multiple Sclerosis. 2004;10:668–74.

Whitehall JS, Patole SK, Campbell P. Recombinant human erythropoietin in anemia of prematurity. Indian Pediatrics. 1999;36:17–27.

Widnell K. Pathophysiology of motor fluctuations in Parkinson's disease. Mov Disord. 2005;20 Suppl 11:S17–22.

Wieringa FT, Dijkhuizen MA. Iron and zinc interactions. Am J Clin Nutr. 2004;80: 787–8.

Wiese J, McPherson S, Odden MC, Shlipak MG. Effect of Opuntia indica on symptoms of the alcohol hangover. Arch Intern Med. 2004;164:1334–40.

Wiggins KJ, Johnson DW. The influence of obesity on the development and survival outcomes of chronic kidney disease. Adv Chronic Kidney Dis. 2005;12:49–55.

Wiggins KL. Guidelines for nutrition care of renal patients. 3rd ed. Chicago (IL): American Dietetic Association; 2002.

Wigmore SJ, Barber MD, Ross JA, et al. Effect of oral eicosapentaenoic acid on weight loss in patients with pancreatic cancer. Nutr Cancer. 2000;36:177–84.

Wigmore SJ, Ross JA, Falconer JS, et al. The effect of polyunsaturated fatty acids on the progress of cachexia in patients with pancreatic cancer. Nutrition. 1996;12(suppl): 27–30.

Wijendran V, Hayes KC. Dietary n-6 and n-3 fatty acid balance and cardiovascular health. Annu Rev Nutr. 2004;24: 597–615.

Wilchanski M, Rivlin J, Cohen S, Augarten A, Blau H, Aviram M, et al. Clinical and genetic risk factors for cystic fibrosis-related liver disease. Pediatrics. 1999;103: 52–7.

Wilcken DE, Dudman NPB, Tyrrell PA, Robertson MR. Folic acid lowers elevated plasma homocysteine in chronic renal insufficiency: possible implications for prevention of vascular disease. Metabolism. 1988;37:697–701.

Wilder RM. Effect of ketonuria on course of epilepsy. Mayo Clinic Proc. 1921;2:307–8.

Wiley JS, Moore MR. Heme biosynthesis and its disorders. Porphyrias and sideroblastic anemias. In: Hoffman R. Hematology, basic principles and practice. 4th ed. Philadelphia (PA): Elsevier Medical Publishing; 2005. p. 499–517.

Wilkinson GR. Drug metabolism and variability among patients in drug response. New Engl J Med. 2005;352: 2211–21.

Willet W. Lessons from dietary studies in Adventists and questions for the future. Am J Clin Nutr. 2003;78: 539S–543S.

Williams R, Olivi S, Li CS, et al. Oral glutamine supplementation decreases resting energy expenditure in children and adolescents with sickle cell anemia. J Pediatr Hematol Oncol. 2004;26:619–25.

Williamson EM. Synergy and other interactions in phytomedicines. Phytomedicine. 2001;8:401–9.

Willison DJ, Keshavjee K, Nair L, Goldsmith C, Holbrook AM. Patients' consent preferences for research uses of information in electronic medical records: interview and survey data. BMJ. 2003;326:373–8.

Willox JC, Corr J, Shaw J, Richardson M, Calman KC, Drennan M. Prednisolone as an appetite stimulant in patients with cancer. BMJ. 1984;288:27.

Wilson IB, Landon BE, Ding L, Zaslavsky AM, Shapiro MF, Bozzette SA, Cleary PD. A national study of the relationship of care site HIV specialization to early adoption of highly active antiretroviral therapy. Med Care. 2005;43: 12–20.

Wilson RE, Krishnamurti L, Kamat D. Management of sickle cell disease in primary care. Clin Pediatr Phila. 2003;42: 753–61.

Wilson TA, Speiser P. Adrenal insufficiency. New York: eMedicine.com, Inc. [updated 2003 Jul 11; cited 2005 Jan 13]. Available from: http://www.emedicine.com.

Wilson, DC, Kalnins D, Stewart C, Hamilton N, Hanna K, Durie PR, et al. Challenges in the dietary treatment of cystic fibrosis related diabetes. Clin Nutr. 2000;19:87–93.

Winkler, MF, Gerrior, SA, Pomp, A. Use of retinol-binding protein and prealbumin as indicators of the response to nutrition therapy. J Am Diet Assoc. 1989;89:684–7.

Winterfeldt E, Bogle M, Ebro L. Dietetics: practice and future trends. 2nd ed. Sudbury (MA): Jones and Bartlett Publishers; 2005.

Witte KK, Clark AL, Cleland JG. Chronic heart failure and micronutrients. J Am Coll Cardiol. 2001;37:1765–74.

Wolf B, Hsia YE, Sweetman L, Gravel R, Harris DJ, Nyhan WL. Propionic acidemia: a clinical update. J Peds. 1981;99:835–46.

Wolfe RR. Regulation of skeletal muscle protein metabolism in catabolic states. Curr Opin Clin Nutr Metabl Care. 2005;8:61–5.

Wolfe S. Handling your health care. Buyer's Market. 1985; 1:2.

Wolff RA, Abbruzzese J, Evans DB. Neoplasms of the exocrine pancreas. In: Holland, Frei, editors. Cancer Medicine. 5th ed. Ontario: BC Decker, Inc.; 2005.

Wolfram RM, Kritz H, Efthimiou Y, Stomatopoulos J, Sinzinger H. Effect of prickly pear (Opuntia robusta) on glucose- and lipid-metabolism in nondiabetes with hyperlipidemia. Wiener klinische Wochenschrift. 2002;114: 840–6.

Wolk R, Swartz R. Nutritional support of patients with acute renal failure. Nutr Support Serv. 1986; 2:38–46.

Wolsko PM, Eisenberg DM, Davis RB, Kessler R, Phillips RS. Patterns and perceptions of care for treatment of back and neck pain: results of a national survey. Spine. 2003;28: 292–7.

Wong KW. Clinical efficacy of n-3 fatty acid supplementation in patients with asthma. J Am Diet Assoc. 2005;105: 98–105.

Wood LS, Gibson PG, Manohar LG. Circulating markers to assess nutritional therapy in cystic fibrosis. Clinica Chimica Acta. 2005;353:13–29.

Wood RJ. Calcium and phosphorus. In: Stipanuk M, editor. Biochemical and physiological aspects of human nutrition. Philadelphia (PA): WB Saunders; 2000. p. 643–70.

Wood RJ, Han O. Recently identified molecular aspects of intestinal iron absorption. J Nutr. 1998;128:1841–4.

Wood RJ, Ronnenberg AG. Iron. In: Shils ME, Shike M, Ross CA, Caballero BC, Cousins RJ, editors. Modern nutrition in health and disease. 10th ed. Philadelphia (PA): Lippincott Williams & Wilkins; 2005. p. 248–70.

Woodman R, Ferrucci L, Guralnik J. Anemia in older adults. Curr Opin Hematol. 2005;12:123–8.

Woods RK, Raven JM, Walters EH, Abramson MJ, Thien FC. Fatty acid levels and risk of asthma in young adults. Thorax 2005;59:105–10.

Wooley JA, Btaiche IF, Good KL. Metabolic and nutritional aspects of acute renal failure in critically ill patients requiring continuous renal replacement therapy. Nutr Clin Pract. 2005;20:176–91.

World Health Organization Department of Noncommunicable Disease Suveillance. Definition, Diagnosis and Classification of Diabetes Mellitus and Its Complications. Geneva: World Health Organization; 1999.

World Health Organization. Adherence to long-term therapies: evidence for action. Geneva: World Health Organization; 2003.

World Health Organization. Diabetes mellitus. Fact sheet #138. 2002. [cited 2006 Feb 11]. Available from: http://www.who.int.

World Health Organization. Energy and protein requirements. Report of a joint FAO/WHO/UNU expert consultation. Technical Report Series 724. Geneva: World Health Organization; 1985.

World Health Organization. HIV/AIDS clinical staging, HIV/AIDS case definitions and use of HIV rapid tests for diagnosis and surveillance: report of a WHO consultation. New Delhi, 1-3 June 2005. Available from: http://w3.whosea.org/LinkFiles/Publications_HIVAIDSClinical_Bookletfinal.pdf.

World Health Organization. Interim WHO clinical staging of HIV/AIDS and HIV/AIDS case definitions for surveillance. 2005. Available from: http://www.who.int/hiv/pub/guidelines/clinicalstaging.pdf.

World Health Organization. International drug monitoring: the role of national centres. WHO technical report series no. 498. Geneva: World Health Organization; 1972.

World Health Organization. International statistical classification of diseases and related health problems. Geneva: World Health Organization; 1989.

World Health Organization. Obesity and overweight facts [monograph on the Internet]. Geneva (Switzerland): World Health Organization; 2005 [cited 2005 Dec].

Available from: http://www.who.int/dietphysicalactivity/publications/facts/obesity/en/index.html.

World Health Organization. WHO International Collaborating Group for the Study of the WHO Staging System. AIDS. 1993;7:711–8.

Worobetz LJ, Hilsden RJ, Shaffer EA, Simon JB, Paré P, Bain VG, et al. The liver. In: Thomson ABR, Shaffer EA, editors. First principles of gastroenterology: the basis of disease and an approach to management. Toronto: Janssen-Ortho; 2005.

Wright CM, Kelly J, Trail A, Parkinson KN, Summerfield G. The diagnosis of borderline iron deficiency: results of a therapeutic trial. Arch Dis Child. 2004;89:1028–31.

Writing Group for the ISCD Position Development Conference. Diagnosis of osteoporosis in men, premenopausal women, and children. J Clin Densitom. 2004;7:17–26.

Writing Group for the ISCD Position Development Conference. Nomenclature and decimal places in bone densitometry. J Clin Densitom. 2004;7:45–9.

Wu CL, Hung CR, Change FY, et al. Involvement of cholecystokinin receptors in the inhibition of gastrointestinal motility by estradiol in ovariectomized rats. Scand J Gastroenterol. 2002, 37:1133–9.

Wu G, Fang YZ, Yang S, Lupton JR, Turner ND. Glutathione metabolism and its implications for health. J Nutr. 2004;134:489–92.

Wynne H. Drug metabolism and ageing. J Br Menopause Soc. 2005;11:51–6.

Xin X, He J, Frontini MG, Ogden LG, Motsamai OI, Whelton PK. Effects of alcohol reduction on blood pressure: a meta-analysis of randomized controlled trials. Hypertension. 2001;38:1112–7.

Yan AT, Yan RT, Tan M, Constance C, Lauzon C, Zaltzman J, et al. Canadian acute coronary syndromes (ACS) Registry investigators. Treatment and one-year outcome of patients with renal dysfunction across the broad spectrum of acute coronary syndromes. Can J Cardiol. 2006;22:115–20.

Yanagawa T, Bunn G, Roberts I, Wentz A. Nutritional support for head-injured patients. The Cochrane Database of Systematic Reviews; 2005:4. The Cochrane Collaboration: John Wiley & Sons, Ltd.

Yang S, Dennehy CE, Tsourounis C. Characterizing adverse events reported to the California poison control system on herbal rememdies and dietary supplements: a pilot study. Journal of Herbal Pharmacotherapy. 2003;2:1–11.

Yanicelli S, Medeiros DM. Elevated plasma phenylalanine concentrations may adversely affect bone status of phenylketonuric mice. J Inherit Metab Dis. 2002;25:347–61.

Yazdanbakhsh M, Kremsner PG, van Ree R. Allergy, parasites, and the hygiene hypothesis. Science. 2002 Apr 19;296(5567):490–4.

Ye H, Losada M, West AB. Diverticulosis coli: update on a "Western" disease. Adv Anat Pathol. 2005;12:74–80.

Yilmaz A, Candan F, Turan M. Coffee phagia and iron-deficiency anemia: a possible association with helicobacter pylori. J Health Popul Nutr. 2005;23:102–3.

Yin S. Making a healthy choice—dieting and weight loss. American Demographics. July 2001.

Yin X, Zhou J, Jie C, Xing D, Zhang Y. Anticancer activity and mechanism of Scutellaria barbata extract on human lung cancer cell line A549. Life Sciences. 2004;75:2233–44.

Yip R. Significance of an abnormally low or high hemoglobin concentration during pregnancy: Special consideration of iron nutrition. Am J Clin Nutr. 2000;72 Suppl:272S–279S.

Young EM, Considine RV, Sattler FR, Deeg MA, Buchanan TA, Degawa-Yamauchi M, et al. Changes in thrombolytic and inflammatory markers after initiation of indinavir- or amprenavir-based antiretroviral therapy. Cardiovasc Toxicol. 2004;4:179–86.

Young IS, Roxborough HE, Woodside JV. Antioxidants and respiratory disease. In: Basu TK, Temple NJ, Garg ML, editors. Antioxidants in human health. New York: CAB Publishing; 1999. p. 293–311.

Young JS. HIV and medical nutrition therapy. J Am Diet Assoc. 1997;97(10 Suppl 2):S161–S166.

Young N. 2005. Aplastic anemia, myelodysplasia, and related bone marrow failure syndromes. In: Kasper DL, Braunwald E, Fauci AS, Hausner SL, Longo DI, Jameson JL, editors. Harrison's principles of internal medicine. 16th ed. New York: McGraw Hill Medical Publishing. p. 617–26.

Young NS, Maciejewski JP. Aplastic anemia. In: Hoffman R. Hematology, basic principles and practice. 4th ed. Philadelphia (PA): Elsevier Medical Publishing; 2005. p. 381–417.

Yousri M, Barri H, Sudhir S. Approach to the patient with renal disease. In: Carpenter G, Griggs R, Loscalzo J, editors. Cecil essentials of medicine. Philadelphia (PA): WB Saunders; 2001. p. 232–7.

Yussman SM, Ryan SA, Auinger P, Weitzman M. Visits to complementary and alternative medicine providers by children and adolescents in the United States. Ambul Ped. 2004;4:429–35.

Yusuf S, Hawken S, Ounpuu S, Bautista L, Franzosi MG, et al. Obesity and the risk of myocardial infarction in 27,000 participants from 52 countries: a case-control study. Lancet. 2005;366:1640–9.

Zachariah R, Teck R, Ascurra O, Humblet P, Harries AD. Targeting CD4 testing to a clinical subgroup of patients could limit unnecessary CD4 measurements, premature antiretroviral treatment and costs in Thyolo Distric, Malawi. Trans R Soc Trop Med Hyg. 2006;100:24–31.

Zachman RD. Role of vitamin A in lung development. J Nutr. 1995;125:1634S–38S.

Zachman, RD, Grummer MA. The interaction of ethanol and vitamin A as a potential mechanism for the pathogenesis of fetal alcohol syndrome. Alcohol Clin Exp Res. 1998;22:1544–56.

Zaloga GP, Roberts P. Permissive underfeeding. New Horizons. 1994;2:257–262.

Zaloga GP, Siddiqui R, Terry C, Marik PE. Arginine: mediator or modulator of sepsis? Nutr Clin Pract. 2004;Jun; 19(3):201–215.

Zamvar V, McClean P, Odeka E, Richards M, Davison S. Hepatitis E virus infection with nonimmune hemolytic anemia. J. Pediatr. Gastroenterol. Nutr.2005;40:223–5.

Zarling E, Parmar J, Mobarhan S, Clapper M. Effect of enteral formula infusion rate, osmolality, and chemical composition upon clinical tolerance and carbohydrate absorption in normal subjects. JPEN. 1986;10:588–590.

Zawada, E. Initiation of dialysis. In: Daugirdas J, Blake P, Ing T, editors. Handbook of dialysis. Philadelphia: Lippincott Williams & Wilkins; 2001. p. 3–11.

Zenuk C, Healey J, Donnelly J, Vaillancourt R, Almalki Y, Smith S. Thiamine deficiency in congestive heart failure patients receiving long term furosemide therapy. Can J Clin Pharmacol. 2003;10:184–8.

Zhang CP, Tian ZB, Liu XS, Zhao QX, Wu J, Liang YX. 2004. Effects of Zhaoyangwan on chronic hepatitis B and posthepatic cirrhosis. World Journal of Gastroenterology, 10:295–298.

Ziegler TR. Glutamine supplementation in bone marrow transplantation. Br. J. Nutr. 2002;87: Suppl 1: S9–15.

Zijp IM, Korver O, Tijburg LB. Effect of tea and other dietary factors on iron absorption. Crit. Rev. Food Sci. Nutr. 2000;40:371–98.

Zimmer JP, Garza C, Heller ME, Butte N, Goldman AS. Postpartum maternal blood helper T (CD3+CD4+) and cytotoxic T (CD3+CD8+) cells: Correlations with iron status, parity, supplement use, and lactation status. Am. J. Clin. Nutr. 1998;67:897–904.

Zimmerman RS, Sirven JL. An overview of surgery for chronic seizures Mayo Clin Proc. 2003;78(1):109–17.

Zimmermann MB, J Köhrle. The impact of iron and selenium deficiencies on iodine and thyroid metabolism: Biochemistry and relevance to public health. Thyroid. 2002; 12:867–78.

Zimmermann MB, Molinari L, Staubli F-Asobayire, et al. 2005. Serum transferrin receptor and zinc protoporphyrin as indicators of iron status in African children. Am. J. Clin. Nutr. 81:615–23.

Zimmermann MB, Wegmueller R, C Zeder, et al. Triple fortification of salt with microcapsules of iodine, iron, and vitamin A. Am. J. Clin. Nutr. 2004;80:1283–90.

Zink EK, McQuilan K. Managing traumatic brain injury. Nursing.2005;35(9):36–43.

Zinna EM, Yarasheski KE. Exercise treatment to counteract protein wasting of chronic diseases. Curr Opin Clin Nutr Metab Care. 2003;6(1):87–93.

Zoorob R, Anderson R, Cefalu C, Sidani M. (2001). Cancer screening guidelines. Am Fam Physician, 63, 1101–12.

Zwischenberger JB, Alpard SK, Bidani A. Early complications, Respiratory failure. Chest Surg Clin N Am. 1999; 9(3):543–64.

Appendix A—General Information

Appendix A

COMMON MEDICAL ABBREVIATIONS

AAL	anterior axillary line
ab lib	at pleasure; as desired (ab libitum)
ACTH	adrenocorticotropic hormone
ac	before meals
AD	Alzheimer's Disease
ADA	American Dietetic Association, American Diabetes Association
ADH	antidiuretic hormone
ad lib	as desired (ad libitum)
ADL	activities of daily living
AGA	antigliadin antibody
AIDS	acquired immunodeficiency syndrome
ALP (Alk phos)	alkaline phosphatase
ALS	amyotrophic lateral sclerosis
ALT	alanine aminotransferase
amp	ampule
ANC	absolute neutrophil count
ANCA	antisacchromyces antibodies
AP	anterior posterior
ARDS	adult respiratory distress syndrome
ARF	acute renal failure, acute respiratory failure
ASA	acetylsalicylic acid, aspirin
ASCA	antineutrophil cytoplasmic antibodies
ASHD	arteriosclerotic heart disease
AV	arteriovenous
BANDS	neutrophils
BCAA	branched chain amino acids
BE	barium enema
BEE	basal energy expenditure
BG	blood glucose
bid	twice a day
bili	bilirubin
BM	bowel movement
BMI	body mass index
BMR	basal metabolic rate
BMT	bone marrow transplant
BP (B/P)	blood pressure
BPD	bronchopulmonary dysplasia
BPH	benign prostate hypertrophy
bpm	beats per minute, breaths per minute
BS	bowel sounds, breath sounds, or blood sugar
BSA	body surface area
BUN	blood urea nitrogen
c	with
c	cup
C	centigrade
CA	cancer; carcinoma
CA1	calcium
CABG	coronary artery bypass graft
CAD	coronary artery disease
CAPD	continuous ambulatory peritoneal dialysis
cath	catheter, catheterize
CAVH	continuous arteriovenous hemofiltration
CBC	complete blood count
cc	cubic centimeter
C.C.E	clubbing, cyanosis, or edema
CCK	cholecystokinin
CCU	coronary care unit
CDAI	Crohn's disease activity index
CDC	Centers for Disease Control
CHD	coronary heart disease
CHF	congestive heart failure
CHI	closed head injury
CHO	carbohydrate
CHOL	cholesterol
cm	centimeter
CNS	central nervous system
c/o	complains of
COPD	chronic obstructive pulmonary disease
CPK	creatinine phosphokinase
Cr	creatinine
CR	complete remission
CSF	cerebrospinal fluid
CT	computed tomography
CVA	cerebrovascular accident
CVD	cardio vascular disease
CVP	central venous pressure
CXR	chest X-ray
DASH	Dietary Approaches to Stop Hypertension
DBW	desirable body weight
d/c	discharge
D/C	discontinue
DCCT	Diabetes Control and Complications Trial

Note: Abbreviations can vary from institution to institution. Although the student will find many of the accepted variations listed in this appendix, other references may be needed to supplement this list.

DKA	diabetic ketoacidosis
dL	deciliter
DM	Diabetes Mellitus
D_5NS	Dextrose, 5% in normal saline
D_5W	Dextrose, 5% in water
DRI	dietary reference intake
DTR	deep tendon reflex
DTs	delirium tremens
DVT	deep vein thrombosis
Dx	diagnosis
ECF	extracellular fluid
ECG/EKG	electrocardiogram
EEG	electroencephalogram
e.g.	for example
EGD	esophagogastroduodenoscopy
ELISA	enzyme-linked immunosorbent assay
EMA	antiendomysial antibody
EMG	electromyography
EOMI	extra occular muscles intact
ER	emergency room
ERT	estrogen replacement therapy
ESR	erythrocyte sedimentation rate
ESRD	end-stage renal disease
ESRF	end-stage renal failure
F	Fahrenheit
FACSM	Fellow American College of Sports Medicine
FBG	fasting blood glucose
FBS	fasting blood sugar
FDA	Food and Drug Administration
FEF	forced mid-expiratory flow
FEV	forced mid-expiratory volume
FFA	free fatty acid
FH	family history
FTT	failure to thrive
FUO	fever of unknown origin
FVC	forced vital capacity
FX	fracture
g	gram
GB	gallbladder
g/dL	grams per deciliter
GERD	gastroesophageal reflux disease
GFR	glomerular filtration rate
GI	gastrointestinal
GM-CSF	granulocyte/macrophage colony stimulating factor
GTF	glucose tolerance factor
GTT	glucose tolerance test
GVHD	graft versus host disease
h	hour
HAV	hepatitis A virus
HBV	hepatitis B virus
HbA_{1c}	glycated hemoglobin
Hct	hematocrit
HC	head circumference
HCV	hepatitis C virus
HDL	high density lipoprotein
HEENT	head, eyes, ears, nose, throat
Hg	mercury
Hgb	hemoglobin
HHNK	hyperosmolar hyperglycemic nonketotic (syndrome)
HIV	human immunodeficiency virus
HLA	human leukocyte antigen
HOB	head of bed
H&P (HPI)	history and physical
HR	heart rate
HS or h.s.	hours of sleep
HTN	hypertension
HX	history
IBD	inflammatory bowel disease
IBS	irritable bowel syndrome
IBW	ideal body weight
ICF	intracranial fluid
ICP	intracranial pressure
ICS	intercostal space
ICU	intensive care unit
i.e.	that is
IGT	impaired glucose tolerance
IM	intramuscularly
inc	incontinent
I&O (I/O)	intake and output
IV	intravenous
IU	international unit
J	joule
K	potassium
kcal	kilocalorie
KCl	potassium chloride
kg	kilogram
KS	Kaposi's sarcoma
KUB	kidney, ureter, bladder
L	liter
LBM	lean body mass
lb	pounds
LCT	long chain triglyceride
LDH	lactic dehydrogenase
LES	lower esophageal sphincter
LFT	liver function test
LIGS	low intermittent gastric suction
LLD	left lateral decubitus position
LLQ	lower left quadrant
LMP	last menstrual period
LOC	level of conciousness
LP	lumbar puncture
LUQ	lower upper quadrant
lytes	electrolytes

MAC	midarm circumference
MAMC	midarm muscle circumference
MAOI	monoamine oxidase inhibitor
MCHC	mean corpuscular hemoglobin concentration
MCL	midclavicular line
MCT	medium chain triglyceride
MCV	mean corpuscular volume
mEq	milliequivalent
mg	milligram
Mg	magnesium
MI	myocardial infarction
mm	millimeter
mmHg	millimeters of mercury
MNT	medical nutrition therapy
MODY	maturity onset diabetes of the young
MOM	Milk of Magnesia
mOsm	milliosmol
MR	mitral regurgitation
MRI	magnetic resonance imaging
MS	multiple sclerosis, morphine sulfate
MVA	motor vehicle accident
MVI	multiple vitamin infusion
N	nitrogen
NG	nasogastric
NH_3	ammonia
NICU	neurointensive care unit, neonatal intensive care unit
NKA	no known allergies
NKDA	no known drug allergies
NPH	neutral protamine Hagedorn insulin
NPO	nothing by mouth
NSAID	nonsteroidal antiinflammatory drug
NTG	nitroglycerin
N/V	nausea and vomiting
O_2	oxygen
OA	osteoarthritis
OC	oral contraceptive
OHA	oral hypoglycemic agent
OR	operating room
ORIF	open reduction internal fixation
OT	occupational therapist
OTC	over the counter
$paco_2$	partial pressure of dissolved carbon dioxide in arterial blood
pao_2	partial pressure of dissolved oxygen in arterial blood
pc	after meals
PCM	protein calorie malnutrition
PD	Parkinson's disease
PE	pulmonary embolus
PED	percutaneous endoscopic duodenostomy
PEEP	positive end expiratory pressure
PEG	percutaneous endoscopic gastrostomy
PEM	protein energy malnutrition
PERRLA	pupils equal, round, and reactive to light and accommodation
pH	hydrogen ion concentration
PKU	phenylketonuria
PMI	point of maximum impulse
PMN	polymorphonuclear
PN	parenteral nutrition
PO	by mouth
PPD	packs per day
PPN	peripheral parenteral nutrition
prn	may be repeated as necessary (pro re nata)
PT	patient, physical therapy, prothrombin time
PTA	prior to admission
PTT	prothromboplastin time
PUD	peptic ulcer disease
PVC	premature ventricular contraction
PVD	peripheral vascular disease
q	every
qd	every day
qh	every hour
qid	four times daily
qns	quantity not sufficient
qod	every other day
RA	rheumatoid arthritis
RBC	red blood cell
RBW	reference body weight
RD	registered dietitian
RDA	recommended dietary allowance
RDS	respiratory distress syndrome
REE	resting energy expenditure
RLL	right lower lobe
RLQ	right lower quadrant
R/O	rule out
ROM	range of motion
ROS	review of systems
RQ	respiratory quotient
RR	respiratory rate
RUL	right upper lobe
RUQ	right upper quadrant
Rx	take, prescribe, or treat
s	without
SBO	small bowel obstruction
SBS	short bowel syndrome
SGOT	serum glutamic oxaloacetic transaminase
SGPT	serum glutamic pyruvic transaminase
SBGM	self blood glucose monitoring
SOB	shortness of breath
S/P	status post
SQ	subcutaneous
ss	half

stat	immediately
susp	suspension
T	temperature
T, tbsp	tablespoon
t, tsp	teaspoon
T&A	tonsillectomy and adenoidectomy
T_3	triiodothyronine
T_4	thyroxine
TB	tuberculosis
TEE	total energy expenditure
TF	tube feeding
TG	triglyceride
TIA	transient ischemic attack
TIBC	total iron binding capacity
tid	three times daily
TKO	to keep open
TLC	total lymphocyte count
TNM	tumor, node, metastasis
TPN	total parenteral nutrition
TSF	triceps skinfold
TSH	thyroid stimulating hormone
TURP	transurethral resection of the prostate
U	unit
UA	urinalysis
UBW	usual body weight
UL	tolerable upper intake level
URI	upper respiratory intake
UTI	urinary tract infection
UUN	urine urea nitrogen
VLCD	very low calorie diet
VOD	venous occlusive disease
VS	vital signs
w.a.	while awake
WBC	white blood cell
WNL	within normal limits
wt	weight
WW	whole wheat
yo	year old

Appendix A2

POSSIBLE NUTRITION ICD 9 CM CODES FOR MEDICAL SERVICES

International Classification of Diseases, 9th Revision, Clinical Modification (ICD-9-CM) is being published by the United States Government[1*]

1. **Infectious and parasitic diseases (001–139)**
   - intestinal infectious diseases (001–009)
   - tuberculosis (010–018)
   - zoonotic bacterial diseases (020–027)
   - other bacterial diseases (030–041)
   - human immunodeficiency virus (hiv) infection (042)
   - poliomyelitis and other non-arthropod-borne viral diseases of central nervous system (045–049)
   - viral diseases accompanied by exanthem (050–057)
   - arthropod-borne viral diseases (060–066)
   - other diseases due to viruses and chlamydiae (070–079)
   - rickettsioses and other arthropod-borne diseases (080–088)
   - syphilis and other venereal diseases (090–099)
   - other spirochetal diseases (100–104)
   - mycoses (110–118)
   - helminthiases (120–129)
   - other infectious and parasitic diseases (130–136)
   - late effects of infectious and parasitic diseases (137–139)

2. **Neoplasms (140–239)**
   - malignant neoplasm of lip, oral cavity, and pharynx (140–149)
   - malignant neoplasm of digestive organs and peritoneum (150–159)
   - malignant neoplasm of respiratory and intrathoracic organs (160–165)
   - malignant neoplasm of bone, connective tissue, skin, and breast (170–176)
   - malignant neoplasm of genitourinary organs (179–189)
   - malignant neoplasm of other and unspecified sites (190–199)
   - benign neoplasms (210–229)
   - carcinoma in situ (230–234)
   - neoplasms of uncertain behavior (235–238)
   - neoplasms of unspecified nature (239)

3. **Endocrine, nutritional and metabolic diseases, and immunity disorders (240–279)**
   - disorders of thyroid gland (240–246)
   - **Diseases of other endocrine glands (250–259)**

     250 Diabetes mellitus

     251 Other disorders of pancreatic internal secretion

     252 Disorders of parathyroid gland

     253 Disorders of the pituitary gland and its hypothalamic control

     254 Diseases of thymus gland

     255 Disorders of adrenal glands

     256 Ovarian dysfunction

     257 Testicular dysfunction

     258 Polyglandular dysfunction and related disorders

     259 Other endocrine disorders
   - **Nutritional deficiencies (260–269)**

     260 Kwashiorkor

     261 Nutritional marasmus

     262 Other severe protein-calorie malnutrition

     263 Other and unspecified protein-calorie malnutrition

     264 Vitamin A deficiency

     265 Thiamine and niacin deficiency states

     266 Deficiency of B-complex components

     267 Ascorbic acid deficiency

     268 Vitamin D deficiency

     269 Other nutritional deficiencies
   - **Other metabolic disorders and immunity disorders (270–279)**

     270 Disorders of amino-acid transport and metabolism

     271 Disorders of carbohydrate transport and metabolism

     272 Disorders of lipoid metabolism

     273 Disorders of plasma protein metabolism

     274 Gout

[1]U.S. DEPARTMENT OF HEALTH AND HUMAN SERVICES Centers for Disease Control and Prevention National Center for Health Statistics, Hyattsville, MD, 20782

*Each year the ICD 9 CMs are updated. Please refer to their website for updates: *http://www.cdc.gov/nchs/about/otheract/icd9/abticd9.htm*

- injury to blood vessels (900–904)
- late effects of injuries, poisonings, toxic effects, and other external causes (905–909)
- superficial injury (910–919)
- contusion with intact skin surface (920–924)
- crushing injury (925–929)
- effects of foreign body entering through orifice (930–939)
- burns (940–949)
- injury to nerves and spinal cord (950–957)
- certain traumatic complications and unspecified injuries (958–959)
- poisoning by drugs, medicinal and biological substances (960–979)
- toxic effects of substances chiefly nonmedicinal as to source (980–989)
- other and unspecified effects of external causes (990–995)
- complications of surgical and medical care, not elsewhere classified (996–999)

Appendix A3

MILLIEQUIVALENTS/MILLIGRAMS OF ELECTROLYTES

Milliequivalents to Milligrams

Cations		Anions	
Milliequivalents	**Milligrams**	**Milliequivalents**	**Milligrams**
1 mEq Potassium (K^+)	39 mg	1 mEq Chloride (Cl^-)	35.5 mg
1 mEq Sodium(Na^{2+})	23 mg	1 mEq Bicarbonate (HCO_3^-)	61 mg
1 mEq Calcium(Ca^{2+})	20 mg	1 mEq Potassium (PO_4^{3-})	31.67 mg
1 mEq Magnesium (Mg^{2+})	12.2 mg		

The equivalent weight of an electrolyte is its molecular weight divided by its valence. Therefore, because the molecular weight of K^+ is 39 and its valance is one, 39/1 is 39 grams. Milliequivalents would be 1/1000 of the equivalents or 39 milligrams. One milliequivalent of Na^+ is 23 milligrams [(23 grams/1) divided by 1000].

Appendix B—Nutrition Assessment

Appendix B1

GROWTH CHARTS

FIGURE B1.1 Weight-for-Age Percentiles: Boys, Birth to 36 Months

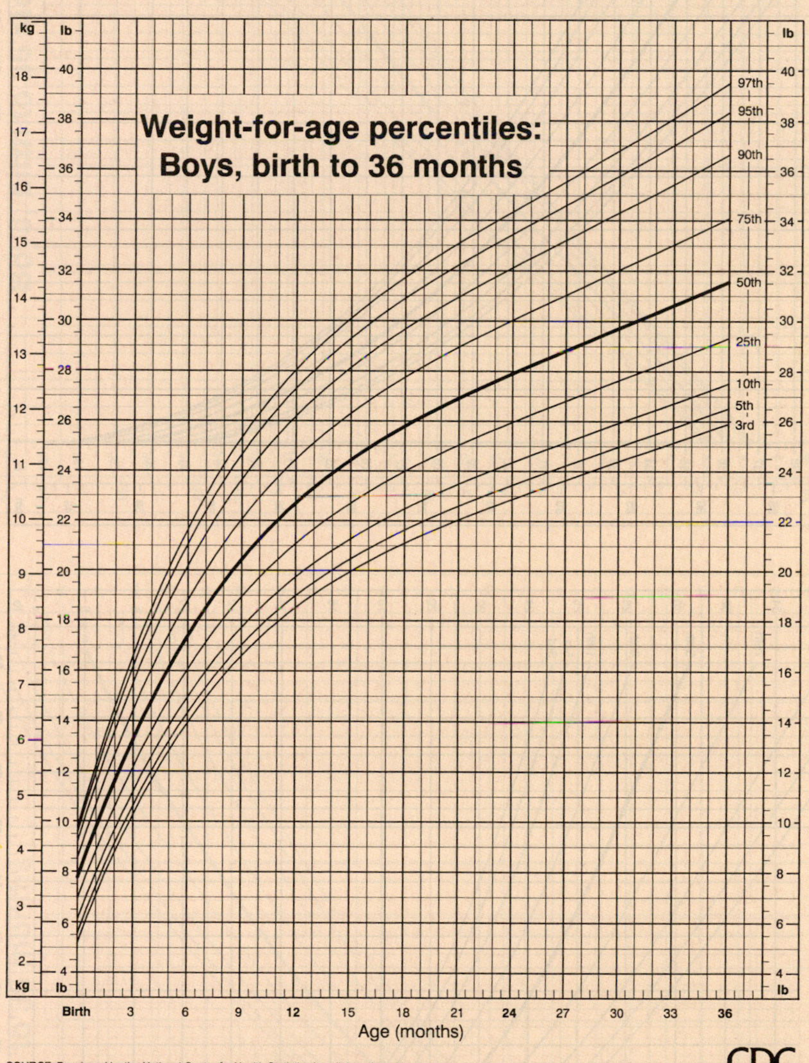

SOURCE: Developed by the National Center for Health Statistics in collaboration with the National Center for Chronic Disease Prevention and Health Promotion (2000).

Figure 1. Weight-for-age percentiles, boys, birth to 36 months, CDC growth charts: United States

FIGURE B1.6 Weight-for-Length Percentiles: Girls, Birth to 36 Months

Weight-for-length percentiles: Girls, birth to 36 months

Length

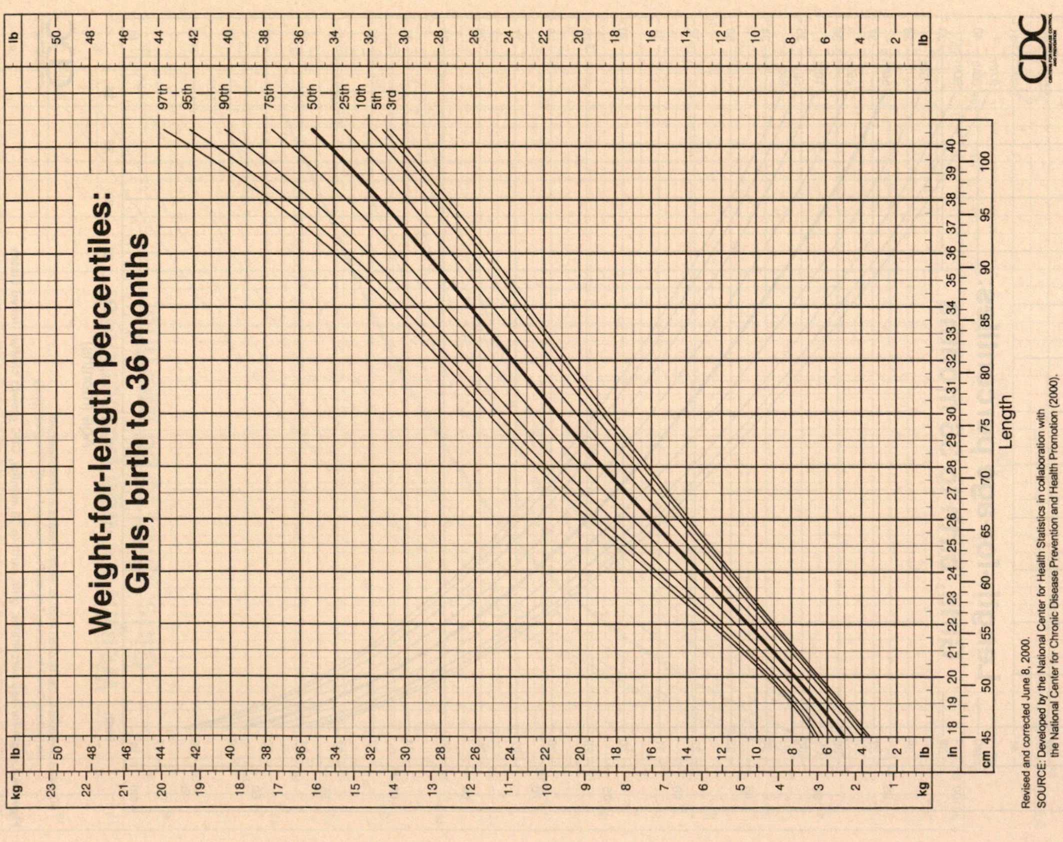

Revised and corrected June 8, 2000.
SOURCE: Developed by the National Center for Health Statistics in collaboration with the National Center for Chronic Disease Prevention and Health Promotion (2000).

CDC

Figure 6. Weight-for-length percentiles, girls, birth to 36 months, CDC growth charts: United States

FIGURE B1.7 Weight-for-Age Percentiles: Boys, 2 to 20 Years

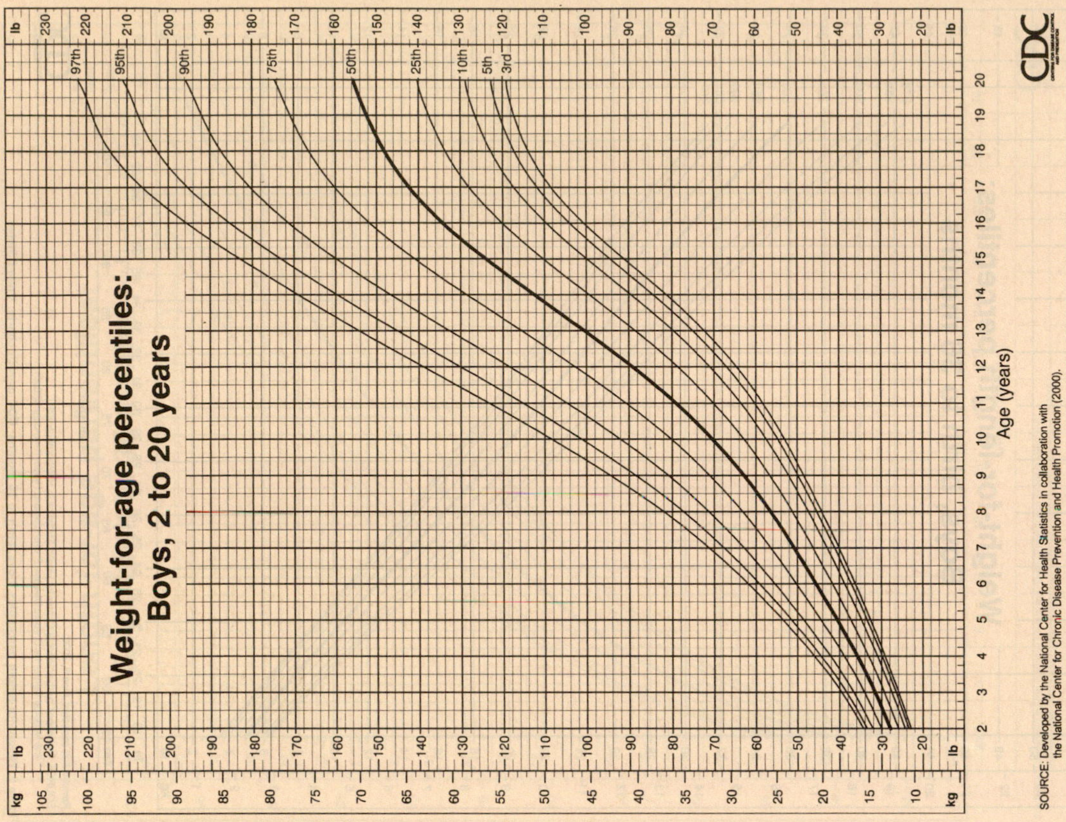

Weight-for-age percentiles: Boys, 2 to 20 years

Age (years)

SOURCE: Developed by the National Center for Health Statistics in collaboration with the National Center for Chronic Disease Prevention and Health Promotion (2000).

CDC

jure 9. Weight-for-age percentiles, boys, 2 to 20 years, CDC growth charts: United States

FIGURE B1.8 Weight-for-Age Percentiles: Girls, 2 to 20 Years

FIGURE B1.9 Stature-for-Age Percentiles: Boys, 2 to 20 Years

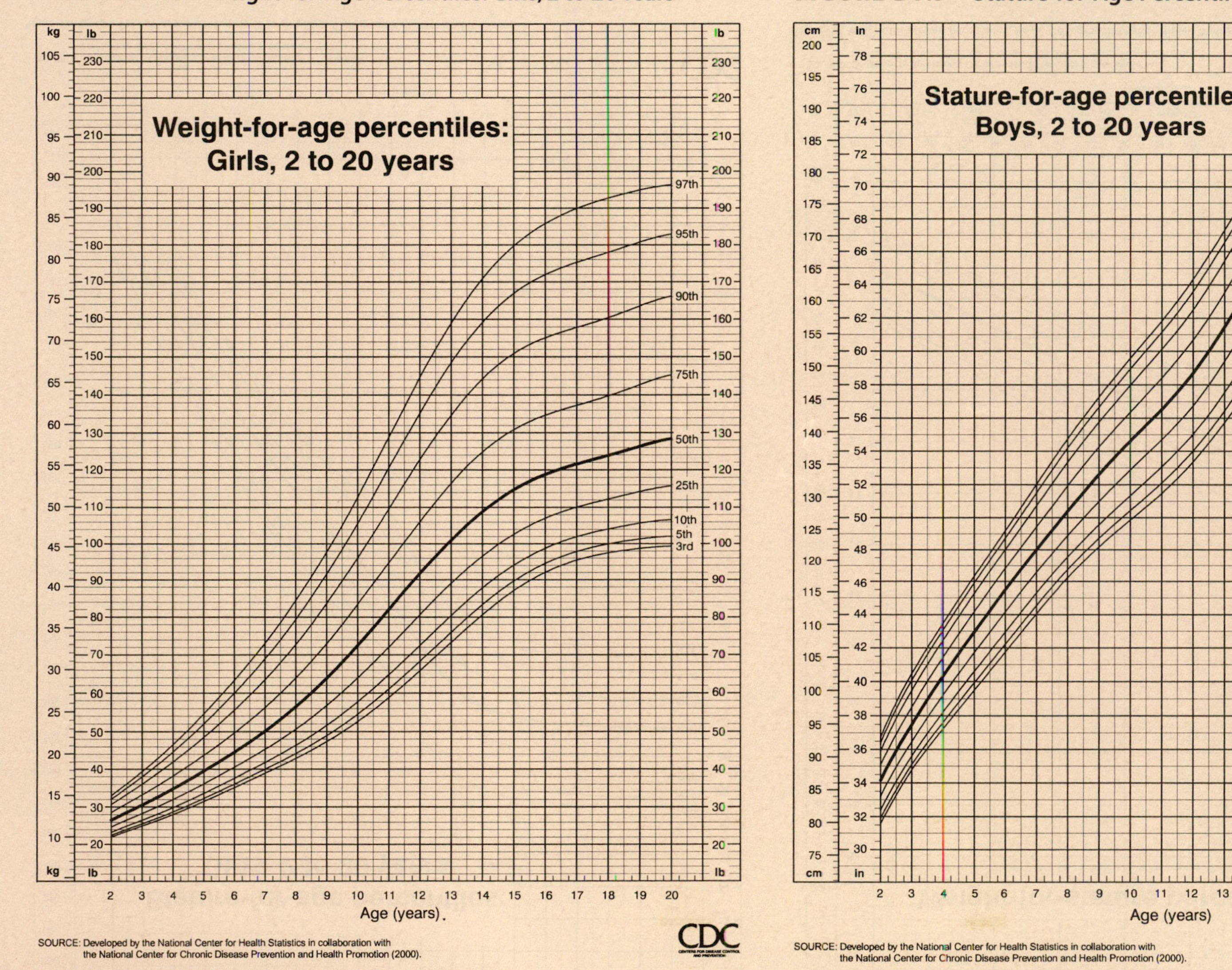

FIGURE B1.10 Stature-for-Age Percentiles: Girls, 2 to 20 Years

FIGURE B1.11 Weight-for-Stature Percentiles: Boys, 2 to 20 Years

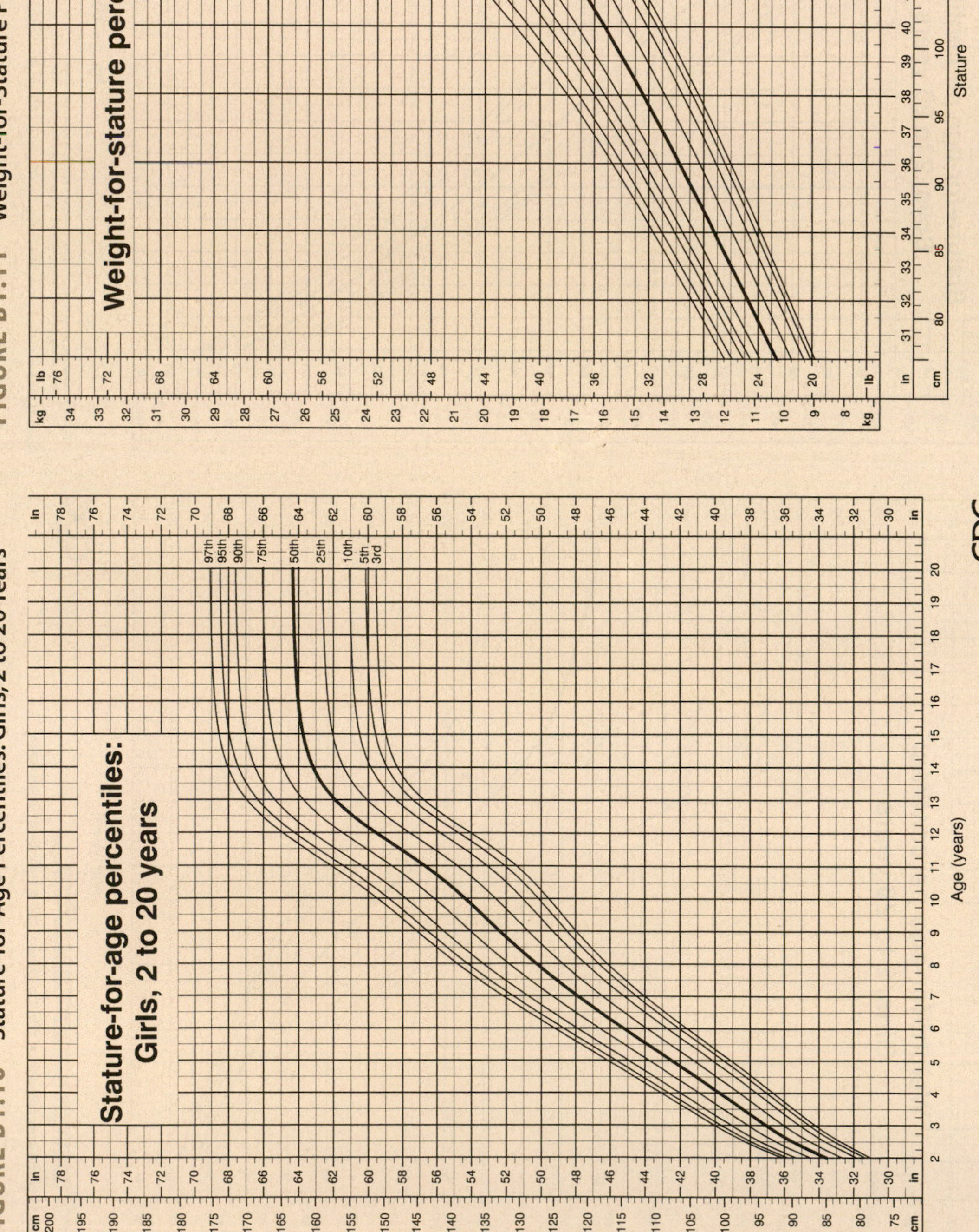

SOURCE: Developed by the National Center for Health Statistics in collaboration with the National Center for Chronic Disease Prevention and Health Promotion (2000).

Figure 12. Stature-for-age percentiles, girls, 2 to 20 years, CDC growth charts: United States

SOURCE: Developed by the National Center for Health Statistics in collaboration with the National Center for Chronic Disease Prevention and Health Promotion (2000).

Figure 13. Weight-for-stature percentiles, boys, CDC growth charts: United States

FIGURE B1.12 Weight-for-Stature Percentiles: Girls, 2 to 20 Years

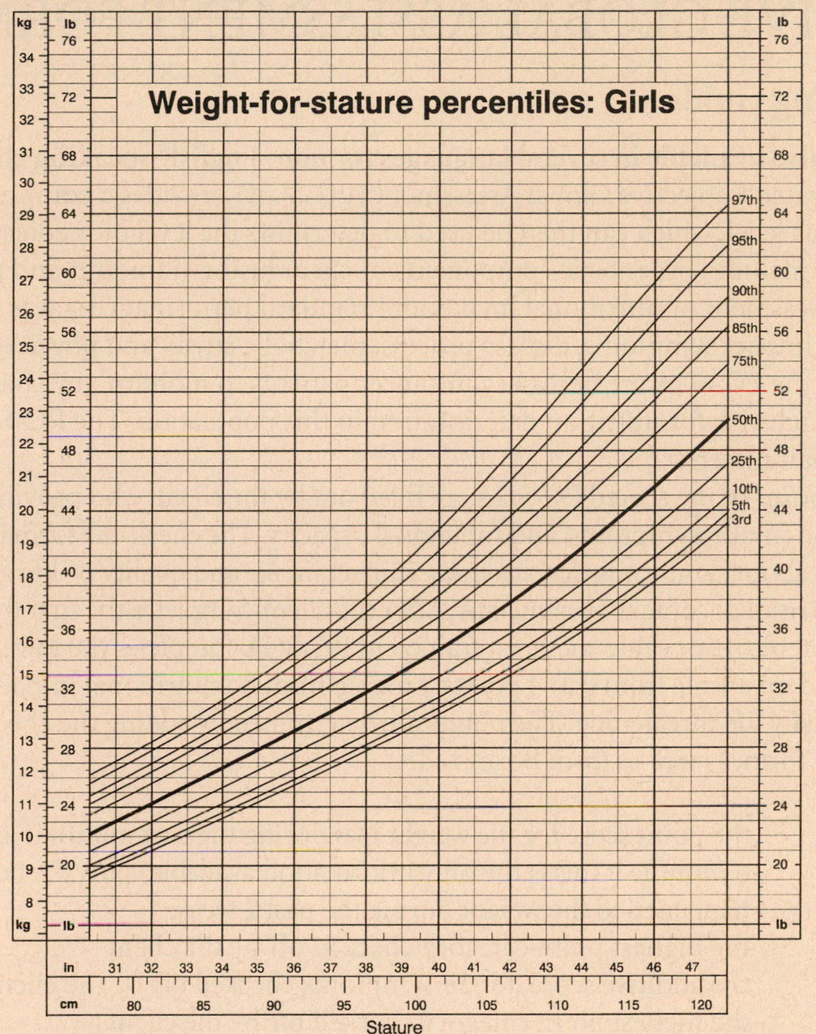

SOURCE: Developed by the National Center for Health Statistics in collaboration with the National Center for Chronic Disease Prevention and Health Promotion (2000).

Figure 14. Weight-for-stature percentiles, girls, CDC growth charts: United States

Appendix B2

SUBJECTIVE GLOBAL ASSESSMENT FORM

PG-SGA Scoring Guide

Note: PG-SGA is also available in several languages for non–English-speaking clients and caregivers. The Patient-Generated Subjective Global Assessment (PG-SGA) provides a comprehensive evaluation of nutritional status level, which can then be used to determine the level of medical nutrition therapy required. The tool includes prognostic components of client history (amount and pattern of weight loss, qualitative assessment of nutritional intake, and standard performance status scales) and clinical history (nutrition impact symptoms, disease process, metabolic stress, and physical examination). Serial assessments using the PG-SGA are necessary in cancer patients to monitor any changes in nutritional status, as there is high risk for nutrition deterioration in this population. The PG-SGA scoring is based on the following parameters.

The first four boxes of the scored PG-SGA are filled out by the client, who provides a current history of weight change, food intake, symptoms, and functional capacity. The check-off format enables clients to be more forthcoming about symptoms that adversely impact intake and quality of life and that are not often thought of in a nutritional context by clinicians. After the client completes the first four boxes, the dietetics professional, doctor, nurse, or other therapist trained in PG-SGA completes the lower section.

Scoring is based on a scale from 0 to 4 points, ranging from no nutritional impact to mild, moderate, severe, and potentially life threatening. The points are determined by adding the checked off points in parentheses on the form, as well as from Boxes 1–4.

Box 1:	the point score for the weight loss during the past month if available (or the past 6 months if this is the only information available) plus the points for what happened to the weight during the past 2 weeks
Box 2:	the highest point category checked off by the client
Box 3:	the additive score, for all symptoms checked off by the client
Box 4:	the highest point category checked off by the client
Disease section:	one point for each diagnosis identified in Box 2
Metabolic section:	a score based on metabolic stressors identified in Box 3
Physical section:	a score based on the physical assessment; refer to Box 4

Once each of these evaluations is made, the trained clinician proficient in nutrition physical assessment determines a global physical scoring (well-nourished or moderately or severely malnourished) using criteria outlined in Box 5. Triaging nutrition intervention is then determined using the information provided in Box 6.

Scored Patient-Generated Subjective Global Assessment (PG-SGA)

Patient ID Information

History

1. Weight:
In summary of my current and recent weight:

I currently weigh about _____ pounds
I am about _____ feet _____ inches tall

One month ago I weighed about _____ pounds
Six months ago I weighed about _____ pounds

During the past two weeks my weight has:
☐ decreased ☐ not changed ☐ increased

☐

2. Food Intake: As compared to my normal, I would rate my food intake during the past month as:
☐ unchanged
☐ more than usual
☐ less than usual
I am now taking:
☐ normal food but less than normal
☐ little solid food
☐ only liquids
☐ only nutritional supplements
☐ very little of anything
☐ only tube feedings or only nutrition by vein

☐

3. Symptoms: I have had the following problems that have kept me from eating enough during the past two weeks (check all that apply):
☐ no problem eating
☐ no appetite, just did not feel like eating
☐ nausea ☐ vomiting
☐ constipation ☐ diarrhea
☐ mouth sores ☐ dry mouth
☐ things taste funny or have no taste ☐ smells bother me
☐ problems swallowing ☐ feel full quickly
☐ pain; where? _____________
☐ other* _______________________________

*Examples: depression, money, or dental problems

☐

4. Activities and Function: Over the past month, I would generally rate my activity as:
☐ normal with no limitations
☐ not my normal self, but able to be up and about with fairly normal activities
☐ not feeling up to most things, but in bed or chair less than half the day
☐ able to do little activity and spend most of the day in bed or chair
☐ pretty much bedridden, rarely out of bed

☐

Additive Score of the Boxes 1–4 ☐ **A**

The remainder of this form will be completed by your doctor, nurse, or therapist. Thank you.

5. Disease and its relation to nutritional requirements
All relevant diagnoses (specify) _______________________________
Primary disease stage (circle if known or appropriate) I II III IV Other _______________
Age _________

6. Metabolic demand
☐ no stress ☐ low stress ☐ moderate stress ☐ high stress

7. Physical

Numerical score from Box 2 ☐
Numerical score from Box 3 ☐
Numerical score from Box 4 ☐

Global Assessment
☐ Well-nourished or anabolic (SGA-A)
☐ Moderate or suspected malnutrition (SGA-B)
☐ Severely malnourished (SGA-C)

Total numerical score of Boxes A+B+C+D ☐
(See triage recommendations below)

Clinician Signature ___________________________ RD RN PA MD DO Other ______ Date _______________

Nutritional Triage Recommendations: Additive score is used to define specific nutritional interventions including patient and family education, symptom management including pharmacologic intervention, and appropriate nutrient intervention (food, nutritional supplements, enteral, or parenteral triage). First line nutrition intervention includes optimal symptom management.

0–1 No intervention required at this time. Reassessment on routine and regular basis during treatment.
2–3 Patient and family education by dietitian, nurse, or other clinician with pharmacologic intervention as indicated by symptom survey (Box 3) and laboratory values as appropriate.
4–8 Requires intervention by dietitian, in conjunction with nurse or physician as indicated by symptoms survey (Box 3).
≥9 Indicates a critical need for improved symptom management and/or nutrient intervention options.

Source: © 2000, American Dietetic Association. "The Clinical Guide to Oncology Nutrition." Used with permission.

TABLE B2.1

Criteria for Scoring Weight Loss

Weight loss in 1 month	Weight loss in 6 months	Points
10% or greater	20% or greater	4
5–9.9%	10–19.9%	3
3–4.9%	6–9.9%	2
2–2.9%	2–5.9%	1
0–1.9%	0–1.9%	0

TABLE B2.2

Scoring Criteria for Diseases or Conditions

Category	Points
Cancer	1
AIDS	1
Pulmonary or cardiac cachexia	1
Presence of decubitus, open wound, or fistula	1
Presence of trauma	1
Age greater than 65 years	1

TABLE B2.3

Scoring of Metabolic Stressors

Stressor	none (0)	low (1)	moderate (2)	high (3)
Fever (°F)	no fever	>99 and <101	≥101 and ≤102	≥102
Fever duration	no fever	<72 hr		>72 hr
Steroids	no steroids	low-dose steroids (<10 mg prednisone equivalents/day)	moderate steroids (≥10, <30 mg prednisone equivalents/day)	high-dose steroids (≥30 mg prednisone equivalents/day)

TABLE B2.4

Components of Quick Physical Examination (none to +++)

Fat status	Muscle status	Fluid status
eyes	temples	skin & skin turgor
triceps fat pinch	shoulders	eyes
anterior lower ribs	clavicle	ankles
	scapula	sacrum
	thumb/index press	abdomen for ascites
	thigh and calf	

TABLE B2.5

PG-SGA Staging Guide

	Stage A	Stage B	Stage C
Category	Well-nourished	Moderately malnourished or suspected of being malnourished	Severely malnourished
Weight	No weight loss or recent non-fluid weight gain	A. Approximately 5% weight loss within 1 month (or 10% in 6 months) B. No weight stabilization or continued weight loss	A. >5% loss in 1 month (or >10% loss in 6 months) B. No weight stabilization or weight gain
Intake	No deficit or significant recent improvement	Definite decrease in intake	Severe deficit in intake
Nutrition impact symptoms	None or significant recent improvement allowing adequate intake	Presence of nutrition impact symptoms (Box 3 of PG-SGA)	Presence of nutrition impact symptoms (Box 3 of PG-SGA)
Functionality	No deficit or significant recent improvement	Moderate functional deficit or recent functional deterioration	Severe functional deficit or recent significant functional deterioration
Physical exam	No deficit or chronic deficit in the face of recent improvement in all history categories listed above	Evidence of mild to moderate loss of subcutaneous fat and/or muscle mass and/or muscle tone on palpation	Obvious signs of malnutrition (eg, severe loss of subcutaneous tissues, possible edema)

TABLE B2.6

Triaging Nutritional Intervention

Additive scores are used to define specific nutritional intervention pathways, including education and/or symptom management, aggressive oral nutrition, and enteral/parenteral triage.

Additive score of 0–1	Indicates that no intervention is required at this time. While these examples include only the client section of the form, the total additive scores include addition of both the client and clinician scores.
Additive score of 2–3	Indicates a need for client education by a dietitian or nurse, with pharmacologic triage by the nurse or physician as indicated by the symptom survey.
Additive score of 4–8	Requires the intervention of the dietitian, working in conjunction with the nurse or physician as indicated by the symptom check-off for pharmacologic management.
Additive score of >9	Indicates a critical need for symptom management and/or nutritional intervention. These clients require an interdisciplinary discussion to address all the aspects that are impacting the nutritional status, as well as the potential need for non-oral nutritional options, including enteral and parenteral nutrition. This decision should be dictated by the presence or absence of GI function.

Reprinted with permission from Ottery FD, Kasenic S, DeBolt S, Rogers K. Volunteer network accrues >1900 patients in 6 months to validate standardized nutritional triage. Abstract 282. Meeting of the American Society of Clinical Oncology, 1998.

Appendix B3

ROUTINE LABORATORY TESTS WITH NUTRITIONAL IMPLICATIONS

TABLE B3.1 Routine Laboratory Tests with Nutritional Implications

This table presents a partial listing of some uses of commonly performed lab tests that have implications for nutritional problems.

Laboratory Test	Acceptable Range	Description
Hematology		
Red blood cell (RBC) count	Male: 4.3–5.7 million/μL Female: 3.8–5.1 million/μL	Number of RBC; aids anemia diagnosis.
Hemoglobin (Hb)	Male: 13.5–17.5 g/dL Female: 12.0–16.0 g/dL	Hemoglobin content of RBC; aids anemia diagnosis.
Hematocrit (Hct)	Male: 39–49% Female: 35–45%	Percentage RBC in total blood volume; aids anemia diagnosis.
Mean corpuscular volume (MCV)	81–99 fL	RBC size, helps to distinguish between microcytic and macrocytic anemias.
Mean corpuscular hemoglobin concentration (MCHC)	31–37% Hb/cell	Hb concentration within RBCs, helps to distinguish iron-deficiency anemia.
White blood cell (WBC) count	4500–11,000 cells/μL	Number of WBC; general assessment of immunity.
Blood Chemistry		
Serum Proteins		
• Total protein	6.4–8.3 g/dL	Protein levels are not specific to disease or highly sensitive; they can reflect poor protein intake, illness or infections, changes in hydration or metabolism, pregnancy, or medications.
• Albumin	3.4–4.8 g/dL	May reflect illness or PEM; slow to respond to improvement or worsening of disease.
• Transferrin	200–400 mg/dL >60 yr: 180–380 mg/dL	May reflect illness, PEM, or iron deficiency; slightly more sensitive to changes than albumin.
• Prealbumin (transthyretin)	10–40 mg/dL	May reflect illness or PEM; more responsive to health status changes than albumin or transferrin.
• C-reactive protein	68–8200 ng/mL	Indicator of inflammation or disease.
Serum Enzymes		
• Creatine kinase (CK)	Male: 38–174 U/L Female: 26–140 U/L	Different forms of CK are found in muscle, brain, and heart. High levels in blood may indicate heart attack, brain tissue damage, or skeletal muscle injury.
• Lactate dehydrogenase (LDH)	208–378 U/L	LDH is found in many tissues. Specific types may be elevated after heart attack, lung damage, or liver disease.
• Alkaline phosphatase	25–100 U/L	Found in many tissues; often measured to evaluate liver function.
• Aspartate aminotransferase (AST, formerly SGOT)	10–30 U/L	Usually monitored to assess liver damage; elevated in most liver diseases. Levels are somewhat increased after muscle injury.
• Alanine aminotransferase (ALT, formerly SGPT)	Male: 10–40 U/L Female: 7–35 U/L	Usually monitored to assess liver damage; elevated in most liver diseases. Levels are somewhat increased after muscle injury.
Serum Electrolytes		
• Sodium	136–146 mEq/L	Helps to evaluate hydration status or neuromuscular, kidney, and adrenal functions.
• Potassium	3.5–5.1 mEq/L	Helps to evaluate acid-base balance and kidney function; can detect potassium imbalances.
• Chloride	98–106 mEq/L	Helps to evaluate hydration status and detect acid-base and electrolyte imbalances.
Other		
• Glucose	74–106 mg/dL	Detects risk of glucose intolerance, diabetes mellitus, and hypoglycemia; helps to monitor diabetes treatment.
• Glycosylated hemoglobin (Hb A_{1c})	5.0–7.5% of Hb	Used to monitor long-term blood glucose control (approximately 1 to 3 months prior).
• Blood urea nitrogen (BUN)	6–20 mg/dL	Primarily used to monitor kidney function; value is altered by liver failure, dehydration, or shock.
• Uric acid	Male: 3.5–7.2 mg/dL Female: 2.6–6.0 mg/dL	Used for detecting gout or changes in kidney function; levels affected by age and diet; varies among different ethnic groups.
• Creatinine (serum or plasma)	Male: 0.7–1.3 mg/dL Female: 0.6–1.1 mg/dL	Used to monitor renal function.

NOTE: μL = microliter; dL = deciliter; fL = femtoliter; ng = nanogram; U/L = units per liter; mEq = milliequivalents.
SOURCE: L. Goldman and J. C. Bennett, eds. *Cecil Textbook of Medicine* (Philadelphia: Saunders, 2000).

Appendix B4

NORMAL VALUES FOR PHYSICAL EXAMINATION

Vital Signs

Temperature
Rectal: C = 37.6°/F = 99.6°
Oral: C = 37°/F = 98.6° (±10°)
Axilla: C = 37.4°/F = 97.6°

Blood Pressure: average 120/80 mmHg

Heart Rate (beats per minute)

Age	At Rest Awake	At Rest Asleep	Exercise or Fever
Newborn	100–180	80–160	<220
1 week–3 months	100–220	80–200	<220
3 months–2 years	80–150	70–120	<200
2–10 years	70–110	60–90	<200
11 years–adult	55–90	50–90	<200

Respiratory Rate (breaths per minute)

Age	Respirations
Newborn	35
1–11 months	30
1–2 years	25
3–4 years	23
5–6 years	21
7–8 years	20
10–11 years	19
12–13 years	19
14–15 years	18
16–17 years	17
17–18 years	16–18
Adult	12–20

(continued on the following page)

Cardiac Exam: carotid pulses equal in rate, rhythm, and strength; normal heart sounds; no murmurs present

HEENT Exam (head, eyes, ears, nose, throat)
Mouth: pink, moist, symmetrical; mucosa pink, soft, moist, smooth
Gums: pink, smooth, moist; may have patchy pigmentation
Teeth: smooth, white, shiny
Tongue: medium red or pink, smooth with free mobility, top surface slightly rough
Eyes: pupils equal, round, reactive to light and accommodation
Ears: tympanic membrane taut, translucent, pearly gray; auricle smooth without lesions; meatus not swollen or occluded; cerumen dry (tan/light yellow) or moist (dark yellow/brown)
Nose: external nose symmetrical, nontender without discharge; mucosa pink; septum at the midline
Pharynx: mucosa pink and smooth
Neck: thyroid gland, lymph nodes not easily palpable or enlarged

Lungs: chest contour symmetrical; spine straight without lateral deviation; no bulging or active movement within the intercostal spaces during breathing; respirations clear to auscultation and percussion

Peripheral Vascular: normal pulse graded at 3+, which indicates that pulse is easy to palpate and not easily obliterated; pulses equal bilaterally and symmetrically

Neurological: normal orientation to people, place, time, with appropriate response and concentration

Skin: warm and dry to touch; should lift easily and return back to original position indicating normal turgor and elasticity

Abdomen: umbilicus flat or concave positioned midway between xyphoid process and symphysis pubis; bowel motility notes normal air and fluid movement every 5–15 seconds; graded as normal, audible, absent, hyperactive, or hypoactive

Appendix B5

NUTRITIONAL DEFICIENCIES REVEALED BY PHYSICAL EXAMINATION

TABLE B5.1

Nutrition Deficiencies Revealed by Physical Examination[3,8,9,12,18]

Deficient Nutrient	Findings
General Survey	
Protein, calories	Loss of weight, muscle, or fat stores; growth retardation, infection
Protein, thiamine	Edema (ankles and feet) (rule out sodium and water retention, pregnancy, protein-losing enteropathy)
Obesity	Excessive fat stores
Vitamin A	Poor growth
Iron	Anemia, fatigue
Skin	
Protein, vitamin C, zinc	Poor wound healing, pressure ulcers
Fat, vitamin A	Xerosis (rule out environmental, lack of hygiene, aging, uremia, hypothyroidism)
	Follicular hyperkeratosis
	Mosaic dermatitis (plaques of skin in center, peeling at periphery on shins)
Vitamin C	Slow wound healing
Niacin	Red, swollen skin lesions
Zinc	Delayed wound healing, acneiform rash, skin lesions, hair loss
Vitamin K or C	Excessive bleeding, petechiae, ecchymoses; small red, purple, black or blue, hemorrhagic spots
Dehydration (fluid)	Poor skin turgor
Nails	
Iron	Koilonychia (rule out cardiopulmonary disease)
Protein deficiency	Dull, lusterless with transverse ridging across nail plate
Vitamin A, C	Pale, poor blanching, irregular, mottled
Protein, calories	Bruising, bleeding
Vitamin C	Splinter hemorrhages
Hair	
Protein	Hair lacks shine, luster (cause may be environmental or chemical)
	Thin, sparse (fine, silky, and sparse with wide gaps between hairs)
Protein, copper	Dyspigmentation (lightening of normal hair color; consider if hair is bleached or dyed)
	Flag sign (alternating bands of light and dark hair in young children): rare
	Easily plucked
Copper	Corkscrew hair (Menkes syndrome)
Face	
Protein	Diffuse depigmentation, swelling
	Moon face (rounded cheeks with pursed mouth, seen in preschoolers)
Calcium	Facial paresthesias

(continued on the following page)

TABLE B5.1 *(continued)*

Nutrition Deficiencies Revealed by Physical Examination

Deficient Nutrient	Findings
Eyes	
Iron, folate, or vitamin B_{12}	Pale conjunctivae (anemia)
Vitam A	Bitot's spots (more common in children)
	Corneal xerosis
	Keratomalacia
Pyridoxine, niacin, riboflavin	Angular palpebritis
Hyperlipidemia	Corneal arcus, xanthelasma
Nose	
Riboflavin, niacin, pyridoxine	Seborrhea on nasolabial area, nose bridge, eyebrows, and backs of ears (rule out poor hygiene)
Lips and mouth	
Niacin, riboflavin	Cheilosis
	Angular scars
Riboflavin, pyridoxine, niacin, iron	Angular stomatitis
Tongue	
Niacin, riboflavin, folic acid, iron, B_{12}	Atrophic filiform papillae
	Glossitis
Zinc	Taste atrophy
Riboflavin	Magenta tongue
Teeth	
Excess sugar, vitamin C	Edentia, caries
Fluorosis	Mottled
Gums	
Vitamin C	Spongy, bleeding, receding
Neck	
Iodine	Enlarged thyroid
Protein, bulimia	Enlarged parotids (bilateral)
Excess fluid	Venous distention, pulsations
Thorax	
Protein, calories	Decreased muscle mass and strength, shortness of breath, fatigue; decreased pulmonary function
Cardiac system	
Thiamine	Heart failure
Gastrointestinal system	
Protein, calories, zinc, vitamin C	Poor wound healing
Protein	Hepatomegaly
Urinary tract	
Dehydration	Dark, concentrated urine
Overhydration	Light, dilute urine

(continued on the following page)

TABLE B5.1 *(continued)*

Musculoskeletal system	
Vitamin D, calcium	Rickets, osteomalacia
Vitamin D	Persistently open anterior fontanel (after age 18 months), craniotabes (softening of skull across back and sides before age 1 year)
	Epiphyseal enlargement (painless) at wrist, knees, and ankles
	Pigeon chest and Harrison's sulcus (horizontal depression on lower chest border)
Protein	Emaciation, muscle wasting, swelling, pain, pale hair patches
Vitamin C	Swollen, painful joints
Thiamine	Pain in thighs, calves
Nervous system	
Protein	Psychomotor changes (listless, apathetic)
	Mental confusion
Thiamin, B_6	Weakness, confusion, depressed reflexes, paresthesias, sensory loss, calf tenderness
Niacin, vitamin B_{12}	Dementia
Calcium, magnesium	Tetany

Appendix B7

TRICEPS SKINFOLD THICKNESS, ARM MUSCLE AREA (AMA), MID-UPPER ARM FAT AREA (AFA): PROCEDURES, COMPUTATIONS, INTERPRETATIONS AND PERCENTILES

BOX B7.1 CLINICAL INSIGHT

Anthropometric Parameters

MEASUREMENT PROCEDURES

Mid-Upper Arm Circumference (AC)

1. Keeping the subject's right arm parallel to the body, bend the elbow 90 degrees.
2. Using either a metallic tape or an insert tape, measure the distance between the acromion (the bony protrusion on the back of the upper shoulder) and the olecranon process (tip) of the elbow. If an insert tape is used, the same number should appear at the top of the shoulder and the elbow, and the midpoint is given by the mark on the tape
3. Mark the midpoint between these two landmarks.
4. Ask the subject to relax the arm, so it hangs loose and parallel to the body.
5. Position the metric tape around the upper arm at the marked midpoint. Make sure that the tape is snug but not so tight as to indent or pinch the skin.
6. Record the AC to the nearest 0.1 cm.

Triceps Skinfold Thickness (TSF)

1. Locate the previously marked midpoint on the posterior or back side of the right upper arm.
2. With the subject's arm hanging down loosely at the side, palpate the measurement site at the midpoint to become familiar with distinguishing muscle from adipose soft tissue.
3. From 1 cm above the midpoint, grasp a vertical pinch of skin and only the subcutaneous fat layer between the thumb and index finger. The skinfold should be gently pulled away from the underlying muscle.
4. Place the skinfold calipers at the midpoint and release the jaw pressure slowly while maintaining a grasp of the skinfold. Three readings should be taken in quick succession, and the average of the three recorded to the nearest 0.1 mm. Each reading should be taken as soon as the jaws of the caliper come into contact with the skin after the pressure is completely released and the dial reading has stabilized (about 4 seconds).

COMPUTATION OF DERIVED ANTHROPOMETRIC PARAMETERS

Calculations of mid-upper arm fat area (AFA) and arm muscle area (AMA) are based on measurements of AC of TSF. Equations for estimating AMA corrected for bone area are presented because they provide more accurate assessments of bone-free muscle area. The computational steps are outlined below. These examples assume that AC is 30 cm and TSF is 35 mm (2.5 cm). (It is essential to convert TSF to centimeters for the following computations.)

1. Determine total upper arm area (TAA):
$$\text{TAA (cm}^2) = \text{AC}^2/4 \times \pi$$
$$\text{TAA} = 30^2/12.57 = 71.6 \text{ cm}^2$$
2. Determine Uncorrected AMA:
$$\text{AMA (cm}^2) = [\text{AC} - (\text{TSF} \times \pi)]^2/4 \times \pi$$
$$\text{AMA} = [30 - (2.5 \times 3.1416)]^2/12.57$$
$$\text{AMA} = 490.44/12.57$$
$$\text{AMA} = 39 \text{ cm}^2$$
3. Determine corrected AMA (AMA_C):
$$\text{Males} = \text{AMA}_C - 10 \text{ cm}^2 = 29.0 \text{ cm}^2$$
$$\text{Females} = \text{AMA}_C - 6.5 \text{ cm}^2 = 32.5 \text{ cm}^2$$
4. Determine the AFA as the difference between TAA and *uncorrected* AMA:
$$\text{AFA (cm}^2) = \text{TAA} - \text{AMA}$$
$$\text{AFA} = 71.6 - 39.0 = 32.6 \text{ cm}^2$$

Note: Arm fat index (AFI) or percent fat area (%FA) can be determined as follows:
$$\text{AFI or \%FA} = (\text{AFA/TAA}) \times 100$$
$$\text{AFI or \%FA} = (32.6/71.6) \times 100 = 45.5\%$$

INTERPRETATION OF ANTHROPOMETRIC PARAMETERS

Comparing the percentile ranking of a specific individual on the various anthropometric measurements with a classification scheme is the basis for interpreting these values. Reference data in percentiles for TSF, AMA, and AFA appear in the appendices. Because the reference data are those compiled by Frisancho from the NHANES I and II data, it is appropriate to use the classification categories derived statistically from these data. The table below displays the percentile categories and their interpretation for arm muscle and arm fat areas as well as total body weight.

Percentile Rank	AMA	AFA	Total Body Weight
<5	Muscle deficit	Fat deficit	Total body wasting
5.1–15	Below average	Below average	Below average
15.1–85	Average	Average	Average
>85	Above average musculature	Excess fat	Excess total body weight

BOX B7.2 CLINICAL INSIGHT

Frame Size

MEASUREMENT PROCEDURE

Elbow breadth is most accurately measured using either sliding or spreading calipers and the following procedure:

1. Raise the subject's right arm so that the forearm is parallel to the body and flexed to a 90-degree angle.
2. Facing the subject, palpate the lateral and medial epicondyles of the humerus (the two prominent bones on either side of the elbow) and place the caliper jaws parallel or slightly at a slant to these two sites.
3. Measure the greatest bony width across the elbow joint twice to the nearest 0.1 cm; take the average of the two measurements.

Note: If sliding calipers are not available, elbow breadth can be estimated by placing the thumb and index finger of one hand parallel to the body on the epicondyles. The distance between the tips of the thumb and index finger is measured as the elbow breadth.

MEASUREMENT INTERPETATION

The National Health and Nutrition Examination Survey (NHANES) classification of frame size as small, medium, or large is based on the Frame Index 2 value, which is derived thus:

$$\text{Frame index 2} = \text{elbow breadth (mm)/stature (cm)} \times 100$$

This index accommodates age-related changes in weight and stature. The computation is made after converting the units of elbow breadth from centimeters to millimeters (by multiplying by 10). The derived Frame Index 2 is compared with the reference values for small, medium, and large frames, as defined below for the appropriate age and gender of the subject.

	Male			Female		
Age (yrs)	**Small**	**Medium**	**Large**	**Small**	**Medium**	**Large**
18.0–24.9	<38.4	38.4 to 41.6	>41.6	<35.2	35.2 to 38.6	>38.6
25.0–29.9	<38.6	38.6 to 41.8	>41.8	<35.7	35.7 to 38.7	>38.7
30.0–34.9	<38.6	38.6 to 42.1	>42.1	<35.7	35.7 to 39.0	>39.0
35.0–39.9	<39.1	39.1 to 42.4	>42.4	<36.2	36.2 to 39.8	>39.8
40.0–44.9	<39.3	39.3 to 42.5	>42.5	<36.7	36.7 to 40.2	>40.2
45.0–49.9	<39.6	39.6 to 43.0	>43.0	<37.2	37.2 to 40.7	>40.7
50.0–54.9	<39.9	39.9 to 43.3	>43.3	<37.2	37.2 to 41.6	>41.6
55.0–59.9	<40.2	40.2 to 43.8	>43.8	<37.8	37.8 to 41.9	>41.9
60.0–64.9	<40.2	40.2 to 43.6	>43.6	<38.2	38.2 to 41.8	>41.8
65.0–69.9	<40.2	40.2 to 43.6	>43.6	<38.2	38.2 to 41.8	>41.8
70.0–74.9	<40.2	40.2 to 43.6	>43.6	<38.2	38.2 to 41.8	>41.8

Adapted from Frisancho AR: *Anthropometric standards for the assessment of growth and nutritional status,* Ann Arbor, MI, 1990, The University of Michigan Press, p. 28.

TABLE B7.1

Means, Standard Deviations, and Percentiles of Triceps Skinfold Thickness (mm) by Age for Males and Females of 1 to 74 Years

				Percentiles								
Age (yr)	N	Mean	SD	*5*	*10*	*15*	*25*	*50*	*75*	*85*	*90*	*95*
MALES												
1.0–1.9	681	10.4	2.9	6.5	7.0	7.5	8.0	10.0	12.0	13.0	14.0	15.5
2.0–2.9	677	10.0	2.9	6.0	6.5	7.0	8.0	10.0	12.0	13.0	14.0	15.0
3.0–3.9	717	9.9	2.7	6.0	7.0	7.0	8.0	9.5	11.5	12.5	13.5	15.0
4.0–4.9	708	9.2	2.7	5.5	6.5	7.0	7.5	9.0	11.0	12.0	12.5	14.0
5.0–5.9	677	8.9	3.1	5.0	6.0	6.0	7.0	8.0	10.0	11.5	13.0	14.5
6.0–6.9	298	8.9	3.8	5.0	5.5	6.0	6.5	8.0	10.0	12.0	13.0	16.0
7.0–7.9	312	9.0	4.0	4.5	5.0	6.0	6.0	8.0	10.5	12.5	14.0	16.0
8.0–8.9	296	9.6	4.4	5.0	5.5	6.0	7.0	8.5	11.0	13.0	16.0	19.0
9.0–9.9	322	10.2	5.1	5.0	5.5	6.0	6.5	9.0	12.5	15.5	17.0	20.0
10.0–10.9	334	11.5	5.7	5.0	6.0	6.0	7.5	10.0	14.0	17.0	20.0	24.0
11.0–11.9	324	12.5	7.0	5.0	6.0	6.5	7.5	10.0	16.0	19.5	23.0	27.0
12.0–12.9	348	12.2	6.8	4.5	6.0	6.0	7.5	10.5	14.5	18.0	22.5	27.5
13.0–13.9	350	11.0	6.7	4.5	5.0	5.5	7.0	9.0	13.0	17.0	20.5	25.0
14.0–14.9	358	10.4	6.5	4.0	5.0	5.0	6.0	8.5	12.5	15.0	18.0	23.5
15.0–15.9	356	9.8	6.5	5.0	5.0	5.0	6.0	7.5	11.0	15.0	18.0	23.5
16.0–16.9	350	10.0	5.9	4.0	5.0	5.1	6.0	8.0	12.0	14.0	17.0	23.0
17.0–17.9	337	9.1	5.3	4.0	5.0	5.0	6.0	7.0	11.0	13.5	16.0	19.5
18.0–24.9	1752	11.3	6.4	4.0	5.0	5.5	6.5	10.0	14.5	17.5	20.0	23.5
25.0–29.9	1251	12.2	6.7	4.0	5.0	6.0	7.0	11.0	15.5	19.0	21.5	25.0
30.0–34.9	941	13.1	6.7	4.5	6.0	6.5	8.0	12.0	16.5	20.0	22.0	25.0
35.0–39.9	832	12.9	6.2	4.5	6.0	7.0	8.5	12.0	16.0	18.5	20.5	24.5
40.0–44.9	828	13.0	6.6	5.0	6.0	6.9	8.0	12.0	16.0	19.0	21.5	26.0
45.0–49.9	867	12.9	6.4	5.0	6.0	7.0	8.0	12.0	16.0	19.0	21.0	25.0
50.0–54.9	879	12.6	6.1	5.0	6.0	7.0	8.0	11.5	15.0	18.5	20.8	25.0
55.0–59.9	807	12.4	6.0	5.0	6.0	6.5	8.0	11.5	15.0	18.0	20.5	25.0
60.0–64.9	1259	12.5	6.0	5.0	6.0	7.0	8.0	11.5	15.5	18.5	20.5	24.0
65.0–69.9	1774	12.1	5.9	4.5	5.0	6.5	8.0	11.0	15.0	18.0	20.0	23.5
70.0–74.9	1251	12.0	5.8	4.5	6.0	6.5	8.0	11.0	15.0	17.0	19.0	23.0

(continued on the following page)

TABLE B7.1 *(continued)*

Means, Standard Deviations, and Percentiles of Triceps Skinfold Thickness (mm) by Age for Males and Females of 1 to 74 Years

				Percentiles								
Age (yr)	N	Mean	SD	*5*	*10*	*15*	*25*	*50*	*75*	*85*	*90*	*95*
FEMALES												
1.0–1.9	622	10.4	3.1	6.0	7.0	7.0	8.0	10.0	12.0	13.0	14.0	16.0
2.0–2.9	614	10.5	2.9	6.0	7.0	7.5	8.5	10.0	12.0	13.5	14.5	16.0
3.0–3.9	652	10.4	2.9	6.0	7.0	7.5	8.5	10.0	12.0	13.0	14.0	16.0
4.0–4.9	681	10.3	3.0	6.0	7.0	7.5	8.0	10.0	12.0	13.0	14.0	15.5
5.0–5.9	673	10.4	3.5	5.5	7.0	7.0	8.0	10.0	12.0	13.5	15.0	17.0
6.0–6.9	296	10.4	3.7	6.0	6.5	7.0	8.0	10.0	12.0	13.0	15.0	17.0
7.0–7.9	330	11.1	4.2	6.0	7.0	7.0	8.0	10.5	12.5	15.0	16.0	19.0
8.0–8.9	276	21.1	5.4	6.0	7.0	7.5	8.5	11.0	14.5	17.0	18.0	22.0
9.0–9.9	322	13.4	5.9	6.5	7.0	8.0	9.0	12.0	16.0	19.0	21.0	25.0
10.0–10.9	329	13.9	6.1	7.0	8.0	8.0	9.0	12.5	17.5	20.0	22.5	27.0
11.0–11.9	302	15.0	6.8	7.0	8.0	8.5	10.0	13.0	18.0	21.5	24.0	29.0
12.0–12.9	323	15.1	6.3	7.0	8.0	9.0	11.0	14.0	18.5	21.5	24.0	27.5
13.0–13.9	360	16.4	7.4	7.0	8.0	9.0	11.0	15.0	20.0	24.0	25.0	30.0
14.0–14.9	370	17.1	7.3	8.0	9.0	10.0	11.5	16.0	21.0	23.5	26.5	32.0
15.0–15.9	309	17.3	7.4	8.0	9.5	10.5	12.0	16.5	20.5	23.0	26.0	32.5
16.0–16.9	343	19.2	7.0	10.5	11.5	12.0	14.0	18.0	23.0	26.0	29.0	32.5
17.0–17.9	291	19.1	8.0	9.0	10.0	12.0	13.0	18.0	24.0	26.5	29.0	34.5
18.0–24.9	2588	20.0	8.2	9.0	1.1	12.0	14.0	18.5	24.5	28.5	31.0	36.0
25.0–29.9	1921	21.7	8.8	10.0	12.0	13.0	15.0	20.0	26.5	31.0	34.0	38.0
30.0–34.9	1619	23.7	9.2	10.5	13.0	15.0	17.0	22.5	29.5	33.0	35.5	41.5
35.0–39.9	1453	24.7	9.3	11.0	13.0	15.5	18.0	23.5	30.0	35.0	37.0	41.0
40.0–44.9	1391	25.1	9.0	12.0	14.0	16.0	19.0	24.5	30.5	35.0	37.0	41.0
45.0–49.9	962	26.1	9.3	12.0	14.5	16.5	19.5	25.5	32.0	35.5	38.0	42.5
50.0–54.9	1006	26.5	9.0	12.0	15.0	17.5	20.5	25.5	32.0	36.0	38.5	42.0
55.0–59.9	880	26.6	9.4	12.0	15.0	17.0	20.5	26.0	32.0	36.0	39.0	42.5
60.0–64.9	1389	26.6	8.8	12.5	16.0	17.5	20.5	26.0	32.0	35.5	38.0	42.5
65.0–69.9	1946	25.1	8.5	12.0	14.5	16.0	19.0	25.0	30.0	33.5	36.0	40.0
70.0–74.9	1462	42.0	8.5	11.0	13.5	15.5	18.0	24.0	29.5	32.0	35.0	38.5

From Frisancho AR: *Anthropometric standards for the assessment of growth and nutritional status*, Ann Arbor, 1990, The University of Michigan Press, p 54.

TABLE B7.2

Means, Standard Deviations, and Percentiles of Upper Arm Muscle Area (cm^2) by Height (cm) for Boys and Girls of 2 to 17 years

				Percentiles								
Height (cm)	N	Mean	SD	*5*	*10*	*15*	*25*	*50*	*75*	*85*	*90*	*95*
BOYS: 2 TO 11 YR												
87–092	94	12.9	2.2	9.3	10.4	10.6	11.2	12.9	14.2	15.0	15.8	16.5
93–098	373	13.7	2.4	10.2	10.9	11.2	12.1	13.5	15.3	15.9	16.5	17.0
99–104	587	14.6	3.1	10.9	11.7	12.2	13.0	14.5	15.9	16.5	17.1	18.4
105–110	587	15.7	3.1	12.0	12.8	13.3	14.1	15.4	17.0	17.8	18.6	19.8
111–116	588	16.7	2.9	12.6	13.6	14.3	15.0	16.6	18.1	18.9	19.6	20.7
117–122	496	18.1	3.5	14.1	14.5	15.0	16.1	17.7	19.7	20.7	21.6	23.4
123–128	376	19.5	3.6	15.0	15.9	16.3	17.4	19.2	21.2	22.3	23.2	24.2
129–134	359	21.6	4.3	16.1	17.3	18.4	19.3	21.1	23.2	34.7	25.3	27.9
135–140	354	22.9	4.2	17.2	18.1	18.9	20.4	22.6	24.9	26.1	27.2	30.2
141–146	325	25.1	5.2	19.3	20.1	20.8	21.9	24.0	27.2	29.0	30.5	34.0
147–152	266	27.5	4.8	21.2	22.4	23.2	24.8	27.0	29.8	31.8	32.9	34.4
153–158	150	29.8	7.2	22.3	23.2	24.3	25.4	28.4	32.2	34.8	36.9	40.1
159–164	65	32.5	6.5	23.7	24.5	25.3	27.5	31.9	35.6	39.7	41.4	44.5
BOYS: 12 TO 17 YR												
141–146	31	26.8	4.7	20.7	21.4	22.7	24.1	25.6	30.3	32.8	33.9	36.3
147–152	90	28.2	4.1	22.4	23.4	24.1	25.6	27.5	30.2	33.1	34.2	36.1
153–158	181	31.4	6.4	22.7	24.9	26.1	27.5	30.4	34.1	36.4	39.1	41.5
159–164	218	35.0	7.7	23.7	26.7	27.8	30.2	34.1	38.6	41.5	44.3	48.4
165–170	323	40.8	9.3	28.1	29.7	31.5	34.2	40.0	45.6	49.0	52.9	58.9
171–176	431	46.6	9.9	32.8	35.2	36.6	39.5	45.8	52.6	56.0	59.1	66.0
177–182	431	50.3	9.3	36.1	38.7	40.8	43.4	49.5	56.4	59.4	62.6	65.9
183–188	269	53.4	11.2	38.3	41.3	42.8	46.1	52.6	57.8	63.0	67.5	74.3
189–194	99	55.4	9.9	41.4	44.2	45.7	48.9	53.9	60.3	65.0	68.5	74.0
GIRLS: 2 TO 10 YR												
87–092	154	12.6	2.1	9.5	10.1	10.5	11.0	12.6	14.2	14.8	15.5	16.2
93–098	384	13.2	2.1	10.1	10.7	11.0	11.8	13.2	14.4	15.3	15.8	16.9
99–104	533	14.1	2.3	10.6	11.2	11.7	12.5	14.0	15.5	16.4	16.9	18.0
105–110	550	14.8	2.4	11.3	11.9	12.4	13.2	14.6	16.3	17.3	17.9	18.9
111–116	543	15.9	2.8	12.3	13.0	13.5	14.2	15.7	17.4	18.4	19.1	20.3
117–122	465	17.0	2.8	13.0	13.9	14.4	15.2	16.7	18.5	19.6	20.3	21.4
123–128	372	18.2	2.8	14.2	15.0	15.5	16.2	17.9	19.6	20.8	21.6	22.9
129–134	333	20.1	4.6	15.3	16.1	16.8	17.6	19.7	21.7	22.9	23.8	25.4
135–140	303	21.6	4.2	16.1	17.4	18.1	19.2	21.1	23.8	24.8	26.3	27.9
141–146	258	23.3	4.0	17.6	18.5	19.5	20.5	23.0	25.8	27.9	28.8	30.6
147–152	161	25.2	5.2	18.5	20.0	20.7	21.7	24.4	27.8	30.0	31.2	32.9
153–158	66	26.7	6.7	19.4	20.1	22.4	23.0	25.4	29.2	31.8	34.0	38.2

(continued on the following page)

TABLE B7.2 *(continued)*

Means, Standard Deviations, and Percentiles of Upper Arm Muscle Area (cm^2) by Height (cm) for Boys and Girls of 2 to 17 years

				Percentiles								
Height (cm)	N	Mean	SD	*5*	*10*	*15*	*25*	*50*	*75*	*85*	*90*	*95*
GIRLS: 11 TO 17 YR												
141–146	53	23.8	4.4	17.1	19.3	19.5	21.0	23.4	25.5	27.9	28.6	33.4
147–152	119	25.2	4.6	18.5	19.7	20.9	22.0	24.3	28.0	29.8	30.3	34.4
153–158	305	29.1	6.4	20.8	22.0	23.0	24.7	28.3	32.9	35.0	37.5	39.2
159–164	587	32.2	7.1	23.3	24.8	26.0	27.7	31.2	35.7	38.0	40.1	43.5
165–170	715	34.2	7.4	25.0	26.6	27.8	29.5	33.2	37.6	40.2	42.8	46.9
171–176	367	34.9	8.0	25.9	27.1	28.0	29.9	33.7	38.0	41.2	43.5	47.6
177–182	113	37.8	8.4	28.6	29.5	30.5	31.7	35.9	41.1	45.9	47.8	58.2

From Frisancho AR: *Anthropometric standards for the assessment of growth and nutritional status*, Ann Arbor, 1990, The University of Michigan Press, p 51.

TABLE B7.4

Means, Standard Deviations, and Percentiles of Upper Arm Fat Area (cm^2) by Age for Males and Females 1 to 74 Years

				Percentiles								
Age (yr)	N	Mean	SD	5	10	15	25	50	75	85	90	95
MALES												
1.0–1.9	681	7.5	2.2	4.5	4.9	5.3	5.9	7.4	8.9	9.6	10.3	11.7
2.0–2.9	672	7.4	2.3	4.2	4.8	5.1	5.8	7.3	8.6	9.7	10.6	11.6
3.0–3.9	715	7.6	2.4	4.5	5.0	5.4	5.9	7.2	8.6	9.7	10.6	11.8
4.0–4.9	707	7.3	2.5	4.1	4.7	5.2	5.7	6.9	8.5	9.3	10.0	11.4
5.0–5.9	676	7.4	3.1	4.0	4.5	4.9	5.5	6.7	8.3	9.8	10.9	12.7
6.0–6.9	298	7.7	4.1	3.7	4.3	4.6	5.2	6.7	8.6	10.3	11.2	15.2
7.0–7.9	312	8.1	4.2	3.8	4.3	4.7	5.4	7.1	9.6	11.6	12.8	15.5
8.0–8.9	296	8.9	5.0	4.1	4.8	5.1	5.8	7.6	10.4	12.4	15.6	18.6
9.0–9.9	322	10.1	6.2	4.2	4.8	5.4	6.1	8.3	11.8	15.8	18.2	21.7
10.0–10.9	333	12.0	7.3	4.7	5.3	5.7	6.9	9.8	14.7	18.3	21.5	37.0
11.0–11.9	324	13.6	9.4	4.9	5.5	6.2	7.3	10.4	16.9	22.3	26.0	32.5
12.0–12.9	348	13.9	9.6	4.7	5.6	6.3	7.6	11.3	15.8	21.1	27.3	35.0
13.0–13.9	350	13.0	9.2	4.7	5.7	6.3	7.6	10.1	14.9	21.2	25.4	32.1
14.0–14.9	358	13.3	10.2	4.6	5.6	6.3	7.4	10.1	15.9	19.5	25.5	31.8
15.0–15.9	356	12.8	9.0	5.6	6.1	6.5	7.3	9.6	14.6	20.2	24.5	31.3
16.0–16.9	350	13.9	9.5	5.6	6.1	6.9	8.3	10.5	16.6	20.6	24.8	33.5
17.0–17.9	337	12.9	8.9	5.4	6.1	6.7	7.4	9.9	15.6	19.7	23.7	28.9
18.0–24.9	1752	16.9	10.8	5.5	6.9	7.7	9.2	13.9	21.5	26.8	30.7	37.2
25.0–29.9	1250	18.8	11.6	6.0	7.3	8.4	10.2	16.3	23.9	29.7	33.3	40.4
30.0–34.9	940	20.4	11.4	6.2	8.4	9.7	11.9	18.4	25.6	31.6	34.8	41.9
35.0–39.9	832	20.1	10.5	6.5	8.1	9.6	12.8	18.8	25.2	29.6	33.4	39.4
40.0–44.9	828	20.4	11.2	7.1	8.7	9.9	12.4	18.0	25.3	30.1	35.3	42.1
45.0–49.9	867	20.1	11.0	7.4	9.0	10.2	12.3	18.1	24.9	29.7	33.7	40.4
50.0–54.9	879	19.4	10.3	7.0	8.6	10.1	12.3	17.3	23.9	29.0	32.4	40.0
55.0–59.9	807	19.2	10.2	6.4	8.2	9.7	12.3	17.4	23.8	28.4	33.3	39.1
60.0–64.9	1259	19.1	10.2	6.9	8.7	9.9	12.1	17.0	23.5	28.3	31.8	38.7
65.0–69.9	1773	18.0	9.8	5.8	7.4	8.5	10.9	16.5	22.8	27.2	30.7	36.3
70.0–74.9	1250	17.5	9.4	6.0	7.5	8.9	11.0	15.9	22.0	25.7	29.1	34.9
FEMALES												
1.0–1.9	622	7.3	2.3	4.1	4.6	5.0	5.6	7.1	8.6	9.5	10.4	11.7
2.0–2.9	614	7.7	2.3	4.4	5.0	5.4	6.1	7.5	9.0	10.0	10.8	12.0
3.0–3.9	651	7.8	2.5	4.3	5.0	5.4	6.1	7.6	9.2	10.2	10.8	12.2
4.0–4.9	680	8.0	2.6	4.3	4.9	5.4	6.2	7.7	9.3	10.4	11.3	12.8
5.0–5.9	672	8.5	3.4	4.4	5.0	5.4	6.3	7.8	9.8	11.3	12.5	14.5
6.0–6.9	296	8.7	3.9	4.5	5.0	5.6	6.2	8.1	10.0	11.2	13.3	16.5
7.0–7.9	329	9.8	4.5	4.8	5.5	6.0	7.0	8.8	11.0	13.2	14.7	19.0
8.0–8.9	275	11.3	6.5	5.2	5.7	6.4	7.2	9.8	13.3	15.8	18.0	23.7
9.0–9.9	321	13.1	7.3	5.4	6.2	6.8	8.1	11.5	15.6	18.8	22.0	27.5
10.0–10.9	329	14.1	7.7	6.1	6.9	7.2	8.4	11.9	18.0	21.5	25.3	29.9

(continued on the following page)

TABLE B7.4 *(continued)*

Means, Standard Deviations, and Percentiles of Upper Arm Fat Area (cm²) by Age for Males and Females 1 to 74 Years

				Percentiles								
Age (yr)	N	Mean	SD	*5*	*10*	*15*	*25*	*50*	*75*	*85*	*90*	*95*
FEMALES												
11.0–11.9	302	16.3	9.7	6.6	7.5	8.2	9.8	13.1	19.9	24.4	28.2	36.8
12.0–12.9	323	16.9	8.9	6.7	8.0	8.8	10.8	14.8	20.8	24.8	29.4	34.0
13.0–13.9	360	19.1	11.0	6.7	7.7	9.4	11.6	16.5	23.7	28.7	32.7	40.8
14.0–14.9	370	20.4	11.0	8.3	9.6	10.9	12.4	17.7	25.1	29.5	34.6	41.2
15.0–15.9	309	20.7	11.4	8.6	10.0	11.4	12.8	18.2	24.4	29.2	32.9	44.3
16.0–16.9	343	23.5	10.9	11.3	12.8	13.7	15.9	20.5	28.0	32.7	37.0	46.0
17.0–17.9	291	23.9	13.0	9.5	11.7	13.0	14.6	21.0	29.5	33.5	38.0	51.6
18.0–24.9	2588	25.2	13.4	10.0	12.0	13.5	16.1	21.9	30.6	37.2	42.0	51.6
25.0–29.9	1921	28.1	14.7	11.0	13.3	15.1	17.7	24.5	34.8	42.1	47.1	57.5
30.0–34.9	1619	31.6	16.1	12.2	14.8	17.2	20.4	28.2	39.0	46.8	52.3	64.5
35.0–39.9	1453	33.6	16.8	13.0	15.8	18.0	21.8	29.7	41.7	49.2	55.5	64.9
40.0–44.9	1390	34.3	16.2	13.8	16.7	19.2	23.0	31.3	42.6	51.0	56.3	64.5
45.0–49.9	961	36.0	17.2	13.6	17.1	19.8	24.3	33.0	44.4	52.3	58.4	68.8
50.0–54.9	1004	36.7	15.9	14.3	18.3	21.4	25.7	34.1	45.6	53.9	57.7	65.7
55.0–59.9	879	37.6	17.7	13.7	18.2	20.7	26.0	34.5	46.4	53.9	59.1	69.7
60.0–64.9	1389	37.1	16.0	15.3	19.1	21.9	26.0	34.8	45.7	51.7	58.3	68.3
65.0–69.9	1946	34.7	15.1	13.9	17.6	20.0	24.1	32.7	42.7	49.2	53.6	62.4
70.0–74.9	1463	32.9	14.6	13.0	16.2	18.8	22.7	31.2	41.0	46.4	51.4	57.7

From Frisancho AR: *Anthropometric standards for the assessment of growth and nutritional status*, Ann Arbor, 1990, The University of Michigan Press, p 52.

Appendix C—Parenteral and Enteral Formulas

Appendix C1

PARENTERAL FORMULAS

Intravenous Fat Emulsion

Product	Oil(%)		Fatty Acid Content (%)					Egg Yolk Phospholipids (%)	Glycerin (%)	Kcals/ml	Osmolality (mOsm/L)
	Safflower	Soybean	Linoleic	Oleic	Palmitic	Linolenic	Stearic				
Intralipid 10%		10	50	26	26	9	305	1.2	2.25	1.1	260
Intralipid 20%		20	50	26	10	9	3.5	1.2	2.25	2	260
Liposyn II 10%	5	5	65.8	17.7	8.8	4.2	3.4	1.2	2.5	1.1	276
Liposyn II 20%	10	10	65.8	17.7	8.8	4.2	3.4	1.2	2.5	2.0	258
Liposyn III 10%		10	54.5	22.4	10.5	8.3	4.2	1.2	2.5	1.1	284
Liposyn III 20%		20	54.5	22.4	10.5	8.3	4.2	1.2	2.5	2.0	292
Liposyn III 30%		30	54.5	22.4	10.5	8.3	4.2	1.2	2.5	2.9	293

Crystalline Amino Acid Infusions

	Aminosyn 3.5% (Abbott)	Aminosyn B 3.5% (Abbott)	Aminosyn 5% (Abbott)	Aminosyn B 5% (Abbott)	Travasol 5.5% (Clintec)	TrophAmine 6% (McGaw)
Amino Acid Concentration	3.5%	3.5%	5%	5%	5.5%	6%
Nitrogen (g/100 mL)	0.55	0.54	0.79	0.77	0.925	0.93
Amino Acids (Essential) (mg/100 mL)						
Isoleucine	252	231	360	300	263	490
Leucine	329	350	470	500	340	840
Lysine	252	368	360	525	318	490
Methionine	140	60	200	86	318	200
Phenylalanine	154	104	220	149	340	290
Threonine	182	140	260	200	230	250
Tryptophan	56	70	80	100	99	129
Valine	290	175	400	250	252	470
Amino Acids (Nonessential) (mg/100 mL)						
Alanine	448	348	640	497	1140	320
Arginine	343	356	490	509	570	730
Histidine[a]	105	105	150	150	241	290
Proline	300	253	430	361	230	410
Serine	147	186	210	265		230
Taurine						15
Tyrosine	31	95	44	135	22	140
Aminoacetic Acid (Glycine)	448	175	640	250	1140	220
Glutamic Acid		258		369		300
Aspartic Acid		245		350		190
Cysteine						< 14
Electrolytes (mEq/L)						
Sodium	7	16.3		19.3		5
Potassium			5.4			
Chloride					22	< 3
Acetate	46	25.2	86	35.9	48	56
Phosphate (mM/L)						
Osmolarity (mOsm/L)	357	308	500	438	575	526
Supplied in (mL)	1000[b]	1000[c]	500[d]	500[c]	500[e]	500
			1000[d]	1000[c]	1000[e]	
					2000[e]	
Labeled Indications						
Peripheral Parenteral Nutrition	Yes	Yes	Yes	Yes	Yes	Yes
Central TPH	No	No	Yes	Yes	Yes	Yes
Protein Sparing	Yes	Yes	Yes	Yes	Yes	No

	Aminosyn 7% (Abbott)	Aminosyn-OF 7% (Abbott)	Aminosyn B 7% (Abbott)	Aminosyn 8.5% (Abbott)
Amino Acid Concentration	7%	7%	7%	8.5%
Nitrogen (g/100 mL)	1.1	1.07	1.07	1.34
Amino Acides (Essential) (mg/100 mL)				
Isoleucine	510	534	462	620
Leucine	660	831	700	810
Lysine	510	475	735	624
Methionine	280	125	120	340
Phenylalanine	310	300	209	380
Threonine	370	360	280	460
Tryptophan	120	125	140	150
Valine	560	452	350	680
Amino Acids (Nonessential) (mg/100 mL)				
Alanine	900	490	695	110
Arginine	690	861	713	850
Histidine[a]	210	220	210	260
Proline	610	570	505	750
Serine	300	347	371	370
Taurine		50		
Tyrosine	44	44	189	44
Aminoacetic Acid (Glycine)	900	270	350	110
Glutamic Acid		576	517	
Aspartic Acid		370	490	
Cysteine				
Electrolytes (mEq/L)				
Sodium		3.4	31.3	
Potassium	5.4			5.4
Chloride				35
Acetate	105	32.5	50.3	90
Phosphate (mM/L)				
Osmolarity (mOsm/L)	700	586	612	860
Supplied in (mL)	500[d]	250[g]	500[c]	500[d]
		500[g]		1000[d]
Labeled Indications				
Peripheral Parenteral Nutrition	Yes	Yes	Yes	Yes
Central TPH	Yes	Yes	Yes	Yes
Protein Sparing	Yes	No	Yes	Yes

(continued on the following page)

Crystalline Amino Acid Infusions *(continued)*

	Aminosyn II 8.5% (Abbott)	Travasol 8.5% without electrolytes (Colintec)	FreAmine III 8.5% (B. Braun)
Amino Acid Concentration	8.5%	8.5%	8.5%
Nitrogen (g/100 mL)	1.3	1.43	
Amino Acides (Essential) (mg/100 mL)			
Isoleucine	561	406	590
Leucine	850	526	770
Lysine	893	492	620
Methionine	146	492	450
Phenylalanine	253	526	480
Threonine	340	356	340
Tryptophan	170	152	130
Valine	425	390	560
Amino Acids (Nonessential) (mg/100 mL)			
Alanine	844	1760	600
Arginine	865	880	810
Histidine[a]	255	372	240
Proline	614	356	950
Serine	450		500
Taurine			
Tyrosine	230	34	
Aminoacetic Acid (Glycine)	425	1760	1190
Glutamic Acid	627		
Aspartic Acid	595		
Cysteine			< 20
Electrolytes (mEq/L)			
Sodium	33.3		10
Potassium			
Chloride		34	< 3
Acetate	61.1	73	72
Phosphate (mM/L)			10
Osmolarity (mOsm/L)	742	890	810
Supplied in (mL)	500[c] 1000[c]	500[h] 1000[h] 2000[h]	500[i] 1000[i]
Labeled Indications			
Peripheral Parenteral Nutrition	Yes	Yes	Yes
Central TPH	Yes	Yes	Yes
Protein Sparing	Yes	No	Yes

	TrophAmine 10% (McGaw)	Aminosyn 10% (Abbott)	Aminosyn-OF 10% (Abbott)	Aminosyn II 10% (Abbott)
Amino Acid Concentration	10%	10%	10%	10%
Nitrogen (g/100 mL)	1.55	1.57	1.52	1.53
Amino Acides (Essential) (mg/100 mL)				
Isoleucine	820	720	760	660
Leucine	1400	940	1200	1000
Lysine	820	720	677	1050
Methionine	340	400	180	172
Phenylalanine	480	440	427	298
Threonine	420	520	512	400
Tryptophan	200	160	180	200
Valine	780	800	673	500
Amino Acids (Nonessential) (mg/100 mL)				
Alanine	540	1280	698	993
Arginine	1200	980	1227	1018
Histidine[a]	480	300	312	300
Proline	680	860	812	722
Serine	380	420	495	530
Taurine	25		70	
Tyrosine	240	44	40	270
Aminoacetic Acid (Glycine)	360	1280	385	500
Glutamic Acid	500		620	738
Aspartic Acid	320		527	700
Cysteine	< 16			
Electrolytes (mEq/L)				
Sodium	5		3.4	45.3
Potassium		5.4		
Chloride	< 3			
Acetate	97	148	46.3	71.8
Phosphate (mM/L)				
Osmolarity (mOsm/L)	875	1000	829	873
Supplied in (mL)	500[f]	500[d] 1000[d]	1000[i]	500[c] 1000[c]
Labeled Indications				
Peripheral Parenteral Nutrition	Yes	Yes	Yes	Yes
Central TPH	Yes	Yes	Yes	Yes
Protein Sparing	No	Yes	No	Yes

(continued on the following page)

(continued)

	Travasol 10% (Clintec)	FreAmine III 10% (McGaw)	Novamine (Clintec)	Aminosyn 15% (Clintec)	Aminosyn II15% (Abbott)
Amino Acid Concentration	10%	10%	11.4%	15%	15%
Nitrogen (g/100 mL)	1.65	1.53	1.8	2.37	2.3
Amino Acides (Essential) (mg/100 mL)					
Isoleucine	600	690	570	749	990
Leucine	730	910	790	1040	1500
Lysine	580	730	900	1180	1575
Methionine	400	530	570	749	258
Phenylalanine	560	560	790	1040	447
Threonine	420	400	570	749	600
Tryptophan	180	150	190	250	300
Valine	580	660	730	960	750
Amino Acids (Nonessential) (mg/100 mL)					
Alanine	2070	710	1650	2170	1490
Arginine	1150	950	1120	1470	1527
Histidine[a]	480	280	680	894	450
Proline	680	1120	680	894	1083
Serine	500	590	450	592	795
Taurine					
Tyrosine	40		30	39	405
Aminoacetic Acid (Glycine)	1030	1400	790	1040	750
Glutamic Acid			570	749	1107
Aspartic Acid			660	434	1050
Cysteine		< 24			

	Travasol 10% (Clintec)	FreAmine III 10% (McGaw)	Novamine (Clintec)	Aminosyn 15% (Clintec)	Aminosyn II15% (Abbott)
Electrolytes (mEq/L)					
Sodium		10			62.7
Potassium					
Chloride	40	< 3			
Acetate	87	≈89	114	151	107.6
Phosphate (mM/L)		10			
Osmolarity (mOsm/L)	1000	≈ 950	1057	1388	1300
Supplied in (mL)	250[l,m] 500[l,m] 1000[l,m] 2000[l,m]	500[i] 1000[i]	500[n] 1000[n]	500[n] 1000[n]	2000[o]
Labeled Indications					
Peripheral Parenteral Nutrition	Yes	Yes	Yes	Yes	Yes
Central TPH	Yes	Yes	Yes	Yes	Yes
Protein Sparing	Yes	No	Yes	No	No

[a] Histidine is considered an essential amino acid in infants and in renal failure.
[b] With 7 mEq/L sodium from the antioxidant sodium hydrosulfite.
[c] Includes 20 mg/dL sodium hydrosulfite.
[d] Includes 5.4 mEq/L potassium from the antioxidant potassium metabisulfite.
[e] With ≈ 3 mEq/L sodium bisulfite.
[f] With < 50 mg sodium metabisulfite per 100 mL.
[g] From the antioxidant sodium hydrosulfite.
[h] With 3 mEq/L sodium bisulfite.
[i] With < 0.1 g sodium bisulfite per 100 mL.
[j] With 230 mg sodium hydrosulfite per 100 m L.
[k] Potassium derived from the antioxidant potassium metabisulfite.
[l] Acetate in Viaflex container ≈ 60 mEq/L; osmolarity is 970 mOsm/L.
[m] Sizes also come in Viaflex containers.
[n] With 30 mg sodium metabisulfite.
[o] With 60 mg sodium hydrosulfite per 100 mL.

TABLE C2.1

Standard Formulas

Product[a]	Volume to Meet 100% RDI[b] (mL)	Energy (kcal/mL)	Protein or Amino Acids (g/L)	Carbohydrate (g/L)	Fat (g/L)	Osmolality[c] (mOsm/kg)	Notes
Lactose-Free, Standard Formulas							
Isocal®	1890	1.06	34	135	44	270	20% fat from MCT
Isosource® Standard	1165	1.20	43	170	39	490	50% fat from MCT
Nutren® 1.0	1500	1.00	40	127	38	315	25% fat from MCT
Osmolite®	2000	1.06	37	151	35	300	20% fat from MCT
Lactose-Free, Fiber-Containing Formulas							
Jevity®	1321	1.06	44	155	35	300	14 g fiber/L
Nutren® Fiber	1500	1.00	40	127	38	330	14 g fiber/L
ProBalance®	1000	1.20	54	156	41	350	10 g fiber/L
Promote® with Fiber	1000	1.00	63	138	28	380	14 g fiber/L
Lactose-Free, High-kCalorie Formulas							
Comply®	830	1.50	60	180	61	460	20% fat from MCT
Deliver® 2.0	1000	2.00	75	200	101	640	30% fat from MCT
Nutren® 1.5	1000	1.50	60	169	68	430	50% fat from MCT
Nutren® 2.0	750	2.00	80	196	104	745	75% fat from MCT
Lactose-Free, High-Protein Formulas							
Isocal® HN	1180	1.06	44	124	45	270	Low residue
Isocal® HN Plus	1000	1.20	54	156	40	400	Low residue
Promote®	1000	1.00	63	130	26	340	20% fat from MCT low residue
Ultracal® HN Plus	1000	1.20	54	156	40	370	30% fat from MCT, 10 g fiber/L
Special-Use Formulas: Pediatric (1 to 10 years)							
Compleat® Pediatric	Varies[d]	1.00	38	130	39	380	Blenderized formula, 6.8 g fiber/L
Kindercal® TF	Varies[d]	1.06	30	135	44	345	12% fat from MCT
Nutren Junior®	Varies[d]	1.00	30	110	50	350	21% fat from MCT
PediaSure®	Varies[d]	1.00	30	110	50	430	
Special-Use Formulas: Glucose Intolerance							
Choice DM® TF	1120	1.06	45	119	51	300	14 g fiber/L
Glucerna®	1420	1.00	42	96	54	355	14 g fiber/L
Glytrol®	1400	1.00	45	100	48	380	15 g fiber/L; 20% fat from MCT
Resource® Diabetic	1180	1.06	63	100	47	300	13 g fiber/L
Special-Use Formulas: Immune System Support							
Impact®	1500	1.00	56	130	28	375	Enriched with arginine, nucleic acids, and omega-3 fatty acids
Impact® 1.5	1250	1.50	84	140	69	550	Same as above
Impact® Glutamine	1000	1.30	78	150	43	630	Same as above and enriched with glutamine; 10 g fiber/L

NOTE: MCT Medium-chain triglycerides.

[a] Formulas come in ready-to-use (liquid) form unless specified under "Notes."

[b] RDI Reference Daily Intakes, which are labeling standards for vitamins, minerals, and protein. Consuming 100 percent of the RDI will meet the nutrient needs of most people using the product.

[c] Osmolality may vary, depending on the flavorings added to a product.

[d] Depends on age of child.

(continued on the following page)

TABLE C2.1 *(continued)*

Standard Formulas

Product[a]	Volume to Meet 100% RDI[b] (mL)	Energy (kcal/mL)	Protein or Amino Acids (g/L)	Carbohydrate (g/L)	Fat (g/L)	Osmolalityc (mOsm/kg)	Notes
Special-Use Formulas: Renal Failure							
Magnacal® Renal	1000	2.00	75	200	101	570	20% fat from MCT; intended for use once hemodialysis has been instituted
Nepro®	947	2.00	70	222	96	665	High-calcium, low-phosphorus; intended for use once dialysis has been instituted
Novasource® Renal	1000	2.00	74	200	100	700	Low in electrolytes; intended for use once dialysis has been instituted
NutriRenal®	750	2.00	70	205	104	650	50% fat from MCT; enriched with vitamins C and B6, olate, zinc, and selenium; intended for use once dialysis as been instituted
Special-Use Formulas: Respiratory Insufficiency							
Novasource® Pulmonary	933	1.50	75	150	68	650	8 g fiber/L
NutriVent®	1000	1.50	68	100	94	330	55% kcal from fat, 40% fat from MCT
Oxepa	1420	1.50	63	106	94	493	55% kcal from fat, enriched with antioxidant nutrients
Pulmocare®	1420	1.50	63	106	93	475	55% kcal from fat, 20% fat from MCT, enriched with antioxidant nutrients
Special-Use Formulas: Wound Healing							
Protain XL®	1250	1.00	57	145	30	340	9 g fiber/L, 20% fat from MCT, enriched with vitamins A and C and zinc
Replete®	1000	1.00	62	113	34	300	Enriched with vitamins A and C and zinc; 25% fat from MCT

TABLE C2.2

Hydrolyzed Protein Formulas

Product[a]	Volume to Meet 100% RDIb (mL)	Energy (kcal/mL)	Protein or Amino Acids (g/L)	Carbohydrate (g/L)	Fat (g/L)	Osmolalityc (mOsm/kg)	Notes
Special-Use Hydrolyzed Formulas: Hepatic Insufficiency							
NutriHep®	1000	1.50	40	290	21	790	Free amino acids, high in branched chain amino acids, low in aromatic amino acids
Special-Use Hydrolyzed Formulas: HIV Infection or AIDS							
Advera®	1184	1.28	60	216	23	680	78% hydrolyzed and 22% intact protein, low fat, fiber added, enriched with vitamins E, C, B6, B12, and folate
Special-Use Hydrolyzed Formulas: Immune System Support							
Alitraq®	1500	1.00	53	165	16	575	Powder form; 47% free amino acids, 42% small peptides, enriched with glutamine and arginine
Crucial®	1000	1.50	94	135	68	490	Enriched with arginine, glutamine, antioxidant nutrients, and zinc
Perative®	1500	1.30	67	180	37	460	Enriched with arginine and beta-carotene
Vivonex® Plus	1800	1.00	45	190	7	650	Powder form; 100% free amino acids, enriched with glutamine, arginine, and branched-chain amino acids
Special-Use Hydrolyzed Formulas: Malabsorption							
Criticare HN®	1890	1.06	38	220	5	650	Mix of free amino acids and small peptides
Optimental®	1422	1.00	51	139	28	540	Contains MCT and arginine; enriched with vitamins C and E and beta-carotene
Peptamen®	1500	1.00	40	127	39	270	70% fat from MCT
Vivonex® T.E.N.	2000	1.00	38	210	3	630	Powder form; 100% free amino acids, enriched with glutamine
Special-Use Hydrolyzed Formulas: Pediatric (1 to 10 years)							
Peptamen Junior®	Varies[c]	1.0	30	138	39	260	60% fat from MCT; contains glutamine
Vivonex® Pediatric	Varies[c]	0.8	24	130	24	360	Powder form; 100% free amino acids

[a] RDI = Reference Daily Intakes, which are labeling standards for vitamins, minerals, and protein. Consuming 100 percent of the RDI will meeet the nutrient needs of most people using the product.

[b] Osmolality may vary depending on the flavorings added to a product.

[c] Depends on age of child.

TABLE C2.3

Protein Modules

Product	Form	Major Protein Source	Energy (kcal/g)	Protein (g/100 g)
Casec®	Powder	Calcium caseinate	3.8	90
ProMod®	Powder	Whey protein	4.2	75

TABLE C2.4

Carbohydrate Modules

Product	Form	Major Carbohydrate Source	Energy (kcal/mL or g)
Polycose Liquid®	Liquid	Hydrolyzed cornstarch	2.0 kcal/ml
Polycose Powder®	Powder	Hydrolyzed cornstarch	3.8 kcal/g

TABLE C2.5

Fat Modules

Product	Form	Major Fat Source	Energy (kcal/mL)	Protein (g/100 mL)
MCT Oil®	Liquid	Medium-chain triglycerides	7.7	86
Microlipid®	Liquid	Safflower oil	4.5	51

Appendix D—Nutrient Tables

Appendix D1

SATURATED FAT, TOTAL FAT, CHOLESTEROL, AND OMEGA-3 CONTENT OF MEAT, FISH, AND POULTRY (3 OZ. PORTIONS)

Saturated Fat, Total Fat, Cholesterol, and Omega-3 Content of Meat, Fish, and Poultry in 3-Ounce Portions Cooked Without Added Fat

Source	Saturated Fat g/3 oz	Total Fat g/3 oz	Cholesterol mg/3 oz	Omega-3 g/3 oz
Lean Red Meats				
Beef (rump roast, shank, bottom round, sirloin)	1.4	4.2	71	–
Lamb (shank roast, sirloin roast, shoulder roast, loin chops, sirloin chops, center leg chop)	2.8	7.8	78	–
Pork (sirloin cutlet, loin roast, sirloin roast, center roast, butterfly chops, loin chops)	3.0	8.6	71	–
Veal (blade roast, sirloin chops, shoulder roast, loin chops, rump roast, shank)	2.0	4.9	93	–
Organ Meats				
Liver				
Beef	1.6	4.2	331	–
Calf	2.2	5.9	477	–
Chicken	1.6	4.6	537	–
Sweetbread	7.3	21.3	250	–
Kidney	0.9	2.9	329	–
Brains	2.5	10.7	1,747	–
Heart	1.4	4.8	164	–
Poultry				
Chicken (without skin)				
Light (roasted)	1.1	3.8	72	–
Dark (roasted)	2.3	8.3	71	–
Turkey (without skin)				
Light (roasted)	0.9	2.7	59	–
Dark (roasted)	2.0	6.1	72	–
Fish				
Haddock	0.1	0.8	63	0.22
Flounder	0.3	1.3	58	0.47
Salmon	1.7	7.0	54	1.88
Tuna, light, canned in water	0.2	0.7	25	0.24

(continued on the following page)

(continued)

Saturated Fat, Total Fat, Cholesterol, and Omega-3 Content of Meat, Fish, and Poultry in 3-Ounce Portions Cooked Without Added Fat

Source	Saturated Fat g/3 oz	Total Fat g/3 oz	Cholesterol mg/3 oz	Omega-3 g/3 oz
Shellfish				
Crustaceans				
Lobster	0.1	0.5	61	0.07
Crab meat				
Alaskan King Crab	0.1	1.3	45	0.38
Blue Crab	0.2	1.5	85	0.45
Shrimp	0.2	0.9	166	0.28
Mollusks				
Abalone	0.3	1.3	144	0.15
Clams	0.2	1.7	57	0.33
Mussels	0.7	3.8	48	0.70
Oysters	1.3	4.2	93	1.06
Scallops	0.1	1.2	56	0.36
Squid	0.6	2.4	400	0.84

Appendix D2

HEME AND NON-HEME IRON CONTENT OF SELECTED FOODS

Selected Food Sources of Heme Iron

Food	Milligrams per serving	% DV*
Chicken liver, cooked, 3½ ounces	12.8	70
Oysters, breaded and fried, 6 pieces	4.5	25
Beef, chuck, lean only, braised, 3 ounces	3.2	20
Clams, breaded, fried, ¾ cup	3.0	15
Beef, tenderloin, roasted, 3 ounces	3.0	15
Turkey, dark meat, roasted, 3½ ounces	2.3	10
Beef, eye of round, roasted, 3 ounces	2.2	10
Turkey, light meat, roasted, 3½ ounces	1.6	8
Chicken, leg, meat only, roasted, 3½ ounces	1.3	6
Tuna, fresh bluefin, cooked, dry heat, 3 ounces	1.1	6
Chicken, breast, roasted, 3 ounces	1.1	6
Halibut, cooked, dry heat, 3 ounces	0.9	6
Crab, blue crab, cooked, moist heat, 3 ounces	0.8	4
Pork, loin, broiled, 3 ounces	0.8	4
Tuna, white, canned in water, 3 ounces	0.8	4
Shrimp, mixed species, cooked, moist heat, 4 large	0.7	4

Selected Food Sources of Nonheme Iron

Food	Milligrams per serving	% DV*
Ready-to-eat cereal, 100% iron fortified, ¾ cup	18.0	100
Oatmeal, instant, fortified, prepared with water, 1 cup	10.0	60
Soybeans, mature, boiled, 1 cup	8.8	50
Lentils, boiled, 1 cup	6.6	35
Beans, kidney, mature, boiled, 1 cup	5.2	25
Beans, lima, large, mature, boiled, 1 cup	4.5	25
Beans, navy, mature, boiled, 1 cup	4.5	25
Ready-to-eat cereal, 25% iron fortified, ¾ cup	4.5	25
Beans, black, mature, boiled, 1 cup	3.6	20
Beans, pinto, mature, boiled, 1 cup	3.6	20
Molasses, blackstrap, 1 tablespoon	3.5	20
Tofu, raw, firm, ½ cup	3.4	20
Spinach, boiled, drained, ½ cup	3.2	20

(continued on the following page)

(continued)

Selected Food Sources of Nonheme Iron

Food	Milligrams per serving	% DV*
Spinach, canned, drained solids ½ cup	2.5	10
Black-eyed peas (cowpeas), boiled, 1 cup	1.8	10
Spinach, frozen, chopped, boiled ½ cup	1.9	10
Grits, white, enriched, quick, prepared with water, 1 cup	1.5	8
Raisins, seedless, packed, ½ cup	1.5	8
Whole wheat bread, 1 slice	0.9	6
White bread, enriched, 1 slice	0.9	6

*DV = Daily Value. DVs are reference numbers developed by the Food and Drug Administration (FDA) to help consumers determine if a food contains a lot or a little of a specific nutrient. The FDA requires all food labels to include the percent DV (%DV) for iron. The percent DV tells you what percent of the DV is provided in one serving. The DV for iron is 18 milligrams (mg). A food providing 5% of the DV or less is a low source while a food that provides 10–19% of the DV is a good source. A food that provides 20% or more of the DV is high in that nutrient. It is important to remember that foods that provide lower percentages of the DV also contribute to a healthful diet. For foods not listed in this table, please refer to the U.S. Department of Agriculture's Nutrient Database Web site: http://www.nal.usda.gov/fnic/cgi-bin/nut_search.pl.

Sources: U.S. Department of Agriculture, Agricultural Research Service. 2003. USDA Nutrient Database for Standard Reference, Release 16. Nutrient Data Laboratory Home Page, http://www.nal.usda.gov/fnic/foodcomp. Office of Dietary Supplements, National Institutes of Health. Dietary Supplement Fact Sheet: Iron [monograph on the Internet]. Bethesda (MD): Office of Dietary Supplements; 2005. Available from http://ods.od.nih.gov/factsheets/iron.asp#en10.

Appendix D3

OXALATE CONTENT OF SELECTED FOODS

High Content (> 10 mg/serving)	Moderately High Content (2–10 mg/serving)	Low Content (< 2 mg/serving)
Beans in tomato sauce	Apple and apple juice	Avocado
Beets*	Apricots	Bananas
Blackberries	Asparagus	Broccoli
Raspberries	Bacon	Beef (lean)
Blueberries	Bottled beer	Bing cherries
Celery	Carrots	Brussels sprouts
Chard	Coffee (8 oz)	Cabbage
Chocolate*	Corn	Cheese
Cocoa*	Lettuce	Chicken noodle soup
Collard greens	Lima beans	Cola Sodas
Currants, red	Mushrooms	Cucumber
Dandelion greens	Onions	Eggs
Eggplant	Oranges	Grapes, Thompson
Escarole	Peaches	Grapefruit
Fruit salad (canned)	Sardines	Jelly
Fruit juice with berries	Sponge cake	Lamb (lean)
Green bell peppers	Tomatoes and tomato juice	Lemonade/Limeade
Grapes, Concord	Turnip	Melons
Grits (white corn)	White bread	Milk
Kale		Nectarine
Leeks		Noodles
Lemon/Lime peel		Oatmeal
Marmalade		Oils
Nuts (peanuts/almonds)*		Orange juice (4 oz)
Okra		Pears
Parsley		Peas
Parsnips		Pineapple
Pepper (> 2 tsp/day)		Pork (lean)
Pokeweed		Poultry
Rhubarb*		Prunes and Plums
Rutabaga		Radishes
Soybeans & all products		Rice
Spinach*		Seafood
Strawberries*		Spaghetti
Summer Squash		Wine
Sweet potatoes		Yogurt
Tea*		
Tofu		
Watercress		
Wheat germ*/bran		

* Have been documented to raise urinary oxalate excretion.

Appendix D4

DIETARY AND SUPPLEMENT CALCIUM GUIDE

Food	Serving Size	Approximate Calcium Content (mg)*
Calcium fortified orange juice	1 cup	290–300
Milk	1 cup	285–300
Yogurt	1 cup	275–450
Salmon, canned, with bones	3 oz	205
Cheese	1 oz	175–275
Tofu, firm	½ cup	155–260
Ice cream	½ cup	90–135
Frozen yogurt	½ cup	105
Turnip greens	½ cup	100
Dried figs	3	80
Broccoli	½ cup	45
Soy milk**	1 cup	varies widely by brand (check label)
Fortified cereals	1 serving	varies widely by brand (check label)
Multivitamin with minerals	1 dose	0–210

* Calcium content of foods may vary; read labels to determine the actual calcium content of a certain food.

** The nutrient content of soy milk varies greatly depending on the manufacturing process and whether the product is fortified with nutrients such as calcium.

Reading Labels

The food label on most products will list its calcium content. The amount of calcium in a product is expressed as a percentage (%). This percentage is based on the recommended calcium intake for many adults of 1000 mg per day.

Example: A product label says the food contains 30% of the daily need for calcium:

30% of 1000 mg per day = 300 mg of calcium

These percentages are meant to be used as a guide; however, since individuals may have different calcium needs, they may not be exact for all adults.

Calcium needs (mg)	Calcium needed per day based on food labels
1000	100%
1200	120%
1500	150%

Calcium Supplements

If you are unable to take enough calcium through the diet, calcium supplements are available to help meet calcium needs. The most common types of calcium supplements are calcium carbonate and calcium citrate.

Calcium is best absorbed in doses of 500 mg or less. If you need to take more than 500 mg of calcium supplements per day, consider taking several smaller doses throughout the day.

Commonly Available Calcium Supplements

Brand	Calcium (mg) per tablet	Vitamin D (IU) per tablet	Approximate cost per tablet	Calcium Source
TumsÃ®	200	0	$ .02	Calcium Carbonate
Extra Strength Tums®	300	0	$ .04	Calcium Carbonate
Oscal® 500	500	0	$ .10	Calcium Carbonate
Oscal® 500 + D	500	200	$ .11	Calcium Carbonate
Caltrate® 600	600	0	$ .09	Calcium Carbonate
Caltrate® 600 + D	600	200	$ .09	Calcium Carbonate
Caltrate® 600 Plus® *	600	200	$ .11	Calcium Carbonate
Viactiv® *	**500**	**100**	**$ .11**	**Calcium Carbonate**
Citrical®	200	0	$ .07	Calcium Citrate
Citrical® + D	315	200	$ .11	Calcium Citrate

* Also contain additional vitamins &/or minerals; Viactiv® contains 20 calories per piece

Reprinted from: University of Virginia Health System Digestive Health Center of Excellence. Calcium and Vitamin D [monograph on the Internet]. Charlottesville (VA): University of Virginia; 2005. Available from: http://www.healthsystem.virginia.edu/internet/digestive-health/nutrition/calcium1.cfm.

Appendix D5

RESISTANT STARCH GUIDE

Types of Resistant Starch

(Cargill Company—http://www.cargilltexturizing.com/1218.html 2004)

RS1: Physically inaccessible starch
- e.g. partly milled grains and seeds and legumes

RS2: Granular starch/Native starch granule
- e.g. native uncooked potato starch and green banana

RS3: Retrograded starch, mainly retrograded amylose
- e.g. cooled-cooked potato, bread, corn flakes

Resistant Starch in Selected Foods

Starchy Foods	%RS (d.b.)	RS Type
Haricot beans (boiled 40 min)	40.0	RS1-**RS3**
Peas (frozen, boiled 5 min)	26.3	RS1-**RS3**
Lentils (boiled 20 min, cold)	16.4	RS1-**RS3**
Boiled potato (cold)	13.5	RS3
Pear barley (boiled 60 min, cold)	12.3	RS3
Millet (boiled 20 min, cold)	7.9	RS3
Boiled potato (hot)	6.7	RS3
Spaghetti (freshly cooked)	6.3	RS1
Ryvita crispbread	4.9	RS3
Cornflakes	3.8	RS3
Porridge oat	3.1	RS1-**RS3**
Abbey crunch biscuit	2.8	RS1-**RS3**
Digestive biscuit	2.1	RS1-**RS3**
Wholemeal bread	1.6	RS3
Instant potato	1.4	RS3
White bread	1.3	RS3
Rice Krispies	1.3	RS3
Banana (green-ripe)	0–60	RS2

Appendix E—Menu Planning

Appendix E1

EXCHANGE LIST FOR MEAL PLANNING

United States: Exchange Lists

CONTENTS

The Exchange Groups and Lists
Combining Food Group Plans and Exchange Lists

The Exchange Groups and Lists

The exchange system sorts foods into three main groups by their proportions of carbohydrate, fat, and protein. These three groups—the carbohydrate group, the fat group, and the meat and meat substitutes group (protein)—organize foods into several exchange lists (see Table E1.1). Then any food on a list can be "exchanged" for any other on that same list. The carbohydrate group covers these exchange lists:

- Starch (cereals, grains, pasta, breads, crackers, snacks, starchy vegetables, and dried beans, peas, and lentils).
- Fruit.
- Milk (fat-free, reduced fat, and whole).
- Other carbohydrates (desserts and snacks with added sugars and fats).
- Vegetables.

The fat group covers this exchange list:

- Fats.

The meat and meat substitutes group (protein) covers these exchange lists:

- Meat and meat substitutes (very lean, lean, medium-fat, and high-fat).

Portion Sizes The exchange system helps people control their energy intakes by paying close attention to portion sizes. The portion sizes have been carefully adjusted and defined so that a portion of any food on a given list provides roughly the same amount of carbohydrate, fat, and protein and, therefore, total kcalories. Any food on a list can then be exchanged, or traded, for any other food on that same list without significantly affecting the diet's balance or total kcalories. For example, a person may select either 17 small grapes or ½ large grapefruit as one fruit portion, and either

TABLE E1.1 The Exchange Groups and Lists

Group/Lists	Typical Item/Portion Size	Carbohydrate (g)	Protein (g)	Fat (g)	Energy[a] (kcal)
Carbohydrate Group					
Starch[b]	1 slice bread	15	3	0–1	80
Fruit	1 small apple	15	—	—	60
Milk					
Fat-free, low-fat	1 c fat-free milk	12	8	0–3	90
Reduced-fat	1 c reduced-fat milk	12	8	5	120
Whole	1 c whole milk	12	8	8	150
Other carbohydrates[c]	2 small cookies	15	varies	varies	varies
Vegetable (nonstarchy)	½ c cooked carrots	5	2	—	25
Meat and Meat Substitute Group [d]					
Meat					
Very lean	1 oz chicken (white meat, no skin)	—	7	0–1	35
Lean	1 oz lean beef	—	7	3	55
Medium-fat	1 oz ground beef	—	7	5	75
High-fat	1 oz pork sausage	—	7	8	100
Fat Group					
Fat	1 tsp butter	—	—	5	45

[a]The energy value for each exchange list represents an approximate average for the group and does not reflect the precise number of grams of carbohydrate, protein, and fat. For example, a slice of bread contains 15 grams of carbohydrate (that's 60 kcalories), 3 grams protein (that's another 12 kcalories), and a little fat—rounded to 80 kcalories for ease in calculating. A half-cup of vegetables (not including starchy vegetables) contains 5 grams carbohydrate (20 kcalories) and 2 grams protein (8 more), which has been rounded down to 25 kcalories.
[b]The starch list includes cereals, grains, breads, crackers, snacks, starchy vegetables (such as corn, peas, and potatoes), and legumes (dried beans, peas, and lentils).
[c]The other carbohydrates list includes foods that contain added sugars and fats such as cakes, cookies, doughnuts, ice cream, potato chips, pudding, syrup, and frozen yogurt.
[d]The meat and meat substitutes list includes legumes, cheeses, and peanut butter.

choice would provide roughly 60 kcalories. A whole grapefruit, however, would count as 2 portions.

A *portion* in the exchange system is not always the same as a *serving* in the Daily Food Guide, especially when it comes to meats. The exchange system lists meats and most cheeses in single ounces; that is, one *portion* (or *exchange*) of meat is 1 ounce, whereas one *serving* in the Daily Food Guide is 2 to 3 ounces. Calculating meat by the ounce encourages a person to keep close track of the exact amounts eaten. This in turn helps control energy and fat intakes. Be aware, too, that most people do not serve foods in carefully measured portions, nor do the amounts reflect the exchange system or Daily Food Guide serving sizes. Many restaurants, for example, offer 8- to 16-ounce steaks that are the equivalent to four or five (2- to 3-ounce) *servings* of meat. Similarly, a bakery may sell muffins or bagels that are two to three times the size of a typical bread serving.

To apply the system successfully, users must become familiar with portion sizes. A convenient way to remember the portion sizes and energy values is to keep in mind a typical item from each list (review Table E1.1).

The Foods on the Lists Foods do not always appear on the exchange list where you might first expect to find them. They are grouped according to their energy-nutrient contents rather than by their source (such as milks), their outward appearance, or their vitamin and mineral contents. Notice, for example, that cheeses are grouped with meats (not milk) because, like meats, cheeses contribute energy from protein and fat but provide negligible carbohydrate. Similarly, starchy vegetables such as potatoes are found on the starch list with breads and cereals, not with the vegetables, and bacon is with the fats and oils, not with the meats.

Users of the exchange lists learn to view mixtures of foods, such as casseroles and soups, as combinations of foods from different exchange lists. They also learn to interpret food labels with the exchange system in mind (see Figure E1.1).

Controlling Energy and Fat By assigning items like bacon to the fat list, the exchange system alerts consumers to foods that are unexpectedly high in fat. Even the starch list specifies which grain products contain added fat (such as biscuits, muffins, and waffles). In addition, the exchange system encourages users to think of fat-free milk as milk and of whole milk as milk with added fat, and to think of very lean meats as meats and of lean, medium-fat, and high-fat meats as meats with added fat. To that end, foods on the milk and meat lists are separated into categories based on their fat contents. The milk group is classed as fat-free, reduced-fat, and whole; the meat group as very lean, lean, medium-fat, and high-fat.

Control of food energy and fat intake can be highly successful with the exchange system. Exchange plans do not, however, guarantee adequate intakes of vitamins and minerals. Food group plans work better from that standpoint because the food groupings are based on similarities in vitamin-mineral content. In the exchange system, for example, meats are grouped with cheeses, yet the meats are iron-rich and calcium-poor, whereas the cheeses are iron-poor and calcium-rich. To take advantage of the strengths of both food group plans and exchange patterns, and to compensate for their weaknesses, diet planners often combine these two diet-planning tools.

FIGURE E1.1 Seeing Exchanges on a Food Label

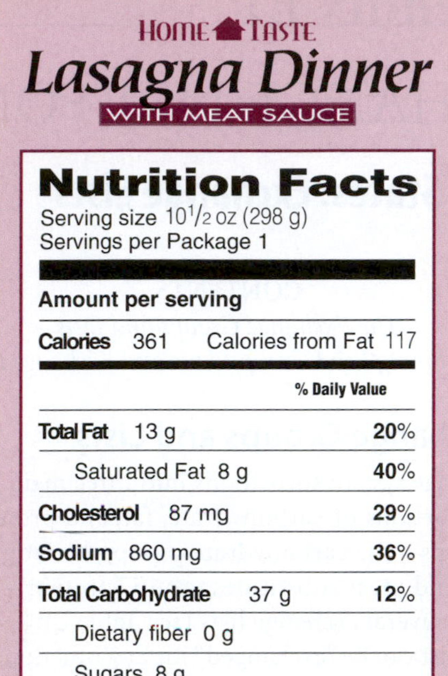

Nutrition Facts
Serving size $10^1/_2$ oz (298 g)
Servings per Package 1

Amount per serving		
Calories 361	Calories from Fat 117	
		% Daily Value
Total Fat 13 g		**20%**
Saturated Fat 8 g		**40%**
Cholesterol 87 mg		**29%**
Sodium 860 mg		**36%**
Total Carbohydrate 37 g		**12%**
Dietary fiber 0 g		
Sugars 8 g		
Protein 26 g		

Can you "see" these exchanges in the label above?

Exchange	Carbohydrate	Protein	Fat
2 starches	30 g	6 g	—
1 vegetable	5 g	2 g	—
3 medium-fat meats	—	21 g	15 g
Exchange totals	**35**	**29**	**15**
Label totals	**37**	**26**	**13**

Knowing that foods on the starch list provide 15 grams of carbohydrate and those on the vegetable list provide 5, you can count a lasagna dinner that provides 37 grams of carbohydrate as "2 starches and 1 vegetable"; knowing that foods on the meat list provide 7 grams of protein, you might count it as "3 meats"; the grams of fat suggest that the meat (and cheese) is probably medium-fat.

Combining Food Group Plans and Exchange Lists

A person may find that using a food group plan together with the exchange lists eases the task of choosing foods that provide all the nutrients. The food group plan ensures the all classes of nutritious foods are included, thus promoting adequacy, balance, and variety. The exchange system classifies the food selections by their energy-yielding nutrients, thus controlling energy and fat intakes.

Table E1.2 shows how to use the Daily Food Guide plan together with the exchange lists to plan a diet. The Daily Food Guide ensures that a certain number of servings is chosen from each of the five food groups (see the first column of the table). The second column translates the number of servings (using the midpoint) into exchanges. With the addition of a small amount of fat, this sample diet plan provides abut 1750 kcalories. Most people can meet their needs for all the nutrients within this reasonable energy allowance. The next step in diet planning is to assign the exchanges to meals and snacks. The final plan might look like the one in Table E1.3.

Next, a person could begin to fill in the plan with real foods to create a menu (use Tables E1.4 through E1.12). For example, the breakfast plan calls for 2 starch exchanges, 1 fruit exchange, and 1 fat-free milk exchange. A person might select a bowl of shredded wheat with banana slices and milk:

1 cup shredded wheat 2 starch exchanges.
1 small banana 1 fruit exchange.
1 cup fat-free milk 1 milk exchange.

Or half a bagel and a bowl of cantaloupe pieces topped with yogurt:

½ bagel 2 starch exchanges
⅓ cantaloupe melon 1 fruit exchange
⅔ cup fat-free plain yogurt 1 milk exchange.

Then the person could move on to complete the menu for lunch, dinner, and snacks. (Table E1.3 includes a sample menu.) As you can see, we all make countless food-related decisions daily—whether we have a plan or not. Following a plan, like the Daily Food Guide, that incorporates health recommendations and diet-planning principles helps a person to make wise decisions.

TABLE E1.2 Diet Planning with the Exchange System Using the Daily Food Guide Pattern

Patterns from Daily Food Guide Plan	Selections Made Using the Exchange System	Energy Cost (kcal)
Grains (breads and cereals)—6 to 11 servings	Starch list—select 9 exchanges	720
Vegetables—3 to 5 servings	Vegetable list—select 4 exchanges	100
Fruits—2 to 4 servings	Fruit list—select 3 exchanges	180
Meat—2 to 3 servings[a]	Meat list—select 6 lean exchanges	330
Milk—2 servings	Milk list—select 2 fat-free exchanges	180
	Fat list—select 5 exchanges	225
Total		1735

[a]In the food group plan, 1 serving is 2 to 3 ounces; in the exchange system, 1 exchange is 1 ounce. The Daily Food Guide suggests that amounts should total 5 to 7 ounces of meat daily.

TABLE E1.3 A Sample Diet Plan and Menu

This diet plan is one of many possibilities. It follows the number of servings suggested by the Daily Food Guide and meets dietary recommendations to provide 45 to 65 percent of its kcalories from carbohydrate, 10 to 35 percent from protein, and 20 to 35 percent from fat.

Exchange	Breakfast	Lunch	Snack	Dinner	Snack
9 starch	2	2	1	3	1
4 vegetables				4	
3 fruit	1	1	1		
6 lean meat		2		4	
2 fat-free milk	1				1
5 fat		1		4	

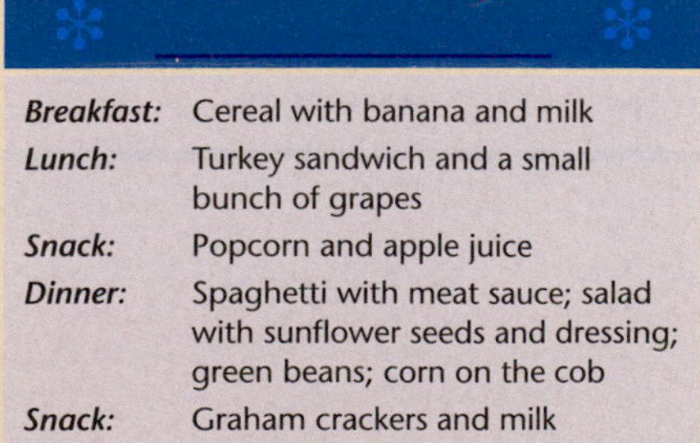

Breakfast: Cereal with banana and milk
Lunch: Turkey sandwich and a small bunch of grapes
Snack: Popcorn and apple juice
Dinner: Spaghetti with meat sauce; salad with sunflower seeds and dressing; green beans; corn on the cob
Snack: Graham crackers and milk

TABLE E1.4 U.S. Exchange System: Starch List

1 starch exchange = 15 g carbohydrate, 3 g protein, 0–1 g fat, and 80 kcal
NOTE: In general, one starch exchange is ½ c cooked cereal, grain, or starchy vegetable; ⅓ c cooked rice or pasta; 1 oz of bread; ¾ to 1 oz snack food.

Serving Size	Food
Bread	
¼ (1 oz)	Bagel, 4 oz
2 slices (1½ oz)	Bread, reduced-kcalorie
1 slice (1 oz)	Bread, white (including French and Italian), whole-wheat, pumpernickel, rye
4 (⅔ oz)	Bread sticks, crisp, 4″ x ½″
½	English muffin
½ (1 oz)	Hot dog or hamburger bun
¼	Naan, 8″ x 2″
1	Pancake, 4″ across, ¼″ thick
½	Pita, 6″ across
1 (1 oz)	Plain roll, small
1 slice (1 oz)	Raisin bread, unfrosted
1	Tortilla, corn, 6″ across
1	Tortilla, flour, 6″ across
⅓	Tortilla, flour, 10″ across
1	Waffle, 4″ square or across, reduced-fat
Cereals and Grains	
½ c	Bran cereals
½ c	Bulgur, cooked
½ c	Cereals, cooked
¾ c	Cereals, unsweetened, ready-to-eat
3 tbs	Cornmeal (dry)
⅓ c	Couscous
3 tbs	Flour (dry)
¼ c	Granola, low-fat
¼ c	Grape nuts
½ c	Grits, cooked
½ c	Kasha
⅓ c	Millet
¼ c	Muesli
½ c	Oats
⅓ c	Pasta, cooked
1½ c	Puffed cereals
⅓ c	Rice, white or brown, cooked
½ c	Shredded wheat
½ c	Sugar-frosted cereal
3 tbs	Wheat germ
Starchy Vegetables	
⅓ c	Baked beans
½ c	Corn
½ cob (5 oz)	Corn on cob, large
1 c	Mixed vegetables with corn, peas, or pasta
½ c	Peas, green
½ c	Plantains
½ medium (3 oz) or ½ c	Potato, boiled
¼ large (3 oz)	Potato, baked with skin
½ c	Potatoes, mashed
1 c	Squash, winter (acorn, butternut, pumpkin)
½ c	Yams, sweet potatoes, plain
Crackers and Snacks	
8	Animal crackers
3	Graham crackers, 2½″ square
¾ oz	Matzoh
4 slices	Melba toast
24	Oyster crackers
3 c	Popcorn (popped, no fat added or low-fat microwave)
¾ oz	Pretzels
2	Rice cakes, 4″ across
6	Saltine-type crackers
15–20 (¾ oz)	Snack chips, fat-free or baked (tortilla, potato)
2–5 (¾ oz)	Whole-wheat crackers, no fat added
Beans, Peas, and Lentils (count as 1 starch + 1 very lean meat)	
½ c	Beans and peas, cooked (garbanzo, lentils, pinto, kidney, white, split, black-eyed)
⅔ c	Lima beans
3 tbs	Miso
Starchy Foods Prepared with Fat (count as 1 starch + 1 fat)	
1	Biscuit, 2½″ across
½ c	Chow mein noodles
1 (2 oz)	Cornbread, 2″ cube
6	Crackers, round butter type
1 c	Croutons
1 c (2 oz)	French-fried potatoes (oven baked)
¼ c	Granola
⅓ c	Hummus
⅕ (1 oz)	Muffin, 5 oz
3 c	Popcorn, microwave
3	Sandwich crackers, cheese or peanut butter filling
9–13 (¾ oz)	Snack chips (potato, tortilla)
⅓ c	Stuffing, bread (prepared)
2	Taco shells, 6″ across
1	Waffle, 4½″ square or across
4–6 (1 oz)	Whole-wheat crackers, fat added

= 400 mg or more of sodium per serving.

TABLE E1.5 U.S. Exchange System: Fruit List

1 fruit exchange = 15 g carbohydrate and 60 kcal
NOTE: In general, one fruit exchange is 1 small fresh fruit; ½ c canned or fresh fruit or unsweetened fruit juice; ¼ c dried fruit.

Serving Size	Food
1 (4 oz)	Apple, unpeeled, small
½ c	Applesauce, unsweetened
4 rings	Apples, dried
4 whole (5½ oz)	Apricots, fresh
8 halves	Apricots, dried
½ c	Apricots, canned
1 (4 oz)	Banana, small
¾ c	Blackberries
¾ c	Blueberries
⅓ melon (11 oz) or 1 c cubes	Cantaloupe, small
12 (3 oz)	Cherries, sweet, fresh
½ c	Cherries, sweet, canned
3	Dates
1½ large or 2 medium (3½ oz)	Figs, fresh
1½	Figs, dried
½ c	Fruit cocktail
½ (11 oz)	Grapefruit, large
¾ c	Grapefruit sections, canned
17 (3 oz)	Grapes, small
1 slice (10 oz) or 1 c cubes	Honeydew melon
1 (3½ oz)	Kiwi
¾ c	Mandarin oranges, canned
½ (5½ oz) or ½ c	Mango, small
1 (5 oz)	Nectarine, small
1 (6½ oz)	Orange, small
½ (8 oz) or 1 c cubes	Papaya
1 (4 oz)	Peach, medium, fresh
½ c	Peaches, canned
½ (4 oz)	Pear, large, fresh
½ c	Pears, canned
¾ c	Pineapple, fresh
½ c	Pineapple, canned
2 (5 oz)	Plums, small
½ c	Plums, canned
3	Plums, dried (prunes)
2 tbs	Raisins
1 c	Raspberries
1¼ c whole berries	Strawberries
2 (8 oz)	Tangerines, small
1 slice (13½ oz) or 1¼ c cubes	Watermelon
Fruit Juice, unsweetened	
½ c	Apple juice/cider
⅓ c	Cranberry juice cocktail
1 c	Cranberry juice cocktail, reduced-kcalorie
⅓ c	Fruit juice blends, 100% juice
⅓ c	Grape juice
½ c	Grapefruit juice
½ c	Orange juice
½ c	Pineapple juice
⅓ c	Prune juice

TABLE E1.6 U.S. Exchange System: Milk List

NOTE: In general, one milk exchange is 1 c milk or yogurt.

Serving Size	Food
Fat-Free and Low-Fat Milk	
1 fat-free/low-fat milk exchange = 12 g carbohydrate, 8 g protein, 0–3 g fat, 90 kcal	
1 c	Fat-free milk
1 c	½% milk
1 c	1% milk
1 c	Fat-free or low-fat buttermilk
½ c	Evaporated fat-free milk
⅓ c dry	Fat-free dry milk
1 c	Soy milk, low-fat or fat-free
⅔ c (6 oz)	Yogurt, fat-free or low-fat, flavored, sweetened with nonnutritive sweetener and fructose
⅔ c (6 oz)	Yogurt, plain fat-free
Reduced-Fat Milk	
1 reduced-fat milk exchange = 12 g carbohydrate, 8 g protein, 5 g fat, 120 kcal	
1 c	2% milk
1 c	Soy milk
1 c	Sweet acidophilus milk
¾ c	Yogurt, plain low-fat
Whole Milk	
1 whole milk exchange = 12 g carbohydrate, 8 g protein, 8 g fat, 150 kcal	
1 c	Whole milk
½ c	Evaporated whole milk
1 c	Goat's milk
1 c	Kefir
¾ c	Yogurt, plain (made from whole milk)

TABLE E1.7 U.S. Exchange System: Sweets, Desserts, and Other Carbohydrates List

1 other carbohydrate exchange = 15 g carbohydrate, or 1 starch, or 1 fruit, or 1 milk exchange

Food	Serving Size	Exchanges per Serving
Angel food cake, unfrosted	1⁄12 cake (2 oz)	2 carbohydrates
Brownies, small, unfrosted	2″ square (1 oz)	1 carbohydrate, 1 fat
Cake, unfrosted	2″ square (1 oz)	1 carbohydrate, 1 fat
Cake, frosted	2″ square (2 oz)	2 carbohydrates, 1 fat
Cookies or sandwich cookies with creme filling	2 small (⅔ oz)	1 carbohydrate, 1 fat
Cookies, sugar-free	3 small or 1 large (¾–1 oz)	1 carbohydrate, 1–2 fats
Cranberry sauce, jellied	¼ c	1½ carbohydrates
Cupcake, frosted	1 small (2 oz)	2 carbohydrates, 1 fat
Doughnut, plain cake	1 medium (1½ oz)	1½ carbohydrates, 2 fats
Doughnut, glazed	3¾″ across (2 oz)	2 carbohydrates, 2 fats
Energy, sport, or breakfast bar	1 bar (1⅓ oz)	1½ carbohydrates, 0–1 fat
Energy, sport, or breakfast bar	1 bar (2 oz)	2 carbohydrates, 1 fat
Fruit cobbler	½ c (3½ oz)	3 carbohydrates, 1 fat
Fruit juice bar, frozen, 100% juice	1 bar (3 oz)	1 carbohydrate
Fruit snacks, chewy (pureed fruit concentrate)	1 roll (¾ oz)	1 carbohydrate
Fruit spreads, 100% fruit	1½ tbs	1 carbohydrate
Gelatin, regular	½ c	1 carbohydrate
Gingersnaps	3	1 carbohydrate
Granola or snack bar, regular or low-fat	1 bar (1 oz)	1½ carbohydrates
Honey	1 tbs	1 carbohydrate
Ice cream	½ c	1 carbohydrate, 2 fats
Ice cream, light	½ c	1 carbohydrate, 1 fat
Ice cream, low-fat	½ c	1½ carbohydrates
Ice cream, fat-free, no sugar added	½ c	1 carbohydrate
Jam or jelly, regular	1 tbs	1 carbohydrate
Milk, chocolate, whole	1 c	2 carbohydrates, 1 fat
Pie, fruit, 2 crusts	⅙ of 8″ commercially prepared pie	3 carbohydrates, 2 fats
Pie, pumpkin or custard	⅛ of 8″ commercially prepared pie	2 carbohydrates, 2 fats
Pudding, regular (made with reduced-fat milk)	½ c	2 carbohydrates
Pudding, sugar-free (made with fat-free milk)	½ c	1 carbohydrate
Reduced-calorie meal replacement (shake)	1 can (10–11 oz)	1½ carbohydrates, 0–1 fats
Rice milk, low-fat or fat-free, plain	1 c	1 carbohydrate
Rice milk, low-fat, flavored	1 c	1½ carbohydrates
Salad dressing, fat-free	¼ c	1 carbohydrate
Sherbet, sorbet	½ c	2 carbohydrates
Spaghetti or pasta sauce, canned	½ c	1 carbohydrate, 1 fat
Sports drinks	8 oz (1 c)	1 carbohydrate
Sugar	1 tbs	1 carbohydrate
Sweet roll or danish	1 (2½ oz)	2½ carbohydrates, 2 fats
Syrup, light	2 tbs	1 carbohydrate
Syrup, regular	1 tbs	1 carbohydrate
Syrup, regular	¼ c	4 carbohydrates
Vanilla wafers	5	1 carbohydrate, 1 fat
Yogurt, frozen	½ c	1 carbohydrate, 0–1 fat
Yogurt, frozen, fat-free	⅓ c	1 carbohydrate
Yogurt, low-fat with fruit	1 c	3 carbohydrates, 0–1 fat

= 400 mg or more sodium per exchange.

TABLE E1.8 U.S. Exchange System: Nonstarchy Vegetable List

1 vegetable exchange = 5 g carbohydrate, 2 g protein, 0 g fat, and 25 kcal
NOTE: In general, one vegetable exchange is ½ c cooked vegetables or vegetable juice; 1 c raw vegetables. Starchy vegetables such as corn, peas, and potatoes are on the starch list (Table G-4).

Artichokes	Mushrooms
Artichoke hearts	Okra
Asparagus	Onions
Beans (green, wax, Italian)	Pea pods
Bean sprouts	Peppers (all varieties)
Beets	Radishes
Broccoli	Salad greens (endive, escarole, lettuce, romaine, spinach)
Brussels sprouts	Sauerkraut 🧂
Cabbage	Spinach
Carrots	Summer squash (crookneck)
Cauliflower	Tomatoes
Celery	Tomatoes, canned
Cucumbers	Tomato sauce 🧂
Eggplant	Tomato/vegetable juice 🧂
Green onions or scallions	Turnips
Greens (collard, kale, mustard, turnip)	Water chestnuts
Kohlrabi	Watercress
Leeks	Zucchini
Mixed vegetables (without corn, peas, or pasta)	

🧂 = 400 mg or more sodium per exchange.

TABLE E1.9 U.S. Exchange System: Meat and Meat Substitutes List

NOTE: In general, a meat exchange is 1 oz meat, poultry, or cheese; ½ c dried beans (weigh meat and poultry and measure beans after cooking).

Serving Size	Food
Very Lean Meat and Substitutes	
1 very lean meat exchange = 7 g protein, 0–1 g fat, 35 kcal	
1 oz	Poultry: Chicken or turkey (white meat, no skin), Cornish hen (no skin)
1 oz	Fish: Fresh or frozen cod, flounder, haddock, halibut, trout, lox (smoked salmon) 🧂, tuna, fresh or canned in water
1 oz	Shellfish: Clams, crab, lobster, scallops, shrimp, imitation shellfish
1 oz	Game: Duck or pheasant (no skin), venison, buffalo, ostrich
	Cheese with ≤1g fat/oz:
¼ c	Fat-free or low-fat cottage cheese
1 oz	Fat-free cheese
1 oz	Processed sandwich meats with ≤1 g fat/oz (such as deli thin, shaved meats, chipped beef 🧂, turkey ham)
2	Egg whites
¼ c	Egg substitutes, plain
1 oz	Hot dogs with ≤1 g fat/oz 🧂
1 oz	Kidney (high in cholesterol)
1 oz	Sausage with ≤1 g fat/oz
Count as 1 very lean meat + 1 starch exchange:	
½ c	Beans, peas, lentils (cooked)
Lean Meat and Substitutes	
1 lean meat exchange = 7 g protein, 3 g fat, 55 kcal	
1 oz	Beef: USDA Select or Choice grades of lean beef trimmed of fat (round, sirloin, and flank steak); tenderloin; roast (rib, chuck, rump); steak (T-bone, porterhouse, cubed), ground round
1 oz	Pork: Lean pork (fresh ham); canned, cured, or boiled ham; Canadian bacon 🧂; tenderloin, center loin chop
1 oz	Lamb: Roast, chop, leg
1 oz	Veal: Lean chop, roast
1 oz	Poultry: Chicken, turkey (dark meat, no skin), chicken (white meat, with skin), domestic duck or goose (well drained of fat, no skin)
	Fish:
1 oz	Herring (uncreamed or smoked)
6 medium	Oysters
1 oz	Salmon (fresh or canned), catfish
2 medium	Sardines (canned)
1 oz	Tuna (canned in oil, drained)
1 oz	Game: Goose (no skin), rabbit
	Cheese:
¼ c	4.5%-fat cottage cheese
2 tbs	Grated Parmesan
1 oz	Cheeses with ≤3 g fat/oz
1½ oz	Hot dogs with ≤3 g fat/oz 🧂
1 oz	Processed sandwich meat with ≤3 g fat/oz (turkey pastrami or kielbasa)
1 oz	Liver, heart (high in cholesterol)
Medium-Fat Meat and Substitutes	
1 medium-fat meat exchange = 7 g protein, 5 g fat, and 75 kcal	
1 oz	Beef: Most beef products (ground beef, meatloaf, corned beef, short ribs, Prime grades of meat trimmed of fat, such as prime rib)
1 oz	Pork: Top loin, chop, Boston butt, cutlet
1 oz	Lamb: Rib roast, ground
1 oz	Veal: Cutlet (ground or cubed, unbreaded)
1 oz	Poultry: Chicken (dark meat, with skin), ground turkey or ground chicken, fried chicken (with skin)
1 oz	Fish: Any fried fish product
	Cheese with ≤5 g fat/oz:
1 oz	Feta
1 oz	Mozzarella
¼ c (2 oz)	Ricotta
1	Egg (high in cholesterol, limit to 3/week)
1 oz	Sausage with ≤5 g fat/oz
1 c	Soy milk
¼ c	Tempeh
4 oz or ½ c	Tofu
High-Fat Meat and Substitutes	
1 high-fat meat exchange = 7 g protein, 8 g fat, 100 kcal	
1 oz	Pork: Spareribs, ground pork, pork sausage
1 oz	Cheese: All regular cheeses (American 🧂, cheddar, Monterey Jack, swiss)
1 oz	Processed sandwich meats with ≤8 g fat/oz (bologna, pimento loaf, salami)
1 oz	Sausage (bratwurst, Italian, knockwurst, Polish, smoked)
1 (10/lb)	Hot dog (turkey or chicken) 🧂
3 slices (20 slices/lb)	Bacon
1 tbs	Peanut butter (contains unsaturated fat)
Count as 1 high-fat meat + 1 fat exchange:	
1 (10/lb)	Hot dog (beef, pork, or combination) 🧂

🧂 = 400 mg or more of sodium per serving.

TABLE E1.10 U.S. Exchange System: Fat List

1 fat exchange = 5 g fat and 45 kcal
NOTE: In general, one fat exchange is 1 tsp regular butter, margarine, or vegetable oil; 1 tbs regular salad dressing. Many fat-free and reduced fat foods are on the Free Foods List (Table G-11).

Serving Size	Food
Monounsaturated Fats	
2 tbs (1 oz)	Avocado
1 tsp	Oil (canola, olive, peanut)
8 large	Olives, ripe (black)
10 large	Olives, green, stuffed
6 nuts	Almonds, cashews
6 nuts	Mixed nuts (50% peanuts)
10 nuts	Peanuts
4 halves	Pecans
½ tbs	Peanut butter, smooth or crunchy
1 tbs	Sesame seeds
2 tsp	Tahini or sesame paste
Polyunsaturated Fats	
4 halves	English walnuts
1 tsp	Margarine, stick, tub, or squeeze
1 tbs	Margarine, lower-fat spread (30% to 50% vegetable oil)
1 tsp	Mayonnaise, regular
1 tbs	Mayonnaise, reduced-fat
1 tsp	Oil (corn, safflower, soybean)
1 tbs	Salad dressing, regular
2 tbs	Salad dressing, reduced-fat
2 tsp	Mayonnaise type salad dressing, regular
1 tbs	Mayonnaise type salad dressing, reduced-fat
1 tbs	Seeds (pumpkin, sunflower)
Saturated Fats*	
1 slice (20 slices/lb)	Bacon, cooked
1 tsp	Bacon, grease
1 tsp	Butter, stick
2 tsp	Butter, whipped
1 tbs	Butter, reduced-fat
2 tbs (½ oz)	Chitterlings, boiled
2 tbs	Coconut, sweetened, shredded
1 tbs	Coconut milk
2 tbs	Cream, half and half
1 tbs (½ oz)	Cream cheese, regular
1½ tbs (¾ oz)	Cream cheese, reduced-fat
	Fatback or salt pork†
1 tsp	Shortening or lard
2 tbs	Sour cream, regular
3 tbs	Sour cream, reduced-fat

= 400 mg or more sodium per exchange
*Saturated fats can raise blood cholesterol levels.
†Use a piece 1″ × 1″ × ¼″ if you plan to eat the fatback cooked with vegetables. Use a piece 2″ × 1″ × ½″ when eating only the vegetables with the fatback removed.

TABLE E1.11 U.S. Exchange System: Free Foods List

NOTE: A serving of free food contains less than 20 kcalories or no more than 5 grams of carbohydrate; those with serving sizes should be limited to 3 servings a day whereas those without serving sizes can be eaten freely.

Serving Size	Food
Fat-Free or Reduced-Fat Foods	
1 tbs (½ oz)	Cream cheese, fat-free
1 tbs	Creamers, nondairy, liquid
2 tsp	Creamers, nondairy, powdered
4 tbs	Margarine spread, fat-free
1 tsp	Margarine spread, reduced-fat
1 tbs	Mayonnaise, fat-free
1 tsp	Mayonnaise, reduced-fat
1 tbs	Mayonnaise type salad dressing, fat-free
1 tsp	Mayonnaise type salad dressing, reduced-fat
	Nonstick cooking spray
1 tbs	Salad dressing, fat-free or low-fat
2 tbs	Salad dressing, fat-free, Italian
1 tbs	Sour cream, fat-free, reduced-fat
1 tbs	Whipped topping, regular
2 tbs	Whipped topping, light or fat-free
Sugar-Free Foods	
1 piece	Candy, hard, sugar-free
	Gelatin dessert, sugar-free
	Gelatin, unflavored
	Gum, sugar-free
2 tsp	Jam or jelly, light
	Sugar substitutes
2 tbs	Syrup, sugar-free
Drinks	
	Bouillon, broth, consommé 🧂
	Bouillon or broth, low-sodium
	Carbonated or mineral water
	Club soda
1 tbs	Cocoa powder, unsweetened
	Coffee
	Diet soft drinks, sugar-free
	Drink mixes, sugar-free
	Tea
	Tonic water, sugar-free
Condiments	
1 tbs	Catsup
	Horseradish
	Lemon juice
	Lime juice
	Mustard
1 tbs	Pickle relish
1½ medium	Pickles, dill 🧂
2 slices	Pickles, sweet (bread and butter)
¾ oz	Pickles, sweet (gherkin)
¼ c	Salsa
1 tbs	Soy sauce, regular or light 🧂
1 tbs	Taco sauce
	Vinegar
2 tbs	Yogurt
Seasonings	
	Flavoring extracts
	Garlic
	Herbs, fresh or dried
	Hot pepper sauces
	Pimento
	Spices
	Wine, used in cooking
	Worcestershire sauce

🧂 = 400 mg or more of sodium per serving.

TABLE E1.12 U.S. Exchange System: Combination Foods List

Food	Serving Size	Exchanges per Serving
Entrées		
Tuna noodle casserole, lasagna, spaghetti with meatballs, chili with beans, macaroni and cheese	1 c (8 oz)	2 carbohydrates, 2 medium-fat meats
Chow mein (without noodles or rice)	2 c (16 oz)	1 carbohydrate, 2 lean meats
Tuna or chicken salad	½ c (3½ oz)	½ carbohydrate, 2 lean meats, 1 fat
Frozen Entrées and Meals		
Dinner-type meal	Generally 14–17 oz	3 carbohydrates, 3 medium-fat meats, 3 fats
Entrée or meal with <340 kcal	About 8–11 oz	2–3 carbohydrates, 1–2 lean meats
Meatless burger, soy based	3 oz	½ carbohydrate, 2 lean meats
Meatless burger, vegetable and starch based	3 oz	1 carbohydrate, 1 lean meat
Pizza, cheese, thin crust	¼ of 12″ (6 oz)	2 carbohydrates, 2 medium-fat meats, 1 fat
Pizza, meat topping, thin crust	¼ of 12″ (6 oz)	2 carbohydrates, 2 medium-fat meats, 2 fats
Pot pie	1 (7 oz)	2½ carbohydrates, 1 medium-fat meat, 3 fats
Soups		
Bean	1 c	1 carbohydrate, 1 very lean meat
Cream (made with water)	1 c (8 oz)	1 carbohydrate, 1 fat
Instant	6 oz prepared	1 carbohydrate
Instant with beans/lentils	8 oz prepared	2½ carbohydrates, 1 very lean meat
Split pea (made with water)	½ c (4 oz)	1 carbohydrate
Tomato (made with water)	1 c (8 oz)	1 carbohydrate
Vegetable beef, chicken noodle, or other broth-type	1 c (8 oz)	1 carbohydrate
Fast Foods		
Burrito with beef	1 (5–7 oz)	3 carbohydrates, 1 medium-fat meat, 1 fat
Chicken nuggets	6	1 carbohydrate, 2 medium-fat meats, 1 fat
Chicken breast and wing, breaded and fried	1 each	1 carbohydrate, 4 medium-fat meats, 2 fats
Chicken sandwich, grilled	1	2 carbohydrates, 3 very lean meats
Chicken wings, hot	6 (5 oz)	1 carbohydrate, 3 medium-fat meats, 4 fats
Fish sandwich/tartar sauce	1	3 carbohydrates, 1 medium-fat meat, 3 fats
French fries	1 medium serving (5 oz)	4 carbohydrates, 4 fats
Hamburger, regular	1	2 carbohydrates, 2 medium-fat meats
Hamburger, large	1	2 carbohydrates, 3 medium-fat meats, 1 fat
Hot dog with bun	1	1 carbohydrate, 1 high-fat meat, 1 fat
Individual pan pizza	1	5 carbohydrates, 3 medium-fat meats, 3 fats
Pizza, cheese, thin crust	¼ of 12″ (about 6 oz)	2½ carbohydrates, 2 medium-fat meats
Pizza, meat, thin crust	¼ of 12″ (about 6 oz)	2½ carbohydrates, 2 medium-fat meats, 1 fat
Soft serve cone	1 small (5 oz)	2½ carbohydrates, 1 fat
Submarine sandwich	1 sub (6″)	3 carbohydrates, 1 vegetable, 2 medium-fat meats, 1 fat
Submarine sandwich (<6 g fat)	1 sub (6″)	2½ carbohydrates, 2 lean meats
Taco, hard or soft shell	1 (3–3½ oz)	1 carbohydrate, 1 medium-fat meat, 1 fat

= 400 mg or more sodium per exchange.

Appendix E2

EXCHANGE CALCULATION FORM

Diet Rx:_______ KCals _______ % CHO (_______ g CHO) _________% Protein (________ g Pro) _______% Fat (__________ g Fat)

	Exchange	No. of Ex	g CHO	Total g CHO	g Prot	Total g Protein	g Fat	Total g Fat	Cal/Exch	Total Exchange Calories
	FF, LF MILK		12		8		1		90	
	RF MILK		12		8		5		120	
	WH MILK		12		8		8		150	
	VEG		5		2		0		25	
	FRUIT		15		0		0		60	
Rx g CHO less Total g CHO so far ÷ 15 = Starch Ex										
	STARCH		15		3		1		80	
Rx g Prot less Total g Prot so far ÷ 7 = Meat Ex										
	VL MEAT		0		7		1		35	
	L MEAT		0		7		3		55	
	MF MEAT		0		7		5.2		75	
	HF MEAT		0		7		8		100	
Rx g Fat less Total g Fat so far ÷ 5 = Fat Ex										
	FAT		0		0		5		45	
TOTALS										

Compare total grams of carbohydrate, protein, and fat to the original diet prescription. Do the number of exchanges need changing to more closely fit the original diet prescription?

Appendix E3

DIABETIC MEDICAL NUTRITION THERAPY EXCHANGE/MENU FORM

Meal Plan

			Grams	Percent
Patient's name: ________	Date: ________	Carbohydrate	______	______
		Protein	______	______
Dietitian: ________	Phone: ________	Fat	______	______
		Calories	______	______

Time	Number of Exchanges/Choices	Menu Ideas	Menu Ideas
	______ Carbohydrate group ______ Starch ______ Fruit ______ Milk ______ ______ Meat group ______ ______ Fat group ______		
	______ ______ ______ ______		
	______ Carbohydrate group ______ Starch ______ Fruit ______ Milk ______ ______ Vegetables ______ Meat group ______ ______ Fat group ______		
	______ ______ ______ ______ ______ ______		
	______ Carbohydrate group ______ Starch ______ Fruit ______ Milk ______ ______ Vegetables ______ Meat group ______ ______ Fat group ______		
	______ ______ ______ ______ ______ ______		

Source: © 1993, American Dietetic Association. "A Healthy Food Guide: Kidney Disease." Used with permission.

Appendix E4

MEAL PLANNING TOOLS FOR INDIVIDUALS WITH DIABETES

Meal-Planning Tool	Key Messages	Appropriate Population	Strengths	Weaknesses
Healthful eating guidelines				
The First Step in Diabetes Meal Planning	• Food pyramid concept • Serving sizes • General guidelines for servings per day • Healthful food choices	• Individuals newly diagnosed with diabetes • Individuals with fixed medication/insulin plans • Individuals who need basic concepts or prefer visual approach to learning • Individuals following a low-fat meal plan • Individuals with poor math skills • Individuals who are unable or unwilling to use a more complex meal planning system	• Easy-to-follow format, simple concept • Can be used to assess nutritional adequacy and quality of individual's usual eating pattern • Basic healthful lifestyle tips • Tips for choosing healthful, low-fat foods • General guidelines for servings per day (could be given to patients before their first RD visit) • Space for individualized mail plan and personal goals • Food-group nutrient composition provided (allows teaching of introductory carbohydrate concepts) • Available in Spanish	• Limited number of foods listed • No space for sample menu • Does not specifically discuss carbohydrate control • Does not review carbohydrate label reading • Does not cover combination and fast foods
Healthy Food Choices	• Healthful lifestyle including physical activity • Food groups • Portion control • Healthful food choices • Basic carbohydrate concepts	• Individuals newly diagnosed with diabetes • Individuals with pre-diabetes (piece does not say diabetes in text) • Individuals with fixed medication/insulin plans • Individuals following a low-fat meal plan • Individuals who are not ready or do not require detailed carbohydrate counting information	• Can be used to assess nutritional adequacy and quality of individual's usual eating pattern • Detailed healthful lifestyle tips: exercise, variety, low fat, fiber, weight, alcohol • Space for individualized meal plan and sample menu • General guidelines for servings per day and nutrient composition • Information on combination foods and fast foods • Basic carbohydrate and meal timing concepts • Goal setting tips and space for personal goals • Available in Spanish	• Less intuitive format • Does not specifically discuss carbohydrate counting • Does not review carbohydrate label reading

(continued on the following page)

(continued)

Plate method				
	• General portion control • Consistency • Basic food categories	• Individuals newly diagnosed with diabetes • Individuals in long-term care • Individuals with poor reading or math skills • Individuals with fixed medication/insulin regimens • Individuals who are unable or unwilling to use more complex meal planning systems • Individuals who do not require detailed carbohydrate information	• Simple and intuitive • Does not require measurement of foods • Easy to use and teach • Useful when counseling non-English speaking individuals • Good tool for visual learners	• Broad portion control method • Portions may vary • Need to adapt for people who do not drink milk • May not emphasize healthful foods • May not cover carbohydrate concepts • Does not review label reading
Menus				
Individualized menus	• Portion control • Meal spacing	• Individuals who have routine eating habits and eat at consistent places • Individuals with fixed medication/insulin regimens • Individuals who are unable or do not wish to use a meal planning system • Individuals with multiple nutrition restrictions and limited ability to integrate the various guidelines	• Easy to use and teach • Does not require meal planning skills • Can model healthful eating and variety • Can be individualized to individual's needs for fat, protein, sodium intake, meal pattern, and carbohydrate intake	• Highly structured • Does not teach guidelines for selecting healthful foods • Does not cover carbohydrate concepts
Exchange system				
Eating Healthy with Diabetes: An Easy Read Guide	• Exchange system food groups • Low-fat, low-sugar food choices • Combination foods	• Individuals with fixed medication/insulin regimens • Individuals who need basic concepts • Individuals with poor math skills • Individuals who are unable or unwilling to use more complex meal planning systems	• Many pictures of foods, food groups, and combination foods • Simplified exchange system • Pictures showing how to count combination foods • Space for personal meal plan by meal	• High degree of structure • Limited food lists • Does not cover carbohydrate concepts or label reading • Does not review healthful lifestyle • Does not show meal plan for day on one page
Exchange Lists for Meal Planning	• Exchange system • Macronutrient information and calories • Portion sizes • Healthful food choices and tips	• Individuals with good reading and math skills • Individuals with fixed medication/insulin regimens • Individuals who are willing to measure foods • Individuals with a desire to have increased flexibility in food choices	• Can be used for nutrition assessment • Can be used for calculation of macronutrient content and calorie level of meal plan or recipes • Detailed food lists (including combination and fast foods) and nutrient information	• High degree of structure • Requires multiple visits for instruction • Information too detailed or complex for some people • Requires good reading and math skills • Does not specifically discuss carbohydrate counting

(continued on the following page)

(continued)

Meal-Planning Tool	Key Messages	Appropriate Population	Strengths	Weaknesses
		• Individuals who need to limit total fat, saturated fat, or calories • Individuals who are already familiar with exchange-based meal planning	• Space for individual meal plan and detailed menu ideas • Detailed information for people limiting fat and calorie intake • Allows more flexibility in food choices • Basic carbohydrate concepts • High-sodium foods identified • Reviews label reading • Basic meal timing information • Available in Spanish	
Carbohydrate Counting				
Basic Carbohydrate Counting	• Basic carbohydrate concepts • Basic blood glucose management concepts • Consistent carbohydrate intake • Portion control	• Individuals with moderate reading and math skills • Individuals who need a focused education message • Individuals who do little medication and insulin adjustment • Individuals with a desire to have increased flexibility in food choices • Individuals who eat out or use foods with labels frequently	• Space for individualized meal plan • Information on combination foods and fast foods • Single nutrient, focused education topic • Consistent carbohydrate and meal spacing concepts • Limited space for setting personal goals	• No general guidelines for servings per day for food groups • Little space for menu ideas • Limited information on healthful food choices
Advanced Carbohydrate Counting	• Advanced carbohydrate counting concepts • Advanced blood glucose and pattern management • Record keeping • Portion control	• Individuals with good reading and math skills • Individuals who are willing and able to monitor blood glucose levels frequently, keep detailed records, and measure or weigh foods • Individuals who use flexible insulin plans, understand insulin action profiles, and feel confident with insulin dose adjustment • Individuals with variable meal times or food intake • Individuals who want maximal flexibility in food choices • Individuals who eat out or use foods with labels frequently • Individuals who already understand basic carbohydrate counting and label reading	• Single nutrient-focused education topic • Detailed review of record keeping • Basic review of blood glucose monitoring and postprandial blood glucose • Reviews how to calculate insulin-to-carbohydrate ratios, insulin sensitivity factors, and pattern management with practice exercises • Basic troubleshooting for weight gain and hypoglycemia • Insulin adjustment for fiber and fat	• Requires multiple visits for instruction • Concepts too complex for many people • Requires good reading and math skills • Does not include any food lists, so must be used with another resource that provides nutrient composition of foods • Does not cover label reading • Does not review insulin action • Limited information on healthful food choices

Reprinted from: Pastors JG, Waslaski J, Gunderson H. Diabetes meal-planning strategies. In Ross TA, Boucher JL, O'Connell BS (eds) *American Dietetic Association Guide to Diabetes: Medical Nutrition Therapy and Education*. Chicago: American Dietetic Association, 2005, Table 18.3, page 213.

Appendix E5

RENAL MEDICAL NUTRITION THERAPY CALCULATION FORM

Patient's name: ______________________ ____________ grams protein
Date: ______________________ ____________ calories
Your dietitian is: ______________________ ____________ milligrams phosphorus
Telephone number: ______________________ ____________ milligrams sodium

Your Daily Meal Plan

Breakfast		**Sample Menu**	**Snack**		**Sample Menu**
Milk	____ choices	________		____ choices	________
Nondairy milk substitute	____ choices	________		____ choices	________
Meat	____ choices	________			
Starch	____ choices	________			
Fruit	____ choices	________	**Dinner**		
Fat	____ choices	________	Milk	____ choices	________
High-calorie	____ choices	________	Nondairy milk substitute	____ choices	________
Salt	____ choices	________	Meat	____ choices	________
			Starch	____ choices	________
			Vegetable	____ choices	________
Snack			Fruit	____ choices	________
	____ choices	________	Fat	____ choices	________
	____ choices	________	High-calorie	____ choices	________
			Salt	____ choices	________
Lunch			**Snack**		
Milk	____ choices	________			
Nondairy milk substitute	____ choices	________		____ choices	________
Meat	____ choices	________		____ choices	________
Starch	____ choices	________			
Vegetable	____ choices	________			
Fruit	____ choices	________			
Fat	____ choices	________			
High-calorie	____ choices	________			
Salt	____ choices	________			

Source: Allen JC, Watters, C. *Lactations: Physiology, Nutrition & Breastfeeding,* 49–102, 1983. Reprinted by permission of Kluwer Academy.

Appendix E6

DIETARY RECOMMENDATIONS FOR HYPOGLYCEMIA

To relieve reactive hypoglycemia, some health professionals recommend taking the following steps:

- eat small meals and snacks about every 3 hours
- exercise regularly
- eat a variety of foods, including meat, poultry, fish, or nonmeat sources of protein; starchy foods such as whole-grain bread, rice, and potatoes; fruits; vegetables; and dairy products
- choose high-fiber foods
- avoid or limit foods high in sugar, especially on an empty stomach

Alcohol

Drinking, especially binge drinking, can cause hypoglycemia because your body's breakdown of alcohol interferes with your liver's efforts to raise blood glucose. Hypoglycemia caused by excessive drinking can be very serious and even fatal.

Reprinted from: National Diabetes Information Clearinghouse. Hypoglycemia (NIH Publication No. 03-3926). Bethesda (MD): National Diabetes Information Clearinghouse; 2003. Available from: http://diabetes.niddk.nih.gov/dm/pubs/hypoglycemia/#nodiabetes.

Appendix E7

GLUTEN-FREE DIET BY FOOD GROUPS[1]

Food Category	Foods Allowed (a)	Foods to Question (b)	Foods to Avoid (c)
Milk & Dairy	Milk, cream, most ice cream, buttermilk, plain yogurt, cheese, cream cheese, processed cheese, processed cheese foods, cottage cheese	Flavored yogurt, frozen yogurt, cheese sauces, cheese spreads, seasoned (flavored) shredded cheese	Malted milk, ice cream made with ingredients not allowed
Grains & Starches	**BREADS, BAKED PRODUCTS AND OTHER ITEMS:** Made with amaranth, arrowroot, buckwheat, corn bran, corn flour, cornmeal, corn starch, flax, legume flours (bean, garbanzo or chickpea, garfava, lentil, pea), mesquite flour, millet, Montina flour (Indian ricegrass), nut flours (almond, chestnut, hazelnut), potato flour, potato starch, pure uncontaminated oat products (oat flour, oat groat, oatmeal)*, quinoa, rice bran, rice flours (brown, glutinous, sweet, white), sago, sorghum flour, soy flour, sweet potato flour, tapioca (cassava, manioc), taro, teff	Items made with buckwheat flour	Items made with wheat bran, wheat farina, wheat flour, wheat germ, wheat-based semolina, wheat starch, durum flour, gluten flour, graham flour, atta, bulgur, einkorn, emmer, farro, kamut, spelt, barley, rye, triticale, commercial oat products (oat bran, oat flour, oat groats, oatmeal)*
	CEREALS: **Hot**: Puffed amaranth, cornmeal, cream of buckwheat, cream of rice (brown, white), hominy grits, quinoa, rice flakes, soy flakes, soy grits **Cold**: Puffed (amaranth, buckwheat, corn, millet, rice), rice crisps or corn flakes (with no barley malt flavoring), rice flakes, soy cereals	Rice and corn cereals, rice and soy pablum	Cereals made from wheat, rye, triticale, barley and commercial oats* Cereals made with added malt extract/malt syrup or malt flavoring
	PASTAS: Macaroni, spaghetti and noodles from beans, corn, peas, potato, quinoa, rice, soy, wild rice	Buckwheat pasta	Pastas made from wheat, wheat starch and other ingredients not allowed (e.g., orzo)
	RICE: Plain (e.g., basmati, brown, jasmine, white, wild)	Seasoned or flavored rice mixes	
	MISCELLANEOUS: Corn tacos, corn tortillas, rice tortillas	Multi-grain or flavored rice crackers, rice cakes and corn cakes	Wheat flour tacos and tortillas Matzoh, matzoh meal, matzoh balls, couscous, tabouli
	Plain rice crackers, rice cakes and corn cakes	Low gluten communion wafers	Regular communion wafers
	Gluten-free communion wafers		
	oat products as they are often cross contaminated with wheat and/or barley.		
Meats & Alternatives	**MEAT, FISH, POULTRY:** Plain (fresh or frozen)	Deli or luncheon meats (e.g., bologna, salami), wieners frankfurters, sausages, pate, meat and sandwich spread, frozen burgers (meat, fish, chicken), meat loaf, ham (ready to cook), seasoned/to cook), seasoned/flavored fish in pouches, imitation fish products (e.g., surimi), meat substitutes, meat product extenders	Canned fish in vegetable brot containing hydrolyzed wheat protein Frozen turkey basted or injected with hydrolyzed wheat protein. Frozen or fresh turkey with bread stuffing. Frozen chicken breasts containing chicken broth (made with ingredients not allowed) Meat, poultry or fish breaded in ingredients not allowed
	Eggs: Fresh, liquid, dried or powdered	Flavored egg products (liquid or frozen)	

[1]Table adapted and revised October 2000 by S. Case, M. Molloy and M. Zarakadas from *Celiac Disease Needs a Diet for Life Handbook*, Canadian Celiac Association. Further revisions made by S. Case for *Gluten-Free Diet: A Comprehensive Resource Guide*, May 2001, April 2002, July 2003, May 2005, January 2005 and March 2006.

(continued on the following page)

(continued)

Notes on Foods Allowed (a)

Category	Food Products	Notes
	Vinegars	Produced from various ingredients: Balsamic (grapes), cider (apples), rice (rice wine), white distilled (corn, wheat or both), wine (red wine). All these vinegars are gluten-free (including distilled white derived from wheat as the distillation process removes the gluten from the final purified product).
	Vanilla	Pure vanilla and pure vanilla extract are derived from the vanilla bean pods of a climbing orchid grown in tropical locations. The vanilla beans are chopped and soaked in alcohol and water; aged and then filtered. It must contain at least 35% ethyl alcohol by volume. The pure vanilla is bottled or the pure extract can be mixed with sugar and a stabilizer and then bottled.
	Natural Vanilla Flavor	Derived from vanilla beans but contains less than 35% ethyl alcohol. May also contain sugar and a stabilizer.
	Artificial (Imitation, Synthetic) Vanilla-Vanillin Extract/Flavoring	Made from a by-product of the pulp and paper industry or a coal-tar derivative that is chemically treated to mimic the flavor of vanilla. Also contains alcohol, water, color and a stabilizer.
	Baker's Yeast	A type of yeast grown on sugar beet molasses. It is available as active dry yeast granules (sold in packets or jars) or compressed yeast (also known as wet yeast, cake yeast or fresh yeast) which must be refrigerated.
	Autolyzed Yeast/Autolyzed Yeast Extract	A special process that causes yeast to be broken down by its own enzymes resulting in the production production of various compounds that can be used as flavoring agents. Autolyzed yeast is almost always derived from baker's yeast.
	Torula Yeast	A yeast grown on wood sugars (a by-product of waste products from the pulp and paper industry). Used as a flavoring agent that has a hickory smoke characteristic.
	Nutritional Yeast	A specific strain of an inactive form of baker's yeast that is grown on a mixture of sugar beet molasses which is fermented, washed, pasteurized and dried at high temperatures. Used as a dietary supplement as it contains protein, fiber, vitamins and minerals. Available in pills, flakes or powder.
	Xanthan Gum	It is produced from the fermentation of corn sugar. This powder is used to thicken sauces and salad dressings, Also used in gluten-free baked products to improvethe structure and texture.
	Guar Gum	A gum extracted from the seed of an East Indian plant. Available as a powder that is used as a thickener and stabilizer. Can be substituted for xanthan gum in gluten-free baked products. It is high in fiber and may have a laxative effect if consumed in large amounts.

Notes on Foods to Question (b)

Category	Food Products	Notes
Milk & Dairy	Cheese Spreads, Cheese Sauces (e.g., Nacho), Seasoned (flavored) Shredded Cheese	May be thickened with wheat flour or wheat starch. Seasonings may contain hydrolyzed wheat protein, wheat flour or wheat starch.
	Flavored Yogurt, Frozen Yogurt	May contain granola, cookie crumbs or wheat bran.
Grains & Starches	Buckwheat Flour	Pure buckwheat flour is gluten-free, however, some buckwheat flour may be mixed with wheat flour.
	Rice and Corn Cereals	May contain barley malt, barley malt extract, barley malt flavoring.
	Buckwheat Pasta	Also called Japanese Soba noodles. Some Soba pasta contains pure buckwheat flour which is gluten-free but others may also containwheat flour.
	Seasoned or Flavored Rice Mixes	Seasonings may contain hydrolyzed wheat protein, wheat flour or wheat starch or have added soy sauce that contains wheat.
	Multi-grain or Flavored Rice Crackers, Rice Cakes and Corn Cakes	Multi-grain products may contain barley and/or oats. Some contain soy sauce (made from wheat), seasonings containing hydrolyzed wheat protein, wheat flour or wheat starch.

(continued on the following page)

(continued)

	Low Gluten Communion Wafers	The Catholic Cannon Law, code 924.2 requires the presence of some wheat in communion wafers and will not accept the gluten-free hosts made with other grains. A very low-gluten host made with a small amount of specially processed wheat starch from the Benedictine Sisters of Perpetual Hope is available. The level of gluten in these hosts is extremely small (less than 37 micrograms or 0.037 milligrams per wafer). The Italian Celiac Association's scientific committee approved the use of the low gluten host. Many health professionals allow the use of this host. Some recommend only consuming only ¼ of a wafer per week. The decision of whether to use this host should be discussed with your health professional. The hosts can be purchased by contacting 1-800-223-2772 or email: altarbreads@benedictinesisters.org or write to Benedictine Sisters Altar Bread Department, 31970 State Highway P, Cyde, MO, 64432, USA. More information for Catholics with celiac disease can be found at www.catholicceliacs.org
Meats &Alternatives	Deli/Luncheon Meats, Hot Dogs and Sausages	May contain fillers made from wheat. Seasonings may contain hydrolyzed wheat protein, wheat flour or wheat starch.
	Meat and Sandwich Spreads	Products such as pâte may contain wheat flour or seasonings containing hydrolyzed wheat protein, wheat flour or wheat starch.
	Frozen Burgers (Meat, Poultry and Fish) and Meat Loaf	May contain fillers (wheat flour, wheat starch, bread crumbs). Seasonings may contain hydrolyzed wheat protein, wheat flour or wheat starch.
	Ham (ready to cook)	Glaze may contain hydrolyzed wheat protein, wheat flour or wheat starch.
	Seasoned/Flavored Fish in Pouches	May contain wheat or barley.
	Imitation Fish Products	Imitation crab/seafood sticks may contain fillers such as wheat starch.
	Meat Substitutes (e.g., vegetarian burgers, sausages, roasts, nuggets, textured vegetable protein)	Often contain hydrolyzed wheat protein, wheat gluten, wheat starch or barley malt.
	Flavored Egg Products (frozen or liquid)	May contain hydrolyzed wheat protein.
	Baked Beans	Some are thickened with wheat flour.
	Seasoned or Dry Roasted Nuts, Pumpkin or Sunflower Seeds	May contain hydrolyzed wheat protein, wheat flour or wheat starch.
	Flavored Tofu	May contain soy sauce (made from wheat) or other seasonings that contain hydrolyzed wheat protein, wheat flour or wheat starch.
	Tempeh	A meat substitute made from fermented soybeans and millet or rice. Often seasoned with soy sauce (made from wheat).
	Miso	A condiment used in Oriental cooking made from fermented soybeans and/or barley, wheat or rice. Wheat or barley are the most common grains used.
Fruits & Vegetables	Dates	Chopped, diced or extruded dates are packaged with oat flour, dextrose or rice flour. Oat flour or dextrose are the most common sources used.
	French Fried Potatoes	Often cooked in oil where gluten-containing foods (e.g., breaded fish and chicken fingers) resulting in cross contamination.
Soups	Canned Soups, Dried Soup Mixes, Soup Bases and Bouillon Cubes	May contain noodles or barley. Cream soups are often thickened with wheat flour. Seasonings may contain hydrolyzed wheat protein, wheat flour or wheat starch.
Fats	Salad Dressings	May contain wheat flour, malt vinegar or soy sauce (made from wheat). Seasonings may contain hydrolyzed wheat protein, wheat flour or wheat starch.
	Suet	The hard fat around the loins and kidneys of beef and sheep. Flour may be added to packaged suet. Suet can be used to make mincemeat, steamed Christmas pudding and Haggis (a traditional Scottish dish).
Desserts	Cake Icing & Frostings	May contain wheat flour or wheat starch.
Sweets	Honey Powder	This commercial powder is used in glazes, seasonings mixes, dry mixes and sauces. May contain wheat flour or wheat starch.

(continued on the following page)

Appendix E8

DIETARY MODIFICATION TO DECREASE SATURATED FAT AND INCREASE POLYUNSATURATED FAT INTAKE

Food Group	Choose	Decrease
Lean meat, poultry, and fish	Beef, pork, lamb – lean cuts well trimmed before cooking	Regular hamburger, fatty cuts of beef, spare ribs, organ meats
	Poultry w/o skin	Poultry with skin, fried chicken
	Fish, shellfish	Fried fish, fried shellfish
	Processed meats prepared from lean meats, e.g., lean ham, lean frankfurters, lean meat with soy protein or carrageen	Regular luncheon meat (bologna, salami, sausage, frankfurters)
Eggs	Egg whites, cholesterol-free egg whites	Egg yolks (if more than the recommended); includes eggs used in baking and cooking
Low-fat dairy products	Milk – skim ½% or 1% fat (fluid, powdered, evaporated, buttermilk)	Whole milk, regular yogurt (fluid, evaporated, condensed), 2% milk, imitation milk
	Yogurt – non-fat or low-fat yogurt or yogurt beverages	Whole milk yogurt
Dairy products	Cheese – low-fat natural or processed cheese	Regular cheeses (American blue, Brie, cheddar, Colby, Edam, Monterey Jack, whole-milk mozzarella, Parmesan, Swiss), cream cheese, Neufchatel cheese
	Low-fat or nonfat varieties, e.g., cottage cheese-low-fat, nonfat, or dry curd 0% to 2%)	Cottage cheese (4% milkfat)
	Frozen dairy dessert – ice milk, frozen yogurt (low-fat or nonfat)	Ice cream
	Low-fat coffee creamer	Cream, half & half, whipping cream
	Low-fat or nonfat sour cream	Non-dairy creamer, whipped topping, sour cream
Fats and Oils	Unsaturated oils – safflower, sunflower, corn, soybean, cottonseed, canola, olive, peanut	Coconut oil, palm kernel oil, palm oil
	Margarines – made from unsaturated oils listed above, especially soft or liquid forms, low in trans fatty acids	Butter, lard, shortening, bacon fat, hard margarine high in trans fatty acids
	Salad dressings – made with unsaturated oils, low-fat or fat-free	Dressings – made with egg yolk, cheese, sour cream, whole milk
	Seeds and nuts – peanut butter, other nut butters	Coconut
	Cocoa powder	Milk chocolate
Breads and cereals	Breads – whole-grain bread, English muffins, bagels, buns, corn or flour tortilla	Bread in which eggs, fat, and/or butter are a major ingredient; croissants
	Cereal – oat, wheat, corn, multi-grain	Most granolas
	Pasta	High-fat crackers
	Rice	Commercial baked pastries, muffins, biscuits
	Dry beans and peas	
	Crackers, low-fat – animal type, graham, soda crackers, breadsticks, melba toast	
	Homemade baked goods using unsaturated oil, skim or 1% milk, and egg substitute – quick breads, biscuits, cornbread muffins, bran muffins, pancakes, waffles	

(continued on the following page)

(continued)

Soups	Reduced – or low-fat and reduced sodium varieties, e.g., chicken or beef noodle, minestrone, tomato, vegetable, potato, reduced-fat soups made with skim milk	Soup containing whole milk, cream, meat fat, poultry fat, or poultry skin
Vegetables	Fresh, frozen, or canned, without added fat or sauce	Vegetables fried or prepared with butter, cheese, or cheese sauce
Fruits	Fruit – fresh, frozen, canned or dried Fruit juice – fresh, frozen or canned	Fried fruit or fruit served with butter or cream sauce
Sweets and modified fat desserts	Beverages – fruit – fruit-flavored drinks, lemonade, fruit punch	Candy made with milk chocolate, coconut oil, palm kernel oil, palm oil
	Sweets – sugar, syrup, honey, jam, preserves, candy made without added fat (candy corn, gumdrops, hard candy), fruit flavored gelatin	Ice cream and frozen treats made with ice cream
	Frozen dessert – low-fat and nonfat yogurt, ice milk, sherbet, sorbet, fruit ice, popsicles	Commercial baked pies, cakes, doughnuts, high-fat cookies, cream pies
	Cookies, cake, pie, pudding – prepared with egg whites, egg substitutes, skim milk or 1% milk, and unsaturated oil or margarine; ginger snaps, fig and other fruit bar cookies, fat-free cookies, angel food cake	

Reprinted from: National Institutes of Health. From: National Cholesterol Education Program. *Second Report of the National Cholesterol Education Program Expert Panel on Detection, Evaluation, and Treatment of High Blood Cholesterol in Adults (Adult Treatment Panel II).* Bethesda, Md: National Institutes of Health, National Heart, Lung, and Blood Institute; 1993. NIH Publication 93-3095. Available from: http://www.nhlbi.nih.gov/guidelines/choleterol/atp3full.pdf.

Appendix E9

GUIDE TO THERAPEUTIC LIFESTYLE CHANGES (TLC)

Healthy Lifestyle Recommendations for a Healthy Heart

Food Items to Choose More Often

Breads and Cereals

≥6 servings per day, adjusted to caloric needs

Breads, cereals, especially whole grain; pasta; rice; potatoes; dry beans and peas; low fat crackers and cookies

Vegetables

3–5 servings per day fresh, frozen, or canned, without added fat, sauce, or salt

Fruits

2–4 servings per day fresh, frozen, canned, dried

Dairy Products

2–3 servings per day

Fat-free, ½%, 1% milk, buttermilk, yogurt, cottage cheese; fat-free & low-fat cheese

Eggs

≤2 egg yolks per week

Egg whites or egg substitute

Meat, Poultry, Fish

≤5 oz per day

Lean cuts loin, leg, round; extra lean hamburger; cold cuts made with lean meat or soy protein; skinless poultry; fish

Fats and Oils

Amount adjusted to caloric level: Unsaturated oils; soft or liquid margarines and vegetable oil spreads, salad dressings, seeds, and nuts

TLC Diet Options

Stanol/sterol-containing margarines; viscous fiber food sources: barley, oats, psyllium, apples, bananas, berries, citrus fruits, nectarines, peaches, pears, plums, prunes, broccoli, brussels sprouts, carrots, dry beans, peas, soy products (tofu, miso)

Food Items to Choose Less Often

Breads and Cereals

Many bakery products, including doughnuts, biscuits, butter rolls, muffins, croissants, sweet rolls, Danish, cakes, pies, coffee cakes, cookies

Many grain-based snacks, including chips, cheese puffs, snack mix, regular crackers, buttered popcorn

Vegetables

Vegetables fried or prepared with butter, cheese, or cream sauce

Fruits

Fruits fried or served with butter or cream

Dairy Products

Whole milk/2% milk, whole-milk yogurt, ice cream, cream, cheese

Eggs

Egg yolks, whole eggs

Meat, Poultry, Fish

Higher fat meat cuts: ribs, t-bone steak, regular ham- burger, bacon, sausage; cold cuts: salami, bologna, hot dogs; organ meats: liver, brains, sweetbreads; poultry with skin; fried meat; fried poultry; fried fish

Fats and Oils

Butter, shortening, stick margarine, chocolate, coconut

Recommendations for Weight Reduction

Weigh Regularly

Record weight, BMI, & waist circumference

Lose Weight Gradually

Goal: lose 10% of body weight in 6 months. Lose ½ to 1 lb per week

Develop Healthy Eating Patterns

- Choose healthy foods (see Column 1)
- Reduce intake of foods in Column 2
- Limit number of eating occasions
- Select sensible portion sizes
- Avoid second helpings
- Identify and reduce hidden fat by reading food labels to choose products lower in saturated fat and calories, and ask about ingredients in ready-to-eat foods prepared away from home
- Identify and reduce sources of excess carbohydrates such as fat-free and regular crackers; cookies and other desserts; snacks; and sugar-containing beverages

Recommendations for Increased Physical Activity

Make Physical Activity Part of Daily Routines

- Reduce sedentary time
- Walk, wheel, or bike-ride more, drive less; Take the stairs instead of an eleva-tor; Get off the bus a few stops early and walk the remaining distance; Mow the lawn with a push mower; Rake leaves; Garden; Push a stroller; Clean the house; Do exercises or pedal a stationary bike while watching television; Play actively with children; Take a brisk 10-minute walk or wheel before work, during your work break, and after dinner

Make Physical Activity Part of Exercise or Recreational Activities

- Walk, wheel, or jog; Bicycle or use an arm pedal bicycle; Swim or do water aerobics; Play basketball; Join a sports team; Play wheelchair sports; Golf (pull cart or carry clubs); Canoe; Cross-country ski; Dance; Take part in an exercise program at work, home, school, or gym

Appendix E10

DASH EATING PLAN

BOX E10.1 FOLLOWING THE DASH EATING PLAN

The DASH eating plan shown below is based on 2,000 calories a day. The number of daily servings in a food group may vary from those listed, depending on your caloric needs. Use this chart to help you plan your menus or take it with you when you go to the store.

Food Group	Daily Servings (except as noted)	Serving Sizes	Examples and Notes	Significance of Each Food Group to the DASH Eating Plan
Grains and grain products	7–8	1 slice bread 1 oz dry cereal* ½ cup cooked rice, pasta, or cereal	Whole wheat bread, English muffin, pita bread, bagel, cereals, grits, oatmeal, crackers, unsalted pretzels and popcorn	Major sources of energy and fiber
Vegetables	4–5	1 cup raw leafy vegetable ½ cup cooked vegetable 6 oz vegetable juice	Tomatoes, potatoes, carrots, green peas, squash, broccoli, turnip greens, collards, kale, spinach, artichokes, green beans, lima beans, sweet potatoes	Rich sources of potassium, magnesium, and fiber
Fruits	4–5	6 oz fruit juice 1 medium fruit ¼ cup dried fruit ½ cup fresh, frozen, or canned fruit	Apricots, bananas, dates, grapes, oranges, orange juice, grapefruit, grapefruit juice, mangoes, melons, peaches, pineapples, prunes, raisins, strawberries, tangerines	Important sources of potassium, magnesium, and fiber
Lowfat or fat free dairy foods	2–3	8 oz milk 1 cup yogurt 1½ oz cheese	Fat free (skim) or lowfat (1%) milk, fat free or lowfat buttermilk, fat free or lowfat regular or frozen yogurt, lowfat and fat free cheese	Major sources of calcium and protein
Meats, poultry, and fish	2 or less	3 oz cooked meats, poultry, or fish	Select only lean; trim away visible fats; broil, roast, or boil, instead of frying; remove skin from poultry	Rich sources of protein and magnesium
Nuts, seeds, and dry beans	4–5 per week	⅓ cup or 1½ oz nuts 2 Tbsp or ½ oz seeds ½ cup cooked dry beans peas	Almonds, filberts, mixed nuts, peanuts, walnuts, sunflower seeds, kidney beans, lentils,	Rich sources of energy, magnesium, potassium, protein, and fiber
Fats and oils†	2–3	1 tsp soft margarine 1 Tbsp lowfatmayonnaise 2 Tbsp light salad dressing 1 tsp vegetable oil	Soft margarine, lowfat mayonnaise, light salad dressing, vegetable oil (such as olive, corn, canola, or safflower)	DASH has 27 percent of calories as fat, including fat in or added to foods
Sweets	5 per week	1 Tbsp sugar 1 Tbsp jelly or jam ½ oz jelly beans 8 oz lemonade	Maple syrup, sugar, jelly, jam, fruit-flavored gelatin, jelly beans, hard candy, fruit punch, sorbet, ices	Sweets should be low in fat

*Equals ½–1¼ cups, depending on cereal type. Check the product's Nutrition Facts Label.

†Fat content changes serving counts for fats and oils: For example, 1 Tbsp of regular salad dressing equals 1 serving; 1 Tbsp of a lowfat dressing equals 1/2 serving; 1 Tbsp of a fat free dressing equals 0 servings.

BOX E10.2 HOW TO LOWER CALORIES ON THE DASH EATING PLAN

The DASH eating plan was not designed to promote weight loss. But it is rich in lower calorie foods, such as fruits and vegetables. You can make it lower in calories by replacing higher calorie foods with more fruits and vegetables—and that also will make it easier for you to reach your DASH goals. Here are some examples:

To increase fruits—

- Eat a medium apple instead of four shortbread cookies. *You'll save 80 calories.*
- Eat ¼ cup of dried apricots instead of a 2-ounce bag of pork rinds. *You'll save 230 calories.*

To increase vegetables—

- Have a hamburger that's 3 ounces of meat instead of 6 ounces. Add ½ cup serving of carrots and ½ cup serving of spinach. *You'll save more than 200 calories.*
- Instead of 5 ounces of chicken, have a stir-fry with 2 ounces of chicken and 1½ cups of raw vegetables. Use a small amount of vegetable oil. *You'll save 50 calories.*

To increase lowfat or fat free dairy products—

- Have a ½ cup serving of lowfat frozen yogurt instead of a 1½-ounce milk chocolate bar. *You'll save about 110 calories.*

And don't forget these calorie-saving tips—

- Use lowfat or fat free condiments.
- Use half as much vegetable oil, soft or liquid margarine, or salad dressing, or choose fat free versions.
- Eat smaller portions—cut back gradually.
- Choose lowfat or fat free dairy products to reduce total fat intake.
- Check the food labels to compare fat content in packaged foods—items marked lowfat or fat free are not always lower in calories than their regular versions.
- Limit foods with lots of added sugar, such as pies, flavored yogurts, candy bars, ice cream, sherbet, regular soft drinks, and fruit drinks.
- Eat fruits canned in their own juice.
- Add fruit to plain yogurt.
- Snack on fruit, vegetable sticks, unbuttered and unsalted popcorn, or bread sticks.
- Drink water or club soda.

Appendix E11

DIETARY RECOMMENDATIONS FOR INCREASING FIBER INTAKE

- To increase your fiber intake:
  - Eat Whole-grain cereals that contain >9 g fiber per serving fro breakfast
  - Eat raw vegetables
  - Eat fruits (such as pears) and vegetables (such as potatoes) with their skins
  - Add legumes to soups, salads, and casseroles
  - Eat fresh and dried fruit for snacks

Food Group	Increase these foods	Limit these food
Protein Foods	Dried beans and legumes (refried beans, baked beans, black-eye peas, lentils, black, pinto, northern, navy, kidney and garbanzo, etc)	
Grains	Brown rice, whole wheat flour and foods made with whole wheat) quinoa, wheat berries, wheat germ, amaranth, millet, and buckwheat, high fiber cereals (bran, granola, shredded wheat, oatmeal, etc.), corn	Refined grains (white rice, white wheat flour)
Diary		
Fruit	Whole fruit with edible, dried fruit,	Fruit juice
Vegetables	Vegetables	Vegetable juice
Fat	Nuts; flaxseed, coconut	

Fiber in Selected Foods

Grains Group

Whole-grain products provide about 1 to 2 grams (or more) of fiber per serving:

- 1 slice whole-wheat, pumpernickel, rye bread.
- 1 oz ready-to-eat cereal (100% bran cereals contain 10 grams or more).
- ½ c cooked barley, bulgur, grits, oatmeal.
- 4 whole-wheat crackers.
- 1 bagel, 3.5″ diameter.
- 1 muffin—bran, blueberry, cornmeal, or English.
- 2 T bran or wheat germ.

Vegetable Group

Most vegetables contain about 2 to 3 grams of fiber per serving:

- 1 c raw bean sprouts.
- ½ c cooked beets, broccoli, Brussels sprouts, cabbage, carrots, cauliflower, collards, corn, eggplant, green beans, green peas, kale, mushrooms, okra, parsnips, potatoes, pumpkin, spinach, summer squash, sweet potatoes, Swiss chard, tomatoes, turnip greens, winter squash, zucchini.
- ½ c chopped raw carrots, peppers.

Fruit Group

Fresh, frozen, and dried fruits have about 2 grams of fiber per serving:

- 1 medium apple, banana, kiwi, nectarine, orange, peach, pear, plum, tangerine.
- ½ grapefruit.
- ½ c applesauce, blackberries, blueberries, cherries, raspberries, strawberries.
- 3 dates, 2 figs, ¼ cup raisins.
- ½ cup canned cherries, peaches, pears, fruit cocktail.

Source: Whitney EN, Rolfes SR. *Understanding Nutrition*. 10th ed. Belmont (CA): Thomson-Wadsworth; 2004. p. 126, modification of Table 4–3 on p. 127.

Appendix E14

LOW-LACTOSE DIETARY RECOMMENDATIONS

Type of Food	Allowed Items	Excluded Items
Beverages	Water, lactose-free carbonated beverages, fruit-flavored drinks, fruit punches, lemonade, limeade, nondairy product drinks, low-lactose milk, acidophilus milk, coffee, and tea	Artificial fruit drinks containing lactose, all beverages and nutritional supplements made with milk and milk products with the exception of buttermilk, low-lactose milk, and yogurt
Bread	Any cooked or dry cereal not containing lactose. Read label for presence of lactose	Instant hot cereals, high-protein cereals, all cereals with added milk or lactose
Flours	All	None
Cheeses	Fermented cheeses (cheddar and any cheese aged with bacteria)	All others
Desserts	Fruit ices; gelatins; angel food cake; desserts made with nondairy products, buttermilk, or sour cream	Ice cream, puddings, and other desserts containing milk or milk products
Eggs	All except raw eggs and eggs prepared with milk or milk products	Creamed, scrambled, omelets, or other eggs prepared with milk; raw eggs
Fats	Margarine not containing milk solids, vegetable oils, mayonnaise, shortening	All others: cream, half-and-half, table and whipping cream, butter.
Fruits, fruit juices	All fresh, canned, or frozen fruit juices; fruits not processed with lactose	Any canned or frozen fruits and fruit juices processed with lactose
Meat, poultry, fish, legumes, nuts	Any except those specifically excluded. Soy milk. Almond milk;	Creamed or breaded fish, poultry or meat; cold cuts, hot dogs, liver, sausage, or other processed meats containing milk or lactose; gravies made with milk
Milk, milk products	Fermented milk products such as acidophilus milk, buttermilk, yogurt, and sour cream; low-lactose products; "lactose-digesting" pills or caplets	All milk, milk products except those allowed
Potatoes, rice, pasta	White or sweet potatoes, macaroni, noodles, spaghetti or other pasta, rice; rice milk	Any prepared with milk, such as commercially prepared creamed or scalloped potato products containing dried milk
Soups	Broth-based soups	Cream soups, chowders, commercially prepared soups that contain milk or milk products
Sweets	Honey, jams, preserves, syrups, molasses	Candy containing lactose, milk, or cocoa; butterscotch candies, caramels, chocolates (Read all labels carefully.)
Vegetables	All vegetables except those prepared with milk	Any prepared with milk, such as creamed, scalloped, or any processed vegetables containing milk, cheese or lactose
Miscellaneous	Catsup, chili sauce, horseradish, olives, pickles, vinegar, gravies prepared without milk, mustard, all herbs and spices, peanut butter, unbuttered popcorn, non diary creams or whipped toppings.	Chocolate, cocoa, milk gravies, cream sauces, chewing gum, instant coffee, powdered soft drinks, artificial juices containing milk, or lactose

Source: National Cancer Institute [homepage on the Internet]. Bethesda (MD): National Cancer Institute; c2006. Available from: http://www.cancer.gov.

Appendix E15

NUTRITION THERAPY FOR CONSTIPATION

For *some* individuals with constipation, nutrition therapy which increases both fiber and fluid may help with the relief of the symptoms of constipation. Increasing these nutrients should proceed slowly so that symptoms are not exacerbated while the gastrointestinal tract adjusts to the higher fiber content.

Fiber

Slowly increase fiber to 25–35 grams per day. Include a variety of grains: wheat, rye, barley, oat, farro, kamut, couscous, soy, and quinoa. Choose foods from the fiber list (Appendix E11) to provide 4 grams or more per serving.

Fluid

Consume at least 6–8 cups of fluid per day. Count a caffeinated beverage as two thirds of its serving size since caffeine is a mild diuretic.

Appendix E16

NUTRITION THERAPY FOR PEPTIC ULCER

Food Groups	Foods Recommended	Food Not Recommended if Symptomatic
Beverages	Non-cola carbonated beverages, Postum®, herbal teas	Cola, coffee, tea, cocoa, alcohol
Milk and milk products	Skim, 1 %, buttermilk, low-fat yogurt	2% or whole milk, crea, high fat yogurt, chocolate milk
Eggs	Poached, hard boiled, or scrambled using low fat cooking methods	Fried or scrambled using high fat cooking methods
Cereals	All ready-to-eat or cooked	None
Meats and Protein Sources	Baked, roasted, broiled, grilled, stewed; trimmed of visible fat beef, veal, lamb, pork, poultry, fish, low-fat cottage cheese, low-fat cheese, peanut butter	Fried meats, bacon, sausage, pepperoni, salami, bologna, frankfurters/hot dogs
Potatoes/Rice/Pasta	All except fried	None except fried
Vegetables	All	Only individual tolerance
Fruits	All	Only individual tolerance
Fat	As tolerated consistent with current dietary guidelines.	As tolerated consistent with current dietary guidelines.
Dessert	All except those considered high fat or those fried such as pastries, doughnuts.	Those considered high fat or those fried such as pastries, doughnuts.
Miscellaneous	All except pepper	Pepper

Appendix E19

NUTRITION THERAPY FOR OSTOMY

The side effects or symptoms that nutrition therapy can address are gas, odor, consistency of the stool, and obstruction.

General Recommendations

- Add one new food at a time; if this food is not tolerated, try it again in several weeks.
- Take small bites of foods and chew thoroughly.
- Eat foods at a regular time each day. Smaller, more frequent meals may be better tolerated. Eating the largest meal in the middle of the day may assist with decreasing stool output at night.
- Consume adequate fluids with at least 8–10 cups of liquid per day, which may be increased during hot weather or other environmental conditions that would result in excessive fluid loss.

Gas

- Avoid practices that may contribute to swallowed air and gas formation, such as the following: chewing gum, use of drinking straws, carbonated beverages, smoking and chewing tobacco, eating quickly.
- Avoid foods that cause excessive gas production and possibly odor, such as the following:
  - Protein

    Fish, eggs
  - Vegetables

    Asparagus, beets, broccoli, Brussels sprouts, cabbage, carrots, cauliflower, corn, cucumbers, garlic, leeks, lettuce, onions, parsley, peppers (sweet)
  - Legumes

    Black-eyed peas, bog beans, broad beans, chickpeas, filed beans, lentils, lima beans, navy beans, mung beans, peanuts, peas, pinto beans, red kidney beans, soybeans, white beans
  - Grains/Cereals/Seeds/Nuts

    Barley, breakfast cereals, granola, oats bran, oat flour, pistachios, rice bran, sesame flour, sorghum (grain), wheat bran, whole wheat flour.
  - Other

    Alcohol (especially beer), bagels, baked beans, bean salad, chili, lentil soup, pasta, peanut butter, soy milk, split pea soup, stir-fried vegetables, tofu, whole grain breads

Odor

- Use foods that may decrease odor, such as the following: buttermilk, parsley yogurt, kefir, cranberry juice.

Consistency of Stool

- Add foods that may thicken stool, such as the following: banana flakes, applesauce, pectin used to thicken jams and jelly, pasta, potatoes, smooth peanut butter, rice, cheese.
- Avoid foods that may cause obstruction, including the following: corn, cabbage (coleslaw), celery, coconut, dried fruits, green peppers, lettuce, nuts, peas, pineapple, popcorn, spinach, turnip greens.
- Avoid foods that may cause diarrhea, including the following: grape juice, high-sugar foods, prune juice, spicy or high-fat foods.

Kidney Stones

- Consider avoiding foods high in oxalate if you have a potential risk to develop oxalate kidney stones. Foods high in oxalate include the following: beans, beer, beets, carob, chocolate, cocoa, dark, leafy greens, instant tea and coffee, nuts, rhubarb, spinach, sweet potato, tofu, wheat bran, whole wheat flour, wheat germ (also see Appendix D3).

Appendix E20

DIETARY RECOMMENDATIONS FOR PANCREATITIS

Eat a variety of nutrient-dense foods from all food groups while avoiding those highest in fat.

Food Group	Foods Recommended	Servings Per Day	Foods not Recommended
Beverages			Alcohol, wine, beer, liquors.
Grains and grain products	Whole-wheat and enriched breads, bagels, pita bread, crackers without added fat, rice, couscous, pasta, cereals, grits, oatmeal, cream of wheat and rice.	7–8	Breads that have added fat in preparation such as pancakes and waffles. Rice, pasta which is fried.
Vegetables	All vegetables except those that are fried.	4–5	Potatoes, and vegetables which are breaded or fried.
Fruits	All fruits and juices except avocado.	4–5	Avocado.
Meats, poultry, eggs, and fish	Lean cuts that are baked, grilled roasted, or broiled. Remove skin from poultry and trim all visible fat.	2–3 oz	Luncheon meats, prime cut beef, duck, goose, and any meat that is fried.
Dairy products	Skim or reduced-fat milk, skim and reduced-fat cheeses and yogurt.	2–3	Whole milk, cream, and dairy products made from whole milk.
Fats and oils	Margarine, mayonnaise, vegetable oils, salad dressings.- Reduced fat and fat-free products will allow more choices. Amounts may vary according to individual needs.	2–3 tsp	Oils, margarine, butter, lard, shortening, mayonnaise, salad dressings, sour cream in amounts more than 3–5 tsp. per day. Snacks and crackers that have added fat in preparation such as potato chips. Olives.
Nuts, Seeds	Any type.	Amounts may be limited according to individual fat allowances.	
Sweets	Any except those with added fat.	Amounts may vary according to individual needs.	Desserts, pies, cookies that are made with whole milk, eggs, and added fat.

Appendix E21

FOOD SAFETY GUIDELINES FOR PATIENTS WITH LOW IMMUNE FUNCTION OR WHO ARE NEUTROPENIC

Safe food handling can help to decrease a person's risk of food-borne illness. People with weakened immune systems must take extra caution to avoid putting themselves at risk to become infected by a food-borne pathogen. it is important to handle food safely, starting with the buying process, through to eating, and on to storing leftovers.

Shopping

- Shop for groceries when you can take food home right away; do not leave food sitting in the car.
- Avoid cans of food that are dented, leaking, or bulging.
- Do not purchase food in cracked glass jars.
- Ensure that safety buttons on metal lids are down and do not make a clicking noise when pushed. Make sure that tamper-resistant safety seals are intact.
- Avoid food in torn or punctured packaging.
- Pick up perishable foods (e.g., meat, eggs, milk) last.
- Place packaged meat, poultry, or fish in separate plastic bags to prevent meat juices from dripping onto other groceries or other meats.
- Make sure the "sell by" or "use by" date has not passed.
- Do not buy any food that has been displayed in any unclean or unsafe manner (e.g., meat allowed to sit outside of refrigeration, cooked shrimp displayed next to raw shrimp).
- When ordering in the deli department, make sure the clerk washes his or her hands between handling raw food and cooked food.

Storage

- Keep your refrigerator and freezer clean.
- Use a refrigerator thermometer to make sure the temperature inside is 40°F or below.
- Make sure the temperature inside the freezer is 0°F.
- Upon arriving home from the store, immediately refrigerate and freeze appropriate foods.
- Leave eggs in their carton, do not place in refrigerator door.
- Store raw meat, poultry, and fish on the bottom shelf of the refrigerator to avoid their juices dripping onto other foods. Raw ground meat, poultry, and fish may be stored for one to two days; other red meat may be stored for three to five days.
- Store canned foods and other shelf-stable products in a cool, dry place. Avoid hot garages and damp basements.

Preparation

- Wash hands before, during, and after food preparation and service.
- Use plastic or glass surfaces for cutting raw meat and poultry. Use a separate cutting board for preparing other foods, such as fruits, vegetables, and bread.
- Wash cutting boards with hot, soapy water after each use. Cutting boards (except those that are made with laminated wood) can all be washed in the dishwasher.
- After handling raw meat, poultry, and fish, wash hands, work surfaces, and utensils with hot soapy water.
- Wash all fruits and vegetables before cutting, cooking, or eating them raw.
- Defrost frozen food on a place in the refrigerator or in the microwave. Cook food immediately after thawing.
- Use different utensils and dishes for cooked foods than you used for raw foods.
- Wash kitchen towels and clothes often in hot water in a washing machine.
- A sanitizing solution can be made with one teaspoon of liquid chlorine bleach mixed with one quart of water. Using solution on countertops and other work surfaces. Do not rinse. Allow surfaces to air dry.

Cooking

- Keep hot foods hot at 140°F or higher and cold foods cold at 40°F or lower
- Do not leave perishable foods out for more than two hours.
- Promptly refrigerate or freeze leftovers in shallow containers or wrapped tightly in bags.
- Use leftovers within three to four days.
- When reheating foods in the microwave, cover and rotate or stir foods once or twice during cooking. The food should be steaming hot.
- Do not eat foods past their expiration date.
- Follow the handling and preparation instructions on produce labels to ensure top quality and safety.

Meat, Poultry, and fish

- Do not eat raw or undercooked meat.
- Cook all meat and poultry until it is no longer pink in the middle
- Fish should be cooked until it is flaky, not rubbery.
- The temperature inside the meat should be more than 165°F.
- Cook poultry to an internal temperature of 180°–185°F.
- Cook fish to 160°F.
- Co not eat stuffing cooked inside poultry. Instead, cook separately to 165°F.
- Cook only shellfish that are closed. Discard any shellfish that do not open during cooking.

Dairy

- Eat or drink only pasteurized milk or dairy products.

Eggs

- Cook eggs until the yolk and white are solid, not runny.
- Do not eat foods that may contain raw eggs, such as Caesar salad dressing or cookie dough.
- If eating fried eggs, be sure eggs are fried on both sides.

Fruits and Vegetables

- Raw fruits and vegetables are safe to eat if washed carefully first.
- Discard any fruits or vegetables with mold.
- Wash fruits and vegetables well under cool running water.
- Do not let cut fruits or vegetables sit unrefrigerated.
- Discard the outermost leaves of a head of lettuce or cabbage.

Water

- Do not drink water straight from lakes, rivers, streams, or springs.
- Always check with your local health department and water company to learn if they have issued any special notices for people with weakened immune systems.
- Water bottles and ice trays should be cleaned with soap and water before use.

Other

- Home canned foods: Use within one year of canning. Cook food for 10 minutes before eating.
- Commercially canned foods: Safe to eat without any further cooking.
- Condiments: Use a clean utensil when dipping into jars. Keep jars refrigerated. Do not use homemade mayonnaise.
- Baby food: Use a clean utensil to remove amount needed from jar. Store open jars in the refrigerator.

Eating Out

- Avoid the same foods when eating out as you would at home (e.g., raw meats, undercooked eggs).
- If the food arrives undercooked, send it back.
- Avoid foods that may contain raw eggs, such as Caesar salad dressing or hollandaise sauce.
- If you are not sure about the ingredients in a dish, ask your waiter before you order.
- Do not order any raw or lightly steamed fish or shellfish, such as oysters, clams, mussels, sushi, or sashimi.

Traveling

- Do not eat uncooked fruits and vegetables unless you can peel them.
- Avoid salads.
- Eat cooked foods while they are still hot.
- Boil all water before drinking it.
- Use only ice made from boiled water.
- Drink only canned or bottled drinks or beverages made with boiled water.
- Steaming hot foods, fruits you peel yourself, bottled and canned processed drinks, and hot coffee or tea should be safe.
- Talk with your healthcare provider about other advice on travel abroad.

Note: Based on information from Centers for Disease Control and Prevention, 2003a, 2003b, 2003c; United States Department of Agriculture, 200.

References

American Cancer Society. (2002). Nutrition for the person with cancer: A guide for patients and families. Atlanta, GA: Author.

Centers for Disease Control and Prevention. (2003a). An ounce of prevention: Keeps the germs away. Retrieved March 13, 2004, from http://www.cdc.gov/ncidod/op/food.htm

Centers for Disease Control and Prevention. (2003b). Safe food and water: A guide for people with HIV infection. Retrieved March 13, 2004, from http://www.cdc.gov/hiv/pubs/brochure/food.htm

Centers for Disease Control and Prevention. (2003c). What can consumers do to protect themselves from foodborne illness? Retrieved March 13, 2004, from http://www.cdc.gov/ncidod/dbmd/diseaseinfo/foodborneinfections_g.htm#consumersprotect

Charney, P., & Malone, A. (Eds.). (2004). ADA pocket guide to nutrition assessment. Chicago: American Dietetic Association.

Dempsey, D., & Mullen, J. (1985). Macronutrient requirements in the malnourished cancer patient: How much of what and why? Cancer, 55, 290–294.

Appendix F—Contemporary and Alternative Medicine Tables

See Chapter 4 for detailed descriptions and cautionary information.

TABLE F1

Complementary and Alternative Medicine Remedies Used in Diseases of the Upper Gastrointestinal System

Remedy	Scientific Name	CAM Use	AYR	TCM	FOLK	NAT	MEG
Aloe vera; kumari; sabila	*Aloe* spp.	Stomach tonic; treat colic, gastrointestinal reflux disorder, peptic ulcers; promote fat/sugar digestion	X		X	X	
Arjuna, arjun	*Terminalia arjuna*	Treat digestive disorders	X				
Atractylodes; bai zhu; cang zhu	*Atractylodes* spp.	Promote digestion		X			
Bilberry	*Vaccinium myrtillus*	Treat mouth, throat problems; dyspepsia			X	X	
Bloodroot, red root, red puccoon	*Sanguinaria Canadensis*	Emetic; stomach "cleansing;" treat peptic ulcers			X		
Castor oil plant; eranda, vatari	*Ricinus communis*	Treat dyspepsia, belching, vomiting; promote absorption of nutrients	X				
Dong quai; angelica	*Angelica sinensis*	Treat peptic ulcers		X			
Gentian, bitter root; long dan cao	*Gentiana scabra*	Treat nausea, vomiting		X		X	
Guava leaf	*Psidium guajava*	Treat digestive disorders			X	X	
Guggula; bedellium	*Commiphora mukal*	Treat dyspepsia	X			X	
Hare's ear; chai hu	*Bupleurum chinense*	Treat dyspepsia		X			
Licorice root; mulethi; gan cao	*Glycyrrhiza glabra*	Treat stomach "weakness," pain, dyspepsia, peptic ulcers	X	X	X	X	
Malabar nut: adhosa, vasaka	*Adhatoda vasika*	Treat nausea, vomiting	X				
Mandarin, tangerine; chen pi	*Citrus reticulata*	Treat dyspepsia, vomiting, peptic ulcers		X		X	
Mandrake, mayapple	*Podophyllum peltatum*	Emetic; treat stomach disorders			X		
Mistletoe	*Phorandendron leucarpum*	Treat stomach disorders			X		
Neem; nimb	*Azadirachta India*	Promote digestion; treat nausea, vomiting, peptic ulcers	X			X	
Peppermint	*Mentha piperita*	Treat nausea, vomiting, morning sickness			X	X	
Pinella; ban shen	*Pinella ternatae*	Treat nausea, vomiting, abdominal bloating		X			
Poke, inkberry; fitolaca	*Phytolacca americana*	Treat stomach disorders			X		

(continued on the following page)

TABLE F1 *(continued)*

Complementary and Alternative Medicine Remedies Used in Diseases of the Upper Gastrointestinal System

Remedy	Scientific Name	CAM Use	AYR	TCM	FOLK	NAT	MEG
Pyridoxine/Vitamin B_6		Treat morning sickness					X
Raspberry	*Rubus* spp.	Induce vomiting			X		
Shilajit; mineral pitch, fulvic acid, "sweat of the rocks"		Treat peptic ulcers	X				
Turmeric; haridra	*Curcuma longa*	Treat dyspepsia; improve protein digestion; improve intestinal flora	X		X		
Water plantain, ze xie	*Alisma plantago-aquaticae*	Treat abdominal bloating		X			
Yellowroot	*Xanthorrhizia simplicissma*	Treat stomach disorders			X		

AYR- Ayurvedic Medicine TCM- Traditional Chinese Medicine FOLK- Folk Remedy NAT- Natural Product MEG- Vitamin/Mineral Megadose

TABLE F2

Complementary and Alternative Medicine Remedies Used in Diseases of the Lower Gastrointestinal System

Remedy	Scientific Name	CAM Use	AYR	TCM	FOLK	NAT	MEG
Amalaka, amla; Indian gooseberry	*Emblica officinalis*	Cleanse intestines, colon; treat bowel disorders, inc. constipation, colitis, hemorrhoids	X			X	
Aloe vera; kumari; sabila	*Aloe* spp.	Treat diarrhea, diverticulitis, hemorrhoids	X		X	X	
Arjuna, arjun	*Terminalia arjuna*	Treat diarrhea	X				
Astragalus, milk vetch; huang qi	*Astragalus membranaceus*	Treat diarrhea		X		X	
Atractylodes; bai zhu; cang zhu	*Atractylodes* spp	Treat irritable bowel syndrome		X			
Bilberry	*Vaccinium myrtillus*	Treat diarrhea			X	X	
Bitter root, dogbane	*Apocynum* spp.	Purgative, treat constipation			X		
Black cohosh	*Cimicifuga racemosa*	Treat diarrhea			X	X	
Castor oil plant; eranda, vatari	*Ricinus communis*	Purgative; treat constipation; hemorrhoids; dysentery; infantile diarrhea	X				
Dong quai; angelica	*Angelica sinensis*	Treat colitis, constipation		X		X	
Gentian, bitter root; long dan cao	*Gentiana scabra*	Treat flatulence, diarrhea		X		X	
Gotu kola; brahmi	*Centella asiatica*	Treat bowel disorders	X			X	
Guava leaf	*Psidium guajava*	Treat diarrhea, dysentery			X	X	
Guggula; bedellium	*Commiphora mukal*	Treat hemorrhoids	X			X	
Hare's ear; chai hu	*Bupleurum chinense*	Treat diarrhea, constipation, colitis, hemorrhoids		X			
Licorice root; mulethi; gan cao	*Glycyrrhiza glabra*	Purgative; treat constipation	X	X	X	X	
Malabar nut: adhosa, vasaka	*Adhatoda vasika*	Treat diarrhea, dysentery	X				
Mandarin, tangerine; chen pi	*Citrus reticulata*	Treat flatulence, diarrhea		X		X	
Mandrake, mayapple	*Podophyllum peltatum*	Purgative; treat constipation, parasites			X		
Mistletoe	*Phorandendron leucarpum*	Treat diarrhea			X		
Morning glory; rinona	*Ipomoea* spp.	Treat diarrhea			X		
Neem; nimb	*Azadirachta india*	Treat parasites	X			X	
Pellitory; akarakara	*Anacyclus pyrethrum*	Treat bowel disorders	X				
Peppermint	*Mentha piperita*	Treat irritable bowel syndrome; diverticulitis			X	X	
Pinella; ban shen	*Pinella ternatae*	Treat diarrhea		X			
Poke, inkberry; fitolaca	*Phytolacca americana*	Purgative; treat constipation			X		
Raspberry	*Rubus* spp.	Treat diarrhea			X		

(continued on the following page)

TABLE F2 *(continued)*

Complementary and Alternative Medicine Remedies Used in Diseases of the Lower Gastrointestinal System

Remedy	Scientific Name	CAM Use	AYR	TCM	FOLK	NAT	MEG
Shilajit; mineral pitch, fulvic acid, "sweat of the rocks"		Treat hemorrhoids, parasites	X				
Snakeroot; sarp-gandha	*Rauwolfia serpentine*	Treat bowel disorders	X				
Turmeric; haridra	*Curcuma longa*	Treat diarrhea, dysentery, hemorrhoids	X		X		
Water plantain, ze xie	*Alisma plantago-aquaticae*	Treat diarrhea, dysentery		X			
Willow	*Salix* spp.	Treat diarrhea			X		
Yellowroot	*Xanthorhiza simplicissma*	Treat dysentery			X		

AYR- Ayurvedic Medicine TCM- Traditional Chinese Medicine FOLK- Folk Remedy NAT- Natural Product MEG- Vitamin/Mineral Megadose

TABLE F3

Complementary and Alternative Medicine Remedies Used in the Diseases of the Hepatobiliary System

Remedy	Scientific Name	CAM Use	AYR	TCM	FOLK	NAT	MEG
Amalaka, amla; Indian gooseberry	*Emblica officinalis*	Treat liver "weakness"	X				
Arjuna, arjun	*Terminalia arjuna*	Treat cirrhosis; treat bile problems	X				
Ashwagandha; winter cherry, Indian ginseng	*Withania somnifera*	Treat alcoholism	X			X	
Bitter root, dogbane	*Apocynum* spp.	Treat liver ailments, gallstones			X		
Bloodroot	*Sanguinaria Canadensis*	Treat liver ailments			X		
Castor oil plant; eranda, vatari	*Ricinus communis*	Treat enlarged liver; jaundice	X				
Chinese, Baikal skullcap; huang qin	*Scutellaria baicalensis*	Treat jaundice, gallstones		X			
Evening Primrose Oil	*Oenthera biennis*	Treat symptoms of alcohol withdrawal			X	X	
Gentian, bitter root; long dan cao	*Gentiana scabra*	Treat jaundice		X		X	
Hare's ear; chai hu	*Bupleurum chinense*	Liver tonic; treat liver ailments, inc. hepatitis, cirrhosis; treat gallstones; inflammation of gallbladder		X			
Licorice root; mulethi; gan cao	*Glycyrrhiza glabra*	Treat liver ailments	X	X	X	X	
Mandrake, mayapple	*Podophyllum peltatum*	Treat liver ailments			X		
Mango; vimang; amra	*Mangifera indica*	Treat liver ailments	X		X	X	
Neem; nimb	*Azadirachta india*	Cleanse liver	X			X	
Noni, Indian mulberry; ashyulka	*Morinda citirolia*	Treat liver ailments	X		X	X	
Peppermint	*Mentha piperita*	Dissolve gallstones			X	X	
Poke, inkberry; fitolaca	*Phytolacca americana*	Treat liver ailments			X		
Rhemannia; Chinese foxglove; sheng di huang; shu di huang	*Rhemannia glutinosa*	Liver tonic		X		X	
Reishi; ling zhi	*Ganoderma lucidum*	Treat hepatitis		X	X	X	
Schizandra; magnolia vine; gomeishi; wu wei zi	*Schizandra chinensis*	Treat hepatitis		X		X	
Shilajit; mineral pitch, fulvic acid, "sweat of the rocks"		Treat gallstones	X				
Turmeric; haridra	*Curcuma longa*	Treat jaundice, hepatitis	X		X		
Yellowroot	*Xanthorhiza simplicissma*	Treat liver ailments			X		
Valerian	*Valeriana* spp.	Treat alcoholism			X		

AYR- Ayurvedic Medicine TCM- Traditional Chinese Medicine FOLK- Folk Remedy NAT- Natural Product MEG- Vitamin/Mineral Megadose

TABLE F4

Complementary and Alternative Medicine Remedies Used in Diseases of the Hematological System

Remedy	Scientific Name	CAM Use	AYR	TCM	FOLK	NAT	MEG
Aconite, monkshood; fu zi	*Aconitum carmichaeli*	Improve spleen function		X			
Amalaka, amla; Indian gooseberry	*Emblica officinalis*	Strengthen blood, treat anemia, treat hemorrhaging	X			X	
Ascorbic acid/Vitamin C		Improve iron absorption					X
Astragalus, milk vetch; huang qi	*Astragalus membranaceus*	Blood, spleen tonic; treat anemia		X		X	
Bloodroot, red root, red puccoon	*Sanguinaria Canadensis*	Treat "weak" blood			X		
Burdock; niu bang zi	*Arctium lappa*	Cleanse blood toxins		X	X	X	
Castor oil plant; eranda, vatari	*Ricinus communis*	Treat enlarged spleen	X				
Cat's claw; gambir; gao teng	*Uncaria rhynchophylla*	Purify blood, esp. during pregnancy, birth	X	X	X	X	
Cobalamine/Vitamin B12		Treat anemia					X
Copper		Treat anemia					X
Dong quai; angelica	*Angelica sinensis*	Purify blood; treat anemia		X		X	
Folic acid		Treat anemia					X
Gotu kola; brahmi	*Centella asiatica*	Purify blood	X			X	
Gudmar, sarpadarushtrika	*Gymnema sylvere*	Improve circulation	X				
Hawthorn	*Crategus oxyacantha*	Treat blood disorders			X	X	
Mistletoe	*Phoradendron leucarpum*	Treat blood problems, hemorrhaging			X		
Neem; nimb	*Azadirachta india*	Purify blood	X			X	
Poke, inkberry; fitolaca	*Phytolacca americana*	Treat "weak" blood			X		
Turmeric; haridra	*Curcuma longa*	Purify blood	X		X		
Raspberry	*Rubus* spp.	Treat anemia; treat hemorrhaging			X		
Rehmannia; Chinese foxglove; sheng di huang; shu di huang	*Rehmannia glutinosa*	Treat anemia, treat hemorrhaging		X		X	

AYR- Ayurvedic Medicine TCM- Traditional Chinese Medicine FOLK- Folk Remedy NAT- Natural Product MEG- Vitamin/Mineral Megadose

TABLE F5

Complementary and Alternative Medicine Remedies Used in Diseases of the Neurological System

Remedy	Scientific Name	CAM Use	AYR	TCM	FOLK	NAT	MEG
Aconite, monkshood; fuzi	*Aconitum carmichaeli*	Treat pain		X			
Amalaka, amla; Indian gooseberry	*Emblica officinalis*	Treat irritability, insomnia	X			X	
Ascorbic acid/vitamin C		Treat depression					X
Ashwagandha; winter cherry, Indian ginseng	*Withania somnifera*	Treat insomnia; treat Alzheimer's disease, memory loss	X			X	
Bitter root, Dogbane	*Apocynum* spp.	Treat headache			X		
Black nightshade, zhoa ia	*Solanum nigrum*	Promote sleep; treat pain			X		
Burdock; niu bang zi	*Arctium lappa*	Treat back pain		X	X	X	
L-Carnitine		Treat chronic fatigue syndrome; reduce memory loss				X	
Castor oil plant; eranda, vatari	*Ricinus communis*	Treat headache, back pain, sciatica	X				
Cat's claw; gambir; gao teng	*Uncaria rhynchophylla*	Treat tremors, seizures, convulsions; treat chronic pain	X	X	X	X	
Chinese, Baikal skullcap; huang qin	*Scutellaria baicalensis*	Treat irritability		X			
Coenzyme Q_{10}		Alleviate fibromyalgia symptoms; treat chronic fatigue syndrome, Parkinsonism					
Cobalamin/vitamin B_{12}		Treat depression, psychosis, Alzheimer's disease, memory loss; assist with myelin regeneration; treat diabetic neuropathy					X
DHEA (dehydro-epiandrosterone)		Alleviate fibromyalgia symptoms; treat chronic fatigue syndrome				X	
Dong quai; angelica	*Angelica sinensis*	Treat headache		X		X	
Echinacea, purple coneflower	*Echinacea purpurea, E. angustifolia*	Treat chronic fatigue syndrome			X	X	
Ephedrine; ephedra; ma huang	*Ephedra sinica*	Treat multiple sclerosis		X		X	
Evening Primrose Oil	*Oenothera biennis*	Provide for source of omega-3 fatty acids; Treat attention deficit hyperactivity disorder, memory loss				X	
Folic acid		Treat depression, insomnia, irritability, dementia					X
Gingko biloba; ying xing	*Gingko biloba*	Optimize brain function; treat headache, dizziness, depression, anxiety, Alzheimer's disease; reduce memory loss; treat diabetic neuropathy		X		X	

(continued on the following page)

TABLE F5 *(continued)*

Complementary and Alternative Medicine Remedies Used in Diseases of the Neurological System

Remedy	Scientific Name	CAM Use	AYR	TCM	FOLK	NAT	MEG
Ginseng; ren shen; fivefinger; tartar root; redberry; sang	*Panax ginseng, P. quinquefolium*	Improve mental performance, reduce memory loss		X	X	X	
Glucosamine		Treat back pain				X	
Goldenseal; yellow root; yellow puccoon; yellow Indian paint; ground raspberry	*Hydrastis canadensis*	Treat chronic fatigue syndrome			X	X	
Gokshura, puncture vine	*Trigulis territris*	Treat back pain, sciatica, neuropathy	X			X	
Gotu kola; brahmi	*Centella asiatica*	Rejuvenate brain cells, nerves; treat nerve disorders, inc. convulsions, epilepsy, tetanus, dementia	X			X	
Grape Seed Extract	*Vitis vinifera, V. coignetiae*	Alleviate fibromyalgia symptoms				X	
Hawthorn	*Crategus oxyacantha*	Treat insomnia			X	X	
Kava kava, 'awa	*Piper methysticum*	Treat headache, back pain; treat obsessive compulsive disorder			X	X	
Magnesium		Treat autism, attention deficit hyperactivity disorder					X
Malabar nut: adhosa, vasaka	*Adhatoda vasika*	Treat neuralgia; treat epilepsy; treat hysteria	X				
Melatonin		Treat seasonal affective disorder, insomnia					
Mistletoe	*Phorandendron leucarpum*	Ease anxiety, panic disorder			X		
Niacin/Vitamin B_3		Treat mania, depression, anxiety, dementia, memory loss					
Pantothenic acid		Treat depression, irritability					X
Pellitory; akarakara	*Anacyclus pyrethrum*	Nerve tonic; treat epilepsy, paralysis	X				
Peppermint	*Mentha piperita*	Treat headache, chronic pain				X	
Pinella; ban shen	*Pinella ternatae*	Treat headache, dizziness		X			
Pyridoxine/vitamin B_6		Treat depression, autism					X
Reishi; ling zhi	*Ganoderma lucidum*	Treat neuralgia		X	X	X	
Rhemannia; Chinese foxglove; sheng di huang; shu di huang	*Rhemannia glutinosa*	Treat irritability, dizziness, insomnia		X		X	
Riboflavin/vitamin B_2		Reduce migraine severity, frequency; treat depression					X

(continued on the following page)

TABLE F5 *(continued)*

St. John's Wort	*Hypericum perforatum*	Alleviate fibromyalgia symptoms, treat chronic fatigue syndrome; reduce memory loss				X	
Salvia; dan shen	*Salvia miltirrhiza*	Treat chronic fatigue syndrome; treat chronic pain, insomnia		X			
Schizandra, magnolia vine; gomeishi; wu wei zi	*Schizandra chinensis*	Treat insomnia		X		X	
Selfheal; all heal; xi ku cao	*Prunella vulgaris*	Treat headache, dizziness		X	X		
Shilajit; mineral pitch, fulvic acid, "sweat of the rocks"		Treat epilepsy, mental disorders	X				
Snakeroot; sarp-gandha	*Rauwolfia serpentine*	Treat central nervous system disorders; treat hypochondria; treat violent mental disorders	X				
Thiamin/vitamin B_1		Treat depression, psychosis, irritability, Alzheimer's disease, memory loss					X
Turmeric; haridra	*Curcuma longa*	Treat chronic pain	X		X		
Valerian	*Valerian* spp.	Treat headache; ease anxiety, panic disorder, obsessive compulsive disorder				X	
Vitamin E		Treat depression; treat Alzheimer's disease, memory loss					X
Willow	*Salix* spp.	Treat headache, chronic pain					
Zinc		Treat autism					X

AYR- Ayurvedic Medicine TCM- Traditional Chinese Medicine FOLK- Folk Remedy NAT- Natural Product MEG- Vitamin/Mineral Megadose

TABLE F6

Complementary and Alternative Medicine Remedies Used in Neoplastic Diseases

Remedy	Scientific Name	CAM Use	AYR	TCM	FOLK	NAT	MEG
Aloe vera; kumari; sabila	*Aloe* spp.	Treat cervical, lung cancers	X		X	X	
Amalaka, amla; Indian gooseberry	*Emblica officinalis*	Prevent/treat cancer	X			X	
Ascorbic acid/Vitamin C		Prevent/treat cancer; antioxidant					X
Ashwagandha; winter cherry, Indian ginseng	*Withania somnifera*	Prevent/treat cancer	X			X	
Astragalus, milk vetch; huang qi	*Astragalus membranaceus*	Adjunct to radiation, chemotherapy; treat melanoma, bladder, bone, breast, cervical, colorectal, endometrial, kidney, liver, lung, ovarian cancers; see Box 24.3 for information on efficacy and potential for adverse reactions		X		X	
Beta-carotene		Reduce cancer risk, esp. cervical cancer					X
Burdock; niu bang zi	*Arctium lappa*	Prevent/treat cancer		X	X	X	
Chinese, Baikal skullcap; huang qin	*Scutellaria baicalensis*	Treat bone, liver cancers		X			
Coenzyme Q_{10}		Reduce cancer risk; treat fibrocystic disease				X	
Copper		Prevent/treat cancer					X
Echinacea, purple coneflower	*Echinacea purpurea, E. angustifolia*	Improve immune system during radiation, chemotherapy			X	X	
Folic acid		Reduce risk cervical, colon cancers					X
Dong quai; angelica	*Angelica sinensis*	Treat esophageal, liver cancers		X		X	
Garlic; ajo; lasuna	*Allium sativum*	Reduce risk of colon, esophageal, lung, stomach cancers			X	X	
Goldenseal; yellow root; yellow puccoon; yellow Indian paint; ground raspberry	*Hydrasitis canadensis*	Inhibit liver cancer growth; see Box 24.3 for information on efficacy and potential for adverse reactions			X	X	
Gotu kola; brahmi	*Centella asiatica*	Treat cancer	X			X	
Grape Seed Extract	*Vitis vinifera, V. coignetiae*	Prevent/treat cancer				X	
Hare's ear; chai hu	*Bupleurum chinense*	Treat bone cancer		X			
Licorice root; mulethi; gan cao	*Glycyrrhiza glabra*	Treat kidney tumors	X	X	X	X	
Magnesium		Reduce cancer risk					X
Mistletoe	*Phorandendron leucarpum*	Treat cancer, esp. breast, ovarian, prostate; see Box 24.3 for information on efficacy and potential for adverse reactions			X		
Neem; nimb	*Azadirachta india*	Prevent/treat cancer	X			X	

(continued on the following page)

TABLE F6 *(continued)*

Melatonin		Treat cancer, adjunct to chemotherapy, radiation				X	
Noni, Indian mulberry, ashyulka	*Morinda citirolia*	Prevent/treat cancer	X		X	X	
Pinella; ban shen	*Pinella ternatae*	Treat esophageal cancer		X			
Poke, inkberry; fiolaca	*Phytolacca Americana*	Treat cancer			X		
Pyridoxine/Vitamin B_6		Reduce risk of cervical cancer					X
Reishi; ling zhi	*Ganoderma lucidum*	Prevent/treat cancer, esp. colorectal, kidney, liver cancers; treat fibrocystic disease		X	X	X	
St. John's Wort	*Hypericum perforatum*	Prevent infiltration of chest wall in breast cancer				X	
Salvia; dan shen	*Salvia miltirrhiza*	Treat abdominal masses; fibrocystic disease		X			
Saw palmetto	*Serenoa repens, Sabal serrulata*	Treat prostate cancer			X	X	
Selenium		Prevent/treat cancer					X
Selfheal; all heal; xi ku cao	*Prunella vulgaris*	Treat cancer		X	X		
Turmeric; haridra	*Curcuma longa*	Treat breast, cervical, colorectal, lung, prostate cancers	X		X		
Vitamin A		Prevent/treat cancer					X
Vitamin E		Reduce colon cancer risk; antioxidant					
Zinc		Prevent/treat cancer, esp. prostate cancer					X

AYR- Ayurvedic Medicine TCM- Traditional Chinese Medicine FOLK- Folk Remedy NAT- Natural Product MEG- Vitamin/Mineral Megadose

TABLE F7

Complementary and Alternative Medicine Remedies Used in Diseases of the Musculoskeletal System

Remedy	Scientific Name	CAM Use	AYR	TCM	FOLK	NAT	MEG
Amalaka, amla; Indian gooseberry	*Emblica officinalis*	Treat osteoporosis	X			X	
Arjuna; arjun	*Terminalia arjuana*	Treat bone fractures	X				
Ashwagandha; winter cherry, Indian ginseng	*Withania somnifera*	Treat arthritis, rheumatism, paralysis	X			X	
Atractylodes; bai zhu; cang zhu	*Atractylodes* spp.	Treat arthritis		X			
Beta-carotene		Reduce pain in osteoarthritis					X
Bitter root, dogbane	*Apocynum spp.*	Treat gout			X		
Black cohosh	*Cimicifuga racemosa*	Treat arthritis			X	X	
Bloodroot, red root, red puccoon	*Sanguinaria Canadensis*	Treat rheumatism			X		
Boron		Improve bone density					X
Burdock; niu bang zi	*Arctium lappa*	Treat gout		X	X	X	
Cat's claw; gambir; gao teng	*Uncaria rhynchophylla*	Treat arthritis	X	X	X	X	
Castor oil plant; eranda, vatari	*Ricinus communis*	Treat arthritis, rheumatism, back pain	X				
Chondroitin		Treat osteoarthritis				X	
Chromium		Increase lean muscle					X
Coenzyme Q_{10}		Alleviate fibromylagia symptoms; treat Parkinsonism				X	
Copper		Treat osteoarthritis					X
DHEA (dehydro-epiandrosterone)		Alleviate fibromylagia symptoms				X	
Dong quai; angelica	*Angelica sinensis*	Treat arthritis		X		X	
Evening primrose oil	*Oenothera biennis*	Treat osteoarthritis			X	X	
Glucosamine		Treat osteoarthritis; alleviate back, joint pain				X	
Gokshura, puncture vine	*Trigulis terristris*	Treat back pain, rheumatism; increase muscle mass	X			X	
Gotu kola; brahmi	*Centella asiatica*	Treat rheumatism, tetnus	X			X	
Guggula; bedellium	*Commiphora mukal*	Treat arthritis, rheumatism, gout	X			X	
Grape Seed Extract	*Vitis vinifera, V. coignetiae*	Alleviate fibromylagia symptoms				X	
Kava kava, 'awa	*Piper methysticum*	Treat back pain			X	X	
Licorice root; mulethi; gan cao	*Glycyrrhiza glabra*	Treat rheumatism	X	X	X	X	

(continued on the following page)

TABLE F7 *(continued)*

Malabar nut: adhosa, vasaka	*Adhatoda vasika*	Treat rheumatism	X				
Mandrake, mayapple	*Podophyllum peltatum*	Treat arthritis, rheumatism			X		
Neem; nimb	*Azadirachta india*	Treat arthritis, rheumatism	X			X	
Pellitory; akarakara	*Anacyclus pyrethrum*	Treat rheumatism; paralysis	X				
Poke, inkberry; fiolaca	*Phytolacca Americana*	Treat arthritis, rheumatism, bursitis			X		
Potassium		Prevent muscle cramps					X
Selenium		Treat osteoarthritis					X
St. John's Wort	*Hypericum perforatum*	Alleviate fibromylagia symptoms				X	
Turmeric; haridra	*Curcuma longa*	Treat arthritis	X		X		
Vitamin A		Treat osteoarthritis					X
Willow	*Salix* spp.	Treat osteoarthritis, rheumatism			X		
Zinc		Treat osteoarthritis					X

AYR- Ayurvedic Medicine TCM- Traditional Chinese Medicine FOLK- Folk Remedy NAT- Natural Product MEG- Vitamin/Mineral Megadose

TABLE F8

Complementary and Alternative Medicine Remedies Used in Underweight and Overweight

Remedy	Scientific Name	CAM Use	AYR	TCM	FOLK	NAT	MEG
Anemarrhena; zhi mu	*Anemarrhena asphodeloidis*	Treat anorexia		X			
Astragalus, milk vetch; huang qi	*Astragalus membranaceus*	Promote weight loss; treat hyperthyroidism		X		X	
Atractylodes; bai zhu; cang zhu	*Atractylodes* spp.	Treat anorexia; treat addiction to sweet, fatty foods, weight loss		X			
L-Carnitine		Increase fat metabolism, promote weight loss				X	
Chromium		Promote weight loss; increase lean muscle					X
Coenzyme Q_{10}		Promote weight loss				X	
DHEA (dehydroepiandrosterone)		Promote weight loss				X	
Ephedrine; ephedra; ma huang	*Ephedra sinica*	Promote weight loss		X		X	
Gentian; bitter root; long dan cao	*Gentiana* spp.	Treat anorexia		X		X	
Gokshura; puncture vine	*Tribulis terrestris*	Increase lean muscle	X			X	
Guggula; bedellium	*Commiphora mukal*	Promote weight loss	X			X	
Hare's ear; chai hu	*Bupleurum chinense*	Treat anorexia; promote weight loss		X			
Malabar nut; adhosa; vasaka	*Adhatoda vasika*	Treat wasting	X				
Mandarin, tangerine; chen pi	*Citrus reticulata*	Treat anorexia; promote weight loss		X		X	
Mandrake, mayapple	*Podophyllum peltatum*	Treat anorexia			X		
Neem; nimb	*Azadirachta india*	Promote weight loss	X			X	
Noni; Indian mulberry; ashyulka	*Morinda citirolia*	Treat anorexia	X		X	X	
Raspberry	*Rubus* spp.	Treat anorexia			X		
St. John's Wort	*Hypericum perfoatum*	Promote weight loss				X	
Schizandra; magnolia vine; gomeishi; wu wei zi	*Schizandra chinensis*	Treat wasting		X		X	
Shilajit; mineral pitch, fulvic acid, "sweat of the rocks"		Promote weight loss	X				
Turmeric; haridra	*Curcuma longa*	Treat anorexia	X		X		

AYR- Ayurvedic Medicine TCM- Traditional Chinese Medicine FOLK- Folk Remedy NAT- Natural Product MEG- Vitamin/Mineral Megadose

(continued on the following page)

TABLE F9 *(continued)*

Reishi; ling zhi	*Ganoderma lucidum*	Lower serum cholesterol, triglyceride levels; treat hypertension		X	X	X	
St. John's Wort	*Hypericum perforatum*	Reduce serum cholesterol levels				X	
Salvia; dan shen	*Salvia miltirrhiza*	Heart, blood tonic; treat angina, arrhythmia, artherosclerosis; treat stroke		X			
Snakeroot; sarp-gandha	*Rauwolfia serpentine*	Treat hypertension	X				
Schizandra; magnolia vine; gomeishi; wu wei zi	*Schizandra chinensis*	Treat arrhythmia		X		X	
Selenium		Reduce risk of cardiovascular, cerebrovascular disease; reduce oxidation of LDL cholesterol					X
Selfheal; all heal; xi ku cao	*Prunella vulgaris*	Treat hypertension		X	X		
Schizandra; magnolia vine; gomeishi; wu wei zi	*Schizandra chinensis*	Treat arrhythmia		X		X	
Vitamin A		Treat cardiovascular disease					X
Vitamin E		Reduce oxidation of LDL cholesterol; increase HDL cholesterol; improve circulation					X
Yellowroot	*Xanthorhiza simplicissma*	Treat hypertension			X		

AYR- Ayruvedic Medicine TCM- Traditional Chinese Medicine FOLK- Folk Remedy NAT- Natural Product MEG- Vitamin/Mineral Megadose

Appendix G—Answers to Case Study Questions

Chapter 4

1. Ms. Thomson's medications and supplements include:
   Lipitor: Antihyperlipidemic agent; HMG Co-A reductase inhibitor; lowers LDL, raises HDL.
   Lisinopril: Cardiovascular agent; ACE inhibitor.
   Peruvian bark: herbal preparation containing quinine and used for alleviating leg cramps.
   Glucosasmine sulfate: amino-acid compound used to treat osteoarthritis, back, and joint pain.
   Flax seed: contains essential fatty acids that may promote an anti-inflammatory effect and contribute to alleviating joint pain.
   Evening primrose oil: herbal preparation used for PMS, menopause, and osteoarthritis, among others.
   St. John's wort: herbal preparation used to treat depression, treat PMS, and reduce cholesterol, among others.
   Dong quai (angelica): Considered yang in Chinese medicine; used to treat menstrual problems, hypertension, and arthritis, among others.
   Black cohosh: used to treat menstrual problems, hypertension, and arthritis, among others.

2. The evidence supporting the efficacy of each of these supplements is inconclusive. According to the National Women's Health Network, very little research supporting the efficacy of these supplements is available. A review in *Menopause* (Huntley and Ernst 2003) states, "There is no convincing evidence for any herbal medical product in the treatment of menopausal symptoms." One concern would be that a common side effect of both lipitor and lisinopril is muscle and joint pain. Perhaps if she stopped taking it, those symptoms would lessen. Additionally, several of these herbs (black cohosh, evening primrose) may exert a cholesterol-lowering effect.

Huntley, A. L. and Ernst, E. (2003). Herbs and menopause. *Menopause, 10*(5): 465–476.

3. In Chinese medicine, energy, or *qi,* travels along meridians in the body. Circulation of *qi* can be enhanced with the insertion of thin needles at specific acupoints. It is believed that bringing harmony and equilibrium to the body will relieve disease and therefore will relieve the pain of aching joints and muscles.

4. The evidence is inconclusive that these treatments will work to alleviate menopausal symptoms or joint pain. If, however, you see improvement and don't find the cocktail to be too expensive, it probably won't do any harm in the dosage recommended. Self-monitor for changes in mood, heart palpitations, GI irritation, or any other symptoms/side effects you experience. Make sure to tell your doctor about all supplements you are taking if he or she prescribes any new medication, especially anticoagulants. If after six months these products are not working for you, they most likely never will and it is a good idea to discontinue their use.

   Advice on avoiding foods from the "nightshade" group (eggplant, tomatoes, peppers, potatoes) is partly based on the history of these foods in European and American colonial culture. Nightshade vegetables were widely viewed as toxic or potentially poisonous, being related to the deadly nightshade plant and to other herbs commonly used in witchcraft. There is also scientific validity to this claim; these vegetables are related to the tobacco plant and contain some amount of a nicotine derivative that acts as a nervous system stimulant. There is no evidence, however, that eating these foods will aggravate menopausal symptoms.

Chapter 6

Subjective information for this case will be:
Diet related:

- Dislikes sweets and fats, rarely eats vegetables.
- Eats 2 large meals a day.
- Never eats breakfast.

Lifestyle:

- Drinks 8 cups of fruit juice/day in addition to coffee and tea (which has added sugar).
- Is fairly inactive.

Medical history: Family history of diabetes.
Learning/motivation: Does want to lose weight because of family history of diabetes.

Objective information:
A—52-year-old male, 5′7″/195 lbs, BMI: 30.7 kg/m^2, IBW 148 lbs, % IBW: 131%.
C—Dx: Obesity
D—Consumes approximately 2800 kcal/day, diet high in sugar (sugar-containing beverages), low in fiber.

Sample PES statement: Excessive energy intake *related to* high consumption of sugar-containing beverages and erratic eating pattern *as evidenced by* average daily kilocalorie intake much greater than estimated recommendations.

Sample assessment:
Current: Obesity, poor quality of diet
Potential: Diabetes
Calories: 1786 kcal; Protein: 70 g

reduced saturated and total fat intake, and a reduction in sodium intake to < 2.4 g/day. Specific nutrition problems that can be identified are: low intake of fruits and vegetables, inadequate fiber intake, inadequate calcium/dairy intake, and high sodium intake.

Chapter 16

1. *Helicobacter pylori* is a spiral-shaped, flagellated, gram-negative rod, which lives under the mucous layer of the stomach. These organisms break down urea to produce ammonia, which helps to neutralize acid in the immediate vicinity of these bacteria and enhances their survival. The *H. pylori* organisms subsequently produce various proteins that damage mucosal cells, attracting lymphocytes and causing persistent inflammation. The by-products produced by the organism result in damage to the epithelium and impair the mucous barrier within the stomach. When it penetrates the stomach, excess acid can irritate the stomach and duodenum, eventually causing an ulcer.
2. Abnormal lab values:

Laboratory Values	High/ Low	Normal values	Relationship to Dx & Nutritional Status
Total Protein 5.9 g/dL	Low	6–8 g/dL	Due to ulcerations of the gastric mucosa, extensive tissue damage and malabsorption which leads to protein deficiency. The Pt. also eats less to avoid pain.
Albumin 3.4 g/dL	Low	3.6–5 g/dL	
Hgb 11.5 g/dL	Low	12–16 g/dL	Indicates active bleeding and may lead to anemia.
Hct 36 %.	Low	37–47 %	

3. Prescribed medications:

Drug	Action
Metronidazole	Metronidazole is an antibiotic that eliminates bacteria and other microorganisms that cause infections of the reproductive system, gastrointestinal tract, skin, vagina, and other areas of the body. Metronidazole selectively blocks some of the cell functions in these microorganisms, resulting in their demise.
Tetracycline	Tetracycline, an antibiotic, eliminates bacteria that cause infections, in this case *Helicobacter pylori*.
Bismuth subsalicylate	It fights infection by *Helicobacter pylori* bacteria, which often occurs with ulcers. Binds or neutralizes the toxins of some bacteria, rendering them nontoxic. It decreases intestinal inflammation and increases the activity of intestinal muscles and lining. Bismuth subsalicylate is considered to have an important role in relapsing gastric ulcers. Bismuth subsalicylate helps to coat the ulcer, allowing it to heal.
Omeprazole	Omeprazole is a proton-pump inhibitor, and provides not only symptom relief, but also symptom resolution in most cases, including those involving more significant ulcers and/or damage to the esophagus. Because proton-pump inhibitors offer the most effective means of impeding acid production, they are useful in treating serious ulcer conditions.

Current Recommendation: Therapy for *H. pylori* infection consists of 10 days to 2 weeks of one or two effective antibiotics, such as amoxicillin, tetracycline (not to be used for children <12 yrs.), metronidazole, or clarithromycin, plus either ranitidine bismuth citrate, bismuth subsalicylate, or a proton-pump inhibitor. Acid suppression by the H2 blocker or proton-pump inhibitor in conjunction with the antibiotics helps alleviate ulcer-related symptoms (i.e., abdominal pain, nausea), helps heal gastric mucosal inflammation, and may enhance efficacy of the antibiotics against *H. pylori* at the gastric mucosal surface. Currently, eight *H. pylori* treatment regimens are approved by the Food and Drug Administration (FDA); however, several other combinations have been used successfully. Antibiotic resistance and patient noncompliance are the two major reasons for treatment failure. Eradication rates of the eight FDA-approved regimens range from 61% to 94% depending on the regimen used. Overall, triple therapy regimens have shown better eradication rates than dual therapy. Longer length of treatment (14 days versus 10 days) results in better eradication rates (http://www.cdc.gov/ulcer/md.htm#treatment).

Drug-nutrient interactions:

Drug	Drug-nutrient interaction
Metronidazole	Flushing, fast heartbeats, nausea, and vomiting may occur when alcohol is ingested during metronidazole therapy. Antibiotics may cause mild side effects such as nausea, vomiting, diarrhea, dark stools, metallic taste in the mouth, dizziness, headache, and yeast infections in women.
Tetracycline	Antibiotics like tetracycline bind to calcium and magnesium found in food, milk, and some other dairy products. This may decrease the absorption of the antibiotic. Antacids, calcium supplements, iron products, and laxatives containing magnesium interfere with tetracycline, making it less effective.
Bismuth subsalicylate	Avoid taking this medication with dairy products or milk. Can cause constipation or diarrhea depending on the combination of salts.
Omeprazole	Proton-pump inhibitors can interfere with vitamin B_{12} absorption from food by slowing the release of gastric acid into the stomach. This is a concern because acid is needed to release vitamin B_{12} from food prior to absorption.

4. Nutrition problems:

Nutrition Problems	Etiology	Signs & Symptoms
Not consuming enough	Ulcerations/inflamation of the gastric mucosa or duodenum caused by *H. pylori*	Chronic indigestion and abdominal pain due to food intake, especially spicy foods, fried foods, coffee, and chocolate. Poor appetite and extensive weight loss. Low HGB & HCT,

Nutrition Problems	Etiology	Signs & Symptoms
PEM/anemia	Extensive tissue damage and malabsorption due to ulcer; active bleeding	Low albumin and total protein levels; extensive weight loss (25% Wt. loss). Low HGB and HCT indicate bleeding and may lead to anemia.

Chapter 17

1. This test evaluates digestion of fats by determining excessive excretion of lipids in patients exhibiting signs of malabsorption, such as weight loss, abdominal distention, and scaly skin. All stool is collected over a 72-hour period and then analyzed for the presence of fat and meat fibers. The presence of fat may indicate a malabsorption problem. Her result indicates fat malabsorption and steatorrhea.
2. When the small intestinal mucosa is exposed to certain sequences of amino acids found in the prolamin fraction of wheat (gliadin), rye (secalin), and barley (hordein), there appears to be both a toxic and inflammatory response. This response damages villi; height is reduced, and they are flattened in appearance. Lack of surface area and reduction of enzyme production cause both malabsorption and maldigestion. Celiac disease (CD), previously referred to as gluten-sensitive enteropathy, gluten intolerance, or non-tropical sprue, is a complex disease whose etiology originates from both genetic and autoimmune factors. In this disease, exposure to gluten results in damage to the intestinal mucosa. Other diseases or conditions associated with CD are type 1 diabetes mellitus and thyroid dysfunction, as well as dermatitis, muscle, and joint pain. Persons with CD may also be at higher risk for lymphoma and other malignancies
3. It is well understood that damage to the intestinal mucosa, which is observed in CD, occurs when the small intestine is exposed to the prolamin fraction—α-gliadin and other protein components of gluten. Gluten is found in wheat, rye, malt, barley, and, in smaller amounts, in oats.

   More recent research indicates damage to the intestinal mucosa is accompanied by an infiltration of white blood cells into the mucosa. This inflammatory response is also reflected in production of IgA antigliadin and antiendomysial antibodies. These antibodies, which now serve as components of the diagnostic procedures for CD, reflect the autoimmune nature of this disease.

   Genetic markers for CD have not been completely established, but as previously stated, evidence is significantly convincing that a genetic component exists. Current research indicates CD is strongly associated with a group of genes on Chromosome 6. These genes for HLA class II antigens are involved in regulation of the body's immune response to gluten protein fractions.
4. Fat malabsorption and protein-energy malnutrition are two nutritional problems as evidenced by steatorrhea and decreased labs (total protein, albumin, and prealbumin). In this disease, exposure to gluten results in damage to the intestinal mucosa. Decreased villous height and, decreased enzyme production have caused malabsorption and/or maldigestion of fat. Dysfunction of the accessory organs of digestion (liver, pancreas, and gallbladder) may also contribute to steatorrhea and general malabsorption. Malabsorption and decreased protein and kcal intake have resulted in catabolism as evidenced by the biochemical labs. In addition, she may have an iron-deficiency anemia since her hematocrit is low (35%).

Chapter 18

1. Acute pancreatitis is acute inflammation of the pancreas. The onset of the disease is rapid and involves pain that may radiate to the back. Pancreatic cells are damaged and destroyed, releasing pancreatic enzymes into the blood. Pain can be intensified with food ingestion, especially in large amounts.
2. The signs and symptoms that support acute pancreatitis are: acute abdominal pain extending into the lower back, severe nausea and vomiting, epigastric pain, upset stomach with fried foods, elevated serum lipase and amylase levels, and low albumin and total protein levels.
3. The factors consistent with risk for pancreatitis are high alcohol intake and epigastric pain for the previous 3 months.
4. First, the patient will be NPO to allow for the pancreas to rest. Next, enteral feeding will be initiated through a nasoenteric tube. The formula will most likely start at 25 mL/hour, and will progress to around 25 kcal/kg of body weight in the next 24 to 48 hours. The formula will be either almost fat free, or composed of mostly medium-chain triglycerides. Finally, oral feeding will be initiated once the patient exhibits decreasing amylase and lipase levels, and pain has ceased for 24 hours.
5. His primary nutrition diagnosis will be "Acute pancreatitis (Altered GI function − NC 1.4) resulting in inadequate oral intake (NI −2.1); signs and symptoms: elevated serum lipase and amylase."
6. Nutrition therapy focuses on providing adequate kcal and protein, resolving negative nitrogen balance by minimizing nitrogen excretion, balancing glucose levels, controlling hypertriglyceridemia and micronutrient imbalances, and finally minimizing stimulation of the pancreas.

Chapter 19

1. Type 1 DM: little or no insulin produced by pancreatic beta cells, meaning that patient is dependent on exogenous insulin; an abrupt onset, usually before age 30.

4. The patient has experienced an overall loss of appetite, possibly due to the inability to swallow properly and the fear of choking. It would appear that the patient might be deficient in both energy and most nutrients because her intake is so limited.
5. Because dysphagia is present, the patient may need to consider tube-feeding placement in order to ensure the adequate intake of nutrients and energy.

Chapter 23

1. What risk factors for emphysema/COPD are present in Stella's history?

   The primary risk factor for the development of COPD is smoking. Other risk factors are Hx of bronchitis and respiratory infections. Females are more than twice as likely to be diagnosed with chronic bronchitis as males.

2. Describe the causes for the following symptoms present in Stella's history and physical exam: anorexia, dyspnea, fatigue, early satiety, bloating.

   Anorexia: Severe dyspnea makes it difficult to eat. Chewing and swallowing may be impaired since both activities change breathing patterns and reduce oxygen uptake. Chronic mouth breathing or certain medications may also cause changes in taste perceptions. She complains of having no appetite and says that food doesn't taste good to her. Her dentures no longer fit properly.

   Fatigue: Her body is not getting enough oxygen to function properly. She might not be consuming enough kcal, which could cause fatigue. Individuals with COPD often complain that they tire easily when eating or experience dyspnea during eating and drinking.

   Early satiety and bloating: Patients with COPD often suffer from hyperinflation of the lungs with accompanying flattening of the diaphragm and reduced abdominal volume, leading to unnecessary fullness and bloating at mealtime. Aerophagia is often seen in COPD and may also cause bloating.

3. Look at Stella's laboratory data. What do they tell you about her nutrition status? Define each of the following blood gases and interpret her values:

   Stella's albumin is low, which indicates malnutrition. Sodium is on the low side, but still in the normal range. Her potassium is also within the normal range.

Blood Gas	Stella's Value	Normal Values
Partial pressure of oxygen (PaO_2)	77.7 mmHg	75 to 100 mm Hg
Partial pressure of carbon dioxide ($PaCO_2$)	50.9 mmHg	35 to 45 mm Hg
pH	7.29	7.35 to 7.45
Oxygen saturation (O_2Sat)	92%	94% to 100%
Bicarbonate - (HCO_3)	29.6 mEq/L	22 to 26 mEq/liter

   Changes in pulmonary function are reflected by changes in the partial pressure of dissolved carbon dioxide ($PaCO_2$), and by changes of partial pressure of dissolved oxygen (PaO_2). Changes in the $PaCO_2$ measure how well carbon dioxide is able to move out of the blood into the airspaces of the lung, and out in the exhaled air. Changes in PaO_2 measure how well oxygen is able to move from the air into the lungs. Oxygen saturation is the measure of the amount of oxygen carried by the red blood cells and can be calculated using the partial pressure of dissolved oxygen (PaO_2). In patients with pulmonary disease, fewer red blood cells carry the usual load of oxygen, and oxygen saturation is decreased.

   The pH is a measure of hydrogen ion concentration ($H+$) in blood. Respiratory acidosis, caused by decreased ventilation, results in carbon dioxide retention, while respiratory alkalosis, caused by increased ventilation, results in loss of carbon dioxide. Carbon dioxide dissolves more readily in the blood than oxygen and forms bicarbonate and smaller amounts of carbonic acid. When present in normal amounts, the ratio of carbonic acid to bicarbonate helps to keep the body pH normal. In situations were acidosis occurs, respiratory activity is increased and the lungs quickly compensate to excrete excess CO_2. If alkalosis is present, respiratory activity automatically decreases and CO_2 is retained, producing a compensatory respiratory acidosis. Stella's increased $PaCO_2$ caused respiratory acidosis, which is indicated by her low pH. Her increased HCO_3, which causes alkalosis, is inconsistent with her acidosis.

4. Evaluate Stella's current weight, usual body weight, ideal body weight, and BMI. How does her 1+ bilateral pitting edema affect your evaluation of her weight?

   UBW: 145–150 lbs.; IBW: 115 lbs.; BMI: 21. Although Stella's current weight is close to her ideal body weight and her BMI is normal, her edema is giving her a false weight due to water. She is probably underweight, and she reports having lost a lot of weight in the past 5 years.

5. What factors can you identify from her nutrition history which probably contribute to her poor food intake?

   Low total intake, low fruit and vegetable intake, low calcium intake, high consumption of sugar in the form of soda, ill-fitting dentures, poor appetite, bad taste, and bloating.

6. Based on the information presented above, what recommendations would you make to enhance Stella's nutrition status? Be specific.

   For her unintentional weight loss, I would tell her to eat high-kcal foods, her favorite foods first. I would suggest she add butter, sauces, or gravies to increase kcal.

   For early satiety I would also suggest she eat her high-kcal foods first, and limit fluids with meals. For mealtime dyspnea and fatigue, I would suggest she eat more slowly, and rest before meals. For bloating, I would suggest small, frequent meals and avoidance of gas-producing foods. And I would refer her to a dentist for her denture problems.

She should eat about 5–6 small meals per day to compensate for some of her satiety and bloating problems in an attempt to get sufficient kcal. She should try to increase her consumption of fruits and vegetables, dairy products, and whole grains. Stella should also begin to take a vitamin and mineral supplement if tolerated.

Energy needs: Based on 25–30 kcal/kg and caloric needs of patients with COPD = ~140% above REE, Stella's caloric requirements would be between 1300 and 1500 kcal/day based on her current weight. Even though she has edema, and her weight is not accurate, she has been unintentionally losing weight. Improving her nutritional status is a goal, so even though 1300–1500 may not be based on her dry weight, I think it is an appropriate intake goal.

Protein needs: To halt wasting, patients are recommended to eat 1.2–1.7 g protein/kg body weight, so Stella's protein needs are ~70g/day.

Chapter 24

1. There are many types of lung cancer, but most can be categorized into two basic types, "small cell" and "non-small cell." Small cell lung cancer is generally faster growing than non-small cell, but more likely to respond to chemotherapy. Smoking is the usual cause of this and all types of lung cancers, although non-smokers do fall victim to this disease.
2. Localized radiation therapy destroys cancer cells by altering their DNA, so they cannot replicate. Chemotherapy drugs are designed to target rapidly dividing cancer cells that are present in malignant tumors. They reduce the size of the cell mass by interrupting the different stages of cell division. Chemotherapy and radiation alter cancer cells so that they do not replicate and metastasize.
3. Radiation therapy targets only the cancerous organs, so increased nutritional demands occur specific to the diagnosis. Increased metabolic demand occurs in all cases of RT and nutritional effects are especially evident with cancers of the head, neck, and abdominal areas, which can lead to feeding difficulties and impaired digestion. Chemotherapy targets the rapidly dividing cells of the gastrointestinal tract. Nausea, vomiting and diarrhea are common. Specialized nutritional support may be needed in both treatments.
4. (P) Inadequate oral food/beverage intake due to (E) self feeding difficulty contributing to (S) weight loss.
5. (P) Increased nutrient needs due to (E) impaired nutrient utilization contributing to (S) malnutrition.

Chapter 25

1. REE = (10 × 54.4 kg) + (6.25 × 165 cm) − (5 × 19) − 161 = 1320 kcal

   Injured: REE × 1.4 = 1850 kcal/day

   Protein: Normal: 43.5 g/day; Injured: 87 g/day
2. Formula given: Novartis (1.0 kcal/mL, 53% CHO, 22% protein, 25% lipid). Given in a volume of 1900 mL/day in feedings initiated at 10–40 mL/hr. The rate is advanced in increments of 10–25 mL/hr every 4–8 hours until goal rate is established.
3. Lisa's diagnosis of a moderate closed head injury indicates that she is in metabolic stress and that her nutritional needs are increased both for energy and protein. In addition, she has suffered from burns on 15% of her body surface area, indicating again that her needs are increased. Because of her diagnosis of two different stress factors, she is at increased nutritional risk for wasting, poor wound healing, and negative nitrogen balance.
4. The most common nutrition diagnosis would be hypermetabolism or increased energy expenditure. Although this patient was not malnourished before the trauma occurred, her BMI was 19, the lowest range for normal weight, indicating that she may be impacted greatly by her increased needs. Also, her diagnosis of a GCS = 10 indicates that she may have suffered some brain damage, which could cause physical, cognitive or behavioral impairments that might affect her nutritional status and intake.
5. (P) Hypermetabolism (E) leading to increased protein and kcalorie need (S) as a result of metabolic stress.

Chapter 26

1. Insulin resistance often occurs side by side with fat redistribution, especially abdominal fat accumulation. This increased visceral fat is not only uncomfortable, but it is also a risk factor for cardiovascular disease. Insulin resistance in combination with HIV medications that he is taking may be responsible for the fat accumulation and weight gain that he is experiencing. This increase in weight and abdominal girth in turn may be responsible for his heartburn.
2. The zidovudine in combovir has been known to cause or exacerbate lipodystrophy. The protease inhibitor combination of lopinavir/ritonavir that he was previously taking has been known to cause insulin resistance as well as body fat redistribution.

3. Consumption of high-fat snacks and large meals, being physically inactive, and a high intake of caffeinated beverages might be responsible for aggravating his lipodystrophy and heartburn.
4. Nutrition-related problems: Over weight, heartburn, a high-fat diet, low fiber intake, high intake of colas/coffee.
5. Eating smaller, low-fat meals at regular intervals and spacing carbohydrate intake throughout the day may be beneficial for this patient. Since efavirenz has been known to cause an increase in serum cholesterol levels, a low-cholesterol diet might also be indicated and for lopinavir/ritonavi's effect on blood sugar, spacing carbohydrates throughout the day would recommended. Eating a balanced diet containing adequate fiber from whole grains, fruits, and vegetables will ensure adequate intake.
6. Although HIV medications are implicated in causing heartburn, there may be many other causes of heartburn including other medications and conditions such as pancreatitis and esophageal reflux. Therefore, he should be advised to see a gastrointestinal doctor to rule out other causes of heartburn. Medications, such as protein pump inhibitors and H_2 blockers, are taken for heartburn along with antacids. He should be advised to keep a journal of his symptoms and any changes in lifestyle, diet, or medication in order to explore the factors responsible for exacerbating his condition. Simple strategies for preventing heartburn include:
   a) Eat small, frequent meals
   b) Avoid lying down after meals for two to three hours
   c) Elevate bed a few inches so that the head is above the stomach
   d) Avoid foods known to cause heartburn including citrus fruits and juices, coffee, cola, and fried and spicy foods.
7. Symptoms of heartburn and distress; body weight and abdominal girth; diet and physical activity levels; and blood glucose and triglyceride levels.

Chapter 27

1. Osteopenia is a condition in which the bone mineral density is low but not low enough to be diagnosed as osteoporosis. The T-score for osteopenia is between 1.0 to 2.5 standard deviations below the mean BMD of the young adult mean value. If osteopenia is not treated, it may result in osteoporosis.
2. The risk factors that this female presents are that she is postmenopausal, is over age 50, has a poor dietary intake of calcium-rich foods, and has a fracture. She also has a family history of osteoporosis.

   More information is needed about her diet history, including levels of caffeine intake, as well as information regarding her level of physical activity including any weight-bearing exercises, any past successes in weight loss, alcohol consumption, any history of smoking, serum 25-hydroxyvitamin D concentration to detect vitamin D deficiency, and current thyroid status.
3. Calcium is a primary bone-forming mineral required in adequate amounts throughout the entire lifecycle to achieve peak bone mass, maintain bone mass, minimize bone mineral loss, and reduce the incidence of osteoporosis-related fracture. Inadequate calcium intake during growth will result in sub-optimal development of peak bone mass and increased risk of osteoporosis in the latter decades of life. During adulthood, a calcium intake insufficient to maintain serum calcium levels leads to increased secretion of parathyroid hormone which, in turn, stimulates the resorption of bone in order to raise serum calcium levels.
4. After menopause, the calcium recommendation is 1000 mg to 1500 mg/day, which is the Adequate Intake. Exceeding the AI will not have any benefits for the treatment of the condition. Therefore, in order to reach her AI the patient should be encouraged to consume calcium-rich foods along with calcium supplements. This patient should take supplements in small doses (500 mg), two or more times per day with meals, because calcium absorption is enhanced at smaller doses and if it is spread throughout the day. In addition, she should not exceed her calcium intake UL of 2,500 mg/day (from foods and supplements). Vitamin D is essential for the absorption of calcium. Therefore, she should make sure that she has adequate intake of vitamin D through diet and/or supplementation.

   Calcium carbonate and calcium citrate are two supplements that provide more elemental calcium than a regular multivitamin. Calcium carbonate products available by prescription or over the counter are TUMS, Os-Cal, and Caltrate. The calcium citrate product on the market is Citracal.
5. Vitamin D plays an important role in maintaining calcium homeostasis and promoting bone health. Vitamin D facilitates the absorption of calcium from the small intestine. It also increases the blood concentrations of calcium and phosphorus by promoting their reabsorption by the kidney. Furthermore, it stimulates osteoclast formation, and thus decreases bone resorption and the release of calcium and phosphorus from bone. Without vitamin D, the bones can become thin and brittle and prone to fractures. This leads to rickets in children and ostomalacia in adults. Vitamin D deficiency can result from inadequate sun exposure, low dietary intake, or malabsorption of vitamin D due to biliary tract or intestinal diseases that impair the absorption of fat and fat-soluble vitamins.

It is not necessary to begin vitamin D supplementation for this woman, because at present her dietary vitamin D intake is low due to lack of knowledge. Thus, the primary goal is to educate her to take in adequate food sources of vitamin D such as skim milk, low-fat cheeses, fat-free yogurts, etc. and also to acquire adequate exposure to the sun.

6. Two types of medications that the physician might consider are estrogen (hormone replacement therapy), and selective estrogen receptor modulators (SERMS). Estrogens reduce bone turnover, prevent bone loss, increase BMD by 5 to 10% in the hip, spine, and total body, significantly reduce fracture risk, and are effective in relieving the hot flushes and night sweats that often accompany menopause. However, they increase the risk of developing breast cancer.

   SERMs are non-hormonal agents that have tissue-specific effects in estrogen-responsive target tissues. In the estrogen-responsive tissues of the skeletal and cardiovascular system, the effects of SERMs are similar to that of estrogen while in breast and uterine tissue SERMs have no estrogen-like effects. Raloxifene is a SERM approved by the FDA and TPD for the prevention and treatment of osteoporosis in postmenopausal women. When raloxifene binds to estrogen receptors in bone, it has an estrogen-like effect that increases BMD and reduces spine fracture risk by 30%. Biophosphanates and PTH hormone are recommended if the patient has osteoporosis.

Chapter 28

1. The screening for PKU is normally done within 12 hours after an infant's birth in a hospital. However, with different birthing practices in the United States, the American Academy of Pediatrics has recommended that the test should be performed from 24 hours to seven days after birth. In the United States, the newborn screening for PKU is done with the Guthrie inhibition assay or the McCammon-Robins fluorometric tests. These procedures require a blood spot sample from the baby's heel and are performed right after the baby's feeding. If the initial PKU tests positive then follow-up tests are performed to confirm the diagnosis. Diagnosis is confirmed when the tests show consecutive PKU levels of 10 mg/dL.
2. Newborn screening allows for early detection/diagnosis and treatment for PKU. Accumulation of phenylketones rises soon after birth when there is absence of phenylalanine hydroxylase enzyme to metabolize phenylalanine. It is important to screen for PKU soon after birth because untreated PKU or late-detected PKU, by contributing to high blood concentrations of phenylalanine, will result in irreversible neurological defects, coma, and even death. The symptoms of PKU are gradual, progressive, and often overlooked until it has reached a severe stage. With early screening, strict and lifelong treatments can be instituted early, thus allowing the individual to have a healthy, productive life.
3. The main principle of dietary treatment for PKU is to allow the maximum amount of dietary protein in the diet in order to promote growth, while still maintaining the levels of phenylalanine in the desired range of 2–6 mg/dL for children ages 12 years and younger, as per the 2000 guidelines from the National Institute of Health consensus guidelines. For individuals above 12 years the diet becomes more relaxed with recommendations of 2–15 mg/dL. The amount of protein that is allowed in an individual's diet is based on the level of activity of the deficient enzyme.
4. Phenex-1 is a powder formula that contains all proteins except for phenylalanine and is specially designed for infants with PKU. Phenex contains a carbohydrate source, a fat source including essential fatty acids, and vitamins and minerals. It is also fortified with tyrosine, which is an essential amino acid for individuals with PKU. Phenex also contains carnitine (found in breast milk) which can scavenge toxic metabolites in the blood. Similac and other standard infant formulas contain significant amounts of all proteins required for the development of a healthy infant. Since tyrosine is not an essential amino acid for normal infants, it is not present in standard formulas.
5. Untreated PKU results in the accumulation of phenylketones including phenylpyruvate, phenyllactic acid, phenylacetic acid, and phenylacetylglutamine which are highly toxic to the brain. As a result, individuals will develop severe mental retardation and seizures that would require lifelong institutionalization. As children with PKU grow older they will have an abnormally small head—microcephaly—and will have stunted growth as well as delayed mental and social skills. Other complications include musty odor due to high levels of phenylketones in the skin and urine and decreased pigmentation of skin and hair due to the retardation of the tyrosinase enzyme activity by phenylalanine.
6. Individuals with PKU have to maintain the PKU diet throughout their lives in order to ensure normal cognitive development in childhood and maintain psychosocial abilities in adulthood. Research studies indicate that adults who do not maintain adequate diet control have degenerated white matter in the brain as shown in brain scans. It has also been found that individuals who do not consume adequate amounts of tyrosine fail to produce dopamine, an important neurotransmitter. These individuals therefore have impaired cognitive function and often suffer from depression.
7. As an infant and growing child, her diet would be based on strict phenylalanine restriction in order to avoid the harmful defects brought about by hyperphenylalanine.

and diagnostic reasoning. When solving a problem, the dietitian must identify and define a problem or situation, assess all options, weigh each option using criteria, test those options, consider the consequences of the decision, and finally make a decision that will potentially solve the problem. Evidence-based dietetics practice requires looking at current research and integrating the information into nutritional practice. The dietitian has to use critical thinking skills to determine how to effectively evaluate and use current research, and must refer to research methodology skills he/she acquired during school.

5. Answers will vary.
6. Outcomes research is part of the ethical practice of dietetics. Data should be collected and dietitians need to report on the method of care taken and the ensuing outcome. Did the patient actually improve, or were there undesired results? As dietitians assess more outcomes, these results can be published for use by other dietitians and health professionals. This can result in updated practices and allows health care professionals to give the best care possible, supported through others' experience.

Chapter 3

1. **Internal Factors:**

   Human Biology Factors (determine nutrient requirements—normal, increased, decreased, change in form, etc.)

   - Biological factors (age, sex, genetics)
   - Physiological phases (growth, pregnancy, lactation, aging)
   - Pathological phases (disease, trauma, altered organ function or metabolism)

   Lifestyle Factors (determine food, physical activity and related choices)

   - Attitudes/beliefs
   - Knowledge
   - Behaviors

   Food and Nutrient Factors (determine the type and amount of nutrients available for use by the body)

   - Intake/composition (quantity, quality, and feeding route)

   External Factors:

   Environmental Factors (external influences that impact consumption and lifestyle)

   - Social (cultural food practices and beliefs, parenting, peer influences)
   - Economic (household finances, economy of the community/country)
   - Food safety and sanitation
   - Food availability/access

   System Factors (external influences that impact on delivery and services)

   - Health care system
   - Educational system
   - Food supply system (industry, agriculture, institutions)

2. The purpose of providing nutrition care is to restore a state of nutritional balance by influencing whatever factors are contributing to the imbalance or altered state of nutritional status. Because of the wide variety of and interaction among the many variables, identifying the underlying causes of a nutritional status imbalance can be a complex process. It is also important to determine what factors are contributing to the cause of a problem, because the type of intervention and/or education that will be provided depends significantly on the underlying cause of the problem.
3. The four steps of ADA's standardized Nutrition Care process (NCP) are as follows:

   1) **Nutrition Assessment**: a systematic process of obtaining, verifying and interpreting data in order to make decisions about the nature and cause of nutrition-related problems. It includes past medical and family history, laboratory data, anthropometric data, physical activity, and dietary history summary. It also includes estimated kcal, protein, and fluid intake.

   2) **Nutrition Diagnosis:** the identification and descriptive labeling of an actual occurrence of a nutrition problem that dietetics professional are responsible for treating independently. PES—Problem, Etiology and Signs and Symptoms—are used in this step. This is the format used in the NCP to write a nutrition diagnosis. It clarifies a specific nutrition problem and logically links the nutrition diagnosis to nutrition intervention, monitoring, and evaluation. When these three parts are used to form the nutrition diagnostic statement, it is generally stated in the following way: The problem (P) *related to* the etiology (E) *as evidenced by* the signs and symptoms (S).

   3) **Nutrition Intervention:** a specific set of activities and associated materials used to address a (nutrition-related) problem. This includes establishment of goals. These are determined by consulting evidence-based practice guides and discussing expectations with the client. They are measurable and realistic, and establish the type of outcome to be tracked over time.

   4) **Nutrition Monitoring and Evaluation:** an active commitment to measuring and recording the appropriate outcome indicators relevant to a nutrition diagnosis in order to determine the degree to which progress is being made and whether or not the client's goals are being met. A dietetics professional may contact the client to provide support and clarify any questions regarding the plan. This is done in order to determine if the plan

is being implemented and whether or not the client fully understands the information provided. Outcome measures may include direct nutrition outcome, clinical and health status outcomes, and patient/client-centered outcomes. For evaluation, baseline data will be compared to changes in the outcome data that is tracked over time. Progress will be discussed with the client and any problems or barriers that are identified will be used to revise PES statements, modify interventions, and/or establish new goals.

4. Standardized nutrition diagnostic terminology refers to a uniform terminology that is used to describe practice. The lack of a standardized nutrition language and common terminology has made it very difficult for dietetics professionals to communicate with each other and other health professionals. Because of this lack of agreement for nutrition language, there was no easy way to classify, measure, and report on the outcomes of nutrition interventions in various patient populations. The lack of specific uniform terminology used in dietetics practice made it impossible to gather data needed for research, education, and reimbursement justification via outcomes analysis. Most notably missing was language that described the specific nutrition problems. Therefore, a Standardized Language Task Force was formed in May of 2003, immediately following the adoption of the NCP, to develop nutrition standardized language. Dietetics professionals can now use these terms to clearly describe specific types of nutrition problems that contribute to a person's nutritional imbalance. Nutrition diagnoses give purpose and focus to the assessment step. They are the missing link in nutrition care and a critical step in the nutrition care process. The standardized nutrition diagnostic terminology now allows dietetics professionals to make explicit that which was implicit in the past.

5. The nutrition diagnostic terms are grouped into three domains: Intake, Clinical, and Behavioral-Environmental.

   Intake domain—domain which contains standardized nutrition diagnostic terms that describe actual problems related to the intake of energy, nutrients, fluids, and bioactive substances through oral diet or nutrition support (enteral or parenteral nutrition). Labels such as inadequate, excessive, or inappropriate are used to describe the specific nutrient or substance that is altered.

   Clinical domain—domain which contains standardized nutrition diagnostic terms that describe nutritional problems that relate to medical or physical conditions.

   Behavioral-Environmental domain—domain which contains standardized nutrition diagnostic terms that describe nutrition problems related to knowledge, attitudes/beliefs, physical environment, access to food, and food safety.

6. **PES**—Problem, Etiology, and Signs and Symptoms. This is the format used in the NCP to write a nutrition diagnosis. It clarifies a specific nutrition problem and logically links the nutrition diagnosis to nutrition intervention and monitoring and evaluation.

   The problem (P) is also referred to as the diagnostic label. It describes in a general way an alteration in the client's nutritional status. Words like excessive, inadequate, and inappropriate are frequently found in these labels. The related factors or etiology (E) are those factors that contribute to the cause or existence of a particular problem. Finally, the signs and symptoms (S) are the defining characteristics obtained from the subjective and objective nutrition assessment data. These data provide evidence that a problem exists and describe the severity of the problem. When these three parts are used to form the nutrition diagnostic statement, it is generally stated in the following way: The problem (P) *related to* the etiology (E) *as evidenced by* the signs and symptoms (S)

7. Example diagnosis:

   - "Inadequate energy intake (P) related to changes in taste and appetite (E) as evidenced by average daily kcal intake 50% less than estimated recommendations (S)."
   - "Involuntary weight loss (P) related to inadequate energy intake (E) as evidenced by 8-lb. weight loss within 4 weeks (S)."

   Let's examine how these diagnoses were made. A comprehensive nutrition assessment reveals the following data:

   - Client is undergoing chemotherapy for cancer treatment (client's medical history).
   - Complaints of meats tasting bitter and most beverages too sweet (food/nutrient intake history).
   - Client states, "I have no appetite and no desire to eat" (food/nutrient intake history).
   - 3-day food records reveal average kcal approximately 50% of estimated needs (dietary intake data compared to estimated needs).
   - Weight loss of 8 lbs. since last out-patient visit 1 month ago (anthropometric data).

8. A nutrition diagnosis is the missing t link between nutrition assessment and nutrition intervention. An accurate nutrition diagnosis is generated from a focused nutrition assessment and sets the stage for the next two steps of the NCP: Step 3 Nutrition Intervention and Step 4 Nutrition Monitoring and Evaluation.

   - The signs and symptoms or defining characteristics represent data obtained from the nutrition assessment in Step 1. These data are used to formulate a PES statement in Step 2, Nutrition Diagnosis.

- Lucena, M. I., Andrade, R. J., de la Cruz, J. P., Rodriquez-Mendizabal, M., Blanco, E., Sánchez de la Cuesta, F. (2002). Effects of silymarin MZ-80 on oxidative stress in patients with alcoholic liver cirrhosis. International Journal of Clinical Pharmacology and Therapeutics, 40(1): 2–8.

  Findings: Silymarin, the active component of milk thistle, produces a small increase in glutathione levels in patients with alcoholic liver disease and also produces a decrease in lipid peroxidation on red blood cells. No actual effect on the liver itself can be determined by this study nor were any changes in routine liver function tests found.

7. Yes, I have used yoga not only as exercise but also for relief of muscular pain as well as for relaxation and stress reduction. I have also explored Ayurveda as diet therapy and used chiropractic and massage for back pain. The reason for choosing these therapies over conventional biomedicine is for the holistic benefits to mind, body, and spirit that are not normally found with biomedicine.

8. It is essential that practitioners be familiar with CAM in order to provide the best possible treatment. The practitioner should honor the client's treatment decisions and reserve judgment or criticism. Understanding different treatment modalities and their purported effectiveness, historical use, and scientific evidence will enable practitioners to identify truly harmful, beneficial, or impotent therapies and make recommendations accordingly.

Chapter 5

1. **Nutritional status** evaluates the body's current nutrient stores. Nutritional status focuses on whether the nutrient intake meets the body's requirements. Change in nutritional status can be assessed with dietary intake, biochemical levels, and clinical assessment.

   Nutritional risk assessment requires the use of clinical judgment to assess the client's current problem(s) in order to project potential nutritional problem(s). Different disease states or treatments may lead to altered nutrient requirements. Nutritional risk is assessed based on the pathophysiology, treatment, and clinical course of a disease or diagnosis.

2. **Nutrition screening** is less involved than nutrition assessment, and is used to quickly identify patients at risk for nutritional problems. Nutrition screening only gathers key pieces of information that have been correlated to nutrition risk. It can be performed by dietetic technicians or other trained personnel rather than a dietitian.

   Nutrition assessment gathers data on the client, and analyze the data to determine the client's current and potential nutritional problems. The data gathered may be both subjective and objective in nature. Areas to gather data on are as follows: medical and social history; dietary history; physical examination; anthropometrics and body composition; biochemical data; and estimation of energy, protein, and fluid requirements.

3. **Subjective data** is usually obtained directly from the patient, family, or other caregivers. Subjective data reflect the client's perception of his or her condition. Subjective data include: food allergies/aversion; appetite and digestion problems; ethnicity and religion (degree of observance); usual pattern of food intake; support systems; health promotion and exercise practices; and learning style/problem-solving abilities.

   Objective data are information obtained from verifiable sources such as medical records, measurements, or laboratory results. Some examples of objective data are: anthropometric data such as weight, height, and age; biochemical laboratory results such as albumin; clinical findings such as diagnosis; and dietary information such as analysis of diet quality.

4. In a **24-hour recall**, a clinician guides the client to recall all the food and drink the client consumed in the previous 24-hour period. This dietary assessment method is easy to conduct, can be completed in a short time, has very little cost involved, and poses little risk for client. However, a 24-hour recall may not reflect typical eating patterns, a client may report information they feel the clinician wants to hear, or a client may over- or underreport. Since this method depends on the client's memory, it may not be accurate.

   Food record/food dairy requires the client to record their dietary intake as it occurs over a specific period of time. This method does not rely on the client's memory and may be more accurate. However, underreporting is common. This method is also more time consuming for the client. The act of recording may also affect a client's food choices, so that they may not reflect the client's normal dietary habits.

   In **food frequency**, a client is asked to identify how often and how much he/she consumes a specific food or food group. This method can be self-administered. It is inexpensive and quick to complete. Since it is self-administered, the response rate may be lower. In addition, the food list provided may not contain all the food choices that the client consumed.

   Observation of food intake/"calorie count" measures actual food intake by kcalorie or kcalorie-protein count; or weighing food before and after a meal is served. This is the most accurate assessment. This is also the most involved and time-consuming method that requires trained personnel.

5. **Analysis based on USDA's MyPyramid** organizes food into food groups and compares the client's dietary intake

in each food group with suggested amounts based on the USDA MyPyramid. This method gives an overview of adequacy but does not reflect the quality of the diet.

Analysis based on exchanges/carbohydrate counting uses the dietary exchanges to estimate macronutrient intake in the diet.

Specific nutrient analysis uses food composition data from the USDA and other published data on food to calculate dietary intake of both macronutrients and various micronutrients.

Computerized dietary analysis uses computerized data analysis programs to analyze dietary intake. It is similar to specific nutrient analysis with the exception that food composition data is stored in a database and analyzed with the use of an electronic device.

6. **Height/Stature** is the measurement of supine or standing height. It can also be estimated with arm span or knee height measurement. Height is used to interpret weight, measure growth for children, and calculate energy requirements and creatinine-height index. However, height has been noted to be one of the most inaccurate measures.

   Weight is measured using a variety of scales. Weight is relatively easy to measure and is used to track growth, development, and health. However, weight is a gross measurement that does not distinguish body composition or fluid shifts.

   Height or weight alone does not accurately determine body composition and/or health status. Using height and weight together with growth charts and body mass index give a better estimate.

7. **Albumin** has been the subject of much nutrition research and thus serves as a good diagnostic tool. Decreased albumin level is correlated with increased morbidity, mortality, and length of hospital stay. However, albumin has a long half life (approximately 20 days), which decreases its sensitivity to short-term changes in protein status or short term interventions. Albumin level is also affected by hydration level, stress, inflammatory response, renal function, liver function, protein-losing enteropathy, and steroid hormone use.

   Prealbumin has been confirmed by clinical research as a sensitive marker for identifying nutritional risk. Prealbumin has a very short half-life (approximately 2 days). Changes in prealbumin levels are indicative of nutrition support during critical illness. As an assessment tool, it is more expensive. Prealbumin level is affected by renal disease, Hodgkin's disease, liver disease, malabsorption, and hyperthyroidism.

   Transferrin is sensitive to acute changes in protein intake or requirement and serves as a good indicator of protein status. However, transferrin level is affected by iron stores, hepatic and renal disease, inflammation, and congestive heart failure.

   Retinol Binding Protein (RBP) has the shortest half-life (approximately 12 hours). This makes RBP one of the most sensitive indicators of protein status. It will reflect short-term changes and responses to nutrition support interventions. However, RBP levels are affected by renal failure, hyperthyroidism, cystic fibrosis, liver failure, vitamin A deficiency, zinc deficiency, and metabolic stress.

8. Energy requirements can be estimated by using a variety of established **equations** such as Harris-Benedict, Mifflin-St Jeor, Owen, and World Health Organization/Food and Agriculture Organization/United Nations University. These equations estimate energy requirements with information on weight, height, age, and activity level. In some situations, energy requirements can be measured by **indirect calorimetry**. Indirect calorimetry calculates energy use by measuring the amount of oxygen used and carbon dioxide expired.

   Stress from disease, infection and trauma tends to modify patients' energy requirements. Depending on the patient's condition, hospitalized patients may become hypermetabolic. The change to energy requirements is accommodated by adding a stress factor appropriate for the patient's condition to the energy requirements calculation.

9. Hemoglobin, hematocrit, mean corpuscular volume, mean corpuscular hemoglobin concentration, mean corpuscular hemoglobin, and total iron binding capacity.

Chapter 6

1. The primary purpose of the medical record in the clinical setting is to provide a basis for determining patient care, for documenting communication among health professionals dedicated to that individual patient's care, and for keeping a clear and comprehensive record of all that is done for the patient for legal reasons.

   A medical record also provides data used by accrediting agencies for health care facilities (including Joint Commission for Accreditation of Healthcare Organizations/JCAHO), for continuous quality improvement programs, and for insurance reimbursement for medical care.

2. Standardized language and abbreviations should be used because a medical record is a legal document, which serves as a description of exactly what happened during the medical care. Also, clients frequently have the right to request and read the copies of their medical records. It thus becomes important to adopt a standardized language to assure the accuracy of a medical record.

lization of nutrients by the body. This way the dietitian knows whether to increase caloric content of the enteral feeding or decrease it, depending on weight fluctuations. A second factor to monitor is blood glucose, in case hyperglycemia and its ensuing complications develop. Finally, a third factor to monitor is skin turgor, for the determination of either edema or dehydration.

8. The caloric content is 1020 kcals, protein amount equals 42.5 g., and osmolarity is 1675 mOsm. If 250 mL of a 20% fat emulsion is added, this provides 500 more kcal.

Chapter 8

1. Electrolytes are substances that separate into charged particles (ions) when dissolved in water or other solvents and thus become capable of conducting an electric current. Anions are ions with negative charges, while cations have positive charges.
2. Sodium, potassium, and chloride are electrolytes primarily found in intracellular and extracellular fluid. The normal serum values are as follows: Sodium: 135–142 mEq/L, Potassium: 3.8–5.0 mEq/L. Chloride: 95–102 mEq/L.
3. Osmolarity is defined as the number of osmols (standard unit of osmotic pressure) per liter of solution (mOsm/L), while osmolality is defined as the number of osmols per kilogram of solvent (water) (mOsm/Kg).
4. 1. Molecular size—smaller molecules transport across more easily than larger molecules.

   2. Method of transport—Solutes transported across the membrane by active transport move more easily than those transported by facilitated diffusion or simple diffusion.

   3. Electrical charge of a solute can dictate its affinity for a specific active transport.
5. Thirst, renal function, and hormone influence are mechanisms that regulate movement of fluids and solutes.
6. A decrease in blood volume will decrease hydostatic pressure, thus stimulating the release of the hormone renin from the kidney. Renin stimulates conversion of angiotensinogen to angiotensin I and a second activation converts angiotensin I to angiotensin II. Increasing amounts of angiotensin II stimulate release of aldosterone from the adrenal cortex. Aldosterone directly influences the kidney to retain Na^+. When Na^+ levels increase, increased osmotic pressure will pull fluid back into the blood; thus, blood volume will increase back to its normal range.
7. Aldosterone directly influences the kidney to retain Na^+. Subsequently, osmotic pressure increases and urine volume decreases. Arginine vasopressin causes fluid to be reabsorbed in the tubules of the kidney. This decreases urine volume, resulting in increased blood volume.
8. Calcium and phosphorus balance is maintained in the body by the hormonal influence over intestinal absorption, exchange between extracellular fluid and bone, and renal excretion of these minerals. When plasma levels of calcium are low, parathyroid hormone (PTH) is secreted from the parathyroid glands. PTH works to raise serum calcium levels by pulling calcium from the bone and decreasing renal excretion of calcium. PTH also stimulates absorption of calcium in the small intestine by activation of vitamin D. When necessary, PTH also acts to increase phosphorus excretion. Calcitonin, another hormone, originates from the thyroid gland. It is capable of decreasing serum calcium levels, and acts by inhibiting osteoclast activity.
9. Hyper- or hypovolemia describe conditions involving an abnormal volume of circulating blood.

   Hypovolemia is almost always related to renal or extrarenal loss of fluids. Extrarenal losses include any excess loss of fluid outside of renal excretion, including gastrointestinal losses, such as in vomiting or diarrhea. Losses through the skin occur during exposure to heat such as increasing body temperature (fever) or increased environmental heat. Excess loss through the skin can also occur through burns or draining wounds.

   The most common cause of hypervolemia is a decrease in urinary output such as seen in acute renal failure. Excess intravenous fluids or the failure of the kidney to accommodate a rapid ingestion of fluids quickly enough may also cause hypervolemia.
10. Hypervolemia is excess extracellular fluid and is usually due to a decrease in urinary output. Hyponatremia is either a decrease in the amount of sodium, an increase in the amount of water in the ECF, or a combination of both. It can be caused by the combination of a sodium restriction used for nutrition therapy with the use of diuretics.
11. Hypernatremia. Cellular dehydration results in an increasing severity of neurological symptoms ranging from lethargy and agitation to seizures and coma. Body temperature can be elevated, skin is flushed, and mucous membranes are dry.
12. Hypokalemia can result from inadequate nutritional intake of potassium, increased renal loss of potassium (use of loop diuretics), or increased loss from the gastrointestinal tract (vomiting, nasogastric suction, and diarrhea). Hypokalemia also results from a shift of potassium from the ECF to the ICF. Signs and symptoms include muscle weakness, diminished deep tendon reflexes, and cardiovascular dysrhythmias which can lead to cardiac arrest. The most common cause of hyperkalemia is inadequate excretion of potassium, commonly found with acute renal failure and chronic kidney disease. Shifts in potassium from the ICF to the ECF can also result in hyperkalemia. Signs and symptoms are a result of the neuromuscular effects of altered potassium

levels—muscle weakness, paralysis, paresthesias, and cardiac dysrhythmias which can lead to cardiac arrest.

13. Hydrogen ions are excreted to correct acidosis, potassium ions are retained, and hyperkalemia may develop.

Chapter 9

1. pH is a measure of the acidity or alkalinity of a solution. It is numerically equal to 7 for neutral solutions, increasing with increasing alkalinity and decreasing with increasing acidity. Volatile acids can be converted to gaseous form and eliminated by the lungs. Non-volatile acids occur through metabolism of carbohydrate, protein, and lipid, and cannot be eliminated by the lungs. A buffer is a substance that reacts with acids or bases to decrease their effect on the pH of a solution.
2. The lungs have the ability to change respiratory rate and depth of breathing to control either release or retention of pCO_2 in the blood. The kidneys reabsorb the majority of HCO_3^- in the blood to maintain a normal physiological pH.
3. Respiratory acidosis occurs when there is an inability of the lungs to expire CO_2. The retention of carbon dioxide causes an excess of acid in relation to base. Respiratory alkalosis generally occurs as a result of conditions causing hyperventilation. Rapid breathing results in a decreased pCO_2, causing an excess of base in relationship to acid.
4. Respiratory acid-base disorders are caused by abnormal CO_2 levels, and CO_2 is a volatile acid. Metabolic acid-base disorders are a result of excessive loss/gain of base (HCO_3) or an excessive loss/gain of fixed (non-volatile) acids.
5. Sleep apnea, cardiac arrest, myasthenia gravis, extreme obesity, injury or trauma to the chest wall, chronic obstructive pulmonary disease, and pneumonia can result in respiratory acidosis.
6. Diarrhea, fistula drainage, end-stage renal failure, ketoacidosis (due to diabetes mellitus, alcoholism, or starvation), or lactic acidosis (due to diabetes mellitus or salicylate overdose) can result in metabolic acidosis.
7. The anion gap is considered a very useful tool to distinguish between the main types of metabolic acidosis. It is calculated by subtracting the chloride and bicarbonate levels from the sodium plus potassium levels.

   Anion gap $= ([Na^+] + [K^+]) - ([Cl^-] + [HCO_3^-])$

   As sodium and potassium are the main extracellular cations, and chloride and bicarbonate are the main anions, the result should reflect the remaining anions. Normally, this concentration is about 8–16 mmol/L. An elevated anion gap (i.e., $>$16 mmol/L) can indicate particular types of metabolic acidosis, particularly certain poisons, lactate acidosis, and ketoacidosis.
8. Respiratory mechanisms such as respiratory rate, tidal volume, ventilation, and rate of CO_2 removal decrease to compensate for metabolic alkalosis. Rate of carbonic acid formation and rate of H^+ generation from CO_2 increase to re-establish acid-base balance. The need for oxygenation of blood is the major limitation of respiratory compensation.

Chapter 10

1. b. epidemiology
2. b. acquired

   Example of a multifactorial disease: atherosclerosis

   Example of a genetic disease: cystic fibrosis
3. b. prognosis
4. a. remission
5. Applying cold to the twisted ankle will result in vasoconstriction to that area. This will limit blood flow and reduce swelling and pain associated with vasodilation and increased vascular permeability. Elevating the ankle will assist in treatment of these same symptoms. It will decrease blood flow as well and will assist in reducing swelling.
6. Nonsteroidal anti-inflammatory drugs treat the acute inflammatory process by blocking/inhibiting, in one or more steps, chemical messengers that signal physiological events for inflammation. Histamine, nitric oxide, prostaglandins, and serotonin are a few such chemical messengers that are blocked by the NSAID. Ibuprofen is an example of an NSAID.

Chapter 11

1. A genome is considered to be the entire set of genes of a given organism. It is the blueprint for approximately 30,000 proteins within the human body. The knowledge of genetic variations within the population regarding diet-related chronic disease may be of great importance for the creation of individual dietary recommendations.
2. The set of specific variants of a gene present in the two alleles in an individual which can result in specific traits or disorders is considered the genotype. The actual expression of an inherited gene can vary, and this determines the phenotype. A haplotype is a group of gene variants that associate together and these may work in concert to produce a specific phenotype. Epigenotype refers to the to the pattern of gene expression regulated by modification to DNA.
3. **autosomal dominant**—an inheritance pattern of a dominant allele on an autosome.

   autosomal recessive—an inheritance pattern of a recessive allele on an autosome.

   X-linked dominant—an inheritance pattern of a dominant allele on the X chromosome. Such disorders are relatively rare.

X-linked recessive—an inheritance pattern of a recessive allele on the X chromosome. Related disorders are more common in males, who carry only one X chromosome.

Y-linked—inheritance based on Y chromosome. Disorders are extremely rare and occur only in males.

heterozygous alleles—having two different alleles or variants of a given gene.

homozygous alleles—having two identical alleles or variants of a given gene.

Phenylketonuria is an autosomal recessive disorder. Familial hypercholesterolemia is an autosomal dominant disorder. Hemophilia is an X-linked recessive disorder.

4. A monogenic disorder arises from a single gene, whereas a polygenic disorder arises from multiple genes interacting with each other.

5. **Single nucleotide polymorphisms** (**SNPs**)—situations in which one nucleotide is replaced by another in a gene, potentially leading to altered function. SNPs are generally identified by the gene name, the location of the affected nucleotide within the gene sequence, the common nucleotide in that position, and an arrow indicating that a less common nucleotide is present.

   MTHFR 667C→T (ala→val) indicates that there is a SNP at nucleotide number 667 in the methylene tetrahydrofolate reductase gene characterized by a thymine in place of the more common cytosine.

6. Epigenic regulation is the inheritance of information based on gene expression levels rather than gene sequence, regulated by genomic modifications such as DNA methylation, histone methylation, acetylation. or phosphorylation, and transcription factors.

   Methyl groups are derived in the diet from sources including folate, choline, methionine, and vitamin B_{12}. A deficiency of methyl groups related to the lack of these nutrients means that as cells divide, methylation may be reduced and some of that transcriptional regulation is lost. Impaired methylation of DNA is related strongly to impaired fetal development and cancer.

7. **Obesity**—the gene encoding for the protein perilipin. Perilipins are localized on the surface of fat droplets inside adipocytes and play a regulatory role, primarily by blocking release of stored triglycerides, and thereby helping to preserve stored fat.

   Type 2 diabetes—the gene encoding for peroxisome proliferator-activated receptor gamma (PPARγ), which is a receptor on the cell nucleus that plays a central role in adipocyte development and function. It's activation is also associated with greater insulin sensitivity.

   Colon Cancer—genes encoding for the glutathione *S*-transferase enzymes. Isothiocyanates derived from cruciferous vegetables are known to induce phase II detoxification enzymes which, like the phase I cytochrome P450 enzymes, are involved in metabolism and removal of potential carcinogens. The glutathione *S*-transferase (GST) family of enzymes is among the most important in this regard.

8. Nutrient deprivation *in utero* may result in fetal adaptation to the deprived environment by increased efficiency of nutrient uptake and usage. This fetal adaptation can cause abnormal carbohydrate metabolism, putting the child at high risk for the eventual development of type 2 diabetes.

Chapter 12

1. Factors that can influence an individual's susceptibility to infectious disease include:
   - *Gender.* Although gender sometimes plays a role, in many cases the underlying mechanism is differential exposure due to occupational and recreational activities.
   - *Age.* The immune system takes time to develop; therefore, the young do not have the full spectrum of immunological defenses available to the adult. As humans grow older, several immune mechanisms decrease, including secretion of mucous and sebaceous glands and the production of cytokines including an interferon. However, natural killer cells that attack infected cells and tumor cells increase.
   - *Nutritional status.* Malnutrition is a major cause of immunodeficiency.
   - *Hormones.* Levels of various hormones play a role in an individual's susceptibility to infectious disease. Individuals with diabetes have an increased risk of fungal and staphylococcal infections, while women with low estrogen have a higher vaginal pH and thus are more susceptible to vaginal infections.
   - *Stress.* Stress activates the flight or fight response, resulting in several physiological changes that impact the immune response. Short-term stress boosts the immune system and long-term and repetitive stress decreases immune responses.
2. The surface of mucous membranes can glue or trap microorganisms so they cannot continue their movement into the body, thus providing a form of natural resistance. For example, the mucus blanket in the respiratory tract can keep organisms from reaching the lungs. Cilia in respiratory mucosa create the ciliary escalator that helps bring the organisms, which are trapped in mucus, to the surface so they can be coughed out.
3. An antigen is a structure that can combine with a cell of the immune system or an antibody, but not necessarily induce activation of the cell or formation of an antibody. An immunogen is an antigen that can induce an im-

mune response. The key difference between an antigen and an immunogen is that the immunogen is foreign to the host producing the response. Haptens are molecules that are too small to be antigenic, but can have antigenic activity if coupled to larger molecules.

4. The immune system is divided into two arms: humoral and cellular immunity. The humoral arm of the immune system refers to antibodies that appear in serum and B cells that ultimately become plasma cells that produce antibodies. The cellular part of the immune system refers to T cells, macrophages, monocytes, and polymorphonuclear leukocytes (a.k.a. PMNs, microphages, granulocytes) that interact with potential pathogens at the cellular level.

   The immune system is also divided into two branches, non-specific and specific. Cells of the non-specific immune system, macrophages, monocytes, natural killer cells and polymorphonuclear leukocytes, react with any antigen; therefore, they can react immediately. In the specific immune response, each B and T cell is programmed to attack one specific antigen, but can interact with others that are closely related or very similar. The specific immune system takes time to respond initially, but improves with additional exposures and responds more rapidly on subsequent encounters with the organism; thus, it normally protects the human from re-infection.

5. Monocytes and macrophages are important in removing pathogens both by themselves and after the pathogen has been targeted by cells of the specific immune system. They are highly specialized to ingest and destroy particulate matter such as bacteria, aged cells, and neoplastic cells by phagocytosis.

   Polymorphonuclear leukocytes (PMN), also called microphages, granulocytes, or polys, are a second group of cells involved in the non-specific immune response. PMNs move easily between blood and tissues, so their number in blood increases or deceases in infections, depending on the type of organism involved.

   Lymphocytes, also called T cells and B cells, are involved in specific immune response.

6. Mast cells produce histamine, which is associated with allergic responses in humans.

7. Descriptions of T helper cells, Th1 and Th2 cells, cytotoxic, CD4 cells, and CD8 cells:

   - T helper cells trigger B cells to make antibodies, activate macrophages, and promote the differentiation of other T cells.
   - T1 helper cells activate the cellular immune system, while Th2 cells increase the production of antibodies.
   - Cytotoxic T cells are capable of killing targeted infected, tumor, or transplant cells directly.
   - CD4 is a marker found predominately on helper T cells that interacts with MHC class II molecules on antigen presenting cells.
   - CD8 is a marker found predominately on cytotoxic T cells that interacts with MHC class I molecules on target cells.

8. Plasma cells are mature B cells that produce protein antibodies.

   After recovery from an illness, B cells produce memory cells that attack the disease-causing organism if it invades again. The second response is quicker than the first and prevents the occurrence of disease symptoms.

   IgG—Fetal immunity, binds pathogens.

   IgA—Helps fight pathogens that contact the body surface, are ingested, or are inhaled.

   IgM—It is the first antibody found in new infections, where its 10 binding sites help ensure that once it binds to the antigen it will stay attached. Acts as part of B cell receptors.

   IgD—Function is currently unknown. It may function as a regulatory antigen receptor.

   IgE—Plays a role in immediate hypersensitivity and the defense against parasites.

9. An antigen presenting cell is a cell that displays a foreign antigen from a pathogen joined to MHC molecules on its surface to be recognized by a T cell. The three major types of antigen presenting cells are dendritic cells, macrophages, and B cells.

10. Negative selection is the process in which B and T cells that react to self molecules are deleted or functionally inactivated during their development. The process of negative selection eliminates self-reactive cells that produce antibodies capable of reacting with a person's own tissue antigens, which could lead to autoimmune diseases.

11. The major immune functions of the lymphatic system are: to concentrate antigens from all over the body into a few lymphoid organs; to circulate lymphocytes through lymphoid organs to allow antigens to interact with antigen-specific cells; and to carry antibody and effector cells to the bloodstream and tissues.

12. Soluble mediators:

    - Complement proteins are activated by other elements of the immune response and are involved in destroying infected cells and some pathogens.
    - Cytokines mediate communication among the cells of the immune system, and between the cells of the immune system and other body systems including the nervous system.

13. Activation of T cells requires that an antigen is "presented" to the T cells attached to the MHC

molecules on the surface of the APC. The T cell is not activated unless its receptor recognizes both the antigen and the MHC molecule.

14. In order to activate a T helper cell, a T cell receptor must bind with both the antigen and MHC II. Upon activation, T helper cells activate the cellular immune system or increase the production of antibodies.
15. Use of an antibody to deliver toxins, drugs, and radioisotopes to tumor cells allows reagents that can be harmful to healthy cells to be concentrated on the tumor. These cells differ from regular antibodies because they are clones produced by an immortal B cell line that reacts with a single antigenic determinate.
16. It is critical to match MHC antigens for tissues used in transplantation because the presence of a MHC antigen on a transplanted organ or tissue that is different from the MHC antigens on the recipient's tissues will initiate an immune response. Unmatched MHC antigens will result in transplant rejection because the immune system attacks the transplant at MHC antigens that are different from those found on the recipient's tissues.
17. Passive immunization is when serum from an individual who has recovered from a disease is given to a person with the disease in an effort to counter the infection. In active immunization, the individual is exposed to an antigen in a harmless form, and produces antibodies as well as activated T and B cells and memory cells to prevent infection.
18. Protein-energy malnutrition in infancy and early childhood has adverse effects on the thymus, including significant reduction in thymic weight, lowered thymic hormone levels, and fewer maturing T cells as well as alterations in the thymic microenvironment and peripheral T-cell function. Lowered helper T cell function will have a negative impact on both the cellular and humoral branches of the immune system.
19. Milk and dairy, peanuts and tree nuts, wheat, soy, eggs, fish, and shellfish.

Chapter 13

1. a. Sublingual, b. Intramuscular, c. Intravenous, d. Inhalation
2. Excipients are substances that are added during the production of medication and that can affect dissolution of a medicine. Binders, lubricants, and coating agents can decrease dissolution. Coloring and flavoring additives and the way the tablet is formulated can also affect dissolution. For example, hard, round and large tablets dissolve more slowly.

   The pH of the stomach and gastric emptying rate are two of the most important nutrition-related factors impacting drug dissolution. Any disease, injury, or surgery that affects oral intake or gastric function can affect dissolution of medications.
3. Distribution of the drug is defined as: The movement of the drug through the body to target tissues. Distribution can vary due to circulation and binding factors. Blood flow (choice a) is the major physiological factor that can affect this.
4. Liver.
5. Age, hepatic function, presence of disease or injury, genetics, and use of other medications.
6. Most drugs are removed through either urinary or biliary excretion. Some can be excreted through the lungs or bowel.

   Kidney disease will effect the excretion of a drug. If the pH of urine changes, this could affect certain types of drugs. Urinary flow can also effect excretion, so a decrease in kidney function could affect this. Another variable is competition for transport across the renal tubule. Other medications are nephrotoxic and could themselves change kidney function.
7. Polypharmacy means: Taking many drugs at one time, which can increase the risk of interactions. Polypharmacy can also include using medications without cause, using more than one medication for the same ailment, using medications that can interact with each other, and using a medication to treat the side effects of an initial medicine.

   The elderly are at risk due to failing health, which can necessitate multiple medications; altered GI function, which can affect absorption; and decreased mental capacity (memory), which can cause accidental misuse of medicine.
8. a. Appetite can be affected by taste, which could result in decreased intake, which could in turn lead to inadequate nutrition.
   b. The antacids could render the phosphorus unavailable for absorption, which could lead to a deficiency and could affect long-term bone health.
   c. This could lead to a potassium deficiency and electrolyte imbalances.
   d. Long-term use may result in megaloblastic anemia secondary to folate deficiency.

Chapter 14

1. The energy of food is generally determined by measuring the CHO, protein, fat, and alcohol in a food using techniques in a lab, and then multiplying the grams of each macronutrient by its energy value (i.e., 4 kcal for CHO and protein, 9 kcal for fat, 7 kcal for alcohol). A kilojoule (or 1000 joules) is a measure of energy that is based on the metric system. One joule is defined as the amount of work

done when 1 Newton of force moves 1 meter. It differs from kilocalories, which is a measure of the amount of heat required to raise the temperature of water, in that it measures work energy rather than heat energy. A kilojoule is also about ¼ of the energy of 1 kilocalorie.

2. The three main components of energy expenditure are resting energy expenditure, thermic effect of food, and activity-related energy expenditure. Resting energy expenditure is the amount of energy required to sustain life when a body is at rest, such as the energy needed for breathing and organ functioning. The thermic effect of food is the energy that is required to utilize food, and can be observed in the increased energy expenditure after a meal. Activity-related energy expenditure is the amount of energy expended during any type of physical activity.

   While resting energy expenditure is the energy used to sustain life, basal energy expenditure is the lowest rate of energy a person will expend while lying flat, having no food for 12 hours, and being in a room that is neither cold nor warm.

3. A person's energy requirements may be determined by using a formula. Formulas are derived by looking at large groups of people and their measured requirements and often incorporate age, sex, weight, height, and activity level as variables. Indirect calorimetery is another method that may be used to determine a person's energy requirements. With indirect calorimetery, a patient at rest breaths into a mask or hood while oxygen consumption and carbon dioxide production are measured. The measurements are then used along with body height and weight to determine REE. Direct calorimetery, on the other hand, is a direct measure of energy that is used by a person. A patient is placed in a chamber that measures heat of evaporation, convection, and radiation over a 24-hour period to determine how much energy they expend in a day, and thus their needs.

4. Appetite is affected by many substances that are produced by the body, including insulin, glucagon, cholecystokinin (CCK), glucagon-like-peptide-1 (GLP), and ghrelin. In particular, the hormone CCK is produced by the pancreas and released into the blood after a meal. It works to signal fullness and decrease the appetite.

   Ghrelin is another hormone and is produced by the cells of the stomach lining. It increases before a meal and will stimulate the appetite.

5. Excess adipose tissue poses different risks depending on the location of the fat tissue. Adipose tissue that is distributed centrally (abdominal fat) is associated with a greater risk of type 2 diabetes, HTN, dyslipidemia, and CHD even with a normal BMI. Fat that is located in the lower body such as in the hips and thighs is not associated with these risks. The best way to measure the amount of fat and where it is located is to use a waist-to-hip ratio, because this method is practical and predicts the risk of CHD well.

6. Several factors which contribute to obesity include specific medical disorders, genetics, dietary factors, energy expenditure, and the environment. Energy expenditure would be a factor that I would address with a client who needed to lose weight, and who had the ability to be physically active, because it would be a factor that she or he would have some ability to control. However, there would be some situations in which addressing this factor would not be appropriate or effective, such as if the client were not ambulatory and were confined to bed-rest.

7. The key elements of nutrition therapy for weight loss include a restriction of kcal of about 500–1000 kcal/day, a modification of the diet to reduce risk factors for CVD, an incremental modification of a client's lifestyle and preferences, and goals that seem achievable to the client.

   Increased physical activity will not only help weight loss and maintenance, but will lower the loss of lean body mass, lower LDL while raising HDL, improve insulin resistance, and improve fitness.

8. For long-term weight loss, sibutramine and orlistat have been approved, and for short-term treatment, mazindol, diethylpropion, benzphetamine, phendimetrazine, and phentermine have been approved.

   Orlistat works by inhibiting the action of gastrointestinal lipase and reducing the digestion of fat. It is supposed to be taken with low-fat meals, and there is increased risk of GI problems if meals contain more than 20 g of fat. Side effects may include flatus with discharge, oily stools, and fecal urgency. The drug also inhibits the absorption of fat-soluble vitamins, so supplements may be needed.

   One surgery that is performed for weight loss is the Roux-en-Y gastric bypass. The stomach, duodenum, and part of the jejunum are bypassed by connecting a small pouch that is formed at the bottom of the esophagus to the jejunum. The body can therefore hold and absorb less food and so weight is lost.

   Several complications may arise, including pulmonary embolism, anastomotic leaks, injury to the spleen, and wound infection. There may also be nausea and vomiting if too much is eaten at once. Another potential complication is dumping syndrome. There may be an increased need for vitamin B_{12}, iron, calcium, and magnesium.

Chapter 15

1. **Systolic:** The force of blood exerted on the walls of blood vessels during the contraction of the ventricles.

   Diastolic: The force of blood exerted on the walls of blood vessels during relaxation of the ventricles.

Stroke Volume: The amount of blood ejected with each contraction of the left ventricle.

Cardiac Output: Equal to heart rate multiplied by stroke volume.

2. Stroke volume is regulated by end-diastolic volume (EDV), mean aortic blood pressure, and strength of ventricular contraction. EDV, often referred to as the preload, is the amount of blood in the ventricles at the end of diastole, which is mostly determined by venous return. Greater EDV means the ventricles will be increasingly stretched. The stretching of the ventricles increases their force of contraction, leading to a greater amount of blood that will be ejected. The fraction of EDV that is ejected from the heart by contraction of the ventricles is termed the ejection fraction (EF). The mean aortic blood pressure or mean arterial pressure (MAP), termed afterload, also affects stroke volume. It is the average force exerted against the walls of the arteries, by the blood, over a cardiac cycle. MAP represents the resistance the ventricles must contract against in order to eject blood; therefore, greater resistance means the ventricles will eject less blood. Finally, the strength of ventricular contraction is affected by the hormones epinephrine and norepinephrine, and the sympathetic stimulation of the heart. These mechanisms increase contractility by increasing the amount of calcium available to the myocardial cells.

   Mean arterial pressure (MAP) is determined through a combination of cardiac output and total peripheral resistance. The MAP must be high enough to force blood through systemic circulation without causing vascular damage. The regulation of MAP involves the sympathetic nervous system, the rennin-angiotensin system and renal function. All three affect blood pressure. Heart rate is dependent upon the balance between parasympathetic activity, which decreases heart rate, and sympathetic activity, which increases heart rate, and both are regulated by the cardiovascular control center. Total peripheral resistance is dependent upon the radium of arterioles (most importantly) and blood viscosity (primarily determined by the number of red blood cells). The radius is controlled by local metabolic controls in skeletal muscle, which may cause vasodilation and increase blood flow to muscles in order to meet metabolic needs. This would decrease resistance by increasing the radius of the vessel. Vasoconstriction will decrease vessel radius and increase resistance, and is caused by sympathetic activity and epinephrine. The hormones vasopressin and angiotensin II also cause vasoconstriction. They affect blood pressure in other ways, as well. When vasopressin (antidiuretic hormone) is released, there is an increase in the reabsorption of water, which increases blood volume, thus increasing blood pressure. Angiotensin II causes aldosterone to be secreted by the adrenal cortex, which causes an increase in sodium and chloride reabsorption, leading to retention of water and increased blood pressure. Additionally, the circulatory system contains baroreceptors, which constantly monitor blood pressure through both short-term and long-term adjustments. The carotid sinus baroreceptor and aortic arch baroreceptor monitor MAP and pulse pressure. If either cardiac output or peripheral resistance increases without a compensating fall in the other, blood pressure increases.

3. Hypertension refers to a chronic elevation in blood pressure. Lifestyle factors that contribute to the development of HTN include diet, exercise, smoking, stress, and weight management. Because development of HTN has such a strong genetic component, poor lifestyle choices may simply exacerbate the problem. Dietary factors also play a role in the development of hypertension.

   Smoking causes acute and chronic elevations in blood pressure. The relationship may be partially explained by the endothelial relaxation that is impaired by cigarette smoking because of its interference with the action of nitrous oxide. The relationship between excess body weight and blood pressure may possibly be attributed to insulin resistance, hyperinsulemia, activation of the sympathetic nervous system, activation of the rennin-angiotensin systems, and physical changes in the kidney. When caloric intake in increased, elevated blood insulin may result, which can cause increased renal sodium reabsorption, leading to increased blood pressure. Studies show that dietary sodium intake is strongly related to an increase in systolic blood pressure. Also, high salt intake has been associated with hypertensive target organ disease.

4. **Loop Diuretics:** act by inhibiting sodium, chloride, and potassium reabsorption in the loop of Henle; increase prostaglandins, resulting in vasodilation.

   Thiazides: inhibit the resabsorption of sodium, chloride and potassium; primarily act in the distal tubule and the ascending loop of Henle.

   Carbonic Anhydrase Inhibitors: block carbonic anhydrase, preventing the exchange of hydrogen ions with sodium and water.

   Potassium-sparing Diuretics: prevent sodium-potassium exchange within the collecting and convoluted tubules; reduce aldosterone stimulation.

   ACE-inhibitors: competitively block the conversion of angiotensin I into angiotensin II; results in vasodilation, decrease in vasopressin release; increase kinin levels that lead to vasodilations; increase release of fibrinolytic substances.

   Beta-adrenergic Blocking Agents: block β-receptors in the heart to decrease rate and cardiac output.

Alpha-receptor Antagonists: block the vascular muscle action that normally responds to sympathetic stimulation; this reduces stroke volume.

Calcium Channel Blocking Agents: affect the movement of calcium, causing the blood vessels to relax, reducing vasoconstriction.

Aldosterone Antagonists: suppress the actions of aldosterone.

The DASH Diet focuses on the use of a variety of foods that not only reduce sodium intake but increase potassium, magnesium, calcium, and fiber within a moderate energy intake. The eating plan is low in saturated fat, cholesterol, and total fat, and emphasizes fruits, vegetables, and low-fat dairy foods. It also includes whole grain products, fish, poultry, and nuts. It is limited in red meat, sweets, and sugar-containing beverages, and rich in magnesium, potassium, calcium, protein, and fiber.

At 2000 kcal a day, the DASH-Sodium Diet provides approximately 4700 mg (120 mEq) potassium, 500 mg magnesium, 1240 mg calcium, 90 g protein, 30 g fiber, and 2400 mg (100 mEq) sodium.

5. The risk factors for development of atherosclerosis are: family history, age, sex, obesity, dyslipidemia, hypertension, diabetes, physical inactivity, and cigarette smoking. Those that are alterable include: obesity, dyslipidemia, hypertension, physical inactivity, atherogenic diet, and cigarette smoking. Dietary changes in the ATP III guidelines include: dietary fat intake within 25–35% of total caloric intake; reducing saturated fat intake to <7% of total kcal or less than 16 grams for an individual on a 2000-kcal.day diet; reduce cholesterol to <200 mg/day; consider increased viscous (soluble) fiber (10–25 g/day) and plant stanols/sterols (2 g/day); eating for weight management; if TG levels are ≥500 mg/dL, then reduce fat intake to ≤15% of total kcal.

6. The four following mechanisms can initiate an MI: sudden blockage of a coronary artery, hemorrhage into an atheromatous plaque, arterial spasm, and increase in myocardial oxygen demand. The rupture or tearing of the fibrous cap (which separates the core and the plaque) exposes the plaque to the blood, allowing blood to push into the fissure. Therefore, hemorrhaging occurs, and clotting ensues. The result may be thrombotic occlusion of the artery. In some cases, occlusion will occur because blood seeping into the plaque causes enlargement, which may obstruct the coronary artery.

7. Peripheral arterial disease (PAD) is used to describe occlusion of blood flow in non-coronary arteries, especially in the lower extremities (pelvis and legs). PAD has been used by some to describe all non-coronary atherosclerotic disease including involvement in the carotid arteries. Atherosclerosis is the development of plaque in the vascular wall that will occlude the lumen of the vessel, creating ischemic conditions. The plaque begins as a fatty and fibrous growth, which may calcify over time. An atherosclerotic plaque in the leg can result in PAD.

   Similarities and differences are found among the risk factors for these diseases. Risk factors for the development of atherosclerosis are important to the development of PAD. However, the impact of the individual risk factors on atherosclerotic disease development and progression in the periphery are not the same as in the coronary arteries. The most influential independent risk factors for PAD development and progression are cigarette smoking and diabetes mellitus. While dyslipidemia and hypertension are important and do increase risk, they do not appear to be as influential as diabetes and cigarette smoking, as in atherosclerosis development. Also, elevated LDL, low HDL, and elevated serum triglycerides do impact risk of PAD, similar to AS. Finally, the effect of hypertension on PAD appears to be much more subdued than the effect in the cerebral or coronary arteries.

   The pathophysiology in PAD is similar to AS, in that an inflammatory response precedes the plaque rupture with subsequent embolus formation. However, there are some small differences between the pathophysiology of PAD and AS. Thrombosis is known to play a critical role in all acute ischemic incidents, but it is postulated that it plays a more critical role in acute ischemic events in those with PAD. Therefore, fibrinolytic therapy in PAD patients is more effective in reducing cardiovascular events. The importance of CRP (a disease risk biomarker) to the progression of PAD is unknown at this time, but, as with AS, it is suspected that the severity of acute events and symptoms are associated with increased CRP levels. Finally, PAD is a major risk factor for lower extremity amputation because ischemic conditions can lead to gangrene. If AS has progressed to the point that it causes occlusion in the vasculature of the leg, then this same progression is likely in the coronary, carotid, and cerebral arteries.

   One symptom (complication) associated with the ischemic conditions of PAD is intermittent claudication, which is a cramp like pain, most common in the calf, associated with activity. Strenuous activity leads to earlier onset and greater pain intensity. As stated above, PAD is a major risk factor for lower extremity amputation because ischemic conditions can lead to gangrene of the distal tissues.

8. The underlying causes of heart failure can be either structural or functional in nature. The primary causes are IHD, hypertension, and dilated cardiomyopathy. Heart failure is a process beginning with an injury to the heart or left ventricle hypertrophy that impairs overall function of the

heart. Heart failure is described as a progressive disorder because even if there is no additional injury to the myocardium there will be a continued deterioration in function. As compensation for the impairment in function, the renin-angiotensin-aldosterone system initiates changes in blood pressure that exacerbate the dysfunction. As a result, the heart becomes weakened and dilated, myocardial fibrosis limits the ability of the walls to respond to stress, and oxidative damage further impairs contractility. Therefore, the overall structure of the heart does not allow for proper functioning.

Following MI, the left ventricle will undergo a change in structure and geometry. The change is referred to as cardiac remodeling. It will result in a hypertrophied and/or dilated left ventricular chamber. Cardiac remodeling is mediated to some extent by neurohormonal systems. Heart failure patients typically have elevated levels of norepinephrine, angiotensin II, aldosterone, endothelin, vasopressin, and cytokines, which can have adverse effects on cardiac structure. Sodium retention and peripheral vasoconstriction result in increases in arterial blood pressure, which increase myocardial workload. Other substances mediate oxidative stress, causing myocardial cell damage and myocardial fibrosis and further altering the structure of the heart, and therefore cardiac remodeling.

9. Nutritional care during congestive heart failure is difficult. A diet restricted in both sodium and fluid is crucial to control acute symptoms and may help reduce the cardiac workload. Also, individuals with heart failure have difficulty eating and many experience a syndrome of malnutrition called cardiac cachexia, characterized by extreme skeletal muscle wasting, fatigue, and anorexia. Further complications from heart failure contributing to nutrition problems include decreased blood flow to the GI tract causing slowed peristalsis, early satiety, and impaired nutrient absorption, and side effects from drugs such as nausea, vomiting, and anorexia. Nutrient deficiencies are also common side effects from the use of diuretics and other medications.

Chapter 16

1. The four basic functions of the GI tract are motility, secretion, digestion, and absorption.

   Motility is the movement of the food consumed along the GI tract. Both propulsive contractions and mixing movements serve not only to move foodstuffs toward sites of digestion and absorption, but to mix foods with digestive secretions and maximize potential absorption.

   Secretions of the GI tract include water, electrolytes, enzymes, bile salts, and mucus.

   Digestion is the breakdown of complex molecules to their simplest form that can be absorbed by cells. Carbohydrates are digested from their most complex form of polysaccharides to the monosaccharides: glucose, fructose, and galactose. Proteins are converted from polypeptides to single amino acids, di- and tri-peptides. Lipids are digested to their simplest forms: free fatty acids, monoglycerides, glycerol, phospholipids, and cholesterol.

   After digestion, these basic molecules are **absorbed** along with water, electrolytes, vitamins, and minerals to provide essential nutrients to every cell. Absorption is the uptake of chemicals into the body.

2. In cases of xerostomia, normal saliva production is interrupted due to infection, damage, a tumor, or medication. So, when xerostomia occurs . . . (1) Food is not moistened and lubricated to help in swallowing; (2) enzymes are not produced to initiate the digestion of carbohydrate; (3) antibacterial protection provided by lysozyme decreases; (4) there is a decrease in taste because there is less interaction of saliva with the taste buds; (5) the chances of dental caries increases, because there is no buffer to neutralize acids, and because xerostomia may also lead to decreased oral hygiene; (6) speech may also be impaired because there is less movement of the lips and tongue.

3. The band of muscle tissue called the lower esophageal sphincter (LES) is responsible for closing and opening the lower end of the esophagus and is essential for maintaining a pressure barrier against contents from the stomach. It is a complex area of smooth muscles and various hormones. If it weakens and loses tone, the LES cannot close up completely after food empties into the stomach. In such cases, acid from the stomach backs up into the esophagus. Many factors have been identified which can lower LES pressure and thus contribute to LES incompetence and GERD. These include:
   a) the increased secretion of hormones: gastrin, estrogen and progesterone;
   b) the presence of other medical conditions such as hiatal hernia or scleroderma; cigarette smoking;
   c) the use of medications including dopamine, morphine, and theophylline; and
   d) specific foods. Foods high in fat, chocolate, spearmint, peppermint, alcohol, and caffeine all may decrease LES pressure.

4. Forceful vomiting can rupture either the esophagus (Boerhaave's syndrome) or tear the lower esophageal sphincter (Mallory-Weiss tear). Bleeding or hematemesis is a serious outcome of these injuries. If gastric contents are aspirated into the lungs, aspiration pneumonia is a likely result.

   Nausea and vomiting can result in inadequate nutrient intake, dehydration, and acid-base imbalances, and over time can lead to learned food aversions. This is similar to anticipatory nausea and vomiting. When a negative consequence is linked to a particular food, most people want to avoid eating that food. Malnutrition can be a long-term consequence for the patient if he or she is not able to ingest an adequate diet for a prolonged amount of time.

5. One cause of peptic ulcer is bacterial infection, and the type of bacteria responsible for this is called *Helicobacter pylori* (*H. pylori*). *H. pylori* weakens the protective mucous coating of the stomach and duodenum, which allows acid to get through to the sensitive lining beneath. Both the acid and the bacteria irritate the lining and cause a sore, or ulcer. *H. pylori* is able to survive in stomach acid because it secretes enzymes that neutralize the acid. This mechanism allows *H. pylori* to make its way to the "safe" area—the protective mucous lining. Once there, the bacterium's spiral shape helps it burrow through the lining.
6. 1) Have small and frequent meals with good nutritional status throughout the day to avoid the delay of gastric emptying and decrease the risk of reflux, and also to help decrease *H. pylori.*
   2) Decrease the intake of black and red pepper, coffee, and alcohol since they are identified as stimulants for gastric acid production. Irritating foods identified by the patient also need to be avoided.
   3) Decrease foods like chocolate, mint and foods high in fat as they tend to lower LES pressure. Increase the intake of n-3 and n-6 fatty acids as they are involved in inflammatory, immune, and cytoprotective physiology of the GI mucosa.
7. **Dumping syndrome** is a complex physiological response to the presence of larger than normal amounts of food and liquid in the proximal small intestine. Usually occurs as a result of a loss of normal regulation of gastric emptying and the GI tract's systemic response to meal.

   Alimentary hypoglycemia occurs in patients who have had upper GI surgical procedures (gastrectomy, gastrojejunostomy, vagotomy, pyloroplasty) and allows rapid glucose entry and absorption in the intestine, provoking excessive insulin response to a meal. This may occur within 1 to 3 hours after a meal.

   Malabsorption and steatorrhea—excessive fat in the feces may also occur.
8. Patients who are symptomatic after gastric surgery often **lose weight**, which may be attributed to inadequate intake. Malabsorption, loss of gastric lipase, or pancreatic or biliary insufficiency all lead to nutritional complications. The patient may become **lactose intolerant** initially due to lactase deficiency. **Anemia, osteoporosis, and selected vitamin and mineral deficiencies** may occur as a result of long-term malabsorption or inadequate intake. **Iron deficiency** may be attributable to loss of acid secretion, which normally facilitates the reduction of iron compounds. If the amount of gastric mucosa is reduced, intrinsic factor may not be produced in quantities adequate to allow for complete **vitamin B_{12}** absorption and may result in pernicious anemia.
9. Dumping syndrome is classified as early or late, depending on the timing of onset of symptoms after ingestion of a meal. **Early dumping syndrome** occurs within 10 to 30 minutes of ingestion of a meal and involves both gastrointestinal and vasomotor complaints. The symptoms of late dumping syndrome, which are mainly vasomotor in nature, occur 2 to 3 hours postprandially. The rapid emptying of the hyperosmolar chyme into the small intestine results in large fluid shifts from the intravascular compartment into the bowel lumen, causing gastrointestinal complaints of bloating and diarrhea. **Late dumping syndrome** is related to the inappropriate release of insulin, which results from rapid absorption of glucose and leads to late hypoglycemia.

Chapter 17

1. Prebiotics are substances in food that stimulate the beneficial flora of the large intestine. An example of such foods is inulin, which is a fructooligosaccharide derived from chicory and oligosaccharides.

   Probiotics are products containing microorganisms that are manufactured and sold as food products such as yogurt with live cultures and dietary supplements. Research is currently underway to determine their role in the promotion of a beneficial environment for the health of the colon and in prevention and treatment of disease.

   Lactate and short-chain fatty acids that result from their fermentation can be absorbed from the colon and utilized elsewhere in the body or utilized by the colon for support of its own tissue growth.
2. Diarrhea can be classified in several different ways. First, diarrhea can be either acute or chronic in origin. Acute diarrhea is short-term, whereas diarrhea lasting several weeks is considered chronic, and is usually associated with more health concerns such as electrolyte imbalances, malabsorption, dehydration, and long-term malnutrition. Nutrition implications of diarrhea are dependent on the volume of gastrointestinal losses and the length of the disease course. Large-volume losses can quickly lead to dehydration and electrolyte and acid-base imbalances. Hyponatremia and hypokalemia are both common with diarrhea. Metabolic acidosis may occur due to excessive loss of bicarbonate ions in stool output.

   Diarrhea can also be classified as either osmotic or secretory. When there is an increase in osmotically active particles in the intestine, the body reacts by pulling water

3.) Hepatitis C is transmitted through blood or bodily fluids via an infected person. This is most commonly done through sharing of needles, but may also have occurred in those who received clotting factors before 1987. Infected mothers may transmit the disease to their babies, and hemodialysis patients have an increased risk.

4.) Other types of hepatitis are types D and E.

Viral hepatitis is inflammation of the liver caused by a virus, whereas alcoholic hepatitis is a result of liver injury due to chronic alcohol consumption. Alcoholic hepatitis patients are often prone to infections and require antibiotics.

5. Cirrhosis is a chronic liver disease where healthy liver tissue is being replaced by scar tissue, which leads to decreased liver function. Common causes are alcoholism, a fatty liver, hepatitis C, and possibly malnutrition, and a genetic predisposition increases susceptibility. Common complications of cirrhosis include encephalopathy, ascites, esophageal varices, and other symptoms that develop from portal hypertension. Hypoglycemia results from an increase in renal sodium absorption and increased sodium retention, which increases fluid retention. Hyperglycemia may result from the ammonia accumulation in arterial blood due to muscle wasting, which stimulates glucagon production and gluconeogenesis. Parameters that can be used to nutritionally assess a patient with cirrhosis include low serum protein levels, elevated SGOT and BSP levels, vitamin deficiencies, lower hematocrit and hemoglobin levels, appearance of jaundice, decreased appetite, and diarrhea.

6. There are four major possible biochemical causes for hepatic encephalopathy. The first is due to a shift from ammonia metabolism in the liver to the skeletal muscle. This leads to elevated levels of ammonia in the brain, and an increase in the conversion of glutamate to glutamine. The second biochemical cause is the accumulation of neurotoxins produced by bacteria, but this theory seems to be improbable. The third is that hyperglycemia leads to an increase in skeletal uptake of branched-chain amino acids. This leads to an abnormal ratio of branched-chain amino acids to aromatic amino acids in the plasma, and accumulation of aromatic amino acids in the brain. In turn, there is a higher serotonin level and inhibition of DOPA, which replaces other catecholamines. Finally, the fourth biochemical cause may be that inhibitory neurotransmitters, like GABA, and substances like benzodiazepine bind to receptor sites in the brain, which lead to an altered mental state. The minimal amount of protein a patient should be given is 50 g. per day, whereas 1.2 g/kg. of body weight may be desirable. Protein can come from vegetables and dairy sources, which are low in aromatic amino acid concentrations. Beneficial effects have also been found with branched-chain amino acid supplementation in some patients.

7. The recommended post-operative diet includes oral intake, once the patient is eliciting bowel sounds and the nasogastric tube can be removed. The patient can progress to a regular diet, but may need to increase fiber intake in order to help manage diarrhea.

8. Chronic pancreatitis develops over many years with greater pancreatic destruction, whereas acute pancreatitis is marked by acute attacks and a quick onset of digestion disturbances. The pertinent labs for pancreatitis are serum and urine bilirubin, aminotransferases, alkaline phosphatase, total protein, and prothrombin time. Common nutritional problems associated with chronic pancreatitis include weight loss, macronutrient and fat-soluble vitamin malabsorption, vitamin B_{12} deficiency, and possible glucose intolerance.

Chapter 19

1. There are three chemical classes of hormones: 1) peptides and proteins, 2) amines, and 3) steroids.

   Examples of peptide and protein hormones are:

   - Insulin (pancreas) which is responsible for cellular glucose uptake and fed-state metabolism
   - Glucagon (pancreas), which is responsible for fasted-state metabolism
   - Leptin (adipose tissue), which influences food intake and metabolic rate
   - GI hormones (gastrin, secretin, choecystekinin, motilin), which control the GI tract, liver, pancreas, and gallbladder
   - Parathyroid hormone (PTH), which controls plasma calcium and phosphate

   Examples of amine hormones are:

   - Thyroid hormones (thyroxine, triiodothyronine, calcitonin), which control metabolic rate, growth, brain function and plasma calcium levels
   - Epinephrine and norepinephrine (produced by the adrenal medulla), which control organic metabolism, cardiovascular function, and response to stress
   - Dopamine (produced by the hypothalamus), which inhibits prolactin secretion

   Examples of steroid hormones are:

   - Sex hormones (estrogen, progestin, testosterone), produced in the gonads, which influence growth and reproductive development
   - Cortisol (adrenal cortex), which influences the stress response, the immune system, and metabolism

- Androgens (adrenal cortex), which control sex drive in women
- Aldosterone (adrenal cortex), which controls sodium and potassium excretion by the kidneys

2. Carbohydrate: glucose homeostasis, net uptake of glucose by liver, glycogen synthesis, glycolysis, fat storage

   Lipid: fat storage

   Protein: metabolized in liver to urea and energy, protein synthesis

3. Diabetes mellitus is insulin deficiency resulting in glucose intolerance.

   Type 1 DM generally presents in people below 30 years of age and is the result of limited insulin production or absolute insulin deficiency of pancreatic beta cells. The disease is usually the result of an autoimmune disorder and can be diagnosed with an antibody test.

   Type 2 DM accounts for 90–95% all diabetes cases and generally presents in older adults, but is becoming more prevalent in children due to lifestyle. Risk factors for type 2 DM include obesity, physical inactivity, older age, family history, and belonging to certain ethnic groups (Hispanic, African American, Native American, South or East Asian, Pacific Islander).

   Pathophysiology: Type 2 DM manifests as decreased cellular response to insulin resulting in increased blood glucose levels.

   Clinical Manifestation: T2DM is different from T1DM in that it comes on slowly, whereas type 1 has a sudden onset. Individuals with T2DM may be asymptomatic for years, only to be diagnosed secondary to a complication such as retinopathy or peripheral neuropathy.

   Gestational DM presents during pregnancy and may resolve itself, though there is an increased risk of developing type 2 DM later in life.

   Epidemiology: Approximately 7% of all pregnancies are complicated by GDM, and, women who have had GDM have a 20 to 50% chance of developing diabetes in the next 5 to 10 years. Risk factors for developing GDM include obesity (BMI >30.0), personal history of GDM, glycosuria, strong family history of diabetes (1st degree relative), prior poor obstetrical outcome (stillbirth, birth defects, or baby >9 lbs), and being a member of a high-risk ethnic group (Hispanic, African American, Native American, South or East Asian, Pacific Islander).

   Etiology: During the second or third trimesters of pregnancy, metabolic alterations occur to meet maternal and fetal demands for energy and nutrients. In addition to alterations in insulin secretion, these alterations affect glucose, amino acid, and lipid metabolism.

   Pathophysiology: Similar to T2DM. Alterations in beta-cell function or cellular glucose resistance.

   Diagnosis:

   1. Casual plasma glucose ≥200 mg/dL (≥11.1mmol/L) in addition to certain symptoms (unexplained weight loss, polydipsia, polyuria)
   2. Fasting plasma glucose ≥126 mg/dL (≥7.0 mmol/L)
   3. 2-hour postprandial glucose ≥200 mg/dL (11.1 mmol/L) during an oral glucose tolerance test (OGTT)

4. Glycemic control refers to the body's ability to maintain glucose homeostasis. Poor glycemic control results in acute complications such as hypoglycemia, hyperglycemia, diabetic ketoacidosis, hyperglycemic hyperosmolar nonketotic state, Somogyi effect, and dawn phenomenon. Long-term chronic complications include macrovascular diseases such as dyslipidemia and hypertension; and microvascular diseases such as nephropathy, retinopathy, and neuropathy.

   The best measure of long-term glycemic control is the hemoglobin A_1C test, which measures glucose associated with blood proteins that have a 120-day turnover. Fructosamine is also an indicator of long-term glycemic control.

   Short-term glycemic control is best monitored using the self-monitoring of blood glucose (SMBG) method in which the individual can use a portable monitor to test up to several times a day. Urine glucose and urine ketones are also short-term indicators of glycemic control.

5. **Rapid-acting insulin**

   - lispro (by Eli Lilly & Company) or insulin aspart (by Novo Nordisk)
   - begins to work about 5 minutes after injection
   - peaks in about 1 hour
   - continues to work for 2 to 4 hours

   Regular or Short-acting insulin (human)

   - usually reaches the bloodstream within 30 minutes after injection
   - peaks anywhere from 2 to 3 hours after injection
   - effective for approximately 3 to 6 hours

   Intermediate-acting insulin (human) - NPH (N) or Lente (L)

   - generally reaches the bloodstream about 2 to 4 hours after injection
   - peaks 4 to 12 hours later
   - effective for about 12 to 18 hours

   Long-acting insulin (ultralente)

   - reaches the bloodstream 6 to 10 hours after injection
   - usually effective for 20 to 24 hours

   Very long-acting insulin - glargine (GLAR-jeen) insulin

   - Starts working within **1 hour** after injection
   - keeps working evenly for 24 hours after injection

Mixed Insulin

- There are a large number of combination insulins available on the market.

Insulin dosage can be determined by using mathematical formulas based on body weight and is further adjusted based on blood glucose levels. There are three basic types of insulin administration regimens: fixed (conventional or standard therapy), flexible (intensive insulin therapy), and continuous subcutaneous insulin infusion (CSII).

Conventional vs. Intensive: *Conventional (standard) insulin therapy* consists of a constant dose of basal (or background) insulin combined with short- or rapid-acting (or bolus) insulin. Individuals using conventional therapy must synchronize administration of their insulin and food intake to avoid hypoglycemia, which requires strict adherence to meal plans and patterns. *Flexible (intensive) insulin therapy* requires multiple daily injections (MDIs) of bolus insulin before meals in addition to basal insulin once or twice daily. Insulin can be adjusted to correspond to food intake, thereby mimicking endogenous insulin secretion in a person without diabetes and allowing for more flexibility regarding meal planning, exercise, and response to hyperglycemia.

6. 1.) Healthful Eating Guidelines: These methods teach making healthy food choices as the primary goal. Guidelines are especially useful for people newly diagnosed with diabetes and for those who may be less educated or have low math skills, which are important for more advanced methods such as exchange lists and carbohydrate counting.

   2.) Plate Method: The plate method teaches portion control, consistency, and basic food categories. This method is also best for newly diagnosed patients and those who are not willing or able to use a more complex system.

   3.) Menus: Menus teach portion control and meal spacing. Menu plans are individualized for those who may have special needs. The main drawback is that self-sufficiency is not developed.

   4.) Carbohydrate counting: In this method, the individual plans meals based on the amount of carbohydrate consumed. Food carbohydrate sources are starches, fruits, milk/yogurt, and sweets. (Nonstarchy vegetables do not need to be counted unless eaten in servings containing >15 g of carbohydrates.) There are two ways to count carbohydrates:

   - the amount of food containing 15 g carbohydrate counts as one carbohydrate choice
   - total grams of carbohydrate in a meal or snack can be counted

   The amount of carbohydrate tolerated varies among individuals and can also be matched to insulin dosage or physical activity.

   5. Exchange system: In the exchange system, foods are grouped together in lists of equal macronutrient value so that they may easily be exchanged or substituted for easy meal planning. Exchange lists may be conceptually difficult for some individuals with diabetes to understand well enough to be used effectively.

   - 70-year-old man: Individualized menus, healthful eating guidelines, plate method because of age and low education.
   - 13-year-old athlete: Carbohydrate counting because of age and physical activity.
   - 32-year-old pregnant woman: Healthful eating guidelines and simple carbohydrate counting because of short duration of therapy.

7. 7 classes of oral diabetes medications:
   - *alpha-glucosidase inhibitors (AGIs)* Delays intestinal absorption of glucose.
   - *amylin analogs* Delays gastric emptying, decreases postprandial glucagon release, suppresses appetite.
   - *biguanides* Decrease hepatic glucose production, increases insulin uptake in muscles.
   - *incretin mimetics* Mimics glucose-dependent insulin secretion, suppresses elevated glucagon secretion, delays gastric emptying.
   - *Meglitinides* Stimulates insulin secretion in presence of glucose.
   - *sulfonylurea agents* Stimulates insulin secretion.
   - *thiazolidinediones* Decreases insulin resistance.

8. Overweight and obesity are risk factors for T2DM. Obesity is also an independent risk factor for hypertension, dyslipidemia, and CVD, the major cause of death in those with diabetes. Moderate weight loss improves glycemic control and reduces CVD risk; therefore, weight loss is recommended for individuals with BMIs >25.0 kg/m2.

Chapter 20

1. The 4 risk factors for developing kidney disease are:

   (1) Proteinuria—one of the strongest predictors of renal disease progression and response to anti-hypertensive therapy;

   (2) Ethnicity—African American males with diabetes have a 2–3x higher risk of end-stage renal disease compared to white patients;

   (3) Gender—the incidence is greater in males than females;

   (4) Smoking—associated with proteinuria, IgA nephrophathy, polycystic kidney disease, lupus nephritis, and progression in type 1 and 2 diabetes

The most frequent diseases associated with the development of CKD are:

(1) Diabetes, representing 39% of patients

(2) Hypertension (28%)

(3) Glomerulonephritis (13%).

2. Diabetic nephropathy is the most common cause of CKD in the United States. People with both type 1 and type 2 diabetes are at increased risk. The risk is greater if blood sugars are not controlled. The kidneys act as filters and the resulting fluid eventually forms urine. The earliest detectable change in the course of diabetic nephropathy is a thickening in the glomerulus, thought to be caused by hyperglycemia. At this stage, the kidney may start allowing more protein (albumin) than normal to be excreted in the urine. As diabetic nephropathy progresses, increasing numbers of glomeruli are destroyed and increasing amounts of albumin are excreted. As the number of functioning nephrons declines, each remaining nephron must clear an increasing solute load. Eventually, the kidneys are unable to filter out the nitrogenous waste. Because the progression is slow, the body can partially adapt to the changes.

3. Kidney function is measured by the glomerular filtration rate (GFR), which is reflected in clearance tests that measure the rate at which substances are cleared from the plasma by the glomeruli. The normal GFR is 135–180 liters per day. The most widely used method for estimating GFR is the Cockcroft-Gault equation. This equation considers the effects of age, sex, and body weight on creatinine generation, thereby adjusting serum creatinine values to more accurately reflect creatinine clearance. The formula is as follows:

   $\text{GFR} = [(140 - \text{age}) \times \text{body weight (kg)} \times 0.85 \text{ if female}] \div [72 \times \text{serum creatinine (mg/dL)}]$

   More recently, the Modification of Diet in Renal Disease (MDRD) modified GFR equation has come to be considered to be the "gold standard" of measurement (in addition to incorporating the influence of age and gender, and the effects of race, three biochemical measures are included):

   $\text{GFR} = 170 \times \text{serum creatinine}^{-0.999} \times \text{age}^{-0.176} \times \text{female}^{0.762} \times (1.18 \times \text{black race}) \times \text{SUN}^{-0.17} \times \text{serum albumin}^{0.318}$

4. Chronic kidney disease (CKD) is a syndrome of progressive and irreversible loss of the excretory, endocrine, and metabolic functions of the kidney secondary to kidney damage. CKD progresses slowly over time, and there may be intervals during which kidney functions remain stable. The onset of renal failure is not usually apparent until 50% to 70% of renal function is lost. ARF, on the other hand, is a disorder in which the kidneys suddenly stop functioning, characterized by abrupt cessation or reduction in GFR and accumulation of nitrogenous wastes. The following are conditions that commonly lead to ARF. (1) A sudden drop in the flow of blood to the kidneys. This may be commonly caused by severe blood loss, severe infection, a serious injury, and dehydration. (2) Damage to the kidneys by, for example, nephrotoxic medication. Such medications include some antibiotics such as gentamicin and streptomycin, toxic levels of some common pain medicines such as aspirin and ibuprofen, and the dyes used in some X-ray tests. (3) The other possible cause of ARF is a sudden blockage that prevents excretion of urine. Some common causes of blockages are kidney stones, a tumor, injury, or an enlarged prostate gland.

5. Nephrotic syndrome (NS) is an abnormal condition that is marked by deficiency of albumin in the blood and its excretion in the urine due to altered permeability of the glomerular basement membranes. Superficially, NS is like protein-energy malnutrition. In both NS and protein energy malnutrition, albumin levels are low, plasma volume is expanded, and albumin pools shift from the extravascular space to the vascular space. Muscle wasting is common in those patients with massive and continual proteinuria. The muscle wasting can often be masked by edema. There are two possible mechanisms that may lead to edema in patients with NS. The first is the classic explanation, also known as the "underfill" model. A decrease in plasma albumin leads to a decrease in difference between interstitial and plasma oncotic pressure resulting in plasma volume contraction. Edema occurs when the amount of fluid flowing into the interstitium exceeds maximal lymph flow. The second mechanism suggests that renal disease creates primary sodium and water retention leading to plasma volume expansion and increased capillary hydrostatic pressure.

   Current protein recommendations for patients with NS are .8 to 1.0 g/kg/day. This dietary protein amount is believed to decrease proteinuria without reducing serum albumin. A study conducted in 1997 indicated that patients with nephrotic syndrome consuming .8 g/kg/day of protein over 24 days maintained positive nitrogen balance. There are no long-term studies to suggest the safety of low-protein diets providing less than .8 g/kg/day, and thus such diets are not recommended. Additionally, protein *supplementation* is not warranted, as there appears to be no benefit in NS patients.

6. In 1995, the National Kidney Foundation set out to improve the outcome for dialysis patients by developing the first Dialysis Outcome Quality Initiative (DOQI). K/DOQI Clinical Practice Guidelines for Chronic Kidney Disease: Evaluation, Classification, and Stratification (National Kidney Foundation 2002) describe five stages of chronic kidney disease. K/DOQI Stage 1 and 2 are defined as kidney damage with normal or increased glomerular

filtration rate (GFR) and kidney damage with mild decrease in GFR. Stage 3 is a moderate decrease in GFR (30–59 mL/min/1.73 m^2), and Stage 4 is severe decrease in GFR (15–29 mL/min/1.73 m^2). The prevalence of Stage 3 or 4 CKD in the United States is estimated at 7,000,000 people for Stage 3 and 400,000 for Stage 4. Stage 5 cannot sustain life and renal replacement therapy must be initiated. Patients should be evaluated every 1 to 4 months for Stage 1 through 4, and monthly in Stage 5.

7. The two kidney replacement treatments are hemodialysis and peritoneal dialysis. Both are started in Stage 5 of CKD. Renal replacement therapy requires access to the patient's circulatory system. For hemodialysis (HD), the preferred permanent access site is an arteriovenous fistula (AVF), created surgically by fashioning in the forearm a subcutaneous joining of the radial artery and the cephalic vein. In peritoneal dialysis (PD), access to the patient's blood supply is gained via a catheter, placed surgically into the peritoneal cavity.

   For PD, dialysate is introduced into the peritoneum through the peritoneal catheter. Solutes from the plasma circulating in the vessels and capillaries perfusing the peritoneal wall pass across the peritoneal membrane into the dialysate, which is subsequently removed and discarded. The dialysate for PD is available with a range of dextrose concentrations that alter the osmolality of the dialysate and assist in fluid removal. In addition, the dwell time (i.e., how long the dialysate remains in the peritoneum) and the number of exchanges (i.e., how many bags of dialysate and the total volume of each used in 24 hours) also affect the amount of fluid and solute removal. PD can be performed continuously during the day or only at night while the patient is sleeping.

   In hemodialysis, the patient's blood is pumped through a selective membrane that is exposed to some rinsing fluid (dialysate) composed of varying ions and minerals. Typical length of dialysis is 3 hours and it is performed three times a week.

8. Protein and energy requires are the same for HD and PD patients and are: 1.2 g/kg/day protein, at least 50% high biological value; 30–35 kcal/kg if 60 years of age or older and 35 kcal/kg if less than 60 years of age. There have been discrepancies in the literature regarding the body weight measurement to be used when calculating protein and energy requirements for underweight and obese patients. According to NKF K/DOQI Nutrition Guidelines (National Kidney Foundation 2000), it is recommended that the adjusted edema-free body weight (aBW_{ef}) be used based on the National Health Nutrition Evaluation Survey (NHANES) II data. It should be used for those patients that have an edema-free body weight <95% or >115% of the median standard weight as determined from the NHANES II data. aBW_{ef} should be used for maintenance in HD and PD patients, and should be obtained post-dialysis for HD patients and after drainage of dialysate for PD patients.

   The aBW_{ef} formula: $aBW_{ef} = BW_{ef} + [\{SBW - BW_{ef}\} \times 0.25]$

   BW_{ef} = actual edema-free body weight;

   SBW = standard body weight as determined from the NHANES II data; this is the median body weight of normal Americans of the same height, gender, frame size and age.

   $aBW_{ef =}$ adjusted edema free body weight

9. Once the kidney fails, the patient can no longer prevent the accumulation of fluid, sodium, potassium, and phosphate between dialysis treatments. Excess fluid gain between treatments will result in hypertension and increased dialysis times. Sodium retention will lead to increased thirst, increased fluid intake, and water retention. Hyperkalemia may result in cardiac arrhythmias. Excess phosphate will stimulate parathyroid hormone release, which increases bone breakdown and bone disease.

   Specially designed vitamins are available and should be used exclusively if supplementation is required. In general, the "renal" vitamin contains B vitamins, folic acid and vitamin C. Fat-soluble vitamins and minerals are not included. Preparations containing vitamin A or high doses of vitamin C should be avoided. Current vitamin supplements formulated specifically for dialysis patients provide: 60–100 mg vitamin C, 1.5 mg thiamin, 1.7 mg riboflavin, 20 mg niacin, 10–50 mg vitamin B_6, 6 mcg-1 mg vitamin B_{12}, 800 mcg-5 mg folic acid, 10 mg pantothentic acid, and 150–300 mcg biotin. Because serum vitamin A levels are elevated in HD and PD patients, supplementation is not necessary. Very little is known regarding the long-term effects of vitamin E supplementation. Vitamin K supplementation may be needed for those patients receiving antibiotic therapy, since the antibiotics may destroy the bacteria found in the gastrointestinal tract (these bacteria are a primary source of vitamin K). However, since most HD patients receive anti-coagulation therapy, caution must be taken due to vitamin K's role in promoting clot formation.

10. Cardiovascular disease (CVD) is the main cause of mortality in patients with CKD. Patients who are diagnosed with CKD are more likely to die from CVD than to progress to end-stage disease. In addition to having a greater incidence of the traditional CVD risk factors (diabetes, hypertension, dyslipidemias), a decreased glomerular filtration rate and proteinuria are independent risk factors for CVD. Aggressive CVD risk factor management is recommended, including pharmacological and diet therapy, smoking cessation and physical activity. The following are the dietary recommendations:

Nutrient	Recommended Intake
Saturated Fat	<7% of total kcal
Polyunsaturated fat	Up to 10% of total kcal
Monounsaturated fat	Up to 20% of total kcal
Total fat	25–35% of total kcal
Carbohydrates	50–60% of total kcal
Protein	Approximately 15% of total kcal
Cholesterol	<200 mg/day
Total kcal	Balance energy intake and expenditure to maintain desirable body weight/prevent weight gain
Fiber	20–30 grams/day with 5–10 grams soluble fiber

11. In patients undergoing dialysis, secondary hyperparathyroidism (SHPT) can progress to severe, intractable forms of bone disease. Prolonged exposure to elevations of PTH causes the development of osteitis fibrosa, a form of bone disease characterized by rapid bone turnover with an excess of collagen production and inadequate mineralization. Although bone mass itself may appear constant, bone quality is poor and the bones are more prone to fracture. Damaged kidneys no longer have the ability to convert inactive vitamin D into active calcitriol, thus limiting calcium absorption from the intestine. Additionally, damaged kidneys may not be able to reabsorb calcium and excrete phosphorus, leading to an imbalance in the blood. As a result, the PTH is constantly stimulated with a continued release of PTH leading to bone breakdown and osteitis fibrosa. Dietary recommendations include supplementation with the active form of vitamin D. Calcium intake should not exceed 1500 mg/day from phosphate binders and no more than 2000 mg/day total including dietary sources. Phosphate intake should be 800–1000 mg/day or <17mg/kg.
12. CKD patients are unable to synthesize adequate amounts of endogenous erthyropoietin, which leads to decreased red blood cell production in the bone marrow and also leads to low hemoglobin (Hgb) levels. The discovery of recombinant human erythropoietin (rHuEPO) has improved the quality of life in CKD patients significantly. Patients are treated with rHuEPO and iron. EPO is given intravenously for HD patients and subcutaneously for PD patients. The effectiveness of EPO therapy depends on adequate iron and protein nutritional status.
13. The goal in the acute post-transplant period is to manage the increased metabolic demands of transplant surgery. In addition to achieving optimal nutritional status, the goals of the transplant diet in the long term include the management of obesity, blood pressure, insulin resistance, diabetes, and hyperlipidemia, maintenance of electrolyte balance, and maximized bone health. Nutrition therapy for kidney transplant patients differs between the acute phase (up to 8 weeks following transplant) and chronic phase (beginning the ninth week following transplant).
14. Risk factors for kidney stones include family history and certain medical conditions, such as hypercalciuria, hyperuricosuria, hyperoxaluria, and low urine volume. Hypercalciuria, an inherited condition, is the cause of more than 50% of all kidney stones. It is defined as urinary calcium excretion greater than 300 mg per 24 hours in men or 250 mg per 24 hours in women. Other causes of kidney stones include gout, excess intake of vitamin D, urinary tract infections, and urinary tract blockages.

Chapter 21

1. a. Low hematocrit and hemoglobin, small red blood cells, low serum ferritin, low serum iron level, high iron binding capacity (TIBC) in the blood, low MCH, low MCHC. **Signs/Symptoms:** Pale skin, fatigue, irritability, weakness, SOB, sore tongue, brittle nails, pica, decreased appetite, headache.

   b. Low hematocrit and hemoglobin with elevated MCV (low red blood cell count with large-sized red blood cells), CBC showing low white blood count and low platelets. **Signs/symptoms:** shortness of breath, fatigue, loss of appetite, diarrhea, tingling/numbness in limbs, sore mouth, unsteady gait, tongue problems, impaired sense of smell, bleeding gums, pallor, rapid heart rate.

   c. Elevated indirect billirubin levels, low serum haptoglobin, hemoglobin in the urine, hemosiderin in the urine, increased urine and fecal urobilinogen, elevated absolute reticulocyte count, low red blood cell count (RBC) and hemoglobin, elevated serum LDH. **Signs/Symptoms:** Chills, fatigue, pale skin, SOB, rapid heart rate, yellow skin (jaundice), dark urine, and enlarged spleen.

   d. Decreased hemoglobin, elevated billirubin, high white blood cell count, elevated serum potassium, elevated serum creatinine, blood oxygen saturation may be decreased. **Signs/Symptoms:** Paleness, yellow eyes/skin, fatigue, breathlessness, rapid heart rate, delayed growth and puberty, susceptibility to infections, ulcers, jaundice, and bone pain.

2. a. Increased WBC's. **Signs/symptoms**: Mucositis, ulceration, chronic fatigue, malaise, poor nutritional status delays healing, lethargy, nutrient deficiencies.

   b. Low WBC, low RBC, and low platelets. **Signs/symptoms:** Headache, dizziness, nausea, shortness of breath, bruising, lack of energy or tiring easily (fatigue), abnormal paleness or lack of color of the skin, blood in stool.

   c. Decreased hemoglobin and low RBC counts. **Signs/symptoms**: In severe thalassemia, fatigue, weakness, pale skin or jaundice, protruding abdomen with enlarged spleen and liver, dark urine, abnormal facial bones, and poor growth.

3. Erythropoietin, a hormone produced by the kidneys, regulates red blood cell production. Its synthesis is regulated by the kidney's ability to monitor blood volume and its production is stimulated by low oxygen levels in the blood. Iron availability dictates the rate of heme synthesis. With an iron deficiency, circulating RBC die off without adequate replacement, the oxygen tension falls, and erythropoietin production is increased. However, these homeostatic mechanisms eventually fail to compensate, resulting in anemia.
4. Patients may require supplementation, be educated on foods/cereal fortified with folate, iron, and cyancobalamin, instructed to limit carbonated beverages, and provided lists of foods high in these particular vitamins.
    a. Dietary iron, zinc, copper, all the B vitamins, and vitamins A, K and E are closely tied to normal hematopoiesis. They are involved with WBC production, RBC production, and cell division.
5. With pregnant women, hemodilution in addition to the demands of the fetus results in an increased utilization of iron and depletion of liver stores. Children may have poor Fe absorption, poor iron density and the presence of increased amounts of calcium, which is another divalent cation that can alter iron uptake from the gastrointestinal tract. Additionally, growing children have an increased need for iron, which may not be met by typical dietary iron intake. Obese children may also be at higher risk for anemia due to the increase of the hormone leptin, which alters the transport, absorption, and storage of iron from food. In elderly people, a decreased immune system may increase risk of anemia. In addition, the presence of iron-deficiency anemia is often associated with various chronic diseases.

Chapter 22

1. Epilepsy is a condition characterized by recurrent seizures, which are characterized by episodes of spontaneous, uncontrolled electrical activity among cerebral neurons. Signs and symptoms differ depending on the type of seizure and location of the focal point within the brain. A simple partial seizure may go unnoticed by the patient or by diagnostic tests. But if the focal point involves a particular sensory, motor, or cognitive control, the subsequent signs and symptoms will reflect this. For example, the patient may experience a change in smell, vision or hearing.

    Tonic-clonic seizures, the most common type of generalized seizure, produce significant signs and symptoms including the loss of consciousness and stiffening of the limbs. Other symptoms can include incontinence, salivation, and cyanosis.
2. A ketogenic diet induces a state of ketosis through a diet with the majority of energy from fat (70–90%) and the remaining kcal from protein and carbohydrate. Although the exact mechanism is not fully understood, it is thought that the increased levels of ketones change neuron metabolism, and that the ketone body acts to change the balance of neurotransmitters, resulting in an anticonvulsant effect. For infants, children, and adolescents, impaired ability to consume adequate nutrients, limited food choices (if on a ketogenic diet), and drug-nutrient interactions may interfere with the ability to achieve optimal growth and development (Peterson et al. 2005). Assuring adequate energy, protein, vitamin and mineral intake is the major component of nutrition therapy for these populations (Liu et al. 2003; Hemingway et al. 2001; Stafstrom and Bough 2003).
3. Risk factors for CVA include cardiovascular disease, diabetes, hyperlipidemia, hypertension and excessive alcohol consumption. Reducing the amout of saturated fat and sodium in the diet while increasing linolenic/linoleic fatty acids, whole grains (fiber), and fruits and vegetables will help to reduce the risk of these factors contributing to CVA. Nutritional implications include: impairment of ability to chew, swallow, or self-feed, dysphagia, drooling, coughing and choking, weight loss, and generalized malnutrition due to inadequate nutritional intake. Aspiration or inhalation of oropharyngeal contents into the lungs is also a primary complication.
4. Parkinson's is a neuromuscular, neurodegenerative disease caused by the loss of dopamine-producing cells in the substantia nigra portion of the brain. Classic motor symptoms of Parkinson's disease include resting tremor, rigidity, bradykinesia (slowed movement), stooped posture and postural instability, mask-like facial features, and a shuffling gait. Other symptoms, such as depression, anxiety, sleep disturbances, sensory abnormalities, and pain, are often experienced prior to the motor symptoms. In Parkinson's there is a progressive loss of dopamine, which causes an imbalance between excitatory and inhibitory communication. The loss of dopamine and the resulting imbalance of neurotransmitters cause the myriad symptoms experienced.
5. Two of the most important drug-nutrient interactions involve the medication Levodopa (L-dopa). L-dopa is transported from the small intestine to the blood-brain barrier; and, at both of these sites, protein carriers transport the medication. However, amino acids also utilize these same carriers, and thus compete with L-dopa for transport. As a result a high protein diet is thought to interfere with drug therapy.

Limiting protein intake 0.5–1 g protein/kg per day, evenly spread throughout all meals, is one way to increase the effect of medications. A high-carbohydrate diet may also be beneficial through an increase in insulin production. Vitamin supplements and foods high in pyridoxine may expedite this conversion before the L-dopa reaches the brain and reduce the amount of dopamine actually transported to the brain.

6. The myelin sheath covering the nerves assures rapid, consistent, non-random communication. With MS, demyelization (destruction of the myelin sheath) occurs. In addition, the axon may also be damaged. When these two components of the nerves are damaged, the communication between neurons is altered. Symptoms of MS vary widely because any nerve in the body can be affected. Numbness, tingling (parasthesia), uncoordination (ataxia), and weakness are all common symptoms. Some individuals experience visual problems from optic neuritis such as double vision, blurred vision, or blindness. Other symptoms can include difficulty swallowing, constipation, and bladder dysfunction.

   There is little research to support any particular nutrient or vitamin supplementation; however, it is proposed that omega-3-fatty acids and restriction of saturated fat (Swank diet) may help.

7. Apolipoprotein E (ApoE) is a cell marker for very-low-density lipoproteins, one of the transport proteins for cholesterol, triglycerides, and phospholipids. **Abnormalities in** apolipoprotein E have been found in those with type 2 Alzheimer's disease. Nutrition implications include (Thomas et al. 2000):
   - Aphasia: cannot verbally express preferences,
   - Apraxia: cannot manipulate utensils and food prior to eating, cannot manipulate food within mouth/swallow,
   - Agnosia: cannot recognize utensils, food,
   - Amnesia: forgets having eaten, does not recognize need to eat,
   - Anorexia: lack of desire to eat, possible psychological basis.

   Other nutrition problems common in AD are unexplained weight loss and increased energy requirements.

8. After a traumatic head injury there is a metabolic and inflammatory response that results in hypermetabolism, hyperglycemia and insulin resistance, increased gluconeogenesis, and lipolysis. Catabolism, as evidenced by nitrogen excretion, appears to peak in the second week post-injury and begins to slow after that time period. Energy requirements may be as high as 140% of resting energy expenditure (REE) for nonparalyzed patients and 100% REE for paralyzed adult patients. Infants, children, and adolescents will have slightly higher requirements, and clinical guidelines recommend providing 130–160% of REE. Protein typically makes up 15–20% of total energy intake.

Chapter 23

1. Three major functions of the respiratory system include:
   - Exchange of oxygen and carbon dioxide
   - Protection of the body against infection
   - Acid-base regulation

   Methods to measure pulmonary function include:
   - Initial evaluation through physical examination and listening to lung sounds, which can help determine abnormalities in breathing.
   - Spirometers, which calculate the amount of air the lungs can hold and the rate at which air can be inhaled and exhaled.
   - Evaluation of arterial blood gases (ABGs), which can determine pH and oxygen and carbon dioxide contents of the blood.

2. Malnutrition associated with poor intake appears to have an impact on the strength and endurance of respiratory muscles, particularly the diaphragm. Symptoms of respiratory disease that may affect dietary intake might include early satiety, anorexia, weight loss, and coughing and dyspnea during eating.

   There is evidence that antioxidants such as vitamin C, vitamin E, β-carotene, and selenium are associated with healthy lung functions. Furthermore, antioxidants in the ECF seem to play a role in protecting the lungs from oxidant injury, which can be caused by cigarette smoke. Cigarette smoking is also associated with reduced levels of antioxidants in body fluids. Smokers metabolize vitamin C faster than nonsmokers, and it is suggested that smokers increase their vitamin C intake to 35 mg/day beyond the DRI.

3. Factors to consider include:
   - According to data from NHANES I, II and III, increases in asthma have paralleled increases in obesity. This would be valuable information to extend to patients in regards to weight management.
   - There is also evidence that shows a protective effect of breastfeeding against the development and severity of asthma in children.
   - Leukotrienes are chemical mediators produced by the body that contribute to the development of asthma. Leukotrienes produce tissue edema, mucus secretion, smooth-muscle proliferation, and powerful bronchoconstriction. There is evidence, albeit inconsistent, that suggests that omega-3 fatty acid supplementation can decrease the production of

inflammatory agents, primarily leukotrienes, in asthmatic patients.

4. BPD is pulmonary inflammation and impaired growth and development of the alveoli.

   Etiology of BPD is complex, but can include extreme premature birth (birth weight <1500 g.), perinatal infection, and the presence of patent ductus arteriosus (PDA).

   Poor vitamin A status is another possible factor in the development of BPD. Vitamin A is important in normal alveolar development and surfactant production, and supports the integrity and regeneration of respiratory epithelial cells. Most premature infants are born with low serum vitamin A levels and lower levels of vitamin A transport carrier protein, a retinol-binding protein.

5. Cystic fibrosis is characterized by abnormally thick mucus secretions from the epithelial surfaces of various organ systems, including the respiratory tract, the gastrointestinal tract, the liver, the genitourinary system, and the sweat glands.

   - Pancreatic insufficiency results in poor digestion, poor absorption of fat and fat-soluble vitamins, and loss of bile and bile salts. Pancreatic enzyme supplements might be necessary.
   - CF and deteriorating pulmonary function may lead to anorexia, increased energy requirements, and malnutrition.
   - Individuals with CF are also at risk for osteopenia and osteoporosis because of pancreatic insufficiency, vitamin and mineral malabsorption (Ca, P), and the chronic use of corticosteroid medications.
   - Adequate kcal to support normal growth and development are essential. Energy needs for children with CF without respiratory infection are comparable to those of healthy children (100–110% of RDA). However, if an individual has significant lung disease or malabsorption, energy requirements may be significantly increased (120 to 150% of the RDA).

   Because there is not a guaranteed method for determining caloric needs, a steady rate of weight gain in growing children should be the goal. Children need to be monitored for appropriate growth. Additional kcal need to be added to the diet to support growth in the form of between-meal snacks and nutrient-dense foods, particularly foods high in fat. Low-fat and low-kcal foods should be avoided. Appropriate feeding behaviors should be encouraged at each age. School-aged children (5–10 years) are at higher risk for decreased growth rate. Certain activities may need to be limited to decrease energy expenditure.

   Adequate protein intake is essential to meet growth needs and maintain protein stores. Good nutrition also plays an important role in preparing the individual with CF for potential transplant later in life.

   Deficiency in several vitamins and minerals might be a problem. These include vitamin A, vitamin D, vitamin K, vitamin E, iron, and zinc. Individuals with CF should consume a high-salt diet. Infants and children with CF are at risk for developing hyponatremia. Nocturnal tube feedings can be used to promote normal eating behaviors during the day.

6. Aspiration pneumonia occurs when aspirated material, such as saliva, food, or gastric juices, causes an inflammatory response in the lung.

   Prevention of aspiration associated with tube feeding includes elevating the head of the patient's bed 30–45° to prevent gastroesophageal reflux, and regularly checking the placement of the feeding tube to ensure it is placed properly and does not move upward beyond the pyloric sphincter. The following procedures have also been proposed to prevent aspiration: use small bowel feedings, rather than gastric feedings to prevent gastric reflux; and use of a continuous method versus an intermittent or bolus method. If the patient has diabetes, maintaining rigid blood glucose control may be helpful.

7. Respiratory failure (RF) occurs when the respiratory system is no longer able to perform its normal functions. RF can result from chronic lung disease like COPD or cystic fibrosis, or as a result of an acute injury to the lung.

   The goals for nutrition care for patients with RF are to meet their nutrition needs, preserve and restore lean body mass, and maintain fluid balance.

   Protocol: A complete nutrition assessment to evaluate the patient's individual nutrition needs, including anthropometric and laboratory status, should be completed. Caloric requirements can be determined by using the Mifflin St. Joer equation or by using indirect calorimetry. Once the kcal requirements have been estimated, the patient's pulmonary status, body weight, and fluid balance must be closely monitored to ensure that overfeeding does not occur. Nutrients of specific concern are α-tocopherol, β-carotene, and vitamin C, which when taken at levels higher than the DRI have been associated with substantial increases in serum α-tocopherol and β-carotene, and appear to prevent further oxidative damage. Phosphate is also essential for optimal pulmonary function and normal contractibility of the diaphragm. Enteral feeding may be necessary, and several products, high in fat and low in carbohydrate, have been developed specifically for patients with RF.

Chapter 24

1. Although a strong link between diet and cancer has not been proven, it is known that nutritional factors contribute to carcinogenesis. Genes may be affected by nutritional components such as antioxidants, soy proteins, fat, kcal, and alcohol. Damage to these genes may result in production of carcinogenic cells during cell reproduction.
2. A normal cell is programmed to grow and reproduce at an innate rate, depending on hormonal and genetic factors. A cancer cell grows and reproduces at an uncontrolled rate autonomous from genetic or hormonal controls. It may even secrete its own growth factors.
3. Tumors are assigned stages depending upon their cell type, tissue of origin, function, whether they are benign or malignant, degree of differentiation, and site according to the Tumor Node Metastases (TNM) Staging System. Stages reflect the size and level of invasion of nearby structures on a scale of I-IV. The patient has stage II diffuse, not localized, B cell or pancreatic cancer, which has spread to the lymphatic system.
4. Chemotherapy medications are designed to target rapidly dividing cells that are present in malignant tumors. They reduce the size of the cell mass by interrupting the different stages of cell division. Chemotherapeutic agents also are often used in combinations to reduce the extent of systemic damage and resistance. Radiation therapy destroys cancer cells by altering their DNA. Unlike chemotherapy, radiation's effect is localized to the target area, but can affect normal cells as well as the cancer cells. Surgery is often used in combination with drug therapy. It is used only if it is determined that the tumor can be excised without too much damage to the surrounding areas or if there is a slim chance of the cells metastasizing.
5. Radiation therapy targets rapidly-dividing cells in a localized region of treatment. It will also affect normal cells within the treatment parameters. Common side effects include conditions specifically related to cell destruction in the area of treatment. Chemotherapy targets rapidly dividing cells, so those normal cells that divide rapidly, such as bone marrow, gastrointestinal tract cells, and cells of the hair follicles, may be affected. Thus, many people undergoing chemotherapy experience toxicity effects including thrombocytopenia, neuropenia, anemia, diarrhea, mucositis, and alopecia.
6. Biological therapies use the body's own immune system to combat the cancerous cells. BRMs can control cancer growth directly or indirectly by making cancer cells more recognizable to the body's defense system. These biological by-products including inteferons, interleukins, and cytokines can also be synthesized by drug companies to be used as a complementary therapy to traditional drugs.
7. SGA is a validated form of nutritional assessment based on medical history of weight changes, changes in dietary intake, and gastrointestinal symptoms. It also includes a physical examination with tests for loss of subcutaneous fat, muscle wasting, ankle edema, and ascites. PG-SGA and scored PG-SGA are recent additions to the SGA which allow for malnourished patients to be quickly triaged for nutrition support.
8. Malignancies precipitate many biochemical changes, including insulin resistance, increased glucose synthesis, gluconeogenesis, and Cori cycle activity. This "futile cycling" accounts for increased energy expenditure in the person even before diagnosis and can lead to deficiencies and the development of cachexia.
9. Nausea and vomiting are common side effects of oncologic therapies. Trigger odors such as cooking or perfumes should be avoided and small, frequent, low-fat meals during treatment periods can be helpful. Early satiety may also be a nutritional problem for cancer patients. Small, frequent, nutrient-dense meals should be eaten to avoid nutritional implications of this problem. Beverages with nutrients should also be consumed. Mucositis in various parts of the GI tract may occur in the cancer patient. Oral mucositis may be eased by the consumption of soft, non-fibrous, non-acidic foods. Liquids (not hot) should be frequently consumed to avoid dehydration. Nutritional supplements may also be helpful.
10. Milk thistle, or Silymarin, is commonly used to decrease the activity of cytochrome P450 enzyme system and may aid in the clearance of certain chemotherapeutic agents. Side effects include a mild-to-moderate laxative effect and allergic reaction.

Chapter 25

1. Five different conditions that may cause stress and lead to a hypermetabolic state include: a closed head injury, a burn, severe inflammation, cancer, or sepsis.
2. When the body is responding to starvation, it will increasingly rely on body tissues such as lipid stores for energy, and it will reduce the basal metabolic rate so as to require less energy overall. In metabolic stress, the response is different. Energy requirements are increased and glucose is used as an energy fuel.
3. The ebb phase in the response to stress begins shortly after the injury occurs. During the ebb phase, shock brings about hypovolemia, thus providing less oxygen to tissues. Lower blood volume causes decreased cardiac and urinary output. At this point the goal of treatment is to restore blood flow, get oxygen to tissues, and stop bleeding.

   Once the blood pressure stabilizes, the flow phase begins. During this phase metabolic stress occurs and leads to hypermetabolism, catabolism, and altered immune and hormonal responses. This phase does not end until the metabolic stress is resolved.

4. During stress the body goes into a "flight or fight" response, releasing stress hormones such as glucagon, cortisol, epinephrine, and norepinephrine. These hormones work to mobilize energy stores so that the immediate energy demand can be met. Specifically, glucagon increases glucose production from amino acids, while cortisol increases levels of glucose, AA, and FFA in the blood and decreases the amount of protein synthesis. Epinephrine and norepinephrine stimulate the breakdown of glycogen and also increase FFA release. Insulin is also released but the body is less insulin resistant at this point.

   Overall, these hormones work to increase the breakdown of body stores of carbohydrate and protein in order to use them as an energy source.

5. Cytokines regulate the release of acute phase proteins. During stress, cytokines such as interleukins, leukotrienes, tumor necrosis factor, and interferons are released and act on target cells to produce stress reactions such as anorexia, fever, inflammation, and possibly hyperglycemia or catabolism.

6. Unlike most fats, which have an immunosuppressive role, omega-3 fatty acids seem to play an important role in improving the body's ability to respond to stress.

   Branched-chain amino acids may play a role in providing energy within the skeletal muscle. Using them might spare other substances that are needed for energy by other tissues during stress.

   Although arginine and glutamine are not essential AA, the body cannot synthesize them at a rate to meet the increased needs. Arginine in particular may help inhibit immunosuppression during stress and improve nitrogen balance, but use of it has also been shown to have a negative outcome. Glutamine supplementation might lower infection rate and prevent translocation of bacteria from the GI tract, and is often given orally.

7. Burns are classified by the depth of the wound and the percentage of body surface that is affected. Wounds are described as superficial, superficial partial thickness, deep partial thickness, or full thickness.

   After a burn injury, the body's response is hypermetabolic and catabolic, meaning that a burn victim is at significant nutritional risk. Body protein may be lost and wound healing requires intense nutrition therapy; however, fluid loss, pain, and immobility may complicate intake and make it difficult to maintain nutritional status. As the % of body surface area increases, the kcal and protein needs increase, especially at the peak of injury. Protein requirements are estimated at 1.5–2.0 g/kg/day to help minimize negative nitrogen balance, while the Mifflin-St. Jeor equation can estimate resting energy expenditure when used with an injury factor of 1.3–1.5.

8. After surgery a patient should be advances to solid food as quickly as possible (and as tolerated) in order to help maintain their nutritional status and the function of the GI tract. Advancement to solid food may not be advised when the surgery interrupts normal nutrition processes, such as with bariatric surgery.

9. Sepsis is presently thought to be an immunosuppressive process due to an infection that prevents an adequate response to infection. An infection or trauma triggers sepsis and causes the immune system to release inflammatory mediators such as TNF, interferon, and interleukins. Since the body in not able to mount a proper immune response, the reaction becomes anti-inflammatory.

   During sepsis there are increased nutritional risk and complications due to abnormalities of metabolism, difficulty determining nutritional needs, fluid restrictions, and multi-system organ dysfunctions.

10. SIRS (systemic inflammatory response syndrome) is another classification of sepsis which may not be caused by an infection, but may occur after a major surgery or trauma, or with other conditions such as myocardial infarction. MODS (multi-organ distress syndrome) may result from complications of sepsis or SIRS. With MODS there is an altered function of 2 or more organs in ill patients and thus these patients cannot maintain homeostasis without intervention. Sepsis, unlike SIRS, must be due to an infection, whereas SIRS may be brought on by infection or trauma.

11. A patient with metabolic stress should not be over fed. Although energy needs increase, complications of overfeeding such as hyperglycemia, fatty liver, and elevated CO_2 production can have a severe negative impact on an ill patient.

Chapter 26

1. The HIV is a retrovirus, which means that it carries RNA rather than DNA; RNA must be transcribed prior to integration into the host cell DNA to reproduce. The human immunodeficiency virus (HIV) targets many cells in the body, particularly the immune cells CD4.

2. HIV infection results in the release of cytokines which sets off a protective immune response. Cytokines are also responsible for a loss of appetite which can result in a reduced nutrient intake. Consistently high levels of cytokines have been reported to cause wasting due to a prolonged suppression of appetite, hypermetabolism, and a preferential loss of lean mass even during the early

and asymptomatic stages of HIV infection. Chronic inflammatory disease and high levels of cytokines can also lead to the altered metabolism of nutrients, further contributing to malnutrition.

3. CD4 levels and presence of opportunistic infections are predictive of disease progression. A BMI of 18 or higher in HIV patients indicates a decreased risk for disease progression.

4. About half of all persons infected with HIV will develop diarrhea during the clinical course of their disease. Diarrhea and intestinal malabsorption are the major causes of nutritional problems in these patients. Possible causes of HIV-related diarrhea include several identifiable enteric pathogens as well as nutritional factors such as fat malabsorption and malnutrition.

   Kaposi's sarcoma (KS) is a malignant disease which affects the gastrointestinal tract. KS lesions can occur in the mouth or esophagus as well as in the intestinal tract.

5. **HIV-liver disease**: Liver function may be compromised through the use of highly active antiretroviral therapy and through infection with cytomegalovirus (CMV) and hepatitis B virus, or as a result of malignant diseases such as Kaposi's sarcoma. Hepatitis C is now considered an HIV opportunistic infection. Dietary restrictions consistent with liver disease will be needed.

   Coronary Artery Disease: Several factors are responsible for the increased risk of developing cardiac disease, including chronic inflammation, increased blood lipid levels due to medications, and the presence of co-factor conditions such as insulin resistance and hypertension.

6. HIV-associated wasting is a primary symptom of HIV infection. Wasting consists of involuntary weight loss, reflecting the considerable depletion of body cell mass. Wasting was declared an AIDS-defining illness by the Centers for Disease Control and Prevention (CDC) in 1987. The CDC defines wasting as the involuntary loss of at least 10% of ideal body weight with associated symptoms of chronic fever, weakness, or diarrhea in the absence of other related illnesses that could contribute to the weight loss.

7. Wasting is associated with considerable morbidity and mortality. When a person reaches 66% of ideal body weight or 54% of lean body mass (LBM), the consequences are fatal. Patients with HIV-associated wasting have a decreased rate of survival regardless of other risk factors. Such patients are also at increased risk for hospitalization and suffer from a decreased quality of life. Even a 5% weight loss is associated with increased mortality and the development of opportunistic infections, even after controlling for CD4 cell counts and other predictive factors.

8. Recommendations for taking ARV medications:

Medications to be taken with food Generic (Brand Name)	Medications to be taken on empty stomach Generic (Brand Name)
Atazanavir/ATV (Reyataza)	Amprenavir/APV(Agenerase)
DelaviridineDLV (rescriptor)	Didanosine (Videx)
Lopinavir/LPV (Kaletra)	Efavirenz/EFV (Sustiva)
Nelfinavir NFV (Viracept)	
Tenofovir/TDF (Viread)	
Tipanavir/TPV (Aptivus)	

9. The drugs Delaviridine and Didanosine both cause diarrhea, nausea, and appetite loss, which in turn affect nutritional status.

10. Insulin resistance is commonly seen in protease-inhibitor treated patients. Besides protease inhibitors (PIs), other factors may contribute to the development of DM. Drugs such as megestrol acetate and corticosteroids are the cause for severe hyperglycemia seen in HIV-infected persons. Other factors implicated in the causation of insulin resistance are lipodystophy, peripheral fat wasting, family history of diabetes, obesity, age, concomitant liver disease, and low CD4 cell count.

11. The type of nutritional interventions required would be prioritized according to individual's needs and problems that may be present. The most common nutritional problems in HIV patients include inadequate intake due to nausea and vomiting, diarrhea, and HIV-anorexia. Following are some dietary interventions that may be useful for each condition.

    (a) Nausea and/or Vomiting: Avoid hot or warm foods; cold or room-temperature foods are better tolerated. Eat small, frequent meals.

    (b) Diarrhea: Replace lost fluids and nutrients with a dequate fluids such as juices, gelatin, or sports drinks. Avoid foods that are irritating such as caffeine-containing foods, citrus, and dairy (if lactose is a problem). Reduce fat intake and increase intake of soluble fiber.

    (c) Anorexia: Eat kcal-dense foods. Try small, frequent meals. Avoid strong-flavored foods. Eat in pleasant surroundings. Try using nutritional supplements if needed.

12. (a) Use of appetite stimulants such as Megestrol acetate and dronabinol to control HIV-related anorexia.

    (b) Anabolic agents such as anabolic steroids, nandrolone, and recombinant growth hormone.

    (c) Exercise, especially resistance training, has been shown to increase lean body mass.

    (d) Testosterone levels are usually decreased in HIV patients and may result in a loss of bone density.

Testosterone deficiency in both men and women can be treated with injectable, patch, gel, or oral forms.

13. This condition involves a loss of fat in the face and extremities and an accumulation of excess fat, especially in the abdomen, breasts, and the lower part of the neck and upper back.

    Lipodystrophy is a complex syndrome, including high blood fats, insulin resistance, and fat redistribution. Because the causes of lipodystrophy are not fully understood, optimal treatment strategies—including dietary strategies—are not yet established. Maintaining a healthy weight, regular physical activity, a low-fat, heart-healthy diet which includes adequate nutrients and fiber, and cutting back on processed foods, sugar, and alcohol will all help in achieving optimal nutritional status. A low-fat diet comprised of healthy fats (mostly unsaturated) seems to help in reducing abdominal fat accumulation. Excess fat in the diet may also lead to weight gain and overweight which in itself may cause insulin resistance and increased blood lipid levels.

Chapter 27

1. Osseous tissue is a major structural and supportive connective tissue that forms bones for the skeletal system. It assists in movement, protects vital organs, and stores calcium phosphate. Three primary components of osseous tissue are osteoblasts, osteocytes, and osteoclasts. Osseous matrix is composed of two components: osteocollagenous fibers (organic) and calcium phosphate crystals (inorganic). The inorganic components of the bone matrix mineralize and calcify the organic component. These osteocollagenous fibers help with the flexibility of the bone, while the calcium phosphate crystals provide the rigidness, so the body can be supported.
2. The four types of cells found in osseous tissue and their respective functions are as fololows:
    - **Osteogenic cells** are present on the surface of bones and the inner portions of the periosteum. They are capable of mitotic division and differentiating into osteoblasts, and are the only source of new osteoblasts. Osteogenic cells are active during normal growth of the skeletal system during childhood and adolescence, and in adulthood are activated in response to bone injury such as a fracture and in response to the stress placed on bones during weight-bearing exercise.
    - **Osteoblasts,** or bone-building cells, synthesize, deposit, and then orient the fibrous proteins of the organic matter of the bone matrix (collagen, proteoglycans, and other proteins) and then participate in the calcification or mineralization of the bone matrix, in a process known as bone formation or mineral deposition. Active osteoblasts are found on the surface of newly forming bone. They remove ions of calcium, phosphate, and other minerals from the blood plasma and deposit them within the bone matrix, thus hardening it. In response to bone fracture or the stress of weight-bearing exercise, the osteogenic cells multiply more rapidly and then differentiate to become osteoblasts.
    - **Osteocytes** are mature osteoblasts surrounded and entrapped by the matrix they have synthesized and calcified, and represent the vast majority of cells in bone. Osteocytes reside in tiny, fluid-filled cavities within the calcified matrix called lacunae, which are interconnected by narrow channels called canaliculi. Delicate cytoplasmic processes extending from the osteocytes pass through the canaliculi to connect with those of other osteocytes, thus allowing the osteocytes to chemically signal each other. The fluid-filled canaliculi also serve as channels for the passage of nutrients and metabolites between the osteocytes and nearby blood vessels. Although the osteocytes neither deposit nor remove bone, they are actively involved in maintaining the bony matrix by monitoring the amount of strain (bending) a bone experiences when it is mechanically loaded (for example, by weight-bearing exercise) and then communicating this information to osteoblasts on the bone surface. The osteoblasts can then build up and strengthen the bone where needed in response to the stress.
    - **Osteoclasts** are bone-removing cells which secrete hydrochloric acid (pH of about 4) to dissolve the mineral components of bone matrix and an enzyme called acid phosphatase that digests the collagen and other protein components of the bone matrix. This process is known as bone resorption or mineral resorption. The dissolved minerals are released into the blood and made available for other uses.
3. Several hormones participate in bone metabolism, including cortisol, growth hormone, and the thyroid hormones, but the primary regulators are parathyroid hormone, calcitonin, and vitamin D. Parathyroid hormone (PTH) is secreted by the parathyroid glands. There are two pairs of parathyroid glands located on the posterior surface of the thyroid gland which is located in the neck. When the blood calcium concentration is low, the parathyroid glands release PTH to raise the blood calcium concentration by the following mechanisms. PTH initiates an immediate release of calcium from the canaliculi and bone cells, as well as a more prolonged release of calcium from bone, by increasing the number of osteoclasts and promoting bone resorption. PTH inhibits collagen synthesis by the osteoblasts, which then inhibits bone deposition and promotes calcium reabsorption by the kidneys. It promotes the final step in the body's synthesis of vitamin D_3 [1,25-$(OH)_2D_3$] by the kidneys, thus enhancing the intestinal absorption of calcium by the intestines. Calcitonin (also known as thyro-

calcitonin) is secreted by the parafollicular or "C" cells of the thyroid gland when the blood calcium concentration is abnormally high. Its secretion lowers blood calcium concentration by inhibiting the activity of osteoclasts (and thus bone resorption), stimulating the activity of osteoblasts (and thus bone formation), and reducing the renal reabsorption of calcium and phosphate. The major function of vitamin D is to increase blood concentrations of calcium and phosphorus by promoting their absorption by the GI tract, promoting their reabsorption by the kidney, and stimulating osteoclast formation and thus bone resorption and the release of calcium and phosphorus from bone.

4. Osteoporosis is the most common bone disease in humans and is characterized by the loss of bone mass and deterioration of bone microarchitecture, compromised bone strength, and an increased susceptibility to fracture and painful morbidity. It is a condition of decreased bone density which has been caused by excessive loss of calcium. The risk factors for osteoporosis include:

   - age ≥65 years
   - being female
   - being of a certain race/ethnicity
   - malabsorption syndrome
   - primary hyperthyroidism
   - family history of osteoporosis (especially maternal hip fracture)
   - early menopause (before age 45)
   - underweight

   Bone mineral density (BMD) accounts for approximately 70% of bone strength, and because there is currently no accurate measure of overall bone strength, BMD is frequently used as a surrogate measure of bone strength. BMD is expressed in grams of mineral per area (g/cm^2) and is best measured using dual-energy X-ray absorptiometry (DXA). In DXA, x-ray beams of two different energy levels are projected through the body and received by a detector opposite the x-ray source. As the x-ray beam passes through the body, some of its energy is absorbed by the body's tissues, particularly dense tissue such as bone; this reduction in energy is known as attenuation. Because bone and soft tissue have different densities, they attenuate or absorb x-ray energy differently, and the differences in attenuation between soft tissues and bone at the two different energy levels are used to calculate BMD. There are several other techniques that are used to measure BMD such as peripheral dual-energy X-ray absorptiometry (P-DEXA), dual photon absorptiometry (DPA), quantitative computed tomography (QCT), and ultrasound.

   Yes, BMD is an appropriate and reliable measure of osteoporosis because it indicates that bone loss has occurred.

5. The important strategies and recommendations used to prevent or treat osteoporosis involve behavioral changes that include adequate calcium and vitamin D intake, weight-bearing and muscle-strengthening exercise, fall prevention, smoking cessation, and avoiding excessive alcohol intake.

   Calcium and vitamin D are two dietary elements that are vital for bone health. Prevention of osteoporosis can be achieved by consuming the recommended Adequate Intakes of calcium, which are 1000 mg/d for adults 19–50 years, and 1200 mg/d for adults over 50 years. Sources of calcium are milk (liquid and powdered), milk products (e.g., cheese, yogurt, and kefir), dark green vegetables (mustard and turnip greens, kale, and broccoli), some nuts (e.g., almonds), some seeds (e.g., sesame), tofu (manufactured using calcium sulfate as opposed to magnesium chloride), corn tortillas, and a variety of calcium-fortified foods such as citrus juice and soy beverages. Persons who are lactose intolerant can consume calcium-rich, lactose-free foods such as soy milk, lactose-free milk, and tablets. Vitamin D in fortified in milk products as well exposure to the sun will provide adequate vitamin D status.

   Other dietary factors involved in the treatment of osteoporosis are phosphorus, vitamin K, protein, fruit and vegetable intake, sodium, caffeine, fluoride, and trace minerals. Phosphorus is an essential bone-forming mineral, and throughout life an adequate supply is necessary for optimal bone health. Vitamin K is also essential for calcium regulation and bone formation. Rich sources of vitamin K are dark green leafy vegetables such as broccoli, Brussels spouts, dark green lettuce, collard greens, or kale. Eating one or more servings of these daily should be enough to meet the recommended target of 120 micrograms/day for men and 90 micrograms/day for women. For those individuals who have poor calcium absorption from the diet or vitamin D insufficiency, supplementation may be required. The most common and least expensive calcium supplements are those containing calcium carbonate, which should be taken with meals because calcium carbonate requires acid to make the calcium more soluble and absorbable. Supplements containing calcium citrate may be taken at any time in reference to meals. Calcium carbonate interferes with the absorption of iron and should not be taken at the same time as an iron supplement; however, calcium citrate does not interfere with the absorption of iron from supplements. Vitamin D and calcium supplementation is now advocated as the basic minimum for treating osteoporosis and reducing risk of fracture in older females and males.

hepatitis, and alcoholic liver cirrhosis, but may be the general entity when subentities are not specified.

algorithm a finite set of well-defined instructions for accomplishing a task; given an initial state, an algorithm will terminate in a corresponding recognizable end-point.

alkalemia condition of excess base in the blood consistent with a pH >7.45.

alkalosis conditions that produce excess base in the blood.

allele a copy of a specific gene situated in a given locus on a chromosome.

allergen an antigen that triggers an allergic response.

allergy an inappropriate and harmful immune reaction to a harmless nonpathogenic substance. Also called hypersensitivity.

allograft a tissue/organ graft between two genetically different individuals from the same species.

allopathic referring to modern or conventional medicine.

allopathic medicine modern or conventional biomedicine.

alpha-glucosidase a digestive enzyme found in the brush border cells of the small intestine that cleaves more complex carbohydrates into sugars.

alternate complement pathway a complement activation pathway that does not involve activation by an antibody.

alternative medicine unconventional therapeutic systems used by clients in place of or parallel to conventional biomedicine; typically administered by trained practitioner.

Alzheimer's disease the most common form of dementia, characterized by formation of amyloid plaques in the brain and neurofibrillary tangles within neurons.

amenorrhea the absence of menstrual cycles when they would be expected to occur.

amylin a hormone synthesized by pancreatic β-cells that contributes to glucose control during the postprandial period.

amyloid a starch-like substance present in diseased tissues.

amyloid plaques cellular deposits found between nerve cells.

amyloid precursor protein (APP) protein from which beta-amyloid is formed.

amylopectin the insoluble component of starch.

amytrophic lateral sclerosis a progressive neurological disease that causes destruction of the motor neurons of the nervous system, resulting in muscle weakness, twitching and atrophy; also known as Lou Gehrig's disease.

anabolic refers to building up or synthesis of larger organic molecules from smaller organic molecular subunits.

anaphylactic shock a life-threatening IgE-mediated allergic reaction. In humans, symptoms include swelling (especially of the lips and face), vomiting, diarrhea, difficulty in breathing, and a sudden drop in blood pressure. Also called anaphylaxis.

anasarca generalized edema with accumulation of serum in the connective tissue

anastomosis the surgical connection of body parts, especially hollow tubular parts like those of the GI tract.

anemia abnormal blood constituents resulting from various etiologies. Anemia is a symptom and is often a result of the decrement in blood constituents, although some forms of elevated blood components that are non-functional may be referred to as an anemia.

anergy antigen-specific nonresponsiveness by a T or B cell in which the cell is present but cannot respond.

aneurysm a weakened portion of the blood vessel wall

angina chest pain caused by oxygen deficit to the heart

angiogenesis the formation of new blood vessels and expanded systems for nutrient delivery and waste removal; a result of chemokines and hormonal messages that up-regulate the formation of these processes. Cancer cells can produce various messengers that trigger this up-regulation, thus allowing the rapidly dividing abnormal cells to acquire materials for growth and spread.

anion gap (AG) Anion gap (AG) = (serum Na^+) − (serum Cl^- + HCO_3^-); normal AG = 12 − 14 mEq/L.

ankle brachial index (ABI) ratio of Doppler-recorded systolic blood pressures between upper and lower extremities; a measure of peripheral vascular disease

anorexia lack of appetite.

anorexigenic appetite inhibiting.

anovulation lack of ovulation during the menstrual cycle.

anthropometry the study of the measurement of size and shape of the human body and its constituents (fat, lean tissue, and bone).

antibody a protein molecule found is serum and tissues that is secreted by B cells in response to a specific antigen that can bind to that antigen and neutralize or help destroy it. Also called immunoglobulin.

anticodons tRNA coding sequences; these sequences are complementary to the codons in mRNA and thus serve as anticodons.

antidiuresis inhibition of water losses through the kidney's reaction to hormones and abnormal cell signals, which reduce tubular losses.

anti-emetic a pharmacologic agent that reduces nausea.

antigen a substance that is specifically bound by an antibody or lymphocytes. Used by the immune system to recognize pathogens and altered cells. See immunogen.

antigenic determinant specific part of an immunogen that stimulates a specific immune response and reacts with the resulting antibody or activated T cell. Also called epitope.

antigen-presenting cell (APC) a cell capable of displaying fragments of antigens from pathogens and altered cells joined to MHC molecules on its surface in a manner that can be recognized by T cells.

antiretroviral (ARV) refers to medications targeted to interrupt the retrovirus life cycle.

antiretroviral therapy (ART) refers to the combination of medications that are typically used for controlling and reducing viral load.

antisense strand the non-coding strand of DNA.

antiseptics agents that kill microbes within living tissue.

antitoxin an antibody to an exotoxin.

aplastic anemia idiopathic anemia from abnormal, deficient, or absent red cell production due to bone marrow disorders.

apolipoprotein protein portion of the lipoprotein; provides cellular stability and allows for cellular recognition and binding

apolipoprotein E a protein that carries cholesterol in blood and that appears to play a role in brain function.

apoptosis genetically programmed cell death.

arginine vasopressin (AVP) previously known as anti-diuretic hormone.

aromatic amino acids (AAA) amino acids containing an aromatic side chain (phenylalanine, tyrosine, tryptophan).

arteriosclerosis a general term for thickening of the walls of the blood vessels with a resulting loss of vascular elasticity and narrowed lumen

arteriovenous graft (AVG) a connection of an artery and vein to provide circulatory access for hemodialysis

ascites abnormal accumulation of fluid in the abdominal cavity.

aspiration inspiration of foreign matter into the lung.

aspiration pneumonia aspiration of inhaled materials (saliva, nasal secretions, bacteria, liquids, food, or gastric contents) into the airway below the level of the vocal cords that results in an inflammatory response in the lung.

asthma a chronic inflammatory lung disease triggered by either an IgE allergic reaction or nonallergic factors that results in inflammation of the airway and reversible airway obstruction.

asymmetric muscle weakness muscle weakness occurring unequally in different parts or sides of the body.

atherosclerosis (AS) thickening of the blood vessel walls specifically caused by the presence of plaque

atopic a milder IgE-mediated allergic response.

atrophic gastritis atrophy of the lining of the stomach, which contains the parietal cells that produce intrinsic factor, proteases, and hydrochloric acid. This form of inflammation is often accompanied by bacterial overgrowth due to neutralization of the gastric pH.

atrophy reduction in size of muscle cells.

attenuated refers to an antigen rendered less virulent but still capable of eliciting an immune response.

auscultation a technique used during physical examination in which a stethoscope is used to evaluate the sounds created in body organs.

autoantibody an antibody to self-antigens.

autograft a tissue graft from one area to another on the same individual.

autoimmunity an immune response to one's own tissues.

autonomic division components of the nervous system that control involuntary functions of the body.

autonomic dysreflexia a complication of spinal cord injury and paralysis. Combination of stimuli that result in sudden increase in blood pressure and changes in heart function.

autosomal dominant an inheritance pattern of a dominant allele on an autosome.

autosomal recessive an inheritance pattern of a recessive allele on an autosome.

autosomal recessive inheritance inheritance of a trait as the result of inheriting a recessive gene for a particular trait from each parent.

autosomes non-sex-determining chromosomes. A human has 22 autosomes.

axon part of a neuron that transmits outgoing signals to other neurons.

axon terminals structure at the end of an axon that releases neurotransmitters.

ayurvedic medicine an Asian Indian medical system based on ancient Sanskrit texts; the name comes from "ayur," meaning "longevity," and "vedic," meaning "knowledge or science."

azotemia a build up of nitrogenous waste products such as urea in the blood and body fluids

B cell a lymphocyte derived from the bone marrow which differentiates into a plasma cell that makes an antibody.

bachytherapy a type of radiation therapy in which radioactive materials are placed in direct contact with the tissue being treated.

bacterial overgrowth syndrome malabsorption and malnutrition that result from cross contamination of bacteria from the colon to the small intestine.

balneotherapy treating disease by bathing. It can involve hot water, steam, or application of hot packs to the body; cold water; contrasting hot and cold baths or showers; a steam bath followed by immersion in cold water; or immersion in mineral springs or a hot bath to which various medicinal herbs or minerals have been added.

BALT (bronchial-associated lymphatic tissue) secondary lymphoid organs of the bronchial tree.

baroreceptor in general, any sensor of pressure changes.

Barrett's esophagus a complication of severe chronic GERD involving changes in the cells of the tissue that line the bottom of the esophagus. These esophageal cells become irritated when the contents of the stomach back up and there is a small but definite increased risk of cancer of the esophagus.

basal energy expenditure the minimum level of energy expended by the body to sustain life. It is measured in the morning when a subject is in a postabsorptive state, comfortably lying motionless in a supine position, and in a thermally neutral environment.

basophils polymorphonuclear leukocytes containing granules that stain with basic dyes. They have much in common with mast cells, including the release of histamine and leukotrienes, which contribute to allergic responses and inflammation.

BCR a B-cell receptor made of an antibody molecule and several auxiliary molecules.

Behavioral-Environmental domain domain that contains standardized nutrition diagnostic terms that describe nutrition problems related to knowledge, attitudes/beliefs, physical environment, access to food, and food safety (Nutrition Diagnosis, ADA, 20065).

beta-amyloid a part of the amyloid precursor protein found in the insoluble deposits outside neurons, which forms the core of plaques.

betadine a povidone-iodine containing solution that is used topically to destroy microorganisms.

bile an emulsifying agent produced in the liver and secreted into the duodenum.

biliary cirrhosis liver cirrhosis in which there is interference with intrahepatic bile flow.

bilirubin the breakdown product of hemoglobin molecules; it is normally excreted from the body via bile secretions.

binding site active site of the enzyme that binds to and acts on a particular substrate.

biomarker a biological molecule used as a marker to measure or indicate the effects or progress of a disease or condition.

biotransformation modification of a drug through metabolism.

body cell mass (BCM) kcalorie-using protein stores in the body; primarily muscle and organ tissues.

body mass index (BMI) weight in kilograms divided by height in meters squared ($BMI = kg \div m^2$). Although technically not a body composition assessment technique, it correlates well with estimates of body composition derived from skinfold measurements, and underwater weighing (hydrodensitometry), and can easily be calculated from weight and height. It is also known as **Quetelet's index**, named after its developer, Adolphe Quetelet (1796–1874), a Belgian statistician, astronomer, mathematician, and sociologist. The formula for calculating body mass index is: {body ~mass~index} ~ = ~{weight~\in~kilograms} over {(height~\in~meters)^2}

bolus feedings rapid administration of 250–500 mL of formula several times daily

bone marrow soft tissue in the cavities of bones where stem cells become red and white blood cells.

bone resorption a process whereby osteoclasts destroy an area of bone as the first step in bone remodeling

botanical remedies a comprehensive term covering all therapeutic parts of plants, including roots, bark, stems, gums, sap, leaves, and flowers. Sometimes the whole plant is used because it is thought that the components work synergistically to produce a more effective cure.

bradykinesia delayed or slowed body movements.

brain stem the part of the brain that connects the brain to the spinal cord and controls autonomic body functions.

branched-chain amino acid (BCAA) one of the amino acids that has a branch chain, namely, leucine, isoleucine, and valine.

bronchial hyperreactivity tendency of the smooth muscle of the tracheobronchial tree to narrow in response to a stimulus; present in virtually all symptomatic patients with asthma.

bronchitis a condition characterized by inflammation and eventual scarring of the lining of the bronchial tubes accompanied by restricted airflow, excessive mucus production, and a persistent cough.

bronchopulmonary dysplasia (BPD) a chronic lung disorder that may affect infants who have been exposed to high levels of oxygen therapy and ventilator support.

buccal refers to placement of a drug in the cheek.

CA-125 a protein that is secreted into the blood by ovarian cells and is used to monitor progress in the treatment of ovarian cancer

cachexia weight loss, wasting of muscle, loss of appetite, and general debility that can occur during a chronic disease.

calcitonin a polypeptide hormone secreted by the "C" cells of the thyroid gland when the blood calcium concentration is high. It lowers blood calcium by inhibiting bone resorption, promoting bone formation, and reducing renal reabsorption of calcium and phosphorus.

calculus calcified deposits that have formed around the teeth.

calorimetry measurement of the flow of heat.

cancer a class of diseases characterized by uncontrolled cell division and the ability of these cells to invade other tissues, either by direct growth into adjacent tissue (invasion) or by migration of cells to distant sites (metastasis).

candida yeastlike fungi found in feces and skin, vaginal, and pharyngeal tissues; GI tract is most important source.

capitation a payment or fee of a fixed amount per person.

carcinoembryonic antigen (CEA) a glycoprotein present in fetal gastrointestinal tissue and in the cells or serum of adults having certain types of cancers. It is used clinically to monitor the effectiveness of a treatment, such as for colorectal cancer.

carcinogen substance that causes cancer.

carcinogenesis the multistep process (initiation, promotion, and progression) through which normal cells are transformed into cancer cells.

cardiac cachexia CVD-associated malnutrition/wasting syndrome characterized by extreme skeletal muscle wasting, fatigue, and anorexia

cardiac output the volume of blood ejected from the left ventricle each minute; mathematically defined as heart rate × stroke volume

carnitine substrate needed for the normal metabolism of fat for energy. It is mostly found in beef and lamb but can be synthesized endogenously from the amino acids l-lysine and l-methionine.

carpopedal referring or pertaining to the hand and foot.

casein hydrolysate product made from the breakdown of casein, a milk-based protein, to smaller components, which are then easier to digest.

catecholamines the chemical classification of adrenodedullary hormones.

CD "Cluster Designation"; an international nomenclature system of leukocyte cell surface molecules (CD number).

CD4 a marker found predominantly on helper T cells that interacts with MHC class II molecules on antigen-presenting cells.

CD4 cell immune cell that is one of the primary targets of HIV for infection.

CD8 a marker found predominantly on cytotoxic T cells that interacts with MHC class I molecules on target cells.

celiac disease (CD) inflammation of the small intestine caused by gluten found in various grains, including wheat.

cellular immunity immune protection provided by the action of immune cells, especially T cells, polymorphonuclear leukocytes, and macrophages.

central nervous system the brain, spinal cord and the associated nerves.

central venous catheter (CVC) intravenous access inserted into large veins such as the subclavian, jugular or femoral veins in the center of the body.

cerebellum the part of the brain that is responsible for maintaining the body's balance and coordination.

cerebral cortex the outer layer of nerve tissue surrounding the cerebral hemispheres.

cerebral hemisphere one side of the cerebrum; each side contains four lobes (frontal, parietal, occipital, and temporal).

ceruloplasmin protein used in copper transport.

chagas disease a parasitic disease caused by *Trypanosoma cruzi.*

chelation therapy the introduction of EDTA (ethylene-diamine-tetra-acetic acid) into the body to bind with and remove metal ions.

chest physiotherapy physical therapy that includes a variety of techniques designed to reduce or prevent infection by clearing pooled secretions and/or infected materials from the lungs.

chiropractic medicine a medical system focusing on nonsurgical, drug-free care through manipulation of the spine and optimization of nerve function.

cholecalciferol a naturally occurring form of vitamin D produced in humans when the precursor molecule 7-dehydrocholesterol present in the skin is exposed to sunlight or to ultraviolet radiation. Also known as vitamin D_3.

cholecystectomy surgical removal of the gallbladder.

cholecystitis an inflammation of the gallbladder, usually due to a gallstone.

choledocholithiasis gallstones that are present in the common bile duct but are usually formed in the gallbladder.

cholelithiasis the presence or formation of gallstones.

cholinergic resembling acetylcholine; stimulated by or releasing acetylcholine or a related compound.

chondroblasts cells that are actively forming cartilage.

chondrocytes cells surrounded by cartilage and located inside small cavities known as lacunae.

chondroitin a nutritional supplement used by some to treat osteoarthritis; it is a specific glycosaminoglycan found in the proteoglycans of articular cartilage.

chondroitin sulfate the most common polysaccharide found in the proteoglycan molecules of cartilage.

chromatin the entire complement of DNA plus the histone proteins with which it is associated.

chromosomes units of the genome, each consisting of a long molecule of DNA that encodes numerous genes plus histone proteins. There are 22 autosomes and 2 sex chromosomes located within the nucleus of a human cell.

chronic kidney disease (CKD) kidney damage or GFR< 60 mL/min/1.73m2 for >3 months; kidney damage is defined as pathologic abnormalities or markers of damage

chronic myeloproliferative disease long-term hyperplasia of hematological tissues, with concomitant overproduction of abnormal cells, growth factors, chemokines, cytokines, and hormones involved in hematopoiesis.

chronic obstructive pulmonary disease (COPD) a disease that limits airflow through either inflammation of the lining of the bronchial tubes or destruction of alveoli.

chyme partially digested food in a semifluid state.

cicatrix scar tissue formation with calcification or hardening of the connective tissue used in repair of tissue damage.

cirrhosis any pathological condition where fibrous connective tissue invades any organ, usually as a consequence of inflammation or other injury.

Civilian Health and Medical Program of the Uniformed Services (CHAMPUS) the health plan that serves the dependents of active-duty military personnel and retired military personnel and their dependents.

Class I MHC antigen glycoproteins found on nucleated cells and encoded by the A, B, and C locus of the major histocompatibility complex. They present antigen to cytotoxic (CD8 +) T cells.

Class II MHC antigen glycoproteins found on nucleated cells and encoded by the Dr, Dq, or DP locus of the major histocompatibility complex. They present antigen to helper (CD4 +) T cells.

claudication pain in arms and legs due to inadequate blood flow to those muscles

clear liquid diet diet consisting of liquids that contribute minimal residue to the gastrointestinal tract. Includes fruit juices without pulp, carbonated sodas, broth, tea, coffee, water, popsicles, fruit ice, Jell-O (gelatin), and liquid nutritional supplements (e.g., Boost Breeze®, Resource Fruit Beverage®).

Clinical domain domain that contains standardized nutrition diagnostic terms that describe nutritional problems that relate to medical or physical conditions (Nutrition Diagnosis, ADA 20065).

clinical manifestations unique signs and symptoms.

clonal deletion a process by which contact with an antigen, usually self-antigen, early in lymphocyte differentiation leads to cell death by apoptosis.

clubbing changes in fingers and toes due to hypoxemia; fingers and toes show a curve at a tip of the nail with flattening surface.

codon a series of three nucleotides in mRNA that encodes a specific amino acid.

cofactors vitamins or other nutrients needed for the proper function of certain enzymes.

coinsurance a cost-sharing requirement under some health insurance policies in which the insured person pays some of the costs of covered services.

collagen a fibrous protein found in connective tissue.

colloid osmotic pressure (oncotic pressure) the osmotic pressure attributed to proteins and other macromolecules.

colonoscopy a procedure for evaluating the lining of the colon using a long, flexible tubular video probe that is inserted into the rectum.

colostomy a procedure in which the rectum only is surgically removed, and the end of the colon is attached to the stoma.

combination in the context of cancer treatment, the use of two or more therapeutic agents/processes for the treatment of a neoplasm.

combination chemotherapy the use of two or more antineoplastic agents to achieve maximum kill of malignant cells.

complement a group of serum proteins activated in a cascade that produces compounds that lyse cells and mediate immune reactions.

complementary medicine unconventional modalities used by clients in addition to conventional biomedicine; may involve practitioner, but often self-prescribed.

congestive heart failure (CHF) impairment of the ventricles' capacity to eject blood from the heart or to fill with blood

constant region (C region) the carboxyl-terminal portion of an immunoglobulin or TCR molecule that is similar from molecule to molecule.

constipation a decrease in frequency of bowel movements with straining with defecation and/or hard stools.

continuous feedings administration of formula for 10–24 hours daily, using a pump to control the feeding rate.

continuous renal replacement therapy (CRRT) type of renal replacement therapy used to treat patients in acute renal failure, particularly those with multiple organ failure; the types of patients treated tend to be hemodynamically unstable, have poor cardiac output, and be unable to tolerate hemodialysis

contracture shortening of muscle tissue resulting in immobility.

cor pulmonale an increase in size of the right ventricle of the heart caused by resistance to the passage of blood through the lungs; can lead to heart failure.

coronary artery disease (CAD) general term for all causes of heart disease characterized by narrowing of vessels supplying blood to the heart

cortical bone dense bone that forms the external surfaces of all bones, the shafts of the long bones, and a shell that caps the ends of the long bones.

cost shifting a much-criticized aspect of the existing health care system in which hospitals and other providers bill indemnity (fee-for-service) insurers at higher rates to recover the costs of charity care and to make up for discounts given to HMOs, PPOs, Medicare, and Medicaid.

costimulatory molecules membrane-bound or secreted products required in addition to MHC/TCR interactions for activation of T cells.

c-reactive protein protein released as a response to inflammation.

creatinine clearance rate at which creatinine is filtered through the kidney; often used as a measure of kidney function.

Crohn's disease a chronic inflammatory bowel disease (IBD) that can affect the entire gastrointestinal tract but most commonly affects the ileum and colon.

curling's ulcer ulceration of gastric or duodenal tissue as a result of burn or trauma.

current procedural terminology codes (CMT) coding system established by the Centers for Medicare and Medicaid Services for identifying medical care interventions.

Cushing's syndrome a disorder resulting from prolonged exposure to high levels of glucocorticoid hormones; symptoms include: muscle weakness, thinning of the skin, moon-shaped face, weight gain, and diabetes mellitus.

cyanosis blue-tinged mucous membranes and skin due to inadequate oxygen supply.

cyclosporine immunosuppressant medication that is often prescribed after organ transplant.

CYP 3A4 a specific cytochrome enzyme involved in drug metabolism.

cystic fibrosis (CF) disease characterized by abnormally thick mucus secretions from the epithelial surfaces of various organ systems, including the respiratory tract, the gastrointestinal tract, the liver, the genitourinary system, and the sweat glands.

cytochrome P-450 isoenzymes (CP450) family of enzyme systems responsible for drug metabolism.

cytokines soluble substances secreted by one cell that cause it or other cells to proliferate, differentiate, migrate, or become activated.

cytologic refers to blood tests that help to determine the morphologic features of a cell.

cytotoxic T cells (CTL) T lymphocytes that kill cells infected by viruses or transformed by cancer.

dawn phenomenon an increase in blood glucose in the early morning, most likely due to increased glucose production in the liver after an overnight fast.

debride to remove dead or injured tissue.

decompensation inability to maintain metabolic balance, leading to derangements in biochemical and clinical parameters.

dehiscence separation of wound edges.

dehydration a deficit of water in the body.

delayed-type hypersensitivity a cell-mediated inflammatory allergic reaction in the skin, (e.g., poison ivy) that takes 24–48 hours to appear.

dendrite branches extending out from the neuron that assists in transmission of impulses.

dendritic cells antigen-trapping and antigen-presenting white blood cells with nerve-like processes (e.g., Langerhans cells and interdigitating cells).

dental caries decay of the teeth that begins when acid dissolves the enamel which covers the tooth.

dentin the hard tissue of the tooth surrounding the central core of nerves and blood vessels.

dephosphorylate to remove a phosphate group from an organic molecule.

developed nation a nation that is generally regarded as one with a high standard of living, a high per capita income, a well-developed infrastructure (e.g., public utilities and systems for transport, public health, and public education), high literacy, long life-expectancy, and so on, when compared to the global average.

developing nation a nation that is generally regarded as one with a low standard of living, a low per capita income, a relatively poorly developed infrastructure (e.g., public utilities and systems for transport, public health, and public education), low literacy rates, low life-expectancy, and so on, when compared to the global norm.

dextrans a mono-dextrin limit product of carbohydrate metabolism $(C_6H_{10}O_5)_{11}$ used to expand plasma volume and chelate iron in solution. It is hydrophilic and branched.

diabetes insipidus chronic excretion of very large amounts of pale urine of low specific gravity.

diabetes mellitus a diverse group of disorders that share the primary symptom of hyperglycemia resulting from defective insulin production, insulin action, or both.

diagnosis-related groups (DRGs) groups developed by Medicare that classify a patient's illness(es) according to principal diagnosis and treatment requirements for the purpose of establishing payment rates.

dialysate fluid used by the dialysis procedure to assist in removal of metabolic by products, wastes, and toxins. Composition is determined by individual patient requirements

dialysis renal replacement procedure that removes excessive and toxic byproducts of metabolism from the blood, thus replacing the filtering function of healthy kidneys

diapedesis part of the of the inflammation response involving movement of blood cells across blood vessel walls into tissues.

diarrhea frequent or unusually liquid bowel movements.

diastole relaxation phase of the cardiac cycle; during this phase, ventricles empty and blood fills the atria

diastolic blood pressure pressure that occurs as ventricles relax (diastole phase of the cardiac cycle)

digoxin cardiac glycoside that is prescribed to alter the contractions of the heart.

dinucleotides paired nucleotide sequences.

direct calorimetry a technique to determine energy expenditure using a highly sophisticated chamber capable of measuring the amount of heat released by a subject's body through evaporation, convection, and radiation.

disinfectants agents that kill microbes on inanimate objects or surfaces.

dissolution dissolving of a medication.

diuresis the production of excessive amounts of urine.

diverticulitis an acute inflammation of the diverticula.

diverticulosis an abnormal presence of outpockets or pouches (diverticula) on the surface of the small intestine or colon.

DNA mutation testing blood test that screens for known mutation(s) that have been shown to result in a particular disease.

docosahexanoic acid (DHA) a fatty acid that can be produced from the essential omega-3 fatty acid linolenic acid. This fatty acid can also be found in many types of fish. DHA has been found to be beneficial for proper brain and vision development as well as in the management of hypertriglyceridemia.

doubly labeled water a technique to determine energy expenditure in which subjects drink a known amount of water containing two different stable isotopic forms of water: $H_2{}^{18}O$ and ^2H_2O. The rate that this water disappears from the subject's body is used to calculate the subject's energy expenditure.

dumping syndrome a group of symptoms that occurs with rapid passage of large amounts of food into the small intestine. Symptoms include dizziness, sweating, decreased blood pressure, and diarrhea.

dysgeusia altered taste.

dyspepsia vague upper abdominal symptoms which may include upper abdominal pain, bloating, early satiety, nausea, or belching.

dysphagia difficulty swallowing.

dysphonia difficulty speaking.

dysplasia abnormal cell growth.

dyspnea shortness of breath or difficulty breathing.

ectopia lentis displaced lenses in the eye.

edema the accumulation of excess fluid in cells, tissue, or a cavity, resulting in swelling.

edentulous without any teeth.

efferent carrying blood away from the designated site; for example, the efferent arteriole carries blood away from the glomerulus

ejection fraction the percentage of the LVEDV that is ejected in the systolic phase; in normal, apparently healthy adults, the typical ejection fraction is 50%–60%; defined mathematically as stroke volume ÷ LVEDV

electrolytes those substances that bear an electrical charge (ions).

electroneutrality the sum of the charges of the anions equals the sum of the charges of the cations.

embolus blood clot that breaks from the cellular surface and freely moves through the circulation

emetogenic an agent that causes nausea and/or vomiting.

emphysema a condition characterized by thinning and destruction of the alveoli, resulting in decreased oxygen transfer into the blood stream and shortness of breath.

enamel Hard outer layer of teeth consisting of hydroxyapatite. This mineral is composed of calcium, phosphorous, fluoride, chloride, sodium and magnesium.

encephalopathy degenerative disease of the brain.

endorphins neuropeptides that assist with pain control.

endoscopy examination of the interior of a canal by means of an endoscope.

endotoxins toxins found in bacteria, often as part of the cell wall, that stimulate an immune response.

end-stage renal disease (ESRD) kidney disease in which kidney function declines to 10 to 15% of normal; an old term that is no longer used; CKD is now used instead; ESRD is equivalent to CKD Stage 5 or when a patient requires renal replacement therapy

enteral nutrition (EN) feeding through the gastrointestinal tract using a tube, catheter, or stoma that delivers nutrients distal to (or beyond) the oral cavity.

enterocyte an intestinal cell.

environmental factors social and economic factors (wages, transportation, etc.) that impact both lifestyle choices and the consumption of food and nutrients. Other external factors such as food safety and sanitation determine the quality of food that is consumed, and food availability/access contributes to the amount and type of food consumed.

eosinophil a polymorphonuclear leukocyte containing granules that produce substances that damage parasites and decrease inflammation; these granules stain with acid dyes.

epicondyle a rounded projection at the end of a bone, located on or above a condyle, and usually serving as a place of attachment for ligaments and tendons.

epidemiology the study of the rates of disease within a given population.

epidural refers to placement of a drug into the spinal fluid.

epidural anesthesia anesthetic drug placed into the epidural space of the lumbar or sacral region of the spine, causing loss of sensation from the abdomen and pelvis to the lower limbs.

epigastric referring to the upper abdominal region.

epigenetics inheritance of information based on gene expression levels rather than gene sequence; regulated by genomic modifications such as DNA methylation and histone acetylation.

epilepsy a neurological disorder characterized by recurrent seizures, generally more than 2 unprovoked seizures.

epinephrine a hormone that is secreted from the adrenal medulla; regulates arterial blood pressure and prepares body for "fight or flight" responses; formerly referred to as adrenaline.

ergocalciferol a form of vitamin d produced by exposing the plant steroid ergosterol to ultraviolet irradiation. Also known as vitamin D_2.

erythroblastosis fetalis an antigen-induced hemolytic anemia of the newborn or premature infant, as a result of incompatibility of maternal Rh factors with the neonate.

erythropoiesis production of erythrocytes or red blood cells (RBC).

erythropoietin the hormone produced in the kidney that regulates marrow production of red blood cells.

esophageal phase of swallowing esophageal peristalsis carries the bolus through the esophagus and LES and into the stomach.

esophageal stricture a significant narrowing of the esophagus that may significantly interfere with swallowing.

esophagectomy surgical procedure resecting or removing the esophagus.

estimated energy requirement (**EER**) the average dietary energy intake that is predicted to maintain energy balance in a healthy adult of a defined age, gender, weight, height, and level of physical activity, consistent with good health. In children and pregnant and lactating women, the EER includes the needs associated with the deposition of tissues or the secretion of milk at rates consistent with good health.

etiology the cause of disease.

euglycemia maintenance of normal blood sugar levels.

evidence-based dietetics practice dietetics practice in which systematically reviewed scientific evidence is used to make food and nutrition practice decisions.

excipients those substances added to formulations of medications, such as color or coating agents.

excitatory in context of the neurological system, refers to a stimulus that results in neural response.

exocrine pancreas part of the pancreas that secretes digestive enzymes and bicarbonate into the duodenal lumen.

exons expressed sequences in mRNA; sequences that are translated into the final protein product.

exotoxins toxins produced by bacteria.

expected outcomes the desired change(s) to be achieved over time as a result of nutrition intervention.

extracellular fluid (**ECF**) the interstitial fluid and the plasma, constituting about 20% of the weight of the body. Sometimes used to mean all fluid outside of cells, usually excluding transcellular fluid.

extracorporeal shockwave lithotripsy (**ESWL**) a common procedure used to treat kidney stones whereby shock waves are used to break down the stones into smaller pieces

exudate fluid and cellular debris that seeps from blood vessels, usually as a result of inflammation.

Fab part of the antibody molecule containing one antigen-binding site. Contains the variable ends or N terminus of one light and one heavy chain.

facultative urine excess water that is excreted through urination.

fasciculations involuntary twitching or movement of muscle.

fatty liver yellow discoloration of the liver due to fatty degeneration of liver parenchymal cells.

Fc the part of the antibody without antigen-binding sites made of the C-terminal or constant domains of the immunoglobulin heavy chains.

federal poverty guidelines guidelines published annually by the Department of Health and Human Services to define "poverty" for legislative purposes. Updates to the poverty line can be found at http://aspe.hhs.gov/poverty/index.shtml.

feedback inhibition regulatory mechanism to limit production of certain substrates, which when present in significant amounts will limit the enzyme involved in making it to decrease the amount of this substrate being produced.

ferritin the storage protein for iron.

fibrin a filamentous protein; for blood clotting to occur, fibrinogen must be converted to fibrin.

fibroblasts connective tissue cells capable of forming collagen fibers.

fibromyalgia a chronic musculoskeletal disorder characterized by widespread muscle pain, joint stiffness, disturbed sleep, fatigue, headache, cognitive and memory problems, paresthesias, and numerous tender points. The word comes from the Latin term for fibrous tissue (fibro) and the Greek terms for muscle (myo) and pain (algia).

fibronectin acute-phase glycoprotein involved in the regulation of cell growth and differentiation, wound healing, and vascular integrity.

first-set reaction rejection of a foreign tissue graft due to antibodies and activated cells formed in response to the graft. Usually occurs 1–2 weeks after the tissue is transplanted.

fistula an abnormal opening or passage between 2 internal organs or from an internal organ to the surface of the body.

flatulence perceived excess gas in the intestinal tract.

foam cells macrophage cells containing lipid; found within the fatty streaks in the development of atherosclerosis

focal segmental glomerulosclerosis (**FSGS**) describes scarring in scattered regions of the kidney, typically limited to one part of the glomerulus and to a minority of glomeruli in the affected region; FSGS may result from a systemic disorder, or it may develop as an idiopathic kidney disease, without a known cause

food and nutrient factors the amount and type of foods and nutrients that are consumed and therefore made available to the body.

food frequency dietary assessment method in which the client describes the frequency and quantity of his or her consumption of certain foods/food groups.

food insecurity lack of adequate access by all people, at all times, to sufficient food for an active and healthy life, including at a minimum a readily available supply of nutritionally adequate and safe foods and an assured ability to acquire acceptable foods in a socially acceptable way.

frontal lobe a division of the cerebrum that is responsible for voluntary movement, speech and complex thought.

fructose a disaccharide absorbed by facilitated transport mechanism but not against a concentration gradient. When the concentration of fructose in the small intestine is greater than that of glucose, its rate of absorption slows and the unabsorbed fructose is fermented in the colon, causing diarrhea. Osmotic diarrhea has been reported in persons who have over-consumed sodas sweetened with high-fructose corn syrup or fruit juices.

full liquid diet diet consisting of all beverages allowed on clear liquid diets with addition of milk, ice cream, yogurt, and liquid nutritional supplements (e.g., Ensure®, Boost®).

fulminant refers to a condition that is severe or aggressive.

fulminant hepatic failure the severe impairment of hepatic functions in the absence of preexisting liver disease.

fundoplication a surgical technique used to suture the fundus of the stomach around the esophagus to prevent reflux.

fusion inhibitors medications that interrupt the viral replication cycle by inhibiting fusion of the HIV virus to the target cell.

galactitol sugar alcohol produced from galactose.

GALT (gut-associated lymphatic tissue) lymphoid tissue including Peyer's patches, the appendix, and solitary lymph nodes in the submucosa.

gamma globulins a group of serum proteins, including most antibody molecules, that migrate fastest towards the cathode during electrophoresis.

gangrene tissue death due to lack of blood flow and oxygen.

gastrectomy surgery to resect a portion of or the entire stomach.

gastritis inflammation of the gastric mucosa.

gastroesophageal reflux disease (GERD) chronic or recurrent gastric pain due to reflux of gastric secretions into the lower esophagus.

gastroparesis delayed gastric emptying, usually due to nerve or muscle injury in the stomach.

gastrostomy an opening into the stomach.

gene expression the level of activity of a specific gene in producing mRNA and, subsequently, protein. Expression can be regulated by many variables, including diet.

general anesthesia total loss of sensation and consciousness as a result of anesthesia drug.

genetic imprinting expression of specific genes, which depends on the parent of origin; some genes are expressed only from the maternal allele and others are expressed only from the paternal allele.

genome the entire set of genes of a given organism.

genotype the specific variants of a gene present in the two alleles in an individual that can result in specific traits or disorders.

gentamycin an antibiotic.

gingiva the gums.

glomerular filtration rate (GFR) the filtration ability of the glomerulus; used as an index of kidney function; normal value is approximately 125 mL/min

glomerulonephritis nephritis marked by inflammation of the capillaries of the renal glomeruli and membrane tissue that serves as a filter

glomerulosclerosis development of scar tissue within the glomerulus

glomerulus a network of thin-walled capillaries closely surrounded by a pear-shaped epithelial membrane called the Bowman's capsule

gluconeogenesis metabolic pathway from which glucose is formed from noncarbohydrate sources.

glucosamine a nutritional supplement used by some as a treatment for osteoarthritis; it is an amino sugar and a raw material for synthesizing glycosaminoglycans and proteoglycans, which are important constituents of articular cartilage.

GLUT-4 glucose transporter that transports glucose between blood and cells; it is the only glucose transporter responsive to insulin.

glycemic control control of blood glucose.

glycogenolysis metabolic pathway from which glycogen is converted to glucose.

glycolysis the anaerobic enzymatic conversion of glucose to lactate or pyruvate, which results in the production of energy in the form of ATP.

glycosuria the presence of glucose in the urine.

gout swelling, redness, heat, pain, and stiffness in a joint due to the formation of uric acid crystals in the synovial fluid, resulting in inflammation within the joint and in the surrounding tissues.

graft-versus-host disease (GVHD) a life-threatening reaction in which transplanted immunocompetent cells, usually T cells, attack the tissues of the immunocompromised recipient.

granuloma a mass of macrophages, with some T lymphocytes at the periphery, formed at the site of persisting inflammation due to the continued presence of a foreign body or infection.

group contract insurance health insurance offered through businesses, union trusts, or other groups and associations.

Guillain-Barré an acute peripheral nervous system disease characterized by progressive paralysis.

H_2 blockers medications that interrupt the production of acid in the stomach.

HAART highly active antiretroviral therapy; a combination of ARVs that is able to fully suppress the virus.

haplotype a group of gene variants that occur together.

hapten a non-immunogenic, low-molecular weight molecule that can be recognized by an antibody. It can initiate an immune response if it is conjugated to a "carrier" molecule.

health educator an individual with a bachelor's degree who educates patients about medical practices, self-care, and health promotion/disease prevention.

health insurance financial protection against health care costs associated with treatment of disease or accidental injury.

Health Insurance Portability and Accountability Act (HIPAA) legislation that guarantees that people who lose their group health insurance will have access to individual insurance, regardless of preexisting medical problems. The law also allows employees to secure health insurance from their new employer when they switch jobs even if they have a preexisting medical condition.

health maintenance organization (HMO) a health plan that provides comprehensive medical services to its members for a fixed, prepaid premium. Members must use participating providers.

heart failure (HF) see *congestive heart failure.*

heavy chain (H chain) the larger of the two types of immunoglobulin chains.

helper T cells (TH) a subset of T cells that triggers B cells to make antibodies, activates macrophages, and promotes the differentiation of other T cells.

hematemesis the vomiting of blood.

hematocrit packed RBC volume expressed as a percentage of whole blood upon centrifugation.

hematopoiesis production of blood cells.

hematopoietic stem cell an undifferentiated bone marrow cell that is a precursor for multiple cell types. Also called pluripotential stem cells.

hematuria the presence of blood in the urine

heme iron-containing, non-protein portion of the hemoglobin molecule that contains iron in the ferrous (+3) state.

hemochromatosis iron overload; elevated levels of iron that can cause tissue damage, especially in the liver.

hemoconcentration the decrease in free water circulating in the blood supply, causing increased levels of proteins, electrolytes, wastes, and nutrients per deciliter of blood. Elevations of several laboratory values are present, and dehydration signs and symptoms may be present.

hemodialysis a type of renal replacement therapy whereby wastes or uremic toxins are filtered from the blood by a semipermeable membrane and removed by the dialysis fluid

hemoglobin the four-pyrrole ring compound in red blood cells that contains iron centers and is responsible for the transport of oxygen.

hemoglobinemia excess free hemoglobin build-up in circulation.

hemoglobinuria excessive free hemoglobin spillage into the urine.

hemolytic anemia an anemia brought on by the rapid, premature destruction of red blood cells in circulation, which may be precipitated by vitamin E deficiency.

hemophilia an inherited disorder of blood clotting, with pronounced bleeding upon tissue injury.

hemorrhage bleeding.

hemorrhagic stroke stroke caused by rupture of a blood vessel (e.g. aneurysm).

hemostasis normal blood flow and blood clotting.

Henderson - Hasselback equation $pH = pK_a + [H_2CO_3]/[HCO_3^-]$

hepatic encephalopathy a syndrome characterized by central nervous system dysfunction in association with liver failure.

hepatitis inflammation of the liver and liver disease involving degenerative or necrotic alterations of hepatocytes.

hepatomegaly enlargement of the liver

herbal remedies technically, preparations made from leafy plants without woody stems; "herbal" is frequently used interchangeably with "botanical."

heterozygous having two different alleles or variants of a given gene.

hiatal hernia protrusion of part of the stomach through the diaphragm into the space normally occupied by the esophagus, heart, and lungs.

high-fiber diet a diet high in fiber (6–10 g above the usual recommendation of 20–35 g/day).

histamine a vasoactive amine that contributes to inflammation and IgE-mediated allergic reaction by causing the dilation of local blood vessels and smooth muscle contraction. Histamine release produces some of the symptoms of immediate hypersensitivity reactions.

histological pertaining to the minute structure, composition, and function of the tissues.

histone a protein around which DNA is wrapped.

hives an itchy skin condition with raised red lumps, often due to an allergic reaction; also called urticaria.

HLA antigens antigens specific to the individual that cause rejection reaction in a host receiving transplantation or foreign cells.

holistic medicine a healthcare approach that considers the physical, environmental, mental, emotional, social, and spiritual aspects of human experience and aims to achieve functioning, balance, and well-being (American Holistic Medical Association, 2004).

holotranscobalamin the fraction of metabolically active B_{12} that is composed of cobalamin linked to transcobalamin in circulation.

homeopathic medicine a medical system based on the idea that like cures like, in which minimal (usually diluted) doses of substances known to cause certain symptoms are used to treat those symptoms.

homozygous having two identical alleles or variants of a given gene.

hormones blood-borne chemical messengers that act on target cells located a long distance from the endocrine gland that produces them.

hot-cold a classification system that evolved from humoral medicine. Unlike yin/yang, it is applied principally to diet, and sometimes to illness, but not to all of nature. To maintain health, hot foods must be balanced with cold foods, and hot or cold illnesses are treated with ample foods of the opposite category. It is sometimes combined with other classifications, such as "cool," "heavy or light," or "acidic or nonacidic."

human biological factors conditions that determine a person's nutrient requirements. One's age, gender, and stage of growth and development are used to estimate kcalorie and nutrient needs. Illnesses that alter organ function or metabolism influence not only the amount of nutrients required but also the form of nutrients that the body needs and can tolerate.

human immunodeficiency virus (HIV) a retrovirus that targets many cells in the body, particularly CD4 T-helper cells.

humoral immunity immunity due to soluble factors such as antibodies circulating in the body's fluids, mainly serum and lymph. "Humors" is an old term for body fluids.

hyaline a histological term used to describe tissue injury that has a glassy, pink appearance.

hydrogenation chemical process of introducing hydrogen to monounsaturated and polyunsaturated fats; used to harden fats at room temperature and improve stability in food production

hydrophilic water loving, or attracting water.

hydrotherapy the use of immersion in water, steam baths and saunas, colonic irrigation, and hot or cold compresses to treat health conditions. It can include therapeutic additives such as botanicals, minerals, or essential oils.

hydroxyapatite a crystallized calcium phosphate salt that gives bones their stiffness.

hypercalcemia high serum calcium (>10.2 mg/dL).

hypercalciuria an excess of calcium in the urine

hypercapnia the term used to describe an excess of the blood gas carbon dioxide (CO_2).

hyperemia increased blood flow to a body tissue.

hyperinflation of the lungs results from loss of elasticity of the alveoli, causing air to be trapped; often seen in emphysema.

hyperkalemia high serum potassium.

hypermagnesemia high serum magnesium levels (>1.1 mmol/L).

hypernatremia abnormally high levels of serum sodium.

hyperosmolar having a higher osmolality than body fluids (>300 mOsm/kg).

hyperosmolar hyperglycemic nonketotic syndrome complication of type 2 diabetes mellitus usually developing after a period of hyperglycemia combined with inadequate fluid intake.

hyperoxaluria an excess of oxalate in the urine

hyperphosphatemia high serum phosphorus (>1.45 mmol/L).

hyperplasia increased number of cells.

hyperpnea rapid breathing.

hyperreflexia over-response or exaggeration of response to a neural stimulus (e.g. twitching).

hypersensitivity an inappropriate and harmful immune reaction to a harmless, nonpathogenic substance. Also called allergy.

hypertension condition of chronically elevated blood pressure

hyperthyroidism excess thyroid secretion.

hypertrophic cardiomyopathy a genetic disorder causing abnormal thickening of the left ventricular wall

hypertrophy increase in cell size.

hyperuricosuria a disorder of uric acid metabolism

hypervolemia increased blood volume.

hypocalcemia low serum calcium (<8.7 mg/dL).

hypochondriasis a somatoform disorder (i.e., a physical ailment stemming from a psychological problem) characterized by an unfounded belief that one is suffering from a serious illness.

hypochromic abnormally pale in color upon visual inspection under a microscope.

hypoglycemia a low serum glucose. Generally considered to be <70 mg/dL.

hypokalemia low serum potassium.

hypomagnesemia low serum magnesium levels (<1.1 mmol/L).

hyponatremia abnormally low concentrations of sodium ions in the circulating blood.

hypoperfusion reduced blood. .flow.

hypophosphatemia low serum phosphorus (<1.45 mmol/L).

hyporesponsiveness hormone resistance.

hypotension low blood pressure.

hypothyroidism deficient thyroid secretion.

hypovolemia decreased blood volume.

hypoxemia condition in which there is an inadequate supply of oxygen in the blood.

hypoxic injury cellular injury as a result of oxygen deprivation.

iatrogenic an adverse condition in a patient resulting from treatment, usually by a physician. Iatrogenic literally means, "brought forth by a physician."

ideal goals science-based values intended to control or improve specific health conditions.

IgA (immunoglobulin A) the predominant immunoglobulin in secretions.

IgA nephropathy a form of glomerular disease that results when immunoglobulin A (IgA) forms deposits in the glomeruli, where it creates inflammation

IgD (immunoglobulin D) an immunoglobulin present in the surfaces of B cells.

IgE the immunoglobulin class that is the predominant mediator of immediate hypersensitivity reactions (allergies).

IgG the predominant immunoglobulin class produced during secondary immune responses. The most prevalent immunoglobulin in the blood.

IgM the predominant immunoglobulin class expressed by virgin B lymphocytes and secreted during primary immune responses.

IL-2 interleukin-2; a lymphokine required by activated T cells for growth.

ileostomy a procedure in which the colon and rectum are surgically removed, and the end of the ileum is attached to the stoma.

ileus decreased or absent motility of the bowel and forward movement of bowel contents.

immediate hypersensitivity a hypersensitivity reaction that appears within minutes after the exposure to the allergen.

immune complex a cluster of antibodies bound to antigens.

immunodeficiency decrease in or lack of an immune response due to absence or defect of one or more components of the immune system.

immunogen an antigen capable of inducing an immune response because it is foreign to the host.

implantable port intravenous access that is completely under the skin, is placed in the vein on the upper chest wall, and exits the body near the xyphoid process, axilla, or abdominal wall.

Indian Health Service an agency within the Department of Health and Human Services that operates a comprehensive health service delivery system for American Indians and Alaska Natives.

indirect calorimetry an approach to determine energy expenditure by measuring a subject's oxygen consumption, carbon dioxide production, and minute ventilation (the amount of air a subject breathes in one minute).

infarct cellular necrosis as a result of lack of oxygen

inflammatory bowel disease (IBD) an autoimmune, chronic inflammatory condition of the gastrointestinal tract. IBD is actually the term designating a syndrome consisting of two diagnoses: **ulcerative colitis** and **Crohn's disease**.

limbic lobe component of the brain involved in control of emotions.

lipoatrophy an immune response related to source and purity of insulin resulting in thinning of subcutaneous fat at the injection site, which causes concaving or pitting of fatty tissue.

lipodystrophy syndrome loss or absence of fat, or the abnormal distribution of fat in the body. In HIV infection, these changes are likely hormonally mediated. Subcutaneous fat loss is most apparent in peripheral limbs and facial areas. Fat deposits are most commonly central, located in the dorsocervical area, breast area, and abdominal region.

lipogenesis the synthesis of triglyceride from carbohydrates and proteins.

lipohypertrophy thickening of subcutaneous fat at an insulin injection site.

local anesthesia loss of sensation only in the area where an anesthetic drug is placed.

lock and key model description of communication between two cells; action between two substances within the body; in order for action to occur the two cells must fit together as a lock and key might.

lower esophageal sphincter (LES) the junction between the esophagus and the stomach.

low-residue diet a diet low in fiber and other food constituents that may contribute to bulk in the large intestine.

lymph extracellular fluid containing WBC (mostly lymphocytes) and antibodies that bathes tissues.

lymph nodes small organs of the immune system where mature B and T lymphocytes respond to an antigen. They are distributed widely throughout the body and linked by lymphatic vessels that bring in antigens from surrounding tissue.

lymphatic system a system of vessels through which lymph travels, consisting of lymphatic vessels and lymph nodes at the intersection of vessels.

lymphocyte a small mononuclear cell with a thin rim of cytoplasm that has antigen-specific receptors.

lymphokine a soluble molecule used for communication between lymphocytes and other cells.

lysosomes cytoplasmic granules involved in the digestion of phagocytosed material that contain hydrolytic enzymes.

macrocytic refers to abnormally large cell size.

macrophage a mononuclear, actively phagocytic cell arising from monocytic stem cells in the bone marrow.

macrophage a large phagocytic antigen-presenting cell derived from the blood monocyte and found in tissues.

macrosomia refers to the condition of abnormally large infants whose mothers have diabetes.

major histocompatibility complex (MHC) a cluster of genes encoding polymorphic cell-surface molecules (MHC class I and class II) that help the organism identify pathogens as foreign. They are important in antigen presentation to T cells, play a role in transplantation rejection, and influence the susceptibility to certain autoimmune diseases. MHC antigens are also called HLA antigens.

MALT (mucosa-associated lymphatic tissue) lymphoid tissue found in the surface mucosa of the respiratory, gastrointestinal, and genitourinary tracts.

managed-care system a health care approach in which insurers try to limit the use of health services, reduce costs, or both. These health plans are subject to utilization review (UR), which aims to prevent unnecessary treatment by requiring enrollees to obtain approval for nonemergency hospital care, denying payment for wasteful treatment, and monitoring severely ill patients to ensure that they get cost-effective care.

mast cell a tissue cell found primarily in mucosal and connective tissue that is similar to the basophil (which is found in blood).

mechanical ventilation artificial ventilation using a ventilator or respirator; performed with a piece of equipment designed to intermittently or continuously assist or control pulmonary ventilation.

Medicaid entitlement program that pays for medical assistance for certain individuals and families with low incomes and resources.

medical doctor a health professional who has earned a post-bachelor degree of doctor of medicine (MD) or doctor of osteopathy (DO) and who has passed a licensing examination.

medical foods foods administered under the supervision of a physician and intended for the specific dietary management of a disease for which distinctive nutritional requirements are established.

medical nutrition therapy (MNT) nutritional diagnostic, therapy, and counseling services for the purpose of disease management that are furnished by a registered dietitian or nutrition professional pursuant to a referral by a physician.

medical pluralism the consecutive or concurrent use of multiple health care systems and therapies by clients.

medium-chain triglycerides (MCTs) triglycerides composed of fatty acids with 8 carbons (octanoic and decanoic fatty acids).

megaloblastic refers to an immature, large red blood cell that is oval in shape and abnormal.

meiosis cell division to produce gametes (sperm and ova) that results in the production of cells with half the complement of chromosomes.

membrane attack complex the final product of the complement cascade that forms a pore on the surface of the target cell, which results in lysis of the cell.

membranous nephropathy disease diagnosed when a kidney biopsy reveals unusual deposits of immunoglobulin G and complement C3, substances created by the body's immune system; 75% of cases are idiopathic

memory cells lymphocytes produced on the first encounter with an antigen that produce a rapid, more vigorous response upon subsequent exposures, which often prevents reinfection.

menhaden oil hydrogenated and partially hydrogenated oils from the menhaden fish – a small plankton-feeding fish.

metabolic acidosis condition resulting from either loss of bicarbonate or retention of nonvolatile acid.

metabolic alkalosis condition resulting from either retention of bicarbonate or loss of nonvolatile acid.

metabolic water water that is produced through nutrient metabolism.

metaplasia replacement of one cell type with another.

metastasis spread of cancer from the primary site to nearby or distant areas through the blood or lymph.

methylation the addition of methyl (-CH_3) groups. DNA methylation patterns can be inherited and impact patterns of gene expression.

microarray technology used to measure expression of thousands of genes simultaneously.

microcytic refers to abnormally small cell size

minor histocompatibility antigens cell surface processed peptides not encoded by the MHC that can contribute to graft rejection.

minute ventilation the volume of air per unit time moved into or out of the lungs; measured by collecting expired volume for a fixed time.

mitosis cell division that produces two cells that are genetically identical to the progenitor cell.

monoamine oxidase (MAO) inhibitors group of medications that block the enzyme system that inactivates some neurotransmitters.

monoclonal antibody an antibody produced by an immortal B cell line that reacts with a single antigenic determinate.

monocyte a large mononuclear phagocytic white blood cell that develops into a macrophage when it enters tissue.

monogenic arising from a single gene.

monounsaturated fats sources of fat that have a predominant amount of fatty acids with one carbon-carbon double bond within their chemical structures

morbidity the state of being diseased.

mortality the incidence of death in a population.

mucositis inflammation of a mucous membrane (e.g., mouth sores).

multiple sclerosis a disorder characterized by demyelination of cells within the CNS, inflammation, and development of scar tissue, causing numbness, tingling, uncoordination, weakness, and varying degrees of blindness.

myasthenia gravis a progressive neuromuscular disorder that affects the skeletal muscles and causes muscle weakness, particularly of the face, eyes, arms and legs.

myelin the covering or insulation of the axon that assures proper communication between neurons.

myeloma a tumor composed of cells derived from hemopoietic tissues of the bone marrow; a plasma cell tumor.

myocardial cells cells found in the myocardium; cells of the heart

myocardial infarction (MI) necrosis of the myocardial cells as a result of oxygen deprivation.

myxedematous non-pitting edema.

Na^+/K^+ pump the enzyme-based mechanism that moves potassium ions into and sodium ions out of a cell by active transport.

nadir the lowest point, usually in reference to the white blood cell count.

nasogastric feeding tube a tube that is inserted nasally (through the nose) into the stomach.

nasointestinal feeding tube a tube that is inserted nasally (through the nose) past the stomach into the intestine.

natural killer cells (NK cells) large granular lymphocyte cells that attack tumors and virally infected cells but do not exhibit antigenic specificity. Also called killer cells (K cells) and null cells.

naturopathic medicine a medical system based on the concept of vitalism, which defines life as an autonomous force that cannot be explained by physical or chemical processes; primary treatments are detoxification and nutritional therapy.

necrosis general term referring to cell death.

necrotizing enterocolitis (NEC) a condition that occurs primarily in premature infants or sick newborns, in which intestinal tissue dies. The cause for this disorder is unknown, but it is thought to be due to decreased blood flow to the bowel, which keeps it from producing the normal protective mucus. If an infant is suspected of having necrotizing enterocolitis, feedings are stopped to allow the bowel to rest.

necrotizing fasciitis inflammation of the connective tissue leading to necrosis of the tissue; may be caused by infection, injury, or an autoimmune reaction.

negative feedback a regulatory mechanism in which a change in a controlled variable triggers a response that opposes the change, thus maintaining a relatively steady state for the regulated factor.

negative nitrogen balance net loss of protein in the body.

negative selection the process in which B and T cells that react to self molecules are deleted or functionally inactivated during their development.

neoadjuvant chemotherapy refers to chemotherapy used prior to primary treatment, which is typically surgery.

neoplasm literally means "new growth"; an abnormal mass of tissue, the growth of which exceeds and is uncoordinated with that of normal tissue.

nephritic syndrome a condition of inflammation of the glomerulus, resulting in hematuria, proteinuria, and oliguria

nephrolithiasis kidney stones, a common disorder in the United States

nephron basic functioning unit of the normal kidney; each nephron has two main parts: the glomerulus and the tubule

nephrotic syndrome a clinical condition consisting of losses of protein in the urine exceeding 3.5 g/day, hyperlipidemia, and low albumin levels ($<$3.5 g/dL) with edema

neurofibrillary tangles collections of twisted *tau* found in the cell bodies of neurons in AD.

neuromodulator substance released that will increase or decrease the activity of specific neurotransmitters.

neuron a nerve cell in the brain.

neuropeptide protein messengers within the brain and nervous system that assist in communication between neurons.

neurotransmitter a chemical messenger that communicates between neurons.

neutropenia low white blood cell count.

neutrophil the most numerous polymorphonuclear leukocyte, with granules that stain with acid and basic dyes. It is phagocytic and enters tissues early in inflammation.

nonexercise activity thermogenesis (**NEAT**) the energy expended through physical activity involved in performing the ordinary activities of daily life. It excludes energy expended in activities to obtain physical exercise or involving sports-like activity.

nonsense codon (**stop codon**) the codon in mRNA that signals completion of translation.

non-specific immunity all aspects of immunity not directly mediated by antigen-specific lymphocytes.

norepinephrine a neurotransmitter released form sympathetic postganglionic fibers; formerly referred to as noradrenaline.

normalized protein equivalent of nitrogen appearance (**nPNA**) an assessment of protein catabolic rate

NPO *Nil per Os*, which is Latin meaning "nothing per mouth."

nucleotide the building block of a nucleic acid, consisting of a ribose sugar, a phosphate group, and a nitrogenous base.

nurse a health care worker who has earned at least an associate's degree in nursing, has been licensed by the state, and assists patients in activities related to maintaining or recovering health.

nutrigenomics the interaction between nutrients and other food-derived bioactive substances with an individual's genome.

nutrition assessment analysis of an individual's nutrition status incorporating both subjective and objective data, including information on diet, psychosocial parameters, education, and motivation.

Nutrition Care Process (NCP) a systematic problem solving method developed by the ADA that dietetic professionals use to think critically, make decisions addressing nutrition-related problems, and provide safe, effective, high-quality nutrition care (Lacey and Pritchett 2003).

nutrition diagnosis the identification and descriptive labeling of an actual occurrence of a nutrition problem that dietetics professionals are responsible for treating independently (Lacey and Pritchett 2003).

nutrition insecurity the provision of an environment that encourages and motivates society to make food choices consistent with short- and long-term good health.

nutrition intervention a specific set of activities and associated materials used to address a (nutrition-related) problem (Lacey and Pritchett 2003)

nutrition monitoring and evaluation an active commitment to measuring and recording the appropriate outcome indicators relevant to a nutrition diagnosis in order to determine the degree to which progress is being made and whether or not the client's goals are being met (Lacey and Pritchett 2003).

nutrition screening the process of gathering data known to correlate with nutritional risk in order to identify individuals who are at risk.

obesigenic promoting or encouraging the development of obesity. An obesigenic environment is one that promotes weight gain and the development of obesity by encouraging consumption of energy and discouraging physical activity.

obesity an excess of body fat or adipose tissue. Obesity can be defined as a proportion of body weight that is adipose tissue (percent body fat) that is greater than some standard. Because it is often impractical in the clinical setting to measure in percent of body fat using body composition analysis, obesity is often defined as a BMI $\geq$30.0 kg/m^2. The term obesity comes from the Latin *obesus,* meaning, "one who has become plump through eating."

obligatory urine the amount of fluid necessary for the body to excrete waste products and solutes (approximately 500 mL).

obstruction blockage.

occipital lobe portion of the cerebral cortex controlling vision.

occupational therapist a health professional who has obtained a bachelor's degree and passed a national registration exam, who helps individuals with mentally, physically, developmentally, or emotionally disabling conditions improve their ability to perform tasks in their daily living and working environments.

odynophagia painful swallowing.

oligomenorrhea abnormally infrequent menstrual cycles.

oligosaccharide a carbohydrate that through hydrolysis yields a small number of monosaccharides.

oliguria urine output less than 400 mL, which is the minimum amount of normal urine that can carry away the daily load of metabolic waste products

omeprazole a type of proton pump inhibitor used to treat GERD and peptic ulcer disease.

oncotic pressure pressure exerted by large protein molecules in blood plasma, which usually do not cross the capillaries. These molecules decrease the fluid that can leak out of the capillaries into the tissue.

ophthalmic refers to placement of a drug into the eye.

oral glucose tolerance test (OGTT) timed glucose challenge to examine efficiency of the body in metabolism of glucose

oral preparatory phase tongue, teeth, and mandible involved in chewing of food and preparation of bolus; food is mixed with saliva, pressed against the hard palate and formed into a bolus.

oral transit phase of swallowing tongue moves bolus to back of throat.

orexigenic appetite stimulating.

orogastric feeding tube a tube that is inserted orally (through the mouth) into the stomach.

orthopnea shortness of breath associated with lying in the supine position

osmolality number of water-attracting particles per weight of water in kilograms (expressed as mOsm/kg).

osmolarity number of millimoles of liquid or solid in a liter of solution.

osmotic pressure the pressure that must be applied to a solution to prevent the passage into it of solvent when solution and pure solvent are separated by a membrane permeable only to the solvent.

osseous tissue the group of cells and cell products that collectively form bone; bone tissue.

osteitis fibrosa cystica a form of high turnover bone disease caused by overproduction of parathyroid hormone (PTH), which increases the rate of bone turnover

osteoarthritis a condition involving progressive loss of articular cartilage and inflammation of the tissues composing the joint, resulting in joint pain, stiffness, and limited joint movement.

osteoblasts cells that synthesize, deposit, and then orient the fibrous proteins of the organic matter of the bone matrix.

osteoclasts bone-removing cells that dissolve the mineral component of the bone matrix, playing a major role in bone resorption.

osteocytes mature osteoblasts surrounded and entrapped by the matrix they have synthesized and then calcified.

osteogenic cells stem cells capable of developing into osteoblasts.

osteomalacia a condition in which the organic matrix of the bones of adults is inadequately mineralized, resulting in muscular weakness, bone pain, and, in advanced cases, deformities of the ribs, pelvis, and bones of the legs.

osteopathic medicine a medical system similar to biomedicine but distinguished by a focus on musculoskeletal tension and restriction as causes for conditions. Uses osteopathic manipulative treatment.

osteopenia a term used to describe a bone mineral density that is low but not low enough to meet the diagnostic criterion for osteoporosis.

osteopetrosis death of bone cells through excessive calcification.

osteophyte a bony outgrowth near the joint affected by osteoarthritis. also referred to as a bone spur.

osteoporosis a disease resulting from a decreased amount of bone mineral and organic matrix which weakens bones, making them more susceptible to fracture.

ostomy an artificial opening created by surgical procedure.

otic refers to placement of a drug into the ear.

outcome the measurable consequence of disease.

outcome measures data used to evaluate the success of interventions; includes direct nutrition outcomes, clinical and health status outcomes, patient/client-centered outcomes, and health care utilization and cost outcomes.

outcomes management system a system that evaluates the effectiveness and efficiency of the entire NCP: assessment, diagnosis, implementation, cost, and other factors. It links care processes and resource utilization with outcomes (Lacey and Pritchett 2003).

outcomes research evaluation of care that focuses on the status of participants after receiving care.

overweight an excess of body weight in in relationship to for height. For adults, overweight is generally defined as a body mass index or BMI of 25.0 kg/m^2 to 29.9 kg/m^2. For children and adolescents, overweight can be defined as a BMI-for-age-and-sex at or above the 95th percentile using the CDC growth charts.

oxidative stress the imbalance of pro-oxidant production and the body's antioxidant supplies that yields cell damage.

oxytocin a hormone that stimulates contraction of the uterus during childbirth, and promotes ejection of milk from mammary glands during breast feeding.

palliative a non-curative treatment which reduces symptoms such as pain.

pancreatic function tests tests to measure pancreatic function, including serum amylase or lipase, a test for the amount of fat in the stool, and an x-ray of the anatomical features of the pancreas and common bile ductMN 23.38 plasma prothrombin concentrations – a measure of blood clotting ability

pancreatitis inflammation of the pancreas.

pancytopenia a reduction in the numbers of all the blood elements—white, red, other cells, and proteins.

pannus an abnormal destructive tissue that develops on the synovial membrane of patients with advanced rheumatoid arthritis. Inflammatory cells in pannus secrete enzymes that are destructive to articular cartilage and subchondral bone.

paracentesis a procedure in which fluid is withdrawn from a body cavity via a trocar and cannula, needle, or other hollow instrument.

paracrine a name for a neurotransmitter that is released from a cell that is close to the target cell.

paraplegia paralysis involving the lower body below the umbilicus.

parasympathetic branch division of the autonomic nervous system that is involved in control of gastrointestinal, cardiac and respiratory systems.

parenchymal referring to the essential elements of an organ.

parenteral refers to injection into the body's circulatory system through a blood vessel.

parenteral nutrition (PN) administration of nutrition directly into the circulatory system (also known as total parenteral nutrition (TPN), central venous nutrition (CVN), or intravenous hyperalimentation [IVH]).

paresthesias symptoms of tingling in fingers and toes; often consistent with electrolyte imbalances.

parietal cell one of the gastric gland cells that lies on the basement membrane covered by chief cells, and secretes hydrochloric acid.

parietal lobe portion of the cerebral cortex responsible for the sensations of pain, touch, taste, temperature, and pressure, and related to mathematical and logical thinking.

Parkinson's disease a neuromuscular, neurodegenerative disease caused by the loss of dopamine-producing cells in the substantia nigra portion of the brain, resulting in resting tremor, rigidity, slowed movement, stooped posture and postural instability, mask-like facial features, and a shuffling gait.

passive immunity immunity due to the transfer of antibodies or activated T cells produced by another individual.

pathogenesis the clinical course of disease.

pathophysiology the study of disease.

"pelvic floor" refers to the pelvic diaphragm, the sphincter mechanism of the lower urinary tract, the upper and lower vaginal supports, and the internal and external anal sphincters. It is a network of muscles, ligaments and other tissues that hold up the pelvic organs (vagina, rectum, uterus and bladder). When this system weakens, known as pelvic floor dysfunction, the organs may shift, bulge and push outward or against each other. As a result, women may suffer from urinary or fecal incontinence or obstruction, vaginal prolapse or pain, sexual dysfunction, and other problems. (University Southern California, University Hospital, (http://www.uscuh.com/CWSContent/uscuh/ourServices/medicalServices/USCUniversityHospitalPelvicFloorDisordersProgram/Patient+Education.htm) 2003)

pelvic floor dysfunction a weakening of the pelvic floor system (consisting of the pelvic diaphragm, the sphincter mechanism of the lower urinary tract, the upper and lower vaginal supports, and the internal and external anal sphincters) that allows internal organs to shift, bulge, and push outward and against each other.

peptic ulcer disease ulceration or perforation in the lining of the stomach, duodenum, or esophagus.

percent weight for height percentage used to evaluate a child's growth pattern relative to population standards.

percussion a technique used during physical examination in which the hands are used to strike the body's surface, and the sounds that are transmitted from the underlying tissues and organs are evaluated.

percutaneous endoscopic gastrostomy (PEG) a procedure used by a physician to insert a feeding tube through the skin and into the stomach using an endoscope.

percutaneous nephrolithotomy a surgical procedure where a surgeon makes an incision in the back and creates a tunnel to the kidney to remove a kidney stone

perforation a break in the integrity of the tissue.

perforin molecule produced by cytotoxic T-cells and NK cells that forms a pore in the membrane of the target cell to allow chemicals from the T-cell or NK cell to enter the target cell and induce apoptosis.

periarticular muscles those muscles located near a joint.

periodontal disease a bacterial infection that destroys the attachment fibers and supporting bone that hold the teeth in the mouth.

periorificial acrodermatitis disease of the skin surrounding the mouth area.

peripheral arterial disease (PAD) atherosclerotic heart disease of all vessels except specific coronary vessels; term used interchangeably with peripheral vascular disease

peripheral nervous system all components of the nervous system except for the brain and spinal cord (central nervous system).

peripheral parenteral nutrition (PPN) administration of nutrition into a vein in the arm or back of the hand (also known as peripheral venous nutrition [PVN]).

peripheral vascular disease (PVD) atherosclerotic heart disease of all vessels except specific coronary vessels

peripherally inserted central catheter (PICC) intravenous access inserted into the arm and threaded into the subclavian vein to the vena cava.

peritoneal dialysis a type of renal replacement therapy during which the peritoneal cavity serves as the reservoir for the dialysate and the peritoneum acts as the semipermeable membrane across which excess body fluid and solutes are removed

peritonitis an inflammation of the peritoneum membrane

pernicious anemia the anemia associated with B_{12} deficiency that is slow, aggressive and potentially life threatening. It is specific to gastrointestinal dysfunction, namely, in gastric enterocytic atrophy, with diminished availability of intrinsic factor, HCL and enzymes. Neuropathy (especially peripheral) results from prolonged deficiency and the ability for the nervous system to regenerate and regain feeling and function in the affected areas is low over time.

PES Problem, Etiology, and Signs and Symptoms. This is the format used in the NCP to write a nutrition diagnosis. It clarifies a specific nutrition problem and logically links the nutrition diagnosis to nutrition intervention and to monitoring and evaluation.

Peyer's patches distinct lymphoid nodules in the intestine that are part of the gut-associated lymphoid tissue (GALT).

phagocytosis the engulfment of a particle or a microorganism by leukocytes such as macrophages and neutrophils, normally followed by destruction of the particle.

phago-lysosome intracellular vacuole where the killing and digestion of phagocytosed material occurs. Produced by the fusion of a phagosome and a lysosome in a phagocytic cell.

phagosome the cytoplasmic vesicle that encloses an ingested organism during phagocytosis.

pharmacogenomics the interaction between drugs and an individual's genome that can impact drug efficacy and toxicity.

pharmacology study of drugs, their properties and their effects.

pharmacotherapy use of drugs for treatment of disease and health maintenance.

pharmakinetics study of drug absorption, distribution, metabolism, and excretion.

pharyngeal phase of swallowing the involuntary swallowing reflux begins and the bolus is carried through the pharynx to the top of the esophagus. The entrance to the trachea (larynx) closes, and the soft palate lifts and closes off entrance to the nose.

phase angle calculates a mathematical relationship between resistance and reactance; for use with bioelectrical impedance to calculate body composition. Higher values for phase angle appear to be consistent with greater body muscle mass and lower risk for morbidity and mortality. Values range from 3–12.

phenotype the expressed or physical properties of an organism.

phlebotomy blood removal through a venous puncture; blood draw.

physical activity-related energy expenditure energy expended in voluntary body movement resulting from the daily activities of life, physical exercise, sports, and play, and nonvoluntary behaviors such as spontaneous muscle contractions, maintenance of posture, and fidgeting. It is the most variable component of 24-hour energy expenditure, depending on how physically active a person is.

pica eating of abnormal items, or non-nutritive substances, such as laundry starch, clay, ice, dirt, paint chips, etc.

pK the constant degree of dissociation (the ability of an acid to release its hydrogen ions) for a given solution. This is a constant amount for any given solution.

plaque the noncalcified accumulation of oral microorganisms and their by-products which adhere to the teeth.

plasma the portion of the blood in which blood constituents are dissolved or suspended. It contains water, proteins, electrolytes, gases, non-pertinacious compounds, wastes, and nutrients.

plasma cells large antibody-producing cells that develop from activated B cells. Also call AFC or antigen forming cells.

plasmapheresis treatment that removes blood from the body, separates out certain cells from the plasma, and then returns the blood back to the body.

pleural effusion accumulation of fluid between the two outer membranes surrounding the lungs.

PLHA people living with HIV and AIDS; other acronyms include PLWHA (people living with HIV/AIDS) and PWA (people with AIDS).

pneumonia inflammation of the lungs, usually caused by bacteria, viruses, or fungi.

polydipsia excessive thirst.

polygenic arising from multiple genes interacting with each other.

polyhydramnios excessive accumulation of amniotic fluid.

polymorphisms DNA sequences of specific genes that vary among individuals.

polymorphonuclear leukocytes (PMN) leukocytes with a multilobed nucleus and cytoplasmic granules that take up acid and basic dyes. Also known as granulocytes, PMNs, and polys.

polyunsaturated fats sources of fat that have a predominant amount of fatty acids that contain more than one double bond in their chemical structures

polyuria frequent urination.

porphyria a cluster of blood-related disorders characterized by abnormal porphyrin synthesis or metabolism. These disorders are hereditary and vary greatly depending upon which enzyme in the cascade of reactions is affected.

portal hypertension abnormally increased pressure in the portal venous system; frequently seen in cirrhosis of the liver and in other conditions that cause obstruction of the portal vein.

positive nitrogen balance net accumulation of protein in the body.

positive selection the rescue from apoptosis of T cells in the thymus that can recognize self MHC molecules.

post-menstrual age age of pregnancy starting from the date of the beginning of the last menstrual period (LMP).

posttranscriptional processing the processing of newly transcribed RNA to excise introns, thus creating the final mRNA product prior to translation of mRNA into a protein.

posttranslational modification modification of a newly synthesized protein to its active form through changes such as phosphorylation or cleavage of specific sections.

prebiotics foods or products containing nondigestible oligosaccharides and inulin, which are thought to stimulate the growth and activity of beneficial bacteria in the gut.

pre-diabetes mellitus blood glucose levels that are higher than normal but not yet high enough to be diagnosed as diabetes.

pre-eclampsia development of hypertension, with symptoms of proteinuria and edema, during pregnancy.

preferred provider organization (PPO) a type of insurance in which the managed care company pays a higher percentage of the costs when a preferred (in-plan) provider is used. The participating providers have agreed to provide their services at negotiated discount fees.

pressor agents substances that cause blood pressure to increase.

primary cancer the location or organ/cells from which the cancer originated.

primary HIV infection the time of the initial seroconversion to HIV infection; usually involves a spike in the level of the virus and sometimes is accompanied by a flu-like syndrome.

primary immune response the immune response when the naive lymphocyte first encounters its antigen.

privileged sites non-vascularized locations in the body where foreign grafts are not rejected.

probiotics foods or products containing live bacteria in quantities known to beneficially alter the microflora of the gut.

prognosis expected outcome; expected response to treatment.

progression the third phase in cancer cell development; the orderly transformation of a preneoplastic lesion to a tumor and, ultimately, invasive cancer.

prokinetic a pharmacologic agent that promotes gastric emptying.

prokinetic agents medications that cause the lower esophageal sphincter to close tightly, preventing gastric reflux. They also act to increase transit time (peristalsis) of stomach contents.

Promoter region Regulatory sequence in a gene to which molecules, such as fatty acids, can bind in order to induce expression of that specific gene. Molecules can also bind to the promoter region to suppress transcription of a specific gene.

promotion the second phase in cancer cell development; process induced in a normal cell that has been exposed to a carcinogen to transform into a cancer cell (promoters are not necessarily carcinogenic).

prophylaxis preventative administration of a compound to avoid consequences of a disease state.

prospective payment system (PPS) system that pays hospitals a fixed sum per case according to a schedule of diagnosis-related groups.

prospectively refers to collecting data as it occurs or happens.

protease inhibitor a medication that prevents protein replication; a common class of drug that is used to prevent human immunodeficiency virus replication.

protein turnover rate a combination of the rates of catabolism and anabolism of protein stores in the body.

protein-losing enteropathy increased fecal loss of serum protein, especially albumin, causing hypoproteinemia.

proteinuria the presence of too much protein in the urine

proton pump inhibitors drugs that reduce acid secretion in the stomach.

protoporphyrin the derivative of hemoglobin containing four pyrrole rings without the iron centers.

pulmonary consolidation changes in tissue structure of the lungs; often visualized as opaque components on a chest x-ray.

pyelonephritis inflammation of both the parenchyma of a kidney and the lining of its renal pelvis, especially due to bacterial infection

pyloroplasty enlarging the pyloric sphincter.

pyruvate complex disorders dysfunction in the metabolism of pyruvate, the end product of glycolysis, via either the Krebs cycle or gluconeogenesis, resulting in the production of lactic acid.

qi in Traditional Chinese Medicine, the fundamental essence or life force.

quadriplegia paralysis involving all arms and legs; also known as tetraplegia.

Quetelet's index see *body mass index*.

rales abnormal respiratory sounds when air flows through liquid present in the airways.

refeeding syndrome metabolic alterations that may occur during nutritional repletion of starved patients.

refractory celiac disease initial or subsequent failure of a strict gluten-free diet to restore normal intestinal architecture and function in patients who have celiac-like enteropathy.

regulatory T-cell a T lymphocyte that turns off specific immune responses.

renal osteodystrophy a general term that refers to bone disease related to CKD, caused by over- or underproduction of PTH or by exposure to aluminum

renal threshold a concentration level of glucose in the blood above which the kidneys pass it through into the urine

respiratory acidosis condition resulting from excess acid in the blood secondary to carbon dioxide retention.

respiratory alkalosis condition resulting from excess base in the blood secondary to increased carbon dioxide expiration.

resting energy expenditure energy expended by the body at rest to keep vital organ systems functioning, including the heart, kidneys, brain, liver, and lungs. It accounts for approximately 60% to 75% of 24-hour energy expenditure and is roughly 1 kcal/kg body weight/hour.

reticulocytes immature RBC. Normal ranges for circulating erythrocytes exist, and levels reflect the ability of the bone marrow to produce precursor cells in normal amounts.

retinopathies non-inflammatory diseases in the retina of the eye.

retroperitoneal lying behind the peritoneum (lining of the abdominal cavity)

retrospectively refers to collecting data from events that have already happened.

retrovirus a virus that carries RNA rather than DNA; RNA must be transcribed prior to integrating into the host cell DNA to reproduce.

reverse transcriptase inhibitors medications that interrupt the viral replication cycle by inhibiting reverse transcriptase enzymes that allow the viral RNA to be transcribed to DNA before being integrated into the host cell DNA.

rhabdomyolysis an acute condition of skeletal muscle destruction.

rheumatoid arthritis a chronic inflammatory disease in which the synovial membrane of the joint becomes inflamed, resulting in swelling, stiffness, pain, limited range of motion, joint deformity, and disability.

rheumatologist a medical doctor specializing in diseases of the muscles and joints that are classified as rheumatic diseases.

rickets a condition characterized by inadequate mineralization of the organic matrix in the bones of children usually caused by a deficiency of vitamin D and resulting in bowing of the legs and skeletal deformity of the rib cage.

salt-resistant describes an individual whose body presents resistance to change in blood pressure as a result of salt intake

salt-sensitive describes an individual who experiences an increase in blood pressure as a result of salt intake

salvage additional treatment, used in hope of a cure or to prolong life, in a patient with recurrence of a malignancy following initial treatment.

saturated fats sources of fat that have a predominant amount of fatty acids that contain all single bonds within their chemical structures

screening and referral system a supportive system within the Nutrition Care Process and Model that helps identify those persons who would benefit from nutrition care (Lacey and Pritchett 2003).

secondary hyperparathyroidism High levels of PTH in the circulation that stimulate bone turnover, which may be accompanied by hyperplasia of the parathyroid glands

secondary immune response rapid, more vigorous immunologic response by memory lymphocytes after the first encounter with an antigen. Produced upon subsequent exposures to the antigen; often prevents reinfection.

secondary polycythemia condition in which an excessive number of red blood cells are produced; occurs in response to compensation for chronic hypoxemia.

second-set rejection accelerated rejection of an allograft due to previous exposure to some of the antigens on the graft.

seizure episode of spontaneous, uncontrolled electrical activity in the brain.

sense strand the coding strand of DNA that is transcribed into RNA.

sensible losses fluid loss that can be measured (usually refers to fluid lost via urine excretion).

sepsis systemic inflammatory response and immunosuppressive process that prevents an adequate response to infection or trauma; may result in organ dysfunction or hypoperfusion abnormalities.

serum the fraction of blood containing water after the removal of cellular components.

serum amyloid a family of apolipoproteins associated with high-density lipoprotein (HDL) in plasma; considered to be an acute-phase protein released in response to inflammation.

serum osmolality a measure of the concentration of solute molecules in the blood.

serum sickness a Type III hypersensitivity response following the administration of a passive antibody in foreign serum.

severe combined immune deficiency (SCID) disease due to several mechanisms that produce an early block in differentiation pathways of both B and T lymphocytes, resulting in infants who are born lacking all major immune defenses.

short bowel syndrome (SBS) decreased digestion and absorption that result from a large resection of the small intestine.

sickle cell anemia a hereditary disease of genetically altered red blood cells that have a sickle shape, carry abnormally formed hemoglobin, and have abnormal transport capabilities for oxygen. The disease is thought to confer protection against malaria.

sideroblastic anemia a form of anemia characterized by the appearance of sideroblasts, immature ferritin-containing blast marrow cells in circulation.

signs observable phenomena such as heart or respiratory rate.

silver nitrate colloidal silver used as an antibacterial treatment in burns.

silver sulfadiazine cream sulfa medicine used to prevent and treat bacterial or fungal infections.

single nucleotide polymorphisms (SNPs) situations in which one nucleotide is replaced by another in a gene, potentially leading to altered function.

Sjogren's syndrome a chronic systematic inflammatory disorder, etiology unknown, characterized by dryness of mucous membranes.

skin fibroblasts connective tissue cells found in the skin.

social worker a professional with at least a bachelor's degree in social work who provides persons, families, or vulnerable populations with psychosocial support, advises family caregivers, counsels patients, and helps plan for patients' needs after discharge.

soma major body portion of the neuron.

somatic division portion of the peripheral nervous system that carries messages from the body back to the brain.

somatization the physical manifestation of stress.

somatostatin a hormone and neurotransmitter that inhibits release of peptide hormones in several tissues.

sorbitol a sugar alcohol; it is used as a sugar substitute.

spasticity involuntary muscle contraction that results in rigidity.

specific gravity the weight of a solution (e.g. urine) in comparison to an equal amount of distilled water. This is used to measure concentrating ability of the kidney.

specific immune response immunity mediated by antigen-specific lymphocytes.

speech-language pathologist a health professional who has earned a master's degree and passed a national examination, who assesses, diagnoses, treats, and helps to prevent speech, language, cognitive, communication, voice, swallowing, fluency, and other related disorders.

sphincter a circular muscle that prevents movement or passage through the circle when contracted; sphincter muscles are located throughout the GI tract and are crucial control factors for peristalsis.

spleen a lymphoid organ in the abdominal cavity that filters blood.

splenomegaly enlargement of the spleen

stable angina chest pain associated with increased oxygen demand such as occurs with physical exertion

stadiometer a calibrated device used to measure stature.

standardized language a uniform terminology that is used to describe practice.

State Children's Health Insurance Program (SCHIP) federal children's health insurance initiative that allows each state to offer health insurance for children up to age 19 who are not already insured.

statin a type of medication that is used to treat hyperlipidemias.

stearic acid an 18-carbon saturated fatty acid

steatorrhea excess fat in the stool (>6 g/24 hrs).

steatosis accumulation of fat in the interstitial tissue of an organ.

stem cells nondifferentiated, primitive cells that have the ability both to multiply and to differentiate into more specialized cells that display unique functions.

sterilization process that destroys all living organisms.

stoma an opening.

stomatitis inflammation of the membrane in the mouth.

stop codon (nonsense codon) the codon in mRNA that signals completion of translation.

stroke volume the volume of blood that is ejected from the left ventricle with each systolic phase; defined mathematically as LVEDV-LVESV

stylet wire guide within the enteral tube that assists with insertion.

subchondral bone bone located beneath the articular cartilage of a joint.

subcutaneous refers to injection into the body under the skin.

sublingual refers to placement of a drug under the tongue.

subluxation in chiropractic, a misalignment of the vertebrae.

substrates any substance that an enzyme acts on to make a product.

subunit vaccine a vaccine made of a single component of an infectious agent and not the whole organism or toxin.

superantigen an antigen that activates a large number of T cells by reacting with the TCR and MHC outside of the normal antigen binding sites.

Supplemental Security Income (SSI) a federal income supplement program designed to help aged, blind, and disabled people who have little or no income that provides cash to meet basic needs for food, clothing, and shelter.

suppressor T-cell a T lymphocyte that suppresses (turns off) specific immune responses. This may or may not be a separate subclass of T cells.

surfactant substance secreted by the alveolar cells of the lung that serves to maintain the stability of pulmonary tissue by reducing the surface tension of fluids that coat the lung.

surgical gastrostomy an opening into the stomach that requires a surgical procedure.

sweat chloride test a test to measure the amount of chloride in the sweat by stimulating the skin to produce a large amount of sweat that is then absorbed by a special filter paper and analyzed for chloride content.

sympathetic branch portion of the peripheral nervous system that prepares the body for action; controls flight or fight response.

symptoms complaints verbalized by a patient.

synapse space or gap between nerve cells across which neurotransmitters pass.

synbiotics products that contain both prebiotics and probiotics.

syncope temporary loss of consciousness; fainting.

synovial fluid a protein-rich, slippery fluid contained inside a fibrous capsule that lubricates and nourishes the cartilage covering the ends of bones at their joints

synovial membrane a membrane lining the capsule that encloses synovial joints and secretes synovial fluid, which lubricates and nourishes the cartilage at the end of bones.

system factors external factors (health care, education, and food supply systems) that influence the type of services that are available to individuals and how these services are delivered.

systemic lupus erythematosus (SLE) an autoimmune, chronic inflammatory disease that affects the connective tissue; affects skin, joints, kidneys, central nervous system, and mucous membranes and eventually spreads to all tissues, invoking a systemic reaction with pain, fever, sensitivity to light, and skin lesions.

systole contraction phase of the cardiac cycle; during this phase blood is ejected from the ventricles into the aorta and pulmonary artery

systolic blood pressure pressure exerted when ejected from the ventricles (systole phase of the cardiac cycle).

T cell receptor (TCR) a two-chain structure on T cells that binds antigen and is associated on the cell with the signal transduction molecules.

T cells lymphocytes that differentiate in the thymus.

tandem mass spectroscopy the methodology used to detect a large number of organic acid compounds on a filter paper blood spot for diagnosing an inborn error of metabolism.

tau a protein that is a principal component of the paired helical filaments in neurofibrillary tangles; helps to maintain the structure of microtubules in normal nerve cells.

telomere the end section of a human chromosome.

temporal lobe portion of the cerebral cortex responsible for hearing and memory.

Temporary Assistance for Needy Families (TANF) program that provides assistance and work opportunities to needy families by providing funds to develop the state's own welfare programs. It was previously known as the welfare programs: Aid to Families with Dependent Children (AFDC) and the Job Opportunities and Basic Skills Training (JOBS) programs.

temporomandibular joint (TMJ) syndrome a condition of facial pain in the joints of the lower jaw.

terminal a condition or disease in which there is no cure.

tetrahydrobiopterin the cofactor needed to stabilize the enzyme phenylalanine hydroxylase.

Th1 a subset of the T helper cells that secretes cytokines, which trigger cell-mediated immune responses that promote inflammation and antiviral responses.

Th2 helper T cells that predominate in the response to allergens and parasites and that make cytokines that promote antibody responses.

thalassemia a group of related blood disorders involving abnormal globin subunits in the hemoglobin molecule. These are hereditary and are most common in persons of Mediterranean or southeastern Asian descent.

thermic effect of food energy expended by the body to digest, absorb, and metabolize food. It accounts for about 10% of 24-hour energy expenditure.

"third space" fluid shift of fluid from the intravascular space to a nonfunctional space.

thrombocytes platelets. Essential to blood clotting, these pieces of larger immature cells contribute to the formation of a thrombus (clot) by aggregating (coalescing) upon chemical activation after endothelial wall (blood vessel) tissue injury.

thrombocytosis low number of platelets.

thrombopoietin a stimulatory protein in red bone marrow that responds to the need for more platelets post-injury; causes an increase in the production of new platelets and also signals other systems to speed up the maturation and activation of the new platelets; up-regulates the complex mechanisms in hemostasis under conditions of injury and trauma.

thrombus blood clot

thymus a primary lymphoid organ, in the chest, where T lymphocytes differentiate, proliferate, and are positively and negatively selected.

tolerance nonresponsiveness to a particular antigen or group of antigens produced by prior exposure to the antigen under nonimmunizing conditions.

topically refers to placement of a drug on the skin.

total iron-binding capacity (TIBC) the capacity for the binding of iron by blood constituents; a surrogate measure for transferrin, since it binds the majority of iron.

toxic megacolon a very inflated colon with abdominal distention, and sometimes fever, abdominal pain, or shock.

toxoid chemically or physically modified toxin that is no longer harmful but can still stimulate an immune response.

trabecular bone loosely organized bone having a sponge-like appearance and found at the ends of long bones.

tracheostomy a surgical opening placed in the trachea to assist breathing.

Traditional Chinese Medicine an ancient holistic medical system based on the concept that health is maintained by keeping the body's vital forces in balance.

transcobalamin I-III a group of proteins that are responsible for the transfer of vitamin B_{12}.

transcription the manufacture of RNA from DNA.

transcription factors a protein that activates transcription of a gene or genes by interacting with RNA polymerase in a gene promoter region.

transferrin the protein responsible for the transport of iron.

transferrin saturation the saturation of the carrier protein for iron, which is a sensitive indicator of iron status and stages of anemia.

translation the assembly of a polypeptide chain based on the sequence of mRNA.

transplantation grafting an organ (e.g., kidney or heart) or cells (e.g., bone marrow) from one individual to another.

trichophyton pathogenic fungi causing dermatophytosis; attacks the hair, skin, and nails.

tropic hormone a hormone that regulates secretion of another hormone.

t-score the number of standard deviations that the patient's BMD is either above or below the mean BMD for healthy young adults of the same sex and race; measure that compares a patient's bone mineral density (BMD) to a standard, healthy BMD, which is set at the mean BMD of healthy young adults of the same sex and race as the patient.

"tube feeding syndrome" hyperosmolar-non-ketotic dehydration, over a short 2–4 day period.

tubules component of the nephron responsible for reabsorption and secretion; designated as the proximal convoluted tubule, the loop of Henle, and the distal convoluted tubule

tumor necrosis factor (TNF) a cytokine that induces programmed cell death, primarily in tumor cells but for any cell with a receptor. Also involved in immunoregulation.

Tumor Node Metastases (TNM) Staging System a systematic way of describing the size, location, and spread of a tumor; T describes the primary tumor according to its size, N applies to the lymph nodes and whether cancer cells have spread to them, M refers to metastases and whether the cancer has spread to distant sites.

tunneled catheter intravenous access that is placed in the vein on the upper chest wall and exits the body near the xyphoid process, axilla, or abdominal wall.

ulceration nonhealing break in skin or tissue surface

ulcerative colitis (UC) a chronic inflammatory bowel disease (IBD) primarily located in the colon and rectum.

ultrafiltrate referring to the initial filtration of metabolic byproducts from the filtered blood within the tubule

unstable angina chest pain that occurs at rest

upper respiratory infection (URI) a nonspecific term used to describe acute infections involving the nose, sinuses, pharynx, larynx, trachea, and bronchi; often referred to as the common cold.

uremia (uremic syndrome) a general term used to encompass a cluster of symptoms resulting from disordered biochemical processes as chronic kidney disease progresses; early symptoms include fatigue, delayed thinking and pruritis

ureterorenoscopy a nonsurgical procedure where a surgeon uses a fiberoptic instrument called a ureteroscope to remove a stone lodged in the ureter

vaccine a substance made from the whole organism or parts that contain critical antigenic components or genes for those components. It stimulates a primary response that produces antibodies and memory cells that protect against subsequent infection by that organism.

vagotomy severing of the vagus nerve; often a component of gastric surgery.

vagus nerve tenth cranial nerve. One of its major functions is to coordinate the autonomic nervous system communication between organs of digestion.

validity the quality of producing desired results.

variable region the part of an antibody or TCR that differs from one antibody or TCR to another and produces a binding site for a specific antigen.

vasomotor referring to nerves that innervate smooth muscles in the walls of arteries and veins and can cause their constriction or dilation.

vasopressin the primary endocrine factor that regulates urinary H_2O loss and overall H_2O balance; regulates blood pressure via this hormone's pressor effects on blood vessels; also known as antidiuretic hormone (ADH).

ventricular fibrillation uncontrolled contractions of the ventricle; often associated with myocardial infarction

ventricular tachycardia rapid heartbeat originating from the ventricle

viral load the level of virus or viral markers measured in the blood.

viscosity thickness of a liquid

Voluntary Chronic Care Improvement Programs (CCIP) programs designed to improve the quality of care and life for people living with chronic illnesses, development and testing of which were authorized by the Medicare Modernization Act of 2003 (MMA). Chronic illnesses account for a significant share of Medicare expenditures.

volvulus the twisting of the bowel causing obstruction.

von Willebrand factor a protein on the platelet membrane surface that is sensitive to the chemical signals of an injured cell and causes the platelet to become sticky and adhere to other platelets and blood constituents.

water intoxication uncontrolled, excessive water consumption resulting in dilutional complications.

Wernicke-Korsakoff syndrome a syndrome consisting of Wernicke's encephalopathy in the acute phase and followed by Korsakoff's syndrome, usually associated with severe alcoholism.

workers' compensation insurance coverage that compensates employees for work-related injuries or disabilities, which employers are required to provide by state law.

xenobiotics chemicals that are found in an organism but are not produced by it or expected to be there, such as drugs or pollutants.

xenograft tissue transplantation between individuals from different species.

xerostomia dry mouth, often the result of damage to the salivary glands.

X-linked dominant an inheritance pattern of a dominant allele on the X chromosome. Such disorders are relatively rare.

X-linked recessive an inheritance pattern of a recessive allele on the X chromosome. Related disorders are more common in males, who carry only one X chromosome.

yin/yang a philosophy with roots in Taoism, the way of nature. Yin and yang are the fundamental duality of the universe, opposite and interacting principles of dark (yin) and light (yang).

Y-linked inheritance based on the Y chromosome. Disorders are extremely rare and occur only in males.

Index

A

Abbreviations, medical, 140
  unacceptable, 140t
Absolute neutrophil count, 774
Absorption
  carbohydrate, 461f
  iron, 662
  in large intestine, 463, 465–466
  lipid, 464f
  of nutrients, 463f
  nutrient, summary of major pathways involving, 567f
  protein, 462f
  sites of nutrient, 465f
  in small intestine, 460
  in stomach, 427–428
  vitamin B_{12}, 671
Acanthosis nigricans, 592
Achalasia, 441–442
  nutrition therapy for, 442
  treatment options for, 443f
Achlorhydria, 445, 565
Acid-base balance, 203. *See also* Acid-base disorders
  arterial blood parameters used for analysis of, 209t
  assessment of, 208
  chemical buffer systems, 206
  effect of, on electrolyte balance, 208
  lung regulation of, 718
  overall schema for maintenance of, 204f
  regulation of, 205–208
  renal regulatory control, 207–208
  respiratory regulatory control, 206–207
Acid-base disorders, 208–214. *See also specific disorders*
  assessment of, 214–215
    common, 215t
  mixed, 214
Acidemia, 205
Acidosis, 205
  summary of renal responses to, 208t
Acid perfusion test, 440
Acids, 203–204
Acquired immunodeficiency syndrome (AIDS). *See* AIDS
Acromegaly, 556
Active artificial immunity, 263
Active immunity, 263
Active immunization, 281
Active natural immunity, 263
Activities of daily living (ADL), 127, 130t
Acupuncture, 79
  meridians, qi traveling along, 82f
Acute care, 5
Acute coronary syndrome, 403
Acute lymphoblastic leukemia (ALL), 681
Acute phase proteins, 818
Acute renal failure, 614
  clinical manifestations, 643
  defined, 643
  epidemiology and etiology, 643
  nutrition therapy for, 644–645
  pathophysiology, 643
  treatment, 644
Acute respiratory failure distress syndrome (ARDS), 740–741
  antioxidant status of patients with, 742
  conditions associated with development of, 740t
Adaptive thermogenesis, 330
Addison's disease, 560
Adhesions, 230
Adipoctyes, 331
Adipose tissue, 331–332
  excess of, 332. *See also* Obesity; Overweight
  location or distribution of, 333–335
Adjustable gastric banding, 355
Adjusted edema-free body weight (aBW_{ef}), 630
Adjuvant chemotherapy, 762
Adrenal glands, 553
  abnormalities of adrenal cortex function, 560
  functions of adrenal cortex hormones, 554t
Adrenocortical hormones, 553
Adult Treatment Panel III Guidelines, 392
  maintenance of dietary fat intake recommendations under, 395
  summary of, 393t–394t
Adverse drug reactions, 310. *See also* Drug-nutrient interactions
Aerophagia, 730
Afferent arteriole, 610
Afterload, 374
Ageusia, 430
AIDS
  anemia and, 665
  challenges for nutrition status maintenance presented y, 806
  -defining diagnosis, 817
  drug-nutrient interactions in, 309–310
  HIV and, 805–838
  progression of HIV infection to, lowering of, 814
AIDS-related wasting syndrome (AWS), 817
Alanine shunt, 885
Albumin, 123–124
Alcohol
  adverse effects on body of, 520t
  derivation of kilocalories from, 520t
  and iron overload in men and post-menopausal women, 675
  malnutrition and, 519–525
  metabolism of, 516
  and osteoporosis, 860
  withdrawal syndrome, 516
    characteristics of, 516t
Alcohol abuse, 514
  gender, age, and racial factors related to, 516
Alcoholic hepatitis, 526
  treatment and nutrition therapy, 526
Alcoholic liver disease (ALD), 514
Alcoholic paralysis, 522
Alcoholism
  anemia and, 665
  as cause of thiamin deficiency, 520–521
  deterioration of nervous system in, 522
  diagnosis and epidemiology of chronic, 514, 516
  and malnutrition, 514, 516–524
  mechanisms of malabsorption related to, 518–519
  nutrition implications of, 519–524
  summary of nutritional effects of, 524–525
Alelles, 242
Alkalemia, 205
Alkalosis, 205
  summary of renal responses to, 208t
Allergens, 285
  common food, 289
  respiratory, 274
Allergic reactions, 285
  characteristics of Type I, II, III, and IV, 286t
  classifications of, 285
  to food, 288
  Type II, 290
  Type IV, 290
Allergies, 262, 285
  food. *See* Food allergies
  shots for, 288
  tests for, 287t
  Type III, 290
  Type I or IgE, 285
Allografts, 280
Allopathic medicine, 24, 65
Alpha-glucosidase, 596
Alternate complement pathways, 276
Alternative medical systems, 73–74, 76–79
  complementary therapies associated with, 79, 82–84
Alternative medicine, 65. *See also* Alternative medical systems; CAM
Alveolar-capillary unit, 717f
  gas exchange in, 716
Alveoli, 716
Alzheimer's disease. *See also* Dementia
  clinical manifestations, 707
  epidemiology, 705
  etiology, 705–706
  nutrition goals and interventions for, 708t
  nutrition therapy for, 707
  pathophysiology, 706–707
  treatment, 707
Amenorrhea, 859
American Board of Medical Specialties, 33t

American College of Rheumatology, 872
American Dietetic Association (ADA), 4
  Code of Ethics, 36–37
  "Nutrigenetics and Nutrigenomics" in Strategic Plan of, 257
  pursuit of third-party reimbursement for nutrition services by, 11
  Renal Practice Group (RPG), 618
  Standards of Professional Practice, 36–37
American Heart Association, 385
American Herbal Products Association, 91
American Holistic Medical Association, 72
American Lung Association, 383
American Society for Parenteral and Enteral Nutrition (ASPEN), 778
Amino acid disorders, 886–892
  adjunct therapies, 892
  and affected amino acids, 886t
  epidemiology, etiology, and clinical manifestations, 886
  nutritional concerns, 890, 892
  nutrition interventions, 887–890
Amylin, 596
Amyloid, 221
Amyloid plaques, 706–707
Amyloid precursor protein, 706
Amylopectin, 903t
Amyotrophic lateral sclerosis (ALS)
  clinical manifestations, 702
  defined, 701
  distinct forms of, 702
  epidemiology, 701
  etiology, 701
  nutrition therapy for, 702
  pathophysiology, 701–702
  treatment, 702
Anabolic reactions, 183
Anaphylactic shock, 285–286
Anaphylaxis
  mediators of, 287t
  risk of, 285–286
Anasarca, 614
Anastomosis, 448
Anemia(s)
  age-related, 653
  anorexia nervosa and, 665–666
  in CKD patients, 639
  clinical signs and symptoms of, 659t
  connections among nutritional, 674
  diagnosis of, 126
  environmental conditions impacting on, 666
  in female athletes, 666
  hemolytic, 656
  kidney disease and, 665
  of liver disease, 523t
  non-nutritional, 676–678
  nutritional, 658–674, 659t
  of prematurity, 678
  rare, 678
  traumatic conditions associated with, 664–665
Anemia of chronic disease (ACD), 664
Anergy, 285
Anesthesia, 797, 799
Aneurysm, 406
  defined, 694
  nutrition implications, 696–697
  risk factors for, 695
  ruptured, 694
  treatment goals for, 696
Angina, 401
Angiogenesis, 664
Angiotensin II, 376, 377
Anion gap, 208. *See also* Anions
Anions, 185, 186. *See also* Anion gap
Ankle brachial index (ABI), 409
Anorexia, 751
  in cancer patients, 777–778
  factors influencing development of, 768f
  nutrition therapy for treatment of, 777t
Anorexia nervosa, 358
  anemia and, 665–666
  common characteristics of, 359t
  diagnostic criteria for, 356
  health complications of, 358–360
  nutrition therapy for, 362–363
  physical and diagnostic findings in patients with, 360t
  physical changes and health complications seen in patients with, 361f
  subtypes of, 356
Anorexigenic stimuli, 330
Anovulation, 565
Anterior pituitary, 553
  hormones, functions of, 555f
Anthropometrics, 102, 114–116
  for fat patterning in lipodystrophy, 822, 822f, 823f, 824f, 825, 825f, 826f
Antibodies, 261, 273–274, 276
  monoclonal, 279
  structure of, 275f
Antibody-dependent cell-mediated cytotoxicity (ADCC), 275
Anticodons, 244
Antidiuresis, 679
Antidiuretic hormone (ADH). *See* Arginine vasopressin (AVP)
Anti-emetics, 766
Antigenic determinant, 261
Antigen-presenting cells (APC), 267, 269
Antigen recognition molecules, 272
  antibodies, 273–274
  B cell receptor, 274
  major histocompatibility complex (MHC), 272–273
  T cell receptor, 274
Antigens
  attack on harmless, 285–293
  characteristics of, 262
  defined, 261
  intradermal injection of, 125
  that mimic human cellular antigens, bacterial and viral, 292t
Antioxidants, 719
Antiretroviral (ARV), 812
Antiretroviral therapy (ART), 812
Antisense strand, 244
Antiseptics, 225
Antitoxin, 290
Antroduodenal manometry, 440
Aplastic anemia, 678
Apolipoprotein, 387, 706
Apoptosis, 231
Appetite, 330–331
Appetite stimulants, 153
Arginine vasopressin (AVP), 187
  influence of RAAS and, 188f
Aromatic amino acids (AAA), 531
Arterial blood gases (ABGs), 717
Arteriosclerosis, 384. *See also* Atherosclerosis (AS)
Arteriovenous graft (AVG), 622
  diagram of, 623f
Arthritic conditions. *See also specific diseases and conditions*
  definition and epidemiology, 865
  factors known to increase risk of, 865
Arthritis. *See* Arthritic conditions
Ascending colon, 463
Ascites, 182
  causes of, 512
  as complication of cirrhosis, 527
    nutrition therapy for, 527
  formation in liver disease, theories of etiology of, 529f
Aspiration, 436–437, 696–697
  associated with enteral feeding, 165
    evidence-based guidelines for prevention of, 739t
  pneumonia, 737–739
  risk factors associated with, 738t
Aspiration pneumonia, 737–739
Assessment, diagnosis, intervention/monitoring evaluation (ADIM) format, 144
  sample, 144t
Asthma, 289–290
  clinical manifestations, 721
  defined, 720
  epidemiology, 720
  etiology, 720
  food intolerances in people with, 288
  increase in rates of U.S., 286
  nutrition therapy for, 721–722
  pathophysiology, 720
  treatment, 721
  triggering, 290
Asymmetric muscle weakness, 702
Atherectomy, 396
Atherosclerosis (AS)
  clinical manifestations, 392
  defined, 383–384
  epidemiology, 385
  etiology, 386–391
    age and sex, 386
    diabetes mellitus, 389
    diet, 389
    dyslipidemia, 387–388
    family history, 386
    impaired fasting glucose and metabolic syndrome, 390–391
    obesity, 387
    physical inactivity, 389

markers in assessing risk for, 387
nutrition therapy and, 392
pathophysiology, 391–392
surgical treatment procedures for, 396
treatment, 392
Atopic reactions, 285
Atria, 373
Atrioventricular bundle, 373
Atrioventricular (AV) node, 373
Atrophic gastritis, 671
Atrophy, 222, 222f, 701
Attenuated vaccines, 282
Auscultation, 114, 717
Autoantibodies, 291, 572
diabetes-related, 577–578
Autograft, 279
Autoimmune disease(s), 290–292
damage caused by, 292–293
induction of, 292
major, 291t
Autoimmunity, 262, 290–292
Automatic division, of peripheral nervous system, 688
Autonomic dysreflexia, 710
Autonomic nervous system
parasympathetic branch, 688
sympathetic branch, 688
Autonomic neuropathy, 576–577
Autosomal dominant traits, 245, 246
Autosomal recessive traits, 245, 246, 731
Autosomes, 242
Autotolerance, 284
Axons, 688
Ayurvedic medicine, 73–74
incorporation of humoral theories in, 83
selected remedies, 75t–76t
Azidothymidine (AZT), 806
Azotemia, 617

B

Bacteria, 278. *See also* Microorganisms
living within colon, 465
Bacterial overgrowth syndrome
clinical manifestations, 503
defined, 503
nutrition therapy for, 503
pathophysiology, 503
treatment, 503
Balnotherapy, 873
Bariatric surgery, 354–355
Barium radiology studies, 440, 441f
Baroreceptors, 187, 376
Barrett's esophagus, 435–436, 760
Barrett's metaplasia, 435–436
Basal acid output, 440
Basal energy expenditure (BEE), 130
defined, 325
Basal metabolic rate (BMR), 130
Bases, 204
Basophils, 267
B cell receptor, 274
B cells, 264, 269
macrophages, and helper T cells, interactions among, 277f
Behavioral-environmental domain, 54
Behavior therapy, 352
Beindorff, Mary Ellen, 544, 684
Benedict, F.G., 326
Bernstein test, 440
Beta-amyloid, 706
Betadine, 225
Bile. *See also* Liver
cholesterol in, 535
functions of, 512
Bilirubin, 465, 656
Billroth I, 448, 449f
Billroth II, 448, 449f
Binding site, 882
Binge-eating disorder, 356
research criteria for, 357
Biochemical assessment, 120–127
hematological assessment, 126
immunocompetence, 125–126
protein assessment, 121–125
vitamin and mineral assessment, 127
Bioelectrical impedance analysis (BIA), 119
in adults with HIV infection, 820–821
Biofeedback, 92
The Biology of Human Starvation (Keys), 785, 786
Biomarkers, 753
Biomedicine, 65
clinical applications of, 66
factors in risk-benefit analysis of CAM versus conventional, 95t
providers, and need for open mind about CAM, 95
response to use of CAM by community practicing, 71–73
Biotransformation, 303–304
Blood. *See also* Hematological system
circulation of, 511–512
to and from liver, 511f
clotting and bleeding disorders, 679
characteristics of, 670t
clotting of, 657
components of, 652f
composition of, 652
disorders, factors in, 656
pH, 718
removal of nonessential solutes from, 611–612
Blood clotting, 267, 657
disorders, 679
characteristics of, 670t
function, select indices of, 667t
Blood pressure. *See also* Hypertension
causes for increase in, 377–378
diastolic, 373
factors influencing arterial, 375f
measurement of, using sphygmomanometer, 374f
physical activity and, 383
regulation of, 374–376
relationship of potassium, calcium, and magnesium to reduction in, 382
smoking cessation and, 383
systolic, 373
Blood urea nitrogen (BUN), 643
Body cell mass (BCM), 818
Body composition, 117–118. *See also* Human body
anthropometric assessment of, 114–116
arm muscle area assessment, 118
bioelectrical impedance analysis, 119
dual energy X-ray absorptiometry (DEXA), 120
height/stature, 114–115
hydrodensitometry, 120
near infrared interactance (NIR), 120
obesity, and overweight, 332–335
skinfold measurements, 118, 118f
use of phase angle, as measure of, 119
Body mass index (BMI), 116, 117, 332–333
in adults, interpretation of, 117t
defined, 327
using and calculating, to classify adults, 333
Body weight
energy balance and, 323–364
genetics and, 343, 345
Bolus, 423
Bolus feedings, 161
Bone, 844–848
cortical and trabecular, 846–848, 846f
metabolism, hormonal control of, 848–849
organization of a long, 847f
subchondral, 865
use of term, 844
Bone marrow, 263, 269
failure, 679, 681
stem cells, 653
transplant, 679, 681
Bone marrow transplantation, 679, 681, 765–766
Bone marrow transplant patients
clinical manifestations, 681
nutrition therapy for, 682–683
pathophysiology, 681
treatment for, 681–682
Bone spurs, 865
Botanical remedies, 73–74
preparation methods for, 87
Brachytherapy, 763
Bradykinesia, 699
Brain, 689–690
recording electrical activity of, 692
regions of, 690, 691f
Brain stem, 690
Branched-chain amino acids (BCAA), 531, 793
Breathwork, 94
Bronchi, 716
Bronchial-associated lymphatic tissue (BALT), 269
components of, 270
Bronchial hyperreactivity, 722
Bronchioles, 716
Bronchitis, 726
Bronchopulmonary dysplasia (BPD)
defined, 722
etiology, 722

nutrition therapy for, 723–725
energy and macronutrient needs, 723
feeding practices, 724–725
vitamins/minerals, 724
preterm infant formulas, 724t
treatment, 723
Brown adipose tissue (BAT), 331
Brundtland, Gro Harlem, 24–25
Brush border, 457
enzymes, 460
Buccal administration, of drugs, 302
Buffers, 204
chemical, 205–206
and their primary roles, 205t
foods that act as, 261
Bulimia nervosa, 360–361
common characteristics of, 359t
diagnostic criteria for, 356
health complications of, 361
nutrition therapy for, 363–364
physical and diagnostic findings in patients with, 362t
physical changes and health complications seen in patients with, 363f
subtypes of, 356–357, 360
Bundle of His, 373
Burns
characteristics of, based on depth, 795t
clinical manifestations, 794–795
defined, 794
epidemiology, 794
etiology, 794
nutrition therapy for, 795–797
pathophysiology, 795
treatment, 795
Bush, George W., 11

C

CA-125, 756
Cachexia
cardiac, 414
factors influencing development of, 768f
Calcitonin, 848
Calcium, 187, 855–858
adequate intakes for, 856t
balance, 190f
dairy food sources of, 857t
imbalance, 195–196
intakes of Americans by age and gender, mean estimated daily, 856t
non-dairy food sources of, 857t
simple approach for estimating dietary intake of, 858
Calculus, 428
Calorimetry, 130
CAM, 65–96
age-adjusted percentages of adults who used selected categories of, during past 12 months (2002), 68t–71t
biomedical response to growing public use of, 71–73
disease/condition for most frequent use of, 71f
factors in risk-benefit analysis of conventional biomedical treatment versus, 95t
guidelines for client selection of, 96t
health conditions most commonly treated with, 67
natural products used in. *See* Natural products
prayer use in, 84
preparation methods for botanical remedies, 87
rationale for choosing, 67, 71
research and regulation, 73
typical clients of, 66–67
Cancer, 278–279
adjuvant therapy for, 759
anorexia and, 777–778
biological therapies used to combat, 766
cell reproduction, 755–756
steps in, 756
clinical manifestations, 758
colon, 761
and rectal screening guidelines, 762
combination treatment for, 759
defined, 751
diagnosis, 756
requiring surgery for treatment, 759–762
diarrhea and, 775–776
dietary risk factors for colorectal, 761
dysgeusia and, 776
epidemiology, 752
esophageal, 760
etiology, 752–754
gastric, 760
of head and neck, 759–760
hematopoietic, 681
immune surveillance against, 279f
and intake of fruits and vegetables, 252–253
intestinal, 760–761
major warning signs of, 756t
MTHFR and ADH polymorphisms interact with dietary folate and alcohol, 252
nutrition support, 778–779
nutrition therapy for, 766–771
abnormalities in carbohydrate, protein, and lipid metabolism, 767
determining nutrient requirements, 770–771
nutrition assessment, 768–770
palliative surgery for, 759
pancreatic, 761–762
pathophysiology, 754–756
patients, popular alternative therapies used by, 772–773
prevention guidelines, 755t
primary, 759
progression, 757f
salvage therapy for, 759
screening and prevention, 754
side effects of antineoplastic agents used for treatment of, 764t
from single gene inherited cancers to gene-nutrient interactions, 251
surgical treatment for, 759–760
terminal, 778
therapy, immunological approaches to, 279
treatment, 758–760
nutritional implications of, 767–768
and treatment associated with anemia, 664
variations in xenobiotic metabolisms influence risk, 251–252
xerostomia and, 776–777
Candida, 125
Capillaries, in alveoli, 716
Capitation, 7
Carbohydrate counting, 589
three levels of, 590
Carbohydrate metabolism disorders, 898–905
Carbohydrates
digestion and absorption, 461f
malabsorption of, 479
Carbon dioxide. *See also* Gas exchange
excess of, 209
transport and exchange of, 206f
Carbonic acid, 204
Carcinoembryonic antigen (CEA), 756
Carcinogenesis, 752
Carcinogens, 752
Cardiac cachexia, 414
Cardiac output, 374
Cardiac rehabilitation, 394
Cardiovascular disease
associated with CKD, 638
effect of dietary modification on, 256
hypertension and, 387. *See also* Hypertension
individual variation in response to environmental influences, 255–256
interaction of dietary fats with genotypes influencing outcomes, 256
and iron-deficiency anemia, 664
as leading cause of death, 385
Cardiovascular system. *See also* Cardiovascular disease
anatomy and physiology of, 371–376
cardiac diagnostic procedures, 407
cardiac function, 373–374
diseases of, 371–417. *See also specific diseases*
heart, 371–373
major functions of, 371
regulation of blood pressure, 374–376. *See also* Blood pressure
role of, 371
Carnitine, 885
Carnitine acyl transferase I (CAT I), 152
Carnitine acyl transferase II (CAT II), 152
Carpopedal spasm, 845
Cartilage, 844
Casein hydrolysate, 900
Case, Shelly, 504
Catabolic reactions, 183–184
Catecholamines, 553
Cations, 185, 186
Cavities, 428
CD4 cells, 268, 807
CD8 cells, 269
Celiac disease, 5
anemia and, 665
clinical manifestations, 482
defined, 481
diagnosis, 482
epidemiology, 481–482
etiology, 482
nutrition therapy for, 483

$$310 \text{ g carbohydrate} \times \frac{4 \text{ kcal}}{1 \text{ g carbohydrate}} = 1240 \text{ kcal.}$$

- Find the total kcalories:

  240 + 720 + 1240 = 2200 kcal.

- Find the percentage of total kcalories from each energy nutrient (see Example 3):

  Protein: 240 ÷ 2200 = 0.109 × 100 = 10.9 = 11% of kcal.

  Fat: 720 ÷ 2200 = 0.327 × 100 = 32.7 = 33% of kcal.

  Carbohydrate: 1240 ÷ 2200 = 0.563 × 100 = 56.3 = 56% of kcal.

  Total: 11% + 33% + 56% = 100% of kcal.

In this case, the percentages total 100 percent, but sometimes they total 99 or 101 because of rounding—a reasonable estimate.

Ratios

A ratio is a comparison of two (or three) values in which one of the values is reduced to 1. A ratio compares identical units and so is expressed without units.

Example 6 Suppose your daily intakes of potassium and sodium are 3000 milligrams (mg) and 2500 milligrams, respectively. What is the potassium-to-sodium ratio?

- Divide the potassium milligrams by the sodium milligrams:

  3000 mg potassium ÷ 2500 mg sodium = 1.2.

The potassium-to-sodium ratio is 1.2:1 (read as "one point two to one" or simply "one point two"), which means there are 1.2 milligrams of potassium for every 1 milligram of sodium. A ratio greater than 1 means that the first value (in this case, potassium) is greater than the second (sodium). When the ratio is less than 1, the second value is larger.

Weights and Measures

LENGTH

1 meter (m) = 39 in.
1 centimeter (cm) = 0.4 in.
1 inch (in) = 2.5 cm.
1 foot (ft) = 30 cm.

TEMPERATURE

Celsius*		Fahrenheit	
Steam	100°C	212°F	Steam
Body temperature	37°C	98.6°F	Body temperature
Ice	0°C	32°F	Ice

- To find degrees Fahrenheit (°F) when you know degrees Celsius (°C), multiply by 9/5 and then add 32.
- To find degrees Celsius (°C) when you know degrees Fahrenheit (°F), subtract 32 and then multiply by 5/9.

VOLUME

1 liter (L) = 1000 mL, 0.26 gal, 1.06 qt, or 2.1 pt.
1 milliliter (mL) = 1/1000 L or 0.03 fluid oz.
1 gallon (gal) = 128 oz, 8 c, or 3.8 L.
1 quart (qt) = 32 oz, 4 c, or 0.95 L.
1 pint (pt) = 16 oz, 2 c, or 0.47 L.
1 cup (c) = 8 oz, 16 tbs, about 250 mL, or 0.25 L.
1 ounce (oz) = 30 mL.
1 tablespoon (tbs) = 3 tsp or 15 mL.
1 teaspoon (tsp) = 5 mL.

WEIGHT

1 kilogram (kg) = 1000 g or 2.2 lb.
1 gram (g) = 1/1000 kg, 1000 mg, or 0.035 oz.
1 milligram (mg) = 1/1000 g or 1000 µg.
1 microgram (µg) = 1/1000 mg.
1 pound (lb) = 16 oz, 454 g, or 0.45 kg.
1 ounce (oz) = about 28 g.

ENERGY

1 kilojoule (kJ) = 0.24 kcal.
1 millijoule (mJ) = 240 kcal.
1 kcalorie (kcal) = 4.2 kJ.
1 g carbohydrate = 4 kcal = 17 kJ.
1 g fat = 9 kcal = 37 kJ.
1 g protein = 4 kcal = 17 kJ.
1 g alcohol = 7 kcal = 29 kJ.

*Also known as *centigrade*.

Daily Values for Food Labels

The Daily Values are standard values developed by the Food and Drug Administration (FDA) for use on food labels. The values are based on 2000 kcalories a day for adults and children over 4 years old. Chapter 2 provides more details.

Nutrient	Amount
Protein[a]	50 g
Thiamin	1.5 mg
Riboflavin	1.7 mg
Niacin	20 mg NE
Biotin	300 µg
Pantothenic acid	10 mg
Vitamin B_6	2 mg
Folate	400 µg
Vitamin B_{12}	6 µg
Vitamin C	60 mg
Vitamin A	5000 IU[b]
Vitamin D	400 IU[b]
Vitamin E	30 IU[b]
Vitamin K	80 µg
Calcium	1000 mg
Iron	18 mg
Zinc	15 mg
Iodine	150 µg
Copper	2 mg
Chromium	120 µg
Selenium	70 µg
Molybdenum	75µg
Manganese	2 mg
Chloride	3400 mg
Magnesium	400 mg
Phosphorus	1000 mg

[a]The Daily Values for protein vary for different groups of people: pregnant women, 60 g; nursing mothers, 65 g; infants under 1 year, 14 g; children 1 to 4 years, 16 g.
[b]Equivalent values for nutrients expressed as IU are: vitamin A, 1500 RAE (assumes a mixture of 40% retinol and 60% beta-carotene); vitamin D, 10 µg; vitamin E, 20 mg.

Food Component	Amount	Calculation Factors
Fat	65 g	30% of kcalories
Saturated fat	20 g	10% of kcalories
Cholesterol	300 mg	Same regardless of kcalories
Carbohydrate (total)	300 g	60% of kcalories
Fiber	25 g	11.5 g per 1000 kcalories
Protein	50 g	10% of kcalories
Sodium	2400 mg	Same regardless of kcalories
Potassium	3500 mg	Same regardless of kcalories

Glossary of Nutrient Measures

kcal: kcalories; a unit by which energy is measured (Chapter 1 provides more details).

g: grams; a unit of weight equivalent to about 0.03 ounces.

mg: milligrams; one-thousandth of a gram.

µg: micrograms; one-millionth of a gram.

IU: international units; an old measure of vitamin activity determined by biological methods (as opposed to new measures that are determined by direct chemical analyses). Many fortified foods and supplements use IU on their labels.
- For vitamin A, 1 IU = 0.3 µg retinol, 3.6 µg β-carotene, or 7.2 µg other vitamin A carotenoids.
- For vitamin D, 1 IU = 0.025 µg cholecalciferol.
- For vitamin E, 1 IU = 0.67 natural α-tocopherol (other conversion factors are used for different forms of vitamin E).

mg NE: milligrams niacin equivalents; a measure of niacin activity (Chapter 10 provides more details).
- 1 NE = 1 mg niacin.
  = 60 mg tryptophan (an amino acid).

µg DFE: micrograms dietary folate equivalents; a measure of folate activity (Chapter 10 provides more details).
- 1 µg DFE = 1 µg food folate.
  = 0.6 µg fortified food or supplement folate.
  = 0.5 µg supplement folate taken on an empty stomach.

µg RAE: micrograms retinol activity equivalents; a measure of vitamin A activity (Chapter 11 provides more details).
- 1 µg RAE = 1 µg retinol.
  = 12 µg β-carotene.
  = 24 µg other vitamin A carotenoids.

mmol: millimoles; one-thousanth of a mole, the molecular weight of a substance. To convert mmol to mg, multiply by the atomic weight of the substance.
- For sodium, mmol × 23 = mg Na.
- For chloride, mmol × 35.5 = mg Cl.
- For sodium chloride, mmol × 58.5 = mg NaCl.